Problem-Solving Strategies and Model Boxes

Note for users of the three-volume edition:
Volume 1 (pp. 1–600) includes Chapters 1–21.
Volume 2 (pp. 601–1062) includes Chapters 22–36.
Volume 3 (pp. 1021–1240) includes Chapters 36–42.

Chapters 37–42 are not in the Standard Edition.

Brief Contents

PHYSICS

FOR SCIENTISTS AND ENGINEERS **A STRATEGIC APPROACH** 4/E

RANDALL D. KNIGHT

California Polytechnic State University
San Luis Obispo

PEARSON

Editor-in-Chief:	Jeanne Zalesky
Acquisitions Editor:	Darien Estes
Project Manager:	Martha Steele
Program Manager:	Katie Conley
Senior Development Editor:	Alice Houston, Ph.D.
Art Development Editors:	Alice Houston, Kim Brucker, and Margot Otway
Development Manager:	Cathy Murphy
Program and Project Management Team Lead:	Kristen Flathman
Production Management:	Rose Kernan
Compositor:	Cenveo® Publisher Services
Design Manager:	Mark Ong
Cover Designer:	John Walker
Illustrators:	Rolin Graphics
Rights & Permissions Project Manager:	Maya Gomez
Rights & Permissions Management:	Rachel Youdelman
Photo Researcher:	Eric Schrader
Manufacturing Buyer:	Maura Zaldivar-Garcia
Executive Marketing Manager:	Christy Lesko
Cover Photo Credit:	Thomas Vogel/Getty Images

Library of Congress Cataloging-in-Publication Data On File

ISBN 10: 0-134-08149-8; ISBN 13: 978-0-134-08149-6 (Standard edition)

www.pearsonhighered.com   1 2 3 4 5 6 7 8 9 10—**V011**—18 17 16 15

About the Author

Randy Knight taught introductory physics for 32 years at Ohio State University and California Polytechnic State University, where he is Professor Emeritus of Physics. Professor Knight received a Ph.D. in physics from the University of California, Berkeley and was a post-doctoral fellow at the Harvard-Smithsonian Center for Astrophysics before joining the faculty at Ohio State University. It was at Ohio State that he began to learn about the research in physics education that, many years later, led to *Five Easy Lessons: Strategies for Successful Physics Teaching* and this book, as well as *College Physics: A Strategic Approach,* co-authored with Brian Jones and Stuart Field. Professor Knight's research interests are in the fields of laser spectroscopy and environmental science. When he's not in front of a computer, you can find Randy hiking, sea kayaking, playing the piano, or spending time with his wife Sally and their five cats.

A research-driven approach, fine-tuned for even greater ease-of-use and student success

REVISED COVERAGE AND ORGANIZATION GIVE INSTRUCTORS GREATER CHOICE AND FLEXIBILITY

FIGURE 9.1 A system-environment perspective on energy.

Heat **Environment** **Work**
Energy added

System
The system has energy E_{sys}

Kinetic ⟷ Potential
Thermal ⟷ Chemical
Energy can be transformed

Heat Energy removed **Work**

Environment

NEW! CHAPTER ORGANIZATION allows instructors to more easily present material as needed to complement labs, course schedules, and different teaching styles. Work and energy are now covered before momentum, oscillations are grouped with mechanical waves, and optics appears after electricity and magnetism. Unchanged is Knight's unique approach of working from concrete to abstract, using multiple representations, balancing qualitative with quantitative, and addressing misconceptions.

NEW! ADVANCED TOPICS as optional sections add even more flexibility for instructors' individual courses. Topics include rocket propulsion, gyroscopes and precession, the wave equation (including for electromagnetic waves), the speed of sound in gases, and more details on the interference of light.

11.6 ADVANCED TOPIC Rocket Propulsion

Newton's second law $\vec{F} = m\vec{a}$ applies to objects whose mass does not change. That's an excellent assumption for balls and bicycles, but what about something like a rocket that loses a significant amount of mass as its fuel is burned? Problems of varying mass are solved with momentum rather than acceleration. We'll look at one important example.

FIGURE 11.29 shows a rocket being propelled by the thrust of burning fuel but *not* influenced by gravity or drag. Perhaps it is a rocket in deep space where gravity is very weak in comparison to the rocket's thrust. This may not be highly realistic, but ignoring gravity allows us to understand the essentials of rocket propulsion without making the mathematics too complicated. Rocket propulsion with gravity is a Challenge Problem in the end-of-chapter problems.

The system rocket + exhaust gases is an isolated system, so its total momentum is conserved. The basic idea is simple: As exhaust gases are shot out the back, the rocket "recoils" in the opposite direction. Putting this idea on a mathematical footing is fairly straightforward—it's basically the same as analyzing an explosion—but we have to be extremely careful with signs.

We'll use a before-and-after approach, as we do with all momentum problems. Th...

FIGURE 11.29 A before-and-after pictorial representation of a rocket burning a small amount of fuel.

Before: m v_x

After: m_{fuel} $m + dm$ $v_x + dv_x$
v_{ex}
Relative to rocket

60. ‖ A clever engineer designs a "sprong" that obeys the force law
CALC $F_x = -q(x - x_{eq})^3$, where x_{eq} is the equilibrium position of the end of the sprong and q is the sprong constant. For simplicity, we'll let $x_{eq} = 0$ m. Then $F_x = -qx^3$.
   a. What are the units of q?
   b. Find an expression for the potential energy of a stretched or compressed sprong.
   c. A sprong-loaded toy gun shoots a 20 g plastic ball. What is the launch speed if the sprong constant is 40,000, with the units you found in part a, and the sprong is compressed 10 cm? Assume the barrel is frictionless.

NEW! MORE CALCULUS-BASED PROBLEMS have been added, along with an icon to make these easy to identify. The significantly revised end-of-chapter problem sets, extensively class-tested and both calibrated and improved using MasteringPhysics® data, expand the range of physics and math skills students will use to solve problems.

Built from the ground up on physics education research and crafted using key ideas from learning theory, Knight has set the standard for effective and accessible pedagogical materials in physics. In this fourth edition, Knight continues to refine and expand the instructional techniques to take students further.

NEW AND UPDATED LEARNING TOOLS PROMOTE DEEPER AND BETTER-CONNECTED UNDERSTANDING

NEW! MODEL BOXES enhance the text's emphasis on modeling—analyzing a complex, real-world situation in terms of simple but reasonable idealizations that can be applied over and over in solving problems. These fundamental simplifications are developed in the text and then deployed more explicitly in the worked examples, helping students to recognize when and how to use recurring models, a key critical-thinking skill.

MODEL 2.2

Constant acceleration

For motion with constant acceleration.

- Model the object as a particle moving in a straight line with constant acceleration.

$\vec{a}$
$\vec{v}$

- Mathematically:
  - $v_{fs} = v_{is} + a_s \Delta t$
  - $s_f = s_i + v_{is} \Delta t + \frac{1}{2} a_s (\ )$
  - $v_{fs}^2 = v_{is}^2 + 2a_s \Delta s$
- Limitations: Model fails i acceleration changes.

The acceleration is constant.

Straight line

MODEL 6.3

Friction

The friction force is *parallel* to the surface.

- Static friction: Acts as needed to prevent motion. Can have *any* magnitude up to $f_{s\,max} = \mu_s n$.
- Kinetic friction: Opposes motion with $f_k = \mu_k n$.
- Rolling friction: Opposes motion with $f_r = \mu_r n$.
- Graphically:

Push or pull

Friction

Motion is *relative to the surface.*

Static   Kinetic

$f_{s\,max} = \mu_s n$

$\mu_k n$

The object slips when static friction reaches $f_{s\,max}$.

Kinetic friction is constant as the object moves.

Static friction increases to match the push or pull.

Rest — Moving   Push or pull force

6 Dynamics I: Motion Along a Line

The powerful thrust of the jet engines accelerates this enormous plane to a speed of over 150 mph in less than a mile.

IN THIS CHAPTER, you will learn to solve linear force-and-motion problems.

How are Newton's laws used to solve problems?
Newton's first and second laws are vector equations. To use them,
- Draw a free-body diagram.
- Read the x- and y-components of the forces directly off the free-body diagram.
- Use $\sum F_x = ma_x$ and $\sum F_y = ma_y$.

How are dynamics problems solved?
A net force on an object causes the object to accelerate.
- Identify the forces and draw a free-body diagram.
- Use Newton's second law to find the object's acceleration.
- Use kinematics for velocity and position.
《 LOOKING BACK Sections 2.4–2.6 Kinematics

How are equilibrium problems solved?
An object at rest or moving with constant velocity is in equilibrium with no net force.
- Identify the forces and draw a free-body diagram.
- Use Newton's second law with $a = 0$ to solve for unknown forces.
《 LOOKING BACK Sections 5.1–5.2 Forces

What are mass and weight?
Mass and weight are not the same.
- Mass describes an object's inertia. Loosely speaking, it is the amount of matter in an object. It is the same everywhere.
- Gravity is a force.
- Weight is the result of weighing an object on a scale. It depends on mass, gravity, and acceleration.

How do we model friction and drag?
Friction and drag are complex forces, but we will develop simple models of each.
- Static, kinetic, and rolling friction depend on the coefficients of friction but not on the object's speed.
- Drag depends on the square of an object's speed and on its cross-section area.
- Falling objects reach terminal speed when drag and gravity are balanced.

Normal $\vec{n}$ Gravity $\vec{F}_G$
Friction $\vec{f}$

Kinetic friction

How do we solve problems?
We will develop and use a four-part problem-solving strategy.
- Model the problem, using information about objects and forces.
- Visualize the situation with a pictorial representation.
- Set up and solve the problem with Newton's laws.
- Assess the result to see if it is reasonable.

SUMMARY

...en to learn to solve linear force-and-motion problems.

...IERAL PRINCIPLES

A Problem-Solving Strategy
A four-part strategy applies to both equilibrium and dynamics problems.
MODEL Make simplifying assumptions.
VISUALIZE
Go back and forth between these steps as needed.
- Translate words into symbols.
- Draw a sketch to define the situation.
- Draw a motion diagram.
- Identify forces.
- Draw a free-body diagram.
SOLVE Use Newton's second law:
$\vec{F}_{net} = \sum_i \vec{F}_i = m\vec{a}$
"Read" the vectors from the free-body diagram. Use kinematics to find velocities and positions.
ASSESS Is the result reasonable? Does it have correct units and significant figures?

...RTANT CONCEPTS

Newton's laws are vector expressions. You must write them out by **components**:
$(F_{net})_x = \sum F_x = ma_x$
$(F_{net})_y = \sum F_y = ma_y$
The acceleration is zero in equilibrium and also along an axis perpendicular to the motion.

...APPLICATIONS

A falling object reaches terminal speed
$v_{term} = \sqrt{\frac{2mg}{C\rho A}}$
Terminal speed is reached when the drag force exactly balances the gravitational force: $\vec{a} = 0$.

TERMS AND NOTATION

equilibrium model	weight	rolling friction	drag coefficient, C
constant-force model	coefficient of static friction, μ_s	coefficient of rolling friction, μ_r	terminal speed, v_{term}
flat-earth approximation	coefficient of kinetic friction, μ_k		

REVISED! ENHANCED CHAPTER PREVIEWS, based on the educational psychology concept of an "advance organizer," have been reconceived to address the questions students are most likely to ask themselves while studying the material for the first time. Questions cover the important ideas, and provide a big-picture overview of the chapter's key principles. Each chapter concludes with the visual Chapter Summary, consolidating and structuring understanding.

A STRUCTURED AND CONSISTENT APPROACH BUILDS PROBLEM-SOLVING SKILLS AND CONFIDENCE

With a research-based 4-step problem-solving framework used throughout the text, students learn the importance of making assumptions (in the MODEL step) and gathering information and making sketches (in the VISUALIZE step) before treating the problem mathematically (SOLVE) and then analyzing their results (ASSESS).

Detailed **PROBLEM-SOLVING STRATEGIES** for different topics and categories of problems (circular-motion problems, calorimetry problems, etc.) are developed throughout, each one built on the 4-step framework and carefully illustrated in worked examples.

PROBLEM-SOLVING STRATEGY 10.1

Energy-conservation problems

MODEL Define the system so that there are no external forces or so that any external forces do no work on the system. If there's friction, bring both surfaces into the system. Model objects as particles and springs as ideal.

VISUALIZE Draw a before-and-after pictorial representation and an energy bar chart. A free-body diagram may be needed to visualize forces.

SOLVE If the system is both isolated and nondissipative, then the mechanical energy is conserved:

$$K_i + U_i = K_f + U_f$$

where K is the total kinetic energy of all moving objects and U is the total potential energy of all interactions within the system. If there's friction, then

$$K_i + U_i = K_f + U_f + \Delta E_{th}$$

where the thermal energy increase due to friction is $\Delta E_{th} = f_k \Delta s$.

ASSESS Check that your result has correct units and significant figures, is reasonable, and answers the question.

Exercise 14

TACTICS BOX 26.1

Finding the potential from the electric field

❶ Draw a picture and identify the point at which you wish to find the potential. Call this position f.

❷ Choose the zero point of the potential, often at infinity. Call this position i.

❸ Establish a coordinate axis from i to f along which you already know or can easily determine the electric field component E_s.

❹ Carry out the integration of Equation 26.3 to find the potential.

Exercise 1

TACTICS BOXES give step-by-step procedures for developing specific skills (drawing free-body diagrams, using ray tracing, etc.).

The **REVISED STUDENT WORKBOOK** is tightly integrated with the main text–allowing students to practice skills from the text's Tactics Boxes, work through the steps of Problem-Solving Strategies, and assess the applicability of the Models. The workbook is referenced throughout the text with the icon .

---

18. The graph shows how the magnetic field changes through
PSS a rectangular loop of wire with resistance R. Draw a graph
30.1 of the current in the loop as a function of time. Let a counterclockwise current be positive, a clockwise current be negative.

a. What is the magnetic flux through the loop at $t = 0$?

b. Does this flux *change* between $t = 0$ and $t = t_1$?

c. Is there an induced current in the loop between $t = 0$ and $t = t_1$?

d. What is the magnetic flux through the loop at $t = t_2$?

e. What is the *change* in flux through the loop between t_1 and t_2?

f. What is the time interval between t_1 and t_2?

g. What is the magnitude of the induced emf between t_1 and t_2?

h. What is the magnitude of the induced current between t_1 and t_2?

i. Does the magnetic field point out of or into the loop?

j. Between t_1 and t_2, is the magnetic flux increasing or decreasing?

k. To oppose the *change* in the flux between t_1 and t_2, should the magnetic field of the induced current point out of or into the loop?

l. Is the induced current between t_1 and t_2 positive or negative?

m. Does the flux through the loop change after t_2?

n. Is there an induced current in the loop after t_2?

o. Use all this information to draw a graph of the induced current. Add appropriate labels on the vertical axis.

BEFORE CLASS

NEW! INTERACTIVE PRELECTURE VIDEOS

address the rapidly growing movement toward pre-lecture teaching and flipped classrooms. These whiteboard-style animations provide an introduction to key topics with embedded assessment to help students prepare and professors identify student misconceptions before lecture.

NEW! DYNAMIC STUDY MODULES (DSMs) con-

tinuously assess students' performance in real time to provide personalized question and explanation content until students master the module with confidence. The DSMs cover basic math skills and key definitions and relationships for topics across all of mechanics and electricity and magnetism.

DURING CLASS

NEW! LEARNING CATALYTICS™ is an interactive

classroom tool that uses students' devices to engage them in more sophisticated tasks and thinking. Learning Catalytics enables instructors to generate classroom discussion and promote peer-to-peer learning to help students develop critical-thinking skills. Instructors can take advantage of real-time analytics to find out where students are struggling and adjust their instructional strategy.

AFTER CLASS

NEW! ENHANCED END-OF-CHAPTER QUESTIONS offer stu-

dents instructional support when and where they need it, including links to the eText, math remediation, and wrong-answer feedback for homework assignments.

ADAPTIVE FOLLOW-UPS

are personalized assignments that pair Mastering's powerful content with Knewton's adaptive learning engine to provide individualized help to students before misconceptions take hold. These adaptive follow-ups address topics students struggled with on assigned homework, including core prerequisite topics.

Preface to the Instructor

This fourth edition of *Physics for Scientists and Engineers: A Strategic Approach* continues to build on the research-driven instructional techniques introduced in the first edition and the extensive feedback from thousands of users. From the beginning, the objectives have been:

- To produce a textbook that is more focused and coherent, less encyclopedic.
- To move key results from physics education research into the classroom in a way that allows instructors to use a range of teaching styles.
- To provide a balance of quantitative reasoning and conceptual understanding, with special attention to concepts known to cause student difficulties.
- To develop students' problem-solving skills in a systematic manner.

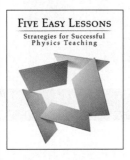

These goals and the rationale behind them are discussed at length in the *Instructor's Guide* and in my small paperback book, *Five Easy Lessons: Strategies for Successful Physics Teaching*. Please request a copy from your local Pearson sales representative if it is of interest to you (ISBN 978-0-805-38702-5).

What's New to This Edition

For this fourth edition, we continue to apply the best results from educational research and to tailor them for this course and its students. At the same time, the extensive feedback we've received from both instructors and students has led to many changes and improvements to the text, the figures, and the end-of-chapter problems. These include:

- **Chapter ordering changes** allow instructors to more easily organize content as needed to accommodate labs, schedules, and different teaching styles. Work and energy are now covered before momentum, oscillations are grouped with mechanical waves, and optics appears after electricity and magnetism.
- **Addition of advanced topics** as optional sections further expands instructors' options. Topics include rocket propulsion, gyroscopes, the wave equation (for mechanical and electromagnetic waves), the speed of sound in gases, and more details on the interference of light.
- **Model boxes** enhance the text's emphasis on modeling—analyzing a complex, real-world situation in terms of simple but reasonable idealizations that can be applied over and over in solving problems. These fundamental simplifications

are developed in the text and then deployed more explicitly in the worked examples, helping students to recognize when and how to use recurring models.

- **Enhanced chapter previews** have been redesigned, with student input, to address the questions students are most likely to ask themselves while studying the material for the first time. The previews provide a big-picture overview of the chapter's key principles.
- **Looking Back pointers** enable students to look back at a previous chapter when it's important to review concepts. Pointers provide the specific section to consult at the exact point in the text where they need to use this material.
- **Focused Part Overviews and Knowledge Structures** consolidate understanding of groups of chapters and give a tighter structure to the book as a whole. Reworked Knowledge Structures provide more targeted detail on overarching themes.
- **Updated visual program** that has been enhanced by revising over 500 pieces of art to increase the focus on key ideas.
- **Significantly revised end-of-chapter problem sets** include more challenging problems to expand the range of physics and math skills students will use to solve problems. A new icon for calculus-based problems has been added.

At the front of this book, you'll find an illustrated walkthrough of the new pedagogical features in this fourth edition.

Textbook Organization

The 42-chapter extended edition (ISBN 978-0-133-94265-1 / 0-133-94265-1) of *Physics for Scientists and Engineers* is intended for a three-semester course. Most of the 36-chapter standard edition (ISBN 978-0-134-08149-6 / 0-134-08149-8), ending with relativity, can be covered in two semesters, although the judicious omission of a few chapters will avoid rushing through the material and give students more time to develop their knowledge and skills.

The full textbook is divided into eight parts: Part I: *Newton's Laws,* Part II: *Conservation Laws,* Part III: *Applications of Newtonian Mechanics,* Part IV: *Oscillations and Waves,* Part V: *Thermodynamics,* Part VI: *Electricity and Magnetism,* Part VII: *Optics,* and Part VIII: *Relativity and Quantum Physics.* Note that covering the parts in this order is by no means essential. Each topic is self-contained, and Parts III–VII can be rearranged to suit an instructor's needs. Part VII: *Optics* does need to follow Part IV: *Oscillations and Waves,* but optics can be taught either before or after electricity and magnetism.

There's a growing sentiment that quantum physics is quickly becoming the province of engineers, not just scientists, and that even a two-semester course should include a reasonable introduction to quantum ideas. The *Instructor's Guide* outlines a couple of routes through the book that allow most of the quantum physics chapters to be included in a two-semester course. I've written the book with the hope that an increasing number of instructors will choose one of these routes.

- **Extended edition,** with modern physics (ISBN 978-0-133-94265-1 / 0-133-94265-1): Chapters 1–42.
- **Standard edition** (ISBN 978-0-134-08149-6 / 0-134-08149-8): Chapters 1-36.
- **Volume 1** (ISBN 978-0-134-11068-4 / 0-134-11068-4) covers mechanics, waves, and thermodynamics: Chapters 1–21.
- **Volume 2** (ISBN 978-0-134-11066-0 / 0-134-11066-8) covers electricity and magnetism and optics, plus relativity: Chapters 22–36.
- **Volume 3** (ISBN 978-0-134-11065-3 / 0-134-11065-X) covers relativity and quantum physics: Chapters 36–42.

The Student Workbook

A key component of *Physics for Scientists and Engineers: A Strategic Approach* is the accompanying *Student Workbook*. The workbook bridges the gap between textbook and home-work problems by providing students the opportunity to learn and practice skills prior to using those skills in quantitative end-of-chapter problems, much as a musician practices technique separately from performance pieces. The workbook exercises, which are keyed to each section of the textbook, focus on developing specific skills, ranging from identifying forces and drawing free-body diagrams to interpreting wave functions.

The workbook exercises, which are generally qualitative and/or graphical, draw heavily upon the physics education research literature. The exercises deal with issues known to cause student difficulties and employ techniques that have proven to be effective at overcoming those difficulties. **New to the fourth edition workbook** are exercises that provide guided practice for the textbook's Model boxes. The workbook exercises can be used in class as part of an active-learning teaching strategy, in recitation sections, or as assigned homework. More information about effective use of the *Student Workbook* can be found in the *Instructor's Guide*.

Instructional Package

Physics for Scientists and Engineers: A Strategic Approach, fourth edition, provides an integrated teaching and learning package of support material for students and instructors. NOTE For convenience, most instructor supplements can be downloaded from the "Instructor Resources" area of MasteringPhysics® and the Instructor Resource Center (www.pearsonhighered.com/educator).

Name of Supplement	Print	Online	Instructor or Student Supplement	Description
MasteringPhysics with Pearson eText ISBN 0-134-08313-X		✓	Instructor and Student Supplement	This product features all of the resources of MasteringPhysics in addition to the new Pearson eText 2.0. Now available on smartphones and tablets, Pearson eText 2.0 comprises the full text, including videos and other rich media. Students can configure reading settings, including resizeable type and night-reading mode, take notes, and highlight, bookmark, and search the text.
Instructor's Solutions Manual ISBN 0-134-09246-5		✓	Instructor Supplement	This comprehensive solutions manual contains complete solutions to all end-of-chapter questions and problems. All problem solutions follow the Model/Visualize/Solve/Assess problem-solving strategy used in the text.
Instructor's Guide ISBN 0-134-09248-1		✓	Instructor Supplement	Written by Randy Knight, this resource provides chapter-by-chapter creative ideas and teaching tips for use in your class. It also contains an extensive review of results of what has been learned from physics education research and provides guidelines for using active-learning techniques in your classroom.
TestGen Test Bank ISBN 0-134-09245-7		✓	Instructor Supplement	The Test Bank contains over 2,000 high-quality conceptual and multiple-choice questions. Test files are provided in both TestGen® and Word format.
Instructor's Resource DVD ISBN 0-134-09247-3	✓	✓	Instructor Supplement	This cross-platform DVD includes an Image Library; editable content for Key Equations, Problem-Solving Strategies, Math Relationship Boxes, Model Boxes, and Tactic Boxes; PowerPoint Lecture Slides and Clicker Questions; Instructor's Guide, Instructor's Solutions Manual; Solutions to Student Workbook exercises; and PhET simulations.
Student Workbook Extended (Ch 1–42) ISBN 0-134-08316-4 Standard (Ch 1–36) ISBN 0-134-08315-6 Volume 1 (Ch 1–21) ISBN 0-134-11064-1 Volume 2 (Ch 22–36) ISBN 0-134-11063-3 Volume 3 (Ch 36–42) ISBN 0-134-11060-9	✓		Student Supplement	For a more detailed description of the *Student Workbook*, see page v.

Acknowledgments

I have relied upon conversations with and, especially, the written publications of many members of the physics education research community. Those whose influence can be seen in these pages include Wendy Adams, the late Arnold Arons, Stuart Field, Uri Ganiel, Ibrahim Halloun, Richard Hake, Ken Heller, Paula Heron, David Hestenes, Brian Jones, the late Leonard Jossem, Jill Larkin, Priscilla Laws, John Mallinckrodt, Kandiah Manivannan, Richard Mayer, Lillian McDermott and members of the Physics Education Research Group at the University of Washington, David Meltzer, Edward "Joe" Redish, Fred Reif, Jeffery Saul, Rachel Scherr, Bruce Sherwood, Josip Slisko, David Sokoloff, Richard Steinberg, Ronald Thornton, Sheila Tobias, Alan Van Heuleven, Carl Wieman, and Michael Wittmann. John Rigden, founder and director of the Introductory University Physics Project, provided the impetus that got me started down this path. Early development of the materials was supported by the National Science Foundation as the Physics for the Year 2000 project; their support is gratefully acknowledged.

I especially want to thank my editors, Jeanne Zalesky and Becky Ruden, development editor Alice Houston, project manager Martha Steele, art development editors Kim Brucker and Margot Otway, and all the other staff at Pearson for their enthusiasm and hard work on this project. Rose Kernan and the team at Cenveo along with photo researcher Eric Schrader get a good deal of the credit for making this complex project all come together. Larry Smith, Larry Stookey, and Michael Ottinger have done an outstanding job of checking the solutions to every end-of-chapter problem and updating the Instructor's Solutions Manual. John Filaseta must be thanked for carefully writing out the solutions to the Student Workbook exercises, and Jason Harlow for putting together the Lecture Slides. In addition to the reviewers and classroom testers listed below, who gave invaluable feedback, I am particularly grateful to Charlie Hibbard for his close scrutiny of every word and figure.

Finally, I am endlessly grateful to my wife Sally for her love, encouragement, and patience, and to our many cats, past and present, who are always ready to suggest "Dinner time?" when I'm in need of a phrase.

Randy Knight, September 2015
rknight@calpoly.edu

Reviewers and Classroom Testers

Gary B. Adams, *Arizona State University*
Ed Adelson, *Ohio State University*
Kyle Altmann, *Elon University*
Wayne R. Anderson, *Sacramento City College*
James H. Andrews, *Youngstown State University*
Kevin Ankoviak, *Las Positas College*
David Balogh, *Fresno City College*
Dewayne Beery, *Buffalo State College*
Joseph Bellina, *Saint Mary's College*
James R. Benbrook, *University of Houston*
David Besson, *University of Kansas*
Matthew Block, *California State University, Sacramento*
Randy Bohn, *University of Toledo*
Richard A. Bone, *Florida International University*
Gregory Boutis, *York College*
Art Braundmeier, *University of Southern Illinois, Edwardsville*
Carl Bromberg, *Michigan State University*
Meade Brooks, *Collin College*
Douglas Brown, *Cabrillo College*
Ronald Brown, *California Polytechnic State University, San Luis Obispo*
Mike Broyles, *Collin County Community College*
Debra Burris, *University of Central Arkansas*
James Carolan, *University of British Columbia*
Michael Chapman, *Georgia Tech University*
Norbert Chencinski, *College of Staten Island*
Tonya Coffey, *Appalachian State University*
Kristi Concannon, *King's College*
Desmond Cook, *Old Dominion University*
Sean Cordry, *Northwestern College of Iowa*
Robert L. Corey, *South Dakota School of Mines*
Michael Crescimanno, *Youngstown State University*
Dennis Crossley, *University of Wisconsin–Sheboygan*
Wei Cui, *Purdue University*
Robert J. Culbertson, *Arizona State University*
Danielle Dalafave, *The College of New Jersey*
Purna C. Das, *Purdue University North Central*
Chad Davies, *Gordon College*
William DeGraffenreid, *California State University–Sacramento*
Dwain Desbien, *Estrella Mountain Community College*
John F. Devlin, *University of Michigan, Dearborn*
John DiBartolo, *Polytechnic University*
Alex Dickison, *Seminole Community College*
Chaden Djalali, *University of South Carolina*
Margaret Dobrowolska, *University of Notre Dame*
Sandra Doty, *Denison University*
Miles J. Dresser, *Washington State University*
Taner Edis, *Truman State University*
Charlotte Elster, *Ohio University*
Robert J. Endorf, *University of Cincinnati*
Tilahun Eneyew, *Embry-Riddle Aeronautical University*
F. Paul Esposito, *University of Cincinnati*
John Evans, *Lee University*
Harold T. Evensen, *University of Wisconsin–Platteville*
Michael R. Falvo, *University of North Carolina*
Abbas Faridi, *Orange Coast College*
Nail Fazleev, *University of Texas–Arlington*
Stuart Field, *Colorado State University*
Daniel Finley, *University of New Mexico*
Jane D. Flood, *Muhlenberg College*
Michael Franklin, *Northwestern Michigan College*

Jonathan Friedman, *Amherst College*

Thomas Furtak, *Colorado School of Mines*

Alina Gabryszewska-Kukawa, *Delta State University*

Lev Gasparov, *University of North Florida*

Richard Gass, *University of Cincinnati*

Delena Gatch, *Georgia Southern University*

J. David Gavenda, *University of Texas, Austin*

Stuart Gazes, *University of Chicago*

Katherine M. Gietzen, *Southwest Missouri State University*

Robert Glosser, *University of Texas, Dallas*

William Golightly, *University of California, Berkeley*

Paul Gresser, *University of Maryland*

C. Frank Griffin, *University of Akron*

John B. Gruber, *San Jose State University*

Thomas D. Gutierrez, *California Polytechnic State University, San Luis Obispo*

Stephen Haas, *University of Southern California*

John Hamilton, *University of Hawaii at Hilo*

Jason Harlow, *University of Toronto*

Randy Harris, *University of California, Davis*

Nathan Harshman, *American University*

J. E. Hasbun, *University of West Georgia*

Nicole Herbots, *Arizona State University*

Jim Hetrick, *University of Michigan–Dearborn*

Scott Hildreth, *Chabot College*

David Hobbs, *South Plains College*

Laurent Hodges, *Iowa State University*

Mark Hollabaugh, *Normandale Community College*

Steven Hubbard, *Lorain County Community College*

John L. Hubisz, *North Carolina State University*

Shane Hutson, *Vanderbilt University*

George Igo, *University of California, Los Angeles*

David C. Ingram, *Ohio University*

Bob Jacobsen, *University of California, Berkeley*

Rong-Sheng Jin, *Florida Institute of Technology*

Marty Johnston, *University of St. Thomas*

Stanley T. Jones, *University of Alabama*

Darrell Judge, *University of Southern California*

Pawan Kahol, *Missouri State University*

Teruki Kamon, *Texas A&M University*

Richard Karas, *California State University, San Marcos*

Deborah Katz, *U.S. Naval Academy*

Miron Kaufman, *Cleveland State University*

Katherine Keilty, *Kingwood College*

Roman Kezerashvili, *New York City College of Technology*

Peter Kjeer, *Bethany Lutheran College*

M. Kotlarchyk, *Rochester Institute of Technology*

Fred Krauss, *Delta College*

Cagliyan Kurdak, *University of Michigan*

Fred Kuttner, *University of California, Santa Cruz*

H. Sarma Lakkaraju, *San Jose State University*

Darrell R. Lamm, *Georgia Institute of Technology*

Robert LaMontagne, *Providence College*

Eric T. Lane, *University of Tennessee–Chattanooga*

Alessandra Lanzara, *University of California, Berkeley*

Lee H. LaRue, *Paris Junior College*

Sen-Ben Liao, *Massachusetts Institute of Technology*

Dean Livelybrooks, *University of Oregon*

Chun-Min Lo, *University of South Florida*

Olga Lobban, *Saint Mary's University*

Ramon Lopez, *Florida Institute of Technology*

Vaman M. Naik, *University of Michigan, Dearborn*

Kevin Mackay, *Grove City College*

Carl Maes, *University of Arizona*

Rizwan Mahmood, *Slippery Rock University*

Mani Manivannan, *Missouri State University*

Mark E. Mattson, *James Madison University*

Richard McCorkle, *University of Rhode Island*

James McDonald, *University of Hartford*

James McGuire, *Tulane University*

Stephen R. McNeil, *Brigham Young University–Idaho*

Theresa Moreau, *Amherst College*

Gary Morris, *Rice University*

Michael A. Morrison, *University of Oklahoma*

Richard Mowat, *North Carolina State University*

Eric Murray, *Georgia Institute of Technology*

Michael Murray, *University of Kansas*

Taha Mzoughi, *Mississippi State University*

Scott Nutter, *Northern Kentucky University*

Craig Ogilvie, *Iowa State University*

Benedict Y. Oh, *University of Wisconsin*

Martin Okafor, *Georgia Perimeter College*

Halina Opyrchal, *New Jersey Institute of Technology*

Derek Padilla, *Santa Rosa Junior College*

Yibin Pan, *University of Wisconsin–Madison*

Georgia Papaefthymiou, *Villanova University*

Peggy Perozzo, *Mary Baldwin College*

Brian K. Pickett, *Purdue University, Calumet*

Joe Pifer, *Rutgers University*

Dale Pleticha, *Gordon College*

Marie Plumb, *Jamestown Community College*

Robert Pompi, *SUNY-Binghamton*

David Potter, *Austin Community College–Rio Grande Campus*

Chandra Prayaga, *University of West Florida*

Kenneth M. Purcell, *University of Southern Indiana*

Didarul Qadir, *Central Michigan University*

Steve Quon, *Ventura College*

Michael Read, *College of the Siskiyous*

Lawrence Rees, *Brigham Young University*

Richard J. Reimann, *Boise State University*

Michael Rodman, *Spokane Falls Community College*

Sharon Rosell, *Central Washington University*

Anthony Russo, *Northwest Florida State College*

Freddie Salsbury, *Wake Forest University*

Otto F. Sankey, *Arizona State University*

Jeff Sanny, *Loyola Marymount University*

Rachel E. Scherr, *University of Maryland*

Carl Schneider, *U.S. Naval Academy*

Bruce Schumm, *University of California, Santa Cruz*
Bartlett M. Sheinberg, *Houston Community College*
Douglas Sherman, *San Jose State University*
Elizabeth H. Simmons, *Boston University*
Marlina Slamet, *Sacred Heart University*
Alan Slavin, *Trent College*
Alexander Raymond Small, *California State Polytechnic University, Pomona*
Larry Smith, *Snow College*
William S. Smith, *Boise State University*
Paul Sokol, *Pennsylvania State University*
LTC Bryndol Sones, *United States Military Academy*
Chris Sorensen, *Kansas State University*
Brian Steinkamp, *University of Southern Indiana*
Anna and Ivan Stern, *AW Tutor Center*
Gay B. Stewart, *University of Arkansas*
Michael Strauss, *University of Oklahoma*
Chin-Che Tin, *Auburn University*
Christos Valiotis, *Antelope Valley College*
Andrew Vanture, *Everett Community College*
Arthur Viescas, *Pennsylvania State University*
Ernst D. Von Meerwall, *University of Akron*
Chris Vuille, *Embry-Riddle Aeronautical University*
Jerry Wagner, *Rochester Institute of Technology*
Robert Webb, *Texas A&M University*
Zodiac Webster, *California State University, San Bernardino*
Robert Weidman, *Michigan Technical University*
Fred Weitfeldt, *Tulane University*
Gary Williams, *University of California, Los Angeles*
Lynda Williams, *Santa Rosa Junior College*
Jeff Allen Winger, *Mississippi State University*
Carey Witkov, *Broward Community College*
Ronald Zammit, *California Polytechnic State University, San Luis Obispo*
Darin T. Zimmerman, *Pennsylvania State University, Altoona*
Fredy Zypman, *Yeshiva University*

Student Focus Groups

California Polytechnic State University, San Luis Obispo
Matthew Bailey
James Caudill
Andres Gonzalez
Mytch Johnson

California State University, Sacramento
Logan Allen
Andrew Fujikawa
Sagar Gupta
Marlene Juarez
Craig Kovac
Alissa McKown
Kenneth Mercado
Douglas Ostheimer
Ian Tabbada
James Womack

Santa Rosa Junior College
Kyle Doughty
Tacho Gardiner
Erik Gonzalez
Joseph Goodwin
Chelsea Irmer
Vatsal Pandya
Parth Parikh
Andrew Prosser
David Reynolds
Brian Swain
Grace Woods

Stanford University
Montserrat Cordero
Rylan Edlin
Michael Goodemote II
Stewart Isaacs
David LaFehr
Sergio Rebeles
Jack Takahashi

Preface to the Student

From Me to You

The most incomprehensible thing about the universe is that it is comprehensible.
 —Albert Einstein

The day I went into physics class it was death.
 —Sylvia Plath, *The Bell Jar*

Let's have a little chat before we start. A rather one-sided chat, admittedly, because you can't respond, but that's OK. I've talked with many of your fellow students over the years, so I have a pretty good idea of what's on your mind.

What's your reaction to taking physics? Fear and loathing? Uncertainty? Excitement? All the above? Let's face it, physics has a bit of an image problem on campus. You've probably heard that it's difficult, maybe impossible unless you're an Einstein. Things that you've heard, your experiences in other science courses, and many other factors all color your *expectations* about what this course is going to be like.

It's true that there are many new ideas to be learned in physics and that the course, like college courses in general, is going to be much faster paced than science courses you had in high school. I think it's fair to say that it will be an *intense* course. But we can avoid many potential problems and difficulties if we can establish, here at the beginning, what this course is about and what is expected of you—and of me!

Just what is physics, anyway? Physics is a way of thinking about the physical aspects of nature. Physics is not better than art or biology or poetry or religion, which are also ways to think about nature; it's simply different. One of the things this course will emphasize is that physics is a human endeavor. The ideas presented in this book were not found in a cave or conveyed to us by aliens; they were discovered and developed by real people engaged in a struggle with real issues.

You might be surprised to hear that physics is not about "facts." Oh, not that facts are unimportant, but physics is far more focused on discovering *relationships* and *patterns* than on learning facts for their own sake.

 For example, the colors of the rainbow appear both when white light passes through a prism and—as in this photo—when white light reflects from a thin film of oil on water. What does this pattern tell us about the nature of light?

Our emphasis on relationships and patterns means that there's not a lot of memorization when you study physics. Some—there are still definitions and equations to learn—but less than in many other courses. Our emphasis, instead, will be on thinking and reasoning. This is important to factor into your expectations for the course.

Perhaps most important of all, *physics is not math!* Physics is much broader. We're going to look for patterns and relationships in nature, develop the logic that relates different ideas, and search for the reasons *why* things happen as they do. In doing so, we're going to stress qualitative reasoning, pictorial and graphical reasoning, and reasoning by analogy. And yes, we will use math, but it's just one tool among many.

It will save you much frustration if you're aware of this physics–math distinction up front. Many of you, I know, want to find a formula and plug numbers into it—that is, to do a math problem. Maybe that worked in high school science courses, but it is *not* what this course expects of you. We'll certainly do many calculations, but the specific numbers are usually the last and least important step in the analysis.

As you study, you'll sometimes be baffled, puzzled, and confused. That's perfectly normal and to be expected. Making mistakes is OK too *if* you're willing to learn from the experience. No one is born knowing how to do physics any more than he or she is born knowing how to play the piano or shoot basketballs. The ability to do physics comes from practice, repetition, and struggling with the ideas until you "own" them and can apply them yourself in new situations. There's no way to make learning effortless, at least for anything worth learning, so expect to have some difficult moments ahead. But also expect to have some moments of excitement at the joy of discovery. There will be instants at which the pieces suddenly click into place and you *know* that you understand a powerful idea. There will be times when you'll surprise yourself by successfully working a difficult problem that you didn't think you could solve. My hope, as an author, is that the excitement and sense of adventure will far outweigh the difficulties and frustrations.

Getting the Most Out of Your Course

Many of you, I suspect, would like to know the "best" way to study for this course. There is no best way. People are different, and what works for one student is less effective for another. But I do want to stress that *reading the text* is vitally important. The basic knowledge for this course is written down on these pages, and your instructor's *number-one expectation* is that you will read carefully to find and learn that knowledge.

Despite there being no best way to study, I will suggest *one* way that is successful for many students.

1. **Read each chapter *before* it is discussed in class.** I cannot stress too strongly how important this step is. Class attendance is much more effective if you are prepared. When you first read a chapter, focus on learning new vocabulary, definitions, and notation. There's a list of terms and notations at the end of each chapter. Learn them! You won't understand

what's being discussed or how the ideas are being used if you don't know what the terms and symbols mean.

2. **Participate actively in class.** Take notes, ask and answer questions, and participate in discussion groups. There is ample scientific evidence that *active participation* is much more effective for learning science than passive listening.

3. **After class, go back for a careful re-reading of the chapter.** In your second reading, pay closer attention to the details and the worked examples. Look for the *logic* behind each example (I've highlighted this to make it clear), not just at what formula is being used. And use the textbook tools that are designed to help your learning, such as the problem-solving strategies, the chapter summaries, and the exercises in the *Student Workbook*.

4. **Finally, apply what you have learned to the homework problems at the end of each chapter.** I strongly encourage you to form a study group with two or three classmates. There's good evidence that students who study regularly with a group do better than the rugged individualists who try to go it alone.

Did someone mention a workbook? The companion *Student Workbook* is a vital part of the course. Its questions and exercises ask you to reason *qualitatively,* to use graphical information, and to give explanations. It is through these exercises that you will learn what the concepts mean and will practice the reasoning skills appropriate to the chapter. You will then have acquired the baseline knowledge and confidence you need *before* turning to the end-of-chapter homework problems. In sports or in music, you would never think of performing before you practice, so why would you want to do so in physics? The workbook is where you practice and work on basic skills.

Many of you, I know, will be tempted to go straight to the homework problems and then thumb through the text looking for a formula that seems like it will work. That approach will not succeed in this course, and it's guaranteed to make you frustrated and discouraged. Very few homework problems are of the "plug and chug" variety where you simply put numbers into a formula. To work the homework problems successfully, you need a better study strategy—either the one outlined above or your own—that helps you learn the concepts and the relationships between the ideas.

Getting the Most Out of Your Textbook

Your textbook provides many features designed to help you learn the concepts of physics and solve problems more effectively.

- ■ **TACTICS BOXES** give step-by-step procedures for particular skills, such as interpreting graphs or drawing special diagrams. Tactics Box steps are explicitly illustrated in subsequent worked examples, and these are often the starting point of a full *Problem-Solving Strategy*.

- ■ **PROBLEM-SOLVING STRATEGIES** are provided for each broad class of problems—problems characteristic of a chapter or group of chapters. The strategies follow a consistent four-step approach to help you develop confidence and proficient problem-solving skills: **MODEL, VISUALIZE, SOLVE, ASSESS**.

- ■ Worked **EXAMPLES** illustrate good problem-solving practices through the consistent use of the four-step problem-solving approach The worked examples are often very detailed and carefully lead you through the *reasoning* behind the solution as well as the numerical calculations.

- ■ **STOP TO THINK** questions embedded in the chapter allow you to quickly assess whether you've understood the main idea of a section. A correct answer will give you confidence to move on to the next section. An incorrect answer will alert you to re-read the previous section.

- ■ Blue annotations on figures help you better understand what the figure is showing. They will help you to interpret graphs; translate between graphs, math, and pictures; grasp difficult concepts through a visual analogy; and develop many other important skills.

The current in a wire is the same at all points.

I = constant

- ■ Schematic *Chapter Summaries* help you organize what you have learned into a hierarchy, from general principles (top) to applications (bottom). Side-by-side pictorial, graphical, textual, and mathematical representations are used to help you translate between these key representations.

- ■ Each part of the book ends with a **KNOWLEDGE STRUCTURE** designed to help you see the forest rather than just the trees.

Now that you know more about what is expected of you, what can you expect of me? That's a little trickier because the book is already written! Nonetheless, the book was prepared on the basis of what I think my students throughout the years have expected—and wanted—from their physics textbook. Further, I've listened to the extensive feedback I have received from thousands of students like you, and their instructors, who used the first three editions of this book.

You should know that these course materials—the text and the workbook—are based on extensive research about how students learn physics and the challenges they face. The effectiveness of many of the exercises has been demonstrated through extensive class testing. I've written the book in an informal style that I hope you will find appealing and that will encourage you to do the reading. And, finally, I have endeavored to make clear not only that physics, as a technical body of knowledge, is relevant to your profession but also that physics is an exciting adventure of the human mind.

I hope you'll enjoy the time we're going to spend together.

Detailed Contents

This text is the Standard edition and contains Chapters 1–36.

Part II   Conservation Laws

Part III   Applications of Newtonian Mechanics

Part IV Oscillations and Waves

Part V Thermodynamics

Part VI   Electricity and Magnetism

Part VII   Optics

Part VIII   Relativity and Quantum Physics

OVERVIEW

Why Things Move

Each of the seven parts of this book opens with an overview to give you a look ahead, a glimpse at where your journey will take you in the next few chapters. It's easy to lose sight of the big picture while you're busy negotiating the terrain of each chapter. In Part I, the big picture, in a word, is *motion*.

There are two big questions we must tackle:

- **How do we describe motion?** It is easy to say that an object moves, but it's not obvious how we should measure or characterize the motion if we want to analyze it mathematically. The mathematical description of motion is called *kinematics,* and it is the subject matter of Chapters 1 through 4.
- **How do we explain motion?** Why do objects have the particular motion they do? Why, when you toss a ball upward, does it go up and then come back down rather than keep going up? Are there "laws of nature" that allow us to predict an object's motion? The explanation of motion in terms of its causes is called *dynamics,* and it is the topic of Chapters 5 through 8.

Two key ideas for answering these questions are *force* (the "cause") and *acceleration* (the "effect"). A variety of pictorial and graphical tools will be developed in Chapters 1 through 5 to help you develop an *intuition* for the connection between force and acceleration. You'll then put this knowledge to use in Chapters 5 through 8 as you analyze motion of increasing complexity.

Another important tool will be the use of *models*. Reality is extremely complicated. We would never be able to develop a science if we had to keep track of every little detail of every situation. A model is a simplified description of reality—much as a model airplane is a simplified version of a real airplane—used to reduce the complexity of a problem to the point where it can be analyzed and understood. We will introduce several important models of motion, paying close attention, especially in these earlier chapters, to where simplifying assumptions are being made, and why.

The "laws of motion" were discovered by Isaac Newton roughly 350 years ago, so the study of motion is hardly cutting-edge science. Nonetheless, it is still extremely important. Mechanics—the science of motion—is the basis for much of engineering and applied science, and many of the ideas introduced here will be needed later to understand things like the motion of waves and the motion of electrons through circuits. Newton's mechanics is the foundation of much of contemporary science, thus we will start at the beginning.

Motion can be slow and steady, or fast and sudden. This rocket, with its rapid acceleration, is responding to forces exerted on it by thrust, gravity, and the air.

1 Concepts of Motion

Motion takes many forms. The ski jumper seen here is an example of translational motion.

IN THIS CHAPTER, you will learn the fundamental concepts of motion.

What is a chapter preview?

Each chapter starts with an overview. Think of it as a roadmap to help you get oriented and make the most of your studying.

«LOOKING BACK A Looking Back reference tells you what material from previous chapters is especially important for understanding the new topics. A quick review will help your learning. You will find additional Looking Back references within the chapter, right at the point they're needed.

What is motion?

Before solving motion problems, we must learn to *describe* motion. We will use

- Motion diagrams
- Graphs
- Pictures

Motion concepts introduced in this chapter include position, velocity, and acceleration.

Known
$x_0 = v_{0x} = t_0 = 0$
$a_x = 2.0 \text{ m/s}^2$
Find
x_1

Why do we need vectors?

Many of the quantities used to describe motion, such as velocity, have both a size and a direction. We use vectors to represent these quantities. This chapter introduces graphical techniques to add and subtract vectors. Chapter 3 will explore vectors in more detail.

Why are units and significant figures important?

Scientists and engineers must communicate their ideas to others. To do so, we have to agree about the *units* in which quantities are measured. In physics we use metric units, called SI units. We also need rules for telling others how accurately a quantity is known. You will learn the rules for using significant figures correctly.

$0.00620 = \boxed{6.20} \times 10^{-3}$

Why is motion important?

The universe is in motion, from the smallest scale of electrons and atoms to the largest scale of entire galaxies. We'll start with the motion of everyday objects, such as cars and balls and people. Later we'll study the motions of waves, of atoms in gases, and of electrons in circuits. Motion is the one theme that will be with us from the first chapter to the last.

1.1 Motion Diagrams

Motion is a theme that will appear in one form or another throughout this entire book. Although we all have intuition about motion, based on our experiences, some of the important aspects of motion turn out to be rather subtle. So rather than jumping immediately into a lot of mathematics and calculations, this first chapter focuses on *visualizing* motion and becoming familiar with the *concepts* needed to describe a moving object. Our goal is to lay the foundations for understanding motion.

FIGURE 1.1 Four basic types of motion.

Linear motion

Circular motion

Projectile motion

Rotational motion

To begin, let's define **motion** as the change of an object's position with time. **FIGURE 1.1** shows four basic types of motion that we will study in this book. The first three—linear, circular, and projectile motion—in which the object moves through space are called **translational motion.** The path along which the object moves, whether straight or curved, is called the object's **trajectory.** Rotational motion is somewhat different because there's movement but the object as a whole doesn't change position. We'll defer rotational motion until later and, for now, focus on translational motion.

Making a Motion Diagram

An easy way to study motion is to make a video of a moving object. A video camera, as you probably know, takes images at a fixed rate, typically 30 every second. Each separate image is called a *frame.* As an example, **FIGURE 1.2** shows four frames from a video of a car going past. Not surprisingly, the car is in a somewhat different position in each frame.

Suppose we edit the video by layering the frames on top of each other, creating the composite image shown in **FIGURE 1.3**. This edited image, showing an object's position at several *equally spaced instants of time,* is called a **motion diagram.** As the examples below show, we can define concepts such as constant speed, speeding up, and slowing down in terms of how an object appears in a motion diagram.

NOTE It's important to keep the camera in a *fixed position* as the object moves by. Don't "pan" it to track the moving object.

FIGURE 1.2 Four frames from a video.

FIGURE 1.3 A motion diagram of the car shows all the frames simultaneously.

The same amount of time elapses between each image and the next.

Examples of motion diagrams

Images that are *equally spaced* indicate an object moving with *constant speed*.

An *increasing distance* between the images shows that the object is *speeding up*.

A *decreasing distance* between the images shows that the object is *slowing down*.

STOP TO THINK 1.1 Which car is going faster, A or B? Assume there are equal intervals of time between the frames of both videos.

Car A   Car B

NOTE Each chapter will have several *Stop to Think* questions. These questions are designed to see if you've understood the basic ideas that have been presented. The answers are given at the end of the book, but you should make a serious effort to think about these questions before turning to the answers.

1.2 Models and Modeling

The real world is messy and complicated. Our goal in physics is to brush aside many of the real-world details in order to discern patterns that occur over and over. For example, a swinging pendulum, a vibrating guitar string, a sound wave, and jiggling atoms in a crystal are all very different—yet perhaps not so different. Each is an example of a system moving back and forth around an equilibrium position. If we focus on understanding a very simple oscillating system, such as a mass on a spring, we'll automatically understand quite a bit about the many real-world manifestations of oscillations.

Stripping away the details to focus on essential features is a process called *modeling*. A **model** is a highly simplified picture of reality, but one that still captures the essence of what we want to study. Thus "mass on a spring" is a simple but realistic model of almost all oscillating systems.

Models allow us to make sense of complex situations by providing a framework for thinking about them. One could go so far as to say that developing and testing models is at the heart of the scientific process. Albert Einstein once said, "Physics should be as simple as possible—but not simpler." We want to find the simplest model that allows us to understand the phenomenon we're studying, but we can't make the model so simple that key aspects of the phenomenon get lost.

We'll develop and use many models throughout this textbook; they'll be one of our most important thinking tools. These models will be of two types:

- *Descriptive models:* What are the essential characteristics and properties of a phenomenon? How do we describe it in the simplest possible terms? For example, the mass-on-a-spring model of an oscillating system is a descriptive model.
- *Explanatory models:* Why do things happen as they do? Explanatory models, based on the laws of physics, have predictive power, allowing us to test—against experimental data—whether a model provides an adequate explanation of our observations.

The Particle Model

For many types of motion, such as that of balls, cars, and rockets, the motion of the object *as a whole* is not influenced by the details of the object's size and shape. All we really need to keep track of is the motion of a single point on the object, so we can treat the object *as if* all its mass were concentrated into this single point. An object that can be represented as a mass at a single point in space is called a **particle.** A particle has no size, no shape, and no distinction between top and bottom or between front and back.

If we model an object as a particle, we can represent the object in each frame of a motion diagram as a simple dot rather than having to draw a full picture. **FIGURE 1.4** shows how much simpler motion diagrams appear when the object is represented as a particle. Note that the dots have been numbered 0, 1, 2, . . . to tell the sequence in which the frames were exposed.

We can model an airplane's takeoff as a particle (a descriptive model) undergoing constant acceleration (a descriptive model) in response to constant forces (an explanatory model).

FIGURE 1.4 Motion diagrams in which the object is modeled as a particle.

(a) Motion diagram of a rocket launch

Numbers show the order in which the frames were exposed.

(b) Motion diagram of a car stopping

The same amount of time elapses between each image and the next.

Treating an object as a particle is, of course, a simplification of reality—but that's what modeling is all about. The **particle model** of motion is a simplification in which we treat a moving object as if all of its mass were concentrated at a single point. The particle model is an excellent approximation of reality for the translational motion of cars, planes, rockets, and similar objects.

Of course, not everything can be modeled as a particle; models have their limits. Consider, for example, a rotating gear. The center doesn't move at all while each tooth is moving in a different direction. We'll need to develop new models when we get to new types of motion, but the particle model will serve us well throughout Part I of this book.

STOP TO THINK 1.2 Three motion diagrams are shown. Which is a dust particle settling to the floor at constant speed, which is a ball dropped from the roof of a building, and which is a descending rocket slowing to make a soft landing on Mars?

(a) 0 ● 1 ● 2 ● 3 ● 4 ● 5 ●

(b) 0 ● 1 ● 2 ● 3 ● 4 ● 5 ●

(c) 0 ● 1 ● 2 ● 3 ● 4 ● 5 ●

1.3 Position, Time, and Displacement

To use a motion diagram, you would like to know *where* the object is (i.e., its *position*) and *when* the object was at that position (i.e., the *time*). Position measurements can be made by laying a coordinate-system grid over a motion diagram. You can then measure the (x, y) coordinates of each point in the motion diagram. Of course, the world does not come with a coordinate system attached. A coordinate system is an artificial grid that *you* place over a problem in order to analyze the motion. You place the origin of your coordinate system wherever you wish, and different observers of a moving object might all choose to use different origins.

Time, in a sense, is also a coordinate system, although you may never have thought of time this way. You can pick an arbitrary point in the motion and label it "$t = 0$ seconds." This is simply the instant you decide to start your clock or stopwatch, so it is the origin of your time coordinate. Different observers might choose to start their clocks at different moments. A video frame labeled "$t = 4$ seconds" was taken 4 seconds after you started your clock.

We typically choose $t = 0$ to represent the "beginning" of a problem, but the object may have been moving before then. Those earlier instants would be measured as negative times, just as objects on the x-axis to the left of the origin have negative values of position. Negative numbers are not to be avoided; they simply locate an event in space or time *relative to an origin*.

To illustrate, **FIGURE 1.5a** shows a sled sliding down a snow-covered hill. **FIGURE 1.5b** is a motion diagram for the sled, over which we've drawn an xy-coordinate system. You can see that the sled's position is $(x_3, y_3) = (15 \text{ m}, 15 \text{ m})$ at time $t_3 = 3$ s. Notice how we've used subscripts to indicate the time and the object's position in a specific frame of the motion diagram.

NOTE The frame at $t = 0$ is frame 0. That is why the fourth frame is labeled 3.

Another way to locate the sled is to draw its **position vector:** an arrow from the origin to the point representing the sled. The position vector is given the symbol $\vec{r}$. Figure 1.5b shows the position vector $\vec{r}_3 = (21 \text{ m}, 45°)$. The position vector $\vec{r}$ does not tell us anything different than the coordinates (x, y). It simply provides the information in an alternative form.

FIGURE 1.5 Motion diagram of a sled with frames made every 1 s.

(a)

(b)

The sled's position in frame 3 can be specified with coordinates.

$t_3 = 3$ s
$(x_3, y_3) = (15 \text{ m}, 15 \text{ m})$

$\vec{r}_3 = (21 \text{ m}, 45°)$

$45°$

Alternatively, the position can be specified by the position vector.

Scalars and Vectors

Some physical quantities, such as time, mass, and temperature, can be described completely by a single number with a unit. For example, the mass of an object is 6 kg and its temperature is 30°C. A single number (with a unit) that describes a physical quantity is called a **scalar.** A scalar can be positive, negative, or zero.

Many other quantities, however, have a directional aspect and cannot be described by a single number. To describe the motion of a car, for example, you must specify not only how fast it is moving, but also the *direction* in which it is moving. A quantity having both a *size* (the "How far?" or "How fast?") and a *direction* (the "Which way?") is called a **vector.** The size or length of a vector is called its *magnitude*. Vectors will be studied thoroughly in Chapter 3, so all we need for now is a little basic information.

We indicate a vector by drawing an arrow over the letter that represents the quantity. Thus $\vec{r}$ and $\vec{A}$ are symbols for vectors, whereas r and A, without the arrows, are symbols for scalars. In handwritten work you must draw arrows over all symbols that represent vectors. This may seem strange until you get used to it, but it is very important because we will often use both r and $\vec{r}$, or both A and $\vec{A}$, in the same problem, and they mean different things! Note that the arrow over the symbol always points to the right, regardless of which direction the actual vector points. Thus we write $\vec{r}$ or $\vec{A}$, never $\overleftarrow{r}$ or $\overleftarrow{A}$.

Displacement

We said that motion is the change in an object's position with time, but how do we show a change of position? A motion diagram is the perfect tool. **FIGURE 1.6** is the motion diagram of a sled sliding down a snow-covered hill. To show how the sled's position changes between, say, $t_3 = 3$ s and $t_4 = 4$ s, we draw a vector arrow between the two dots of the motion diagram. This vector is the sled's **displacement,** which is given the symbol $\Delta\vec{r}$. The Greek letter delta (Δ) is used in math and science to indicate the *change* in a quantity. In this case, as we'll show, the displacement $\Delta\vec{r}$ is the change in an object's position.

NOTE $\Delta\vec{r}$ is a *single* symbol. You cannot cancel out or remove the Δ.

Notice how the sled's position vector $\vec{r}_4$ is a combination of its early position $\vec{r}_3$ with the displacement vector $\Delta\vec{r}$. In fact, $\vec{r}_4$ is the *vector sum* of the vectors $\vec{r}_3$ and $\Delta\vec{r}$. This is written

$$\vec{r}_4 = \vec{r}_3 + \Delta\vec{r} \tag{1.1}$$

Here we're adding vector quantities, not numbers, and vector addition differs from "regular" addition. We'll explore vector addition more thoroughly in Chapter 3, but for now you can add two vectors $\vec{A}$ and $\vec{B}$ with the three-step procedure shown in Tactics Box 1.1.

FIGURE 1.6 The sled undergoes a displacement $\Delta\vec{r}$ from position $\vec{r}_3$ to position $\vec{r}_4$.

The sled's displacement between $t_3 = 3$ s and $t_4 = 4$ s is the vector drawn from one postion to the next.

TACTICS BOX 1.1

Vector addition

To add $\vec{B}$ to $\vec{A}$:

❶ Draw $\vec{A}$.

❷ Place the tail of $\vec{B}$ at the tip of $\vec{A}$.

❸ Draw an arrow from the tail of $\vec{A}$ to the tip of $\vec{B}$. This is vector $\vec{A} + \vec{B}$.

If you examine Figure 1.6, you'll see that the steps of Tactics Box 1.1 are exactly how $\vec{r}_3$ and $\Delta\vec{r}$ are added to give $\vec{r}_4$.

NOTE A vector is not tied to a particular location on the page. You can move a vector around as long as you don't change its length or the direction it points. Vector $\vec{B}$ is not changed by sliding it to where its tail is at the tip of $\vec{A}$.

Equation 1.1 told us that $\vec{r}_4 = \vec{r}_3 + \Delta\vec{r}$. This is easily rearranged to give a more precise definition of displacement: **The displacement $\Delta\vec{r}$ of an object as it moves from an initial position $\vec{r}_i$ to a final position $\vec{r}_f$ is**

$$\Delta\vec{r} = \vec{r}_f - \vec{r}_i \qquad (1.2)$$

Graphically, $\Delta\vec{r}$ is a vector arrow drawn from position $\vec{r}_i$ to position $\vec{r}_f$.

NOTE To be more general, we've written Equation 1.2 in terms of an *initial position* and a *final position*, indicated by subscripts i and f. We'll frequently use i and f when writing general equations, then use specific numbers or values, such as 3 and 4, when working a problem.

This definition of $\Delta\vec{r}$ involves *vector subtraction*. With numbers, subtraction is the same as the addition of a negative number. That is, $5 - 3$ is the same as $5 + (-3)$. Similarly, we can use the rules for vector addition to find $\vec{A} - \vec{B} = \vec{A} + (-\vec{B})$ if we first define what we mean by $-\vec{B}$. As **FIGURE 1.7** shows, the negative of vector $\vec{B}$ is a vector with the same length but pointing in the opposite direction. This makes sense because $\vec{B} - \vec{B} = \vec{B} + (-\vec{B}) = \vec{0}$, where $\vec{0}$, a vector with zero length, is called the **zero vector.**

FIGURE 1.7 The negative of a vector.

Vector $-\vec{B}$ has the same length as $\vec{B}$ but points in the opposite direction.

$\vec{B} + (-\vec{B}) = \vec{0}$ because the sum returns to the starting point.

The zero vector $\vec{0}$ has no length.

TACTICS BOX 1.2 (MP)

Vector subtraction

To subtract $\vec{B}$ from $\vec{A}$:

❶ Draw $\vec{A}$.

❷ Place the tail of $-\vec{B}$ at the tip of $\vec{A}$.

❸ Draw an arrow from the tail of $\vec{A}$ to the tip of $-\vec{B}$. This is vector $\vec{A} - \vec{B}$.

FIGURE 1.8 uses the vector subtraction rules of Tactics Box 1.2 to prove that the displacement $\Delta\vec{r}$ is simply the vector connecting the dots of a motion diagram.

▼ **FIGURE 1.8** Using vector subtraction to find $\Delta\vec{r} = \vec{r}_f - \vec{r}_i$.

(a) Initial and final position vectors

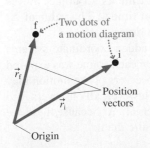

Two dots of a motion diagram

Position vectors

Origin

(b) Procedure for finding the particle's displacement vector $\Delta\vec{r}$

❶ Draw $\vec{r}_f$.

❷ Draw $-\vec{r}_i$ at the tip of $\vec{r}_f$.

❸ Draw $\vec{r}_f - \vec{r}_i$. This is $\Delta\vec{r}$.

Finally, slide $\Delta\vec{r}$ back to the motion diagram. It is a vector from dot i to dot f.

FIGURE 1.9 Motion diagrams with the displacement vectors.

(a) Rocket launch

$\Delta \vec{r}_3$

$\Delta \vec{r}_2$

$\Delta \vec{r}_1$

Start $\Delta \vec{r}_0$

(b) Car stopping

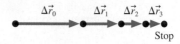

$\Delta \vec{r}_0$ $\Delta \vec{r}_1$ $\Delta \vec{r}_2$ $\Delta \vec{r}_3$

Stop

Motion Diagrams with Displacement Vectors

The first step in analyzing a motion diagram is to determine all of the displacement vectors. As Figure 1.8 shows, the displacement vectors are simply the arrows connecting each dot to the next. Label each arrow with a *vector* symbol $\Delta \vec{r}_n$, starting with $n = 0$. **FIGURE 1.9** shows the motion diagrams of Figure 1.4 redrawn to include the displacement vectors. You do not need to show the position vectors.

NOTE When an object either starts from rest or ends at rest, the initial or final dots are *as close together* as you can draw the displacement vector arrow connecting them. In addition, just to be clear, you should write "Start" or "Stop" beside the initial or final dot. It is important to distinguish stopping from merely slowing down.

Now we can conclude, more precisely than before, that, as time proceeds:

- An object is speeding up if its displacement vectors are increasing in length.
- An object is slowing down if its displacement vectors are decreasing in length.

EXAMPLE 1.1 | Headfirst into the snow

Alice is sliding along a smooth, icy road on her sled when she suddenly runs headfirst into a large, very soft snowbank that gradually brings her to a halt. Draw a motion diagram for Alice. Show and label all displacement vectors.

MODEL The details of Alice and the sled—their size, shape, color, and so on—are not relevant to understanding their overall motion. So we can model Alice and the sled as one particle.

VISUALIZE **FIGURE 1.10** shows a motion diagram. The problem statement suggests that the sled's speed is very nearly constant until it hits the snowbank. Thus the displacement vectors are of equal length as Alice slides along the icy road. She begins slowing when she hits the snowbank, so the displacement vectors then get shorter until the sled stops. We're told that her stop is gradual, so we want the vector lengths to get shorter gradually rather than suddenly.

FIGURE 1.10 The motion diagram of Alice and the sled.

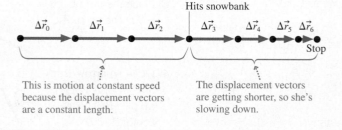

Hits snowbank

$\Delta \vec{r}_0$ $\Delta \vec{r}_1$ $\Delta \vec{r}_2$ $\Delta \vec{r}_3$ $\Delta \vec{r}_4$ $\Delta \vec{r}_5$ $\Delta \vec{r}_6$

Stop

This is motion at constant speed because the displacement vectors are a constant length.

The displacement vectors are getting shorter, so she's slowing down.

Time Interval

It's also useful to consider a *change* in time. For example, the clock readings of two frames of film might be t_1 and t_2. The specific values are arbitrary because they are timed relative to an arbitrary instant that you chose to call $t = 0$. But the **time interval** $\Delta t = t_2 - t_1$ is *not* arbitrary. It represents the elapsed time for the object to move from one position to the next.

The time interval $\Delta t = t_f - t_i$ measures the elapsed time as an object moves from an initial position $\vec{r}_i$ at time t_i to a final position $\vec{r}_f$ at time t_f. The value of Δt is independent of the specific clock used to measure the times.

To summarize the main idea of this section, we have added coordinate systems and clocks to our motion diagrams in order to measure *when* each frame was exposed and *where* the object was located at that time. Different observers of the motion may choose different coordinate systems and different clocks. However, all observers find the *same* values for the displacements $\Delta \vec{r}$ and the time intervals Δt because these are independent of the specific coordinate system used to measure them.

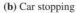

A stopwatch is used to measure a time interval.

1.4 Velocity

It's no surprise that, during a given time interval, a speeding bullet travels farther than a speeding snail. To extend our study of motion so that we can compare the bullet to the snail, we need a way to measure how fast or how slowly an object moves.

One quantity that measures an object's fastness or slowness is its **average speed,** defined as the ratio

$$\text{average speed} = \frac{\text{distance traveled}}{\text{time interval spent traveling}} = \frac{d}{\Delta t} \qquad (1.3)$$

If you drive 15 miles (mi) in 30 minutes ($\frac{1}{2}$ h), your average speed is

$$\text{average speed} = \frac{15 \text{ mi}}{\frac{1}{2} \text{ h}} = 30 \text{ mph} \qquad (1.4)$$

Although the concept of speed is widely used in our day-to-day lives, it is not a sufficient basis for a science of motion. To see why, imagine you're trying to land a jet plane on an aircraft carrier. It matters a great deal to you whether the aircraft carrier is moving at 20 mph (miles per hour) to the north or 20 mph to the east. Simply knowing that the boat's speed is 20 mph is not enough information!

It's the displacement $\Delta \vec{r}$, a vector quantity, that tells us not only the distance traveled by a moving object, but also the *direction* of motion. Consequently, a more useful ratio than $d/\Delta t$ is the ratio $\Delta \vec{r}/\Delta t$. In addition to measuring how fast an object moves, this ratio is a vector that points in the direction of motion.

It is convenient to give this ratio a name. We call it the **average velocity,** and it has the symbol $\vec{v}_{\text{avg}}$. **The average velocity of an object during the time interval Δt, in which the object undergoes a displacement $\Delta \vec{r}$, is the vector**

$$\vec{v}_{\text{avg}} = \frac{\Delta \vec{r}}{\Delta t} \qquad (1.5)$$

An object's average velocity vector points in the same direction as the displacement vector $\Delta \vec{r}$. This is the direction of motion.

> **NOTE** In everyday language we do not make a distinction between speed and velocity, but in physics *the distinction is very important*. In particular, speed is simply "How fast?" whereas velocity is "How fast, and in which direction?" As we go along we will be giving other words more precise meanings in physics than they have in everyday language.

As an example, **FIGURE 1.11a** shows two ships that move 5 miles in 15 minutes. Using Equation 1.5 with $\Delta t = 0.25$ h, we find

$$\vec{v}_{\text{avg A}} = (20 \text{ mph, north})$$
$$\vec{v}_{\text{avg B}} = (20 \text{ mph, east}) \qquad (1.6)$$

Both ships have a speed of 20 mph, but their velocities are different. Notice how the velocity *vectors* in **FIGURE 1.11b** point in the direction of motion.

> **NOTE** Our goal in this chapter is to *visualize* motion with motion diagrams. Strictly speaking, the vector we have defined in Equation 1.5, and the vector we will show on motion diagrams, is the *average* velocity $\vec{v}_{\text{avg}}$. But to allow the motion diagram to be a useful tool, we will drop the subscript and refer to the average velocity as simply $\vec{v}$. Our definitions and symbols, which somewhat blur the distinction between average and instantaneous quantities, are adequate for visualization purposes, but they're not the final word. We will refine these definitions in Chapter 2, where our goal will be to develop the mathematics of motion.

The victory goes to the runner with the highest average speed.

FIGURE 1.11 The displacement vectors and velocities of ships A and B.

(a)

$\Delta \vec{r}_A = (5 \text{ mi, north})$

A

$\Delta \vec{r}_B = (5 \text{ mi, east})$

B

(b)

$\vec{v}_{\text{avg A}} = (20 \text{ mph, north})$

The velocity vectors point in the direction of motion.

$\vec{v}_{\text{avg B}} = (20 \text{ mph, east})$

Motion Diagrams with Velocity Vectors

The velocity vector points in the same direction as the displacement $\Delta\vec{r}$, and the length of $\vec{v}$ is directly proportional to the length of $\Delta\vec{r}$. Consequently, the vectors connecting each dot of a motion diagram to the next, which we previously labeled as displacements, could equally well be identified as velocity vectors.

This idea is illustrated in **FIGURE 1.12**, which shows four frames from the motion diagram of a tortoise racing a hare. The vectors connecting the dots are now labeled as velocity vectors $\vec{v}$. **The length of a velocity vector represents the average speed with which the object moves between the two points.** Longer velocity vectors indicate faster motion. You can see that the hare moves faster than the tortoise.

Notice that the hare's velocity vectors do not change; each has the same length and direction. We say the hare is moving with *constant velocity*. The tortoise is also moving with its own constant velocity.

FIGURE 1.12 Motion diagram of the tortoise racing the hare.

These are average velocity vectors.

$\vec{v}_0$ $\vec{v}_1$ $\vec{v}_2$

Hare

$\vec{v}_0$ $\vec{v}_1$ $\vec{v}_2$

Tortoise

The length of each arrow represents the average speed. The hare moves faster than the tortoise.

EXAMPLE 1.2 | Accelerating up a hill

The light turns green and a car accelerates, starting from rest, up a 20° hill. Draw a motion diagram showing the car's velocity.

MODEL Use the particle model to represent the car as a dot.

VISUALIZE The car's motion takes place along a straight line, but the line is neither horizontal nor vertical. A motion diagram should show the object moving with the correct orientation—in this case, at an angle of 20°. **FIGURE 1.13** shows several frames of the motion diagram, where we see the car speeding up. The car starts from rest, so the first arrow is drawn as short as possible and the first dot is labeled "Start." The displacement vectors have been drawn from each dot to the next, but then they are identified and labeled as average velocity vectors $\vec{v}$.

FIGURE 1.13 Motion diagram of a car accelerating up a hill.

This labels the whole row of vectors as velocity vectors.

$\vec{v}$

The velocity vectors are getting longer, so the car is speeding up.

Start

EXAMPLE 1.3 | A rolling soccer ball

Marcos kicks a soccer ball. It rolls along the ground until stopped by Jose. Draw a motion diagram of the ball.

MODEL This example is typical of how many problems in science and engineering are worded. The problem does not give a clear statement of where the motion begins or ends. Are we interested in the motion of the ball just during the time it is rolling between Marcos and Jose? What about the motion *as* Marcos kicks it (ball rapidly speeding up) or *as* Jose stops it (ball rapidly slowing down)? The point is that *you* will often be called on to make a *reasonable interpretation* of a problem statement. In this problem, the details of kicking and stopping the ball are complex. The motion of the ball across the ground is easier to describe, and it's a motion you might expect to learn about in a physics class. So our *interpretation* is that the motion diagram should start as the ball leaves Marcos's foot (ball already moving) and should end the instant it touches Jose's foot

(ball still moving). In between, the ball will slow down a little. We will model the ball as a particle.

VISUALIZE With this interpretation in mind, **FIGURE 1.14** shows the motion diagram of the ball. Notice how, in contrast to the car of Figure 1.13, the ball is already moving as the motion diagram video begins. As before, the average velocity vectors are found by connecting the dots. You can see that the average velocity vectors get shorter as the ball slows. Each $\vec{v}$ is different, so this is *not* constant-velocity motion.

FIGURE 1.14 Motion diagram of a soccer ball rolling from Marcos to Jose.

The velocity vectors are gradually getting shorter.

$\vec{v}$

Marcos   Jose

STOP TO THINK 1.3   A particle moves from position 1 to position 2 during the time interval Δt. Which vector shows the particle's average velocity?

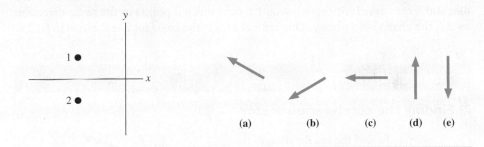

1.5 Linear Acceleration

Position, time, and velocity are important concepts, and at first glance they might appear to be sufficient to describe motion. But that is not the case. Sometimes an object's velocity is constant, as it was in Figure 1.12. More often, an object's velocity changes as it moves, as in Figures 1.13 and 1.14. We need one more motion concept to describe a *change* in the velocity.

Because velocity is a vector, it can change in two possible ways:

1. The magnitude can change, indicating a change in speed; or
2. The direction can change, indicating that the object has changed direction.

We will concentrate for now on the first case, a change in speed. The car accelerating up a hill in Figure 1.13 was an example in which the magnitude of the velocity vector changed but not the direction. We'll return to the second case in Chapter 4.

When we wanted to measure changes in position, the ratio $\Delta \vec{r}/\Delta t$ was useful. This ratio is the *rate of change of position*. By analogy, consider an object whose velocity changes from an initial $\vec{v}_i$ to a final $\vec{v}_f$ during the time interval Δt. Just as $\Delta \vec{r} = \vec{r}_f - \vec{r}_i$ is the change of position, the quantity $\Delta \vec{v} = \vec{v}_f - \vec{v}_i$ is the change of velocity. The ratio $\Delta \vec{v}/\Delta t$ is then the *rate of change of velocity*. It has a large magnitude for objects that speed up quickly and a small magnitude for objects that speed up slowly.

The ratio $\Delta \vec{v}/\Delta t$ is called the **average acceleration,** and its symbol is $\vec{a}_{avg}$. **The average acceleration of an object during the time interval Δt, in which the object's velocity changes by $\Delta \vec{v}$, is the vector**

$$\vec{a}_{avg} = \frac{\Delta \vec{v}}{\Delta t} \tag{1.7}$$

The average acceleration vector points in the same direction as the vector $\Delta \vec{v}$.

Acceleration is a fairly abstract concept. Yet it is essential to develop a good intuition about acceleration because it will be a key concept for understanding why objects move as they do. Motion diagrams will be an important tool for developing that intuition.

The Audi TT accelerates from 0 to 60 mph in 6 s.

NOTE As we did with velocity, we will drop the subscript and refer to the average acceleration as simply $\vec{a}$. This is adequate for visualization purposes, but not the final word. We will refine the definition of acceleration in Chapter 2.

Finding the Acceleration Vectors on a Motion Diagram

Let's look at how we can determine the average acceleration vector $\vec{a}$ from a motion diagram. From its definition, Equation 1.7, we see that $\vec{a}$ **points in the same direction as** $\Delta\vec{v}$, the change of velocity. This critical idea is the basis for a technique to find $\vec{a}$.

TACTICS BOX 1.3   (MP)

Finding the acceleration vector

To find the acceleration as the velocity changes from $\vec{v}_i$ to $\vec{v}_f$, we must determine the *change* of velocity $\Delta\vec{v} = \vec{v}_f - \vec{v}_i$.

❶ Draw the velocity vector $\vec{v}_f$.

❷ Draw $-\vec{v}_i$ at the tip of $\vec{v}_f$.

❸ Draw $\Delta\vec{v} = \vec{v}_f - \vec{v}_i$
   $= \vec{v}_f + (-\vec{v}_i)$
   This is the direction of $\vec{a}$.

❹ Return to the original motion diagram. Draw a vector at the middle dot in the direction of $\Delta\vec{v}$; label it $\vec{a}$. This is the average acceleration at the midpoint between $\vec{v}_i$ and $\vec{v}_f$.

Exercises 21–24

Many Tactics Boxes will refer you to exercises in the ·······
Student Workbook where you can practice the new skill.

Notice that the acceleration vector goes beside the middle dot, not beside the velocity vectors. This is because each acceleration vector is determined by the *difference* between the *two* velocity vectors on either side of a dot. The length of $\vec{a}$ does not have to be the exact length of $\Delta\vec{v}$; it is the direction of $\vec{a}$ that is most important.

The procedure of Tactics Box 1.3 can be repeated to find $\vec{a}$ at each point in the motion diagram. Note that we cannot determine $\vec{a}$ at the first and last points because we have only one velocity vector and can't find $\Delta\vec{v}$.

The Complete Motion Diagram

You've now seen several *Tactics Boxes* that help you accomplish specific tasks. Tactics Boxes will appear in nearly every chapter in this book. We'll also, where appropriate, provide *Problem-Solving Strategies*.

PROBLEM-SOLVING STRATEGY 1.1   (MP)

Motion diagrams

MODEL Determine whether it is appropriate to model the moving object as a particle. Make simplifying assumptions when interpreting the problem statement.

VISUALIZE A complete motion diagram consists of:

- The position of the object in each frame of the video, shown as a dot. Use five or six dots to make the motion clear but without overcrowding the picture. More complex motions may need more dots.
- The average velocity vectors, found by connecting each dot in the motion diagram to the next with a vector arrow. There is *one* velocity vector linking each *two* position dots. Label the row of velocity vectors $\vec{v}$.
- The average acceleration vectors, found using Tactics Box 1.3. There is *one* acceleration vector linking each *two* velocity vectors. Each acceleration vector is drawn at the dot between the two velocity vectors it links. Use $\vec{0}$ to indicate a point at which the acceleration is zero. Label the row of acceleration vectors $\vec{a}$.

STOP TO THINK 1.4   A particle undergoes acceleration $\vec{a}$ while moving from point 1 to point 2. Which of the choices shows the most likely velocity vector $\vec{v}_2$ as the particle leaves point 2?

(a)   (b)   (c)   (d)

Examples of Motion Diagrams

Let's look at some examples of the full strategy for drawing motion diagrams.

EXAMPLE 1.4 | The first astronauts land on Mars

A spaceship carrying the first astronauts to Mars descends safely to the surface. Draw a motion diagram for the last few seconds of the descent.

MODEL The spaceship is small in comparison with the distance traveled, and the spaceship does not change size or shape, so it's reasonable to model the spaceship as a particle. We'll assume that its motion in the last few seconds is straight down. The problem ends as the spacecraft touches the surface.

VISUALIZE FIGURE 1.15 shows a complete motion diagram as the spaceship descends and slows, using its rockets, until it comes to rest on the surface. Notice how the dots get closer together as it slows. The inset uses the steps of Tactics Box 1.3 (numbered circles) to show how the acceleration vector $\vec{a}$ is determined at one point. All the other acceleration vectors will be similar because for each pair of velocity vectors the earlier one is longer than the later one.

FIGURE 1.15 Motion diagram of a spaceship landing on Mars.

EXAMPLE 1.5 | Skiing through the woods

A skier glides along smooth, horizontal snow at constant speed, then speeds up going down a hill. Draw the skier' motion diagram.

MODEL Model the skier as a particle. It's reasonable to assume that the downhill slope is a straight line. Although the motion as a whole is not linear, we can treat the skier's motion as two separate linear motions.

VISUALIZE FIGURE 1.16 shows a complete motion diagram of the skier. The dots are equally spaced for the horizontal motion, indicating constant speed; then the dots get farther apart as the skier speeds up going down the hill. The insets show how the average acceleration vector $\vec{a}$ is determined for the horizontal motion and along the slope. All the other acceleration vectors along the slope will be similar to the one shown because each velocity vector is longer than the preceding one. Notice that we've explicitly written $\vec{0}$ for the acceleration beside the dots where the velocity is constant. The acceleration at the point where the direction changes will be considered in Chapter 4.

FIGURE 1.16 Motion diagram of a skier.

Notice something interesting in Figures 1.15 and 1.16. Where the object is speeding up, the acceleration and velocity vectors point in the *same direction*. Where the object is slowing down, the acceleration and velocity vectors point in *opposite directions*. These results are always true for motion in a straight line. **For motion along a line:**

- **An object is speeding up if and only if $\vec{v}$ and $\vec{a}$ point in the same direction.**
- **An object is slowing down if and only if $\vec{v}$ and $\vec{a}$ point in opposite directions.**
- **An object's velocity is constant if and only if $\vec{a} = \vec{0}$.**

NOTE In everyday language, we use the word *accelerate* to mean "speed up" and the word *decelerate* to mean "slow down." But speeding up and slowing down are both changes in the velocity and consequently, by our definition, *both* are accelerations. In physics, *acceleration* refers to changing the velocity, no matter what the change is, and not just to speeding up.

EXAMPLE 1.6 | Tossing a ball

Draw the motion diagram of a ball tossed straight up in the air.

MODEL This problem calls for some interpretation. Should we include the toss itself, or only the motion after the ball is released? Should we include the ball hitting the ground? It appears that this problem is really concerned with the ball's motion through the air. Consequently, we begin the motion diagram at the instant that the tosser releases the ball and end the diagram at the instant the ball hits the ground. We will consider neither the toss nor the impact. And, of course, we will model the ball as a particle.

VISUALIZE We have a slight difficulty here because the ball retraces its route as it falls. A literal motion diagram would show the upward motion and downward motion on top of each other, leading to confusion. We can avoid this difficulty by horizontally separating the upward motion and downward motion diagrams. This will not affect our conclusions because it does not change any of the vectors.

FIGURE 1.17 shows the motion diagram drawn this way. Notice that the very top dot is shown twice—as the end point of the upward motion and the beginning point of the downward motion.

The ball slows down as it rises. You've learned that the acceleration vectors point opposite the velocity vectors for an object that is slowing down along a line, and they are shown accordingly. Similarly, $\vec{a}$ and $\vec{v}$ point in the same direction as the falling ball speeds up. Notice something interesting: The acceleration vectors point downward both while the ball is rising *and* while it is falling. Both "speeding up" and "slowing down" occur with the *same* acceleration vector. This is an important conclusion, one worth pausing to think about.

Now let's look at the top point on the ball's trajectory. The velocity vectors point upward but are getting shorter as the ball approaches the top. As the ball starts to fall, the velocity vectors point downward and are getting longer. There must be a moment—just an instant as $\vec{v}$ switches from pointing up to pointing down—when the velocity is zero. Indeed, the ball's velocity *is* zero for an instant at the precise top of the motion!

But what about the acceleration at the top? The inset shows how the average acceleration is determined from the last upward velocity before the top point and the first downward velocity. We find that the acceleration at the top is pointing downward, just as it does elsewhere in the motion.

Many people expect the acceleration to be zero at the highest point. But the velocity at the top point *is* changing—from up to down. If the velocity is changing, there *must* be an acceleration. A downward-pointing acceleration vector is needed to turn the velocity vector from up to down. Another way to think about this is to note that zero acceleration would mean no change of velocity. When the ball reached zero velocity at the top, it would hang there and not fall if the acceleration were also zero!

FIGURE 1.17 Motion diagram of a ball tossed straight up in the air.

For clarity, we displace the upward and downward motions. They really occur along the same line.

1.6 Motion in One Dimension

An object's motion can be described in terms of three fundamental quantities: its position $\vec{r}$, velocity $\vec{v}$, and acceleration $\vec{a}$. These are vectors, but for motion in one dimension, the vectors are restricted to point only "forward" or "backward." Consequently, we can describe one-dimensional motion with the simpler quantities x, v_x, and a_x (or y, v_y, and a_y). However, we need to give each of these quantities an explicit *sign,* positive or negative, to indicate whether the position, velocity, or acceleration vector points forward or backward.

Determining the Signs of Position, Velocity, and Acceleration

Position, velocity, and acceleration are measured with respect to a coordinate system, a grid or axis that *you* impose on a problem to analyze the motion. We will find it convenient to use an x-axis to describe both horizontal motion and motion along an inclined plane. A y-axis will be used for vertical motion. A coordinate axis has two essential features:

1. An origin to define zero; and
2. An x or y label (with units) to indicate the positive end of the axis.

In this textbook, we will follow the convention that **the positive end of an x-axis is to the right and the positive end of a y-axis is up.** The signs of position, velocity, and acceleration are based on this convention.

TACTICS BOX 1.4　　　　　　　　　　　　　　　　　　　　　　MP

Determining the sign of the position, velocity, and acceleration

$x > 0$　Position to right of origin.

$x < 0$　Position to left of origin.

$\vec{v}$　$v_x > 0$　Direction of motion is to the right.

$\vec{v}$　$v_x < 0$　Direction of motion is to the left.

$\vec{a}$　$a_x > 0$　Acceleration vector points to the right.

$\vec{a}$　$a_x < 0$　Acceleration vector points to the left.

$y > 0$　Position above origin.　　$y < 0$　Position below origin.

$v_y > 0$　Direction of motion is up.　　$v_y < 0$　Direction of motion is down.

$a_y > 0$　Acceleration vector points up.　　$a_y < 0$　Acceleration vector points down.

- The sign of position (x or y) tells us *where* an object is.

- The sign of velocity (v_x or v_y) tells us *which direction* the object is moving.

- The sign of acceleration (a_x or a_y) tells us which way the acceleration vector points, *not* whether the object is speeding up or slowing down.

Exercises 30–31

Acceleration is where things get a bit tricky. A natural tendency is to think that a positive value of a_x or a_y describes an object that is speeding up while a negative value describes an object that is slowing down (decelerating). However, this interpretation *does not work*.

Acceleration is defined as $\vec{a}_{avg} = \Delta\vec{v}/\Delta t$. The direction of $\vec{a}$ can be determined by using a motion diagram to find the direction of $\Delta\vec{v}$. The one-dimensional acceleration a_x (or a_y) is then positive if the vector $\vec{a}$ points to the right (or up), negative if $\vec{a}$ points to the left (or down).

FIGURE 1.18 shows that this method for determining the sign of a does not conform to the simple idea of speeding up and slowing down. The object in Figure 1.18a has a positive acceleration ($a_x > 0$) not because it is speeding up but because the vector $\vec{a}$ points in the positive direction. Compare this with the motion diagram of Figure 1.18b. Here the object is slowing down, but it still has a positive acceleration ($a_x > 0$) because $\vec{a}$ points to the right.

In the previous section, we found that an object is speeding up if $\vec{v}$ and $\vec{a}$ point in the same direction, slowing down if they point in opposite directions. For one-dimensional motion this rule becomes:

- An object is speeding up if and only if v_x and a_x have the same sign.
- An object is slowing down if and only if v_x and a_x have opposite signs.
- An object's velocity is constant if and only if $a_x = 0$.

Notice how the first two of these rules are at work in Figure 1.18.

FIGURE 1.18 One of these objects is speeding up, the other slowing down, but they both have a positive acceleration a_x.

(a) Speeding up to the right

$\vec{a}$

$\vec{v}$

$x > 0$　$v_x > 0$　$a_x > 0$

(b) Slowing down to the left

$\vec{a}$

$\vec{v}$

$x > 0$　$v_x < 0$　$a_x > 0$

Position-versus-Time Graphs

FIGURE 1.19 is a motion diagram, made at 1 frame per minute, of a student walking to school. You can see that she leaves home at a time we choose to call $t = 0$ min and makes steady progress for a while. Beginning at $t = 3$ min there is a period where the distance traveled during each time interval becomes less—perhaps she slowed down to speak with a friend. Then she picks up the pace, and the distances within each interval are longer.

FIGURE 1.19 The motion diagram of a student walking to school and a coordinate axis for making measurements.

Figure 1.19 includes a coordinate axis, and you can see that every dot in a motion diagram occurs at a specific position. **TABLE 1.1** shows the student's positions at different times as measured along this axis. For example, she is at position $x = 120$ m at $t = 2$ min.

The motion diagram is one way to represent the student's motion. Another is to make a graph of the measurements in Table 1.1. **FIGURE 1.20a** is a graph of x versus t for the student. The motion diagram tells us only where the student is at a few discrete points of time, so this graph of the data shows only points, no lines.

NOTE A graph of "a versus b" means that a is graphed on the vertical axis and b on the horizontal axis. Saying "graph a versus b" is really a shorthand way of saying "graph a as a function of b."

TABLE 1.1 Measured positions of a student walking to school

Time t (min)	Position x (m)	Time t (min)	Position x (m)
0	0	5	220
1	60	6	240
2	120	7	340
3	180	8	440
4	200	9	540

FIGURE 1.20 Position graphs of the student's motion.

However, common sense tells us the following. First, the student was *somewhere specific* at all times. That is, there was never a time when she failed to have a well-defined position, nor could she occupy two positions at one time. (As reasonable as this belief appears to be, it will be severely questioned and found not entirely accurate when we get to quantum physics!) Second, the student moved *continuously* through all intervening points of space. She could not go from $x = 100$ m to $x = 200$ m without passing through every point in between. It is thus quite reasonable to believe that her motion can be shown as a continuous line passing through the measured points, as shown in **FIGURE 1.20b**. A continuous line or curve showing an object's position as a function of time is called a **position-versus-time graph** or, sometimes, just a *position graph*.

NOTE A graph is *not* a "picture" of the motion. The student is walking along a straight line, but the graph itself is not a straight line. Further, we've graphed her position on the vertical axis even though her motion is horizontal. Graphs are *abstract representations* of motion. We will place significant emphasis on the process of interpreting graphs, and many of the exercises and problems will give you a chance to practice these skills.

EXAMPLE 1.7 | Interpreting a position graph

The graph in FIGURE 1.21a represents the motion of a car along a straight road. Describe the motion of the car.

MODEL We'll model the car as a particle with a precise position at each instant.

VISUALIZE As FIGURE 1.21b shows, the graph represents a car that travels to the left for 30 minutes, stops for 10 minutes, then travels back to the right for 40 minutes.

FIGURE 1.21 Position-versus-time graph of a car.

(a)

(b) 1. At $t = 0$ min, the car is 10 km to the right of the origin.

2. The value of x decreases for 30 min, indicating that the car is moving to the left.

5. The car reaches the origin at $t = 80$ min.

3. The car stops for 10 min at a position 20 km to the left of the origin.

4. The car starts moving back to the right at $t = 40$ min.

1.7 Solving Problems in Physics

Physics is not mathematics. Math problems are clearly stated, such as "What is $2 + 2$?" Physics is about the world around us, and to describe that world we must use language. Now, language is wonderful—we couldn't communicate without it—but language can sometimes be imprecise or ambiguous.

The challenge when reading a physics problem is to translate the words into symbols that can be manipulated, calculated, and graphed. **The translation from words to symbols is the heart of problem solving in physics.** This is the point where ambiguous words and phrases must be clarified, where the imprecise must be made precise, and where you arrive at an understanding of exactly what the question is asking.

Using Symbols

Symbols are a language that allows us to talk with precision about the relationships in a problem. As with any language, we all need to agree to use words or symbols in the same way if we want to communicate with each other. Many of the ways we use symbols in science and engineering are somewhat arbitrary, often reflecting historical roots. Nonetheless, practicing scientists and engineers have come to agree on how to use the language of symbols. Learning this language is part of learning physics.

We will use subscripts on symbols, such as x_3, to designate a particular point in the problem. Scientists usually label the starting point of the problem with the subscript "0," not the subscript "1" that you might expect. When using subscripts, make sure that all symbols referring to the same point in the problem have the *same numerical subscript*. To have the same point in a problem characterized by position x_1 but velocity v_{2x} is guaranteed to lead to confusion!

Drawing Pictures

You may have been told that the first step in solving a physics problem is to "draw a picture," but perhaps you didn't know why, or what to draw. The purpose of drawing a picture is to aid you in the words-to-symbols translation. Complex problems have far more information than you can keep in your head at one time. Think of a picture as a "memory extension," helping you organize and keep track of vital information.

Although any picture is better than none, there really is a *method* for drawing pictures that will help you be a better problem solver. It is called the **pictorial representation** of the problem. We'll add other pictorial representations as we go along, but the following procedure is appropriate for motion problems.

TACTICS BOX 1.5

Drawing a pictorial representation

❶ **Draw a motion diagram.** The motion diagram develops your intuition for the motion.

❷ **Establish a coordinate system.** Select your axes and origin to match the motion. For one-dimensional motion, you want either the *x*-axis or the *y*-axis parallel to the motion. The coordinate system determines whether the signs of *v* and *a* are positive or negative.

❸ **Sketch the situation.** Not just any sketch. Show the object at the *beginning* of the motion, at the *end,* and at any point where the character of the motion changes. Show the object, not just a dot, but very simple drawings are adequate.

❹ **Define symbols.** Use the sketch to define symbols representing quantities such as position, velocity, acceleration, and time. *Every* variable used later in the mathematical solution should be defined on the sketch. Some will have known values, others are initially unknown, but all should be given symbolic names.

❺ **List known information.** Make a table of the quantities whose values you can determine from the problem statement or that can be found quickly with simple geometry or unit conversions. Some quantities are implied by the problem, rather than explicitly given. Others are determined by your choice of coordinate system.

❻ **Identify the desired unknowns.** What quantity or quantities will allow you to answer the question? These should have been defined as symbols in step 4. Don't list every unknown, only the one or two needed to answer the question.

It's not an overstatement to say that a well-done pictorial representation of the problem will take you halfway to the solution. The following example illustrates how to construct a pictorial representation for a problem that is typical of problems you will see in the next few chapters.

EXAMPLE 1.8 | Drawing a pictorial representation

Draw a pictorial representation for the following problem: A rocket sled accelerates horizontally at 50 m/s² for 5.0 s, then coasts for 3.0 s. What is the total distance traveled?

VISUALIZE FIGURE 1.22, on the next page, is the pictorial representation. The motion diagram shows an acceleration phase followed by a coasting phase. Because the motion is horizontal, the appropriate coordinate system is an *x*-axis. We've chosen to place the origin at the starting point. The motion has a beginning, an end, and a point where the motion changes from accelerating to coasting, and these are the three sled positions sketched in the figure. The quantities *x*, v_x, and *t* are needed at each of three *points,* so these

have been defined on the sketch and distinguished by subscripts. Accelerations are associated with *intervals* between the points, so only two accelerations are defined. Values for three quantities are given in the problem statement, although we need to use the motion diagram, where $\vec{a}$ points to the right, and our choice of coordinate system to know that $a_{0x} = +50$ m/s² rather than -50 m/s². The values $x_0 = 0$ m and $t_0 = 0$ s are choices we made when setting up the coordinate system. The value $v_{0x} = 0$ m/s is part of our *interpretation* of the problem. Finally, we identify x_2 as the quantity that will answer the question. We now understand quite a bit about the problem and would be ready to start a quantitative analysis.

Continued

FIGURE 1.22 A pictorial representation.

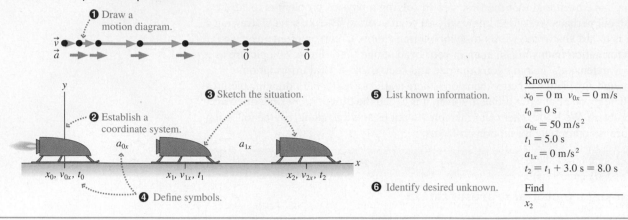

❶ Draw a motion diagram.

❷ Establish a coordinate system.

❸ Sketch the situation.

❹ Define symbols.

a_{0x} a_{1x}

x_0, v_{0x}, t_0 x_1, v_{1x}, t_1 x_2, v_{2x}, t_2

❺ List known information.

Known
$x_0 = 0$ m $v_{0x} = 0$ m/s
$t_0 = 0$ s
$a_{0x} = 50$ m/s^2
$t_1 = 5.0$ s
$a_{1x} = 0$ m/s^2
$t_2 = t_1 + 3.0$ s $= 8.0$ s

❻ Identify desired unknown.

Find
x_2

We didn't *solve* the problem; that is not the purpose of the pictorial representation. The pictorial representation is a systematic way to go about interpreting a problem and getting ready for a mathematical solution. Although this is a simple problem, and you probably know how to solve it if you've taken physics before, you will soon be faced with much more challenging problems. Learning good problem-solving skills at the beginning, while the problems are easy, will make them second nature later when you really need them.

Representations

A picture is one way to *represent* your knowledge of a situation. You could also represent your knowledge using words, graphs, or equations. Each **representation of knowledge** gives us a different perspective on the problem. The more tools you have for thinking about a complex problem, the more likely you are to solve it.

There are four representations of knowledge that we will use over and over:

A new building requires careful planning. The architect's visualization and drawings have to be complete before the detailed procedures of construction get under way. The same is true for solving problems in physics.

1. The *verbal* representation. A problem statement, in words, is a verbal representation of knowledge. So is an explanation that you write.
2. The *pictorial* representation. The pictorial representation, which we've just presented, is the most literal depiction of the situation.
3. The *graphical* representation. We will make extensive use of graphs.
4. The *mathematical* representation. Equations that can be used to find the numerical values of specific quantities are the mathematical representation.

NOTE The mathematical representation is only one of many. Much of physics is more about thinking and reasoning than it is about solving equations.

A Problem-Solving Strategy

One of the goals of this textbook is to help you learn a *strategy* for solving physics problems. The purpose of a strategy is to guide you in the right direction with minimal wasted effort. The four-part problem-solving strategy shown on the next page—**Model, Visualize, Solve, Assess**—is based on using different representations of knowledge. You will see this problem-solving strategy used consistently in the worked examples throughout this textbook, and you should endeavor to apply it to your own problem solving.

Throughout this textbook we will emphasize the first two steps. They are the *physics* of the problem, as opposed to the mathematics of solving the resulting equations. This is not to say that those mathematical operations are always easy—in many cases they are not. But our primary goal is to understand the physics.

GENERAL PROBLEM-SOLVING STRATEGY

MODEL It's impossible to treat every detail of a situation. Simplify the situation with a model that captures the essential features. For example, the object in a mechanics problem is often represented as a particle.

VISUALIZE This is where expert problem solvers put most of their effort.

- Draw a *pictorial representation*. This helps you visualize important aspects of the physics and assess the information you are given. It starts the process of translating the problem into symbols.
- Use a *graphical representation* if it is appropriate for the problem.
- Go back and forth between these representations; they need not be done in any particular order.

SOLVE Only after modeling and visualizing are complete is it time to develop a *mathematical representation* with specific equations that must be solved. All symbols used here should have been defined in the pictorial representation.

ASSESS Is your result believable? Does it have proper units? Does it make sense?

Textbook illustrations are obviously more sophisticated than what you would draw on your own paper. To show you a figure very much like what *you* should draw, the final example of this section is in a "pencil sketch" style. We will include one or more pencil-sketch examples in nearly every chapter to illustrate exactly what a good problem solver would draw.

EXAMPLE 1.9 | Launching a weather rocket

Use the first two steps of the problem-solving strategy to analyze the following problem: A small rocket, such as those used for meteorological measurements of the atmosphere, is launched vertically with an acceleration of 30 m/s². It runs out of fuel after 30 s. What is its maximum altitude?

MODEL We need to do some interpretation. Common sense tells us that the rocket does not stop the instant it runs out of fuel. Instead, it continues upward, while slowing, until it reaches its maximum altitude. This second half of the motion, after running out of fuel, is like the ball that was tossed upward in the first half of Example 1.6. Because the problem does not ask about the rocket's descent, we conclude that the problem ends at the point of maximum altitude. We'll model the rocket as a particle.

VISUALIZE FIGURE 1.23 shows the pictorial representation in pencil-sketch style. The rocket is speeding up during the first half of the motion, so $\vec{a}_0$ points upward, in the positive y-direction. Thus the initial acceleration is $a_{0y} = 30$ m/s². During the second half, as the rocket slows, $\vec{a}_1$ points downward. Thus a_{1y} is a negative number.

This information is included with the known information. Although the velocity v_{2y} wasn't given in the problem statement, it must—just like for the ball in Example 1.6—be zero at the very top of the trajectory. Last, we have identified y_2 as the desired unknown. This, of course, is not the only unknown in the problem, but it is the one we are specifically asked to find.

ASSESS If you've had a previous physics class, you may be tempted to assign a_{1y} the value -9.8 m/s², the free-fall acceleration.

FIGURE 1.23 Pictorial representation for the rocket.

However, that would be true only if there is no air resistance on the rocket. We will need to consider the *forces* acting on the rocket during the second half of its motion before we can determine a value for a_{1y}. For now, all that we can safely conclude is that a_{1y} is negative.

Our task in this section is not to *solve* problems—all that in due time—but to focus on what is happening in a problem. In other words, to make the translation from words to symbols in preparation for subsequent mathematical analysis. Modeling and the pictorial representation will be our most important tools.

1.8 Units and Significant Figures

Science is based upon experimental measurements, and measurements require *units*. The system of units used in science is called *le Système Internationale d'Unités*. These are commonly referred to as **SI units.** In casual speaking we often refer to *metric units*.

All of the quantities needed to understand motion can be expressed in terms of the three basic SI units shown in TABLE 1.2. Other quantities can be expressed as a combination of these basic units. Velocity, expressed in meters per second or m/s, is a ratio of the length unit to the time unit.

TABLE 1.2 The basic SI units

Quantity	Unit	Abbreviation
time	second	s
length	meter	m
mass	kilogram	kg

Time

The standard of time prior to 1960 was based on the *mean solar day*. As time-keeping accuracy and astronomical observations improved, it became apparent that the earth's rotation is not perfectly steady. Meanwhile, physicists had been developing a device called an *atomic clock*. This instrument is able to measure, with incredibly high precision, the frequency of radio waves absorbed by atoms as they move between two closely spaced energy levels. This frequency can be reproduced with great accuracy at many laboratories around the world. Consequently, the SI unit of time—the second—was redefined in 1967 as follows:

One *second* is the time required for 9,192,631,770 oscillations of the radio wave absorbed by the cesium-133 atom. The abbreviation for second is the letter s.

Several radio stations around the world broadcast a signal whose frequency is linked directly to the atomic clocks. This signal is the time standard, and any time-measuring equipment you use was calibrated from this time standard.

An atomic clock at the National Institute of Standards and Technology is the primary standard of time.

Length

The SI unit of length—the meter—was originally defined as one ten-millionth of the distance from the north pole to the equator along a line passing through Paris. There are obvious practical difficulties with implementing this definition, and it was later abandoned in favor of the distance between two scratches on a platinum-iridium bar stored in a special vault in Paris. The present definition, agreed to in 1983, is as follows:

One *meter* is the distance traveled by light in vacuum during 1/299,792,458 of a second. The abbreviation for meter is the letter m.

This is equivalent to defining the speed of light to be exactly 299,792,458 m/s. Laser technology is used in various national laboratories to implement this definition and to calibrate secondary standards that are easier to use. These standards ultimately make their way to your ruler or to a meter stick. It is worth keeping in mind that any measuring device you use is only as accurate as the care with which it was calibrated.

Mass

The original unit of mass, the gram, was defined as the mass of 1 cubic centimeter of water. That is why you know the density of water as 1 g/cm^3. This definition proved to be impractical when scientists needed to make very accurate measurements. The SI unit of mass—the kilogram—was redefined in 1889 as:

One *kilogram* is the mass of the international standard kilogram, a polished platinum-iridium cylinder stored in Paris. The abbreviation for kilogram is kg.

By international agreement, this metal cylinder, stored in Paris, is the definition of the kilogram.

The kilogram is the only SI unit still defined by a manufactured object. Despite the prefix *kilo*, it is the kilogram, not the gram, that is the SI unit.

Using Prefixes

We will have many occasions to use lengths, times, and masses that are either much less or much greater than the standards of 1 meter, 1 second, and 1 kilogram. We will do so by using *prefixes* to denote various powers of 10. TABLE 1.3 lists the common prefixes that will be used frequently throughout this book. Memorize it! Few things in science are learned by rote memory, but this list is one of them. A more extensive list of prefixes is shown inside the front cover of the book.

Although prefixes make it easier to talk about quantities, the SI units are meters, seconds, and kilograms. Quantities given with prefixed units must be converted to SI units before any calculations are done. Unit conversions are best done at the very beginning of a problem, as part of the pictorial representation.

TABLE 1.3 Common prefixes

Prefix	Power of 10	Abbreviation
giga-	10^9	G
mega-	10^6	M
kilo-	10^3	k
centi-	10^{-2}	c
milli-	10^{-3}	m
micro-	10^{-6}	μ
nano-	10^{-9}	n

Unit Conversions

Although SI units are our standard, we cannot entirely forget that the United States still uses English units. Thus it remains important to be able to convert back and forth between SI units and English units. TABLE 1.4 shows several frequently used conversions, and these are worth memorizing if you do not already know them. While the English system was originally based on the length of the king's foot, it is interesting to note that today the conversion 1 in = 2.54 cm is the *definition* of the inch. In other words, the English system for lengths is now based on the meter!

There are various techniques for doing unit conversions. One effective method is to write the conversion factor as a ratio equal to one. For example, using information in Tables 1.3 and 1.4, we have

TABLE 1.4 Useful unit conversions

1 in = 2.54 cm
1 mi = 1.609 km
1 mph = 0.447 m/s
1 m = 39.37 in
1 km = 0.621 mi
1 m/s = 2.24 mph

$$\frac{10^{-6}\ m}{1\ \mu m} = 1 \quad \text{and} \quad \frac{2.54\ cm}{1\ in} = 1$$

Because multiplying any expression by 1 does not change its value, these ratios are easily used for conversions. To convert 3.5 μm to meters we compute

$$3.5\ \mu m \times \frac{10^{-6}\ m}{1\ \mu m} = 3.5 \times 10^{-6}\ m$$

Similarly, the conversion of 2 feet to meters is

$$2.00\ ft \times \frac{12\ in}{1\ ft} \times \frac{2.54\ cm}{1\ in} \times \frac{10^{-2}\ m}{1\ cm} = 0.610\ m$$

Notice how units in the numerator and in the denominator cancel until only the desired units remain at the end. You can continue this process of multiplying by 1 as many times as necessary to complete all the conversions.

Assessment

As we get further into problem solving, you will need to decide whether or not the answer to a problem "makes sense." To determine this, at least until you have more experience with SI units, you may need to convert from SI units back to the English units in which you think. But this conversion does not need to be very accurate. For example, if you are working a problem about automobile speeds and reach an answer of 35 m/s, all you really want to know is whether or not this is a realistic speed for a car. That requires a "quick and dirty" conversion, not a conversion of great accuracy.

TABLE 1.5 Approximate conversion factors. Use these for assessment, not in problem solving.

$1 \text{ cm} \approx \frac{1}{2} \text{ in}$
$10 \text{ cm} \approx 4 \text{ in}$
$1 \text{ m} \approx 1 \text{ yard}$
$1 \text{ m} \approx 3 \text{ feet}$
$1 \text{ km} \approx 0.6 \text{ mile}$
$1 \text{ m/s} \approx 2 \text{ mph}$

TABLE 1.5 shows several approximate conversion factors that can be used to assess the answer to a problem. Using $1 \text{ m/s} \approx 2 \text{ mph}$, you find that 35 m/s is roughly 70 mph, a reasonable speed for a car. But an answer of 350 m/s, which you might get after making a calculation error, would be an unreasonable 700 mph. Practice with these will allow you to develop intuition for metric units.

NOTE These approximate conversion factors are accurate to only one significant figure. This is sufficient to assess the answer to a problem, but do *not* use the conversion factors from Table 1.5 for converting English units to SI units at the start of a problem. Use Table 1.4.

Significant Figures

It is necessary to say a few words about a perennial source of difficulty: significant figures. Mathematics is a subject where numbers and relationships can be as precise as desired, but physics deals with a real world of ambiguity. It is important in science and engineering to state clearly what you know about a situation—no less and, especially, no more. Numbers provide one way to specify your knowledge.

If you report that a length has a value of 6.2 m, the implication is that the actual value falls between 6.15 m and 6.25 m and thus rounds to 6.2 m. If that is the case, then reporting a value of simply 6 m is saying less than you know; you are withholding information. On the other hand, to report the number as 6.213 m is wrong. Any person reviewing your work—perhaps a client who hired you—would interpret the number 6.213 m as meaning that the actual length falls between 6.2125 m and 6.2135 m, thus rounding to 6.213 m. In this case, you are claiming to have knowledge and information that you do not really possess.

The way to state your knowledge precisely is through the proper use of **significant figures.** You can think of a significant figure as being a digit that is reliably known. A number such as 6.2 m has *two* significant figures because the next decimal place—the one-hundredths—is not reliably known. As **FIGURE 1.24** shows, the best way to determine how many significant figures a number has is to write it in scientific notation.

FIGURE 1.24 Determining significant figures.

Leading zeros locate the decimal point. They are not significant.

$$0.00620 = \boxed{6.20} \times 10^{-3}$$

A trailing zero after the decimal place is reliably known. It is significant.

The number of significant figures is the number of digits when written in scientific notation.

- The number of significant figures $\neq$ the number of decimal places.
- Changing units shifts the decimal point but does not change the number of significant figures.

What about numbers like 320 m and 20 kg? Whole numbers with trailing zeros are ambiguous unless written in scientific notation. Even so, writing 2.0×10^1 kg is tedious, and few practicing scientists or engineers would do so. In this textbook, we'll adopt the rule that *whole numbers always have at least two significant figures,* even if one of those is a trailing zero. By this rule, 320 m, 20 kg, and 8000 s each have two significant figures, but 8050 s would have three.

Calculations with numbers follow the "weakest link" rule. The saying, which you probably know, is that "a chain is only as strong as its weakest link." If nine out of ten links in a chain can support a 1000 pound weight, that strength is meaningless if the tenth link can support only 200 pounds. Nine out of the ten numbers used in a calculation might be known with a precision of 0.01%; but if the tenth number is poorly known, with a precision of only 10%, then the result of the calculation cannot possibly be more precise than 10%.

TACTICS BOX 1.6

Using significant figures

❶ When multiplying or dividing several numbers, or taking roots, the number of significant figures in the answer should match the number of significant figures of the *least* precisely known number used in the calculation.

❷ When adding or subtracting several numbers, the number of decimal places in the answer should match the *smallest* number of decimal places of any number used in the calculation.

❸ Exact numbers are perfectly known and do not affect the number of significant figures an answer should have. Examples of exact numbers are the 2 and the π in the formula $C = 2\pi r$ for the circumference of a circle.

❹ It is acceptable to keep one or two extra digits during intermediate steps of a calculation, to minimize rounding error, as long as the final answer is reported with the proper number of significant figures.

❺ Examples and problems in this textbook will normally provide data to either two or three significant figures, as is appropriate to the situation. **The appropriate number of significant figures for the answer is determined by the data provided.**

Exercises 38–39

NOTE Be careful! Many calculators have a default setting that shows two decimal places, such as 5.23. This is dangerous. If you need to calculate 5.23/58.5, your calculator will show 0.09 and it is all too easy to write that down as an answer. By doing so, you have reduced a calculation of two numbers having three significant figures to an answer with only one significant figure. The proper result of this division is 0.0894 or 8.94×10^{-2}. You will avoid this error if you keep your calculator set to display numbers in *scientific notation* with two decimal places.

EXAMPLE 1.10 | Using significant figures

An object consists of two pieces. The mass of one piece has been measured to be 6.47 kg. The volume of the second piece, which is made of aluminum, has been measured to be 4.44×10^{-4} m³. A handbook lists the density of aluminum as 2.7×10^3 kg/m³. What is the total mass of the object?

SOLVE First, calculate the mass of the second piece:

$$m = (4.44 \times 10^{-4} \text{ m}^3)(2.7 \times 10^3 \text{ kg/m}^3)$$
$$= 1.199 \text{ kg} = 1.2 \text{ kg}$$

The number of significant figures of a product must match that of the *least* precisely known number, which is the two-significant-figure density of aluminum. Now add the two masses:

$$\begin{array}{r} 6.47 \text{ kg} \\ + 1.2 \text{ kg} \\ \hline 7.7 \text{ kg} \end{array}$$

The sum is 7.67 kg, but the hundredths place is not reliable because the second mass has no reliable information about this digit. Thus we must round to the one decimal place of the 1.2 kg. The best we can say, with reliability, is that the total mass is 7.7 kg.

Proper use of significant figures is part of the "culture" of science and engineering. We will frequently emphasize these "cultural issues" because you must learn to speak the same language as the natives if you wish to communicate effectively. Most students know the rules of significant figures, having learned them in high school, but many fail to apply them. It is important to understand the reasons for significant figures and to get in the habit of using them properly.

TABLE 1.6 Some approximate lengths

	Length (m)
Altitude of jet planes	10,000
Distance across campus	1000
Length of a football field	100
Length of a classroom	10
Length of your arm	1
Width of a textbook	0.1
Length of a fingernail	0.01

TABLE 1.7 Some approximate masses

	Mass (kg)
Small car	1000
Large human	100
Medium-size dog	10
Science textbook	1
Apple	0.1
Pencil	0.01
Raisin	0.001

Orders of Magnitude and Estimating

Precise calculations are appropriate when we have precise data, but there are many times when a very rough estimate is sufficient. Suppose you see a rock fall off a cliff and would like to know how fast it was going when it hit the ground. By doing a mental comparison with the speeds of familiar objects, such as cars and bicycles, you might judge that the rock was traveling at "about" 20 mph.

This is a one-significant-figure estimate. With some luck, you can distinguish 20 mph from either 10 mph or 30 mph, but you certainly cannot distinguish 20 mph from 21 mph. A one-significant-figure estimate or calculation, such as this, is called an **order-of-magnitude estimate.** An order-of-magnitude estimate is indicated by the symbol $\sim$, which indicates even less precision than the "approximately equal" symbol $\approx$. You would say that the speed of the rock is $v \sim 20$ mph.

A useful skill is to make reliable estimates on the basis of known information, simple reasoning, and common sense. This is a skill that is acquired by practice. Many chapters in this book will have homework problems that ask you to make order-of-magnitude estimates. The following example is a typical estimation problem.

TABLES 1.6 and 1.7 have information that will be useful for doing estimates.

EXAMPLE 1.11 | Estimating a sprinter's speed

Estimate the speed with which an Olympic sprinter crosses the finish line of the 100 m dash.

SOLVE We do need one piece of information, but it is a widely known piece of sports trivia. That is, world-class sprinters run the 100 m dash in about 10 s. Their *average* speed is $v_{avg} \approx (100 \text{ m})/(10 \text{ s}) \approx 10$ m/s. But that's only average. They go slower than average at the beginning, and they cross the finish line at a speed faster than average. How much faster? Twice as fast, 20 m/s, would be ≈ 40 mph. Sprinters don't seem like they're running as fast as a 40 mph car, so this probably is too fast. Let's *estimate* that their final speed is 50% faster than the average. Thus they cross the finish line at $v \sim 15$ m/s.

STOP TO THINK 1.5 Rank in order, from the most to the least, the number of significant figures in the following numbers. For example, if b has more than c, c has the same number as a, and a has more than d, you could give your answer as b > c = a > d.

a. 82   b. 0.0052   c. 0.430   d. 4.321×10^{-10}

SUMMARY

The goal of Chapter 1 has been to learn the fundamental concepts of motion.

GENERAL STRATEGY

Problem Solving

MODEL Make simplifying assumptions.

VISUALIZE Use:

- **Pictorial representation**
- **Graphical representation**

SOLVE Use a **mathematical representation** to find numerical answers.

ASSESS Does the answer have the proper units and correct significant figures? Does it make sense?

Motion Diagrams

- Help visualize motion.
- Provide a tool for finding acceleration vectors.

Dots show positions at equal time intervals.

Velocity vectors go dot to dot.

The acceleration vector points in the direction of $\Delta \vec{v}$.

▶ These are the *average* velocity and acceleration vectors.

IMPORTANT CONCEPTS

The **particle model** represents a moving object as if all its mass were concentrated at a single point.

Position locates an object with respect to a chosen coordinate system. Change in position is called **displacement.**

Velocity is the rate of change of the position vector $\vec{r}$.

Acceleration is the rate of change of the velocity vector $\vec{v}$.

An object has an acceleration if it

- Changes speed and/or
- Changes direction.

Pictorial Representation

❶ Draw a motion diagram.

❷ Establish coordinates.

❸ Sketch the situation.

❹ Define symbols.

❺ List knowns.

❻ Identify desired unknown.

Known
$x_0 = v_{0x} = t_0 = 0$
$a_x = 2.0 \text{ m/s}^2 \quad t_1 = 2.0 \text{ s}$

Find
x_1

APPLICATIONS

For **motion along a line:**

- Speeding up: $\vec{v}$ and $\vec{a}$ point in the same direction, v_x and a_x have the same sign.
- Slowing down: $\vec{v}$ and $\vec{a}$ point in opposite directions, v_x and a_x have opposite signs.
- Constant speed: $\vec{a} = \vec{0}$, $a_x = 0$.

Acceleration a_x is positive if $\vec{a}$ points right, negative if $\vec{a}$ points left. The sign of a_x does *not* imply speeding up or slowing down.

Significant figures are reliably known digits. The number of significant figures for:

- **Multiplication, division, powers** is set by the value with the fewest significant figures.
- **Addition, subtraction** is set by the value with the smallest number of decimal places.

The appropriate number of significant figures in a calculation is determined by the data provided.

TERMS AND NOTATION

motion	particle model	time interval, Δt	representation of knowledge
translational motion	position vector, $\vec{r}$	average speed	SI units
trajectory	scalar	average velocity, $\vec{v}$	significant figures
motion diagram	vector	average acceleration, $\vec{a}$	order-of-magnitude estimate
model	displacement, $\Delta \vec{r}$	position-versus-time graph	
particle	zero vector, $\vec{0}$	pictorial representation	

CONCEPTUAL QUESTIONS

1. How many significant figures does each of the following numbers have?
   a. 0.73   b. 7.30   c. 73   d. 0.073
2. How many significant figures does each of the following numbers have?
   a. 290   b. 2.90×10^4 c. 0.0029   d. 2.90
3. Is the particle in FIGURE Q1.3 speeding up? Slowing down? Or can you tell? Explain.

   FIGURE Q1.3 ● ●   ●   ●   ●

4. Does the object represented in FIGURE Q1.4 have a positive or negative value of a_x? Explain.
5. Does the object represented in FIGURE Q1.5 have a positive or negative value of a_y? Explain.

   FIGURE Q1.4   FIGURE Q1.5

6. Determine the signs (positive, negative, or zero) of the position, velocity, and acceleration for the particle in FIGURE Q1.6.

   FIGURE Q1.6

7. Determine the signs (positive, negative, or zero) of the position, velocity, and acceleration for the particle in FIGURE Q1.7.

   FIGURE Q1.7   FIGURE Q1.8

8. Determine the signs (positive, negative, or zero) of the position, velocity, and acceleration for the particle in FIGURE Q1.8.

EXERCISES AND PROBLEMS

Exercises

Section 1.1 Motion Diagrams

1. | A car skids to a halt to avoid hitting an object in the road. Draw a basic motion diagram, using the images from the video, from the time the skid begins until the car is stopped.
2. | A rocket is launched straight up. Draw a basic motion diagram, using the images from the video, from the moment of liftoff until the rocket is at an altitude of 500 m.
3. | You are watching a jet ski race. A racer speeds up from rest to 70 mph in just a few seconds, then continues at a constant speed. Draw a basic motion diagram of the jet ski, using images from the video, from 10 s before reaching top speed until 10 s afterward.

Section 1.2 Models and Modeling

4. | a. Write a paragraph describing the particle model. What is it, and why is it important?
   b. Give two examples of situations, different from those described in the text, for which the particle model is appropriate.
   c. Give an example of a situation, different from those described in the text, for which it would be inappropriate.

Section 1.3 Position, Time, and Displacement

Section 1.4 Velocity

5. | You drop a soccer ball from your third-story balcony. Use the particle model to draw a motion diagram showing the ball's position and average velocity vectors from the time you release the ball until the instant it touches the ground.

6. | A baseball player starts running to the left to catch the ball as soon as the hit is made. Use the particle model to draw a motion diagram showing the position and average velocity vectors of the player during the first few seconds of the run.
7. | A softball player slides into second base. Use the particle model to draw a motion diagram showing his position and his average velocity vectors from the time he begins to slide until he reaches the base.

Section 1.5 Linear Acceleration

8. | a. FIGURE EX1.8 shows the first three points of a motion diagram. Is the object's average speed between points 1 and 2 greater than, less than, or equal to its average speed between points 0 and 1? Explain how you can tell.
   b. Use Tactics Box 1.3 to find the average acceleration vector at point 1. Draw the completed motion diagram, showing the velocity vectors and acceleration vector.

   0 ●

   1 ●
                            ●   ●   ●   ●   ●
   2 ●   0   1   2   3   4
   FIGURE EX1.8   FIGURE EX1.9

9. | FIGURE EX1.9 shows five points of a motion diagram. Use Tactics Box 1.3 to find the average acceleration vectors at points 1, 2, and 3. Draw the completed motion diagram showing velocity vectors and acceleration vectors.

10. ‖ **FIGURE EX1.10** shows two dots of a motion diagram and vector $\vec{v}_1$. Copy this figure, then add dot 3 and the next velocity vector $\vec{v}_2$ if the acceleration vector $\vec{a}$ at dot 2 (a) points up and (b) points down.

FIGURE EX1.10   FIGURE EX1.11

11. ‖ **FIGURE EX1.11** shows two dots of a motion diagram and vector $\vec{v}_2$. Copy this figure, then add dot 4 and the next velocity vector $\vec{v}_3$ if the acceleration vector $\vec{a}$ at dot 3 (a) points right and (b) points left.

12. | A speed skater accelerates from rest and then keeps skating at a constant speed. Draw a complete motion diagram of the skater.

13. | A car travels to the left at a steady speed for a few seconds, then brakes for a stop sign. Draw a complete motion diagram of the car.

14. | A goose flies toward a pond. It lands on the water and slides for some distance before it comes to a stop. Draw the motion diagram of the goose, starting shortly before it hits the water and assuming the motion is entirely horizontal.

15. | You use a long rubber band to launch a paper wad straight up. Draw a complete motion diagram of the paper wad from the moment you release the stretched rubber band until the paper wad reaches its highest point.

16. | A roof tile falls straight down from a two-story building. It lands in a swimming pool and settles gently to the bottom. Draw a complete motion diagram of the tile.

17. | Your roommate drops a tennis ball from a third-story balcony. It hits the sidewalk and bounces as high as the second story. Draw a complete motion diagram of the tennis ball from the time it is released until it reaches the maximum height on its bounce. Be sure to determine and show the acceleration at the lowest point.

Section 1.6 Motion in One Dimension

18. ‖ **FIGURE EX1.18** shows the motion diagram of a drag racer. The camera took one frame every 2 s.

FIGURE EX1.18

a. Measure the x-value of the racer at each dot. List your data in a table similar to Table 1.1, showing each position and the time at which it occurred.

b. Make a position-versus-time graph for the drag racer. Because you have data only at certain instants, your graph should consist of dots that are not connected together.

19. | Write a short description of the motion of a real object for which **FIGURE EX1.19** would be a realistic position-versus-time graph.

FIGURE EX1.19

20. | Write a short description of the motion of a real object for which **FIGURE EX1.20** would be a realistic position-versus-time graph.

FIGURE EX1.20

Section 1.7 Solving Problems in Physics

21. ‖ Draw a pictorial representation for the following problem. Do *not* solve the problem. The light turns green, and a bicyclist starts forward with an acceleration of 1.5 m/s^2. How far must she travel to reach a speed of 7.5 m/s?

22. ‖ Draw a pictorial representation for the following problem. Do *not* solve the problem. What acceleration does a rocket need to reach a speed of 200 m/s at a height of 1.0 km?

Section 1.8 Units and Significant Figures

23. | How many significant figures are there in the following values?
     a. 0.05×10^{-4} b. 0.00340
     c. 7.2×10^4 d. 103.00

24. ‖ Convert the following to SI units:
     a. 8.0 in   b. 66 ft/s
     c. 60 mph   d. 14 in^2

25. | Convert the following to SI units:
     a. 75 in   b. $3.45 \times 10^6 \text{ yr}$
     c. 62 ft/day   d. $2.2 \times 10^4 \text{ mi}^2$

26. ‖ Using the approximate conversion factors in Table 1.5, convert the following SI units to English units *without* using your calculator.
     a. 30 cm   b. 25 m/s
     c. 5 km   d. 0.5 cm

27. | Using the approximate conversion factors in Table 1.5, convert the following to SI units *without* using your calculator.
     a. 20 ft   b. 60 mi
     c. 60 mph   d. 8 in

28. | Compute the following numbers, applying the significant figure rules adopted in this textbook.
    a. 33.3×25.4 b. $33.3 - 25.4$
    c. $\sqrt{33.3}$ d. $333.3 \div 25.4$

29. | Perform the following calculations with the correct number of significant figures.
    a. 159.31×204.6 b. $5.1125 + 0.67 + 3.2$
    c. $7.662 - 7.425$ d. $16.5/3.45$

30. | Estimate (don't measure!) the length of a typical car. Give your answer in both feet and meters. Briefly describe how you arrived at this estimate.

31. | Estimate the height of a telephone pole. Give your answer in both feet and meters. Briefly describe how you arrived at this estimate.

32. | Estimate the average speed with which the hair on your head
BIO grows. Give your answer in both m/s and μm/hour. Briefly describe how you arrived at this estimate.

33. | Motor neurons in mammals transmit signals from the brain
BIO to skeletal muscles at approximately 25 m/s. Estimate how long in ms it takes a signal to get from your brain to your hand.

Problems

For Problems 34 through 43, draw a complete pictorial representation. **Do *not* solve these problems or do any mathematics.**

34. | A Porsche accelerates from a stoplight at 5.0 m/s² for five seconds, then coasts for three more seconds. How far has it traveled?

35. | A jet plane is cruising at 300 m/s when suddenly the pilot turns the engines up to full throttle. After traveling 4.0 km, the jet is moving with a speed of 400 m/s. What is the jet's acceleration as it speeds up?

36. | Sam is recklessly driving 60 mph in a 30 mph speed zone when he suddenly sees the police. He steps on the brakes and slows to 30 mph in three seconds, looking nonchalant as he passes the officer. How far does he travel while braking?

37. | You would like to stick a wet spit wad on the ceiling, so you toss it straight up with a speed of 10 m/s. How long does it take to reach the ceiling, 3.0 m above?

38. | A speed skater moving across frictionless ice at 8.0 m/s hits a 5.0-m-wide patch of rough ice. She slows steadily, then continues on at 6.0 m/s. What is her acceleration on the rough ice?

39. | Santa loses his footing and slides down a frictionless, snowy roof that is tilted at an angle of 30°. If Santa slides 10 m before reaching the edge, what is his speed as he leaves the roof?

40. | A motorist is traveling at 20 m/s. He is 60 m from a stoplight when he sees it turn yellow. His reaction time, before stepping on the brake, is 0.50 s. What steady deceleration while braking will bring him to a stop right at the light?

41. | A car traveling at 30 m/s runs out of gas while traveling up a 10° slope. How far up the hill will the car coast before starting to roll back down?

42. ‖ Ice hockey star Bruce Blades is 5.0 m from the blue line and gliding toward it at a speed of 4.0 m/s. You are 20 m from the blue line, directly behind Bruce. You want to pass the puck to Bruce. With what speed should you shoot the puck down the ice so that it reaches Bruce exactly as he crosses the blue line?

43. ‖ David is driving a steady 30 m/s when he passes Tina, who is sitting in her car at rest. Tina begins to accelerate at a steady 2.0 m/s² at the instant when David passes. How far does Tina drive before passing David?

Problems 44 through 48 show a motion diagram. For each of these problems, write a one or two sentence "story" about a *real object* that has this motion diagram. Your stories should talk about people or objects by name and say what they are doing. Problems 34 through 43 are examples of motion short stories.

44. |

FIGURE P1.44

45. |

FIGURE P1.45

46. |

FIGURE P1.46

47. |

FIGURE P1.47

48. |

FIGURE P1.48

Problems 49 through 52 show a partial motion diagram. For each:
a. Complete the motion diagram by adding acceleration vectors.
b. Write a physics *problem* for which this is the correct motion diagram. Be imaginative! Don't forget to include enough information to make the problem complete and to state clearly what is to be found.
c. Draw a pictorial representation for your problem.

49.

FIGURE P1.49

50.

FIGURE P1.50

51.

FIGURE P1.51

52.

FIGURE P1.52

53. | A regulation soccer field for international play is a rectangle with a length between 100 m and 110 m and a width between 64 m and 75 m. What are the smallest and largest areas that the field could be?

54. | As an architect, you are designing a new house. A window has a height between 140 cm and 150 cm and a width between 74 cm and 70 cm. What are the smallest and largest areas that the window could be?

55. ‖ A 5.4-cm-diameter cylinder has a length of 12.5 cm. What is the cylinder's volume in SI units?

56. | An intravenous saline drip has 9.0 g of sodium chloride per
BIO liter of water. By definition, 1 mL = 1 cm³. Express the salt concentration in kg/m³.

57. ‖ The quantity called *mass density* is the mass per unit volume of a substance. What are the mass densities in SI units of the following objects?
a. A 215 cm³ solid with a mass of 0.0179 kg.
b. 95 cm³ of a liquid with a mass of 77 g.

58. | **FIGURE P1.58** shows a motion diagram of a car traveling down a street. The camera took one frame every 10 s. A distance scale is provided.

FIGURE P1.58

a. Measure the *x*-value of the car at each dot. Place your data in a table, similar to Table 1.1, showing each position and the instant of time at which it occurred.
b. Make a position-versus-time graph for the car. Because you have data only at certain instants of time, your graph should consist of dots that are not connected together.

59. | Write a short description of a real object for which **FIGURE P1.59** would be a realistic position-versus-time graph.

FIGURE P1.59

60. | Write a short description of a real object for which **FIGURE P1.60** would be a realistic position-versus-time graph.

FIGURE P1.60

2 Kinematics in One Dimension

This Japanese "bullet train" accelerates slowly but steadily until reaching a speed of 300 km/h.

IN THIS CHAPTER, you will learn to solve problems about motion along a straight line.

What is kinematics?

Kinematics is the mathematical description of motion. We begin with motion along a straight line. Our primary tools will be an object's position, velocity, and acceleration.

« LOOKING BACK Sections 1.4–1.6 Velocity, acceleration, and Tactics Box 1.4 about signs

How are graphs used in kinematics?

Graphs are a very important visual representation of motion, and learning to "think graphically" is one of our goals. We'll work with graphs showing how position, velocity, and acceleration change with time. These graphs are related to each other:

- Velocity is the slope of the position graph.
- Acceleration is the slope of the velocity graph.

How is calculus used in kinematics?

Motion is change, and calculus is the mathematical tool for describing a quantity's rate of change. We'll find that

- Velocity is the time derivative of position.
- Acceleration is the time derivative of velocity.

What are models?

A model is a simplified description of a situation that focuses on essential features while ignoring many details. Models allow us to make sense of complex situations by seeing them as variations on a common theme, all with the same underlying physics.

> **MODEL 2.1**
>
> Look for model boxes like this throughout the book.
> - Key figures
> - Key equations
> - Model limitations

What is free fall?

Free fall is motion under the influence of gravity only. Free fall is not literally "falling" because it also applies to objects thrown straight up and to projectiles. Surprisingly, all objects in free fall, *regardless of their mass*, have the same acceleration. Motion on a frictionless inclined plane is closely related to free-fall motion.

How will I use kinematics?

The equations of motion that you learn in this chapter will be used throughout the entire book. In Part I, we'll see how an object's motion is related to forces acting on the object. We'll later apply these kinematic equations to the motion of waves and to the motion of charged particles in electric and magnetic fields.

2.1 Uniform Motion

The simplest possible motion is motion along a straight line at a constant, unvarying speed. We call this **uniform motion.** Because velocity is the combination of speed and direction, **uniform motion is motion with constant velocity.**

FIGURE 2.1 shows the motion diagram of an object in uniform motion. For example, this might be you riding your bicycle along a straight line at a perfectly steady 5 m/s ($\approx$ 10 mph). Notice how all the displacements are exactly the same; this is a characteristic of uniform motion.

If we make a position-versus-time graph—remember that position is graphed on the *vertical* axis—it's a straight line. In fact, an alternative definition is that **an object's motion is uniform if and only if its position-versus-time graph is a straight line.**

≪ Section 1.4 defined an object's **average velocity** as $\Delta\vec{r}/\Delta t$. For one-dimensional motion, this is simply $\Delta x/\Delta t$ (for horizontal motion) or $\Delta y/\Delta t$ (for vertical motion). You can see in Figure 2.1 that Δx and Δt are, respectively, the "rise" and "run" of the position graph. Because rise over run is the slope of a line,

$$v_{\text{avg}} \equiv \frac{\Delta x}{\Delta t} \quad \text{or} \quad \frac{\Delta y}{\Delta t} = \text{slope of the position-versus-time graph} \qquad (2.1)$$

That is, **the average velocity is the slope of the position-versus-time graph.** Velocity has units of "length per time," such as "miles per hour." The SI units of velocity are meters per second, abbreviated m/s.

> **NOTE** The symbol $\equiv$ in Equation 2.1 stands for "is defined as." This is a stronger statement than the two sides simply being equal.

The constant slope of a straight-line graph is another way to see that the velocity is constant for uniform motion. There's no real need to specify "average" for a velocity that doesn't change, so we will drop the subscript and refer to the average velocity as v_x or v_y.

An object's **speed** v is how fast it's going, independent of direction. This is simply $v = |v_x|$ or $v = |v_y|$, the magnitude or absolute value of the object's velocity. Although we will use speed from time to time, our mathematical analysis of motion is based on velocity, not speed. The subscript in v_x or v_y is an essential part of the notation, reminding us that, even in one dimension, the velocity is a vector.

FIGURE 2.1 Motion diagram and position graph for uniform motion.

The displacements between successive frames are the same.

The position graph is a straight line. Its slope is $\Delta x/\Delta t$.

EXAMPLE 2.1 | **Relating a velocity graph to a position graph**

FIGURE 2.2 is the position-versus-time graph of a car.

a. Draw the car's velocity-versus-time graph.

b. Describe the car's motion.

MODEL Model the car as a particle, with a well-defined position at each instant of time.

VISUALIZE Figure 2.2 is the graphical representation.

SOLVE a. The car's position-versus-time graph is a sequence of three straight lines. Each of these straight lines represents uniform motion at a constant velocity. We can determine the car's velocity during each interval of time by measuring the slope of the line.

The position graph starts out sloping downward—a negative slope. Although the car moves a distance of 4.0 m during the first 2.0 s, its *displacement* is

$$\Delta x = x_{\text{at 2.0 s}} - x_{\text{at 0.0 s}} = -4.0\ \text{m} - 0.0\ \text{m} = -4.0\ \text{m}$$

The time interval for this displacement is $\Delta t = 2.0$ s, so the velocity during this interval is

$$v_x = \frac{\Delta x}{\Delta t} = \frac{-4.0\ \text{m}}{2.0\ \text{s}} = -2.0\ \text{m/s}$$

The car's position does not change from $t = 2$ s to $t = 4$ s ($\Delta x = 0$), so $v_x = 0$. Finally, the displacement between $t = 4$ s and $t = 6$ s is $\Delta x = 10.0$ m. Thus the velocity during this interval is

$$v_x = \frac{10.0\ \text{m}}{2.0\ \text{s}} = 5.0\ \text{m/s}$$

These velocities are shown on the velocity-versus-time graph of FIGURE 2.3.

b. The car backs up for 2 s at 2.0 m/s, sits at rest for 2 s, then drives forward at 5.0 m/s for at least 2 s. We can't tell from the graph what happens for $t > 6$ s.

ASSESS The velocity graph and the position graph look completely different. The *value* of the velocity graph at any instant of time equals the *slope* of the position graph.

Continued

FIGURE 2.2 Position-versus-time graph.

FIGURE 2.3 The corresponding velocity-versus-time graph.

Example 2.1 brought out several points that are worth emphasizing.

TACTICS BOX 2.1 (MP)

Interpreting position-versus-time graphs

❶ Steeper slopes correspond to faster speeds.
❷ Negative slopes correspond to negative velocities and, hence, to motion to the left (or down).
❸ The slope is a ratio of intervals, $\Delta x / \Delta t$, not a ratio of coordinates. That is, the slope is *not* simply x/t.

Exercises 1–3

NOTE We are distinguishing between the *actual* slope and the *physically meaningful* slope. If you were to use a ruler to measure the rise and the run of the graph, you could compute the actual slope of the line as drawn on the page. That is not the slope to which we are referring when we equate the velocity with the slope of the line. Instead, we find the *physically meaningful* slope by measuring the rise and run using the scales along the axes. The "rise" Δx is some number of meters; the "run" Δt is some number of seconds. The physically meaningful rise and run include units, and the ratio of these units gives the units of the slope.

The Mathematics of Uniform Motion

The physics of the motion is the same regardless of whether an object moves along the x-axis, the y-axis, or any other straight line. Consequently, it will be convenient to write equations for a "generic axis" that we will call the s-axis. The position of an object will be represented by the symbol s and its velocity by v_s.

NOTE In a specific problem you should use either x or y rather than s.

Consider an object in uniform motion along the s-axis with the linear position-versus-time graph shown in FIGURE 2.4. The object's **initial position** is s_i at time t_i. The term *initial position* refers to the starting point of our analysis or the starting point in a problem; the object may or may not have been in motion prior to t_i. At a later time t_f, the ending point of our analysis, the object's **final position** is s_f.

The object's velocity v_s along the s-axis can be determined by finding the slope of the graph:

$$v_s = \frac{\text{rise}}{\text{run}} = \frac{\Delta s}{\Delta t} = \frac{s_f - s_i}{t_f - t_i} \tag{2.2}$$

FIGURE 2.4 The velocity is found from the slope of the position-versus-time graph.

We will use s as a generic label for position. In practice, s could be either x or y.

The slope of the line is $v_s = \Delta s/\Delta t$.

Equation 2.2 is easily rearranged to give

$$s_f = s_i + v_s \, \Delta t \qquad \text{(uniform motion)} \qquad (2.3)$$

Equation 2.3 tells us that the object's position increases linearly as the elapsed time Δt increases—exactly as we see in the straight-line position graph.

The Uniform-Motion Model

Chapter 1 introduced a *model* as a simplified picture of reality, but one that still captures the essence of what we want to study. When it comes to motion, few real objects move with a precisely constant velocity. Even so, there are many cases in which it is quite reasonable to model their motion as being uniform. That is, uniform motion is a very good approximation of their actual, but more complex, motion. The **uniform-motion model** is a coherent set of representations—words, pictures, graphs, and equations—that allows us to explain an object's motion and to predict where the object will be at a future instant of time.

MODEL 2.1

Uniform motion

For motion with constant velocity.

- Model the object as a particle moving in a straight line at constant speed:

- Mathematically:
  - $v_s = \Delta s / \Delta t$
  - $s_f = s_i + v_s \, \Delta t$

- Limitations: Model fails if the particle has a significant change of speed or direction.

Exercise 4

EXAMPLE 2.2 | Lunch in Cleveland?

Bob leaves home in Chicago at 9:00 A.M. and drives east at 60 mph. Susan, 400 miles to the east in Pittsburgh, leaves at the same time and travels west at 40 mph. Where will they meet for lunch?

MODEL Here is a problem where, for the first time, we can really put all four aspects of our problem-solving strategy into play. To begin, we'll model Bob's and Susan's cars as being in uniform

motion. Their real motion is certainly more complex, but over a long drive it's reasonable to approximate their motion as constant speed along a straight line.

VISUALIZE FIGURE 2.5 shows the pictorial representation. The equal spacings of the dots in the motion diagram indicate that the motion is uniform. In evaluating the given information, we

FIGURE 2.5 Pictorial representation for Example 2.2.

Continued

recognize that the starting time of 9:00 A.M. is not relevant to the problem. Consequently, the initial time is chosen as simply $t_0 = 0$ h. Bob and Susan are traveling in opposite directions, hence one of the velocities must be a negative number. We have chosen a coordinate system in which Bob starts at the origin and moves to the right (east) while Susan is moving to the left (west). Thus Susan has the negative velocity. Notice how we've assigned position, velocity, and time symbols to each point in the motion. Pay special attention to how subscripts are used to distinguish different points in the problem and to distinguish Bob's symbols from Susan's.

One purpose of the pictorial representation is to establish what we need to find. Bob and Susan meet when they have the same position at the same time t_1. Thus we want to find $(x_1)_B$ at the time when $(x_1)_B = (x_1)_S$. Notice that $(x_1)_B$ and $(x_1)_S$ are Bob's and Susan's *positions,* which are equal when they meet, not the distances they have traveled.

SOLVE The goal of the mathematical representation is to proceed from the pictorial representation to a mathematical solution of the problem. We can begin by using Equation 2.3 to find Bob's and Susan's positions at time t_1 when they meet:

$$(x_1)_B = (x_0)_B + (v_x)_B(t_1 - t_0) = (v_x)_B t_1$$
$$(x_1)_S = (x_0)_S + (v_x)_S(t_1 - t_0) = (x_0)_S + (v_x)_S t_1$$

Notice two things. First, we started by writing the *full* statement of Equation 2.3. Only then did we simplify by dropping those terms known to be zero. You're less likely to make accidental errors if you follow this procedure. Second, we replaced the generic symbol s with the specific horizontal-position symbol x, and we replaced the generic subscripts i and f with the specific symbols 0 and 1 that we defined in the pictorial representation. This is also good problem-solving technique.

The condition that Bob and Susan meet is

$$(x_1)_B = (x_1)_S$$

By equating the right-hand sides of the above equations, we get

$$(v_x)_B t_1 = (x_0)_S + (v_x)_S t_1$$

Solving for t_1 we find that they meet at time

$$t_1 = \frac{(x_0)_S}{(v_x)_B - (v_x)_S} = \frac{400 \text{ miles}}{60 \text{ mph} - (-40) \text{ mph}} = 4.0 \text{ hours}$$

Finally, inserting this time back into the equation for $(x_1)_B$ gives

$$(x_1)_B = \left(60 \frac{\text{miles}}{\text{hour}}\right) \times (4.0 \text{ hours}) = 240 \text{ miles}$$

As noted in Chapter 1, this textbook will assume that all data are good to at least two significant figures, even when one of those is a trailing zero. So 400 miles, 60 mph, and 40 mph each have two significant figures, and consequently we've calculated results to two significant figures.

While 240 miles is a number, it is not yet the answer to the question. The phrase "240 miles" by itself does not say anything meaningful. Because this is the value of Bob's *position,* and Bob was driving east, the answer to the question is, "They meet 240 miles east of Chicago."

ASSESS Before stopping, we should check whether or not this answer seems reasonable. We certainly expected an answer between 0 miles and 400 miles. We also know that Bob is driving faster than Susan, so we expect that their meeting point will be *more* than halfway from Chicago to Pittsburgh. Our assessment tells us that 240 miles is a reasonable answer.

FIGURE 2.6 Position-versus-time graphs for Bob and Susan.

It is instructive to look at this example from a graphical perspective. **FIGURE 2.6** shows position-versus-time graphs for Bob and Susan. Notice the negative slope for Susan's graph, indicating her negative velocity. The point of interest is the intersection of the two lines; this is where Bob and Susan have the same position at the same time. Our method of solution, in which we equated $(x_1)_B$ and $(x_1)_S$, is really just solving the mathematical problem of finding the intersection of two lines. This procedure is useful for many problems in which there are two moving objects.

STOP TO THINK 2.1 Which position-versus-time graph represents the motion shown in the motion diagram?

2.2 Instantaneous Velocity

Uniform motion is simple, but objects rarely travel for long with a constant velocity. Far more common is a velocity that changes with time. For example, FIGURE 2.7 shows the motion diagram and position graph of a car speeding up after the light turns green. Notice how the velocity vectors increase in length, causing the graph to curve upward as the car's displacements get larger and larger.

If you were to watch the car's speedometer, you would see it increase from 0 mph to 10 mph to 20 mph and so on. At any instant of time, the speedometer tells you how fast the car is going *at that instant*. If we include directional information, we can define an object's **instantaneous velocity**—speed and direction—as its velocity at a single instant of time.

For uniform motion, the slope of the straight-line position graph is the object's velocity. FIGURE 2.8 shows that there's a similar connection between instantaneous velocity and the slope of a curved position graph.

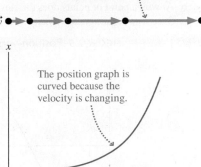

FIGURE 2.7 Motion diagram and position graph of a car speeding up.

The spacing between the dots increases as the car speeds up.

The position graph is curved because the velocity is changing.

FIGURE 2.8 Instantaneous velocity at time t is the slope of the tangent to the curve at that instant.

What is the velocity at time t?

Zoom in on a *very* small segment of the curve centered on the point of interest. This little piece of the curve is essentially a straight line. Its slope $\Delta s/\Delta t$ is the average velocity during the interval Δt.

The little segment of straight line, when extended, is the tangent to the curve at time t. Its slope is the instantaneous velocity at time t.

What we see graphically is that the average velocity $v_{\text{avg}} = \Delta s/\Delta t$ becomes a better and better approximation to the instantaneous velocity v_s as the time interval Δt over which the average is taken gets smaller and smaller. We can state this idea mathematically in terms of the limit $\Delta t \rightarrow 0$:

$$v_s \equiv \lim_{\Delta t \to 0} \frac{\Delta s}{\Delta t} = \frac{ds}{dt} \qquad \text{(instantaneous velocity)} \qquad (2.4)$$

As Δt continues to get smaller, the average velocity $v_{\text{avg}} = \Delta s/\Delta t$ reaches a constant or *limiting* value. That is, **the instantaneous velocity at time t is the average velocity during a time interval Δt, centered on t, as Δt approaches zero.** In calculus, this limit is called *the derivative of s with respect to t*, and it is denoted ds/dt.

Graphically, $\Delta s/\Delta t$ is the slope of a straight line. As Δt gets smaller (i.e., more and more magnification), the straight line becomes a better and better approximation of the curve *at that one point*. In the limit $\Delta t \rightarrow 0$, the straight line is tangent to the curve. As Figure 2.8 shows, **the instantaneous velocity at time t is the slope of the line that is tangent to the position-versus-time graph at time t.** That is,

$$v_s = \text{slope of the position-versus-time graph at time } t \qquad (2.5)$$

The steeper the slope, the larger the magnitude of the velocity.

EXAMPLE 2.3 | **Finding velocity from position graphically**

FIGURE 2.9 shows the position-versus-time graph of an elevator.

a. At which labeled point or points does the elevator have the least velocity?

b. At which point or points does the elevator have maximum velocity?

c. Sketch an approximate velocity-versus-time graph for the elevator.

FIGURE 2.9 Position-versus-time graph.

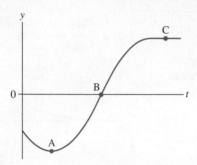

MODEL Model the elevator as a particle.

VISUALIZE Figure 2.9 is the graphical representation.

SOLVE a. At any instant, an object's velocity is the slope of its position graph. **FIGURE 2.10a** shows that the elevator has the least velocity—no velocity at all!—at points A and C where the slope is zero. At point A, the velocity is only instantaneously zero. At point C, the elevator has actually stopped and remains at rest.

b. The elevator has maximum velocity at B, the point of steepest slope.

c. Although we cannot find an exact velocity-versus-time graph, we can see that the slope, and hence v_y, is initially negative, becomes zero at point A, rises to a maximum value at point B, decreases back to zero a little before point C, then remains at zero thereafter.

FIGURE 2.10 The velocity-versus-time graph is found from the slope of the position graph.

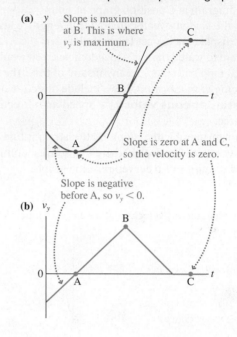

Thus **FIGURE 2.10b** shows, at least approximately, the elevator's velocity-versus-time graph.

ASSESS Once again, the shape of the velocity graph bears no resemblance to the shape of the position graph. You must transfer *slope* information from the position graph to *value* information on the velocity graph.

A Little Calculus: Derivatives

Calculus—invented simultaneously in England by Newton and in Germany by Leibniz—is designed to deal with instantaneous quantities. In other words, it provides us with the tools for evaluating limits such as the one in Equation 2.4.

The notation ds/dt is called *the derivative of s with respect to t*, and Equation 2.4 defines it as the limiting value of a ratio. As Figure 2.8 showed, ds/dt can be interpreted graphically as the slope of the line that is tangent to the position graph.

The most common functions we will use in Parts I and II of this book are powers and polynomials. Consider the function $u(t) = ct^n$, where c and n are constants. The symbol u is a "dummy name" to represent any function of time, such as $x(t)$ or $y(t)$. The following result is proven in calculus:

The derivative of $u = ct^n$ is $\dfrac{du}{dt} = nct^{n-1}$ (2.6)

For example, suppose the position of a particle as a function of time is $s(t) = 2t^2$ m, where t is in s. We can find the particle's velocity $v_s = ds/dt$ by using Equation 2.6 with $c = 2$ and $n = 2$ to calculate

$$v_s = \frac{ds}{dt} = 2 \cdot 2t^{2-1} = 4t$$

This is an expression for the particle's velocity as a function of time.

Scientists and engineers must use calculus to calculate the orbits of satellites.

FIGURE 2.11 shows the particle's position and velocity graphs. It is critically important to understand the relationship between these two graphs. The *value* of the velocity graph at any instant of time, which we can read directly off the vertical axis, is the *slope* of the position graph at that same time. This is illustrated at $t = 3$ s.

A value that doesn't change with time, such as the position of an object at rest, can be represented by the function $u = c =$ constant. That is, the exponent of t^n is $n = 0$. You can see from Equation 2.6 that the derivative of a constant is zero. That is,

$$\frac{du}{dt} = 0 \text{ if } u = c = \text{constant} \qquad (2.7)$$

This makes sense. The graph of the function $u = c$ is simply a horizontal line. The slope of a horizontal line—which is what the derivative du/dt measures—is zero.

The only other information we need about derivatives for now is how to evaluate the derivative of the sum of two functions. Let u and w be two separate functions of time. You will learn in calculus that

$$\frac{d}{dt}(u + w) = \frac{du}{dt} + \frac{dw}{dt} \qquad (2.8)$$

That is, the derivative of a sum is the sum of the derivatives.

NOTE You may have learned in calculus to take the derivative dy/dx, where y is a function of x. The derivatives we use in physics are the same; only the notation is different. We're interested in how quantities change with time, so our derivatives are with respect to t instead of x.

FIGURE 2.11 Position-versus-time graph and the corresponding velocity-versus-time graph.

(a)

(b)

EXAMPLE 2.4 | **Using calculus to find the velocity**

A particle's position is given by the function $x(t) = (-t^3 + 3t)$ m, where t is in s.

a. What are the particle's position and velocity at $t = 2$ s?

b. Draw graphs of x and v_x during the interval -3 s $\leq t \leq 3$ s.

c. Draw a motion diagram to illustrate this motion.

SOLVE

a. We can compute the position directly from the function x:

$$x(\text{at } t = 2 \text{ s}) = -(2)^3 + (3)(2) = -8 + 6 = -2 \text{ m}$$

The velocity is $v_x = dx/dt$. The function for x is the sum of two polynomials, so

$$v_x = \frac{dx}{dt} = \frac{d}{dt}(-t^3 + 3t) = \frac{d}{dt}(-t^3) + \frac{d}{dt}(3t)$$

The first derivative is a power with $c = -1$ and $n = 3$; the second has $c = 3$ and $n = 1$. Using Equation 2.6, we have

$$v_x = (-3t^2 + 3) \text{ m/s}$$

where t is in s. Evaluating the velocity at $t = 2$ s gives

$$v_x(\text{at } t = 2 \text{ s}) = -3(2)^2 + 3 = -9 \text{ m/s}$$

The negative sign indicates that the particle, at this instant of time, is moving to the *left* at a speed of 9 m/s.

b. **FIGURE 2.12** shows the position graph and the velocity graph. You can make graphs like these with a graphing calculator or graphing software. The slope of the position-versus-time graph at $t = 2$ s is -9 m/s; this becomes the *value* that is graphed for the velocity at $t = 2$ s.

FIGURE 2.12 Position and velocity graphs.

Continued

c. Finally, we can interpret the graphs in Figure 2.12 to draw the motion diagram shown in **FIGURE 2.13**.

- The particle is initially to the right of the origin ($x > 0$ at $t = -3$ s) but moving to the left ($v_x < 0$). Its *speed* is slowing ($v = |v_x|$ is decreasing), so the velocity vector arrows are getting shorter.
- The particle passes the origin $x = 0$ m at $t \approx -1.5$ s, but it is still moving to the left.
- The position reaches a minimum at $t = -1$ s; the particle is as far left as it is going. The velocity is *instantaneously* $v_x = 0$ m/s as the particle reverses direction.

- The particle moves back to the right between $t = -1$ s and $t = 1$ s ($v_x > 0$).
- The particle turns around again at $t = 1$ s and begins moving back to the left ($v_x < 0$). It keeps speeding up, then disappears off to the left.

A point in the motion where a particle reverses direction is called a **turning point.** It is a point where the velocity is instantaneously zero while the position is a maximum or minimum. This particle has two turning points, at $t = -1$ s and again at $t = +1$ s. We will see many other examples of turning points.

FIGURE 2.13 Motion diagram for Example 2.4.

STOP TO THINK 2.2 Which velocity-versus-time graph goes with the position-versus-time graph on the left?

(a)   (b)   (c)   (d)

FIGURE 2.14 Approximating a velocity-versus-time graph with a series of constant-velocity steps.

2.3 Finding Position from Velocity

Equation 2.4 allows us to find the instantaneous velocity v_s if we know the position s as a function of time. But what about the reverse problem? Can we use the object's velocity to calculate its position at some future time t? Equation 2.3, $s_f = s_i + v_s \Delta t$, does this for the case of uniform motion with a constant velocity. We need to find a more general expression that is valid when v_s is not constant.

FIGURE 2.14a is a velocity-versus-time graph for an object whose velocity varies with time. Suppose we know the object's position to be s_i at an initial time t_i. Our goal is to find its position s_f at a later time t_f.

Because we know how to handle constant velocities, using Equation 2.3, let's *approximate* the velocity function of Figure 2.14a as a series of constant-velocity steps of width Δt. This is illustrated in **FIGURE 2.14b**. During the first step, from time t_i to time $t_i + \Delta t$, the velocity has the constant value $(v_s)_1$. The velocity during step k has the constant value $(v_s)_k$. Although the approximation shown in the figure is rather rough, with only 11 steps, we can easily imagine that it could be made as accurate as desired by having more and more ever-narrower steps.

The velocity during each step is constant (uniform motion), so we can apply Equation 2.3 to each step. The object's displacement Δs_1 during the first step is simply $\Delta s_1 = (v_s)_1 \Delta t$. The displacement during the second step $\Delta s_2 = (v_s)_2 \Delta t$, and during step k the displacement is $\Delta s_k = (v_s)_k \Delta t$.

The total displacement of the object between t_i and t_f can be approximated as the sum of all the individual displacements during each of the N constant-velocity steps. That is,

$$\Delta s = s_f - s_i \approx \Delta s_1 + \Delta s_2 + \cdots + \Delta s_N = \sum_{k=1}^{N} (v_s)_k \Delta t \qquad (2.9)$$

where Σ (Greek sigma) is the symbol for summation. With a simple rearrangement, the particle's final position is

$$s_f \approx s_i + \sum_{k=1}^{N} (v_s)_k \Delta t \qquad (2.10)$$

Our goal was to use the object's velocity to find its final position s_f. Equation 2.10 nearly reaches that goal, but Equation 2.10 is only approximate because the constant-velocity steps are only an approximation of the true velocity graph. But if we now let $\Delta t \to 0$, each step's width approaches zero while the total number of steps N approaches infinity. In this limit, the series of steps becomes a perfect replica of the velocity-versus-time graph and Equation 2.10 becomes exact. Thus

$$s_f = s_i + \lim_{\Delta t \to 0} \sum_{k=1}^{N} (v_s)_k \Delta t = s_i + \int_{t_i}^{t_f} v_s \, dt \qquad (2.11)$$

The expression on the right is read, "the integral of $v_s \, dt$ from t_i to t_f." Equation 2.11 is the result that we were seeking. It allows us to predict an object's position s_f at a future time t_f.

We can give Equation 2.11 an important geometric interpretation. FIGURE 2.15 shows step k in the approximation of the velocity graph as a tall, thin rectangle of height $(v_s)_k$ and width Δt. The product $\Delta s_k = (v_s)_k \Delta t$ is the area (base × height) of this small rectangle. The sum in Equation 2.11 adds up all of these rectangular areas to give the total area enclosed between the t-axis and the tops of the steps. The limit of this sum as $\Delta t \to 0$ is the total area enclosed between the t-axis and the velocity curve. This is called the "area under the curve." Thus a graphical interpretation of Equation 2.11 is

$$s_f = s_i + \text{area under the velocity curve } v_s \text{ between } t_i \text{ and } t_f \qquad (2.12)$$

NOTE Wait a minute! The displacement $\Delta s = s_f - s_i$ is a length. How can a length equal an area? Recall earlier, when we found that the velocity is the slope of the position graph, we made a distinction between the *actual* slope and the *physically meaningful* slope? The same distinction applies here. We need to measure the quantities we are using, v_s and Δt, by referring to the scales on the axes. Δt is some number of seconds while v_s is some number of meters per second. When these are multiplied together, the *physically meaningful* area has units of meters.

FIGURE 2.15 The total displacement Δs is the "area under the curve."

During step k, the product $\Delta s_k = (v_s)_k \Delta t$ is the area of the shaded rectangle.

During the interval t_i to t_f, the total displacement Δs is the "area under the curve."

EXAMPLE 2.5 | The displacement during a drag race

FIGURE 2.16 shows the velocity-versus-time graph of a drag racer. How far does the racer move during the first 3.0 s?

FIGURE 2.16 Velocity-versus-time graph for Example 2.5.

The line is the function $v_x = 4t$ m/s.

The displacement Δx is the area of the shaded triangle.

MODEL Model the drag racer as a particle with a well-defined position at all times.

VISUALIZE Figure 2.16 is the graphical representation.

SOLVE The question "How far?" indicates that we need to find a displacement Δx rather than a position x. According to Equation 2.12, the car's displacement $\Delta x = x_f - x_i$ between $t = 0$ s and $t = 3$ s is the area under the curve from $t = 0$ s to $t = 3$ s. The curve in this case is an angled line, so the area is that of a triangle:

$$\Delta x = \text{area of triangle between } t = 0 \text{ s and } t = 3 \text{ s}$$
$$= \tfrac{1}{2} \times \text{base} \times \text{height}$$
$$= \tfrac{1}{2} \times 3 \text{ s} \times 12 \text{ m/s} = 18 \text{ m}$$

The drag racer moves 18 m during the first 3 seconds.

ASSESS The "area" is a product of s with m/s, so Δx has the proper units of m.

EXAMPLE 2.6 | Finding the turning point

FIGURE 2.17 is the velocity graph for a particle that starts at $x_i = 30$ m at time $t_i = 0$ s.

a. Draw a motion diagram for the particle.

b. Where is the particle's turning point?

c. At what time does the particle reach the origin?

FIGURE 2.17 Velocity-versus-time graph for the particle of Example 2.6.

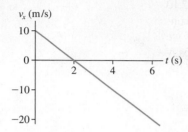

FIGURE 2.18 Motion diagram for the particle whose velocity graph was shown in Figure 2.17.

VISUALIZE The particle is initially 30 m to the right of the origin and moving *to the right* ($v_x > 0$) with a speed of 10 m/s. But v_x is decreasing, so the particle is slowing down. At $t = 2$ s the velocity, just for an instant, is zero before becoming negative. This is the turning point. The velocity is negative for $t > 2$ s, so the particle has reversed direction and moves back toward the origin. At some later time, which we want to find, the particle will pass $x = 0$ m.

SOLVE a. **FIGURE 2.18** shows the motion diagram. The distance scale will be established in parts b and c but is shown here for convenience.

b. The particle reaches the turning point at $t = 2$ s. To learn *where* it is at that time we need to find the displacement during the first two seconds. We can do this by finding the area under the curve between $t = 0$ s and $t = 2$ s:

$$x(\text{at } t = 2 \text{ s}) = x_i + \text{area under the curve between 0 s and 2 s}$$
$$= 30 \text{ m} + \tfrac{1}{2}(2 \text{ s} - 0 \text{ s})(10 \text{ m/s} - 0 \text{ m/s})$$
$$= 40 \text{ m}$$

The turning point is at $x = 40$ m.

c. The particle needs to move $\Delta x = -40$ m to get from the turning point to the origin. That is, the area under the curve from $t = 2$ s to the desired time t needs to be -40 m. Because the curve is below the axis, with negative values of v_x, the area to the right of $t = 2$ s is a *negative* area. With a bit of geometry, you will find that the triangle with a base extending from $t = 2$ s to $t = 6$ s has an area of -40 m. Thus the particle reaches the origin at $t = 6$ s.

A Little More Calculus: Integrals

Taking the derivative of a function is equivalent to finding the slope of a graph of the function. Similarly, evaluating an integral is equivalent to finding the area under a graph of the function. The graphical method is very important for building intuition about motion but is limited in its practical application. Just as derivatives of standard functions can be evaluated and tabulated, so can integrals.

The integral in Equation 2.11 is called a *definite integral* because there are two definite boundaries to the area we want to find. These boundaries are called the lower (t_i) and upper (t_f) *limits of integration*. For the important function $u(t) = ct^n$, the essential result from calculus is that

$$\int_{t_i}^{t_f} u\, dt = \int_{t_i}^{t_f} ct^n\, dt = \frac{ct^{n+1}}{n+1} \bigg|_{t_i}^{t_f} = \frac{ct_f^{n+1}}{n+1} - \frac{ct_i^{n+1}}{n+1} \qquad (n \neq -1) \qquad (2.13)$$

The vertical bar in the third step with subscript t_i and superscript t_f is a shorthand notation from calculus that means—as seen in the last step—the integral evaluated at the upper limit t_f *minus* the integral evaluated at the lower limit t_i. You also need to know that for two functions u and w,

$$\int_{t_i}^{t_f} (u + w)\, dt = \int_{t_i}^{t_f} u\, dt + \int_{t_i}^{t_f} w\, dt \qquad (2.14)$$

That is, the integral of a sum is equal to the sum of the integrals.

EXAMPLE 2.7 | Using calculus to find the position

Use calculus to solve Example 2.6.

SOLVE Figure 2.17 is a linear graph. Its "y-intercept" is seen to be 10 m/s and its slope is −5 (m/s)/s. Thus the velocity can be described by the equation

$$v_x = (10 - 5t) \text{ m/s}$$

where t is in s. We can find the position x at time t by using Equation 2.11:

$$x = x_i + \int_0^t v_x \, dt = 30 \text{ m} + \int_0^t (10 - 5t) \, dt$$

$$= 30 \text{ m} + \int_0^t 10 \, dt - \int_0^t 5t \, dt$$

We used Equation 2.14 for the integral of a sum to get the final expression. The first integral is a function of the form $u = ct^n$ with $c = 10$ and $n = 0$; the second is of the form $u = ct^n$ with $c = 5$ and $n = 1$. Using Equation 2.13, we have

$$\int_0^t 10 \, dt = 10t \Big|_0^t = 10 \cdot t - 10 \cdot 0 = 10t \text{ m}$$

and

$$\int_0^t 5t \, dt = \tfrac{5}{2}t^2 \Big|_0^t = \tfrac{5}{2} \cdot t^2 - \tfrac{5}{2} \cdot 0^2 = \tfrac{5}{2}t^2 \text{ m}$$

Combining the pieces gives

$$x = \left(30 + 10t - \tfrac{5}{2}t^2\right) \text{ m}$$

This is a general result for the position at *any* time t.

The particle's turning point occurs at $t = 2$ s, and its position at that time is

$$x(\text{at } t = 2 \text{ s}) = 30 + (10)(2) - \tfrac{5}{2}(2)^2 = 40 \text{ m}$$

The time at which the particle reaches the origin is found by setting $x = 0$ m:

$$30 + 10t - \tfrac{5}{2}t^2 = 0$$

This quadratic equation has two solutions: $t = -2$ s or $t = 6$ s.

When we solve a quadratic equation, we cannot just arbitrarily select the root we want. Instead, we must decide which is the *meaningful* root. Here the negative root refers to a time before the problem began, so the meaningful one is the positive root, $t = 6$ s.

ASSESS The results agree with the answers we found previously from a graphical solution.

STOP TO THINK 2.3 Which position-versus-time graph goes with the velocity-versus-time graph on the left? The particle's position at $t_i = 0$ s is $x_i = -10$ m.

(a)   (b)   (c)   (d)

2.4 Motion with Constant Acceleration

We need one more major concept to describe one-dimensional motion: acceleration. Acceleration, as we noted in Chapter 1, is a rather abstract concept. Nonetheless, acceleration is the linchpin of mechanics. We will see very shortly that Newton's laws relate the acceleration of an object to the forces that are exerted on it.

Let's conduct a race between a Volkswagen Beetle and a Porsche to see which can achieve a velocity of 30 m/s ($\approx$60 mph) in the shortest time. Both cars are equipped with computers that will record the speedometer reading 10 times each second. This gives a nearly continuous record of the *instantaneous* velocity of each car. **TABLE 2.1** shows some of the data. The velocity-versus-time graphs, based on these data, are shown in **FIGURE 2.19** on the next page.

How can we describe the difference in performance of the two cars? It is not that one has a different velocity from the other; both achieve every velocity between 0 and 30 m/s. The distinction is how long it took each to *change* its velocity from 0 to 30 m/s. The Porsche changed velocity quickly, in 6.0 s, while the VW needed 15 s to make

TABLE 2.1 Velocities of a Porsche and a Volkswagen Beetle

t (s)	v_{Porsche} (m/s)	v_{VW} (m/s)
0.0	0.0	0.0
0.1	0.5	0.2
0.2	1.0	0.4
0.3	1.5	0.6
⋮	⋮	⋮

FIGURE 2.19 Velocity-versus-time graphs for the Porsche and the VW Beetle.

the same velocity change. Because the Porsche had a velocity change $\Delta v_s = 30$ m/s during a time interval $\Delta t = 6.0$ s, the *rate* at which its velocity changed was

$$\text{rate of velocity change} = \frac{\Delta v_s}{\Delta t} = \frac{30 \text{ m/s}}{6.0 \text{ s}} = 5.0 \text{ (m/s)/s} \qquad (2.15)$$

Notice the units. They are units of "velocity per second." A rate of velocity change of 5.0 "meters per second per second" means that the velocity increases by 5.0 m/s during the first second, by another 5.0 m/s during the next second, and so on. In fact, the velocity will increase by 5.0 m/s during any second in which it is changing at the rate of 5.0 (m/s)/s.

Chapter 1 introduced *acceleration* as "the rate of change of velocity." That is, acceleration measures how quickly or slowly an object's velocity changes. In parallel with our treatment of velocity, let's define the **average acceleration** a_{avg} during the time interval Δt to be

$$a_{\text{avg}} \equiv \frac{\Delta v_s}{\Delta t} \qquad \text{(average acceleration)} \qquad (2.16)$$

Equations 2.15 and 2.16 show that the Porsche had the rather large acceleration of 5.0 (m/s)/s.

Because Δv_s and Δt are the "rise" and "run" of a velocity-versus-time graph, we see that a_{avg} can be interpreted graphically as the *slope* of a straight-line velocity-versus-time graph. In other words,

$$a_{\text{avg}} = \text{slope of the velocity-versus-time graph} \qquad (2.17)$$

Figure 2.19 uses this idea to show that the VW's average acceleration is

$$a_{\text{VW avg}} = \frac{\Delta v_s}{\Delta t} = \frac{10 \text{ m/s}}{5.0 \text{ s}} = 2.0 \text{ (m/s)/s}$$

This is less than the acceleration of the Porsche, as expected.

An object whose velocity-versus-time graph is a straight-line graph has a steady and unchanging acceleration. There's no need to specify "average" if the acceleration is constant, so we'll use the symbol a_s as we discuss motion along the s-axis with constant acceleration.

Signs and Units

An important aspect of acceleration is its *sign*. Acceleration $\vec{a}$, like position $\vec{r}$ and velocity $\vec{v}$, is a vector. For motion in one dimension, the sign of a_x (or a_y) is positive if the vector $\vec{a}$ points to the right (or up), negative if it points to the left (or down). This was illustrated in « Figure 1.18 and the very important « Tactics Box 1.4, which you may wish to review. It's particularly important to emphasize that positive and negative values of a_s do *not* correspond to "speeding up" and "slowing down."

EXAMPLE 2.8 | **Relating acceleration to velocity**

a. A bicyclist has a velocity of 6 m/s and a constant acceleration of 2 (m/s)/s. What is her velocity 1 s later? 2 s later?

b. A bicyclist has a velocity of −6 m/s and a constant acceleration of 2 (m/s)/s. What is his velocity 1 s later? 2 s later?

SOLVE

a. An acceleration of 2 (m/s)/s *means* that the velocity increases by 2 m/s every 1 s. If the bicyclist's initial velocity is 6 m/s, then 1 s later her velocity will be 8 m/s. After 2 s, which is 1

additional second later, it will increase by another 2 m/s to 10 m/s. After 3 s it will be 12 m/s. Here a positive a_x is causing the bicyclist to speed up.

b. If the bicyclist's initial velocity is a *negative* −6 m/s but the acceleration is a positive +2 (m/s)/s, then 1 s later his velocity will be −4 m/s. After 2 s it will be −2 m/s, and so on. In this case, a positive a_x is causing the object to *slow down* (decreasing speed v). This agrees with the rule from Tactics Box 1.4: An object is slowing down if and only if v_x and a_x have opposite signs.

NOTE It is customary to abbreviate the acceleration units $(m/s)/s$ as m/s^2. For example, the bicyclists in Example 2.8 had an acceleration of $2 \, m/s^2$. We will use this notation, but keep in mind the *meaning* of the notation as "(meters per second) per second."

EXAMPLE 2.9 | Running the court

A basketball player starts at the left end of the court and moves with the velocity shown in **FIGURE 2.20**. Draw a motion diagram and an acceleration-versus-time graph for the basketball player.

FIGURE 2.20 Velocity-versus-time graph for the basketball player of Example 2.9.

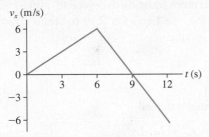

VISUALIZE The velocity is positive (motion to the right) and increasing for the first 6 s, so the velocity arrows in the motion diagram are to the right and getting longer. From $t = 6$ s to 9 s the motion is still to the right (v_x is still positive), but the arrows are getting shorter because v_x is decreasing. There's a turning point at $t = 9$ s, when $v_x = 0$, and after that the motion is to the left (v_x is negative) and getting faster. The motion diagram of **FIGURE 2.21a** shows the velocity and the acceleration vectors.

SOLVE Acceleration is the slope of the velocity graph. For the first 6 s, the slope has the constant value

$$a_x = \frac{\Delta v_x}{\Delta t} = \frac{6.0 \, m/s}{6.0 \, s} = 1.0 \, m/s^2$$

The velocity then decreases by 12 m/s during the 6 s interval from $t = 6$ s to $t = 12$ s, so

$$a_x = \frac{\Delta v_x}{\Delta t} = \frac{-12 \, m/s}{6.0 \, s} = -2.0 \, m/s^2$$

The acceleration graph for these 12 s is shown in **FIGURE 2.21b**. Notice that there is no change in the acceleration at $t = 9$ s, the turning point.

FIGURE 2.21 Motion diagram and acceleration graph for Example 2.9.

ASSESS The *sign* of a_x does *not* tell us whether the object is speeding up or slowing down. The basketball player is slowing down from $t = 6$ s to $t = 9$ s, then speeding up from $t = 9$ s to $t = 12$ s. Nonetheless, his acceleration is negative during this entire interval because his acceleration vector, as seen in the motion diagram, always points to the left.

The Kinematic Equations of Constant Acceleration

Consider an object whose acceleration a_s remains constant during the time interval $\Delta t = t_f - t_i$. At the beginning of this interval, at time t_i, the object has initial velocity v_{is} and initial position s_i. Note that t_i is often zero, but it does not have to be. We would like to predict the object's final position s_f and final velocity v_{fs} at time t_f.

The object's velocity is changing because the object is accelerating. **FIGURE 2.22a** shows the acceleration-versus-time graph, a horizontal line between t_i and t_f. It is not hard to find the object's velocity v_{fs} at a later time t_f. By definition,

$$a_s = \frac{\Delta v_s}{\Delta t} = \frac{v_{fs} - v_{is}}{\Delta t} \tag{2.18}$$

which is easily rearranged to give

$$v_{fs} = v_{is} + a_s \Delta t \tag{2.19}$$

The velocity-versus-time graph, shown in **FIGURE 2.22b**, is a straight line that starts at v_{is} and has slope a_s.

FIGURE 2.22 Acceleration and velocity graphs for constant acceleration.

As you learned in the last section, the object's final position is

$$s_f = s_i + \text{area under the velocity curve } v_s \text{ between } t_i \text{ and } t_f \qquad (2.20)$$

The shaded area in Figure 2.22b can be subdivided into a rectangle of area $v_{is}\,\Delta t$ and a triangle of area $\frac{1}{2}(a_s \Delta t)(\Delta t) = \frac{1}{2}a_s(\Delta t)^2$. Adding these gives

$$s_f = s_i + v_{is}\,\Delta t + \tfrac{1}{2}a_s(\Delta t)^2 \qquad (2.21)$$

where $\Delta t = t_f - t_i$ is the elapsed time. The quadratic dependence on Δt causes the position-versus-time graph for constant-acceleration motion to have a parabolic shape, as shown in Model 2.2.

Equations 2.19 and 2.21 are two of the basic kinematic equations for motion with *constant* acceleration. They allow us to predict an object's position and velocity at a future instant of time. We need one more equation to complete our set, a direct relation between position and velocity. First use Equation 2.19 to write $\Delta t = (v_{fs} - v_{is})/a_s$. Substitute this into Equation 2.21, giving

$$s_f = s_i + v_{is}\left(\frac{v_{fs} - v_{is}}{a_s}\right) + \tfrac{1}{2}a_s\left(\frac{v_{fs} - v_{is}}{a_s}\right)^2 \qquad (2.22)$$

With a bit of algebra, this is rearranged to read

$$v_{fs}^{\,2} = v_{is}^{\,2} + 2a_s\,\Delta s \qquad (2.23)$$

where $\Delta s = s_f - s_i$ is the *displacement* (not the distance!). Equation 2.23 is the last of the three kinematic equations for motion with constant acceleration.

The Constant-Acceleration Model

Few objects with changing velocity have a perfectly constant acceleration, but it is often reasonable to model their acceleration as being constant. We do so by utilizing the **constant-acceleration model.** Once again, a model is a set of words, pictures, graphs, and equations that allows us to explain and predict an object's motion.

MODEL 2.2   (MP)

Constant acceleration

For motion with constant acceleration.

■ Model the object as a particle moving in a straight line with constant acceleration.

■ Mathematically:
- $v_{fs} = v_{is} + a_s\,\Delta t$
- $s_f = s_i + v_{is}\,\Delta t + \tfrac{1}{2}a_s(\Delta t)^2$
- $v_{fs}^2 = v_{is}^2 + 2a_s\,\Delta s$

■ Limitations: Model fails if the particle's acceleration changes.

a_s Horizontal line

The acceleration is constant.

v_s Straight line

The slope is a_s.

s Parabola

The slope is v_s.

Exercise 16

In this text, we'll usually model runners, cars, planes, and rockets as having constant acceleration. Their actual acceleration is often more complicated (for example, a car's acceleration gradually decreases rather than remaining constant until full speed is reached), but the mathematical complexity of dealing with realistic accelerations would detract from the physics we're trying to learn.

The constant-acceleration model is the basis for a problem-solving strategy.

PROBLEM-SOLVING STRATEGY 2.1

Kinematics with constant acceleration

MODEL Model the object as having constant acceleration.

VISUALIZE Use different representations of the information in the problem.

- Draw a *pictorial representation*. This helps you assess the information you are given and starts the process of translating the problem into symbols.
- Use a *graphical representation* if it is appropriate for the problem.
- Go back and forth between these two representations as needed.

SOLVE The mathematical representation is based on the three kinematic equations:

$$v_{fs} = v_{is} + a_s \Delta t$$
$$s_f = s_i + v_{is} \Delta t + \tfrac{1}{2} a_s (\Delta t)^2$$
$$v_{fs}^{\,2} = v_{is}^{\,2} + 2a_s \Delta s$$

- Use x or y, as appropriate to the problem, rather than the generic s.
- Replace i and f with numerical subscripts defined in the pictorial representation.

ASSESS Check that your result has the correct units and significant figures, is reasonable, and answers the question.

NOTE You are strongly encouraged to solve problems on the Dynamics Worksheets found at the back of the Student Workbook. These worksheets will help you use the Problem-Solving Strategy and develop good problem-solving skills.

EXAMPLE 2.10 | The motion of a rocket sled

A rocket sled's engines fire for 5.0 s, boosting the sled to a speed of 250 m/s. The sled then deploys a braking parachute, slowing by 3.0 m/s per second until it stops. What is the total distance traveled?

MODEL We're not given the sled's initial acceleration, while the rockets are firing, but rocket sleds are aerodynamically shaped to minimize air resistance and so it seems reasonable to model the sled as a particle undergoing constant acceleration.

VISUALIZE FIGURE 2.23 shows the pictorial representation. We've made the reasonable assumptions that the sled starts from rest and that the braking parachute is deployed just as the rocket burn ends. There are three points of interest in this problem: the start, the change from propulsion to braking, and the stop. Each of these points has been assigned a position, velocity, and time. Notice that we've replaced the generic subscripts i and f of the kinematic equations with the numerical subscripts 0, 1, and 2. Accelerations are associated not with specific points in the motion but with the intervals between the points, so acceleration a_{0x} is the acceleration between points 0 and 1 while acceleration a_{1x} is the acceleration between points 1 and 2. The acceleration vector $\vec{a}_1$ points to the left, so a_{1x} is negative. The sled stops at the end point, so $v_{2x} = 0$ m/s.

SOLVE We know how long the rocket burn lasts and the velocity at the end of the burn. Because we're modeling the sled as having uniform acceleration, we can use the first kinematic equation of Problem-Solving Strategy 2.1 to write

$$v_{1x} = v_{0x} + a_{0x}(t_1 - t_0) = a_{0x} t_1$$

We started with the complete equation, then simplified by noting which terms were zero. Solving for the boost-phase acceleration, we have

$$a_{0x} = \frac{v_{1x}}{t_1} = \frac{250 \text{ m/s}}{5.0 \text{ s}} = 50 \text{ m/s}^2$$

Notice that we worked algebraically until the last step—a hallmark of good problem-solving technique that minimizes the chances of

FIGURE 2.23 Pictorial representation of the rocket sled.

Known
$x_0 = 0$ m $v_{0x} = 0$ m/s $t_0 = 0$ s
$v_{1x} = 250$ m/s $t_1 = 5.0$ s
$a_{1x} = -3.0$ m/s² $v_{2x} = 0$ m/s

Find
x_2

Continued

calculation errors. Also, in accord with the significant figure rules of Chapter 1, 50 m/s² is considered to have two significant figures.

Now we have enough information to find out how far the sled travels while the rockets are firing. The second kinematic equation of Problem-Solving Strategy 2.1 is

$$x_1 = x_0 + v_{0x}(t_1 - t_0) + \tfrac{1}{2}a_{0x}(t_1 - t_0)^2 = \tfrac{1}{2}a_{0x}t_1^2$$
$$= \tfrac{1}{2}(50 \text{ m/s}^2)(5.0 \text{ s})^2 = 625 \text{ m}$$

The braking phase is a little different because we don't know how long it lasts. But we do know both the initial and final velocities, so we can use the third kinematic equation of Problem-Solving Strategy 2.1:

$$v_{2x}^2 = v_{1x}^2 + 2a_{1x}\Delta x = v_{1x}^2 + 2a_{1x}(x_2 - x_1)$$

Notice that Δx is *not* x_2; it's the displacement $(x_2 - x_1)$ during the braking phase. We can now solve for x_2:

$$x_2 = x_1 + \frac{v_{2x}^2 - v_{1x}^2}{2a_{1x}}$$
$$= 625 \text{ m} + \frac{0 - (250 \text{ m/s})^2}{2(-3.0 \text{ m/s}^2)} = 11{,}000 \text{ m}$$

We kept three significant figures for x_1 at an intermediate stage of the calculation but rounded to two significant figures at the end.

ASSESS The total distance is 11 km ≈ 7 mi. That's large but believable. Using the approximate conversion factor 1 m/s ≈ 2 mph from Table 1.5, we see that the top speed is ≈ 500 mph. It will take a long distance for the sled to gradually stop from such a high speed.

EXAMPLE 2.11 | A two-car race

Fred is driving his Volkswagen Beetle at a steady 20 m/s when he passes Betty sitting at rest in her Porsche. Betty instantly begins accelerating at 5.0 m/s². How far does Betty have to drive to overtake Fred?

MODEL Model the VW as a particle in uniform motion and the Porsche as a particle with constant acceleration.

VISUALIZE FIGURE 2.24 is the pictorial representation. Fred's motion diagram is one of uniform motion, while Betty's shows uniform acceleration. Fred is ahead in frames 1, 2, and 3, but Betty catches up with him in frame 4. The coordinate system shows the cars with the same position at the start and at the end—but with the important difference that Betty's Porsche has an acceleration while Fred's VW does not.

SOLVE This problem is similar to Example 2.2, in which Bob and Susan met for lunch. As we did there, we want to find Betty's position $(x_1)_B$ at the instant t_1 when $(x_1)_B = (x_1)_F$. We know, from the models of uniform motion and uniform acceleration, that Fred's position graph is a straight line but Betty's is a parabola. The position graphs in Figure 2.24 show that we're solving for the intersection point of the line and the parabola.

Fred's and Betty's positions at t_1 are

$$(x_1)_F = (x_0)_F + (v_{0x})_F(t_1 - t_0) = (v_{0x})_F t_1$$
$$(x_1)_B = (x_0)_B + (v_{0x})_B(t_1 - t_0) + \tfrac{1}{2}(a_{0x})_B(t_1 - t_0)^2 = \tfrac{1}{2}(a_{0x})_B t_1^2$$

By equating these,

$$(v_{0x})_F t_1 = \tfrac{1}{2}(a_{0x})_B t_1^2$$

we can solve for the time when Betty passes Fred:

$$t_1\left[\tfrac{1}{2}(a_{0x})_B t_1 - (v_{0x})_F\right] = 0$$
$$t_1 = \begin{cases} 0 \text{ s} \\ 2(v_{0x})_F/(a_{0x})_B = 8.0 \text{ s} \end{cases}$$

Interestingly, there are two solutions. That's not surprising, when you think about it, because the line and the parabola of the position graphs have *two* intersection points: when Fred first passes Betty, and 8.0 s later when Betty passes Fred. We're interested in only the second of these points. We can now use either of the distance equations to find $(x_1)_B = (x_1)_F = 160$ m. Betty has to drive 160 m to overtake Fred.

ASSESS 160 m ≈ 160 yards. Because Betty starts from rest while Fred is moving at 20 m/s ≈ 40 mph, needing 160 yards to catch him seems reasonable.

> **NOTE** The purpose of the Assess step is not to prove that an answer must be right but to rule out answers that, with a little thought, are clearly wrong.

FIGURE 2.24 Pictorial representation for Example 2.11.

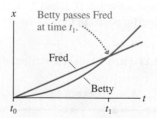

Known		
$(x_0)_F = 0$ m	$(x_0)_B = 0$ m	$t_0 = 0$ s
$(v_{0x})_F = 20$ m/s	$(v_{0x})_B = 0$ m/s	
$(a_{0x})_B = 5.0$ m/s²	$(v_{1x})_F = 20$ m/s	

Find
$(x_1)_B$ at t_1 when $(x_1)_B = (x_1)_F$

2.5 Free Fall

The motion of an object moving under the influence of gravity only, and no other forces, is called **free fall.** Strictly speaking, free fall occurs only in a vacuum, where there is no air resistance. Fortunately, the effect of air resistance is small for "heavy objects," so we'll make only a very slight error in treating these objects *as if* they were in free fall. For very light objects, such as a feather, or for objects that fall through very large distances and gain very high speeds, the effect of air resistance is *not* negligible. Motion with air resistance is a problem we will study in Chapter 6. Until then, we will restrict our attention to "heavy objects" and will make the reasonable assumption that falling objects are in free fall.

Galileo, in the 17th century, was the first to make detailed measurements of falling objects. The story of Galileo dropping different weights from the leaning bell tower at the cathedral in Pisa is well known, although historians cannot confirm its truth. Based on his measurements, wherever they took place, Galileo developed a *model* for motion in the absence of air resistance:

- Two objects dropped from the same height will, if air resistance can be neglected, hit the ground at the same time and with the same speed.
- Consequently, **any two objects in free fall, regardless of their mass, have the same acceleration** $a_{\text{free fall}}$.

FIGURE 2.25a shows the motion diagram of an object that was released from rest and falls freely. **FIGURE 2.25b** shows the object's velocity graph. The motion diagram and graph are identical for a falling pebble and a falling boulder. The fact that the velocity graph is a straight line tells us the motion is one of constant acceleration, and $a_{\text{free fall}}$ is found from the slope of the graph. Careful measurements show that the value of $\vec{a}_{\text{free fall}}$ varies ever so slightly at different places on the earth, due to the slightly nonspherical shape of the earth and to the fact that the earth is rotating. A global average, at sea level, is

$$\vec{a}_{\text{free fall}} = (9.80 \text{ m/s}^2, \text{ vertically downward}) \qquad (2.24)$$

Vertically downward means along a line toward the center of the earth.

The length, or magnitude, of $\vec{a}_{\text{free fall}}$ is known as the **free-fall acceleration,** and it has the special symbol g:

$$g = 9.80 \text{ m/s}^2 \text{ (free-fall acceleration)}$$

Several points about free fall are worthy of note:

- g, by definition, is *always* positive. **There will never be a problem that will use a negative value for g.** But, you say, objects fall when you release them rather than rise, so how can g be positive?
- g is *not* the acceleration $a_{\text{free fall}}$, but simply its magnitude. Because we've chosen the y-axis to point vertically upward, the downward acceleration vector $\vec{a}_{\text{free fall}}$ has the one-dimensional acceleration

$$a_y = a_{\text{free fall}} = -g \qquad (2.25)$$

It is a_y that is negative, not g.

In a vacuum, the apple and feather fall at the same rate and hit the ground at the same time.

FIGURE 2.25 Motion of an object in free fall.

(a)

(b)

■ We can model free fall as motion with constant acceleration, with $a_y = -g$.

■ g is not called "gravity." Gravity is a force, not an acceleration. The symbol g recognizes the influence of gravity, but g is *the free-fall acceleration*.

■ $g = 9.80 \text{ m/s}^2$ only on earth. Other planets have different values of g. You will learn in Chapter 13 how to determine g for other planets.

NOTE Despite the name, free fall is not restricted to objects that are literally falling. Any object moving under the influence of gravity only, and no other forces, is in free fall. This includes objects falling straight down, objects that have been tossed or shot straight up, and projectile motion.

EXAMPLE 2.12 | **A falling rock**

A rock is dropped from the top of a 20-m-tall building. What is its impact velocity?

MODEL A rock is fairly heavy, and air resistance is probably not a serious concern in a fall of only 20 m. It seems reasonable to model the rock's motion as free fall: constant acceleration with $a_y = a_{\text{free fall}} = -g$.

VISUALIZE FIGURE 2.26 shows the pictorial representation. We have placed the origin at the ground, which makes $y_0 = 20 \text{ m}$. Although the rock falls 20 m, it is important to notice that the *displacement* is $\Delta y = y_1 - y_0 = -20 \text{ m}$.

FIGURE 2.26 Pictorial representation of a falling rock.

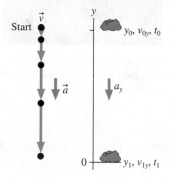

Known
$y_0 = 20 \text{ m}$
$v_{0y} = 0 \text{ m/s}$ $t_0 = 0 \text{ s}$
$y_1 = 0 \text{ m}$
$a_y = -g = -9.80 \text{ m/s}^2$

Find
v_{1y}

SOLVE In this problem we know the displacement but not the time, which suggests that we use the third kinematic equation from Problem-Solving Strategy 2.1:

$$v_{1y}^2 = v_{0y}^2 + 2a_y \Delta y = -2g \Delta y$$

We started by writing the general equation, then noted that $v_{0y} = 0 \text{ m/s}$ and substituted $a_y = -g$. Solving for v_{1y}:

$$v_{1y} = \sqrt{-2g\Delta y} = \sqrt{-2(9.8 \text{ m/s}^2)(-20 \text{ m})} = \pm 20 \text{ m/s}$$

A common error would be to say, "The rock fell 20 m, so $\Delta y = 20 \text{ m}$." That would have you trying to take the square root of a negative number. As noted above, Δy is a *displacement,* not a distance, and in this case $\Delta y = -20 \text{ m}$.

The $\pm$ sign indicates that there are two mathematical solutions; therefore, we have to use physical reasoning to choose between them. The rock does hit with a *speed* of 20 m/s, but the question asks for the impact *velocity*. The velocity vector points down, so the sign of v_{1y} is negative. Thus the impact velocity is -20 m/s.

ASSESS Is the answer reasonable? Well, 20 m is about 60 feet, or about the height of a five- or six-story building. Using $1 \text{ m/s} \approx 2 \text{ mph}$, we see that $20 \text{ m/s} \approx 40 \text{ mph}$. That seems quite reasonable for the speed of an object after falling five or six stories. If we had misplaced a decimal point, though, and found 2.0 m/s, we would be suspicious that this was much too small after converting it to $\approx 4 \text{ mph}$.

EXAMPLE 2.13 | **Finding the height of a leap**

The springbok, an antelope found in Africa, gets its name from its remarkable jumping ability. When startled, a springbok will leap straight up into the air—a maneuver called a "pronk." A springbok goes into a crouch to perform a pronk. It then extends its legs forcefully, accelerating at 35 m/s² for 0.70 m as

its legs straighten. Legs fully extended, it leaves the ground and rises into the air. How high does it go?

MODEL The springbok is changing shape as it leaps, so can we reasonably model it as a particle? We can if we focus on the *body* of the springbok, treating the expanding legs like external springs. Initially, the body of the springbok is driven upward by its legs. We'll model this as a particle—the body—undergoing constant acceleration. Once the springbok's feet leave the ground, we'll model the motion of the springbok's body as a particle in free fall.

VISUALIZE FIGURE 2.27 shows the pictorial representation. This is a problem with a beginning point, an end point, and a point in between where the nature of the motion changes. We've identified these points with subscripts 0, 1, and 2. The motion from 0 to 1 is a rapid upward acceleration until the springbok's feet leave the ground at 1. Even though the springbok is moving upward from 1 to 2, this is free-fall motion because the springbok is now moving under the influence of gravity only.

FIGURE 2.27 Pictorial representation of a startled springbok.

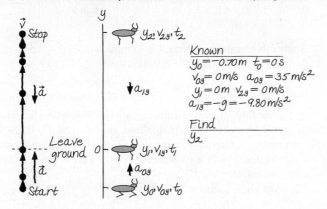

How do we put "How high?" into symbols? The clue is that the very top point of the trajectory is a *turning point,* and we've seen that the instantaneous velocity at a turning point is $v_{2y} = 0$.

This was not explicitly stated but is part of our interpretation of the problem.

SOLVE For the first part of the motion, pushing off, we know a displacement but not a time interval. We can use

$$v_{1y}^2 = v_{0y}^2 + 2a_{0y}\,\Delta y = 2(35 \text{ m/s}^2)(0.70 \text{ m}) = 49 \text{ m}^2/\text{s}^2$$

$$v_{1y} = \sqrt{49 \text{ m}^2/\text{s}^2} = 7.0 \text{ m/s}$$

The springbok leaves the ground with a velocity of 7.0 m/s. This is the starting point for the problem of a projectile launched straight up from the ground. One possible solution is to use the velocity equation to find how long it takes to reach maximum height, then the position equation to calculate the maximum height. But that takes two separate calculations. It is easier to make another use of the velocity-displacement equation:

$$v_{2y}^2 = 0 = v_{1y}^2 + 2a_{1y}\,\Delta y = v_{1y}^2 - 2g(y_2 - y_1)$$

where now the acceleration is $a_{1y} = -g$. Using $y_1 = 0$, we can solve for y_2, the height of the leap:

$$y_2 = \frac{v_{1y}^2}{2g} = \frac{(7.0 \text{ m/s})^2}{2(9.80 \text{ m/s}^2)} = 2.5 \text{ m}$$

ASSESS 2.5 m is a bit over 8 feet, a remarkable vertical jump. But these animals are known for their jumping ability, so the answer seems reasonable. Note that it is especially important in a multipart problem like this to use numerical subscripts to distinguish different points in the motion.

2.6 Motion on an Inclined Plane

FIGURE 2.28a shows a problem closely related to free fall: that of motion down a straight, but frictionless, inclined plane, such as a skier going down a slope on frictionless snow. What is the object's acceleration? Although we're not yet prepared to give a rigorous derivation, we can deduce the acceleration with a plausibility argument.

FIGURE 2.28b shows the free-fall acceleration $\vec{a}_{\text{free fall}}$ the object would have if the incline suddenly vanished. The free-fall acceleration points straight down. This vector can be broken into two pieces: a vector $\vec{a}_\parallel$ that is parallel to the incline and a vector $\vec{a}_\perp$ that is perpendicular to the incline. The surface of the incline somehow "blocks" $\vec{a}_\perp$, through a process we will examine in Chapter 6, but $\vec{a}_\parallel$ is unhindered. It is this piece of $\vec{a}_{\text{free fall}}$, parallel to the incline, that accelerates the object.

By definition, the length, or magnitude, of $\vec{a}_{\text{free fall}}$ is g. Vector $\vec{a}_\parallel$ is opposite angle θ (Greek *theta*), so the length, or magnitude, of $\vec{a}_\parallel$ must be $g \sin \theta$. Consequently, the one-dimensional acceleration along the incline is

$$a_s = \pm g \sin \theta \qquad (2.26)$$

The correct sign depends on the direction in which the ramp is tilted. Examples will illustrate.

Equation 2.26 makes sense. Suppose the plane is perfectly horizontal. If you place an object on a horizontal surface, you expect it to stay at rest with no acceleration. Equation 2.26 gives $a_s = 0$ when $\theta = 0°$, in agreement with our expectations. Now suppose you tilt the plane until it becomes vertical, at $\theta = 90°$. Without friction, an object would simply fall, in free fall, parallel to the vertical surface. Equation 2.26 gives $a_s = -g = a_{\text{free fall}}$ when $\theta = 90°$, again in agreement with our expectations. Equation 2.26 gives the correct result in these *limiting cases*.

FIGURE 2.28 Acceleration on an inclined plane.

(a)

Angle of incline θ

(b)

This piece of $\vec{a}_{\text{free fall}}$ accelerates the object down the incline.

$\vec{a}_\parallel$

$\vec{a}_{\text{free fall}}$ θ $\vec{a}_\perp$

θ

Same angle

EXAMPLE 2.14 | Measuring acceleration

In the laboratory, a 2.00-m-long track has been inclined as shown in FIGURE 2.29. Your task is to measure the acceleration of a cart on the ramp and to compare your result with what you might have expected. You have available five "photogates" that measure the cart's speed as it passes through. You place a gate every 30 cm from a line you mark near the top of the track as the starting line. One run generates the data shown in the table. The first entry isn't a photogate, but it is a valid data point because you know the cart's speed is zero at the point where you release it.

FIGURE 2.29 The experimental setup.

Distance (cm)	Speed (m/s)
0	0.00
30	0.75
60	1.15
90	1.38
120	1.56
150	1.76

NOTE Physics is an experimental science. Our knowledge of the universe is grounded in observations and measurements. Consequently, some examples and homework problems throughout this book will be based on data. Data-based homework problems require the use of a spreadsheet, graphing software, or a graphing calculator in which you can "fit" data with a straight line.

MODEL Model the cart as a particle.

VISUALIZE FIGURE 2.30 shows the pictorial representation. The track and axis are tilted at angle $\theta = \tan^{-1}(20.0 \text{ cm}/180 \text{ cm}) = 6.34°$. This is motion on an inclined plane, so you might expect the cart's acceleration to be $a_x = g \sin\theta = 1.08 \text{ m/s}^2$.

FIGURE 2.30 The pictorial representation of the cart on the track.

Known	
$x_0 = 0$ m	$v_{0x} = 0$ m/s
$t_0 = 0$ s	$\theta = 6.34°$

Find
a_x

SOLVE In analyzing data, we want to use *all* the data. Further, we almost always want to use graphs when we have a series of measurements. We might start by graphing speed versus distance traveled. This is shown in FIGURE 2.31a, where we've converted distances to meters. As expected, speed increases with distance, but the graph isn't linear and that makes it hard to analyze.

Rather than proceeding by trial and error, let's be guided by theory. *If* the cart has constant acceleration—which we don't yet know and need to confirm—the third kinematic equation tells us that velocity and displacement should be related by

$$v_x^2 = v_{0x}^2 + 2a_x \Delta x = 2a_x x$$

The last step was based on starting from rest ($v_{0x} = 0$) at the origin ($\Delta x = x - x_0 = x$).

FIGURE 2.31 Graphs of velocity and of velocity squared.

Rather than graphing v_x versus x, suppose we graph v_x^2 versus x. If we let $y = v_x^2$, the kinematic equation reads

$$y = 2a_x x$$

This is in the form of a linear equation: $y = mx + b$, where m is the slope and b is the y-intercept. In this case, $m = 2a_x$ and $b = 0$. So if the cart really does have constant acceleration, a graph of v_x^2 versus x should be linear with a y-intercept of zero. This is a prediction that we can test.

Thus our analysis has three steps:

1. Graph v_x^2 versus x. If the graph is a straight line with a y-intercept of zero (or very close to zero), then we can conclude that the cart has constant acceleration on the ramp. If not, the acceleration is *not* constant and we cannot use the kinematic equations for constant acceleration.
2. If the graph has the correct shape, we can determine its slope m.
3. Because kinematics predicts $m = 2a_x$, the acceleration must be $a_x = m/2$.

FIGURE 2.31b is the graph of v_x^2 versus x. It does turn out to be a straight line with a y-intercept of zero, and this is the evidence we need that the cart has a constant acceleration on the ramp. To proceed, we want to determine the slope by finding the straight line that is the "best fit" to the data. This is a statistical technique, justified in a statistics class, but one that is implemented in spreadsheets and graphing calculators. The solid line in Figure 2.31b is the best-fit line for this data, and its equation is shown. We see that the slope is $m = 2.06 \text{ m/s}^2$. **Slopes have units,** and the units come not from the fitting procedure but by looking at the axes of the graph. Here the vertical axis is velocity squared, with units of $(\text{m/s})^2$, while the horizontal axis is position, measured in m. Thus the slope, rise over run, has units of m/s^2.

Finally, we can determine that the cart's acceleration was

$$a_x = \frac{m}{2} = 1.03 \text{ m/s}^2$$

This is about 5% less than the 1.08 m/s² we expected. Two possibilities come to mind. Perhaps the distances used to find the tilt angle weren't measured accurately. Or, more likely, the cart rolls with a small bit of friction. The predicted acceleration $a_x = g \sin\theta$ is for a *frictionless* inclined plane; any friction would decrease the acceleration.

ASSESS The acceleration is just slightly less than predicted for a frictionless incline, so the result is reasonable.

Thinking Graphically

A good way to solidify your intuitive understanding of motion is to consider the problem of a hard, smooth ball rolling on a smooth track. The track is made up of several straight segments connected together. Each segment may be either horizontal or inclined. Your task is to analyze the ball's motion graphically.

There are a small number of rules to follow:

1. Assume that the ball passes smoothly from one segment of the track to the next, with no abrupt change of speed and without ever leaving the track.
2. The graphs have no numbers, but they should show the correct *relationships*. For example, the position graph should be steeper in regions of higher speed.
3. The position s is the position measured *along* the track. Similarly, v_s and a_s are the velocity and acceleration parallel to the track.

EXAMPLE 2.15 | From track to graphs

Draw position, velocity, and acceleration graphs for the ball on the smooth track of **FIGURE 2.32**.

FIGURE 2.32 A ball rolling along a track.

VISUALIZE It is often easiest to begin with the velocity. There is no acceleration on the horizontal surface ($a_s = 0$ if $\theta = 0°$), so the velocity remains constant at v_{0s} until the ball reaches the slope. The slope is an inclined plane where the ball has constant acceleration. The velocity increases linearly with time during constant-acceleration motion. The ball returns to constant-velocity motion after reaching the bottom horizontal segment. The middle graph of **FIGURE 2.33** shows the velocity.

We can easily draw the acceleration graph. The acceleration is zero while the ball is on the horizontal segments and has a constant positive value on the slope. These accelerations are consistent with the slope of the velocity graph: zero slope, then positive slope, then a return to zero. The acceleration cannot *really* change instantly from zero to a nonzero value, but the change can be so quick that we do not see it on the time scale of the graph. That is what the vertical dotted lines imply.

Finally, we need to find the position-versus-time graph. The position increases linearly with time during the first segment at constant velocity. It also does so during the third segment of motion, but with a steeper slope to indicate a faster velocity. In between, while the acceleration is nonzero but constant, the position graph has a *parabolic* shape. Notice that the parabolic section blends *smoothly* into the straight lines on either side. An abrupt change of slope (a "kink") would indicate an abrupt change in velocity and would violate rule 1.

FIGURE 2.33 Motion graphs for the ball in Example 2.15.

The position graph changes smoothly, without kinks.

EXAMPLE 2.16 | From graphs to track

FIGURE 2.34 shows a set of motion graphs for a ball moving on a track. Draw a picture of the track and describe the ball's initial condition. Each segment of the track is *straight,* but the segments may be tilted.

VISUALIZE The ball starts with initial velocity $v_{0s} > 0$ and maintains this velocity for awhile; there's no acceleration. Thus the ball must start out rolling to the right on a horizontal track. At the end of the motion, the ball is again rolling on a horizontal track (no acceleration, constant velocity), but it's rolling to the *left* because v_s is negative. Further, the final speed ($|v_s|$) is greater than the initial speed. The middle section of the graph shows us what happens. The ball starts slowing with constant acceleration (rolling uphill), reaches a turning point (s is maximum, $v_s = 0$), then speeds up in the opposite direction (rolling downhill). This is still a negative acceleration because the ball is speeding up in the negative

FIGURE 2.34 Motion graphs of a ball rolling on a track of unknown shape.

Continued

s-direction. It must roll farther downhill than it had rolled uphill before reaching a horizontal section of track. **FIGURE 2.35** shows the track and the initial conditions that are responsible for the graphs of Figure 2.34.

FIGURE 2.35 Track responsible for the motion graphs of Figure 2.34.

This track has a "switch." A ball moving to the right goes up the incline, but a ball rolling downhill goes straight through.

$v_{0s} > 0$

STOP TO THINK 2.5 The ball rolls up the ramp, then back down. Which is the correct acceleration graph?

(a)   (b)   (c)   (d)   (e)

2.7 ADVANCED TOPIC Instantaneous Acceleration

FIGURE 2.36 Velocity and acceleration graphs of a car leaving a stop sign.

(a)   The car speeds up from rest until it reaches a steady cruising speed.

(b) a_x The *slope* of the velocity graph is the *value* of the acceleration.

Although the constant-acceleration model is very useful, real moving objects only rarely have constant acceleration. For example, **FIGURE 2.36a** is a realistic velocity-versus-time graph for a car leaving a stop sign. The graph is not a straight line, so this is *not* motion with constant acceleration.

We can define an instantaneous acceleration much as we defined the instantaneous velocity. The instantaneous velocity at time *t* is the slope of the position-versus-time graph at that time or, mathematically, the derivative of the position with respect to time. By analogy: **The instantaneous acceleration a_s is the slope of the line that is tangent to the velocity-versus-time curve at time *t*.** Mathematically, this is

$$a_s = \frac{dv_s}{dt} = \text{slope of the velocity-versus-time graph at time } t \qquad (2.27)$$

FIGURE 2.36b applies this idea by showing the car's acceleration graph. At each instant of time, the *value* of the car's acceleration is the *slope* of its velocity graph. The initially steep slope indicates a large initial acceleration. The acceleration decreases to zero as the car reaches cruising speed.

The reverse problem—to find the velocity v_s if we know the acceleration a_s at all instants of time—is also important. Again, with analogy to velocity and position, we have

$$v_{fs} = v_{is} + \int_{t_i}^{t_f} a_s \, dt \qquad (2.28)$$

The graphical interpretation of Equation 2.28 is

$$v_{fs} = v_{is} + \text{area under the acceleration curve } a_s \text{ between } t_i \text{ and } t_f \qquad (2.29)$$

EXAMPLE 2.17 | Finding velocity from acceleration

FIGURE 2.37 shows the acceleration graph for a particle with an initial velocity of 10 m/s. What is the particle's velocity at $t = 8$ s?

FIGURE 2.37 Acceleration graph for Example 2.17.

MODEL We're told this is the motion of a particle.

VISUALIZE Figure 2.37 is a graphical representation of the motion.

SOLVE The change in velocity is found as the area under the acceleration curve:

$v_{fs} = v_{is} +$ area under the acceleration curve a_s between t_i and t_f

The area under the curve between $t_i = 0$ s and $t_f = 8$ s can be subdivided into a rectangle (0 s $\leq t \leq 4$ s) and a triangle (4 s $\leq t \leq 8$ s). These areas are easily computed. Thus

$$v_s(\text{at } t = 8 \text{ s}) = 10 \text{ m/s} + (4 \text{ (m/s)/s})(4 \text{ s})$$
$$+ \tfrac{1}{2}(4 \text{ (m/s)/s})(4 \text{ s})$$
$$= 34 \text{ m/s}$$

EXAMPLE 2.18 | A realistic car acceleration

Starting from rest, a car takes T seconds to reach its cruising speed v_{max}. A plausible expression for the velocity as a function of time is

$$v_x(t) = \begin{cases} v_{max}\left(\dfrac{2t}{T} - \dfrac{t^2}{T^2}\right) & t \leq T \\ v_{max} & t \geq T \end{cases}$$

a. Demonstrate that this is a plausible function by drawing velocity and acceleration graphs.

b. Find an expression for the distance traveled at time T in terms of T and the maximum acceleration a_{max}.

c. What are the maximum acceleration and the distance traveled for a car that reaches a cruising speed of 15 m/s in 8.0 s?

MODEL Model the car as a particle.

VISUALIZE **FIGURE 2.38a** shows the velocity graph. It's an inverted parabola that reaches v_{max} at time T and then holds that value. From the slope, we see that the acceleration should start at a maximum value a_{max}, steadily decrease until T, and be zero for $t > T$.

SOLVE

a. We can find an expression for a_x by taking the derivative of v_x. Starting with $t \leq T$, and using Equation 2.6 for the derivatives of polynomials, we find

$$a_x = \frac{dv_x}{dt} = v_{max}\left(\frac{2}{T} - \frac{2t}{T^2}\right) = \frac{2v_{max}}{T}\left(1 - \frac{t}{T}\right) = a_{max}\left(1 - \frac{t}{T}\right)$$

where $a_{max} = 2v_{max}/T$. For $t \geq T$, $a_x = 0$. Altogether,

$$a_x(t) = \begin{cases} a_{max}\left(1 - \dfrac{t}{T}\right) & t \leq T \\ 0 & t \geq T \end{cases}$$

This expression for the acceleration is graphed in **FIGURE 2.38b**. The acceleration decreases linearly from a_{max} to 0 as the car accelerates from rest to its cruising speed.

b. To find the position as a function of time, we need to integrate the velocity (Equation 2.11) using Equation 2.13 for the integrals of polynomials. At time T, when cruising speed is reached,

FIGURE 2.38 Velocity and acceleration graphs for Example 2.18.

(a)

(b)

$$x_T = x_0 + \int_0^T v_x \, dt = 0 + \frac{2v_{max}}{T}\int_0^T t \, dt - \frac{v_{max}}{T^2}\int_0^T t^2 \, dt$$

$$= \frac{2v_{max}}{T}\frac{t^2}{2}\Big|_0^T - \frac{v_{max}}{T^2}\frac{t^3}{3}\Big|_0^T$$

$$= v_{max}T - \tfrac{1}{3}v_{max}T = \tfrac{2}{3}v_{max}T$$

Recalling that $a_{max} = 2v_{max}/T$, we can write the distance traveled as

$$x_T = \tfrac{2}{3}v_{max}T = \tfrac{1}{3}\left(\frac{2v_{max}}{T}\right)T^2 = \tfrac{1}{3}a_{max}T^2$$

If the acceleration stayed constant, the distance would be $\frac{1}{2}aT^2$. We have found a similar expression but, because the acceleration is steadily decreasing, a smaller fraction in front.

c. With $v_{max} = 15$ m/s and $T = 8.0$ s, realistic values for city driving, we find

$$a_{max} = \frac{2v_{max}}{T} = \frac{2(15 \text{ m/s})}{8.0 \text{ s}} = 3.75 \text{ m/s}^2$$

$$x_T = \tfrac{1}{3}a_{max}T^2 = \tfrac{1}{3}(3.75 \text{ m/s}^2)(8.0 \text{ s})^2 = 80 \text{ m}$$

ASSESS 80 m in 8.0 s to reach a cruising speed of 15 m/s $\approx$ 30 mph is very reasonable. This gives us good reason to believe that a car's initial acceleration is $\approx \frac{1}{3}g$.

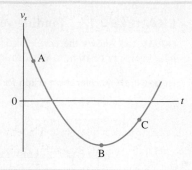
CHALLENGE EXAMPLE 2.19 │ Rocketing along

A rocket sled accelerates along a long, horizontal rail. Starting from rest, two rockets burn for 10 s, providing a constant acceleration. One rocket then burns out, halving the acceleration, but the other burns for an additional 5 s to boost the sled's speed to 625 m/s. How far has the sled traveled when the second rocket burns out?

MODEL Model the rocket sled as a particle with constant acceleration.

VISUALIZE FIGURE 2.39 shows the pictorial representation. This is a two-part problem with a beginning, an end (the second rocket burns out), and a point in between where the motion changes (the first rocket burns out).

FIGURE 2.39 The pictorial representation of the rocket sled.

SOLVE The difficulty with this problem is that there's not enough information to completely analyze either the first or the second part of the motion. A successful solution will require combining information about both parts of the motion, and that can be done only by working algebraically, not worrying about numbers until the end of the problem. A well-drawn pictorial representation and clearly defined symbols are essential.

The first part of the motion, with both rockets firing, has acceleration a_{0x}. The sled's position and velocity when the first rocket burns out are

$$x_1 = x_0 + v_{0x}\,\Delta t + \tfrac{1}{2}a_{0x}(\Delta t)^2 = \tfrac{1}{2}a_{0x}t_1^2$$
$$v_{1x} = v_{0x} + a_{0x}\,\Delta t = a_{0x}t_1$$

where we simplified as much as possible by knowing that the sled started from rest at the origin at $t_0 = 0$ s. We can't compute numerical values, but these are valid algebraic expressions that we can carry over to the second part of the motion.

From t_1 to t_2, the acceleration is a smaller a_{1x}. The velocity when the second rocket burns out is

$$v_{2x} = v_{1x} + a_{1x}\,\Delta t = a_{0x}t_1 + a_{1x}(t_2 - t_1)$$

where for v_{1x} we used the algebraic result from the first part of the motion. Now we have enough information to complete the solution. We know that the acceleration is halved when the first rocket burns out, so $a_{1x} = \tfrac{1}{2}a_{0x}$. Thus

$$v_{2x} = 625 \text{ m/s} = a_{0x}(10 \text{ s}) + \tfrac{1}{2}a_{0x}(5 \text{ s}) = (12.5 \text{ s})a_{0x}$$

Solving, we find $a_{0x} = 50 \text{ m/s}^2$.

With the acceleration now known, we can calculate the position and velocity when the first rocket burns out:

$$x_1 = \tfrac{1}{2}a_{0x}t_1^2 = \tfrac{1}{2}(50 \text{ m/s}^2)(10 \text{ s})^2 = 2500 \text{ m}$$
$$v_{1x} = a_{0x}t_1 = (50 \text{ m/s}^2)(10 \text{ s}) = 500 \text{ m/s}$$

Finally, the position when the second rocket burns out is

$$x_2 = x_1 + v_{1x}\,\Delta t + \tfrac{1}{2}a_{1x}(\Delta t)^2$$
$$= 2500 \text{ m} + (500 \text{ m/s})(5 \text{ s}) + \tfrac{1}{2}(25 \text{ m/s}^2)(5 \text{ s})^2 = 5300 \text{ m}$$

The sled has traveled 5300 m when it reaches 625 m/s at the burnout of the second rocket.

ASSESS 5300 m is 5.3 km, or roughly 3 miles. That's a long way to travel in 15 s! But the sled reaches incredibly high speeds. At the final speed of 625 m/s, over 1200 mph, the sled would travel nearly 10 km in 15 s. So 5.3 km in 15 s for the accelerating sled seems reasonable.

SUMMARY

The goal of Chapter 2 has been to learn to solve problems about motion along a straight line.

GENERAL PRINCIPLES

Kinematics describes motion in terms of position, velocity, and acceleration.

General kinematic relationships are given **mathematically** by:

Instantaneous velocity $\quad v_s = ds/dt =$ slope of position graph

Instantaneous acceleration $\quad a_s = dv_s/dt =$ slope of velocity graph

Final position $\quad s_f = s_i + \int_{t_i}^{t_f} v_s \, dt = s_i + \begin{cases} \text{area under the velocity} \\ \text{curve from } t_i \text{ to } t_f \end{cases}$

Final velocity $\quad v_{fs} = v_{is} + \int_{t_i}^{t_f} a_s \, dt = v_{is} + \begin{cases} \text{area under the acceleration} \\ \text{curve from } t_i \text{ to } t_f \end{cases}$

Solving Kinematics Problems

MODEL Uniform motion or constant acceleration.

VISUALIZE Draw a pictorial representation.

SOLVE

- Uniform motion $s_f = s_i + v_s \Delta t$
- Constant acceleration $v_{fs} = v_{is} + a_s \Delta t$
$$s_f = s_i + v_s \Delta t + \tfrac{1}{2} a_s (\Delta t)^2$$
$$v_{fs}^2 = v_{is}^2 + 2 a_s \Delta s$$

ASSESS Is the result reasonable?

IMPORTANT CONCEPTS

Position, velocity, and acceleration are related graphically.

- The slope of the position-versus-time graph is the value on the velocity graph.

- The slope of the velocity graph is the value on the acceleration graph.

- s is a maximum or minimum at a turning point, and $v_s = 0$.

- Displacement is the area under the velocity curve.

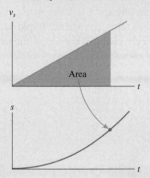

APPLICATIONS

The **sign of v_s** indicates the direction of motion.

- $v_s > 0$ is motion to the right or up.

- $v_s < 0$ is motion to the left or down.

The **sign of a_s** indicates which way $\vec{a}$ points, *not* whether the object is speeding up or slowing down.

- $a_s > 0$ if $\vec{a}$ points to the right or up.

- $a_s < 0$ if $\vec{a}$ points to the left or down.

- The direction of $\vec{a}$ is found with a motion diagram.

An object is **speeding up** if and only if v_s and a_s have the same sign.

An object is **slowing down** if and only if v_s and a_s have opposite signs.

Free fall is constant-acceleration motion with

$$a_y = -g = -9.80 \text{ m/s}^2$$

Motion on an inclined plane has $a_s = \pm g \sin\theta$. The sign depends on the direction of the tilt.

TERMS AND NOTATION

kinematics
uniform motion
average velocity, v_{avg}
speed, v

initial position, s_i
final position, s_f
uniform-motion model
instantaneous velocity, v_s

turning point
average acceleration, a_{avg}
constant-acceleration model
free fall

free-fall acceleration, g
instantaneous acceleration, a_s

CONCEPTUAL QUESTIONS

For Questions 1 through 3, interpret the position graph given in each figure by writing a very short "story" of what is happening. Be creative! Have characters and situations! Simply saying that "a car moves 100 meters to the right" doesn't qualify as a story. Your stories should make *specific reference* to information you obtain from the graph, such as distance moved or time elapsed.

1.

FIGURE Q2.1

2.

FIGURE Q2.2

3.

FIGURE Q2.3

4. **FIGURE Q2.4** shows a position-versus-time graph for the motion of objects A and B as they move along the same axis.
   a. At the instant $t = 1$ s, is the speed of A greater than, less than, or equal to the speed of B? Explain.
   b. Do objects A and B ever have the *same* speed? If so, at what time or times? Explain.

FIGURE Q2.4

FIGURE Q2.5

5. **FIGURE Q2.5** shows a position-versus-time graph for the motion of objects A and B as they move along the same axis.
   a. At the instant $t = 1$ s, is the speed of A greater than, less than, or equal to the speed of B? Explain.
   b. Do objects A and B ever have the *same* speed? If so, at what time or times? Explain.

6. **FIGURE Q2.6** shows the position-versus-time graph for a moving object. At which lettered point or points:
   a. Is the object *moving* the slowest?
   b. Is the object moving the fastest?
   c. Is the object at rest?
   d. Is the object moving to the left?

FIGURE Q2.6

7. **FIGURE Q2.7** shows the position-versus-time graph for a moving object. At which lettered point or points:
   a. Is the object moving the fastest?
   b. Is the object moving to the left?
   c. Is the object speeding up?
   d. Is the object turning around?

FIGURE Q2.7

8. **FIGURE Q2.8** shows six frames from the motion diagrams of two moving cars, A and B.
   a. Do the two cars ever have the same position at one instant of time? If so, in which frame number (or numbers)?
   b. Do the two cars ever have the same velocity at one instant of time? If so, between which two frames?

FIGURE Q2.8

9. You're driving along the highway at a steady speed of 60 mph when another driver decides to pass you. At the moment when the front of his car is exactly even with the front of your car, and you turn your head to smile at him, do the two cars have equal velocities? Explain.

10. A bicycle is traveling east. Can its acceleration vector ever point west? Explain.

11. (a) Give an example of a vertical motion with a positive velocity and a negative acceleration. (b) Give an example of a vertical motion with a negative velocity and a negative acceleration.

12. A ball is thrown straight up into the air. At each of the following instants, is the magnitude of the ball's acceleration greater than g, equal to g, less than g, or 0? Explain.
    a. Just after leaving your hand.
    b. At the very top (maximum height).
    c. Just before hitting the ground.

13. A rock is *thrown* (not dropped) straight down from a bridge into the river below. At each of the following instants, is the magnitude of the rock's acceleration greater than g, equal to g, less than g, or 0? Explain.
    a. Immediately after being released.
    b. Just before hitting the water.

14. **FIGURE Q2.14** shows the velocity-versus-time graph for a moving object. At which lettered point or points:
    a. Is the object speeding up?
    b. Is the object slowing down?
    c. Is the object moving to the left?
    d. Is the object moving to the right?

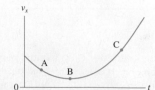

FIGURE Q2.14

EXERCISES AND PROBLEMS

Exercises

Section 2.1 Uniform Motion

1. ‖ Alan leaves Los Angeles at 8:00 A.M. to drive to San Francisco, 400 mi away. He travels at a steady 50 mph. Beth leaves Los Angeles at 9:00 A.M. and drives a steady 60 mph.
   a. Who gets to San Francisco first?
   b. How long does the first to arrive have to wait for the second?

2. ‖ Julie drives 100 mi to Grandmother's house. On the way to Grandmother's, Julie drives half the distance at 40 mph and half the distance at 60 mph. On her return trip, she drives half the time at 40 mph and half the time at 60 mph.
   a. What is Julie's average speed on the way to Grandmother's house?
   b. What is her average speed on the return trip?

3. ‖ Larry leaves home at 9:05 and runs at constant speed to the lamppost seen in **FIGURE EX2.3**. He reaches the lamppost at 9:07, immediately turns, and runs to the tree. Larry arrives at the tree at 9:10.
   a. What is Larry's average velocity, in m/min, during each of these two intervals?
   b. What is Larry's average velocity for the entire run?

FIGURE EX2.3

4. ‖ **FIGURE EX2.4** is the position-versus-time graph of a jogger. What is the jogger's velocity at $t = 10$ s, at $t = 25$ s, and at $t = 35$ s?

FIGURE EX2.4

Section 2.2 Instantaneous Velocity

Section 2.3 Finding Position from Velocity

5. | **FIGURE EX2.5** shows the position graph of a particle.
   a. Draw the particle's velocity graph for the interval 0 s $\le t \le 4$ s.
   b. Does this particle have a turning point or points? If so, at what time or times?

FIGURE EX2.5   **FIGURE EX2.6**

6. ‖ A particle starts from $x_0 = 10$ m at $t_0 = 0$ s and moves with the velocity graph shown in **FIGURE EX2.6**.
   a. Does this particle have a turning point? If so, at what time?
   b. What is the object's position at $t = 2$ s and 4 s?

7. ‖ **FIGURE EX2.7** is a somewhat idealized graph of the velocity
BIO of blood in the ascending aorta during one beat of the heart. Approximately how far, in cm, does the blood move during one beat?

FIGURE EX2.7

8. | **FIGURE EX2.8** shows the velocity graph for a particle having initial position $x_0 = 0$ m at $t_0 = 0$ s. At what time or times is the particle found at $x = 35$ m?

FIGURE EX2.8

Section 2.4 Motion with Constant Acceleration

9. ‖ **FIGURE EX2.9** shows the velocity graph of a particle. Draw the particle's acceleration graph for the interval 0 s $\le t \le 4$ s.

FIGURE EX2.9

10. ‖ **FIGURE EX2.7** showed the velocity graph of blood in the aorta.
BIO What is the blood's acceleration during each phase of the motion, speeding up and slowing down?

11. ‖ **FIGURE EX2.11** shows the velocity graph of a particle moving along the x-axis. Its initial position is $x_0 = 2.0$ m at $t_0 = 0$ s. At $t = 2.0$ s, what are the particle's (a) position, (b) velocity, and (c) acceleration?

FIGURE EX2.11

12. ‖ **FIGURE EX2.12** shows the velocity-versus-time graph for a particle moving along the x-axis. Its initial position is $x_0 = 2.0$ m at $t_0 = 0$ s.
   a. What are the particle's position, velocity, and acceleration at $t = 1.0$ s?
   b. What are the particle's position, velocity, and acceleration at $t = 3.0$ s?

FIGURE EX2.12

13. | a. What constant acceleration, in SI units, must a car have to go from zero to 60 mph in 10 s?
   b. How far has the car traveled when it reaches 60 mph? Give your answer both in SI units and in feet.

14. ‖ A jet plane is cruising at 300 m/s when suddenly the pilot turns the engines up to full throttle. After traveling 4.0 km, the jet is moving with a speed of 400 m/s. What is the jet's acceleration, assuming it to be a constant acceleration?

15. ‖ a. How many days will it take a spaceship to accelerate to the speed of light $(3.0 \times 10^8$ m/s) with the acceleration g?
   b. How far will it travel during this interval?
   c. What fraction of a light year is your answer to part b? A light year is the distance light travels in one year.

   NOTE We know, from Einstein's theory of relativity, that no object can travel at the speed of light. So this problem, while interesting and instructive, is not realistic.

16. ‖ When you sneeze, the air in your lungs accelerates from rest to 150 km/h in approximately 0.50 s. What is the acceleration of the air in m/s^2?
   BIO

17. ‖ A speed skater moving to the left across frictionless ice at 8.0 m/s hits a 5.0-m-wide patch of rough ice. She slows steadily, then continues on at 6.0 m/s. What is her acceleration on the rough ice?

18. ‖ A Porsche challenges a Honda to a 400 m race. Because the Porsche's acceleration of 3.5 m/s^2 is larger than the Honda's 3.0 m/s^2, the Honda gets a 1.0 s head start. Who wins? By how many seconds?

19. ‖ A car starts from rest at a stop sign. It accelerates at 4.0 m/s^2 for 6.0 s, coasts for 2.0 s, and then slows down at a rate of 3.0 m/s^2 for the next stop sign. How far apart are the stop signs?

Section 2.5 Free Fall

20. | Ball bearings are made by letting spherical drops of molten metal fall inside a tall tower—called a *shot tower*—and solidify as they fall.
   a. If a bearing needs 4.0 s to solidify enough for impact, how high must the tower be?
   b. What is the bearing's impact velocity?

21. ‖ A student standing on the ground throws a ball straight up. The ball leaves the student's hand with a speed of 15 m/s when the hand is 2.0 m above the ground. How long is the ball in the air before it hits the ground? (The student moves her hand out of the way.)

22. ‖ A rock is tossed straight up from ground level with a speed of 20 m/s. When it returns, it falls into a hole 10 m deep.
   a. What is the rock's velocity as it hits the bottom of the hole?
   b. How long is the rock in the air, from the instant it is released until it hits the bottom of the hole?

23. ‖ When jumping, a flea accelerates at an astounding 1000 m/s^2,
   BIO but over only the very short distance of 0.50 mm. If a flea jumps straight up, and if air resistance is neglected (a rather poor approximation in this situation), how high does the flea go?

24. ‖ As a science project, you drop a watermelon off the top of the Empire State Building, 320 m above the sidewalk. It so happens that Superman flies by at the instant you release the watermelon. Superman is headed straight down with a speed of 35 m/s. How fast is the watermelon going when it passes Superman?

25. ‖‖ A rock is dropped from the top of a tall building. The rock's displacement in the last second before it hits the ground is 45% of the entire distance it falls. How tall is the building?

Section 2.6 Motion on an Inclined Plane

26. ‖ A skier is gliding along at 3.0 m/s on horizontal, frictionless snow. He suddenly starts down a 10° incline. His speed at the bottom is 15 m/s.
   a. What is the length of the incline?
   b. How long does it take him to reach the bottom?

27. ‖ A car traveling at 30 m/s runs out of gas while traveling up a 10° slope. How far up the hill will it coast before starting to roll back down?

28. ‖ Santa loses his footing and slides down a frictionless, snowy roof that is tilted at an angle of 30°. If Santa slides 10 m before reaching the edge, what is his speed as he leaves the roof?

29. ‖ A snowboarder glides down a 50-m-long, 15° hill. She then glides horizontally for 10 m before reaching a 25° upward slope. Assume the snow is frictionless.
   a. What is her velocity at the bottom of the hill?
   b. How far can she travel up the 25° slope?

30. ‖ A small child gives a plastic frog a big push at the bottom of a slippery 2.0-m-long, 1.0-m-high ramp, starting it with a speed of 5.0 m/s. What is the frog's speed as it flies off the top of the ramp?

Section 2.7 Instantaneous Acceleration

31. ‖ **FIGURE EX2.31** shows the acceleration-versus-time graph of a particle moving along the x-axis. Its initial velocity is $v_{0x} = 8.0$ m/s at $t_0 = 0$ s. What is the particle's velocity at $t = 4.0$ s?

FIGURE EX2.31

FIGURE EX2.32

32. ‖ **FIGURE EX2.32** shows the acceleration graph for a particle that starts from rest at $t = 0$ s. What is the particle's velocity at $t = 6$ s?

33. | A particle moving along the x-axis has its position described by
   CALC the function $x = (2t^3 + 2t + 1)$ m, where t is in s. At $t = 2$ s what are the particle's (a) position, (b) velocity, and (c) acceleration?

34. ‖ A particle moving along the x-axis has its velocity described
CALC by the function $v_x = 2t^2$ m/s, where t is in s. Its initial position is $x_0 = 1$ m at $t_0 = 0$ s. At $t = 1$ s what are the particle's (a) position, (b) velocity, and (c) acceleration?

35. ‖ The position of a particle is given by the function
CALC $x = (2t^3 - 9t^2 + 12)$ m, where t is in s.
   a. At what time or times is $v_x = 0$ m/s?
   b. What are the particle's position and its acceleration at this time(s)?

36. ‖ The position of a particle is given by the function
CALC $x = (2t^3 - 6t^2 + 12)$ m, where t is in s.
   a. At what time does the particle reach its minimum velocity? What is $(v_x)_{min}$?
   b. At what time is the acceleration zero?

Problems

37. ‖ Particles A, B, and C move along the x-axis. Particle C has an initial velocity of 10 m/s. In **FIGURE P2.37**, the graph for A is a position-versus-time graph; the graph for B is a velocity-versus-time graph; the graph for C is an acceleration-versus-time graph. Find each particle's velocity at $t = 7.0$ s.

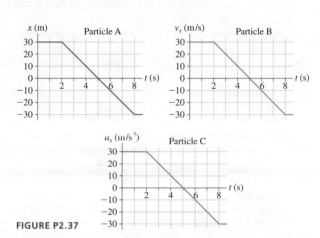

FIGURE P2.37

38. ‖ A block is suspended from a spring, pulled down, and released. The block's position-versus-time graph is shown in **FIGURE P2.38**.
   a. At what times is the velocity zero? At what times is the velocity most positive? Most negative?
   b. Draw a reasonable velocity-versus-time graph.

FIGURE P2.38

39. ‖ A particle's velocity is described by the function
CALC $v_x = (t^2 - 7t + 10)$ m/s, where t is in s.
   a. At what times does the particle reach its turning points?
   b. What is the particle's acceleration at each of the turning points?

40. ‖‖ A particle's velocity is described by the function $v_x = kt^2$ m/s,
CALC where k is a constant and t is in s. The particle's position at $t_0 = 0$ s is $x_0 = -9.0$ m. At $t_1 = 3.0$ s, the particle is at $x_1 = 9.0$ m. Determine the value of the constant k. Be sure to include the proper units.

41. ‖ A particle's acceleration is described by the function
CALC $a_x = (10 - t)$ m/s^2, where t is in s. Its initial conditions are $x_0 = 0$ m and $v_{0x} = 0$ m/s at $t = 0$ s.
   a. At what time is the velocity again zero?
   b. What is the particle's position at that time?

42. ‖ A particle's velocity is given by the function
CALC $v_x = (2.0 \text{ m/s})\sin(\pi t)$, where t is in s.
   a. What is the first time after $t = 0$ s when the particle reaches a turning point?
   b. What is the particle's acceleration at that time?

43. ‖ A ball rolls along the smooth track shown in **FIGURE P2.43**. Each segment of the track is straight, and the ball passes smoothly from one segment to the next without changing speed or leaving the track. Draw three vertically stacked graphs showing position, velocity, and acceleration versus time. Each graph should have the same time axis, and the proportions of the graph should be qualitatively correct. Assume that the ball has enough speed to reach the top.

FIGURE P2.43

44. ‖ Draw position, velocity, and acceleration graphs for the ball shown in **FIGURE P2.44**. See Problem 43 for more information.

FIGURE P2.44

45. ‖ **FIGURE P2.45** shows a set of kinematic graphs for a ball rolling on a track. All segments of the track are straight lines, but some may be tilted. Draw a picture of the track and also indicate the ball's initial condition.

FIGURE P2.45   **FIGURE P2.46**

46. ‖ **FIGURE P2.46** shows a set of kinematic graphs for a ball rolling on a track. All segments of the track are straight lines, but some may be tilted. Draw a picture of the track and also indicate the ball's initial condition.

47. ‖ The takeoff speed for an Airbus A320 jetliner is 80 m/s. Velocity data measured during takeoff are as shown.

t (s)	v_x (m/s)
0	0
10	23
20	46
30	69

a. Is the jetliner's acceleration constant during takeoff? Explain.
b. At what time do the wheels leave the ground?
c. For safety reasons, in case of an aborted takeoff, the runway must be three times the takeoff distance. Can an A320 take off safely on a 2.5-mi-long runway?

48. ‖ You are driving to the grocery store at 20 m/s. You are 110 m from an intersection when the traffic light turns red. Assume that your reaction time is 0.50 s and that your car brakes with constant acceleration. What magnitude braking acceleration will bring you to a stop exactly at the intersection?

49. ‖ You're driving down the highway late one night at 20 m/s when a deer steps onto the road 35 m in front of you. Your reaction time before stepping on the brakes is 0.50 s, and the maximum deceleration of your car is 10 m/s².
a. How much distance is between you and the deer when you come to a stop?
b. What is the maximum speed you could have and still not hit the deer?

50. ‖ Two cars are driving at the same constant speed on a straight road, with car 1 in front of car 2. Car 1 suddenly starts to brake with constant acceleration and stops in 10 m. At the instant car 1 comes to a stop, car 2 begins to brake with the same acceleration. It comes to a halt just as it reaches the back of car 1. What was the separation between the cars before they starting braking?

51. ‖ You are playing miniature golf at the golf course shown in FIGURE P2.51. Due to the fake plastic grass, the ball decelerates at 1.0 m/s² when rolling horizontally and at 6.0 m/s² on the slope. What is the slowest speed with which the ball can leave your golf club if you wish to make a hole in one?

FIGURE P2.51

52. ‖ The minimum stopping distance for a car traveling at a speed of 30 m/s is 60 m, including the distance traveled during the driver's reaction time of 0.50 s. What is the minimum stopping distance for the same car traveling at a speed of 40 m/s?

53. ‖ A cheetah spots a Thomson's gazelle, its preferred prey, and
BIO leaps into action, quickly accelerating to its top speed of 30 m/s, the highest of any land animal. However, a cheetah can maintain this extreme speed for only 15 s before having to let up. The cheetah is 170 m from the gazelle as it reaches top speed, and the gazelle sees the cheetah at just this instant. With negligible reaction time, the gazelle heads directly away from the cheetah, accelerating at 4.6 m/s² for 5.0 s, then running at constant speed. Does the gazelle escape? If so, by what distance is the gazelle in front when the cheetah gives up?

54. ‖ You are at a train station, standing next to the train at the front of the first car. The train starts moving with constant acceleration, and 5.0 s later the back of the first car passes you. How long does it take after the train starts moving until the back of the seventh car passes you? All cars are the same length.

55. ‖‖ A 200 kg weather rocket is loaded with 100 kg of fuel and fired straight up. It accelerates upward at 30 m/s² for 30 s, then runs out of fuel. Ignore any air resistance effects.
a. What is the rocket's maximum altitude?
b. How long is the rocket in the air before hitting the ground?

56. ‖‖ A 1000 kg weather rocket is launched straight up. The rocket motor provides a constant acceleration for 16 s, then the motor stops. The rocket altitude 20 s after launch is 5100 m. You can ignore any effects of air resistance. What was the rocket's acceleration during the first 16 s?

57. ‖‖ A lead ball is dropped into a lake from a diving board 5.0 m above the water. After entering the water, it sinks to the bottom with a constant velocity equal to the velocity with which it hit the water. The ball reaches the bottom 3.0 s after it is released. How deep is the lake?

58. ‖ A hotel elevator ascends 200 m with a maximum speed of 5.0 m/s. Its acceleration and deceleration both have a magnitude of 1.0 m/s².
a. How far does the elevator move while accelerating to full speed from rest?
b. How long does it take to make the complete trip from bottom to top?

59. ‖ A basketball player can jump to a height of 55 cm. How far above the floor can he jump in an elevator that is descending at a constant 1.0 m/s?

60. ‖ You are 9.0 m from the door of your bus, behind the bus, when it pulls away with an acceleration of 1.0 m/s². You instantly start running toward the still-open door at 4.5 m/s.
a. How long does it take for you to reach the open door and jump in?
b. What is the maximum time you can wait before starting to run and still catch the bus?

61. ‖ Ann and Carol are driving their cars along the same straight road. Carol is located at $x = 2.4$ mi at $t = 0$ h and drives at a steady 36 mph. Ann, who is traveling in the same direction, is located at $x = 0.0$ mi at $t = 0.50$ h and drives at a steady 50 mph.
a. At what time does Ann overtake Carol?
b. What is their position at this instant?
c. Draw a position-versus-time graph showing the motion of both Ann and Carol.

62. ‖ Amir starts riding his bike up a 200-m-long slope at a speed of 18 km/h, decelerating at 0.20 m/s² as he goes up. At the same instant, Becky starts down from the top at a speed of 6.0 km/h, accelerating at 0.40 m/s² as she goes down. How far has Amir ridden when they pass?

63. ‖ A very slippery block of ice slides down a smooth ramp tilted at angle θ. The ice is released from rest at vertical height h above the bottom of the ramp. Find an expression for the speed of the ice at the bottom.

64. ‖ Bob is driving the getaway car after the big bank robbery. He's going 50 m/s when his headlights suddenly reveal a nail strip that the cops have placed across the road 150 m in front of him. If Bob can stop in time, he can throw the car into reverse and escape. But if he crosses the nail strip, all his tires will go flat and he will be caught. Bob's reaction time before he can hit the brakes is 0.60 s, and his car's maximum deceleration is 10 m/s². Does Bob stop before or after the nail strip? By what distance?

65. ‖ One game at the amusement park has you push a puck up a long, frictionless ramp. You win a stuffed animal if the puck, at its highest point, comes to within 10 cm of the end of the ramp without going off. You give the puck a push, releasing it with a speed of 5.0 m/s when it is 8.5 m from the end of the ramp. The puck's speed after traveling 3.0 m is 4.0 m/s. How far is it from the end when it stops?

66. ‖ A motorist is driving at 20 m/s when she sees that a traffic light 200 m ahead has just turned red. She knows that this light stays red for 15 s, and she wants to reach the light just as it turns green again. It takes her 1.0 s to step on the brakes and begin slowing. What is her speed as she reaches the light at the instant it turns green?

67. ‖ Nicole throws a ball straight up. Chad watches the ball from a window 5.0 m above the point where Nicole released it. The ball passes Chad on the way up, and it has a speed of 10 m/s as it passes him on the way back down. How fast did Nicole throw the ball?

68. ‖ David is driving a steady 30 m/s when he passes Tina, who is sitting in her car at rest. Tina begins to accelerate at a steady 2.0 m/s^2 at the instant when David passes.
a. How far does Tina drive before passing David?
b. What is her speed as she passes him?

69. ‖ A cat is sleeping on the floor in the middle of a 3.0-m-wide room when a barking dog enters with a speed of 1.50 m/s. As the dog enters, the cat (as only cats can do) immediately accelerates at 0.85 m/s^2 toward an open window on the opposite side of the room. The dog (all bark and no bite) is a bit startled by the cat and begins to slow down at 0.10 m/s^2 as soon as it enters the room. How far is the cat in front of the dog as it leaps through the window?

70. ‖‖ Water drops fall from the edge of a roof at a steady rate. A fifth drop starts to fall just as the first drop hits the ground. At this instant, the second and third drops are exactly at the bottom and top edges of a 1.00-m-tall window. How high is the edge of the roof?

71. ‖‖ I was driving along at 20 m/s, trying to change a CD and not watching where I was going. When I looked up, I found myself 45 m from a railroad crossing. And wouldn't you know it, a train moving at 30 m/s was only 60 m from the crossing. In a split second, I realized that the train was going to beat me to the crossing and that I didn't have enough distance to stop. My only hope was to accelerate enough to cross the tracks before the train arrived. If my reaction time before starting to accelerate was 0.50 s, what minimum acceleration did my car need for me to be here today writing these words?

72. ‖ As an astronaut visiting Planet X, you're assigned to measure the free-fall acceleration. Getting out your meter stick and stop watch, you time the fall of a heavy ball from several heights. Your data are as follows:

Height (m)	Fall time (s)
0.0	0.00
1.0	0.54
2.0	0.72
3.0	0.91
4.0	1.01
5.0	1.17

Analyze these data to determine the free-fall acceleration on Planet X. Your analysis method should involve fitting a straight line to an appropriate graph, similar to the analysis in Example 2.14.

73. ‖ Your goal in laboratory is to launch a ball of mass m straight up so that it reaches exactly height h above the top of the launching tube. You and your lab partners will earn fewer points if the ball goes too high or too low. The launch tube uses compressed air to accelerate the ball over a distance d, and you have a table of data telling you how to set the air compressor to achieve a desired acceleration. Find an expression for the acceleration that will earn you maximum points.

74. ‖ When a 1984 Alfa Romeo Spider sports car accelerates at
CALC the maximum possible rate, its motion during the first 20 s is extremely well modeled by the simple equation

$$v_x^2 = \frac{2P}{m}t$$

where $P = 3.6 \times 10^4$ watts is the car's power output, $m = 1200$ kg is its mass, and v_x is in m/s. That is, the square of the car's velocity increases linearly with time.
a. Find an algebraic expression in terms of P, m, and t for the car's acceleration at time t.
b. What is the car's speed at $t = 2$ s and $t = 10$ s?
c. Evaluate the acceleration at $t = 2$ s and $t = 10$ s.

75. ‖ The two masses in FIGURE P2.75 slide on frictionless wires. They are connected by a pivoting rigid rod of length L. Prove that $v_{2x} = -v_{1y}\tan\theta$.

FIGURE P2.75

In Problems 76 through 79, you are given the kinematic equation or equations that are used to solve a problem. For each of these, you are to:
a. Write a *realistic* problem for which this is the correct equation(s). Be sure that the answer your problem requests is consistent with the equation(s) given.
b. Draw the pictorial representation for your problem.
c. Finish the solution of the problem.

76. $64 \text{ m} = 0 \text{ m} + (32 \text{ m/s})(4 \text{ s} - 0 \text{ s}) + \frac{1}{2}a_x(4 \text{ s} - 0 \text{ s})^2$

77. $(10 \text{ m/s})^2 = v_{0y}^2 - 2(9.8 \text{ m/s}^2)(10 \text{ m} - 0 \text{ m})$

78. $(0 \text{ m/s})^2 = (5 \text{ m/s})^2 - 2(9.8 \text{ m/s}^2)(\sin 10°)(x_1 - 0 \text{ m})$

79. $v_{1x} = 0 \text{ m/s} + (20 \text{ m/s}^2)(5 \text{ s} - 0 \text{ s})$
$x_1 = 0 \text{ m} + (0 \text{ m/s})(5 \text{ s} - 0 \text{ s}) + \frac{1}{2}(20 \text{ m/s}^2)(5 \text{ s} - 0 \text{ s})^2$
$x_2 = x_1 + v_{1x}(10 \text{ s} - 5 \text{ s})$

Challenge Problems

80. ‖‖ A rocket is launched straight up with constant acceleration. Four seconds after liftoff, a bolt falls off the side of the rocket. The bolt hits the ground 6.0 s later. What was the rocket's acceleration?

81. ‖‖ Careful measurements have been made of Olympic sprinters in the 100 meter dash. A simple but reasonably accurate model is that a sprinter accelerates at 3.6 m/s^2 for $3\frac{1}{3}$ s, then runs at constant velocity to the finish line.
a. What is the race time for a sprinter who follows this model?
b. A sprinter could run a faster race by accelerating faster at the beginning, thus reaching top speed sooner. If a sprinter's top speed is the same as in part a, what acceleration would he need to run the 100 meter dash in 9.9 s?
c. By what percent did the sprinter need to increase his acceleration in order to decrease his time by 1%?

82. ||| Careful measurements have been made of Olympic sprinters
CALC in the 100 meter dash. A quite realistic model is that the sprinter's velocity is given by

$$v_x = a(1 - e^{-bt})$$

where t is in s, v_x is in m/s, and the constants a and b are characteristic of the sprinter. Sprinter Carl Lewis's run at the 1987
World Championships is modeled with $a = 11.81$ m/s and
$b = 0.6887$ s^{-1}.
   a. What was Lewis's acceleration at $t = 0$ s, 2.00 s, and 4.00 s?
   b. Find an expression for the distance traveled at time t.
   c. Your expression from part b is a transcendental equation,
      meaning that you can't solve it for t. However, it's not hard to
      use trial and error to find the time needed to travel a specific
      distance. To the nearest 0.01 s, find the time Lewis needed to
      sprint 100.0 m. His official time was 0.01 s more than your
      answer, showing that this model is very good, but not perfect.
83. ||| A sprinter can accelerate with constant acceleration for 4.0 s
    before reaching top speed. He can run the 100 meter dash in 10.0 s.
    What is his speed as he crosses the finish line?
84. ||| A rubber ball is shot straight up from the ground with speed v_0.
    Simultaneously, a second rubber ball at height h directly above
    the first ball is dropped from rest.

   a. At what height above the ground do the balls collide? Your an-
      swer will be an *algebraic expression* in terms of h, v_0, and g.
   b. What is the maximum value of h for which a collision occurs
      before the first ball falls back to the ground?
   c. For what value of h does the collision occur at the instant
      when the first ball is at its highest point?
85. ||| The Starship Enterprise returns from warp drive to ordinary
    space with a forward speed of 50 km/s. To the crew's great sur-
    prise, a Klingon ship is 100 km directly ahead, traveling in the
    same direction at a mere 20 km/s. Without evasive action, the
    Enterprise will overtake and collide with the Klingons in just
    slightly over 3.0 s. The Enterprise's computers react instantly to
    brake the ship. What magnitude acceleration does the Enterprise
    need to just barely avoid a collision with the Klingon ship?
    Assume the acceleration is constant.
    Hint: Draw a position-versus-time graph showing the motions
    of both the Enterprise and the Klingon ship. Let $x_0 = 0$ km be
    the location of the Enterprise as it returns from warp drive. How
    do you show graphically the situation in which the collision is
    "barely avoided"? Once you decide what it looks like graphically,
    express that situation mathematically.

3 Vectors and Coordinate Systems

Wind has both a speed and a direction, hence the motion of the wind is described by a vector.

IN THIS CHAPTER, you will learn how vectors are represented and used.

What is a vector?

A vector is a quantity with both a size—its magnitude—and a direction. Vectors you'll meet in the next few chapters include position, displacement, velocity, acceleration, force, and momentum.

« LOOKING BACK Tactics Boxes 1.1 and 1.2 Vector addition and subtraction

How are vectors added and subtracted?

Vectors are added "tip to tail." The order of addition does not matter. To subtract vectors, turn the subtraction into addition by writing $\vec{A} - \vec{B} = \vec{A} + (-\vec{B})$. The vector $-\vec{B}$ is the same length as $\vec{B}$ but points in the opposite direction.

What are unit vectors?

Unit vectors define what we *mean* by the $+x$- and $+y$-directions in space.

- A unit vector has magnitude 1.
- A unit vector has no units.

Unit vectors simply point.

What are components?

Components of vectors are the pieces of vectors parallel to the coordinate axes—in the directions of the unit vectors. We write

$$\vec{E} = E_x \hat{\imath} + E_y \hat{\jmath}$$

Components simplify vector math.

How are components used?

Components let us do vector math with algebra, which is easier and more precise than adding and subtracting vectors using geometry and trigonometry. Multiplying a vector by a number simply multiplies all of the vector's components by that number.

$$\vec{C} = 2\vec{A} + 3\vec{B}$$

means

$$\begin{cases} C_x = 2A_x + 3B_x \\ C_y = 2A_y + 3B_y \end{cases}$$

How will I use vectors?

Vectors appear everywhere in physics and engineering—from velocities to electric fields and from forces to fluid flows. The tools and techniques you learn in this chapter will be used throughout your studies and your professional career.

3.1 Scalars and Vectors

A quantity that is fully described by a single number (with units) is called a **scalar.** Mass, temperature, volume and energy are all scalars. We will often use an algebraic symbol to represent a scalar quantity. Thus m will represent mass, T temperature, V volume, E energy, and so on.

Our universe has three dimensions, so some quantities also need a direction for a full description. If you ask someone for directions to the post office, the reply "Go three blocks" will not be very helpful. A full description might be, "Go three blocks south." A quantity having both a size and a direction is called a **vector.**

The mathematical term for the length, or size, of a vector is **magnitude,** so we can also say that **a vector is a quantity having a magnitude and a direction.**

FIGURE 3.1 shows that the *geometric representation* of a vector is an arrow, with the tail of the arrow (not its tip!) placed at the point where the measurement is made. An arrow makes a natural representation of a vector because it inherently has both a length and a direction. As you've already seen, we label vectors by drawing a small arrow over the letter that represents the vector: $\vec{r}$ for position, $\vec{v}$ for velocity, $\vec{a}$ for acceleration.

FIGURE 3.1 The velocity vector $\vec{v}$ has both a magnitude and a direction.

Magnitude of vector

Direction of vector

$v = 5$ m/s

$\vec{v}$ — Name of vector

The vector is drawn across the page, but it represents the particle's velocity at this one point.

NOTE Although the vector arrow is drawn across the page, from its tail to its tip, this does *not* indicate that the vector "stretches" across this distance. Instead, the vector arrow tells us the value of the vector quantity only at the one point where the tail of the vector is placed.

The magnitude of a vector can be written using absolute value signs or, more frequently, as the letter without the arrow. For example, the magnitude of the velocity vector in Figure 3.1 is $v = |\vec{v}| = 5$ m/s. This is the object's *speed*. The magnitude of the acceleration vector $\vec{a}$ is written a. **The magnitude of a vector is a scalar.** Note that magnitude of a vector cannot be a negative number; it must be positive or zero, with appropriate units.

It is important to get in the habit of using the arrow symbol for vectors. If you omit the vector arrow from the velocity vector $\vec{v}$ and write only v, then you're referring only to the object's speed, not its velocity. The symbols $\vec{r}$ and r, or $\vec{v}$ and v, do *not* represent the same thing.

FIGURE 3.2 Displacement vectors.

(a)

Sam's actual path

$\vec{S}$

N

200 ft

Sam

Sam's displacement

Displacement is the straight-line connection from the initial to the final position.

(b)

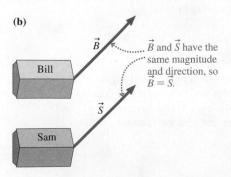

$\vec{B}$

Bill

$\vec{S}$

Sam

$\vec{B}$ and $\vec{S}$ have the same magnitude and direction, so $\vec{B} = \vec{S}$.

3.2 Using Vectors

Suppose Sam starts from his front door, walks across the street, and ends up 200 ft to the northeast of where he started. Sam's displacement, which we will label $\vec{S}$, is shown in **FIGURE 3.2a.** The displacement vector is a *straight-line connection* from his initial to his final position, not necessarily his actual path.

To describe a vector we must specify both its magnitude and its direction. We can write Sam's displacement as $\vec{S} = (200 \text{ ft, northeast})$. The magnitude of Sam's displacement is $S = |\vec{S}| = 200$ ft, the distance between his initial and final points.

Sam's next-door neighbor Bill also walks 200 ft to the northeast, starting from his own front door. Bill's displacement $\vec{B} = (200 \text{ ft, northeast})$ has the same magnitude and direction as Sam's displacement $\vec{S}$. Because vectors are defined by their magnitude and direction, **two vectors are equal if they have the same magnitude and direction.** Thus the two displacements in **FIGURE 3.2b** are equal to each other, and we can write $\vec{B} = \vec{S}$.

NOTE A vector is unchanged if you move it to a different point on the page as long as you don't change its length or the direction it points.

Vector Addition

If you earn $50 on Saturday and $60 on Sunday, your *net* income for the weekend is the sum of $50 and $60. With numbers, the word *net* implies addition. The same is true with vectors. For example, FIGURE 3.3 shows the displacement of a hiker who first hikes 4 miles to the east, then 3 miles to the north. The first leg of the hike is described by the displacement $\vec{A} = (4\text{ mi, east})$. The second leg of the hike has displacement $\vec{B} = (3\text{ mi, north})$. Vector $\vec{C}$ is the *net displacement* because it describes the net result of the hiker's first having displacement $\vec{A}$, then displacement $\vec{B}$.

The net displacement $\vec{C}$ is an initial displacement $\vec{A}$ *plus* a second displacement $\vec{B}$, or

$$\vec{C} = \vec{A} + \vec{B} \tag{3.1}$$

The sum of two vectors is called the **resultant vector.** It's not hard to show that vector addition is commutative: $\vec{A} + \vec{B} = \vec{B} + \vec{A}$. That is, you can add vectors in any order you wish.

« Tactics Box 1.1 on page 6 showed the three-step procedure for adding two vectors, and it's highly recommended that you turn back for a quick review. This tip-to-tail method for adding vectors, which is used to find $\vec{C} = \vec{A} + \vec{B}$ in Figure 3.3, is called **graphical addition.** Any two vectors of the same type—two velocity vectors or two force vectors—can be added in exactly the same way.

The graphical method for adding vectors is straightforward, but we need to do a little geometry to come up with a complete description of the resultant vector $\vec{C}$. Vector $\vec{C}$ of Figure 3.3 is defined by its magnitude C and by its direction. Because the three vectors $\vec{A}$, $\vec{B}$, and $\vec{C}$ form a right triangle, the magnitude, or length, of $\vec{C}$ is given by the Pythagorean theorem:

$$C = \sqrt{A^2 + B^2} = \sqrt{(4\text{ mi})^2 + (3\text{ mi})^2} = 5\text{ mi} \tag{3.2}$$

Notice that Equation 3.2 uses the magnitudes A and B of the vectors $\vec{A}$ and $\vec{B}$. The angle θ, which is used in Figure 3.3 to describe the direction of $\vec{C}$, is easily found for a right triangle:

$$\theta = \tan^{-1}\left(\frac{B}{A}\right) = \tan^{-1}\left(\frac{3\text{ mi}}{4\text{ mi}}\right) = 37° \tag{3.3}$$

Altogether, the hiker's net displacement is $\vec{C} = \vec{A} + \vec{B} = (5\text{ mi, }37°\text{ north of east})$.

NOTE Vector mathematics makes extensive use of geometry and trigonometry. Appendix A, at the end of this book, contains a brief review of these topics.

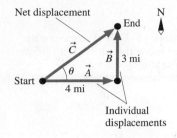

FIGURE 3.3 The net displacement $\vec{C}$ resulting from two displacements $\vec{A}$ and $\vec{B}$.

EXAMPLE 3.1 | Using graphical addition to find a displacement

A bird flies 100 m due east from a tree, then 50 m northwest (that is, 45° north of west). What is the bird's net displacement?

VISUALIZE FIGURE 3.4 shows the two individual displacements, which we've called $\vec{A}$ and $\vec{B}$. The net displacement is the vector sum $\vec{C} = \vec{A} + \vec{B}$, which is found graphically.

FIGURE 3.4 The bird's net displacement is $\vec{C} = \vec{A} + \vec{B}$.

SOLVE The two displacements are $\vec{A} = (100\text{ m, east})$ and $\vec{B} = (50\text{ m, northwest})$. The net displacement $\vec{C} = \vec{A} + \vec{B}$ is found by drawing a vector from the initial to the final position. But

describing $\vec{C}$ is a bit trickier than the example of the hiker because $\vec{A}$ and $\vec{B}$ are not at right angles. First, we can find the magnitude of $\vec{C}$ by using the law of cosines from trigonometry:

$$C^2 = A^2 + B^2 - 2AB\cos 45°$$
$$= (100\text{ m})^2 + (50\text{ m})^2 - 2(100\text{ m})(50\text{ m})\cos 45°$$
$$= 5430\text{ m}^2$$

Thus $C = \sqrt{5430\text{ m}^2} = 74\text{ m}$. Then a second use of the law of cosines can determine angle ϕ (the Greek letter phi):

$$B^2 = A^2 + C^2 - 2AC\cos\phi$$
$$\phi = \cos^{-1}\left[\frac{A^2 + C^2 - B^2}{2AC}\right] = 29°$$

The bird's net displacement is

$$\vec{C} = (74\text{ m, }29°\text{ north of east})$$

It is often convenient to draw two vectors with their tails together, as shown in **FIGURE 3.5a**. To evaluate $\vec{D} + \vec{E}$, you could move vector $\vec{E}$ over to where its tail is on the tip of $\vec{D}$, then use the tip-to-tail rule of graphical addition. That gives vector $F = \vec{D} + \vec{E}$ in **FIGURE 3.5b**. Alternatively, **FIGURE 3.5c** shows that the vector sum $\vec{D} + \vec{E}$ can be found as the diagonal of the parallelogram defined by $\vec{D}$ and $\vec{E}$. This method for vector addition is called the *parallelogram rule* of vector addition.

▶ **FIGURE 3.5** Two vectors can be added using the tip-to-tail rule or the parallelogram rule.

(a) What is $\vec{D} + \vec{E}$?

(b) **Tip-to-tail rule:** Slide the tail of $\vec{E}$ to the tip of $\vec{D}$.

(c) **Parallelogram rule:** Find the diagonal of the parallelogram formed by $\vec{D}$ and $\vec{E}$.

FIGURE 3.6 The net displacement after four individual displacements.

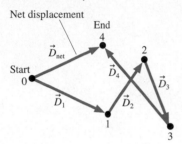

Vector addition is easily extended to more than two vectors. **FIGURE 3.6** shows the path of a hiker moving from initial position 0 to position 1, then position 2, then position 3, and finally arriving at position 4. These four segments are described by displacement vectors $\vec{D}_1, \vec{D}_2, \vec{D}_3,$ and $\vec{D}_4$. The hiker's *net* displacement, an arrow from position 0 to position 4, is the vector $\vec{D}_{net}$. In this case,

$$\vec{D}_{net} = \vec{D}_1 + \vec{D}_2 + \vec{D}_3 + \vec{D}_4 \qquad (3.4)$$

The vector sum is found by using the tip-to-tail method three times in succession.

STOP TO THINK 3.1 Which figure shows $\vec{A}_1 + \vec{A}_2 + \vec{A}_3$?

(a) **(b)** **(c)** **(d)** **(e)**

More Vector Mathematics

In addition to adding vectors, we will need to subtract vectors (« Tactics Box 1.2 on page 7), multiply vectors by scalars, and understand how to interpret the negative of a vector. These operations are illustrated in **FIGURE 3.7**.

FIGURE 3.7 Working with vectors.

The length of $\vec{B}$ is "stretched" by the factor c. That is, $B = cA$.

$\vec{A} = (A, \theta)$

$\vec{B} = c\vec{A} = (cA, \theta)$

$\vec{B}$ points in the same direction as $\vec{A}$.

Multiplication by a scalar

$\vec{A} + (-\vec{A}) = \vec{0}$. The tip of $-\vec{A}$ returns to the starting point.

Vector $-\vec{A}$ is equal in magnitude but opposite in direction to $\vec{A}$.

The **zero vector** $\vec{0}$ has zero length

The negative of a vector

$-2\vec{A}$

Multiplication by a negative scalar

Vector subtraction: What is $\vec{A} - \vec{C}$? Write it as $\vec{A} + (-\vec{C})$ and add!

$\vec{A} - \vec{C}$

Tip-to-tail subtraction using $-\vec{C}$

$\vec{A} - \vec{C}$

Parallelogram subtraction using $-\vec{C}$

EXAMPLE 3.2 | Velocity and displacement

Carolyn drives her car north at 30 km/h for 1 hour, east at 60 km/h for 2 hours, then north at 50 km/h for 1 hour. What is Carolyn's net displacement?

SOLVE Chapter 1 defined average velocity as

$$\vec{v} = \frac{\Delta \vec{r}}{\Delta t}$$

so the displacement $\Delta \vec{r}$ during the time interval Δt is $\Delta \vec{r} = (\Delta t)\vec{v}$. This is multiplication of the vector $\vec{v}$ by the scalar Δt. Carolyn's velocity during the first hour is $\vec{v}_1 = (30 \text{ km/h, north})$, so her displacement during this interval is

$$\Delta \vec{r}_1 = (1 \text{ hour})(30 \text{ km/h, north}) = (30 \text{ km, north})$$

Similarly,

$$\Delta \vec{r}_2 = (2 \text{ hours})(60 \text{ km/h, east}) = (120 \text{ km, east})$$

$$\Delta \vec{r}_3 = (1 \text{ hour})(50 \text{ km/h, north}) = (50 \text{ km, north})$$

In this case, multiplication by a scalar changes not only the length of the vector but also its units, from km/h to km. The direction, however, is unchanged. Carolyn's net displacement is

$$\Delta \vec{r}_{\text{net}} = \Delta \vec{r}_1 + \Delta \vec{r}_2 + \Delta \vec{r}_3$$

FIGURE 3.8 The net displacement is the vector sum $\Delta \vec{r}_{\text{net}} = \Delta \vec{r}_1 + \Delta \vec{r}_2 + \Delta \vec{r}_3$.

This addition of the three vectors is shown in **FIGURE 3.8**, using the tip-to-tail method. $\Delta \vec{r}_{\text{net}}$ stretches from Carolyn's initial position to her final position. The magnitude of her net displacement is found using the Pythagorean theorem:

$$r_{\text{net}} = \sqrt{(120 \text{ km})^2 + (80 \text{ km})^2} = 144 \text{ km}$$

The direction of $\Delta \vec{r}_{\text{net}}$ is described by angle θ, which is

$$\theta = \tan^{-1}\left(\frac{80 \text{ km}}{120 \text{ km}}\right) = 34°$$

Thus Carolyn's net displacement is $\Delta \vec{r}_{\text{net}} = (144 \text{ km}, 34° \text{ north of east})$.

STOP TO THINK 3.2 Which figure shows $2\vec{A} - \vec{B}$?

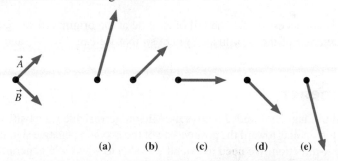

(a)　　　(b)　　　(c)　　　(d)　　　(e)

3.3 Coordinate Systems and Vector Components

Vectors do not require a coordinate system. We can add and subtract vectors graphically, and we will do so frequently to clarify our understanding of a situation. But the graphical addition of vectors is not an especially good way to find quantitative results. In this section we will introduce a *coordinate representation* of vectors that will be the basis of an easier method for doing vector calculations.

Coordinate Systems

The world does not come with a coordinate system attached to it. A coordinate system is an artificially imposed grid that you place on a problem in order to make quantitative measurements. You are free to choose:

- Where to place the origin, and
- How to orient the axes.

Different problem solvers may choose to use different coordinate systems; that is perfectly acceptable. However, some coordinate systems will make a problem easier

A GPS uses satellite signals to find your position in the earth's coordinate system with amazing accuracy.

FIGURE 3.9 A conventional *xy*-coordinate system and the quadrants of the *xy*-plane.

to solve. Part of our goal is to learn how to choose an appropriate coordinate system for each problem.

FIGURE 3.9 shows the *xy*-coordinate system we will use in this book. The placement of the axes is not entirely arbitrary. By convention, the positive *y*-axis is located 90° *counterclockwise* (ccw) from the positive *x*-axis. Figure 3.9 also identifies the four **quadrants** of the coordinate system, I through IV.

Coordinate axes have a positive end and a negative end, separated by zero at the origin where the two axes cross. When you draw a coordinate system, it is important to label the axes. This is done by placing *x* and *y* labels at the *positive* ends of the axes, as in Figure 3.9. The purpose of the labels is twofold:

- To identify which axis is which, and
- To identify the positive ends of the axes.

This will be important when you need to determine whether the quantities in a problem should be assigned positive or negative values.

Component Vectors

FIGURE 3.10 Component vectors $\vec{A}_x$ and $\vec{A}_y$ are drawn parallel to the coordinate axes such that $\vec{A} = \vec{A}_x + \vec{A}_y$.

The *y*-component vector is parallel to the *y*-axis.   The *x*-component vector is parallel to the *x*-axis.

FIGURE 3.10 shows a vector $\vec{A}$ and an *xy*-coordinate system that we've chosen. Once the directions of the axes are known, we can define two new vectors *parallel to the axes* that we call the **component vectors** of $\vec{A}$. You can see, using the parallelogram rule, that $\vec{A}$ is the vector sum of the two component vectors:

$$\vec{A} = \vec{A}_x + \vec{A}_y \tag{3.5}$$

In essence, we have broken vector $\vec{A}$ into two perpendicular vectors that are parallel to the coordinate axes. This process is called the **decomposition** of vector $\vec{A}$ into its component vectors.

> **NOTE** It is not necessary for the tail of $\vec{A}$ to be at the origin. All we need to know is the *orientation* of the coordinate system so that we can draw $\vec{A}_x$ and $\vec{A}_y$ parallel to the axes.

Components

You learned in Chapters 1 and 2 to give the kinematic variable v_x a positive sign if the velocity vector $\vec{v}$ points toward the positive end of the *x*-axis, a negative sign if $\vec{v}$ points in the negative *x*-direction. We need to extend this idea to vectors in general.

Suppose vector $\vec{A}$ has been decomposed into component vectors $\vec{A}_x$ and $\vec{A}_y$ parallel to the coordinate axes. We can describe each component vector with a single number called the **component**. The *x-component* and *y-component* of vector $\vec{A}$, denoted A_x and A_y, are determined as follows:

TACTICS BOX 3.1

Determining the components of a vector

❶ The absolute value $|A_x|$ of the *x*-component A_x is the magnitude of the component vector $\vec{A}_x$.

❷ The sign of A_x is positive if $\vec{A}_x$ points in the positive *x*-direction (right), negative if $\vec{A}_x$ points in the negative *x*-direction (left).

❸ The *y*-component A_y is determined similarly.

Exercises 10–18

In other words, the component A_x tells us two things: how big $\vec{A}_x$ is and, with its sign, which end of the axis $\vec{A}_x$ points toward. **FIGURE 3.11** shows three examples of determining the components of a vector.

FIGURE 3.11 Determining the components of a vector.

$\vec{A}_y$ points in the positive y-direction, so $A_y = +2$ m.

$\vec{A}_x$ points in the positive x-direction, so $A_x = +3$ m.

$\vec{B}_y$ points in the positive y-direction, so $B_y = +2$ m.

$\vec{B}_x$ points in the negative x-direction, so $B_x = -2$ m.

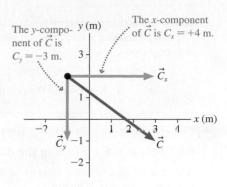

The y-component of $\vec{C}$ is $C_y = -3$ m.

The x-component of $\vec{C}$ is $C_x = +4$ m.

NOTE Beware of the somewhat confusing terminology. $\vec{A}_x$ and $\vec{A}_y$ are called *component vectors,* whereas A_x and A_y are simply called *components.* The components A_x and A_y are just numbers (with units), so make sure you do *not* put arrow symbols over the components.

We will frequently need to decompose a vector into its components. We will also need to "reassemble" a vector from its components. In other words, we need to move back and forth between the geometric and the component representations of a vector. **FIGURE 3.12** shows how this is done.

The magnitude and direction of $\vec{A}$ are found from the components. In this example,

$A = \sqrt{A_x^2 + A_y^2}$ $\theta = \tan^{-1}(A_y/A_x)$

$A_y = A\sin\theta$

$A_x = A\cos\theta$

The components of $\vec{A}$ are found from the magnitude and direction.

The angle is defined differently. In this example, the magnitude and direction are

$B = \sqrt{B_x^2 + B_y^2}$ $\phi = \tan^{-1}(B_x/|B_y|)$

$B_y = -B\cos\phi$

$B_x = B\sin\phi$

Minus signs must be inserted manually, depending on the vector's direction.

◄ **FIGURE 3.12** Moving between the geometric representation and the component representation.

Each decomposition requires that you pay close attention to the direction in which the vector points and the angles that are defined.

- If a component vector points left (or down), you must *manually* insert a minus sign in front of the component, as was done for B_y in Figure 3.12.
- The role of sines and cosines can be reversed, depending upon which angle is used to define the direction. Compare A_x and B_x.
- The angle used to define direction is almost always between 0° and 90°, so you must take the inverse tangent of a positive number. Use absolute values of the components, as was done to find angle ϕ (Greek phi) in Figure 3.12.

EXAMPLE 3.3 | Finding the components of an acceleration vector

Seen from above, a hummingbird's acceleration is (6.0 m/s², 30° south of west). Find the x- and y-components of the acceleration vector $\vec{a}$.

VISUALIZE It's important to *draw* vectors. **FIGURE 3.13** establishes a map-like coordinate system with the x-axis pointing east and the y-axis north. Vector $\vec{a}$ is then decomposed into components parallel to the axes. Notice that the axes are "acceleration axes" with units of acceleration, not xy-axes, because we're measuring an acceleration vector.

▶ **FIGURE 3.13** Decomposition of $\vec{a}$.

a_x is negative.

a_y is negative.

$a = 6.0$ m/s²

Continued

SOLVE The acceleration vector points to the left (negative *x*-direction) and down (negative *y*-direction), so the components a_x and a_y are both negative:

$$a_x = -a \cos 30° = -(6.0 \text{ m/s}^2) \cos 30° = -5.2 \text{ m/s}^2$$

$$a_y = -a \sin 30° = -(6.0 \text{ m/s}^2) \sin 30° = -3.0 \text{ m/s}^2$$

ASSESS The units of a_x and a_y are the same as the units of vector $\vec{a}$. Notice that we had to insert the minus signs manually by observing that the vector points left and down.

EXAMPLE 3.4 | **Finding the direction of motion**

FIGURE 3.14 shows a car's velocity vector $\vec{v}$. Determine the car's speed and direction of motion.

FIGURE 3.14 The velocity vector $\vec{v}$ of Example 3.4.

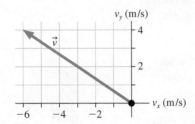

VISUALIZE **FIGURE 3.15** shows the components v_x and v_y and defines an angle θ with which we can specify the direction of motion.

SOLVE We can read the components of $\vec{v}$ directly from the axes: $v_x = -6.0$ m/s and $v_y = 4.0$ m/s. Notice that v_x is negative. This is enough information to find the car's speed v, which is the magnitude of $\vec{v}$:

$$v = \sqrt{v_x^2 + v_y^2} = \sqrt{(-6.0 \text{ m/s})^2 + (4.0 \text{ m/s})^2} = 7.2 \text{ m/s}$$

FIGURE 3.15 Decomposition of $\vec{v}$.

From trigonometry, angle θ is

$$\theta = \tan^{-1}\left(\frac{v_y}{|v_x|}\right) = \tan^{-1}\left(\frac{4.0 \text{ m/s}}{6.0 \text{ m/s}}\right) = 34°$$

The absolute value signs are necessary because v_x is a negative number. The velocity vector $\vec{v}$ can be written in terms of the speed and the direction of motion as

$$\vec{v} = (7.2 \text{ m/s}, 34° \text{ above the negative } x\text{-axis})$$

STOP TO THINK 3.3   What are the *x*- and *y*-components C_x and C_y of vector $\vec{C}$?

3.4 Unit Vectors and Vector Algebra

The vectors (1, +*x*-direction) and (1, +*y*-direction), shown in **FIGURE 3.16**, have some interesting and useful properties. Each has a magnitude of 1, has no units, and is parallel to a coordinate axis. A vector with these properties is called a **unit vector.** These unit vectors have the special symbols

$$\hat{\imath} \equiv (1, \text{ positive } x\text{-direction})$$

$$\hat{\jmath} \equiv (1, \text{ positive } y\text{-direction})$$

FIGURE 3.16 The unit vectors $\hat{\imath}$ and $\hat{\jmath}$.

The unit vectors have magnitude 1, no units, and point in the +*x*-direction and +*y*-direction.

The notation $\hat{\imath}$ (read "i hat") and $\hat{\jmath}$ (read "j hat") indicates a unit vector with a magnitude of 1. Recall that the symbol $\equiv$ means "is defined as."

Unit vectors establish the directions of the positive axes of the coordinate system. Our choice of a coordinate system may be arbitrary, but once we decide to place a coordinate system on a problem we need something to tell us "That direction is the positive *x*-direction." This is what the unit vectors do.

The unit vectors provide a useful way to write component vectors. The component vector $\vec{A}_x$ is the piece of vector $\vec{A}$ that is parallel to the x-axis. Similarly, $\vec{A}_y$ is parallel to the y-axis. Because, by definition, the vector $\hat{i}$ points along the x-axis and $\hat{j}$ points along the y-axis, we can write

$$\vec{A}_x = A_x \hat{i}$$
$$\vec{A}_y = A_y \hat{j}$$

(3.6)

Equations 3.6 separate each component vector into a length and a direction. The full decomposition of vector $\vec{A}$ can then be written

$$\vec{A} = \vec{A}_x + \vec{A}_y = A_x \hat{i} + A_y \hat{j}$$

(3.7)

FIGURE 3.17 shows how the unit vectors and the components fit together to form vector $\vec{A}$.

NOTE In three dimensions, the unit vector along the +z-direction is called $\hat{k}$, and to describe vector $\vec{A}$ we would include an additional component vector $\vec{A}_z = A_z \hat{k}$.

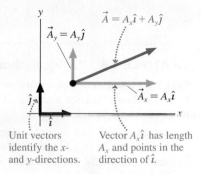

FIGURE 3.17 The decomposition of vector $\vec{A}$ is $A_x \hat{i} + A_y \hat{j}$.

Unit vectors identify the x- and y-directions.   Vector $A_x \hat{i}$ has length A_x and points in the direction of $\hat{i}$.

EXAMPLE 3.5 | Run rabbit run!

A rabbit, escaping a fox, runs 40.0° north of west at 10.0 m/s. A coordinate system is established with the positive x-axis to the east and the positive y-axis to the north. Write the rabbit's velocity in terms of components and unit vectors.

VISUALIZE FIGURE 3.18 shows the rabbit's velocity vector and the coordinate axes. We're showing a velocity vector, so the axes are labeled v_x and v_y rather than x and y.

FIGURE 3.18 The velocity vector $\vec{v}$ is decomposed into components v_x and v_y.

SOLVE 10.0 m/s is the rabbit's *speed*, not its velocity. The velocity, which includes directional information, is

$$\vec{v} = (10.0 \text{ m/s}, 40.0° \text{ north of west})$$

Vector $\vec{v}$ points to the left and up, so the components v_x and v_y are negative and positive, respectively. The components are

$$v_x = -(10.0 \text{ m/s}) \cos 40.0° = -7.66 \text{ m/s}$$
$$v_y = +(10.0 \text{ m/s}) \sin 40.0° = 6.43 \text{ m/s}$$

With v_x and v_y now known, the rabbit's velocity vector is

$$\vec{v} = v_x \hat{i} + v_y \hat{j} = (-7.66 \hat{i} + 6.43 \hat{j}) \text{ m/s}$$

Notice that we've pulled the units to the end, rather than writing them with each component.

ASSESS Notice that the minus sign for v_x was inserted manually. **Signs don't occur automatically; you have to set them after checking the vector's direction.**

Vector Math

You learned in Section 3.2 how to add vectors graphically, but it is a tedious problem in geometry and trigonometry to find precise values for the magnitude and direction of the resultant. The addition and subtraction of vectors become much easier if we use components and unit vectors.

To see this, let's evaluate the vector sum $\vec{D} = \vec{A} + \vec{B} + \vec{C}$. To begin, write this sum in terms of the components of each vector:

$$\vec{D} = D_x \hat{i} + D_y \hat{j} = \vec{A} + \vec{B} + \vec{C}$$
$$= (A_x \hat{i} + A_y \hat{j}) + (B_x \hat{i} + B_y \hat{j}) + (C_x \hat{i} + C_y \hat{j})$$

(3.8)

We can group together all the x-components and all the y-components on the right side, in which case Equation 3.8 is

$$(D_x) \hat{i} + (D_y) \hat{j} = (A_x + B_x + C_x) \hat{i} + (A_y + B_y + C_y) \hat{j}$$

(3.9)

Comparing the x- and y-components on the left and right sides of Equation 3.9, we find:

$$D_x = A_x + B_x + C_x$$
$$D_y = A_y + B_y + C_y$$

(3.10)

Stated in words, Equation 3.10 says that we can perform vector addition by adding the x-components of the individual vectors to give the x-component of the resultant and by adding the y-components of the individual vectors to give the y-component of the resultant. This method of vector addition is called **algebraic addition.**

EXAMPLE 3.6 | **Using algebraic addition to find a displacement**

Example 3.1 was about a bird that flew 100 m to the east, then 50 m to the northwest. Use the algebraic addition of vectors to find the bird's net displacement.

VISUALIZE FIGURE 3.19 shows displacement vectors $\vec{A} = (100 \text{ m},$ east) and $\vec{B} = (50 \text{ m, northwest})$. We draw vectors tip-to-tail to add them graphically, but it's usually easier to draw them all from the origin if we are going to use algebraic addition.

FIGURE 3.19 The net displacement is $\vec{C} = \vec{A} + \vec{B}$.

The net displacement $\vec{C} = \vec{A} + \vec{B}$ is drawn according to the parallelogram rule.

SOLVE To add the vectors algebraically we must know their components. From the figure these are seen to be

$$\vec{A} = 100\,\hat{\imath} \text{ m}$$

$$\vec{B} = (-50 \cos 45° \,\hat{\imath} + 50 \sin 45° \hat{\jmath}) \text{ m} = (-35.3\hat{\imath} + 35.3\hat{\jmath}) \text{ m}$$

Notice that vector quantities must include units. Also notice, as you would expect from the figure, that $\vec{B}$ has a negative x-component. Adding $\vec{A}$ and $\vec{B}$ by components gives

$$\vec{C} = \vec{A} + \vec{B} = 100\hat{\imath} \text{ m} + (-35.3\hat{\imath} + 35.3\hat{\jmath}) \text{ m}$$

$$= (100 \text{ m} - 35.3 \text{ m})\hat{\imath} + (35.3 \text{ m})\hat{\jmath} = (64.7\hat{\imath} + 35.3\hat{\jmath}) \text{ m}$$

This would be a perfectly acceptable answer for many purposes. However, we need to calculate the magnitude and direction of $\vec{C}$ if we want to compare this result to our earlier answer. The magnitude of $\vec{C}$ is

$$C = \sqrt{C_x^2 + C_y^2} = \sqrt{(64.7 \text{ m})^2 + (35.3 \text{ m})^2} = 74 \text{ m}$$

The angle ϕ, as defined in Figure 3.19, is

$$\phi = \tan^{-1}\left(\frac{C_y}{|C_x|}\right) = \tan^{-1}\left(\frac{35.3 \text{ m}}{64.7 \text{ m}}\right) = 29°$$

Thus $\vec{C} = (74 \text{ m}, 29°$ north of west$)$, in perfect agreement with Example 3.1.

Vector subtraction and the multiplication of a vector by a scalar, using components, are very much like vector addition. To find $\vec{R} = \vec{P} - \vec{Q}$ we would compute

$$R_x = P_x - Q_x$$
$$R_y = P_y - Q_y \tag{3.11}$$

Similarly, $\vec{T} = c\vec{S}$ would be

$$T_x = cS_x$$
$$T_y = cS_y \tag{3.12}$$

In other words, a vector equation is interpreted as meaning: Equate the x-components on both sides of the equals sign, then equate the y-components, and then the z-components. Vector notation allows us to write these three equations in a much more compact form.

Tilted Axes and Arbitrary Directions

As we've noted, the coordinate system is entirely your choice. It is a grid that you impose on the problem in a manner that will make the problem easiest to solve. As you've already seen in Chapter 2, it is often convenient to tilt the axes of the coordinate system, such as those shown in FIGURE 3.20. The axes are perpendicular, and the y-axis is oriented correctly with respect to the x-axis, so this is a legitimate coordinate system. There is no requirement that the x-axis has to be horizontal.

Finding components with tilted axes is no harder than what we have done so far. Vector $\vec{C}$ in Figure 3.20 can be decomposed into $\vec{C} = C_x\hat{\imath} + C_y\hat{\jmath}$, where $C_x = C \cos \theta$ and $C_y = C \sin \theta$. Note that the unit vectors $\hat{\imath}$ and $\hat{\jmath}$ correspond to the *axes*, not to "horizontal" and "vertical," so they are also tilted.

Tilted axes are useful if you need to determine component vectors "parallel to" and "perpendicular to" an arbitrary line or surface. This is illustrated in the following example.

FIGURE 3.20 A coordinate system with tilted axes.

The components of $\vec{C}$ are found with respect to the tilted axes.

$$\vec{C} = C_x\hat{\imath} + C_y\hat{\jmath}$$

Unit vectors $\hat{\imath}$ and $\hat{\jmath}$ define the x- and y-axes.

EXAMPLE 3.7 | Muscle and bone

The deltoid—the rounded muscle across the top of your upper arm—allows you to lift your arm away from your side. It does so by pulling on an attachment point on the humerus, the upper arm bone, at an angle of 15° with respect to the humerus. If you hold your arm at an angle 30° below horizontal, the deltoid must pull with a force of 720 N to support the weight of your arm, as shown in **FIGURE 3.21a**. (You'll learn in Chapter 5 that force is a vector quantity measured in units of *newtons*, abbreviated N.) What are the components of the muscle force parallel to and perpendicular to the bone?

VISUALIZE FIGURE 3.21b shows a tilted coordinate system with the *x*-axis parallel to the humerus. The force $\vec{F}$ is shown 15° from the *x*-axis. The component of force parallel to the bone, which we can denote $F_{\parallel}$, is equivalent to the *x*-component: $F_{\parallel} = F_x$. Similarly, the component of force perpendicular to the bone is $F_{\perp} = F_y$.

SOLVE From the geometry of Figure 3.21b, we see that

$$F_{\parallel} = F \cos 15° = (720 \text{ N}) \cos 15° = 695 \text{ N}$$

$$F_{\perp} = F \sin 15° = (720 \text{ N}) \sin 15° = 186 \text{ N}$$

ASSESS The muscle pulls nearly parallel to the bone, so we expected $F_{\parallel} \approx 720 \text{ N}$ and $F_{\perp} \ll F_{\parallel}$. Thus our results seem reasonable.

FIGURE 3.21 Finding the components of force parallel and perpendicular to the humerus.

(a)

(b)

STOP TO THINK 3.4 Angle ϕ that specifies the direction of $\vec{C}$ is given by

a. $\tan^{-1}(|C_x|/C_y)$

b. $\tan^{-1}(C_x/|C_y|)$

c. $\tan^{-1}(|C_x|/|C_y|)$

d. $\tan^{-1}(|C_y|/C_x)$

e. $\tan^{-1}(C_y/|C_x|)$

f. $\tan^{-1}(|C_y|/|C_x|)$

CHALLENGE EXAMPLE 3.8 | Finding the net force

FIGURE 3.22 shows three forces acting at one point. What is the net force $\vec{F}_{net} = \vec{F}_1 + \vec{F}_2 + \vec{F}_3$?

VISUALIZE Figure 3.22 shows the forces and a tilted coordinate system.

SOLVE The vector equation $\vec{F}_{net} = \vec{F}_1 + \vec{F}_2 + \vec{F}_3$ is really two simultaneous equations:

$$(F_{net})_x = F_{1x} + F_{2x} + F_{3x}$$

$$(F_{net})_y = F_{1y} + F_{2y} + F_{3y}$$

The components of the forces are determined with respect to the axes. Thus

$$F_{1x} = F_1 \cos 45° = (50 \text{ N}) \cos 45° = 35 \text{ N}$$

$$F_{1y} = F_1 \sin 45° = (50 \text{ N}) \sin 45° = 35 \text{ N}$$

$\vec{F}_2$ is easier. It is pointing along the *y*-axis, so $F_{2x} = 0 \text{ N}$ and $F_{2y} = 20 \text{ N}$. To find the components of $\vec{F}_3$, we need to recognize—because $\vec{F}_3$ points straight down—that the angle between $\vec{F}_3$ and the *x*-axis is 75°. Thus

$$F_{3x} = F_3 \cos 75° = (57 \text{ N}) \cos 75° = 15 \text{ N}$$

$$F_{3y} = -F_3 \sin 75° = -(57 \text{ N}) \sin 75° = -55 \text{ N}$$

FIGURE 3.22 Three forces.

The minus sign in F_{3y} is critical, and it appears not from some formula but because we recognized—from the figure—that the *y*-component of $\vec{F}_3$, points in the −*y*-direction. Combining the pieces, we have

$$(F_{net})_x = 35 \text{ N} + 0 \text{ N} + 15 \text{ N} = 50 \text{ N}$$

$$(F_{net})_y = 35 \text{ N} + 20 \text{ N} + (-55 \text{ N}) = 0 \text{ N}$$

Thus the net force is $\vec{F}_{net} = 50\hat{\imath} \text{ N}$. It points along the *x*-axis of the tilted coordinate system.

ASSESS Notice that all work was done with reference to the axes of the coordinate system, not with respect to vertical or horizontal.

SUMMARY

The goals of Chapter 3 have been to learn how vectors are represented and used.

IMPORTANT CONCEPTS

A vector is a quantity described by both a magnitude and a direction.

The vector describes the situation at this point.

$\vec{A}$

Direction

The length or magnitude is denoted A. Magnitude is a scalar.

Unit Vectors

Unit vectors have magnitude 1 and no units. Unit vectors $\hat{\imath}$ and $\hat{\jmath}$ define the directions of the x- and y-axes.

USING VECTORS

Components

The component vectors are parallel to the x- and y-axes:

$$\vec{A} = \vec{A}_x + \vec{A}_y = A_x\hat{\imath} + A_y\hat{\jmath}$$

In the figure at the right, for example:

$$A_x = A\cos\theta \qquad A = \sqrt{A_x^2 + A_y^2}$$
$$A_y = A\sin\theta \qquad \theta = \tan^{-1}(A_y/A_x)$$

▶ Minus signs need to be included if the vector points down or left.

$A_x < 0$	$A_x > 0$
$A_y > 0$	$A_y > 0$
$A_x < 0$	$A_x > 0$
$A_y < 0$	$A_y < 0$

The components A_x and A_y are the magnitudes of the component vectors $\vec{A}_x$ and $\vec{A}_y$ *and* a plus or minus sign to show whether the component vector points toward the positive end or the negative end of the axis.

Working Graphically

Addition

Negative

Subtraction

Multiplication

Working Algebraically

Vector calculations are done component by component: $\vec{C} = 2\vec{A} + \vec{B}$ means $\begin{cases} C_x = 2A_x + B_x \\ C_y = 2A_y + B_y \end{cases}$

The magnitude of $\vec{C}$ is then $C = \sqrt{C_x^2 + C_y^2}$ and its direction is found using $\tan^{-1}$.

TERMS AND NOTATION

scalar	resultant vector	quadrants	component
vector	graphical addition	component vector	unit vector, $\hat{\imath}$ or $\hat{\jmath}$
magnitude	zero vector, $\vec{0}$	decomposition	algebraic addition

CONCEPTUAL QUESTIONS

1. Can the magnitude of the displacement vector be more than the distance traveled? Less than the distance traveled? Explain.
2. If $\vec{C} = \vec{A} + \vec{B}$, can $C = A + B$? Can $C > A + B$? For each, show how or explain why not.
3. If $\vec{C} = \vec{A} + \vec{B}$, can $C = 0$? Can $C < 0$? For each, show how or explain why not.
4. Is it possible to add a scalar to a vector? If so, demonstrate. If not, explain why not.
5. How would you define the *zero vector* $\vec{0}$?
6. Can a vector have a component equal to zero and still have nonzero magnitude? Explain.

7. Can a vector have zero magnitude if one of its components is nonzero? Explain.
8. Suppose two vectors have unequal magnitudes. Can their sum be zero? Explain.
9. Are the following statements true or false? Explain your answer.
   a. The magnitude of a vector can be different in different coordinate systems.
   b. The direction of a vector can be different in different coordinate systems.
   c. The components of a vector can be different in different coordinate systems.

EXERCISES AND PROBLEMS

Exercises

Section 3.1 Scalars and Vectors

Section 3.2 Using Vectors

1. | Trace the vectors in **FIGURE EX3.1** onto your paper. Then find (a) $\vec{A} + \vec{B}$ and (b) $\vec{A} - \vec{B}$.

FIGURE EX3.1   **FIGURE EX3.2**

2. | Trace the vectors in **FIGURE EX3.2** onto your paper. Then find (a) $\vec{A} + \vec{B}$ and (b) $\vec{A} - \vec{B}$.

Section 3.3 Coordinate Systems and Vector Components

3. | a. What are the x- and y-components of vector $\vec{E}$ shown in **FIGURE EX3.3** in terms of the angle θ and the magnitude E?
   b. For the same vector, what are the x- and y-components in terms of the angle ϕ and the magnitude E?

FIGURE EX3.3

4. ‖ A velocity vector 40° below the positive x-axis has a y-component of -10 m/s. What is the value of its x-component?
5. | A position vector in the first quadrant has an x-component of 6 m and a magnitude of 10 m. What is the value of its y-component?
6. | Draw each of the following vectors. Then find its x- and y-components.
   a. $\vec{a} = (3.5$ m/s^2, negative x-direction)
   b. $\vec{v} = (440$ m/s, 30° below the positive x-axis)
   c. $\vec{r} = (12$ m, 40° above the positive x-axis)
7. ‖ Draw each of the following vectors. Then find its x- and y-components.
   a. $\vec{v} = (7.5$ m/s, 30° clockwise from the positive y-axis)
   b. $\vec{a} = (1.5$ m/s^2, 30° above the negative x-axis)
   c. $\vec{F} = (50.0$ N, 36.9° counterclockwise from the positive y-axis)

8. | Let $\vec{C} = (3.15$ m, 15° above the negative x-axis) and $\vec{D} = (25.6$ m, 30° to the right of the negative y-axis). Find the x- and y-components of each vector.
9. | A runner is training for an upcoming marathon by running around a 100-m-diameter circular track at constant speed. Let a coordinate system have its origin at the center of the circle with the x-axis pointing east and the y-axis north. The runner starts at $(x, y) = (50$ m, 0 m) and runs 2.5 times around the track in a clockwise direction. What is his displacement vector? Give your answer as a magnitude and direction.

Section 3.4 Unit Vectors and Vector Algebra

10. ‖ Draw each of the following vectors, label an angle that specifies the vector's direction, then find its magnitude and direction.
    a. $\vec{B} = -4.0\hat{\imath} + 4.0\hat{\jmath}$
    b. $\vec{r} = (-2.0\hat{\imath} - 1.0\hat{\jmath})$ cm
    c. $\vec{v} = (-10\hat{\imath} - 100\hat{\jmath})$ m/s
    d. $\vec{a} = (20\hat{\imath} + 10\hat{\jmath})$ m/s^2
11. ‖ Draw each of the following vectors, label an angle that specifies the vector's direction, and then find the vector's magnitude and direction.
    a. $\vec{A} = 3.0\hat{\imath} + 7.0\hat{\jmath}$
    b. $\vec{a} = (-2.0\hat{\imath} + 4.5\hat{\jmath})$ m/s^2
    c. $\vec{v} = (14\hat{\imath} - 11\hat{\jmath})$ m/s
    d. $\vec{r} = (-2.2\hat{\imath} - 3.3\hat{\jmath})$ m
12. ‖ Let $\vec{A} = 2\hat{\imath} + 3\hat{\jmath}$, $\vec{B} = 2\hat{\imath} - 4\hat{\jmath}$, and $\vec{C} = \vec{A} + \vec{B}$.
    a. Write vector $\vec{C}$ in component form.
    b. Draw a coordinate system and on it show vectors $\vec{A}$, $\vec{B}$, and $\vec{C}$.
    c. What are the magnitude and direction of vector $\vec{C}$?
13. | Let $\vec{A} = 4\hat{\imath} - 2\hat{\jmath}$, $\vec{B} = -3\hat{\imath} + 5\hat{\jmath}$, and $\vec{C} = \vec{A} + \vec{B}$.
    a. Write vector $\vec{C}$ in component form.
    b. Draw a coordinate system and on it show vectors $\vec{A}$, $\vec{B}$, and $\vec{C}$.
    c. What are the magnitude and direction of vector $\vec{C}$?
14. | Let $\vec{A} = 4\hat{\imath} - 2\hat{\jmath}$, $\vec{B} = -3\hat{\imath} + 5\hat{\jmath}$, and $\vec{D} = \vec{A} - \vec{B}$.
    a. Write vector $\vec{D}$ in component form.
    b. Draw a coordinate system and on it show vectors $\vec{A}$, $\vec{B}$, and $\vec{D}$.
    c. What are the magnitude and direction of vector $\vec{D}$?
15. | Let $\vec{A} = 4\hat{\imath} - 2\hat{\jmath}$, $\vec{B} = -3\hat{\imath} + 5\hat{\jmath}$, and $\vec{E} = 2\vec{A} + 3\vec{B}$.
    a. Write vector $\vec{E}$ in component form.
    b. Draw a coordinate system and on it show vectors $\vec{A}$, $\vec{B}$, and $\vec{E}$.
    c. What are the magnitude and direction of vector $\vec{E}$?

16. | Let $\vec{A} = 4\hat{\imath} - 2\hat{\jmath}$, $\vec{B} = -3\hat{\imath} + 5\hat{\jmath}$, and $\vec{F} = \vec{A} - 4\vec{B}$.
    a. Write vector $\vec{F}$ in component form.
    b. Draw a coordinate system and on it show vectors $\vec{A}$, $\vec{B}$, and $\vec{F}$.
    c. What are the magnitude and direction of vector $\vec{F}$?
17. | Let $\vec{E} = 2\hat{\imath} + 3\hat{\jmath}$ and $\vec{F} = 2\hat{\imath} - 2\hat{\jmath}$. Find the magnitude of
    a. $\vec{E}$ and $\vec{F}$ b. $\vec{E} + \vec{F}$ c. $-\vec{E} - 2\vec{F}$
18. | Let $\vec{B} = (5.0$ m, $30°$ counterclockwise from vertical). Find the x- and y-components of $\vec{B}$ in each of the two coordinate systems shown in **FIGURE EX3.18**.

FIGURE EX3.18

FIGURE EX3.19

19. | What are the x- and y-components of the velocity vector shown in **FIGURE EX3.19**?
20. || For the three vectors shown in **FIGURE EX3.20**, $\vec{A} + \vec{B} + \vec{C} = 1\hat{\jmath}$. What is vector $\vec{B}$?
    a. Write $\vec{B}$ in component form.
    b. Write $\vec{B}$ as a magnitude and a direction.

FIGURE EX3.20

21. | The *magnetic field* inside an instrument is $\vec{B} = (2.0\hat{\imath} - 1.0\hat{\jmath})$ T where $\vec{B}$ represents the magnetic field vector and T stands for *tesla*, the unit of the magnetic field. What are the magnitude and direction of the magnetic field?

Problems

22. || Let $\vec{A} = (3.0$ m, $20°$ south of east), $\vec{B} = (2.0$ m, north), and $\vec{C} = (5.0$ m, $70°$ south of west).
    a. Draw and label $\vec{A}$, $\vec{B}$, and $\vec{C}$ with their tails at the origin. Use a coordinate system with the x-axis to the east.
    b. Write $\vec{A}$, $\vec{B}$, and $\vec{C}$ in component form, using unit vectors.
    c. Find the magnitude and the direction of $\vec{D} = \vec{A} + \vec{B} + \vec{C}$.
23. || The position of a particle as a function of time is given by
    CALC $\vec{r} = (5.0\hat{\imath} + 4.0\hat{\jmath})t^2$ m, where t is in seconds.
    a. What is the particle's distance from the origin at $t = 0$, 2, and 5 s?
    b. Find an expression for the particle's velocity $\vec{v}$ as a function of time.
    c. What is the particle's speed at $t = 0$, 2, and 5 s?
24. || a. What is the angle ϕ between vectors $\vec{E}$ and $\vec{F}$ in **FIGURE P3.24**?
    b. Use geometry and trigonometry to determine the magnitude and direction of $\vec{G} = \vec{E} + \vec{F}$.
    c. Use components to determine the magnitude and direction of $\vec{G} = \vec{E} + \vec{F}$.

FIGURE P3.24

FIGURE P3.25

25. || **FIGURE P3.25** shows vectors $\vec{A}$ and $\vec{B}$. Find vector $\vec{C}$ such that $\vec{A} + \vec{B} + \vec{C} = \vec{0}$. Write your answer in component form.
26. ||| **FIGURE P3.26** shows vectors $\vec{A}$ and $\vec{B}$. Find $\vec{D} = 2\vec{A} + \vec{B}$. Write your answer in component form.

FIGURE P3.26

27. || Find a vector that points in the same direction as the vector $(\hat{\imath} + \hat{\jmath})$ and whose magnitude is 1.
28. || While vacationing in the mountains you do some hiking. In the morning, your displacement is $\vec{S}_{\text{morning}} = (2000$ m, east) + (3000 m, north) + (200 m, vertical). After lunch, your displacement is $\vec{S}_{\text{afternoon}} = (1500$ m, west) + (2000 m, north) − (300 m, vertical).
    a. At the end of the hike, how much higher or lower are you compared to your starting point?
    b. What is the magnitude of your net displacement for the day?
29. || The minute hand on a watch is 2.0 cm in length. What is the displacement vector of the tip of the minute hand
    a. From 8:00 to 8:20 A.M.?
    b. From 8:00 to 9:00 A.M.?
30. || You go to an amusement park with your friend Betty, who wants to ride the 30-m-diameter Ferris wheel. She starts the ride at the lowest point of a wheel that, as you face it, rotates counter-clockwise. What is her displacement vector when the wheel has rotated by an angle of $60°$? Give your answer as a magnitude and direction.
31. || Ruth sets out to visit her friend Ward, who lives 50 mi north and 100 mi east of her. She starts by driving east, but after 30 mi she comes to a detour that takes her 15 mi south before going east again. She then drives east for 8 mi and runs out of gas, so Ward flies there in his small plane to get her. What is Ward's displacement vector? Give your answer (a) in component form, using a coordinate system in which the y-axis points north, and (b) as a magnitude and direction.
32. | A cannon tilted upward at $30°$ fires a cannonball with a speed of 100 m/s. What is the component of the cannonball's velocity parallel to the ground?
33. | You are fixing the roof of your house when a hammer breaks loose and slides down. The roof makes an angle of $35°$ with the horizontal, and the hammer is moving at 4.5 m/s when it reaches the edge. What are the horizontal and vertical components of the hammer's velocity just as it leaves the roof?
34. | Jack and Jill ran up the hill at 3.0 m/s. The horizontal component of Jill's velocity vector was 2.5 m/s.
    a. What was the angle of the hill?
    b. What was the vertical component of Jill's velocity?

35. | A pine cone falls straight down from a pine tree growing on a 20° slope. The pine cone hits the ground with a speed of 10 m/s. What is the component of the pine cone's impact velocity (a) parallel to the ground and (b) perpendicular to the ground?

36. | Kami is walking through the airport with her two-wheeled suitcase. The suitcase handle is tilted 40° from vertical, and Kami pulls parallel to the handle with a force of 120 N. (Force is measured in *newtons,* abbreviated N.) What are the horizontal and vertical components of her applied force?

37. || Dee is on a swing in the playground. The chains are 2.5 m long, and the tension in each chain is 450 N when Dee is 55 cm above the lowest point of her swing. Tension is a vector directed along the chain, measured in *newtons,* abbreviated N. What are the horizontal and vertical components of the tension at this point in the swing?

38. || Your neighbor Paul has rented a truck with a loading ramp. The ramp is tilted upward at 25°, and Paul is pulling a large crate up the ramp with a rope that angles 10° above the ramp. If Paul pulls with a force of 550 N, what are the horizontal and vertical components of his force? (Force is measured in *newtons,* abbreviated N.)

39. || Tom is climbing a 3.0-m-long ladder that leans against a vertical wall, contacting the wall 2.5 m above the ground. His weight of 680 N is a vector pointing vertically downward. (Weight is measured in *newtons,* abbreviated N.) What are the components of Tom's weight parallel and perpendicular to the ladder?

40. || The treasure map in FIGURE P3.40 gives the following directions to the buried treasure: "Start at the old oak tree, walk due north for 500 paces, then due east for 100 paces. Dig." But when you arrive, you find an angry dragon just north of the tree. To avoid the dragon, you set off along the yellow brick road at an angle 60° east of north. After walking 300 paces you see an opening through the woods. Which direction should you go, and how far, to reach the treasure?

FIGURE P3.40

41. ||| The bacterium *E. coli* is a single-cell organism that lives in the
BIO gut of healthy animals, including humans. When grown in a uniform medium in the laboratory, these bacteria swim along zig-zag paths at a constant speed of 20 μm/s. FIGURE P3.41 shows the trajectory of an *E. coli* as it moves from point A to point E. What are the magnitude and direction of the bacterium's average velocity for the entire trip?

FIGURE P3.41

42. || A flock of ducks is trying to migrate south for the winter, but they keep being blown off course by a wind blowing from the west at 6.0 m/s. A wise elder duck finally realizes that the solution is to fly at an angle to the wind. If the ducks can fly at 8.0 m/s relative to the air, what direction should they head in order to move directly south?

43. || FIGURE P3.43 shows three ropes tied together in a knot. One of your friends pulls on a rope with 3.0 units of force and another pulls on a second rope with 5.0 units of force. How hard and in what direction must you pull on the third rope to keep the knot from moving?

FIGURE P3.43

44. || Four forces are exerted on the object shown in FIGURE P3.44. (Forces are measured in *newtons,* abbreviated N.) The *net force* on the object is $\vec{F}_{net} = \vec{F}_1 + \vec{F}_2 + \vec{F}_3 + \vec{F}_4 = 4.0\hat{\imath}$ N. What are (a) $\vec{F}_3$ and (b) $\vec{F}_4$? Give your answers in component form.

FIGURE P3.44

45. || FIGURE P3.45 shows four electric charges located at the corners of a rectangle. Like charges, you will recall, repel each other while opposite charges attract. Charge B exerts a repulsive force (directly *away from* B) on charge A of 3.0 N. Charge C exerts an attractive force (directly *toward* C) on charge A of 6.0 N. Finally, charge D exerts an attractive force of 2.0 N on charge A. Assuming that forces are vectors, what are the magnitude and direction of the net force $\vec{F}_{net}$ exerted on charge A?

FIGURE P3.45

4 Kinematics in Two Dimensions

The water droplets are following the parabolic trajectories of projectile motion.

IN THIS CHAPTER, you will learn how to solve problems about motion in a plane.

How do objects accelerate in two dimensions?

An object accelerates when it changes velocity. In two dimensions, velocity can change by changing magnitude (speed) or by changing direction. These are represented by acceleration components tangent to and perpendicular to an object's trajectory.

« LOOKING BACK Section 1.5 Finding acceleration vectors on a motion diagram

What is projectile motion?

Projectile motion is two-dimensional free-fall motion under the influence of only gravity. Projectile motion follows a parabolic trajectory. It has uniform motion in the horizontal direction and $a_y = -g$ in the vertical direction.

« LOOKING BACK Section 2.5 Free fall

What is relative motion?

Coordinate systems that move relative to each other are called reference frames. If object C has velocity $\vec{v}_{CA}$ relative to a reference frame A, and if A moves with velocity $\vec{v}_{AB}$ relative to another reference frame B, then the velocity of C in reference frame B is $\vec{v}_{CB} = \vec{v}_{CA} + \vec{v}_{AB}$.

What is circular motion?

An object moving in a circle (or rotating) has an *angular displacement* instead of a linear displacement. Circular motion is described by angular velocity ω (analogous to velocity v_s) and angular acceleration α (analogous to acceleration a_s). We'll study both uniform and accelerated circular motion.

What is centripetal acceleration?

An object in circular motion is always changing direction. The acceleration of changing direction—called centripetal acceleration—points to the center of the circle. All circular motion has a centripetal acceleration. An object also has a *tangential acceleration* if it is changing speed.

Where is two-dimensional motion used?

Linear motion allowed us to introduce the concepts of motion, but most real motion takes place in two or even three dimensions. Balls move along curved trajectories, cars turn corners, planets orbit the sun, and electrons spiral in the earth's magnetic field. Where is two-dimensional motion used? Everywhere!

4.1 Motion in Two Dimensions

Motion diagrams are an important tool for visualizing motion, and we'll continue to use them, but we also need to develop a mathematical description of motion in two dimensions. For convenience, we'll say that any two-dimensional motion is in the *xy*-plane regardless of whether the plane of motion is horizontal or vertical.

FIGURE 4.1 shows a particle moving along a curved path—its *trajectory*—in the *xy*-plane. We can locate the particle in terms of its position vector $\vec{r} = x\hat{\imath} + y\hat{\jmath}$.

NOTE In Chapter 2 we made extensive use of position-versus-time graphs, either *x* versus *t* or *y* versus *t*. Figure 4.1, like many of the graphs we'll use in this chapter, is a graph of *y* versus *x*. In other words, it's an actual *picture* of the trajectory, not an abstract representation of the motion.

FIGURE 4.2a shows the particle moving from position $\vec{r}_1$ at time t_1 to position $\vec{r}_2$ at a later time t_2. The average velocity—pointing in the direction of the displacement $\Delta\vec{r}$—is

$$\vec{v}_{avg} = \frac{\Delta\vec{r}}{\Delta t} = \frac{\Delta x}{\Delta t}\hat{\imath} + \frac{\Delta y}{\Delta t}\hat{\jmath} \tag{4.1}$$

You learned in Chapter 2 that the instantaneous velocity is the limit of $\vec{v}_{avg}$ as $\Delta t \to 0$. As Δt decreases, point 2 moves closer to point 1 until, as FIGURE 4.2b shows, the displacement vector becomes tangent to the curve. Consequently, **the instantaneous velocity vector $\vec{v}$ is tangent to the trajectory**.

Mathematically, the limit of Equation 4.1 gives

$$\vec{v} = \lim_{\Delta t \to 0} \frac{\Delta\vec{r}}{\Delta t} = \frac{d\vec{r}}{dt} = \frac{dx}{dt}\hat{\imath} + \frac{dy}{dt}\hat{\jmath} \tag{4.2}$$

We can also write the velocity vector in terms of its *x*- and *y*-components as

$$\vec{v} = v_x\hat{\imath} + v_y\hat{\jmath} \tag{4.3}$$

Comparing Equations 4.2 and 4.3, you can see that the velocity vector $\vec{v}$ has *x*- and *y*-components

$$v_x = \frac{dx}{dt} \quad \text{and} \quad v_y = \frac{dy}{dt} \tag{4.4}$$

That is, the *x*-component v_x of the velocity vector is the rate dx/dt at which the particle's *x*-coordinate is changing. The *y*-component is similar.

FIGURE 4.2c illustrates another important feature of the velocity vector. If the vector's angle θ is measured from the positive *x*-direction, the velocity vector components are

$$v_x = v\cos\theta$$
$$v_y = v\sin\theta \tag{4.5}$$

where

$$v = \sqrt{v_x^2 + v_y^2} \tag{4.6}$$

is the particle's *speed* at that point. Speed is always a positive number (or zero), whereas the components are *signed* quantities (i.e., they can be positive or negative) to convey information about the direction of the velocity vector. Conversely, we can use the two velocity components to determine the direction of motion:

$$\theta = \tan^{-1}\left(\frac{v_y}{v_x}\right) \tag{4.7}$$

NOTE In Chapter 2, you learned that the *value* of the velocity is the *slope* of the position-versus-time graph. Now we see that the *direction* of the velocity vector $\vec{v}$ is the *tangent* to the *y*-versus-*x* graph of the trajectory. FIGURE 4.3, on the next page, reminds you that these two graphs use different interpretations of the tangent lines. The tangent to the trajectory does not tell us anything about how fast the particle is moving.

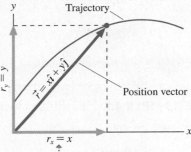

FIGURE 4.1 A particle moving along a trajectory in the *xy*-plane.

The *x*- and *y*-components of $\vec{r}$ are simply *x* and *y*.

FIGURE 4.2 The instantaneous velocity vector is tangent to the trajectory.

(a)

The displacement is $\Delta\vec{r} = \Delta x\hat{\imath} + \Delta y\hat{\jmath}$

The average velocity points in the direction of $\Delta\vec{r}$.

(b)

The instantaneous velocity $\vec{v}$ is tangent to the curve at 1.

As $\Delta t \to 0$, $\Delta\vec{r}$ becomes tangent to the curve at 1.

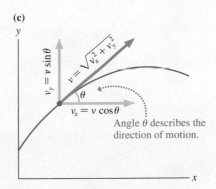

(c)

Angle θ describes the direction of motion.

FIGURE 4.3 Two different uses of tangent lines.

Position-versus-time graph

The *value* of the velocity is the slope of the position graph.

Trajectory

The *direction* of the velocity is tangent to the trajectory.

EXAMPLE 4.1 | **Finding velocity**

A sports car's position on a winding road is given by

$$\vec{r} = (6.0t - 0.10t^2 + 0.00048t^3)\hat{\imath} + (8.0t + 0.060t^2 - 0.00095t^3)\hat{\jmath}$$

where the *y*-axis points north, *t* is in s, and *r* is in m. What are the car's speed and direction at *t* = 120 s?

MODEL Model the car as a particle.

SOLVE Velocity is the derivative of position, so

$$v_x = \frac{dx}{dt} = 6.0 - 2(0.10t) + 3(0.00048t^2)$$

$$v_y = \frac{dy}{dt} = 8.0 + 2(0.060t) - 3(0.00095t^2)$$

Written as a vector, the velocity is

$$\vec{v} = (6.0 - 0.20t + 0.00144t^2)\hat{\imath} + (8.0 + 0.120t - 0.00285t^2)\hat{\jmath}$$

where *t* is in s and *v* is in m/s. At *t* = 120 s, we can calculate $\vec{v} = (2.7\hat{\imath} - 18.6\hat{\jmath})$ m/s. The car's speed at this instant is

$$v = \sqrt{v_x^2 + v_y^2} = \sqrt{(2.7 \text{ m/s})^2 + (-18.6 \text{ m/s})^2} = 19 \text{ m/s}$$

The velocity vector has a negative *y*-component, so the direction of motion is to the right (east) and down (south). The angle below the *x*-axis is

$$\theta = \tan^{-1}\left(\frac{|-18.6 \text{ m/s}|}{2.7 \text{ m/s}}\right) = 82°$$

So, at this instant, the car is headed 82° south of east at a speed of 19 m/s.

STOP TO THINK 4.1   During which time interval or intervals is the particle described by these position graphs at rest? More than one may be correct.

a. 0–1 s

b. 1–2 s

c. 2–3 s

d. 3–4 s

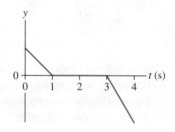

Acceleration Graphically

In ◀ Section 1.5 we defined the *average acceleration* $\vec{a}_{avg}$ of a moving object to be

$$\vec{a}_{avg} = \frac{\Delta\vec{v}}{\Delta t} \tag{4.8}$$

From its definition, we see that $\vec{a}$ **points in the same direction as** $\Delta\vec{v}$, the change of velocity. As an object moves, its velocity vector can change in two possible ways:

1. The magnitude of $\vec{v}$ can change, indicating a change in speed, or
2. The direction of $\vec{v}$ can change, indicating that the object has changed direction.

The kinematics of Chapter 2 considered only the acceleration due to changing speed. Now it's time to look at the acceleration associated with changing direction. Tactics Box 4.1 shows how we can use the velocity vectors on a motion diagram to determine the direction of the average acceleration vector. This is an extension of Tactics Box 1.3, which showed how to find $\vec{a}$ for one-dimensional motion.

TACTICS BOX 4.1

Finding the acceleration vector

To find the acceleration between velocity $\vec{v}_i$ and velocity $\vec{v}_f$:

❶ Draw the velocity vector $\vec{v}_f$.

❷ Draw $-\vec{v}_i$ at the tip of $\vec{v}_f$.

❸ Draw $\Delta\vec{v} = \vec{v}_f - \vec{v}_i$
        $= \vec{v}_f + (-\vec{v}_i)$
This is the direction of $\vec{a}$.

❹ Return to the original motion diagram. Draw a vector at the middle point in the direction of $\Delta\vec{v}$; label it $\vec{a}$. This is the average acceleration between $\vec{v}_i$ and $\vec{v}_f$.

Exercises 1–4

Our everyday use of the word "accelerate" means "speed up." The mathematical definition of acceleration—the rate of change of velocity—also includes slowing down, as you learned in Chapter 2, as well as changing direction. All these are motions that change the velocity.

EXAMPLE 4.2 | Through the valley

A ball rolls down a long hill, through the valley, and back up the other side. Draw a complete motion diagram of the ball.

MODEL Model the ball as a particle.

VISUALIZE FIGURE 4.4 is the motion diagram. Where the particle moves along a *straight line,* it speeds up if $\vec{a}$ and $\vec{v}$ point in the same direction and slows down if $\vec{a}$ and $\vec{v}$ point in opposite directions. This important idea was the basis for the one-dimensional kinematics we developed in Chapter 2. When the direction of $\vec{v}$ changes, as it does when the ball goes through the valley, we need to use vector subtraction to find the direction of $\Delta\vec{v}$ and thus of $\vec{a}$. The procedure is shown at two points in the motion diagram.

FIGURE 4.4 The motion diagram of the ball of Example 4.2.

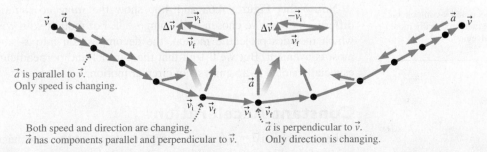

$\vec{a}$ is parallel to $\vec{v}$.
Only speed is changing.

Both speed and direction are changing.
$\vec{a}$ has components parallel and perpendicular to $\vec{v}$.

$\vec{a}$ is perpendicular to $\vec{v}$.
Only direction is changing.

FIGURE 4.5 Analyzing the acceleration vector.

This component of $\vec{a}$ is changing the direction of motion.

This component of $\vec{a}$ is changing the speed of the motion.

FIGURE 4.5 shows that an object's acceleration vector can be decomposed into a component parallel to the velocity—that is, parallel to the direction of motion—and a component perpendicular to the velocity. $\vec{a}_{\parallel}$ **is the piece of the acceleration that causes the object to change speed,** speeding up if $\vec{a}_{\parallel}$ points in the same direction as $\vec{v}$, slowing down if they point in opposite directions. $\vec{a}_{\perp}$ **is the piece of the acceleration that causes the object to change direction.** An object changing direction *always* has a component of acceleration perpendicular to the direction of motion.

Looking back at Example 4.2, we see that $\vec{a}$ is parallel to $\vec{v}$ on the straight portions of the hill where only speed is changing. At the very bottom, where the ball's direction is changing but not its speed, $\vec{a}$ is perpendicular to $\vec{v}$. The acceleration is angled with respect to velocity—having both parallel and perpendicular components—at those points where both speed and direction are changing.

STOP TO THINK 4.2 This acceleration will cause the particle to

a. Speed up and curve upward.
b. Speed up and curve downward.
c. Slow down and curve upward.
d. Slow down and curve downward.
e. Move to the right and down.
f. Reverse direction.

Acceleration Mathematically

FIGURE 4.6 The instantaneous acceleration $\vec{a}$.

(a) The parallel component is associated with a change of speed.

The perpendicular component is associated with a change of direction.

(b)

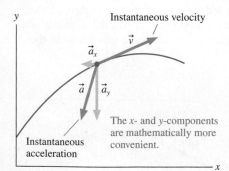

The *x*- and *y*-components are mathematically more convenient.

In Tactics Box 4.1, the average acceleration is found from two velocity vectors separated by the time interval Δt. If we let Δt get smaller and smaller, the two velocity vectors get closer and closer. In the limit $\Delta t \rightarrow 0$, we have the instantaneous acceleration $\vec{a}$ at the same point on the trajectory (and the same instant of time) as the instantaneous velocity $\vec{v}$. This is shown in **FIGURE 4.6**.

By definition, the acceleration vector $\vec{a}$ is the rate at which the velocity $\vec{v}$ is changing at that instant. To show this, Figure 4.6a decomposes $\vec{a}$ into components $\vec{a}_{\parallel}$ and $\vec{a}_{\perp}$ that are parallel and perpendicular to the trajectory. As we just showed, $\vec{a}_{\parallel}$ is associated with a change of speed, and $\vec{a}_{\perp}$ is associated with a change of direction. Both kinds of changes are accelerations. Notice that $\vec{a}_{\perp}$ always points toward the "inside" of the curve because that is the direction in which $\vec{v}$ is changing.

Although the parallel and perpendicular components of $\vec{a}$ convey important ideas about acceleration, it's often more practical to write $\vec{a}$ in terms of the x- and y-components shown in Figure 4.6b. Because $\vec{v} = v_x \hat{\imath} + v_y \hat{\jmath}$, we find

$$\vec{a} = a_x \hat{\imath} + a_y \hat{\jmath} = \frac{d\vec{v}}{dt} = \frac{dv_x}{dt} \hat{\imath} + \frac{dv_y}{dt} \hat{\jmath} \qquad (4.9)$$

from which we see that

$$a_x = \frac{dv_x}{dt} \qquad \text{and} \qquad a_y = \frac{dv_y}{dt} \qquad (4.10)$$

That is, the x-component of $\vec{a}$ is the rate dv_x/dt at which the x-component of velocity is changing.

Notice that Figures 4.6a and 4.6b show the *same* acceleration vector; all that differs is how we've chosen to decompose it. For motion with constant acceleration, which includes projectile motion, the decomposition into x- and y-components is most convenient. But we'll find that the parallel and perpendicular components are especially suited to an analysis of circular motion.

Constant Acceleration

If the acceleration $\vec{a} = a_x \hat{\imath} + a_y \hat{\jmath}$ is constant, then the two components a_x and a_y are both constant. In this case, everything you learned about constant-acceleration kinematics in « Section 2.4 carries over to two-dimensional motion.

Consider a particle that moves with constant acceleration from an initial position $\vec{r}_i = x_i \hat{\imath} + y_i \hat{\jmath}$, starting with initial velocity $\vec{v}_i = v_{ix}\hat{\imath} + v_{iy}\hat{\jmath}$. Its position and velocity at a final point f are

$$x_f = x_i + v_{ix}\,\Delta t + \tfrac{1}{2}a_x(\Delta t)^2 \qquad y_f = y_i + v_{iy}\,\Delta t + \tfrac{1}{2}a_y(\Delta t)^2$$
$$v_{fx} = v_{ix} + a_x\,\Delta t \qquad\qquad v_{fy} = v_{iy} + a_y\,\Delta t \qquad\qquad (4.11)$$

There are *many* quantities to keep track of in two-dimensional kinematics, making the pictorial representation all the more important as a problem-solving tool.

NOTE For constant acceleration, the *x*-component of the motion and the *y*-component of the motion are independent of each other. However, they remain connected through the fact that Δt must be the same for both.

EXAMPLE 4.3 | Plotting a spacecraft trajectory

In the distant future, a small spacecraft is drifting "north" through the galaxy at 680 m/s when it receives a command to return to the starship. The pilot rotates the spacecraft until the nose is pointed 25° north of east, then engages the ion engine. The spacecraft accelerates at 75 m/s². Plot the spacecraft's trajectory for the first 20 s.

MODEL Model the spacecraft as a particle with constant acceleration.

VISUALIZE FIGURE 4.7 shows a pictorial representation in which the *y*-axis points north and the spacecraft starts at the origin. Notice that each point in the motion is labeled with *two* positions (x and y), *two* velocity components (v_x and v_y), and the time t. This will be our standard labeling scheme for trajectory problems.

SOLVE The acceleration vector has both *x*- and *y*-components; their values have been calculated in the pictorial representation. But it is a *constant* acceleration, so we can write

$$x_1 = x_0 + v_{0x}(t_1 - t_0) + \tfrac{1}{2}a_x(t_1 - t_0)^2$$
$$= 34.0t_1^2 \text{ m}$$
$$y_1 = y_0 + v_{0y}(t_1 - t_0) + \tfrac{1}{2}a_y(t_1 - t_0)^2$$
$$= 680t_1 + 15.8t_1^2 \text{ m}$$

where t_1 is in s. Graphing software produces the trajectory shown in FIGURE 4.8. The trajectory is a parabola, which is characteristic of two-dimensional motion with constant acceleration.

FIGURE 4.8 The spacecraft trajectory.

FIGURE 4.7 Pictorial representation of the spacecraft.

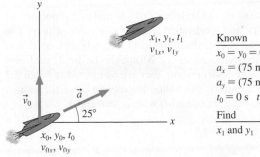

Known
$x_0 = y_0 = 0$ m $v_{0x} = 0$ m/s $v_{0y} = 680$ m/s
$a_x = (75 \text{ m/s}^2)\cos 25° = 68.0 \text{ m/s}^2$
$a_y = (75 \text{ m/s}^2)\sin 25° = 31.6 \text{ m/s}^2$
$t_0 = 0$ s $t_1 = 0$ s to 20 s

Find
x_1 and y_1

4.2 Projectile Motion

Baseballs and tennis balls flying through the air, Olympic divers, and daredevils shot from cannons all exhibit what we call *projectile motion*. A **projectile is an object that moves in two dimensions under the influence of only gravity.** Projectile motion is an extension of the free-fall motion we studied in Chapter 2. We will continue to neglect the influence of air resistance, leading to results that are a good approximation of reality for relatively heavy objects moving relatively slowly over relatively short distances. As we'll see, projectiles in two dimensions follow a *parabolic trajectory* like the one seen in FIGURE 4.9.

The start of a projectile's motion, be it thrown by hand or shot from a gun, is called the *launch,* and the angle θ of the initial velocity $\vec{v}_0$ above the horizontal (i.e., above

FIGURE 4.9 A parabolic trajectory.

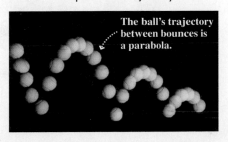

The ball's trajectory between bounces is a parabola.

FIGURE 4.10 A projectile launched with initial velocity $\vec{v}_0$.

the x-axis) is called the **launch angle**. FIGURE 4.10 illustrates the relationship between the initial velocity vector $\vec{v}_0$ and the initial values of the components v_{0x} and v_{0y}. You can see that

$$v_{0x} = v_0 \cos\theta$$
$$v_{0y} = v_0 \sin\theta \qquad (4.12)$$

where v_0 is the initial speed.

NOTE A projectile launched at an angle *below* the horizontal (such as a ball thrown downward from the roof of a building) has *negative* values for θ and v_{0y}. However, the *speed* v_0 is always positive.

Gravity acts downward, and we know that objects released from rest fall straight down, not sideways. Hence a projectile has no horizontal acceleration, while its vertical acceleration is simply that of free fall. Thus

$$a_x = 0$$
$$a_y = -g \qquad \text{(projectile motion)} \qquad (4.13)$$

FIGURE 4.11 The velocity and acceleration vectors of a projectile.

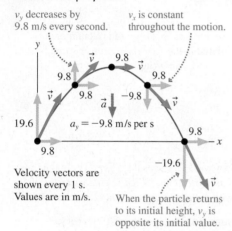

Velocity vectors are shown every 1 s. Values are in m/s.

When the particle returns to its initial height, v_y is opposite its initial value.

In other words, **the vertical component of acceleration a_y is just the familiar $-g$ of free fall, while the horizontal component a_x is zero. Projectiles are in free fall.**

To see how these conditions influence the motion, FIGURE 4.11 shows a projectile launched from $(x_0, y_0) = (0 \text{ m}, 0 \text{ m})$ with an initial velocity $\vec{v}_0 = (9.8\hat{\imath} + 19.6\hat{\jmath})$ m/s. The value of v_x never changes because there's no horizontal acceleration, but v_y decreases by 9.8 m/s every second. This is what it *means* to accelerate at $a_y = -9.8 \text{ m/s}^2 = (-9.8 \text{ m/s})$ per second.

You can see from Figure 4.11 that **projectile motion is made up of two independent motions:** uniform motion at constant velocity in the horizontal direction and free-fall motion in the vertical direction. The kinematic equations that describe these two motions are simply Equations 4.11 with $a_x = 0$ and $a_y = -g$.

EXAMPLE 4.4 | **Don't try this at home!**

A stunt man drives a car off a 10.0-m-high cliff at a speed of 20.0 m/s. How far does the car land from the base of the cliff?

MODEL Model the car as a particle in free fall. Assume that the car is moving horizontally as it leaves the cliff.

VISUALIZE The pictorial representation, shown in FIGURE 4.12, is *very* important because the number of quantities to keep track of is quite large. We have chosen to put the origin at the base of the cliff. The assumption that the car is moving horizontally as it leaves the cliff leads to $v_{0x} = v_0$ and $v_{0y} = 0$ m/s.

FIGURE 4.12 Pictorial representation for the car of Example 4.4.

Known	Find
$x_0 = 0$ m $\quad v_{0y} = 0$ m/s $\quad t_0 = 0$ s	x_1
$y_0 = 10.0$ m $\quad v_{0x} = v_0 = 20.0$ m/s	
$a_x = 0$ m/s^2 $\quad a_y = -g$ $\quad y_1 = 0$ m	

SOLVE Each point on the trajectory has x- and y-components of position, velocity, and acceleration but only *one* value of time. The time needed to move horizontally to x_1 is the *same* time needed to fall vertically through distance y_0. **Although the horizontal and vertical motions are independent, they are connected through the time t.** This is a critical observation for solving projectile motion problems. The kinematics equations with $a_x = 0$ and $a_y = -g$ are

$$x_1 = x_0 + v_{0x}(t_1 - t_0) = v_0 t_1$$
$$y_1 = 0 = y_0 + v_{0y}(t_1 - t_0) - \tfrac{1}{2}g(t_1 - t_0)^2 = y_0 - \tfrac{1}{2}g t_1^2$$

We can use the vertical equation to determine the time t_1 needed to fall distance y_0:

$$t_1 = \sqrt{\frac{2y_0}{g}} = \sqrt{\frac{2(10.0 \text{ m})}{9.80 \text{ m/s}^2}} = 1.43 \text{ s}$$

We then insert this expression for t into the horizontal equation to find the distance traveled:

$$x_1 = v_0 t_1 = (20.0 \text{ m/s})(1.43 \text{ s}) = 28.6 \text{ m}$$

ASSESS The cliff height is ≈ 33 ft and the initial speed is $v_0 \approx 40$ mph. Traveling $x_1 = 29$ m ≈ 95 ft before hitting the ground seems reasonable.

The x- and y-equations of Example 4.4 are parametric equations. It's not hard to eliminate t and write an expression for y as a function of x. From the x_1 equation, $t_1 = x_1/v_0$. Substituting this into the y_1 equation, we find

$$y = y_0 - \frac{g}{2v_0^2}x^2 \tag{4.14}$$

The graph of $y = cx^2$ is a parabola, so Equation 4.14 represents an inverted parabola that starts from height y_0. This proves, as we asserted previously, that a projectile follows a parabolic trajectory.

Reasoning About Projectile Motion

Suppose a heavy ball is launched exactly horizontally at height h above a horizontal field. At the exact instant that the ball is launched, a second ball is simply dropped from height h. Which ball hits the ground first?

It may seem hard to believe, but—if air resistance is neglected—the balls hit the ground *simultaneously*. They do so because the horizontal and vertical components of projectile motion are independent of each other. The initial horizontal velocity of the first ball has *no* influence over its vertical motion. Neither ball has any initial motion in the vertical direction, so both fall distance h in the same amount of time. You can see this in FIGURE 4.13.

FIGURE 4.14a shows a useful way to think about the trajectory of a projectile. Without gravity, a projectile would follow a straight line. Because of gravity, the particle at time t has "fallen" a distance $\frac{1}{2}gt^2$ below this line. The separation grows as $\frac{1}{2}gt^2$, giving the trajectory its parabolic shape.

Use this idea to think about the following "classic" problem in physics:

A hungry bow-and-arrow hunter in the jungle wants to shoot down a coconut that is hanging from the branch of a tree. He points his arrow directly at the coconut, but as luck would have it, the coconut falls from the branch at the *exact* instant the hunter releases the string. Does the arrow hit the coconut?

You might think that the arrow will miss the falling coconut, but it doesn't. Although the arrow travels very fast, it follows a slightly curved parabolic trajectory, not a straight line. Had the coconut stayed on the tree, the arrow would have curved under its target as gravity caused it to fall a distance $\frac{1}{2}gt^2$ below the straight line. But $\frac{1}{2}gt^2$ is also the distance the coconut falls while the arrow is in flight. Thus, as FIGURE 4.14b shows, the arrow and the coconut fall the same distance and meet at the same point!

FIGURE 4.13 A projectile launched horizontally falls in the same time as a projectile that is released from rest.

FIGURE 4.14 A projectile follows a parabolic trajectory because it "falls" a distance $\frac{1}{2}gt^2$ below a straight-line trajectory.

(a)

y

Trajectory without gravity

The distance between the gravity-free trajectory and the actual trajectory grows as the particle "falls" $\frac{1}{2}gt^2$.

$\frac{1}{2}gt^2$

$\frac{1}{2}gt^2$

Actual trajectory

x

(b)

y

Trajectory without gravity

$\frac{1}{2}gt^2$

Actual trajectory of arrow

x

The Projectile Motion Model

Projectile motion is an ideal that's rarely achieved by real objects. Nonetheless, the **projectile motion model** is another important simplification of reality that we can add to our growing list of models.

MODEL 4.1

Projectile motion

For motion under the influence of only gravity.

■ Model the object as a particle launched with speed v_0 at angle θ:

■ Mathematically:

- **Uniform motion** in the horizontal direction with $v_x = v_0 \cos\theta$.

- **Constant acceleration** in the vertical direction with $a_y = -g$.

- Same Δt for both motions.

A projectile follows a parabolic trajectory.

■ Limitations: Model fails if air resistance is significant.

Exercise 9

PROBLEM-SOLVING STRATEGY 4.1

Projectile motion problems

MODEL Is it reasonable to ignore air resistance? If so, use the projectile motion model.

VISUALIZE Establish a coordinate system with the x-axis horizontal and the y-axis vertical. Define symbols and identify what the problem is trying to find. For a launch at angle θ, the initial velocity components are $v_{ix} = v_0 \cos\theta$ and $v_{iy} = v_0 \sin\theta$.

SOLVE The acceleration is known: $a_x = 0$ and $a_y = -g$. Thus the problem is one of two-dimensional kinematics. The kinematic equations are

Horizontal	Vertical
$x_f = x_i + v_{ix}\,\Delta t$	$y_f = y_i + v_{iy}\,\Delta t - \frac{1}{2}g(\Delta t)^2$
$v_{fx} = v_{ix} = \text{constant}$	$v_{fy} = v_{iy} - g\,\Delta t$

Δt is the same for the horizontal and vertical components of the motion. Find Δt from one component, then use that value for the other component.

ASSESS Check that your result has correct units and significant figures, is reasonable, and answers the question.

EXAMPLE 4.5 | Jumping frog contest

Frogs, with their long, strong legs, are excellent jumpers. And thanks to the good folks of Calaveras County, California, who have a jumping frog contest every year in honor of a Mark Twain story, we have very good data on how far a determined frog can jump.

High-speed cameras show that a good jumper goes into a crouch, then rapidly extends his legs by typically 15 cm during a 65 ms push off, leaving the ground at a 30° angle. How far does this frog leap?

MODEL Model the push off as linear motion with uniform acceleration. A bullfrog is fairly heavy and dense, so ignore air resistance and model the leap as projectile motion.

VISUALIZE This is a two-part problem: linear acceleration followed by projectile motion. A key observation is that **the final velocity for pushing off the ground becomes the initial velocity of the projectile motion**. FIGURE 4.15 shows a separate pictorial representation for each part. Notice that we've used different coordinate systems for the two parts; coordinate systems are our choice, and for each part of the motion we've chosen the coordinate system that makes the problem easiest to solve.

SOLVE While pushing off, the frog travels 15 cm = 0.15 m in 65 ms = 0.065 s. We could find his speed at the end of pushing off if we knew the acceleration. Because the initial velocity is zero,

FIGURE 4.15 Pictorial representations of the jumping frog.

Pushing off

x_1, v_{1x}, t_1

x_0, v_{0x}, t_0

$\vec{a}$

Known	Find
$x_0 = 0\ m$ $v_{0x} = 0\ m/s$ $t_0 = 0\ s$	v_{1x}
$x_1 = 0.15\ m$ $t_1 = 0.065\ s$	

In flight

y

$\vec{v}_0$

θ

$x_0, v_{0x}, y_0, v_{0y}, t_0$ $x_1, v_{1x}, y_1, v_{1y}, t_1$

Known	Find
$x_0 = 0\ m$ $y_0 = 0\ m$ $t_0 = 0\ s$	x_1
v_0 $\theta = 30°$ $y_1 = 0\ m$ $a_y = -g$	

we can find the acceleration from the position-acceleration-time kinematic equation:

$$x_1 = x_0 + v_{0x}\,\Delta t + \tfrac{1}{2}a_x(\Delta t)^2 = \tfrac{1}{2}a_x(\Delta t)^2$$

$$a_x = \frac{2x_1}{(\Delta t)^2} = \frac{2(0.15\ m)}{(0.065\ s)^2} = 71\ m/s^2$$

This is a substantial acceleration, but it doesn't last long. At the end of the 65 ms push off, the frog's velocity is

$$v_{1x} = v_{0x} + a_x\,\Delta t = (71\ m/s^2)(0.065\ s) = 4.62\ m/s$$

We'll keep an extra significant figure here to avoid round-off error in the second half of the problem.

The end of the push off is the beginning of the projectile motion, so the second part of the problem is to find the distance of a projectile launched with velocity $\vec{v}_0 = (4.62\ m/s, 30°)$. The initial x- and y-components of the launch velocity are

$$v_{0x} = v_0\cos\theta \qquad v_{0y} = v_0\sin\theta$$

The kinematic equations of projectile motion, with $a_x = 0$ and $a_y = -g$, are

$$x_1 = x_0 + v_{0x}\,\Delta t \qquad\qquad y_1 = y_0 + v_{0y}\,\Delta t - \tfrac{1}{2}g(\Delta t)^2$$
$$= (v_0\cos\theta)\,\Delta t \qquad\qquad = (v_0\sin\theta)\,\Delta t - \tfrac{1}{2}g(\Delta t)^2$$

We can find the time of flight from the vertical equation by setting $y_1 = 0$:

$$0 = (v_0\sin\theta)\,\Delta t - \tfrac{1}{2}g(\Delta t)^2 = (v_0\sin\theta - \tfrac{1}{2}g\,\Delta t)\,\Delta t$$

and thus

$$\Delta t = 0 \qquad \text{or} \qquad \Delta t = \frac{2v_0\sin\theta}{g}$$

Both are legitimate solutions. The first corresponds to the instant when $y = 0$ at the launch, the second to when $y = 0$ as the frog hits the ground. Clearly, we want the second solution. Substituting this expression for Δt into the equation for x_1 gives

$$x_1 = (v_0\cos\theta)\frac{2v_0\sin\theta}{g} = \frac{2v_0^2\sin\theta\cos\theta}{g}$$

We can simplify this result with the trigonometric identity $2\sin\theta\cos\theta = \sin(2\theta)$. Thus the distance traveled by the frog is

$$x_1 = \frac{v_0^2\sin(2\theta)}{g}$$

Using $v_0 = 4.62\ m/s$ and $\theta = 30°$, we find that the frog leaps a distance of 1.9 m.

ASSESS 1.9 m is about 6 feet, or about 10 times the frog's body length. That's pretty amazing, but true. Jumps of 2.2 m have been recorded in the lab. And the Calaveras County record holder, Rosie the Ribeter, covered 6.5 m—21 feet—in three jumps!

The distance a projectile travels is called its *range*. As Example 4.5 found, a projectile that lands at the same elevation from which it was launched has

$$\text{range} = \frac{v_0^2\sin(2\theta)}{g} \qquad\qquad (4.15)$$

The maximum range occurs for $\theta = 45°$, where $\sin(2\theta) = 1$. But there's more that we can learn from this equation. Because $\sin(180° - x) = \sin x$, it follows that $\sin(2(90° - \theta)) = \sin(2\theta)$. Consequently, a projectile launched either at angle θ *or* at angle $(90° - \theta)$ will travel the same distance *over level ground*. FIGURE 4.16 shows several trajectories of projectiles launched with the same initial speed.

NOTE Equation 4.15 is *not* a general result. It applies *only* in situations where the projectile lands at the same elevation from which it was fired.

FIGURE 4.16 Trajectories of a projectile launched at different angles with a speed of 99 m/s.

y (m)

Launch angles of θ and $90° - \theta$ give the same range.

Maximum range is achieved at 45°.

75° 60° 45° 30° 15°

x (m)

$v_0 = 99\ m/s$

4.3 Relative Motion

FIGURE 4.17 Velocities in Amy's reference frame.

FIGURE 4.17 shows Amy and Bill watching Carlos on his bicycle. According to Amy, Carlos's velocity is $v_x = 5$ m/s. Bill sees the bicycle receding in his rearview mirror, in the *negative x*-direction, getting 10 m farther away from him every second. According to Bill, Carlos's velocity is $v_x = -10$ m/s. Which is Carlos's *true* velocity?

Velocity is not a concept that can be true or false. Carlos's velocity *relative to Amy* is $(v_x)_{CA} = 5$ m/s, where the subscript notation means "C relative to A." Similarly, Carlos's velocity *relative to Bill* is $(v_x)_{CB} = -10$ m/s. These are both valid descriptions of Carlos's motion.

It's not hard to see how to combine the velocities for one-dimensional motion:

The first subscript is the same on both sides.   The last subscript is the same on both sides.

$$(v_x)_{CB} = (v_x)_{CA} + (v_x)_{AB} \qquad (4.16)$$

The inner subscripts "cancel."

We'll justify this relationship later in this section and then extend it to two-dimensional motion.

Equation 4.16 tells us that the velocity of C relative to B is the velocity of C relative to A *plus* the velocity of A relative to B. Note that

$$(v_x)_{AB} = -(v_x)_{BA} \qquad (4.17)$$

because if B is moving to the right relative to A, then A is moving to the left relative to B. In Figure 4.17, Bill is moving to the right relative to Amy with $(v_x)_{BA} = 15$ m/s, so $(v_x)_{AB} = -15$ m/s. Knowing that Carlos's velocity relative to Amy is 5 m/s, we find that Carlos's velocity relative to Bill is, as expected, $(v_x)_{CB} = (v_x)_{CA} + (v_x)_{AB} = 5$ m/s $+ (-15)$ m/s $= -10$ m/s.

EXAMPLE 4.6 | A speeding bullet

The police are chasing a bank robber. While driving at 50 m/s, they fire a bullet to shoot out a tire of his car. The police gun shoots bullets at 300 m/s. What is the bullet's speed as measured by a TV camera crew parked beside the road?

MODEL Assume that all motion is in the positive *x*-direction. The bullet is the object that is observed from both the police car and the ground.

SOLVE The bullet B's velocity relative to the gun G is $(v_x)_{BG} = 300$ m/s. The gun, inside the car, is traveling relative to the TV crew C at $(v_x)_{GC} = 50$ m/s. We can combine these values to find that the bullet's velocity relative to the TV crew on the ground is

$$(v_x)_{BC} = (v_x)_{BG} + (v_x)_{GC} = 300 \text{ m/s} + 50 \text{ m/s} = 350 \text{ m/s}$$

ASSESS It should be no surprise in this simple situation that we simply add the velocities.

Reference Frames

A coordinate system in which an experimenter (possibly with the assistance of helpers) makes position and time measurements of physical events is called a **reference frame.** In Figure 4.17, Amy and Bill each had their own reference frame (where they were at rest) in which they measured Carlos's velocity.

More generally, **FIGURE 4.18** shows two reference frames, A and B, and an object C. It is assumed that the reference frames are moving with respect to each other. At this instant of time, the position vector of C in reference frame A is $\vec{r}_{CA}$, meaning "the position of C relative to the origin of frame A." Similarly, $\vec{r}_{CB}$ is the position vector of C in reference frame B. Using vector addition, you can see that

$$\vec{r}_{CB} = \vec{r}_{CA} + \vec{r}_{AB} \tag{4.18}$$

where $\vec{r}_{AB}$ locates the origin of A relative to the origin of B.

In general, object C is moving relative to both reference frames. To find its velocity in each reference frame, take the time derivative of Equation 4.18:

$$\frac{d\vec{r}_{CB}}{dt} = \frac{d\vec{r}_{CA}}{dt} + \frac{d\vec{r}_{AB}}{dt} \tag{4.19}$$

By definition, $d\vec{r}/dt$ is a velocity. The first derivative is $\vec{v}_{CB}$, the velocity of C relative to B. Similarly, the second derivative is the velocity of C relative to A, $\vec{v}_{CA}$. The last derivative is slightly different because it doesn't refer to object C. Instead, this is the velocity $\vec{v}_{AB}$ of reference frame A relative to reference frame B. As we noted in one dimension, $\vec{v}_{AB} = -\vec{v}_{BA}$.

Writing Equation 4.19 in terms of velocities, we have

$$\vec{v}_{CB} = \vec{v}_{CA} + \vec{v}_{AB} \tag{4.20}$$

This relationship between velocities in different reference frames was recognized by Galileo in his pioneering studies of motion, hence it is known as the **Galilean transformation of velocity.** If you know an object's velocity in one reference frame, you can *transform* it into the velocity that would be measured in a different reference frame. Just as in one dimension, the velocity of C relative to B is the velocity of C relative to A plus the velocity of A relative to B, *but* you must add the velocities as vectors for two-dimensional motion.

As we've seen, the Galilean velocity transformation is pretty much common sense for one-dimensional motion. The real usefulness appears when an object travels in a *medium* moving with respect to the earth. For example, a boat moves relative to the water. What is the boat's net motion if the water is a flowing river? Airplanes fly relative to the air, but the air at high altitudes often flows at high speed. Navigation of boats and planes requires knowing both the motion of the vessel in the medium and the motion of the medium relative to the earth.

FIGURE 4.18 Two reference frames.

Object C can be located relative to A or to B.

Reference frame A

Reference frame B

EXAMPLE 4.7 | Flying to Cleveland I

Cleveland is 300 miles east of Chicago. A plane leaves Chicago flying due east at 500 mph. The pilot forgot to check the weather and doesn't know that the wind is blowing to the south at 50 mph. What is the plane's ground speed? Where is the plane 0.60 h later, when the pilot expects to land in Cleveland?

MODEL Establish a coordinate system with the *x*-axis pointing east and the *y*-axis north. The plane P flies in the air, so its velocity relative to the air A is $\vec{v}_{PA} = 500\hat{\imath}$ mph. Meanwhile, the air is moving relative to the ground G at $\vec{v}_{AG} = -50\hat{\jmath}$ mph.

FIGURE 4.19 The wind causes a plane flying due east in the air to move to the southeast relative to the ground.

$\vec{v}_{PA}$ of plane relative to air

Chicago

Cleveland

$\vec{v}_{AG}$ of air

$\vec{v}_{PG}$ of plane relative to ground

SOLVE The velocity equation $\vec{v}_{PG} = \vec{v}_{PA} + \vec{v}_{AG}$ is a vector-addition equation. **FIGURE 4.19** shows graphically what happens. Although the nose of the plane points east, the wind carries the plane in a direction somewhat south of east. The plane's velocity relative to the ground is

$$\vec{v}_{PG} = \vec{v}_{PA} + \vec{v}_{AG} = (500\hat{\imath} - 50\hat{\jmath}) \text{ mph}$$

The plane's ground speed is

$$v = \sqrt{(v_x)_{PG}^2 + (v_y)_{PG}^2} = 502 \text{ mph}$$

After flying for 0.60 h at this velocity, the plane's location (relative to Chicago) is

$$x = (v_x)_{PG}t = (500 \text{ mph})(0.60 \text{ h}) = 300 \text{ mi}$$
$$y = (v_y)_{PG}t = (-50 \text{ mph})(0.60 \text{ h}) = -30 \text{ mi}$$

The plane is 30 mi due south of Cleveland! Although the pilot thought he was flying to the east, his actual heading has been $\tan^{-1}(50 \text{ mph}/500 \text{ mph}) = \tan^{-1}(0.10) = 5.71°$ south of east.

EXAMPLE 4.8 | Flying to Cleveland II

A wiser pilot flying from Chicago to Cleveland on the same day plots a course that will take her directly to Cleveland. In which direction does she fly the plane? How long does it take to reach Cleveland?

MODEL Establish a coordinate system with the *x*-axis pointing east and the *y*-axis north. The air is moving relative to the ground at $\vec{v}_{AG} = -50\hat{\jmath}$ mph.

SOLVE The objective of navigation is to move between two points on the earth's surface. The wiser pilot, who knows that the wind will affect her plane, draws the vector picture of **FIGURE 4.20**. She sees that she'll need $(v_y)_{PG} = 0$, in order to fly due east to Cleveland. This will require turning the nose of the plane at an angle θ north of east, making $\vec{v}_{PA} = (500\cos\theta\,\hat{\imath} + 500\sin\theta\,\hat{\jmath})$ mph.

The velocity equation is $\vec{v}_{PG} = \vec{v}_{PA} + \vec{v}_{AG}$. The desired heading is found from setting the *y*-component of this equation to zero:

$$(v_y)_{PG} = (v_y)_{PA} + (v_y)_{AG} = (500\sin\theta - 50)\text{ mph} = 0\text{ mph}$$

$$\theta = \sin^{-1}\left(\frac{50\text{ mph}}{500\text{ mph}}\right) = 5.74°$$

FIGURE 4.20 To travel due east in a south wind, a pilot has to point the plane somewhat to the northeast.

The plane's velocity relative to the ground is then $\vec{v}_{PG} = (500\text{ mph}) \times \cos 5.74°\,\hat{\imath} = 497\hat{\imath}$ mph. This is slightly slower than the speed relative to the air. The time needed to fly to Cleveland at this speed is

$$t = \frac{300\text{ mi}}{497\text{ mph}} = 0.604\text{ h}$$

It takes 0.004 h = 14 s longer to reach Cleveland than it would on a day without wind.

ASSESS A boat crossing a river or an ocean current faces the same difficulties. These are exactly the kinds of calculations performed by pilots of boats and planes as part of navigation.

STOP TO THINK 4.4 A plane traveling horizontally to the right at 100 m/s flies past a helicopter that is going straight up at 20 m/s. From the helicopter's perspective, the plane's direction and speed are

a. Right and up, less than 100 m/s.
b. Right and up, 100 m/s.
c. Right and up, more than 100 m/s.
d. Right and down, less than 100 m/s.
e. Right and down, 100 m/s.
f. Right and down, more than 100 m/s.

4.4 Uniform Circular Motion

FIGURE 4.21 A particle in uniform circular motion.

The velocity is tangent to the circle.
The velocity vectors are all the same length.

FIGURE 4.21 shows a particle moving around a circle of radius *r*. The particle might be a satellite in an orbit, a ball on the end of a string, or even just a dot painted on the side of a rotating wheel. Circular motion is another example of motion in a plane, but it is quite different from projectile motion.

To begin the study of circular motion, consider a particle that moves at *constant speed* around a circle of radius *r*. This is called **uniform circular motion.** Regardless of what the particle represents, its velocity vector $\vec{v}$ is always tangent to the circle. The particle's speed *v* is constant, so vector $\vec{v}$ is always the same length.

The time interval it takes the particle to go around the circle once, completing one revolution (abbreviated rev), is called the **period** of the motion. Period is represented by the symbol *T*. It's easy to relate the particle's period *T* to its speed *v*. For a particle moving with constant speed, speed is simply distance/time. In one period, the particle moves once around a circle of radius *r* and travels the circumference $2\pi r$. Thus

$$v = \frac{1\text{ circumference}}{1\text{ period}} = \frac{2\pi r}{T} \tag{4.21}$$

EXAMPLE 4.9 | **A rotating crankshaft**

A 4.0-cm-diameter crankshaft turns at 2400 rpm (revolutions per minute). What is the speed of a point on the surface of the crankshaft?

SOLVE We need to determine the time it takes the crankshaft to make 1 rev. First, we convert 2400 rpm to revolutions per second:

$$\frac{2400 \text{ rev}}{1 \text{ min}} \times \frac{1 \text{ min}}{60 \text{ s}} = 40 \text{ rev/s}$$

If the crankshaft turns 40 times in 1 s, the time for 1 rev is

$$T = \frac{1}{40} \text{ s} = 0.025 \text{ s}$$

Thus the speed of a point on the surface, where $r = 2.0 \text{ cm} = 0.020 \text{ m}$, is

$$v = \frac{2\pi r}{T} = \frac{2\pi(0.020 \text{ m})}{0.025 \text{ s}} = 5.0 \text{ m/s}$$

Angular Position

Rather than using xy-coordinates, it will be more convenient to describe the position of a particle in circular motion by its distance r from the center of the circle and its angle θ from the positive x-axis. This is shown in **FIGURE 4.22**. The angle θ is the **angular position** of the particle.

We can distinguish a position above the x-axis from a position that is an equal angle below the x-axis by *defining* θ to be positive when measured *counterclockwise* (ccw) from the positive x-axis. An angle measured clockwise (cw) from the positive x-axis has a negative value. "Clockwise" and "counterclockwise" in circular motion are analogous, respectively, to "left of the origin" and "right of the origin" in linear motion, which we associated with negative and positive values of x. A particle 30° below the positive x-axis is equally well described by either $\theta = -30°$ or $\theta = +330°$. We could also describe this particle by $\theta = \frac{11}{12}$ rev, where *revolutions* are another way to measure the angle.

Although degrees and revolutions are widely used measures of angle, mathematicians and scientists usually find it more useful to measure the angle θ in Figure 4.22 by using the **arc length** s that the particle travels along the edge of a circle of radius r. We define the angular unit of **radians** such that

$$\theta(\text{radians}) \equiv \frac{s}{r} \tag{4.22}$$

The radian, which is abbreviated rad, is the SI unit of angle. An angle of 1 rad has an arc length s exactly equal to the radius r.

The arc length completely around a circle is the circle's circumference $2\pi r$. Thus the angle of a full circle is

$$\theta_{\text{full circle}} = \frac{2\pi r}{r} = 2\pi \text{ rad}$$

This relationship is the basis for the well-known conversion factors

$$1 \text{ rev} = 360° = 2\pi \text{ rad}$$

As a simple example of converting between radians and degrees, let's convert an angle of 1 rad to degrees:

$$1 \text{ rad} = 1 \text{ rad} \times \frac{360°}{2\pi \text{ rad}} = 57.3°$$

FIGURE 4.22 A particle's position is described by distance r and angle θ.

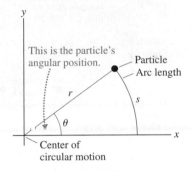

This is the particle's angular position.

Particle
Arc length
r
s
θ
Center of circular motion

Circular motion is one of the most common types of motion.

Thus a rough approximation is 1 rad $\approx$ 60°. We will often specify angles in degrees, but keep in mind that the SI unit is the radian.

An important consequence of Equation 4.22 is that the arc length spanning angle θ is

$$s = r\theta \qquad (\text{with } \theta \text{ in rad}) \tag{4.23}$$

This is a result that we will use often, but it is valid *only* if θ is measured in radians and not in degrees. This very simple relationship between angle and arc length is one of the primary motivations for using radians.

NOTE Units of angle are often troublesome. Unlike the kilogram or the second, for which we have standards, the radian is a *defined* unit. It's really just a *name* to remind us that we're dealing with an angle. Consequently, the radian unit sometimes appears or disappears without warning. This seems rather mysterious until you get used to it. This textbook will call your attention to such behavior the first few times it occurs. With a little practice, you'll soon learn when the rad unit is needed and when it's not.

Angular Velocity

FIGURE 4.23 A particle moves with angular velocity ω.

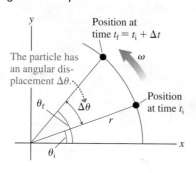

FIGURE 4.23 shows a particle moving in a circle from an initial angular position θ_i at time t_i to a final angular position θ_f at a later time t_f. The change $\Delta\theta = \theta_f - \theta_i$ is called the **angular displacement.** We can measure the particle's circular motion in terms of the rate of change of θ, just as we measured the particle's linear motion in terms of the rate of change of its position s.

In analogy with linear motion, let's define the *average angular velocity* to be

$$\text{average angular velocity} \equiv \frac{\Delta\theta}{\Delta t} \tag{4.24}$$

As the time interval Δt becomes very small, $\Delta t \to 0$, we arrive at the definition of the instantaneous **angular velocity:**

$$\omega \equiv \lim_{\Delta t \to 0} \frac{\Delta\theta}{\Delta t} = \frac{d\theta}{dt} \qquad (\text{angular velocity}) \tag{4.25}$$

The symbol ω is a lowercase Greek omega, *not* an ordinary w. The SI unit of angular velocity is rad/s, but °/s, rev/s, and rev/min are also common units. Revolutions per minute is abbreviated rpm.

Angular velocity is the *rate* at which a particle's angular position is changing as it moves around a circle. A particle that starts from $\theta = 0$ rad with an angular velocity of 0.5 rad/s will be at angle $\theta = 0.5$ rad after 1 s, at $\theta = 1.0$ rad after 2 s, at $\theta = 1.5$ rad after 3 s, and so on. Its angular position is increasing at the *rate* of 0.5 radian per second. **A particle moves with uniform circular motion if and only if its angular velocity ω is constant and unchanging**.

FIGURE 4.24 Positive and negative angular velocities.

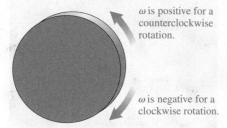

ω is positive for a counterclockwise rotation.

ω is negative for a clockwise rotation.

Angular velocity, like the velocity v_s of one-dimensional motion, can be positive or negative. The signs shown in FIGURE 4.24 are based on the fact that θ was defined to be positive for a counterclockwise rotation. Because the definition $\omega = d\theta/dt$ for circular motion parallels the definition $v_s = ds/dt$ for linear motion, the graphical relationships we found between v_s and s in Chapter 2 apply equally well to ω and θ:

$$\omega = \text{slope of the } \theta\text{-versus-}t \text{ graph at time } t$$
$$\theta_f = \theta_i + \text{area under the } \omega\text{-versus-}t \text{ curve between } t_i \text{ and } t_f \tag{4.26}$$
$$= \theta_i + \omega \Delta t$$

You will see many more instances where circular motion is analogous to linear motion with angular variables replacing linear variables. Thus much of what you learned about linear kinematics carries over to circular motion.

EXAMPLE 4.10 | A graphical representation of circular motion

FIGURE 4.25 shows the angular position of a painted dot on the edge of a rotating wheel. Describe the wheel's motion and draw an ω-versus-t graph.

FIGURE 4.25 Angular position graph for the wheel of Example 4.10.

SOLVE Although circular motion seems to "start over" every revolution (every 2π rad), the angular position θ continues to increase. $\theta = 6\pi$ rad corresponds to three revolutions. This wheel makes 3 ccw rev (because θ is getting more positive) in 3 s, immediately reverses direction and makes 1 cw rev in 2 s, then stops at $t = 5$ s

and holds the position $\theta = 4\pi$ rad. The angular velocity is found by measuring the slope of the graph:

$t = 0$–3 s	slope $= \Delta\theta/\Delta t = 6\pi$ rad/3 s $= 2\pi$ rad/s
$t = 3$–5 s	slope $= \Delta\theta/\Delta t = -2\pi$ rad/2 s $= -\pi$ rad/s
$t > 5$ s	slope $= \Delta\theta/\Delta t = 0$ rad/s

These results are shown as an ω-versus-t graph in **FIGURE 4.26**. For the first 3 s, the motion is uniform circular motion with $\omega = 2\pi$ rad/s. The wheel then changes to a different uniform circular motion with $\omega = -\pi$ rad/s for 2 s, then stops.

FIGURE 4.26 ω-versus-t graph for the wheel of Example 4.10.

NOTE In physics, we nearly always want to give results as numerical values. Example 4.9 had a π in the equation, but we used its numerical value to compute $v = 5.0$ m/s. However, angles in radians are an exception to this rule. It's okay to leave a π in the value of θ or ω, and we have done so in Example 4.10.

Not surprisingly, the angular velocity ω is closely related to the period and speed of the motion. As a particle goes around a circle one time, its angular displacement is $\Delta\theta = 2\pi$ rad during the interval $\Delta t = T$. Thus, using the definition of angular velocity, we find

$$|\omega| = \frac{2\pi \text{ rad}}{T} \quad \text{or} \quad T = \frac{2\pi \text{ rad}}{|\omega|} \quad\quad (4.27)$$

The period alone gives only the absolute value of $|\omega|$. You need to know the direction of motion to determine the sign of ω.

EXAMPLE 4.11 | At the roulette wheel

A small steel roulette ball rolls ccw around the inside of a 30-cm-diameter roulette wheel. The ball completes 2.0 rev in 1.20 s.

a. What is the ball's angular velocity?

b. What is the ball's position at $t = 2.0$ s? Assume $\theta_i = 0$.

MODEL Model the ball as a particle in uniform circular motion.

SOLVE a. The period of the ball's motion, the time for 1 rev, is $T = 0.60$ s. Angular velocity is positive for ccw motion, so

$$\omega = \frac{2\pi \text{ rad}}{T} = \frac{2\pi \text{ rad}}{0.60 \text{ s}} = 10.47 \text{ rad/s}$$

b. The ball starts at $\theta_i = 0$ rad. After $\Delta t = 2.0$ s, its position is

$$\theta_f = 0 \text{ rad} + (10.47 \text{ rad/s})(2.0 \text{ s}) = 20.94 \text{ rad}$$

where we've kept an extra significant figure to avoid round-off error. Although this is a mathematically acceptable answer, an observer would say that the ball is always located somewhere between 0° and 360°. Thus it is common practice to subtract an integer number of 2π rad, representing the completed revolutions. Because $20.94/2\pi = 3.333$, we can write

$$\begin{aligned}\theta_f &= 20.94 \text{ rad} = 3.333 \times 2\pi \text{ rad} \\ &= 3 \times 2\pi \text{ rad} + 0.333 \times 2\pi \text{ rad} \\ &= 3 \times 2\pi \text{ rad} + 2.09 \text{ rad}\end{aligned}$$

In other words, at $t = 2.0$ s the ball has completed 3 rev and is 2.09 rad = 120° into its fourth revolution. An observer would say that the ball's position is $\theta_f = 120°$.

As Figure 4.21 showed, the velocity vector $\vec{v}$ is always tangent to the circle. In other words, the velocity vector has only a *tangential component,* which we will designate v_t. The tangential velocity is positive for ccw motion, negative for cw motion.

Combining $v = 2\pi r/T$ for the speed with $\omega = 2\pi/T$ for the angular velocity—but keeping the sign of ω to indicate the direction of motion—we see that the tangential velocity and the angular velocity are related by

$$v_t = \omega r \qquad (\text{with } \omega \text{ in rad/s}) \tag{4.28}$$

Because v_t is the only nonzero component of $\vec{v}$, the particle's speed is $v = |v_t| = |\omega|r$. We'll sometimes write this as $v = \omega r$ if there's no ambiguity about the sign of ω.

NOTE While it may be convenient in some problems to measure ω in rev/s or rpm, you must convert to SI units of rad/s before using Equation 4.28.

As a simple example, a particle moving cw at 2.0 m/s in a circle of radius 40 cm has angular velocity

$$\omega = \frac{v_t}{r} = \frac{-2.0 \text{ m/s}}{0.40 \text{ m}} = -5.0 \text{ rad/s}$$

where v_t and ω are negative because the motion is clockwise. Notice the units. Velocity divided by distance has units of s^{-1}. But because the division, in this case, gives us an angular quantity, we've inserted the *dimensionless* unit rad to give ω the appropriate units of rad/s.

STOP TO THINK 4.5 A particle moves cw around a circle at constant speed for 2.0 s. It then reverses direction and moves ccw at half the original speed until it has traveled through the same angle. Which is the particle's angle-versus-time graph?

    (a)   (b)   (c)   (d)

FIGURE 4.27 Using Tactics Box 4.1 to find Maria's acceleration on the Ferris wheel.

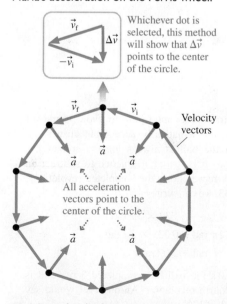

Maria's acceleration is an acceleration of changing direction, not of changing speed.

4.5 Centripetal Acceleration

FIGURE 4.27 shows a motion diagram of Maria riding a Ferris wheel at the amusement park. Maria has constant speed but *not* constant velocity because her velocity vector is changing direction. She may not be speeding up, but Maria *is* accelerating because her velocity is changing. The inset to Figure 4.27 applies the rules of Tactics Box 4.1 to find that—at every point—**Maria's acceleration vector points toward the center of the circle.** This is an acceleration due to changing direction rather than changing speed. Because the instantaneous velocity is tangent to the circle, $\vec{v}$ and $\vec{a}$ are perpendicular to each other at all points on the circle.

The acceleration of uniform circular motion is called **centripetal acceleration,** a term from a Greek root meaning "center seeking." Centripetal acceleration is not a new type of acceleration; all we are doing is *naming* an acceleration that corresponds to a particular type of motion. The magnitude of the centripetal acceleration is constant because each successive $\Delta\vec{v}$ in the motion diagram has the same length.

The motion diagram tells us the direction of $\vec{a}$, but it doesn't give us a value for a. To complete our description of uniform circular motion, we need to find a quantitative relationship between a and the particle's speed v. **FIGURE 4.28** shows the velocity $\vec{v}_i$

at one instant of motion and the velocity $\vec{v}_f$ an infinitesimal amount of time dt later. During this small interval of time, the particle has moved through the infinitesimal angle $d\theta$ and traveled distance $ds = r\, d\theta$.

By definition, the acceleration is $\vec{a} = d\vec{v}/dt$. We can see from the inset to Figure 4.28 that $d\vec{v}$ points toward the center of the circle—that is, $\vec{a}$ is a centripetal acceleration. To find the magnitude of $\vec{a}$, we can see from the isosceles triangle of velocity vectors that, if $d\theta$ is in radians,

$$dv = |d\vec{v}| = v\, d\theta \tag{4.29}$$

For uniform circular motion at constant speed, $v = ds/dt = r\, d\theta/dt$ and thus the time to rotate through angle $d\theta$ is

$$dt = \frac{r\, d\theta}{v} \tag{4.30}$$

Combining Equations 4.29 and 4.30, we see that the acceleration has magnitude

$$a = |\vec{a}| = \frac{|d\vec{v}|}{dt} = \frac{v\, d\theta}{r\, d\theta/v} = \frac{v^2}{r}$$

In vector notation, we can write

$$\vec{a} = \left(\frac{v^2}{r},\ \text{toward center of circle}\right) \qquad \text{(centripetal acceleration)} \tag{4.31}$$

Using Equation 4.28, $v = \omega r$, we can also express the magnitude of the centripetal acceleration in terms of the angular velocity ω as

$$a = \omega^2 r \tag{4.32}$$

NOTE Centripetal acceleration is not a constant acceleration. The magnitude of the centripetal acceleration is constant during uniform circular motion, but the direction of $\vec{a}$ is constantly changing. **Thus the constant-acceleration kinematics equations of Chapter 2 do *not* apply to circular motion.**

The Uniform Circular Motion Model

The **uniform circular motion model** is especially important because it applies not only to particles moving in circles but also to the uniform rotation of solid objects.

MODEL 4.2

Uniform circular motion

For motion with constant angular velocity ω.

- Applies to a particle moving along a circular trajectory at constant speed or to points on a solid object rotating at a steady rate.

- Mathematically:
  - The tangential velocity is $v_t = \omega r$.
  - The centripetal acceleration is v^2/r or $\omega^2 r$.
  - ω and v_t are positive for ccw rotation, negative for cw rotation.

- Limitations: Model fails if rotation isn't steady.

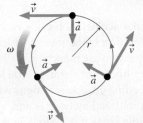

The velocity is tangent to the circle. The acceleration points to the center.

Exercise 20

FIGURE 4.28 Finding the acceleration of circular motion.

$d\vec{v}$ is the arc of a circle with arc length $dv = v\, d\theta$.

These are the velocities at times t and $t + dt$.

Same angle

EXAMPLE 4.12 | **The acceleration of a Ferris wheel**

A typical carnival Ferris wheel has a radius of 9.0 m and rotates 4.0 times per minute. What speed and acceleration do the riders experience?

MODEL Model the rider as a particle in uniform circular motion.

SOLVE The period is $T = \frac{1}{4}$ min = 15 s. From Equation 4.21, a rider's speed is

$$v = \frac{2\pi r}{T} = \frac{2\pi (9.0 \text{ m})}{15 \text{ s}} = 3.77 \text{ m/s}$$

Consequently, the centripetal acceleration has magnitude

$$a = \frac{v^2}{r} = \frac{(3.77 \text{ m/s})^2}{9.0 \text{ m}} = 1.6 \text{ m/s}^2$$

ASSESS This was not intended to be a profound problem, merely to illustrate how centripetal acceleration is computed. The acceleration is enough to be noticed and make the ride interesting, but not enough to be scary.

STOP TO THINK 4.6 Rank in order, from largest to smallest, the centripetal accelerations a_a to a_e of particles a to e.

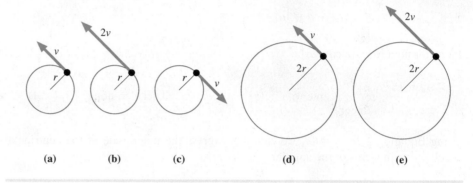

(a) (b) (c) (d) (e)

4.6 Nonuniform Circular Motion

A roller coaster car doing a loop-the-loop slows down as it goes up one side, speeds up as it comes back down the other. The ball in a roulette wheel gradually slows until it stops. Circular motion with a changing speed is called **nonuniform circular motion.** As you'll see, nonuniform circular motion is analogous to accelerated linear motion.

FIGURE 4.29 shows a point speeding up as it moves around a circle. This might be a car speeding up around a curve or simply a point on a solid object that is rotating faster and faster. The key feature of the motion is a *changing angular velocity*. For linear motion, we defined acceleration as $a_x = dv_x/dt$. By analogy, let's define the **angular acceleration** α (Greek alpha) of a rotating object, or a point on the object, to be

$$\alpha \equiv \frac{d\omega}{dt} \quad \text{(angular acceleration)} \tag{4.33}$$

Angular acceleration is the *rate* at which the angular velocity ω changes, just as linear acceleration is the rate at which the linear velocity v_x changes. The units of angular acceleration are rad/s².

For linear acceleration, you learned that a_x and v_x have the same sign when an object is speeding up, opposite signs when it is slowing down. The same rule applies to circular and rotational motion: ω and α have the same sign when the rotation is speeding up, opposite signs if it is slowing down. These ideas are illustrated in **FIGURE 4.30**.

NOTE Be careful with the sign of α. You learned in Chapter 2 that positive and negative values of the acceleration can't be interpreted as simply "speeding up" and "slowing down." Similarly, positive and negative values of angular acceleration can't be interpreted as a rotation that is speeding up or slowing down.

FIGURE 4.29 Circular motion with a changing angular velocity.

The angular velocity is changing.

FIGURE 4.30 The signs of angular velocity and acceleration. The rotation is speeding up if ω and α have the same sign, slowing down if they have opposite signs.

Initial angular velocity

$\omega > 0$	$\omega > 0$	$\omega < 0$	$\omega < 0$
$\alpha > 0$	$\alpha < 0$	$\alpha > 0$	$\alpha < 0$
Speeding up ccw	Slowing down ccw	Slowing down cw	Speeding up cw

Angular position, angular velocity, and angular acceleration are defined exactly the same as linear position, velocity, and acceleration—simply starting with an angular rather than a linear measurement of position. Consequently, **the graphical interpretation and the kinematic equations of circular/rotational motion with constant angular acceleration are exactly the same as for linear motion with constant acceleration.** This is shown in the **constant angular acceleration model** below. All the problem-solving techniques you learned in Chapter 2 for linear motion carry over to circular and rotational motion.

MODEL 4.3

Constant angular acceleration

For motion with constant angular acceleration α.

- Applies to particles with circular trajectories and to rotating solid objects.

- Mathematically: The graphs and equations for this circular/rotational motion are analogous to linear motion with constant acceleration.

  - Analogs: $s \rightarrow \theta \quad v_s \rightarrow \omega \quad a_s \rightarrow \alpha$

ω is the slope of θ

α is the slope of ω

Rotational kinematics	**Linear kinematics**
$\omega_f = \omega_i + \alpha \Delta t$	$v_{fs} = v_{is} + a_s \Delta t$
$\theta_f = \theta_i + \omega_i \Delta t + \frac{1}{2}\alpha(\Delta t)^2$	$s_f = s_i + v_{is}\Delta t + \frac{1}{2}a_s(\Delta t)^2$
$\omega_f^2 = \omega_i^2 + 2\alpha \Delta\theta$	$v_{fs}^2 = v_{is}^2 + 2a_s \Delta s$

EXAMPLE 4.13 | A rotating wheel

FIGURE 4.31a is a graph of angular velocity versus time for a rotating wheel. Describe the motion and draw a graph of angular acceleration versus time.

SOLVE This is a wheel that starts from rest, gradually speeds up *counterclockwise* until reaching top speed at t_1, maintains a constant angular velocity until t_2, then gradually slows down until stopping at t_3. The motion is always ccw because ω is always positive. The angular acceleration graph of **FIGURE 4.32b** is based on the fact that α is the slope of the ω-versus-t graph.

Conversely, the initial linear increase of ω can be seen as the increasing area under the α-versus-t graph as t increases from 0 to t_1. The angular velocity doesn't change from t_1 to t_2 when the area under the α-versus-t is zero.

▶ **FIGURE 4.31** ω-versus-t graph and the corresponding α-versus-t graph for a rotating wheel.

(a) ω

Constant positive slope, so α is positive.   Zero slope, so α is zero.   Constant negative slope, so α is negative.

(b) α

EXAMPLE 4.14 | **A slowing fan**

A ceiling fan spinning at 60 rpm coasts to a stop 25 s after being turned off. How many revolutions does it make while stopping?

MODEL Model the fan as a rotating object with constant angular acceleration.

SOLVE We don't know which direction the fan is rotating, but the fact that the rotation is slowing tells us that ω and α have opposite signs. We'll assume that ω is positive. We need to convert the initial angular velocity to SI units:

$$\omega_i = 60 \, \frac{\text{rev}}{\text{min}} \times \frac{1 \, \text{min}}{60 \, \text{s}} \times \frac{2\pi \, \text{rad}}{1 \, \text{rev}} = 6.28 \, \text{rad/s}$$

We can use the first rotational kinematics equation in Model 4.3 to find the angular acceleration:

$$\alpha = \frac{\omega_f - \omega_i}{\Delta t} = \frac{0 \, \text{rad/s} - 6.28 \, \text{rad/s}}{25 \, \text{s}} = -0.25 \, \text{rad/s}^2$$

Then, from the second rotational kinematic equation, the angular displacement during these 25 s is

$$\Delta\theta = \omega_i \Delta t + \tfrac{1}{2}\alpha(\Delta t)^2$$

$$= (6.28 \, \text{rad/s})(25 \, \text{s}) + \tfrac{1}{2}(-0.25 \, \text{rad/s}^2)(25 \, \text{s})^2$$

$$= 78.9 \, \text{rad} \times \frac{1 \, \text{rev}}{2\pi \, \text{rad}} = 13 \, \text{rev}$$

The kinematic equation returns an angle in rad, but the question asks for revolutions, so the last step was a unit conversion.

ASSESS Turning through 13 rev in 25 s while stopping seems reasonable. Notice that the problem is solved just like the linear kinematics problems you learned to solve in Chapter 2.

Tangential Acceleration

FIGURE 4.32 Acceleration in nonuniform circular motion.

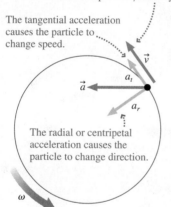

The velocity is always tangent to the circle, so the radial component v_r is always zero.

The tangential acceleration causes the particle to change speed.

The radial or centripetal acceleration causes the particle to change direction.

FIGURE 4.32 shows a particle in nonuniform circular motion. *Any* circular motion, whether uniform or nonuniform, has a centripetal acceleration because the particle is changing direction; this was the acceleration component $\vec{a}_\perp$ of Figure 4.6. As a vector component, the centripetal acceleration, which points radially toward the center of the circle, is the **radial acceleration** a_r. The expression $a_r = v_t^2/r = \omega^2 r$ is still valid in nonuniform circular motion.

For a particle to speed up or slow down as it moves around a circle, it needs—in addition to the centripetal acceleration—an acceleration parallel to the trajectory or, equivalently, parallel to $\vec{v}$. This is the acceleration component $\vec{a}_\parallel$ associated with changing speed. We'll call this the **tangential acceleration** a_t because, like the velocity v_t, it is always tangent to the circle. Because of the tangential acceleration, **the acceleration vector $\vec{a}$ of a particle in nonuniform circular motion does not point toward the center of the circle**. It points "ahead" of center for a particle that is speeding up, as in Figure 4.32, but it would point "behind" center for a particle slowing down. You can see from Figure 4.32 that the magnitude of the acceleration is

$$a = \sqrt{a_r^2 + a_t^2} \tag{4.34}$$

If a_t is constant, then the arc length s traveled by the particle around the circle and the tangential velocity v_t are found from constant-acceleration kinematics:

$$s_f = s_i + v_{it}\Delta t + \tfrac{1}{2}a_t(\Delta t)^2$$
$$v_{ft} = v_{it} + a_t\Delta t \tag{4.35}$$

Because tangential acceleration is the rate at which the tangential velocity changes, $a_t = dv_t/dt$, and we already know that the tangential velocity is related to the angular velocity by $v_t = \omega r$, it follows that

$$a_t = \frac{dv_t}{dt} = \frac{d(\omega r)}{dt} = \frac{d\omega}{dt} r = \alpha r \tag{4.36}$$

Thus $v_t = \omega r$ and $a_t = \alpha r$ are analogous equations for the tangential velocity and acceleration. In Example 4.14, where we found the fan to have angular acceleration $\alpha = -0.25 \, \text{rad/s}^2$, a blade tip 65 cm from the center would have tangential acceleration

$$a_t = \alpha r = (-0.25 \, \text{rad/s}^2)(0.65 \, \text{m}) = -0.16 \, \text{m/s}^2$$

EXAMPLE 4.15 | Analyzing rotational data

You've been assigned the task of measuring the start-up characteristics of a large industrial motor. After several seconds, when the motor has reached full speed, you know that the angular acceleration will be zero, but you hypothesize that the angular acceleration may be constant during the first couple of seconds as the motor speed increases. To find out, you attach a shaft encoder to the 3.0-cm-diameter axle. A shaft encoder is a device that converts the angular position of a shaft or axle to a signal that can be read by a computer. After setting the computer program to read four values a second, you start the motor and acquire the following data:

Time (s)	Angle (°)	Time (s)	Angle (°)
0.00	0	1.00	267
0.25	16	1.25	428
0.50	69	1.50	620
0.75	161		

a. Do the data support your hypothesis of a constant angular acceleration? If so, what is the angular acceleration? If not, is the angular acceleration increasing or decreasing with time?
b. A 76-cm-diameter blade is attached to the motor shaft. At what time does the acceleration of the tip of the blade reach 10 m/s²?

MODEL The axle is rotating with nonuniform circular motion. Model the tip of the blade as a particle.

VISUALIZE FIGURE 4.33 shows that the blade tip has both a tangential and a radial acceleration.

FIGURE 4.33 Pictorial representation of the axle and blade.

SOLVE a. *If* the motor starts up with constant angular acceleration, with $\theta_i = 0$ and $\omega_i = 0$ rad/s, the angle-time equation of rotational kinematics is $\theta = \frac{1}{2}\alpha t^2$. This can be written as a linear equation $y = mx + b$ if we let $\theta = y$ and $t^2 = x$. That is, constant angular acceleration predicts that a graph of θ versus t^2 should be a straight line with slope $m = \frac{1}{2}\alpha$ and y-intercept $b = 0$. We can test this.

FIGURE 4.34 is the graph of θ versus t^2, and it confirms our hypothesis that the motor starts up with constant angular acceleration. The

FIGURE 4.34 Graph of θ versus t^2 for the motor shaft.

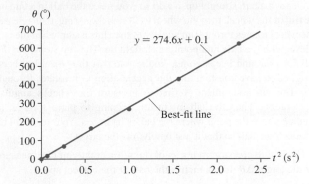

best-fit line, found using a spreadsheet, gives a slope of 274.6°/s². The units come not from the spreadsheet but by looking at the units of rise (°) over run (s² because we're graphing t^2 on the x-axis). Thus the angular acceleration is

$$\alpha = 2m = 549.2°/s^2 \times \frac{\pi \text{ rad}}{180°} = 9.6 \text{ rad/s}^2$$

where we used $180° = \pi$ rad to convert to SI units of rad/s².
c. The magnitude of the linear acceleration is

$$a = \sqrt{a_r^2 + a_t^2}$$

The tangential acceleration of the blade tip is

$$a_t = \alpha r = (9.6 \text{ rad/s}^2)(0.38 \text{ m}) = 3.65 \text{ m/s}^2$$

We were careful to use the blade's radius, not its diameter, and we kept an extra significant figure to avoid round-off error. The radial (centripetal) acceleration increases as the rotation speed increases, and the total acceleration reaches 10 m/s² when

$$a_r = \sqrt{a^2 - a_t^2} = \sqrt{(10 \text{ m/s}^2)^2 - (3.65 \text{ m/s}^2)^2} = 9.31 \text{ m/s}^2$$

Radial acceleration is $a_r = \omega^2 r$, so the corresponding angular velocity is

$$\omega = \sqrt{\frac{a_r}{r}} = \sqrt{\frac{9.31 \text{ m/s}^2}{0.38 \text{ m}}} = 4.95 \text{ rad/s}$$

For constant angular acceleration, $\omega = \alpha t$, so this angular velocity is achieved at

$$t = \frac{\omega}{\alpha} = \frac{4.95 \text{ rad/s}}{9.6 \text{ rad/s}^2} = 0.52 \text{ s}$$

Thus it takes 0.52 s for the acceleration of the blade tip to reach 10 m/s².

ASSESS The acceleration at the tip of a long blade is likely to be large. It seems plausible that the acceleration would reach 10 m/s² in ≈ 0.5 s.

STOP TO THINK 4.7 The fan blade is slowing down. What are the signs of ω and α?

a. ω is positive and α is positive.
b. ω is positive and α is negative.
c. ω is negative and α is positive.
d. ω is negative and α is negative.

CHALLENGE EXAMPLE 4.16 | Hit the target!

One day when you come into lab, you see a spring-loaded wheel that can launch a ball straight up. To do so, you place the ball in a cup on the rim of the wheel, turn the wheel to stretch the spring, then release. The wheel rotates through an angle $\Delta\theta$, then hits a stop when the cup is level with the axle and pointing straight up. The cup stops, but the ball flies out and keeps going. You're told that the wheel has been designed to have constant angular acceleration as it rotates through $\Delta\theta$. The lab assignment is first to measure the wheel's angular acceleration. Then the lab instructor is going to place a target at height h above the point where the ball is launched. Your task will be to launch the ball so that it just barely hits the target.

a. Find an expression in terms of quantities that you can measure for the angle $\Delta\theta$ that launches the ball at the correct speed.

b. Evaluate $\Delta\theta$ if the wheel's diameter is 62 cm, you've determined that its angular acceleration is 200 rad/s^2, the mass of the ball is 25 g, and the instructor places the target 190 cm above the launch point.

MODEL Model the ball as a particle. It first undergoes circular motion that we'll model as having constant angular acceleration. We'll then ignore air resistance and model the vertical motion as free fall.

VISUALIZE FIGURE 4.35 is a pictorial representation. This is a two-part problem, with the speed at the end of the angular acceleration being the launch speed for the vertical motion. We've chosen to call the wheel radius R and the target height h. These and the angular

FIGURE 4.35 Pictorial representation of the ball launcher.

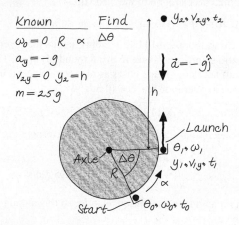

acceleration α are considered "known" because we will measure them, but we don't have numerical values at this time.

SOLVE

a. The circular motion problem and the vertical motion problem are connected through the ball's speed: The final speed of the angular acceleration is the launch speed of the vertical motion. We don't know anything about time intervals, which suggests using the kinematic equations that relate distance and acceleration (for the vertical motion) and angle and angular acceleration (for the circular motion). For the angular acceleration, with $\omega_0 = 0$ rad/s,

$$\omega_1^2 = \omega_0^2 + 2\alpha\,\Delta\theta = 2\alpha\,\Delta\theta$$

The final speed of the ball and cup, when the wheel hits the stop, is

$$v_1 = \omega_1 R = R\sqrt{2\alpha\,\Delta\theta}$$

Thus the vertical-motion problem begins with the ball being shot upward with velocity $v_{1y} = R\sqrt{2\alpha\,\Delta\theta}$. How high does it go? The highest point is the point where $v_{2y} = 0$, so the free-fall equation is

$$v_{2y}^2 = 0 = v_{1y}^2 - 2g\,\Delta y = R^2 \cdot 2\alpha\,\Delta\theta - 2gh$$

Rather than solve for height h, we need to solve for the angle that produces a given height. This is

$$\Delta\theta = \frac{gh}{\alpha R^2}$$

Once we've determined the properties of the wheel and then measured the height at which our instructor places the target, we'll quickly be able to calculate the angle through which we should pull back the wheel to launch the ball.

b. For the values given in the problem statement, $\Delta\theta = 0.969$ rad $= 56°$. Don't forget that equations involving angles need values in radians and return values in radians.

ASSESS The angle needed to be less than 90° or else the ball would fall out of the cup before launch. And an angle of only a few degrees would seem suspiciously small. Thus 56° seems to be reasonable. Notice that the mass was not needed in this problem. Part of becoming a better problem solver is evaluating the information you have to see what is relevant. Some homework problems will help you develop this skill by providing information that isn't necessary.

SUMMARY

The goal of Chapter 4 has been to learn how to solve problems about motion in a plane.

GENERAL PRINCIPLES

The **instantaneous velocity**

$$\vec{v} = d\vec{r}/dt$$

is a vector tangent to the trajectory.

The **instantaneous acceleration** is

$$\vec{a} = d\vec{v}/dt$$

$\vec{a}_\parallel$, the component of $\vec{a}$ parallel to $\vec{v}$, is responsible for change of *speed*. $\vec{a}_\perp$, the component of $\vec{a}$ perpendicular to $\vec{v}$, is responsible for change of *direction*.

Relative Motion

If object C moves relative to reference frame A with velocity $\vec{v}_{CA}$, then it moves relative to a different reference frame B with velocity

$$\vec{v}_{CB} = \vec{v}_{CA} + \vec{v}_{AB}$$

where $\vec{v}_{AB}$ is the velocity of A relative to B. This is the Galilean transformation of velocity.

Object C moves relative to both A and B.

Reference frame A

Reference frame B

IMPORTANT CONCEPTS

Uniform Circular Motion

Angular velocity $\omega = d\theta/dt$.
v_t and ω are constant:

$$v_t = \omega r$$

The centripetal acceleration points toward the center of the circle:

$$a = \frac{v^2}{r} = \omega^2 r$$

It changes the particle's direction but not its speed.

Nonuniform Circular Motion

Angular acceleration $\alpha = d\omega/dt$.
The radial acceleration

$$a_r = \frac{v^2}{r} = \omega^2 r$$

changes the particle's direction. The tangential component

$$a_t = \alpha r$$

changes the particle's speed.

APPLICATIONS

Kinematics in two dimensions

If $\vec{a}$ is constant, then the x- and y-components of motion are independent of each other.

$$x_f = x_i + v_{ix}\,\Delta t + \tfrac{1}{2}a_x(\Delta t)^2$$
$$y_f = y_i + v_{iy}\,\Delta t + \tfrac{1}{2}a_y(\Delta t)^2$$
$$v_{fx} = v_{ix} + a_x\,\Delta t$$
$$v_{fy} = v_{iy} + a_y\,\Delta t$$

Projectile motion is motion under the influence of only gravity.

MODEL Model as a particle launched with speed v_0 at angle θ.

VISUALIZE Use coordinates with the x-axis horizontal and the y-axis vertical.

SOLVE The horizontal motion is uniform with $v_x = v_0\cos\theta$. The vertical motion is free fall with $a_y = -g$. The x and y kinematic equations have the *same* value for Δt.

The trajectory is a parabola.

Circular motion kinematics

Period $T = \dfrac{2\pi r}{v} = \dfrac{2\pi}{\omega}$

Angular position $\theta = \dfrac{s}{r}$

Constant angular acceleration

$$\omega_f = \omega_i + \alpha\,\Delta t$$
$$\theta_f = \theta_i + \omega_i\,\Delta t + \tfrac{1}{2}\alpha(\Delta t)^2$$
$$\omega_f^2 = \omega_i^2 + 2\alpha\,\Delta\theta$$

Circular motion graphs and kinematics are analogous to linear motion with constant acceleration.

Angle, angular velocity, and angular acceleration are related graphically.

- The angular velocity is the slope of the angular position graph.
- The angular acceleration is the slope of the angular velocity graph.

TERMS AND NOTATION

projectile	uniform circular motion	angular velocity, ω	angular acceleration, α
launch angle, θ	period, T	centripetal acceleration	constant angular acceleration
projectile motion model	angular position, θ	uniform circular motion	model
reference frame	arc length, s	model	radial acceleration, a_r
Galilean transformation	radians	nonuniform circular	tangential acceleration, a_t
of velocity	angular displacement, $\Delta\theta$	motion	

CONCEPTUAL QUESTIONS

1. a. At this instant, is the particle in **FIGURE Q4.1** speeding up, slowing down, or traveling at constant speed?
   b. Is this particle curving to the right, curving to the left, or traveling straight?

FIGURE Q4.1   FIGURE Q4.2

2. a. At this instant, is the particle in **FIGURE Q4.2** speeding up, slowing down, or traveling at constant speed?
   b. Is this particle curving upward, curving downward, or traveling straight?
3. Tarzan swings through the jungle by hanging from a vine.
   a. Immediately after stepping off a branch to swing over to another tree, is Tarzan's acceleration $\vec{a}$ zero or not zero? If not zero, which way does it point? Explain.
   b. Answer the same question at the lowest point in Tarzan's swing.
4. A projectile is launched at an angle of 30°.
   a. Is there any point on the trajectory where $\vec{v}$ and $\vec{a}$ are parallel to each other? If so, where?
   b. Is there any point where $\vec{v}$ and $\vec{a}$ are perpendicular to each other? If so, where?
5. For a projectile, which of the following quantities are constant during the flight: $x, y, r, v_x, v_y, v, a_x, a_y$? Which of these quantities are zero throughout the flight?
6. A cart that is rolling at constant velocity on a level table fires a ball straight up.
   a. When the ball comes back down, will it land in front of the launching tube, behind the launching tube, or directly in the tube? Explain.
   b. Will your answer change if the cart is accelerating in the forward direction? If so, how?
7. A rock is thrown from a bridge at an angle 30° below horizontal. Immediately after the rock is released, is the magnitude of its acceleration greater than, less than, or equal to g? Explain.
8. Anita is running to the right at 5 m/s in **FIGURE Q4.8**. Balls 1 and 2 are thrown toward her by friends standing on the ground. According to Anita, both balls are approaching her at 10 m/s.

Which ball was thrown at a faster speed? Or were they thrown with the same speed? Explain.

FIGURE Q4.8

9. An electromagnet on the ceiling of an airplane holds a steel ball. When a button is pushed, the magnet releases the ball. The experiment is first done while the plane is parked on the ground, and the point where the ball hits the floor is marked with an X. Then the experiment is repeated while the plane is flying level at a steady 500 mph. Does the ball land slightly in front of the X (toward the nose of the plane), on the X, or slightly behind the X (toward the tail of the plane)? Explain.
10. Zack is driving past his house in **FIGURE Q4.10**. He wants to toss his physics book out the window and have it land in his driveway. If he lets go of the book exactly as he passes the end of the driveway, should he direct his throw outward and toward the front of the car (throw 1), straight outward (throw 2), or outward and toward the back of the car (throw 3)? Explain.

FIGURE Q4.10   **FIGURE Q4.11**

11. In **FIGURE Q4.11**, Yvette and Zack are driving down the freeway side by side with their windows down. Zack wants to toss his physics book out the window and have it land in Yvette's front seat. Ignoring air resistance, should he direct his throw outward and toward the front of the car (throw 1), straight outward (throw 2), or outward and toward the back of the car (throw 3)? Explain.

12. In uniform circular motion, which of the following quantities are constant: speed, instantaneous velocity, the tangential component of velocity, the radial component of acceleration, the tangential component of acceleration? Which of these quantities are zero throughout the motion?

13. **FIGURE Q4.13** shows three points on a steadily rotating wheel.
    a. Rank in order, from largest to smallest, the angular velocities ω_1, ω_2, and ω_3 of these points. Explain.
    b. Rank in order, from largest to smallest, the speeds v_1, v_2, and v_3 of these points. Explain.

FIGURE Q4.13

14. **FIGURE Q4.14** shows four rotating wheels. For each, determine the signs ($+$ or $-$) of ω and α.

(a) Speeding up (b) Slowing down (c) Slowing down (d) Speeding up

FIGURE Q4.14

15. **FIGURE Q4.15** shows a pendulum at one end point of its arc.
    a. At this point, is ω positive, negative, or zero? Explain.
    b. At this point, is α positive, negative, or zero? Explain.

FIGURE Q4.15

EXERCISES AND PROBLEMS

Exercises

Section 4.1 Motion in Two Dimensions

Problems 1 and 2 show a partial motion diagram. For each:
a. Complete the motion diagram by adding acceleration vectors.
b. Write a physics *problem* for which this is the correct motion diagram. Be imaginative! Don't forget to include enough information to make the problem complete and to state clearly what is to be found.

1. |

FIGURE EX4.1

2. |

Top view of motion in a horizontal plane

Circular arc

FIGURE EX4.2

Answer Problems 3 through 5 by choosing one of the eight labeled acceleration vectors or selecting option I: $\vec{a} = \vec{0}$.

3. || At this instant, the particle has steady speed and is curving to the right. What is the direction of its acceleration?

FIGURE EX4.3

4. || At this instant, the particle is speeding up and curving upward. What is the direction of its acceleration?

FIGURE EX4.4

5. || At this instant, the particle is speeding up and curving downward. What is the direction of its acceleration?

FIGURE EX4.5

6. || A rocket-powered hockey puck moves on a horizontal frictionless table. **FIGURE EX4.6** shows graphs of v_x and v_y, the x- and y-components of the puck's velocity. The puck starts at the origin.
    a. In which direction is the puck moving at $t = 2$ s? Give your answer as an angle from the x-axis.
    b. How far from the origin is the puck at $t = 5$ s?

FIGURE EX4.6

7. || A rocket-powered hockey puck moves on a horizontal frictionless table. **FIGURE EX4.7** shows graphs of v_x and v_y, the x- and y-components of the puck's velocity. The puck starts at the origin. What is the magnitude of the puck's acceleration at $t = 5$ s?

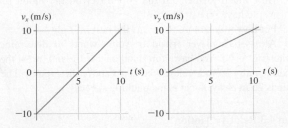

FIGURE EX4.7

8. ‖ A particle's trajectory is described by $x = \left(\frac{1}{2}t^3 - 2t^2\right)$ m and
CALC $y = \left(\frac{1}{2}t^2 - 2t\right)$ m, where t is in s.
   a. What are the particle's position and speed at $t = 0$ s and $t = 4$ s?
   b. What is the particle's direction of motion, measured as an angle from the x-axis, at $t = 0$ s and $t = 4$ s?

9. ‖ A particle moving in the xy-plane has velocity $\vec{v} = (2t\hat{\imath} + (3 - t^2)\hat{\jmath})$ m/s, where t is in s. What is the particle's acceleration vector at $t = 4$ s?
CALC

10. ‖ You have a remote-controlled car that has been programmed to
CALC have velocity $\vec{v} = (-3t\hat{\imath} + 2t^2\hat{\jmath})$ m/s, where t is in s. At $t = 0$ s, the car is at $\vec{r}_0 = (3.0\hat{\imath} + 2.0\hat{\jmath})$ m. What are the car's (a) position vector and (b) acceleration vector at $t = 2.0$ s?

Section 4.2 Projectile Motion

11. ‖ A ball thrown horizontally at 25 m/s travels a horizontal distance of 50 m before hitting the ground. From what height was the ball thrown?

12. | A physics student on Planet Exidor throws a ball, and it follows the parabolic trajectory shown in **FIGURE EX4.12**. The ball's position is shown at 1 s intervals until $t = 3$ s. At $t = 1$ s, the ball's velocity is $\vec{v} = (2.0\hat{\imath} + 2.0\hat{\jmath})$ m/s.
   a. Determine the ball's velocity at $t = 0$ s, 2 s, and 3 s.
   b. What is the value of g on Planet Exidor?
   c. What was the ball's launch angle?

FIGURE EX4.12

13. ‖ A supply plane needs to drop a package of food to scientists working on a glacier in Greenland. The plane flies 100 m above the glacier at a speed of 150 m/s. How far short of the target should it drop the package?

14. ‖ A rifle is aimed horizontally at a target 50 m away. The bullet hits the target 2.0 cm below the aim point.
   a. What was the bullet's flight time?
   b. What was the bullet's speed as it left the barrel?

15. ‖ In the Olympic shotput event, an athlete throws the shot with an initial speed of 12.0 m/s at a 40.0° angle from the horizontal. The shot leaves her hand at a height of 1.80 m above the ground. How far does the shot travel?

16. ‖ On the Apollo 14 mission to the moon, astronaut Alan Shepard hit a golf ball with a 6 iron. The free-fall acceleration on the moon is 1/6 of its value on earth. Suppose he hit the ball with a speed of 25 m/s at an angle 30° above the horizontal.
   a. How much farther did the ball travel on the moon than it would have on earth?
   b. For how much more time was the ball in flight?

17. ‖‖ A baseball player friend of yours wants to determine his pitching speed. You have him stand on a ledge and throw the ball horizontally from an elevation 4.0 m above the ground. The ball lands 25 m away. What is his pitching speed?

Section 4.3 Relative Motion

18. ‖ A boat takes 3.0 hours to travel 30 km down a river, then 5.0 hours to return. How fast is the river flowing?

19. ‖ When the moving sidewalk at the airport is broken, as it often seems to be, it takes you 50 s to walk from your gate to baggage claim. When it is working and you stand on the moving sidewalk the entire way, without walking, it takes 75 s to travel the same distance. How long will it take you to travel from the gate to baggage claim if you walk while riding on the moving sidewalk?

20. | Mary needs to row her boat across a 100-m-wide river that is flowing to the east at a speed of 1.0 m/s. Mary can row with a speed of 2.0 m/s.
   a. If Mary points her boat due north, how far from her intended landing spot will she be when she reaches the opposite shore?
   b. What is her speed with respect to the shore?

21. | A kayaker needs to paddle north across a 100-m-wide harbor. The tide is going out, creating a tidal current that flows to the east at 2.0 m/s. The kayaker can paddle with a speed of 3.0 m/s.
   a. In which direction should he paddle in order to travel straight across the harbor?
   b. How long will it take him to cross?

22. ‖ Susan, driving north at 60 mph, and Trent, driving east at 45 mph, are approaching an intersection. What is Trent's speed relative to Susan's reference frame?

Section 4.4 Uniform Circular Motion

23. ‖ **FIGURE EX4.23** shows the angular-velocity-versus-time graph for a particle moving in a circle. How many revolutions does the object make during the first 4 s?

FIGURE EX4.23

FIGURE EX4.24

24. | **FIGURE EX4.24** shows the angular-position-versus-time graph for a particle moving in a circle. What is the particle's angular velocity at (a) $t = 1$ s, (b) $t = 4$ s, and (c) $t = 7$ s?

25. ‖ **FIGURE EX4.25** shows the angular-velocity-versus-time graph for a particle moving in a circle, starting from $\theta_0 = 0$ rad at $t = 0$ s. Draw the angular-position-versus-time graph. Include an appropriate scale on both axes.

FIGURE EX4.25

26. ‖ The earth's radius is about 4000 miles. Kampala, the capital of Uganda, and Singapore are both nearly on the equator. The distance between them is 5000 miles. The flight from Kampala to Singapore takes 9.0 hours. What is the plane's angular velocity with respect to the earth's surface? Give your answer in °/h.

27. | An old-fashioned single-play vinyl record rotates on a turntable at 45 rpm. What are (a) the angular velocity in rad/s and (b) the period of the motion?

28. ‖ As the earth rotates, what is the speed of (a) a physics student in Miami, Florida, at latitude 26°, and (b) a physics student in Fairbanks, Alaska, at latitude 65°? Ignore the revolution of the earth around the sun. The radius of the earth is 6400 km.

29. | How fast must a plane fly along the earth's equator so that the sun stands still relative to the passengers? In which direction must the plane fly, east to west or west to east? Give your answer in both km/h and mph. The earth's radius is 6400 km.

30. ‖ A 3000-m-high mountain is located on the equator. How much faster does a climber on top of the mountain move than a surfer at a nearby beach? The earth's radius is 6400 km.

Section 4.5 Centripetal Acceleration

31. | Peregrine falcons are known for their maneuvering ability. In a
BIO tight circular turn, a falcon can attain a centripetal acceleration 1.5 times the free-fall acceleration. What is the radius of the turn if the falcon is flying at 25 m/s?

32. | To withstand "g-forces" of up to 10 g's, caused by suddenly
BIO pulling out of a steep dive, fighter jet pilots train on a "human centrifuge." 10 g's is an acceleration of 98 m/s². If the length of the centrifuge arm is 12 m, at what speed is the rider moving when she experiences 10 g's?

33. ‖ The radius of the earth's very nearly circular orbit around the sun is 1.5×10^{11} m. Find the magnitude of the earth's (a) velocity, (b) angular velocity, and (c) centripetal acceleration as it travels around the sun. Assume a year of 365 days.

34. ‖ A speck of dust on a spinning DVD has a centripetal acceleration of 20 m/s².
   a. What is the acceleration of a different speck of dust that is twice as far from the center of the disk?
   b. What would be the acceleration of the first speck of dust if the disk's angular velocity was doubled?

35. ‖ Your roommate is working on his bicycle and has the bike upside down. He spins the 60-cm-diameter wheel, and you notice that a pebble stuck in the tread goes by three times every second. What are the pebble's speed and acceleration?

Section 4.6 Nonuniform Circular Motion

36. | FIGURE EX4.36 shows the angular velocity graph of the crankshaft in a car. What is the crankshaft's angular acceleration at (a) $t = 1$ s, (b) $t = 3$ s, and (c) $t = 5$ s?

FIGURE EX4.36

FIGURE EX4.37

37. ‖ FIGURE EX4.37 shows the angular acceleration graph of a turntable that starts from rest. What is the turntable's angular velocity at (a) $t = 1$ s, (b) $t = 2$ s, and (c) $t = 3$ s?

38. ‖ FIGURE EX4.38 shows the angular-velocity-versus-time graph for a particle moving in a circle. How many revolutions does the object make during the first 4 s?

FIGURE EX4.38

39. ‖ A wheel initially rotating at 60 rpm experiences the angular acceleration shown in FIGURE EX4.39. What is the wheel's angular velocity, in rpm, at $t = 3.0$ s?

FIGURE EX4.39

40. ‖ A 5.0-m-diameter merry-go-round is initially turning with a 4.0 s period. It slows down and stops in 20 s.
   a. Before slowing, what is the speed of a child on the rim?
   b. How many revolutions does the merry-go-round make as it stops?

41. ‖ An electric fan goes from rest to 1800 rpm in 4.0 s. What is its angular acceleration?

42. ‖ A bicycle wheel is rotating at 50 rpm when the cyclist begins to pedal harder, giving the wheel a constant angular acceleration of 0.50 rad/s².
   a. What is the wheel's angular velocity, in rpm, 10 s later?
   b. How many revolutions does the wheel make during this time?

43. ‖ Starting from rest, a DVD steadily accelerates to 500 rpm in 1.0 s, rotates at this angular speed for 3.0 s, then steadily decelerates to a halt in 2.0 s. How many revolutions does it make?

Problems

44. ‖‖ A spaceship maneuvering near Planet Zeta is located at $\vec{r} = (600\hat{\imath} - 400\hat{\jmath} + 200\hat{k}) \times 10^3$ km, relative to the planet, and traveling at $\vec{v} = 9500\hat{\imath}$ m/s. It turns on its thruster engine and accelerates with $\vec{a} = (40\hat{\imath} - 20\hat{k})$ m/s² for 35 min. What is the spaceship's position when the engine shuts off? Give your answer as a position vector measured in km.

45. ‖‖ A particle moving in the xy-plane has velocity $\vec{v}_0 = v_{0x}\hat{\imath} + v_{0y}\hat{\jmath}$
CALC at $t = 0$. It undergoes acceleration $\vec{a} = bt\hat{\imath} - cv_y\hat{\jmath}$, where b and c are constants. Find an expression for the particle's velocity at a later time t.

46. ‖ A projectile's horizontal range over level ground is $v_0^2 \sin 2\theta/g$. At what launch angle or angles will the projectile land at half of its maximum possible range?

47. ‖ a. A projectile is launched with speed v_0 and angle θ. Derive an expression for the projectile's maximum height h.
   b. A baseball is hit with a speed of 33.6 m/s. Calculate its height and the distance traveled if it is hit at angles of 30.0°, 45.0°, and 60.0°.

48. ‖‖ A projectile is launched from ground level at angle θ and speed
CALC v_0 into a headwind that causes a constant horizontal acceleration of magnitude a opposite the direction of motion.
   a. Find an expression in terms of a and g for the launch angle that gives maximum range.
   b. What is the angle for maximum range if a is 10% of g?

49. ‖‖ A gray kangaroo can bound across level ground with each jump
BIO carrying it 10 m from the takeoff point. Typically the kangaroo leaves the ground at a 20° angle. If this is so:
   a. What is its takeoff speed?
   b. What is its maximum height above the ground?

50. ‖ A ball is thrown toward a cliff of height h with a speed of 30 m/s and an angle of 60° above horizontal. It lands on the edge of the cliff 4.0 s later.
   a. How high is the cliff?
   b. What was the maximum height of the ball?
   c. What is the ball's impact speed?

51. ‖ A tennis player hits a ball 2.0 m above the ground. The ball leaves his racquet with a speed of 20.0 m/s at an angle 5.0° above the horizontal. The horizontal distance to the net is 7.0 m, and the net is 1.0 m high. Does the ball clear the net? If so, by how much? If not, by how much does it miss?

52. ‖ You are target shooting using a toy gun that fires a small ball at a speed of 15 m/s. When the gun is fired at an angle of 30° above horizontal, the ball hits the bull's-eye of a target at the same height as the gun. Then the target distance is halved. At what angle must you aim the gun to hit the bull's-eye in its new position? (Mathematically there are two solutions to this problem; the physically reasonable answer is the smaller of the two.)

53. ‖ A 35 g steel ball is held by a ceiling-mounted electromagnet 3.5 m above the floor. A compressed-air cannon sits on the floor, 4.0 m to one side of the point directly under the ball. When a button is pressed, the ball drops and, simultaneously, the cannon fires a 25 g plastic ball. The two balls collide 1.0 m above the floor. What was the launch speed of the plastic ball?

54. ‖ You are watching an archery tournament when you start wondering how fast an arrow is shot from the bow. Remembering your physics, you ask one of the archers to shoot an arrow parallel to the ground. You find the arrow stuck in the ground 60 m away, making a 3.0° angle with the ground. How fast was the arrow shot?

55. ‖ You're 6.0 m from one wall of the house seen in FIGURE P4.55. You want to toss a ball to your friend who is 6.0 m from the opposite wall. The throw and catch each occur 1.0 m above the ground.
    a. What minimum speed will allow the ball to clear the roof?
    b. At what angle should you toss the ball?

FIGURE P4.55   FIGURE P4.56

56. ‖ Sand moves without slipping at 6.0 m/s down a conveyer that is tilted at 15°. The sand enters a pipe 3.0 m below the end of the conveyer belt, as shown in FIGURE P4.56. What is the horizontal distance d between the conveyer belt and the pipe?

57. ‖ A stunt man drives a car at a speed of 20 m/s off a 30-m-high cliff. The road leading to the cliff is inclined upward at an angle of 20°.
    a. How far from the base of the cliff does the car land?
    b. What is the car's impact speed?

58. ‖‖ A javelin thrower standing at rest holds the center of the javelin
BIO behind her head, then accelerates it through a distance of 70 cm as she throws. She releases the javelin 2.0 m above the ground traveling at an angle of 30° above the horizontal. Top-rated javelin throwers do throw at about a 30° angle, not the 45° you might have expected, because the biomechanics of the arm allow them to throw the javelin much faster at 30° than they would be able to at 45°. In this throw, the javelin hits the ground 62 m away. What was the acceleration of the javelin during the throw? Assume that it has a constant acceleration.

59. ‖ A rubber ball is dropped onto a ramp that is tilted at 20°, as shown in FIGURE P4.59. A bouncing ball obeys the "law of reflection," which says that the ball leaves the surface at the same angle it approached the surface. The ball's next bounce is 3.0 m to the right of its first bounce. What is the ball's rebound speed on its first bounce?

FIGURE P4.59   FIGURE P4.60

60. ‖‖ You are asked to consult for the city's research hospital, where
BIO a group of doctors is investigating the bombardment of cancer tumors with high-energy ions. As FIGURE P4.60 shows, ions are fired directly toward the center of the tumor at speeds of 5.0×10^6 m/s. To cover the entire tumor area, the ions are deflected sideways by passing them between two charged metal plates that accelerate the ions perpendicular to the direction of their initial motion. The acceleration region is 5.0 cm long, and the ends of the acceleration plates are 1.5 m from the target. What sideways acceleration is required to deflect an ion 2.0 cm to one side?

61. ‖ Ships A and B leave port together. For the next two hours, ship A travels at 20 mph in a direction 30° west of north while ship B travels 20° east of north at 25 mph.
    a. What is the distance between the two ships two hours after they depart?
    b. What is the speed of ship A as seen by ship B?

62. ‖ While driving north at 25 m/s during a rainstorm you notice that the rain makes an angle of 38° with the vertical. While driving back home moments later at the same speed but in the opposite direction, you see that the rain is falling straight down. From these observations, determine the speed and angle of the raindrops relative to the ground.

63. ‖ You've been assigned the task of using a shaft encoder—a device that measures the angle of a shaft or axle and provides a signal to a computer—to analyze the rotation of an engine crankshaft under certain conditions. The table lists the crankshaft's angles over a 0.6 s interval.

Time (s)	Angle (rad)
0.0	0.0
0.1	2.0
0.2	3.2
0.3	4.3
0.4	5.3
0.5	6.1
0.6	7.0

Is the crankshaft rotating with uniform circular motion? If so, what is its angular velocity in rpm? If not, is the angular acceleration positive or negative?

64. ‖ A circular track has several concentric rings where people can run at their leisure. Phil runs on the outermost track with radius r_P while Annie runs on an inner track with radius $r_A = 0.80r_P$. The runners start side by side, along a radial line, and run at the same speed in a counterclockwise direction. How many revolutions has Annie made when Annie's and Phil's velocity vectors point in opposite directions for the first time?

65. ||| A typical laboratory centrifuge rotates at 4000 rpm. Test tubes have to be placed into a centrifuge very carefully because of the very large accelerations.

 a. What is the acceleration at the end of a test tube that is 10 cm from the axis of rotation?

 b. For comparison, what is the magnitude of the acceleration a test tube would experience if dropped from a height of 1.0 m and stopped in a 1.0-ms-long encounter with a hard floor?

66. || Astronauts use a centrifuge to simulate the acceleration of a
BIO rocket launch. The centrifuge takes 30 s to speed up from rest to its top speed of 1 rotation every 1.3 s. The astronaut is strapped into a seat 6.0 m from the axis.

 a. What is the astronaut's tangential acceleration during the first 30 s?

 b. How many g's of acceleration does the astronaut experience when the device is rotating at top speed? Each 9.8 m/s² of acceleration is 1 g.

67. || Communications satellites are placed in a circular orbit where they stay directly over a fixed point on the equator as the earth rotates. These are called *geosynchronous orbits*. The radius of the earth is 6.37×10^6 m, and the altitude of a geosynchronous orbit is 3.58×10^7 m ($\approx 22{,}000$ miles). What are (a) the speed and (b) the magnitude of the acceleration of a satellite in a geosynchronous orbit?

68. || A computer hard disk 8.0 cm in diameter is initially at rest. A small dot is painted on the edge of the disk. The disk accelerates at 600 rad/s² for $\frac{1}{2}$s, then coasts at a steady angular velocity for another $\frac{1}{2}$s.

 a. What is the speed of the dot at $t = 1.0$ s?

 b. Through how many revolutions has the disk turned?

69. || A high-speed drill rotating ccw at 2400 rpm comes to a halt in 2.5 s.

 a. What is the magnitude of the drill's angular acceleration?

 b. How many revolutions does it make as it stops?

70. || A turbine is spinning at 3800 rpm. Friction in the bearings is so low that it takes 10 min to coast to a stop. How many revolutions does the turbine make while stopping?

71. || Your 64-cm-diameter car tire is rotating at 3.5 rev/s when suddenly you press down hard on the accelerator. After traveling 200 m, the tire's rotation has increased to 6.0 rev/s. What was the tire's angular acceleration? Give your answer in rad/s².

72. || The angular velocity of a process control motor is
CALC $\omega = \left(20 - \frac{1}{2}t^2\right)$ rad/s, where t is in seconds.

 a. At what time does the motor reverse direction?

 b. Through what angle does the motor turn between $t = 0$ s and the instant at which it reverses direction?

73. || A Ferris wheel of radius R speeds up with angular acceleration α starting from rest. Find an expression for the (a) velocity and (b) centripetal acceleration of a rider after the Ferris wheel has rotated through angle $\Delta\theta$.

74. || A 6.0-cm-diameter gear rotates with angular velocity $\omega =$
CALC $\left(2.0 + \frac{1}{2}t^2\right)$ rad/s, where t is in seconds. At $t = 4.0$ s, what are:

 a. The gear's angular acceleration?

 b. The tangential acceleration of a tooth on the gear?

75. || A painted tooth on a spinning gear has angular acceleration
CALC $\alpha = (20 - t)$ rad/s², where t is in s. Its initial angular velocity, at $t = 0$ s, is 300 rpm. What is the tooth's angular velocity in rpm at $t = 20$ s?

76. ||| A car starts from rest on a curve with a radius of 120 m and accelerates tangentially at 1.0 m/s². Through what angle will the car have traveled when the magnitude of its total acceleration is 2.0 m/s²?

77. ||| A long string is wrapped around a 6.0-cm-diameter cylinder, initially at rest, that is free to rotate on an axle. The string is then pulled with a constant acceleration of 1.5 m/s² until 1.0 m of string has been unwound. If the string unwinds without slipping, what is the cylinder's angular speed, in rpm, at this time?

In Problems 78 through 80 you are given the equations that are used to solve a problem. For each of these, you are to

 a. Write a realistic problem for which these are the correct equations. Be sure that the answer your problem requests is consistent with the equations given.

 b. Finish the solution of the problem, including a pictorial representation.

78. $100 \text{ m} = 0 \text{ m} + (50\cos\theta \text{ m/s})t_1$

 $0 \text{ m} = 0 \text{ m} + (50\sin\theta \text{ m/s})t_1 - \frac{1}{2}(9.80 \text{ m/s}^2)t_1^2$

79. $v_x = -(6.0\cos 45°) \text{ m/s} + 3.0 \text{ m/s}$

 $v_y = (6.0\sin 45°) \text{ m/s} + 0 \text{ m/s}$

 $100 \text{ m} = v_y t_1, x_1 = v_x t_1$

80. $2.5 \text{ rad} = 0 \text{ rad} + \omega_i(10 \text{ s}) + \big((1.5 \text{ m/s}^2)/2(50 \text{ m})\big)(10 \text{ s})^2$

 $\omega_f = \omega_i + \big((1.5 \text{ m/s}^2)/(50 \text{ m})\big)(10 \text{ s})$

Challenge Problems

81. ||| In one contest at the county fair, seen in **FIGURE CP4.81**, a spring-loaded plunger launches a ball at a speed of 3.0 m/s from one corner of a smooth, flat board that is tilted up at a 20° angle. To win, you must make the ball hit a small target at the adjacent corner, 2.50 m away. At what angle θ should you tilt the ball launcher?

FIGURE CP4.81   **FIGURE CP4.82**

82. ||| An archer standing on a 15° slope shoots an arrow 20° above the horizontal, as shown in **FIGURE CP4.82**. How far down the slope does the arrow hit if it is shot with a speed of 50 m/s from 1.75 m above the ground?

83. ||| A skateboarder starts up a 1.0-m-high, 30° ramp at a speed of 7.0 m/s. The skateboard wheels roll without friction. At the top she leaves the ramp and sails through the air. How far from the end of the ramp does the skateboarder touch down?

84. ||| A cannon on a train car fires a projectile to the right with speed
CALC v_0, relative to the train, from a barrel elevated at angle θ. The cannon fires just as the train, which had been cruising to the right along a level track with speed v_{train}, begins to accelerate with acceleration a, which can be either positive (speeding up) or negative (slowing down). Find an expression for the angle at which the projectile should be fired so that it lands as far as possible from the cannon. You can ignore the small height of the cannon above the track.

85. ||| A child in danger of drowning in a river is being carried downstream by a current that flows uniformly with a speed of 2.0 m/s. The child is 200 m from the shore and 1500 m upstream of the boat dock from which the rescue team sets out. If their boat speed is 8.0 m/s with respect to the water, at what angle from the shore should the pilot leave the shore to go directly to the child?

5 Force and Motion

These ice boats are a memorable example of the connection between force and motion.

IN THIS CHAPTER, you will learn about the connection between force and motion.

What is a force?

The fundamental concept of **mechanics** is force.

- A force is a push or a pull.
- A force acts on an object.
- A force requires an agent.
- A force is a vector.

How do we identify forces?

A force can be a contact force or a long-range force.

- Contact forces occur at points where the environment touches the object.
- Contact forces disappear the instant contact is lost. Forces have no memory.
- Long-range forces include gravity and magnetism.

Thrust force $\vec{F}_{thrust}$

Gravity $\vec{F}_G$ Normal force $\vec{n}$

How do we show forces?

Forces can be displayed on a free-body diagram. You'll draw all forces—both pushes and pulls—as vectors with their tails on the particle. A well-drawn free-body diagram is an essential step in solving problems, as you'll see in the next chapter.

What do forces do?

A net force causes an object to *accelerate* with an acceleration directly proportional to the size of the force. This is Newton's second law, the most important statement in mechanics. For a particle of mass m,

$$\vec{a} = \frac{1}{m} \vec{F}_{net}$$

« LOOKING BACK Sections 1.4, 2.4, and 3.2 Acceleration and vector addition

What is Newton's first law?

Newton's first law—an object at rest stays at rest and an object in motion continues moving at constant speed in a straight line if and only if the net force on the object is zero—helps us define what a force *is*. It is also the basis for identifying the reference frames—called inertial reference frames—in which Newton's laws are valid.

What good are forces?

Kinematics describes *how* an object moves. For the more important tasks of knowing *why* an object moves and being able to predict its position and orientation at a future time, we have to know the forces acting on the object. Relating force to motion is the subject of **dynamics,** and it is one of the most important underpinnings of all science and engineering.

5.1 Force

The two major issues that this chapter will examine are:

- What is a force?
- What is the connection between force and motion?

We begin with the first of these questions in the table below.

What is a force?

A force is a push or a pull.

Our commonsense idea of a **force** is that it is a *push* or a *pull*. We will refine this idea as we go along, but it is an adequate starting point. Notice our careful choice of words: We refer to "*a* force," rather than simply "force." We want to think of a force as a very specific *action,* so that we can talk about a single force or perhaps about two or three individual forces that we can clearly distinguish. Hence the concrete idea of "a force" acting on an object.

A force acts on an object.

Implicit in our concept of force is that a **force acts on an object.** In other words, pushes and pulls are applied *to* something—an object. From the object's perspective, it has a force *exerted* on it. Forces do not exist in isolation from the object that experiences them.

A force requires an agent.

Every force has an **agent,** something that acts or exerts power. That is, a force has a specific, identifiable *cause.* As you throw a ball, it is your hand, while in contact with the ball, that is the agent or the cause of the force exerted on the ball. *If* a force is being exerted on an object, you must be able to identify a specific cause (i.e., the agent) of that force. Conversely, a force is not exerted on an object *unless* you can identify a specific cause or agent. Although this idea may seem to be stating the obvious, you will find it to be a powerful tool for avoiding some common misconceptions about what is and is not a force.

A force is a vector.

If you push an object, you can push either gently or very hard. Similarly, you can push either left or right, up or down. To quantify a push, we need to specify both a magnitude *and* a direction. It should thus come as no surprise that force is a vector. The general symbol for a force is the vector symbol $\vec{F}$. The size or strength of a force is its magnitude F.

A force can be either a contact force …

There are two basic classes of forces, depending on whether the agent touches the object or not. **Contact forces** are forces that act on an object by touching it at a point of contact. The bat must touch the ball to hit it. A string must be tied to an object to pull it. The majority of forces that we will examine are contact forces.

… or a long-range force.

Long-range forces are forces that act on an object without physical contact. Magnetism is an example of a long-range force. You have undoubtedly held a magnet over a paper clip and seen the paper clip leap up to the magnet. A coffee cup released from your hand is pulled to the earth by the long-range force of gravity.

NOTE In the particle model, objects cannot exert forces on themselves. A force on an object will always have an agent or cause external to the object. Now, there are certainly objects that have internal forces (think of all the forces inside the engine of your car!), but the particle model is not valid if you need to consider those internal forces. If you are going to treat your car as a particle and look only at the overall motion of the car as a whole, that motion will be a consequence of external forces acting on the car.

Force Vectors

We can use a simple diagram to visualize how forces are exerted on objects.

TACTICS BOX 5.1

Drawing force vectors

❶ Model the object as a particle.

❷ Place the *tail* of the force vector on the particle.

❸ Draw the force vector as an arrow pointing in the proper direction and with a length proportional to the size of the force.

❹ Give the vector an appropriate label.

$\vec{F}$

Step 2 may seem contrary to what a "push" should do, but recall that moving a vector does not change it as long as the length and angle do not change. The vector $\vec{F}$ is the same regardless of whether the tail or the tip is placed on the particle. FIGURE 5.1 shows three examples of force vectors.

FIGURE 5.1 Three examples of forces and their vector representations.

The rope is the agent.

The spring is the agent.

Box

Long-range force of gravity

Box

Pulling force of rope

Box

Pushing force of spring

Earth is the agent.

Combining Forces

FIGURE 5.2 Two forces applied to a box.

(a)

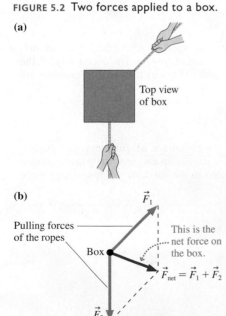

Top view of box

(b)

Pulling forces of the ropes

Box

$\vec{F}_1$

This is the net force on the box.

$\vec{F}_{net} = \vec{F}_1 + \vec{F}_2$

$\vec{F}_2$

FIGURE 5.2a shows a box being pulled by two ropes, each exerting a force on the box. How will the box respond? Experimentally, we find that when several forces $\vec{F}_1$, $\vec{F}_2$, $\vec{F}_3$, ... are exerted on an object, they combine to form a **net force** given by the *vector* sum of *all* the forces:

$$\vec{F}_{net} \equiv \sum_{i=1}^{N} \vec{F}_i = \vec{F}_1 + \vec{F}_2 + \cdots + \vec{F}_N \tag{5.1}$$

Recall that $\equiv$ is the symbol meaning "is defined as." Mathematically, this summation is called a **superposition of forces.** FIGURE 5.2b shows the net force on the box.

STOP TO THINK 5.1 Two of the three forces exerted on an object are shown. The net force points to the left. Which is the missing third force?

$\vec{F}_1$

$\vec{F}_2$

Two of the three forces exerted on an object

$\vec{F}_3$

(a)

$\vec{F}_3$

(b)

$\vec{F}_3$

(c)

$\vec{F}_3$

(d)

5.2 A Short Catalog of Forces

There are many forces we will deal with over and over. This section will introduce you to some of them. Many of these forces have special symbols. As you learn the major forces, be sure to learn the symbol for each.

Gravity

Gravity—the only long-range force we will encounter in the next few chapters—keeps you in your chair and the planets in their orbits around the sun. We'll have a thorough look at gravity in Chapter 13. For now we'll concentrate on objects on or near the surface of the earth (or other planet).

The pull of a planet on an object on or near the surface is called the **gravitational force.** The agent for the gravitational force is the *entire planet*. Gravity acts on *all* objects, whether moving or at rest. The symbol for gravitational force is $\vec{F}_G$. **The gravitational force vector always points vertically downward,** as shown in FIGURE 5.3.

> **NOTE** We often refer to "the weight" of an object. For an object at rest on the surface of a planet, its weight is simply the magnitude F_G of the gravitational force. However, weight and gravitational force are not the same thing, nor is weight the same as mass. We will briefly examine mass later in the chapter, and we'll explore the rather subtle connections among gravity, weight, and mass in Chapter 6.

FIGURE 5.3 Gravity.

The gravitational force pulls the box down.

$\vec{F}_G$

Ground

Spring Force

Springs exert one of the most common contact forces. **A spring can either push (when compressed) or pull (when stretched).** FIGURE 5.4 shows the **spring force,** for which we use the symbol $\vec{F}_{Sp}$. In both cases, pushing and pulling, the tail of the force vector is placed on the particle in the force diagram.

Although you may think of a spring as a metal coil that can be stretched or compressed, this is only one type of spring. Hold a ruler, or any other thin piece of wood or metal, by the ends and bend it slightly. It flexes. When you let go, it "springs" back to its original shape. This is just as much a spring as is a metal coil.

FIGURE 5.4 The spring force.

(a) A compressed spring exerts a pushing force on an object.

$\vec{F}_{Sp}$

(b) A stretched spring exerts a pulling force on an object.

$\vec{F}_{Sp}$

Tension Force

When a string or rope or wire pulls on an object, it exerts a contact force that we call the **tension force,** represented by a capital $\vec{T}$. **The direction of the tension force is always along the direction of the string or rope,** as you can see in FIGURE 5.5. The commonplace reference to "the tension" in a string is an informal expression for T, the size or magnitude of the tension force.

> **NOTE** Tension is represented by the symbol T. This is logical, but there's a risk of confusing the tension T with the identical symbol T for the period of a particle in circular motion. The number of symbols used in science and engineering is so large that some letters are used several times to represent different quantities. The use of T is the first time we've run into this problem, but it won't be the last. You must be alert to the *context* of a symbol's use to deduce its meaning.

FIGURE 5.5 Tension.

The rope exerts a tension force on the sled.

$\vec{T}$

We can obtain a deeper understanding of some forces and interactions with a picture of what's happening at the atomic level. You'll recall from chemistry that matter consists of *atoms* that are attracted to each other by *molecular bonds*. Although the details are complex, governed by quantum physics, we can often use a simple **ball-and-spring model** of a solid to get an idea of what's happening at the atomic level.

MODEL 5.1

Ball-and-spring model of solids

Solids consist of atoms held together by molecular bonds.

- Represent the solid as **an array of balls connected by springs.**
- Pulling on or pushing on a solid causes the bonds to be stretched or compressed. **Stretched or compressed bonds exert spring forces.**
- There are an immense number of bonds. The force of one bond is very tiny, but the combined force of all bonds can be very large.
- Limitations: Model fails for liquids and gases.

Ball-like atoms

Spring-like bonds

In the case of tension, pulling on the ends of a string or rope stretches the spring-like molecular bonds ever so slightly. What we call "tension" is then the net spring force being exerted by trillions and trillions of microscopic springs.

Normal Force

If you sit on a bed, the springs in the mattress compress and, as a consequence of the compression, exert an upward force on you. Stiffer springs would show less compression but still exert an upward force. The compression of extremely stiff springs might be measurable only by sensitive instruments. Nonetheless, the springs would compress ever so slightly and exert an upward spring force on you.

FIGURE 5.6 shows an object resting on top of a sturdy table. The table may not visibly flex or sag, but—just as you do to the bed—the object compresses the spring-like molecular bonds in the table. The size of the compression is very small, but it is not zero. As a consequence, the compressed "molecular springs" *push upward* on the object. We say that "the table" exerts the upward force, but it is important to understand that the pushing is *really* done by molecular bonds.

We can extend this idea. Suppose you place your hand on a wall and lean against it. Does the wall exert a force on your hand? As you lean, you compress the molecular bonds in the wall and, as a consequence, they push outward against your hand. So the answer is yes, the wall does exert a force on you.

The force the table surface exerts is vertical; the force the wall exerts is horizontal. In all cases, the force exerted on an object that is pressing against a surface is in a direction *perpendicular* to the surface. Mathematicians refer to a line that is perpendicular to a surface as being *normal* to the surface. In keeping with this terminology, we define the **normal force** as **the force exerted perpendicular to a surface (the agent) against an object that is pressing against the surface.** The symbol for the normal force is $\vec{n}$.

We're not using the word *normal* to imply that the force is an "ordinary" force or to distinguish it from an "abnormal force." A surface exerts a force *perpendicular* (i.e., normal) to itself as the molecular springs press *outward*. **FIGURE 5.7** shows an object on an inclined surface, a common situation.

In essence, the normal force is just a spring force, but one exerted by a vast number of microscopic springs acting at once. The normal force is responsible for the "solidness" of solids. It is what prevents you from passing right through the chair you are sitting in and what causes the pain and the lump if you bang your head into a door.

Friction

Friction, like the normal force, is exerted by a surface. But whereas the normal force is perpendicular to the surface, **the friction force is always *parallel* to the surface.** It is useful to distinguish between two kinds of friction:

FIGURE 5.6 The table exerts an upward force on the book.

The compressed molecular bonds push upward on the object.

FIGURE 5.7 The normal force.

$\vec{n}$

The surface pushes outward against the bottom of the frog.

■ *Kinetic friction,* denoted $\vec{f}_k$, appears as an object slides across a surface. This is a force that "opposes the motion," meaning that the friction force vector $\vec{f}_k$ points in a direction opposite the velocity vector $\vec{v}$ (i.e., "the motion").

■ *Static friction,* denoted $\vec{f}_s$, is the force that keeps an object "stuck" on a surface and prevents its motion. Finding the direction of $\vec{f}_s$ is a little trickier than finding it for $\vec{f}_k$. Static friction points opposite the direction in which the object *would* move if there were no friction. That is, it points in the direction necessary to *prevent* motion.

FIGURE 5.8 shows examples of kinetic and static friction.

> **NOTE** A surface exerts a kinetic friction force when an object moves *relative to* the surface. A package on a conveyor belt is in motion, but it does not experience a kinetic friction force because it is not moving relative to the belt. So to be precise, we should say that the kinetic friction force points opposite to an object's motion *relative to* a surface.

FIGURE 5.8 Kinetic and static friction.

Kinetic friction opposes the motion.

Static friction acts in the direction that prevents slipping.

Drag

Friction at a surface is one example of a *resistive force,* a force that opposes or resists motion. Resistive forces are also experienced by objects moving through fluids—gases and liquids. The resistive force of a fluid is called **drag,** with symbol $\vec{F}_{drag}$. **Drag, like kinetic friction, points opposite the direction of motion.** **FIGURE 5.9** shows an example.

Drag can be a significant force for objects moving at high speeds or in dense fluids. Hold your arm out the window as you ride in a car and feel how the air resistance against it increases rapidly as the car's speed increases. Drop a lightweight object into a beaker of water and watch how slowly it settles to the bottom.

For objects that are heavy and compact, that move in air, and whose speed is not too great, the drag force of air resistance is fairly small. To keep things as simple as possible, **you can neglect air resistance in all problems unless a problem explicitly asks you to include it.**

FIGURE 5.9 Air resistance is an example of drag.

Air resistance points opposite the direction of motion.

Thrust

A jet airplane obviously has a force that propels it forward during takeoff. Likewise for the rocket being launched in **FIGURE 5.10**. This force, called **thrust,** occurs when a jet or rocket engine expels gas molecules at high speed. Thrust is a contact force, with the exhaust gas being the agent that pushes on the engine. The process by which thrust is generated is rather subtle, and we will postpone a full discussion until we study Newton's third law in Chapter 7. For now, we will treat thrust as a force opposite the direction in which the exhaust gas is expelled. There's no special symbol for thrust, so we will call it $\vec{F}_{thrust}$.

FIGURE 5.10 Thrust force on a rocket.

Thrust force is exerted on a rocket by exhaust gases.

Electric and Magnetic Forces

Electricity and magnetism, like gravity, exert long-range forces. We will study electric and magnetic forces in detail in Part VI. For now, it is worth noting that the forces holding molecules together—the molecular bonds—are not actually tiny springs. Atoms and molecules are made of charged particles—electrons and protons—and what we call a molecular bond is really an electric force between these particles. So when we say that the normal force and the tension force are due to "molecular springs," or that friction is due to atoms running into each other, what we're really saying is that these forces, at the most fundamental level, are actually electric forces between the charged particles in the atoms.

5.3 Identifying Forces

A typical physics problem describes an object that is being pushed and pulled in various directions. Some forces are given explicitly; others are only implied. In order to proceed, it is necessary to determine all the forces that act on the object. The procedure for identifying forces will become part of the *pictorial representation* of the problem.

Force	Notation
General force	$\vec{F}$
Gravitational force	$\vec{F}_G$
Spring force	$\vec{F}_{Sp}$
Tension	$\vec{T}$
Normal force	$\vec{n}$
Static friction	$\vec{f}_s$
Kinetic friction	$\vec{f}_k$
Drag	$\vec{F}_{drag}$
Thrust	$\vec{F}_{thrust}$

TACTICS BOX 5.2

Identifying forces

❶ **Identify the object of interest.** This is the object you wish to study.

❷ **Draw a picture of the situation.** Show the object of interest and all other objects—such as ropes, springs, or surfaces—that touch it.

❸ **Draw a closed curve around the object.** Only the object of interest is inside the curve; everything else is outside.

❹ **Locate every point on the boundary of this curve where other objects touch the object of interest.** These are the points where *contact forces* are exerted on the object.

❺ **Name and label each contact force acting on the object.** There is at least one force at each point of contact; there may be more than one. When necessary, use subscripts to distinguish forces of the same type.

❻ **Name and label each long-range force acting on the object.** For now, the only long-range force is the gravitational force.

Exercises 3–8

EXAMPLE 5.1 | Forces on a bungee jumper

A bungee jumper has leapt off a bridge and is nearing the bottom of her fall. What forces are being exerted on the jumper?

VISUALIZE   FIGURE 5.11 Forces on a bungee jumper.

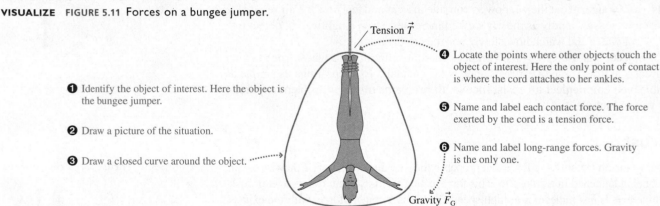

❶ Identify the object of interest. Here the object is the bungee jumper.

❷ Draw a picture of the situation.

❸ Draw a closed curve around the object.

Tension $\vec{T}$

❹ Locate the points where other objects touch the object of interest. Here the only point of contact is where the cord attaches to her ankles.

❺ Name and label each contact force. The force exerted by the cord is a tension force.

❻ Name and label long-range forces. Gravity is the only one.

Gravity $\vec{F}_G$

EXAMPLE 5.2 | Forces on a skier

A skier is being towed up a snow-covered hill by a tow rope. What forces are being exerted on the skier?

VISUALIZE   FIGURE 5.12 Forces on a skier.

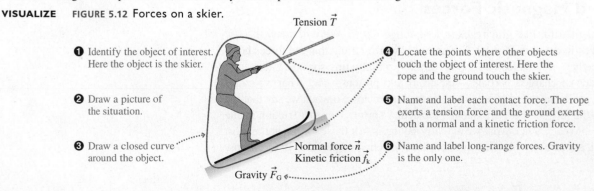

Tension $\vec{T}$

❶ Identify the object of interest. Here the object is the skier.

❷ Draw a picture of the situation.

❸ Draw a closed curve around the object.

❹ Locate the points where other objects touch the object of interest. Here the rope and the ground touch the skier.

❺ Name and label each contact force. The rope exerts a tension force and the ground exerts both a normal and a kinetic friction force.

❻ Name and label long-range forces. Gravity is the only one.

Normal force $\vec{n}$
Kinetic friction $\vec{f}_k$
Gravity $\vec{F}_G$

NOTE You might have expected two friction forces and two normal forces in Example 5.2, one on each ski. Keep in mind, however, that we're working within the particle model, which represents the skier by a single point. A particle has only one contact with the ground, so there is one normal force and one friction force.

EXAMPLE 5.3 | **Forces on a rocket**

A rocket is being launched to place a new satellite in orbit. Air resistance is not negligible. What forces are being exerted on the rocket?

VISUALIZE This drawing is much more like the sketch you would make when identifying forces as part of solving a problem.

▶ FIGURE 5.13 Forces on a rocket.

STOP TO THINK 5.2 You've just kicked a rock, and it is now sliding across the ground 2 m in front of you. Which of these forces act on the rock? List all that apply.

a. Gravity, acting downward.   b. The normal force, acting upward.
c. The force of the kick, acting in the direction of motion.
d. Friction, acting opposite the direction of motion.

5.4 What Do Forces Do?

Having learned to identify forces, we ask the next question: How does an object move when a force is exerted on it? The only way to answer this question is to do experiments. Let's conduct a "virtual experiment," one you can easily visualize. Imagine using your fingers to stretch a rubber band to a certain length—say 10 centimeters—that you can measure with a ruler, as shown in FIGURE 5.14. You know that a stretched rubber band exerts a force—a spring force—because your fingers *feel* the pull. Furthermore, this is a reproducible force; the rubber band exerts the same force every time you stretch it to this length. We'll call this the *standard force F*. Not surprisingly, two identical rubber bands exert twice the pull of one rubber band, and N side-by-side rubber bands exert N times the standard force: $F_{net} = NF$.

Now attach one rubber band to a 1 kg block and stretch it to the standard length. The object experiences the same force F as did your finger. The rubber band gives us a way of applying a known and reproducible force to an object. Then imagine using the rubber band to pull the block across a horizontal, frictionless table. (We can imagine a frictionless table since this is a virtual experiment, but in practice you could nearly eliminate friction by supporting the object on a cushion of air.)

If you stretch the rubber band and then release the object, the object moves toward your hand. But as it does so, the rubber band gets shorter and the pulling force decreases. To keep the pulling force constant, you must *move your hand* at just the right speed to keep the length of the rubber band from changing! FIGURE 5.15a shows the experiment being carried out. Once the motion is complete, you can use motion diagrams and kinematics to analyze the object's motion.

FIGURE 5.14 **A reproducible force.**

One rubber band stretched the standard length exerts the standard force F.

Two rubber bands stretched the standard length exert twice the standard force.

FIGURE 5.15 Measuring the motion of a 1 kg block that is pulled with a constant force.

(a)

Maintain constant stretch.

Rubber band
Pull
Frictionless surface
Motion diagram

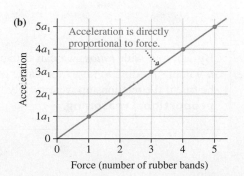

(b)

Acceleration is directly proportional to force.

Force (number of rubber bands)

The first important finding of this experiment is that **an object pulled with a constant force moves with a constant acceleration.** That is, the answer to the question What does a force do? is: A force causes an object to accelerate, and a constant force produces a constant acceleration.

What happens if you increase the force by using several rubber bands? To find out, use two rubber bands, then three rubber bands, then four, and so on. With N rubber bands, the force on the block is NF. **FIGURE 5.15b** shows the results of this experiment. You can see that doubling the force causes twice the acceleration, tripling the force causes three times the acceleration, and so on. The graph reveals our second important finding: **The acceleration is directly proportional to the force.** This result can be written as

$$a = cF \qquad\qquad (5.2)$$

where c, called the *proportionality constant,* is the slope of the graph.

MATHEMATICAL ASIDE

Proportionality and proportional reasoning

The concept of **proportionality** arises frequently in physics. A quantity symbolized by u is *proportional* to another quantity symbolized by v if

$$u = cv$$

where c (which might have units) is called the **proportionality constant**. This relationship between u and v is often written

$$u \propto v$$

where the symbol $\propto$ means "is proportional to."

If v is doubled to $2v$, then u doubles to $c(2v) = 2(cv) = 2u$. In general, if v is changed by any factor f, then u changes by the same factor. This is the essence of what we *mean* by proportionality.

A graph of u versus v is a straight line *passing through the origin* (i.e., the vertical intercept is zero) with slope $= c$. Notice that proportionality is a much more specific relationship between u and v than mere linearity. The linear equation $u = cv + b$ has a straight-line graph, but it doesn't pass through the origin (unless b happens to be zero) and doubling v does not double u.

The slope is c.
The graph passes through the origin.

u is proportional to v.

If $u \propto v$, then $u_1 = cv_1$ and $u_2 = cv_2$. Dividing the second equation by the first, we find

$$\frac{u_2}{u_1} = \frac{v_2}{v_1}$$

By working with *ratios,* we can deduce information about u without needing to know the value of c. (This would not be true if the relationship were merely linear.) This is called **proportional reasoning.**

Proportionality is not limited to being linearly proportional. The graph on the left shows that u is clearly not proportional to w. But a graph of u versus $1/w^2$ is a straight line passing through the origin; thus, in this case, u is proportional to $1/w^2$, or $u \propto 1/w^2$. We would say that "u is proportional to the inverse square of w."

Is not proportional

Is proportional

u is proportional to the inverse square of w.

EXAMPLE u is proportional to the inverse square of w. By what factor does u change if w is tripled?

SOLUTION This is an opportunity for proportional reasoning; we don't need to know the proportionality constant. If u is proportional to $1/w^2$, then

$$\frac{u_2}{u_1} = \frac{1/w_2^2}{1/w_1^2} = \frac{w_1^2}{w_2^2} = \left(\frac{w_1}{w_2}\right)^2$$

Tripling w, with $w_2/w_1 = 3$, and thus $w_1/w_2 = \frac{1}{3}$, changes u to

$$u_2 = \left(\frac{w_1}{w_2}\right)^2 u_1 = \left(\frac{1}{3}\right)^2 u_1 = \frac{1}{9}u_1$$

Tripling w causes u to become $\frac{1}{9}$ of its original value.

Many *Student Workbook* and end-of-chapter homework questions will require proportional reasoning. It's an important skill to learn.

The final question for our virtual experiment is: How does the acceleration depend on the mass of the object being pulled? To find out, apply the *same force*—for example, the standard force of one rubber band—to a 2 kg block, then a 3 kg block, and so on, and for each measure the acceleration. Doing so gives you the results shown in **FIGURE 5.16**. An object with twice the mass of the original block has only half the acceleration when both are subjected to the same force.

Mathematically, the graph of Figure 5.16 is one of *inverse proportionality*. That is, **the acceleration is inversely proportional to the object's mass.** We can combine these results —that the acceleration is directly proportional to the force applied and inversely proportional to the object's mass—into the single statement

$$a = \frac{F}{m} \tag{5.3}$$

if we define the basic unit of force as the force that causes a 1 kg mass to accelerate at 1 m/s^2. That is,

$$1 \text{ basic unit of force} \equiv 1 \text{ kg} \times 1 \frac{\text{m}}{\text{s}^2} = 1 \frac{\text{kg m}}{\text{s}^2}$$

This basic unit of force is called a newton:

One **newton** is the force that causes a 1 kg mass to accelerate at 1 m/s^2. The abbreviation for newton is N. Mathematically, $1 \text{ N} = 1 \text{ kg m/s}^2$.

TABLE 5.1 lists some typical forces. As you can see, "typical" forces on "typical" objects are likely to be in the range 0.01–10,000 N.

Mass

We've been using the term *mass* without a clear definition. As we learned in Chapter 1, the SI unit of mass, the kilogram, is based on a particular metal block kept in a vault in Paris. This suggests that *mass* is the amount of matter an object contains, and that is certainly our everyday concept of mass. Now we see that a more precise way of defining an object's mass is in terms of its acceleration in response to a force. Figure 5.16 shows that an object with twice the amount of matter accelerates only half as much in response to the same force. The more matter an object has, the more it *resists* accelerating in response to a force. You're familiar with this idea: Your car is much harder to push than your bicycle. The tendency of an object to resist a *change* in its velocity (i.e., to resist acceleration) is called **inertia.** Consequently, the mass used in Equation 5.3, a measure of an object's resistance to changing its motion, is called **inertial mass.** We'll meet a different concept of mass, *gravitational mass,* when we study Newton's law of gravity in Chapter 13.

FIGURE 5.16 Acceleration is inversely proportional to mass.

TABLE 5.1 Approximate magnitude of some typical forces

Force	Approximate magnitude (newtons)
Weight of a U.S. quarter	0.05
Weight of 1/4 cup sugar	0.5
Weight of a 1 pound object	5
Weight of a house cat	50
Weight of a 110 pound person	500
Propulsion force of a car	5,000
Thrust force of a small jet engine	50,000

STOP TO THINK 5.3 Two rubber bands stretched to the standard length cause an object to accelerate at 2 m/s^2. Suppose another object with twice the mass is pulled by four rubber bands stretched to the standard length. The acceleration of this second object is

a. 1 m/s^2
b. 2 m/s^2
c. 4 m/s^2
d. 8 m/s^2
e. 16 m/s^2

Hint: Use proportional reasoning.

5.5 Newton's Second Law

Equation 5.3 is an important finding, but our experiment was limited to looking at an object's response to a single applied force. Realistically, an object is likely to be subjected to several distinct forces $\vec{F}_1$, $\vec{F}_2$, $\vec{F}_3$,... that may point in different directions. What happens then? In that case, it is found experimentally that the acceleration is determined by the *net* force.

Newton was the first to recognize the connection between force and motion. This relationship is known today as **Newton's second law**.

Newton's second law An object of mass m subjected to forces $\vec{F}_1$, $\vec{F}_2$, $\vec{F}_3$,... will undergo an acceleration $\vec{a}$ given by

$$\vec{a} = \frac{\vec{F}_{net}}{m} \tag{5.4}$$

where the net force $\vec{F}_{net} = \vec{F}_1 + \vec{F}_2 + \vec{F}_3 + \cdots$ is the vector sum of all forces acting on the object. The acceleration vector $\vec{a}$ points in the same direction as the net force vector $\vec{F}_{net}$.

The significance of Newton's second law cannot be overstated. There was no reason to suspect that there should be any simple relationship between force and acceleration. Yet there it is, a simple but exceedingly powerful equation relating the two. The critical idea is that **an object accelerates in the direction of the net force vector $\vec{F}_{net}$.**

We can rewrite Newton's second law in the form

$$\vec{F}_{net} = m\vec{a} \tag{5.5}$$

which is how you'll see it presented in many textbooks. Equations 5.4 and 5.5 are mathematically equivalent, but Equation 5.4 better describes the central idea of New-tonian mechanics: A force applied to an object causes the object to accelerate.

It's also worth noting that **the object responds only to the forces acting on it** *at this instant.* The object has no memory of forces that may have been exerted at earlier times. This idea is sometimes called **Newton's zeroth law.**

> **NOTE** Be careful not to think that one force "overcomes" the others to determine the motion. Forces are not in competition with each other! It is $\vec{F}_{net}$, the sum of *all* the forces, that determines the acceleration $\vec{a}$.

As an example, **FIGURE 5.17a** shows a box being pulled by two ropes. The ropes exert tension forces $\vec{T}_1$ and $\vec{T}_2$ on the box. **FIGURE 5.17b** represents the box as a particle, shows the forces acting on the box, and adds them graphically to find the net force $\vec{F}_{net}$. The box will accelerate in the direction of $\vec{F}_{net}$ with acceleration

$$\vec{a} = \frac{\vec{F}_{net}}{m} = \frac{\vec{T}_1 + \vec{T}_2}{m}$$

> **NOTE** The acceleration is *not* $(T_1 + T_2)/m$. You must add the forces as *vectors,* not merely add their magnitudes as scalars.

Forces Are Interactions

There's one more important aspect of forces. If you push against a door (the object) to close it, the door pushes back against your hand (the agent). If a tow rope pulls on a car (the object), the car pulls back on the rope (the agent). In general, if an agent exerts a force on an object, the object exerts a force on the agent. We really need to think of a force as an *interaction* between two objects. This idea is captured in **Newton's third law**—that for every action there is an equal but opposite reaction.

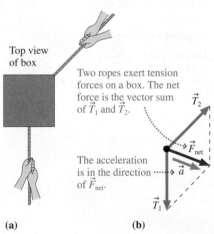

FIGURE 5.17 Acceleration of a pulled box.

Top view of box

Two ropes exert tension forces on a box. The net force is the vector sum of $\vec{T}_1$ and $\vec{T}_2$.

$\vec{T}_2$

$\vec{F}_{net}$

The acceleration is in the direction of $\vec{F}_{net}$.

$\vec{a}$

$\vec{T}_1$

(a)   (b)

Although the interaction perspective is a more exact way to view forces, it adds complications that we would like to avoid for now. Our approach will be to start by focusing on how a single object responds to forces exerted on it. Then, in Chapter 7, we'll return to Newton's third law and the larger issue of how two or more objects interact with each other.

STOP TO THINK 5.4 Three forces act on an object. In which direction does the object accelerate?

5.6 Newton's First Law

For 2000 years, scientists and philosophers thought that the "natural state" of an object is to be at rest. An object at rest requires no explanation. A moving object, though, is not in its natural state and thus requires an explanation: Why is this object moving? What keeps it going?

Galileo, in around 1600, was one of the first scientists to carry out controlled experiments. Many careful measurements in which he minimized the influence of friction led Galileo to conclude that in the absence of friction or air resistance, a moving object would continue to move along a straight line forever with no loss of speed. In other words, the natural state of an object—its behavior if free of external influences—is not rest but is *uniform motion* with constant velocity! "At rest" has no special significance in Galileo's view of motion; it is simply uniform motion that happens to have $\vec{v} = \vec{0}$.

It was left to Newton to generalize this result, and today we call it **Newton's first law** of motion.

> **Newton's first law** An object that is at rest will remain at rest, or an object that is moving will continue to move in a straight line with constant velocity, if and only if the net force acting on the object is zero.

Newton's first law is also known as the *law of inertia*. If an object is at rest, it has a tendency to stay at rest. If it is moving, it has a tendency to continue moving with the *same velocity*.

NOTE The first law refers to *net* force. An object can remain at rest, or can move in a straight line with constant velocity, even though forces are exerted on it as long as the *net* force is zero.

Notice the "if and only if" aspect of Newton's first law. If an object is at rest or moves with constant velocity, then we can conclude that there is no net force acting on it. Conversely, if no net force is acting on it, we can conclude that the object will have constant velocity, not just constant speed. The direction remains constant, too!

An object on which the net force is zero—and thus is either at rest or moving in a straight line with constant velocity—is said to be in **mechanical equilibrium**. As **FIGURE 5.18** shows, objects in mechanical equilibrium have no acceleration: $\vec{a} = \vec{0}$.

FIGURE 5.18 Two examples of mechanical equilibrium.

An object at rest is in equilibrium: $\vec{F}_{net} = \vec{0}$.

$\vec{v} = \vec{0}$
$\vec{a} = \vec{0}$

$\vec{a} = \vec{0}$

An object moving in a straight line at constant velocity is also in equilibrium: $\vec{F}_{net} = \vec{0}$.

What Good Is Newton's First Law?

So what causes an object to move? Newton's first law says **no cause is needed for an object to move!** Uniform motion is the object's natural state. Nothing at all is required for it to remain in that state. The proper question, according to Newton, is: What causes an object to *change* its velocity? Newton, with Galileo's help, also gave us the answer. **A *force* is what causes an object to change its velocity.**

The preceding paragraph contains the essence of Newtonian mechanics. This new perspective on motion, however, is often contrary to our common experience. We all know perfectly well that you must keep pushing an object—exerting a force on it—to keep it moving. Newton is asking us to change our point of view and to consider motion *from the object's perspective* rather than from our personal perspective. As far as the object is concerned, our push is just one of several forces acting on it. Others might include friction, air resistance, or gravity. Only by knowing the *net* force can we determine the object's motion.

Newton's first law may seem to be merely a special case of Newton's second law. After all, the equation $\vec{F}_{net} = m\vec{a}$ tells us that an object moving with constant velocity $(\vec{a} = \vec{0})$ has $\vec{F}_{net} = \vec{0}$. The difficulty is that the second law assumes that we already know what force is. The purpose of the first law is to *identify* a force as something that disturbs a state of equilibrium. The second law then describes how the object responds to this force. Thus from a *logical* perspective, the first law really is a separate statement that must precede the second law. But this is a rather formal distinction. From a pedagogical perspective it is better—as we have done—to use a commonsense understanding of force and start with Newton's second law.

This guy thinks there's a force hurling him into the windshield. What a dummy!

FIGURE 5.19 Reference frames.

(a) Cruising at constant speed.

The ball stays in place; the airplane is an inertial reference frame.

(b) Accelerating during takeoff.

The ball accelerates toward the back even though there are no horizontal forces; the airplane is *not* an inertial reference frame.

Inertial Reference Frames

If a car stops suddenly, you may be "thrown" into the windshield if you're not wearing your seat belt. You have a very real forward acceleration *relative to the car,* but is there a force pushing you forward? A force is a push or a pull caused by an identifiable agent in contact with the object. Although you *seem* to be pushed forward, there's no agent to do the pushing.

The difficulty—an acceleration without an apparent force—comes from using an inappropriate reference frame. Your acceleration measured in a reference frame attached to the car is not the same as your acceleration measured in a reference frame attached to the ground. Newton's second law says $\vec{F}_{net} = m\vec{a}$. But which $\vec{a}$? Measured in which reference frame?

We define an **inertial reference frame** as a reference frame in which Newton's first law is valid. If $\vec{a} = \vec{0}$ (an object is at rest or moving with constant velocity) only when $\vec{F}_{net} = \vec{0}$, then the reference frame in which $\vec{a}$ is measured is an inertial reference frame.

Not all reference frames are inertial reference frames. **FIGURE 5.19a** shows a physics student cruising at constant velocity in an airplane. If the student places a ball on the floor, it stays there. There are no horizontal forces, and the ball remains at rest relative to the airplane. That is, $\vec{a} = \vec{0}$ in the airplane's reference frame when $\vec{F}_{net} = \vec{0}$. Newton's first law is satisfied, so this airplane is an inertial reference frame.

The physics student in **FIGURE 5.19b** conducts the same experiment during takeoff. He carefully places the ball on the floor just as the airplane starts to accelerate down the runway. You can imagine what happens. The ball rolls to the back of the plane as the passengers are being pressed back into their seats. Nothing exerts a horizontal contact force on the ball, yet the ball accelerates *in the plane's reference frame*. This violates Newton's first law, so the plane is *not* an inertial reference frame during takeoff.

In the first example, the plane is traveling with constant velocity. In the second, the plane is accelerating. **Accelerating reference frames are not inertial reference frames.** Consequently, Newton's laws are not valid in an accelerating reference frame.

The earth is not exactly an inertial reference frame because the earth rotates on its axis and orbits the sun. However, the earth's acceleration is so small that violations of Newton's laws can be measured only in very careful experiments. We will treat the earth and laboratories attached to the earth as inertial reference frames, an approximation that is exceedingly well justified.

To understand the motion of the passengers in a braking car, you need to measure velocities and accelerations *relative to the ground*. From the perspective of an observer on the ground, the body of a passenger in a braking car tries to continue moving forward with constant velocity, exactly as we would expect on the basis of Newton's first law, while his immediate surroundings are decelerating. The passenger is not "thrown" into the windshield. Instead, the windshield runs into the passenger!

Thinking About Force

It is important to identify correctly all the forces acting on an object. It is equally important not to include forces that do not really exist. We have established a number of criteria for identifying forces; the three critical ones are:

- A force has an agent. Something tangible and identifiable causes the force.
- Forces exist at the point of contact between the agent and the object experiencing the force (except for the few special cases of long-range forces).
- Forces exist due to interactions happening *now*, not due to what happened in the past.

Consider a bowling ball rolling along on a smooth floor. It is very tempting to think that a horizontal "force of motion" keeps it moving in the forward direction. But *nothing contacts the ball* except the floor. No agent is giving the ball a forward push. According to our definition, then, there is *no* forward "force of motion" acting on the ball. So what keeps it going? Recall our discussion of the first law: *No* cause is needed to keep an object moving at constant velocity. It continues to move forward simply because of its inertia.

A related problem occurs if you throw a ball. A pushing force was indeed required to accelerate the ball *as it was thrown*. But that force disappears the instant the ball loses contact with your hand. The force does not stick with the ball as the ball travels through the air. Once the ball has acquired a velocity, *nothing* is needed to keep it moving with that velocity.

There's no "force of motion" or any other forward force on this arrow. It continues to move because of inertia.

5.7 Free-Body Diagrams

Having discussed at length what is and is not a force, we are ready to assemble our knowledge about force and motion into a single diagram called a *free-body diagram*. You will learn in the next chapter how to write the equations of motion directly from the free-body diagram. Solution of the equations is a mathematical exercise—possibly a difficult one, but nonetheless an exercise that could be done by a computer. The *physics* of the problem, as distinct from the purely calculational aspects, are the steps that lead to the free-body diagram.

A **free-body diagram,** part of the *pictorial representation* of a problem, represents the object as a particle and shows *all* of the forces acting on the object.

TACTICS BOX 5.3   (MP)

Drawing a free-body diagram

❶ **Identify all forces acting on the object.** This step was described in Tactics Box 5.2.

❷ **Draw a coordinate system.** Use the axes defined in your pictorial representation.

❸ **Represent the object as a dot at the origin of the coordinate axes.** This is the particle model.

❹ **Draw vectors representing each of the identified forces.** This was described in Tactics Box 5.1. Be sure to label each force vector.

❺ **Draw and label the *net force* vector $\vec{F}_{net}$.** Draw this vector beside the diagram, not on the particle. Or, if appropriate, write $\vec{F}_{net} = \vec{0}$. Then check that $\vec{F}_{net}$ points in the same direction as the acceleration vector $\vec{a}$ on your motion diagram.

Exercises 24–29

EXAMPLE 5.4 | An elevator accelerates upward

An elevator, suspended by a cable, speeds up as it moves upward from the ground floor. Identify the forces and draw a free-body diagram of the elevator.

MODEL Model the elevator as a particle.

VISUALIZE

FIGURE 5.20 Free-body diagram of an elevator accelerating upward.

ASSESS The coordinate axes, with a vertical y-axis, are the ones we would use in a pictorial representation of the motion. The elevator is accelerating upward, so $\vec{F}_{net}$ must point upward. For this to be true, the magnitude of $\vec{T}$ must be larger than the magnitude of $\vec{F}_G$. The diagram has been drawn accordingly.

EXAMPLE 5.5 | An ice block shoots across a frozen lake

Bobby straps a small model rocket to a block of ice and shoots it across the smooth surface of a frozen lake. Friction is negligible. Draw a pictorial representation of the block of ice.

MODEL Model the block of ice as a particle. The pictorial representation consists of a motion diagram to determine $\vec{a}$, a force-identification picture, and a free-body diagram. The statement of the situation implies that friction is negligible.

VISUALIZE

FIGURE 5.21 Pictorial representation for a block of ice shooting across a frictionless frozen lake.

ASSESS The motion diagram tells us that the acceleration is in the positive x-direction. According to the rules of vector addition, this can be true only if the upward-pointing $\vec{n}$ and the downward-pointing $\vec{F}_G$ are equal in magnitude and thus cancel each other. The vectors have been drawn accordingly, and this leaves the net force vector pointing toward the right, in agreement with $\vec{a}$ from the motion diagram.

EXAMPLE 5.6 | A skier is pulled up a hill

A tow rope pulls a skier up a snow-covered hill at a constant speed. Draw a pictorial representation of the skier.

MODEL This is Example 5.2 again with the additional information that the skier is moving at constant speed. The skier will be modeled as a particle in *mechanical equilibrium.* If we were doing a kinematics problem, the pictorial representation would use a tilted coordinate system with the *x*-axis parallel to the slope, so we use these same tilted coordinate axes for the free-body diagram.

VISUALIZE

FIGURE 5.22 Pictorial representation for a skier being towed at a constant speed.

ASSESS We have shown $\vec{T}$ pulling parallel to the slope and $\vec{f}_k$, which opposes the direction of motion, pointing down the slope. $\vec{n}$ is perpendicular to the surface and thus along the *y*-axis. Finally, and this is important, the gravitational force $\vec{F}_G$ is *vertically* downward, *not* along the negative *y*-axis. In fact, you should convince yourself from the geometry that the angle θ between the $\vec{F}_G$ vector and the negative *y*-axis is the same as the angle θ of the incline above the horizontal. The skier moves in a straight line with constant speed, so $\vec{a} = \vec{0}$ and, from Newton's first law, $\vec{F}_{net} = \vec{0}$. Thus we have drawn the vectors such that the *y*-component of $\vec{F}_G$ is equal in magnitude to $\vec{n}$. Similarly, $\vec{T}$ must be large enough to match the negative *x*-components of both $\vec{f}_k$ and $\vec{F}_G$.

Free-body diagrams will be our major tool for the next several chapters. Careful practice with the workbook exercises and homework in this chapter will pay immediate benefits in the next chapter. Indeed, it is not too much to assert that a problem is half solved, or even more, when you complete the free-body diagram.

STOP TO THINK 5.5 An elevator suspended by a cable is moving upward and slowing to a stop. Which free-body diagram is correct?

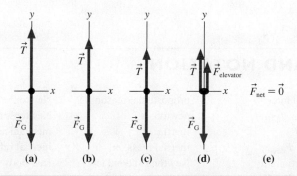

SUMMARY

The goal of Chapter 5 has been to learn about the connection between force and motion.

GENERAL PRINCIPLES

Newton's First Law

An object at rest will remain at rest, or an object that is moving will continue to move in a straight line with constant velocity, if and only if the net force on the object is zero.

$$\vec{F}_{net} = \vec{0}$$

$$\vec{a} = \vec{0}$$

The first law tells us that no "cause" is needed for motion. Uniform motion is the "natural state" of an object.

Newton's laws are valid only in inertial reference frames.

Newton's Zeroth Law

An object responds only to forces acting on it *at this instant.*

Newton's Second Law

An object with mass m has acceleration

$$\vec{a} = \frac{1}{m}\vec{F}_{net}$$

where $\vec{F}_{net} = \vec{F}_1 + \vec{F}_2 + \vec{F}_3 + \cdots$ is the vector sum of all the individual forces acting on the object.

$$\vec{F}_{net}$$

$$\vec{a}$$

The second law tells us that a net force causes an object to accelerate. This is the connection between force and motion.

IMPORTANT CONCEPTS

Acceleration is the link to kinematics.

From $\vec{F}_{net}$, find $\vec{a}$.

From a, find v and x.

$\vec{a} = \vec{0}$ is the condition for **equilibrium**.

An object **at rest** is in equilibrium.

So is an object with **constant velocity**.

Equilibrium occurs if and only if $\vec{F}_{net} = \vec{0}$.

Mass is the resistance of an object to acceleration. It is an intrinsic property of an object.

Mass is the inverse of the slope. Larger mass, smaller slope.

Force is a push or a pull on an object.

- Force is a vector, with a magnitude and a direction.
- Force requires an agent.
- Force is either a contact force or a long-range force.

KEY SKILLS

Identifying Forces

Forces are identified by locating the points where other objects touch the object of interest. These are points where contact forces are exerted. In addition, objects with mass feel a long-range gravitational force.

Thrust force $\vec{F}_{thrust}$

Gravity $\vec{F}_G$ Normal force $\vec{n}$

Free-Body Diagrams

A free-body diagram represents the object as a particle at the origin of a coordinate system. Force vectors are drawn with their tails on the particle. The net force vector is drawn beside the diagram.

TERMS AND NOTATION

mechanics	net force, $\vec{F}_{net}$	normal force, $\vec{n}$	proportional reasoning	Newton's third law
dynamics	superposition of forces	friction, $\vec{f}_k$ or $\vec{f}_s$	newton, N	Newton's first law
force, $\vec{F}$	gravitational force, $\vec{F}_G$	drag, $\vec{F}_{drag}$	inertia	mechanical equilibrium
agent	spring force, $\vec{F}_{Sp}$	thrust, $\vec{F}_{thrust}$	inertial mass, m	inertial reference frame
contact force	tension force, $\vec{T}$	proportionality	Newton's second law	free-body diagram
long-range force	ball-and-spring model	proportionality constant	Newton's zeroth law	

CONCEPTUAL QUESTIONS

1. An elevator suspended by a cable is descending at constant velocity. How many force vectors would be shown on a free-body diagram? Name them.

2. A compressed spring is pushing a block across a rough horizontal table. How many force vectors would be shown on a free-body diagram? Name them.

3. A brick is falling from the roof of a three-story building. How many force vectors would be shown on a free-body diagram? Name them.

4. In FIGURE Q5.4, block B is falling and dragging block A across a table. How many force vectors would be shown on a free-body diagram of block A? Name them.

FIGURE Q5.4

5. You toss a ball straight up in the air. Immediately after you let go of it, what force or forces are acting on the ball? For each force you name, (a) state whether it is a contact force or a long-range force and (b) identify the agent of the force.

6. A constant force applied to A causes A to accelerate at 5 m/s². The same force applied to B causes an acceleration of 3 m/s². Applied to C, it causes an acceleration of 8 m/s².
   a. Which object has the largest mass? Explain.
   b. Which object has the smallest mass?
   c. What is the ratio m_A/m_B of the mass of A to the mass of B?

7. An object experiencing a constant force accelerates at 10 m/s². What will the acceleration of this object be if
   a. The force is doubled? Explain.
   b. The mass is doubled?
   c. The force is doubled *and* the mass is doubled?

8. An object experiencing a constant force accelerates at 8 m/s². What will the acceleration of this object be if
   a. The force is halved? Explain.
   b. The mass is halved?
   c. The force is halved *and* the mass is halved?

9. If an object is at rest, can you conclude that there are no forces acting on it? Explain.

10. If a force is exerted on an object, is it possible for that object to be moving with constant velocity? Explain.

11. Is the statement "An object always moves in the direction of the net force acting on it" true or false? Explain.

12. Newton's second law says $\vec{F}_{net} = m\vec{a}$. So is $m\vec{a}$ a force? Explain.

13. Is it possible for the friction force on an object to be in the direction of motion? If so, give an example. If not, why not?

14. Suppose you press your physics book against a wall hard enough to keep it from moving. Does the friction force on the book point (a) into the wall, (b) out of the wall, (c) up, (d) down, or (e) is there no friction force? Explain.

15. FIGURE Q5.15 shows a hollow tube forming three-quarters of a circle. It is lying flat on a table. A ball is shot through the tube at high speed. As the ball emerges from the other end, does it follow path A, path B, or path C? Explain.

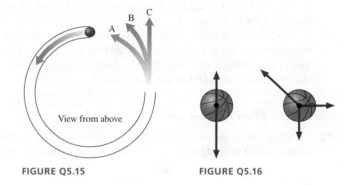

FIGURE Q5.15   FIGURE Q5.16

16. Which, if either, of the basketballs in FIGURE Q5.16 are in equilibrium? Explain.

17. Which of the following are inertial reference frames? Explain.
   a. A car driving at steady speed on a straight and level road.
   b. A car driving at steady speed up a 10° incline.
   c. A car speeding up after leaving a stop sign.
   d. A car driving at steady speed around a curve.

EXERCISES AND PROBLEMS

Exercises

Section 5.3 Identifying Forces

1. | A car is parked on a steep hill. Identify the forces on the car.

2. | A chandelier hangs from a chain in the middle of a dining room. Identify the forces on the chandelier.

3. | A baseball player is sliding into second base. Identify the forces on the baseball player.

4. ‖ A jet plane is speeding down the runway during takeoff. Air resistance is not negligible. Identify the forces on the jet.

5. ‖ An arrow has just been shot from a bow and is now traveling horizontally. Air resistance is not negligible. Identify the forces on the arrow.

Section 5.4 What Do Forces Do?

6. | Two rubber bands cause an object to accelerate with acceleration *a*. How many rubber bands are needed to cause an object with half the mass to accelerate three times as quickly?

7. | Two rubber bands pulling on an object cause it to accelerate at 1.2 m/s².
   a. What will be the object's acceleration if it is pulled by four rubber bands?
   b. What will be the acceleration of two of these objects glued together if they are pulled by two rubber bands?

8. ‖ FIGURE EX5.8 shows acceleration-versus-force graphs for two objects pulled by rubber bands. What is the mass ratio m_1/m_2?

FIGURE EX5.8

FIGURE EX5.9

9. ‖ FIGURE EX5.9 shows an acceleration-versus-force graph for three objects pulled by rubber bands. The mass of object 2 is 0.20 kg. What are the masses of objects 1 and 3? Explain your reasoning.

10. ‖ For an object starting from rest and accelerating with constant acceleration, distance traveled is proportional to the square of the time. If an object travels 2.0 furlongs in the first 2.0 s, how far will it travel in the first 4.0 s?

11. ‖ You'll learn in Chapter 25 that the *potential energy* of two electric charges is inversely proportional to the distance between them. Two charges 30 nm apart have 1.0 J of potential energy. What is their potential energy if they are 20 nm apart?

Section 5.5 Newton's Second Law

12. | FIGURE EX5.12 shows an acceleration-versus-force graph for a 200 g object. What force values go in the blanks on the horizontal scale?

FIGURE EX5.12

FIGURE EX5.13

13. | FIGURE EX5.13 shows an acceleration-versus-force graph for a 500 g object. What acceleration values go in the blanks on the vertical scale?

14. ‖ FIGURE EX5.14 shows the acceleration of objects of different mass that experience the same force. What is the magnitude of the force?

FIGURE EX5.14

FIGURE EX5.15

15. | FIGURE EX5.15 shows an object's acceleration-versus-force graph. What is the object's mass?

16. | Based on the information in Table 5.1, *estimate*
   a. The weight of a laptop computer.
   b. The propulsion force of a bicycle.

Section 5.6 Newton's First Law

Exercises 17 through 19 show two of the three forces acting on an object in equilibrium. Redraw the diagram, showing all three forces. Label the third force $\vec{F}_3$.

FIGURE EX5.17   FIGURE EX5.18   FIGURE EX5.19

Section 5.7 Free-Body Diagrams

Exercises 20 through 22 show a free-body diagram. For each, write a short description of a real object for which this would be the correct free-body diagram. Use Examples 5.4, 5.5, and 5.6 as examples of what a description should be like.

FIGURE EX5.20   FIGURE EX5.21   FIGURE EX5.22

Exercises 23 through 27 describe a situation. For each, identify all forces acting on the object and draw a free-body diagram of the object.

23. | A cat is sitting on a window sill.
24. | An ice hockey puck glides across frictionless ice.
25. | Your physics textbook is sliding across the table.
26. | A steel beam, suspended by a single cable, is being lowered by a crane at a steadily decreasing speed.
27. | A jet plane is accelerating down the runway during takeoff. Friction is negligible, but air resistance is not.

Problems

28. | Redraw the two motion diagrams shown in FIGURE P5.28, then draw a vector beside each one to show the direction of the net force acting on the object. Explain your reasoning.

FIGURE P5.28   (a)   (b)

29. | A single force with x-component F_x acts on a 2.0 kg object as it moves along the x-axis. The object's acceleration graph (a_x versus t) is shown in FIGURE P5.29. Draw a graph of F_x versus t.

FIGURE P5.29   FIGURE P5.30

30. ‖ A single force with x-component F_x acts on a 500 g object as it moves along the x-axis. The object's acceleration graph (a_x versus t) is shown in FIGURE P5.30. Draw a graph of F_x versus t.

31. | A single force with x-component F_x acts on a 2.0 kg object as it moves along the x-axis. A graph of F_x versus t is shown in FIGURE P5.31. Draw an acceleration graph (a_x versus t) for this object.

FIGURE P5.31   FIGURE P5.32

32. ‖ A single force with x-component F_x acts on a 500 g object as it moves along the x-axis. A graph of F_x versus t is shown in FIGURE P5.32. Draw an acceleration graph (a_x versus t) for this object.

33. | A constant force is applied to an object, causing the object to accelerate at 8.0 m/s^2. What will the acceleration be if
   a. The force is doubled?
   b. The object's mass is doubled?
   c. The force and the object's mass are both doubled?
   d. The force is doubled and the object's mass is halved?

34. | A constant force is applied to an object, causing the object to accelerate at 10 m/s^2. What will the acceleration be if
   a. The force is halved?
   b. The object's mass is halved?
   c. The force and the object's mass are both halved?
   d. The force is halved and the object's mass is doubled?

Problems 35 through 40 show a free-body diagram. For each:
   a. Identify the direction of the acceleration vector $\vec{a}$ and show it as a vector next to your diagram. Or, if appropriate, write $\vec{a} = \vec{0}$.
   b. If possible, identify the direction of the velocity vector $\vec{v}$ and show it as a labeled vector.
   c. Write a short description of a real object for which this is the correct free-body diagram. Use Examples 5.4, 5.5, and 5.6 as models of what a description should be like.

FIGURE P5.35   FIGURE P5.36

FIGURE P5.37   FIGURE P5.38

FIGURE P5.39   FIGURE P5.40

41. ‖ In lab, you propel a cart with four known forces while using an ultrasonic motion detector to measure the cart's acceleration. Your data are as follows:

Force (N)	Acceleration (m/s^2)
0.25	0.5
0.50	0.8
0.75	1.3
1.00	1.8

   a. How should you graph these data so as to determine the mass of the cart from the slope of the line? That is, what values should you graph on the horizontal axis and what on the vertical axis?
   b. Is there another data point that would be reasonable to add, even though you made no measurements? If so, what is it?
   c. What is your best determination of the cart's mass?

Problems 42 through 52 describe a situation. For each, draw a motion diagram, a force-identification diagram, and a free-body diagram.
42. | An elevator, suspended by a single cable, has just left the tenth floor and is speeding up as it descends toward the ground floor.
43. ‖ A rocket is being launched straight up. Air resistance is not negligible.
44. | A Styrofoam ball has just been shot straight up. Air resistance is not negligible.
45. | You are a rock climber going upward at a steady pace on a vertical wall.

46. ‖   You've slammed on the brakes and your car is skidding to a stop while going down a 20° hill.

47. | You've just kicked a rock on the sidewalk and it is now sliding along the concrete.

48. ‖   You've jumped down from a platform. Your feet are touching the ground and your knees are flexing as you stop.

49. ‖   You are bungee jumping from a high bridge. You are moving downward while the bungee cord is stretching.

50. ‖   Your friend went for a loop-the-loop ride at the amusement park. Her car is upside down at the top of the loop.

51. ‖   A spring-loaded gun shoots a plastic ball. The trigger has just been pulled and the ball is starting to move down the barrel. The barrel is horizontal.

52. ‖   A person on a bridge throws a rock straight down toward the water. The rock has just been released.

53. ‖   The leaf hopper, champion jumper of the insect world, can
BIO   jump straight up at 4 m/s^2. The jump itself lasts a mere 1 ms before the insect is clear of the ground.
   a. Draw a free-body diagram of this mighty leaper while the jump is taking place.
   b. While the jump is taking place, is the force of the ground on the leaf hopper greater than, less than, or equal to the force of gravity on the leaf hopper? Explain.

54. ‖   A bag of groceries is on the seat of your car as you stop for a stop light. The bag does not slide. Draw a motion diagram, a force-identification diagram, and a free-body diagram for the bag.

55. ‖   A heavy box is in the back of a truck. The truck is accelerating to the right. Draw a motion diagram, a force-identification diagram, and a free-body diagram for the box.

56. ‖   If a car stops suddenly, you feel "thrown forward." We'd like to understand what happens to the passengers as a car stops.

Imagine yourself sitting on a *very* slippery bench inside a car. This bench has no friction, no seat back, and there's nothing for you to hold onto.
   a. Draw a picture and identify all of the forces acting on you as the car travels at a perfectly steady speed on level ground.
   b. Draw your free-body diagram. Is there a net force on you? If so, in which direction?
   c. Repeat parts a and b with the car slowing down.
   d. Describe what happens to you as the car slows down.
   c. Use Newton's laws to explain why you seem to be "thrown forward" as the car stops. Is there really a force pushing you forward?
   f. Suppose now that the bench is not slippery. As the car slows down, you stay on the bench and don't slide off. What force is responsible for your deceleration? In which direction does this force point? Include a free-body diagram as part of your answer.

57. ‖   A rubber ball bounces. We'd like to understand *how* the ball bounces.
   a. A rubber ball has been dropped and is bouncing off the floor. Draw a motion diagram of the ball during the brief time interval that it is in contact with the floor. Show 4 or 5 frames as the ball compresses, then another 4 or 5 frames as it expands. What is the direction of $\vec{a}$ during each of these parts of the motion?
   b. Draw a picture of the ball in contact with the floor and identify all forces acting on the ball.
   c. Draw a free-body diagram of the ball during its contact with the ground. Is there a net force acting on the ball? If so, in which direction?
   d. Write a paragraph in which you describe what you learned from parts a to c and in which you answer the question: How does a ball bounce?

6 Dynamics I: Motion Along a Line

The powerful thrust of the jet engines accelerates this enormous plane to a speed of over 150 mph in less than a mile.

IN THIS CHAPTER, you will learn to solve linear force-and-motion problems.

How are Newton's laws used to solve problems?

Newton's first and second laws are vector equations. To use them,

- Draw a free-body diagram.
- Read the x- and y-components of the forces directly off the free-body diagram.
- Use $\sum F_x = ma_x$ and $\sum F_y = ma_y$.

How are dynamics problems solved?

A net force on an object causes the object to accelerate.

- Identify the forces and draw a free-body diagram.
- Use Newton's second law to find the object's acceleration.
- Use kinematics for velocity and position.

Normal $\vec{n}$
Friction $\vec{f_s}$ Gravity $\vec{F_G}$

« LOOKING BACK Sections 2.4–2.6 Kinematics

How are equilibrium problems solved?

An object at rest or moving with constant velocity is in equilibrium with no net force.

- Identify the forces and draw a free-body diagram.
- Use Newton's second law with $a = 0$ to solve for unknown forces.

« LOOKING BACK Sections 5.1–5.2 Forces

What are mass and weight?

Mass and weight are not the same.

- Mass describes an object's inertia. Loosely speaking, it is the amount of matter in an object. It is the same everywhere.
- Gravity is a force.
- Weight is the result of weighing an object on a scale. It depends on mass, gravity, and acceleration.

How do we model friction and drag?

Friction and drag are complex forces, but we will develop simple models of each.

Kinetic friction

- Static, kinetic, and rolling friction depend on the coefficients of friction but not on the object's speed.
- Drag depends on the *square* of an object's speed and on its cross-section area.
- Falling objects reach terminal speed when drag and gravity are balanced.

How do we solve problems?

We will develop and use a four-part problem-solving strategy:

- Model the problem, using information about objects and forces.
- Visualize the situation with a pictorial representation.
- Set up and solve the problem with Newton's laws.
- Assess the result to see if it is reasonable.

6.1 The Equilibrium Model

Kinematics is a description of *how* an object moves. But our goal is deeper: We would like an explanation for *why* an object moves as it does. Galileo and Newton discovered that motion is determined by forces. In the absence of a net force, an object is at rest or moves with constant velocity. **Its acceleration is zero,** and this is the basis for our first explanatory model: the **equilibrium model.**

MODEL 6.1

Mechanical equilibrium

For objects on which the net force is zero.

- Model the object as a particle with no acceleration.
  - A particle at rest is in equilibrium.
  - A particle moving in a straight line at constant speed is also in equilibrium.
- Mathematically: $\vec{a} = \vec{0}$ in equilibrium; thus
  - **Newton's second law** is $\vec{F}_{\text{net}} = \sum_i \vec{F}_i = \vec{0}$.
  - The forces are "read" from the free-body diagram,
- Limitations: Model fails if the forces aren't balanced.

$\vec{F}_{\text{net}} = \vec{0}$
$\vec{a} = \vec{0}$

The object is at rest or moves with constant velocity.

Newton's laws are *vector equations.* The requirement for equilibrium, $\vec{F}_{\text{net}} = \vec{0}$ and thus $\vec{a} = \vec{0}$, is a shorthand way of writing two simultaneous equations:

$$(F_{\text{net}})_x = \sum_i (F_i)_x = 0 \quad \text{and} \quad (F_{\text{net}})_y = \sum_i (F_i)_y = 0 \tag{6.1}$$

In other words, the sum of all x-components and the sum of all y-components must simultaneously be zero. Although real-world situations often have forces pointing in three dimensions, thus requiring a third equation for the z-component of $\vec{F}_{\text{net}}$, we will restrict ourselves for now to problems that can be analyzed in two dimensions.

NOTE The equilibrium condition of Equations 6.1 applies only to particles, which cannot rotate. Equilibrium of an extended object, which can rotate, requires an additional condition that we will study in Chapter 12.

Equilibrium problems occur frequently. Let's look at a couple of examples.

The concept of equilibrium is essential for the engineering analysis of stationary objects such as bridges.

EXAMPLE 6.1 | Finding the force on the kneecap

Your kneecap (patella) is attached by a tendon to your quadriceps muscle. This tendon pulls at a 10° angle relative to the femur, the bone of your upper leg. The patella is also attached to your lower leg (tibia) by a tendon that pulls parallel to the leg. To balance these forces, the end of your femur pushes outward on the patella. Bending your knee increases the tension in the tendons, and both have a tension of 60 N when the knee is bent to make a 70° angle between the upper and lower leg. What force does the femur exert on the kneecap in this position?

MODEL Model the kneecap as a particle in equilibrium.

FIGURE 6.1 Pictorial representation of the kneecap in equilibrium.

VISUALIZE FIGURE 6.1 shows how to draw a pictorial representation. We've chosen to align the x-axis with the femur. The three forces—shown on the free-body diagram—are labeled $\vec{T}_1$ and $\vec{T}_2$ for the tensions and $\vec{F}$ for the femur's push. Notice that we've *defined* angle θ to indicate the direction of the femur's force on the kneecap.

SOLVE This is an equilibrium problem, with three forces on the kneecap that must sum to zero. For $\vec{a} = 0$, Newton's second law, written in component form, is

$$(F_{net})_x = \sum_i (F_i)_x = T_{1x} + T_{2x} + F_x = 0$$

$$(F_{net})_y = \sum_i (F_i)_y = T_{1y} + T_{2y} + F_y = 0$$

NOTE You might have been tempted to write $-T_{1x}$ in the equation since $\vec{T}_1$ points to the left. But the net force, by definition, is the *sum* of all the individual forces. The fact that $\vec{T}_1$ points to the left will be taken into account when we *evaluate* the components.

The components of the force vectors can be evaluated directly from the free-body diagram:

$$T_{1x} = -T_1 \cos 10° \qquad T_{1y} = T_1 \sin 10°$$

$$T_{2x} = -T_2 \cos 70° \qquad T_{2y} = -T_2 \sin 70°$$

$$F_x = F \cos \theta \qquad F_y = F \sin \theta$$

This is where signs enter, with T_{1x} being assigned a negative value because $\vec{T}_1$ points to the left. Similarly, $\vec{T}_2$ points both to the left and down, so both T_{2x} and T_{2y} are negative. With these components, Newton's second law becomes

$$-T_1 \cos 10° - T_2 \cos 70° + F \cos \theta = 0$$

$$T_1 \sin 10° - T_2 \sin 70° + F \sin \theta = 0$$

These are two simultaneous equations for the two unknowns F and θ. We will encounter equations of this form on many occasions,

so make a note of the method of solution. First, rewrite the two equations as

$$F \cos \theta = T_1 \cos 10° + T_2 \cos 70°$$

$$F \sin \theta = -T_1 \sin 10° + T_2 \sin 70°$$

Next, divide the second equation by the first to eliminate F:

$$\frac{F \sin \theta}{F \cos \theta} = \tan \theta = \frac{-T_1 \sin 10° + T_2 \sin 70°}{T_1 \cos 10° + T_2 \cos 70°}$$

Then solve for θ:

$$\theta = \tan^{-1}\left(\frac{-T_1 \sin 10° + T_2 \sin 70°}{T_1 \cos 10° + T_2 \cos 70°}\right)$$

$$= \tan^{-1}\left(\frac{-(60 \text{ N}) \sin 10° + (60 \text{ N}) \sin 70°}{(60 \text{ N}) \cos 10° + (60 \text{ N}) \cos 70°}\right) = 30°$$

Finally, use θ to find F:

$$F = \frac{T_1 \cos 10° + T_2 \cos 70°}{\cos \theta}$$

$$= \frac{(60 \text{ N}) \cos 10° + (60 \text{ N}) \cos 70°}{\cos 30°} = 92 \text{ N}$$

The question asked What force? and force is a vector, so we must specify both the magnitude and the direction. With the knee in this position, the femur exerts a force $\vec{F} = (92 \text{ N}, 30° \text{ above the femur})$ on the kneecap.

ASSESS The magnitude of the force would be 0 N if the leg were straight, 120 N if the knee could be bent 180° so that the two tendons pull in parallel. The knee is closer to fully bent than to straight, so we would expect a femur force between 60 N and 120 N. Thus the calculated magnitude of 92 N seems reasonable.

EXAMPLE 6.2 | Towing a car up a hill

A car with a weight of 15,000 N is being towed up a 20° slope at constant velocity. Friction is negligible. The tow rope is rated at 6000 N maximum tension. Will it break?

MODEL Model the car as a particle in equilibrium.

VISUALIZE Part of our analysis of the problem statement is to determine which quantity or quantities allow us to answer the yes-or-no

question. In this case, we need to calculate the tension in the rope. FIGURE 6.2 shows the pictorial representation. Note the similarities to Examples 5.2 and 5.6 in Chapter 5, which you may want to review.

We noted in Chapter 5 that the weight of an object at rest is the magnitude F_G of the gravitational force acting on it, and that information has been listed as known.

FIGURE 6.2 Pictorial representation of a car being towed up a hill.

Continued

SOLVE The free-body diagram shows forces $\vec{T}$, $\vec{n}$, and $\vec{F}_G$ acting on the car. Newton's second law with $\vec{a} = \vec{0}$ is

$$(F_{net})_x = \sum F_x = T_x + n_x + (F_G)_x = 0$$

$$(F_{net})_y = \sum F_y = T_y + n_y + (F_G)_y = 0$$

From here on, we'll use $\sum F_x$ and $\sum F_y$, without the label i, as a simple shorthand notation to indicate that we're adding all the x-components and all the y-components of the forces.

We can find the components directly from the free-body diagram:

$$T_x = T \qquad\qquad T_y = 0$$

$$n_x = 0 \qquad\qquad n_y = n$$

$$(F_G)_x = -F_G \sin\theta \qquad (F_G)_y = -F_G \cos\theta$$

NOTE The gravitational force has both x- and y-components in this coordinate system, both of which are negative due to the direction of the vector $\vec{F}_G$. You'll see this situation often, so be sure you understand where $(F_G)_x$ and $(F_G)_y$ come from.

With these components, the second law becomes

$$T - F_G \sin\theta = 0$$

$$n - F_G \cos\theta = 0$$

The first of these can be rewritten as

$$T = F_G \sin\theta = (15,000 \text{ N}) \sin 20° = 5100 \text{ N}$$

Because $T < 6000$ N, we conclude that the rope will *not* break. It turned out that we did not need the y-component equation in this problem.

ASSESS Because there's no friction, it would not take *any* tension force to keep the car rolling along a horizontal surface ($\theta = 0°$). At the other extreme, $\theta = 90°$, the tension force would need to equal the car's weight ($T = 15,000$ N) to lift the car straight up at constant velocity. The tension force for a 20° slope should be somewhere in between, and 5100 N is a little less than half the weight of the car. That our result is reasonable doesn't prove it's right, but we have at least ruled out careless errors that give unreasonable results.

6.2 Using Newton's Second Law

The essence of Newtonian mechanics, introduced in « Section 5.4, can be expressed in two steps:

- The forces acting on an object determine its acceleration $\vec{a} = \vec{F}_{net}/m$.
- The object's trajectory can be determined by using $\vec{a}$ in the equations of kinematics.

These two ideas are the basis of a problem-solving strategy.

PROBLEM-SOLVING STRATEGY 6.1

Newtonian mechanics

MODEL Model the object as a particle. Make other simplifications depending on what kinds of forces are acting.

VISUALIZE Draw a **pictorial representation**.

- Show important points in the motion with a sketch, establish a coordinate system, define symbols, and identify what the problem is trying to find.
- Use a motion diagram to determine the object's acceleration vector $\vec{a}$. The acceleration is zero for an object in equilibrium.
- Identify all forces acting on the object *at this instant* and show them on a free-body diagram.
- It's OK to go back and forth between these steps as you visualize the situation.

SOLVE The mathematical representation is based on Newton's second law:

$$\vec{F}_{net} = \sum_i \vec{F}_i = m\vec{a}$$

The forces are "read" directly from the free-body diagram. Depending on the problem, either

- Solve for the acceleration, then use kinematics to find velocities and positions; or
- Use kinematics to determine the acceleration, then solve for unknown forces.

ASSESS Check that your result has correct units and significant figures, is reasonable, and answers the question.

Exercise 23

Newton's second law is a vector equation. To apply the step labeled Solve, you must write the second law as two simultaneous equations:

$$(F_{net})_x = \sum F_x = ma_x$$
$$(F_{net})_y = \sum F_y = ma_y \qquad (6.2)$$

The primary goal of this chapter is to illustrate the use of this strategy.

EXAMPLE 6.3 | Speed of a towed car

A 1500 kg car is pulled by a tow truck. The tension in the tow rope is 2500 N, and a 200 N friction force opposes the motion. If the car starts from rest, what is its speed after 5.0 seconds?

MODEL Model the car as an accelerating particle. We'll assume, as part of our *interpretation* of the problem, that the road is horizontal and that the direction of motion is to the right.

VISUALIZE FIGURE 6.3 shows the pictorial representation. We've established a coordinate system and defined symbols to represent kinematic quantities. We've identified the speed v_1, rather than the velocity v_{1x}, as what we're trying to find.

SOLVE We begin with Newton's second law:

$$(F_{net})_x = \sum F_x = T_x + f_x + n_x + (F_G)_x = ma_x$$
$$(F_{net})_y = \sum F_y = T_y + f_y + n_y + (F_G)_y = ma_y$$

All four forces acting on the car have been included in the vector sum. The equations are perfectly general, with $+$ signs everywhere, because the four vectors are *added* to give $\vec{F}_{net}$. We can now "read" the vector components from the free-body diagram:

$$T_x = +T \qquad T_y = 0 \qquad n_x = 0 \qquad n_y = +n$$
$$f_x = -f \qquad f_y = 0 \qquad (F_G)_x = 0 \qquad (F_G)_y = -F_G$$

The signs, which we had to insert by hand, depend on which way the vectors point. Substituting these into the second-law equations and dividing by m give

$$a_x = \frac{1}{m}(T - f)$$
$$= \frac{1}{1500 \text{ kg}}(2500 \text{ N} - 200 \text{ N}) = 1.53 \text{ m/s}^2$$
$$a_y = \frac{1}{m}(n - F_G)$$

NOTE Newton's second law has allowed us to determine a_x exactly but has given only an algebraic expression for a_y. However, we know *from the motion diagram* that $a_y = 0$! That is, the motion is purely along the x-axis, so there is *no* acceleration along the y-axis. The requirement $a_y = 0$ allows us to conclude that $n = F_G$.

Because a_x is a constant 1.53 m/s², we can finish by using constant-acceleration kinematics to find the velocity:

$$v_{1x} = v_{0x} + a_x \, \Delta t$$
$$= 0 + (1.53 \text{ m/s}^2)(5.0 \text{ s}) = 7.7 \text{ m/s}$$

The problem asked for the *speed* after 5.0 s, which is $v_1 = 7.7$ m/s.

ASSESS 7.7 m/s ≈ 15 mph, a quite reasonable speed after 5 s of acceleration.

FIGURE 6.3 Pictorial representation of a car being towed.

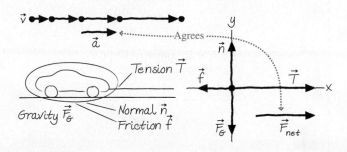

If all the forces acting on an object are constant, as in the last example, then the object moves with constant acceleration and we can deploy the uniform-acceleration model of kinetics. Now not all forces are constant—you will later meet forces that vary with position or time—but in many situations it is reasonable to model the motion as being due to constant forces. The **constant-force model** will be our most important dynamics model for the next several chapters.

MODEL 6.2

Constant force

For objects on which the net force is constant.

- Model the object as a particle with uniform acceleration.
  - The particle accelerates in the direction of the net force.
- Mathematically:
  - **Newton's second law** is $\vec{F}_{net} = \sum_i \vec{F}_i = m\vec{a}$.
  - Use the kinematics of constant acceleration.
- Limitations: Model fails if the forces aren't constant.

$$\vec{a} = \frac{\vec{F}_{net}}{m}$$

The object undergoes uniform acceleration.

EXAMPLE 6.4 | Altitude of a rocket

A 500 g model rocket with a weight of 4.90 N is launched straight up. The small rocket motor burns for 5.00 s and has a steady thrust of 20.0 N. What maximum altitude does the rocket reach?

MODEL We'll model the rocket as a particle acted on by constant forces by neglecting the velocity-dependent air resistance (rockets have very aerodynamic shapes) and neglecting the mass loss of the burned fuel.

VISUALIZE The pictorial representation of FIGURE 6.4 finds that this is a two-part problem. First, the rocket accelerates straight up. Second, the rocket continues going up as it slows down, a free-fall situation. The maximum altitude is at the end of the second part of the motion.

SOLVE We now know what the problem is asking, have established relevant symbols and coordinates, and know what the forces are.

We begin the mathematical representation by writing Newton's second law, in component form, as the rocket accelerates upward. The free-body diagram shows two forces, so

$$(F_{net})_x = \sum F_x = (F_{thrust})_x + (F_G)_x = ma_{0x}$$
$$(F_{net})_y = \sum F_y = (F_{thrust})_y + (F_G)_y = ma_{0y}$$

The fact that vector $\vec{F}_G$ points downward—and which might have tempted you to use a minus sign in the y-equation—will be taken into account when we *evaluate* the components. None of the vectors in this problem has an x-component, so only the y-component of the second law is needed. We can use the free-body diagram to see that

$$(F_{thrust})_y = +F_{thrust}$$
$$(F_G)_y = -F_G$$

FIGURE 6.4 Pictorial representation of a rocket launch.

Known
$y_0 = 0$ m	$F_{thrust} = 20.0$ N
$v_{0y} = 0$ m/s	$F_G = 4.90$ N
$t_0 = 0$ s	
$t_1 = 5.00$ s	$a_{1y} = -9.80$ m/s^2
$v_{2y} = 0$ (top)	$m = 500$ g $= 0.500$ kg

Find

y_2

This is the point at which the directional information about the force vectors enters. The y-component of the second law is then

$$a_{0y} = \frac{1}{m}(F_{\text{thrust}} - F_G)$$

$$= \frac{20.0 \text{ N} - 4.90 \text{ N}}{0.500 \text{ kg}} = 30.2 \text{ m/s}^2$$

Notice that we converted the mass to SI units of kilograms before doing any calculations and that, because of the definition of the newton, the division of newtons by kilograms automatically gives the correct SI units of acceleration.

The acceleration of the rocket is constant until it runs out of fuel, so we can use constant-acceleration kinematics to find the altitude and velocity at burnout ($\Delta t = t_1 = 5.00$ s):

$$y_1 = y_0 + v_{0y}\,\Delta t + \tfrac{1}{2}a_{0y}(\Delta t)^2$$

$$= \tfrac{1}{2}a_{0y}(\Delta t)^2 = 377 \text{ m}$$

$$v_{1y} = v_{0y} + a_{0y}\,\Delta t = a_{0y}\,\Delta t = 151 \text{ m/s}$$

The only force on the rocket after burnout is gravity, so the second part of the motion is free fall. We do not know how long it takes to reach the top, but we do know that the final velocity is $v_{2y} = 0$. Constant-acceleration kinematics with $a_{1y} = -g$ gives

$$v_{2y}^2 = 0 = v_{1y}^2 - 2g\,\Delta y = v_{1y}^2 - 2g(y_2 - y_1)$$

which we can solve to find

$$y_2 = y_1 + \frac{v_{1y}^2}{2g} = 377 \text{ m} + \frac{(151 \text{ m/s})^2}{2(9.80 \text{ m/s}^2)}$$

$$= 1540 \text{ m} = 1.54 \text{ km}$$

ASSESS The maximum altitude reached by this rocket is 1.54 km, or just slightly under one mile. While this does not seem unreasonable for a high-acceleration rocket, the neglect of air resistance was probably not a terribly realistic assumption.

STOP TO THINK 6.1 A Martian lander is approaching the surface. It is slowing its descent by firing its rocket motor. Which is the correct free-body diagram?

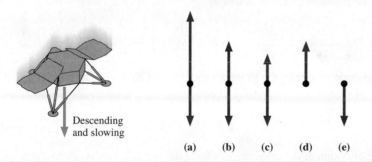

Descending and slowing

(a)   (b)   (c)   (d)   (e)

6.3 Mass, Weight, and Gravity

Ordinary language does not make a large distinction between mass and weight. However, these are separate and distinct concepts in science and engineering. We need to understand how they differ, and how they're related to gravity, if we're going to think clearly about force and motion.

Mass: An Intrinsic Property

Mass, you'll recall from « Section 5.4, is a scalar quantity that describes an object's inertia. Loosely speaking, it also describes the amount of matter in an object. **Mass is an intrinsic property of an object.** It tells us something about the object, regardless of where the object is, what it's doing, or whatever forces may be acting on it.

A *pan balance*, shown in FIGURE 6.5, is a device for measuring mass. Although a pan balance requires gravity to function, it does not depend on the strength of gravity. Consequently, the pan balance would give the same result on another planet.

FIGURE 6.5 A pan balance measures mass.

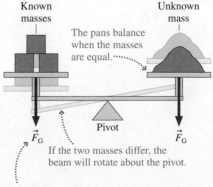

Known masses

Unknown mass

The pans balance when the masses are equal.

$\vec{F}_G$ Pivot $\vec{F}_G$

If the two masses differ, the beam will rotate about the pivot.

Both pans are pulled down by the force of gravity.

Gravity: A Force

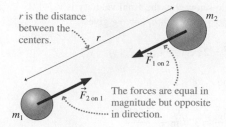

FIGURE 6.6 Newton's law of gravity.

r is the distance between the centers.

$\vec{F}_{1 \text{ on } 2}$

$\vec{F}_{2 \text{ on } 1}$

The forces are equal in magnitude but opposite in direction.

The idea of gravity has a long and interesting history intertwined with our evolving ideas about the solar system. It was Newton who—along with discovering his three laws of motion—first recognized that **gravity is an attractive, long-range force between** *any* **two objects.**

FIGURE 6.6 shows two objects with masses m_1 and m_2 separated by distance r. Each object pulls on the other with a force given by *Newton's law of gravity:*

$$F_{1 \text{ on } 2} = F_{2 \text{ on } 1} = \frac{Gm_1m_2}{r^2} \qquad \text{(Newton's law of gravity)} \qquad (6.3)$$

where $G = 6.67 \times 10^{-11}$ N m²/kg², called the *gravitational constant,* is one of the basic constants of nature. Notice that gravity is *not* a constant force—the force gets weaker as the distance between the objects increases.

The gravitational force between two human-sized objects is minuscule, completely insignificant in comparison with other forces. That's why you're not aware of being tugged toward everything around you. Only when one or both objects are planet-sized or larger does gravity become an important force. Indeed, Chapter 13 will explore in detail the application of Newton's law of gravity to the orbits of satellites and planets.

For objects moving near the surface of the earth (or other planet), things like balls and cars and planes that we'll be studying in the next few chapters, we can make the **flat-earth approximation** shown in FIGURE 6.7. That is, if the object's height above the surface is very small in comparison with the size of the planet, then the curvature of the surface is not noticeable and there's virtually no difference between r and the planet's radius R. Consequently, a very good approximation for the gravitational force of the planet on mass m is simply

FIGURE 6.7 Gravity near the surface of a planet.

$h \ll R$

$\vec{F}_G$

The surface of the planet seems to be flat for objects very near the surface, with $h \ll R$.

Center of planet of mass M

$$\vec{F}_G = \vec{F}_{\text{planet on } m} = \left(\frac{GMm}{R^2}, \text{ straight down} \right) = (mg, \text{ straight down}) \qquad (6.4)$$

The magnitude or size of the gravitational force is $F_G = mg$, where the quantity g—a property of the planet—is defined to be

$$g = \frac{GM}{R^2} \qquad (6.5)$$

Also, the direction of the gravitational force defines what we *mean* by "straight down."

But why did we choose to call it g, a symbol we've already used for free-fall acceleration? To see the connection, recall that free fall is motion under the influence of gravity only. FIGURE 6.8 shows the free-body diagram of an object in free fall near the surface of a planet. With $\vec{F}_{\text{net}} = \vec{F}_G$, Newton's second law predicts the acceleration to be

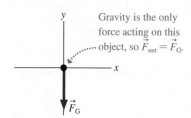

FIGURE 6.8 The free-body diagram of an object in free fall.

Gravity is the only force acting on this object, so $\vec{F}_{\text{net}} = \vec{F}_G$.

$\vec{F}_G$

$$\vec{a}_{\text{free fall}} = \frac{\vec{F}_{\text{net}}}{m} = \frac{\vec{F}_G}{m} = (g, \text{ straight down}) \qquad (6.6)$$

Because g is a property of the planet, independent of the object, **all objects on the same planet, regardless of mass, have the same free-fall acceleration.** We introduced this idea in Chapter 2 as an experimental discovery of Galileo, but now we see that the mass independence of $\vec{a}_{\text{free fall}}$ is a prediction of Newton's law of gravity.

But does Newton's law predict the correct value, which we know from experiment to be $g = |a_{\text{free fall}}| = 9.80$ m/s²? We can use the average radius ($R_{\text{earth}} = 6.37 \times 10^6$ m) and mass ($M_{\text{earth}} = 5.98 \times 10^{24}$ kg) of the earth to calculate

$$g_{\text{earth}} = \frac{GM_{\text{earth}}}{(R_{\text{earth}})^2} = \frac{(6.67 \times 10^{-11} \text{ N m}^2/\text{kg}^2)(5.98 \times 10^{24} \text{ kg})}{(6.37 \times 10^6 \text{ m})^2} = 9.83 \text{ N/kg}$$

You should convince yourself that N/kg is equivalent to m/s², so $g_{\text{earth}} = 9.83$ m/s².

NOTE Astronomical data are provided inside the back cover of the book.

Newton's prediction is very close, but it's not quite right. The free-fall acceleration *would* be 9.83 m/s^2 on a stationary earth, but, in reality, the earth is rotating on its axis. The "missing" 0.03 m/s^2 is due to the earth's rotation, a claim we'll justify when we study circular motion in Chapter 8. Because we're on the outside of a rotating sphere, rather like being on the outside edge of a merry-go-round, the effect of rotation is to "weaken" gravity.

Strictly speaking, Newton's laws of motion are not valid in an earth-based reference frame because it is rotating and thus is not an inertial reference frame. Fortunately, we can use Newton's laws to analyze motion near the earth's surface, and we can use $F_G = mg$ for the gravitational force *if* we use $g = |a_{\text{free fall}}| = 9.80$ m/s^2 rather than $g = g_{\text{earth}}$. (This assertion is proved in more advanced classes.) In our rotating reference frame, $\vec{F}_G$ is the *effective gravitational force,* the true gravitational force given by Newton's law of gravity plus a small correction due to our rotation. This is the force to show on free-body diagrams and use in calculations.

Weight: A Measurement

When you weigh yourself, you stand on a *spring scale* and compress a spring. The reading of a spring scale, such as the one shown in FIGURE 6.9, is F_{Sp}, the magnitude of the upward force the spring is exerting.

With that in mind, let's define the **weight** of an object to be the reading F_{Sp} of a calibrated spring scale when the object is at rest relative to the scale. That is, **weight is a measurement, the result of "weighing" an object.** Because F_{Sp} is a force, weight is measured in newtons.

If the object and scale in Figure 6.9 are stationary, then the object being weighed is in equilibrium. $\vec{F}_{\text{net}} = \vec{0}$ only if the upward spring force exactly balances the downward gravitational force of magnitude mg:

$$F_{\text{Sp}} = F_G = mg \qquad (6.7)$$

Because we defined weight as the reading F_{Sp} of a spring scale, the weight of a stationary object is

$$w = mg \qquad \text{(weight of a stationary object)} \qquad (6.8)$$

Note that the scale does not "know" the weight of the object. All it can do is to measure how much its spring is compressed. On earth, a student with a mass of 70 kg has weight $w = (70 \text{ kg})(9.80 \text{ m/s}^2) = 686$ N *because* he compresses a spring until the spring pushes upward with 686 N. On a different planet, with a different value for g, the compression of the spring would be different and the student's weight would be different.

NOTE **Mass and weight are not the same thing.** Mass, in kg, is an intrinsic property of an object; its value is unique and always the same. Weight, in N, depends on the object's mass, but it also depends on the situation—the strength of gravity and, as we will see, whether or not the object is accelerating. Weight is *not* a property of the object, and thus weight does not have a unique value.

Surprisingly, you cannot directly feel or sense gravity. Your *sensation*—how heavy you feel—is due to contact forces pressing against you, forces that touch you and activate nerve endings in your skin. As you read this, your sensation of weight is due to the normal force exerted on you by the chair in which you are sitting. When you stand, you feel the contact force of the floor pushing against your feet.

But recall the sensations you feel while accelerating. You feel "heavy" when an elevator suddenly accelerates upward, but this sensation vanishes as soon as the elevator reaches a steady speed. Your stomach seems to rise a little and you feel lighter than normal as the upward-moving elevator brakes to a halt or a roller coaster goes over the top. Has your weight actually changed?

FIGURE 6.9 A spring scale measures weight.

FIGURE 6.10 A man weighing himself in an accelerating elevator.

Spring scale

To answer this question, **FIGURE 6.10** shows a man weighing himself on a spring scale in an accelerating elevator. The only forces acting on the man are the upward spring force of the scale and the downward gravitational force. This seems to be the same situation as Figure 6.9, but there's one big difference: The man is accelerating, hence there must be a net force on the man in the direction of $\vec{a}$.

For the net force $\vec{F}_{\text{net}}$ to point upward, the magnitude of the spring force must be *greater* than the magnitude of the gravitational force. That is, $F_{\text{Sp}} > mg$. Looking at the free-body diagram in Figure 6.10, we see that the y-component of Newton's second law is

$$(F_{\text{net}})_y = (F_{\text{Sp}})_y + (F_{\text{G}})_y = F_{\text{Sp}} - mg = ma_y \tag{6.9}$$

where m is the man's mass.

We defined weight as the reading F_{sp} of a calibrated spring scale *when the object is at rest relative to the scale*. That is the case here as the scale and man accelerate upward together. Thus the man's weight as he accelerates vertically is

$$w = \text{scale reading } F_{\text{Sp}} = mg + ma_y = mg\left(1 + \frac{a_y}{g}\right) \tag{6.10}$$

If an object is either at rest or moving with constant velocity, then $a_y = 0$ and $w = mg$. That is, the weight of a stationary object is the magnitude of the (effective) gravitational force acting on it. But its weight differs if it has a vertical acceleration.

You *do* weigh more when accelerating upward $(a_y > 0)$ because the reading of a scale—a weighing—increases. Similarly, your weight is less when the acceleration vector $\vec{a}$ points downward $(a_y < 0)$ because the scale reading goes down. Weight, as we've defined it, corresponds to your sensation of heaviness or lightness.*

We found Equation 6.10 by considering a person in an accelerating elevator, but it applies to any object with a vertical acceleration. Further, an object doesn't really have to be on a scale to have a weight; an object's weight is the magnitude of the contact force supporting it. It makes no difference whether this is the spring force of the scale or simply the normal force of the floor.

> **NOTE** Informally, we sometimes say "This object weighs such and such" or "The weight of this object is . . . ." We'll interpret these expressions as meaning *mg*, the weight of an object of mass m at rest $(a_y = 0)$ on the surface of the earth or some other astronomical body.

Astronauts are weightless as they orbit the earth.

Weightlessness

Suppose the elevator cable breaks and the elevator, along with the man and his scale, plunges straight down in free fall! What will the scale read? When the free-fall acceleration $a_y = -g$ is used in Equation 6.10, we find $w = 0$. In other words, *the man has no weight!*

Suppose, as the elevator falls, the man inside releases a ball from his hand. In the absence of air resistance, as Galileo discovered, both the man and the ball would fall at the same rate. From the man's perspective, the ball would appear to "float" beside him. Similarly, the scale would float beneath him and not press against his feet. He is what we call *weightless*. Gravity is still pulling down on him—that's why he's falling—but he has no *sensation* of weight as everything floats around him in free fall.

But isn't this exactly what happens to astronauts orbiting the earth? If an astronaut tries to stand on a scale, it does not exert any force against her feet and reads zero. She is said to be weightless. But if the criterion to be weightless is to be in free fall, and if astronauts orbiting the earth are weightless, does this mean that they are in free fall? This is a very interesting question to which we shall return in Chapter 8.

---

* Surprisingly, there is no universally agreed-upon definition of *weight*. Some textbooks define weight as the gravitational force on an object, $\vec{w} = (mg, \text{down})$. In that case, the scale reading of an accelerating object, and your sensation of weight, is often called *apparent weight*. This textbook prefers the definition of *weight* as being what a scale reads, the result of a weighing measurement.

STOP TO THINK 6.2 An elevator that has descended from the 50th floor is coming to a halt at the 1st floor. As it does, your weight is

 a. More than mg.   b. Less than mg.   c. Equal to mg.   d. Zero.

6.4 Friction

Friction is absolutely essential for many things we do. Without friction you could not walk, drive, or even sit down (you would slide right off the chair!). Although friction is a complicated force, many aspects of friction can be described with a simple model.

Static Friction

« Section 5.2 defined *static friction* $\vec{f}_s$ as the force on an object that keeps it from slipping. **FIGURE 6.11** shows a rope pulling on a box that, due to static friction, isn't moving. The box is in equilibrium, so the static friction force must exactly balance the tension force:

$$f_s = T \qquad (6.11)$$

To determine the direction of $\vec{f}_s$, decide which way the object would move if there were no friction. The static friction force $\vec{f}_s$ points in the *opposite* direction to prevent the motion.

Unlike the gravitational force, which has the precise and unambiguous magnitude $F_G = mg$, the size of the static friction force depends on how hard you push or pull. The harder the rope in Figure 6.11 pulls, the harder the floor pulls back. Reduce the tension, and the static friction force will automatically be reduced to match. Static friction acts in *response* to an applied force. **FIGURE 6.12** illustrates this idea.

But there's clearly a limit to how big f_s can get. If you pull hard enough, the object slips and starts to move. In other words, the static friction force has a *maximum* possible size $f_{s\,max}$.

- An object remains at rest as long as $f_s < f_{s\,max}$.
- The object slips when $f_s = f_{s\,max}$.
- A static friction force $f_s > f_{s\,max}$ is not physically possible.

Experiments with friction show that $f_{s\,max}$ is proportional to the magnitude of the normal force. That is,

$$f_{s\,max} = \mu_s n \qquad (6.12)$$

where the proportionality constant μ_s is called the **coefficient of static friction.** The coefficient is a dimensionless number that depends on the materials of which the object and the surface are made. **TABLE 6.1** on the next page shows some typical coefficients of friction. It is to be emphasized that these are only approximate; the exact value of the coefficient depends on the roughness, cleanliness, and dryness of the surfaces.

> **NOTE** The static friction force is *not* given by Equation 6.12; this equation is simply the maximum possible static friction. The static friction force is not found with an equation but by determining how much force is needed to maintain equilibrium.

Kinetic Friction

Once the box starts to slide, as in **FIGURE 6.13**, the static friction force is replaced by a kinetic friction force $\vec{f}_k$. Experiments show that kinetic friction, unlike static friction, has a nearly *constant* magnitude. Furthermore, the size of the kinetic friction force

FIGURE 6.11 Static friction keeps an object from slipping.

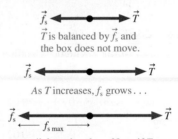

The direction of static friction is opposite to the pull, preventing motion.

FIGURE 6.12 Static friction acts in response to an applied force.

$\vec{T}$ is balanced by $\vec{f}_s$ and the box does not move.

As T increases, f_s grows . . .

. . . until f_s reaches $f_{s\,max}$. Now, if T gets any bigger, the object will start to move.

FIGURE 6.13 The kinetic friction force is opposite the direction of motion.

The direction of kinetic friction is opposite to the motion.

is *less* than the maximum static friction, $f_k < f_{s\,max}$, which explains why it is easier to keep the box moving than it was to start it moving. The direction of $\vec{f}_k$ is always opposite to the direction in which an object slides across the surface.

The kinetic friction force is also proportional to the magnitude of the normal force:

$$f_k = \mu_k n \tag{6.13}$$

where μ_k is called the **coefficient of kinetic friction.** Table 6.1 includes typical values of μ_k. You can see that $\mu_k < \mu_s$, causing the kinetic friction to be less than the maximum static friction.

Rolling Friction

If you slam on the brakes hard enough, your car tires slide against the road surface and leave skid marks. This is kinetic friction. A wheel *rolling* on a surface also experiences friction, but not kinetic friction. As FIGURE 6.14 shows, the portion of the wheel that contacts the surface is stationary with respect to the surface, not sliding. The interaction of this contact area with the surface causes **rolling friction.** The force of rolling friction can be calculated in terms of a **coefficient of rolling friction** μ_r:

$$f_r = \mu_r n \tag{6.14}$$

Rolling friction acts very much like kinetic friction, but values of μ_r (see Table 6.1) are much lower than values of μ_k. This is why it is easier to roll an object on wheels than to slide it.

A Model of Friction

The friction equations are not "laws of nature" on a level with Newton's laws. Instead, they provide a reasonably accurate, but not perfect, description of how friction forces act. That is, they are a *model* of friction. And because we characterize friction in terms of constant forces, this model of friction meshes nicely with our model of dynamics with constant force.

FIGURE 6.14 Rolling friction is also opposite the direction of motion

$\vec{f}_r$

The wheel is stationary across the area of contact, not sliding.

TABLE 6.1 Coefficients of friction

Materials	Static μ_s	Kinetic μ_k	Rolling μ_r
Rubber on dry concrete	1.00	0.80	0.02
Rubber on wet concrete	0.30	0.25	0.02
Steel on steel (dry)	0.80	0.60	0.002
Steel on steel (lubricated)	0.10	0.05	
Wood on wood	0.50	0.20	
Wood on snow	0.12	0.06	
Ice on ice	0.10	0.03	

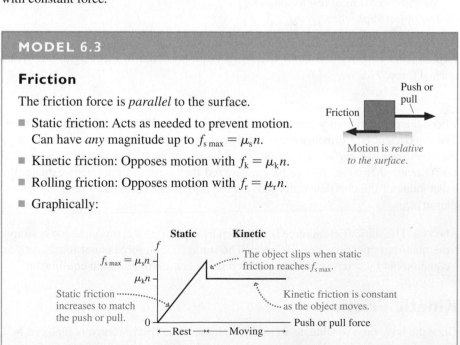

MODEL 6.3

Friction

The friction force is *parallel* to the surface.

- Static friction: Acts as needed to prevent motion. Can have *any* magnitude up to $f_{s\,max} = \mu_s n$.
- Kinetic friction: Opposes motion with $f_k = \mu_k n$.
- Rolling friction: Opposes motion with $f_r = \mu_r n$.
- Graphically:

Push or pull

Friction

Motion is *relative to the surface.*

Static · Kinetic

f

$f_{s\,max} = \mu_s n$

$\mu_k n$

The object slips when static friction reaches $f_{s\,max}$.

Static friction increases to match the push or pull.

Kinetic friction is constant as the object moves.

Push or pull force

0

← Rest → ← Moving →

STOP TO THINK 6.3 Rank in order, from largest to smallest, the sizes of the friction forces $\vec{f}_a$ to $\vec{f}_e$ in these 5 different situations. The box and the floor are made of the same materials in all situations.

EXAMPLE 6.5 | How far does a box slide?

Carol pushes a 25 kg wood box across a wood floor at a steady speed of 2.0 m/s. How much force does Carol exert on the box? If she stops pushing, how far will the box slide before coming to rest?

MODEL This situation can be modeled as dynamics with constant force—one of the forces being friction. Notice that this is a two-part problem: first while Carol is pushing the box, then as it slides after she releases it.

VISUALIZE This is a fairly complex situation, one that calls for careful visualization. **FIGURE 6.15** shows the pictorial representation both while Carol pushes, when $\vec{a} = \vec{0}$, and after she stops. We've placed $x = 0$ at the point where she stops pushing because this is the point where the kinematics calculation for How far? will begin. Notice that each part of the motion needs its own free-body diagram. The box is moving until the very instant that the problem ends, so only kinetic friction is relevant.

SOLVE We'll start by finding how hard Carol has to push to keep the box moving at a steady speed. The box is in equilibrium (constant velocity, $\vec{a} = \vec{0}$), and Newton's second law is

$$\sum F_x = F_{push} - f_k = 0$$
$$\sum F_y = n - F_G = n - mg = 0$$

where we've used $F_G = mg$ for the gravitational force. The negative sign occurs in the first equation because $\vec{f}_k$ points to the left and thus the *component* is negative: $(f_k)_x = -f_k$. Similarly, $(F_G)_y = -F_G$ because the gravitational force vector—with magnitude mg—points down. In addition to Newton's laws, we also have our model of kinetic friction:

$$f_k = \mu_k n$$

Altogether we have three simultaneous equations in the three unknowns F_{push}, f_k, and n. Fortunately, these equations are easy to solve. The y-component of Newton's second law tells us that $n = mg$. We can then find the friction force to be

$$f_k = \mu_k mg$$

We substitute this into the x-component of the second law, giving

$$F_{push} = f_k = \mu_k mg$$
$$= (0.20)(25 \text{ kg})(9.80 \text{ m/s}^2) = 49 \text{ N}$$

Carol pushes this hard to keep the box moving at a steady speed.

The box is not in equilibrium after Carol stops pushing it. Our strategy for the second half of the problem is to use Newton's second law to find the acceleration, then use constant-acceleration kinematics to find how far the box moves before stopping. We know

FIGURE 6.15 Pictorial representation of a box sliding across a floor.

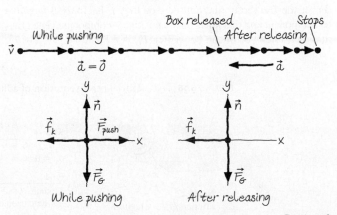

Continued

from the motion diagram that $a_y = 0$. Newton's second law, applied to the second free-body diagram of Figure 6.15, is

$$\sum F_x = -f_k = ma_x$$

$$\sum F_y = n - mg = ma_y = 0$$

We also have our model of friction,

$$f_k = \mu_k n$$

We see from the y-component equation that $n = mg$, and thus $f_k = \mu_k mg$. Using this in the x-component equation gives

$$ma_x = -f_k = -\mu_k mg$$

This is easily solved to find the box's acceleration:

$$a_x = -\mu_k g = -(0.20)(9.80 \text{ m/s}^2) = -1.96 \text{ m/s}^2$$

The acceleration component a_x is negative because the acceleration vector $\vec{a}$ points to the left, as we see from the motion diagram.

Now we are left with a problem of constant-acceleration kinematics. We are interested in a distance, rather than a time interval, so the easiest way to proceed is

$$v_{1x}^2 = 0 = v_{0x}^2 + 2a_x \Delta x = v_{0x}^2 + 2a_x x_1$$

from which the distance that the box slides is

$$x_1 = \frac{-v_{0x}^2}{2a_x} = \frac{-(2.0 \text{ m/s})^2}{2(-1.96 \text{ m/s}^2)} = 1.0 \text{ m}$$

ASSESS Carol was pushing at 2 m/s ≈ 4 mph, which is fairly fast. The box slides 1.0 m, which is slightly over 3 feet. That sounds reasonable.

NOTE Example 6.5 needed both the horizontal and the vertical components of the second law even though the motion was entirely horizontal. This need is typical when friction is involved because we must find the normal force before we can evaluate the friction force.

EXAMPLE 6.6 | Dumping a file cabinet

A 50 kg steel file cabinet is in the back of a dump truck. The truck's bed, also made of steel, is slowly tilted. What is the size of the static friction force on the cabinet when the bed is tilted 20°? At what angle will the file cabinet begin to slide?

MODEL Model the file cabinet as a particle in equilibrium. We'll also use the model of static friction. The file cabinet will slip when the static friction force reaches its maximum value $f_{s\,max}$.

VISUALIZE FIGURE 6.16 shows the pictorial representation when the truck bed is tilted at angle θ. We can make the analysis easier if we tilt the coordinate system to match the bed of the truck.

SOLVE The file cabinet is in equilibrium. Newton's second law is

$$(F_{net})_x = \sum F_x = n_x + (F_G)_x + (f_s)_x = 0$$

$$(F_{net})_y = \sum F_y = n_y + (F_G)_y + (f_s)_y = 0$$

From the free-body diagram we see that f_s has only a *negative* x-component and that n has only a positive y-component. The gravitational force vector can be written $\vec{F}_G = +F_G \sin\theta\, \hat{i} - F_G \cos\theta\, \hat{j}$,

so $\vec{F}_G$ has both x- and y-components in this coordinate system. Thus the second law becomes

$$\sum F_x = F_G \sin\theta - f_s = mg \sin\theta - f_s = 0$$

$$\sum F_y = n - F_G \cos\theta = n - mg \cos\theta = 0$$

where we've used $F_G = mg$.

You might be tempted to solve the y-component equation for n, then to use Equation 6.12 to calculate the static friction force as $\mu_s n$. **But Equation 6.12 does *not* say $f_s = \mu_s n$.** Equation 6.12 gives only the maximum possible static friction force $f_{s\,max}$, the point at which the object slips. In nearly all situations, the actual static friction force is less than $f_{s\,max}$. In this problem, we can use the x-component equation—which tells us that static friction has to exactly balance the component of the gravitational force along the incline—to find the size of the static friction force:

$$f_s = mg \sin\theta = (50 \text{ kg})(9.80 \text{ m/s}^2) \sin 20°$$

$$= 170 \text{ N}$$

FIGURE 6.16 The pictorial representation of a file cabinet in a tilted dump truck.

Known
$\mu_s = 0.80$ $m = 50$ kg
$\mu_k = 0.60$

Find
f_s where $\theta = 20°$
θ where cabinet slips

The coefficients of friction are found in Table 6.1.

Normal $\vec{n}$
Friction $\vec{f}_s$ Gravity $\vec{F}_G$

To prevent slipping, static friction must point *up* the slope.

Slipping occurs when the static friction reaches its maximum value

$$f_s = f_{s\,max} = \mu_s n$$

From the y-component of Newton's law we see that $n = mg\cos\theta$. Consequently,

$$f_{s\,max} = \mu_s mg\cos\theta$$

Substituting this into the x-component of the first law gives

$$mg\sin\theta - \mu_s mg\cos\theta = 0$$

The mg in both terms cancels, and we find

$$\frac{\sin\theta}{\cos\theta} = \tan\theta = \mu_s$$

$$\theta = \tan^{-1}\mu_s = \tan^{-1}(0.80) = 39°$$

ASSESS Steel doesn't slide all that well on unlubricated steel, so a fairly large angle is not surprising. The answer seems reasonable.

> **NOTE** A common error is to use simply $n = mg$. Be sure to evaluate the normal force within the context of each specific problem. In this example, $n = mg\cos\theta$.

Causes of Friction

It is worth a brief pause to look at the *causes* of friction. All surfaces, even those quite smooth to the touch, are very rough on a microscopic scale. When two objects are placed in contact, they do not make a smooth fit. Instead, as FIGURE 6.17 shows, the high points on one surface become jammed against the high points on the other surface, while the low points are not in contact at all. The amount of contact depends on how hard the surfaces are pushed together, which is why friction forces are proportional to n.

At the points of actual contact, the atoms in the two materials are pressed closely together and molecular bonds are established between them. These bonds are the "cause" of the static friction force. For an object to slip, you must push it hard enough to break these molecular bonds between the surfaces. Once they are broken, and the two surfaces are sliding against each other, there are still attractive forces between the atoms on the opposing surfaces as the high points of the materials push past each other. However, the atoms move past each other so quickly that they do not have time to establish the tight bonds of static friction. That is why the kinetic friction force is smaller. Friction can be minimized with lubrication, a very thin film of liquid between the surfaces that allows them to "float" past each other with many fewer points in actual contact.

FIGURE 6.17 An atomic-level view of friction.

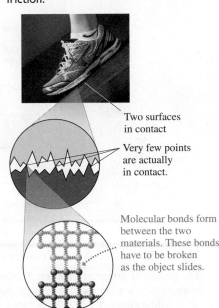

Two surfaces in contact

Very few points are actually in contact.

Molecular bonds form between the two materials. These bonds have to be broken as the object slides.

6.5 Drag

The air exerts a drag force on objects as they move through the air. You experience drag forces every day as you jog, bicycle, ski, or drive your car. The drag force $\vec{F}_{drag}$

- Is opposite in direction to $\vec{v}$.
- Increases in magnitude as the object's speed increases.

FIGURE 6.18 illustrates the drag force.

Drag is a more complex force than ordinary friction because drag depends on the object's speed. Drag also depends on the object's shape and on the density of the medium through which it moves. Fortunately, we can use a fairly simple *model* of drag if the following three conditions are met:

- The object is moving through the air near the earth's surface.
- The object's size (diameter) is between a few millimeters and a few meters.
- The object's speed is less than a few hundred meters per second.

These conditions are usually satisfied for balls, people, cars, and many other objects in our everyday world. Under these conditions, the drag force on an object moving with speed v can be written

$$\vec{F}_{drag} = \left(\tfrac{1}{2}C\rho Av^2,\ \text{direction opposite the motion}\right) \qquad (6.15)$$

FIGURE 6.18 The drag force on a high-speed motorcyclist is significant.

$\vec{F}_{drag}$

The symbols in Equation 6.15 are:

- A is the *cross-section area* of the object as it "faces into the wind," as illustrated in **FIGURE 6.19**.
- ρ is the density of the air, which is 1.3 kg/m^3 at atmospheric pressure and 0°C, a common reference point of pressure and temperature.
- C is the **drag coefficient.** It is smaller for aerodynamically shaped objects, larger for objects presenting a flat face to the wind. Figure 6.19 gives approximate values for a sphere and two cylinders. C is dimensionless; it has no units.

Notice that the drag force is proportional to the *square* of the object's speed. So drag is *not* a constant force (unless v is constant) and you cannot use constant-acceleration kinematics.

This model of drag fails for objects that are very small (such as dust particles), very fast (such as bullets), or that move in liquids (such as water). Motion in a liquid will be considered in Challenge Problems 6.76 and 6.77, but otherwise we'll leave these situations to more advanced textbooks.

FIGURE 6.19 Cross-section areas for objects of different shape.

A falling sphere $C \approx 0.5$

The cross section is an equatorial circle. $A = \pi r^2$

A cylinder falling end down $C \approx 0.8$

The cross section is a circle. $A = \pi r^2$

A cylinder falling side down $C \approx 1.1$

The cross section is a rectangle. $A = 2rL$

EXAMPLE 6.7 | Air resistance compared to rolling friction

The profile of a typical 1500 kg passenger car, as seen from the front, is 1.6 m wide and 1.4 m high. Aerodynamic body shaping gives a drag coefficient of 0.35. At what speed does the magnitude of the drag equal the magnitude of the rolling friction?

MODEL Model the car as a particle. Use the models of rolling friction and drag. Note that this is *not* a constant-force situation.

VISUALIZE FIGURE 6.20 shows the car and a free-body diagram. A full pictorial representation is not needed because we won't be doing any kinematics calculations.

FIGURE 6.20 A car experiences both rolling friction and drag.

Drag due to air resistance

Car's cross-section area A

Propulsion due to car engine

Rolling friction due to road

SOLVE Drag is less than friction at low speeds, where air resistance is negligible. But drag increases as v increases, so there will be a speed at which the two forces are equal in size. Above this speed, drag is more important than rolling friction.

There's no motion and no acceleration in the vertical direction, so we can see from the free-body diagram that $n = F_G = mg$. Thus $f_r = \mu_r mg$. Equating friction and drag, we have

$$\tfrac{1}{2}C\rho A v^2 = \mu_r mg$$

Solving for v, we find

$$v = \sqrt{\frac{2\mu_r mg}{C\rho A}} = \sqrt{\frac{2(0.02)(1500 \text{ kg})(9.80 \text{ m/s}^2)}{(0.35)(1.3 \text{ kg/m}^3)\,(1.4 \text{ m} \times 1.6 \text{ m})}} = 24 \text{ m/s}$$

where the value of μ_r for rubber on concrete was taken from Table 6.1.

ASSESS 24 m/s is approximately 50 mph, a reasonable result. This calculation shows that we can reasonably ignore air resistance for car speeds less than 30 or 40 mph. Calculations that neglect drag will be increasingly inaccurate as speeds go above 50 mph.

Terminal Speed

The drag force increases as an object falls and gains speed. If the object falls far enough, it will eventually reach a speed, shown in **FIGURE 6.21**, at which $F_{drag} = F_G$. That is, the drag force will be equal and opposite to the gravitational force. The net force at this speed is $\vec{F}_{net} = \vec{0}$, so there is no further acceleration and the object falls with a *constant* speed. The speed at which the exact balance between the upward drag force and the downward gravitational force causes an object to fall without acceleration is called the **terminal speed** v_{term}. Once an object has reached terminal speed, it will continue falling at that speed until it hits the ground.

It's not hard to compute the terminal speed. It is the speed, by definition, at which $F_{drag} = F_G$ or, equivalently, $\frac{1}{2}C\rho Av^2 = mg$. This speed is

$$v_{term} = \sqrt{\frac{2mg}{C\rho A}} \qquad (6.16)$$

A more massive object has a larger terminal speed than a less massive object of equal size and shape. A 10-cm-diameter lead ball, with a mass of 6 kg, has a terminal speed of 160 m/s, while a 10-cm-diameter Styrofoam ball, with a mass of 50 g, has a terminal speed of only 15 m/s.

A popular use of Equation 6.16 is to find the terminal speed of a skydiver. A skydiver is rather like the cylinder of Figure 6.19 falling "side down," for which we see that $C \approx 1.1$. A typical skydiver is 1.8 m long and 0.40 m wide ($A = 0.72$ m²) and has a mass of 75 kg. His terminal speed is

$$v_{term} = \sqrt{\frac{2mg}{C\rho A}} = \sqrt{\frac{2(75 \text{ kg})(9.8 \text{ m/s}^2)}{(1.1)(1.3 \text{ kg/m}^3)(0.72 \text{ m}^2)}} = 38 \text{ m/s}$$

This is roughly 90 mph. A higher speed can be reached by falling feet first or head first, which reduces the area A and the drag coefficient.

Although we've focused our analysis on objects moving vertically, the same ideas apply to objects moving horizontally. If an object is thrown or shot horizontally, $\vec{F}_{drag}$ causes the object to slow down. An airplane reaches its maximum speed, which is analogous to the terminal speed, when the drag is equal and opposite to the thrust: $F_{drag} = F_{thrust}$. The net force is then zero and the plane cannot go any faster. The maximum speed of a passenger jet is about 550 mph.

FIGURE 6.21 An object falling at terminal speed.

Terminal speed is reached when the drag force exactly balances the gravitational force: $\vec{a} = \vec{0}$.

STOP TO THINK 6.4   The terminal speed of a Styrofoam ball is 15 m/s. Suppose a Styrofoam ball is shot straight down from a high tower with an initial speed of 30 m/s. Which velocity graph is correct?

(a)

(b)

(c)

(d)

(e)

6.6 More Examples of Newton's Second Law

We will finish this chapter with four additional examples in which we use the problem-solving strategy in more complex scenarios.

EXAMPLE 6.8 | Stopping distances

A 1500 kg car is traveling at a speed of 30 m/s when the driver slams on the brakes and skids to a halt. Determine the stopping distance if the car is traveling up a 10° slope, down a 10° slope, or on a level road.

MODEL Model the car's motion as dynamics with constant force and use the model of kinetic friction. We want to solve the problem only once, not three separate times, so we'll leave the slope angle θ unspecified until the end.

VISUALIZE FIGURE 6.22 shows the pictorial representation. We've shown the car sliding uphill, but these representations work equally well for a level or downhill slide if we let θ be zero or negative, respectively. We've used a tilted coordinate system so that the motion is along one of the axes. We've *assumed* that the car is traveling to the right, although the problem didn't state this. You could equally well make the opposite assumption, but you would have to be careful with negative values of x and v_x. The car *skids* to a halt, so we've taken the coefficient of *kinetic* friction for rubber on concrete from Table 6.1.

SOLVE Newton's second law and the model of kinetic friction are

$$\sum F_x = n_x + (F_G)_x + (f_k)_x$$
$$= -mg\sin\theta - f_k = ma_x$$
$$\sum F_y = n_y + (F_G)_y + (f_k)_y$$
$$= n - mg\cos\theta = ma_y = 0$$
$$f_k = \mu_k n$$

We've written these equations by "reading" the motion diagram and the free-body diagram. Notice that both components of the gravitational force vector $\vec{F}_G$ are negative. $a_y = 0$ because the motion is entirely along the x-axis.

The second equation gives $n = mg\cos\theta$. Using this in the friction model, we find $f_k = \mu_k mg\cos\theta$. Inserting this result back into the first equation then gives

$$ma_x = -mg\sin\theta - \mu_k mg\cos\theta$$
$$= -mg(\sin\theta + \mu_k\cos\theta)$$
$$a_x = -g(\sin\theta + \mu_k\cos\theta)$$

This is a constant acceleration. Constant-acceleration kinematics gives

$$v_{1x}^2 = 0 = v_{0x}^2 + 2a_x(x_1 - x_0) = v_{0x}^2 + 2a_x x_1$$

which we can solve for the stopping distance x_1:

$$x_1 = -\frac{v_{0x}^2}{2a_x} = \frac{v_{0x}^2}{2g(\sin\theta + \mu_k\cos\theta)}$$

Notice how the minus sign in the expression for a_x canceled the minus sign in the expression for x_1. Evaluating our result at the three different angles gives the stopping distances:

$$x_1 = \begin{cases} 48\text{ m} & \theta = 10° & \text{uphill} \\ 57\text{ m} & \theta = 0° & \text{level} \\ 75\text{ m} & \theta = -10° & \text{downhill} \end{cases}$$

The implications are clear about the danger of driving downhill too fast!

ASSESS 30 m/s $\approx$ 60 mph and 57 m $\approx$ 180 feet on a level surface. This is similar to the stopping distances you learned when you got your driver's license, so the results seem reasonable. Additional confirmation comes from noting that the expression for a_x becomes $-g\sin\theta$ if $\mu_k = 0$. This is what you learned in Chapter 2 for the acceleration on a frictionless inclined plane.

FIGURE 6.22 Pictorial representation of a skidding car.

This representation works for a downhill slide if we let θ be negative.

Known
$x_0 = 0$ m $t_0 = 0$ s $v_{0x} = 30$ m/s
$m = 1500$ kg $v_{1x} = 0$ m/s
$\mu_k = 0.80$
$\theta = -10°, 0°, 10°$

Find
$\Delta x = x_1 - x_0 = x_1$

EXAMPLE 6.9 | Measuring the tension pulling a cart

Your instructor has set up a lecture demonstration in which a 250 g cart can roll along a level, 2.00-m-long track while its velocity is measured with a motion detector. First, the instructor simply gives the cart a push and measures its velocity as it rolls down the track. The data below show that the cart slows slightly before reaching the end of the track. Then, as **FIGURE 6.23** shows, the instructor attaches a string to the cart and uses a falling weight to pull the cart. She then asks you to determine the tension in the string. For extra credit, find the coefficient of rolling friction.

Time (s)	Rolled velocity (m/s)	Pulled velocity (m/s)
0.00	1.20	0.00
0.25	1.17	0.36
0.50	1.15	0.80
0.75	1.12	1.21
1.00	1.08	1.52
1.25	1.04	1.93
1.50	1.02	2.33

FIGURE 6.23 The experimental arrangement.

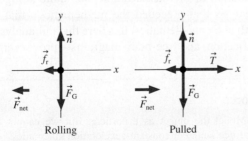

MODEL Model the cart as a particle acted on by constant forces.

VISUALIZE The cart changes velocity—it accelerates—when both pulled and rolled. Consequently, there must be a net force for both motions. For rolling, force identification finds that the only horizontal force is rolling friction, a force that opposes the motion and slows the cart. There is no "force of motion" or "force of the hand" because the hand is no longer in contact with the cart. (Recall Newton's "zeroth law": The cart responds only to forces applied *at this instant.*) Pulling adds a tension force in the direction of motion. The two free-body diagrams are shown in **FIGURE 6.24**.

FIGURE 6.24 Pictorial representations of the cart.

SOLVE The cart's acceleration when pulled, which we can find from the velocity data, will allow us to find the net force. Isolating the tension force will require knowing the friction force, but we can determine that from the rolling motion. For the rolling motion, Newton's second law can be written by "reading" the free-body diagram on the left:

$$\sum F_x = (f_r)_x = -f_r = ma_x = ma_{roll}$$
$$\sum F_y = n_y + (F_G)_y = n - mg = 0$$

Make sure you understand where the signs come from and how we used our knowledge that $\vec{a}$ has only an x-component, which we called a_{roll}. The magnitude of the friction force, which is all we'll need to determine the tension, is found from the x-component equation:

$$f_r = -ma_{roll} = -m \times \text{slope of the rolling-velocity graph}$$

But we'll need to do a bit more analysis to get the coefficient of rolling friction. The y-component equation tells us that $n - mg$. Using this in the model of rolling friction, $f_r = \mu_r n = \mu_r mg$, we see that the coefficient of rolling friction is

$$\mu_r = \frac{f_r}{mg}$$

The x-component equation of Newton's second law when the cart is pulled is

$$\sum F_x = T + (f_r)_x = T - f_r = ma_x = ma_{pulled}$$

Thus the tension that we seek is

$$T = f_r + ma_{pulled} = f_r + m \times \text{slope of the pulled-velocity graph}$$

FIGURE 6.25 shows the graphs of the velocity data. The accelerations are the slopes of these lines, and from the equations of the best-fit lines we find $a_{roll} = -0.124 \text{ m/s}^2$ and $a_{pulled} = 1.55 \text{ m/s}^2$. Thus the friction force is

$$f_r = -ma_{roll} = -(0.25 \text{ kg})(-0.124 \text{ m/s}^2) = 0.031 \text{ N}$$

Knowing this, we find that the string tension pulling the cart is

$$T = f_r + ma_{pulled} = 0.031 \text{ N} + (0.25 \text{ kg})(1.55 \text{ m/s}^2) = 0.42 \text{ N}$$

and the coefficient of rolling friction is

$$\mu_r = \frac{f_r}{mg} = \frac{0.031 \text{ N}}{(0.25 \text{ kg})(9.80 \text{ m/s}^2)} = 0.013$$

FIGURE 6.25 The velocity graphs of the rolling and pulled motion. The slopes of these graphs are the cart's acceleration.

ASSESS The coefficient of rolling friction is very small, but it's similar to the values in Table 6.1 and thus believable. That gives us confidence that our value for the tension is also correct. It's reasonable that the tension needed to accelerate the cart is small because the cart is light and there's very little friction.

EXAMPLE 6.10 | **Make sure the cargo doesn't slide**

A 100 kg box of dimensions 50 cm × 50 cm × 50 cm is in the back of a flatbed truck. The coefficients of friction between the box and the bed of the truck are $\mu_s = 0.40$ and $\mu_k = 0.20$. What is the maximum acceleration the truck can have without the box slipping?

MODEL This is a somewhat different problem from any we have looked at thus far. Let the box, which we'll model as a particle, be the object of interest. It contacts other objects only where it touches the truck bed, so only the truck can exert contact forces on the box. If the box does *not* slip, then there is no motion of the box *relative to the truck* and the box must accelerate *with the truck*: $a_{box} = a_{truck}$. As the box accelerates, it must, according to Newton's second law, have a net force acting on it. But from what?

Imagine, for a moment, that the truck bed is frictionless. The box would slide backward (as seen in the truck's reference frame) as the truck accelerates. The force that prevents sliding is *static friction*, so the truck must exert a static friction force on the box to "pull" the box along with it and prevent the box from sliding *relative to the truck.*

VISUALIZE This situation is shown in **FIGURE 6.26**. There is only one horizontal force on the box, $\vec{f}_s$, and it points in the *forward* direction to accelerate the box. Notice that we're solving the problem with the ground as our reference frame. Newton's laws are not valid in the accelerating truck because it is not an inertial reference frame.

SOLVE Newton's second law, which we can "read" from the free-body diagram, is

$$\sum F_x = f_s = ma_x$$
$$\sum F_y = n - F_G = n - mg = ma_y = 0$$

Now, static friction, you will recall, can be *any* value between 0 and $f_{s\,max}$. If the truck accelerates slowly, so that the box doesn't slip, then $f_s < f_{s\,max}$. However, we're interested in the acceleration a_{max} at which the box begins to slip. This is the acceleration at which f_s reaches its maximum possible value

$$f_s = f_{s\,max} = \mu_s n$$

The y-equation of the second law and the friction model combine to give $f_{s\,max} = \mu_s mg$. Substituting this into the x-equation, and noting that a_x is now a_{max}, we find

$$a_{max} = \frac{f_{s\,max}}{m} = \mu_s g = 3.9 \text{ m/s}^2$$

The truck must keep its acceleration less than 3.9 m/s² if slipping is to be avoided.

ASSESS 3.9 m/s² is about one-third of g. You may have noticed that items in a car or truck are likely to *tip over* when you start or stop, but they slide only if you really floor it and accelerate very quickly. So this answer seems reasonable. Notice that neither the dimensions of the crate nor μ_k was needed. Real-world situations rarely have exactly the information you need, no more and no less. Many problems in this textbook will require you to assess the information in the problem statement in order to learn which is relevant to the solution.

FIGURE 6.26 Pictorial representation for the box in a flatbed truck.

Known
$m = 100$ kg
Box dimensions 50 cm × 50 cm × 50 cm
$\mu_s = 0.40$ $\mu_k = 0.20$

Find
Acceleration at which box slips

Gravity $\vec{F}_G$ Normal $\vec{n}$ Static friction $\vec{f}_s$

The mathematical representation of this last example was quite straightforward. The challenge was in the analysis that preceded the mathematics—that is, in the *physics* of the problem rather than the mathematics. It is here that our analysis tools—motion diagrams, force identification, and free-body diagrams—prove their value.

CHALLENGE EXAMPLE 6.11 | **Acceleration from a variable force**

Force $F_x = c\sin(\pi t/T)$, where c and T are constants, is applied to an object of mass m that moves on a horizontal, frictionless surface. The object is at rest at the origin at $t = 0$.

a. Find an expression for the object's velocity. Graph your result for $0 \le t \le T$.

b. What is the maximum velocity of a 500 g object if $c = 2.5$ N and $T = 1.0$ s?

MODEL Model the object as a particle. But we cannot use the constant-force model or constant-acceleration kinematics.

VISUALIZE The sine function is 0 at $t = 0$ and again at $t = T$, when the value of the argument is π rad. Over the interval $0 \le t \le T$, the force grows from 0 to c and then returns to 0, always pointing in the positive x-direction. **FIGURE 6.27** shows a graph of the force and a pictorial representation.

FIGURE 6.27 Pictorial representation for a variable force.

SOLVE The object's acceleration increases between 0 and $T/2$ as the force increases. You might expect the object to slow down between $T/2$ and T as the force decreases. However, *there's still a net force in the positive x-direction, so there must be an acceleration in the positive x-direction*. The object continues to speed up, only more slowly as the acceleration decreases. Maximum velocity is reached at $t = T$.

a. This is not constant-acceleration motion, so we cannot use the familiar equations of constant-acceleration kinematics. Instead, we must use the definition of acceleration as the rate of change—the time derivative—of velocity. With no friction, we need only the x-component equation of Newton's second law:

$$a_x = \frac{dv_x}{dt} = \frac{F_{net}}{m} = \frac{c}{m} \sin\left(\frac{\pi t}{T}\right)$$

First we rewrite this as

$$dv_x = \frac{c}{m} \sin\left(\frac{\pi t}{T}\right) dt$$

Then we integrate both sides from the initial conditions ($v_x = v_{0x} = 0$ at $t = t_0 = 0$) to the final conditions (v_x at the later time t):

$$\int_0^{v_x} dv_x = \frac{c}{m} \int_0^t \sin\left(\frac{\pi t}{T}\right) dt$$

The fraction c/m is a constant that we could take outside the integral. The integral on the right side is of the form

$$\int \sin(bx)\, dx = -\frac{1}{b} \cos(bx)$$

Using this, and integrating both sides of the equation, we find

$$v_x \Big|_0^{v_x} = v_x - 0 = -\frac{cT}{\pi m} \cos\left(\frac{\pi t}{T}\right)\Big|_0^t = -\frac{cT}{\pi m}\left(\cos\left(\frac{\pi t}{T}\right) - 1\right)$$

Simplifying, we find the object's velocity at time t is

$$v_x = \frac{cT}{\pi m}\left(1 - \cos\left(\frac{\pi t}{T}\right)\right)$$

This expression is graphed in **FIGURE 6.28**, where we see that, as predicted, maximum velocity is reached at $t = T$.

FIGURE 6.28 The object's velocity as a function of time.

b. Maximum velocity, at $t = T$, is

$$v_{max} = \frac{cT}{\pi m}(1 - \cos\pi) = \frac{2cT}{\pi m} = \frac{2(2.5\ \text{N})(1.0\ \text{s})}{\pi(0.50\ \text{kg})} = 3.2\ \text{m/s}$$

ASSESS A steady 2.5 N force would cause a 0.5 kg object to accelerate at 5 m/s² and reach a speed of 5 m/s in 1 s. A variable force with a maximum of 2.5 N will produce less acceleration, so a top speed of 3.2 m/s seems reasonable.

SUMMARY

The goal of Chapter 6 has been to learn to solve linear force-and-motion problems.

GENERAL PRINCIPLES

Two Explanatory Models

An object on which there is no net force is in **mechanical equilibrium**.

- Objects at rest.
- Objects moving with constant velocity.
- Newton's second law applies with $\vec{a} = \vec{0}$.

$\vec{F}_{net} = \vec{0}$

An object on which the net force is constant undergoes **dynamics with constant force**.

- The object accelerates.
- The kinematic model is that of constant acceleration.
- Newton's second law applies.

$\vec{a}$

Go back and forth between these steps as needed.

A Problem-Solving Strategy

A four-part strategy applies to both equilibrium and dynamics problems.

MODEL Make simplifying assumptions.

VISUALIZE

- Translate words into symbols.
- Draw a sketch to define the situation.
- Draw a motion diagram.
- Identify forces.
- Draw a free-body diagram.

SOLVE Use Newton's second law:

$$\vec{F}_{net} = \sum_i \vec{F}_i = m\vec{a}$$

"Read" the vectors from the free-body diagram. Use kinematics to find velocities and positions.

ASSESS Is the result reasonable? Does it have correct units and significant figures?

IMPORTANT CONCEPTS

Specific information about three important descriptive models:

Gravity $\vec{F}_G = (mg, \text{ downward})$

Friction $\vec{f}_s = (0 \text{ to } \mu_s n, \text{ direction as necessary to prevent motion})$

$\vec{f}_k = (\mu_k n, \text{ direction opposite the motion})$

$\vec{f}_r = (\mu_r n, \text{ direction opposite the motion})$

Drag $\vec{F}_{drag} = \left(\frac{1}{2}C\rho A v^2, \text{ direction opposite the motion}\right)$

Newton's laws are vector expressions. You must write them out by **components**:

$$(F_{net})_x = \sum F_x = ma_x$$

$$(F_{net})_y = \sum F_y = ma_y$$

The acceleration is zero in equilibrium and also along an axis perpendicular to the motion.

APPLICATIONS

Mass is an intrinsic property of an object that describes the object's inertia and, loosely speaking, its quantity of matter.

The **weight** of an object is the reading of a spring scale when the object is at rest relative to the scale. Weight is the result of weighing. An object's weight depends on its mass, its acceleration, and the strength of gravity. An object in free fall is weightless.

A falling object reaches **terminal speed**

$$v_{term} = \sqrt{\frac{2mg}{C\rho A}}$$

Terminal speed is reached when the drag force exactly balances the gravitational force: $\vec{a} = \vec{0}$.

TERMS AND NOTATION

equilibrium model
constant-force model
flat-earth approximation

weight
coefficient of static friction, μ_s
coefficient of kinetic friction, μ_k

rolling friction
coefficient of rolling friction, μ_r

drag coefficient, C
terminal speed, v_{term}

CONCEPTUAL QUESTIONS

1. Are the objects described here in equilibrium while at rest, in equilibrium while in motion, or not in equilibrium at all? Explain.
   a. A 200 pound barbell is held over your head.
   b. A girder is lifted at constant speed by a crane.
   c. A girder is being lowered into place. It is slowing down.
   d. A jet plane has reached its cruising speed and altitude.
   e. A box in the back of a truck doesn't slide as the truck stops.

2. A ball tossed straight up has $v = 0$ at its highest point. Is it in equilibrium? Explain.

3. Kat, Matt, and Nat are arguing about why a physics book on a table doesn't fall. According to Kat, "Gravity pulls down on it, but the table is in the way so it can't fall." "Nonsense," says Matt. "An upward force simply overcomes the downward force to prevent it from falling." "But what about Newton's first law?" counters Nat. "It's not moving, so there can't be any forces acting on it." None of the statements is exactly correct. Who comes closest, and how would you change his or her statement to make it correct?

4. If you know all of the forces acting on a moving object, can you tell the direction the object is moving? If yes, explain how. If no, give an example.

5. An elevator, hanging from a single cable, moves upward at constant speed. Friction and air resistance are negligible. Is the tension in the cable greater than, less than, or equal to the gravitational force on the elevator? Explain. Include a free-body diagram as part of your explanation.

6. An elevator, hanging from a single cable, moves downward and is slowing. Friction and air resistance are negligible. Is the tension in the cable greater than, less than, or equal to the gravitational force on the elevator? Explain. Include a free-body diagram as part of your explanation.

7. Are the following statements true or false? Explain.
   a. The mass of an object depends on its location.
   b. The weight of an object depends on its location.
   c. Mass and weight describe the same thing in different units.

8. An astronaut takes his bathroom scale to the moon and then stands on it. Is the reading of the scale his weight? Explain.

9. The four balls in FIGURE Q6.9 have been thrown straight up. They have the same size, but different masses. Air resistance is negligible. Rank in order, from largest to smallest, the magnitude of the net force acting on each ball. Some may be equal. Give your answer in the form a > b > c = d and explain your ranking.

FIGURE Q6.9

10. Suppose you attempt to pour out 100 g of salt, using a pan balance for measurements, while in a rocket accelerating upward. Will the quantity of salt be too much, too little, or the correct amount? Explain.

11. An astronaut orbiting the earth is handed two balls that have identical outward appearances. However, one is hollow while the other is filled with lead. How can the astronaut determine which is which? Cutting or altering the balls is not allowed.

12. A hand presses down on the book in FIGURE Q6.12. Is the normal force of the table on the book larger than, smaller than, or equal to mg?

FIGURE Q6.12 Book of mass m

13. Boxes A and B in FIGURE Q6.13 both remain at rest. Is the friction force on A larger than, smaller than, or equal to the friction force on B? Explain.

FIGURE Q6.13

14. Suppose you push a hockey puck of mass m across frictionless ice for 1.0 s, starting from rest, giving the puck speed v after traveling distance d. If you repeat the experiment with a puck of mass $2m$, pushing with the same force,
   a. How long will you have to push for the puck to reach the same speed v?
   b. How long will you have to push for the puck to travel the same distance d?

15. A block pushed along the floor with velocity v_{0x} slides a distance d after the pushing force is removed.
   a. If the mass of the block is doubled but its initial velocity is not changed, what distance does the block slide before stopping?
   b. If the initial velocity is doubled to $2v_{0x}$ but the mass is not changed, what distance does the block slide before stopping?

16. A crate of fragile dishes is in the back of a pickup truck. The truck accelerates north from a stop sign, and the crate moves without slipping. Does the friction force on the crate point north or south? Or is the friction force zero? Explain.

17. Five balls move through the air as shown in FIGURE Q6.17. All five have the same size and shape. Air resistance is not negligible. Rank in order, from largest to smallest, the magnitudes of the accelerations a_a to a_e. Some may be equal. Give your answer in the form a > b = c > d > e and explain your ranking.

FIGURE Q6.17

EXERCISES AND PROBLEMS

Exercises

Section 6.1 The Equilibrium Model

1. ‖ The three ropes in **FIGURE EX6.1** are tied to a small, very light ring. Two of these ropes are anchored to walls at right angles with the tensions shown in the figure. What are the magnitude and direction of the tension $\vec{T}_3$ in the third rope?

FIGURE EX6.1

2. | The three ropes in **FIGURE EX6.2** are tied to a small, very light ring. Two of the ropes are anchored to walls at right angles, and the third rope pulls as shown. What are T_1 and T_2, the magnitudes of the tension forces in the first two ropes?

FIGURE EX6.2

3. | A football coach sits on a sled while two of his players build their strength by dragging the sled across the field with ropes. The friction force on the sled is 1000 N, the players have equal pulls, and the angle between the two ropes is 20°. How hard must each player pull to drag the coach at a steady 2.0 m/s?

4. ‖ A 20 kg loudspeaker is suspended 2.0 m below the ceiling by two 3.0-m-long cables that angle outward at equal angles. What is the tension in the cables?

5. | A 65 kg gymnast wedges himself between two closely spaced vertical walls by pressing his hands and feet against the walls. What is the magnitude of the friction force on each hand and foot? Assume they are all equal.

6. ‖ A construction worker with a weight of 850 N stands on a roof that is sloped at 20°. What is the magnitude of the normal force of the roof on the worker?

7. ‖ In an electricity experiment, a 1.0 g plastic ball is suspended on a 60-cm-long string and given an electric charge. A charged rod brought near the ball exerts a horizontal electrical force $\vec{F}_{\text{elec}}$ on it, causing the ball to swing out to a 20° angle and remain there.
   a. What is the magnitude of $\vec{F}_{\text{elec}}$?
   b. What is the tension in the string?

Section 6.2 Using Newton's Second Law

8. | The forces in **FIGURE EX6.8** act on a 2.0 kg object. What are the values of a_x and a_y, the x- and y-components of the object's acceleration?

FIGURE EX6.8

9. | The forces in **FIGURE EX6.9** act on a 2.0 kg object. What are the values of a_x and a_y, the x- and y-components of the object's acceleration?

FIGURE EX6.9

10. | **FIGURE EX6.10** shows the velocity graph of a 2.0 kg object as it moves along the x-axis. What is the net force acting on this object at $t = 1$ s? At 4 s? At 7 s?

FIGURE EX6.10

11. ‖ **FIGURE EX6.11** shows the force acting on a 2.0 kg object as it moves along the x-axis. The object is at rest at the origin at $t = 0$ s. What are its acceleration and velocity at $t = 6$ s?

FIGURE EX6.11

12. | A horizontal rope is tied to a 50 kg box on frictionless ice. What is the tension in the rope if:
   a. The box is at rest?
   b. The box moves at a steady 5.0 m/s?
   c. The box has $v_x = 5.0$ m/s and $a_x = 5.0$ m/s^2?

13. | A 50 kg box hangs from a rope. What is the tension in the rope if:
   a. The box is at rest?
   b. The box moves up at a steady 5.0 m/s?
   c. The box has $v_y = 5.0$ m/s and is speeding up at 5.0 m/s^2?
   d. The box has $v_y = 5.0$ m/s and is slowing down at 5.0 m/s^2?

14. | A 2.0×10^7 kg train applies its brakes with the intent of slowing down at a 1.2 m/s^2 rate. What magnitude force must its brakes provide?

15. || A 8.0×10^4 kg spaceship is at rest in deep space. Its thrusters provide a force of 1200 kN. The spaceship fires its thrusters for 20 s, then coasts for 12 km. How long does it take the spaceship to coast this distance?

16. || The position of a 2.0 kg mass is given by $x = (2t^3 - 3t^2)$ m,
CALC where t is in seconds. What is the net horizontal force on the mass at (a) $t = 0$ s and (b) $t = 1$ s?

Section 6.3 Mass, Weight, and Gravity

17. | A woman has a mass of 55 kg.
    a. What is her weight while standing on earth?
    b. What are her mass and her weight on Mars, where $g = 3.76$ m/s^2?

18. | It takes the elevator in a skyscraper 4.0 s to reach its cruising speed of 10 m/s. A 60 kg passenger gets aboard on the ground floor. What is the passenger's weight
    a. Before the elevator starts moving?
    b. While the elevator is speeding up?
    c. After the elevator reaches its cruising speed?

19. || Zach, whose mass is 80 kg, is in an elevator descending at 10 m/s. The elevator takes 3.0 s to brake to a stop at the first floor.
    a. What is Zach's weight before the elevator starts braking?
    b. What is Zach's weight while the elevator is braking?

20. || **FIGURE EX6.20** shows the velocity graph of a 75 kg passenger in an elevator. What is the passenger's weight at $t = 1$ s? At 5 s? At 9 s?

FIGURE EX6.20

21. | What thrust does a 200 g model rocket need in order to have a vertical acceleration of 10 m/s^2
    a. On earth?
    b. On the moon, where $g = 1.62$ m/s^2?

22. || A 20,000 kg rocket has a rocket motor that generates 3.0×10^5 N of thrust. Assume no air resistance.
    a. What is the rocket's initial upward acceleration?
    b. At an altitude of 5000 m the rocket's acceleration has increased to 6.0 m/s^2. What mass of fuel has it burned?

23. || The earth is 1.50×10^{11} m from the sun. The earth's mass is 5.98×10^{24} kg, while the mass of the sun is 1.99×10^{30} kg. What is earth's acceleration toward the sun?

Section 6.4 Friction

24. | Bonnie and Clyde are sliding a 300 kg bank safe across the floor to their getaway car. The safe slides with a constant speed if Clyde pushes from behind with 385 N of force while Bonnie pulls forward on a rope with 350 N of force. What is the safe's coefficient of kinetic friction on the bank floor?

25. | A stubborn, 120 kg mule sits down and refuses to move. To drag the mule to the barn, the exasperated farmer ties a rope around the mule and pulls with his maximum force of 800 N. The coefficients of friction between the mule and the ground are $\mu_s = 0.8$ and $\mu_k = 0.5$. Is the farmer able to move the mule?

26. || A 10 kg crate is placed on a horizontal conveyor belt. The materials are such that $\mu_s = 0.5$ and $\mu_k = 0.3$.
    a. Draw a free-body diagram showing all the forces on the crate if the conveyer belt runs at constant speed.
    b. Draw a free-body diagram showing all the forces on the crate if the conveyer belt is speeding up.
    c. What is the maximum acceleration the belt can have without the crate slipping?

27. || Bob is pulling a 30 kg filing cabinet with a force of 200 N, but the filing cabinet refuses to move. The coefficient of static friction between the filing cabinet and the floor is 0.80. What is the magnitude of the friction force on the filing cabinet?

28. || A rubber-wheeled 50 kg cart rolls down a 15° concrete incline. What is the cart's acceleration if rolling friction is (a) neglected and (b) included?

29. || A 4000 kg truck is parked on a 15° slope. How big is the friction force on the truck? The coefficient of static friction between the tires and the road is 0.90.

30. | A 1500 kg car skids to a halt on a wet road where $\mu_k = 0.50$. How fast was the car traveling if it leaves 65-m-long skid marks?

31. || A 50,000 kg locomotive is traveling at 10 m/s when its engine and brakes both fail. How far will the locomotive roll before it comes to a stop? Assume the track is level.

32. || You and your friend Peter are putting new shingles on a roof pitched at 25°. You're sitting on the very top of the roof when Peter, who is at the edge of the roof directly below you, 5.0 m away, asks you for the box of nails. Rather than carry the 2.5 kg box of nails down to Peter, you decide to give the box a push and have it slide down to him. If the coefficient of kinetic friction between the box and the roof is 0.55, with what speed should you push the box to have it gently come to rest right at the edge of the roof?

33. || An Airbus A320 jetliner has a takeoff mass of 75,000 kg. It reaches its takeoff speed of 82 m/s (180 mph) in 35 s. What is the thrust of the engines? You can neglect air resistance but not rolling friction.

Section 6.5 Drag

34. || A medium-sized jet has a 3.8-m-diameter fuselage and a loaded mass of 85,000 kg. The drag on an airplane is primarily due to the cylindrical fuselage, and aerodynamic shaping gives it a drag coefficient of 0.37. How much thrust must the jet's engines provide to cruise at 230 m/s at an altitude where the air density is 1.0 kg/m^3?

35. ||| A 75 kg skydiver can be modeled as a rectangular "box" with dimensions 20 cm × 40 cm × 180 cm. What is his terminal speed if he falls feet first? Use 0.8 for the drag coefficient.

36. ||| A 6.5-cm-diameter ball has a terminal speed of 26 m/s. What is the ball's mass?

Problems

37. || A 2.0 kg object initially at rest at the origin is subjected to the time-varying force shown in **FIGURE P6.37**. What is the object's velocity at $t = 4$ s?

FIGURE P6.37

38. ‖ A 5.0 kg object initially at rest at the origin is subjected to the time-varying force shown in **FIGURE P6.38**. What is the object's velocity at $t = 6$ s?

FIGURE P6.38

39. ‖ The 1000 kg steel beam in **FIGURE P6.39** is supported by two ropes. What is the tension in each?

FIGURE P6.39

40. ‖ Henry, whose mass is 95 kg, stands on a bathroom scale in an elevator. The scale reads 830 N for the first 3.0 s after the elevator starts moving, then 930 N for the next 3.0 s. What is the elevator's velocity 6.0 s after starting?

41. ‖ An accident victim with a broken
BIO leg is being placed in traction. The patient wears a special boot with a pulley attached to the sole. The foot and boot together have a mass of 4.0 kg, and the doctor has decided to hang a 6.0 kg mass from the rope. The boot is held suspended by the ropes, as shown in **FIGURE P6.41**, and does not touch the bed.

FIGURE P6.41

   a. Determine the amount of tension in the rope by using Newton's laws to analyze the hanging mass.
   b. The net traction force needs to pull straight out on the leg. What is the proper angle θ for the upper rope?
   c. What is the net traction force pulling on the leg?
   Hint: If the pulleys are frictionless, which we will assume, the tension in the rope is constant from one end to the other.

42. ‖ Seat belts and air bags save lives by reducing the forces exerted
BIO on the driver and passengers in an automobile collision. Cars are designed with a "crumple zone" in the front of the car. In the event of an impact, the passenger compartment decelerates over a distance of about 1 m as the front of the car crumples. An occupant restrained by seat belts and air bags decelerates with the car. By contrast, an unrestrained occupant keeps moving forward with no loss of speed (Newton's first law!) until hitting the dashboard or windshield. These are unyielding surfaces, and the unfortunate occupant then decelerates over a distance of only about 5 mm.
   a. A 60 kg person is in a head-on collision. The car's speed at impact is 15 m/s. Estimate the net force on the person if he or she is wearing a seat belt and if the air bag deploys.
   b. Estimate the net force that ultimately stops the person if he or she is not restrained by a seat belt or air bag.

43. ‖ The piston of a machine exerts a constant force on a ball as it moves horizontally through a distance of 15 cm. You use a motion detector to measure the speed of five different balls as they come off the piston; the data are shown in the table. Use theory to find two quantities that, when graphed, should give a straight line. Then use the graph to find the size of the piston's force.

Mass (g)	Speed (m/s)
200	9.4
400	6.3
600	5.2
800	4.9
1000	4.0

44. ‖‖ Compressed air is used to fire a 50 g ball vertically upward from a 1.0-m-tall tube. The air exerts an upward force of 2.0 N on the ball as long as it is in the tube. How high does the ball go above the top of the tube?

45. ‖ a. A rocket of mass m is launched straight up with thrust $\vec{F}_{thrust}$. Find an expression for the rocket's speed at height h if air resistance is neglected.
   b. The motor of a 350 g model rocket generates 9.5 N thrust. If air resistance can be neglected, what will be the rocket's speed as it reaches a height of 85 m?

46. ‖ A rifle with a barrel length of 60 cm fires a 10 g bullet with a horizontal speed of 400 m/s. The bullet strikes a block of wood and penetrates to a depth of 12 cm.
   a. What resistive force (assumed to be constant) does the wood exert on the bullet?
   b. How long does it take the bullet to come to rest?

47. ‖ A truck with a heavy load has a total mass of 7500 kg. It is climbing a 15° incline at a steady 15 m/s when, unfortunately, the poorly secured load falls off! Immediately after losing the load, the truck begins to accelerate at 1.5 m/s². What was the mass of the load? Ignore rolling friction.

48. ‖ An object of mass m is at rest at the top of a smooth slope of height h and length L. The coefficient of kinetic friction between the object and the surface, μ_k, is small enough that the object will slide down the slope after being given a very small push to get it started. Find an expression for the object's speed at the bottom of the slope.

49. ‖ Sam, whose mass is 75 kg, takes off across level snow on his jet-powered skis. The skis have a thrust of 200 N and a coefficient of kinetic friction on snow of 0.10. Unfortunately, the skis run out of fuel after only 10 s.
   a. What is Sam's top speed?
   b. How far has Sam traveled when he finally coasts to a stop?

50. ‖ A baggage handler drops your 10 kg suitcase onto a conveyor belt running at 2.0 m/s. The materials are such that $\mu_s = 0.50$ and $\mu_k = 0.30$. How far is your suitcase dragged before it is riding smoothly on the belt?

51. ‖ A 2.0 kg wood block is launched up a wooden ramp that is inclined at a 30° angle. The block's initial speed is 10 m/s.
   a. What vertical height does the block reach above its starting point?
   b. What speed does it have when it slides back down to its starting point?

52. ‖ It's a snowy day and you're pulling a friend along a level road on a sled. You've both been taking physics, so she asks what you think the coefficient of friction between the sled and the snow is. You've been walking at a steady 1.5 m/s, and the rope pulls up on the sled at a 30° angle. You estimate that the mass of the sled, with your friend on it, is 60 kg and that you're pulling with a force of 75 N. What answer will you give?

53. ‖ A large box of mass M is pulled across a horizontal, frictionless surface by a horizontal rope with tension T. A small box of mass m sits on top of the large box. The coefficients of static and kinetic friction between the two boxes are μ_s and μ_k, respectively. Find an expression for the maximum tension T_{max} for which the small box rides on top of the large box without slipping.

54. ‖ A large box of mass M is moving on a horizontal surface at speed v_0. A small box of mass m sits on top of the large box. The coefficients of static and kinetic friction between the two boxes are μ_s and μ_k, respectively. Find an expression for the shortest distance d_{min} in which the large box can stop without the small box slipping.

55. ‖ You're driving along at 25 m/s with your aunt's valuable antiques in the back of your pickup truck when suddenly you see a giant hole in the road 55 m ahead of you. Fortunately, your foot is right beside the brake and your reaction time is zero!

a. Can you stop the truck before it falls into the hole?

b. If your answer to part a is yes, can you stop without the antiques sliding and being damaged? Their coefficients of friction are $\mu_s = 0.60$ and $\mu_k = 0.30$.

Hint: You're not trying to stop in the shortest possible distance. What's your best strategy for avoiding damage to the antiques?

56. ‖ The 2.0 kg wood box in FIGURE P6.56 slides down a vertical wood wall while you push on it at a 45° angle. What magnitude of force should you apply to cause the box to slide down at a constant speed?

FIGURE P6.56

57. ‖ A 1.0 kg wood block is pressed against a vertical wood wall by the 12 N force shown in FIGURE P6.57. If the block is initially at rest, will it move upward, move downward, or stay at rest?

FIGURE P6.57

58. ‖ A person with compromised pinch strength in his fingers can exert a force of only 6.0 N to either side of a pinch-held object, such as the book shown in FIGURE P6.58. What is the heaviest book he can hold vertically before it slips out of his fingers? The coefficient of static friction between his fingers and the book cover is 0.80.

FIGURE P6.58

59. ‖ A ball is shot from a compressed-air gun at twice its terminal speed.

a. What is the ball's initial acceleration, as a multiple of g, if it is shot straight up?

b. What is the ball's initial acceleration, as a multiple of g, if it is shot straight down?

60. ‖‖‖ Starting from rest, a 2500 kg helicopter accelerates straight up at a constant 2.0 m/s². What is the helicopter's height at the moment its blades are providing an upward force of 26 kN? The helicopter can be modeled as a 2.6-m-diameter sphere.

61. ‖ Astronauts in space "weigh" themselves by oscillating on a
CALC spring. Suppose the position of an oscillating 75 kg astronaut is given by $x = (0.30 \text{ m}) \sin((\pi \text{ rad/s}) \times t)$, where t is in s. What force does the spring exert on the astronaut at (a) $t = 1.0$ s and (b) 1.5 s? Note that the angle of the sine function is in radians.

62. ‖ A particle of mass m moving along the x-axis experiences the
CALC net force $F_x = ct$, where c is a constant. The particle has velocity v_{0x} at $t = 0$. Find an algebraic expression for the particle's velocity v_x at a later time t.

63. ‖ At $t = 0$, an object of mass m is at rest at $x = 0$ on a horizontal,
CALC frictionless surface. A horizontal force $F_x = F_0(1 - t/T)$, which decreases from F_0 at $t = 0$ to zero at $t = T$, is exerted on the object. Find an expression for the object's (a) velocity and (b) position at time T.

64. ‖ At $t = 0$, an object of mass m is at rest at $x = 0$ on a hori-
CALC zontal, frictionless surface. Starting at $t = 0$, a horizontal force $F_x = F_0 e^{-t/T}$ is exerted on the object.

a. Find and graph an expression for the object's velocity at an arbitrary later time t.

b. What is the object's velocity after a very long time has elapsed?

65. ‖‖‖ Large objects have inertia and tend to keep moving—Newton's
BIO first law. Life is very different for small microorganisms that swim through water. For them, drag forces are so large that they instantly stop, without coasting, if they cease their swimming motion. To swim at constant speed, they must exert a constant propulsion force by rotating corkscrew-like flagella or beating hair-like cilia. The quadratic model of drag of Equation 6.15 fails for very small particles. Instead, a small object moving in a liquid experiences a *linear* drag force, $\vec{F}_{drag} = (bv,$ direction opposite the motion), where b is a constant. For a sphere of radius R, the drag constant can be shown to be $b = 6\pi\eta R$, where η is the *viscosity* of the liquid. Water at 20°C has viscosity 1.0×10^{-3} N s/m².

a. A *paramecium* is about 100 μm long. If it's modeled as a sphere, how much propulsion force must it exert to swim at a typical speed of 1.0 mm/s? How about the propulsion force of a 2.0-μm-diameter *E. coli* bacterium swimming at 30 μm/s?

b. The propulsion forces are very small, but so are the organisms. To judge whether the propulsion force is large or small *relative to the organism*, compute the acceleration that the propulsion force could give each organism if there were no drag. The density of both organisms is the same as that of water, 1000 kg/m³.

66. ‖‖‖ A 60 kg skater is gliding across frictionless ice at 4.0 m/s. Air
CALC resistance is not negligible. You can model the skater as a 170-cm-tall, 36-cm-diameter cylinder. What is the skater's speed 2.0 s later?

67. ‖‖‖ Very small objects, such as dust particles, experience a *linear* drag force, $\vec{F}_{drag} = (bv,$ direction opposite the motion), where b is a constant. That is, the quadratic model of drag of Equation 6.15 fails for very small particles. For a sphere of radius R, the drag constant can be shown to be $b = 6\pi\eta R$, where η is the *viscosity* of the gas.

a. Find an expression for the terminal speed v_{term} of a spherical particle of radius R and mass m falling through a gas of viscosity η.

b. Suppose a gust of wind has carried a 50-μm-diameter dust particle to a height of 300 m. If the wind suddenly stops, how long will it take the dust particle to settle back to the ground? Dust has a density of 2700 kg/m³, the viscosity of 25°C air is 2.0×10^{-5} N s/m², and you can assume that the falling dust particle reaches terminal speed almost instantly.

Problems 68 and 69 show a free-body diagram. For each:

a. Write a realistic dynamics problem for which this is the correct free-body diagram. Your problem should ask a question that can be answered with a value of position or velocity (such as "How far?" or "How fast?"), and should give sufficient information to allow a solution.

b. Solve your problem!

68.

69.

FIGURE P6.68   FIGURE P6.69

In Problems 70 through 72 you are given the dynamics equations that are used to solve a problem. For each of these, you are to

a. Write a realistic problem for which these are the correct equations.

b. Draw the free-body diagram and the pictorial representation for your problem.

c. Finish the solution of the problem.

70. $-0.80n = (1500 \text{ kg})a_x$

$n - (1500 \text{ kg})(9.80 \text{ m/s}^2) = 0$

71. $T - 0.20n - (20 \text{ kg})(9.80 \text{ m/s}^2) \sin 20°$

$\quad = (20 \text{ kg})(2.0 \text{ m/s}^2)$

$n - (20 \text{ kg})(9.80 \text{ m/s}^2) \cos 20° = 0$

72. $(100 \text{ N}) \cos 30° - f_k = (20 \text{ kg})a_x$

$n + (100 \text{ N}) \sin 30° - (20 \text{ kg})(9.80 \text{ m/s}^2) = 0$

$f_k = 0.20n$

Challenge Problems

73. ⫿ A block of mass m is at rest at the origin at $t = 0$. It is pushed
CALC with constant force F_0 from $x = 0$ to $x = L$ across a horizontal surface whose coefficient of kinetic friction is $\mu_k = \mu_0(1 - x/L)$. That is, the coefficient of friction decreases from μ_0 at $x = 0$ to zero at $x = L$.

a. Use what you've learned in calculus to prove that

$$a_x = v_x \frac{dv_x}{dx}$$

b. Find an expression for the block's speed as it reaches position L.

74. ⫿ A spring-loaded toy gun exerts a variable force on a plastic ball as
CALC the spring expands. Consider a horizontal spring and a ball of mass m whose position when barely touching a fully expanded spring is $x = 0$. The ball is pushed to the left, compressing the spring. You'll learn in Chapter 9 that the spring force on the ball, when the ball is at position x (which is negative), can be written as $(F_{Sp})_x = -kx$, where k is called the *spring constant*. The minus sign is needed to make the x-component of the force positive. Suppose the ball is initially pushed to $x_0 = -L$, then released and shot to the right.

a. Use what you've learned in calculus to prove that

$$a_x = v_x \frac{dv_x}{dx}$$

b. Find an expression, in terms of m, k, and L, for the speed of the ball as it comes off the spring at $x = 0$.

75. ⫿ **FIGURE CP6.75** shows an
CALC *accelerometer*, a device for measuring the horizontal acceleration of cars and airplanes. A ball is free to roll on a parabolic track described by the equation $y = x^2$, where both x and y are in meters. A scale along the bottom is used to measure the ball's horizontal position x.

FIGURE CP6.75

a. Find an expression that allows you to use a measured position x (in m) to compute the acceleration a_x (in m/s²). (For example, $a_x = 3x$ is a possible expression.)

b. What is the acceleration if $x = 20$ cm?

76. ⫿ An object moving in a liquid experiences a *linear* drag
CALC force: $\vec{F}_{drag} = (bv$, direction opposite the motion), where b is a constant called the *drag coefficient*. For a sphere of radius R, the drag constant can be computed as $b = 6\pi\eta R$, where η is the *viscosity* of the liquid.

a. Find an algebraic expression for $v_x(t)$, the x-component of velocity as a function of time, for a spherical particle of radius R and mass m that is shot horizontally with initial speed v_0 through a liquid of viscosity η.

b. Water at 20°C has viscosity $\eta = 1.0 \times 10^{-3}$ N s/m². Suppose a 4.0-cm-diameter, 33 g ball is shot horizontally into a tank of 20°C water. How long will it take for the horizontal speed to decrease to 50% of its initial value?

77. ⫿ An object moving in a liquid experiences a *linear* drag
CALC force: $\vec{F}_{drag} = (bv$, direction opposite the motion), where b is a constant called the *drag coefficient*. For a sphere of radius R, the drag constant can be computed as $b = 6\pi\eta R$, where η is the *viscosity* of the liquid.

a. Use what you've learned in calculus to prove that

$$a_x = v_x \frac{dv_x}{dx}$$

b. Find an algebraic expression for $v_x(x)$, the x-component of velocity as a function of distance traveled, for a spherical particle of radius R and mass m that is shot horizontally with initial speed v_0 through a liquid of viscosity η.

c. Water at 20°C has viscosity $\eta = 1.0 \times 10^{-3}$ N s/m². Suppose a 1.0-cm-diameter, 1.0 g marble is shot horizontally into a tank of 20°C water at 10 cm/s. How far will it travel before stopping?

78. ⫿ An object with cross section A is shot horizontally across
CALC frictionless ice. Its initial velocity is v_{0x} at $t_0 = 0$ s. Air resistance is not negligible.

a. Show that the velocity at time t is given by the expression

$$v_x = \frac{v_{0x}}{1 + C\rho A v_{0x} t/2m}$$

b. A 1.6-m-wide, 1.4-m-high, 1500 kg car with a drag coefficient of 0.35 hits a very slick patch of ice while going 20 m/s. If friction is neglected, how long will it take until the car's speed drops to 10 m/s? To 5 m/s?

c. Assess whether or not it is reasonable to neglect kinetic friction.

7 Newton's Third Law

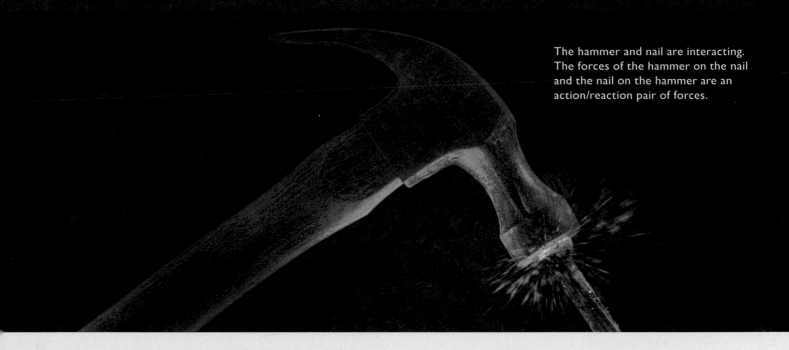

The hammer and nail are interacting. The forces of the hammer on the nail and the nail on the hammer are an action/reaction pair of forces.

IN THIS CHAPTER, you will use Newton's third law to understand how objects interact.

What is an interaction?

All forces are interactions in which objects exert forces on each other. If A pushes on B, then B pushes back on A. These two forces form an action/reaction pair of forces. One can't exist without the other.

Action/reaction pair

« LOOKING BACK Section 5.5 Forces, interactions, and Newton's second law

What is an interaction diagram?

We will often analyze a problem by defining a system—the objects of interest—and the larger environment that acts on the system. An interaction diagram is a key visual tool for identifying action/reaction forces of interaction *inside* the system and external forces from agents in the environment.

How do we model ropes and pulleys?

A common way that two objects interact is to be connected via a rope or cable or string. Pulleys change the direction of the tension forces. We will often model

- Ropes and strings as massless;
- Pulleys as massless and frictionless.

The objects' accelerations are constrained to have the same magnitude.

What is Newton's third law?

Newton's third law governs interactions:

- Every force is a member of an action/reaction pair.
- The two members of a pair act on different objects.
- The two members of a pair are equal in magnitude but opposite in direction.

How is Newton's third law used?

The dynamics problem-solving strategy of Chapter 6 is still our primary tool.

- Draw a free-body diagram for *each object*.
- Identify and show action/reaction pairs.
- Use Newton's second law for *each object*.
- Relate forces with Newton's third law.

« LOOKING BACK Section 6.2 Problem-Solving Strategy 6.1

Why is Newton's third law important?

We started our study of dynamics with only the first two of Newton's laws in order to practice identifying and using forces. But objects in the real world don't exist in isolation—they *interact* with each other. Newton's third law gives us a much more complete view of mechanics. The third law is also an essential tool in the practical application of physics to problems in engineering and technology.

7.1 Interacting Objects

FIGURE 7.1 The hammer and nail are interacting with each other.

The force of the nail on the hammer...

The force of the hammer on the nail

FIGURE 7.1 shows a hammer hitting a nail. The hammer exerts a force on the nail as it drives the nail forward. At the same time, the nail exerts a force on the hammer. If you're not sure that it does, imagine hitting the nail with a glass hammer. It's the force of the nail on the hammer that would cause the glass to shatter.

In fact, any time an object A pushes or pulls on another object B, B pushes or pulls back on A. When you pull someone with a rope in a tug-of-war, that person pulls back on you. Your chair pushes up on you (the normal force) as you push down on the chair. These are examples of an **interaction**, the mutual influence of two objects on each other.

To be more specific, if object A exerts a force $\vec{F}_{A\,on\,B}$ on object B, then object B exerts a force $\vec{F}_{B\,on\,A}$ on object A. This pair of forces, shown in **FIGURE 7.2**, is called an **action/reaction pair.** Two objects interact by exerting an action/reaction pair of forces on each other. Notice the very explicit subscripts on the force vectors. The first letter is the *agent;* the second letter is the object on which the force acts. $\vec{F}_{A\,on\,B}$ is a force exerted *by* A *on* B.

FIGURE 7.2 An action/reaction pair of forces.

$\vec{F}_{B\,on\,A}$ $\vec{F}_{A\,on\,B}$

A   B

Action/reaction pair

NOTE The name "action/reaction pair" is somewhat misleading. The forces occur simultaneously, and we cannot say which is the "action" and which the "reaction." **An action/reaction pair of forces exists as a pair, or not at all.**

The hammer and nail interact through contact forces. Does the same idea hold true for long-range forces such as gravity? Newton was the first to realize that it does. His evidence was the tides. Astronomers had known since antiquity that the tides depend on the phase of the moon, but Newton was the first to understand that tides are the ocean's response to the gravitational pull of the moon on the earth.

Objects, Systems, and the Environment

Chapters 5 and 6 considered forces acting on a single object that we modeled as a particle. **FIGURE 7.3a** shows a diagrammatic representation of single-particle dynamics. We can use Newton's second law, $\vec{a} = \vec{F}_{net}/m$, to determine the particle's acceleration.

We now want to extend the particle model to situations in which two or more objects, each represented as a particle, interact with each other. For example, **FIGURE 7.3b** shows three objects interacting via action/reaction pairs of forces. The forces can be given labels such as $\vec{F}_{A\,on\,B}$ and $\vec{F}_{B\,on\,A}$. How do these particles move?

We will often be interested in the motion of some of the objects, say objects A and B, but not of others. For example, objects A and B might be the hammer and the nail, while object C is the earth. The earth interacts with both the hammer and the nail via gravity, but in a practical sense the earth remains "at rest" while the hammer and nail move. Let's define the **system** as those objects whose motion we want to analyze and the **environment** as objects external to the system.

FIGURE 7.3c is a new kind of diagram, an **interaction diagram**, in which we've enclosed the objects of the system in a box and represented interactions as lines

FIGURE 7.3 Single-particle dynamics and a model of interacting objects.

(a) Single-particle dynamics

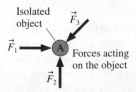

Isolated object

$\vec{F}_3$

$\vec{F}_1$

A

Forces acting on the object

$\vec{F}_2$

This is a force diagram.

(b) Interacting objects

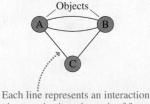

Objects

A   B

C

Each line represents an interaction via an action/reaction pair of forces.

(c) System and environment

System

A   B

C

External forces

Internal interactions

Environment

This is an *interaction diagram.*

connccting objects. This is a rather abstract, schematic diagram, but it captures the essence of the interactions. Notice that interactions with objects in the environment are called **external forces.** For the hammer and nail, the gravitational force on each—an interaction with the earth—is an external force.

NOTE *Every* force is one member of an action/reaction pair, so there is no such thing as a true "external force." What we call an external force is simply an interaction between an object of interest, one we've chosen to place inside the system, and an object whose motion is not of interest.

The bat and the ball are interacting with each other.

7.2 Analyzing Interacting Objects

TACTICS BOX 7.1 (MP)

Analyzing interacting objects

❶ **Represent each object as a circle** with a name and label. Place each in the correct position relative to other objects. The surface of the earth (label S; contact forces) and the entire earth (label EE; long-range forces) should be considered separate objects.

❷ **Identify interactions.** Draw *one* connecting line between relevant circles to represent each interaction.

- Every interaction line connects two and only two objects.
- A surface can have two interactions: friction (parallel to the surface) and a normal force (perpendicular to the surface).
- The entire earth interacts only by the long-range gravitational force.

❸ **Identify the system.** Identify the objects of interest; draw and label a box enclosing them. This completes the interaction diagram.

❹ **Draw a free-body diagram for each object in the system.** Include only the forces acting *on* each object, not forces exerted by the object.

- Every interaction line crossing the system boundary is one external force acting on an object. The usual symbols, such as $\vec{n}$ and $\vec{T}$ can be used.
- Every interaction line within the system represents an action/reaction pair of forces. There is one force vector on *each* of the objects, and these forces point in opposite directions. Use labels like $\vec{F}_{\text{A on B}}$ and $\vec{F}_{\text{B on A}}$.
- Connect the two action/reaction forces—which must be on *different* free-body diagrams—with a dashed line.

Exercises 1–7

We'll illustrate these ideas with two concrete examples. The first cxample will be much longer than usual because we'll go carefully through all the steps in the reasoning.

EXAMPLE 7.1 | **Pushing a crate**

FIGURE 7.4 shows a person pushing a large crate across a rough surface. Identify all interactions, show them on an interaction diagram, then draw free-body diagrams of the person and the crate.

FIGURE 7.4 A person pushes a crate across a rough floor.

VISUALIZE The interaction diagram of **FIGURE 7.5** on the next page starts by representing every object as a circle in the correct position but separated from all other objects. The person and the crate are obvious objects. The earth is also an object that both exerts and experiences forces, but it's necessary to distinguish between the surface, which exerts contact forces, and thc cntire earth, which exerts the long-range gravitational force.

Figure 7.5 also identifies the various interactions. Some, like the pushing interaction between the person and the crate, are fairly

Continued

FIGURE 7.5 The interaction diagram.

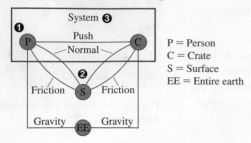

P = Person
C = Crate
S = Surface
EE = Entire earth

FIGURE 7.6 Free-body diagrams of the person and the crate.

obvious. The interactions with the earth are a little trickier. Gravity, a long-range force, is an interaction between each object and the earth as a whole. Friction forces and normal forces are contact interactions between each object and the earth's surface. These are two different interactions, so two interaction lines connect the crate to the surface and the person to the surface. Finally, we've enclosed the person and crate in a box labeled System. These are the objects whose motion we wish to analyze.

NOTE Interactions are between two *different* objects. None of the interactions are between an object and itself.

We can now draw free-body diagrams for the objects in the system, the crate and the person. **FIGURE 7.6** correctly locates the crate's free-body diagram to the right of the person's free-body diagram. For each, three interaction lines cross the system boundary and thus represent external forces. These are the gravitational force from the entire earth, the upward normal force from the surface, and a friction force from the surface. We can use familiar labels such as $\vec{n}_P$ and $\vec{f}_C$, but **it's very important to distinguish different forces with subscripts.** There's now more than one normal force. If you call both simply $\vec{n}$, you're almost certain to make mistakes when you start writing out the second-law equations.

The directions of the normal forces and the gravitational forces are clear, but we have to be careful with friction. Friction force $\vec{f}_C$ is kinetic friction of the crate sliding across the surface, so it

points left, opposite the motion. But what about friction between the person and the surface? It is tempting to draw force $\vec{f}_P$ pointing to the left. After all, friction forces are supposed to be in the direction opposite the motion. But if we did so, the person would have two forces to the left, $\vec{F}_{C \text{ on } P}$ and $\vec{f}_P$, and none to the right, causing the person to accelerate *backward!* That is clearly not what happens, so what is wrong?

Imagine pushing a crate to the right across loose sand. Each time you take a step, you tend to kick the sand to the *left,* behind you. Thus friction force $\vec{f}_{P \text{ on } S}$, the force of the person pushing against the earth's surface, is to the *left.* In reaction, the force of the earth's surface against the person is a friction force to the *right.* It is force $\vec{f}_{S \text{ on } P}$, which we've shortened to $\vec{f}_P$, that causes the person to accelerate in the forward direction. Further, as we'll discuss more below, this is a *static* friction force; your foot is planted on the ground, not sliding across the surface.

Finally, we have one internal interaction. The crate is pushed with force $\vec{F}_{P \text{ on } C}$. If A pushes or pulls on B, then B pushes or pulls back on A, so the reaction to force $\vec{F}_{P \text{ on } C}$ is $\vec{F}_{C \text{ on } P}$, the crate pushing back against the person's hands. Force $\vec{F}_{P \text{ on } C}$ is a force exerted on the crate, so it's shown on the crate's free-body diagram. Force $\vec{F}_{C \text{ on } P}$ is exerted on the person, so it is drawn on the person's free-body diagram. **The two forces of an action/reaction pair never occur on the same free-body diagram.** We've connected forces $\vec{F}_{P \text{ on } C}$ and $\vec{F}_{C \text{ on } P}$ with a dashed line to show that they are an action/reaction pair.

Propulsion

The friction force $\vec{f}_P$ (force of surface on person) is an example of **propulsion.** It is the force that a system with an internal source of energy uses to drive itself forward. Propulsion is an important feature not only of walking or running but also of the forward motion of cars, jets, and rockets. Propulsion is somewhat counterintuitive, so it is worth a closer look.

If you try to walk across a frictionless floor, your foot slips and slides *backward.* In order for you to walk, the floor needs to have friction so that your foot *sticks* to the floor as you straighten your leg, moving your body forward. The friction that prevents slipping is *static* friction. Static friction, you will recall, acts in the direction that prevents slipping. The static friction force $\vec{f}_P$ has to point in the *forward* direction to prevent your foot from slipping backward. It is this forward-directed static friction force that propels you forward! The force of your foot on the floor, the other half of the action/reaction pair, is in the opposite direction.

The distinction between you and the crate is that you have an *internal source of energy* that allows you to straighten your leg by pushing backward against the surface.

In essence, you walk by pushing the earth away from you. The earth's surface responds by pushing you forward. These are static friction forces. In contrast, all the crate can do is slide, so *kinetic* friction opposes the motion of the crate.

FIGURE 7.7 shows how propulsion works. A car uses its motor to spin the tires, causing the tires to push backward against the ground. This is why dirt and gravel are kicked backward, not forward. The earth's surface responds by pushing the car forward. These are also *static* friction forces. The tire is rolling, but the bottom of the tire, where it contacts the road, is instantaneously at rest. If it weren't, you would leave one giant skid mark as you drove and would burn off the tread within a few miles.

FIGURE 7.7 Examples of propulsion.

The person pushes backward against the earth.
The earth pushes forward on the person.
Static friction

The car pushes backward against the earth.
The earth pushes forward on the car.
Static friction

The rocket pushes the hot gases backward.
The gases push the rocket forward.
Thrust

EXAMPLE 7.2 | Towing a car

A tow truck uses a rope to pull a car along a horizontal road, as shown in **FIGURE 7.8**. Identify all interactions, show them on an interaction diagram, then draw free-body diagrams of each object in the system.

FIGURE 7.8 A truck towing a car.

VISUALIZE The interaction diagram of **FIGURE 7.9** represents the objects as separate circles, but with the correct relative positions. The rope is shown as a separate object. Many of the interactions are identical to those in Example 7.1. The system—the objects in motion—consists of the truck, the rope, and the car.

The three objects in the system require three free-body diagrams, shown in **FIGURE 7.10**. Gravity, friction, and normal forces at the surface are all interactions that cross the system boundary and are shown as external forces. The car is an

FIGURE 7.9 The interaction diagram.

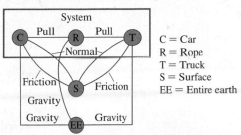

C = Car
R = Rope
T = Truck
S = Surface
EE = Entire earth

inert object rolling along. It would slow and stop if the rope were cut, so the surface must exert a rolling friction force $\vec{f}_C$ to the left. The truck, however, has an internal source of energy. The truck's drive wheels push the ground to the left with force $\vec{f}_{T \text{ on S}}$. In reaction, the ground propels the truck forward, to the right, with force $\vec{f}_T$.

We next need to identify the forces between the car, the truck, and the rope. The rope pulls on the car with a tension force $\vec{T}_{R \text{ on C}}$. You might be tempted to put the reaction force on the truck because we say that "the truck pulls the car," but the truck is not in contact

FIGURE 7.10 Free-body diagrams of Example 7.2.

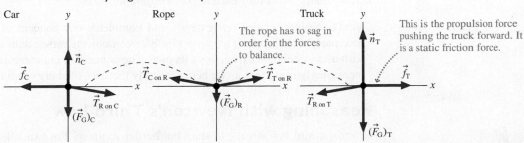

Continued

with the car. The truck pulls on the rope, then the rope pulls on the car. Thus the reaction to $\vec{T}_{\text{R on C}}$ is a force on the *rope:* $\vec{T}_{\text{C on R}}$. These are an action/reaction pair. At the other end, $\vec{T}_{\text{T on R}}$ and $\vec{T}_{\text{R on T}}$ are also an action/reaction pair.

> **NOTE** Drawing an interaction diagram helps you avoid mistakes because it shows very clearly what is interacting with what.

Notice that the tension forces of the rope *cannot* be horizontal. If they were, the rope's free-body diagram would show a net downward force, because of its weight, and the rope would accelerate downward.

The tension forces $\vec{T}_{\text{T on R}}$ and $\vec{T}_{\text{C on R}}$ have to angle slightly upward to balance the gravitational force, so any real rope has to sag at least a little in the center.

ASSESS Make sure you avoid the common error of considering $\vec{n}$ and $\vec{F}_{\text{G}}$ to be an action/reaction pair. These are both forces on the *same* object, whereas the two forces of an action/reaction pair are always on two *different* objects that are interacting with each other. The normal and gravitational forces are often equal in magnitude, as they are in this example, but that doesn't make them an action/reaction pair of forces.

STOP TO THINK 7.1   A rope of negligible mass pulls a crate across the floor. The rope and crate are the system; the hand pulling the rope is part of the environment. What, if anything, is wrong with the free-body diagrams?

7.3 Newton's Third Law

Newton was the first to recognize how the two members of an action/reaction pair of forces are related to each other. Today we know this as **Newton's third law:**

> **Newton's third law** Every force occurs as one member of an action/reaction pair of forces.
> - The two members of an action/reaction pair act on two *different* objects.
> - The two members of an action/reaction pair are equal in magnitude but opposite in direction: $\vec{F}_{\text{A on B}} = -\vec{F}_{\text{B on A}}$.

We deduced most of the third law in Section 7.2. There we found that the two members of an action/reaction pair are always opposite in direction (see Figures 7.6 and 7.10). According to the third law, this will always be true. But the most significant portion of the third law, which is by no means obvious, is that the two members of an action/reaction pair have *equal* magnitudes. That is, $F_{\text{A on B}} = F_{\text{B on A}}$. This is the quantitative relationship that will allow you to solve problems of interacting objects.

Newton's third law is frequently stated as "For every action there is an equal but opposite reaction." While this is indeed a catchy phrase, it lacks the preciseness of our preferred version. In particular, it fails to capture an essential feature of action/reaction pairs—that they each act on a *different* object.

> **NOTE** Newton's third law extends and completes our concept of *force*. We can now recognize force as an *interaction* between objects rather than as some "thing" with an independent existence of its own. The concept of an interaction will become increasingly important as we begin to study the laws of energy and momentum.

Reasoning with Newton's Third Law

Newton's third law is easy to state but harder to grasp. For example, consider what happens when you release a ball. Not surprisingly, it falls down. But if the ball and the earth exert equal and opposite forces on each other, as Newton's third law alleges, why doesn't the earth "fall up" to meet the ball?

The key to understanding this and many similar puzzles is that **the forces are equal but the accelerations are not.** Equal causes can produce very unequal effects. **FIGURE 7.11** shows equal-magnitude forces on the ball and the earth. The force on ball B is simply the gravitational force of Chapter 6:

$$\vec{F}_{\text{earth on ball}} = (\vec{F}_{\text{G}})_{\text{B}} = -m_{\text{B}}g\hat{j} \qquad (7.1)$$

where m_{B} is the mass of the ball. According to Newton's second law, this force gives the ball an acceleration

$$\vec{a}_{\text{B}} = \frac{(\vec{F}_{\text{G}})_{\text{B}}}{m_{\text{B}}} = -g\hat{j} \qquad (7.2)$$

This is just the familiar free-fall acceleration.

According to Newton's third law, the ball pulls up on the earth with force $\vec{F}_{\text{ball on earth}}$. Because $\vec{F}_{\text{earth on ball}}$ and $\vec{F}_{\text{ball on earth}}$ are an action/reaction pair, $\vec{F}_{\text{ball on earth}}$ must be equal in magnitude and opposite in direction to $\vec{F}_{\text{earth on ball}}$. That is,

$$\vec{F}_{\text{ball on earth}} = -\vec{F}_{\text{earth on ball}} = -(\vec{F}_{\text{G}})_{\text{B}} = +m_{\text{B}}g\hat{j} \qquad (7.3)$$

Using this result in Newton's second law, we find the upward acceleration of the earth as a whole is

$$\vec{a}_{\text{E}} = \frac{\vec{F}_{\text{ball on earth}}}{m_{\text{E}}} = \frac{m_{\text{B}}g\hat{j}}{m_{\text{E}}} = \left(\frac{m_{\text{B}}}{m_{\text{E}}}\right)g\hat{j} \qquad (7.4)$$

The upward acceleration of the earth is less than the downward acceleration of the ball by the factor $m_{\text{B}}/m_{\text{E}}$. If we assume a 1 kg ball, we can estimate the magnitude of $\vec{a}_{\text{E}}$:

$$\vec{a}_{\text{E}} = \frac{\vec{F}_{\text{ball on earth}}}{m_{\text{E}}} = \frac{m_{\text{B}}g\hat{j}}{m_{\text{E}}} = \left(\frac{m_{\text{B}}}{m_{\text{E}}}\right)g\hat{j}$$

With this incredibly small acceleration, it would take the earth 8×10^{15} years, approximately 500,000 times the age of the universe, to reach a speed of 1 mph! So we certainly would not expect to see or feel the earth "fall up" after we drop a ball.

NOTE Newton's third law equates the size of two forces, not two accelerations. The acceleration continues to depend on the mass, as Newton's second law states. **In an interaction between two objects of different mass, the lighter mass will do essentially all of the accelerating even though the forces exerted on the two objects are equal.**

FIGURE 7.11 The action/reaction forces of a ball and the earth are equal in magnitude.

The earth pulls on the ball. $\vec{F}_{\text{earth on ball}}$

The ball pulls equally hard on the earth. $\vec{F}_{\text{ball on earth}}$

EXAMPLE 7.3 | The forces on accelerating boxes

The hand shown in **FIGURE 7.12** pushes boxes A and B to the right across a frictionless table. The mass of B is larger than the mass of A.

a. Draw free-body diagrams of A, B, and the hand H, showing only the *horizontal* forces. Connect action/reaction pairs with dashed lines.

b. Rank in order, from largest to smallest, the horizontal forces shown on your free-body diagrams.

FIGURE 7.12 Hand H pushes boxes A and B.

$m_{\text{B}} > m_{\text{A}}$

H

Frictionless surface   A   B $\vec{a}$

VISUALIZE a. The hand H pushes on box A, and A pushes back on H. Thus $\vec{F}_{\text{H on A}}$ and $\vec{F}_{\text{A on H}}$ are an action/reaction pair. Similarly, A pushes on B and B pushes back on A. **The hand H does not touch box B, so there is no interaction between them.** There is no friction. **FIGURE 7.13** on the next page shows five horizontal forces and identifies two action/reaction pairs. Notice that each force is shown on the free-body diagram of the object that it acts *on*.

b. According to Newton's third law, $F_{\text{A on H}} = F_{\text{H on A}}$ and $F_{\text{A on B}} = F_{\text{B on A}}$. But the third law is not our only tool. The boxes are *accelerating* to the right, because there's no friction, so Newton's *second* law tells us that box A must have a net force to the right. Consequently, $F_{\text{H on A}} > F_{\text{B on A}}$. Similarly, $F_{\text{arm on H}} > F_{\text{A on H}}$ is needed to accelerate the hand. Thus

$$F_{\text{arm on H}} > F_{\text{A on H}} = F_{\text{H on A}} > F_{\text{B on A}} = F_{\text{A on B}}$$

Continued

ASSESS You might have expected $F_{\text{A on B}}$ to be larger than $F_{\text{H on A}}$ because $m_{\text{B}} > m_{\text{A}}$. It's true that the *net* force on B is larger than the *net* force on A, but we have to reason more closely to judge the individual forces. Notice how we used both the second and the third laws to answer this question.

FIGURE 7.13 The free-body diagrams, showing only the horizontal forces.

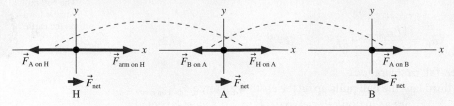

STOP TO THINK 7.2 A small car is pushing a larger truck that has a dead battery. The mass of the truck is larger than the mass of the car. Which of the following statements is true?

a. The car exerts a force on the truck, but the truck doesn't exert a force on the car.
b. The car exerts a larger force on the truck than the truck exerts on the car.
c. The car exerts the same amount of force on the truck as the truck exerts on the car.
d. The truck exerts a larger force on the car than the car exerts on the truck.
e. The truck exerts a force on the car, but the car doesn't exert a force on the truck.

Acceleration Constraints

FIGURE 7.14 The car and the truck have the same acceleration.

The rope is under tension.

Newton's third law is one quantitative relationship you can use to solve problems of interacting objects. In addition, we frequently have other information about the motion in a problem. For example, if two objects A and B move together, their accelerations are *constrained* to be equal: $\vec{a}_{\text{A}} = \vec{a}_{\text{B}}$. A well-defined relationship between the accelerations of two or more objects is called an **acceleration constraint**. It is an independent piece of information that can help solve a problem.

In practice, we'll express acceleration constraints in terms of the x- and y-components of $\vec{a}$. Consider the car being towed in FIGURE 7.14. This is one-dimensional motion, so we can write the acceleration constraint as

$$a_{\text{C}x} = a_{\text{T}x} = a_x$$

Because the accelerations of both objects are equal, we can drop the subscripts C and T and call both of them a_x.

Don't assume the accelerations of A and B will always have the same sign. Consider blocks A and B in FIGURE 7.15. The blocks are connected by a string, so they are constrained to move together and their accelerations have equal magnitudes. But A has a positive acceleration (to the right) in the x-direction while B has a negative acceleration (downward) in the y-direction. Thus the acceleration constraint is

FIGURE 7.15 The string constrains the two objects to accelerate together.

The accelerations have the same magnitude.

$$a_{\text{A}x} = -a_{\text{B}y}$$

This relationship does *not* say that $a_{\text{A}x}$ is a negative number. It is simply a relational statement, saying that $a_{\text{A}x}$ is (-1) times whatever $a_{\text{B}y}$ happens to be. The acceleration $a_{\text{B}y}$ in Figure 7.15 is a negative number, so $a_{\text{A}x}$ is positive. In some problems, the signs of $a_{\text{A}x}$ and $a_{\text{B}y}$ may not be known until the problem is solved, but the *relationship* is known from the beginning.

A Revised Strategy for Interacting-Objects Problems

Problems of interacting objects can be solved with a few modifications to the problem-solving strategy we developed in « Section 6.2.

PROBLEM-SOLVING STRATEGY 7.1

Interacting-objects problems

MODEL Identify which objects are part of the system and which are part of the environment. Make simplifying assumptions.

VISUALIZE Draw a pictorial representation.

- Show important points in the motion with a sketch. You may want to give each object a separate coordinate system. Define symbols, list acceleration constraints, and identify what the problem is trying to find.
- Draw an interaction diagram to identify the forces on each object and all action/reaction pairs.
- Draw a *separate* free-body diagram for each object showing only the forces acting *on* that object, not forces exerted by the object. Connect the force vectors of action/reaction pairs with dashed lines.

SOLVE Use Newton's second and third laws.

- Write the equations of Newton's second law for *each* object, using the force information from the free-body diagrams.
- Equate the magnitudes of action/reaction pairs.
- Include the acceleration constraints, the friction model, and other quantitative information relevant to the problem.
- Solve for the acceleration, then use kinematics to find velocities and positions.

ASSESS Check that your result has the correct units and significant figures, is reasonable, and answers the question.

You might be puzzled that the Solve step calls for the use of the third law to equate just the *magnitudes* of action/reaction forces. What about the "opposite in direction" part of the third law? You have already used it! Your free-body diagrams should show the two members of an action/reaction pair to be opposite in direction, and that information will have been utilized in writing the second-law equations. Because the directional information has already been used, all that is left is the magnitude information.

EXAMPLE 7.4 | Keep the crate from sliding

You and a friend have just loaded a 200 kg crate filled with priceless art objects into the back of a 2000 kg truck. As you press down on the accelerator, force $\vec{F}_{\text{surface on truck}}$ propels the truck forward. To keep things simple, call this just $\vec{F}_T$. What is the maximum magnitude $\vec{F}_T$ can have without the crate sliding? The static and kinetic coefficients of friction between the crate and the bed of the truck are 0.80 and 0.30. Rolling friction of the truck is negligible.

MODEL The crate and the truck are separate objects that form the system. We'll model them as particles. The earth and the road surface are part of the environment.

VISUALIZE The sketch in **FIGURE 7.16** on the next page establishes a coordinate system, lists the known information, and—new

to problems of interacting objects—identifies the acceleration constraint. As long as the crate doesn't slip, it must accelerate *with* the truck. Both accelerations are in the positive x-direction, so the acceleration constraint in this problem is $a_{Cx} = a_{Tx} = a_x$.

The interaction diagram of Figure 7.16 shows the crate interacting twice with the truck—a friction force parallel to the surface of the truck bed and a normal force perpendicular to this surface. The truck interacts similarly with the road surface, but notice that the crate does not interact with the ground; there's no contact between them. The two interactions within the system are each an action/reaction pair, so this is a total of four forces. You can also see four external forces crossing the system boundary, so the free-body diagrams should show a total of eight forces.

Continued

FIGURE 7.16 Pictorial representation of the crate and truck in Example 7.4.

Sketch

Known
$m_T = 2000$ kg
$m_C = 200$ kg
$\mu_s = 0.80$
$\mu_k = 0.30$

Acceleration constraint
$a_{Cx} = a_{Tx} = a_x$

Find
$(F_T)_{max}$ without slipping

Interaction diagram

Free-body diagrams

Finally, the interaction information is transferred to the free-body diagrams, where we see friction between the crate and truck as an action/reaction pair and the normal forces (the truck pushes up on the crate, the crate pushes down on the truck) as another action/reaction pair. It's easy to overlook forces such as $\vec{f}_{C \text{ on } T}$, but you won't make this mistake if you first identify action/reaction pairs on an interaction diagram. Note that $\vec{f}_{C \text{ on } T}$ and $\vec{f}_{T \text{ on } C}$ are *static* friction forces because they are forces that prevent slipping; force $\vec{f}_{T \text{ on } C}$ must point forward to prevent the crate from sliding out the back of the truck.

SOLVE Now we're ready to write Newton's second law. For the crate:

$$\sum (F_{\text{on crate}})_x = f_{T \text{ on } C} = m_C a_{Cx} = m_C a_x$$
$$\sum (F_{\text{on crate}})_y = n_{T \text{ on } C} - (F_G)_C = n_{T \text{ on } C} - m_C g = 0$$

For the truck:

$$\sum (F_{\text{on truck}})_x = F_T - f_{C \text{ on } T} = m_T a_{Tx} = m_T a_x$$
$$\sum (F_{\text{on truck}})_y = n_T - (F_G)_T - n_{C \text{ on } T}$$
$$= n_T - m_T g - n_{C \text{ on } T} = 0$$

Be sure you agree with all the signs, which are based on the free-body diagrams. The net force in the y-direction is zero because there's no motion in the y-direction. It may seem like a lot of effort to write all the subscripts, but it is very important in problems with more than one object.

Notice that we've already used the acceleration constraint $a_{Cx} = a_{Tx} = a_x$. Another important piece of information is Newton's third law, which tells us that $f_{C \text{ on } T} = f_{T \text{ on } C}$ and $n_{C \text{ on } T} = n_{T \text{ on } C}$. Finally, we know that the maximum value of F_T will occur when the static friction on the crate reaches its maximum value:

$$f_{T \text{ on } C} = f_{s \text{ max}} = \mu_s n_{T \text{ on } C}$$

The friction depends on the normal force on the crate, not the normal force on the truck.

Now we can assemble all the pieces. From the y-equation of the crate, $n_{T \text{ on } C} = m_C g$. Thus

$$f_{T \text{ on } C} = \mu_s n_{T \text{ on } C} = \mu_s m_C g$$

Using this in the x-equation of the crate, we find that the acceleration is

$$a_x = \frac{f_{T \text{ on } C}}{m_C} = \mu_s g$$

This is the crate's maximum acceleration without slipping. Now use this acceleration *and* the fact that $f_{C \text{ on } T} = f_{T \text{ on } C} = \mu_s m_C g$ in the x-equation of the truck to find

$$F_T - f_{C \text{ on } T} = F_T - \mu_s m_C g = m_T a_x = m_T \mu_s g$$

Solving for F_T, we find the maximum propulsion without the crate sliding is

$$(F_T)_{max} = \mu_s(m_T + m_C)g$$
$$= (0.80)(2200 \text{ kg})(9.80 \text{ m/s}^2) = 17{,}000 \text{ N}$$

ASSESS This is a hard result to assess. Few of us have any intuition about the size of forces that propel cars and trucks. Even so, the fact that the forward force on the truck is a significant fraction (80%) of the combined weight of the truck and the crate seems plausible. We might have been suspicious if F_T had been only a tiny fraction of the weight or much greater than the weight.

As you can see, there are many equations and many pieces of information to keep track of when solving a problem of interacting objects. These problems are not inherently harder than the problems you learned to solve in Chapter 6, but they do require a high level of organization. Using the systematic approach of the problem-solving strategy will help you solve similar problems successfully.

STOP TO THINK 7.3 Boxes A and B are sliding to the right across a frictionless table. The hand H is slowing them down. The mass of A is larger than the mass of B. Rank in order, from largest to smallest, the *horizontal* forces on A, B, and H.

a. $F_{B \text{ on } H} = F_{H \text{ on } B} = F_{A \text{ on } B} = F_{B \text{ on } A}$
b. $F_{B \text{ on } H} = F_{H \text{ on } B} > F_{A \text{ on } B} = F_{B \text{ on } A}$
c. $F_{B \text{ on } H} = F_{H \text{ on } B} < F_{A \text{ on } B} = F_{B \text{ on } A}$
d. $F_{H \text{ on } B} = F_{H \text{ on } A} > F_{A \text{ on } B}$

7.4 Ropes and Pulleys

Many objects are connected by strings, ropes, cables, and so on. In single-particle dynamics, we defined *tension* as the force exerted on an object by a rope or string. Now we need to think more carefully about the string itself. Just what do we mean when we talk about the tension "in" a string?

Tension Revisited

FIGURE 7.17 shows a heavy safe hanging from a rope, placing the rope under tension. If you cut the rope, the safe and the lower portion of the rope will fall. Thus there must be a force *within* the rope by which the upper portion of the rope pulls upward on the lower portion to prevent it from falling.

Chapter 5 introduced an atomic-level model in which tension is due to the stretching of spring-like molecular bonds within the rope. Stretched springs exert pulling forces, and the combined pulling force of billions of stretched molecular springs in a string or rope is what we call *tension*.

An important aspect of tension is that it pulls equally *in both directions*. To gain a mental picture, imagine holding your arms outstretched and having two friends pull on them. You'll remain at rest—but "in tension"—as long as they pull with equal strength in opposite directions. But if one lets go, analogous to the breaking of molecular bonds if a rope breaks or is cut, you'll fly off in the other direction!

FIGURE 7.17 Tension forces within the rope are due to stretching the spring-like molecular bonds.

Stretched molecular bonds

Safe

Tension forces pull in both directions.

EXAMPLE 7.5 | **Pulling a rope**

FIGURE 7.18a shows a student pulling horizontally with a 100 N force on a rope that is attached to a wall. In FIGURE 7.18b, two students in a tug-of-war pull on opposite ends of a rope with 100 N each. Is the tension in the second rope larger than, smaller than, or the same as that in the first rope?

FIGURE 7.18 Which rope has a larger tension?

(a)

T = ?

100 N   Rope 1

(b)

T = ?

100 N   Rope 2   100 N

SOLVE Surely pulling on a rope from both ends causes more tension than pulling on one end. Right? Before jumping to conclusions, let's analyze the situation carefully.

FIGURE 7.19a shows the first student, the rope, and the wall as separate, interacting objects. Force $\vec{F}_{\text{S on R}}$ is the student pulling on the rope, so it has magnitude 100 N. Forces $\vec{F}_{\text{S on R}}$ and $\vec{F}_{\text{R on S}}$ are an action/reaction pair and must have equal magnitudes. Similarly for forces $\vec{F}_{\text{W on R}}$ and $\vec{F}_{\text{R on W}}$. Finally, because the rope is in static equilibrium, *force $\vec{F}_{\text{W on R}}$ has to balance force $\vec{F}_{\text{S on R}}$*. Thus

$$F_{\text{R on W}} = F_{\text{W on R}} = F_{\text{S on R}} = F_{\text{R on S}} = 100 \text{ N}$$

The first and third equalities are Newton's third law; the second equality is Newton's first law for the rope.

Forces $\vec{F}_{\text{R on S}}$ and $\vec{F}_{\text{R on W}}$ are the pulling forces exerted by the rope and are what we mean by "the tension in the rope." Thus the tension in the first rope is 100 N.

FIGURE 7.19b repeats the analysis for the rope pulled by two students. Each student pulls with 100 N, so $F_{\text{S1 on R}} = 100$ N and $F_{\text{S2 on R}} = 100$ N. Just as before, there are two action/reaction pairs and the rope is in static equilibrium. Thus

$$F_{\text{R on S2}} = F_{\text{S2 on R}} = F_{\text{S1 on R}} = F_{\text{R on S1}} = 100 \text{ N}$$

The tension in the rope—the pulling forces $\vec{F}_{\text{R on S1}}$ and $\vec{F}_{\text{R on S2}}$—is still 100 N!

You may have *assumed* that the student on the right in Figure 7.18b is doing something to the rope that the wall in Figure 7.18a does not do. But our analysis finds that the wall, just like the student, pulls to the right with 100 N. The rope doesn't care whether it's pulled by a wall or a hand. It experiences the same forces in both cases, so the rope's tension is the same in both.

ASSESS Ropes and strings exert forces at *both* ends. The force with which they pull—and thus the force pulling on them at each end—*is* the tension in the rope. Tension is not the sum of the pulling forces.

FIGURE 7.19 Analysis of tension forces.

(a)

The rope is in equilibrium.

$\vec{F}_{\text{S on R}}$ $\vec{F}_{\text{W on R}}$

$\vec{F}_{\text{R on S}}$ 100 N pull $\vec{F}_{\text{R on W}}$

Student   Wall

(b)

The rope is in equilibrium.

$\vec{F}_{\text{S1 on R}}$ $\vec{F}_{\text{S2 on R}}$

$\vec{F}_{\text{R on S1}}$ 100 N pull $\vec{F}_{\text{R on S2}}$

Student 1   Student 2

STOP TO THINK 7.4 All three 50 kg blocks are at rest. Is the tension in rope 2 greater than, less than, or equal to the tension in rope 1?

The Massless String Approximation

FIGURE 7.20 Tension pulls forward on block A, backward on block B.

(a)

(b)

The tension is constant throughout a rope that is in equilibrium, but what happens if the rope is accelerating? For example, FIGURE 7.20a shows two connected blocks being pulled by force $\vec{F}$. Is the string's tension at the right end, where it pulls back on B, the same as the tension at the left end, where it pulls on A?

FIGURE 7.20b shows the horizontal forces acting on the blocks and the string. The only horizontal forces acting on the string are $\vec{T}_{\text{A on S}}$ and $\vec{T}_{\text{B on S}}$, so Newton's second law *for the string* is

$$(F_{\text{net}})_x = T_{\text{B on S}} - T_{\text{A on S}} = m_S a_x \tag{7.5}$$

where m_S is the mass of the string. If the string is accelerating, then the tensions at the two ends can *not* be the same. The tension at the "front" of the string must be greater than the tension at the "back" in order to accelerate the string!

Often in physics and engineering problems the mass of the string or rope is much less than the masses of the objects that it connects. In such cases, we can adopt the **massless string approximation.** In the limit $m_S \to 0$, Equation 7.5 becomes

$$T_{\text{B on S}} = T_{\text{A on S}} \quad \text{(massless string approximation)} \tag{7.6}$$

In other words, **the tension in a massless string is constant.** This is nice, but it isn't the primary justification for the massless string approximation.

Look again at Figure 7.20b. If $T_{\text{B on S}} = T_{\text{A on S}}$, then

$$\vec{T}_{\text{S on A}} = -\vec{T}_{\text{S on B}} \tag{7.7}$$

FIGURE 7.21 The massless string approximation allows objects A and B to act *as if* they are directly interacting.

We can omit the string if we assume it is massless.

That is, the force on block A is equal and opposite to the force on block B. Forces $\vec{T}_{\text{S on A}}$ and $\vec{T}_{\text{S on B}}$ act *as if* they are an action/reaction pair of forces. Thus we can draw the simplified diagram of FIGURE 7.21 in which the string is missing and blocks A and B interact directly with each other through forces that we can call $\vec{T}_{\text{A on B}}$ and $\vec{T}_{\text{B on A}}$.

In other words, **if objects A and B interact with each other through a massless string, we can omit the string and treat forces $\vec{F}_{\text{A on B}}$ and $\vec{F}_{\text{B on A}}$ as if they are an action/reaction pair.** This is not literally true because A and B are not in contact. Nonetheless, all a massless string does is transmit a force from A to B without changing the magnitude of that force. This is the real significance of the massless string approximation.

> **NOTE** For problems in this book, you can assume that any strings or ropes are massless unless the problem explicitly states otherwise. The simplified view of Figure 7.21 is appropriate under these conditions. But if the string has a mass, it must be treated as a separate object.

EXAMPLE 7.6 | Comparing two tensions

Blocks A and B in **FIGURE 7.22** are connected by massless string 2 and pulled across a frictionless table by massless string 1. B has a larger mass than A. Is the tension in string 2 larger than, smaller than, or equal to the tension in string 1?

FIGURE 7.22 Blocks A and B are pulled across a frictionless table by massless strings.

$$m_B > m_A$$

MODEL The massless string approximation allows us to treat A and B *as if* they interact directly with each other. The blocks are accelerating because there's a force to the right and no friction.

SOLVE B has a larger mass, so it may be tempting to conclude that string 2, which pulls B, has a greater tension than string 1, which pulls A. The flaw in this reasoning is that Newton's second law tells us only about the *net* force. The net force on B *is* larger than the net force on A, but the net force on A is *not* just the tension $\vec{T}_1$ in the forward direction. The tension in string 2 also pulls *backward* on A!

FIGURE 7.23 shows the horizontal forces in this frictionless situation. Because the string is massless, forces $\vec{T}_{A\text{ on }B}$ and $\vec{T}_{B\text{ on }A}$ act *as if* they are an action/reaction pair.

FIGURE 7.23 The horizontal forces on blocks A and B.

From Newton's third law,

$$T_{A\text{ on }B} = T_{B\text{ on }A} = T_2$$

where T_2 is the tension in string 2. From Newton's second law, the net force on A is

$$(F_{A\text{ net}})_x = T_1 - T_{B\text{ on }A} = T_1 - T_2 = m_A a_{Ax}$$

The net force on A is the *difference* in tensions. The blocks are accelerating to the right, making $a_{Ax} > 0$, so

$$T_1 > T_2$$

The tension in string 2 is *smaller* than the tension in string 1.

ASSESS This is not an intuitively obvious result. A careful study of the reasoning in this example is worthwhile. An alternative analysis would note that $\vec{T}_1$ accelerates *both* blocks, of combined mass $(m_A + m_B)$, whereas $\vec{T}_2$ accelerates only block B. Thus string 1 must have the larger tension.

Pulleys

Strings and ropes often pass over pulleys. The application might be as simple as lifting a heavy weight or as complex as the internal cable-and-pulley arrangement that precisely moves a robot arm.

FIGURE 7.24a shows a simple situation in which block B, as it falls, drags block A across a table. As the string moves, static friction between the string and pulley causes the pulley to turn. If we assume that

- The string *and* the pulley are both massless, and
- There is no friction where the pulley turns on its axle,

then no net force is needed to accelerate the string or turn the pulley. Thus **the tension in a massless string remains constant as it passes over a massless, frictionless pulley.**

Because of this, we can draw the simplified free-body diagram of **FIGURE 7.24b**, in which the string and pulley are omitted. Forces $\vec{T}_{A\text{ on }B}$ and $\vec{T}_{B\text{ on }A}$ act *as if* they are an action/reaction pair, even though they are not opposite in direction because the tension force gets "turned" by the pulley.

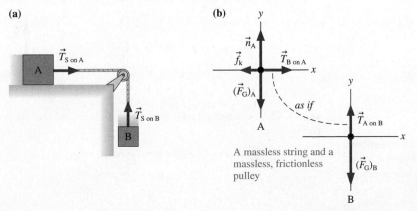

◄ **FIGURE 7.24** Blocks A and B are connected by a string that passes over a pulley.

TACTICS BOX 7.2

Working with ropes and pulleys

For massless ropes or strings and massless, frictionless pulleys:

- If a force pulls on one end of a rope, the tension in the rope equals the magnitude of the pulling force.
- If two objects are connected by a rope, the tension is the same at both ends.
- If the rope passes over a pulley, the tension in the rope is unaffected.

Exercises 17–22

STOP TO THINK 7.5 In Figure 7.24, on the previous page, is the tension in the string greater than, less than, or equal to the gravitational force acting on block B?

7.5 Examples of Interacting-Objects Problems

We will conclude this chapter with three extended examples. Although the mathematics will be more involved than in any of our work up to this point, we will continue to emphasize the *reasoning* one uses in approaching problems such as these. The solutions will be based on Problem-Solving Strategy 7.1. In fact, these problems are now reaching such a level of complexity that, for all practical purposes, it becomes impossible to work them unless you are following a well-planned strategy. Our earlier emphasis on forces and free-body diagrams will now really begin to pay off!

EXAMPLE 7.7 | **Placing a leg in traction**

Serious fractures of the leg often need a stretching force to keep contracting leg muscles from forcing the broken bones together too hard. This is done using *traction*, an arrangement of a rope, a weight, and pulleys as shown in **FIGURE 7.25**. The rope must make the same angle on both sides of the pulley so that the net force on the leg is horizontal, but the angle can be adjusted to control the amount of traction. The doctor has specified 50 N of traction for this patient with a 4.2 kg hanging mass. What is the proper angle?

FIGURE 7.25 A leg in traction.

4.2 kg

MODEL Model the leg and the weight as particles. The other point where forces are applied is the pulley attached to the patient's foot, which we'll treat as a separate object. We'll assume massless ropes and a massless, frictionless pulley.

VISUALIZE **FIGURE 7.26** shows three free-body diagrams. Forces $\vec{T}_{P1}$ and $\vec{T}_{P2}$ are the tension forces of the rope as it pulls on the pulley. The pulley is in equilibrium, so these forces are balanced by $\vec{F}_{L \text{ on } P}$, which forms an action/reaction pair with the 50 N traction force $\vec{F}_{P \text{ on } L}$. Our model of the rope and pulley makes the tension force constant, $T_{P1} = T_{P2} = T_W$, so we'll call it simply T.

FIGURE 7.26 The free-body diagrams.

SOLVE The x-component equation of Newton's second law for the pulley is

$$\sum (F_{\text{on P}})_x = T_{P1} \cos\theta + T_{P2} \cos\theta - F_{L \text{ on } P}$$
$$= 2T\cos\theta - F_{L \text{ on } P} = 0$$

Thus the correct angle for the ropes is

$$\theta = \cos^{-1}\left(\frac{F_{\text{L on P}}}{2T}\right)$$

We know, from Newton's third law, that $F_{\text{L on P}} = F_{\text{P on L}} = 50$ N. We can determine the tension force by analyzing the weight. It also is in equilibrium, so the upward tension force exactly balances the downward gravitational force:

$$T = (F_G)_\text{w} = m_\text{w}g = (4.2 \text{ kg})(9.80 \text{ m/s}^2) = 41 \text{ N}$$

Thus the proper angle is

$$\theta = \cos^{-1}\left(\frac{50 \text{ N}}{2(41 \text{ N})}\right) = 52°$$

ASSESS The traction force would approach 82 N if angle θ approached zero because the two ropes would pull in parallel. Conversely, the traction would approach 0 N if θ approached 90°. The desired traction is roughly midway between these two extremes, so an angle near 45° seems reasonable.

EXAMPLE 7.8 | The show must go on!

A 200 kg set used in a play is stored in the loft above the stage. The rope holding the set passes up and over a pulley, then is tied backstage. The director tells a 100 kg stagehand to lower the set. When he unties the rope, the set falls and the unfortunate man is hoisted into the loft. What is the stagehand's acceleration?

MODEL The system is the stagehand M and the set S, which we will model as particles. Assume a massless rope and a massless, frictionless pulley.

VISUALIZE FIGURE 7.27 shows the pictorial representation. The man's acceleration a_{My} is positive, while the set's acceleration a_{Sy} is negative. These two accelerations have the same magnitude because the two objects are connected by a rope, but they have opposite signs. Thus the acceleration constraint is $a_{\text{Sy}} = -a_{\text{My}}$. Forces $\vec{T}_{\text{M on S}}$ and $\vec{T}_{\text{S on M}}$ are not literally an action/reaction pair, but they act *as if* they are because the rope is massless and the pulley is massless and frictionless. Notice that the pulley has "turned" the tension force so that $\vec{T}_{\text{M on S}}$ and $\vec{T}_{\text{S on M}}$ are *parallel* to each other rather than opposite, as members of a true action/reaction pair would have to be.

SOLVE Newton's second law for the man and the set is

$$\sum(F_{\text{on M}})_y = T_{\text{S on M}} - m_\text{M}g = m_\text{M}a_{\text{My}}$$

$$\sum(F_{\text{on S}})_y = T_{\text{M on S}} - m_\text{S}g = m_\text{S}a_{\text{Sy}} = -m_\text{S}a_{\text{My}}$$

Only the y-equations are needed. Notice that we used the acceleration constraint in the last step. Newton's third law is

$$T_{\text{M on S}} = T_{\text{S on M}} = T$$

where we can drop the subscripts and call the tension simply T. With this substitution, the two second-law equations can be written

$$T - m_\text{M}g = m_\text{M}a_{\text{My}}$$
$$T - m_\text{S}g = -m_\text{S}a_{\text{My}}$$

These are simultaneous equations in the two unknowns T and a_{My}. We can eliminate T by subtracting the second equation from the first to give

$$(m_\text{S} - m_\text{M})g = (m_\text{S} + m_\text{M})a_{\text{My}}$$

Finally, we can solve for the hapless stagehand's acceleration:

$$a_{\text{My}} = \frac{m_\text{S} - m_\text{M}}{m_\text{S} + m_\text{M}}g = \frac{100 \text{ kg}}{300 \text{ kg}}9.80 \text{ m/s}^2 = 3.27 \text{ m/s}^2$$

This is also the acceleration with which the set falls. If the rope's tension was needed, we could now find it from $T = m_\text{M}a_{\text{My}} + m_\text{M}g$.

ASSESS If the stagehand weren't holding on, the set would fall with free-fall acceleration g. The stagehand acts as a *counterweight* to reduce the acceleration.

FIGURE 7.27 Pictorial representation for Example 7.8.

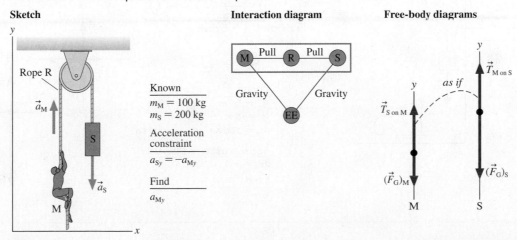

Sketch

Rope R

Known
$m_\text{M} = 100$ kg
$m_\text{S} = 200$ kg

Acceleration constraint
$a_{\text{Sy}} = -a_{\text{My}}$

Find
a_{My}

Interaction diagram

Pull   Pull

Gravity   Gravity

Free-body diagrams

as if

CHALLENGE EXAMPLE 7.9 | A not-so-clever bank robbery

Bank robbers have pushed a 1000 kg safe to a second-story floor-to-ceiling window. They plan to break the window, then lower the safe 3.0 m to their truck. Not being too clever, they stack up 500 kg of furniture, tie a rope between the safe and the furniture, and place the rope over a pulley. Then they push the safe out the window. What is the safe's speed when it hits the truck? The coefficient of kinetic friction between the furniture and the floor is 0.50.

MODEL This is a continuation of the situation that we analyzed in Figures 7.15 and 7.24, which are worth reviewing. The system is the safe S and the furniture F, which we will model as particles. We will assume a massless rope and a massless, frictionless pulley.

VISUALIZE The safe and the furniture are tied together, so their accelerations have the same magnitude. The safe has a y-component of acceleration a_{Sy} that is negative because the safe accelerates in the negative y-direction. The furniture has an x-component a_{Fx} that is positive. Thus the acceleration constraint is

$$a_{Fx} = -a_{Sy}$$

The free-body diagrams shown in **FIGURE 7.28** are modeled after Figure 7.24 but now include a kinetic friction force on the furniture. Forces $\vec{T}_{F \text{ on } S}$ and $\vec{T}_{S \text{ on } F}$ act *as if* they are an action/reaction pair, so they have been connected with a dashed line.

SOLVE We can write Newton's second law directly from the free-body diagrams. For the furniture,

$$\sum (F_{\text{on F}})_x = T_{S \text{ on } F} - f_k = T - f_k = m_F a_{Fx} = -m_F a_{Sy}$$

$$\sum (F_{\text{on F}})_y = n - m_F g = 0$$

And for the safe,

$$\sum (F_{\text{on S}})_y = T - m_S g = m_S a_{Sy}$$

Notice how we used the acceleration constraint in the first equation. We also went ahead and made use of Newton's third law:

$T_{F \text{ on } S} = T_{S \text{ on } F} = T$. We have one additional piece of information, the model of kinetic friction:

$$f_k = \mu_k n = \mu_k m_F g$$

where we used the y-equation of the furniture to deduce that $n = m_F g$. Substitute this result for f_k into the x-equation of the furniture, then rewrite the furniture's x-equation and the safe's y-equation:

$$T - \mu_k m_F g = -m_F a_{Sy}$$
$$T - m_S g = m_S a_{Sy}$$

We have succeeded in reducing our knowledge to two simultaneous equations in the two unknowns a_{Sy} and T. Subtract the second equation from the first to eliminate T:

$$(m_S - \mu_k m_F)g = -(m_S + m_F)a_{Sy}$$

Finally, solve for the safe's acceleration:

$$a_{Sy} = -\left(\frac{m_S - \mu_k m_F}{m_S + m_F}\right)g$$

$$= -\left(\frac{1000 \text{ kg} - (0.50)(500 \text{ kg})}{1000 \text{ kg} + 500 \text{ kg}}\right)9.80 \text{ m/s}^2 = -4.9 \text{ m/s}^2$$

Now we need to calculate the kinematics of the falling safe. Because the time of the fall is not known or needed, we can use

$$v_{1y}^2 = v_{0y}^2 + 2a_{Sy}\Delta y = 0 + 2a_{Sy}(y_1 - y_0) = -2a_{Sy}y_0$$

$$v_1 = \sqrt{-2a_{Sy}y_0} = \sqrt{-2(-4.9 \text{ m/s}^2)(3.0 \text{ m})} = 5.4 \text{ m/s}$$

ASSESS The value of v_{1y} is negative, but we only needed to find the speed so we took the absolute value. This is about 12 mph, so it seems unlikely that the truck will survive the impact of the 1000 kg safe!

FIGURE 7.28 Pictorial representation for Challenge Example 7.9.

SUMMARY

The goal of Chapter 7 has been to use Newton's third law to understand how objects interact.

GENERAL PRINCIPLES

Newton's Third Law

Every force occurs as one member of an **action/reaction pair** of forces. The two members of an action/reaction pair:
- Act on two *different* objects.
- Are equal in magnitude but opposite in direction:

$$\vec{F}_{A \text{ on } B} = -\vec{F}_{B \text{ on } A}$$

Action/reaction pair

Solving Interacting-Objects Problems

MODEL Identify which objects form the system.

VISUALIZE Draw a pictorial representation.

- Define symbols and coordinates.
- Identify acceleration constraints.
- Draw an interaction diagram.
- Draw a separate free-body diagram for each object.
- Connect action/reaction pairs with dashed lines.

SOLVE Write Newton's second law for each object.

- Use the free-body diagrams.
- Equate the magnitudes of action/reaction pairs.
- Include acceleration constraints and friction.

ASSESS Is the result reasonable?

IMPORTANT CONCEPTS

Objects, systems, and the environment

Objects whose motion is of interest are the system.

Objects whose motion is not of interest form the environment.

The objects of interest interact with the environment, but those interactions can be considered external forces.

Interaction diagram

APPLICATIONS

Acceleration constraints

Objects that are constrained to move together must have accelerations of equal magnitude: $a_A = a_B$. This must be expressed in terms of components, such as $a_{Ax} = -a_{By}$.

Strings and pulleys

The tension in a string or rope pulls in both directions. The tension is constant in a string if the string is:

- Massless, or
- In equilibrium

Objects connected by massless strings passing over massless, frictionless pulleys act *as if* they interact via an action/reaction pair of forces.

TERMS AND NOTATION

interaction	environment	propulsion	acceleration constraint
action/reaction pair	interaction diagram	Newton's third law	massless string approximation
system	external force		

CONCEPTUAL QUESTIONS

1. You find yourself in the middle of a frozen lake with a surface so slippery ($\mu_s = \mu_k = 0$) you cannot walk. However, you happen to have several rocks in your pocket. The ice is extremely hard. It cannot be chipped, and the rocks slip on it just as much as your feet do. Can you think of a way to get to shore? Use pictures, forces, and Newton's laws to explain your reasoning.

2. How does a sprinter sprint? What is the forward force on a sprinter as she accelerates? Where does that force come from? Your explanation should include an interaction diagram and a free-body diagram.

3. How does a rocket take off? What is the upward force on it? Your explanation should include an interaction diagram and free-body diagrams of the rocket and of the parcel of gas being exhausted.

4. How do basketball players jump straight up into the air? Your explanation should include an interaction diagram and a free-body diagram.

5. A mosquito collides head-on with a car traveling 60 mph. Is the force of the mosquito on the car larger than, smaller than, or equal to the force of the car on the mosquito? Explain.

6. A mosquito collides head-on with a car traveling 60 mph. Is the magnitude of the mosquito's acceleration larger than, smaller than, or equal to the magnitude of the car's acceleration? Explain.

7. A small car is pushing a large truck. They are speeding up. Is the force of the truck on the car larger than, smaller than, or equal to the force of the car on the truck?

8. A very smart 3-year-old child is given a wagon for her birthday. She refuses to use it. "After all," she says, "Newton's third law says that no matter how hard I pull, the wagon will exert an equal but opposite force on me. So I will never be able to get it to move forward." What would you say to her in reply?

9. Teams red and blue are having a tug-of-war. According to Newton's third law, the force with which the red team pulls on the blue team exactly equals the force with which the blue team pulls on the red team. How can one team ever win? Explain.

10. Will hanging a magnet in front of the iron cart in FIGURE Q7.10 make it go? Explain.

FIGURE Q7.10

11. FIGURE Q7.11 shows two masses at rest. The string is massless and the pulley is frictionless. The spring scale reads in kg. What is the reading of the scale?

FIGURE Q7.11

12. FIGURE Q7.12 shows two masses at rest. The string is massless and the pulleys are frictionless. The spring scale reads in kg. What is the reading of the scale?

FIGURE Q7.12

13. The hand in FIGURE Q7.13 is pushing on the back of block A. Blocks A and B, with $m_B > m_A$, are connected by a massless string and slide on a frictionless surface. Is the force of the string on B larger than, smaller than, or equal to the force of the hand on A? Explain.

FIGURE Q7.13

14. Blocks A and B in FIGURE Q7.14 are connected by a massless string over a massless, frictionless pulley. The blocks have just been released from rest. Will the pulley rotate clockwise, counter-clockwise, or not at all? Explain.

FIGURE Q7.14

15. In case a in FIGURE Q7.15, block A is accelerated across a frictionless table by a hanging 10 N weight (1.02 kg). In case b, block A is accelerated across a frictionless table by a steady 10 N tension in the string. The string is massless, and the pulley is massless and frictionless. Is A's acceleration in case b greater than, less than, or equal to its acceleration in case a? Explain.

FIGURE Q7.15

EXERCISES AND PROBLEMS

Exercises

Section 7.2 Analyzing Interacting Objects

For Exercises 1 through 5:
  a. Draw an interaction diagram.
  b. Identify the "system" on your interaction diagram.
  c. Draw a free-body diagram for each object in the system. Use dashed lines to connect members of an action/reaction pair.

1. | A soccer ball and a bowling ball have a head-on collision at this instant. Rolling friction is negligible.

2. | A weightlifter stands up at constant speed from a squatting position while holding a heavy barbell across his shoulders.

3. | A steel cable with mass is lifting a girder. The girder is speeding up.

4. || Block A in FIGURE EX7.4 is heavier than block B and is sliding down the incline. All surfaces have friction. The rope is massless, and the massless pulley turns on frictionless bearings. The rope and the pulley are among the interacting objects, but you'll have to decide if they're part of the system.

FIGURE EX7.4   FIGURE EX7.5

5. || Block A in FIGURE EX7.5 is sliding down the incline. The rope is massless, and the massless pulley turns on frictionless bearings, but the surface is not frictionless. The rope and the pulley are among the interacting objects, but you'll have to decide if they're part of the system.

Section 7.3 Newton's Third Law

6. | a. How much force does an 80 kg astronaut exert on his chair while sitting at rest on the launch pad?
   b. How much force does the astronaut exert on his chair while accelerating straight up at 10 m/s²?

7. | Block B in FIGURE EX7.7 rests on a surface for which the static and kinetic coefficients of friction are 0.60 and 0.40, respectively. The ropes are massless. What is the maximum mass of block A for which the system remains in equilibrium?

FIGURE EX7.7

8. || A 1000 kg car pushes a 2000 kg truck that has a dead battery. When the driver steps on the accelerator, the drive wheels of the car push against the ground with a force of 4500 N. Rolling friction can be neglected.
   a. What is the magnitude of the force of the car on the truck?
   b. What is the magnitude of the force of the truck on the car?

9. || Blocks with masses of 1 kg, 2 kg, and 3 kg are lined up in a row on a frictionless table. All three are pushed forward by a 12 N force applied to the 1 kg block.
   a. How much force does the 2 kg block exert on the 3 kg block?
   b. How much force does the 2 kg block exert on the 1 kg block?

10. | A 3000 kg meteorite falls toward the earth. What is the magnitude of the *earth's* acceleration just before impact? The earth's mass is 5.98×10^{24} kg.

11. | The foot of a 55 kg sprinter is on the ground for 0.25 s while her body accelerates from rest to 2.0 m/s.
   a. Is the friction between her foot and the ground static friction or kinetic friction?
   b. What is the magnitude of the friction force?

12. || A steel cable lying flat on the floor drags a 20 kg block across a horizontal, frictionless floor. A 100 N force applied to the cable causes the block to reach a speed of 4.0 m/s in a distance of 2.0 m. What is the mass of the cable?

13. || An 80 kg spacewalking astronaut pushes off a 640 kg satellite, exerting a 100 N force for the 0.50 s it takes him to straighten his arms. How far apart are the astronaut and the satellite after 1.0 min?

14. || The sled dog in FIGURE EX7.14 drags sleds A and B across the snow. The coefficient of friction between the sleds and the snow is 0.10. If the tension in rope 1 is 150 N, what is the tension in rope 2?

FIGURE EX7.14

15. || Two-thirds of the weight of a 1500 kg car rests on the drive wheels. What is the maximum acceleration of this car on a concrete surface?

Section 7.4 Ropes and Pulleys

16. || FIGURE EX7.16 shows two 1.0 kg blocks connected by a rope. A second rope hangs beneath the lower block. Both ropes have a mass of 250 g. The entire assembly is accelerated upward at 3.0 m/s² by force $\vec{F}$.
   a. What is F?
   b. What is the tension at the top end of rope 1?
   c. What is the tension at the bottom end of rope 1?
   d. What is the tension at the top end of rope 2?

FIGURE EX7.16   FIGURE EX7.17

17. || What is the tension in the rope of FIGURE EX7.17?

18. ‖ A 2.0-m-long, 500 g rope pulls a 10 kg block of ice across a horizontal, frictionless surface. The block accelerates at 2.0 m/s². How much force pulls forward on (a) the ice, (b) the rope? Assume that the rope is perfectly horizontal.

19. | A woman living in a third-story apartment is moving out. Rather than carrying everything down the stairs, she decides to pack her belongings into crates, attach a frictionless pulley to her balcony railing, and lower the crates by rope. How hard must she pull on the horizontal end of the rope to lower a 25 kg crate at steady speed?

20. ‖ Two blocks are attached to opposite ends of a massless rope that goes over a massless, frictionless, stationary pulley. One of the blocks, with a mass of 6.0 kg, accelerates downward at $\frac{3}{4}g$. What is the mass of the other block?

21. ‖ The cable cars in San Francisco are pulled along their tracks by an underground steel cable that moves along at 9.5 mph. The cable is driven by large motors at a central power station and extends, via an intricate pulley arrangement, for several miles beneath the city streets. The length of a cable stretches by up to 100 ft during its lifetime. To keep the tension constant, the cable passes around a 1.5-m-diameter "tensioning pulley" that rolls back and forth on rails, as shown in FIGURE EX7.21. A 2000 kg block is attached to the tensioning pulley's cart, via a rope and pulley, and is suspended in a deep hole. What is the tension in the cable car's cable?

FIGURE EX7.21

22. ‖ A 2.0 kg rope hangs from the ceiling. What is the tension at the midpoint of the rope?

23. ‖ A mobile at the art museum has a 2.0 kg steel cat and a 4.0 kg steel dog suspended from a lightweight cable, as shown in FIGURE EX7.23. It is found that $\theta_1 = 20°$ when the center rope is adjusted to be perfectly horizontal. What are the tension and the angle of rope 3?

FIGURE EX7.23

24. ‖ The 1.0 kg block in FIGURE EX7.24 is tied to the wall with a rope. It sits on top of the 2.0 kg block. The lower block is pulled to the right with a tension force of 20 N. The coefficient of kinetic friction at both the lower and upper surfaces of the 2.0 kg block is $\mu_k = 0.40$.
   a. What is the tension in the rope attached to the wall?
   b. What is the acceleration of the 2.0 kg block?

FIGURE EX7.24   **FIGURE EX7.25**

25. ‖ The 100 kg block in FIGURE EX7.25 takes 6.0 s to reach the floor after being released from rest. What is the mass of the block on the left? The pulley is massless and frictionless.

Problems

26. ‖ FIGURE P7.26 shows two strong magnets on opposite sides of a small table. The long-range attractive force between the magnets keeps the lower magnet in place.
   a. Draw an interaction diagram and draw free-body diagrams for both magnets and the table. Use dashed lines to connect the members of an action/reaction pair.
   b. The lower magnet is being pulled upward against the bottom of the table. Suppose that each magnet's weight is 2.0 N and that the magnetic force of the lower magnet on the upper magnet is 6.0 N. How hard does the lower magnet push against the table?

FIGURE P7.26   **FIGURE P7.27**

27. ‖ FIGURE P7.27 shows a 6.0 N force pushing two gliders along an air track. The 200 g spring between the gliders is compressed. How much force does the spring exert on (a) glider A and (b) glider B? The spring is firmly attached to the gliders, and it does not sag.

28. ‖ A rope of length L and mass m is suspended from the ceiling. Find an expression for the tension in the rope at position y, measured upward from the free end of the rope.

29. ‖ While driving to work last year, I was holding my coffee mug in my left hand while changing the CD with my right hand. Then the cell phone rang, so I placed the mug on the flat part of my dashboard. Then, believe it or not, a deer ran out of the woods and on to the road right in front of me. Fortunately, my reaction time was zero, and I was able to stop from a speed of 20 m/s in a mere 50 m, just barely avoiding the deer. Later tests revealed that the static and kinetic coefficients of friction of the coffee mug on the dash are 0.50 and 0.30, respectively; the coffee and mug had a mass of 0.50 kg; and the mass of the deer was 120 kg. Did my coffee mug slide?

30. ‖ A Federation starship (2.0×10^6 kg) uses its tractor beam to pull a shuttlecraft (2.0×10^4 kg) aboard from a distance of 10 km away. The tractor beam exerts a constant force of 4.0×10^4 N on the shuttlecraft. Both spacecraft are initially at rest. How far does the starship move as it pulls the shuttlecraft aboard?

31. ‖ Your forehead can withstand a force of about 6.0 kN before
BIO fracturing, while your cheekbone can withstand only about 1.3 kN. Suppose a 140 g baseball traveling at 30 m/s strikes your head and stops in 1.5 ms.
   a. What is the magnitude of the force that stops the baseball?
   b. What force does the baseball exert on your head? Explain.
   c. Are you in danger of a fracture if the ball hits you in the forehead? On the cheek?

32. ‖ Bob, who has a mass of 75 kg, can throw a 500 g rock with a speed of 30 m/s. The distance through which his hand moves as he accelerates the rock from rest until he releases it is 1.0 m.
   a. What constant force must Bob exert on the rock to throw it with this speed?
   b. If Bob is standing on frictionless ice, what is his recoil speed after releasing the rock?

33. ⫼ Two packages at UPS start sliding down the 20° ramp shown in **FIGURE P7.33**. Package A has a mass of 5.0 kg and a coefficient of friction of 0.20. Package B has a mass of 10 kg and a coefficient of friction of 0.15. How long does it take package A to reach the bottom?

FIGURE P7.33   **FIGURE P7.34**

34. ⫼ The two blocks in **FIGURE P7.34** are sliding down the incline. What is the tension in the massless string?

35. ‖ The coefficient of static friction is 0.60 between the two blocks in **FIGURE P7.35**. The coefficient of kinetic friction between the lower block and the floor is 0.20. Force $\vec{F}$ causes both blocks to cross a distance of 5.0 m, starting from rest. What is the least amount of time in which this motion can be completed without the top block sliding on the lower block?

FIGURE P7.35   **FIGURE P7.36**

36. ⫼ The block of mass M in **FIGURE P7.36** slides on a frictionless surface. Find an expression for the tension in the string.

37. ‖ The 10.2 kg block in **FIGURE P7.37** is held in place by a force applied to a rope passing over two massless, frictionless pulleys. Find the tensions T_1 to T_5 and the magnitude of force $\vec{F}$.

FIGURE P7.37   **FIGURE P7.38**

38. ‖ The coefficient of kinetic friction between the 2.0 kg block in **FIGURE P7.38** and the table is 0.30. What is the acceleration of the 2.0 kg block?

39. ⫼ **FIGURE P7.39** shows a block of mass m resting on a 20° slope. The block has coefficients of friction $\mu_s = 0.80$ and $\mu_k = 0.50$ with the surface. It is connected via a massless string over a massless, frictionless pulley to a hanging block of mass 2.0 kg.
    a. What is the minimum mass m that will stick and not slip?
    b. If this minimum mass is nudged ever so slightly, it will start being pulled up the incline. What acceleration will it have?

FIGURE P7.39

40. ‖ A 4.0 kg box is on a frictionless 35° slope and is connected via a massless string over a massless, frictionless pulley to a hanging 2.0 kg weight. The picture for this situation is similar to **FIGURE P7.39**.
    a. What is the tension in the string if the 4.0 kg box is *held* in place, so that it cannot move?
    b. If the box is then released, which way will it move on the slope?
    c. What is the tension in the string once the box begins to move?

41. ‖ The 1.0 kg physics book in **FIGURE P7.41** is connected by a string to a 500 g coffee cup. The book is given a push up the slope and released with a speed of 3.0 m/s. The coefficients of friction are $\mu_s = 0.50$ and $\mu_k = 0.20$.
    a. How far does the book slide?
    b. At the highest point, does the book stick to the slope, or does it slide back down?

FIGURE P7.41

42. ‖ The 2000 kg cable car shown in **FIGURE P7.42** descends a 200-m-high hill. In addition to its brakes, the cable car controls its speed by pulling an 1800 kg counterweight up the other side of the hill. The rolling friction of both the cable car and the counterweight are negligible.
    a. How much braking force does the cable car need to descend at constant speed?
    b. One day the brakes fail just as the cable car leaves the top on its downward journey. What is the runaway car's speed at the bottom of the hill?

FIGURE P7.42

43. ‖ The century-old *ascensores* in Valparaiso, Chile, are picturesque cable cars built on stilts to keep the passenger compartments level as they go up and down the steep hillsides. As **FIGURE P7.43** shows, one car ascends as the other descends. The cars use a two-cable arrangement to compensate for friction; one cable passing around a large pulley connects the cars, the second is pulled by a small motor. Suppose the mass of both cars (with passengers) is 1500 kg, the coefficient of rolling friction is 0.020, and the cars move at constant speed. What is the tension in (a) the connecting cable and (b) the cable to the motor?

FIGURE P7.43

44. ‖ A 3200 kg helicopter is flying horizontally. A 250 kg crate is suspended from the helicopter by a massless cable that is constantly 20° from vertical. What propulsion force $\vec{F}_{prop}$ is being provided by the helicopter's rotor? Air resistance can be ignored. Give your answer in component form in a coordinate system where $\hat{\imath}$ points in the direction of motion and $\hat{\jmath}$ points upward.

45. ‖ A house painter uses the chair-and-pulley arrangement of **FIGURE P7.45** to lift himself up the side of a house. The painter's mass is 70 kg and the chair's mass is 10 kg. With what force must he pull down on the rope in order to accelerate upward at 0.20 m/s²?

FIGURE P7.45

46. ‖ A long, 1.0 kg rope hangs from a support that breaks, causing the rope to fall, if the pull exceeds 40 N. A student team has built a 2.0 kg robot "mouse" that runs up and down the rope. What maximum acceleration can the robot have—both magnitude and direction—without the rope falling?

47. ‖ A 50-cm-diameter, 400 g beach ball is dropped with a 4.0 mg ant riding on the top. The ball experiences air resistance, but the ant does not. What is the magnitude of the normal force exerted on the ant when the ball's speed is 2.0 m/s?

48. ‖ A 70 kg tightrope walker stands at the center of a rope. The rope supports are 10 m apart and the rope sags 10° at each end. The tightrope walker crouches down, then leaps straight up with an acceleration of 8.0 m/s² to catch a passing trapeze. What is the tension in the rope as he jumps?

49. ‖ Find an expression for the magnitude of the horizontal force F in **FIGURE P7.49** for which m_1 does not slip either up or down along the wedge. All surfaces are frictionless.

FIGURE P7.49

50. ‖ A rocket burns fuel at a rate of 5.0 kg/s, expelling the exhaust gases at a speed of 4.0 km/s *relative to the rocket*. We would like to find the thrust of the rocket engine.
    a. Model the fuel burning as a steady ejection of small pellets, each with the small mass Δm. Suppose it takes a short time Δt to accelerate a pellet (at constant acceleration) to the exhaust speed v_{ex}. Further, suppose the rocket is clamped down so that it can't recoil. Find an expression for the magnitude of the force that one pellet exerts on the rocket during the short time while the pellet is being expelled.
    b. If the rocket is moving, v_{ex} is no longer the pellet's speed through space but it is still the pellet's speed *relative to the rocket*. By considering the limiting case of Δm and Δt approaching zero, in which case the rocket is now burning fuel continuously, calculate the rocket thrust for the values given above.

Problems 51 and 52 show the free-body diagrams of two interacting systems. For each of these, you are to
  a. Write a realistic problem for which these are the correct free-body diagrams. Be sure that the answer your problem requests is consistent with the diagrams shown.
  b. Finish the solution of the problem.

51.

52.

FIGURE P7.51   **FIGURE P7.52**

Challenge Problems

53. ‖‖ The lower block in **FIGURE CP7.53** is pulled on by a rope with a tension force of 20 N. The coefficient of kinetic friction between the lower block and the surface is 0.30. The coefficient of kinetic friction between the lower block and the upper block is also 0.30. What is the acceleration of the 2.0 kg block?

FIGURE CP7.53

54. ‖‖ In **FIGURE CP7.54**, find an expression for the acceleration of m_1. The pulleys are massless and frictionless.
    Hint: Think carefully about the acceleration constraint.

FIGURE CP7.54   **FIGURE CP7.55**

55. ‖‖ What is the acceleration of the 3.0 kg block in **FIGURE CP7.55** across the frictionless table?
    Hint: Think carefully about the acceleration constraint.

56. ▌▌▌ A 40-cm-diameter, 50-cm-tall, 15 kg hollow cylinder is placed on top of a 40-cm-diameter, 30-cm-tall, 100 kg cylinder of solid aluminum, then the two are sent sliding across frictionless ice. The static and kinetic coefficients of friction between the cylinders are 0.45 and 0.25, respectively. Air resistance cannot be neglected. What is the maximum speed the cylinders can have without the top cylinder sliding off?

57. ▌▌▌ **FIGURE CP7.57** shows a 200 g hamster sitting on an 800 g wedge-shaped block. The block, in turn, rests on a spring scale. An extra-fine lubricating oil having $\mu_s = \mu_k = 0$ is sprayed on the top surface of the block, causing the hamster to slide down. Friction between the block and the scale is large enough that the block does *not* slip on the scale. What does the scale read, in grams, as the hamster slides down?

40°

FIGURE CP7.57   Scale

58. ▌▌▌ **FIGURE CP7.58** shows three hanging masses connected by massless strings over two massless, frictionless pulleys.

a. Find the acceleration constraint for this system. It is a single equation relating a_{1y}, a_{2y}, and a_{3y}.
   Hint: y_A isn't constant.

b. Find an expression for the tension in string A.
   Hint: You should be able to write four second-law equations. These, plus the acceleration constraint, are five equations in five unknowns.

c. Suppose: $m_1 = 2.5$ kg, $m_2 = 1.5$ kg, and $m_3 = 4.0$ kg. Find the acceleration of each.

d. The 4.0 kg mass would appear to be in equilibrium. Explain why it accelerates.

FIGURE CP7.58

8 Dynamics II: Motion in a Plane

Why doesn't the roller coaster fall off the track at the top of the loop?

IN THIS CHAPTER, you will learn to solve problems about motion in two dimensions.

Are Newton's laws different in two dimensions?

No. Newton's laws are vector equations, and they work equally well in two and three dimensions. For motion in a plane, we'll focus on how a force *tangent* to a particle's trajectory changes its speed, while a force *perpendicular* to the trajectory changes the particle's direction.

Force of changing speed

$\vec{v}$

$\vec{F}$

Force of changing direction

« LOOKING BACK Chapter 4 Kinematics of projectile and circular motion

How do we analyze projectile-like motion?

For linear motion, one component of the acceleration was always zero. Motion in a plane generally has acceleration along two axes. If the accelerations are independent, we can use *x*- and *y*-coordinates and we will find motions analogous to the projectile motion we studied in Chapter 4.

Projectile motion with drag

y

45°

30°

x

How do we analyze circular motion?

Circular motion must have a force component toward the center of the circle to create the centripetal acceleration. In this case the acceleration components are radial and, perhaps, tangential. We'll use a different coordinate system, *rtz* coordinates, to study the dynamics of circular motion

$\vec{v}$

$\vec{a}$

$\vec{F}$

Does this analysis apply to orbits?

Yes, it does. The circular orbit of a satellite or planet is motion in which the force of gravity is creating the inward centripetal acceleration. You'll see that an orbiting projectile is in free fall.

$\vec{F}_G$
Planet

$\vec{F}_G$ $\vec{F}_G$

« LOOKING BACK Section 6.3 Gravity and weight

Why doesn't the water fall out of the bucket?

How can you swing a bucket of water over your head without the water falling out? Why doesn't a car going around a loop-the-loop fall off at the top? Circular motion is not always intuitive, but you'll strengthen your ability to use Newtonian reasoning by thinking about some of these problems.

Why is planar motion important?

By starting with linear motion, we were able to develop the ideas and tools of Newtonian mechanics with minimal distractions. But planes and rockets move in a plane. Satellites and electrons orbit in a plane. The points on a rotating hard drive move in a plane. In fact, much of this chapter is a prelude to Chapter 12, where we will study rotational motion. This chapter gives you the tools you need to analyze more complex—and more realistic—forms of motion.

8.1 Dynamics in Two Dimensions

Newton's second law, $\vec{a} = \vec{F}_{net}/m$, determines an object's acceleration; it makes no distinction between linear motion and two-dimensional motion in a plane. We began with motion along a line, in order to focus on the essential physics, but now we turn our attention to the motion of projectiles, satellites, and other objects that move in two dimensions. We'll continue to follow « Problem-Solving Strategy 6.1, which is well worth a review, but we'll find that we need to think carefully about the appropriate coordinate system for each problem.

EXAMPLE 8.1 | Rocketing in the wind

A small rocket for gathering weather data has a mass of 30 kg and generates 1500 N of thrust. On a windy day, the wind exerts a 20 N horizontal force on the rocket. If the rocket is launched straight up, what is the shape of its trajectory, and by how much has it been deflected sideways when it reaches a height of 1.0 km? Because the rocket goes much higher than this, assume there's no significant mass loss during the first 1.0 km of flight.

MODEL Model the rocket as a particle. We need to find the *function* $y(x)$ describing the curve the rocket follows. Because rockets have aerodynamic shapes, we'll assume no vertical air resistance.

VISUALIZE FIGURE 8.1 shows a pictorial representation. We've chosen a coordinate system with a vertical y-axis. Three forces act on the rocket: two vertical and one horizontal.

FIGURE 8.1 Pictorial representation of the rocket launch.

Known
$x_i = y_i = 0$ m
$v_{ix} = v_{iy} = 0$ m/s
$y_f = 1000$ m
$m = 30$ kg
$F_{thrust} = 1500$ N
$F_{wind} = 20$ N

Find
x_f

SOLVE In this problem, the vertical and horizontal forces are independent of each other. Newton's second law is

$$a_x = \frac{(F_{net})_x}{m} = \frac{F_{wind}}{m}$$

$$a_y = \frac{(F_{net})_y}{m} = \frac{F_{thrust} - mg}{m}$$

The primary difference from the linear-motion problems you've been solving is that the rocket accelerates along both axes. However, both accelerations are constant, so we can use kinematics to find

$$x = \tfrac{1}{2}a_x(\Delta t)^2 = \frac{F_{wind}}{2m}(\Delta t)^2$$

$$y = \tfrac{1}{2}a_y(\Delta t)^2 = \frac{F_{thrust} - mg}{2m}(\Delta t)^2$$

where we used the fact that all initial positions and velocities are zero. From the x-equation, $(\Delta t)^2 = 2mx/F_{wind}$. Substituting this into the y-equation, we find

$$y(x) = \left(\frac{F_{thrust} - mg}{F_{wind}} \right) x$$

This is the equation of the rocket's trajectory. It is a linear equation. Somewhat surprisingly, given that the rocket has both vertical and horizontal accelerations, its trajectory is a *straight line*. We can rearrange this result to find the deflection at height y:

$$x = \left(\frac{F_{wind}}{F_{thrust} - mg} \right) y$$

From the data provided, we can calculate a deflection of 17 m at a height of 1000 m.

ASSESS The solution depended on the fact that the time parameter Δt is the *same* for both components of the motion.

Projectile Motion

We found in Chapter 6 that the gravitational force on an object near the surface of a planet is $\vec{F}_G = (mg, \text{down})$. For a coordinate system with a vertical y-axis,

$$\vec{F}_G = -mg\hat{\jmath} \qquad (8.1)$$

Consequently, from Newton's second law, the acceleration is

$$a_x = \frac{(F_G)_x}{m} = 0$$

$$a_y = \frac{(F_G)_y}{m} = -g \qquad (8.2)$$

Equations 8.2 justify the analysis of projectile motion in « Section 4.2—a downward acceleration $a_y = -g$ with no horizontal acceleration—where we found that a drag-free

FIGURE 8.2 A projectile is affected by drag. These are trajectories of a plastic ball launched at different angles.

projectile follows a parabolic trajectory. The fact that the two components of acceleration are independent of each other allows us to solve for the vertical and horizontal motions.

However, the situation is quite different for a low-mass projectile, where the effects of drag are too large to ignore. We'll leave it as a homework problem for you to show that the acceleration of a projectile subject to drag is

$$a_x = -\frac{\rho CA}{2m} v_x \sqrt{v_x^2 + v_y^2}$$

$$a_y = -g - \frac{\rho CA}{2m} v_y \sqrt{v_x^2 + v_y^2}$$

(8.3)

Here the components of acceleration are neither constant nor independent of each other because a_x depends on v_y and vice versa. It turns out that these two equations cannot be solved exactly for the trajectory, but they can be solved numerically. FIGURE 8.2 shows the numerical solution for the motion of a 5 g plastic ball that's been hit with an initial speed of 25 m/s. It doesn't travel very far (the maximum distance without drag would be more than 60 m), and the maximum range is no longer reached for a launch angle of 45°. Notice that the trajectories are not at all parabolic.

STOP TO THINK 8.1 This force will cause the particle to

a. Speed up and curve upward.
b. Speed up and curve downward.
c. Slow down and curve upward.
d. Slow down and curve downward.
e. Move to the right and down.
f. Reverse direction.

8.2 Uniform Circular Motion

FIGURE 8.3 Uniform circular motion and the *rtz*-coordinate system.

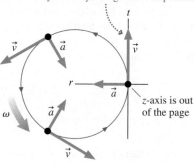

Velocity has only a tangential component.

Acceleration has only a radial component.

Circle seen edge-on

The kinematics of uniform circular motion were introduced in « Sections 4.4–4.5, and a review is *highly* recommended. Now we're ready to study *dynamics*—how forces *cause* circular motion. FIGURE 8.3 reminds you that the particle's velocity is tangent to the circle, and its acceleration—a *centripetal acceleration*—points toward the center. If the particle has angular velocity ω and speed $v = \omega r$, its centripetal acceleration is

$$\vec{a} = \left(\frac{v^2}{r}, \text{ toward center of circle}\right) = (\omega^2 r, \text{ toward center of circle})$$

(8.4)

An *xy*-coordinate system is not a good choice to analyze circular motion because the *x*- and *y*-components of the acceleration are not constant. Instead, as Figure 8.3 shows, we'll use a coordinate system whose axes are defined as follows:

- The origin is at the point where the particle is located.
- The *r*-axis (radial axis) points *from* the particle *toward* the center of the circle.
- The *t*-axis (tangential axis) is tangent to the circle, pointing in the counterclockwise direction.
- The *z*-axis is perpendicular to the plane of motion.

These three mutually perpendicular axes form the **rtz-coordinate system.**

You can see that the *rtz*-components of $\vec{v}$ and $\vec{a}$ are

$$v_r = 0 \qquad a_r = \frac{v^2}{r} = \omega^2 r$$

$$v_t = \omega r \qquad a_t = 0$$

$$v_z = 0 \qquad a_z = 0$$

(8.5)

where the angular velocity $\omega = d\theta/dt$ must be in rad/s. For uniform circular motion, **the velocity vector has only a tangential component and the acceleration vector has only a radial component.** Now you can begin to see the advantages of the *rtz*-coordinate system.

NOTE Recall that ω and v_t are positive for a counterclockwise (ccw) rotation, negative for a clockwise (cw) rotation. The particle's speed is $v = |v_t|$.

Dynamics of Uniform Circular Motion

A particle in uniform circular motion is clearly not traveling at constant velocity in a straight line. Consequently, according to Newton's first law, the particle *must* have a net force acting on it. We've already determined the acceleration of a particle in uniform circular motion—the centripetal acceleration of Equation 8.4. Newton's second law tells us exactly how much net force is needed to cause this acceleration:

$$\vec{F}_{net} = m\vec{a} = \left(\frac{mv^2}{r}, \text{ toward center of circle}\right) \quad (8.6)$$

In other words, a particle of mass m moving at constant speed v around a circle of radius r *must* have a net force of magnitude mv^2/r pointing toward the center of the circle. Without such a force, the particle would move off in a straight line tangent to the circle.

FIGURE 8.4 shows the net force $\vec{F}_{net}$ acting on a particle as it undergoes uniform circular motion. You can see that $\vec{F}_{net}$, like $\vec{a}$, **points along the radial axis of the *rtz*-coordinate system, toward the center of the circle.** The tangential and perpendicular components of $\vec{F}_{net}$ are zero.

NOTE The force described by Equation 8.6 is not a *new* force. The force itself must have an identifiable agent and will be one of our familiar forces, such as tension, friction, or the normal force. Equation 8.6 simply tells us how the force needs to act—how strongly and in which direction—to cause the particle to move with speed v in a circle of radius r.

The usefulness of the *rtz*-coordinate system becomes apparent when we write Newton's second law, Equation 8.6, in terms of the *r*-, *t*-, and *z*-components:

$$(F_{net})_r = \sum F_r = ma_r = \frac{mv^2}{r} = m\omega^2 r$$
$$(F_{net})_t = \sum F_t = ma_t = 0 \quad (8.7)$$
$$(F_{net})_z = \sum F_z = ma_z = 0$$

For uniform circular motion, the sum of the forces along the *t*-axis and along the *z*-axis must equal zero, and the sum of the forces along the *r*-axis *must* equal ma_r, where a_r is the centripetal acceleration.

On banked curves, the normal force of the road assists in providing the centripetal acceleration of the turn.

FIGURE 8.4 The net force points in the radial direction, toward the center of the circle.

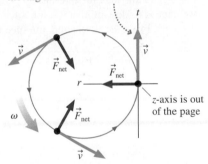

With no force, the particle would continue moving in the direction of $\vec{v}$.

z-axis is out of the page

EXAMPLE 8.2 | Spinning in a circle

An energetic father places his 20 kg child on a 5.0 kg cart to which a 2.0-m-long rope is attached. He then holds the end of the rope and spins the cart and child around in a circle, keeping the rope parallel to the ground. If the tension in the rope is 100 N, how many revolutions per minute (rpm) does the cart make? Rolling friction between the cart's wheels and the ground is negligible.

MODEL Model the child in the cart as a particle in uniform circular motion.

VISUALIZE FIGURE 8.5 on the next page shows the pictorial representation. A circular-motion problem usually does not have starting and ending points like a projectile problem, so numerical subscripts such as x_1 or y_2 are usually not needed. Here we need to define the cart's speed v and the radius r of the circle. Further, a

motion diagram is not needed for uniform circular motion because we already know the acceleration $\vec{a}$ points to the center of the circle.

The essential part of the pictorial representation is the free-body diagram. **For uniform circular motion we'll draw the free-body diagram in the *rz*-plane, looking at the edge of the circle, because this is the plane of the forces.** The contact forces acting on the cart are the normal force of the ground and the tension force of the rope. The normal force is perpendicular to the plane of the motion and thus in the *z*-direction. The direction of $\vec{T}$ is determined by the statement that the rope is parallel to the ground. In addition, there is the long-range gravitational force $\vec{F}_G$.

SOLVE We defined the *r*-axis to point toward the center of the circle, so $\vec{T}$ points in the positive *r*-direction and has *r*-component $T_r = T$.

Continued

FIGURE 8.5 Pictorial representation of a cart spinning in a circle.

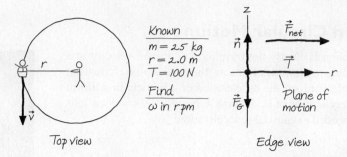

Known
$m = 25$ kg
$r = 2.0$ m
$T = 100$ N

Find
ω in rpm

Top view   Edge view

Newton's second law, using the rtz-components of Equations 8.7, is

$$\sum F_r = T = \frac{mv^2}{r}$$

$$\sum F_z = n - mg = 0$$

We've taken the r- and z-components of the forces directly from the free-body diagram, as you learned to do in Chapter 6. Then we've *explicitly* equated the sums to $a_r = v^2/r$ and $a_z = 0$. This is the basic strategy for all uniform circular-motion problems. From the z-equation we can find that $n = mg$. This would be useful if we needed to determine a friction force, but it's not needed in this problem. From the r-equation, the speed of the cart is

$$v = \sqrt{\frac{rT}{m}} = \sqrt{\frac{(2.0 \text{ m})(100 \text{ N})}{25 \text{ kg}}} = 2.83 \text{ m/s}$$

The cart's angular velocity is then found from Equations 8.5:

$$\omega = \frac{v_t}{r} = \frac{v}{r} = \frac{2.83 \text{ m/s}}{2.0 \text{ m}} = 1.41 \text{ rad/s}$$

This is another case where we inserted the radian unit because ω is specifically an *angular* velocity. Finally, we need to convert ω to rpm:

$$\omega = \frac{1.41 \text{ rad}}{1 \text{ s}} \times \frac{1 \text{ rev}}{2\pi \text{ rad}} \times \frac{60 \text{ s}}{1 \text{ min}} = 14 \text{ rpm}$$

ASSESS 14 rpm corresponds to a period $T \approx 4$ s. This result is reasonable.

The Central-Force Model

A force that is always directed toward the same point is called a **central force.** The tension in the rope of the last example is a central force, as is the gravitational force acting on an orbiting satellite. An object acted on by an attractive central force can undergo uniform circular motion around the central point. More complicated trajectories can occur in some situations—such as satellites following elliptical orbits—but for now we'll focus on circular motion, or motion with constant r. This **central-force model** is another important model of motion.

MODEL 8.1

Central force with constant *r*

For objects on which a constant net force points toward a central point.

- Model the object as a particle.
- The force causes a centripetal acceleration.
  - The motion is uniform circular motion.
- Mathematically:
  - Newton's second law is

$$\vec{F}_{\text{net}} = \left(\frac{mv^2}{r} \text{ or } m\omega^2 r, \text{ toward center} \right)$$

  - Use the kinematics of uniform circular motion.
- Limitations: Model fails if the force has a tangential component or if r changes.

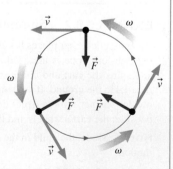

The object undergoes uniform circular motion.

Exercise 10

Let's look at more examples of the central-force model in action.

EXAMPLE 8.3 | Turning the corner I

What is the maximum speed with which a 1500 kg car can make a left turn around a curve of radius 50 m on a level (unbanked) road without sliding?

MODEL The car doesn't complete a full circle, but it is in uniform circular motion for a quarter of a circle while turning. We can model the car as a particle subject to a central force. Assume that rolling friction is negligible.

VISUALIZE FIGURE 8.6 shows the pictorial representation. The issue we must address is *how* a car turns a corner. What force or forces cause the direction of the velocity vector to change? Imagine driving on a completely frictionless road, such as a very icy road. You would not be able to turn a corner. Turning the steering wheel would be of no use; the car would slide straight ahead, in accordance with both Newton's first law and the experience of anyone who has ever driven on ice! So it must be *friction* that somehow allows the car to turn.

Figure 8.6 shows the top view of a tire as it turns a corner. If the road surface were frictionless, the tire would slide straight ahead. The force that prevents an object from sliding across a surface is *static friction*. Static friction $\vec{f}_s$ pushes *sideways* on the tire, toward the center of the circle. How do we know the direction is sideways? If $\vec{f}_s$ had a component either parallel to $\vec{v}$ or opposite to $\vec{v}$, it would cause the car to speed up or slow down. Because the car changes direction but not speed, static friction must be perpendicular to $\vec{v}$. $\vec{f}_s$ causes the centripetal acceleration of circular motion around the curve, and thus the free-body diagram, drawn from behind the car, shows the static friction force pointing toward the center of the circle.

SOLVE The maximum turning speed is reached when the static friction force reaches its maximum $f_{s\,max} = \mu_s n$. If the car enters the curve at a speed higher than the maximum, static friction will not

be large enough to provide the necessary centripetal acceleration and the car will slide.

The static friction force points in the positive r-direction, so its radial component is simply the magnitude of the vector: $(f_s)_r = f_s$. Newton's second law in the rtz-coordinate system is

$$\sum F_r = f_s = \frac{mv^2}{r}$$

$$\sum F_z = n - mg = 0$$

The only difference from Example 8.2 is that the tension force toward the center has been replaced by a static friction force toward the center. From the radial equation, the speed is

$$v = \sqrt{\frac{rf_s}{m}}$$

The speed will be a maximum when f_s reaches its maximum value:

$$f_s = f_{s\,max} = \mu_s n = \mu_s mg$$

where we used $n = mg$ from the z-equation. At that point,

$$v_{max} = \sqrt{\frac{rf_{s\,max}}{m}} = \sqrt{\mu_s rg}$$

$$= \sqrt{(1.0)(50\ \text{m})(9.80\ \text{m/s}^2)} = 22\ \text{m/s}$$

where the coefficient of static friction was taken from Table 6.1.

ASSESS 22 m/s $\approx$ 45 mph, a reasonable answer for how fast a car can take an unbanked curve. Notice that the car's mass canceled out and that the final equation for v_{max} is quite simple. This is another example of why it pays to work algebraically until the very end.

FIGURE 8.6 Pictorial representation of a car turning a corner.

Because μ_s depends on road conditions, the maximum safe speed through turns can vary dramatically. Wet roads, in particular, lower the value of μ_s and thus lower the speed of turns. Icy conditions are even worse. The corner you turn every day at 45 mph will require a speed of no more than 15 mph if the coefficient of static friction drops to 0.1.

EXAMPLE 8.4 | Turning the corner II

A highway curve of radius 70 m is banked at a 15° angle. At what speed v_0 can a car take this curve without assistance from friction?

MODEL Model the car as a particle subject to a central force.

VISUALIZE Having just discussed the role of friction in turning corners, it is perhaps surprising to suggest that the same turn can also be accomplished without friction. Example 8.3 considered a level roadway, but real highway curves are *banked* by being tilted

Continued

up at the outside edge of the curve. The angle is modest on ordinary highways, but it can be quite large on high-speed racetracks. The purpose of banking becomes clear if you look at the free-body diagram in **FIGURE 8.7**. The normal force $\vec{n}$ is perpendicular to the road, so tilting the road causes $\vec{n}$ to have a component toward the center of the circle. **The radial component n_r is the central force that causes the centripetal acceleration needed to turn the car.** Notice that we are *not* using a tilted coordinate system, although this looks rather like an inclined-plane problem. The center of the circle is in the same horizontal plane as the car, and for circular-motion problems we need the r-axis to pass through the center. Tilted axes are for *linear* motion along an incline.

FIGURE 8.7 Pictorial representation of a car on a banked curve.

SOLVE Without friction, $n_r = n \sin \theta$ is the only component of force in the radial direction. It is this inward component of the normal force on the car that causes it to turn the corner. Newton's second law is

$$\sum F_r = n \sin \theta = \frac{mv_0^2}{r}$$

$$\sum F_z = n \cos \theta - mg = 0$$

where θ is the angle at which the road is banked and we've assumed that the car is traveling at the correct speed v_0. From the z-equation,

$$n = \frac{mg}{\cos \theta}$$

Substituting this into the r-equation and solving for v_0 give

$$\frac{mg}{\cos \theta} \sin \theta = mg \tan \theta = \frac{mv_0^2}{r}$$

$$v_0 = \sqrt{rg \tan \theta} = 14 \text{ m/s}$$

ASSESS This is ≈ 28 mph, a reasonable speed. Only at this very specific speed can the turn be negotiated without reliance on friction forces.

It's interesting to explore what happens at other speeds on a banked curve. **FIGURE 8.8** shows that the car will need to rely on both the banking *and* friction if it takes the curve at a speed faster or slower than v_0.

FIGURE 8.8 Free-body diagrams for a car going around a banked curve at speeds slower and faster than the friction-free speed v_0.

EXAMPLE 8.5 | A rock in a sling

A Stone Age hunter places a 1.0 kg rock in a sling and swings it in a horizontal circle around his head on a 1.0-m-long vine. If the vine breaks at a tension of 200 N, what is the maximum angular speed, in rpm, with which he can swing the rock?

MODEL Model the rock as a particle in uniform circular motion.

VISUALIZE This problem appears, at first, to be essentially the same as Example 8.2, where the father spun his child around on

a rope. However, the lack of a normal force from a supporting surface makes a *big* difference. In this case, the *only* contact force on the rock is the tension in the vine. Because the rock moves in a horizontal circle, you may be tempted to draw a free-body diagram like **FIGURE 8.9a**, where $\vec{T}$ is directed along the r-axis. You will quickly run into trouble, however, because this diagram has a net force in the z-direction and it is impossible to satisfy $\sum F_z = 0$. The

FIGURE 8.9 Pictorial representation of a rock in a sling.

(a)

Wrong diagram!

$\vec{T}$

r

$\vec{F}_G$

z

(b)

L

θ

r

Center of circle

Known
$m = 1.0$ kg
$L = 1.0$ m
$T_{max} = 200$ N

Find
ω_{max}

$\vec{T}$

z

θ

r

$\vec{F}_G$

gravitational force $\vec{F}_G$ certainly points vertically downward, so the difficulty must be with $\vec{T}$.

As an experiment, tie a small weight to a string, swing it over your head, and check the *angle* of the string. You will quickly discover that the string is *not* horizontal but, instead, is angled downward. The sketch of **FIGURE 8.9b** labels the angle θ. Notice that the rock moves in a *horizontal* circle, so the center of the circle is *not* at his hand. The r-axis points to the center of the circle, but the tension force is directed along the vine. Thus the correct free-body diagram is the one in Figure 8.9b.

SOLVE The free-body diagram shows that the downward gravitational force is balanced by an upward component of the tension, leaving the radial component of the tension to cause the centripetal acceleration. Newton's second law is

$$\sum F_r = T\cos\theta = \frac{mv^2}{r}$$

$$\sum F_z = T\sin\theta - mg = 0$$

where θ is the angle of the vine below horizontal. From the z-equation we find

$$\sin\theta = \frac{mg}{T}$$

$$\theta_{max} = \sin^{-1}\left(\frac{(1.0\text{ kg})(9.8\text{ m/s}^2)}{200\text{ N}}\right) = 2.81°$$

where we've evaluated the angle at the maximum tension of 200 N. The vine's angle of inclination is small but not zero.

Turning now to the r-equation, we find the rock's speed is

$$v_{max} = \sqrt{\frac{rT\cos\theta_{max}}{m}}$$

Careful! The radius r of the circle is *not* the length L of the vine. You can see in Figure 8.9b that $r = L\cos\theta$. Thus

$$v_{max} = \sqrt{\frac{LT\cos^2\theta_{max}}{m}} = \sqrt{\frac{(1.0\text{ m})(200\text{ N})(\cos 2.81°)^2}{1.0\text{ kg}}} = 14.1\text{ m/s}$$

We can now find the maximum angular speed, the value of ω that brings the tension to the breaking point:

$$\omega_{max} = \frac{v_{max}}{r} = \frac{v_{max}}{L\cos\theta_{max}} = \frac{14.1\text{ rad}}{1\text{ s}} \times \frac{1\text{ rev}}{2\pi\text{ rad}} \times \frac{60\text{ s}}{1\text{ min}} = 135\text{ rpm}$$

STOP TO THINK 8.2 A block on a string spins in a horizontal circle on a frictionless table. Rank in order, from largest to smallest, the tensions T_a to T_e acting on blocks a to e.

$r = 100$ cm
$\omega = 50$ rpm

(a)

$r = 100$ cm
$\omega = 100$ rpm

(b)

$r = 50$ cm
$\omega = 100$ rpm

(c)

$r = 50$ cm
$\omega = 200$ rpm

(d)

$r = 25$ cm
$\omega = 200$ rpm

(e)

8.3 Circular Orbits

Satellites orbit the earth, the earth orbits the sun, and our entire solar system orbits the center of the Milky Way galaxy. Not all orbits are circular, but in this section we'll limit our analysis to circular orbits.

How does a satellite orbit the earth? What forces act on it? To answer these important questions, let's return, for a moment, to projectile motion. Projectile motion occurs when the only force on an object is gravity. Our analysis of projectiles assumed that

FIGURE 8.10 Projectiles being launched at increasing speeds from height h on a smooth, airless planet.

A Projectile motion

B The ground is curving away from the projectile.

C

D This projectile "falls" all the way around the planet because the curvature of its trajectory matches the planet's curvature.

FIGURE 8.11 The "real" gravitational force is always directed toward the center of the planet.

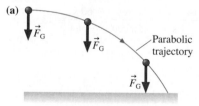

(a)

$\vec{F}_G$

$\vec{F}_G$

Parabolic trajectory

$\vec{F}_G$

$\vec{F}_G$

Flat-earth approximation

(b)

Circular orbit

$\vec{F}_G$
Planet

$\vec{F}_G$ $\vec{F}_G$

Spherical planet

the earth is flat and that the acceleration due to gravity is everywhere straight down. This is an acceptable approximation for projectiles of limited range, such as baseballs or cannon balls, but there comes a point where we can no longer ignore the curvature of the earth.

FIGURE 8.10 shows a perfectly smooth, spherical, airless planet with one tower of height h. A projectile is launched from this tower parallel to the ground ($\theta = 0°$) with speed v_0. If v_0 is very small, as in trajectory A, the "flat-earth approximation" is valid and the problem is identical to Example 8.4 in which a car drove off a cliff. The projectile simply falls to the ground along a parabolic trajectory.

As the initial speed v_0 is increased, the projectile begins to notice that the ground is curving out from beneath it. It is falling the entire time, always getting closer to the ground, but the distance that the projectile travels before finally reaching the ground—that is, its range—increases because the projectile must "catch up" with the ground that is curving away from it. Trajectories B and C are of this type. The actual calculation of these trajectories is beyond the scope of this textbook, but you should be able to understand the factors that influence the trajectory.

If the launch speed v_0 is sufficiently large, there comes a point where the curve of the trajectory and the curve of the earth are parallel. In this case, the projectile "falls" but it never gets any closer to the ground! This is the situation for trajectory D. A closed trajectory around a planet or star, such as trajectory D, is called an **orbit.**

The most important point of this qualitative analysis is that **an orbiting projectile is in free fall.** This is, admittedly, a strange idea, but one worth careful thought. An orbiting projectile is really no different from a thrown baseball or a car driving off a cliff. The only force acting on it is gravity, but its tangential velocity is so large that the curvature of its trajectory matches the curvature of the earth. When this happens, the projectile "falls" under the influence of gravity but never gets any closer to the surface.

In the flat-earth approximation, shown in **FIGURE 8.11a**, the gravitational force acting on an object of mass m is

$$\vec{F}_G = (mg, \text{ vertically downward}) \qquad \text{(flat-earth approximation)} \qquad (8.8)$$

But since stars and planets are actually spherical (or very close to it), the "real" force of gravity acting on an object is directed toward the *center* of the planet, as shown in **FIGURE 8.11b**. In this case the gravitational force is

$$\vec{F}_G = (mg, \text{ toward center}) \qquad \text{(spherical planet)} \qquad (8.9)$$

That is, gravity is a central force causing the centripetal acceleration of uniform circular motion. Thus the gravitational force causes the object in Figure 8.11b to have acceleration

$$\vec{a} = \frac{\vec{F}_{net}}{m} = (g, \text{ toward center}) \qquad (8.10)$$

An object moving in a circle of radius r at speed v_{orbit} will have this centripetal acceleration if

$$a_r = \frac{(v_{orbit})^2}{r} = g \qquad (8.11)$$

That is, if an object moves parallel to the surface with the speed

$$v_{orbit} = \sqrt{rg} \qquad (8.12)$$

then the free-fall acceleration is exactly the centripetal acceleration needed for a circular orbit of radius r. An object with any other speed will not follow a circular orbit.

The earth's radius is $r = R_e = 6.37 \times 10^6$ m. (A table of useful astronomical data is inside the back cover of this book.) The orbital speed of a projectile just skimming the surface of an airless, bald earth is

$$v_{orbit} = \sqrt{rg} = \sqrt{(6.37 \times 10^6 \text{ m})(9.80 \text{ m/s}^2)} = 7900 \text{ m/s} \approx 16,000 \text{ mph}$$

Even if there were no trees and mountains, a real projectile moving at this speed would burn up from the friction of air resistance.

Satellites

Suppose, however, that we launched the projectile from a tower of height $h = 230$ mi $\approx 3.8 \times 10^5$ m, just above the earth's atmosphere. This is approximately the height of the International Space Station and other low-earth-orbit satellites. Note that $h \ll R_e$, so the radius of the orbit $r = R_e + h = 6.75 \times 10^6$ m is only 5% greater than the earth's radius. Many people have a mental image that satellites orbit far above the earth, but in fact many satellites come pretty close to skimming the surface. Our calculation of v_{orbit} thus turns out to be quite a good estimate of the speed of a satellite in low earth orbit.

We can use v_{orbit} to calculate the period of a satellite orbit:

$$T = \frac{2\pi r}{v_{orbit}} = 2\pi \sqrt{\frac{r}{g}} \tag{8.13}$$

For a low earth orbit, with $r = R_e + 230$ miles, we find $T = 5210$ s $= 87$ min. The period of the International Space Station at an altitude of 230 mi is, indeed, close to 87 minutes. (The actual period is 93 min. The difference, you'll learn in Chapter 13, arises because g is slightly less at a satellite's altitude.)

When we discussed *weightlessness* in Chapter 6, we discovered that it occurs during free fall. We asked the question, at the end of « Section 6.3, whether astronauts and their spacecraft were in free fall. We can now give an affirmative answer: They are, indeed, in free fall. They are falling continuously around the earth, under the influence of only the gravitational force, but never getting any closer to the ground because the earth's surface curves beneath them. Weightlessness in space is no different from the weightlessness in a free-falling elevator. It does *not* occur from an absence of gravity. Instead, the astronaut, the spacecraft, and everything in it are weightless because they are all falling together.

The International Space Station is in free fall.

8.4 Reasoning About Circular Motion

Some aspects of circular motion are puzzling and counterintuitive. Examining a few of these will give us a chance to practice Newtonian reasoning.

Centrifugal Force?

If the car turns a corner quickly, you feel "thrown" against the door. But there's really no such force because there is no agent exerting it. FIGURE 8.12 shows a bird's-eye view of you riding in a car as it makes a left turn. You try to continue moving in a straight line, obeying Newton's first law, when—without having been provoked—the door suddenly turns in front of you and runs into you! You do, indeed, then feel the force of the door because it is now the normal force of the door, pointing *inward* toward the center of the curve, causing you to turn the corner. But you were not "thrown" into the door; the door ran into you.

The "force" that seems to push an object to the outside of a circle is commonly known as the *centrifugal force*. Despite having a name, the centrifugal force is fictitious. It describes your experience *relative to a noninertial reference frame,* but there really is no such force. **You must always use Newton's laws in an inertial reference frame.** There are no centrifugal forces in an inertial reference frame.

> **NOTE** You might wonder if the *rtz*-coordinate system is an inertial reference frame. It is. We're using the *rtz*-coordinates to establish directions for decomposing vectors, but we're not making measurements in the *rtz*-system. That is, velocities and accelerations are measured in the laboratory reference frame. The particle would always be at rest ($\vec{v} = \vec{0}$) if we measured velocities in a reference frame attached to the particle.

FIGURE 8.12 Bird's-eye view of a passenger as a car turns a corner.

You try to keep moving straight ahead.

The door provides the central force that makes you move in a circle.

Gravity on a Rotating Earth

There is one small problem with the admonition that you must use Newton's laws in an inertial reference frame: A reference frame attached to the ground isn't truly inertial because of the earth's rotation. Fortunately, we can make a simple correction that allows us to continue using Newton's laws on the earth's surface.

FIGURE 8.13 shows an object being weighed by a spring scale on the earth's equator. An observer hovering above the north pole sees two forces on the object: the gravitational force $\vec{F}_{M \text{ on } m}$, given by Newton's law of gravity, and the outward spring force $\vec{F}_{\text{Sp}}$. The object moves in a circle as the earth rotates, so Newton's second law is

$$\sum F_r = F_{M \text{ on } m} - F_{\text{Sp}} = m\omega^2 R$$

where ω is the angular speed of the rotating earth. The spring-scale reading $F_{\text{Sp}} = F_{M \text{ on } m} - m\omega^2 R$ is *less* than it would be on a nonrotating earth.

The blow-up in Figure 8.13 shows how we see things in a noninertial, flat-earth reference frame. For us the object is at rest, in equilibrium, hence the upward spring force must be exactly balanced by a downward gravitational force $\vec{F}_{\text{G}}$. Thus $F_{\text{Sp}} = F_{\text{G}}$.

Now both we and the hovering, inertial observer see the same reading on the scale. If F_{Sp} is the same for both of us, then

$$F_{\text{G}} = F_{M \text{ on } m} - m\omega^2 R \tag{8.14}$$

In other words, force $\vec{F}_{\text{G}}$—what we called the *effective* gravitational force in Chapter 6—is slightly less than the true gravitational force $\vec{F}_{M \text{ on } m}$ because of the earth's rotation. In essence, $m\omega^2 R$ is the centrifugal force, a fictitious force trying—from our perspective in a noninertial reference frame—to "throw" us off the rotating platform. There really is no such force, but—this is the important point—**we can continue to use Newton's laws in our rotating reference frame if we pretend there is.**

Because $F_{\text{G}} = mg$ for an object at rest, the effect of the centrifugal term in Equation 8.14 is to make g a little smaller than it would be on a nonrotating earth:

$$g = \frac{F_{\text{G}}}{m} = \frac{F_{M \text{ on } m} - m\omega^2 R}{m} = \frac{GM}{R^2} - \omega^2 R = g_{\text{earth}} - \omega^2 R \tag{8.15}$$

We calculated $g_{\text{earth}} = 9.83 \text{ m/s}^2$ in Chapter 6. Using $\omega = 1$ rev/day (which must be converted to SI units) and $R = 6370$ km, we find $\omega^2 R = 0.033 \text{ m/s}^2$ at the equator. Thus the free-fall acceleration—what we actually measure in our rotating reference frame—is about 9.80 m/s^2, exactly what we measure in the laboratory.

Things are a little more complicated at other latitudes, but the bottom line is that we can safely use Newton's laws in our rotating, noninertial reference frame on the earth's surface if we calculate the gravitational force—as we've been doing—as $F_{\text{G}} = mg$ with g the measured free-fall value, a value that compensates for our rotation, rather than the purely gravitational g_{earth}.

Why Does the Water Stay in the Bucket?

If you swing a bucket of water over your head quickly, the water stays in, but you'll get a shower if you swing too slowly. Why? We'll answer this question by starting with an equivalent situation, a roller coaster doing a loop-the-loop.

FIGURE 8.14 shows a roller-coaster car going around a vertical loop-the-loop of radius r. Why doesn't the car fall off at the top of the circle? Now, motion in a vertical circle is *not* uniform circular motion; the car slows down as it goes up one side and speeds up as it comes back down the other. But at the very top and very bottom points, only the car's direction is changing, not its speed, so at those points the acceleration is purely centripetal. Thus **there must be a net force toward the center of the circle.**

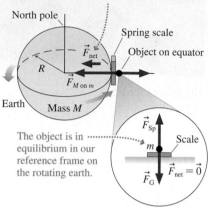

FIGURE 8.13 The earth's rotation affects the measured value of g.

The object is in circular motion on a rotating earth, so there is a net force toward the center.

North pole

Spring scale

Object on equator

$\vec{F}_{\text{net}}$

R

$\vec{F}_{M \text{ on } m}$

Earth   Mass M

The object is in equilibrium in our reference frame on the rotating earth.

$\vec{F}_{\text{Sp}}$

m Scale

$\vec{F}_{\text{G}}$ $\vec{F}_{\text{net}} = \vec{0}$

FIGURE 8.14 A roller-coaster car going around a loop-the-loop.

$\vec{v}_{\text{top}}$

$\vec{F}_{\text{net}}$ $\vec{n}$ $\vec{F}_{\text{G}}$

r

The r-axis points toward the center.

r

$\vec{F}_{\text{net}}$ $\vec{n}$

$\vec{v}_{\text{bot}}$

$\vec{F}_{\text{G}}$

First consider the very bottom of the loop. To have a net force toward the center—upward at this point—requires $n > F_G$. The normal force has to *exceed* the gravitational force to provide the net force needed to "turn the corner" at the bottom of the circle. This is why you "feel heavy" at the bottom of the circle or at the bottom of a valley on a roller coaster.

We can analyze the situation quantitatively by writing the r-component of Newton's second law. At the bottom of the circle, with the r-axis pointing upward, we have

$$\sum F_r = n_r + (F_G)_r = n - mg = ma_r = \frac{m(v_{\text{bot}})^2}{r} \tag{8.16}$$

From Equation 8.16 we find

$$n = mg + \frac{m(v_{\text{bot}})^2}{r} \tag{8.17}$$

The normal force at the bottom is *larger* than *mg*.

Things are a little trickier as the roller-coaster car crosses the top of the loop. Whereas the normal force of the track pushes up when the car is at the bottom of the circle, it *presses down* when the car is at the top and the track is above the car. Think about the free-body diagram to make sure you agree.

The car is still moving in a circle, so there *must* be a net force toward the center of the circle. The r-axis, which points toward the center of the circle, now points *downward*. Consequently, both forces have *positive* components. Newton's second law at the top of the circle is

$$\sum F_r = n_r + (F_G)_r = n + mg = \frac{m(v_{\text{top}})^2}{r} \tag{8.18}$$

Thus at the top the normal force of the track on the car is

$$n = \frac{m(v_{\text{top}})^2}{r} - mg \tag{8.19}$$

The normal force at the top can exceed *mg* if v_{top} is large enough. Our interest, however, is in what happens as the car goes slower and slower. As v_{top} decreases, there comes a point when n reaches zero. "No normal force" means "no contact," so at that speed, the track is *not* pushing against the car. Instead, the car is able to complete the circle because gravity alone provides sufficient centripetal acceleration.

The speed at which $n = 0$ is called the *critical speed* v_c:

$$v_c = \sqrt{\frac{rmg}{m}} = \sqrt{rg} \tag{8.20}$$

The critical speed is the slowest speed at which the car can complete the circle. Equation 8.19 would give a negative value for n if $v < v_c$, but that is physically impossible. The track can push against the wheels of the car ($n > 0$), but it can't pull on them. If $v < v_c$, the car cannot turn the full loop but, instead, comes off the track and becomes a projectile! **FIGURE 8.15** summarizes our reasoning.

Water stays in a bucket swung over your head for the same reason: Circular motion requires a net force toward the center of the circle. At the top of the circle—if you swing the bucket fast enough—the bucket adds to the force of gravity by pushing *down* on the water, just like the downward normal force of the track on the roller-coaster car. As long as the bucket is pushing against the water, the bucket and the water are in contact and thus the water is "in" the bucket. As you swing slower and slower, requiring the water to have less and less centripetal acceleration, the bucket-on-water normal force decreases until it becomes zero at the critical speed. At the critical speed, gravity alone provides sufficient centripetal acceleration. Below the critical speed, gravity provides *too much* downward force for circular motion, so the water leaves the bucket and becomes a projectile following a parabolic trajectory toward your head!

FIGURE 8.15 A roller-coaster car at the top of the loop.

The normal force adds to gravity to make a large enough force for the car to turn the circle.

$v > v_c$

At v_c, gravity alone is enough force for the car to turn the circle. $\vec{n} = \vec{0}$ at the top point.

v_c

The gravitational force is too large for the car to stay in the circle!

Normal force became zero here.

$v < v_c$

Parabolic trajectory

8.5 Nonuniform Circular Motion

Many interesting examples of circular motion involve objects whose speed changes. As we've already noted, a roller-coaster car doing a loop-the-loop slows down as it goes up one side, speeds up as it comes back down the other side. Circular motion with a changing speed is called *nonuniform circular motion*.

FIGURE 8.16 shows a particle moving in a circle of radius r. In addition to a radial force component—required for all circular motion—this particle experiences a *tangential* force component $(F_{net})_t$ and hence a tangential acceleration

$$a_t = \frac{dv_t}{dt} \qquad (8.21)$$

Now v_t is the particle's velocity *around* the circle, with speed $v = |v_t|$, so a tangential force component causes the particle to change speed. That is, the particle is undergoing nonuniform circular motion. Note that $(F_{net})_t$, like v_t, is positive when ccw, negative when cw.

Force and acceleration are still related to each other through Newton's second law:

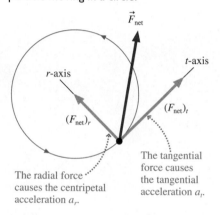

FIGURE 8.16 Net force $\vec{F}_{net}$ is applied to a particle moving in a circle.

The radial force causes the centripetal acceleration a_r.

The tangential force causes the tangential acceleration a_t.

$$(F_{net})_r = \sum F_r = ma_r = \frac{mv_t^2}{r} = m\omega^2 r$$

$$(F_{net})_t = \sum F_t = ma_t \qquad (8.22)$$

$$(F_{net})_z = \sum F_z = 0$$

If the tangential force is constant, you can apply what you know about constant-acceleration kinematics to solve for v_t at a later time.

NOTE Equations 8.22 differ from Equations 8.7 for uniform circular motion only in the fact that a_t is no longer zero.

EXAMPLE 8.6 | **Sliding out of the curve**

A 1500 kg car drives around a flat, 50-m-diameter track, starting from rest. The drive wheels supply a small but steady 525 N force in the forward direction. The coefficient of static friction between the car tires and the road is 0.90. How many revolutions of the track have been made when the car slides out of the curve?

MODEL Model the car as a particle in nonuniform circular motion. Assume that rolling friction and air resistance can be neglected.

VISUALIZE FIGURE 8.17 shows a pictorial representation. As in earlier examples, it's static friction, perpendicular to the tires, that causes the centripetal acceleration of circular motion. The propulsion force is a tangential force. For the first time, we need a free-body diagram showing forces in three dimensions.

FIGURE 8.17 Pictorial representation of a car speeding up around a circle.

Known	Find
$r = 50$ m $m = 1500$ kg	$\Delta\theta$
$\mu_s = 0.90$ $F_{wheels} = 525$ N	

SOLVE At slow speeds, static friction in the radial direction keeps the car moving in a circle. But there's an upper limit to the size of the static friction force, and the car will begin to slide out of the curve when that limit is reached. The r-component of Newton's second law is

$$\sum F_r = f_s = \frac{mv_t^2}{r}$$

That is, static friction increases proportional to v_t^2 until the car reaches a velocity v_{max} at which the static friction is $f_{s\,max}$.

Recall that the maximum possible static friction is $f_{s\,max} = \mu_s n$. We can find the normal force from the z-component of Newton's second law:

$$\sum F_z = n - F_G = 0$$

Thus $n = F_G = mg$ and $f_{s\,max} = \mu_s mg$. Combining these two equations, we see that the mass cancels and we have

$$v_{max}^2 = \mu_s rg$$

How far does the car have to travel to reach this speed? We can find the car's tangential acceleration from the t-component of the second law: $a_t = F_{wheels}/m$. This is a constant acceleration, so we can use constant-acceleration kinematics. Let s measure the distance around the circle—the arc length. Thus, because the initial velocity is $v_0 = 0$, we have

$$v_t^2 = v_0^2 + 2a_s s = 2a_s s = \frac{2sF_{wheels}}{m}$$

You'll recall that the angular displacement, measured in radians, is $\Delta\theta = s/r$. So when the car reaches velocity v_t, it has revolved through an angle

$$\Delta\theta = \frac{s}{r} = \frac{mv_t^2}{2rF_{wheels}}$$

Using the maximum speed before sliding, we find that the car slides out of the curve after revolving through an angle

$$\Delta\theta_{max} = \frac{mv_{max}^2}{2rF_{wheels}} = \frac{m}{2rF_{wheels}} \times \mu_s rg = \frac{\mu_s mg}{2F_{wheels}}$$

For the car in this problem,

$$\Delta\theta_{max} = \frac{(0.90)(1500\ \text{kg})(9.80\ \text{m/s}^2)}{2(525\ \text{N})}$$

$$= 12.6\ \text{rad} \times \frac{1\ \text{rev}}{2\pi\ \text{rad}} = 2.0\ \text{rev}$$

It completes 2.0 revolutions before it starts to slide.

ASSESS A 525 N force on a 1500 kg car causes a tangential acceleration of $a_t \approx 0.3\ \text{m/s}^2$. That's a quite modest acceleration, so it seems reasonable that the car would complete 2 rev before gaining enough speed to start sliding.

We've come a long way since our first dynamics problems in Chapter 6, but our basic strategy has not changed.

PROBLEM-SOLVING STRATEGY 8.1

Circular-motion problems

MODEL Model the object as a particle and make other simplifying assumptions.

VISUALIZE Draw a pictorial representation. Use rtz-coordinates.

■ Establish a coordinate system with the r-axis pointing toward the center of the circle.
■ Show important points in the motion on a sketch. Define symbols and identify what the problem is trying to find.
■ Identify the forces and show them on a free-body diagram.

SOLVE Newton's second law is

$$(F_{net})_r = \sum F_r = ma_r = \frac{mv_t^2}{r} = m\omega^2 r$$

$$(F_{net})_t = \sum F_t = ma_t$$

$$(F_{net})_z = \sum F_z = 0$$

■ Determine the force components from the free-body diagram. Be careful with signs.
■ The tangential acceleration for uniform circular motion is $a_t = 0$.
■ Solve for the acceleration, then use kinematics to find velocities and positions.

ASSESS Check that your result has the correct units and significant figures, is reasonable, and answers the question.

Exercise 11

STOP TO THINK 8.4 A ball on a string is swung in a vertical circle. The string happens to break when it is parallel to the ground and the ball is moving up. Which trajectory does the ball follow?

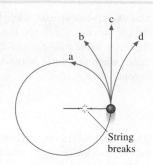

CHALLENGE EXAMPLE 8.7 | Swinging on two strings

The 250 g ball shown in **FIGURE 8.18** revolves in a horizontal plane as the vertical shaft spins. What is the critical angular speed, in rpm, that the shaft must exceed to keep both strings taut?

FIGURE 8.18 A ball revolving on two strings.

MODEL Model the ball as a particle in uniform circular motion. For both strings to be straight, as shown, both must be under tension. If the angular speed is slowly decreased, eventually the lower string will go slack and the ball will sag. The critical angular speed ω_c is the angular speed at which the tension in the lower string reaches zero. We need to find an expression for the tension in the lower string, then determine when that tension becomes zero.

VISUALIZE **FIGURE 8.19** is the ball's free-body diagram with the r-axis pointing toward the center of the circle. The ball is acted on by two tension forces, at equal angles above and below horizontal, and by gravity. The free-body diagram is similar to Example 8.5, the rock in the sling, but with an additional tension force.

FIGURE 8.19 Free-body diagram of the ball.

SOLVE This is uniform circular motion, so we need to consider only the r- and z-components of Newton's second law. All the information is on the free-body diagram, where we see that gravity has only a z-component but the tensions have both r- and z-components. The two equations are

$$\sum F_r = T_{\text{top}} \cos\theta + T_{\text{bot}} \cos\theta = m\omega^2 r$$

$$\sum F_z = T_{\text{top}} \sin\theta - T_{\text{bot}} \sin\theta - mg = 0$$

Factoring out the $\cos\theta$ and $\sin\theta$ terms, we have two simultaneous equations:

$$T_{\text{top}} + T_{\text{bot}} = \frac{m\omega^2 r}{\cos\theta}$$

$$T_{\text{top}} - T_{\text{bot}} = \frac{mg}{\sin\theta}$$

Subtracting the second equation from the first will eliminate T_{top}:

$$2T_{\text{bot}} = \frac{m\omega^2 r}{\cos\theta} - \frac{mg}{\sin\theta}$$

and thus

$$T_{\text{bot}} = \frac{m}{2}\left(\frac{\omega^2 r}{\cos\theta} - \frac{g}{\sin\theta}\right)$$

You can see that this expression becomes negative—a physically impossible situation—if ω is too small. The angular speed at which the tension reaches zero—the critical angular speed—is found by setting the expression in parentheses equal to zero. This gives

$$\omega_c = \sqrt{\frac{g}{r\tan\theta}}$$

For our situation,

$$r = \sqrt{(1.0 \text{ m})^2 - (0.50 \text{ m})^2} = 0.866 \text{ m}$$

$$\theta = \sin^{-1}[(0.50 \text{ m})/(1.0 \text{ m})] = 30°$$

Thus the critical angular speed is

$$\omega_c = \sqrt{\frac{9.80 \text{ m/s}^2}{(0.866 \text{ m})\tan 30°}} = 4.40 \text{ rad/s}$$

Converting to rpm:

$$\omega_c = 4.40 \text{ rad/s} \times \frac{1 \text{ rev}}{2\pi \text{ rad}} \times \frac{60 \text{ s}}{1 \text{ min}} = 42 \text{ rpm}$$

ASSESS ω_c is the *minimum* angular speed needed to keep both strings taut. For a ball attached to meter-long strings, 42 rpm—a bit less than 1 revolution per second—seems plausible. Your intuition probably suggests that the bottom string wouldn't be taut if the shaft spun at only a few rpm, and hundreds of rpm seems much too high. Remember that the goal of Assess is not to prove that an answer is correct but to rule out answers that, with a little thought, are clearly wrong.

SUMMARY

The goal of Chapter 8 has been to learn to solve problems about motion in two dimensions.

GENERAL PRINCIPLES

Newton's Second Law

Expressed in x- and y-component form:

$$(F_{net})_x = \sum F_x = ma_x$$

$$(F_{net})_y = \sum F_y = ma_y$$

Expressed in rtz-component form:

$$(F_{net})_r = \sum F_r = ma_r = \frac{mv_t^2}{r} = m\omega^2 r$$

$$(F_{net})_t = \sum F_t = \begin{cases} 0 & \text{uniform circular motion} \\ ma_t & \text{nonuniform circular motion} \end{cases}$$

$$(F_{net})_z = \sum F_z = 0$$

Uniform Circular Motion

- Speed is constant.
- $\vec{F}_{net}$ points toward the center of the circle.
- The **centripetal acceleration** $\vec{a}$ points toward the center of the circle. It changes the particle's direction but not its speed.

Nonuniform Circular Motion

- Speed changes.
- $\vec{F}_{net}$ and $\vec{a}$ have both radial and tangential components.
- The radial component changes the particle's direction.
- The tangential component changes the particle's speed.

IMPORTANT CONCEPTS

rtz-coordinates

- The r-axis points toward the center of the circle.
- The t-axis is tangent, pointing counterclockwise.

Plane of motion

Projectile motion

- With no drag, the x- and y-components of acceleration are independent. The trajectory is a parabola.
- With drag, the trajectory is not a parabola. Maximum range is achieved for an angle less than $45°$.

Projectile motion with drag

APPLICATIONS

Orbits

An object acted on only by gravity has a circular orbit of radius r if its speed is

$$v = \sqrt{rg}$$

The object is in free fall.

Circular motion on surfaces

Circular motion requires a net force pointing to the center. n must be > 0 for the object to be in contact with a surface.

TERMS AND NOTATION

rtz-coordinate system	central force	central-force model	orbit

CONCEPTUAL QUESTIONS

1. In uniform circular motion, which of the following are constant: speed, velocity, angular velocity, centripetal acceleration, magnitude of the net force?

2. A car runs out of gas while driving down a hill. It rolls through the valley and starts up the other side. At the very bottom of the valley, which of the free-body diagrams in **FIGURE Q8.2** is correct? The car is moving to the right, and drag and rolling friction are negligible.

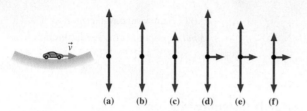

FIGURE Q8.2

3. **FIGURE Q8.3** is a bird's-eye view of particles on strings moving in horizontal circles on a tabletop. All are moving at the same speed. Rank in order, from largest to smallest, the tensions T_a to T_d. Give your answer in the form a > b = c > d and explain your ranking.

FIGURE Q8.3

4. Tarzan swings through the jungle on a massless vine. At the lowest point of his swing, is the tension in the vine greater than, less than, or equal to the gravitational force on Tarzan? Explain.

5. **FIGURE Q8.5** shows two balls of equal mass moving in vertical circles. Is the tension in string A greater than, less than, or equal to the tension in string B if the balls travel over the top of the circle (a) with equal speed and (b) with equal angular velocity?

FIGURE Q8.5

6. Ramon and Sally are observing a toy car speed up as it goes around a circular track. Ramon says, "The car's speeding up, so there must be a net force parallel to the track." "I don't think so," replies Sally. "It's moving in a circle, and that requires centripetal acceleration. The net force has to point to the center of the circle." Do you agree with Ramon, Sally, or neither? Explain.

7. A jet plane is flying on a level course at constant speed. The engines are at full throttle.
   a. What is the net force on the plane? Explain.
   b. Draw a free-body diagram of the plane as seen from the side with the plane flying to the right. Name (don't just label) any and all forces shown on your diagram.
   c. Airplanes bank when they turn. Draw a free-body diagram of the plane as seen from behind as it makes a right turn.
   d. *Why* do planes bank as they turn? Explain.

8. A small projectile is launched parallel to the ground at height $h = 1$ m with sufficient speed to orbit a completely smooth, airless planet. A bug rides inside a small hole inside the projectile. Is the bug weightless? Explain.

9. You can swing a ball on a string in a vertical circle if you swing it fast enough. But if you swing too slowly, the string goes slack as the ball nears the top. Explain *why* there's a minimum speed to keep the ball moving in a circle.

10. A golfer starts with the club over her head and swings it to reach maximum speed as it contacts the ball. Halfway through her swing, when the golf club is parallel to the ground, does the acceleration vector of the club head point (a) straight down, (b) parallel to the ground, approximately toward the golfer's shoulders, (c) approximately toward the golfer's feet, or (d) toward a point above the golfer's head? Explain.

EXERCISES AND PROBLEMS

Problems labeled ▓ integrate material from earlier chapters.

Exercises

Section 8.1 Dynamics in Two Dimensions

1. ‖ As a science fair project, you want to launch an 800 g model rocket straight up and hit a horizontally moving target as it passes 30 m above the launch point. The rocket engine provides a constant thrust of 15.0 N. The target is approaching at a speed of 15 m/s. At what horizontal distance between the target and the rocket should you launch?

2. ‖ A 500 g model rocket is on a cart that is rolling to the right at a speed of 3.0 m/s. The rocket engine, when it is fired, exerts an 8.0 N vertical thrust on the rocket. Your goal is to have the rocket pass through a small horizontal hoop that is 20 m above the ground. At what horizontal distance left of the hoop should you launch?

3. ‖ A 4.0×10^{10} kg asteroid is heading directly toward the center of the earth at a steady 20 km/s. To save the planet, astronauts strap a giant rocket to the asteroid perpendicular to its direction of travel. The rocket generates 5.0×10^9 N of thrust. The rocket is fired when the asteroid is 4.0×10^6 km away from earth. You can ignore the earth's gravitational force on the asteroid and their rotation about the sun.

a. If the mission fails, how many hours is it until the asteroid impacts the earth?

b. The radius of the earth is 6400 km. By what minimum angle must the asteroid be deflected to just miss the earth?

c. What is the actual angle of deflection if the rocket fires at full thrust for 300 s before running out of fuel?

4. || A 55 kg astronaut who weighs 180 N on a distant planet is pondering whether she can leap over a 3.5-m-wide chasm without falling in. If she leaps at a 15° angle, what initial speed does she need to clear the chasm?

Section 8.2 Uniform Circular Motion

5. | A 1500 kg car drives around a flat 200-m-diameter circular track at 25 m/s. What are the magnitude and direction of the net force on the car? What causes this force?

6. | A 1500 kg car takes a 50-m-radius unbanked curve at 15 m/s. What is the size of the friction force on the car?

7. || A 200 g block on a 50-cm-long string swings in a circle on a horizontal, frictionless table at 75 rpm.
a. What is the speed of the block?
b. What is the tension in the string?

8. || In the Bohr model of the hydrogen atom, an electron (mass $m = 9.1 \times 10^{-31}$ kg) orbits a proton at a distance of 5.3×10^{-11} m. The proton pulls on the electron with an electric force of 8.2×10^{-8} N. How many revolutions per second does the electron make?

9. | Suppose the moon were held in its orbit not by gravity but by a massless cable attached to the center of the earth. What would be the tension in the cable? Use the table of astronomical data inside the back cover of the book.

10. || A highway curve of radius 500 m is designed for traffic moving at a speed of 90 km/h. What is the correct banking angle of the road?

11. || It is proposed that future space stations create an artificial gravity by rotating. Suppose a space station is constructed as a 1000-m-diameter cylinder that rotates about its axis. The inside surface is the deck of the space station. What rotation period will provide "normal" gravity?

12. || A 5.0 g coin is placed 15 cm from the center of a turntable. The coin has static and kinetic coefficients of friction with the turntable surface of $\mu_s = 0.80$ and $\mu_k = 0.50$. The turntable very slowly speeds up to 60 rpm. Does the coin slide off?

13. || Mass m_1 on the frictionless table of **FIGURE EX8.13** is connected by a string through a hole in the table to a hanging mass m_2. With what speed must m_1 rotate in a circle of radius r if m_2 is to remain hanging at rest?

FIGURE EX8.13

Section 8.3 Circular Orbits

14. | A satellite orbiting the moon very near the surface has a period of 110 min. What is free-fall acceleration on the surface of the moon?

15. || What is free-fall acceleration toward the sun at the distance of the earth's orbit? Astronomical data are inside the back cover of the book.

16. || A 9.4×10^{21} kg moon orbits a distant planet in a circular orbit of radius 1.5×10^8 m. It experiences a 1.1×10^{19} N gravitational pull from the planet. What is the moon's orbital period in earth days?

17. || Communications satellites are placed in circular orbits where they stay directly over a fixed point on the equator as the earth rotates. These are called *geosynchronous orbits*. The altitude of a geosynchronous orbit is 3.58×10^7 m ($\approx 22,000$ miles).
a. What is the period of a satellite in a geosynchronous orbit?
b. Find the value of g at this altitude.
c. What is the weight of a 2000 kg satellite in a geosynchronous orbit?

Section 8.4 Reasoning About Circular Motion

18. | A car drives over the top of a hill that has a radius of 50 m. What maximum speed can the car have at the top without flying off the road?

19. || The weight of passengers on a roller coaster increases by 50% as the car goes through a dip with a 30 m radius of curvature. What is the car's speed at the bottom of the dip?

20. || A roller coaster car crosses the top of a circular loop-the-loop at twice the critical speed. What is the ratio of the normal force to the gravitational force?

21. || The normal force equals the magnitude of the gravitational force as a roller coaster car crosses the top of a 40-m-diameter loop-the-loop. What is the car's speed at the top?

22. || A student has 65-cm-long arms. What is the minimum angular velocity (in rpm) for swinging a bucket of water in a vertical circle without spilling any? The distance from the handle to the bottom of the bucket is 35 cm.

23. | While at the county fair, you decide to ride the Ferris wheel. Having eaten too many candy apples and elephant ears, you find the motion somewhat unpleasant. To take your mind off your stomach, you wonder about the motion of the ride. You estimate the radius of the big wheel to be 15 m, and you use your watch to find that each loop around takes 25 s.
a. What are your speed and the magnitude of your acceleration?
b. What is the ratio of your weight at the top of the ride to your weight while standing on the ground?
c. What is the ratio of your weight at the bottom of the ride to your weight while standing on the ground?

24. || A 500 g ball swings in a vertical circle at the end of a 1.5-m-long string. When the ball is at the bottom of the circle, the tension in the string is 15 N. What is the speed of the ball at that point?

25. || A 500 g ball moves in a vertical circle on a 102-cm-long string. If the speed at the top is 4.0 m/s, then the speed at the bottom will be 7.5 m/s. (You'll learn how to show this in Chapter 10.)
a. What is the gravitational force acting on the ball?
b. What is the tension in the string when the ball is at the top?
c. What is the tension in the string when the ball is at the bottom?

26. || A heavy ball with a weight of 100 N ($m = 10.2$ kg) is hung from the ceiling of a lecture hall on a 4.5-m-long rope. The ball is pulled to one side and released to swing as a pendulum, reaching a speed of 5.5 m/s as it passes through the lowest point. What is the tension in the rope at that point?

Section 8.5 Nonuniform Circular Motion

27. || A toy train rolls around a horizontal 1.0-m-diameter track. The coefficient of rolling friction is 0.10. How long does it take the train to stop if it's released with an angular speed of 30 rpm?

28. || A new car is tested on a 200-m-diameter track. If the car speeds up at a steady 1.5 m/s², how long after starting is the magnitude of its centripetal acceleration equal to the tangential acceleration?

29. ‖ An 85,000 kg stunt plane performs a loop-the-loop, flying in a 260-m-diameter vertical circle. At the point where the plane is flying straight down, its speed is 55 m/s and it is speeding up at a rate of 12 m/s per second.
   a. What is the magnitude of the net force on the plane? You can neglect air resistance.
   b. What angle does the net force make with the horizontal? Let an angle above horizontal be positive and an angle below horizontal be negative.

30. ‖ Three cars are driving at 25 m/s along the road shown in FIGURE EX8.30. Car B is at the bottom of a hill and car C is at the top. Both hills have a 200 m radius of curvature. Suppose each car suddenly brakes hard and starts to skid. What is the tangential acceleration (i.e., the acceleration parallel to the road) of each car? Assume $\mu_k = 1.0$.

FIGURE EX8.30

Problems

31. ‖ Derive Equations 8.3 for the acceleration of a projectile subject to drag.

32. ‖ A 100 g bead slides along a frictionless wire with the parabolic
CALC shape $y = (2 \text{ m}^{-1})x^2$.
   a. Find an expression for a_y, the vertical component of acceleration, in terms of x, v_x, and a_x.
   Hint: Use the basic definitions of velocity and acceleration.
   b. Suppose the bead is released at some negative value of x and has a speed of 2.3 m/s as it passes through the lowest point of the parabola. What is the net force on the bead at this instant? Write your answer in component form.

33. ‖ Space scientists have a large test chamber from which all the air can be evacuated and in which they can create a horizontal uniform electric field. The electric field exerts a constant horizontal force on a charged object. A 15 g charged projectile is launched with a speed of 6.0 m/s at an angle 35° above the horizontal. It lands 2.9 m in front of the launcher. What is the magnitude of the electric force on the projectile?

34. ‖ A 5000 kg interceptor rocket is launched at an angle of 44.7°. The thrust of the rocket motor is 140,700 N.
   a. Find an equation $y(x)$ that describes the rocket's trajectory.
   b. What is the shape of the trajectory?
   c. At what elevation does the rocket reach the speed of sound, 330 m/s?

35. ‖ A motorcycle daredevil plans to ride up a 2.0-m-high, 20° ramp, sail across a 10-m-wide pool filled with hungry crocodiles, and land at ground level on the other side. He has done this stunt many times and approaches it with confidence. Unfortunately, the motorcycle engine dies just as he starts up the ramp. He is going 11 m/s at that instant, and the rolling friction of his rubber tires (coefficient 0.02) is not negligible. Does he survive, or does he become crocodile food? Justify your answer by calculating the distance he travels through the air after leaving the end of the ramp.

36. ‖‖ A rocket-powered hockey puck has a thrust of 2.0 N and a total mass of 1.0 kg. It is released from rest on a frictionless table, 4.0 m from the edge of a 2.0 m drop. The front of the rocket is pointed directly toward the edge. How far does the puck land from the base of the table?

37. ‖ A 500 g model rocket is resting horizontally at the top edge of a 40-m-high wall when it is accidentally bumped. The bump pushes it off the edge with a horizontal speed of 0.5 m/s and at the same time causes the engine to ignite. When the engine fires, it exerts a constant 20 N horizontal thrust away from the wall.
   a. How far from the base of the wall does the rocket land?
   b. Describe the rocket's trajectory as it travels to the ground.

38. ‖ A 2.0 kg projectile with initial velocity $\vec{v} = 8.0\,\hat{i}$ m/s experi-
CALC ences the variable force $\vec{F} = -2.0t\,\hat{i} + 4.0t^2\,\hat{j}$ N, where t is in s.
   a. What is the projectile's speed at $t = 2.0$ s?
   b. At what instant of time is the projectile moving parallel to the y-axis?

39. ‖ A 75 kg man weighs himself at the north pole and at the equator. Which scale reading is higher? By how much? Assume the earth is spherical.

40. ‖ A concrete highway curve of radius 70 m is banked at a 15° angle. What is the maximum speed with which a 1500 kg rubber-tired car can take this curve without sliding?

41. ‖ a. An object of mass m swings in a horizontal circle on a string of length L that tilts downward at angle θ. Find an expression for the angular velocity ω.
   b. A student ties a 500 g rock to a 1.0-m-long string and swings it around her head in a horizontal circle. At what angular speed, in rpm, does the string tilt down at a 10° angle?

42. ‖‖‖ You've taken your neighbor's young child to the carnival to ride the rides. She wants to ride The Rocket. Eight rocket-shaped cars hang by chains from the outside edge of a large steel disk. A vertical axle through the center of the ride turns the disk, causing the cars to revolve in a circle. You've just finished taking physics, so you decide to figure out the speed of the cars while you wait. You estimate that the disk is 5 m in diameter and the chains are 6 m long. The ride takes 10 s to reach full speed, then the cars swing out until the chains are 20° from vertical. What is the cars' speed?

43. ‖ A 4.4-cm-diameter, 24 g plastic ball is attached to a 1.2-m-long string and swung in a vertical circle. The ball's speed is 6.1 m/s at the point where it is moving straight up. What is the magnitude of the net force on the ball? Air resistance is not negligible.

44. ‖ A charged particle of mass m moving with speed v in a plane perpendicular to a magnetic field experiences a force $\vec{F} = (qvB, \text{perpendicular to } \vec{v})$, where q is the amount of charge and B is the magnetic field strength. Because the force is always perpendicular to the particle's velocity, the particle undergoes uniform circular motion. Find an expression for the period of the motion. Gravity can be neglected.

45. ‖‖‖ Two wires are tied to the 2.0 kg sphere shown in FIGURE P8.45. The sphere revolves in a horizontal circle at constant speed.
   a. For what speed is the tension the same in both wires?
   b. What is the tension?

FIGURE P8.45

46. ||| Two wires are tied to the 300 g sphere shown in **FIGURE P8.46**. The sphere revolves in a horizontal circle at a constant speed of 7.5 m/s. What is the tension in each of the wires?

FIGURE P8.46

47. || A conical pendulum is formed by attaching a ball of mass m to a string of length L, then allowing the ball to move in a horizontal circle of radius r. **FIGURE P8.47** shows that the string traces out the surface of a cone, hence the name.
   a. Find an expression for the tension T in the string.
   b. Find an expression for the ball's angular speed ω.
   c. What are the tension and angular speed (in rpm) for a 500 g ball swinging in a 20-cm-radius circle at the end of a 1.0-m-long string?

FIGURE P8.47   **FIGURE P8.48**

48. || The 10 mg bead in **FIGURE P8.48** is free to slide on a frictionless wire loop. The loop rotates about a vertical axis with angular velocity ω. If ω is less than some critical value ω_c, the bead sits at the bottom of the spinning loop. When $\omega > \omega_c$, the bead moves out to some angle θ.
   a. What is ω_c in rpm for the loop shown in the figure?
   b. At what value of ω, in rpm, is $\theta = 30°$?

49. ||| In an old-fashioned amusement park ride, passengers stand inside a 5.0-m-diameter hollow steel cylinder with their backs against the wall. The cylinder begins to rotate about a vertical axis. Then the floor on which the passengers are standing suddenly drops away! If all goes well, the passengers will "stick" to the wall and not slide. Clothing has a static coefficient of friction against steel in the range 0.60 to 1.0 and a kinetic coefficient in the range 0.40 to 0.70. A sign next to the entrance says "No children under 30 kg allowed." What is the minimum angular speed, in rpm, for which the ride is safe?

50. || The ultracentrifuge is an important tool for separating and analyzing proteins. Because of the enormous centripetal accelerations, the centrifuge must be carefully balanced, with each sample matched by a sample of identical mass on the opposite side. Any difference in the masses of opposing samples creates a net force on the shaft of the rotor, potentially leading to a catastrophic failure of the apparatus. Suppose a scientist makes a slight error in sample preparation and one sample has a mass 10 mg larger than the opposing sample. If the samples are 12 cm from the axis of the rotor and the ultracentrifuge spins at 70,000 rpm, what is the magnitude of the net force on the rotor due to the unbalanced samples?

51. || In an amusement park ride called The Roundup, passengers stand inside a 16-m-diameter rotating ring. After the ring has acquired sufficient speed, it tilts into a vertical plane, as shown in **FIGURE P8.51**.
   a. Suppose the ring rotates once every 4.5 s. If a rider's mass is 55 kg, with how much force does the ring push on her at the top of the ride? At the bottom?
   b. What is the longest rotation period of the wheel that will prevent the riders from falling off at the top?

FIGURE P8.51

52. || Suppose you swing a ball of mass m in a vertical circle on a string of length L. As you probably know from experience, there is a minimum angular velocity ω_{min} you must maintain if you want the ball to complete the full circle without the string going slack at the top.
   a. Find an expression for ω_{min}.
   b. Evaluate ω_{min} in rpm for a 65 g ball tied to a 1.0-m-long string.

53. || A 30 g ball rolls around a 40-cm-diameter L-shaped track, shown in **FIGURE P8.53**, at 60 rpm. What is the magnitude of the *net* force that the track exerts on the ball? Rolling friction can be neglected.
   Hint: The track exerts more than one force on the ball.

FIGURE P8.53   **FIGURE P8.54**

54. ||| **FIGURE P8.54** shows a small block of mass m sliding around the inside of an L-shaped track of radius r. The bottom of the track is frictionless; the coefficient of kinetic friction between the block and the wall of the track is μ_k. The block's speed is v_0 at $t_0 = 0$. Find an expression for the block's speed at a later time t.

55. || The physics of circular motion sets an upper limit to the speed of human walking. (If you need to go faster, your gait changes from a walk to a run.) If you take a few steps and watch what's happening, you'll see that your body pivots in circular motion over your forward foot as you bring your rear foot forward for the next step. As you do so, the normal force of the ground on your foot decreases and your body tries to "lift off" from the ground.
   a. A person's center of mass is very near the hips, at the top of the legs. Model a person as a particle of mass m at the top of a leg of length L. Find an expression for the person's maximum walking speed v_{max}.
   b. Evaluate your expression for the maximum walking speed of a 70 kg person with a typical leg length of 70 cm. Give your answer in both m/s and mph, then comment, based on your experience, as to whether this is a reasonable result. A "normal" walking speed is about 3 mph.

56. || A 100 g ball on a 60-cm-long string is swung in a vertical circle about a point 200 cm above the floor. The tension in the string when the ball is at the very bottom of the circle is 5.0 N. A very sharp knife is suddenly inserted, as shown in **FIGURE P8.56**, to cut the string directly below the point of support. How far to the right of where the string was cut does the ball hit the floor?

FIGURE P8.56　　　　　　　**FIGURE P8.57**

57. || A 60 g ball is tied to the end of a 50-cm-long string and swung in a vertical circle. The center of the circle, as shown in **FIGURE P8.57**, is 150 cm above the floor. The ball is swung at the minimum speed necessary to make it over the top without the string going slack. If the string is released at the instant the ball is at the top of the loop, how far to the right does the ball hit the ground?

58. || Elm Street has a pronounced dip at the bottom of a steep hill before going back uphill on the other side. Your science teacher has asked everyone in the class to measure the radius of curvature of the dip. Some of your classmates are using surveying equipment, but you decide to base your measurement on what you've learned in physics. To do so, you sit on a spring scale, drive through the dip at different speeds, and for each speed record the scale's reading as you pass through the bottom of the dip. Your data are as follows:

Speed (m/s)	Scale reading (N)
5	599
10	625
15	674
20	756
25	834

Sitting on the scale while the car is parked gives a reading of 588 N. Analyze your data, using a graph, to determine the dip's radius of curvature.

59. || A 100 g ball on a 60-cm-long string is swung in a vertical circle about a point 200 cm above the floor. The string suddenly breaks when it is parallel to the ground and the ball is moving upward. The ball reaches a height 600 cm above the floor. What was the tension in the string an instant before it broke?

60. || Scientists design a new particle accelerator in which protons (mass 1.7×10^{-27} kg) follow a circular trajectory given by $\vec{r} = c\cos(kt^2)\hat{\imath} + c\sin(kt^2)\hat{\jmath}$, where $c = 5.0$ m and $k = 8.0 \times 10^4$ rad/s^2 are constants and t is the elapsed time.
   a. What is the radius of the circle?
   b. What is the proton's speed at $t = 3.0$ s?
   c. What is the force on the proton at $t = 3.0$ s? Give your answer in component form.

61. ||| A 1500 kg car starts from rest and drives around a flat 50-m-diameter circular track. The forward force provided by the car's drive wheels is a constant 1000 N.
   a. What are the magnitude and direction of the car's acceleration at $t = 10$ s? Give the direction as an angle from the r-axis.
   b. If the car has rubber tires and the track is concrete, at what time does the car begin to slide out of the circle?

62. ||| A 500 g steel block rotates on a steel table while attached to a 2.0-m-long massless rod. Compressed air fed through the rod is ejected from a nozzle on the back of the block, exerting a thrust force of 3.5 N. The nozzle is 70° from the radial line, as shown in **FIGURE P8.62**. The block starts from rest.
   a. What is the block's angular velocity after 10 rev?
   b. What is the tension in the rod after 10 rev?

FIGURE P8.62

63. ||| A 2.0 kg ball swings in a vertical circle on the end of an 80-cm-long string. The tension in the string is 20 N when its angle from the highest point on the circle is $\theta = 30°$.
   a. What is the ball's speed when $\theta = 30°$?
   b. What are the magnitude and direction of the ball's acceleration when $\theta = 30°$?

In Problems 64 and 65 you are given the equation used to solve a problem. For each of these, you are to
   a. Write a realistic problem for which this is the correct equation. Be sure that the answer your problem requests is consistent with the equation given.
   b. Finish the solution of the problem.

64. $60 \text{ N} = (0.30 \text{ kg})\omega^2(0.50 \text{ m})$

65. $(1500 \text{ kg})(9.8 \text{ m/s}^2) - 11{,}760 \text{ N} = (1500 \text{ kg})\,v^2/(200 \text{ m})$

Challenge Problems

66. ||| Sam (75 kg) takes off up a 50-m-high, 10° frictionless slope on his jet-powered skis. The skis have a thrust of 200 N. He keeps his skis tilted at 10° after becoming airborne, as shown in **FIGURE CP8.66**. How far does Sam land from the base of the cliff?

FIGURE CP8.66

67. ||| In the absence of air resistance, a projectile that lands at the elevation from which it was launched achieves maximum range when launched at a 45° angle. Suppose a projectile of mass m is launched with speed v_0 into a headwind that exerts a constant, horizontal retarding force $\vec{F}_{\text{wind}} = -F_{\text{wind}}\hat{\imath}$.
   a. Find an expression for the angle at which the range is maximum.
   b. By what percentage is the maximum range of a 0.50 kg ball reduced if $F_{\text{wind}} = 0.60$ N?

68. ||| The father of Example 8.2 stands at the summit of a conical hill as he spins his 20 kg child around on a 5.0 kg cart with a 2.0-m-long rope. The sides of the hill are inclined at 20°. He again keeps the rope parallel to the ground, and friction is negligible. What rope tension will allow the cart to spin with the same 14 rpm it had in the example?

69. ||| A small bead slides around a horizontal circle at height y inside the cone shown in **FIGURE CP8.69**. Find an expression for the bead's speed in terms of a, h, y, and g.

FIGURE CP8.69

FIGURE CP8.70

70. ||| A 500 g steel block rotates on a steel table while attached to a 1.2-m-long hollow tube as shown in **FIGURE CP8.70**. Compressed air fed through the tube and ejected from a nozzle on the back of the block exerts a thrust force of 4.0 N perpendicular to the tube.

The maximum tension the tube can withstand without breaking is 50 N. If the block starts from rest, how many revolutions does it make before the tube breaks?

71. ||| If a vertical cylinder of water (or any other liquid) rotates about its axis, as shown in **FIGURE CP8.71**, the surface forms a smooth curve. Assuming that the water rotates as a unit (i.e., all the water rotates with the same angular velocity), show that the shape of the surface is a parabola described by the equation $z = (\omega^2/2g)r^2$.

Hint: Each particle of water on the surface is subject to only two forces: gravity and the normal force due to the water underneath it. The normal force, as always, acts perpendicular to the surface.

FIGURE CP8.71

Newton's Laws

KEY FINDINGS   What are the overarching findings of Part I?

- **Kinematics** is the description of motion. Motion can be described

  - Visually   - Graphically   - Mathematically

- **Forces** cause objects to *change* their motion—that is, to accelerate.
- Objects **interact** by exerting equal but opposite forces on each other.

LAWS   What laws of physics govern motion?

Newton's first law   An object will remain at rest or will continue to move with constant velocity if and only if $\vec{F}_{net} = \vec{0}$. The object is in **mechanical equilibrium**.

Newton's second law $\vec{F}_{net} = m\vec{a}$ A net force on an object causes the object to accelerate.

Newton's third law $\vec{F}_{A\,on\,B} = -\vec{F}_{B\,on\,A}$ For every action, there is an equal but opposite reaction.

MODELS   What are the most common models for applying the laws of physics to moving objects?

Constant force/Uniform acceleration

- Model the object as a particle.

  - Acceleration is in the direction of the net force and is constant.

- Mathematically:

  - Newton's second law is

  $$\vec{F}_{net} = \sum_i \vec{F}_i = m\vec{a}$$

  - Use *xy*-coordinates.

  - Constant-acceleration kinematics:

  $$v_{fs} = v_{is} + a_s\Delta t$$
  $$s_f = s_i + v_{is}\Delta t + \tfrac{1}{2}a_s(\Delta t)^2$$
  $$v_{fs}^2 = v_{is}^2 + 2a_s\Delta s$$

- Special cases:

  - Uniform motion: $a_s = 0$. The displacement graph is a straight line with slope v_s.

  - Projectile motion: The only force is gravity. Horizontal motion is uniform; vertical motion has constant $a_y = -g$.

a_s Horizontal line

0 ——————————— t

The acceleration is constant.

v_s Straight line

v_{is} ————————— t

The slope is a_s.

s Parabola

s_i ————————— t

The slope is v_s.

Central force/Uniform circular motion

- Model the object as a particle.

  - The force causes a constant centripetal acceleration. The particle moves around a circle at constant speed and with constant angular velocity.

- Mathematically: Newton's second law is

  - $\vec{F}_{net} = (mv^2/r$ or $m\omega^2 r$, toward the center)

  - Use *rtz*-coordinates.

  - Uniform-circular-motion kinematics:

    - The tangential velocity is $v_t = \omega r$.

    - The centripetal acceleration is v^2/r or $\omega^2 r$.

    - ω and v_t are positive for a ccw rotation, negative for a cw rotation.

- General case: Accelerated circular/rotational motion. Angular acceleration is contant.

  $$\omega_f = \omega_i + \alpha\Delta t$$
  $$\theta_f = \theta_i + \omega_i\Delta t + \tfrac{1}{2}\alpha(\Delta t)^2$$
  $$\omega_f^2 = \omega_i^2 + 2\alpha\Delta\theta$$

  These equations are analogous to constant-acceleration kinematics.

TOOLS   What are the most important tools for analyzing the physics of motion?

- The particle model and motion diagrams

- Vectors

- Free-body diagrams

- Interaction diagrams

- Calculus and graphical analysis

  $v_s = ds/dt =$ slope of the position graph

  $a_s = dv_s/dt =$ slope of the velocity graph

  $$v_{fs} = v_{is} + \int_{t_i}^{t_f} a_s\,dt = v_{is} +\ \text{area under acceleration curve}$$

  $$s_f = s_i + \int_{t_i}^{t_f} v_s\,dt = s_i +\ \text{area under the velocity curve}$$

OVERVIEW

Why Some Things Don't Change

Part I of this textbook was about *change*. One particular type of change—motion—is governed by Newton's second law. Although Newton's second law is a very powerful statement, it isn't the whole story. Part II will now focus on things that *stay the same* as other things around them change.

Consider, for example, an explosive chemical reaction taking place inside a closed, sealed box. No matter how violent the explosion, the total mass of the products—the final mass M_f —is the same as the initial mass M_i of the reactants. In other words, matter cannot be created or destroyed, only rearranged. This is an important and powerful statement about nature.

A quantity that stays the same throughout an interaction is said to be *conserved*. The most important such quantity is *energy*. If a system of interacting objects is *isolated*—an important qualification—then the energy of the system never changes no matter how complex the interactions. This description of how nature behaves, called the *law of conservation of energy*, is perhaps the most important physical law ever discovered.

But what is energy? How do you determine the energy of a system? These are not easy questions. Energy is an abstract idea, not as tangible or easy to picture as mass or force. Our modern concept of energy wasn't fully formulated until the middle of the 19th century, two hundred years after Newton, when the relationship between *energy* and *heat* was finally understood. That is a topic we will take up in Part V, where the concept of energy will be found to be the basis of thermodynamics. But all that in due time. In Part II we will be content to introduce the concept of energy and show how energy can be a useful problem-solving tool. We'll also meet another quantity—*momentum*—that is conserved under the proper circumstances.

Conservation laws give us a new and different perspective on motion. This is not insignificant. You've seen optical illusions where a figure appears first one way, then another, even though the information has not changed. Likewise with motion. Some situations are most easily analyzed from the perspective of Newton's laws; others make more sense from a conservation-law perspective. An important goal of Part II is to learn which is better for a given problem.

Energy is the lifeblood of modern society. These photovoltaic panels transform solar energy into electrical energy and, unavoidably, increased thermal energy.

9 Work and Kinetic Energy

The bow may be very contemporary, but it's still the bow string doing work on the arrow that makes the arrow fly.

IN THIS CHAPTER, you will begin your study of how energy is transferred and transformed.

How should we think about energy?

Chapters 9 and 10 will develop the basic energy model, a powerful set of ideas for using energy. A key distinction is between the system, which has energy, and the environment. Energy can be transferred between the system and the environment or transformed within the system.

《 LOOKING BACK Section 7.1 Interacting objects

What are some important forms of energy?

Three important forms of energy:

- Potential energy is energy associated with an object's *position*.
- Kinetic energy is energy associated with an object's *motion*.
- Thermal energy is the energy of the random motion of *atoms* within an object.

Energy is measured in joules.

Potential energy

Kinetic energy

What is work?

A process that *changes* the energy of a system by mechanical means—pushing or pulling on it—is called work.

Work W is done when a force pushes or pulls a particle through a displacement, thus changing the particle's kinetic energy.

Force does work on the particle

What laws govern energy?

Working with energy is very much like accounting: A system's energy E changes by the amount of work done on the system. The mathematical statement of this idea is called the energy principle:

$$\Delta E_{\text{sys}} = W_{\text{ext}}$$

What is power?

Power is the rate at which energy is transferred or transformed. For machines, power is the rate at which they do work. For electricity, power is the rate at which electric energy is transformed into heat, sound, or light. Power is measured in watts, where 1 watt is a rate of 1 joule per second.

Why is energy important?

Energy is one of the most important concepts in science, engineering, and society. Some would say it is *the* most important. All life depends on energy, transformed from solar energy to chemical energy to us. Society depends on energy, from industry and transportation to heating and cooling our buildings. Using energy wisely and efficiently is a key concern of the 21st century.

9.1 Energy Overview

Energy. It's a word you hear all the time, and everyone has some sense of what *energy* means. Moving objects have energy; energy is the ability to make things happen; energy is associated with heat and with electricity; we're constantly told to conserve energy; living organisms need energy; and engineers harness energy to do useful things. Some scientists consider the *law of conservation of energy* to be the most important of all the laws of nature. But all that in due time—first we have to start with the basic ideas.

Just what is energy? The concept of energy has grown and changed with time, and it is not easy to define in a general way just what energy is. Rather than starting with a formal definition, we're going to let the concept of energy expand slowly over the course of several chapters. Our goal is to understand the characteristics of energy, how energy is used, and how energy is transformed from one form into another. It's a complex story, so we'll take it step by step until all the pieces are in place.

Some important forms of energy

Kinetic energy K	Potential energy U	Thermal energy E_{th}
Kinetic energy is the energy of motion. All moving objects have kinetic energy. The more massive an object or the faster it moves, the larger its kinetic energy.	Potential energy is stored energy associated with an object's position. The roller coaster's gravitational potential energy depends on its height above the ground.	Thermal energy is the sum of the microscopic kinetic and potential energies of all the atoms and bonds that make up the object. An object has more thermal energy when hot than when cold.

The Energy Principle

« Section 7.1 introduced *interaction diagrams* and the very important distinction between the **system,** those objects whose motion and interactions we wish to analyze, and the **environment,** objects external to the system but exerting forces on the system. The most important step in an energy analysis is to clearly define the system. Why? Because energy is not some disembodied, ethereal substance; it's the energy *of something.* Specifically, it's *the energy of a system.*

FIGURE 9.1 illustrates the idea pictorially. The system *has energy,* the **system energy,** which we'll designate E_{sys}. There are many kinds or forms of energy: kinetic energy K, potential energy U, thermal energy E_{th}, chemical energy, and so on. We'll introduce these one by one as we go along. Within the system, energy can be *transformed without loss.* Chemical energy can be transformed into kinetic energy, which is then transformed into thermal energy. As long as the system is not interacting with the environment, the total energy of the system is unchanged. You'll recognize this idea as an initial statement of the *law of conservation of energy.*

But systems often do interact with their environment. Those interactions *change* the energy of the system, either increasing it (energy added) or decreasing it (energy removed). We say that interactions with the environment *transfer* energy into or out of the system. Interestingly, there are only two ways to transfer energy. One is by mechanical means, using forces to push and pull on the system. A process that

FIGURE 9.1 A system-environment perspective on energy.

transfers energy to or from a system by mechanical means is called **work,** with the symbol *W*. We'll have a lot to say about work in this chapter. The second is by thermal means when the environment is hotter or colder than the system. A process that transfers energy to or from a system by thermal means is called **heat.** We'll defer a discussion of heat until Chapter 18, but we wanted to mention it now in order to gain an overview of what the energy story is all about.

Some energy transfers ...   **... and transformations**

Putting a shot
System: The shot
Transfer: $W \rightarrow K$
The athlete (the environment) does work pushing the shot to give it kinetic energy.

A falling diver
System: The diver and the earth
Transformation: $U \rightarrow K$
The diver is speeding up as gravitational potential energy is transformed into kinetic energy.

Pulling a slingshot
System: The slingshot
Transfer: $W \rightarrow U$
The boy (the environment) does work by stretching the rubber band to give it potential energy.

A speeding meteor
System: The meteor and the air
Transformation: $K \rightarrow E_{\text{th}}$
The meteor and the air get hot enough to glow as the meteor's kinetic energy is transformed into thermal energy.

The key ideas are **energy transfer** between the environment and the system and **energy transformation** within the system. This is much like what happens with money. You may have several accounts at the bank—perhaps a checking account and a couple of savings accounts. You can move money back and forth between the accounts, thus transforming it without changing the total amount of money. Of course, you can also transfer money into or out of your accounts by making deposits or withdrawals. If we treat a withdrawal as a negative deposit—which is exactly what accountants do— simple accounting tells you that

$$\Delta(\text{balance}) = \text{ net deposit}$$

That is, the change in your bank balance is simply the sum of all your deposits.

Energy accounting works the same way. Transformations of energy within the system move the energy around but don't change the total energy of the system. *Change* occurs only when there's a transfer of energy between the system and the environment. If we treat incoming energy as a positive transfer and outgoing energy as a negative transfer, and with work being the only energy-transfer process that we consider for now, we can write

$$\Delta E_{\text{sys}} = W_{\text{ext}} \tag{9.1}$$

where the subscript on *W* refers to external work done by the environment. This very simple looking statement, which is just a statement of energy accounting, is called the **energy principle.** But don't let the simplicity fool you; this will turn out to be an incredibly powerful tool for analyzing physical situations and solving problems.

The Basic Energy Model

We'll complete our energy overview—a roadmap of the next two chapters—with the **basic energy model.**

MODEL 9.1

Basic energy model

Energy is a property of the system.

- Energy is *transformed* within the system without loss.
- Energy is *transferred* to and from the system by forces from the environment.
  - The forces do *work* on the system.
  - $W > 0$ for energy added.
  - $W < 0$ for energy removed.
- The energy of an *isolated system*—one that doesn't interact with its environment—does not change. We say it is *conserved*.
- The energy principle is $\Delta E_{\text{sys}} = W_{\text{ext}}$.
- Limitations: Model fails if there is energy transfer via thermal processes (heat).

Exercise 1

We call this model *basic* because, for now, the only forms of energy we'll consider are kinetic energy, potential energy, and thermal energy, and the only energy-transfer process we'll consider is work. This is an excellent model for a mechanical process, but it's not complete. We'll expand the model when we get to thermodynamics by adding chemical energy, another form of energy, and heat, another energy-transfer process. And this model, although basic, still has many complexities, so we'll be developing it piece by piece in this chapter and the next.

9.2 Work and Kinetic Energy for a Single Particle

Let's start our investigation of energy with the simplest possible situation: One particle of mass m is acted on by one constant force $\vec{F}$ that acts parallel to the direction of motion, pushing or pulling on the particle as it undergoes a displacement Δs. We define the particle to be the system—a one-particle system—while the agent of the force is in the environment. FIGURE 9.2 shows both an interaction diagram and a new kind of pictorial representation, a **before-and-after representation,** in which we show an object *before* and *after* an interaction and, as usual, establish a coordinate system and define appropriate symbols.

You know what's going to happen. If the force is in the direction of motion—the situation shown in the figure—the particle will speed up and its "energy of motion" will increase. Conversely, if the force opposes the motion, the particle will slow down and lose energy. Our goal is to make this idea precise by discovering exactly how the changing energy is related to the applied force.

We'll start by writing Newton's second law for the particle:

$$F_s = ma_s = m\frac{dv_s}{dt} \tag{9.2}$$

Newton's second law tells us how the particle's velocity changes with time. But suppose we want to know how the velocity changes with position. To answer that question, we can use the chain rule that you've learned in calculus:

$$\frac{dv_s}{dt} = \frac{dv_s}{ds}\frac{ds}{dt} = v_s\frac{dv_s}{ds} \tag{9.3}$$

where in the last step we used $ds/dt = v_s$. With this, we can write Newton's second law as

$$F_s = mv_s\frac{dv_s}{ds} \tag{9.4}$$

FIGURE 9.2 The interaction diagram and before-and-after representation for a one-particle system.

Interaction diagram

A before-and-after representation shows the object's position and velocity before and after an interaction.

Before-and-after representation

Now we have an alternative version of Newton's second law in terms of the rate of change of velocity with position.

To use Equation 9.4, we first rewrite it as

$$mv_s\, dv_s = F_s\, ds \tag{9.5}$$

Now we can integrate. This is going to be a definite integral over just the motion shown in the before-and-after representation. That is, the right side will be an integral over position s from the initial position s_i to the final position s_f. The left side will be an integral over velocity v_s, and its limits have to match the limits of the right-hand integral: from v_{is} at s_i to v_{fs} at s_f. Thus we have

$$\int_{v_{is}}^{v_{fs}} mv_s\, dv_s = \int_{s_i}^{s_f} F_s\, ds \tag{9.6}$$

We have two integrals to examine, and we'll do them one by one. We can start with the integral on the left, which is of the form $\int x\, dx$. Factoring out m, which is a constant, we find

$$m \int_{v_{is}}^{v_{fs}} v_s\, dv_s = m \left[\tfrac{1}{2}v_s^2\right]_{v_{is}}^{v_{fs}} = \tfrac{1}{2}mv_{fs}^2 - \tfrac{1}{2}mv_{is}^2 = \Delta\left(\tfrac{1}{2}mv^2\right) = \Delta K \tag{9.7}$$

You'll notice that we dropped the subscript s in the next-to-last step. v_s is a vector component, with a sign to indicate direction, but the sign makes no difference after v_s is squared. All that matters is the particle's *speed* v.

The last step in Equation 9.7 introduces a new quantity

$$K = \tfrac{1}{2}mv^2 \qquad \text{(kinetic energy)} \tag{9.8}$$

which is called the **kinetic energy** of the particle. **Kinetic energy is energy of motion.** It depends on the particle's mass and speed but not on its position. Furthermore, kinetic energy is a property or characteristic of the system. So what we've calculated with the left-hand integral is $\Delta K = K_f - K_i$, the *change* in the system's kinetic energy as the force pushes the particle through the displacement Δs. ΔK is positive if the particle speeds up (gain of kinetic energy), negative if it slows down (loss of kinetic energy).

> **NOTE** By its definition, kinetic energy can *never* be negative. Finding a negative value for K while solving a problem is an indication that you've made a mistake somewhere.

The unit of kinetic energy is mass multiplied by velocity squared. In SI units, this is $kg\, m^2/s^2$. Because energy is so important, the unit of energy is given its own name, the **joule.** We define

$$1\text{ joule} = 1\text{ J} = 1\text{ kg}\, m^2/s^2$$

All other forms of energy are also measured in joules.

To give you an idea about the size of a joule, consider a 0.5 kg mass (≈ 1 lb on earth) moving at 4 m/s (≈ 10 mph). Its kinetic energy is

$$K = \tfrac{1}{2}mv^2 = \tfrac{1}{2}(0.5\text{ kg})(4\text{ m/s})^2 = 4\text{ J}$$

This suggests that everyday objects moving at ordinary speeds will have energies from a fraction of a joule up to, perhaps, a few thousand joules. A running person has $K \approx 1000$ J, while a high-speed truck might have $K \approx 10^6$ J.

> **NOTE** You *must* have masses in kilograms and velocities in m/s before doing energy calculations.

STOP TO THINK 9.1 A 1000 kg car has a speed of 20 m/s. A 2000 kg truck has a speed of 10 m/s. Which has more kinetic energy?

a. The car.   b. The truck.   c. Their kinetic energies are the same.

Work

Now let's turn to the integral on the right-hand side of Equation 9.6. This integral is telling us *by how much* the kinetic energy changes due to the force. That is, it is the energy transferred to or from the system by the force. Earlier we said that a process that transfers energy to or from a system by mechanical means—by forces—is called work. So the integral on the right-hand side of Equation 9.6 must be the *work W done by force* $\vec{F}$.

Having identified the left side of Equation 9.6 with the changing kinetic energy of the system and the right side with the work done on the system, we can rewrite Equation 9.6 as

Change in the system's kinetic energy

Amount of work done by an external force

$$\Delta K = K_f - K_i = W \tag{9.9}$$

Final kinetic energy ⋯   ⋯ Initial kinetic energy

(Energy principle for a one-particle system)

This is our first version of the energy principle. Notice that it's a cause-and-effect statement: **The work done on a one-particle system causes the system's kinetic energy to change.**

We'll study work thoroughly in the next section, but for now we're considering only the simplest case of a constant force parallel to the direction of motion (the *s*-axis). A constant force can be factored out of the integral, giving

$$W = \int_{s_i}^{s_f} F_s \, ds = F_s \int_{s_i}^{s_f} ds = F_s s \Big|_{s_i}^{s_f} = F_s(s_f - s_i)$$
$$= F_s \, \Delta s \tag{9.10}$$

The unit of work, that of force multiplied by distance, is the N m. Recall that $1 \, \text{N} = 1 \, \text{kg m/s}^2$. Thus

$$1 \, \text{N m} = 1 \, (\text{kg m/s}^2) \, \text{m} = 1 \, \text{kg m}^2/\text{s}^2 = 1 \, \text{J}$$

Thus the unit of work is really the unit of energy. This is consistent with the idea that work is a transfer of energy. Rather than use N m, we will measure work in joules.

EXAMPLE 9.1 | Firing a cannonball

A 5.0 kg cannonball is fired straight up at 35 m/s. What is its speed after rising 45 m?

MODEL Let the system consist of only the cannonball, which we model as a particle. Assume that air resistance is negligible.

VISUALIZE FIGURE 9.3 is a before-and-after pictorial representation. Because before-and-after representations are usually simpler than

FIGURE 9.3 Before-and-after representation of the cannonball.

the pictorial representation used in dynamics problems, you can include known information right on the diagram instead of making a Known table.

SOLVE It isn't necessary to use work and energy to solve this problem. You could solve it as a free-fall problem. Or you might have previously learned to solve problems like this using potential energy, a topic we'll take up in the next chapter. But using work and energy emphasizes how these two key ideas are related, and it gives us a simple example of the *problem-solving process* before we get to more complex problems. The energy principle is $\Delta K = W$, where work is done by the force of gravity. The cannonball is rising, so its displacement Δy is positive. But the force vector points down, with component $F_y = -mg$. Thus gravity does work

$$W = F_y \Delta y = -mg \, \Delta y = -(5.0 \, \text{kg})(9.80 \, \text{m/s}^2)(45 \, \text{m}) = -2210 \, \text{J}$$

as the cannonball rises 45 m. A negative work means that the system is losing energy, which is what we expect as the cannonball slows.

Continued

The cannonball's change of kinetic energy is $\Delta K = K_f - K_i$. The initial kinetic energy is

$$K_i = \tfrac{1}{2}mv_i^2 = \tfrac{1}{2}(5.0 \text{ kg})(35 \text{ m/s})^2 = 3060 \text{ J}$$

Using the energy principle, we find the final kinetic energy to be $K_f = K_i + W = 3060 \text{ J} - 2210 \text{ J} = 850 \text{ J}$. Then

$$v_f = \sqrt{\frac{2K_f}{m}} = \sqrt{\frac{2(850 \text{ J})}{5.0 \text{ kg}}} = 18 \text{ m/s}$$

ASSESS 35 m/s ≈ 70 mph. A cannonball fired upward at that speed is going to go fairly high. To have lost half its speed at a height of 45 m ≈ 150 ft seems reasonable.

Signs of Work

FIGURE 9.4 How to determine the sign of W.

Work is *positive* when the force acts in the *same direction* as the displacement.

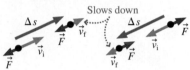

Work is *negative* when the force and displacement are in *opposite* directions.

Work is *zero* if the particle doesn't move (no displacement).

Work can be either positive or negative, but some care is needed to get the sign right when calculating work. The key is to remember that work is an energy transfer. If the force causes the particle to speed up, then the work done by that force is positive. Similarly, negative work means that the force is causing the object to slow and lose energy.

The sign of W is *not* determined by the direction the force vector points. That's only half the issue. The displacement Δs also has a sign, so you have to consider both the force direction *and* the displacement direction. As **FIGURE 9.4** shows, **work is positive when the force acts in the direction of the displacement** (causing the particle to speed up). Similarly, **work is negative when force and displacement are in opposite directions** (causing the particle to slow). And there's no work at all ($W = 0$) if the particle doesn't move!

STOP TO THINK 9.2 A rock falls to the bottom of a deep canyon. Is the work done on the rock by gravity positive, negative, or zero?

Extending the Model

Our initial model has been of a single particle acted on by a constant force parallel to the displacement. We can easily make some straightforward extensions of this model to slightly more complex—and interesting—situations:

■ **Force perpendicular to the displacement:** A force parallel to a particle's displacement causes the particle to speed up or slow down, changing its energy. But a force *perpendicular* to the displacement does *not* change the particle's speed; it is neither speeding up nor slowing down. Its energy is not changing, so no work is being done on it. **A force perpendicular to the displacement does no work.**

■ **Multiple forces:** If multiple forces act on a system, their works add. That is, $\Delta K = W_{tot}$, where the total work done is

$$W_{tot} = W_1 + W_2 + W_3 + \cdots \qquad (9.11)$$

■ **Multiparticle systems:** If a system has more than one particle, the system's energy is the total kinetic energy of all the particles:

$$E_{sys} = K_{tot} = K_1 + K_2 + K_3 + \cdots \qquad (9.12)$$

K_{tot} is truly a *system* energy, not the energy of any one particle. How does K_{tot} change when work is done? You can see from its definition that ΔK_{tot} is the sum of all the individual kinetic-energy changes, and each of those changes is the work done on that particular particle. Thus

$$\Delta K_{tot} = W_{tot} \qquad (9.13)$$

where now W_{tot} is the total work done on *all* the particles in the system.

NOTE You might expect $W_{tot} = (F_{net})_s \Delta s$, where $\vec{F}_{net}$ is the net work on the system. This is true for a one-particle system (if all the forces are constant), but in general it is *not* true for a multiparticle system because each particle undergoes a different displacement. You must find the work done on each particle, then sum those to find the total work done on the system.

Two equal-mass pucks on frictionless ice are pushed toward each other by two equal but opposite forces. Is the total work positive, negative, or zero?

9.3 Calculating the Work Done

Section 9.2 introduced two key ideas: (1) a system has energy and (2) work is a mechanical process that changes the system's energy. Now we're ready to look more closely at how to calculate the work done in different situations. Although we'll be focusing on the mathematical techniques of calculating work, it's important to keep in mind that our real goal is to learn how the energy of a system changes when forces are applied to it.

"Work" is a common word in the English language, with many meanings. Work might refer to physical exertion, to your job or occupation, or even to a work of art. But set aside those ideas about work because they are *not* what work means in physics. Work, as we'll use the word, is a *process*. Specifically, it is a process that changes a system's energy by mechanical means—pushing or pulling on it with forces. We say that work *transfers* energy between the environment and the system.

Equation 9.6 defined work as

$$W = \int_{s_i}^{s_f} F_s \, ds \tag{9.14}$$

(work done by force $\vec{F}$ as a particle is displaced from s_i to s_f)

where, to remind you, F_s is the component of $\vec{F}$ in the direction of motion (the s-direction). We began by looking at a force that was parallel to the displacement, but such a restriction is not required because any force component perpendicular to the motion does no work. Equation 9.14 is, in fact, a general definition of work.

We'll start by learning how to calculate work for constant forces, and we'll introduce a new mathematical idea, the *dot product* of two vectors, that will allow us to write the work in a compact notation. Then we'll consider the work done by a variable force that changes as the particle moves.

Constant Force

FIGURE 9.5 shows a particle moving in a straight line. A constant force $\vec{F}$, which makes an angle θ with respect to the particle's displacement $\Delta \vec{r}$, acts on the particle throughout its motion. We've established an s-axis in the direction of motion, and you can see that the force component along the direction of motion is $F_s = F \cos\theta$. According to Equation 9.14, the work done on the particle by this force is

$$W = \int_{s_i}^{s_f} F_s \, ds = \int_{s_i}^{s_f} F \cos\theta \, ds \tag{9.15}$$

Both F and $\cos\theta$ are constant, so they can be taken outside the integral. Thus

$$W = F \cos\theta \int_{s_i}^{s_f} ds = F \cos\theta (s_f - s_i) \tag{9.16}$$

Now s_i and s_f are specific to this coordinate system, but their *difference* $s_f - s_i$ is Δr, the magnitude of the particle's displacement vector. Thus we can write more generally, independent of any specific coordinate system, that the work done by the constant force $\vec{F}$ is

$$W = F(\Delta r) \cos\theta \quad \text{(work done by a constant force)} \tag{9.17}$$

where θ is the angle between the force and the particle's displacement $\Delta \vec{r}$.

NOTE You may have learned in an earlier physics course that work is "force times distance." This is *not* the definition of work, merely a special case. Work is "force times distance" only if the force is constant *and* parallel to the displacement ($\theta = 0°$).

FIGURE 9.5 Work being done by a constant force.

A *constant* force acts on the particle as it moves.

F_s is the component of $\vec{F}$ in the direction of motion. It causes the particle to speed up or slow down.

EXAMPLE 9.2 | **Pulling a suitcase**

A strap inclined upward at a 45° angle pulls a suitcase 100 m through the airport. The tension in the strap is 20 N. How much work does the tension force do on the suitcase?

MODEL Let the system consist of only the suitcase, which we model as a particle.

VISUALIZE FIGURE 9.6 is a before-and-after pictorial representation.

SOLVE The motion is along the x-axis, so in this case $\Delta r = \Delta x$. We can use Equation 9.17 to find that the tension does work:

$$W = T(\Delta x)\cos\theta = (20\text{ N})(100\text{ m})\cos 45° = 1400\text{ J}$$

FIGURE 9.6 Pictorial representation of the suitcase.

ASSESS Because a person pulls the strap, we would say informally that the person does 1400 J of work on the suitcase.

According to the basic energy model, work can be either positive or negative to indicate energy transfer into or out of the system. The quantities F and Δr are always positive, so **the sign of W is determined entirely by the angle θ between the force $\vec{F}$ and the displacement $\Delta\vec{r}$.**

TACTICS BOX 9.1 (MP)

Calculating the work done by a constant force

Force and displacement	θ	Work W	Sign of W	Energy transfer
	0°	$F(\Delta r)$	+	
	< 90°	$F(\Delta r)\cos\theta$	+	Energy is transferred into the system. The particle speeds up. K increases.
	90°	0	0	No energy is transferred. Speed and K are constant.
	> 90°	$F(\Delta r)\cos\theta$	−	
	180°	$-F(\Delta r)$	−	Energy is transferred out of the system. The particle slows down. K decreases.

Exercises 3–9

NOTE The sign of W depends on the angle between the force vector and the displacement vector, *not* on the coordinate axes. A force to the left does *positive* work if it pushes a particle to the left (the force and the displacement are in the same direction) even though the force component F_x is negative. Think about whether the force is trying to increase the particle's speed ($W > 0$) or decrease the particle's speed ($W < 0$).

EXAMPLE 9.3 | **Launching a rocket**

A 150,000 kg rocket is launched straight up. The rocket motor generates a thrust of 4.0×10^6 N. What is the rocket's speed at a height of 500 m?

MODEL Let the system consist of only the rocket, which we model as a particle. Thrust and gravity are constant forces that do work on the rocket. We'll ignore air resistance and any slight mass loss.

VISUALIZE FIGURE 9.7 shows a before-and-after representation and a free-body diagram.

FIGURE 9.7 Before-and-after representation and free-body diagram of a rocket launch.

SOLVE We can solve this problem with the energy principle, $\Delta K = W_{tot}$. Both forces do work on the rocket. The thrust is in the direction of motion, with $\theta = 0°$, and thus

$$W_{thrust} = F_{thrust}(\Delta r) = (4.0 \times 10^6 \text{ N})(500 \text{ m}) = 2.00 \times 10^9 \text{ J}$$

The gravitational force points downward, opposite the displacement $\Delta \vec{r}$, so $\theta = 180°$. Thus the work done by gravity is

$$W_{grav} = -F_G(\Delta r) = -mg(\Delta r)$$
$$= -(1.5 \times 10^5 \text{ kg})(9.8 \text{ m/s}^2)(500 \text{ m}) = -0.74 \times 10^9 \text{ J}$$

The work done by the thrust is positive. By itself, the thrust would cause the rocket to speed up. The work done by gravity is negative, not because $\vec{F}_G$ points down but because $\vec{F}_G$ is opposite the displacement. By itself, gravity would cause the rocket to slow down. The energy principle, using $K_i = 0$, is

$$\Delta K = \tfrac{1}{2}mv_1^2 - 0 = W_{tot} = W_{thrust} + W_{grav} = 1.26 \times 10^9 \text{ J}$$

Solving for the speed, we find

$$v_1 = \sqrt{\frac{2W_{tot}}{m}} = 130 \text{ m/s}$$

ASSESS The total work is positive, meaning that energy is transferred *to* the rocket. In response, the rocket speeds up.

STOP TO THINK 9.4 A crane uses a single cable to lower a steel girder into place. The girder moves with constant speed. The cable tension does work W_T and gravity does work W_G. Which statement is true?

a. W_T is positive and W_G is positive.
b. W_T is positive and W_G is negative.
c. W_T is negative and W_G is positive.
d. W_T is negative and W_G is negative.
e. W_T and W_G are both zero.

Work as a Dot Product of Two Vectors

There's something different about the quantity $F(\Delta r)\cos\theta$ in Equation 9.17. We've spent many chapters adding vectors, but this is the first time we've *multiplied* two vectors. Multiplying vectors is not like multiplying scalars. In fact, there is more than one way to multiply vectors. We will introduce one way now, the *dot product*.

FIGURE 9.8 shows two vectors, $\vec{A}$ and $\vec{B}$, with angle α between them. We define the **dot product** of $\vec{A}$ and $\vec{B}$ as

$$\vec{A} \cdot \vec{B} = AB\cos\alpha \qquad (9.18)$$

A dot product *must have* the dot symbol · between the vectors. The notation $\vec{A}\vec{B}$, without the dot, is *not* the same thing as $\vec{A} \cdot \vec{B}$. The dot product is also called the **scalar product** because the value is a scalar. Later, when we need it, we'll introduce a different way to multiply vectors called the *cross product*.

The dot product of two vectors depends on the orientation of the vectors. FIGURE 9.9 shows five different situations, including the three "special cases" where $\alpha = 0°$, $90°$, and $180°$.

FIGURE 9.8 Vectors $\vec{A}$ and $\vec{B}$, with angle α between them.

NOTE The dot product of a vector with itself is well defined. If $\vec{B} = \vec{A}$ (i.e., $\vec{B}$ is a copy of $\vec{A}$), then $\alpha = 0°$. Thus $\vec{A} \cdot \vec{A} = A^2$.

FIGURE 9.9 The dot product $\vec{A} \cdot \vec{B}$ as α ranges from 0° to 180°.

EXAMPLE 9.4 | **Calculating a dot product**

Compute the dot product of the two vectors in **FIGURE 9.10**.

SOLVE The angle between the vectors is $\alpha = 30°$, so

$$\vec{A} \cdot \vec{B} = AB\cos\alpha = (3)(4)\cos 30° = 10.4$$

FIGURE 9.10 Vectors $\vec{A}$ and $\vec{B}$ of Example 9.4.

FIGURE 9.11 The unit vectors $\hat{\imath}$ and $\hat{\jmath}$.

Magnitude = 1

Like vector addition and subtraction, calculating the dot product of two vectors is often performed most easily using vector components. **FIGURE 9.11** reminds you of the unit vectors $\hat{\imath}$ and $\hat{\jmath}$ that point in the positive x-direction and positive y-direction. The two unit vectors are perpendicular to each other, so their dot product is $\hat{\imath} \cdot \hat{\jmath} = 0$. Furthermore, because the magnitudes of $\hat{\imath}$ and $\hat{\jmath}$ are 1, $\hat{\imath} \cdot \hat{\imath} = 1$ and $\hat{\jmath} \cdot \hat{\jmath} = 1$.

In terms of components, we can write the dot product of vectors $\vec{A}$ and $\vec{B}$ as

$$\vec{A} \cdot \vec{B} = (A_x\hat{\imath} + A_y\hat{\jmath}) \cdot (B_x\hat{\imath} + B_y\hat{\jmath})$$

Multiplying this out, and using the results for the dot products of the unit vectors:

$$\vec{A} \cdot \vec{B} = A_xB_x\hat{\imath} \cdot \hat{\imath} + (A_xB_y + A_yB_x)\hat{\imath} \cdot \hat{\jmath} + A_yB_y\hat{\jmath} \cdot \hat{\jmath}$$
$$= A_xB_x + A_yB_y \tag{9.19}$$

That is, the **dot product is the sum of the products of the components**.

EXAMPLE 9.5 | **Calculating a dot product using components**

Compute the dot product of $\vec{A} = 3\hat{\imath} + 3\hat{\jmath}$ and $\vec{B} = 4\hat{\imath} - \hat{\jmath}$.

SOLVE FIGURE 9.12 shows vectors $\vec{A}$ and $\vec{B}$. We could calculate the dot product by first doing the geometry needed to find the angle between the vectors and then using Equation 9.18. But calculating the dot product from the vector components is much easier. It is

$$\vec{A} \cdot \vec{B} = A_xB_x + A_yB_y = (3)(4) + (3)(-1) = 9$$

FIGURE 9.12 Vectors $\vec{A}$ and $\vec{B}$.

Looking at Equation 9.17, the work done by a constant force, you should recognize that it is the dot product of the force vector and the displacement vector:

$$W = \vec{F} \cdot \Delta\vec{r} \quad \text{(work done by a constant force)} \tag{9.20}$$

This definition of work is valid for a constant force.

EXAMPLE 9.6 | Calculating work using the dot product

A 70 kg skier is gliding at 2.0 m/s when he starts down a very slippery 50-m-long, 10° slope. What is his speed at the bottom?

MODEL Model the skier as a particle and interpret "very slippery" to mean frictionless. Use the energy principle to find his final speed.

VISUALIZE FIGURE 9.13 shows a pictorial representation.

FIGURE 9.13 Pictorial representation of the skier.

Before:
$x_0 = 0$ m
$v_0 = 2.0$ m/s
$m = 70$ kg

After:
$x_1 = 50$ m
v_1

Find: v_1

SOLVE The only forces on the skier are $\vec{F}_G$ and $\vec{n}$. The normal force is perpendicular to the motion and thus does no work. The work done by gravity is easily calculated as a dot product:

$$W = \vec{F}_G \cdot \Delta \vec{r} = mg(\Delta r)\cos \alpha$$
$$= (70 \text{ kg})(9.8 \text{ m/s}^2)(50 \text{ m})\cos 80° = 5960 \text{ J}$$

Notice that the angle *between* the vectors is 80°, not 10°. Then, from the energy principle, we find

$$\Delta K = \tfrac{1}{2}mv_1^2 - \tfrac{1}{2}mv_0^2 = W$$

$$v_1 = \sqrt{v_0^2 + \frac{2W}{m}} = \sqrt{(2.0 \text{ m/s})^2 + \frac{2(5960 \text{ J})}{70 \text{ kg}}} = 13 \text{ m/s}$$

NOTE While in the midst of the mathematics of calculating work, do not lose sight of what the energy principle is all about. It is a statement about *energy transfer:* Work causes a particle's kinetic energy to either increase or decrease.

STOP TO THINK 9.5 Which force does the most work as a particle undergoes displacement $\Delta \vec{r}$?

a. The 10 N force.
b. The 8 N force.
c. The 6 N force.
d. They all do the same amount of work.

Zero-Work Situations

There are three common situations where *no* work is done. The most obvious is when the object doesn't move ($\Delta s = 0$). If you were to hold a 200 lb weight over your head, you might break out in a sweat and your arms would tire. You might feel that you had done a lot of work, but you would have done *zero* work in the physics sense because the weight was not displaced and thus you transferred no energy to it. **A force acting on a particle does no work unless the particle is displaced.**

FIGURE 9.14 shows a particle moving in uniform circular motion. As you learned in Chapter 8, uniform circular motion requires a force pointing toward the center of the circle. How much work does this force do?

Zero! You can see that the force is everywhere perpendicular to the small displacement $d\vec{s}$, so the dot product is zero and the force does *no* work on the particle. This shouldn't be surprising. The particle's speed, and hence its kinetic energy, doesn't change in uniform circular motion, so no energy is transferred to or from the system. **A force everywhere perpendicular to the motion does no work.** The friction force on a car turning a corner does no work. Neither does the tension force when a ball on a string is in circular motion.

Last, consider the roller skater in FIGURE 9.15 who straightens her arms and pushes off from a wall. She applies a force to the wall and thus, by Newton's third law, the wall applies a force $\vec{F}_{\text{W on S}}$ to her. How much work does this force do?

Surprisingly, zero. The reason is subtle but worth discussing because it gives us insight into how energy is transferred and transformed. The skater differs from suitcases and rockets in two important ways. First, the skater, as she extends her arms,

FIGURE 9.14 A perpendicular force does no work.

The force is everywhere perpendicular to the displacement, so it does no work.

FIGURE 9.15 Does the wall do work on the skater?

How much work does this force do as she pushes away from the wall?

is a *deformable object*. We cannot use the particle model for a deformable object. Second, the skater has an *internal source of energy*. Because she's a living object, she has an internal store of chemical energy that is available through metabolic processes.

Although the skater's center of mass is displaced, *the palms of her hands—where the force is exerted—are not*. The particles on which force $F_{\text{W on S}}$ acts have no displacement, and we've just seen that there's no work without displacement. The force acts, but the force doesn't push any physical thing through a displacement. Hence no work is done.

But the skater indisputably gains kinetic energy. How? Recall, from the energy overview that started this chapter, that the full energy principle is $\Delta E_{\text{sys}} = W_{\text{ext}}$. A system can gain kinetic energy without any work being done *if* it can transform some other energy into kinetic energy. In this case, the skater transforms chemical energy into kinetic energy. The same is true if you jump straight up from the ground. The ground applies an upward force to your feet, but that force does no work because the point of application—the soles of your feet—has no displacement while you're jumping. Instead, your increased kinetic energy comes via a decrease in your body's chemical energy. A brick cannot jump or push off from a wall because it cannot deform and has no usable source of internal energy.

STOP TO THINK 9.6 A car accelerates smoothly away from a stop sign. Is the work done on the car positive, negative, or zero?

Variable Force

We've learned how to calculate the work done on an object by a constant force, but what about a force that changes as the object moves? Equation 9.14, the definition of work, is all we need:

$$W = \int_{s_i}^{s_f} F_s\, ds = \text{area under the force-versus-position graph} \qquad (9.21)$$

$$\text{(work done by a variable force)}$$

The integral sums up the small amounts of work $F_s\, ds$ done in each step along the trajectory. The only new feature, because F_s now varies with position, is that we cannot take F_s outside the integral. We must evaluate the integral either geometrically by finding the area under the curve (which we'll do in the next example) or by actually doing the integration (which we'll do in the next section).

EXAMPLE 9.7 | Using work to find the speed of a car

A 1500 kg car is towed, starting from rest. **FIGURE 9.16** shows the tension force in the tow rope as the car travels from $x = 0$ m to $x = 200$ m. What is the car's speed after being pulled 200 m?

MODEL Let the system consist of only the car, which we model as a particle. We'll neglect rolling friction. Two vertical forces, the normal force and gravity, are perpendicular to the motion and thus do no work.

FIGURE 9.16 Force-versus-position graph for a car.

SOLVE We can solve this problem with the energy principle, $\Delta K = K_f - K_i = W$, where W is the work done by the tension force, but the force is not constant so we have to use the full definition of work as an integral. In this case, we can do the integral graphically:

$$W = \int_{0\text{ m}}^{200\text{ m}} T_x\, dx$$

$$= \text{area under the force curve from 0 m to 200 m}$$

$$= \tfrac{1}{2}(5000\text{ N})(200\text{ m}) = 500{,}000\text{ J}$$

The initial kinetic energy is zero, so the final kinetic energy is simply the energy transferred to the system by the work of the tension: $K_f = W = 500{,}000$ J. Then, from the definition of kinetic energy,

$$v_f = \sqrt{\frac{2K_f}{m}} = \sqrt{\frac{2(500{,}000\text{ J})}{1500\text{ kg}}} = 26\text{ m/s}$$

ASSESS 26 m/s $\approx$ 55 mph is a reasonable final speed after being towed 200 m.

9.4 Restoring Forces and the Work Done by a Spring

If you stretch a rubber band, a force tries to pull the rubber band back to its equilibrium, or unstretched, length. A force that restores a system to an equilibrium position is called a **restoring force.** Objects that exert restoring forces are called **elastic.** The most basic examples of elasticity are things like springs and rubber bands, but other examples of elasticity and restoring forces abound. For example, the steel beams flex slightly as you drive your car over a bridge, but they are restored to equilibrium after your car passes by. Nearly everything that stretches, compresses, flexes, bends, or twists exhibits a restoring force and can be called elastic.

We didn't introduce restoring forces in Part I of this textbook because we didn't have the mathematical tools to deal with them. But now—using work and energy—we do. We're going to use a simple spring as our model of elasticity. Suppose you have a spring whose **equilibrium length** is L_0. This is the length of the spring when it is neither pushing nor pulling. If you stretch (or compress) the spring, how hard does it pull (or push) back? Measurements show that

- The force is *opposite the displacement*. This is what we *mean* by a restoring force.
- If you don't stretch or compress the spring too much, the force is *proportional to the displacement from equilibrium*. The farther you push or pull, the larger the force.

FIGURE 9.17 shows a spring along a generic s-axis exerting force $\vec{F}_{Sp}$. Notice that s_{eq} is the position, or coordinate, of the free end of the spring, *not* the spring's equilibrium length L_0. When the spring is stretched, the **spring displacement** $\Delta s = s - s_{eq}$ is positive while $(F_{Sp})_s$, the s-component of the restoring force, is negative. Similarly, compressing the spring makes $\Delta s < 0$ and $(F_{Sp})_s > 0$. The graph of force versus displacement is a straight line with negative slope, showing that the spring force is proportional to but *opposite* the displacement.

The equation of the straight-line graph passing through the origin is

$$(F_{Sp})_s = -k\,\Delta s \qquad \text{(Hooke's law)} \qquad (9.22)$$

The minus sign is the mathematical indication of a *restoring* force, and the constant k—the absolute value of the slope of the line—is called the **spring constant** of the spring. The units of the spring constant are N/m. This relationship between the force and displacement of a spring was discovered by Robert Hooke, a contemporary (and sometimes bitter rival) of Newton. **Hooke's law** is not a true "law of nature," in the sense that Newton's laws are, but is actually just a *model* of a restoring force. It works well for *small* displacements from equilibrium, but Hooke's law will fail for any real spring that is compressed or stretched too far. A hypothetical massless spring for which Hooke's law is true at all displacements is called an **ideal spring.**

> **NOTE** The force does not depend on the spring's physical length L but, instead, on the *displacement* Δs of the end of the spring.

The spring constant k is a property that characterizes a spring, just as mass m characterizes a particle. For a given spring, k is a constant—it does not change as the spring is stretched or compressed. If k is large, it takes a large pull to cause a significant stretch, and we call the spring a "stiff" spring. A spring with small k can be stretched with very little force, and we call it a "soft" spring.

> **NOTE** In an earlier physics course, you may have learned Hooke's law as $F_{Sp} = -kx$ rather than as $-k\,\Delta s$. This can be misleading, and it is a common source of errors. The restoring force is $-kx$ *only* if the coordinate system in the problem is chosen such that the origin is at the equilibrium position of the free end of the spring. That is, $x = \Delta s$ only if $x_{eq} = 0$. This choice of origin is often made, but in some problems it will be more convenient to locate the origin of the coordinate system elsewhere.

FIGURE 9.17 Properties of a spring.

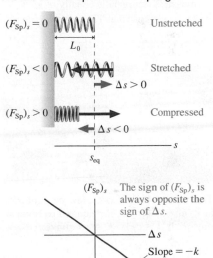

$(F_{Sp})_s = 0$ Unstretched
L_0

$(F_{Sp})_s < 0$ Stretched
$\Delta s > 0$

$(F_{Sp})_s > 0$ Compressed
$\Delta s < 0$

s_{eq}

$(F_{Sp})_s$ The sign of $(F_{Sp})_s$ is always opposite the sign of Δs.

Δs
Slope $= -k$

EXAMPLE 9.8 | Pull until it slips

FIGURE 9.18 shows a spring attached to a 2.0 kg block. The other end of the spring is pulled by a motorized toy train that moves forward at 5.0 cm/s. The spring constant is 50 N/m, and the coefficient of static friction between the block and the surface is 0.60. The spring is at its equilibrium length at $t = 0$ s when the train starts to move. When does the block slip?

FIGURE 9.18 A toy train stretches the spring until the block slips.

5.0 cm/s

2.0 kg

MODEL Model the block as a particle and the spring as an ideal spring obeying Hooke's law.

FIGURE 9.19 The free-body diagram.

VISUALIZE FIGURE 9.19 is a free-body diagram for the block.

SOLVE Recall that the tension in a massless string pulls equally at *both* ends of the string. The same is true

for the spring force: It pulls (or pushes) equally at *both* ends. This is the key to solving the problem. As the right end of the spring moves, stretching the spring, the spring pulls backward on the train *and* forward on the block with equal strength. As the spring stretches, the static friction force on the block increases in magnitude to keep the block at rest. The block is in static equilibrium, so

$$\sum (F_{net})_x = (F_{Sp})_x + (f_s)_x = F_{Sp} - f_s = 0$$

where F_{Sp} is the *magnitude* of the spring force. The magnitude is $F_{Sp} = k\,\Delta x$, where $\Delta x = v_x t$ is the distance the train has moved. Thus

$$f_s = F_{Sp} = k\,\Delta x$$

The block slips when the static friction force reaches its maximum value $f_{s\,max} = \mu_s n = \mu_s mg$. This occurs when the train has moved

$$\Delta x = \frac{f_{s\,max}}{k} = \frac{\mu_s mg}{k} = \frac{(0.60)(2.0\,\text{kg})(9.80\,\text{m/s}^2)}{50\,\text{N/m}}$$

$$= 0.235\,\text{m} = 23.5\,\text{cm}$$

The time at which the block slips is

$$t = \frac{\Delta x}{v_x} = \frac{23.5\,\text{cm}}{5.0\,\text{cm/s}} = 4.7\,\text{s}$$

The slip can range from a few centimeters in a relatively small earthquake to several meters in a very large earthquake.

This example illustrates a class of motion called *stick-slip motion*. Once the block slips, it will shoot forward some distance, then stop and stick again. As the train continues, there will be a recurring sequence of stick, slip, stick, slip, stick. . . .

Earthquakes are an important example of stick-slip motion. The large tectonic plates making up the earth's crust are attempting to slide past each other, but friction causes the edges of the plates to stick together. You may think of rocks as rigid and brittle, but large masses of rock are somewhat elastic and can be "stretched." Eventually the elastic force of the deformed rocks exceeds the friction force between the plates. An earthquake occurs as the plates slip and lurch forward. Once the tension is released, the plates stick together again and the process starts all over.

STOP TO THINK 9.7 The graph shows the force magnitude versus displacement for three springs. Rank in order, from largest to smallest, the spring constants k_a, k_b, and k_c.

F_{Sp}

a

b

c

Δs

Work Done by Springs

The primary goal of this section is to calculate the work done by a spring. FIGURE 9.20 shows a spring acting on an object as it moves from s_i to s_f. The spring force on the object varies as the object moves, but we can calculate the spring's work by using Equation 9.21 for a variable force. Hooke's law for the spring is $(F_{Sp})_s = -k\,\Delta s = -k(s - s_{eq})$. Thus

$$W = \int_{s_i}^{s_f} (F_{Sp})_s\,ds = -k \int_{s_i}^{s_f} (s - s_{eq})\,ds \qquad (9.23)$$

This is an integration best carried out with a change of variables. Define $u = s - s_{eq}$, in which case $ds = du$. This changes the integrand from $(s - s_{eq}) ds$ to $u\, du$. When we change variables, we also have to change the integration limits. At the lower limit, where $s = s_i$, the new variable u is $s_i - s_{eq} = \Delta s_i$. The lower limit becomes the initial *displacement*. Similarly, $s = s_f$ makes $u = s_f - s_{eq} = \Delta s_f$ at the upper limit. With these changes, the integral is

$$W = -k \int_{\Delta s_i}^{\Delta s_f} u\, ds = -\tfrac{1}{2} k u^2 \Big|_{\Delta s_i}^{\Delta s_f} = -\tfrac{1}{2} k (\Delta s_f)^2 + \tfrac{1}{2} k (\Delta s_i)^2 \qquad (9.24)$$

With a small rearrangement of the right side, we see that the work done by a spring is

$$W = -\left(\tfrac{1}{2} k (\Delta s_f)^2 - \tfrac{1}{2} k (\Delta s_i)^2 \right) \qquad \text{(work done by a spring)} \qquad (9.25)$$

Because the displacements are squared, it makes no difference whether the initial and final displacements are stretches or compressions.

The work done by a spring is energy transferred to the object by the force of the spring. We can use this—and the energy principle—to solve problems that we were unable to solve with a direct application of Newton's laws.

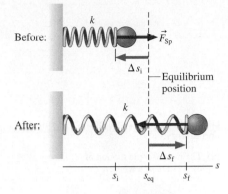

FIGURE 9.20 The spring does work on the object.

EXAMPLE 9.9 | Using the energy principle for a spring

The "pincube machine" was an ill-fated predecessor of the pinball machine. A 100 g cube is launched by pulling a spring back 12 cm and releasing it. The spring's spring constant is 65 N/m. What is the cube's launch speed as it leaves the spring? Assume that the surface is frictionless.

MODEL Let the system consist of only the cube, which we model as a particle. Two vertical forces, the normal force and gravity, are perpendicular to the cube's displacement, and we've seen that perpendicular forces do no work. Only the spring force does work.

VISUALIZE FIGURE 9.21 is a before-and-after pictorial representation in which, for horizontal motion, we've replaced the generic s-axis with an x-axis. Notice that for problem solving we use numerical subscripts in place of the generic i and f.

SOLVE We can solve this problem with the energy principle, $\Delta K = K_1 - K_0 = W$, where W is the work done by the spring. The initial displacement is $\Delta x_0 = -0.12$ m. The cube will separate from the spring when the spring has expanded back to its equilibrium length, so the final displacement is $\Delta x_1 = 0$ m. From Equation 9.25, the spring does work

$$W = -\left(\tfrac{1}{2} k (\Delta x_1)^2 - \tfrac{1}{2} k (\Delta x_0)^2 \right) = \tfrac{1}{2}(65 \text{ N/m})(-0.12 \text{ m})^2 - 0$$
$$= 0.468 \text{ J}$$

FIGURE 9.21 Pictorial representation of the pincube machine.

The initial kinetic energy is zero, so the final kinetic energy is simply the energy transferred to the system by the work of the spring: $K_1 = W = 0.468$ J. Then, from the definition of kinetic energy,

$$v_1 = \sqrt{\frac{2 K_1}{m}} = \sqrt{\frac{2(0.468 \text{ J})}{0.10 \text{ kg}}} = 3.1 \text{ m/s}$$

ASSESS 3.1 m/s ≈ 6 mph seems a reasonable final speed for a small, spring-launched cube.

STOP TO THINK 9.8 A block is attached to a spring, the spring is stretched, and the block is released at the position shown. As the block moves to the right, is the work done by the wall positive, negative, or zero?

9.5 Dissipative Forces and Thermal Energy

Suppose you drag a heavy sofa across the floor at a steady speed. You are doing work, but the sofa is not gaining kinetic energy. And when you stop pulling, the sofa almost instantly stops moving. Where is the energy going that you're adding to the system? And what happens to the sofa's kinetic energy when you stop pulling?

You know that rubbing things together raises their temperature, in extreme cases making them hot enough to start a fire. As the sofa slides across the floor, friction causes the bottom of the sofa and the floor to get hotter. An increasing temperature is associated with increasing *thermal energy,* so in this situation the work done by pulling is increasing the system's thermal energy instead of its kinetic energy. Our goal in this section is to understand what thermal energy is and how it is related to *dissipative forces.*

Energy at the Microscopic Level

FIGURE 9.22 Two perspectives of motion and energy.

(a) The macroscopic motion of the system as a whole

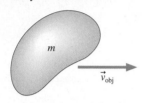

(b) The microscopic motion of the atoms inside

Atoms in motion; each has kinetic energy.

Molecular bonds stretch and compress; each has potential energy.

FIGURE 9.22 shows two different perspectives of an object. In the macrophysics perspective of Figure 9.22a you see an object of mass m moving as a whole with velocity v_{obj}. As a consequence of its motion, the object has macroscopic kinetic energy $K_{macro} = \frac{1}{2}mv_{obj}^2$.

NOTE You recognize the prefix *micro,* meaning "small." You may not be familiar with *macro,* which means "large." Everyday objects, which consist of vast numbers of particle-like atoms, are *macroscopic objects.* We will use the term **macrophysics** to refer to the motion and dynamics of an object as a whole and **microphysics** to refer to the motions of atoms within an object.

Figure 9.22b is a microphysics view of the same object, where now we see a *system of particles.* Each of these atoms is jiggling about and has kinetic energy. As the atoms move, they stretch and compress the spring-like bonds between them. We'll study potential energy in Chapter 10, but you'll recall from the energy overview at the beginning of this chapter that potential energy is *stored* energy. Stretched and compressed springs store energy, so the bonds have potential energy.

The kinetic energy of one atom is exceedingly small, but there are enormous numbers of atoms in a macroscopic object. The total kinetic energy of all the atoms is what we call the *microscopic kinetic energy,* K_{micro}. The total potential energy of all the bonds is the *microscopic potential energy,* U_{micro}. These energies are distinct from the macroscopic energy of the object as a whole.

The combined microscopic kinetic and potential energy of the atoms—the energy of the jiggling atoms and stretching bonds—is called the **thermal energy** of the system:

$$E_{th} = K_{micro} + U_{micro} \tag{9.26}$$

This energy is hidden from view in our macrophysics perspective, but it is quite real. We will discover later, when we reach thermodynamics, that the thermal energy is related to the *temperature* of the system. Raising the temperature causes the atoms to move faster and the bonds to stretch more, giving the system more thermal energy.

With the inclusion of thermal energy, the system has both macroscopic kinetic energy *and* thermal energy: $E_{sys} = K + E_{th}$. K is understood to be the total macroscopic kinetic energy; we'll use the subscript "macro" only if there's a need to distinguish macroscopic energy from microscopic energy. With this, the energy principle becomes

$$\Delta E_{sys} = \Delta K + \Delta E_{th} = W_{ext} \tag{9.27}$$

Work done on the system might increase the system's kinetic energy, its thermal energy, or both. Or, in the absence of work, kinetic energy can be transformed into thermal energy as long as the total energy change is zero. Recognizing thermal energy greatly expands the range of problems we can analyze with the energy principle.

NOTE The microscopic energy of atoms is *not* called "heat." As we've already seen, heat is a *process,* similar to work, for transferring energy between the system and the environment. We'll have a lot more to say about heat in future chapters. For the time being we want to use the correct term "thermal energy" to describe the random, thermal motions of the atoms in a system. If the temperature of a system goes up (i.e., it gets hotter), it is because the system's thermal energy has increased.

Dissipative Forces

Forces such as friction and drag cause the macroscopic kinetic energy of a system to be *dissipated* as thermal energy. Hence these are called **dissipative forces**. FIGURE 9.23 shows how microscopic interactions are responsible for transforming macroscopic kinetic energy into thermal energy when two objects slide against each other. Because friction causes *both* objects to get warmer, with increased thermal energy, **we must define the system to include both objects whose temperature changes**—both the sofa *and* the floor.

For example, FIGURE 9.24 shows a box being pulled at constant speed across a horizontal surface with friction. As you can imagine, both the surface and the box are getting warmer—increasing thermal energy—but the kinetic energy is not changing. If we define the system to be box + surface, then the increasing thermal energy of the system is entirely due to the work being done on the system by tension in the rope: $\Delta E_{th} = W_{tension}$.

The work done by tension in pulling the box through distance Δs is simply $W_{tension} = T\Delta s$; thus $\Delta E_{th} = T\Delta s$. Because the box is moving with constant velocity, and thus no net force, the tension force has to exactly balance the friction force: $T = f_k$. Consequently, the increase in thermal energy due to the dissipative force of friction is

$$\Delta E_{th} = f_k \Delta s \qquad (9.28)$$

Notice that the increase in thermal energy is directly proportional to the total distance of sliding. **Dissipative forces always increase the thermal energy; they never decrease it.**

You might wonder why we didn't simply calculate the work done by friction. The rather subtle reason is that work is defined only for forces acting on a *particle*. There is work being done on individual atoms at the boundary as they are pulled this way and that, but we would need a detailed knowledge of atomic-level friction forces to calculate this work. The friction force $\vec{f}_k$ is an average force on the object as a whole; it is not a force on any particular particle, so we cannot use it to calculate work. Furthermore, increasing thermal energy is not an energy transfer—the definition of work—from the box to the surface or from the surface to the box; both the box *and* the surface are gaining thermal energy. **The techniques used to calculate the work done on a particle cannot be used to calculate the work done by dissipative forces.**

> **NOTE** The considerations that led to Equation 9.28 allow us to calculate the total increase in thermal energy of the entire system, but we cannot determine what fraction of ΔE_{th} goes to the box and what fraction goes to the surface.

FIGURE 9.23 Motion with friction leads to thermal energy.

Atoms at the interface push and pull on each other.

The spring-like bonds stretch and store potential energy.

The potential energy is transformed into kinetic energy when the bonds break.

FIGURE 9.24 Work done by tension is dissipated as thermal energy.

The system is the box *plus* the surface.

Work done by tension increases the thermal energy of the box *and* the surface.

EXAMPLE 9.10 | Increasing kinetic and thermal energy

A rope with 30 N of tension pulls a 10 kg crate 3.0 m across a horizontal floor, starting from rest. The coefficient of friction between the crate and the floor is 0.20. What is the increase in thermal energy? What is the crate's final speed?

MODEL Let the system consist of both the crate and the floor. The tension in the rope does work on the system, but the vertical normal force and gravitational force do not.

SOLVE The energy principle, Equation 9.27, is $\Delta K + \Delta E_{th} = W_{ext}$. The friction force on an object moving on a horizontal surface is $f_k = \mu_k n = \mu_k mg$, so the increase in thermal energy, given by Equation 9.28, is

$$\Delta E_{th} = f_k \Delta s = \mu_k mg \Delta s$$
$$= (0.20)(10\,kg)(9.80\,m/s^2)(3.0\,m) = 59\,J$$

The tension force does work $W_{ext} = T\Delta s = (30\,N)(3.0\,m) = 90\,J$. 59 J of this goes to increasing the thermal energy, so $\Delta K = 31\,J$ is the crate's change of kinetic energy. Because $K_i = 0\,J$, $K_f = \Delta K = 31\,J$. Using the definition of kinetic energy, we find that the crate's final speed is

$$v_f = \sqrt{\frac{2K_f}{m}} = \sqrt{\frac{2(31\,J)}{10\,kg}} = 2.5\,m/s$$

ASSESS The thermal energy of the crate *and* floor increases by 59 J. We cannot determine ΔE_{th} for the crate (or floor) alone.

9.6 Power

Work is a transfer of energy between the environment and a system. In many situations we would like to know *how quickly* the energy is transferred. Does the force act quickly and transfer the energy very rapidly, or is it a slow and lazy transfer of energy? If you need to buy a motor to lift 1000 kg of bricks up 20 m, it makes a *big* difference whether the motor has to do this in 30 s or 30 min!

The question How quickly? implies that we are talking about a *rate*. For example, the velocity of an object—how quickly it is moving—is the *rate of change* of position. So when we raise the issue of how quickly the energy is transferred, we are talking about the *rate of transfer* of energy. The rate at which energy is transferred or transformed is called the **power** *P*, and it is defined as

$$P = \frac{dE_{sys}}{dt} \tag{9.29}$$

The unit of power is the **watt,** which is defined as 1 watt = 1 W = 1 J/s. Common prefixes used with power are mW (milliwatts), kW (kilowatts), and MW (megawatts).

For example, the rope in Example 9.10 pulled with a tension of 30 N and, by doing work, transferred 90 J of energy to the system. If it took 10 s to drag the crate 3.0 m, then energy was being transferred at the rate of 9 J/s. We would say that whatever was supplying this energy—whether a human or a motor—has a "power output" of 9 W.

The idea of power as a *rate* of energy transfer applies no matter what the form of energy. **FIGURE 9.25** shows three examples of the idea of power. For now, we want to focus primarily on *work* as the source of energy transfer. Within this more limited scope, power is simply the **rate of doing work:** $P = dW/dt$. If a particle moves through a small displacement $d\vec{r}$ while acted on by force $\vec{F}$, the force does a small amount of work dW given by

$$dW = \vec{F} \cdot d\vec{r}$$

Dividing both sides by dt, to give a rate of change, yields

$$\frac{dW}{dt} = \vec{F} \cdot \frac{d\vec{r}}{dt}$$

But $d\vec{r}/dt$ is the velocity $\vec{v}$, so we can write the power as

$$P = \vec{F} \cdot \vec{v} = Fv \cos\theta \tag{9.30}$$

In other words, the power delivered to a particle by a force acting on it is the dot product of the force and the particle's velocity. These ideas will become clearer with some examples.

The English unit of power is the *horsepower.* The conversion factor to watts is

> 1 horsepower = 1 hp = 746 W

Many common appliances, such as motors, are rated in hp.

FIGURE 9.25 Examples of power.

100 W
Lightbulb

Electric energy ⟶ light
and thermal energy at 100 J/s

Athlete

Chemical energy of glucose
and fat ⟶ mechanical
energy at ≈350 J/s ≈ ½ hp

½ hp

20 kW
Gas furnace

Chemical energy of gas ⟶
thermal energy at 20,000 J/s

EXAMPLE 9.11 | **Power output of a motor**

A factory uses a motor and a cable to drag a 300 kg machine to the proper place on the factory floor. What power must the motor supply to drag the machine at a speed of 0.50 m/s? The coefficient of friction between the machine and the floor is 0.60.

SOLVE The force applied by the motor, through the cable, is the tension force $\vec{T}$. This force does work on the machine with power $P = Tv$. The machine is in equilibrium, because the motion is at

constant velocity, hence the tension in the rope balances the friction and is

$$T = f_k = \mu_k mg$$

The motor's power output is

$$P = Tv = \mu_k mgv = 882 \text{ W}$$

EXAMPLE 9.12 | **Power output of a car engine**

A 1500 kg car has a front profile that is 1.6 m wide by 1.4 m high and a drag coefficient of 0.50. The coefficient of rolling friction is 0.02. What power must the engine provide to drive at a steady 30 m/s (≈65 mph) if 25% of the power is "lost" before reaching the drive wheels?

SOLVE The net force on a car moving at a steady speed is zero. The motion is opposed both by rolling friction and by air resistance. The forward force on the car $\vec{F}_{car}$ (recall that this is really $\vec{F}_{ground\ on\ car}$, a

reaction to the drive wheels pushing backward on the ground with $\vec{F}_{car\ on\ ground}$) exactly balances the two opposing forces:

$$F_{car} = f_r + F_{drag}$$

where $\vec{F}_{drag}$ is the drag due to the air. Using the results of Chapter 6, where both rolling friction and drag were introduced, this becomes

$$F_{car} = \mu_r mg + \tfrac{1}{2}C\rho Av^2 = 294\ \text{N} + 655\ \text{N} = 949\ \text{N}$$

Here $A = (1.6\ \text{m}) \times (1.4\ \text{m})$ is the front cross-section area of the car, and we used $1.3\ \text{kg/m}^3$ as the density of air. The power required to push the car forward at 30 m/s is

$$P_{car} = F_{car}v = (949\ \text{N})(30\ \text{m/s}) = 28{,}500\ \text{W} = 38\ \text{hp}$$

This is the power *needed* at the drive wheels to push the car against the dissipative forces of friction and air resistance. The power output

of the engine is larger because some energy is used to run the water pump, the power steering, and other accessories. In addition, energy is lost to friction in the drive train. If 25% of the power is lost (a typical value), leading to $P_{car} = 0.75P_{engine}$, the engine's power output is

$$P_{engine} = \frac{P_{car}}{0.75} = 38{,}000\ \text{W} = 51\ \text{hp}$$

ASSESS Automobile engines are typically rated at ≈ 200 hp. Most of that power is reserved for fast acceleration and climbing hills.

STOP TO THINK 9.9 Four students run up the stairs in the time shown. Rank in order, from largest to smallest, their power outputs P_a to P_d.

(a)   (b)   (c)   (d)

CHALLENGE EXAMPLE 9.13 | Stopping a brick

A 25.0-cm-long spring stands vertically on the ground, with its lower end secured in a base. A 1.5 kg brick is held 40 cm directly above the spring and dropped onto the spring. The spring compresses to a length of 17.0 cm before starting to launch the brick back upward. What is the spring's spring constant?

MODEL Let the system consist of only the brick, which is modeled as a particle. Assume that the spring is ideal. Both gravity and, after contact, the spring do work on the brick.

VISUALIZE FIGURE 9.26 is a before-and-after pictorial representation. We've chosen to place the origin of the y-axis at the equilibrium position of the spring's upper end. The length of the spring is not relevant, but the *difference* between its before and after lengths—8.0 cm—is the spring's maximum compression. The point of maximum compression is a turning point for the brick, as it reverses direction and starts moving back up, so the brick's instantaneous velocity is zero.

SOLVE At first, this seems like a two-part problem: free fall until hitting the spring, then deceleration as the spring compresses. But by using the energy principle, $\Delta E_{sys} = \Delta K = W_{tot} = W_G + W_{Sp}$, we can do it in one step. Interestingly, $\Delta K = 0$ because the brick is instantaneously at rest at the beginning and again at the point of maximum spring compression. Consequently, $W_{tot} = 0$. That's not a difficulty because gravity does positive work (the downward gravitational force is in the direction of the brick's displacement) while

FIGURE 9.26 Pictorial representation of the brick and spring.

the spring does negative work (the upward spring force is opposite the displacement).

The work done by gravity is

$$W_G = (F_G)_y \Delta y_{brick} = -mg\,\Delta y_{brick}$$

where Δy_{brick} is the brick's total displacement. The negative sign comes from $(F_G)_y = -mg$ but W_G is positive because $\Delta y_{brick} = y_1 - y_0 = -0.48$ m is also negative. It may seem strange that calculating the work done by gravity is so simple when the brick first accelerates, then slows quickly after hitting the spring. But work depends on only the displacement, not how fast or slow the object is moving.

We have to be careful with the spring because its displacement is not the same as the brick's displacement. The spring begins compressing only when contact is made, so $\Delta y_{Sp\,0} = 0$ m at the instant the brick is released. The work done by the spring then continues until maximum compression, when the spring's displacement is $\Delta y_{Sp\,1} = -0.080$ m. The work done by the spring, Equation 9.25, is thus

$$W_{Sp} = -\left(\tfrac{1}{2}k(\Delta y_{Sp\,1})^2 - \tfrac{1}{2}k(\Delta y_{Sp\,0})^2\right) = -\tfrac{1}{2}k(\Delta y_{Sp\,1})^2$$

With this information about the two works, the energy principle is

$$\Delta K = 0 = W_G + W_{Sp} = -mg\,\Delta y_{brick} - \tfrac{1}{2}k(\Delta y_{Sp\,1})^2$$

Solving for the spring constant gives

$$k = -\frac{2mg\Delta y_{brick}}{(\Delta y_{Sp\,1})^2} = -\frac{2(1.5\ \text{kg})(9.80\ \text{m/s}^2)(-0.48\ \text{m})}{(-0.080\ \text{m})^2}$$

$$= 2200\ \text{N/m}$$

ASSESS 2200 N/m is a fairly large spring constant, but that's to be expected for a spring that's going to stop a falling ≈ 3 pound brick. The complexity of this problem was not the math, which was fairly simple, but the reasoning. It's a good illustration of how to apply energy reasoning to other problems.

SUMMARY

The goal of Chapter 9 has been to begin your study of how energy is transferred and transformed.

GENERAL PRINCIPLES

Basic Energy Model

- Energy is a property of the system.
- Energy is *transformed* within the system without loss.
- Energy is *transferred* to and from the system by forces that do work W.
- $W > 0$ for energy added.
- $W < 0$ for energy removed.

The Energy Principle

Doing work on a system changes the **system energy**:

$$\Delta E_{sys} = W_{ext}$$

For systems containing only particles, no interactions, $E_{sys} = K + E_{th}$. All forces are external forces, so

$$\Delta K + \Delta E_{th} = W_{tot}$$

where W_{tot} is the total work done on all particles.

IMPORTANT CONCEPTS

Kinetic energy is an energy of motion: $K = \frac{1}{2}mv^2$

Potential energy is stored energy.

Thermal energy is the **microscopic** energy of moving atoms and stretched bonds.

Dissipative forces, such as friction and drag, transform **macroscopic** energy into thermal energy. For friction:

$$\Delta E_{th} = f_k \, \Delta s$$

The **work** done by a force on a particle as it moves from s_i to s_f is

$$W = \int_{s_i}^{s_f} F_s \, ds = \text{area under the force curve}$$

The work done by a constant force is

$$W = \vec{F} \cdot \Delta \vec{r}$$

The work done by a spring is

$$W = -\left(\frac{1}{2}k(\Delta s_f)^2 - \frac{1}{2}k(\Delta s_i)^2 \right)$$

where Δs is the displacement of the end of the spring.

APPLICATIONS

Hooke's law

The **restoring force** of an ideal spring is

$$(F_{Sp})_s = -k \, \Delta s$$

where k is the **spring constant** and Δs is the displacement of the end of the spring from equilibrium.

Power is the rate at which energy is transferred or transformed:

$$P = dE_{sys}/dt$$

For a particle with velocity $\vec{v}$, the power delivered to the particle by force $\vec{F}$ is $P = \vec{F} \cdot \vec{v} = Fv\cos\theta$.

Dot product

$$\vec{A} \cdot \vec{B} = AB\cos\theta = A_xB_x + A_yB_y$$

TERMS AND NOTATION

energy	energy transformation	scalar product	ideal spring
system	energy principle	restoring force	macrophysics
environment	basic energy model	elastic	microphysics
system energy, E_{sys}	before-and-after representation	equilibrium length, L_0	thermal energy, E_{th}
work, W	kinetic energy, K	spring displacement, Δs	dissipative force
heat	joule, J	spring constant, k	power, P
energy transfer	dot product	Hooke's law	watt, W

CONCEPTUAL QUESTIONS

1. If a particle's speed increases by a factor of 3, by what factor does its kinetic energy change?
2. Particle A has half the mass and eight times the kinetic energy of particle B. What is the speed ratio v_A/v_B?
3. An elevator held by a single cable is ascending but slowing down. Is the work done by tension positive, negative, or zero? What about the work done by gravity? Explain.
4. The rope in FIGURE Q9.4 pulls the box to the left across a rough surface. Is the work done by tension positive, negative, or zero? Explain.

FIGURE Q9.4

5. A 0.2 kg plastic cart and a 20 kg lead cart both roll without friction on a horizontal surface. Equal forces are used to push both carts forward a distance of 1 m, starting from rest. After traveling 1 m, is the kinetic energy of the plastic cart greater than, less than, or equal to the kinetic energy of the lead cart? Explain.
6. A particle moving to the left is slowed by a force pushing to the right. Is the work done on the particle positive or negative? Or is there not enough information to tell? Explain.
7. A particle moves in a vertical plane along the *closed* path seen in FIGURE Q9.7, starting at A and eventually returning to its starting point. Is the work done by gravity positive, negative, or zero? Explain.

FIGURE Q9.7

8. You need to raise a heavy block by pulling it with a massless rope. You can either (a) pull the block straight up height h, or (b) pull it up a long, frictionless plane inclined at a 15° angle until its height has increased by h. Assume you will move the block at constant speed either way. Will you do more work in case a or case b? Or is the work the same in both cases? Explain.
9. A ball on a string travels once around a circle with a circumference of 2.0 m. The tension in the string is 5.0 N. How much work is done by tension?
10. A sprinter accelerates from rest. Is the work done on the sprinter positive, negative, or zero? Explain.
11. A spring has an unstretched length of 10 cm. It exerts a restoring force F when stretched to a length of 11 cm.
    a. For what length of the spring is its restoring force $3F$?
    b. At what compressed length is the restoring force $2F$?
12. The left end of a spring is attached to a wall. When Bob pulls on the right end with a 200 N force, he stretches the spring by 20 cm. The same spring is then used for a tug-of-war between Bob and Carlos. Each pulls on his end of the spring with a 200 N force. How far does the spring stretch? Explain.
13. The driver of a car traveling at 60 mph slams on the brakes, and the car skids to a halt. What happened to the kinetic energy the car had just before stopping?
14. The motor of a crane uses power P to lift a steel beam. By what factor must the motor's power increase to lift the beam twice as high in half the time?

EXERCISES AND PROBLEMS

Problems labeled ▨ integrate material from earlier chapters.

Exercises

Section 9.2 Work and Kinetic Energy for a Single Particle

1. | Which has the larger kinetic energy, a 10 g bullet fired at 500 m/s or a 75 kg student running at 5.5 m/s?
2. || At what speed does a 1000 kg compact car have the same kinetic energy as a 20,000 kg truck going 25 km/h?
3. || A mother has four times the mass of her young son. Both are running with the same kinetic energy. What is the ratio v_{son}/v_{mother} of their speeds?
4. | A horizontal rope with 15 N tension drags a 25 kg box 2.0 m to the left across a horizontal surface. How much work is done by (a) tension and (b) gravity?
5. | A 25 kg box sliding to the left across a horizontal surface is brought to a halt in a distance of 35 cm by a horizontal rope pulling to the right with 15 N tension. How much work is done by (a) tension and (b) gravity?
6. | A 2.0 kg book is lying on a 0.75-m-high table. You pick it up and place it on a bookshelf 2.25 m above the floor.
    a. How much work does gravity do on the book?
    b. How much work does your hand do on the book?
7. || A 20 g particle is moving to the left at 30 m/s. A force on the particle causes it to move the right at 30 m/s. How much work is done by the force?

8. || FIGURE EX9.8 is the kinetic-energy graph for a 2.0 kg object moving along the x-axis. Determine the work done on the object during each of the four intervals AB, BC, CD, and DE.

FIGURE EX9.8

9. | You throw a 5.5 g coin straight down at 4.0 m/s from a 35-m-high bridge.
    a. How much work does gravity do as the coin falls to the water below?
    b. What is the speed of the coin just as it hits the water?
10. || The cable of a crane is lifting a 750 kg girder. The girder increases its speed from 0.25 m/s to 0.75 m/s in a distance of 3.5 m.
    a. How much work is done by gravity?
    b. How much work is done by tension?

Section 9.3 Calculating the Work Done

11. | Evaluate the dot product $\vec{A} \cdot \vec{B}$ if
    a. $\vec{A} = 4\hat{\imath} - 2\hat{\jmath}$ and $\vec{B} = -2\hat{\imath} - 3\hat{\jmath}$.
    b. $\vec{A} = -4\hat{\imath} + 2\hat{\jmath}$ and $\vec{B} = 2\hat{\imath} + 4\hat{\jmath}$.

12. | Evaluate the dot product $\vec{A} \cdot \vec{B}$ if
   a. $\vec{A} = 3\hat{\imath} + 4\hat{\jmath}$ and $\vec{B} = 2\hat{\imath} - 6\hat{\jmath}$.
   b. $\vec{A} = 3\hat{\imath} - 2\hat{\jmath}$ and $\vec{B} = 6\hat{\imath} + 4\hat{\jmath}$.

13. | What is the angle θ between vectors $\vec{A}$ and $\vec{B}$ in each part of Exercise 12?

14. | Evaluate the dot product of the three pairs of vectors in FIGURE EX9.14.

FIGURE EX9.14

15. | Evaluate the dot product of the three pairs of vectors in FIGURE EX9.15.

FIGURE EX9.15

16. ‖ A 25 kg air compressor is dragged up a rough incline from $\vec{r}_1 = (1.3\hat{\imath} + 1.3\hat{\jmath})$ m to $\vec{r}_2 = (8.3\hat{\imath} + 2.9\hat{\jmath})$ m, where the y-axis is vertical. How much work does gravity do on the compressor during this displacement?

17. ‖ A 45 g bug is hovering in the air. A gust of wind exerts a force $\vec{F} = (4.0\hat{\imath} - 6.0\hat{\jmath}) \times 10^{-2}$ N on the bug.
   a. How much work is done by the wind as the bug undergoes displacement $\Delta\vec{r} = (2.0\hat{\imath} - 2.0\hat{\jmath})$ m?
   b. What is the bug's speed at the end of this displacement? Assume that the speed is due entirely to the wind.

18. ‖ The two ropes seen in FIGURE EX9.18 are used to lower a 255 kg piano 5.00 m from a second-story window to the ground. How much work is done by each of the three forces?

FIGURE EX9.18   **FIGURE EX9.19**

19. ‖ The three ropes shown in the bird's-eye view of FIGURE EX9.19 are used to drag a crate 3.0 m across the floor. How much work is done by each of the three forces?

20. | FIGURE EX9.20 is the force-versus-position graph for a particle moving along the x-axis. Determine the work done on the particle during each of the three intervals 0–1 m, 1–2 m, and 2–3 m.

FIGURE EX9.20

21. ‖ A 500 g particle moving along the x-axis experiences the force shown in FIGURE EX9.21. The particle's velocity is 2.0 m/s at $x = 0$ m. What is its velocity at $x = 3$ m?

FIGURE EX9.21   **FIGURE EX9.22**

22. ‖ A 2.0 kg particle moving along the x-axis experiences the force shown in FIGURE EX9.22. The particle's velocity is 4.0 m/s at $x = 0$ m. What is its velocity at $x = 2$ m and $x = 4$ m?

23. ‖ A particle moving on the x-axis experiences a force given by
CALC $F_x = qx^2$, where q is a constant. How much work is done on the particle as it moves from $x = 0$ to $x = d$?

24. ‖ A 150 g particle at $x = 0$ is moving at 2.00 m/s in the
CALC $+x$-direction. As it moves, it experiences a force given by $F_x = (0.250 \text{ N}) \sin(x/2.00 \text{ m})$. What is the particle's speed when it reaches $x = 3.14$ m?

Section 9.4 Restoring Forces and the Work Done by a Spring

25. | A horizontal spring with spring constant 750 N/m is attached to a wall. An athlete presses against the free end of the spring, compressing it 5.0 cm. How hard is the athlete pushing?

26. | A 35-cm-long vertical spring has one end fixed on the floor. Placing a 2.2 kg physics textbook on the spring compresses it to a length of 29 cm. What is the spring constant?

27. ‖ A 10-cm-long spring is attached to the ceiling. When a 2.0 kg mass is hung from it, the spring stretches to a length of 15 cm.
   a. What is the spring constant?
   b. How long is the spring when a 3.0 kg mass is suspended from it?

28. ‖ A 60 kg student is standing atop a spring in an elevator as it accelerates upward at 3.0 m/s². The spring constant is 2500 N/m. By how much is the spring compressed?

29. ‖ A 5.0 kg mass hanging from a spring scale is slowly lowered onto a vertical spring, as shown in FIGURE EX9.29. The scale reads in newtons.
   a. What does the spring scale read just before the mass touches the lower spring?
   b. The scale reads 20 N when the lower spring has been compressed by 2.0 cm. What is the value of the spring constant for the lower spring?
   c. At what compression will the scale read zero?

FIGURE EX9.29

30. ‖ A horizontal spring with spring constant 85 N/m extends outward from a wall just above floor level. A 1.5 kg box sliding across a frictionless floor hits the end of the spring and compresses it 6.5 cm before the spring expands and shoots the box back out. How fast was the box going when it hit the spring?

Section 9.5 Dissipative Forces and Thermal Energy

31. | One mole $(6.02 \times 10^{23}$ atoms) of helium atoms in the gas phase has 3700 J of microscopic kinetic energy at room temperature. If we assume that all atoms move with the same speed, what is that speed? The mass of a helium atom is 6.68×10^{-27} kg.

32. | A 55 kg softball player slides into second base, generating 950 J of thermal energy in her legs and the ground. How fast was she running?

33. ‖ A baggage handler throws a 15 kg suitcase along the floor of an airplane luggage compartment with a speed of 1.2 m/s. The suitcase slides 2.0 m before stopping. Use work and energy to find the suitcase's coefficient of kinetic friction on the floor.

34. ‖‖ An 8.0 kg crate is pulled 5.0 m up a 30° incline by a rope angled 18° above the incline. The tension in the rope is 120 N, and the crate's coefficient of kinetic friction on the incline is 0.25.
    a. How much work is done by tension, by gravity, and by the normal force?
    b. What is the increase in thermal energy of the crate and incline?

35. ‖ Justin, with a mass of 30 kg, is going down an 8.0-m-high water slide. He starts at rest, and his speed at the bottom is 11 m/s. How much thermal energy is created by friction during his descent?

Section 9.6 Power

36. | a. How much work does an elevator motor do to lift a 1000 kg elevator a height of 100 m?
    b. How much power must the motor supply to do this in 50 s at constant speed?

37. | a. How much work must you do to push a 10 kg block of steel across a steel table at a steady speed of 1.0 m/s for 3.0 s?
    b. What is your power output while doing so?

38. ‖ How much energy is consumed by (a) a 1.2 kW hair dryer used for 10 min and (b) a 10 W night light left on for 24 h?

39. | At midday, solar energy strikes the earth with an intensity of about 1 kW/m². What is the area of a solar collector that could collect 150 MJ of energy in 1 h? This is roughly the energy content of 1 gallon of gasoline.

40. ‖‖ A 50 kg sprinter, starting from rest, runs 50 m in 7.0 s at constant acceleration.
    a. What is the magnitude of the horizontal force acting on the sprinter?
    b. What is the sprinter's power output at 2.0 s, 4.0 s, and 6.0 s?

41. ‖ A 70 kg human sprinter can accelerate from rest to 10 m/s in
BIO 3.0 s. During the same time interval, a 30 kg greyhound can go from rest to 20 m/s. What is the average power output of each? *Average* power over a time interval Δt is $\Delta E / \Delta t$.

42. ‖ The human heart pumps the average adult's 6.0 L (6000 cm³)
BIO of blood through the body every minute. The heart must do work to overcome frictional forces that resist blood flow. The average adult blood pressure is 1.3×10^4 N/m².
    a. How much work does the heart do to move the 6.0 L of blood completely through the body?
    b. What power output must the heart have to do this task once a minute?
    Hint: When the heart contracts, it applies force to the blood. Pressure is force/area. Model the circulatory system as a single closed tube, with cross-section area A and volume $V = 6.0$ L, filled with blood to which the heart applies a force.

Problems

43. ‖ A 1000 kg elevator accelerates upward at 1.0 m/s² for 10 m, starting from rest.
    a. How much work does gravity do on the elevator?
    b. How much work does the tension in the elevator cable do on the elevator?
    c. What is the elevator's kinetic energy after traveling 10 m?

44. ‖ a. Starting from rest, a crate of mass m is pushed up a frictionless slope of angle θ by a *horizontal* force of magnitude F. Use work and energy to find an expression for the crate's speed v when it is at height h above the bottom of the slope.
    b. Doug uses a 25 N horizontal force to push a 5.0 kg crate up a 2.0-m-high, 20° frictionless slope. What is the speed of the crate at the top of the slope?

45. ‖‖ Susan's 10 kg baby brother Paul sits on a mat. Susan pulls the mat across the floor using a rope that is angled 30° above the floor. The tension is a constant 30 N and the coefficient of friction is 0.20. Use work and energy to find Paul's speed after being pulled 3.0 m.

46. ‖ A particle of mass m moving along the x-axis has velocity $v_x = v_0 \sin(\pi x / 2L)$. How much work is done on the particle as it moves (a) from $x = 0$ to $x = L$ and (b) from $x = 0$ to $x = 2L$?

47. | A ball shot straight up with kinetic energy K_0 reaches height h. What height will it reach if the initial kinetic energy is doubled?

48. ‖ A pile driver lifts a 250 kg weight and then lets it fall onto the end of a steel pipe that needs to be driven into the ground. A fall of 1.5 m drives the pipe in 35 cm. What is the average force exerted on the pipe?

49. ‖ A 50 kg ice skater is gliding along the ice, heading due north at 4.0 m/s. The ice has a small coefficient of static friction, to prevent the skater from slipping sideways, but $\mu_k = 0$. Suddenly, a wind from the northeast exerts a force of 4.0 N on the skater.
    a. Use work and energy to find the skater's speed after gliding 100 m in this wind.
    b. What is the minimum value of μ_s that allows her to continue moving straight north?

50. | You're fishing from a tall pier and have just caught a 1.5 kg fish. As it breaks the surface, essentially at rest, the tension in the vertical fishing line is 16 N. Use work and energy to find the fish's upward speed after you've lifted it 2.0 m.

51. ‖ Hooke's law describes an ideal spring. Many real springs are
CALC better described by the restoring force $(F_{Sp})_s = -k\Delta s - q(\Delta s)^3$, where q is a constant. Consider a spring with $k = 250$ N/m and $q = 800$ N/m³.
    a. How much work must you do to compress this spring 15 cm? Note that, by Newton's third law, the work you do *on* the spring is the negative of the work done *by* the spring.
    b. By what percent has the cubic term increased the work over what would be needed to compress an ideal spring?
    Hint: Let the spring lie along the s-axis with the equilibrium position of the end of the spring at $s = 0$. Then $\Delta s = s$.

52. ‖ The force acting on a particle is $F_x = F_0 e^{-x/L}$. How much work
CALC does this force do as the particle moves along the x-axis from $x = 0$ to $x = L$?

53. ‖ The gravitational attraction between two objects with masses
CALC m_1 and m_2, separated by distance x, is $F = Gm_1m_2/x^2$, where G is the *gravitational constant*.
    a. How much work is done by gravity when the separation changes from x_1 to x_2? Assume $x_2 < x_1$.
    b. If one mass is much greater than the other, the larger mass stays essentially at rest while the smaller mass moves toward it. Suppose a 1.5×10^{13} kg comet is passing the orbit of Mars, heading straight for the sun at a speed of 3.5×10^4 m/s. What will its speed be when it crosses the orbit of Mercury? Astronomical data are given in the tables at the back of the book, and $G = 6.67 \times 10^{-11}$ N m²/kg².

54. || An *electric dipole* consists of two equal but opposite electric
CALC charges, $+q$ and $-q$, separated by a small distance d. If another
charge Q is distance x from the center of the dipole, in the plane
perpendicular to the line between $+q$ and $-q$, and if $x \gg d$, the
dipole exerts an electric force $F = KqQd/x^3$ on charge Q, where
K is a constant. How much work does the electric force do if
charge Q moves from distance x_1 to distance x_2?

55. || A 50 g rock is placed in a
slingshot and the rubber band
is stretched. The magnitude of
the force of the rubber band
on the rock is shown by the
graph in FIGURE P9.55. The
rubber band is stretched 30 cm
and then released. What is the
speed of the rock?

FIGURE P9.55

56. || When a 65 kg cheerleader stands on a vertical spring, the spring
compresses by 5.5 cm. When a second cheerleader stands on the
shoulders of the first, the spring compresses an additional 4.5 cm.
What is the mass of the second cheerleader?

57. || Two identical horizontal springs are attached to opposite sides
of a box that sits on a frictionless table. The outer ends of the springs
are clamped while the springs are at their equilibrium lengths.
Then a 2.0 N force applied to the box, parallel to the springs,
compresses one spring by 3.0 cm while stretching the other by
the same amount. What is the spring constant of the springs?

58. || A spring of equilibrium length L_1 and spring constant k_1 hangs
from the ceiling. Mass m_1 is suspended from its lower end. Then a
second spring, with equilibrium length L_2 and spring constant k_2,
is hung from the bottom of m_1. Mass m_2 is suspended from this
second spring. How far is m_2 below the ceiling?

59. || A horizontal spring with spring constant 250 N/m is com-
pressed by 12 cm and then used to launch a 250 g box across the
floor. The coefficient of kinetic friction between the box and the
floor is 0.23. What is the box's launch speed?

60. | A 90 kg firefighter needs to climb the stairs of a 20-m-tall
building while carrying a 40 kg backpack filled with gear. How
much power does he need to reach the top in 55 s?

61. || A hydroelectric power plant uses spinning turbines to transform
the kinetic energy of moving water into electric energy with 80%
efficiency. That is, 80% of the kinetic energy becomes electric
energy. A small hydroelectric plant at the base of a dam generates
50 MW of electric power when the falling water has a speed of
18 m/s. What is the water flow rate—kilograms of water per
second—through the turbines?

62. ||| When you ride a bicycle at constant speed, nearly all the energy
BIO you expend goes into the work you do against the drag force of the
air. Model a cyclist as having cross-section area 0.45 m² and, because the human body is not aerodynamically shaped, a drag coefficient of 0.90.
   a. What is the cyclist's power output while riding at a steady
   7.3 m/s (16 mph)?
   b. *Metabolic power* is the rate at which your body "burns" fuel
   to power your activities. For many activities, your body is
   roughly 25% efficient at converting the chemical energy of
   food into mechanical energy. What is the cyclist's metabolic
   power while cycling at 7.3 m/s?
   c. The food calorie is equivalent to 4190 J. How many calories
   does the cyclist burn if he rides over level ground at 7.3 m/s
   for 1 h?

63. || A farmer uses a tractor to pull a 150 kg bale of hay up a 15°
incline to the barn at a steady 5.0 km/h. The coefficient of kinetic
friction between the bale and the ramp is 0.45. What is the tractor's
power output?

64. || A Porsche 944 Turbo has a rated engine power of 217 hp. 30%
of the power is lost in the drive train, and 70% reaches the wheels.
The total mass of the car and driver is 1480 kg, and two-thirds of
the weight is over the drive wheels.
   a. What is the maximum acceleration of the Porsche on a
   concrete surface where $\mu_s = 1.00$?
   Hint: What force pushes the car forward?
   b. If the Porsche accelerates at a_{max}, what is its speed when it
   reaches maximum power output?
   c. How long does it take the Porsche to reach the maximum
   power output?

65. || Astronomers using a 2.0-m-diameter telescope observe a
distant supernova—an exploding star. The telescope's detector
records 9.1×10^{-11} J of light energy during the first 10 s. It's
known that this type of supernova has a visible-light power
output of 5.0×10^{37} W for the first 10 s of the explosion. How
distant is the supernova? Give your answer in *light years*, where
one light year is the distance light travels in one year. The speed
of light is 3.0×10^8 m/s.

66. || Six dogs pull a two-person sled with a total mass of 220 kg.
The coefficient of kinetic friction between the sled and the snow is
0.080. The sled accelerates at 0.75 m/s² until it reaches a cruising
speed of 12 km/h. What is the team's (a) maximum power
output during the acceleration phase and (b) power output during
the cruising phase?

In Problems 67 through 69 you are given the equation(s) used to solve
a problem. For each of these, you are to
   a. Write a realistic problem for which this is the correct equation(s).
   b. Draw a pictorial representation.
   c. Finish the solution of the problem.

67. $\frac{1}{2}(2.0 \text{ kg})(4.0 \text{ m/s})^2 + 0$
$+ (0.15)(2.0 \text{ kg})(9.8 \text{ m/s}^2)(2.0 \text{ m}) = 0 + 0 + T(2.0 \text{ m})$

68. $F_{push} - (0.20)(30 \text{ kg})(9.8 \text{ m/s}^2) = 0$
$75 \text{ W} = F_{push}v$

69. $T - (1500 \text{ kg})(9.8 \text{ m/s}^2) = (1500 \text{ kg})(1.0 \text{ m/s}^2)$
$P = T(2.0 \text{ m/s})$

Challenge Problems

70. ||| A 12 kg weather rocket generates a thrust of 200 N. The
rocket, pointing upward, is clamped to the top of a vertical spring.
The bottom of the spring, whose spring constant is 550 N/m, is
anchored to the ground.
   a. Initially, before the engine is ignited, the rocket sits at rest on
   top of the spring. How much is the spring compressed?
   b. After the engine is ignited, what is the rocket's speed when
   the spring has stretched 40 cm?

71. ||| A gardener pushes a 12 kg lawnmower whose handle is tilted
up 37° above horizontal. The lawnmower's coefficient of rolling
friction is 0.15. How much power does the gardener have to supply
to push the lawnmower at a constant speed of 1.2 m/s? Assume his
push is parallel to the handle.

72. ||| A uniform solid bar with mass m and length L rotates with
CALC angular velocity ω about an axle at one end of the bar. What is the
bar's kinetic energy?

10 Interactions and Potential Energy

These windmills are transforming the kinetic energy of the wind into electric energy.

IN THIS CHAPTER, you will develop a better understanding of energy and its conservation.

How do interactions affect energy?

We continue our investigation of energy by allowing interactions to be part of the system, rather than external forces. You will learn that interactions can store energy within the system. Further, this interaction energy can be transformed—via the interaction forces—into kinetic energy.

Environment

System

The interaction is part of the system.

What is potential energy?

Interaction energy is usually called potential energy. There are many kinds of potential energy, each associated with *position*.

- Gravitational potential energy changes with height.
- Elastic potential energy changes with stretching.

« LOOKING BACK Section 9.1 Energy overview

When is energy conserved?

- If a system is isolated, its total energy is conserved.
- If a system both is isolated and has *no dissipative forces*, its mechanical energy, $K + U$, is conserved.

Energy bar charts are a tool for visualizing energy conservation.

$$K_i + U_i = K_f + U_f$$

What is an energy diagram?

An energy diagram is a graphical representation of how the energy of a particle changes as it moves. Turning points occur where the total energy line crosses the potential-energy curve. And potential-energy minima are points of stable equilibrium.

How is force related to potential energy?

Only certain types of forces, called conservative forces, are associated with a potential energy. For these forces,

- The work done changes the potential energy by $\Delta U = -W$.
- Force is the negative of the slope of the potential-energy curve.

$F_s = -\text{slope}$

Where are we now in our study of energy?

Energy is a big topic, not one that can be presented in a single chapter. Chapters 9 and 10 are primarily about mechanical energy and the mechanical transfer of energy via work. And we've touched on thermal energy because it's unavoidable in realistic mechanical systems with friction. These are related by the energy principle:

$$\Delta E_{sys} = \Delta K + \Delta U + \Delta E_{th} = W_{ext}$$

Part V of this book, Thermodynamics, will expand our energy ideas to include heat and a deeper understanding of thermal energy. Then we'll add another form of energy—electric energy—in Part VI.

10.1 Potential Energy

If you press a ball against a spring and release it, the ball shoots forward. It certainly seems like the spring had a supply of stored energy that was transferred to the ball. Or imagine tossing the ball straight up. Where does its kinetic energy go as it slows? And from where does it acquire kinetic energy as it falls? There's again a sense that the energy is stored somewhere as the ball rises, then released as the ball falls. But is energy really stored? And if so, where? And how? Answering these questions is key to expanding our understanding of the basic energy model.

Chapter 9 emphasized the importance of the *system* and the *environment*. The system has energy E_{sys}, and forces from the environment—external forces—change the system's energy by doing work on the system. In Chapter 9, we considered only systems of particles, and all forces originated in the environment. But that's not the only way to define the system. What happens if we bring some of the interactions inside the system?

FIGURE 10.1 shows two particle-like objects A and B that interact with each other and nothing else. For example, these might be two objects connected by a spring, two masses exerting gravitational forces on each other, or two charged particles exerting electric forces on each other. Regardless of what the interaction is, this is an action/reaction pair of forces that obeys Newton's third law. There are two ways to define a system.

System 1 has been chosen to consist of only the two particles; the forces are external forces. This is exactly the analysis we did in Chapter 9, so we know that the energy principle for system 1 is

$$\Delta E_{sys\,1} = \Delta K_{tot} = W_{ext} = W_A + W_B \tag{10.1}$$

where K_{tot} is the combined kinetic energies of A and B. Work W_A is the work done on A by force $\vec{F}_{B\,on\,A}$, and similarly W_B is the work done on B by force $\vec{F}_{A\,on\,B}$. The work of these two forces changes the system's kinetic energy.

Now consider the same two particles but with a different choice of system, system 2, where we've included the interaction within the system. It's important to recognize that a system is not a physical thing. It's an analysis tool that we can define however we wish, and our choice doesn't change the behavior of physical objects. Objects A and B are oblivious to our choice of system, so ΔK_{tot} for system 2 is exactly the same as for system 1. But W_{ext} *has* changed. System 2 has no external forces to transfer energy to or from the system, so $W_{ext} = 0$. Consequently, the energy principle for system 2 is

$$\Delta E_{sys\,2} = W_{ext} = 0 \tag{10.2}$$

Now the fact that $\Delta E_{sys\,1} \neq \Delta E_{sys\,2}$ is not an issue; after all, they are different systems. But we know that system 2 has a changing kinetic energy, so how can $\Delta E_{sys\,2} = 0$?

Because system 2 has an interaction inside the system that system 1 lacks, let's postulate that system 2 has an additional form of energy associated with the interaction. That is, system 1 has $E_{sys\,1} = K_{tot}$, because particles have only kinetic energy, but system 2 has $E_{sys\,2} = K_{tot} + U$, where U, called **potential energy,** is the energy of the interaction.

If this is true, we can combine $\Delta E_{sys\,2} = 0$, from Equation 10.2, with our knowledge of ΔK_{tot} from Equation 10.1 to write

$$\Delta E_{sys\,2} = \Delta K_{tot} + \Delta U = (W_A + W_B) + \Delta U = 0 \tag{10.3}$$

That is, system 2 can have $\Delta E_{sys} = 0$ if it has a potential energy that changes by

$$\Delta U = -(W_A + W_B) = -W_{int} \tag{10.4}$$

where W_{int} is the total work done *inside the system* by the interaction forces.

Equation 10.3 tells us that the system's kinetic energy can increase ($\Delta K > 0$) if its potential energy decreases ($\Delta U < 0$) by the same amount. In effect, **the interaction stores energy inside the system** with the *potential* to be converted to kinetic energy (or, in other situations, thermal energy)—hence the name *potential energy*. This

FIGURE 10.1 Two choices of the system and the environment.

Action/reaction pair of forces

External forces do work on the system.

System 1:

The interaction is part of the system.

System 2:

idea will become more concrete as we start looking at specific examples. And, since we *postulated* the existence of an energy associated with interactions, we'll need to investigate the types of interactions for which this is true.

> **NOTE** Kinetic energy is the energy of an object. In contrast, potential energy is the energy of an interaction. You can say "The ball has kinetic energy" but not "The ball has potential energy." We'll look at the best way to describe potential energy when we get to specific examples.

Systems Matter

When solving a problem, *you* get to define the system. But your choice has consequences! E_{sys} is the energy *of* the system, so a different system will have a different energy. Similarly, W_{ext} is the work done on the system by forces originating in the environment, and that will depend on the boundary between the system and the environment.

In Figure 10.1, system 1 is a restricted system of just the particles, so system 1 has only kinetic energy. All the interaction forces are external forces that do work. Thus system 1 obeys

$$\Delta E_{sys} = \Delta K_{tot} = W_A + W_B$$

System 2 includes the interaction, so system 2 has both kinetic and potential energy. But the choice of the system boundary means that no work is done by external forces. So for system 2,

$$\Delta E_{sys} = \Delta K_{tot} + \Delta U = 0$$

Both mathematical statements are true because they refer to different systems. Notice that, for system 2, kinetic energy can be transformed into potential energy, or vice versa, but **the total energy of the system does not change.** This is our first glimpse of the idea of *conservation of energy*.

The point to remember is that **any choice of system is acceptable, but you can't mix and match.** You can define the system so that you have to calculate work, or you can define the system where you use potential energy, but using both work *and* potential energy is incorrect because it double counts the contribution of the interaction. Thus the most critical step in an energy analysis is to clearly define the system you're working with.

10.2 Gravitational Potential Energy

We'll start our exploration of potential energy with **gravitational potential energy,** the interaction energy associated with the gravitational interaction between two masses. The symbol for gravitational potential energy is U_G. We'll restrict ourselves to the "flat-earth approximation" $\vec{F}_G = -mg\hat{j}$. The gravitational potential energy of two astronomical bodies will be taken up in Chapter 13.

FIGURE 10.2 shows a ball of mass m moving upward from an initial vertical position y_i to a final vertical position y_f. The earth exerts force $\vec{F}_{E \text{ on } B}$ on the ball and, by Newton's third law, the ball exerts an equal-but-opposite force $\vec{F}_{B \text{ on } E}$ on the earth.

We could define the system to consist of only the ball, in which case the force of gravity is an external force that does work on the ball, changing its kinetic energy. We did exactly this in Chapter 9. Now let's define the system to be ball + earth. This brings the interaction inside the system, so (ignoring any gravitational forces from distant astronomical bodies) there's no external work. Instead, we have an energy of interaction—the gravitational potential energy—described by Equation 10.4:

$$\Delta U_G = -(W_B + W_E) \tag{10.5}$$

where W_B is the work gravity does on the ball and W_E is the work gravity does on the earth. The latter, practically speaking, is zero. $\vec{F}_{E \text{ on } B}$ and $\vec{F}_{B \text{ on } E}$ have equal magnitudes, by Newton's third law, but the earth's displacement is completely insignificant

FIGURE 10.2 The ball + earth system has a gravitational potential energy.

compared to the ball's displacement. Because work is a product of force and displacement, the work done on the earth is essentially zero and we can write

$$\Delta U_G = -W_B \qquad (10.6)$$

You learned in Chapter 9 to compute the work of gravity on the ball: $W_B = (F_G)_y \Delta y = -mg \, \Delta y$. So if the ball changes its vertical position by Δy, the gravitational potential energy changes by

$$\Delta U_G = -W_B = mg \, \Delta y \qquad (10.7)$$

Notice that increasing the ball's height ($\Delta y > 0$) increases the gravitational potential energy ($\Delta U_G > 0$), as we would expect.

Our energy analysis has given us an expression for ΔU_G, the *change* in potential energy, but not an expression for U_G itself. If we write out what the Δ in Equation 10.7 means—final value minus initial value—we have

$$U_{Gf} - U_{Gi} = mgy_f - mgy_i \qquad (10.8)$$

Consequently, we define the gravitational potential energy to be

$$U_G = mgy \qquad \text{(gravitational potential energy)} \qquad (10.9)$$

Notice that **gravitational potential energy is an energy of position.** It depends on the object's position but not on its speed. You should convince yourself that the units of mass times acceleration times position are joules, the unit of energy.

STOP TO THINK 10.1 Rank in order, from largest to smallest, the gravitational potential energies of balls a to d.

EXAMPLE 10.1 | Launching a pebble

Rafael uses a slingshot to shoot a 25 g pebble straight up at 17 m/s. How high does the pebble go?

MODEL Let the system consist of both the earth and the pebble, which we model as a particle. Assume that air resistance is negligible. There are no external forces to do work, but the system does have gravitational potential energy.

VISUALIZE FIGURE 10.3 is a before-and-after pictorial representation. The before-and-after representation will continue to be our primary visualization tool.

SOLVE The energy principle for the pebble + earth system is

$$\Delta E_{sys} = \Delta K + \Delta U_G = W_{ext} = 0$$

That is, the system energy does not change at all. Instead, kinetic energy is transformed into potential energy without loss inside the system. In principle, the kinetic energy is that of the ball plus kinetic energy of the earth. But as we just noted, the enormous mass difference means that the earth is effectively at rest while the pebble does all the moving, so the only kinetic energy we need to consider is that of the pebble. Thus we have

$$0 = \Delta K + \Delta U_G = \left(\tfrac{1}{2}mv_1^2 - \tfrac{1}{2}mv_0^2\right) + (mgy_1 - mgy_0)$$

FIGURE 10.3 Pictorial representation of the pebble + earth system.

We know that $v_1 = 0$ m/s, and we chose a coordinate system in which $y_0 = 0$ m, so we're left with

$$y_1 = \frac{v_0^2}{2g} = \frac{(17 \text{ m/s})^2}{2(9.80 \text{ m/s}^2)} = 15 \text{ m}$$

The answer did not depend on the pebble's mass, which is not surprising after our earlier practice with free-fall problems.

ASSESS A height of 15 m ≈ 45 ft seems reasonable for a slingshot.

The Zero of Potential Energy

Our expression for the gravitational potential energy $U_G = mgy$ seems straightforward. But you might notice, on further reflection, that the value of U_G depends on where you choose to put the origin of your coordinate system. Consider FIGURE 10.4, where Amber and Carlos are attempting to determine the potential energy when a 1 kg rock is 1 m above the ground. Amber chooses to put the origin of her coordinate system on the ground, measures $y_{rock} = 1$ m, and quickly computes $U_G = mgy = 9.8$ J. Carlos, on the other hand, read Chapter 1 very carefully and recalls that it is entirely up to him where to locate the origin of his coordinate system. So he places his origin next to the rock, measures $y_{rock} = 0$ m, and declares that $U_G = mgy = 0$ J!

How can the potential energy have two different values? The source of this apparent difficulty comes from our interpretation of Equation 10.7. Our energy analysis found that the potential energy *changes* by $\Delta U_G = mg(y_f - y_i)$. Our claim that $U_G = mgy$ is consistent with this finding, but so also would be a claim that $U_G = mgy + C$, where C is any constant.

In other words, potential energy does not have a uniquely defined value. Adding or subtracting the same constant from all potential energies in a problem has no physical consequences because our analysis uses only ΔU_G, the *change* in the potential energy, never the actual value of U_G. In practice, we work with potential energies by setting a *reference point* or *reference level* where $U_G = 0$. This is the **zero of potential energy.** Where you place the reference point is entirely up to you; it makes no difference as long as every potential energy in the problem uses the same reference point. For gravitational potential energy, we choose the reference level by placing the origin of the y-axis at that point. Where $y = 0$, $U_G = 0$. In Figure 10.4, Amber has placed her zero of potential energy at the ground, whereas Carlos has set a reference level 1 m above the ground. Either is perfectly acceptable as long as Amber and Carlos use their reference levels consistently.

But what happens when the rock falls? When it gets to the ground, Amber measures $y = 0$ m and computes $U_G = 0$ J. No problem. But Carlos measures $y = -1$ m and thus computes $U_G = -9.8$ J. A negative potential energy may seem surprising, but it's not wrong; it simply means that the potential energy is less than at the reference point. The potential energy with the rock on the ground is certainly less than when the rock was 1 m above the ground, so for Carlos—with an elevated reference level—the potential energy is negative. The important point is that both Amber and Carlos agree that the gravitational potential energy *changes* by $\Delta U_G = -9.8$ J as the rock falls.

FIGURE 10.4 Amber and Carlos use different coordinate systems to determine the gravitational potential energy.

$U_G = 9.8$ J

$U_G = 0$ J

−1 m   0 m−

1 kg rock

0 m   −1 m

Amber's coordinate system   Carlos's coordinate system

Energy Bar Charts

If an object of mass m interacts with the earth (or other astronomical body) and there are no other forces, the energy principle for the object + earth system, Equation 10.3, is

$$\Delta K + \Delta U_G = (K_f - K_i) + (U_{Gf} - U_{Gi}) = 0 \qquad (10.10)$$

We can rewrite this as

$$K_i + U_{Gi} = K_f + U_{Gf} \qquad (10.11)$$

The quantity $E_{mech} = K_{tot} + U_{int}$, the total macroscopic kinetic and potential energy, is called the **mechanical energy** of the system. Equation 10.11 is telling us that—in this situation—**the mechanical energy does not change** as the object undergoes vertical motion. Whatever initial mechanical energy the system had before the vertical motion, it has exactly the same mechanical energy after the motion. Kinetic energy may be transformed into potential energy during the motion, or vice versa, but their sum remains unchanged.

A quantity that is unchanged during an interaction is said to be *conserved*, and Equation 10.11 is our first statement of the *law of conservation of energy*. Now this is a highly restricted situation: only the gravitational force, no other forces, and only vertical motion. We'll explore energy conservation thoroughly later in this chapter and see to what extent these restrictions can be lifted, but we're already beginning to see the power of thinking about mechanical systems in terms of energy.

Equation 10.11, which is really just energy accounting, can be represented graphically with an **energy bar chart.** For example, FIGURE 10.5 is a bar chart showing how energy is transformed when a ball is tossed straight up. Kinetic energy is gradually transformed into potential energy as the ball rises, then potential energy is transformed into kinetic energy as it falls, but **the combined height of the bars does not change.** That is, the mechanical energy of the ball + earth system is conserved.

FIGURE 10.5 Energy bar charts for a ball tossed into the air.

As it rises, the ball loses kinetic energy and gains potential energy.

$K = 0$

As it falls, the ball loses potential energy and gains kinetic energy.

$U_G = 0$ $U_G = 0$

The sum $K + U_G$ remains constant.

$K \ U_G$ $K \ U_G$ $K \ U_G$ $K \ U_G$ $K \ U_G$

NOTE Most bar charts have no numbers. Their purpose is to think about the relative changes—what's increasing, what's decreasing, and what remains constant; there's no significance to how tall a bar is.

EXAMPLE 10.2 | Dropping a watermelon

A 5.0 kg watermelon is dropped from a third-story balcony, 11 m above the street. Unfortunately, the water department forgot to replace the cover on a manhole, and the watermelon falls into a 3.0-m-deep hole. How fast is the watermelon going when it hits bottom?

MODEL Let the system consist of both the earth and the watermelon, which we model as a particle. Assume that air resistance is negligible. There are no external forces, and the motion is vertical, so the system's mechanical energy is conserved.

VISUALIZE FIGURE 10.6 shows both a before-and-after pictorial representation and an energy bar chart. Initially the system has gravitational potential energy but no kinetic energy. Potential

FIGURE 10.6 Pictorial representation and energy bar chart of the watermelon + earth system.

System

y
y_0
Before:
$y_0 = 11$ m
$v_0 = 0$ m/s
$m = 5.0$ kg

0

After:
$y_1 = -3.0$ m
v_1
y_1
Find: v_1

The combined height of the two bars is unchanged.

+

0

−

$K_i \ + \ U_{Gi} \ = \ K_f \ + \ U_{Gf}$

energy is transformed into kinetic energy as the watermelon falls. Our choice of the y-axis origin has placed the zero of potential energy at ground level, so the potential energy is negative when the watermelon reaches the bottom of the hole. Even so, the combined height of the two bars has not changed.

SOLVE The energy principle for the watermelon + earth system, written as a conservation statement, is

$$K_i + U_{Gi} = 0 + mgy_0 = K_f + U_{Gf} = \tfrac{1}{2}mv_1^2 + mgy_1$$

Solving for the impact speed, we find

$$v_1 = \sqrt{2g(y_0 - y_1)}$$
$$= \sqrt{2(9.80 \text{ m/s}^2)(11.0 \text{ m} - (-3.0 \text{ m}))}$$
$$= 17 \text{ m/s}$$

ASSESS A speed of $17 \text{ m/s} \approx 35$ mph seems reasonable for the watermelon after falling ≈ 4 stories. In thinking about this problem, you might be concerned that, once below ground level, potential energy continues being transformed into kinetic energy even though the potential energy is "less than none." Keep in mind that the actual value of U is not relevant because we can place the zero of potential energy anywhere we wish, so a negative potential energy is just a number with no implication that it's "less than none." There's no "storehouse" of potential energy that might run dry. As long as the interaction acts, potential energy can continue being transformed into kinetic energy.

Digging Deeper into Gravitational Potential Energy

The concept of gravitational potential energy would be of little interest or use if it applied only to vertical free fall. Let's begin to expand the idea. FIGURE 10.7 shows a particle of mass m moving at an angle while acted on by gravity. How much work does gravity do?

Gravity is a constant force. In Chapter 9 you learned that, in general, the work done by a constant force is $W = \vec{F} \cdot \Delta \vec{r}$. If we write both $\vec{F}_G$ and $\Delta \vec{r}$ in terms of components, and use the Chapter 9 result for calculating the dot product with components, we find that the work done by gravity is

$$W_{\text{by grav}} = \vec{F}_G \cdot \Delta \vec{r} = (F_G)_x(\Delta r_x) + (F_G)_y(\Delta r_y) = 0 + (-mg)(\Delta y)$$

$$= -mg\,\Delta y \qquad (10.12)$$

Because $\vec{F}_G$ has no x-component, the work depends only on the vertical displacement Δy.

Consequently, **the change in gravitational potential energy depends only on an object's vertical displacement.** This is true not only for motion along a straight line, as in Figure 10.7, but also for motion along a *curved* trajectory because a curve can be represented as the limit of a very large number of very short straight-line segments.

For example, FIGURE 10.8 shows an object sliding down a curved, frictionless surface. The change in gravitational potential energy of the object + earth system depends only on Δy, the distance the object descends, *not* on the shape of the curve. But now there's an additional force—the normal force of the surface. Does this force affect the system's energy? No! The normal force is always perpendicular to the box's instantaneous displacement, and you learned in Chapter 9 that forces perpendicular to the displacement do no work. **Forces always perpendicular to the motion do not affect the system's energy.** They can be ignored during an energy analysis.

FIGURE 10.7 Gravity does work on a particle moving at an angle.

FIGURE 10.8 For motion on any frictionless surface, only the vertical displacement Δy affects the energy.

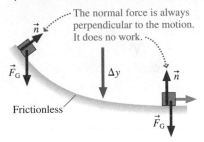

STOP TO THINK 10.2 Two identical projectiles are fired with the same speed but at different angles. Neglect air resistance. At the elevation shown as a dashed line,

a. The speed of A is greater than the speed of B.
b. The speed of A is the same as the speed of B.
c. The speed of A is less than the speed of B.

EXAMPLE 10.3 | The speed of a sled

Christine runs forward with her sled at 2.0 m/s. She hops onto the sled at the top of a 5.0-m-high, very slippery slope. What is her speed at the bottom?

MODEL Let the system consist of the earth and the sled, which we model as a particle. Because the slope is "very slippery," we'll assume that friction is negligible. The slope exerts a normal force on the sled, but it is always perpendicular to the motion and does not affect the energy.

VISUALIZE FIGURE 10.9a shows a before-and-after pictorial representation. We are not told the angle of the slope, or even if it is a straight slope, but the change in potential energy depends only on the vertical distance Christine descends and *not* on the shape of the hill. FIGURE 10.9b is an energy bar chart in which we see an initial

FIGURE 10.9 Pictorial representation and energy bar chart of the sled + earth system.

Continued

kinetic *and* potential energy being transformed into entirely kinetic energy as Christine goes down the slope.

SOLVE The energy analysis is just like in Example 10.2; the fact that the object is moving on a curved surface hasn't changed anything. The energy principle, written as a conservation statement, is

$$K_i + U_{Gi} = \tfrac{1}{2}mv_0^2 + mgy_0$$
$$= K_f + U_{Gf} = \tfrac{1}{2}mv_1^2 + 0$$

Her speed at the bottom is

$$v_1 = \sqrt{v_0^2 + 2gy_0}$$
$$= \sqrt{(2.0 \text{ m/s})^2 + 2(9.80 \text{ m/s}^2)(5.0 \text{ m})}$$
$$= 10 \text{ m/s}$$

ASSESS $10 \text{ m/s} \approx 20 \text{ mph}$ is fast but believable for a $5 \text{ m} \approx 15 \text{ ft}$ descent.

STOP TO THINK 10.3 A small child slides down the four frictionless slides a–d. Each has the same height. Rank in order, from largest to smallest, her speeds v_a to v_d at the bottom.

Motion with Gravity and Friction

What if there's friction? You learned in **« Section 9.5** that friction increases the thermal energy of the system—defined to include *both* objects—by $\Delta E_{th} = f_k \Delta s$. For a system with both gravitational potential energy and friction, the energy principle becomes

$$\Delta K + \Delta U_G + \Delta E_{th} = 0 \tag{10.13}$$

or, equivalently,

$$K_i + U_{Gi} = K_f + U_{Gf} + \Delta E_{th} \tag{10.14}$$

Mechanical energy $K + U_G$ is *not* conserved if there is friction. Because $\Delta E_{th} > 0$ (friction always makes surfaces hotter, never cooler), the final mechanical energy is less than the initial mechanical energy. That is, some fraction of the initial kinetic and potential energy is transformed into thermal energy during the motion. Friction causes objects to slow down, and motion ceases when all the mechanical energy has been transformed into thermal energy. Mechanical energy is conserved only when there are no dissipative forces and thus $\Delta E_{th} = 0$.

> **NOTE** We can write the energy principle in terms of initial and final values of the kinetic energy and the potential energy, but *not* the thermal energy. Objects always have thermal energy—the atoms are constantly in motion—but we have no way to know how much. All we can calculate is the *change* in thermal energy.

Although mechanical energy is not conserved, *the system's energy is*. Equation 10.13 tells us that the sum of kinetic, potential, and thermal energy—the energy of the system—does not change as the object moves on a surface with friction. The initial mechanical energy does not disappear; it's merely transformed into an equal amount of thermal energy.

EXAMPLE 10.4 | Skateboarding up a ramp

During the skateboard finals, Isabella encounters a 6.0-m-long, 15° upward ramp. Isabella's mass, including the skateboard, is 55 kg, and the coefficient of rolling friction between her wheels and the ramp is 0.025. With what speed must she start up the ramp to reach the top at 2.5 m/s? What percentage of her mechanical energy is "lost" to friction?

MODEL Let the system consist of the earth (including the ramp) and Isabella on the skateboard.

VISUALIZE FIGURE 10.10 shows a before-and-after pictorial representation.

FIGURE 10.10 Pictorial representation of Isabella on the ramp.

SOLVE Isabella's kinetic energy is transformed into potential energy as she gains height, but some of her kinetic energy is also transformed into increased thermal energy of her wheels and the ramp because of rolling friction. The energy principle including friction is

$$K_i + U_{Gi} = \tfrac{1}{2}mv_0^2 + 0$$
$$= K_f + U_{Gf} + \Delta E_{th} = \tfrac{1}{2}mv_1^2 + mgy_1 + f_r \Delta s$$

where we've used rolling friction f_r rather than the kinetic friction of sliding. Rolling friction is $f_r = \mu_r n$, and recall—from Chapter 6 —that the normal force of an object on a slope is $n = mg\cos\theta$. (Draw a free-body diagram if you're not sure.) Thus

$$\tfrac{1}{2}mv_0^2 = \tfrac{1}{2}mv_1^2 + mgy_1 + \mu_r mg\,\Delta s\cos\theta$$

The mass cancels. Solving for Isabella's speed at the bottom of the ramp, we find

$$v_0 = \sqrt{v_1^2 + 2gy_1 + 2\mu_r g\,\Delta s\cos\theta} = 6.3\ \text{m/s}$$

Isabella's initial mechanical energy is entirely kinetic energy: $K_0 = \tfrac{1}{2}mv_0^2 = 1090$ J. The thermal energy of the ramp and her wheels increases by $\Delta E_{th} = \mu_r mg\,\Delta s\cos\theta = 78$ J. Thus the percentage of mechanical energy transformed into thermal energy as Isabella ascends the ramp is

$$\frac{78\ \text{J}}{1090\ \text{J}} \times 100 = 7.2\%$$

This energy is not truly lost—it's still in the system—but it's no longer available for motion.

ASSESS The ramp is $1.55\ \text{m} \approx 5$ ft high. Starting up the ramp at $6.3\ \text{m/s} \approx 12$ mph in order to reach the top at $2.5\ \text{m/s} \approx 5$ mph seems reasonable.

STOP TO THINK 10.4 A skier glides down a gentle slope at constant speed. What energy transformation is taking place?

a. $U_G \to K$
b. $U_G \to K + E_{th}$
c. $U_G \to E_{th}$
d. $K \to E_{th}$
e. No energy transformation is occurring.

10.3 Elastic Potential Energy

Much of what you've just learned about gravitational potential energy carries over to the *elastic potential energy* of a spring. FIGURE 10.11 shows a spring exerting a force on a block while the block moves on a frictionless, horizontal surface. In Chapter 9, we analyzed this problem by defining the system to consist of only the block, and we calculated the work of the spring on the block. Now let's define the system to be block + spring + wall. That is, the system is the spring and the objects connected by the spring. The surface and the earth exert forces on the block—the normal force and gravity—but those forces are always perpendicular to the displacement and do not transfer any energy to the system.

We'll assume the spring to be massless, so it has no kinetic energy. Instead, the spring is the *interaction* between the block and the wall. Because the interaction is inside the system, it has an interaction energy, the **elastic potential energy,** given by

$$\Delta U_{Sp} = -(W_B + W_W) \qquad (10.15)$$

where W_B is the work the spring does on the block and W_W is the work done on the wall. But the wall is rigid and has no displacement, so $W_W = 0$ and thus $\Delta U_{Sp} = -W_B$.

We calculated the work done by an ideal spring—one that produces a linear restoring force for all displacement—in « Section 9.4. If the block moves from an initial position

FIGURE 10.11 The block + spring + wall system has an elastic potential energy.

The system is the spring and the objects the spring is attached to.

Frictionless

s_i, where the spring's displacement is $\Delta s_i = s_i - s_{eq}$, to a final position s_f with displacement $\Delta s_f = s_f - s_{eq}$, the spring does work

$$W_B = -\left(\tfrac{1}{2}k(\Delta s_f)^2 - \tfrac{1}{2}k(\Delta s_i)^2\right) \tag{10.16}$$

With the minus sign of Equation 10.15, we have

$$\Delta U_{Sp} = U_f - U_i = -W_B = \tfrac{1}{2}k(\Delta s_f)^2 - \tfrac{1}{2}k(\Delta s_i)^2 \tag{10.17}$$

Thus the elastic potential energy is

$$U_{Sp} = \tfrac{1}{2}k(\Delta s)^2 \quad \text{(elastic potential energy)} \tag{10.18}$$

where Δs is the displacement of the spring from its equilibrium length. Elastic potential energy, like gravitational potential energy, is an *energy of position*. It depends on where the block is, not on how fast the block is moving. Although we derived Equation 10.18 for a spring, it applies to *any* linear restoring force if k is the appropriate "spring constant" for that force.

The energy principle for a system with elastic potential energy and no external interactions is either $\Delta E_{sys} = \Delta K + \Delta U_{Sp} = 0$ or, recognizing that mechanical energy is again conserved,

$$K_i + U_{Sp\,i} = K_f + U_{Sp\,f} \tag{10.19}$$

EXAMPLE 10.5 | An air-track glider compresses a spring

In a laboratory experiment, your instructor challenges you to figure out how fast a 500 g air-track glider is traveling when it collides with a horizontal spring attached to the end of the track. He pushes the glider, and you notice that the spring compresses 2.7 cm before the glider rebounds. After discussing the situation with your lab partners, you decide to hang the spring on a hook and suspend the glider from the bottom end of the spring. This stretches the spring by 3.5 cm. Based on your measurements, how fast was the glider moving?

MODEL Let the system consist of the track, the spring, and the glider. The spring is inside the system, so the elastic interaction will be treated as a potential energy. Gravity and the normal force of the track on the glider are perpendicular to the glider's displacement, so they do no work and do not enter into an energy analysis. An air track is essentially frictionless, and there are no other external forces.

VISUALIZE FIGURE 10.12 shows a before-and-after pictorial representation of the collision, an energy bar chart, and a free-body diagram of the suspended glider.

SOLVE The glider's kinetic energy is gradually transformed into elastic potential energy as it compresses the spring. The point of maximum compression—After in Figure 10.12—is a turning point in the motion. The velocity is instantaneously zero, the glider's kinetic energy is zero, and thus—as the bar chart shows—all the energy has been transformed into potential energy. The spring will expand and cause the glider to rebound, but that's not part of this problem. The energy principle with elastic potential energy, in conservation form, is

$$K_i + U_{Sp\,i} = \tfrac{1}{2}mv_0^2 + 0 = K_f + U_{Sp\,f} = 0 + \tfrac{1}{2}k(\Delta x_1)^2$$

where we utilized our knowledge that the initial elastic potential energy and the final kinetic energy, at the turning point, are zero. Solving for the glider's initial speed, we find

$$v_0 = \sqrt{\frac{k}{m}}\,\Delta x_1$$

FIGURE 10.12 Pictorial representation of the experiment.

It was at this point that you and your lab partners realized you needed to determine the spring constant k. One way to do so is to measure the stretch caused by a suspended mass. The hanging glider is in equilibrium with no net force, and the free-body diagram shows that the upward spring force exactly balances the downward gravitational force. From Hooke's law, the *magnitude* of the spring force is $F_{Sp} = k|\Delta y|$. Thus Newton's first law for the suspended glider is

$$F_{Sp} = k|\Delta y| = F_G = mg$$

from which the spring's spring constant is

$$k = \frac{mg}{|\Delta y|} = \frac{(0.50 \text{ kg})(9.80 \text{ m/s}^2)}{0.035 \text{ m}} = 140 \text{ N/m}$$

Knowing k, you can now find that the glider's speed was

$$v_0 = \sqrt{\frac{k}{m}} \Delta x_1 = \sqrt{\frac{140 \text{ N/m}}{0.50 \text{ kg}}} (0.027 \text{ m}) = 0.45 \text{ m/s}$$

ASSESS A speed of ≈ 0.5 m/s is typical for gliders on an air track.

Including Gravity

Now that we see how the basic energy model works, it's easy to extend it to new situations. If a problem has both a spring *and* a vertical displacement, we define the system so that both the gravitational interaction and the elastic interaction are inside the system. Then we have both elastic *and* gravitational potential energy. That is,

$$U = U_G + U_{Sp} \tag{10.20}$$

You have to be careful with the energy accounting because there are more ways that energy can be transformed, but nothing fundamental has changed by having two potential energies rather than one.

And we know how to include the increased thermal energy if there's friction. Thus for a system that has gravitational interactions, elastic interactions, and friction, but no external forces that do work, the energy principle is

$$\Delta E_{sys} = \Delta K + \Delta U_G + \Delta U_{Sp} + \Delta E_{th} = 0 \tag{10.21}$$

or, in conservation form,

$$K_i + U_{Gi} + U_{Spi} = K_f + U_{Gf} + U_{Spf} + \Delta E_{th} \tag{10.22}$$

This is looking a bit more complex as we have more and more energies to keep track of, but the message of Equations 10.21 and 10.22 is both simple and profound: **For a system that has no other interactions with its environment, the total energy of the system does not change.** It can be transformed in many ways by the interactions, but the total does not change.

EXAMPLE 10.6 | **A spring-launched block**

Your lab assignment for the week is to devise an innovative method to determine the spring constant of a spring. You see several small blocks of different mass lying around, so you decide to measure how high the compressed spring will launch each of the blocks. You and your lab partners realize that you need to compress the spring the same amount each time, so that only the mass is varying, and you choose to use a compression of 4.0 cm. You decide to measure height from the point on the compressed spring at which the block is released. Four launches generate the data in the table:

Mass (g)	Height (m)
50	2.07
100	1.11
150	0.65
200	0.51

What value will you report for the spring constant?

MODEL Let the system consist of the earth, the block, the spring, and the floor, so there will be two potential energies. There's no friction and we'll assume no drag; hence the mechanical energy of this system is conserved. Model the spring as ideal.

VISUALIZE FIGURE 10.13 shows a pictorial representation, including an energy bar chart. We've chosen to place the origin of the

FIGURE 10.13 Pictorial representation of the experiment.

Continued

coordinate system at the point of launch. The projectile reaches height $y_1 = h$, at which point $v_1 = 0$ m/s.

SOLVE You might think we would need to find the block's speed as it leaves the spring. That would be true if we were solving this problem with Newton's laws of motion. But with an energy analysis, we can compare the system's pre-launch energy to its energy when the block reaches its highest point, completely bypassing the launch speed. The block certainly has kinetic energy *during* the motion, but the net energy transfer, shown on the energy bar chart, is from elastic potential energy to gravitational potential energy.

With both elastic and gravitational potential energy included, the energy principle is

$$K_i + U_{Gi} + U_{Sp\,i} = 0 + 0 + \tfrac{1}{2}k(\Delta y_0)^2$$
$$= K_f + U_{Gf} + U_{Sp\,f} = 0 + mgy_1 + 0$$

The block travels to position y_1, but the end of the spring does not! Be careful in spring problems not to mistake the position of an object for the position of the end of the spring; sometimes they are the same, but not always. Here the final elastic potential energy is that of an empty, unstretched spring: zero. Solving for the height, we find

$$y_1 = h = \frac{k(\Delta y_0)^2}{2mg} = \frac{k(\Delta y_0)^2}{2g} \times \frac{1}{m}$$

The first expression is correct as an algebraic expression, but here we want to use the result to analyze an experiment in which we measure h as m is varied. By isolating the mass term, we see that plotting h versus $1/m$ (that is, using $1/m$ as the x-variable) should yield a straight line with slope $k(\Delta y_0)^2/2g$.

FIGURE 10.14 is a graph of h versus $1/m$, with masses first converted to kg. The graph is linear and the best-fit line has a y-intercept very near zero, confirming our analysis of the situation. The experimentally determined slope is 0.105 m kg, with the units determined by rise over run. Thus the experimental value of the spring constant is

$$k = \frac{2g}{(\Delta y_0)^2} \times \text{slope} = 1290 \text{ N/m}$$

FIGURE 10.14 Graph of the block height versus the inverse of its mass.

ASSESS 1290 N/m is a reasonably stiff spring, but that's to be expected if you're launching blocks a meter or more into the air.

STOP TO THINK 10.5 A spring-loaded pop gun shoots a plastic ball with a speed of 4 m/s. If the spring is compressed twice as far, the ball's speed will be

a. 2 m/s   b. 4 m/s

c. 8 m/s   d. 16 m/s

10.4 Conservation of Energy

One of the most powerful statements in physics is the **law of conservation of energy:**

> **Law of conservation of energy** The total energy $E_{sys} = E_{mech} + E_{th}$ of an isolated system is a constant. The kinetic, potential, and thermal energy within the system can be transformed into each other, but their sum cannot change. Further, the mechanical energy $E_{mech} = K + U$ is conserved if the system is both isolated and nondissipative.

FIGURE 10.15 The basic energy model for an isolated system.

The system's total energy E_{sys} is conserved.

Energy can still be transformed within the system.

The key is that energy is conserved for an **isolated system,** a system that does not exchange energy with its environment either because it has no interactions with the environment or because those interactions do no work. **FIGURE 10.15** shows our basic energy model for an isolated system.

It Depends on the System

As significant as the law of conservation of energy is, it's critical to notice that the law does *not* say "Energy is always conserved." The law of conservation of energy refers to the energy *of a system*—hence our emphasis on systems in Chapters 9 and 10. Energy is conserved for some choices of system, but not others. For example, energy is conserved for a projectile moving near the earth if you define the system to be projectile + earth, but not if you define the system to be only the projectile.

In addition, the law of conservation of energy comes with an important qualification: Is the system isolated? Energy is certainly not conserved if an external force does work on the system. Thus the answer to the question "Is energy conserved?" is "It depends on the system."

A Strategy for Energy Problems

This is a good place to summarize the problem-solving strategy we've been developing for using the law of conservation of energy.

PROBLEM-SOLVING STRATEGY 10.1

Energy-conservation problems

MODEL Define the system so that there are no external forces or so that any external forces do no work on the system. If there's friction, bring both surfaces into the system. Model objects as particles and springs as ideal.

VISUALIZE Draw a before-and-after pictorial representation and an energy bar chart. A free-body diagram may be needed to visualize forces.

SOLVE If the system is both isolated and nondissipative, then the mechanical energy is conserved:

$$K_i + U_i = K_f + U_f$$

where K is the total kinetic energy of all moving objects and U is the total potential energy of all interactions within the system. If there's friction, then

$$K_i + U_i = K_f + U_f + \Delta E_{th}$$

where the thermal energy increase due to friction is $\Delta E_{th} = f_k \Delta s$.

ASSESS Check that your result has correct units and significant figures, is reasonable, and answers the question.

Exercise 14

EXAMPLE 10.7 | The speed of a pendulum

A pendulum is created by attaching one end of a 78-cm-long string to the ceiling and tying a 150 g steel ball to the other end. The ball is pulled back until the string is 60° from vertical, then released. What is the speed of the ball at its lowest point?

MODEL Let the system consist of the earth and the ball. The tension force, like a normal force, is always perpendicular to the motion and does no work, so this is an isolated system with no friction. Its mechanical energy is conserved.

VISUALIZE FIGURE 10.16 shows a before-and-after pictorial representation, where we've placed the zero of potential energy at the lowest point of the ball's swing. Trigonometry is needed to determine the ball's initial height.

SOLVE Conservation of mechanical energy is

$$K_i + U_{Gi} = 0 + mgy_0 = K_f + U_{Gf} = \tfrac{1}{2}mv_1^2 + 0$$

Thus the ball's speed at the bottom is

$$v_1 = \sqrt{2gy_0} = \sqrt{2(9.80 \text{ m/s}^2)(0.39 \text{ m})} = 2.8 \text{ m/s}$$

The speed is exactly the same as if the ball had simply fallen 0.39 m.

FIGURE 10.16 Pictorial representation of a pendulum.

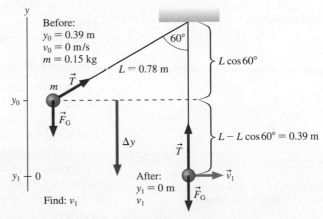

ASSESS To solve this problem directly from Newton's laws of motion requires advanced mathematics because of the complex way the net force changes with angle. But we can solve it in one line with an energy analysis!

Where Is Potential Energy?

Kinetic energy is the energy of a moving object. The basic energy model says that kinetic energy can be transformed into potential energy without loss, but where *is* the potential energy? If energy is real, not just an accounting fiction, what is it that has potential energy?

Potential energy is stored in *fields*. We've not yet introduced fields in this textbook, although we'll have a lot to say about electric and magnetic fields in later chapters. Even so, you've no doubt heard of magnetic fields and gravitational fields. Our modern understanding of the fundamental forces of nature, the long-range forces such as gravitational and electric forces, is that they are mediated by fields. How do two masses exert forces on each other through empty space? Or two electric charges? Through their fields!

When two masses move apart, the gravitational field changes to a new configuration that can store more energy. Thus the phrase "kinetic energy is transformed into gravitational potential energy" really means that the energy of a moving object is transformed into the energy of the gravitational field. At a later time, the field's energy can be transformed back into kinetic energy. The same holds true for the energy of charges and electric fields, a topic we'll take up in Part VII.

What about elastic potential energy? Remember that all solids, including springs, are held together by molecular bonds. Although quantum physics is needed for a complete understanding of bonds, they are essentially electric forces between neighboring atoms. When a solid is placed under tension, a vast number of molecular bonds stretch just a little and more energy is stored in their electric fields. What we call elastic potential energy at the macroscopic level is really energy stored in the electric fields of molecular bonds.

The theory of field energy is an advanced topic in physics. Nonetheless, this brief discussion helps complete our picture of what energy is and how it's associated with physical objects.

10.5 Energy Diagrams

Potential energy is an energy of position. The gravitational potential energy depends on the height of an object, and the elastic potential energy depends on a spring's displacement. Other potential energies you will meet in the future will depend in some way on position. Functions of position are easy to represent as graphs. A graph showing a system's potential energy and total energy as a function of position is called an **energy diagram.** Energy diagrams allow you to visualize motion based on energy considerations.

FIGURE 10.17 is the energy diagram of a particle in free fall. The gravitational potential energy $U_G = mgy$ is graphed as a line through the origin with slope mg. The *potential-energy curve* is labeled PE. The line labeled TE is the *total energy line*, $E = K + U_G$. It is horizontal because mechanical energy is conserved, meaning that the object's mechanical energy E has the same value at every position.

Suppose the particle is at position y_1. By definition, the distance from the axis to the potential-energy curve is the system's potential energy U_{G1} at that position. Because $K_1 = E - U_{G1}$, the distance between the potential-energy curve and the total energy line is the particle's kinetic energy.

The four-frame "movie" of **FIGURE 10.18** illustrates how an energy diagram is used to visualize motion. The first frame shows a particle projected upward from $y_a = 0$ with kinetic energy K_a. Initially the energy is entirely kinetic, with $U_{Ga} = 0$. A pictorial representation and an energy bar chart help to illustrate what the energy diagram is showing.

In the second frame, the particle has gained height but lost speed. The potential energy U_{Gb} is larger, and the distance K_b between the potential-energy curve and the total energy line is less. The particle continues rising and slowing until, in the third frame, it reaches the y-value where the total energy line crosses the potential-energy

FIGURE 10.17 The energy diagram of a particle in free fall.

K and U_G change as the particle moves from y_1 to y_2, but their sum is always E.

FIGURE 10.18 A four-frame movie of a particle in free fall.

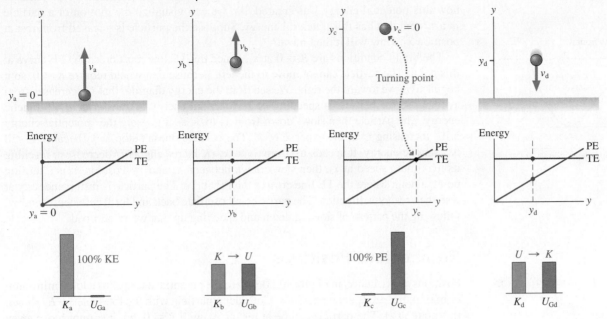

curve. This point, where $K = 0$ and the energy is entirely potential, is a *turning point* where the particle reverses direction. Finally, we see the particle speeding up as it falls.

A particle with this amount of total energy would need negative kinetic energy to be to the right of the point, at y_c, where the total energy line crosses the potential-energy curve. Negative K is not physically possible, so **the particle cannot be at positions with $U > E$.** Now, it's certainly true that you could make the particle reach a larger value of y simply by throwing it harder. But that would increase E and move the total energy line higher.

NOTE It's important to realize that the TE line is under your control. If you project an object with a different speed, or drop it from a different height, you're giving it a different total energy. You can give the object different *initial conditions* and use the energy diagram to explore how it will move with that amount of total energy.

FIGURE 10.19 shows the energy diagram of a mass on a horizontal spring, where x has been measured from the wall where the spring is attached. The equilibrium length of the spring is L_0 and the displacement of the end of the spring is $\Delta x = x - L_0$, so the spring's potential energy is $U_{Sp} = \frac{1}{2}k(\Delta x)^2 = \frac{1}{2}k(x - L_0)^2$. The potential-energy curve, a graph of U_{Sp} versus x, is a parabola centered at the equilibrium position. You can't change the PE curve—it's determined by the spring constant—but you can set the TE to any height you wish simply by stretching the spring to the proper length. The figure shows one possible TE line.

If you pull the mass out to position x_R and release it, the initial mechanical energy is entirely potential. As the restoring force of the spring pulls the mass to the left, the kinetic energy increases as the potential energy decreases. The mass has maximum speed at $x = L_0$, where $U_{Sp} = 0$, and then it slows down as the spring starts to compress. You should be able to visualize that x_L, where the PE curve crosses the TE line, is a turning point. It's the point of maximum compression where the mass instantaneously has $K = 0$. The mass will reverse direction, speed up until $x = L_0$, then slow down until reaching x_R, where it started. This is another turning point, so it will reverse direction again and start the process over. In other words, the mass will *oscillate* back and forth between the left and right turning points at x_L and x_R where the TE line crosses the PE curve. We'll study oscillations in Chapter 15, but we can already see from the energy diagram that a mass on a spring undergoes oscillatory motion.

FIGURE 10.19 The energy diagram of a mass on a horizontal spring.

The height of the TE line is determined by how far you stretch or compress the spring.

The PE curve is a parabola determined by the spring constant.

FIGURE 10.20 A more general energy diagram.

FIGURE 10.21 Points of stable and unstable equilibrium.

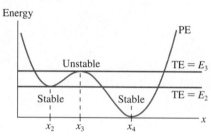

FIGURE 10.20 applies these ideas to a more general energy diagram. We don't know how this potential energy was created, but we can visualize the motion of a particle in a system that has this potential energy. Suppose the particle is released from rest at position x_1. How will it then move?

The initial conditions are $K = 0$ at x_1, hence the TE line must cross the PE curve at this point. The particle cannot move to the left, because that would require $K < 0$, so it begins to move toward the right. We see from the energy diagram that U decreases from x_1 to x_2, so the particle is speeding up as potential energy is transformed into kinetic energy. The particle then slows down from x_2 to x_3 as it goes up the "potential-energy hill," increasing U at the expense of K. The particle doesn't stop at x_3 because it still has kinetic energy. It speeds up from x_3 to x_4 (K increasing as U decreases), reaching its maximum speed at x_4, then slows down between x_4 and x_5. Position x_5 is a turning point, a point where the TE line crosses the PE curve. The particle is instantaneously at rest, then reverses direction. The particle will oscillate back and forth between x_1 and x_5, following the pattern of slowing down and speeding up that we've outlined.

Equilibrium Positions

Positions x_2, x_3, and x_4 in Figure 10.20, where the potential energy has a local minimum or maximum, are special positions. Consider a particle with the total energy E_2 shown in **FIGURE 10.21**. The particle can be at rest at x_2, with $K = 0$, but it cannot move away from x_2. In other words, a particle with energy E_2 is in *equilibrium* at x_2. If you disturb the particle, giving it a small kinetic energy and a total energy just *slightly* larger than E_2, the particle will undergo a very small oscillation centered on x_2, like a marble in the bottom of a bowl. An equilibrium for which small disturbances cause small oscillations is called a point of **stable equilibrium.** You should recognize that *any* minimum in the PE curve is a point of stable equilibrium. Position x_4 is also a point of stable equilibrium, in this case for a particle with $E = 0$.

Figure 10.21 also shows a particle with energy E_3 that is tangent to the curve at x_3. If a particle is placed *exactly* at x_3, it will stay there at rest ($K = 0$). But if you disturb the particle at x_3, giving it an energy only slightly more than E_3, it will speed up as it moves away from x_3. This is like trying to balance a marble on top of a hill. The slightest displacement will cause the marble to roll down the hill. A point of equilibrium for which a small disturbance causes the particle to move away is called a point of **unstable equilibrium.** Any maximum in the PE curve, such as x_3, is a point of unstable equilibrium.

We can summarize these lessons as follows:

TACTICS BOX 10.1

Interpreting an energy diagram

❶ The distance from the axis to the PE curve is the particle's potential energy. The distance from the PE curve to the TE line is its kinetic energy. These are transformed as the position changes, causing the particle to speed up or slow down, but the sum $K + U$ doesn't change.

❷ A point where the TE line crosses the PE curve is a turning point. The particle reverses direction.

❸ The particle cannot be at a point where the PE curve is above the TE line.

❹ The PE curve is determined by the properties of the system—mass, spring constant, and the like. You cannot change the PE curve. However, you can raise or lower the TE line simply by changing the initial conditions to give the particle more or less total energy.

❺ A minimum in the PE curve is a point of stable equilibrium. A maximum in the PE curve is a point of unstable equilibrium.

Exercises 15–17 ✎

EXAMPLE 10.8 | Balancing a mass on a spring

A spring of length L_0 and spring constant k is standing on one end. A block of mass m is placed on the spring, compressing it. What is the length of the compressed spring?

MODEL Assume an ideal spring obeying Hooke's law. The block + earth + spring system has both gravitational potential energy U_G and elastic potential energy U_{Sp}. The block sitting on top of the spring is at a point of stable equilibrium (small disturbances cause the block to oscillate slightly around the equilibrium position), so we can solve this problem by looking at the energy diagram.

VISUALIZE FIGURE 10.22a is a pictorial representation. We've used a coordinate system with the origin at ground level, so the displacement of the spring is $y - L_0$.

SOLVE FIGURE 10.22b shows the two potential energies separately and also shows the total potential energy:

$$U_{tot} = U_G + U_{Sp}$$

$$= mgy + \frac{1}{2}k(y - L_0)^2$$

The equilibrium position (the minimum of U_{tot}) has shifted from L_0 to a smaller value of y, closer to the ground. We can find the equilibrium by locating the position of the minimum in the PE curve. You know from calculus that the minimum of a function is at the point where the derivative (or slope) is zero. The derivative of U_{tot} is

$$\frac{dU_{tot}}{dy} = mg + k(y - L_0)$$

The derivative is zero at the point y_{eq}, so we can easily find

$$mg + k(y_{eq} - L_0) = 0$$

$$y_{eq} = L_0 - \frac{mg}{k}$$

The block compresses the spring by the length mg/k from its original length L_0, giving it a new equilibrium length $L_0 - mg/k$.

FIGURE 10.22 The block + earth + spring system has both gravitational and elastic potential energy.

(a)

(b)

STOP TO THINK 10.6 A particle with the potential energy shown in the graph is moving to the right. It has 1 J of kinetic energy at $x = 1$ m. Where is the particle's turning point?

10.6 Force and Potential Energy

As you've seen, we can find the energy of an interaction—potential energy—by calculating the work the interaction force does inside the system. Can we reverse this procedure? That is, if we know a system's potential energy, can we find the interaction force?

We defined the change in potential energy to be $\Delta U = -W_{int}$. Suppose that an object undergoes a *very small* displacement Δs, so small that the interaction force $\vec{F}$ is essentially constant. The work done by a constant force is $W = F_s \Delta s$, where F_s is the force component parallel to the displacement. During this small displacement, the system's potential energy changes by

$$\Delta U = -W_{int} = -F_s \Delta s \qquad (10.23)$$

FIGURE 10.23 Relating force to the PE curve.

F_s is the *negative* of the slope of the potential-energy curve.

which we can rewrite as

$$F_s = -\frac{\Delta U}{\Delta s} \qquad (10.24)$$

In the limit $\Delta s \to 0$, the force on the object is

$$F_s = \lim_{\Delta s \to 0}\left(-\frac{\Delta U}{\Delta s}\right) = -\frac{dU}{ds} \qquad (10.25)$$

That is, the interaction force on an object is the *negative* of the derivative of the potential energy with respect to position.

Graphically, as FIGURE 10.23 shows, force is the negative of the slope, at position s, of the potential-energy curve in an energy diagram:

$$F_s = -\frac{dU}{ds} = \text{the negative of the slope of the PE curve at } s \qquad (10.26)$$

FIGURE 10.24 Elastic potential energy and force graphs.

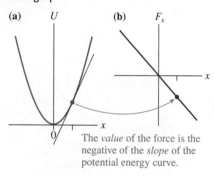

(a) U (b) F_x

The *value* of the force is the negative of the *slope* of the potential energy curve.

In practice, of course, we'll usually use either $F_x = -dU/dx$ or $F_y = -dU/dy$. Thus

- A positive slope corresponds to a negative force: to the left or downward.
- A negative slope corresponds to a positive force: to the right or upward.
- The steeper the slope, the larger the force.

As an example, consider the elastic potential energy $U_{Sp} = \frac{1}{2}kx^2$ for a horizontal spring with $x_{eq} = 0$ so that $\Delta x = x$. FIGURE 10.24a shows that the potential-energy curve is a parabola, with changing slope. If an object attached to the spring is at position x, the force on the object is

$$F_x = -\frac{dU_{Sp}}{dx} = -\frac{d}{dx}\left(\tfrac{1}{2}kx^2\right) = -kx$$

This is just Hooke's law for an ideal spring, with the minus sign indicating that Hooke's law is a restoring force. FIGURE 10.24b is a graph of force versus x. At each position x, the *value* of the force is equal to the negative of the *slope* of the PE curve.

We already knew Hooke's law, of course, so the point of this particular exercise was to illustrate the meaning of Equation 10.26. But if we had *not* known the force, we see that it's possible to find the force from the PE curve. For example, you'll learn in Part VIII that FIGURE 10.25a is a possible potential-energy function for a charged particle, one that we could create with suitably shaped electrodes. What force does the particle experience in this region of space? We find out by measuring the slope of the PE curve. The result is shown in FIGURE 10.25b. On the left side of this region of space ($x < 2$ m), a negative slope, and thus a positive value of F_x, means that the force pushes the particle to the right. A negative force on the right side ($x > 3$ m) tells us that F_x pushes the particle to the left. And there's no force at all in the center. This is a *restoring force* because a particle trying to leave this region is pushed back toward the center, but it's not a linear restoring force like that of a spring.

FIGURE 10.25 A potential-energy curve and the associated force curve.

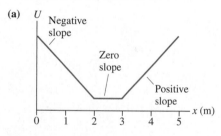

(a) U Negative slope Zero slope Positive slope x (m)

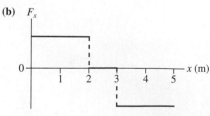

(b) F_x x (m)

EXAMPLE 10.9 | **Finding equilibrium positions**

A system's potential energy is given by $U(x) = (2x^3 - 3x^2)$ J, where x is a particle's position in m. Where are the equilibrium positions for this system, and are they stable or unstable equilibria?

SOLVE You learned in Chapter 6 that a particle in equilibrium has $\vec{F}_{net} = \vec{0}$. Then, in the previous section, you learned that the maxima and minima of the PE curve are points of equilibrium.

These may seem to be two different criteria for equilibrium, but actually they are identical. The interaction force on the particle is $F_x = -dU/dx$. The force is zero—equilibrium—at positions where the derivative is zero. But you've learned in calculus that positions where the derivative of a function is zero are the maxima and minima of the function. At either a maximum or minimum of the PE curve, the slope is zero and hence the force is zero.

For this potential-energy function,

$$F_x = -\frac{dU}{dx} = (-6x^2 + 6x) \text{ N}$$

when x is in m. The force is zero—minima or maxima of U—when $6x_{eq}^2 = 6x_{eq}$. This has two solutions:

$$x_{eq} = 0 \text{ m} \quad \text{and} \quad x_{eq} = 1 \text{ m}$$

These are positions of equilibrium, where a particle at rest will remain at rest. But how do we know if these are positions of stable equilibrium or unstable equilibrium?

A *minimum* in the PE curve is a stable equilibrium. Recall, from calculus, that a minimum of a function has a first derivative equal to zero *and* a second derivative that's positive. Similarly, a maximum of a function has a first derivative equal to zero and a second derivative that's negative. The second derivative of U is

$$\frac{d^2U}{dx^2} = \frac{d}{dx}(6x^2 - 6x) = (12x - 6) \text{ N/m}$$

The second derivative evaluated at $x = 0$ m is -6 N/m < 0, so $x = 0$ m is a maximum of the PE curve. At $x = 1$ m, the second derivative is $+6$ N/m > 0, hence a minimum in the PE curve. Thus this system has an unstable equilibrium if the particle is at $x = 0$ m and a stable equilibrium if it is at $x = 1$ m. There's a force on the particle at all other positions.

STOP TO THINK 10.7   A particle moves along the x-axis with the potential energy shown. The x-component of the force on the particle when it is at $x = 4$ m is

a. 4 N
c. 1 N
e. −2 N

b. 2 N
d. −4 N
f. −1 N

10.7 Conservative and Nonconservative Forces

A system in which particles interact gravitationally or elastically or, as we'll discover later, electrically has a potential energy. But do all forces have potential energies? Is there a "tension potential energy" or a "friction potential energy"? If not, what's special about the gravitational and elastic forces? What conditions must an interaction satisfy to have an associated potential energy?

FIGURE 10.26 shows a particle that can move from point A to point B along two possible paths while a force $\vec{F}$ acts on it. In general, the force experienced along path 1 is not the same as the force experienced along path 2. The force changes the particle's speed, so the particle's kinetic energy when it arrives at B differs from the kinetic energy it had when it left A.

Let's assume that there is a potential energy U associated with force $\vec{F}$ just as the gravitational potential energy $U_G = mgy$ is associated with the gravitational force $\vec{F}_G = -mg\hat{\jmath}$. What restrictions does this assumption place on $\vec{F}$? There are three steps in the logic:

1. Potential energy is an energy of position. U depends only on where the particle is, not on how it got there. The system has one value of potential energy when the particle is at A, a different value when the particle is at B. Thus the change in potential energy, $\Delta U = U_B - U_A$, is the same whether the particle moves along path 1 or path 2.

FIGURE 10.26 A particle can move from A to B along either of two paths.

The particle can move along either path from A to B.

Potential energy U_B

B

Path 1

$\vec{F}$

$\vec{F}$

Path 2

A

Potential energy U_A

The force does work on the particle as it moves from A to B, changing the particle's kinetic energy.

2. Potential energy is transformed into kinetic energy, with $\Delta K = -\Delta U$. If ΔU is independent of the path followed, then ΔK is also independent of the path. The particle has the same kinetic energy at B no matter which path it follows.

3. According to the energy principle, the change in a particle's kinetic energy is equal to the work done on the particle by force $\vec{F}$. That is, $\Delta K = W$. Because ΔK is independent of the path followed, it *must* be the case that **the work done by force $\vec{F}$ as the particle moves from A to B is independent of the path followed.**

A force for which the work done on a particle as it moves from an initial to a final position is independent of the path followed is called a **conservative force.** The importance of conservative forces is that **a potential energy can be associated with any conservative force.** Specifically, the potential-energy difference between an initial position i and a final position f is

$$\Delta U = -W_c(i \rightarrow f) \tag{10.27}$$

where the notation $W_c(i \rightarrow f)$ is the work done by a conservative force as the particle moves along *any* path from i to f. Equation 10.27 is a general definition of the potential energy associated with a conservative force.

A force for which we can define a potential energy is called *conservative* because the mechanical energy $K + U$ is conserved for a system in which this is the only interaction. We've already shown that the gravitational force is a conservative force by showing that ΔU_G depends only on the vertical displacement, not on the path followed; hence mechanical energy is conserved when two masses interact gravitationally. Similarly, mechanical energy is conserved for a mass on a spring—an elastic interaction—if there are no other forces. Conservative forces do not contribute to any loss of mechanical energy.

Nonconservative Forces

A characteristic of a conservative force is that **an object returning to its starting point will return with no loss of kinetic energy** because $\Delta U = 0$ if the initial and final points are the same. If a ball is tossed into the air, energy is transformed from kinetic into potential and back such that the ball's kinetic energy is unchanged when it returns to its initial height. The same is true for a puck sliding up and back down a frictionless slope.

But not all forces are conservative forces. If the slope has friction, then the puck returns with *less* kinetic energy. Part of its kinetic energy is transformed into gravitational potential energy as it slides up, but part is transformed into some other form of energy—thermal energy—that lacks the "potential" to be transformed back into kinetic energy. A force for which we cannot define a potential energy is called a **nonconservative force.** Friction and drag, which transform mechanical energy into thermal energy, are nonconservative forces, so there is no "friction potential energy."

Similarly, forces like tension and thrust are nonconservative. If you pull an object with a rope, the work done by tension is proportional to the distance traveled. More work is done along a longer path between two points than along a shorter path, so tension fails the "Work is independent of the path followed" test and does not have a potential energy.

All in all, most forces are *not* conservative forces. Gravitational forces, linear restoring forces, and, later, electric forces turn out to be fairly special because they are among the few forces for which we *can* define a potential energy. Fortunately, these are some of the most important forces in nature, so the energy principle is powerful and useful despite there being only a small number of conservative forces.

10.8 The Energy Principle Revisited

We opened Chapter 9 by introducing the energy principle—basically a statement of energy accounting—but noted that we would need to develop many new ideas to make sense of energy. We've now explored kinetic energy, potential energy, work, conservative and nonconservative forces, and much more. It's time to return to the basic energy model and start pulling together the many ideas introduced in Chapters 9 and 10.

FIGURE 10.27 shows a system of three objects that interact with each other and are acted on by external forces from the environment. These forces cause the system's kinetic energy K to change. By how much? Kinetic energy is energy of motion, and the kinetic energy would be the same if we had defined the system—as we did in Chapter 9—to consist of only the objects, not the interactions. Thus $\Delta K = W_{tot} = W_c + W_{nc}$, where in the second step we've divided the total work done by all forces into the work W_c done by conservative forces and the work W_{nc} done by nonconservative forces.

Now let's make a further distinction by dividing the nonconservative forces into *dissipative* forces and *external* forces. Dissipative forces, like friction and drag, transform mechanical energy into thermal energy.

To illustrate what we mean by an external force, suppose you pick up a box at rest on the floor and place it at rest on a table. The box + earth system gains gravitational potential energy, but $\Delta K = 0$ and $\Delta E_{th} = 0$. So where did the energy come from? Or consider pulling the box across the table with a string. The box gains kinetic energy and possibly thermal energy, but not by transforming potential energy. The force of your hand and the tension of the string are forces that "reach in" from the environment to change the system. Thus they are *external forces*. They are nonconservative forces, with no potential energy, but they change the system's mechanical energy rather than its thermal energy.

With this distinction, the system's change in kinetic energy is

$$\Delta K = W_{tot} = W_c + W_{nc} = W_c + W_{diss} + W_{ext} \qquad (10.28)$$

When we bring the conservative interactions inside the system, the work done by conservative forces becomes potential energy: $W_c = -\Delta U$. And, as we learned in Chapter 9, the work done by dissipative forces becomes thermal energy: $W_{diss} = -\Delta E_{th}$. With these substitutions, Equation 10.28 becomes

$$\Delta K = -\Delta U - \Delta E_{th} + W_{ext} \qquad (10.29)$$

Separating energy terms from the work, we can write Equation 10.29 as

$$\Delta K + \Delta U + \Delta E_{th} = \Delta E_{mech} + \Delta E_{th} = \Delta E_{sys} = W_{ext} \qquad (10.30)$$

where $E_{sys} = K + U + E_{th} = E_{mech} + E_{th}$ is the energy of the system. Equation 10.30, the energy principle but with all the terms now defined, is our most general statement about how the energy of a mechanical system changes.

In Section 10.4 we defined an *isolated system* as a system that does not exchange energy with its environment. That is, an isolated system is one on which no work is done by external forces: $W_{ext} = 0$. Thus an immediate conclusion from Equation 10.30 is that **the total energy E_{sys} of an isolated system is conserved.** If, in addition, the system is nondissipative (i.e., no friction forces), then $\Delta E_{th} = 0$. In that case, the mechanical energy E_{mech} is conserved. You'll recognize this as the *law of conservation of energy* from Section 10.4. The law of conservation of energy is one of the most powerful statements in physics.

FIGURE 10.28 reproduces the basic energy model of Chapter 9. Now you can see that this is a pictorial representation of Equation 10.30. E_{sys}, the total energy of the system, changes only if external forces transfer energy into or out of the system by doing work on the system. The kinetic, potential, and thermal energies within the system can be

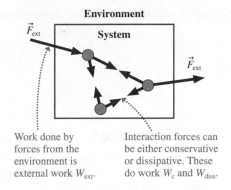

FIGURE 10.27 A system with both internal interactions and external forces.

Work done by forces from the environment is external work W_{ext}.

Interaction forces can be either conservative or dissipative. These do work W_c and W_{diss}.

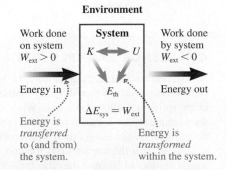

FIGURE 10.28 The basic energy model.

Work done on system $W_{ext} > 0$

Work done by system $W_{ext} < 0$

Energy in

Energy out

$\Delta E_{sys} = W_{ext}$

Energy is *transferred* to (and from) the system.

Energy is *transformed* within the system.

transformed into each other by forces within the system. And E_{sys} is conserved in the absence of interactions with the environment.

Energy Bar Charts Expanded

Energy bar charts can now be expanded to include the thermal energy and the work done by external forces. The energy principle, Equation 10.30, can be rewritten as

$$K_i + U_i + W_{ext} = K_f + U_f + \Delta E_{th} \qquad (10.31)$$

The initial mechanical energy $(K_i + U_i)$ plus any energy added to or removed from the system (W_{ext}) becomes, without loss, the final mechanical energy $(K_f + U_f)$ plus any increase in the system's thermal energy (ΔE_{th}). Remember that we have no way to determine $E_{th\,i}$ or $E_{th\,f}$, only the *change* in thermal energy. ΔE_{th} is always positive when the system contains dissipative forces.

EXAMPLE 10.10 | Hauling up supplies

A mountain climber uses a rope to drag a bag of supplies up a slope at constant speed. Show the energy transfers and transformations on an energy bar chart.

MODEL Let the system consist of the earth, the bag of supplies, and the slope.

SOLVE The tension in the rope is an external force that does work on the bag of supplies. This is an energy transfer into the system. The bag has kinetic energy, but it moves at a steady speed and so K is not *changing*. Instead, the energy transfer into the system increases both gravitational potential energy (the bag is gaining height) and thermal energy (the bag and the slope are getting warmer due to friction). The overall process is $W_{ext} \rightarrow U + E_{th}$. This is shown in **FIGURE 10.29**.

FIGURE 10.29 The energy bar chart for Example 10.10.

STOP TO THINK 10.8 A weight attached to a rope is released from rest. As the weight falls, picking up speed, the rope spins a generator that causes a lightbulb to glow. Define the system to be the weight and the earth. In this situation,

 a. $U \rightarrow K + W_{ext}$. E_{mech} is not conserved but E_{sys} is.
 b. $U + W_{ext} \rightarrow K$. Both E_{mech} and E_{sys} are conserved.
 c. $U \rightarrow K + E_{th}$. E_{mech} is not conserved but E_{sys} is.
 d. $U \rightarrow K + W_{ext}$. Neither E_{mech} nor E_{sys} is conserved.
 e. $W_{ext} \rightarrow K + U$. E_{mech} is not conserved but E_{sys} is.

CHALLENGE EXAMPLE 10.11 | A spring workout

An exercise machine at the gym has a 5.0 kg weight attached to one end of a horizontal spring with spring constant 80 N/m. The other end of the spring is anchored to a wall. When a woman working out on the machine pushes her arms forward, a cable stretches the spring by dragging the weight along a track with a coefficient of kinetic friction of 0.30. What is the woman's power output at the moment when the weight has moved 50 cm if the cable tension is a constant 100 N?

MODEL This is a complex situation, but one that we can analyze. First, identify the weight, the spring, the wall, and the track as the system. We need to have the track inside the system because friction increases the temperature of both the weight *and* the track. The tension in the cable is an external force. The work W_{ext} done by the cable's tension transfers energy into the system, causing K, U_{Sp}, and E_{th} all to increase.

FIGURE 10.30 Pictorial representation and energy bar chart for Challenge Example 10.11.

(a)

$m = 5.0$ kg

$k = 80$ N/m

$\vec{T}$

$\vec{f_k}$

0

$\Delta r = 0.50$ m

Before: $x_0 = 0$ m
$v_0 = 0$ m/s
$T = 100$ N
$\mu_k = 0.30$

After: $x_1 = 0.50$ m
v_1

Find: v_1 and P_1

(b)

$K_i + U_i + W_{ext} = K_f + U_f + \Delta E_{th}$

VISUALIZE **FIGURE 10.30a** is a before-and-after pictorial representation. The energy transfers and transformations are shown in the energy bar chart of **FIGURE 10.30b**.

SOLVE You learned in ❮ Section 9.6 that power is the *rate* at which work is done and that the power delivered by force $\vec{F}$ to an object moving with velocity $\vec{v}$ is $P = \vec{F} \cdot \vec{v}$. Here the tension $\vec{T}$ pulls parallel to the weight's velocity, so the power being supplied when the weight has velocity $\vec{v}$ is $P = Tv$. We know the cable's tension, so we need to use energy considerations to find the weight's speed v_1 after the spring has been stretched to $\Delta x_1 = 50$ cm.

The energy principle $K_i + U_i + W_{ext} = K_f + U_f + \Delta E_{th}$ is

$$\tfrac{1}{2}mv_0^2 + \tfrac{1}{2}k(\Delta x_0)^2 + W_{ext} = \tfrac{1}{2}mv_1^2 + \tfrac{1}{2}k(\Delta x_1)^2 + \Delta E_{th}$$

The initial displacement is $\Delta x_0 = 0$ m and we know that $v_0 = 0$ m/s, so the energy principle simplifies to

$$\tfrac{1}{2}mv_1^2 = W_{ext} - \tfrac{1}{2}k(\Delta x_1)^2 - \Delta E_{th}$$

The external work done by the cable's tension is

$$W_{ext} = T\Delta r = (100 \text{ N})(0.50 \text{ m}) = 50.0 \text{ J}$$

From Chapter 9, the increase in thermal energy due to friction is

$$\Delta E_{th} = f_k \Delta r = \mu_k mg \Delta r$$
$$= (0.30)(5.0 \text{ kg})(9.80 \text{ m/s}^2)(0.50 \text{ m}) = 7.4 \text{ J}$$

Solving for the speed v_1, when the spring's displacement is $\Delta x_1 = 50$ cm $= 0.50$ m, we have

$$v_1 = \sqrt{\frac{2(W_{ext} - \tfrac{1}{2}k(\Delta x_1)^2 - \Delta E_{th})}{m}} = 3.6 \text{ m/s}$$

The power being supplied at this instant to keep stretching the spring is

$$P = Tv_1 = (100 \text{ N})(3.6 \text{ m/s}) = 360 \text{ W}$$

ASSESS The work done by the cable's tension is energy transferred to the system. Part of the energy increases the speed of the weight, part increases the potential energy stored in the spring, and part is transformed into increased thermal energy, thus increasing the temperature. We had to bring all these energy ideas together to solve this problem.

SUMMARY

The goal of Chapter 10 has been to develop a better understanding of energy and its conservation.

GENERAL PRINCIPLES

The Energy Principle Revisited

- Energy is *transformed* within the system.
- Energy is *transferred* to and from the system by work W.

Two variations of the energy principle are

$$\Delta E_{sys} = \Delta K + \Delta U + \Delta E_{th} = W_{ext}$$
$$K_i + U_i + W_{ext} = K_f + U_f + \Delta E_{th}$$

Basic energy model

Solving Energy Problems

MODEL Define the system.

VISUALIZE Draw a before-and-after pictorial representation and an energy bar chart.

SOLVE Use the energy principle:

$$K_i + U_i + W_{ext} = K_f + U_f + \Delta E_{th}$$

ASSESS Is the result reasonable?

Law of Conservation of Energy

- **Isolated system:** $W_{ext} = 0$. The total system energy $E_{sys} = K + U + E_{th}$ is conserved. $\Delta E_{sys} = 0$.
- **Isolated, nondissipative system:** $W_{ext} = 0$ and $W_{diss} = 0$. The **mechanical energy** $E_{mech} = K + U$ is conserved: $K_i + U_i = K_f + U_f$.

IMPORTANT CONCEPTS

Potential energy, or *interaction energy*, is energy stored inside a system via interaction forces. The energy is stored in *fields*.

- Potential energy is associated only with **conservative forces** for which the work done is independent of the path.
- Work W_{int} by the interaction forces causes $\Delta U = -W_{int}$.
- Force $F_s = -dU/ds = -$(slope of the potential energy curve).
- Potential energy is an energy of the system, not an energy of a specific object.

Energy diagrams show the potential-energy curve PE and the total mechanical energy line TE.

- From the axis to the curve is U. From the curve to the TE line is K.
- **Turning points** occur where the TE line crosses the PE curve.
- Minima and maxima in the PE curve are, respectively, positions of **stable** and **unstable equilibrium.**

APPLICATIONS

Gravitational potential energy is an energy of the earth + object system:

$$U_G = mgy$$

Elastic potential energy is an energy of the spring + attached objects system:

$$U_{Sp} = \tfrac{1}{2}k(\Delta s)^2$$

Energy bar charts show the energy principle in graphical form.

$$K_i + U_i + W_{ext} = K_f + U_f + \Delta E_{th}$$

TERMS AND NOTATION

potential energy, U
gravitational potential energy, U_G
zero of potential energy
mechanical energy, E_{mech}

energy bar chart
elastic potential energy, U_{Sp}
law of conservation of energy

isolated system
energy diagram
stable equilibrium

unstable equilibrium
conservative force
nonconservative force

CONCEPTUAL QUESTIONS

1. Upon what basic quantity does kinetic energy depend? Upon what basic quantity does potential energy depend?

2. Can kinetic energy ever be negative? Can gravitational potential energy ever be negative? For each, give a plausible *reason* for your answer without making use of any equations.

3. A roller-coaster car rolls down a frictionless track, reaching speed v_0 at the bottom. If you want the car to go twice as fast at the bottom, by what factor must you increase the height of the track? Explain.

4. The three balls in **FIGURE Q10.4**, which have equal masses, are fired with equal speeds from the same height above the ground. Rank in order, from largest to smallest, their speeds v_a, v_b, and v_c as they hit the ground. Explain.

FIGURE Q10.4   **FIGURE Q10.5**

5. Rank in order, from most to least, the elastic potential energy $(U_{Sp})_a$ to $(U_{Sp})_d$ stored in the springs of **FIGURE Q10.5**. Explain.

6. A spring is compressed 1.0 cm. How far must you compress a spring with twice the spring constant to store the same amount of energy?

7. A spring gun shoots out a plastic ball at speed v_0. The spring is then compressed twice the distance it was on the first shot. By what factor is the ball's speed increased? Explain.

8. A particle with the potential energy shown in **FIGURE Q10.8** is moving to the right at $x = 5$ m with total energy E.
   a. At what value or values of x is this particle's speed a maximum?
   b. Does this particle have a turning point or points in the range of x covered by the graph? If so, where?
   c. If E is changed appropriately, could the particle remain at rest at any point or points in the range of x covered by the graph? If so, where?

FIGURE Q10.8

9. A compressed spring launches a block up an incline. Which objects should be included within the system in order to make an energy analysis as easy as possible?

10. A process occurs in which a system's potential energy decreases while the system does work on the environment. Does the system's kinetic energy increase, decrease, or stay the same? Or is there not enough information to tell? Explain.

11. A process occurs in which a system's potential energy increases while the environment does work on the system. Does the system's kinetic energy increase, decrease, or stay the same? Or is there not enough information to tell? Explain.

12. **FIGURE Q10.12** is the energy bar chart for a firefighter sliding down a fire pole from the second floor to the ground. Let the system consist of the firefighter, the pole, and the earth. What are the bar heights of W_{ext}, K_f, and U_{Gf}?

FIGURE Q10.12

13. a. If the force on a particle at some point in space is zero, must its potential energy also be zero at that point? Explain.
    b. If the potential energy of a particle at some point in space is zero, must the force on it also be zero at that point? Explain.

EXERCISES AND PROBLEMS

Problems labeled ▓ integrate material from earlier chapters.

Exercises

Section 10.1 Potential Energy

1. | Object A is stationary while objects B and C are in motion. Forces from object A do 10 J of work on object B and –5 J of work on object C. Forces from the environment do 4 J of work on object B and 8 J of work on object C. Objects B and C do not interact. What are ΔK_{tot} and ΔU_{int} if (a) objects A, B, and C are defined as separate systems and (b) one system is defined to include objects A, B, and C and their interactions?

2. | A system of two objects has $\Delta K_{tot} = 7$ J and $\Delta U_{int} = -5$ J.
   a. How much work is done by interaction forces?
   b. How much work is done by external forces?

Section 10.2 Gravitational Potential Energy

3. | The lowest point in Death Valley is 85 m below sea level. The summit of nearby Mt. Whitney has an elevation of 4420 m. What is the change in potential energy when an energetic 65 kg hiker makes it from the floor of Death Valley to the top of Mt. Whitney?

4. | a. What is the kinetic energy of a 1500 kg car traveling at a speed of 30 m/s (≈ 65 mph)?
   b. From what height would the car have to be dropped to have this same amount of kinetic energy just before impact?

5. | a. With what minimum speed must you toss a 100 g ball straight up to just touch the 10-m-high roof of the gymnasium if you release the ball 1.5 m above the ground? Solve this problem using energy.
   b. With what speed does the ball hit the ground?

6. | What height does a frictionless playground slide need so that a 35 kg child reaches the bottom at a speed of 4.5 m/s?

7. | A 55 kg skateboarder wants to just make it to the upper edge of a "quarter pipe," a track that is one-quarter of a circle with a radius of 3.0 m. What speed does he need at the bottom?

8. | What minimum speed does a 100 g puck need to make it to the top of a 3.0-m-long, 20° frictionless ramp?

9. || A pendulum is made by tying a 500 g ball to a 75-cm-long string. The pendulum is pulled 30° to one side, then released. What is the ball's speed at the lowest point of its trajectory?

10. || A 20 kg child is on a swing that hangs from 3.0-m-long chains. What is her maximum speed if she swings out to a 45° angle?

11. || A 1500 kg car traveling at 10 m/s suddenly runs out of gas while approaching the valley shown in **FIGURE EX10.11**. The alert driver immediately puts the car in neutral so that it will roll. What will be the car's speed as it coasts into the gas station on the other side of the valley?

FIGURE EX10.11

12. | The maximum energy a bone can absorb without breaking is
BIO surprisingly small. Experimental data show that the leg bones of a healthy, 60 kg human can absorb about 200 J. From what maximum height could a 60 kg person jump and land rigidly upright on both feet without breaking his legs? Assume that all energy is absorbed by the leg bones in a rigid landing.

13. || A cannon tilted up at a 30° angle fires a cannon ball at 80 m/s from atop a 10-m-high fortress wall. What is the ball's impact speed on the ground below?

14. || In a hydroelectric dam, water falls 25 m and then spins a turbine to generate electricity.
   a. What is ΔU_G of 1.0 kg of water?
   b. Suppose the dam is 80% efficient at converting the water's potential energy to electrical energy. How many kilograms of water must pass through the turbines each second to generate 50 MW of electricity? This is a typical value for a small hydroelectric dam.

Section 10.3 Elastic Potential Energy

15. | How far must you stretch a spring with $k = 1000$ N/m to store 200 J of energy?

16. || A stretched spring stores 2.0 J of energy. How much energy will be stored if the spring is stretched three times as far?

17. | A student places her 500 g physics book on a frictionless table. She pushes the book against a spring, compressing the spring by 4.0 cm, then releases the book. What is the book's speed as it slides away? The spring constant is 1250 N/m.

18. | A block sliding along a horizontal frictionless surface with speed v collides with a spring and compresses it by 2.0 cm. What will be the compression if the same block collides with the spring at a speed of $2v$?

19. | A 10 kg runaway grocery cart runs into a spring with spring constant 250 N/m and compresses it by 60 cm. What was the speed of the cart just before it hit the spring?

20. || As a 15,000 kg jet plane lands on an aircraft carrier, its tail hook snags a cable to slow it down. The cable is attached to a spring with spring constant 60,000 N/m. If the spring stretches 30 m to stop the plane, what was the plane's landing speed?

21. || The elastic energy stored in your tendons can contribute up to
BIO 35% of your energy needs when running. Sports scientists find that (on average) the knee extensor tendons in sprinters stretch 41 mm while those of nonathletes stretch only 33 mm. The spring constant of the tendon is the same for both groups, 33 N/mm. What is the difference in maximum stored energy between the sprinters and the nonathletes?

22. || The spring in **FIGURE EX10.22a** is compressed by Δx. It launches the block across a frictionless surface with speed v_0. The two springs in **FIGURE EX10.22b** are identical to the spring of Figure EX10.22a. They are compressed by the same Δx and used to launch the same block. What is the block's speed now?

FIGURE EX10.22

23. ||| The spring in **FIGURE EX10.23a** is compressed by Δx. It launches the block across a frictionless surface with speed v_0. The two springs in **FIGURE EX10.23b** are identical to the spring of Figure EX10.23a. They are compressed the same *total* Δx and used to launch the same block. What is the block's speed now?

FIGURE EX10.23

Section 10.4 Conservation of Energy

Section 10.5 Energy Diagrams

24. || **FIGURE EX10.24** is the potential-energy diagram for a 20 g particle that is released from rest at $x = 1.0$ m.
   a. Will the particle move to the right or to the left?
   b. What is the particle's maximum speed? At what position does it have this speed?
   c. Where are the turning points of the motion?

FIGURE EX10.24

FIGURE EX10.25

25. || **FIGURE EX10.25** is the potential-energy diagram for a 500 g particle that is released from rest at A. What are the particle's speeds at B, C, and D?

26. ‖ In **FIGURE EX10.26**, what is the maximum speed of a 2.0 g particle that oscillates between $x = 2.0$ mm and $x = 8.0$ mm?

FIGURE EX10.26

FIGURE EX10.27

27. ‖ a. In **FIGURE EX10.27**, what minimum speed does a 100 g particle need at point A to reach point B?
    b. What minimum speed does a 100 g particle need at point B to reach point A?

28. ‖ **FIGURE EX10.28** shows the potential energy of a 500 g particle as it moves along the x-axis. Suppose the particle's mechanical energy is 12 J.
    a. Where are the particle's turning points?
    b. What is the particle's speed when it is at $x = 4.0$ m?
    c. What is the particle's maximum speed? At what position or positions does this occur?
    d. Suppose the particle's energy is lowered to 4.0 J. Can the particle ever be at $x = 2.0$ m? At $x = 4.0$ m?

FIGURE EX10.28

29. ‖ In **FIGURE EX10.28**, what is the maximum speed a 200 g particle could have at $x = 2.0$ m and never reach $x = 6.0$ m?

Section 10.6 Force and Potential Energy

30. ‖ A system in which only one particle can move has the potential energy shown in **FIGURE EX10.30**. What is the x-component of the force on the particle at $x = 5$, 15, and 25 cm?

FIGURE EX10.30

FIGURE EX10.31

31. ‖ A system in which only one particle can move has the potential energy shown in **FIGURE EX10.31**. What is the y-component of the force on the particle at $y = 0.5$ m and 4 m?

32. ‖ A particle moving along the y-axis is in a system with potential energy $U = 4y^3$ J, where y is in m. What is the y-component of the force on the particle at $y = 0$ m, 1 m, and 2 m?
CALC

33. ‖ A particle moving along the x-axis is in a system with potential energy $U = 10/x$ J, where x is in m. What is the x-component of the force on the particle at $x = 2$ m, 5 m, and 8 m?
CALC

34. ‖ **FIGURE EX10.34** shows the potential energy of a system in which a particle moves along the x-axis. Draw a graph of the force F_x as a function of position x.

FIGURE EX10.34

Section 10.7 Conservative and Nonconservative Forces

35. ‖ A particle moves from A to D in **FIGURE EX10.35** while experiencing force $\vec{F} = (6\hat{\imath} + 8\hat{\jmath})$ N. How much work does the force do if the particle follows path (a) ABD, (b) ACD, and (c) AD? Is this a conservative force? Explain.

FIGURE EX10.35

36. ‖ A force does work on a 50 g particle as the particle moves along the following straight paths in the xy-plane: 25 J from $(0 \text{ m}, 0 \text{ m})$ to $(5 \text{ m}, 0 \text{ m})$; 35 J from $(0 \text{ m}, 0 \text{ m})$ to $(0 \text{ m}, 5 \text{ m})$; -5 J from $(5 \text{ m}, 0 \text{ m})$ to $(5 \text{ m}, 5 \text{ m})$; -15 J from $(0 \text{ m}, 5 \text{ m})$ to $(5 \text{ m}, 5 \text{ m})$; and 20 J from $(0 \text{ m}, 0 \text{ m})$ to $(5 \text{ m}, 5 \text{ m})$.
    a. Is this a conservative force?
    b. If the zero of potential energy is at the origin, what is the potential energy at $(5 \text{ m}, 5 \text{ m})$?

Section 10.8 The Energy Principle Revisited

37. ‖ A system loses 400 J of potential energy. In the process, it does 400 J of work on the environment and the thermal energy increases by 100 J. Show this process on an energy bar chart.

38. ‖ What is the final kinetic energy of the system for the process shown in **FIGURE EX10.38**?

FIGURE EX10.38

FIGURE EX10.39

39. ‖ How much work is done by the environment in the process shown in **FIGURE EX10.39**? Is energy transferred to the system or from the system to the environment?

40. ‖ A cable with 20.0 N of tension pulls straight up on a 1.50 kg block that is initially at rest. What is the block's speed after being lifted 2.00 m? Solve this problem using work and energy.

Problems

41. ‖ A very slippery ice cube slides in a *vertical* plane around the inside of a smooth, 20-cm-diameter horizontal pipe. The ice cube's speed at the bottom of the circle is 3.0 m/s. What is the ice cube's speed at the top?

42. ‖ A 50 g ice cube can slide up and down a frictionless 30° slope. At the bottom, a spring with spring constant 25 N/m is compressed 10 cm and used to launch the ice cube up the slope. How high does it go above its starting point?

43. ‖ You have been hired to design a spring-launched roller coaster that will carry two passengers per car. The car goes up a 10-m-high hill, then descends 15 m to the track's lowest point. You've determined that the spring can be compressed a maximum of 2.0 m and that a loaded car will have a maximum mass of 400 kg. For safety reasons, the spring constant should be 10% larger than the minimum needed for the car to just make it over the top.
    a. What spring constant should you specify?
    b. What is the maximum speed of a 350 kg car if the spring is compressed the full amount?

44. ‖ It's been a great day of new, frictionless snow. Julie starts at the top of the 60° slope shown in FIGURE P10.44. At the bottom, a circular arc carries her through a 90° turn, and she then launches off a 3.0-m-high ramp. How far horizontally is her touchdown point from the end of the ramp?

25 m

90°

60° ↕3.0 m

FIGURE P10.44

45. ‖ A block of mass m slides down a frictionless track, then around the inside of a circular loop-the-loop of radius R. From what minimum height h must the block start to make it around without falling off? Give your answer as a multiple of R.

46. ‖‖ A 1000 kg safe is 2.0 m above a heavy-duty spring when the rope holding the safe breaks. The safe hits the spring and compresses it 50 cm. What is the spring constant of the spring?

47. ‖‖ You have a ball of unknown mass, a spring with spring constant 950 N/m, and a meter stick. You use various compressions of the spring to launch the ball vertically, then use the meter stick to measure the ball's maximum height above the launch point. Your data are as follows:

Compression (cm)	Height (cm)
2.0	32
3.0	65
4.0	115
5.0	189

Use an appropriate graph of the data to determine the ball's mass.

48. ‖ Sam, whose mass is 75 kg, straps on his skis and starts down a 50-m-high, 20° frictionless slope. A strong headwind exerts a *horizontal* force of 200 N on him as he skies. Use work and energy to find Sam's speed at the bottom.

49. ‖ A horizontal spring with spring constant 100 N/m is compressed 20 cm and used to launch a 2.5 kg box across a frictionless, horizontal surface. After the box travels some distance, the surface becomes rough. The coefficient of kinetic friction of the box on the surface is 0.15. Use work and energy to find how far the box slides across the rough surface before stopping.

50. ‖ Truck brakes can fail if they get too hot. In some mountainous areas, ramps of loose gravel are constructed to stop runaway trucks that have lost their brakes. The combination of a slight upward slope and a large coefficient of rolling resistance as the truck tires sink into the gravel brings the truck safely to a halt. Suppose a gravel ramp slopes upward at 6.0° and the coefficient of rolling friction is 0.40. Use work and energy to find the length of a ramp that will stop a 15,000 kg truck that enters the ramp at 35 m/s (≈ 75 mph).

51. ‖ A freight company uses a compressed spring to shoot 2.0 kg packages up a 1.0-m-high frictionless ramp into a truck, as FIGURE P10.51 shows. The spring constant is 500 N/m and the spring is compressed 30 cm.
    a. What is the speed of the package when it reaches the truck?
    b. A careless worker spills his soda on the ramp. This creates a 50-cm-long sticky spot with a coefficient of kinetic friction 0.30. Will the next package make it into the truck?

1.0 m

50-cm-long sticky spot

FIGURE P10.51   **FIGURE P10.52**

52. ‖ Use work and energy to find an expression for the speed of the block in FIGURE P10.52 just before it hits the floor if (a) the coefficient of kinetic friction for the block on the table is μ_k and (b) the table is frictionless.

53. ‖ a. A 50 g ice cube can slide without friction up and down a 30° slope. The ice cube is pressed against a spring at the bottom of the slope, compressing the spring 10 cm. The spring constant is 25 N/m. When the ice cube is released, what total distance will it travel up the slope before reversing direction?
    b. The ice cube is replaced by a 50 g plastic cube whose coefficient of kinetic friction is 0.20. How far will the plastic cube travel up the slope? Use work and energy.

54. ‖ The spring shown in FIGURE P10.54 is compressed 50 cm and used to launch a 100 kg physics student. The track is frictionless until it starts up the incline. The student's coefficient of kinetic friction on the 30° incline is 0.15.
    a. What is the student's speed just after losing contact with the spring?
    b. How far up the incline does the student go?

$k = 80,000$ N/m

$m = 100$ kg

10 m

30°

FIGURE P10.54

55. ||| Protons and neutrons (together called *nucleons*) are held
CALC together in the nucleus of an atom by a force called the *strong force*. At very small separations, the strong force between two nucleons is larger than the repulsive electrical force between two protons—hence its name. But the strong force quickly weakens as the distance between the protons increases. A well-established model for the potential energy of two nucleons interacting via the strong force is

$$U = U_0[1 - e^{-x/x_0}]$$

where x is the distance between the centers of the two nucleons, x_0 is a constant having the value $x_0 = 2.0 \times 10^{-15}$ m, and $U_0 = 6.0 \times 10^{-11}$ J.

   Quantum effects are essential for a proper understanding of nucleons, but let us innocently consider two neutrons as if they were small, hard, electrically neutral spheres of mass 1.67×10^{-27} kg and diameter 1.0×10^{-15} m. Suppose you hold two neutrons 5.0×10^{-15} m apart, measured between their centers, then release them. What is the speed of each neutron as they crash together? Keep in mind that *both* neutrons are moving.

56. || A 2.6 kg block is attached to a horizontal rope that exerts a
CALC variable force $F_x = (20 - 5x)$ N, where x is in m. The coefficient of kinetic friction between the block and the floor is 0.25. Initially the block is at rest at $x = 0$ m. What is the block's speed when it has been pulled to $x = 4.0$ m?

57. || A system has potential energy
CALC

$$U(x) = x + \sin((2 \text{ rad/m})x)$$

as a particle moves over the range $0 \text{ m} \leq x \leq \pi \text{ m}$.
   a. Where are the equilibrium positions in this range?
   b. For each, is it a point of stable or unstable equilibrium?

58. || A particle that can move along the x-axis is part of a system
CALC with potential energy

$$U(x) = \frac{A}{x^2} - \frac{B}{x}$$

where A and B are positive constants.
   a. Where are the particle's equilibrium positions?
   b. For each, is it a point of stable or unstable equilibrium?

59. || A 100 g particle experiences the one-dimensional, conservative force F_x shown in FIGURE P10.59.

FIGURE P10.59

   a. Let the zero of potential energy be at $x = 0$ m. What is the potential energy at $x = 1.0, 2.0, 3.0,$ and 4.0 m?
      Hint: Use the definition of potential energy and the geometric interpretation of work.
   b. Suppose the particle is shot to the right from $x = 1.0$ m with a speed of 25 m/s. Where is its turning point?

60. || A clever engineer designs a "sprong" that obeys the force law
CALC $F_x = -q(x - x_{eq})^3$, where x_{eq} is the equilibrium position of the end of the sprong and q is the sprong constant. For simplicity, we'll let $x_{eq} = 0$ m. Then $F_x = -qx^3$.

   a. What are the units of q?
   b. Find an expression for the potential energy of a stretched or compressed sprong.
   c. A sprong-loaded toy gun shoots a 20 g plastic ball. What is the launch speed if the sprong constant is 40,000, with the units you found in part a, and the sprong is compressed 10 cm? Assume the barrel is frictionless.

61. || The potential energy for a particle that can move along the
CALC x-axis is $U = Ax^2 + B\sin(\pi x/L)$, where $A, B,$ and L are constants. What is the force on the particle at (a) $x = 0$, (b) $x = L/2$, and (c) $x = L$?

62. || A particle that can move along the x-axis experiences an
CALC interaction force $F_x = (3x^2 - 5x)$ N, where x is in m. Find an expression for the system's potential energy.

63. || An object moving in the xy-plane is subjected to the force
CALC $\vec{F} = (2xy\,\hat{\imath} + x^2\hat{\jmath})$ N, where x and y are in m.
   a. The particle moves from the origin to the point with coordinates (a, b) by moving first along the x-axis to $(a, 0)$, then parallel to the y-axis. How much work does the force do?
   b. The particle moves from the origin to the point with coordinates (a, b) by moving first along the y-axis to $(0, b)$, then parallel to the x-axis. How much work does the force do?
   c. Is this a conservative force?

64. || An object moving in the xy-plane is subjected to the force
CALC $\vec{F} = (2xy\hat{\imath} + 3y\hat{\jmath})$ N, where x and y are in m.
   a. The particle moves from the origin to the point with coordinates (a, b) by moving first along the x-axis to $(a, 0)$, then parallel to the y-axis. How much work does the force do?
   b. The particle moves from the origin to the point with coordinates (a, b) by moving first along the y-axis to $(0, b)$, then parallel to the x-axis. How much work does the force do?
   c. Is this a conservative force?

65. Write a realistic problem for which the energy bar chart shown in FIGURE P10.65 correctly shows the energy at the beginning and end of the problem.

FIGURE P10.65 $K_i + U_{Gi} + U_{Spi} = K_f + U_{Gf} + U_{Spf}$

In Problems 66 through 68 you are given the equation used to solve a problem. For each of these, you are to
   a. Write a realistic problem for which this is the correct equation.
   b. Draw the before-and-after pictorial representation.
   c. Finish the solution of the problem.

66. $\frac{1}{2}(1500 \text{ kg})(5.0 \text{ m/s})^2 + (1500 \text{ kg})(9.80 \text{ m/s}^2)(10 \text{ m})$
   $= \frac{1}{2}(1500 \text{ kg})v_i^2 + (1500 \text{ kg})(9.80 \text{ m/s}^2)(0 \text{ m})$

67. $\frac{1}{2}(0.20 \text{ kg})(2.0 \text{ m/s})^2 + \frac{1}{2}k(0 \text{ m})^2$
   $= \frac{1}{2}(0.20 \text{ kg})(0 \text{ m/s})^2 + \frac{1}{2}k(-0.15 \text{ m})^2$

68. $\frac{1}{2}(0.50 \text{ kg})v_f^2 + (0.50 \text{ kg})(9.80 \text{ m/s}^2)(0 \text{ m})$
   $+ \frac{1}{2}(400 \text{ N/m})(0 \text{ m})^2 = \frac{1}{2}(0.50 \text{ kg})(0 \text{ m/s})^2$
   $+ (0.50 \text{ kg})(9.80 \text{ m/s}^2)((-0.10 \text{ m}) \sin 30°)$
   $+ \frac{1}{2}(400 \text{ N/m})(-0.10 \text{ m})^2$

Challenge Problems

69. ⫴ A pendulum is formed from a small ball of mass m on a string of length L. As **FIGURE CP10.69** shows, a peg is height $h = L/3$ above the pendulum's lowest point. From what minimum angle θ must the pendulum be released in order for the ball to go over the top of the peg without the string going slack?

FIGURE CP10.69

70. ⫴ In a physics lab experiment, a compressed spring launches a 20 g metal ball at a 30° angle. Compressing the spring 20 cm causes the ball to hit the floor 1.5 m below the point at which it leaves the spring after traveling 5.0 m horizontally. What is the spring constant?

71. ⫴ It's your birthday, and to celebrate you're going to make your first bungee jump. You stand on a bridge 100 m above a raging river and attach a 30-m-long bungee cord to your harness. A bungee cord, for practical purposes, is just a long spring, and this cord has a spring constant of 40 N/m. Assume that your mass is 80 kg. After a long hesitation, you dive off the bridge. How far are you above the water when the cord reaches its maximum elongation?

72. ⫴ A 10 kg box slides 4.0 m down the frictionless ramp shown
CALC in **FIGURE CP10.72**, then collides with a spring whose spring constant is 250 N/m.
   a. What is the maximum compression of the spring?
   b. At what compression of the spring does the box have its maximum speed?

FIGURE CP10.72   **FIGURE CP10.73**

73. ⫴ The spring in **FIGURE CP10.73** has a spring constant of 1000 N/m. It is compressed 15 cm, then launches a 200 g block. The horizontal surface is frictionless, but the block's coefficient of kinetic friction on the incline is 0.20. What distance d does the block sail through the air?

74. ⫴ A sled starts from rest at the top of the frictionless, hemispherical, snow-covered hill shown in **FIGURE CP10.74**.
   a. Find an expression for the sled's speed when it is at angle ϕ.
   b. Use Newton's laws to find the maximum speed the sled can have at angle ϕ without leaving the surface.
   c. At what angle ϕ_{max} does the sled "fly off" the hill?

FIGURE CP10.74

11 Impulse and Momentum

An exploding firework is a dramatic event. Nonetheless, the explosion obeys some simple laws of physics.

IN THIS CHAPTER, you will learn to use the concepts of impulse and momentum.

What is momentum?

An object's momentum is the product of its mass and velocity. An object can have a large momentum by having a large mass or a large velocity. Momentum is a vector, and it is especially important to pay attention to the *signs* of the components of momentum.

Same momentum:

Large mass

Large velocity

What is impulse?

A force of short duration is an impulsive force. The impulse J_x that this force delivers to an object is the area under the force-versus-time graph. For time-dependent forces, impulse and momentum are often more useful than Newton's laws.

F_x Impulse = area

$\leftarrow\Delta t\rightarrow$ t

How are impulse and momentum related?

Working with momentum is similar to working with energy. It's important to clearly define the system. The momentum principle says that a system's momentum changes when an impulse is delivered:

$+$

0

$-$ p_{ix} $+$ J_x $=$ p_{fx}

$$\Delta p_x = J_x$$

A momentum bar chart, similar to an energy bar chart, shows this principle graphically.

« LOOKING BACK Section 9.1 Energy overview

Is momentum conserved?

The total momentum of an isolated system is conserved. The particles of an isolated system interact with each other but not with the environment. Regardless of how intense the interactions are, the final momentum equals the initial momentum.

System

Interactions

Environment

« LOOKING BACK Section 10.4 Energy conservation

How does momentum apply to collisions?

One important application of momentum conservation is the study of collisions.

■ In a totally inelastic collision, the objects stick together. Momentum is conserved.

■ In a perfectly elastic collision, the objects bounce apart. Both momentum and energy are conserved.

Where else is momentum used?

This chapter looks at two other important applications of momentum conservation:

■ An explosion is a short interaction that drives two or more objects apart.

■ In rocket propulsion the object's mass is changing continuously.

A tennis ball collides with a racket. Notice that the left side of the ball is flattened.

FIGURE 11.1 A collision.

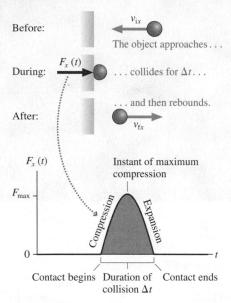

11.1 Momentum and Impulse

A **collision** is a short-duration interaction between two objects. The collision between a tennis ball and a racket, or a baseball and a bat, may seem instantaneous to your eye, but that is a limitation of your perception. A high-speed photograph reveals that the side of the ball is significantly flattened during the collision. It takes time to compress the ball, and more time for the ball to re-expand as it leaves the racket or bat.

The duration of a collision depends on the materials from which the objects are made, but 1 to 10 ms (0.001 to 0.010 s) is fairly typical. This is the time during which the two objects are in contact with each other. The harder the objects, the shorter the contact time. A collision between two steel balls lasts less than 200 microseconds.

FIGURE 11.1 shows an object colliding with a wall. The object approaches with an initial horizontal velocity v_{ix}, experiences a force of duration Δt, and leaves with final velocity v_{fx}. Notice that the object, as in the photo above, *deforms* during the collision. A particle cannot be deformed, so we cannot model colliding objects as particles. Instead, we model a colliding object as an *elastic object* that compresses and then expands, much like a spring. Indeed, that's exactly what happens during a collision at the microscopic level: Molecular bonds compress, store elastic potential energy, then transform some or all of that potential energy back into the kinetic energy of the rebounding object. We'll examine the energy issues of collisions later in this chapter.

The force of a collision is usually very large in comparison to other forces exerted on the object. A large force exerted for a small interval of time is called an **impulsive force.** The graph of Figure 11.1 shows how a typical impulsive force behaves, rapidly growing to a maximum at the instant of maximum compression, then decreasing back to zero. The force is zero before contact begins and after contact ends. Because an impulsive force is a function of time, we will write it as $F_x(t)$.

NOTE Both v_x and F_x are components of vectors and thus have *signs* indicating which way the vectors point.

We can use Newton's second law to find how the object's velocity changes as a result of the collision. Acceleration in one dimension is $a_x = dv_x/dt$, so the second law is

$$ma_x = m\frac{dv_x}{dt} = F_x(t)$$

After multiplying both sides by dt, we can write the second law as

$$m\,dv_x = F_x(t)\,dt \tag{11.1}$$

The force is nonzero only during an interval of time from t_i to $t_f = t_i + \Delta t$, so let's integrate Equation 11.1 over this interval. The velocity changes from v_{ix} to v_{fx} during the collision; thus

$$m\int_{v_i}^{v_f} dv_x = mv_{fx} - mv_{ix} = \int_{t_i}^{t_f} F_x(t)\,dt \tag{11.2}$$

We need some new tools to help us make sense of Equation 11.2.

Momentum

The product of a particle's mass and velocity is called the *momentum* of the particle:

$$\textbf{momentum} = \vec{p} \equiv m\vec{v} \tag{11.3}$$

Momentum, like velocity, is a vector. The units of momentum are kg m/s. The plural of "momentum" is "momenta," from its Latin origin.

The momentum vector $\vec{p}$ is parallel to the velocity vector $\vec{v}$. **FIGURE 11.2** shows that $\vec{p}$, like any vector, can be decomposed into x- and y-components. Equation 11.3, which is a vector equation, is a shorthand way to write the simultaneous equations

$$p_x = mv_x$$
$$p_y = mv_y$$

FIGURE 11.2 The momentum $\vec{p}$ can be decomposed into x- and y-components.

Momentum is a vector pointing in the same direction as the object's velocity.

NOTE One of the most common errors in momentum problems is a failure to use the appropriate signs. The momentum component p_x has the same sign as v_x. Momentum is *negative* for a particle moving to the left (on the x-axis) or down (on the y-axis).

An object can have a large momentum by having either a small mass but a large velocity or a small velocity but a large mass. For example, a 5.5 kg (12 lb) bowling ball rolling at a modest 2 m/s has momentum of magnitude $p = (5.5 \text{ kg})(2 \text{ m/s}) = 11 \text{ kg m/s}$. This is almost exactly the same momentum as a 9 g bullet fired from a high-speed rifle at 1200 m/s.

Newton actually formulated his second law in terms of momentum rather than acceleration:

$$\vec{F} = m\vec{a} = m\frac{d\vec{v}}{dt} = \frac{d(m\vec{v})}{dt} = \frac{d\vec{p}}{dt} \tag{11.4}$$

This statement of the second law, saying that **force is the rate of change of momentum,** is more general than our earlier version $\vec{F} = m\vec{a}$. It allows for the possibility that the mass of the object might change, such as a rocket that is losing mass as it burns fuel.

Returning to Equation 11.2, you can see that mv_{ix} and mv_{fx} are p_{ix} and p_{fx}, the x-component of the particle's momentum before and after the collision. Further, $p_{fx} - p_{ix}$ is Δp_x, the *change* in the particle's momentum. In terms of momentum, Equation 11.2 is

$$\Delta p_x = p_{fx} - p_{ix} = \int_{t_i}^{t_f} F_x(t)\, dt \tag{11.5}$$

Now we need to examine the right-hand side of Equation 11.5.

Impulse

Equation 11.5 tells us that the particle's change in momentum is related to the time integral of the force. Let's define a quantity J_x called the *impulse* to be

$$\textbf{impulse } = J_x \equiv \int_{t_i}^{t_f} F_x(t)\, dt \tag{11.6}$$
$$= \text{area under the } F_x(t) \text{ curve between } t_i \text{ and } t_f$$

Strictly speaking, impulse has units of N s, but you should be able to show that N s are equivalent to kg m/s, the units of momentum.

The interpretation of the integral in Equation 11.6 as an area under a curve is especially important. **FIGURE 11.3a** portrays the impulse graphically. Because the force changes in a complicated way during a collision, it is often useful to describe the collision in terms of an *average* force F_{avg}. As **FIGURE 11.3b** shows, F_{avg} is the height of a rectangle that has the same area, and thus the same impulse, as the real force curve. The impulse exerted during the collision is

$$J_x = F_{avg}\, \Delta t \tag{11.7}$$

Equation 11.2, which we found by integrating Newton's second law, can now be rewritten in terms of impulse and momentum as

$$\Delta p_x = J_x \qquad \text{(momentum principle)} \tag{11.8}$$

This result, called the **momentum principle,** says that **an impulse delivered to an object causes the object's momentum to change.** The momentum p_{fx} "after" an interaction, such as a collision or an explosion, is equal to the momentum p_{ix} "before" the interaction *plus* the impulse that arises from the interaction:

$$p_{fx} = p_{ix} + J_x \tag{11.9}$$

FIGURE 11.3 Looking at the impulse graphically.

(a)

(b)

FIGURE 11.4 The momentum principle helps us understand a rubber ball bouncing off a wall.

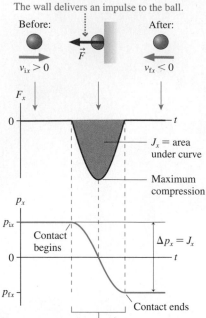

The wall delivers an impulse to the ball.

The impulse changes the ball's momentum.

FIGURE 11.4 illustrates the momentum principle for a rubber ball bouncing off a wall. Notice the signs; they are very important. The ball is initially traveling toward the right, so v_{ix} and p_{ix} are positive. After the bounce, v_{fx} and p_{fx} are negative. The force *on the ball* is toward the left, so F_x is also negative. The graphs show how the force and the momentum change with time.

Although the interaction is very complex, the impulse—the area under the force graph—is all we need to know to find the ball's velocity as it rebounds from the wall. The final momentum is

$$p_{fx} = p_{ix} + J_x = p_{ix} + \text{area under the force curve} \qquad (11.10)$$

and the final velocity is $v_{fx} = p_{fx}/m$. In this example, the area has a negative value.

An Analogy with the Energy Principle

You've probably noticed that there is a similarity between the momentum principle and the energy principle of Chapters 9 and 10. For a system of one object acted on by a force:

$$\text{energy principle:} \qquad \Delta K = W = \int_{x_i}^{x_f} F_x \, dx$$

$$\text{momentum principle:} \qquad \Delta p_x = J_x = \int_{t_i}^{t_f} F_x \, dt \qquad (11.11)$$

In both cases, a force acting on an object changes the state of the system. If the force acts over the spatial interval from x_i to x_f, it does *work* that changes the object's kinetic energy. If the force acts over a time interval from t_i to t_f, it delivers an *impulse* that changes the object's momentum. **FIGURE 11.5** shows that the geometric interpretation of work as the area under the F-versus-x graph parallels an interpretation of impulse as the area under the F-versus-t graph.

FIGURE 11.5 Impulse and work are both the area under a force curve, but it's very important to know what the horizontal axis is.

This does not mean that a force *either* creates an impulse *or* does work but does not do both. Quite the contrary. A force acting on a particle *both* creates an impulse *and* does work, changing both the momentum and the kinetic energy of the particle. Whether you use the energy principle or the momentum principle depends on the question you are trying to answer.

In fact, we can express the kinetic energy in terms of momentum as

$$K = \tfrac{1}{2}mv^2 = \frac{(mv)^2}{2m} = \frac{p^2}{2m} \qquad (11.12)$$

You cannot change a particle's kinetic energy without also changing its momentum.

Momentum Bar Charts

The momentum principle tells us that **impulse transfers momentum to an object.** If an object has 2 kg m/s of momentum, a 1 kg m/s impulse delivered to the object increases its momentum to 3 kg m/s. That is, $p_{fx} = p_{ix} + J_x$.

Just as we did with energy, we can represent this "momentum accounting" with a **momentum bar chart.** For example, the bar chart of **FIGURE 11.6** represents the ball colliding with a wall in Figure 11.4. Momentum bar charts are a tool for visualizing an interaction.

NOTE The vertical scale of a momentum bar chart has no numbers; it can be adjusted to match any problem. However, be sure that all bars in a given problem use a consistent scale.

FIGURE 11.6 A momentum bar chart.

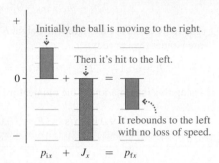

Initially the ball is moving to the right.
Then it's hit to the left.
It rebounds to the left with no loss of speed.

$$p_{ix} + J_x = p_{fx}$$

STOP TO THINK 11.1 The cart's change of momentum is

a. -30 kg m/s
b. -20 kg m/s
c. 0 kg m/s
d. 10 kg m/s
e. 20 kg m/s
f. 30 kg m/s

Before: 10 kg   After:
2 m/s   1 m/s

Solving Impulse and Momentum Problems

Impulse and momentum problems, like energy problems, relate the situation before an interaction to the situation afterward. Consequently, the *before-and-after pictorial representation* remains our primary visualization tool. Let's look at an example.

EXAMPLE 11.1 | Hitting a baseball

A 150 g baseball is thrown with a speed of 20 m/s. It is hit straight back toward the pitcher at a speed of 40 m/s. The interaction force between the ball and the bat is shown in **FIGURE 11.7**. What *maximum* force F_{max} does the bat exert on the ball? What is the *average* force of the bat on the ball?

FIGURE 11.7 The interaction force between the baseball and the bat.

F_{max}

3.0 ms

MODEL Model the baseball as an elastic object and the interaction as a collision.

VISUALIZE **FIGURE 11.8** is a before-and-after pictorial representation. Because F_x is positive (a force to the right), we know the ball was initially moving toward the left and is hit back toward the right. Thus we converted the statements about *speeds* into information about *velocities,* with v_{ix} negative.

SOLVE The momentum principle is

$$\Delta p_x = J_x = \text{area under the force curve}$$

We know the velocities before and after the collision, so we can calculate the ball's momenta:

$$p_{ix} = mv_{ix} = (0.15 \text{ kg})(-20 \text{ m/s}) = -3.0 \text{ kg m/s}$$
$$p_{fx} = mv_{fx} = (0.15 \text{ kg})(40 \text{ m/s}) = 6.0 \text{ kg m/s}$$

FIGURE 11.8 A before-and-after pictorial representation.

Before:
$v_{ix} = -20$ m/s
$m = 0.15$ kg

After:
$v_{fx} = 40$ m/s
Find: F_{max} and F_{avg}

Initially the ball is moving to the left.
Then it's hit to the right.
It rebounds to the right at a higher speed.

$$p_{ix} + J_x = p_{fx}$$

Continued

Thus the *change* in momentum is

$$\Delta p_x = p_{fx} - p_{ix} = 9.0 \text{ kg m/s}$$

The force curve is a triangle with height F_{max} and width 3.0 ms. The area under the curve is

$$J_x = \text{area} = \tfrac{1}{2}(F_{max})(0.0030 \text{ s}) = (F_{max})(0.0015 \text{ s})$$

Using this information in the momentum principle, we have

$$9.0 \text{ kg m/s} = (F_{max})(0.0015 \text{ s})$$

Thus the *maximum* force is

$$F_{max} = \frac{9.0 \text{ kg m/s}}{0.0015 \text{ s}} = 6000 \text{ N}$$

The *average* force, which depends on the collision duration $\Delta t = 0.0030$ s, has the smaller value:

$$F_{avg} = \frac{J_x}{\Delta t} = \frac{\Delta p_x}{\Delta t} = \frac{9.0 \text{ kg m/s}}{0.0030 \text{ s}} = 3000 \text{ N}$$

ASSESS F_{max} is a large force, but quite typical of the impulsive forces during collisions. The main thing to focus on is our new perspective: An impulse changes the momentum of an object.

Other forces often act on an object during a collision or other brief interaction. In Example 11.1, for instance, the baseball is also acted on by gravity. Usually these other forces are *much* smaller than the interaction forces. The 1.5 N weight of the ball is vastly less than the 6000 N force of the bat on the ball. We can reasonably neglect these small forces *during* the brief time of the impulsive force by using what is called the **impulse approximation.**

When we use the impulse approximation, p_{ix} and p_{fx} (and v_{ix} and v_{fx}) are the momenta (and velocities) *immediately* before and *immediately* after the collision. For example, the velocities in Example 11.1 are those of the ball just before and after it collides with the bat. We could then do a follow-up problem, including gravity and drag, to find the ball's speed a second later as the second baseman catches it. We'll look at some two-part examples later in the chapter.

STOP TO THINK 11.2 A 10 g rubber ball and a 10 g clay ball are thrown at a wall with equal speeds. The rubber ball bounces, the clay ball sticks. Which ball delivers a larger impulse to the wall?

a. The clay ball delivers a larger impulse because it sticks.
b. The rubber ball delivers a larger impulse because it bounces.
c. They deliver equal impulses because they have equal momenta.
d. Neither delivers an impulse to the wall because the wall doesn't move.

11.2 Conservation of Momentum

The momentum principle was derived from Newton's second law and is really just an alternative way of looking at single-particle dynamics. To discover the real power of momentum for problem solving, we need also to invoke Newton's third law, which will lead us to one of the most important principles in physics: conservation of momentum.

FIGURE 11.9 shows two objects with initial velocities $(v_{ix})_1$ and $(v_{ix})_2$. The objects collide, then bounce apart with final velocities $(v_{fx})_1$ and $(v_{fx})_2$. The forces during the collision, as the objects are interacting, are the action/reaction pair $\vec{F}_{1 \text{ on } 2}$ and $\vec{F}_{2 \text{ on } 1}$. For now, we'll continue to assume that the motion is one dimensional along the x-axis.

NOTE The notation, with all the subscripts, may seem excessive. But there are two objects, and each has an initial and a final velocity, so we need to distinguish among four different velocities.

Newton's second law for each object *during* the collision is

$$\frac{d(p_x)_1}{dt} = (F_x)_{2 \text{ on } 1}$$

$$\frac{d(p_x)_2}{dt} = (F_x)_{1 \text{ on } 2} = -(F_x)_{2 \text{ on } 1} \tag{11.13}$$

We made explicit use of Newton's third law in the second equation.

FIGURE 11.9 A collision between two objects.

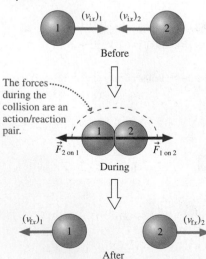

The forces during the collision are an action/reaction pair.

Although Equations 11.13 are for two different objects, suppose—just to see what happens—we were to *add* these two equations. If we do, we find that

$$\frac{d(p_x)_1}{dt} + \frac{d(p_x)_2}{dt} = \frac{d}{dt}\left[(p_x)_1 + (p_x)_2 \right] = (F_x)_{2 \text{ on } 1} + (-(F_x)_{2 \text{ on } 1}) = 0 \quad (11.14)$$

If the time derivative of the quantity $(p_x)_1 + (p_x)_2$ is zero, it must be the case that

$$(p_x)_1 + (p_x)_2 = \text{constant} \quad (11.15)$$

Equation 11.15 is a conservation law! If $(p_x)_1 + (p_x)_2$ is a constant, then the sum of the momenta *after* the collision equals the sum of the momenta *before* the collision. That is,

$$(p_{fx})_1 + (p_{fx})_2 = (p_{ix})_1 + (p_{ix})_2 \quad (11.16)$$

Furthermore, this equality is independent of the interaction force. We don't need to know *anything* about $\vec{F}_{1 \text{ on } 2}$ and $\vec{F}_{2 \text{ on } 1}$ to make use of Equation 11.16.

As an example, **FIGURE 11.10** is a before-and-after pictorial representation of two equal-mass train cars colliding and coupling. Equation 11.16 relates the momenta of the cars after the collision to their momenta before the collision:

$$m_1(v_{fx})_1 + m_2(v_{fx})_2 = m_1(v_{ix})_1 + m_2(v_{ix})_2$$

Initially, car 1 is moving with velocity $(v_{ix})_1 = v_i$ while car 2 is at rest. Afterward, they roll together with the common final velocity v_f. Furthermore, $m_1 = m_2 = m$. With this information, the momentum equation is

$$mv_f + mv_f = mv_i + 0$$

The mass cancels, and we find that the train cars' final velocity is $v_f = \frac{1}{2}v_i$. That is, we can predict that the final speed is exactly half the initial speed of car 1 without knowing anything about the complex interaction between the two cars as they collide.

FIGURE 11.10 Two colliding train cars.

Before:
$(v_{ix})_1 = v_i$ $(v_{ix})_2 = 0$
$m_1 = m$ $m_2 = m$

After:
$(v_{fx})_1 = (v_{fx})_2 = v_f$
m m

Systems of Particles

Equation 11.16 illustrates the idea of a conservation law for momentum, but it was derived for the specific case of two particles colliding in one dimension. Our goal is to develop a more general law of conservation of momentum, a law that will be valid in three dimensions and that will work for any type of interaction. The next few paragraphs are fairly mathematical, so you might want to begin by looking ahead to Equations 11.24 and the statement of the law of conservation of momentum to see where we're heading.

Our study of energy in the last two chapters has emphasized the importance of having a clearly defined system. The same is true for momentum. Consider a system consisting of N particles. **FIGURE 11.11** shows a simple case where $N = 3$, but N could be anything. The particles might be large entities (cars, baseballs, etc.), or they might be the microscopic atoms in a gas. We can identify each particle by an identification number k. Every particle in the system *interacts* with every other particle via action/reaction pairs of forces $\vec{F}_{j \text{ on } k}$ and $\vec{F}_{k \text{ on } j}$. In addition, every particle is subjected to possible *external forces* $\vec{F}_{\text{ext on } k}$ from agents outside the system.

If particle k has velocity $\vec{v}_k$, its momentum is $\vec{p}_k = m_k\vec{v}_k$. We define the **total momentum** $\vec{P}$ of the system as the vector sum

$$\vec{P} = \text{total momentum} = \vec{p}_1 + \vec{p}_2 + \vec{p}_3 + \cdots + \vec{p}_N = \sum_{k=1}^{N} \vec{p}_k \quad (11.17)$$

That is the total momentum *of the system* is the vector sum of the individual momenta.

FIGURE 11.11 A system of particles.

External force

External force

The total momentum of the rocket + gases system is conserved, so the rocket accelerates forward as the gases are expelled backward.

The time derivative of $\vec{P}$ tells us how the total momentum of the system changes with time:

$$\frac{d\vec{P}}{dt} = \sum_k \frac{d\vec{p}_k}{dt} = \sum_k \vec{F}_k \qquad (11.18)$$

where we used Newton's second law for each particle in the form $\vec{F}_k = d\vec{p}_k/dt$, which was Equation 11.4.

The net force acting on particle k can be divided into *external forces,* from outside the system, and *interaction forces* due to all the other particles in the system:

$$\vec{F}_k = \sum_{j \neq k} \vec{F}_{j\,\text{on}\,k} + \vec{F}_{\text{ext on}\,k} \qquad (11.19)$$

The restriction $j \neq k$ expresses the fact that particle k does not exert a force on itself. Using this in Equation 11.18 gives the rate of change of the total momentum $\vec{P}$ of the system:

$$\frac{d\vec{P}}{dt} = \sum_k \sum_{j \neq k} \vec{F}_{j\,\text{on}\,k} + \sum_k \vec{F}_{\text{ext on}\,k} \qquad (11.20)$$

The double sum on $\vec{F}_{j\,\text{on}\,k}$ adds *every* interaction force within the system. But the interaction forces come in action/reaction pairs, with $\vec{F}_{k\,\text{on}\,j} = -\vec{F}_{j\,\text{on}\,k}$, so $\vec{F}_{k\,\text{on}\,j} + \vec{F}_{j\,\text{on}\,k} = \vec{0}$. Consequently, **the sum of all the interaction forces is zero.** As a result, Equation 11.20 becomes

$$\frac{d\vec{P}}{dt} = \sum_k \vec{F}_{\text{ext on}\,k} = \vec{F}_{\text{net}} \qquad (11.21)$$

where $\vec{F}_{\text{net}}$ is the net force exerted on the system by agents outside the system. But this is just Newton's second law written for the system as a whole! That is, **the rate of change of the total momentum of the system is equal to the net force applied to the system.**

Equation 11.21 has two very important implications. First, we can analyze the motion of the system as a whole without needing to consider interaction forces between the particles that make up the system. In fact, we have been using this idea all along as an *assumption* of the particle model. When we treat cars and rocks and baseballs as particles, we assume that the internal forces between the atoms—the forces that hold the object together—do not affect the motion of the object as a whole. Now we have *justified* that assumption.

Isolated Systems

The second implication of Equation 11.21, and the more important one from the perspective of this chapter, applies to an isolated system. In Chapter 10, we defined an *isolated system* as one that is not influenced or altered by external forces from the environment. For momentum, that means a system on which the *net* external force is zero: $\vec{F}_{\text{net}} = \vec{0}$. That is, an isolated system is one on which there are *no* external forces or for which the external forces are balanced and add to zero.

For an isolated system, Equation 11.21 is simply

$$\frac{d\vec{P}}{dt} = \vec{0} \qquad \text{(isolated system)} \qquad (11.22)$$

In other words, **the *total* momentum of an isolated system does not change.** The total momentum $\vec{P}$ remains constant, *regardless* of whatever interactions are going on *inside* the system. The importance of this result is sufficient to elevate it to a law of nature, alongside Newton's laws.

Law of conservation of momentum The total momentum $\vec{P}$ of an isolated system is a constant. Interactions within the system do not change the system's total momentum. Mathematically, the law of conservation of momentum is

$$\vec{P}_f = \vec{P}_i \qquad (11.23)$$

The total momentum *after* an interaction is equal to the total momentum *before* the interaction. Because Equation 11.23 is a vector equation, the equality is true for each of the components of the momentum vector. That is,

$$(p_{fx})_1 + (p_{fx})_2 + (p_{fx})_3 + \cdots = (p_{ix})_1 + (p_{ix})_2 + (p_{ix})_3 + \cdots$$
$$(p_{fy})_1 + (p_{fy})_2 + (p_{fy})_3 + \cdots = (p_{iy})_1 + (p_{iy})_2 + (p_{iy})_3 + \cdots \qquad (11.24)$$

The *x*-equation is an extension of Equation 11.16 to *N* interacting particles.

NOTE It is worth emphasizing the critical role of Newton's third law. The law of conservation of momentum is a direct consequence of the fact that interactions within an isolated system are action/reaction pairs.

EXAMPLE 11.2 | A glider collision

A 250 g air-track glider is pushed across a level track toward a 500 g glider that is at rest. **FIGURE 11.12** shows a position-versus-time graph of the 250 g glider as recorded by a motion detector. Best-fit lines have been found. What is the speed of the 500 g glider after the collision?

FIGURE 11.12 Position graph of the 250 g glider.

MODEL Let the system be the two gliders. The gliders interact with each other, but the external forces (normal force and gravity) balance to make $\vec{F}_{net} = \vec{0}$. Thus the gliders form an isolated system and their total momentum is conserved.

VISUALIZE FIGURE 11.13 is a before-and-after pictorial representation. The graph of Figure 11.12 tells us that the 250 g glider initially moves to the right, collides at $t = 1.0$ s, then rebounds to the left (decreasing *x*). Note that the best-fit lines are written as a generic $y = \ldots$, which is what you would see in data-analysis software.

SOLVE Conservation of momentum for this one-dimensional problem requires that the final momentum equal the initial momentum: $P_{fx} = P_{ix}$. In terms of the individual components, conservation of momentum is

$$(p_{fx})_1 + (p_{fx})_2 = (p_{ix})_1 + (p_{ix})_2$$

Each momentum is mv_x, so conservation of momentum in terms of velocities is

FIGURE 11.13 Before-and-after representation of a collision.

$$m_1(v_{fx})_1 + m_2(v_{fx})_2 = m_1(v_{ix})_1 + m_2(v_{ix})_2 = m_1(v_{ix})_1$$

where, in the last step, we used $(v_{ix})_2 = 0$ for the 500 g glider. Solving for the heavier glider's final velocity gives

$$(v_{fx})_2 = \frac{m_1}{m_2}\left[(v_{ix})_1 - (v_{fx})_1\right]$$

From Chapter 2 kinematics, the velocities of the 250 g glider before and after the collision are the slopes of the position-versus-time graph. Referring to Figure 11.12, we see that $(v_{ix})_1 = 0.75$ m/s and $(v_{fx})_1 = -0.21$ m/s. The latter is negative because the rebound motion is to the left. Thus

$$(v_{fx})_2 = \frac{250 \text{ g}}{500 \text{ g}}\left[0.75 \text{ m/s} - (-0.21 \text{ m/s})\right] = 0.48 \text{ m/s}$$

The 500 g glider moves away from the collision at 0.48 m/s.

ASSESS The 500 g glider has twice the mass of the glider that was pushed, so a somewhat smaller speed seems reasonable. Paying attention to the *signs*—which are positive and which negative—was very important for reaching a correct answer. We didn't convert the masses to kilograms because only the mass *ratio* of 0.50 was needed.

A Strategy for Conservation of Momentum Problems

PROBLEM-SOLVING STRATEGY 11.1

Conservation of momentum

MODEL Clearly define the *system*.

■ If possible, choose a system that is isolated ($\vec{F}_{net} = \vec{0}$) or within which the interactions are sufficiently short and intense that you can ignore external forces for the duration of the interaction (the impulse approximation). Momentum is conserved.

■ If it's not possible to choose an isolated system, try to divide the problem into parts such that momentum is conserved during one segment of the motion. Other segments of the motion can be analyzed using Newton's laws or conservation of energy.

VISUALIZE Draw a before-and-after pictorial representation. Define symbols that will be used in the problem, list known values, and identify what you're trying to find.

SOLVE The mathematical representation is based on the law of conservation of momentum: $\vec{P}_f = \vec{P}_i$. In component form, this is

$$(p_{fx})_1 + (p_{fx})_2 + (p_{fx})_3 + \cdots = (p_{ix})_1 + (p_{ix})_2 + (p_{ix})_3 + \cdots$$
$$(p_{fy})_1 + (p_{fy})_2 + (p_{fy})_3 + \cdots = (p_{iy})_1 + (p_{iy})_2 + (p_{iy})_3 + \cdots$$

ASSESS Check that your result has correct units and significant figures, is reasonable, and answers the question.

Exercise 17

EXAMPLE 11.3 | Rolling away

Bob sees a stationary cart 8.0 m in front of him. He decides to run to the cart as fast as he can, jump on, and roll down the street. Bob has a mass of 75 kg and the cart's mass is 25 kg. If Bob accelerates at a steady 1.0 m/s², what is the cart's speed just after Bob jumps on?

MODEL This is a two-part problem. First Bob accelerates across the ground. Then Bob lands on and sticks to the cart, a "collision" between Bob and the cart. The interaction forces between Bob and the cart (i.e., friction) act only over the fraction of a second it takes Bob's feet to become stuck to the cart. Using the impulse approximation allows the system Bob + cart to be treated as an isolated system during the brief interval of the "collision," and thus the total momentum of Bob + cart is conserved during this interaction. But the system Bob + cart is *not* an isolated system for the entire problem because Bob's initial acceleration has nothing to do with the cart.

VISUALIZE Our strategy is to divide the problem into an *acceleration* part, which we can analyze using kinematics, and a *collision* part, which we can analyze with momentum conservation. The pictorial representation of **FIGURE 11.14** includes information about both parts. Notice that Bob's velocity $(v_{1x})_B$ at the end of his run is his "before" velocity for the collision.

SOLVE The first part of the mathematical representation is kinematics. We don't know how long Bob accelerates, but we do know his acceleration and the distance. Thus

$$(v_{1x})_B^2 = (v_{0x})_B^2 + 2a_x \Delta x = 2a_x x_1$$

His velocity after accelerating for 8.0 m is

$$(v_{1x})_B = \sqrt{2a_x x_1} = 4.0 \text{ m/s}$$

The second part of the problem, the collision, uses conservation of momentum: $P_{2x} = P_{1x}$. Equation 11.24 is

$$m_B(v_{2x})_B + m_C(v_{2x})_C = m_B(v_{1x})_B + m_C(v_{1x})_C = m_B(v_{1x})_B$$

where we've used $(v_{1x})_C = 0$ m/s because the cart starts at rest. In this problem, Bob and the cart move together at the end with a common velocity, so we can replace both $(v_{2x})_B$ and $(v_{2x})_C$ with simply v_{2x}. Solving for v_{2x}, we find

$$v_{2x} = \frac{m_B}{m_B + m_C}(v_{1x})_B = \frac{75 \text{ kg}}{100 \text{ kg}} \times 4.0 \text{ m/s} = 3.0 \text{ m/s}$$

The cart's speed is 3.0 m/s immediately after Bob jumps on.

FIGURE 11.14 Pictorial representation of Bob and the cart.

Notice how easy this was! No forces, no acceleration constraints, no simultaneous equations. Why didn't we think of this before? Conservation laws are indeed powerful, but they can answer only certain questions. Had we wanted to know how far Bob slid across the cart before sticking to it, how long the slide took, or what the cart's acceleration was during the collision, we would not have been able to answer such questions on the basis of the conservation law. There is a price to pay for finding a simple connection between before and after, and that price is the loss of information about the details of the interaction. If we are satisfied with knowing only about before and after, then conservation laws are a simple and straightforward way to proceed. But many problems *do* require us to understand the interaction, and for these there is no avoiding Newton's laws.

It Depends on the System

The first step in the problem-solving strategy asks you to clearly define the system. **The goal is to choose a system whose momentum will be conserved.** Even then, it is the *total* momentum of the system that is conserved, not the momenta of the individual objects within the system.

As an example, consider what happens if you drop a rubber ball and let it bounce off a hard floor. Is momentum conserved? You might be tempted to answer yes because the ball's rebound speed is very nearly equal to its impact speed. But there are two errors in this reasoning.

First, momentum depends on *velocity*, not speed. The ball's velocity and momentum change sign during the collision. Even if their magnitudes are equal, the ball's momentum after the collision is *not* equal to its momentum before the collision.

But more important, we haven't defined the system. The momentum of what? Whether or not momentum is conserved depends on the system. FIGURE 11.15 shows two different choices of systems. In Figure 11.15a, where the ball itself is chosen as the system, the gravitational force of the earth on the ball is an external force. This force causes the ball to accelerate toward the earth, changing the ball's momentum. When the ball hits, the force of the floor on the ball is also an external force. The impulse of $\vec{F}_{\text{floor on ball}}$ changes the ball's momentum from "down" to "up" as the ball bounces. The momentum of this system is most definitely *not* conserved.

Figure 11.15b shows a different choice. Here the system is ball + earth. Now the gravitational forces and the impulsive forces of the collision are interactions *within* the system. This is an isolated system, so the *total* momentum $\vec{P} = \vec{p}_{\text{ball}} + \vec{p}_{\text{earth}}$ is conserved.

In fact, the total momentum (in this reference frame) is $\vec{P} = \vec{0}$ because both the ball and the earth are initially at rest. The ball accelerates toward the earth after you release it, while the earth—due to Newton's third law—accelerates toward the ball in such a way that their individual momenta are always equal but opposite.

FIGURE 11.15 Whether or not momentum is conserved as a ball falls to earth depends on your choice of the system.

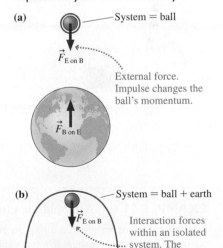

(a)   System = ball

$\vec{F}_{\text{E on B}}$

External force. Impulse changes the ball's momentum.

$\vec{F}_{\text{B on E}}$

(b)   System = ball + earth

$\vec{F}_{\text{E on B}}$

Interaction forces within an isolated system. The system's total momentum is conserved.

$\vec{F}_{\text{B on E}}$

Why don't we notice the earth "leaping up" toward us each time we drop something? Because of the earth's enormous mass relative to everyday objects—roughly 10^{25} times larger. Momentum is the product of mass and velocity, so the earth would need an "upward" speed of only about 10^{-25} m/s to match the momentum of a typical falling object. At that speed, it would take 300 million years for the earth to move the diameter of an atom! The earth does, indeed, have a momentum equal and opposite to that of the ball, but we'll never notice it.

STOP TO THINK 11.3 Objects A and C are made of different materials, with different "springiness," but they have the same mass and are initially at rest. When ball B collides with object A, the ball ends up at rest. When ball B is thrown with the same speed and collides with object C, the ball rebounds to the left. Compare the velocities of A and C after the collisions. Is v_A greater than, equal to, or less than v_C?

11.3 Collisions

Collisions can have different possible outcomes. A rubber ball dropped on the floor bounces, but a ball of clay sticks to the floor without bouncing. A golf club hitting a golf ball causes the ball to rebound away from the club, but a bullet striking a block of wood embeds itself in the block.

FIGURE 11.16 An inelastic collision.

Inelastic Collisions

A collision in which the two objects stick together and move with a common final velocity is called a **perfectly inelastic collision.** The clay hitting the floor and the bullet embedding itself in the wood are examples of perfectly inelastic collisions. Other examples include railroad cars coupling together upon impact and darts hitting a dart board. As **FIGURE 11.16** shows, the key to analyzing a perfectly inelastic collision is the fact that **the two objects have a common final velocity.**

A system consisting of the two colliding objects is isolated, so its total momentum is conserved. However, mechanical energy is *not* conserved because some of the initial kinetic energy is transformed into thermal energy during the collision.

EXAMPLE 11.4 | An inelastic glider collision

In a laboratory experiment, a 200 g air-track glider and a 400 g air-track glider are pushed toward each other from opposite ends of the track. The gliders have Velcro tabs on the front and will stick together when they collide. The 200 g glider is pushed with an initial speed of 3.0 m/s. The collision causes it to reverse direction at 0.40 m/s. What was the initial speed of the 400 g glider?

MODEL Define the system to be the two gliders. This is an isolated system, so its total momentum is conserved in the collision. The gliders stick together, so this is a perfectly inelastic collision.

VISUALIZE **FIGURE 11.17** shows a pictorial representation. We've chosen to let the 200 g glider (glider 1) start out moving to the right, so $(v_{ix})_1$ is a positive 3.0 m/s. The gliders move to the left after the collision, so their common final velocity is $v_{fx} = -0.40$ m/s.

SOLVE The law of conservation of momentum, $P_{fx} = P_{ix}$, is

$$(m_1 + m_2)v_{fx} = m_1(v_{ix})_1 + m_2(v_{ix})_2$$

where we made use of the fact that the combined mass $m_1 + m_2$ moves together after the collision. We can easily solve for the initial velocity of the 400 g glider:

$$(v_{ix})_2 = \frac{(m_1 + m_2)v_{fx} - m_1(v_{ix})_1}{m_2}$$

$$= \frac{(0.60 \text{ kg})(-0.40 \text{ m/s}) - (0.20 \text{ kg})(3.0 \text{ m/s})}{0.40 \text{ kg}} = -2.1 \text{ m/s}$$

The negative sign indicates that the 400 g glider started out moving to the left. The initial *speed* of the glider, which we were asked to find, is 2.1 m/s.

FIGURE 11.17 The before-and-after pictorial representation of an inelastic collision.

Find: $(v_{ix})_2$

EXAMPLE 11.5 | A ballistic pendulum

A 10 g bullet is fired into a 1200 g wood block hanging from a 150-cm-long string. The bullet embeds itself into the block, and the block then swings out to an angle of 40°. What was the speed of the bullet? (This is called a *ballistic pendulum.*)

MODEL This is a two-part problem. Part one, the impact of the bullet on the block, is an inelastic collision. For this part, we define the system to be bullet + block. Momentum is conserved, but mechanical energy is not because some of the energy is transformed into thermal energy. For part two, the subsequent swing, mechanical energy is conserved for the system bullet + block + earth (there's no friction). The *total* momentum is conserved, including the momentum of the earth, but that's not helpful. The momentum of the block with the bullet—which is all that we can calculate—is not conserved because the block is acted on by the external forces of tension and gravity.

VISUALIZE FIGURE 11.18 is a pictorial representation in which we've identified before-and-after quantities for both the collision and the swing.

FIGURE 11.18 A ballistic pendulum is used to measure the speed of a bullet.

SOLVE The momentum conservation equation $P_f = P_i$ applied to the inelastic collision gives

$$(m_W + m_B)v_{1x} = m_W(v_{0x})_W + m_B(v_{0x})_B$$

The wood block is initially at rest, with $(v_{0x})_W = 0$, so the bullet's velocity is

$$(v_{0x})_B = \frac{m_W + m_B}{m_B}v_{1x}$$

where v_{1x} is the velocity of the block + bullet *immediately* after the collision, as the pendulum begins to swing. If we can determine v_{1x} from an analysis of the swing, then we will be able to calculate the speed of the bullet. Turning our attention to the swing, the energy conservation equation $K_f + U_{Gf} = K_i + U_{Gi}$ is

$$\tfrac{1}{2}(m_W + m_B)v_2^2 + (m_W + m_B)gy_2 = \tfrac{1}{2}(m_W + m_B)v_1^2 + (m_W + m_B)gy_1$$

We used the *total* mass $(m_W + m_B)$ of the block and embedded bullet, but notice that it cancels out. We also dropped the x-subscript on v_1 because for energy calculations we need only speed, not velocity. The speed is zero at the top of the swing ($v_2 = 0$), and we've defined the y-axis such that $y_1 = 0$ m. Thus

$$v_1 = \sqrt{2gy_2}$$

The initial speed is found simply from the maximum height of the swing. You can see from the geometry of Figure 11.18 that

$$y_2 = L - L\cos\theta = L(1 - \cos\theta) = 0.351 \text{ m}$$

With this, the initial velocity of the pendulum, immediately after the collision, is

$$v_{1x} = v_1 = \sqrt{2gy_2} = \sqrt{2(9.80 \text{ m/s}^2)(0.351 \text{ m})} = 2.62 \text{ m/s}$$

Having found v_{1x} from an energy analysis of the swing, we can now calculate that the speed of the bullet was

$$(v_{0x})_B = \frac{m_W + m_B}{m_B}v_{1x} = \frac{1.210 \text{ kg}}{0.010 \text{ kg}} \times 2.62 \text{ m/s} = 320 \text{ m/s}$$

ASSESS It would have been very difficult to solve this problem using Newton's laws, but it yielded to a straightforward analysis based on the concepts of momentum and energy.

STOP TO THINK 11.4 The two particles are both moving to the right. Particle 1 catches up with particle 2 and collides with it. The particles stick together and continue on with velocity v_f. Which of these statements is true?

a. v_f is greater than v_1.
b. $v_f = v_1$
c. v_f is greater than v_2 but less than v_1.
d. $v_f = v_2$
e. v_f is less than v_2.
f. Can't tell without knowing the masses.

Before   After

A perfectly elastic collision conserves both momentum and mechanical energy.

FIGURE 11.19 A perfectly elastic collision.

Before: K_i

Energy is stored in compressed bonds, then released as the bonds re-expand.

During: ①②

After: $K_f = K_i$

Elastic Collisions

In an inelastic collision, some of the mechanical energy is dissipated inside the objects as thermal energy and not all of the kinetic energy is recovered. We're now interested in "perfect bounce" collisions in which kinetic energy is stored as elastic potential energy in compressed molecular bonds, and then *all* of the stored energy is transformed back into the post-collision kinetic energy of the objects. A collision in which mechanical energy is conserved is called a **perfectly elastic collision.** A perfectly elastic collision is an idealization, like a frictionless surface, but collisions between two very hard objects, such as two billiard balls or two steel balls, come close to being perfectly elastic.

FIGURE 11.19 shows a head-on, perfectly elastic collision of a ball of mass m_1, having initial velocity $(v_{ix})_1$, with a ball of mass m_2 that is initially at rest. The balls' velocities after the collision are $(v_{fx})_1$ and $(v_{fx})_2$. These are velocities, not speeds, and have signs. Ball 1, in particular, might bounce backward and have a negative value for $(v_{fx})_1$.

The collision must obey two conservation laws: conservation of momentum (obeyed in any collision) and conservation of mechanical energy (because the collision is perfectly elastic). Although the energy is transformed into potential energy during the collision, the mechanical energy before and after the collision is purely kinetic energy. Thus

momentum conservation: $\quad m_1(v_{fx})_1 + m_2(v_{fx})_2 = m_1(v_{ix})_1 \quad$ (11.25)

energy conservation: $\quad \frac{1}{2}m_1(v_{fx})_1^2 + \frac{1}{2}m_2(v_{fx})_2^2 = \frac{1}{2}m_1(v_{ix})_1^2 \quad$ (11.26)

Momentum conservation alone is not sufficient to analyze the collision because there are two unknowns: the two final velocities. Energy conservation provides the additional information that we need. Isolating $(v_{fx})_1$ in Equation 11.25 gives

$$(v_{fx})_1 = (v_{ix})_1 - \frac{m_2}{m_1}(v_{fx})_2 \quad (11.27)$$

We substitute this into Equation 11.26:

$$\frac{1}{2}m_1\left[(v_{ix})_1 - \frac{m_2}{m_1}(v_{fx})_2\right]^2 + \frac{1}{2}m_2(v_{fx})_2^2 = \frac{1}{2}m_1(v_{ix})_1^2$$

With a bit of algebra, this can be rearranged to give

$$(v_{fx})_2\left[\left(1 + \frac{m_2}{m_1}\right)(v_{fx})_2 - 2(v_{ix})_1\right] = 0 \quad (11.28)$$

One possible solution to this equation is seen to be $(v_{fx})_2 = 0$. However, this solution is of no interest; it is the case where ball 1 misses ball 2. The other solution is

$$(v_{fx})_2 = \frac{2m_1}{m_1 + m_2}(v_{ix})_1$$

which, finally, can be substituted back into Equation 11.27 to yield $(v_{fx})_1$. The complete solution is

$$(v_{fx})_1 = \frac{m_1 - m_2}{m_1 + m_2}(v_{ix})_1 \qquad (v_{fx})_2 = \frac{2m_1}{m_1 + m_2}(v_{ix})_1 \qquad (11.29)$$

(perfectly elastic collision with ball 2 initially at rest)

Equations 11.29 allow us to compute the final velocity of each ball. These equations are a little difficult to interpret, so let us look at the three special cases shown in FIGURE 11.20.

Case a: $m_1 = m_2$. This is the case of one billiard ball striking another of equal mass. For this case, Equations 11.29 give

$$v_{f1} = 0 \qquad v_{f2} = v_{i1}$$

Case b: $m_1 \gg m_2$. This is the case of a bowling ball running into a Ping-Pong ball. We do not want an exact solution here, but an approximate solution for the limiting case that $m_1 \to \infty$. Equations 11.29 in this limit give

$$v_{f1} \approx v_{i1} \qquad v_{f2} \approx 2v_{i1}$$

Case c: $m_1 \ll m_2$. Now we have the reverse case of a Ping-Pong ball colliding with a bowling ball. Here we are interested in the limit $m_1 \to 0$, in which case Equations 11.29 become

$$v_{f1} \approx -v_{i1} \qquad v_{f2} \approx 0$$

These cases agree well with our expectations and give us confidence that Equations 11.29 accurately describe a perfectly elastic collision.

Using Reference Frames

Equations 11.29 assumed that ball 2 was at rest prior to the collision. Suppose, however, you need to analyze the perfectly elastic collision that is just about to take place in FIGURE 11.21. What are the direction and speed of each ball after the collision? You could solve the simultaneous momentum and energy equations, but the mathematics becomes quite messy when both balls have an initial velocity. Fortunately, there's an easier way.

You already know the answer—Equations 11.29—when ball 2 is initially at rest. And in Chapter 4 you learned the Galilean transformation of velocity. This transformation relates an object's velocity as measured in one reference frame to its velocity in a different reference frame that moves with respect to the first. The Galilean transformation provides an elegant and straightforward way to analyze the collision of Figure 11.21.

TACTICS BOX 11.1   (MP)

Analyzing elastic collisions

❶ Use the Galilean transformation to transform the initial velocities of balls 1 and 2 from the "lab frame" to a reference frame in which ball 2 is at rest.
❷ Use Equations 11.29 to determine the outcome of the collision in the frame where ball 2 is initially at rest.
❸ Transform the final velocities back to the "lab frame."

FIGURE 11.22a on the next page shows the situation, just before the collision, in the lab frame L. Ball 1 has initial velocity $(v_{ix})_{1L} = 2.0$ m/s. Recall from Chapter 4 that the subscript notation means "velocity of ball 1 relative to the lab frame L." Because ball 2 is moving to the left, it has $(v_{ix})_{2L} = -3.0$ m/s. We would like to observe the collision from a reference frame in which ball 2 is at rest. That will be true if we choose a moving reference frame M that travels alongside ball 2 with the same velocity: $(v_x)_{ML} = -3.0$ m/s.

FIGURE 11.20 Three special elastic collisions.

Case a: $m_1 = m_2$

Ball 1 stops. Ball 2 goes forward with $v_{f2} = v_{i1}$.

Case b: $m_1 \gg m_2$

Ball 1 hardly slows down. Ball 2 is knocked forward at $v_{f2} \approx 2v_{i1}$.

Case c: $m_1 \ll m_2$

Ball 1 bounces off ball 2 with almost no loss of speed. Ball 2 hardly moves.

FIGURE 11.21 A perfectly elastic collision in which both balls have an initial velocity.

2.0 m/s   3.0 m/s

$m_1 = 200$ g $m_2 = 100$ g

FIGURE 11.22 The collision seen in two reference frames: the lab frame L and a moving frame M in which ball 2 is initially at rest.

(a)

$(v_{ix})_{2L} = -3.0$ m/s

$(v_{ix})_{1L} = 2.0$ m/s

Frame M moves with ball 2.

$(v_x)_{ML} = -3.0$ m/s

The collision seen in the lab frame L.

(b)

Ball 2 is at rest in M.

$(v_{ix})_{1M} = 5.0$ m/s $(v_{ix})_{2M} = 0$ m/s

The collision seen in moving frame M.

We first need to transform the balls' velocities from the lab frame to the moving reference frame. From Chapter 4, the Galilean transformation of velocity for an object O is

$$(v_x)_{OM} = (v_x)_{OL} + (v_x)_{LM} \qquad (11.30)$$

That is, O's velocity in reference frame M is its velocity in reference frame L plus the velocity of frame L relative to frame M. Because reference frame M is moving to the left relative to L with $(v_x)_{ML} = -3.0$ m/s, reference frame L is moving to the right relative to M with $(v_x)_{LM} = +3.0$ m/s. Applying the transformation to the two initial velocities gives

$$(v_{ix})_{1M} = (v_{ix})_{1L} + (v_x)_{LM} = 2.0 \text{ m/s} + 3.0 \text{ m/s} = 5.0 \text{ m/s}$$
$$(v_{ix})_{2M} = (v_{ix})_{2L} + (v_x)_{LM} = -3.0 \text{ m/s} + 3.0 \text{ m/s} = 0 \text{ m/s} \qquad (11.31)$$

$(v_{ix})_{2M} = 0$ m/s, as expected, because we chose a moving reference frame in which ball 2 would be at rest.

FIGURE 11.22b now shows a situation—with ball 2 initially at rest—in which we can use Equations 11.29 to find the post-collision velocities in frame M:

$$(v_{fx})_{1M} = \frac{m_1 - m_2}{m_1 + m_2} (v_{ix})_{1M} = 1.7 \text{ m/s}$$

$$(v_{fx})_{2M} = \frac{2m_1}{m_1 + m_2} (v_{ix})_{1M} = 6.7 \text{ m/s} \qquad (11.32)$$

Reference frame M hasn't changed—it's still moving to the left in the lab frame at 3.0 m/s—but the collision has changed both balls' velocities in frame M.

To finish, we need to transform the post-collision velocities in frame M back to the lab frame L. We can do so with another application of the Galilean transformation:

$$(v_{fx})_{1L} = (v_{fx})_{1M} + (v_x)_{ML} = 1.7 \text{ m/s} + (-3.0 \text{ m/s}) = -1.3 \text{ m/s}$$
$$(v_{fx})_{2L} = (v_{fx})_{2M} + (v_x)_{ML} = 6.7 \text{ m/s} + (-3.0 \text{ m/s}) = 3.7 \text{ m/s} \qquad (11.33)$$

FIGURE 11.23 The post-collision velocities in the lab frame.

$(v_{fx})_{1L} = -1.3$ m/s $(v_{fx})_{2L} = 3.7$ m/s

FIGURE 11.23 shows the outcome of the collision in the lab frame. It's not hard to confirm that these final velocities do, indeed, conserve both momentum and energy.

Two Collision Models

No collision is perfectly elastic, although collisions between two very hard objects (metal spheres) or between two springs (such as a collision on an air track) come close. Collisions can be perfectly inelastic, although many real-world inelastic collisions exhibit a small residual bounce. Thus perfectly elastic and perfectly inelastic collisions are *models* of collisions in which we simplify reality in order to gain understanding without getting bogged down in the messy details of real collisions.

MODEL 11.1

Collisions

For two colliding objects.

- Represent the objects as elastic objects moving in a straight line.

- In a **perfectly inelastic collision**, the objects stick and move together. Kinetic energy is transformed into thermal energy.
  Mathematically:

Before:
After:

$$(m_1 + m_2)v_{fx} = m_1(v_{ix})_1 + m_2(v_{ix})_2$$

- In a **perfectly elastic collision**, the objects bounce apart with no loss of energy.
  Mathematically:

Before:
After:

  - If object 2 is initially at rest, then

$$(v_{fx})_1 = \frac{m_1 - m_2}{m_1 + m_2}(v_{ix})_1 \qquad (v_{fx})_2 = \frac{2m_1}{m_1 + m_2}(v_{ix})_1$$

  - If both objects are moving, use the Galilean transformation to transform the velocities to a reference frame in which object 2 is at rest.

- Limitations: Model fails if the collision is not head-on or cannot reasonably be approximated as a "thud" or as a "perfect bounce."

Exercise 22

11.4 Explosions

An **explosion,** where the particles of the system move apart from each other after a brief, intense interaction, is the opposite of a collision. The explosive forces, which could be from an expanding spring or from expanding hot gases, are *internal* forces. If the system is isolated, its total momentum during the explosion will be conserved.

EXAMPLE 11.6 | Recoil

A 10 g bullet is fired from a 3.0 kg rifle with a speed of 500 m/s. What is the recoil speed of the rifle?

MODEL The rifle causes a small mass of gunpowder to explode, and the expanding gas then exerts forces on *both* the bullet and the rifle. Let's define the system to be bullet + gas + rifle. The forces due to the expanding gas during the explosion are internal forces, within the system. Any friction forces between the bullet and the rifle as the bullet travels down the barrel are also internal forces. Gravity is balanced by the upward force of the person holding the rifle, so $\vec{F}_{net} = \vec{0}$. This is an isolated system and the law of conservation of momentum applies.

VISUALIZE FIGURE 11.24 shows a pictorial representation before and after the bullet is fired.

FIGURE 11.24 Before-and-after pictorial representation of a rifle firing a bullet.

Before: $m_B = 0.010$ kg
$(v_{ix})_B = 0$ m/s
System
$m_R = 3.0$ kg
$(v_{ix})_R = 0$ m/s

After:
$(v_{fx})_R$ $(v_{fx})_B = 500$ m/s

Find: $(v_{fx})_R$

Continued

SOLVE The x-component of the total momentum is $P_x = (p_x)_B + (p_x)_R + (p_x)_{gas}$. Everything is at rest before the trigger is pulled, so the initial momentum is zero. After the trigger is pulled, the momentum of the expanding gas is the sum of the momenta of all the molecules in the gas. For every molecule moving in the forward direction with velocity v and momentum mv there is, on average, another molecule moving in the opposite direction with velocity $-v$ and thus momentum $-mv$. When the values are summed over the enormous number of molecules in the gas, we will be left with $p_{gas} \approx 0$. In addition, the mass of the gas is much less than that of the rifle or bullet. For both reasons, we can reasonably neglect the momentum of the gas. The law of conservation of momentum is thus

$$P_{fx} = m_B (v_{fx})_B + m_R (v_{fx})_R = P_{ix} = 0$$

Solving for the rifle's velocity, we find

$$(v_{fx})_R = -\frac{m_B}{m_R}(v_{fx})_B = -\frac{0.010 \text{ kg}}{3.0 \text{ kg}} \times 500 \text{ m/s} = -1.7 \text{ m/s}$$

The minus sign indicates that the rifle's recoil is to the left. The recoil *speed* is 1.7 m/s.

We would not know where to begin to solve a problem such as this using Newton's laws. But Example 11.6 is a simple problem when approached from the before-and-after perspective of a conservation law. The selection of bullet + gas + rifle as "the system" was the critical step. For momentum conservation to be a useful principle, we had to select a system in which the complicated forces due to expanding gas and friction were all internal forces. The rifle by itself is *not* an isolated system, so its momentum is *not* conserved.

EXAMPLE 11.7 | Radioactivity

A ^{238}U uranium nucleus is radioactive. It spontaneously disintegrates into a small fragment that is ejected with a measured speed of 1.50×10^7 m/s and a "daughter nucleus" that recoils with a measured speed of 2.56×10^5 m/s. What are the atomic masses of the ejected fragment and the daughter nucleus?

MODEL The notation ^{238}U indicates the isotope of uranium with an atomic mass of 238 u, where u is the abbreviation for the *atomic mass unit*. The nucleus contains 92 protons (uranium is atomic number 92) and 146 neutrons. The disintegration of a nucleus is, in essence, an explosion. Only *internal* nuclear forces are involved, so the total momentum is conserved in the decay.

VISUALIZE FIGURE 11.25 shows the pictorial representation. The mass of the daughter nucleus is m_1 and that of the ejected fragment is m_2. Notice that we converted the speed information to velocity information, giving $(v_{fx})_1$ and $(v_{fx})_2$ opposite signs.

FIGURE 11.25 Before-and-after pictorial representation of the decay of a ^{238}U nucleus.

Before: ^{238}U $m = 238$ u $v_{ix} = 0$ m/s

After: m_1 ① m_2 ② $(v_{fx})_2 = 1.50 \times 10^7$ m/s

$(v_{fx})_1 = -2.56 \times 10^5$ m/s

Find: m_1 and m_2

SOLVE The nucleus was initially at rest, hence the total momentum is zero. The momentum after the decay is still zero if the two pieces fly apart in opposite directions with momenta equal in magnitude but opposite in sign. That is,

$$P_{fx} = m_1 (v_{fx})_1 + m_2 (v_{fx})_2 = P_{ix} = 0$$

Although we know both final velocities, this is not enough information to find the two unknown masses. However, we also have another conservation law, conservation of mass, that requires

$$m_1 + m_2 = 238 \text{ u}$$

Combining these two conservation laws gives

$$m_1 (v_{fx})_1 + (238 \text{ u} - m_1)(v_{fx})_2 = 0$$

The mass of the daughter nucleus is

$$m_1 = \frac{(v_{fx})_2}{(v_{fx})_2 - (v_{fx})_1} \times 238 \text{ u}$$

$$= \frac{1.50 \times 10^7 \text{ m/s}}{(1.50 \times 10^7 - (-2.56 \times 10^5)) \text{ m/s}} \times 238 \text{ u} = 234 \text{ u}$$

With m_1 known, the mass of the ejected fragment is $m_2 = 238 - m_1 = 4$ u.

ASSESS All we learn from a momentum analysis is the masses. Chemical analysis shows that the daughter nucleus is the element thorium, atomic number 90, with two fewer protons than uranium. The ejected fragment carried away two protons as part of its mass of 4 u, so it must be a particle with two protons and two neutrons. This is the nucleus of a helium atom, ^{4}He, which in nuclear physics is called an *alpha particle* α. Thus the radioactive decay of ^{238}U can be written as ^{238}U $\rightarrow$ ^{234}Th $+ \alpha$.

Much the same reasoning explains how a rocket or jet aircraft accelerates. **FIGURE 11.26** shows a rocket with a parcel of fuel on board. Burning converts the fuel to hot gases that are expelled from the rocket motor. If we choose rocket + gases to be the system, the burning and expulsion are both internal forces. There are no other forces, so the total momentum of the rocket + gases system must be conserved. The rocket gains forward velocity and momentum as the exhaust gases are shot out the back, but the *total* momentum of the system remains zero.

Section 11.6 looks at rocket propulsion in more detail, but even without the details you should be able to understand that jet and rocket propulsion is a consequence of momentum conservation.

FIGURE 11.26 Rocket propulsion is an example of conservation of momentum.

STOP TO THINK 11.5 An explosion in a rigid pipe shoots out three pieces. A 6 g piece comes out the right end. A 4 g piece comes out the left end with twice the speed of the 6 g piece. From which end, left or right, does the third piece emerge?

11.5 Momentum in Two Dimensions

The law of conservation of momentum $\vec{P}_f = \vec{P}_i$ is not restricted to motion along a line. Many interesting examples of collisions and explosions involve motion in a plane, and for these both the magnitude *and the direction* of the total momentum vector are unchanged. The total momentum is the vector sum of the individual momenta, so the total momentum is conserved only if each component is conserved:

$$(p_{fx})_1 + (p_{fx})_2 + (p_{fx})_3 + \cdots = (p_{ix})_1 + (p_{ix})_2 + (p_{ix})_3 + \cdots$$
$$(p_{fy})_1 + (p_{fy})_2 + (p_{fy})_3 + \cdots = (p_{iy})_1 + (p_{iy})_2 + (p_{iy})_3 + \cdots$$
(11.34)

Let's look at some examples of momentum conservation in two dimensions.

Collisions and explosions often involve motion in two dimensions.

EXAMPLE 11.8 | **A peregrine falcon strike**

Peregrine falcons often grab their prey from above while both falcon and prey are in flight. A 0.80 kg falcon, flying at 18 m/s, swoops down at a 45° angle from behind a 0.36 kg pigeon flying horizontally at 9.0 m/s. What are the speed and direction of the falcon (now holding the pigeon) immediately after impact?

MODEL The two birds, modeled as particles, are the system. This is a perfectly inelastic collision because after the collision the falcon and pigeon move at a common final velocity. The birds are not a perfectly isolated system because of external forces of the air, but during the brief collision the external impulse delivered by the air resistance will be negligible. Within this approximation, the total momentum of the falcon + pigeon system is conserved during the collision.

VISUALIZE **FIGURE 11.27** is a before-and-after pictorial representation. We've used angle ϕ to label the post-collision direction.

SOLVE The initial velocity components of the falcon are $(v_{ix})_F = v_F\cos\theta$ and $(v_{iy})_F = -v_F\sin\theta$. The pigeon's initial velocity is entirely along the x-axis. After the collision, when the falcon and pigeon have the common velocity $\vec{v}_f$, the components are $v_{fx} = v_f\cos\phi$ and $v_{fy} = -v_f\sin\phi$. Conservation of momentum in two

FIGURE 11.27 Pictorial representation of a falcon catching a pigeon.

dimensions requires conservation of both the x- and y-components of momentum. This gives two conservation equations:

$$(m_F + m_P)v_{fx} = (m_F + m_P)v_f\cos\phi$$
$$= m_F(v_{ix})_F + m_P(v_{ix})_P = m_Fv_F\cos\theta + m_Pv_P$$
$$(m_F + m_P)v_{fy} = -(m_F + m_P)v_f\sin\phi$$
$$= m_F(v_{iy})_F + m_P(v_{iy})_P = -m_Fv_F\sin\theta$$

Continued

The unknowns are v_f and ϕ. Dividing both equations by the total mass gives

$$v_f \cos \phi = \frac{m_F v_F \cos \theta + m_P v_P}{m_F + m_P} = 11.6 \text{ m/s}$$

$$v_f \sin \phi = \frac{m_F v_F \sin \theta}{m_F + m_P} = 8.78 \text{ m/s}$$

We can eliminate v_f by dividing the second equation by the first to give

$$\frac{v_f \sin \phi}{v_f \cos \phi} = \tan \phi = \frac{8.78 \text{ m/s}}{11.6 \text{ m/s}} = 0.757$$

$$\phi = \tan^{-1}(0.757) = 37°$$

Then $v_f = (11.6 \text{ m/s})/\cos(37°) = 15 \text{ m/s}$. Immediately after impact, the falcon, with its meal, is traveling at 15 m/s at an angle 37° below the horizontal.

ASSESS It makes sense that the falcon would slow down after grabbing the slower-moving pigeon. And Figure 11.27 tells us that the total momentum is at an angle between 0° (the pigeon's momentum) and 45° (the falcon's momentum). Thus our answer seems reasonable.

EXAMPLE 11.9 | A three-piece explosion

A 10.0 g projectile is traveling east at 2.0 m/s when it suddenly explodes into three pieces. A 3.0 g fragment is shot due west at 10 m/s while another 3.0 g fragment travels 40° north of east at 12 m/s. What are the speed and direction of the third fragment?

MODEL Although many complex forces are involved in the explosion, they are all internal to the system. There are no external forces, so this is an isolated system and its total momentum is conserved.

VISUALIZE FIGURE 11.28 shows a before-and-after pictorial representation. We'll use uppercase M and V to distinguish the initial object from the three pieces into which it explodes.

FIGURE 11.28 Before-and-after pictorial representation of the three-piece explosion.

Find: v_3 and θ

SOLVE The system is the initial object and the subsequent three pieces. Conservation of momentum requires

$$m_1 (v_{fx})_1 + m_2 (v_{fx})_2 + m_3 (v_{fx})_3 = MV_{ix}$$

$$m_1 (v_{fy})_1 + m_2 (v_{fy})_2 + m_3 (v_{fy})_3 = MV_{iy}$$

Conservation of mass implies that

$$m_3 = M - m_1 - m_2 = 4.0 \text{ g}$$

Neither the original object nor m_2 has any momentum along the y-axis. We can use Figure 11.28 to write out the x- and y-components of $\vec{v}_1$ and $\vec{v}_3$, leading to

$$m_1 v_1 \cos 40° - m_2 v_2 + m_3 v_3 \cos \theta = MV$$

$$m_1 v_1 \sin 40° - m_3 v_3 \sin \theta = 0$$

where we used $(v_{fx})_2 = -v_2$ because m_2 is moving in the negative x-direction. Inserting known values in these equations gives us

$$-2.42 + 4v_3 \cos \theta = 20$$

$$23.14 - 4v_3 \sin \theta = 0$$

We can leave the masses in grams in this situation because the conversion factor to kilograms appears on both sides of the equation and thus cancels out. To solve, first use the second equation to write $v_3 = 5.79/\sin \theta$. Substitute this result into the first equation, noting that $\cos \theta / \sin \theta = 1/\tan \theta$, to get

$$-2.42 + 4\left(\frac{5.79}{\sin \theta}\right) \cos \theta = -2.42 + \frac{23.14}{\tan \theta} = 20$$

Now solve for θ:

$$\tan \theta = \frac{23.14}{20 + 2.42} = 1.03$$

$$\theta = \tan^{-1}(1.03) = 45.8°$$

Finally, use this result in the earlier expression for v_3 to find

$$v_3 = \frac{5.79}{\sin 45.8°} = 8.1 \text{ m/s}$$

The third fragment, with a mass of 4.0 g, is shot 46° south of east at a speed of 8.1 m/s.

An object traveling to the right with $\vec{p} = 2\,\hat{\imath}$ kg m/s suddenly explodes into two pieces. Piece 1 has the momentum $\vec{p}_1$ shown in the figure. What is the momentum $\vec{p}_2$ of the second piece?

11.6 ADVANCED TOPIC Rocket Propulsion

Newton's second law $\vec{F} = m\vec{a}$ applies to objects whose mass does not change. That's an excellent assumption for balls and bicycles, but what about something like a rocket that loses a significant amount of mass as its fuel is burned? Problems of varying mass are solved with momentum rather than acceleration. We'll look at one important example.

FIGURE 11.29 shows a rocket being propelled by the thrust of burning fuel but *not* influenced by gravity or drag. Perhaps it is a rocket in deep space where gravity is very weak in comparison to the rocket's thrust. This may not be highly realistic, but ignoring gravity allows us to understand the essentials of rocket propulsion without making the mathematics too complicated. Rocket propulsion with gravity is a Challenge Problem in the end-of-chapter problems.

The system rocket + exhaust gases is an isolated system, so its total momentum is conserved. The basic idea is simple: As exhaust gases are shot out the back, the rocket "recoils" in the opposite direction. Putting this idea on a mathematical footing is fairly straightforward—it's basically the same as analyzing an explosion—but we have to be extremely careful with signs.

We'll use a before-and-after approach, as we do with all momentum problems. The Before state is a rocket of mass m (including all onboard fuel) moving with velocity v_x and having initial momentum $P_{ix} = mv_x$. During a small interval of time dt, the rocket burns a small mass of fuel m_{fuel} and expels the resulting gases from the back of the rocket at an exhaust speed v_{ex} *relative to the rocket*. That is, a space cadet on the rocket sees the gases leaving the rocket at speed v_{ex} regardless of how fast the rocket is traveling through space.

After this little packet of burned fuel has been ejected, the rocket has new velocity $v_x + dv_x$ and new mass $m + dm$. Now you're probably thinking that this can't be right; the rocket *loses* mass rather than gaining mass. But that's *our* understanding of the physical situation. The mathematical analysis knows only that the mass changes, not whether it increases or decreases. Saying that the mass is $m + dm$ at time $t + dt$ is a formal statement that the mass has changed, and that's how analysis of change is done in calculus. The fact that the rocket's mass is decreasing means that dm has a negative value. That is, the minus goes with the value of dm, not with the statement that the mass has changed.

After the gas has been ejected, both the rocket and the gas have momentum. Conservation of momentum tells us that

$$P_{fx} = m_{rocket}(v_x)_{rocket} + m_{fuel}(v_x)_{fuel} = P_{ix} = mv_x \qquad (11.35)$$

The mass of this little packet of burned fuel is the mass *lost* by the rocket: $m_{fuel} = -dm$. Mathematically, the minus sign tells us that the mass of the burned fuel (the gases) and the rocket mass are changing in opposite directions. Physically, we know that $dm < 0$, so the exhaust gases have a positive mass.

FIGURE 11.29 A before-and-after pictorial representation of a rocket burning a small amount of fuel.

Before:

After:

Relative to rocket

The gases are ejected toward the left at speed v_{ex} relative to the rocket. If the rocket's velocity is v_x, then the gas's velocity *through space* is $v_x - v_{ex}$. When we assemble all these pieces of information, the momentum conservation equation is

$$(m + dm)(v_x + dv_x) + (-dm)(v_x - v_{ex}) = mv_x \qquad (11.36)$$

Multiplying this out gives

$$mv_x + v_x dm + m\,dv_x + dm\,dv_x - v_x dm + v_{ex} dm = mv_x \qquad (11.37)$$

You can see that several terms cancel, leading to $m\,dv_x + v_{ex} dm + dm\,dv_x = 0$. We can drop the third term; it is the product of two infinitesimal terms and thus is negligible compared to the first two terms. With one final algebraic rearrangement, we're left with

$$dv_x = -v_{ex}\frac{dm}{m} \qquad (11.38)$$

Remember that dm is negative—it's the mass *lost* by the rocket when a small amount of fuel is burned—and so dv_x is positive. Physically, Equation 11.38 is telling us the amount by which the rocket's velocity increases when it burns a small amount of fuel. Not surprisingly, a lighter rocket (smaller m) gains more velocity than a heavier rocket (larger m).

There are a couple of ways to use Equation 11.38. First, divide both sides by the small interval of time dt in which the fuel is burned, which will make this a rate equation:

$$\frac{dv_x}{dt} = \frac{-v_{ex}dm/dt}{m} = \frac{v_{ex}R}{m} \qquad (11.39)$$

where $R = |dm/dt|$ is the rate—in kg/s—at which fuel is burned. The fuel burn rate is reasonably constant for most rocket engines.

The left side of Equation 11.39 is the rocket's acceleration: $a_x = dv_x/dt$. Thus from Newton's second law, $a_x = F_x/m$, the numerator on the right side of Equation 11.39 must be a force. This is the *thrust* of the rocket engine:

$$F_{thrust} = v_{ex}R \qquad (11.40)$$

So Equation 11.39 is just Newton's second law, $a = F_{thrust}/m$, for the instantaneous acceleration, which will change as the rocket's mass m changes. But now we know how the thrust force is related to physical properties of the rocket engine.

Returning to Equation 11.38, we can find out how the rocket's velocity changes as fuel is burned by integrating. Suppose the rocket starts from rest ($v_x = 0$) with mass $m_0 = m_R + m_{F0}$, where m_R is the mass of the empty rocket and m_{F0} is the initial mass of the fuel. At a later time, when the mass has been reduced to m, the velocity is v.

Integrating between this Before and After, we find

$$\int_0^v dv_x = v = -v_{ex}\int_{m_0}^m \frac{dm}{m} = -v_{ex}\ln m \Big|_{m_0}^m \qquad (11.41)$$

where $\ln m$ is the *natural logarithm* (logarithm with base e) of m. Evaluating this between the limits, and using the properties of logarithms, gives

$$-v_{ex}\ln m \Big|_{m_0}^m = -v_{ex}(\ln m - \ln m_0) = -v_{ex}\ln\left(\frac{m}{m_0}\right) = v_{ex}\ln\left(\frac{m_0}{m}\right) \qquad (11.42)$$

Thus the rocket's velocity when its mass has decreased to m is

$$v = v_{ex}\ln\left(\frac{m_0}{m}\right) \qquad (11.43)$$

Initially, when $m = m_0$, $v = 0$ because $\ln 1 = 0$. The maximum speed occurs when the fuel is completely gone and $m = m_R$. This is

$$v_{max} = v_{ex}\ln\left(\frac{m_R + m_{F0}}{m_R}\right) \qquad (11.44)$$

Notice that the rocket speed can exceed v_{ex} if the fuel-mass-to-rocket-mass ratio is large enough. Also, because v_{max} is greatly improved by reducing m_R, we can see why rockets that need enough speed to go into orbit are usually *multistage rockets*, dropping off the mass of the lower stages when their fuel is depleted.

EXAMPLE 11.10 | Firing a rocket

Sounding rockets are small rockets used to gather weather data and do atmospheric research. One of the most popular sounding rockets has been the fairly small (10-in-diameter, 16-ft-long) Black Brant III. It is loaded with 210 kg of fuel, has a launch mass of 290 kg, and generates 49 kN of thrust for 9.0 s. What would be the maximum speed of a Black Brant III if launched from rest in deep space?

MODEL We define the system to be the rocket and its exhaust gases. This is an isolated system, its total momentum is conserved, and the rocket's maximum speed is given by Equation 11.44.

SOLVE We're given $m_{F0} = 210$ kg. Knowing that the launch mass is 290 kg, we can deduce that the mass of the empty rocket is $m_R = 80$ kg. Because the rocket burns 210 kg of fuel in 9.0 s, the fuel burn rate is

$$R = \frac{210 \text{ kg}}{9.0 \text{ s}} = 23.3 \text{ kg/s}$$

Knowing the burn rate and the thrust, we can use Equation 11.40 to calculate the exhaust velocity:

$$v_{ex} = \frac{F_{thrust}}{R} = \frac{49{,}000 \text{ N}}{23.3 \text{ kg/s}} = 2100 \text{ m/s}$$

Thus the rocket's maximum speed in deep space would be

$$v_{max} = v_{ex} \ln\left(\frac{m_R + m_{F0}}{m_R}\right) = (2100 \text{ m/s})\ln\left(\frac{290 \text{ kg}}{80 \text{ kg}}\right) = 2700 \text{ m/s}$$

ASSESS An actual sounding rocket doesn't reach this speed because it's affected both by gravity and by drag. Even so, the rocket's acceleration is so large that gravity plays a fairly minor role. A Black Brant III launched into the earth's atmosphere achieves a maximum speed of 2100 m/s and, because it continues to coast upward long after the fuel is exhausted, reaches a maximum altitude of 175 km (105 mi).

CHALLENGE EXAMPLE 11.11 | A rebounding pendulum

A 200 g steel ball hangs on a 1.0-m-long string. The ball is pulled sideways so that the string is at a 45° angle, then released. At the very bottom of its swing the ball strikes a 500 g steel paperweight that is resting on a frictionless table. To what angle does the ball rebound?

MODEL We can divide this problem into three parts. First the ball swings down as a pendulum. Second, the ball and paperweight have a collision. Steel balls bounce off each other very well, so we will model the collision as perfectly elastic. Third, the ball, after it bounces off the paperweight, swings back up as a pendulum.

For Parts 1 and 3, we define the system as the ball and the earth. This is an isolated, nondissipative system, so its mechanical energy is conserved. In Part 2, let the system consist of the ball and the paperweight, which have a perfectly elastic collision.

VISUALIZE FIGURE 11.30 shows four distinct moments of time: as the ball is released, an instant before the collision, an instant after the collision but before the ball and paperweight have had time to move, and as the ball reaches its highest point on the rebound. Call the ball A and the paperweight B, so $m_A = 0.20$ kg and $m_B = 0.50$ kg.

FIGURE 11.30 Four moments in the collision of a pendulum with a paperweight.

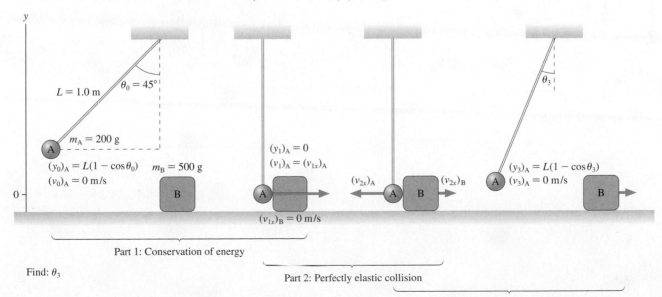

Part 1: Conservation of energy

Find: θ_3

Part 2: Perfectly elastic collision

Part 3: Conservation of energy

Continued

SOLVE Part 1: The first part involves the ball only. Its initial height is

$$(y_0)_A = L - L\cos\theta_0 = L(1 - \cos\theta_0) = 0.293 \text{ m}$$

We can use conservation of mechanical energy to find the ball's velocity at the bottom, just before impact on the paperweight:

$$\tfrac{1}{2}m_A(v_1)_A^2 + m_A g(y_1)_A = \tfrac{1}{2}m_A(v_0)_A^2 + m_A g(y_0)_A$$

We know $(v_0)_A = 0$. Solving for the velocity at the bottom, where $(y_1)_A = 0$, gives

$$(v_1)_A = \sqrt{2g(y_0)_A} = 2.40 \text{ m/s}$$

Part 2: The ball and paperweight undergo a perfectly elastic collision in which the paperweight is initially at rest. These are the conditions for which Equations 11.29 were derived. The velocities *immediately* after the collision, prior to any further motion, are

$$(v_{2x})_A = \frac{m_A - m_B}{m_A + m_B}(v_{1x})_A = -1.03 \text{ m/s}$$

$$(v_{2x})_B = \frac{2m_A}{m_A + m_B}(v_{1x})_A = +1.37 \text{ m/s}$$

The ball rebounds toward the left with a speed of 1.03 m/s while the paperweight moves to the right at 1.37 m/s. Kinetic energy has been conserved (you might want to check this), but it is now shared between the ball and the paperweight.

Part 3: Now the ball is a pendulum with an initial speed of 1.03 m/s. Mechanical energy is again conserved, so we can find its maximum height at the point where $(v_3)_A = 0$:

$$\tfrac{1}{2}m_A(v_3)_A^2 + m_A g(y_3)_A = \tfrac{1}{2}m_A(v_2)_A^2 + m_A g(y_2)_A$$

Solving for the maximum height gives

$$(y_3)_A = \frac{(v_2)_A^2}{2g} = 0.0541 \text{ m}$$

The height $(y_3)_A$ is related to angle θ_3 by $(y_3)_A = L(1 - \cos\theta_3)$. This can be solved to find the angle of rebound:

$$\theta_3 = \cos^{-1}\left(1 - \frac{(y_3)_A}{L}\right) = 19°$$

The paperweight speeds away at 1.37 m/s and the ball rebounds to an angle of 19°.

ASSESS The ball and the paperweight aren't hugely different in mass, so we expect the ball to transfer a significant fraction of its energy to the paperweight when they collide. Thus a rebound to roughly half the initial angle seems reasonable.

SUMMARY

The goals of Chapter 11 have been to learn to use the concepts of impulse and momentum.

GENERAL PRINCIPLES

Law of Conservation of Momentum

The total momentum $\vec{P} = \vec{p}_1 + \vec{p}_2 + \cdots$ of an isolated system is a constant. Thus

$$\vec{P}_\text{f} = \vec{P}_\text{i}$$

Newton's Second Law

In terms of momentum, Newton's second law is

$$\vec{F} = \frac{d\vec{p}}{dt}$$

Solving Momentum Conservation Problems

MODEL Choose an isolated system or a system that is isolated during at least part of the problem.

VISUALIZE Draw a pictorial representation of the system before and after the interaction.

SOLVE Write the law of conservation of momentum in terms of vector components:

$$(p_{\text{f}x})_1 + (p_{\text{f}x})_2 + \cdots = (p_{\text{i}x})_1 + (p_{\text{i}x})_2 + \cdots$$
$$(p_{\text{f}y})_1 + (p_{\text{f}y})_2 + \cdots = (p_{\text{i}y})_1 + (p_{\text{i}y})_2 + \cdots$$

ASSESS Is the result reasonable?

IMPORTANT CONCEPTS

Momentum $\vec{p} = m\vec{v}$

Impulse $J_x = \displaystyle\int_{t_\text{i}}^{t_\text{f}} F_x(t)\, dt = $ area under force curve

Impulse and momentum are related by the **momentum principle**

$$\Delta p_x = J_x$$

The impulse delivered to an object causes the object's momentum to change. This is an alternative statement of Newton's second law.

Momentum bar charts display the momentum principle $p_{\text{f}x} = p_{\text{i}x} + J_x$ in graphical form.

System A group of interacting particles.

Isolated system A system on which there are no external forces or the net external force is zero.

APPLICATIONS

Collisions In a **perfectly inelastic collision,** two objects stick together and move with a common final velocity. In a **perfectly elastic collision,** they bounce apart and conserve mechanical energy as well as momentum.

Explosions Two or more objects fly apart from each other. Their total momentum is conserved.

Two dimensions The same ideas apply in two dimensions. Both the x- and y-components of $\vec{P}$ must be conserved. This gives two simultaneous equations to solve.

Rockets The momentum of the exhaust-gas + rocket system is conserved. **Thrust** is the product of the exhaust speed and the rate at which fuel is burned.

TERMS AND NOTATION

collision	impulse, J_x	impulse approximation	perfectly inelastic collision
impulsive force	momentum principle	total momentum, $\vec{P}$	perfectly elastic collision
momentum, $\vec{p}$	momentum bar chart	law of conservation of momentum	explosion

CONCEPTUAL QUESTIONS

1. Rank in order, from largest to smallest, the momenta $(p_x)_a$ to $(p_x)_e$ of the objects in **FIGURE Q11.1**.

FIGURE Q11.1

2. A 2 kg object is moving to the right with a speed of 1 m/s when it experiences an impulse of 4 N s. What are the object's speed and direction after the impulse?

3. A 2 kg object is moving to the right with a speed of 1 m/s when it experiences an impulse of −4 N s. What are the object's speed and direction after the impulse?

4. A 0.2 kg plastic cart and a 20 kg lead cart can both roll without friction on a horizontal surface. Equal forces are used to push both carts forward for a time of 1 s, starting from rest. After the force is removed at $t = 1$ s, is the momentum of the plastic cart greater than, less than, or equal to the momentum of the lead cart? Explain.

5. A 0.2 kg plastic cart and a 20 kg lead cart can both roll without friction on a horizontal surface. Equal forces are used to push both carts forward for a distance of 1 m, starting from rest. After traveling 1 m, is the momentum of the plastic cart greater than, less than, or equal to the momentum of the lead cart? Explain.

6. Angie, Brad, and Carlos are discussing a physics problem in which two identical bullets are fired with equal speeds at equal-mass wood and steel blocks resting on a frictionless table. One bullet bounces off the steel block while the second becomes embedded in the wood block. "All the masses and speeds are the same," says Angie, "so I think the blocks will have equal speeds after the collisions." "But what about momentum?" asks Brad. "The bullet hitting the wood block transfers all its momentum and energy to the block, so the wood block should end up going faster

than the steel block." "I think the bounce is an important factor," replies Carlos. "The steel block will be faster because the bullet bounces off it and goes back the other direction." Which of these three do you agree with, and why?

7. It feels better to catch a hard ball while wearing a padded glove than to catch it bare handed. Use the ideas of this chapter to explain why.

8. Automobiles are designed with "crumple zones" intended to collapse in a collision. Use the ideas of this chapter to explain why.

9. A golf club continues forward after hitting the golf ball. Is momentum conserved in the collision? Explain, making sure you are careful to identify "the system."

10. Suppose a rubber ball collides head-on with a more massive steel ball traveling in the opposite direction with equal speed. Which ball, if either, receives the larger impulse? Explain.

11. Two particles collide, one of which was initially moving and the other initially at rest.
    a. Is it possible for *both* particles to be at rest after the collision? Give an example in which this happens, or explain why it can't happen.
    b. Is it possible for *one* particle to be at rest after the collision? Give an example in which this happens, or explain why it can't happen.

12. Two ice skaters, Paula and Ricardo, push off from each other. Ricardo weighs more than Paula.
    a. Which skater, if either, has the greater momentum after the push-off? Explain.
    b. Which skater, if either, has the greater speed after the push-off? Explain.

13. Two balls of clay of known masses hang from the ceiling on massless strings of equal length. They barely touch when both hang at rest. One ball is pulled back until its string is at 45°, then released. It swings down, collides with the second ball, and they stick together. To determine the angle to which the balls swing on the opposite side, would you invoke (a) conservation of momentum, (b) conservation of mechanical energy, (c) both, (d) either but not both, or (e) these laws alone are not sufficient to find the angle? Explain.

EXERCISES AND PROBLEMS

Problems labeled ▨ integrate material from earlier chapters.

Exercises

Section 11.1 Momentum and Impulse

1. | At what speed do a bicycle and its rider, with a combined mass of 100 kg, have the same momentum as a 1500 kg car traveling at 5.0 m/s?

2. | What is the magnitude of the momentum of
   a. A 3000 kg truck traveling at 15 m/s?
   b. A 200 g baseball thrown at 40 m/s?

3. ‖ What impulse does the force shown in **FIGURE EX11.3** exert on a 250 g particle?

FIGURE EX11.3

FIGURE EX11.4

4. ‖ What is the impulse on a 3.0 kg particle that experiences the force shown in **FIGURE EX11.4**?

5. ‖ In **FIGURE EX11.5**, what value of F_{max} gives an impulse of 6.0 N s?

FIGURE EX11.5

FIGURE EX11.6

6. ‖ **FIGURE EX11.6** is an incomplete momentum bar chart for a collision that lasts 10 ms. What are the magnitude and direction of the average collision force exerted on the object?

7. ‖ **FIGURE EX11.7** is an incomplete momentum bar chart for a 50 g particle that experiences an impulse lasting 10 ms. What were the speed and direction of the particle before the impulse?

FIGURE EX11.7

FIGURE EX11.8

8. | A 2.0 kg object is moving to the right with a speed of 1.0 m/s when it experiences the force shown in **FIGURE EX11.8**. What are the object's speed and direction after the force ends?

9. | A 2.0 kg object is moving to the right with a speed of 1.0 m/s when it experiences the force shown in **FIGURE EX11.9**. What are the object's speed and direction after the force ends?

FIGURE EX11.9

10. | A sled slides along a horizontal surface on which the coefficient of kinetic friction is 0.25. Its velocity at point A is 8.0 m/s and at point B is 5.0 m/s. Use the momentum principle to find how long the sled takes to travel from A to B.

11. | Far in space, where gravity is negligible, a 425 kg rocket traveling at 75 m/s fires its engines. **FIGURE EX11.11** shows the thrust force as a function of time. The mass lost by the rocket during these 30 s is negligible.
   a. What impulse does the engine impart to the rocket?
   b. At what time does the rocket reach its maximum speed? What is the maximum speed?

FIGURE EX11.11

12. ‖ A 600 g air-track glider collides with a spring at one end of the track. **FIGURE EX11.12** shows the glider's velocity and the force exerted on the glider by the spring. How long is the glider in contact with the spring?

FIGURE EX11.12

13. ‖ A 250 g ball collides with a wall. **FIGURE EX11.13** shows the ball's velocity and the force exerted on the ball by the wall. What is v_{fx}, the ball's rebound velocity?

FIGURE EX11.13

Section 11.2 Conservation of Momentum

14. | A 5000 kg open train car is rolling on frictionless rails at 22 m/s when it starts pouring rain. A few minutes later, the car's speed is 20 m/s. What mass of water has collected in the car?

15. | A 10,000 kg railroad car is rolling at 2.0 m/s when a 4000 kg load of gravel is suddenly dropped in. What is the car's speed just after the gravel is loaded?

16. ‖ A 10-m-long glider with a mass of 680 kg (including the passengers) is gliding horizontally through the air at 30 m/s when a 60 kg skydiver drops out by releasing his grip on the glider. What is the glider's velocity just after the skydiver lets go?

17. | Three identical train cars, coupled together, are rolling east at speed v_0. A fourth car traveling east at $2v_0$ catches up with the three and couples to make a four-car train. A moment later, the train cars hit a fifth car that was at rest on the tracks, and it couples to make a five-car train. What is the speed of the five-car train?

Section 11.3 Collisions

18. | A 300 g bird flying along at 6.0 m/s sees a 10 g insect heading straight toward it at a speed of 30 m/s. The bird opens its mouth wide and enjoys a nice lunch. What is the bird's speed immediately after swallowing?

19. | The parking brake on a 2000 kg Cadillac has failed, and it is rolling slowly, at 1.0 mph, toward a group of small children. Seeing the situation, you realize you have just enough time to drive your 1000 kg Volkswagen head-on into the Cadillac and save the children. With what speed should you impact the Cadillac to bring it to a halt?

20. | A 1500 kg car is rolling at 2.0 m/s. You would like to stop the car by firing a 10 kg blob of sticky clay at it. How fast should you fire the clay?

21. ‖ Fred (mass 60 kg) is running with the football at a speed of 6.0 m/s when he is met head-on by Brutus (mass 120 kg), who is moving at 4.0 m/s. Brutus grabs Fred in a tight grip, and they fall to the ground. Which way do they slide, and how far? The coefficient of kinetic friction between football uniforms and Astroturf is 0.30.

22. ‖ A 50 g marble moving at 2.0 m/s strikes a 20 g marble at rest. What is the speed of each marble immediately after the collision?

23. | A proton is traveling to the right at 2.0×10^7 m/s. It has a head-on perfectly elastic collision with a carbon atom. The mass of the carbon atom is 12 times the mass of the proton. What are the speed and direction of each after the collision?

24. ‖ A 50 g ball of clay traveling at speed v_0 hits and sticks to a 1.0 kg brick sitting at rest on a frictionless surface.
   a. What is the speed of the brick after the collision?
   b. What percentage of the mechanical energy is lost in this collision?

25. ‖ A package of mass m is released from rest at a warehouse loading dock and slides down the 3.0-m-high, frictionless chute of FIGURE EX11.25 to a waiting truck. Unfortunately, the truck driver went on a break without having removed the previous package, of mass $2m$, from the bottom of the chute.
   a. Suppose the packages stick together. What is their common speed after the collision?
   b. Suppose the collision between the packages is perfectly elastic. To what height does the package of mass m rebound?

FIGURE EX11.25

Section 11.4 Explosions

26. | A 50 kg archer, standing on frictionless ice, shoots a 100 g arrow at a speed of 100 m/s. What is the recoil speed of the archer?

27. ‖ A 70.0 kg football player is gliding across very smooth ice at 2.00 m/s. He throws a 0.450 kg football straight forward. What is the player's speed afterward if the ball is thrown at
   a. 15.0 m/s relative to the ground?
   b. 15.0 m/s relative to the player?

28. ‖ Dan is gliding on his skateboard at 4.0 m/s. He suddenly jumps backward off the skateboard, kicking the skateboard forward at 8.0 m/s. How fast is Dan going as his feet hit the ground? Dan's mass is 50 kg and the skateboard's mass is 5.0 kg.

29. | Two ice skaters, with masses of 50 kg and 75 kg, are at the center of a 60-m-diameter circular rink. The skaters push off against each other and glide to opposite edges of the rink. If the heavier skater reaches the edge in 20 s, how long does the lighter skater take to reach the edge?

30. | A ball of mass m and another ball of mass $3m$ are placed inside a smooth metal tube with a massless spring compressed between them. When the spring is released, the heavier ball flies out of one end of the tube with speed v_0. With what speed does the lighter ball emerge from the other end?

Section 11.5 Momentum in Two Dimensions

31. ‖ Two particles collide and bounce apart. FIGURE EX11.31 shows the initial momenta of both and the final momentum of particle 2. What is the final momentum of particle 1? Write your answer using unit vectors.

FIGURE EX11.31

32. ‖ An object at rest explodes into three fragments. FIGURE EX11.32 shows the momentum vectors of two of the fragments. What is the momentum of the third fragment? Write your answer using unit vectors.

FIGURE EX11.32

33. ‖ A 20 g ball of clay traveling east at 3.0 m/s collides with a 30 g ball of clay traveling north at 2.0 m/s. What are the speed and the direction of the resulting 50 g ball of clay? Give your answer as an angle north of east.

34. ‖ At the center of a 50-m-diameter circular ice rink, a 75 kg skater traveling north at 2.5 m/s collides with and holds on to a 60 kg skater who had been heading west at 3.5 m/s.
   a. How long will it take them to glide to the edge of the rink?
   b. Where will they reach it? Give your answer as an angle north of west.

Section 11.6 Rocket Propulsion

35. ‖ A small rocket with 15 kN thrust burns 250 kg of fuel in 30 s. What is the exhaust speed of the hot gases?

36. ‖ A rocket in deep space has an empty mass of 150 kg and exhausts the hot gases of burned fuel at 2500 m/s. What mass of fuel is needed to reach a top speed of 4000 m/s?

37. ‖ A rocket in deep space has an exhaust-gas speed of 2000 m/s. When the rocket is fully loaded, the mass of the fuel is five times the mass of the empty rocket. What is the rocket's speed when half the fuel has been burned?

Problems

38. ‖ A tennis player swings her 1000 g racket with a speed of 10 m/s. She hits a 60 g tennis ball that was approaching her at a speed of 20 m/s. The ball rebounds at 40 m/s.
   a. How fast is her racket moving immediately after the impact? You can ignore the interaction of the racket with her hand for the brief duration of the collision.
   b. If the tennis ball and racket are in contact for 10 ms, what is the average force that the racket exerts on the ball? How does this compare to the gravitational force on the ball?

39. ‖ A 60 g tennis ball with an initial speed of 32 m/s hits a wall and rebounds with the same speed. **FIGURE P11.39** shows the force of the wall on the ball during the collision. What is the value of F_{max}, the maximum value of the contact force during the collision?

FIGURE P11.39

40. ‖ A 500 g cart is released from rest 1.00 m from the bottom of a frictionless, 30.0° ramp. The cart rolls down the ramp and bounces off a rubber block at the bottom. **FIGURE P11.40** shows the force during the collision. After the cart bounces, how far does it roll back up the ramp?

FIGURE P11.40

FIGURE P11.41

41. ‖‖‖ A 200 g ball is dropped from a height of 2.0 m, bounces on a hard floor, and rebounds to a height of 1.5 m. **FIGURE P11.41** shows the impulse received from the floor. What maximum force does the floor exert on the ball?

42. ‖ The flowers of the bunchberry plant open with astonishing
BIO force and speed, causing the pollen grains to be ejected out of the flower in a mere 0.30 ms at an acceleration of 2.5×10^4 m/s². If the acceleration is constant, what impulse is delivered to a pollen grain with a mass of 1.0×10^{-7} g?

43. ‖ A particle of mass m is at rest at $t = 0$. Its momentum for $t > 0$
CALC is given by $p_x = 6t^2$ kg m/s, where t is in s. Find an expression for $F_x(t)$, the force exerted on the particle as a function of time.

44. ‖ Air-track gliders with masses 300 g, 400 g, and 200 g are lined up and held in place with lightweight springs compressed between them. All three are released at once. The 200 g glider flies off to the right while the 300 g glider goes left. Their position-versus-time graphs, as measured by motion detectors, are shown in **FIGURE P11.44**. What are the direction (right or left) and speed of the 400 g glider that was in the middle?

FIGURE P11.44

45. ‖ Most geologists believe that the dinosaurs became extinct 65 million years ago when a large comet or asteroid struck the earth, throwing up so much dust that the sun was blocked out for a period of many months. Suppose an asteroid with a diameter of 2.0 km and a mass of 1.0×10^{13} kg hits the earth (6.0×10^{24} kg) with an impact speed of 4.0×10^4 m/s.
  a. What is the earth's recoil speed after such a collision? (Use a reference frame in which the earth was initially at rest.)
  b. What percentage is this of the earth's speed around the sun? The earth orbits the sun at a distance of 1.5×10^{11} m.

46. ‖ Squids rely on jet propulsion to move around. A 1.50 kg squid
BIO (including the mass of water inside the squid) drifting at 0.40 m/s suddenly ejects 0.10 kg of water to get itself moving at 2.50 m/s. If drag is ignored over the small interval of time needed to expel the water (the impulse approximation), what is the water's ejection speed relative to the squid?

47. ‖ A firecracker in a coconut blows the coconut into three pieces. Two pieces of equal mass fly off south and west, perpendicular to each other, at speed v_0. The third piece has twice the mass as the other two. What are the speed and direction of the third piece? Give the direction as an angle east of north.

48. ‖ One billiard ball is shot east at 2.0 m/s. A second, identical billiard ball is shot west at 1.0 m/s. The balls have a glancing collision, not a head-on collision, deflecting the second ball by 90° and sending it north at 1.41 m/s. What are the speed and direction of the first ball after the collision? Give the direction as an angle south of east.

49. ‖ a. A bullet of mass m is fired into a block of mass M that is at rest. The block, with the bullet embedded, slides distance d across a horizontal surface. The coefficient of kinetic friction is μ_k. Find an expression for the bullet's speed v_{bullet}.
  b. What is the speed of a 10 g bullet that, when fired into a 10 kg stationary wood block, causes the block to slide 5.0 cm across a wood table?

50. ‖ You are part of a search-and-rescue mission that has been called out to look for a lost explorer. You've found the missing explorer, but, as **FIGURE P11.50** shows, you're separated from him by a 200-m-high cliff and a 30-m-wide raging river. To save his life, you need to get a 5.0 kg package of emergency supplies across the river. Unfortunately, you can't throw the package hard enough to make it across. Fortunately, you happen to have a 1.0 kg rocket intended for launching flares. Improvising quickly, you attach a sharpened stick to the front of the rocket, so that it will impale itself into the package of supplies, then fire the rocket at ground level toward the supplies. What minimum speed must the rocket have just before impact in order to save the explorer's life?

FIGURE P11.50

51. ‖ An object at rest on a flat, horizontal surface explodes into two fragments, one seven times as massive as the other. The heavier fragment slides 8.2 m before stopping. How far does the lighter fragment slide? Assume that both fragments have the same coefficient of kinetic friction.

52. || A 1500 kg weather rocket accelerates upward at 10 m/s². It explodes 2.0 s after liftoff and breaks into two fragments, one twice as massive as the other. Photos reveal that the lighter fragment traveled straight up and reached a maximum height of 530 m. What were the speed and direction of the heavier fragment just after the explosion?

53. || In a ballistics test, a 25 g bullet traveling horizontally at 1200 m/s goes through a 30-cm-thick 350 kg stationary target and emerges with a speed of 900 m/s. The target is free to slide on a smooth horizontal surface. What is the target's speed just after the bullet emerges?

54. || Two 500 g blocks of wood are 2.0 m apart on a frictionless table. A 10 g bullet is fired at 400 m/s toward the blocks. It passes all the way through the first block, then embeds itself in the second block. The speed of the first block immediately afterward is 6.0 m/s. What is the speed of the second block after the bullet stops in it?

55. ||| A 100 g granite cube slides down a 40° frictionless ramp. At the bottom, just as it exits onto a horizontal table, it collides with a 200 g steel cube at rest. How high above the table should the granite cube be released to give the steel cube a speed of 150 cm/s?

56. ||| You have been asked to design a "ballistic spring system" to measure the speed of bullets. A spring whose spring constant is k is suspended from the ceiling. A block of mass M hangs from the spring. A bullet of mass m is fired vertically upward into the bottom of the block and stops in the block. The spring's maximum compression d is measured.
    a. Find an expression for the bullet's speed v_B in terms of m, M, k, and d.
    b. What was the speed of a 10 g bullet if the block's mass is 2.0 kg and if the spring, with $k = 50$ N/m, was compressed by 45 cm?

57. || In FIGURE P11.57, a block of mass m slides along a frictionless track with speed v_m. It collides with a stationary block of mass M. Find an expression for the minimum value of v_m that will allow the second block to circle the loop-the-loop without falling off if the collision is
    (a) perfectly inelastic or (b) perfectly elastic.

FIGURE P11.57

58. || The stoplight had just changed and a 2000 kg Cadillac had entered the intersection, heading north at 3.0 m/s, when it was struck by a 1000 kg eastbound Volkswagen. The cars stuck together and slid to a halt, leaving skid marks angled 35° north of east. How fast was the Volkswagen going just before the impact?

59. || Ann (mass 50 kg) is standing at the left end of a 15-m-long, 500 kg cart that has frictionless wheels and rolls on a frictionless track. Initially both Ann and the cart are at rest. Suddenly, Ann starts running along the cart at a speed of 5.0 m/s relative to the cart. How far will Ann have run *relative to the ground* when she reaches the right end of the cart?

60. ||| Force $F_x = (10 \text{ N}) \sin(2\pi t/4.0 \text{ s})$ is exerted on a 250 g particle during the interval $0 \text{ s} \le t \le 2.0$ s. If the particle starts from rest, what is its speed at $t = 2.0$ s?
CALC

61. ||| A 500 g particle has velocity $v_x = -5.0$ m/s at $t = -2$ s. Force $F_x = (4 - t^2)$ N, where t is in s, is exerted on the particle between $t = -2$ s and $t = 2$ s. This force increases from 0 N at $t = -2$ s to 4 N at $t = 0$ s and then back to 0 N at $t = 2$ s. What is the particle's velocity at $t = 2$ s?
CALC

62. || A 30 ton rail car and a 90 ton rail car, initially at rest, are connected together with a giant but massless compressed spring between them. When released, the 30 ton car is pushed away at a speed of 4.0 m/s relative to the 90 ton car. What is the speed of the 30 ton car relative to the ground?

63. ||| A 20 g ball is fired horizontally with speed v_0 toward a 100 g ball hanging motionless from a 1.0-m-long string. The balls undergo a head-on, perfectly elastic collision, after which the 100 g ball swings out to a maximum angle $\theta_{max} = 50°$. What was v_0?

64. || A 100 g ball moving to the right at 4.0 m/s catches up and collides with a 400 g ball that is moving to the right at 1.0 m/s. If the collision is perfectly elastic, what are the speed and direction of each ball after the collision?

65. || A 100 g ball moving to the right at 4.0 m/s collides head-on with a 200 g ball that is moving to the left at 3.0 m/s.
    a. If the collision is perfectly elastic, what are the speed and direction of each ball after the collision?
    b. If the collision is perfectly inelastic, what are the speed and direction of the combined balls after the collision?

66. || Old naval ships fired 10 kg cannon balls from a 200 kg cannon. It was very important to stop the recoil of the cannon, since otherwise the heavy cannon would go careening across the deck of the ship. In one design, a large spring with spring constant 20,000 N/m was placed behind the cannon. The other end of the spring braced against a post that was firmly anchored to the ship's frame. What was the speed of the cannon ball if the spring compressed 50 cm when the cannon was fired?

67. || A proton (mass 1 u) is shot toward an unknown target nucleus at a speed of 2.50×10^6 m/s. The proton rebounds with its speed reduced by 25% while the target nucleus acquires a speed of 3.12×10^5 m/s. What is the mass, in atomic mass units, of the target nucleus?

68. || The nucleus of the polonium isotope ²¹⁴Po (mass 214 u) is radioactive and decays by emitting an alpha particle (a helium nucleus with mass 4 u). Laboratory experiments measure the speed of the alpha particle to be 1.92×10^7 m/s. Assuming the polonium nucleus was initially at rest, what is the recoil speed of the nucleus that remains after the decay?

69. || A neutron is an electrically neutral subatomic particle with a mass just slightly greater than that of a proton. A free neutron is radioactive and decays after a few minutes into other subatomic particles. In one experiment, a neutron at rest was observed to decay into a proton (mass 1.67×10^{-27} kg) and an electron (mass 9.11×10^{-31} kg). The proton and electron were shot out back-to-back. The proton speed was measured to be 1.0×10^5 m/s and the electron speed was 3.0×10^7 m/s. No other decay products were detected.
    a. Did momentum seem to be conserved in the decay of this neutron?

    NOTE Experiments such as this were first performed in the 1930s and seemed to indicate a failure of the law of conservation of momentum. In 1933, Wolfgang Pauli postulated that the neutron might have a *third* decay product that is virtually impossible to detect. Even so, it can carry away just enough momentum to keep the total momentum conserved. This proposed particle was named the *neutrino*, meaning "little neutral one." Neutrinos were, indeed, discovered nearly 20 years later.

    b. If a neutrino was emitted in the above neutron decay, in which direction did it travel? Explain your reasoning.
    c. How much momentum did this neutrino "carry away" with it?

70. ‖ A 20 g ball of clay traveling east at 2.0 m/s collides with a 30 g ball of clay traveling 30° south of west at 1.0 m/s. What are the speed and direction of the resulting 50 g blob of clay?

71. ‖ FIGURE P11.71 shows a collision between three balls of clay. The three hit simultaneously and stick together. What are the speed and direction of the resulting blob of clay?

FIGURE P11.71

72. ‖ A 2100 kg truck is traveling east through an intersection at 2.0 m/s when it is hit simultaneously from the side and the rear. (Some people have all the luck!) One car is a 1200 kg compact traveling north at 5.0 m/s. The other is a 1500 kg midsize traveling east at 10 m/s. The three vehicles become entangled and slide as one body. What are their speed and direction just after the collision?

73. ‖ A rocket in deep space has an empty mass of 150 kg and exhausts the hot gases of burned fuel at 2500 m/s. It is loaded with 600 kg of fuel, which it burns in 30 s. What is the rocket's speed 10 s, 20 s, and 30 s after launch?

74. ‖ a. To understand why rockets often have multiple stages, first consider a single-stage rocket with an empty mass of 200 kg, 800 kg of fuel, and a 2000 m/s exhaust speed. If fired in deep space, what is the rocket's maximum speed?

   b. Now divide the rocket into two stages, each with an empty mass of 100 kg, 400 kg of fuel, and a 2000 m/s exhaust speed. The first stage is released after it runs out of fuel. What is the top speed of the second stage? You'll need to consider how the equation for v_{max} should be altered when a rocket is not starting from rest.

In Problems 75 through 78 you are given the equation(s) used to solve a problem. For each of these, you are to
   a. Write a realistic problem for which this is the correct equation(s).
   b. Finish the solution of the problem, including a pictorial representation.

75. $(0.10\,\text{kg})(40\,\text{m/s}) - (0.10\,\text{kg})(-30\,\text{m/s}) = \frac{1}{2}(1400\,\text{N})\Delta t$

76. $(600\,\text{g})(4.0\,\text{m/s}) = (400\,\text{g})(3.0\,\text{m/s}) + (200\,\text{g})(v_{ix})_2$

77. $(3000\,\text{kg})v_{fx} = (2000\,\text{kg})(5.0\,\text{m/s}) + (1000\,\text{kg})(-4.0\,\text{m/s})$

78. $(0.10\,\text{kg} + 0.20\,\text{kg})v_{1x} = (0.10\,\text{kg})(3.0\,\text{m/s})$

$\frac{1}{2}(0.30\,\text{kg})(0\,\text{m/s})^2 + \frac{1}{2}(3.0\,\text{N/m})(\Delta x_2)^2$

$= \frac{1}{2}(0.30\,\text{kg})(v_{1x})^2 + \frac{1}{2}(3.0\,\text{N/m})(0\,\text{m})^2$

Challenge Problems

79. ‖‖ A 1000 kg cart is rolling to the right at 5.0 m/s. A 70 kg man is standing on the right end of the cart. What is the speed of the cart if the man suddenly starts running to the left with a speed of 10 m/s relative to the cart?

80. ‖‖ A spaceship of mass 2.0×10^6 kg is cruising at a speed of 5.0×10^6 m/s when the antimatter reactor fails, blowing the ship into three pieces. One section, having a mass of 5.0×10^5 kg, is blown straight backward with a speed of 2.0×10^6 m/s. A second piece, with mass 8.0×10^5 kg, continues forward at 1.0×10^6 m/s. What are the direction and speed of the third piece?

81. ‖‖ A 20 kg wood ball hangs from a 2.0-m-long wire. The maximum tension the wire can withstand without breaking is 400 N. A 1.0 kg projectile traveling horizontally hits and embeds itself in the wood ball. What is the greatest speed this projectile can have without causing the wire to break?

82. ‖‖ A two-stage rocket is traveling at 1200 m/s with respect to the earth when the first stage runs out of fuel. Explosive bolts release the first stage and push it backward with a speed of 35 m/s relative to the second stage. The first stage is three times as massive as the second stage. What is the speed of the second stage after the separation?

83. ‖‖ The air-track carts in FIGURE CP11.83 are sliding to the right at 1.0 m/s. The spring between them has a spring constant of 120 N/m and is compressed 4.0 cm. The carts slide past a flame that burns through the string holding them together. Afterward, what are the speed and direction of each cart?

FIGURE P11.83

84. ‖‖ Section 11.6 found an equation for v_{max} of a rocket fired in deep space. What is v_{max} for a rocket fired vertically from the surface of an airless planet with free-fall acceleration g? Referring to Section 11.6, you can write an equation for ΔP_y, the change of momentum in the vertical direction, in terms of dm and dv_y. ΔP_y is no longer zero because now gravity delivers an impulse. Rewrite the momentum equation by including the impulse due to gravity during the time dt during which the mass changes by dm. Pay attention to signs! Your equation will have three differentials, but two are related through the fuel burn rate R. Use this relationship—again pay attention to signs; m is decreasing—to write your equation in terms of dm and dv_y. Then integrate to find a modified expression for v_{max} at the instant all the fuel has been burned.

   a. What is v_{max} for a vertical launch from an airless planet? Your answer will be in terms of m_R, the empty rocket mass; m_{F0}, the initial fuel mass; v_{ex}, the exhaust speed; R, the fuel burn rate; and g.

   b. A rocket with a total mass of 330,000 kg when fully loaded burns all 280,000 kg of fuel in 250 s. The engines generate 4.1 MN of thrust. What is this rocket's speed at the instant all the fuel has been burned if it is launched in deep space? If it is launched vertically from the earth?

CALC

Conservation Laws

KEY FINDINGS   What are the overarching findings of Part II?

■ **Work** causes a system's energy to change. It is a *transfer* of energy to or from the environment.

■ **Energy** can be *transformed* within a system, but the **total energy** of an isolated system does not change.

■ An **impulse** causes a system's momentum to change.

■ Momentum can be *exchanged* within a system, but the **total momentum** of an isolated system does not change.

LAWS   What laws of physics govern energy and momentum?

Energy principle $\Delta E_{sys} = \Delta K + \Delta U + \Delta E_{th} = W_{ext}$ or $K_i + U_i + W_{ext} = K_f + U_f + \Delta E_{th}$

Conservation of energy   For an isolated system ($W_{ext} = 0$), the total energy $E_{sys} = K + U + E_{th}$ is conserved. $\Delta E_{sys} = 0$.
                             For an isolated, nondissipative system, the **mechanical energy** $E_{mech} = K + U$ is conserved.

Momentum principle $\Delta \vec{p} = \vec{J}$

Conservation of momentum   For an isolated system ($\vec{J} = \vec{0}$), the total momentum is conserved. $\Delta \vec{P} = \vec{0}$.

MODELS   What are the most common models for using conservation laws?

Basic energy model

■ Energy is a property of the system.

■ Energy is *transformed* within the system without loss.

■ Energy is *transferred* to and from the system by forces that do *work*.

   ▪ $W > 0$ for energy added.

   ▪ $W < 0$ for energy removed.

   ▪ $\Delta E_{sys} = \Delta K + \Delta U + \Delta E_{th} = W_{ext}$

Other models and approximations

■ An isolated system does not interact with its environment.

   ▪ For energy, an isolated system has no work done on it.

   ▪ For momentum, an isolated system experiences no impulse.

■ Thermal energy is the microscopic energy of moving atoms and stretched bonds.

Collision model

■ Perfectly inelastic collision: Objects stick together and move with a common final velocity.

   ▪ Momentum is conserved.

   ▪ $(m_1 + m_2)v_{fx} = m_1(v_{ix})_1 + m_2(v_{ix})_2$

■ Perfectly elastic collision: Objects bounce apart with no loss of energy.

   ▪ Momentum and energy are both conserved.

■ An ideal spring obeys **Hooke's law** for all displacements: $(F_{Sp})_s = -k\,\Delta s$.

■ The impulse approximation ignores forces that are small compared to impulsive forces during the brief time of a collision or explosion.

TOOLS   What are the most important tools for using energy and momentum?

■ Kinetic energy:
  $K = \frac{1}{2}mv^2$

   ▪ Kinetic energy is an energy of *motion*.

■ Gravitational potential energy:
  $U_G = mgy$

■ Elastic potential energy:
  $U_{Sp} = \frac{1}{2}k(\Delta s)^2$

   ▪ Potential energy is an energy of *position*.

■ Momentum: $\vec{p} = m\vec{v}$.

■ Force acting through a displacement does work.

   ▪ Constant force:
     $W = \vec{F} \cdot \Delta \vec{r} = F(\Delta r)\cos\theta$

   ▪ Variable force:

     $W = \int_{s_i}^{s_f} F_s\,ds$

     = area under the F_s-versus-s curve

■ Force acting over time delivers an impulse:

   $J_x = \int_{t_i}^{t_f} F_x\,dt$

   = area under the F_x-versus-t curve

■ Energy diagrams

   ▪ Visualize speed changes and turning points.
     $F_x =$ negative of the slope of the curve.

■ Bar charts

   ▪ Momentum and energy conservation.

■ Before-and-after pictorial representation.

   ▪ Important problem-solving tool.

OVERVIEW

Power Over Our Environment

Early humans had to endure whatever nature provided. Only within the last few thousand years have agriculture and technology provided some level of control over the environment. And it has been a mere couple of centuries since machines, and later electronics, began to do much of our work and provide us with "creature comforts."

It's no coincidence that machines began to appear about a century after Galileo, Newton, and others ignited what we now call the *scientific revolution*. The machines and other devices we take for granted today are direct consequences of scientific knowledge and the scientific method.

Parts I and II have established Newton's theory of motion, the foundation of modern science. Most of the applications will be developed in other science and engineering courses, but we're now in a good position to examine a few of the more practical aspects of Newtonian mechanics.

Our goal for Part III is to apply our newfound theory to three important topics:

■ **Rotation.** Rotation is a very important form of motion, but to understand rotational motion we'll need to introduce a new model—the *rigid-body model*. We'll then be able to study rolling wheels and spinning space stations. Rotation will also lead to the law of conservation of angular momentum.

■ **Gravity.** By adding one more law, Newton's law of gravity, we'll be able to understand much about the physics of the space station, communication satellites, the solar system, and interplanetary travel.

■ **Fluids.** Liquids and gases *flow*. Surprisingly, it takes no new physics to understand the basic mechanical properties of fluids. By applying our understanding of force, we'll be able to understand what pressure is, how a steel ship can float, and how fluids flow through pipes.

Newton's laws of motion and the conservation laws, especially conservation of energy, will be the tools that allow us to analyze and understand a variety of interesting and practical applications.

A hurricane is a fluid—the air—moving on a rotating sphere—the earth—under the influence of gravity. Understanding hurricanes is very much an application of Newtonian mechanics.

12 Rotation of a Rigid Body

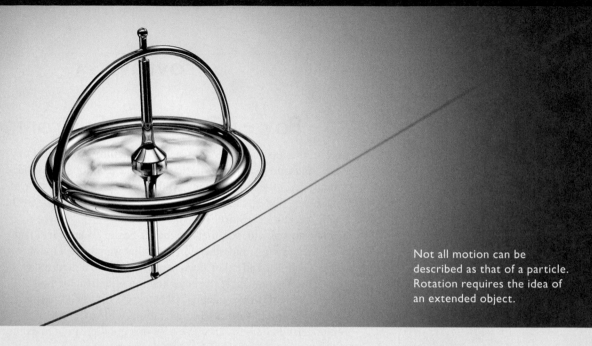

Not all motion can be described as that of a particle. Rotation requires the idea of an extended object.

IN THIS CHAPTER, you will learn to understand and apply the physics of rotation.

What is a rigid body?

An object whose size and shape don't change as it moves is called a rigid body. A rigid body is characterized by its moment of inertia I, which is the rotational equivalent of mass. We'll consider

- Rotation about an axle.
- Rolling without slipping.

«LOOKING BACK Section 6.1 Equilibrium

What is torque?

Torque is the tendency or ability of a force to rotate an object about a pivot point. You'll learn that torque depends on both the force *and* where the force is applied. A longer wrench provides a larger torque.

What does torque do?

Torque is to rotation what force is to linear motion. Torque τ causes an object to have angular acceleration. Newton's second law for rotation is $\alpha = \tau/I$. Much of rotational dynamics will look familiar because it is analogous to linear dynamics.

«LOOKING BACK Section 6.2 Newton's second law

What is angular momentum?

Angular momentum is to rotation what momentum is to linear motion. Angular momentum is an object's tendency to "keep rotating." Angular momentum $\vec{L}$ is a *vector* pointing along the rotation axis. You'll use angular momentum to understand the precession of a spinning top or gyroscope.

What is conserved in rotational motion?

The mechanical energy of a rotating object includes its rotational kinetic energy $\frac{1}{2}I\omega^2$. This is analogous to linear kinetic energy.

- Mechanical energy is conserved for frictionless, rotating systems.
- Angular momentum is conserved for isolated systems.

«LOOKING BACK Section 10.4 Conservation of energy; Section 11.2 Conservation of momentum

Why is rigid-body motion important?

The world is full of rotating objects, from windmill turbines to the gyroscopes used in navigation. The wheels on your bicycle or car roll without slipping. Scientists investigate rotating molecules and rotating galaxies. No understanding of motion is complete without understanding rotational motion, and this chapter will develop the tools you need. We'll also expand our understanding of equilibrium by exploring the conditions under which objects *don't* rotate.

12.1 Rotational Motion

Thus far, our study of motion has focused almost exclusively on the *particle model*, in which an object is represented as a mass at a single point in space. As we expand our study of motion to rotation, we need to consider *extended objects* whose size and shape *do* matter. Thus this chapter will be based on the **rigid-body model:**

MODEL 12.1

Rigid-body model

A **rigid body** is an extended object whose size and shape do not change as it moves.

- Particle-like atoms are held together by rigid massless rods.
- A rigid body cannot be stretched, compressed, or deformed. All points on the body have the same angular velocity and angular acceleration.
- Limitations: Model fails if an object changes shape or is deformed.

Atoms are particles.

Bonds are rigid, massless rods.

Exercise 1

FIGURE 12.1 illustrates the three basic types of motion of a rigid body: **translational motion, rotational motion,** and **combination motion.**

FIGURE 12.1 Three basic types of motion of a rigid body.

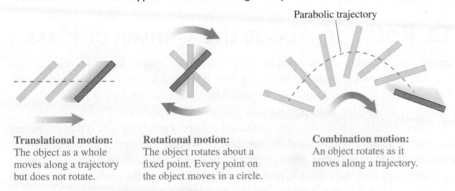

Parabolic trajectory

Translational motion:
The object as a whole moves along a trajectory but does not rotate.

Rotational motion:
The object rotates about a fixed point. Every point on the object moves in a circle.

Combination motion:
An object rotates as it moves along a trajectory.

Brief Review of Rotational Kinematics

Rotation is an extension of circular motion, so we begin with a brief summary of Chapter 4. **A review of « Sections 4.4–4.6 is highly recommended.** FIGURE 12.2 shows a wheel rotating on an axle. Its angular velocity

$$\omega = \frac{d\theta}{dt} \tag{12.1}$$

is the rate at which the wheel rotates. The SI units of ω are radians per second (rad/s), but revolutions per second (rev/s) and revolutions per minute (rpm) are frequently used. Notice that all points have equal angular velocities, so we can refer to the angular velocity ω *of the wheel.*

If the wheel is speeding up or slowing down, its angular acceleration is

$$\alpha = \frac{d\omega}{dt} \tag{12.2}$$

The units of angular acceleration are rad/s^2. Angular acceleration is the *rate* at which the angular velocity ω changes, just as the linear acceleration is the rate at which the

FIGURE 12.2 Rotational motion.

Every point on the wheel rotates with the *same* angular velocity ω.

All points have a tangential velocity and a radial (centripetal) acceleration. They have a tangential acceleration if the wheel has angular acceleration.

TABLE 12.1 Rotational kinematics for constant angular acceleration

$$\omega_f = \omega_i + \alpha \, \Delta t$$

$$\theta_f = \theta_i + \omega_i \, \Delta t + \tfrac{1}{2} \alpha (\Delta t)^2$$

$$\omega_f^2 = \omega_i^2 + 2\alpha \, \Delta\theta$$

linear velocity v changes. **TABLE 12.1** summarizes the kinematic equations for rotation with constant angular acceleration.

FIGURE 12.3 reminds you of the sign conventions for angular velocity and acceleration. They will be especially important in the present chapter. Be careful with the sign of α. Just as with linear acceleration, positive and negative values of α can't be interpreted as simply "speeding up" and "slowing down."

FIGURE 12.3 The signs of angular velocity and angular acceleration.

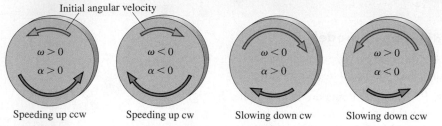

Initial angular velocity

| $\omega > 0$ $\alpha > 0$ | $\omega < 0$ $\alpha < 0$ | $\omega < 0$ $\alpha > 0$ | $\omega > 0$ $\alpha < 0$ |

Speeding up ccw   Speeding up cw   Slowing down cw   Slowing down ccw

The rotation is speeding up if ω and α have the same sign, slowing down if they have opposite signs.

A point at distance r from the rotation axis has instantaneous velocity and acceleration, shown in Figure 12.2, given by

$$v_r = 0 \qquad a_r = \frac{v_t^2}{r} = \omega^2 r$$

$$v_t = r\omega \qquad a_t = r\alpha \tag{12.3}$$

The sign convention for ω implies that v_t and a_t are positive if they point in the counterclockwise (ccw) direction, negative if they point in the clockwise (cw) direction.

12.2 Rotation About the Center of Mass

FIGURE 12.4 Rotation about the center of mass.

(a)

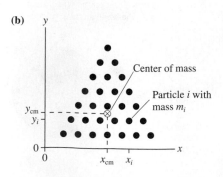

The object rotates about this point, which is the center of mass.

(b)

Center of mass

Particle i with mass m_i

Imagine yourself floating in a space capsule deep in space. Suppose you take an object like that shown in **FIGURE 12.4a** and spin it so that it simply rotates but has no translational motion as it floats beside you. *About what point does it rotate?* That is the question we need to answer.

An unconstrained object (i.e., one not on an axle or a pivot) on which there is no net force rotates about a point called the **center of mass.** The center of mass remains motionless while every other point in the object undergoes circular motion around it. You need not go deep into space to demonstrate rotation about the center of mass. If you have an air table, a flat object rotating on the air table rotates about its center of mass.

To locate the center of mass, **FIGURE 12.4b** models the object as a set of particles numbered $i = 1, 2, 3, \ldots$. Particle i has mass m_i and is located at position (x_i, y_i). We'll prove later in this section that the center of mass is located at position

$$x_{cm} = \frac{1}{M} \sum_i m_i x_i = \frac{m_1 x_1 + m_2 x_2 + m_3 x_3 + \cdots}{m_1 + m_2 + m_3 + \cdots}$$

$$y_{cm} = \frac{1}{M} \sum_i m_i y_i = \frac{m_1 y_1 + m_2 y_2 + m_3 y_3 + \cdots}{m_1 + m_2 + m_3 + \cdots} \tag{12.4}$$

where $M = m_1 + m_2 + m_3 + \cdots$ is the object's total mass.

Let's see if Equations 12.4 make sense. Suppose you have an object consisting of N particles, all with the same mass m. That is, $m_1 = m_2 = \cdots = m_N = m$. We can factor the m out of the numerator, and the denominator becomes simply Nm. The m cancels, and the x-coordinate of the center of mass is

$$x_{cm} = \frac{x_1 + x_2 + \cdots + x_N}{N} = x_{average}$$

In this case, x_{cm} is simply the *average* x-coordinate of all the particles. Likewise, y_{cm} will be the average of all the y-coordinates.

This *does* make sense! If the particle masses are all the same, the center of mass should be at the center of the object. And the "center of the object" is the average position of all the particles. To allow for *unequal* masses, Equations 12.4 are called *weighted averages*. Particles of higher mass count more than particles of lower mass, but the basic idea remains the same. **The center of mass is the mass-weighted center of the object.**

EXAMPLE 12.1 | **The center of mass of a barbell**

A barbell consists of a 500 g ball and a 2.0 kg ball connected by a massless 50-cm-long rod.

a. Where is the center of mass?

b. What is the speed of each ball if they rotate about the center of mass at 40 rpm?

MODEL Model the barbell as a rigid body.

VISUALIZE FIGURE 12.5 shows the two masses. We've chosen a coordinate system in which the masses are on the x-axis with the 2.0 kg mass at the origin.

SOLVE a. We can use Equations 12.4 to calculate that the center of mass is

$$x_{cm} = \frac{m_1 x_1 + m_2 x_2}{m_1 + m_2}$$

$$= \frac{(2.0\ \text{kg})(0.0\ \text{m}) + (0.50\ \text{kg})(0.50\ \text{m})}{2.0\ \text{kg} + 0.50\ \text{kg}} = 0.10\ \text{m}$$

$y_{cm} = 0$ because all the masses are on the x-axis. The center of mass is 20% of the way from the 2.0 kg ball to the 0.50 kg ball.

FIGURE 12.5 Finding the center of mass.

b. Each ball rotates about the center of mass. The radii of the circles are $r_1 = 0.10$ m and $r_2 = 0.40$ m. The tangential velocities are $(v_i)_t = r_i \omega$, but this equation requires ω to be in rad/s. The conversion is

$$\omega = 40\ \frac{\text{rev}}{\text{min}} \times \frac{1\ \text{min}}{60\ \text{s}} \times \frac{2\pi\ \text{rad}}{1\ \text{rev}} = 4.19\ \text{rad/s}$$

Consequently,

$$(v_1)_t = r_1\omega = (0.10\ \text{m})(4.19\ \text{rad/s}) = 0.42\ \text{m/s}$$

$$(v_2)_t = r_2\omega = (0.40\ \text{m})(4.19\ \text{rad/s}) = 1.68\ \text{m/s}$$

ASSESS The center of mass is closer to the heavier ball than to the lighter ball. We expected this because x_{cm} is a mass-weighted average of the positions. But the lighter mass moves faster because it is farther from the rotation axis.

Finding the Center of Mass by Integration

For any realistic object, carrying out the summations of Equations 12.4 over all the atoms in the object is not practical. Instead, as **FIGURE 12.6** shows, we can divide an extended object into many small cells or boxes, each with the same very small mass Δm. We will number the cells 1, 2, 3, ... , just as we did the particles. Cell i has coordinates (x_i, y_i) and mass $m_i = \Delta m$. The center-of-mass coordinates are then

$$x_{cm} = \frac{1}{M} \sum_i x_i\, \Delta m \quad \text{and} \quad y_{cm} = \frac{1}{M} \sum_i y_i\, \Delta m$$

Now, as you might expect, we'll let the cells become smaller and smaller, with the total number increasing. As each cell becomes infinitesimally small, we can replace Δm with dm and the sum by an integral. Then

$$x_{cm} = \frac{1}{M} \int x\, dm \quad \text{and} \quad y_{cm} = \frac{1}{M} \int y\, dm \qquad (12.5)$$

FIGURE 12.6 Calculating the center of mass of an extended object.

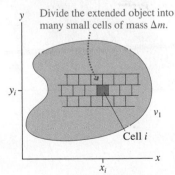

Equations 12.5 are a formal definition of the center of mass, but they are *not* ready to integrate in this form. First, integrals are carried out over *coordinates*, not over masses. Before we can integrate, we must replace dm by an equivalent expression involving a coordinate differential such as dx or dy. Second, no limits of integration have been specified. The procedure for using Equations 12.5 is best shown with an example.

EXAMPLE 12.2 | The center of mass of a rod

Find the center of mass of a thin, uniform rod of length L and mass M. Use this result to find the tangential acceleration of one tip of a 1.60-m-long rod that rotates about its center of mass with an angular acceleration of 6.0 rad/s^2.

VISUALIZE **FIGURE 12.7** shows the rod. We've chosen a coordinate system such that the rod lies along the x-axis from 0 to L. Because the rod is "thin," we'll assume that $y_{cm} = 0$.

FIGURE 12.7 Finding the center of mass of a long, thin rod.

A small cell of width dx at position x has mass $dm = (M/L)dx$.

SOLVE Our first task is to find x_{cm}, which lies somewhere on the x-axis. To do this, we divide the rod into many small cells of mass dm. One such cell, at position x, is shown. The cell's width is dx. Because the rod is *uniform,* the mass of this little cell is the *same fraction* of the total mass M that dx is of the total length L. That is,

$$\frac{dm}{M} = \frac{dx}{L}$$

Consequently, we can express dm in terms of the coordinate differential dx as

$$dm = \frac{M}{L} dx$$

NOTE The change of variables from dm to the differential of a coordinate is *the* key step in calculating the center of mass.

With this expression for dm, Equation 12.5 for x_{cm} becomes

$$x_{cm} = \frac{1}{M}\left(\frac{M}{L}\int x \, dx\right) = \frac{1}{L}\int_0^L x \, dx$$

where in the last step we've noted that summing "all the mass in the rod" means integrating from $x = 0$ to $x = L$. This is a straightforward integral to carry out, giving

$$x_{cm} = \frac{1}{L}\left[\frac{x^2}{2}\right]_0^L = \frac{1}{L}\left[\frac{L^2}{2} - 0\right] = \tfrac{1}{2}L$$

The center of mass is at the center of the rod, as you probably guessed. For a 1.60-m-long rod, each tip of the rod rotates in a circle with $r = \tfrac{1}{2}L = 0.80$ m. The tangential acceleration, the rate at which the tip is speeding up, is

$$a_t = r\alpha = (0.80 \text{ m})(6.0 \text{ rad/s}^2) = 4.8 \text{ m/s}^2$$

NOTE For any symmetrical object of uniform density, the center of mass is at the physical center of the object.

FIGURE 12.8 Finding the center of mass.

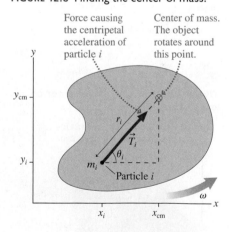

Force causing the centripetal acceleration of particle i

Center of mass. The object rotates around this point.

To see where the center-of-mass equations come from, **FIGURE 12.8** shows an object rotating about its center of mass. Particle i is moving in a circle, so it *must* have a centripetal acceleration. Acceleration requires a force, and this force is due to tension in the molecular bonds that hold the object together. Force $\vec{T}_i$ on particle i has magnitude

$$T_i = m_i(a_i)_r = m_i r_i \omega^2 \tag{12.6}$$

where we used Equation 12.3 for a_r. All points in a rigid rotating object have the *same* angular velocity, so ω doesn't need a subscript.

At every instant of time, the internal tension forces are all paired as action/reaction forces, equal in magnitude but opposite in direction, so the sum of all the tension forces must be zero. That is, $\sum \vec{T}_i = \vec{0}$. The x-component of this sum is

$$\sum_i (T_i)_x = \sum_i T_i \cos\theta_i = \sum_i (m_i r_i \omega^2)\cos\theta_i = 0 \tag{12.7}$$

You can see from Figure 12.8 that $\cos\theta_i = (x_{cm} - x_i)/r_i$. Thus

$$\sum_i (T_i)_x = \sum_i (m_i r_i \omega^2)\frac{x_{cm} - x_i}{r_i} = \left(\sum_i m_i x_{cm} - \sum_i m_i x_i\right)\omega^2 = 0 \tag{12.8}$$

This equation will be true if the term in parentheses is zero. x_{cm} is a constant, so we can bring it outside the summation to write

$$\sum_i m_i x_{cm} - \sum_i m_i x_i = \left(\sum_i m_i\right)x_{cm} - \sum_i m_i x_i = M x_{cm} - \sum_i m_i x_i = 0 \tag{12.9}$$

where we used the fact that $\sum m_i$ is simply the object's total mass M. Solving for x_{cm}, we find the x-coordinate of the object's center of mass to be

$$x_{cm} = \frac{1}{M}\sum_i m_i x_i = \frac{m_1 x_1 + m_2 x_2 + m_3 x_3 + \cdots}{m_1 + m_2 + m_3 + \cdots} \tag{12.10}$$

This was Equation 12.4. The y-equation is found similarly.

STOP TO THINK 12.1 A baseball bat is cut into two pieces at its center of mass. Which end is heavier?

a. The handle end (left end).
b. The hitting end (right end).
c. The two ends weigh the same.

12.3 Rotational Energy

A rotating rigid body—whether it's rotating freely about its center of mass or constrained to rotate on an axle—has kinetic energy because all atoms in the object are in motion. The kinetic energy due to rotation is called **rotational kinetic energy.**

FIGURE 12.9 shows a few of the particles making up a solid object that rotates with angular velocity ω. Particle i, which rotates in a circle of radius r_i, moves with speed $v_i = r_i\omega$. The object's rotational kinetic energy is the sum of the kinetic energies of each of the particles:

$$K_{rot} = \tfrac{1}{2}m_1v_1^2 + \tfrac{1}{2}m_2v_2^2 + \cdots \qquad (12.11)$$

$$= \tfrac{1}{2}m_1r_1^2\omega^2 + \tfrac{1}{2}m_2r_2^2\omega^2 + \cdots = \tfrac{1}{2}\left(\sum_i m_ir_i^2\right)\omega^2$$

The quantity $\sum m_ir_i^2$ is called the object's **moment of inertia** I:

$$I = m_1r_1^2 + m_2r_2^2 + m_3r_3^2 + \cdots = \sum_i m_ir_i^2 \qquad (12.12)$$

The units of moment of inertia are $\text{kg}\,\text{m}^2$. **An object's moment of inertia depends on the axis of rotation.** Once the axis is specified, allowing the values of r_i to be determined, the moment of inertia *about that axis* can be calculated from Equation 12.12.

Written using the moment of inertia I, the rotational kinetic energy is

$$K_{rot} = \tfrac{1}{2}I\omega^2 \qquad (12.13)$$

NOTE Rotational kinetic energy is *not* a new form of energy. It is the familiar kinetic energy of motion, simply expressed in a form that is convenient for rotational motion. Notice the analogy with the familiar $\tfrac{1}{2}mv^2$.

FIGURE 12.9 Rotational kinetic energy is due to the motion of the particles.

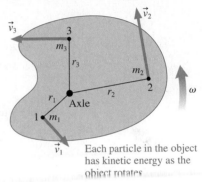

Each particle in the object has kinetic energy as the object rotates.

EXAMPLE 12.3 | **A rotating widget**

Students participating in an engineering project design the triangular widget seen in FIGURE 12.10. The three masses, held together by lightweight plastic rods, rotate in the plane of the page about an axle passing through the right-angle corner. At what angular velocity does the widget have 100 mJ of rotational energy?

FIGURE 12.10 The rotating widget.

150 g

6.0 cm

Axle

250 g

8.0 cm   300 g

MODEL Model the widget as a rigid body consisting of three particles connected by massless rods.

SOLVE Rotational energy is $K = \tfrac{1}{2}I\omega^2$. The moment of inertia is measured about the rotation axis, thus

$$I = \sum_i m_ir_i^2 = (0.25\,\text{kg})(0.080\,\text{m})^2 + (0.15\,\text{kg})(0.060\,\text{m})^2$$
$$+ (0.30\,\text{kg})(0\,\text{m})^2$$
$$= 2.14 \times 10^{-3}\,\text{kg}\,\text{m}^2$$

The largest mass makes no contribution to I because it is on the rotation axis with $r = 0$. With I known, the angular velocity is

$$\omega = \sqrt{\frac{2K}{I}} = \sqrt{\frac{2(0.10\,\text{J})}{2.14 \times 10^{-3}\,\text{kg}\,\text{m}^2}}$$

$$= 9.67\,\text{rad/s} \times \frac{1\,\text{rev}}{2\pi\,\text{rad}} = 1.54\,\text{rev/s} = 92\,\text{rpm}$$

ASSESS The moment of inertia depends on the distance of each mass from the rotation axis. The moment of inertia would be different for an axle passing through either of the other two masses, and thus the required angular velocity would be different.

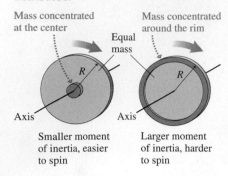

FIGURE 12.11 Moment of inertia depends on both the mass and how the mass is distributed.

Mass concentrated at the center

Mass concentrated around the rim

Equal mass

Axis

Axis

Smaller moment of inertia, easier to spin

Larger moment of inertia, harder to spin

Before rushing to calculate moments of inertia, let's get a better understanding of the meaning. First, notice that **moment of inertia is the rotational equivalent of mass**. It plays the same role in Equation 12.13 as mass m in the now-familiar $K = \frac{1}{2}mv^2$. Recall that the quantity we call *mass* was actually defined as the *inertial mass*. Objects with larger mass have a larger *inertia,* meaning that they're harder to accelerate. Similarly, an object with a larger moment of inertia is harder to rotate. The fact that *moment of inertia* retains the word "inertia" reminds us of this.

Consider the two wheels shown in FIGURE 12.11. They have the same total mass M and the same radius R. As you probably know from experience, it's much easier to spin the wheel whose mass is concentrated at the center than to spin the one whose mass is concentrated around the rim. This is because having the mass near the center (smaller values of r_i) lowers the moment of inertia.

Moments of inertia for many solid objects are tabulated and found online. You would need to compute I yourself only for an object of unusual shape. TABLE 12.2 is a short list of common moments of inertia. We'll see in the next section where these come from, but do notice how I depends on the rotation axis.

TABLE 12.2 Moments of inertia of objects with uniform density

Object and axis	Picture	I	Object and axis	Picture	I
Thin rod, about center		$\frac{1}{12}ML^2$	Cylinder or disk, about center		$\frac{1}{2}MR^2$
Thin rod, about end		$\frac{1}{3}ML^2$	Cylindrical hoop, about center		MR^2
Plane or slab, about center		$\frac{1}{12}Ma^2$	Solid sphere, about diameter		$\frac{2}{5}MR^2$
Plane or slab, about edge		$\frac{1}{3}Ma^2$	Spherical shell, about diameter		$\frac{2}{3}MR^2$

If the rotation axis is not through the center of mass, then rotation may cause the center of mass to move up or down. In that case, the object's gravitational potential energy $U_G = Mgy_{cm}$ will change. If there are no dissipative forces (i.e., if the axle is frictionless) and if no work is done by external forces, then the mechanical energy

$$E_{mech} = K_{rot} + U_G = \frac{1}{2}I\omega^2 + Mgy_{cm} \qquad (12.14)$$

is a conserved quantity.

EXAMPLE 12.4 | The speed of a rotating rod

A 1.0-m-long, 200 g rod is hinged at one end and connected to a wall. It is held out horizontally, then released. What is the speed of the tip of the rod as it hits the wall?

MODEL The mechanical energy is conserved if we assume the hinge is frictionless. The rod's gravitational potential energy is transformed into rotational kinetic energy as it "falls."

FIGURE 12.12 A before-and-after pictorial representation of the rod.

y

Hinge

$L = 1.0$ m

Before: $y_{cm0} = 0$ m
$\omega_0 = 0$ rad/s
$m = 0.20$ kg

After: $y_{cm1} = -\frac{1}{2}L$

$\vec{v}_{tip}$

Find: $v_{tip} = \omega_1 L$

VISUALIZE FIGURE 12.12 is a familiar before-and-after pictorial representation of the rod.

SOLVE Mechanical energy is conserved, so we can equate the rod's final mechanical energy to its initial mechanical energy:

$$\tfrac{1}{2}I\omega_1^2 + Mgy_{cm1} = \tfrac{1}{2}I\omega_0^2 + Mgy_{cm0}$$

The initial conditions are $\omega_0 = 0$ and $y_{cm0} = 0$. The center of mass moves to $y_{cm1} = -\frac{1}{2}L$ as the rod hits the wall. From Table 12.2 we find $I = \frac{1}{3}ML^2$ for a rod rotating about one end. Thus

$$\tfrac{1}{2}I\omega_1^2 + Mgy_{cm1} = \tfrac{1}{6}ML^2\omega_1^2 - \tfrac{1}{2}MgL = 0$$

We can solve this for the rod's angular velocity as it hits the wall:

$$\omega_1 = \sqrt{\frac{3g}{L}}$$

The tip of the rod is moving in a circle with radius $r = L$. Its final speed is

$$v_{tip} = \omega_1 L = \sqrt{3gL} = 5.4 \text{ m/s}$$

ASSESS 5.4 m/s = 12 mph, which seems plausible for a meter-long rod swinging through 90°.

STOP TO THINK 12.2 A solid cylinder and a cylindrical shell, each with radius R and mass M, rotate about their axes with the same angular velocity ω. Which has more kinetic energy?

a. The solid cylinder.
b. The cylindrical shell.
c. They have the same kinetic energy.
d. Neither has kinetic energy because they are only rotating, not moving.

12.4 Calculating Moment of Inertia

The equation for rotational energy is easy to write, but we can't make use of it without knowing an object's moment of inertia. Unlike mass, we can't measure moment of inertia by putting an object on a scale. And while we can guess that the center of mass of a symmetrical object is at the physical center of the object, we can *not* guess the moment of inertia of even a simple object. To find I, we really must carry through the calculation.

Equation 12.12 defines the moment of inertia as a sum over all the particles in the system. As we did for the center of mass, we can replace the individual particles with cells 1, 2, 3, … of mass Δm. Then the moment of inertia summation can be converted to an integration:

$$I = \sum_i r_i^2 \Delta m \xrightarrow[\Delta m \to 0]{} I = \int r^2 \, dm \qquad (12.15)$$

where r is the distance from the rotation axis. If we let the rotation axis be the z-axis, then we can write the moment of inertia as

$$I = \int (x^2 + y^2) \, dm \qquad \text{(rotation about the } z\text{-axis)} \qquad (12.16)$$

NOTE You *must* replace dm by an equivalent expression involving a coordinate differential such as dx or dy before you can carry out the integration.

You can use any coordinate system to calculate the coordinates x_{cm} and y_{cm} of the center of mass. But the moment of inertia is defined for rotation about a particular axis, and r is measured from that axis. Thus the coordinate system used for moment-of-inertia calculations *must* have its origin at the pivot point.

EXAMPLE 12.5 | Moment of inertia of a rod about a pivot at one end

Find the moment of inertia of a thin, uniform rod of length L and mass M that rotates about a pivot at one end.

MODEL An object's moment of inertia depends on the axis of rotation. In this case, the rotation axis is at the end of the rod.

VISUALIZE FIGURE 12.13 defines an x-axis with the origin at the pivot point.

FIGURE 12.13 Setting up the integral to find the moment of inertia of a rod.

A small cell of width dx at position x has mass $dm = (M/L)dx$.

SOLVE Because the rod is thin, we can assume that $y \approx 0$ for all points on the rod. Thus

$$I = \int x^2 \, dm$$

The small amount of mass dm in the small length dx is $dm = (M/L) \, dx$, as we found in Example 12.2. The rod extends from $x = 0$ to $x = L$, so the moment of inertia about one end is

$$I_{end} = \frac{M}{L} \int_0^L x^2 \, dx = \frac{M}{L} \left[\frac{x^3}{3} \right]_0^L = \tfrac{1}{3} ML^2$$

ASSESS The moment of inertia involves a product of the total mass M with the *square* of a length, in this case L. All moments of inertia have a similar form, although the fraction in front will vary. This is the result shown earlier in Table 12.2.

EXAMPLE 12.6 | Moment of inertia of a circular disk about an axis through the center

Find the moment of inertia of a circular disk of radius R and mass M that rotates on an axis passing through its center.

VISUALIZE FIGURE 12.14 shows the disk and defines distance r from the axis.

FIGURE 12.14 Setting up the integral to find the moment of inertia of a disk.

A narrow ring of width dr has mass $dm = (M/A)dA$. Its area is dA = width $\times$ circumference = $2\pi r \, dr$.

SOLVE This is a situation of great practical importance. To solve this problem, we need to use a two-dimensional integration scheme that you learned in calculus. Rather than dividing the disk into little boxes, let's divide it into narrow *rings* of mass dm. Figure 12.14 shows one such ring, of radius r and width dr. Let dA represent the area of this ring. The mass dm in this ring is the same fraction of the total mass M as dA is of the total area A. That is,

$$\frac{dm}{M} = \frac{dA}{A}$$

Thus the mass in the small area dA is

$$dm = \frac{M}{A} \, dA$$

This is the reasoning we used to find the center of mass of the rod in Example 12.2, only now we're using it in two dimensions.

The total area of the disk is $A = \pi R^2$, but what is dA? If we imagine unrolling the little ring, it would form a long, thin rectangle of length $2\pi r$ and height dr. Thus the *area* of this little ring is $dA = 2\pi r \, dr$. With this information we can write

$$dm = \frac{M}{\pi R^2} (2\pi r \, dr) = \frac{2M}{R^2} r \, dr$$

Now we have an expression for dm in terms of a coordinate differential dr, so we can proceed to carry out the integration for I. Using Equation 12.15, we find

$$I_{disk} = \int r^2 \, dm = \int r^2 \left(\frac{2M}{R^2} r \, dr \right) = \frac{2M}{R^2} \int_0^R r^3 \, dr$$

where in the last step we have used the fact that the disk extends from $r = 0$ to $r = R$. Performing the integration gives

$$I_{disk} = \frac{2M}{R^2} \left[\frac{r^4}{4} \right]_0^R = \tfrac{1}{2} MR^2$$

ASSESS Once again, the moment of inertia involves a product of the total mass M with the *square* of a length, in this case R.

If a complex object can be divided into simpler pieces 1, 2, 3, . . . whose moments of inertia $I_1, I_2, I_3 \ldots$ are already known, the moment of inertia of the entire object is

$$I_{object} = I_1 + I_2 + I_3 + \cdots \tag{12.17}$$

The Parallel-Axis Theorem

The moment of inertia depends on the rotation axis. Suppose you need to know the moment of inertia for rotation about the off-center axis in FIGURE 12.15. You can find this quite easily if you know the moment of inertia for rotation around a *parallel axis* through the center of mass.

If the axis of interest is distance d from a parallel axis through the center of mass, the moment of inertia is

$$I = I_{cm} + Md^2 \qquad (12.18)$$

Equation 12.18 is called the **parallel-axis theorem.** We'll give a proof for the one-dimensional object shown in FIGURE 12.16.

The x-axis has its origin at the rotation axis, and the x'-axis has its origin at the center of mass. You can see that the coordinates of dm along these two axes are related by $x = x' + d$. By definition, the moment of inertia about the rotation axis is

$$I = \int x^2 \, dm = \int (x' + d)^2 \, dm = \int (x')^2 \, dm + 2d \int x' \, dm + d^2 \int dm \qquad (12.19)$$

The first of the three integrals on the right, by definition, is the moment of inertia I_{cm} about the center of mass. The third is simply Md^2 because adding up (integrating) all the dm gives the total mass M.

If you refer back to Equations 12.5, the definition of the center of mass, you'll see that the middle integral on the right is equal to Mx'_{cm}. But $x'_{cm} = 0$ because we specifically chose the x'-axis to have its origin at the center of mass. Thus the second integral is zero and we end up with Equation 12.18. The proof in two dimensions is similar.

FIGURE 12.15 An off-center axis.

An off-center rotation axis

Mass M

Axis through the center of mass

The moment of inertia about this axis is $I = I_{cm} + Md^2$.

FIGURE 12.16 Proving the parallel-axis theorem.

Use this axis for calculating I about the pivot.

Pivot $x = x' + d$ dm

$\otimes$ cm

d x'

Use this axis for calculating I_{cm}.

EXAMPLE 12.7 | **The moment of inertia of a thin rod**

Find the moment of inertia of a thin rod with mass M and length L about an axis one-third of the length from one end.

SOLVE From Table 12.2 we know the moment of inertia about the center of mass is $\frac{1}{12}ML^2$. The center of mass is at the center of the rod. An axis $\frac{1}{3}L$ from one end is $d = \frac{1}{6}L$ from the center of mass. Using the parallel-axis theorem, we have

$$I = I_{cm} + Md^2 = \tfrac{1}{12}ML^2 + M\left(\tfrac{1}{6}L\right)^2 = \tfrac{1}{9}ML^2$$

STOP TO THINK 12.3 Four Ts are made from two identical rods of equal mass and length. Rank in order, from largest to smallest, the moments of inertia I_a to I_d for rotation about the dashed line.

(a)   (b)   (c)   (d)

12.5 Torque

Consider the common experience of pushing open a door. FIGURE 12.17 is a top view of a door hinged on the left. Four pushing forces are shown, all of equal strength. Which of these will be most effective at opening the door?

Force $\vec{F}_1$ will open the door, but force $\vec{F}_2$, which pushes straight at the hinge, will not. Force $\vec{F}_3$ will open the door, but not as easily as $\vec{F}_1$. What about $\vec{F}_4$? It is perpendicular to the door, it has the same magnitude as $\vec{F}_1$, but you know from

FIGURE 12.17 The four forces have different effects on the swinging door.

Top view of door $\vec{F}_2$

Hinge $\vec{F}_4$ $\vec{F}_3$ $\vec{F}_1$

experience that pushing close to the hinge is not as effective as pushing at the outer edge of the door.

The ability of a force to cause a rotation depends on three factors:

1. The magnitude F of the force.
2. The distance r from the point of application to the pivot.
3. The angle at which the force is applied.

We can incorporate these three factors into a single quantity called the *torque*. **FIGURE 12.18** shows a force $\vec{F}$ trying to rotate the wrench and nut about a *pivot point*—the axis about which the nut will rotate. We say that this force exerts a **torque** τ (Greek tau), which we define as

$$\tau \equiv rF\sin\phi \qquad (12.20)$$

Torque depends on the three properties we just listed: the magnitude of the force, its distance from the pivot, and its angle. Loosely speaking, τ measures the "effectiveness" of the force at causing an object to rotate about a pivot. **Torque is the rotational equivalent of force.**

NOTE Angle ϕ is measured *counterclockwise* from the dashed line that extends outward along the radial line. This is consistent with our sign convention for the angular position θ.

The SI units of torque are newton-meters, abbreviated N m. Although we defined 1 N m = 1 J during our study of energy, torque is not an energy-related quantity and so we do *not* use joules as a measure of torque.

Torque, like force, has a sign. A torque that tries to rotate the object in a ccw direction is positive while a negative torque gives a cw rotation. **FIGURE 12.19** summarizes the signs. Notice that a force pushing straight toward the pivot or pulling straight out from the pivot exerts *no* torque.

FIGURE 12.18 Force $\vec{F}$ exerts a torque about the pivot point.

$\vec{F}$ exerts a torque about the pivot point.

Angle ϕ is measured ccw from the radial line.

Point where force is applied

Rigid body

Pivot point

r is measured from the pivot to the point where the force is applied.

FIGURE 12.19 Signs and strengths of the torque.

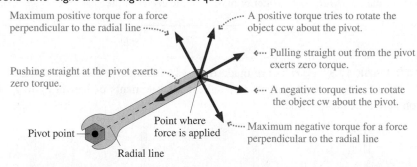

Maximum positive torque for a force perpendicular to the radial line

A positive torque tries to rotate the object ccw about the pivot.

Pulling straight out from the pivot exerts zero torque.

Pushing straight at the pivot exerts zero torque.

A negative torque tries to rotate the object cw about the pivot.

Point where force is applied

Pivot point

Maximum negative torque for a force perpendicular to the radial line

Radial line

NOTE Torque differs from force in a very important way. Torque is calculated or measured *about a pivot point*. To say that a torque is 20 N m is meaningless. You need to say that the torque is 20 N m about a particular point. Torque can be calculated about any pivot point, but its value depends on the point chosen because this choice determines r and ϕ.

Returning to the door of Figure 12.17, you can see that $\vec{F}_1$ is most effective at opening the door because $\vec{F}_1$ exerts the largest torque *about the pivot point*. $\vec{F}_3$ has equal magnitude, but it is applied at an angle less than 90° and thus exerts less torque. $\vec{F}_2$, pushing straight at the hinge with $\phi = 0°$, exerts no torque at all. And $\vec{F}_4$, with a smaller value for r, exerts less torque than $\vec{F}_1$.

Your foot exerts a torque that rotates the crank.

Interpreting Torque

Torque can be interpreted from two perspectives, as shown in FIGURE 12.20. First, the quantity $F\sin\phi$ is the tangential force component F_t. Consequently, the torque is

$$\tau = rF_t \qquad (12.21)$$

In other words, torque is the product of r with the force component F_t that is tangent to the circular path followed by this point on the wrench. This interpretation makes sense because the radial component of $\vec{F}$ points straight at the pivot point and cannot exert a torque.

A second perspective, widely used in applications, is based on the idea of a *moment arm*. Figure 12.20 shows the **line of action,** the line along which the force acts. The minimum distance between the pivot point and the line of action—the length of a line drawn *perpendicular to the line of action*—is called the **moment arm** (or the *lever arm*) d. Because $\sin(180° - \phi) = \sin\phi$, it is easy to see that $d = r\sin\phi$. Thus the torque $rF\sin\phi$ can also be written

$$|\tau| = dF \qquad (12.22)$$

NOTE Equation 12.22 gives only $|\tau|$, the magnitude of the torque; the sign has to be supplied by observing the direction in which the torque acts.

FIGURE 12.20 Two useful interpretations of the torque.

1. Torque is due to the tangential component of force: $\tau = rF_t$.

$F_t = F\sin\phi$

2. Torque is the force multiplied by the moment arm: $|\tau| = dF$.

EXAMPLE 12.8 | Applying a torque

Luis uses a 20-cm-long wrench to turn a nut. The wrench handle is tilted 30° above the horizontal, and Luis pulls straight down on the end with a force of 100 N. How much torque does Luis exert on the nut?

VISUALIZE FIGURE 12.21 shows the situation. The angle is a negative $\phi = -120°$ because it is *clockwise* from the radial line.

FIGURE 12.21 A wrench being used to turn a nut.

$r = 20$ cm

$\phi = -120°$

F_t

30°

Luis's pull (100 N)

Line of action

Moment arm d

SOLVE The tangential component of the force is

$$F_t = F\sin\phi = -86.6 \text{ N}$$

According to our sign convention, F_t is negative because it points in a cw direction. The torque, from Equation 12.21, is

$$\tau = rF_t = (0.20 \text{ m})(-86.6 \text{ N}) = -17 \text{ N m}$$

Alternatively, Figure 12.21 has drawn the *line of action* by extending the force vector forward and backward. The *moment arm*, the distance between the pivot point and the line of action, is

$$d = r\sin(60°) = 0.17 \text{ m}$$

Inserting the moment arm in Equation 12.22 gives

$$|\tau| = dF = (0.17 \text{ m})(100 \text{ N}) = 17 \text{ N m}$$

The torque acts to give a cw rotation, so we insert a minus sign to end up with

$$\tau = -17 \text{ N m}$$

ASSESS The largest possible torque, if Luis pulled perpendicular to the 20-cm-long wrench, would have a magnitude of 20 N m. Pulling at an angle will reduce this, so 17 N m is a reasonable answer.

STOP TO THINK 12.4 Rank in order, from largest to smallest, the five torques τ_a to τ_e. The rods all have the same length and are pivoted at the dot.

2 N

2 N

2 N

2 N

4 N

45°   4 N

(a)   (b)   (c)   (d)   (e)

FIGURE 12.22 The forces exert a net torque about the pivot point.

The axle exerts a force on the crank to keep $\vec{F}_{net} = 0$. This force does not exert a torque.

Net Torque

FIGURE 12.22 shows the forces acting on the crankset of a bicycle. The crankset is free to rotate about the axle, but the axle prevents it from having any translational motion relative to the bike frame. It does so by exerting force $\vec{F}_{axle}$ on the crankset to balance the other forces and keep $\vec{F}_{net} = 0$.

Forces $\vec{F}_1, \vec{F}_2, \vec{F}_3, \ldots$ exert torques $\tau_1, \tau_2, \tau_3, \ldots$ on the crankset, but $\vec{F}_{axle}$ does *not* exert a torque because it is applied at the pivot point and has zero moment arm. Thus the *net* torque about the axle is the sum of the torques due to the applied forces:

$$\tau_{net} = \tau_1 + \tau_2 + \tau_3 + \cdots = \sum_i \tau_i \qquad (12.23)$$

Gravitational Torque

Gravity exerts a torque on many objects. If the object in **FIGURE 12.23** is released, a torque due to gravity will cause it to rotate around the axle. To calculate the torque about the axle, we start with the fact that gravity acts on *every* particle in the object, exerting a downward force of magnitude $F_i = m_i g$ on particle i. The *magnitude* of the gravitational torque on particle i is $|\tau_i| = d_i m_i g$, where d_i is the moment arm. But we need to be careful with signs.

FIGURE 12.23 Gravitational torque.

(a)

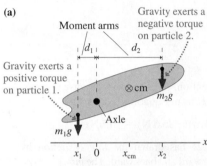

Moment arms

Gravity exerts a negative torque on particle 2.

Gravity exerts a positive torque on particle 1.

(b)

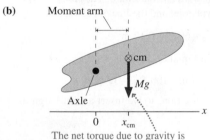

Moment arm

The net torque due to gravity is found by pretending the object's entire mass is at the center of mass.

A moment arm must be a positive number because it's a distance. If we establish a coordinate system with the origin at the axle, then you can see from Figure 12.23a that the moment arm d_i of particle i is $|x_i|$. A particle to the right of the axle (positive x_i) experiences a *negative* torque because gravity tries to rotate this particle in a clockwise direction. Similarly, a particle to the left of the axle (negative x_i) has a positive torque. The torque is opposite in sign to x_i, so we can get the sign right by writing

$$\tau_i = -x_i m_i g = -(m_i x_i)g \qquad (12.24)$$

The net torque due to gravity is found by summing Equation 12.24 over all particles:

$$\tau_{grav} = \sum_i \tau_i = \sum_i (-m_i x_i g) = -\left(\sum_i m_i x_i\right)g \qquad (12.25)$$

But according to the definition of center of mass, Equations 12.4, $\sum m_i x_i = M x_{cm}$. Thus the torque due to gravity is

$$\tau_{grav} = -M g x_{cm} \qquad (12.26)$$

where x_{cm} is the position of the center of mass *relative to the axis of rotation*.

Equation 12.26 has the simple interpretation shown in Figure 12.23b. Mg is the net gravitational force on the entire object, and x_{cm} is the moment arm between the rotation axis and the center of mass. The gravitational torque on an extended object of mass M is equivalent to the torque of a *single* force vector $\vec{F}_G = -Mg\hat{j}$ acting at the object's center of mass.

In other words, **the gravitational torque is found by treating the object as if all its mass were concentrated at the center of mass.** This is the basis for the well-known technique of finding an object's center of mass by balancing it. An object will balance on a pivot, as shown in **FIGURE 12.24**, only if the center of mass is directly above the pivot point. If the pivot is *not* under the center of mass, the gravitational torque will cause the object to rotate.

FIGURE 12.24 An object balances on a pivot that is directly under the center of mass.

The line of action passes through the pivot.

Pivot

Mg

NOTE The point at which gravity acts is also called the *center of gravity*. As long as gravity is uniform over the object—always true for earthbound objects—there's no difference between center of mass and center of gravity.

EXAMPLE 12.9 | The gravitational torque on a beam

The 4.00-m-long, 500 kg steel beam shown in FIGURE 12.25 is supported 1.20 m from the right end. What is the gravitational torque about the support?

FIGURE 12.25 A steel beam supported at one point.

MODEL The center of mass of the beam is at the midpoint. $x_{cm} = -0.80$ m is measured from the pivot point.

SOLVE This is a straightforward application of Equation 12.26. The gravitational torque is

$$\tau_{grav} = -Mgx_{cm} = -(500 \text{ kg})(9.80 \text{ m/s}^2)(-0.80 \text{ m}) = 3920 \text{ N m}$$

ASSESS The torque is positive because gravity tries to rotate the beam ccw. Notice that the beam in Figure 12.25 is *not* in equilibrium. It will fall over unless other forces, not shown, are supporting it.

12.6 Rotational Dynamics

What does a torque do? **A torque causes an angular acceleration.** To see why, FIGURE 12.26 shows a rigid body undergoing *pure rotational motion* about a fixed and unmoving axis. This might be a rotation about the object's center of mass, such as we considered in Section 12.2. Or it might be an object, such as a turbine, rotating on an axle.

The forces $\vec{F}_1$, $\vec{F}_2$, $\vec{F}_3$,... in Figure 12.26 are external forces acting on particles of masses $m_1, m_2, m_3,...$ that are part of the rigid body. These forces exert torques τ_1, τ_2, τ_3,... about the rotation axis. The *net* torque on the object is the sum of the torques:

$$\tau_{net} = \sum_i \tau_i \tag{12.27}$$

Focus on particle i, which is acted on by force $\vec{F}_i$ and undergoes circular motion with radius r_i. In Chapter 8, we found that the radial component of $\vec{F}_i$ is responsible for the centripetal acceleration of circular motion, while the tangential component $(F_i)_t$ causes the particle to speed up or slow down with a tangential acceleration $(a_i)_t$. Newton's second law is

$$(F_i)_t = m_i(a_i)_t = m_i r_i \alpha \tag{12.28}$$

where in the last step we used the relationship between tangential and angular acceleration: $a_t = r\alpha$. The angular acceleration α does not have a subscript because *all particles in the object have the same angular acceleration.* That is, α is the angular acceleration of the entire object.

Multiplying both sides by r_i gives

$$r_i(F_i)_t = m_i r_i^2 \alpha \tag{12.29}$$

But $r_i(F_i)_t$ is the torque τ_i about the axis on particle i; hence Newton's second law for a single particle in the object is

$$\tau_i = m_i r_i^2 \alpha \tag{12.30}$$

Returning now to Equation 12.27, we see that the net torque on the object is

$$\tau_{net} = \sum_i \tau_i = \sum_i m_i r_i^2 \alpha = \left(\sum_i m_i r_i^2\right)\alpha \tag{12.31}$$

In the last step, we factored out α by using the key idea that every particle in a rotating rigid body has the *same* angular acceleration.

You'll recognize the quantity in parentheses as the moment of inertia. Substituting I into Equation 12.31 puts the final piece of the puzzle into place. An object that experiences a net torque τ_{net} about the axis of rotation undergoes an angular acceleration

$$\alpha = \frac{\tau_{net}}{I} \quad \text{(Newton's second law for rotational motion)} \tag{12.32}$$

where I is the object's moment of inertia *about the rotation axis*. This result, Newton's second law for rotation, is the fundamental equation of rigid-body dynamics.

FIGURE 12.26 The external forces exert a torque about the rotation axis.

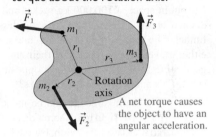

A net torque causes the object to have an angular acceleration.

In practice we often write $\tau_{net} = I\alpha$, but Equation 12.32 better conveys the idea that **torque is the cause of angular acceleration.** In the absence of a net torque ($\tau_{net} = 0$), the object either does not rotate ($\omega = 0$) or rotates with *constant* angular velocity ($\omega = $ constant).

TABLE 12.3 summarizes the analogies between linear and rotational dynamics.

TABLE 12.3 Rotational and linear dynamics

Rotational dynamics		Linear dynamics	
torque (N m)	τ_{net}	force (N)	$\vec{F}_{net}$
moment of inertia (kg m^2)	I	mass (kg)	m
angular acceleration (rad/s^2)	α	acceleration (m/s^2)	$\vec{a}$
second law	$\alpha = \tau_{net}/I$	second law	$\vec{a} = \vec{F}_{net}/m$

EXAMPLE 12.10 | Rotating rockets

Far out in space, a 100,000 kg rocket and a 200,000 kg rocket are docked at opposite ends of a motionless 90-m-long connecting tunnel. The tunnel is rigid and its mass is much less than that of either rocket. The rockets start their engines simultaneously, each generating 50,000 N of thrust in opposite directions. What is the structure's angular velocity after 30 s?

MODEL The entire structure can be modeled as two masses at the ends of a massless, rigid rod. There's no net force, so the structure does not undergo translational motion, but the thrusts do create torques that will give the structure angular acceleration and cause it to rotate. We'll assume the thrust forces are perpendicular to the connecting tunnel. This is an unconstrained rotation, so the structure will rotate about its center of mass.

VISUALIZE FIGURE 12.27 shows the rockets and defines distances r_1 and r_2 from the center of mass.

FIGURE 12.27 The thrusts exert a torque on the structure.

SOLVE Our strategy will be to use Newton's second law to find the angular acceleration, followed by rotational kinematics to find ω. We'll need to determine the moment of inertia, and that requires knowing the distances of the two rockets from the rotation axis. As we did in Example 12.1, we choose a coordinate system in which the masses are on the *x*-axis and in which m_1 is at the origin. Then

$$x_{cm} = \frac{m_1 x_1 + m_2 x_2}{m_1 + m_2}$$

$$= \frac{(100,000 \text{ kg})(0 \text{ m}) + (200,000 \text{ kg})(90 \text{ m})}{100,000 \text{ kg} + 200,000 \text{ kg}} = 60 \text{ m}$$

The structure's center of mass is $r_1 = 60$ m from the 100,000 kg rocket and $r_2 = 30$ m from the 200,000 kg rocket. The moment of inertia about the center of mass is

$$I = m_1 r_1^2 + m_2 r_2^2 = 540,000,000 \text{ kg m}^2$$

The two rocket thrusts exert net torque

$$\tau_{net} = r_1 F_1 + r_2 F_2 = (60 \text{ m})(50,000 \text{ N}) + (30 \text{ m})(50,000 \text{ N})$$

$$= 4,500,000 \text{ N m}$$

With I and τ_{net} now known, we can use Newton's second law to find the angular acceleration:

$$\alpha = \frac{\tau}{I} = \frac{4,500,000 \text{ N m}}{540,000,000 \text{ kg m}^2} = 0.00833 \text{ rad/s}^2$$

After 30 seconds, the structure's angular velocity is

$$\omega = \alpha \, \Delta t = 0.25 \text{ rad/s}$$

STOP TO THINK 12.5 Rank in order, from largest to smallest, the angular accelerations α_a to α_d.

(a)

(b)

(c)

(d)

12.7 Rotation About a Fixed Axis

In this section we'll look at rigid bodies that rotate about a fixed axis. The problem-solving strategy for rotational dynamics is very similar to that for linear dynamics.

PROBLEM-SOLVING STRATEGY 12.1

Rotational dynamics problems

MODEL Model the object as a rigid body.

VISUALIZE Draw a pictorial representation to clarify the situation, define coordinates and symbols, and list known information.

- Identify the axis about which the object rotates.
- Identify forces and determine their distances from the axis. For most problems it will be useful to draw a free-body diagram.
- Identify any torques caused by the forces and the signs of the torques.

SOLVE The mathematical representation is based on Newton's second law for rotational motion:

$$\tau_{net} = I\alpha \qquad \text{or} \qquad \alpha = \frac{\tau_{net}}{I}$$

- Find the moment of inertia in Table 12.2 or, if needed, calculate it as an integral or by using the parallel-axis theorem.
- Use rotational kinematics to find angles and angular velocities.

ASSESS Check that your result has correct units and significant figures, is reasonable, and answers the question.

Exercise 28

EXAMPLE 12.11 | **Starting an airplane engine**

The engine in a small airplane is specified to have a torque of 60 N m. This engine drives a 2.0-m-long, 40 kg propeller. On start-up, how long does it take the propeller to reach 200 rpm?

MODEL The propeller can be modeled as a rigid rod that rotates about its center. The engine exerts a torque on the propeller.

VISUALIZE FIGURE 12.28 shows the propeller and the rotation axis.

FIGURE 12.28 A rotating airplane propeller.

The torque from the engine rotates the propeller.

$M = 40$ kg

$L = 2.0$ m

Axis

SOLVE The moment of inertia of a rod rotating about its center is found from Table 12.2:

$$I = \tfrac{1}{12}ML^2 = \tfrac{1}{12}(40 \text{ kg})(2.0 \text{ m})^2 = 13.33 \text{ kg m}^2$$

The 60 N m torque of the engine causes an angular acceleration

$$\alpha = \frac{\tau}{I} = \frac{60 \text{ N m}}{13.33 \text{ kg m}^2} = 4.50 \text{ rad/s}^2$$

The time needed to reach $\omega_f = 200$ rpm $= 3.33$ rev/s $= 20.9$ rad/s is

$$\Delta t = \frac{\Delta\omega}{\alpha} = \frac{\omega_f - \omega_i}{\alpha} = \frac{20.9 \text{ rad/s} - 0 \text{ rad/s}}{4.5 \text{ rad/s}^2} = 4.6 \text{ s}$$

ASSESS We've assumed a constant angular acceleration, which is reasonable for the first few seconds while the propeller is still turning slowly. Eventually, air resistance and friction will cause opposing torques and the angular acceleration will decrease. At full speed, the negative torque due to air resistance and friction cancels the torque of the engine. Then $\tau_{net} = 0$ and the propeller turns at *constant* angular velocity with no angular acceleration.

Constraints Due to Ropes and Pulleys

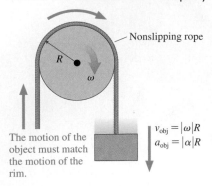

Nonslipping rope

R

ω

The motion of the object must match the motion of the rim.

$v_{\text{obj}} = |\omega|R$
$a_{\text{obj}} = |\alpha|R$

Many important applications of rotational dynamics involve objects, such as pulleys, that are connected via ropes or belts to other objects. **FIGURE 12.29** shows a rope passing over a pulley and connected to an object in linear motion. If the rope does not slip as the pulley rotates, then the rope's speed v_{rope} must exactly match the speed of the rim of the pulley, which is $v_{\text{rim}} = |\omega|R$. If the pulley has an angular acceleration, the rope's acceleration a_{rope} must match the *tangential* acceleration of the rim of the pulley, $a_t = |\alpha|R$.

The object attached to the other end of the rope has the same speed and acceleration as the rope. Consequently, an object connected to a pulley of radius R by a rope that does not slip must obey the constraints

$$v_{\text{obj}} = |\omega|R$$
$$a_{\text{obj}} = |\alpha|R \qquad \text{(motion constraints for a nonslipping rope)} \qquad (12.33)$$

These constraints are very similar to the acceleration constraints introduced in Chapter 7 for two objects connected by a string or rope.

> **NOTE** The constraints are given as magnitudes. Specific problems will need to introduce signs that depend on the direction of motion and on the choice of coordinate system.

The Constant-Torque Model

If all the torques exerted on an object are constant, the object rotates with constant angular acceleration. Even if the torques aren't perfectly constant, there are many situations where it's reasonable to model them as if they were. The **constant-torque model,** analogous to the constant-force model of Section 6.2, is the most important model of rotational dynamics.

MODEL 12.2

Constant torque

For objects on which the net torque is constant.

- Model the object as a rigid body with constant angular acceleration.
- Take into account constraints due to ropes and pulleys.
- Mathematically:
  - **Newton's second law** is $\tau_{\text{net}} = I\alpha$.
  - Use the kinematics of constant angular acceleration.
- Limitations: Model fails if the torque isn't constant.

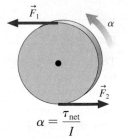

$\vec{F}_1$

α

$\vec{F}_2$

$$\alpha = \frac{\tau_{\text{net}}}{I}$$

The object has constant angular acceleration.

EXAMPLE 12.12 | Lowering a bucket

A 2.0 kg bucket is attached to a massless string that is wrapped around a 1.0 kg, 4.0-cm-diameter cylinder, as shown in **FIGURE 12.30a**. The cylinder rotates on an axle through the center. The bucket is released from rest 1.0 m above the floor. How long does it take to reach the floor?

MODEL Assume the string does not slip.

VISUALIZE **FIGURE 12.30b** shows the free-body diagram for the cylinder and the bucket. The string tension exerts an upward force on the bucket and a downward force on the outer edge of the cylinder. The string is massless, so these two tension forces act as if they are an action/reaction pair: $T_b = T_c = T$.

SOLVE Newton's second law applied to the linear motion of the bucket is

$$ma_y = T - mg$$

where, as usual, the y-axis points upward. What about the cylinder? The only torque comes from the string tension. The moment arm for the tension is $d = R$, and the torque is positive because the string turns the cylinder ccw. Thus $\tau_{\text{string}} = TR$ and Newton's second law for the rotational motion is

$$\alpha = \frac{\tau_{\text{net}}}{I} = \frac{TR}{\frac{1}{2}MR^2} = \frac{2T}{MR}$$

FIGURE 12.30 The falling bucket turns the cylinder.

(a)

(b)

$$a_y = -\alpha R$$

Using α from the cylinder's equation in the constraint, we find

$$a_y = -\alpha R = -\frac{2T}{MR}R = -\frac{2T}{M}$$

Thus the tension is $T = -\frac{1}{2}Ma_y$. If we use this value of the tension in the bucket's equation, we can solve for the acceleration:

$$ma_y = -\tfrac{1}{2}Ma_y - mg$$

$$a_y = -\frac{g}{(1 + M/2m)} = -7.84 \text{ m/s}^2$$

The time to fall through $\Delta y = -1.0$ m is found from kinematics:

$$\Delta y = \tfrac{1}{2}a_y(\Delta t)^2$$

$$\Delta t = \sqrt{\frac{2\,\Delta y}{a_y}} = \sqrt{\frac{2(-1.0 \text{ m})}{-7.84 \text{ m/s}^2}} = 0.50 \text{ s}$$

The moment of inertia of a cylinder rotating about a center axis was taken from Table 12.2.

The last piece of information we need is the constraint due to the fact that the string doesn't slip. Equation 12.33 relates only the absolute values, but in this problem α is positive (ccw acceleration) while a_y is negative (downward acceleration). Hence

ASSESS The expression for the acceleration gives $a_y = -g$ if $M = 0$. This makes sense because the bucket would be in free fall if there were no cylinder. When the cylinder has mass, the downward force of gravity on the bucket has to accelerate the bucket *and* spin the cylinder. Consequently, the acceleration is reduced and the bucket takes longer to fall.

12.8 Static Equilibrium

An extended object that is completely stationary is in *static equilibrium*. It has no linear acceleration ($\vec{a} = \vec{0}$) and no angular acceleration ($\alpha = 0$). Thus, from Newton's laws, the conditions for static equilibrium are no net force *and* no net torque. These two rules are the basis for a branch of engineering, called *statics*, that analyzes buildings, dams, bridges, and other structures in static equilibrium.

Section 6.1 introduced the model of mechanical equilibrium for objects that can be represented as particles. For extended objects, we have the **static equilibrium model.**

Structures such as bridges are analyzed in engineering statics.

MODEL 12.3

Static equilibrium

For extended objects at rest.

- Model the object as a rigid body with no acceleration.
- Mathematically:
  - No net force: $\vec{F}_{\text{net}} = \sum \vec{F}_i = \vec{0}$, and
  - No net torque: $\tau_{\text{net}} = \sum \tau_i = 0$
- The torque is zero about *every* point, so use any point that is convenient for the pivot point.
- Limitations: Model fails if either the forces or the torques aren't balanced.

$\vec{a} = \vec{0} \quad \alpha = 0$

The object is at rest.

For any point you choose, an object that is not rotating is not rotating about that point. This seems to be a trivial statement, but it has an important implication: For a rigid body in static equilibrium, the net torque is zero about *any* point. You can use any point you wish as the pivot point for calculating torque. Even so, some choices are better than others for problem solving. As the examples will show, it's often best to choose a point at which several forces act because the torques exerted by those forces will be zero.

EXAMPLE 12.13 | Lifting weights

Weightlifting can exert extremely large forces on the body's joints and tendons. In the *strict curl* event, a standing athlete uses both arms to lift a barbell by moving only his forearms, which pivot at the elbows. The record weight lifted in the strict curl is over 200 pounds (about 900 N). **FIGURE 12.31** shows the arm bones and the biceps, the main lifting muscle when the forearm is horizontal. What is the tension in the tendon connecting the biceps muscle to the bone while a 900 N barbell is held stationary in this position?

FIGURE 12.31 An arm holding a barbell.

MODEL Model the arm as two rigid rods connected by a hinge. We'll ignore the arm's weight because it is so much less than that of the barbell. Although the tendon pulls at a slight angle, it is close enough to vertical that we'll treat it as such.

VISUALIZE **FIGURE 12.32** shows the forces acting on our simplified model of the forearm. The biceps pulls the forearm up against the upper arm at the elbow, so the force $\vec{F}_{elbow}$ *on* the forearm at the elbow—a force due to the upper arm—is a downward force.

SOLVE Static equilibrium requires both the net force *and* the net torque on the forearm to be zero. Only the y-component of force is relevant, and setting it to zero gives a first equation:

$$\sum F_y = F_{tendon} - F_{elbow} - F_{barbell} = 0$$

FIGURE 12.32 A pictorial representation of the forces involved.

Because each arm supports half the weight of the barbell, $F_{barbell} = 450$ N. We don't know either F_{tendon} or F_{elbow}, nor does the force equation give us enough information to find them. But the fact that the net torque also must be zero gives us that extra information. The torque is zero about *every* point, so we can choose any point we wish to calculate the torque. The elbow joint is a convenient point because force $\vec{F}_{elbow}$ exerts no torque about this point; its moment arm is zero. Thus the torque equation is

$$\tau_{net} = d_{tendon}F_{tendon} - d_{arm}F_{barbell} = 0$$

The tension in the tendon tries to rotate the arm ccw, so it produces a positive torque. Similarly, the torque due to the barbell is negative. We can solve the torque equation for F_{tendon} to find

$$F_{tendon} = F_{barbell}\frac{d_{arm}}{d_{tendon}} = (450\text{ N})\frac{35\text{ cm}}{4.0\text{ cm}} = 3900\text{ N}$$

ASSESS The short distance d_{tendon} from the tendon to the elbow joint means that the force supplied by the biceps has to be very large to counter the torque generated by a force applied at the opposite end of the forearm. Although we ended up not needing the force equation in this problem, we could now use it to calculate that the force exerted at the elbow is $F_{elbow} = 3450$ N. These large forces can easily damage the tendon or the elbow.

EXAMPLE 12.14 | Walking the plank

Adrienne (50 kg) and Bo (90 kg) are playing on a 100 kg rigid plank resting on the supports seen in **FIGURE 12.33**. If Adrienne stands on the left end, can Bo walk all the way to the right end without the plank tipping over? If not, how far can he get past the support on the right?

FIGURE 12.33 Adrienne and Bo on the plank.

MODEL Model the plank as a uniform rigid body with its center of mass at the center.

VISUALIZE **FIGURE 12.34** shows the forces acting on the plank. Both supports exert upward forces. $\vec{n}_A$ and $\vec{n}_B$ are the normal forces of Adrienne's and Bo's feet pushing down on the board.

SOLVE Because the plank is resting on the supports, not held down, forces $\vec{n}_1$ and $\vec{n}_2$ must point upward. (The supports could pull down if the plank were nailed to them, but that's not the case here.) Force $\vec{n}_1$ will decrease as Bo moves to the right, and the tipping point occurs when $n_1 = 0$. The plank remains in static equilibrium right up to the tipping point, so both the net force and the net torque on it are zero. The force equation is

$$\sum F_y = n_1 + n_2 - n_A - n_B - Mg$$
$$= n_1 + n_2 - m_Ag - m_Bg - Mg = 0$$

Adrienne is at rest, with zero net force, so her downward force on the board, an action/reaction pair with the upward normal force of the board on her, equals her weight: $n_A = m_Ag$. Bo's center of

FIGURE 12.34 A pictorial representation of the forces on the plank.

Calculate torques about this point.

Known
$m_A = 50$ kg $d_A = 2.0$ m
$m_B = 90$ kg $M = 100$ kg
$d_2 = 3.0$ m $d_M = 2.5$ m

Find
d_B for which $n_I = 0$

mass oscillates up and down as he walks, so he's *not* in equilibrium and, strictly speaking, $n_B \neq m_B g$. But we'll assume that he edges out onto the board slowly, with minimal bouncing, in which case $n_B = m_B g$ is a reasonable approximation.

We can again choose any point we wish for calculating torque. Let's use the support on the left. Adrienne and the support on the right exert positive torques about this point; the other forces exert negative torques. Force $\vec{n}_1$ exerts no torque because it acts at the pivot point. Thus the torque equation is

$$\tau_{\text{net}} = d_A m_A g - d_B m_B g - d_M M g + d_2 n_2 = 0$$

At the tipping point, where $n_1 = 0$, the force equation gives $n_2 = (m_A + m_B + M)g$. Substituting this into the torque equation and then solving for Bo's position give

$$d_B = \frac{d_A m_A - d_M M + d_2(m_A + m_B + M)}{m_B} = 6.3 \text{ m}$$

Bo doesn't quite make it to the end. The plank tips when he's 6.3 m past the left support, our pivot point, and thus 3.3 m past the support on the right.

ASSESS We could have solved this problem somewhat more simply had we chosen the support on the right for calculating the torques. However, you might not recognize the "best" point for calculating the torques in a problem. The point of this example is that it doesn't matter which point you choose.

EXAMPLE 12.15 | Will the ladder slip?

A 3.0-m-long ladder leans against a frictionless wall at an angle of 60°. What is the minimum value of μ_s, the coefficient of static friction with the ground, that prevents the ladder from slipping?

MODEL The ladder is a rigid rod of length L. To not slip, it must be in both translational equilibrium ($\vec{F}_{\text{net}} = \vec{0}$) and rotational equilibrium ($\tau_{\text{net}} = 0$).

VISUALIZE FIGURE 12.35 shows the ladder and the forces acting on it.

FIGURE 12.35 A ladder in total equilibrium.

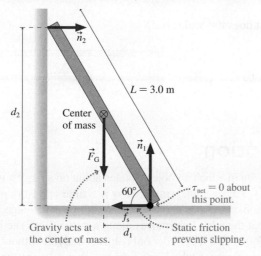

Gravity acts at the center of mass.

Static friction prevents slipping.

SOLVE The x- and y-components of $\vec{F}_{\text{net}} = \vec{0}$ are

$$\sum F_x = n_2 - f_s = 0$$
$$\sum F_y = n_1 - Mg = 0$$

The net torque is zero about *any* point, so which should we choose? The bottom corner of the ladder is a good choice because two forces pass through this point and have no torque about it. The torque about the bottom corner is

$$\tau_{\text{net}} = d_1 F_G - d_2 n_2 = \tfrac{1}{2}(L\cos 60°)Mg - (L\sin 60°)n_2 = 0$$

The signs are based on the observation that $\vec{F}_G$ would cause the ladder to rotate ccw while $\vec{n}_2$ would cause it to rotate cw. All together, we have three equations in the three unknowns n_1, n_2, and f_s. If we solve the third for n_2,

$$n_2 = \frac{\tfrac{1}{2}(L\cos 60°)Mg}{L\sin 60°} = \frac{Mg}{2\tan 60°}$$

we can then substitute this into the first to find

$$f_s = \frac{Mg}{2\tan 60°}$$

Our model of friction is $f_s \leq f_{s\,\text{max}} = \mu_s n_1$. We can find n_1 from the second equation: $n_1 = Mg$. Using this, the model of static friction tells us that

$$f_s \leq \mu_s Mg$$

Comparing these two expressions for f_s, we see that μ_s must obey

$$\mu_s \geq \frac{1}{2\tan 60°} = 0.29$$

Thus the minimum value of the coefficient of static friction is 0.29.

ASSESS You know from experience that you can lean a ladder or other object against a wall if the ground is "rough," but it slips if the surface is too smooth. 0.29 is a "medium" value for the coefficient of static friction, which is reasonable.

Balance and Stability

If you tilt a box up on one edge by a small amount and let go, it falls back down. If you tilt it too much, it falls over. And if you tilt "just right," you can get the box to balance on its edge. What determines these three possible outcomes?

FIGURE 12.36 illustrates the idea with a car, but the results are general and apply in many situations. As long as the object's center of mass remains over the base of support, torque due to gravity will rotate the object back to its equilibrium position.

FIGURE 12.36 Stability depends on the position of the center of mass.

(a) The torque due to gravity will bring the car back down as long as the center of mass is above the base of support.

Base of support

(b) The vehicle is at the critical angle θ_c when its center of mass is exactly over the pivot.

Pivot

(c) Now the center of mass is outside the base of support. Torque due to gravity will cause the car to roll over.

Base of support

A *critical angle* θ_c is reached when the center of mass is directly over the pivot point. This is the point of balance, with no net torque. For vehicles, the distance between the tires is called the track width t. If the height of the center of mass is h, you can see from Figure 12.36b that the critical angle is

$$\theta_c = \tan^{-1}\left(\frac{t}{2h}\right)$$

For passenger cars with $h \approx 0.33t$, the critical angle is $\theta_c \approx 57°$. But for a sport utility vehicle (SUV) with $h \approx 0.47t$, a higher center of mass, the critical angle is only $\theta_c \approx 47°$. Various automobile safety groups have determined that a vehicle with $\theta_c > 50°$ is unlikely to roll over in an accident. A rollover becomes increasingly likely when θ_c is reduced below 50°. The general rule is that **a wider base of support and/or a lower center of mass improve stability.**

This dancer balances *en pointe* by having her center of mass directly over her toes, her base of support.

STOP TO THINK 12.6 What does the scale read?

a. 500 N
b. 1000 N
c. 2000 N
d. 4000 N

Massless rod

Scale

1000 N

12.9 Rolling Motion

Rolling is a *combination motion* in which an object rotates about an axis that is moving along a straight-line trajectory. For example, FIGURE 12.37 is a time-exposure photo of a rolling wheel with one lightbulb on the axis and a second lightbulb at the edge. The axis light moves straight ahead, but the edge light moves along a curve. Let's see if we can understand this interesting motion. We'll consider only objects that roll without slipping.

FIGURE 12.38 shows a round object—a wheel or a sphere—that rolls forward exactly one revolution. The point that had been on the bottom follows the curve you saw in Figure 12.37 to the top and back to the bottom. *Because the object doesn't slip,* the center of mass moves forward exactly one circumference: $\Delta x_{cm} = 2\pi R$.

We can also write the distance traveled in terms of the velocity of the center of mass: $\Delta x_{cm} = v_{cm} \Delta t$. But Δt, the time it takes the object to make one complete revolution, is nothing other than the rotation period T. In other words, $\Delta x_{cm} = v_{cm}T$.

FIGURE 12.37 The trajectories of the center of a wheel and of a point on the rim are seen in a time-exposure photograph.

Path of wheel rim

Path of center of wheel

Path followed by the point on the rim

Object rolls one revolution without slipping.

R

$\vec{v}_{cm}$ $\vec{v}_{cm}$ R $\vec{v}_{cm}$

$\Delta x_{cm} = v_{cm}\Delta t = 2\pi R$

◀ **FIGURE 12.38** An object rolling through one revolution.

These two expressions for Δx_{cm} come from two perspectives on the motion: one looking at the rotation and the other looking at the translation of the center of mass. But it's the same distance no matter how you look at it, so these two expressions must be equal. Consequently,

$$\Delta x_{cm} = 2\pi R = v_{cm}T \qquad (12.34)$$

If we divide by T, we can write the center-of-mass velocity as

$$v_{cm} = \frac{2\pi}{T} R \qquad (12.35)$$

But $2\pi/T$ is the angular velocity ω, as you learned in Chapter 4, leading to

$$v_{cm} = R\omega \qquad (12.36)$$

Equation 12.36 is the **rolling constraint,** the basic link between translation and rotation for objects that roll without slipping.

Let's look carefully at a particle in the rolling object. As **FIGURE 12.39a** shows, the position vector $\vec{r}_i$ for particle i is the vector sum $\vec{r}_i = \vec{r}_{cm} + \vec{r}_{i,\,rel}$. Taking the time derivative of this equation, we can write the velocity of particle i as

$$\vec{v}_i = \vec{v}_{cm} + \vec{v}_{i,\,rel} \qquad (12.37)$$

In other words, the velocity of particle i can be divided into two parts: the velocity $\vec{v}_{cm}$ of the object as a whole plus the velocity $\vec{v}_{i,\,rel}$ of particle i relative to the center of mass (i.e., the velocity that particle i would have if the object were only rotating and had no translational motion).

FIGURE 12.39b applies this idea to point P at the very bottom of the rolling object, the point of contact between the object and the surface. This point is moving around the center of the object at angular velocity ω, so $v_{i,\,rel} = -R\omega$. The negative sign indicates that the motion is cw. At the same time, the center-of-mass velocity, Equation 12.36, is $v_{cm} = R\omega$. Adding these, we find that the velocity of point P, the lowest point, is $v_i = 0$. In other words, **the point on the bottom of a rolling object is instantaneously at rest.**

Although this seems surprising, it is really what we mean by "rolling without slipping." If the bottom point had a velocity, it would be moving horizontally relative to the surface. In other words, it would be slipping or sliding across the surface. To roll without slipping, the bottom point, the point touching the surface, must be at rest.

FIGURE 12.40 shows how the velocity vectors at the top, center, and bottom of a rotating wheel are found by adding the rotational velocity vectors to the center-of-mass velocity. You can see that $v_{bottom} = 0$ and that $v_{top} = 2R\omega = 2v_{cm}$.

FIGURE 12.39 The motion of a particle in the rolling object.

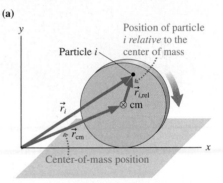

(a)

Particle i

Position of particle i *relative* to the center of mass

$\vec{r}_{i,rel}$

$\otimes$ cm

$\vec{r}_i$

$\vec{r}_{cm}$

Center-of-mass position

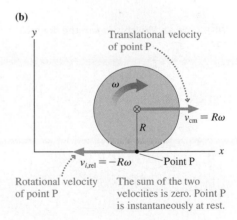

(b)

Translational velocity of point P

ω

$\otimes$

R

$v_{cm} = R\omega$

$v_{i,rel} = -R\omega$ Point P

Rotational velocity of point P

The sum of the two velocities is zero. Point P is instantaneously at rest.

FIGURE 12.40 Rolling without slipping is a combination of translation and rotation.

Translation   +   Rotation   =   Rolling

v_{cm} $R\omega$ $v = 2v_{cm} = 2R\omega$

$\otimes$ $+$ $\otimes$ 0 $=$ $\otimes$ $v = v_{cm} = R\omega$

v_{cm}

v_{cm} $-R\omega$ $v = 0$

Kinetic Energy of a Rolling Object

We found earlier that the rotational kinetic energy of a rigid body in pure rotational motion is $K_{rot} = \frac{1}{2}I\omega^2$. Now we would like to find the kinetic energy of an object that rolls without slipping, a combination of rotational and translation motion.

We begin with the observation that the bottom point in FIGURE 12.41 is instantaneously at rest. Consequently, we can think of an axis through P as an *instantaneous axis of rotation*. The idea of an instantaneous axis of rotation seems a little far-fetched, but it is confirmed by looking at the instantaneous velocities of the center point and the top point. We found these in Figure 12.40 and they are shown again in Figure 12.41. They are exactly what you would expect as the tangential velocity $v_t = r\omega$ for rotation about P at distances R and $2R$.

From this perspective, the object's motion is pure rotation about point P. Thus the kinetic energy is that of pure rotation:

$$K = K_{\text{rotation about P}} = \frac{1}{2}I_P\omega^2 \tag{12.38}$$

I_P is the moment of inertia for rotation about point P. We can use the parallel-axis theorem to write I_P in terms of the moment of inertia I_{cm} about the center of mass. Point P is displaced by distance $d = R$; thus

$$I_P = I_{cm} + MR^2$$

Using this expression in Equation 12.38 gives us the kinetic energy:

$$K = \frac{1}{2}I_{cm}\omega^2 + \frac{1}{2}M(R\omega)^2 \tag{12.39}$$

We know from the rolling constraint that $R\omega$ is the center-of-mass velocity v_{cm}. Thus the kinetic energy of a rolling object is

$$K_{\text{rolling}} = \frac{1}{2}I_{cm}\omega^2 + \frac{1}{2}Mv_{cm}^2 = K_{rot} + K_{cm} \tag{12.40}$$

In other words, **the rolling motion of a rigid body can be described as a translation of the center of mass (with kinetic energy K_{cm}) plus a rotation about the center of mass (with kinetic energy K_{rot}).**

The Great Downhill Race

FIGURE 12.42 shows a contest in which a sphere, a cylinder, and a circular hoop, all of mass M and radius R, are placed at height h on a slope of angle θ. All three are released from rest at the same instant of time and roll down the ramp without slipping. To make things more interesting, they are joined by a particle of mass M that slides down the ramp without friction. Which one will win the race to the bottom of the hill? Does rotation affect the outcome?

An object's initial gravitational potential energy is transformed into kinetic energy as it rolls (or slides, in the case of the particle). The kinetic energy, as we just discovered, is a combination of translational and rotational kinetic energy. If we choose the bottom of the ramp as the zero point of potential energy, the statement of energy conservation $K_f = U_i$ can be written

$$\frac{1}{2}I_{cm}\omega^2 + \frac{1}{2}Mv_{cm}^2 = Mgh \tag{12.41}$$

The translational and rotational velocities are related by $\omega = v_{cm}/R$. In addition, notice from Table 12.2 that the moments of inertia of all the objects can be written in the form

$$I_{cm} = cMR^2 \tag{12.42}$$

where c is a constant that depends on the object's geometry. For example, $c = \frac{2}{5}$ for a sphere but $c = 1$ for a circular hoop. Even the particle can be represented by $c = 0$, which eliminates the rotational kinetic energy.

With this information, Equation 12.41 becomes

$$\frac{1}{2}(cMR^2)\left(\frac{v_{cm}}{R}\right)^2 + \frac{1}{2}Mv_{cm}^2 = \frac{1}{2}M(1 + c)v_{cm}^2 = Mgh$$

FIGURE 12.41 Rolling motion is an instantaneous rotation about point P.

Instantaneous rotation about point P

ω

$v = 2R\omega$

$v = R\omega$

P

Point P, which is instantaneously at rest, is the pivot point for the entire object.

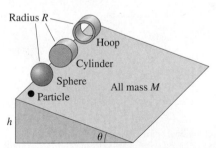

FIGURE 12.42 Which will win the downhill race?

Radius R

Hoop

Cylinder

Sphere

Particle

All mass M

h

θ

Thus the finishing speed of an object with $I = cMR^2$ is

$$v_{cm} = \sqrt{\frac{2gh}{1+c}} \qquad (12.43)$$

The final speed is independent of both M and R, but it does depend on the *shape* of the rolling object. The particle, with the smallest value of c, will finish with the highest speed, while the circular hoop, with the largest c, will be the slowest. In other words, the rolling aspect of the motion *does* matter!

We can use Equation 12.43 to find the acceleration a_{cm} of the center of mass. The objects move through distance $\Delta x = h/\sin\theta$, so we can use constant-acceleration kinematics to find

$$v_{cm}^2 = 2a_{cm}\,\Delta x$$
$$a_{cm} = \frac{v_{cm}^2}{2\Delta x} = \frac{2gh/(1+c)}{2h/\sin\theta} = \frac{g\sin\theta}{1+c} \qquad (12.44)$$

Recall, from Chapter 2, that $a_{particle} = g\sin\theta$ is the acceleration of a particle sliding down a frictionless incline. We can use this fact to write Equation 12.44 in an interesting form:

$$a_{cm} = \frac{a_{particle}}{1+c} \qquad (12.45)$$

This analysis leads us to the conclusion that **the acceleration of a rolling object is less—in some cases significantly less—than the acceleration of a particle.** The reason is that the energy has to be shared between translational kinetic energy and rotational kinetic energy. A particle, by contrast, can put all its energy into translational kinetic energy.

FIGURE 12.43 shows the results of the race. The simple particle wins by a fairly wide margin. Of the solid objects, the sphere has the largest acceleration. Even so, its acceleration is only 71% the acceleration of a particle. The acceleration of the circular hoop, which comes in last, is a mere 50% that of a particle.

NOTE The objects having the largest acceleration are those whose mass is most concentrated near the center. Placing the mass far from the center, as in the hoop, increases the moment of inertia. Thus it requires a larger effort to get a hoop rolling than to get a sphere of equal mass rolling.

12.10 The Vector Description of Rotational Motion

Rotation about a fixed axis, such as an axle, can be described in terms of a scalar angular velocity ω and a scalar torque τ, using a plus or minus sign to indicate the direction of rotation. This is very much analogous to the one-dimensional kinematics of Chapter 2. For more general rotational motion, angular velocity, torque, and other quantities must be treated as *vectors*. We won't go into much detail because the subject rapidly gets very complicated, but we will sketch some important basic ideas.

The Angular Velocity Vector

FIGURE 12.44 shows a rotating rigid body. We can define an angular velocity vector $\vec{\omega}$ as follows:

- The magnitude of $\vec{\omega}$ is the object's angular velocity ω.
- $\vec{\omega}$ points along the axis of rotation in the direction given by the *right-hand rule* illustrated in Figure 12.44.

FIGURE 12.43 And the winner is ...

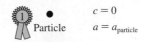
① Particle $c = 0$
$a = a_{particle}$

Solid sphere $c = \frac{2}{5}$
$a_{cm} = \frac{5}{7}a_{particle}$
$= 0.71a_{particle}$

Solid cylinder $c = \frac{1}{2}$
$a_{cm} = \frac{2}{3}a_{particle}$
$= 0.67a_{particle}$

Circular hoop $c = 1$
$a_{cm} = \frac{1}{2}a_{particle}$
$= 0.50a_{particle}$

FIGURE 12.44 The angular velocity vector $\vec{\omega}$ is found using the right-hand rule.

1. Using your right hand, curl your fingers in the direction of rotation with your thumb along the rotation axis.
2. Your thumb is then pointing in the direction of $\vec{\omega}$.

Rotation axis

If the object rotates in the xy-plane, the vector $\vec{\omega}$ points along the z-axis. The scalar angular velocity $\omega = v_t/r$ that we've been using is now seen to be ω_z, the z-component of the vector $\vec{\omega}$. You should convince yourself that the sign convention for ω (positive for ccw rotation, negative for cw rotation) is equivalent to having the vector $\vec{\omega}$ pointing in the positive z-direction or the negative z-direction.

The Cross Product of Two Vectors

We defined the torque exerted by force $\vec{F}$ to be $\tau = rF \sin\phi$. The quantity F is the magnitude of the force vector $\vec{F}$, and the distance r is really the magnitude of the position vector $\vec{r}$. Hence torque looks very much like a product of the two vectors $\vec{r}$ and $\vec{F}$. Previously, in conjunction with the definition of work, we introduced the dot product of two vectors: $\vec{A} \cdot \vec{B} = AB \cos\alpha$, where α is the angle between the vectors. $\tau = rF \sin\phi$ is a different way of multiplying vectors that depends on the *sine* of the angle between them.

FIGURE 12.45 shows two vectors, $\vec{A}$ and $\vec{B}$, with angle α between them. We define the **cross product** of $\vec{A}$ and $\vec{B}$ as the vector

$$\vec{A} \times \vec{B} \equiv (AB \sin\alpha, \text{ in the direction given by the right-hand rule}) \quad (12.46)$$

The symbol $\times$ between the vectors is *required* to indicate a cross product. The cross product is also called the **vector product** because the result is a vector.

The **right-hand rule,** which specifies the direction of $\vec{A} \times \vec{B}$, can be stated in three different but equivalent ways:

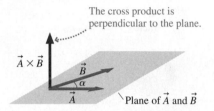

FIGURE 12.45 The cross product $\vec{A} \times \vec{B}$, is a vector perpendicular to the plane of vectors $\vec{A}$ and $\vec{B}$.

The cross product is perpendicular to the plane.

Plane of $\vec{A}$ and $\vec{B}$

Using the right-hand rule

Spread your *right* thumb and index finger apart by angle α. Bend your middle finger so that it is *perpendicular* to your thumb and index finger. Orient your hand so that your thumb points in the direction of $\vec{A}$ and your index finger in the direction of $\vec{B}$. Your middle finger now points in the direction of $\vec{A} \times \vec{B}$.

Make a loose fist with your *right* hand with your thumb extended outward. Orient your hand so that your thumb is perpendicular to the plane of $\vec{A}$ and $\vec{B}$ and your fingers are curling *from* the line of vector $\vec{A}$ *toward* the line of vector $\vec{B}$. Your thumb now points in the direction of $\vec{A} \times \vec{B}$.

Imagine using a screwdriver to turn the slot in the head of a screw from the direction of $\vec{A}$ to the direction of $\vec{B}$. The screw will move either "in" or "out." The direction in which the screw moves is the direction of $\vec{A} \times \vec{B}$.

These methods are easier to demonstrate than to describe in words! Your instructor will show you how they work. Some individuals find one method of thinking about the direction of the cross product easier than the others, but they all work, and you'll soon find the method that works best for you.

Referring back to Figure 12.45, you should use the right-hand rule to convince yourself that the cross product $\vec{A} \times \vec{B}$ is a vector that points *upward,* perpendicular to the plane of $\vec{A}$ and $\vec{B}$. FIGURE 12.46 shows that the cross product, like the dot product, depends on the angle between the two vectors. Notice the two special cases: $\vec{A} \times \vec{B} = \vec{0}$ when $\alpha = 0°$ (parallel vectors) and $\vec{A} \times \vec{B}$ has its maximum magnitude AB when $\alpha = 90°$ (perpendicular vectors).

FIGURE 12.46 The magnitude of the cross-product vector increases from 0 to AB as α increases from 0° to 90°.

The cross product is zero when $\vec{A}$ and $\vec{B}$ are parallel.

As α increases from 0° to 90°, the length of $\vec{A} \times \vec{B}$ increases.

The cross product is maximum when $\vec{A}$ and $\vec{B}$ are perpendicular.

EXAMPLE 12.16 | **Calculating a cross product**

FIGURE 12.47 shows vectors $\vec{C}$ and $\vec{D}$ in the plane of the page. What is the cross product $\vec{E} = \vec{C} \times \vec{D}$?

FIGURE 12.47 Vectors $\vec{C}$ and $\vec{D}$.

SOLVE The angle between the two vectors is $\alpha = 110°$. Consequently, the magnitude of the cross product is

$$E = CD \sin \alpha = (2 \text{ m})(1 \text{ m}) \sin(110°) = 1.88 \text{ m}^2$$

The direction of $\vec{E}$ is given by the right-hand rule. To curl your right fingers from $\vec{C}$ to $\vec{D}$, you have to point your thumb *into* the page. Alternatively, if you turned a screwdriver from $\vec{C}$ to $\vec{D}$ you would be driving a screw *into* the page. Thus

$$\vec{E} = (1.88 \text{ m}^2, \text{ into page})$$

ASSESS Notice that $\vec{E}$ has units of m^2.

The cross product has three important properties:

1. The product $\vec{A} \times \vec{B}$ is *not* equal to the product $\vec{B} \times \vec{A}$. That is, the cross product does not obey the commutative rule $ab = ba$ that you know from arithmetic. In fact, you can see from the right-hand rule that the product $\vec{B} \times \vec{A}$ points in exactly the opposite direction from $\vec{A} \times \vec{B}$. Thus, as **FIGURE 12.48a** shows,

$$\vec{B} \times \vec{A} = -\vec{A} \times \vec{B}$$

2. In a *right-handed coordinate system*, which is the standard coordinate system of science and engineering, the z-axis is oriented relative to the xy-plane such that the unit vectors obey $\hat{\imath} \times \hat{\jmath} = \hat{k}$. This is shown in **FIGURE 12.48b.** You can also see from this figure that $\hat{\jmath} \times \hat{k} = \hat{\imath}$ and $\hat{k} \times \hat{\imath} = \hat{\jmath}$.

3. The derivative of a cross product is

$$\frac{d}{dt}(\vec{A} \times \vec{B}) = \frac{d\vec{A}}{dt} \times \vec{B} + \vec{A} \times \frac{d\vec{B}}{dt} \qquad (12.47)$$

Torque

Now let's return to torque. As a concrete example, **FIGURE 12.49** on the next page shows a long wrench being used to loosen the nuts holding a car wheel on. We've established a right-handed coordinate system with its origin at the nut, so force $\vec{F}$ exerts a torque about the origin. Let's define a *torque vector*

$$\vec{\tau} \equiv \vec{r} \times \vec{F} \qquad (12.48)$$

If we place the vector tails together in order to use the right-hand rule, we see that the torque vector is perpendicular to the plane of $\vec{r}$ and $\vec{F}$. The angle between the vectors is ϕ, so the magnitude of the torque is $\tau = rF|\sin \phi|$.

FIGURE 12.48 Properties of the cross product.

(a)

(b)

$\hat{\imath} \times \hat{\jmath} = \hat{k}$
$\hat{\jmath} \times \hat{k} = \hat{\imath}$
$\hat{k} \times \hat{\imath} = \hat{\jmath}$

FIGURE 12.49 The torque vector.

This is a right-handed coordinate system.

Force $\vec{F}$ exerts a torque.

$\vec{\tau} = \vec{r} \times \vec{F}$

The torque vector is perpendicular to the plane of $\vec{r}$ and $\vec{F}$.

The force vector is redrawn with its tail on the tail of $\vec{r}$.

You can see that the scalar torque $\tau = rF\sin\phi$ we've been using is really the component along the rotation axis—in this case τ_x—of the vector $\vec{\tau}$. This is the basis for our earlier sign convention for τ. In Figure 12.49, where the force causes a ccw rotation, the torque vector points in the positive x-direction, and thus τ_x is positive.

EXAMPLE 12.17 | **Wrench torque revisited**

Example 12.8 found the torque that Luis exerts on a nut by pulling on the end of a wrench. What is the torque vector?

VISUALIZE FIGURE 12.50 shows the position vector $\vec{r}$, drawn from the pivot point to the point where the force is applied. The figure

FIGURE 12.50 Calculating the torque vector.

The torque vector points into the page.

The force vector is redrawn at the pivot point in order to use the right-hand rule.

also redraws the force vector $\vec{F}$ at the pivot point, not because force is applied there but because it's easiest to use the right-hand rule if the vectors are drawn with their tails together.

SOLVE We already know the magnitude of the torque, 17 N m, from Example 12.8. Now we need to apply the right-hand rule. If you place your right thumb along $\vec{r}$ and your index finger along $\vec{F}$, which is somewhat awkward, you'll see that your middle finger points into the page. Alternatively, make a loose fist of your right hand, then orient your fist so that your fingers curl *from $\vec{r}$ toward $\vec{F}$*. Doing so requires your thumb to point into the page. Using either method, we conclude that

$$\vec{\tau} = (17\ \text{N m, into page})$$

12.11 Angular Momentum

FIGURE 12.51 shows a particle that, at this instant, is located at position $\vec{r}$ and is moving with momentum $\vec{p} = m\vec{v}$. Together, $\vec{r}$ and $\vec{p}$ define the *plane of motion*. We define the particle's **angular momentum** $\vec{L}$ relative to the origin to be the vector

$$\vec{L} \equiv \vec{r} \times \vec{p} = (mrv\sin\beta, \text{ direction of right-hand rule}) \qquad (12.49)$$

Because of the cross product, **the angular momentum vector is perpendicular to the plane of motion.** The units of angular momentum are kg m²/s.

NOTE Angular momentum is the rotational equivalent of linear momentum in much the same way that torque is the rotational equivalent of force. Notice that the vector definitions are parallel: $\vec{\tau} \equiv \vec{r} \times \vec{F}$ and $\vec{L} \equiv \vec{r} \times \vec{p}$.

FIGURE 12.51 The angular momentum vector $\vec{L}$.

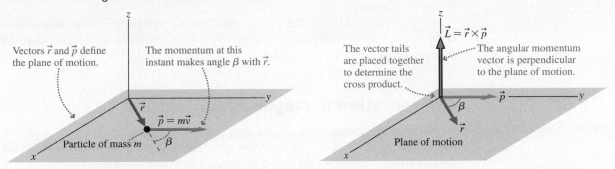

Angular momentum, like torque, is *about* the point from which $\vec{r}$ is measured. A different origin would yield a different angular momentum. Angular momentum is especially simple for a particle in circular motion. As FIGURE 12.52 shows, the angle β between $\vec{p}$ (or $\vec{v}$) and $\vec{r}$ is always 90° if we make the obvious choice of measuring $\vec{r}$ from the center of the circle. For motion in the xy-plane, the angular momentum vector $\vec{L}$—which must be perpendicular to the plane of motion—is entirely along the z-axis:

$$L_z = mrv_t \qquad \text{(particle in circular motion)} \qquad (12.50)$$

where v_t is the tangential component of velocity. Our sign convention for v_t makes L_z, like ω, positive for a ccw rotation, negative for a cw rotation.

In Chapter 11, we found that Newton's second law for a particle can be written $\vec{F}_{net} = d\vec{p}/dt$. There's a similar connection between torque and angular momentum. To show this, we take the time derivative of $\vec{L}$:

$$\frac{d\vec{L}}{dt} = \frac{d}{dt}(\vec{r} \times \vec{p}) = \frac{d\vec{r}}{dt} \times \vec{p} + \vec{r} \times \frac{d\vec{p}}{dt} \qquad (12.51)$$
$$= \vec{v} \times \vec{p} + \vec{r} \times \vec{F}_{net}$$

where we used Equation 12.47 for the derivative of a cross product. We also used the definitions $\vec{v} = d\vec{r}/dt$ and $\vec{F}_{net} = d\vec{p}/dt$.

Vectors $\vec{v}$ and $\vec{p}$ are parallel, and the cross product of two parallel vectors is $\vec{0}$. Thus the first term in Equation 12.51 vanishes. The second term $\vec{r} \times \vec{F}_{net}$ is the net torque, $\vec{\tau}_{net} = \vec{\tau}_1 + \vec{\tau}_2 + \cdots$, so we arrive at

$$\frac{d\vec{L}}{dt} = \vec{\tau}_{net} \qquad (12.52)$$

Equation 12.52, which says **a net torque causes the particle's angular momentum to change**, is the rotational equivalent of $d\vec{p}/dt = \vec{F}_{net}$.

Angular Momentum of a Rigid Body

Equation 12.52 is the angular momentum of a single particle. The angular momentum of a rigid body composed of particles with individual angular momenta $\vec{L}_1, \vec{L}_2, \vec{L}_3, \ldots$ is the vector sum

$$\vec{L} = \vec{L}_1 + \vec{L}_2 + \vec{L}_3 + \cdots = \sum_i \vec{L}_i \qquad (12.53)$$

We can combine Equations 12.52 and 12.53 to find the rate of change of the system's angular momentum:

$$\frac{d\vec{L}}{dt} = \sum_i \frac{d\vec{L}_i}{dt} = \sum_i \vec{\tau}_i = \vec{\tau}_{net} \qquad (12.54)$$

Because any internal forces are action/reaction pairs of forces, acting with the same strength in opposite directions, the net torque due to internal forces is zero. Thus the

FIGURE 12.52 Angular momentum of circular motion.

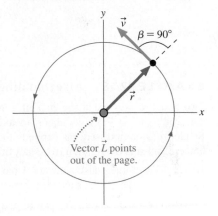

only forces that contribute to the net torque are external forces exerted on the system by the environment.

For a system of particles, **the rate of change of the system's angular momentum is the net torque on the system.** Equation 12.54 is analogous to the Chapter 11 result $d\vec{P}/dt = \vec{F}_{net}$, which says that the rate of change of a system's total linear momentum is the net force on the system.

Conservation of Angular Momentum

A net torque on a rigid body causes its angular momentum to change. Conversely, the angular momentum does *not* change—it is *conserved*—for a system with no net torque. This is the basis of the law of conservation of angular momentum.

> **Law of conservation of angular momentum** The angular momentum $\vec{L}$ of an isolated system ($\vec{\tau}_{net} = \vec{0}$) is conserved. The final angular momentum $\vec{L}_f$ is equal to the initial angular momentum $\vec{L}_i$. Both the magnitude *and* the direction of $\vec{L}$ are unchanged.

EXAMPLE 12.18 | An expanding rod

Two equal masses are at the ends of a massless 50-cm-long rod. The rod spins at 2.0 rev/s about an axis through its midpoint. Suddenly, a compressed gas expands the rod out to a length of 160 cm. What is the rotation frequency after the expansion?

MODEL The forces push outward from the pivot and exert no torques. Thus the system's angular momentum is conserved.

VISUALIZE FIGURE 12.53 is a before-and-after pictorial representation. The angular momentum vectors $\vec{L}_i$ and $\vec{L}_f$ are perpendicular to the plane of motion.

SOLVE The particles are moving in circles, so each has angular momentum $L = mrv_t = mr^2\omega = \frac{1}{4}ml^2\omega$, where we used $r = \frac{1}{2}l$. Thus the initial angular momentum of the system is

$$L_i = \tfrac{1}{4}ml_i^2\omega_i + \tfrac{1}{4}ml_i^2\omega_i = \tfrac{1}{2}ml_i^2\omega_i$$

Similarly, the angular momentum after the expansion is $L_f = \frac{1}{2}ml_f^2\omega_f$. Angular momentum is conserved as the rod expands, thus

$$\tfrac{1}{2}ml_f^2\omega_f = \tfrac{1}{2}ml_i^2\omega_i$$

Solving for ω_f, we find

$$\omega_f = \left(\frac{l_i}{l_f}\right)^2 \omega_i = \left(\frac{50\text{ cm}}{160\text{ cm}}\right)^2 (2.0\text{ rev/s}) = 0.20\text{ rev/s}$$

ASSESS The values of the masses weren't needed. All that matters is the ratio of the lengths.

FIGURE 12.53 The system before and after the rod expands.

Before:
$\omega_i = 2$ rev/s
$\vec{L}_i = \vec{L}_{1i} + \vec{L}_{2i}$
$\vec{p}_{2i}$
2
$\vec{p}_{1i}$
$l_i = 50$ cm

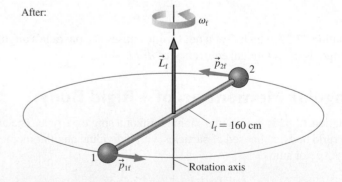

After:
ω_f
$\vec{L}_f$
$\vec{p}_{2f}$
2
$l_f = 160$ cm
1
$\vec{p}_{1f}$
Rotation axis

Angular Momentum and Angular Velocity

The analogy between linear and rotational motion has been so consistent that you might expect one more. The Chapter 9 result $\vec{P} = M\vec{v}$, which we can now write as $M\vec{v}_{cm}$ because it is translational motion of the object as a whole, might give us reason to anticipate that angular momentum and angular velocity are related by $\vec{L} = I\vec{\omega}$. Unfortunately, the analogy breaks down here. For an arbitrarily shaped object, the angular momentum vector and the angular velocity vector don't necessarily point in the same direction. The general relationship between $\vec{L}$ and $\vec{\omega}$ is beyond the scope of this text.

The good news is that the analogy *does* continue to hold in two important situations: the rotation of a *symmetrical* object about the symmetry axis and the rotation of any object about a fixed axle. For example, the axis of a cylinder or disk is a symmetry axis, as is any diameter through a sphere. In these two situations, the angular momentum and angular velocity are related by

$$\vec{L} = I\vec{\omega} \quad \text{(rotation about a fixed axle or axis of symmetry)} \quad (12.55)$$

This relationship is shown for a spinning disk in FIGURE 12.54. Equation 12.55 is particularly important for applying the law of conservation of angular momentum.

If an object's angular momentum is conserved, its angular speed is inversely proportional to its moment of inertia. The rotation of the rod in Example 12.18 slowed dramatically as it expanded because its moment of inertia increased. Similarly, the ice skater in FIGURE 12.55 uses her moment of inertia to control her spin. She spins faster if she pulls in her arms, decreasing her moment of inertia. Similarly, extending her arms increases her moment of inertia, and her angular velocity drops until she can skate out of the spin. It's all a matter of conserving angular momentum.

TABLE 12.4 summarizes the analogies between linear and angular quantities.

FIGURE 12.54 The angular momentum vector about an axis of symmetry.

Angular velocity and angular momentum vectors point along the rotation axis in the direction determined by the right-hand rule.

TABLE 12.4 Angular and linear momentum and energy

Angular motion	Linear motion
$K_{rot} = \frac{1}{2}I\omega^2$	$K_{cm} = \frac{1}{2}Mv_{cm}^2$
$\vec{L} = I\vec{\omega}$ *	$\vec{P} = M\vec{v}_{cm}$
$d\vec{L}/dt = \vec{\tau}_{net}$	$d\vec{P}/dt = \vec{F}_{net}$
The angular momentum of a system is conserved if there is no net torque.	The linear momentum of a system is conserved if there is no net force.

*Rotation about an axis of symmetry.

FIGURE 12.55 An ice skater's rotation speed depends on her moment of inertia.

Large moment of inertia; slow spin

Small moment of inertia; fast spin

EXAMPLE 12.19 | Two interacting disks

A 20-cm-diameter, 2.0 kg solid disk is rotating at 200 rpm. A 20-cm-diameter, 1.0 kg circular loop is dropped straight down onto the rotating disk. Friction causes the loop to accelerate until it is "riding" on the disk. What is the final angular velocity of the combined system?

MODEL The friction between the two objects creates torques that speed up the loop and slow down the disk. But these torques are internal to the combined disk + loop system, so $\tau_{net} = 0$ and the *total* angular momentum of the disk + loop system is conserved.

VISUALIZE FIGURE 12.56 is a before-and-after pictorial representation. Initially only the disk is rotating, at angular velocity $\vec{\omega}_i$. The rotation is about an axis of symmetry, so the angular momentum $\vec{L} = I\vec{\omega}$ is parallel to $\vec{\omega}$. At the end of the problem, $\vec{\omega}_{disk} = \vec{\omega}_{loop} = \vec{\omega}_f$.

SOLVE Both angular momentum vectors point along the rotation axis. Conservation of angular momentum tells us that the magnitude of $\vec{L}$ is unchanged. Thus

$$L_f = I_{disk}\omega_f + I_{loop}\omega_f = L_i = I_{disk}\omega_i$$

Solving for ω_f gives

$$\omega_f = \frac{I_{disk}}{I_{disk} + I_{loop}}\omega_i$$

FIGURE 12.56 The circular loop drops onto the rotating disk.

Before: $\omega_i = 200$ rpm
20 cm L_i
$M_{loop} = 1.0$ kg
$M_{disk} = 2.0$ kg
Symmetry axis

After: ω_f
L_f

The moments of inertia for a disk and a loop can be found in Table 12.2, leading to

$$\omega_f = \frac{\frac{1}{2}M_{disk}R^2}{\frac{1}{2}M_{disk}R^2 + M_{loop}R^2}\omega_i = 100 \text{ rpm}$$

ASSESS The angular velocity has been reduced to half its initial value, which seems reasonable.

STOP TO THINK 12.7 Two buckets spin around in a horizontal circle on frictionless bearings. Suddenly, it starts to rain. As a result,

a. The buckets continue to rotate at constant angular velocity because the rain is falling vertically while the buckets move in a horizontal plane.

b. The buckets continue to rotate at constant angular velocity because the total mechanical energy of the bucket + rain system is conserved.

c. The buckets speed up because the potential energy of the rain is transformed into kinetic energy.

d. The buckets slow down because the angular momentum of the bucket + rain system is conserved.

e. Both a and b.   f. None of the above.

12.12 ADVANCED TOPIC Precession of a Gyroscope

Rotating objects can exhibit surprising and unexpected behaviors. For example, a common lecture demonstration makes use of a bicycle wheel with two handles along the axis. The wheel is spun, then handed to an unsuspecting student who is asked to turn the spinning wheel 90°. Surprisingly, this is *very hard to do*. The reason is that the angular momentum is a *vector*, so the wheel's rotation axis—the direction of $\vec{L}$—is highly resistant to change. If the wheel is spinning fast, a *large* torque is required to turn the wheel's axis.

We'll look at a related example: the precession of a gyroscope. A **gyroscope**—whether it's a toy or a precision instrument used for navigation—is a rapidly spinning wheel or disk whose axis of rotation can assume any orientation. As it spins, it has angular momentum $\vec{L} = I\vec{\omega}$ along the rotation axis. A navigation gyroscope is mounted in gimbals that allow it to spin with virtually no torque from the environment. Once its axis is pointed north, conservation of angular momentum will ensure that the axis continues to point north no matter how the ship or plane moves.

We want to consider a horizontal gyroscope, with the disk spinning in a vertical plane, that is supported at only one end of its axle, as shown in **FIGURE 12.57**. You would expect it to simply fall over—but it doesn't. Instead, the axle remains horizontal, parallel to the ground, while the entire gyroscope slowly rotates in a horizontal plane. This steady change in the orientation of the rotation axis is called **precession,** and we say that the gyroscope precesses about its point of support. The **precession frequency** Ω (capital Greek omega) is much less than the disk's rotation frequency ω. Note that Ω, like ω, is in rad/s.

You might object that angular momentum is not conserved during precession. This is true. The *magnitude* of $\vec{L}$ is constant, but its *direction* is changing. However, angular momentum is conserved only for an isolated system, one on which there is no net torque. The spinning gyroscope is *not* an isolated system because gravity is exerting a torque on it. Indeed, understanding the relationship between the gravitational torque and the angular momentum is the key to understanding why the gyroscope precesses.

FIGURE 12.58a shows a gyroscope that is *not* spinning. When released, it most definitely falls over by rotating about the point of support until the disk hits the table. Because the motion is *rotation*, rather than the translational motion of a gyroscope that is simply dropped, we can analyze it using the concepts of torque and angular momentum.

There are two forces acting on the gyroscope: gravity pulling downward at the disk's center of mass (we'll assume that the axle is massless) and the normal force of the support pushing upward. The normal force exerts no torque about the pivot point because it acts at the pivot point, so **the net torque on the gyroscope is entirely a gravitational torque:**

$$\vec{\tau} = \vec{r} \times \vec{F}_{\text{G}} = Mgd\,\hat{\imath} \tag{12.56}$$

FIGURE 12.57 A spinning gyroscope precesses in a horizontal plane.

The gyroscope precesses around a horizontal circle at frequency Ω.

The gyroscope spins with angular velocity ω.

where $d = |\vec{r}|$ is the distance from the pivot to the center of the disk. To evaluate the cross product, we redrew $\vec{F}_G$ at the pivot and established a coordinate system with the z-axis along the axle. Vectors $\vec{r}$ and $\vec{F}_G$ are perpendicular ($\sin\alpha = 1$), and by using the right-hand rule we see that $\vec{\tau}$ points along the x-axis.

We found in the previous section that a torque causes the angular momentum to change. In particular,

$$\frac{d\vec{L}}{dt} = \vec{\tau} \qquad (12.57)$$

So in a small interval of time dt, the torque causes the gyroscope's angular momentum about the point of support to change by $d\vec{L} = \vec{\tau}\, dt$.

FIGURE 12.58b shows graphically what happens. Initially, when the gyroscope is first released, $\vec{L} = \vec{0}$. After a small interval of time, the gyroscope acquires a small amount of angular momentum $d\vec{L}$ in the direction of $\vec{\tau}$, the $\hat{\imath}$ direction. An angular momentum along the x-axis means that the gyroscope is rotating in the yz-plane—which is exactly what it does as it starts to fall. During the next interval of time, $\vec{L}$ increases a bit more in the $\hat{\imath}$ direction, and then a bit more. This is what we expect as the falling gyroscope picks up speed, increasing its angular momentum.

Now the magnitude of $\vec{\tau}$ does not remain constant—the angle between $\vec{r}$ and $\vec{F}_G$ changes as the gyroscope falls, changing the cross product—so integrating Equation 12.57 symbolically is very difficult. Nonetheless, the *direction* of $\vec{\tau}$ is always the $\hat{\imath}$ direction, so we can see that the angular momentum keeps increasing in the $\hat{\imath}$ direction as the gyroscope falls.

What's different about a spinning gyroscope that causes it to precess rather than fall? In **FIGURE 12.59a** we've again just released the gyroscope, its axle is again along the z-axis, but now it's spinning with angular velocity $\vec{\omega} = \omega\,\hat{k}$. Consequently, the gyroscope has initial angular momentum $\vec{L} = I\vec{\omega} = I\omega\,\hat{k}$ along the z-axis. The torque is exactly as we calculated above, and that torque again causes the angular momentum to change by $d\vec{L} = \vec{\tau}\, dt$. **The only difference is that the gyroscope starts with initial angular momentum**—but that makes all the difference.

FIGURE 12.59b, looking down from above, shows the initial angular momentum $\vec{L}$. A very small time interval dt after we release the gyroscope, its angular momentum will have changed to $\vec{L} + d\vec{L}$. The small *change* in angular momentum, $d\vec{L}$, is parallel to the torque and thus *perpendicular* to the spinning gyroscope's angular momentum $\vec{L}$. Because we're adding vectors, not scalars, the "new" angular momentum has rotated to a new position but *not* increased in magnitude. So during dt, the angular momentum—and thus the entire gyroscope—rotates through a small angle $d\phi$ in the horizontal plane.

The torque vector is always perpendicular to the axle, and thus $d\vec{L}$ is always perpendicular to $\vec{L}$. With each subsequent time interval dt, the gyroscope rotates through another small angle $d\phi$ while the magnitude of the angular momentum (and hence the disk's angular velocity ω) is unchanged. The gyroscope is precessing in the horizontal plane!

You've encountered a similar situation previously. If a ball is initially at rest, pulling on it with a string causes the ball to accelerate (increasing $\vec{v}$) in the direction of the pull. That is, $\vec{v}$ increases in magnitude but doesn't change direction. But if a ball on a string is in uniform circular motion, a force directed to the center—the string tension—has a very different effect. $d\vec{v}$ points toward the center, because that's the direction of the centripetal acceleration, but now $d\vec{v}$ is perpendicular to $\vec{v}$. Adding them as vectors to get $\vec{v} + d\vec{v}$ changes the *direction* of the velocity vector but not its magnitude. Then, as now, having an initial vector ($\vec{v}$ or $\vec{L}$) leads to very different behavior than not having an initial vector.

The small horizontal rotation $d\phi$ is a small piece of the precession. Because it occurs during the small time interval dt, the *rate* of horizontal rotation—the precession frequency—is

$$\Omega = \frac{d\phi}{dt} \qquad (12.58)$$

FIGURE 12.58 The gravitational torque on a nonspinning gyroscope causes it to fall over.

(a)

The torque is perpendicular to $\vec{r}$.

The gyroscope falls.

(b)

The initial angular momentum is zero.

The angular momentum increases in the direction of the torque.

FIGURE 12.59 For a spinning gyroscope, the gravitational torque changes the direction but not the magnitude of the angular momentum.

(a)

The torque hasn't changed.

The gyroscope precesses.

(b)

View from above

The angular momentum changes direction but not magnitude.

$d\vec{L}$ is always perpendicular to $\vec{L}$.

In Figure 12.59b the small length dL is the arc length spanned by $d\phi$, hence $d\phi = dL/L$ and the precession frequency is

$$\Omega = \frac{dL/L}{dt} = \frac{dL/dt}{L} \qquad (12.59)$$

From Equations 12.56 and 12.57, $dL/dt = \tau = Mgd$. Further, the gyroscope's angular momentum has magnitude $L = I\omega$, where I is the moment of inertia of the disk rotating about the axle. Thus the precession frequency of the gyroscope—in rad/s—is

$$\Omega = \frac{Mgd}{I\omega} \qquad (12.60)$$

Because the spin angular velocity ω is in the denominator, a very rapidly spinning gyroscope precesses very slowly. As the gyroscope runs down, due to any little bit of friction, it begins to precess faster and faster.

We have made one tacit assumption. As the gyroscope precesses, the precessional motion has its own angular momentum along the vertical axis. The gyroscope's angular momentum $\vec{L}$ is not simply the angular momentum of the spinning disk, as we assumed, but the vector sum $\vec{L}_{\text{spin}} + \vec{L}_{\text{precess}}$. As long as the gyroscope precesses slowly, with $\Omega \ll \omega$, the precessional angular momentum is very small compared to the spin angular momentum and our assumption is well justified. But toward the end of the gyroscope's motion, as ω decreases and Ω increases, our model of precession breaks down and the gyroscope's motion becomes more complex.

EXAMPLE 12.20 | **A precessing gyroscope**

A gyroscope used in a lecture demonstration consists of a 120 g, 7.0-cm-diameter solid disk that rotates on a lightweight axle. From the center of the disk to the end of the axle is 5.0 cm. When spun, placed on a stand, and released, the gyroscope is observed to precess with a period of 1.0 s. How fast, in rpm, is it spinning?

SOLVE The precession frequency is given by Equation 12.60. The moment of inertia of a disk of mass M and radius R about an axis through its center is $I = \frac{1}{2}MR^2$. Inserting this into Equation 12.60, we see that the precession frequency

$$\Omega = \frac{Mgd}{I\omega} = \frac{Mgd}{\frac{1}{2}MR^2\omega} = \frac{2gd}{\omega R^2}$$

is actually independent of the gyroscope's mass. Solving for ω gives

$$\omega = \frac{2gd}{\Omega R^2}$$

A precession period of 1.0 s corresponds to the precession frequency

$$\Omega = \frac{2\pi \text{ rad}}{1.0 \text{ s}} = 6.28 \text{ rad/s}$$

Thus the gyroscope's spin angular velocity is

$$\omega = \frac{2(9.80 \text{ m/s}^2)(0.050 \text{ m})}{(6.28 \text{ rad/s})(0.035 \text{ m})^2} = 127 \text{ rad/s}$$

Converting to rpm gives

$$\omega = 127 \text{ rad/s} \times \frac{1 \text{ rev}}{2\pi \text{ rad}} \times \frac{60 \text{ s}}{1 \text{ min}} = 1200 \text{ rpm}$$

ASSESS 1200 rpm is 20 rev/s. That seems reasonable for a spinning top or gyroscope. And $\Omega \ll \omega$, so our precession model of the gyroscope is valid.

CHALLENGE EXAMPLE 12.21 | **The ballistic pendulum revisited**

A 2.0 kg block hangs from the end of a 1.5 kg, 1.0-m-long rod, together forming a pendulum that swings from a frictionless pivot at the top end of the rod. A 10 g bullet is fired horizontally into the block, where it sticks, causing the pendulum to swing out to a 30° angle. What was the speed of the bullet?

MODEL Model the rod as a uniform rod that can rotate around one end, and assume the block is small enough to model as a particle. There are no external torques on the bullet + block + rod system, so angular momentum is conserved in the inelastic

collision. Further, the mechanical energy of the system is conserved after (but not during) the collision as the pendulum swings outward.

VISUALIZE FIGURE 12.60 is a pictorial representation. This is a two-part problem, so we've separated the collision's before-and-after from the pendulum swing's before-and-after. The end of the collision is the beginning of the swing.

SOLVE This is a *ballistic pendulum*. Example 11.5 considered a simpler ballistic pendulum with a mass on a string, rather than on

FIGURE 12.60 Pictorial representation of the bullet hitting the pendulum.

Collision: Angular momentum

Swing: Mechanical energy

a rod, and a review of that example is highly recommended. The key to both is that a different conservation law applies to each part of the problem.

Angular momentum is conserved in the collision, thus $L_1 = L_0$. Before the collision, the angular momentum—which we'll measure about the pendulum's pivot point—is entirely that of the bullet. The angular momentum of a particle is $L = mrv \sin \beta$. An instant before the collision, just as the bullet reaches the block, $r = d$ and, because $\vec{v}$ is perpendicular to $\vec{r}$ at that instant, $\beta = 90°$. Thus $L_0 = m_b d v_{0b}$. (This is the magnitude of the angular momentum; from the right-hand rule, the angular momentum vector points out of the page.)

An instant after the collision, but before the pendulum has had time to move, the rod has angular velocity ω_1 and the block, with the embedded bullet, is moving in a circle with speed $v_1 = \omega_1 r = \omega_1 d$. The angular momentum of the block + bullet system is that of a particle, still with $\beta = 90°$, while that of the rod—an object rotating on a fixed axle—is $I_{rod}\omega_1$. Thus the post-collision angular momentum is

$$L_1 = (m_B + m_b)v_1 r + I_{rod}\omega_1 = (m_B + m_b)d^2\omega_1 + \tfrac{1}{3}m_R d^2\omega_1$$

The moment of inertia of the rod was taken from Table 12.2.

Equating the before-and-after angular momenta, then solving for v_{0b}, gives

$$m_b d v_{0b} = (m_B + m_b)d^2\omega_1 + \tfrac{1}{3}m_R d^2\omega_1$$

$$v_{0b} = \frac{m_B + m_b + \tfrac{1}{3}m_R}{m_b}d\omega_1 = 251 d\omega_1$$

Once we know ω_1, which we'll find from energy conservation in the swing, we'll be able to compute the bullet's speed.

Mechanical energy is conserved during the swing, but you must be careful to include all the energies. The kinetic energy has two components: the translational kinetic energy of the block + bullet system and the rotational kinetic energy of the rod. The gravitational potential energy also has two components: the potential energy of the block + bullet system and the potential energy of the rod. The latter changes because the center of mass moves upward as the rod swings. Thus the energy conservation statement is

$$\tfrac{1}{2}(m_B + m_b)v_2^2 + \tfrac{1}{2}I_{rod}\omega_2^2 + (m_B + m_b)gy_2 + m_R gy_{cm2} =$$
$$\tfrac{1}{2}(m_B + m_b)v_1^2 + \tfrac{1}{2}I_{rod}\omega_1^2 + (m_B + m_b)gy_1 + m_R gy_{cm1}$$

Although this looks very complicated, you should convince yourself that we've done nothing more than add up two kinetic energies and two potential energies before and after the swing.

We know that $v_2 = 0$ and $\omega_2 = 0$ at the end of the swing, and that $v_1 = d\omega_1$ at the beginning. We also know the moment of inertia of a rod pivoted at one end. Combining the potential energy terms and using $\Delta y = y_f - y_i$, we thus have

$$\tfrac{1}{2}(m_B + m_b + \tfrac{1}{3}m_R)d^2\omega_1^2 = (m_B + m_b)g\,\Delta y + m_R g\,\Delta y_{cm}$$

We see from Figure 12.60 that the block, at its highest point, is distance $d\cos\theta$ below the pivot. It started distance d below the pivot, so the bullet + block system gained height $\Delta y = d - d\cos\theta = d(1 - \cos\theta)$. The rod's center of mass started distance $d/2$ below the pivot and rises only half as much as the block, so $\Delta y_{cm} = \tfrac{1}{2}d(1 - \cos\theta)$. With these, the energy equation becomes

$$\tfrac{1}{2}(m_B + m_b + \tfrac{1}{3}m_R)d^2\omega_1^2 = (m_B + m_b + \tfrac{1}{2}m_R)gd(1 - \cos\theta)$$

We can now solve for ω_1:

$$\omega_1 = \sqrt{\frac{m_B + m_b + \tfrac{1}{2}m_R}{m_B + m_b + \tfrac{1}{3}m_R}\frac{2g(1 - \cos\theta)}{d}} = 1.70 \text{ rad/s}$$

and with that

$$v_{0b} = 251 d\omega_1 = 430 \text{ m/s}$$

ASSESS 430 m/s seems a reasonable speed for a bullet. This was a challenging problem, but one that you can solve if you focus on the problem-solving strategies—drawing a careful pictorial representation, defining the system, and thinking about which conservation laws apply—rather than hunting for the "right" equation.

SUMMARY

The goal of Chapter 12 has been to understand and apply the physics of rotation.

GENERAL PRINCIPLES

Solving Rotational Dynamics Problems

MODEL Model the object as a **rigid body**.

VISUALIZE Draw a pictorial representation.

SOLVE Use **Newton's second law** for rotational motion:

$$\alpha = \frac{\tau_{net}}{I}$$

Use rotational kinematics to find angles and angular velocities.

ASSESS Is the result reasonable?

Conservation Laws

Energy is conserved for an isolated system.

- Pure rotation $E = K_{rot} + U_G = \frac{1}{2}I\omega^2 + Mgy_{cm}$
- Rolling $E = K_{rot} + K_{cm} + U_G = \frac{1}{2}I\omega^2 + \frac{1}{2}Mv_{cm}^2 + Mgy_{cm}$

Angular momentum is conserved if $\vec{\tau}_{net} = \vec{0}$.

- Particle $\vec{L} = \vec{r} \times \vec{p}$
- Rotation about a symmetry axis or fixed axle $\vec{L} = I\vec{\omega}$

IMPORTANT CONCEPTS

Torque is the rotational equivalent of force:

$$\tau = rF \sin\phi = rF_t = dF$$

The vector description of torque is

$$\vec{\tau} = \vec{r} \times \vec{F}$$

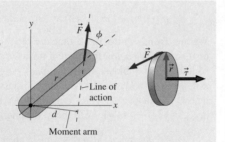

Vector description of rotation

Angular velocity $\vec{\omega}$ points along the rotation axis in the direction of the right-hand rule.

For a rigid body rotating about a fixed axle or an axis of symmetry, the angular momentum is $\vec{L} = I\vec{\omega}$.

Newton's second law is $\dfrac{d\vec{L}}{dt} = \vec{\tau}_{net}$.

A system of particles on which there is no net force undergoes unconstrained rotation about the **center of mass**:

$$x_{cm} = \frac{1}{M}\int x\,dm \qquad y_{cm} = \frac{1}{M}\int y\,dm$$

The gravitational torque on a body can be found by treating the body as a particle with all the mass M concentrated at the center of mass.

The **moment of inertia**

$$I = \sum_i m_i r_i^2 = \int r^2\,dm$$

is the rotational equivalent of mass. The moment of inertia depends on how the mass is distributed around the axis. If I_{cm} is known, I about a parallel axis distance d away is given by the **parallel-axis theorem:** $I = I_{cm} + Md^2$.

APPLICATIONS

Rigid-body model

- Size and shape do not change as the object moves.
- The object is modeled as particle-like atoms connected by massless, rigid rods.

Rigid-body equilibrium

An object is in total equilibrium only if both $\vec{F}_{net} = \vec{0}$ and $\vec{\tau}_{net} = \vec{0}$.

No rotational or translational motion

Rolling motion

For an object that rolls without slipping

$$v_{cm} = R\omega$$
$$K = K_{rot} + K_{cm}$$

TERMS AND NOTATION

rigid-body model	rotational kinetic energy, K_{rot}	constant-torque model	angular momentum, $\vec{L}$
rigid body	moment of inertia, I	static equilibrium model	law of conservation of angular
translational motion	parallel-axis theorem	rolling constraint	momentum
rotational motion	torque, τ	cross product	gyroscope
combination motion	line of action	vector product	precession
center of mass	moment arm, d	right-hand rule	precession frequency

CONCEPTUAL QUESTIONS

1. Is the center of mass of the dumbbell in **FIGURE Q12.1** at point a, b, or c? Explain.

FIGURE Q12.1   **FIGURE Q12.2**

2. If the angular velocity ω is held constant, by what *factor* must R change to double the rotational kinetic energy of the dumbbell in **FIGURE Q12.2**?

3. **FIGURE Q12.3** shows three rotating disks, all of equal mass. Rank in order, from largest to smallest, their rotational kinetic energies K_a to K_c.

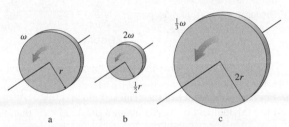

FIGURE Q12.3

4. Must an object be rotating to have a moment of inertia? Explain.

5. The moment of inertia of a uniform rod about an axis through its center is $\frac{1}{12}mL^2$. The moment of inertia about an axis at one end is $\frac{1}{3}mL^2$. Explain *why* the moment of inertia is larger about the end than about the center.

6. You have two solid steel spheres. Sphere 2 has twice the radius of sphere 1. By what *factor* does the moment of inertia I_2 of sphere 2 exceed the moment of inertia I_1 of sphere 1?

7. The professor hands you two spheres. They have the same mass, the same radius, and the same exterior surface. The professor claims that one is a solid sphere and the other is hollow. Can you determine which is which without cutting them open? If so, how? If not, why not?

8. Six forces are applied to the door in **FIGURE Q12.8**. Rank in order, from largest to smallest, the six torques τ_a to τ_f about the hinge on the left. Explain.

FIGURE Q12.8

9. A student gives a quick push to a ball at the end of a mass-less, rigid rod, as shown in **FIGURE Q12.9**, causing the ball to rotate clockwise in a *horizontal* circle. The rod's pivot is frictionless.

FIGURE Q12.9

   a. As the student is pushing, is the torque about the pivot positive, negative, or zero?

   b. After the push has ended, does the ball's angular velocity (i) steadily increase; (ii) increase for awhile, then hold steady; (iii) hold steady; (iv) decrease for awhile, then hold steady; or (v) steadily decrease? Explain.

   c. Right after the push has ended, is the torque positive, negative, or zero?

10. Rank in order, from largest to smallest, the angular accelerations α_a to α_d in **FIGURE Q12.10**. Explain.

FIGURE Q12.10

11. The solid cylinder and cylindrical shell in **FIGURE Q12.11** have the same mass, same radius, and turn on frictionless, horizontal axles. (The cylindrical shell has light-weight spokes connecting the shell to the axle.) A rope is wrapped around each cylinder and tied to a block. The blocks have the same mass and are held the same height above the ground. Both blocks are released simultaneously. Which hits the ground first? Or is it a tie? Explain.

FIGURE Q12.11

12. A diver in the pike position (legs straight, hands on ankles) usually makes only one or one-and-a-half rotations. To make two or three rotations, the diver goes into a tuck position (knees bent, body curled up tight). Why?

13. Is the angular momentum of disk a in **FIGURE Q12.13** larger than, smaller than, or equal to the angular momentum of disk b? Explain.

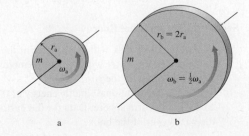

FIGURE Q12.13

EXERCISES AND PROBLEMS

Problems labeled � integrate material from earlier chapters.

Exercises

Section 12.1 Rotational Motion

1. ‖ A high-speed drill reaches 2000 rpm in 0.50 s.
   a. What is the drill's angular acceleration?
   b. Through how many revolutions does it turn during this first 0.50 s?
2. ‖ A skater holds her arms outstretched as she spins at 180 rpm. What is the speed of her hands if they are 140 cm apart?
3. ‖ A ceiling fan with 80-cm-diameter blades is turning at 60 rpm. Suppose the fan coasts to a stop 25 s after being turned off.
   a. What is the speed of the tip of a blade 10 s after the fan is turned off?
   b. Through how many revolutions does the fan turn while stopping?
4. ‖‖ An 18-cm-long bicycle crank arm, with a pedal at one end, is attached to a 20-cm-diameter sprocket, the toothed disk around which the chain moves. A cyclist riding this bike increases her pedaling rate from 60 rpm to 90 rpm in 10 s.
   a. What is the tangential acceleration of the pedal?
   b. What length of chain passes over the top of the sprocket during this interval?

Section 12.2 Rotation About the Center of Mass

5. | How far from the center of the earth is the center of mass of the earth + moon system? Data for the earth and moon can be found inside the back cover of the book.
6. | The three masses shown in **FIGURE EX12.6** are connected by massless, rigid rods. What are the coordinates of the center of mass?

FIGURE EX12.6

FIGURE EX12.7

7. | The three masses shown in **FIGURE EX12.7** are connected by massless, rigid rods. What are the coordinates of the center of mass?
8. ‖ A 100 g ball and a 200 g ball are connected by a 30-cm-long, massless, rigid rod. The balls rotate about their center of mass at 120 rpm. What is the speed of the 100 g ball?

Section 12.3 Rotational Energy

9. ‖ A thin, 100 g disk with a diameter of 8.0 cm rotates about an axis through its center with 0.15 J of kinetic energy. What is the speed of a point on the rim?
10. ‖ What is the rotational kinetic energy of the earth? Assume the earth is a uniform sphere. Data for the earth can be found inside the back cover of the book.

11. ‖ The three 200 g masses in **FIGURE EX12.11** are connected by massless, rigid rods.
    a. What is the triangle's moment of inertia about the axis through the center?
    b. What is the triangle's kinetic energy if it rotates about the axis at 5.0 rev/s?

FIGURE EX12.11

12. ‖ A drum major twirls a 96-cm-long, 400 g baton about its center of mass at 100 rpm. What is the baton's rotational kinetic energy?

Section 12.4 Calculating Moment of Inertia

13. ‖ The four masses shown in **FIGURE EX12.13** are connected by massless, rigid rods.
    a. Find the coordinates of the center of mass.
    b. Find the moment of inertia about an axis that passes through mass A and is perpendicular to the page.

FIGURE EX12.13

14. ‖ The four masses shown in **FIGURE EX12.13** are connected by massless, rigid rods.
    a. Find the coordinates of the center of mass.
    b. Find the moment of inertia about a diagonal axis that passes through masses B and D.
15. | The three masses shown in **FIGURE EX12.15** are connected by massless, rigid rods.
    a. Find the coordinates of the center of mass.
    b. Find the moment of inertia about an axis that passes through mass A and is perpendicular to the page.
    c. Find the moment of inertia about an axis that passes through masses B and C.

FIGURE EX12.15

16. ‖ A 12-cm-diameter CD has a mass of 21 g. What is the CD's moment of inertia for rotation about a perpendicular axis (a) through its center and (b) through the edge of the disk?
17. ‖ A 25 kg solid door is 220 cm tall, 91 cm wide. What is the door's moment of inertia for (a) rotation on its hinges and (b) rotation about a vertical axis inside the door, 15 cm from one edge?

Section 12.5 Torque

18. ‖ In **FIGURE EX12.18**, what is the net torque about the axle?

FIGURE EX12.18

19. ‖ In **FIGURE EX12.19**, what magnitude force provides 5.0 N m net torque about the axle?

FIGURE EX12.19

20. ‖ The 20-cm-diameter disk in **FIGURE EX12.20** can rotate on an axle through its center. What is the net torque about the axle?

FIGURE EX12.20

FIGURE EX12.21

21. ‖ The axle in **FIGURE EX12.21** is half the distance from the center to the rim. What is the net torque about the axle?

22. ‖ A 4.0-m-long, 500 kg steel beam extends horizontally from the point where it has been bolted to the framework of a new building under construction. A 70 kg construction worker stands at the far end of the beam. What is the magnitude of the torque about the bolt due to the worker and the weight of the beam?

23. ‖ An athlete at the gym holds a 3.0 kg steel ball in his hand. His
BIO arm is 70 cm long and has a mass of 3.8 kg, with the center of mass at 40% of the arm length. What is the magnitude of the torque about his shoulder due to the ball and the weight of his arm if he holds his arm
   a. Straight out to his side, parallel to the floor?
   b. Straight, but 45° below horizontal?

Section 12.6 Rotational Dynamics

Section 12.7 Rotation About a Fixed Axis

24. ‖ An object's moment of inertia is 2.0 kg m². Its angular velocity is increasing at the rate of 4.0 rad/s per second. What is the net torque on the object?

25. ‖ An object whose moment of inertia is 4.0 kg m² experiences the torque shown in **FIGURE EX12.25**. What is the object's angular velocity at $t = 3.0$ s? Assume it starts from rest.

FIGURE EX12.25

26. ‖‖ A 1.0 kg ball and a 2.0 kg ball are connected by a 1.0-m-long rigid, massless rod. The rod is rotating cw about its center of mass at 20 rpm. What net torque will bring the balls to a halt in 5.0 s?

27. ‖ Starting from rest, a 12-cm-diameter compact disk takes 3.0 s to reach its operating angular velocity of 2000 rpm. Assume that the angular acceleration is constant. The disk's moment of inertia is 2.5×10^{-5} kg m².
   a. How much net torque is applied to the disk?
   b. How many revolutions does it make before reaching full speed?

28. ‖ A 4.0 kg, 36-cm-diameter metal disk, initially at rest, can rotate on an axle along its axis. A steady 5.0 N tangential force is applied to the edge of the disk. What is the disk's angular velocity, in rpm, 4.0 s later?

Section 12.8 Static Equilibrium

29. ‖ The two objects in **FIGURE EX12.29** are balanced on the pivot. What is distance d?

FIGURE EX12.29   **FIGURE EX12.30**

30. ‖ The object shown in **FIGURE EX12.30** is in equilibrium. What are the magnitudes of $\vec{F}_1$ and $\vec{F}_2$?

31. ‖ The 3.0-m-long, 100 kg rigid beam of **FIGURE EX12.31** is supported at each end. An 80 kg student stands 2.0 m from support 1. How much upward force does each support exert on the beam?

FIGURE EX12.31   Support 1   Support 2

32. ‖ A 5.0 kg cat and a 2.0 kg bowl of tuna fish are at opposite ends of the 4.0-m-long seesaw of **FIGURE EX12.32**. How far to the left of the pivot must a 4.0 kg cat stand to keep the seesaw balanced?

FIGURE EX12.32

Section 12.9 Rolling Motion

33. ‖ A car tire is 60 cm in diameter. The car is traveling at a speed of 20 m/s.
   a. What is the tire's angular velocity, in rpm?
   b. What is the speed of a point at the top edge of the tire?
   c. What is the speed of a point at the bottom edge of the tire?

34. ‖ A 500 g, 8.0-cm-diameter can is filled with uniform, dense food. It rolls across the floor at 1.0 m/s. What is the can's kinetic energy?

35. ‖ An 8.0-cm-diameter, 400 g solid sphere is released from rest at the top of a 2.1-m-long, 25° incline. It rolls, without slipping, to the bottom.
   a. What is the sphere's angular velocity at the bottom of the incline?
   b. What fraction of its kinetic energy is rotational?

36. ‖ A solid sphere of radius R is placed at a height of 30 cm on a 15° slope. It is released and rolls, without slipping, to the bottom. From what height should a circular hoop of radius R be released on the same slope in order to equal the sphere's speed at the bottom?

Section 12.10 The Vector Description of Rotational Motion

37. | Evaluate the cross products $\vec{A} \times \vec{B}$ and $\vec{C} \times \vec{D}$.

FIGURE EX12.37

38. | Evaluate the cross products $\vec{A} \times \vec{B}$ and $\vec{C} \times \vec{D}$.

FIGURE EX12.38

39. | Vector $\vec{A} = 3\hat{\imath} + \hat{\jmath}$ and vector $\vec{B} = 3\hat{\imath} - 2\hat{\jmath} + 2\hat{k}$. What is the cross product $\vec{A} \times \vec{B}$?

40. || Force $\vec{F} = -10\hat{\jmath}$ N is exerted on a particle at $\vec{r} = (5\hat{\imath} + 5\hat{\jmath})$ m. What is the torque on the particle about the origin?

41. || A 1.3 kg ball on the end of a lightweight rod is located at $(x, y) = (3.0 \text{ m}, 2.0 \text{ m})$, where the y-axis is vertical. The other end of the rod is attached to a pivot at $(x, y) = (0 \text{ m}, 3.0 \text{ m})$. What is the torque about the pivot? Write your answer using unit vectors.

Section 12.11 Angular Momentum

42. || What are the magnitude and direction of the angular momentum relative to the origin of the 200 g particle in **FIGURE EX12.42**?

FIGURE EX12.42

43. || What is the angular momentum vector of the 2.0 kg, 4.0-cm-diameter rotating disk in **FIGURE EX12.43**?

FIGURE EX12.43   **FIGURE EX12.44**

44. || What is the angular momentum vector of the 500 g rotating bar in **FIGURE EX12.44**?

45. || How fast, in rpm, would a 5.0 kg, 22-cm-diameter bowling ball have to spin to have an angular momentum of 0.23 kg m²/s?

46. || A 2.0 kg, 20-cm-diameter turntable rotates at 100 rpm on frictionless bearings. Two 500 g blocks fall from above, hit the turntable simultaneously at opposite ends of a diameter, and stick. What is the turntable's angular velocity, in rpm, just after this event?

Section 12.12 Precession of a Gyroscope

47. || A 75 g, 6.0-cm-diameter solid spherical top is spun at 1200 rpm on an axle that extends 1.0 cm past the edge of the sphere. The tip of the axle is placed on a support. What is the top's precession frequency in rpm?

48. || A toy gyroscope has a ring of mass M and radius R attached to the axle by lightweight spokes. The end of the axle is distance R from the center of the ring. The gyroscope is spun at angular velocity ω, then the end of the axle is placed on a support that allows the gyroscope to process.
    a. Find an expression for the precession frequency Ω in terms of M, R, ω, and g.
    b. A 120 g, 8.0-cm-diameter gyroscope is spun at 1000 rpm and allowed to process. What is the precession period?

Problems

49. || A 300 g ball and a 600 g ball are connected by a 40-cm-long massless, rigid rod. The structure rotates about its center of mass at 100 rpm. What is its rotational kinetic energy?

50. || An 800 g steel plate has the shape of the isosceles triangle
CALC shown in **FIGURE P12.50**. What are the x- and y-coordinates of the center of mass?

Hint: Divide the triangle into vertical strips of width dx, then relate the mass dm of a strip at position x to the values of x and dx.

FIGURE P12.50   **FIGURE P12.51**

51. || Determine the moment of inertia about the axis of the object shown in **FIGURE P12.51**.

52. || What is the moment of inertia of a 2.0 kg, 20-cm-diameter disk for rotation about an axis (a) through the center, and (b) through the edge of the disk?

53. || Calculate by direct integration the moment of inertia for a thin
CALC rod of mass M and length L about an axis located distance d from one end. Confirm that your answer agrees with Table 12.2 when $d = 0$ and when $d = L/2$.

54. || Calculate the moment
CALC of inertia of the rectangular plate in **FIGURE P12.54** for rotation about a perpendicular axis through the center.

FIGURE P12.54

55. || a. A disk of mass M and radius R has a hole of radius r centered
CALC on the axis. Calculate the moment of inertia of the disk.
    b. Confirm that your answer agrees with Table 12.2 when $r = 0$ and when $r = R$.
    c. A 4.0-cm-diameter disk with a 3.0-cm-diameter hole rolls down a 50-cm-long, 20° ramp. What is its speed at the bottom? What percent is this of the speed of a particle sliding down a frictionless ramp?

56. ||| Consider a solid cone of radius R, height H, and mass M. The
CALC volume of a cone is $\frac{1}{3}\pi HR^2$.
   a. What is the distance from the apex (the point) to the center of mass?
   b. What is the moment of inertia for rotation about the axis of the cone?
   Hint: The moment of inertia can be calculated as the sum of the moments of inertia of lots of small pieces.

57. || A person's center of mass is easily found by having the per-
BIO son lie on a *reaction board*. A horizontal, 2.5-m-long, 6.1 kg reaction board is supported only at the ends, with one end resting on a scale and the other on a pivot. A 60 kg woman lies on the reaction board with her feet over the pivot. The scale reads 25 kg. What is the distance from the woman's feet to her center of mass?

58. || A 3.0-m-long ladder, as shown in Figure 12.35, leans against a frictionless wall. The coefficient of static friction between the ladder and the floor is 0.40. What is the minimum angle the ladder can make with the floor without slipping?

59. || In **FIGURE P12.59**, an 80 kg construction worker sits down 2.0 m from the end of a 1450 kg steel beam to eat his lunch. What is the tension in the cable?

FIGURE P12.59

Cable

30°

6.0 m

60. || A 40 kg, 5.0-m-long beam is supported by, but not attached to, the two posts in **FIGURE P12.60**. A 20 kg boy starts walking along the beam. How close can he get to the right end of the beam without it falling over?

3.0 m

FIGURE P12.60

L

d

FIGURE P12.61

61. || Your task in a science contest is to stack four identical uniform bricks, each of length L, so that the top brick is as far to the right as possible without the stack falling over. Is it possible, as **FIGURE P12.61** shows, to stack the bricks such that no part of the top brick is over the table? Answer this question by determining the maximum possible value of d.

62. || A 120-cm-wide sign hangs from a 5.0 kg, 200-cm-long pole. A cable of negligible mass supports the end of the rod as shown in **FIGURE P12.62**. What is the maximum mass of the sign if the maximum tension in the cable without breaking is 300 N?

Cable

250 cm

200 cm

80 cm

PHYSICS SHOPPE

FIGURE P12.62

63. || A piece of modern sculpture consists of an 8.0-m-long, 150 kg stainless steel bar passing diametrically through a 50 kg copper sphere. The center of the sphere is 2.0 m from one end of the bar. To be mounted for display, the bar is oriented vertically, with the copper sphere at the lower end, then tilted 35° from vertical and held in place by one horizontal steel cable attached to the bar 2.0 m from the top end. What is the tension in the cable?

64. ||| Flywheels are large, massive wheels used to store energy. They can be spun up slowly, then the wheel's energy can be released quickly to accomplish a task that demands high power. An industrial flywheel has a 1.5 m diameter and a mass of 250 kg. Its maximum angular velocity is 1200 rpm.
   a. A motor spins up the flywheel with a constant torque of 50 N m. How long does it take the flywheel to reach top speed?
   b. How much energy is stored in the flywheel?
   c. The flywheel is disconnected from the motor and connected to a machine to which it will deliver energy. Half the energy stored in the flywheel is delivered in 2.0 s. What is the average power delivered to the machine?
   d. How much torque does the flywheel exert on the machine?

65. || Blocks of mass m_1 and m_2 are connected by a massless string that passes over the pulley in **FIGURE P12.65**. The pulley turns on frictionless bearings. Mass m_1 slides on a horizontal, frictionless surface. Mass m_2 is released while the blocks are at rest.
   a. Assume the pulley is massless. Find the acceleration of m_1 and the tension in the string. This is a Chapter 7 review problem.
   b. Suppose the pulley has mass m_p and radius R. Find the acceleration of m_1 and the tensions in the upper and lower portions of the string. Verify that your answers agree with part a if you set $m_p = 0$.

m_1

m_2

ω

Brake

FIGURE P12.65

FIGURE P12.66

66. || The 2.0 kg, 30-cm-diameter disk in **FIGURE P12.66** is spinning at 300 rpm. How much friction force must the brake apply to the rim to bring the disk to a halt in 3.0 s?

67. || A 30-cm-diameter, 1.2 kg solid turntable rotates on a 1.2-cm-diameter, 450 g shaft at a constant 33 rpm. When you hit the stop switch, a brake pad presses against the shaft and brings the turntable to a halt in 15 seconds. How much friction force does the brake pad apply to the shaft?

68. || Your engineering team has been assigned the task of measuring the properties of a new jet-engine turbine. You've previously determined that the turbine's moment of inertia is 2.6 kg m². The next job is to measure the frictional torque of the bearings. Your plan is to run the turbine up to a predetermined rotation speed, cut the power, and time how long it takes the turbine to reduce its rotation speed by 50%. Your data are given in the table. Draw an appropriate graph of the data and, from the slope of the best-fit line, determine the frictional torque.

Rotation (rpm)	Time (s)
1500	19
1800	22
2100	25
2400	30
2700	34

69. ‖ A hollow sphere is rolling along a horizontal floor at 5.0 m/s when it comes to a 30° incline. How far up the incline does it roll before reversing direction?

70. ‖ A 750 g disk and a 760 g ring, both 15 cm in diameter, are rolling along a horizontal surface at 1.5 m/s when they encounter a 15° slope. How far up the slope does each travel before rolling back down?

71. ‖‖ A cylinder of radius R, length L, and mass M is released from rest on a slope inclined at angle θ. It is oriented to roll straight down the slope. If the slope were frictionless, the cylinder would *slide* down the slope without rotating. What minimum coefficient of static friction is needed for the cylinder to roll down without slipping?

72. ‖ The 5.0 kg, 60-cm-diameter disk in FIGURE P12.72 rotates on an axle passing through one edge. The axle is parallel to the floor. The cylinder is held with the center of mass at the same height as the axle, then released.

    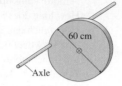
    FIGURE P12.72

    a. What is the cylinder's initial angular acceleration?
    b. What is the cylinder's angular velocity when it is directly below the axle?

73. ‖ A thin, uniform rod of length L and mass M is placed vertically on a horizontal table. If tilted ever so slightly, the rod will fall over.
    a. What is the speed of the center of mass just as the rod hits the table if there's so much friction that the bottom tip of the rod does not slide?
    b. What is the speed of the center of mass just as the rod hits the table if the table is frictionless?

74. ‖ A long, thin rod of mass M and length L is standing straight up on a table. Its lower end rotates on a frictionless pivot. A very slight push causes the rod to fall over. As it hits the table, what are (a) the angular velocity and (b) the speed of the tip of the rod?

75. ‖‖ The marble rolls down the track shown in FIGURE P12.75 and around a loop-the-loop of radius R. The marble has mass m and radius r. What minimum height h must the track have for the marble to make it around the loop-the-loop without falling off?

FIGURE P12.75

FIGURE P12.76

76. ‖ The sphere of mass M and radius R in FIGURE P12.76 is rigidly attached to a thin rod of radius r that passes through the sphere at distance $\frac{1}{2}R$ from the center. A string wrapped around the rod pulls with tension T. Find an expression for the sphere's angular acceleration. The rod's moment of inertia is negligible.

77. ‖ A satellite follows the elliptical orbit shown in FIGURE P12.77. The only force on the satellite is the gravitational attraction of the planet. The satellite's speed at point a is 8000 m/s.
    a. Does the satellite experience any torque about the center of the planet? Explain.
    b. What is the satellite's speed at point b?
    c. What is the satellite's speed at point c?

FIGURE P12.77

78. ‖‖ A 10 g bullet traveling at 400 m/s strikes a 10 kg, 1.0-m-wide door at the edge opposite the hinge. The bullet embeds itself in the door, causing the door to swing open. What is the angular velocity of the door just after impact?

79. ‖ A 200 g, 40-cm-diameter turntable rotates on frictionless bearings at 60 rpm. A 20 g block sits at the center of the turntable. A compressed spring shoots the block radially outward along a frictionless groove in the surface of the turntable. What is the turntable's rotation angular velocity when the block reaches the outer edge?

80. ‖ Luc, who is 1.80 m tall and weighs 950 N, is standing at the center of a playground merry-go-round with his arms extended, holding a 4.0 kg dumbbell in each hand. The merry-go-round can be modeled as a 4.0-m-diameter disk with a weight of 1500 N. Luc's body can be modeled as a uniform 40-cm-diameter cylinder with massless arms extending to hands that are 85 cm from his center. The merry-go-round is coasting at a steady 35 rpm when Luc brings his hands in to his chest. Afterward, what is the angular velocity, in rpm, of the merry-go-round?

81. ‖‖ A merry-go-round is a common piece of playground equipment. A 3.0-m-diameter merry-go-round with a mass of 250 kg is spinning at 20 rpm. John runs tangent to the merry-go-round at 5.0 m/s, in the same direction that it is turning, and jumps onto the outer edge. John's mass is 30 kg. What is the merry-go-round's angular velocity, in rpm, after John jumps on?

82. ‖‖ A 45 kg figure skater is spinning on the toes of her skates at 1.0 rev/s. Her arms are outstretched as far as they will go. In this orientation, the skater can be modeled as a cylindrical torso (40 kg, 20 cm average diameter, 160 cm tall) plus two rod-like arms (2.5 kg each, 66 cm long) attached to the outside of the torso. The skater then raises her arms straight above her head, where she appears to be a 45 kg, 20-cm-diameter, 200-cm-tall cylinder. What is her new angular velocity, in rev/s?

83. ‖‖ During most of its lifetime, a star maintains an equilibrium size in which the inward force of gravity on each atom is balanced by an outward pressure force due to the heat of the nuclear reactions in the core. But after all the hydrogen "fuel" is consumed by nuclear fusion, the pressure force drops and the star undergoes a *gravitational collapse* until it becomes a *neutron star*. In a neutron star, the electrons and protons of the atoms are squeezed together by gravity until they fuse into neutrons. Neutron stars spin very rapidly and emit intense pulses of radio and light waves, one pulse per rotation. These "pulsing stars" were discovered in the 1960s and are called *pulsars*.
    a. A star with the mass ($M = 2.0 \times 10^{30}$ kg) and size ($R = 7.0 \times 10^8$ m) of our sun rotates once every 30 days. After undergoing gravitational collapse, the star forms a pulsar that is observed by astronomers to emit radio pulses every 0.10 s. By treating the neutron star as a solid sphere, deduce its radius.
    b. What is the speed of a point on the equator of the neutron star? Your answers will be somewhat too large because a star cannot be accurately modeled as a solid sphere. Even so, you will be able to show that a star, whose mass is 10^6 larger than the earth's, can be compressed by gravitational forces to a size smaller than a typical state in the United States!

84. ‖ The earth's rotation axis, which is tilted 23.5° from the plane of the earth's orbit, today points to Polaris, the north star. But Polaris has not always been the north star because the earth, like a spinning gyroscope, precesses. That is, a line extending along the earth's rotation axis traces out a 23.5° cone as the earth precesses with a period of 26,000 years. This occurs because the earth is not a perfect sphere. It has an *equatorial bulge*, which allows both the moon and the sun to exert a gravitational torque on the earth. Our expression for the precession frequency of a gyroscope can be written $\Omega = \tau/I\omega$. Although we derived this equation for a specific situation, it's a valid result, differing by at most a constant close to 1, for the precession of any rotating object. What is the average gravitational torque on the earth due to the moon and the sun?

Challenge Problems

85. ‖‖‖ The bunchberry flower has the fastest-moving parts ever
BIO observed in a plant. Initially, the stamens are held by the petals in a bent position, storing elastic energy like a coiled spring. When the petals release, the tips of the stamen act like medieval catapults, flipping through a 60° angle in just 0.30 ms to launch pollen from anther sacs at their ends. The human eye just sees a burst of pollen; only high-speed photography reveals the details. As FIGURE CP12.85 shows, we can model the stamen tip as a 1.0-mm-long, 10 μg rigid rod with a 10 μg anther sac at the end. Although oversimplifying, we'll assume a constant angular acceleration.
   a. How large is the "straightening torque"?
   b. What is the speed of the anther sac as it releases its pollen?

FIGURE CP12.85

86. ‖‖‖ The two blocks in FIGURE CP12.86 are connected by a massless rope that passes over a pulley. The pulley is 12 cm in diameter and has a mass of 2.0 kg. As the pulley turns, friction at the axle exerts a torque of magnitude 0.50 N m. If the blocks are released from rest, how long does it take the 4.0 kg block to reach the floor?

FIGURE CP12.86

87. ‖‖‖ A rod of length L and mass M has a nonuniform mass distribu-
CALC tion. The linear mass density (mass per length) is $\lambda = cx^2$, where x is measured from the *center* of the rod and c is a constant.
   a. What are the units of c?
   b. Find an expression for c in terms of L and M.
   c. Find an expression in terms of L and M for the moment of inertia of the rod for rotation about an axis through the center.

88. ‖‖‖ In FIGURE CP12.88, a 200 g toy car is placed on a narrow 60-cm-diameter track with wheel grooves that keep the car going in a circle. The 1.0 kg track is free to turn on a frictionless, vertical axis. The spokes have negligible mass. After the car's switch is turned on, it soon reaches a steady speed of 0.75 m/s relative to the track. What then is the track's angular velocity, in rpm?

FIGURE CP12.88   FIGURE CP12.89

89. ‖‖‖ FIGURE CP12.89 shows a cube of mass m sliding without friction at speed v_0. It undergoes a perfectly elastic collision with the bottom tip of a rod of length d and mass $M = 2m$. The rod is pivoted about a frictionless axle through its center, and initially it hangs straight down and is at rest. What is the cube's velocity— both speed and direction—after the collision?

90. ‖‖‖ A 75 g, 30-cm-long rod hangs vertically on a frictionless, horizontal axle passing through its center. A 10 g ball of clay traveling horizontally at 2.5 m/s hits and sticks to the very bottom tip of the rod. To what maximum angle, measured from vertical, does the rod (with the attached ball of clay) rotate?

13 Newton's Theory of Gravity

The Andromeda Galaxy—seen here in a false-color image showing ultraviolet (blue) and infrared (red) light—consists of billions of stars orbiting the galactic center.

IN THIS CHAPTER, you will understand the motion of satellites and planets.

What are Kepler's laws?

Before Galileo and the telescope, Kepler used naked-eye measurements to make three major discoveries:

- The planets move in elliptical orbits.
- Planets "sweep out" equal areas in equal times.
- For circular orbits, the square of a planet's period is proportional to the cube of the orbit's radius.

Kepler's discoveries were the first solid proof of Copernicus's assertion that earth and the other planets orbit the sun.

What is Newton's theory?

Newton proposed that *any* two masses M and m are attracted toward each other by a gravitational force of magnitude

$$F_{M \text{ on } m} = F_{m \text{ on } M} = \frac{GMm}{r^2}$$

where r is the distance between the masses and G is the gravitational constant.

- Newton's law is an inverse-square law.
- Newton's law predicts the value of g.

In addition to a specific force law for gravity, Newton's three laws of motion apply to all objects in the universe.

What is gravitational energy?

The gravitational potential energy of two masses is

$$U_G = -\frac{GMm}{r}$$

Gravitational potential energy is negative, with a zero point at infinity. The $U_G = mgy$ that you learned in Chapter 10 is a special case for objects very near the surface of a planet. Gravitationally interacting stars, planets, and satellites are always isolated systems, so mechanical energy is conserved.

« LOOKING BACK Chapter 10 Potential energy and energy conservation

What does the theory say about orbits?

Kepler's laws can be *derived* from Newton's theory of gravity:

- Orbits can be circular or elliptical.
- Orbits conserve energy and angular momentum.
- Geosynchronous orbits have the same period as the rotating planet.

« LOOKING BACK Sections 8.2–8.3 Circular motion

Why is the theory of gravity important?

Satellites, space stations, the GPS system, and future missions to planets all depend on Newton's theory of gravity. Our modern understanding of the cosmos—from stars and galaxies to the Big Bang—is based on understanding gravity. Newton's theories of motion and gravity were the beginnings of modern science.

13.1 A Little History

The study of the structure of the universe is called **cosmology.** The ancient Greeks developed a cosmological model, with the earth at the center of the universe while the moon, the sun, the planets, and the stars were points of light turning about the earth on large "celestial spheres." This viewpoint was further expanded by the second-century Egyptian astronomer Ptolemy (the P is silent). He developed an elaborate mathematical model of the solar system that quite accurately predicted the complex planetary motions.

Then, in 1543, the medieval world was turned on its head with the publication of Nicholas Copernicus's *De Revolutionibus.* Copernicus argued that it is not the earth at rest in the center of the universe—it is the sun! Furthermore, Copernicus asserted that all of the planets revolve about the sun (hence his title) in circular orbits.

Tycho and Kepler

The greatest medieval astronomer was Tycho Brahe. For 30 years, from 1570 to 1600, Tycho compiled the most accurate astronomical observations the world had known. The invention of the telescope was still to come, but Tycho developed ingenious mechanical sighting devices that allowed him to determine the positions of stars and planets in the sky with unprecedented accuracy.

Tycho had a young mathematical assistant named Johannes Kepler. Kepler had become one of the first outspoken defenders of Copernicus, and his goal was to find evidence for circular planetary orbits in Tycho's records. To appreciate the difficulty of this task, keep in mind that Kepler was working before the development of graphs or of calculus—and certainly before calculators! His mathematical tools were algebra, geometry, and trigonometry, and he was faced with thousands upon thousands of individual observations of planetary positions measured as angles above the horizon.

Many years of work led Kepler to discover that the orbits are not circles, as Copernicus claimed, but *ellipses.* Furthermore, the speed of a planet is not constant but varies as it moves around the ellipse.

Kepler's laws, as we call them today, state that

1. Planets move in elliptical orbits, with the sun at one focus of the ellipse.
2. A line drawn between the sun and a planet sweeps out equal areas during equal intervals of time.
3. The square of a planet's orbital period is proportional to the cube of the semimajor-axis length.

FIGURE 13.1a shows that an ellipse has two *foci* (plural of *focus*), and the sun occupies one of these. The long axis of the ellipse is the *major axis,* and half the length of this axis is called the *semimajor-axis length.* As the planet moves, a line drawn from the sun to the planet "sweeps out" an area. **FIGURE 13.1b** shows two such areas. Kepler's discovery that the areas are equal for equal Δt implies that the planet moves faster when near the sun, slower when farther away.

All the planets except Mercury have elliptical orbits that are only very slightly distorted circles. As **FIGURE 13.2** shows, a circle is an ellipse in which the two foci move to the center, effectively making one focus, and the semimajor-axis length becomes the radius. Because the mathematics of ellipses is difficult, this chapter will focus on circular orbits.

Kepler published the first two of his laws in 1609, the same year in which Galileo first turned a telescope to the heavens. Through his telescope Galileo could *see* moons orbiting Jupiter, just as Copernicus had suggested the planets orbit the sun. He could *see* that Venus has phases, like the moon, which implied its orbital motion about the sun. By the time of Galileo's death in 1642, the Copernican revolution was complete.

FIGURE 13.1 The elliptical orbit of a planet about the sun.

(a) The planet moves in an elliptical orbit with the sun at one focus.

Minor axis · Planet · Semiminor-axis length · Sun · Major axis · Foci · Semimajor-axis length

(b) The line between the sun and the planet sweeps out equal areas during equal intervals of time.

Slower · Faster

FIGURE 13.2 A circular orbit is a special case of an elliptical orbit.

Kepler's second law: Equal areas in equal times imply the speed is constant.

Kepler's first law: The sun is at the center.

The motion is uniform circular motion.

Sun · r

Circular orbit

Kepler's third law: The square of the period is proportional to r^3.

Isaac Newton, 1642–1727.

13.2 Isaac Newton

A popular image has Newton thinking of the idea of gravity after an apple fell on his head. This amusing story is at least close to the truth. Newton himself said that the "notion of gravitation" came to him as he "sat in a contemplative mood" and "was occasioned by the fall of an apple." It occurred to him that, perhaps, the apple was attracted to the *center* of the earth but was prevented from getting there by the earth's surface. And if the apple was so attracted, why not the moon? In other words, perhaps gravitation is a *universal* force between all objects in the universe! This is not shocking today, but no one before Newton had ever thought that the mundane motion of objects on earth had any connection at all with the stately motion of the planets through the heavens.

Suppose the moon's circular motion around the earth is due to the pull of the earth's gravity. Then, as you learned in Chapter 8 and is shown in **FIGURE 13.3**, the moon must be in *free fall* with the free-fall acceleration $g_{\text{at moon}}$.

> **NOTE** The symbol g_{moon} is the free-fall acceleration caused by the *moon's* gravity—that is, the acceleration of a falling object on the moon. Here we're interested in the acceleration *of* the moon by the earth's gravity, which we'll call $g_{\text{at moon}}$.

FIGURE 13.3 The moon is in free fall around the earth.

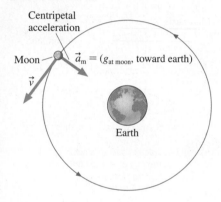

The centripetal acceleration of an object in uniform circular motion is

$$a_r = g_{\text{at moon}} = \frac{v_m^2}{r_m} \tag{13.1}$$

The moon's speed is related to the radius r_m and period T_m of its orbit by $v_m = $ circumference/period $= 2\pi r_m/T_m$. Combining these, Newton found

$$g_{\text{at moon}} = \frac{4\pi^2 r_m}{T_m^2} = \frac{4\pi^2(3.84 \times 10^8 \text{ m})}{(2.36 \times 10^6 \text{ s})^2} = 0.00272 \text{ m/s}^2$$

Astronomical measurements had established a reasonably good value for r_m by the time of Newton, and the period $T_m = 27.3$ days (2.36×10^6 s) was quite well known.

The moon's centripetal acceleration is significantly less than the free-fall acceleration on the earth's surface. In fact,

$$\frac{g_{\text{at moon}}}{g_{\text{on earth}}} = \frac{0.00272 \text{ m/s}^2}{9.80 \text{ m/s}^2} = \frac{1}{3600}$$

This is an interesting result, but it was Newton's next step that was critical. He compared the radius of the moon's orbit to the radius of the earth:

$$\frac{r_m}{R_e} = \frac{3.84 \times 10^8 \text{ m}}{6.37 \times 10^6 \text{ m}} = 60.2$$

> **NOTE** We'll use a lowercase r, as in r_m, to indicate the radius of an orbit. We'll use an uppercase R, as in R_e, to indicate the radius of a star or planet.

I deduced that the forces which keep the planets in their orbs must be reciprocally as the squares of their distances from the centers about which they revolve; and thereby compared the force requisite to keep the Moon in her orb with the force of gravity at the surface of the Earth; and found them answer pretty nearly.

Isaac Newton

Newton recognized that $(60.2)^2$ is almost exactly 3600. Thus, he reasoned:

- If g has the value 9.80 at the earth's surface, and
- If the force of gravity and g decrease in size depending inversely on the square of the distance from the center of the earth,
- Then g will have exactly the value it needs at the distance of the moon to cause the moon to orbit the earth with a period of 27.3 days.

His two ratios were not identical, but he found them to "answer pretty nearly" and knew that he had to be on the right track.

STOP TO THINK 13.1 A satellite orbits the earth with constant speed at a height above the surface equal to the earth's radius. The magnitude of the satellite's acceleration is

a. $4g_{\text{on earth}}$ b. $2g_{\text{on earth}}$ c. $g_{\text{on earth}}$

d. $\frac{1}{2}g_{\text{on earth}}$ e. $\frac{1}{4}g_{\text{on earth}}$ f. 0

13.3 Newton's Law of Gravity

Newton proposed that *every* object in the universe attracts *every other* object with a force that is

1. Inversely proportional to the square of the distance between the objects.
2. Directly proportional to the product of the masses of the two objects.

To make these ideas more specific, FIGURE 13.4 shows masses m_1 and m_2 separated by distance r. Each mass exerts an attractive force on the other, a force that we call the **gravitational force.** These two forces form an action/reaction pair, so $\vec{F}_{1 \text{ on } 2}$ is equal and opposite to $\vec{F}_{2 \text{ on } 1}$. The magnitude of the forces is given by Newton's law of gravity.

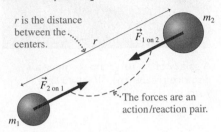

FIGURE 13.4 The gravitational forces on masses m_1 and m_2.

r is the distance between the centers.

$\vec{F}_{1 \text{ on } 2}$

$\vec{F}_{2 \text{ on } 1}$

The forces are an action/reaction pair.

m_2

m_1

> **Newton's law of gravity** If two objects with masses m_1 and m_2 are a distance r apart, the objects exert attractive forces on each other of magnitude
>
> $$F_{1 \text{ on } 2} = F_{2 \text{ on } 1} = \frac{Gm_1 m_2}{r^2} \tag{13.2}$$
>
> The forces are directed along the straight line joining the two objects.

The constant G, called the **gravitational constant,** is a proportionality constant necessary to relate the masses, measured in kilograms, to the force, measured in newtons. In the SI system of units, G has the value

$$G = 6.67 \times 10^{-11} \text{ N m}^2/\text{kg}^2$$

FIGURE 13.5 is a graph of the gravitational force as a function of the distance between the two masses. As you can see, an inverse-square force decreases rapidly.

Strictly speaking, Equation 13.2 is valid only for particles. However, Newton was able to show that this equation also applies to spherical objects, such as planets, if r is the distance between their centers. Our intuition and common sense suggest this to us, as they did to Newton. The rather difficult proof is not essential, so we will omit it.

FIGURE 13.5 The gravitational force is an inverse-square force.

Gravitational force

F

Doubling the distance between the masses causes the force to decrease by a factor of 4.

$F/4$
$F/9$

Separation of masses

$1r$ $2r$ $3r$

Gravitational Force and Weight

Knowing G, we can calculate the size of the gravitational force. Consider two 1.0 kg masses that are 1.0 m apart. According to Newton's law of gravity, these two masses exert an attractive gravitational force on each other of magnitude

$$F_{1 \text{ on } 2} = F_{2 \text{ on } 1} = \frac{Gm_1 m_2}{r^2}$$

$$= \frac{(6.67 \times 10^{-11} \text{ N m}^2/\text{kg}^2)(1.0 \text{ kg})(1.0 \text{ kg})}{(1.0 \text{ m})^2} = 6.67 \times 10^{-11} \text{ N}$$

This is an exceptionally tiny force, especially when compared to the gravitational force of the entire earth on each mass: $F_G = mg = 9.8 \text{ N}$.

The fact that the gravitational force between two ordinary-size objects is so small is the reason we are not aware of it. You are being attracted to every object around you, but the forces are so tiny in comparison to the normal forces and friction forces acting on you that they are completely undetectable. Only when one (or both) of the masses is exceptionally large—planet-size—does the force of gravity become important.

We find a more respectable result if we calculate the force *of the earth* on a 1.0 kg mass at the earth's surface:

$$F_{\text{earth on 1 kg}} = \frac{GM_e m_{1 \text{ kg}}}{R_e^{\,2}}$$

$$= \frac{(6.67 \times 10^{-11} \text{ N m}^2/\text{kg}^2)(5.98 \times 10^{24} \text{ kg})(1.0 \text{ kg})}{(6.37 \times 10^6 \text{ m})^2} = 9.8 \text{ N}$$

where the distance between the mass and the center of the earth is the earth's radius.

The dynamics of stellar motions, spanning many thousands of light years, are governed by Newton's law of gravity.

A galaxy of $\approx 10^{11}$ stars spanning a distance greater than 100,000 light years.

The earth's mass M_e and radius R_e were taken from Table 13.2 in Section 13.6. This table, which is also printed inside the back cover of the book, contains astronomical data that will be used for examples and homework.

The force $F_{\text{earth on 1 kg}} = 9.8$ N is exactly the weight of a stationary 1.0 kg mass: $F_G = mg = 9.8$ N. Is this a coincidence? Of course not. Weight—the upward force of a spring scale—exactly balances the downward gravitational force, so numerically they must be equal.

Although weak, gravity is a *long-range* force. No matter how far apart two objects may be, there is a gravitational attraction between them given by Equation 13.2. Consequently, gravity is the most ubiquitous force in the universe. It not only keeps your feet on the ground, it also keeps the earth orbiting the sun, the solar system orbiting the center of the Milky Way galaxy, and the entire Milky Way galaxy performing an intricate orbital dance with other galaxies making up what is called the "local cluster" of galaxies.

The Principle of Equivalence

Newton's law of gravity depends on a rather curious assumption. The concept of *mass* was introduced in Chapter 5 by considering the relationship between force and acceleration. The *inertial mass* of an object, which is the mass that appears in Newton's second law, is found by measuring the object's acceleration a in response to force F:

$$m_{\text{inert}} = \text{inertial mass} = \frac{F}{a} \tag{13.3}$$

Gravity plays no role in this definition of mass.

The quantities m_1 and m_2 in Newton's law of gravity are being used in a very different way. Masses m_1 and m_2 govern the strength of the gravitational attraction between two objects. The mass used in Newton's law of gravity is called the **gravitational mass.** The gravitational mass of an object can be determined by measuring the attractive force exerted on it by another mass M a distance r away:

$$m_{\text{grav}} = \text{gravitational mass} = \frac{r^2 F_{M\,\text{on}\,m}}{GM} \tag{13.4}$$

Acceleration does not enter into the definition of the gravitational mass.

These are two very different concepts of mass. Yet Newton, in his theory of gravity, asserts that the inertial mass in his second law is the very same mass that governs the strength of the gravitational attraction between two objects. The assertion that $m_{\text{grav}} = m_{\text{inert}}$ is called the **principle of equivalence.** It says that inertial mass is *equivalent to* gravitational mass.

As a hypothesis about nature, the principle of equivalence is subject to experimental verification or disproof. Many exceptionally clever experiments have looked for any difference between the gravitational mass and the inertial mass, and they have shown that any difference, if it exists at all, is less than 10 parts in a trillion! As far as we know today, the gravitational mass and the inertial mass are exactly the same thing.

But why should a quantity associated with the dynamics of motion, relating force to acceleration, have anything at all to do with the gravitational attraction? This is a question that intrigued Einstein and eventually led to his general theory of relativity, the theory about curved spacetime and black holes. General relativity is beyond the scope of this textbook, but it explains the principle of equivalence as a property of space itself.

Newton's Theory of Gravity

Newton's theory of gravity is more than just Equation 13.2. The *theory* of gravity consists of:

1. A specific force law for gravity, given by Equation 13.2, *and*
2. The principle of equivalence, *and*
3. An assertion that Newton's three laws of motion are universally applicable. These laws are as valid for heavenly bodies, the planets and stars, as for earthly objects.

Consequently, everything we have learned about forces, motion, and energy is relevant to the dynamics of satellites, planets, and galaxies.

STOP TO THINK 13.2 The figure shows a binary star system. The mass of star 2 is twice the mass of star 1. Compared to $\vec{F}_{1\text{ on }2}$, the magnitude of the force $\vec{F}_{2\text{ on }1}$ is

a. Four times as big.
b. Twice as big.
c. The same size.
d. Half as big.
e. One-quarter as big.

13.4 Little *g* and Big *G*

The familiar equation $F_G = mg$ works well when an object is on the surface of a planet, but mg will not help us find the force exerted on the same object if it is in orbit around the planet. Neither can we use mg to find the force of attraction between the earth and the moon. Newton's law of gravity provides a more fundamental starting point because it describes a *universal* force that exists between all objects.

To illustrate the connection between Newton's law of gravity and the familiar $F_G = mg$, **FIGURE 13.6** shows an object of mass m on the surface of Planet X. Planet X inhabitant Mr. Xhzt, standing on the surface, finds that the downward gravitational force is $F_G = mg_X$, where g_X is the free-fall acceleration on Planet X.

We, taking a more cosmic perspective, reply, "Yes, that is the force *because* of a universal force of attraction between your planet and the object. The size of the force is determined by Newton's law of gravity."

We and Mr. Xhzt are both correct. Whether you think locally or globally, we and Mr. Xhzt must arrive at the *same numerical value* for the magnitude of the force. Suppose an object of mass m is on the surface of a planet of mass M and radius R. The local gravitational force is

$$F_G = mg_{\text{surface}} \tag{13.5}$$

where g_{surface} is the free-fall acceleration at the planet's surface. The force of gravitational attraction for an object on the surface $(r = R)$, as given by Newton's law of gravity, is

$$F_{M\text{ on }m} = \frac{GMm}{R^2} \tag{13.6}$$

Because these are two names and two expressions for the same force, we can equate the right-hand sides to find that

$$g_{\text{surface}} = \frac{GM}{R^2} \tag{13.7}$$

We have used Newton's law of gravity to *predict* the value of g at the surface of a planet. The value depends on the mass and radius of the planet as well as on the value of G, which establishes the overall strength of the gravitational force.

The expression for g_{surface} in Equation 13.7 is valid for any planet or star. Using the mass and radius of Mars, we can predict the Martian value of g:

$$g_{\text{Mars}} = \frac{GM_{\text{Mars}}}{R_{\text{Mars}}^2} = \frac{(6.67 \times 10^{-11}\,\text{N}\,\text{m}^2/\text{kg}^2)(6.42 \times 10^{23}\,\text{kg})}{(3.37 \times 10^6\,\text{m})^2} = 3.8\,\text{m/s}^2$$

NOTE We noted in Chapter 6 that measured values of g are very slightly smaller on a rotating planet. We'll ignore rotation in this chapter.

FIGURE 13.6 Weighing an object of mass m on Planet X.

Planetary perspective:
$F = mg_X$

Planet X

Universal perspective:
$F = \dfrac{GM_Xm}{R_X^2}$

Planet X

Decrease of g with Distance

Equation 13.7 gives $g_{surface}$ at the surface of a planet. More generally, imagine an object of mass m at distance $r > R$ from the center of a planet. Further, suppose that gravity from the planet is the only force acting on the object. Then its acceleration, the free-fall acceleration, is given by Newton's second law:

$$g = \frac{F_{M \text{ on } m}}{m} = \frac{GM}{r^2} \qquad (13.8)$$

This more general result agrees with Equation 13.7 if $r = R$, but it allows us to determine the "local" free-fall acceleration at distances $r > R$. Equation 13.8 expresses Newton's discovery, with regard to the moon, that g decreases inversely with the square of the distance.

FIGURE 13.7 shows a satellite orbiting at height h above the earth's surface. Its distance from the center of the earth is $r = R_e + h$. Most people have a mental image that satellites orbit "far" from the earth, but in reality h is typically 200 miles $\approx 3 \times 10^5$ m, while $R_e = 6.37 \times 10^6$ m. Thus the satellite is barely "skimming" the earth at a height only about 5% of the earth's radius!

The value of g at height h above the earth (i.e., above sea level) is

$$g = \frac{GM_e}{(R_e + h)^2} = \frac{GM_e}{R_e^2(1 + h/R_e)^2} = \frac{g_{earth}}{(1 + h/R_e)^2} \qquad (13.9)$$

where $g_{earth} = 9.83$ m/s^2 is the value calculated from Equation 13.7 for $h = 0$ on a nonrotating earth. **TABLE 13.1** shows the value of g evaluated at several values of h.

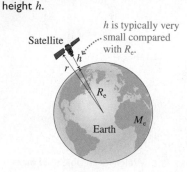

FIGURE 13.7 A satellite orbits the earth at height h.

h is typically very small compared with R_e.

Satellite

h

r

R_e

Earth

M_e

TABLE 13.1 Variation of g with height above the ground

Height h	Example	g (m/s^2)
0 m	ground	9.83
4500 m	Mt. Whitney	9.82
10,000 m	jet airplane	9.80
300,000 m	space station	8.90
35,900,000 m	communications satellite	0.22

NOTE The free-fall acceleration of a satellite such as the space station is only slightly less than the ground-level value. An object in orbit is not "weightless" because there is no gravity in space but because it is in free fall, as you learned in Chapter 8.

Weighing the Earth

We can predict g if we know the earth's mass. But how do we know the value of M_e? We cannot place the earth on a giant pan balance, so how is its mass known? Furthermore, how do we know the value of G?

Newton did not know the value of G. He could say that the gravitational force is proportional to the product m_1m_2 and inversely proportional to r^2, but he had no means of knowing the value of the proportionality constant.

Determining G requires a *direct* measurement of the gravitational force between two known masses at a known separation. The small size of the gravitational force between ordinary-size objects makes this quite a feat. Yet the English scientist Henry Cavendish came up with an ingenious way of doing so with a device called a *torsion balance*. Two fairly small masses m, typically about 10 g, are placed on the ends of a lightweight rod. The rod is hung from a thin fiber, as shown in **FIGURE 13.8a**, and allowed to reach equilibrium.

If the rod is then rotated slightly and released, a *restoring force* will return it to equilibrium. This is analogous to displacing a spring from equilibrium, and in fact the restoring force and the angle of displacement obey a version of Hooke's law:

FIGURE 13.8 Cavendish's experiment to measure G.

(a)

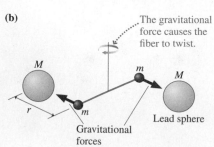

Fiber

$\Delta\theta$

Rod with small masses

The force needed to rotate the rod by $\Delta\theta$ is $k\Delta\theta$.

(b)

The gravitational force causes the fiber to twist.

M

m

M

r

m

Gravitational forces

Lead sphere

$F_{restore} = k\,\Delta\theta$. The "torsion constant" k can be determined by timing the period of oscillations. Once k is known, a force that twists the rod slightly away from equilibrium can be measured by the product $k\,\Delta\theta$. It is possible to measure very small angular deflections, so this device can be used to determine very small forces.

Two larger masses M (typically lead spheres with $M \approx 10$ kg) are then brought close to the torsion balance, as shown in **FIGURE 13.8b**. The gravitational attraction that they exert on the smaller hanging masses causes a very small but measurable twisting of the balance, enough to measure $F_{M\,on\,m}$. Because m, M, and r are all known, Cavendish was able to determine G from

$$G = \frac{F_{M\,on\,m}\,r^2}{Mm} \tag{13.10}$$

His first results were not highly accurate, but improvements over the years in this and similar experiments have produced the value of G accepted today.

With an independently determined value of G, we can return to Equation 13.7 to find

$$M_e = \frac{g_{earth}R_e^{\,2}}{G} \tag{13.11}$$

We have weighed the earth! The value of g_{earth} at the earth's surface is known with great accuracy from kinematics experiments. The earth's radius R_e is determined by surveying techniques. Combining our knowledge from these very different measurements has given us a way to determine the mass of the earth.

The gravitational constant G is what we call a *universal constant*. Its value establishes the strength of one of the fundamental forces of nature. As far as we know, the gravitational force between two masses would be the same anywhere in the universe. Universal constants tell us something about the most basic and fundamental properties of nature. You will soon meet other universal constants.

The free-fall acceleration varies slightly due to mountains and to variation in the density of the earth's crust. This map shows the *gravitational anomaly*, with red regions of slightly stronger gravity and blue regions of slightly weaker gravity. The variation is tiny, less than 0.001 m/s^2.

STOP TO THINK 13.3 A planet has four times the mass of the earth, but the acceleration due to gravity on the planet's surface is the same as on the earth's surface. The planet's radius is

a. $4R_e$ b. $2R_e$ c. R_e d. $\frac{1}{2}R_e$ e. $\frac{1}{4}R_e$

13.5 Gravitational Potential Energy

Gravitational problems are ideal for the conservation-law tools we developed in Chapters 9 through 11. Because gravity is the only force, and it is a conservative force, both the momentum and the mechanical energy of the system $m_1 + m_2$ are conserved. To employ conservation of energy, however, we need to determine an appropriate form for the gravitational potential energy for two interacting masses.

The definition of potential energy that we developed in Chapter 11 is

$$\Delta U = U_f - U_i = -W_c(i \rightarrow f) \tag{13.12}$$

where $W_c(i \rightarrow f)$ is the work done by a conservative force as a particle moves from position i to position f. For a flat earth, we used $F = -mg$ and the choice that $U = 0$ at the surface ($y = 0$) to arrive at the now-familiar $U_G = mgy$. This result for U_G is valid only for $y \ll R_e$, when the earth's curvature and size are not apparent. We now need to find an expression for the gravitational potential energy of masses that interact over *large* distances.

FIGURE 13.9 shows two particles of mass m_1 and m_2. Let's calculate the work done on mass m_2 by the conservative force $\vec{F}_{1\,on\,2}$ as m_2 moves from an initial position at distance r to a final position very far away. The force, which points to the left, is opposite the displacement; hence this force does *negative* work. Consequently, due to the minus sign in Equation 13.12, ΔU is *positive*. A pair of masses *gains* potential energy as the masses move farther apart, just as a particle near the earth's surface gains potential energy as it moves to a higher altitude.

FIGURE 13.9 Calculating the work done by the gravitational force as mass m_2 moves from r to ∞.

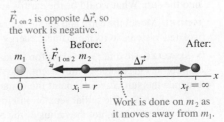

$\vec{F}_{1\,on\,2}$ is opposite $\Delta\vec{r}$, so the work is negative.

Work is done on m_2 as it moves away from m_1.

We can establish a coordinate system with m_1 at the origin and m_2 moving along the x-axis. The gravitational force is a variable force, so we need the full definition of work:

$$W(\text{i} \rightarrow \text{f}) = \int_{x_\text{i}}^{x_\text{f}} F_x \, dx \tag{13.13}$$

$\vec{F}_{1 \text{ on } 2}$ points toward the left, so its x-component is $(F_{1 \text{ on } 2})_x = -Gm_1m_2/x^2$. As mass m_2 moves from $x_\text{i} = r$ to $x_\text{f} = \infty$, the potential energy changes by

$$\Delta U = U_{\text{at} \infty} - U_{\text{at} r} = -\int_r^\infty (F_{1 \text{ on } 2})_x \, dx = -\int_r^\infty \left(\frac{-Gm_1m_2}{x^2} \right) dx$$

$$= +Gm_1m_2 \int_r^\infty \frac{dx}{x^2} = -\left. \frac{Gm_1m_2}{x} \right|_r^\infty = \frac{Gm_1m_2}{r} \tag{13.14}$$

FIGURE 13.10 The gravitational potential-energy curve.

U_G

U approaches 0 as $r \rightarrow \infty$.

U approaches $-\infty$ as $r \rightarrow 0$, but this is not physically significant because two objects can't occupy the same point.

FIGURE 13.11 Two masses gain kinetic energy as their separation decreases.

The system loses potential energy and gains kinetic energy.

Before

After

Energy

The kinetic energy K is the distance from the potential-energy curve to the total energy line.

PE

NOTE We chose to integrate along the x-axis, but the fact that gravity is a conservative force means that ΔU will have this value if m_2 moves from r to ∞ along *any* path.

To proceed further, we need to choose the point where $U = 0$. We would like our choice to be valid for any star or planet, regardless of its mass and radius. This will be the case if we set $U = 0$ at the point where the interaction between the masses vanishes. According to Newton's law of gravity, the strength of the interaction is zero only when $r = \infty$. Two masses infinitely far apart will have no tendency, or potential, to move together, so we will *choose* to place the zero point of potential energy at $r = \infty$. That is, $U_{\text{at} \infty} = 0$.

This choice gives us the gravitational potential energy of masses m_1 and m_2:

$$U_\text{G} = -\frac{Gm_1m_2}{r} \tag{13.15}$$

This is the potential energy of masses m_1 and m_2 when their *centers* are separated by distance r. **FIGURE 13.10** is a graph of U_G as a function of the distance r between the masses. Notice that it asymptotically approaches 0 as $r \rightarrow \infty$.

NOTE Although Equation 13.15 looks rather similar to Newton's law of gravity, it depends only on $1/r$, *not* on $1/r^2$.

It may seem disturbing that the potential energy is negative, but we encountered similar situations in Chapter 10. All a negative potential energy means is that the potential energy of the two masses at separation r is *less* than their potential energy at infinite separation. Only the *change* in U has physical significance, and the change will be the same no matter where we place the zero of potential energy.

To illustrate, suppose two masses a distance r_1 apart are released from rest. How will they move? From a force perspective, you would note that each mass experiences an attractive force and accelerates toward the other. The energy perspective of **FIGURE 13.11** tells the same thing. By moving toward smaller r (that is, $r_1 \rightarrow r_2$), the system *loses* potential energy and *gains* kinetic energy while conserving E_mech. The system is "falling downhill," although in a more general sense than we think about on a flat earth.

EXAMPLE 13.1 | Crashing into the sun

Suppose the earth suddenly came to a halt and ceased revolving around the sun. The gravitational force would then pull it directly into the sun. What would be the earth's speed as it crashed?

MODEL Model the earth and the sun as spherical masses. This is an isolated system, so its mechanical energy is conserved.

VISUALIZE FIGURE 13.12 is a before-and-after pictorial representation for this gruesome cosmic event. The "crash" occurs as the earth touches the sun, at which point the distance between their centers is $r_2 = R_\text{s} + R_\text{e}$. The initial separation r_1 is the radius of the earth's *orbit* about the sun, not the radius of the earth.

FIGURE 13.12 Before-and-after pictorial representation of the earth crashing into the sun (not to scale).

Before:

R_e

$v_1 = 0$ m/s

Earth

After:

v_2 R_s

Sun

$r_2 = R_\text{s} + R_\text{e} = 7.02 \times 10^8$ m

$r_1 = 1.50 \times 10^{11}$ m

SOLVE Strictly speaking, the kinetic energy is the sum $K = K_{earth} + K_{sun}$. However, the sun is so much more massive than the earth that the lightweight earth does almost all of the moving. It is a reasonable approximation to consider the sun as remaining at rest. In that case, the energy conservation equation $K_2 + U_2 = K_1 + U_1$ is

$$\tfrac{1}{2}M_e v_2{}^2 - \frac{GM_s M_e}{R_s + R_e} = 0 - \frac{GM_s M_e}{r_1}$$

This is easily solved for the earth's speed at impact. Using data from Table 13.2, we find

$$v_2 = \sqrt{2GM_s\left(\frac{1}{R_s + R_e} - \frac{1}{r_1}\right)} = 6.13 \times 10^5 \text{ m/s}$$

ASSESS The earth would be really flying along at over 1 million miles per hour as it crashed into the sun! It is worth noting that we do not have the mathematical tools to solve this problem using Newton's second law because the acceleration is not constant. But the solution is straightforward when we use energy conservation.

EXAMPLE 13.2 | Escape speed

A 1000 kg rocket is fired straight away from the surface of the earth. What speed does the rocket need to "escape" from the gravitational pull of the earth and never return? Assume a nonrotating earth.

MODEL In a simple universe, consisting of only the earth and the rocket, an insufficient launch speed will cause the rocket eventually to fall back to earth. Once the rocket finally slows to a halt, gravity will ever so slowly pull it back. The only way the rocket can escape is to never stop ($v = 0$) and thus never have a turning point! That is, the rocket must continue moving away from the earth forever. The *minimum* launch speed for escape, which is called the **escape speed,** will cause the rocket to stop ($v = 0$) only as it reaches $r = \infty$. Now ∞, of course, is not a "place," so a statement like this means that we want the rocket's speed to approach $v = 0$ asymptotically as $r \to \infty$.

VISUALIZE FIGURE 13.13 is a before-and-after pictorial representation.

SOLVE Energy conservation $K_2 + U_2 = K_1 + U_1$ is

$$0 + 0 = \tfrac{1}{2}mv_1{}^2 - \frac{GM_e m}{R_e}$$

FIGURE 13.13 Pictorial representation of a rocket launched with sufficient speed to escape the earth's gravity.

where we used the fact that both the kinetic and potential energy are zero at $r = \infty$. Thus the escape speed is

$$v_{escape} = v_1 = \sqrt{\frac{2GM_e}{R_e}} = 11,200 \text{ m/s} \approx 25,000 \text{ mph}$$

ASSESS It is difficult to assess the answer to a problem with which we have no direct experience, although we certainly expected the escape speed to be very large. Notice that the problem was mathematically easy; the difficulty was deciding how to interpret it. That is why—as you have now seen many times—the "physics" of a problem consists of thinking, interpreting, and modeling. We will see variations on this problem in the future, with both gravity and electricity, so you might want to review the *reasoning* involved.

The Flat-Earth Approximation

How does Equation 13.15 for the gravitational potential energy relate to our previous use of $U_G = mgy$ on a flat earth? FIGURE 13.14 shows an object of mass m located at height y above the surface of the earth. The object's distance from the earth's center is $r = R_e + y$ and its gravitational potential energy is

$$U_G = -\frac{GM_e m}{r} = -\frac{GM_e m}{R_e + y} = -\frac{GM_e m}{R_e(1 + y/R_e)} \quad (13.16)$$

where, in the last step, we factored R_e out of the denominator.

Suppose the object is very close to the earth's surface ($y \ll R_e$). In that case, the ratio $y/R_e \ll 1$. There is an approximation you will learn about in calculus, called the *binomial approximation,* that says

$$(1 + x)^n \approx 1 + nx \qquad \text{if } x \ll 1 \quad (13.17)$$

As an illustration, you can easily use your calculator to find that $1/1.01 = 0.9901$, to four significant figures. But suppose you wrote $1.01 = 1 + 0.01$. You could then use the binomial approximation to calculate

$$\frac{1}{1.01} = \frac{1}{1 + 0.01} = (1 + 0.01)^{-1} \approx 1 + (-1)(0.01) = 0.9900$$

You can see that the approximate answer is off by only 0.01%.

FIGURE 13.14 Gravity on a flat earth.

For a spherical earth:
$$U_G = -\frac{GM_e m}{R_e + y}$$

We can treat the earth as flat if $y \ll R_e$:
$$U_G = mgy$$

If we call $y/R_e = x$ in Equation 13.16 and use the binomial approximation, with $n = -1$, we find

$$U_G(\text{if } y \ll R_e) \approx -\frac{GM_e m}{R_e}\left(1 - \frac{y}{R_e}\right) = -\frac{GM_e m}{R_e} + m\left(\frac{GM_e}{R_e^2}\right)y \qquad (13.18)$$

Now the first term is just the gravitational potential energy U_0 when the object is at ground level ($y = 0$). In the second term, you can recognize $GM_e/R_e^2 = g_{\text{earth}}$ from the definition of g in Equation 13.7. Thus we can write Equation 13.18 as

$$U_G(\text{if } y \ll R_e) = U_0 + mg_{\text{earth}}y \qquad (13.19)$$

Although we chose U_G to be zero when $r = \infty$, we are always free to change our minds. If we change the zero point of potential energy to be $U_0 = 0$ at the surface, which is the choice we made in Chapter 10, then Equation 13.19 becomes

$$U_G(\text{if } y \ll R_e) = mg_{\text{earth}}y \qquad (13.20)$$

We can sleep easier knowing that Equation 13.15 for the gravitational potential energy is consistent with our earlier "flat-earth" expression for the potential energy.

EXAMPLE 13.3 | The speed of a satellite

A less-than-successful inventor wants to launch small satellites into orbit by launching them straight up from the surface of the earth at very high speed.

a. With what speed should he launch the satellite if it is to have a speed of 500 m/s at a height of 400 km? Ignore air resistance.

b. By what percentage would your answer be in error if you used a flat-earth approximation?

MODEL Mechanical energy is conserved if we ignore drag.

VISUALIZE FIGURE 13.15 shows a pictorial representation.

FIGURE 13.15 Pictorial representation of a satellite launched straight up.

SOLVE a. Although the height is exaggerated in the figure, 400 km = 400,000 m is high enough that we cannot ignore the earth's spherical shape. The energy conservation equation $K_2 + U_2 = K_1 + U_1$ is

$$\tfrac{1}{2}mv_2^2 - \frac{GM_e m}{R_e + y_2} = \tfrac{1}{2}mv_1^2 - \frac{GM_e m}{R_e + y_1}$$

where we've written the distance between the satellite and the earth's center as $r = R_e + y$. The initial height is $y_1 = 0$. Notice that the satellite mass m cancels and is not needed. Solving for the launch speed, we have

$$v_1 = \sqrt{v_2^2 + 2GM_e\left(\frac{1}{R_e} - \frac{1}{R_e + y_2}\right)} = 2770 \text{ m/s}$$

This is about 6000 mph, much less than the escape speed.

b. The calculation is the same in the flat-earth approximation except that we use $U_G = mgy$. Thus

$$\tfrac{1}{2}mv_2^2 + mgy_2 = \tfrac{1}{2}mv_1^2 + mgy_1$$
$$v_1 = \sqrt{v_2^2 + 2gy_2} = 2840 \text{ m/s}$$

The flat-earth value of 2840 m/s is 70 m/s too big. The error, as a percentage of the correct 2770 m/s, is

$$\text{error} = \frac{70}{2770} \times 100 = 2.5\%$$

ASSESS The true speed is less than the flat-earth approximation because the force of gravity decreases with height. Launching a rocket against a decreasing force takes less effort than it would with the flat-earth force of mg at all heights.

STOP TO THINK 13.4 Rank in order, from largest to smallest, the absolute values of the gravitational potential energies of these pairs of masses. The numbers give the relative masses and distances.

(a) $m_1 = 2$ ---- $r = 4$ ---- $m_2 = 2$

(b) $m_1 = 1$ -- $r = 1$ -- $m_2 = 1$

(c) $m_1 = 1$ -- $r = 2$ -- $m_2 = 1$

(d) $m_1 = 1$ ---- $r = 4$ ---- $m_2 = 4$

(e) $m_1 = 4$ -------- $r = 8$ -------- $m_2 = 4$

13.6 Satellite Orbits and Energies

Solving Newton's second law to find the trajectory of a mass moving under the influence of gravity is mathematically beyond this textbook. It turns out that the solution is a set of elliptical orbits, which is Kepler's first law. Kepler had no *reason* why orbits should be ellipses rather than some other shape. Newton was able to show that ellipses are a *consequence* of his theory of gravity.

The mathematics of ellipses is rather difficult, so we will restrict most of our analysis to the limiting case in which an ellipse becomes a circle. Most planetary orbits differ only very slightly from being circular. The earth's orbit, for example has a (semiminor axis/semimajor axis) ratio of 0.99986—very close to a true circle!

FIGURE 13.16 shows a massive body M, such as the earth or the sun, with a lighter body m orbiting it. The lighter body is called a **satellite,** even though it may be a planet orbiting the sun. For circular motion, where the gravitational force provides the centripetal acceleration v^2/r, Newton's second law for the satellite is

$$F_{M \text{ on } m} = \frac{GMm}{r^2} = ma_r = \frac{mv^2}{r} \tag{13.21}$$

Thus the speed of a satellite in a circular orbit is

$$v = \sqrt{\frac{GM}{r}} \tag{13.22}$$

A satellite must have this specific speed in order to have a circular orbit of radius r about the larger mass M. If the velocity differs from this value, the orbit will become elliptical rather than circular. Notice that the orbital speed does *not* depend on the satellite's mass m. This is consistent with our previous discovery, for motion on a flat earth, that motion due to gravity is independent of the mass.

FIGURE 13.16 The orbital motion of a satellite due to the force of gravity.

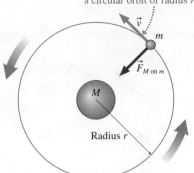

The satellite must have speed $\sqrt{GM/r}$ to maintain a circular orbit of radius r.

EXAMPLE 13.4 | **The speed of the space station**

A small supply satellite needs to dock with the International Space Station. The ISS is in a near-circular orbit at a height of 420 km. What are the speeds of the ISS and the supply satellite in this orbit?

SOLVE Despite their different masses, the satellite and the ISS travel side by side with the same speed. They are simply in free fall together. Using $r = R_e + h$ with $h = 420$ km $= 4.20 \times 10^5$ m, we find the speed

$$v = \sqrt{\frac{(6.67 \times 10^{-11} \text{ N m}^2/\text{kg}^2)(5.98 \times 10^{24} \text{ kg})}{6.79 \times 10^6 \text{ m}}}$$

$$= 7660 \text{ m/s} \approx 17,000 \text{ mph}$$

ASSESS The answer depends on the mass of the earth but *not* on the mass of the satellite.

Kepler's Third Law

An important parameter of circular motion is the *period*. Recall that the period T is the time to complete one full orbit. The relationship among speed, radius, and period is

$$v = \frac{\text{circumference}}{\text{period}} = \frac{2\pi r}{T} \tag{13.23}$$

We can find a relationship between a satellite's period and the radius of its orbit by using Equation 13.22 for v:

$$v = \frac{2\pi r}{T} = \sqrt{\frac{GM}{r}} \tag{13.24}$$

Squaring both sides and solving for T give

$$T^2 = \left(\frac{4\pi^2}{GM}\right)r^3 \tag{13.25}$$

The International Space Station appears to be floating, but it's actually traveling at nearly 8000 m/s as it orbits the earth.

In other words, the *square* of the period is proportional to the *cube* of the radius. This is Kepler's third law. You can see that Kepler's third law is a direct consequence of Newton's law of gravity.

TABLE 13.2 contains astronomical information about the solar system. We can use these data to check the validity of Equation 13.25. **FIGURE 13.17** is a graph of log T versus log r for all the planets in Table 13.2 except Mercury. Notice that the scales on each axis are increasing logarithmically—by *factors* of 10—rather than linearly. (Also, the vertical axis has converted T to the SI units of s.) As you can see, the graph is a straight line with a best-fit equation

$$\log T = 1.500 \log r - 9.264$$

Taking the logarithm of both sides of Equation 13.25, and using the logarithm properties $\log a^n = n \log a$ and $\log(ab) = \log a + \log b$, we have

$$\log T = \tfrac{3}{2} \log r + \tfrac{1}{2} \log\!\left(\frac{4\pi^2}{GM}\right)$$

In other words, theory predicts that the slope of a log T-versus-log r graph should be exactly $\tfrac{3}{2}$. As Figure 13.17 shows, the solar-system data agree to an impressive four significant figures. A homework problem will let you use the y-intercept of the graph to determine the mass of the sun.

FIGURE 13.17 The graph of log T versus log r for the planetary data of Table 13.2.

TABLE 13.2 Useful astronomical data

Planetary body	Mean distance from sun (m)	Period (years)	Mass (kg)	Mean radius (m)
Sun	—	—	1.99×10^{30}	6.96×10^8
Moon	3.84×10^8*	27.3 days	7.36×10^{22}	1.74×10^6
Mercury	5.79×10^{10}	0.241	3.18×10^{23}	2.43×10^6
Venus	1.08×10^{11}	0.615	4.88×10^{24}	6.06×10^6
Earth	1.50×10^{11}	1.00	5.98×10^{24}	6.37×10^6
Mars	2.28×10^{11}	1.88	6.42×10^{23}	3.37×10^6
Jupiter	7.78×10^{11}	11.9	1.90×10^{27}	6.99×10^7
Saturn	1.43×10^{12}	29.5	5.68×10^{26}	5.85×10^7
Uranus	2.87×10^{12}	84.0	8.68×10^{25}	2.33×10^7
Neptune	4.50×10^{12}	165	1.03×10^{26}	2.21×10^7

*Distance from earth.

A particularly interesting application of Equation 13.25 is to communications satellites that are in **geosynchronous orbits** above the earth. These satellites have a period of 24 h = 86,400 s, making their orbital motion synchronous with the earth's rotation. As a result, a satellite in such an orbit appears to remain stationary over one point on the earth's equator. Equation 13.25 allows us to compute the radius of an orbit with this period:

$$r_{\text{geo}} = R_{\text{e}} + h_{\text{geo}} = \left[\left(\frac{GM}{4\pi^2}\right)T^2\right]^{1/3}$$

$$= \left[\left(\frac{(6.67 \times 10^{-11}\,\text{N m}^2/\text{kg}^2)(5.98 \times 10^{24}\,\text{kg})}{4\pi^2}\right)(86{,}400\,\text{s})^2\right]^{1/3}$$

$$= 4.225 \times 10^7\,\text{m}$$

The height of the orbit is

$$h_{\text{geo}} = r_{\text{geo}} - R_{\text{e}} = 3.59 \times 10^7\,\text{m} = 35{,}900\,\text{km} \approx 22{,}300\,\text{mi}$$

NOTE When you use Equation 13.25, the period *must* be in SI units of s.

Geosynchronous orbits are much higher than the low-earth orbits used by the space station and remote-sensing satellites, where $h \approx 300$ km. Communications satellites in geosynchronous orbits were first proposed in 1948 by science fiction writer Arthur C. Clarke, 10 years before the first artificial satellite of any type!

EXAMPLE 13.5 | **Extrasolar planets**

In recent years, astronomers have discovered thousands of planets orbiting nearby stars. These are called *extrasolar planets*. Suppose a planet is observed to have a 1200 day period as it orbits a star at the same distance that Jupiter is from the sun. What is the mass of the star in solar masses? (1 *solar mass* is defined to be the mass of the sun.)

SOLVE Here "day" means earth days, as used by astronomers to measure the period. Thus the planet's period in SI units is $T = 1200$ days $= 1.037 \times 10^8$ s. The orbital radius is that of Jupiter,

which we can find in Table 13.2 to be $r = 7.78 \times 10^{11}$ m. Solving Equation 13.25 for the mass of the star gives

$$M = \frac{4\pi^2 r^3}{GT^2} = 2.59 \times 10^{31} \text{ kg} \times \frac{1 \text{ solar mass}}{1.99 \times 10^{30} \text{ kg}}$$

$$= 13 \text{ solar masses}$$

ASSESS This is a large, but not extraordinary, star.

STOP TO THINK 13.5 Two planets orbit a star. Planet 1 has orbital radius r_1 and planet 2 has $r_2 = 4r_1$. Planet 1 orbits with period T_1. Planet 2 orbits with period

a. $T_2 = 8T_1$ b. $T_2 = 4T_1$ c. $T_2 = 2T_1$

d. $T_2 = \frac{1}{2}T_1$ e. $T_2 = \frac{1}{4}T_1$ f. $T_2 = \frac{1}{8}T_1$

Kepler's Second Law

FIGURE 13.18a shows a planet moving in an elliptical orbit. In Chapter 12 we defined a particle's *angular momentum* to be

$$L = mrv \sin \beta \tag{13.26}$$

where β is the angle between $\vec{r}$ and $\vec{v}$. For a circular orbit, where β is always 90°, this reduces to simply $L = mrv$.

The only force on the satellite, the gravitational force, points directly toward the star or planet that the satellite is orbiting and exerts no torque; thus **the satellite's angular momentum is conserved as it orbits.**

The satellite moves forward a small distance $\Delta s = v\,\Delta t$ during the small interval of time Δt. This motion defines the triangle of area ΔA shown in **FIGURE 13.18b**. ΔA is the area "swept out" by the satellite during Δt. You can see that the height of the triangle is $h = \Delta s \sin \beta$, so the triangle's area is

$$\Delta A = \tfrac{1}{2} \times \text{base} \times \text{height} = \tfrac{1}{2} \times r \times \Delta s \sin \beta = \tfrac{1}{2}rv \sin \beta \, \Delta t \tag{13.27}$$

The *rate* at which the area is swept out by the satellite as it moves is

$$\frac{\Delta A}{\Delta t} = \tfrac{1}{2}rv \sin \beta = \frac{mrv \sin \beta}{2m} = \frac{L}{2m} \tag{13.28}$$

The angular momentum L is conserved, so it has the same value at every point in the orbit. Consequently, the rate at which the area is swept out by the satellite is constant. This is Kepler's second law, which says that a line drawn between the sun and a planet sweeps out equal areas during equal intervals of time. We see that Kepler's second law is a consequence of the conservation of angular momentum.

Another consequence of angular momentum is that the orbital speed is constant only for a circular orbit. Consider the "ends" of an elliptical orbit, where r is a minimum or maximum. At these points, $\beta = 90°$ and thus $L = mrv$. Because L is constant, the satellite's speed at the farthest point must be less than its speed at the nearest point. In general, a satellite slows as r increases, then speeds up as r decreases, to keep its angular momentum (and its energy) constant.

FIGURE 13.18 Angular momentum is conserved for a planet in an elliptical orbit.

(a)

The gravitational force points straight at the sun and exerts no torque.

(b)

Area ΔA is swept out during Δt.

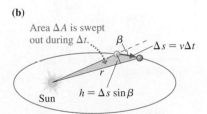

$\Delta s = v\Delta t$

$h = \Delta s \sin\beta$

Kepler's laws summarize observational data about the motions of the planets. They were an outstanding achievement, but they did not form a theory. Newton put forward a *theory,* a specific set of relationships between force and motion that allows *any* motion to be understood and calculated. Newton's theory of gravity has allowed us to *deduce* Kepler's laws and, thus, to understand them at a more fundamental level.

Orbital Energetics

Let us conclude this chapter by thinking about the energetics of orbital motion. We found, with Equation 13.24, that a satellite in a circular orbit must have $v^2 = GM/r$. A satellite's speed is determined entirely by the size of its orbit. The satellite's kinetic energy is thus

$$K = \tfrac{1}{2}mv^2 = \frac{GMm}{2r} \tag{13.29}$$

But $-GMm/r$ is the potential energy, U_G, so

$$K = -\tfrac{1}{2}U_G \tag{13.30}$$

This is an interesting result. In all our earlier examples, the kinetic and potential energy were two independent parameters. In contrast, a satellite can move in a circular orbit *only* if there is a very specific relationship between K and U. It is not that K and U *have* to have this relationship, but if they do not, the trajectory will be elliptical rather than circular.

Equation 13.30 gives us the mechanical energy of a satellite in a circular orbit:

$$E_{\text{mech}} = K + U_G = \tfrac{1}{2}U_G \tag{13.31}$$

The gravitational potential energy is negative, hence the *total* mechanical energy is also negative. Negative total energy is characteristic of a **bound system,** a system in which the satellite is bound to the central mass by the gravitational force and cannot get away. The total energy of an unbound system must be ≥ 0 because the satellite can reach infinity, where $U = 0$, while still having kinetic energy. A negative value of E_{mech} tells us that the satellite is unable to escape the central mass.

FIGURE 13.19 shows the energies of a satellite in a circular orbit as a function of the orbit's radius. Notice how $E_{\text{mech}} = \tfrac{1}{2}U_G$. This figure can help us understand the energetics of transferring a satellite from one orbit to another. Suppose a satellite is in an orbit of radius r_1 and we'd like it to be in a larger orbit of radius r_2. The kinetic energy at r_2 is less than at r_1 (the satellite moves more slowly in the larger orbit), but you can see that the total energy *increases* as r increases. Consequently, transferring a satellite to a larger orbit requires a net energy increase $\Delta E > 0$. Where does this increase of energy come from?

Artificial satellites are raised to higher orbits by firing their rocket motors to create a forward thrust. This force does work on the satellite, and the energy principle of Chapter 10 tells us that this work increases the satellite's energy by $\Delta E_{\text{mech}} = W_{\text{ext}}$. Thus the energy to "lift" a satellite into a higher orbit comes from the chemical energy stored in the rocket fuel.

FIGURE 13.19 The kinetic, potential, and total energy of a satellite in a circular orbit.

Energy ΔE must be added to move a satellite from an orbit with radius r_1 to radius r_2.

EXAMPLE 13.6 | Raising a satellite

How much work must be done to boost a 1000 kg communications satellite from a low earth orbit with $h = 300$ km to a geosynchronous orbit?

SOLVE The required work is $W_{\text{ext}} = \Delta E_{\text{mech}}$, and from Equation 13.31 we see that $\Delta E_{\text{mech}} = \tfrac{1}{2}\Delta U_G$. The initial orbit has radius $r_{\text{low}} = R_e + h = 6.67 \times 10^6$ m. We earlier found the radius of a geosynchronous orbit to be 4.22×10^7 m. Thus

$$W_{\text{ext}} = \Delta E_{\text{mech}} = \tfrac{1}{2}\Delta U_G = \tfrac{1}{2}(-GM_em)\left(\frac{1}{r_{\text{geo}}} - \frac{1}{r_{\text{low}}}\right) = 2.52 \times 10^{10} \text{ J}$$

ASSESS It takes a lot of energy to boost satellites to high orbits!

You might think that the way to get a satellite into a larger orbit would be to point the thrusters toward the earth and blast outward. That would work fine *if* the satellite were initially at rest and moved straight out along a linear trajectory. But an orbiting satellite is already moving and has significant inertia. A force directed straight outward would *change* the satellite's velocity vector in that direction but would not cause it to *move* along that line. (Remember all those earlier motion diagrams for motion along curved trajectories.) In addition, a force directed outward would be almost at right angles to the motion and would do essentially zero work. Navigating in space is not as easy as it appears in *Star Wars!*

To move the satellite in FIGURE 13.20 from the orbit with radius r_1 to the larger circular orbit of radius r_2, the thrusters are turned on at point 1 to apply a brief *forward* thrust force in the direction of motion, *tangent* to the circle. This force does a significant amount of work because the force is parallel to the displacement, so the satellite quickly gains kinetic energy ($\Delta K > 0$). But $\Delta U_G = 0$ because the satellite does not have time to change its distance from the earth during a thrust of short duration. With the kinetic energy increased, but not the potential energy, the satellite no longer meets the requirement $K = -\frac{1}{2}U_G$ for a circular orbit. Instead, it goes into an elliptical orbit.

In the elliptical orbit, the satellite moves "uphill" toward point 2 by transforming kinetic energy into potential energy. At point 2, the satellite has arrived at the desired distance from earth and has the "right" value of the potential energy, but its kinetic energy is now *less* than needed for a circular orbit. (The analysis is more complex than we want to pursue here. It will be left for a homework Challenge Problem.) If no action is taken, the satellite will continue on its elliptical orbit and "fall" back to point 1. But another *forward* thrust at point 2 increases its kinetic energy, without changing U_G, until the kinetic energy reaches the value $K = -\frac{1}{2}U_G$ required for a circular orbit. Presto! The second burn kicks the satellite into the desired circular orbit of radius r_2. The work $W_{ext} = \Delta E_{mech}$ is the *total* work done in both burns. It takes a more extended analysis to see how the work has to be divided between the two burns, but even without those details you now have enough knowledge about orbits and energy to understand the ideas that are involved.

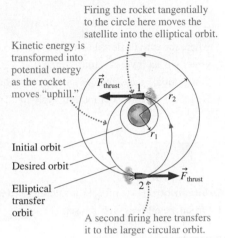

FIGURE 13.20 Transferring a satellite to a larger circular orbit.

Firing the rocket tangentially to the circle here moves the satellite into the elliptical orbit.

Kinetic energy is transformed into potential energy as the rocket moves "uphill."

$\vec{F}_{thrust}$ 1 r_2

r_1

Initial orbit

Desired orbit

Elliptical transfer orbit

$\vec{F}_{thrust}$ 2

A second firing here transfers it to the larger circular orbit.

CHALLENGE EXAMPLE 13.7 | A binary star system

Astronomers discover a binary star system with a period of 90 days. Both stars have a mass twice that of the sun. How far apart are the two stars?

MODEL Model the stars as spherical masses exerting gravitational forces on each other.

VISUALIZE *An isolated system rotates around its center of mass.* FIGURE 13.21 shows the orbits and the forces. If r is the distance of each star to the center of mass—the radius of that star's orbit—then the distance between the stars is $d = 2r$.

FIGURE 13.21 The binary star system.

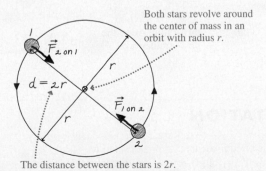

Both stars revolve around the center of mass in an orbit with radius r.

$\vec{F}_{2\,on\,1}$

r

$d = 2r$

$\vec{F}_{1\,on\,2}$

r

2

The distance between the stars is $2r$.

SOLVE Star 2 has only one force acting on it, $\vec{F}_{1\,on\,2}$, and that force has to provide the centripetal acceleration v^2/r of circular motion. Newton's second law for star 2 is

$$F_{1\,on\,2} = \frac{GM_1 M_2}{d^2} = \frac{GM^2}{4r^2} = Ma_r = \frac{Mv^2}{r}$$

where we used $M_1 = M_2 = M$. The equation for star 1 is identical. The star's speed is related to the period and the circumference of its orbit by $v = 2\pi r/T$. With this, the force equation becomes

$$\frac{GM^2}{4r^2} = \frac{4\pi^2 Mr}{T^2}$$

Solving for r gives

$$r = \left[\frac{GMT^2}{16\pi^2}\right]^{1/3}$$

$$= \left[\frac{(6.67 \times 10^{-11}\,\text{N}\,\text{m}^2/\text{kg}^2)(2 \times 1.99 \times 10^{30}\,\text{kg})(7.78 \times 10^6\,\text{s})^2}{16\pi^2}\right]^{1/3}$$

$$= 4.67 \times 10^{10}\,\text{m}$$

The distance between the stars is $d = 2r = 9.3 \times 10^{10}$ m.

ASSESS The result is in the range of solar-system distances and thus is reasonable.

SUMMARY

The goal of Chapter 13 has been to understand the motion of satellites and planets.

GENERAL PRINCIPLES

Newton's Theory of Gravity

1. Two objects with masses M and m a distance r apart exert attractive **gravitational forces** on each other of magnitude

$$F_{M \text{ on } m} = F_{m \text{ on } M} = \frac{GMm}{r^2}$$

where the **gravitational constant** is $G = 6.67 \times 10^{-11} \text{ N m}^2/\text{kg}^2$. Gravity is an **inverse-square** force.

2. Gravitational mass and inertial mass are equivalent.

3. Newton's three laws of motion apply to all objects in the universe.

The forces are an action/reaction pair.

IMPORTANT CONCEPTS

Orbital motion of a planet (or satellite) is described by **Kepler's laws**:

1. Orbits are ellipses with the sun (or planet) at one focus.

2. A line between the sun and the planet sweeps out equal areas during equal intervals of time.

3. The square of the planet's period T is proportional to the cube of the orbit's semimajor axis.

Circular orbits are a special case of an ellipse. For a circular orbit around a mass M,

$$v = \sqrt{\frac{GM}{r}} \qquad \text{and} \qquad T^2 = \left(\frac{4\pi^2}{GM}\right) r^3$$

Conservation of angular momentum

The angular momentum $L = mrv \sin \beta$ remains constant throughout the orbit. Kepler's second law is a consequence of this law.

Orbital energetics

A satellite's mechanical energy $E_{\text{mech}} = K + U_G$ is conserved, where the gravitational potential energy is

$$U_G = -\frac{GMm}{r}$$

For circular orbits, $K = -\frac{1}{2} U_G$ and $E_{\text{mech}} = \frac{1}{2} U_G$. Negative total energy is characteristic of a **bound system.**

APPLICATIONS

For a planet of mass M and radius R:

- Free-fall acceleration on the surface is $g_{\text{surface}} = \dfrac{GM}{R^2}$

- **Escape speed** is $v_{\text{escape}} = \sqrt{\dfrac{2GM}{R}}$

- Radius of a **geosynchronous orbit** is $r_{\text{geo}} = \left(\dfrac{GM}{4\pi^2} T^2\right)^{1/3}$

To move a satellite to a larger orbit:

- Forward thrust to move to an elliptical transfer orbit

- A second forward thrust to move to the new circular orbit

TERMS AND NOTATION

cosmology	Newton's law of gravity	principle of equivalence	satellite
Kepler's laws	gravitational constant, G	Newton's theory of gravity	geosynchronous orbit
gravitational force	gravitational mass	escape speed	bound system

CONCEPTUAL QUESTIONS

1. Is the earth's gravitational force on the sun larger than, smaller than, or equal to the sun's gravitational force on the earth? Explain.
2. The gravitational force of a star on orbiting planet 1 is F_1. Planet 2, which is twice as massive as planet 1 and orbits at twice the distance from the star, experiences gravitational force F_2. What is the ratio F_1/F_2?
3. A 1000 kg satellite and a 2000 kg satellite follow exactly the same orbit around the earth.
   a. What is the ratio F_1/F_2 of the gravitational force on the first satellite to that on the second satellite?
   b. What is the ratio a_1/a_2 of the acceleration of the first satellite to that of the second satellite?
4. How far away from the earth must an orbiting spacecraft be for the astronauts inside to be weightless? Explain.
5. A space station astronaut is working outside the station as it orbits the earth. If he drops a hammer, will it fall to earth? Explain why or why not.
6. The free-fall acceleration at the surface of planet 1 is 20 m/s². The radius and the mass of planet 2 are twice those of planet 1. What is g on planet 2?
7. *Why* is the gravitational potential energy of two masses negative? Note that saying "because that's what the equation gives" is *not* an explanation.
8. The escape speed from Planet X is 10,000 m/s. Planet Y has the same radius as Planet X but is twice as dense. What is the escape speed from Planet Y?
9. The mass of Jupiter is 300 times the mass of the earth. Jupiter orbits the sun with $T_{Jupiter} = 11.9$ yr in an orbit with $r_{Jupiter} = 5.2 r_{earth}$. Suppose the earth could be moved to the distance of Jupiter and placed in a circular orbit around the sun. Which of the following describes the earth's new period? Explain.
   a. 1 yr
   b. Between 1 yr and 11.9 yr
   c. 11.9 yr
   d. More than 11.9 yr
   e. It would depend on the earth's speed.
   f. It's impossible for a planet of earth's mass to orbit at the distance of Jupiter.
10. Satellites in near-earth orbit experience a very slight drag due to the extremely thin upper atmosphere. These satellites slowly but surely spiral inward, where they finally burn up as they reach the thicker lower levels of the atmosphere. The radius decreases so slowly that you can consider the satellite to have a circular orbit at all times. As a satellite spirals inward, does it speed up, slow down, or maintain the same speed? Explain.

EXERCISES AND PROBLEMS

Problems labeled ▒ integrate material from earlier chapters.

Exercises

Section 13.3 Newton's Law of Gravity

1. | What is the ratio of the sun's gravitational force on you to the earth's gravitational force on you?
2. || What is the ratio of the sun's gravitational force on the moon to the earth's gravitational force on the moon?
3. || The centers of a 10 kg lead ball and a 100 g lead ball are separated by 10 cm.
   a. What gravitational force does each exert on the other?
   b. What is the ratio of this gravitational force to the gravitational force of the earth on the 100 g ball?
4. | What is the force of attraction between a 50 kg woman and a 70 kg man sitting 1.0 m apart?
5. || The International Space Station orbits 300 km above the surface of the earth. What is the gravitational force on a 1.0 kg sphere inside the International Space Station?
6. || Two 65 kg astronauts leave earth in a spacecraft, sitting 1.0 m apart. How far are they from the center of the earth when the gravitational force between them is as strong as the gravitational force of the earth on one of the astronauts?
7. || A 20 kg sphere is at the origin and a 10 kg sphere is at $x = 20$ cm. At what position on the x-axis could you place a small mass such that the net gravitational force on it due to the spheres is zero?

Section 13.4 Little g and Big G

8. | a. What is the free-fall acceleration at the surface of the sun?
   b. What is the free-fall acceleration toward the sun at the distance of the earth?

9. || What is the free-fall acceleration at the surface of (a) the moon and (b) Jupiter?
10. || A sensitive gravimeter at a mountain observatory finds that the free-fall acceleration is 0.0075 m/s² less than that at sea level. What is the observatory's altitude?
11. | Saturn's moon Titan has a mass of 1.35×10^{23} kg and a radius of 2580 km. What is the free-fall acceleration on Titan?
12. || A newly discovered planet has a radius twice as large as earth's and a mass five times as large. What is the free-fall acceleration on its surface?
13. | Suppose we could shrink the earth without changing its mass. At what fraction of its current radius would the free-fall acceleration at the surface be three times its present value?
14. || Planet Z is 10,000 km in diameter. The free-fall acceleration on Planet Z is 8.0 m/s².
   a. What is the mass of Planet Z?
   b. What is the free-fall acceleration 10,000 km above Planet Z's north pole?

Section 13.5 Gravitational Potential Energy

15. | An astronaut on earth can throw a ball straight up to a height of 15 m. How high can he throw the ball on Mars?
16. || What is the escape speed from Jupiter?
17. || A rocket is launched straight up from the earth's surface at a speed of 15,000 m/s. What is its speed when it is very far away from the earth?
18. | A space station orbits the sun at the same distance as the earth but on the opposite side of the sun. A small probe is fired away from the station. What minimum speed does the probe need to escape the solar system?

19. ‖ A proposed *space elevator* would consist of a cable stretching from the earth's surface to a satellite, orbiting far in space, that would keep the cable taut. A motorized climber could slowly carry rockets to the top, where they could be launched away from the earth using much less energy. What would be the escape speed for a craft launched from a space elevator at a height of 36,000 km? Ignore the earth's rotation.

20. ‖ Nothing can escape the *event horizon* of a black hole, not even light. You can think of the event horizon as being the distance from a black hole at which the escape speed is the speed of light, 3.00×10^8 m/s, making all escape impossible. What is the radius of the event horizon for a black hole with a mass 5.0 times the mass of the sun? This distance is called the *Schwarzschild radius*.

21. ‖ You have been visiting a distant planet. Your measurements have determined that the planet's mass is twice that of earth but the free-fall acceleration at the surface is only one-fourth as large.
    a. What is the planet's radius?
    b. To get back to earth, you need to escape the planet. What minimum speed does your rocket need?

22. ‖ Two meteoroids are heading for earth. Their speeds as they cross the moon's orbit are 2.0 km/s.
    a. The first meteoroid is heading straight for earth. What is its speed of impact?
    b. The second misses the earth by 5000 km. What is its speed at its closest point?

23. ‖ A binary star system has two stars, each with the same mass as our sun, separated by 1.0×10^{12} m. A comet is very far away and essentially at rest. Slowly but surely, gravity pulls the comet toward the stars. Suppose the comet travels along a trajectory that passes through the midpoint between the two stars. What is the comet's speed at the midpoint?

Section 13.6 Satellite Orbits and Energies

24. ‖ The *asteroid belt* circles the sun between the orbits of Mars and Jupiter. One asteroid has a period of 5.0 earth years. What are the asteroid's orbital radius and speed?

25. ‖ You are the science officer on a visit to a distant solar system. Prior to landing on a planet you measure its diameter to be 1.8×10^7 m and its rotation period to be 22.3 hours. You have previously determined that the planet orbits 2.2×10^{11} m from its star with a period of 402 earth days. Once on the surface you find that the free-fall acceleration is 12.2 m/s². What is the mass of (a) the planet and (b) the star?

26. ‖ Three satellites orbit a planet of radius R, as shown in FIGURE EX13.26. Satellites S_1 and S_3 have mass m. Satellite S_2 has mass $2m$. Satellite S_1 orbits in 250 minutes and the force on S_1 is 10,000 N.
    a. What are the periods of S_2 and S_3?
    b. What are the forces on S_2 and S_3?
    c. What is the kinetic-energy ratio K_1/K_3 for S_1 and S_3?

FIGURE EX13.26

27. ‖ A satellite orbits the sun with a period of 1.0 day. What is the radius of its orbit?

28. ‖ A new planet is discovered orbiting the star Vega in a circular orbit. The planet takes 55 earth years to complete one orbit around the star. Vega's mass is 2.1 times the sun's mass. What is the radius of the planet's orbit? Give your answer as a multiple of the radius of the earth's orbit.

29. ‖ A small moon orbits its planet in a circular orbit at a speed of 7.5 km/s. It takes 28 hours to complete one full orbit. What is the mass of the planet?

30. ‖ An earth satellite moves in a circular orbit at a speed of 5500 m/s. What is its orbital period?

31. ‖ What are the speed and altitude of a geosynchronous satellite orbiting Mars? Mars rotates on its axis once every 24.8 hours.

32. ‖ a. At what height above the earth is the free-fall acceleration 10% of its value at the surface?
    b. What is the speed of a satellite orbiting at that height?

33. ‖ In 2000, NASA placed a satellite in orbit around an asteroid. Consider a spherical asteroid with a mass of 1.0×10^{16} kg and a radius of 8.8 km.
    a. What is the speed of a satellite orbiting 5.0 km above the surface?
    b. What is the escape speed from the asteroid?

34. ‖ Pluto moves in a fairly elliptical orbit around the sun. Pluto's speed at its closest approach of 4.43×10^9 km is 6.12 km/s. What is Pluto's speed at the most distant point in its orbit, where it is 7.30×10^9 km from the sun?

Problems

35. ‖ FIGURE P13.35 shows three masses. What are the magnitude and the direction of the net gravitational force on (a) the 20.0 kg mass and (b) the 5.0 kg mass? Give the direction as an angle cw or ccw from the *y*-axis.

FIGURE P13.35   FIGURE P13.36

36. ‖ What are the magnitude and direction of the net gravitational force on the 20.0 kg mass in FIGURE P13.36?

37. ‖ What is the total gravitational potential energy of the three masses in FIGURE P13.35?

38. ‖ What is the total gravitational potential energy of the three masses in FIGURE P13.36?

39. ‖ Two spherical objects have a combined mass of 150 kg. The gravitational attraction between them is 8.00×10^{-6} N when their centers are 20 cm apart. What is the mass of each?

40. ‖ Two 100 kg lead spheres are suspended from 100-m-long massless cables. The tops of the cables have been carefully anchored *exactly* 1 m apart. By how much is the distance between the centers of the spheres less than 1 m?

41. ‖ A rogue black hole with a mass 12 times the mass of the sun drifts into the solar system on a collision course with earth. How far is the black hole from the center of the earth when objects on the earth's surface begin to lift into the air and "fall" up into the black hole? Give your answer as a multiple of the earth's radius.

42. ||| An object of mass m is dropped from height h above a planet of mass M and radius R. Find an expression for the object's speed as it hits the ground.

43. ||| A projectile is shot straight up from the earth's surface at a speed of 10,000 km/h. How high does it go?

44. || The unexplored planet Alpha Centauri III has a radius of 7.0×10^6 m. A visiting astronaut drops a rock, from rest, into a 100-m-deep crevasse. She records that it takes 6.0 s for the rock to reach the bottom. What is the mass of Alpha Centauri III?

45. || An astronaut circling the earth at an altitude of 400 km is horrified to discover that a cloud of space debris is moving in the exact same orbit as his spacecraft, but in the opposite direction. The astronaut detects the debris when it is 25 km away. How much time does he have to fire his rockets and change orbits?

46. || Suppose that on earth you can jump straight up a distance of 45 cm. Asteroids are made of material with mass density 2800 kg/m³. What is the maximum diameter of a spherical asteroid from which you could escape by jumping?

47. ||| A rogue band of colonists on the moon declares war and
CALC prepares to use a catapult to launch large boulders at the earth. Assume that the boulders are launched from the point on the moon nearest the earth. For this problem you can ignore the rotation of the two bodies and the orbiting of the moon.
   a. What is the minimum speed with which a boulder must be launched to reach the earth?
   Hint: The minimum speed is not the escape speed. You need to analyze a three-body system.
   b. Ignoring air resistance, what is the impact speed on earth of a boulder launched at this minimum speed?

48. ||| Two spherical asteroids have the same radius R. Asteroid 1 has mass M and asteroid 2 has mass $2M$. The two asteroids are released from rest with distance $10R$ between their centers. What is the speed of each asteroid just before they collide?
   Hint: You will need to use two conservation laws.

49. || A starship is circling a distant planet of radius R. The astronauts find that the free-fall acceleration at their altitude is half the value at the planet's surface. How far above the surface are they orbiting? Your answer will be a multiple of R.

50. ||| The two stars in a binary star system have masses 2.0×10^{30} kg and 6.0×10^{30} kg. They are separated by 2.0×10^{12} m. What are
   a. The system's rotation period, in years?
   b. The speed of each star?

51. ||| A 4000 kg lunar lander is in orbit 50 km above the surface of the moon. It needs to move out to a 300-km-high orbit in order to link up with the mother ship that will take the astronauts home. How much work must the thrusters do?

52. ||| The 75,000 kg space shuttle used to fly in a 250-km-high circular orbit. It needed to reach a 610-km-high circular orbit to service the Hubble Space Telescope. How much energy was required to boost it to the new orbit?

53. ||| How much energy would be required to move the earth into a circular orbit with a radius 1.0 km larger than its current radius?

54. || NASA would like to place a satellite in orbit around the moon such that the satellite always remains in the same position over the lunar surface. What is the satellite's altitude?

55. || In 2014, the European Space Agency placed a satellite in orbit around comet 67P/Churyumov-Gerasimenko and then landed a probe on the surface. The actual orbit was elliptical, but we'll approximate it as a 50-km-diameter circular orbit with a period of 11 days.
   a. What was the satellite's orbital speed around the comet? (Both the comet and the satellite are orbiting the sun at a much higher speed.)

b. What is the mass of the comet?
   c. The lander was pushed from the satellite, toward the comet, at a speed of 70 cm/s, and it then fell—taking about 7 hours—to the surface. What was its landing speed? The comet's shape is irregular, but on average it has a diameter of 3.6 km.

56. || A satellite orbiting the earth is directly over a point on the equator at 12:00 midnight every two days. It is not over that point at any time in between. What is the radius of the satellite's orbit?

57. ||| FIGURE P13.57 shows two planets of mass m orbiting a star of mass M. The planets are in the same orbit, with radius r, but are always at opposite ends of a diameter. Find an exact expression for the orbital period T.
   Hint: Each planet feels two forces.

FIGURE P13.57

58. || Figure 13.17 showed a graph of log T versus log r for the planetary data given in Table 13.2. Such a graph is called a *log-log graph*. The scales in Figure 13.17 are logarithmic, not linear, meaning that each division along the axis corresponds to a *factor* of 10 increase in the value. Strictly speaking, the "correct" labels on the y-axis should be 7, 8, 9, and 10 because these are the logarithms of $10^7, \ldots, 10^{10}$.
   a. Consider two quantities u and v that are related by the expression $v^p = Cu^q$, where C is a constant. The exponents p and q are not necessarily integers. Define $x = \log u$ and $y = \log v$. Find an expression for y in terms of x.
   b. What *shape* will a graph of y versus x have? Explain.
   c. What *slope* will a graph of y versus x have? Explain.
   d. Use the experimentally determined "best-fit" line in Figure 13.17 to find the mass of the sun.

59. || Large stars can explode as they finish burning their nuclear fuel, causing a *supernova*. The explosion blows away the outer layers of the star. According to Newton's third law, the forces that push the outer layers away have *reaction forces* that are inwardly directed on the core of the star. These forces compress the core and can cause the core to undergo a *gravitational collapse*. The gravitational forces keep pulling all the matter together tighter and tighter, crushing atoms out of existence. Under these extreme conditions, a proton and an electron can be squeezed together to form a neutron. If the collapse is halted when the neutrons all come into contact with each other, the result is an object called a *neutron star*, an entire star consisting of solid nuclear matter. Many neutron stars rotate about their axis with a period of ≈ 1 s and, as they do so, send out a pulse of electromagnetic waves once a second. These stars were discovered in the 1960s and are called *pulsars*.
   a. Consider a neutron star with a mass equal to the sun, a radius of 10 km, and a rotation period of 1.0 s. What is the speed of a point on the equator of the star?
   b. What is g at the surface of this neutron star?
   c. A stationary 1.0 kg mass has a weight on earth of 9.8 N. What would be its weight on the star?
   d. How many revolutions per minute are made by a satellite orbiting 1.0 km above the surface?
   e. What is the radius of a geosynchronous orbit?

60. || The solar system is 25,000 light years from the center of our Milky Way galaxy. One *light year* is the distance light travels in one year at a speed of 3.0×10^8 m/s. Astronomers have determined that the solar system is orbiting the center of the galaxy at a speed of 230 km/s.
   a. Assuming the orbit is circular, what is the period of the solar system's orbit? Give your answer in years.

b. Our solar system was formed roughly 5 billion years ago. How many orbits has it completed?

c. The gravitational force on the solar system is the net force due to all the matter inside our orbit. Most of that matter is concentrated near the center of the galaxy. Assume that the matter has a spherical distribution, like a giant star. What is the approximate mass of the galactic center?

d. Assume that the sun is a typical star with a typical mass. If galactic matter is made up of stars, approximately how many stars are in the center of the galaxy?

Astronomers have spent many years trying to determine how many stars there are in the Milky Way. The number of stars seems to be only about 10% of what you found in part d. In other words, about 90% of the mass of the galaxy appears to be in some form other than stars. This is called the *dark matter* of the universe. No one knows what the dark matter is. This is one of the outstanding scientific questions of our day.

61. ‖ Three stars, each with the mass of our sun, form an equilateral triangle with sides 1.0×10^{12} m long. (This triangle would just about fit within the orbit of Jupiter.) The triangle has to rotate, because otherwise the stars would crash together in the center. What is the period of rotation?

62. ‖ Comets move around the sun in very elliptical orbits. At its closet approach, in 1986, Comet Halley was 8.79×10^7 km from the sun and moving with a speed of 54.6 km/s. What was the comet's speed when it crossed Neptune's orbit in 2006?

63. ‖‖ A 55,000 kg space capsule is in a 28,000-km-diameter circular orbit around the moon. A brief but intense firing of its engine in the forward direction suddenly decreases its speed by 50%. This causes the space capsule to go into an elliptical orbit. What are the space capsule's (a) maximum and (b) minimum distances from the center of the moon in its new orbit?

Hint: You will need to use two conservation laws.

In Problems 64 through 66 you are given the equation(s) used to solve a problem. For each of these, you are to

a. Write a realistic problem for which this is the correct equation(s).

b. Draw a pictorial representation.

c. Finish the solution of the problem.

64. $\dfrac{(6.67 \times 10^{-11}\,\text{N}\,\text{m}^2/\text{kg}^2)(5.68 \times 10^{26}\,\text{kg})}{r^2}$

$= \dfrac{(6.67 \times 10^{-11}\,\text{N}\,\text{m}^2/\text{kg}^2)(5.98 \times 10^{24}\,\text{kg})}{(6.37 \times 10^6\,\text{m})^2}$

65. $\dfrac{(6.67 \times 10^{-11}\,\text{N}\,\text{m}^2/\text{kg}^2)(5.98 \times 10^{24}\,\text{kg})(1000\,\text{kg})}{r^2}$

$= \dfrac{(1000\,\text{kg})(1997\,\text{m/s})^2}{r}$

66. $\frac{1}{2}(100\,\text{kg})v_2^2$

$- \dfrac{(6.67 \times 10^{-11}\,\text{N}\,\text{m}^2/\text{kg}^2)(7.36 \times 10^{22}\,\text{kg})(100\,\text{kg})}{1.74 \times 10^6\,\text{m}}$

$= 0 - \dfrac{(6.67 \times 10^{-11}\,\text{N}\,\text{m}^2/\text{kg}^2)(7.36 \times 10^{22}\,\text{kg})(100\,\text{kg})}{3.48 \times 10^6\,\text{m}}$

Challenge Problems

67. ‖‖ Two Jupiter-size planets are released from rest 1.0×10^{11} m apart. What are their speeds as they crash together?

68. ‖‖ A satellite in a circular orbit of radius r has period T. A satellite in a nearby orbit with radius $r + \Delta r$, where $\Delta r \ll r$, has the very slightly different period $T + \Delta T$.

a. Show that

$$\frac{\Delta T}{T} = \frac{3}{2}\frac{\Delta r}{r}$$

b. Two earth satellites are in parallel orbits with radii 6700 km and 6701 km. One day they pass each other, 1 km apart, along a line radially outward from the earth. How long will it be until they are again 1 km apart?

69. ‖‖ While visiting Planet Physics, you toss a rock straight up at 11 m/s and catch it 2.5 s later. While you visit the surface, your cruise ship orbits at an altitude equal to the planet's radius every 230 min. What are the (a) mass and (b) radius of Planet Physics?

70. ‖‖ A moon lander is orbiting the moon at an altitude of 1000 km. By what percentage must it decrease its speed so as to just graze the moon's surface one-half period later?

71. ‖‖ Let's look in more detail at how a satellite is moved from one circular orbit to another. FIGURE CP13.71 shows two circular orbits, of radii r_1 and r_2, and an elliptical orbit that connects them. Points 1 and 2 are at the ends of the semimajor axis of the ellipse.

FIGURE CP13.71

a. A satellite moving along the elliptical orbit has to satisfy two conservation laws. Use these two laws to prove that the velocities at points 1 and 2 are

$$v_1' = \sqrt{\frac{2GM(r_2/r_1)}{r_1 + r_2}} \qquad \text{and} \qquad v_2' = \sqrt{\frac{2GM(r_1/r_2)}{r_1 + r_2}}$$

The prime indicates that these are the velocities on the elliptical orbit. Both reduce to Equation 13.22 if $r_1 = r_2 = r$.

b. Consider a 1000 kg communications satellite that needs to be boosted from an orbit 300 km above the earth to a geosynchronous orbit 35,900 km above the earth. Find the velocity v_1 on the inner circular orbit and the velocity v_1' at the low point on the elliptical orbit that spans the two circular orbits.

c. How much work must the rocket motor do to transfer the satellite from the circular orbit to the elliptical orbit?

d. Now find the velocity v_2' at the high point of the elliptical orbit and the velocity v_2 of the outer circular orbit.

e. How much work must the rocket motor do to transfer the satellite from the elliptical orbit to the outer circular orbit?

f. Compute the total work done and compare your answer to the result of Example 13.6.

72. ‖‖ FIGURE CP13.72 shows a particle of mass m at distance x from the center of a very thin cylinder of mass M and length L. The particle is outside the cylinder, so $x > L/2$.

a. Calculate the gravitational potential energy of these two masses.

b. Use what you know about the relationship between force and potential energy to find the magnitude of the gravitational force on m when it is at position x.

FIGURE CP13.72

14 Fluids and Elasticity

A defining characteristic of any fluid is that—like this waterfall—it flows.

IN THIS CHAPTER, you will learn about systems that flow or deform.

What is a fluid?

A fluid is a substance that flows. Both gases and liquids are fluids.

- A gas is compressible. The molecules move freely with few interactions.
- A liquid is incompressible. The molecules are weakly bound to one another.

Gas

Liquid

What is pressure?

Fluids exert forces on the walls of their containers. Pressure is the force-to-area ratio F/A.

- Pressure in liquids, called hydrostatic pressure, is due to gravity. Pressure increases with depth.
- Pressure in gases is primarily thermal. Pressure is constant in a container.

What is buoyancy?

Buoyancy is the upward force a fluid exerts on an object.

- Archimedes' principle says that the buoyant force equals the weight of the displaced fluid.
- An object floats if the buoyant force is sufficient to balance the object's weight.

« LOOKING BACK Section 6.1 Equilibrium

How does a fluid flow?

An ideal fluid—an incompressible, nonviscous fluid flowing smoothly—flows along streamlines. Bernoulli's equation, a statement of energy conservation, relates the pressures, speeds, and heights at two points on a streamline.

What is elasticity?

Elasticity describes how objects deform under stress. An object's

- Young's modulus characterizes how it stretches when pulled.
- Bulk modulus tells us how much it is compressed by pressure.

« LOOKING BACK Section 9.4 Restoring forces

Why are fluids important?

Gases and liquids are two of the three common states of matter. Scientists study atmospheres and oceans, while engineers use the controlled flow of fluids in a vast number of applications. This chapter will let you see how Newton's laws can be applied to systems that can flow or deform.

14.1 Fluids

Quite simply, a **fluid** is a substance that flows. Because they flow, fluids take the shape of their container rather than retaining a shape of their own. You may think that gases and liquids are quite different, but both are fluids, and their similarities are often more important than their differences.

The detailed behavior of gases and, especially, liquids can be complex. Fortunately, the essential characteristics of gases and liquids are captured in simple molecular models that are related to the ball-and-spring model of solids that we presented in « Section 5.2.

MODEL 14.1

Molecular model of gases and liquids

Gases and liquids are fluids—they flow and exert pressure.

■ **Gases**
- Molecules move freely through space.
- Molecules do not interact except for occasional collisions with each other or the walls.
- Molecules are far apart, so a gas is *compressible*.

A gas fills the container.

■ **Liquids**
- Molecules are weakly bound and stay close together.
- A liquid is *incompressible* because the molecules can't get any closer.
- Weak bonds allow the molecules to move around.

A liquid has a surface.

Volume and Density

One important parameter that characterizes a macroscopic system is its volume V, the amount of space the system occupies. The SI unit of volume is m^3. Nonetheless, both cm^3 and, to some extent, liters (L) are widely used metric units of volume. In most cases, you *must* convert these to m^3 before doing calculations.

While it is true that $1\ m = 100\ cm$, it is *not* true that $1\ m^3 = 100\ cm^3$. FIGURE 14.1 shows that the volume conversion factor is $1\ m^3 = 10^6\ cm^3$. A liter is $1000\ cm^3$, so $1\ m^3 = 10^3\ L$. A milliliter (1 mL) is the same as $1\ cm^3$.

A system is also characterized by its *density*. Suppose you have several blocks of copper, each of different size. Each block has a different mass m and a different volume V. Nonetheless, all the blocks are copper, so there should be some quantity that has the *same* value for all the blocks, telling us, "This is copper, not some other material." The most important such parameter is the *ratio* of mass to volume, which we call the **mass density** ρ (lowercase Greek rho):

$$\rho = \frac{m}{V} \quad \text{(mass density)} \tag{14.1}$$

Conversely, an object of density ρ has mass $m = \rho V$.

The SI units of mass density are kg/m^3. Nonetheless, units of g/cm^3 are widely used. You need to convert these to SI units before doing most calculations. You must convert both the grams to kilograms and the cubic centimeters to cubic meters. The net result is the conversion factor

$$1\ g/cm^3 = 1000\ kg/m^3$$

The mass density is usually called simply "the density" if there is no danger of confusion. However, we will meet other types of density as we go along, and sometimes it is important to be explicit about which density you are using. TABLE 14.1 provides a

FIGURE 14.1 There are $10^6\ cm^3$ in $1\ m^3$.

Subdivide the $1\ m \times 1\ m \times 1\ m$ cube into little cubes 1 cm on a side. You will get 100 subdivisions along each edge.

There are $100 \times 100 \times 100 = 10^6$ little $1\ cm^3$ cubes in the big $1\ m^3$ cube.

short list of mass densities of various fluids. Notice the enormous difference between the densities of gases and liquids. Gases have lower densities because the molecules in gases are farther apart than in liquids.

What does it *mean* to say that the density of gasoline is 680 kg/m³ or, equivalently, 0.68 g/cm³? Density is a mass-to-volume ratio. It is often described as the "mass per unit volume," but for this to make sense you have to know what is meant by "unit volume." Regardless of which system of length units you use, a **unit volume** is one of those units cubed. For example, if you measure lengths in meters, a unit volume is 1 m³. But 1 cm³ is a unit volume if you measure lengths in centimeters, and 1 mi³ is a unit volume if you measure lengths in miles.

Density is the mass of one unit of volume, whatever the units happen to be. To say that the density of gasoline is 680 kg/m³ is to say that the mass of 1 m³ of gasoline is 680 kg. The mass of 1 cm³ of gasoline is 0.68 g, so the density of gasoline in those units is 0.68 g/cm³.

The mass density is independent of the object's size. Mass and volume are parameters that characterize a *specific piece* of some substance—say copper—whereas the mass density characterizes the substance itself. All pieces of copper have the same mass density, which differs from the mass density of any other substance.

TABLE 14.1 Densities of fluids at standard temperature (0°C) and pressure (1 atm)

Substance	ρ (kg/m³)
Helium gas	0.18
Air	1.29
Gasoline	680
Ethyl alcohol	790
Benzene	880
Oil (typical)	900
Water	1000
Seawater	1030
Glycerin	1260
Mercury	13,600

EXAMPLE 14.1 | **Weighing the air**

What is the mass of air in a living room with dimensions 4.0 m × 6.0 m × 2.5 m?

MODEL Table 14.1 gives air density at a temperature of 0°C. The air density doesn't vary significantly over a small range of temperatures (we'll study this issue in a later chapter), so we'll use this value even though most people keep their living room warmer than 0°C.

SOLVE The room's volume is

$$V = (4.0 \text{ m}) \times (6.0 \text{ m}) \times (2.5 \text{ m}) = 60 \text{ m}^3$$

The mass of the air is

$$m = \rho V = (1.29 \text{ kg/m}^3)(60 \text{ m}^3) = 77 \text{ kg}$$

ASSESS This is perhaps more mass than you might have expected from a substance that hardly seems to be there. For comparison, a swimming pool this size would contain 60,000 kg of water.

STOP TO THINK 14.1 A piece of glass is broken into two pieces of different size. Rank in order, from largest to smallest, the mass densities of pieces a, b, and c.

14.2 Pressure

"Pressure" is a word we all know and use. You probably have a commonsense idea of what pressure is. For example, you feel the effects of varying pressure against your eardrums when you swim underwater or take off in an airplane. Cans of whipped cream are "pressurized" to make the contents squirt out when you press the nozzle. It's hard to open a "vacuum sealed" jar of jelly the first time, but easy after the seal is broken.

You've probably seen water squirting out of a hole in the side of a container, as in FIGURE 14.2. Notice that the water emerges at greater speed from a hole at greater depth. And you've probably felt the air squirting out of a hole in a bicycle tire or inflatable air mattress. These observations suggest that

- "Something" pushes the water or air *sideways,* out of the hole.
- In a liquid, the "something" is larger at greater depths. In a gas, the "something" appears to be the same everywhere.

Our goal is to turn these everyday observations into a precise definition of pressure.

FIGURE 14.2 Water pressure pushes the water *sideways*, out of the holes.

FIGURE 14.3 The fluid presses against area A with force $\vec{F}$.

The fluid pushes with force $\vec{F}$ against area A.

FIGURE 14.3 shows a fluid—either a liquid or a gas—pressing against a small area A with force $\vec{F}$. This is the force that pushes the fluid out of a hole. In the absence of a hole, $\vec{F}$ pushes against the wall of the container. Let's define the **pressure** at this point in the fluid to be the ratio of the force to the area on which the force is exerted:

$$p = \frac{F}{A} \tag{14.2}$$

Notice that pressure is a scalar, not a vector. You can see, from Equation 14.2, that a fluid exerts a force of magnitude

$$F = pA \tag{14.3}$$

on a surface of area A. The force is *perpendicular* to the surface.

> **NOTE** Pressure itself is *not* a force, even though we sometimes talk informally about "the force exerted by the pressure." The correct statement is that the *fluid* exerts a force on a surface.

From its definition, pressure has units of N/m^2. The SI unit of pressure is the **pascal,** defined as

$$1 \text{ pascal} = 1 \text{ Pa} \equiv 1 \text{ N/m}^2$$

This unit is named for the 17th-century French scientist Blaise Pascal, who was one of the first to study fluids. Large pressures are often given in kilopascals, where $1 \text{ kPa} = 1000 \text{ Pa}$.

Equation 14.2 is the basis for the simple pressure-measuring device shown in **FIGURE 14.4a**. Because the spring constant k and the area A are known, we can determine the pressure by measuring the compression of the spring. Once we've built such a device, we can place it in various liquids and gases to learn about pressure. **FIGURE 14.4b** shows what we can learn from a series of simple experiments.

FIGURE 14.4 Learning about pressure.

(a) Piston attached to spring

Vacuum; no fluid force is exerted on the piston from this side.

1. The fluid exerts force $\vec{F}$ on a piston with surface area A.

2. The force compresses the spring. Because the spring constant k is known, we can use the spring's compression to find F.

3. Because A is known, we can find the pressure from $p = F/A$.

(b) Pressure-measuring device in fluid

1. There is pressure *everywhere* in a fluid, not just at the bottom or at the walls of the container.

2. The pressure at one point in the fluid is the same whether you point the pressure-measuring device up, down, or sideways. The fluid pushes up, down, and sideways with equal strength.

3. In a *liquid*, the pressure increases with depth below the surface. In a *gas*, the pressure is nearly the same at all points (at least in laboratory-size containers).

The first statement in Figure 14.4b is especially important. Pressure exists at *all* points within a fluid, not just at the walls of the container. You may recall that tension exists at *all* points in a string, not only at its ends where it is tied to an object. We understood tension as the different parts of the string *pulling* against each other. Pressure is an analogous idea, except that the different parts of a fluid are *pushing* against each other.

Causes of Pressure

Gases and liquids are both fluids, but they have some important differences. Liquids are nearly incompressible; gases are highly compressible. The molecules in a liquid attract each other via molecular bonds; the molecules in a gas do not interact other than through occasional collisions. These differences affect how we think about pressure in gases and liquids.

Imagine that you have two sealed jars, each containing a small amount of mercury and nothing else. All the air has been removed from the jars. Suppose you take the two jars into orbit on the space station, where they are weightless. One jar you keep cool, so that the mercury is a liquid. The other you heat until the mercury becomes a gas. What can we say about the pressure in these two jars?

As **FIGURE 14.5** shows, molecular bonds hold the liquid mercury together. It might quiver like Jello, but it remains a cohesive drop floating in the center of the jar. The liquid drop exerts no forces on the walls, so there's *no* pressure in the jar containing the liquid. (If we actually did this experiment, a very small fraction of the mercury would be in the vapor phase and create what is called *vapor pressure.*)

The gas is different. The gas molecules collide with the wall of the container, and each bounce exerts a force on the wall. The force from any one collision is incredibly small, but there are an extraordinarily large number of collisions every second. These collisions cause the gas to have a pressure. We will do the calculation in Chapter 20.

FIGURE 14.6 shows the jars back on earth. Because of gravity, the liquid now fills the bottom of the jar and exerts a force on the bottom and the sides. Liquid mercury is incompressible, so the volume of liquid in Figure 14.6 is the same as in Figure 14.5. There is still no pressure on the top of the jar (other than the very small vapor pressure).

At first glance, the situation in the gas-filled jar seems unchanged from Figure 14.5. However, the earth's gravitational pull causes the gas density to be *slightly* more at the bottom of the jar than at the top. Because the pressure due to collisions is proportional to the density, the pressure is *slightly* larger at the bottom of the jar than at the top.

Thus there appear to be two contributions to the pressure in a container of fluid:

1. A *gravitational contribution* that arises from gravity pulling down on the fluid. Because a fluid can flow, forces are exerted on both the bottom and sides of the container. The gravitational contribution depends on the strength of the gravitational force.
2. A *thermal contribution* due to the collisions of freely moving gas molecules with the walls. The thermal contribution depends on the absolute temperature of the gas.

A detailed analysis finds that these two contributions are not entirely independent of each other, but the distinction is useful for a basic understanding of pressure. Let's see how these two contributions apply to different situations.

Pressure in Gases

The pressure in a laboratory-size container of gas is due almost entirely to the thermal contribution. A container would have to be $\approx$ 100 m tall for gravity to cause the pressure at the top to be even 1% less than the pressure at the bottom. Laboratory-size containers are much less than 100 m tall, so we can quite reasonably assume that p has the *same* value at all points in a laboratory-size container of gas.

Decreasing the number of molecules in a container decreases the gas pressure simply because there are fewer collisions with the walls. If a container is completely empty, with no atoms or molecules, then the pressure is $p = 0$ Pa. This is a *perfect vacuum*. No perfect vacuum exists in nature, not even in the most remote depths of outer space, because it is impossible to completely remove every atom from a region of space. In practice, a **vacuum** is an enclosed space in which $p \ll 1$ atm. Using $p = 0$ Pa is then a very good approximation.

Atmospheric Pressure

The earth's atmosphere is *not* a laboratory-size container. The height of the atmosphere is such that the gravitational contribution to pressure *is* important. As **FIGURE 14.7** shows, the density of air slowly decreases with increasing height until approaching

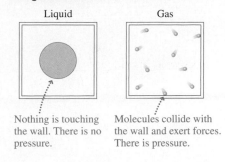

FIGURE 14.5 A liquid and a gas in a weightless environment.

Liquid   Gas

Nothing is touching the wall. There is no pressure.

Molecules collide with the wall and exert forces. There is pressure.

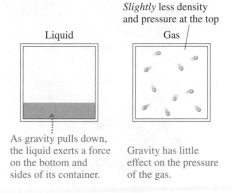

FIGURE 14.6 Gravity affects the pressure of the fluids.

Slightly less density and pressure at the top

Liquid   Gas

As gravity pulls down, the liquid exerts a force on the bottom and sides of its container.

Gravity has little effect on the pressure of the gas.

FIGURE 14.7 The pressure and density decrease with increasing height in the atmosphere.

Space

3. The density and pressure approach zero in outer space.

Walls of an imaginary container

2. Because of gravity, the density and pressure decrease with increasing height.

Air

1. The air's density and pressure are greatest at the earth's surface.

Earth

zero in the vacuum of space. Consequently, the pressure of the air, what we call the *atmospheric pressure* p_{atmos}, decreases with height. The air pressure is less in Denver than in Miami.

The atmospheric pressure *at sea level* varies slightly with the weather, but the global average sea-level pressure is 101,300 Pa. Consequently, we define the **standard atmosphere** as

$$1 \text{ standard atmosphere} = 1 \text{ atm} \equiv 101,300 \text{ Pa} = 101.3 \text{ kPa}$$

The standard atmosphere, usually referred to simply as "atmospheres," is a commonly used unit of pressure. But it is not an SI unit, so you must convert atmospheres to pascals before doing most calculations with pressure.

> **NOTE** Unless you happen to live right at sea level, the atmospheric pressure around you is somewhat less than 1 atm. Pressure experiments use a barometer to determine the actual atmospheric pressure. For simplicity, this textbook will always assume that the pressure of the air is $p_{atmos} = 1$ atm unless stated otherwise.

Given that the pressure of the air at sea level is 101.3 kPa, you might wonder why the weight of the air doesn't crush your forearm when you rest it on a table. Your forearm has a surface area of $\approx 200 \text{ cm}^2 = 0.02 \text{ m}^2$, so the force of the air pressing against it is $\approx 2000 \text{ N}$ (≈ 450 pounds). How can you even lift your arm?

The reason, as **FIGURE 14.8** shows, is that a fluid exerts pressure forces in *all* directions. There *is* a downward force of $\approx 2000 \text{ N}$ on your forearm, but the air underneath your arm exerts an upward force of the same magnitude. The *net* force is very close to zero. (To be accurate, there is a net *upward* force called the buoyant force. We'll study buoyancy in Section 14.4. The buoyant force of the air is usually too small to notice.)

But, you say, there isn't any air under my arm if I rest it on a table. Actually, there is. There would be a *vacuum* under your arm if there were no air. Imagine placing your arm on the top of a large vacuum cleaner suction tube. What happens? You feel a downward force as the vacuum cleaner "tries to suck your arm in." However, the downward force you feel is not a *pulling* force from the vacuum cleaner. It is the *pushing* force of the air above your arm *when the air beneath your arm is removed and cannot push back*. Air molecules do not have hooks! They have no ability to "pull" on your arm. The air can only push.

Vacuum cleaners, suction cups, and other similar devices are powerful examples of how strong atmospheric pressure forces can be *if* the air is removed from one side of an object so as to produce an unbalanced force. The fact that we are *surrounded* by the fluid allows us to move around in the air, just as we swim underwater, oblivious of these strong forces.

FIGURE 14.8 Pressure forces in a fluid push with equal strength in all directions.

The forces of a fluid push in *all* directions.

EXAMPLE 14.2 | A suction cup

A 10.0-cm-diameter suction cup is pushed against a smooth ceiling. What is the maximum mass of an object that can be suspended from the suction cup without pulling it off the ceiling? The mass of the suction cup is negligible.

MODEL Pushing the suction cup against the ceiling pushes the air out. We'll assume that the volume enclosed between the suction cup and the ceiling is a perfect vacuum with $p = 0$ Pa. We'll also assume that the pressure in the room is 1 atm.

VISUALIZE FIGURE 14.9 shows a free-body diagram of the suction cup stuck to the ceiling. The downward normal force of the ceiling is distributed around the rim of the suction cup, but in the particle model we can show this as a single force vector.

FIGURE 14.9 A suction cup is held to the ceiling by air pressure pushing upward on the bottom.

SOLVE The suction cup remains stuck to the ceiling, in static equilibrium, as long as $F_{air} = n + F_G$. The magnitude of the upward force exerted by the air is

$$F_{air} = pA = p\pi r^2 = (101{,}300 \text{ Pa})\pi(0.050 \text{ m})^2 = 796 \text{ N}$$

There is no downward force from the air in this case because there is no air inside the cup. Increasing the hanging mass decreases the normal force n by an equal amount. The maximum weight has been reached when n is reduced to zero. Thus

$$(F_G)_{max} = mg = F_{air} = 796 \text{ N}$$

$$m = \frac{796 \text{ N}}{g} = 81 \text{ kg}$$

ASSESS The suction cup can support a mass of up to 81 kg if all the air is pushed out, leaving a perfect vacuum inside. A real suction cup won't achieve a perfect vacuum, but suction cups can hold substantial weight.

Pressure in Liquids

Gravity causes a liquid to fill the bottom of a container. Thus it's not surprising that the pressure in a liquid is due almost entirely to the gravitational contribution. We'd like to determine the pressure at depth d below the surface of the liquid. We will assume that the liquid is at rest; flowing liquids will be considered later in this chapter.

The shaded cylinder of liquid in **FIGURE 14.10** extends from the surface to depth d. This cylinder, like the rest of the liquid, is in static equilibrium with $\vec{F}_{net} = \vec{0}$. Three forces act on this cylinder: the gravitational force mg on the liquid in the cylinder, a downward force $p_0 A$ due to the pressure p_0 at the surface of the liquid, and an upward force pA due to the liquid beneath the cylinder pushing up on the bottom of the cylinder. This third force is a consequence of our earlier observation that different parts of a fluid push against each other. Pressure p, which is what we're trying to find, is the pressure at the bottom of the cylinder.

The upward force balances the two downward forces, so

$$pA = p_0 A + mg \qquad (14.4)$$

The liquid is a cylinder of cross-section area A and height d. Its volume is $V = Ad$ and its mass is $m = \rho V = \rho Ad$. Substituting this expression for the mass of the liquid into Equation 14.4, we find that the area A cancels from all terms. The pressure at depth d in a liquid is

$$p = p_0 + \rho gd \qquad \text{(hydrostatic pressure at depth } d) \qquad (14.5)$$

where ρ is the liquid's density. Because the fluid is at rest, the pressure given by Equation 14.5 is called the **hydrostatic pressure.** The fact that g appears in Equation 14.5 reminds us that this is a gravitational contribution to the pressure.

As expected, $p = p_0$ at the surface, where $d = 0$. Pressure p_0 is often due to the air or other gas above the liquid. $p_0 = 1$ atm $= 101.3$ kPa for a liquid that is open to the air. However, p_0 can also be the pressure due to a piston or a closed surface pushing down on the top of the liquid.

NOTE Equation 14.5 assumes that the liquid is *incompressible;* that is, its density ρ doesn't increase with depth. This is an excellent assumption for liquids, but not a good one for a gas, which *is* compressible.

FIGURE 14.10 Measuring the pressure at depth d in a liquid.

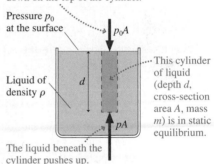

Whatever is above the liquid pushes down on the top of the cylinder.

Pressure p_0 at the surface → $p_0 A$

Liquid of density ρ · · · d

This cylinder of liquid (depth d, cross-section area A, mass m) is in static equilibrium.

The liquid beneath the cylinder pushes up. Pressure at depth d is p. ↗ pA

Free-body diagram of the column of liquid

$\vec{F}_{net} = \vec{0}$

pA (up), mg (down), $p_0 A$ (down)

EXAMPLE 14.3 | The pressure on a submarine

A submarine cruises at a depth of 300 m. What is the pressure at this depth? Give the answer in both pascals and atmospheres.

SOLVE The density of seawater, from Table 14.1, is $\rho = 1030 \text{ kg/m}^3$. The pressure at depth $d = 300$ m is found from Equation 14.5 to be

$$p = p_0 + \rho gd = 1.013 \times 10^5 \text{ Pa}$$
$$+ (1030 \text{ kg/m}^3)(9.80 \text{ m/s}^2)(300 \text{ m}) = 3.13 \times 10^6 \text{ Pa}$$

Converting the answer to atmospheres gives

$$p = 3.13 \times 10^6 \text{ Pa} \times \frac{1 \text{ atm}}{1.013 \times 10^5 \text{ Pa}} = 30.9 \text{ atm}$$

ASSESS The pressure deep in the ocean is very large. Windows on submersibles must be very thick to withstand the large forces.

FIGURE 14.11 Some properties of a liquid in hydrostatic equilibrium are not what you might expect.

(a)

(b)

The hydrostatic pressure in a liquid depends only on the depth and the pressure at the surface. This observation has some important implications. **FIGURE 14.11a** shows two connected tubes. It's certainly true that the larger volume of liquid in the wide tube weighs more than the liquid in the narrow tube. You might think that this extra weight would push the liquid in the narrow tube higher than in the wide tube. But it doesn't. If d_1 were larger than d_2, then, according to the hydrostatic pressure equation, the pressure at the bottom of the narrow tube would be higher than the pressure at the bottom of the wide tube. This *pressure difference* would cause the liquid to *flow* from right to left until the heights were equal.

Thus a first conclusion: **A connected liquid in hydrostatic equilibrium rises to the same height in all open regions of the container.**

FIGURE 14.11b shows two connected tubes of different shape. The conical tube holds more liquid above the dotted line, so you might think that $p_1 > p_2$. But it isn't. Both points are at the same depth, thus $p_1 = p_2$. If p_1 were larger than p_2, the pressure at the bottom of the left tube would be larger than the pressure at the bottom of the right tube. This would cause the liquid to flow until the pressures were equal.

Thus a second conclusion: **The pressure is the same at all points on a horizontal line through a connected liquid in hydrostatic equilibrium.**

> **NOTE** Both of these conclusions are restricted to liquids in hydrostatic equilibrium. The situation is different for flowing fluids, as we'll see later in the chapter.

EXAMPLE 14.4 | **Pressure in a closed tube**

Water fills the tube shown in **FIGURE 14.12**. What is the pressure at the top of the closed tube?

MODEL This is a liquid in hydrostatic equilibrium. The closed tube is not an open region of the container, so the water cannot rise to an equal height. Nevertheless, the pressure is still the same at all points on a horizontal line. In particular, the pressure at the top of the closed tube equals the

FIGURE 14.12 A water-filled tube.

100 cm

Closed

40 cm

pressure in the open tube at the height of the dashed line. Assume $p_0 = 1.00$ atm.

SOLVE A point 40 cm above the bottom of the open tube is at a depth of 60 cm. The pressure at this depth is

$$p = p_0 + \rho g d$$
$$= 1.013 \times 10^5 \text{ Pa} + (1000 \text{ kg/m}^3)(9.80 \text{ m/s}^2)(0.60 \text{ m})$$
$$= 1.072 \times 10^5 \text{ Pa} = 1.06 \text{ atm}$$

This is the pressure at the top of the closed tube.

ASSESS The water in the open tube *pushes* the water in the closed tube up against the top of the tube, which is why the pressure is greater than 1 atm.

We can draw one more conclusion from the hydrostatic pressure equation $p = p_0 + \rho g d$. If we change the pressure p_0 at the surface to p_1, the pressure at depth d becomes $p' = p_1 + \rho g d$. The *change* in pressure $\Delta p = p_1 - p_0$ is the same at all points in the fluid, independent of the size or shape of the container. This idea, that **a change in the pressure at one point in an incompressible fluid appears undiminished at all points in the fluid**, was first recognized by Blaise Pascal and is called **Pascal's principle.**

For example, if we compressed the air above the open tube in Example 14.4 to a pressure of 1.5 atm, an increase of 0.5 atm, the pressure at the top of the closed tube would increase to 1.56 atm. Pascal's principle is the basis for hydraulic systems, as we'll see in the next section.

STOP TO THINK 14.2 Water is slowly poured into the container until the water level has risen into tubes A, B, and C. The water doesn't overflow from any of the tubes. How do the water depths in the three columns compare to each other?

a. $d_A > d_B > d_C$
b. $d_A < d_B < d_C$
c. $d_A = d_B = d_C$
d. $d_A = d_C > d_B$
e. $d_A = d_C < d_B$

Water

14.3 Measuring and Using Pressure

The pressure in a fluid is measured with a *pressure gauge*. The fluid pushes against some sort of spring, and the spring's displacement is registered by a pointer on a dial.

Many pressure gauges, such as tire gauges and the gauges on air tanks, measure not the actual or absolute pressure p but what is called **gauge pressure.** The gauge pressure, denoted p_g, is the pressure *in excess* of 1 atm. That is,

$$p_g = p - 1 \text{ atm} \qquad (14.6)$$

You must add 1 atm = 101.3 kPa to the reading of a pressure gauge to find the absolute pressure p that you need for doing most science or engineering calculations: $p = p_g + 1$ atm.

A tire-pressure gauge reads the gauge pressure p_g, not the absolute pressure p.

EXAMPLE 14.5 | **An underwater pressure gauge**

An underwater pressure gauge reads 60 kPa. What is its depth?

MODEL The gauge reads gauge pressure, not absolute pressure.

SOLVE The hydrostatic pressure at depth d, with $p_0 = 1$ atm, is $p = 1$ atm $+ \rho g d$. Thus the gauge pressure is

$$p_g = p - 1 \text{ atm} = (1 \text{ atm} + \rho g d) - 1 \text{ atm} = \rho g d$$

The term $\rho g d$ is the pressure *in excess* of atmospheric pressure and thus *is* the gauge pressure. Solving for d, we find

$$d = \frac{60,000 \text{ Pa}}{(1000 \text{ kg/m}^3)(9.80 \text{ m/s}^2)} = 6.1 \text{ m}$$

Solving Hydrostatic Problems

We now have enough information to formulate a set of rules for thinking about hydrostatic problems.

TACTICS BOX 14.1 (MP)

Hydrostatics

❶ **Draw a picture.** Show open surfaces, pistons, boundaries, and other features that affect pressure. Include height and area measurements and fluid densities. Identify the points at which you need to find the pressure.

❷ **Determine the pressure at surfaces.**

- **Surface open to the air:** $p_0 = p_{atmos}$, usually 1 atm.
- **Surface covered by a gas:** $p_0 = p_{gas}$.
- **Closed surface:** $p = F/A$, where F is the force the surface, such as a piston, exerts on the fluid.

❸ **Use horizontal lines.** Pressure in a connected fluid is the same at any point along a horizontal line.

❹ **Allow for gauge pressure.** Pressure gauges read $p_g = p - 1$ atm.

❺ **Use the hydrostatic pressure equation.** $p = p_0 + \rho g d$.

Exercises 4–13

Manometers and Barometers

Gas pressure is sometimes measured with a device called a *manometer*. A manometer, shown in **FIGURE 14.13**, is a U-shaped tube connected to the gas at one end and open to the air at the other end. The tube is filled with a liquid—usually mercury—of density ρ. The liquid is in static equilibrium. A scale allows the user to measure the height h of the right side above the left side.

Steps 1–3 from Tactics Box 14.1 lead to the conclusion that the pressures p_1 and p_2 must be equal. Pressure p_1, at the surface on the left, is simply the gas pressure: $p_1 = p_{gas}$. Pressure p_2 is the hydrostatic pressure at depth $d = h$ in the liquid on the right: $p_2 = 1$ atm $+ \rho g h$. Equating these two pressures gives

$$p_{gas} = 1 \text{ atm} + \rho g h \qquad (14.7)$$

FIGURE 14.13 A manometer is used to measure gas pressure.

❶ Draw a picture.

❷ This is an open surface, so $p_0 = 1$ atm.

Liquid, density ρ

This is a surface covered by a gas, so $p_0 = p_{gas}$.

Gas at pressure p_{gas}

h

1 2

p_1 p_2

❸ Points 1 and 2 are on a horizontal line, so $p_1 = p_2$.

Figure 14.13 assumed $p_{gas} > 1$ atm, so the right side of the liquid is higher than the left. Equation 14.7 is also valid for $p_{gas} < 1$ atm if the distance of the right side *below* the left side is considered to be a negative value of h.

EXAMPLE 14.6 | Using a manometer

The pressure of a gas cell is measured with a mercury manometer. The mercury is 36.2 cm higher in the outside arm than in the arm connected to the gas cell.

a. What is the gas pressure?

b. What is the reading of a pressure gauge attached to the gas cell?

SOLVE a. From Table 14.1, the density of mercury is $\rho = 13,600$ kg/m³. Equation 14.7 with $h = 0.362$ m gives

$$p_{gas} = 1 \text{ atm} + \rho gh = 149.5 \text{ kPa}$$

We had to change 1 atm to 101,300 Pa before adding. Converting the result to atmospheres, we have $p_{gas} = 1.476$ atm.

b. The pressure gauge reads gauge pressure: $p_g = p - 1$ atm $= 0.476$ atm or 48.2 kPa.

FIGURE 14.14 A barometer.

(a) Seal and invert tube.

Liquid, density ρ

(b)

Vacuum (zero pressure)

$p_2 = \rho gh$

$p_1 = p_{atmos}$

h

1 2

Another important pressure-measuring instrument is the *barometer,* which is used to measure the atmospheric pressure p_{atmos}. **FIGURE 14.14a** shows a glass tube, sealed at the bottom, that has been completely filled with a liquid. If we temporarily seal the top end, we can invert the tube and place it in a beaker of the same liquid. When the temporary seal is removed, some, but not all, of the liquid runs out, leaving a liquid column in the tube that is a height h above the surface of the liquid in the beaker. This device, shown in **FIGURE 14.14b**, is a barometer. What does it measure? And why doesn't *all* the liquid in the tube run out?

We can analyze the barometer much as we did the manometer. Points 1 and 2 in Figure 14.14b are on a horizontal line drawn even with the surface of the liquid. The liquid is in hydrostatic equilibrium, so the pressure at these two points must be equal. Liquid runs out of the tube only until a balance is reached between the pressure at the base of the tube and the pressure of the air.

You can think of a barometer as rather like a seesaw. If the pressure of the atmosphere increases, it presses down on the liquid in the beaker. This forces liquid up the tube until the pressures at points 1 and 2 are equal. If the atmospheric pressure falls, liquid has to flow out of the tube to keep the pressures equal at these two points.

The pressure at point 2 is the pressure due to the weight of the liquid in the tube plus the pressure of the gas above the liquid. But in this case there is no gas above the liquid! Because the tube had been completely full of liquid when it was inverted, the space left behind when the liquid ran out is a vacuum (ignoring a very slight *vapor pressure* of the liquid, negligible except in extremely precise measurements). Thus pressure p_2 is simply $p_2 = \rho gh$.

Equating p_1 and p_2 gives

$$p_{atmos} = \rho gh \qquad (14.8)$$

Thus we can measure the atmosphere's pressure by measuring the height of the liquid column in a barometer.

The average air pressure at sea level causes a column of mercury in a mercury barometer to stand 760 mm above the surface. Knowing that the density of mercury is 13,600 kg/m³ (at 0°C), we can use Equation 14.8 to find that the average atmospheric pressure is

$$p_{atmos} = \rho_{Hg}gh = (13,600 \text{ kg/m}^3)(9.80 \text{ m/s}^2)(0.760 \text{ m})$$
$$= 1.013 \times 10^5 \text{ Pa} = 101.3 \text{ kPa}$$

This is the value given earlier as "one standard atmosphere."

The barometric pressure varies slightly from day to day as the weather changes. Weather systems are called *high-pressure systems* or *low-pressure systems,* depending on whether the local sea-level pressure is higher or lower than one standard atmosphere. Higher pressure is usually associated with fair weather, while lower pressure portends rain.

Pressure Units

In practice, pressure is measured in several different units. This plethora of units and abbreviations has arisen historically as scientists and engineers working on different subjects (liquids, high-pressure gases, low-pressure gases, weather, etc.) developed what seemed to them the most convenient units. These units continue in use through tradition, so it is necessary to become familiar with converting back and forth between them. TABLE 14.2 gives the basic conversions.

TABLE 14.2 Pressure units

Unit	Abbreviation	Conversion to 1 atm	Uses
pascal	Pa	101.3 kPa	SI unit: $1\ Pa = 1\ N/m^2$
atmosphere	atm	1 atm	general
millimeters of mercury	mm of Hg	760 mm of Hg	gases and barometric pressure
inches of mercury	in	29.92 in	barometric pressure in U.S. weather forecasting
pounds per square inch	psi	14.7 psi	engineering and industry

Blood Pressure

The last time you had a medical checkup, the doctor may have told you something like "Your blood pressure is 120 over 80." What does that mean?

About every 0.8 s, assuming a pulse rate of 75 beats per minute, your heart "beats." The heart muscles contract and push blood out into your aorta. This contraction, like squeezing a balloon, raises the pressure in your heart. The pressure increase, in accordance with Pascal's principle, is transmitted through all your arteries.

FIGURE 14.15 is a pressure graph showing how blood pressure changes during one cycle of the heartbeat. The medical condition of *high blood pressure* usually means that your resting systolic pressure is higher than necessary for blood circulation. The high pressure causes undue stress and strain on your entire circulatory system, often leading to serious medical problems. Low blood pressure can cause you to get dizzy if you stand up quickly because the pressure isn't adequate to pump the blood up to your brain.

Blood pressure is measured with a cuff that goes around your arm. The doctor or nurse pressurizes the cuff, places a stethoscope over the artery in your arm, then slowly releases the pressure while watching a pressure gauge. Initially, the cuff squeezes the artery shut and cuts off the blood flow. When the cuff pressure drops below the systolic pressure, the pressure pulse during each beat of your heart forces the artery open briefly and a squirt of blood goes through. You can feel this, and the doctor or nurse records the pressure when she hears the blood start to flow. This is your systolic pressure.

This pulsing of the blood through your artery lasts until the cuff pressure reaches the diastolic pressure. Then the artery remains open continuously and the blood flows smoothly. This transition is easily heard in the stethoscope, and the doctor or nurse records your diastolic pressure.

Blood pressure is measured in millimeters of mercury. And it is a gauge pressure, the pressure in excess of 1 atm. A fairly typical blood pressure of a healthy young adult is 120/80, meaning that the systolic pressure is $p_g = 120$ mm of Hg (absolute pressure $p = 880$ mm of Hg) and the diastolic pressure is 80 mm of Hg.

FIGURE 14.15 Blood pressure during one cycle of a heartbeat.

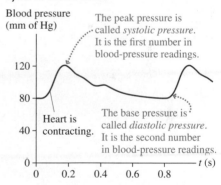

The Hydraulic Lift

The use of pressurized liquids to do useful work is a technology known as **hydraulics.** Pascal's principle is the fundamental idea underlying hydraulic devices. If you increase the pressure at one point in a liquid by pushing a piston in, that pressure increase is transmitted to all points in the liquid. A second piston at some other point in the fluid can then push outward and do useful work.

The brake system in your car is a hydraulic system. Stepping on the brake pushes a piston into the *master brake cylinder* and increases the pressure in the *brake fluid*. The fluid itself hardly moves, but the pressure increase is transmitted to the four wheels where it pushes the brake pads against the spinning brake disk. You've used a pressurized liquid to achieve the useful goal of stopping your car.

One advantage of hydraulic systems over simple mechanical linkages is the possibility of *force multiplication*. To see how this works, we'll analyze a *hydraulic lift*, such as the one that lifts your car at the repair shop. **FIGURE 14.16a** shows force $\vec{F}_2$, due to the weight of the car, pressing down on a liquid via a piston of area A_2. A much smaller force $\vec{F}_1$ presses down on a piston of area A_1. Can this system possibly be in equilibrium?

As you now know, the hydrostatic pressure is the same at all points along a horizontal line through a fluid. Consider the line passing through the liquid/piston interface on the left in Figure 14.16a. Pressures p_1 and p_2 must be equal, thus

$$p_0 + \frac{F_1}{A_1} = p_0 + \frac{F_2}{A_2} + \rho g h \tag{14.9}$$

The atmosphere presses equally on both sides, so p_0 cancels. The system is in static equilibrium if

$$F_2 = \frac{A_2}{A_1} F_1 - \rho g h A_2 \tag{14.10}$$

NOTE Force $\vec{F}_2$ is the force of the heavy object pushing *down* on the liquid. According to Newton's third law, the liquid pushes *up* on the object with a force of equal magnitude. Thus F_2 in Equation 14.10 is the "lifting force."

Suppose we need to lift the car higher. If piston 1 is pushed down distance d_1, as in **FIGURE 14.16b**, it displaces volume $V_1 = A_1 d_1$ of liquid. Because the liquid is incompressible, V_1 must equal the volume $V_2 = A_2 d_2$ added beneath piston 2 as it rises distance d_2. That is,

$$d_2 = \frac{d_1}{A_2/A_1} \tag{14.11}$$

The distance is *divided* by the same factor as that by which force is multiplied. A small force may be able to support a heavy weight, but you have to push the small piston a large distance to raise the heavy weight by a small amount.

This conclusion is really just a statement of energy conservation. Work is done *on* the liquid by a small force pushing the liquid through a large displacement. Work is done *by* the liquid when it lifts the heavy weight through a small distance. A full analysis must consider the fact that the gravitational potential energy of the liquid is also changing, so we can't simply equate the output work to the input work, but you can see that energy considerations require piston 1 to move farther than piston 2.

FIGURE 14.16 A hydraulic lift.

(a)

Pressure p_1 is due to atmospheric pressure p_0 *plus* pressure F_1/A_1, due to $\vec{F}_1$.

Pressure p_2 is p_0 *plus* F_2/A_2 *plus* $\rho g h$ from the liquid column of height h.

(b)

Because the fluid is incompressible, $A_1 d_1 = A_2 d_2$.

EXAMPLE 14.7 | Lifting a car

The hydraulic lift at a car repair shop is filled with oil. The car rests on a 25-cm-diameter piston. To lift the car, compressed air is used to push down on a 6.0-cm-diameter piston. What does the pressure gauge read when a 1300 kg car is 2.0 m above the compressed-air piston?

MODEL Assume that the oil is incompressible. Its density, from Table 14.1, is 900 kg/m³.

SOLVE F_2 is the weight of the car pressing down on the piston: $F_2 = mg = 12{,}700$ N. The piston areas are $A_1 = \pi(0.030 \text{ m})^2 = 0.00283 \text{ m}^2$ and $A_2 = \pi(0.125 \text{ m})^2 = 0.0491 \text{ m}^2$. The force required to hold the car at height $h = 2.0$ m is found by solving Equation 14.10 for F_1:

$$F_1 = \frac{A_1}{A_2} F_2 + \rho g h A_1$$

$$= \frac{0.00283 \text{ m}^2}{0.0491 \text{ m}^2} \times 12{,}700 \text{ N} + (900 \text{ kg/m}^3)(9.8 \text{ m/s}^2)(2.0 \text{ m})(0.00283 \text{ m}^2)$$

$$= 782 \text{ N}$$

The pressure applied to the fluid by the compressed-air piston is

$$\frac{F_1}{A_1} = \frac{782 \text{ N}}{0.00283 \text{ m}^2} = 2.76 \times 10^5 \text{ Pa} = 2.7 \text{ atm}$$

This is the pressure *in excess* of atmospheric pressure, which is what a pressure gauge measures, so the gauge reads, depending on its units, 276 kPa or 2.7 atm.

ASSESS 782 N is roughly the weight of an average adult man. The multiplication factor $A_2/A_1 = 17$ makes it quite easy for this much force to lift the car.

STOP TO THINK 14.3 Rank in order, from largest to smallest, the magnitudes of the forces $\vec{F}_a$, $\vec{F}_b$, and $\vec{F}_c$ required to balance the masses. The masses are in kilograms, the three smaller cylinders have the same diameter, and the four larger cylinders have the same diameter.

14.4 Buoyancy

A rock, as you know, sinks like a rock. Wood floats on the surface of a lake. A penny with a mass of a few grams sinks, but a massive steel aircraft carrier floats. How can we understand these diverse phenomena?

An air mattress floats effortlessly on the surface of a swimming pool. But if you've ever tried to push an air mattress underwater, you know it is nearly impossible. As you push down, the water pushes up. This net upward force of a fluid is called the **buoyant force.**

The basic reason for the buoyant force is easy to understand. **FIGURE 14.17** shows a cylinder submerged in a liquid. The pressure in the liquid increases with depth, so the pressure at the bottom of the cylinder is larger than at the top. Both cylinder ends have equal area, so force $\vec{F}_{up}$ is larger than force $\vec{F}_{down}$. (Remember that pressure forces push in *all* directions.) Consequently, the pressure in the liquid exerts a *net upward force* on the cylinder of magnitude $F_{net} = F_{up} - F_{down}$. This is the buoyant force.

The submerged cylinder illustrates the idea in a simple way, but the result is not limited to cylinders or to liquids. Suppose we isolate a parcel of fluid of arbitrary shape and volume by drawing an imaginary boundary around it, as shown in **FIGURE 14.18a** on the next page. This parcel is in static equilibrium. Consequently, the gravitational force pulling down on the parcel must be balanced by an upward force. The upward force, which is exerted on this parcel of fluid by the surrounding fluid, is the buoyant force $\vec{F}_B$. The buoyant force matches the weight of the fluid: $F_B = mg$.

Imagine that we could somehow remove this parcel of fluid and instantaneously replace it with an object of exactly the same shape and size, as shown in **FIGURE 14.18b**. Because the buoyant force is exerted by the *surrounding* fluid, and the surrounding

FIGURE 14.17 The buoyant force arises because the fluid pressure at the bottom of the cylinder is larger than at the top.

The net force of the fluid on the cylinder is the buoyant force $\vec{F}_B$.

$F_{up} > F_{down}$ because the pressure increases with depth. Hence the fluid exerts a net upward force.

FIGURE 14.18 The buoyant force.

(a) Imaginary boundary around a parcel of fluid

$\vec{F}_B$

$\vec{F}_G$ These are equal because the parcel is in static equilibrium.

(b) Real object with same size and shape as the parcel of fluid

$\vec{F}_B$

The buoyant force on the object is the same as on the parcel of fluid because the *surrounding* fluid has not changed.

fluid hasn't changed, the buoyant force on this new object is *exactly the same* as the buoyant force on the parcel of fluid that we removed.

When an object (or a portion of an object) is immersed in a fluid, it *displaces* fluid that would otherwise fill that region of space. This fluid is called the **displaced fluid.** The displaced fluid's volume is exactly the volume of the portion of the object that is immersed in the fluid. Figure 14.18 leads us to conclude that the magnitude of the upward buoyant force matches the weight of this displaced fluid.

This idea was first recognized by the ancient Greek mathematician and scientist Archimedes, perhaps the greatest scientist of antiquity, and today we know it as *Archimedes' principle.*

Archimedes' principle A fluid exerts an upward buoyant force $\vec{F}_B$ on an object immersed in or floating on the fluid. The magnitude of the buoyant force equals the weight of the fluid displaced by the object.

Suppose the fluid has density ρ_f and the object displaces volume V_f of fluid. The mass of the displaced fluid is $m_f = \rho_f V_f$ and so its weight is $m_f g = \rho_f V_f g$. Thus Archimedes' principle in equation form is

$$F_B = \rho_f V_f g \qquad (14.12)$$

NOTE It is important to distinguish the density and volume of the displaced fluid from the density and volume of the object. To do so, we'll use subscript f for the fluid and o for the object.

EXAMPLE 14.8 | **Holding a block of wood underwater**

A 10 cm $\times$ 10 cm $\times$ 10 cm block of wood with density 700 kg/m³ is held underwater by a string tied to the bottom of the container. What is the tension in the string?

MODEL The buoyant force is given by Archimedes' principle.

VISUALIZE FIGURE 14.19 shows the forces acting on the wood.

FIGURE 14.19 The forces acting on the submerged wood.

The buoyant force pushes up on the block.

$\vec{F}_B$

Block of wood

$\vec{F}_G$

String $\vec{T}$

SOLVE The block is in equilibrium, so

$$\sum F_y = F_B - T - m_o g = 0$$

Thus the tension is $T = F_B - m_o g$. The mass of the block is $m_o = \rho_o V_o$, and the buoyant force, given by Equation 14.12, is $F_B = \rho_f V_f g$. Thus

$$T = \rho_f V_f g - \rho_o V_o g = (\rho_f - \rho_o) V_o g$$

where we've used the fact that $V_f = V_o$ for a completely submerged object. The volume is $V_o = 1000 \text{ cm}^3 = 1.0 \times 10^{-3} \text{ m}^3$, and hence the tension in the string is

$$T = ((1000 \text{ kg/m}^3) - (700 \text{ kg/m}^3))$$
$$\times (1.0 \times 10^{-3} \text{ m}^3)(9.8 \text{ m/s}^2) = 2.9 \text{ N}$$

ASSESS The tension depends on the *difference* in densities. The tension would vanish if the wood density matched the water density.

Float or Sink?

If you *hold* an object underwater and then release it, it floats to the surface, sinks, or remains "hanging" in the water. How can we predict which it will do? The net force on the object an instant after you release it is $\vec{F}_{net} = (F_B - m_o g)\hat{k}$. Whether it heads for the surface or the bottom depends on whether the buoyant force F_B is larger or smaller than the object's weight $m_o g$.

The magnitude of the buoyant force is $\rho_f V_f g$. The weight of a uniform object, such as a block of steel, is simply $\rho_o V_o g$. But a compound object, such as a scuba diver, may have pieces of varying density. If we define the **average density** to be $\rho_{avg} = m_o/V_o$, the weight of a compound object is $\rho_{avg} V_o g$.

Comparing $\rho_f V_f g$ to $\rho_{avg} V_o g$, and noting that $V_f = V_o$ for an object that is fully submerged, we see that an object floats or sinks depending on whether the fluid density ρ_f is larger or smaller than the object's average density ρ_{avg}. If the densities are equal, the object is in static equilibrium and hangs motionless. This is called **neutral buoyancy.** These conditions are summarized in Tactics Box 14.2.

TACTICS BOX 14.2

Finding whether an object floats or sinks

❶ Object sinks

$\rho_{avg} > \rho_f$

$m_o g$

An object sinks if it weighs more than the fluid it displaces—that is, if its average density is greater than the density of the fluid.

❷ Object floats

$\rho_{avg} < \rho_f$

$m_o g$

An object floats on the surface if it weighs less than the fluid it displaces—that is, if its average density is less than the density of the fluid.

❸ Neutral buoyancy

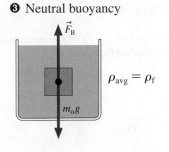

$\rho_{avg} = \rho_f$

$m_o g$

An object hangs motionless if it weighs exactly the same as the fluid it displaces—that is, if its average density equals the density of the fluid.

Exercises 14–18

As an example, steel is denser than water, so a chunk of steel sinks. Oil is less dense than water, so oil floats on water. Fish use *swim bladders* filled with air and scuba divers use weighted belts to adjust their average density to match the density of water. Both are examples of neutral buoyancy.

If you release a block of wood underwater, the net upward force causes the block to shoot to the surface. Then what? Let's begin with a *uniform* object such as the block shown in FIGURE 14.20. This object contains nothing tricky, like indentations or voids. Because it's floating, it must be the case that $\rho_o < \rho_f$.

Now that the object is floating, it's in static equilibrium. The upward buoyant force, given by Archimedes' principle, exactly balances the downward weight of the object. That is,

$$F_B = \rho_f V_f g = m_o g = \rho_o V_o g \qquad (14.13)$$

In this case, the volume of the displaced fluid is *not* the same as the volume of the object. In fact, we can see from Equation 14.13 that the volume of fluid displaced by a floating object of uniform density is

$$V_f = \frac{\rho_o}{\rho_f} V_o < V_o \qquad (14.14)$$

You've often heard it said that "90% of an iceberg is underwater." Equation 14.14 is the basis for that statement. Most icebergs break off glaciers and are fresh-water ice with a density of 917 kg/m³. The density of seawater is 1030 kg/m³. Thus

$$V_f = \frac{917 \text{ kg/m}^3}{1030 \text{ kg/m}^3} V_o = 0.89 V_o$$

V_f, the displaced water, is the volume of the iceberg that is underwater. You can see that, indeed, 89% of the volume of an iceberg is underwater.

> **NOTE** Equation 14.14 applies only to *uniform* objects. It does not apply to boats, hollow spheres, or other objects of nonuniform composition.

FIGURE 14.20 A floating object is in static equilibrium.

An object of density ρ_o and volume V_o is floating on a fluid of density ρ_f.

$\vec{F}_B$

$m_o g$

Fluid density ρ_f

The submerged volume of the object is equal to the volume V_f of displaced fluid.

About 90% of an iceberg is underwater.

EXAMPLE 14.9 | **Measuring the density of an unknown liquid**

You need to determine the density of an unknown liquid. You notice that a block floats in this liquid with 4.6 cm of the side of the block submerged. When the block is placed in water, it also floats but with 5.8 cm submerged. What is the density of the unknown liquid?

MODEL The block is an object of uniform composition.

VISUALIZE FIGURE 14.21 shows the block and defines the cross-section area A and submerged lengths h_u in the unknown liquid and h_w in water.

FIGURE 14.21 More of the block is submerged in water than in an unknown liquid.

SOLVE The block is floating, so Equation 14.14 applies. The block displaces volume $V_u = Ah_u$ of the unknown liquid. Thus

$$V_u = Ah_u = \frac{\rho_o}{\rho_u} V_o$$

Similarly, the block displaces volume $V_w = Ah_w$ of the water, leading to

$$V_w = Ah_w = \frac{\rho_o}{\rho_w} V_o$$

Because there are two fluids, we've used subscripts w for water and u for the unknown in place of the fluid subscript f. The product $\rho_o V_o$ appears in both equations; hence

$$\rho_u Ah_u = \rho_w Ah_w$$

The area A cancels, and the density of the unknown liquid is

$$\rho_u = \frac{h_w}{h_u} \rho_w = \frac{5.8 \text{ cm}}{4.6 \text{ cm}} \times 1000 \text{ kg/m}^3 = 1260 \text{ kg/m}^3$$

ASSESS Comparison with Table 14.1 shows that the unknown liquid is likely to be glycerin.

Boats

FIGURE 14.22 A physicist's boat.

We'll conclude by designing a boat. FIGURE 14.22 is a physicist's idea of a boat. Four massless but rigid walls are attached to a solid steel plate of mass m_o and area A. As the steel plate settles down into the water, the sides allow the boat to displace a volume of water much larger than that displaced by the steel alone. The boat will float if the weight of the displaced water equals the weight of the boat.

In terms of density, the boat will float if $\rho_{avg} < \rho_f$. If the sides of the boat are height h, the boat's volume is $V_o = Ah$ and its average density is $\rho_{avg} = m_o/V_o = m_o/Ah$. The boat will float if

$$\rho_{avg} = \frac{m_o}{Ah} < \rho_f \tag{14.15}$$

Thus the minimum height of the sides, a height that would allow the boat to float (in perfectly still water!) with water right up to the rails, is

$$h_{min} = \frac{m_o}{\rho_f A} \tag{14.16}$$

As a quick example, a 5 m × 10 m steel "barge" with a 2-cm-thick floor has an area of 50 m² and a mass of 7900 kg. The minimum height of the massless walls, as given by Equation 14.16, is 16 cm.

Real ships and boats are more complicated, but the same idea holds true. Whether it's made of concrete, steel, or lead, **a boat will float if its geometry allows it to displace enough water to equal the weight of the boat.**

STOP TO THINK 14.4 An ice cube is floating in a glass of water that is filled entirely to the brim. When the ice cube melts, the water level will

a. Fall.
b. Stay the same, right at the brim.
c. Rise, causing the water to spill.

14.5 Fluid Dynamics

The wind blowing through your hair, a white-water river, and oil gushing from an oil well are examples of fluids in motion. We've focused thus far on fluid statics, but it's time to turn our attention to fluid dynamics.

Fluid flow is a complex subject. Many aspects, especially turbulence and the formation of eddies, are still not well understood and are areas of current science and engineering research. We will avoid these difficulties by using a simplified model. The **ideal-fluid model** provides a good, though not perfect, description of fluid flow in many situations. It captures the essence of fluid flow while eliminating unnecessary details.

MODEL 14.2

Ideal fluid

Applies to liquids and gases. A fluid can be considered *ideal* if

- The fluid is *incompressible*.
- The fluid is *nonviscous*.
  - **Viscosity,** a fluid's resistance to flow, is analogous to kinetic friction.
  - Nonviscous flow is analogous to friction-free motion.
- The flow is *laminar*.
  - **Laminar flow** occurs when the fluid velocity at each point in the fluid is constant. The flow is smooth; it doesn't change or fluctuate.
- Limitations: The model fails if
  - the fluid has significant viscosity.
  - the flow is *turbulent* rather than laminar.

Exercise 20

FIGURE 14.23 Rising smoke changes from laminar flow to turbulent flow.

The rising smoke in the photograph of FIGURE 14.23 begins as laminar flow, recognizable by the smooth contours, but at some point undergoes a transition to turbulent flow. A laminar-to-turbulent transition is not uncommon in fluid flow. The ideal-fluid model can be applied to the laminar flow, but not to the turbulent flow.

The Equation of Continuity

FIGURE 14.24a shows smoke being used to help engineers visualize the airflow around a car in a wind tunnel. The smoothness of the flow tells us this is laminar flow. But notice also how the individual smoke trails retain their identity. They don't cross or get mixed together. Each smoke trail represents a **streamline** in the fluid. FIGURE 14.24b illustrates three important properties of streamlines.

FIGURE 14.24 Particles in an ideal fluid move along streamlines.

(a) Streamline

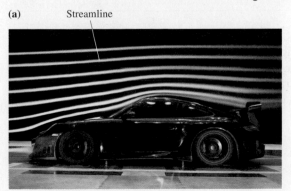

(b) 1. Streamlines never cross.

$\vec{v}$

2. Fluid particle velocity is tangent to the streamline.

3. The speed is higher where the streamlines are closer together.

FIGURE 14.25 The flow speed changes as a tube's cross-section area changes.

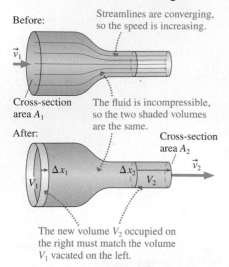

Before:

Streamlines are converging, so the speed is increasing.

$\vec{v}_1$

Cross-section area A_1

The fluid is incompressible, so the two shaded volumes are the same.

After:

Cross-section area A_2

Δx_1 Δx_2 $\vec{v}_2$

V_1 V_2

The new volume V_2 occupied on the right must match the volume V_1 vacated on the left.

FIGURE 14.26 Narrowing the cross section increases the water's speed.

A key observation is that the flow speed changes as the streamlines get closer together or farther apart. In **FIGURE 14.25**, where an ideal fluid flows through a tube, the streamlines get closer together as the tube gets narrower, so the flow speed must increase. To understand quantitatively how the speed changes, focus on the shaded volume of fluid as it moves through the tube.

An ideal fluid is incompressible, so this shaded portion of the fluid moves forward *without changing volume*. If the flow speed on the left side of the tube is v_1, then during a time interval Δt the shaded fluid moves forward distance $\Delta x_1 = v_1 \Delta t$ and "vacates" volume $V_1 = A_1 \Delta x_1 = A_1 v_1 \Delta t$. (New fluid moves in to fill this volume, but we're focusing on the shaded portion of the fluid.) On the right side of the tube, where the flow speed is v_2, the shaded fluid moves forward distance $\Delta x_2 = v_2 \Delta t$ and occupies a new volume $V_2 = A_2 \Delta x_2 = A_2 v_2 \Delta t$. These two volumes must be equal, leading to the conclusion that

$$v_1 A_1 = v_2 A_2 \tag{14.17}$$

Equation 14.17 is called the **equation of continuity**, and it is one of two important equations for the flow of an ideal fluid. The equation of continuity says that **flow is faster in narrower parts of a tube, slower in wider parts.** You're familiar with this conclusion from many everyday observations. For example, water is shot from the narrow nozzle in **FIGURE 14.26** much faster than it's flowing through the wider hose.

The quantity

$$Q = vA \tag{14.18}$$

is called the **volume flow rate.** The SI units of Q are m^3/s, although in practice Q may be measured in cm^3/s, liters per minute, or, in the United States, gallons per minute. Another way to express the meaning of the equation of continuity is to say that **the volume flow rate is constant at all points in a tube.**

EXAMPLE 14.10 | **River rafting**

A 20-m-wide, 4.0-m-deep river with a triangular cross section flows at a lazy 1.0 m/s. As it makes its way to the sea, the river enters a gorge where rocky walls confine it to a width of 4.0 m and a depth of 6.0 m, again with a triangular cross section. What is the river's volume flow rate? And what is its speed through the gorge?

MODEL Model the flowing river as an ideal fluid.

SOLVE A triangular river cross section of width w and depth d has cross-section area $A = \frac{1}{2}wd$. In the first part of the river,

$$A_1 = \frac{1}{2}(20 \text{ m})(4.0 \text{ m}) = 40 \text{ m}^2$$

and thus the river's flow rate is

$$Q = v_1 A_1 = (1.0 \text{ m/s})(40 \text{ m}^2) = 40 \text{ m}^3/\text{s}$$

The volume flow rate is the same at all points along the river, including the gorge. The gorge has cross-section area

$$A_2 = \frac{1}{2}(4.0 \text{ m})(6.0 \text{ m}) = 12 \text{ m}^2$$

Thus the river's speed through the gorge is

$$v_2 = \frac{Q}{A_2} = \frac{40 \text{ m}^3/\text{s}}{12 \text{ m}^2} = 3.3 \text{ m/s}$$

ASSESS A river speed of 3.3 m/s, or about 7 mph, is typical of a Class III whitewater-rafting river. For a 6-m-deep river, some whitewater and waves on the surface don't greatly affect our assumption of smooth flow through a perfectly triangular cross section.

STOP TO THINK 14.5 The figure shows volume flow rates (in cm^3/s) for all but one tube. What is the volume flow rate through the unmarked tube? Is the flow direction in or out?

Flows in cm^3/s

Bernoulli's Equation

The equation of continuity is one of two important relationships for ideal fluids. The other is a statement of energy conservation. The general statement of the energy principle that you learned in Chapter 10 is

$$\Delta K + \Delta U = W_{ext} \qquad (14.19)$$

where W_{ext} is the work done by any external forces.

Let's see how this applies to the fluid flowing through the tube of **FIGURE 14.27**. This is the situation we considered in Figure 14.25, but now the tube changes height. The more darkly shaded volume of fluid is the system to which we will apply Equation 14.19. Work is done on this system by pressure forces from the *surrounding* fluid in the tube. The unseen fluid to the left of our system exerts force $\vec{F}_1 = (p_1 A_1$, to the right), where p_1 is the fluid pressure at this point in the tube and A_1 is the cross-section area. During a small time interval Δt, this force pushes the fluid through displacement $\Delta\vec{r}_1 = (\Delta x_1$, to the right) and does work

$$W_1 = \vec{F}_1 \cdot \Delta\vec{r}_1 = F_1 \Delta x_1 = (p_1 A_1)\Delta x_1 = p_1(A_1 \Delta x_1) = p_1 V \qquad (14.20)$$

The A_1 and Δx_1 enter the equation from different terms, but they conveniently combine to give $V = A_1 \Delta x_1$, the volume "vacated" by the shaded fluid as it's pushed forward.

The situation is much the same on the right edge of the system, where the surrounding fluid exerts a pressure force $\vec{F}_2 = (p_2 A_2$, to the left). However, force $\vec{F}_2$ is *opposite* the displacement $\Delta\vec{r}_2$, which introduces a minus sign into the dot product for the work, giving

$$W_2 = \vec{F}_2 \cdot \Delta\vec{r}_2 = -F_2 \Delta x_2 = -(p_2 A_2)\Delta x_2 = -p_2(A_2 \Delta x_2) = -p_2 V \qquad (14.21)$$

Because the fluid is incompressible, the volume $V = A_2 \Delta x_2$ "gained" on the right side is exactly the same as that lost on the left. Altogether, the work done on the system by the surrounding fluid is

$$W_{ext} = W_1 + W_2 = (p_1 - p_2)V \qquad (14.22)$$

The work depends on the *pressure difference* $p_1 - p_2$.

Now let's see what happens to the system's potential and kinetic energy. Most of the system does not change during time interval Δt; it's fluid at the same height moving at the same speed. All we need to consider are the volumes V at the two ends. On the right, the system *gains* kinetic and gravitational potential energy as the fluid moves into volume V. Simultaneously, the system *loses* kinetic and gravitational potential energy on the left as fluid vacates volume V.

The mass of fluid in volume V is $m = \rho V$, where ρ is the fluid density. Thus the net change in the system's gravitational potential energy during Δt is

$$\Delta U_G = mgy_2 - mgy_1 = \rho Vgy_2 - \rho Vgy_1 \qquad (14.23)$$

Similarly, the system's change in kinetic energy is

$$\Delta K = \tfrac{1}{2}mv_2^2 - \tfrac{1}{2}mv_1^2 = \tfrac{1}{2}\rho Vv_2^2 - \tfrac{1}{2}\rho Vv_1^2 \qquad (14.24)$$

Combining Equations 14.22, 14.23, and 14.24 gives us the energy equation for the fluid in the flow tube:

$$\tfrac{1}{2}\rho Vv_2^2 - \tfrac{1}{2}\rho Vv_1^2 + \rho Vgy_2 - \rho Vgy_1 = p_1 V - p_2 V \qquad (14.25)$$

The volume V cancels out of all the terms. Regrouping terms, we have

$$p_1 + \tfrac{1}{2}\rho v_1^2 + \rho gy_1 = p_2 + \tfrac{1}{2}\rho v_2^2 + \rho gy_2 \qquad (14.26)$$

Equation 14.26 is called **Bernoulli's equation.** It is named for the 18th-century Swiss scientist Daniel Bernoulli, who made some of the earliest studies of fluid dynamics.

FIGURE 14.27 Energy analysis of fluid flow through a tube.

Bernoulli's equation is really nothing more than a statement about work and energy. It is sometimes useful to express Bernoulli's equation in the alternative form

$$p + \tfrac{1}{2}\rho v^2 + \rho g y = \text{constant} \qquad (14.27)$$

This version of Bernoulli's equation tells us that the quantity $p + \tfrac{1}{2}\rho v^2 + \rho g y$ remains constant along a streamline.

> **NOTE** Using Bernoulli's equation is very much like using the law of conservation of energy. Rather than identifying a "before" and "after," you want to identify two points on a streamline. As the following examples show, Bernoulli's equation is often used in conjunction with the equation of continuity.

EXAMPLE 14.11 | An irrigation system

Water flows through the pipes shown in FIGURE 14.28. The water's speed through the lower pipe is 5.0 m/s and a pressure gauge reads 75 kPa. What is the reading of the pressure gauge on the upper pipe?

FIGURE 14.28 The water pipes of an irrigation system.

MODEL Treat the water as an ideal fluid obeying Bernoulli's equation. Consider a streamline connecting point 1 in the lower pipe with point 2 in the upper pipe.

SOLVE Bernoulli's equation, Equation 14.26, relates the pressure, fluid speed, and heights at points 1 and 2. It is easily solved for the pressure p_2 at point 2:

$$p_2 = p_1 + \tfrac{1}{2}\rho v_1^2 - \tfrac{1}{2}\rho v_2^2 + \rho g y_1 - \rho g y_2$$
$$= p_1 + \tfrac{1}{2}\rho(v_1^2 - v_2^2) + \rho g(y_1 - y_2)$$

All quantities on the right are known except v_2, and that is where the equation of continuity will be useful. The cross-section areas and water speeds at points 1 and 2 are related by

$$v_1 A_1 = v_2 A_2$$

from which we find

$$v_2 = \frac{A_1}{A_2} v_1 = \frac{r_1^2}{r_2^2} v_1 = \frac{(0.030 \text{ m})^2}{(0.020 \text{ m})^2}(5.0 \text{ m/s}) = 11.25 \text{ m/s}$$

The pressure at point 1 is $p_1 = 75 \text{ kPa} + 1 \text{ atm} = 176{,}300 \text{ Pa}$. We can now use the above expression for p_2 to calculate $p_2 = 105{,}900 \text{ Pa}$. This is the absolute pressure; the pressure gauge on the upper pipe will read

$$p_2 = 105{,}900 \text{ Pa} - 1 \text{ atm} = 4.6 \text{ kPa}$$

ASSESS Reducing the pipe size decreases the pressure because it makes $v_2 > v_1$. Gaining elevation also reduces the pressure.

EXAMPLE 14.12 | Hydroelectric power

Small hydroelectric plants in the mountains sometimes bring the water from a reservoir down to the power plant through enclosed tubes. In one such plant, the 100-cm-diameter intake tube in the base of the dam is 50 m below the reservoir surface. The water drops 200 m through the tube before flowing into the turbine through a 50-cm-diameter nozzle.

a. What is the water speed into the turbine?

b. By how much does the inlet pressure differ from the hydrostatic pressure at that depth?

MODEL Treat the water as an ideal fluid obeying Bernoulli's equation. Consider a streamline that begins at the surface of the reservoir and ends at the exit of the nozzle. The pressure at the surface is $p_1 = p_{\text{atmos}}$ and $v_1 \approx 0$ m/s. The water discharges into air, so $p_3 = p_{\text{atmos}}$ at the exit.

VISUALIZE FIGURE 14.29 is a pictorial representation of the situation.

FIGURE 14.29 Pictorial representation of the water flow to a hydroelectric plant.

SOLVE a. Bernoulli's equation, with $v_1 = 0$ m/s and $y_3 = 0$ m, is

$$p_{atmos} + \rho g y_1 = p_{atmos} + \tfrac{1}{2}\rho v_3^2$$

The power plant is in the mountains, where $p_{atmos} < 1$ atm, but p_{atmos} occurs on both sides of Bernoulli's equation and cancels. Solving for v_3 gives

$$v_3 = \sqrt{2gy_1} = \sqrt{2(9.80 \text{ m/s}^2)(250 \text{ m})} = 70 \text{ m/s}$$

b. You might expect the pressure p_2 at the intake to be the hydrostatic pressure $p_{atmos} + \rho g d$ at depth d. But the water is *flowing* into the intake tube, so it's not in static equilibrium. We can find the intake speed v_2 from the equation of continuity:

$$v_2 = \frac{A_3}{A_2}v_3 = \frac{r_3^2}{r_2^2}\sqrt{2gy_1}$$

The intake is along the streamline between points 1 and 3, so we can apply Bernoulli's equation to points 1 and 2:

$$p_{atmos} + \rho g y_1 = p_2 + \tfrac{1}{2}\rho v_2^2 + \rho g y_2$$

Solving this equation for p_2, and noting that $y_1 - y_2 = d$, we find

$$p_2 = p_{atmos} + \rho g(y_1 - y_2) - \tfrac{1}{2}\rho v_2^2$$

$$= p_{atmos} + \rho g d - \tfrac{1}{2}\rho \left(\frac{r_3}{r_2}\right)^4 (2gy_1)$$

$$= p_{static} - \rho g y_1 \left(\frac{r_3}{r_2}\right)^4$$

The intake pressure is less than hydrostatic pressure by the amount

$$\rho g y_1 \left(\frac{r_3}{r_2}\right)^4 = 153{,}000 \text{ Pa} = 1.5 \text{ atm}$$

ASSESS The water's exit speed from the nozzle is the same as if it fell 250 m from the surface of the reservoir. This isn't surprising because we've assumed a nonviscous (i.e., frictionless) liquid. "Real" water would have less speed but still flow very fast.

Two Applications

The speed of a flowing gas is often measured with a device called a **Venturi tube.** Venturi tubes measure gas speeds in environments as different as chemistry laboratories, wind tunnels, and jet engines.

FIGURE 14.30 shows gas flowing through a tube that changes from cross-section area A_1 to area A_2. A U-shaped glass tube containing liquid of density ρ_{liq} connects the two segments of the flow tube. When gas flows through the horizontal tube, the liquid stands height h higher in the side of the U tube connected to the narrow segment of the flow tube.

Figure 14.30 shows how a Venturi tube works. We can make this analysis quantitative and determine the gas-flow speed from the liquid height h. Two pieces of information we have to work with are Bernoulli's equation

$$p_1 + \tfrac{1}{2}\rho v_1^2 + \rho g y_1 = p_2 + \tfrac{1}{2}\rho v_2^2 + \rho g y_2 \tag{14.28}$$

and the equation of continuity

$$v_2 A_2 = v_1 A_1 \tag{14.29}$$

In addition, the hydrostatic equation for the liquid tells us that the pressure p_2 above the right tube differs from the pressure p_1 above the left tube by $\rho_{liq}gh$. That is,

$$p_2 = p_1 - \rho_{liq}gh \tag{14.30}$$

First we use Equations 14.29 and 14.30 to eliminate v_2 and p_2 in Bernoulli's equation:

$$p_1 + \tfrac{1}{2}\rho v_1^2 = (p_1 - \rho_{liq}gh) + \tfrac{1}{2}\rho \left(\frac{A_1}{A_2}\right)^2 v_1^2 \tag{14.31}$$

The potential energy terms have disappeared because $y_1 = y_2$ for a horizontal tube. Equation 14.31 can now be solved for v_1, then v_2 is obtained from Equation 14.29. We'll skip a few algebraic steps and go right to the result:

$$v_1 = A_2 \sqrt{\frac{2\rho_{liq}gh}{\rho(A_1^2 - A_2^2)}}$$

$$\tag{14.32}$$

$$v_2 = A_1 \sqrt{\frac{2\rho_{liq}gh}{\rho(A_1^2 - A_2^2)}}$$

Equations 14.32 are reasonably accurate as long as the flow speeds are much less than the speed of sound, about 340 m/s. The Venturi tube is an example of the power of Bernoulli's equation.

FIGURE 14.30 A Venturi tube measures gas-flow speeds.

1. As the gas flows into a smaller cross section, it speeds up (equation of continuity). As it speeds up, the pressure decreases (Bernoulli's equation).

Area A_1 Gas of density ρ

$\vec{v}_2$ Area A_2

$\vec{v}_1$

Pressure p_2

Pressure p_1

h

Liquid of density ρ_{liq}

2. The U tube acts like a manometer. The liquid level is higher on the side where the pressure is lower.

FIGURE 14.31 Airflow over a wing generates lift by creating unequal pressures above and below.

1. The streamlines in the flow tube are compressed, indicating that the air speeds up as it flows over the top of the wing. This lowers the pressure to $p < p_{atmos}$.

$\vec{F}_{lift}$

2. The pressure difference exerts an upward force on the wing.

$p \approx p_{atmos}$ beneath wing

As a final example, we can use Bernoulli's equation to understand, at least qualitatively, how airplane wings generate *lift*. FIGURE 14.31 shows the cross section of an airplane wing. This shape is called an *airfoil*.

Although you usually think of an airplane moving through the air, in the airplane's reference frame it is the air that flows across a stationary wing. As it does, the streamlines must separate. The bottom of the wing does not significantly alter the streamlines going under the wing. But the streamlines going over the top of the wing get bunched together. As we've seen, with the equation of continuity, the flow speed has to increase when streamlines get closer together. Consequently, the air speed increases as it flows across the top of the wing.

If the air speed increases, then, from Bernoulli's equation, the air pressure must decrease. And if the air pressure above the wing is less than the air pressure below, the air will exert a net *upward* force on the wing. The upward force of the air due to the pressure difference across the wing is called **lift.** A full understanding of lift in aerodynamics involves other, more complicated factors, such as the creation of vortices on the trailing edge of the wing, but our introduction to fluid dynamics has given you enough tools to at least begin to understand how airplanes stay aloft.

STOP TO THINK 14.6 Rank in order, from highest to lowest, the liquid heights h_a to h_d. The airflow is from left to right.

Direction of airflow

Air pump

h_a h_b h_c h_d

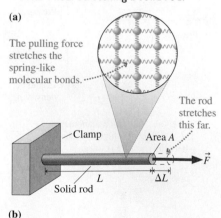

FIGURE 14.32 Stretching a solid rod.

(a)

The pulling force stretches the spring-like molecular bonds.

The rod stretches this far.

Clamp   Area A

$\vec{F}$

L ΔL

Solid rod

(b)

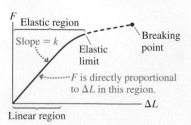

F
Elastic region

Slope $= k$

Elastic limit

Breaking point

F is directly proportional to ΔL in this region.

ΔL

Linear region

14.6 Elasticity

The final subject to explore in this chapter is elasticity. Although elasticity applies primarily to solids rather than fluids, you will see that similar ideas come into play.

Tensile Stress and Young's Modulus

Suppose you clamp one end of a solid rod while using a strong machine to pull on the other with force $\vec{F}$. FIGURE 14.32a shows the experimental arrangement. We usually think of solids as being, well, solid. But any material, be it plastic, concrete, or steel, will stretch as the spring-like molecular bonds expand.

FIGURE 14.32b shows graphically the amount of force needed to stretch the rod by the amount ΔL. This graph contains several regions of interest. First is the *elastic region,* ending at the *elastic limit.* As long as ΔL is less than the elastic limit, the rod will return to its initial length L when the force is removed. Just such a reversible stretch is what we mean when we say a material is *elastic.* A stretch beyond the elastic limit will permanently deform the object; it will not return to its initial length when the force is removed. And, not surprisingly, there comes a point when the rod breaks.

For most materials, the graph begins with a *linear region,* which is where we will focus our attention. If ΔL is within the linear region, the force needed to stretch the rod is

$$F = k \, \Delta L \qquad (14.33)$$

where k is the slope of the graph. You'll recognize Equation 14.33 as none other than Hooke's law.

The difficulty with Equation 14.33 is that the proportionality constant k depends both on the composition of the rod—whether it is, say, steel or aluminum—and on the rod's length and cross-section area. It would be useful to characterize the elastic properties of steel in general, or aluminum in general, without needing to know the dimensions of a specific rod.

We can meet this goal by thinking about Hooke's law at the atomic scale. The elasticity of a material is directly related to the spring constant of the molecular bonds between neighboring atoms. As **FIGURE 14.33** shows, the force pulling each bond is proportional to the quantity F/A. This force causes each bond to stretch by an amount proportional to $\Delta L/L$. We don't know what the proportionality constants are, but we don't need to. Hooke's law applied to a molecular bond tells us that the force pulling on a bond is proportional to the amount that the bond stretches. Thus F/A must be proportional to $\Delta L/L$. We can write their proportionality as

$$\frac{F}{A} = Y\frac{\Delta L}{L} \qquad (14.34)$$

The proportionality constant Y is called **Young's modulus.** It is directly related to the spring constant of the molecular bonds, so it depends on the material from which the object is made but *not* on the object's geometry.

A comparison of Equations 14.33 and 14.34 shows that Young's modulus can be written as $Y = kL/A$. This is not a definition of Young's modulus but simply an expression for making an experimental determination of the value of Young's modulus. This k is the spring constant of the rod seen in Figure 14.32. It is a quantity easily measured in the laboratory.

The quantity F/A, where A is the cross-section area, is called **tensile stress.** Notice that it is essentially the same definition as pressure. Even so, tensile stress differs in that the stress is applied in a particular direction whereas pressure forces are exerted in all directions. Another difference is that stress is measured in N/m^2 rather than pascals. The quantity $\Delta L/L$, the fractional increase in the length, is called **strain.** Strain is dimensionless. The numerical values of strain are always very small because solids cannot be stretched very much before reaching the breaking point.

With these definitions, Equation 14.34 can be written

$$\text{stress} = Y \times \text{strain} \qquad (14.35)$$

Because strain is dimensionless, Young's modulus Y has the same dimensions as stress, namely N/m^2. **TABLE 14.3** gives values of Young's modulus for several common materials. Large values of Y characterize materials that are stiff and rigid. "Softer" materials, at least relatively speaking, have smaller values of Y. You can see that steel has a larger Young's modulus than aluminum.

FIGURE 14.33 A material's elasticity is directly related to the spring constant of the molecular bonds.

The number of bonds is proportional to area A. If the rod is pulled with force F, the force pulling on each bond is proportional to F/A.

Area A

Length L

The number of bonds along the rod is proportional to length L. If the rod stretches by ΔL, the stretch of each bond is proportional to $\Delta L/L$.

TABLE 14.3 Elastic properties of various materials

Substance	Young's modulus (N/m^2)	Bulk modulus (N/m^2)
Steel	20×10^{10}	16×10^{10}
Copper	11×10^{10}	14×10^{10}
Aluminum	7×10^{10}	7×10^{10}
Concrete	3×10^{10}	–
Wood (Douglas fir)	1×10^{10}	–
Plastic (polystyrene)	0.3×10^{10}	–
Mercury	–	3×10^{10}
Water	–	0.2×10^{10}

Concrete is a widely used building material because it is relatively inexpensive and, with its large Young's modulus, it has tremendous compressional strength.

We introduced Young's modulus by considering how materials stretch. But Equation 14.35 and Young's modulus also apply to the compression of materials. Compression is particularly important in engineering applications, where beams, columns, and support foundations are compressed by the load they bear. Concrete is often compressed, as in columns that support highway overpasses, but rarely stretched.

> **NOTE** Whether the rod is stretched or compressed, Equation 14.35 is valid only in the linear region of the graph in Figure 14.32b. The breaking point is usually well outside the linear region, so you can't use Young's modulus to compute the maximum possible stretch or compression.

EXAMPLE 14.13 | **Stretching a wire**

A 2.0-m-long, 1.0-mm-diameter wire is suspended from the ceiling. Hanging a 4.5 kg mass from the wire stretches the wire's length by 1.0 mm. What is Young's modulus for this wire? Can you identify the material?

MODEL The hanging mass creates tensile stress in the wire.

SOLVE The force pulling on the wire, which is simply the weight of the hanging mass, produces tensile stress

$$\frac{F}{A} = \frac{mg}{\pi r^2} = \frac{(4.5\ \text{kg})(9.80\ \text{m/s}^2)}{\pi (0.0005\ \text{m})^2} = 5.6 \times 10^7\ \text{N/m}^2$$

The resulting stretch of 1.0 mm is a strain of $\Delta L/L = (1.0\ \text{mm})/(2000\ \text{mm}) = 5.0 \times 10^{-4}$. Thus Young's modulus for the wire is

$$Y = \frac{F/A}{\Delta L/L} = 11 \times 10^{10}\ \text{N/m}^2$$

Referring to Table 14.3, we see that the wire is made of copper.

Volume Stress and the Bulk Modulus

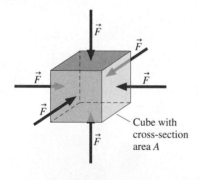

FIGURE 14.34 An object is compressed by pressure forces pushing equally on all sides.

Cube with cross-section area A

Young's modulus characterizes the response of an object to being pulled in one direction. **FIGURE 14.34** shows an object being squeezed in all directions. For example, objects under water are squeezed from all sides by the water pressure. The force per unit area F/A applied to *all* surfaces of an object is called the **volume stress.** Because the force pushes equally on all sides, the volume stress (unlike the tensile stress) really is the same as pressure p.

No material is perfectly rigid. A volume stress applied to an object compresses its volume slightly. The **volume strain** is defined as $\Delta V/V$. The volume strain is a *negative* number because the volume stress *decreases* the volume.

Volume stress, or pressure, is linearly proportional to the volume strain, much as the tensile stress is linearly proportional to the strain in a rod. That is,

$$\frac{F}{A} = p = -B\frac{\Delta V}{V} \tag{14.36}$$

where B is called the **bulk modulus.** The negative sign in Equation 14.36 ensures that the pressure is a positive number. Table 14.3 gives values of the bulk modulus for several materials. Smaller values of B correspond to materials that are more easily compressed. Both solids and liquids can be compressed and thus have a bulk modulus, whereas Young's modulus applies only to solids.

EXAMPLE 14.14 | **Compressing a sphere**

A 1.00-m-diameter solid steel sphere is lowered to a depth of 10,000 m in a deep ocean trench. By how much does its diameter shrink?

MODEL The water pressure applies a volume stress to the sphere.

SOLVE The water pressure at $d = 10{,}000$ m is

$$p = p_0 + \rho g d = 1.01 \times 10^8\ \text{Pa}$$

where we used the density of seawater. The bulk modulus of steel, taken from Table 14.3, is $16 \times 10^{10}\ \text{N/m}^2$. Thus the volume strain is

$$\frac{\Delta V}{V} = -\frac{p}{B} = -\frac{1.01 \times 10^8\ \text{Pa}}{16 \times 10^{10}\ \text{Pa}} = -6.3 \times 10^{-4}$$

The volume of a sphere is $V = \frac{4}{3}\pi r^3$. If the radius changes by the infinitesimal amount dr, we can use calculus to find that the volume changes by

$$dV = \tfrac{4}{3}\pi d(r^3) = \tfrac{4}{3}\pi \times 3r^2 dr = 4\pi r^2 dr$$

Thus $\Delta V = 4\pi r^2 \Delta r$ is a quite good approximation for very small changes in the sphere's radius and volume. Using this, the volume strain is

$$\frac{\Delta V}{V} = \frac{4\pi r^2 \Delta r}{\frac{4}{3}\pi r^3} = \frac{3\Delta r}{r} = -6.3 \times 10^{-4}$$

Solving for Δr gives $\Delta r = -1.05 \times 10^{-4} \text{ m} = -0.105 \text{ mm}$. The diameter changes by twice this, decreasing 0.21 mm.

ASSESS The immense pressure of the deep ocean causes only a tiny change in the sphere's diameter. You can see that treating solids and liquids as incompressible is an excellent approximation under nearly all circumstances.

CHALLENGE EXAMPLE 14.15 | Draining a cone

A conical tank of radius R and height H, pointed end down, is full of water. A small hole of radius r is opened at the bottom of the tank, with $r \ll R$ so that the tank drains slowly. Find an expression for the time T it takes to drain the tank completely.

MODEL Model the water as an ideal fluid. We can use Bernoulli's equation to relate the flow speed from the hole to the height of the water in the cone.

VISUALIZE FIGURE 14.35 is a pictorial representation. Because the tank drains slowly, we've assumed that the water velocity at the top surface is always very close to zero: $v_1 = 0$. The pressure at the surface is $p_1 = p_{atmos}$. The water discharges into air, so we also have $p_2 = p_{atmos}$ at the exit.

FIGURE 14.35 Pictorial representation of water draining from a tank.

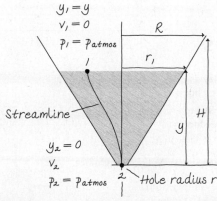

SOLVE As the tank drains, the water height y decreases from H to 0. If we can find an expression for dy/dt, the *rate* at which the water height changes, we'll be able to find T by integrating from "full tank" at $t = 0$ to "empty tank" at $t = T$. Our starting point is the rate at which water flows out of the hole at the bottom—the volume flow rate $Q = v_2 A_2 = \pi r^2 v_2$, where v_2 is the exit speed. The volume of water inside the tank is changing at the rate

$$\frac{dV_{water}}{dt} = -Q = -\pi r^2 v_2$$

where the minus sign shows that the volume is *decreasing* with time.

We need to relate both V_{water} and v_2 to the height y of the water surface. The volume of a cone is $V = \frac{1}{3} \times \text{base} \times \text{height}$, so the cone of water has volume $V_{water} = \frac{1}{3}\pi r_1^2 y$. Based on the similar triangles in Figure 14.35, $r_1/R = y/H$. Thus $r_1 = (R/H)y$ and

$$V_{water} = \frac{\pi R^2}{3H^2} y^3$$

Taking the time derivative, we find

$$\frac{dV_{water}}{dt} = \frac{d}{dt}\left[\frac{\pi R^2}{3H^2} y^3\right] = \frac{\pi R^2}{H^2} y^2 \frac{dy}{dt}$$

This relates the rate at which the volume changes to the rate at which the height changes.

We can next relate v_2 to the water height y by using Bernoulli's equation to connect the conditions at the surface (point 1) to conditions at the exit (point 2):

$$p_1 + \tfrac{1}{2}\rho v_1^2 + \rho g y_1 = p_2 + \tfrac{1}{2}\rho v_2^2 + \rho g y_2$$

With $p_1 = p_2$, $v_1 = 0$, $y_1 = y$, and $y_2 = 0$ at the bottom, Bernoulli's equation simplifies to $\rho g y = \frac{1}{2}\rho v_2^2$. Thus the exit speed of the water is

$$v_2 = \sqrt{2gy}$$

The exit speed decreases as the water height drops because the pressure at the bottom is less.

With this information, our equation for the rate at which the volume is changing becomes

$$\frac{dV_{water}}{dt} = \frac{\pi R^2}{H^2} y^2 \frac{dy}{dt} = -\pi r^2 v_2 = -\pi r^2 \sqrt{2gy}$$

In preparation for integration, we need to get all the y's on one side of the equation and dt on the other. Rearranging gives

$$dt = -\frac{R^2}{r^2 H^2 \sqrt{2g}} y^{3/2}\, dy$$

We need to integrate this from the beginning, with $y = H$ at $t = 0$, to the moment the tank is empty, with $y = 0$ at $t = T$:

$$\int_0^T dt = T = -\frac{R^2}{r^2 H^2 \sqrt{2g}} \int_H^0 y^{3/2}\, dy = \frac{R^2}{r^2 H^2 \sqrt{2g}} \int_0^H y^{3/2}\, dy$$

The minus sign was eliminated by reversing the integration limits. Performing the integration gives us the desired result for the time to drain the tank:

$$T = \frac{R^2}{r^2 H^2 \sqrt{2g}} \int_0^H y^{3/2}\, dy = \frac{R^2}{r^2 H^2 \sqrt{2g}} \left[\frac{2}{5} y^{5/2}\right]_0^H$$

$$= \frac{2}{5}\frac{R^2}{r^2}\sqrt{\frac{H}{2g}}$$

ASSESS Making the tank larger by increasing R or H increases the time needed to drain. Making the hole at the bottom larger—a larger value of r—decreases the time. These are as we would have expected, giving us confidence in our result.

SUMMARY

The goal of Chapter 14 has been to learn about systems that flow or deform.

GENERAL PRINCIPLES

Fluid Statics

Gases

- Freely moving particles
- Compressible
- Pressure primarily thermal
- Pressure is constant in a laboratory-size container

Liquids

- Loosely bound particles
- Incompressible
- Pressure primarily gravitational
- Hydrostatic pressure at depth d is $p = p_0 + \rho g d$

Fluid Dynamics

Ideal-fluid model

- Incompressible
- Nonviscous
- Smooth, laminar flow

The fluid moves along streamlines.

Equation of continuity

$$v_1 A_1 = v_2 A_2$$

Bernoulli's equation

$$p_1 + \tfrac{1}{2}\rho v_1^2 + \rho g y_1 = p_2 + \tfrac{1}{2}\rho v_2^2 + \rho g y_2$$

Bernoulli's equation is a statement of energy conservation.

IMPORTANT CONCEPTS

Density $\rho = m/V$, where m is mass and V is volume.

Pressure $p = F/A$, where F is the magnitude of the fluid force and A is the area on which the force acts.

- Pressure exists at all points in a fluid.
- Pressure pushes equally in all directions.
- Pressure is constant along a horizontal line.
- Gauge pressure is $p_g = p - 1$ atm.

APPLICATIONS

Buoyancy is the upward force of a fluid on an object.

Archimedes' principle

The magnitude of the buoyant force equals the weight of the fluid displaced by the object.

Sink	$\rho_{avg} > \rho_f$	$F_B < m_o g$
Rise to surface	$\rho_{avg} < \rho_f$	$F_B > m_o g$
Neutrally buoyant	$\rho_{avg} = \rho_f$	$F_B = m_o g$

Elasticity describes the deformation of solids and liquids under stress.

Linear stretch and compression

$$(F/A) = Y(\Delta L/L)$$

Tensile stress   Young's modulus   Strain

Volume compression

$$p = -B(\Delta V/V)$$

Bulk modulus   Volume strain

TERMS AND NOTATION

fluid	hydrostatic pressure	neutral buoyancy	Venturi tube
gas	Pascal's principle	ideal-fluid model	lift
liquid	gauge pressure, p_g	viscosity	Young's modulus, Y
mass density, ρ	hydraulics	laminar flow	tensile stress
unit volume	buoyant force	streamline	strain
pressure, p	displaced fluid	equation of continuity	volume stress
pascal, Pa	Archimedes' principle	volume flow rate, Q	volume strain
vacuum	average density, ρ_{avg}	Bernoulli's equation	bulk modulus, B
standard atmosphere, atm			

CONCEPTUAL QUESTIONS

1. An object has density ρ.
   a. Suppose each of the object's three dimensions is increased by a factor of 2 without changing the material of which the object is made. Will the density change? If so, by what factor? Explain.
   b. Suppose each of the object's three dimensions is increased by a factor of 2 without changing the object's mass. Will the density change? If so, by what factor? Explain.

2. Rank in order, from largest to smallest, the pressures at a, b, and c in **FIGURE Q14.2**. Explain.

FIGURE Q14.2

3. Rank in order, from largest to smallest, the pressures at d, e, and f in **FIGURE Q14.2**. Explain.

4. **FIGURE Q14.4** shows two rectangular tanks, A and B, full of water. They have equal depths and equal thicknesses (the dimension into the page) but different widths.

FIGURE Q14.4

   a. Compare the forces the water exerts on the bottoms of the tanks. Is F_A larger than, smaller than, or equal to F_B? Explain.
   b. Compare the forces the water exerts on the sides of the tanks. Is F_A larger than, smaller than, or equal to F_B? Explain.

5. In **FIGURE Q14.5**, is p_A larger than, smaller than, or equal to p_B? Explain.

FIGURE Q14.5

FIGURE Q14.6

6. Rank in order, from largest to smallest, the densities of blocks a, b, and c in **FIGURE Q14.6**. Explain.

7. Blocks a, b, and c in **FIGURE Q14.7** have the same volume. Rank in order, from largest to smallest, the sizes of the buoyant forces F_a, F_b, and F_c on a, b, and c. Explain.

FIGURE Q14.7

8. Blocks a, b, and c in **FIGURE Q14.7** have the same density. Rank in order, from largest to smallest, the sizes of the buoyant forces F_a, F_b, and F_c on a, b, and c. Explain.

9. The two identical beakers in **FIGURE Q14.9** are filled to the same height with water. Beaker B has a plastic sphere floating in it. Which beaker, with all its contents, weighs more? Or are they equal? Explain.

FIGURE Q14.9   **FIGURE Q14.10**

10. Gas flows through the pipe of **FIGURE Q14.10**. You can't see into the pipe to know how the inner diameter changes. Rank in order, from largest to smallest, the gas speeds v_a, v_b, and v_c at points a, b, and c. Explain.

11. Wind blows over the house in **FIGURE Q14.11**. A window on the ground floor is open. Is there an airflow through the house? If so, does the air flow in the window and out the chimney, or in the chimney and out the window? Explain.

FIGURE Q14.11

12. A 2000 N force stretches a wire by 1 mm. A second wire of the same material is twice as long and has twice the diameter. How much force is needed to stretch it by 1 mm? Explain.

13. A wire is stretched right to the breaking point by a 5000 N force. A longer wire made of the same material has the same diameter. Is the force that will stretch it right to the breaking point larger than, smaller than, or equal to 5000 N? Explain.

EXERCISES AND PROBLEMS

Problems labeled ▮ integrate material from earlier chapters.

Exercises

Section 14.1 Fluids

1. ‖ What is the volume in mL of 55 g of a liquid with density 1100 kg/m³?

2. ‖ Cylinders A and B have equal heights. Cylinder A is filled with helium gas at 1.0 atm pressure and 0°C. The diameter of cylinder B is half that of cylinder A, and cylinder B is filled with glycerin. What is the ratio of the fluid mass in cylinder B to that in cylinder A?

3. ‖ a. 50 g of gasoline are mixed with 50 g of water. What is the average density of the mixture?
   b. 50 cm³ of gasoline are mixed with 50 cm³ of water. What is the average density of the mixture?

4. ‖ A 6.0 m × 12.0 m swimming pool slopes linearly from a 1.0 m depth at one end to a 3.0 m depth at the other. What is the mass of water in the pool?

Section 14.2 Pressure

5. ‖ A 1.0-m-diameter vat of liquid is 2.0 m deep. The pressure at the bottom of the vat is 1.3 atm. What is the mass of the liquid in the vat?

6. | The deepest point in the ocean is 11 km below sea level, deeper than Mt. Everest is tall. What is the pressure in atmospheres at this depth?

7. | A 3.0-cm-diameter tube is held upright and filled to the top with mercury. The mercury pressure at the bottom of the tube—the pressure in excess of atmospheric pressure—is 50 kPa. How tall is the tube?

8. ‖ a. What volume of water has the same mass as 8.0 m³ of ethyl alcohol?
   b. If this volume of water is in a cubic tank, what is the pressure at the bottom?

9. ‖ A 50-cm-thick layer of oil floats on a 120-cm-thick layer of water. What is the pressure at the bottom of the water layer?

10. ‖ A research submarine has a 20-cm-diameter window 8.0 cm thick. The manufacturer says the window can withstand forces up to 1.0×10^6 N. What is the submarine's maximum safe depth? The pressure inside the submarine is maintained at 1.0 atm.

11. ‖ A 20-cm-diameter circular cover is placed over a 10-cm-diameter hole that leads into an evacuated chamber. The pressure in the chamber is 20 kPa. How much force is required to pull the cover off?

12. ‖ The container shown in **FIGURE EX14.12** is filled with oil. It is open to the atmosphere on the left.
   a. What is the pressure at point A?
   b. What is the pressure difference between points A and B? Between points A and C?

FIGURE EX14.12

Section 14.3 Measuring and Using Pressure

13. | What is the height of a water barometer at atmospheric pressure?

14. ‖ What is the minimum hose diameter of an ideal vacuum cleaner that could lift a 10 kg (22 lb) dog off the floor?

15. ‖ How far must a 2.0-cm-diameter piston be pushed down into one cylinder of a hydraulic lift to raise an 8.0-cm-diameter piston by 20 cm?

Section 14.4 Buoyancy

16. | A 6.00-cm-diameter sphere with a mass of 89.3 g is neutrally buoyant in a liquid. Identify the liquid.

17. ‖ A 2.0 cm × 2.0 cm × 6.0 cm block floats in water with its long axis vertical. The length of the block above water is 2.0 cm. What is the block's mass density?

18. ‖ Astronauts visiting a new planet find a lake filled with an unknown liquid. They have with them a plastic cube, 6.0 cm on each side, with a density of 840 kg/m³. First they weigh the cube with a spring scale, measuring a weight of 21 N. Then they float the cube in the lake and find that two-thirds of the cube is submerged. At what depth in the lake will the pressure be twice the atmospheric pressure of 85 kPa?

19. ‖ A sphere completely submerged in water is tethered to the bottom with a string. The tension in the string is one-third the weight of the sphere. What is the density of the sphere?

20. ‖ A 5.0 kg rock whose density is 4800 kg/m³ is suspended by a string such that half of the rock's volume is under water. What is the tension in the string?

21. ‖ What is the tension of the string in **FIGURE EX14.21**?

100 cm³ of aluminum, density $\rho_{Al} = 2700$ kg/m³

Ethyl alcohol

FIGURE EX14.21

22. ‖ A 10-cm-diameter, 20-cm-tall steel cylinder ($\rho_{steel} = 7900$ kg/m³) floats in mercury. The axis of the cylinder is perpendicular to the surface. What length of steel is above the surface?

23. | You need to determine the density of a ceramic statue. If you suspend it from a spring scale, the scale reads 28.4 N. If you then lower the statue into a tub of water, so that it is completely submerged, the scale reads 17.0 N. What is the statue's density?

24. ‖ Styrofoam has a density of 150 kg/m³. What is the maximum mass that can hang without sinking from a 50-cm-diameter Styrofoam sphere in water? Assume the volume of the mass is negligible compared to that of the sphere.

25. ‖ You and your friends are playing in the swimming pool with a 60-cm-diameter beach ball. How much force would be needed to push the ball completely under water?

Section 14.5 Fluid Dynamics

26. | Water flows through a 2.5-cm-diameter hose at 3.0 m/s. How long, in minutes, will it take to fill a 600 L child's wading pool?

27. ‖ A long horizontal tube has a square cross section with sides of width L. A fluid moves through the tube with speed v_0. The tube then changes to a circular cross section with diameter L. What is the fluid's speed in the circular part of the tube?

28. ‖ A 1.0-cm-diameter pipe widens to 2.0 cm, then narrows to 5.0 mm. Liquid flows through the first segment at a speed of 4.0 m/s.
   a. What is the speed in the second and third segments?
   b. What is the volume flow rate through the pipe?

29. | A bucket is filled with water to a height of 23 cm, then a plug is removed from a 4.0-mm-diameter hole in the bottom of the bucket. As the water begins to pour out of the hole, how fast is it moving?

30. ‖ What does the top pressure gauge read in **FIGURE EX14.30**?

FIGURE EX14.30

31. ‖‖‖ A 2.0 mL syringe has an inner diameter of 6.0 mm, a needle
BIO inner diameter of 0.25 mm, and a plunger pad diameter (where you place your finger) of 1.2 cm. A nurse uses the syringe to inject medicine into a patient whose blood pressure is 140/100.
   a. What is the minimum force the nurse needs to apply to the syringe?
   b. The nurse empties the syringe in 2.0 s. What is the flow speed of the medicine through the needle?

Section 14.6 Elasticity

32. ‖ A 70 kg mountain climber dangling in a crevasse stretches a 50-m-long, 1.0-cm-diameter rope by 8.0 cm. What is Young's modulus for the rope?

33. ‖ An 80-cm-long, 1.0-mm-diameter steel guitar string must be tightened to a tension of 2000 N by turning the tuning screws. By how much is the string stretched?

34. ‖ A 3.0-m-tall, 50-cm-diameter concrete column supports a 200,000 kg load. By how much is the column compressed?

35. ‖ a. What is the pressure at a depth of 5000 m in the ocean?
   b. What is the fractional volume change $\Delta V/V$ of seawater at this pressure?
   c. What is the density of seawater at this pressure?

36. ‖ A large 10,000 L aquarium is supported by four wood posts (Douglas fir) at the corners. Each post has a square 4.0 cm × 4.0 cm cross section and is 80 cm tall. By how much is each post compressed by the weight of the aquarium?

37. ‖ A 5.0-m-diameter solid aluminum sphere is launched into space. By how much does its diameter increase? Give your answer in μm.

Problems

38. ‖ a. In **FIGURE P14.38**, how much force does the fluid exert on the end of the cylinder at A?
   b. How much force does the fluid exert on the end of the cylinder at B?

FIGURE P14.38   **FIGURE P14.39**

39. ‖ The two 60-cm-diameter cylinders in **FIGURE P14.39**, closed at one end, open at the other, are joined to form a single cylinder, then the air inside is removed.
   a. How much force does the atmosphere exert on the flat end of each cylinder?
   b. Suppose one cylinder is bolted to a sturdy ceiling. How many 100 kg football players would need to hang from the lower cylinder to pull the two cylinders apart?

40. ‖ *Postural hypotension* is the occurrence of low systolic blood
BIO pressure when a person stands up too quickly from a reclining position. A brain blood pressure lower than 90 mm of Hg can cause fainting or lightheadedness. In a healthy adult, the automatic constriction and expansion of blood vessels keep the brain blood pressure constant while posture is changing, but disease or aging can weaken this response. If the blood pressure in your brain is 118 mm of Hg while lying down, what would it be when you stand up if this automatic response failed? Assume your brain is 40 cm from your heart and the density of blood is 1060 kg/m³.

41. ‖ A friend asks you how much pressure is in your car tires. You know that the tire manufacturer recommends 30 psi, but it's been a while since you've checked. You can't find a tire gauge in the car, but you do find the owner's manual and a ruler. Fortunately, you've just finished taking physics, so you tell your friend, "I don't know, but I can figure it out." From the owner's manual you find that the car's mass is 1500 kg. It seems reasonable to assume that each tire supports one-fourth of the weight. With the ruler you find that the tires are 15 cm wide and the flattened segment of the tire in contact with the road is 13 cm long. What answer—in psi—will you give your friend?

42. ‖ a. The 70 kg student in **FIGURE P14.42** balances a 1200 kg elephant on a hydraulic lift. What is the diameter of the piston the student is standing on?
   b. When a second student joins the first, the piston sinks 35 cm. What is the second student's mass?

FIGURE P14.42

43. ‖‖‖ A 55 kg cheerleader uses an oil-filled hydraulic lift to hold four 110 kg football players at a height of 1.0 m. If her piston is 16 cm in diameter, what is the diameter of the football players' piston?

44. ‖ A U-shaped tube, open to the air on both ends, contains mercury. Water is poured into the left arm until the water column is 10.0 cm deep. How far upward from its initial position does the mercury in the right arm rise?

45. ‖ Geologists place *tiltmeters* on the sides of volcanoes to measure the displacement of the surface as magma moves inside the volcano. Although most tiltmeters today are electronic, the traditional tiltmeter, used for decades, consisted of two or more water-filled metal cans placed some distance apart and connected by a hose. **FIGURE P14.45** shows two such cans, each having a window to measure the water height. Suppose the cans are placed so that the water level in both is initially at the 5.0 cm mark. A week later, the water level in can 2 is at the 6.5 cm mark.
   a. Did can 2 move up or down relative to can 1? By what distance?
   b. Where is the water level now in can 1?

FIGURE P14.45   100 m

46. ‖ Glycerin is poured into an open U-shaped tube until the height in both sides is 20 cm. Ethyl alcohol is then poured into one arm until the height of the alcohol column is 20 cm. The two liquids do not mix. What is the difference in height between the top surface of the glycerin and the top surface of the alcohol?

47. ‖ An aquarium of length L, width (front to back) W, and depth
CALC D is filled to the top with liquid of density ρ.
   a. Find an expression for the force of the liquid on the bottom of the aquarium.
   b. Find an expression for the force of the liquid on the front window of the aquarium.
   c. Evaluate the forces for a 100-cm-long, 35-cm-wide, 40-cm-deep aquarium filled with water.

48. ‖ It's possible to use the ideal-gas law to show that the density of
CALC the earth's atmosphere decreases exponentially with height. That is, $\rho = \rho_0 \exp(-z/z_0)$, where z is the height above sea level, ρ_0 is the density at sea level (you can use the Table 14.1 value), and z_0 is called the *scale height* of the atmosphere.
   a. Determine the value of z_0.
   Hint: What is the weight of a column of air?
   b. What is the density of the air in Denver, at an elevation of 1600 m? What percent of sea-level density is this?

49. ‖ The average density of the body of a fish is 1080 kg/m³. To
BIO keep from sinking, a fish increases its volume by inflating an internal air bladder, known as a swim bladder, with air. By what percent must the fish increase its volume to be neutrally buoyant in fresh water? The density of air at 20°C is 1.19 kg/m³.

50. ‖ A cylinder with cross-section area A floats with its long axis
CALC vertical in a liquid of density ρ.
   a. Pressing down on the cylinder pushes it deeper into the liquid. Find an expression for the force needed to push the cylinder distance x deeper into the liquid and hold it there.
   b. A 4.0-cm-diameter cylinder floats in water. How much work must be done to push the cylinder 10 cm deeper into the water?

51. ‖ A less-dense liquid of density ρ_1 floats on top of a more-dense liquid of density ρ_2. A uniform cylinder of length l and density ρ, with $\rho_1 < \rho < \rho_2$, floats at the interface with its long axis vertical. What fraction of the length is in the more-dense liquid?

52. ‖ A 30-cm-tall, 4.0-cm-diameter plastic tube has a sealed bottom. 250 g of lead pellets are poured into the bottom of the tube, whose mass is 30 g, then the tube is lowered into a liquid. The tube floats with 5.0 cm extending above the surface. What is the density of the liquid?

53. ‖ One day when you come into physics lab you find several plastic hemispheres floating like boats in a tank of fresh water. Each lab group is challenged to determine the heaviest rock that can be placed in the bottom of a plastic boat without sinking it. You get one try. Sinking the boat gets you no points, and the maximum number of points goes to the group that can place the heaviest rock without sinking. You begin by measuring one of the hemispheres, finding that it has a mass of 21 g and a diameter of 8.0 cm. What is the mass of the heaviest rock that, in perfectly still water, won't sink the plastic boat?

54. ‖‖‖ A spring with spring constant 35 N/m is attached to the ceiling, and a 5.0-cm-diameter, 1.0 kg metal cylinder is attached to its lower end. The cylinder is held so that the spring is neither stretched nor compressed, then a tank of water is placed underneath with the surface of the water just touching the bottom of the cylinder. When released, the cylinder will oscillate a few times but, damped by the water, quickly reach an equilibrium position. When in equilibrium, what length of the cylinder is submerged?

55. ‖‖‖ A plastic "boat" with a 25 cm² square cross section floats in a liquid. One by one, you place 50 g masses inside the boat and measure how far the boat extends below the surface. Your data are as follows:

Mass added (g)	Depth (cm)
50	2.9
100	5.0
150	6.6
200	8.6

Draw an appropriate graph of the data and, from the slope and intercept of the best-fit line, determine the mass of the boat and the density of the liquid.

56. ‖‖‖ A 355 mL soda can is 6.2 cm in diameter and has a mass of 20 g. Such a soda can half full of water is floating upright in water. What length of the can is above the water level?

57. ‖ A nuclear power plant draws 3.0×10^6 L/min of cooling water from the ocean. If the water is drawn in through two parallel, 3.0-m-diameter pipes, what is the water speed in each pipe?

58. ‖ a. A liquid of density ρ flows at speed v_0 through a horizontal pipe that expands smoothly from diameter d_0 to a larger diameter d_1. The pressure in the narrower section is p_0. Find an expression for the pressure p_1 in the wider section.
   b. A pressure gauge reads 50 kPa as water flows at 10.0 m/s through a 16.8-cm-diameter horizontal pipe. What is the reading of a pressure gauge after the pipe has expanded to 20.0 cm in diameter?

59. ‖‖‖ A tree loses water to the air by the process of *transpiration* at
BIO the rate of 110 g/h. This water is replaced by the upward flow of sap through vessels in the trunk. If the trunk contains 2000 vessels, each 100 μm in diameter, what is the upward speed of the sap in each vessel? The density of tree sap is 1040 kg/m³.

60. ‖ Water flows from the pipe shown in **FIGURE P14.60** with a speed of 4.0 m/s.
   a. What is the water pressure as it exits into the air?
   b. What is the height h of the standing column of water?

FIGURE P14.60   **FIGURE P14.61**

61. ‖ The 1.0-m-tall cylinder in **FIGURE P14.61** contains air at a pressure of 1 atm. A very thin, frictionless piston of negligible mass is placed at the top of the cylinder, to prevent any air from escaping, then mercury is slowly poured into the cylinder until no more can be added without the cylinder overflowing. What is the height h of the column of compressed air?
   Hint: Boyle's law, which you learned in chemistry, says $p_1V_1 = p_2V_2$ for a gas compressed at constant temperature, which we will assume to be the case.

62. ‖‖‖ Water flowing out of a 16-mm-diameter faucet fills a 2.0 L bottle in 10 s. At what distance below the faucet has the water stream narrowed to 10 mm diameter?

63. ‖ Water from a vertical pipe emerges as a 10-cm-diameter cylinder and falls straight down 7.5 m into a bucket. The water exits the pipe with a speed of 2.0 m/s. What is the diameter of the column of water as it hits the bucket?

64. ‖ A 2.0-m-diameter vertical cylinder, open at the top, is filled with ethyl alcohol to the 2.5 m mark. A worker pulls a plug from a hole 55 cm from the bottom, causing the fluid level to begin dropping at a rate of 3.4 mm/min. What is the diameter of the hole?

65. | A hurricane wind blows across a 6.0 m × 15.0 m flat roof at a speed of 130 km/h.
    a. Is the air pressure above the roof higher or lower than the pressure inside the house? Explain.
    b. What is the pressure difference?
    c. How much force is exerted on the roof? If the roof cannot withstand this much force, will it "blow in" or "blow out"?

66. ‖ Air flows through the tube shown in **FIGURE P14.66** at a rate of 1200 cm³/s. Assume that air is an ideal fluid. What is the height h of mercury in the right side of the U-tube?

FIGURE P14.66   **FIGURE P14.67**

67. ‖ Air flows through the tube shown in **FIGURE P14.67**. Assume that air is an ideal fluid.
    a. What are the air speeds v_1 and v_2 at points 1 and 2?
    b. What is the volume flow rate?

68. ‖ A water tank of height h has a small hole at height y. The water is replenished to keep h from changing. The water squirting from the hole has range x. The range approaches zero as $y \rightarrow 0$ because the water squirts right onto the ground. The range also approaches zero as $y \rightarrow h$ because the horizontal velocity becomes zero. Thus there must be some height y between 0 and h for which the range is a maximum.
    a. Find an algebraic expression for the flow speed v with which the water exits the hole at height y.
    b. Find an algebraic expression for the range of a particle shot horizontally from height y with speed v.
    c. Combine your expressions from parts a and b. Then find the maximum range x_{max} and the height y of the hole. "Real" water won't achieve quite this range because of viscosity, but it will be close.

69. ‖ a. A cylindrical tank of radius R, filled to the top with a liquid, has a small hole in the side, of radius r, at distance d below the surface. Find an expression for the volume flow rate through the hole.
    b. A 4.0-mm-diameter hole is 1.0 m below the surface of a 2.0-m-diameter tank of water. What is the rate, in mm/min, at which the water level will initially drop if the water is not replenished?

70. ‖ There is a disk of cartilage between each pair of vertebrae in
BIO your spine. Young's modulus for cartilage is 1.0×10^6 N/m².
Suppose a relaxed disk is 4.0 cm in diameter and 5.0 mm thick. If a disk in the lower spine supports half the weight of a 66 kg person, by how many mm does the disk compress?

Challenge Problems

71. ‖‖ The bottom of a steel "boat" is a 5.0 m × 10 m × 2.0 cm piece of steel ($\rho_{steel} = 7900$ kg/m³). The sides are made of 0.50-cm-thick steel. What minimum height must the sides have for this boat to float in perfectly calm water?

72. ‖‖ The tank shown in **FIGURE CP14.72** is completely filled with a
CALC liquid of density ρ. The right face is not permanently attached to the tank but, instead, is held against a rubber seal by the tension in a spring. To prevent leakage, the spring must both pull with sufficient strength *and* prevent a torque from pushing the bottom of the right face out.
    a. What minimum spring tension is needed?
    b. If the spring has the minimum tension, at what height d from the bottom must it be attached?

FIGURE CP14.72   **FIGURE CP14.73**

73. ‖‖ In **FIGURE CP14.73**, a cone of density ρ_0 and total height l floats in a liquid of density ρ_f. The height of the cone above the liquid is h. What is the ratio h/l of the exposed height to the total height?

74. ‖‖ Disk brakes, such as those in your car, operate by using pressurized oil to push outward on a piston. The piston, in turn, presses brake pads against a spinning rotor or wheel, as seen in **FIGURE CP14.74**. Consider a 15 kg industrial grinding wheel, 26 cm in diameter, spinning at 900 rpm. The brake pads are actuated by 2.0-cm-diameter pistons, and they contact the wheel an average distance 12 cm from the axis. If the coefficient of kinetic friction between the brake pad and the wheel is 0.60, what oil pressure is needed to stop the wheel in 5.0 s?

FIGURE CP14.74

75. ‖‖ In addition to the buoyant force, an object moving in a liquid
CALC experiences a *linear* drag force $\vec{F}_{drag} = (bv$, direction opposite the motion), where b is a constant. For a sphere of radius R, the drag constant can be shown to be $b = 6\pi\eta R$, where η is the *viscosity* of the liquid. Consider a sphere of radius R and density ρ that is released from rest at the surface of a liquid with density ρ_f.
    a. Find an expression in terms of R, η, g, and the densities for the sphere's *terminal speed* v_{term} as it falls through the liquid.
    b. Solve Newton's second law to find an expression for $v_y(t)$, the sphere's vertical velocity as a function of time as it falls. Pay careful attention to signs!
    c. Water at 20°C has viscosity $\eta = 1.0 \times 10^{-3}$ Pa s. Aluminum has density 2700 kg/m³. If a 3.0-mm-diameter aluminum pellet is dropped into water, what is its terminal speed, and how long does it take to reach 90% of its terminal speed?

PART III

KNOWLEDGE STRUCTURE

Applications of Newtonian Mechanics

KEY FINDINGS   What are the overarching findings of Part III?

- Newton's laws of motion together with the conservation laws for energy and momentum form what is called **Newtonian mechanics.** Part III has introduced relatively little new physics and instead has focused on applying Newtonian mechanics to new situations.

- Rotational motion is analogous to linear motion.
- With Newton's law of gravity, we can use Newtonian mechanics to understand the motions of satellites and planets.
- Liquids and gases cannot be modeled as particles. Even so, fluids still obey Newtonian laws of statics and motion.

LAWS   What laws of physics govern these applications?

Newton's second law for rotation   A net torque causes an extended object to have angular acceleration: $\tau_{net} = I\alpha$

Conservation of angular momentum   For an isolated system ($\tau_{net} = 0$), the total angular momentum is conserved: $\Delta \vec{L} = \vec{0}$.

Newton's law of gravity   The attractive force between two objects separated by r is $F_{m \text{ on } M} = F_{M \text{ on } m} = \dfrac{GMm}{r^2}$.

Bernoulli's equation   For two points along a streamline of a flowing fluid, $p_1 + \frac{1}{2}\rho v_1^2 + \rho g y_1 = p_2 + \frac{1}{2}\rho v_2^2 + \rho g y_2$.

MODELS   What are the most important models of Part III?

Rotation models

- A **rigid body** is an extended object whose size and shape do not change.
  - Particle-like atoms held together by bonds that are massless, rigid rods.

- A rigid body that experiences a **constant torque** undergoes constant angular acceleration.
  - Obeys Newton's second law for rotation.

- An extended object is in **static equilibrium** if and only if there are no net force *and* no net torque.

Fluid models

- A **gas** fills its container.
  - Molecules move freely.
  - Molecules are far apart.
  - A gas is compressible.

- A **liquid** has a well-defined surface.
  - Molecules are loosely bound.
  - Molecules are close together.
  - A liquid is incompressible.

- An **ideal fluid** obeys Bernoulli's equation.
  - Incompressible
  - Nonviscous
  - Smooth, laminar flow

TOOLS   What are the most important tools introduced in Part III?

Rotation

- Rotational/linear analogs:
Moment of inertia I	Mass m
Torque τ	Force F
Angular acceleration α	Acceleration a
Angular momentum $\vec{L}$	Momentum $\vec{p}$

- Rotational kinetic energy: $K = \frac{1}{2}I\omega^2$

- **Torque:** $\tau = rF\sin\phi = rF_\parallel = dF$
  - d is the moment arm or lever arm.
  - Vector torque is $\vec{\tau} = \vec{r} \times \vec{F}$.

- **Angular momentum:** $\vec{L} = I\vec{\omega}$

- Rolling motion:
  - Rolling without slipping: $v_{cm} = R\omega$
  - $K = K_{cm} + K_{rot}$

Gravity

- Newton's theory of gravity predicts Kepler's three observational laws. We can summarize them as:
  - Planets and satellites move in elliptical orbits.
  - Angular momentum is conserved.
  - For circular orbits, the square of the period is proportional to the cube of the orbit's radius.

- Gravitational potential energy:

$$U_G = -\frac{GMm}{r}$$

  - The potential energy is negative with a zero at infinity
  - **Escape speed** is the minimum speed needed to reach infinity.

Fluids

- **Pressure** $p = F/A$ exists at all points in a fluid.
  - Gas pressure is primarily thermal.
  - Liquid pressure is primarily gravitational.
  - Hydrostatic pressure at depth d:
    $$p = p_0 + \rho g d$$

- **Archimedes' principle** says that the upward buoyant force equals the weight of the displaced liquid.
  - An object floats if its average density is less than the fluid density.

- **Equation of continuity:** $v_1 A_1 = v_2 A_2$
  - Relates two points on a streamline.
  - v is the flow speed.
  - A is the cross-section area.

OVERVIEW

The Wave Model

Parts I–III of this text have been primarily about the physics of particles. You've seen that systems ranging from balls and rockets to planets can be thought of as particles or as systems of particles. A *particle* is one of the two fundamental models of classical physics. The other, to which we now turn our attention, is a *wave*.

Waves are ubiquitous in nature. Familiar examples include

- Undulating ripples on a pond.
- The swaying ground of an earthquake.
- A vibrating guitar string.
- The sweet sound of a flute.
- The colors of the rainbow.

The physics of oscillations and waves is the subject of Part IV. The most basic oscillation, called *simple harmonic motion,* is still the motion of a particle. But oscillators are the sources of waves, and the mathematical description of oscillation will carry over to the mathematical description of waves.

A wave, in contrast with a particle, is diffuse, spread out, not to be found at a single point in space. We will start with waves traveling outward through some medium, like the spreading ripples after a pebble hits a pool of water. These are called *traveling waves.* An investigation of what happens when waves travel through each other will lead us to *standing waves,* which are essential for understanding phenomena ranging from those as common as musical instruments to as complex as lasers and the electrons in atoms. We'll also study one of the most important defining characteristics of waves—their ability to exhibit *interference.*

Our exploration of wave phenomena will call upon sound waves, light waves, and vibrating strings for examples, but our goal will be to emphasize the unity and coherence that are common to *all* types of waves. Later, in Part VII, we will devote three chapters to light and optics, perhaps the most important application of waves. Although light is an electromagnetic wave, your understanding of those chapters will depend on nothing more than the "waviness" of light. If you wish, you can proceed to those chapters immediately after finishing Part IV. The electromagnetic aspects of light waves will be taken up in Chapter 31.

The song of a humpback whale can travel hundreds of kilometers underwater. This graph uses a procedure called wavelet analysis to study the frequency structure of a humpback whale song.

15 Oscillations

A pendulum swinging back and forth is one of the most common examples of simple harmonic motion.

IN THIS CHAPTER, you will learn about systems that oscillate in simple harmonic motion.

What are oscillations?

Oscillatory motion is a repetitive motion back and forth around an equilibrium position. We'll describe oscillations in terms of their amplitude, period, and frequency. The most important oscillation is simple harmonic motion (SHM), where the position and velocity graphs are sinusoidal.

What things undergo SHM?

The prototype of SHM is a mass oscillating on a spring. Lessons learned from this system apply to all SHM.

- A pendulum is a classic example of SHM.
- Any system with a linear restoring force undergoes SHM.

« LOOKING BACK Section 9.4 Restoring forces

How is SHM related to circular motion?

The projection of uniform circular motion onto a line oscillates back and forth in SHM.

- This link to circular motion will help us develop the mathematics of SHM.
- A phase constant, based on the angle on a circle, will describe the initial conditions.

« LOOKING BACK Section 4.4 Circular motion

Is energy conserved in SHM?

If there is no friction or other dissipative force, the mechanical energy of an oscillating system is conserved. Energy is transformed back and forth between kinetic and potential energy. Energy conservation is an important problem-solving strategy.

« LOOKING BACK Sections 10.3–10.5 Elastic potential energy and energy diagrams

What if there's friction?

If there's dissipation, the system "runs down." This is called a damped oscillation. The oscillation amplitude undergoes exponential decay. But the amplitude can grow very large when an oscillatory system is *driven* at its natural frequency. This is called resonance.

Why is SHM important?

Simple harmonic motion is one of the most common and important motions in science and engineering.

- Oscillations and vibrations occur in mechanical, electrical, chemical, and atomic systems. Understanding how a system might oscillate is an important part of engineering design.
- More complex oscillations can be understood in terms of SHM.
- Oscillations are the sources of waves, which we'll study in the next two chapters.

15.1 Simple Harmonic Motion

One of the most common and important types of motion is **oscillatory motion**—a repetitive motion back and forth around an equilibrium position. Swinging chandeliers, vibrating guitar strings, the electrons in cell phone circuits, and even atoms in solids are all undergoing oscillatory motion. In addition, oscillations are the sources of *waves,* our subject in the following two chapters. Oscillations play a major role in all fields of science and engineering.

FIGURE 15.1 shows position-versus-time graphs for two different oscillating systems. The shape of the graph depends on the details of the oscillator, but all oscillators have two things in common:

1. The oscillations take place around an equilibrium position.
2. The motion is *periodic,* repeating at regular intervals of time.

The time to complete one full cycle, or one oscillation, is called the **period** of the oscillation. Period is represented by the symbol T.

A system can oscillate in many ways, but the most fundamental oscillation is the smooth *sinusoidal* oscillation (i.e., like a sine or cosine) of Figure 15.1. This sinusoidal oscillation is called **simple harmonic motion,** abbreviated SHM. You'll learn in more advanced courses that *any* oscillation can be represented as a sum of sinusoidal oscillations, so SHM is the basis for understanding all oscillatory motion.

The prototype of simple harmonic motion is a mass oscillating on a spring. **FIGURE 15.2** shows an air-track glider attached to a spring. If the glider is pulled out a few centimeters and released, it oscillates back and forth. The graph shows an actual air-track measurement in which the glider's position was recorded 20 times per second. This is a position-versus-time graph that has been rotated 90° from its usual orientation to match the motion of the glider. You can see that it's a sinusoidal oscillation—simple harmonic motion.

As Figures 15.1 and 15.2 show, an oscillator moves back and forth between $x = -A$ and $x = +A$, where A, the **amplitude** of the motion, is the maximum displacement from equilibrium. Notice that the amplitude is the distance from the *axis* to a maximum or minimum, *not* the distance from the minimum to the maximum.

Period and amplitude are two important characteristics of oscillatory motion. A third is the **frequency,** f, which is the number of cycles or oscillations completed per second. If one cycle takes T seconds, the oscillator can complete $1/T$ cycles each second. That is, period and frequency are inverses of each other:

$$f = \frac{1}{T} \quad \text{or} \quad T = \frac{1}{f} \tag{15.1}$$

The units of frequency are **hertz,** abbreviated Hz, named in honor of the German physicist Heinrich Hertz, who produced the first artificially generated radio waves in 1887. By definition,

$$1 \text{ Hz} \equiv 1 \text{ cycle per second} = 1 \text{ s}^{-1}$$

We will often deal with very rapid oscillations and make use of the units shown in **TABLE 15.1.** For example, electrons oscillating back and forth at 101 MHz in an FM radio circuit have an oscillation period $T = 1/(101 \times 10^6 \text{ Hz}) = 9.9 \times 10^{-9} \text{ s} = 9.9 \text{ ns}$.

NOTE Uppercase and lowercase letters *are* important. 1 MHz is 1 megahertz = 10^6 Hz, but 1 mHz is 1 millihertz = 10^{-3} Hz!

Kinematics of Simple Harmonic Motion

We'll start by *describing* simple harmonic motion mathematically—that is, with kinematics. Then in Section 15.4 we'll take up the dynamics of how forces *cause* simple harmonic motion.

FIGURE 15.1 Examples of oscillatory motion.

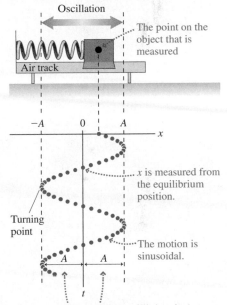

FIGURE 15.2 A prototype simple-harmonic-motion experiment.

TABLE 15.1 Units of frequency

Frequency	Period
10^3 Hz = 1 kilohertz = 1 kHz	1 ms
10^6 Hz = 1 megahertz = 1 MHz	1 μs
10^9 Hz = 1 gigahertz = 1 GHz	1 ns

FIGURE 15.3 Position and velocity graphs for simple harmonic motion.

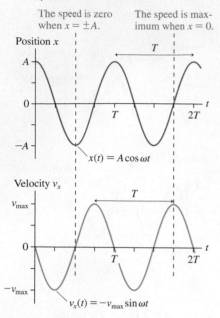

The speed is zero when $x = \pm A$.

The speed is maximum when $x = 0$.

Position x

$x(t) = A\cos\omega t$

Velocity v_x

$v_x(t) = -v_{max}\sin\omega t$

FIGURE 15.3 shows a SHM position graph—such as the one generated by the air-track glider—in its "normal" position. For the moment we'll assume that the oscillator starts at maximum displacement $(x = +A)$ at $t = 0$. Also shown is the oscillator's velocity-versus-time graph, which we can deduce from the slope of the position graph.

- The instantaneous velocity is zero at the instants when $x = \pm A$ because the slope of the position graph is zero. These are the *turning points* in the motion.
- The position graph has maximum slope when $x = 0$, so these are points of maximum speed. When $x = 0$ with a positive slope—maximum speed to the right—the instantaneous velocity is $v_x = +v_{max}$, where v_{max} is the amplitude of the velocity curve. Similarly, $v_x = -v_{max}$ when $x = 0$ with a negative slope—maximum speed to the left.

Although these are empirical observations (we don't yet have any "theory" of oscillation), we can see that the position graph, with a maximum at $t = 0$, is a cosine function with amplitude A and period T. We can write this as

$$x(t) = A\cos\left(\frac{2\pi t}{T}\right) \tag{15.2}$$

where the notation $x(t)$ indicates that x is a *function* of time t.

NOTE The arguments of sine and cosine functions are in *radians*. This will be true throughout our study of oscillations and waves. Be sure to set your calculator to radian mode before doing calculations.

Because $\cos(0 \text{ rad}) = \cos(2\pi \text{ rad}) = 1$, we can see that $x = A$ at $t = 0$ and again at $t = T$. In other words, this is a cosine function with amplitude A and period T. Notice that x passes through zero at $t = \frac{1}{4}T$ and $t = \frac{3}{4}T$ because $\cos\left(\frac{1}{2}\pi\right) = \cos\left(\frac{3}{2}\pi\right) = 0$.

We can write Equation 15.2 in two alternative forms. First, because the oscillation frequency is $f = 1/T$, we can write

$$x(t) = A\cos(2\pi f t) \tag{15.3}$$

Second, recall from Chapter 4 that a particle in circular motion has *angular velocity* ω that is related to the period by $\omega = 2\pi/T$, where ω is in rad/s. For oscillations and waves, ω is called the **angular frequency.** We can write the position in terms of ω as

$$x(t) = A\cos(\omega t) \tag{15.4}$$

Most of our work with oscillations and waves will be in terms of the angular frequency.

Now that we've defined $f = 1/T$, we see that ω, f, and T are related by

$$\omega \text{ (in rad/s)} = \frac{2\pi}{T} = 2\pi f \text{ (in Hz)} \tag{15.5}$$

Be careful! Both f and ω are frequencies, but they're not the same and they're not interchangeable. Frequency f is the true frequency, in cycles per second, and it's always measured in Hz. Angular frequency ω is useful because it's related to the angle of the cosine function and, as you'll learn in the next section, to a circular-motion analog of SHM, but it's always in rad/s.

Just as the position graph is a cosine function, you can see that the velocity graph in Figure 15.3 is an "upside-down" sine function with the same period. We can write the velocity function as

$$v_x(t) = -v_{max}\sin\left(\frac{2\pi t}{T}\right) \tag{15.6}$$

where the minus sign inverts the graph. This function is zero at $t = 0$ and again at $t = T$.

NOTE v_{max} is the maximum *speed* and thus is a *positive* number.

We deduced Equation 15.6 from the experimental results, but we could equally well find it from the position function of Equation 15.2. After all, velocity is the time derivative of position. **TABLE 15.2** reminds you of the derivatives of the sine and cosine functions. Using the derivative of the position function, we find

$$v_x(t) = \frac{dx}{dt} = -\frac{2\pi A}{T} \sin\left(\frac{2\pi t}{T}\right) = -2\pi f A \sin(2\pi f t) = -\omega A \sin \omega t \qquad (15.7)$$

Comparing Equation 15.7, the mathematical definition of velocity, to Equation 15.6, the empirical description, we see that the maximum speed of an oscillation is

$$v_{max} = \frac{2\pi A}{T} = 2\pi f A = \omega A \qquad (15.8)$$

Not surprisingly, the object has a greater maximum speed if you stretch the spring farther and give the oscillation a larger amplitude.

TABLE 15.2 Derivatives of sine and cosine functions

$$\frac{d}{dt}\big(a\sin(bt + c)\big) = +ab\cos(bt + c)$$

$$\frac{d}{dt}\big(a\cos(bt + c)\big) = -ab\sin(bt + c)$$

EXAMPLE 15.1 | A system in simple harmonic motion

An air-track glider is attached to a spring, pulled 20.0 cm to the right, and released at $t = 0$ s. It makes 15 oscillations in 10.0 s.

a. What is the period of oscillation?
b. What is the object's maximum speed?
c. What are the position and velocity at $t = 0.800$ s?

MODEL An object oscillating on a spring is in SHM.

SOLVE a. The oscillation frequency is

$$f = \frac{15 \text{ oscillations}}{10.0 \text{ s}} = 1.50 \text{ oscillations/s} = 1.50 \text{ Hz}$$

Thus the period is $T = 1/f = 0.667$ s.
b. The oscillation amplitude is $A = 0.200$ m. Thus

$$v_{max} = \frac{2\pi A}{T} = \frac{2\pi(0.200 \text{ m})}{0.667 \text{ s}} = 1.88 \text{ m/s}$$

c. The object starts at $x = +A$ at $t = 0$ s. This is exactly the oscillation described by Equations 15.2 and 15.6. The position at $t = 0.800$ s is

$$x = A\cos\left(\frac{2\pi t}{T}\right) = (0.200 \text{ m})\cos\left(\frac{2\pi(0.800 \text{ s})}{0.667 \text{ s}}\right)$$

$$= (0.200 \text{ m})\cos(7.54 \text{ rad}) = 0.0625 \text{ m} = 6.25 \text{ cm}$$

The velocity at this instant of time is

$$v_x = -v_{max}\sin\left(\frac{2\pi t}{T}\right) = -(1.88 \text{ m/s})\sin\left(\frac{2\pi(0.800 \text{ s})}{0.667 \text{ s}}\right)$$

$$= -(1.88 \text{ m/s})\sin(7.54 \text{ rad}) = -1.79 \text{ m/s} = -179 \text{ cm/s}$$

At $t = 0.800$ s, which is slightly more than one period, the object is 6.25 cm to the right of equilibrium and moving to the *left* at 179 cm/s. Notice the use of radians in the calculations.

EXAMPLE 15.2 | Finding the time

A mass oscillating in simple harmonic motion starts at $x = A$ and has period T. At what time, as a fraction of T, does the object first pass through $x = \frac{1}{2}A$?

SOLVE Figure 15.3 showed that the object passes through the equilibrium position $x = 0$ at $t = \frac{1}{4}T$. This is one-quarter of the total distance in one-quarter of a period. You might expect it to take $\frac{1}{8}T$ to reach $\frac{1}{2}A$, but this is not the case because the SHM graph is not linear between $x = A$ and $x = 0$. We need to use $x(t) = A\cos(2\pi t/T)$. First, we write the equation with $x = \frac{1}{2}A$:

$$x = \frac{A}{2} = A\cos\left(\frac{2\pi t}{T}\right)$$

Then we solve for the time at which this position is reached:

$$t = \frac{T}{2\pi}\cos^{-1}\left(\frac{1}{2}\right) = \frac{T}{2\pi}\frac{\pi}{3} = \frac{1}{6}T$$

ASSESS The motion is slow at the beginning and then speeds up, so it takes longer to move from $x = A$ to $x = \frac{1}{2}A$ than it does to move from $x = \frac{1}{2}A$ to $x = 0$. Notice that the answer is independent of the amplitude A.

STOP TO THINK 15.1 An object moves with simple harmonic motion. If the amplitude and the period are both doubled, the object's maximum speed is

a. Quadrupled.   b. Doubled.   c. Unchanged.
d. Halved.   e. Quartered.

FIGURE 15.4 A projection of the circular motion of a rotating ball matches the simple harmonic motion of an object on a spring.

Simple harmonic motion of block

15.2 SHM and Circular Motion

The graphs of Figure 15.3 and the position function $x(t) = A\cos\omega t$ are for an oscillation in which the object just happened to be at $x_0 = A$ at $t = 0$. But you will recall that $t = 0$ is an arbitrary choice, the instant of time when you or someone else starts a stopwatch. What if you had started the stopwatch when the object was at $x_0 = -A$, or when the object was somewhere in the middle of an oscillation? In other words, what if the oscillator had different *initial conditions?* The position graph would still show an oscillation, but neither Figure 15.3 nor $x(t) = A\cos\omega t$ would describe the motion correctly.

To learn how to describe the oscillation for other initial conditions it will help to turn to a topic you studied in Chapter 4—circular motion. There's a very close connection between simple harmonic motion and circular motion.

Imagine you have a turntable with a small ball glued to the edge. FIGURE 15.4a shows how to make a "shadow movie" of the ball by projecting a light past the ball and onto a screen. The ball's shadow oscillates back and forth as the turntable rotates. This is certainly periodic motion, with the same period as the turntable, but is it simple harmonic motion?

To find out, you could place a real object on a real spring directly below the shadow, as shown in FIGURE 15.4b. If you did so, and if you adjusted the turntable to have the same period as the spring, you would find that the shadow's motion exactly matches the simple harmonic motion of the object on the spring. **Uniform circular motion projected onto one dimension is simple harmonic motion.**

To understand this, consider the particle in FIGURE 15.5. It is in uniform circular motion, moving *counterclockwise* in a circle with radius A. As in Chapter 4, we can locate the particle by the angle ϕ measured counterclockwise (ccw) from the x-axis. Projecting the ball's shadow onto a screen in Figure 15.4 is equivalent to observing just the x-component of the particle's motion. Figure 15.5 shows that the x-component, when the particle is at angle ϕ, is

$$x = A\cos\phi \tag{15.9}$$

FIGURE 15.5 A particle in uniform circular motion with radius A and angular velocity ω.

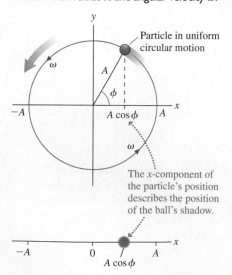

The x-component of the particle's position describes the position of the ball's shadow.

Recall that the particle's *angular velocity,* in rad/s, is

$$\omega = \frac{d\phi}{dt} \tag{15.10}$$

This is the rate at which the angle ϕ is increasing. If the particle starts from $\phi_0 = 0$ at $t = 0$, its angle at a later time t is simply

$$\phi = \omega t \tag{15.11}$$

As ϕ increases, the particle's x-component is

$$x(t) = A\cos\omega t \tag{15.12}$$

This is identical to Equation 15.4 for the position of a mass on a spring! Thus the x-component of a particle in uniform circular motion is simple harmonic motion.

NOTE When used to describe oscillatory motion, ω is called the *angular frequency* rather than the angular velocity. The angular frequency of an oscillator has the same numerical value, in rad/s, as the angular velocity of the corresponding particle in circular motion.

The names and units can be a bit confusing until you get used to them. It may help to notice that *cycle* and *oscillation* are not true units. Unlike the "standard meter" or the "standard kilogram," to which you could compare a length or a mass, there is no "standard cycle" to which you can compare an oscillation. Cycles and oscillations are simply counted events. Thus the frequency f has units of hertz, where $1\ \text{Hz} = 1\ \text{s}^{-1}$. We may *say* "cycles per second" just to be clear, but the actual units are only "per second."

The radian is the SI unit of angle. However, the radian is a *defined* unit. Further, its definition as a ratio of two lengths ($\theta = s/r$) makes it a pure number without dimensions. As we noted in Chapter 4, the unit of angle, be it radians or degrees, is really just a *name* to remind us that we're dealing with an angle. The 2π in the equation $\omega = 2\pi f$ (and in similar situations), which is stated without units, *means* 2π rad/cycle. When multiplied by the frequency f in cycles/s, it gives the frequency in rad/s. That is why, in this context, ω is called the angular *frequency*.

NOTE *Hertz* is specifically "cycles per second" or "oscillations per second." It is used for f but *not* for ω. We'll always be careful to use rad/s for ω, but you should be aware that many books give the units of ω as simply s^{-1}.

Initial Conditions: The Phase Constant

Now we're ready to consider the issue of other initial conditions. The particle in Figure 15.5 started at $\phi_0 = 0$. This was equivalent to an oscillator starting at the far right edge, $x_0 = A$. **FIGURE 15.6** shows a more general situation in which the initial angle ϕ_0 can have any value. The angle at a later time t is then

$$\phi = \omega t + \phi_0 \tag{15.13}$$

In this case, the particle's projection onto the x-axis at time t is

$$x(t) = A\cos(\omega t + \phi_0) \tag{15.14}$$

If Equation 15.14 describes the particle's projection, then it must also be the position of an oscillator in simple harmonic motion. The oscillator's velocity v_x is found by taking the derivative dx/dt. The resulting equations,

$$x(t) = A\cos(\omega t + \phi_0)$$
$$v_x(t) = -\omega A\sin(\omega t + \phi_0) = -v_{max}\sin(\omega t + \phi_0) \tag{15.15}$$

are the two primary kinematic equations of simple harmonic motion.

The quantity $\phi = \omega t + \phi_0$, which steadily increases with time, is called the **phase** of the oscillation. The phase is simply the *angle* of the circular-motion particle whose shadow matches the oscillator. The constant ϕ_0 is called the **phase constant.** It is determined by the *initial conditions* of the oscillator.

To see what the phase constant means, set $t = 0$ in Equations 15.15:

$$x_0 = A\cos\phi_0$$
$$v_{0x} = -\omega A\sin\phi_0 \tag{15.16}$$

The position x_0 and velocity v_{0x} at $t = 0$ are the initial conditions. **Different values of the phase constant correspond to different starting points on the circle and thus to different initial conditions.**

The cosine function of Figure 15.3 and the equation $x(t) = A\cos\omega t$ are for an oscillation with $\phi_0 = 0$ rad. You can see from Equations 15.16 that $\phi_0 = 0$ rad implies $x_0 = A$ and $v_0 = 0$. That is, the particle starts from rest at the point of maximum displacement.

FIGURE 15.7, on the next page, illustrates these ideas by looking at three values of the phase constant: $\phi_0 = \pi/3$ rad (60°), $-\pi/3$ rad ($-60°$), and π rad (180°). Notice that $\phi_0 = \pi/3$ rad and $\phi_0 = -\pi/3$ rad have the same starting position, $x_0 = \frac{1}{2}A$. This is a property of the cosine function in Equation 15.16. But these are *not* the same initial conditions. In one case the oscillator starts at $\frac{1}{2}A$ while moving to the left, in the other case it starts at $\frac{1}{2}A$ while moving to the right. You can distinguish between the two by visualizing the motion.

All values of the phase constant ϕ_0 between 0 and π rad correspond to a particle in the upper half of the circle and *moving to the left*. Thus v_{0x} is negative. All values of the phase constant ϕ_0 between π and 2π rad (or, as they are usually stated, between $-\pi$ and 0 rad) have the particle in the lower half of the circle and *moving to the right*. Thus v_{0x} is positive. If you're told that the oscillator is at $x = \frac{1}{2}A$ and moving to the right at $t = 0$, then the phase constant must be $\phi_0 = -\pi/3$ rad, not $+\pi/3$ rad.

FIGURE 15.6 A particle in uniform circular motion with initial angle ϕ_0.

Angle at time t is
$\phi = \omega t + \phi_0$.

Initial position of particle at $t = 0$

$x_0 = A\cos\phi_0$

The initial x-component of the particle's position can be anywhere between $-A$ and A, depending on ϕ_0.

FIGURE 15.7 Different initial conditions are described by different values of the phase constant.

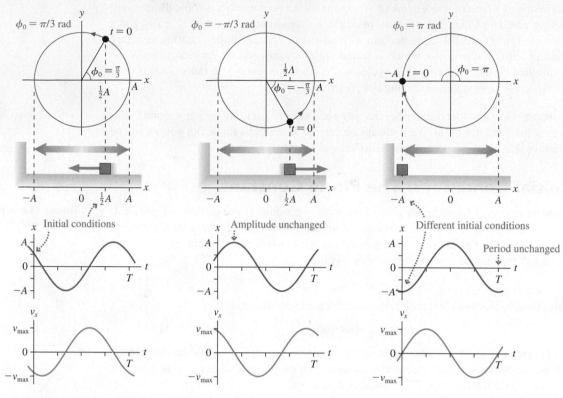

EXAMPLE 15.3 | Using the initial conditions

An object on a spring oscillates with a period of 0.80 s and an amplitude of 10 cm. At $t = 0$ s, it is 5.0 cm to the left of equilibrium and moving to the left. What are its position and direction of motion at $t = 2.0$ s?

MODEL An object oscillating on a spring is in simple harmonic motion.

SOLVE We can find the phase constant ϕ_0 from the initial condition $x_0 = -5.0$ cm $= A \cos \phi_0$. This condition gives

$$\phi_0 = \cos^{-1}\left(\frac{x_0}{A}\right) = \cos^{-1}\left(-\frac{1}{2}\right) = \pm \tfrac{2}{3}\pi \text{ rad} = \pm 120°$$

Because the oscillator is moving to the *left* at $t = 0$, it is in the upper half of the circular-motion diagram and must have a phase constant between 0 and π rad. Thus ϕ_0 is $\tfrac{2}{3}\pi$ rad. The angular frequency is

$$\omega = \frac{2\pi}{T} = \frac{2\pi}{0.80 \text{ s}} = 7.85 \text{ rad/s}$$

Thus the object's position at time $t = 2.0$ s is

$$x(t) = A \cos(\omega t + \phi_0)$$
$$= (10 \text{ cm}) \cos\left((7.85 \text{ rad/s})(2.0 \text{ s}) + \tfrac{2}{3}\pi\right)$$
$$= (10 \text{ cm}) \cos(17.8 \text{ rad}) = 5.0 \text{ cm}$$

The object is now 5.0 cm to the right of equilibrium. But which way is it moving? There are two ways to find out. The direct way is to calculate the velocity at $t = 2.0$ s:

$$v_x = -\omega A \sin(\omega t + \phi_0) = +68 \text{ cm/s}$$

The velocity is positive, so the motion is to the right. Alternatively, we could note that the phase at $t = 2.0$ s is $\phi = 17.8$ rad. Dividing by π, you can see that

$$\phi = 17.8 \text{ rad} = 5.67\pi \text{ rad} = (4\pi + 1.67\pi) \text{ rad}$$

The 4π rad represents two complete revolutions. The "extra" phase of 1.67π rad falls between π and 2π rad, so the particle in the circular-motion diagram is in the lower half of the circle and moving to the right.

NOTE The inverse-cosine function $\cos^{-1}$ is a *two-valued* function. Your calculator returns a single value, an angle between 0 rad and π rad. But the negative of this angle is also a solution. As Example 15.3 demonstrates, you must use additional information to choose between them.

STOP TO THINK 15.2 The figure shows four oscillators at $t = 0$. Which one has the phase constant $\phi_0 = \pi/4$ rad?

15.3 Energy in SHM

We've begun to develop the mathematical language of simple harmonic motion, but thus far we haven't included any physics. We've made no mention of the mass of the object or the spring constant of the spring. An energy analysis, using the tools of Chapter 10, is a good starting place.

FIGURE 15.8 shows an object oscillating on a spring, our prototype of simple harmonic motion. Now we'll specify that the object has mass m, the spring has spring constant k, and the motion takes place on a frictionless surface. You learned in Chapter 10 that the elastic potential energy when the object is at position x is $U_{Sp} = \frac{1}{2}k(\Delta x)^2$, where $\Delta x = x - x_{eq}$ is the displacement from the equilibrium position x_{eq}. In this chapter we'll always use a coordinate system in which $x_{eq} = 0$, making $\Delta x = x$. There's no chance for confusion with gravitational potential energy, so we can omit the subscript Sp and write the elastic potential energy as

$$U = \tfrac{1}{2}kx^2 \tag{15.17}$$

Thus the mechanical energy of an object oscillating on a spring is

$$E = K + U = \tfrac{1}{2}mv^2 + \tfrac{1}{2}kx^2 \tag{15.18}$$

The lower portion of Figure 15.8 is an energy diagram, showing the parabolic potential-energy curve $U = \frac{1}{2}kx^2$ and the kinetic energy $K = E - U$. Recall that a particle oscillates between the *turning points* where the total energy line E crosses the potential-energy curve. The left turning point is at $x = -A$, and the right turning point is at $x = +A$. To go beyond these points would require a negative kinetic energy, which is physically impossible.

You can see that **the particle has purely potential energy at $x = \pm A$ and purely kinetic energy as it passes through the equilibrium point at $x = 0$.** At maximum displacement, with $x = \pm A$ and $v = 0$, the energy is

$$E(\text{at } x = \pm A) = U = \tfrac{1}{2}kA^2 \tag{15.19}$$

At $x = 0$, where $v = \pm v_{max}$, the energy is

$$E(\text{at } x = 0) = K = \tfrac{1}{2}m(v_{max})^2 \tag{15.20}$$

The system's mechanical energy is conserved because the surface is frictionless and there are no external forces, so the energy at maximum displacement and the energy at maximum speed, Equations 15.19 and 15.20, must be equal. That is

$$\tfrac{1}{2}m(v_{max})^2 = \tfrac{1}{2}kA^2 \tag{15.21}$$

FIGURE 15.8 Energy transformations during SHM.

Energy is transformed between kinetic and potential, but the total mechanical energy E doesn't change.

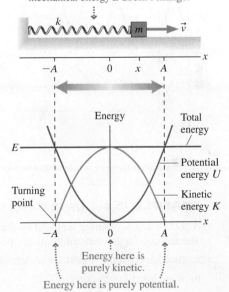

Thus the maximum speed is related to the amplitude by

$$v_{max} = \sqrt{\frac{k}{m}}\,A \qquad (15.22)$$

This is a relationship based on the physics of the situation.

Earlier, using kinematics, we found that

$$v_{max} = \frac{2\pi A}{T} = 2\pi f A = \omega A \qquad (15.23)$$

Comparing Equations 15.22 and 15.23, we see that frequency and period of an oscillating spring are determined by the spring constant k and the object's mass m:

$$\omega = \sqrt{\frac{k}{m}} \qquad f = \frac{1}{2\pi}\sqrt{\frac{k}{m}} \qquad T = 2\pi\sqrt{\frac{m}{k}} \qquad (15.24)$$

These three expressions are really only one equation. They say the same thing, but each expresses it in slightly different terms.

Equations 15.24 tell us that the period and frequency are related to the object's mass m and the spring constant k. It is perhaps surprising, but **the period and frequency do not depend on the amplitude A**. A small oscillation and a large oscillation have the same period.

Conservation of Energy

Because energy is conserved, we can combine Equations 15.18, 15.19, and 15.20 to write

$$E = \tfrac{1}{2}mv^2 + \tfrac{1}{2}kx^2 = \tfrac{1}{2}kA^2 = \tfrac{1}{2}m(v_{max})^2 \quad \text{(conservation of energy)} \qquad (15.25)$$

Any pair of these expressions may be useful, depending on the known information. For example, you can use the amplitude A to find the speed at any point x by combining the first and second expressions for E. The speed v at position x is

$$v = \sqrt{\frac{k}{m}\,(A^2 - x^2)} = \omega\sqrt{A^2 - x^2} \qquad (15.26)$$

FIGURE 15.9 shows graphically how the kinetic and potential energy change with time. They both oscillate but remain *positive* because x and v are squared. Energy is continuously being transformed back and forth between the kinetic energy of the moving block and the stored potential energy of the spring, but their sum remains constant. Notice that K and U both oscillate *twice* each period; make sure you understand why.

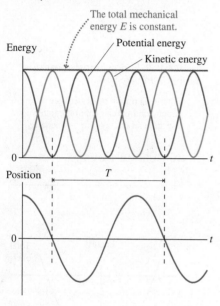

FIGURE 15.9 Kinetic energy, potential energy, and the total mechanical energy for simple harmonic motion.

The total mechanical energy E is constant.

Energy

Potential energy

Kinetic energy

Position

T

EXAMPLE 15.4 | Using conservation of energy

A 500 g block on a spring is pulled a distance of 20 cm and released. The subsequent oscillations are measured to have a period of 0.80 s.

a. At what position or positions is the block's speed 1.0 m/s?

b. What is the spring constant?

MODEL The motion is SHM. Energy is conserved.

SOLVE a. The block starts from the point of maximum displacement, where $E = U = \tfrac{1}{2}kA^2$. At a later time, when the position is x and the speed is v, energy conservation requires

$$\tfrac{1}{2}mv^2 + \tfrac{1}{2}kx^2 = \tfrac{1}{2}kA^2$$

Solving for x, we find

$$x = \sqrt{A^2 - \frac{mv^2}{k}} = \sqrt{A^2 - \left(\frac{v}{\omega}\right)^2}$$

where we used $k/m = \omega^2$ from Equation 15.24. The angular frequency is easily found from the period: $\omega = 2\pi/T = 7.85$ rad/s. Thus

$$x = \sqrt{(0.20\text{ m})^2 - \left(\frac{1.0\text{ m/s}}{7.85\text{ rad/s}}\right)^2} = \pm 0.15\text{ m} = \pm 15\text{ cm}$$

There are two positions because the block has this speed on either side of equilibrium.

b. Although part a did not require that we know the spring constant, it is straightforward to find from Equation 15.24:

$$T = 2\pi\sqrt{\frac{m}{k}}$$

$$k = \frac{4\pi^2 m}{T^2} = \frac{4\pi^2(0.50\text{ kg})}{(0.80\text{ s})^2} = 31\text{ N/m}$$

STOP TO THINK 15.3 The four springs shown here have been compressed from their equilibrium position at $x = 0$ cm. When released, the attached mass will start to oscillate. Rank in order, from highest to lowest, the maximum speeds of the masses.

15.4 The Dynamics of SHM

Our analysis thus far has been based on the experimental observation that the oscillation of a spring "looks" sinusoidal. It's time to look at force and acceleration and to see that Newton's second law *predicts* sinusoidal motion.

A motion diagram will help us visualize the object's acceleration. FIGURE 15.10 shows one cycle of the motion, separating motion to the left and motion to the right to make the diagram clear. As you can see, the object's velocity is large as it passes through the equilibrium point at $x = 0$, but $\vec{v}$ is *not changing* at that point. Acceleration measures the *change* of the velocity; hence $\vec{a} = \vec{0}$ at $x = 0$.

FIGURE 15.10 Motion diagram of simple harmonic motion. The left and right motions are separated vertically for clarity but really occur along the same line.

In contrast, the velocity is changing rapidly at the turning points. At the right turning point, $\vec{v}$ changes from a right-pointing vector to a left-pointing vector. Thus the acceleration $\vec{a}$ at the right turning point is large and *to the left*. In one-dimensional motion, the acceleration component a_x has a large *negative* value at the right turning point. Similarly, the acceleration $\vec{a}$ at the left turning point is large and *to the right*. Consequently, a_x has a large positive value at the left turning point.

Our motion-diagram analysis suggests that the acceleration a_x is most positive when the displacement is most negative, most negative when the displacement is a maximum, and zero when $x = 0$. This is confirmed by taking the derivative of the velocity:

$$a_x = \frac{dv_x}{dt} = \frac{d}{dt}(-\omega A \sin \omega t) = -\omega^2 A \cos \omega t \qquad (15.27)$$

then graphing it.

FIGURE 15.11 shows the position graph that we started with in Figure 15.3 and the corresponding acceleration graph. Comparing the two, you can see that the acceleration

FIGURE 15.11 Position and acceleration graphs for an oscillating spring. We've chosen $\phi_0 = 0$.

graph looks like an upside-down position graph. In fact, because $x = A \cos \omega t$, Equation 15.27 for the acceleration can be written

$$a_x = -\omega^2 x \qquad (15.28)$$

That is, **the acceleration is proportional to the negative of the displacement.** The acceleration is, indeed, most positive when the displacement is most negative and is most negative when the displacement is most positive.

Recall that the acceleration is related to the net force by Newton's second law. Consider again our prototype mass on a spring, shown in **FIGURE 15.12**. This is the simplest possible oscillation, with no distractions due to friction or gravitational forces. We will assume the spring itself to be massless.

You learned in Chapter 9 that the spring force is given by Hooke's law:

$$(F_{Sp})_x = -k\,\Delta x \qquad (15.29)$$

The minus sign indicates that the spring force is a **restoring force,** a force that always points back toward the equilibrium position. If we place the origin of the coordinate system at the equilibrium position, as we've done throughout this chapter, then $\Delta x = x$ and Hooke's law is simply $(F_{Sp})_x = -kx$.

The x-component of Newton's second law for the object attached to the spring is

$$(F_{net})_x = (F_{Sp})_x = -kx = ma_x \qquad (15.30)$$

Equation 15.30 is easily rearranged to read

$$a_x = -\frac{k}{m}x \qquad (15.31)$$

You can see that Equation 15.31 is identical to Equation 15.28 if the system oscillates with angular frequency $\omega = \sqrt{k/m}$. We previously found this expression for ω from an energy analysis. Our experimental observation that the acceleration is proportional to the *negative* of the displacement is exactly what Hooke's law would lead us to expect. That's the good news.

The bad news is that a_x is not a constant. As the object's position changes, so does the acceleration. Nearly all of our kinematic tools have been based on constant acceleration. We can't use those tools to analyze oscillations, so we must go back to the very definition of acceleration:

$$a_x = \frac{dv_x}{dt} = \frac{d^2x}{dt^2}$$

Acceleration is the second derivative of position with respect to time. If we use this definition in Equation 15.31, it becomes

$$\frac{d^2x}{dt^2} = -\frac{k}{m}x \quad \text{(equation of motion for a mass on a spring)} \qquad (15.32)$$

Equation 15.32, which is called the **equation of motion,** is a second-order differential equation. Unlike other equations we've dealt with, Equation 15.32 cannot be solved by direct integration. We'll need to take a different approach.

Solving the Equation of Motion

The solution to an algebraic equation such as $x^2 = 4$ is a number. The solution to a differential equation is a *function.* The x in Equation 15.32 is really $x(t)$, the position as a function of time. The solution to this equation is a function $x(t)$ whose second derivative is the function itself multiplied by $(-k/m)$.

One important property of differential equations that you will learn about in math is that the solutions are *unique.* That is, there is only *one* solution to Equation 15.32 that satisfies the initial conditions. If we were able to *guess* a solution, the uniqueness property would tell us that we had found the *only* solution. That might seem a rather

FIGURE 15.12 The prototype of simple harmonic motion: a mass oscillating on a horizontal spring without friction.

strange way to solve equations, but in fact differential equations are frequently solved by using your knowledge of what the solution needs to look like to guess an appropriate function. Let us give it a try!

We know from experimental evidence that the oscillatory motion of a spring appears to be sinusoidal. Let us *guess* that the solution to Equation 15.32 should have the functional form

$$x(t) = A\cos(\omega t + \phi_0) \tag{15.33}$$

where A, ω, and ϕ_0 are unspecified constants that we can adjust to any values that might be necessary to satisfy the differential equation.

If you were to guess that a solution to the algebraic equation $x^2 = 4$ is $x = 2$, you would verify your guess by substituting it into the original equation to see if it works. We need to do the same thing here: Substitute our guess for $x(t)$ into Equation 15.32 to see if, for an appropriate choice of the three constants, it works. To do so, we need the second derivative of $x(t)$. That is straightforward:

An optical technique called *interferometry* reveals the bell-like vibrations of a wine glass.

$$x(t) = A\cos(\omega t + \phi_0)$$

$$\frac{dx}{dt} = -\omega A \sin(\omega t + \phi_0) \tag{15.34}$$

$$\frac{d^2x}{dt^2} = -\omega^2 A \cos(\omega t + \phi_0)$$

If we now substitute the first and third of Equations 15.34 into Equation 15.32, we find

$$-\omega^2 A \cos(\omega t + \phi_0) = -\frac{k}{m} A \cos(\omega t + \phi_0) \tag{15.35}$$

Equation 15.35 will be true at all instants of time if and only if $\omega^2 = k/m$. There do not seem to be any restrictions on the two constants A and ϕ_0—they are determined by the initial conditions.

So we have found—by guessing!—that *the* solution to the equation of motion for a mass oscillating on a spring is

$$x(t) = A\cos(\omega t + \phi_0) \tag{15.36}$$

where the angular frequency

$$\omega = 2\pi f = \sqrt{\frac{k}{m}} \tag{15.37}$$

is determined by the mass and the spring constant.

NOTE Once again we see that the oscillation frequency is independent of the amplitude A.

Equations 15.36 and 15.37 seem somewhat anticlimactic because we've been using these results for the last several pages. But keep in mind that we had been *assuming* $x = A\cos\omega t$ simply because the experimental observations "looked" like a cosine function. We've now justified that assumption by showing that Equation 15.36 really is the solution to Newton's second law for a mass on a spring. **The *theory* of oscillation, based on Hooke's law for a spring and Newton's second law, is in good agreement with the experimental observations.**

EXAMPLE 15.5 | **Analyzing an oscillator**

At $t = 0$ s, a 500 g block oscillating on a spring is observed moving to the right at $x = 15$ cm. It reaches a maximum displacement of 25 cm at $t = 0.30$ s.

a. Draw a position-versus-time graph for one cycle of the motion.

b. What is the maximum force on the block, and what is the first time at which this occurs?

Continued

MODEL The motion is simple harmonic motion.

SOLVE a. The position equation of the block is $x(t) = A\cos(\omega t + \phi_0)$. We know that the amplitude is $A = 0.25$ m and that $x_0 = 0.15$ m. From these two pieces of information we obtain the phase constant:

$$\phi_0 = \cos^{-1}\left(\frac{x_0}{A}\right) = \cos^{-1}(0.60) = \pm 0.927 \text{ rad}$$

The object is initially moving to the right, which tells us that the phase constant must be between $-\pi$ and 0 rad. Thus $\phi_0 = -0.927$ rad. The block reaches its maximum displacement $x_{max} = A$ at time $t = 0.30$ s. At that instant of time

$$x_{max} = A = A\cos(\omega t + \phi_0)$$

This equation can be true only if $\cos(\omega t + \phi_0) = 1$, which requires $\omega t + \phi_0 = 0$. Thus

$$\omega = \frac{-\phi_0}{t} = \frac{-(-0.927 \text{ rad})}{0.30 \text{ s}} = 3.09 \text{ rad/s}$$

Now that we know ω, it is straightforward to compute the period:

$$T = \frac{2\pi}{\omega} = 2.0 \text{ s}$$

FIGURE 15.13 graphs $x(t) = (25 \text{ cm})\cos(3.09t - 0.927)$, where t is in s, from $t = 0$ s to $t = 2.0$ s.

FIGURE 15.13 Position-versus-time graph for the oscillator of Example 15.5.

b. We found the acceleration of SHM to be $a_x = -\omega^2 x$, so the force on the block at position x is $F_x = -m\omega^2 x$. The force will be a maximum, $F_{max} = m\omega^2 A$, when x reaches its minimum displacement $x = -A$. For the block,

$$F_{max} = (0.50 \text{ kg})(3.09 \text{ rad/s})^2(0.25 \text{ m}) = 1.2 \text{ N}$$

This occurs exactly half a period (1.0 s) after the block reaches its maximum displacement, thus at $t = 1.3$ s.

ASSESS A 2 s period is a modest oscillation, so we don't expect the block's acceleration to be extreme. A maximum force of 1.2 N on a 0.5 kg block causes a maximum acceleration of 2.4 m/s^2, which seems reasonable.

STOP TO THINK 15.4 This is the position graph of a mass on a spring. What can you say about the velocity and the force at the instant indicated by the dashed line?

a. Velocity positive; force to the right.
b. Velocity negative; force to the right.
c. Velocity zero; force to the right.
d. Velocity positive; force to the left.
e. Velocity negative; force to the left.
f. Velocity zero; force to the left.
g. Velocity and force both zero.

15.5 Vertical Oscillations

We have focused our analysis on a horizontally oscillating spring. But the typical demonstration you'll see in class is a mass bobbing up and down on a spring hung vertically from a support. Is it safe to assume that a vertical oscillation has the same mathematical description as a horizontal oscillation? Or does the additional force of gravity change the motion? Let us look at this more carefully.

FIGURE 15.14 shows a block of mass m hanging from a spring of spring constant k. An important fact to notice is that the equilibrium position of the block is *not* where the spring is at its unstretched length. At the equilibrium position of the block, where it hangs motionless, the spring has stretched by ΔL.

Finding ΔL is an equilibrium problem in which the upward spring force balances the downward gravitational force on the block. The y-component of the spring force is given by Hooke's law:

$$(F_{Sp})_y = -k\,\Delta y = +k\,\Delta L \tag{15.38}$$

FIGURE 15.14 Gravity stretches the spring.

Unstretched spring

ΔL

$\vec{F}_{Sp}$

m

$\vec{F}_G$

The block hanging at rest has stretched the spring by ΔL.

Equation 15.38 makes a distinction between ΔL, which is simply a *distance* and is a positive number, and the displacement Δy. The block is displaced downward, so $\Delta y = -\Delta L$. Newton's first law for the block in equilibrium is

$$(F_{net})_y = (F_{Sp})_y + (F_G)_y = k\,\Delta L - mg = 0 \qquad (15.39)$$

from which we can find

$$\Delta L = \frac{mg}{k} \qquad (15.40)$$

This is the distance the spring stretches when the block is attached to it.

Let the block oscillate around this equilibrium position, as shown in FIGURE 15.15. We've now placed the origin of the y-axis at the block's equilibrium position in order to be consistent with our analyses of oscillations throughout this chapter. If the block moves upward, as the figure shows, the spring gets shorter compared to its equilibrium length, but the spring is still *stretched* compared to its unstretched length in Figure 15.14. When the block is at position y, the spring is stretched by an amount $\Delta L - y$ and hence exerts an *upward* spring force $F_{Sp} = k(\Delta L - y)$. The net force on the block at this point is

$$(F_{net})_y = (F_{Sp})_y + (F_G)_y = k(\Delta L - y) - mg = (k\,\Delta L - mg) - ky \qquad (15.41)$$

But $k\,\Delta L - mg$ is zero, from Equation 15.40, so the net force on the block is simply

$$(F_{net})_y = -ky \qquad (15.42)$$

Equation 15.42 for vertical oscillations is *exactly* the same as Equation 15.30 for horizontal oscillations, where we found $(F_{net})_x = -kx$. That is, the restoring force for vertical oscillations is identical to the restoring force for horizontal oscillations. The role of gravity is to determine where the equilibrium position is, but it doesn't affect the oscillatory motion around the equilibrium position.

Because the net force is the same, Newton's second law has exactly the same oscillatory solution:

$$y(t) = A\cos(\omega t + \phi_0) \qquad (15.43)$$

with, again, $\omega = \sqrt{k/m}$. **The vertical oscillations of a mass on a spring are the same simple harmonic motion as those of a block on a horizontal spring.** This is an important finding because it was not obvious that the motion would still be simple harmonic motion when gravity was included.

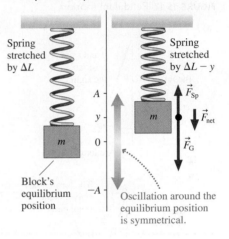

FIGURE 15.15 The block oscillates around the equilibrium position.

Spring stretched by ΔL

Spring stretched by $\Delta L - y$

Block's equilibrium position

Oscillation around the equilibrium position is symmetrical.

EXAMPLE 15.6 | Bungee oscillations

An 83 kg student hangs from a bungee cord with spring constant 270 N/m. The student is pulled down to a point where the cord is 5.0 m longer than its unstretched length, then released. Where is the student, and what is his velocity 2.0 s later?

MODEL A bungee cord can be modeled as a spring. Vertical oscillations on the bungee cord are SHM.

VISUALIZE FIGURE 15.16 shows the situation.

SOLVE Although the cord is stretched by 5.0 m when the student is released, this is *not* the amplitude of the oscillation. Oscillations occur around the equilibrium position, so we have to begin by finding the equilibrium point where the student hangs motionless. The cord stretch at equilibrium is given by Equation 15.40:

$$\Delta L = \frac{mg}{k} = 3.0 \text{ m}$$

Stretching the cord 5.0 m pulls the student 2.0 m below the equilibrium point, so $A = 2.0$ m. That is, the student oscillates with amplitude $A = 2.0$ m about a point 3.0 m beneath the bungee cord's

FIGURE 15.16 A student on a bungee cord oscillates about the equilibrium position.

270 N/m

The bungee cord is modeled as a spring.

5.0 m

Equilibrium

Release

Oscillation

83 kg

original end point. The student's position as a function of time, as measured from the equilibrium position, is

$$y(t) = (2.0 \text{ m})\cos(\omega t + \phi_0)$$

where $\omega = \sqrt{k/m} = 1.80$ rad/s.

Continued

The initial condition

$$y_0 = A\cos\phi_0 = -A$$

requires the phase constant to be $\phi_0 = \pi$ rad. At $t = 2.0$ s the student's position and velocity are

$$y = (2.0\text{ m})\cos\big((1.80\text{ rad/s})(2.0\text{ s}) + \pi\text{ rad}\big) = 1.8\text{ m}$$
$$v_y = -\omega A\sin(\omega t + \phi_0) = -1.6\text{ m/s}$$

The student is 1.8 m *above* the equilibrium position, or 1.2 m *below* the original end of the cord. Because his velocity is negative, he's passed through the highest point and is heading down.

15.6 The Pendulum

Now let's look at another very common oscillator: a pendulum. **FIGURE 15.17a** shows a mass m attached to a string of length L and free to swing back and forth. The pendulum's position can be described by the arc of length s, which is zero when the pendulum hangs straight down. Because angles are measured ccw, s and θ are positive when the pendulum is to the right of center, negative when it is to the left.

Two forces are acting on the mass: the string tension $\vec{T}$ and gravity $\vec{F}_G$. As we did with circular motion, it will be useful to divide the forces into tangential components, parallel to the motion, and radial components parallel to the string. These are shown on the free-body diagram of **FIGURE 15.17b**.

Newton's second law for the tangential component, parallel to the motion, is

$$(F_{\text{net}})_t = \sum F_t = (F_G)_t = -mg\sin\theta = ma_t \tag{15.44}$$

Using $a_t = d^2s/dt^2$ for acceleration "around" the circle, and noting that the mass cancels, we can write Equation 15.44 as

$$\frac{d^2s}{dt^2} = -g\sin\theta \tag{15.45}$$

This is the equation of motion for an oscillating pendulum. The sine function makes this equation more complicated than the equation of motion for an oscillating spring.

The Small-Angle Approximation

Suppose we restrict the pendulum's oscillations to *small angles* of less than about 10°. This restriction allows us to make use of an interesting and important piece of geometry.

FIGURE 15.18 shows an angle θ and a circular arc of length $s = r\theta$. A right triangle has been constructed by dropping a perpendicular from the top of the arc to the axis. The height of the triangle is $h = r\sin\theta$. Suppose that the angle θ is "small," which, in practice, means $\theta \ll 1$ rad. In that case there is very little difference between h and s. If $h \approx s$, then $r\sin\theta \approx r\theta$. It follows that

$$\sin\theta \approx \theta \quad \text{if} \quad \theta \ll 1\text{ rad}$$

The result that $\sin\theta \approx \theta$ for small angles is called the **small-angle approximation.** We can similarly note that $l \approx r$ for small angles. Because $l = r\cos\theta$, it follows that

$$\cos\theta \approx 1 \quad \text{if} \quad \theta \ll 1\text{ rad}$$

Finally, we can take the ratio of sine and cosine to find $\tan\theta \approx \sin\theta \approx \theta$. We will have other occasions to use the small-angle approximation throughout the remainder of this text.

NOTE The small-angle approximation is valid *only* if angle θ is in radians!

How small does θ have to be to justify using the small-angle approximation? It's easy to use your calculator to find that the small-angle approximation is good to three significant figures, an error of $\leq 0.1\%$, up to angles of ≈ 0.10 rad ($\approx 5°$). In practice, we will use the approximation up to about 10°, but for angles any larger it rapidly loses validity and produces unacceptable results.

FIGURE 15.17 Pendulum motion

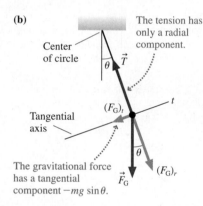

(a)

θ and s are negative on the left.
θ and s are positive on the right.

L

m

s

0 Arc length

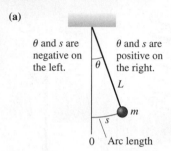

(b)

Center of circle

The tension has only a radial component.

$\vec{T}$

$(F_G)_t$

Tangential axis

The gravitational force has a tangential component $-mg\sin\theta$.

$\vec{F}_G$

$(F_G)_r$

FIGURE 15.18 The geometrical basis of the small-angle approximation.

r

$h = r\sin\theta$

$s = r\theta$

θ

$l = r\cos\theta$

If we restrict the pendulum to $\theta < 10°$, we can use $\sin\theta \approx \theta$. In that case, Equation 15.44 for the net force on the mass is

$$(F_{net})_t = -mg\sin\theta \approx -mg\theta = -\frac{mg}{L}s$$

where, in the last step, we used the fact that angle θ is related to the arc length by $\theta = s/L$. Then the equation of motion becomes

$$\frac{d^2s}{dt^2} = \frac{(F_{net})_t}{m} = -\frac{g}{L}s \qquad (15.46)$$

This is *exactly* the same as Equation 15.32 for a mass oscillating on a spring. The names are different, with x replaced by s and k/m by g/L, but that does not make it a different equation.

Because we know the solution to the spring problem, we can immediately write the solution to the pendulum problem just by changing variables and constants:

$$s(t) = A\cos(\omega t + \phi_0) \qquad \text{or} \qquad \theta(t) = \theta_{max}\cos(\omega t + \phi_0) \qquad (15.47)$$

The angular frequency

$$\omega = 2\pi f = \sqrt{\frac{g}{L}} \qquad (15.48)$$

is determined by the length of the string. The pendulum is interesting in that **the frequency, and hence the period, is independent of the mass.** It depends only on the length of the pendulum. The amplitude A and the phase constant ϕ_0 are determined by the initial conditions, just as they were for an oscillating spring.

EXAMPLE 15.7 | The maximum angle of a pendulum

A 300 g mass on a 30-cm-long string oscillates as a pendulum. It has a speed of 0.25 m/s as it passes through the lowest point. What maximum angle does the pendulum reach?

MODEL Assume that the angle remains small, in which case the motion is simple harmonic motion.

SOLVE The angular frequency of the pendulum is

$$\omega = \sqrt{\frac{g}{L}} = \sqrt{\frac{9.8 \text{ m/s}^2}{0.30 \text{ m}}} = 5.72 \text{ rad/s}$$

The speed at the lowest point is $v_{max} = \omega A$, so the amplitude is

$$A = s_{max} = \frac{v_{max}}{\omega} = \frac{0.25 \text{ m/s}}{5.72 \text{ rad/s}} = 0.0437 \text{ m}$$

The maximum angle, at the maximum arc length s_{max}, is

$$\theta_{max} = \frac{s_{max}}{L} = \frac{0.0437 \text{ m}}{0.30 \text{ m}} = 0.146 \text{ rad} = 8.3°$$

ASSESS Because the maximum angle is less than 10°, our analysis based on the small-angle approximation is reasonable.

EXAMPLE 15.8 | The gravimeter

Deposits of minerals and ore can alter the local value of the free-fall acceleration because they tend to be denser than surrounding rocks. Geologists use a *gravimeter*—an instrument that accurately measures the local free-fall acceleration—to search for ore deposits. One of the simplest gravimeters is a pendulum. To achieve the highest accuracy, a stopwatch is used to time 100 oscillations of a pendulum of different lengths. At one location in the field, a geologist makes the following measurements:

Length (m)	Time (s)
0.500	141.7
1.000	200.6
1.500	245.8
2.000	283.5

What is the local value of g?

MODEL Assume the oscillation angle is small, in which case the motion is simple harmonic motion with a period independent of the mass of the pendulum. Because the data are known to four significant figures (± 1 mm on the length and ± 0.1 s on the timing, both of which are easily achievable), we expect to determine g to four significant figures.

SOLVE From Equation 15.48, using $f = 1/T$, we find

$$T^2 = \left(2\pi\sqrt{\frac{L}{g}}\right)^2 = \frac{4\pi^2}{g}L$$

That is, the square of a pendulum's period is proportional to its length. Consequently, a graph of T^2 versus L should be a straight line passing through the origin with slope $4\pi^2/g$. We can use the

Continued

experimentally measured slope to determine g. **FIGURE 15.19** is a graph of the data, with the period found by dividing the measured time by 100.

As expected, the graph is a straight line passing through the origin. The slope of the best-fit line is 4.021 s²/m. Consequently,

$$g = \frac{4\pi^2}{\text{slope}} = \frac{4\pi^2}{4.021 \text{ s}^2/\text{m}} = 9.818 \text{ m/s}^2$$

ASSESS The fact that the graph is linear and passes through the origin confirms our model of the situation. Had this *not* been the case, we would have had to conclude either that our model of the pendulum as a simple, small-angle pendulum was not valid or that our measurements were bad. This is an important reason for having multiple data points rather than using only one length.

FIGURE 15.19 Graph of the square of the pendulum's period versus its length.

$y = 4.021x + 0.001$

Best-fit line

The Simple-Harmonic-Motion Model

You can begin to see how, in a sense, we have solved *all* simple-harmonic-motion problems once we have solved the problem of the horizontal spring. The restoring force of a spring, $F_{Sp} = -kx$, is directly proportional to the displacement x from equilibrium. The pendulum's restoring force, in the small-angle approximation, is directly proportional to the displacement s. A restoring force that is directly proportional to the displacement from equilibrium is called a **linear restoring force.** For *any* linear restoring force, the equation of motion is identical to the spring equation (other than perhaps using different symbols). Consequently, **any system with a linear restoring force will undergo simple harmonic motion around the equilibrium position.**

This is why an oscillating spring is the prototype of SHM. Everything that we learn about an oscillating spring can be applied to the oscillations of any other linear restoring force, ranging from the vibration of airplane wings to the motion of electrons in electric circuits.

MODEL 15.1

Simple harmonic motion

For any system with a restoring force that's linear or can be well approximated as linear.

- Motion is SHM around the equilibrium position.
- Frequency and period are independent of the amplitude.
- Mathematically:

  - For an appropriate position variable u, the equation of motion can be written
    $$d^2u/dt^2 = -Cu$$
    where C is a collection of constants.
  - The angular frequency is $\omega = \sqrt{C}$.
  - The position and velocity are
    $$u = A\cos(\omega t + \phi_0) \quad v_u = -v_{max}\sin(\omega t + \phi_0)$$
    where A and ϕ_0 are determined by the initial conditions.
  - Mechanical energy is conserved.
- Limitations: Model fails if the restoring force deviates significantly from linear.

Exercise 22

The Physical Pendulum

A mass on a string is often called a *simple pendulum*. But you can also make a pendulum from any solid object that swings back and forth on a pivot under the influence of gravity. This is called a *physical pendulum*.

FIGURE 15.20 shows a physical pendulum of mass M for which the distance between the pivot and the center of mass is l. The moment arm of the gravitational force acting at the center of mass is $d = l\sin\theta$, so the gravitational torque is

$$\tau = -Mgd = -Mgl\sin\theta$$

The torque is negative because, for positive θ, it's causing a clockwise rotation. If we restrict the angle to being small $(\theta < 10°)$, as we did for the simple pendulum, we can use the small-angle approximation to write

$$\tau = -Mgl\theta \tag{15.49}$$

Gravity exerts a linear restoring torque on the pendulum—that is, the torque is directly proportional to the angular displacement θ—so we expect the physical pendulum to undergo SHM.

From Chapter 12, Newton's second law for rotational motion is

$$\alpha = \frac{d^2\theta}{dt^2} = \frac{\tau}{I}$$

where I is the object's moment of inertia about the pivot point. Using Equation 15.49 for the torque, we find

$$\frac{d^2\theta}{dt^2} = \frac{-Mgl}{I}\theta \tag{15.50}$$

The equation of motion is of the form $d^2\theta/dt^2 = -C\theta$, so the model for simple harmonic motion tells us that the motion is SHM with angular frequency

$$\omega = 2\pi f = \sqrt{\frac{Mgl}{I}} \tag{15.51}$$

It appears that the frequency depends on the mass of the pendulum, but recall that the moment of inertia is directly proportional to M. Thus M cancels and the frequency of a physical pendulum, like that of a simple pendulum, is independent of mass.

FIGURE 15.20 A physical pendulum.

Distance from pivot to center of mass

l

θ

d

Moment arm of gravitational torque Mg

EXAMPLE 15.9 | A swinging leg as a pendulum

A student in a biomechanics lab measures the length of his leg, from hip to heel, to be 0.90 m. What is the frequency of the pendulum motion of the student's leg? What is the period?

MODEL We can model a human leg reasonably well as a rod of uniform cross section, pivoted at one end (the hip) to form a physical pendulum. For small-angle oscillations it will undergo SHM. The center of mass of a uniform leg is at the midpoint, so $l = L/2$.

SOLVE The moment of inertia of a rod pivoted about one end is $I = \frac{1}{3}ML^2$, so the pendulum frequency is

$$f = \frac{1}{2\pi}\sqrt{\frac{Mgl}{I}} = \frac{1}{2\pi}\sqrt{\frac{Mg(L/2)}{ML^2/3}} = \frac{1}{2\pi}\sqrt{\frac{3g}{2L}} = 0.64 \text{ Hz}$$

The corresponding period is $T = 1/f = 1.6$ s. Notice that we didn't need to know the mass.

ASSESS As you walk, your legs do swing as physical pendulums as you bring them forward. The frequency is fixed by the length of your legs and their distribution of mass; it doesn't depend on amplitude. Consequently, you don't increase your walking speed by taking more rapid steps—changing the frequency is difficult. You simply take longer strides, changing the amplitude but not the frequency.

STOP TO THINK 15.5 One person swings on a swing and finds that the period is 3.0 s. A second person of equal mass joins him. With two people swinging, the period is

a. 6.0 s
b. >3.0 s but not necessarily 6.0 s
c. 3.0 s
d. <3.0 s but not necessarily 1.5 s
e. 1.5 s
f. Can't tell without knowing the length

The shock absorbers in cars and trucks are heavily damped springs. The vehicle's vertical motion, after hitting a rock or a pothole, is a damped oscillation.

15.7 Damped Oscillations

A pendulum left to itself gradually slows down and stops. The sound of a ringing bell gradually dies away. All real oscillators do run down—some very slowly but others quite quickly—as friction or other dissipative forces transform their mechanical energy into the thermal energy of the oscillator and its environment. An oscillation that runs down and stops is called a **damped oscillation.**

There are many possible reasons for the dissipation of energy, such as air resistance, friction, and internal forces within a metal spring as it flexes. The forces involved in dissipation are complex, but a simple *linear drag* model gives a quite accurate description of most damped oscillations. That is, we'll assume a drag force that depends linearly on the velocity as

$$\vec{F}_{drag} = -b\vec{v} \quad \text{(model of the drag force)} \tag{15.52}$$

where the minus sign is the mathematical statement that the force is always opposite in direction to the velocity in order to slow the object.

The **damping constant** b depends in a complicated way on the shape of the object *and* on the viscosity of the air or other medium in which the particle moves. The damping constant plays the same role in our model of drag that the coefficient of friction does in our model of friction.

The units of b need to be such that they will give units of force when multiplied by units of velocity. As you can confirm, these units are kg/s. A value $b = 0$ kg/s corresponds to the limiting case of no resistance, in which case the mechanical energy is conserved. A typical value of b for a spring or a pendulum in air is ≤ 0.10 kg/s. Objects moving in a liquid can have significantly larger values of b.

FIGURE 15.21 shows a mass oscillating on a spring in the presence of a drag force. With the drag included, Newton's second law is

$$(F_{net})_x = (F_{Sp})_x + (F_{drag})_x = -kx - bv_x = ma_x \tag{15.53}$$

Using $v_x = dx/dt$ and $a_x = d^2x/dt^2$, we can write Equation 15.53 as

$$\frac{d^2x}{dt^2} + \frac{b}{m}\frac{dx}{dt} + \frac{k}{m}x = 0 \tag{15.54}$$

Equation 15.54 is the equation of motion of a damped oscillator. If you compare it to Equation 15.32, the equation of motion for a block on a frictionless surface, you'll see that it differs by the inclusion of the term involving dx/dt.

Equation 15.54 is another second-order differential equation. We will simply assert (and, as a homework problem, you can confirm) that the solution is

$$x(t) = Ae^{-bt/2m}\cos(\omega t + \phi_0) \quad \text{(damped oscillator)} \tag{15.55}$$

where the angular frequency is given by

$$\omega = \sqrt{\frac{k}{m} - \frac{b^2}{4m^2}} = \sqrt{\omega_0{}^2 - \frac{b^2}{4m^2}} \tag{15.56}$$

Here $\omega_0 = \sqrt{k/m}$ is the angular frequency of an undamped oscillator ($b = 0$). The constant e is the base of natural logarithms, so $e^{-bt/2m}$ is an *exponential function*. Because $e^0 = 1$, Equation 15.55 reduces to our previous $x(t) = A\cos(\omega t + \phi_0)$ when $b = 0$. This makes sense and gives us confidence in Equation 15.55.

Lightly Damped Oscillators

A *lightly damped* oscillator, which oscillates many times before stopping, is one for which $b/2m \ll \omega_0$. In that case, $\omega \approx \omega_0$ is a good approximation. That is, light damping does not affect the oscillation frequency.

FIGURE 15.22 is a graph of the position $x(t)$ for a lightly damped oscillator, as given by Equation 15.55. To keep things simple, we've assumed that the phase constant is

FIGURE 15.21 An oscillating mass in the presence of a drag force.

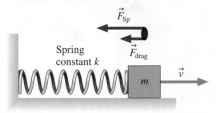

FIGURE 15.22 Position-versus-time graph for a lightly damped oscillator.

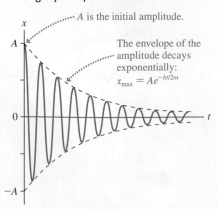

zero. Notice that the term $Ae^{-bt/2m}$, which is shown by the dashed line, acts as a slowly varying amplitude:

$$x_{max}(t) = Ae^{-bt/2m} \tag{15.57}$$

where A is the *initial* amplitude, at $t = 0$. The oscillation keeps bumping up against this line, slowly dying out with time.

A slowly changing line that provides a border to a rapid oscillation is called the **envelope** of the oscillations. In this case, the oscillations have an *exponentially decaying envelope*. Make sure you study Figure 15.22 long enough to see how both the oscillations and the decaying amplitude are related to Equation 15.55.

Changing the amount of damping, by changing the value of b, affects how quickly the oscillations decay. **FIGURE 15.23** shows just the envelope $x_{max}(t)$ for several oscillators that are identical except for the value of the damping constant b. (You need to imagine a rapid oscillation within each envelope, as in Figure 15.22.) Increasing b causes the oscillations to damp more quickly, while decreasing b makes them last longer.

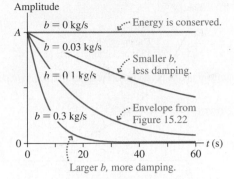

FIGURE 15.23 Oscillation envelopes for several values of b, mass 1.0 kg.

MATHEMATICAL ASIDE

Exponential decay

Exponential decay occurs in a vast number of physical systems of importance in science and engineering. Mechanical vibrations, electric circuits, and nuclear radioactivity all exhibit exponential decay.

The mathematical analysis of physical systems frequently leads to solutions of the form

$$u = Ae^{-v/v_0} = A \exp(-v/v_0)$$

where exp is the *exponential function*. The number $e = 2.71828\ldots$ is the base of natural logarithms in the same way that 10 is the base of ordinary logarithms.

A graph of u illustrates what we mean by exponential decay. It starts with $u = A$ at $v = 0$ (because $e^0 = 1$) and then steadily decays, asymptotically approaching zero. The quantity v_0 is called the *decay constant*. When $v = v_0$, $u = e^{-1}A = 0.37A$. When $v = 2v_0$, $u = e^{-2}A = 0.13A$.

The decay constant v_0 must have the same units as v. If v represents position, then v_0 is a length; if v represents time, then v_0 is a time interval. In a specific situation, v_0 is often called the *decay length* or the *decay time*. It is the length or time in which the quantity decays to 37% of its initial value.

No matter what the process is or what u represents, **a quantity that decays exponentially decays to 37% of its initial value when one decay constant has passed.** Thus exponential decay is a universal behavior. The decay curve always looks exactly like the figure shown here. Once you've learned the properties of exponential decay, you'll immediately know how to apply this knowledge to a new situation.

The mechanical energy of a damped oscillator is *not* conserved because of the drag force. We previously found the energy of an undamped oscillator to be $E = \frac{1}{2}kA^2$. This is still valid for a lightly damped oscillator if we replace A with the slowly decaying amplitude x_{max}. Thus

$$E(t) = \frac{1}{2}k(x_{max})^2 = \frac{1}{2}k(Ae^{-bt/2m})^2 = \frac{1}{2}kA^2e^{-bt/m} \tag{15.58}$$

Here A is the initial amplitude, so $\frac{1}{2}kA^2$ is the initial energy, which we call E_0. Let's define the **time constant** τ (also called the *decay constant* or the *decay time*) to be

$$\tau = \frac{m}{b} \tag{15.59}$$

FIGURE 15.24 Energy decay of a lightly damped oscillator.

Because b has units of kg/s, τ has units of seconds. With this, we can write the energy decay as

$$E(t) = E_0 e^{-t/\tau} \qquad (15.60)$$

In other words, **a lightly damped oscillator's mechanical energy decays exponentially with time constant τ.**

As **FIGURE 15.24** shows, the time constant is the amount of time needed for the energy to decay to e^{-1}, or 37%, of its initial value. We say that the time constant τ measures the "characteristic time" during which the energy of the oscillation is dissipated. Roughly two-thirds of the initial energy is gone after one time constant has elapsed, and nearly 90% has dissipated after two time constants have gone by.

For practical purposes, we can speak of the time constant as the *lifetime* of the oscillation—about how long it lasts. Mathematically, there is never a time when the oscillation is "over." The decay approaches zero asymptotically, but it never gets there in any finite time. The best we can do is define a characteristic time when the motion is "almost over," and that is what the time constant τ does.

EXAMPLE 15.10 | **A damped pendulum**

A 500 g mass swings on a 60-cm-string as a pendulum. The amplitude is observed to decay to half its initial value after 35 oscillations.

a. What is the time constant for this oscillator?

b. At what time will the *energy* have decayed to half its initial value?

MODEL The motion is a damped oscillation.

SOLVE a. The initial amplitude at $t = 0$ is $x_{max} = A$. After 35 oscillations the amplitude is $x_{max} = \frac{1}{2}A$. The period of the pendulum is

$$T = 2\pi\sqrt{\frac{L}{g}} = 2\pi\sqrt{\frac{0.60 \text{ m}}{9.8 \text{ m/s}^2}} = 1.55 \text{ s}$$

so 35 oscillations have occurred at $t = 54.2$ s.

The amplitude of oscillation at time t is given by Equation 15.57: $x_{max}(t) = Ae^{-bt/2m} = Ae^{-t/2\tau}$. In this case,

$$\tfrac{1}{2}A = Ae^{-(54.2 \text{ s})/2\tau}$$

Notice that we do not need to know A itself because it cancels out. To solve for τ, we take the natural logarithm of both sides of the equation:

$$\ln\left(\tfrac{1}{2}\right) = -\ln 2 = \ln e^{-(54.2 \text{ s})/2\tau} = -\frac{54.2 \text{ s}}{2\tau}$$

This is easily rearranged to give

$$\tau = \frac{54.2 \text{ s}}{2 \ln 2} = 39 \text{ s}$$

If desired, we could now determine the damping constant to be $b = m/\tau = 0.013$ kg/s.

b. The energy at time t is given by

$$E(t) = E_0 e^{-t/\tau}$$

The time at which an exponential decay is reduced to $\frac{1}{2}E_0$, half its initial value, has a special name. It is called the **half-life** and given the symbol $t_{1/2}$. The concept of the half-life is widely used in applications such as radioactive decay. To relate $t_{1/2}$ to τ, we first write

$$E(\text{at } t = t_{1/2}) = \tfrac{1}{2}E_0 = E_0 e^{-t_{1/2}/\tau}$$

The E_0 cancels, giving

$$\tfrac{1}{2} = e^{-t_{1/2}/\tau}$$

Again, we take the natural logarithm of both sides:

$$\ln\left(\tfrac{1}{2}\right) = -\ln 2 = \ln e^{-t_{1/2}/\tau} = -t_{1/2}/\tau$$

Finally, we solve for $t_{1/2}$:

$$t_{1/2} = \tau \ln 2 = 0.693\tau$$

This result that $t_{1/2}$ is 69% of τ is valid for any exponential decay. In this particular problem, half the energy is gone at

$$t_{1/2} = (0.693)(39 \text{ s}) = 27 \text{ s}$$

ASSESS The oscillator loses energy faster than it loses amplitude. This is what we should expect because the energy depends on the *square* of the amplitude.

STOP TO THINK 15.6 Rank in order, from largest to smallest, the time constants τ_a to τ_d of the decays shown in the figure. All the graphs have the same scale.

15.8 Driven Oscillations and Resonance

Thus far we have focused on the free oscillations of an isolated system. Some initial disturbance displaces the system from equilibrium, and it then oscillates freely until its energy is dissipated. These are very important situations, but they do not exhaust the possibilities. Another important situation is an oscillator that is subjected to a periodic external force. Its motion is called a **driven oscillation.**

A simple example of a driven oscillation is pushing a child on a swing, where your push is a periodic external force applied to the swing. A more complex example is a car driving over a series of equally spaced bumps. Each bump causes a periodic upward force on the car's shock absorbers, which are big, heavily damped springs. The electromagnetic coil on the back of a loudspeaker cone provides a periodic magnetic force to drive the cone back and forth, causing it to send out sound waves. Air turbulence moving across the wings of an aircraft can exert periodic forces on the wings and other aerodynamic surfaces, causing them to vibrate if they are not properly designed.

As these examples suggest, driven oscillations have many important applications. However, driven oscillations are a mathematically complex subject. We will simply hint at some of the results, saving the details for more advanced classes.

Consider an oscillating system that, when left to itself, oscillates at a frequency f_0. We will call this the **natural frequency** of the oscillator. The natural frequency for a mass on a spring is $\sqrt{k/m}/2\pi$, but it might be given by some other expression for another type of oscillator. Regardless of the expression, f_0 is simply the frequency of the system if it is displaced from equilibrium and released.

Suppose that this system is subjected to a *periodic* external force of frequency f_{ext}. This frequency, which is called the **driving frequency,** is completely independent of the oscillator's natural frequency f_0. Somebody or something in the environment selects the frequency f_{ext} of the external force, causing the force to push on the system f_{ext} times every second.

Although it is possible to solve Newton's second law with an external driving force, we will be content to look at a graphical representation of the solution. The most important result is that the oscillation amplitude depends very sensitively on the frequency f_{ext} of the driving force. The response to the driving frequency is shown in **FIGURE 15.25** for a system with $m = 1.0$ kg, a natural frequency $f_0 = 2.0$ Hz, and a damping constant $b = 0.20$ kg/s. This graph of amplitude versus driving frequency, called the **response curve,** occurs in many different applications.

When the driving frequency is substantially different from the oscillator's natural frequency, at the right and left edges of Figure 15.25, the system oscillates but the amplitude is very small. The system simply does not respond well to a driving frequency that differs much from f_0. As the driving frequency gets closer and closer to the natural frequency, the amplitude of the oscillation rises dramatically. After all, f_0 is the frequency at which the system "wants" to oscillate, so it is quite happy to respond to a driving frequency near f_0. Hence the amplitude reaches a maximum when the driving frequency exactly matches the system's natural frequency: $f_{ext} = f_0$.

The amplitude can become exceedingly large when the frequencies match, especially if the damping constant is very small. **FIGURE 15.26** shows the same oscillator with three different values of the damping constant. There's very little response if the damping constant is increased to 0.80 kg/s, but the amplitude for $f_{ext} = f_0$ becomes very large when the damping constant is reduced to 0.08 kg/s. This large-amplitude response to a driving force whose frequency matches the natural frequency of the system is a phenomenon called **resonance.** The condition for resonance is

$$f_{ext} = f_0 \quad \text{(resonance condition)} \tag{15.61}$$

Within the context of driven oscillations, the natural frequency f_0 is often called the **resonance frequency.**

An important feature of Figure 15.26 is how the amplitude and width of the resonance depend on the damping constant. A heavily damped system responds fairly

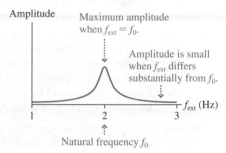

FIGURE 15.25 The response curve of a driven oscillator at frequencies near its natural frequency of 2.0 Hz.

FIGURE 15.26 The resonance amplitude becomes higher and narrower as the damping constant decreases.

A singer or musical instrument can shatter a crystal goblet by matching the goblet's natural oscillation frequency.

little, even at resonance, but it responds to a wide range of driving frequencies. Very lightly damped systems can reach exceptionally high amplitudes, but notice that the range of frequencies to which the system responds becomes narrower and narrower as b decreases.

This allows us to understand why a few singers can break crystal goblets but not inexpensive, everyday glasses. An inexpensive glass gives a "thud" when tapped, but a fine crystal goblet "rings" for several seconds. In physics terms, the goblet has a much longer time constant than the glass. That, in turn, implies that the goblet is very lightly damped while the ordinary glass is heavily damped (because the internal forces within the glass are not those of a high-quality crystal structure).

The singer causes a sound wave to impinge on the goblet, exerting a small driving force at the frequency of the note she is singing. If the singer's frequency matches the natural frequency of the goblet—resonance! Only the lightly damped goblet, like the top curve in Figure 15.26, can reach amplitudes large enough to shatter. The restriction, though, is that its natural frequency has to be matched very precisely. The sound also has to be very loud.

CHALLENGE **EXAMPLE 15.11** | A swinging pendulum

A pendulum consists of a massless, rigid rod with a mass at one end. The other end is pivoted on a frictionless pivot so that the rod can rotate in a complete circle. The pendulum is inverted, with the mass directly above the pivot point, then released. The speed of the mass as it passes through the lowest point is 5.0 m/s. If the pendulum later undergoes small-amplitude oscillations at the bottom of the arc, what will its frequency be?

MODEL This is a simple pendulum because the rod is massless. However, our analysis of a pendulum used the small-angle approximation. It applies only to the small-amplitude oscillations at the end, *not* to the pendulum swinging down from the inverted position. Fortunately, energy is conserved throughout, so we can analyze the big swing using conservation of mechanical energy.

VISUALIZE FIGURE 15.27 is a pictorial representation of the pendulum swinging down from the inverted position. The pendulum length is L, so the initial height is $2L$.

FIGURE 15.27 Before-and-after pictorial representation of the pendulum swinging down from an inverted position.

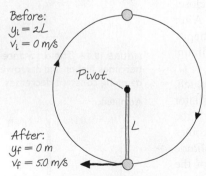

Before:
$y_i = 2L$
$v_i = 0$ m/s

Pivot

L

After:
$y_f = 0$ m
$v_f = 5.0$ m/s

SOLVE The frequency of a simple pendulum is $f = \sqrt{g/L}\,/2\pi$. We're not given L, but we can find it by analyzing the pendulum's swing down from an inverted position. Mechanical energy is conserved, and the only potential energy is gravitational potential energy. Conservation of mechanical energy $K_f + U_{Gf} = K_i + U_{Gi}$, with $U_G = mgy$, is

$$\tfrac{1}{2}mv_f^2 + mgy_f = \tfrac{1}{2}mv_i^2 + mgy_i$$

The mass cancels, which is good since we don't know it, and two terms are zero. Thus

$$\tfrac{1}{2}v_f^2 = g(2L) = 2gL$$

Solving for L, we find

$$L = \frac{v_f^2}{4g} = \frac{(5.0 \text{ m/s})^2}{4(9.80 \text{ m/s}^2)} = 0.638 \text{ m}$$

Now we can calculate the frequency:

$$f = \frac{1}{2\pi}\sqrt{\frac{g}{L}} = \frac{1}{2\pi}\sqrt{\frac{9.80 \text{ m/s}^2}{0.638 \text{ m}}} = 0.62 \text{ Hz}$$

ASSESS The frequency corresponds to a period of about 1.5 s, which seems reasonable.

SUMMARY

The goal of Chapter 15 has been to learn about systems that oscillate in simple harmonic motion.

GENERAL PRINCIPLES

Dynamics

SHM occurs when a **linear restoring force** acts to return a system to an equilibrium position.

Horizontal spring

$(F_{net})_x = -kx$

Vertical spring

The origin is at the equilibrium position $\Delta L = mg/k$.

$(F_{net})_y = -ky$

Both: $\omega = \sqrt{\dfrac{k}{m}}$ $T = 2\pi\sqrt{\dfrac{m}{k}}$

Simple pendulum

$\omega = \sqrt{\dfrac{g}{L}}$

$T = 2\pi\sqrt{\dfrac{L}{g}}$

Physical pendulum

$\omega = \sqrt{\dfrac{Mgl}{I}}$

$T = 2\pi\sqrt{\dfrac{I}{Mgl}}$

Energy

If there is no friction or dissipation, kinetic and potential energy are alternately transformed into each other, but the total mechanical energy $E = K + U$ is conserved.

$$E = \tfrac{1}{2}mv^2 + \tfrac{1}{2}kx^2$$
$$= \tfrac{1}{2}m(v_{max})^2$$
$$= \tfrac{1}{2}kA^2$$

The energy of a lightly damped oscillator decays exponentially

$$E = E_0 e^{-t/\tau}$$

where τ is the **time constant**.

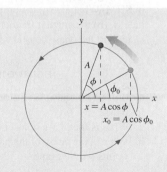

IMPORTANT CONCEPTS

Simple harmonic motion (SHM) is a sinusoidal oscillation with period T and amplitude A.

Frequency $f = \dfrac{1}{T}$

Angular frequency

$$\omega = 2\pi f = \dfrac{2\pi}{T}$$

Position $x(t) = A\cos(\omega t + \phi_0)$

$$= A\cos\left(\dfrac{2\pi t}{T} + \phi_0\right)$$

Velocity $v_x(t) = -v_{max}\sin(\omega t + \phi_0)$ with maximum speed $v_{max} = \omega A$

Acceleration $a_x(t) = -\omega^2 x(t) = -\omega^2 A\cos(\omega t + \phi_0)$

SHM is the projection onto the x-axis of **uniform circular motion.**

$\phi = \omega t + \phi_0$ is the **phase**

The position at time t is

$$x(t) = A\cos\phi$$
$$= A\cos(\omega t + \phi_0)$$

The **phase constant** ϕ_0 is determined by the initial conditions:

$$x_0 = A\cos\phi_0 \qquad v_{0x} = -\omega A\sin\phi_0$$

APPLICATIONS

Resonance

When a system is driven by a periodic external force, it responds with a large-amplitude oscillation if $f_{ext} \approx f_0$, where f_0 is the system's natural oscillation frequency, or **resonant frequency.**

Damping

If there is a drag force $\vec{F}_{drag} = -b\vec{v}$, where b is the damping constant, then (for lightly damped systems)

$$x(t) = Ae^{-bt/2m}\cos(\omega t + \phi_0)$$

The time constant for energy loss is $\tau = m/b$.

TERMS AND NOTATION

oscillatory motion	angular frequency, ω	damped oscillation	natural frequency, f_0
period, T	phase, ϕ	damping constant, b	driving frequency, f_{ext}
simple harmonic motion, SHM	phase constant, ϕ_0	envelope	response curve
	restoring force	exponential decay	resonance
amplitude, A	equation of motion	time constant, τ	resonance frequency, f_0
frequency, f	small-angle approximation	half-life, $t_{1/2}$	
hertz, Hz	linear restoring force	driven oscillation	

CONCEPTUAL QUESTIONS

1. A block oscillating on a spring has period $T = 2$ s. What is the period if:
   a. The block's mass is doubled? Explain. Note that you do not know the value of either m or k, so do *not* assume any particular values for them. The required analysis involves thinking about ratios.
   b. The value of the spring constant is quadrupled?
   c. The oscillation amplitude is doubled while m and k are unchanged?

2. A pendulum on Planet X, where the value of g is unknown, oscillates with a period $T = 2$ s. What is the period of this pendulum if:
   a. Its mass is doubled? Explain. Note that you do not know the value of m, L, or g, so do not assume any specific values. The required analysis involves thinking about ratios.
   b. Its length is doubled?
   c. Its oscillation amplitude is doubled?

3. **FIGURE Q15.3** shows a position-versus-time graph for a particle in SHM. What are (a) the amplitude A, (b) the angular frequency ω, and (c) the phase constant ϕ_0?

FIGURE Q15.3

4. **FIGURE Q15.4** shows a position-versus-time graph for a particle in SHM.
   a. What is the phase constant ϕ_0? Explain.
   b. What is the phase of the particle at each of the three numbered points on the graph?

FIGURE Q15.4

5. Equation 15.25 states that $\frac{1}{2}kA^2 = \frac{1}{2}m(v_{max})^2$. What does this mean? Write a couple of sentences explaining how to interpret this equation.

6. A block oscillating on a spring has an amplitude of 20 cm. What will the amplitude be if the total energy is doubled? Explain.

7. A block oscillating on a spring has a maximum speed of 20 cm/s. What will the block's maximum speed be if the total energy is doubled? Explain.

8. The solid disk and circular hoop in **FIGURE Q15.8** have the same radius and the same mass. Each can swing back and forth as a pendulum from a pivot at the top edge. Which, if either, has the larger period of oscillation?

FIGURE Q15.8

9. **FIGURE Q15.9** shows the potential-energy diagram and the total energy line of a particle oscillating on a spring.
   a. What is the spring's equilibrium length?
   b. Where are the turning points of the motion? Explain.
   c. What is the particle's maximum kinetic energy?
   d. What will be the turning points if the particle's total energy is doubled?

FIGURE Q15.9

10. Suppose the damping constant b of an oscillator increases.
    a. Is the medium more resistive or less resistive?
    b. Do the oscillations damp out more quickly or less quickly?
    c. Is the time constant τ increased or decreased?

11. a. Describe the difference between τ and T. Don't just *name* them; say what is different about the physical concepts they represent.
    b. Describe the difference between τ and $t_{1/2}$.

12. What is the difference between the driving frequency and the natural frequency of an oscillator?

EXERCISES AND PROBLEMS

Problems labeled ▓ integrate material from earlier chapters.

Exercises

Section 15.1 Simple Harmonic Motion

1. | An air-track glider attached to a spring oscillates between the 10 cm mark and the 60 cm mark on the track. The glider completes 10 oscillations in 33 s. What are the (a) period, (b) frequency, (c) angular frequency, (d) amplitude, and (e) maximum speed of the glider?

2. ‖ An air-track glider is attached to a spring. The glider is pulled to the right and released from rest at $t = 0$ s. It then oscillates with a period of 2.0 s and a maximum speed of 40 cm/s.
   a. What is the amplitude of the oscillation?
   b. What is the glider's position at $t = 0.25$ s?

3. | When a guitar string plays the note "A," the string vibrates at 440 Hz. What is the period of the vibration?

4. ‖ An object in SHM oscillates with a period of 4.0 s and an amplitude of 10 cm. How long does the object take to move from $x = 0.0$ cm to $x = 6.0$ cm?

Section 15.2 SHM and Circular Motion

5. | What are the (a) amplitude, (b) frequency, and (c) phase constant of the oscillation shown in FIGURE EX15.5?

FIGURE EX15.5

FIGURE EX15.6

6. ‖ What are the (a) amplitude, (b) frequency, and (c) phase constant of the oscillation shown in FIGURE EX15.6?

7. ‖ FIGURE EX15.7 is the position-versus-time graph of a particle in simple harmonic motion.
   a. What is the phase constant?
   b. What is the velocity at $t = 0$ s?
   c. What is v_{max}?

FIGURE EX15.7

FIGURE EX15.8

8. ‖ FIGURE EX15.8 is the velocity-versus-time graph of a particle in simple harmonic motion.
   a. What is the amplitude of the oscillation?
   b. What is the phase constant?
   c. What is the position at $t = 0$ s?

9. ‖ An object in simple harmonic motion has an amplitude of 4.0 cm, a frequency of 2.0 Hz, and a phase constant of $2\pi/3$ rad. Draw a position graph showing two cycles of the motion.

10. ‖ An object in simple harmonic motion has an amplitude of 8.0 cm, a frequency of 0.25 Hz, and a phase constant of $-\pi/2$ rad. Draw a position graph showing two cycles of the motion.

11. | An object in simple harmonic motion has amplitude 4.0 cm and frequency 4.0 Hz, and at $t = 0$ s it passes through the equilibrium point moving to the right. Write the function $x(t)$ that describes the object's position.

12. | An object in simple harmonic motion has amplitude 8.0 cm and frequency 0.50 Hz. At $t = 0$ s it has its most negative position. Write the function $x(t)$ that describes the object's position.

13. ‖ An air-track glider attached to a spring oscillates with a period of 1.5 s. At $t = 0$ s the glider is 5.00 cm left of the equilibrium position and moving to the right at 36.3 cm/s.
   a. What is the phase constant?
   b. What is the phase at $t = 0$ s, 0.5 s, 1.0 s, and 1.5 s?

Section 15.3 Energy in SHM

Section 15.4 The Dynamics of SHM

14. | A block attached to a spring with unknown spring constant oscillates with a period of 2.0 s. What is the period if
   a. The mass is doubled?
   b. The mass is halved?
   c. The amplitude is doubled?
   d. The spring constant is doubled?
   Parts a to d are independent questions, each referring to the initial situation.

15. ‖ A 200 g air-track glider is attached to a spring. The glider is pushed in 10 cm and released. A student with a stopwatch finds that 10 oscillations take 12.0 s. What is the spring constant?

16. ‖ A 200 g mass attached to a horizontal spring oscillates at a frequency of 2.0 Hz. At $t = 0$ s, the mass is at $x = 5.0$ cm and has $v_x = -30$ cm/s. Determine:
   a. The period.   b. The angular frequency.
   c. The amplitude.   d. The phase constant.
   e. The maximum speed.   f. The maximum acceleration.
   g. The total energy.   h. The position at $t = 0.40$ s.

17. ‖ The position of a 50 g oscillating mass is given by $x(t) = (2.0 \text{ cm})\cos(10t - \pi/4)$, where t is in s. Determine:
   a. The amplitude.   b. The period.
   c. The spring constant.   d. The phase constant.
   e. The initial conditions. f. The maximum speed.
   g. The total energy.   h. The velocity at $t = 0.40$ s.

18. ‖ A 1.0 kg block is attached to a spring with spring constant 16 N/m. While the block is sitting at rest, a student hits it with a hammer and almost instantaneously gives it a speed of 40 cm/s. What are
   a. The amplitude of the subsequent oscillations?
   b. The block's speed at the point where $x = \frac{1}{2}A$?

19. ‖ A student is bouncing on a trampoline. At her highest point, her feet are 55 cm above the trampoline. When she lands, the trampoline sags 15 cm before propelling her back up. For how long is she in contact with the trampoline?

20. || **FIGURE EX15.20** is a kinetic-energy graph of a mass oscillating on a *very* long horizontal spring. What is the spring constant?

FIGURE EX15.20

Section 15.5 Vertical Oscillations

21. | A spring is hanging from the ceiling. Attaching a 500 g physics book to the spring causes it to stretch 20 cm in order to come to equilibrium.
    a. What is the spring constant?
    b. From equilibrium, the book is pulled down 10 cm and released. What is the period of oscillation?
    c. What is the book's maximum speed?

22. || A spring with spring constant 15 N/m hangs from the ceiling. A ball is attached to the spring and allowed to come to rest. It is then pulled down 6.0 cm and released. If the ball makes 30 oscillations in 20 s, what are its (a) mass and (b) maximum speed?

23. || A spring is hung from the ceiling. When a block is attached to its end, it stretches 2.0 cm before reaching its new equilibrium length. The block is then pulled down slightly and released. What is the frequency of oscillation?

Section 15.6 The Pendulum

24. | A grandfather clock ticks each time the pendulum passes through the lowest point. If the pendulum is modeled as a simple pendulum, how long must it be for the ticks to occur once a second?

25. || A 200 g ball is tied to a string. It is pulled to an angle of 8.0° and released to swing as a pendulum. A student with a stopwatch finds that 10 oscillations take 12 s. How long is the string?

26. | A mass on a string of unknown length oscillates as a pendulum with a period of 4.0 s. What is the period if
    a. The mass is doubled?
    b. The string length is doubled?
    c. The string length is halved?
    d. The amplitude is doubled?
    Parts a to d are independent questions, each referring to the initial situation.

27. | What is the length of a pendulum whose period on the moon matches the period of a 2.0-m-long pendulum on the earth?

28. | What is the period of a 1.0-m-long pendulum on (a) the earth and (b) Venus?

29. | Astronauts on the first trip to Mars take along a pendulum that has a period on earth of 1.50 s. The period on Mars turns out to be 2.45 s. What is the free-fall acceleration on Mars?

30. ||| A 100 g mass on a 1.0-m-long string is pulled 8.0° to one side and released. How long does it take for the pendulum to reach 4.0° on the opposite side?

31. || A uniform steel bar swings from a pivot at one end with a period of 1.2 s. How long is the bar?

Section 15.7 Damped Oscillations

Section 15.8 Driven Oscillations and Resonance

32. | A 2.0 g spider is dangling at the end of a silk thread. You can make the spider bounce up and down on the thread by tapping lightly on his feet with a pencil. You soon discover that you can give the spider the largest amplitude on his little bungee cord if you tap exactly once every second. What is the spring constant of the silk thread?

33. || The amplitude of an oscillator decreases to 36.8% of its initial value in 10.0 s. What is the value of the time constant?

34. || In a science museum, a 110 kg brass pendulum bob swings at the end of a 15.0-m-long wire. The pendulum is started at exactly 8:00 A.M. every morning by pulling it 1.5 m to the side and releasing it. Because of its compact shape and smooth surface, the pendulum's damping constant is only 0.010 kg/s. At exactly 12:00 noon, how many oscillations will the pendulum have completed and what is its amplitude?

35. | Vision is blurred if the head is vibrated at 29 Hz because the
BIO vibrations are resonant with the natural frequency of the eyeball in its socket. If the mass of the eyeball is 7.5 g, a typical value, what is the effective spring constant of the musculature that holds the eyeball in the socket?

36. || A 350 g mass on a 45-cm-long string is released at an angle of 4.5° from vertical. It has a damping constant of 0.010 kg/s. After 25 s, (a) how many oscillations has it completed and (b) how much energy has been lost?

Problems

37. || The motion of a particle is given by $x(t) = (25 \text{ cm})\cos(10t)$, where t is in s. What is the first time at which the kinetic energy is twice the potential energy?

38. || a. When the displacement of a mass on a spring is $\frac{1}{2}A$, what fraction of the energy is kinetic energy and what fraction is potential energy?
    b. At what displacement, as a fraction of A, is the energy half kinetic and half potential?

39. || For a particle in simple harmonic motion, show that $v_{max} = (\pi/2)v_{avg}$, where v_{avg} is the average speed during one cycle of the motion.

40. || A 100 g block attached to a spring with spring constant 2.5 N/m oscillates horizontally on a frictionless table. Its velocity is 20 cm/s when $x = -5.0$ cm.
    a. What is the amplitude of oscillation?
    b. What is the block's maximum acceleration?
    c. What is the block's position when the acceleration is maximum?
    d. What is the speed of the block when $x = 3.0$ cm?

41. || A 0.300 kg oscillator has a speed of 95.4 cm/s when its displacement is 3.00 cm and 71.4 cm/s when its displacement is 6.00 cm. What is the oscillator's maximum speed?

42. || An ultrasonic transducer, of the type used in medical ultra-
BIO sound imaging, is a very thin disk ($m = 0.10$ g) driven back and forth in SHM at 1.0 MHz by an electromagnetic coil.
    a. The maximum restoring force that can be applied to the disk without breaking it is 40,000 N. What is the maximum oscillation amplitude that won't rupture the disk?
    b. What is the disk's maximum speed at this amplitude?

43. ‖ Astronauts in space cannot weigh themselves by standing on a
BIO bathroom scale. Instead, they determine their mass by oscillating
on a large spring. Suppose an astronaut attaches one end of a large
spring to her belt and the other end to a hook on the wall of the
space capsule. A fellow astronaut then pulls her away from the
wall and releases her. The spring's length as a function of time is
shown in FIGURE P15.43.
a. What is her mass if the spring constant is 240 N/m?
b. What is her speed when the spring's length is 1.2 m?

FIGURE P15.43

44. ‖ Your lab instructor has asked you to measure a spring constant
using a dynamic method—letting it oscillate—rather than a static
method of stretching it. You and your lab partner suspend the
spring from a hook, hang different masses on the lower end, and
start them oscillating. One of you uses a meter stick to measure
the amplitude, the other uses a stopwatch to time 10 oscillations.
Your data are as follows:

Mass (g)	Amplitude (cm)	Time (s)
100	6.5	7.8
150	5.5	9.8
200	6.0	10.9
250	3.5	12.4

Use the best-fit line of an appropriate graph to determine the spring
constant.

45. ‖ A 5.0 kg block hangs from a spring with spring constant
2000 N/m. The block is pulled down 5.0 cm from the equilibrium
position and given an initial velocity of 1.0 m/s back toward equilibrium. What are the (a) frequency, (b) amplitude, and (c) total
mechanical energy of the motion?

46. ‖‖ A 200 g block hangs from a spring with spring constant 10 N/m.
At $t = 0$ s the block is 20 cm below the equilibrium point and
moving upward with a speed of 100 cm/s. What are the block's
a. Oscillation frequency?
b. Distance from equilibrium when the speed is 50 cm/s?
c. Distance from equilibrium at $t = 1.0$ s?

47. ‖ A block hangs in equilibrium from a vertical spring. When a
second identical block is added, the original block sags by 5.0 cm.
What is the oscillation frequency of the two-block system?

48. ‖ A 75 kg student jumps off a bridge with a 12-m-long bungee
cord tied to his feet. The massless bungee cord has a spring
constant of 430 N/m.
a. How far below the bridge is the student's lowest point?
b. How long does it take the student to reach his lowest point?
You can assume that the bungee cord exerts no force until it
begins to stretch.

49. ‖ Scientists are measuring the properties of a newly discovered
elastic material. They create a 1.5-m-long, 1.6-mm-diameter cord,
attach an 850 g mass to the lower end, then pull the mass down
2.5 mm and release it. Their high-speed video camera records 36
oscillations in 2.0 s. What is Young's modulus of the material?

50. ‖‖‖ A mass hanging from a spring oscillates with a period of 0.35 s.
Suppose the mass and spring are swung in a horizontal circle, with
the free end of the spring at the pivot. What rotation frequency, in
rpm, will cause the spring's length to stretch by 15%?

51. ‖ A compact car has a mass of 1200 kg. Assume that the car has
one spring on each wheel, that the springs are identical, and that
the mass is equally distributed over the four springs.
a. What is the spring constant of each spring if the empty car
bounces up and down 2.0 times each second?
b. What will be the car's oscillation frequency while carrying
four 70 kg passengers?

52. ‖ The two blocks in FIGURE P15.52 oscillate on a frictionless surface with a period of 1.5 s. The upper block just begins to slip
when the amplitude is increased to 40 cm. What is the coefficient
of static friction between the two blocks?

FIGURE P15.52

53. ‖ A 1.00 kg block is attached to a horizontal spring with spring
constant 2500 N/m. The block is at rest on a frictionless surface. A
10 g bullet is fired into the block, in the face opposite the spring,
and sticks. What was the bullet's speed if the subsequent oscillations have an amplitude of 10.0 cm?

54. ‖‖‖ It has recently become possible to "weigh" DNA molecules
BIO by measuring the influence of their mass on a nano-oscillator.
FIGURE P15.54 shows a thin rectangular cantilever etched out of
silicon (density 2300 kg/m^3) with a small gold dot (not visible)
at the end. If pulled down and released, the end of the cantilever
vibrates with SHM, moving up and down like a diving board after
a jump. When bathed with DNA molecules whose ends have been
modified to bind with gold, one or more molecules may attach to
the gold dot. The addition of their mass causes a very slight—but
measurable—decrease in the oscillation frequency.

FIGURE P15.54

A vibrating cantilever of mass M can be modeled as a block of
mass $\frac{1}{3}M$ attached to a spring. (The factor of $\frac{1}{3}$ arises from the moment of inertia of a bar pivoted at one end.) Neither the mass nor
the spring constant can be determined very accurately— perhaps
to only two significant figures—but the oscillation frequency can
be measured with very high precision simply by counting the oscillations. In one experiment, the cantilever was initially vibrating
at exactly 12 MHz. Attachment of a DNA molecule caused the
frequency to decrease by 50 Hz. What was the mass of the DNA?

55. ‖ It is said that Galileo discovered a basic principle of the pendulum—that the period is independent of the amplitude—by using
his pulse to time the period of swinging lamps in the cathedral
as they swayed in the breeze. Suppose that one oscillation of a
swinging lamp takes 5.5 s.
a. How long is the lamp chain?
b. What maximum speed does the lamp have if its maximum
angle from vertical is 3.0°?

56. || Orangutans can move by *brachiation,* swinging like a pendu
BIO lum beneath successive handholds. If an orangutan has arms that
are 0.90 m long and repeatedly swings to a 20° angle, taking one
swing after another, estimate its speed of forward motion in m/s.
While this is somewhat beyond the range of validity of the
small-angle approximation, the standard results for a pendulum
are adequate for making an estimate.

57. || The pendulum shown in FIGURE P15.57
is pulled to a 10° angle on the left side and
released.
   a. What is the period of this pendulum?
   b. What is the pendulum's maximum angle
   on the right side?

12 cm

Peg

12 cm

FIGURE P15.57

58. || A uniform rod of mass M and length L swings as a pendulum
on a pivot at distance $L/4$ from one end of the rod. Find an expression for the frequency f of small-angle oscillations.

59. || Interestingly, there have been several studies using cadavers to
BIO determine the moments of inertia of human body parts, information that is important in biomechanics. In one study, the center of
mass of a 5.0 kg lower leg was found to be 18 cm from the knee.
When the leg was allowed to pivot at the knee and swing freely as
a pendulum, the oscillation frequency was 1.6 Hz. What was the
moment of inertia of the lower leg about the knee joint?

60. || A 500 g air-track glider attached to a spring with spring con-
stant 10 N/m is sitting at rest on a frictionless air track. A 250 g
glider is pushed toward it from the far end of the track at a speed of
120 cm/s. It collides with and sticks to the 500 g glider. What are
the amplitude and period of the subsequent oscillations?

61. || A 200 g block attached to a horizontal spring is oscillating with
an amplitude of 2.0 cm and a frequency of 2.0 Hz. Just as it passes
through the equilibrium point, moving to the right, a sharp blow
directed to the left exerts a 20 N force for 1.0 ms. What are the new
(a) frequency and (b) amplitude?

62. || FIGURE P15.62 is a top view of an object of mass m connected
between two stretched rubber bands of length L. The object rests
on a frictionless surface. At equilibrium, the tension in each rubber band is T. Find an expression for the frequency of oscillations *perpendicular* to the rubber bands. Assume the amplitude is
sufficiently small that the magnitude of the tension in the rubber
bands is essentially unchanged as the mass oscillates.

L L

Rubber bands

FIGURE P15.62

Potential energy (J)

4×10^{-19}
3×10^{-19}
2×10^{-19}
1×10^{-19}

0.08 0.10 0.12 0.14 0.16
Bond length (nm)

FIGURE P15.63

63. || A molecular bond can be modeled as a spring between two
atoms that vibrate with simple harmonic motion. FIGURE P15.63
shows an SHM approximation for the potential energy of an HCl
molecule. Because the chlorine atom is so much more massive
than the hydrogen atom, it is reasonable to assume that the hydrogen atom ($m = 1.67 \times 10^{-27}$ kg) vibrates back and forth while
the chlorine atom remains at rest. Use the graph to estimate the
vibrational frequency of the HCl molecule.

64. || A penny rides on top of a piston as it undergoes vertical sim-
ple harmonic motion with an amplitude of 4.0 cm. If the frequency is low, the penny rides up and down without difficulty. If the
frequency is steadily increased, there comes a point at which the
penny leaves the surface.
   a. At what point in the cycle does the penny first lose contact
   with the piston?
   b. What is the maximum frequency for which the penny just
   barely remains in place for the full cycle?

65. || On your first trip to Planet X you happen to take along a 200 g
mass, a 40-cm-long spring, a meter stick, and a stopwatch. You're
curious about the free-fall acceleration on Planet X, where ordinary
tasks seem easier than on earth, but you can't find this information
in your Visitor's Guide. One night you suspend the spring from
the ceiling in your room and hang the mass from it. You find that
the mass stretches the spring by 31.2 cm. You then pull the mass
down 10.0 cm and release it. With the stopwatch you find that 10
oscillations take 14.5 s. Based on this information, what is g?

66. ||| Suppose a large spherical object, such as a planet, with radius R
and mass M has a narrow tunnel passing diametrically through it. A
particle of mass m is inside the tunnel at a distance $x \leq R$ from the
center. It can be shown that the net gravitational force on the particle is
due entirely to the sphere of mass with radius $r \leq x$; there is no net
gravitational force from the mass in the spherical shell with $r > x$.
   a. Find an expression for the gravitational force on the particle,
   assuming the object has uniform density. Your expression will
   be in terms of x, R, m, M, and any necessary constants.
   b. You should have found that the gravitational force is a linear
   restoring force. Consequently, in the absence of air resistance,
   objects in the tunnel will oscillate with SHM. Suppose an in-
   trepid astronaut exploring a 150-km-diameter, 3.5×10^{18} kg
   asteroid discovers a tunnel through the center. If she jumps
   into the hole, how long will it take her to fall all the way
   through the asteroid and emerge on the other side?

67. || The 15 g head of a bobble-head doll oscillates in SHM at a
frequency of 4.0 Hz.
   a. What is the spring constant of the spring on which the head
   is mounted?
   b. The amplitude of the head's oscillations decreases to 0.5 cm
   in 4.0 s. What is the head's damping constant?

68. || An oscillator with a mass of 500 g and a period of 0.50 s has
an amplitude that decreases by 2.0% during each complete oscillation. If the initial amplitude is 10 cm, what will be the amplitude
after 25 oscillations?

69. ||| A spring with spring constant 15.0 N/m hangs from the ceiling.
A 500 g ball is attached to the spring and allowed to come to
rest. It is then pulled down 6.0 cm and released. What is the time
constant if the ball's amplitude has decreased to 3.0 cm after
30 oscillations?

70. ||| A captive James Bond is strapped to a table beneath a huge
pendulum made of a 2.0-m-diameter uniform circular metal blade
rigidly attached, at its top edge, to a 6.0-m-long, massless rod.
The pendulum is set swinging with a 10° amplitude when its lower
edge is 3.0 m above the prisoner, then the table slowly starts ascending at 1.0 mm/s. After 25 minutes, the pendulum's amplitude
has decreased to 7.0°. Fortunately, the prisoner is freed with a
mere 30 s to spare. What was the speed of the lower edge of the
blade as it passed over him for the last time?

71. || A 250 g air-track glider is attached to a spring with spring
constant 4.0 N/m. The damping constant due to air resistance is
0.015 kg/s. The glider is pulled out 20 cm from equilibrium and
released. How many oscillations will it make during the time in
which the amplitude decays to e^{-1} of its initial value?

72. ‖ A 200 g oscillator in a vacuum chamber has a frequency of 2.0 Hz. When air is admitted, the oscillation decreases to 60% of its initial amplitude in 50 s. How many oscillations will have been completed when the amplitude is 30% of its initial value?

73. ‖ Prove that the expression for $x(t)$ in Equation 15.55 is a solution to the equation of motion for a damped oscillator, Equation 15.54, if and only if the angular frequency ω is given by the expression in Equation 15.56.

74. ‖ A block on a frictionless table is connected as shown in FIGURE P15.74 to two springs having spring constants k_1 and k_2. Show that the block's oscillation frequency is given by

$$f = \sqrt{f_1^2 + f_2^2}$$

where f_1 and f_2 are the frequencies at which it would oscillate if attached to spring 1 or spring 2 alone.

FIGURE P15.74

75. ‖ A block on a frictionless table is connected as shown in FIGURE P15.75 to two springs having spring constants k_1 and k_2. Find an expression for the block's oscillation frequency f in terms of the frequencies f_1 and f_2 at which it would oscillate if attached to spring 1 or spring 2 alone.

FIGURE P15.75

Challenge Problems

76. ‖‖ A 15-cm-long, 200 g rod is pivoted at one end. A 20 g ball of clay is stuck on the other end. What is the period if the rod and clay swing as a pendulum?

77. ‖‖ A solid sphere of mass M and radius R is suspended from a thin rod, as shown in FIGURE CP15.77. The sphere can swing back and forth at the bottom of the rod. Find an expression for the frequency f of small-angle oscillations.

FIGURE CP15.77

78. ‖‖ A uniform rod of length L oscillates as a pendulum about a
CALC pivot that is a distance x from the center.
   a. For what value of x, in terms of L, is the oscillation period a minimum?
   b. What is the minimum oscillation period of a 15 kg, 1.0-m-long steel bar?

79. ‖‖ A spring is standing upright on a table with its bottom end fastened to the table. A block is dropped from a height 3.0 cm above the top of the spring. The block sticks to the top end of the spring and then oscillates with an amplitude of 10 cm. What is the oscillation frequency?

80. ‖‖ The analysis of a simple pendulum assumed that the mass was a particle, with no size. A realistic pendulum is a small, uniform sphere of mass M and radius R at the end of a massless string, with L being the distance from the pivot to the center of the sphere.
   a. Find an expression for the period of this pendulum.
   b. Suppose $M = 25$ g, $R = 1.0$ cm, and $L = 1.0$ m, typical values for a real pendulum. What is the ratio T_{real}/T_{simple}, where T_{real} is your expression from part a and T_{simple} is the expression derived in this chapter?

81. ‖‖ FIGURE CP15.81 shows a 200 g uniform rod pivoted at one end. The other end is attached to a horizontal spring. The spring is neither stretched nor compressed when the rod hangs straight down. What is the rod's oscillation period? You can assume that the rod's angle from vertical is always small.

FIGURE CP15.81

16 Traveling Waves

This surfer is "catching a wave." At the same time, he's seeing light waves and hearing sound waves.

IN THIS CHAPTER, you will learn the basic properties of traveling waves.

What is a wave?

A wave is a disturbance traveling through a medium. In a transverse wave, the displacement is perpendicular to the direction of travel. In a longitudinal wave, the displacement is parallel to the direction of travel.

Transverse wave

Longitudinal wave

What are some wave properties?

A wave is characterized by:

- Wave speed: How fast it travels through the medium.
- Wavelength: The distance between two neighboring crests.
- Frequency: The number of oscillations per second.
- Amplitude: The maximum displacement.

« LOOKING BACK Sections 15.1–15.2 Properties of simple harmonic motion

Wavelength λ

Are sound and light waves?

Yes! Very important waves.

- Sound waves are longitudinal waves.
- Light waves are transverse waves.

The colors of visible light correspond to different wavelengths.

Visible light

700 600 500 400
Wavelength (nm)

Do waves carry energy?

They do. The rate at which a wave delivers energy to a surface is the intensity of the wave. For sound waves, we'll use a logarithmic decibel scale to characterize the loudness of a sound.

Intensity is decreasing

What is the Doppler effect?

The frequency and wavelength of a wave are shifted if there is relative motion between the source and the observer of the waves. This is called the Doppler effect. It explains why the pitch of an ambulance siren drops as it races past you.

Reduced wavelength

Increased wavelength

How will I use waves?

Waves are literally everywhere. Communications systems from radios to cell phones to fiber optics use waves. Sonar and radar and medical ultrasound use waves. Music and musical instruments are all about waves. Waves are present in the oceans, the atmosphere, and the earth. This chapter and the next will allow you to understand and work with a wide variety of waves that you may meet in your career.

16.1 An Introduction to Waves

From sound and light to ocean waves and seismic waves, we're surrounded by waves. Understanding musical instruments, cell phones, or lasers requires a knowledge of waves. With this chapter we shift our focus from the particle model to a new **wave model** that emphasizes those aspects of wave behavior common to all waves.

The wave model is built around the idea of a **traveling wave,** which is an organized disturbance traveling with a well-defined wave speed. We'll begin our study of traveling waves by looking at two distinct wave motions.

Two types of traveling waves

A transverse wave

Up/down   Motion of wave at speed v ⟶

Displacement

A longitudinal wave

Push/pull   Motion of wave at speed v ⟶

Displacement ⟶

A **transverse wave** is a wave in which the displacement is *perpendicular* to the direction in which the wave travels. For example, a wave travels along a string in a horizontal direction while the particles that make up the string oscillate vertically. Electromagnetic waves are also transverse waves because the electromagnetic fields oscillate perpendicular to the direction in which the wave travels.

In a **longitudinal wave,** the particles in the medium are displaced *parallel* to the direction in which the wave travels. Here we see a chain of masses connected by springs. If you give the first mass in the chain a sharp push, a disturbance travels down the chain by compressing and expanding the springs. Sound waves in gases and liquids are the most well known examples of longitudinal waves.

We can also classify waves on the basis of what is "waving":

1. **Mechanical waves** travel only within a material *medium,* such as air or water. Two familiar mechanical waves are sound waves and water waves.
2. **Electromagnetic waves,** from radio waves to visible light to x rays, are a self-sustaining oscillation of the *electromagnetic field.* Electromagnetic waves require no material medium and can travel through a vacuum.

The **medium** of a mechanical wave is the substance through or along which the wave moves. For example, the medium of a water wave is the water, the medium of a sound wave is the air, and the medium of a wave on a stretched string is the string. A medium must be *elastic.* That is, a restoring force of some sort brings the medium back to equilibrium after it has been displaced or disturbed. The tension in a stretched string pulls the string back straight after you pluck it. Gravity restores the level surface of a lake after the wave generated by a boat has passed by.

As a wave passes through a medium, the atoms of the medium—we'll simply call them the particles of the medium—are displaced from equilibrium. This is a **disturbance** of the medium. The water ripples of **FIGURE 16.1** are a disturbance of the water's surface. A pulse traveling down a string is a disturbance, as are the wake of a boat and the sonic boom created by a jet traveling faster than the speed of sound. **The disturbance of a wave is an *organized* motion of the particles in the medium,** in contrast to the *random* molecular motions of thermal energy.

FIGURE 16.1 Ripples on a pond are a traveling wave.

The disturbance is the rippling of the water's surface.

The water is the medium.

Wave Speed

A wave disturbance is created by a *source.* The source of a wave might be a rock thrown into water, your hand plucking a stretched string, or an oscillating loudspeaker cone pushing on the air. Once created, the disturbance travels outward through the medium at the **wave speed** v. This is the speed with which a ripple moves across the water or a pulse travels down a string.

NOTE The disturbance propagates through the medium, but the medium as a whole does not move! The ripples on the pond (the disturbance) move outward from the splash of the rock, but there is no outward flow of water from the splash. **A wave transfers energy, but it does not transfer any material or substance outward from the source.**

As an example, we'll prove in Section 16.4 that the wave speed on a string stretched with tension T_s is

$$v_{string} = \sqrt{\frac{T_s}{\mu}} \qquad \text{(wave speed on a stretched string)} \qquad (16.1)$$

where μ is the string's **linear density,** its mass-to-length ratio:

$$\mu = \frac{m}{L} \qquad (16.2)$$

The SI unit of linear density is kg/m. A fat string has a larger value of μ than a skinny string made of the same material. Similarly, a steel wire has a larger value of μ than a plastic string of the same diameter. We'll assume that strings are *uniform,* meaning the linear density is the same everywhere along the length of the string.

NOTE The subscript s on the symbol T_s for the string's tension distinguishes it from the symbol T for the *period* of oscillation.

Equation 16.1 is the wave *speed,* not the wave velocity, so v_{string} always has a positive value. Every point on a wave travels with this speed. You can increase the wave speed either by *increasing* the string's tension (make it tighter) or by *decreasing* the string's linear density (make it skinnier).

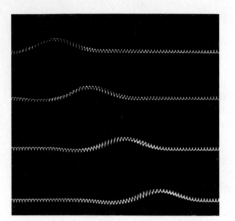

This sequence of photographs shows a wave pulse traveling along a spring.

EXAMPLE 16.1 | Measuring the linear density

A 2.00-m-long metal wire is attached to a motion sensor, stretched horizontally to a pulley 1.50 m away, then connected to a 2.00 kg hanging mass that provides tension. A mechanical pick plucks a horizontal segment of wire right at the pulley, creating a small wave pulse that travels along the wire. The plucking motion starts a timer that is stopped by the motion sensor when the pulse reaches the end of the wire. What is the wire's linear density if the pulse takes 18.0 ms to travel the length of the wire?

MODEL Model the pulse as a traveling wave and the pulley as frictionless.

VISUALIZE FIGURE 16.2 is a pictorial representation. The free-body diagram is for the hanging mass.

SOLVE The wave speed on a wire is determined by the wire's linear density μ and tension T_s. By measuring the wave speed and the tension, we can determine the linear density. The hanging mass is in equilibrium, with no net force, so we see from the free-body diagram that the tension throughout the wire (because the pulley is frictionless) is $T_s = F_G = Mg = 19.6$ N. Because the wave pulse travels 1.50 m in 18.0 ms, its speed is

$$v = \frac{1.50 \text{ m}}{0.0180 \text{ s}} = 83.3 \text{ m/s}$$

FIGURE 16.2 A wave pulse on a wire.

Thus, from Equation 16.1, the wire's linear density is

$$\mu = \frac{T_s}{v^2} = \frac{19.6 \text{ N}}{(83.3 \text{ m/s})^2} = 2.82 \times 10^{-3} \text{ kg/m} = 2.82 \text{ g/m}$$

Linear densities of strings are often stated in g/m, although these are not SI units. You must use kg/m in any calculations.

ASSESS A meter of thin wire is likely to have a mass of a few grams, so a linear density of a few g/m is reasonable. Note that the total length of the wire was not relevant.

The wave speed on a string is a property of the string—its tension and linear density. In general, **the wave speed is a property *of the medium.*** The wave speed depends on the restoring forces within the medium but not at all on the shape or size of the pulse, how the pulse was generated, or how far it has traveled.

STOP TO THINK 16.1 Which of the following actions would make a pulse travel faster along a stretched string? More than one answer may be correct. If so, give all that are correct.

a. Move your hand up and down more quickly as you generate the pulse.
b. Move your hand up and down a larger distance as you generate the pulse.
c. Use a heavier string of the same length, under the same tension.
d. Use a lighter string of the same length, under the same tension.
e. Stretch the string tighter to increase the tension.
f. Loosen the string to decrease the tension.
g. Put more force into the wave.

16.2 One-Dimensional Waves

To understand waves we must deal with functions of *two* variables. Until now, we have been concerned with quantities that depend only on time, such as $x(t)$ or $v(t)$. Functions of the one variable t are appropriate for a particle because a particle is only in one place at a time, but a wave is not localized. It is spread out through space at each instant of time. To describe a wave mathematically requires a function that specifies not only an instant of time (when) but also a point in space (where).

Rather than leaping into mathematics, we will start by thinking about waves graphically. Consider the wave pulse shown moving along a stretched string in **FIGURE 16.3**. (We will consider somewhat artificial triangular and square-shaped pulses in this section to make clear where the edges of the pulse are.) The graph shows the string's displacement Δy at a particular instant of time t_1 as a function of position x along the string. This is a "snapshot" of the wave, much like what you might make with a camera whose shutter is opened briefly at t_1. A graph that shows the wave's displacement as a function of position at a single instant of time is called a **snapshot graph.** For a wave on a string, a snapshot graph is literally a picture of the wave at this instant.

FIGURE 16.4 shows a sequence of snapshot graphs as the wave of Figure 16.3 continues to move. These are like successive frames from a video. Notice that the wave pulse moves forward distance $\Delta x = v\,\Delta t$ during the time interval Δt. That is, the wave moves with constant speed.

A snapshot graph tells only half the story. It tells us *where* the wave is and how it varies with position, but only at one instant of time. It gives us no information about how the wave *changes* with time. As a different way of portraying the wave, suppose we follow the dot marked on the string in Figure 16.4 and produce a graph showing how the displacement of this dot changes with time. The result, shown in **FIGURE 16.5**, is a displacement-versus-time graph at a single position in space. A graph that shows the wave's displacement as a function of time at a single position in space is called a **history graph.** It tells the history of that particular point in the medium.

FIGURE 16.3 A snapshot graph of a wave pulse on a string.

FIGURE 16.4 A sequence of snapshot graphs shows the wave in motion.

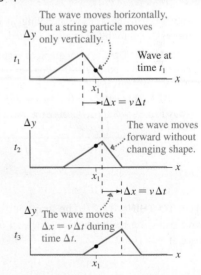

FIGURE 16.5 A history graph for the dot on the string in Figure 16.4.

FIGURE 16.6 An alternative look at a traveling wave.

You might think we have made a mistake; the graph of Figure 16.5 is reversed compared to Figure 16.4. It is not a mistake, but it requires careful thought to see why. As the wave moves toward the dot, the steep **leading edge** causes the dot to rise quickly. On the displacement-versus-time graph, *earlier* times (smaller values of *t*) are to the *left* and later times (larger *t*) to the right. Thus the leading edge of the wave is on the *left* side of the Figure 16.5 history graph. As you move to the right on Figure 16.5 you see the slowly falling **trailing edge** of the wave as it moves past the dot at later times.

The snapshot graph of Figure 16.3 and the history graph of Figure 16.5 portray complementary information. The snapshot graph tells us how things look throughout all of space, but at only one instant of time. The history graph tells us how things look at all times, but at only one position in space. We need them both to have the full story of the wave. An alternative representation of the wave is the series of graphs in **FIGURE 16.6**, where we can get a clearer sense of the wave moving forward. But graphs like these are essentially impossible to draw by hand, so it is necessary to move back and forth between snapshot graphs and history graphs.

EXAMPLE 16.2 | Finding a history graph from a snapshot graph

FIGURE 16.7 is a snapshot graph at $t = 0$ s of a wave moving to the right at a speed of 2.0 m/s. Draw a history graph for the position $x = 8.0$ m.

FIGURE 16.7 A snapshot graph at $t = 0$ s.

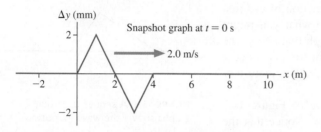

MODEL This is a wave traveling at constant speed. The pulse moves 2.0 m to the right every second.

VISUALIZE The snapshot graph of Figure 16.7 shows the wave at all points on the *x*-axis at $t = 0$ s. You can see that nothing is happening at $x = 8.0$ m at this instant of time because the wave has not yet reached $x = 8.0$ m. In fact, at $t = 0$ s the leading edge of the wave is still 4.0 m away from $x = 8.0$ m. Because the wave is traveling at 2.0 m/s, it will

take 2.0 s for the leading edge to reach $x = 8.0$ m. Thus the history graph for $x = 8.0$ m will be zero until $t = 2.0$ s. The first part of the wave causes a *downward* displacement of the medium, so immediately after $t = 2.0$ s the displacement at $x = 8.0$ m will be negative. The negative portion of the wave pulse is 2.0 m wide and takes 1.0 s to pass $x = 8.0$ m, so the midpoint of the pulse reaches $x = 8.0$ m at $t = 3.0$ s. The positive portion takes another 1.0 s to go past, so the trailing edge of the pulse arrives at $t = 4.0$ s. You could also note that the trailing edge was initially 8.0 m away from $x = 8.0$ m and needed 4.0 s to travel that distance at 2.0 m/s. The displacement at $x = 8.0$ m returns to zero at $t = 4.0$ s and remains zero for all later times. This information is all portrayed on the history graph of **FIGURE 16.8**.

FIGURE 16.8 The corresponding history graph at $x = 8.0$ m.

STOP TO THINK 16.2 The graph at the right is the history graph at $x = 4.0$ m of a wave traveling to the right at a speed of 2.0 m/s. Which is the history graph of this wave at $x = 0$ m?

(a)

(b)

(c)

(d)

Longitudinal Waves

For a wave on a string, a transverse wave, the snapshot graph is literally a picture of the wave. Not so for a longitudinal wave, where the particles in the medium are displaced parallel to the direction in which the wave is traveling. Thus the displacement is Δx rather than Δy, and a snapshot graph is a graph of Δx versus x.

FIGURE 16.9a is a snapshot graph of a longitudinal wave, such as a sound wave. It's purposefully drawn to have the same shape as the string wave in Example 16.2. Without practice, it's not clear what this graph tells us about the particles in the medium.

FIGURE 16.9 Visualizing a longitudinal wave.

(a)

(b)

1. Draw a series of equally spaced vertical lines to represent the equilibrium positions of particles before the wave arrives.

2. Use information from the graph to displace the particles in the medium to the right or left.

3. The wave propagates to the right at 1.0 cm/s.

To help you find out, **FIGURE 16.9b** provides a tool for visualizing longitudinal waves. In the second row, we've used information from the graph to displace the particles in the medium to the right or to the left of their equilibrium positions. For example, the particle at $x = 1.0$ cm has been displaced 0.5 cm to the right because the snapshot graph shows $\Delta x = 0.5$ cm at $x = 1.0$ cm. We now have a picture of the longitudinal wave pulse at $t_1 = 0$ s. You can see that the medium is compressed to higher density at the center of the pulse and, to compensate, expanded to lower density at the leading and trailing edges. Two more lines show the medium at $t_2 = 1$ s and $t_3 = 2$ s so that you can see the wave propagating through the medium at 1.0 cm/s.

The Displacement

A traveling wave causes the particles of the medium to be displaced from their equilibrium positions. Because one of our goals is to develop a mathematical representation to describe all types of waves, we'll use the generic symbol D to stand for the *displacement* of a wave of any type. But what do we mean by a "particle" in the medium? And what about electromagnetic waves, for which there is no medium?

For a string, where the atoms stay fixed relative to each other, you can think of either the atoms themselves or very small segments of the string as being the particles of the medium. D is then the perpendicular displacement Δy of a point on the string. For a sound wave, D is the longitudinal displacement Δx of a small volume of fluid. For any other mechanical wave, D is the appropriate displacement. Even electromagnetic waves can be described within the same mathematical representation if D is interpreted as a yet-undefined *electromagnetic field strength,* a "displacement" in a more abstract sense as an electromagnetic wave passes through a region of space.

You've probably seen or participated in "the wave" at a sporting event. The wave moves around the stadium, but the people (the medium) simply undergo small displacements from their equilibrium positions.

Because the displacement of a particle in the medium depends both on *where* the particle is (position *x*) and on *when* you observe it (time *t*), *D* must be a function of the two variables *x* and *t*. That is,

$$D(x, t) = \text{the displacement at time } t \text{ of a particle at position } x$$

The values of *both* variables—where and when—must be specified before you can evaluate the displacement *D*.

16.3 Sinusoidal Waves

A wave source that oscillates with simple harmonic motion (SHM) generates a **sinusoidal wave.** For example, a loudspeaker cone that oscillates in SHM radiates a sinusoidal sound wave. The sinusoidal electromagnetic waves broadcast by television and FM radio stations are generated by electrons oscillating back and forth in the antenna wire with SHM. **The frequency *f* of the wave is the frequency of the oscillating source.**

FIGURE 16.10 shows a sinusoidal wave moving through a medium. To understand how this wave propagates, let's look at history and snapshot graphs. FIGURE 16.11a is a history graph, showing the displacement of the medium at one point in space. Each particle in the medium undergoes simple harmonic motion with frequency *f*, so this graph of SHM is identical to the graphs you learned to work with in Chapter 15. The *period* of the wave, shown on the graph, is the time interval for one cycle of the motion. The period is related to the wave frequency *f* by

$$T = \frac{1}{f} \tag{16.3}$$

exactly as in simple harmonic motion. The **amplitude** *A* of the wave is the maximum value of the displacement. The crests of the wave have displacement $D_{\text{crest}} = A$ and the troughs have displacement $D_{\text{trough}} = -A$.

FIGURE 16.10 A sinusoidal wave moving along the *x*-axis.

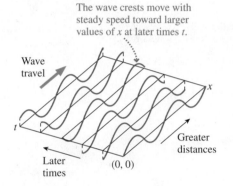

The wave crests move with steady speed toward larger values of *x* at later times *t*.

FIGURE 16.11 History and snapshot graphs for a sinusoidal wave.

(a) A history graph at one point in space

(b) A snapshot graph at one instant of time

Displacement versus time is only half the story. FIGURE 16.11b shows a snapshot graph for the same wave at one instant in time. Here we see the wave stretched out in space, moving to the right with speed *v*. An important characteristic of a sinusoidal wave is that it is *periodic in space* as well as in time. As you move from left to right along the "frozen" wave in the snapshot graph, the disturbance repeats itself over and over. The distance spanned by one cycle of the motion is called the **wavelength** of the wave. Wavelength is symbolized by λ (lowercase Greek lambda) and, because it is a length, it is measured in units of meters. The wavelength is shown in Figure 16.11b as the distance between two crests, but it could equally well be the distance between two troughs.

NOTE Wavelength is the spatial analog of period. The period *T* is the *time* in which the disturbance at a single point in space repeats itself. The wavelength λ is the *distance* in which the disturbance at one instant of time repeats itself.

The Fundamental Relationship for Sinusoidal Waves

There is an important relationship between the wavelength and the period of a wave. FIGURE 16.12 shows this relationship through five snapshot graphs of a sinusoidal wave at time increments of one-quarter of the period T. One full period has elapsed between the first graph and the last, which you can see by observing the motion at a fixed point on the x-axis. Each point in the medium has undergone exactly one complete oscillation.

The critical observation is that the wave crest marked by an arrow has moved one full wavelength between the first graph and the last. That is, **during a time interval of exactly one period T, each crest of a sinusoidal wave travels forward a distance of exactly one wavelength λ**. Because speed is distance divided by time, the wave speed must be

$$v = \frac{\text{distance}}{\text{time}} = \frac{\lambda}{T} \tag{16.4}$$

Because $f = 1/T$, it is customary to write Equation 16.4 in the form

$$v = \lambda f \tag{16.5}$$

Although Equation 16.5 has no special name, it is *the* fundamental relationship for periodic waves. When using it, keep in mind the *physical* meaning that a wave moves forward a distance of one wavelength during a time interval of one period.

NOTE Wavelength and period are defined only for *periodic* waves, so Equations 16.4 and 16.5 apply only to periodic waves. A wave pulse has a wave speed, but it doesn't have a wavelength or a period. Hence Equations 16.4 and 16.5 cannot be applied to wave pulses.

Because the wave speed is a property of the medium while the wave frequency is a property of the oscillating source, it is often useful to write Equation 16.5 as

$$\lambda = \frac{v}{f} = \frac{\text{property of the medium}}{\text{property of the source}} \tag{16.6}$$

The wavelength is a *consequence* of a wave of frequency f traveling through a medium in which the wave speed is v.

FIGURE 16.12 A series of snapshot graphs at time increments of one-quarter of the period T.

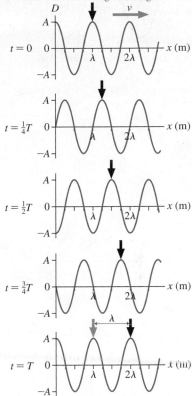

During a time interval of exactly one period, the crest has moved forward exactly one wavelength.

STOP TO THINK 16.3 What is the frequency of this traveling wave?

a. 0.1 Hz
b. 0.2 Hz
c. 2 Hz
d. 5 Hz
e. 10 Hz
f. 500 Hz

The Mathematics of Sinusoidal Waves

FIGURE 16.13 on the next page shows a snapshot graph at $t = 0$ of a sinusoidal wave. The sinusoidal function that describes the displacement of this wave is

$$D(x, t = 0) = A \sin\left(2\pi \frac{x}{\lambda} + \phi_0\right) \tag{16.7}$$

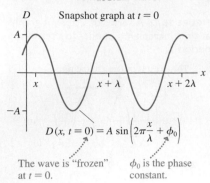

FIGURE 16.13 A sinusoidal wave.

The wave is "frozen" at $t = 0$.

ϕ_0 is the phase constant.

$$D(x, t = 0) = A \sin\left(2\pi \frac{x}{\lambda} + \phi_0\right)$$

where the notation $D(x, t = 0)$ means that we've frozen the time at $t = 0$ to make the displacement a function of only x. The term ϕ_0 is a *phase constant* that characterizes the initial conditions. (We'll return to the phase constant momentarily.)

The function of Equation 16.7 is periodic with period λ. We can see this by writing

$$D(x + \lambda) = A \sin\left(2\pi \frac{(x + \lambda)}{\lambda} + \phi_0\right) = A \sin\left(2\pi \frac{x}{\lambda} + \phi_0 + 2\pi \text{ rad}\right)$$

$$= A \sin\left(2\pi \frac{x}{\lambda} + \phi_0\right) = D(x)$$

where we used the fact that $\sin(a + 2\pi \text{ rad}) = \sin a$. In other words, the disturbance created by the wave at $x + \lambda$ is exactly the same as the disturbance at x.

The next step is to set the wave in motion. We can do this by replacing x in Equation 16.7 with $x - vt$. To see why this works, recall that the wave moves distance vt during time t. In other words, whatever displacement the wave has at position x at time t, the wave must have had that same displacement at position $x - vt$ at the earlier time $t = 0$. Mathematically, this idea can be captured by writing

$$D(x, t) = D(x - vt, t = 0) \tag{16.8}$$

Make sure you understand how this statement describes a wave moving in the positive x-direction at speed v.

This is what we were looking for. $D(x, t)$ is the general function describing the traveling wave. It's found by taking the function that describes the wave at $t = 0$—the function of Equation 16.7—and replacing x with $x - vt$. Thus the displacement equation of a sinusoidal wave traveling in the positive x-direction at speed v is

$$D(x, t) = A \sin\left(2\pi \frac{x - vt}{\lambda} + \phi_0\right) = A \sin\left(2\pi \left(\frac{x}{\lambda} - \frac{t}{T}\right) + \phi_0\right) \tag{16.9}$$

In the last step we used $v = \lambda f = \lambda/T$ to write $v/\lambda = 1/T$. The function of Equation 16.9 is not only periodic in space with period λ, it is also periodic in time with period T. That is, $D(x, t + T) = D(x, t)$.

It will be useful to introduce two new quantities. First, recall from simple harmonic motion the *angular frequency*

$$\omega = 2\pi f = \frac{2\pi}{T} \tag{16.10}$$

The units of ω are rad/s, although many textbooks use simply s^{-1}.

You can see that ω is 2π times the reciprocal of the period in time. This suggests that we define an analogous quantity, called the **wave number** k, that is 2π times the reciprocal of the period in space:

$$k = \frac{2\pi}{\lambda} \tag{16.11}$$

The units of k are rad/m, although many textbooks use simply m^{-1}.

NOTE The wave number k is *not* a spring constant, even though it uses the same symbol. This is a most unfortunate use of symbols, but every major textbook and professional tradition uses the same symbol k for these two very different meanings, so we have little choice but to follow along.

We can use the fundamental relationship $v = \lambda f$ to find an analogous relationship between ω and k:

$$v = \lambda f = \frac{2\pi}{k} \frac{\omega}{2\pi} = \frac{\omega}{k} \tag{16.12}$$

which is usually written

$$\omega = vk \qquad (16.13)$$

Equation 16.13 contains no new information. It is a variation of Equation 16.5, but one that is convenient when working with k and ω.

If we use the definitions of Equations 16.10 and 16.11, Equation 16.9 for the displacement can be written

$$D(x, t) = A \sin(kx - \omega t + \phi_0) \qquad (16.14)$$
(sinusoidal wave traveling in the positive x-direction)

A sinusoidal wave traveling in the negative x-direction is $A \sin(kx + \omega t + \phi_0)$. Equation 16.14 is graphed versus x and t in **FIGURE 16.14**.

You learned in **« Section 15.2** that the initial conditions of an oscillator can be characterized by a phase constant. The same is true for a sinusoidal wave. At $(x, t) = (0 \text{ m}, 0 \text{ s})$ Equation 16.14 becomes

$$D(0 \text{ m}, 0 \text{ s}) = A \sin \phi_0 \qquad (16.15)$$

Different values of ϕ_0 describe different initial conditions for the wave.

FIGURE 16.14 Interpreting the equation of a sinusoidal traveling wave.

If x is fixed, $D(x_1, t) = A \sin(kx_1 - \omega t + \phi_0)$ gives a sinusoidal history graph at one point in space, x_1. It repeats every T s.

If t is fixed, $D(x, t_1) = A \sin(kx - \omega t_1 + \phi_0)$ gives a sinusoidal snapshot graph at one instant of time, t_1. It repeats every λ m.

EXAMPLE 16.3 | Analyzing a sinusoidal wave

A sinusoidal wave with an amplitude of 1.00 cm and a frequency of 100 Hz travels at 200 m/s in the positive x-direction. At $t = 0$ s, the point $x = 1.00$ m is on a crest of the wave.

a. Determine the values of A, v, λ, k, f, ω, T, and ϕ_0 for this wave.
b. Write the equation for the wave's displacement as it travels.
c. Draw a snapshot graph of the wave at $t = 0$ s.

VISUALIZE The snapshot graph will be sinusoidal, but we must do some numerical analysis before we know how to draw it.

SOLVE a. There are several numerical values associated with a sinusoidal traveling wave, but they are not all independent. From the problem statement itself we learn that

$$A = 1.00 \text{ cm} \qquad v = 200 \text{ m/s} \qquad f = 100 \text{ Hz}$$

We can then find:

$$\lambda = v/f = 2.00 \text{ m}$$
$$k = 2\pi/\lambda = \pi \text{ rad/m or } 3.14 \text{ rad/m}$$
$$\omega = 2\pi f = 628 \text{ rad/s}$$
$$T = 1/f = 0.0100 \text{ s} = 10.0 \text{ ms}$$

The phase constant ϕ_0 is determined by the initial conditions. We know that a wave crest, with displacement $D = A$, is passing $x_0 = 1.00$ m at $t_0 = 0$ s. Equation 16.14 at x_0 and t_0 is

$$D(x_0, t_0) = A = A \sin[k(1.00 \text{ m}) + \phi_0]$$

This equation is true only if $\sin[k(1.00 \text{ m}) + \phi_0] = 1$, which requires

$$k(1.00 \text{ m}) + \phi_0 = \frac{\pi}{2} \text{ rad}$$

Solving for the phase constant gives

$$\phi_0 = \frac{\pi}{2} \text{ rad} - (\pi \text{ rad/m})(1.00 \text{ m}) = -\frac{\pi}{2} \text{ rad}$$

b. With the information gleaned from part a, the wave's displacement is

$$D(x, t) = 1.00 \text{ cm} \times$$
$$\sin[(3.14 \text{ rad/m})x - (628 \text{ rad/s})t - \pi/2 \text{ rad}]$$

Notice that we included units with A, k, ω, and ϕ_0.

c. We know that $x = 1.00$ m is a wave crest at $t = 0$ s and that the wavelength is $\lambda = 2.00$ m. Because the origin is $\lambda/2$ away from the crest at $x = 1.00$ m, we expect to find a wave trough at $x = 0$. This is confirmed by calculating $D(0 \text{ m}, 0 \text{ s}) = (1.00 \text{ cm}) \sin(-\pi/2 \text{ rad}) = -1.00$ cm. **FIGURE 16.15** is a snapshot graph that portrays this information.

FIGURE 16.15 A snapshot graph at $t = 0$ s of the sinusoidal wave of Example 16.3.

The Velocity of a Particle in the Medium

As a sinusoidal wave travels along the x-axis with speed v, the particles of the medium oscillate back and forth in SHM. For a transverse wave, such as a wave on a string, the oscillation is in the y-direction. For a longitudinal sound wave, the particles oscillate

FIGURE 16.16 A snapshot graph of a wave on a string with vectors showing the velocity *of the string* at various points.

The velocity of the wave ➡

The velocity of a particle on the string

At a turning point, the particle has zero velocity.

A particle's velocity is maximum at zero displacement.

in the *x*-direction, parallel to the propagation. We can use the displacement equation, Equation 16.14, to find the velocity of a particle in the medium.

At time *t*, the displacement of the medium at position *x* is

$$D(x, t) = A \sin(kx - \omega t + \phi_0) \tag{16.16}$$

The velocity of the medium—**which is not the same as the velocity of the wave along the string**—is the time derivative of Equation 16.16:

$$v = \frac{dD}{dt} = -\omega A \cos(kx - \omega t + \phi_0) \tag{16.17}$$

Thus the maximum speed of particles in the medium is $v_{max} = \omega A$. This is the same result we found for simple harmonic motion because the motion of the medium *is* simple harmonic motion. **FIGURE 16.16** shows velocity vectors *of the particles* at different points along a string as a sinusoidal wave moves from left to right.

NOTE Creating a wave of larger amplitude increases the speed of particles in the medium, but it does *not* change the speed of the wave *through* the medium.

16.4 ADVANCED TOPIC The Wave Equation on a String

Why do waves propagate along a string? We've described wave motion—the kinematics of waves—but not explained why it occurs. The motion of a string, like that of a baseball, is governed by Newton's second law. But a baseball can be modeled as a moving particle. To explain waves, we need to see how Newton's laws apply to a *continuous* object that is spread out in space.

This section will be significantly more mathematical than any analysis that we've done so far, so it will be good to get an overview of where we're going. We have two primary goals:

■ To use Newton's second law to find an *equation of motion* for displacements on a string. This is called the *wave equation*. We'll see that Equation 16.14, the displacement of a sinusoidal wave, is a solution to the wave equation.
■ To predict the wave speed on a string.

Although we'll derive the wave equation for a string, the equation itself occurs in many other contexts in science and engineering. Wherever this equation arises, the solutions are traveling waves.

FIGURE 16.17 shows a small piece of string that is displaced from its equilibrium position. This piece is at position *x* and has a small horizontal width Δx. We're going to apply Newton's second law, the familiar $F_{net} = ma$, to this little piece of string. Notice that it's curved, so the tension forces at the ends are *not* opposite each other. This is essential in order for there to be a net force.

We'll begin by making the realistic assumption that the wave amplitude *A* is much smaller than the wavelength λ. On average, the string "rises" distance *A* over a "run" of $\lambda/4$. If $A \ll \lambda$, then the slope of the string is always very small. That is, the string itself and the tension vectors are always very close to being horizontal. (Our drawings greatly exaggerate the amplitude for clarity.)

This assumption has two immediate implications. First, a small-amplitude wave doesn't noticeably increase the length of the string. With no additional stretching, the string tension T_s is not altered by the wave. Second, because the string is always very close to being horizontal, there's virtually no difference between the actual length of the piece of string in Figure 16.17 and its horizontal width Δx. Thus the mass of this little piece of string is $m = \mu \, \Delta x$, where, you'll recall, μ is the string's linear density (mass per unit length).

Because we know the mass, let's start with the *ma* side of Newton's second law. For a transverse wave, this little piece of string oscillates perpendicular to the direction

FIGURE 16.17 Apply Newton's second law to this small piece of string.

$\vec{T}_s(x + \Delta x)$

$\vec{a}$

$\vec{T}_s(x)$

$m = \mu \Delta x$

Apply Newton's second law to this small piece of string.

x $x + \Delta x$

of wave propagation. For wave motion along the x-axis, the string accelerates in the y-direction. If we had *only* this little piece of string, not an entire string, we would model it as a particle and write its acceleration as $a_y = dv_y/dt = d^2y/dt^2$. The acceleration of a particle, as you've learned, is the second derivative of its position with respect to time.

But the string isn't a particle. At any instant of time, different pieces of the string have different accelerations. To find the acceleration at a specific point on the string, we want to know how the displacement varies with t at that specific value of x. Or, because the displacement $D(x, t)$ is a function of two variables, we want to know the rate at which $D(x, t)$ changes with respect to t at a specific value of x. In multivariable calculus, the rate of change of a function with respect to one variable while all other variables are held fixed is called a **partial derivative.** Partial, because by holding other variables constant we're only partially examining the many ways in which the function could change.

Partial derivatives have a special notation, using a "curly d." The velocity of our little piece of string is written

$$v_y = \frac{\partial D}{\partial t} \tag{16.18}$$

and its acceleration is

$$a_y = \frac{\partial^2 D}{\partial t^2} \tag{16.19}$$

Don't panic if you've not reached partial derivatives in calculus. They are evaluated exactly like regular derivatives, but the partial-derivative notation means "treat all the other variables as if they were constants." Using the partial derivative, we find the first half of Newton's second law for our little piece of string is

$$ma_y = \mu\,\Delta x \frac{\partial^2 D}{\partial t^2} \tag{16.20}$$

Now we can turn our attention to finding the net force on this little piece of string. The type of analysis we're going to do may be new to you, but it is widely used in more advanced science and engineering courses. **FIGURE 16.18** shows our little piece of the string, this time with the tension forces—one at each end of the string—resolved into x- and y-components.

Strictly speaking, the tension force T_s is tangent to the string. However, our small-amplitude assumption, requiring this piece of string to be almost horizontal, means there is virtually no distinction between T_s and its horizontal component. (This is the small-angle approximation $\cos\theta \approx 1$ if $\theta \ll 1$ rad.) Thus we've identified the two horizontal components as T_s. Because they are equal but opposite, the net horizontal force is zero. This has to be true because each piece of this transverse wave accelerates only in the y-direction.

The net force on this little piece of string in the transverse direction (the y-direction) is

$$F_{\text{net }y} = T_y(x + \Delta x) + T_y(x) \tag{16.21}$$

The notation $T_y(x)$ means "the y-component of the string tension at position x on the string." And we're adding, not subtracting, because this is a formal statement that the net force is the sum of all forces.

You can see, from the force triangle in Figure 16.18, that the ratio $T_y(x + \Delta x)/T_s$—rise over run—is the *slope of the string* at position $x + \Delta x$. This is a key part of the analysis, so make sure you understand it. The slope is the derivative of the string's displacement with respect to x at *this specific instant of time.* We're holding t constant while looking at the spatial variation of the string, so this is another partial derivative:

$$\text{string slope } = \frac{\partial D}{\partial x} \tag{16.22}$$

FIGURE 16.18 Finding the net force on the string.

At the right end of this little piece of string, at position $x + \Delta x$, the y-component of the tension is

$$T_y(x + \Delta x) = (\text{slope at } x + \Delta x) \times T_s = T_s \frac{\partial D}{\partial x}\bigg|_{x+\Delta x} \tag{16.23}$$

where the subscript on the partial derivative means to evaluate the slope at $x + \Delta x$. The same analysis holds at the left end, position x, with one change: Because T_s points toward the left, a "negative run," the ratio $T_y(x)/T_s$ is the *negative* of the string slope. Thus

$$T_y(x) = -(\text{slope at } x) \times T_s = -T_s \frac{\partial D}{\partial x}\bigg|_x \tag{16.24}$$

Combining Equations 16.23 and 16.24, we find that the net force on this little piece of string is

$$F_{\text{net } y} = T_y(x + \Delta x) + T_y(x) = T_s\left[\frac{\partial D}{\partial x}\bigg|_{x+\Delta x} - \frac{\partial D}{\partial x}\bigg|_x\right] \tag{16.25}$$

If this little piece of string were straight, the two slopes would be the same and there would be no net force. As we noted above, the string *has* to have a curvature to have a net force.

We're almost done. We know the net force on this little piece of string (Equation 16.25) and we know its mass and acceleration (Equation 16.20). Because $F_{\text{net } y} = ma_y$, we can equate these two results:

$$T_s\left[\frac{\partial D}{\partial x}\bigg|_{x+\Delta x} - \frac{\partial D}{\partial x}\bigg|_x\right] = \mu\,\Delta x \frac{\partial^2 D}{\partial t^2} \tag{16.26}$$

Dividing by $\mu\,\Delta x$, we have

$$\frac{\partial^2 D}{\partial t^2} = \frac{T_s}{\mu} \times \frac{\dfrac{\partial D}{\partial x}\bigg|_{x+\Delta x} - \dfrac{\partial D}{\partial x}\bigg|_x}{\Delta x} \tag{16.27}$$

Recall, from calculus, that the derivative of the function $f(x)$ is defined as

$$\frac{df}{dx} = \lim_{\Delta x \to 0}\left[\frac{f(x + \Delta x) - f(x)}{\Delta x}\right]$$

This is exactly what we have on the right side of Equation 16.27 if we let the width of our little piece of string approach zero: $\Delta x \to 0$. The function for which we're evaluating the difference between $x + \Delta x$ and x is the partial derivative $\partial D/\partial x$, and the derivative of a derivative is a second derivative.

Thus in the limit $\Delta x \to 0$, Equation 16.27 becomes

$$\frac{\partial^2 D}{\partial t^2} = \frac{T_s}{\mu}\frac{\partial^2 D}{\partial x^2} \quad \text{(wave equation for a string)} \tag{16.28}$$

Equation 16.28 is the *wave equation for a string*. It's really Newton's second law in disguise, but written for a continuous object where the displacement is a function of both position and time. Just like Newton's second law for a particle, it governs the dynamics of motion on a string.

Traveling Wave Solutions

The equation of motion for a simple harmonic oscillator turned out to be a second-order differential equation. Although there are systematic ways to solve differential equations, we noted that—because solutions are unique—we can sometimes use what we know about a situation to *guess* the solution. The same is true for Equation 16.28, which is a *partial differential equation*. We have reason to think that sinusoidal waves can travel on stretched strings, so let's guess that a solution to Equation 16.28 is

$$D(x, t) = A \sin(kx - \omega t + \phi_0) \tag{16.29}$$

where the minus sign gives a wave traveling in the $+x$-direction and—from Equation 16.13—the wave speed is $v = \omega/k$.

To evaluate this possible solution we need its second partial derivatives. With respect to position we have

$$\frac{\partial D}{\partial x} = kA \cos(kx - \omega t + \phi_0)$$

$$\frac{\partial^2 D}{\partial x^2} = \frac{\partial}{\partial x}\left(\frac{\partial D}{\partial x}\right) = -k^2 A \sin(kx - \omega t + \phi_0) \tag{16.30}$$

and with respect to time we have

$$\frac{\partial D}{\partial t} = -\omega A \cos(kx - \omega t + \phi_0)$$

$$\frac{\partial^2 D}{\partial t^2} = \frac{\partial}{\partial t}\left(\frac{\partial D}{\partial t}\right) = -\omega^2 A \sin(kx - \omega t + \phi_0) \tag{16.31}$$

Substituting the second partial derivatives into the wave equation, Equation 16.28, gives

$$-\omega^2 A \sin(kx - \omega t + \phi_0) = \frac{T_s}{\mu}(-k^2 A \sin(kx - \omega t + \phi_0)) \tag{16.32}$$

This will be true only if

$$\omega^2 = \frac{T_s}{\mu} k^2 \tag{16.33}$$

But ω/k is the wave speed v, so what we've found is that the sinusoidal wave of Equation 16.1 *is* a solution to the wave equation, but only if the wave travels with speed

$$v = \frac{\omega}{k} = \sqrt{\frac{T_s}{\mu}} \tag{16.34}$$

You should be able to convince yourself that we would have arrived at the same result if we had started with $D(x, t) = A \sin(kx + \omega t + \phi_0)$ for a wave traveling in the $-x$-direction.

Let's summarize. We used Newton's second law for a small piece of the string to come up with an equation for the dynamics of motion on a string. We then showed that a solution to this equation is a sinusoidal traveling wave, and we made a specific prediction for the wave speed in terms of two properties or characteristics of the string: its tension and its mass density. Thus the answer to the question with which we started this section—*why* do waves propagate along a string—is that wave motion is simply a consequence of Newton's second law, the relationship between force and acceleration, when applied to a continuous object.

With Equation 16.9 in hand, we can write Equation 16.28 as

$$\frac{\partial^2 D}{\partial t^2} = v^2 \frac{\partial^2 D}{\partial x^2} \quad \text{(the general wave equation)} \tag{16.35}$$

We derived Equation 16.28 specifically for a string, but *any* physical system that obeys Equation 16.35 for some type of displacement D will have sinusoidal waves traveling with speed v. Equation 16.35 is called the **wave equation,** and it occurs over and over in science and engineering. We will see it again in this chapter in our analysis of sound waves. And much later, in Chapter 31, we'll discover that electromagnetic fields also obey this equation. Thus electromagnetic waves exist, and we'll be able to predict that all electromagnetic waves, regardless of wavelength, travel through vacuum with the same speed—the speed of light.

EXAMPLE 16.4 | Generating a sinusoidal wave

A very long string with $\mu = 2.0$ g/m is stretched along the x-axis with a tension of 5.0 N. At $x = 0$ m it is tied to a 100 Hz simple harmonic oscillator that vibrates perpendicular to the string with an amplitude of 2.0 mm. The oscillator is at its maximum positive displacement at $t = 0$ s.

a. Write the displacement equation for the traveling wave on the string.

b. At $t = 5.0$ ms, what is the string's displacement at a point 2.7 m from the oscillator?

MODEL The oscillator generates a sinusoidal traveling wave on a string. The displacement of the wave has to match the displacement of the oscillator at $x = 0$ m.

SOLVE a. The equation for the displacement is

$$D(x, t) = A \sin(kx - \omega t + \phi_0)$$

with A, k, ω, and ϕ_0 to be determined. The wave amplitude is the same as the amplitude of the oscillator that generates the wave, so $A = 2.0$ mm. The oscillator has its maximum displacement $y_{osc} = A = 2.0$ mm at $t = 0$ s, thus

$$D(0 \text{ m}, 0 \text{ s}) = A \sin(\phi_0) = A$$

This requires the phase constant to be $\phi_0 = \pi/2$ rad. The wave's frequency is $f = 100$ Hz, the frequency of the source; therefore

the angular frequency is $\omega = 2\pi f = 200\pi$ rad/s. We still need $k = 2\pi/\lambda$, but we do not know the wavelength. However, we have enough information to determine the wave speed, and we can then use either $\lambda = v/f$ or $k = \omega/v$. The speed is

$$v = \sqrt{\frac{T_s}{\mu}} = \sqrt{\frac{5.0 \text{ N}}{0.0020 \text{ kg/m}}} = 50 \text{ m/s}$$

Using v, we find $\lambda = 0.50$ m and $k = 2\pi/\lambda = 4\pi$ rad/m. Thus the wave's displacement equation is

$$D(x, t) = (2.0 \text{ mm}) \times$$
$$\sin[2\pi((2.0 \text{ m}^{-1})x - (100 \text{ s}^{-1})t) + \pi/2 \text{ rad}]$$

Notice that we have separated out the 2π. This step is not essential, but for some problems it makes subsequent steps easier.

b. The wave's displacement at $t = 5.0$ ms $= 0.0050$ s is

$$D(x, t = 5.0 \text{ ms}) = (2.0 \text{ mm}) \sin(4\pi x - \pi \text{ rad} + \pi/2 \text{ rad})$$
$$= (2.0 \text{ mm}) \sin(4\pi x - \pi/2 \text{ rad})$$

At $x = 2.7$ m (calculator set to radians!), the displacement is

$$D(2.7 \text{ m}, 5.0 \text{ ms}) = 1.6 \text{ mm}$$

16.5 Sound and Light

Although there are many kinds of waves in nature, two are especially significant for us as humans. These are sound waves and light waves, the basis of hearing and seeing.

Sound Waves

FIGURE 16.19 A sound wave is a sequence of compressions and rarefactions.

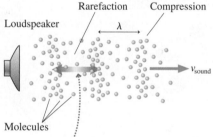

Individual molecules oscillate back and forth. As they do so, the compressions propagate forward at speed v_{sound}. Compressions are regions of higher pressure, so a sound wave is a pressure wave.

We usually think of sound waves traveling in air, but sound can travel through any gas, through liquids, and even through solids. **FIGURE 16.19** shows a loudspeaker cone vibrating back and forth in a fluid such as air or water. Each time the cone moves forward, it collides with the molecules and pushes them closer together. A half cycle later, as the cone moves backward, the fluid has room to expand and the density decreases a little. These regions of higher and lower density (and thus higher and lower pressure) are called **compressions** and **rarefactions.**

This periodic sequence of compressions and rarefactions travels outward from the loudspeaker as a longitudinal sound wave. When the wave reaches your ear, the oscillating pressure causes your eardrum to vibrate. These vibrations are transferred into your inner ear and perceived as sound.

The speed of sound waves depends on the compressibility of the medium. As **TABLE 16.1** shows, the speed is faster in liquids and solids (relatively incompressible) than in gases (highly compressible). For sound waves in air, the speed at temperature T (in °C) is

TABLE 16.1 The speed of sound

Medium	Speed (m/s)
Air (0°C)	331
Air (20°C)	343
Helium (0°C)	970
Ethyl alcohol	1170
Water (20°C)	1480
Granite	6000
Aluminum	6420

$$v_{\text{sound in air}} = 331 \text{ m/s} \times \sqrt{\frac{T(°C) + 273}{273}} \tag{16.36}$$

We'll derive this result in Section 16.6, but recall from chemistry that adding 273 to a Celsius temperature converts it to an absolute temperature in kelvins. The speed of sound increases with increasing temperature but, interestingly, does *not* depend on the air pressure. For air at room temperature (20°C),

$$v_{\text{sound in air}} = 343 \text{ m/s} \qquad \text{(sound speed in air at 20°C)}$$

This is the value you should use when solving problems unless you're given temperature information.

A speed of 343 m/s is high, but not extraordinarily so. A distance as small as 100 m is enough to notice a slight delay between when you see something, such as a person hammering a nail, and when you hear it. The time required for sound to travel 1 km is $t = (1000 \text{ m})/(343 \text{ m/s}) \approx 3$ s. You may have learned to estimate the distance to a bolt of lightning by timing the number of seconds between when you see the flash and when you hear the thunder. Because sound takes 3 s to travel 1 km, the time divided by 3 gives the distance in kilometers. Or, in English units, the time divided by 5 gives the distance in miles.

Your ears are able to detect sound waves with frequencies between roughly 20 Hz and 20,000 Hz, or 20 kHz. You can use the fundamental relationship $v_{\text{sound}} = \lambda f$ to calculate that a 20 Hz sound wave has a 17-m-long wavelength, while the wavelength of a 20 kHz note is a mere 17 mm. Low frequencies are perceived as "low-pitch" bass notes, while high frequencies are heard as "high-pitch" treble notes. Your high-frequency range of hearing can deteriorate with age (10 kHz is the average upper limit at age 65) or as a result of exposure to very loud sounds.

Sound waves exist at frequencies well above 20 kHz, even though humans can't hear them. These are called *ultrasonic* frequencies. Oscillators vibrating at frequencies of many MHz generate the ultrasonic waves used in ultrasound medical imaging. A 3 MHz wave traveling through water (which is basically what your body is) at a sound speed of 1480 m/s has a wavelength of about 0.5 mm. It is this very small wavelength that allows ultrasound to image very small objects. We'll see why when we study *diffraction* in Chapter 33.

This ultrasound image is an example of using high-frequency sound waves to "see" within the human body.

Electromagnetic Waves

A light wave is an *electromagnetic wave,* a self-sustaining oscillation of the electromagnetic field. Other electromagnetic waves, such as radio waves, microwaves, and ultraviolet light, have the same physical characteristics as light waves even though we cannot sense them with our eyes. It is easy to demonstrate that light will pass unaffected through a container from which all the air has been removed, and light reaches us from distant stars through the vacuum of interstellar space. Such observations raise interesting but difficult questions. If light can travel through a region in which there is no matter, then what is the *medium* of a light wave? What is it that is waving?

It took scientists over 50 years, most of the 19th century, to answer this question. We will examine the answers in more detail in Chapter 31 after we introduce the ideas of electric and magnetic fields. For now we can say that light waves are a "self-sustaining oscillation of the electromagnetic field." That is, the displacement D is an electric or magnetic field. Being self-sustaining means that electromagnetic waves require *no material medium* in order to travel; hence electromagnetic waves are not mechanical waves. Fortunately, we can learn about the wave properties of light without having to understand electromagnetic fields.

It was predicted theoretically in the late 19th century, and has been subsequently confirmed, that all electromagnetic waves travel through vacuum with the same speed, called the *speed of light*. The value of the speed of light is

$$v_{\text{light}} = c = 299{,}792{,}458 \text{ m/s} \qquad \text{(electromagnetic wave speed in vacuum)}$$

where the special symbol c is used to designate the speed of light. (This is the c in Einstein's famous formula $E = mc^2$.) Now *this* is really moving—about one million times faster than the speed of sound in air!

NOTE $c = 3.00 \times 10^8$ m/s is the appropriate value to use in calculations.

The wavelengths of light are extremely small. You will learn in Chapter 33 how these wavelengths are determined, but for now we will note that visible light

White light passing through a prism is spread out into a band of colors called the *visible spectrum*.

is an electromagnetic wave with a wavelength (in air) in the range of roughly 400 nm (400×10^{-9} m) to 700 nm (700×10^{-9} m). Each wavelength is perceived as a different color, with the longer wavelengths seen as orange or red light and the shorter wavelengths seen as blue or violet light. A prism is able to spread the different wavelengths apart, from which we learn that "white light" is all the colors, or wavelengths, combined. The spread of colors seen with a prism, or seen in a rainbow, is called the *visible spectrum*.

If the wavelengths of light are unbelievably small, the oscillation frequencies are unbelievably large. The frequency for a 600 nm wavelength of light (orange) is

$$f = \frac{v}{\lambda} = \frac{3.00 \times 10^8 \text{ m/s}}{600 \times 10^{-9} \text{ m}} = 5.00 \times 10^{14} \text{ Hz}$$

The frequencies of light waves are roughly a factor of a trillion (10^{12}) higher than sound frequencies.

Electromagnetic waves exist at many frequencies other than the rather limited range that our eyes detect. One of the major technological advances of the 20th century was learning to generate and detect electromagnetic waves at many frequencies, ranging from low-frequency radio waves to the extraordinarily high frequencies of x rays. **FIGURE 16.20** shows that the visible spectrum is a small slice of the much broader **electromagnetic spectrum.**

FIGURE 16.20 The electromagnetic spectrum from 10^6 Hz to 10^{18} Hz.

EXAMPLE 16.5 | Traveling at the speed of light

A satellite exploring Jupiter transmits data to the earth as a radio wave with a frequency of 200 MHz. What is the wavelength of the electromagnetic wave, and how long does it take the signal to travel 800 million kilometers from Jupiter to the earth?

SOLVE Radio waves are sinusoidal electromagnetic waves traveling with speed c. Thus

$$\lambda = \frac{c}{f} = \frac{3.00 \times 10^8 \text{ m/s}}{2.00 \times 10^8 \text{ Hz}} = 1.5 \text{ m}$$

The time needed to travel 800×10^6 km $= 8.0 \times 10^{11}$ m is

$$\Delta t = \frac{\Delta x}{c} = \frac{8.0 \times 10^{11} \text{ m}}{3.00 \times 10^8 \text{ m/s}} = 2700 \text{ s} = 45 \text{ min}$$

The Index of Refraction

Light waves travel with speed c in a vacuum, but they slow down as they pass through transparent materials such as water or glass or even, to a very slight extent, air. The slowdown is a consequence of interactions between the electromagnetic field of the wave and the electrons in the material. The speed of light in a material is characterized by the material's **index of refraction** n, defined as

$$n = \frac{\text{speed of light in a vacuum}}{\text{speed of light in the material}} = \frac{c}{v} \tag{16.37}$$

The index of refraction of a material is always greater than 1 because $v < c$. A vacuum has $n = 1$ exactly. **TABLE 16.2** shows the index of refraction for several materials. You can see that liquids and solids have larger indices of refraction than gases.

TABLE 16.2 Typical indices of refraction

Material	Index of refraction
Vacuum	1 exactly
Air	1.0003
Water	1.33
Glass	1.50
Diamond	2.42

NOTE An accurate value for the index of refraction of air is relevant only in very precise measurements. We will assume $n_{air} = 1.00$ in this text.

If the speed of a light wave changes as it enters into a transparent material, such as glass, what happens to the light's frequency and wavelength? Because $v = \lambda f$, either λ or f or both have to change when v changes.

As an analogy, think of a sound wave in the air as it impinges on the surface of a pool of water. As the air oscillates back and forth, it periodically pushes on the surface of the water. These pushes generate the compressions of the sound wave that continues on into the water. Because each push of the air causes one compression of the water, the frequency of the sound wave in the water must be *exactly the same* as the frequency of the sound wave in the air. In other words, **the frequency of a wave is the frequency of the source. It does not change as the wave moves from one medium to another.**

The same is true for electromagnetic waves; the frequency does not change as the wave moves from one material to another.

FIGURE 16.21 shows a light wave passing through a transparent material with index of refraction n. As the wave travels through vacuum it has wavelength λ_{vac} and frequency f_{vac} such that $\lambda_{vac} f_{vac} = c$. In the material, $\lambda_{mat} f_{mat} = v = c/n$. The frequency does not change as the wave enters ($f_{mat} = f_{vac}$), so the wavelength must. The wavelength in the material is

$$\lambda_{mat} = \frac{v}{f_{mat}} = \frac{c}{nf_{mat}} = \frac{c}{nf_{vac}} = \frac{\lambda_{vac}}{n} \qquad (16.38)$$

The wavelength in the transparent material is less than the wavelength in vacuum. This makes sense. Suppose a marching band is marching at one step per second at a speed of 1 m/s. Suddenly they slow their speed to $\frac{1}{2}$ m/s but maintain their march at one step per second. The only way to go slower while marching at the same pace is to take *smaller steps*. When a light wave enters a material, the only way it can go slower while oscillating at the same frequency is to have a *smaller wavelength*.

FIGURE 16.21 Light passing through a transparent material with index of refraction n.

A transparent material in which light travels slower, at speed $v = c/n$

Vacuum $n = 1$ Index n $n = 1$

λ_{vac} $\lambda = \lambda_{vac}/n$

The wavelength inside the material decreases, but the frequency doesn't change.

EXAMPLE 16.6 | **Light traveling through glass**

Orange light with a wavelength of 600 nm is incident upon a 1.00-mm-thick glass microscope slide.

a. What is the light speed in the glass?

b. How many wavelengths of the light are inside the slide?

SOLVE a. From Table 16.2 we see that the index of refraction of glass is $n_{glass} = 1.50$. Thus the speed of light in glass is

$$v_{glass} = \frac{c}{n_{glass}} = \frac{3.00 \times 10^8 \text{ m/s}}{1.50} = 2.00 \times 10^8 \text{ m/s}$$

b. The wavelength inside the glass is

$$\lambda_{glass} = \frac{\lambda_{vac}}{n_{glass}} = \frac{600 \text{ nm}}{1.50} = 400 \text{ nm} = 4.00 \times 10^{-7} \text{ m}$$

N wavelengths span a distance $d = N\lambda$, so the number of wavelengths in $d = 1.00$ mm is

$$N = \frac{d}{\lambda} = \frac{1.00 \times 10^{-3} \text{ m}}{4.00 \times 10^{-7} \text{ m}} = 2500$$

ASSESS The fact that 2500 wavelengths fit within 1 mm shows how small the wavelengths of light are.

STOP TO THINK 16.4 A light wave travels from left to right through three transparent materials of equal thickness. Rank in order, from largest to smallest, the indices of refraction n_a, n_b, and n_c.

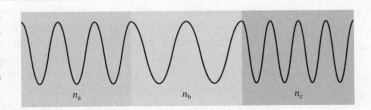

The Wave Model

We introduced the concept of a *wave model* at the beginning of this chapter. Now we're in a position to articulate what this means.

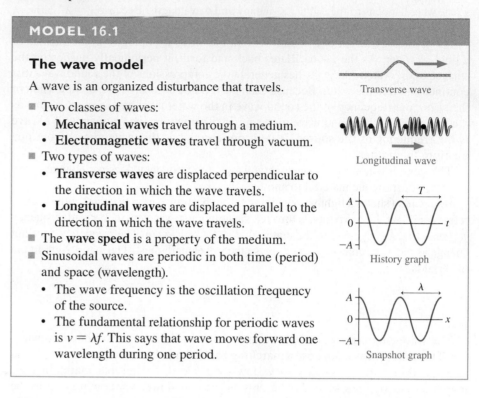

MODEL 16.1

The wave model

A wave is an organized disturbance that travels.

- Two classes of waves:
  - **Mechanical waves** travel through a medium.
  - **Electromagnetic waves** travel through vacuum.
- Two types of waves:
  - **Transverse waves** are displaced perpendicular to the direction in which the wave travels.
  - **Longitudinal waves** are displaced parallel to the direction in which the wave travels.
- The **wave speed** is a property of the medium.
- Sinusoidal waves are periodic in both time (period) and space (wavelength).
  - The wave frequency is the oscillation frequency of the source.
  - The fundamental relationship for periodic waves is $v = \lambda f$. This says that wave moves forward one wavelength during one period.

Transverse wave

Longitudinal wave

History graph

Snapshot graph

16.6 ADVANCED TOPIC The Wave Equation in a Fluid

In Section 16.4 we used Newton's second law to show that traveling waves can propagate on a stretched string *and* to predict the wave speed in terms of properties of the string. Now we wish to do the same for sound waves—longitudinal waves propagating through a fluid.

A sound wave is a sequence of compressions and rarefactions in which the fluid is alternately compressed and expanded. A substance's compressibility is characterized by its *bulk modulus B*, which you met in « Section 14.6 when we looked at the elastic properties of materials. If excess pressure p is applied to an object of volume V, then the fractional change in volume—the fraction by which it's compressed—is

$$\frac{\Delta V}{V} = -\frac{p}{B} \tag{16.39}$$

The minus sign indicates that the volume *decreases* when pressure is applied. Gases are much more compressible than liquids, so gases have much smaller values of B than liquids.

Let's apply this to a fluid—either a liquid or a gas. **FIGURE 16.22** shows a small cylindrical piece of fluid with equilibrium pressure p_0 located between positions x and $x + \Delta x$. The initial length and volume of this little piece of fluid are $L_i = \Delta x$ and $V_i = aL_i = a \Delta x$. Notice that we use a for area in this chapter so that there is no conflict with A for amplitude. Suppose the pressure changes to $p_0 + p$. The volume of this little piece of fluid will either decrease (compression) or increase (expansion), depending on whether p is positive or negative.

The volume changes only if the ends of the cylinder undergo *different* displacements. (Equal displacements would shift the cylinder but not change its volume.) In

FIGURE 16.22 An element of fluid changes volume as the pressure changes.

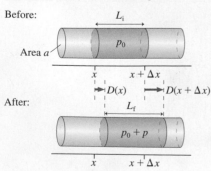

the bottom half of Figure 16.22 we see that the left end of the cylinder has undergone displacement $D(x, t)$ while the displacement at the right end is $D(x + \Delta x, t)$. Now the cylinder has length

$$L_f = L_i + (D(x + \Delta x, t) - D(x, t)) \qquad (16.40)$$

and consequently its volume has *changed* by

$$\Delta V = a(L_f - L_i) = a(D(x + \Delta x, t) - D(x, t)) \qquad (16.41)$$

Substituting both the initial volume and the volume change into Equation 16.39, we have

$$\frac{\Delta V}{V} = \frac{a(D(x + \Delta x, t) - D(x, t))}{a \Delta x} = -\frac{p}{B} \qquad (16.42)$$

After canceling the a, you can see that, just as in Section 16.4, we're left—in the limit $\Delta x \rightarrow 0$—with the definition of the derivative of D with respect to x. It is again a *partial* derivative because we're holding the variable t constant. Thus we find that the fluid pressure (or, to be exact, the pressure deviation from p_0) at position x is related to the displacement of the medium by

$$p(x, t) = -B\frac{\partial D}{\partial x} \qquad (16.43)$$

The pressure depends on how rapidly the fluid's displacement changes with position.

We anticipate that we'll discover sinusoidal sound waves later in this section, so a displacement wave of amplitude A,

$$D(x, t) = A\sin(kx - \omega t + \phi_0) \qquad (16.44)$$

is associated with a pressure wave

$$p(x, t) = -B\frac{\partial D}{\partial x} = -kBA\cos(kx - \omega t + \phi_0)$$
$$= -p_{max}\cos(kx - \omega t + \phi_0) \qquad (16.45)$$

The *pressure amplitude,* or maximum pressure, is

$$p_{max} = kBA = \frac{2\pi fBA}{v_{sound}} \qquad (16.46)$$

where we used $\omega = 2\pi f = vk$ (Equation 16.13) in the last step to write the result in terms of the wave's speed and frequency. In other words, a sound wave is not just a traveling wave of molecular displacement. **A sound wave is also a traveling pressure wave.**

As an example, a quite loud 100 decibel, 500 Hz sound wave in air has a pressure amplitude of 2 Pa. That is, the pressure varies around atmospheric pressure by ± 2 Pa. You can use Equation 16.46 and $B_{air} = 1.42 \times 10^5$ Pa to find that the amplitude of the oscillating air molecules is a microscopic 1.5 μm.

FIGURE 16.23 uses Equations 16.44 and 16.45 to draw snapshot graphs of displacement and pressure for a sound wave propagating to the right. Positive displacement pushes molecules to the right while negative displacement pushes them to the left, so molecules pile up (a compression) at points where the displacement is changing from positive to negative. These are the points where the displacement has the *most negative slope* and thus, from Equation 16.43, the greatest pressure.

In general, the pressure wave has a maximum or minimum at points where the displacement wave is zero, and vice versa. This observation will help us understand standing sound waves in Chapter 17.

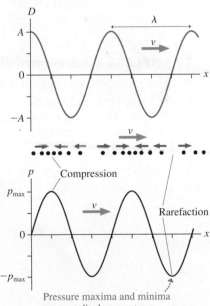

FIGURE 16.23 Snapshot graphs of the sound displacement and pressure.

Pressure maxima and minima occur at displacement zeros.

Predicting the Speed of Sound

Let's return to our small, cylindrical piece of fluid and this time, in **FIGURE 16.24** on the next page, apply Newton's second law to it. The fluid pressure at position x is $p(x, t)$,

FIGURE 16.24 Fluid pressure exerts a net force on the cylinder.

and this pressure pushes the cylinder to the right with force $ap(x, t)$. At the same time, the fluid at position $x + \Delta x$ has pressure $p(x + \Delta x, t)$, and it pushes the cylinder to the left with force $ap(x + \Delta x, t)$. The net force is

$$F_{\text{net }x} = ap(x, t) - ap(x + \Delta x, t) = -a(p(x + \Delta x, t) - p(x, t)) \qquad (16.47)$$

The minus sign arises because the pressure forces push in opposite directions.

Newton's second law is $F_{\text{net }x} = ma_x$. The cylinder's mass is $m = \rho V = \rho a \, \Delta x$, where ρ is the fluid density. (Don't confuse area a with acceleration a_x!) The acceleration, just as in our analysis of a string, is the second partial derivative of displacement with respect to time. Thus the second law for our little cylinder of fluid is

$$F_{\text{net }x} = -a(p(x + \Delta x, t) - p(x, t)) = ma_x = \rho a \, \Delta x \, \frac{\partial^2 D}{\partial t^2} \qquad (16.48)$$

The area a cancels, and a slight rearrangement gives

$$\frac{\partial^2 D}{\partial t^2} = -\frac{1}{\rho} \frac{p(x + \Delta x, t) - p(x, t)}{\Delta x} \rightarrow -\frac{1}{\rho} \frac{\partial p}{\partial x} \qquad (16.49)$$

where, once again, we have a derivative in the limit $\Delta x \rightarrow 0$.

Fortunately, we already found, in Equation 16.43, that $p = -B \, \partial D/\partial x$. Substituting this for p in Equation 16.49 gives

$$\frac{\partial^2 D}{\partial t^2} = \frac{B}{\rho} \frac{\partial^2 D}{\partial x^2} \qquad (16.50)$$

Equation 16.50 is a wave equation! Just as in our analysis of a string, applying Newton's second law to a small piece of the medium has led to a wave equation. We've already shown that a sinusoidal traveling wave, Equation 16.44, is a solution, so we don't need to prove it again. Further, by comparing Equation 16.50 to the general wave equation, Equation 16.35, we can predict the speed of sound in a fluid:

$$v_{\text{sound}} = \sqrt{\frac{B}{\rho}} \qquad (16.51)$$

TABLE 16.3 gives values of the bulk modulus for several common fluids.

TABLE 16.3 Bulk moduli of common fluids

Medium	B (Pa)
Mercury (20°C)	2.85×10^{10}
Water (20°C)	2.18×10^9
Ethyl alcohol (20°C)	1.06×10^9
Helium (0°C, 1 atm)	1.688×10^5
Air (0°C, 1 atm)	1.418×10^5

EXAMPLE 16.7 | **The speed of sound in water**

Predict the speed of sound in water at 20°C.

SOLVE From Table 16.3, the bulk modulus of water at 20°C is 2.18×10^9 Pa. The density of water is usually given as 1000 kg/m³, but this is at 4°C. To three significant figures, the density at 20°C is 998 kg/m³. Thus we predict

$$v_{\text{sound}} = \sqrt{\frac{2.18 \times 10^9 \text{ Pa}}{998 \text{ kg/m}^3}} = 1480 \text{ m/s}$$

This is exactly the value given earlier in Table 16.1.

For gases, both B and ρ are proportional to the pressure, so their ratio is independent of pressure. At 0°C and 1 atm, the density of air is $\rho_0 = 1.292$ kg/m³. Thus the speed of sound in air at 0°C is

$$v_{\text{sound in air}} = \sqrt{\frac{B_0}{\rho_0}} = \sqrt{\frac{1.418 \times 10^5 \text{ Pa}}{1.292 \text{ kg/m}^3}} = 331 \text{ m/s} \qquad \text{(at 0°C)} \qquad (16.52)$$

exactly as shown in Table 16.1.

You can use the ideal-gas law to show that the density (at constant pressure) of a gas is inversely proportional to its absolute temperature T in kelvins. If the density at 0°C and 1 atm is ρ_0, then the density at temperature T is

$$\rho_T = \rho_0 \frac{273}{T(\text{K})} = \rho_0 \frac{273}{T(°\text{C}) + 273} \qquad (16.53)$$

where we used $0°C = 273$ K to convert kelvin to °C. Thus a general expression for the speed of sound in air is

$$v_{\text{sound in air}} = \sqrt{\frac{B}{\rho}} = \sqrt{\frac{B_0}{\rho_0} \frac{T(°C) + 273}{273}} = 331 \text{ m/s} \times \sqrt{\frac{T(°C) + 273}{273}} \quad (16.54)$$

This was the expression given without proof in Section 16.5. Now we see that it comes from the wave equation for sound with a little help from the ideal-gas law.

16.7 Waves in Two and Three Dimensions

Suppose you were to take a photograph of ripples spreading on a pond. If you mark the location of the *crests* on the photo, your picture would look like FIGURE 16.25a. The lines that locate the crests are called **wave fronts,** and they are spaced precisely one wavelength apart. The diagram shows only a single instant of time, but you can imagine a movie in which you would see the wave fronts moving outward from the source at speed v. A wave like this is called a **circular wave.** It is a two-dimensional wave that spreads across a surface.

Although the wave fronts are circles, you would hardly notice the curvature if you observed a small section of the wave front very, very far away from the source. The wave fronts would appear to be parallel lines, still spaced one wavelength apart and traveling at speed v. A good example is an ocean wave reaching a beach. Ocean waves are generated by storms and wind far out at sea, hundreds or thousands of miles away. By the time they reach the beach where you are working on your tan, the crests appear to be straight lines. An aerial view of the ocean would show a wave diagram like FIGURE 16.25b.

Many waves of interest, such as sound waves or light waves, move in three dimensions. For example, loudspeakers and lightbulbs emit **spherical waves.** That is, the crests of the wave form a series of concentric spherical shells separated by the wavelength λ. In essence, the waves are three-dimensional ripples. It will still be useful to draw wave-front diagrams such as Figure 16.25, but now the circles are slices through the spherical shells locating the wave crests.

If you observe a spherical wave very, very far from its source, the small piece of the wave front that you can see is a little patch on the surface of a very large sphere. If the radius of the sphere is sufficiently large, you will not notice the curvature and this little patch of the wave front appears to be a plane. FIGURE 16.26 illustrates the idea of a **plane wave.**

To visualize a plane wave, imagine standing on the x-axis facing a sound wave as it comes toward you from a very distant loudspeaker. Sound is a longitudinal wave, so the particles of medium oscillate toward you and away from you. If you were to locate all of the particles that, at one instant of time, were at their maximum displacement toward you, they would all be located in a plane perpendicular to the travel direction. This is one of the wave fronts in Figure 16.26, and all the particles in this plane are doing exactly the same thing at that instant of time. This plane is moving toward you at speed v. There is another plane one wavelength behind it where the molecules are also at maximum displacement, yet another two wavelengths behind the first, and so on.

Because a plane wave's displacement depends on x but not on y or z, the displacement function $D(x, t)$ describes a plane wave just as readily as it does a one-dimensional wave. Once you specify a value for x, the displacement is the same at every point in the yz-plane that slices the x-axis at that value (i.e., one of the planes shown in Figure 16.26).

NOTE There are no perfect plane waves in nature, but many waves of practical interest can be modeled as plane waves.

We can describe a circular wave or a spherical wave by changing the mathematical description from $D(x, t)$ to $D(r, t)$, where r is the radial distance measured outward from the source. Then the displacement of the medium will be the same at every point

FIGURE 16.25 The wave fronts of a circular or spherical wave.

(a)   Wave fronts are the crests of the wave. They are spaced one wavelength apart.

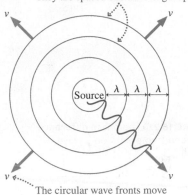

The circular wave fronts move outward from the source at speed v.

(b)

Very far away from the source, small sections of the wave fronts appear to be straight lines.

FIGURE 16.26 A plane wave.

Very far from the source, small segments of spherical wave fronts appear to be planes. The wave is cresting at every point in these planes.

on a spherical surface. In particular, a sinusoidal spherical wave with wave number k and angular frequency ω is written

$$D(r, t) = A(r)\sin(kr - \omega t + \phi_0) \tag{16.55}$$

Other than the change of x to r, the only difference is that the amplitude is now a function of r. A one-dimensional wave propagates with no change in the wave amplitude. But circular and spherical waves spread out to fill larger and larger volumes of space. To conserve energy, an issue we'll look at later in the chapter, the wave's amplitude has to decrease with increasing distance r. This is why sound and light decrease in intensity as you get farther from the source. We don't need to specify exactly how the amplitude decreases with distance, but you should be aware that it does.

Phase and Phase Difference

« Section 15.2 introduced the concept of *phase* for an oscillator in simple harmonic motion. Phase is also important for waves. The **phase** of a sinusoidal wave, denoted ϕ, is the quantity $(kx - \omega t + \phi_0)$. Phase will be an important concept in Chapter 17, where we will explore the consequences of adding various waves together. For now, we can note that the wave fronts seen in Figures 16.25 and 16.26 are "surfaces of constant phase." To see this, write the displacement as simply $D(x, t) = A\sin\phi$. Because each point on a wave front has the same displacement, the phase must be the same at every point.

It will be useful to know the *phase difference* $\Delta\phi$ between two different points on a sinusoidal wave. FIGURE 16.27 shows two points on a sinusoidal wave at time t. The phase difference between these points is

$$\Delta\phi = \phi_2 - \phi_1 = (kx_2 - \omega t + \phi_0) - (kx_1 - \omega t + \phi_0)$$
$$= k(x_2 - x_1) = k\,\Delta x = 2\pi\frac{\Delta x}{\lambda} \tag{16.56}$$

That is, **the phase difference between two points on a wave depends on only the ratio of their separation Δx to the wavelength λ.** For example, two points on a wave separated by $\Delta x = \frac{1}{2}\lambda$ have a phase difference $\Delta\phi = \pi$ rad.

An important consequence of Equation 16.56 is that **the phase difference between two adjacent wave fronts is $\Delta\phi = 2\pi$ rad.** This follows from the fact that two adjacent wave fronts are separated by $\Delta x = \lambda$. This is an important idea. Moving from one crest of the wave to the next corresponds to changing the *distance* by λ and changing the *phase* by 2π rad.

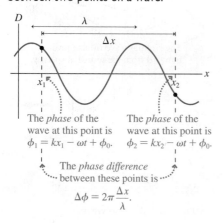

FIGURE 16.27 The phase difference between two points on a wave.

The *phase* of the wave at this point is $\phi_1 = kx_1 - \omega t + \phi_0$.

The *phase* of the wave at this point is $\phi_2 = kx_2 - \omega t + \phi_0$.

The *phase difference* between these points is

$\Delta\phi = 2\pi\dfrac{\Delta x}{\lambda}$.

EXAMPLE 16.8 | **The phase difference between two points on a sound wave**

A 100 Hz sound wave travels with a wave speed of 343 m/s.

a. What is the phase difference between two points 60.0 cm apart along the direction the wave is traveling?

b. How far apart are two points whose phase differs by 90°?

MODEL Treat the wave as a plane wave traveling in the positive x-direction.

SOLVE a. The phase difference between two points is

$$\Delta\phi = 2\pi\frac{\Delta x}{\lambda}$$

In this case, $\Delta x = 60.0$ cm $= 0.600$ m. The wavelength is

$$\lambda = \frac{v}{f} = \frac{343 \text{ m/s}}{100 \text{ Hz}} = 3.43 \text{ m}$$

and thus

$$\Delta\phi = 2\pi\frac{0.600 \text{ m}}{3.43 \text{ m}} = 0.350\pi \text{ rad} = 63.0°$$

b. A phase difference $\Delta\phi = 90°$ is $\pi/2$ rad. This will be the phase difference between two points when $\Delta x/\lambda = \frac{1}{4}$, or when $\Delta x = \lambda/4$. Here, with $\lambda = 3.43$ m, $\Delta x = 85.8$ cm.

ASSESS The phase difference increases as Δx increases, so we expect the answer to part b to be larger than 60 cm.

STOP TO THINK 16.5 What is the phase difference between the crest of a wave and the adjacent trough?

a. -2π rad   b. 0 rad   c. $\pi/4$ rad
d. $\pi/2$ rad   e. π rad   f. 3π rad

16.8 Power, Intensity, and Decibels

A traveling wave transfers energy from one point to another. The sound wave from a loudspeaker sets your eardrum into motion. Light waves from the sun warm the earth. The *power* of a wave is the rate, in joules per second, at which the wave transfers energy. As you learned in Chapter 9, power is measured in watts. A loudspeaker might emit 2 W of power, meaning that energy in the form of sound waves is radiated at the rate of 2 joules per second.

A focused light, like that of a projector, is more *intense* than the diffuse light that goes in all directions. Similarly, a loudspeaker that beams its sound forward into a small area produces a louder sound in that area than a speaker of equal power that radiates the sound in all directions. Quantities such as brightness and loudness depend not only on the rate of energy transfer, or power, but also on the *area* that receives that power.

FIGURE 16.28 shows a wave impinging on a surface of area a. The surface is perpendicular to the direction in which the wave is traveling. This might be a real, physical surface, such as your eardrum or a photovoltaic cell, but it could equally well be a mathematical surface in space that the wave passes right through. If the wave has power P, we define the **intensity** I of the wave to be

$$I = \frac{P}{a} = \text{power-to-area ratio} \qquad (16.57)$$

The SI units of intensity are W/m^2. Because intensity is a power-to-area ratio, a wave focused into a small area will have a larger intensity than a wave of equal power that is spread out over a large area.

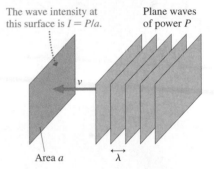

FIGURE 16.28 Plane waves of power P impinge on area a with intensity $I = P/a$.

The wave intensity at this surface is $I = P/a$.

Plane waves of power P

Area a λ v

EXAMPLE 16.9 | The intensity of a laser beam

A typical red laser pointer emits 1.0 mW of light power into a 1.0-mm-diameter laser beam. What is the intensity of the laser beam?

MODEL The laser beam is a light wave.

SOLVE The light waves of the laser beam pass through a mathematical surface that is a circle of diameter 1.0 mm. The intensity of the laser beam is

$$I = \frac{P}{a} = \frac{P}{\pi r^2} = \frac{0.0010 \text{ W}}{\pi (0.00050 \text{ m})^2} = 1300 \text{ W/m}^2$$

ASSESS This is roughly the intensity of sunlight at noon on a summer day. The difference between the sun and a small laser is not their intensities, which are about the same, but their powers. The laser has a small power of 1 mW. It can produce a very intense wave only because the area through which the wave passes is very small. The sun, by contrast, radiates a total power $P_{sun} \approx 4 \times 10^{26}$ W. This immense power is spread through *all* of space, producing an intensity of 1400 W/m^2 at a distance of 1.5×10^{11} m, the radius of the earth's orbit.

If a source of spherical waves radiates uniformly in all directions, then, as **FIGURE 16.29** on the next page shows, the power at distance r is spread uniformly over the surface of a sphere of radius r. The surface area of a sphere is $a = 4\pi r^2$, so the intensity of a uniform spherical wave is

$$I = \frac{P_{source}}{4\pi r^2} \qquad \text{(intensity of a uniform spherical wave)} \qquad (16.58)$$

The inverse-square dependence of r is really just a statement of energy conservation. The source emits energy at the rate P joules per second. The energy is spread over a

FIGURE 16.29 A source emitting uniform spherical waves.

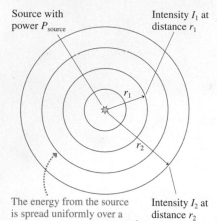

Source with power P_{source}

Intensity I_1 at distance r_1

The energy from the source is spread uniformly over a spherical surface of area $4\pi r^2$.

Intensity I_2 at distance r_2

larger and larger area as the wave moves outward. Consequently, the energy *per unit area* must decrease in proportion to the surface area of a sphere.

If the intensity at distance r_1 is $I_1 = P_{source}/4\pi r_1^2$ and the intensity at r_2 is $I_2 = P_{source}/4\pi r_2^2$, then you can see that the intensity *ratio* is

$$\frac{I_1}{I_2} = \frac{r_2^2}{r_1^2} \tag{16.59}$$

You can use Equation 16.59 to compare the intensities at two distances from a source without needing to know the power of the source.

> **NOTE** Wave intensities are strongly affected by reflections and absorption. Equations 16.58 and 16.59 apply to situations such as the light from a star or the sound from a firework exploding high in the air. Indoor sound does *not* obey a simple inverse-square law because of the many reflecting surfaces.

For a sinusoidal wave, each particle in the medium oscillates back and forth in simple harmonic motion. You learned in Chapter 15 that a particle in SHM with amplitude A has energy $E = \frac{1}{2}kA^2$, where k is the spring constant of the medium, not the wave number. It is this oscillatory energy of the medium that is transferred, particle to particle, as the wave moves through the medium.

Because a wave's intensity is proportional to the rate at which energy is transferred through the medium, and because the oscillatory energy in the medium is proportional to the *square* of the amplitude, we can infer that

$$I \propto A^2 \tag{16.60}$$

That is, **the intensity of a wave is proportional to the square of its amplitude.** If you double the amplitude of a wave, you increase its intensity by a factor of 4.

Sound Intensity Level

Human hearing spans an extremely wide range of intensities, from the *threshold of hearing* at $\approx 1 \times 10^{-12}$ W/m² (at midrange frequencies) to the *threshold of pain* at ≈ 10 W/m². If we want to make a scale of loudness, it's convenient and logical to place the zero of our scale at the threshold of hearing. To do so, we define the **sound intensity level,** expressed in **decibels** (dB), as

$$\beta = (10 \text{ dB}) \log_{10}\left(\frac{I}{I_0}\right) \tag{16.61}$$

where $I_0 = 1.0 \times 10^{-12}$ W/m². The symbol β is the Greek letter beta. Notice that β is computed as a base-10 logarithm, not a natural logarithm.

The decibel is named after Alexander Graham Bell, inventor of the telephone. Sound intensity level is actually dimensionless because it's formed from the ratio of two intensities, so decibels are just a *name* to remind us that we're dealing with an intensity *level* rather than a true intensity.

Right at the threshold of hearing, where $I = I_0$, the sound intensity level is

$$\beta = (10 \text{ dB}) \log_{10}\left(\frac{I_0}{I_0}\right) = (10 \text{ dB}) \log_{10}(1) = 0 \text{ dB}$$

Note that 0 dB doesn't mean no sound; it means that, for most people, no sound is heard. Dogs have more sensitive hearing than humans, and most dogs can easily perceive a 0 dB sound. The sound intensity level at the pain threshold is

$$\beta = (10 \text{ dB}) \log_{10}\left(\frac{10 \text{ W/m}^2}{10^{-12} \text{ W/m}^2}\right) = (10 \text{ dB}) \log_{10}(10^{13}) = 130 \text{ dB}$$

The major point to notice is that the sound intensity level increases by 10 dB each time the actual intensity increases by a *factor* of 10. For example, the sound intensity level increases from 70 dB to 80 dB when the sound intensity increases from 10^{-5} W/m^2 to 10^{-4} W/m^2. Perception experiments find that sound is perceived as "twice as loud" when the intensity increases by a factor of 10. In terms of decibels, we can say that the perceived loudness of a sound doubles with each increase in the sound intensity level by 10 dB.

TABLE 16.4 gives the sound intensity levels for a number of sounds. Although 130 dB is the threshold of pain, quieter sounds can damage your hearing. A fairly short exposure to 120 dB can cause damage to the hair cells in the ear, but lengthy exposure to sound intensity levels of over 85 dB can produce damage as well.

TABLE 16.4 Sound intensity levels of common sounds

Sound	β (dB)
Threshold of hearing	0
Person breathing, at 3 m	10
A whisper, at 1 m	20
Quiet room	30
Outdoors, no traffic	40
Quiet restaurant	50
Normal conversation, at 1 m	60
Busy traffic	70
Vacuum cleaner, for user	80
Niagara Falls, at viewpoint	90
Snowblower, at 2 m	100
Stereo, at maximum volume	110
Rock concert	120
Threshold of pain	130
Loudest football stadium	140

EXAMPLE 16.10 | Blender noise

The blender making a smoothie produces a sound intensity level of 83 dB. What is the intensity of the sound? What will the sound intensity level be if a second blender is turned on?

SOLVE We can solve Equation 16.61 for the sound intensity, finding $I = I_0 \times 10^{\beta/10 \text{ dB}}$. Here we used the fact that 10 raised to a power is an "antilogarithm." In this case,

$$I = (1.0 \times 10^{-12} \text{ W/m}^2) \times 10^{8.3} = 2.0 \times 10^{-4} \text{ W/m}^2$$

A second blender doubles the sound power and thus raises the intensity to $I = 4.0 \times 10^{-4}$ W/m^2. The new sound intensity level is

$$\beta = (10 \text{ dB}) \log_{10}\left(\frac{4.0 \times 10^{-4} \text{ W/m}^2}{1.0 \times 10^{-12} \text{ W/m}^2}\right) = 86 \text{ dB}$$

ASSESS In general, doubling the actual sound intensity increases the decibel level by 3 dB.

STOP TO THINK 16.6 Four trumpet players are playing the same note. If three of them suddenly stop, the sound intensity level decreases by

a. 40 dB   b. 12 dB   c. 6 dB   d. 4 dB

16.9 The Doppler Effect

Our final topic for this chapter is an interesting effect that occurs when you are in motion relative to a wave source. It is called the *Doppler effect*. You've likely noticed that the pitch of an ambulance's siren drops as it goes past you. Why?

FIGURE 16.30a on the next page shows a source of sound waves moving away from Pablo and toward Nancy at a steady speed v_s. The subscript s indicates that this is the speed of the source, not the speed of the waves. The source is emitting sound waves of frequency f_0 as it travels. The figure is a motion diagram showing the position of the source at times $t = 0$, T, $2T$, and $3T$, where $T = 1/f_0$ is the period of the waves.

Nancy measures the frequency of the wave emitted by the *approaching source* to be f_+. At the same time, Pablo measures the frequency of the wave emitted by the *receding source* to be f_-. Our task is to relate f_+ and f_- to the source frequency f_0 and speed v_s.

After a wave crest leaves the source, its motion is governed by the properties of the medium. That is, the motion of the source cannot affect a wave that has already been emitted. Thus each circular wave front in FIGURE 16.30b is centered on the point from which it was emitted. The wave crest from point 3 was emitted just as this figure was made, but it hasn't yet had time to travel any distance.

FIGURE 16.30 A motion diagram showing the wave fronts emitted by a source as it moves to the right at speed v_s.

(a) Motion of the source

Pablo   Nancy

The dots are the positions of the source at $t = 0$, T, $2T$, and $3T$. The source emits frequency f_0.

v_s

Pablo sees the source receding at speed v_s.

Nancy sees the source approaching at speed v_s.

(b) Snapshot at time $3T$

Behind the source, the wavelength is expanded to λ_-.

In front of the source, the wavelength is compressed to λ_+.

Distance d

Pablo detects frequency f_-.

Nancy detects frequency f_+.

$2\lambda_0$

$3\lambda_0$

λ_0

Crest 0 was emitted at $t = 0$. The wave front is a circle centered on point 0.

Crest 1 was emitted at $t = T$. The wave front is a circle centered on point 1.

Crest 2 was emitted at $t = 2T$. The wave front is a circle centered on point 2.

The wave crests are bunched up in the direction the source is moving, stretched out behind it. The distance between one crest and the next is one wavelength, so the wavelength λ_+ Nancy measures is *less* than the wavelength $\lambda_0 = v/f_0$ that would be emitted if the source were at rest. Similarly, λ_- behind the source is larger than λ_0.

These crests move through the medium at the wave speed v. Consequently, the frequency $f_+ = v/\lambda_+$ detected by the observer whom the source is approaching is *higher* than the frequency f_0 emitted by the source. Similarly, $f_- = v/\lambda_-$ detected behind the source is *lower* than frequency f_0. This change of frequency when a source moves relative to an observer is called the **Doppler effect.**

The distance labeled d in Figure 16.30b is the difference between how far the wave has moved and how far the source has moved at time $t = 3T$. These distances are

$$\Delta x_{\text{wave}} = vt = 3vT$$
$$\Delta x_{\text{source}} = v_s t = 3v_s T \tag{16.62}$$

The distance d spans three wavelengths; thus the wavelength of the wave emitted by an approaching source is

$$\lambda_+ = \frac{d}{3} = \frac{\Delta x_{\text{wave}} - \Delta x_{\text{source}}}{3} = \frac{3vT - 3v_s T}{3} = (v - v_s)T \tag{16.63}$$

You can see that our arbitrary choice of three periods was not relevant because the 3 cancels. The frequency detected in Nancy's direction is

$$f_+ = \frac{v}{\lambda_+} = \frac{v}{(v - v_s)T} = \frac{v}{(v - v_s)}f_0 \tag{16.64}$$

where $f_0 = 1/T$ is the frequency of the source and is the frequency you would detect if the source were at rest. We'll find it convenient to write the detected frequency as

$$f_+ = \frac{f_0}{1 - v_s/v} \qquad \text{(Doppler effect for an approaching source)}$$
$$\tag{16.65}$$
$$f_- = \frac{f_0}{1 + v_s/v} \qquad \text{(Doppler effect for a receding source)}$$

Doppler weather radar uses the Doppler shift of reflected radar signals to measure wind speeds and thus better gauge the severity of a storm.

Proof of the second version, for the frequency f_- of a receding source, is similar. You can see that $f_+ > f_0$ in front of the source, because the denominator is less than 1, and $f_- < f_0$ behind the source.

EXAMPLE 16.11 | **How fast are the police traveling?**

A police siren has a frequency of 550 Hz as the police car approaches you, 450 Hz after it has passed you and is receding. How fast are the police traveling? The temperature is 20°C.

MODEL The siren's frequency is altered by the Doppler effect. The frequency is f_+ as the car approaches and f_- as it moves away.

SOLVE To find v_s, we rewrite Equations 16.65 as

$$f_0 = (1 + v_s/v)f_-$$
$$f_0 = (1 - v_s/v)f_+$$

We subtract the second equation from the first, giving

$$0 = f_- - f_+ + \frac{v_s}{v}(f_- + f_+)$$

This is easily solved to give

$$v_s = \frac{f_+ - f_-}{f_+ + f_-} v = \frac{100 \text{ Hz}}{1000 \text{ Hz}} \times 343 \text{ m/s} = 34.3 \text{ m/s}$$

ASSESS If you now solve for the siren frequency when at rest, you will find $f_0 = 495$ Hz. Surprisingly, the at-rest frequency is not halfway between f_- and f_+.

A Stationary Source and a Moving Observer

Suppose the police car in Example 16.11 is at rest while you drive toward it at 34.3 m/s. You might think that this is equivalent to having the police car move toward you at 34.3 m/s, but it isn't. Mechanical waves move through a medium, and the Doppler effect depends not just on how the source and the observer move with respect to each other but also on how they move with respect to the medium. We'll omit the proof, but it's not hard to show that the frequencies heard by an observer moving at speed v_o relative to a stationary source emitting frequency f_0 are

$$f_+ = (1 + v_o/v)f_0 \quad \text{(observer approaching a source)}$$
$$f_- = (1 - v_o/v)f_0 \quad \text{(observer receding from a source)}$$

(16.66)

A quick calculation shows that the frequency of the police siren as you approach it at 34.3 m/s is 545 Hz, not the 550 Hz you heard as it approached you at 34.3 m/s.

The Doppler Effect for Light Waves

The Doppler effect is observed for all types of waves, not just sound waves. If a source of light waves is receding from you, the wavelength λ_- that you detect is longer than the wavelength λ_0 emitted by the source.

Although the reason for the Doppler shift for light is the same as for sound waves, there is one fundamental difference. We derived Equations 16.65 for the Doppler-shifted frequencies by measuring the wave speed v relative to the medium. For electromagnetic waves in empty space, there is no medium. Consequently, we need to turn to Einstein's theory of relativity to determine the frequency of light waves from a moving source. The result, which we state without proof, is

$$\lambda_- = \sqrt{\frac{1 + v_s/c}{1 - v_s/c}} \, \lambda_0 \quad \text{(receding source)}$$

(16.67)

$$\lambda_+ = \sqrt{\frac{1 - v_s/c}{1 + v_s/c}} \, \lambda_0 \quad \text{(approaching source)}$$

Here v_s is the speed of the source *relative to* the observer.

The light waves from a receding source are shifted to longer wavelengths ($\lambda_- > \lambda_0$). Because the longest visible wavelengths are perceived as the color red, the light from a receding source is **red shifted.** That is *not* to say that the light is red, simply that its wavelength is shifted toward the red end of the spectrum. If $\lambda_0 = 470$ nm (blue) light emitted by a rapidly receding source is detected at $\lambda_- = 520$ nm (green), we would say that the light has been red shifted. Similarly, light from an approaching source is **blue shifted,** meaning that the detected wavelengths are shorter than the emitted wavelengths ($\lambda_+ < \lambda_0$) and thus are shifted toward the blue end of the spectrum.

EXAMPLE 16.12 | Measuring the velocity of a galaxy

Hydrogen atoms in the laboratory emit red light with wavelength 656 nm. In the light from a distant galaxy, this "spectral line" is observed at 691 nm. What is the speed of this galaxy relative to the earth?

MODEL The observed wavelength is longer than the wavelength emitted by atoms at rest with respect to the observer (i.e., red shifted), so we are looking at light emitted from a galaxy that is receding from us.

SOLVE Squaring the expression for λ_- in Equations 16.67 and solving for v_s give

$$v_s = \frac{(\lambda_-/\lambda_0)^2 - 1}{(\lambda_-/\lambda_0)^2 + 1} c$$

$$= \frac{(691 \text{ nm}/656 \text{ nm})^2 - 1}{(691 \text{ nm}/656 \text{ nm})^2 + 1} c$$

$$= 0.052c = 1.56 \times 10^7 \text{ m/s}$$

ASSESS The galaxy is moving away from the earth at about 5% of the speed of light!

FIGURE 16.31 A Hubble Space Telescope picture of a quasar.

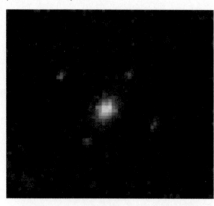

In the 1920s, an analysis of the red shifts of many galaxies led the astronomer Edwin Hubble to the conclusion that the galaxies of the universe are *all* moving apart from each other. Extrapolating backward in time must bring us to a point when all the matter of the universe—and even space itself, according to the theory of relativity—began rushing out of a primordial fireball. Many observations and measurements since have given support to the idea that the universe began in a *Big Bang* about 14 billion years ago.

As an example, **FIGURE 16.31** is a Hubble Space Telescope picture of a *quasar*, short for *quasistellar object*. Quasars are extraordinarily powerful sources of light and radio waves. The light reaching us from quasars is highly red shifted, corresponding in some cases to objects that are moving away from us at greater than 90% of the speed of light. Astronomers have determined that some quasars are 10 to 12 *billion* light years away from the earth, hence the light we see was emitted when the universe was only about 25% of its present age. Today, the red shifts of distant quasars and supernovae (exploding stars) are being used to refine our understanding of the structure and evolution of the universe.

STOP TO THINK 16.7 Amy and Zack are both listening to the source of sound waves that is moving to the right. Compare the frequencies each hears.

a. $f_{\text{Amy}} > f_{\text{Zack}}$
b. $f_{\text{Amy}} = f_{\text{Zack}}$
c. $f_{\text{Amy}} < f_{\text{Zack}}$

Amy   10 m/s f_0 10 m/s   10 m/s   Zack

CHALLENGE EXAMPLE 16.13 | Decreasing the sound

The loudspeaker on a homecoming float—mounted on a pole—is stuck playing an annoying 210 Hz tone. When the speaker is 10 m away, you measure the sound to be a loud 95 dB at 208 Hz. How long will it take for the sound intensity level to drop to a tolerable 55 dB?

MODEL The source is on a pole, so model the sound waves as uniform spherical waves. Assume a temperature of 20°C.

SOLVE The 208 Hz frequency you measure is less than the 210 Hz frequency that was emitted, so the float must be moving away from you. The Doppler effect for a receding source is

$$f_- = \frac{f_0}{1 + v_s/v}$$

We can solve this to find the speed of the float:

$$v_s = \left(\frac{f_0}{f_-} - 1\right)v = \left(\frac{210 \text{ Hz}}{208 \text{ Hz}} - 1\right) \times 343 \text{ m/s} = 3.3 \text{ m/s}$$

The sound intensity of a spherical wave decreases with the inverse square of the distance from the source. A sound intensity level β corresponds to an intensity $I = I_0 \times 10^{\beta/10 \text{ dB}}$, where $I_0 = 1.0 \times 10^{-12} \text{ W/m}^2$. At the initial 95 dB, the intensity is

$$I_1 = I_0 \times 10^{9.5} = 3.2 \times 10^{-3} \text{ W/m}^2$$

At the desired 55 dB, the intensity will have dropped to

$$I_2 = I_0 \times 10^{5.5} = 3.2 \times 10^{-7} \text{ W/m}^2$$

The intensity ratio is related to the distances by

$$\frac{I_1}{I_2} = \frac{r_2^2}{r_1^2}$$

Thus the sound will have dropped to 55 dB when the distance to the speaker is

$$r_2 = \sqrt{\frac{I_1}{I_2}} r_1 = \sqrt{10^4} \times 10 \text{ m} = 1000 \text{ m}$$

The float has to travel $\Delta x = 990$ m, which will take

$$\Delta t = \frac{\Delta x}{v_s} = \frac{990 \text{ m}}{3.3 \text{ m/s}} = 300 \text{ s} = 5.0 \text{ min}$$

ASSESS To drop the sound intensity level by 40 dB requires decreasing the intensity by a factor of 10^4. And with the intensity depending on the inverse square of the distance, that requires increasing the distance by a factor of 100. Floats don't move very fast—3.3 m/s is about 7 mph—so needing several minutes to travel the ≈ 1000 m seems reasonable.

SUMMARY

The goal of Chapter 16 has been to learn the basic properties of traveling waves.

GENERAL PRINCIPLES

The Wave Model

This model is based on the idea of a **traveling wave**, which is an organized disturbance traveling at a well-defined **wave speed** v.

- In **transverse waves** the displacement is perpendicular to the direction in which the wave travels.

- In **longitudinal waves** the particles of the medium are displaced parallel to the direction in which the wave travels.

A wave transfers **energy,** but no material or substance is transferred outward from the source.

Two basic classes of waves:

- **Mechanical waves** travel through a material medium such as water or air.

- **Electromagnetic waves** require no material medium and can travel through a vacuum.

For mechanical waves, such as sound waves and waves on strings, the speed of the wave is a property of the medium. Speed does not depend on the size or shape of the wave.

IMPORTANT CONCEPTS

The **displacement** D of a wave is a function of both position (where) and time (when).

- A **snapshot graph** shows the wave's displacement as a function of position at a single instant of time.

- A **history graph** shows the wave's displacement as a function of time at a single point in space.

For a transverse wave on a string, the snapshot graph is a picture of the wave. The displacement of a longitudinal wave is parallel to the motion; thus the snapshot graph of a longitudinal sound wave is *not* a picture of the wave.

Sinusoidal waves are periodic in both time (period T) and space (wavelength λ):

$$D(x, t) = A \sin\left[2\pi(x/\lambda - t/T) + \phi_0\right]$$
$$= A \sin(kx - \omega t + \phi_0)$$

where A is the **amplitude,** $k = 2\pi/\lambda$ is the **wave number,** $\omega = 2\pi f = 2\pi/T$ is the **angular frequency,** and ϕ_0 is the **phase constant** that describes initial conditions.

One-dimensional waves

Two- and three-dimensional waves

The fundamental relationship for any sinusoidal wave is $v = \lambda f$.

APPLICATIONS

- **String** (transverse): $v = \sqrt{T_s/\mu}$

- **Sound** (longitudinal): $v = \sqrt{B/\rho} = 343$ m/s in 20°C air

- **Light** (transverse): $v = c/n$, where $c = 3.00 \times 10^8$ m/s is the speed of light in a vacuum and n is the material's **index of refraction**

The wave intensity is the power-to-area ratio: $I = P/a$

For a circular or spherical wave: $I = P_{\text{source}}/4\pi r^2$

The sound intensity level is

$$\beta = (10 \text{ dB}) \log_{10}(I/1.0 \times 10^{-12} \text{ W/m}^2)$$

The **Doppler effect** occurs when a wave source and detector are moving with respect to each other: the frequency detected differs from the frequency f_0 emitted.

Approaching source

$$f_+ = \frac{f_0}{1 - v_s/v}$$

Receding source

$$f_- = \frac{f_0}{1 + v_s/v}$$

Observer approaching a source

$$f_+ = (1 + v_o/v)f_0$$

Observer receding from a source

$$f_- = (1 - v_o/v)f_0$$

The Doppler effect for light uses a result derived from the theory of relativity.

TERMS AND NOTATION

wave model	linear density, μ	partial derivative	plane wave
traveling wave	snapshot graph	wave equation	phase, ϕ
transverse wave	history graph	compression	intensity, I
longitudinal wave	leading edge	rarefaction	sound intensity level, β
mechanical wave	trailing edge	electromagnetic spectrum	decibels
electromagnetic wave	sinusoidal wave	index of refraction, n	Doppler effect
medium	amplitude, A	wave front	red shifted
disturbance	wavelength, λ	circular wave	blue shifted
wave speed, v	wave number, k	spherical wave	

CONCEPTUAL QUESTIONS

1. The three wave pulses in **FIGURE Q16.1** travel along the same stretched string. Rank in order, from largest to smallest, their wave speeds v_a, v_b, and v_c. Explain.

FIGURE Q16.1

2. A wave pulse travels along a stretched string at a speed of 200 cm/s. What will be the speed if:
   a. The string's tension is doubled?
   b. The string's mass is quadrupled (but its length is unchanged)?
   c. The string's length is quadrupled (but its mass is unchanged)?
   Note: Each part is independent and refers to changes made to the original string.

3. **FIGURE Q16.3** is a history graph showing the displacement as a function of time at one point on a string. Did the displacement at this point reach its maximum of 2 mm *before* or *after* the interval of time when the displacement was a constant 1 mm?

FIGURE Q16.3

4. **FIGURE Q16.4** shows a snapshot graph *and* a history graph for a wave pulse on a stretched string. They describe the same wave from two perspectives.
   a. In which direction is the wave traveling? Explain.
   b. What is the speed of this wave?

FIGURE Q16.4

5. Rank in order, from largest to smallest, the wavelengths λ_a, λ_b, and λ_c for sound waves having frequencies $f_a = 100$ Hz, $f_b = 1000$ Hz, and $f_c = 10,000$ Hz. Explain.

6. A sound wave with wavelength λ_0 and frequency f_0 moves into a new medium in which the speed of sound is $v_1 = 2v_0$. What are the new wavelength λ_1 and frequency f_1? Explain.

7. What are the amplitude, wavelength, frequency, and phase constant of the traveling wave in **FIGURE Q16.7**?

FIGURE Q16.7

FIGURE Q16.8

8. **FIGURE Q16.8** is a snapshot graph of a sinusoidal wave at $t = 1.0$ s. What is the phase constant of this wave?

9. **FIGURE Q16.9** shows the wave fronts of a circular wave. What is the phase difference between (a) points A and B, (b) points C and D, and (c) points E and F?

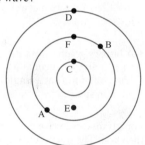

FIGURE Q16.9

10. Sound wave A delivers 2 J of energy in 2 s. Sound wave B delivers 10 J of energy in 5 s. Sound wave C delivers 2 mJ of energy in 1 ms. Rank in order, from largest to smallest, the sound powers P_A, P_B, and P_C of these three sound waves. Explain.

11. One physics professor talking produces a sound intensity level of 52 dB. It's a frightening idea, but what would be the sound intensity level of 100 physics professors talking simultaneously?

12. You are standing at $x = 0$ m, listening to a sound that is emitted at frequency f_0. The graph of **FIGURE Q16.12** shows the frequency you hear during a 4-second interval. Which of the following describes the sound source? Explain your choice.

FIGURE Q16.12

   A. It moves from left to right and passes you at $t = 2$ s.
   B. It moves from right to left and passes you at $t = 2$ s.
   C. It moves toward you but doesn't reach you. It then reverses direction at $t = 2$ s.
   D. It moves away from you until $t = 2$ s. It then reverses direction and moves toward you but doesn't reach you.

EXERCISES AND PROBLEMS

Problems labeled ▓ integrate material from earlier chapters.

Exercises

Section 16.1 An Introduction to Waves

1. | The wave speed on a string under tension is 200 m/s. What is the speed if the tension is halved?

2. | The wave speed on a string is 150 m/s when the tension is 75 N. What tension will give a speed of 180 m/s?

3. ‖ A 25 g string is under 20 N of tension. A pulse travels the length of the string in 50 ms. How long is the string?

Section 16.2 One-Dimensional Waves

4. ‖ Draw the history graph $D(x = 4.0 \text{ m}, t)$ at $x = 4.0$ m for the wave shown in FIGURE EX16.4.

Snapshot graph of a wave at $t = 2$ s

FIGURE EX16.4

Snapshot graph of a wave at $t = 0$ s

FIGURE EX16.5

5. ‖ Draw the history graph $D(x = 0 \text{ m}, t)$ at $x = 0$ m for the wave shown in FIGURE EX16.5.

6. ‖ Draw the snapshot graph $D(x, t = 0 \text{ s})$ at $t = 0$ s for the wave shown in FIGURE EX16.6.

History graph of a wave at $x = 2$ m
Wave moving to the right at 1.0 m/s

FIGURE EX16.6

History graph of a wave at $x = 0$ m
Wave moving to the left at 1.0 m/s

FIGURE EX16.7

7. ‖ Draw the snapshot graph $D(x, t = 1.0 \text{ s})$ at $t = 1.0$ s for the wave shown in FIGURE EX16.7.

8. ‖ FIGURE EX16.8 is a picture at $t = 0$ s of the particles in a medium as a longitudinal wave is passing through. The equilibrium spacing between the particles is 1.0 cm. Draw the snapshot graph $D(x, t = 0 \text{ s})$ of this wave at $t = 0$ s.

FIGURE EX16.8

9. ‖ FIGURE EX16.9 is the snapshot graph at $t = 0$ s of a *longitudinal* wave. Draw the corresponding picture of the particle positions, as was done in Figure 16.9b. Let the equilibrium spacing between the particles be 1.0 cm.

FIGURE EX16.9

Section 16.3 Sinusoidal Waves

10. | A wave has angular frequency 30 rad/s and wavelength 2.0 m. What are its (a) wave number and (b) wave speed?

11. | A wave travels with speed 200 m/s. Its wave number is 1.5 rad/m. What are its (a) wavelength and (b) frequency?

12. | The displacement of a wave traveling in the negative y-direction is $D(y, t) = (5.2 \text{ cm}) \sin(5.5y + 72t)$, where y is in m and t is in s. What are the (a) frequency, (b) wavelength, and (c) speed of this wave?

13. | The displacement of a wave traveling in the positive x-direction is $D(x, t) = (3.5 \text{ cm}) \sin(2.7x - 124t)$, where x is in m and t is in s. What are the (a) frequency, (b) wavelength, and (c) speed of this wave?

14. ‖ What are the amplitude, frequency, and wavelength of the wave in FIGURE EX16.14?

History graph at $x = 0$ m
Wave traveling left at 2.0 m/s

FIGURE EX16.14

Section 16.4 The Wave Equation on a String

15. ‖ Show that the displacement $D(x, t) = cx^2 + dt^2$, where c and
CALC d are constants, is a solution to the wave equation. Then find an expression in terms of c and d for the wave speed.

16. ‖ Show that the displacement $D(x, t) = \ln(ax + bt)$, where a
CALC and b are constants, is a solution to the wave equation. Then find an expression in terms of a and b for the wave speed.

Section 16.5 Sound and Light

17. ‖ a. What is the wavelength of a 2.0 MHz ultrasound wave traveling through aluminum?
   b. What frequency of electromagnetic wave would have the same wavelength as the ultrasound wave of part a?

18. | a. What is the frequency of an electromagnetic wave with a wavelength of 20 cm?
   b. What would be the wavelength of a sound wave in water with the same frequency as the electromagnetic wave of part a?

19. | a. What is the frequency of blue light that has a wavelength of 450 nm?
   b. What is the frequency of red light that has a wavelength of 650 nm?
   c. What is the index of refraction of a material in which the red-light wavelength is 450 nm?

20. | a. An FM radio station broadcasts at a frequency of 101.3 MHz. What is the wavelength?
   b. What is the frequency of a sound source that produces the same wavelength in 20°C air?

21. | a. Telephone signals are often transmitted over long distances by microwaves. What is the frequency of microwave radiation with a wavelength of 3.0 cm?
   b. Microwave signals are beamed between two mountaintops 50 km apart. How long does it take a signal to travel from one mountaintop to the other?

22. ‖ A hammer taps on the end of a 4.00-m-long metal bar at room temperature. A microphone at the other end of the bar picks up two pulses of sound, one that travels through the metal and one that travels through the air. The pulses are separated in time by 9.00 ms. What is the speed of sound in this metal?

23. | Cell phone conversations are transmitted by high-frequency radio waves. Suppose the signal has wavelength 35 cm while traveling through air. What are the (a) frequency and (b) wavelength as the signal travels through 3-mm-thick window glass into your room?

24. || a. How long does it take light to travel through a 3.0-mm-thick piece of window glass?
    b. Through what thickness of water could light travel in the same amount of time?

25. | A light wave has a 670 nm wavelength in air. Its wavelength in a transparent solid is 420 nm.
    a. What is the speed of light in this solid?
    b. What is the light's frequency in the solid?

26. | A 440 Hz sound wave in 20°C air propagates into the water of a swimming pool. What are the wave's (a) frequency and (b) wavelength in the water?

Section 16.6 The Wave Equation in a Fluid

27. | What is the speed of sound in air (a) on a cold winter day in Minnesota when the temperature is −25°F, and (b) on a hot summer day in Death Valley when the temperature is 125°F?

28. | The density of mercury is 13,600 kg/m³. What is the speed of sound in mercury at 20°C?

Section 16.7 Waves in Two and Three Dimensions

29. | A circular wave travels outward from the origin. At one instant of time, the phase at $r_1 = 20$ cm is 0 rad and the phase at $r_2 = 80$ cm is 3π rad. What is the wavelength of the wave?

30. || A spherical wave with a wavelength of 2.0 m is emitted from the origin. At one instant of time, the phase at $r = 4.0$ m is π rad. At that instant, what is the phase at $r = 3.5$ m and at $r = 4.5$ m?

31. || A loudspeaker at the origin emits a 120 Hz tone on a day when the speed of sound is 340 m/s. The phase difference between two points on the x-axis is 5.5 rad. What is the distance between these two points?

32. || A sound source is located somewhere along the x-axis. Experiments show that the same wave front simultaneously reaches listeners at $x = -7.0$ m and $x = +3.0$ m.
    a. What is the x-coordinate of the source?
    b. A third listener is positioned along the positive y-axis. What is her y-coordinate if the same wave front reaches her at the same instant it does the first two listeners?

Section 16.8 Power, Intensity, and Decibels

33. || A sound wave with intensity 2.0×10^{-3} W/m² is perceived to be modestly loud. Your eardrum is 6.0 mm in diameter. How much energy will be transferred to your eardrum while listening to this sound for 1.0 min?
BIO

34. || The intensity of electromagnetic waves from the sun is 1.4 kW/m² just above the earth's atmosphere. Eighty percent of this reaches the surface at noon on a clear summer day. Suppose you think of your back as a 30 cm × 50 cm rectangle. How many joules of solar energy fall on your back as you work on your tan for 1.0 h?

35. || A concert loudspeaker suspended high above the ground emits 35 W of sound power. A small microphone with a 1.0 cm² area is 50 m from the speaker.
    a. What is the sound intensity at the position of the microphone?
    b. How much sound energy impinges on the microphone each second?

36. || During takeoff, the sound intensity level of a jet engine is 140 dB at a distance of 30 m. What is the sound intensity level at a distance of 1.0 km?

37. | The sun emits electromagnetic waves with a power of 4.0×10^{26} W. Determine the intensity of electromagnetic waves from the sun just outside the atmospheres of Venus, the earth, and Mars.

38. | What are the sound intensity levels for sound waves of intensity (a) 3.0×10^{-6} W/m² and (b) 3.0×10^{-2} W/m²?

39. || A loudspeaker on a tall pole broadcasts sound waves equally in all directions. What is the speaker's power output if the sound intensity level is 90 dB at a distance of 20 m?

40. || The sound intensity level 5.0 m from a large power saw is 100 dB. At what distance will the sound be a more tolerable 80 dB?

Section 16.9 The Doppler Effect

41. | A friend of yours is loudly singing a single note at 400 Hz while racing toward you at 25.0 m/s on a day when the speed of sound is 340 m/s.
    a. What frequency do you hear?
    b. What frequency does your friend hear if you suddenly start singing at 400 Hz?

42. | An opera singer in a convertible sings a note at 600 Hz while cruising down the highway at 90 km/h. What is the frequency heard by
    a. A person standing beside the road in front of the car?
    b. A person on the ground behind the car?

43. || A bat locates insects by emitting ultrasonic "chirps" and then listening for echoes from the bugs. Suppose a bat chirp has a frequency of 25 kHz. How fast would the bat have to fly, and in what direction, for you to just barely be able to hear the chirp at 20 kHz?
BIO

44. || A mother hawk screeches as she dives at you. You recall from biology that female hawks screech at 800 Hz, but you hear the screech at 900 Hz. How fast is the hawk approaching?

Problems

45. || **FIGURE P16.45** is a history graph at $x = 0$ m of a wave traveling in the positive x-direction at 4.0 m/s.
    a. What is the wavelength?
    b. What is the phase constant of the wave?
    c. Write the displacement equation for this wave.

History graph at $x = 0$ m
Wave traveling right at 4.0 m/s

FIGURE P16.45

Snapshot graph at $t = 0$ s

FIGURE P16.46

46. || **FIGURE P16.46** is a snapshot graph at $t = 0$ s of a 5.0 Hz wave traveling to the left.
    a. What is the wave speed?
    b. What is the phase constant of the wave?
    c. Write the displacement equation for this wave.

47. | String 1 in **FIGURE P16.47** has linear density 2.0 g/m and string 2 has linear density 4.0 g/m. A student sends pulses in both directions by quickly pulling up on the knot, then releasing it.

FIGURE P16.47

What should the string lengths L_1 and L_2 be if the pulses are to reach the ends of the strings simultaneously?

48. || Oil explorers set off explosives to make loud sounds, then listen for the echoes from underground oil deposits. Geologists suspect that there is oil under 500-m-deep Lake Physics. It's known that Lake Physics is carved out of a granite basin. Explorers detect a weak echo 0.94 s after exploding dynamite at the lake surface. If it's really oil, how deep will they have to drill into the granite to reach it?

49. || One cue your hearing system uses to localize a sound (i.e., to
BIO tell where a sound is coming from) is the slight difference in the
arrival times of the sound at your ears. Your ears are spaced
approximately 20 cm apart. Consider a sound source 5.0 m from
the center of your head along a line 45° to your right. What is the
difference in arrival times? Give your answer in microseconds.
Hint: You are looking for the difference between two numbers that
are nearly the same. What does this near equality imply about the
necessary precision during intermediate stages of the calculation?

50. || A helium-neon laser beam has a wavelength in air of 633 nm. It
takes 1.38 ns for the light to travel through 30 cm of an unknown
liquid. What is the wavelength of the laser beam in the liquid?

51. || Earthquakes are essentially sound waves—called seismic
waves—traveling through the earth. Because the earth is solid,
it can support both longitudinal and transverse seismic waves.
The speed of longitudinal waves, called P waves, is 8000 m/s.
Transverse waves, called S waves, travel at a slower 4500 m/s.
A seismograph records the two waves from a distant earthquake.
If the S wave arrives 2.0 min after the P wave, how far away was
the earthquake? You can assume that the waves travel in straight
lines, although actual seismic waves follow more complex routes.

52. || Helium (density 0.18 kg/m^3 at 0°C and 1 atm pressure)
remains a gas until the extraordinarily low temperature of 4.2 K.
What is the speed of sound in helium at 5 K?

53. || A 20.0-cm-long, 10.0-cm-diameter cylinder with a piston at
one end contains 1.34 kg of an unknown liquid. Using the piston
to compress the length of the liquid by 1.00 mm increases the
pressure by 41.0 atm. What is the speed of sound in the liquid?

54. || A sound wave is described by $D(y, t) = (0.0200 \text{ mm}) \times$
$\sin[(8.96 \text{ rad/m})y + (3140 \text{ rad/s})t + \pi/4 \text{ rad}]$, where y is in m
and t is in s.
   a. In what direction is this wave traveling?
   b. Along which axis is the air oscillating?
   c. What are the wavelength, the wave speed, and the period of
   oscillation?

55. || A wave on a string is described by $D(x, t) = (3.0 \text{ cm}) \times$
$\sin[2\pi(x/(2.4 \text{ m}) + t/(0.20 \text{ s}) + 1)]$, where x is in m and t is in s.
   a. In what direction is this wave traveling?
   b. What are the wave speed, the frequency, and the wave number?
   c. At $t = 0.50$ s, what is the displacement of the string at
   $x = 0.20$ m?

56. || A wave on a string is described by $D(x, t) = (2.00 \text{ cm}) \times$
$\sin[(12.57 \text{ rad/m})x - (638 \text{ rad/s})t]$, where x is in m and t in s.
The linear density of the string is 5.00 g/m. What are
   a. The string tension?
   b. The maximum displacement of a point on the string?
   c. The maximum speed of a point on the string?

57. || **FIGURE P16.57** shows a snapshot graph of a wave traveling
to the right along a string at 45 m/s. At this instant, what is the
velocity of points 1, 2, and 3 on the string?

FIGURE P16.57   FIGURE P16.58

58. ||| **FIGURE P16.58** shows two masses hanging from a steel wire.
The mass of the wire is 60.0 g. A wave pulse travels along the
wire from point 1 to point 2 in 24.0 ms. What is mass m?

59. || A wire is made by welding
together two metals having different densities. **FIGURE P16.59**
shows a 2.00-m-long section of
wire centered on the junction,

FIGURE P16.59

but the wire extends much farther in both directions. The wire is
placed under 2250 N tension, then a 1500 Hz wave with an amplitude
of 3.00 mm is sent down the wire. How many wavelengths (complete
cycles) of the wave are in this 2.00-m-long section of the wire?

60. || The string in **FIGURE P16.60** has linear
density μ. Find an expression in terms of
M, μ, and θ for the speed of waves on the
string.

FIGURE P16.60

61. || A string that is under 50.0 N of tension has linear density
5.0 g/m. A sinusoidal wave with amplitude 3.0 cm and wavelength
2.0 m travels along the string. What is the maximum speed of a
particle on the string?

62. || The G string on a guitar is a 0.46-mm-diameter steel string
with a linear density of 1.3 g/m. When the string is properly
tuned to 196 Hz, the wave speed on the string is 250 m/s. Tuning
is done by turning the tuning screw, which slowly tightens—and
stretches—the string. By how many mm does a 75-cm-long G
string stretch when it's first tuned?

63. || A sinusoidal wave travels along a stretched string. A particle on the
string has a maximum speed of 2.0 m/s and a maximum acceleration
of 200 m/s^2. What are the frequency and amplitude of the wave?

64. ||| Is the displacement $D(x, t) = (0.10 - 0.10x^2 + xt - 2.5t^2)$ m,
CALC where x is in m and t is in s, a possible traveling wave? If so, what
is the wave speed?

65. ||| Is the displacement $D(x, t) = (3.0 \text{ mm}) e^{i(2.0x + 8.0t + 5.0)}$, where
CALC x is in m, t is in s, and $i = \sqrt{-1}$, a possible traveling wave? If so,
what is the wave speed? *Complex exponentials* are often used to
represent waves in more advanced treatments.

66. || An AM radio station broadcasts with a power of 25 kW at a
frequency of 920 kHz. Estimate the intensity of the radio wave at
a point 10 km from the broadcast antenna.

67. || LASIK eye surgery uses pulses of laser light to shave off tissue
BIO from the cornea, reshaping it. A typical LASIK laser emits a
1.0-mm-diameter laser beam with a wavelength of 193 nm. Each
laser pulse lasts 15 ns and contains 1.0 mJ of light energy.
   a. What is the power of one laser pulse?
   b. During the very brief time of the pulse, what is the intensity
   of the light wave?

68. || The sound intensity 50 m from a wailing tornado siren is
0.10 W/m^2.
   a. What is the intensity at 1000 m?
   b. The weakest intensity likely to be heard over background
   noise is $\approx 1 \ \mu\text{W/m}^2$. Estimate the maximum distance at
   which the siren can be heard.

69. || A distant star system is discovered in which a planet with
twice the radius of the earth and rotating 3.0 times as fast as the
earth orbits a star with a total power output of 6.8×10^{29} W.
   a. If the star's radius is 6.0 times that of the sun, what is the
   electromagnetic wave intensity at the surface? Astronomers
   call this the *surface flux*. Astronomical data are provided
   inside the back cover of the book.
   b. Every planet-day (one rotation), the planet receives 9.4×10^{22} J
   of energy. What is the planet's distance from its star? Give
   your answer in *astronomical units* (AU), where 1 AU is the
   distance of the earth from the sun.

70. ‖ A compact sound source radiates 25 W of sound energy uniformly in all directions. What is the ratio of the sound intensity at a distance of 1.0 m to that at 5.0 m in (a) a two-dimensional universe, (b) our normal three-dimensional universe, and (c) a hypothetical four-dimensional universe?

71. ‖ A loudspeaker, mounted on a tall pole, is engineered to emit 75% of its sound energy into the forward hemisphere, 25% toward the back. You measure an 85 dB sound intensity level when standing 3.5 m in front of and 2.5 m below the speaker. What is the speaker's power output?

72. ‖ Your ears are sensitive to differences in pitch, but they are not
BIO very sensitive to differences in intensity. You are not capable of detecting a difference in sound intensity level of less than 1 dB. By what factor does the sound intensity increase if the sound intensity level increases from 60 dB to 61 dB?

73. ‖ The intensity of a sound source is described by an inverse-square law only if the source is very small (a point source) and only if the waves can travel unimpeded in all directions. For an extended source or in a situation where obstacles absorb or reflect the waves, the intensity at distance r can often be expressed as $I = cP_{source}/r^x$, where c is a constant and the exponent x—which would be 2 for an ideal spherical wave—depends on the situation. In one such situation, you use a sound meter to measure the sound intensity level at different distances from a source, acquiring the data in the table. Use the best-fit line of an appropriate graph to determine the exponent x that characterizes this sound source.

Distance (m)	Intensity level (dB)
1	100
3	93
10	85
30	78
100	70

74. ‖ A physics professor demonstrates the Doppler effect by tying a 600 Hz sound generator to a 1.0-m-long rope and whirling it around her head in a horizontal circle at 100 rpm. What are the highest and lowest frequencies heard by a student in the classroom?

75. ‖ An avant-garde composer wants to use the Doppler effect in his new opera. As the soprano sings, he wants a large bat to fly toward her from the back of the stage. The bat will be outfitted with a microphone to pick up the singer's voice and a loudspeaker to rebroadcast the sound toward the audience. The composer wants the sound the audience hears from the bat to be, in musical terms, one half-step higher in frequency than the note they are hearing from the singer. Two notes a half-step apart have a frequency ratio of $2^{1/12} = 1.059$. With what speed must the bat fly toward the singer?

76. ‖ A loudspeaker on a pole is radiating 100 W of sound energy in all
CALC directions. You are walking directly toward the speaker at 0.80 m/s. When you are 20 m away, what are (a) the sound intensity level and (b) the rate (dB/s) at which the sound intensity level is increasing?
Hint: Use the chain rule and the relationship $\log_{10} x = \ln x/\ln 10$.

77. ‖ Show that the Doppler frequency f_- of a receding source is $f_- = f_0/(1 + v_s/v)$.

78. ‖ A starship approaches its home planet at a speed of $0.10c$. When it is 54×10^6 km away, it uses its green laser beam ($\lambda = 540$ nm) to signal its approach.
a. How long does the signal take to travel to the home planet?
b. At what wavelength is the signal detected on the home planet?

79. ‖ Wavelengths of light from a distant galaxy are found to be 0.50% longer than the corresponding wavelengths measured in a terrestrial laboratory. Is the galaxy approaching or receding from the earth? At what speed?

80. ‖ You have just been pulled over for running a red light, and the police officer has informed you that the fine will be $250. In desperation, you suddenly recall an idea that your physics professor recently discussed in class. In your calmest voice, you tell the officer that the laws of physics prevented you from knowing that the light was red. In fact, as you drove toward it, the light was Doppler shifted to where it appeared green to you. "OK," says the officer, "Then I'll ticket you for speeding. The fine is $1 for every 1 km/h over the posted speed limit of 50 km/h." How big is your fine? Use 650 nm as the wavelength of red light and 540 nm as the wavelength of green light.

Challenge Problems

81. ‖‖ One way to monitor global warming is to measure the average temperature of the ocean. Researchers are doing this by measuring the time it takes sound pulses to travel underwater over large distances. At a depth of 1000 m, where ocean temperatures hold steady near 4°C, the average sound speed is 1480 m/s. It's known from laboratory measurements that the sound speed increases 4.0 m/s for every 1.0°C increase in temperature. In one experiment, where sounds generated near California are detected in the South Pacific, the sound waves travel 8000 km. If the smallest time change that can be reliably detected is 1.0 s, what is the smallest change in average temperature that can be measured?

82. ‖‖ A rope of mass m and length L hangs from a ceiling.
CALC a. Show that the wave speed on the rope a distance y above the lower end is $v = \sqrt{gy}$.
b. Show that the time for a pulse to travel the length of the string is $\Delta t = 2\sqrt{L/g}$.

83. ‖‖ A communications truck with a 44-cm-diameter dish receiver
CALC on the roof starts out 10 km from its base station. It drives directly away from the base station at 50 km/h for 1.0 h, keeping the receiver pointed at the base station. The base station antenna broadcasts continuously with 2.5 kW of power, radiated uniformly in all directions. How much electromagnetic energy does the truck's dish receive during that 1.0 h?

84. ‖‖ Some modern optical devic-
CALC es are made with glass whose index of refraction changes with distance from the front surface. **FIGURE CP16.84** shows the index of refraction as a function of the distance into a slab of glass of thickness L. The index of refraction increases linearly from n_1 at the front surface to n_2 at the rear surface.
a. Find an expression for the time light takes to travel through this piece of glass.
b. Evaluate your expression for a 1.0-cm-thick piece of glass for which $n_1 = 1.50$ and $n_2 = 1.60$.

FIGURE CP16.84

85. ‖‖ A water wave is a *shallow-water wave* if the water depth d is less
CALC than $\approx \lambda/10$. It is shown in hydrodynamics that the speed of a shallow-water wave is $v = \sqrt{gd}$, so waves slow down as they move into shallower water. Ocean waves, with wavelengths of typically 100 m, are shallow-water waves when the water depth is less than ≈ 10 m. Consider a beach where the depth increases linearly with distance from the shore until reaching a depth of 5.0 m at a distance of 100 m. How long does it take a wave to move the last 100 m to the shore? Assume that the waves are so small that they don't break before reaching the shore.

17 Superposition

This swirl of colors is due to a very thin layer of oil. Oil is clear. The colors arise from the interference of reflected light waves.

IN THIS CHAPTER, you will understand and use the ideas of superposition.

What is superposition?

Waves can pass through each other. When they do, their displacements add together at each point. This is called the principle of superposition. It is a property of waves but not of particles.

《 LOOKING BACK Sections 16.1–16.4 Properties of traveling waves

What is a standing wave?

A standing wave is created when two waves travel in opposite directions between two boundaries.

- Standing waves have well-defined patterns called modes.
- Some points on the wave, called nodes, do not oscillate at all.

How are standing waves related to music?

The notes played by musical instruments are standing waves.

- Guitars have string standing waves.
- Flutes have pressure standing waves.

Changing the length of a standing wave changes its frequency and the note played.

《 LOOKING BACK Section 16.5 Sound waves

What is interference?

When two sources emit waves with the same wavelength, the overlapped waves create an interference pattern.

- Constructive interference (red) occurs where waves add to produce a wave with a larger amplitude.
- Destructive interference (black) occurs where waves cancel.

What are beats?

The superposition of two waves with slightly different frequencies produces a loud-soft-loud-soft modulation of the intensity called beats. Beats have important applications in music, ultrasonics, and telecommunications.

Why is superposition important?

Superposition and standing waves occur often in the world around us, especially when there are reflections. Musical instruments, microwave systems, and lasers all depend on standing waves. Standing waves are also important for large structures such as buildings and bridges. Superposition of light waves causes interference, which is used in electro-optic devices and precision measuring techniques.

17.1 The Principle of Superposition

FIGURE 17.1a shows two baseball players, Alan and Bill, at batting practice. Unfortunately, someone has turned the pitching machines so that pitching machine A throws baseballs toward Bill while machine B throws toward Alan. If two baseballs are launched at the same time, and with the same speed, they collide at the crossing point. Two particles cannot occupy the same point of space at the same time.

FIGURE 17.1 Unlike particles, two waves can pass directly through each other.

But waves, unlike particles, can pass directly through each other. In **FIGURE 17.1b** Alan and Bill are listening to the stereo system in the locker room after practice. Because both hear the music quite well, the sound wave that travels from loudspeaker A toward Bill must pass through the wave traveling from loudspeaker B toward Alan.

What happens to the medium at a point where two waves are present simultaneously? If wave 1 displaces a particle in the medium by D_1 and wave 2 *simultaneously* displaces it by D_2, the net displacement of the particle is simply $D_1 + D_2$. This is a very important idea because it tells us how to combine waves. It is known as the *principle of superposition*.

> **Principle of superposition** When two or more waves are *simultaneously* present at a single point in space, the displacement of the medium at that point is the sum of the displacements due to each individual wave.

Mathematically, the net displacement of a particle in the medium is

$$D_{net} = D_1 + D_2 + \cdots = \sum_i D_i \tag{17.1}$$

where D_i is the displacement that would be caused by wave i alone. We will make the simplifying assumption that the displacements of the individual waves are along the same line so that we can add displacements as scalars rather than vectors.

To use the principle of superposition you must know the displacement caused by each wave if traveling alone. Then you go through the medium *point by point* and add the displacements due to each wave *at that point* to find the net displacement at that point.

To illustrate, **FIGURE 17.2** shows snapshot graphs taken 1 s apart of two waves traveling at the same speed (1 m/s) in opposite directions. The principle of superposition comes into play wherever the waves overlap. The solid line is the sum *at each point* of the two displacements at that point. This is the displacement that you would actually observe as the two waves pass through each other.

Notice how two overlapping positive displacements add to give a displacement twice that of the individual waves. This is called *constructive interference*. Similarly, *destructive interference* is occurring at the points where positive and negative displacements add to give a superposition with zero displacement. We will defer the main discussion until later in this chapter, but you can already see that *interference is a consequence of superposition*.

FIGURE 17.2 The superposition of two waves as they pass through each other.

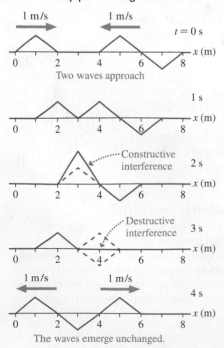

STOP TO THINK 17.1   Two pulses on a string approach each other at speeds of 1 m/s. What is the shape of the string at $t = 6$ s?

Approaching waves at $t = 0$ s

(a)

(b)

(c)

(d)

17.2 Standing Waves

FIGURE 17.3 is a time-lapse photograph of a *standing wave* on a vibrating string. It's not obvious from the photograph, but this is actually a superposition of two waves. To understand this, consider two sinusoidal waves **with the same frequency, wavelength, and amplitude** traveling in opposite directions. For example, **FIGURE 17.4a** shows two waves on a string, and **FIGURE 17.4b** shows nine snapshot graphs, at intervals of $\frac{1}{8}T$. The dots identify two of the crests to help you visualize the wave movement.

At *each point*, the net displacement—the superposition—is found by adding the red displacement and the green displacement. **FIGURE 17.4c** shows the result. It is the wave you would actually observe. The blue dot shows that the blue wave is moving neither right nor left. The wave of Figure 17.4c is called a **standing wave** because the crests and troughs "stand in place" as the wave oscillates.

FIGURE 17.3 A vibrating string is an example of a standing wave.

FIGURE 17.4 The superposition of two sinusoidal waves traveling in opposite directions.

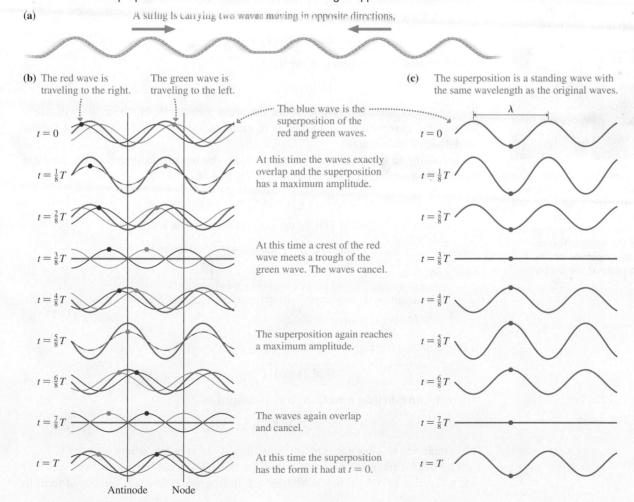

FIGURE 17.5 The intensity of a standing wave is maximum at the antinodes, zero at the nodes.

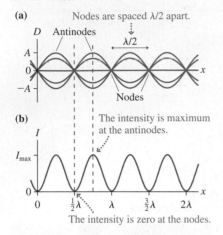

(a) Nodes are spaced λ/2 apart.

(b) The intensity is maximum at the antinodes.

The intensity is zero at the nodes.

This photograph shows the Tacoma Narrows suspension bridge on the day in 1940 when it experienced a wind-induced standing-wave oscillation that led to its collapse. The red line shows the original line of the deck of the bridge. You can clearly see the large amplitude of the oscillation and the node at the center of the span.

Nodes and Antinodes

FIGURE 17.5a has collapsed the nine graphs of Figure 17.4b into a single graphical representation of a standing wave. Compare this to the Figure 17.3 photograph of a vibrating string. A striking feature of a standing-wave pattern is the existence of **nodes,** points that *never move!* **The nodes are spaced λ/2 apart.** Halfway between the nodes are the points where the particles in the medium oscillate with maximum displacement. These points of maximum amplitude are called **antinodes,** and you can see that they are also spaced λ/2 apart.

It seems surprising and counterintuitive that some particles in the medium have no motion at all. To understand this, look closely at the two traveling waves in Figure 17.4a. You will see that the nodes occur at points where at *every instant* of time the displacements of the two traveling waves have equal magnitudes but *opposite signs.* That is, nodes are points of destructive interference where the net displacement is always zero. In contrast, antinodes are points of constructive interference where two displacements of the same sign always add to give a net displacement larger than that of the individual waves.

In Chapter 16 you learned that the *intensity* of a wave is proportional to the square of the amplitude: $I \propto A^2$. You can see in **FIGURE 17.5b** that maximum intensity occurs at the antinodes and that the intensity is zero at the nodes. If this is a sound wave, the loudness is maximum at the antinodes and zero at the nodes. A standing light wave is bright at the antinodes, dark at the nodes. The key idea is that **the intensity is maximum at points of constructive interference and zero at points of destructive interference.**

The Mathematics of Standing Waves

A sinusoidal wave traveling to the right along the x-axis with angular frequency $\omega = 2\pi f$, wave number $k = 2\pi/\lambda$, and amplitude a is

$$D_R = a\sin(kx - \omega t) \tag{17.2}$$

An equivalent wave traveling to the left is

$$D_L = a\sin(kx + \omega t) \tag{17.3}$$

We previously used the symbol A for the wave amplitude, but here we will use a lowercase a to represent the amplitude of each individual wave and reserve A for the amplitude of the net wave.

According to the principle of superposition, the net displacement of the medium when both waves are present is the sum of D_R and D_L:

$$D(x, t) = D_R + D_L = a\sin(kx - \omega t) + a\sin(kx + \omega t) \tag{17.4}$$

We can simplify Equation 17.4 by using the trigonometric identity

$$\sin(\alpha \pm \beta) = \sin\alpha\cos\beta \pm \cos\alpha\sin\beta$$

Doing so gives

$$D(x, t) = a(\sin kx \cos \omega t - \cos kx \sin \omega t) + a(\sin kx \cos \omega t + \cos kx \sin \omega t)$$
$$= (2a\sin kx)\cos \omega t \tag{17.5}$$

It is useful to write Equation 17.5 as

$$D(x, t) = A(x)\cos \omega t \tag{17.6}$$

where the **amplitude function** $A(x)$ is defined as

$$A(x) = 2a\sin kx \tag{17.7}$$

The amplitude reaches a maximum value $A_{max} = 2a$ at points where $\sin kx = 1$.

The displacement $D(x, t)$ given by Equation 17.6 is neither a function of $x - vt$ nor a function of $x + vt$; hence it is *not* a traveling wave. Instead, the $\cos \omega t$ term in

Equation 17.6 describes a medium in which each point oscillates in simple harmonic motion with frequency $f = \omega/2\pi$. The function $A(x) = 2a \sin kx$ gives the amplitude of the oscillation for a particle at position x.

FIGURE 17.6 graphs Equation 17.6 at several different instants of time. Notice that the graphs are identical to those of Figure 17.5a, showing us that Equation 17.6 is the mathematical description of a standing wave.

The nodes of the standing wave are the points at which the amplitude is zero. They are located at positions x for which

$$A(x) = 2a \sin kx = 0 \tag{17.8}$$

The sine function is zero if the angle is an integer multiple of π rad, so Equation 17.8 is satisfied if

$$kx_m = \frac{2\pi x_m}{\lambda} = m\pi \qquad m = 0, 1, 2, 3, \ldots \tag{17.9}$$

Thus the position x_m of the mth node is

$$x_m = m\frac{\lambda}{2} \qquad m = 0, 1, 2, 3, \ldots \tag{17.10}$$

You can see that the spacing between two adjacent nodes is $\lambda/2$, in agreement with Figure 17.5b. The nodes are *not* spaced by λ, as you might have expected.

FIGURE 17.6 The net displacement resulting from two counter-propagating sinusoidal waves.

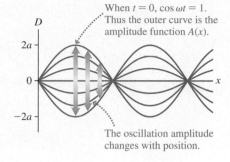

When $t = 0$, $\cos \omega t = 1$. Thus the outer curve is the amplitude function $A(x)$.

The oscillation amplitude changes with position.

EXAMPLE 17.1 | **Node spacing on a string**

A very long string has a linear density of 5.0 g/m and is stretched with a tension of 8.0 N. 100 Hz waves with amplitudes of 2.0 mm are generated at the ends of the string.

a. What is the node spacing along the resulting standing wave?

b. What is the maximum displacement of the string?

MODEL Two counter-propagating waves of equal frequency create a standing wave.

VISUALIZE The standing wave will look like Figure 17.5a.

SOLVE a. The speed of the waves on the string is

$$v = \sqrt{\frac{T_s}{\mu}} = \sqrt{\frac{8.0 \text{ N}}{0.0050 \text{ kg/m}}} = 40 \text{ m/s}$$

and the wavelength is

$$\lambda = \frac{v}{f} = \frac{40 \text{ m/s}}{100 \text{ Hz}} = 0.40 \text{ m} = 40 \text{ cm}$$

Thus the spacing between adjacent nodes is $\lambda/2 = 20$ cm.

b. The maximum displacement is $A_{max} = 2a = 4.0$ mm.

17.3 Standing Waves on a String

Wiggling both ends of a very long string is not a practical way to generate standing waves. Instead, as in the photograph in Figure 17.3, standing waves are usually seen on a string that is fixed at both ends. To understand why this condition causes standing waves, we need to examine what happens when a traveling wave encounters a discontinuity.

FIGURE 17.7a shows a *discontinuity* between a string with a larger linear density and one with a smaller linear density. The tension is the same in both strings, so the wave speed is slower on the left, faster on the right. Whenever a wave encounters a discontinuity, some of the wave's energy is *transmitted* forward and some is *reflected*.

FIGURE 17.7 A wave reflects when it encounters a discontinuity or a boundary.

(a) Discontinuity where the wave speed increases

Before:

After:

String with slower wave speed   String with faster wave speed

(b) Discontinuity where the wave speed decreases

Before:

After:

The reflected pulse is inverted.

(c) Boundary

Before:

After:

The reflected pulse is inverted and its amplitude is unchanged.

Light waves exhibit an analogous behavior when they encounter a piece of glass. Most of the light wave's energy is transmitted through the glass, which is why glass is transparent, but a small amount of energy is reflected. That is how you see your reflection dimly in a storefront window.

In FIGURE 17.7b, an incident wave encounters a discontinuity at which the wave speed decreases. In this case, the reflected pulse is *inverted*. A positive displacement of the incident wave becomes a negative displacement of the reflected wave. Because $\sin(\phi + \pi) = -\sin\phi$, we say that the reflected wave has a *phase change of π upon reflection*. This aspect of reflection will be important later in the chapter when we look at the interference of light waves.

The wave in FIGURE 17.7c reflects from a *boundary*. This is like Figure 17.7b in the limit that the string on the right becomes infinitely massive. Thus the reflection in Figure 17.7c looks like that of Figure 17.7b with one exception: Because there is no transmitted wave, *all* the wave's energy is reflected. Hence **the amplitude of a wave reflected from a boundary is unchanged.**

Creating Standing Waves

FIGURE 17.8 shows a string of length L tied at $x = 0$ and $x = L$. If you wiggle the string in the middle, sinusoidal waves travel outward in both directions and soon reach the boundaries. Because the speed of a reflected wave does not change, **the wavelength and frequency of a reflected sinusoidal wave are unchanged.** Consequently, reflections at the ends of the string cause two waves of *equal amplitude and wavelength* to travel in opposite directions along the string. As we've just seen, these are the conditions that cause a standing wave!

To connect the mathematical analysis of standing waves in Section 17.2 with the physical reality of a string tied down at the ends, we need to impose *boundary conditions*. A **boundary condition** is a mathematical statement of any constraint that *must* be obeyed at the boundary or edge of a medium. Because the string is tied down at the ends, the displacements at $x = 0$ and $x = L$ must be zero at all times. Thus the standing-wave boundary conditions are $D(x = 0, t) = 0$ and $D(x = L, t) = 0$. Stated another way, we require nodes at both ends of the string.

We found that the displacement of a standing wave is $D(x, t) = (2a \sin kx) \cos \omega t$. This equation already satisfies the boundary condition $D(x = 0, t) = 0$. That is, the origin has already been located at a node. The second boundary condition, at $x = L$, requires $D(x = L, t) = 0$. This condition will be met at all times if

$$2a \sin kL = 0 \qquad \text{(boundary condition at } x = L\text{)} \qquad (17.11)$$

Equation 17.11 will be true if $\sin kL = 0$, which in turn requires

$$kL = \frac{2\pi L}{\lambda} = m\pi \qquad m = 1, 2, 3, 4, \ldots \qquad (17.12)$$

kL must be a multiple of $m\pi$, but $m = 0$ is excluded because L can't be zero.

For a string of fixed length L, the only quantity in Equation 17.12 that can vary is λ. That is, the boundary condition is satisfied only if the wavelength has one of the values

$$\lambda_m = \frac{2L}{m} \qquad m = 1, 2, 3, 4, \ldots \qquad (17.13)$$

A standing wave can exist on the string *only* if its wavelength is one of the values given by Equation 17.13. The mth possible wavelength $\lambda_m = 2L/m$ is just the right size so that its mth node is located at the end of the string (at $x = L$).

NOTE Other wavelengths, which would be perfectly acceptable wavelengths for a traveling wave, cannot exist as a *standing* wave of length L because they cannot meet the boundary conditions requiring a node at each end of the string.

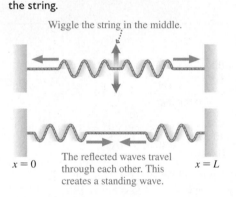

FIGURE 17.8 Reflections at the two boundaries cause a standing wave on the string.

Wiggle the string in the middle.

$x = 0$

The reflected waves travel through each other. This creates a standing wave.

$x = L$

If standing waves are possible only for certain wavelengths, then only a few specific oscillation frequencies are allowed. Because $\lambda f = v$ for a sinusoidal wave, the oscillation frequency corresponding to wavelength λ_m is

$$f_m = \frac{v}{\lambda_m} = \frac{v}{2L/m} = m\frac{v}{2L} \qquad m = 1, 2, 3, 4, \ldots \qquad (17.14)$$

The lowest allowed frequency

$$f_1 = \frac{v}{2L} \qquad \text{(fundamental frequency)} \qquad (17.15)$$

which corresponds to wavelength $\lambda_1 = 2L$, is called the **fundamental frequency** of the string. The allowed frequencies can be written in terms of the fundamental frequency as

$$f_m = mf_1 \qquad m = 1, 2, 3, 4, \ldots \qquad (17.16)$$

The allowed standing-wave frequencies are all integer multiples of the fundamental frequency. The higher-frequency standing waves are called **harmonics,** with the $m = 2$ wave at frequency f_2 called the *second harmonic,* the $m = 3$ wave called the *third harmonic,* and so on.

FIGURE 17.9 graphs the first four possible standing waves on a string of fixed length L. These possible standing waves are called the **modes** of the string, or sometimes the *normal modes.* Each mode, numbered by the integer m, has a unique wavelength and frequency. Keep in mind that these drawings simply show the *envelope,* or outer edge, of the oscillations. The string is continuously oscillating at all positions between these edges, as we showed in more detail in Figure 17.5a.

There are three things to note about the modes of a string.

1. m is the number of *antinodes* on the standing wave, not the number of nodes. You can tell a string's mode of oscillation by counting the number of antinodes.
2. The *fundamental mode,* with $m = 1$, has $\lambda_1 = 2L$, not $\lambda_1 = L$. Only half of a wavelength is contained between the boundaries, a direct consequence of the fact that the spacing between nodes is $\lambda/2$.
3. The frequencies of the normal modes form a series: $f_1, 2f_1, 3f_1, 4f_1, \ldots$. The fundamental frequency f_1 can be found as the *difference* between the frequencies of any two adjacent modes. That is, $f_1 = \Delta f = f_{m+1} - f_m$.

FIGURE 17.9 The first four modes for standing waves on a string of length L.

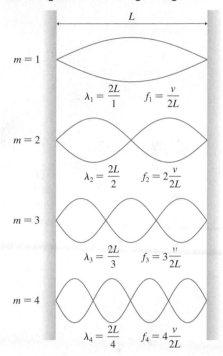

$$m = 1 \qquad \lambda_1 = \frac{2L}{1} \qquad f_1 = \frac{v}{2L}$$

$$m = 2 \qquad \lambda_2 = \frac{2L}{2} \qquad f_2 = 2\frac{v}{2L}$$

$$m = 3 \qquad \lambda_3 = \frac{2L}{3} \qquad f_3 = 3\frac{v}{2L}$$

$$m = 4 \qquad \lambda_4 = \frac{2L}{4} \qquad f_4 = 4\frac{v}{2L}$$

EXAMPLE 17.2 | **Measuring g**

Standing-wave frequencies can be measured very accurately. Consequently, standing waves are often used in experiments to make accurate measurements of other quantities. One such experiment, shown in **FIGURE 17.10**, uses standing waves to measure the free-fall acceleration g. A heavy mass is suspended from a 1.65-m-long, 5.85 g steel wire; then an oscillating magnetic field (because steel is magnetic) is used to excite the $m = 3$ standing wave on the wire. Measuring the frequency for different masses yields the data given in the table. Analyze these data to determine the local value of g.

FIGURE 17.10 An experiment to measure g.

Electromagnet

L

M

Mass (kg)	f_3 (Hz)
2.00	68
4.00	97
6.00	117
8.00	135
10.00	152

MODEL The hanging mass creates tension in the wire. This establishes the wave speed along the wire and thus the frequencies of standing waves. Masses of a few kg might stretch the wire a mm or so, but that doesn't change the length L until the third decimal place. The mass of the wire itself is insignificant in comparison to that of the hanging mass. We'll be justified in determining g to three significant figures.

SOLVE The frequency of the third harmonic is

$$f_3 = \frac{3}{2}\frac{v}{L}$$

The wave speed on the wire is

$$v = \sqrt{\frac{T_s}{\mu}} = \sqrt{\frac{Mg}{m/L}} = \sqrt{\frac{MgL}{m}}$$

where Mg is the weight of the hanging mass, and thus the tension in the wire, while m is the mass of the wire. Combining these two equations, we have

Continued

$$f_3 = \frac{3}{2}\sqrt{\frac{Mg}{mL}} = \frac{3}{2}\sqrt{\frac{g}{mL}}\sqrt{M}$$

Squaring both sides gives

$$f_3^2 = \frac{9g}{4mL}M$$

A graph of the square of the standing-wave frequency versus mass M should be a straight line passing through the origin with slope $9g/4mL$. We can use the experimental slope to determine g.

FIGURE 17.11 is a graph of f_3^2 versus M. The slope of the best-fit line is 2289 $\text{kg}^{-1}\,\text{s}^{-2}$, from which we find

$$g = \text{slope} \times \frac{4mL}{9}$$

$$= 2289\ \text{kg}^{-1}\,\text{s}^{-2} \times \frac{4(0.00585\ \text{kg})(1.65\ \text{m})}{9} = 9.82\ \text{m/s}^2$$

FIGURE 17.11 Graph of the data.

ASSESS The fact that the graph is linear and passes through the origin confirms our model. This is an important reason for having multiple data points rather than using only one mass.

STOP TO THINK 17.2 A standing wave on a string vibrates as shown at the right. Suppose the string tension is quadrupled while the frequency and the length of the string are held constant. Which standing-wave pattern is produced?

Original standing wave

(a)   (b)   (c)   (d)

Standing Electromagnetic Waves

FIGURE 17.12 A laser contains a standing light wave between two parallel mirrors.

Laser cavity

Standing light wave

Full reflector

Laser beam

Partial reflector

Because electromagnetic waves are transverse waves, a standing electromagnetic wave is very much like a standing wave on a string. Standing electromagnetic waves can be established between two parallel mirrors that reflect light back and forth. The mirrors are boundaries, analogous to the boundaries at the ends of a string. In fact, this is exactly how a laser operates. The two facing mirrors in **FIGURE 17.12** form what is called a *laser cavity*.

Because the mirrors act like the points to which a string is tied, the light wave must have a node at the surface of each mirror. One of the mirrors is only partially reflective, to allow some light to escape and form the laser beam, but this doesn't affect the boundary condition.

Because the boundary conditions are the same, Equations 17.13 and 17.14 for λ_m and f_m apply to a laser just as they do to a vibrating string. The primary difference is the size of the wavelength. A typical laser cavity has a length $L \approx 30$ cm, and visible light has a wavelength $\lambda \approx 600$ nm. The standing light wave in a laser cavity has a mode number m that is approximately

$$m = \frac{2L}{\lambda} \approx \frac{2 \times 0.30\ \text{m}}{6.00 \times 10^{-7}\ \text{m}} = 1{,}000{,}000$$

In other words, the standing light wave inside a laser cavity has approximately one million antinodes! This is a consequence of the very short wavelength of light.

EXAMPLE 17.3 | **The standing light wave inside a laser**

Helium-neon lasers emit the red laser light commonly used in classroom demonstrations and supermarket checkout scanners. A helium-neon laser operates at a wavelength of precisely 632.9924 nm when the spacing between the mirrors is 310.372 mm.

a. In which mode does this laser operate?

b. What is the next longest wavelength that could form a standing wave in this laser cavity?

MODEL The light wave forms a standing wave between the two mirrors.

VISUALIZE The standing wave looks like Figure 17.12.

SOLVE a. We can use $\lambda_m = 2L/m$ to find that m (the mode) is

$$m = \frac{2L}{\lambda_m} = \frac{2(0.310372 \text{ m})}{6.329924 \times 10^{-7} \text{ m}} = 980{,}650$$

There are 980,650 antinodes in the standing light wave.

b. The next longest wavelength that can fit in this laser cavity will have one fewer node. It will be the $m = 980{,}649$ mode and its wavelength will be

$$\lambda = \frac{2L}{m} = \frac{2(0.310372 \text{ m})}{980{,}649} = 632.9930 \text{ nm}$$

ASSESS The wavelength increases by a mere 0.0006 nm when the mode number is decreased by 1.

Microwaves, with a wavelength of a few centimeters, can also set up standing waves. This is not always good. If the microwaves in a microwave oven form a standing wave, there are nodes where the electromagnetic field intensity is always zero. These nodes cause cold spots where the food does not heat. Although designers of microwave ovens try to prevent standing waves, ovens usually do have cold spots spaced $\lambda/2$ apart at nodes in the microwave field. A turntable in a microwave oven keeps the food moving so that no part of your dinner remains at a node.

17.4 Standing Sound Waves and Musical Acoustics

A long, narrow column of air, such as the air in a tube or pipe, can support a *longitudinal* standing sound wave. Longitudinal waves are somewhat trickier than string waves because a graph—showing displacement *parallel* to the tube—is not a picture of the wave.

To illustrate the ideas, **FIGURE 17.13** is a series of three graphs and pictures that show the $m = 2$ standing wave inside a column of air closed at both ends. We call this a *closed-closed tube*. The air at the closed ends cannot oscillate because the air molecules are pressed up against the wall, unable to move; hence **a closed end of a column of air must be a displacement node.** Thus the boundary conditions—nodes at the ends—are the same as for a standing wave on a string.

FIGURE 17.13 The $m = 2$ standing sound wave in a closed-closed tube of air.

FIGURE 17.14 The $m = 2$ longitudinal standing wave can be represented as a displacement wave or as a pressure wave.

The closed end is a displacement node and a pressure antinode.

Air molecules undergo longitudinal oscillations. This is a displacement antinode and a pressure node.

The displacement and pressure nodes and antinodes are interchanged.

The pressure is oscillating around atmospheric pressure p_{atmos}.

Although the graph looks familiar, it is now a graph of *longitudinal* displacement. At $t = 0$, positive displacements in the left half and negative displacements in the right half cause all the air molecules to converge at the center of the tube. The density and pressure rise at the center and fall at the ends—a *compression* and *rarefaction* in the terminology of Chapter 16. A half cycle later, the molecules have rushed to the ends of the tube. Now the pressure is maximum at the ends, minimum in the center. Try to visualize the air molecules sloshing back and forth this way.

FIGURE 17.14 combines these illustrations into a single picture showing where the molecules are oscillating (antinodes) and where they're not (nodes). A graph of the displacement Δx looks just like the $m = 2$ graph of a standing wave on a string. Because the boundary conditions are the same, the possible wavelengths and frequencies of standing waves in a closed-closed tube are the same as for a string of the same length.

It is often useful to think of sound as a *pressure wave* rather than a displacement wave, and the bottom graph in Figure 17.14 shows the $m = 2$ pressure standing wave in a closed-closed tube. Notice that the pressure is oscillating around p_{atmos}, its equilibrium value.

We showed in « Section 16.6 that the pressure in a sound wave is minimum or maximum at points where the displacement is zero, and vice versa. Consequently, **the nodes and antinodes of the pressure wave are interchanged with those of the displacement wave.** You can see in Figure 17.13 that the gas molecules are alternately pushed up against the ends of the tube, then pulled away, causing the pressure at the closed ends to oscillate with maximum amplitude—an antinode—at a point where the displacement is a node.

EXAMPLE 17.4 | **Singing in the shower**

A shower stall is 2.45 m (8 ft) tall. For what frequencies less than 500 Hz are there standing sound waves in the shower stall?

MODEL The shower stall, to a first approximation, is a column of air 2.45 m long. It is closed at the ends by the ceiling and floor. Assume a 20°C speed of sound.

VISUALIZE A standing sound wave will have nodes at the ceiling and the floor. The $m = 2$ mode will look like Figure 17.14 rotated 90°.

SOLVE The fundamental frequency for a standing sound wave in this air column is

$$f_1 = \frac{v}{2L} = \frac{343 \text{ m/s}}{2(2.45 \text{ m})} = 70 \text{ Hz}$$

The possible standing-wave frequencies are integer multiples of the fundamental frequency. These are 70 Hz, 140 Hz, 210 Hz, 280 Hz, 350 Hz, 420 Hz, and 490 Hz.

ASSESS The many possible standing waves in a shower cause the sound to *resonate,* which helps explain why some people like to sing in the shower. Our approximation of the shower stall as a one-dimensional tube is actually a bit too simplistic. A three-dimensional analysis would find additional modes, making the "sound spectrum" even richer.

Tubes with Openings

Air columns closed at both ends are of limited interest unless, as in Example 17.4, you are inside the column. Columns of air that *emit* sound are open at one or both ends. Many musical instruments fit this description. For example, a flute is a tube of air open at both ends. The flutist blows across one end to create a standing wave inside the tube, and a note of that frequency is emitted from both ends of the flute. (The blown end of a flute is open on the side, rather than across the tube. That is necessary for practical reasons of how flutes are played, but from a physics perspective this is the "end" of the tube because it opens the tube to the atmosphere.) A trumpet, however, is open at the bell end but is *closed* by the player's lips at the other end.

You saw earlier that a wave is partially transmitted and partially reflected at a discontinuity. When a sound wave traveling through a tube of air reaches an open end, some of the wave's energy is transmitted out of the tube to become the sound that you hear and some portion of the wave is reflected back into the tube. These reflections, analogous to the reflection of a string wave from a boundary, allow standing sound waves to exist in a tube of air that is open at one or both ends.

Not surprisingly, the *boundary condition* at the open end of a column of air is not the same as the boundary condition at a closed end. The air pressure at the open end of a tube is constrained to match the atmospheric pressure of the surrounding air. Consequently, the open end of a tube must be a pressure node. Because pressure nodes and antinodes are interchanged with those of the displacement wave, **an open end of an air column is required to be a displacement antinode.** (A careful analysis shows that the antinode is actually just outside the open end, but for our purposes we'll assume the antinode is exactly at the open end.)

FIGURE 17.15 The first three standing sound wave modes in columns of air with different boundary conditions.

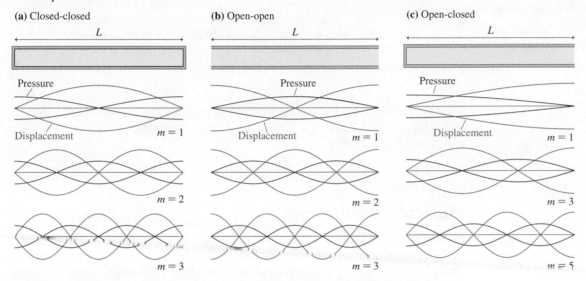

(a) Closed-closed

(b) Open-open

(c) Open-closed

FIGURE 17.15 shows displacement and pressure graphs of the first three standing-wave modes of a tube closed at both ends (a *closed-closed tube*), a tube open at both ends (an *open-open tube*), and a tube open at one end but closed at the other (an *open-closed tube*), all with the same length L. Notice the pressure and displacement boundary conditions. The standing wave in the open-open tube looks like the closed-closed tube except that the positions of the nodes and antinodes are interchanged. In both cases there are m half-wavelength segments between the ends; thus the wavelengths and frequencies of an open-open tube and a closed-closed tube are the same as those of a string tied at both ends:

$$\lambda_m = \frac{2L}{m} \qquad m = 1, 2, 3, 4, \ldots$$
$$f_m = m\frac{v}{2L} = mf_1 \qquad \text{(open-open or closed-closed tube)} \qquad (17.17)$$

The open-closed tube is different. The fundamental mode has only one-quarter of a wavelength in a tube of length L; hence the $m = 1$ wavelength is $\lambda_1 = 4L$. This is twice the λ_1 wavelength of an open-open or a closed-closed tube. Consequently, **the fundamental frequency of an open-closed tube is half that of an open-open or a closed-closed tube of the same length.** It will be left as a homework problem for you to show that the possible wavelengths and frequencies of an open-closed tube of length L are

$$\lambda_m = \frac{4L}{m} \qquad m = 1, 3, 5, 7, \ldots$$
$$f_m = m\frac{v}{4L} = mf_1 \qquad \text{(open-closed tube)} \qquad (17.18)$$

Notice that m in Equation 17.18 takes on only *odd* values.

EXAMPLE 17.5 | Resonances of the ear canal

The eardrum, which transmits sound vibrations to the sensory organs of the inner ear, lies at the end of the ear canal. For adults, the ear canal is about 2.5 cm in length. What frequency standing waves can occur in the ear canal that are within the range of human hearing? The speed of sound in the warm air of the ear canal is 350 m/s.

MODEL The ear canal is open to the air at one end, closed by the eardrum at the other. We can model it as an open-closed tube. The standing waves will be those of Figure 17.15c.

SOLVE The lowest standing-wave frequency is the fundamental frequency for a 2.5-cm-long open-closed tube:

$$f_1 = \frac{v}{4L} = \frac{350 \text{ m/s}}{4(0.025 \text{ m})} = 3500 \text{ Hz}$$

Standing waves also occur at the harmonics, but an open-closed tube has only odd harmonics. These are

$$f_3 = 3f_1 = 10{,}500 \text{ Hz}$$
$$f_5 = 5f_1 = 17{,}500 \text{ Hz}$$

Higher harmonics are beyond the range of human hearing, as discussed in Section 16.5.

ASSESS The ear canal is short, so we expected the standing-wave frequencies to be relatively high. The air in your ear canal responds readily to sounds at these frequencies—what we call a *resonance* of the ear canal—and transmits theses sounds to the eardrum. Consequently, your ear actually is slightly more sensitive to sounds with frequencies around 3500 Hz and 10,500 Hz than to sounds at nearby frequencies.

STOP TO THINK 17.3 An open-open tube of air supports standing waves at frequencies of 300 Hz and 400 Hz and at no frequencies between these two. The second harmonic of this tube has frequency

a. 100 Hz   b. 200 Hz   c. 400 Hz   d. 600 Hz   e. 800 Hz

Musical Instruments

An important application of standing waves is to musical instruments. Instruments such as the guitar, the piano, and the violin have strings fixed at the ends and tightened to create tension. A disturbance generated on the string by plucking, striking, or bowing it creates a standing wave on the string.

The fundamental frequency of a vibrating string is

$$f_1 = \frac{v}{2L} = \frac{1}{2L} \sqrt{\frac{T_s}{\mu}}$$

where T_s is the tension in the string and μ is its linear density. The fundamental frequency is the musical note you hear when the string is sounded. Increasing the tension in the string raises the fundamental frequency, which is how stringed instruments are tuned.

NOTE v is the wave speed *on the string,* not the speed of sound in air.

For the guitar or the violin, the strings are all the same length and under approximately the same tension. Were that not the case, the neck of the instrument would tend to twist toward the side of higher tension. The strings have different frequencies because they differ in linear density: The lower-pitched strings are "fat" while the higher-pitched strings are "skinny." This difference changes the frequency by changing the wave speed. *Small* adjustments are then made in the tension to bring each string to the exact desired frequency. Once the instrument is tuned, you play it by using your fingertips to alter the effective length of the string. As you shorten the string's length, the frequency and pitch go up.

A piano covers a much wider range of frequencies than a guitar or violin. This range cannot be produced by changing only the linear densities of the strings. The high end would have strings too thin to use without breaking, and the low end would have solid rods rather than flexible wires! So a piano is tuned through a combination of changing the linear density *and* the length of the strings. The bass note strings are not only fatter, they are also longer.

The strings on a harp vibrate as standing waves. Their frequencies determine the notes that you hear.

With a wind instrument, blowing into the mouthpiece creates a standing sound wave inside a tube of air. The player changes the notes by using her fingers to cover holes or open valves, changing the length of the tube and thus its frequency. The fact that the holes are on the side makes very little difference; the first open hole becomes an antinode because the air is free to oscillate in and out of the opening.

A wind instrument's frequency depends on the speed of sound *inside* the instrument. But the speed of sound depends on the temperature of the air. When a wind player first blows into the instrument, the air inside starts to rise in temperature. This increases the sound speed, which in turn raises the instrument's frequency for each note until the air temperature reaches a steady state. Consequently, wind players must "warm up" before tuning their instrument.

Many wind instruments have a "buzzer" at one end of the tube, such as a vibrating reed on a saxophone or vibrating lips on a trombone. Buzzers generate a continuous range of frequencies rather than single notes, which is why they sound like a "squawk" if you play on just the mouthpiece without the rest of the instrument. When a buzzer is connected to the body of the instrument, most of those frequencies cause no response of the air molecules. But the frequency from the buzzer that matches the fundamental frequency of the instrument causes the buildup of a large-amplitude response at just that frequency—a standing-wave resonance. This is the energy input that generates and sustains the musical note.

EXAMPLE 17.6 | Flutes and clarinets

A clarinet is 66.0 cm long. A flute is nearly the same length, with 63.6 cm between the hole the player blows across and the end of the flute. What are the frequencies of the lowest note and the next higher harmonic on a flute and on a clarinet? The speed of sound in warm air is 350 m/s.

MODEL The flute is an open-open tube, open at the end as well as at the hole the player blows across. A clarinet is an open-closed tube because the player's lips and the reed seal the tube at the upper end.

SOLVE The lowest frequency is the fundamental frequency. For the flute, an open-open tube, this is

$$f_1 = \frac{v}{2L} = \frac{350 \text{ m/s}}{2(0.636 \text{ m})} = 275 \text{ Hz}$$

The clarinet, an open-closed tube, has

$$f_1 = \frac{v}{4L} = \frac{350 \text{ m/s}}{4(0.660 \text{ m})} = 133 \text{ Hz}$$

The next higher harmonic on the flute's open-open tube is $m = 2$ with frequency $f_2 = 2f_1 = 550$ Hz. An open-closed tube has only odd harmonics, so the next higher harmonic of the clarinet is $f_3 = 3f_1 = 399$ Hz.

ASSESS The clarinet plays a much lower note than the flute—musically, about an octave lower—because it is an open-closed tube. It's worth noting that neither of our fundamental frequencies is exactly correct because our open-open and open-closed tube models are a bit too simplified to adequately describe a real instrument. However, both calculated frequencies are close because our models do capture the essence of the physics.

A vibrating string plays the musical note corresponding to the fundamental frequency f_1, so stringed instruments must use several strings to obtain a reasonable range of notes. In contrast, wind instruments can sound at the second or third harmonic of the tube of air (f_2 or f_3). These higher frequencies are sounded by *overblowing* (flutes, brass instruments) or with keys that open small holes to encourage the formation of an antinode at that point (clarinets, saxophones). The controlled use of these higher harmonics gives wind instruments a wide range of notes.

17.5 Interference in One Dimension

One of the most basic characteristics of waves is the ability of two waves to combine into a single wave whose displacement is given by the principle of superposition. The pattern resulting from the superposition of two waves is often called **interference**. A standing wave is the interference pattern produced when two waves of equal frequency travel in opposite directions. In this section we will look at the interference of two waves traveling in the *same* direction.

FIGURE 17.16 Two overlapped waves travel along the x-axis.

(a) Two overlapped light waves

(b) Two overlapped sound waves

FIGURE 17.16a shows two light waves impinging on a partially silvered mirror. Such a mirror partially transmits and partially reflects each wave, causing two *overlapped* light waves to travel along the x-axis to the right of the mirror. Or consider the two loudspeakers in **FIGURE 17.16b**. The sound wave from loudspeaker 2 passes just to the side of loudspeaker 1; hence two overlapped sound waves travel to the right along the x-axis. We want to find out what happens when two overlapped waves travel in the same direction along the same axis.

Figure 17.16b shows a point on the x-axis where the overlapped waves are detected, either by your ear or by a microphone. This point is distance x_1 from loudspeaker 1 and distance x_2 from loudspeaker 2. (We will use loudspeakers and sound waves for most of our examples, but our analysis is valid for any wave.) What is the amplitude of the combined waves at this point?

Throughout this section, **we will assume that the waves are sinusoidal, have the same frequency and amplitude, and travel to the right along the x-axis.** Thus we can write the displacements of the two waves as

$$D_1(x_1, t) = a\sin(kx_1 - \omega t + \phi_{10}) = a\sin\phi_1$$
$$D_2(x_2, t) = a\sin(kx_2 - \omega t + \phi_{20}) = a\sin\phi_2$$

(17.19)

where ϕ_1 and ϕ_2 are the *phases* of the waves. Both waves have the same wave number $k = 2\pi/\lambda$ and the same angular frequency $\omega = 2\pi f$.

The phase constants ϕ_{10} and ϕ_{20} are characteristics of *the sources,* not the medium. **FIGURE 17.17** shows snapshot graphs at $t = 0$ of waves emitted by three sources with phase constants $\phi_0 = 0$ rad, $\phi_0 = \pi/2$ rad, and $\phi_0 = \pi$ rad. You can see that **the phase constant tells us what the source is doing at $t = 0$.** For example, a loudspeaker at its center position and moving backward at $t = 0$ has $\phi_0 = 0$ rad. Looking back at Figure 17.16b, you can see that loudspeaker 1 has phase constant $\phi_{10} = 0$ rad and loudspeaker 2 has $\phi_{20} = \pi$ rad.

NOTE We will often consider *identical sources,* by which we mean that $\phi_{20} = \phi_{10}$. That is, the sources oscillate in phase.

Let's examine overlapped waves graphically before diving into the mathematics. **FIGURE 17.18** shows two important situations. In part a, the crests of the two waves are aligned as they travel along the x-axis. In part b, the crests of one wave align with the troughs of the other wave. The graphs and the wave fronts are slightly displaced from each other so that you can see what each wave is doing, but the *physical situation* is one in which the waves are traveling *on top of* each other.

The two waves of Figure 17.18a have the same displacement at every point: $D_1(x) = D_2(x)$. Two waves that are aligned crest to crest and trough to trough are said to be **in phase.** Waves that are in phase march along "in step" with each other.

When we combine two in-phase waves, using the principle of superposition, the net displacement at each point is twice the displacement of each individual wave. The superposition of two waves to create a traveling wave with an amplitude *larger* than either individual wave is called **constructive interference.** When the waves are exactly in phase, giving $A = 2a$, we have *maximum constructive interference.*

FIGURE 17.17 Waves from three sources having phase constants $\phi_0 = 0$ rad, $\phi_0 = \pi/2$ rad, and $\phi_0 = \pi$ rad.

(a) Snapshot graph at $t = 0$ for $\phi_0 = 0$ rad

When this crest was emitted, a quarter-cycle ago, the speaker cone was all the way forward.

Now this speaker cone, at $x = 0$, is centered and moving backward.

(b) Snapshot graph at $t = 0$ for $\phi_0 = \pi/2$ rad

This speaker cone is all the way forward.

(c) Snapshot graph at $t = 0$ for $\phi_0 = \pi$ rad

This speaker cone is centered and moving forward.

When this trough was emitted, a quarter-cycle ago, the speaker cone was all the way back.

In Figure 17.18b, where the crests of one wave align with the troughs of the other, the waves march along "out of step" with $D_1(x) = -D_2(x)$ at every point. Two waves that are aligned crest to trough are said to be *180° out of phase* or, more generally, just **out of phase.** A superposition of two waves to create a wave with an amplitude smaller than either individual wave is called **destructive interference.** In this case, because $D_1 = -D_2$, the net displacement is *zero* at *every point* along the axis. The combination of two waves that cancel each other to give no wave is called *perfect destructive interference.*

NOTE Perfect destructive interference occurs only if the two waves have exactly equal amplitudes, as we're assuming. A 180° phase difference always produces *maximum destructive interference,* but the cancellation won't be perfect if there is any difference in the amplitudes.

The Phase Difference

To understand interference, we need to focus on the *phases* of the two waves, which are

$$\phi_1 = kx_1 - \omega t + \phi_{10}$$
$$\phi_2 = kx_2 - \omega t + \phi_{20} \tag{17.20}$$

The difference between the two phases ϕ_1 and ϕ_2, called the **phase difference** $\Delta\phi$, is

$$\Delta\phi = \phi_2 - \phi_1 = (kx_2 - \omega t + \phi_{20}) - (kx_1 - \omega t + \phi_{10})$$
$$= k(x_2 - x_1) + (\phi_{20} - \phi_{10}) \tag{17.21}$$
$$= 2\pi \frac{\Delta x}{\lambda} + \Delta\phi_0$$

You can see that there are two contributions to the phase difference. $\Delta x = x_2 - x_1$, the distance between the two sources, is called **path-length difference.** It is the extra distance traveled by wave 2 on the way to the point where the two waves are combined. $\Delta\phi_0 = \phi_{20} - \phi_{10}$ is the *inherent phase difference* between the sources.

The condition of being in phase, where crests are aligned with crests and troughs with troughs, is $\Delta\phi = 0, 2\pi, 4\pi$, or any integer multiple of 2π rad. Thus the condition for maximum constructive interference is

> Maximum constructive interference:
> $$\Delta\phi = 2\pi \frac{\Delta x}{\lambda} + \Delta\phi_0 = m \cdot 2\pi \text{ rad} \qquad m = 0, 1, 2, 3, \ldots \tag{17.22}$$

For identical sources, which have $\Delta\phi_0 = 0$ rad, maximum constructive interference occurs when $\Delta x = m\lambda$. That is, **two identical sources produce maximum constructive interference when the path-length difference is an integer number of wavelengths.**

FIGURE 17.19 shows two identical sources (i.e., the two loudspeakers are doing the same thing at the same time), so $\Delta\phi_0 = 0$ rad. The path-length difference Δx is the extra distance traveled by the wave from loudspeaker 2 before it combines with loudspeaker 1. In this case, $\Delta x = \lambda$. Because a wave moves forward exactly one wavelength during one period, loudspeaker 1 emits a crest exactly as a crest of wave 2 passes by. The two waves are "in step," with $\Delta\phi = 2\pi$ rad, so the two waves interfere constructively to produce a wave of amplitude $2a$.

Maximum destructive interference, where the crests of one wave are aligned with the troughs of the other, occurs when two waves are *out of phase,* meaning that $\Delta\phi = \pi, 3\pi, 5\pi$, or any odd multiple of π rad. Thus the condition for maximum destructive interference is

> Maximum destructive interference:
> $$\Delta\phi = 2\pi \frac{\Delta x}{\lambda} + \Delta\phi_0 = \left(m + \tfrac{1}{2}\right) \cdot 2\pi \text{ rad} \qquad m = 0, 1, 2, 3, \ldots \tag{17.23}$$

FIGURE 17.18 Interference of two waves traveling along the x-axis.

(a) Maximum constructive interference

These two waves are in phase. Their crests are aligned.

Their superposition produces a traveling wave with amplitude $2a$.

(b) Perfect destructive interference

These two waves are out of phase. The crests of one wave are aligned with the troughs of the other.

Their superposition produces a wave with zero amplitude.

FIGURE 17.19 Two identical sources one wavelength apart.

Speaker 2

This crest is emitted as a crest from speaker 2 passes by.

Identical sources $\Delta\phi_0 = 0$

Speaker 1

$\Delta x = \lambda$
Path-length difference

The two waves are in phase ($\Delta\phi = 2\pi$ rad) and interfere constructively.

For identical sources, which have $\Delta\phi_0 = 0$ rad, maximum destructive interference occurs when $\Delta x = (m + \frac{1}{2})\lambda$. **That is, two identical sources produce maximum destructive interference when the path-length difference is a half-integer number of wavelengths.**

Two waves can be out of phase because the sources are located at different positions, because the sources themselves are out of phase, or because of a combination of these two. **FIGURE 17.20** illustrates these ideas by showing three different ways in which two waves interfere destructively. Each of these three arrangements creates waves with $\Delta\phi = \pi$ rad.

FIGURE 17.20 Destructive interference three ways.

(a) The sources are out of phase.

$\Delta\phi_0 = \pi$ rad

$\Delta x = 0$

(b) Identical sources are separated by half a wavelength.

$\Delta\phi_0 = 0$ rad

$\Delta x = \frac{1}{2}\lambda$

(c) The sources are both separated and partially out of phase.

$\Delta\phi_0 = \frac{\pi}{2}$ rad

$\Delta x = \frac{1}{4}\lambda$

NOTE Don't confuse the phase difference of the waves ($\Delta\phi$) with the phase difference of the sources ($\Delta\phi_0$). It is $\Delta\phi$, the phase difference of the waves, that governs interference.

EXAMPLE 17.7 | Interference between two sound waves

You are standing in front of two side-by-side loudspeakers playing sounds of the same frequency. Initially there is almost no sound at all. Then one of the speakers is moved slowly away from you. The sound intensity increases as the separation between the speakers increases, reaching a maximum when the speakers are 0.75 m apart. Then, as the speaker continues to move, the intensity starts to decrease. What is the distance between the speakers when the sound intensity is again a minimum?

MODEL The changing sound intensity is due to the interference of two overlapped sound waves.

VISUALIZE Moving one speaker relative to the other changes the phase difference between the waves.

SOLVE A minimum sound intensity implies that the two sound waves are interfering destructively. Initially the loudspeakers are side by side, so the situation is as shown in Figure 17.20a with $\Delta x = 0$ and $\Delta\phi_0 = \pi$ rad. That is, the speakers themselves are out of phase. Moving one of the speakers does not change $\Delta\phi_0$, but it does change the path-length difference Δx and thus increases the overall phase difference $\Delta\phi$. Constructive interference, causing maximum intensity, is reached when

$$\Delta\phi = 2\pi \frac{\Delta x}{\lambda} + \Delta\phi_0 = 2\pi \frac{\Delta x}{\lambda} + \pi = 2\pi \text{ rad}$$

where we used $m = 1$ because this is the first separation giving constructive interference. The speaker separation at which this occurs is $\Delta x = \lambda/2$. This is the situation shown in **FIGURE 17.21**.

FIGURE 17.21 Two out-of-phase sources generate waves that are in phase if the sources are one half-wavelength apart.

The sources are out of phase, $\Delta\phi_0 = \pi$ rad.

$\Delta x = \frac{1}{2}\lambda$

The sources are separated by half a wavelength.

As a result, the waves are in phase.

Because $\Delta x = 0.75$ m is $\lambda/2$, the sound's wavelength is $\lambda = 1.50$ m. The next point of destructive interference, with $m = 1$, occurs when

$$\Delta\phi = 2\pi \frac{\Delta x}{\lambda} + \Delta\phi_0 = 2\pi \frac{\Delta x}{\lambda} + \pi = 3\pi \text{ rad}$$

Thus the distance between the speakers when the sound intensity is again a minimum is

$$\Delta x = \lambda = 1.50 \text{ m}$$

ASSESS A separation of λ gives constructive interference for two *identical* speakers ($\Delta\phi_0 = 0$). Here the phase difference of π rad between the speakers (one is pushing forward as the other pulls back) gives destructive interference at this separation.

STOP TO THINK 17.4 Two loudspeakers emit waves with $\lambda = 2.0$ m. Speaker 2 is 1.0 m in front of speaker 1. What, if anything, can be done to cause constructive interference between the two waves?

a. Move speaker 1 forward (to the right) 1.0 m.
b. Move speaker 1 forward (to the right) 0.5 m.
c. Move speaker 1 backward (to the left) 0.5 m.
d. Move speaker 1 backward (to the left) 1.0 m.
e. Nothing. The situation shown already causes constructive interference.
f. Constructive interference is not possible for any placement of the speakers.

17.6 The Mathematics of Interference

Let's look more closely at the superposition of two waves. As two waves of equal amplitude and frequency travel together along the x-axis, the net displacement of the medium is

$$D = D_1 + D_2 = a\sin(kx_1 - \omega t + \phi_{10}) + a\sin(kx_2 - \omega t + \phi_{20})$$
$$= a\sin\phi_1 + a\sin\phi_2 \qquad (17.24)$$

where the phases ϕ_1 and ϕ_2 were defined in Equations 17.20.

A useful trigonometric identity is

$$\sin\alpha + \sin\beta = 2\cos\left[\tfrac{1}{2}(\alpha - \beta)\right]\sin\left[\tfrac{1}{2}(\alpha + \beta)\right] \qquad (17.25)$$

This identity is certainly not obvious, although it is easily proven by working backward from the right side. We can use this identity to write the net displacement of Equation 17.24 as

$$D = \left[2a\cos\left(\frac{\Delta\phi}{2}\right)\right]\sin(kx_{\text{avg}} - \omega t + (\phi_0)_{\text{avg}}) \qquad (17.26)$$

where $\Delta\phi = \phi_2 - \phi_1$ is the phase difference between the two waves, exactly as in Equation 17.21. $x_{\text{avg}} = (x_1 + x_2)/2$ is the average distance to the two sources and $(\phi_0)_{\text{avg}} = (\phi_{10} + \phi_{20})/2$ is the average phase constant of the sources.

The sine term shows that the superposition of the two waves is still a traveling wave. An observer would see a sinusoidal wave moving along the x-axis with the *same* wavelength and frequency as the original waves.

But how *big* is this wave compared to the two original waves? They each had amplitude a, but the amplitude of their superposition is

$$A = \left|2a\cos\left(\frac{\Delta\phi}{2}\right)\right| \qquad (17.27)$$

where we have used an absolute value sign because amplitudes must be positive. Depending upon the phase difference of the two waves, the amplitude of their superposition can be anywhere from zero (perfect destructive interference) to $2a$ (maximum constructive interference).

The amplitude has its maximum value $A = 2a$ if $\cos(\Delta\phi/2) = \pm 1$. This occurs when

$$\Delta\phi = m \cdot 2\pi \qquad \text{(maximum amplitude } A = 2a\text{)} \qquad (17.28)$$

where m is an integer. Similarly, the amplitude is zero if $\cos(\Delta\phi/2) = 0$, which occurs when

$$\Delta\phi = \left(m + \tfrac{1}{2}\right) \cdot 2\pi \qquad \text{(minimum amplitude } A = 0\text{)} \qquad (17.29)$$

Equations 17.28 and 17.29 are identical to the conditions of Equations 17.22 and 17.23 for constructive and destructive interference. We initially found these conditions by

FIGURE 17.22 The interference of two waves for three different values of the phase difference.

For $\Delta\phi = 40°$, the interference is constructive but not maximum constructive.

For $\Delta\phi = 160°$, the interference is destructive but not perfect destructive.

considering the alignment of the crests and troughs. Now we have confirmed them with an algebraic addition of the waves.

It is entirely possible, of course, that the two waves are neither exactly in phase nor exactly out of phase. Equation 17.27 allows us to calculate the amplitude of the superposition for any value of the phase difference. As an example, **FIGURE 17.22** on the previous page shows the calculated interference of two waves that differ in phase by 40°, by 90°, and by 160°.

EXAMPLE 17.8 | **More interference of sound waves**

Two loudspeakers emit 500 Hz sound waves with an amplitude of 0.10 mm. Speaker 2 is 1.00 m behind speaker 1, and the phase difference between the speakers is 90°. What is the amplitude of the sound wave at a point 2.00 m in front of speaker 1?

MODEL The amplitude is determined by the interference of the two waves. Assume that the speed of sound has a room-temperature (20°C) value of 343 m/s.

SOLVE The amplitude of the sound wave is

$$A = |2a\cos(\Delta\phi/2)|$$

where $a = 0.10$ mm and the phase difference between the waves is

$$\Delta\phi = \phi_2 - \phi_1 = 2\pi\frac{\Delta x}{\lambda} + \Delta\phi_0$$

The sound's wavelength is

$$\lambda = \frac{v}{f} = \frac{343 \text{ m/s}}{500 \text{ Hz}} = 0.686 \text{ m}$$

Distances $x_1 = 2.00$ m and $x_2 = 3.00$ m are measured from the speakers, so the path-length difference is $\Delta x = 1.00$ m. We're given that the inherent phase difference between the speakers is $\Delta\phi_0 = \pi/2$ rad. Thus the phase difference at the observation point is

$$\Delta\phi = 2\pi\frac{\Delta x}{\lambda} + \Delta\phi_0 = 2\pi\frac{1.00 \text{ m}}{0.686 \text{ m}} + \frac{\pi}{2} \text{ rad} = 10.73 \text{ rad}$$

and the amplitude of the wave at this point is

$$A = \left|2a\cos\left(\frac{\Delta\phi}{2}\right)\right| = \left|(0.200 \text{ mm})\cos\left(\frac{10.73}{2}\right)\right| = 0.121 \text{ mm}$$

ASSESS The interference is constructive because $A > a$, but less than maximum constructive interference.

Application: Thin-Film Optical Coatings

The shimmering colors of soap bubbles and oil slicks, as seen in the photo at the beginning of the chapter, are due to the interference of light waves. In fact, the idea of light-wave interference in one dimension has an important application in the optics industry, namely the use of **thin-film optical coatings.** These films, less than 1 μm (10^{-6} m) thick, are placed on glass surfaces, such as lenses, to control reflections from the glass. Antireflection coatings on the lenses in cameras, microscopes, and other optical equipment are examples of thin-film coatings.

FIGURE 17.23 shows a light wave of wavelength λ approaching a piece of glass that has been coated with a transparent film of thickness d whose index of refraction is n. The air-film boundary is a discontinuity at which the wave speed suddenly decreases, and you saw earlier, in Figure 17.7, that a discontinuity causes a reflection. Most of the light is transmitted into the film, but a little bit is reflected.

Furthermore, you saw in Figure 17.7 that the wave reflected from a discontinuity at which the speed decreases is *inverted* with respect to the incident wave. For a sinusoidal wave, which we're now assuming, the inversion is represented mathematically as a phase shift of π rad. The speed of a light wave decreases when it enters a material with a *larger* index of refraction. Thus **a light wave that reflects from a boundary at which the index of refraction increases has a phase shift of π rad.** There is no phase shift for the reflection from a boundary at which the index of refraction decreases. The reflection in Figure 17.23 is from a boundary between air ($n_{\text{air}} = 1.00$) and a transparent film with $n_{\text{film}} > n_{\text{air}}$, so the reflected wave is inverted due to the phase shift of π rad.

When the transmitted wave reaches the glass, most of it continues on into the glass but a portion is reflected back to the left. We'll assume that the index of refraction of the glass is larger than that of the film, $n_{\text{glass}} > n_{\text{film}}$, so this reflection also has a phase shift of π rad. This second reflection, after traveling back through the film, passes back into the air. There are now *two* equal-frequency reflected waves traveling to the left, and these waves will interfere. If the two reflected waves are *in phase*, they will interfere constructively to cause a *strong reflection*. If the two reflected waves are *out of*

FIGURE 17.23 The two reflections, one from the coating and one from the glass, interfere.

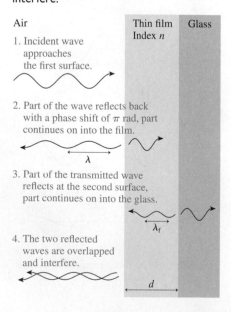

Air | Thin film Index n | Glass

1. Incident wave approaches the first surface.

2. Part of the wave reflects back with a phase shift of π rad, part continues on into the film.

   λ

3. Part of the transmitted wave reflects at the second surface, part continues on into the glass.

   λ_f

4. The two reflected waves are overlapped and interfere.

   d

phase, they will interfere destructively to cause a *weak reflection* or, if their amplitudes are equal, *no reflection* at all.

This suggests practical uses for thin-film optical coatings. The reflections from glass surfaces, even if weak, are often undesirable. For example, reflections degrade the performance of optical equipment. These reflections can be eliminated by coating the glass with a film whose thickness is chosen to cause *destructive* interference of the two reflected waves. This is an *antireflection coating*.

The amplitude of the reflected light depends on the phase difference between the two reflected waves. This phase difference is

$$\Delta\phi = \phi_2 - \phi_1 = (kx_2 + \phi_{20} + \pi \text{ rad}) - (kx_1 + \phi_{10} + \pi \text{ rad})$$

$$= 2\pi \frac{\Delta x}{\lambda_f} + \Delta\phi_0 \tag{17.30}$$

Antireflection coatings use the interference of light waves to nearly eliminate reflections from glass surfaces.

where we explicitly included the reflection phase shift of each wave. In this case, because *both* waves had a phase shift of π rad, the reflection phase shifts cancel.

The wavelength λ_f is the wavelength *in the film* because that's where the path-length difference Δx occurs. You learned in Chapter 16 that the wavelength in a transparent material with index of refraction n is $\lambda_f = \lambda/n$, where the unsubscripted λ is the wavelength in vacuum or air. That is, λ is the wavelength that we measure on "our" side of the air-film boundary.

The path-length difference between the two waves is $\Delta x = 2d$ because wave 2 travels through the film *twice* before rejoining wave 1. The two waves have a common origin—the initial division of the incident wave at the front surface of the film—so the inherent phase difference is $\Delta\phi_0 = 0$. Thus the phase difference of the two reflected waves is

$$\Delta\phi = 2\pi \frac{2d}{\lambda/n} = 2\pi \frac{2nd}{\lambda} \tag{17.31}$$

The interference is constructive, causing a strong reflection, when $\Delta\phi = m \cdot 2\pi$ rad. So when both reflected waves have a phase shift of π rad, constructive interference occurs for wavelengths

$$\lambda_C = \frac{2nd}{m} \qquad m = 1, 2, 3, \ldots \qquad \text{(constructive interference)} \tag{17.32}$$

You will notice that m starts with 1, rather than 0, in order to give meaningful results. Destructive interference, with minimum reflection, requires $\Delta\phi = (m - \frac{1}{2}) \cdot 2\pi$ rad. This—again, when both waves have a phase shift of π rad—occurs for wavelengths

$$\lambda_D = \frac{2nd}{m - \frac{1}{2}} \qquad m = 1, 2, 3, \ldots \qquad \text{(destructive interference)} \tag{17.33}$$

We've used $m - \frac{1}{2}$, rather than $m + \frac{1}{2}$, so that m can start with 1 to match the condition for constructive interference.

NOTE The exact condition for constructive or destructive interference is satisfied for only a few discrete wavelengths λ. Nonetheless, reflections are strongly enhanced (nearly constructive interference) for a range of wavelengths near λ_C. Likewise, there is a range of wavelengths near λ_D for which the reflection is nearly canceled.

EXAMPLE 17.9 | Designing an antireflection coating

Magnesium fluoride (MgF_2) is used as an antireflection coating on lenses. The index of refraction of MgF_2 is 1.39. What is the thinnest film of MgF_2 that works as an antireflection coating at $\lambda = 510$ nm, near the center of the visible spectrum?

MODEL Reflection is minimized if the two reflected waves interfere destructively.

SOLVE The film thicknesses that cause destructive interference at wavelength λ are

$$d = \left(m - \frac{1}{2}\right)\frac{\lambda}{2n}$$

The thinnest film has $m = 1$. Its thickness is

$$d = \frac{\lambda}{4n} = \frac{510 \text{ nm}}{4(1.39)} = 92 \text{ nm}$$

The film thickness is significantly less than the wavelength of visible light!

Continued

as the wavelength moves away from 510 nm. It is the increasing reflection at the ends of the visible spectrum ($\lambda \approx 400$ nm and $\lambda \approx 700$ nm), where $\Delta\phi$ deviates significantly from π rad, that gives a reddish-purple tinge to the lenses on cameras and binoculars. Homework problems will let you explore situations where only one of the two reflections has a reflection phase shift of π rad.

17.7 Interference in Two and Three Dimensions

FIGURE 17.24 A circular or spherical wave.

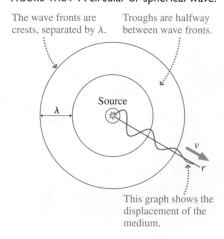

The wave fronts are crests, separated by λ.

Troughs are halfway between wave fronts.

Source

λ

v

r

This graph shows the displacement of the medium.

FIGURE 17.25 The overlapping ripple patterns of two sources. Several points of constructive and destructive interference are noted.

Two in-phase sources emit circular or spherical waves.

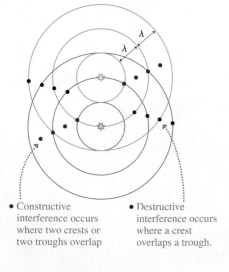

λ

λ

• Constructive interference occurs where two crests or two troughs overlap

• Destructive interference occurs where a crest overlaps a trough.

Ripples on a lake move in two dimensions. The glow from a lightbulb spreads outward as a spherical wave. A circular or spherical wave, illustrated in **FIGURE 17.24**, can be written

$$D(r, t) = a\sin(kr - \omega t + \phi_0) \tag{17.34}$$

where r is the distance measured outward from the source. Equation 17.34 is our familiar wave equation with the one-dimensional coordinate x replaced by a more general radial coordinate r. Recall that the wave fronts represent the *crests* of the wave and are spaced by the wavelength λ.

What happens when two circular or spherical waves overlap? For example, imagine two paddles oscillating up and down on the surface of a pond. We will assume that the two paddles oscillate with the same frequency and amplitude and that they are in phase. **FIGURE 17.25** shows the wave fronts of the two waves. The ripples overlap as they travel, and, as was the case in one dimension, this causes interference. An important difference, though, is that amplitude decreases with distance as waves spread out in two or three dimensions—a consequence of energy conservation—so the two overlapped waves generally do *not* have equal amplitudes. Consequently, destructive interference rarely produces perfect cancellation.

Maximum constructive interference occurs where two crests align or two troughs align. Several locations of constructive interference are marked in Figure 17.25. Intersecting wave fronts are points where two crests are aligned. It's a bit harder to visualize, but two troughs are aligned when a midpoint between two wave fronts is overlapped with another midpoint between two wave fronts. Maximum, but usually not perfect, destructive interference occurs where the crest of one wave aligns with a trough of the other wave. Several points of destructive interference are also indicated in Figure 17.25.

A picture on a page is static, but **the wave fronts are in motion**. Try to imagine the wave fronts of Figure 17.25 expanding outward as new circular rings are born at the sources. The waves will move forward half a wavelength during half a period, causing the crests in Figure 17.25 to be replaced by troughs while the troughs become crests.

The important point to recognize is that **the motion of the waves does not affect the points of constructive and destructive interference.** Points in the figure where two crests overlap will become points where two troughs overlap, but this overlap is still constructive interference. Similarly, points in the figure where a crest and a trough overlap will become a point where a trough and a crest overlap—still destructive interference.

The mathematical description of interference in two or three dimensions is very similar to that of one-dimensional interference. The net displacement of a particle in the medium is

$$D = D_1 + D_2 = a_1\sin(kr_1 - \omega t + \phi_{10}) + a_2\sin(kr_2 - \omega t + \phi_{20}) \tag{17.35}$$

The only differences between Equation 17.35 and the earlier one-dimensional Equation 17.24 are that the linear coordinates have been changed to radial coordinates and we've allowed the amplitudes to differ. These changes do not affect the phase difference, which, with x replaced by r, is now

$$\Delta\phi = 2\pi\frac{\Delta r}{\lambda} + \Delta\phi_0 \qquad (17.36)$$

The term $2\pi(\Delta r/\lambda)$ is the phase difference that arises when the waves travel different distances from the sources to the point at which they combine. Δr itself is the *path-length difference*. As before, $\Delta\phi_0$ is any inherent phase difference of the sources themselves.

Maximum constructive interference occurs, just as in one dimension, at those points where $\cos(\Delta\phi/2) = \pm 1$. Similarly, maximum destructive interference occurs at points where $\cos(\Delta\phi/2) = 0$. The conditions for constructive and destructive interference are

> Maximum constructive interference:
>
> $$\Delta\phi = 2\pi\frac{\Delta r}{\lambda} + \Delta\phi_0 = m\cdot 2\pi$$
>
> $$m = 0, 1, 2, \ldots \qquad (17.37)$$
>
> Maximum destructive interference:
>
> $$\Delta\phi = 2\pi\frac{\Delta r}{\lambda} + \Delta\phi_0 = (m + \tfrac{1}{2})\cdot 2\pi$$

Two overlapping water waves create an interference pattern.

Identical Sources

For two identical sources (i.e., sources that oscillate in phase with $\Delta\phi_0 = 0$), the conditions for constructive and destructive interference are simple:

Constructive: $\Delta r = m\lambda$

$$\text{(identical sources)} \qquad (17.38)$$

Destructive: $\Delta r = \left(m + \dfrac{1}{2}\right)\lambda$

The waves from two identical sources interfere constructively at points where the path-length difference is an integer number of wavelengths because, for these values of Δr, crests are aligned with crests and troughs with troughs. **The waves interfere destructively at points where the path-length difference is a half-integer number of wavelengths** because, for these values of Δr, crests are aligned with troughs. These two statements are the essence of interference.

> **NOTE** Equation 17.38 applies only if the sources are in phase. If the sources are not in phase, you must use the more general Equation 17.37 to locate the points of constructive and destructive interference.

Wave fronts are spaced exactly one wavelength apart; hence we can measure the distances r_1 and r_2 simply by counting the rings in the wave-front pattern. In **FIGURE 17.26**, which is based on Figure 17.25, point A is distance $r_1 = 3\lambda$ from the first source and $r_2 = 2\lambda$ from the second. The path-length difference is $\Delta r_A = 1\lambda$, the condition for the maximum constructive interference of identical sources. Point B has $\Delta r_B = \frac{1}{2}\lambda$, so it is a point of maximum destructive interference.

> **NOTE** Interference is determined by Δr, the path-length *difference*, rather than by r_1 or r_2.

FIGURE 17.26 For identical sources, the path-length difference Δr determines whether the interference at a particular point is constructive or destructive.

- At A, $\Delta r_A = \lambda$, so this is a point of constructive interference.

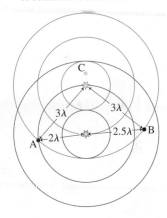

- At B, $\Delta r_B = \frac{1}{2}\lambda$, so this is a point of destructive interference.

STOP TO THINK 17.5 The interference at point C in Figure 17.26 is

a. Maximum constructive.
b. Constructive, but less than maximum.
c. Maximum destructive.
d. Destructive, but less than maximum.
e. There is no interference at point C.

We can now locate the points of maximum constructive interference by drawing a line through *all* the points at which $\Delta r = 0$, another line through all the points at which $\Delta r = \lambda$, and so on. These lines, shown in red in FIGURE 17.27, are called **antinodal lines.** They are analogous to the antinodes of a standing wave, hence the name. An antinode is a *point* of maximum constructive interference; for circular waves, oscillation at maximum amplitude occurs along a continuous *line*. Similarly, destructive interference occurs along lines called **nodal lines.** The amplitude is a minimum along a nodal line, usually close to zero, just as it is at a node in a standing-wave pattern.

FIGURE 17.27 The points of constructive and destructive interference fall along antinodal and nodal lines.

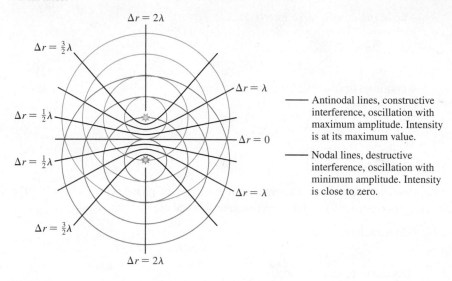

Antinodal lines, constructive interference, oscillation with maximum amplitude. Intensity is at its maximum value.

Nodal lines, destructive interference, oscillation with minimum amplitude. Intensity is close to zero.

A Problem-Solving Strategy for Interference Problems

The information in this section is the basis of a strategy for solving interference problems. This strategy applies equally well to interference in one dimension if you use Δx instead of Δr.

PROBLEM-SOLVING STRATEGY 17.1

Interference of two waves

MODEL Model the waves as linear, circular, or spherical.

VISUALIZE Draw a picture showing the sources of the waves and the point where the waves interfere. Give relevant dimensions. Identify the distances r_1 and r_2 from the sources to the point. Note any phase difference $\Delta\phi_0$ between the two sources.

SOLVE The interference depends on the path-length difference $\Delta r = r_2 - r_1$ and the source phase difference $\Delta\phi_0$.

Constructive: $\quad \Delta\phi = 2\pi\dfrac{\Delta r}{\lambda} + \Delta\phi_0 = m \cdot 2\pi$

$$m = 0, 1, 2, \ldots$$

Destructive: $\quad \Delta\phi = 2\pi\dfrac{\Delta r}{\lambda} + \Delta\phi_0 = \left(m + \tfrac{1}{2}\right) \cdot 2\pi$

For identical sources ($\Delta\phi_0 = 0$), the interference is maximum constructive if $\Delta r = m\lambda$, maximum destructive if $\Delta r = (m + \tfrac{1}{2})\lambda$.

ASSESS Check that your result has correct units and significant figures, is reasonable, and answers the question.

Exercise 18

EXAMPLE 17.10 | Two-dimensional interference between two loudspeakers

Two loudspeakers in a plane are 2.0 m apart and in phase with each other. Both emit 700 Hz sound waves into a room where the speed of sound is 341 m/s. A listener stands 5.0 m in front of the loudspeakers and 2.0 m to one side of the center. Is the interference at this point maximum constructive, maximum destructive, or in between? How will the situation differ if the loudspeakers are out of phase?

MODEL The two speakers are sources of in-phase, spherical waves. The overlap of these waves causes interference.

VISUALIZE FIGURE 17.28 shows the loudspeakers and defines the distances r_1 and r_2 to the point of observation. The figure includes dimensions and notes that $\Delta\phi_0 = 0$ rad.

FIGURE 17.28 Pictorial representation of the interference between two loudspeakers.

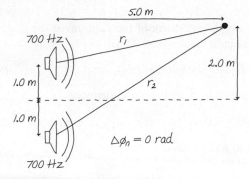

SOLVE It's not r_1 and r_2 that matter, but the *difference* Δr between them. From the geometry of the figure we can calculate that

$$r_1 = \sqrt{(5.0 \text{ m})^2 + (1.0 \text{ m})^2} = 5.10 \text{ m}$$
$$r_2 = \sqrt{(5.0 \text{ m})^2 + (3.0 \text{ m})^2} = 5.83 \text{ m}$$

Thus the path-length difference is $\Delta r = r_2 - r_1 = 0.73$ m. The wavelength of the sound waves is

$$\lambda = \frac{v}{f} = \frac{341 \text{ m/s}}{700 \text{ Hz}} = 0.487 \text{ m}$$

In terms of wavelengths, the path-length difference is $\Delta r/\lambda = 1.50$, or

$$\Delta r = \tfrac{3}{2}\lambda$$

Because the sources are in phase ($\Delta\phi_0 = 0$), this is the condition for *destructive* interference. If the sources were out of phase ($\Delta\phi_0 = \pi$ rad), then the phase difference of the waves at the listener would be

$$\Delta\phi = 2\pi \frac{\Delta r}{\lambda} + \Delta\phi_0 = 2\pi\left(\frac{3}{2}\right) + \pi \text{ rad} = 4\pi \text{ rad}$$

This is an integer multiple of 2π rad, so in this case the interference would be *constructive*.

ASSESS Both the path-length difference and any inherent phase difference of the sources must be considered when evaluating interference.

STOP TO THINK 17.6 These two loudspeakers are in phase. They emit equal-amplitude sound waves with a wavelength of 1.0 m. At the point indicated, is the interference maximum constructive, maximum destructive, or something in between?

17.8 Beats

Thus far we have looked at the superposition of sources having the same wavelength and frequency. We can also use the principle of superposition to investigate a phenomenon that is easily demonstrated with two sources of slightly different frequency.

If you listen to two sounds with very different frequencies, such as a high note and a low note, you hear two distinct tones. But if the frequency difference is very small, just one or two hertz, then you hear a single tone whose intensity is *modulated* once or twice every second. That is, the sound goes up and down in volume, loud, soft, loud, soft, ... , making a distinctive sound pattern called **beats.**

Consider two sinusoidal waves traveling along the x-axis with angular frequencies $\omega_1 = 2\pi f_1$ and $\omega_2 = 2\pi f_2$. The two waves are

$$D_1 = a\sin(k_1 x - \omega_1 t + \phi_{10})$$
$$D_2 = a\sin(k_2 x - \omega_2 t + \phi_{20})$$

(17.39)

where the subscripts 1 and 2 indicate that the frequencies, wave numbers, and phase constants of the two waves may be different.

To simplify the analysis, let's make several assumptions:

1. The two waves have the same amplitude a,
2. A detector, such as your ear, is located at the origin $(x = 0)$,
3. The two sources are in phase $(\phi_{10} = \phi_{20})$, and
4. The source phases happen to be $\phi_{10} = \phi_{20} = \pi$ rad.

None of these assumptions is essential to the outcome. All could be otherwise and we would still come to basically the same conclusion, but the mathematics would be far messier. Making these assumptions allows us to emphasize the physics with the least amount of mathematics.

With these assumptions, the two waves as they reach the detector at $x = 0$ are

$$D_1 = a\sin(-\omega_1 t + \pi) = a\sin\omega_1 t$$
$$D_2 = a\sin(-\omega_2 t + \pi) = a\sin\omega_2 t \tag{17.40}$$

where we've used the trigonometric identity $\sin(\pi - \theta) = \sin\theta$. The principle of superposition tells us that the *net* displacement of the medium at the detector is the sum of the displacements of the individual waves. Thus

$$D = D_1 + D_2 = a(\sin\omega_1 t + \sin\omega_2 t) \tag{17.41}$$

Earlier, for interference, we used the trigonometric identity

$$\sin\alpha + \sin\beta = 2\cos\left[\tfrac{1}{2}(\alpha - \beta)\right]\sin\left[\tfrac{1}{2}(\alpha + \beta)\right]$$

We can use this identity again to write Equation 17.41 as

$$D = 2a\cos\left[\tfrac{1}{2}(\omega_1 - \omega_2)t\right]\sin\left[\tfrac{1}{2}(\omega_1 + \omega_2)t\right]$$
$$= \left[2a\cos(\omega_{mod}t)\right]\sin(\omega_{avg}t) \tag{17.42}$$

where $\omega_{avg} = \tfrac{1}{2}(\omega_1 + \omega_2)$ is the *average* angular frequency and $\omega_{mod} = \tfrac{1}{2}|\omega_1 - \omega_2|$ is called the *modulation frequency*. We've used the absolute value because the modulation depends only on the frequency *difference* between the sources, not on which has the larger frequency.

We are interested in the situation when the two frequencies are very nearly equal: $\omega_1 \approx \omega_2$. In that case, ω_{avg} hardly differs from either ω_1 or ω_2 while ω_{mod} is very near to—but not exactly—zero. When ω_{mod} is very small, the term $\cos(\omega_{mod}t)$ oscillates *very* slowly. We have grouped it with the $2a$ term because, together, they provide a slowly changing "amplitude" for the rapid oscillation at frequency ω_{avg}.

FIGURE 17.29 is a history graph of the wave at the detector $(x = 0)$. It shows the oscillation of the air against your eardrum at frequency $f_{avg} = \omega_{avg}/2\pi = \tfrac{1}{2}(f_1 + f_2)$. This oscillation determines the note you hear; it differs little from the two notes at frequencies f_1 and f_2. We are especially interested in the time-dependent amplitude, shown as a dashed line, that is given by the term $2a\cos(\omega_{mod}t)$. This periodically varying amplitude is called a **modulation** of the wave, which is where ω_{mod} gets its name.

As the amplitude rises and falls, the sound alternates as loud, soft, loud, soft, and so on. But that is exactly what you hear when you listen to beats! The alternating loud and soft sounds arise from the two waves being alternately in phase and out of phase, causing constructive and then destructive interference.

Imagine two people walking side by side at just slightly different paces. Initially both of their right feet hit the ground together, but after a while they get out of step. A little bit later they are back in step and the process alternates. The sound waves are doing the same. Initially the crests of each wave, of amplitude a, arrive together at your ear and the net displacement is doubled to $2a$. But after a while the two waves, being of slightly different frequency, get out of step and a crest of one arrives with a

FIGURE 17.29 Beats are caused by the superposition of two waves of nearly identical frequency.

The medium oscillates rapidly at frequency f_{avg}.

The amplitude is slowly modulated as $2a\cos(\omega_{mod}t)$.

Loud Soft Loud Soft Loud Soft Loud

trough of the other. When this happens, the two waves cancel each other to give a net displacement of zero. This process alternates over and over, loud and soft.

Notice, in Figure 17.29, that the sound intensity rises and falls *twice* during one cycle of the modulation envelope. Each "loud-soft-loud" is one beat, so the **beat frequency** f_{beat}, which is the number of beats per second, is *twice* the modulation frequency $f_{mod} = \omega_{mod}/2\pi$. From the above definition of ω_{mod}, the beat frequency is

$$f_{beat} = 2f_{mod} = 2\frac{\omega_{mod}}{2\pi} = 2 \cdot \frac{1}{2}\left(\frac{\omega_1}{2\pi} - \frac{\omega_2}{2\pi}\right) = |f_1 - f_2| \qquad (17.43)$$

The beat frequency is simply the *difference* between the two individual frequencies.

EXAMPLE 17.11 | Detecting bats with beats

The little brown bat is a common species in North America. It emits echolocation pulses at a frequency of 40 kHz, well above the range of human hearing. To allow researchers to "hear" these bats, the bat detector shown in **FIGURE 17.30** combines the bat's sound wave at frequency f_1 with a wave of frequency f_2 from a tunable oscillator. The resulting beat frequency is then amplified and sent to a loudspeaker. To what frequency should the tunable oscillator be set to produce an audible beat frequency of 3 kHz?

SOLVE Combining two waves with different frequencies gives a beat frequency

$$f_{beat} = |f_1 - f_2|$$

A beat frequency will be generated at 3 kHz if the oscillator frequency and the bat frequency *differ* by 3 kHz. An oscillator frequency of either 37 kHz or 43 kHz will work nicely.

ASSESS The electronic circuitry of radios, televisions, and cell phones makes extensive use of *mixers* to generate difference frequencies.

FIGURE 17.30 The operation of a bat detector.

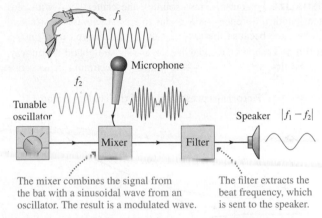

The mixer combines the signal from the bat with a sinusoidal wave from an oscillator. The result is a modulated wave.

The filter extracts the beat frequency, which is sent to the speaker.

Beats aren't limited to sound waves. **FIGURE 17.31** shows a graphical example of beats. Two "fences" of slightly different frequencies are superimposed on each other. The difference in the two frequencies is two lines per inch. You can confirm, with a ruler, that the figure has two "beats" per inch, in agreement with Equation 17.43.

Beats are important in many other situations. For example, you have probably seen movies where rotating wheels seem to turn slowly backward. Why is this? Suppose the movie camera is shooting at 30 frames per second but the wheel is rotating 32 times per second. The combination of the two produces a "beat" of 2 Hz, meaning that the wheel appears to rotate only twice per second. The same is true if the wheel is rotating 28 times per second, but in this case, where the wheel frequency slightly lags the camera frequency, it appears to rotate *backward* twice per second!

FIGURE 17.31 A graphical example of beats.

STOP TO THINK 17.7 You hear three beats per second when two sound tones are generated. The frequency of one tone is 610 Hz. The frequency of the other is

a. 604 Hz   b. 607 Hz   c. 613 Hz
d. 616 Hz   e. Either a or d.   f. Either b or c.

CHALLENGE EXAMPLE 17.12 | An airplane landing system

Your firm has been hired to design a system that allows airplane pilots to make instrument landings in rain or fog. You've decided to place two radio transmitters 50 m apart on either side of the runway. These two transmitters will broadcast the same frequency, but out of phase with each other. This will cause a nodal line to extend straight off the end of the runway. As long as the airplane's receiver is silent, the pilot knows she's directly in line with the runway. If she drifts to one side or the other, the radio will pick up a signal and sound a warning beep. To have sufficient accuracy, the first intensity maxima need to be 60 m on either side of the nodal line at a distance of 3.0 km. What frequency should you specify for the transmitters?

MODEL The two transmitters are sources of out-of-phase, circular waves. The overlap of these waves produces an interference pattern.

VISUALIZE For out-of-phase sources, the center line—with zero path-length difference—is a nodal line of maximum destructive interference because the two signals always arrive out of phase. FIGURE 17.32 shows the nodal line, extending straight off the runway, and the first antinodal line—the points of maximum constructive

FIGURE 17.32 Pictorial representation of the landing system.

interference—on either side. Comparing this to Figure 17.27, where the two sources were in phase, you can see that the nodal and antinodal lines have been reversed.

SOLVE Point P, 60 m to the side at a distance of 3000 m, needs to be a point of maximum constructive interference. The distances are

$$r_1 = \sqrt{(3000 \text{ m})^2 + (60 \text{ m} - 25 \text{ m})^2} = 3000.204 \text{ m}$$
$$r_2 = \sqrt{(3000 \text{ m})^2 + (60 \text{ m} + 25 \text{ m})^2} = 3001.204 \text{ m}$$

We needed to keep several extra significant figures because we're looking for the difference between two numbers that are almost the same. The path-length difference at P is

$$\Delta r = r_2 - r_1 = 1.000 \text{ m}$$

We know, for out-of-phase transmitters, that the phase difference of the sources is $\Delta\phi_0 = \pi$ rad. The first maximum will occur where the phase difference between the waves is $\Delta\phi = 1 \cdot 2\pi$ rad. Thus the condition that we must satisfy at P is

$$\Delta\phi = 2\pi \text{ rad} = 2\pi \frac{\Delta r}{\lambda} + \pi \text{ rad}$$

Solving for λ, we find

$$\lambda = 2\,\Delta r = 2.00 \text{ m}$$

Consequently, the required frequency is

$$f = \frac{c}{\lambda} = \frac{3.00 \times 10^8 \text{ m/s}}{2.00 \text{ m}} = 1.50 \times 10^8 \text{ Hz} = 150 \text{ MHz}$$

ASSESS 150 MHz is slightly higher than the frequencies of FM radio (≈ 100 MHz) but is well within the radio frequency range. Notice that the condition to be satisfied at P is that the path-length difference must be $\frac{1}{2}\lambda$. This makes sense. A path-length difference of $\frac{1}{2}\lambda$ contributes π rad to the phase difference. When combined with the π rad from the out-of-phase sources, the total phase difference of 2π rad creates constructive interference.

SUMMARY

The goal of Chapter 17 has been to understand and use the idea of superposition.

GENERAL PRINCIPLES

Principle of Superposition

The displacement of a medium when more than one wave is present is the sum at each point of the displacements due to each individual wave.

IMPORTANT CONCEPTS

Standing Waves

Standing waves are due to the superposition of two traveling waves moving in opposite directions.

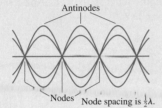

Antinodes

Nodes   Node spacing is $\frac{1}{2}\lambda$.

The amplitude at position x is

$$A(x) = 2a \sin kx$$

where a is the amplitude of each wave.

The boundary conditions determine which standing-wave frequencies and wavelengths are allowed. The allowed standing waves are **modes** of the system.

$m = 1$

$m = 2$

$m = 3$

Standing waves on a string

Solving Interference Problems

Maximum constructive interference occurs where crests are aligned with crests and troughs with troughs. The waves are in phase.

Maximum destructive interference occurs where crests are aligned with troughs. The waves are out of phase.

MODEL Model the wave as linear, circular, or spherical.

VISUALIZE Find distances to the sources.

SOLVE Interference depends on the **phase difference** $\Delta\phi$ between the waves:

Constructive: $\Delta\phi = 2\pi \dfrac{\Delta r}{\lambda} + \Delta\phi_0 = m \cdot 2\pi$

Destructive: $\Delta\phi = 2\pi \dfrac{\Delta r}{\lambda} + \Delta\phi_0 = \left(m + \frac{1}{2}\right) \cdot 2\pi$

Δr is the path-length difference of the two waves, and $\Delta\phi_0$ is any phase difference between the sources. For identical (in-phase) sources:

Constructive: $\Delta r = m\lambda$ Destructive: $\Delta r = \left(m + \frac{1}{2}\right)\lambda$

ASSESS Is the result reasonable?

Antinodal lines, maximum constructive interference.

Nodal lines, maximum destructive interference.

APPLICATIONS

Boundary conditions

Strings, electromagnetic waves, and sound waves in closed-closed tubes must have nodes at both ends:

$$\lambda_m = \frac{2L}{m} \qquad f_m = m\frac{v}{2L} = mf_1 \qquad m = 1, 2, 3, \ldots$$

The frequencies and wavelengths are the same for a sound wave in an open-open tube, which has antinodes at both ends.

A sound wave in an open-closed tube must have a node at the closed end but an antinode at the open end. This leads to

$$\lambda_m = \frac{4L}{m} \qquad f_m = m\frac{v}{4L} = mf_1 \qquad m = 1, 3, 5, 7, \ldots$$

Beats (loud-soft-loud-soft modulations of intensity) occur when two waves of slightly different frequency are superimposed.

Soft   Loud   Soft   Loud   Soft

The beat frequency between waves of frequencies f_1 and f_2 is

$$f_{\text{beat}} = |f_1 - f_2|$$

TERMS AND NOTATION

principle of superposition	fundamental frequency, f_1	out of phase	antinodal line
standing wave	harmonic	destructive interference	nodal line
node	mode	phase difference, $\Delta\phi$	beats
antinode	interference	path-length difference, Δx	modulation
amplitude function, $A(x)$	in phase	or Δr	beat frequency, f_{beat}
boundary condition	constructive interference	thin-film optical coating	

CONCEPTUAL QUESTIONS

1. **FIGURE Q17.1** shows a standing wave oscillating on a string at frequency f_0.

   FIGURE Q17.1

   a. What mode (m-value) is this?
   b. How many antinodes will there be if the frequency is doubled to $2f_0$?

2. If you take snapshots of a standing wave on a string, there are certain instants when the string is totally flat. What has happened to the energy of the wave at those instants?

3. **FIGURE Q17.3** shows the displacement of a standing sound wave in a 32-cm-long horizontal tube of air open at both ends.

   a. What mode (m-value) is this?
   b. Are the air molecules moving horizontally or vertically? Explain.
   c. At what distances from the left end of the tube do the molecules oscillate with maximum amplitude?
   d. At what distances from the left end of the tube does the air pressure oscillate with maximum amplitude?

   FIGURE Q17.3

4. An organ pipe is tuned to exactly 384 Hz when the room temperature is 20°C. If the room temperature later increases to 22°C, does the pipe's frequency increase, decrease, or stay the same? Explain.

5. If you pour liquid into a tall, narrow glass, you may hear sound with a steadily rising pitch. What is the source of the sound? And why does the pitch rise as the glass fills?

6. A flute filled with helium will, until the helium escapes, play notes at a much higher pitch than normal. Why?

7. In music, two notes are said to be an *octave* apart when one note is exactly twice the frequency of the other. Suppose you have a guitar string playing frequency f_0. To increase the frequency by an octave, to $2f_0$, by what factor would you have to (a) increase the tension or (b) decrease the length?

8. **FIGURE Q17.8** is a snapshot graph of two plane waves passing through a region of space. Each wave has a 2.0 mm amplitude and the same wavelength. What is the net displacement of the medium at points a, b, and c?

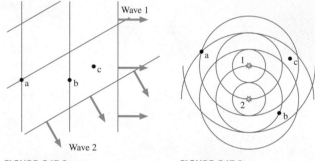

FIGURE Q17.8   **FIGURE Q17.9**

9. **FIGURE Q17.9** shows the circular waves emitted by two in-phase sources. Are a, b, and c points of maximum constructive interference, maximum destructive interference, or in between?

10. A trumpet player hears 5 beats per second when she plays a note and simultaneously sounds a 440 Hz tuning fork. After pulling her tuning valve out to slightly increase the length of her trumpet, she hears 3 beats per second against the tuning fork. Was her initial frequency 435 Hz or 445 Hz? Explain.

EXERCISES AND PROBLEMS

Problems labeled ▓ integrate material from earlier chapters.

Exercises

Section 17.1 The Principle of Superposition

1. | **FIGURE EX17.1** is a snapshot graph at $t = 0$ s of two waves approaching each other at 1.0 m/s. Draw six snapshot graphs, stacked vertically, showing the string at 1 s intervals from $t = 1$ s to $t = 6$ s.

   FIGURE EX17.1

2. | **FIGURE EX17.2** is a snapshot graph at $t = 0$ s of two waves approaching each other at 1.0 m/s. Draw six snapshot graphs, stacked vertically, showing the string at 1 s intervals from $t = 1$ s to $t = 6$ s.

   FIGURE EX17.2

3. ‖ **FIGURE EX17.3a** is a snapshot graph at $t = 0$ s of two waves approaching each other at 1.0 m/s. At what time was the snapshot graph in **FIGURE EX17.3b** taken?

(a)

(b)

FIGURE EX17.3

Section 17.2 Standing Waves

Section 17.3 Standing Waves on a String

4. ‖ **FIGURE EX17.4** is a snapshot graph at $t = 0$ s of two waves moving to the right at 1.0 m/s. The string is fixed at $x = 8.0$ m. Draw four snapshot graphs, stacked vertically, showing the string at $t = 2, 4, 6,$ and 8 s.

FIGURE EX17.4

5. ‖ **FIGURE EX17.5** shows a standing wave oscillating at 100 Hz on a string. What is the wave speed?

FIGURE EX17.5

FIGURE EX17.6

6. ‖ **FIGURE EX17.6** shows a standing wave on a 2.0-m-long string that has been fixed at both ends and tightened until the wave speed is 40 m/s. What is the frequency?

7. | **FIGURE EX17.7** shows a standing wave on a string that is oscillating at 100 Hz.

FIGURE EX17.7

   a. How many antinodes will there be if the frequency is increased to 200 Hz?
   b. If the tension is increased by a factor of 4, at what frequency will the string continue to oscillate as a standing wave that looks like the one in the figure?

8. | a. What are the three longest wavelengths for standing waves on a 240-cm-long string that is fixed at both ends?
   b. If the frequency of the second-longest wavelength is 50 Hz, what is the frequency of the third-longest wavelength?

9. | Standing waves on a 1.0-m-long string that is fixed at both ends are seen at successive frequencies of 36 Hz and 48 Hz.
   a. What are the fundamental frequency and the wave speed?
   b. Draw the standing-wave pattern when the string oscillates at 48 Hz.

10. | The two highest-pitch strings on a violin are tuned to 440 Hz (the A string) and 659 Hz (the E string). What is the ratio of the mass of the A string to that of the E string? Violin strings are all the same length and under essentially the same tension.

11. ‖ A heavy piece of hanging sculpture is suspended by a 90-cm-long, 5.0 g steel wire. When the wind blows hard, the wire hums at its fundamental frequency of 80 Hz. What is the mass of the sculpture?

12. | A carbon dioxide laser is an infrared laser. A CO_2 laser with a cavity length of 53.00 cm oscillates in the $m = 100,000$ mode. What are the wavelength and frequency of the laser beam?

13. | Microwaves pass through a small hole into the "microwave cavity" of **FIGURE EX17.13**. What frequencies between 10 GHz and 20 GHz will create standing waves in the cavity?

FIGURE EX17.13

Section 17.4 Standing Sound Waves and Musical Acoustics

14. | What are the three longest wavelengths for standing sound waves in a 121-cm-long tube that is (a) open at both ends and (b) open at one end, closed at the other?

15. | **FIGURE EX17.15** shows a standing sound wave in an 80-cm-long tube. The tube is filled with an unknown gas. What is the speed of sound in this gas?

FIGURE EX17.15

16. | The fundamental frequency of an open-open tube is 1500 Hz when the tube is filled with 0°C helium. What is its frequency when filled with 0°C air?

17. | We can make a simple model of the human vocal tract as an
BIO open-closed tube extending from the opening of the mouth to the diaphragm. What is the length of this tube if its fundamental frequency equals a typical speech frequency of 250 Hz? The speed of sound in the warm air is 350 m/s.

18. ‖ The lowest note on a grand piano has a frequency of 27.5 Hz. The entire string is 2.00 m long and has a mass of 400 g. The vibrating section of the string is 1.90 m long. What tension is needed to tune this string properly?

19. | A bass clarinet can be modeled as a 120-cm-long open-closed tube. A bass clarinet player starts playing in a 20°C room, but soon the air inside the clarinet warms to where the speed of sound is 352 m/s. Does the fundamental frequency increase or decrease? By how much?

20. ‖ A violin string is 30 cm long. It sounds the musical note A (440 Hz) when played without fingering. How far from the end of the string should you place your finger to play the note C (523 Hz)?

21. | A longitudinal standing wave can be created in a long, thin aluminum rod by stroking the rod with very dry fingers. This is often done as a physics demonstration, creating a high-pitched, very annoying whine. From a wave perspective, the standing wave is equivalent to a sound standing wave in an open-open tube. As **FIGURE EX17.21** shows, both ends of the rod are antinodes. What is the fundamental frequency of a 2.0-m-long aluminum rod?

FIGURE EX17.21

Section 17.5 Interference in One Dimension

Section 17.6 The Mathematics of Interference

22. | Two loudspeakers emit sound waves along the x-axis. The sound has maximum intensity when the speakers are 20 cm apart. The sound intensity decreases as the distance between the speakers is increased, reaching zero at a separation of 60 cm.
    a. What is the wavelength of the sound?
    b. If the distance between the speakers continues to increase, at what separation will the sound intensity again be a maximum?

23. || Two loudspeakers in a 20°C room emit 686 Hz sound waves along the x-axis.
    a. If the speakers are in phase, what is the smallest distance between the speakers for which the interference of the sound waves is maximum destructive?
    b. If the speakers are out of phase, what is the smallest distance between the speakers for which the interference of the sound waves is maximum constructive?

24. | Noise-canceling headphones are an application of destructive interference. Each side of the headphones uses a microphone to pick up noise, delays it slightly, then rebroadcasts the noise next to your ear where it can interfere with the incoming sound wave of the noise. Suppose you are sitting 1.8 m from an annoying, 110 Hz buzzing sound. What is the minimum headphone delay, in ms, that will cancel this noise?

25. | What is the thinnest film of MgF_2 ($n = 1.39$) on glass that produces a strong reflection for orange light with a wavelength of 600 nm?

26. || A very thin oil film ($n = 1.25$) floats on water ($n = 1.33$). What is the thinnest film that produces a strong reflection for green light with a wavelength of 500 nm?

Section 17.7 Interference in Two and Three Dimensions

27. || FIGURE EX17.27 shows the circular wave fronts emitted by two wave sources.
    a. Are these sources in phase or out of phase? Explain.
    b. Make a table with rows labeled P, Q, and R and columns labeled r_1, r_2, Δr, and C/D. Fill in the table for points P, Q, and R, giving the distances as multiples of λ and indicating, with a C or a D, whether the interference at that point is constructive or destructive.

FIGURE EX17.27

FIGURE EX17.28

28. || FIGURE EX17.28 shows the circular wave fronts emitted by two wave sources.
    a. Are these sources in phase or out of phase? Explain.
    b. Make a table with rows labeled P, Q, and R and columns labeled r_1, r_2, Δr, and C/D. Fill in the table for points P, Q, and R, giving the distances as multiples of λ and indicating, with a C or a D, whether the interference at that point is constructive or destructive.

29. || Two in-phase loudspeakers, which emit sound in all directions, are sitting side by side. One of them is moved sideways by 3.0 m, then forward by 4.0 m. Afterward, constructive interference is observed $\frac{1}{4}$ and $\frac{3}{4}$ of the distance between the speakers along the line that joins them. What is the maximum possible wavelength of the sound waves?

30. || Two in-phase speakers 2.0 m apart in a plane are emitting 1800 Hz sound waves into a room where the speed of sound is 340 m/s. Is the point 4.0 m in front of one of the speakers, perpendicular to the plane of the speakers, a point of maximum constructive interference, maximum destructive interference, or something in between?

31. || Two out-of-phase radio antennas at $x = \pm 300$ m on the x-axis are emitting 3.0 MHz radio waves. Is the point $(x, y) = (300 \text{ m}, 800 \text{ m})$ a point of maximum constructive interference, maximum destructive interference, or something in between?

Section 17.8 Beats

32. | Two strings are adjusted to vibrate at exactly 200 Hz. Then the tension in one string is increased slightly. Afterward, three beats per second are heard when the strings vibrate at the same time. What is the new frequency of the string that was tightened?

33. | A flute player hears four beats per second when she compares her note to a 523 Hz tuning fork (the note C). She can match the frequency of the tuning fork by pulling out the "tuning joint" to lengthen her flute slightly. What was her initial frequency?

34. | Traditional Indonesian music uses an ensemble called a *gamelan* that is based on tuned percussion instruments somewhat like gongs. In Bali, the gongs are often grouped in pairs that are slightly out of tune with each other. When both are played at once, the beat frequency lends a distinctive vibrating quality to the music. Suppose a pair of gongs are tuned to produce notes at 151 Hz and 155 Hz. How many beats are heard if the gongs are struck together and both ring for 2.5 s?

35. || Two microwave signals of nearly equal wavelengths can generate a beat frequency if both are directed onto the same microwave detector. In an experiment, the beat frequency is 100 MHz. One microwave generator is set to emit microwaves with a wavelength of 1.250 cm. If the second generator emits the longer wavelength, what is that wavelength?

Problems

36. | A 2.0-m-long string vibrates at its second-harmonic frequency with a maximum amplitude of 2.0 cm. One end of the string is at $x = 0$ cm. Find the oscillation amplitude at $x = 10, 20, 30, 40,$ and 50 cm.

37. || A string vibrates at its third-harmonic frequency. The amplitude at a point 30 cm from one end is half the maximum amplitude. How long is the string?

38. || Tendons are, essentially, elastic cords stretched between two BIO fixed ends. As such, they can support standing waves. A woman has a 20-cm-long Achilles tendon—connecting the heel to a muscle in the calf—with a cross-section area of 90 mm². The density of tendon tissue is 1100 kg/m³. For a reasonable tension of 500 N, what will be the fundamental frequency of her Achilles tendon?

39. || Biologists think that some spiders "tune" strands of their web BIO to give enhanced response at frequencies corresponding to those at which desirable prey might struggle. Orb spider web silk has a typical diameter of 20 μm, and spider silk has a density of

1300 kg/m^3. To have a fundamental frequency at 100 Hz, to what tension must a spider adjust a 12-cm-long strand of silk?

40. ‖ A particularly beautiful note reaching your ear from a rare Stradivarius violin has a wavelength of 39.1 cm. The room is slightly warm, so the speed of sound is 344 m/s. If the string's linear density is 0.600 g/m and the tension is 150 N, how long is the vibrating section of the violin string?

41. ‖ A violinist places her finger so that the vibrating section of a 1.0 g/m string has a length of 30 cm, then she draws her bow across it. A listener nearby in a 20°C room hears a note with a wavelength of 40 cm. What is the tension in the string?

42. ‖ A steel wire is used to stretch the spring of **FIGURE P17.42**. An oscillating magnetic field drives the steel wire back and forth. A standing wave with three antinodes is created when the spring is stretched 8.0 cm. What stretch of the spring produces a standing wave with two antinodes?

FIGURE P17.42   Spring

43. ‖ Astronauts visiting Planet X have a 250-cm-long string whose mass is 5.00 g. They tie the string to a support, stretch it horizontally over a pulley 2.00 m away, and hang a 4.00 kg mass on the free end. Then the astronauts begin to excite standing waves on the horizontal portion of the string. Their data are as follows:

m	Frequency (Hz)
1	31
2	66
3	95
4	130
5	162

Use the best-fit line of an appropriate graph to determine the value of g, the free-fall acceleration on Planet X.

44. ‖ A 75 g bungee cord has an equilibrium length of 1.20 m. The cord is stretched to a length of 1.80 m, then vibrated at 20 Hz. This produces a standing wave with two antinodes. What is the spring constant of the bungee cord?

45. ‖ A metal wire under tension T_0 vibrates at its fundamental frequency. For what tension will the second-harmonic frequency be the same as the fundamental frequency at tension T_0?

46. ‖‖ In a laboratory experiment, one end of a horizontal string is tied to a support while the other end passes over a frictionless pulley and is tied to a 1.5 kg sphere. Students determine the frequencies of standing waves on the horizontal segment of the string, then they raise a beaker of water until the hanging 1.5 kg sphere is completely submerged. The frequency of the fifth harmonic with the sphere submerged exactly matches the frequency of the third harmonic before the sphere was submerged. What is the diameter of the sphere?

47. ‖‖ A vibrating standing wave on a string radiates a sound wave with intensity proportional to the square of the standing-wave amplitude. When a piano key is struck and held down, so that the string continues to vibrate, the sound level decreases by 8.0 dB in 1.0 s. What is the string's damping time constant τ?

48. ‖ What is the fundamental frequency of the steel wire in **FIGURE P17.48**?

75 g steel wire

4.0 kg bar

45°

2.0 m   8.0 kg

FIGURE P17.48

49. ‖ The two strings in **FIGURE P17.49** are of equal length and are being driven at equal frequencies. The linear density of the left string is 5.0 g/m. What is the linear density of the right string?

Stretched spring

FIGURE P17.49

50. ‖ Western music uses a musical scale with *equal temperament* tuning, which means that any two adjacent notes have the same frequency ratio r. That is, notes n and $n + 1$ are related by $f_{n+1} = rf_n$ for all n. In this system, the frequency doubles every 12 notes—an interval called an *octave*.
   a. What is the value of r?
   b. Orchestras tune to the note A, which has a frequency of 440 Hz. What is the frequency of the next note of the scale (called A-sharp)?

51. ‖ An open-open organ pipe is 78.0 cm long. An open-closed pipe has a fundamental frequency equal to the third harmonic of the open-open pipe. How long is the open-closed pipe?

52. ‖ Deep-sea divers often breathe a mixture of helium and oxygen
BIO to avoid getting the "bends" from breathing high-pressure nitrogen. The helium has the side effect of making the divers' voices sound odd. Although your vocal tract can be roughly described as an open-closed tube, the way you hold your mouth and position your lips greatly affects the standing-wave frequencies of the vocal tract. This is what allows different vowels to sound different. The "ee" sound is made by shaping your vocal tract to have standing-wave frequencies at, normally, 270 Hz and 2300 Hz. What will these frequencies be for a helium-oxygen mixture in which the speed of sound at body temperature is 750 m/s? The speed of sound in air at body temperature is 350 m/s.

53. ‖ In 1866, the German scientist Adolph Kundt developed a technique for accurately measuring the speed of sound in various gases. A long glass tube, known today as a Kundt's tube, has a vibrating piston at one end and is closed at the other. Very finely ground particles of cork are sprinkled in the bottom of the tube before the piston is inserted. As the vibrating piston is slowly moved forward, there are a few positions that cause the cork particles to collect in small, regularly spaced piles along the bottom. **FIGURE P17.53** shows an experiment in which the tube is filled with pure oxygen and the piston is driven at 400 Hz. What is the speed of sound in oxygen?

Piston   Glass   Piles of cork
            tube   particles

400 Hz

123 cm

FIGURE P17.53

54. || The 40-cm-long tube of FIG-
URE P17.54 has a 40-cm-long insert
that can be pulled in and out. A vibrating tuning fork is held next to the
tube. As the insert is slowly pulled
out, the sound from the tuning fork
creates standing waves in the tube

FIGURE P17.54

when the total length L is 42.5 cm, 56.7 cm, and 70.9 cm. What is
the frequency of the tuning fork? Assume $v_{sound} = 343$ m/s.

55. || A 1.0-m-tall vertical tube is filled with 20°C water. A tuning
fork vibrating at 580 Hz is held just over the top of the tube as the
water is slowly drained from the bottom. At what water heights,
measured from the bottom of the tube, will there be a standing
wave in the tube above the water?

56. ||| A 44-cm-diameter water tank is filled with 35 cm of water. A
CALC 3.0-mm-diameter spigot at the very bottom of the tank is opened
and water begins to flow out. The water falls into a 2.0-cm-diameter,
40-cm-tall glass cylinder. As the water falls and creates noise, resonance causes the column of air in the cylinder to produce a tone
at the column's fundamental frequency. What are (a) the frequency
and (b) the rate at which the frequency is changing (Hz/s) when the
cylinder has been filling for 4.0 s? You can assume that the height
of the water in the tank does not appreciably change in 4.0 s.

57. || A 25-cm-long wire with a linear density of 20 g/m passes
across the open end of an 85-cm-long open-closed tube of air. If
the wire, which is fixed at both ends, vibrates at its fundamental
frequency, the sound wave it generates excites the second vibrational mode of the tube of air. What is the tension in the wire?
Assume $v_{sound} = 340$ m/s.

58. || An old mining tunnel disappears into a hillside. You would like
to know how long the tunnel is, but it's too dangerous to go inside.
Recalling your recent physics class, you decide to try setting up
standing-wave resonances inside the tunnel. Using your subsonic
amplifier and loudspeaker, you find resonances at 4.5 Hz and 6.3 Hz,
and at no frequencies between these. It's rather chilly inside the
tunnel, so you estimate the sound speed to be 335 m/s. Based on
your measurements, how far is it to the end of the tunnel?

59. || Two in-phase loudspeakers emit identical 1000 Hz sound
waves along the x-axis. What distance should one speaker be placed
behind the other for the sound to have an amplitude 1.5 times that
of each speaker alone?

60. || Analyze the standing sound waves in an open-closed tube to
show that the possible wavelengths and frequencies are given by
Equation 17.18.

61. || Two loudspeakers emit sound waves of the same frequency
along the x-axis. The amplitude of each wave is a. The sound
intensity is minimum when speaker 2 is 10 cm behind speaker 1.
The intensity increases as speaker 2 is moved forward and first
reaches maximum, with amplitude $2a$, when it is 30 cm in front
of speaker 1. What is
a. The wavelength of the sound?
b. The phase difference between the two loudspeakers?
c. The amplitude of the sound (as a multiple of a) if the speakers
are placed side by side?

62. || Two loudspeakers emit sound waves along the x-axis. A lis-
tener in front of both speakers hears a maximum sound intensity
when speaker 2 is at the origin and speaker 1 is at $x = 0.50$ m. If
speaker 1 is slowly moved forward, the sound intensity decreases
and then increases, reaching another maximum when speaker 1 is
at $x = 0.90$ m.
a. What is the frequency of the sound? Assume $v_{sound} = 340$ m/s.
b. What is the phase difference between the speakers?

63. || A sheet of glass is coated with a 500-nm-thick layer of oil
($n = 1.42$).
a. For what *visible* wavelengths of light do the reflected waves
interfere constructively?
b. For what *visible* wavelengths of light do the reflected waves
interfere destructively?
c. What is the color of reflected light? What is the color of transmitted light?

64. || A manufacturing firm has hired your company, Acoustical Con-
sulting, to help with a problem. Their employees are complaining
about the annoying hum from a piece of machinery. Using a frequency meter, you quickly determine that the machine emits a rather
loud sound at 1200 Hz. After investigating, you tell the owner that
you cannot solve the problem entirely, but you can at least improve
the situation by eliminating reflections of this sound from the walls.
You propose to do this by installing mesh screens in front of the
walls. A portion of the sound will reflect from the mesh; the rest will
pass through the mesh and reflect from the wall. How far should the
mesh be placed in front of the wall for this scheme to work?

65. || A soap bubble is essentially a very thin film of water ($n =
1.33$) surrounded by air. The colors that you see in soap bubbles
are produced by interference.
a. Derive an expression for the wavelengths λ_C for which constructive interference causes a strong reflection from a soap
bubble of thickness d.
Hint: Think about the reflection phase shifts at both boundaries.
b. What visible wavelengths of light are strongly reflected from
a 390-nm-thick soap bubble? What color would such a soap
bubble appear to be?

66. || Engineers are testing a new thin-film coating whose index of re-
fraction is less than that of glass. They deposit a 560-nm-thick layer on
glass, then shine lasers on it. A red laser with a wavelength of 640 nm
has no reflection at all, but a violet laser with a wavelength of 400 nm
has a maximum reflection. How the coating behaves at other wavelengths is unknown. What is the coating's index of refraction?

67. || Scientists are testing a transparent material whose in-
dex of refraction for visible light varies with wavelength as
$n = 30.0 \text{ nm}^{1/2}/\lambda^{1/2}$, where λ is in nm. If a 295-nm-thick coating
is placed on glass ($n = 1.50$) for what visible wavelengths will
the reflected light have maximum constructive interference?

68. || You are standing 2.5 m
directly in front of one of the
two loudspeakers shown in
FIGURE P17.68. They are 3.0
m apart and both are playing a
686 Hz tone in phase. As you
begin to walk directly away
from the speaker, at what distances from the speaker do you
hear a *minimum* sound intensity?
The room temperature is 20°C.

FIGURE P17.68

69. || Two loudspeakers in a plane, 5.0 m apart, are playing the same
frequency. If you stand 12.0 m in front of the plane of the speakers, centered between them, you hear a sound of maximum intensity. As you walk parallel to the plane of the speakers, staying
12.0 m in front of them, you first hear a minimum of sound intensity when you are directly in front of one of the speakers. What is
the frequency of the sound? Assume a sound speed of 340 m/s.

70. || Two identical loudspeakers separated by distance Δx each
emit sound waves of wavelength λ and amplitude a along the
x-axis. What is the minimum value of the ratio $\Delta x/\lambda$ for which
the amplitude of their superposition is also a?

71. ‖ The three identical loudspeakers in **FIGURE P17.71** play a 170 Hz tone in a room where the speed of sound is 340 m/s. You are standing 4.0 m in front of the middle speaker. At this point, the amplitude of the wave from each speaker is a.

3.0 m

3.0 m

4.0 m

FIGURE P17.71

   a. What is the amplitude at this point?
   b. How far must speaker 2 be moved to the left to produce a maximum amplitude at the point where you are standing?
   c. When the amplitude is maximum, by what factor is the sound intensity greater than the sound intensity from a single speaker?

72. ‖ Piano tuners tune pianos by listening to the beats between the *harmonics* of two different strings. When properly tuned, the note A should have a frequency of 440 Hz and the note E should be at 659 Hz.
   a. What is the frequency difference between the third harmonic of the A and the second harmonic of the E?
   b. A tuner first tunes the A string very precisely by matching it to a 440 Hz tuning fork. She then strikes the A and E strings simultaneously and listens for beats between the harmonics. What beat frequency indicates that the E string is properly tuned?
   c. The tuner starts with the tension in the E string a little low, then tightens it. What is the frequency of the E string when she hears four beats per second?

73. ‖ A flutist assembles her flute in a room where the speed of sound is 342 m/s. When she plays the note A, it is in perfect tune with a 440 Hz tuning fork. After a few minutes, the air inside her flute has warmed to where the speed of sound is 346 m/s.
   a. How many beats per second will she hear if she now plays the note A as the tuning fork is sounded?
   b. How far does she need to extend the "tuning joint" of her flute to be in tune with the tuning fork?

74. ‖ You have two small, identical boxes that generate 440 Hz
CALC notes. While holding one, you drop the other from a 20-m-high balcony. How many beats will you hear before the falling box hits the ground? You can ignore air resistance.

75. ‖ Two loudspeakers emit 400 Hz notes. One speaker sits on the ground. The other speaker is in the back of a pickup truck. You hear eight beats per second as the truck drives away from you. What is the truck's speed?

Challenge Problems

76. ‖‖ Two radio antennas are separated by 2.0 m. Both broadcast identical 750 MHz waves. If you walk around the antennas in a circle of radius 10 m, how many maxima will you detect?

77. ‖‖ A 280 Hz sound wave is directed into one end of the trombone slide seen in **FIG-URE CP17.77**. A microphone is placed at the other end to record the intensity of sound waves that are transmitted through the tube. The straight sides of the slide are 80 cm in length and 10 cm apart with a semicircular bend at the end. For what slide extensions s will the microphone detect a maximum of sound intensity?

80 cm

10 cm   80 cm

s

FIGURE CP17.77

78. ‖‖ As the captain of the scientific team sent to Planet Physics, one of your tasks is to measure g. You have a long, thin wire labeled 1.00 g/m and a 1.25 kg weight. You have your accurate space cadet chronometer but, unfortunately, you seem to have forgotten a meter stick. Undeterred, you first find the midpoint of the wire by folding it in half. You then attach one end of the wire to the wall of your laboratory, stretch it horizontally to pass over a pulley at the midpoint of the wire, then tie the 1.25 kg weight to the end hanging over the pulley. By vibrating the wire, and measuring time with your chronometer, you find that the wire's second-harmonic frequency is 100 Hz. Next, with the 1.25 kg weight still tied to one end of the wire, you attach the other end to the ceiling to make a pendulum. You find that the pendulum requires 314 s to complete 100 oscillations. Pulling out your trusty calculator, you get to work. What value of g will you report back to headquarters?

79. ‖‖ When mass M is tied to the bottom of a long, thin wire suspended from the ceiling, the wire's second-harmonic frequency is 200 Hz. Adding an additional 1.0 kg to the hanging mass increases the second-harmonic frequency to 245 Hz. What is M?

80. ‖‖ Ultrasound has many medical applications, one of which is to mon-
BIO itor fetal heartbeats by reflecting ultrasound off a fetus in the womb.
   a. Consider an object moving at speed v_o toward an at-rest source that is emitting sound waves of frequency f_0. Show that the reflected wave (i.e., the echo) that returns to the source has a Doppler-shifted frequency

$$f_{echo} = \left(\frac{v + v_o}{v - v_o}\right) f_0$$

   where v is the speed of sound in the medium.
   b. Suppose the object's speed is much less than the wave speed: $v_o \ll v$. Then $f_{echo} \approx f_0$, and a microphone that is sensitive to these frequencies will detect a beat frequency if it listens to f_0 and f_{echo} simultaneously. Use the binomial approximation and other appropriate approximations to show that the beat frequency is $f_{beat} \approx (2v_o/v)f_0$.
   c. The reflection of 2.40 MHz ultrasound waves from the surface of a fetus's beating heart is combined with the 2.40 MHz wave to produce a beat frequency that reaches a maximum of 65 Hz. What is the maximum speed of the surface of the heart? The speed of ultrasound waves within the body is 1540 m/s.
   d. Suppose the surface of the heart moves in simple harmonic motion at 90 beats/min. What is the amplitude in mm of the heartbeat?

81. ‖‖ A water wave is called a *deep-water wave* if the water's depth is more than one-quarter of the wavelength. Unlike the waves we've considered in this chapter, the speed of a deep-water wave depends on its wavelength:

$$v = \sqrt{\frac{g\lambda}{2\pi}}$$

Longer wavelengths travel faster. Let's apply this to standing waves. Consider a diving pool that is 5.0 m deep and 10.0 m wide. Standing water waves can set up across the width of the pool. Because water sloshes up and down at the sides of the pool, the boundary conditions require antinodes at $x = 0$ and $x = L$. Thus a standing water wave resembles a standing sound wave in an open-open tube.
   a. What are the wavelengths of the first three standing-wave modes for water in the pool? Do they satisfy the condition for being deep-water waves?
   b. What are the wave speeds for each of these waves?
   c. Derive a general expression for the frequencies f_m of the possible standing waves. Your expression should be in terms of m, g, and L.
   d. What are the oscillation *periods* of the first three standing wave modes?

Oscillations and Waves

KEY FINDINGS What are the overarching findings of Part IV?

- Particles are
  - Localized
  - Discrete
  - Two particles cannot occupy the same point in space.

- Waves are
  - Diffuse
  - Spread out
  - Two waves can pass through each other.

LAWS What laws of physics govern oscillations and waves?

Newton's second law   SHM: $d^2x/dt^2 = -\omega^2 x$ Wave equation: $\partial^2 D/\partial t^2 = v^2 \partial^2 D/\partial x^2$

Conservation of energy   For SHM: $E = \frac{1}{2}mv^2 + \frac{1}{2}kx^2 = \frac{1}{2}m(v_{max})^2 = \frac{1}{2}kA^2$

Fundamental relationship for sinusoidal waves $v = \lambda f = \omega/k$

Principle of superposition   The net displacement is the sum of the displacements due to each wave.

MODELS What are the most important models of Part IV?

Simple harmonic motion

- Any object with a **linear restoring force** can undergo SHM. This is sinusoidal motion with

  $x = A\cos(\omega t + \phi_0)$

  $v = -v_{max}\sin(\omega t + \phi_0)$

- Mechanical energy is conserved if there's no friction.

- With friction, the oscillations are damped. A simple model of **damping** predicts oscillations that decay exponentially with time.

- **Resonance** is a large-amplitude response when an oscillator is driven at its natural frequency.

Waves

- A wave is a disturbance that travels.
  - **Mechanical waves** travel through a medium.
  - **Electromagnetic waves** travel through a vacuum.
  - Waves can be **transverse** or **longitudinal.**

- Wave speed is a property of the medium.

- Sinusoidal waves are periodic in both time (period) and space (wavelength). They obey $v = \lambda f$.

- Waves obey the principle of superposition.

Transverse wave

Longitudinal wave

History graph

Snapshot graph

TOOLS What are the most important tools introduced in Part IV?

Oscillation period
- Frequency $f = 1/T$
- Angular frequency $\omega = 2\pi f = 2\pi/T$
- Spring $T = 2\pi\sqrt{m/k}$
- Pendulum $T = 2\pi\sqrt{L/g}$
- Wave $f = v/\lambda$

Sinusoidal wave
- Displacement is a function of x and t:

  $D(x,t) = A\sin(kx \mp \omega t + \phi_0)$

- $-\omega t$ for motion to the right
- $+\omega t$ for motion to the left
- The *wave number* is $k = 2\pi/\lambda$

Sound intensity level

$\beta = (10\text{ dB})\log_{10}(I/10^{-12}\text{ W/m}^2)$

Wave speed
- String $v = \sqrt{T_s/\mu}$
- Sound $v = \sqrt{B/\rho}$

Phase
- The quantity $\omega t + \phi_0$ is called the phase ϕ of SHM.
- The quantity $k - \omega t + \phi_0$ is the phase ϕ of a sinusoidal wave.
- The *phase constant* ϕ_0 is given by the initial conditions.

Doppler effect

A frequency shift when the source moves relative to an observer:

  $f = f_0/(1 \mp v_s/v)$

for an approaching/receding source.

Standing waves

Two waves traveling in opposite directions.

- Strings, open-open tubes, and closed-closed tubes:

  $f = mf_1$

  $f_1 = v/2L$ = fundamental frequency
- Standing waves have points that never move called *nodes.*

$m = 1$

$m = 2$

Interference
- *Constructive interference* if *in phase*

  $\Delta\phi = m \cdot 2\pi$ rad
- *Destructive interference* if *out of phase*

  $\Delta\phi = \left(m + \frac{1}{2}\right) \cdot 2\pi$ rad
- *Beats* $f_{beat} = |f_1 - f_2|$ if frequencies differ.

PART

V Thermodynamics

OVERVIEW

It's All About Energy

Thermodynamics—the science of energy in its broadest context—arose hand in hand with the industrial revolution as the systematic study of converting heat energy into mechanical motion and work. Hence the name *thermo + dynamics*. Indeed, the analysis of engines and generators of various kinds remains the focus of engineering thermodynamics. But thermodynamics, as a science, now extends to all forms of energy conversions, including those involving living organisms. For example:

- **Engines** convert the energy of a fuel into the mechanical energy of moving pistons, gears, and wheels.
- **Fuel cells** convert chemical energy into electrical energy.
- **Photovoltaic cells** convert the electromagnetic energy of light into electrical energy.
- **Lasers** convert electrical energy into the electromagnetic energy of light.
- **Organisms** convert the chemical energy of food into a variety of other forms of energy, including kinetic energy, sound energy, and thermal energy.

The major goals of Part V are to understand both *how* energy transformations such as these take place and *how efficient* they are. We'll discover that the laws of thermodynamics place limits on the efficiency of energy transformations, and understanding these limits is essential for analyzing the very real energy needs of society in the 21st century.

Our ultimate destination in Part V is an understanding of the thermodynamics of *heat engines*. A heat engine is a device, such as a power plant or an internal combustion engine, that transforms heat energy into useful work. These are the devices that power our modern society.

Understanding how to transform heat into work will be a significant achievement, but we first have many steps to take along the way. We need to understand the concepts of temperature and pressure. We need to learn about the properties of solids, liquids, and gases. Most important, we need to expand our view of energy to include *heat*, the energy that is transferred between two systems at different temperatures.

At a deeper level, we need to see how these concepts are connected to the underlying microphysics of randomly moving molecules. We will find that the familiar concepts of thermodynamics, such as temperature and pressure, have their roots in atomic-level motion and collisions. This *micro/macro connection* will lead to the second law of thermodynamics, one of the most subtle but also one of the most profound and far-reaching statements in physics.

Only after all these steps have been taken will we be able to analyze a real heat engine. It is an ambitious goal, but one we can achieve.

Smoke particles allow us to visualize *convection,* one of the ways in which heat is transferred from one place to another.

18 A Macroscopic Description of Matter

The phrase "solid as a rock" is cast in doubt when rocks melt, as they do in this flowing lava.

IN THIS CHAPTER, you will learn some of the characteristics of macroscopic systems.

What are the phases of matter?

Most materials can exist as a solid, a liquid, or a gas. These are the most common phases of matter.

Starting with this chapter and continuing through Part V you will come to understand that the macroscopic properties of matter, such as volume, density, pressure, and temperature, can often be understood in terms of the microscopic motions of their atoms and molecules. This micro/macro connection is an important part of our modern understanding of matter.

« LOOKING BACK Sections 14.1–14.3 Fluids and pressure

What is temperature?

You're familiar with temperature, but what does it actually measure? We'll start with the simple idea that temperature measures "hotness" and "coldness," but we'll come to recognize that temperature measures a system's thermal energy. We'll study the well-known fact that objects expand or contract when the temperature changes.

What is an ideal gas?

We'll model a gas as consisting of tiny, hard spheres that occasionally collide but otherwise do not interact with each other. This ideal gas obeys a law relating four state variables—the ideal-gas law:

$$pV = nRT$$

You will use the ideal-gas law to analyze what happens when a gas changes state.

What is an ideal-gas process?

Heating or compressing a gas is a *process* that changes the state of the gas. An ideal-gas process can be shown as a trajectory through a pV diagram. We'll study three basic processes:

- Constant-volume process
- Constant-pressure process
- Constant-temperature process

Why are macroscopic properties important?

Physicists, chemists, biologists, and engineers all work with matter at the macroscopic level. Everything from basic science to engineering design depends on knowing how materials respond when they are heated, compressed, melted, or otherwise changed by factors in their environment. Changing states of matter underlie devices ranging from car engines to power plants to spacecraft.

18.1 Solids, Liquids, and Gases

In Part V we will study the properties of matter itself, as opposed to the motion of matter. We will focus on a *macroscopic description* of large quantities of matter. Even so, part of our modern understanding of matter is that macroscopic properties, such as pressure and temperature, have their basis in the microscopic motions of atoms and molecules, and we'll spend some time exploring this **micro/macro connection.**

As you know, each of the elements and most compounds can exist as a solid, liquid, or gas—the three most common **phases** of matter. The change between liquid and solid (freezing or melting) or between liquid and gas (boiling or condensing) is called a **phase change.** Water is the only substance for which all three phases—ice, liquid, and steam—are everyday occurrences.

NOTE This use of the word "phase" has no relationship at all to the *phase* or *phase constant* of simple harmonic motion and waves.

MODEL 18.1

Solids, liquids, and gases

Atoms vibrate around equilibrium positions.

Atoms are held close together by weak molecular bonds, but they can slide around each other.

Atoms are far apart and travel freely through space except for occasional collisions.

Solid ⟷ Melt / Freeze ⟷ Liquid ⟷ Boil / Condense ⟷ Gas

A **solid** is a rigid macroscopic system consisting of particle-like atoms connected by spring-like molecular bonds. Solids are nearly *incompressible,* which tells us that the atoms in a solid are just about as close together as they can get.

The solid shown here is a **crystal,** meaning that the atoms are arranged in a periodic array. Elements and many compounds have a crystal structure in their solid phase.

A **liquid** is a system in which the molecules are loosely held together by weak molecular bonds. The bonds are strong enough that the molecules never get far apart but not strong enough to prevent the molecules from sliding around each other.

A liquid is more complicated than either a solid or a gas. Like a solid, a liquid is nearly *incompressible.* Like a gas, a liquid flows and deforms to fit the shape of its container.

A **gas** is a system in which each molecule moves through space as a free, noninteracting particle until, on occasion, it collides with another molecule or with the wall of the container. A gas is a *fluid.* A gas is also highly *compressible,* which tells us that there is lots of space between the molecules.

Gases are fairly simple macroscopic systems; hence many of our examples in Part V will be based on gases.

State Variables

The parameters used to characterize or describe a macroscopic system are known as **state variables** because, taken all together, they describe the *state* of the macroscopic system. You met some state variables in earlier chapters: volume, pressure, mass, mass density, and thermal energy. We'll soon introduce several new state variables.

One important state variable, the mass density, is defined as the ratio of two other state variables:

$$\rho = \frac{M}{V} \quad \text{(mass density)} \tag{18.1}$$

In this chapter we'll use an uppercase M for the system mass and a lowercase m for the mass of an atom. **TABLE 18.1** is a short list of mass densities.

A system is said to be in **thermal equilibrium** if its state variables are constant and not changing. As an example, a gas is in thermal equilibrium if it has been left undisturbed long enough for p, V, and T to reach steady values.

TABLE 18.1 Densities of materials

Substance	ρ (kg/m^3)
Air at STP*	1.29
Ethyl alcohol	790
Water (solid)	920
Water (liquid)	1000
Aluminum	2700
Copper	8920
Gold	19,300
Iron	7870
Lead	11,300
Mercury	13,600
Silicon	2330

*$T = 0°C$, $p = 1$ atm

EXAMPLE 18.1 | **The mass of a lead pipe**

A project on which you are working uses a cylindrical lead pipe with outer and inner diameters of 4.0 cm and 3.5 cm, respectively, and a length of 50 cm. What is its mass?

SOLVE The mass density of lead is $\rho_{lead} = 11{,}300 \text{ kg/m}^3$. The volume of a circular cylinder of length l is $V = \pi r^2 l$. In this case we need to find the volume of the outer cylinder, of radius r_2, *minus*

the volume of air in the inner cylinder, of radius r_1. The volume of the pipe is

$$V = \pi r_2^2 l - \pi r_1^2 l = \pi(r_2^2 - r_1^2)l = 1.47 \times 10^{-4} \text{ m}^3$$

Hence the pipe's mass is

$$M = \rho_{lead} V = 1.7 \text{ kg}$$

STOP TO THINK 18.1   The pressure in a system is measured to be 60 kPa. At a later time the pressure is 40 kPa. The value of Δp is

a. 60 kPa   b. 40 kPa   c. 20 kPa   d. −20 kPa

18.2 Atoms and Moles

The mass of a macroscopic system is directly related to the total number of atoms or molecules in the system, denoted N. Because N is determined simply by counting, it is a number with no units. A typical macroscopic system has $N \sim 10^{25}$ atoms, an incredibly large number.

The symbol $\sim$, if you are not familiar with it, stands for "has the order of magnitude." It means that the number is known only to within a factor of 10 or so. The statement $N \sim 10^{25}$, which is read "N is of order 10^{25}," implies that N is somewhere in the range 10^{24} to 10^{26}. It is far less precise than the "approximately equal" symbol $\approx$. Saying $N \sim 10^{25}$ gives us a rough idea of how large N is and allows us to know that it differs significantly from 10^5 or even 10^{15}.

It is often useful to know the number of atoms or molecules per cubic meter in a system. We call this quantity the **number density.** It characterizes how densely the atoms are packed together within the system. In an N-atom system that fills volume V, the number density is

$$\frac{N}{V} \quad \text{(number density)} \tag{18.2}$$

The SI units of number density are m^{-3}. The number density of atoms in a solid is $(N/V)_{solid} \sim 10^{29} \text{ m}^{-3}$. The number density of a gas depends on the pressure, but is usually less than 10^{27} m^{-3}. As **FIGURE 18.1** shows, **the value of N/V in a *uniform* system is independent of the volume V.** That is, the number density is the same whether you look at the whole system or just a portion of it.

> **NOTE** While we might say "There are 100 tennis balls per cubic meter," or "There are 10^{29} atoms per cubic meter," tennis balls and atoms are not units. The units of N/V are simply m^{-3}.

FIGURE 18.1 The number density of a uniform system is independent of the volume.

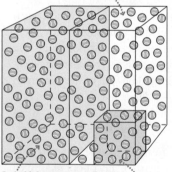

A 100 m^3 room contains 10,000 tennis balls. The number density of balls in the room is
$N/V = 10{,}000/100 \text{ m}^3 = 100 \text{ m}^{-3}$

In half the room, we find 5000 balls in 50 m^3:
$N/V = 5000/50 \text{ m}^3 = 100 \text{ m}^{-3}$

In one-tenth of the room, we find 1000 balls in 10 m^3:
$N/V = 1000/10 \text{ m}^3 = 100 \text{ m}^{-3}$

Atomic Mass and Atomic Mass Number

You will recall from chemistry that atoms of different elements have different masses. The mass of an atom is determined primarily by its most massive constituents, the protons and neutrons in its nucleus. The *sum* of the number of protons and neutrons is called the **atomic mass number** A:

$$A = \text{number of protons} + \text{number of neutrons}$$

A, which by definition is an integer, is written as a leading superscript on the atomic symbol. For example, the common isotope of hydrogen, with one proton and no neutrons, is ^1H. The "heavy hydrogen" isotope called *deuterium*, which includes one neutron, is ^{2}H. The primary isotope of carbon, with six protons (which makes it carbon) and six neutrons, is ^{12}C. The radioactive isotope ^{14}C, used for carbon dating of archeological finds, contains six protons and eight neutrons.

The **atomic mass** scale is established by defining the mass of ^{12}C to be exactly 12 u, where u is the symbol for the **atomic mass unit.** That is, $m(^{12}C) = 12$ u. The atomic mass of any other atom is its mass relative to ^{12}C. For example, careful experiments with hydrogen find that the mass *ratio* $m(^1H)/m(^{12}C)$ is 1.0078/12. Thus the atomic mass of hydrogen is $m(^1H) = 1.0078$ u.

The numerical value of the atomic mass of ^{1}H is close to, but not exactly, its atomic mass number $A = 1$. For our purposes, it will be sufficient to overlook the slight difference and **use the integer atomic mass numbers as the values of the atomic mass.** That is, we'll use $m(^1H) = 1$ u, $m(^4He) = 4$ u, and $m(^{16}O) = 16$ u. For molecules, the **molecular mass** is the sum of the atomic masses of the atoms forming the molecule. Thus the molecular mass of O_2, the constituent of oxygen gas, is $m(O_2) = 32$ u.

NOTE An element's atomic mass number is *not* the same as its atomic number. The *atomic number,* the element's position in the periodic table, is the number of protons in the nucleus.

TABLE 18.2 shows the atomic mass numbers of some of the elements that we'll use for examples and homework problems. A complete periodic table of the elements, including atomic masses, is found in Appendix B.

Moles and Molar Mass

One way to specify the amount of substance in a macroscopic system is to give its mass. Another is to measure the amount of substance in *moles.* By definition, one **mole** of matter, be it solid, liquid, or gas, is the amount of substance containing as many basic particles as there are atoms in 0.012 kg (12 g) of ^{12}C. Many ingenious experiments have determined that there are 6.02×10^{23} atoms in 0.012 kg of ^{12}C, so we can say that 1 mole of substance, abbreviated 1 mol, is 6.02×10^{23} basic particles.

The basic particle depends on the substance. Helium is a **monatomic gas,** meaning that the basic particle is the helium atom. Thus 6.02×10^{23} helium atoms are 1 mol of helium. But oxygen gas is a **diatomic gas** because the basic particle is the two-atom diatomic molecule O_2. 1 mol of oxygen gas contains 6.02×10^{23} *molecules* of O_2 and thus $2 \times 6.02 \times 10^{23}$ oxygen atoms. **TABLE 18.3** lists the monatomic and diatomic gases that we will use for examples and homework problems.

The number of basic particles per mole of substance is called **Avogadro's number,** N_A. The value of Avogadro's number is

$$N_A = 6.02 \times 10^{23} \text{ mol}^{-1}$$

Despite its name, Avogadro's number is not simply "a number"; it has units. Because there are N_A particles per mole, the number of moles in a substance containing N basic particles is

$$n = \frac{N}{N_A} \quad \text{(moles of substance)} \qquad (18.3)$$

Avogadro's number allows us to determine atomic masses in kilograms. Knowing that N_A ^{12}C atoms have a mass of 0.012 kg, the mass of one ^{12}C atom must be

$$m(^{12}C) = \frac{0.012 \text{ kg}}{6.02 \times 10^{23}} = 1.993 \times 10^{-26} \text{ kg}$$

We defined the atomic mass scale such that $m(^{12}C) = 12$ u. Thus the conversion factor between atomic mass units and kilograms is

$$1 \text{ u} = \frac{m(^{12}C)}{12} = 1.66 \times 10^{-27} \text{ kg}$$

One mole of helium, sulfur, copper, and mercury.

TABLE 18.2 Some atomic mass numbers

Element		A
^{1}H	Hydrogen	1
^{4}He	Helium	4
^{12}C	Carbon	12
^{14}N	Nitrogen	14
^{16}O	Oxygen	16
^{20}Ne	Neon	20
^{27}Al	Aluminum	27
^{40}Ar	Argon	40
^{207}Pb	Lead	207

TABLE 18.3 Monatomic and diatomic gases

Monatomic		Diatomic	
He	Helium	H_2	Hydrogen
Ne	Neon	N_2	Nitrogen
Ar	Argon	O_2	Oxygen

This conversion factor allows us to calculate the mass in kg of any atom. For example, a ^{20}Ne atom has atomic mass $m(^{20}\text{Ne}) = 20$ u. Multiplying by 1.66×10^{-27} kg/u gives $m(^{20}\text{Ne}) = 3.32 \times 10^{-26}$ kg. If the atomic mass is specified in kilograms, the number of atoms in a system of mass M can be found from

$$N = \frac{M}{m} \qquad (18.4)$$

The **molar mass** of a substance is the mass of 1 mol of substance. The molar mass, which we'll designate M_{mol}, has units kg/mol. By definition, the molar mass of ^{12}C is 0.012 kg/mol. For other substances, whose atomic or molecular masses are given relative to ^{12}C, the numerical value of the molar mass is the numerical value of the atomic or molecular mass divided by 1000. For example, the molar mass of He, with $m = 4$ u, is $M_{\text{mol}}(\text{He}) = 0.004$ kg/mol and the molar mass of diatomic O_2 is $M_{\text{mol}}(O_2) = 0.032$ kg/mol.

Equation 18.4 uses the atomic mass to find the number of atoms in a system. Similarly, you can use the molar mass to determine the number of moles. For a system of mass M consisting of atoms or molecules with molar mass M_{mol},

$$n = \frac{M}{M_{\text{mol}}} \qquad (18.5)$$

EXAMPLE 18.2 | Moles of oxygen

100 g of oxygen gas is how many moles of oxygen?

SOLVE We can do the calculation two ways. First, let's determine the number of molecules in 100 g of oxygen. The diatomic oxygen molecule O_2 has molecular mass $m = 32$ u. Converting this to kg, we get the mass of one molecule:

$$m = 32 \text{ u} \times \frac{1.66 \times 10^{-27} \text{ kg}}{1 \text{ u}} = 5.31 \times 10^{-26} \text{ kg}$$

Thus the number of molecules in 100 g = 0.100 kg is

$$N = \frac{M}{m} = \frac{0.100 \text{ kg}}{5.31 \times 10^{-26} \text{ kg}} = 1.88 \times 10^{24}$$

Knowing the number of molecules gives us the number of moles:

$$n = \frac{N}{N_A} = 3.13 \text{ mol}$$

Alternatively, we can use Equation 18.5 to find

$$n = \frac{M}{M_{\text{mol}}} = \frac{0.100 \text{ kg}}{0.032 \text{ kg/mol}} = 3.13 \text{ mol}$$

STOP TO THINK 18.2 Which system contains more atoms: 5 mol of helium ($A = 4$) or 1 mol of neon ($A = 20$)?

a. Helium.　　　　　b. Neon.　　　　　c. They have the same number of atoms.

Thermal expansion of the liquid in the thermometer tube pushes it higher in the hot water than in the ice water.

18.3 Temperature

We are all familiar with the idea of temperature. Mass is a measure of the amount of substance in a system. Velocity is a measure of how fast a system moves. What physical property of the system have you determined if you measure its temperature?

We will begin with the commonsense idea that temperature is a measure of how "hot" or "cold" a system is. As we develop these ideas, we'll find that **temperature** T is related to a system's *thermal energy*. We defined thermal energy in Chapter 9 as the kinetic and potential energy of the atoms and molecules in a system as they vibrate (a solid) or move around (a gas). A system has more thermal energy when it is "hot" than when it is "cold." In Chapter 20, we'll replace these vague notions of hot and cold with a precise relationship between temperature and thermal energy.

To start, we need a means to measure the temperature of a system. This is what a *thermometer* does. A thermometer can be any small macroscopic system that undergoes a measurable change as it exchanges thermal energy with its surroundings. It is placed in contact with a larger system whose temperature it will measure. In a common glass-tube thermometer, for example, a small volume of mercury or alcohol expands or contracts when placed in contact with a "hot" or "cold" object. The object's temperature is determined by the length of the column of liquid.

A thermometer needs a *temperature scale* to be a useful measuring device. In 1742, the Swedish astronomer Anders Celsius sealed mercury into a small capillary tube and observed how it moved up and down the tube as the temperature changed. He selected two temperatures that anyone could reproduce, the freezing and boiling points of pure water, and labeled them 0 and 100. He then marked off the glass tube into one hundred equal intervals between these two reference points. By doing so, he invented the temperature scale that we today call the *Celsius scale*. The units of the Celsius temperature scale are "degrees Celsius," which we abbreviate °C. Note that the degree symbol ° is part of the unit, not part of the number.

The *Fahrenheit scale,* still widely used in the United States, is related to the Celsius scale by

$$T_F = \tfrac{9}{5}T_C + 32° \tag{18.6}$$

FIGURE 18.2 shows several temperatures measured on the Celsius and Fahrenheit scales and also on the Kelvin scale.

FIGURE 18.2 Temperatures measured with different scales.

	°F	°C	K
Water boils	212	100	373
Normal body temp	99	37	310
Room temperature	68	20	293
Water freezes	32	0	273
CO_2 sublimates	−109	−78	195
Nitrogen boils	−321	−196	77
Absolute zero	−460	−273	0

Absolute Zero and Absolute Temperature

Any physical property that changes with temperature can be used as a thermometer. In practice, the most useful thermometers have a physical property that changes *linearly* with temperature. One of the most important scientific thermometers is the **constant-volume gas thermometer** shown in FIGURE 18.3a. This thermometer depends on the fact that the *absolute* pressure (not the gauge pressure) of a gas in a sealed container increases linearly as the temperature increases.

A gas thermometer is first calibrated by recording the pressure at two reference temperatures, such as the boiling and freezing points of water. These two points are plotted on a pressure-versus-temperature graph and a straight line is drawn through them. The gas bulb is then brought into contact with the system whose temperature is to be measured. The pressure is measured, then the corresponding temperature is read off the graph.

FIGURE 18.3b shows the pressure-temperature relationship for three different gases. Notice two important things about this graph.

1. There is a *linear* relationship between temperature and pressure.
2. All gases extrapolate to *zero pressure* at the same temperature: $T_0 = −273°C$. No gas actually gets that cold without condensing, although helium comes very close, but it is surprising that you get the same zero-pressure temperature for any gas and any starting pressure.

The pressure in a gas is due to collisions of the molecules with each other and the walls of the container. A pressure of zero would mean that all motion, and thus all collisions, had ceased. If there were no atomic motion, the system's thermal energy would be zero. The temperature at which all motion would cease, and at which $E_{th} = 0$, is called **absolute zero.** Because temperature is related to thermal energy, absolute zero is the lowest temperature that has physical meaning. We see from the gas-thermometer data that $T_0 = −273°C$.

It is useful to have a temperature scale with the zero point at absolute zero. Such a temperature scale is called an **absolute temperature scale.** Any system whose temperature is measured on an absolute scale will have $T > 0$. The absolute temperature scale having the same unit size as the Celsius scale is called the *Kelvin scale*. It is the SI scale of temperature. The units of the Kelvin scale are *kelvins,* abbreviated as K. The conversion between the Celsius scale and the Kelvin scale is

$$T_K = T_C + 273 \tag{18.7}$$

FIGURE 18.3 The pressure in a constant-volume gas thermometer extrapolates to zero at $T_0 = −273°C$.

(a)   Pressure gauge reading absolute pressure

Rigid gas-filled sphere

System whose temperature is to be measured

(b)

Each gas thermometer is calibrated at 0°C and 100°C.

Gas 1
Gas 2
Gas 3

Condensation points

$T_0 = −273°C$

On the Kelvin scale, absolute zero is 0 K, the freezing point of water is 273 K, and the boiling point of water is 373 K.

NOTE The units are simply "kelvins," *not* "degrees Kelvin."

STOP TO THINK 18.3 The temperature of a glass of water increases from 20°C to 30°C. What is ΔT?

a. 10 K   b. 283 K   c. 293 K   d. 303 K

18.4 Thermal Expansion

FIGURE 18.4 An object's length changes when the temperature changes.

Before: Temperature T

After: Temperature $T + \Delta T$

L ΔL

Objects expand when heated. This **thermal expansion** is why the liquid rises in a thermometer and why pipes, highways, and bridges have expansion joints. **FIGURE 18.4** shows an object of length L that changes by ΔL when the temperature is changed from T to $T + \Delta T$. For most solids, the *fractional* change in length, $\Delta L/L$, is proportional to the temperature change ΔT with a proportionality coefficient that depends on the material. That is,

$$\frac{\Delta L}{L} = \alpha \, \Delta T \qquad (18.8)$$

where α (Greek alpha) is the material's **coefficient of linear expansion.** Equation 18.8 characterizes both expansion ($\Delta L > 0$ if the temperature increases) and contraction ($\Delta L < 0$ if the temperature falls).

TABLE 18.4 gives the coefficients of linear expansion for a few common materials and also for a metal alloy called *invar* that is specifically designed to have extremely low thermal expansion. Because the fractional change in length is dimensionless, α has units of °C^{-1}, read as "per degree Celsius." The units may also be written K^{-1}, because a temperature *change* ΔT is the same in °C and K, but practical measurements are made in °C, not K.

TABLE 18.4 Coefficients of linear and volume expansion

Material	α (°C^{-1})
Aluminum	2.3×10^{-5}
Brass	1.9×10^{-5}
Concrete	1.2×10^{-5}
Steel	1.1×10^{-5}
Invar	0.09×10^{-5}

Material	β (°C^{-1})
Gasoline	9.6×10^{-4}
Mercury	1.8×10^{-4}
Ethyl alcohol	1.1×10^{-4}

EXAMPLE 18.3 | **An expanding pipe**

A 55-m-long steel pipe runs from one side of a refinery to the other. By how much does the pipe expand on a 5°C winter day when 155°C oil is pumped through it?

SOLVE The expansion is given by Equation 18.8, with the coefficient of linear expansion for steel taken from Table 18.4:

$$\Delta L = \alpha L \, \Delta T = (1.1 \times 10^{-5} \, °\text{C}^{-1})(55 \text{ m})(150°\text{C})$$
$$= 0.091 \text{ m} = 9.1 \text{ cm}$$

ASSESS 9.1 cm is a very small fraction of 55 m, so the pipe as a whole has expanded very little. Nevertheless, 9.1 cm $\approx$ 3.5 in is a huge expansion in terms of the engineering of the pipe. Pipes like this have to be designed with flexible expansion joints that allow for thermal expansion and contraction.

Volume expansion is treated the same way. If an object's volume changes by ΔV during a temperature change ΔT, the *fractional* change in volume is

$$\frac{\Delta V}{V} = \beta \, \Delta T \qquad (18.9)$$

where β (Greek beta) is the material's **coefficient of volume expansion**.

Liquids are constrained by the shape of their container, so liquids are characterized by a volume-expansion coefficient but not by a linear-expansion coefficient. A few values are given in Table 18.4, where you can see that β, like α, has units of °C^{-1}.

Solids expand linearly in all three directions, as given by Equation 18.8, and in the process a solid changes its volume. Imagine a cube of edge length L and volume $V = L^3$. If the edge length changes by a very small amount dL, the volume changes by

$$dV = 3L^2 dL \qquad (18.10)$$

If we divide both sides by $V = L^3$, we see that

$$\frac{dV}{V} = 3\frac{dL}{L} \qquad (18.11)$$

That is, the fractional change in volume is three times the fractional change in the length of the sides. Consequently, a solid's coefficient of volume expansion is

$$\beta_{\text{solid}} = 3\alpha \qquad (18.12)$$

You may have noticed that Table 18.4 does not include water. Water is a very common substance, but—due to its molecular structure—water has some unusual properties that set it apart from other liquids. If you lower the temperature of water, its volume contracts as expected. But only until the temperature reaches 4°C. If you continue cooling water, from 4°C down to the freezing point at 0°C, the volume expands! Water has maximum density at 4°C, with slightly less density—due to the expanded volume—at both higher and lower temperatures. Consequently, the thermal expansion of water cannot be characterized by a single coefficient of thermal expansion.

STOP TO THINK 18.4   A steel plate has a 2.000-cm-diameter hole through it. If the plate is heated, does the diameter of the hole increase or decrease?

18.5 Phase Changes

The temperature inside the freezer compartment of a refrigerator is typically about −20°C. Suppose you were to remove a few ice cubes from the freezer, place them in a sealed container with a thermometer, then heat them, as FIGURE 18.5a shows. We'll assume that the heating is done so slowly that the inside of the container always has a single, well-defined temperature.

FIGURE 18.5b shows the temperature as a function of time. After steadily rising from the initial −20°C, the temperature remains fixed at 0°C for an extended period of time. This is the interval of time during which the ice melts. As it's melting, the ice temperature is 0°C and the liquid water temperature is 0°C. Even though the system is being heated, the liquid water temperature doesn't begin to rise until all the ice has melted. If you were to turn off the flame at any point, the system would remain a mixture of ice and liquid water at 0°C.

NOTE In everyday language, the three phases of water are called *ice, water,* and *steam.* That is, the term *water* implies the liquid phase. Scientifically, these are the solid, liquid, and gas phases of the compound called *water.* To be clear, we'll use the term *water* in the scientific sense of a collection of H_2O molecules. We'll say either *liquid* or *liquid water* to denote the liquid phase.

The thermal energy of a solid is the kinetic energy of the vibrating atoms plus the potential energy of the stretched and compressed molecular bonds. Melting occurs when the thermal energy gets so large that molecular bonds begin to break, allowing the atoms to move around. The temperature at which a solid becomes a liquid or, if the thermal energy is reduced, a liquid becomes a solid is called the **melting point** or the **freezing point.** Melting and freezing are *phase changes.*

A system at the melting point is in **phase equilibrium,** meaning that any amount of solid can coexist with any amount of liquid. Raise the temperature ever so slightly

FIGURE 18.5 Water is transformed from solid to liquid to gas.

(a)

(b)

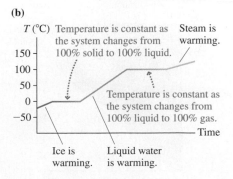

and the entire system becomes liquid. Lower it slightly and it all becomes solid. But exactly at the melting point the system has no tendency to move one way or the other. That is why the temperature remains constant at the melting point until the phase change is complete.

You can see the same thing happening in Figure 18.5b at 100°C, the boiling point. This is a phase equilibrium between the liquid phase and the gas phase, and any amount of liquid can coexist with any amount of gas at this temperature. Above this temperature, the thermal energy is too large for bonds to be established between molecules, so the system is a gas. If the thermal energy is reduced, the molecules begin to bond with each other and stick together. In other words, the gas condenses into a liquid. The temperature at which a gas becomes a liquid or, if the thermal energy is increased, a liquid becomes a gas is called the **condensation point** or the **boiling point.**

NOTE Liquid water becomes solid ice at 0°C, but that doesn't mean the temperature of ice is always 0°C. Ice reaches the temperature of its surroundings. If the air temperature in a freezer is −20°C, then the ice temperature is −20°C. Likewise, steam can be heated to temperatures above 100°C. That doesn't happen when you boil water on the stove because the steam escapes, but steam can be heated far above 100°C in a sealed container.

Phase Diagrams

A **phase diagram** is used to show how the phases and phase changes of a substance vary with both temperature and pressure. FIGURE 18.6 shows the phase diagrams for water and carbon dioxide. You can see that each diagram is divided into three regions corresponding to the solid, liquid, and gas phases. The boundary lines separating the regions indicate the phase transitions. The system is in phase equilibrium at a pressure-temperature point that falls on one of these lines.

Phase diagrams contain a great deal of information. Notice on the water phase diagram that the dashed line at $p = 1$ atm crosses the solid-liquid boundary at 0°C and the liquid-gas boundary at 100°C. These well-known melting and boiling point temperatures of water apply only at standard atmospheric pressure. You can see that in Denver, where $p_{atmos} < 1$ atm, water melts at slightly above 0°C and boils at a temperature below 100°C. A *pressure cooker* works by allowing the pressure inside to exceed 1 atm. This raises the boiling point, so foods that are in boiling water are at a temperature above 100°C and cook faster.

Crossing the solid-liquid boundary corresponds to melting or freezing while crossing the liquid-gas boundary corresponds to boiling or condensing. But there's another possibility—crossing the solid-gas boundary. The phase change in which a solid becomes a gas is called **sublimation.** It is not an everyday experience with water, but you probably are familiar with the sublimation of dry ice. Dry ice is solid carbon dioxide. You can see on the carbon dioxide phase diagram that the dashed line at $p = 1$ atm crosses the solid-*gas* boundary, rather than the solid-liquid boundary, at $T = -78$°C. This is the *sublimation temperature* of dry ice.

Liquid carbon dioxide does exist, but only at pressures greater than 5 atm and temperatures greater than −56°C. A CO_2 fire extinguisher contains *liquid* carbon dioxide under high pressure. (You can hear the liquid slosh if you shake a CO_2 fire extinguisher.)

One important difference between the water and carbon dioxide phase diagrams is the slope of the solid-liquid boundary. For most substances, the solid phase is denser than the liquid phase and the liquid is denser than the gas. Pressurizing the substance compresses it and increases the density. If you start compressing CO_2 gas at room temperature, thus moving upward through the phase diagram along a vertical line, you'll first condense it to a liquid and eventually, if you keep compressing, change it into a solid.

FIGURE 18.6 Phase diagrams (not to scale) for water and carbon dioxide.

Water

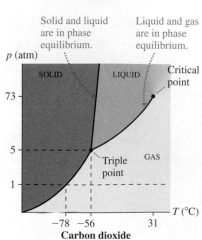

Carbon dioxide

Water is a very unusual substance in that the density of ice is *less* than the density of liquid water. That is why ice floats. If you compress ice, making it denser, you eventually cause a phase transition in which the ice turns to liquid water! Consequently, the solid-liquid boundary for water slopes to the left.

The liquid-gas boundary ends at a point called the **critical point.** Below the critical point, liquid and gas are clearly distinct and there is a phase change if you go from one to the other. But there is no clear distinction between liquid and gas at pressures or temperatures above the critical point. The system is a *fluid,* but it can be varied continuously between high density and low density without a phase change.

The final point of interest on the phase diagram is the **triple point** where the phase boundaries meet. Two phases are in phase equilibrium along the boundaries. The triple point is the *one* value of temperature and pressure for which all three phases can coexist in phase equilibrium. That is, any amounts of solid, liquid, and gas can happily coexist at the triple point. For water, the triple point occurs at $T_3 = 0.01°C$ and $p_3 = 0.006$ atm.

The significance of the triple point of water is its connection to the Kelvin temperature scale. The Celsius scale required two *reference points,* the boiling and melting points of water. We can now see that these are not very satisfactory reference points because their values vary as the pressure changes. In contrast, there's only one temperature at which ice, liquid water, and water vapor will coexist in equilibrium. If you produce this equilibrium in the laboratory, then you *know* the system is at the triple-point temperature.

The triple-point temperature of water is an ideal reference point, hence the Kelvin temperature scale is *defined* to be a linear temperature scale starting from 0 K at absolute zero and passing through 273.16 K at the triple point of water. Because $T_3 = 0.01°C$, absolute zero on the Celsius scale is $T_0 = -273.15°C$.

NOTE To be consistent with our use of significant figures, $T_0 = -273$ K is the appropriate value to use in calculations *unless* you know other temperatures with an accuracy of better than 1°C.

Food takes longer to cook at high altitudes because the boiling point of water is less than 100°C.

STOP TO THINK 18.5 For which is there a sublimation temperature that is higher than a melting temperature?

a. Water   b. Carbon dioxide   c. Both   d. Neither

18.6 Ideal Gases

We noted earlier in the chapter that solids and liquids are nearly incompressible, an observation suggesting that atoms are fairly hard and cannot be pressed together once they come into contact with each other. Based on this observation, suppose we were to model atoms as "hard spheres" that do not interact except for occasional elastic collisions when two atoms come into contact and bounce apart.

This is a *model* of an atom—what we might call the *ideal atom*—because it ignores the weak attractive interactions that hold liquids and solids together. A gas of these noninteracting atoms is called an **ideal gas.** It is a gas of small, hard, randomly moving atoms that bounce off each other and the walls of their container but otherwise do not interact. The ideal gas is a somewhat simplified description of a real gas, but experiments show that the ideal-gas model is quite good for real gases if two conditions are met:

1. The density is low (i.e., the atoms occupy a volume much smaller than that of the container), and
2. The temperature is well above the condensation point.

If the density gets too high, or the temperature too low, then the attractive forces between the atoms begin to play an important role and our model, which ignores those attractive forces, fails. These are the forces that are responsible, under the right conditions, for the gas condensing into a liquid.

We've been using the term "atoms," but many gases, as you know, consist of molecules rather than atoms. Only helium, neon, argon, and the other inert elements in the far-right column of the periodic table of the elements form monatomic gases. Hydrogen (H_2), nitrogen (N_2), and oxygen (O_2) are diatomic gases. As far as translational motion is concerned, the ideal-gas model does not distinguish between a monatomic gas and a diatomic gas; both are considered as simply small, hard spheres. Hence the terms "atoms" and "molecules" can be used interchangeably to mean the basic constituents of the gas.

The Ideal-Gas Law

Section 18.1 introduced the idea of *state variables,* those parameters that describe the state of a macroscopic system. The state variables for an ideal gas are the volume V of its container, the number of moles n of the gas present in the container, the temperature T of the gas and its container, and the pressure p that the gas exerts on the walls of the container. These four state parameters are not independent of each other. If you change the value of one—by, say, raising the temperature—then one or more of the others will change as well. Each change of the parameters is a *change of state* of the system.

Experiments during the 17th and 18th centuries found a very specific relationship between the four state variables. Suppose you change the state of a gas, by heating it or compressing it or doing something else to it, and measure p, V, n, and T. Repeat this many times, changing the state of the gas each time, until you have a large table of p, V, n, and T values.

Then make a graph on which you plot pV, the product of the pressure and volume, on the vertical axis and nT, the product of the number of moles and temperature (in kelvins), on the horizontal axis. The very surprising result is that for *any* gas, whether it is hydrogen or helium or oxygen or methane, **you get exactly the same graph,** the linear graph shown in FIGURE 18.7. In other words, nothing about the graph indicates what gas was used because all gases give the same result.

> **NOTE** No real gas could extend to $nT = 0$ because it would condense. But an ideal gas never condenses because the only interactions among the molecules are hard-sphere collisions.

As you can see, there is a very clear proportionality between the quantity pV and the quantity nT. If we designate the slope of the line in this graph as R, then we can write the relationship as

$$pV = R \times (nT)$$

It is customary to write this relationship in a slightly different form, namely

$$pV = nRT \qquad \text{(ideal-gas law)} \qquad (18.13)$$

Equation 18.13 is the **ideal-gas law. The ideal-gas law is a relationship among the four state variables—p, V, n, and T—that characterize a gas in thermal equilibrium.**

The constant R, which is determined experimentally as the slope of the graph in Figure 18.7, is called the **universal gas constant.** Its value, in SI units, is

$$R = 8.31 \text{ J/mol K}$$

The units of R seem puzzling. The denominator mol K is clear because R multiplies nT. But what about the joules? The left side of the ideal-gas law, pV, has units

$$\text{Pa m}^3 = \frac{\text{N}}{\text{m}^2} \text{m}^3 = \text{N m} = \text{joules}$$

The product pV has units of joules, as shown on the vertical axis in Figure 18.7.

FIGURE 18.7 A graph of pV versus nT for an ideal gas.

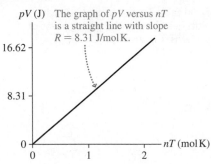

pV (J)   The graph of pV versus nT is a straight line with slope $R = 8.31$ J/mol K.

16.62

8.31

0

0   1   2

nT (mol K)

NOTE You perhaps learned in chemistry to work gas problems using units of atmospheres and liters. To do so, you had a different numerical value of R expressed in those units. In physics, however, we always work gas problems in SI units. Pressures *must* be in Pa, volumes in m^3, and temperatures in K.

The surprising fact, and one worth commenting upon, is that *all* gases have the *same* graph and the *same* value of R. There is no obvious reason a very simple atomic gas such as helium should have the same slope as a more complex gas such as methane (CH_4). Nonetheless, both turn out to have the same value for R. The ideal-gas law, within its limits of validity, describes *all* gases with a single value of the constant R.

EXAMPLE 18.4 | Calculating a gas pressure

100 g of oxygen gas is distilled into an evacuated 600 cm³ container. What is the gas pressure at a temperature of 150°C?

MODEL The gas can be treated as an ideal gas. Oxygen is a diatomic gas of O_2 molecules.

SOLVE From the ideal-gas law, the pressure is $p = nRT/V$. In Example 18.2 we calculated the number of moles in 100 g of O_2 and found $n = 3.13$ mol. Gas problems typically involve several conversions to get quantities into the proper units, and this example is no exception. The SI units of V and T are m^3 and K, respectively, thus

$$V = (600 \text{ cm}^3)\left(\frac{1 \text{ m}}{100 \text{ cm}}\right)^3 = 6.00 \times 10^{-4} \text{ m}^3$$

$$T = (150 + 273) \text{ K} = 423 \text{ K}$$

With this information, the pressure is

$$p = \frac{nRT}{V} = \frac{(3.13 \text{ mol})(8.31 \text{ J/mol K})(423 \text{ K})}{6.00 \times 10^{-4} \text{ m}^3}$$

$$= 1.83 \times 10^7 \text{ Pa} = 181 \text{ atm}$$

In this text we will consider only gases in sealed containers. The number of moles (and number of molecules) will not change during a problem. In that case,

$$\frac{pV}{T} = nR = \text{constant} \tag{18.14}$$

If the gas is initially in state i, characterized by the state variables p_i, V_i, and T_i, and at some later time in a final state f, the state variables for these two states are related by

$$\frac{p_f V_f}{T_f} = \frac{p_i V_i}{T_i} \quad \text{(ideal gas in a sealed container)} \tag{18.15}$$

This before-and-after relationship between the two states, reminiscent of a conservation law, will be valuable for many problems.

EXAMPLE 18.5 | Calculating a gas temperature

A cylinder of gas is at 0°C. A piston compresses the gas to half its original volume and three times its original pressure. What is the final gas temperature?

MODEL Treat the gas as an ideal gas in a sealed container.

SOLVE The before-and-after relationship of Equation 18.15 can be written

$$T_2 = T_1 \frac{p_2}{p_1} \frac{V_2}{V_1}$$

In this problem, the compression of the gas results in $V_2/V_1 = \frac{1}{2}$ and $p_2/p_1 = 3$. The initial temperature is $T_1 = 0°C = 273$ K. With this information,

$$T_2 = 273 \text{ K} \times 3 \times \tfrac{1}{2} = 409 \text{ K} = 136°C$$

ASSESS We did not need to know actual values of the pressure and volume, just the *ratios* by which they change.

We will often want to refer to the number of molecules N in a gas rather than the number of moles n. This is an easy change to make. Because $n = N/N_A$, the ideal-gas law in terms of N is

$$pV = nRT = \frac{N}{N_A} RT = N \frac{R}{N_A} T \tag{18.16}$$

R/N_A, the ratio of two known constants, is known as **Boltzmann's constant** k_B:

$$k_B = \frac{R}{N_A} = 1.38 \times 10^{-23} \text{ J/K}$$

The subscript B distinguishes Boltzmann's constant from a spring constant or other uses of the symbol k.

Ludwig Boltzmann was an Austrian physicist who did some of the pioneering work in statistical physics during the mid-19th century. Boltzmann's constant k_B can be thought of as the "gas constant per molecule," whereas R is the "gas constant per mole." With this definition, the ideal-gas law in terms of N is

$$pV = Nk_BT \quad \text{(ideal-gas law)} \tag{18.17}$$

Equations 18.13 and 18.17 are both the ideal-gas law, just expressed in terms of different state variables.

Recall that the number density (molecules per m³) was defined as N/V. A rearrangement of Equation 18.17 gives the number density as

$$\frac{N}{V} = \frac{p}{k_BT} \tag{18.18}$$

This is a useful consequence of the ideal-gas law, but keep in mind that the pressure *must* be in SI units of pascals and the temperature *must* be in SI units of kelvins.

EXAMPLE 18.6 | **The distance between molecules**

"Standard temperature and pressure," abbreviated **STP**, are $T = 0°C$ and $p = 1$ atm. Estimate the average distance between gas molecules at STP.

MODEL Consider the gas to be an ideal gas.

SOLVE Suppose a container of volume V holds N molecules at STP. How do we estimate the distance between them? Imagine placing an imaginary sphere around each molecule, separating it from its neighbors. This divides the total volume V into N little spheres of volume v_i, where $i = 1$ to N. The spheres of two neighboring molecules touch each other, like a crate full of Ping-Pong balls of somewhat different sizes all touching their neighbors, so the distance between two molecules is the sum of the radii of their two spheres. Each of these spheres is somewhat different, but a reasonable *estimate* of the distance between molecules would be twice the *average* radius of a sphere.

The average volume of one of these little spheres is

$$v_{avg} = \frac{V}{N} = \frac{1}{N/V}$$

That is, the average volume per molecule (m³ per molecule) is the inverse of the number density, the number of molecules per m³. This is not the volume of the molecule itself, which is much smaller, but the average volume of space that each molecule can claim as its own. We can use Equation 18.18 to calculate the number density:

$$\frac{N}{V} = \frac{p}{k_BT} = \frac{1.01 \times 10^5 \text{ Pa}}{(1.38 \times 10^{-23} \text{ J/K})(273 \text{ K})}$$
$$= 2.69 \times 10^{25} \text{ molecules/m}^3$$

where we used the definition of STP in SI units. Thus the average volume per molecule is

$$v_{avg} = \frac{1}{N/V} = 3.72 \times 10^{-26} \text{ m}^3$$

The volume of a sphere is $\frac{4}{3}\pi r^3$, so the average radius of a sphere is

$$r_{avg} = \left(\frac{3}{4\pi} v_{avg}\right)^{1/3} = 2.1 \times 10^{-9} \text{ m} = 2.1 \text{ nm}$$

The average distance between two molecules, with their spheres touching, is twice r_{avg}. Thus

$$\text{average distance} = 2r_{avg} \approx 4 \text{ nm}$$

This is a simple estimate, so we've given the answer with only one significant figure.

ASSESS One of the assumptions of the ideal-gas model is that atoms or molecules are "far apart" in comparison to the sizes of atoms and molecules. Chemistry experiments find that small molecules, such as N_2 and O_2, are roughly 0.3 nm in diameter. For a gas at STP, we see that the average distance between molecules is more than 10 times the size of a molecule. Thus the ideal-gas model works very well for a gas at STP.

STOP TO THINK 18.6 You have two containers of equal volume. One is full of helium gas. The other holds an equal mass of nitrogen gas. Both gases have the same pressure. How does the temperature of the helium compare to the temperature of the nitrogen?

a. $T_{helium} > T_{nitrogen}$ b. $T_{helium} = T_{nitrogen}$ c. $T_{helium} < T_{nitrogen}$

18.7 Ideal-Gas Processes

The ideal-gas law is the connection between the state variables pressure, temperature, and volume. If the state variables change, as they would from heating or compressing the gas, the state of the gas changes. An **ideal-gas process** is the means by which the gas changes from one state to another.

> **NOTE** Even in a sealed container, the ideal-gas law is a relationship among *three* variables. In general, *all three change* during an ideal-gas process. As a result, thinking about cause and effect can be rather tricky. Don't make the mistake of thinking that one variable is constant unless you're sure, beyond a doubt, that it is.

The *p*V Diagram

It will be very useful to represent ideal-gas processes on a graph called a **pV diagram.** This is nothing more than a graph of pressure versus volume. The important idea behind the *pV* diagram is that *each point* on the graph represents a single, unique state of the gas. That seems surprising at first, because a point on the graph only directly specifies the values of *p* and *V*. But knowing *p* and *V*, and assuming that *n* is known for a sealed container, we can find the temperature by using the ideal-gas law. Thus each point actually represents a triplet of values (p, V, T) specifying the state of the gas.

For example, FIGURE 18.8 is a *pV* diagram showing three states of a system consisting of 1 mol of gas. The values of *p* and *V* can be read from the axes for each point, then the temperature at that point determined from the ideal-gas law.

An ideal-gas *process* is a "trajectory" in the *pV* diagram showing all the intermediate states through which the gas passes. Figure 18.8 shows two different processes by which the gas can be changed from state 1 to state 3.

There are infinitely many ways to change the gas from state 1 to state 3. Although the initial and final states are the same for each of them, the particular process by which the gas changes—that is, the particular trajectory—will turn out to have very real consequences. The *pV* diagram is an important graphical representation of the process.

FIGURE 18.8 The state of the gas and ideal-gas processes can be shown on a *pV* diagram.

Quasi-Static Processes

Strictly speaking, the ideal-gas law applies only to gases in *thermal equilibrium,* meaning that the state variables are constant and not changing. But, by definition, an ideal-gas process causes some of the state variables to change. The gas is *not* in thermal equilibrium while the process of changing from state 1 to state 2 is under way.

To use the ideal-gas law throughout, we will assume that the process occurs *so slowly* that the system is never far from equilibrium. In other words, the values of *p*, *V*, and *T* at any point in the process are essentially the same as the equilibrium values they would assume if we stopped the process at that point. A process that is essentially in thermal equilibrium at all times is called a **quasi-static process.** It is an idealization, like a frictionless surface, but one that is a very good approximation in many real situations.

An important characteristic of a quasi-static process is that the trajectory through the *pV* diagram can be *reversed.* If you quasi-statically expand a gas by slowly pulling a piston out, as shown in FIGURE 18.9a, you can reverse the process by slowly pushing the piston in. The gas retraces its *pV* trajectory until it has returned to its initial state. Contrast this with what happens when the membrane bursts in FIGURE 18.9b. That is a sudden process, not at all quasi-static. The expanding gas is *not* in thermal equilibrium until some later time when it has completely filled the larger container, so the *irreversible* process of Figure 18.9b cannot be represented on a *pV* diagram.

The critical question is: How slow must a process be to qualify as quasi-static? That is a difficult question to answer. This textbook will always assume that processes are quasi-static. It turns out to be a reasonable assumption for the types of examples and homework problems we will look at. Irreversible processes will be left to more advanced courses.

FIGURE 18.9 The slow motion of the piston is a quasi-static process. The bursting of the membrane is not.

Constant-Volume Process

Many important gas processes take place in a container of constant, unchanging volume. A constant-volume process is called an **isochoric process,** where *iso* is a prefix meaning "constant" or "equal" while *choric* is from a Greek root meaning "volume." An isochoric process is one for which

$$V_f = V_i \qquad (18.19)$$

For example, suppose that you have a gas in the closed, rigid container shown in FIGURE 18.10a. Warming the gas with a Bunsen burner will raise its pressure without changing its volume. This process is shown as the vertical line $1 \rightarrow 2$ on the pV diagram of FIGURE 18.10b. A constant-volume cooling, by placing the container on a block of ice, would lower the pressure and be represented as the vertical line from 2 to 1. **Any isochoric process appears on a pV diagram as a vertical line.**

FIGURE 18.10 A constant-volume (isochoric) process.

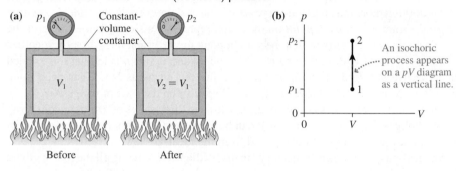

Before   After

EXAMPLE 18.7 | A constant-volume gas thermometer

A constant-volume gas thermometer is placed in contact with a reference cell containing water at the triple point. After reaching equilibrium, the gas pressure is recorded as 55.78 kPa. The thermometer is then placed in contact with a sample of unknown temperature. After the thermometer reaches a new equilibrium, the gas pressure is 65.12 kPa. What is the temperature of this sample?

MODEL The thermometer's volume doesn't change, so this is an isochoric process.

SOLVE The temperature at the triple point of water is $T_1 = 0.01°C = 273.16$ K. The ideal-gas law for a closed system is $p_2V_2/T_2 = p_1V_1/T_1$. The volume doesn't change, so $V_2/V_1 = 1$. Thus

$$T_2 = T_1 \frac{V_2}{V_1}\frac{p_2}{p_1} = T_1 \frac{p_2}{p_1} = (273.16 \text{ K}) \frac{65.12 \text{ kPa}}{55.78 \text{ kPa}}$$
$$= 318.90 \text{ K} = 45.75°C$$

The temperature *must* be in kelvins to do this calculation, although it is common to convert the final answer to °C. The fact that the pressures were given to four significant figures justified using $T_K = T_C + 273.15$ rather than the usual $T_C + 273$.

ASSESS $T_2 > T_1$, which we expected from the increase in pressure.

Constant-Pressure Process

Other gas processes take place at a constant, unchanging pressure. A constant-pressure process is called an **isobaric process,** where *baric* is from the same root as "barometer" and means "pressure." An isobaric process is one for which

$$p_f = p_i \qquad (18.20)$$

FIGURE 18.11a shows one method of changing the state of a gas while keeping the pressure constant. A cylinder of gas has a tight-fitting piston of mass M that can slide up and down but seals the container so that no atoms enter or escape. As the free-body diagram of FIGURE 18.11b shows, the piston and the air press down with force $p_{atmos}A + Mg$ while the gas inside pushes up with force $p_{gas}A$. In equilibrium, the gas pressure inside the cylinder is

$$p_{gas} = p_{atmos} + \frac{Mg}{A} \qquad (18.21)$$

In other words, the gas pressure is determined by the requirement that the gas must support both the mass of the piston and the air pressing inward. **This pressure is independent of the temperature of the gas or the height of the piston, so it stays constant as long as M is unchanged.**

If the cylinder is warmed, the gas will expand and push the piston up. But the pressure, determined by mass M, will not change. This process is shown on the pV diagram of FIGURE 18.11c as the horizontal line $1 \rightarrow 2$. We call this an *isobaric expansion*. An *isobaric compression* occurs if the gas is cooled, lowering the piston. **Any isobaric process appears on a pV diagram as a horizontal line.**

FIGURE 18.11 A constant-pressure (isobaric) process.

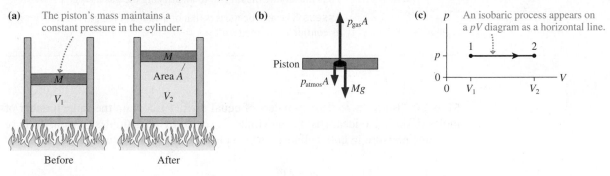

EXAMPLE 18.8 | **Comparing pressure**

The two cylinders in FIGURE 18.12 contain ideal gases at 20°C. Each cylinder is sealed by a frictionless piston of mass M.

a. How does the pressure of gas 2 compare to that of gas 1? Is it larger, smaller, or the same?

b. Suppose gas 2 is warmed to 80°C. Describe what happens to the pressure and volume.

FIGURE 18.12 Compare the pressures of the two gases.

MODEL Treat the gases as ideal gases.

SOLVE a. The pressure in the gas is determined by the requirement that the piston be in mechanical equilibrium. The pressure of the gas inside pushes up on the piston; the air pressure and the weight of the piston press down. The gas pressure $p = p_{atmos} + Mg/A$ depends on the mass of the piston, but not at all on how high the piston is or what type of gas is inside the cylinder. Thus both pressures are the same.

b. Neither does the pressure depend on temperature. Warming the gas increases the temperature, but the pressure—determined by the mass and area of the piston—is unchanged. Because $pV/T = $ constant, and p is constant, it must be true that $V/T = $ constant. As T increases, the volume V also must increase to keep V/T unchanged. In other words, increasing the gas temperature causes the volume to expand—the piston goes up—but with no change in pressure. This is an isobaric process.

EXAMPLE 18.9 | **Identifying a gas**

Your lab assistant distilled 50 g of a gas into a cylinder, but he left without writing down what kind of gas it is. The cylinder has a pressure regulator that adjusts a piston to keep the pressure at a constant 2.00 atm. To identify the gas, you measure the cylinder volume at several different temperatures, acquiring the data shown at the right. What is the gas?

T (°C)	V (L)
−50	11.6
0	14.0
50	16.2
100	19.4
150	21.8

MODEL The pressure doesn't change, so heating the gas is an isobaric process.

SOLVE The ideal-gas law is $pV = nRT$. Writing this as

$$V = \frac{nR}{p} T$$

we see that a graph of V versus T should be a straight line passing through the origin. Further, we can use the slope of the graph, nR/p, to measure the number of moles of gas, and from that we can identify the gas by determining its molar mass.

FIGURE 18.13 on the next page is a graph of the data, with the volumes and temperatures converted to SI units of m^3 ($1 \ m^3 = 1000 \ L$) and kelvins. The y-intercept of the graph is

Continued

FIGURE 18.13 A graph of the gas volume versus its temperature.

essentially zero, confirming the behavior of the gas as ideal, and the slope of the best-fit line is 5.16×10^{-5} m³/K. The number of moles of gas is

$$n = \frac{p}{R} \times \text{slope} = \frac{2 \times 101{,}300 \text{ Pa}}{8.31 \text{ J/mol K}} \times 5.16 \times 10^{-5} \text{ m}^3/\text{K} = 1.26 \text{ mol}$$

From this, the molar mass is

$$M = \frac{0.050 \text{ kg}}{1.26 \text{ mol}} = 0.040 \text{ kg/mol}$$

Thus the atomic mass is 40 u, identifying the gas as argon.

ASSESS The atomic mass is that of a well-known gas, which gives us confidence in the result.

STOP TO THINK 18.7 Two cylinders of equal diameter contain the same number of moles of the same ideal gas. Each cylinder is sealed by a frictionless piston. To have the same pressure in both cylinders, which piston would you use in cylinder 2?

FIGURE 18.14 A constant-temperature (isothermal) process.

(a)

Constant-Temperature Process

The last process we wish to look at for now is one that takes place at a constant temperature. A constant-temperature process is called an **isothermal process.** An isothermal process is one for which $T_f = T_i$. Because $pV = nRT$, a constant-temperature process in a closed system (constant n) is one for which the product pV doesn't change. Thus

$$p_f V_f = p_i V_i \tag{18.22}$$

in an isothermal process.

One possible isothermal process is illustrated in **FIGURE 18.14a**, where a piston is being pushed down to compress a gas. If the piston is pushed *slowly,* then heat energy transfer through the walls of the cylinder keeps the gas at the same temperature as the surrounding liquid. This is an *isothermal compression.* The reverse process, with the piston slowly pulled out, would be an *isothermal expansion.*

(b)

Representing an isothermal process on the pV diagram is a little more complicated than the two previous processes because both p and V change. As long as T remains fixed, we have the relationship

$$p = \frac{nRT}{V} = \frac{\text{constant}}{V} \tag{18.23}$$

(c)

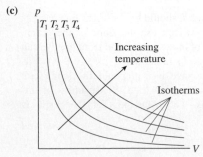

The inverse relationship between p and V causes the graph of an isothermal process to be a _hyperbola_. As one state variable goes up, the other goes down.

The process shown as $1 \rightarrow 2$ in **FIGURE 18.14b** represents the *isothermal compression* shown in Figure 18.14a. An *isothermal expansion* would move in the opposite direction along the hyperbola.

The location of the hyperbola depends on the value of T. A lower-temperature process is represented by a hyperbola closer to the origin than a higher-temperature process. FIGURE 18.14c shows four hyperbolas representing the temperatures T_1 to T_4, where $T_4 > T_3 > T_2 > T_1$. These are called **isotherms**. A gas undergoing an isothermal process moves along the isotherm of the appropriate temperature.

EXAMPLE 18.10 | Compressing air in the lungs

An ocean snorkeler takes a deep breath at the surface, filling his lungs with 4.0 L of air. He then descends to a depth of 5.0 m. At this depth, what is the volume of air in the snorkeler's lungs?

MODEL At the surface, the pressure in the lungs is 1.00 atm. Because the body cannot sustain large pressure differences between inside and outside, the air pressure in the lungs rises—and the volume decreases—to match the surrounding water pressure as he descends.

SOLVE The ideal-gas law for a sealed container is

$$V_2 = \frac{p_1}{p_2} \frac{T_2}{T_1} V_1$$

Air is quickly warmed to body temperature as it enters through the nose and mouth, and it remains at body temperature as the snorkeler dives, so $T_2/T_1 = 1$. We know $p_1 = 1.00$ atm $= 101{,}300$ Pa at the surface. We can find p_2 from the hydrostatic pressure equation, using the density of seawater:

$$p_2 = p_1 + \rho g d = 101{,}300 \text{ Pa} + (1030 \text{ kg/m}^3)(9.80 \text{ m/s}^2)(5.0 \text{ m})$$
$$= 151{,}800 \text{ Pa}$$

With this, the volume of the lungs at a depth of 5.0 m is

$$V_2 = \frac{101{,}300 \text{ Pa}}{151{,}800 \text{ Pa}} \times 1 \times 4.0 \text{ L} = 2.7 \text{ L}$$

ASSESS The air inside your lungs does compress—significantly—as you dive below the surface.

EXAMPLE 18.11 | A multistep process

A gas at 2.0 atm pressure and a temperature of 200°C is first expanded isothermally until its volume has doubled. It then undergoes an isobaric compression until it returns to its original volume. First show this process on a pV diagram. Then find the final temperature (in °C) and pressure.

MODEL The final state of the isothermal expansion is the initial state for an isobaric compression.

VISUALIZE FIGURE 18.15 shows the process. As the gas expands isothermally, it moves downward along an isotherm until it reaches volume $V_2 = 2V_1$. The gas is then compressed at constant pressure p_2 until its final volume V_3 equals its original volume V_1. State 3 is on an isotherm closer to the origin, so we expect to find $T_3 < T_1$.

SOLVE $T_2/T_1 = 1$ during the isothermal expansion and $V_2 = 2V_1$, so the pressure at point 2 is

$$p_2 = p_1 \frac{T_2}{T_1} \frac{V_1}{V_2} = p_1 \frac{V_1}{2V_1} = \tfrac{1}{2} p_1 = 1.0 \text{ atm}$$

We have $p_3/p_2 = 1$ during the isobaric compression and $V_3 = V_1 = \tfrac{1}{2} V_2$, so

FIGURE 18.15 A pV diagram for the process of Example 18.11.

$$T_3 = T_2 \frac{p_3}{p_2} \frac{V_3}{V_2} = T_2 \frac{\tfrac{1}{2} V_2}{V_2} = \tfrac{1}{2} T_2 = 236.5 \text{ K} = -36.5°\text{C}$$

where we converted T_2 to 473 K before doing calculations and then converted T_3 back to °C. The final state, with $T_3 = -36.5°$C and $p_3 = 1.0$ atm, is one in which both the pressure and the absolute temperature are half their original values.

STOP TO THINK 18.8 What is the ratio T_2/T_1 for this process?

a. $\frac{1}{4}$

b. $\frac{1}{2}$

c. 1 (no change)

d. 2

e. 4

f. There's not enough information to tell.

CHALLENGE EXAMPLE 18.12 | Depressing a piston

A large, 50.0-cm-diameter metal cylinder filled with air supports a 20.0 kg piston that can slide up and down without friction. The piston is 100.0 cm above the bottom when the temperature is 20°C. An 80.0 kg student then stands on the piston. After several minutes have elapsed, by how much has the piston been depressed?

MODEL The metal walls of the cylinder are a good thermal conductor, so after several minutes the gas temperature—even if it initially changed—will return to room temperature. The final temperature matches the initial temperature. Assume that the atmospheric pressure is 1 atm.

VISUALIZE FIGURE 18.16 shows the cylinder before and after the student stands on it. The volume of the cylinder is $V = Ah$, and only h changes.

SOLVE The ideal-gas law for a sealed container is

$$\frac{p_2 A h_2}{T_2} = \frac{p_1 A h_1}{T_1}$$

Because $T_2 = T_1$, the final height of the piston is

$$h_2 = \frac{p_1}{p_2} h_1$$

The pressure of the gas is determined by the mass of the piston (and anything on the piston) and the pressure of the air above. In equilibrium,

$$p = p_{\text{atmos}} + \frac{Mg}{A} = \begin{cases} 1.023 \times 10^5 \text{ Pa} & \text{piston only} \\ 1.063 \times 10^5 \text{ Pa} & \text{piston and student} \end{cases}$$

where we used $p_{\text{atmos}} = 1 \text{ atm} = 1.013 \times 10^5 \text{ Pa}$ and $A = \pi r^2 = 0.196 \text{ m}^2$. The final height of the piston is

FIGURE 18.16 The student compresses the gas.

$$h_2 = \frac{1.023 \times 10^5 \text{ Pa}}{1.063 \times 10^5 \text{ Pa}} \times 100.0 \text{ cm} = 96.2 \text{ cm}$$

The question, however, was by how much the piston is depressed. This is $h_1 - h_2 = 3.8 \text{ cm}$.

ASSESS Neither the piston nor the student increases the gas pressure to much above 1 atm, so it's not surprising that the added weight of the student doesn't push the piston down very far.

SUMMARY

The goal of Chapter 18 has been to learn some of the characteristics of macroscopic systems.

GENERAL PRINCIPLES

Three Common Phases of Matter

Solid Rigid, definite shape. Nearly incompressible.

Liquid Molecules loosely held together by molecular bonds, but able to move around. Nearly incompressible.

Gas Molecules moving freely through space. Compressible.

The different phases exist for different conditions of temperature T and pressure p. The boundaries separating the regions of a **phase diagram** are lines of phase equilibrium. Any amounts of the two phases can coexist in equilibrium. The **triple point** is the one value of temperature and pressure at which all three phases can coexist in equilibrium.

IMPORTANT CONCEPTS

Ideal-Gas Model

- Atoms and molecules are small, hard spheres that travel freely through space except for occasional collisions with each other or the walls.
- The model is valid when the density is low and the temperature well above the condensation point.

Ideal-Gas Law

The **state variables** of an ideal gas are related by the ideal-gas law

$$pV = nRT \quad \text{or} \quad pV = Nk_BT$$

where $R = 8.31$ J/mol K is the universal gas constant and $k_B = 1.38 \times 10^{-23}$ J/K is Boltzmann's constant. p, V, and T *must* be in SI units of Pa, m^3, and K.

For a gas in a sealed container, with constant n:

$$\frac{p_2V_2}{T_2} = \frac{p_1V_1}{T_1}$$

Counting atoms and moles

A macroscopic sample of matter consists of N atoms (or molecules), each of mass m (the **atomic** or **molecular mass**):

$$N = \frac{M}{m}$$

Alternatively, we can state that the sample consists of n **moles**:

$$n = \frac{N}{N_A} \quad \text{or} \quad \frac{M}{M_{mol}}$$

where $N_A = 6.02 \times 10^{23} \text{ mol}^{-1}$ is **Avogadro's number.**

The molar mass M_{mol}, in kg/mol, is the numerical value of the atomic or molecular mass in u divided by 1000. The atomic or molecular mass, in atomic mass units u, is well approximated by the **atomic mass number** A. The atomic mass unit is

$$1 \text{ u} = 1.66 \times 10^{-27} \text{ kg}$$

The **number density** of the sample is $\frac{N}{V}$.

APPLICATIONS

Temperature scales

The Kelvin scale has absolute zero at $T_0 = 0$ K and the triple point of water at $T_3 = 273.16$ K.

$$T_K = T_C + 273$$

Thermal expansion

For a temperature change ΔT,

$$\Delta L/L = \alpha \Delta T \qquad \Delta V/V = \beta \Delta T$$

For a solid, $\beta = 3\alpha$.

Three basic gas processes

Ideal-gas processes are shown as trajectories through the pV diagram.

1. **Isochoric,** or constant volume
2. **Isobaric,** or constant pressure
3. **Isothermal,** or constant temperature

pV diagram

TERMS AND NOTATION

micro/macro connection	atomic mass unit, u	thermal expansion	ideal gas
phase	molecular mass	coefficient of linear expansion, α	ideal-gas law
phase change	mole, n	coefficient of volume expansion, β	universal gas constant, R
solid	monatomic gas	melting point	Boltzmann's constant, k_B
crystal	diatomic gas	freezing point	STP
liquid	Avogadro's number, N_A	phase equilibrium	ideal-gas process
gas	molar mass, M_{mol}	condensation point	pV diagram
state variable	temperature, T	boiling point	quasi-static process
thermal equilibrium	constant-volume gas	phase diagram	isochoric process
number density, N/V	thermometer	sublimation	isobaric process
atomic mass number, A	absolute zero, T_0	critical point	isothermal process
atomic mass	absolute temperature scale	triple point	isotherm

CONCEPTUAL QUESTIONS

1. Rank in order, from highest to lowest, the temperatures $T_1 = 0$ K, $T_2 = 0°$C, and $T_3 = 0°$F.

2. The sample in an experiment is initially at 10°C. If the sample's temperature is doubled, what is the new temperature in °C?

3. a. Is there a highest temperature at which ice can exist? If so, what is it? If not, why not?
   b. Is there a lowest temperature at which water vapor can exist? If so, what is it? If not, why not?

4. The cylinder in FIGURE Q18.4 is divided into two compartments by a frictionless piston that can slide back and forth. If the piston is in equilibrium, is the pressure on the left side greater than, less than, or equal to the pressure on the right? Explain.

FIGURE Q18.4

5. A gas is in a sealed container. By what factor does the gas temperature change if:
   a. The volume is doubled and the pressure is tripled?
   b. The volume is halved and the pressure is tripled?

6. A gas is in a sealed container. The gas pressure is tripled and the temperature is doubled.
   a. Does the number of moles of gas in the container increase, decrease, or stay the same?
   b. By what factor does the number density of the gas increase?

7. An aquanaut lives in an underwater apartment 100 m beneath the surface of the ocean. Compare the freezing and boiling points of water in the aquanaut's apartment to their values at the surface. Are they higher, lower, or the same? Explain.

8. a. A sample of water vapor in an enclosed cylinder has an initial pressure of 500 Pa at an initial temperature of −0.01°C. A piston squeezes the sample smaller and smaller, without limit. Describe what happens to the water as the squeezing progresses.
   b. Repeat part a if the initial temperature is 0.03°C warmer.

9. A gas is in a sealed container. By what factor does the gas pressure change if:
   a. The volume is doubled and the temperature is tripled?
   b. The volume is halved and the temperature is tripled?

10. A gas undergoes the process shown in FIGURE Q18.10. By what factor does the temperature change?

FIGURE Q18.10   FIGURE Q18.11

11. The temperature increases from 300 K to 1200 K as a gas undergoes the process shown in FIGURE Q18.11. What is the final pressure?

12. A student is asked to sketch a pV diagram for a gas that goes through a cycle consisting of (a) an isobaric expansion, (b) a constant-volume reduction in temperature, and (c) an isothermal process that returns the gas to its initial state. The student draws the diagram shown in FIGURE Q18.12. What, if anything, is wrong with the student's diagram?

FIGURE Q18.12

EXERCISES AND PROBLEMS

Problems labeled ▓ integrate material from earlier chapters.

Exercises

Section 18.1 Solids, Liquids, and Gases

1. ‖ What volume of water has the same mass as 100 cm³ of gold?
2. | The nucleus of a uranium atom has a diameter of 1.5×10^{-14} m and a mass of 4.0×10^{-25} kg. What is the density of the nucleus?
3. ‖ What is the diameter of a copper sphere that has the same mass as a 10 cm × 10 cm × 10 cm cube of aluminum?
4. ‖ A hollow aluminum sphere with outer diameter 10.0 cm has a mass of 690 g. What is the sphere's inner diameter?

Section 18.2 Atoms and Moles

5. ‖ How many moles are in a 2.0 cm × 2.0 cm × 2.0 cm cube of copper?
6. ‖ How many atoms are in a 2.0 cm × 2.0 cm × 2.0 cm cube of aluminum?
7. ‖ What is the number density of (a) aluminum and (b) lead?
8. ‖ An element in its solid phase has mass density 1750 kg/m³ and number density 4.39×10^{28} atoms/m³. What is the element's atomic mass number?
9. ‖ 1.0 mol of gold is shaped into a sphere. What is the sphere's diameter?
10. ‖ What volume of aluminum has the same number of atoms as 10 cm³ of mercury?

Section 18.3 Temperature

11. | The lowest and highest natural temperatures ever recorded on earth are −129°F in Antarctica and 134°F in Death Valley. What are these temperatures in °C and in K?
12. ‖ At what temperature does the numerical value in °F match the numerical value in °C?
13. ‖ A demented scientist creates a new temperature scale, the "Z scale." He decides to call the boiling point of nitrogen 0°Z and the melting point of iron 1000°Z.
    a. What is the boiling point of water on the Z scale?
    b. Convert 500°Z to degrees Celsius and to kelvins

Section 18.4 Thermal Expansion

Section 18.5 Phase Changes

14. | A concrete bridge is built of 325-cm-long concrete slabs with an expansion joint between them. The slabs just touch on a 115°F day, the hottest day for which the bridge is designed. What is the gap between the slabs when the temperature is 0°F?
15. | A surveyor has a steel measuring tape that is calibrated to be 100.000 m long (i.e., accurate to ± 1 mm) at 20°C. If she measures the distance between two stakes to be 65.175 m on a 3°C day, does she need to add or subtract a correction factor to get the true distance? How large, in mm, is the correction factor?
16. | Two students each build a piece of scientific equipment that uses a 655-mm-long metal rod. One student uses a brass rod, the other an invar rod. If the temperature increases by 5.0°C, how much more does the brass rod expand than the invar rod?

17. | A 60.00 L fuel tank is filled with gasoline on a −10°C day, then rolled into a storage shed where the temperature is 20°C. If the tank is not vented, what minimum volume needs to be left empty at filling time so that the tank doesn't rupture as it warms?
18. ‖ What is the temperature in °F and the pressure in Pa at the triple point of (a) water and (b) carbon dioxide?

Section 18.6 Ideal Gases

19. | A cylinder contains nitrogen gas. A piston compresses the gas to half its initial volume. Afterward,
    a. Has the mass density of the gas changed? If so, by what factor? If not, why not?
    b. Has the number of moles of gas changed? If so, by what factor? If not, why not?
20. | 3.0 mol of gas at a temperature of −120°C fills a 2.0 L container. What is the gas pressure?
21. ‖ A rigid container holds 2.0 mol of gas at a pressure of 1.0 atm and a temperature of 30°C.
    a. What is the container's volume?
    b. What is the pressure if the temperature is raised to 130°C?
22. ‖ A gas at 100°C fills volume V_0. If the pressure is held constant, what is the volume if (a) the Celsius temperature is doubled and (b) the Kelvin temperature is doubled?
23. ‖ The total lung capacity of a typical adult is 5.0 L. Approximately
BIO 20% of the air is oxygen. At sea level and at a body temperature of 37°C, how many oxygen molecules do the lungs contain at the end of a strong inhalation?
24. ‖‖ A 20-cm-diameter cylinder that is 40 cm long contains 50 g of oxygen gas at 20°C.
    a. How many moles of oxygen are in the cylinder?
    b. How many oxygen molecules are in the cylinder?
    c. What is the number density of the oxygen?
    d. What is the reading of a pressure gauge attached to the tank?
25. | The solar corona is a very hot atmosphere surrounding the visible surface of the sun. X-ray emissions from the corona show that its temperature is about 2×10^6 K. The gas pressure in the corona is about 0.03 Pa. Estimate the number density of particles in the solar corona.
26. ‖ A gas at temperature T_0 and atmospheric pressure fills a cylinder. The gas is transferred to a new cylinder with three times the volume, after which the pressure is half the original pressure. What is the new temperature of the gas?

Section 18.7 Ideal-Gas Processes

27. | A gas with initial state variables p_1, V_1, and T_1 expands isothermally until $V_2 = 2V_1$. What are (a) T_2 and (b) p_2?
28. | A gas with initial state variables p_1, V_1, and T_1 is cooled in an isochoric process until $p_2 = \frac{1}{3}p_1$. What are (a) V_2 and (b) T_2?
29. ‖ A rigid, hollow sphere is submerged in boiling water in a room where the air pressure is 1.0 atm. The sphere has an open valve with its inlet just above the water level. After a long period of time has elapsed, the valve is closed. What will be the pressure inside the sphere if it is then placed in (a) a mixture of ice and water and (b) an insulated box filled with dry ice?
30. | A rigid container holds hydrogen gas at a pressure of 3.0 atm and a temperature of 20°C. What will the pressure be if the temperature is lowered to −20°C?

31. ‖ A 24-cm-diameter vertical cylinder is sealed at the top by a frictionless 20 kg piston. The piston is 84 cm above the bottom when the gas temperature is 303°C. The air above the piston is at 1.00 atm pressure.
    a. What is the gas pressure inside the cylinder?
    b. What will the height of the piston be if the temperature is lowered to 15°C?

32. ‖ 0.10 mol of argon gas is admitted to an evacuated 50 cm³ container at 20°C. The gas then undergoes an isochoric heating to a temperature of 300°C.
    a. What is the final pressure of the gas?
    b. Show the process on a pV diagram. Include a proper scale on both axes.

33. ‖ 0.10 mol of argon gas is admitted to an evacuated 50 cm³ container at 20°C. The gas then undergoes an isobaric heating to a temperature of 300°C.
    a. What is the final volume of the gas?
    b. Show the process on a pV diagram. Include a proper scale on both axes.

34. ‖ 0.10 mol of argon gas is admitted to an evacuated 50 cm³ container at 20°C. The gas then undergoes an isothermal expansion to a volume of 200 cm³.
    a. What is the final pressure of the gas?
    b. Show the process on a pV diagram. Include a proper scale on both axes.

35. ‖ 0.0040 mol of gas undergoes the process shown in FIGURE EX18.35.
    a. What type of process is this?
    b. What are the initial and final temperatures in °C?

FIGURE EX18.35   **FIGURE EX18.36**

36. ‖ A gas with an initial temperature of 900°C undergoes the process shown in FIGURE EX18.36.
    a. What type of process is this?
    b. What is the final temperature in °C?
    c. How many moles of gas are there?

37. ‖ 0.020 mol of gas undergoes the process shown in FIGURE EX18.37.
    a. What type of process is this?
    b. What is the final temperature in °C?
    c. What is the final volume V_2?

FIGURE EX18.37

38. ‖ 0.0050 mol of gas undergoes the process $1 \rightarrow 2 \rightarrow 3$ shown in FIGURE P16.38. What are (a) temperature T_1, (b) pressure p_2, and (c) volume V_3?

FIGURE P16.38

39. ‖ An ideal gas starts with pressure p_1 and volume V_1. Draw a pV diagram showing the process in which the gas undergoes an isochoric process that doubles the pressure, then an isobaric process that doubles the volume, followed by an isothermal process that doubles the volume again. Label each of the three processes.

40. ‖ An ideal gas starts with pressure p_1 and volume V_1. Draw a pV diagram showing the process in which the gas undergoes an isothermal process during which the volume is halved, then an isochoric process during which the pressure is halved, followed by an isobaric process during which the volume is doubled. Label each of the three processes.

Problems

41. ‖ The atomic mass number of copper is $A = 64$. Assume that atoms in solid copper form a cubic crystal lattice. To envision this, imagine that you place atoms at the centers of tiny sugar cubes, then stack the little sugar cubes to form a big cube. If you dissolve the sugar, the atoms left behind are in a cubic crystal lattice. What is the smallest distance between two copper atoms?

42. ‖ The molecular mass of water (H_2O) is $A = 18$. How many protons are there in 1.0 L of liquid water?

43. ‖ A brass ring with inner diameter 2.00 cm and outer diameter 3.00 cm needs to fit over a 2.00-cm-diameter steel rod, but at 20°C the hole through the brass ring is 50 μm too small in diameter. To what temperature, in °C, must the rod *and* ring be heated so that the ring just barely slips over the rod?

44. ‖ A 15°C, 2.0-cm-diameter aluminum bar just barely slips between two rigid steel walls 10.0 cm apart. If the bar is warmed to 25°C, how much force does it exert on each wall?

45. ‖ The semiconductor industry manufactures integrated circuits in large vacuum chambers where the pressure is 1.0×10^{-10} mm of Hg.
    a. What fraction is this of atmospheric pressure?
    b. At $T = 20$°C, how many molecules are in a cylindrical chamber 40 cm in diameter and 30 cm tall?

46. ‖‖ A 6.0-cm-diameter, 10-cm-long cylinder contains 100 mg of oxygen (O_2) at a pressure less than 1 atm. The cap on one end of the cylinder is held in place only by the pressure of the air. One day when the atmospheric pressure is 100 kPa, it takes a 184 N force to pull the cap off. What is the temperature of the gas?

47. ‖ A nebula—a region of the galaxy where new stars are forming—contains a very tenuous gas with 100 atoms/cm³. This gas is heated to 7500 K by ultraviolet radiation from nearby stars. What is the gas pressure in atm?

48. ‖ On average, each person in the industrialized world is responsible for the emission of 10,000 kg of carbon dioxide (CO_2) every year. This includes CO_2 that you generate directly, by burning fossil fuels to operate your car or your furnace, as well as CO_2 generated on your behalf by electric generating stations and manufacturing plants. CO_2 is a greenhouse gas that contributes to global warming. If you were to store your yearly CO_2 emissions in a cube at STP, how long would each edge of the cube be?

49. ‖ To determine the mass of neon contained in a rigid, 2.0 L cylinder, you vary the cylinder's temperature while recording the reading of a pressure gauge. Your data are as follows:

Temperature (°C)	Pressure gauge (atm)
100	6.52
150	7.80
200	8.83
250	9.59

Use the best-fit line of an appropriate graph to determine the mass of the neon.

50. ‖ The 3.0-m-long pipe in FIGURE P18.50 is closed at the top end. It is slowly pushed straight down into the water until the top end of the pipe is level with the water's surface. What is the length L of the trapped volume of air?

FIGURE P18.50

51. ‖ A diving bell is a 3.0-m-tall cylinder closed at the upper end but open at the lower end. The temperature of the air in the bell is 20°C. The bell is lowered into the ocean until its lower end is 100 m deep. The temperature at that depth is 10°C.
   a. How high does the water rise in the bell after enough time has passed for the air inside to reach thermal equilibrium?
   b. A compressed-air hose from the surface is used to expel all the water from the bell. What minimum air pressure is needed to do this?

52. ‖ An electric generating plant boils water to produce high-pressure steam. The steam spins a turbine that is connected to the generator.
   a. How many liters of water must be boiled to fill a 5.0 m³ boiler with 50 atm of steam at 400°C?
   b. The steam has dropped to 2.0 atm pressure at 150°C as it exits the turbine. How much volume does it now occupy?

53. ‖ On a cool morning, when the temperature is 15°C, you measure the pressure in your car tires to be 30 psi. After driving 20 mi on the freeway, the temperature of your tires is 45°C. What pressure will your tire gauge now show?

54. ‖ The air temperature and pressure in a laboratory are 20°C and 1.0 atm. A 1.0 L container is open to the air. The container is then sealed and placed in a bath of boiling water. After reaching thermal equilibrium, the container is opened. How many moles of air escape?

55. ‖ 10,000 cm³ of 200°C steam at a pressure of 20 atm is cooled until it condenses. What is the volume of the liquid water? Give your answer in cm³.

56. ‖ The mercury manometer shown in FIGURE P18.56 is attached to a gas cell. The mercury height h is 120 mm when the cell is placed in an ice-water mixture. The mercury height drops to 30 mm when the device is carried into an industrial freezer. What is the freezer temperature?
   Hint: The right tube of the manometer is much narrower than the left tube. What reasonable assumption can you make about the gas volume?

FIGURE P18.56   FIGURE P18.57

57. ‖ The U-shaped tube in FIGURE P18.57 has a total length of 1.0 m. It is open at one end, closed at the other, and is initially filled with air at 20°C and 1.0 atm pressure. Mercury is poured slowly into the open end without letting any air escape, thus compressing the air. This is continued until the open side of the tube is completely filled with mercury. What is the length L of the column of mercury?

58. ‖ The 50 kg circular piston shown in FIGURE P18.58 floats on 0.12 mol of compressed air.
   a. What is the piston height h if the temperature is 30°C?
   b. How far does the piston move if the temperature is increased by 100°C?

FIGURE P18.58

59. ‖ A diver 50 m deep in 10°C fresh water exhales a 1.0-cm-diameter bubble. What is the bubble's diameter just as it reaches the surface of the lake, where the water temperature is 20°C?
   Hint: Assume that the air bubble is always in thermal equilibrium with the surrounding water.

60. ‖ 8.0 g of helium gas follows the process $1 \rightarrow 2 \rightarrow 3$ shown in FIGURE P18.60. Find the values of V_1, V_3, p_2, and T_3.

FIGURE P18.60   FIGURE P18.61

61. ‖ FIGURE P18.61 shows two different processes by which 1.0 g of nitrogen gas moves from state 1 to state 2. The temperature of state 1 is 25°C. What are (a) pressure p_1 and (b) temperatures (in °C) T_2, T_3, and T_4?

62. || **FIGURE P18.62** shows two different processes by which 80 mol of gas move from state 1 to state 2. The dashed line is an isotherm.
   a. What is the temperature of the isothermal process?
   b. What maximum temperature is reached along the straight-line process?

FIGURE P18.62   **FIGURE P18.63**

63. || 4.0 g of oxygen gas, starting at 820°C, follow the process $1 \rightarrow 2$ shown in **FIGURE P18.63**. What is temperature T_2 (in °C)?
64. || 10 g of dry ice (solid CO_2) is placed in a 10,000 cm^3 container, then all the air is quickly pumped out and the container sealed. The container is warmed to 0°C, a temperature at which CO_2 is a gas.
   a. What is the gas pressure? Give your answer in atm.
   The gas then undergoes an isothermal compression until the pressure is 3.0 atm, immediately followed by an isobaric compression until the volume is 1000 cm^3.
   b. What is the final temperature of the gas (in °C)?
   c. Show the process on a pV diagram.
65. || A container of gas at 2.0 atm pressure and 127°C is compressed at constant temperature until the volume is halved. It is then further compressed at constant pressure until the volume is halved again.
   a. What are the final pressure and temperature of the gas?
   b. Show this process on a pV diagram.
66. || Five grams of nitrogen gas at an initial pressure of 3.0 atm and at 20°C undergo an isobaric expansion until the volume has tripled.
   a. What is the gas volume after the expansion?
   b. What is the gas temperature after the expansion (in °C)?
   The gas pressure is then decreased at constant volume until the original temperature is reached.
   c. What is the gas pressure after the decrease?
   Finally, the gas is isothermally compressed until it returns to its initial volume.
   d. What is the final gas pressure?
   e. Show the full three-step process on a pV diagram. Use appropriate scales on both axes.

In Problems 67 through 70 you are given the equation(s) used to solve a problem. For each of these, you are to
   a. Write a realistic problem for which this is the correct equation(s).
   b. Draw a pV diagram.
   c. Finish the solution of the problem.

67. $p_2 = \dfrac{300 \text{ cm}^3}{100 \text{ cm}^3} \times 1 \times 2 \text{ atm}$

68. $(T_2 + 273) \text{ K} = \dfrac{200 \text{ kPa}}{500 \text{ kPa}} \times 1 \times (400 + 273) \text{ K}$

69. $V_2 = \dfrac{(400 + 273) \text{ K}}{(50 + 273) \text{ K}} \times 1 \times 200 \text{ cm}^3$

70. $(2.0 \times 101{,}300 \text{ Pa})(100 \times 10^{-6} \text{ m}^3) = n(8.31 \text{ J/mol K})T_1$

   $n = \dfrac{0.12 \text{ g}}{20 \text{ g/mol}}$

   $T_2 = \dfrac{200 \text{ cm}^3}{100 \text{ cm}^3} \times 1 \times T_1$

Challenge Problems

71. ||| An inflated bicycle inner tube is 2.2 cm in diameter and 200 cm in circumference. A small leak causes the gauge pressure to decrease from 110 psi to 80 psi on a day when the temperature is 20°C. What mass of air is lost? Assume the air is pure nitrogen.
72. ||| The cylinder in **FIGURE CP18.72** has a moveable piston attached to a spring. The cylinder's cross-section area is 10 cm^2, it contains 0.0040 mol of gas, and the spring constant is 1500 N/m. At 20°C the spring is neither compressed nor stretched. How far is the spring compressed if the gas temperature is raised to 100°C?

FIGURE CP18.72   **FIGURE CP18.73**

73. ||| Containers A and B in **FIGURE CP18.73** hold the same gas. The volume of B is four times the volume of A. The two containers are connected by a thin tube (negligible volume) and a valve that is closed. The gas in A is at 300 K and pressure of 1.0×10^5 Pa. The gas in B is at 400 K and pressure of 5.0×10^5 Pa. Heaters will maintain the temperatures of A and B even after the valve is opened. After the valve is opened, gas will flow one way or the other until A and B have equal pressure. What is this final pressure?
74. ||| The closed cylinder of **FIGURE CP18.74** has a tight-fitting but frictionless piston of mass M. The piston is in equilibrium when the left chamber has pressure p_0 and length L_0 while the spring on the right is compressed by ΔL.
   a. What is ΔL in terms of $p_0, L_0, A, M,$ and k?
   b. Suppose the piston is moved a small distance x to the right. Find an expression for the net force $(F_x)_\text{net}$ on the piston. Assume all motions are slow enough for the gas to remain at the same temperature as its surroundings.
   c. If released, the piston will oscillate around the equilibrium position. Assuming $x \ll L_0$ find an expression for the oscillation period T.
   Hint: Use the binomial approximation.

FIGURE CP18.74

19 Work, Heat, and the First Law of Thermodynamics

This false-color thermal image—an infrared photo—shows the hotter and cooler parts of the anvil.

IN THIS CHAPTER, you will learn and apply the first law of thermodynamics.

What is the first law of thermodynamics?

The first law of thermodynamics is a very general statement about energy and its conservation. A system's thermal energy changes if energy is transferred into or out of the system as *work* or as *heat*. That is,

$$\Delta E_{th} = W + Q$$

« LOOKING BACK Sections 10.4 and 10.8 Conservation of energy

How is work done on a gas?

Work W is the transfer of energy in a mechanical interaction when forces push or pull. Work is done on a gas by changing its volume.

- $W > 0$ (energy added) in a compression.
- $W < 0$ (energy extracted) in an expansion.

« LOOKING BACK Sections 9.2–9.3 Work

What is heat?

Heat Q is the transfer of energy in a thermal interaction when the system and its environment are at different temperatures.

- $Q > 0$ (energy added) when the environment is hotter than the system.
- $Q < 0$ (energy extracted) when the system is hotter than the environment.

How is heat transferred?

Heat energy can be transferred between a system and its environment by

- Conduction
- Convection
- Radiation
- Evaporation

What are some thermal properties of matter?

Heat can cause either a temperature change or a phase change. A material's response to heat is governed by its specific heat, its heat of fusion, its heat of vaporization, and its thermal conductivity. These are thermal properties of matter. An important application is calorimetry—determining the final temperature when two or more systems interact thermally.

Why is the first law important?

The first law of thermodynamics is a very general statement that energy cannot be created or destroyed—merely moved from one place to another. Much of the machinery of modern society, from automobile engines to electric power plants to spacecraft, depends on and is an application of the first law. We'll look at some of these applications in more detail in Chapter 21.

19.1 It's All About Energy

In « Section 10.8 we wrote the *energy principle* as

$$\Delta K = W_c + W_{diss} + W_{ext} \tag{19.1}$$

Recall that energy is about *systems* and their interactions. Equation 19.1 tells us that the kinetic energy of a system of particles is changed when forces do work on the particles by pushing or pulling them through a distance. Here

1. W_c is the work done by conservative forces. This work can be represented as a change in the system's potential energy: $\Delta U = -W_c$.
2. W_{diss} is the work done by friction-like dissipative forces within the system. This work increases the system's thermal energy: $\Delta E_{th} = -W_{diss}$.
3. W_{ext} is the work done by external forces that originate in the environment. The push of a piston rod would be an external force.

With these definitions, Equation 19.1 becomes

$$\Delta K + \Delta U + \Delta E_{th} = W_{ext} \tag{19.2}$$

The system's *mechanical energy* was defined as $E_{mech} = K + U$. **FIGURE 19.1** reminds you that the mechanical energy is the *macroscopic* energy of the system as a whole, while E_{th} is the *microscopic* energy of the atoms and molecules within the system. This led to our final energy statement of Chapter 10:

$$\Delta E_{sys} = \Delta E_{mech} + \Delta E_{th} = W_{ext} \tag{19.3}$$

Thus the total energy of an *isolated system,* for which $W_{ext} = 0$, is constant.

The emphasis in Chapters 9 and 10 was on isolated systems. There we were interested in learning how kinetic and potential energy were *transformed* into each other and, where there is friction, into thermal energy. Now we want to focus on how energy is *transferred* between the system and its environment when W_{ext} is *not* zero.

NOTE Strictly speaking, Equation 19.3 should use the *internal energy* E_{int} rather than the thermal energy E_{th}, where $E_{int} = E_{th} + E_{chem} + E_{nuc} + \cdots$ includes all the various kinds of energies that can be stored inside a system. This textbook will focus on simple thermodynamics systems in which the internal energy is entirely thermal: $E_{int} = E_{th}$. We'll leave other forms of internal energy to more advanced courses.

Energy Transfer

Doing work on a system can have very different consequences. **FIGURE 19.2a** shows an object being lifted at steady speed by a rope. The rope's tension is an external force doing work W_{ext} on the system. In this case, the energy transferred into the system goes entirely to increasing the system's macroscopic potential energy U_{grav}, part of the mechanical energy. The energy-transfer process $W_{ext} \rightarrow E_{mech}$ is shown graphically in the energy bar chart of Figure 19.2a.

Contrast this with **FIGURE 19.2b**, where the same rope with the same tension now drags the object at steady speed across a rough surface. The tension does the same amount of work, but the mechanical energy does not change. Instead, friction increases the thermal energy of the object + surface system. The energy-transfer process $W_{ext} \rightarrow E_{th}$ is shown in the energy bar chart of Figure 19.2b.

The point of this example is that the energy transferred to a system can go entirely to the system's mechanical energy, entirely to its thermal energy, or (imagine dragging the object up an incline) some combination of the two. The energy isn't lost, but where it ends up depends on the circumstances.

FIGURE 19.1 The total energy of a system.

The *macroscopic* energy of the system as a whole is its mechanical energy E_{mech}.

$$E_{sys} = E_{mech} + E_{th}$$

The *microscopic* energy of the atoms and molecules is the system's thermal energy E_{th}.

FIGURE 19.2 The work done by tension can have very different consequences.

(a) Lift at steady speed

The tension does work on the system.

$$K_i + U_i + W_{ext} = K_f + U_f + \Delta E_{th}$$

Energy transferred to the system goes entirely to mechanical energy.

(b) Drag at steady speed

The tension does work on the system.

$$K_i + U_i + W_{ext} = K_f + U_f + \Delta E_{th}$$

Energy transferred to the system goes entirely to thermal energy.

The Missing Piece: Heat

You can transfer energy into a system by the mechanical process of doing work on the system. But that can't be all there is to energy transfer. What happens when you place a pan of water on the stove and light the burner? The water temperature increases, so $\Delta E_{th} > 0$. But no work is done ($W_{ext} = 0$) and there is no change in the water's mechanical energy ($\Delta E_{mech} = 0$). This process clearly violates the energy equation $\Delta E_{mech} + \Delta E_{th} = W_{ext}$. What's wrong?

Nothing is wrong. The energy equation is correct as far as it goes, but it is incomplete. Work is energy transferred in a mechanical interaction, but that is not the only way a system can interact with its environment. Energy can also be transferred between the system and the environment if they have a *thermal interaction*. The energy transferred in a thermal interaction is called *heat*.

The symbol for heat is Q. When heat is included, the energy equation becomes

$$\Delta E_{sys} = \Delta E_{mech} + \Delta E_{th} = W + Q \qquad (19.4)$$

Heat and work are both energy transferred between the system and the environment.

> **NOTE** We've dropped the subscript "ext" from W. The work that we consider in thermodynamics is *always* the work done by the environment on the system. We won't need to distinguish this work from W_c or W_{diss}, so the subscript is superfluous.

We'll return to Equation 19.4 in Section 19.4 after we look at how work is calculated for ideal-gas processes and at what heat is.

STOP TO THINK 19.1 A gas cylinder and piston are covered with heavy insulation. The piston is pushed into the cylinder, compressing the gas. In this process the gas temperature

a. Increases.
b. Decreases.
c. Doesn't change.
d. There's not sufficient information to tell.

19.2 Work in Ideal-Gas Processes

We introduced the idea of *work* in Chapter 9. **Work** is the energy transferred between a system and the environment when a net force acts on the system over a distance. The process itself is a **mechanical interaction,** meaning that the system and the environment interact via macroscopic pushes and pulls. Loosely speaking, we say that the environment (or a particular force from the environment) "does work" on the system. A system is in **mechanical equilibrium** if there is no net force on the system.

FIGURE 19.3 on the next page reminds you that work can be either positive or negative. **The sign of the work is *not* just an arbitrary convention, nor does it have anything to do with the choice of coordinate system.** The sign of the work tells us which way energy is being transferred.

In contrast to the mechanical energy or the thermal energy, **work is not a state variable.** That is, work is not a number characterizing the system. Instead, work is the amount of energy that moves between the system and the environment during a mechanical interaction. We can measure the *change* in a state variable, such as a temperature change $\Delta T = T_f - T_i$, but it would make no sense to talk about a "change of work." Consequently, work always appears as W, never as ΔW.

The pistons in a car engine do work on the air-fuel mixture by compressing it.

FIGURE 19.3 The sign of work.

Before   After

Work is *positive* when the force
is in the direction of motion.
- The force causes the object to speed up.
- Energy is transferred from the
  environment to the system.

Before   After

Work is *negative* when the force
is opposite to the motion.
- The force causes the object to slow down.
- Energy is transferred from the system to
  the environment.

You learned in « Section 9.3 how to calculate work. The small amount of work dW done by force $\vec{F}$ as a system moves through the small displacement $d\vec{s}$ is $dW = \vec{F} \cdot d\vec{s}$. If we restrict ourselves to situations where $\vec{F}$ is either parallel or opposite to $d\vec{s}$, then the total work done on the system as it moves from s_i to s_f is

$$W = \int_{s_i}^{s_f} F_s \, ds \qquad (19.5)$$

FIGURE 19.4 The external force does work on the gas as the piston moves.

(a) The gas pushes
with force $\vec{F}_{gas}$. To keep the piston in place,
an external force must be
equal and opposite to $\vec{F}_{gas}$.

$\vec{F}_{gas}$ $\vec{F}_{ext}$

Pressure p

0   x

Piston area A

(b) As the piston moves dx, the
external force does work $(F_{ext})_x \, dx$
on the gas.

dx $\vec{F}_{ext}$

The volume changes by $dV = A \, dx$
as the piston moves dx.

Let's apply this definition to calculate the work done when a gas is expanded or compressed. FIGURE 19.4a shows a gas of pressure p whose volume can be changed by a moving piston. The *system* is the gas in the cylinder, which exerts a force $\vec{F}_{gas}$ of magnitude pA on the left side of the piston. To prevent the pressure from blowing the piston out of the cylinder, there must be an equal but opposite external force $\vec{F}_{ext}$ pushing on the right side of the piston. In practice, this would likely be the force exerted by a piston rod. Using the coordinate system of Figure 19.4a, we have

$$(F_{ext})_x = -(F_{gas})_x = -pA \qquad (19.6)$$

Suppose the piston moves a small distance dx, as shown in FIGURE 19.4b. As it does so, the external force (i.e., the environment) does work

$$dW_{ext} = (F_{ext})_x \, dx = -pA \, dx \qquad (19.7)$$

If dx is positive (expanding gas), then dW_{ext} is negative. This is to be expected because the force is opposite the displacement. dW_{ext} is positive if the gas is slightly compressed ($dx < 0$).

If the piston moves dx, the gas volume changes by $dV = A \, dx$. Thus the work is

$$dW_{ext} = -p \, dV \qquad (19.8)$$

If we allow the gas to change in a slow, quasi-static process from an initial volume V_i to a final volume V_f, the total work done by the environment is found by integrating Equation 19.8:

$$W_{ext} = -\int_{V_i}^{V_f} p \, dV \quad \text{(work done on a gas)} \qquad (19.9)$$

Equation 19.9 is a key result of thermodynamics. Although we used a cylinder to derive it, it turns out to be true for a container of any shape.

NOTE The pressure of a gas usually changes as the gas expands or contracts. Consequently, p is *not* a constant that can be brought outside the integral. You need to know how the pressure changes with volume before you can carry out the integration.

What is this work done *on*? Not, as you might think, the piston. Work is an energy transfer, but the piston's energy is not changing. It is in equilibrium, with equal forces exerted on both faces. Further, the motion of the piston is very slow—it's a quasi-static process—so the piston has negligible kinetic energy. The piston is merely a movable boundary of the gas, so **the work is done by the environment on the gas.** When a gas expands or contracts, energy is transferred between the gas and the environment.

We could also look at the moving piston from the perspective of the gas. The gas is exerting a force on the piston as it's displaced, so the gas does work W_{gas}. In fact, because $\vec{F}_{gas} = -\vec{F}_{ext}$, the work done by the gas is $W_{gas} = -W_{ext}$. This is really just another way of saying that there's no net work done on the piston. However, the work that appears in the energy principle, and now in the laws of thermodynamics, is W_{ext}, the work done by the environment—either positive or negative—*on* the system, and that is the work that will appear (without a subscript) in our equations.

But some care is needed. Both the environment *and* the gas do work, with opposite signs. It is *not* an either-or situation. When a gas is compressed, it is often said that "Work is done *on* the gas." This is a situation with $W_{ext} > 0$ and $W_{gas} < 0$. When a gas expands, it is often said that "Work is done *by* the gas." But this doesn't mean that $W_{ext} = 0$! It simply means that $W_{gas} > 0$ and so, in this case, you would use $W_{ext} < 0$.

We can give the work done on a gas a nice geometric interpretation. You learned in Chapter 18 how to represent an ideal-gas process as a curve in the pV diagram. **FIGURE 19.5** shows that the work done on a gas is the negative of the area under the pV curve as the volume changes from V_i to V_f. That is,

> $W =$ the negative of the area under the pV curve between V_i and V_f

Figure 19.5a shows a process in which a gas *expands* from V_i to a larger volume V_f. The area under the curve is positive, so the environment does a negative amount of work on an expanding gas. Figure 19.5b shows a process in which a gas is compressed to a smaller volume. This one is a little trickier because we have to integrate "backward" along the V-axis. You learned in calculus that integrating from a larger limit to a smaller limit gives a negative result, so the area in Figure 19.5b is a negative area. Consequently, as the minus sign in Equation 19.9 indicates, the environment does positive work on a gas to compress it.

(a)

Expanding gas ($V_f > V_i$): Integration direction to right; Area under pV curve positive; Environment does *negative* work.

$W = -$area

Integration direction

(b)

Compressed gas ($V_f < V_i$): Integration direction to left; Area under pV curve negative; Environment does *positive* work.

$W = -$area

Integration direction

EXAMPLE 19.1 | **The work done on an expanding gas**

How much work is done in the ideal-gas process of **FIGURE 19.6**?

FIGURE 19.6 The ideal-gas process of Example 19.1.

$(500 \times 10^{-6} \text{ m}^3)(300,000 \text{ Pa}) = 150 \text{ J}$

$\frac{1}{2}(500 \times 10^{-6} \text{ m}^3)(200,000 \text{ Pa}) = 50 \text{ J}$

$(500 \times 10^{-6} \text{ m}^3)(100,000 \text{ Pa}) = 50 \text{ J}$

MODEL The work done on a gas is the negative of the area under the pV curve. The gas is *expanding*, so we expect the work to be negative.

SOLVE As Figure 19.6 shows, the area under the curve can be divided into two rectangles and a triangle. Volumes *must* be converted to SI units of m^3. The total area under the curve is 250 J, so the work done on the gas as it expands is

$$W = -(\text{area under the } pV \text{ curve}) = -250 \text{ J}$$

ASSESS We noted previously that the product $Pa \, m^3$ is equivalent to joules. The work is negative, as expected, because the external force pushing on the piston is opposite the direction of the piston's displacement.

Equation 19.9 is the basis for a problem-solving strategy.

Isochoric Process

FIGURE 19.7 Calculating the work done during ideal-gas processes.

(a)

For an *isochoric* process, the area under the pV curve is zero. No work is done.

The isochoric process in **FIGURE 19.7a** is one in which the volume does not change. Consequently,

$$W = 0 \qquad \text{(isochoric process)} \qquad (19.10)$$

An isochoric process is the *only* ideal-gas process in which no work is done.

Isobaric Process

(b)

For an *isobaric* process, the area is $p\Delta V$. The work done on the gas is $-p\Delta V$.

FIGURE 19.7b shows an isobaric process in which the volume changes from V_i to V_f. The rectangular area under the curve is $p\,\Delta V$, so the work done during this process is

$$W = -p\,\Delta V \qquad \text{(isobaric process)} \qquad (19.11)$$

where $\Delta V = V_f - V_i$. ΔV is positive if the gas expands $(V_f > V_i)$, so W is negative. ΔV is *negative* if the gas is compressed $(V_f < V_i)$, making W positive.

Isothermal Process

FIGURE 19.8 An isothermal process.

Isotherm for temperature T

For an *isothermal* process, the work done on the gas is the negative of the area under the hyperbola.

FIGURE 19.8 shows an isothermal process. Here we need to know the pressure as a function of volume before we can integrate Equation 19.9. From the ideal-gas law, $p = nRT/V$. Thus the work on the gas as the volume changes from V_i to V_f is

$$W = -\int_{V_i}^{V_f} p \, dV = -\int_{V_i}^{V_f} \frac{nRT}{V} \, dV = -nRT \int_{V_i}^{V_f} \frac{dV}{V} \qquad (19.12)$$

where we could take the T outside the integral because temperature is constant during an isothermal process. This is a straightforward integration, giving

$$W = -nRT \int_{V_i}^{V_f} \frac{dV}{V} = -nRT \ln V \Big|_{V_i}^{V_f}$$

$$= -nRT(\ln V_f - \ln V_i) = -nRT \ln\left(\frac{V_f}{V_i}\right) \qquad (19.13)$$

Because $nRT = p_iV_i = p_fV_f$ during an isothermal process, the work is:

$$W = -nRT \ln\left(\frac{V_f}{V_i}\right) = -p_iV_i \ln\left(\frac{V_f}{V_i}\right) = -p_fV_f \ln\left(\frac{V_f}{V_i}\right) \qquad (19.14)$$
$$\text{(isothermal process)}$$

Which version of Equation 19.14 is easiest to use will depend on the information you're given. The pressure, volume, and temperature *must* be in SI units.

EXAMPLE 19.2 | **The work of an isothermal compression**

A cylinder contains 7.0 g of nitrogen gas. How much work must be done to compress the gas at a constant temperature of 80°C until the volume is halved?

MODEL This is an isothermal ideal-gas process.

SOLVE Nitrogen gas is N_2, with molar mass $M_{mol} = 28$ g/mol, so 7.0 g is 0.25 mol of gas. The temperature is $T = 353$ K. Although we don't know the actual volume, we do know that $V_f = \frac{1}{2}V_i$. The volume ratio is all we need to calculate the work:

$$W = -nRT \ln\left(\frac{V_f}{V_i}\right)$$
$$= -(0.25 \text{ mol})(8.31 \text{ J/mol K})(353 \text{ K})\ln(1/2) = 508 \text{ J}$$

ASSESS The work is positive because a force from the environment pushes the piston inward to compress the gas.

Work Depends on the Path

FIGURE 19.9a shows two different processes that take a gas from an initial state i to a final state f. Although the initial and final states are the same, the work done during these two processes is *not* the same. **The work done during an ideal-gas process depends on the path followed through the pV diagram.**

You may recall that "work is independent of the path," but that referred to a different situation. In Chapter 10, we found that the work done by a conservative force is independent of the physical path of the object through space. For an ideal-gas process, the "path" is a sequence of thermodynamic states on a pV diagram. It is a figurative path because we can draw a picture of it on a pV diagram, but it is not a literal path.

The path dependence of work has an important implication for multistep processes. For a process $1 \rightarrow 2 \rightarrow 3$, the work has to be calculated separately for each step in the process, giving $W_{tot} = W_{1 \text{ to } 2} + W_{2 \text{ to } 3}$. In most cases, calculating the work for a process that goes directly from 1 to 3 will given an incorrect answer. The initial and final states are the same, but the work is *not* the same because work depends on the path followed through the pV diagram.

FIGURE 19.9 The work done during an ideal-gas process depends on the path.

The area under the process A curve is larger than the area under the process B curve. Thus $|W_A| > |W_B|$.

STOP TO THINK 19.2 Two processes take an ideal gas from state 1 to state 3. Compare the work done by process A to the work done by process B.

a. $W_A = W_B = 0$
b. $W_A = W_B$ but neither is zero
c. $W_A > W_B$
d. $W_A < W_B$

19.3 Heat

Heat is a more elusive concept than work. We use the word "heat" very loosely in the English language, often as synonymous with *hot*. We might say on a very hot day, "This heat is oppressive." If your apartment is cold, you may say, "Turn up the heat." These expressions date to a time long ago when it was thought that heat was a *substance* with fluid-like properties.

FIGURE 19.10 Joule's experiments to show the equivalence of heat and work.

Water

The flame heats the water. The temperature increases.

The paddle does work on the water. The temperature increases.

Our concept of heat changed with the work of British physicist James Joule in the 1840s. Joule was the first to carry out careful experiments to learn how it is that systems change their temperature. Using experiments like those shown in FIGURE 19.10, Joule found that you can raise the temperature of a beaker of water by two entirely different means:

1. Heating it with a flame, or
2. Doing work on it with a rapidly spinning paddle wheel.

The final state of the water is *exactly* the same in both cases. This implies that heat and work are essentially equivalent. In other words, heat is not a substance. Instead, **heat is energy.** Heat and work, which previously had been regarded as two completely different phenomena, were now seen to be simply two different ways of transferring energy to or from a system.

Thermal Interactions

To be specific, **heat** is the energy transferred between a system and the environment as a consequence of a *temperature difference* between them. Unlike a mechanical interaction in which work is done, heat requires no macroscopic motion of the system. Instead (we'll look at the details in Chapter 20), energy is transferred when the *faster* molecules in the hotter object collide with the *slower* molecules in the cooler object. On average, these collisions cause the faster molecules to lose energy and the slower molecules to gain energy. The net result is that energy is transferred from the hotter object to the colder object. The process itself, whereby energy is transferred between the system and the environment via atomic-level collisions, is called a **thermal interaction.**

When you place a pan of water on the stove, heat is the energy transferred *from* the hotter flame *to* the cooler water. If you place the water in a freezer, heat is the energy transferred from the warmer water to the colder air in the freezer. A system is in **thermal equilibrium** with the environment, or two systems are in thermal equilibrium with each other, if there is no temperature difference.

Like work, **heat is not a state variable. That is, heat is not a property of the system.** Instead, heat is the amount of energy that moves between the system and the environment during a thermal interaction. It would not be meaningful to talk about a "change of heat." Thus heat appears in the energy equation simply as a value Q, never as ΔQ. FIGURE 19.11 shows how to interpret the sign of Q.

FIGURE 19.11 The sign of heat.

(a) Positive heat

$Q > 0$

System

$T_{env} > T_{sys}$

(b) Negative heat

$Q < 0$

System

$T_{sys} > T_{env}$

(c) Thermal equilibrium

$Q = 0$

System

$T_{sys} = T_{env}$

NOTE For both heat and work, a positive value indicates energy being transferred from the environment to the system. TABLE 19.1 summarizes the similarities and differences between work and heat.

TABLE 19.1 Understanding work and heat

	Work	Heat
Interaction:	Mechanical	Thermal
Requires:	Force and displacement	Temperature difference
Process:	Macroscopic pushes and pulls	Microscopic collisions
Positive value:	$W > 0$ when a gas is compressed. Energy is transferred in.	$Q > 0$ when the environment is at a higher temperature than the system. Energy is transferred in.
Negative value:	$W < 0$ when a gas expands. Energy is transferred out.	$Q < 0$ when the system is at a higher temperature than the environment. Energy is transferred out.
Equilibrium:	A system is in mechanical equilibrium when there is no net force or torque on it.	A system is in thermal equilibrium when it is at the same temperature as the environment.

Units of Heat

Heat is energy transferred between the system and the environment. Consequently, the SI unit of heat is the joule. Historically, before the connection between heat and work had been recognized, a unit for measuring heat, the calorie, had been defined as

$$1 \text{ calorie} = 1 \text{ cal} = \text{the quantity of heat needed to change the}$$
$$\text{temperature of 1 g of water by 1°C}$$

Once Joule established that heat is energy, it was apparent that the calorie is really a unit of energy. In today's SI units, the conversion is

$$1 \text{ cal} = 4.186 \text{ J}$$

The calorie you know in relation to food is not the same as the heat calorie. The *food calorie*, abbreviated Cal with a capital C, is

$$1 \text{ food calorie} = 1 \text{ Cal} = 1000 \text{ cal} = 1 \text{ kcal} = 4186 \text{ J}$$

We will not use calories in this textbook, but there are some fields of science and engineering where calories are still widely used. All the calculations you learn to do with joules can equally well be done with calories.

Heat, Temperature, and Thermal Energy

It is important to distinguish among *heat, temperature,* and *thermal energy.* These three ideas are related, but the distinctions among them are crucial. In brief,

- Thermal energy is an energy *of the system* due to the motion of its atoms and molecules and the stretching/compressing of spring-like molecular bonds. It is a *form* of energy. Thermal energy is a state variable, and it makes sense to talk about how E_{th} changes during a process. The system's thermal energy continues to exist even if the system is isolated and not interacting thermally with its environment.
- Heat is energy transferred *between the system* and the environment as they interact. Heat is *not* a particular form of energy, nor is it a state variable. It makes no sense to talk about how heat changes. $Q = 0$ if a system does not interact thermally with its environment. Heat may cause the system's thermal energy to change, but that doesn't make heat and thermal energy the same.
- Temperature is a state variable that quantifies the "hotness" or "coldness" of a system. We haven't yet given a precise definition of temperature, but it is related to the thermal energy *per molecule.* A temperature difference is a requirement for a thermal interaction in which heat energy is transferred between the system and the environment.

It is especially important not to associate an observed temperature increase with heat. Heating a system is one way to change its temperature, but, as Joule showed, not the only way. You can also change the system's temperature by doing work on the system or, as is the case with friction, transforming mechanical energy into thermal energy. **Observing the system tells us nothing about the process by which energy enters or leaves the system.**

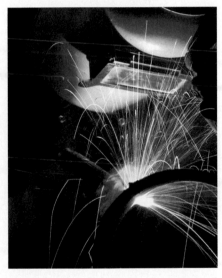

Heat is the energy transferred in a thermal interaction.

STOP TO THINK 19.3 Which one or more of the following processes involves heat?

a. The brakes in your car get hot when you stop.
b. A steel block is held over a candle.
c. You push a rigid cylinder of gas across a frictionless surface.
d. You push a piston into a cylinder of gas, increasing the temperature of the gas.
e. You place a cylinder of gas in hot water. The gas expands, causing a piston to rise and lift a weight. The temperature of the gas does not change.

19.4 The First Law of Thermodynamics

Heat was the missing piece that we needed to arrive at a completely general statement of the law of conservation of energy. Restating Equation 19.4, we have

$$\Delta E_{sys} = \Delta E_{mech} + \Delta E_{th} = W + Q$$

Work and heat, two ways of transferring energy between a system and the environment, cause the system's energy to change.

At this point in the text we are not interested in systems that have a macroscopic motion of the system as a whole. Moving macroscopic systems were important to us for many chapters, but now, as we investigate the thermal properties of a system, we would like the system as a whole to rest peacefully on the laboratory bench while we study it. So we will assume, throughout the remainder of Part V, that $\Delta E_{mech} = 0$.

With this assumption clearly stated, the law of conservation of energy becomes

$$\Delta E_{th} = W + Q \qquad \text{(first law of thermodynamics)} \qquad (19.15)$$

The energy equation, in this form, is called the **first law of thermodynamics** or simply "the first law." The first law is a very general statement about energy.

Chapters 9 and 10 introduced the basic energy model—*basic* because it included work but not heat. The first law of thermodynamics is the basis for a more general energy model, the **thermodynamic energy model,** in which work and heat are on an equal footing.

MODEL 19.1

Thermodynamic energy model

Thermal energy is a property of the system.

- **Work** and **heat** are energies transferred between the system and the environment.
  - Work is energy transferred in a mechanical interaction.
  - Heat is energy transferred in a thermal interaction.
- The **first law of thermodynamics** says that transferring energy to the system changes the system's thermal energy.
  - $\Delta E_{th} = W + Q$
  - W and $Q > 0$ for energy added.
  - W and $Q < 0$ for energy removed.
- Limitations: Model fails if the system's mechanical energy also changes.

Keep in mind that thermal energy isn't the only thing that changes. Work or heat that changes the system's thermal energy also changes other state variables, such as pressure, volume, or temperature. The first law tells us only about ΔE_{th}. Other laws and relationships, such as the ideal-gas law, are needed to learn how the other state variables change.

Three Special Ideal-Gas Processes

The ideal-gas law relates the state variables p, V, and T, but it tells us nothing about how these variables change if we *do* something to the gas. The first law of thermodynamics is an additional law that gases must obey, one focused specifically on the process by which a gas is changed. In general, you need to consider both laws to solve thermodynamics problems about gases.

There are three ideal-gas processes in which one of the three terms in the first law—ΔE_{th}, W, or Q—is zero:

- **Isothermal process ($\Delta E_{th} = 0$):** If the temperature of a gas doesn't change, neither does its thermal energy. In this case, the first law is $W + Q = 0$. For example, a gas can be compressed ($W > 0$) without a temperature increase if all the energy is returned to the environment as heat ($Q < 0$ for heat energy leaving the system). An isothermal process exchanges work and heat (one entering the system, the other leaving) without changing the gas temperature. But pressure and volume *do* change. Equation 19.14 for the work done in an isothermal process, together with the ideal-gas law, will allow you to calculate what happens to p and V. FIGURE 19.12a shows an energy bar chart for an isothermal process.

- **Isochoric process ($W = 0$):** We saw in Section 19.2 that work is done on (or by) a gas when its volume changes. An isochoric process has $\Delta V = 0$; thus no work is done and the first law can be written $\Delta E_{th} = Q$. This is a process in which the system is *mechanically isolated* from its environment. Heating (or cooling) the gas increases (or decreases) its thermal energy, so the temperature goes up (or down). You'll learn later in this chapter how to calculate the temperature change ΔT. After using the first law to find ΔT, *then* you can use the ideal-gas law to calculate the pressure change. FIGURE 19.12b is an energy bar chart for a cooling process that lowers the temperature.

- **Adiabatic process ($Q = 0$):** A process in which no heat is transferred—perhaps the system is extremely well insulated—is called an **adiabatic process.** The system is *thermally isolated* from its environment. With $Q = 0$, the first law is $\Delta E_{th} = W$. But just because there's no heat does not mean that the temperature remains constant. Compressing an insulated gas—a mechanical interaction with $W > 0$—raises its temperature! Similarly, an adiabatic expansion (one with no heat exchanged) lowers the gas temperature. $Q = 0$ does *not* mean $\Delta T = 0$. We'll examine adiabatic processes and their pV curves later in the chapter, but notice that *all three* state variables p, V, and T change during an adiabatic process. FIGURE 19.12c is an energy bar chart for an adiabatic process with an increasing temperature.

FIGURE 19.12 First-law bar charts for three special ideal-gas processes.

(a) Isothermal process: $\Delta E_{th} = 0$

$E_{th\,i} + W + Q = E_{th\,f}$

Energy that enters the system as heat leaves as work. The thermal energy is unchanged.

(b) Isochoric process: $W = 0$

$E_{th\,i} + W + Q = E_{th\,f}$

Thermal energy has decreased by the amount of energy that left the system as heat.

(c) Adiabatic process: $Q = 0$

$E_{th\,i} + W + Q = E_{th\,f}$

Energy that enters the system as work increases the thermal energy —and thus the temperature.

STOP TO THINK 19.4 Which first-law bar chart describes the process shown in the pV diagram?

$E_{th\,i} + W + Q = E_{th\,f}$

(a)

$E_{th\,i} + W + Q = E_{th\,f}$

(b)

$E_{th\,i} + W + Q = E_{th\,f}$

(c)

$E_{th\,i} + W + Q = E_{th\,f}$

(d)

19.5 Thermal Properties of Matter

Heat and work are equivalent in the sense that the change of the system is *exactly the same* whether you transfer heat energy to it or do an equal amount of work on it. Adding energy to the system, or removing it, changes the system's thermal energy.

What happens to a system when you change its thermal energy? In this section we'll consider two distinct possibilities:

- The temperature of the system changes.
- The system undergoes a phase change, such as melting or freezing.

Temperature Change and Specific Heat

Suppose you do an experiment in which you add energy to water, either by doing work on it or by transferring heat to it. Either way, you will find that adding 4190 J of energy raises the temperature of 1 kg of water by 1 K. If you were fortunate enough to have 1 kg of gold, you would need to add only 129 J of energy to raise its temperature by 1 K.

The amount of energy that raises the temperature of 1 kg of a substance by 1 K is called the **specific heat** of that substance. The symbol for specific heat is c. Water has specific heat $c_{\text{water}} = 4190$ J/kg K. The specific heat of gold is $c_{\text{gold}} = 129$ J/kg K. Specific heat depends only on the material from which an object is made. **TABLE 19.2** provides some specific heats for common liquids and solids.

> **NOTE** The term *specific heat* does not use the word "heat" in the way that we have defined it. Specific heat is an old idea, dating back to the days of the caloric theory when heat was thought to be a substance contained in the object. The term has continued in use even though our understanding of heat has changed.

If energy c is required to raise the temperature of 1 kg of a substance by 1 K, then energy Mc is needed to raise the temperature of mass M by 1 K and $(Mc)\Delta T$ is needed to raise the temperature of mass M by ΔT. In other words, the thermal energy of the system changes by

$$\Delta E_{\text{th}} = Mc\,\Delta T \qquad \text{(temperature change)} \qquad (19.16)$$

when its temperature changes by ΔT. ΔE_{th} can be either positive (thermal energy increases as the temperature goes up) or negative (thermal energy decreases as the temperature goes down). Recall that uppercase M is used for the mass of an entire system while lowercase m is reserved for the mass of an atom or molecule.

> **NOTE** In practice, ΔT is usually measured in °C. But the Kelvin and the Celsius temperature scales have the same step size, so ΔT in K has exactly the same numerical value as ΔT in °C. Thus

- You do not need to convert temperatures from °C to K if you need only a temperature change ΔT.
- You do need to convert anytime you need the actual temperature T.

The first law of thermodynamics, $\Delta E_{\text{th}} = W + Q$, allows us to write Equation 19.16 as $Mc\,\Delta T = W + Q$. In other words, **we can change the system's temperature either by heating it or by doing an equivalent amount of work on it.** In working with solids and liquids, we almost always change the temperature by heating. If $W = 0$, which we will assume for the rest of this section, then the heat energy needed to bring about a temperature change ΔT is

$$Q = Mc\,\Delta T \qquad \text{(temperature change)} \qquad (19.17)$$

Because $\Delta T = \Delta E_{\text{th}}/Mc$, it takes more energy to change the temperature of a substance with a large specific heat than to change the temperature of a substance with a small

TABLE 19.2 Specific heats and molar specific heats of solids and liquids

Substance	c (J/kg K)	C (J/mol K)
Solids		
Aluminum	900	24.3
Copper	385	24.4
Iron	449	25.1
Gold	129	25.4
Lead	128	26.5
Ice	2090	37.6
Liquids		
Ethyl alcohol	2400	110.4
Mercury	140	28.1
Water	4190	75.4

specific heat. You can think of specific heat as measuring the *thermal inertia* of a substance. Metals, with small specific heats, warm up and cool down quickly. A piece of aluminum foil can be safely held within seconds of removing it from a hot oven. Water, with a very large specific heat, is slow to warm up and slow to cool down. This is fortunate for us. The large thermal inertia of water is essential for the biological processes of life. We wouldn't be here studying physics if water had a small specific heat!

EXAMPLE 19.3 | **Running a fever**

A 70 kg student catches the flu, and his body temperature increases from 37.0°C (98.6°F) to 39.0°C (102.2°F). How much energy is required to raise his body's temperature? The specific heat of a mammalian body is 3400 J/kg K, nearly that of water because mammals are mostly water.

MODEL Energy is supplied by the chemical reactions of the body's metabolism. These exothermic reactions transfer heat to the body. Normal metabolism provides enough heat energy to offset energy losses (radiation, evaporation, etc.) while maintaining a normal

body temperature of 37°C. We need to calculate the additional energy needed to raise the body's temperature by 2.0°C, or 2.0 K.

SOLVE The necessary heat energy is

$$Q = Mc\,\Delta T = (70\text{ kg})(3400\text{ J/kg K})(2.0\text{ K}) = 4.8 \times 10^5\text{ J}$$

ASSESS This appears to be a lot of energy, but a joule is actually a very small amount of energy. It is only 110 Cal, approximately the energy gained by eating an apple.

The **molar specific heat** is the amount of energy that raises the temperature of 1 mol of a substance by 1 K. We'll use an uppercase C for the molar specific heat. The heat energy needed to bring about a temperature change ΔT of n moles of substance is

$$Q = nC\,\Delta T \qquad (19.18)$$

Molar specific heats are listed in Table 19.2. Look at the five elemental solids (excluding ice). All have C very near 25 J/mol K. If we were to expand the table, we would find that most elemental solids have $C \approx 25$ J/mol K. This can't be a coincidence, but what is it telling us? This is a puzzle we will address in Chapter 20, where we will explore thermal energy at the atomic level.

Phase Change and Heat of Transformation

Suppose you start with a system in its solid phase and heat it at a steady rate. FIGURE 19.13, which you saw in Chapter 18, shows how the system's temperature changes. At first, the temperature increases linearly. This is not hard to understand because Equation 19.17 can be written

$$\text{slope of the } T\text{-versus-}Q \text{ graph} = \frac{\Delta T}{Q} = \frac{1}{Mc} \qquad (19.19)$$

The slope of the graph depends inversely on the system's specific heat. A constant specific heat implies a constant slope and thus a linear graph. In fact, you can measure c from such a graph.

| **NOTE** The different slopes indicate that the solid, liquid, and gas phases of a substance have different specific heats.

But there are times, shown as horizontal line segments, during which heat is being transferred to the system but the temperature isn't changing. These are *phase changes*. The thermal energy continues to increase during a phase change, but the additional energy goes into breaking molecular bonds rather than speeding up the molecules. **A phase change is characterized by a change in thermal energy without a change in temperature.**

The amount of heat energy that causes 1 kg of a substance to undergo a phase change is called the **heat of transformation** of that substance. For example, laboratory experiments show that 333,000 J of heat are needed to melt 1 kg of ice at 0°C.

FIGURE 19.13 The temperature of a system that is heated at a steady rate.

The system's thermal energy is increasing as heat energy is added, but it's causing a phase change rather than a temperature change.

Lava—molten rock—undergoes a phase change when it contacts the much colder water. This is one way in which new islands are formed.

The symbol for heat of transformation is L. The heat required for the entire system of mass M to undergo a phase change is

$$Q = ML \qquad \text{(phase change)} \qquad (19.20)$$

Heat of transformation is a generic term that refers to any phase change. Two specific heats of transformation are the **heat of fusion** L_f, the heat of transformation between a solid and a liquid, and the **heat of vaporization** L_v, the heat of transformation between a liquid and a gas. The heat needed for these phase changes is

$$Q = \begin{cases} \pm ML_f & \text{melt/freeze} \\ \pm ML_v & \text{boil/condense} \end{cases} \qquad (19.21)$$

where the $\pm$ indicates that heat must be *added* to the system during melting or boiling but *removed* from the system during freezing or condensing. **You must explicitly include the minus sign when it is needed.**

TABLE 19.3 gives the heats of transformation of a few substances. Notice that the heat of vaporization is always much larger than the heat of fusion. We can understand this. Melting breaks just enough molecular bonds to allow the system to lose rigidity and flow. Even so, the molecules in a liquid remain close together and loosely bonded. Vaporization breaks all bonds completely and sends the molecules flying apart. This process requires a larger increase in the thermal energy and thus a larger quantity of heat.

TABLE 19.3 Melting/boiling temperatures and heats of transformation

Substance	T_m (°C)	L_f (J/kg)	T_b (°C)	L_v (J/kg)
Nitrogen (N_2)	−210	0.26×10^5	−196	1.99×10^5
Ethyl alcohol	−114	1.09×10^5	78	8.79×10^5
Mercury	−39	0.11×10^5	357	2.96×10^5
Water	0	3.33×10^5	100	22.6×10^5
Lead	328	0.25×10^5	1750	8.58×10^5

EXAMPLE 19.4 | Melting wax

An insulated jar containing 200 g of solid candle wax is placed on a hot plate that supplies heat energy to the wax at the rate of 220 J/s. The wax temperature is measured every 30 s, yielding the following data:

Time (s)	Temperature (°C)	Time (s)	Temperature (°C)
0	20.0	180	70.5
30	31.7	210	70.5
60	42.2	240	70.6
90	55.0	270	70.5
120	64.7	300	70.4
150	70.4	330	74.5

What are the specific heat of the solid wax, the melting point, and the wax's heat of fusion?

MODEL The wax is in an insulated jar, so assume that heat loss to the environment is negligible.

VISUALIZE Heat energy is being supplied at the rate of 220 J/s, so the total heat energy that has been transferred into the wax at time t is $Q = 220t$ J. FIGURE 19.14 shows the temperature graphed

FIGURE 19.14 The heating curve of the wax.

against the cumulative heat Q, although notice that the horizontal axis is in kJ, not J. The initial linear slope corresponds to raising the wax's temperature to the melting point. Temperature remains constant during a phase change, even though the sample is still being heated, so the horizontal section of the graph is when the wax is melting. The temperature increase at the end shows that the temperature of the liquid wax is beginning to rise after melting is complete.

SOLVE From $Q = Mc\,\Delta T$, the slope of the T-versus-Q graph is $\Delta T/Q = 1/Mc$. The experimental slope of the best-fit line is $1.708°C/kJ = 0.001708$ K/J. Thus the specific heat of the solid wax is

$$c = \frac{1}{M \times \text{slope}} = \frac{1}{(0.200 \text{ kg})(0.001708 \text{ K/J})} = 2930 \text{ J/kg K}$$

From the table, we see that the melting temperature—which remains constant during the phase change—is 70.5°C. The heat required for the phase change is $Q = ML_f$, so the heat of fusion is $L_f = Q/M$. With data recorded only every 30 s, it's not exactly clear when the melting began and when it ended. The extension of the initial slope shows that the temperature reached the melting point about halfway between 120 s and 150 s, so the melting started at about 135 s. We'll assume it was complete about halfway between 300 s and 330 s, or at about 315 s. Thus the melting took 180 s, during which, at 220 J/s, 39,600 J of heat energy was transferred from the hot plate to the wax. With this value of Q, the heat of fusion is

$$L_f = \frac{Q}{M} = \frac{39,600 \text{ J}}{0.200 \text{ kg}} = 2.0 \times 10^5 \text{ J/kg}$$

ASSESS Both the specific heat and the heat of fusion are similar to values in Tables 19.2 and 19.3, which gives us confidence in our results.

STOP TO THINK 19.5 Objects A and B are brought into close thermal contact with each other, but they are well isolated from their surroundings. Initially $T_A = 0°C$ and $T_B = 100°C$. The specific heat of A is less than the specific heat of B. The two objects will soon reach a common final temperature T_f. The final temperature is

a. $T_f > 50°C$ b. $T_f = 50°C$ c. $T_f < 50°C$

19.6 Calorimetry

At one time or another you've probably put an ice cube into a hot drink to cool it quickly. You were engaged, in a somewhat trial-and-error way, in a practical aspect of heat transfer known as **calorimetry.**

FIGURE 19.15 shows two systems thermally interacting with each other but isolated from everything else. Suppose they start at different temperatures T_1 and T_2. As you know from experience, heat energy will be transferred from the hotter to the colder system until they reach a common final temperature T_f. The systems will then be in thermal equilibrium and the temperature will not change further.

The insulation prevents any heat energy from being transferred to or from the environment, so energy conservation tells us that any energy leaving the hotter system must enter the colder system. That is, the systems *exchange* energy with no net loss or gain. The concept is straightforward, but to state the idea mathematically we need to be careful with signs.

Let Q_1 be the energy transferred to system 1 as heat. Similarly, Q_2 is the energy transferred to system 2. The fact that the systems are merely exchanging energy can be written $|Q_1| = |Q_2|$. The energy *lost* by the hotter system is the energy *gained* by the colder system, so Q_1 and Q_2 have opposite signs: $Q_1 = -Q_2$. No energy is exchanged with the environment, hence it makes more sense to write this relationship as

$$Q_{net} = Q_1 + Q_2 = 0 \qquad (19.22)$$

This idea is not limited to the interaction of only two systems. If three or more systems are combined in isolation from the rest of their environment, each at a different initial temperature, they will all come to a common final temperature that can be found from the relationship

$$Q_{net} = Q_1 + Q_2 + Q_3 + \cdots = 0 \qquad (19.23)$$

NOTE The signs are very important in calorimetry problems. ΔT is always $T_f - T_i$, so ΔT and Q are negative for any system whose temperature decreases. The proper sign of Q for any phase change must be supplied *by you*, depending on the direction of the phase change.

FIGURE 19.15 Two systems interact thermally.

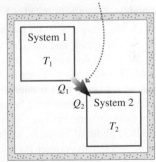

Heat energy is transferred from system 1 to system 2.

Energy conservation requires
$$|Q_1| = |Q_2|$$

PROBLEM-SOLVING STRATEGY 19.2

Calorimetry problems

MODEL Model the systems as interacting with each other but isolated from the larger environment.

VISUALIZE List known information and identify what you need to find. Convert all quantities to SI units.

SOLVE The mathematical representation, a statement of energy conservation, is

$$Q_{net} = Q_1 + Q_2 + \cdots = 0$$

- For systems that undergo a temperature change, $Q = Mc(T_f - T_i)$. Be sure to have the temperatures T_i and T_f in the correct order.
- For systems that undergo a phase change, $Q = \pm ML$. Supply the correct sign by observing whether energy enters or leaves the system.
- Some systems may undergo a temperature change *and* a phase change. Treat the changes separately. The heat energy is $Q = Q_{\Delta T} + Q_{phase}$.

ASSESS Is the final temperature in the middle? T_f that is higher or lower than all initial temperatures is an indication that something is wrong, usually a sign error.

Exercise 15

NOTE You may have learned to solve calorimetry problems in other courses by writing $Q_{gained} = Q_{lost}$. That is, by balancing heat gained with heat lost. That approach works in simple problems, but it has two drawbacks. First, you often have to "fudge" the signs to make them work. Second, and more serious, you can't extend this approach to a problem with three or more interacting systems. Using $Q_{net} = 0$ is much preferred.

EXAMPLE 19.5 | Calorimetry with a phase change

Your 500 mL soda is at 20°C, room temperature, so you add 100 g of ice from the −20°C freezer. Does all the ice melt? If so, what is the final temperature? If not, what fraction of the ice melts? Assume that you have a well-insulated cup.

MODEL We have a thermal interaction between the soda, which is essentially water, and the ice. We need to distinguish between three possible outcomes. If all the ice melts, then $T_f > 0°C$. It's also possible that the soda will cool to 0°C before all the ice has melted, leaving the ice and liquid in equilibrium at 0°C. A third possibility is that the soda will freeze solid before the ice warms up to 0°C. That seems unlikely here, but there are situations, such as the pouring of molten metal out of furnaces, when all the liquid does solidify. We need to distinguish between these before knowing how to proceed.

VISUALIZE All the initial temperatures, masses, and specific heats are known. The final temperature of the combined soda + ice system is unknown.

SOLVE Let's first calculate the heat needed to melt all the ice and leave it as liquid water at 0°C. To do so, we must warm the ice to 0°C, then change it to water. The heat input for this two-stage process is

$$Q_{melt} = M_i c_i (20\ K) + M_i L_f = 37,500\ J$$

where L_f is the heat of fusion of water. It is used as a *positive* quantity because we must *add* heat to melt the ice. Next, let's calculate how much heat energy will leave the soda if it cools all the way to 0°C. The volume is $V = 500\ mL = 5.00 \times 10^{-4}\ m^3$ and thus the mass is $M_s = \rho V = 0.500\ kg$. The heat loss is

$$Q_{cool} = M_s c_w (-20\ K) = -41,900\ J$$

where $\Delta T = -20\ K$ because the temperature decreases. Because $|Q_{cool}| > Q_{melt}$, the soda has sufficient energy to melt all the ice. Hence the final state will be all liquid at $T_f > 0$. (Had we found $|Q_{cool}| < Q_{melt}$, then the final state would have been an ice-liquid mixture at 0°C.)

Energy conservation requires $Q_{ice} + Q_{soda} = 0$. The heat Q_{ice} consists of three terms: warming the ice to 0°C, melting the ice to water at 0°C, then warming the 0°C water to T_f. The mass will still be M_i in the last of these steps because it is the "ice system," but we need to use the specific heat of *liquid water*. Thus

$$Q_{ice} + Q_{soda} = [M_i c_i (20\ K) + M_i L_f + M_i c_w (T_f - 0°C)]$$
$$+ M_s c_w (T_f - 20°C) = 0$$

We've already done part of the calculation, allowing us to write

$$37,500\ J + M_i c_w (T_f - 0°C) + M_s c_w (T_f - 20°C) = 0$$

Solving for T_f gives

$$T_f = \frac{20 M_s c_w - 37,500}{M_i c_w + M_s c_w} = 1.7°C$$

ASSESS As expected, the soda has been cooled to nearly the freezing point.

EXAMPLE 19.6 | Three interacting systems

A 200 g piece of iron at 120°C and a 150 g piece of copper at −50°C are dropped into an insulated beaker containing 300 g of ethyl alcohol at 20°C. What is the final temperature?

MODEL Here you can't use a simple $Q_{gained} = Q_{lost}$ approach because you don't know whether the alcohol is going to warm up or cool down. In principle, the alcohol could freeze or boil, but the masses and temperatures of the metals suggest that the temperature will not change greatly. We will assume that the alcohol remains a liquid.

VISUALIZE All the initial temperatures, masses, and specific heats are known. We need to find the final temperature.

SOLVE Energy conservation requires

$$Q_i + Q_c + Q_e = M_i c_i (T_f - 120°C) + M_c c_c (T_f - (-50°C))$$
$$+ M_e c_e (T_f - 20°C) = 0$$

Solving for T_f gives

$$T_f = \frac{120 M_i c_i - 50 M_c c_c + 20 M_e c_e}{M_i c_i + M_c c_c + M_e c_e} = 26°C$$

ASSESS The temperature is between the initial iron and copper temperatures, as expected, and well below the boiling temperature of the alcohol, validating our assumption of no phase change. It turns out that the alcohol warms up ($Q_e > 0$), but we had no way to know this without doing the calculation.

19.7 The Specific Heats of Gases

Specific heats are given in Table 19.2 for solids and liquids. Gases are harder to characterize because the heat required to cause a specified temperature change depends on the *process* by which the gas changes state.

FIGURE 19.16 shows two isotherms on the pV diagram for a gas. Processes A and B, which start on the T_i isotherm and end on the T_f isotherm, have the *same* temperature change $\Delta T = T_f - T_i$. But process A, which takes place at constant volume, requires a *different* amount of heat than does process B, which occurs at constant pressure. The reason is that work is done in process B but not in process A. This is a situation that we are now equipped to analyze.

It is useful to define two different versions of the specific heat of gases, one for constant-volume (isochoric) processes and one for constant-pressure (isobaric) processes. We will define these as molar specific heats because we usually do gas calculations using moles instead of mass. The quantity of heat needed to change the temperature of n moles of gas by ΔT is

$$Q = n C_V \Delta T \quad \text{(temperature change at constant volume)}$$
$$Q = n C_P \Delta T \quad \text{(temperature change at constant pressure)} \quad (19.24)$$

where C_V is the **molar specific heat at constant volume** and C_P is the **molar specific heat at constant pressure**. **TABLE 19.4** gives the values of C_V and C_P for a few common monatomic and diatomic gases. The units are J/mol K. Molar specific heats do vary somewhat with temperature—we'll look at an example in the next chapter—but the values in the table are adequate for temperatures from 200 K to 800 K.

NOTE Equations 19.24 apply to two specific ideal-gas processes. In a general gas process, for which neither p nor V is constant, we have no direct way to relate Q to ΔT. In that case, the heat must be found indirectly from the first law as $Q = \Delta E_{th} - W$.

FIGURE 19.16 Processes A and B have the same ΔT and the same ΔE_{th}, but they require different amounts of heat.

TABLE 19.4 Molar specific heats of gases (J/mol K) at $T = 0°C$

Gas	C_P	C_V	$C_P - C_V$
Monatomic Gases			
He	20.8	12.5	8.3
Ne	20.8	12.5	8.3
Ar	20.8	12.5	8.3
Diatomic Gases			
H_2	28.7	20.4	8.3
N_2	29.1	20.8	8.3
O_2	29.2	20.9	8.3

EXAMPLE 19.7 | Heating and cooling a gas

Three moles of O_2 gas are at 20.0°C. 600 J of heat energy are transferred to the gas at constant pressure, then 600 J are removed at constant volume. What is the final temperature? Show the process on a pV diagram.

MODEL O_2 is a diatomic ideal gas. The gas is heated as an isobaric process, then cooled as an isochoric process.

SOLVE The heat transferred during the constant-pressure process causes a temperature rise

$$\Delta T = T_2 - T_1 = \frac{Q}{nC_P} = \frac{600 \text{ J}}{(3.0 \text{ mol})(29.2 \text{ J/mol K})} = 6.8°C$$

Continued

where C_P for oxygen was taken from Table 19.4. Heating leaves the gas at temperature $T_2 = T_1 + \Delta T = 26.8°C$. The temperature then falls as heat is removed during the constant-volume process:

$$\Delta T = T_3 - T_2 = \frac{Q}{nC_V} = \frac{-600 \text{ J}}{(3.0 \text{ mol})(20.9 \text{ J/mol K})} = -9.5°C$$

We used a *negative* value for Q because heat energy is transferred from the gas to the environment. The final temperature of the gas is $T_3 = T_2 + \Delta T = 17.3°C$. **FIGURE 19.17** shows the process on a pV diagram. The gas expands (moves horizontally on the diagram) as heat is added, then cools at constant volume (moves vertically on the diagram) as heat is removed.

ASSESS The final temperature is lower than the initial temperature because $C_P > C_V$.

FIGURE 19.17 The pV diagram for Example 19.7.

EXAMPLE 19.8 | **Calorimetry with a gas and a solid**

The interior volume of a 200 g hollow aluminum box is 800 cm³. The box contains nitrogen gas at STP. A 20 cm³ block of copper at a temperature of 300°C is placed inside the box, then the box is sealed. What is the final temperature?

MODEL This example has three interacting systems: the aluminum box, the nitrogen gas, and the copper block. They must all come to a common final temperature T_f.

VISUALIZE The box and gas have the same initial temperature: $T_{Al} = T_{N2} = 0°C$. The box doesn't change size, so this is a constant-volume process. The final temperature is unknown.

SOLVE Although one of the systems is now a gas, the calorimetry equation $Q_{net} = Q_{Al} + Q_{N2} + Q_{Cu} = 0$ is still appropriate. In this case,

$$Q_{net} = m_{Al}c_{Al}(T_f - T_{Al}) + n_{N2}C_V(T_f - T_{N2})$$

$$+ m_{Cu}c_{Cu}(T_f - T_{Cu}) = 0$$

Notice that we used masses and specific heats for the solids but moles and the molar specific heat for the gas. We used C_V because this is a constant-volume process. Solving for T_f gives

$$T_f = \frac{m_{Al}c_{Al}T_{Al} + n_{N2}C_V T_{N2} + m_{Cu}c_{Cu}T_{Cu}}{m_{Al}c_{Al} + n_{N2}C_V + m_{Cu}c_{Cu}}$$

The specific heat values are found in Tables 19.2 and 19.4. The mass of the copper is

$$m_{Cu} = \rho_{Cu}V_{Cu} = (8920 \text{ kg/cm}^3)(20 \times 10^{-6} \text{ m}^3) = 0.178 \text{ kg}$$

The number of moles of the gas is found from the ideal-gas law, using the initial conditions. Notice that inserting the copper block *displaces* 20 cm³ of gas; hence the gas volume is only $V = 780 \text{ cm}^3 = 7.80 \times 10^{-4} \text{ m}^3$. Thus

$$n_{N2} = \frac{pV}{RT} = 0.0348 \text{ mol}$$

Computing the final temperature gives $T_f = 83°C$.

C_P and C_V

You may have noticed two curious features in Table 19.4. First, the molar specific heats of monatomic gases are *all alike*. And the molar specific heats of diatomic gases, while different from monatomic gases, are again *very nearly alike*. We saw a similar feature in Table 19.2 for the molar specific heats of solids. Second, the *difference* $C_P - C_V = 8.3$ J/mol K is the same in every case. And, most puzzling of all, the value of $C_P - C_V$ appears to be equal to the universal gas constant R! Why should this be?

The relationship between C_V and C_P hinges on one crucial idea: ΔE_{th}, **the change in the thermal energy of a gas, is the same for *any* two processes that have the same ΔT.** The thermal energy of a gas is associated with temperature, so any process that changes the gas temperature from T_i to T_f has the same ΔE_{th} as any other process that goes from T_i to T_f. Furthermore, the first law $\Delta E_{th} = Q + W$ tells us that a gas cannot distinguish between heat and work. The system's thermal energy changes in response to energy added to or removed from the system, but the response of the gas is the same whether you heat the system, do work on the system, or do some combination of both. Thus **any two processes that change the thermal energy of the gas by ΔE_{th} will cause the same temperature change ΔT.**

With that in mind, look back at Figure 19.16. Both gas processes have the same ΔT, so both have the same value of ΔE_{th}. Process A is an isochoric process in which no work is done (the piston doesn't move), so the first law for this process is

$$(\Delta E_{th})_A = W + Q = 0 + Q_{const\ vol} = nC_V\,\Delta T \qquad (19.25)$$

Process B is an isobaric process. You learned earlier that the work done on the gas during an isobaric process is $W = -p\,\Delta V$. Thus

$$(\Delta E_{th})_B = W + Q = -p\,\Delta V + Q_{const\ press} = -p\,\Delta V + nC_P\,\Delta T \qquad (19.26)$$

$(\Delta E_{th})_B = (\Delta E_{th})_A$ because both have the same ΔT, so we can equate the right sides of Equations 19.25 and 19.26:

$$-p\,\Delta V + nC_P\,\Delta T = nC_V\,\Delta T \qquad (19.27)$$

For the final step, we can use the ideal-gas law $pV = nRT$ to relate ΔV and ΔT during process B. For any gas process,

$$\Delta(pV) = \Delta(nRT) \qquad (19.28)$$

For a constant-pressure process, where p is constant, Equation 19.28 becomes

$$p\,\Delta V = nR\,\Delta T \qquad (19.29)$$

Substituting this expression for $p\,\Delta V$ into Equation 19.27 gives

$$-nR\,\Delta T + nC_P\,\Delta T = nC_V\,\Delta T \qquad (19.30)$$

The $n\,\Delta T$ cancels, and we are left with

$$C_P = C_V + R \qquad (19.31)$$

This result, which applies to all ideal gases, is exactly what we see in the data of Table 19.4.

But that's not the only conclusion we can draw. Equation 19.25 found that $\Delta E_{th} = nC_V\,\Delta T$ for a constant-volume process. But we had just noted that ΔE_{th} is the same for *all* gas processes that have the same ΔT. Consequently, this expression for ΔE_{th} is equally true for any other process. That is

$$\Delta E_{th} = nC_V\,\Delta T \qquad \text{(any ideal-gas process)} \qquad (19.32)$$

Compare this result to Equations 19.24. We first made a distinction between constant-volume and constant-pressure processes, but now we're saying that Equation 19.32 is true for any process. Are we contradicting ourselves? No, the difference lies in what you need to calculate.

- The change in thermal energy when the temperature changes by ΔT is the same for any process. That's Equation 19.32.
- The *heat* required to bring about the temperature change depends on what the process is. That's Equations 19.24. An isobaric process requires more heat than an isochoric process that produces the same ΔT.

The reason for the difference is seen by writing the first law as $Q = \Delta E_{th} - W$. In an isochoric process, where $W = 0$, *all* the heat input is used to increase the gas temperature. But in an isobaric process, some of the energy that enters the system as heat then leaves the system as work ($W < 0$) done by the expanding gas. Thus more heat is needed to produce the same ΔT.

Heat Depends on the Path

Consider the two ideal-gas processes shown in FIGURE 19.18. Even though the initial and final states are the same, the heat added during these two processes is *not* the same. We can use the first law $\Delta E_{th} = W + Q$ to see why.

FIGURE 19.18 Is the heat input along these two paths the same or different?

The thermal energy is a state variable. That is, its value depends on the state of the gas, not the process by which the gas arrived at that state. Thus $\Delta E_{th} = E_{th\,f} - E_{th\,i}$ is the same for both processes. If ΔE_{th} is the same for processes A and B, then $W_A + Q_A = W_B + Q_B$.

You learned in Section 19.2 that the work done during an ideal-gas process depends on the path in the pV diagram. There's more area under the process B curve, so $|W_B| > |W_A|$. Both values of W are negative because the gas expands, so W_B is more negative than W_A. Consequently, $W_A + Q_A$ can equal $W_B + Q_B$ only if $Q_B > Q_A$. **The heat added or removed during an ideal-gas process depends on the path followed through the pV diagram.**

Adiabatic Processes

Section 19.4 introduced the idea of an *adiabatic process,* a process in which no heat energy is transferred ($Q = 0$). **FIGURE 19.19** compares an adiabatic process with isothermal and isochoric processes. We're now prepared to look at adiabatic processes in more detail.

In practice, there are two ways that an adiabatic process can come about. First, a gas cylinder can be completely surrounded by thermal insulation, such as thick pieces of Styrofoam. The environment can interact mechanically with the gas by pushing or pulling on the insulated piston, but there is no thermal interaction.

Second, the gas can be expanded or compressed very rapidly in what we call an *adiabatic expansion* or an *adiabatic compression*. In a rapid process there is essentially no time for heat to be transferred between the gas and the environment. We've already alluded to the idea that heat is transferred via atomic-level collisions. These collisions take time. If you stick one end of a copper rod into a flame, the other end will eventually get too hot to hold—but not instantly. Some amount of time is required for heat to be transferred from one end to the other. A process that takes place faster than the heat can be transferred is adiabatic.

> **NOTE** You may recall reading in Chapter 18 that we are going to study only quasi-static processes, processes that proceed slowly enough to remain essentially in equilibrium at all times. Now we're proposing to study processes that take place very rapidly. Isn't this a contradiction? Yes, to some extent it is. What we need to establish are the appropriate time scales. How slow must a process go to be quasi-static? How fast must it go to be adiabatic? These types of calculations must be deferred to a more advanced course. It turns out—fortunately!—that many practical applications, such as the compression strokes in gasoline and diesel engines, are fast enough to be adiabatic yet slow enough to still be considered quasi-static.

For an adiabatic process, with $Q = 0$, the first law of thermodynamics is $\Delta E_{th} = W$. Compressing a gas adiabatically ($W > 0$) increases the thermal energy. Thus an adiabatic compression raises the temperature of a gas. A gas that expands adiabatically ($W < 0$) gets colder as its thermal energy decreases. Thus an adiabatic expansion lowers the temperature of a gas. You can use an adiabatic process to change the gas temperature without using heat!

The work done in an adiabatic process goes entirely to changing the thermal energy of the gas. But we just found that $\Delta E_{th} = nC_V \Delta T$ for *any* process. Thus

$$W = nC_V \Delta T \qquad \text{(adiabatic process)} \qquad (19.33)$$

Equation 19.33 joins with the equations we derived earlier for the work done in isochoric, isobaric, and isothermal processes.

Gas processes can be represented as trajectories in the pV diagram. For example, a gas moves along a hyperbola during an isothermal process. How does an adiabatic process appear in a pV diagram? The result is more important than the derivation,

FIGURE 19.19 The relationship of three important processes to the first law of thermodynamics.

An *isothermal* process has $\Delta E_{th} = 0$, so $W = -Q$.

An *isochoric* process has $W = 0$, so $\Delta E_{th} = Q$.

$$\Delta E_{th} = W + Q$$

An *adiabatic* process has $Q = 0$, so $\Delta E_{th} = W$.

which is a bit tedious, so we'll begin with the answer and then, at the end of this section, show where it comes from.

First, we define the **specific heat ratio** γ (lowercase Greek gamma) to be

$$\gamma = \frac{C_P}{C_V} = \begin{cases} 1.67 & \text{monatomic gas} \\ 1.40 & \text{diatomic gas} \end{cases} \qquad (19.34)$$

The specific heat ratio has many uses in thermodynamics. Notice that γ is dimensionless.

An adiabatic process is one in which

$$pV^\gamma = \text{constant} \qquad \text{or} \qquad p_f V_f^\gamma = p_i V_i^\gamma \qquad (19.35)$$

This is similar to the isothermal $pV = \text{constant}$, but somewhat more complex due to the exponent γ.

The curves found by graphing $p = \text{constant}/V^\gamma$ are called **adiabats.** In FIGURE 19.20 you see that the two adiabats are steeper than the hyperbolic isotherms. An adiabatic process moves along an adiabat in the same way that an isothermal process moves along an isotherm. You can see that the temperature falls during an adiabatic expansion and rises during an adiabatic compression.

If we use the ideal-gas-law expression $p = nRT/V$ in the adiabatic equation $pV^\gamma = \text{constant}$, we see that $TV^{\gamma-1}$ is also constant during an adiabatic process. Thus another useful equation for adiabatic processes is

$$T_f V_f^{\gamma-1} = T_i V_i^{\gamma-1} \qquad (19.36)$$

FIGURE 19.20 An adiabatic process moves along pV curves called *adiabats*.

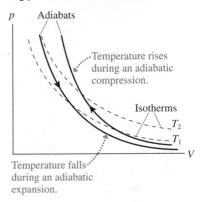

EXAMPLE 19.9 | An adiabatic compression

Air containing gasoline vapor is admitted into the cylinder of an internal combustion engine at 1.00 atm pressure and 30°C. The piston rapidly compresses the gas from 500 cm³ to 50 cm³, a *compression ratio* of 10.

a. What are the final temperature and pressure of the gas?
b. Show the compression process on a pV diagram.
c. How much work is done to compress the gas?

MODEL The compression is rapid, with insufficient time for heat to be transferred from the gas to the environment, so we will model it as an adiabatic compression. We'll treat the gas as if it were 100% air.

SOLVE a. We know the initial pressure and volume, and we know the volume after the compression. For an adiabatic process, where pV^γ remains constant, the final pressure is

$$p_f = p_i \left(\frac{V_i}{V_f}\right)^\gamma = (1.00 \text{ atm})(10)^{1.40} = 25.1 \text{ atm}$$

Air is a mixture of N_2 and O_2, diatomic gases, so we used $\gamma = 1.40$. We can now find the temperature by using the ideal-gas law:

$$T_f = T_i \frac{p_f V_f}{p_i V_i} = (303 \text{ K})(25.1)\left(\frac{1}{10}\right) = 761 \text{ K} = 488°C$$

Temperature *must* be in kelvins for doing gas calculations such as these.

b. FIGURE 19.21 shows the pV diagram. The 30°C and 488°C isotherms are included to show how the temperature changes during the process.

FIGURE 19.21 The adiabatic compression of the gas in an internal combustion engine.

c. The work done is $W = nC_V \Delta T$, with $\Delta T = 458$ K. The number of moles is found from the ideal-gas law and the initial conditions:

$$n = \frac{p_i V_i}{RT_i} = 0.0201 \text{ mol}$$

Thus the work done to compress the gas is

$$W = nC_V \Delta T = (0.0201 \text{ mol})(20.8 \text{ J/mol K})(458 \text{ K}) = 192 \text{ J}$$

ASSESS The temperature rises dramatically during the compression stroke of an engine. But the higher temperature has nothing to do with heat! **The temperature and thermal energy of the gas are increased not by heating the gas but by doing work on it.** This is an important idea to understand.

Proof of Equation 19.35

Now let's see where Equation 19.35 comes from. Consider an adiabatic process in which an infinitesimal amount of work dW done on a gas causes an infinitesimal change in the thermal energy. For an adiabatic process, with $dQ = 0$, the first law of thermodynamics is

$$dE_{th} = dW \tag{19.37}$$

We can use Equation 19.32, which is valid for *any* gas process, to write $dE_{th} = nC_V\,dT$. Earlier in the chapter we found that the work done during a small volume change is $dW = -p\,dV$. With these substitutions, Equation 19.37 becomes

$$nC_V\,dT = -p\,dV \tag{19.38}$$

The ideal-gas law can now be used to write $p = nRT/V$. The n cancels, and the C_V can be moved to the other side of the equation to give

$$\frac{dT}{T} = -\frac{R}{C_V}\frac{dV}{V} \tag{19.39}$$

We're going to integrate Equation 19.39, but anticipating the need for $\gamma = C_P/C_V$ we can first use the fact that $C_P = C_V + R$ to write

$$\frac{R}{C_V} = \frac{C_P - C_V}{C_V} = \frac{C_P}{C_V} - 1 = \gamma - 1 \tag{19.40}$$

Now we integrate Equation 19.39 from the initial state i to the final state f:

$$\int_{T_i}^{T_f} \frac{dT}{T} = -(\gamma - 1)\int_{V_i}^{V_f} \frac{dV}{V} \tag{19.41}$$

Carrying out the integration gives

$$\ln\left(\frac{T_f}{T_i}\right) = \ln\left(\frac{V_i}{V_f}\right)^{\gamma-1} \tag{19.42}$$

where we used the logarithm properties $\log a - \log b = \log(a/b)$ and $c\log a = \log(a^c)$.
Taking the exponential of both sides now gives

$$\left(\frac{T_f}{T_i}\right) = \left(\frac{V_i}{V_f}\right)^{\gamma-1}$$
$$T_f V_f^{\gamma-1} = T_i V_i^{\gamma-1} \tag{19.43}$$

This was Equation 19.36. Writing $T = pV/nR$ and canceling $1/nR$ from both sides of the equation give Equation 19.35:

$$p_f V_f^{\gamma} = p_i V_i^{\gamma} \tag{19.44}$$

This was a lengthy derivation, but it is good practice at seeing how the ideal-gas law and the first law of thermodynamics can work together to yield results of great importance.

STOP TO THINK 19.6 For the two processes shown, which of the following is true:

a. $Q_A > Q_B$
b. $Q_A = Q_B$
c. $Q_A < Q_B$

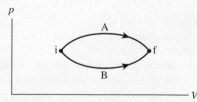

19.8 Heat-Transfer Mechanisms

You feel warmer when the sun is shining on you, colder when sitting on a metal bench or when the wind is blowing, especially if your skin is wet. This is due to the transfer of heat. Although we've talked about heat a lot in this chapter, we haven't said much about *how* heat is transferred from a hotter object to a colder object. There are four basic mechanisms by which objects exchange heat with their surroundings. Evaporation was treated in Section 19.5; in this section, we will consider the other mechanisms.

Heat-transfer mechanisms

When two objects are in direct contact, such as the soldering iron and the circuit board, heat is transferred by *conduction.*

Air currents near a lighted candle rise, taking thermal energy with them in a process known as *convection.*

The lamp at the top shines on the lambs huddled below, warming them. The energy is transferred by *radiation.*

Blowing on a hot cup of tea or coffee cools it by *evaporation.*

Conduction

FIGURE 19.22 shows an object sandwiched between a higher temperature T_H and a lower temperature T_C. The temperature *difference* causes heat energy to be transferred from the hot side to the cold side in a process known as **conduction.**

It is not surprising that more heat is transferred if the temperature difference ΔT is larger. A material with a larger cross section A (a fatter pipe) transfers more heat, while a thicker material, increasing the distance L between the hot and cold sources, decreases the rate of heat transfer.

These observations about heat conduction can be summarized in a single formula. If a small amount of heat dQ is transferred in a small time interval dt, the *rate* of heat transfer is dQ/dt. For a material of cross-section area A and length L, spanning a temperature difference $\Delta T = T_H - T_C$, the rate of heat transfer is

$$\frac{dQ}{dt} = k\frac{A}{L}\Delta T \qquad (19.45)$$

The quantity k, which characterizes whether the material is a good conductor of heat or a poor conductor, is called the **thermal conductivity** of the material. Because the heat-transfer rate J/s is a *power,* measured in watts, the units of k are W/m K. Values of k for common materials are given in **TABLE 19.5**; a material with a larger value of k is a better conductor of heat.

NOTE Heat conductivity is yet *another* use of the symbol k. Whenever you see a k, be very alert to the context in which it is used.

Most good heat conductors are metals, which are also good conductors of electricity. One exception is diamond, in which the strong bonds among atoms that make diamond such a hard material lead to a rapid transfer of thermal energy. Air and other gases are poor conductors of heat because there are no bonds between adjacent molecules.

FIGURE 19.22 Conduction of heat through a solid.

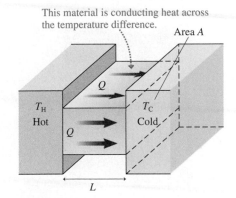

This material is conducting heat across the temperature difference.

Area A

T_H Hot

T_C Cold

Q

L

TABLE 19.5 Thermal conductivities

Material	k (W/m K)
Diamond	2000
Silver	430
Copper	400
Aluminum	240
Iron	80
Stainless steel	14
Ice	1.7
Concrete	0.8
Glass	0.8
Styrofoam	0.035
Air (20°C, 1 atm)	0.023

Some of our perceptions of hot and cold have more to do with thermal conductivity than with temperature. For example, a metal chair feels colder to your bare skin than a wooden chair not because it has a lower temperature—both are at room temperature—but because it has a much larger thermal conductivity that conducts heat away from your body at a much higher rate.

EXAMPLE 19.10 | Keeping a freezer cold

A 1.8-m-wide by 1.0-m-tall by 0.65-m-deep home freezer is insulated with 5.0-cm-thick Styrofoam insulation. At what rate must the compressor remove heat from the freezer to keep the inside at $-20°C$ in a room where the air temperature is $25°C$?

MODEL Heat is transferred through each of the six sides by conduction. The compressor must remove heat at the same rate it enters to maintain a steady temperature inside. The heat conduction is determined primarily by the thick insulation, so we'll neglect the thin inner and outer panels.

SOLVE Each of the six sides is a slab of Styrofoam with cross-section area A_i and thickness $L = 5.0$ cm. The total rate of heat transfer is

$$\frac{dQ}{dt} = \sum_{i=1}^{6} k \frac{A_i}{L} \Delta T = \frac{k \Delta T}{L} \sum_{i=1}^{6} A_i = \frac{k \Delta T}{L} A_{\text{total}}$$

The total surface area is

$$A_{\text{total}} = 2 \times (1.8 \text{ m} \times 1.0 \text{ m} + 1.8 \text{ m} \times 0.65 \text{ m}$$
$$+ 1.0 \text{ m} \times 0.65 \text{ m}) = 7.24 \text{ m}^2$$

Using $k = 0.035$ W/m K from Table 19.5, we find

$$\frac{dQ}{dt} = \frac{k \Delta T}{L} A_{\text{total}} = \frac{(0.035 \text{ W/m K})(45 \text{ K})(7.24 \text{ m}^2)}{0.050 \text{ m}} = 230 \text{ W}$$

Heat enters the freezer through the walls at the rate 230 J/s; thus the compressor must remove 230 J of heat energy every second to keep the temperature at $-20°C$.

ASSESS We'll learn in Chapter 19 how the compressor does this and how much work it must do. A typical freezer uses electric energy at a rate of about 150 W, so our result seems reasonable.

Warm water (colored) moves by convection.

Convection

Air is a poor conductor of heat, but thermal energy is easily transferred through air, water, and other fluids because the air and water can flow. A pan of water on the stove is heated at the bottom. This heated water expands, becomes less dense than the water above it, and thus rises to the surface, while cooler, denser water sinks to take its place. The same thing happens to air. This transfer of thermal energy by the motion of a fluid—the well-known idea that "heat rises"—is called **convection.**

Convection is usually the main mechanism for heat transfer in fluid systems. On a small scale, convection mixes the pan of water that you heat on the stove; on a large scale, convection is responsible for making the wind blow and ocean currents circulate. Air is a very poor thermal conductor, but it is very effective at transferring energy by convection. To use air for thermal insulation, it is necessary to trap the air in small pockets to limit convection. And that's exactly what feathers, fur, double-paned windows, and fiberglass insulation do. Convection is much more rapid in water than in air, which is why people can die of hypothermia in 68°F (20°C) water but can live quite happily in 68°F air.

Because convection involves the often-turbulent motion of fluids, there is no simple equation for energy transfer by convection. Our description must remain qualitative.

Radiation

The sun *radiates* energy to earth through the vacuum of space. Similarly, you feel the warmth from the glowing red coals in a fireplace.

All objects emit energy in the form of **radiation,** electromagnetic waves generated by oscillating electric charges in the atoms that form the object. These waves transfer energy from the object that emits the radiation to the object that absorbs it. Electromagnetic waves carry energy from the sun; this energy is absorbed when sunlight falls on your skin, warming you by increasing your thermal energy. Your skin also emits electromagnetic radiation, helping to keep your body cool by decreasing your thermal energy. Radiation is a significant part of the *energy balance* that keeps your body at the proper temperature.

NOTE The word "radiation" comes from "radiate," meaning "to beam." Radiation can refer to x rays or to the radioactive decay of nuclei, but it also can refer simply to light and other forms of electromagnetic waves that "beam" from an object. Here we are using this second meaning of the term.

You are familiar with radiation from objects hot enough to glow "red hot" or, at a high enough temperature, "white hot." The sun is simply a very hot ball of glowing gas, and the white light from an incandescent lightbulb is radiation emitted by a thin wire filament heated to a very high temperature by an electric current. Objects at lower temperatures also radiate, but at infrared wavelengths. You can't see this radiation (although you can sometimes feel it), but infrared-sensitive detectors can measure it and are used to make thermal images.

The energy radiated by an object depends strongly on temperature. If a small amount of heat energy dQ is radiated during a small time interval dt by an object with surface area A and absolute temperature T, the *rate* of heat transfer is found to be

This satellite image shows radiation emitted by the ocean waters off the east coast of the United States. You can clearly see the warm waters of the Gulf Stream, a large-scale convection that transfers heat to northern latitudes.

$$\frac{dQ}{dt} = e\sigma AT^4 \qquad (19.46)$$

Because the rate of energy transfer is power ($1 \text{ J/s} = 1 \text{ W}$), dQ/dt is often called the *radiated power*. Notice the very strong fourth-power dependence on temperature. Doubling the absolute temperature of an object increases the radiated power by a factor of 16!

The parameter e in Equation 19.46 is the **emissivity** of the surface, a measure of how effectively it radiates. The value of e ranges from 0 to 1. σ is a constant, known as the Stefan-Boltzmann constant, with the value

$$\sigma = 5.67 \times 10^{-8} \text{ W/m}^2 \text{ K}^4$$

NOTE Just as in the ideal-gas law, the temperature in Equation 19.46 *must* be in kelvins.

Objects not only emit radiation, they also *absorb* radiation emitted by their surroundings. Suppose an object at temperature T is surrounded by an environment at temperature T_0. The *net* rate at which the object radiates heat energy—that is, radiation emitted minus radiation absorbed—is

$$\frac{dQ_{\text{net}}}{dt} = e\sigma A(T^4 - T_0^4) \qquad (19.47)$$

This makes sense. An object should have no *net* radiation if it's in thermal equilibrium ($T = T_0$) with its surroundings.

Notice that the emissivity e appears for absorption as well as emission; good emitters are also good absorbers. A perfect absorber ($e = 1$), one absorbing all light and radiation impinging on it but reflecting none, would appear completely black. Thus a perfect absorber is sometimes called a **black body.** But a perfect absorber would also be a perfect emitter, so thermal radiation from an ideal emitter is called **black-body radiation.** It seems strange that black objects are perfect emitters, but think of black charcoal glowing bright red in a fire. At room temperature, it "glows" equally bright with infrared.

EXAMPLE 19.11 | **Taking the sun's temperature**

The radius of the sun is 6.96×10^8 m. At the distance of the earth, 1.50×10^{11} m, the intensity of solar radiation (measured by satellites above the atmosphere) is 1370 W/m². What is the temperature of the sun's surface?

MODEL Assume the sun to be an ideal radiator with $e = 1$.

SOLVE The total power radiated by the sun is the power per m² multiplied by the surface area of a sphere extending to the earth:

$$P = \frac{1370 \text{ W}}{1 \text{ m}^2} \times 4\pi (1.50 \times 10^{11} \text{ m})^2 = 3.87 \times 10^{26} \text{ W}$$

That is, the sun radiates energy at the rate $dQ/dt = 3.87 \times 10^{26}$ J/s. That's a lot of power! This energy is radiated from the surface of a sphere of radius R_S. Using this information in Equation 19.46, we find that the sun's surface temperature is

$$T = \left[\frac{dQ/dt}{e\sigma(4\pi R_S^2)} \right]^{1/4}$$

$$= \left[\frac{3.87 \times 10^{26} \text{ W}}{(1)(5.67 \times 10^{-8} \text{ W/m}^2 \text{ K}^4)4\pi(6.96 \times 10^8 \text{ m})^2} \right]^{1/4}$$

$$= 5790 \text{ K}$$

ASSESS This temperature is confirmed by measurements of the solar spectrum, a topic we'll explore in Part VIII.

Thermal radiation plays a prominent role in climate and global warming. The earth as a whole is in thermal equilibrium. Consequently, it must radiate back into space exactly as much energy as it receives from the sun. The incoming radiation from the hot sun is mostly visible light. The earth's atmosphere is transparent to visible light, so this radiation reaches the surface and is absorbed. The cooler earth radiates infrared radiation, but the atmosphere is *not* completely transparent to infrared. Some components of the atmosphere, notably water vapor and carbon dioxide, are strong absorbers of infrared radiation. They hinder the emission of radiation and, rather like a blanket, keep the earth's surface warmer than it would be without these gases in the atmosphere.

The **greenhouse effect,** as it's called, is a natural part of the earth's climate. The earth would be much colder and mostly frozen were it not for naturally occurring carbon dioxide in the atmosphere. But carbon dioxide also results from the burning of fossil fuels, and human activities since the beginning of the industrial revolution have increased the atmospheric concentration of carbon dioxide by nearly 50%. This human contribution has amplified the greenhouse effect and is the primary cause of global warming.

STOP TO THINK 19.7 Suppose you are an astronaut in space, hard at work in your sealed spacesuit. The only way that you can transfer excess heat to the environment is by

a. Conduction. b. Convection. c. Radiation. d. Evaporation.

CHALLENGE EXAMPLE 19.12 | **Boiling water**

400 mL of water is poured into a covered 8.0-cm-diameter, 150 g glass beaker with a 2.0-mm-thick bottom; then the beaker is placed on a 400°C hot plate. Once the water reaches the boiling point, how long will it take to boil away all the water?

MODEL The bottom of the beaker is a heat-conducting material transferring heat energy from the 400°C hot plate to the 100°C boiling water. The temperature of both the water and the beaker remains constant until the water has boiled away. We'll assume that heat losses due to convection and radiation are negligible, in which case the heat energy entering the system is used entirely for the phase change of the water. The beaker's mass isn't relevant because its temperature isn't changing.

SOLVE The heat energy required to boil mass M of water is

$$Q = ML_v$$

where $L_v = 2.26 \times 10^6$ J/kg is the heat of vaporization. The heat energy transferred through the bottom of the beaker during a time interval Δt is

$$Q = k\frac{A}{L}\Delta T \Delta t$$

where $k = 0.80$ W/m K is the thermal conductivity of glass. Because the heat transferred by conduction is used entirely for boiling the water, we can combine these two expressions:

$$k\frac{A}{L}\Delta T \Delta t = ML_v$$

and then solve for Δt:

$$\Delta t = \frac{MLL_v}{kA\,\Delta T} = \frac{(0.40 \text{ kg})(0.0020 \text{ m})(2.26 \times 10^6 \text{ J/kg})}{(0.80 \text{ W/m K})(0.0050 \text{ m}^2)(300 \text{ K})}$$

$$= 1500 \text{ s} = 25 \text{ min}$$

We used the density of water to find that $M = 400$ g $= 0.40$ kg and calculated $A = \pi r^2 = 0.0050$ m² as the area through which heat conduction occurs.

ASSESS 400 mL is roughly 2 cups, a small hot plate can bring 2 cups of water to a boil in 5 min or so, and boiling the water away takes quite a bit longer than bringing it to a boil. 25 min is a slight underestimate since we neglected energy losses due to convection and radiation, but it seems reasonable. A stove could boil the water away much faster because the burner temperature (gas flame or red-hot heating coil) is much higher.

SUMMARY

The goal of Chapter 19 has been to learn and apply the first law of thermodynamics.

GENERAL PRINCIPLES

First Law of Thermodynamics

$$\Delta E_{th} = W + Q$$

The first law is a general statement of energy conservation.

Work W and heat Q depend on the process by which the system is changed.

Energy

Thermal energy E_{th} Microscopic energy of moving molecules and stretched molecular bonds. ΔE_{th} depends on the initial/final states but is independent of the process.

Work W Energy transferred to the system by forces in a mechanical interaction.

Heat Q Energy transferred to the system via atomic-level collisions in a thermal interaction.

IMPORTANT CONCEPTS

Solving Problems of Work on an Ideal Gas

The work done on a gas is

$$W = -\int_{V_i}^{V_f} p\, dV$$

$$= -(\text{area under the } pV \text{ curve})$$

Solving Calorimetry Problems

When two or more systems interact thermally, they come to a common final temperature determined by

$$Q_{net} = Q_1 + Q_2 + \cdots = 0$$

An **adiabatic process** has $Q = 0$. Gases move along an **adiabat** for which $pV^\gamma = $ constant, where $\gamma = C_P/C_V$ is the **specific heat ratio**. An adiabatic process changes the temperature of the gas without heating it.

The **heat of transformation L** is the energy needed to cause 1 kg of substance to undergo a phase change

$$Q = \pm ML$$

The **specific heat c** of a substance is the energy needed to raise the temperature of 1 kg by 1 K:

$$Q = Mc\,\Delta T$$

The **molar specific heat C** is the energy needed to raise the temperature of 1 mol by 1 K:

$$Q = nC\,\Delta T$$

The molar specific heat of gases depends on the *process* by which the temperature is changed:

$C_V = $ molar specific heat at **constant volume**

$C_P = C_V + R = $ molar specific heat at **constant pressure**

Heat is transferred by **conduction, convection, radiation,** and **evaporation.**

Conduction: $dQ/dt = (kA/L)\Delta T$

Radiation: $dQ/dt = e\sigma A T^4$

SUMMARY OF BASIC GAS PROCESSES

Process	Definition	Stays constant	Work	Heat
Isochoric	$\Delta V = 0$	V and p/T	$W = 0$	$Q = nC_V\,\Delta T$
Isobaric	$\Delta p = 0$	p and V/T	$W = -p\,\Delta V$	$Q = nC_P\,\Delta T$
Isothermal	$\Delta T = 0$	T and pV	$W = -nRT\ln(V_f/V_i)$	$\Delta E_{th} = 0$
Adiabatic	$Q = 0$	pV^γ	$W = \Delta E_{th}$	$Q = 0$
All gas processes	First law $\Delta E_{th} = W + Q = nC_V\,\Delta T$		Ideal-gas law $pV = nRT$	

TERMS AND NOTATION

work, W	adiabatic process	molar specific heat at constant	thermal conductivity, k
mechanical interaction	specific heat, c	volume, C_V	convection
mechanical equilibrium	molar specific heat, C	molar specific heat at constant	radiation
heat, Q	heat of transformation, L	pressure, C_P	emissivity, e
thermal interaction	heat of fusion, L_f	specific heat ratio, γ	black body
thermal equilibrium	heat of vaporization, L_v	adiabat	black-body radiation
first law of thermodynamics	calorimetry	conduction	greenhouse effect
thermodynamic energy model			

CONCEPTUAL QUESTIONS

1. When a space capsule returns to earth, its surfaces get very hot as it passes through the atmosphere at high speed. Has the space capsule been heated? If so, what was the source of the heat? If not, why is it hot?

2. Do (a) temperature, (b) heat, and (c) thermal energy describe a property of a system, an interaction of the system with its environment, or both? Explain.

3. Two containers hold equal masses of nitrogen gas at equal temperatures. You supply 10 J of heat to container A while not allowing its volume to change, and you supply 10 J of heat to container B while not allowing its pressure to change. Afterward, is temperature T_A greater than, less than, or equal to T_B? Explain.

4. You need to raise the temperature of a gas by 10°C. To use the least amount of heat energy, should you heat the gas at constant pressure or at constant volume? Explain.

5. *Why* is the molar specific heat of a gas at constant pressure larger than the molar specific heat at constant volume?

6. **FIGURE Q19.6** shows an adiabatic process.
   a. Is the final temperature higher than, lower than, or equal to the initial temperature?
   b. Is any heat energy added to or removed from the system in this process? Explain.

FIGURE Q19.6

FIGURE Q19.7

7. **FIGURE Q19.7** shows two different processes taking an ideal gas from state i to state f. Is the work done on the gas in process A greater than, less than, or equal to the work done in process B? Explain.

8. **FIGURE Q19.8** shows two different processes taking an ideal gas from state i to state f.
   a. Is the temperature *change* ΔT during process A larger than, smaller than, or equal to the change during process B? Explain.
   b. Is the heat energy added during process A greater than, less than, or equal to the heat added during process B? Explain.

FIGURE Q19.8

FIGURE Q19.9

9. The gas cylinder in **FIGURE Q19.9** is a rigid container that is well insulated except for the bottom surface, which is in contact with a block of ice. The initial gas temperature is $> 0°C$.
   a. During the process that occurs until the gas reaches a new equilibrium, are (i) ΔT, (ii) W, and (iii) Q greater than, less than, or equal to zero? Explain.
   b. Draw a pV diagram showing the process.

10. The gas cylinder in **FIGURE Q19.10** is well insulated except for the bottom surface, which is in contact with a block of ice. The piston can slide without friction. The initial gas temperature is $> 0°C$.
    a. During the process that occurs until the gas reaches a new equilibrium, are (i) ΔT, (ii) W, and (iii) Q greater than, less than, or equal to zero? Explain.
    b. Draw a pV diagram showing the process.

FIGURE Q19.10

11. The gas cylinder in **FIGURE Q19.11** is well insulated on all sides. The piston can slide without friction. Many small masses on top of the piston are removed one by one until the total mass is reduced by 50%.
    a. During this process, are (i) ΔT, (ii) W, and (iii) Q greater than, less than, or equal to zero? Explain.
    b. Draw a pV diagram showing the process.

FIGURE Q19.11

EXERCISES AND PROBLEMS

Problems labeled ▉ integrate material from earlier chapters.

Exercises

Section 19.1 It's All About Energy

Section 19.2 Work in Ideal-Gas Processes

1. ‖ How much work is done on the gas in the process shown in **FIGURE EX19.1**?

FIGURE EX19.1

FIGURE EX19.2

2. ‖ How much work is done on the gas in the process shown in **FIGURE EX19.2**?

3. ‖ 80 J of work are done on the gas in the process shown in **FIGURE EX19.3**. What is V_1 in cm^3?

FIGURE EX19.3

4. ‖ A 2000 cm^3 container holds 0.10 mol of helium gas at 300°C. How much work must be done to compress the gas to 1000 cm^3 at (a) constant pressure and (b) constant temperature?

5. ‖ 500 J of work must be done to compress a gas to half its initial volume at constant temperature. How much work must be done to compress the gas by a factor of 10, starting from its initial volume?

Section 19.3 Heat

Section 19.4 The First Law of Thermodynamics

6. ❘ Draw a first-law bar chart (see Figure 19.12) for the gas process in **FIGURE EX19.6**.

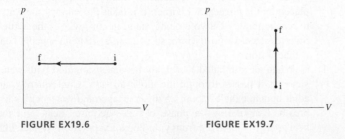

FIGURE EX19.6   **FIGURE EX19.7**

7. ❘ Draw a first-law bar chart (see Figure 19.12) for the gas process in **FIGURE EX19.7**.

8. ❘ Draw a first-law bar chart (see Figure 19.12) for the gas process in **FIGURE EX19.8**.

FIGURE EX19.8

FIGURE EX19.9

9. ❘ Draw a first-law bar chart (see Figure 19.12) for the gas process in **FIGURE EX19.9**.

10. ❘ A gas is compressed from 600 cm^3 to 200 cm^3 at a constant pressure of 400 kPa. At the same time, 100 J of heat energy is transferred out of the gas. What is the change in thermal energy of the gas during this process?

11. ❘ 500 J of work are done on a system in a process that decreases the system's thermal energy by 200 J. How much heat energy is transferred to or from the system?

Section 19.5 Thermal Properties of Matter

12. ‖ How much heat energy must be added to a 6.0-cm-diameter copper sphere to raise its temperature from −50°C to 150°C?

13. ‖ A rapidly spinning paddle wheel raises the temperature of 200 mL of water from 21°C to 25°C. How much (a) heat is transferred and (b) work is done in this process?

14. ❘ a. 100 J of heat energy are transferred to 20 g of mercury. By how much does the temperature increase?
    b. How much heat is needed to raise the temperature of 20 g of water by the same amount?

15. ‖ How much heat is needed to change 20 g of mercury at 20°C into mercury vapor at the boiling point?

16. ‖ What is the maximum mass of ethyl alcohol you could boil with 1000 J of heat, starting from 20°C?

17. ❘ One way you keep from overheating is by perspiring.
BIO Evaporation—a phase change—requires heat, and the heat energy is removed from your body. Evaporation is much like boiling, only water's heat of vaporization at 35°C is a somewhat larger 24×10^5 J/kg because at lower temperatures more energy is required to break the molecular bonds. Very strenuous activity can cause an adult human to produce 30 g of perspiration per minute. If all the perspiration evaporates, rather than dripping off, at what rate (in J/s) is it possible to exhaust heat by perspiring?

18. ❘ A scientist whose scale is broken but who has a working 2.5 kW heating coil and a thermometer decides to improvise to determine the mass of a block of aluminum she has recently acquired. She heats the aluminum for 30 s and finds that its temperature increases from 20°C to 35°C. What is the mass of the aluminum?

19. ‖ Two cars collide head-on while each is traveling at 80 km/h. Suppose all their kinetic energy is transformed into the thermal energy of the wrecks. What is the temperature increase of each car? You can assume that each car's specific heat is that of iron.

20. ‖ An experiment measures the temperature of a 500 g substance while steadily supplying heat to it. **FIGURE EX19.20** shows the results of the experiment. What are the (a) specific heat of the solid phase, (b) specific heat of the liquid phase, (c) melting and boiling temperatures, and (d) heats of fusion and vaporization?

FIGURE EX19.20

Section 19.6 Calorimetry

21. ‖ 30 g of copper pellets are removed from a 300°C oven and immediately dropped into 100 mL of water at 20°C in an insulated cup. What will the new water temperature be?

22. ‖ A 750 g aluminum pan is removed from the stove and plunged into a sink filled with 10.0 L of water at 20.0°C. The water temperature quickly rises to 24.0°C. What was the initial temperature of the pan in °C and in °F?

23. ‖ A 50.0 g thermometer is used to measure the temperature of 200 mL of water. The specific heat of the thermometer, which is mostly glass, is 750 J/kg K, and it reads 20.0°C while lying on the table. After being completely immersed in the water, the thermometer's reading stabilizes at 71.2°C. What was the actual water temperature before it was measured?

24. ‖ A 500 g metal sphere is heated to 300°C, then dropped into a beaker containing 300 cm^3 of mercury at 20.0°C. A short time later the mercury temperature stabilizes at 99.0°C. Identify the metal.

25. ‖‖ A 65 cm^3 block of iron is removed from an 800°C furnace and immediately dropped into 200 mL of 20°C water. What fraction of the water boils away?

Section 19.7 The Specific Heats of Gases

26. ‖ A container holds 1.0 g of argon at a pressure of 8.0 atm.
   a. How much heat is required to increase the temperature by 100°C at constant volume?
   b. How much will the temperature increase if this amount of heat energy is transferred to the gas at constant pressure?

27. ‖ A container holds 1.0 g of oxygen at a pressure of 8.0 atm.
   a. How much heat is required to increase the temperature by 100°C at constant pressure?
   b. How much will the temperature increase if this amount of heat energy is transferred to the gas at constant volume?

28. ‖ The volume of a gas is halved during an adiabatic compression that increases the pressure by a factor of 2.5.
   a. What is the specific heat ratio γ?
   b. By what factor does the temperature increase?

29. ‖ A gas cylinder holds 0.10 mol of O_2 at 150°C and a pressure of 3.0 atm. The gas expands adiabatically until the pressure is halved. What are the final (a) volume and (b) temperature?

30. ‖ A gas cylinder holds 0.10 mol of O_2 at 150°C and a pressure of 3.0 atm. The gas expands adiabatically until the volume is doubled. What are the final (a) pressure and (b) temperature?

31. ‖ 0.10 mol of nitrogen gas follow the two processes shown in **FIGURE EX19.31**. How much heat is required for each?

FIGURE EX19.31   **FIGURE EX19.32**

32. ‖ A monatomic gas follows the process $1 \rightarrow 2 \rightarrow 3$ shown in **FIGURE EX19.32**. How much heat is needed for (a) process $1 \rightarrow 2$ and (b) process $2 \rightarrow 3$?

Section 19.8 Heat-Transfer Mechanisms

33. ‖ The ends of a 20-cm-long, 2.0-cm-diameter rod are maintained at 0°C and 100°C by immersion in an ice-water bath and boiling water. Heat is conducted through the rod at 4.5×10^4 J per hour. Of what material is the rod made?

34. ‖ A 10 m × 14 m house is built on a 12-cm-thick concrete slab. What is the heat-loss rate through the slab if the ground temperature is 5°C while the interior of the house is 22°C?

35. ‖ You are boiling pasta and absentmindedly grab a copper stirring spoon rather than your wooden spoon. The copper spoon has a 20 mm × 1.5 mm rectangular cross section, and the distance from the boiling water to your 35°C hand is 18 cm. How long does it take the spoon to transfer 25 J of energy to your hand?

36. ‖ What maximum power can be radiated by a 10-cm-diameter solid lead sphere? Assume an emissivity of 1.

37. ‖‖ Radiation from the head is a major source of heat loss from the human body. Model a head as a 20-cm-diameter, 20-cm-tall cylinder with a flat top. If the body's surface temperature is 35°C, what is the net rate of heat loss on a chilly 5°C day? All skin, regardless of color, is effectively black in the infrared where the radiation occurs, so use an emissivity of 0.95.

BIO

Problems

38. ‖ A 5.0 g ice cube at −20°C is in a rigid, sealed container from which all the air has been evacuated. How much heat is required to change this ice cube into steam at 200°C? Steam has $C_V = 1500$ J/kg K and $C_P = 1960$ J/kg K.

39. ‖ A 5.0-m-diameter garden pond is 30 cm deep. Solar energy is incident on the pond at an average rate of 400 W/m^2. If the water absorbs all the solar energy and does not exchange energy with its surroundings, how many hours will it take to warm from 15°C to 25°C?

40. ‖ The burner on an electric stove has a power output of 2.0 kW. A 750 g stainless steel teakettle is filled with 20°C water and placed on the already hot burner. If it takes 3.0 min for the water to reach a boil, what volume of water, in cm^3, was in the kettle? Stainless steel is mostly iron, so you can assume its specific heat is that of iron.

41. ‖ When air is inhaled, it quickly becomes saturated with water vapor as it passes through the moist airways. Consequently, an adult human exhales about 25 mg of evaporated water with each breath. Evaporation—a phase change—requires heat, and the heat energy is removed from your body. Evaporation is much like boiling, only water's heat of vaporization at 35°C is a somewhat larger 24×10^5 J/kg because at lower temperatures more energy is required to break the molecular bonds. At 12 breaths/min, on a

BIO

dry day when the inhaled air has almost no water content, what is the body's rate of energy loss (in J/s) due to exhaled water? (For comparison, the energy loss from radiation, usually the largest loss on a cool day, is about 100 J/s.)

42. ‖ 512 g of an unknown metal at a temperature of 15°C is dropped into a 100 g aluminum container holding 325 g of water at 98°C. A short time later, the container of water and metal stabilizes at a new temperature of 78°C. Identify the metal.

43. ‖ A 150 L (≈ 40 gal) electric hot-water tank has a 5.0 kW heater. How many minutes will it take to raise the water temperature from 65°F to 140°F?

44. ‖ The specific heat of most solids is nearly constant over a wide CALC temperature range. Not so for diamond. Between 200 K and 600 K, the specific heat of diamond is reasonably well described by $c = 2.8T - 350$ J/kg K, where T is in K. For gemstone diamonds, 1 carat = 200 mg. How much heat energy is needed to raise the temperature of a 3.5 carat diamond from −50°C to 250°C?

45. ‖ A lava flow is threatening to engulf a small town. A 400-m-wide, 35-cm-thick tongue of 1200°C lava is advancing at the rate of 1.0 m per minute. The mayor devises a plan to stop the lava in its tracks by flying in large quantities of 20°C water and dousing it. The lava has density 2500 kg/m³, specific heat 1100 J/kg K, melting temperature 800°C, and heat of fusion 4.0×10^5 J/kg. How many liters of water per minute, at a minimum, will be needed to save the town?

46. ‖ Suppose you take and hold a deep breath on a chilly day, inhaling BIO 3.0 L of air at 0°C and 1 atm.
   a. How much heat must your body supply to warm the air to your internal body temperature of 37°C?
   b. By how much does the air's volume increase as it warms?

47. ‖ Your 300 mL cup of coffee is too hot to drink when served at 90°C. What is the mass of an ice cube, taken from a −20°C freezer, that will cool your coffee to a pleasant 60°C?

48. ‖ A typical nuclear reactor generates 1000 MW (1000 MJ/s) of electrical energy. In doing so, it produces 2000 MW of "waste heat" that must be removed from the reactor to keep it from melting down. Many reactors are sited next to large bodies of water so that they can use the water for cooling. Consider a reactor where the intake water is at 18°C. State regulations limit the temperature of the output water to 30°C so as not to harm aquatic organisms. How many liters of cooling water have to be pumped through the reactor each minute?

49. ‖ 2.0 mol of gas are at 30°C and a pressure of 1.5 atm. How much work must be done on the gas to compress it to one third of its initial volume at (a) constant temperature and (b) constant pressure? (c) Show both processes on a single pV diagram.

50. ‖ A 6.0-cm-diameter cylinder of nitrogen gas has a 4.0-cm-thick movable copper piston. The cylinder is oriented vertically, as shown in FIGURE P19.50, and the air above the piston is evacuated. When the gas temperature is 20°C, the piston floats 20 cm above the bottom of the cylinder.
   a. What is the gas pressure?
   b. How many gas molecules are in the cylinder?
   Then 2.0 J of heat energy are transferred to the gas.
   c. What is the new equilibrium temperature of the gas?
   d. What is the final height of the piston?
   e. How much work is done on the gas as the piston rises?

FIGURE P19.50

51. ‖ A 560 kg concrete table needs to be supported at the four corners by compressed-air cylinders. Each cylinder is 25 cm in diameter and has a 1.20 m initial length when the pressure inside is 1.0 atm. A hoist lowers the table very slowly, compressing the cylinders while allowing them to stay in thermal equilibrium with their surroundings. How much work has been done on the gas of the four cylinders when the table reaches its equilibrium position?

52. ‖ An ideal-gas process is described by $p = cV^{1/2}$, where c is a CALC constant.
   a. Find an expression for the work done on the gas in this process as the volume changes from V_1 to V_2.
   b. 0.033 mol of gas at an initial temperature of 150°C is compressed, using this process, from 300 cm³ to 200 cm³. How much work is done on the gas?
   c. What is the final temperature of the gas in °C?

53. ‖ A 10-cm-diameter cylinder contains argon gas at 10 atm pressure and a temperature of 50°C. A piston can slide in and out of the cylinder. The cylinder's initial length is 20 cm. 2500 J of heat are transferred to the gas, causing the gas to expand at constant pressure. What are (a) the final temperature and (b) the final length of the cylinder?

54. ‖‖ A cube 20 cm on each side contains 3.0 g of helium at 20°C. 1000 J of heat energy are transferred to this gas. What are (a) the final pressure if the process is at constant volume and (b) the final volume if the process is at constant pressure? (c) Show and label both processes on a single pV diagram.

55. ‖ An 8.0-cm-diameter, well-insulated vertical cylinder containing nitrogen gas is sealed at the top by a 5.1 kg frictionless piston. The air pressure above the piston is 100 kPa.
   a. What is the gas pressure inside the cylinder?
   b. Initially, the piston height above the bottom of the cylinder is 26 cm. What will be the piston height if an additional 3.5 kg are placed on top of the piston?

56. ‖ n moles of an ideal gas at temperature T_1 and volume V_1 expand isothermally until the volume has doubled. In terms of n, T_1, and V_1, what are (a) the final temperature, (b) the work done on the gas, and (c) the heat energy transferred to the gas?

57. ‖‖ 5.0 g of nitrogen gas at 20°C and an initial pressure of 3.0 atm undergo an isobaric expansion until the volume has tripled.
   a. What are the gas volume and temperature after the expansion?
   b. How much heat energy is transferred to the gas to cause this expansion?
   The gas pressure is then decreased at constant volume until the original temperature is reached.
   c. What is the gas pressure after the decrease?
   d. What amount of heat energy is transferred from the gas as its pressure decreases?
   e. Show the total process on a pV diagram. Provide an appropriate scale on both axes.

58. ‖ 0.10 mol of nitrogen gas follow the two processes shown in FIGURE P19.58. How much heat is required for each?

FIGURE P19.58

59. ||| You come into lab one day and find a well-insulated 2000 mL thermos bottle containing 500 mL of boiling liquid nitrogen. The remainder of the thermos has nitrogen gas at a pressure of 1.0 atm. The gas and liquid are in thermal equilibrium. While waiting for lab to start, you notice a piece of iron on the table with "197 g" written on it. Just for fun, you drop the iron into the thermos and seal the cap tightly so that no gas can escape. After a few seconds have passed, what is the pressure inside the thermos? The density of liquid nitrogen is 810 kg/m³.

60. ||| Your laboratory assignment for the week is to measure the specific heat ratio γ of carbon dioxide. The gas is contained in a cylinder with a movable piston and a thermometer. When the piston is withdrawn as far as possible, the cylinder's length is 20 cm. You decide to push the piston in very rapidly by various amounts and, for each push, to measure the temperature of the carbon dioxide. Before each push, you withdraw the piston all the way and wait several minutes for the gas to come to the room temperature of 21°C. Your data are as follows:

Push (cm)	Temperature (°C)
5	35
10	68
13	110
15	150

Use the best-fit line of an appropriate graph to determine γ for carbon dioxide.

61. || Two cylinders each contain 0.10 mol of a diatomic gas at 300 K and a pressure of 3.0 atm. Cylinder A expands isothermally and cylinder B expands adiabatically until the pressure of each is 1.0 atm.
   a. What are the final temperature and volume of each?
   b. Show both processes on a single pV diagram. Use an appropriate scale on both axes.

62. ||| FIGURE P19.62 shows a thermodynamic process followed by 120 mg of helium.
   a. Determine the pressure (in atm), temperature (in °C), and volume (in cm³) of the gas at points 1, 2, and 3. Put your results in a table for easy reading.
   b. How much work is done on the gas during each of the three segments?
   c. How much heat energy is transferred to or from the gas during each of the three segments?

FIGURE P19.62

63. || Two containers of a diatomic gas have the same initial conditions. One container, heated at constant pressure, has a temperature increase of 20°C. The other container receives the same quantity of heat energy, but at constant volume. What is its temperature increase?

64. || 14 g of nitrogen gas at STP are adiabatically compressed to a pressure of 20 atm. What are (a) the final temperature, (b) the work done on the gas, (c) the heat transfer to the gas, and (d) the compression ratio V_{max}/V_{min} ? (e) Show the process on a pV diagram, using proper scales on both axes.

65. || 14 g of nitrogen gas at STP are pressurized in an isochoric process to a pressure of 20 atm. What are (a) the final temperature, (b) the work done on the gas, (c) the heat transfer to the gas, and (d) the pressure ratio p_{max}/p_{min} ? (e) Show the process on a pV diagram, using proper scales on both axes.

66. || A cylinder with a movable piston contains n moles of gas at
CALC a temperature higher than that of the surrounding environment. An external force on the piston keeps the pressure constant while the gas cools as $\Delta T = (\Delta T)_0 e^{-t/\tau}$, where ΔT is the temperature *difference* between the gas and the environment, $(\Delta T)_0$ is the initial temperature difference, and τ is the time constant.
   a. Find an expression for the *rate* at which the environment does work on the gas. Recall that the rate of doing work is *power*.
   b. What power is initially supplied by the environment if 0.15 mol of gas are initially 12°C warmer than the surroundings and cool with a time constant of 60 s?

67. || When strong winds rapidly carry air down from mountains to a lower elevation, the air has no time to exchange heat with its surroundings. The air is compressed as the pressure rises, and its temperature can increase dramatically. These warm winds are called Chinook winds in the Rocky Mountains and Santa Ana winds in California. Suppose the air temperature high in the mountains behind Los Angeles is 0°C at an elevation where the air pressure is 60 kPa. What will the air temperature be, in °C and °F, when the Santa Ana winds have carried this air down to an elevation near sea level where the air pressure is 100 kPa?

68. || You would like to put a solar hot water system on your roof, but you're not sure it's feasible. A reference book on solar energy shows that the ground-level solar intensity in your city is 800 W/m² for at least 5 hours a day throughout most of the year. Assuming that a completely black collector plate loses energy only by radiation, and that the air temperature is 20°C, what is the equilibrium temperature of a collector plate directly facing the sun? Note that while a plate has two sides, only the side facing the sun will radiate because the opposite side will be well insulated.

69. || A cubical box 20 cm on a side is constructed from 1.2-cm-thick concrete panels. A 100 W lightbulb is sealed inside the box. What is the air temperature inside the box when the light is on if the surrounding air temperature is 20°C?

70. || A cylindrical copper rod and an iron rod with exactly the same dimensions are welded together end to end. The outside end of the copper rod is held at 100°C, and the outside end of the iron rod is held at 0°C. What is the temperature at the midpoint where the rods are joined together?

71. ||| Most stars are *main-sequence* stars, a group of stars for which size, mass, surface temperature, and radiated power are closely related. The sun, for instance, is a yellow main-sequence star with a surface temperature of 5800 K. For a main-sequence star whose mass M is more than twice that of the sun, the total radiated power, relative to the sun, is approximately $P/P_{sun} = 1.5(M/M_{sun})^{3.5}$. The star Regulus A is a bluish main-sequence star with mass $3.8M_{sun}$ and radius $3.1R_{sun}$. What is the surface temperature of Regulus A?

72. ||| A satellite to reflect radar is a 2.0-m-diameter, 2.0-mm-thick
CALC spherical copper shell. While orbiting the earth, the satellite absorbs sunlight and is warmed to 50°C. When it passes into the

earth's shadow, the satellite radiates energy to deep space. The temperature of deep space is actually 3 K, as a result of the Big Bang 14 billion years ago, but it is so much colder than the satellite that you can assume a deep-space temperature of 0 K. If the satellite's emissivity is 0.75, to what temperature, in °C, will it drop during the 45 minutes it takes to move through the earth's shadow?

73. ‖ The sun's intensity at the distance of the earth is 1370 W/m². 30% of this energy is reflected by water and clouds; 70% is absorbed. What would be the earth's average temperature (in °C) if the earth had no atmosphere? The emissivity of the surface is very close to 1. (The actual average temperature of the earth, about 15°C, is higher than your calculation because of the greenhouse effect.)

In Problems 74 through 76 you are given the equation used to solve a problem. For each of these, you are to

   a. Write a realistic problem for which this is the correct equation.
   b. Finish the solution of the problem.

74. $50 \text{ J} = -n(8.31 \text{ J/mol K})(350 \text{ K})\ln\left(\frac{1}{3}\right)$

75. $(200 \times 10^{-6} \text{ m}^3)(13{,}600 \text{ kg/m}^3)$
    $\times (140 \text{ J/kg K})(90°\text{C} - 15°\text{C})$
    $+ (0.50 \text{ kg})(449 \text{ J/kg K})(90°\text{C} - T_i) = 0$

76. $(10 \text{ atm})V_2^{1.40} = (1.0 \text{ atm})V_1^{1.40}$

Challenge Problems

77. ‖‖ 10 g of aluminum at 200°C and 20 g of copper are dropped into 50 cm³ of ethyl alcohol at 15°C. The temperature quickly comes to 25°C. What was the initial temperature of the copper?

78. ‖‖ A beaker with a metal bottom is filled with 20 g of water at 20°C. It is brought into good thermal contact with a 4000 cm³ container holding 0.40 mol of a monatomic gas at 10 atm pressure. Both containers are well insulated from their surroundings.

What is the gas pressure after a long time has elapsed? You can assume that the containers themselves are nearly massless and do not affect the outcome.

FIGURE CP19.78

FIGURE CP19.79

79. ‖‖‖ **FIGURE CP19.79** shows a thermodynamic process followed by 0.015 mol of hydrogen. How much heat energy is transferred to the gas?

80. ‖‖‖ One cylinder in the diesel engine of a truck has an initial volume of 600 cm³. Air is admitted to the cylinder at 30°C and a pressure of 1.0 atm. The piston rod then does 400 J of work to rapidly compress the air. What are its final temperature and volume?

81. ‖‖‖ 0.020 mol of a diatomic gas, with initial temperature 20°C,
CALC are compressed from 1500 cm³ to 500 cm³ in a process in which $pV^2 = $ constant. How much heat energy is added during this process?

82. ‖‖‖ A monatomic gas fills the left end of the cylinder in
CALC **FIGURE CP19.82**. At 300 K, the gas cylinder length is 10.0 cm and the spring is compressed by 2.0 cm. How much heat energy must be added to the gas to expand the cylinder length to 16.0 cm?

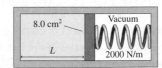

FIGURE CP19.82

20 The Micro/Macro Connection

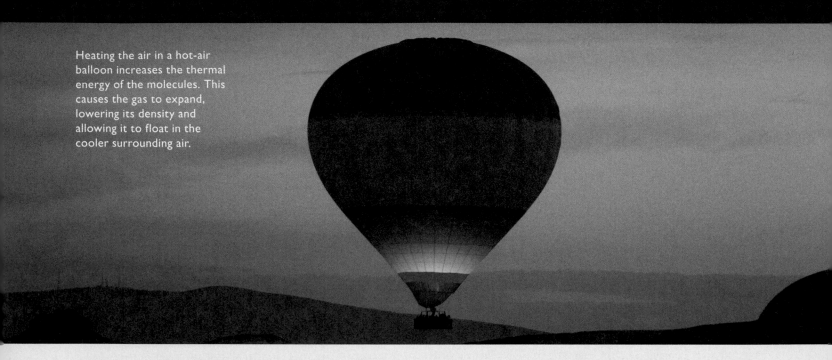

Heating the air in a hot-air balloon increases the thermal energy of the molecules. This causes the gas to expand, lowering its density and allowing it to float in the cooler surrounding air.

IN THIS CHAPTER, you will see how macroscopic properties depend on the motion of atoms.

What is the micro/macro connection?

We've discovered several puzzles in the last two chapters:

- Why does the ideal-gas law work for every gas?
- Why is the molar specific heat the same for every monatomic gas? And for every diatomic gas? And for every elemental solid?
- Just what does temperature actually measure?

We can resolve these puzzles and understand many properties of macroscopic systems by studying the microscopic behavior of the system's atoms and molecules. This is the micro/macro connection.

« LOOKING BACK Sections 19.3–19.5 Heat, the first law, and specific heats

Why do gases have pressure?

Gases have pressure due to the collisions of the molecules with the walls of the container. We'll find that we can calculate an average molecular speed, the root-mean-square speed, by relating the ideal-gas law to a microscopic calculation of the gas pressure.

What is temperature?

At a microscopic level, temperature measures the average translational kinetic energy of moving atoms and molecules. We will use this discovery to explain why all monatomic (and diatomic) gases have exactly the same molar specific heat.

« LOOKING BACK Section 19.7 The specific heats of gases

How do interacting systems reach equilibrium?

Two thermally interacting systems reach a common final temperature because they exchange energy via collisions. On average, more-energetic atoms transfer energy to less-energetic atoms until both systems have the same *average* translational kinetic energy.

What is the second law of thermodynamics?

The second law of thermodynamics governs how systems evolve in time. One statement of the second law is that heat energy is transferred spontaneously from a hotter system to a colder system, but never from colder to hotter. Heat transfer is an irreversible process. The concept of entropy will help us see that irreversible processes occur because some macroscopic states are vastly more likely to occur than others.

20.1 Molecular Speeds and Collisions

If matter really consists of atoms and molecules, then the macroscopic properties of matter, such as temperature, pressure, and specific heat, ought to be related to the microscopic motion of those atoms and molecules. Our goal in this chapter is to explore this *micro/macro connection*. Let's begin by looking at a gas. Do all molecules in the gas move with the same speed, or is there a range of speeds?

FIGURE 20.1 shows an experiment to measure the speeds of molecules in a gas. The two rotating disks form a *velocity selector*. Once every revolution, the slot in the first disk allows a small pulse of molecules to pass through. By the time these molecules reach the second disk, the slots have rotated. The molecules can pass through the second slot and be detected *only* if they have exactly the right speed $v = L/\Delta t$ to travel between the two disks during time interval Δt it takes the axle to complete one revolution. Molecules having any other speed are blocked by the second disk. By changing the rotation speed of the axle, this apparatus can measure how many molecules have each of many possible speeds.

FIGURE 20.2 shows the results for nitrogen gas (N_2) at $T = 20°C$. The data are presented in the form of a **histogram,** a bar chart in which the height of each bar tells how many (or, in this case, what percentage) of the molecules have a speed in the *range* of speeds shown below the bar. For example, 16% of the molecules have speeds in the range from 600 m/s to 700 m/s. All the bars sum to 100%, showing that this histogram describes *all* of the molecules leaving the source.

It turns out that the molecules have what is called a *distribution* of speeds, ranging from as low as $\approx$ 100 m/s to as high as $\approx$ 1200 m/s. But not all speeds are equally likely; there is a *most likely speed* of $\approx$ 550 m/s. This is really fast, $\approx$ 1200 mph! Changing the temperature or changing to a different gas changes the most likely speed, as we'll learn later in the chapter, but it does not change the *shape* of the distribution.

If you were to repeat the experiment, you would again find the most likely speed to be $\approx$ 550 m/s and that 16% of the molecules have speeds between 600 m/s and 700 m/s. This is an important lesson. Although a gas consists of a vast number of molecules, each moving randomly, *averages,* such as the average number of molecules in the speed range 600 to 700 m/s, have precise, predictable values. **The micro/macro connection is built on the idea that the macroscopic properties of a system, such as temperature or pressure, are related to the *average* behavior of the atoms and molecules.**

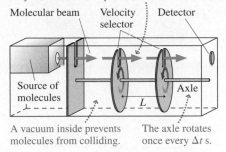

FIGURE 20.1 An experiment to measure the speeds of molecules in a gas.

Only molecules with speed $L/\Delta t$ reach the detector.

Molecular beam · Velocity selector · Detector

Source of molecules · Axle · L

A vacuum inside prevents molecules from colliding.

The axle rotates once every Δt s.

FIGURE 20.2 The distribution of molecular speeds in a sample of nitrogen gas.

% of molecules

Most likely speed

16% of the molecules have speeds between 600 m/s and 700 m/s.

N_2 molecules at 20°C

Speed range (m/s)

Mean Free Path

Imagine someone opening a bottle of strong perfume a few feet away from you. If molecular speeds are hundreds of meters per second, you might expect to smell the perfume almost instantly. But that isn't what happens. As you know, it takes many seconds for the molecules to *diffuse* across the room. Let's see why this is.

FIGURE 20.3 shows a "movie" of one molecule. Instead of zipping along in a straight line, as it would in a vacuum, the molecule follows a convoluted zig-zag path in which it frequently collides with other molecules. A question we could ask is: What is the *average* distance between collisions? If a molecule has N_{coll} collisions as it travels distance L, the average distance between collisions, which is called the **mean free path** λ (lowercase Greek lambda), is

$$\lambda = \frac{L}{N_{coll}} \tag{20.1}$$

FIGURE 20.4a on the next page shows two molecules approaching each other. We will assume that the molecules are spherical and of radius r. We will also continue the ideal-gas assumption that the molecules undergo hard-sphere collisions, like billiard balls. In that case, the molecules will collide if the distance between their *centers* is less than $2r$. They will miss if the distance is greater than $2r$.

FIGURE 20.3 A single molecule follows a zig-zag path through a gas.

The molecule changes direction and speed with each collision.

It moves freely between collisions.

Initial position

Later position

FIGURE 20.4 A "sample" molecule collides with "target" molecules.

(a)

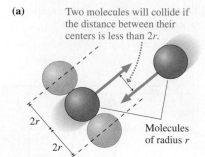

Two molecules will collide if the distance between their centers is less than $2r$.

$2r$
$2r$

Molecules of radius r

(b)

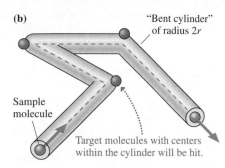

"Bent cylinder" of radius $2r$

Sample molecule

Target molecules with centers within the cylinder will be hit.

In **FIGURE 20.4b** we've drawn a cylinder of radius $2r$ centered on the trajectory of a "sample" molecule. The sample molecule collides with any "target" molecule whose center is located within the cylinder, causing the cylinder to bend at that point. Hence the number of collisions N_{coll} is equal to the number of molecules in a cylindrical volume of length L.

The volume of a cylinder is $V_{cyl} = AL = \pi(2r)^2 L$. If the number density of the gas is N/V particles per m³, then the number of collisions along a trajectory of length L is

$$N_{coll} = \frac{N}{V} V_{cyl} = \frac{N}{V} \pi(2r)^2 L = 4\pi \frac{N}{V} r^2 L \qquad (20.2)$$

Thus the mean free path between collisions is

$$\lambda = \frac{L}{N_{coll}} = \frac{1}{4\pi(N/V)r^2}$$

We made a tacit assumption in this derivation that the target molecules are at rest. While the general idea behind our analysis is correct, a more detailed calculation with all the molecules moving introduces an extra factor of $\sqrt{2}$, giving

$$\lambda = \frac{1}{4\sqrt{2}\,\pi(N/V)r^2} \qquad \text{(mean free path)} \qquad (20.3)$$

Laboratory measurements are necessary to determine atomic and molecular radii, but a reasonable rule of thumb is to assume that atoms in a monatomic gas have $r \approx 0.5 \times 10^{-10}$ m and diatomic molecules have $r \approx 1.0 \times 10^{-10}$ m.

EXAMPLE 20.1 | **The mean free path at room temperature**

What is the mean free path of a nitrogen molecule at 1.0 atm pressure and room temperature (20°C)?

SOLVE Nitrogen is a diatomic molecule, so $r \approx 1.0 \times 10^{-10}$ m. We can use the ideal-gas law in the form $pV = Nk_BT$ to determine the number density:

$$\frac{N}{V} = \frac{p}{k_BT} = \frac{101,300\ \text{Pa}}{(1.38 \times 10^{-23}\ \text{J/K})(293\ \text{K})} = 2.5 \times 10^{25}\ \text{m}^{-3}$$

Thus the mean free path is

$$\lambda = \frac{1}{4\sqrt{2}\,\pi(N/V)r^2}$$

$$= \frac{1}{4\sqrt{2}\,\pi(2.5 \times 10^{25}\ \text{m}^{-3})(1.0 \times 10^{-10}\ \text{m})^2}$$

$$= 2.3 \times 10^{-7}\ \text{m} = 230\ \text{nm}$$

ASSESS You learned in Example 18.6 that the average separation between gas molecules at STP is ≈ 4 nm. It seems that any given molecule can slip between its neighbors, which are spread out in three dimensions, and travel—on average—about 60 times the average spacing before it collides with another molecule.

STOP TO THINK 20.1 The table shows the properties of four gases, each having the same number of molecules. Rank in order, from largest to smallest, the mean free paths λ_A to λ_D of molecules in these gases.

Gas	A	B	C	D
Volume	V	$2V$	V	V
Atomic mass	m	m	$2m$	m
Atomic radius	r	r	r	$2r$

20.2 Pressure in a Gas

Why does a gas have pressure? In Chapter 14, where pressure was introduced, we suggested that the pressure in a gas is due to collisions of the molecules with the walls of its container. The force due to one such collision may be unmeasurably tiny, but the steady rain of a vast number of molecules striking a wall each second exerts a measurable macroscopic force. The gas pressure is the force per unit area ($p = F/A$) resulting from these molecular collisions.

Our task in this section is to calculate the pressure by doing the appropriate averaging over molecular motions and collisions. This task can be divided into three main pieces:

1. Calculate the momentum change of a single molecule during a collision.
2. Find the force due to all collisions.
3. Introduce an appropriate average speed.

Force Due to Collisions

FIGURE 20.5 shows a molecule with an x-component of velocity v_x having a perfectly elastic collision with a wall and rebounding with its velocity changed to $-v_x$. This one molecule has a momentum *change*

$$\Delta p_x = m(-v_x) - mv_x = -2mv_x \qquad (20.4)$$

Suppose there are N_{coll} collisions with the wall during a small interval of time Δt. Further, suppose that all the molecules have the same speed. (This latter assumption isn't really necessary, and we'll soon remove this constraint, but it helps us focus on the physics without getting lost in the math.) Then the total momentum change of the gas during Δt is

$$\Delta P_x = N_{coll}\Delta p_x = -2N_{coll}mv_x \qquad (20.5)$$

You learned in Section 11.1 that the momentum principle can be written as

$$\Delta P_x = (F_{avg})_x \Delta t \qquad (20.6)$$

Thus the average force of the wall *on the gas* is

$$(F_{on\,gas})_x = \frac{\Delta P_x}{\Delta t} = -\frac{2N_{coll}mv_x}{\Delta t} \qquad (20.7)$$

Equation 20.7 has a negative sign because, as we've set it up, the collision force of the wall on the gas molecules is to the left. But Newton's third law is $(F_{on\,gas})_x = -(F_{on\,wall})_x$, so the force *on the wall* due to these collisions is

$$(F_{on\,wall})_x = \frac{2N_{coll}mv_x}{\Delta t} \qquad (20.8)$$

We need to determine how many collisions occur during Δt. Assume that Δt is smaller than the mean time between collisions, so no collisions alter the molecular speeds during this interval. **FIGURE 20.6** has shaded a volume of the gas of length $\Delta x = v_x \Delta t$. *Every one* of the molecules in this shaded region *that is moving to the right* will reach and collide with the wall during Δt. Molecules outside this region will not reach the wall and will not collide.

The shaded region has volume $A\Delta x$, where A is the area of the wall. If the gas has number density N/V, the number of molecules in the shaded region is $(N/V)A\Delta x = (N/V)Av_x\Delta t$. But only half these molecules are moving to the right, so the number of collisions during Δt is

$$N_{coll} = \frac{1}{2}\frac{N}{V}Av_x\Delta t \qquad (20.9)$$

Substituting Equation 20.9 into Equation 20.8, we see that Δt cancels and that the force of the molecules on the wall is

$$F_{on\,wall} = \frac{N}{V}mv_x^2A \qquad (20.10)$$

Notice that this expression for $F_{on\,wall}$ does not depend on any details of the collisions.

We can relax the assumption that all molecules have the same speed by replacing the squared velocity v_x^2 in Equation 20.10 with its average value. That is,

$$F_{on\,wall} = \frac{N}{V}m(v_x^2)_{avg}A \qquad (20.11)$$

where $(v_x^2)_{avg}$ is the quantity v_x^2 averaged over all the molecules in the container.

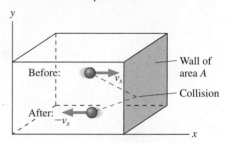

FIGURE 20.5 A molecule colliding with the wall exerts an impulse on it.

Before:

After:

Wall of area A

Collision

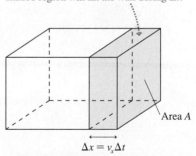

FIGURE 20.6 Determining the rate of collisions.

Only molecules moving to the right in the shaded region will hit the wall during Δt.

Area A

$\Delta x = v_x\Delta t$

The Root-Mean-Square Speed

We need to be somewhat careful when averaging velocities. The velocity component v_x has a sign. At any instant of time, half the molecules in a container move to the right and have positive v_x while the other half move to the left and have negative v_x. Thus the *average velocity* is $(v_x)_{avg} = 0$. If this weren't true, the entire container of gas would move away!

The speed of a molecule is $v = (v_x^2 + v_y^2 + v_z^2)^{1/2}$. Thus the average of the speed squared is

$$(v^2)_{avg} = (v_x^2 + v_y^2 + v_z^2)_{avg} = (v_x^2)_{avg} + (v_y^2)_{avg} + (v_z^2)_{avg} \quad (20.12)$$

The square root of $(v^2)_{avg}$ is called the **root-mean-square speed** v_{rms}:

$$v_{rms} = \sqrt{(v^2)_{avg}} \quad \text{(root-mean-square speed)} \quad (20.13)$$

This is usually called the *rms speed*. You can remember its definition by noting that its name is the *opposite* of the sequence of operations: First you square all the speeds, then you average the squares (find the mean), then you take the square root. Because the square root "undoes" the square, v_{rms} must, in some sense, give an average speed.

NOTE We could compute a true average speed v_{avg}, but that calculation is difficult. More important, the root-mean-square speed tends to arise naturally in many scientific and engineering calculations. It turns out that v_{rms} differs from v_{avg} by less than 10%, so for practical purposes we can interpret v_{rms} as being essentially the average speed of a molecule in a gas.

There's nothing special about the x-axis. The coordinate system is something that *we* impose on the problem, so *on average* it must be the case that

$$(v_x^2)_{avg} = (v_y^2)_{avg} = (v_z^2)_{avg} \quad (20.14)$$

Hence we can use Equation 20.12 and the definition of v_{rms} to write

$$v_{rms}^2 = (v_x^2)_{avg} + (v_y^2)_{avg} + (v_z^2)_{avg} = 3(v_x^2)_{avg} \quad (20.15)$$

Consequently, $(v_x^2)_{avg}$ is

$$(v_x^2)_{avg} = \tfrac{1}{3}v_{rms}^2 \quad (20.16)$$

Using this result in Equation 20.11 gives us the net force on the wall of the container:

$$F_{on\ wall} = \frac{1}{3}\frac{N}{V}mv_{rms}^2 A \quad (20.17)$$

Thus the pressure on the wall of the container due to all the molecular collisions is

$$p = \frac{F_{on\ wall}}{A} = \frac{1}{3}\frac{N}{V}mv_{rms}^2 \quad (20.18)$$

We have met our goal. Equation 20.18 expresses the macroscopic pressure in terms of the microscopic physics. The pressure depends on the number density of molecules in the container and on how fast, on average, the molecules are moving.

EXAMPLE 20.2 | **The rms speed of helium atoms**

A container holds helium at a pressure of 200 kPa and a temperature of 60.0°C. What is the rms speed of the helium atoms?

SOLVE The rms speed can be found from the pressure and the number density. Using the ideal-gas law gives us the number density:

$$\frac{N}{V} = \frac{p}{k_B T} = \frac{200,000\ \text{Pa}}{(1.38 \times 10^{-23}\ \text{J/K})(333\ \text{K})} = 4.35 \times 10^{25}\ \text{m}^{-3}$$

The mass of a helium atom is $m = 4\ \text{u} = 6.64 \times 10^{-27}$ kg. Thus

$$v_{rms} = \sqrt{\frac{3p}{(N/V)m}} = 1440\ \text{m/s}$$

ASSESS We found in Chapter 17 that the speed of sound in helium is roughly 1000 m/s. Individual atoms probably move somewhat faster than a wave front, so 1440 m/s seems quite reasonable.

STOP TO THINK 20.2 The speed of every molecule in a gas is suddenly increased by a factor of 4. As a result, v_{rms} increases by a factor of

a. 2.   b. <4 but not necessarily 2.
c. 4.   d. >4 but not necessarily 16.
e. 16.   f. v_{rms} doesn't change.

20.3 Temperature

A molecule of mass m and velocity v has translational kinetic energy

$$\epsilon = \tfrac{1}{2}mv^2 \qquad (20.19)$$

We'll use ϵ (lowercase Greek epsilon) to distinguish the energy of a molecule from the system energy E. Thus the average translational kinetic energy is

ϵ_{avg} = average translational kinetic energy of a molecule

$$= \tfrac{1}{2}m(v^2)_{avg} = \tfrac{1}{2}mv_{rms}^2 \qquad (20.20)$$

We've included the word "translational" to distinguish ϵ from rotational kinetic energy, which we will consider later in this chapter.

We can write the gas pressure, Equation 20.18, in terms of the average translational kinetic energy as

$$p = \frac{2}{3}\frac{N}{V}(\tfrac{1}{2}mv_{rms}^2) = \frac{2}{3}\frac{N}{V}\epsilon_{avg} \qquad (20.21)$$

The pressure is directly proportional to the average molecular translational kinetic energy. This makes sense. More-energetic molecules will hit the walls harder as they bounce and thus exert more force on the walls.

It's instructive to write Equation 20.21 as

$$pV = \tfrac{2}{3}N\epsilon_{avg} \qquad (20.22)$$

We know, from the ideal-gas law, that

$$pV = Nk_BT \qquad (20.23)$$

Comparing these two equations, we reach the significant conclusion that the average translational kinetic energy per molecule is

$$\epsilon_{avg} = \tfrac{3}{2}k_BT \qquad \text{(average translational kinetic energy)} \qquad (20.24)$$

where the temperature T is in kelvins. For example, the average translational kinetic energy of a molecule at room temperature (20°C) is

$$\epsilon_{avg} = \tfrac{3}{2}(1.38 \times 10^{-23} \text{ J/K})(293 \text{ K}) = 6.1 \times 10^{-21} \text{ J}$$

NOTE A molecule's average translational kinetic energy depends *only* on the temperature, not on the molecule's mass. If two gases have the same temperature, their molecules have the same average translational kinetic energy.

Equation 20.24 is especially satisfying because it finally gives real meaning to the concept of temperature. Writing it as

$$T = \frac{2}{3k_B}\epsilon_{avg} \qquad (20.25)$$

we can see that, for a gas, this **thing we call** *temperature* **measures the average translational kinetic energy.** A higher temperature corresponds to a larger value of ϵ_{avg} and thus to higher molecular speeds. This concept of temperature also gives meaning to *absolute zero* as the temperature at which $\epsilon_{avg} = 0$ and all molecular motion

FIGURE 20.7 The micro/macro connection for pressure and temperature.

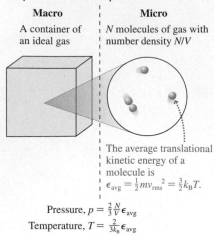

Macro	Micro
A container of an ideal gas	N molecules of gas with number density N/V

The average translational kinetic energy of a molecule is $\epsilon_{avg} = \frac{1}{2}mv_{rms}^2 = \frac{3}{2}k_BT$.

Pressure, $p = \frac{2}{3}\frac{N}{V}\epsilon_{avg}$

Temperature, $T = \frac{2}{3k_B}\epsilon_{avg}$

ceases. (Quantum effects at very low temperatures prevent the motions from actually stopping, but our classical theory predicts that they would.) **FIGURE 20.7** summarizes what we've learned thus far about the micro/macro connection.

We can now justify our assumption that molecular collisions are perfectly elastic. Suppose they were not. If kinetic energy was lost in collisions, the average translational kinetic energy ϵ_{avg} of the gas would decrease and we would see a steadily decreasing temperature. But that doesn't happen. The temperature of an isolated system remains constant, indicating that ϵ_{avg} is not changing with time. Consequently, the collisions must be perfectly elastic.

EXAMPLE 20.3 | **Total microscopic kinetic energy**

What is the total translational kinetic energy of the molecules in 1.0 mol of gas at STP?

SOLVE The average translational kinetic energy of each molecule is

$$\epsilon_{avg} = \frac{3}{2}k_BT = \frac{3}{2}(1.38 \times 10^{-23} \text{ J/K})(273 \text{ K})$$
$$= 5.65 \times 10^{-21} \text{ J}$$

1.0 mol of gas contains N_A molecules; hence the total kinetic energy is

$$K_{micro} = N_A\epsilon_{avg} = 3400 \text{ J}$$

ASSESS The energy of any one molecule is incredibly small. Nonetheless, a macroscopic system has substantial thermal energy because it consists of an incredibly large number of molecules.

By definition, $\epsilon_{avg} = \frac{1}{2}mv_{rms}^2$. Using the ideal-gas law, we found $\epsilon_{avg} = \frac{3}{2}k_BT$. By equating these expressions we find that the rms speed of molecules in a gas is

$$v_{rms} = \sqrt{\frac{3k_BT}{m}} \tag{20.26}$$

The rms speed depends on the square root of the temperature and inversely on the square root of the molecular mass. For example, room-temperature nitrogen (molecular mass 28 u) has rms speed

$$v_{rms} = \sqrt{\frac{3(1.38 \times 10^{-23} \text{ J/K})(293 \text{ K})}{28(1.66 \times 10^{-27} \text{ kg})}} = 509 \text{ m/s}$$

This is in excellent agreement with the experimental results of Figure 20.2.

EXAMPLE 20.4 | **Mean time between collisions**

Estimate the mean time between collisions for a nitrogen molecule at 1.0 atm pressure and room temperature (20°C).

MODEL Because v_{rms} is essentially the average molecular speed, the *mean time between collisions* is simply the time needed to travel distance λ, the mean free path, at speed v_{rms}.

SOLVE We found $\lambda = 2.3 \times 10^{-7}$ m in Example 20.1 and $v_{rms} = 509$ m/s above. Thus the mean time between collisions is

$$(\Delta t)_{avg} = \frac{\lambda}{v_{rms}} = \frac{2.3 \times 10^{-7} \text{ m}}{509 \text{ m/s}} = 4.5 \times 10^{-10} \text{ s}$$

ASSESS The air molecules around us move very fast, they collide with their neighbors about two billion times every second, and they manage to move, on average, only about 230 nm between collisions.

STOP TO THINK 20.3 The speed of every molecule in a gas is suddenly increased by a factor of 4. As a result, the temperature T increases by a factor of

a. 2.

b. <4 but not necessarily 2.

c. 4.

d. >4 but not necessarily 16.

e. 16.

f. T doesn't change.

20.4 Thermal Energy and Specific Heat

We defined the thermal energy of a system to be $E_{th} = K_{micro} + U_{micro}$, where K_{micro} is the microscopic kinetic energy of the moving molecules and U_{micro} is the potential energy of the stretched and compressed molecular bonds. We're now ready to take a microscopic look at thermal energy.

Monatomic Gases

FIGURE 20.8 shows a monatomic gas such as helium or neon. The atoms in an ideal gas have no molecular bonds with their neighbors; hence $U_{micro} = 0$. Furthermore, the kinetic energy of a monatomic gas particle is entirely translational kinetic energy ϵ. Thus the thermal energy of a monatomic gas of N atoms is

$$E_{th} = K_{micro} = \epsilon_1 + \epsilon_2 + \epsilon_3 + \cdots + \epsilon_N = N\epsilon_{avg} \qquad (20.27)$$

where ϵ_i is the translational kinetic energy of atom i. We found that $\epsilon_{avg} = \frac{3}{2}k_B T$; hence the thermal energy is

$$E_{th} = \tfrac{3}{2}Nk_B T = \tfrac{3}{2}nRT \qquad \text{(thermal energy of a monatomic gas)} \qquad (20.28)$$

where we used $N = nN_A$ and the definition of Boltzmann's constant, $k_B = R/N_A$.

We've noted for the last two chapters that thermal energy is associated with temperature. Now we have an explicit result for a monatomic gas: E_{th} is directly proportional to the temperature. Notice that E_{th} is independent of the atomic mass. Any two monatomic gases will have the same thermal energy if they have the same temperature and the same number of atoms (or moles).

If the temperature of a monatomic gas changes by ΔT, its thermal energy changes by

$$\Delta E_{th} = \tfrac{3}{2}nR\,\Delta T \qquad (20.29)$$

In Chapter 19 we found that the change in thermal energy for *any* ideal-gas process is related to the molar specific heat at constant volume by

$$\Delta E_{th} = nC_V\,\Delta T \qquad (20.30)$$

Equation 20.29 is a microscopic result that we obtained by relating the temperature to the average translational kinetic energy of the atoms. Equation 20.30 is a macroscopic result that we arrived at from the first law of thermodynamics. We can make a micro/macro connection by combining these two equations. Doing so gives us a *prediction* for the molar specific heat:

$$C_V = \tfrac{3}{2}R = 12.5 \text{ J/mol K} \qquad \text{(monatomic gas)} \qquad (20.31)$$

This was exactly the value of C_V for all three monatomic gases in Table 19.4. The perfect agreement of theory and experiment is strong evidence that gases really do consist of moving, colliding molecules.

The Equipartition Theorem

The particles of a monatomic gas are atoms. Their energy consists exclusively of their translational kinetic energy. A particle's translational kinetic energy can be written

$$\epsilon = \tfrac{1}{2}mv^2 = \tfrac{1}{2}mv_x^2 + \tfrac{1}{2}mv_y^2 + \tfrac{1}{2}mv_z^2 = \epsilon_x + \epsilon_y + \epsilon_z \qquad (20.32)$$

where we have written separately the energy associated with translational motion along the three axes. Because each axis in space is independent, we can think of ϵ_x, ϵ_y, and ϵ_z as independent *modes* of storing energy within the system.

FIGURE 20.8 The atoms in a monatomic gas have only translational kinetic energy.

Atom i has translational kinetic energy ϵ_i but no potential energy or rotational kinetic energy.

N atoms in a gas at temperature T

The thermal energy of the gas is $E_{th} = \epsilon_1 + \epsilon_2 + \epsilon_3 + \cdots = N\epsilon_{avg}$.

Other systems have additional modes of energy storage. For example,

- Two atoms joined by a spring-like molecular bond can vibrate back and forth. Both kinetic and potential energy are associated with this vibration.
- A diatomic molecule, in addition to translational kinetic energy, has rotational kinetic energy if it rotates end-over-end like a dumbbell.

We define the number of **degrees of freedom** as the number of distinct and independent modes of energy storage. A monatomic gas has three degrees of freedom, the three modes of translational kinetic energy. Systems that can vibrate or rotate have more degrees of freedom.

An important result of statistical physics says that the energy in a system is distributed so that all modes of energy storage have equal amounts of energy. This conclusion is known as the *equipartition theorem,* meaning that the energy is equally divided. The proof is beyond what we can do in this textbook, so we will state the theorem without proof:

> **Equipartition theorem** The thermal energy of a system of particles is equally divided among all the possible degrees of freedom. For a system of N particles at temperature T, the energy stored in each mode (each degree of freedom) is $\frac{1}{2}Nk_BT$ or, in terms of moles, $\frac{1}{2}nRT$.

A monatomic gas has three degrees of freedom and thus, as we found above, $E_{th} = \frac{3}{2}Nk_BT$.

Solids

FIGURE 20.9 A simple model of a solid.

Each atom has microscopic translational kinetic energy *and* microscopic potential energy along all three axes.

FIGURE 20.9 reminds you of our "bedspring model" of a solid with particle-like atoms connected by a lattice of spring-like molecular bonds. How many degrees of freedom does a solid have? Three degrees of freedom are associated with the kinetic energy, just as in a monatomic gas. In addition, the molecular bonds can be compressed or stretched independently along the x-, y-, and z-axes. Three additional degrees of freedom are associated with these three modes of potential energy. Altogether, a solid has six degrees of freedom.

The energy stored in each of these six degrees of freedom is $\frac{1}{2}Nk_BT$. The thermal energy of a solid is the total energy stored in all six modes, or

$$E_{th} = 3Nk_BT = 3nRT \qquad \text{(thermal energy of a solid)} \qquad (20.33)$$

We can use this result to predict the molar specific heat of a solid. If the temperature changes by ΔT, then the thermal energy changes by

$$\Delta E_{th} = 3nR\,\Delta T \qquad (20.34)$$

In Chapter 19 we defined the molar specific heat of a solid such that

$$\Delta E_{th} = nC\,\Delta T \qquad (20.35)$$

By comparing Equations 20.34 and 20.35 we can predict that the molar specific heat of a solid is

$$C = 3R = 25.0 \text{ J/mol K} \quad \text{(solid)} \qquad (20.36)$$

Not bad. The five elemental solids in Table 19.2 had molar specific heats clustered right around 25 J/mol K. They ranged from 24.3 J/mol K for aluminum to 26.5 J/mol K for lead. There are two reasons the agreement between theory and experiment isn't quite as perfect as it was for monatomic gases. First, our simple bedspring model of a solid isn't quite as accurate as our model of a monatomic gas. Second, quantum effects are beginning to make their appearance. More on this shortly. Nonetheless, our ability to predict C to within a few percent from a simple model of a solid is further evidence for the atomic structure of matter.

Diatomic Molecules

Diatomic molecules are a bigger challenge. How many degrees of freedom does a diatomic molecule have? FIGURE 20.10 shows a diatomic molecule, such as molecular nitrogen N_2, oriented along the x-axis. Three degrees of freedom are associated with the molecule's translational kinetic energy. The molecule can have a dumbbell-like end-over-end rotation about either the y-axis or the z-axis. It can also rotate about its own axis. These are three rotational degrees of freedom. The two atoms can also vibrate back and forth, stretching and compressing the molecular bond. This vibrational motion has both kinetic and potential energy—thus two more degrees of freedom.

Altogether, then, a diatomic molecule has eight degrees of freedom, and we would expect the thermal energy of a gas of diatomic molecules to be $E_{th} = 4k_BT$. The analysis we followed for a monatomic gas would then lead to the prediction $C_V = 4R = 33.2$ J/mol K. As compelling as this reasoning seems to be, this is *not* the experimental value of C_V that was reported for diatomic gases in Table 19.4. Instead, we found $C_V = 20.8$ J/mol K.

Why should a theory that works so well for monatomic gases and solids fail so miserably for diatomic molecules? To see what's going on, notice that 20.8 J/mol K $= \frac{5}{2}R$. A monatomic gas, with three degrees of freedom, has $C_V = \frac{3}{2}R$. A solid, with six degrees of freedom, has $C = 3R$. A diatomic gas would have $C_V = \frac{5}{2}R$ if it had five degrees of freedom, not eight.

This discrepancy was a major conundrum as statistical physics developed in the late 19th century. Although it was not recognized as such at the time, we are here seeing our first evidence for the breakdown of classical Newtonian physics. Classically, a diatomic molecule has eight degrees of freedom. The equipartition theorem doesn't distinguish between them; all eight should have the same energy. But atoms and molecules are not classical particles. It took the development of quantum theory in the 1920s to accurately characterize the behavior of atoms and molecules. We don't yet have the tools needed to see why, but quantum effects prevent three of the modes—the two vibrational modes and the rotation of the molecule about its own axis—from being active at room temperature.

FIGURE 20.11 shows C_V as a function of temperature for hydrogen gas. C_V is right at $\frac{5}{2}R$ for temperatures from ≈ 200 K up to ≈ 800 K. But at very low temperatures C_V drops to the monatomic-gas value $\frac{3}{2}R$. The two rotational modes become "frozen out" and the nonrotating molecule has only translational kinetic energy. Quantum physics can explain this, but not Newtonian physics. You can also see that the two vibrational modes *do* become active at very high temperatures, where C_V rises to $\frac{7}{2}R$. Thus the real answer to What's wrong? is that Newtonian physics is not the right physics for describing atoms and molecules. We are somewhat fortunate that Newtonian physics is adequate to understand monatomic gases and solids, at least at room temperature.

Accepting the quantum result that a diatomic gas has only five degrees of freedom at commonly used temperatures (the translational degrees of freedom and the two end-over-end rotations), we find

$$E_{th} = \frac{5}{2}Nk_BT = \frac{5}{2}nRT$$
$$C_V = \frac{5}{2}R = 20.8 \text{ J/mol K}$$

(diatomic gases) (20.37)

A diatomic gas has more thermal energy than a monatomic gas at the same temperature because the molecules have rotational as well as translational kinetic energy.

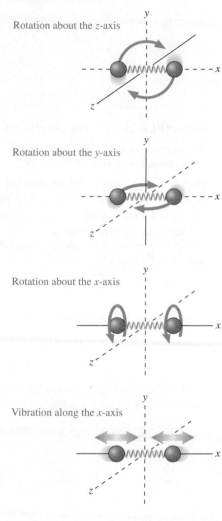

FIGURE 20.10 A diatomic molecule can rotate or vibrate.

Rotation about the z-axis

Rotation about the y-axis

Rotation about the x-axis

Vibration along the x-axis

FIGURE 20.11 Hydrogen molar specific heat at constant volume as a function of temperature.

C_V (J/mol K)

Vibration

Rotation

Translation

The temperature scale is logarithmic.

While the micro/macro connection firmly establishes the atomic structure of matter, it also heralds the need for a new theory of matter at the atomic level. That is a task we will take up in Part VIII. For now, **TABLE 20.1** summarizes what we have learned from kinetic theory about thermal energy and molar specific heats.

TABLE 20.1 Kinetic theory predictions for the thermal energy and the molar specific heat

System	Degrees of freedom	E_{th}	C_V
Monatomic gas	3	$\frac{3}{2}Nk_BT = \frac{3}{2}nRT$	$\frac{3}{2}R = 12.5$ J/mol K
Diatomic gas	5	$\frac{5}{2}Nk_BT = \frac{5}{2}nRT$	$\frac{5}{2}R = 20.8$ J/mol K
Elemental solid	6	$3Nk_BT = 3nRT$	$3R = 25.0$ J/mol K

EXAMPLE 20.5 | The rotational frequency of a molecule

The nitrogen molecule N_2 has a bond length of 0.12 nm. Estimate the rotational frequency of N_2 at 20°C.

MODEL The molecule can be modeled as a rigid dumbbell of length $L = 0.12$ nm rotating about its center.

SOLVE The rotational kinetic energy of the molecule is $\epsilon_{rot} = \frac{1}{2}I\omega^2$, where I is the moment of inertia about the center. Because we have two point masses each moving in a circle of radius $r = L/2$, the moment of inertia is

$$I = mr^2 + mr^2 = 2m\left(\frac{L}{2}\right)^2 = \frac{mL^2}{2}$$

Thus the rotational kinetic energy is

$$\epsilon_{rot} = \frac{1}{2}\frac{mL^2}{2}\omega^2 = \frac{mL^2\omega^2}{4} = \pi^2 mL^2 f^2$$

where we used $\omega = 2\pi f$ to relate the rotational frequency f to the angular frequency ω. From the equipartition theorem, the energy

associated with this mode is $\frac{1}{2}Nk_BT$, so the *average* rotational kinetic energy per molecule is

$$(\epsilon_{rot})_{avg} = \frac{1}{2}k_BT$$

Equating these two expressions for ϵ_{rot} gives us

$$\pi^2 mL^2 f^2 = \frac{1}{2}k_BT$$

Thus the rotational frequency is

$$f = \sqrt{\frac{k_BT}{2\pi^2 mL^2}} = 7.8 \times 10^{11} \text{ rev/s}$$

We evaluated f at $T = 293$ K, using $m = 14$ u $= 2.34 \times 10^{-26}$ kg for each *atom*.

ASSESS This is a *very* high frequency, but these values are typical of molecular rotations.

STOP TO THINK 20.4 How many degrees of freedom does a bead on a rigid rod have?

a. 1   b. 2   c. 3   d. 4   e. 5   f. 6

20.5 Thermal Interactions and Heat

We can now look in more detail at what happens when two systems at different temperatures interact with each other. **FIGURE 20.12** shows a rigid, insulated container divided into two sections by a very thin membrane. The left side, which we'll call system 1, has N_1 atoms at an initial temperature T_{1i}. System 2 on the right has N_2 atoms at an initial temperature T_{2i}. The membrane is so thin that atoms can collide at the boundary as if the membrane were not there, yet it is a barrier that prevents atoms from moving from one side to the other. The situation is analogous, on an atomic scale, to basketballs colliding through a shower curtain.

Suppose that system 1 is initially at a higher temperature: $T_{1i} > T_{2i}$. This is not an equilibrium situation. The temperatures will change with time until the systems eventually reach a common final temperature T_f. If you *watch* the gases as one warms and the other cools, you see nothing happening. This interaction is quite

FIGURE 20.12 Two gases can interact thermally through a very thin barrier.

Insulation prevents heat from entering or leaving the container.

A thin barrier prevents atoms from moving from system 1 to 2 but still allows them to collide. The barrier is clamped in place and cannot move.

different from a mechanical interaction in which, for example, you might see a piston move from one side toward the other. The only way in which the gases can interact is via molecular collisions at the boundary. This is a *thermal interaction,* and our goal is to understand how thermal interactions bring the systems to thermal equilibrium.

System 1 and system 2 begin with thermal energies

$$E_{1i} = \tfrac{3}{2}N_1 k_B T_{1i} = \tfrac{3}{2}n_1 R T_{1i}$$
$$E_{2i} = \tfrac{3}{2}N_2 k_B T_{2i} = \tfrac{3}{2}n_2 R T_{2i} \tag{20.38}$$

We've written the energies for monatomic gases; you could do the same calculation if one or both of the gases is diatomic by replacing the $\tfrac{3}{2}$ with $\tfrac{5}{2}$. Notice that we've omitted the subscript "th" to keep the notation manageable.

The total energy of the combined systems is $E_{tot} = E_{1i} + E_{2i}$. As systems 1 and 2 interact, their individual thermal energies E_1 and E_2 can change but their sum E_{tot} remains constant. The system will have reached thermal equilibrium when the individual thermal energies reach final values E_{1f} and E_{2f} that no longer change.

The Systems Exchange Energy

FIGURE 20.13 shows a fast atom and a slow atom approaching the barrier from opposite sides. They undergo a perfectly elastic collision at the barrier. Although no net energy is lost in a perfectly elastic collision, in most such collisions the more-energetic atom loses energy while the less-energetic atom gains energy. In other words, there's an energy *transfer* from the more-energetic atom's side to the less-energetic atom's side.

The average translational kinetic energy per atom is directly proportional to the temperature: $\epsilon_{avg} = \tfrac{3}{2}k_B T$. Because $T_{1i} > T_{2i}$, the atoms in system 1 are, on average, more energetic than the atoms in system 2. Thus *on average* the collisions transfer energy from system 1 to system 2. Not in every collision: sometimes a fast atom in system 2 collides with a slow atom in system 1, transferring energy from 2 to 1. But the net energy transfer, from all collisions, is from the warmer system 1 to the cooler system 2. In other words, **heat is the energy transferred *via collisions* between the more-energetic (warmer) atoms on one side and the less-energetic (cooler) atoms on the other.**

How do the systems "know" when they've reached thermal equilibrium? Energy transfer continues until the atoms on both sides of the barrier have the *same average translational kinetic energy*. Once the average translational kinetic energies are the same, there is no tendency for energy to flow in either direction. This is the state of thermal equilibrium, so the condition for thermal equilibrium is

$$\left(\epsilon_1\right)_{avg} = \left(\epsilon_2\right)_{avg} \quad \text{(thermal equilibrium)} \tag{20.39}$$

where, as before, ϵ is the translational kinetic energy of an atom.

Because the average energies are directly proportional to the final temperatures, $\epsilon_{avg} = \tfrac{3}{2}k_B T_f$, thermal equilibrium is characterized by the macroscopic condition

$$T_{1f} = T_{2f} = T_f \quad \text{(thermal equilibrium)} \tag{20.40}$$

In other words, **two thermally interacting systems reach a common final temperature *because* they exchange energy via collisions until the atoms on each side have, on average, equal translational kinetic energies.** This is a very important idea.

Equation 20.40 can be used to determine the equilibrium thermal energies. Because these are monatomic gases, $E_{th} = N\epsilon_{avg}$. Thus the equilibrium condition $\left(\epsilon_1\right)_{avg} = \left(\epsilon_2\right)_{avg} = \left(\epsilon_{tot}\right)_{avg}$ implies

$$\frac{E_{1f}}{N_1} = \frac{E_{2f}}{N_2} = \frac{E_{tot}}{N_1 + N_2} \tag{20.41}$$

FIGURE 20.13 On average, collisions transfer energy from more-energetic atoms to less-energetic atoms.

FIGURE 20.14 Equilibrium is reached when the atoms on each side have, on average, equal energies.

Collisions transfer energy from the warmer system to the cooler system as more-energetic atoms lose energy to less-energetic atoms.

Thermal equilibrium occurs when the systems have the same average translational kinetic energy and thus the same temperature.

In general, the thermal energies E_{1f} and E_{2f} are *not* equal.

from which we can conclude

$$E_{1f} = \frac{N_1}{N_1 + N_2} E_{tot} = \frac{n_1}{n_1 + n_2} E_{tot}$$

$$E_{2f} = \frac{N_2}{N_1 + N_2} E_{tot} = \frac{n_2}{n_1 + n_2} E_{tot} \quad (20.42)$$

where in the last step we used moles rather than molecules.

Notice that $E_{1f} + E_{2f} = E_{tot}$, verifying that energy has been conserved even while being redistributed between the systems.

No work is done on either system because the barrier has no macroscopic displacement, so the first law of thermodynamics is

$$Q_1 = \Delta E_1 = E_{1f} - E_{1i}$$

$$Q_2 = \Delta E_2 = E_{2f} - E_{2i} \quad (20.43)$$

As a homework problem you can show that $Q_1 = -Q_2$, as required by energy conservation. That is, the heat lost by one system is gained by the other. $|Q_1|$ is the quantity of heat that is transferred from the warmer gas to the cooler gas during the thermal interaction.

> **NOTE** In general, the equilibrium thermal energies of the system are *not* equal. That is, $E_{1f} \neq E_{2f}$. They will be equal only if $N_1 = N_2$. Equilibrium is reached when the average translational kinetic energies in the two systems are equal—that is, when $(\epsilon_1)_{avg} = (\epsilon_2)_{avg}$, not when $E_{1f} = E_{2f}$. The distinction is important. **FIGURE 20.14** summarizes these ideas.

EXAMPLE 20.6 | A thermal interaction

A sealed, insulated container has 2.0 g of helium at an initial temperature of 300 K on one side of a barrier and 10.0 g of argon at an initial temperature of 600 K on the other side.

a. How much heat energy is transferred, and in which direction?

b. What is the final temperature?

MODEL The systems start with different temperatures, so they are not in thermal equilibrium. Energy will be transferred via collisions from the argon to the helium until both systems have the same average molecular energy.

SOLVE a. Let the helium be system 1. Helium has molar mass $M_{mol} = 0.004$ kg/mol, so $n_1 = M/M_{mol} = 0.50$ mol. Similarly, argon has $M_{mol} = 0.040$ kg/mol, so $n_2 = 0.25$ mol. The initial thermal energies of the two monatomic gases are

$$E_{1i} = \tfrac{3}{2} n_1 R T_{1i} = 225R = 1870 \text{ J}$$

$$E_{2i} = \tfrac{3}{2} n_2 R T_{2i} = 225R = 1870 \text{ J}$$

The systems start with *equal* thermal energies, but they are not in thermal equilibrium. The total energy is $E_{tot} = 3740$ J. In equilibrium, this energy is distributed between the two systems as

$$E_{1f} = \frac{n_1}{n_1 + n_2} E_{tot} = \frac{0.50}{0.75} \times 3740 \text{ J} = 2493 \text{ J}$$

$$E_{2f} = \frac{n_2}{n_1 + n_2} E_{tot} = \frac{0.25}{0.75} \times 3740 \text{ J} = 1247 \text{ J}$$

The heat entering or leaving each system is

$$Q_1 = Q_{He} = E_{1f} - E_{1i} = 623 \text{ J}$$

$$Q_2 = Q_{Ar} = E_{2f} - E_{2i} = -623 \text{ J}$$

The helium and the argon interact thermally via collisions at the boundary, causing 623 J of heat to be transferred from the warmer argon to the cooler helium.

b. These are constant-volume processes, thus $Q = n C_V \Delta T$. $C_V = \tfrac{3}{2} R$ for monatomic gases, so the temperature changes are

$$\Delta T_{He} = \frac{Q_{He}}{\tfrac{3}{2} n R} = \frac{623 \text{ J}}{1.5(0.50 \text{ mol})(8.31 \text{ J/mol K})} = 100 \text{ K}$$

$$\Delta T_{Ar} = \frac{Q_{Ar}}{\tfrac{3}{2} n R} = \frac{-623 \text{ J}}{1.5(0.25 \text{ mol})(8.31 \text{ J/mol K})} = -200 \text{ K}$$

Both gases reach the common final temperature $T_f = 400$ K.

ASSESS $E_{1f} = 2 E_{2f}$ because there are twice as many atoms in system 1.

The main idea of this section is that two systems reach a common final temperature not by magic or by a prearranged agreement but simply from the energy exchange of vast numbers of molecular collisions. Real interacting systems, of course, are separated by walls rather than our unrealistic thin membrane. As the systems interact, the energy is first transferred via collisions from system 1 into the wall and subsequently, as the cooler molecules collide with a warm wall, into system 2. That is, the energy transfer is $E_1 \rightarrow E_{\text{wall}} \rightarrow E_2$. This is still heat because the energy transfer is occurring via molecular collisions rather than mechanical motion.

STOP TO THINK 20.5 Systems A and B are interacting thermally. At this instant of time,

a. $T_A > T_B$
b. $T_A = T_B$
c. $T_A < T_B$

A	B
$N = 1000$	$N = 2000$
$\epsilon_{\text{avg}} = 1.0 \times 10^{-20}\,\text{J}$	$\epsilon_{\text{avg}} = 0.5 \times 10^{-20}\,\text{J}$
$E_{\text{th}} = 1.0 \times 10^{-17}\,\text{J}$	$E_{\text{th}} = 1.0 \times 10^{-17}\,\text{J}$

20.6 Irreversible Processes and the Second Law of Thermodynamics

The preceding section looked at the thermal interaction between a warm gas and a cold gas. Heat energy is transferred from the warm gas to the cold gas until they reach a common final temperature. But why isn't heat transferred from the cold gas to the warm gas, making the cold side colder and the warm side warmer? Such a process could still conserve energy, but it never happens. The transfer of heat energy from hot to cold is an example of an **irreversible process,** a process that can happen only in one direction.

Examples of irreversible processes abound. Stirring the cream in your coffee mixes the cream and coffee together. No amount of stirring ever unmixes them. If you shake a jar that has red marbles on the top and blue marbles on the bottom, the two colors are quickly mixed together. No amount of shaking ever separates them again. If you watched a movie of someone shaking a jar and saw the red and blue marbles separating, you would be certain that the movie was running backward. In fact, a reasonable definition of an irreversible process is one for which a backward-running movie shows a physically impossible process.

FIGURE 20.15a is a two-frame movie of a collision between two particles, perhaps two gas molecules. Suppose that sometime after the collision is over we could reach in and reverse the velocities of both particles. That is, replace vector $\vec{v}$ with vector $-\vec{v}$. Then, as in a movie playing backward, the collision would happen in reverse. This is the movie of **FIGURE 20.15b**.

FIGURE 20.15 Molecular collisions are reversible.

(a) Forward movie
Before: After:

(b) The backward movie is equally plausible.
Before: After:

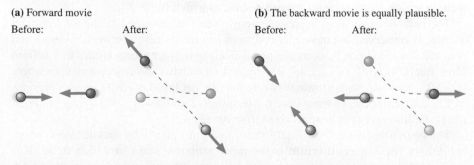

You cannot tell, just by looking at the two movies, which is really going forward and which is being played backward. Maybe Figure 20.15b was the original collision and Figure 20.15a is the backward version. Nothing in either collision looks wrong, and no measurements you might make on either would reveal any violations of Newton's laws. **Interactions between molecules are reversible processes.**

FIGURE 20.16 A car crash is irreversible.

(a) Forward movie

(b) The backward movie is physically impossible.

FIGURE 20.17 Two interacting systems. Balls are chosen at random and moved to the other box.

Balls are chosen at random and moved from one box to the other.

Box 1
N_1 balls

Box 2
N_2 balls

Tossing all heads, while not impossible, is extremely unlikely, and the probability of doing so rapidly decreases as the number of coins increases.

Contrast this with the two-frame car crash movies in **FIGURE 20.16**. Past and future are clearly distinct in an irreversible process, and the backward movie of Figure 20.16b is obviously wrong. But what has been violated in the backward movie? To have the crumpled car spring away from the wall would not violate any laws of physics we have so far discovered. It would simply require transforming the thermal energy of the car and wall back into the macroscopic center-of-mass energy of the car as a whole.

The paradox stems from our assertion that macroscopic phenomena can be understood on the basis of microscopic molecular motions. If the microscopic motions are all reversible, how can the macroscopic phenomena end up being irreversible? If reversible collisions can cause heat to be transferred from hot to cold, why do they never cause heat to be transferred from cold to hot? There must be another law of physics preventing it. The law we seek must, in some sense, be able to distinguish the past from the future.

Which Way to Equilibrium?

Stated another way, how do two systems initially at different temperatures "know" which way to go to reach equilibrium? Perhaps an analogy will help.

FIGURE 20.17 shows two boxes, numbered 1 and 2, containing identical balls. Box 1 starts with more balls than box 2, so $N_{1i} > N_{2i}$. Once every second, one ball is chosen at random and moved to the other box. This is a reversible process because a ball can move from box 2 to box 1 just as easily as from box 1 to box 2. What do you expect to see if you return several hours later?

Because balls are chosen at random, and because $N_{1i} > N_{2i}$, it's initially more likely that a ball will move from box 1 to box 2 than from box 2 to box 1. Sometimes a ball will move "backward" from box 2 to box 1, but overall there's a net movement of balls from box 1 to box 2. The system will evolve until $N_1 \approx N_2$. This is a stable situation—equilibrium!—with an equal number of balls moving in both directions.

But couldn't it go the other way, with N_1 getting even larger while N_2 decreases? In principle, any possible arrangement of the balls is possible in the same way that any number of heads are possible if you throw N coins in the air and let them fall. If you throw four coins, the odds are 1 in 2^4, or 1 in 16, of getting four heads. With four balls, the odds are 1 in 16 that, at a randomly chosen instant of time, you would find $N_1 = 4$. You wouldn't find that to be terribly surprising.

With 10 balls, the probability that $N_1 = 10$ is $0.5^{10} \approx 1/1000$. With 100 balls, the probability that $N_1 = 100$ has dropped to $\approx 10^{-30}$. With 10^{20} balls, the odds of finding all of them, or even most of them, in one box are so staggeringly small that it's safe to say it will "never" happen. Although each transfer is reversible, **the statistics of large numbers make it overwhelmingly more likely that the system will evolve toward a state in which $N_1 \approx N_2$ than toward a state in which $N_1 > N_2$.**

The balls in our analogy represent energy. The total energy, like the total number of balls, is conserved, but molecular collisions can move energy between system 1 and system 2. Each collision is reversible, just as likely to transfer energy from 1 to 2 as from 2 to 1. But if $(\epsilon_{1i})_{avg} > (\epsilon_{2i})_{avg}$, and if we're dealing with two macroscopic systems where $N > 10^{20}$, then it's overwhelmingly likely that the net result of many, many collisions will be to transfer energy from system 1 to system 2 until $(\epsilon_{1f})_{avg} = (\epsilon_{2f})_{avg}$—in other words, for heat energy to be transferred from hot to cold.

The system reaches thermal equilibrium not by any plan or by outside intervention, but simply because **equilibrium is the *most probable* state in which to be**. It is *possible* that the system will move away from equilibrium, with heat moving from cold to hot, but remotely improbable in any realistic system. The consequence of a vast number of random events is that the system evolves in one direction, toward equilibrium, and not the other. **Reversible microscopic events lead to irreversible macroscopic behavior because some macroscopic states are vastly more probable than others.**

Order, Disorder, and Entropy

FIGURE 20.18 shows three different systems. At the top is a group of atoms arranged in a crystal-like lattice. This is a highly ordered and nonrandom system, with each atom's position precisely specified. Contrast this with the system on the bottom, where there is no order at all. The position of every atom was assigned entirely at random.

It is extremely improbable that the atoms in a container would *spontaneously* arrange themselves into the ordered pattern of the top picture. In a system of, say, 10^{20} atoms, the probability of this happening is similar to the probability that 10^{20} tossed coins will all be heads. We can safely say that it will never happen. By contrast, there are a vast number of arrangements like the one on the bottom that randomly fill the container.

The middle picture of Figure 20.18 is an in-between situation. This situation might arise as a solid melts. The positions of the atoms are clearly not completely random, so the system preserves some degree of order. This in-between situation is more likely to occur spontaneously than the highly ordered lattice on the top, but is less likely to occur than the completely random system on the bottom.

Scientists and engineers use a state variable called **entropy** to measure the probability that a macroscopic state will occur spontaneously. The ordered lattice, which has a very small probability of spontaneous occurrence, has a very low entropy. The entropy of the randomly filled container is high. The entropy of the middle picture is somewhere in between. It is often said that entropy measures the amount of *disorder* in a system. The entropy in Figure 20.18 increases as you move from the ordered system on the top to the disordered system on the bottom.

Similarly, two thermally interacting systems with different temperatures have a low entropy. These systems are ordered in the sense that the faster atoms are on one side of the barrier, the slower atoms on the other. The most random possible distribution of energy, and hence the least ordered system, corresponds to the situation where the two systems are in thermal equilibrium with equal temperatures. Entropy increases as two systems with initially different temperatures move toward equilibrium. Entropy would decrease if heat energy moved from cold to hot, making the hot system hotter and the cold system colder.

Entropy can be calculated, but we'll leave that to more advanced courses. For our purposes, the *concept* of entropy as a measure of the disorder in a system, or of the probability that a macroscopic state will occur, is more important than a numerical value.

FIGURE 20.18 Ordered and disordered systems.

Increasing order
Decreasing entropy
Decreasing probability

Increasing randomness
Increasing entropy
Increasing probability

The Second Law of Thermodynamics

The fact that macroscopic systems evolve irreversibly toward equilibrium is a statement about nature that is not contained in any of the laws of physics we have encountered. It is, in fact, a new law of physics, one known as the **second law of thermodynamics.**

The formal statement of the second law of thermodynamics is given in terms of entropy:

Second law, formal statement The entropy of an isolated system (or group of systems) never decreases. The entropy either increases, until the system reaches equilibrium, or, if the system began in equilibrium, stays the same.

The qualifier "isolated" is most important. We can order the system by reaching in from the outside, perhaps using tiny tweezers to place the atoms in a lattice. Similarly, we can transfer heat from cold to hot by using a refrigerator. The second law is about what a system can or cannot do *spontaneously,* on its own, without outside intervention.

The second law of thermodynamics tells us that an isolated system evolves such that

- Order turns into disorder and randomness.
- Information is lost rather than gained.
- The system "runs down."

An isolated system never spontaneously generates order out of randomness. It is not that the system "knows" about order or randomness, but rather that there are vastly more states corresponding to randomness than there are corresponding to order. As collisions occur at the microscopic level, the laws of probability dictate that the system will, on average, move inexorably toward the most probable and thus most random macroscopic state.

The second law of thermodynamics is often stated in several equivalent but more informal versions. One of these, and the one most relevant to our discussion, is

> **Second law, informal statement #1** When two systems at different temperatures interact, heat energy is transferred spontaneously from the hotter to the colder system, never from the colder to the hotter.

This version of the second law will be used in Chapter 21 to understand the thermodynamics of engines.

The second law of thermodynamics is an independent statement about nature, separate from the first law. The first law is a precise statement about energy conservation. The second law, by contrast, is a *probabilistic* statement, based on the statistics of very large numbers. While it is conceivable that heat could spontaneously move from cold to hot, it will never occur in any realistic macroscopic system.

The irreversible evolution from less-likely macroscopic states to more-likely macroscopic states is what gives us a macroscopic direction of time. Stirring blends your coffee and cream; it never unmixes them. Friction causes an object to stop while increasing its thermal energy; the random atomic motions of thermal energy never spontaneously organize themselves into a macroscopic motion of the entire object. A plant in a sealed jar dies and decomposes to carbon and various gases; the gases and carbon never spontaneously assemble themselves into a flower. These are all examples of irreversible processes. They each show a clear direction of time, a distinct difference between past and future.

Thus another statement of the second law is

> **Second law, informal statement #2** The time direction in which the entropy of an isolated macroscopic system increases is "the future."

Establishing the "arrow of time" is one of the most profound implications of the second law of thermodynamics.

STOP TO THINK 20.6 Two identical boxes each contain 1,000,000 molecules. In box A, 750,000 molecules happen to be in the left half of the box while 250,000 are in the right half. In box B, 499,900 molecules happen to be in the left half of the box while 500,100 are in the right half. At this instant of time,

a. The entropy of box A is larger than the entropy of box B.
b. The entropy of box A is equal to the entropy of box B.
c. The entropy of box A is smaller than the entropy of box B.

SUMMARY

The goal of Chapter 20 has been to see how macroscopic properties depend on the motion of atoms.

GENERAL PRINCIPLES

The **micro/macro connection** relates the macroscopic properties of a system to the motion and collisions of its atoms and molecules.

The Equipartition Theorem

Tells us how collisions distribute the energy in the system. The energy stored in each mode of the system (each **degree of freedom**) is $\frac{1}{2}Nk_BT$ or, in terms of moles, $\frac{1}{2}nRT$.

The Second Law of Thermodynamics

Tells us how collisions move a system toward equilibrium. The entropy of an isolated system can only increase or, in equilibrium, stay the same.

- Order turns into disorder and randomness.
- Systems run down.
- Heat energy is transferred spontaneously from a hotter to a colder system, never from colder to hotter.

IMPORTANT CONCEPTS

Pressure is due to the force of the molecules colliding with the walls:

$$p = \frac{1}{3}\frac{N}{V}mv_{rms}^2 = \frac{2}{3}\frac{N}{V}\epsilon_{avg}$$

The **average translational kinetic energy** of a molecule is $\epsilon_{avg} = \frac{3}{2}k_BT$. The temperature of the gas $T = 2\epsilon_{avg}/3k_B$ measures the average translational kinetic energy.

Entropy measures the probability that a macroscopic state will occur or, equivalently, the amount of disorder in a system.

Increasing entropy ➡

The **thermal energy** of a system is

E_{th} = translational kinetic energy + rotational kinetic energy + vibrational energy

- **Monatomic gas** $E_{th} = \frac{3}{2}Nk_BT = \frac{3}{2}nRT$
- **Diatomic gas** $E_{th} = \frac{5}{2}Nk_BT = \frac{5}{2}nRT$
- **Elemental solid** $E_{th} = 3Nk_BT = 3nRT$

Heat is energy transferred via collisions from more-energetic molecules on one side to less-energetic molecules on the other. Equilibrium is reached when $(\epsilon_1)_{avg} = (\epsilon_2)_{avg}$, which implies $T_{1f} = T_{2f}$.

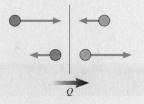

APPLICATIONS

The **root-mean-square speed** v_{rms} is the square root of the average of the squares of the molecular speeds:

$$v_{rms} = \sqrt{(v^2)_{avg}}$$

For molecules of mass m at temperature T, $v_{rms} = \sqrt{\dfrac{3k_BT}{m}}$

Molar specific heats can be predicted from the thermal energy because $\Delta E_{th} = nC\,\Delta T$.

- **Monatomic gas** $C_V = \frac{3}{2}R$
- **Diatomic gas** $C_V = \frac{5}{2}R$
- **Elemental solid** $C = 3R$

TERMS AND NOTATION

histogram	root-mean-square speed, v_{rms}	equipartition theorem	entropy
mean free path, λ	degrees of freedom	irreversible process	second law of thermodynamics

CONCEPTUAL QUESTIONS

1. Solids and liquids resist being compressed. They are not totally incompressible, but it takes large forces to compress them even slightly. If it is true that matter consists of atoms, what can you infer about the microscopic nature of solids and liquids from their incompressibility?
2. Gases, in contrast with solids and liquids, are very compressible. What can you infer from this observation about the microscopic nature of gases?
3. The density of air at STP is about $\frac{1}{1000}$ the density of water. How does the average distance between air molecules compare to the average distance between water molecules? Explain.
4. The mean free path of molecules in a gas is 200 nm.
   a. What will be the mean free path if the pressure is doubled while the temperature is held constant?
   b. What will be the mean free path if the absolute temperature is doubled while the pressure is held constant?
5. If the pressure of a gas is really due to the *random* collisions of molecules with the walls of the container, why do pressure gauges—even very sensitive ones—give perfectly steady readings? Shouldn't the gauge be continually jiggling and fluctuating? Explain.
6. Suppose you could suddenly increase the speed of every molecule in a gas by a factor of 2.
   a. Would the rms speed of the molecules increase by a factor of $2^{1/2}$, 2, or 2^2? Explain.
   b. Would the gas pressure increase by a factor of $2^{1/2}$, 2, or 2^2? Explain.
7. Suppose you could suddenly increase the speed of every molecule in a gas by a factor of 2.
   a. Would the temperature of the gas increase by a factor of $2^{1/2}$, 2, or 2^2? Explain.

b. Would the molar specific heat at constant volume change? If so, by what factor? If not, why not?
8. The two containers of gas in FIGURE Q20.8 are in good thermal contact with each other but well insulated from the environment. They have been in contact for a long time and are in thermal equilibrium.
   a. Is v_{rms} of helium greater than, less than, or equal to v_{rms} of argon? Explain.
   b. Does the helium have more thermal energy, less thermal energy, or the same amount of thermal energy as the argon? Explain.

FIGURE Q20.8

9. Suppose you place an ice cube in a beaker of room-temperature water, then seal them in a rigid, well-insulated container. No energy can enter or leave the container.
   a. If you open the container an hour later, will you find a beaker of water slightly cooler than room temperature, or a large ice cube and some 100°C steam?
   b. Finding a large ice cube and some 100°C steam would not violate the first law of thermodynamics. $W = 0$ J and $Q = 0$ J because the container is sealed, and $\Delta E_{th} = 0$ J because the increase in thermal energy of the water molecules that became steam is offset by the decrease in thermal energy of the water molecules that turned to ice. Energy would be conserved, yet we never see an outcome like this. Why not?

EXERCISES AND PROBLEMS

Problems labeled ▓ integrate material from earlier chapters.

Exercises

Section 20.1 Molecular Speeds and Collisions

1. ‖ The number density of an ideal gas at STP is called the *Loschmidt number*. Calculate the Loschmidt number.
2. ‖ A 1.0 m × 1.0 m × 1.0 m cube of nitrogen gas is at 20°C and 1.0 atm. Estimate the number of molecules in the cube with a speed between 700 m/s and 1000 m/s.
3. ‖ At what pressure will the mean free path in room-temperature (20°C) nitrogen be 1.0 m?
4. ‖‖ Integrated circuits are manufactured in vacuum chambers in which the air pressure is 1.0×10^{-10} mm of Hg. What are (a) the number density and (b) the mean free path of a molecule? Assume $T = 20°C$.
5. ‖ The mean free path of a molecule in a gas is 300 nm. What will the mean free path be if the gas temperature is doubled at (a) constant volume and (b) constant pressure?

6. ‖ A lottery machine uses blowing air to keep 2000 Ping-Pong balls bouncing around inside a 1.0 m × 1.0 m × 1.0 m box. The diameter of a Ping-Pong ball is 3.0 cm. What is the mean free path between collisions? Give your answer in cm.
7. ‖ For a monatomic gas, what is the ratio of the volume per atom (V/N) to the volume *of* an atom when the mean free path is ten times the atomic diameter?

Section 20.2 Pressure in a Gas

8. ‖ The molecules in a six-particle gas have velocities

   $$\vec{v}_1 = (20\hat{i} - 30\hat{j}) \text{ m/s} \qquad \vec{v}_4 = 30\hat{i} \text{ m/s}$$
   $$\vec{v}_2 = (40\hat{i} + 70\hat{j}) \text{ m/s} \qquad \vec{v}_5 = (40\hat{i} - 40\hat{j}) \text{ m/s}$$
   $$\vec{v}_3 = (-80\hat{i} + 20\hat{j}) \text{ m/s} \qquad \vec{v}_6 = (-50\hat{i} - 20\hat{j}) \text{ m/s}$$

   Calculate (a) $\vec{v}_{avg}$, (b) v_{avg}, and (c) v_{rms}.
9. ‖ Eleven molecules have speeds 15, 16, 17, ... , 25 m/s. Calculate (a) v_{avg} and (b) v_{rms}.

10. | FIGURE EX20.10 is a histogram showing the speeds of the molecules in a very small gas. What are (a) the most probable speed, (b) the average speed, and (c) the rms speed?

FIGURE EX20.10

11. ‖ The number density in a container of neon gas is 5.00×10^{25} m^{-3}. The atoms are moving with an rms speed of 660 m/s. What are (a) the temperature and (b) the pressure inside the container?

12. ‖‖ At 100°C the rms speed of nitrogen molecules is 576 m/s. Nitrogen at 100°C and a pressure of 2.0 atm is held in a container with a 10 cm × 10 cm square wall. Estimate the rate of molecular collisions (collisions/s) on this wall.

13. ‖ A cylinder contains gas at a pressure of 2.0 atm and a number density of 4.2×10^{25} m^{-3}. The rms speed of the atoms is 660 m/s. Identify the gas.

Section 20.3 Temperature

14. | A gas consists of a mixture of neon and argon. The rms speed of the neon atoms is 400 m/s. What is the rms speed of the argon atoms?

15. ‖ What are the rms speeds of (a) argon atoms and (b) hydrogen molecules at 800°C?

16. ‖ 1.5 m/s is a typical walking speed. At what temperature (in °C) would nitrogen molecules have an rms speed of 1.5 m/s?

17. ‖ At what temperature (in °C) do hydrogen molecules have the same rms speed as nitrogen molecules at 100°C?

18. | The rms speed of molecules in a gas is 600 m/s. What will be the rms speed if the gas pressure and volume are both halved?

19. ‖ By what factor does the rms speed of a molecule change if the temperature is increased from 10°C to 1000°C?

20. ‖ Atoms can be "cooled" to incredibly low temperatures by letting them interact with a laser beam. Various novel quantum phenomena appear at these temperatures. What is the rms speed of cesium atoms that have been cooled to a temperature of 100 nK?

21. | At STP, what is the total translational kinetic energy of the molecules in 1.0 mol of (a) hydrogen, (b) helium, and (c) oxygen?

22. | Suppose you double the temperature of a gas at constant volume. Do the following change? If so, by what factor?
    a. The average translational kinetic energy of a molecule.
    b. The rms speed of a molecule.
    c. The mean free path.

23. ‖ During a physics experiment, helium gas is cooled to a temperature of 10 K at a pressure of 0.10 atm. What are (a) the mean free path in the gas, (b) the rms speed of the atoms, and (c) the average energy per atom?

24. | What are (a) the average kinetic energy and (b) the rms speed of a proton in the center of the sun, where the temperature is 2.0×10^7 K?

25. ‖ The atmosphere of the sun consists mostly of hydrogen *atoms* (not molecules) at a temperature of 6000 K. What are (a) the average translational kinetic energy per atom and (b) the rms speed of the atoms?

Section 20.4 Thermal Energy and Specific Heat

26. ‖ A 10 g sample of neon gas has 1700 J of thermal energy. Estimate the average speed of a neon atom.

27. ‖ The rms speed of the atoms in a 2.0 g sample of helium gas is 700 m/s. What is the thermal energy of the gas?

28. ‖ A 6.0 m × 8.0 m × 3.0 m room contains air at 20°C. What is the room's thermal energy?

29. | The thermal energy of 1.0 mol of a substance is increased by 1.0 J. What is the temperature change if the system is (a) a monatomic gas, (b) a diatomic gas, and (c) a solid?

30. ‖ What is the thermal energy of 100 cm³ of aluminum at 100°C?

31. | 1.0 mol of a monatomic gas interacts thermally with 1.0 mol of an elemental solid. The gas temperature decreases by 50°C at constant volume. What is the temperature change of the solid?

32. ‖ A cylinder of nitrogen gas has a volume of 15,000 cm³ and a pressure of 100 atm.
    a. What is the thermal energy of this gas at room temperature (20°C)?
    b. What is the mean free path in the gas?
    c. The valve is opened and the gas is allowed to expand slowly and isothermally until it reaches a pressure of 1.0 atm. What is the change in the thermal energy of the gas?

33. ‖ A rigid container holds 0.20 g of hydrogen gas. How much heat is needed to change the temperature of the gas
    a. From 50 K to 100 K?
    b. From 250 K to 300 K?
    c. From 2250 K to 2300 K?

Section 20.5 Thermal Interactions and Heat

34. | 2.0 mol of monatomic gas A initially has 5000 J of thermal energy. It interacts with 3.0 mol of monatomic gas B, which initially has 8000 J of thermal energy.
    a. Which gas has the higher initial temperature?
    b. What is the final thermal energy of each gas?

35. ‖‖ 4.0 mol of monatomic gas A interacts with 3.0 mol of monatomic gas B. Gas A initially has 9000 J of thermal energy, but in the process of coming to thermal equilibrium it transfers 1000 J of heat energy to gas B. How much thermal energy did gas B have initially?

Section 20.6 Irreversible Processes and the Second Law of Thermodynamics

36. ‖ Two containers hold several balls. Once a second, one of the balls is chosen at random and switched to the other container. After a long time has passed, you record the number of balls in each container every second. In 10,000 s, you find 80 times when all the balls were in one container (either one) and the other container was empty.
    a. How many balls are there?
    b. What is the most likely number of balls to be found in one of the containers?

Problems

37. ‖ The pressure inside a tank of neon is 150 atm. The temperature is 25°C. On average, how many atomic diameters does a neon atom move between collisions?

38. ‖‖ From what height must an oxygen molecule fall in a vacuum so that its kinetic energy at the bottom equals the average energy of an oxygen molecule at 300 K?

39. ‖ Dust particles are $\approx 10 \ \mu m$ in diameter. They are pulverized rock, with $\rho \approx 2500 \ kg/m^3$. If you treat dust as an ideal gas, what is the rms speed of a dust particle at 20°C?

40. ‖ A mad engineer builds a cube, 2.5 m on a side, in which 6.2-cm-diameter rubber balls are constantly sent flying in random directions by vibrating walls. He will award a prize to anyone who can figure out how many balls are in the cube without entering it or taking out any of the balls. You decide to shoot 6.2-cm-diameter plastic balls into the cube, through a small hole, to see how far they get before colliding with a rubber ball. After many shots, you find they travel an average distance of 1.8 m. How many rubber balls do you think are in the cube?

41. ‖ Photons of light scatter off molecules, and the distance you can see through a gas is proportional to the mean free path of photons through the gas. Photons are not gas molecules, so the mean free path of a photon is not given by Equation 20.3, but its dependence on the number density of the gas and on the molecular radius is the same. Suppose you are in a smoggy city and can barely see buildings 500 m away.
    a. How far would you be able to see if all the molecules around you suddenly doubled in volume?
    b. How far would you be able to see if the temperature suddenly rose from 20°C to a blazing hot 1500°C with the pressure unchanged?

42. ‖ Interstellar space, far from any stars, is filled with a very low density of hydrogen atoms (H, not H_2). The number density is about 1 atom/cm^3 and the temperature is about 3 K.
    a. Estimate the pressure in interstellar space. Give your answer in Pa and in atm.
    b. What is the rms speed of the atoms?
    c. What is the edge length L of an $L \times L \times L$ cube of gas with 1.0 J of thermal energy?

43. ‖ You are watching a science fiction movie in which the hero shrinks down to the size of an atom and fights villains while jumping from air molecule to air molecule. In one scene, the hero's molecule is about to crash head-on into the molecule on which a villain is riding. The villain's molecule is initially 50 molecular radii away and, in the movie, it takes 3.5 s for the molecules to collide. Estimate the air temperature required for this to be possible. Assume the molecules are nitrogen molecules, each traveling at the rms speed. Is this a plausible temperature for air?

44. ‖ CALC
    a. Find an expression for the v_{rms} of gas molecules in terms of p, V, and the total mass of the gas M.
    b. A gas cylinder has a piston at one end that is moving outward at speed v_{piston} during an isobaric expansion of the gas. Find an expression for the *rate* at which v_{rms} is changing in terms of v_{piston}, the instantaneous value of v_{rms}, and the instantaneous value L of the length of the cylinder.
    c. A cylindrical sample chamber has a piston moving outward at 0.50 m/s during an isobaric expansion. The rms speed of the gas molecules is 450 m/s at the instant the chamber length is 1.5 m. At what rate is v_{rms} changing?

45. ‖ Equation 20.3 is the mean free path of a particle through a gas of identical particles of equal radius. An electron can be thought of as a point particle with zero radius.
    a. Find an expression for the mean free path of an electron through a gas.
    b. Electrons travel 3 km through the Stanford Linear Accelerator. In order for scattering losses to be negligible, the pressure inside the accelerator tube must be reduced to the point where the mean free path is at least 50 km. What is the maximum possible pressure inside the accelerator tube, assuming $T = 20°C$? Give your answer in both Pa and atm.

46. ‖ Uranium has two naturally occurring isotopes. ^{238}U has a natural abundance of 99.3% and ^{235}U has an abundance of 0.7%. It is the rarer ^{235}U that is needed for nuclear reactors. The isotopes are separated by forming uranium hexafluoride, UF_6, which is a gas, then allowing it to diffuse through a series of porous membranes. $^{235}UF_6$ has a slightly larger rms speed than $^{238}UF_6$ and diffuses slightly faster. Many repetitions of this procedure gradually separate the two isotopes. What is the ratio of the rms speed of $^{235}UF_6$ to that of $^{238}UF_6$?

47. ‖ On earth, STP is based on the average atmospheric pressure at the surface and on a phase change of water that occurs at an easily produced temperature, being only slightly cooler than the average air temperature. The atmosphere of Venus is almost entirely carbon dioxide (CO_2), the pressure at the surface is a staggering 93 atm, and the average temperature is 470°C. Venusian scientists, if they existed, would certainly use the surface pressure as part of their definition of STP. To complete the definition, they would seek a phase change that occurs near the average temperature. Conveniently, the melting point of the element tellurium is 450°C. What are (a) the rms speed and (b) the mean free path of carbon dioxide molecules at Venusian STP based on this phase change in tellurium? The radius of a CO_2 molecule is 1.5×10^{-10} m.

48. ‖‖‖ 5.0×10^{23} nitrogen molecules collide with a 10 cm^2 wall each second. Assume that the molecules all travel with a speed of 400 m/s and strike the wall head-on. What is the pressure on the wall?

49. ‖‖‖ A 10 cm × 10 cm × 10 cm box contains 0.010 mol of nitrogen at 20°C. What is the rate of collisions (collisions/s) on one wall of the box?

50. ‖ **FIGURE P20.50** shows the thermal energy of 0.14 mol of gas as a function of temperature. What is C_V for this gas?

FIGURE P20.50

51. ‖ A 100 cm^3 box contains helium at a pressure of 2.0 atm and a temperature of 100°C. It is placed in thermal contact with a 200 cm^3 box containing argon at a pressure of 4.0 atm and a temperature of 400°C.
    a. What is the initial thermal energy of each gas?
    b. What is the final thermal energy of each gas?
    c. How much heat energy is transferred, and in which direction?
    d. What is the final temperature?
    e. What is the final pressure in each box?

52. ‖ 2.0 g of helium at an initial temperature of 300 K interacts thermally with 8.0 g of oxygen at an initial temperature of 600 K.
    a. What is the initial thermal energy of each gas?
    b. What is the final thermal energy of each gas?
    c. How much heat energy is transferred, and in which direction?
    d. What is the final temperature?

53. ‖ A gas of 1.0×10^{20} atoms or molecules has 1.0 J of thermal energy. Its molar specific heat at constant pressure is 20.8 J/mol K. What is the temperature of the gas?

54. || Scientists studying the behavior of hydrogen at low temperatures need to lower the temperature of 0.50 mol of hydrogen gas from 300 K to 30 K. How much thermal energy must they remove from the gas?

55. || A water molecule has its three atoms arranged in a "V" shape, so it has rotational kinetic energy around any of three mutually perpendicular axes. However, like diatomic molecules, its vibrational modes are not active at temperatures below 1000 K. What is the thermal energy of 2.0 mol of steam at a temperature of 160°C?

56. || A monatomic gas and a diatomic gas have equal numbers of moles and equal temperatures. Both are heated at constant pressure until their volume doubles. What is the ratio $Q_{diatomic}/Q_{monatomic}$?

57. || In the discussion following Equation 20.43 it was said that $Q_1 = -Q_2$. Prove that this is so.

58. || A monatomic gas is adiabatically compressed to $\frac{1}{8}$ of its initial volume. Does each of the following quantities change? If so, does it increase or decrease, and by what factor? If not, why not?
    a. The rms speed.
    b. The mean free path.
    c. The thermal energy of the gas.
    d. The molar specific heat at constant volume.

59. || n moles of a diatomic gas with $C_V = \frac{5}{2}R$ has initial pressure p_i and volume V_i. The gas undergoes a process in which the pressure is directly proportional to the volume until the rms speed of the molecules has doubled.
    a. Show this process on a pV diagram.
    b. How much heat does this process require? Give your answer in terms of n, p_i, and V_i.

60. || The 2010 Nobel Prize in Physics was awarded for the discovery of graphene, a two-dimensional form of carbon in which the atoms form a two-dimensional crystal-lattice sheet only one atom thick. Predict the molar specific heat of graphene. Give your answer as a multiple of R.

61. ||| The rms speed of the molecules in 1.0 g of hydrogen gas is 1800 m/s.
    a. What is the total translational kinetic energy of the gas molecules?
    b. What is the thermal energy of the gas?
    c. 500 J of work are done to compress the gas while, in the same process, 1200 J of heat energy are transferred from the gas to the environment. Afterward, what is the rms speed of the molecules?

62. || At what temperature does the rms speed of (a) a nitrogen molecule and (b) a hydrogen molecule equal the escape speed from the earth's surface? (c) You'll find that these temperatures are very high, so you might think that the earth's gravity could easily contain both gases. But not all molecules move with v_{rms}. There is a distribution of speeds, and a small percentage of molecules have speeds several times v_{rms}. Bit by bit, a gas can

slowly leak out of the atmosphere as its fastest molecules escape. A reasonable rule of thumb is that the earth's gravity can contain a gas only if the average translational kinetic energy per molecule is less than 1% of the kinetic energy needed to escape. Use this rule to show why the earth's atmosphere contains nitrogen but not hydrogen, even though hydrogen is the most abundant element in the universe.

63. || n_1 moles of a monatomic gas and n_2 moles of a diatomic gas are mixed together in a container.
    a. Derive an expression for the molar specific heat at constant volume of the mixture.
    b. Show that your expression has the expected behavior if $n_1 \rightarrow 0$ or $n_2 \rightarrow 0$.

64. || A 1.0 kg ball is at rest on the floor in a 2.0 m × 2.0 m × 2.0 m room of air at STP. Air is 80% nitrogen (N_2) and 20% oxygen (O_2) by volume.
    a. What is the thermal energy of the air in the room?
    b. What fraction of the thermal energy would have to be conveyed to the ball for it to be spontaneously launched to a height of 1.0 m?
    c. By how much would the air temperature have to decrease to launch the ball?
    d. Your answer to part c is so small as to be unnoticeable, yet this event never happens. Why not?

Challenge Problems

65. ||| An experiment you're designing needs a gas with $\gamma = 1.50$. You recall from your physics class that no individual gas has this value, but it occurs to you that you could produce a gas with $\gamma = 1.50$ by mixing together a monatomic gas and a diatomic gas. What fraction of the molecules need to be monatomic?

66. ||| Consider a container like that shown in Figure 20.12, with n_1 moles of a monatomic gas on one side and n_2 moles of a diatomic gas on the other. The monatomic gas has initial temperature T_{1i}. The diatomic gas has initial temperature T_{2i}.
    a. Show that the equilibrium thermal energies are

$$E_{1f} = \frac{3n_1}{3n_1 + 5n_2}(E_{1i} + E_{2i})$$

$$E_{2f} = \frac{5n_2}{3n_1 + 5n_2}(E_{1i} + E_{2i})$$

    b. Show that the equilibrium temperature is

$$T_f = \frac{3n_1 T_{1i} + 5n_2 T_{2i}}{3n_1 + 5n_2}$$

    c. 2.0 g of helium at an initial temperature of 300 K interacts thermally with 8.0 g of oxygen at an initial temperature of 600 K. What is the final temperature? How much heat energy is transferred, and in which direction?

21 Heat Engines and Refrigerators

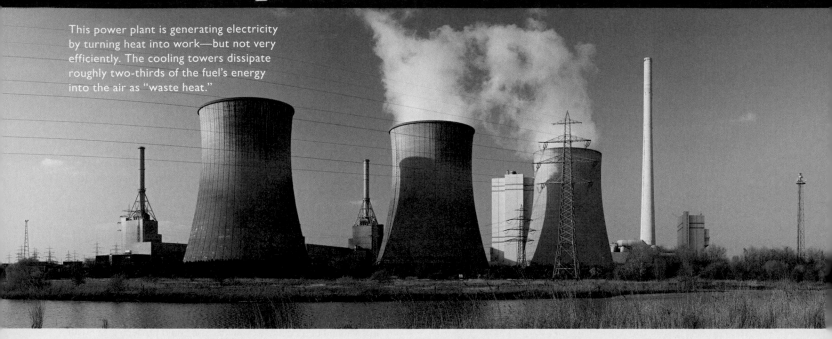

This power plant is generating electricity by turning heat into work—but not very efficiently. The cooling towers dissipate roughly two-thirds of the fuel's energy into the air as "waste heat."

IN THIS CHAPTER, you will study the principles that govern heat engines and refrigerators.

What is a heat engine?

A heat engine is a device for transforming heat energy into useful work. Heat engines

- Follow a cyclical process that can be shown on a pV diagram or an energy-transfer diagram.
- Require not only a source of heat but also a source of cooling. These are called the hot reservoir and the cold reservoir.
- Are governed by the first and second laws of thermodynamics.

《 LOOKING BACK Sections 19.2–19.4 and 19.7 Work, heat, the first law of thermodynamics, and the specific heats of gases

What is a refrigerator?

A refrigerator is any device—including air conditioners—in which external work is used to "pump energy uphill" from cold to hot. A refrigerator is essentially a heat engine running in reverse. A refrigerator's efficiency is given by its coefficient of performance.

How is the efficiency of an engine determined?

How good is an engine at transforming heat energy into work? We'll define an engine's thermal efficiency as

$$\text{efficiency} = \frac{\text{work done}}{\text{heat required}}$$

Conservation of energy—the first law of thermodynamics—says that no engine can have an efficiency greater than one.

Is there a maximum possible efficiency?

Yes. And, surprisingly, it's set by the second law's prohibition of spontaneous heat flow from cold to hot. We'll find that a *perfectly reversible heat engine*—a Carnot engine—has the maximum efficiency allowed by the laws of thermodynamics. The Carnot efficiency depends only on the temperatures of the hot and cold reservoirs.

《 LOOKING BACK Section 20.6 The second law of thermodynamics

Why are heat engines important?

Modern society is powered by devices that transform fuel energy into useful work. Examples include engines for propelling cars and airplanes, power plants for generating electricity, and a wide variety of machines used in industry and manufacturing. These are all heat engines, and all are governed by the laws of thermodynamics. Maximizing efficiency—an important engineering and societal goal—requires a good understanding of the underlying physical principles.

21.1 Turning Heat into Work

Thermodynamics is the branch of physics that studies the transformation of energy. Many practical devices are designed to transform energy from one form, such as the heat from burning fuel, into another, such as work. Chapters 19 and 20 established two laws of thermodynamics that any such device must obey:

> **First law** Energy is conserved; that is, $\Delta E_{th} = W + Q$.
>
> **Second law** Most macroscopic processes are irreversible. In particular, heat energy is transferred spontaneously from a hotter system to a colder system but never from a colder system to a hotter system.

Our goal in this chapter is to discover what these two laws, especially the second law, imply about devices that turn heat into work. In particular:

- How does a practical device transform heat into work?
- What are the limitations and restrictions on these energy transformations?

Work Done by the System

The work W in the first law is the work done *on* the system by external forces from the environment. However, it makes more sense in "practical thermodynamics" to use the work done *by* the system. For example, you want to know how much useful work you can obtain from an expanding gas. The work done by the system is called W_s.

Work done by the environment and work done by the system are not mutually exclusive. In fact, they are very simply related by $W_s = -W$. In **FIGURE 21.1**, force $\vec{F}_{gas}$ due to the gas pressure does work when the piston moves. This is W_s, the work done *by* the system. At the same time, some object in the environment, such as a piston rod, must be pushing inward with force $\vec{F}_{ext} = -\vec{F}_{gas}$ to keep the gas pressure from blowing the piston out. This force does the work W *on* the system, work that you've learned is the negative of the area under the pV curve of the process.

Because the forces are equal but opposite, we see that

$$W_s = -W = \text{the area under the } pV \text{ curve} \qquad (21.1)$$

When a gas expands and pushes the piston out, transferring energy out of the system, we say "the system does work on the environment." While this may seem to imply that the environment is doing no work on the system, all the phrase means is that W_s is positive and W is negative.

Similarly, "the environment does work on the system" means that $W > 0$ (energy is transferred into the system) and thus $W_s < 0$. Whether we use W or W_s is a matter of convenience. They are always opposite to each other rather than one being zero.

The first law of thermodynamics $\Delta E_{th} = W + Q$ can be written in terms of W_s as

$$Q = W_s + \Delta E_{th} \qquad \text{(first law of thermodynamics)} \qquad (21.2)$$

Any energy transferred into a system as heat is either used to do work or stored within the system as an increased thermal energy.

Energy-Transfer Diagrams

Suppose you drop a hot rock into the ocean. Heat is transferred from the rock to the ocean until the rock and ocean are at the same temperature. Although the ocean warms up ever so slightly, ΔT_{ocean} is so small as to be completely insignificant. For all practical purposes, the ocean is infinite and unchangeable.

An **energy reservoir** is an object or a part of the environment so large that **its temperature does not change** when heat is transferred between the system and the reservoir. A reservoir at a higher temperature than the system is called a *hot reservoir*.

FIGURE 21.1 Forces $\vec{F}_{gas}$ and $\vec{F}_{ext}$ both do work as the piston moves.

(a)

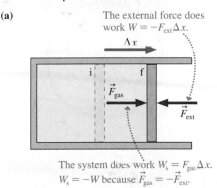

The external force does work $W = -F_{ext}\Delta x$.

The system does work $W_s = F_{gas}\Delta x$. $W_s = -W$ because $\vec{F}_{gas} = -\vec{F}_{ext}$.

(b)

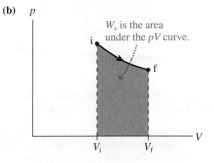

W_s is the area under the pV curve.

FIGURE 21.2 Energy-transfer diagrams.

(a)

Heat is transferred from hot to cold.

Copper bar

Q_H Q_C

Fire   Ice
Hot reservoir at T_H Cold reservoir at T_C

(b) Heat energy is transferred from a hot reservoir to a cold reservoir. Energy conservation requires $Q_C = Q_H$.

Hot reservoir T_H

Q_H The copper bar is the system.

System

Q_C

Cold reservoir T_C

(c) Heat is never spontaneously transferred from a colder object to a hotter object.

Hot reservoir T_H

Q_H

System

Q_C

Cold reservoir T_C

FIGURE 21.3 Work can be transformed into heat with **100% efficiency.**

Hot reservoir T_H

W System

Q_C

Cold reservoir T_C
Energy-transfer diagrams show work entering or leaving from the side.

A vigorously burning flame is a hot reservoir for small objects placed in the flame. A reservoir at a lower temperature than the system is called a *cold reservoir*. The ocean is a cold reservoir for the hot rock. We will use T_H and T_C to designate the temperatures of the hot and cold reservoirs.

Hot and cold reservoirs are idealizations, in the same category as frictionless surfaces and massless strings. No real object can maintain a perfectly constant temperature as heat is transferred in or out. Even so, an object can be modeled as a reservoir if it is much larger than the system that thermally interacts with it.

Heat energy is transferred between a system and a reservoir if they have different temperatures. We will define

$$Q_H = \text{amount of heat transferred to or from a hot reservoir}$$

$$Q_C = \text{amount of heat transferred to or from a cold reservoir}$$

By definition, Q_H and Q_C are *positive* quantities. The direction of heat transfer, which determines the sign of Q in the first law, will always be clear as we deal with thermodynamic devices.

FIGURE 21.2a shows a heavy copper bar between a hot reservoir (at temperature T_H) and a cold reservoir (at temperature T_C). Heat Q_H is transferred from the hot reservoir into the copper and heat Q_C is transferred from the copper to the cold reservoir. **FIGURE 21.2b** is an **energy-transfer diagram** for this process. The hot reservoir is always drawn at the top, the cold reservoir at the bottom, and the system—the copper bar in this case—between them. Figure 21.2b shows heat Q_H being transferred into the system and Q_C being transferred out.

The first law of thermodynamics $Q = W_s + \Delta E_{th}$ refers to the *system*. Q is the *net* heat to the system, which, in this case, is $Q = Q_H - Q_C$. The copper bar does no work, so $W_s = 0$. The bar warms up when first placed between the two reservoirs, but it soon comes to a steady state where its temperature no longer changes. Then $\Delta E_{th} = 0$. Thus the first law tells us that $Q = Q_H - Q_C = 0$, from which we conclude that $Q_C = Q_H$.

In other words, all of the heat transferred into the hot end of the rod is subsequently transferred out of the cold end. This isn't surprising. After all, we know that heat is transferred spontaneously from a hotter object to a colder object. Even so, there has to be some *means* by which the heat energy gets from the hotter object to the colder. The copper bar provides a route for the energy transfer, and $Q_C = Q_H$ is the statement that energy is conserved as it moves through the bar.

Contrast Figure 21.2b with **FIGURE 21.2c**. Figure 21.2c shows a system in which heat is being transferred from the cold reservoir to the hot reservoir. The first law of thermodynamics is not violated, because $Q_H = Q_C$, but the second law is. If there were such a system, it would allow the spontaneous (i.e., with no outside input or assistance) transfer of heat from a colder object to a hotter object. The process of Figure 21.2c is forbidden by the second law of thermodynamics.

Work into Heat and Heat into Work

Turning work into heat is easy—just rub two objects together. Work from the friction force increases the objects' thermal energy and their temperature. Heat energy is then transferred from the warmer objects to the cooler environment. **FIGURE 21.3** is the energy-transfer diagram for this process. The conversion of work into heat is 100% efficient in that *all* the energy supplied to the system as work is ultimately transferred to the environment as heat. Notice that the objects have returned to their initial state at the end of this process, ready to repeat the process for as long as there's a source of motion.

The reverse—transforming heat into work—isn't so easy. Heat can be transformed into work in a one-time process, such as an isothermal expansion of a gas, but at the end the system is not restored to its initial state. To be practical, **a device that transforms heat into work must return to its initial state at the end of the process and be ready for continued use.** You want your car engine to turn over and over for as long as there's fuel.

Interestingly, no one has ever invented a "perfect engine" that transforms heat into work with 100% efficiency *and returns to its initial state* so that it can continue to do work as long as there is fuel. Of course, that such a device has not been invented is not a proof that it can't be done. We'll provide a proof shortly, but for now we'll make the hypothesis that the process of FIGURE 21.4 is somehow forbidden.

Notice the asymmetry between Figures 21.3 and 21.4. The perfect transformation of work into heat is permitted, but the perfect transformation of heat into work is forbidden. This asymmetry parallels the asymmetry of the two processes in Figure 21.2. In fact, we'll soon see that the "perfect engine" of Figure 21.4 is forbidden for exactly the same reason: the second law of thermodynamics.

21.2 Heat Engines and Refrigerators

The steam generator at your local electric power plant works by boiling water to produce high-pressure steam that spins a turbine (which then spins a generator to produce electricity). That is, the steam pressure is doing work. The steam is then condensed to liquid water and pumped back to the boiler to start the process again. There are two crucial ideas here. First, the device works in a cycle, with the water returning to its initial conditions once a cycle. Second, heat is transferred to the water in the boiler, but heat is transferred *out* of the water in the condenser.

Car engines and steam generators are examples of what we call *heat engines*. A **heat engine** is any closed-cycle device that extracts heat Q_H from a hot reservoir, does useful work, and exhausts heat Q_C to a cold reservoir. A **closed-cycle device** is one that periodically *returns to its initial conditions,* repeating the same process over and over. That is, all state variables (pressure, temperature, thermal energy, and so on) return to their initial values once every cycle. Consequently, a heat engine can continue to do useful work for as long as it is attached to the reservoirs.

FIGURE 21.5 is the energy-transfer diagram of a heat engine. Unlike the forbidden "perfect engine" of Figure 21.4, a heat engine is connected to both a hot reservoir *and* a cold reservoir. You can think of a heat engine as "siphoning off" some of the heat that moves from the hot reservoir to the cold reservoir and transforming that heat into work—*some* of the heat, but not all.

Because the temperature and thermal energy of a heat engine return to their initial values at the end of each cycle, there is no *net* change in E_{th}:

$$(\Delta E_{th})_{net} = 0 \qquad \text{(any heat engine, over one full cycle)} \qquad (21.3)$$

Consequently, the first law of thermodynamics *for a full cycle* of a heat engine is $(\Delta E_{th})_{net} = Q - W_s = 0$.

Let's define W_{out} to be the useful work done *by* the heat engine *per cycle*. The first law applied to a heat engine is

$$W_{out} = Q_{net} = Q_H - Q_C \quad \text{(work per cycle done by a heat engine)} \qquad (21.4)$$

This is just energy conservation. The energy-transfer diagram of Figure 21.5 is a pictorial representation of Equation 21.4.

> **NOTE** Equations 21.3 and 21.4 apply only to a *full cycle* of the heat engine. They are *not* valid for any of the individual processes that make up a cycle.

For practical reasons, we would like an engine to do the maximum amount of work with the minimum amount of fuel. We can measure the performance of a heat engine in terms of its **thermal efficiency** η (lowercase Greek eta), defined as

$$\eta = \frac{W_{out}}{Q_H} = \frac{\text{what you get}}{\text{what you had to pay}} \qquad (21.5)$$

FIGURE 21.4 There are no perfect engines that turn heat into work with 100% efficiency.

The steam turbine in a modern power plant is an enormous device. Expanding steam does work by spinning the turbine.

FIGURE 21.5 The energy-transfer diagram of a heat engine.

1. Heat energy Q_H is transferred from the hot reservoir (typically burning fuel) to the system.

2. Part of the energy is used to do useful work W_{out}.

3. The remaining energy $Q_C = Q_H - W_{out}$ is exhausted to the cold reservoir (cooling water or the air) as waste heat.

Using Equation 21.4 for W_{out}, we can also write the thermal efficiency as

$$\eta = 1 - \frac{Q_C}{Q_H} \tag{21.6}$$

FIGURE 21.6 illustrates the idea of thermal efficiency.

A *perfect* heat engine would have $\eta_{perfect} = 1$. That is, it would be 100% efficient at converting heat from the hot reservoir (the burning fuel) into work. You can see from Equation 21.6 that a perfect engine would have no exhaust ($Q_C = 0$) and would not need a cold reservoir. Figure 21.4 has already suggested that there are no perfect heat engines, that an engine with $\eta = 1$ is impossible. A heat engine *must* exhaust **waste heat** to a cold reservoir. It is energy that was extracted from the hot reservoir but *not* transformed to useful work.

Practical heat engines, such as car engines and steam generators, have thermal efficiencies in the range $\eta \approx 0.1 - 0.5$. This is not large. Can a clever designer do better, or is this some kind of physical limitation?

STOP TO THINK 21.1 Rank in order, from largest to smallest, the work W_{out} performed by these four heat engines.

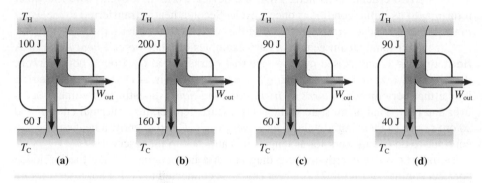

A Heat-Engine Example

To illustrate how these ideas actually work, FIGURE 21.7 shows a simple engine that converts heat into the work of lifting mass M.

FIGURE 21.7 A simple heat engine transforms heat into work.

(a) Heat is transferred into the gas from the burning fuel.

(b) The gas does work by lifting the mass in an isobaric expansion.

(c) The piston is locked and the mass is removed. The heat is turned off.

(d) The gas cools back to room temperature at constant volume. Then the piston is unlocked.

(e) A steadily increasing external force steadily raises the pressure in an isothermal compression until the initial pressure has been restored.

Isobaric heating and expansion   Constant-volume cooling   Isothermal compression

The net effect of this multistep process is to convert some of the fuel's energy into the useful work of lifting the mass. There has been no net change in the gas, which has returned to its initial pressure, volume, and temperature at the end of step (e). We can start the whole process over again and continue lifting masses (doing work) as long as we have fuel.

FIGURE 21.8 shows the heat-engine process on a *pV* diagram. It is a *closed-cycle* because the gas returns to its initial conditions. No work is done during the isochoric process, and, as you can see from the areas under the curve, the work done *by* the gas to lift the mass is greater than the work the environment must do *on* the gas to recompress it. Thus this heat engine, by burning fuel, does *net* work per cycle:

$$W_{net} = W_{lift} - W_{ext} = (W_s)_{1 \to 2} + (W_s)_{3 \to 1}.$$

Notice that the cyclical process of Figure 21.8 involves two cooling processes in which heat is transferred *from* the gas to the environment. Heat energy is transferred from hotter objects to colder objects, so the system *must* be connected to a cold reservoir with $T_C < T_{gas}$ during these two processes. A key to understanding heat engines is that they require both a heat source (burning fuel) *and* a heat sink (cooling water, the air, or something at a lower temperature than the system).

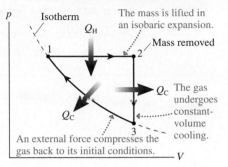

FIGURE 21.8 The closed-cycle *pV* diagram for the heat engine of Figure 21.7.

EXAMPLE 21.1 | Analyzing a heat engine I

Analyze the heat engine of **FIGURE 21.9** to determine (a) the net work done per cycle, (b) the engine's thermal efficiency, and (c) the engine's power output if it runs at 600 rpm. Assume the gas is monatomic.

FIGURE 21.9 The heat engine of Example 21.1.

MODEL The gas follows a closed cycle consisting of three distinct processes, each of which was studied in Chapters 18 and 19. For each of the three we need to determine the work done and the heat transferred.

SOLVE To begin, we can use the initial conditions at state 1 and the ideal-gas law to determine the number of moles of gas:

$$n = \frac{p_1 V_1}{RT_1} = \frac{(200 \times 10^3 \text{ Pa})(2.0 \times 10^{-4} \text{ m}^3)}{(8.31 \text{ J/mol K})(300 \text{ K})} = 0.0160 \text{ mol}$$

Process $1 \to 2$: The work done *by* the gas in the isobaric expansion is

$$(W_s)_{12} = p \Delta V = (200 \times 10^3 \text{ Pa})\left((6.0 - 2.0) \times 10^{-4} \text{ m}^3\right) = 80 \text{ J}$$

We can use the ideal-gas law at constant pressure to find $T_2 = (V_2/V_1)T_1 = 3T_1 = 900$ K. The heat transfer during a constant-pressure process is

$$Q_{12} = nC_P \Delta T$$
$$= (0.0160 \text{ mol})(20.8 \text{ J/mol K})(900 \text{ K} - 300 \text{ K}) = 200 \text{ J}$$

where we used $C_P = \frac{5}{2}R$ for a monatomic ideal gas.

Process $2 \to 3$: No work is done in an isochoric process, so $(W_s)_{23} = 0$. The temperature drops back to 300 K, so the heat transfer, with $C_V = \frac{3}{2}R$, is

$$Q_{23} = nC_V \Delta T$$
$$= (0.0160 \text{ mol})(12.5 \text{ J/mol K})(300 \text{ K} - 900 \text{ K}) = -120 \text{ J}$$

Process $3 \to 1$: The gas returns to its initial state with volume V_1. The work done *by* the gas during an isothermal process is

$$(W_s)_{31} = nRT \ln\left(\frac{V_1}{V_3}\right)$$
$$= (0.0160 \text{ mol})(8.31 \text{ J/mol K})(300 \text{ K}) \ln\left(\frac{1}{3}\right) = -44 \text{ J}$$

W_s is negative because the environment does work on the gas to compress it. An isothermal process has $\Delta E_{th} = 0$ and hence, from the first law,

$$Q_{31} = (W_s)_{31} = -44 \text{ J}$$

Q is negative because the gas must be cooled as it is compressed to keep the temperature constant.

a. The *net* work done by the engine during one cycle is

$$W_{out} = (W_s)_{12} + (W_s)_{23} + (W_s)_{31} = 36 \text{ J}$$

As a consistency check, notice that the net heat transfer is

$$Q_{net} = Q_{12} + Q_{23} + Q_{31} = 36 \text{ J}$$

Equation 21.4 told us that a heat engine *must* have $W_{out} = Q_{net}$, and we see that it does.

b. The efficiency depends not on the net heat transfer but on the heat Q_H transferred into the engine from the flame. Heat enters during process $1 \to 2$, where Q is positive, and exits during processes $2 \to 3$ and $3 \to 1$, where Q is negative. Thus

$$Q_H = Q_{12} = 200 \text{ J}$$
$$Q_C = |Q_{23}| + |Q_{31}| = 164 \text{ J}$$

Notice that $Q_H - Q_C = 36 \text{ J} = W_{out}$. In this heat engine, 200 J of heat from the hot reservoir does 36 J of useful work. Thus the thermal efficiency is

$$\eta = \frac{W_{out}}{Q_H} = \frac{36 \text{ J}}{200 \text{ J}} = 0.18 \text{ or } 18\%$$

This heat engine is far from being a perfect engine!

Continued

c. An engine running at 600 rpm goes through 10 cycles per second. The power output is the work done *per second*:

$$P_{out} = (\text{work per cycle}) \times (\text{cycles per second})$$
$$= 360 \text{ J/s} = 360 \text{ W}$$

ASSESS Although we didn't need Q_{net}, verifying that $Q_{net} = W_{out}$ was a check of self-consistency. Heat-engine analysis requires many calculations and offers many opportunities to get signs wrong. However, there are a sufficient number of self-consistency checks so that you can almost always spot calculational errors *if you check for them*.

Let's think about this example a bit more before going on. We've said that a heat engine operates between a hot reservoir and a cold reservoir. Figure 21.9 doesn't explicitly show the reservoirs. Nonetheless, we know that heat is transferred from a hotter object to a colder object. Heat Q_H is transferred into the system during process $1 \rightarrow 2$ as the gas warms from 300 K to 900 K. For this to be true, the hot-reservoir temperature T_H must be ≥ 900 K. Likewise, heat Q_C is transferred from the system to the cold reservoir as the temperature drops from 900 K to 300 K in process $2 \rightarrow 3$. For this to be true, the cold-reservoir temperature T_C must be ≤ 300 K.

So, while we really don't know what the reservoirs are or their exact temperatures, we can say with certainty that the hot-reservoir temperature T_H must exceed the highest temperature reached by the system and the cold-reservoir temperature T_C must be less than the coldest system temperature.

Refrigerators

Your house or apartment has a refrigerator. Very likely it has an air conditioner. The purpose of these devices is to make air that is cooler than its environment even colder. The first does so by blowing hot air out into a warm room, the second by blowing it out to the hot outdoors. You've probably felt the hot air exhausted by an air conditioner compressor or coming out from beneath the refrigerator.

At first glance, a refrigerator or air conditioner may seem to violate the second law of thermodynamics. After all, doesn't the second law forbid heat from being transferred from a colder object to a hotter object? Not quite: The second law says that heat is not *spontaneously* transferred from a colder to a hotter object. A refrigerator or air conditioner requires electric power to operate. They do cause heat to be transferred from cold to hot, but the transfer is "assisted" rather than spontaneous.

A **refrigerator** is any closed-cycle device that uses external work W_{in} to remove heat Q_C from a cold reservoir and exhaust heat Q_H to a hot reservoir. **FIGURE 21.10** is the energy-transfer diagram of a refrigerator. The cold reservoir is the air inside the refrigerator or the air inside your house on a summer day. To keep the air cold, in the face of inevitable "heat leaks," the refrigerator or air conditioner compressor continuously removes heat from the cold reservoir and exhausts heat into the room or outdoors. You can think of a refrigerator as "pumping heat uphill," much as a water pump lifts water uphill.

Because a refrigerator, like a heat engine, is a cyclical device, $\Delta E_{th} = 0$. Conservation of energy requires

$$Q_H = Q_C + W_{in} \tag{21.7}$$

To move energy from a colder to a hotter reservoir, a refrigerator must exhaust *more* heat to the outside than it removes from the inside. This has significant implications for whether or not you can cool a room by leaving the refrigerator door open.

The thermal efficiency of a heat engine was defined as "what you get (useful work W_{out})" versus "what you had to pay (fuel to supply Q_H)." By analogy, we define the **coefficient of performance** K of a refrigerator to be

$$K = \frac{Q_C}{W_{in}} = \frac{\text{what you get}}{\text{what you had to pay}} \tag{21.8}$$

What you get, in this case, is the removal of heat from the cold reservoir. But you have to pay the electric company for the work needed to run the refrigerator. A better

This air conditioner transfers heat energy *from* the cool indoors *to* the hot exterior.

FIGURE 21.10 The energy-transfer diagram of a refrigerator.

The amount of heat exhausted to the hot reservoir is larger than the amount of heat extracted from the cold reservoir.

External work is used to remove heat from a cold reservoir and exhaust heat to a hot reservoir.

refrigerator will require less work to remove a given amount of heat, thus having a larger coefficient of performance.

A perfect refrigerator would require no work ($W_{in} = 0$) and would have $K_{perfect} = \infty$. But if Figure 21.10 had no work input, it would look like Figure 21.2c. That device was forbidden by the second law of thermodynamics because, with no work input, heat would move *spontaneously* from cold to hot.

We noted in Chapter 20 that the second law of thermodynamics can be stated several different but equivalent ways. We can now give a third informal statement:

> **Second law, informal statement #3** There are no perfect refrigerators with coefficient of performance $K = \infty$.

Any real refrigerator or air conditioner *must* use work to move energy from the cold reservoir to the hot reservoir, hence $K < \infty$.

No Perfect Heat Engines

We hypothesized above that there are no perfect heat engines—that is, no heat engines like the one shown in Figure 21.4 with $Q_C = 0$ and $\eta = 1$. Now we're ready to prove this hypothesis. **FIGURE 21.11** shows a hot reservoir at temperature T_H and a cold reservoir at temperature T_C. An ordinary refrigerator, one that obeys all the laws of physics, is operating between these two reservoirs.

�lm◀ **FIGURE 21.11** A perfect engine driving an ordinary refrigerator would be able to violate the second law of thermodynamics.

Suppose we had a perfect heat engine, one that takes in heat Q_H from the high-temperature reservoir and transforms that energy entirely into work W_{out}. If we had such a heat engine, we could use its output to provide the work input to the refrigerator. The two devices combined have no connection to the external world. That is, there's no net input or net output of work.

If we built a box around the heat engine and refrigerator, so that you couldn't see what was inside, the only thing you would observe is heat being transferred *with no outside assistance* from the cold reservoir to the hot reservoir. But a spontaneous or unassisted transfer of heat from a colder to a hotter object is exactly what the second law of thermodynamics forbids. Consequently, our assumption of a perfect heat engine must be wrong. Hence another statement of the second law of thermodynamics is:

> **Second law, informal statement #4** There are no perfect heat engines with efficiency $\eta = 1$.

Any real heat engine *must* exhaust waste heat Q_C to a cold reservoir.

STOP TO THINK 21.2 It's a hot day and your air conditioner is broken. Your roommate says, "Let's open the refrigerator door and cool this place off." Will this work?

a. Yes.   b. No.   c. It might, but it will depend on how hot the room is.

21.3 Ideal-Gas Heat Engines

We will focus on heat engines that use a gas as the *working substance*. The gasoline or diesel engine in your car is an engine that alternately compresses and expands a gaseous fuel-air mixture. A discussion of engines such as steam generators that rely on phase changes will be deferred to more advanced courses.

A gas heat engine can be represented by a closed-cycle trajectory in the pV diagram, such as the one shown in FIGURE 21.12a. This observation leads to an important geometric interpretation of the work done by the system during one full cycle. You learned in Section 21.1 that the work done *by* the system is the area under the curve of a pV trajectory. As FIGURE 21.12b shows, the net work done during a full cycle is

$$W_{out} = W_{expand} - |W_{compress}| = \text{area } \textit{inside} \text{ the closed curve} \qquad (21.9)$$

FIGURE 21.12 The work W_{out} done by the system during one full cycle is the area enclosed within the curve.

(a) A typical heat-engine cycle

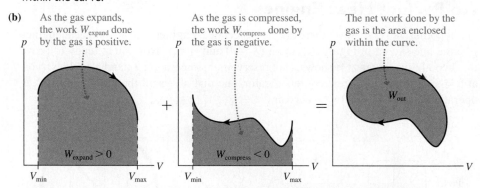

(b) As the gas expands, the work W_{expand} done by the gas is positive.

$W_{expand} > 0$

As the gas is compressed, the work $W_{compress}$ done by the gas is negative.

$W_{compress} < 0$

The net work done by the gas is the area enclosed within the curve.

W_{out}

You can see that **the net work done by a gas heat engine during one full cycle is the area enclosed by the pV curve for the cycle.** A thermodynamic cycle with a larger enclosed area does more work than one with a smaller enclosed area. Notice that the gas must go around the pV trajectory in a *clockwise* direction for W_{out} to be positive. We'll see later that a refrigerator uses a counterclockwise (ccw) cycle.

Ideal-Gas Summary

We've learned a lot about ideal gases in the last three chapters. All gas processes obey the ideal-gas law $pV = nRT$ and the first law of thermodynamics $\Delta E_{th} = Q - W_s$. TABLE 21.1 summarizes the results for specific gas processes. This table shows W_s, the work done *by* the system, so the signs are opposite those in Chapter 19.

TABLE 21.1 Summary of ideal-gas processes

Process	Gas law	Work W_s	Heat Q	Thermal energy
Isochoric	$p_i/T_i = p_f/T_f$	0	$nC_V \Delta T$	$\Delta E_{th} = Q$
Isobaric	$V_i/T_i = V_f/T_f$	$p \Delta V$	$nC_P \Delta T$	$\Delta E_{th} = Q - W_s$
Isothermal	$p_i V_i = p_f V_f$	$nRT \ln(V_f/V_i)$	$Q = W_s$	$\Delta E_{th} = 0$
		$pV \ln(V_f/V_i)$		
Adiabatic	$p_i V_i^\gamma = p_f V_f^\gamma$	$(p_f V_f - p_i V_i)/(1 - \gamma)$	0	$\Delta E_{th} = -W_s$
	$T_i V_i^{\gamma-1} = T_f V_f^{\gamma-1}$	$-nC_V \Delta T$		
Any	$p_i V_i/T_i = p_f V_f/T_f$	area under curve		$\Delta E_{th} = nC_V \Delta T$

There is one entry in this table that you haven't seen before. The expression

$$W_s = \frac{p_f V_f - p_i V_i}{1 - \gamma} \quad \text{(work in an adiabatic process)} \qquad (21.10)$$

for the work done in an adiabatic process follows from writing $W_s = -\Delta E_{th} = -nC_V\Delta T$, which you learned in Chapter 19, then using $\Delta T = \Delta(pV)/nR$ and the definition of γ. The proof will be left for a homework problem.

You learned in Chapter 20 that the thermal energy of an ideal gas depends only on its temperature. **TABLE 21.2** lists the thermal energy, molar specific heats, and specific heat ratio $\gamma = C_P/C_V$ for monatomic and diatomic gases.

TABLE 21.2 Properties of monatomic and diatomic gases

	Monatomic	Diatomic
E_{th}	$\frac{3}{2}nRT$	$\frac{5}{2}nRT$
C_V	$\frac{3}{2}R$	$\frac{5}{2}R$
C_P	$\frac{5}{2}R$	$\frac{7}{2}R$
γ	$\frac{5}{3} = 1.67$	$\frac{7}{5} = 1.40$

A Strategy for Heat-Engine Problems

The engine of Example 21.1 was not a realistic heat engine, but it did illustrate the kinds of reasoning and computations involved in the analysis of a heat engine.

PROBLEM-SOLVING STRATEGY 21.1   (MP)

Heat-engine problems

MODEL Identify each process in the cycle.

VISUALIZE Draw the pV diagram of the cycle.

SOLVE There are several steps in the mathematical analysis.

- Use the ideal-gas law to complete your knowledge of n, p, V, and T at one point in the cycle.
- Use the ideal-gas law and equations for specific gas processes to determine p, V, and T at the beginning and end of each process.
- Calculate Q, W_s, and ΔE_{th} for each process.
- Find W_{out} by adding W_s for each process in the cycle. If the geometry is simple, you can confirm this value by finding the area enclosed within the pV curve.
- Add just the positive values of Q to find Q_H.
- Verify that $(\Delta E_{th})_{net} = 0$. This is a self-consistency check to verify that you haven't made any mistakes.
- Calculate the thermal efficiency η and any other quantities you need to complete the solution.

ASSESS Is $(\Delta E_{th})_{net} = 0$? Do all the signs of W_s and Q make sense? Does η have a reasonable value? Have you answered the question?

EXAMPLE 21.2 | Analyzing a heat engine II

A heat engine with a diatomic gas as the working substance uses the closed cycle shown in **FIGURE 21.13**. How much work does this engine do per cycle, and what is its thermal efficiency?

FIGURE 21.13 The pV diagram for the heat engine of Example 21.2.

MODEL Processes $1 \rightarrow 2$ and $3 \rightarrow 4$ are isobaric. Processes $2 \rightarrow 3$ and $4 \rightarrow 1$ are isochoric.

VISUALIZE The pV diagram has already been drawn.

SOLVE We know the pressure, volume, and temperature at state 4. The number of moles of gas in the heat engine is

$$n = \frac{p_4 V_4}{RT_4} = \frac{(101{,}300\ \text{Pa})(1.0\ \text{m}^3)}{(8.31\ \text{J/mol K})(300\ \text{K})} = 40.6\ \text{mol}$$

$p/T = $ constant during an isochoric process and $V/T = $ constant during an isobaric process. These allow us to find that $T_1 = T_3 = 900$ K and $T_2 = 2700$ K. This completes our knowledge of the state variables at all four corners of the diagram.

Process $1 \rightarrow 2$ is an isobaric expansion, so

$$(W_s)_{12} = p\,\Delta V = (3.0 \times 101{,}300\ \text{Pa})(2.0\ \text{m}^3) = 6.08 \times 10^5\ \text{J}$$

Continued

where we converted the pressure to pascals. The heat transfer during an isobaric expansion is

$$Q_{12} = nC_P \Delta T = (40.6 \text{ mol})(29.1 \text{ J/mol K})(1800 \text{ K})$$

$$= 21.27 \times 10^5 \text{ J}$$

where $C_P = \frac{7}{2}R$ for a diatomic gas. Then, using the first law,

$$\Delta E_{12} = Q_{12} - (W_s)_{12} = 15.19 \times 10^5 \text{ J}$$

Process $2 \rightarrow 3$ is an isochoric process, so $(W_s)_{23} = 0$ and

$$\Delta E_{23} = Q_{23} = nC_V \Delta T = -15.19 \times 10^5 \text{ J}$$

Notice that ΔT is *negative*.

Process $3 \rightarrow 4$ is an isobaric compression. Now ΔV is negative, so

$$(W_s)_{34} = p \Delta V = -2.03 \times 10^5 \text{ J}$$

and

$$Q_{34} = nC_P \Delta T = -7.09 \times 10^5 \text{ J}$$

Then $\Delta E_{th} = Q_{34} - (W_s)_{34} = -5.06 \times 10^5 \text{ J}$.

Process $4 \rightarrow 1$ is another constant-volume process, so again $(W_s)_{41} = 0$ and

$$\Delta E_{41} = Q_{41} = nC_V \Delta T = 5.06 \times 10^5 \text{ J}$$

The results of all four processes are shown in **TABLE 21.3**. The net results for W_{out}, Q_{net}, and $(\Delta E_{th})_{net}$ are found by summing the columns. As expected, $W_{out} = Q_{net}$ and $(\Delta E_{th})_{net} = 0$.

TABLE 21.3 Energy transfers in Example 21.2. All energies $\times 10^5$ J

Process	W_s	Q	ΔE_{th}
$1 \rightarrow 2$	6.08	21.27	15.19
$2 \rightarrow 3$	0	−15.19	−15.19
$3 \rightarrow 4$	−2.03	−7.09	−5.06
$4 \rightarrow 1$	0	5.06	5.06
Net	4.05	4.05	0

The work done during one cycle is $W_{out} = 4.05 \times 10^5$ J. Heat enters the system from the hot reservoir during processes $1 \rightarrow 2$ and $4 \rightarrow 1$, where Q is positive. Summing these gives $Q_H = 26.33 \times 10^5$ J. Thus the thermal efficiency of this engine is

$$\eta = \frac{W_{out}}{Q_H} = \frac{4.05 \times 10^5 \text{ J}}{26.33 \times 10^5 \text{ J}} = 0.15 = 15\%$$

ASSESS The verification that $W_{out} = Q_{net}$ and $(\Delta E_{th})_{net} = 0$ gives us great confidence that we didn't make any calculational errors. This engine may not seem very efficient, but η is quite typical of many real engines.

We noted in Example 21.1 that a heat engine's hot-reservoir temperature T_H must exceed the highest temperature reached by the system and the cold-reservoir temperature T_C must be less than the coldest system temperature. Although we don't know what the reservoirs are in Example 21.2, we can be sure that $T_H > 2700$ K and $T_C < 300$ K.

STOP TO THINK 21.3 What is the thermal efficiency of this heat engine?

a. 0.10
b. 0.50
c. 0.25
d. 4
e. Can't tell without knowing Q_C

The Brayton Cycle

The heat engines of Examples 21.1 and 21.2 have been educational but not realistic. As an example of a more realistic heat engine we'll look at the thermodynamic cycle known as the *Brayton cycle*. It is a reasonable model of a *gas turbine engine*. Gas turbines are used for electric power generation and as the basis for jet engines in aircraft and rockets. The *Otto cycle,* which describes the gasoline internal combustion engine, and the *Diesel cycle,* which, not surprisingly, describes the diesel engine, will be the subject of homework problems.

FIGURE 21.14a is a schematic look at a gas turbine engine, and **FIGURE 21.14b** is the corresponding pV diagram. To begin the Brayton cycle, air at an initial pressure p_1 is rapidly compressed in a *compressor*. This is an *adiabatic process,* with $Q = 0$,

A jet engine uses a modified Brayton cycle.

because there is no time for heat to be exchanged with the surroundings. Recall that an adiabatic compression raises the temperature of a gas by doing work on it, not by heating it, so the air leaving the compressor is very hot.

The hot gas flows into a combustion chamber. Fuel is continuously admitted to the combustion chamber where it mixes with the hot gas and is ignited, transferring heat to the gas at constant pressure and raising the gas temperature yet further. The high-pressure gas then expands, spinning a turbine that does some form of useful work. This adiabatic expansion, with $Q = 0$, drops the temperature and pressure of the gas. The pressure at the end of the expansion through the turbine is back to p_1, but the gas is still quite hot. The gas completes the cycle by flowing through a device called a **heat exchanger** that transfers heat energy to a cooling fluid. Large power plants are often sited on rivers or oceans in order to use the water for the cooling fluid in the heat exchanger.

This thermodynamic cycle, called a Brayton cycle, has two adiabatic processes—the compression and the expansion through the turbine—plus a constant-pressure heating and a constant-pressure cooling. There's no heat transfer during the adiabatic processes. The hot-reservoir temperature must be $T_H \geq T_3$ for heat to be transferred into the gas during process $2 \rightarrow 3$. Similarly, the heat exchanger will remove heat from the gas only if $T_C \leq T_1$.

The thermal efficiency of any heat engine is

$$\eta = \frac{W_{out}}{Q_H} = 1 - \frac{Q_C}{Q_H}$$

Heat is transferred into the gas only during process $2 \rightarrow 3$. This is an isobaric process, so $Q_H = nC_P \Delta T = nC_P(T_3 - T_2)$. Similarly, heat is transferred out only during the isobaric process $4 \rightarrow 1$.

We have to be careful with signs. Q_{41} is negative because the temperature decreases, but Q_C was defined as the *amount* of heat exchanged with the cold reservoir, a positive quantity. Thus

$$Q_C = |Q_{41}| = |nC_P(T_1 - T_4)| = nC_P(T_4 - T_1) \tag{21.11}$$

With these expressions for Q_H and Q_C, the thermal efficiency is

$$\eta_{Brayton} = 1 - \frac{T_4 - T_1}{T_3 - T_2} \tag{21.12}$$

This expression isn't useful unless we compute all four temperatures. Fortunately, we can cast Equation 21.12 into a more useful form.

You learned in Chapter 19 that $pV^\gamma =$ constant during an adiabatic process, where $\gamma = C_P/C_V$ is the specific heat ratio. If we use $V = nRT/p$ from the ideal-gas law, $V^\gamma = (nR)^\gamma T^\gamma p^{-\gamma}$. $(nR)^\gamma$ is a constant, so we can write $pV^\gamma =$ constant as

$$p^{(1-\gamma)}T^\gamma = \text{constant} \tag{21.13}$$

Equation 21.13 is a pressure-temperature relationship for an adiabatic process. Because $(T^\gamma)^{1/\gamma} = T$, we can simplify Equation 21.13 by raising both sides to the power $1/\gamma$. Doing so gives

$$p^{(1-\gamma)/\gamma}T = \text{constant} \tag{21.14}$$

during an adiabatic process.

Process $1 \rightarrow 2$ is an adiabatic process; hence

$$p_1^{(1-\gamma)/\gamma}T_1 = p_2^{(1-\gamma)/\gamma}T_2 \tag{21.15}$$

Isolating T_1 gives

$$T_1 = \frac{p_2^{(1-\gamma)/\gamma}}{p_1^{(1-\gamma)/\gamma}}T_2 = \left(\frac{p_2}{p_1}\right)^{(1-\gamma)/\gamma}T_2 = \left(\frac{p_{max}}{p_{min}}\right)^{(1-\gamma)/\gamma}T_2 \tag{21.16}$$

If we define the **pressure ratio** r_p as $r_p = p_{max}/p_{min}$, then T_1 and T_2 are related by

$$T_1 = r_p^{(1-\gamma)/\gamma}T_2 \tag{21.17}$$

(a)

(b)

FIGURE 21.15 The efficiency of a Brayton cycle as a function of the pressure ratio r_p.

The efficiency first grows quickly as the pressure ratio is increased, reaching $\approx$ 50% at $r_p = 10$, then levels off.

The algebra of getting to Equation 21.17 was a bit tricky, but the final result is fairly simple. Process $3 \rightarrow 4$ is also an adiabatic process. The same reasoning leads to

$$T_4 = r_p^{(1-\gamma)/\gamma} T_3 \qquad (21.18)$$

If we substitute these expressions for T_1 and T_4 into Equation 21.12, the efficiency is

$$\eta_B = 1 - \frac{T_4 - T_1}{T_3 - T_2} = 1 - \frac{r_p^{(1-\gamma)/\gamma} T_3 - r_p^{(1-\gamma)/\gamma} T_2}{T_3 - T_2} = 1 - \frac{r_p^{(1-\gamma)/\gamma}(T_3 - T_2)}{T_3 - T_2}$$

$$= 1 - r_p^{(1-\gamma)/\gamma}$$

Remarkably, all the temperatures cancel and we're left with an expression that depends only on the pressure ratio. Noting that $(1 - \gamma)$ is negative, we can make one final change and write

$$\eta_B = 1 - \frac{1}{r_p^{(\gamma-1)/\gamma}} \qquad (21.19)$$

FIGURE 21.15 is a graph of the efficiency of the Brayton cycle as a function of the pressure ratio, assuming $\gamma = 1.40$ for a diatomic gas such as air.

21.4 Ideal-Gas Refrigerators

Suppose we were to operate a Brayton heat engine backward, going ccw rather than cw in the pV diagram. FIGURE 21.16a (which you should compare to Figure 21.14a) shows a device for doing this. FIGURE 21.16b is its pV diagram, and FIGURE 21.16c is the energy-transfer diagram. Starting from point 4, the gas is adiabatically compressed to increase its temperature and pressure. It then flows through a high-temperature heat exchanger where the gas *cools* at constant pressure from temperature T_3 to T_2. The gas then expands adiabatically, leaving it significantly colder at T_1 than it started at T_4. It completes the cycle by flowing through a low-temperature heat exchanger, where it *warms* back to its starting temperature.

FIGURE 21.16 A refrigerator that extracts heat from the cold reservoir and exhausts heat to the hot reservoir.

(a)

(b)

(c)

Suppose that the low-temperature heat exchanger is a closed container of air surrounding a pipe through which the engine's cold gas is flowing. The heat-exchange process $1 \rightarrow 4$ *cools* the air in the container as it warms the gas flowing through the pipe. If you were to place eggs and milk inside this closed container, you would call it a refrigerator!

Going around a closed pV cycle in a ccw direction reverses the sign of W for each process in the cycle. Consequently, the area inside the curve of Figure 21.16b is W_{in}, the work done *on* the system. Here work is used to extract heat Q_C from the cold reservoir and exhaust a larger amount of heat $Q_H = Q_C + W_{in}$ to the hot reservoir. But where, in this situation, are the energy reservoirs?

Understanding a refrigerator is a little harder than understanding a heat engine. The key is to remember that **heat is always transferred from a hotter object to a colder object.** In particular,

- The gas in a refrigerator can extract heat Q_C *from* the cold reservoir only if the gas temperature is *lower* than the cold-reservoir temperature T_C. Heat energy is then transferred *from* the cold reservoir *into* the colder gas.
- The gas in a refrigerator can exhaust heat Q_H *to* the hot reservoir only if the gas temperature is *higher* than the hot-reservoir temperature T_H. Heat energy is then transferred *from* the warmer gas *into* the hot reservoir.

These two requirements place severe constraints on the thermodynamics of a refrigerator. Because there is no reservoir colder than T_C, the gas cannot reach a temperature lower than T_C by heat exchange. The gas in a refrigerator *must* use an adiabatic expansion ($Q = 0$) to lower the temperature below T_C. Likewise, a gas refrigerator requires an adiabatic compression to raise the gas temperature above T_H.

EXAMPLE 21.3 | **Analyzing a refrigerator**

A refrigerator using helium gas operates on a reversed Brayton cycle with a pressure ratio of 5.0. Prior to compression, the gas occupies 100 cm³ at a pressure of 150 kPa and a temperature of −23°C. Its volume at the end of the expansion is 80 cm³. What are the refrigerator's coefficient of performance and its power input if it operates at 60 cycles per second?

MODEL The Brayton cycle has two adiabatic processes and two isobaric processes. The work per cycle needed to run the refrigerator is $W_{in} = Q_H - Q_C$; hence we can determine both the coefficient of performance and the power requirements from Q_H and Q_C. Heat energy is transferred only during the two isobaric processes.

VISUALIZE FIGURE 21.17 shows the pV cycle. We know from the pressure ratio of 5.0 that the maximum pressure is 750 kPa. Neither V_2 nor V_3 is known.

FIGURE 21.17 A Brayton-cycle refrigerator.

SOLVE To calculate heat we're going to need the temperatures at the four corners of the cycle. First, we can use the conditions of state 4 to find the number of moles of helium:

$$n = \frac{p_4 V_4}{R T_4} = 0.00722 \text{ mol}$$

Process $1 \rightarrow 4$ is isobaric; hence temperature T_1 is

$$T_1 = \frac{V_1}{V_4} T_4 = (0.80)(250 \text{ K}) = 200 \text{ K} = -73°C$$

With Equation 21.14 we found that the quantity $p^{(1-\gamma)/\gamma}T$ remains constant during an adiabatic process. Helium is a monatomic gas with $\gamma = \frac{5}{3}$, so $(1-\gamma)/\gamma = -\frac{2}{5} = -0.40$. For the adiabatic compression $4 \rightarrow 3$,

$$p_3^{-0.40} T_3 = p_4^{-0.40} T_4$$

Solving for T_3 gives

$$T_3 = \left(\frac{p_4}{p_3}\right)^{-0.40} T_4 = \left(\frac{1}{5}\right)^{-0.40} (250 \text{ K}) = 476 \text{ K} = 203°C$$

The same analysis applied to the $2 \rightarrow 1$ adiabatic expansion gives

$$T_2 = \left(\frac{p_1}{p_2}\right)^{-0.40} T_1 = \left(\frac{1}{5}\right)^{-0.40} (200 \text{ K}) = 381 \text{ K} = 108°C$$

Now we can use $C_P = \frac{5}{2} R = 20.8$ J/mol K for a monatomic gas to compute the heat transfers:

$$Q_H = |Q_{32}| = n C_P (T_3 - T_2)$$
$$= (0.00722 \text{ mol})(20.8 \text{ J/mol K})(95 \text{ K}) = 14.3 \text{ J}$$
$$Q_C = |Q_{14}| = n C_P (T_4 - T_1)$$
$$= (0.00722 \text{ mol})(20.8 \text{ J/mol K})(50 \text{ K}) = 7.5 \text{ J}$$

Thus the work *input* to the refrigerator is $W_{in} = Q_H - Q_C = 6.8$ J. During each cycle, 6.8 J of work are done *on* the gas to extract 7.5 J of heat from the cold reservoir. Then 14.3 J of heat are exhausted into the hot reservoir.

The refrigerator's coefficient of performance is

$$K = \frac{Q_C}{W_{in}} = \frac{7.5 \text{ J}}{6.8 \text{ J}} = 1.1$$

The power input needed to run the refrigerator is

$$P_{in} = 6.8 \frac{\text{J}}{\text{cycle}} \times 60 \frac{\text{cycles}}{\text{s}} = 410 \frac{\text{J}}{\text{s}} = 410 \text{ W}$$

ASSESS These are fairly realistic values for a kitchen refrigerator. You pay your electric company for providing the work W_{in} that operates the refrigerator. The cold reservoir is the freezer compartment. The cold temperature T_C must be higher than T_4 ($T_C > -23°C$) in order for heat to be transferred *from* the cold reservoir *to* the gas. A typical freezer temperature is $-15°C$, so this condition is satisfied. The hot reservoir is the air in the room. The back and underside of a refrigerator have heat-exchanger coils where the hot gas, after compression, transfers heat to the air. The hot temperature T_H must be less than T_2 ($T_H < 108°C$) in order for heat to be transferred *from* the gas *to* the air. An air temperature $\approx 25°C$ under a refrigerator satisfies this condition.

STOP TO THINK 21.4 What, if anything, is wrong with this refrigerator?

21.5 The Limits of Efficiency

Thermodynamics has its historical roots in the development of the steam engine and other machines of the early industrial revolution. Early steam engines, built on the basis of experience rather than scientific understanding, were not very efficient at converting fuel energy into work. The first major theoretical analysis of heat engines was published by the French engineer Sadi Carnot in 1824. The question that Carnot raised was: Can we make a heat engine whose thermal efficiency η approaches 1, or is there an upper limit η_{max} that cannot be exceeded? To frame the question more clearly, imagine we have a hot reservoir at temperature T_H and a cold reservoir at T_C. What is the most efficient heat engine (maximum η) that can operate between these two energy reservoirs? Similarly, what is the most efficient refrigerator (maximum K) that can operate between the two reservoirs?

We just saw that a refrigerator is, in some sense, a heat engine running backward. We might thus suspect that the most efficient heat engine is related to the most efficient refrigerator. Suppose we have a heat engine that we can turn into a refrigerator by reversing the direction of operation, thus changing the direction of the energy transfers, and with *no other changes*. In particular, the heat engine and the refrigerator operate between the same two energy reservoirs at temperatures T_H and T_C.

FIGURE 21.18a shows such a heat engine and its corresponding refrigerator. Notice that the refrigerator has *exactly the same* work and heat transfer as the heat engine, only in the opposite directions. A device that can be operated as either a heat engine or a refrigerator between the same two energy reservoirs and with the same energy transfers, with only their direction changed, is called a **perfectly reversible engine.** A perfectly reversible engine is an idealization, as was the concept of a perfectly elastic collision. Nonetheless, it will allow us to establish limits that no real engine can exceed.

FIGURE 21.18 If a perfectly reversible heat engine is used to operate a perfectly reversible refrigerator, the two devices exactly cancel each other.

(a)

Perfectly reversible heat engine   Perfectly reversible refrigerator

(b)

Perfectly reversible heat engine + Perfectly reversible refrigerator = No work done and no heat transferred

Suppose we have a perfectly reversible heat engine and a perfectly reversible refrigerator (the same device running backward) operating between a hot reservoir at temperature T_H and a cold reservoir at temperature T_C. Because the work W_{in} needed to operate the refrigerator is exactly the same as the useful work W_{out} done by the heat engine, we can use the heat engine, as shown in **FIGURE 21.18b**, to drive the refrigerator. The heat Q_C the engine exhausts to the cold reservoir is exactly the same as the heat Q_C the refrigerator extracts from the cold reservoir. Similarly, the heat Q_H the engine extracts from the hot reservoir matches the heat Q_H the refrigerator exhausts to the hot reservoir. Consequently, there is no net heat transfer in either direction. The refrigerator exactly replaces all the heat energy that had been transferred out of the hot reservoir by the heat engine.

You may want to compare the reasoning used here with the reasoning we used with Figure 21.11. There we tried to use the output of a "perfect" heat engine to run a refrigerator but did *not* succeed.

A Perfectly Reversible Engine Has Maximum Efficiency

Now we've arrived at the critical step in the reasoning. Suppose I claim to have a heat engine that can operate between temperatures T_H and T_C with *more* efficiency than a perfectly reversible engine. **FIGURE 21.19** shows the output of this heat engine operating the same perfectly reversible refrigerator that we used in Figure 21.18b.

FIGURE 21.19 A heat engine more efficient than a perfectly reversible engine could be used to violate the second law of thermodynamics.

Recall that the thermal efficiency and the work of a heat engine are

$$\eta = \frac{W_{out}}{Q_H} \quad \text{and} \quad W_{out} = Q_H - Q_C$$

If the new heat engine is more efficient than the perfectly reversible engine it replaces, it needs *less* heat Q_H from the hot reservoir to perform the *same* work W_{out}. If Q_H is less while W_{out} is the same, then Q_C must also be less. That is, the new heat engine exhausts less heat to the cold reservoir than does the perfectly reversible heat engine.

When this new heat engine drives the perfectly reversible refrigerator, the heat it exhausts to the cold reservoir is *less* than the heat extracted from the cold reservoir by the refrigerator. Similarly, this engine extracts *less* heat from the hot reservoir than the refrigerator exhausts. Thus the net result of using this superefficient heat engine to operate a perfectly reversible refrigerator is that heat is transferred from the cold reservoir to the hot reservoir *without outside assistance*.

But this can't happen. It would violate the second law of thermodynamics. Hence we have to conclude that no heat engine operating between reservoirs at temperatures T_H and T_C can be more efficient than a perfectly reversible engine. This very important conclusion is another version of the second law:

Second law, informal statement #5 No heat engine operating between reservoirs at temperatures T_H and T_C can be more efficient than a perfectly reversible engine operating between these temperatures.

The answer to our question "Is there a maximum η that cannot be exceeded?" is a clear Yes! The maximum possible efficiency η_{max} is that of a perfectly reversible engine. Because the perfectly reversible engine is an idealization, any real engine will have an efficiency less than η_{max}.

A similar argument shows that no refrigerator can be more efficient than a perfectly reversible refrigerator. If we had such a refrigerator, and if we ran it with the output of a perfectly reversible heat engine, we could transfer heat from cold to hot with no outside assistance. Thus:

> **Second law, informal statement #6** No refrigerator operating between reservoirs at temperatures T_H and T_C can have a coefficient of performance larger than that of a perfectly reversible refrigerator operating between these temperatures.

Conditions for a Perfectly Reversible Engine

This argument tells us that η_{max} and K_{max} exist, but it doesn't tell us what they are. Our final task will be to "design" and analyze a perfectly reversible engine. Under what conditions is an engine reversible?

An engine transfers energy by both mechanical and thermal interactions. Mechanical interactions are pushes and pulls. The environment does work on the system, transferring energy into the system by pushing in on a piston. The system transfers energy back to the environment by pushing out on the piston.

The energy transferred by a moving piston is perfectly reversible, returning the system to its initial state, with no change of temperature or pressure, only if the motion is *frictionless*. The slightest bit of friction will prevent the mechanical transfer of energy from being perfectly reversible.

The circumstances under which heat transfer can be *completely* reversed aren't quite so obvious. After all, Chapter 20 emphasized the *irreversible* nature of heat transfer. If objects A and B are in thermal contact, with $T_A > T_B$, then heat energy is transferred from A to B. But the second law of thermodynamics prohibits a heat transfer from B back to A. Heat transfer through a temperature *difference* is an irreversible process.

But suppose $T_A = T_B$. With no temperature difference, any heat that is transferred from A to B can, at a later time, be transferred from B back to A. This transfer wouldn't violate the second law, which prohibits only heat transfer from a colder object to a hotter object. Now you might object, and rightly so, that heat *can't* move from A to B if they are at the same temperature because heat, by definition, is the energy transferred between two objects at different temperatures.

This is true, but consider a heat engine in which, during one part of the cycle, $T_H = T_{engine} + dT$. That is, the hot reservoir is only infinitesimally hotter than the engine. Heat Q_H can be transferred to the heat engine, but *very slowly*. If you later, with a reversed cycle, try to make the heat move from the engine back to the hot reservoir, the second law will prevent you from doing so with perfect precision. But, because the temperature difference is infinitesimal, you'll be missing only an infinitesimal amount of heat. You can transfer heat reversibly in the limit $dT \to 0$ (making this an isothermal process), but you must be prepared to spend an infinite amount of time doing so. Similarly, the engine can reversibly exchange heat with a cold reservoir at temperature $T_C = T_{engine} - dT$.

Thus the thermal transfer of energy is reversible *if the heat is transferred infinitely slowly in an isothermal process*. This is an idealization, but so are completely frictionless processes. Nonetheless, we can now say that a perfectly reversible engine must use only two types of processes:

1. Frictionless mechanical interactions with no heat transfer ($Q = 0$), and
2. Thermal interactions in which heat is transferred in an isothermal process ($\Delta E_{th} = 0$).

Any engine that uses only these two types of processes is called a **Carnot engine.** A Carnot engine is a perfectly reversible engine; thus it has the maximum possible thermal efficiency $\eta_{\max}$ and, if operated as a refrigerator, the maximum possible coefficient of performance $K_{\max}$.

21.6 The Carnot Cycle

The definition of a Carnot engine does not specify whether the engine's working substance is a gas or a liquid. It makes no difference. Our argument that a perfectly reversible engine is the most efficient possible heat engine depended only on the engine's reversibility. Consequently, **any Carnot engine operating between T_H and T_C must have exactly the same efficiency as any other Carnot engine operating between the same two energy reservoirs.** If we can determine the thermal efficiency of one Carnot engine, we'll know the efficiency of all Carnot engines. Because liquids and phase changes are complicated, we'll analyze a Carnot engine that uses an ideal gas.

Designing a Carnot Engine

The **Carnot cycle** is an ideal-gas cycle that consists of the two adiabatic processes ($Q = 0$) and two isothermal processes ($\Delta E_{\text{th}} = 0$) shown in **FIGURE 21.20**. These are the two types of processes allowed in a perfectly reversible gas engine. As a Carnot cycle operates,

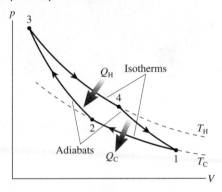

FIGURE 21.20 The Carnot cycle is perfectly reversible.

1. The gas is isothermally compressed while in thermal contact with the cold reservoir at temperature T_C. Heat energy $Q_C = |Q_{12}|$ is removed from the gas as it is compressed in order to keep the temperature constant. The compression must take place extremely slowly because there can be only an infinitesimal temperature difference between the gas and the reservoir.
2. The gas is adiabatically compressed while thermally isolated from the environment. This compression increases the gas temperature until it matches temperature T_H of the hot reservoir. No heat is transferred during this process.
3. After reaching maximum compression, the gas expands isothermally at temperature T_H. Heat $Q_H = Q_{34}$ is transferred from the hot reservoir into the gas as it expands in order to keep the temperature constant.
4. Finally, the gas expands adiabatically, with $Q = 0$, until the temperature decreases back to T_C.

Work is done in all four processes of the Carnot cycle, but heat is transferred only during the two isothermal processes.

The thermal efficiency of any heat engine is

$$\eta = \frac{W_{\text{out}}}{Q_H} = 1 - \frac{Q_C}{Q_H}$$

We can determine η_{Carnot} by finding the heat transfer in the two isothermal processes.

Process $1 \rightarrow 2$: Table 21.1 gives us the heat transfer in an isothermal process at temperature T_C:

$$Q_{12} = (W_s)_{12} = nRT_C \ln\left(\frac{V_2}{V_1}\right) = -nRT_C \ln\left(\frac{V_1}{V_2}\right) \tag{21.20}$$

$V_1 > V_2$, so the logarithm on the right is positive. Q_{12} is negative because heat is transferred out of the system, but Q_C is simply the *amount* of heat transferred to the cold reservoir:

$$Q_C = |Q_{12}| = nRT_C \ln\left(\frac{V_1}{V_2}\right) \tag{21.21}$$

Process $3 \rightarrow 4$: Similarly, the heat transferred in the isothermal expansion at temperature T_H is

$$Q_H = Q_{34} = (W_s)_{34} = nRT_H \ln\left(\frac{V_4}{V_3}\right) \tag{21.22}$$

Thus the thermal efficiency of the Carnot cycle is

$$\eta_{\text{Carnot}} = 1 - \frac{Q_C}{Q_H} = 1 - \frac{T_C \ln(V_1/V_2)}{T_H \ln(V_4/V_3)} \tag{21.23}$$

We can simplify this expression. During the two adiabatic processes,

$$T_C V_2^{\gamma-1} = T_H V_3^{\gamma-1} \quad \text{and} \quad T_C V_1^{\gamma-1} = T_H V_4^{\gamma-1} \tag{21.24}$$

An algebraic rearrangement gives

$$V_2 = V_3 \left(\frac{T_H}{T_C}\right)^{1/(\gamma-1)} \quad \text{and} \quad V_1 = V_4 \left(\frac{T_H}{T_C}\right)^{1/(\gamma-1)} \tag{21.25}$$

from which it follows that

$$\frac{V_1}{V_2} = \frac{V_4}{V_3} \tag{21.26}$$

Consequently, the two logarithms in Equation 21.23 cancel and we're left with the result that the thermal efficiency of a Carnot engine operating between a hot reservoir at temperature T_H and a cold reservoir at temperature T_C is

$$\eta_{\text{Carnot}} = 1 - \frac{T_C}{T_H} \quad \text{(Carnot thermal efficiency)} \tag{21.27}$$

This remarkably simple result, an efficiency that depends only on the ratio of the temperatures of the hot and cold reservoirs, is Carnot's legacy to thermodynamics.

NOTE Temperatures T_H and T_C are *absolute* temperatures.

EXAMPLE 21.4 | A Carnot engine

A Carnot engine is cooled by water at $T_C = 10°C$. What temperature must be maintained in the hot reservoir of the engine to have a thermal efficiency of 70%?

MODEL The efficiency of a Carnot engine depends only on the temperatures of the hot and cold reservoirs.

SOLVE The thermal efficiency $\eta_{\text{Carnot}} = 1 - T_C/T_H$ can be rearranged to give

$$T_H = \frac{T_C}{1 - \eta_{\text{Carnot}}} = 943 \text{ K} = 670°C$$

where we used $T_C = 283$ K.

ASSESS A "real" engine would need a higher temperature than this to provide 70% efficiency because no real engine will match the Carnot efficiency.

EXAMPLE 21.5 | A real engine

The heat engine of Example 21.2 had a highest temperature of 2700 K, a lowest temperature of 300 K, and a thermal efficiency of 15%. What is the efficiency of a Carnot engine operating between these two temperatures?

SOLVE The Carnot efficiency is

$$\eta_{\text{Carnot}} = 1 - \frac{T_C}{T_H} = 1 - \frac{300 \text{ K}}{2700 \text{ K}} = 0.89 = 89\%$$

ASSESS The thermodynamic cycle used in Example 21.2 doesn't come anywhere close to the Carnot efficiency.

The Maximum Efficiency

In Section 21.2 we tried to invent a perfect engine with $\eta = 1$ and $Q_C = 0$. We found that we could not do so without violating the second law, so no engine can have $\eta = 1$. However, that example didn't rule out an engine with $\eta = 0.9999$. Further analysis has now shown that no heat engine operating between energy reservoirs at temperatures T_H and T_C can be more efficient than a perfectly reversible engine operating between these temperatures.

We've now reached the endpoint of this line of reasoning by establishing an exact result for the thermal efficiency of a perfectly reversible engine, the Carnot engine. We can summarize our conclusions:

> **Second law, informal statement #7** No heat engine operating between energy reservoirs at temperatures T_H and T_C can exceed the Carnot efficiency
>
> $$\eta_{\text{Carnot}} = 1 - \frac{T_C}{T_H}$$

As Example 21.5 showed, real engines usually fall well short of the Carnot limit.

We also found that no refrigerator can exceed the coefficient of performance of a perfectly reversible refrigerator. We'll leave the proof as a homework problem, but an analysis very similar to that above shows that the coefficient of performance of a Carnot refrigerator is

$$K_{\text{Carnot}} = \frac{T_C}{T_H - T_C} \qquad \text{(Carnot coefficient of performance)} \qquad (21.28)$$

Thus we can state:

> **Second law, informal statement #8** No refrigerator operating between energy reservoirs at temperatures T_H and T_C can exceed the Carnot coefficient of performance
>
> $$K_{\text{Carnot}} = \frac{T_C}{T_H - T_C}$$

EXAMPLE 21.6 | **Brayton versus Carnot**

The Brayton-cycle refrigerator of Example 21.3 had coefficient of performance $K = 1.1$. Compare this to the limit set by the second law of thermodynamics.

SOLVE Example 21.3 found that the reservoir temperatures had to be $T_C \geq 250$ K and $T_H \leq 381$ K. A Carnot refrigerator operating between 250 K and 381 K has

$$K_{\text{Carnot}} = \frac{T_C}{T_H - T_C} = \frac{250 \text{ K}}{381 \text{ K} - 250 \text{ K}} = 1.9$$

ASSESS This is the minimum value of K_{Carnot}. It will be even higher if $T_C > 250$ K or $T_H < 381$ K. The coefficient of performance of the reasonably realistic refrigerator of Example 21.3 is less than 60% of the limiting value.

Statements #7 and #8 of the second law are a major result of this chapter, one with profound implications. The efficiency limit of a heat engine is set by the temperatures of the hot and cold reservoirs. High efficiency requires $T_C/T_H \ll 1$ and thus $T_H \gg T_C$. However, practical realities often prevent T_H from being significantly larger than T_C, in which case the engine cannot possibly have a large efficiency. This limit on the efficiency of heat engines is a consequence of the second law of thermodynamics.

EXAMPLE 21.7 | Generating electricity

An electric power plant boils water to produce high-pressure steam at 400°C. The high-pressure steam spins a turbine as it expands, then the turbine spins the generator. The steam is then condensed back to water in an ocean-cooled heat exchanger at 25°C. What is the *maximum* possible efficiency with which heat energy can be converted to electric energy?

MODEL The maximum possible efficiency is that of a Carnot engine operating between these temperatures.

SOLVE The Carnot efficiency depends on absolute temperatures, so we must use $T_H = 400°C = 673$ K and $T_C = 25°C = 298$ K. Then

$$\eta_{max} = 1 - \frac{298}{673} = 0.56 = 56\%$$

ASSESS This is an upper limit. Real coal-, oil-, gas-, and nuclear-heated steam generators actually operate at $\approx 35\%$ thermal efficiency, converting only about one-third of the fuel energy to electric energy while exhausting about two-thirds of the energy to the environment as waste heat. (The heat *source* has nothing to do with the efficiency. All it does is boil water.) Not much can be done to alter the low-temperature limit. The high-temperature limit is determined by the maximum temperature and pressure the boiler and turbine can withstand. The efficiency of electricity generation is far less than most people imagine, but it is an unavoidable consequence of the second law of thermodynamics.

A limit on the efficiency of heat engines was not expected. We are used to thinking in terms of energy conservation, so it comes as no surprise that we cannot make an engine with $\eta > 1$. But the limits arising from the second law were not anticipated, nor are they obvious. Nonetheless, they are a very real fact of life and a very real constraint on any practical device. No one has ever invented a machine that exceeds the second-law limits, and we have seen that the maximum efficiency for realistic engines is surprisingly low.

STOP TO THINK 21.5 Could this heat engine be built?

a. Yes.
b. No.
c. It's impossible to tell without knowing what kind of cycle it uses.

CHALLENGE EXAMPLE 21.8 | Calculating efficiency

A heat engine using a monatomic ideal gas goes through the following closed cycle:

- Isochoric heating until the pressure is doubled.
- Isothermal expansion until the pressure is restored to its initial value.
- Isobaric compression until the volume is restored to its initial value.

What is the thermal efficiency of this heat engine? What would be the thermal efficiency of a Carnot engine operating between the highest and lowest temperatures reached by this engine?

MODEL The cycle consists of three familiar processes; we'll need to analyze each. The amount of work and heat will depend on the quantity of gas, which we don't know, but efficiency is a work-to-heat ratio that is independent of the amount of gas.

VISUALIZE FIGURE 21.21 shows the cycle. The initial pressure, volume, and temperature are p, V, and T. The isochoric process increases the pressure to $2p$ and, because the ratio p/T is constant

FIGURE 21.21 The pV cycle of the heat engine.

in an isochoric process, increases the temperature to $2T$. The isothermal expansion is along the $2T$ isotherm. The product pV is constant in an isothermal process, so the volume doubles to $2V$ as the pressure returns to p.

SOLVE We know, symbolically, the state variables at each corner of the pV diagram. That is sufficient for calculating W_s, Q, and ΔE_{th}.

Process $1 \rightarrow 2$: An isochoric process has $W_s = 0$ and

$$Q = \Delta E_{th} = nC_V \Delta T = \tfrac{3}{2}nRT$$

where we used $C_V = \tfrac{3}{2}R$ for a monatomic gas and $\Delta T = 2T - T = T$.

Process $2 \rightarrow 3$: An isothermal process has $\Delta E_{th} = 0$ and

$$Q = W_s = nR(2T)\ln\left(\frac{2V}{V}\right) = (2\ln 2)nRT$$

Here we used the Table 21.1 result for the work done in an isothermal process.

Process $3 \rightarrow 1$: The work done by the gas is the area under the curve, which is negative because $\Delta V = V - 2V = -V$ in the compression:

$$W_s = \text{area} = p\,\Delta V = -pV = -nRT$$

We used the ideal-gas law in the last step to express the result in terms of n and T. The heat transfer is also negative because $\Delta T = T - 2T = -T$:

$$Q = nC_P \Delta T = -\tfrac{5}{2}nRT$$

where we used $C_P = \tfrac{5}{2}R$ for a monatomic gas. Based on the first law, $\Delta E_{th} = Q - W_s = -\tfrac{3}{2}nRT$.

Summing over the three processes, we see that $(\Delta E_{th})_{net} = 0$, as expected, and

$$W_{out} = (2\ln 2 - 1)nRT$$

Heat energy is supplied to the gas $(Q > 0)$ in processes $1 \rightarrow 2$ and $2 \rightarrow 3$, so

$$Q_H = (2\ln 2 + \tfrac{3}{2})nRT$$

Thus the thermal efficiency of this heat engine is

$$\eta = \frac{W_{out}}{Q_H} = \frac{(2\ln 2 - 1)nRT}{(2\ln 2 + \tfrac{3}{2})nRT} = 0.134 = 13.4\%$$

A Carnot engine would be able to operate between a high temperature $T_H = 2T$ and a low temperature $T_C = T$. Its efficiency would be

$$\eta_{Carnot} = 1 - \frac{T_C}{T_H} = 1 - \frac{T}{2T} = 0.500 = 50.0\%$$

ASSESS As we anticipated, the thermal efficiency depends on the *shape* of the pV cycle but not on the quantity of gas or even on the values of p, V, or T. The heat engine's 13.4% efficiency is considerably less than the 50% maximum possible efficiency set by the second law of thermodynamics.

SUMMARY

The goal of Chapter 21 has been to study the principles that govern heat engines and refrigerators.

GENERAL PRINCIPLES

Heat Engines

Devices that transform heat into work. They require two energy reservoirs at different temperatures.

Thermal efficiency

$$\eta = \frac{W_{out}}{Q_H} = \frac{\text{what you get}}{\text{what you pay}}$$

Second-law limit:

$$\eta \le 1 - \frac{T_C}{T_H}$$

Refrigerators

Devices that use work to transfer heat from a colder object to a hotter object.

Coefficient of performance

$$K = \frac{Q_C}{W_{in}} = \frac{\text{what you get}}{\text{what you pay}}$$

Second-law limit:

$$K \le \frac{T_C}{T_H - T_C}$$

IMPORTANT CONCEPTS

A **perfectly reversible engine** (a **Carnot engine**) can be operated as either a heat engine or a refrigerator between the same two energy reservoirs by reversing the cycle and with no other changes.

- A **Carnot heat engine** has the maximum possible thermal efficiency of any heat engine operating between T_H and T_C:

$$\eta_{Carnot} = 1 - \frac{T_C}{T_H}$$

- A **Carnot refrigerator** has the maximum possible coefficient of performance of any refrigerator operating between T_H and T_C:

$$K_{Carnot} = \frac{T_C}{T_H - T_C}$$

The **Carnot cycle** for a gas engine consists of two isothermal processes and two adiabatic processes.

An **energy reservoir** is a part of the environment so large in comparison to the system that its temperature doesn't change as the system extracts heat energy from or exhausts heat energy to the reservoir. All heat engines and refrigerators operate between two energy reservoirs at different temperatures T_H and T_C.

The **work** W_s done *by* the system has the opposite sign to the work done *on* the system.

$$W_s = \text{area under } pV \text{ curve}$$

APPLICATIONS

To analyze a heat engine or refrigerator:

MODEL Identify each process in the cycle.

VISUALIZE Draw the pV diagram of the cycle.

SOLVE There are several steps:

- Determine p, V, and T at the beginning and end of each process.
- Calculate ΔE_{th}, W_s, and Q for each process.
- Determine W_{in} or W_{out}, Q_H, and Q_C.
- Calculate $\eta = W_{out}/Q_H$ or $K = Q_C/W_{in}$.

ASSESS Verify $(\Delta E_{th})_{net} = 0$. Check signs.

TERMS AND NOTATION

thermodynamics	closed-cycle device	coefficient of performance, K	Carnot engine
energy reservoir	thermal efficiency, η	heat exchanger	Carnot cycle
energy-transfer diagram	waste heat	pressure ratio, r_p	
heat engine	refrigerator	perfectly reversible engine	

CONCEPTUAL QUESTIONS

1. In going from i to f in each of the three processes of FIGURE Q21.1, is work done *by* the system ($W < 0$, $W_s > 0$), is work done *on* the system ($W > 0$, $W_s < 0$), or is *no* net work done?

FIGURE Q21.1

2. Rank in order, from largest to smallest, the amount of work $(W_s)_1$ to $(W_s)_4$ done by the gas in each of the cycles shown in FIGURE Q21.2. Explain.

FIGURE Q21.2

3. Could you have a heat engine with $\eta > 1$? Explain.

4. FIGURE Q21.4 shows the pV diagram of a heat engine. During which stage or stages is (a) heat added to the gas, (b) heat removed from the gas, (c) work done on the gas, and (d) work done by the gas?

FIGURE Q21.4

5. Rank in order, from largest to smallest, the thermal efficiencies η_1 to η_4 of the four heat engines in FIGURE Q21.5. Explain.

FIGURE Q21.5

6. FIGURE Q21.6 shows the thermodynamic cycles of two heat engines. Which heat engine has the larger thermal efficiency? Or are they the same? Explain.

FIGURE Q21.6

7. A heat engine satisfies $W_{out} = Q_{net}$. Why is there no ΔE_{th} term in this relationship?

8. Do the energy-transfer diagrams in FIGURE Q21.8 represent possible heat engines? If not, what is wrong?

FIGURE Q21.8

9. Do the energy-transfer diagrams in **FIGURE Q21.9** represent possible refrigerators? If not, what is wrong?

FIGURE Q21.9

10. It gets pretty hot in your apartment. In browsing the Internet, you find a company selling small "room air conditioners." You place the air conditioner on the floor, plug it in, and—the advertisement says—it will lower the room temperature up to 10°F. Should you order one? Explain.

11. The first and second laws of thermodynamics are sometimes stated as "You can't win" and "You can't even break even." Do these sayings accurately characterize the laws of thermodynamics as applied to heat engines? Why or why not?

EXERCISES AND PROBLEMS

Problems labeled ▩ integrate material from earlier chapters.

Exercises

Section 21.1 Turning Heat into Work

Section 21.2 Heat Engines and Refrigerators

1. ‖ A heat engine does 200 J of work per cycle while exhausting 400 J of waste heat. What is the engine's thermal efficiency?

2. | A heat engine with a thermal efficiency of 40% does 100 J of work per cycle. How much heat is (a) extracted from the hot reservoir and (b) exhausted to the cold reservoir per cycle?

3. | A heat engine extracts 55 kJ of heat from the hot reservoir each cycle and exhausts 40 kJ of heat. What are (a) the thermal efficiency and (b) the work done per cycle?

4. | 50 J of work are done per cycle on a refrigerator with a coefficient of performance of 4.0. How much heat is (a) extracted from the cold reservoir and (b) exhausted to the hot reservoir per cycle?

5. ‖ A refrigerator requires 200 J of work and exhausts 600 J of heat per cycle. What is the refrigerator's coefficient of performance?

6. ‖ A 32%-efficient electric power plant produces 900 MW of electric power and discharges waste heat into 20°C ocean water. Suppose the waste heat could be used to heat homes during the winter instead of being discharged into the ocean. A typical American house requires an average of 20 kW for heating. How many homes could be heated with the waste heat of this one power plant?

7. ‖ The power output of a car engine running at 2400 rpm is 500 kW. How much (a) work is done and (b) heat is exhausted per cycle if the engine's thermal efficiency is 20%? Give your answers in kJ.

8. ‖ 1.0 L of 20°C water is placed in a refrigerator. The refrigerator's motor must supply an extra 8.0 W of power to chill the water to 5°C in 1.0 h. What is the refrigerator's coefficient of performance?

Section 21.3 Ideal-Gas Heat Engines

Section 21.4 Ideal-Gas Refrigerators

9. ‖ The cycle of **FIGURE EX21.9** consists of three processes. Make a table with rows labeled A–C and columns labeled ΔE_{th}, W_s, and Q. Fill each box in the table with $+$, $-$, or 0 to indicate whether the quantity increases, decreases, or stays the same during that process.

FIGURE EX21.9

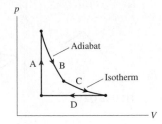

FIGURE EX21.10

10. ‖ The cycle of **FIGURE EX21.10** consists of four processes. Make a table with rows labeled A to D and columns labeled ΔE_{th}, W_s, and Q. Fill each box in the table with $+$, $-$, or 0 to indicate whether the quantity increases, decreases, or stays the same during that process.

11. ‖ How much work is done per cycle by a gas following the pV trajectory of **FIGURE EX21.11**?

FIGURE EX21.11

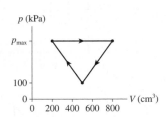

FIGURE EX21.12

12. ‖ A gas following the pV trajectory of **FIGURE EX21.12** does 60 J of work per cycle. What is p_{max}?

13. ‖ What are (a) W_{out} and Q_H and (b) the thermal efficiency for the heat engine shown in **FIGURE EX21.13**?

FIGURE EX21.13   **FIGURE EX21.14**

14. ‖ What are (a) W_{out} and Q_H and (b) the thermal efficiency for the heat engine shown in **FIGURE EX21.14**?

15. ‖ What are (a) the thermal efficiency and (b) the heat extracted from the hot reservoir for the heat engine shown in **FIGURE EX21.15**?

FIGURE EX21.15   **FIGURE EX21.16**

16. ‖ How much heat is exhausted to the cold reservoir by the heat engine shown in **FIGURE EX21.16**?

17. ‖ A heat engine uses a diatomic gas in a Brayton cycle. What is the engine's thermal efficiency if the gas volume is halved during the adiabatic compression?

18. | At what pressure ratio does a Brayton cycle using a monatomic gas have an efficiency of 50%?

19. ‖ The coefficient of performance of a refrigerator is 6.0. The refrigerator's compressor uses 115 W of electric power and is 95% efficient at converting electric power into work. What are (a) the rate at which heat energy is removed from inside the refrigerator and (b) the rate at which heat energy is exhausted into the room?

20. ‖ What are (a) the heat extracted from the cold reservoir and (b) the coefficient of performance for the refrigerator shown in **FIGURE EX21.20**?

FIGURE EX21.20

21. ‖ An air conditioner removes 5.0×10^5 J/min of heat from a house and exhausts 8.0×10^5 J/min to the hot outdoors.
   a. How much power does the air conditioner's compressor require?
   b. What is the air conditioner's coefficient of performance?

Section 21.5 The Limits of Efficiency

Section 21.6 The Carnot Cycle

22. | Which, if any, of the heat engines in **FIGURE EX21.22** violate (a) the first law of thermodynamics or (b) the second law of thermodynamics? Explain.

FIGURE EX21.22

23. | Which, if any, of the refrigerators in **FIGURE EX21.23** violate (a) the first law of thermodynamics or (b) the second law of thermodynamics? Explain.

FIGURE EX21.23

24. ‖ At what cold-reservoir temperature (in °C) would a Carnot engine with a hot-reservoir temperature of 427°C have an efficiency of 60%?

25. ‖ A heat engine does 10 J of work and exhausts 15 J of waste heat during each cycle.
   a. What is the engine's thermal efficiency?
   b. If the cold-reservoir temperature is 20°C, what is the minimum possible temperature in °C of the hot reservoir?

26. ‖ a. A heat engine does 200 J of work per cycle while exhausting 600 J of heat to the cold reservoir. What is the engine's thermal efficiency?
   b. A Carnot engine with a hot-reservoir temperature of 400°C has the same thermal efficiency. What is the cold-reservoir temperature in °C?

27. | A Carnot engine operating between energy reservoirs at temperatures 300 K and 500 K produces a power output of 1000 W. What are (a) the thermal efficiency of this engine, (b) the rate of heat input, in W, and (c) the rate of heat output, in W?

28. ‖ A Carnot engine whose hot-reservoir temperature is 400°C has a thermal efficiency of 40%. By how many degrees should the temperature of the cold reservoir be decreased to raise the engine's efficiency to 60%?

29. ‖ The ideal gas in a Carnot engine extracts 1000 J of heat energy during the isothermal expansion at 300°C. How much heat energy is exhausted during the isothermal compression at 50°C?

30. ‖ A heat engine operating between energy reservoirs at 20°C and 600°C has 30% of the maximum possible efficiency. How much energy must this engine extract from the hot reservoir to do 1000 J of work?

31. ‖ A heat engine operating between a hot reservoir at 500°C and a cold reservoir at 0°C is 60% as efficient as a Carnot engine. If this heat engine and the Carnot engine do the same amount of work, what is the ratio $Q_H/(Q_H)_{Carnot}$?

32. ‖ A Carnot refrigerator operating between −20°C and +20°C extracts heat from the cold reservoir at the rate 200 J/s. What are (a) the coefficient of performance of this refrigerator, (b) the rate at which work is done on the refrigerator, and (c) the rate at which heat is exhausted to the hot side?

33. ‖ The coefficient of performance of a refrigerator is 5.0. The compressor uses 10 J of energy per cycle.
    a. How much heat energy is exhausted per cycle?
    b. If the hot-reservoir temperature is 27°C, what is the lowest possible temperature in °C of the cold reservoir?

34. ‖ A Carnot heat engine with thermal efficiency $\frac{1}{3}$ is run backward as a Carnot refrigerator. What is the refrigerator's coefficient of performance?

Problems

35. ‖ **FIGURE P21.35** shows a heat engine going through one cycle. The gas is diatomic. The masses are such that when the pin is removed, in steps 3 and 6, the piston does not move.
    a. Draw the pV diagram for this heat engine.
    b. How much work is done per cycle?
    c. What is this engine's thermal efficiency?

FIGURE P21.35

36. ‖ The engine that powers a crane burns fuel at a flame temperature of 2000°C. It is cooled by 20°C air. The crane lifts a 2000 kg steel girder 30 m upward. How much heat energy is transferred to the engine by burning fuel if the engine is 40% as efficient as a Carnot engine?

37. ‖ A heat engine with 50% of the Carnot efficiency operates between reservoirs at 20°C and 200°C. The engine inputs heat energy at an average rate of 63 W while compressing a spring 22 cm in 0.50 s. What is the spring constant?

38. ‖ Prove that the work done in an adiabatic process i → f is $W_s = (p_f V_f - p_i V_i)/(1 - \gamma)$.

39. ‖ A Carnot refrigerator operates between reservoirs at −20°C and 50°C in a 25°C room. The refrigerator is a 40 cm × 40 cm × 40 cm box. Five of the walls are perfect insulators, but the sixth is a 1.0-cm-thick piece of stainless steel. What electric power does the refrigerator require to maintain the inside temperature at −20°C?

40. ‖ Prove that the coefficient of performance of a Carnot refrigerator is $K_{\text{Carnot}} = T_C/(T_H - T_C)$.

41. ‖‖ An ideal refrigerator utilizes a Carnot cycle operating between 0°C and 25°C. To turn 10 kg of liquid water at 0°C into 10 kg of ice at 0°C, (a) how much heat is exhausted into the room and (b) how much energy must be supplied to the refrigerator?

42. ‖ A freezer with a coefficient of performance 30% that of a Carnot refrigerator keeps the inside temperature at −22°C in a 25°C room. 3.0 L of water at 20°C are placed in the freezer. How long does it take for the water to freeze if the freezer's compressor does work at the rate of 200 W while the water is freezing?

43. ‖ There has long been an interest in using the vast quantities of thermal energy in the oceans to run heat engines. A heat engine needs a temperature *difference,* a hot side and a cold side. Conveniently, the ocean surface waters are warmer than the deep ocean waters. Suppose you build a floating power plant in the tropics where the surface water temperature is ≈ 30°C. This would be the hot reservoir of the engine. For the cold reservoir, water would be pumped up from the ocean bottom where it is always ≈ 5°C. What is the maximum possible efficiency of such a power plant?

44. ‖ A Carnot heat engine operates between reservoirs at 182°C and 0°C. If the engine extracts 25 J of energy from the hot reservoir per cycle, how many cycles will it take to lift a 10 kg mass a height of 10 m?

45. ‖ A Carnot engine operates between temperatures of 5°C and 500°C. The output is used to run a Carnot refrigerator operating between −5°C and 25°C. How many joules of heat energy does the refrigerator exhaust into the room for each joule of heat energy used by the heat engine?

46. ‖ **FIGURE P21.46** shows a Carnot heat engine driving a Carnot refrigerator.
    a. Determine Q_2, Q_3, and Q_4.
    b. Is Q_3 greater than, less than, or equal to Q_1?
    c. Do these two devices, when operated together in this way, violate the second law?

FIGURE P21.46

47. ‖‖‖ A Carnot heat engine and an ordinary refrigerator with coefficient of performance 2.00 operate between reservoirs at 350 K and 250 K. The work done by the Carnot heat engine drives the refrigerator. If the heat engine extracts 10.0 J of energy from the hot reservoir, how much energy does the refrigerator exhaust to the hot reservoir?

48. ‖ A heat engine running backward is called a refrigerator if its purpose is to extract heat from a cold reservoir. The same engine running backward is called a *heat pump* if its purpose is to exhaust warm air into the hot reservoir. Heat pumps are widely used for home heating. You can think of a heat pump as a refrigerator that is cooling the already cold outdoors and, with its exhaust heat Q_H, warming the indoors. Perhaps this seems a little silly, but

consider the following. Electricity can be directly used to heat a home by passing an electric current through a heating coil. This is a direct, 100% conversion of work to heat. That is, 15 kW of electric power (generated by doing work at the rate of 15 kJ/s at the power plant) produces heat energy inside the home at a rate of 15 kJ/s. Suppose that the neighbor's home has a heat pump with a coefficient of performance of 5.0, a realistic value. Note that "what you get" with a heat pump is heat delivered, Q_H, so a heat pump's coefficient of performance is defined as $K = Q_H/W_{in}$.

   a. How much electric power (in kW) does the heat pump use to deliver 15 kJ/s of heat energy to the house?

   b. An average price for electricity is about 40 MJ per dollar. A furnace or heat pump will run typically 250 hours per month during the winter. What does one month's heating cost in the home with a 15 kW electric heater and in the home of the neighbor who uses a heat pump?

49. ‖ A car's internal combustion engine can be modeled as a heat engine operating between a combustion temperature of 1500°C and an air temperature of 20°C with 30% of the Carnot efficiency. The heat of combustion of gasoline is 47 kJ/g. What mass of gasoline is burned to accelerate a 1500 kg car from rest to a speed of 30 m/s?

50. ‖ Consider a 1.0 MW power plant (this is the useful output in the form of electric energy) that operates between 30°C and 450°C at 65% of the Carnot efficiency. This is enough electric energy for about 750 homes. One way to use energy more efficiently would be to use the 30°C "waste" energy to heat the homes rather than releasing that heat energy into the environment. This is called *cogeneration,* and it is used in some parts of Europe but rarely in the United States. The average home uses 70 GJ of energy per year for heating. For estimating purposes, assume that all the power plant's exhaust energy can be transported to homes without loss and that home heating takes place at a steady rate for half a year each year. How many homes could be heated by the power plant?

51. ‖ A typical coal-fired power plant burns 300 metric tons of coal *every hour* to generate 750 MW of electricity. 1 metric ton = 1000 kg. The density of coal is 1500 kg/m³ and its heat of combustion is 28 MJ/kg. Assume that *all* heat is transferred from the fuel to the boiler and that *all* the work done in spinning the turbine is transformed into electric energy.

   a. Suppose the coal is piled up in a 10 m × 10 m room. How tall must the pile be to operate the plant for one day?

   b. What is the power plant's thermal efficiency?

52. ‖ A nuclear power plant generates 3000 MW of heat energy from nuclear reactions in the reactor's core. This energy is used to boil water and produce high-pressure steam at 300°C. The steam spins a turbine, which produces 1000 MW of electric power, then the steam is condensed and the water is cooled to 25°C before starting the cycle again.

   a. What is the maximum possible thermal efficiency of the power plant?

   b. What is the plant's actual efficiency?

   c. Cooling water from a river flows through the condenser (the low-temperature heat exchanger) at the rate of 1.2×10^8 L/h ($\approx$ 30 million gallons per hour). If the river water enters the condenser at 18°C, what is its exit temperature?

53. ‖ The electric output of a power plant is 750 MW. Cooling water flows through the power plant at the rate 1.0×10^8 L/h. The cooling water enters the plant at 16°C and exits at 27°C. What is the power plant's thermal efficiency?

54. ‖ **CALC** Engineers testing the efficiency of an electric generator gradually vary the temperature of the hot steam used to power it while leaving the temperature of the cooling water at a constant 20°C. They find that the generator's efficiency increases at a rate of 3.5×10^{-4} K^{-1} at steam temperatures near 300°C. What is the ratio of the generator's efficiency to the efficiency of a Carnot engine?

55. ‖ A heat engine using 1.0 mol of a monatomic gas follows the cycle shown in **FIGURE P21.55**. 3750 J of heat energy is transferred to the gas during process $1 \rightarrow 2$.

   a. Determine W_s, Q, and ΔE_{th} for each of the four processes in this cycle. Display your results in a table.

   b. What is the thermal efficiency of this heat engine?

FIGURE P21.55   **FIGURE P21.56**

56. ‖ A heat engine using a diatomic gas follows the cycle shown in **FIGURE P21.56**. Its temperature at point 1 is 20°C.

   a. Determine W_s, Q, and ΔE_{th} for each of the three processes in this cycle. Display your results in a table.

   b. What is the thermal efficiency of this heat engine?

   c. What is the power output of the engine if it runs at 500 rpm?

57. ‖ **FIGURE P21.57** shows the cycle for a heat engine that uses a gas having $\gamma = 1.25$. The initial temperature is $T_1 = 300$ K, and this engine operates at 20 cycles per second.

   a. What is the power output of the engine?

   b. What is the engine's thermal efficiency?

FIGURE P21.57   **FIGURE P21.58**

58. ‖ A heat engine using a monatomic gas follows the cycle shown in **FIGURE P21.58**.

   a. Find W_s, Q, and ΔE_{th} for each process in the cycle. Display your results in a table.

   b. What is the thermal efficiency of this heat engine?

59. ‖ A heat engine uses a diatomic gas that follows the pV cycle in **FIGURE P21.59**.

   a. Determine the pressure, volume, and temperature at point 2.

   b. Determine ΔE_{th}, W_s, and Q for each of the three processes. Put your results in a table for easy reading.

   c. How much work does this engine do per cycle and what is its thermal efficiency?

FIGURE P21.59

60. ‖ FIGURE P21.60 is the pV diagram of Example 21.2, but now the device is operated in reverse.
   a. During which processes is heat transferred into the gas?
   b. Is this Q_H, heat extracted from a hot reservoir, or Q_C, heat extracted from a cold reservoir? Explain.
   c. Determine the values of Q_H and Q_C.
   Hint: The calculations have been done in Example 21.2 and do not need to be repeated. Instead, you need to determine which processes now contribute to Q_H and which to Q_C.
   d. Is the area inside the curve W_{in} or W_{out}? What is its value?
   e. The device is now being operated in a ccw cycle. Is it a refrigerator? Explain.

FIGURE P21.60

FIGURE P21.61

61. ‖ A heat engine using 120 mg of helium as the working substance follows the cycle shown in FIGURE P21.61.
   a. Determine the pressure, temperature, and volume of the gas at points 1, 2, and 3.
   b. What is the engine's thermal efficiency?
   c. What is the maximum possible efficiency of a heat engine that operates between T_{max} and T_{min}?
62. ‖‖ The heat engine shown in FIGURE P21.62 uses 2.0 mol of a monatomic gas as the working substance.
   a. Determine T_1, T_2, and T_3.
   b. Make a table that shows ΔE_{th}, W_s, and Q for each of the three processes.
   c. What is the engine's thermal efficiency?

FIGURE P21.62

FIGURE P21.63

63. ‖ The heat engine shown in FIGURE P21.63 uses 0.020 mol of a diatomic gas as the working substance.
   a. Determine T_1, T_2, and T_3.
   b. Make a table that shows ΔE_{th}, W_s, and Q for each of the three processes.
   c. What is the engine's thermal efficiency?
64. ‖‖ A heat engine with 0.20 mol of a monatomic ideal gas initially fills a 2000 cm³ cylinder at 600 K. The gas goes through the following closed cycle:

■ Isothermal expansion to 4000 cm³.
■ Isochoric cooling to 300 K.
■ Isothermal compression to 2000 cm³.
■ Isochoric heating to 600 K.

How much work does this engine do per cycle and what is its thermal efficiency?

In Problems 65 through 68 you are given the equation(s) used to solve a problem. For each of these, you are to
   a. Write a realistic problem for which this is the correct equation(s).
   b. Finish the solution of the problem.

65. $0.80 = 1 - (0°C + 273)/(T_H + 273)$
66. $4.0 = Q_C/W_{in}$
    $Q_H = 100$ J
67. $0.20 = 1 - Q_C/Q_H$
    $W_{out} = Q_H - Q_C = 20$ J
68. $400\text{ kJ} = \frac{1}{2}(p_{max} - 100\text{ kPa})(3.0\text{ m}^3 - 1.0\text{ m}^3)$

Challenge Problems

69. ‖‖‖ 100 mL of water at 15°C is placed in the freezer compartment of a refrigerator with a coefficient of performance of 4.0. How much heat energy is exhausted into the room as the water is changed to ice at −15°C?
70. ‖‖‖ FIGURE CP21.70 shows two insulated compartments separated by a thin wall. The left side contains 0.060 mol of helium at an initial temperature of 600 K and the right side contains 0.030 mol of helium at an initial temperature of 300 K. The compartment on the right is attached to a vertical cylinder, above which the air pressure is 1.0 atm. A 10-cm-diameter, 2.0 kg piston can slide without friction up and down the cylinder. Neither the cylinder diameter nor the volumes of the compartments are known.
   a. What is the final temperature?
   b. How much heat is transferred from the left side to the right side?
   c. How high is the piston lifted due to this heat transfer?
   d. What fraction of the heat is converted into work?

FIGURE CP21.70

71. ‖‖‖ A refrigerator using helium gas operates on the reversed cycle shown in FIGURE CP21.71. What are the refrigerator's (a) coefficient of performance and (b) power input if it operates at 60 cycles per second?

FIGURE CP21.71

72. ||| A heat engine using a diatomic ideal gas goes through the following closed cycle:

- Isothermal compression until the volume is halved.
- Isobaric expansion until the volume is restored to its initial value.
- Isochoric cooling until the pressure is restored to its initial value.

What are the thermal efficiencies of (a) this heat engine and (b) a Carnot engine operating between the highest and lowest temperatures reached by this engine?

73. ||| The gasoline engine in your car can be modeled as the Otto cycle shown in FIGURE CP21.73. A fuel-air mixture is sprayed into the cylinder at point 1, where the piston is at its farthest distance from the spark plug. This mixture is compressed as the piston moves toward the spark plug during the adiabatic *compression stroke*. The spark plug fires at point 2, releasing heat energy that had been stored in the gasoline. The fuel burns so quickly that the piston doesn't have time to move, so the heating is an isochoric process. The hot, high-pressure gas then pushes the piston outward during the *power stroke*. Finally, an exhaust value opens to allow the gas temperature and pressure to drop back to their initial values before starting the cycle over again.

FIGURE CP21.73

a. Analyze the Otto cycle and show that the work done per cycle is

$$W_{out} = \frac{nR}{1-\gamma}(T_2 - T_1 + T_4 - T_3)$$

b. Use the adiabatic connection between T_1 and T_2 and also between T_3 and T_4 to show that the thermal efficiency of the Otto cycle is

$$\eta = 1 - \frac{1}{r^{(\gamma-1)}}$$

where $r = V_{max}/V_{min}$ is the engine's *compression ratio*.

c. Graph η versus r out to $r = 30$ for a diatomic gas.

74. ||| FIGURE CP21.74 shows the Diesel cycle. It is similar to the Otto cycle (see Problem 21.73), but there are two important differences. First, the fuel is not admitted until the air is fully compressed at point 2. Because of the high temperature at the end of an adiabatic compression, the fuel begins to burn spontaneously. (There are no spark plugs in a diesel engine!) Second, combustion takes place more slowly, with fuel continuing to be injected. This makes the ignition stage a constant-pressure process. The cycle shown, for one cylinder of a diesel engine, has a *displacement* $V_{max} - V_{min}$ of 1000 cm³ and a compression ratio $r = V_{max}/V_{min} = 21$. These are typical values for a diesel truck. The engine operates with intake air ($\gamma = 1.40$) at 25°C and 1.0 atm pressure. The quantity of fuel injected into the cylinder has a heat of combustion of 1000 J.

FIGURE CP21.74

a. Find p, V, and T at each of the four corners of the cycle. Display your results in a table.

b. What is the net work done by the cylinder during one full cycle?

c. What is the thermal efficiency of this engine?

d. What is the power output in kW and horsepower (1 hp = 746 W) of an eight-cylinder diesel engine running at 2400 rpm?

Thermodynamics

KEY FINDINGS   What are the overarching findings of Part V?

Thermodynamics is an expanded view of systems and energy.

- A system exchanges energy with its environment via both **work** and **heat.** These are energy transfers.
- In a **heat engine,** heat energy is transformed into useful work when a system follows a cyclical process.

- Some processes are **irreversible.** A reversible **Carnot engine** is a heat engine with the maximum possible efficiency.
- Many macroscopic thermal properties of materials can be understood in terms of the motions of atoms and molecules.

LAWS   What laws of physics govern thermodynamics?

First law of thermodynamics   Energy is conserved: $\Delta E_{th} = W + Q$.

Second law of thermodynamics   Heat is not spontaneously transferred from a colder object to a hotter object.

Ideal-gas law $pV = nRT$ or $pV = Nk_BT$

Equipartition theorem   The energy stored in each degree of freedom is $\frac{1}{2}Nk_BT$ or $\frac{1}{2}nRT$.

MODELS   What are the most important models of Part V?

Thermodynamic energy model

- **Work** and **heat** are energies transferred between the system and the environment.
  - Work is a mechanical interaction.
  - Heat is a thermal interaction.
- Transferring energy changes the system's thermal energy as given by the first law: $\Delta E_{th} = W + Q$.

Ideal-gas model

- For low densities and temperatures not too close to the condensation point, all gases, regardless of composition, obey the ideal-gas law with the same value of the gas constant R.

Phases of matter

- Solid: Rigid, definite shape, nearly incompressible.
- Liquid: Fluid, takes shape of container, nearly incompressible.
- Gas: Non-interacting particles, highly compressible.

Carnot engine

- A perfectly reversible heat engine has the maximum possible efficiency of any heat engine operating between T_H and T_C.
- The efficiency depends only on the reservoir temperatures, not on any details of the engine: $\eta_{Carnot} = 1 - T_C/T_H$.

TOOLS   What are the most important tools introduced in Part V?

pV diagrams show

- States
- Processes

Four fundamental gas processes

- Isochoric: $\Delta V = 0$ and $W = 0$
- Isobaric: $\Delta p = 0$
- Isothermal: $\Delta T = 0$ and $\Delta E_{th} = 0$
- Adiabatic: $Q = 0$

Work in gas processes

- The work done *on* a gas is
$$W = -\int_{V_i}^{V_f} p\, dV$$
= −area under the pV curve
- For a closed cycle, the work done *by* a gas is W_s = area enclosed.

Heat and thermal energy

- Heat is energy transferred in a thermal process when there is a temperature difference.
- Thermal energy is the microscopic energy of moving atoms.
- Heat, thermal energy, and temperature are related but not the same.

Heat is transferred by

- Conduction
- Convection
- Radiation
- Evaporation

Heating and cooling

- The heat energy needed for a temperature change is $Q = mc\,\Delta T$ or $Q = nC\,\Delta T$.
- For thermally isolated systems
$$Q_{net} = Q_1 + Q_2 + \cdots = 0$$

Heat engines and refrigerators

Heat engines and refrigerators require both a **hot reservoir** and a **cold reservoir.**

- A heat engine transforms heat energy from a hot reservoir into work while exhausting energy to the cold reservoir.

- A refrigerator uses external work to "pump" heat energy from a cold reservoir to the hot reservoir.
- A Carnot engine is the best possible engine or refrigerator.

OVERVIEW

Forces and Fields

Amber, or fossilized tree resin, has long been prized for its beauty. It has been known since antiquity that a piece of amber rubbed with fur can attract feathers or straw—seemingly magical powers to a pre-scientific society. It was also known to the ancient Greeks that certain stones from the region they called *Magnesia* could pick up pieces of iron. It is from these humble beginnings that we today have high-speed computers, lasers, and magnetic resonance imaging as well as such mundane modern-day miracles as the lightbulb.

The basic phenomena of electricity and magnetism are not as familiar as those of mechanics. You have spent your entire life exerting forces on objects and watching them move, but your experience with electricity and magnetism is probably much more limited. We will deal with this lack of experience by placing a large emphasis on the *phenomena* of electricity and magnetism.

We will begin by looking in detail at *electric charge* and the process of *charging* an object. It is easy to make systematic observations of how charges behave, and we will consider the forces between charges and how charges behave in different materials. Similarly, we will begin our study of magnetism by observing how magnets stick to some metals but not others and how magnets affect compass needles. But our most important observation will be that an electric current affects a compass needle in exactly the same way as a magnet. This observation, suggesting a close connection between electricity and magnetism, will eventually lead us to the discovery of electromagnetic waves.

Our goal in Part VI is to develop a theory to explain the phenomena of electricity and magnetism. The linchpin of our theory will be the entirely new concept of a *field*. Electricity and magnetism are about the long-range interactions of charges, both static charges and moving charges, and the field concept will help us understand how these interactions take place. We will want to know how fields are created by charges and how charges, in return, respond to the fields. Bit by bit, we will assemble a theory—based on the new concepts of electric and magnetic fields—that will allow us to understand, explain, and predict a wide range of electromagnetic behavior.

The story of electricity and magnetism is vast. The 19th-century formulation of the theory of electromagnetism, which led to sweeping revolutions in science and technology, has been called by no less than Einstein "the most important event in physics since Newton's time." Not surprisingly, all we can do in this text is develop some of the basic ideas and concepts, leaving many details and applications to later courses. Even so, our study of electricity and magnetism will explore some of the most exciting and important topics in physics.

These bright loops above the surface of the sun—called *coronal loops*—are an extremely hot gas (>10^6 K) of charged particles moving along field lines of the sun's magnetic field.

22 Electric Charges and Forces

Electricity is one of the fundamental forces of nature. Lightning is a vivid manifestation of electric charges and forces.

IN THIS CHAPTER, you will learn that electric phenomena are based on charges, forces, and fields.

What is electric charge?

Electric phenomena depend on charge.

- There are two kinds of charge, called positive and negative.
- Electrons and protons—the constituents of atoms—are the basic charges of ordinary matter.
- Charging is the transfer of electrons from one object to another.

How do charges behave?

Charges have well-established behaviors:

- Two charges of the same kind repel; two opposite charges attract.
- Small neutral objects are attracted to a charge of either sign.
- Charge can be transferred from one object to another.
- Charge is conserved.

What are conductors and insulators?

There are two classes of materials with very different electrical properties:

- Conductors are materials through or along which charge moves easily.
- Insulators are materials on or in which charge is immobile.

What is Coulomb's law?

Coulomb's law is the fundamental law for the electric force between two charged particles. Coulomb's law, like Newton's law of gravity, is an inverse-square law: The electric force is inversely proportional to the *square* of the distance between charges.

« LOOKING BACK Sections 3.2–3.4 Vector addition
« LOOKING BACK Sections 13.2–13.4 Gravity

What is an electric field?

How is a long-range force transmitted from one charge to another? We'll develop the idea that charges create an electric field, and the electric field of one charge is the agent that exerts a force on another charge. That is, charges interact via electric fields. The electric field is present at all points in space.

Why are electric charges important?

Computers, cell phones, and optical fiber communications may seem to have little in common with the fact that you can get a shock when you touch a doorknob after walking across a carpet. But the physics of electric charges—how objects get charged and how charges interact with each other—is the foundation for all modern electronic devices and communications technology. Electricity and magnetism is a very large and very important topic, and it starts with simple observations of electric charges and forces.

22.1 The Charge Model

You can receive a mildly unpleasant shock and produce a little spark if you touch a metal doorknob after walking across a carpet. Vigorously brushing your freshly washed hair makes all the hairs fly apart. A plastic comb that you've run through your hair will pick up bits of paper and other small objects, but a metal comb won't.

The common factor in these observations is that two objects are *rubbed* together. Why should rubbing an object cause forces and sparks? What kind of forces are these? Why do metallic objects behave differently from nonmetallic? These are the questions with which we begin our study of electricity.

Our first goal is to develop a model for understanding electric phenomena in terms of *charges* and *forces*. We will later use our contemporary knowledge of atoms to understand electricity on a microscopic level, but the basic concepts of electricity make *no* reference to atoms or electrons. The theory of electricity was well established long before the electron was discovered.

Experimenting with Charges

Let us enter a laboratory where we can make observations of electric phenomena. The major tools in the lab are plastic, glass, and metal rods; pieces of wool and silk; and small metal spheres on wood stands. Let's see what we can learn with these tools.

Discovering electricity I

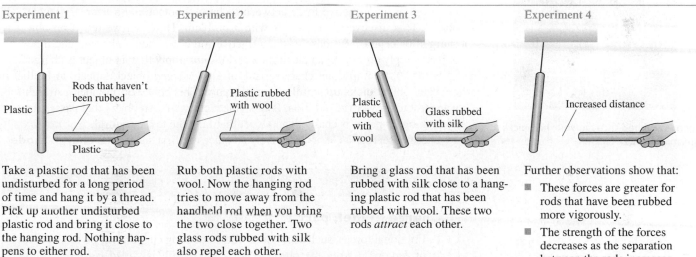

Experiment 1	Experiment 2	Experiment 3	Experiment 4

Take a plastic rod that has been undisturbed for a long period of time and hang it by a thread. Pick up another undisturbed plastic rod and bring it close to the hanging rod. Nothing happens to either rod.

Rub both plastic rods with wool. Now the hanging rod tries to move away from the handheld rod when you bring the two close together. Two glass rods rubbed with silk also repel each other.

Bring a glass rod that has been rubbed with silk close to a hanging plastic rod that has been rubbed with wool. These two rods *attract* each other.

Further observations show that:
- These forces are greater for rods that have been rubbed more vigorously.
- The strength of the forces decreases as the separation between the rods increases.

No forces were observed in Experiment 1. We will say that the original objects are **neutral**. Rubbing the rods (Experiments 2 and 3) somehow causes forces to be exerted between them. We will call the rubbing process **charging** and say that a rubbed rod is *charged*. For now, these are simply descriptive terms. The terms don't tell us anything about the process itself.

Experiments 2 and 3 show that there is a *long-range repulsive force,* requiring no contact, between two identical objects that have been charged in the *same* way. Furthermore, Experiment 4 shows that the force between two charged objects depends on the distance between them. This is the first long-range force we've encountered since gravity was introduced in Chapter 5. It is also the first time we've observed a repulsive force, so right away we see that new ideas will be needed to understand electricity.

Experiment 3 is a puzzle. Two rods *seem* to have been charged in the same way, by rubbing, but these two rods *attract* each other rather than repel. Why does the outcome of Experiment 3 differ from that of Experiment 2? Back to the lab.

Discovering electricity II

Experiment 5	Experiment 6	Experiment 7	Experiment 8

Hold a charged (i.e., rubbed) plastic rod over small pieces of paper. The pieces of paper leap up and stick to the rod. A charged glass rod does the same. However, a neutral rod has no effect on the pieces of paper.

Hang charged plastic and glass rods. Both are attracted to a *neutral* (i.e., unrubbed) plastic rod. Both are also attracted to a *neutral* glass rod. In fact, the charged rods are attracted to *any* neutral object, such as a finger or a piece of paper.

Rub a hanging plastic rod with wool and then hold the *wool* close to the rod. The rod is weakly *attracted* to the wool. The plastic rod is *repelled* by a piece of silk that has been used to rub glass.

Further experiments show that there appear to be *no* objects that, after being rubbed, pick up pieces of paper and attract *both* the charged plastic and glass rods.

A comb rubbed through your hair picks up small pieces of paper.

Our first set of experiments found that charged objects exert forces on each other. The forces are sometimes attractive, sometimes repulsive. Experiments 5 and 6 show that there is an attractive force between a charged object and a *neutral* (uncharged) object. This discovery presents us with a problem: How can we tell if an object is charged or neutral? Because of the attractive force between a charged and a neutral object, simply observing an electric force does *not* imply that an object is charged.

However, an important characteristic of any *charged* object appears to be that **a charged object picks up small pieces of paper.** This behavior provides a straightforward test to answer the question, Is this object charged? An object that passes the test by picking up paper is charged; an object that fails the test is neutral.

These observations let us tentatively advance the first stages of a **charge model.**

MODEL 22.1

Charge model, part I

1. Frictional forces, such as rubbing, add something called **charge** to an object or remove it from the object. The process itself is called *charging*. More vigorous rubbing produces a larger quantity of charge.

2. There are two and only two kinds of charge. For now we will call these "plastic charge" and "glass charge." Other objects can sometimes be charged by rubbing, but the charge they receive is either "plastic charge" or "glass charge."

3. Two **like charges** (plastic/plastic or glass/glass) exert repulsive forces on each other. Two **opposite charges** (plastic/glass) attract each other.

4. The force between two charges is a long-range force. The size of the force increases as the quantity of charge increases and decreases as the distance between the charges increases.

5. *Neutral* objects have an *equal mixture* of both "plastic charge" and "glass charge." The rubbing process somehow manages to separate the two.

Postulate 2 is based on Experiment 8. If an object is charged (i.e., picks up paper), it always attracts one charged rod and repels the other. That is, it acts either "like plastic" or "like glass." If there were a third kind of charge, different from the first two, an object with that charge should pick up paper and attract *both* the charged plastic and glass rods. No such objects have ever been found.

The basis for postulate 5 is the observation in Experiment 7 that a charged plastic rod is attracted to the wool used to rub it but repelled by silk that has rubbed glass. It appears that rubbing glass causes the silk to acquire "plastic charge." The easiest way to explain this is to hypothesize that the silk starts out with equal amounts of "glass charge" and "plastic charge" and that the rubbing somehow transfers "glass charge" from the silk to the rod. This leaves an excess of "glass charge" on the rod and an excess of "plastic charge" on the silk.

While the charge model is *consistent* with the observations, it is by no means proved. We still have some large unexplained puzzles, such as why charged objects exert attractive forces on neutral objects.

Electric Properties of Materials

We still need to clarify how different types of materials respond to charges.

Discovering electricity III

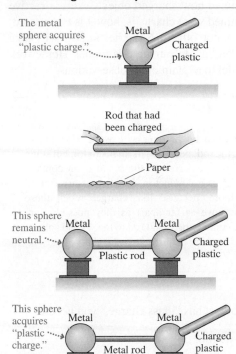

The metal sphere acquires "plastic charge."
Metal
Charged plastic

Rod that had been charged
Paper

This sphere remains neutral.
Metal   Metal
Plastic rod
Charged plastic

This sphere acquires "plastic charge."
Metal   Metal
Metal rod
Charged plastic

Experiment 9

Charge a plastic rod by rubbing it with wool. Touch a neutral metal sphere with the rubbed area of the rod. The metal sphere then picks up small pieces of paper and repels a charged, hanging plastic rod. The metal sphere appears to have acquired "plastic charge."

Experiment 10

Charge a plastic rod, then run your finger along it. After you've done so, the rod no longer picks up small pieces of paper or repels a charged, hanging plastic rod. Similarly, the metal sphere of Experiment 9 no longer repels the plastic rod after you touch it with your finger.

Experiment 11

Place two metal spheres close together with a plastic rod connecting them. Charge a second plastic rod, by rubbing, and touch it to one of the metal spheres. Afterward, the metal sphere that was touched picks up small pieces of paper and repels a charged, hanging plastic rod. The other metal sphere does neither.

Experiment 12

Repeat Experiment 11 with a metal rod connecting the two metal spheres. Touch one metal sphere with a charged plastic rod. Afterward, *both* metal spheres pick up small pieces of paper and repel a charged, hanging plastic rod.

Our final set of experiments has shown that

- Charge can be *transferred* from one object to another, but only when the objects *touch*. Contact is required. Removing charge from an object, which you can do by touching it, is called **discharging.**
- There are two types or classes of materials with very different electric properties. We call these *conductors* and *insulators*.

Experiment 12, in which a metal rod is used, is in sharp contrast to Experiment 11. Charge somehow *moves through* or along a metal rod, from one sphere to the other, but remains *fixed in place* on a plastic or glass rod. Let us define **conductors** as those materials through or along which charge easily moves and **insulators** as those materials on or in which charges remain immobile. Glass and plastic are insulators; metal is a conductor.

This information lets us add two more postulates to our charge model:

> ## MODEL 22.1
>
> ### Charge model, part II
> 6. There are two types of materials. Conductors are materials through or along which charge easily moves. Insulators are materials on or in which charges remain fixed in place.
> 7. Charge can be transferred from one object to another by contact.

NOTE Both insulators and conductors can be charged. They differ in the *mobility* of the charge.

We have by no means exhausted the number of experiments and observations we might try. Early scientific investigators were faced with all of these results, plus many others. How should we make sense of it all? The charge model seems promising, but certainly not proven. We have not yet explained how charged objects exert attractive forces on *neutral* objects, nor have we explained what charge is, how it is transferred, or *why* it moves through some objects but not others. Nonetheless, we will take advantage of our historical hindsight and continue to pursue this model. Homework problems will let you practice using the model to explain other observations.

EXAMPLE 22.1 | Transferring charge

In Experiment 12, touching one metal sphere with a charged plastic rod caused a second metal sphere to become charged with the same type of charge as the rod. Use the postulates of the charge model to explain this.

SOLVE We need the following postulates from the charge model:

7. Charge is transferred upon contact.
6. Metal is a conductor, and charge moves through a conductor
3. Like charges repel.

The plastic rod was charged by rubbing with wool. The charge doesn't move around on the rod, because it is an insulator, but some of the "plastic charge" is transferred to the metal upon contact. Once in the metal, which is a conductor, the charges are free to move around. Furthermore, because like charges repel, these plastic charges quickly move as far apart as they possibly can. Some move through the connecting metal rod to the second sphere. Consequently, the second sphere acquires "plastic charge."

STOP TO THINK 22.1 To determine if an object has "glass charge," you need to

a. See if the object attracts a charged plastic rod.
b. See if the object repels a charged glass rod.
c. Do both a and b.
d. Do either a or b.

22.2 Charge

As you probably know, the modern names for the two types of charge are *positive charge* and *negative charge*. You may be surprised to learn that the names were coined by Benjamin Franklin.

So what is positive and what is negative? It's entirely up to us! Franklin established the convention that **a glass rod that has been rubbed with silk is *positively* charged.** That's it. Any other object that repels a charged glass rod is also positively charged. Any charged object that attracts a charged glass rod is negatively charged. Thus **a plastic rod rubbed with wool is negative.** It was only long afterward, with the discovery of electrons and protons, that electrons were found to be attracted to a charged glass rod while protons were repelled. Thus *by convention* electrons have a negative charge and protons a positive charge.

Atoms and Electricity

Now let's fast forward to the 21st century. The theory of electricity was developed without knowledge of atoms, but there is no reason for us to continue to overlook this important part of our contemporary perspective. FIGURE 22.1 shows that an atom consists of a very small and dense *nucleus* (diameter $\sim 10^{-14}$ m) surrounded by much less massive orbiting *electrons*. The electron orbital frequencies are so enormous ($\sim 10^{15}$ revolutions per second) that the electrons seem to form an **electron cloud** of diameter $\sim 10^{-10}$ m, a factor 10^4 larger than the nucleus.

Experiments at the end of the 19th century revealed that electrons are particles with both mass and a negative charge. The nucleus is a composite structure consisting of *protons,* positively charged particles, and neutral *neutrons.* The atom is held together by the attractive electric force between the positive nucleus and the negative electrons.

One of the most important discoveries is that **charge, like mass, is an inherent property of electrons and protons.** It's no more possible to have an electron without charge than it is to have an electron without mass. As far as we know today, electrons and protons have charges of opposite sign but *exactly* equal magnitude. (Very careful experiments have never found any difference.) This atomic-level unit of charge, called the **fundamental unit of charge,** is represented by the symbol e. TABLE 22.1 shows the masses and charges of protons and electrons. We need to define a unit of charge, which we will do in Section 22.4, before we can specify how much charge e is.

The Micro/Macro Connection

Electrons and protons are the basic charges of ordinary matter. Consequently, the various observations we made in Section 22.1 need to be explained in terms of electrons and protons.

> **NOTE** Electrons and protons are particles of matter. Their motion is governed by Newton's laws. Electrons can move from one object to another when the objects are in contact, but neither electrons nor protons can leap through the air from one object to another. An object does not become charged simply from being close to a charged object.

Charge is represented by the symbol q (or sometimes Q). A macroscopic object, such as a plastic rod, has charge

$$q = N_p e - N_e e = (N_p - N_e)e \tag{22.1}$$

where N_p and N_e are the number of protons and electrons contained in the object. An object with an equal number of protons and electrons has no *net* charge (i.e., $q = 0$) and is said to be *electrically neutral.*

> **NOTE** *Neutral* does *not* mean "no charges" but, instead, no *net* charge.

A charged object has an unequal number of protons and electrons. An object is positively charged if $N_p > N_e$. It is negatively charged if $N_p < N_e$. Notice that an object's charge is always an integer multiple of e. That is, the amount of charge on an object varies by small but discrete steps, not continuously. This is called **charge quantization.**

In practice, objects acquire a positive charge not by gaining protons, as you might expect, but by losing electrons. Protons are *extremely* tightly bound within the nucleus and cannot be added to or removed from atoms. Electrons, on the other hand, are bound rather loosely and can be removed without great difficulty. The process of removing an electron from the electron cloud of an atom is called **ionization.** An atom that is missing an electron is called a *positive ion.* Its *net* charge is $q = +e$.

Some atoms can accommodate an *extra* electron and thus become a *negative ion* with net charge $q = -e$. A saltwater solution is a good example. When table salt (the chemical sodium chloride, NaCl) dissolves, it separates into positive sodium ions Na^+ and negative chlorine ions Cl^-. FIGURE 22.2 shows positive and negative ions.

FIGURE 22.1 An atom.

The nucleus, exaggerated for clarity, contains positive protons.

$\sim 10^{-10}$ m

The electron cloud is negatively charged.

TABLE 22.1 Protons and electrons

Particle	Mass (kg)	Charge
Proton	1.67×10^{-27}	$+e$
Electron	9.11×10^{-31}	$-e$

FIGURE 22.2 Positive and negative ions.

Positive ion   Negative ion

The atom has lost one electron, giving it a net positive charge.

The atom has gained one electron, giving it a net negative charge.

FIGURE 22.3 Charging by friction usually creates molecular ions as bonds are broken.

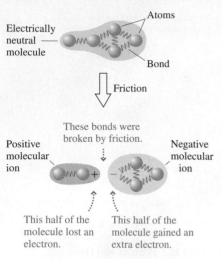

All the charging processes we observed in Section 22.1 involved rubbing and friction. The forces of friction cause molecular bonds at the surface to break as the two materials slide past each other. Molecules are electrically neutral, but **FIGURE 22.3** shows that *molecular ions* can be created when one of the bonds in a large molecule is broken. The positive molecular ions remain on one material and the negative ions on the other, so one of the objects being rubbed ends up with a net positive charge and the other with a net negative charge. This is the way in which a plastic rod is charged by rubbing with wool or a comb is charged by passing through your hair.

Charge Conservation and Charge Diagrams

One of the most important discoveries about charge is the **law of conservation of charge:** Charge is neither created nor destroyed. Charge can be transferred from one object to another as electrons and ions move about, but the *total* amount of charge remains constant. For example, charging a plastic rod by rubbing it with wool transfers electrons from the wool to the plastic as the molecular bonds break. The wool is left with a positive charge equal in magnitude but opposite in sign to the negative charge of the rod: $q_{wool} = -q_{plastic}$. The *net* charge remains zero.

Diagrams are going to be an important tool for understanding and explaining charges and the forces on charged objects. As you begin to use diagrams, it will be important to make explicit use of charge conservation. The net number of plusses and minuses drawn on your diagrams should *not* change as you show them moving around.

STOP TO THINK 22.2 Rank in order, from most positive to most negative, the charges q_a to q_e of these five systems.

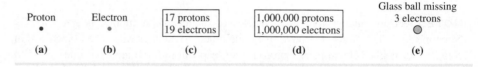

22.3 Insulators and Conductors

FIGURE 22.4 A microscopic look at insulators and conductors.

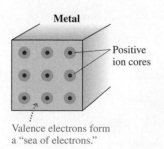

You have seen that there are two classes of materials as defined by their electrical properties: insulators and conductors. **FIGURE 22.4** looks inside an insulator and a metallic conductor. The electrons in the insulator are all tightly bound to the positive nuclei and not free to move around. Charging an insulator by friction leaves patches of molecular ions on the surface, but these patches are immobile.

In metals, the outer atomic electrons (called the *valence electrons* in chemistry) are only weakly bound to the nuclei. As the atoms come together to form a solid, these outer electrons become detached from their parent nuclei and are free to wander about through the entire solid. The solid *as a whole* remains electrically neutral, because we have not added or removed any electrons, but the electrons are now rather like a negatively charged gas or liquid—what physicists like to call a **sea of electrons**—permeating an array of positively charged **ion cores.**

The primary consequence of this structure is that electrons in a metal are highly mobile. They can quickly and easily move through the metal in response to electric forces. The motion of charges through a material is what we will later call a **current,** and the charges that physically move are called the **charge carriers.** The charge carriers in metals are electrons.

Metals aren't the only conductors. Ionic solutions, such as salt water, are also good conductors. But the charge carriers in an ionic solution are the ions, not electrons. We'll focus on metallic conductors because of their importance in applications of electricity.

Charging

Insulators are often charged by rubbing. The charge diagrams of FIGURE 22.5 show that the charges on the rod are on the surface and that charge is conserved. The charge can be transferred to another object upon contact, but it doesn't move around on the rod.

FIGURE 22.5 An insulating rod is charged by rubbing.

Rub the plastic rod with a piece of wool.

Wool

Plastic

This end is still neutral.

Negative charges are immobile on the rod's surface.

The positive charge on the wool is equal to the negative charge on the rod.

Metals usually cannot be charged by rubbing, but Experiment 9 showed that a metal sphere can be charged by contact with a charged plastic rod. FIGURE 22.6 gives a pictorial explanation. An essential idea is that **the electrons in a conductor are free to move.** Once charge is transferred to the metal, repulsive forces between the negative charges cause the electrons to move apart from each other.

Note that the newly added electrons do not themselves need to move to the far corners of the metal. Because of the repulsive forces, the newcomers simply "shove" the entire electron sea a little to the side. The electron sea takes an extremely short time to adjust itself to the presence of the added charge, typically less than 10^{-9} s. For all practical purposes, a conductor responds *instantaneously* to the addition or removal of charge.

Other than this very brief interval during which the electron sea is adjusting, the charges in an *isolated* conductor are in static equilibrium. That is, the charges are at rest (i.e., static) and there is no net force on any charge. This condition is called **electrostatic equilibrium.** If there *were* a net force on one of the charges, it would quickly move to an equilibrium point at which the force was zero.

Electrostatic equilibrium has an important consequence:

In an isolated conductor, any excess charge is located on the surface of the conductor.

To see this, suppose there *were* an excess electron in the interior of an isolated conductor. The extra electron would upset the electrical neutrality of the interior and exert forces on nearby electrons, causing them to move. But their motion would violate the assumption of static equilibrium, so we're forced to conclude that there cannot be any excess electrons in the interior. Any excess electrons push each other apart until they're all on the surface.

FIGURE 22.6 A conductor is charged by contact with a charged plastic rod.

Plastic

Metal

Charge is transferred to the metal upon contact.

These charges repel each other.

Very fast

Charge spreads over the surface of the metal.

EXAMPLE 22.2 | Charging an electroscope

Many electricity demonstrations are carried out with the help of an *electroscope* like the one shown in FIGURE 22.7. Touching the sphere at the top of an electroscope with a charged plastic rod causes the leaves to fly apart and remain hanging at an angle. Use charge diagrams to explain why.

MODEL We'll use the charge model and the model of a conductor as a material through which electrons move.

VISUALIZE FIGURE 22.8 on the next page uses a series of charge diagrams to show the charging of an electroscope.

FIGURE 22.7 A charged electroscope.

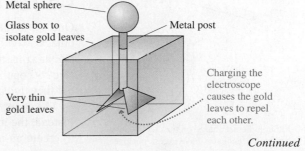

Metal sphere

Glass box to isolate gold leaves

Metal post

Very thin gold leaves

Charging the electroscope causes the gold leaves to repel each other.

Continued

FIGURE 22.8 The process by which an electroscope is charged.

Plastic

Electroscope

Very fast

1. Negative charges (i.e., electrons) are transferred from the rod to the metal sphere upon contact.

2. Metal is a conductor. Therefore charge *very quickly* spreads throughout the entire electroscope.

3. Like charges repel. The negatively charged leaves exert repulsive forces on each other, causing them to spread apart.

FIGURE 22.9 Touching a charged metal discharges it.

The metal is positively charged

+ + + +
+ Metal +
+ + + +

Touch

Charges spread through the metal + human system. Very little charge is left on the metal.

Discharging

The human body consists largely of salt water. Pure water is not a terribly good conductor, but salt water, with its Na^+ and Cl^- ions, is. Consequently, and occasionally tragically, humans are reasonably good conductors. This fact allows us to understand how it is that *touching* a charged object discharges it, as we observed in Experiment 10. As **FIGURE 22.9** shows, the net effect of touching a charged metal is that it and the conducting human together become a much larger conductor than the metal alone. Any excess charge that was initially confined to the metal can now spread over the larger metal + human conductor. This may not entirely discharge the metal, but in typical circumstances, where the human is much larger than the metal, the residual charge remaining on the metal is much reduced from the original charge. The metal, for most practical purposes, is discharged. In essence, two conductors in contact "share" the charge that was originally on just one of them.

Moist air is a conductor, although a rather poor one. Charged objects in air slowly lose their charge as the object shares its charge with the air. The earth itself is a giant conductor because of its water, moist soil, and a variety of ions. Any object that is physically connected to the earth through a conductor is said to be **grounded.** The effect of being grounded is that the object shares any excess charge it has with the entire earth! But the earth is so enormous that any conductor attached to the earth will be completely discharged.

The purpose of *grounding* objects, such as circuits and appliances, is to prevent the buildup of any charge on the objects. The third prong on appliances and electronics that have a three-prong plug is the ground connection. The building wiring physically connects that third wire deep into the ground somewhere just outside the building, often by attaching it to a metal water pipe that goes underground.

FIGURE 22.10 A charged rod held close to an electroscope causes the leaves to repel each other.

The electroscope is polarized by the charged rod. The sea of electrons shifts toward the positive rod.

Although the net charge on the electroscope is still zero, the leaves have excess positive charge and repel each other.

Charge Polarization

One observation from Section 22.1 still needs an explanation. How do charged objects exert an attractive force on a *neutral* object? To begin answering this question, **FIGURE 22.10** shows a positively charged rod held close to—but not touching—a *neutral* electroscope. The leaves move apart and stay apart as long as you hold the rod near, but they quickly collapse when it is removed.

The charged rod doesn't touch the electroscope, so no charge is added or removed. Instead, the metal's sea of electrons is attracted to the positive rod and shifts slightly to create an excess of negative charge on the side near the rod. The far side of the electroscope now has a deficit of electrons—an excess positive charge. We say that the electroscope has been *polarized*. **Charge polarization** is a slight separation of the positive and negative charges in a neutral object. Because there's no net charge, the electron sea quickly readjusts when the rod is removed.

Why don't *all* the electrons rush to the side near the positive charge? Once the electron sea shifts slightly, the stationary positive ions begin to exert a force, a restoring force, pulling the electrons back to the right. The equilibrium position for the sea of electrons shifts just enough that the forces due to the external charge and the positive ions are in balance. In practice, the displacement of the electron sea is usually *less than 10^{-15} m!*

Charge polarization is the key to understanding how a charged object exerts an attractive force on a neutral object. **FIGURE 22.11** shows a positively charged rod near a neutral piece of metal. Because the electric force decreases with distance, **the attractive force on the electrons at the top surface is slightly greater than the repulsive force on the ions at the bottom.** The net force toward the charged rod is called a **polarization force.** The polarization force arises because the charges in the metal are separated, *not* because the rod and metal are oppositely charged.

A negatively charged rod would push the electron sea slightly away, polarizing the metal to have a positive upper surface charge and a negative lower surface charge. Once again, these are the conditions for the charge to exert a *net attractive force* on the metal. Thus our charge model explains how a charged object of *either* sign attracts neutral pieces of metal.

The Electric Dipole

Polarizing a conductor is one thing, but why does a charged rod pick up paper, which is an insulator? Consider what happens when we bring a positive charge near an atom. As **FIGURE 22.12** shows, the charge polarizes the atom. The electron cloud doesn't move far, because the force from the positive nucleus pulls it back, but the center of positive charge and the center of negative charge are now slightly separated.

FIGURE 22.12 A neutral atom is polarized by and attracted toward an external charge.

The external charge attracts the atom's negative charge, pulling the negative charge slightly toward it.

The atom's negative charge is closer to the external charge than its positive charge, so the atom is *attracted* toward the external charge.

Two opposite charges with a slight separation between them form what is called an **electric dipole.** (The actual distortion from a perfect sphere is minuscule, nothing like the distortion shown in the figure.) The attractive force on the dipole's near end *slightly* exceeds the repulsive force on its far end because the near end is closer to the external charge. The net force, an *attractive* force between the charge and the atom, is another example of a polarization force.

An insulator has no sea of electrons to shift if an external charge is brought close. Instead, as **FIGURE 22.13** shows, all the individual atoms inside the insulator become polarized. The polarization force acting *on each atom* produces a net polarization force toward the external charge. This solves the puzzle. A charged rod picks up pieces of paper by

- Polarizing the atoms in the paper,
- Then exerting an attractive polarization force on each atom.

This is important. Make sure you understand all the steps in the reasoning.

FIGURE 22.11 The polarization force on a neutral piece of metal is due to the slight charge separation.

1. The charged rod polarizes the neutral metal.

2. The nearby negative charge is attracted to the rod more strongly than the distant positive charge is repelled, resulting in a net upward force.

FIGURE 22.13 The atoms in an insulator are polarized by an external charge.

Electric dipoles

External charge

Insulator

Net force

STOP TO THINK 22.3 An electroscope is positively charged by *touching* it with a positive glass rod. The electroscope leaves spread apart and the glass rod is removed. Then a negatively charged plastic rod is brought close to the top of the electroscope, but it doesn't touch. What happens to the leaves?

a. The leaves get closer together.
b. The leaves spread farther apart.
c. One leaf moves higher, the other lower.
d. The leaves don't move.

Charging by Induction

Charge polarization is responsible for an interesting and counterintuitive way of charging an electroscope. **FIGURE 22.14** shows a positively charged glass rod held near an electroscope but not touching it, while a person touches the electroscope with a finger. Unlike what happens in Figure 22.10, the electroscope leaves hardly move.

FIGURE 22.14 A positive rod can charge an electroscope negatively by induction.

1. The charged rod polarizes the electroscope + person conductor. The leaves repel slightly due to polarization.

2. The negative charge on the electroscope is isolated when contact is broken.

3. When the rod is removed, the leaves first collapse as the polarization vanishes, then repel as the excess negative charge spreads out.

Charge polarization occurs, as it did in Figure 22.10, but this time in the much larger electroscope + person conductor. If the person removes his or her finger while the system is polarized, the electroscope is left with a net *negative* charge and the person has a net positive charge. The electroscope has been charged *opposite to the rod* in a process called **charging by induction.**

22.4 Coulomb's Law

The first three sections have established a *model* of charges and electric forces. This model has successfully provided a qualitative explanation of electric phenomena; now it's time to become quantitative. Experiment 4 in Section 22.1 found that the electric force decreases with distance. The force law that describes this behavior is known as *Coulomb's law.*

Charles Coulomb was one of many scientists investigating electricity in the late 18th century. Coulomb had the idea of studying electric forces using the torsion balance scheme by which Cavendish had measured the value of the gravitational constant G (see Section 13.4). This was a difficult experiment. Despite many obstacles, Coulomb announced in 1785 that the electric force obeys an *inverse-square law* analogous to Newton's law of gravity. Today we know it as **Coulomb's law.**

Coulomb's law

1. If two charged particles having charges q_1 and q_2 are a distance r apart, the particles exert forces on each other of magnitude

$$F_{1 \text{ on } 2} = F_{2 \text{ on } 1} = \frac{K|q_1||q_2|}{r^2} \qquad (22.2)$$

where K is called the **electrostatic constant.** These forces are an action/reaction pair, equal in magnitude and opposite in direction.
2. The forces are directed along the line joining the two particles. The forces are *repulsive* for two like charges and *attractive* for two opposite charges.

We sometimes speak of the "force between charge q_1 and charge q_2," but keep in mind that we are really dealing with charged *objects* that also have a mass, a size, and other properties. Charge is not some disembodied entity that exists apart from matter. Coulomb's law describes the force between charged *particles,* which are also called **point charges.** A charged particle, which is an extension of the particle model we used in Part I, has a mass and a charge but has no size.

Coulomb's law looks much like Newton's law of gravity, but there is one important difference: The charge q can be either positive or negative. Consequently, the absolute value signs in Equation 22.2 are especially important. The first part of Coulomb's law gives only the *magnitude* of the force, which is always positive. The direction must be determined from the second part of the law. FIGURE 22.15 shows the forces between different combinations of positive and negative charges.

Units of Charge

Coulomb had no *unit* of charge, so he was unable to determine a value for K, whose numerical value depends on the units of both charge and distance. The SI unit of charge, the **coulomb** (C), is derived from the SI unit of *current,* so we'll have to await the study of current in Chapter 27 before giving a precise definition. For now we'll note that the fundamental unit of charge e has been measured to have the value

$$e = 1.60 \times 10^{-19} \text{ C}$$

This is a very small amount of charge. Stated another way, 1 C is the net charge of roughly 6.25×10^{18} protons.

NOTE The amount of charge produced by friction is typically in the range 1 nC (10^{-9} C) to 100 nC. This is an excess or deficit of 10^{10} to 10^{12} electrons.

Once the unit of charge is established, torsion balance experiments such as Coulomb's can be used to measure the electrostatic constant K. In SI units

$$K = 8.99 \times 10^9 \text{ N m}^2/\text{C}^2$$

It is customary to round this to $K = 9.0 \times 10^9 \text{ N m}^2/\text{C}^2$ for all but extremely precise calculations, and we will do so.

Surprisingly, we will find that Coulomb's law is not explicitly used in much of the theory of electricity. While it *is* the basic force law, most of our future discussion and calculations will be of things called *fields* and *potentials.* It turns out that we can make many future equations easier to use if we rewrite Coulomb's law in a somewhat more complicated way. Let's define a new constant, called the **permittivity constant** ϵ_0 (pronounced "epsilon zero" or "epsilon naught"), as

$$\epsilon_0 = \frac{1}{4\pi K} = 8.85 \times 10^{-12} \text{ C}^2/\text{N m}^2$$

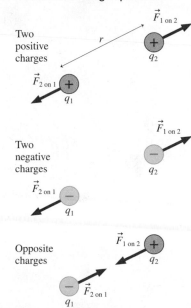

FIGURE 22.15 Attractive and repulsive forces between charged particles.

Two positive charges

Two negative charges

Opposite charges

Rewriting Coulomb's law in terms of ϵ_0 gives us

$$F = \frac{1}{4\pi\epsilon_0} \frac{|q_1||q_2|}{r^2} \qquad (22.3)$$

It will be easiest when using Coulomb's law directly to use the electrostatic constant K. However, in later chapters we will switch to the second version with ϵ_0.

Using Coulomb's Law

Coulomb's law is a force law, and forces are vectors. It has been many chapters since we made much use of vectors and vector addition, but these mathematical techniques will be essential in our study of electricity and magnetism.

There are two important observations regarding Coulomb's law:

1. **Coulomb's law applies only to point charges.** A point charge is an idealized material object with charge and mass but with no size or extension. For practical purposes, two charged objects can be modeled as point charges if they are much smaller than the separation between them.
2. **Electric forces, like other forces, can be superimposed.** If multiple charges 1, 2, 3, … are present, the *net* electric force on charge j due to all other charges is

$$\vec{F}_{\text{net}} = \vec{F}_{1\,\text{on}\,j} + \vec{F}_{2\,\text{on}\,j} + \vec{F}_{3\,\text{on}\,j} + \cdots \qquad (22.4)$$

where each of the $\vec{F}_{i\,\text{on}\,j}$ is given by Equation 22.2 or 22.3.

These conditions are the basis of a strategy for using Coulomb's law to solve electrostatic force problems.

PROBLEM-SOLVING STRATEGY 22.1

Electrostatic forces and Coulomb's law

MODEL Identify point charges or model objects as point charges.

VISUALIZE Use a *pictorial representation* to establish a coordinate system, show the positions of the charges, show the force vectors on the charges, define distances and angles, and identify what the problem is trying to find. This is the process of translating words to symbols.

SOLVE The mathematical representation is based on Coulomb's law:

$$F_{1\,\text{on}\,2} = F_{2\,\text{on}\,1} = \frac{K|q_1||q_2|}{r^2}$$

- Show the directions of the forces—repulsive for like charges, attractive for opposite charges—on the pictorial representation.
- When possible, do graphical vector addition on the pictorial representation. While not exact, it tells you the type of answer you should expect.
- Write each force vector in terms of its x- and y-components, then add the components to find the net force. Use the pictorial representation to determine which components are positive and which are negative.

ASSESS Check that your result has correct units and significant figures, is reasonable, and answers the question.

Exercise 26

EXAMPLE 22.3 | Lifting a glass bead

A small plastic sphere charged to −10 nC is held 1.0 cm above a small glass bead at rest on a table. The bead has a mass of 15 mg and a charge of +10 nC. Will the glass bead "leap up" to the plastic sphere?

MODEL Model the plastic sphere and glass bead as point charges.

VISUALIZE FIGURE 22.16 establishes a y-axis, identifies the plastic sphere as q_1 and the glass bead as q_2, and shows a free-body diagram.

FIGURE 22.16 A pictorial representation of the charges and forces.

accelerate upward from the table. Using the values provided, we have

$$F_{1 \text{ on } 2} = \frac{K|q_1||q_2|}{r^2}$$

$$= \frac{(9.0 \times 10^9 \,\text{N}\,\text{m}^2/\text{C}^2)(10 \times 10^{-9}\text{C})(10 \times 10^{-9}\text{C})}{(0.010 \,\text{m})^2}$$

$$= 9.0 \times 10^{-3} \,\text{N}$$

$$F_G = m_{\text{bead}}g = 1.5 \times 10^{-4} \,\text{N}$$

$F_{1 \text{ on } 2}$ exceeds $m_{\text{bead}}g$ by a factor of 60, so the glass bead will leap upward.

ASSESS The values used in this example are realistic for spheres ≈ 2 mm in diameter. In general, as in this example, electric forces are *significantly* larger than gravitational forces. Consequently, we can neglect gravity when working electric-force problems unless the particles are fairly massive.

SOLVE If $F_{1 \text{ on } 2}$ is less than the gravitational force $F_G = m_{\text{bead}}g$, then the bead will remain at rest on the table with $\vec{F}_{1 \text{ on } 2} + \vec{F}_G + \vec{n} = \vec{0}$. But if $F_{1 \text{ on } 2}$ is greater than $m_{\text{bead}}g$, the glass bead will

---

EXAMPLE 22.4 | The point of zero force

Two positively charged particles q_1 and $q_2 = 3q_1$ are 10.0 cm apart on the x-axis. Where (other than at infinity) could a third charge q_3 be placed so as to experience no net force?

MODEL Model the charged particles as point charges.

VISUALIZE FIGURE 22.17 establishes a coordinate system with q_1 at the origin and q_2 at $x = d$. We have no information about the sign of q_3, so apparently the position we're looking for will work for either sign. Suppose q_3 is off the x-axis, such as at point A. The two repulsive (or attractive) electric forces on q_3 cannot possibly add to zero, so q_3 must be somewhere on the x-axis. At point B, outside the two charges, the two forces on q_3 will always be in the same direction and, again, cannot add to zero. The only possible location is at point C, on the x-axis *between* the charges where the two forces are in opposite directions.

FIGURE 22.17 A pictorial representation of the charges and forces.

Only if q_3 is somewhere along the line between q_1 and q_2 can the forces add to zero.

SOLVE The mathematical problem is to find the position for which the forces $\vec{F}_{1 \text{ on } 3}$ and $\vec{F}_{2 \text{ on } 3}$ are equal in magnitude. If q_3 is distance x from q_1, it is distance $d - x$ from q_2. The *magnitudes* of the forces are

$$F_{1 \text{ on } 3} = \frac{Kq_1|q_3|}{r_{13}^2} = \frac{Kq_1|q_3|}{x^2}$$

$$F_{2 \text{ on } 3} = \frac{Kq_2|q_3|}{r_{23}^2} = \frac{K(3q_1)|q_3|}{(d-x)^2}$$

Charges q_1 and q_2 are positive and do not need absolute value signs. Equating the two forces gives

$$\frac{Kq_1|q_3|}{x^2} = \frac{3Kq_1|q_3|}{(d-x)^2}$$

The term $Kq_1|q_3|$ cancels. Multiplying by $x^2(d-x)^2$ gives

$$(d-x)^2 = 3x^2$$

which can be rearranged into the quadratic equation

$$2x^2 + 2dx - d^2 = 2x^2 + 20x - 100 = 0$$

where we used $d = 10$ cm and x is in cm. The solutions to this equation are

$$x = +3.66 \text{ cm and } -13.66 \text{ cm}$$

Both are points where the *magnitudes* of the two forces are equal, but $x = -13.66$ cm is a point where the magnitudes are equal while the directions are the same. The solution we want, which is between the charges, is $x = 3.66$ cm. Thus the point to place q_3 is 3.66 cm from q_1 along the line joining q_1 and q_2.

ASSESS q_1 is smaller than q_2, so we expect the point at which the forces balance to be closer to q_1 than to q_2. The solution seems reasonable. Note that the problem statement has no coordinates, so "$x = 3.66$ cm" is *not* an acceptable answer. You need to describe the position relative to q_1 and q_2.

EXAMPLE 22.5 | Three charges

Three charged particles with $q_1 = -50$ nC, $q_2 = +50$ nC, and $q_3 = +30$ nC are placed on the corners of the 5.0 cm $\times$ 10.0 cm rectangle shown in **FIGURE 22.18**. What is the net force on charge q_3 due to the other two charges? Give your answer both in component form and as a magnitude and direction.

FIGURE 22.18 The three charges.

MODEL Model the charged particles as point charges.

VISUALIZE The pictorial representation of **FIGURE 22.19** establishes a coordinate system. q_1 and q_3 are opposite charges, so force vector $\vec{F}_{1 \text{ on } 3}$ is an attractive force toward q_1. q_2 and q_3 are like charges, so force vector $\vec{F}_{2 \text{ on } 3}$ is a repulsive force away from q_2. q_1 and q_2 have equal magnitudes, but $\vec{F}_{2 \text{ on } 3}$ has been drawn shorter than $\vec{F}_{1 \text{ on } 3}$ because q_2 is farther from q_3. Vector addition has been used to draw the net force vector $\vec{F}_3$ and to define its angle ϕ.

FIGURE 22.19 A pictorial representation of the charges and forces.

SOLVE The question asks for a *force*, so our answer will be the *vector* sum $\vec{F}_3 = \vec{F}_{1 \text{ on } 3} + \vec{F}_{2 \text{ on } 3}$. We need to write $\vec{F}_{1 \text{ on } 3}$ and $\vec{F}_{2 \text{ on } 3}$ in component form. The magnitude of force $\vec{F}_{1 \text{ on } 3}$ can be found using Coulomb's law:

$$F_{1 \text{ on } 3} = \frac{K|q_1||q_3|}{r_{13}^2}$$
$$= \frac{(9.0 \times 10^9 \,\text{N m}^2/\text{C}^2)(50 \times 10^{-9}\,\text{C})(30 \times 10^{-9}\,\text{C})}{(0.100 \,\text{m})^2}$$
$$= 1.35 \times 10^{-3}\,\text{N}$$

where we used $r_{13} = 10.0$ cm.

The pictorial representation shows that $\vec{F}_{1 \text{ on } 3}$ points downward, in the negative y-direction, so

$$\vec{F}_{1 \text{ on } 3} = -1.35 \times 10^{-3}\hat{\jmath}\,\text{N}$$

To calculate $\vec{F}_{2 \text{ on } 3}$ we first need the distance r_{23} between the charges:

$$r_{23} = \sqrt{(5.0 \,\text{cm})^2 + (10.0 \,\text{cm})^2} = 11.2 \,\text{cm}$$

The magnitude of $\vec{F}_{2 \text{ on } 3}$ is thus

$$F_{2 \text{ on } 3} = \frac{K|q_2||q_3|}{r_{23}^2}$$
$$= \frac{(9.0 \times 10^9 \,\text{N m}^2/\text{C}^2)(50 \times 10^{-9}\,\text{C})(30 \times 10^{-9}\,\text{C})}{(0.112 \,\text{m})^2}$$
$$= 1.08 \times 10^{-3}\,\text{N}$$

This is only a magnitude. The *vector* $\vec{F}_{2 \text{ on } 3}$ is

$$\vec{F}_{2 \text{ on } 3} = -F_{2 \text{ on } 3}\cos\theta\,\hat{\imath} + F_{2 \text{ on } 3}\sin\theta\,\hat{\jmath}$$

where angle θ is defined in the figure and the signs (negative x-component, positive y-component) were determined from the pictorial representation. From the geometry of the rectangle,

$$\theta = \tan^{-1}\left(\frac{10.0 \,\text{cm}}{5.0 \,\text{cm}}\right) = \tan^{-1}(2.0) = 63.4°$$

Thus $\vec{F}_{2 \text{ on } 3} = (-4.83\hat{\imath} + 9.66\hat{\jmath}) \times 10^{-4}\,\text{N}$. Now we can add $\vec{F}_{1 \text{ on } 3}$ and $\vec{F}_{2 \text{ on } 3}$ to find

$$\vec{F}_3 = \vec{F}_{1 \text{ on } 3} + \vec{F}_{2 \text{ on } 3} = (-4.83\hat{\imath} - 3.84\hat{\jmath}) \times 10^{-4}\,\text{N}$$

This would be an acceptable answer for many problems, but sometimes we need the net force as a magnitude and direction. With angle ϕ as defined in the figure, these are

$$F_3 = \sqrt{F_{3x}^2 + F_{3y}^2} = 6.2 \times 10^{-4}\,\text{N}$$

$$\phi = \tan^{-1}\left|\frac{F_{3y}}{F_{3x}}\right| = 38°$$

Thus $\vec{F}_3 = (6.2 \times 10^{-4}\,\text{N},\ 38°$ below the negative x-axis).

ASSESS The forces are not large, but they are typical of electrostatic forces. Even so, you'll soon see that these forces can produce very large accelerations because the masses of the charged objects are usually very small.

STOP TO THINK 22.4 Charged spheres A and B exert repulsive forces on each other. $q_A = 4q_B$. Which statement is true?

A   B

a. $F_{\text{A on B}} > F_{\text{B on A}}$ b. $F_{\text{A on B}} = F_{\text{B on A}}$ c. $F_{\text{A on B}} < F_{\text{B on A}}$

22.5 The Electric Field

Electric and magnetic forces, like gravity, are *long-range* forces; no contact is required for one charged particle to exert a force on another. But this raises some troubling issues. For example, consider the charged particles A and B in **FIGURE 22.20**. If A suddenly starts moving, as shown by the arrow, the force vector on B must pivot to follow A. Does this happen *instantly*? Or is there some *delay* between when A moves and when the force $\vec{F}_{A\,on\,B}$ responds?

Neither Coulomb's law nor Newton's law of gravity is dependent on time, so the answer from the perspective of Newtonian physics has to be "instantly." Yet most scientists found this troubling. What if A is 100,000 light years from B? Will B respond *instantly* to an event 100,000 light years away? The idea of instantaneous transmission of forces had become unbelievable to most scientists by the beginning of the 19th century. But if there is a delay, how long is it? How does the information to "change force" get sent from A to B? These were the issues when a young Michael Faraday appeared on the scene.

Michael Faraday is one of the most interesting figures in the history of science. Because of the late age at which he started his education—he was a teenager—he never became fluent in mathematics. In place of equations, Faraday's brilliant and insightful mind developed many ingenious *pictorial* methods for thinking about and describing physical phenomena. By far the most important of these was the field.

The Concept of a Field

Faraday was particularly impressed with the pattern that iron filings make when sprinkled around a magnet, as seen in **FIGURE 22.21**. The pattern's regularity and the curved lines suggested to Faraday that the *space itself* around the magnet is filled with some kind of magnetic influence. Perhaps the magnet in some way alters the space around it. In this view, a piece of iron near the magnet responds not directly to the magnet but, instead, to the alteration of space caused by the magnet. This space alteration, whatever it is, is the *mechanism* by which the long-range force is exerted.

FIGURE 22.22 illustrates Faraday's idea. The Newtonian view was that A and B interact directly. In Faraday's view, A first alters or modifies the space around it, and particle B then comes along and interacts with this altered space. The alteration of space becomes the *agent* by which A and B interact. Furthermore, this alteration could easily be imagined to take a finite time to propagate outward from A, perhaps in a wave-like fashion. If A changes, B responds only when the new alteration of space reaches it. The interaction between B and this alteration of space is a *local* interaction, rather like a contact force.

Faraday's idea came to be called a **field.** The term "field," which comes from mathematics, describes a function that assigns a vector to every point in space. When used in physics, a field conveys the idea that the physical entity exists at every point in space. That is, indeed, what Faraday was suggesting about how long-range forces operate. The charge makes an alteration *everywhere* in space. Other charges then respond to the alteration at their position. The alteration of the space around a mass is called the *gravitational field.* Similarly, the space around a charge is altered to create the **electric field.**

> **NOTE** The concept of a field is in sharp contrast to the concept of a particle. A particle exists at *one* point in space. The purpose of Newton's laws of motion is to determine how the particle moves from point to point along a trajectory. A field exists simultaneously at *all* points in space.

Faraday's idea was not taken seriously at first; it seemed too vague and nonmathematical to scientists steeped in the Newtonian tradition of particles and forces. But the significance of the concept of field grew as electromagnetic theory developed during the first half of the 19th century. What seemed at first a pictorial "gimmick" came to be seen as more and more essential for understanding electric and magnetic forces.

The Field Model

The basic idea is that **the electric field is the agent that exerts an electric force on a charged particle.** Or, if you prefer, that charged particles interact via the electric field. We postulate:

FIGURE 22.20 If charge A moves, how long does it take the force vector on B to respond?

FIGURE 22.21 Iron filings sprinkled around the ends of a magnet suggest that the influence of the magnet extends into the space around it.

FIGURE 22.22 Newton's and Faraday's ideas about long-range forces.

In the Newtonian view, A exerts a force directly on B.

In Faraday's view, A alters the space around it. (The wavy lines are poetic license. We don't know what the alteration looks like.)

Particle B then responds to the altered space. The altered space is the agent that exerts the force on B.

1. Some set of charges, which we call the **source charges,** alters the space around them by creating an electric field $\vec{E}$ at all points in space.
2. A separate charge q *in* the electric field experiences force $\vec{F} = q\vec{E}$ exerted *by the field.* The force on a positive charge is in the direction of $\vec{E}$; the force on a negative charge is directed opposite to $\vec{E}$.

Thus the source charges exert an electric force on q through the electric field that they've created.

We can learn about the electric field by measuring the force on a *probe charge q*. If, as in **FIGURE 22.23a**, we place a probe charge at position (x, y, z) and measure force $\vec{F}_{\text{on } q}$, then the electric field at that point is

$$\vec{E}(x, y, z) = \frac{\vec{F}_{\text{on } q} \text{ at } (x, y, z)}{q} \qquad (22.5)$$

We're *defining* the electric field as a force-to-charge ratio; hence the units of electric field are newtons per coulomb, or N/C. The magnitude E of the electric field is called the **electric field strength.**

It is important to recognize that probe charge q allows us to *observe* the field, but q is not responsible for creating the field. The field was already there, created by the source charges. **FIGURE 22.23b** shows the field at this point in space after probe charge q has been removed. You could imagine "mapping out" the field by moving charge q all through space.

Because q appears in Equation 22.5, you might think that the electric field depends on the size of the charge used to probe it. It doesn't. Coulomb's law tells us that $\vec{F}_{\text{on } q}$ is proportional to q, so the electric field defined in Equation 22.5 is *independent* of the charge that probes it. The electric field depends only on the source charges that create it.

We can summarize these important ideas with the **field model** of charge interactions:

FIGURE 22.23 Charge q is a probe of the electric field.

(a) If the probe charge feels an electric force . . .

q Position (x, y, z)

(b)

$\vec{E}$

. . . then there's an electric field at this point in space.

MODEL 22.2

Electric field

Charges interact via the electric field.

- The electric force on a charge is exerted by the electric field.
- The electric field is created by other charges, the **source charges.**
  - The electric force is a vector.
  - The field exists at all points in space.
  - A charge does not feel its own field.
- If the electric field at a point in space is $\vec{E}$, a particle with charge q experiences an electric force $\vec{F}_{\text{on } q} = q\vec{E}$.
  - The force on a positive charge is in the direction of the field.
  - The force on a negative charge is opposite the direction of the field.

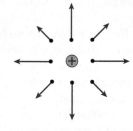

EXAMPLE 22.6 | **Electric forces in a cell**

Every cell in your body is electrically active in various ways. For example, nerve propagation occurs when large electric fields in the cell membranes of neurons cause ions to move through the cell walls. The field strength in a typical cell membrane is 1.0×10^7 N/C. What is the magnitude of the electric force on a singly charged calcium ion?

MODEL The ion is a point charge in an electric field. A singly charged ion is missing one electron and has net charge $q = +e$.

SOLVE A charged particle in an electric field experiences an electric force $\vec{F}_{\text{on } q} = q\vec{E}$. In this case, the magnitude of the force is

$$F = eE = (1.6 \times 10^{-19}\,\text{C})(1.0 \times 10^7\,\text{N/C}) = 1.6 \times 10^{-12}\,\text{N}$$

ASSESS This may seem like an incredibly tiny force, but it is applied to a particle with mass $m \sim 10^{-26}$ kg. The ion would have an unimaginable acceleration ($F/m \sim 10^{14}$ m/s^2) were it not for resistive forces as it moves through the membrane. Even so, an ion can cross the cell wall in less than 1 μs.

STOP TO THINK 22.5 An electron is placed at the position marked by the dot. The force on the electron is

a. Zero.
b. To the right.
c. To the left.
d. There's not enough information to tell.

The Electric Field of a Point Charge

We will begin to put the definition of the electric field to full use in the next chapter. For now, to develop the ideas, we will determine the electric field of a single point charge q. **FIGURE 22.24a** shows charge q and a point in space at which we would like to know the electric field. To do so, we use a second charge q' as a probe of the electric field.

For the moment, assume both charges are positive. The force on q', which is repulsive and directed straight away from q, is given by Coulomb's law:

$$\vec{F}_{\text{on } q'} = \left(\frac{1}{4\pi\epsilon_0} \frac{qq'}{r^2}, \text{ away from } q \right) \tag{22.6}$$

It's customary to use $1/4\pi\epsilon_0$ rather than K for field calculations. Equation 22.5 defined the electric field in terms of the force on a probe charge; thus the electric field at this point is

$$\vec{E} = \frac{\vec{F}_{\text{on } q'}}{q'} = \left(\frac{1}{4\pi\epsilon_0} \frac{q}{r^2}, \text{ away from } q \right) \tag{22.7}$$

The electric field is shown in **FIGURE 22.24b**.

> **NOTE** The expression for the electric field is similar to Coulomb's law. To distinguish the two, remember that Coulomb's law has a product of two charges in the numerator. It describes the force between *two* charges. The electric field has a single charge in the numerator. It is the field of *a* charge.

If we calculate the field at a sufficient number of points in space, we can draw a **field diagram** such as the one shown in **FIGURE 22.25**. Notice that the field vectors all point straight away from charge q. Also notice how quickly the arrows decrease in length due to the inverse-square dependence on r.

Keep these three important points in mind when using field diagrams:

1. The diagram is just a representative sample of electric field vectors. The field exists at all the other points. A well-drawn diagram can tell you fairly well what the field would be like at a neighboring point.
2. The arrow indicates the direction and the strength of the electric field *at the point to which it is attached*—that is, at the point where the *tail* of the vector is placed. In this chapter, we indicate the point at which the electric field is measured with a dot. The length of any vector is significant only relative to the lengths of other vectors.
3. Although we have to draw a vector across the page, from one point to another, an electric field vector is *not* a spatial quantity. It does not "stretch" from one point to another. Each vector represents the electric field at *one point* in space.

Unit Vector Notation

Equation 22.7 is precise, but it's not terribly convenient. Furthermore, what happens if the source charge q is negative? We need a more concise notation to write the electric field, a notation that will allow q to be either positive or negative.

The basic need is to express "away from q" in mathematical notation. "Away from q" is a *direction* in space. To guide us, recall that we already have a notation for

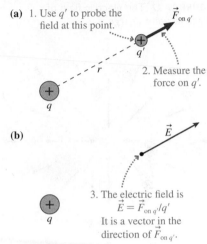

FIGURE 22.24 Charge q' is used to probe the electric field of point charge q.

(a) 1. Use q' to probe the field at this point.
$\vec{F}_{\text{on } q'}$
q'
2. Measure the force on q'.

(b)
$\vec{E}$
3. The electric field is $\vec{E} = \vec{F}_{\text{on } q'}/q'$
It is a vector in the direction of $\vec{F}_{\text{on } q'}$.

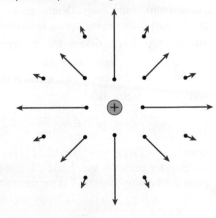

FIGURE 22.25 The electric field of a positive point charge.

FIGURE 22.26 Using the unit vector $\hat{r}$.

Electric field at point 1 is in the direction of $\hat{r}_1$.

The unit vectors specify the directions to the points.

$\vec{E}_2$ is in the direction of $\hat{r}_2$.

FIGURE 22.27 The electric field of a negative point charge.

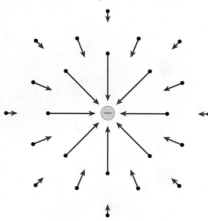

expressing certain directions—namely, the unit vectors $\hat{i}$, $\hat{j}$, and $\hat{k}$. For example, unit vector $\hat{i}$ means "in the direction of the positive x-axis." With a minus sign, $-\hat{i}$ means "in the direction of the negative x-axis." Unit vectors, with a magnitude of 1 and no units, provide purely directional information.

With this in mind, let's define the unit vector $\hat{r}$ to be a vector of length 1 that points from the origin to a point of interest. Unit vector $\hat{r}$ provides no information at all about the distance to the point; it merely specifies the direction.

FIGURE 22.26 shows unit vectors $\hat{r}_1$, $\hat{r}_2$, and $\hat{r}_3$ pointing toward points 1, 2, and 3. Unlike $\hat{i}$ and $\hat{j}$, unit vector $\hat{r}$ does not have a fixed direction. Instead, unit vector $\hat{r}$ specifies the direction "straight outward from this point." But that's just what we need to describe the electric field vector, which is shown at points 1, 2, and 3 due to a positive charge at the origin. No matter which point you choose, the electric field at that point is "straight outward" from the charge. In other words, the electric field $\vec{E}$ points in the direction of the unit vector $\hat{r}$.

With this notation, the electric field at distance r from a point charge q is

$$\vec{E} = \frac{1}{4\pi\epsilon_0} \frac{q}{r^2} \hat{r} \qquad \text{(electric field of a point charge)} \qquad (22.8)$$

where $\hat{r}$ is the unit vector from the charge toward the point at which we want to know the field. Equation 22.8 is identical to Equation 22.7, but written in a notation in which the unit vector $\hat{r}$ expresses the idea "away from q."

Equation 22.8 works equally well if q is negative. A negative sign in front of a vector simply reverses its direction, so the unit vector $-\hat{r}$ points *toward* charge q. **FIGURE 22.27** shows the electric field of a negative point charge. It looks like the electric field of a positive point charge except that the vectors point inward, toward the charge, instead of outward.

We'll end this chapter with three examples of the electric field of a point charge. Chapter 23 will expand these ideas to the electric fields of multiple charges and of extended objects.

EXAMPLE 22.7 | **Calculating the electric field**

A -1.0 nC charged particle is located at the origin. Points 1, 2, and 3 have (x, y) coordinates $(1\,\text{cm}, 0\,\text{cm})$, $(0\,\text{cm}, 1\,\text{cm})$, and $(1\,\text{cm}, 1\,\text{cm})$, respectively. Determine the electric field $\vec{E}$ at these points, then show the vectors on an electric field diagram.

MODEL The electric field is that of a negative point charge.

VISUALIZE The electric field points straight *toward* the origin. It will be weaker at $(1\,\text{cm}, 1\,\text{cm})$, which is farther from the charge.

SOLVE The electric field is

$$\vec{E} = \frac{1}{4\pi\epsilon_0} \frac{q}{r^2} \hat{r}$$

where $q = -1.0\,\text{nC} = -1.0 \times 10^{-9}\,\text{C}$. The distance r is $1.0\,\text{cm} = 0.010\,\text{m}$ for points 1 and 2 and $(\sqrt{2} \times 1.0\,\text{cm}) = 0.0141\,\text{m}$ for point 3. The *magnitude* of $\vec{E}$ at the three points is

$$E_1 = E_2 = \frac{1}{4\pi\epsilon_0} \frac{|q|}{r_1^2}$$

$$= \frac{(9.0 \times 10^9\,\text{N m}^2/\text{C}^2)(1.0 \times 10^{-9}\,\text{C})}{(0.010\,\text{m})^2} = 90{,}000\,\text{N/C}$$

$$E_3 = \frac{1}{4\pi\epsilon_0} \frac{|q|}{r_3^2}$$

$$= \frac{(9.0 \times 10^9\,\text{N m}^2/\text{C}^2)(1.0 \times 10^{-9}\,\text{C})}{(0.0141\,\text{m})^2} = 45{,}000\,\text{N/C}$$

Because q is negative, the field at each of these positions points directly at charge q. The electric field vectors, in component form, are

$$\vec{E}_1 = -90{,}000\,\hat{i}\,\text{N/C}$$

$$\vec{E}_2 = -90{,}000\,\hat{j}\,\text{N/C}$$

$$\vec{E}_3 = -E_3 \cos 45°\hat{i} - E_3 \sin 45°\hat{j} = (-31{,}800\,\hat{i} - 31{,}800\,\hat{j})\,\text{N/C}$$

These vectors are shown on the electric field diagram of **FIGURE 22.28**.

FIGURE 22.28 The electric field diagram of a -1.0 nC charged particle.

EXAMPLE 22.8 | The electric field of a proton

The electron in a hydrogen atom orbits the proton at a radius of 0.053 nm.

a. What is the proton's electric field strength at the position of the electron?

b. What is the magnitude of the electric force on the electron?

SOLVE a. The proton's charge is $q = e$. Its electric field strength at the distance of the electron is

$$E = \frac{1}{4\pi\epsilon_0} \frac{e}{r^2} = \frac{1}{4\pi\epsilon_0} \frac{1.6 \times 10^{-19}\,\text{C}}{(5.3 \times 10^{-11}\,\text{m})^2} = 5.1 \times 10^{11}\,\text{N/C}$$

Note how large this field is compared to the field of Example 22.7.

b. We could use Coulomb's law to find the force on the electron, but the whole point of knowing the electric field is that we can use it directly to find the force on a charge in the field. The magnitude of the force on the electron is

$$F_{\text{on elec}} = |q_e| E_{\text{of proton}}$$
$$= (1.60 \times 10^{-19}\,\text{C})(5.1 \times 10^{11}\,\text{N/C})$$
$$= 8.2 \times 10^{-8}\,\text{N}$$

STOP TO THINK 22.6 Rank in order, from largest to smallest, the electric field strengths E_a to E_d at points a to d.

CHALLENGE EXAMPLE 22.9 | A charge in static equilibrium

A horizontal electric field causes the charged ball in FIGURE 22.29 to hang at a 15° angle, as shown. The spring is plastic, so it doesn't discharge the ball, and in its equilibrium position the spring extends only to the vertical dashed line. What is the electric field strength?

FIGURE 22.29 A charged ball hanging in static equilibrium.

MODEL Model the ball as a point charge in static equilibrium. The electric force on the ball is $\vec{F}_E = q\vec{E}$. The charge is positive, so the force is in the same direction as the field.

VISUALIZE FIGURE 22.30 is a free-body diagram for the ball.

SOLVE The ball is in equilibrium, so the net force on the ball must be zero. With the field applied, the spring is stretched by $\Delta x = L \sin\theta = (0.60\,\text{m})(\sin 15°) = 0.16\,\text{m}$, where L is the string length, and exerts a pulling force $F_{\text{Sp}} = k\,\Delta x$ to the left. Newton's first law, with $\vec{a} = \vec{0}$ for equilibrium, is

$$\sum F_x = F_E - F_{\text{Sp}} - T\sin\theta = 0$$

$$\sum F_y = T\cos\theta - F_G = T\cos\theta - mg = 0$$

FIGURE 22.30 The free-body diagram.

From the y-equation,

$$T = \frac{mg}{\cos\theta}$$

The x-equation is then

$$qE - k\,\Delta x - mg\tan\theta = 0$$

We can solve this for the electric field strength:

$$E = \frac{mg\tan\theta + k\,\Delta x}{q}$$
$$= \frac{(0.0030\,\text{kg})(9.8\,\text{m/s}^2)\tan 15° + (0.050\,\text{N/m})(0.16\,\text{m})}{20 \times 10^{-9}\,\text{C}}$$
$$= 7.9 \times 10^5\,\text{N/C}$$

ASSESS We don't yet have a way of judging whether this is a reasonable field strength, but we'll see in the next chapter that this is typical of the electric field strength near an object that has been charged by rubbing.

SUMMARY

The goal of Chapter 22 has been to learn that electric phenomena are based on charges, forces, and fields.

GENERAL PRINCIPLES

Coulomb's Law

The forces between two charged particles q_1 and q_2 separated by distance r are

$$F_{1 \text{ on } 2} = F_{2 \text{ on } 1} = \frac{K|q_1||q_2|}{r^2}$$

The forces are repulsive for two like charges, attractive for two opposite charges.

To solve electrostatic force problems:

MODEL Model objects as point charges.

VISUALIZE Draw a picture showing charges and force vectors.

SOLVE Use Coulomb's law and the vector addition of forces.

ASSESS Is the result reasonable?

IMPORTANT CONCEPTS

The Charge Model

There are two kinds of charge, positive and negative.

- Fundamental charges are protons and electrons, with charge $\pm e$ where $e = 1.60 \times 10^{-19}$ C.

- Objects are charged by adding or removing electrons.

- The amount of charge is $q = (N_p - N_e)e$.

- An object with an equal number of protons and electrons is **neutral**, meaning no *net* charge.

Charged objects exert electric forces on each other.

- Like charges repel, opposite charges attract.

- The force increases as the charge increases.

- The force decreases as the distance increases.

There are two types of material, insulators and conductors.

- Charge remains fixed in or on an insulator.

- Charge moves easily through or along conductors.

- Charge is transferred by contact between objects.

Charged objects attract neutral objects.

- Charge polarizes metal by shifting the electron sea.

- Charge polarizes atoms, creating electric dipoles.

- The **polarization** force is always an attractive force.

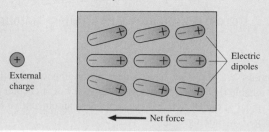

The Field Model

Charges interact with each other via the **electric field** $\vec{E}$.

- Charge A alters the space around it by creating an electric field.

- The field is the agent that exerts a force. The force on charge q_B is $\vec{F}_{\text{on B}} = q_B\vec{E}$.

An electric field is identified and measured in terms of the force on a **probe charge** q:

$$\vec{E} = \vec{F}_{\text{on } q}/q$$

- The electric field exists at all points in space.

- An electric field vector shows the field only at one point, the point at the tail of the vector.

The electric field of a **point charge** is

$$\vec{E} = \frac{1}{4\pi\epsilon_0} \frac{q}{r^2} \hat{r}$$

Unit vector $\hat{r}$ indicates "away from q."

TERMS AND NOTATION

neutral	electron cloud	electrostatic equilibrium	coulomb, C
charging	fundamental unit of charge, e	grounded	permittivity constant, ϵ_0
charge model	charge quantization	charge polarization	field
charge, q or Q	ionization	polarization force	electric field, $\vec{E}$
like charges	law of conservation of charge	electric dipole	source charge
opposite charges	sea of electrons	charging by induction	electric field strength, E
discharging	ion core	Coulomb's law	field model
conductor	current	electrostatic constant, K	field diagram
insulator	charge carrier	point charge	

CONCEPTUAL QUESTIONS

1. Can an insulator be charged? If so, how would you charge an insulator? If not, why not?

2. Can a conductor be charged? If so, how would you charge a conductor? If not, why not?

3. Four lightweight balls A, B, C, and D are suspended by threads. Ball A has been touched by a plastic rod that was rubbed with wool. When the balls are brought close together, without touching, the following observations are made:
   - Balls B, C, and D are attracted to ball A.
   - Balls B and D have no effect on each other.
   - Ball B is attracted to ball C.

   What are the charge states (glass, plastic, or neutral) of balls A, B, C, and D? Explain.

4. Charged plastic and glass rods hang by threads.
   a. An object repels the plastic rod. Can you predict what it will do to the glass rod? If so, what? If not, why not?
   b. A different object attracts the plastic rod. Can you predict what it will do to the glass rod? If so, what? If not, why not?

5. A lightweight metal ball hangs by a thread. When a charged rod is held near, the ball moves toward the rod, touches the rod, then quickly "flies away" from the rod. Explain this behavior.

6. A plastic balloon that has been rubbed with wool will stick to a wall. Can you conclude that the wall is charged? If so, where does the charge come from? If not, why does the balloon stick?

7. Suppose there exists a third type of charge in addition to the two types we've called glass and plastic. Call this third type X charge. What experiment or series of experiments would you use to test whether an object has X charge? State clearly how each possible outcome of the experiments is to be interpreted.

8. The two oppositely charged metal spheres in FIGURE Q22.8 have equal quantities of charge. They are brought into contact with a neutral metal rod. What is the final charge state of each sphere and of the rod? Use both charge diagrams and words to explain.

FIGURE Q22.8

FIGURE Q22.9

9. Metal sphere A in FIGURE Q22.9 has 4 units of negative charge and metal sphere B has 2 units of positive charge. The two spheres are brought into contact. What is the final charge state of each sphere? Explain.

10. A negatively charged electroscope has separated leaves.
    a. Suppose you bring a negatively charged rod close to the top of the electroscope, but not touching. How will the leaves respond? Use both charge diagrams and words to explain.
    b. How will the leaves respond if you bring a positively charged rod close to the top of the electroscope, but not touching? Use both charge diagrams and words to explain.

11. Metal spheres A and B in FIGURE Q22.11 are initially neutral and are touching. A positively charged rod is brought near A, but not touching. Is A now positive, negative, or neutral? Use both charge diagrams and words to explain.

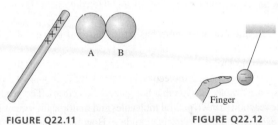

FIGURE Q22.11   FIGURE Q22.12

12. If you bring your finger near a lightweight, negatively charged hanging ball, the ball swings over toward your finger as shown in FIGURE Q22.12. Use charge diagrams and words to explain this observation.

13. Reproduce FIGURE Q22.13 on your paper. Then draw a dot (or dots) on the figure to show the position (or positions) where an electron would experience no net force.

FIGURE Q22.13

14. Charges A and B in FIGURE Q22.14 are equal. Each charge exerts a force on the other of magnitude F. Suppose the charge of B is increased by a factor of 4, but everything else is unchanged. In terms of F, (a) what is the magnitude of the force on A, and (b) what is the magnitude of the force on B?

FIGURE Q22.14

15. The electric force on a charged particle in an electric field is F. What will be the force if the particle's charge is tripled and the electric field strength is halved?

EXERCISES AND PROBLEMS

Problems labeled ▆▆ integrate material from earlier chapters.

Exercises

Section 22.1 The Charge Model

Section 22.2 Charge

1. | A glass rod is charged to +8.0 nC by rubbing.
   a. Have electrons been removed from the rod or protons added?
   b. How many electrons have been removed or protons added?
2. | A plastic rod is charged to −12 nC by rubbing.
   a. Have electrons been added to the rod or protons removed?
   b. How many electrons have been added or protons removed?
3. | A plastic rod that has been charged to −15 nC touches a metal sphere. Afterward, the rod's charge is −10 nC.
   a. What kind of charged particle was transferred between the rod and the sphere, and in which direction? That is, did it move from the rod to the sphere or from the sphere to the rod?
   b. How many charged particles were transferred?
4. | A glass rod that has been charged to +12 nC touches a metal sphere. Afterward, the rod's charge is +8.0 nC.
   a. What kind of charged particle was transferred between the rod and the sphere, and in which direction? That is, did it move from the rod to the sphere or from the sphere to the rod?
   b. How many charged particles were transferred?
5. ‖ What is the total charge of all the electrons in 1.0 L of liquid water?
6. ‖ What mass of aluminum has a total nuclear charge of 1.0 C? Aluminum has atomic number 13.
7. | A chemical reaction takes place among 3 molecular ions that have each lost 2 electrons, 2 molecular ions that have each gained 3 electrons, and 1 molecular ion that has gained 2 electrons. The products of the reaction are two neutral molecules and multiple molecular ions that each have a charge of magnitude e. How many molecular ions are produced, and are they charged positively or negatively?
8. | A *linear accelerator* uses alternating electric fields to accelerate electrons to close to the speed of light. A small number of the electrons collide with a target, but a large majority pass through the target and impact a *beam dump* at the end of the accelerator. In one experiment the beam dump measured charge accumulating at a rate of −2.0 nC/s. How many electrons traveled down the accelerator during the 2.0 h run?

Section 22.3 Insulators and Conductors

9. | Figure 22.8 showed how an electroscope becomes negatively charged. The leaves will also repel each other if you touch the electroscope with a positively charged glass rod. Use a series of charge diagrams to explain what happens and why the leaves repel each other.
10. | Two neutral metal spheres on wood stands are touching. A negatively charged rod is held directly above the top of the left sphere, not quite touching it. While the rod is there, the right sphere is moved so that the spheres no longer touch. Then the rod is withdrawn. Afterward, what is the charge state of each sphere? Use charge diagrams to explain your answer.
11. ‖ You have two neutral metal spheres on wood stands. Devise a procedure for charging the spheres so that they will have like charges of *exactly* equal magnitude. Use charge diagrams to explain your procedure.
12. ‖ You have two neutral metal spheres on wood stands. Devise a procedure for charging the spheres so that they will have opposite charges of *exactly* equal magnitude. Use charge diagrams to explain your procedure.

Section 22.4 Coulomb's Law

13. | Two 1.0 kg masses are 1.0 m apart (center to center) on a frictionless table. Each has +10 μC of charge.
   a. What is the magnitude of the electric force on one of the masses?
   b. What is the initial acceleration of this mass if it is released and allowed to move?
14. ‖ Two small plastic spheres each have a mass of 2.0 g and a charge of −50.0 nC. They are placed 2.0 cm apart (center to center).
   a. What is the magnitude of the electric force on each sphere?
   b. By what factor is the electric force on a sphere larger than its weight?
15. ‖ A small glass bead has been charged to +20 nC. A small metal ball bearing 1.0 cm above the bead feels a 0.018 N downward electric force. What is the charge on the ball bearing?
16. | Two protons are 2.0 fm apart.
   a. What is the magnitude of the electric force on one proton due to the other proton?
   b. What is the magnitude of the gravitational force on one proton due to the other proton?
   c. What is the ratio of the electric force to the gravitational force?
17. ‖ What is the net electric force on charge A in **FIGURE EX22.17**?

FIGURE EX22.17   **FIGURE EX22.18**

18. ‖ What is the net electric force on charge B in **FIGURE EX22.18**?
19. ‖ What is the force $\vec{F}$ on the 1.0 nC charge in **FIGURE EX22.19**? Give your answer as a magnitude and a direction.

FIGURE EX22.19   **FIGURE EX22.20**

20. ‖ What is the force $\vec{F}$ on the 1.0 nC charge in **FIGURE EX22.20**? Give your answer as a magnitude and a direction.

21. | Object A, which has been charged to +4.0 nC, is at the origin. Object B, which has been charged to −8.0 nC, is at $(x, y) = (0.0 \text{ cm}, 2.0 \text{ cm})$. Determine the electric force on each object. Write each force vector in component form.

22. | A small plastic bead has been charged to −15 nC. What are the magnitude and direction of the acceleration of (a) a proton and (b) an electron that is 1.0 cm from the center of the bead?

23. | A 2.0 g plastic bead charged to −4.0 nC and a 4.0 g glass bead charged to +8.0 nC are 2.0 cm apart and free to move. What are the accelerations of (a) the plastic bead and (b) the glass bead?

24. ‖ Two positive point charges q and $4q$ are at $x = 0$ and $x = L$, respectively, and free to move. A third charge is placed so that the entire three-charge system is in static equilibrium. What are the magnitude, sign, and x-coordinate of the third charge?

25. ‖ A massless spring is attached to a support at one end and has a 2.0 μC charge glued to the other end. A −4.0 μC charge is slowly brought near. The spring has stretched 1.2 cm when the charges are 2.6 cm apart. What is the spring constant of the spring?

Section 22.5 The Electric Field

26. | What are the strength and direction of the electric field 1.0 mm from (a) a proton and (b) an electron?

27. ‖ The electric field at a point in space is $\vec{E} = (400\hat{\imath} + 100\hat{\jmath})$ N/C.
   a. What is the electric force on a proton at this point? Give your answer in component form.
   b. What is the electric force on an electron at this point? Give your answer in component form.
   c. What is the magnitude of the proton's acceleration?
   d. What is the magnitude of the electron's acceleration?

28. | What are the strength and direction of the electric field 4.0 cm from a small plastic bead that has been charged to −8.0 nC?

29. ‖ What magnitude charge creates a 1.0 N/C electric field at a point 1.0 m away?

30. ‖ What are the strength and direction of an electric field that will balance the weight of a 1.0 g plastic sphere that has been charged to −3.0 nC?

31. ‖ The electric field 2.0 cm from a small object points away from the object with a strength of 270,000 N/C. What is the object's charge?

32. ‖ A +12 nC charge is located at the origin.
   a. What are the electric fields at the positions $(x, y) = $ (5.0 cm, 0 cm), (−5.0 cm, 5.0 cm), and (−5.0 cm, −5.0 cm)? Write each electric field vector in component form.
   b. Draw a field diagram showing the electric field vectors at these points.

33. ‖ A −12 nC charge is located at $(x, y) = (1.0 \text{ cm}, 0 \text{ cm})$. What are the electric fields at the positions $(x, y) = $ (5.0 cm, 0 cm), (−5.0 cm, 0 cm), and (0 cm, 5.0 cm)? Write each electric field vector in component form.

34. | A 0.10 g honeybee acquires a charge of +23 pC while flying.
BIO   a. The earth's electric field near the surface is typically (100 N/C, downward). What is the ratio of the electric force on the bee to the bee's weight?
   b. What electric field (strength and direction) would allow the bee to hang suspended in the air?

Problems

35. ‖ Pennies today are copper-covered zinc, but older pennies are 3.1 g of solid copper. What are the total positive charge and total negative charge in a solid copper penny that is electrically neutral?

36. ‖ Two 1.0 g spheres are charged equally and placed 2.0 cm apart. When released, they begin to accelerate at 150 m/s². What is the magnitude of the charge on each sphere?

37. ‖ The nucleus of a ^{125}Xe atom (an isotope of the element xenon with mass 125 u) is 6.0 fm in diameter. It has 54 protons and charge $q = +54e$.
   a. What is the electric force on a proton 2.0 fm from the surface of the nucleus?
   b. What is the proton's acceleration?
   Hint: Treat the spherical nucleus as a point charge.

38. ‖ A *Van de Graaff generator* is a device that accumulates electrons on a large metal sphere until the large amount of charge causes sparks. As you'll learn in Chapter 23, the electric field of a charged sphere is exactly the same as if the charge were a point charge at the center of the sphere. Suppose that a 25-cm-diameter sphere has accumulated 1.0×10^{13} extra electrons and that a small ball 50 cm from the edge of the sphere feels the force $\vec{F} = (9.2 \times 10^{-4} \text{ N, away from sphere})$. What is the charge on the ball?

39. ‖ A smart phone charger delivers charge to the phone, in the form of electrons, at a rate of −0.75 C/s. How many electrons are delivered to the phone during 30 min of charging?

40. ‖ Objects A and B are both positively charged. Both have a mass of 100 g, but A has twice the charge of B. When A and B are placed 10 cm apart, B experiences an electric force of 0.45 N.
   a. What is the charge on A?
   b. If the objects are released, what is the initial acceleration of A?

41. ‖ What is the force $\vec{F}$ on the −10 nC charge in **FIGURE P22.41**? Give your answer as a magnitude and an angle measured cw or ccw (specify which) from the +x-axis.

FIGURE P22.41   FIGURE P22.42

42. ‖ What is the force $\vec{F}$ on the −10 nC charge in **FIGURE P22.42**? Give your answer as a magnitude and an angle measured cw or ccw (specify which) from the +x-axis.

43. ‖ What is the force $\vec{F}$ on the 5.0 nC charge in **FIGURE P22.43**? Give your answer as a magnitude and an angle measured cw or ccw (specify which) from the +x-axis.

FIGURE P22.43   FIGURE P22.44

44. ‖ What is the force $\vec{F}$ on the −1.0 nC charge in the middle of **FIGURE P22.44** due to the four other charges? Give your answer in component form.

45. ‖ What is the force $\vec{F}$ on the 1.0 nC charge at the bottom in FIGURE P22.45? Give your answer in component form.

FIGURE P22.45   **FIGURE P22.46**

46. ‖ What is the force $\vec{F}$ on the 1.0 nC charge at the bottom in FIGURE P22.46? Give your answer in component form.

47. ‖‖ A +2.0 nC charge is at the origin and a −4.0 nC charge is at $x = 1.0$ cm.
    a. At what x-coordinate could you place a proton so that it would experience no net force?
    b. Would the net force be zero for an electron placed at the same position? Explain.

48. ‖ The net force on the 1.0 nC charge in FIGURE P22.48 is zero. What is q?

FIGURE P22.48   **FIGURE P22.49**

49. ‖ Charge q_2 in FIGURE P22.49 is in equilibrium. What is q_1?

50. ‖ A positive point charge Q is located at $x = a$ and a negative point charge $-Q$ is at $x = -a$. A positive charge q can be placed anywhere on the y-axis. Find an expression for $(F_{net})_x$, the x-component of the net force on q.

51. ‖ FIGURE P22.51 shows four charges at the corners of a square of side L. What is the magnitude of the net force on q?

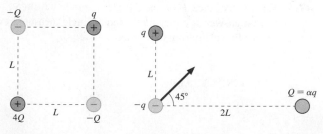

FIGURE P22.51   **FIGURE P22.52**

52. ‖ FIGURE P22.52 shows three charges and the net force on charge $-q$. Charge Q is some multiple α of q. What is α?

53. ‖‖ Suppose the magnitude of the proton charge differs from the magnitude of the electron charge by a mere 1 part in 10^9.
    a. What would be the force between two 2.0-mm-diameter copper spheres 1.0 cm apart? Assume that each copper atom has an equal number of electrons and protons.
    b. Would this amount of force be detectable? What can you conclude from the fact that no such forces are observed?

54. ‖ In a simple model of the hydrogen atom, the electron moves in a circular orbit of radius 0.053 nm around a stationary proton. How many revolutions per second does the electron make?

55. ‖ You have two small, 2.0 g balls that have been given equal but opposite charges, but you don't know the magnitude of the charge. To find out, you place the balls distance d apart on a slippery horizontal surface, release them, and use a motion detector to measure the initial acceleration of one of the balls toward the other. After repeating this for several different separation distances, your data are as follows:

Distance (cm)	Acceleration (m/s^2)
2.0	0.74
3.0	0.30
4.0	0.19
5.0	0.10

Use an appropriate graph of the data to determine the magnitude of the charge.

56. ‖ A 2.0 g metal cube and a 4.0 g metal cube are 6.0 cm apart, measured between their centers, on a horizontal surface. For both, the coefficient of static friction is 0.65. Both cubes, initially neutral, are charged at a rate of 7.0 nC/s. How long after charging begins does one cube begin to slide away? Which cube moves first?

57. ‖ Space explorers discover an 8.7×10^{17} kg asteroid that happens to have a positive charge of 4400 C. They would like to place their 3.3×10^5 kg spaceship in orbit around the asteroid. Interestingly, the solar wind has given their spaceship a charge of −1.2 C. What speed must their spaceship have to achieve a 7500-km-diameter circular orbit?

58. ‖ Two equal point charges 2.5 cm apart, both initially neutral,
CALC are being charged at the rate of 5.0 nC/s. At what rate (N/s) is the force between them increasing 1.0 s after charging begins?

59. ‖ You have a lightweight spring whose unstretched length is 4.0 cm. First, you attach one end of the spring to the ceiling and hang a 1.0 g mass from it. This stretches the spring to a length of 5.0 cm. You then attach two small plastic beads to the opposite ends of the spring, lay the spring on a frictionless table, and give each plastic bead the same charge. This stretches the spring to a length of 4.5 cm. What is the magnitude of the charge (in nC) on each bead?

60. ‖ An electric dipole consists of two opposite charges $\pm q$ separated by a small distance s. The product $p = qs$ is called the *dipole moment*. FIGURE P22.60 shows an electric dipole perpendicular to an electric field $\vec{E}$. Find an expression in terms of p and E for the magnitude of the torque that the electric field exerts on the dipole.

FIGURE P22.60   Electric dipole

61. ‖ You sometimes create a spark when you touch a doorknob after shuffling your feet on a carpet. Why? The air always has a few free electrons that have been kicked out of atoms by cosmic rays. If an electric field is present, a free electron is accelerated until it collides with an air molecule. Most such collisions are elastic, so the electron collides, accelerates, collides, accelerates, and so on, gradually gaining speed. But if the electron's kinetic energy just before a collision is 2.0×10^{-18} J or more, it has sufficient energy to kick an electron out of the molecule it hits. Where there was one free electron, now there are two! Each of these can then accelerate,

hit a molecule, and kick out another electron. Then there will be four free electrons. In other words, as **FIGURE P22.61** shows, a sufficiently strong electric field causes a "chain reaction" of electron production. This is called a *breakdown* of the air. The current of moving electrons is what gives you the shock, and a spark is generated when the electrons recombine with the positive ions and give off excess energy as a burst of light.

a. The average distance between ionizing collisions is 2.0 μm. (The electron's mean free path is less than this, but most collisions are elastic collisions in which the electron bounces with no loss of energy.) What acceleration must an electron have to gain 2.0×10^{-18} J of kinetic energy in this distance?

b. What force must act on an electron to give it the acceleration found in part a?

c. What strength electric field will exert this much force on an electron? This is the *breakdown field strength*. **Note:** The measured breakdown field strength is a little less than your calculated value because our model of the process is a bit too simple. Even so, your calculated value is close.

d. Suppose a free electron in air is 1.0 cm away from a point charge. What minimum charge is needed to cause a breakdown and create a spark as the electron moves toward the point charge?

Breakdown of the air   2.0 μm

FIGURE P22.61

62. ‖ Two 5.0 g point charges on 1.0-m-long threads repel each other after being charged to +100 nC, as shown in **FIGURE P22.62**. What is the angle θ? You can assume that θ is a small angle.

FIGURE P22.62

63. ‖ What are the electric fields at points 1, 2, and 3 in **FIGURE P22.63**? Give your answer in component form.

FIGURE P22.63

FIGURE P22.64

64. ‖ What are the electric fields at points 1, 2, and 3 in **FIGURE P22.64**? Give your answer in component form.

65. ‖‖ A 10.0 nC charge is located at position $(x, y) = (1.0$ cm, 2.0 cm). At what (x, y) position(s) is the electric field
   a. $-225,000\,\hat{\imath}$ N/C?
   b. $(161,000\,\hat{\imath} + 80,500\,\hat{\jmath})$ N/C?
   c. $(21,600\,\hat{\imath} - 28,800\,\hat{\jmath})$ N/C?

66. ‖ Three 1.0 nC charges are placed as shown in **FIGURE P22.66**. Each of these charges creates an electric field $\vec{E}$ at a point 3.0 cm in front of the middle charge.

FIGURE P22.66

   a. What are the three fields $\vec{E}_1$, $\vec{E}_2$, and $\vec{E}_3$ created by the three charges? Write your answer for each as a vector in component form.
   b. Do you think that electric fields obey a principle of superposition? That is, is there a "net field" at this point given by $\vec{E}_{net} = \vec{E}_1 + \vec{E}_2 + \vec{E}_3$? Use what you learned in this chapter and previously in our study of forces to argue why this is or is not true.
   c. If it is true, what is $\vec{E}_{net}$?

67. ‖ An electric field $\vec{E} = 100,000\,\hat{\imath}$ N/C causes the 5.0 g point charge in **FIGURE P22.67** to hang at a 20° angle. What is the charge on the ball?

FIGURE P22.67   **FIGURE P22.68**

68. ‖ An electric field $\vec{E} = 200,000\,\hat{\imath}$ N/C causes the point charge in **FIGURE P25.68** to hang at an angle. What is θ?

In Problems 69 through 72 you are given the equation(s) used to solve a problem. For each of these,
   a. Write a realistic problem for which this is the correct equation(s).
   b. Finish the solution of the problem.

69. $$\frac{(9.0 \times 10^9 \text{ N m}^2/\text{C}^2) \times N \times (1.60 \times 10^{-19} \text{ C})}{(1.0 \times 10^{-6} \text{ m})^2}$$

$$= 1.5 \times 10^6 \text{ N/C}$$

70. $$\frac{(9.0 \times 10^9 \text{ N m}^2/\text{C}^2)q^2}{(0.0150 \text{ m})^2} = 0.020 \text{ N}$$

71. $$\frac{(9.0 \times 10^9 \text{ N m}^2/\text{C}^2)(15 \times 10^{-9} \text{ C})}{r^2} = 54,000 \text{ N/C}$$

72. $$\sum F_x = 2 \times \frac{(9.0 \times 10^9 \text{ N m}^2/\text{C}^2)(1.0 \times 10^{-9} \text{ C})q}{((0.020 \text{ m})/\sin 30°)^2} \times \cos 30°$$

$$= 5.0 \times 10^{-5} \text{ N}$$

$$\sum F_y = 0 \text{ N}$$

Challenge Problems

73. ⫼ Two 3.0 g point charges on 1.0-m-long threads repel each other after being equally charged, as shown in **FIGURE CP22.73**. What is the magnitude of the charge q?

FIGURE CP22.73

74. ⫼ Three 3.0 g balls are tied to 80-cm-long threads and hung from a *single* fixed point. Each of the balls is given the same charge q. At equilibrium, the three balls form an equilateral triangle in a horizontal plane with 20 cm sides. What is q?

75. ⫼ The identical small spheres shown in **FIGURE CP22.75** are charged to $+100$ nC and -100 nC. They hang as shown in a 100,000 N/C electric field. What is the mass of each sphere?

FIGURE CP22.75

76. ⫼ The force on the -1.0 nC charge is as shown in **FIGURE CP22.76**. What is the magnitude of this force?

FIGURE CP22.76 10 nC

77. ⫼ In Section 22.3 we claimed that a charged object exerts a net attractive force on an electric dipole. Let's investigate this. **FIGURE CP22.77** shows a *permanent* electric dipole consisting of charges $+q$ and $-q$ separated by the fixed distance s. Charge $+Q$ is distance r from the center of the dipole. We'll assume, as is usually the case in practice, that $s \ll r$.

a. Write an expression for the net force exerted on the dipole by charge $+Q$.

b. Is this force toward $+Q$ or away from $+Q$? Explain.

c. Use the *binomial approximation* $(1 + x)^{-n} \approx 1 - nx$ if $x \ll 1$ to show that your expression from part a can be written $F_{net} = 2KqQs/r^3$.

d. How can an electric force have an inverse-cube dependence? Doesn't Coulomb's law say that the electric force depends on the inverse square of the distance? Explain.

FIGURE CP22.77

23 The Electric Field

In a plasma ball, electrons follow the electric field lines outward from the center electrode. The streamers appear where gas atoms emit light after the high-speed electrons collide with them.

IN THIS CHAPTER, you will learn how to calculate and use the electric field.

Where do electric fields come from?

Electric fields are created by charges.

- Electric fields add. The field due to several point charges is the sum of the fields due to each charge.
- Electric fields are *vectors*. Summing electric fields is vector addition.
- Two equal but opposite charges form an electric dipole.
- Electric fields can be represented by electric field vectors or electric field lines.

Field of a dipole

« LOOKING BACK Section 22.5 The electric field of a point charge

What if the charge is continuous?

For macroscopic charged objects, like rods or disks, we can think of the charge as having a continuous distribution.

- A charged object is characterized by its charge density—the charge per length, area, or volume.
- We'll divide objects into small point charge-like pieces ΔQ.
- The summation of their electric fields will become an integral.
- We'll calculate the electric fields of charged rods, loops, disks, and planes.

Total charge Q

ΔQ

L

Linear charge density $\lambda = Q/L$

What fields are especially important?

We will develop and use four important electric field models.

Point charge   Line of charge   Plane of charge   Sphere of charge

What is a parallel-plate capacitor?

Two parallel conducting plates with equal but opposite charges form a parallel-plate capacitor. You'll learn that the electric field between the plates is a uniform electric field, the same at every point. Capacitors are also important elements of circuits, as you'll see in Chapter 26.

How do charges respond to fields?

Electric fields exert forces on charges.

- Charged particles *accelerate*. Acceleration depends on the charge-to-mass ratio.
- A charged particle in a uniform field follows a parabolic trajectory.
- A dipole in an electric field feels a torque that aligns the dipole with the field.

« LOOKING BACK Section 4.2 Projectiles

23.1 Electric Field Models

Chapter 22 introduced the key idea that **charged particles interact via the electric field.** In this chapter you will learn how to calculate the electric field of a set of charges. We will start with discrete point charges, then move on to calculating the electric field of a continuous distribution of charge. The latter will let you practice the mathematical skills you've been learning in calculus. Only at the end of the chapter will we look at what happens to charges that find themselves *in* an electric field.

The electric fields used in science and engineering are often caused by fairly complex distributions of charge. Sometimes these fields require exact calculations, but much of the time we can understand the physics by using simplified models of the electric field. A *model,* you'll recall, is a highly simplified picture of reality, one that captures the essence of what we want to study without unnecessary complications.

Four common electric field models are the basis for understanding a wide variety of electric phenomena. We present them here together as a reference; the first half of this chapter will then be devoted to justifying and explaining these results.

MODEL 23.1

Four key electric fields

A point charge:
- Small charged objects

$$\vec{E} = \frac{1}{4\pi\epsilon_0} \frac{q}{r^2} \hat{r}$$

An infinitely long line of charge:
- Wires

$$\vec{E} = \left(\frac{1}{4\pi\epsilon_0} \frac{2|\lambda|}{r}, \begin{cases} \text{away if } + \\ \text{toward if } - \end{cases} \right)$$

An infinitely wide plane of charge:
- Capacitors

$$\vec{E} = \left(\frac{\eta}{2\epsilon_0}, \begin{cases} \text{away if } + \\ \text{toward if } - \end{cases} \right)$$

A sphere of charge:
- Electrodes

$$\vec{E} = \frac{1}{4\pi\epsilon_0} \frac{Q}{r^2} \hat{r} \text{ for } r > R$$

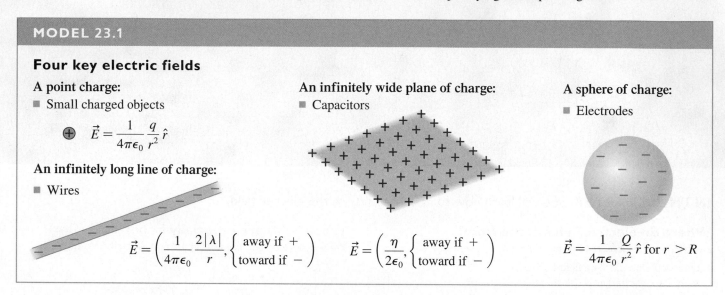

23.2 The Electric Field of Point Charges

Our starting point, from « Section 22.5, is the electric field of a point charge *q:*

$$\vec{E} = \frac{1}{4\pi\epsilon_0} \frac{q}{r^2} \hat{r} \qquad \text{(electric field of a point charge)} \tag{23.1}$$

where $\hat{r}$ is a unit vector pointing away from *q* and $\epsilon_0 = 8.85 \times 10^{-12}$ C^2/N m^2 is the permittivity constant. **FIGURE 23.1** reminds you of the electric fields of point charges. Although we have to give each vector we draw a length, keep in mind that each arrow represents the electric field *at a point.* The electric field is not a spatial quantity that "stretches" from one end of the arrow to the other.

▶ **FIGURE 23.1** The electric field of a positive and a negative point charge.

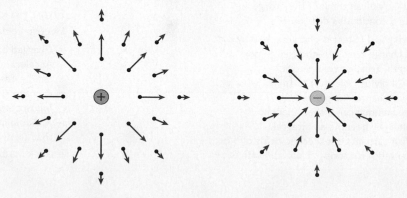

Multiple Point Charges

What happens if there is more than one charge? The electric field was defined as $\vec{E} = \vec{F}_{\text{on }q}/q$, where $\vec{F}_{\text{on }q}$ is the electric force on charge q. Forces add as vectors, so the net force on q due to a group of point charges is the vector sum

$$\vec{F}_{\text{on }q} = \vec{F}_{1\text{ on }q} + \vec{F}_{2\text{ on }q} + \cdots$$

Consequently, the net electric field due to a group of point charges is

$$\vec{E}_{\text{net}} = \frac{\vec{F}_{\text{on }q}}{q} = \frac{\vec{F}_{1\text{ on }q}}{q} + \frac{\vec{F}_{2\text{ on }q}}{q} + \cdots = \vec{E}_1 + \vec{E}_2 + \cdots = \sum_i \vec{E}_i \qquad (23.2)$$

where $\vec{E}_i$ is the field from point charge i. That is, **the net electric field is the *vector sum* of the electric fields due to each charge.** In other words, electric fields obey the *principle of superposition*.

Thus vector addition is the key to electric field calculations.

PROBLEM-SOLVING STRATEGY 23.1   (MP)

The electric field of multiple point charges

MODEL Model charged objects as point charges.

VISUALIZE For the pictorial representation:

- Establish a coordinate system and show the locations of the charges.
- Identify the point P at which you want to calculate the electric field.
- Draw the electric field of each charge at P.
- Use symmetry to determine if any components of $\vec{E}_{\text{net}}$ are zero.

SOLVE The mathematical representation is $\vec{E}_{\text{net}} = \sum \vec{E}_i$.

- For each charge, determine its distance from P and the angle of $\vec{E}_i$ from the axes.
- Calculate the field strength of each charge's electric field.
- Write each vector $\vec{E}_i$ in component form.
- Sum the vector components to determine $\vec{E}_{\text{net}}$.

ASSESS Check that your result has correct units and significant figures, is reasonable (see **TABLE 23.1**), and agrees with any known limiting cases.

TABLE 23.1 Typical electric field strengths

Field location	Field strength (N/C)
Inside a current-carrying wire	10^{-3}–10^{-1}
Near the earth's surface	10^2–10^4
Near objects charged by rubbing	10^3–10^6
Electric breakdown in air, causing a spark	3×10^6
Inside an atom	10^{11}

EXAMPLE 23.1 | The electric field of three equal point charges

Three equal point charges q are located on the y-axis at $y = 0$ and at $y = \pm d$. What is the electric field at a point on the x-axis?

MODEL This problem is a step along the way to understanding the electric field of a charged wire. We'll assume that q is positive when drawing pictures, but the solution should allow for the possibility that q is negative. The question does not ask about any specific point, so we will be looking for a symbolic expression in terms of the unspecified position x.

VISUALIZE **FIGURE 23.2** shows the charges, the coordinate system, and the three electric field vectors $\vec{E}_1$, $\vec{E}_2$, and $\vec{E}_3$. Each of these fields points *away from* its source charge because of the assumption that q is positive. We need to find the vector sum $\vec{E}_{\text{net}} = \vec{E}_1 + \vec{E}_2 + \vec{E}_3$.

Before rushing into a calculation, we can make our task *much* easier by first thinking qualitatively about the situation. For example, the fields $\vec{E}_1$, $\vec{E}_2$, and $\vec{E}_3$ all lie in the xy-plane, hence we can conclude without any calculations that $(E_{\text{net}})_z = 0$. Next, look

FIGURE 23.2 Calculating the electric field of three equal point charges.

at the y-components of the fields. The fields $\vec{E}_1$ and $\vec{E}_3$ have equal magnitudes and are tilted away from the x-axis by the same angle θ. Consequently, the y-components of $\vec{E}_1$ and $\vec{E}_3$ will *cancel* when added. $\vec{E}_2$ has no y-component, so we can conclude that $(E_{\text{net}})_y = 0$. The only component we need to calculate is $(E_{\text{net}})_x$.

Continued

SOLVE We're ready to calculate. The x-component of the field is

$$(E_{net})_x = (E_1)_x + (E_2)_x + (E_3)_x = 2(E_1)_x + (E_2)_x$$

where we used the fact that fields $\vec{E}_1$ and $\vec{E}_3$ have *equal* x-components. Vector $\vec{E}_2$ has *only* the x-component

$$(E_2)_x = E_2 = \frac{1}{4\pi\epsilon_0}\frac{q_2}{r_2^2} = \frac{1}{4\pi\epsilon_0}\frac{q}{x^2}$$

where $r_2 = x$ is the distance from q_2 to the point at which we are calculating the field. Vector $\vec{E}_1$ is at angle θ from the x-axis, so its x-component is

$$(E_1)_x = E_1\cos\theta = \frac{1}{4\pi\epsilon_0}\frac{q_1}{r_1^2}\cos\theta$$

where r_1 is the distance from q_1. This expression for $(E_1)_x$ is correct, but it is not yet sufficient. Both the distance r_1 and the angle θ vary with the position x and need to be expressed as functions of x. From the Pythagorean theorem, $r_1 = (x^2 + d^2)^{1/2}$. Thus

$$\cos\theta = \frac{x}{r_1} = \frac{x}{(x^2 + d^2)^{1/2}}$$

By combining these pieces, we see that $(E_1)_x$ is

$$(E_1)_x = \frac{1}{4\pi\epsilon_0}\frac{q}{x^2 + d^2}\frac{x}{(x^2 + d^2)^{1/2}} = \frac{1}{4\pi\epsilon_0}\frac{xq}{(x^2 + d^2)^{3/2}}$$

This expression is a bit complex, but notice that the dimensions of $x/(x^2 + d^2)^{3/2}$ are $1/m^2$, as they *must* be for the field of a point charge. Checking dimensions is a good way to verify that you haven't made algebra errors.

We can now combine $(E_1)_x$ and $(E_2)_x$ to write

$$(E_{net})_x = 2(E_1)_x + (E_2)_x = \frac{q}{4\pi\epsilon_0}\left[\frac{1}{x^2} + \frac{2x}{(x^2 + d^2)^{3/2}}\right]$$

The other two components of $\vec{E}_{net}$ are zero, hence the electric field of the three charges at a point on the x-axis is

$$\vec{E}_{net} = \frac{q}{4\pi\epsilon_0}\left[\frac{1}{x^2} + \frac{2x}{(x^2 + d^2)^{3/2}}\right]\hat{\imath}$$

ASSESS This is the electric field only at points *on the x-axis*. Furthermore, this expression is valid only for $x > 0$. The electric field to the left of the charges points in the opposite direction, but our expression doesn't change sign for negative x. (This is a consequence of how we wrote $(E_2)_x$.) We would need to modify this expression to use it for negative values of x. The good news, though, is that our expression *is* valid for both positive and negative q. A negative value of q makes $(E_{net})_x$ negative, which would be an electric field pointing to the left, toward the negative charges.

Limiting Cases

There are two cases for which we know what the result should be. First, let x become really small. As the point in Figure 23.2 approaches the origin, the fields $\vec{E}_1$ and $\vec{E}_3$ become opposite to each other and cancel. Thus as $x \to 0$, the field *should* be that of the single point charge q at the origin, a field we already know. Is it? Notice that

$$\lim_{x\to 0}\frac{2x}{(x^2 + d^2)^{3/2}} = 0 \tag{23.3}$$

Thus $E_{net} \to q/4\pi\epsilon_0 x^2$ as $x \to 0$, the expected field of a single point charge.

Now consider the opposite situation, where x becomes extremely large. From very far away, the three source charges will seem to merge into a single charge of size $3q$, just as three very distant lightbulbs appear to be a single light. Thus the field for $x \gg d$ *should* be that of a point charge $3q$. Is it?

The field is zero in the limit $x \to \infty$. That doesn't tell us much, so we don't want to go *that* far away. We simply want x to be very large in comparison to the spacing d between the source charges. If $x \gg d$, then the denominator of the second term of $\vec{E}_{net}$ is well approximated by $(x^2 + d^2)^{3/2} \approx (x^2)^{3/2} = x^3$. Thus

$$\lim_{x\gg d}\left[\frac{1}{x^2} + \frac{2x}{(x^2 + d^2)^{3/2}}\right] = \frac{1}{x^2} + \frac{2x}{x^3} = \frac{3}{x^2} \tag{23.4}$$

Consequently, the net electric field far from the source charges is

$$\vec{E}_{net}(x \gg d) = \frac{1}{4\pi\epsilon_0}\frac{(3q)}{x^2}\hat{\imath} \tag{23.5}$$

As expected, this is the field of a point charge $3q$. These checks of limiting cases provide confidence in the result of the calculation.

FIGURE 23.3 is a graph of the field strength E_{net} for the three charges of Example 23.1. Although we don't have any numerical values, we can specify x as a multiple of the charge separation d. Notice how the graph matches the field of a single point charge when $x \ll d$ and matches the field of a point charge $3q$ when $x \gg d$.

FIGURE 23.3 The electric field strength along a line perpendicular to three equal point charges.

The electric field matches that of a single point charge q when $x \ll d$.

Electric field of point charge $3q$

Electric field of point charge q

The electric field matches that of point charge $3q$ when $x \gg d$.

The Electric Field of a Dipole

Two equal but opposite charges separated by a small distance form an *electric dipole.* **FIGURE 23.4** shows two examples. In a *permanent electric dipole,* such as the water molecule, the oppositely charged particles maintain a small permanent separation. We can also create an electric dipole, as you learned in Chapter 22, by polarizing a neutral atom with an external electric field. This is an *induced electric dipole.*

FIGURE 23.5 shows that we can represent an electric dipole, whether permanent or induced, by two opposite charges $\pm q$ separated by the small distance s. The dipole has zero net charge, but it *does* have an electric field. Consider a point on the positive y-axis. This point is slightly closer to $+q$ than to $-q$, so the fields of the two charges do not cancel. You can see in the figure that $\vec{E}_{\text{dipole}}$ points in the positive y-direction. Similarly, vector addition shows that $\vec{E}_{\text{dipole}}$ points in the negative y-direction at points along the x-axis.

Let's calculate the electric field of a dipole at a point on the axis of the dipole. This is the y-axis in Figure 23.5. The point is distance $r_+ = y - s/2$ from the positive charge and $r_- = y + s/2$ from the negative charge. The net electric field at this point has only a y-component, and the sum of the fields of the two point charges gives

$$(E_{\text{dipole}})_y = (E_+)_y + (E_-)_y = \frac{1}{4\pi\epsilon_0}\frac{q}{(y-\frac{1}{2}s)^2} + \frac{1}{4\pi\epsilon_0}\frac{(-q)}{(y+\frac{1}{2}s)^2}$$

$$= \frac{q}{4\pi\epsilon_0}\left[\frac{1}{(y-\frac{1}{2}s)^2} - \frac{1}{(y+\frac{1}{2}s)^2}\right] \quad (23.6)$$

Combining the two terms over a common denominator, we find

$$(E_{\text{dipole}})_y = \frac{q}{4\pi\epsilon_0}\left[\frac{2ys}{(y-\frac{1}{2}s)^2(y+\frac{1}{2}s)^2}\right] \quad (23.7)$$

We omitted some of the algebraic steps, but be sure you can do this yourself. Some of the homework problems will require similar algebra.

In practice, we almost always observe the electric field of a dipole at distances $y \gg s$—that is, for distances much larger than the charge separation. In such cases, the denominator can be approximated $(y-\frac{1}{2}s)^2(y+\frac{1}{2}s)^2 \approx y^4$. With this approximation, Equation 23.7 becomes

$$(E_{\text{dipole}})_y \approx \frac{1}{4\pi\epsilon_0}\frac{2qs}{y^3} \quad (23.8)$$

It is useful to define the **dipole moment** $\vec{p}$, shown in **FIGURE 23.6**, as the vector

$$\vec{p} = (qs, \text{ from the negative to the positive charge}) \quad (23.9)$$

The direction of $\vec{p}$ identifies the orientation of the dipole, and the dipole-moment magnitude $p = qs$ determines the electric field strength. The SI units of the dipole moment are $C\,m$.

We can use the dipole moment to write a succinct expression for the electric field at a point on the axis of a dipole:

$$\vec{E}_{\text{dipole}} \approx \frac{1}{4\pi\epsilon_0}\frac{2\vec{p}}{r^3} \quad \text{(on the axis of an electric dipole)} \quad (23.10)$$

where r is the distance measured from the *center* of the dipole. We've switched from y to r because we've now specified that Equation 23.10 is valid only along the axis of the dipole. Notice that the electric field along the axis points in the direction of the dipole moment $\vec{p}$.

A homework problem will let you calculate the electric field in the plane that bisects the dipole. This is the field shown on the x-axis in Figure 23.5, but it could equally well be the field on the z-axis as it comes out of the page. The field, for $r \gg s$, is

$$\vec{E}_{\text{dipole}} \approx -\frac{1}{4\pi\epsilon_0}\frac{\vec{p}}{r^3} \quad \text{(bisecting plane)} \quad (23.11)$$

FIGURE 23.4 Permanent and induced electric dipoles.

A water molecule is a *permanent* dipole because the negative electrons spend more time with the oxygen atom.

This dipole is *induced,* or stretched, by the electric field acting on the + and − charges.

FIGURE 23.5 The dipole electric field at two points.

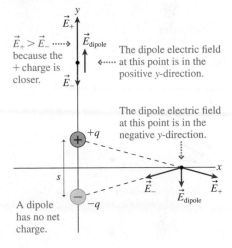

$\vec{E}_+ > \vec{E}_-$ because the + charge is closer.

The dipole electric field at this point is in the positive y-direction.

The dipole electric field at this point is in the negative y-direction.

A dipole has no net charge.

FIGURE 23.6 The dipole moment.

The dipole moment $\vec{p}$ is a vector pointing from the negative to the positive charge with magnitude qs.

This field is *opposite* to $\vec{p}$, and it is only half the strength of the on-axis field at the same distance.

> **NOTE** Do these inverse-cube equations violate Coulomb's law? Not at all. Coulomb's law describes the force between two *point charges,* and from Coulomb's law we found that the electric field of a *point charge* varies with the inverse square of the distance. But a dipole is not a point charge. The field of a dipole decreases more rapidly than that of a point charge, which is to be expected because the dipole is, after all, electrically neutral.

EXAMPLE 23.2 | **The electric field of a water molecule**

The water molecule H_2O has a permanent dipole moment of magnitude 6.2×10^{-30} C m. What is the electric field strength 1.0 nm from a water molecule at a point on the dipole's axis?

MODEL The size of a molecule is ≈ 0.1 nm. Thus $r \gg s$, and we can use Equation 23.10 for the on-axis electric field of the molecule's dipole moment.

SOLVE The on-axis electric field strength at $r = 1.0$ nm is

$$E \approx \frac{1}{4\pi\epsilon_0}\frac{2p}{r^3} = (9.0 \times 10^9\,\text{N m}^2/\text{C}^2)\frac{2(6.2 \times 10^{-30}\,\text{C m})}{(1.0 \times 10^{-9}\,\text{m})^3}$$

$$= 1.1 \times 10^8\,\text{N/C}$$

ASSESS By referring to Table 23.1 you can see that the field strength is "strong" compared to our everyday experience with charged objects but "weak" compared to the electric field inside the atoms themselves. This seems reasonable.

Electric Field Lines

We can't see the electric field. Consequently, we need pictorial tools to help us visualize it in a region of space. One method, introduced in Chapter 22, is to picture the electric field by drawing electric field vectors at various points in space.

Another common way to picture the field is to draw **electric field lines.** As **FIGURE 23.7** shows,

FIGURE 23.7 Electric field lines.

- Electric field lines are *continuous* curves tangent to the electric field vectors.
- Closely spaced field lines indicate a greater field strength; widely spaced field lines indicate a smaller field strength.
- Electric field lines start on positive charges and end on negative charges.
- Electric field lines never cross.

The third bullet point follows from the fact that electric fields are created by charge. However, we will have to modify this idea in Chapter 30 when we find another way to create an electric field.

FIGURE 23.8 shows three electric fields represented by electric field lines. Notice that the electric field of a dipole points in the direction of $\vec{p}$ (left to right) on both sides of the dipole, but points opposite to $\vec{p}$ in the bisecting plane.

FIGURE 23.8 The electric field lines of (a) a positive point charge, (b) a negative point charge, and (c) a dipole.

(a)

(b)

(c)

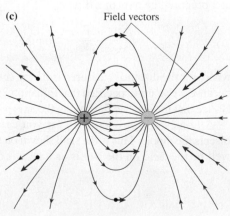

Neither field-vector diagrams nor field-line diagrams are perfect pictorial representations of an electric field. The field vectors are somewhat harder to draw, and they show the field at only a few points, but they do clearly indicate the direction and strength of the electric field at those points. Field-line diagrams perhaps look more elegant, and they're sometimes easier to sketch, but there's no formula for knowing where to draw the lines. We'll use both field-vector diagrams and field-line diagrams, depending on the circumstances.

STOP TO THINK 23.1   At the dot, the electric field points

a. Left.
b. Right.
c. Up.
d. Down.
e. The electric field is zero.

 •

23.3 The Electric Field of a Continuous Charge Distribution

Ordinary objects—tables, chairs, beakers of water—seem to our senses to be continuous distributions of matter. There is no obvious evidence for atoms, even though we have good reasons to believe that we would find atoms if we subdivided the matter sufficiently far. Thus it is easier, for many practical purposes, to consider matter to be continuous and to talk about the *density* of matter. Density—the number of kilograms of matter per cubic meter—allows us to describe the distribution of matter *as if* the matter were continuous rather than atomic.

Much the same situation occurs with charge. If a charged object contains a large number of excess electrons—for example, 10^{12} extra electrons on a metal rod—it is not practical to track every electron. It makes more sense to consider the charge to be *continuous* and to describe how it is distributed over the object.

FIGURE 23.9a shows an object of length L, such as a plastic rod or a metal wire, with charge Q spread uniformly along it. (We will use an uppercase Q for the total charge of an object, reserving lowercase q for individual point charges.) The **linear charge density** λ is defined to be

$$\lambda = \frac{Q}{L} \tag{23.12}$$

Linear charge density, which has units of C/m, is the amount of charge *per meter* of length. The linear charge density of a 20-cm-long wire with 40 nC of charge is 2.0 nC/cm or 2.0×10^{-7} C/m.

NOTE The linear charge density λ is analogous to the linear mass density μ that you used in Chapter 16 to find the speed of a wave on a string.

We'll also be interested in charged surfaces. **FIGURE 23.9b** shows a two-dimensional distribution of charge across a surface of area A. We define the **surface charge density** η (lowercase Greek eta) to be

$$\eta = \frac{Q}{A} \tag{23.13}$$

Surface charge density, with units of C/m², is the amount of charge *per square meter*. A 1.0 mm × 1.0 mm square on a surface with $\eta = 2.0 \times 10^{-4}$ C/m² contains 2.0×10^{-10} C or 0.20 nC of charge. (The volume charge density $\rho = Q/V$, measured in C/m³, will be used in Chapter 24.)

FIGURE 23.9 One-dimensional and two-dimensional continuous charge distributions.

(a) Charge Q on a rod of length L. The linear charge density is $\lambda = Q/L$.

L

ΔL

The charge in a small length ΔL is $\Delta Q = \lambda \Delta L$.

(b) Charge Q on a surface of area A. The surface charge density is $\eta = Q/A$.

Area A

The charge in a small area ΔA is $\Delta Q = \eta \Delta A$.

Figure 23.9 and the definitions of Equations 23.12 and 23.13 assume that the object is **uniformly charged,** meaning that the charges are evenly spread over the object. We will assume objects are uniformly charged unless noted otherwise.

> **NOTE** Some textbooks represent the surface charge density with the symbol σ. Because σ is also used to represent *conductivity,* an idea we'll introduce in Chapter 27, we've selected a different symbol for surface charge density.

STOP TO THINK 23.2 A piece of plastic is uniformly charged with surface charge density η_a. The plastic is then broken into a large piece with surface charge density η_b and a small piece with surface charge density η_c. Rank in order, from largest to smallest, the surface charge densities η_a to η_c.

Integration Is Summation

Calculating the electric field of a continuous charge distribution usually requires setting up and evaluating integrals—a skill you've been learning in calculus. It is common to think that "an integral is the area under a curve," an idea we used in our study of kinematics.

But integration is much more than a tool for finding areas. More generally, **integration is summation.** That is, an integral is a sophisticated way to add an infinite number of infinitesimally small pieces. The area under a curve happens to be a special case in which you're summing the small areas $y(x)\,dx$ of an infinite number of tall, narrow boxes, but the idea of integration as summation has many other applications.

Suppose, for example, that a charged object is divided into a large number of small pieces numbered $i = 1, 2, 3, \ldots, N$ having small quantities of charge $\Delta Q_1, \Delta Q_2, \Delta Q_3, \ldots, \Delta Q_N$. Figure 23.9 showed small pieces of charge for a charged rod and a charged sheet, but the object could have any shape. The total charge on the object is found by *summing* all the small charges:

$$Q = \sum_{i=1}^{N} \Delta Q_i \tag{23.14}$$

If we now let $\Delta Q_i \to 0$ and $N \to \infty$, then we *define* the integral:

$$Q = \lim_{\Delta Q \to 0} \sum_{i=1}^{N} \Delta Q_i = \int_{\text{object}} dQ \tag{23.15}$$

That is, integration is the summing of an infinite number of infinitesimally small pieces of charge. This use of integration has nothing to do with the area under a curve.

Although Equation 23.15 is a formal statement of "add up all the little pieces," it's not yet an expression that can actually be integrated with the tools of calculus. Integrals are carried out over coordinates, such as dx or dy, and we also need coordinates to specify what is meant by "integrate over the object." This is where densities come in.

Suppose we want to find the total charge of a thin, charged rod of length L. First, we establish an x-axis with the origin at one end of the rod, as shown in **FIGURE 23.10**. Then we divide the rod into lots of tiny segments of length dx. Each of these little segments has a charge dQ, and the total charge on the rod is the sum of all the dQ values—that's what Equation 23.15 is saying. Now the critical step: The rod has some linear charge density λ. Consequently, the charge of a small segment of the rod is $dQ = \lambda\,dx$. **Densities are the link between quantities and coordinates.** Finally, "integrate over the rod" means to integrate from $x = 0$ to $x = L$. Thus the total charge on the rod is

$$Q = \int_{\text{rod}} dQ = \int_{0}^{L} \lambda\,dx \tag{23.16}$$

FIGURE 23.10 Setting up an integral to calculate the charge on a rod.

The rod's linear charge density is λ.

The charge in a small length dx is $dQ = \lambda\,dx$.

Now we have an expression that can be integrated. If λ is constant, as it is for a uniformly charged rod, we can take it outside the integral to find $Q = \lambda L$. But we could also use Equation 23.16 for a nonuniformly charged rod where λ is some function of x.

This discussion reveals two key ideas that will be needed for calculating electric fields:

- Integration is the tool for summing a vast number of small pieces.
- Density is the connection between quantities and coordinates.

A Problem-Solving Strategy

Our goal is to find the electric field of a continuous distribution of charge, such as a charged rod or a charged disk. We have two basic tools to work with:

- The electric field of a point charge, and
- The principle of superposition.

We can apply these tools with a three-step strategy:

1. Divide the total charge Q into many small point-like charges ΔQ.
2. Use our knowledge of the electric field of a point charge to find the electric field of each ΔQ.
3. Calculate the net field $\vec{E}_{\text{net}}$ by summing the fields of all the ΔQ.

As you've no doubt guessed, we'll let the sum become an integral.

We will go step by step through several examples to illustrate the procedures. However, we first need to flesh out the steps of the problem-solving strategy. The aim of this problem-solving strategy is to break a difficult problem down into small steps that are individually manageable.

PROBLEM-SOLVING STRATEGY 23.2

The electric field of a continuous distribution of charge

MODEL Model the charge distribution as a simple shape.

VISUALIZE For the pictorial representation:

- Draw a picture, establish a coordinate system, and identify the point P at which you want to calculate the electric field.
- Divide the total charge Q into small pieces of charge ΔQ, using shapes for which you *already know* how to determine $\vec{E}$. This is often, but not always, a division into point charges.
- Draw the electric field vector at P for one or two small pieces of charge. This will help you identify distances and angles that need to be calculated.

SOLVE The mathematical representation is $\vec{E}_{\text{net}} = \sum \vec{E}_i$.

- Write an algebraic expression for *each* of the three components of $\vec{E}$ (unless you are sure one or more is zero) at point P. Let the (x, y, z) coordinates of the point remain variables.
- Replace the small charge ΔQ with an equivalent expression involving a charge density and a coordinate, such as dx. **This is the critical step in making the transition from a sum to an integral** because you need a coordinate to serve as the integration variable.
- Express all angles and distances in terms of the coordinates.
- Let the sum become an integral. The integration limits for this variable must "cover" the entire charged object.

ASSESS Check that your result is consistent with any limits for which you know what the field should be.

Exercise 16

EXAMPLE 23.3 | The electric field of a line of charge

FIGURE 23.11 shows a thin, uniformly charged rod of length L with total charge Q. Find the electric field strength at radial distance r in the plane that bisects the rod.

FIGURE 23.11 A thin, uniformly charged rod.

MODEL The rod is thin, so we'll assume the charge lies along a line and forms what we call a *line of charge*. The rod's linear charge density is $\lambda = Q/L$.

VISUALIZE FIGURE 23.12 illustrates the steps of the problem-solving strategy. We've chosen a coordinate system in which the rod lies along the y-axis and point P, in the bisecting plane, is on the x-axis. We've then divided the rod into N small segments of charge ΔQ, each of which is small enough to model as a point charge. For every ΔQ in the bottom half of the wire with a field that points to the right and up, there's a matching ΔQ in the top half whose field points to the right and down. The y-components of these two fields cancel, hence **the net electric field on the x-axis points straight away from the rod.** The only component we need to calculate is E_x. (This is the same reasoning on the basis of symmetry that we used in Example 23.1.)

FIGURE 23.12 Calculating the electric field of a line of charge.

Draw field vectors at P to establish distances and angles. Note how the y-components from symmetrically opposed charge segments cancel.

SOLVE Each of the little segments of charge can be modeled as a point charge. We know the electric field of a point charge, so we can write the x-component of $\vec{E}_i$, the electric field of segment i, as

$$(E_i)_x = E_i \cos\theta_i = \frac{1}{4\pi\epsilon_0} \frac{\Delta Q}{r_i^2} \cos\theta_i$$

where r_i is the distance from charge i to point P. You can see from the figure that $r_i = (y_i^2 + r^2)^{1/2}$ and $\cos\theta_i = r/r_i = r/(y_i^2 + r^2)^{1/2}$. With these, $(E_i)_x$ is

$$(E_i)_x = \frac{1}{4\pi\epsilon_0} \frac{\Delta Q}{y_i^2 + r^2} \frac{r}{\sqrt{y_i^2 + r^2}}$$

$$= \frac{1}{4\pi\epsilon_0} \frac{r\,\Delta Q}{(y_i^2 + r^2)^{3/2}}$$

Compare this result to the very similar calculation we did in Example 23.1. If we now sum this expression over all the charge segments, the net x-component of the electric field is

$$E_x = \sum_{i=1}^{N} (E_i)_x = \frac{1}{4\pi\epsilon_0} \sum_{i=1}^{N} \frac{r\,\Delta Q}{(y_i^2 + r^2)^{3/2}}$$

This is the same superposition we did for the $N = 3$ case in Example 23.1. The only difference is that we have now written the result as an explicit summation so that N can have any value. We want to let $N \rightarrow \infty$ and to replace the sum with an integral, but we can't integrate over Q; it's not a geometric quantity. This is where the linear charge density enters. The quantity of charge in each segment is related to its length Δy by $\Delta Q = \lambda\,\Delta y = (Q/L)\Delta y$. In terms of the linear charge density, the electric field is

$$E_x = \frac{Q/L}{4\pi\epsilon_0} \sum_{i=1}^{N} \frac{r\,\Delta y}{(y_i^2 + r^2)^{3/2}}$$

Now we're ready to let the sum become an integral. If we let $N \rightarrow \infty$, then each segment becomes an infinitesimal length $\Delta y \rightarrow dy$ while the discrete position variable y_i becomes the continuous integration variable y. The sum from $i = 1$ at the bottom end of the line of charge to $i = N$ at the top end will be replaced with an integral from $y = -L/2$ to $y = +L/2$. Thus in the limit $N \rightarrow \infty$,

$$E_x = \frac{Q/L}{4\pi\epsilon_0} \int_{-L/2}^{L/2} \frac{r\,dy}{(y^2 + r^2)^{3/2}}$$

This is a standard integral that you have learned to do in calculus and that can be found in Appendix A. Note that r is a *constant* as far as this integral is concerned. Integrating gives

$$E_x = \frac{Q/L}{4\pi\epsilon_0} \frac{y}{r\sqrt{y^2 + r^2}} \Big|_{-L/2}^{L/2}$$

$$= \frac{Q/L}{4\pi\epsilon_0} \left[\frac{L/2}{r\sqrt{(L/2)^2 + r^2}} - \frac{-L/2}{r\sqrt{(-L/2)^2 + r^2}} \right]$$

$$= \frac{1}{4\pi\epsilon_0} \frac{Q}{r\sqrt{r^2 + (L/2)^2}}$$

Because E_x is the *only* component of the field, the electric field strength E_{rod} at distance r from the center of a charged rod is

$$E_{\text{rod}} = \frac{1}{4\pi\epsilon_0} \frac{|Q|}{r\sqrt{r^2 + (L/2)^2}}$$

The field strength must be positive, so we added absolute value signs to Q to allow for the possibility that the charge could be negative. The only restriction is to remember that this is the electric field at a point in the plane that bisects the rod.

ASSESS Suppose we are at a point *very* far from the rod. If $r \gg L$, the length of the rod is not relevant and the rod appears to be a point charge Q in the distance. Thus in the *limiting case $r \gg L$*, we expect the rod's electric field to be that of a point charge. If $r \gg L$, the square root becomes $(r^2 + (L/2)^2)^{1/2} \approx (r^2)^{1/2} = r$ and the electric field strength at distance r becomes $E_{\text{rod}} \approx Q/4\pi\epsilon_0 r^2$, the field of a point charge. The fact that our expression of E_{rod} has the correct limiting behavior gives us confidence that we haven't made any mistakes in its derivation.

An Infinite Line of Charge

What if the rod or wire becomes very long, becoming a **line of charge,** while the linear charge density λ remains constant? To answer this question, we can rewrite the expression for E_{rod} by factoring $(L/2)^2$ out of the denominator:

$$E_{rod} = \frac{1}{4\pi\epsilon_0} \frac{|Q|}{r \cdot L/2} \frac{1}{\sqrt{1 + 4r^2/L^2}} = \frac{1}{4\pi\epsilon_0} \frac{2|\lambda|}{r} \frac{1}{\sqrt{1 + 4r^2/L^2}}$$

where $|\lambda| = |Q|/L$ is the magnitude of the linear charge density. If we now let $L \to \infty$, the last term becomes simply 1 and we're left with

$$\vec{E}_{line} = \left(\frac{1}{4\pi\epsilon_0} \frac{2|\lambda|}{r}, \begin{cases} \text{away from line if charge } + \\ \text{toward line if charge } - \end{cases} \right) \quad \text{(line of charge)} \quad (23.17)$$

where we've now included the field's direction. **FIGURE 23.13** shows the electric field vectors of an infinite line of positive charge. The vectors would point inward for a negative line of charge.

NOTE Unlike a point charge, for which the field decreases as $1/r^2$, the field of an infinitely long charged wire decreases more slowly—as only $1/r$.

The infinite line of charge is the second of our important electric field models. Although no real wire is infinitely long, the field of a realistic finite-length wire is well approximated by Equation 23.17 except at points near the end of the wire.

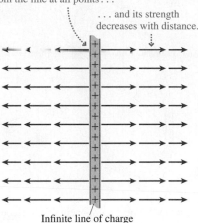

FIGURE 23.13 The electric field of an infinite line of charge.

The field points straight away from the line at all points...

...and its strength decreases with distance.

Infinite line of charge

STOP TO THINK 23.3 Which of the following actions will increase the electric field strength at the position of the dot?

a. Make the rod longer without changing the charge.
b. Make the rod shorter without changing the charge.
c. Make the rod wider without changing the charge.
d. Make the rod narrower without changing the charge.
e. Add charge to the rod.
f. Remove charge from the rod.
g. Move the dot farther from the rod.
h. Move the dot closer to the rod.

Charged rod

23.4 The Electric Fields of Rings, Disks, Planes, and Spheres

In this section we'll derive the electric fields for several important charge distributions.

EXAMPLE 23.4 | The electric field of a ring of charge

A thin ring of radius R is uniformly charged with total charge Q. Find the electric field at a point on the axis of the ring (perpendicular to the ring).

MODEL Because the ring is thin, we'll assume the charge lies along a circle of radius R. You can think of this as a line of charge of length $2\pi R$ wrapped into a circle. The linear charge density along the ring is $\lambda = Q/2\pi R$.

VISUALIZE FIGURE 23.14 on the next page shows the ring and illustrates the steps of the problem-solving strategy. We've chosen a coordinate system in which the ring lies in the xy-plane and point P is on the z-axis. We've then divided the ring into N small segments of charge ΔQ, each of which can be modeled as a point charge. As you can see from the figure, the component of the field perpendicular to the axis cancels for two diametrically opposite segments. Thus we need to calculate only the z-component E_z.

Continued

FIGURE 23.14 Calculating the on-axis electric field of a ring of charge.

Divide the ring into segments.

Segment i with charge ΔQ

The field perpendicular to the axis cancels for two diametrically opposed segments.

SOLVE The z-component of the electric field due to segment i is

$$(E_i)_z = E_i \cos\theta_i = \frac{1}{4\pi\epsilon_0} \frac{\Delta Q}{r_i^2} \cos\theta_i$$

You can see from the figure that *every* segment of the ring, independent of i, has

$$r_i = \sqrt{z^2 + R^2}$$

$$\cos\theta_i = \frac{z}{r_i} = \frac{z}{\sqrt{z^2 + R^2}}$$

Consequently, the field of segment i is

$$(E_i)_z = \frac{1}{4\pi\epsilon_0} \frac{\Delta Q}{z^2 + R^2} \frac{z}{\sqrt{z^2 + R^2}} = \frac{1}{4\pi\epsilon_0} \frac{z}{(z^2 + R^2)^{3/2}} \Delta Q$$

The net electric field is found by summing $(E_i)_z$ due to all N segments:

$$E_z = \sum_{i=1}^{N} (E_i)_z = \frac{1}{4\pi\epsilon_0} \frac{z}{(z^2 + R^2)^{3/2}} \sum_{i=1}^{N} \Delta Q$$

We were able to bring all terms involving z to the front because z is a constant as far as the summation is concerned. Surprisingly, we don't need to convert the sum to an integral to complete this calculation. The sum of all the ΔQ around the ring is simply the ring's total charge, $\sum \Delta Q = Q$, hence the field on the axis is

$$(E_{\text{ring}})_z = \frac{1}{4\pi\epsilon_0} \frac{zQ}{(z^2 + R^2)^{3/2}}$$

This expression is valid for both positive and negative z (i.e., on either side of the ring) and for both positive and negative charge.

ASSESS It will be left as a homework problem to show that this result gives the expected limit when $z \gg R$.

FIGURE 23.15 The on-axis electric field of a ring of charge.

(a)

The field is zero in the center.

Maximum field strength

(b)

$(E_{\text{ring}})_z$

$-2R$　$-R$　R　$2R$　z

FIGURE 23.15 shows two representations of the on-axis electric field of a positively charged ring. Figure 23.15a shows that the electric field vectors point away from the ring, increasing in length until reaching a maximum when $|z| \approx R$, then decreasing. The graph of $(E_{\text{ring}})_z$ in Figure 23.15b confirms that the field strength has a maximum on either side of the ring. Notice that the electric field at the center of the ring is zero, even though this point is surrounded by charge. You might want to spend a minute thinking about why this has to be the case.

A Disk of Charge

FIGURE 23.16 shows a disk of radius R that is uniformly charged with charge Q. This is a mathematical disk, with no thickness, and its surface charge density is

$$\eta = \frac{Q}{A} = \frac{Q}{\pi R^2} \tag{23.18}$$

We would like to calculate the on-axis electric field of this disk. Our problem-solving strategy tells us to divide a continuous charge into segments for which we already know how to find $\vec{E}$. Because we now know the on-axis electric field of a ring of charge, let's divide the disk into N very narrow rings of radius r and width Δr. One such ring, with radius r_i and charge ΔQ_i, is shown.

We need to be careful with notation. The R in Example 23.4 was the radius of the ring. Now we have many rings, and the radius of ring i is r_i. Similarly, Q was the charge on the ring. Now the charge on ring i is ΔQ_i, a small fraction of the total charge on the disk. With these changes, the electric field of ring i, with radius r_i, is

$$(E_i)_z = \frac{1}{4\pi\epsilon_0} \frac{z\,\Delta Q_i}{(z^2 + r_i^2)^{3/2}} \tag{23.19}$$

The on-axis electric field of the charged disk is the sum of the electric fields of all of the rings:

$$(E_{\text{disk}})_z = \sum_{i=1}^{N} (E_i)_z = \frac{z}{4\pi\epsilon_0} \sum_{i=1}^{N} \frac{\Delta Q_i}{(z^2 + r_i^2)^{3/2}} \tag{23.20}$$

The critical step, as always, is to relate ΔQ to a coordinate. Because we now have a surface, rather than a line, the charge in ring i is $\Delta Q = \eta \, \Delta A_i$, where ΔA_i is the area of ring i. We can find ΔA_i, as you've learned to do in calculus, by "unrolling" the ring to form a narrow rectangle of length $2\pi r_i$ and height Δr. Thus the area of ring i is $\Delta A_i = 2\pi r_i \, \Delta r$ and the charge is $\Delta Q_i = 2\pi \eta r_i \, \Delta r$. With this substitution, Equation 23.20 becomes

$$(E_{\text{disk}})_z = \frac{\eta z}{2\epsilon_0} \sum_{i=1}^{N} \frac{r_i \, \Delta r}{(z^2 + r_i^2)^{3/2}} \qquad (23.21)$$

As $N \to \infty$, $\Delta r \to dr$ and the sum becomes an integral. Adding all the rings means integrating from $r = 0$ to $r = R$; thus

$$(E_{\text{disk}})_z = \frac{\eta z}{2\epsilon_0} \int_0^R \frac{r \, dr}{(z^2 + r^2)^{3/2}} \qquad (23.22)$$

All that remains is to carry out the integration. This is straightforward if we make the variable change $u = z^2 + r^2$. Then $du = 2r \, dr$ or, equivalently, $r \, dr = \frac{1}{2} du$. At the lower integration limit $r = 0$, our new variable is $u = z^2$. At the upper limit $r = R$, the new variable is $u = z^2 + R^2$.

> **NOTE** When changing variables in a definite integral, you *must* also change the limits of integration.

With this variable change the integral becomes

$$(E_{\text{disk}})_z = \frac{\eta z}{2\epsilon_0} \frac{1}{2} \int_{z^2}^{z^2+R^2} \frac{du}{u^{3/2}} = \frac{\eta z}{4\epsilon_0} \frac{-2}{u^{1/2}} \bigg|_{z^2}^{z^2+R^2} = \frac{\eta z}{2\epsilon_0} \left[\frac{1}{z} - \frac{1}{\sqrt{z^2 + R^2}} \right] \qquad (23.23)$$

If we multiply through by z, the on-axis electric field of a charged disk with surface charge density $\eta = Q/\pi R^2$ is

$$(E_{\text{disk}})_z = \frac{\eta}{2\epsilon_0} \left[1 - \frac{z}{\sqrt{z^2 + R^2}} \right] \qquad (23.24)$$

> **NOTE** This expression is valid only for $z > 0$. The field for $z < 0$ has the same magnitude but points in the opposite direction.

Limiting Cases

It's a bit difficult to see what Equation 23.24 is telling us, so let's compare it to what we already know. First, you can see that the quantity in square brackets is dimensionless. The surface charge density $\eta = Q/A$ has the same units as q/r^2, so $\eta/2\epsilon_0$ has the same units as $q/4\pi\epsilon_0 r^2$. This tells us that $\eta/2\epsilon_0$ really is an electric field.

Next, let's move very far away from the disk. At distance $z \gg R$, the disk appears to be a point charge Q in the distance and the field of the disk should approach that of a point charge. If we let $z \to \infty$ in Equation 23.24, so that $z^2 + R^2 \approx z^2$, we find $(E_{\text{disk}})_z \to 0$. This is true, but not quite what we wanted. We need to let z be very large in comparison to R, but not so large as to make E_{disk} vanish. That requires a little more care in taking the limit.

We can cast Equation 23.24 into a somewhat more useful form by factoring the z^2 out of the square root to give

$$(E_{\text{disk}})_z = \frac{\eta}{2\epsilon_0} \left[1 - \frac{1}{\sqrt{1 + R^2/z^2}} \right] \qquad (23.25)$$

Now $R^2/z^2 \ll 1$ if $z \gg R$, so the second term in the square brackets is of the form $(1 + x)^{-1/2}$ where $x \ll 1$. We can then use the *binomial approximation*

$$(1 + x)^n \approx 1 + nx \quad \text{if} \quad x \ll 1 \qquad \text{(binomial approximation)}$$

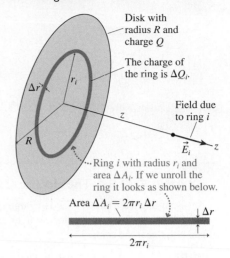

FIGURE 23.16 Calculating the on-axis field of a charged disk.

Disk with radius R and charge Q

The charge of the ring is ΔQ_i.

Field due to ring i

$\vec{E}_i$

Ring i with radius r_i and area ΔA_i. If we unroll the ring it looks as shown below.

Area $\Delta A_i = 2\pi r_i \, \Delta r$

Δr

$2\pi r_i$

to simplify the expression in square brackets:

$$1 - \frac{1}{\sqrt{1 + R^2/z^2}} = 1 - (1 + R^2/z^2)^{-1/2} \approx 1 - \left(1 + \left(-\frac{1}{2}\right)\frac{R^2}{z^2}\right) = \frac{R^2}{2z^2} \qquad (23.26)$$

This is a good approximation when $z \gg R$. Substituting this approximation into Equation 23.25, we find that the electric field of the disk for $z \gg R$ is

$$(E_{\text{disk}})_z \approx \frac{\eta}{2\epsilon_0}\frac{R^2}{2z^2} = \frac{Q/\pi R^2}{4\epsilon_0}\frac{R^2}{z^2} = \frac{1}{4\pi\epsilon_0}\frac{Q}{z^2} \qquad \text{if } z \gg R \qquad (23.27)$$

This is, indeed, the field of a point charge Q, giving us confidence in Equation 23.24 for the on-axis electric field of a disk of charge.

EXAMPLE 23.5 | **The electric field of a charged disk**

A 10-cm-diameter plastic disk is charged uniformly with an extra 10^{11} electrons. What is the electric field 1.0 mm above the surface at a point near the center?

MODEL Model the plastic disk as a uniformly charged disk. We are seeking the on-axis electric field. Because the charge is negative, the field will point *toward* the disk.

SOLVE The total charge on the plastic square is $Q = N(-e) = -1.60 \times 10^{-8}$ C. The surface charge density is

$$\eta = \frac{Q}{A} = \frac{Q}{\pi R^2} = \frac{-1.60 \times 10^{-8}\text{ C}}{\pi (0.050\text{ m})^2} = -2.04 \times 10^{-6}\text{ C/m}^2$$

The electric field at $z = 0.0010$ m, given by Equation 23.25, is

$$E_z = \frac{\eta}{2\epsilon_0}\left[1 - \frac{1}{\sqrt{1 + R^2/z^2}}\right] = -1.1 \times 10^5\text{ N/C}$$

The minus sign indicates that the field points *toward*, rather than away from, the disk. As a vector,

$$\vec{E} = (1.1 \times 10^5\text{ N/C, toward the disk})$$

ASSESS The total charge, -16 nC, is typical of the amount of charge produced on a small plastic object by rubbing or friction. Thus 10^5 N/C is a typical electric field strength near an object that has been charged by rubbing.

A Plane of Charge

Many electronic devices use charged, flat surfaces—disks, squares, rectangles, and so on—to steer electrons along the proper paths. These charged surfaces are called **electrodes.** Although any real electrode is finite in extent, we can often model an electrode as an infinite **plane of charge.** As long as the distance z to the electrode is small in comparison to the distance to the edges, we can reasonably treat the edges *as if* they are infinitely far away.

The electric field of a plane of charge is found from the on-axis field of a charged disk by letting the radius $R \rightarrow \infty$. That is, a disk with infinite radius is an infinite plane. From Equation 23.24, we see that the electric field of a plane of charge with surface charge density η is

$$E_{\text{plane}} = \frac{\eta}{2\epsilon_0} = \text{constant} \qquad (23.28)$$

This is a simple result, but what does it tell us? First, the field strength is directly proportional to the charge density η: more charge, bigger field. Second, and more interesting, the field strength is the same at *all* points in space, independent of the distance z. The field strength 1000 m from the plane is the same as the field strength 1 mm from the plane.

How can this be? It seems that the field should get weaker as you move away from the plane of charge. But remember that we are dealing with an *infinite* plane of charge. What does it mean to be "close to" or "far from" an infinite object? For a disk of finite radius R, whether a point at distance z is "close to" or "far from" the disk is a comparison of z to R. If $z \ll R$, the point is close to the disk. If $z \gg R$, the point is far from the disk. But as $R \rightarrow \infty$, we have no *scale* for distinguishing near and far. In essence, *every* point in space is "close to" a disk of infinite radius.

No real plane is infinite in extent, but we can interpret Equation 23.28 as saying that the field of a surface of charge, regardless of its shape, is a constant $\eta/2\epsilon_0$ for those points whose distance z to the surface is much smaller than their distance to the edge.

We do need to note that the derivation leading to Equation 23.28 considered only $z > 0$. For a positively charged plane, with $\eta > 0$, the electric field points *away from* the plane on both sides of the plane. This requires $E_z < 0$ ($\vec{E}$ pointing in the negative z-direction) on the side with $z < 0$. Thus a complete description of the electric field, valid for both sides of the plane and for either sign of η, is

$$\vec{E}_{\text{plane}} = \left(\frac{|\eta|}{2\epsilon_0}, \begin{cases} \text{away from plane if charge } + \\ \text{toward plane if charge } - \end{cases} \right) \quad \text{(plane of charge)} \quad (23.29)$$

The infinite plane of charge is the third of our important electric field models.

FIGURE 23.17 shows two views of the electric field of a positively charged plane. All the arrows would be reversed for a negatively charged plane. It would have been very difficult to anticipate this result from Coulomb's law or from the electric field of a single point charge, but step by step we have been able to use the concept of the electric field to look at increasingly complex distributions of charge.

A Sphere of Charge

The one last charge distribution for which we need to know the electric field is a **sphere of charge.** This problem is analogous to wanting to know the gravitational field of a spherical planet or star. The procedure for calculating the field of a sphere of charge is the same as we used for lines and planes, but the integrations are significantly more difficult. We will skip the details of the calculations and, for now, simply assert the result without proof. In Chapter 24 we'll use an alternative procedure to find the field of a sphere of charge.

A sphere of charge Q and radius R, be it a uniformly charged sphere or just a spherical shell, has an electric field *outside* the sphere ($r \geq R$) that is exactly the same as that of a point charge Q located at the center of the sphere:

$$\vec{E}_{\text{sphere}} = \frac{Q}{4\pi\epsilon_0 r^2}\hat{r} \quad \text{for } r \geq R \quad (23.30)$$

This assertion is analogous to our earlier assertion that the gravitational force between stars and planets can be computed as if all the mass is at the center.

FIGURE 23.18 shows the electric field of a sphere of positive charge. The field of a negative sphere would point inward.

STOP TO THINK 23.4 Rank in order, from largest to smallest, the electric field strengths E_a to E_e at these five points near a plane of charge.

23.5 The Parallel-Plate Capacitor

FIGURE 23.19 shows two electrodes, one with charge $+Q$ and the other with $-Q$, placed face-to-face a distance d apart. This arrangement of two electrodes, charged equally but oppositely, is called a **parallel-plate capacitor.** Capacitors play important roles in many electric circuits. Our goal is to find the electric field both inside the capacitor (i.e., between the plates) and outside the capacitor.

NOTE The *net* charge of a capacitor is zero. Capacitors are charged by transferring electrons from one plate to the other. The plate that gains N electrons has charge $-Q = N(-e)$. The plate that loses electrons has charge $+Q$.

FIGURE 23.17 Two views of the electric field of a plane of charge.

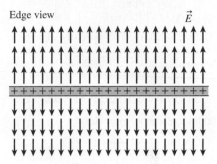

FIGURE 23.18 The electric field of a sphere of positive charge.

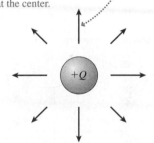

FIGURE 23.19 A parallel-plate capacitor.

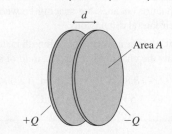

FIGURE 23.20 The electric fields inside and outside a parallel-plate capacitor.

The capacitor's charge resides on the inner surfaces as planes of charge.

Edge view of electrodes

$\vec{E}_+$ $\vec{E}_-$
$\vec{E}_{\text{net}} = \vec{0}$

$\vec{E}_+$
$\vec{E}_-$
$\vec{E}_{\text{net}}$

$\vec{E}_-$ $\vec{E}_+$
$\vec{E}_{\text{net}} = \vec{0}$

d

Outside the capacitor, $\vec{E}_+$ and $\vec{E}_-$ are opposite, so the net field is zero.

Inside the capacitor, $\vec{E}_+$ and $\vec{E}_-$ are parallel, so the net field is large.

FIGURE 23.21 Ideal versus real capacitors.

(a) Ideal capacitor—edge view

The field is uniform

(b) Real capacitor—edge view

FIGURE 23.20 is an enlarged view of the capacitor plates, seen from the side. Because opposite charges attract, all of the charge is on the *inner* surfaces of the two plates. Thus the inner surfaces of the plates can be modeled as two planes of charge with equal but opposite surface charge densities. As you can see from the figure, at all points in space the electric field $\vec{E}_+$ of the positive plate points *away from* the plane of positive charges. Similarly, the field $\vec{E}_-$ of the negative plate everywhere points *toward* the plane of negative charges.

NOTE You might think the right capacitor plate would somehow "block" the electric field created by the positive plate and prevent the presence of an $\vec{E}_+$ field to the right of the capacitor. To see that it doesn't, consider an analogous situation with gravity. The strength of gravity above a table is the same as its strength below it. Just as the table doesn't block the earth's gravitational field, intervening matter or charges do not alter or block an object's electric field.

Outside the capacitor, $\vec{E}_+$ and $\vec{E}_-$ point in opposite directions and, because the field of a plane of charge is independent of the distance from the plane, have equal magnitudes. Consequently, the fields $\vec{E}_+$ and $\vec{E}_-$ add to zero outside the capacitor plates. There's *no* electric field outside an ideal parallel-plate capacitor.

Inside the capacitor, between the electrodes, field $\vec{E}_+$ points from positive to negative and has magnitude $\eta/2\epsilon_0 = Q/2\epsilon_0 A$, where A is the surface area of each electrode. Field $\vec{E}_-$ *also* points from positive to negative and *also* has magnitude $Q/2\epsilon_0 A$, so the inside field $\vec{E}_+ + \vec{E}_-$ is twice that of a plane of charge. Thus the electric field of a parallel-plate capacitor is

$$\vec{E}_{\text{capacitor}} = \begin{cases} \left(\dfrac{Q}{\epsilon_0 A}, \text{from positive to negative} \right) & \text{inside} \\ \vec{0} & \text{outside} \end{cases} \quad (23.31)$$

FIGURE 23.21a shows the electric field—this time using field lines—of an ideal parallel-plate capacitor. Now, it's true that no real capacitor is infinite in extent, but the ideal parallel-plate capacitor is a very good approximation for all but the most precise calculations as long as the electrode separation d is much smaller than the electrodes' size. **FIGURE 23.21b** shows that the interior field of a real capacitor is virtually identical to that of an ideal capacitor but that the exterior field isn't quite zero. This weak field outside the capacitor is called the **fringe field.** We will keep things simple by always assuming the plates are very close together and using Equation 23.31 for the field inside a parallel-plate capacitor.

NOTE The shape of the electrodes—circular or square or any other shape—is not relevant as long as the electrodes are very close together.

EXAMPLE 23.6 | **The electric field inside a capacitor**

Two $1.0 \text{ cm} \times 2.0 \text{ cm}$ rectangular electrodes are 1.0 mm apart. What charge must be placed on each electrode to create a uniform electric field of strength 2.0×10^6 N/C? How many electrons must be moved from one electrode to the other to accomplish this?

MODEL The electrodes can be modeled as an ideal parallel-plate capacitor because the spacing between them is much smaller than their lateral dimensions.

SOLVE The electric field strength inside the capacitor is $E = Q/\epsilon_0 A$. Thus the charge to produce a field of strength E is

$$Q = (8.85 \times 10^{-12} \text{ C}^2/\text{N m}^2)(2.0 \times 10^{-4} \text{ m}^2)(2.0 \times 10^6 \text{ N/C})$$
$$= 3.5 \times 10^{-9} \text{ C} = 3.5 \text{ nC}$$

The positive plate must be charged to $+3.5$ nC and the negative plate to -3.5 nC. In practice, the plates are charged by using a *battery* to move electrons from one plate to the other. The number of electrons in 3.5 nC is

$$N = \frac{Q}{e} = \frac{3.5 \times 10^{-9} \text{ C}}{1.60 \times 10^{-19} \text{ C/electron}} = 2.2 \times 10^{10} \text{ electrons}$$

Thus 2.2×10^{10} electrons are moved from one electrode to the other. Note that the capacitor *as a whole* has no net charge.

ASSESS The plate spacing does not enter the result. As long as the spacing is much smaller than the plate dimensions, as is true in this example, the field is independent of the spacing.

Uniform Electric Fields

FIGURE 23.22 shows an electric field that is the *same*—in strength and direction—at every point in a region of space. This is called a **uniform electric field.** A uniform electric field is analogous to the uniform gravitational field near the surface of the earth. Uniform fields are of great practical significance because, as you will see in the next section, computing the trajectory of a charged particle moving in a uniform electric field is a straightforward process.

The easiest way to produce a uniform electric field is with a parallel-plate capacitor, as you can see in Figure 23.21a. Indeed, our interest in capacitors is due in large measure to the fact that the electric field is uniform. Many electric field problems refer to a uniform electric field. Such problems carry an implicit assumption that the action is taking place *inside* a parallel-plate capacitor.

FIGURE 23.22 A uniform electric field.

STOP TO THINK 23.5 Rank in order, from largest to smallest, the forces F_a to F_e a proton would experience if placed at points a to e in this parallel-plate capacitor.

23.6 Motion of a Charged Particle in an Electric Field

Our motivation for introducing the concept of the electric field was to understand the long-range electric interaction of charges. Some charges, the *source charges,* create an electric field. Other charges then respond to that electric field. The first five sections of this chapter have focused on the electric field of the source charges. Now we turn our attention to the second half of the interaction.

FIGURE 23.23 shows a particle of charge q and mass m at a point where an electric field $\vec{E}$ has been produced by *other* charges, the source charges. The electric field exerts a force

$$\vec{F}_{\text{on }q} = q\vec{E}$$

on the charged particle. Notice that the force on a negatively charged particle is *opposite* in direction to the electric field vector. Signs are important!

FIGURE 23.23 The electric field exerts a force on a charged particle.

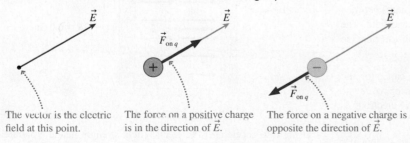

The vector is the electric field at this point.   The force on a positive charge is in the direction of $\vec{E}$.   The force on a negative charge is opposite the direction of $\vec{E}$.

If $\vec{F}_{\text{on }q}$ is the only force acting on q, it causes the charged particle to accelerate with

$$\vec{a} = \frac{\vec{F}_{\text{on }q}}{m} = \frac{q}{m}\vec{E} \tag{23.32}$$

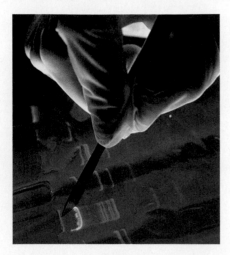

The technique of *gel electrophoresis* uses an electric field to measure "DNA fingerprints." DNA fragments are charged, and fragments with different charge-to-mass ratios are separated by the field.

This acceleration is the *response* of the charged particle to the source charges that created the electric field. The ratio q/m is especially important for the dynamics of charged-particle motion. It is called the **charge-to-mass ratio.** Two *equal* charges, say a proton and a Na^+ ion, will experience *equal* forces $\vec{F} = q\vec{E}$ if placed at the same point in an electric field, but their accelerations will be *different* because they have different masses and thus different charge-to-mass ratios. Two particles with different charges and masses *but* with the same charge-to-mass ratio will undergo the same acceleration and follow the same trajectory.

Motion in a Uniform Field

The motion of a charged particle in a *uniform* electric field is especially important for its basic simplicity and because of its many valuable applications. A uniform field is *constant* at all points—constant in both magnitude and direction—within the region of space where the charged particle is moving. It follows, from Equation 23.32, that **a charged particle in a uniform electric field will move with constant acceleration.** The magnitude of the acceleration is

$$a = \frac{qE}{m} = \text{constant} \tag{23.33}$$

where E is the electric field strength, and the direction of $\vec{a}$ is parallel or antiparallel to $\vec{E}$, depending on the sign of q.

Identifying the motion of a charged particle in a uniform field as being one of constant acceleration brings into play all the kinematic machinery that we developed in Chapters 2 and 4 for constant-acceleration motion. The basic trajectory of a charged particle in a uniform field is a *parabola,* analogous to the projectile motion of a mass in the near-earth uniform gravitational field. In the special case of a charged particle moving parallel to the electric field vectors, the motion is one-dimensional, analogous to the one-dimensional vertical motion of a mass tossed straight up or falling straight down.

NOTE The gravitational acceleration $\vec{a}_{grav}$ always points straight down. The electric field acceleration $\vec{a}_{elec}$ can point in *any* direction. You must determine the electric field $\vec{E}$ in order to learn the direction of $\vec{a}$.

EXAMPLE 23.7 | An electron moving across a capacitor

Two 6.0-cm-diameter electrodes are spaced 5.0 mm apart. They are charged by transferring 1.0×10^{11} electrons from one electrode to the other. An electron is released from rest just above the surface of the negative electrode. How long does it take the electron to cross to the positive electrode? What is its speed as it collides with the positive electrode? Assume the space between the electrodes is a vacuum.

MODEL The electrodes form a parallel-plate capacitor. The charges *on* the electrodes cannot escape, but any additional charges *between* the capacitor plates will be accelerated by the electric field. The electric field inside a parallel-plate capacitor is a uniform field, so the electron will have constant acceleration.

VISUALIZE FIGURE 23.24 shows an edge view of the capacitor and the electron. The force on the negative electron is *opposite* the electric field, so the electron is repelled by the negative electrode as it accelerates across the gap of width d.

SOLVE The electrodes are not point charges, so we cannot use Coulomb's law to find the force on the electron. Instead, we must analyze the electron's motion in terms of the electric field inside the capacitor. The field is the agent that exerts the force on the electron,

FIGURE 23.24 An electron accelerates across a capacitor (plate separation exaggerated).

The capacitor was charged by transferring 10^{11} electrons from the right electrode to the left electrode.

causing it to accelerate. The electric field strength inside a parallel-plate capacitor with charge $Q = Ne$ is

$$E = \frac{\eta}{\epsilon_0} = \frac{Q}{\epsilon_0 A} = \frac{Ne}{\epsilon_0 \pi R^2} = 639{,}000 \text{ N/C}$$

The electron's acceleration in this field is

$$a = \frac{eE}{m} = 1.1 \times 10^{17} \text{ m/s}^2$$

where we used the electron mass $m = 9.11 \times 10^{-31}$ kg. This is an enormous acceleration compared to accelerations we're familiar with for macroscopic objects. We can use one-dimensional kinematics, with $x_i = 0$ and $v_i = 0$, to find the time required for the electron to cross the capacitor:

$$x_f = d = \tfrac{1}{2}a(\Delta t)^2$$

$$\Delta t = \sqrt{\frac{2d}{a}} = 3.0 \times 10^{-10} \text{ s} = 0.30 \text{ ns}$$

The electron's speed as it reaches the positive electrode is

$$v = a\,\Delta t = 3.3 \times 10^7 \text{ m/s}$$

ASSESS We used e rather than $-e$ to find the acceleration because we already knew the direction; we needed only the magnitude. The electron's speed, after traveling a mere 5 mm, is approximately 10% the speed of light.

Parallel electrodes such as those in Example 23.7 are often used to accelerate charged particles. If the positive plate has a small hole in the center, a *beam* of electrons will pass through the hole and emerge with a speed of 3.3×10^7 m/s. This is the basic idea of the *electron gun* used until quite recently in *cathode-ray tube* (CRT) devices such as televisions and computer display terminals. (A negatively charged electrode is called a *cathode*, so the physicists who first learned to produce electron beams in the late 19th century called them *cathode rays*.)

EXAMPLE 23.8 | Deflecting an electron beam

An electron gun creates a beam of electrons moving horizontally with a speed of 3.3×10^7 m/s. The electrons enter a 2.0-cm-long gap between two parallel electrodes where the electric field is $\vec{E} = (5.0 \times 10^4 \text{ N/C, down})$. In which direction, and by what angle, is the electron beam deflected by these electrodes?

MODEL The electric field between the electrodes is uniform. Assume that the electric field outside the electrodes is zero.

VISUALIZE FIGURE 23.25 shows an electron moving through the electric field. The electric field points down, so the force on the (negative) electrons is upward. The electrons will follow a parabolic trajectory, analogous to that of a ball thrown horizontally, except that the electrons "fall up" rather than down.

FIGURE 23.25 The deflection of an electron beam in a uniform electric field.

Deflection plates

SOLVE This is a two-dimensional motion problem. The electron enters the capacitor with velocity *vector* $\vec{v}_0 = v_{0x}\hat{\imath} = 3.3 \times 10^7\,\hat{\imath}$ m/s and leaves with velocity $\vec{v}_1 = v_{1x}\hat{\imath} + v_{1y}\hat{\jmath}$. The electron's angle of travel upon leaving the electric field is

$$\theta = \tan^{-1}\left(\frac{v_{1y}}{v_{1x}}\right)$$

This is the *deflection angle*. To find θ we must compute the final velocity vector $\vec{v}_1$.

There is no horizontal force on the electron, so $v_{1x} = v_{0x} = 3.3 \times 10^7$ m/s. The electron's upward acceleration has magnitude

$$a = \frac{eE}{m} = \frac{(1.60 \times 10^{-19} \text{ C})(5.0 \times 10^4 \text{ N/C})}{9.11 \times 10^{-31} \text{ kg}}$$

$$= 8.78 \times 10^{15} \text{ m/s}^2$$

We can use the fact that the horizontal velocity is constant to determine the time interval Δt needed to travel length 2.0 cm:

$$\Delta t = \frac{L}{v_{0x}} = \frac{0.020 \text{ m}}{3.3 \times 10^7 \text{ m/s}} = 6.06 \times 10^{-10} \text{ s}$$

Vertical acceleration will occur during this time interval, resulting in a final vertical velocity

$$v_{1y} = v_{0y} + a\,\Delta t = 5.3 \times 10^6 \text{ m/s}$$

The electron's velocity as it leaves the capacitor is thus

$$\vec{v}_1 = (3.3 \times 10^7\,\hat{\imath} + 5.3 \times 10^6\,\hat{\jmath}) \text{ m/s}$$

and the deflection angle θ is

$$\theta = \tan^{-1}\left(\frac{v_{1y}}{v_{1x}}\right) = 9.1°$$

ASSESS We know that the electron beam in a cathode-ray tube can be deflected enough to cover the screen, so a deflection angle of 9° seems reasonable. Our neglect of the gravitational force is seen to be justified because the acceleration of the electrons is enormous in comparison to the free-fall acceleration g.

By using two sets of deflection plates—one for vertical deflection and one for horizontal—a cathode-ray tube could steer the electrons to any point on the screen. Electrons striking a phosphor coating on the inside of the screen would then make a dot of light.

STOP TO THINK 23.6   Which electric field is responsible for the proton's trajectory?

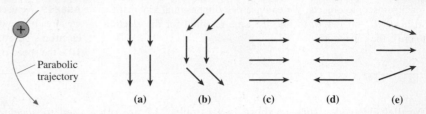

23.7 Motion of a Dipole in an Electric Field

Let us conclude this chapter by returning to one of the more striking puzzles we faced when making the observations at the beginning of Chapter 22. There you found that charged objects of *either* sign exert forces on neutral objects, such as when a comb used to brush your hair picks up pieces of paper. Our qualitative understanding of the *polarization force* was that it required two steps:

- The charge polarizes the neutral object, creating an induced electric dipole.
- The charge then exerts an attractive force on the near end of the dipole that is slightly stronger than the repulsive force on the far end.

We are now in a position to make that understanding more quantitative.

Dipoles in a Uniform Field

FIGURE 23.26a shows an electric dipole in a *uniform* external electric field $\vec{E}$ that has been created by source charges we do not see. That is, $\vec{E}$ is *not* the field of the dipole but, instead, is a field to which the dipole is responding. In this case, because the field is uniform, the dipole is presumably inside an unseen parallel-plate capacitor.

The net force on the dipole is the sum of the forces on the two charges forming the dipole. Because the charges $\pm q$ are equal in magnitude but opposite in sign, the two forces $\vec{F}_+ = +q\vec{E}$ and $\vec{F}_- = -q\vec{E}$ are also equal but opposite. Thus the net force on the dipole is

$$\vec{F}_{net} = \vec{F}_+ + \vec{F}_- = \vec{0} \tag{23.34}$$

There is no net force on a dipole in a uniform electric field.

There may be no net force, but the electric field *does* affect the dipole. Because the two forces in Figure 23.26a are in opposite directions but not aligned with each other, the electric field exerts a *torque* on the dipole and causes the dipole to *rotate*.

The torque rotates the dipole until it is aligned with the electric field, as shown in **FIGURE 23.26b**. In this position, the dipole experiences not only no net force but also no torque. Thus Figure 23.26b represents the *equilibrium position* for a dipole in a uniform electric field. Notice that the positive end of the dipole is in the direction in which $\vec{E}$ points.

FIGURE 23.27 shows a sample of permanent dipoles, such as water molecules, in an external electric field. All the dipoles rotate until they are aligned with the electric field. This is the mechanism by which the sample becomes *polarized*. Once the dipoles are aligned, there is an excess of positive charge at one end of the sample and an excess of negative charge at the other end. The excess charges at the ends of the sample are the basis of the polarization forces we discussed in Section 22.3.

FIGURE 23.26 A dipole in a uniform electric field.

(a)   The electric field exerts a torque on this dipole.

(b)   This dipole is in equilibrium.

FIGURE 23.27 A sample of permanent dipoles is *polarized* in an electric field.

The dipoles align with the electric field.

Excess negative charge on this surface   Excess positive charge on this surface

It's not hard to calculate the torque. Recall from Chapter 12 that the magnitude of a torque is the product of the force and the moment arm. FIGURE 23.28 shows that there are two forces of the same magnitude ($F_+ = F_- = qE$), each with the same moment arm ($d = \frac{1}{2}s\sin\theta$). Thus the torque on the dipole is

$$\tau = 2 \times dF_+ = 2(\tfrac{1}{2}s\sin\theta)(qE) = pE\sin\theta \tag{23.35}$$

where $p = qs$ was our definition of the dipole moment. The torque is zero when the dipole is aligned with the field, making $\theta = 0$.

Also recall from Chapter 12 that the torque can be written in a compact mathematical form as the cross product between two vectors. The terms p and E in Equation 23.35 are the magnitudes of vectors, and θ is the angle between them. Thus in vector notation, the torque exerted on a dipole moment $\vec{p}$ by an electric field $\vec{E}$ is

$$\vec{\tau} = \vec{p} \times \vec{E} \tag{23.36}$$

The torque is greatest when $\vec{p}$ is perpendicular to $\vec{E}$, zero when $\vec{p}$ is aligned with or opposite to $\vec{E}$.

FIGURE 23.28 The torque on a dipole.

In terms of vectors, $\vec{\tau} = \vec{p} \times \vec{E}$.

EXAMPLE 23.9 | The angular acceleration of a dipole dumbbell

Two 1.0 g balls are connected by a 2.0-cm-long insulating rod of negligible mass. One ball has a charge of $+10$ nC, the other a charge of -10 nC. The rod is held in a 1.0×10^4 N/C uniform electric field at an angle of 30° with respect to the field, then released. What is its initial angular acceleration?

MODEL The two oppositely charged balls form an electric dipole. The electric field exerts a torque on the dipole, causing an angular acceleration.

VISUALIZE FIGURE 23.29 shows the dipole in the electric field.

FIGURE 23.29 The dipole of Example 23.9.

SOLVE The dipole moment is $p = qs = (1.0 \times 10^{-8}\,\text{C}) \times (0.020\,\text{m}) = 2.0 \times 10^{-10}$ C m. The torque exerted on the dipole moment by the electric field is

$$\tau = pE\sin\theta = (2.0 \times 10^{-10}\,\text{C m})(1.0 \times 10^4\,\text{N/C})\sin 30°$$
$$= 1.0 \times 10^{-6}\,\text{N m}$$

You learned in Chapter 12 that a torque causes an angular acceleration $\alpha = \tau/I$, where I is the moment of inertia. The dipole rotates about its center of mass, which is at the center of the rod, so the moment of inertia is

$$I = m_1 r_1^2 + m_2 r_2^2 = 2m(\tfrac{1}{2}s)^2 = \tfrac{1}{2}ms^2 = 2.0 \times 10^{-7}\,\text{kg m}^2$$

Thus the rod's angular acceleration is

$$\alpha = \frac{\tau}{I} = \frac{1.0 \times 10^{-6}\,\text{N m}}{2.0 \times 10^{-7}\,\text{kg m}^2} = 5.0\,\text{rad/s}^2$$

ASSESS This value of α is the initial angular acceleration, when the rod is first released. The torque and the angular acceleration will decrease as the rod rotates toward alignment with $\vec{E}$.

Dipoles in a Nonuniform Field

Suppose that a dipole is placed in a nonuniform electric field, one in which the field strength changes with position. For example, FIGURE 23.30 shows a dipole in the nonuniform field of a point charge. The first response of the dipole is to rotate until it is aligned with the field, with the dipole's positive end pointing in the same direction as the field. Now, however, there is a *slight difference* between the forces acting on the two ends of the dipole. This difference occurs because the electric field, which depends on the distance from the point charge, is stronger at the end of the dipole nearest the charge. This causes a net force to be exerted on the dipole.

Which way does the force point? Once the dipole is aligned, the leftward attractive force on its negative end is slightly stronger than the rightward repulsive force on its positive end. This causes a net force *toward* the point charge.

In fact, for any nonuniform electric field, **the net force on a dipole is toward the direction of the strongest field.** Because any finite-size charged object, such as a charged rod or a charged disk, has a field strength that increases as you get closer to the object, we can conclude that **a dipole will experience a net force toward any charged object.**

FIGURE 23.30 An aligned dipole is drawn toward a point charge.

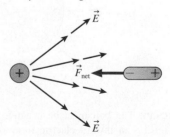

EXAMPLE 23.10 | The force on a water molecule

The water molecule H_2O has a permanent dipole moment of magnitude 6.2×10^{-30} C m. A water molecule is located 10 nm from a Na^+ ion in a saltwater solution. What force does the ion exert on the water molecule?

VISUALIZE FIGURE 23.31 shows the ion and the dipole. The forces are an action/reaction pair.

FIGURE 23.31 The interaction between an ion and a permanent dipole.

$\vec{F}_{\text{dipole on ion}}$ $\vec{F}_{\text{ion on dipole}}$

Na^+ ion   Water molecule

$r = 10$ nm

SOLVE A Na^+ ion has charge $q = +e$. The electric field of the ion aligns the water's dipole moment and exerts a net force on it. We could calculate the net force on the dipole as the small difference between the attractive force on its negative end and the repulsive force on its positive end. Alternatively, we know from Newton's

third law that the force $\vec{F}_{\text{dipole on ion}}$ has the same magnitude as the force $\vec{F}_{\text{ion on dipole}}$ that we are seeking. We calculated the on-axis field of a dipole in Section 23.2. An ion of charge $q = e$ will experience a force of magnitude $F = qE_{\text{dipole}} = eE_{\text{dipole}}$ when placed in that field. The dipole's electric field, which we found in Equation 23.10, is

$$E_{\text{dipole}} = \frac{1}{4\pi\epsilon_0} \frac{2p}{r^3}$$

The force on the ion at distance $r = 1.0 \times 10^{-8}$ m is

$$F_{\text{dipole on ion}} = eE_{\text{dipole}} = \frac{1}{4\pi\epsilon_0} \frac{2ep}{r^3} = 1.8 \times 10^{-14} \text{ N}$$

Thus the force on the water molecule is $F_{\text{ion on dipole}} = 1.8 \times 10^{-14}$ N.

ASSESS While 1.8×10^{-14} N may seem like a very small force, it is $\approx 10^{11}$ times larger than the size of the earth's gravitational force on these atomic particles. Forces such as these cause water molecules to cluster around any ions that are in solution. This clustering plays an important role in the microscopic physics of solutions studied in chemistry and biochemistry.

CHALLENGE EXAMPLE 23.11 | An orbiting proton

In a vacuum chamber, a proton orbits a 1.0-cm-diameter metal ball 1.0 mm above the surface with a period of 1.0 μs. What is the charge on the ball?

MODEL Model the ball as a charged sphere. The electric field of a charged sphere is the same as that of a point charge at the center, so the radius of the ball is irrelevant. Assume that the gravitational force on the proton is extremely small compared to the electric force and can be neglected.

VISUALIZE FIGURE 23.32 shows the orbit and the force on the proton.

FIGURE 23.32 An orbiting proton.

SOLVE The ball must be negative, with an inward electric field exerting an inward electric force on the positive proton. This is

exactly the necessary condition for uniform circular motion. Recall from Chapter 8 that Newton's second law for uniform circular motion is $(F_{\text{net}})_r = mv^2/r$. Here the only radial force has magnitude $F_{\text{elec}} = eE$, so the proton will move in a circular orbit if

$$eE = \frac{mv^2}{r}$$

The electric field strength of a sphere of charge Q at distance r is $E = Q/4\pi\epsilon_0 r^2$. From Chapter 4, orbital speed and period are related by $v = \text{circumference/period} = 2\pi r/T$. With these substitutions, Newton's second law becomes

$$\frac{eQ}{4\pi\epsilon_0 r^2} = \frac{4\pi^2 m}{T^2} r$$

Solving for Q, we find

$$Q = \frac{16\pi^3 \epsilon_0 m r^3}{eT^2} = 9.9 \times 10^{-12} \text{ C}$$

where we used $r = 6.0$ mm as the radius of the proton's orbit. Q is the *magnitude* of the charge on the ball. Including the sign, we have

$$Q_{\text{ball}} = -9.9 \times 10^{-12} \text{ C}$$

ASSESS This is not a lot of charge, but it shouldn't take much charge to affect the motion of something as light as a proton.

SUMMARY

The goal of Chapter 23 has been to learn how to calculate and use the electric field.

GENERAL PRINCIPLES

Sources of $\vec{E}$

Electric fields are created by charges.

Multiple point charges

MODEL Model objects as point charges.

VISUALIZE Establish a coordinate system and draw field vectors.

SOLVE Use superposition: $\vec{E} = \vec{E}_1 + \vec{E}_2 + \vec{E}_3 + \cdots$

Continuous distribution of charge

MODEL Model objects as simple shapes.

VISUALIZE
- Establish a coordinate system.
- Divide the charge into small segments ΔQ.
- Draw a field vector for one or two pieces of charge.

SOLVE
- Find the field of each ΔQ.
- Write $\vec{E}$ as the sum of the fields of all ΔQ. Don't forget that it's a *vector* sum; use components.
- Use the charge density (λ or η) to replace ΔQ with an integration coordinate, then integrate.

Consequences of $\vec{E}$

The electric field exerts a force on a charged particle:

$$\vec{F} = q\vec{E}$$

The force causes acceleration:

$$\vec{a} = (q/m)\vec{E}$$

Trajectories of charged particles are calculated with kinematics.

The electric field exerts a torque on a dipole:

$$\tau = pE\sin\theta$$

The torque tends to align the dipoles with the field.

In a nonuniform electric field, a dipole has a net force in the direction of increasing field strength.

APPLICATIONS

Four Key Electric Field Models

Point charge with charge q

$$\vec{E} = \frac{1}{4\pi\epsilon_0}\frac{q}{r^2}\hat{r}$$

Infinite line of charge with linear charge density λ

$$\vec{E}_{\text{line}} = \left(\frac{1}{4\pi\epsilon_0}\frac{2|\lambda|}{r}, \begin{cases} \text{away if } + \\ \text{toward if } - \end{cases}\right)$$

Infinite plane of charge with surface charge density η

$$\vec{E}_{\text{plane}} = \left(\frac{|\eta|}{2\epsilon_0}, \begin{cases} \text{away if } + \\ \text{toward if } - \end{cases}\right)$$

Sphere of charge with total charge Q

Same as a point charge Q for $r > R$

Electric dipole

The electric dipole moment is

$$\vec{p} = (qs, \text{ from negative to positive})$$

Field on axis: $\vec{E} = \dfrac{1}{4\pi\epsilon_0}\dfrac{2\vec{p}}{r^3}$

Field in bisecting plane: $\vec{E} = -\dfrac{1}{4\pi\epsilon_0}\dfrac{\vec{p}}{r^3}$

Parallel-plate capacitor

The electric field inside an ideal capacitor is a **uniform electric field**:

$$\vec{E} = \left(\frac{\eta}{\epsilon_0}, \text{ from positive to negative}\right)$$

TERMS AND NOTATION

dipole moment, $\vec{p}$	uniformly charged	plane of charge	fringe field
electric field line	line of charge	sphere of charge	uniform electric field
linear charge density, λ	electrode	parallel-plate capacitor	charge-to-mass ratio, q/m
surface charge density, η			

CONCEPTUAL QUESTIONS

1. You've been assigned the task of determining the magnitude and direction of the electric field at a point in space. Give a step-by-step procedure of how you will do so. List any objects you will use, any measurements you will make, and any calculations you will need to perform. Make sure that your measurements do not disturb the charges that are creating the field.

2. Reproduce **FIGURE Q23.2** on your paper. For each part, draw a dot or dots on the figure to show any position or positions (other than infinity) where $\vec{E} = \vec{0}$.

(a)

(b)

FIGURE Q23.2

3. Rank in order, from largest to smallest, the electric field strengths E_1 to E_4 at points 1 to 4 in **FIGURE Q23.3**. Explain.

FIGURE Q23.3   **FIGURE Q23.4**

4. A small segment of wire in **FIGURE Q23.4** contains 10 nC of charge.
   a. The segment is shrunk to one-third of its original length. What is the ratio λ_f/λ_i, where λ_i and λ_f are the initial and final linear charge densities?
   b. A proton is very far from the wire. What is the ratio F_f/F_i of the electric force on the proton after the segment is shrunk to the force before the segment was shrunk?
   c. Suppose the original segment of wire is stretched to 10 times its original length. How much charge must be *added* to the wire to keep the linear charge density unchanged?

5. An electron experiences a force of magnitude F when it is 1 cm from a very long, charged wire with linear charge density λ. If the charge density is doubled, at what distance from the wire will a proton experience a force of the same magnitude F?

6. **FIGURE Q23.6** shows a hollow soda straw that has been uniformly charged with positive charge. What is the electric field at the center (inside) of the straw? Explain.

Inside straw

FIGURE Q23.6

7. The irregularly shaped area of charge in **FIGURE Q23.7** has surface charge density η_i. Each dimension (x and y) of the area is reduced by a factor of 3.163.
   a. What is the ratio η_f/η_i, where η_f is the final surface charge density?
   b. An electron is very far from the area. What is the ratio F_f/F_i of the electric force on the electron after the area is reduced to the force before the area was reduced?

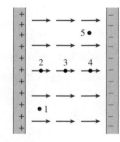

FIGURE Q23.7

8. A circular disk has surface charge density 8 nC/cm². What will the surface charge density be if the radius of the disk is doubled?

9. A sphere of radius R has charge Q. The electric field strength at distance $r > R$ is E_i. What is the ratio E_f/E_i of the final to initial electric field strengths if (a) Q is halved, (b) R is halved, and (c) r is halved (but is still $> R$)? Each part changes only one quantity; the other quantities have their initial values.

10. The ball in **FIGURE Q23.10** is suspended from a large, uniformly charged positive plate. It swings with period T. If the ball is discharged, will the period increase, decrease, or stay the same? Explain.

Mass m
Charge q

FIGURE Q23.10   **FIGURE Q23.11**

11. Rank in order, from largest to smallest, the electric field strengths E_1 to E_5 at the five points in **FIGURE Q23.11**. Explain.

12. A parallel-plate capacitor consists of two square plates, size $L \times L$, separated by distance d. The plates are given charge $\pm Q$. What is the ratio E_f/E_i of the final to initial electric field strengths if (a) Q is doubled, (b) L is doubled, and (c) d is doubled? Each part changes only one quantity; the other quantities have their initial values.

13. A small object is released at point 3 in the center of the capacitor in **FIGURE Q23.11**. For each situation, does the object move to the right, to the left, or remain in place? If it moves, does it accelerate or move at constant speed?
   a. A positive object is released from rest.
   b. A neutral but polarizable object is released from rest.
   c. A negative object is released from rest.

14. A proton and an electron are released from rest in the center of a capacitor.
   a. Is the force ratio F_p/F_e greater than 1, less than 1, or equal to 1? Explain.
   b. Is the acceleration ratio a_p/a_e greater than 1, less than 1, or equal to 1? Explain.

15. Three charges are placed at the corners of the triangle in **FIGURE Q23.15**. The ++ charge has twice the quantity of charge of the two − charges; the net charge is zero. Is the triangle in equilibrium? If so, explain why. If not, draw the equilibrium orientation.

FIGURE Q23.15

EXERCISES AND PROBLEMS

Problems labeled   integrate material from earlier chapters.

Exercises

Section 23.2 The Electric Field of Point Charges

1. ‖ What are the strength and direction of the electric field at the position indicated by the dot in **FIGURE EX23.1**? Specify the direction as an angle above or below horizontal.

FIGURE EX23.1   **FIGURE EX23.2**

2. ‖ What are the strength and direction of the electric field at the position indicated by the dot in **FIGURE EX23.2**? Specify the direction as an angle above or below horizontal.

3. ‖ What are the strength and direction of the electric field at the position indicated by the dot in **FIGURE EX23.3**? Specify the direction as an angle above or below horizontal.

FIGURE EX23.3   **FIGURE EX23.4**

4. ‖ What are the strength and direction of the electric field at the position indicated by the dot in **FIGURE EX23.4**? Specify the direction as an angle above or below horizontal.

5. ‖ An electric dipole is formed from two charges, $\pm q$, spaced 1.0 cm apart. The dipole is at the origin, oriented along the y-axis. The electric field strength at the point $(x, y) = (0 \text{ cm}, 10 \text{ cm})$ is 360 N/C.
   a. What is the charge q? Give your answer in nC.
   b. What is the electric field strength at the point $(x, y) = (10 \text{ cm}, 0 \text{ cm})$?

6. ‖ An electric dipole is formed from ± 1.0 nC charges spaced 2.0 mm apart. The dipole is at the origin, oriented along the x-axis. What is the electric field strength at the points (a) $(x, y) = (10 \text{ cm}, 0 \text{ cm})$ and (b) $(x, y) = (0 \text{ cm}, 10 \text{ cm})$?

7. ‖ An *electret* is similar to a magnet, but rather than being permanently magnetized, it has a permanent electric dipole moment. Suppose a small electret with electric dipole moment 1.0×10^{-7} C m is 25 cm from a small ball charged to +25 nC, with the ball on the axis of the electric dipole. What is the magnitude of the electric force on the ball?

Section 23.3 The Electric Field of a Continuous Charge Distribution

8. ‖ The electric field strength 10.0 cm from a very long charged wire is 2000 N/C. What is the electric field strength 5.0 cm from the wire?

9. ‖ A 10-cm-long thin glass rod uniformly charged to +10 nC and a 10-cm-long thin plastic rod uniformly charged to −10 nC are placed side by side, 4.0 cm apart. What are the electric field strengths E_1 to E_3 at distances 1.0 cm, 2.0 cm, and 3.0 cm from the glass rod along the line connecting the midpoints of the two rods?

10. ‖ Two 10-cm-long thin glass rods uniformly charged to +10 nC are placed side by side, 4.0 cm apart. What are the electric field strengths E_1 to E_3 at distances 1.0 cm, 2.0 cm, and 3.0 cm to the right of the rod on the left along the line connecting the midpoints of the two rods?

11. ‖ A small glass bead charged to + 6.0 nC is in the plane that bisects a thin, uniformly charged, 10-cm-long glass rod and is 4.0 cm from the rod's center. The bead is repelled from the rod with a force of 840 μN. What is the total charge on the rod?

12. ‖‖ The electric field 5.0 cm from a very long charged wire is (2000 N/C, toward the wire). What is the charge (in nC) on a 1.0-cm-long segment of the wire?

13. ‖‖ A 12-cm-long thin rod has the *nonuniform* charge density
   CALC $\lambda(x) = (2.0 \text{ nC/cm})e^{-|x|/(6.0 \text{ cm})}$, where x is measured from the center of the rod. What is the total charge on the rod?
   Hint: This exercise requires an integration. Think about how to handle the absolute value sign.

Section 23.4 The Electric Fields of Rings, Disks, Planes, and Spheres

14. ‖ Two 10-cm-diameter charged rings face each other, 20 cm apart. The left ring is charged to −20 nC and the right ring is charged to +20 nC.
   a. What is the electric field $\vec{E}$, both magnitude and direction, at the midpoint between the two rings?
   b. What is the force on a proton at the midpoint?

15. || Two 10-cm-diameter charged rings face each other, 20 cm apart. Both rings are charged to +20 nC. What is the electric field strength at (a) the midpoint between the two rings and (b) the center of the left ring?

16. || Two 10-cm-diameter charged disks face each other, 20 cm apart. The left disk is charged to −50 nC and the right disk is charged to +50 nC.
   a. What is the electric field $\vec{E}$, both magnitude and direction, at the midpoint between the two disks?
   b. What is the force $\vec{F}$ on a −1.0 nC charge placed at the midpoint?

17. || The electric field strength 2.0 cm from the surface of a 10-cm-diameter metal ball is 50,000 N/C. What is the charge (in nC) on the ball?

18. || A 20 cm × 20 cm horizontal metal electrode is uniformly charged to +80 nC. What is the electric field strength 2.0 mm above the center of the electrode?

19. || Two 2.0-cm-diameter insulating spheres have a 6.0 cm space between them. One sphere is charged to +10 nC, the other to −15 nC. What is the electric field strength at the midpoint between the two spheres?

20. || You've hung two very large sheets of plastic facing each other with distance d between them, as shown in FIGURE EX23.20. By rubbing them with wool and silk, you've managed to give one sheet a uniform surface charge density $\eta_1 = -\eta_0$ and the other a uniform surface charge density $\eta_2 = +3\eta_0$. What are the electric field vectors at points 1, 2, and 3?

FIGURE EX23.20

21. | A 2.0 m × 4.0 m flat carpet acquires a uniformly distributed charge of −10 μC after you and your friends walk across it several times. A 2.5 μg dust particle is suspended in midair just above the center of the carpet. What is the charge on the dust particle?

Section 23.5 The Parallel-Plate Capacitor

22. || Two circular disks spaced 0.50 mm apart form a parallel-plate capacitor. Transferring 3.0×10^9 electrons from one disk to the other causes the electric field strength to be 2.0×10^5 N/C. What are the diameters of the disks?

23. || A parallel-plate capacitor is formed from two 6.0-cm-diameter electrodes spaced 2.0 mm apart. The electric field strength inside the capacitor is 1.0×10^6 N/C. What is the charge (in nC) on each electrode?

24. ||| Air "breaks down" when the electric field strength reaches 3.0×10^6 N/C, causing a spark. A parallel-plate capacitor is made from two 4.0 cm × 4.0 cm electrodes. How many electrons must be transferred from one electrode to the other to create a spark between the electrodes?

25. || Two parallel plates 1.0 cm apart are equally and oppositely charged. An electron is released from rest at the surface of the negative plate and simultaneously a proton is released from rest at the surface of the positive plate. How far from the negative plate is the point at which the electron and proton pass each other?

Section 23.6 Motion of a Charged Particle in an Electric Field

26. || Two 2.0-cm-diameter disks face each other, 1.0 mm apart. They are charged to ± 10 nC.

   a. What is the electric field strength between the disks?
   b. A proton is shot from the negative disk toward the positive disk. What launch speed must the proton have to just barely reach the positive disk?

27. | Honeybees acquire a charge while flying due to friction with the air. A 100 mg bee with a charge of +23 pC experiences an electric force in the earth's electric field, which is typically 100 N/C, directed downward.
BIO
   a. What is the ratio of the electric force on the bee to the bee's weight?
   b. What electric field strength and direction would allow the bee to hang suspended in the air?

28. || An electron traveling parallel to a uniform electric field increases its speed from 2.0×10^7 m/s to 4.0×10^7 m/s over a distance of 1.2 cm. What is the electric field strength?

29. || The surface charge density on an infinite charged plane is -2.0×10^{-6} C/m². A proton is shot straight away from the plane at 2.0×10^6 m/s. How far does the proton travel before reaching its turning point?

30. || An electron in a vacuum chamber is fired with a speed of 8300 km/s toward a large, uniformly charged plate 75 cm away. The electron reaches a closest distance of 15 cm before being repelled. What is the plate's surface charge density?

31. ||| A 1.0-μm-diameter oil droplet (density 900 kg/m³) is negatively charged with the addition of 25 extra electrons. It is released from rest 2.0 mm from a very wide plane of positive charge, after which it accelerates toward the plane and collides with a speed of 3.5 m/s. What is the surface charge density of the plane?

Section 23.7 Motion of a Dipole in an Electric Field

32. | The permanent electric dipole moment of the water molecule (H_2O) is 6.2×10^{-30} C m. What is the maximum possible torque on a water molecule in a 5.0×10^8 N/C electric field?

33. || A point charge Q is distance r from a dipole consisting of charges $\pm q$ separated by distance s. The dipole is initially oriented so that Q is in the plane bisecting the dipole. Immediately after the dipole is released, what are (a) the magnitude of the force and (b) the magnitude of the torque on the dipole? You can assume $r \gg s$.

34. || An ammonia molecule (NH_3) has a permanent electric dipole moment 5.0×10^{-30} C m. A proton is 2.0 nm from the molecule in the plane that bisects the dipole. What is the electric force of the molecule on the proton?

Problems

35. || What are the strength and direction of the electric field at the position indicated by the dot in FIGURE P23.35? Give your answer (a) in component form and (b) as a magnitude and angle measured cw or ccw (specify which) from the positive x-axis.

FIGURE P23.35   FIGURE P23.36

36. ‖ What are the strength and direction of the electric field at the position indicated by the dot in **FIGURE P23.36**? Give your answer (a) in component form and (b) as a magnitude and angle measured cw or ccw (specify which) from the positive x-axis.

37. ‖ What are the strength and direction of the electric field at the position indicated by the dot in **FIGURE P23.37**? Give your answer (a) in component form and (b) as a magnitude and angle measured cw or ccw (specify which) from the positive x-axis.

FIGURE P23.37

FIGURE P23.38

38. ‖ **FIGURE P23.38** shows three charges at the corners of a square. Write the electric field at point P in component form.

39. ‖ Charges $-q$ and $+2q$ in **FIGURE P23.39** are located at $x = \pm a$. Determine the electric field at points 1 to 4. Write each field in component form.

FIGURE P23.39

40. ‖ Derive Equation 23.11 for the field $\vec{E}_{dipole}$ in the plane that bisects an electric dipole.

41. ‖ **FIGURE P23.41** is a cross section of two infinite lines of charge that extend out of the page. Both have linear charge density λ. Find an expression for the electric field strength E at height y above the midpoint between the lines.

FIGURE P23.41   **FIGURE P23.42**

42. ‖‖ **FIGURE P23.42** is a cross section of two infinite lines of charge that extend out of the page. The linear charge densities are $\pm \lambda$. Find an expression for the electric field strength E at height y above the midpoint between the lines.

43. ‖ **FIGURE P23.43** shows a thin rod of length L with total charge Q.
CALC a. Find an expression for the electric field strength at point P on the axis of the rod at distance r from the center.
b. Verify that your expression has the expected behavior if $r \gg L$.
c. Evaluate E at $r = 3.0$ cm if $L = 5.0$ cm and $Q = 3.0$ nC.

FIGURE P23.43   **FIGURE P23.44**

44. ‖ **FIGURE P23.44** shows a thin rod of length L with total charge
CALC Q. Find an expression for the electric field $\vec{E}$ at point P. Give your answer in component form.

45. ‖ Show that the on-axis electric field of a ring of charge has the expected behavior when $z \ll R$ and when $z \gg R$.

46. ‖ A ring of radius R has total charge Q.
CALC a. At what distance along the z-axis is the electric field strength a maximum?
b. What is the electric field strength at this point?

47. ‖ Charge Q is uniformly distributed along a thin, flexible rod
CALC of length L. The rod is then bent into the semicircle shown in **FIGURE P23.47**.
a. Find an expression for the electric field $\vec{E}$ at the center of the semicircle.
Hint: A small piece of arc length Δs spans a small angle $\Delta \theta = \Delta s/R$, where R is the radius.
b. Evaluate the field strength if $L = 10$ cm and $Q = 30$ nC.

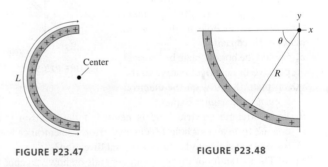

FIGURE P23.47   **FIGURE P23.48**

48. ‖ A plastic rod with linear charge density λ is bent into the
CALC quarter circle shown in **FIGURE P23.48**. We want to find the electric field at the origin.
a. Write expressions for the x- and y-components of the electric field at the origin due to a small piece of charge at angle θ.
b. Write, but do not evaluate, definite integrals for the x- and y-components of the net electric field at the origin.
c. Evaluate the integrals and write $\vec{E}_{net}$ in component form.

49. ‖ An infinite plane of charge with surface charge density $3.2 \ \mu C/m^2$ has a 20-cm-diameter circular hole cut out of it. What is the electric field strength directly over the center of the hole at a distance of 12 cm?
Hint: Can you create this charge distribution as a superposition of charge distributions for which you know the electric field?

50. ‖ A sphere of radius R and surface charge density η is positioned with its center distance $2R$ from an infinite plane with surface charge density η. At what distance from the plane, along a line toward the center of the sphere, is the electric field zero?

51. ‖ A parallel-plate capacitor has 2.0 cm × 2.0 cm electrodes with surface charge densities $\pm 1.0 \times 10^{-6}$ C/m². A proton traveling parallel to the electrodes at 1.0×10^6 m/s enters the center of the gap between them. By what distance has the proton been deflected sideways when it reaches the far edge of the capacitor? Assume the field is uniform inside the capacitor and zero outside the capacitor.

52. ‖ An electron is launched at a 45° angle and a speed of 5.0×10^6 m/s from the positive plate of the parallel-plate capacitor shown in **FIGURE P23.52**. The electron lands 4.0 cm away.
   a. What is the electric field strength inside the capacitor?
   b. What is the smallest possible spacing between the plates?

FIGURE P23.52

FIGURE P23.53

53. ‖‖ The two parallel plates in **FIGURE P23.53** are 2.0 cm apart and the electric field strength between them is 1.0×10^4 N/C. An electron is launched at a 45° angle from the positive plate. What is the maximum initial speed v_0 the electron can have without hitting the negative plate?

54. ‖ A problem of practical interest is to make a beam of electrons turn a 90° corner. This can be done with the parallel-plate capacitor shown in **FIGURE P23.54**. An electron with kinetic energy 3.0×10^{-17} J enters through a small hole in the bottom plate of the capacitor.
   a. Should the bottom plate be charged positive or negative relative to the top plate if you want the electron to turn to the right? Explain.
   b. What strength electric field is needed if the electron is to emerge from an exit hole 1.0 cm away from the entrance hole, traveling at right angles to its original direction?
   Hint: The difficulty of this problem depends on how you choose your coordinate system.
   c. What minimum separation d_{min} must the capacitor plates have?

FIGURE P23.54

55. ‖‖ A *positron* is an elementary particle identical to an electron except that its charge is $+e$. An electron and a positron can rotate about their center of mass as if they were a dumbbell connected by a massless rod. What is the orbital frequency for an electron and a positron 1.0 nm apart?

56. ‖‖ Your physics assignment is to figure out a way to use electricity to launch a small 6.0-cm-long plastic drink stirrer. You decide that you'll charge the little plastic rod by rubbing it with fur, then hold it near a long, charged wire, as shown in **FIGURE P23.56**. When you let go, the electric

FIGURE P23.56

force of the wire on the plastic rod will shoot it away. Suppose you can uniformly charge the plastic stirrer to 10 nC and that the linear charge density of the long wire is 1.0×10^{-7} C/m. What is the net electric force on the plastic stirrer if the end closest to the wire is 2.0 cm away?
Hint: The stirrer cannot be modeled as a point charge; an integration is required.

57. ‖‖‖ The combustion of fossil fuels produces micron-sized particles of soot, one of the major components of air pollution. The terminal speeds of these particles are extremely small, so they remain suspended in air for very long periods of time. Furthermore, very small particles almost always acquire small amounts of charge from cosmic rays and various atmospheric effects, so their motion is influenced not only by gravity but also by the earth's weak electric field. Consider a small spherical particle of radius r, density ρ, and charge q. A small sphere moving with speed v experiences a drag force $F_{drag} = 6\pi\eta rv$, where η is the viscosity of the air. (This differs from the drag force you learned in Chapter 6 because there we considered macroscopic rather than microscopic objects.)
   a. A particle falling at its terminal speed v_{term} is in equilibrium with no net force. Write Newton's first law for this particle falling in the presence of a *downward* electric field of strength E, then solve to find an expression for v_{term}.
   b. Soot is primarily carbon, and carbon in the form of graphite has a density of 2200 kg/m³. In the absence of an electric field, what is the terminal speed in mm/s of a 1.0-μm-diameter graphite particle? The viscosity of air at 20°C is 1.8×10^{-5} kg/m s.
   c. The earth's electric field is typically (150 N/C, downward). In this field, what is the terminal speed in mm/s of a 1.0-μm-diameter graphite particle that has acquired 250 extra electrons?

58. ‖ A 2.0-mm-diameter glass sphere has a charge of +1.0 nC. What speed does an electron need to orbit the sphere 1.0 mm above the surface?

59. ‖ In a classical model of the hydrogen atom, the electron orbits the proton in a circular orbit of radius 0.053 nm. What is the orbital frequency? The proton is so much more massive than the electron that you can assume the proton is at rest.

60. ‖ An electric field can *induce* an electric dipole in a neutral atom or molecule by pushing the positive and negative charges in opposite directions. The dipole moment of an induced dipole is directly proportional to the electric field. That is, $\vec{p} = \alpha \vec{E}$, where α is called the *polarizability* of the molecule. A bigger field stretches the molecule farther and causes a larger dipole moment.
   a. What are the units of α?
   b. An ion with charge q is distance r from a molecule with polarizability α. Find an expression for the force $\vec{F}_{ion\ on\ dipole}$.

61. ‖ Show that an infinite line of charge with linear charge density λ exerts an attractive force on an electric dipole with magnitude $F = 2\lambda p / 4\pi\epsilon_0 r^2$. Assume that r, the distance from the line, is much larger than the charge separation in the dipole.

62. ‖‖‖ The ozone molecule O_3 has a permanent dipole moment of 1.8×10^{-30} C m. Although the molecule is very slightly bent—which is why it has a dipole moment—it can be modeled as a uniform rod of length 2.5×10^{-10} m with the dipole moment perpendicular to the axis of the rod. Suppose an ozone molecule is in a 5000 N/C uniform electric field. In equilibrium, the dipole moment is aligned with the electric field. But if the molecule is rotated by a *small* angle and released, it will oscillate back and forth in simple harmonic motion. What is the frequency f of oscillation?

In Problems 63 through 66 you are given the equation(s) used to solve a problem. For each of these

a. Write a realistic problem for which this is the correct equation(s).
b. Finish the solution of the problem.

63. $(9.0 \times 10^9 \, \text{N m}^2/\text{C}^2) \dfrac{(2.0 \times 10^{-9} \, \text{C}) \, s}{(0.025 \, \text{m})^3} = 1150 \, \text{N/C}$

64. $(9.0 \times 10^9 \, \text{N m}^2/\text{C}^2) \dfrac{2(2.0 \times 10^{-7} \, \text{C/m})}{r} = 25,000 \, \text{N/C}$

65. $\dfrac{\eta}{2\epsilon_0} \left[1 - \dfrac{z}{\sqrt{z^2 + R^2}} \right] = \dfrac{1}{2} \dfrac{\eta}{2\epsilon_0}$

66. $2.0 \times 10^{12} \, \text{m/s}^2 = \dfrac{(1.60 \times 10^{-19} \, \text{C}) E}{(1.67 \times 10^{-27} \, \text{kg})}$

$E = \dfrac{Q}{(8.85 \times 10^{-12} \, \text{C}^2/\text{N m}^2)(0.020 \, \text{m})^2}$

Challenge Problems

67. ||| A rod of length L lies along the y-axis with its center at the
CALC origin. The rod has a nonuniform linear charge density $\lambda = a|y|$, where a is a constant with the units C/m^2.
   a. Draw a graph of λ versus y over the length of the rod.
   b. Determine the constant a in terms of L and the rod's total charge Q.
   c. Find the electric field strength of the rod at distance x on the x-axis.

68. ||| a. An infinitely long *sheet* of charge of width L lies in the xy-
CALC   plane between $x = -L/2$ and $x = L/2$. The surface charge density is η. Derive an expression for the electric field $\vec{E}$ at height z above the centerline of the sheet.
   b. Verify that your expression has the expected behavior if $z \ll L$ and if $z \gg L$.
   c. Draw a graph of field strength E versus z.

69. ||| a. An infinitely long *sheet* of charge of width L lies in the xy-
CALC   plane between $x = -L/2$ and $x = L/2$. The surface charge density is η. Derive an expression for the electric field $\vec{E}$ along the x-axis for points outside the sheet $(x > L/2)$.
   b. Verify that your expression has the expected behavior if $x \gg L$.
   Hint: $\ln(1 + u) \approx u$ if $u \ll 1$.
   c. Draw a graph of field strength E versus x for $x > L/2$.

70. ||| A thin cylindrical shell of radius R and length L, like a soda
CALC straw, is uniformly charged with surface charge density η. What is the electric field strength at the center of one end of the cylinder?

71. ||| One type of ink-jet printer, called an electrostatic ink-jet printer, forms the letters by using deflecting electrodes to steer charged ink drops up and down vertically as the ink jet sweeps horizontally across the page. The ink jet forms 30-μm-diameter drops of ink, charges them by spraying 800,000 electrons on the surface, and shoots them toward the page at a speed of 20 m/s. Along the way, the drops pass through two horizontal, parallel electrodes that are 6.0 mm long, 4.0 mm wide, and spaced 1.0 mm apart. The distance from the center of the electrodes to the paper is 2.0 cm. To form the tallest letters, which have a height of 6.0 mm, the drops need to be deflected upward (or downward) by 3.0 mm. What electric field strength is needed between the electrodes to achieve this deflection? Ink, which consists of dye particles suspended in alcohol, has a density of 800 kg/m^3.

72. ||| A proton orbits a long charged wire, making 1.0×10^6 revolutions per second. The radius of the orbit is 1.0 cm. What is the wire's linear charge density?

73. ||| You have a summer intern position with a company that designs and builds nanomachines. An engineer with the company is designing a microscopic oscillator to help keep time, and you've been assigned to help him analyze the design. He wants to place a negative charge at the center of a very small, positively charged metal ring. His claim is that the negative charge will undergo simple harmonic motion at a frequency determined by the amount of charge on the ring.
   a. Consider a negative charge near the center of a positively charged ring centered on the z-axis. Show that there is a restoring force on the charge if it moves along the z-axis but stays close to the center of the ring. That is, show there's a force that tries to keep the charge at $z = 0$.
   b. Show that for *small* oscillations, with amplitude $\ll R$, a particle of mass m with charge $-q$ undergoes simple harmonic motion with frequency

$$f = \frac{1}{2\pi} \sqrt{\frac{qQ}{4\pi\epsilon_0 m R^3}}$$

   R and Q are the radius and charge of the ring.
   c. Evaluate the oscillation frequency for an electron at the center of a 2.0-μm-diameter ring charged to 1.0×10^{-13} C.

24 Gauss's Law

An electric field image of blood plasma from healthy blood. The wire in the center creates the electric field. Variations in the shape and color of the pattern can give early warning of cancer.

IN THIS CHAPTER, you will learn about and apply Gauss's law.

What is Gauss's law?

Gauss's law is a general statement about the nature of electric fields. It is more fundamental than Coulomb's law and is the first of what we will later call Maxwell's equations, the governing equations of electricity and magnetism.

Gauss's law says that the electric flux through a closed surface is proportional to the amount of charge Q_{in} enclosed within the surface. This seemingly abstract statement will be the basis of a powerful strategy for finding the electric fields of charge distributions that have a high degree of symmetry.

« LOOKING BACK Section 22.5 The electric field of a point charge Section 23.2 Electric field lines

What good is symmetry?

For charge distributions with a high degree of symmetry, the symmetry of the electric field must match the symmetry of the charge distribution. Important symmetries are planar symmetry, cylindrical symmetry, and spherical symmetry. The concept of symmetry plays an important role in math and science.

Cylindrical symmetry

What is electric flux?

The amount of electric field passing through a surface is called the electric flux. Electric flux is analogous to the amount of air or water flowing through a loop. You will learn to calculate the flux through open and closed surfaces.

« LOOKING BACK Section 9.3 Vector dot products

How is Gauss's law used?

Gauss's law is easier to use than superposition for finding the electric field both inside and outside of charged spheres, cylinders, and planes. To use Gauss's law, you calculate the electric flux through a Gaussian surface surrounding the charge. This will turn out to be much easier than it sounds!

Gaussian surface

What can we learn about conductors?

Gauss's law can be used to establish several properties of conductors in electrostatic equilibrium. In particular:

- Any excess charge is all on the surface.
- The interior electric field is zero.
- The external field is perpendicular to the surface.

24.1 Symmetry

To continue our exploration of electric fields, suppose we knew only two things:

1. The field points away from positive charges, toward negative charges, and
2. An electric field exerts a force on a charged particle.

From this information alone, what can we deduce about the electric field of the infinitely long charged cylinder shown in **FIGURE 24.1**?

We don't know if the cylinder's diameter is large or small. We don't know if the charge density is the same at the outer edge as along the axis. All we know is that the charge is positive and the charge distribution has *cylindrical symmetry*. We say that a charge distribution is **symmetric** if there is a group of *geometric transformations* that don't cause any *physical* change.

To make this idea concrete, suppose you close your eyes while a friend transforms a charge distribution in one of the following three ways. He or she can

- *Translate* (that is, displace) the charge parallel to an axis,
- *Rotate* the charge about an axis, or
- *Reflect* the charge in a mirror.

When you open your eyes, will you be able to tell if the charge distribution has been changed? You might tell by observing a visual difference in the distribution. Or the results of an experiment with charged particles could reveal that the distribution has changed. If nothing you can see or do reveals any change, then we say that the charge distribution is symmetric under that particular transformation.

FIGURE 24.2 shows that the charge distribution of Figure 24.1 is symmetric with respect to

- Translation parallel to the cylinder axis. Shifting an infinitely long cylinder by 1 mm or 1000 m makes no noticeable or measurable change.
- Rotation by any angle about the cylinder axis. Turning a cylinder about its axis by 1° or 100° makes no detectable change.
- Reflections in any plane containing or perpendicular to the cylinder axis. Exchanging top and bottom, front and back, or left and right makes no detectable change.

A charge distribution that is symmetric under these three groups of geometric transformations is said to be *cylindrically symmetric*. Other charge distributions have other types of symmetries. Some charge distributions have no symmetry at all.

Our interest in symmetry can be summed up in a single statement:

> **The symmetry of the electric field must match the symmetry of the charge distribution.**

If this were not true, you could use the electric field to test whether the charge distribution had undergone a transformation.

Now we're ready to see what we can learn about the electric field in Figure 24.1. Could the field look like **FIGURE 24.3a**? (Imagine this picture rotated about the axis.) That is, is this a *possible* field? This field looks the same if it's translated parallel to the

FIGURE 24.1 A charge distribution with cylindrical symmetry.

Infinitely long
charged cylinder

FIGURE 24.2 Transformations that don't change an infinite cylinder of charge.

Original
cylinder

Translation
parallel to
the axis

Rotation
about the
axis

Reflection
in plane
containing
the axis

Reflection
perpendicular
to the axis

FIGURE 24.3 Could the field of a cylindrical charge distribution look like this?

(a) Is this a possible electric field of an infinitely long charged cylinder? Suppose the charge and the field are reflected in a plane perpendicular to the axis.

Reflection plane
$\vec{E}$
$\vec{E}$
Reflect

(b) The charge distribution is not changed by the reflection, but the field is. This field doesn't match the symmetry of the cylinder, so the cylinder's field can't look like this.

$\vec{E}$
$\vec{E}$

FIGURE 24.4 Or might the field of a cylindrical charge distribution look like this?

(a)

End view of cylinder — Reflection plane

The charge distribution is not changed by reflecting it in a plane containing the axis.

$\vec{E}$

Reflect

(b)

This field *is* changed. It doesn't match the symmetry of the cylinder, so the field can't look like this.

$\vec{E}$

cylinder axis, if up and down are exchanged by reflecting the field in a plane coming out of the page, or if you rotate the cylinder about its axis.

But the proposed field fails one test: reflection in a plane perpendicular to the axis, a reflection that exchanges left and right. This reflection, which would *not* make any change in the charge distribution itself, produces the field shown in **FIGURE 24.3b**. This change in the field is detectable because a positively charged particle would now have a component of motion to the left instead of to the right.

The field of Figure 24.3a, which makes a distinction between left and right, is not cylindrically symmetric and thus is *not* a possible field. In general, **the electric field of a cylindrically symmetric charge distribution cannot have a component parallel to the cylinder axis.**

Well then, what about the electric field shown in **FIGURE 24.4a**? Here we're looking down the axis of the cylinder. The electric field vectors are restricted to planes perpendicular to the cylinder and thus do not have any component parallel to the cylinder axis. This field is symmetric for rotations about the axis, but it's *not* symmetric for a reflection in a plane containing the axis.

The field of **FIGURE 24.4b**, after this reflection, is easily distinguishable from the field of Figure 24.4a. Thus **the electric field of a cylindrically symmetric charge distribution cannot have a component tangent to the circular cross section.**

FIGURE 24.5 shows the only remaining possible field shape. The electric field is radial, pointing straight out from the cylinder like the bristles on a bottle brush. This is the one electric field shape matching the symmetry of the charge distribution.

FIGURE 24.5 This is the only shape for the electric field that matches the symmetry of the charge distribution.

What Good Is Symmetry?

Given how little we assumed about Figure 24.1—that the charge distribution is cylindrically symmetric and that electric fields point away from positive charges— we've been able to deduce a great deal about the electric field. In particular, we've deduced the *shape* of the electric field.

Now, shape is not everything. We've learned nothing about the strength of the field or how strength changes with distance. Is E constant? Does it decrease like $1/r$ or $1/r^2$? We don't yet have a complete description of the field, but knowing what shape the field *has* to have will make finding the field strength a much easier task.

That's the good of symmetry. Symmetry arguments allow us to *rule out* many conceivable field shapes as simply being incompatible with the symmetry of the charge distribution. Knowing what doesn't happen, or can't happen, is often as useful as knowing what does happen. By the process of elimination, we're led to the one and only shape the field can possibly have. Reasoning on the basis of symmetry is a sometimes subtle but always powerful means of reasoning.

Three Fundamental Symmetries

Three fundamental symmetries appear frequently in electrostatics. The first row of **FIGURE 24.6** shows the simplest form of each symmetry. The second row shows a more complex, but more realistic, situation with the same symmetry.

FIGURE 24.6 Three fundamental symmetries.

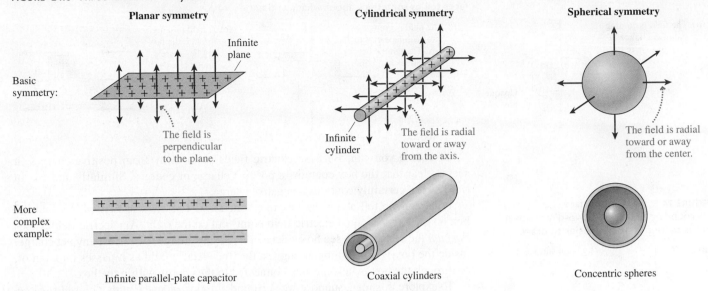

NOTE Figures must be finite in extent, but the planes and cylinders in Figure 24.6 are assumed to be infinite.

Objects do exist that are extremely close to being perfect spheres, but no real cylinder or plane can be infinite in extent. Even so, the fields of infinite planes and cylinders are good models for the fields of finite planes and cylinders at points not too close to an edge or an end. The fields that we'll study in this chapter, even if idealized, have many important applications.

STOP TO THINK 24.1 A uniformly charged rod has a *finite* length L. The rod is symmetric under rotations about the axis and under reflection in any plane containing the axis. It is *not* symmetric under translations or under reflections in a plane perpendicular to the axis unless that plane bisects the rod. Which field shape or shapes match the symmetry of the rod?

24.2 The Concept of Flux

FIGURE 24.7a on the next page shows an opaque box surrounding a region of space. We can't see what's in the box, but there's an electric field vector coming out of each face of the box. Can you figure out what's in the box?

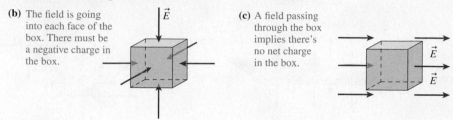

FIGURE 24.7 Although we can't see into the boxes, the electric fields passing through the faces tell us something about what's in them.

(a) The field is coming out of each face of the box. There must be a positive charge in the box.

Opaque box

(b) The field is going into each face of the box. There must be a negative charge in the box.

(c) A field passing through the box implies there's no net charge in the box.

FIGURE 24.8 Gaussian surface surrounding a charge. A two-dimensional cross section is usually easier to draw.

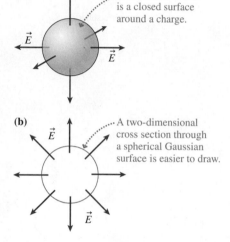

(a) A Gaussian surface is a closed surface around a charge.

(b) A two-dimensional cross section through a spherical Gaussian surface is easier to draw.

Of course you can. Because electric fields point away from positive charges, it seems clear that the box contains a positive charge or charges. Similarly, the box in **FIGURE 24.7b** certainly contains a negative charge.

What can we tell about the box in **FIGURE 24.7c**? The electric field points into the box on the left. An equal electric field points out on the right. An electric field passes *through* the box, but we see no evidence there's any charge (or at least any net charge) inside the box. These examples suggest that the electric field as it passes into, out of, or through the box is in some way connected to the charge within the box.

To explore this idea, suppose we surround a region of space with a *closed surface,* a surface that divides space into distinct inside and outside regions. Within the context of electrostatics, a closed surface through which an electric field passes is called a **Gaussian surface,** named after the 19th-century mathematician Karl Gauss. This is an imaginary, mathematical surface, not a physical surface, although it might coincide with a physical surface. For example, **FIGURE 24.8a** shows a spherical Gaussian surface surrounding a charge.

A closed surface must, of necessity, be a surface in three dimensions. But three-dimensional pictures are hard to draw, so we'll often look at two-dimensional cross sections through a Gaussian surface, such as the one shown in **FIGURE 24.8b**. Now we can tell from the *spherical symmetry* of the electric field vectors poking through the surface that the positive charge inside must be spherically symmetric and centered at the *center* of the sphere.

FIGURE 24.9 A Gaussian surface is most useful when it matches the shape of the field.

(a)

Cylindrical Gaussian surface

(b)

Two-dimensional cross sections of a Gaussian surface

A Gaussian surface is most useful when it matches the shape and symmetry of the field. For example, **FIGURE 24.9a** shows a cylindrical Gaussian surface—a *closed* cylinder—surrounding some kind of cylindrical charge distribution, such as a charged wire. **FIGURE 24.9b** simplifies the drawing by showing two-dimensional end and side views. Because the Gaussian surface matches the symmetry of the charge distribution, the electric field is everywhere *perpendicular* to the side wall and *no* field passes through the top and bottom surfaces.

For contrast, consider the spherical surface in **FIGURE 24.10**. This is also a Gaussian surface, and the protruding electric field tells us there's a positive charge inside. It might be a point charge located on the left side, but we can't really say. A Gaussian surface that doesn't match the symmetry of the charge distribution isn't terribly useful.

These examples lead us to two conclusions:

FIGURE 24.10 Not every surface is useful for learning about charge.

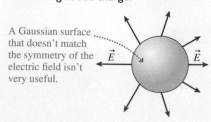

A Gaussian surface that doesn't match the symmetry of the electric field isn't very useful.

1. The electric field, in some sense, "flows" *out of* a closed surface surrounding a region of space containing a net positive charge and *into* a closed surface surrounding a net negative charge. The electric field may flow *through* a closed surface surrounding a region of space in which there is no net charge, but the *net flow* is zero.
2. The electric field pattern through the surface is particularly simple if the closed surface matches the symmetry of the charge distribution inside.

The electric field doesn't really flow like a fluid, but the metaphor is a useful one. The Latin word for flow is *flux*, and the amount of electric field passing through a surface is called the **electric flux.** Our first conclusions, stated in terms of electric flux, are

- There is an outward flux through a closed surface around a net positive charge.
- There is an inward flux through a closed surface around a net negative charge.
- There is no net flux through a closed surface around a region of space in which there is no net charge.

This chapter has been entirely qualitative thus far as we've established pictorially what we mean by symmetry, flux, and the fact that the electric flux through a closed surface has something to do with the charge inside. In the next two sections you'll learn how to calculate the electric flux through a surface and how the flux is related to the enclosed charge. That relationship, Gauss's law, will allow us to determine the electric fields of some interesting and useful charge distributions.

STOP TO THINK 24.2   This box contains

a. A positive charge.
b. A negative charge.
c. No charge.
d. A net positive charge.
e. A net negative charge.
f. No net charge.

24.3 Calculating Electric Flux

Let's start with a brief overview of where this section will take us. We'll begin with a definition of flux that is easy to understand, then we'll turn that simple definition into a formidable-looking integral. We need the integral because the simple definition applies only to uniform electric fields and flat surfaces. Those are good starting points, but we'll soon need to calculate the flux of nonuniform fields through curved surfaces.

Mathematically, the flux of a nonuniform field through a curved surface is described by a special kind of integral called a *surface integral*. It's quite possible that you have not yet encountered surface integrals in your calculus course, and the "novelty factor" contributes to making this integral look worse than it really is. We will emphasize over and over the idea that an integral is just a fancy way of doing a sum, in this case the sum of the small amounts of flux through many small pieces of a surface.

The good news is that *every* surface integral we need to evaluate in this chapter, or that you will need to evaluate for the homework problems, is either zero or is so easy that you will be able to do it in your head. This seems like an astounding claim, but you will soon see it is true. The key will be to make effective use of the *symmetry* of the electric field.

The Basic Definition of Flux

Imagine holding a rectangular wire loop of area A in front of a fan. As **FIGURE 24.11** on the next page shows, the volume of air flowing through the loop each second depends on the angle between the loop and the direction of flow. The flow is maximum through a loop that is perpendicular to the airflow; no air goes through the same loop if it lies parallel to the flow.

FIGURE 24.11 The amount of air flowing through a loop depends on the angle between $\vec{v}$ and $\hat{n}$.

(a)

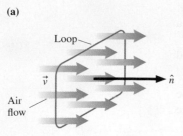

Loop

Air flow

The air flowing through the loop is maximum when $\theta = 0°$.

(b)

Unit vector normal to loop

No air flows through the loop when $\theta = 90°$.

(c) The loop is tilted by angle θ.

$v_\perp = v\cos\theta$ is the component of the air velocity perpendicular to the loop.

The flow direction is identified by the velocity vector $\vec{v}$. We can identify the loop's orientation by defining a unit vector $\hat{n}$ normal to the plane of the loop. Angle θ is then the angle between $\vec{v}$ and $\hat{n}$. The loop per pendicular to the flow in Figure 24.11a has $\theta = 0°$; the loop parallel to the flow in Figure 24.11b has $\theta = 90°$. You can think of θ as the angle by which a loop has been tilted away from perpendicular.

NOTE A surface has two sides, so $\hat{n}$ could point either way. We'll choose the side that makes $\theta \leq 90°$.

You can see from Figure 24.11c that the velocity vector $\vec{v}$ can be decomposed into components $v_\perp = v\cos\theta$ perpendicular to the loop and $v_\parallel = v\sin\theta$ parallel to the loop. Only the perpendicular component $v_\perp$ carries air *through* the loop. Consequently, the volume of air flowing through the loop each second is

$$\text{volume of air per second (m}^3\text{/s)} = v_\perp A = vA\cos\theta \tag{24.1}$$

$\theta = 0°$ is the orientation for maximum flow through the loop, as expected, and no air flows through the loop if it is tilted 90°.

An electric field doesn't flow in a literal sense, but we can apply the same idea to an electric field passing through a surface. **FIGURE 24.12** shows a surface of area A in a uniform electric field $\vec{E}$. Unit vector $\hat{n}$ is normal to the surface, and θ is the angle between $\hat{n}$ and $\vec{E}$. Only the component $E_\perp = E\cos\theta$ passes *through* the surface.

With this in mind, and using Equation 24.1 as an analog, let's define the *electric flux* Φ_e (uppercase Greek phi) as

$$\Phi_e = E_\perp A = EA\cos\theta \tag{24.2}$$

The electric flux measures the amount of electric field passing through a surface of area A if the normal to the surface is tilted at angle θ from the field.

Equation 24.2 looks very much like a vector dot product: $\vec{E} \cdot \vec{A} = EA\cos\theta$. For this idea to work, let's define an **area vector** $\vec{A} = A\hat{n}$ to be a vector in the direction of $\hat{n}$—that is, *perpendicular* to the surface—with a magnitude A equal to the area of the surface. Vector $\vec{A}$ has units of m². **FIGURE 24.13a** shows two area vectors.

FIGURE 24.12 An electric field passing through a surface.

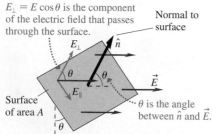

$E_\perp = E\cos\theta$ is the component of the electric field that passes through the surface.

Normal to surface

Surface of area A

θ is the angle between $\hat{n}$ and $\vec{E}$.

FIGURE 24.13 The electric flux can be defined in terms of the area vector $\vec{A}$.

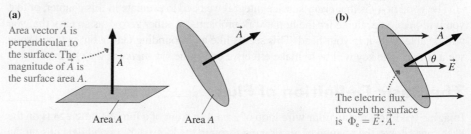

(a)

Area vector $\vec{A}$ is perpendicular to the surface. The magnitude of $\vec{A}$ is the surface area A.

Area A

Area A

(b)

The electric flux through the surface is $\Phi_e = \vec{E} \cdot \vec{A}$.

FIGURE 24.13b shows an electric field passing through a surface of area A. The angle between vectors $\vec{A}$ and $\vec{E}$ is the same angle used in Equation 24.2 to define the electric flux, so Equation 24.2 really is a dot product. We can define the electric flux more concisely as

$$\Phi_e = \vec{E} \cdot \vec{A} \quad \text{(electric flux of a constant electric field)} \qquad (24.3)$$

Writing the flux as a dot product helps make clear how angle θ is defined: θ is the angle between the electric field and a line *perpendicular* to the plane of the surface.

NOTE Figure 24.13b shows a circular area, but the shape of the surface is not relevant. However, Equation 24.3 is restricted to a *constant* electric field passing through a *planar* surface.

EXAMPLE 24.1 | **The electric flux inside a parallel-plate capacitor**

Two 100 cm^2 parallel electrodes are spaced 2.0 cm apart. One is charged to $+5.0$ nC, the other to -5.0 nC. A 1.0 cm $\times$ 1.0 cm surface between the electrodes is tilted to where its normal makes a $45°$ angle with the electric field. What is the electric flux through this surface?

MODEL Assume the surface is located near the center of the capacitor where the electric field is uniform. The electric flux doesn't depend on the shape of the surface.

VISUALIZE The surface is square, rather than circular, but otherwise the situation looks like Figure 24.13b.

SOLVE In Chapter 23, we found the electric field inside a parallel-plate capacitor to be

$$E = \frac{Q}{\epsilon_0 A_{\text{plates}}} = \frac{5.0 \times 10^{-9} \text{ C}}{(8.85 \times 10^{-12} \text{ C}^2/\text{N m}^2)(1.0 \times 10^{-2} \text{ m}^2)}$$

$$= 5.65 \times 10^4 \text{ N/C}$$

A 1.0 cm $\times$ 1.0 cm surface has $A = 1.0 \times 10^{-4}$ m^2. The electric flux through this surface is

$$\Phi_e = \vec{E} \cdot \vec{A} = EA \cos\theta$$

$$= (5.65 \times 10^4 \text{ N/C})(1.0 \times 10^{-4} \text{ m}^2) \cos 45°$$

$$= 4.0 \text{ N m}^2/\text{C}$$

ASSESS The units of electric flux are the product of electric field and area units: $\text{N m}^2/\text{C}$.

The Electric Flux of a Nonuniform Electric Field

Our initial definition of the electric flux assumed that the electric field $\vec{E}$ was constant over the surface. How should we calculate the electric flux if $\vec{E}$ varies from point to point on the surface? We can answer this question by returning to the analogy of air flowing through a loop. Suppose the airflow varies from point to point. We can still find the total volume of air passing through the loop each second by dividing the loop into many small areas, finding the flow through each small area, then adding them. Similarly, **the electric flux through a surface can be calculated as the sum of the fluxes through smaller pieces of the surface.** Because flux is a scalar, adding fluxes is easier than adding electric fields.

FIGURE 24.14 shows a surface in a nonuniform electric field. Imagine dividing the surface into many small pieces of area δA. Each little area has an area vector $\delta\vec{A}$ perpendicular to the surface. Two of the little pieces are shown in the figure. The electric fluxes through these two pieces differ because the electric fields are different.

Consider the small piece i where the electric field is $\vec{E}_i$. The small electric flux $\delta\Phi_i$ through area $(\delta\vec{A})_i$ is

$$\delta\Phi_i = \vec{E}_i \cdot (\delta\vec{A})_i \qquad (24.4)$$

The flux through every other little piece of the surface is found the same way. The total electric flux through the entire surface is then the sum of the fluxes through each of the small areas:

$$\Phi_e = \sum_i \delta\Phi_i = \sum_i \vec{E}_i \cdot (\delta\vec{A})_i \qquad (24.5)$$

Now let's go to the limit $\delta\vec{A} \rightarrow d\vec{A}$. That is, the little areas become infinitesimally small, and there are infinitely many of them. Then the sum becomes an integral, and the electric flux through the surface is

$$\Phi_e = \int_{\text{surface}} \vec{E} \cdot d\vec{A} \qquad (24.6)$$

The integral in Equation 24.6 is called a **surface integral.**

FIGURE 24.14 A surface in a nonuniform electric field.

Piece j $(\delta\vec{A})_j$

$(\delta\vec{A})_i$ $\vec{E}_j$

$\vec{E}_i$

Piece i The total area A can be divided into many small pieces of area δA. $\vec{E}$ may be different at each piece.

Equation 24.6 may look rather frightening if you haven't seen surface integrals before. Despite its appearance, a surface integral is no more complicated than integrals you know from calculus. After all, what does $\int f(x)\,dx$ really mean? This expression is a shorthand way to say "Divide the x-axis into many little segments of length δx, evaluate the function $f(x)$ in each of them, then add up $f(x)\,\delta x$ for all the segments along the line." The integral in Equation 24.6 differs only in that we're dividing a surface into little pieces instead of a line into little segments. In particular, we're summing the fluxes through a vast number of very tiny pieces.

You may be thinking, "OK, I understand the idea, but I don't know what to *do*. In calculus, I learned formulas for evaluating integrals such as $\int x^2\,dx$. How do I evaluate a surface integral?" This is a good question. We'll deal with evaluation shortly, and it will turn out that the surface integrals in electrostatics are quite easy to evaluate. But don't confuse *evaluating* the integral with understanding what the integral *means*. The surface integral in Equation 24.6 is simply a shorthand notation for the summation of the electric fluxes through a vast number of very tiny pieces of a surface.

The electric field might be different at every point on the surface, but suppose it isn't. That is, suppose a flat surface is in a uniform electric field $\vec{E}$. A field that is the same at every single point on a surface is a constant as far as the integration of Equation 24.6 is concerned, so we can take it outside the integral. In that case,

$$\Phi_e = \int_{\text{surface}} \vec{E} \cdot d\vec{A} = \int_{\text{surface}} E\cos\theta\,dA = E\cos\theta \int_{\text{surface}} dA \tag{24.7}$$

The integral that remains in Equation 24.7 tells us to add up all the little areas into which the full surface was subdivided. But the sum of all the little areas is simply the area of the surface:

$$\int_{\text{surface}} dA = A \tag{24.8}$$

This idea—that the surface integral of dA is the area of the surface—is one we'll use to evaluate most of the surface integrals of electrostatics. If we substitute Equation 24.8 into Equation 24.7, we find that the electric flux in a uniform electric field is $\Phi_e = EA\cos\theta$. We already knew this, from Equation 24.2, but it was important to see that the surface integral of Equation 24.6 gives the correct result for the case of a uniform electric field.

The Flux Through a Curved Surface

Most of the Gaussian surfaces we considered in the last section were curved surfaces. FIGURE 24.15 shows an electric field passing through a curved surface. How do we find the electric flux through this surface? Just as we did for a flat surface!

Divide the surface into many small pieces of area δA. For each, define the area vector $\delta\vec{A}$ perpendicular to the surface *at that point*. Compared to Figure 24.14, the only difference that the curvature of the surface makes is that the $\delta\vec{A}$ are no longer parallel to each other. Find the small electric flux $\delta\Phi_i = \vec{E}_i \cdot (\delta\vec{A})_i$ through each little area, then add them all up. The result, once again, is

$$\Phi_e = \int_{\text{surface}} \vec{E} \cdot d\vec{A} \tag{24.9}$$

We *assumed*, in deriving this expression the first time, that the surface was flat and that all the $\delta\vec{A}$ were parallel to each other. But that assumption wasn't necessary. The *meaning* of Equation 24.9—a summation of the fluxes through a vast number of very tiny pieces—is unchanged if the pieces lie on a curved surface.

FIGURE 24.15 A curved surface in an electric field.

The flux through this little piece is $\delta\Phi_i = \vec{E}_i \cdot (\delta\vec{A})_i$.

Curved surface of total area A

We seem to be getting more and more complex, using surface integrals first for nonuniform fields and now for curved surfaces. But consider the two situations shown in FIGURE 24.16. The electric field $\vec{E}$ in Figure 24.16a is everywhere tangent, or parallel, to the curved surface. We don't need to know the magnitude of $\vec{E}$ to recognize that $\vec{E} \cdot d\vec{A}$ is *zero at every point* on the surface because $\vec{E}$ is perpendicular to $d\vec{A}$ at every point. Thus $\Phi_e = 0$. A tangent electric field never pokes through the surface, so it has no flux through the surface.

The electric field in Figure 24.16b is everywhere perpendicular to the surface *and* has the same magnitude E at every point. $\vec{E}$ differs in direction at different points on a curved surface, but at any particular point $\vec{E}$ is parallel to $d\vec{A}$ and $\vec{E} \cdot d\vec{A}$ is simply $E\,dA$. In this case,

$$\Phi_e = \int_{\text{surface}} \vec{E} \cdot d\vec{A} = \int_{\text{surface}} E\,dA = E \int_{\text{surface}} dA = EA \qquad (24.10)$$

As we evaluated the integral, the fact that E has the same magnitude at every point on the surface allowed us to bring the constant value outside the integral. We then used the fact that the integral of dA over the surface is the surface area A.

We can summarize these two situations with a Tactics Box.

(a)

$\vec{E}$ is everywhere tangent to the surface. The flux is zero.

(b)

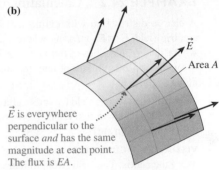

$\vec{E}$ is everywhere perpendicular to the surface *and* has the same magnitude at each point. The flux is EA.

TACTICS BOX 24.1   (MP)

Evaluating surface integrals

❶ If the electric field is everywhere tangent to a surface, the electric flux through the surface is $\Phi_e = 0$.

❷ If the electric field is everywhere perpendicular to a surface *and* has the same magnitude E at every point, the electric flux through the surface is $\Phi_e = EA$.

These two results will be of immeasurable value for using Gauss's law because *every* flux we'll need to calculate will be one of these situations. This is the basis for our earlier claim that the evaluation of surface integrals is not going to be difficult.

The Electric Flux Through a Closed Surface

Our final step, to calculate the electric flux through a closed surface such as a box, a cylinder, or a sphere, requires nothing new. We've already learned how to calculate the electric flux through flat and curved surfaces, and a closed surface is nothing more than a surface that happens to be closed.

However, the mathematical notation for the surface integral over a closed surface differs slightly from what we've been using. It is customary to use a little circle on the integral sign to indicate that the surface integral is to be performed over a closed surface. With this notation, the electric flux through a closed surface is

$$\Phi_e = \oint \vec{E} \cdot d\vec{A} \qquad (24.11)$$

Only the notation has changed. The electric flux is still the summation of the fluxes through a vast number of tiny pieces, pieces that now cover a closed surface.

NOTE A closed surface has a distinct inside and outside. The area vector $d\vec{A}$ is defined to always point *toward the outside*. This removes an ambiguity that was present for a single surface, where $d\vec{A}$ could point to either side.

Now we're ready to calculate the flux through a closed surface.

EXAMPLE 24.2 | Calculating the electric flux through a closed cylinder

A charge distribution with cylindrical symmetry has created the electric field $\vec{E} = E_0(r^2/r_0^2)\hat{r}$, where E_0 and r_0 are constants and where unit vector $\hat{r}$ lies in the xy-plane. Calculate the electric flux through a closed cylinder of length L and radius R that is centered along the z-axis.

MODEL The electric field extends radially outward from the z-axis with cylindrical symmetry. The z-component is $E_z = 0$. The cylinder is a Gaussian surface.

VISUALIZE FIGURE 24.17a is a view of the electric field looking along the z-axis. The field strength increases with increasing radial distance, and it's symmetric about the z-axis. FIGURE 24.17b is the closed Gaussian surface for which we need to calculate the electric flux. We can place the cylinder anywhere along the z-axis because the electric field extends forever in that direction.

FIGURE 24.17 The electric field and the closed surface through which we will calculate the electric flux.

(a)
Electric field, looking along the z-axis

(b)
Radius R
The field is everywhere perpendicular to the wall.
Gaussian surface
There is no field through the end.

SOLVE To calculate the flux, we divide the closed cylinder into three surfaces: the top, the bottom, and the cylindrical wall. The electric field is tangent to the surface at every point on the top and bottom surfaces. Hence, according to step 1 in Tactics Box 24.1, the flux through those two surfaces is zero. For the cylindrical wall, the electric field is perpendicular to the surface at every point *and* has the constant magnitude $E = E_0(R^2/r_0^2)$ at every point on the surface. Thus, from step 2 in Tactics Box 24.1,

$$\Phi_{\text{wall}} = EA_{\text{wall}}$$

If we add the three pieces, the net flux through the closed surface is

$$\Phi_e = \oint \vec{E} \cdot d\vec{A} = \Phi_{\text{top}} + \Phi_{\text{bottom}} + \Phi_{\text{wall}} = 0 + 0 + EA_{\text{wall}}$$
$$= EA_{\text{wall}}$$

We've evaluated the surface integral, using the two steps in Tactics Box 24.1, and there was nothing to it! To finish, all we need to recall is that the surface area of a cylindrical wall is circumference × height, or $A_{\text{wall}} = 2\pi RL$. Thus

$$\Phi_e = \left(E_0 \frac{R^2}{r_0^2}\right)(2\pi RL) = \frac{2\pi LR^3}{r_0^2} E_0$$

ASSESS LR^3/r_0^2 has units of m², an area, so this expression for Φ_e has units of N m²/C. These are the correct units for electric flux, giving us confidence in our answer. Notice the important role played by symmetry. The electric field was perpendicular to the wall and of constant value at every point on the wall *because* the Gaussian surface had the same symmetry as the charge distribution. We would not have been able to evaluate the surface integral in such an easy way for a surface of any other shape. Symmetry is the key.

STOP TO THINK 24.3 The total electric flux through this box is

a. 0 N m²/C
b. 1 N m²/C
c. 2 N m²/C
d. 4 N m²/C
e. 6 N m²/C
f. 8 N m²/C

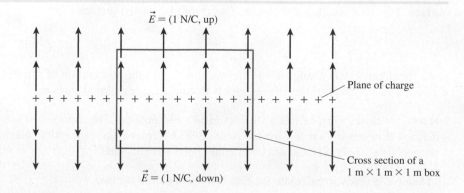

$\vec{E} = (1 \text{ N/C, up})$

Plane of charge

$\vec{E} = (1 \text{ N/C, down})$

Cross section of a 1 m × 1 m × 1 m box

24.4 Gauss's Law

The last section was long, but knowing how to calculate the electric flux through a closed surface is essential for the main topic of this chapter: Gauss's law. Gauss's law is equivalent to Coulomb's law for static charges, although Gauss's law will look very different.

The purpose of learning Gauss's law is twofold:

■ Gauss's law allows the electric fields of some continuous distributions of charge to be found much more easily than does Coulomb's law.

■ Gauss's law is valid for *moving* charges, but Coulomb's law is not (although it's a very good approximation for velocities that are much less than the speed of light). Thus Gauss's law is ultimately a more fundamental statement about electric fields than is Coulomb's law.

Let's start with Coulomb's law for the electric field of a point charge. FIGURE 24.18 shows a spherical Gaussian surface of radius r centered on a positive charge q. Keep in mind that this is an imaginary, mathematical surface, not a physical surface. There is a net flux through this surface because the electric field points outward at every point on the surface. To evaluate the flux, given formally by the surface integral of Equation 24.11, notice that the electric field is perpendicular to the surface at every point on the surface *and*, from Coulomb's law, it has the same magnitude $E = q/4\pi\epsilon_0 r^2$ at every point on the surface. This simple situation arises because **the Gaussian surface has the same symmetry as the electric field.**

Thus we know, without having to do any hard work, that the flux integral is

$$\Phi_e = \oint \vec{E} \cdot d\vec{A} = EA_{\text{sphere}} \qquad (24.12)$$

The surface area of a sphere of radius r is $A_{\text{sphere}} = 4\pi r^2$. If we use A_{sphere} and the Coulomb-law expression for E in Equation 24.12, we find that the electric flux through the spherical surface is

$$\Phi_e = \frac{q}{4\pi\epsilon_0 r^2} 4\pi r^2 = \frac{q}{\epsilon_0} \qquad (24.13)$$

You should examine the logic of this calculation closely. We really did evaluate the surface integral of Equation 24.11, although it may appear, at first, as if we didn't do much. The integral was easily evaluated, we reiterate for emphasis, because the closed surface on which we performed the integration matched the *symmetry* of the charge distribution.

NOTE We found Equation 24.13 for a positive charge, but it applies equally to negative charges. According to Equation 24.13, Φ_e is negative if q is negative. And that's what we would expect from the basic definition of flux, $\vec{E} \cdot \vec{A}$. The electric field of a negative charge points inward, while the area vector of a closed surface points outward, making the dot product negative.

Electric Flux Is Independent of Surface Shape and Radius

Notice something interesting about Equation 24.13. The electric flux depends on the amount of charge but *not* on the radius of the sphere. Although this may seem a bit surprising, it's really a direct consequence of what we *mean* by flux. Think of the fluid analogy with which we introduced the term "flux." If fluid flows outward from a central point, all the fluid crossing a small-radius spherical surface will, at some later time, cross a large-radius spherical surface. No fluid is lost along the way, and no new fluid is created. Similarly, the point charge in FIGURE 24.19 is the only source of electric field. Every electric field line passing through a small-radius spherical surface also passes through a large-radius spherical surface. Hence the electric flux is independent of r.

FIGURE 24.18 A spherical Gaussian surface surrounding a point charge.

Cross section of a Gaussian sphere of radius r. This is a mathematical surface, not a physical surface.

Point charge q

The electric field is everywhere perpendicular to the surface *and* has the same magnitude at every point.

FIGURE 24.19 The electric flux is the same through *every* sphere centered on a point charge.

Every field line passes through the smaller *and* the larger sphere. The flux through the two spheres is the same.

NOTE This argument hinges on the fact that Coulomb's law is an inverse-square force law. The electric field strength, which is proportional to $1/r^2$, decreases with distance. But the surface area, which increases in proportion to r^2, exactly compensates for this decrease. Consequently, the electric flux of a point charge through a spherical surface is independent of the radius of the sphere.

This conclusion about the flux has an extremely important generalization. **FIGURE 24.20a** shows a point charge and a closed Gaussian surface of arbitrary shape and dimensions. All we know is that the charge is *inside* the surface. What is the electric flux through this arbitrary surface?

One way to answer the question is to approximate the surface as a patchwork of spherical and radial pieces. The spherical pieces are centered on the charge and the radial pieces lie along lines extending outward from the charge. (Figure 24.20 is a two-dimensional drawing so you need to imagine these arcs as actually being pieces of a spherical shell.) The figure, of necessity, shows fairly large pieces that don't match the actual surface all that well. However, we can make this approximation as good as we want by letting the pieces become sufficiently small.

The electric field is everywhere tangent to the radial pieces. Hence the electric flux through the radial pieces is zero. The spherical pieces, although at varying distances from the charge, form a *complete sphere*. That is, any line drawn radially outward from the charge will pass through exactly one spherical piece, and no radial lines can avoid passing through a spherical piece. You could even imagine, as **FIGURE 24.20b** shows, sliding the spherical pieces in and out *without changing the angle they subtend* until they come together to form a complete sphere.

Consequently, the electric flux through these spherical pieces that, when assembled, form a complete sphere must be exactly the same as the flux q/ϵ_0 through a spherical Gaussian surface. In other words, **the flux through *any* closed surface surrounding a point charge q is**

$$\Phi_e = \oint \vec{E} \cdot d\vec{A} = \frac{q}{\epsilon_0} \tag{24.14}$$

This surprisingly simple result is a consequence of the fact that Coulomb's law is an inverse-square force law. Even so, the reasoning that got us to Equation 24.14 is rather subtle and well worth reviewing.

Charge Outside the Surface

The closed surface shown in **FIGURE 24.21a** has a point charge q outside the surface but no charges inside. Now what can we say about the flux? By approximating this surface with spherical and radial pieces *centered on the charge,* as we did in Figure 24.20, we can reassemble the surface into the equivalent surface of **FIGURE 24.21b**. This closed

FIGURE 24.20 An arbitrary Gaussian surface can be approximated with spherical and radial pieces.

(a)

The spherical pieces are centered on the charge.

Point charge

Gaussian surface of arbitrary shape

The radial pieces are along lines extending out from the charge. There's no flux through these.

(b)

The spherical pieces can slide in or out to form a complete sphere. Hence the flux through the pieces is the same as the flux through a sphere.

FIGURE 24.21 A point charge outside a Gaussian surface.

(a)

The flux is negative on some pieces of the surface.

The flux is positive on some pieces of the surface.

Closed surface

Point charge outside surface

$\vec{E}$

Approximating this surface with spherical and radial pieces allows it to be reassembled as the surface in part (b) that has the same flux.

(b)

$\vec{A}$ is parallel to $\vec{E}$, so the flux is positive.

$\vec{A}$

Two-dimensional cross section

$\vec{A}$ is opposite to $\vec{E}$, so the flux is negative.

$\vec{E}$

The fluxes through these surfaces are equal but opposite. The net flux is zero.

surface consists of sections of two spherical shells, and it is equivalent in the sense that the electric flux through this surface is the same as the electric flux through the original surface of Figure 24.21a.

If the electric field were a fluid flowing outward from the charge, all the fluid *entering* the closed region through the first spherical surface would later *exit* the region through the second spherical surface. There is no *net* flow into or out of the closed region. Similarly, every electric field line entering this closed volume through one spherical surface exits through the other spherical surface.

Mathematically, the electric fluxes through the two spherical surfaces have the same magnitude because Φ_e is independent of r. But they have *opposite signs* because the outward-pointing area vector $\vec{A}$ is parallel to $\vec{E}$ on one surface but opposite to $\vec{E}$ on the other. The sum of the fluxes through the two surfaces is zero, and we are led to the conclusion that **the net electric flux is zero through a closed surface that does not contain any net charge.** Charges outside the surface do not produce a net flux through the surface.

This isn't to say that the flux through a small piece of the surface is zero. In fact, as Figure 24.21a shows, nearly every piece of the surface has an electric field either entering or leaving and thus has a nonzero flux. But some of these are positive and some are negative. When summed over the *entire* surface, the positive and negative contributions exactly cancel to give no *net* flux.

Multiple Charges

Finally, consider an arbitrary Gaussian surface and a group of charges $q_1, q_2, q_3, \ldots$ such as those shown in **FIGURE 24.22**. What is the net electric flux through the surface?

By definition, the net flux is

$$\Phi_e = \oint \vec{E} \cdot d\vec{A}$$

From the principle of superposition, the electric field is $\vec{E} = \vec{E}_1 + \vec{E}_2 + \vec{E}_3 + \cdots$, where $\vec{E}_1, \vec{E}_2, \vec{E}_3, \ldots$ are the fields of the individual charges. Thus the flux is

$$\Phi_e = \oint \vec{E}_1 \cdot d\vec{A} + \oint \vec{E}_2 \cdot d\vec{A} + \oint \vec{E}_3 \cdot d\vec{A} + \cdots \tag{24.15}$$

$$= \Phi_1 + \Phi_2 + \Phi_3 + \cdots$$

where $\Phi_1, \Phi_2, \Phi_3, \ldots$ are the fluxes through the Gaussian surface due to the individual charges. That is, the net flux is the sum of the fluxes due to individual charges. But we know what those are: q/ϵ_0 for the charges inside the surface and zero for the charges outside. Thus

$$\Phi_e = \left(\frac{q_1}{\epsilon_0} + \frac{q_2}{\epsilon_0} + \cdots + \frac{q_i}{\epsilon_0} \text{ for all charges inside the surface} \right) \tag{24.16}$$

$$+ (0 + 0 + \cdots + 0 \text{ for all charges outside the surface})$$

We define

$$Q_{in} = q_1 + q_2 + \cdots + q_i \text{ for all charges inside the surface} \tag{24.17}$$

as the total charge enclosed *within* the surface. With this definition, we can write our result for the net electric flux in a very neat and compact fashion. For any *closed* surface enclosing total charge Q_{in}, the net electric flux through the surface is

$$\Phi_e = \oint \vec{E} \cdot d\vec{A} = \frac{Q_{in}}{\epsilon_0} \tag{24.18}$$

This result for the electric flux is known as **Gauss's law.**

FIGURE 24.22 Charges both inside and outside a Gaussian surface.

The fluxes due to charges outside the surface are all zero.

q_3

q_2

q_1

Two-dimensional cross section of a Gaussian surface

Total charge inside is Q_{in}.

The fluxes due to charges inside the surface add.

What Does Gauss's Law Tell Us?

In one sense, Gauss's law doesn't say anything new or anything that we didn't already know from Coulomb's law. After all, we derived Gauss's law from Coulomb's law. But in another sense, Gauss's law is more important than Coulomb's law. Gauss's law states a very general property of electric fields—namely, that charges create electric fields in just such a way that the net flux of the field is the same through *any* surface surrounding the charges, no matter what its size and shape may be. This fact may have been implied by Coulomb's law, but it was by no means obvious. And Gauss's law will turn out to be particularly useful later when we combine it with other electric and magnetic field equations.

Gauss's law is the mathematical statement of our observations in Section 24.2. There we noticed a net "flow" of electric field out of a closed surface containing charges. Gauss's law quantifies this idea by making a specific connection between the "flow," now measured as electric flux, and the amount of charge.

But is it useful? Although to some extent Gauss's law is a formal statement about electric fields, not a tool for solving practical problems, there are exceptions: Gauss's law will allow us to find the electric fields of some very important and very practical charge distributions much more easily than if we had to rely on Coulomb's law. We'll consider some examples in the next section.

STOP TO THINK 24.4 These are two-dimensional cross sections through three-dimensional closed spheres and a cube. Rank in order, from largest to smallest, the electric fluxes Φ_a to Φ_e through surfaces a to e.

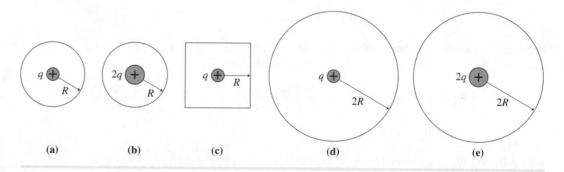

(a) (b) (c) (d) (e)

24.5 Using Gauss's Law

In this section, we'll use Gauss's law to determine the electric fields of several important charge distributions. Some of these you already know, from Chapter 23; others will be new. Three important observations can be made about using Gauss's law:

1. Gauss's law applies only to a *closed* surface, called a Gaussian surface.
2. A Gaussian surface is not a physical surface. It need not coincide with the boundary of any physical object (although it could if we wished). It is an imaginary, mathematical surface in the space surrounding one or more charges.
3. We can't find the electric field from Gauss's law alone. We need to apply Gauss's law in situations where, from symmetry and superposition, we already can guess the *shape* of the field.

These observations and our previous discussion of symmetry and flux lead to the following strategy for solving electric field problems with Gauss's law.

PROBLEM-SOLVING STRATEGY 24.1

Gauss's law

MODEL Model the charge distribution as a distribution with symmetry.

VISUALIZE Draw a picture of the charge distribution.
- Determine the symmetry of its electric field.
- Choose and draw a Gaussian surface with the *same symmetry*.
- You need not enclose all the charge within the Gaussian surface.
- Be sure every part of the Gaussian surface is either tangent to or perpendicular to the electric field.

SOLVE The mathematical representation is based on Gauss's law

$$\Phi_e = \oint \vec{E} \cdot d\vec{A} = \frac{Q_{in}}{\epsilon_0}$$

- Use Tactics Boxes 24.1 and 24.2 to evaluate the surface integral.

ASSESS Check that your result has correct units and significant figures, is reasonable, and answers the question.

Exercise 19

EXAMPLE 24.3 | Outside a sphere of charge

In Chapter 23 we asserted, without proof, that the electric field outside a sphere of total charge Q is the same as the field of a point charge Q at the center. Use Gauss's law to prove this result.

MODEL The charge distribution within the sphere need not be uniform (i.e., the charge density might increase or decrease with r), but it must have spherical symmetry in order for us to use Gauss's law. We will assume that it does.

VISUALIZE FIGURE 24.23 shows a sphere of charge Q and radius R. We want to find $\vec{E}$ outside this sphere, for distances $r > R$. The spherical symmetry of the charge distribution tells us that the electric field must point *radially outward* from the sphere. Although Gauss's law is true for any surface surrounding the charged sphere, it is useful only if we choose a Gaussian surface to match the spherical symmetry of the charge distribution and the field. Thus a spherical surface of radius $r > R$ concentric with the charged sphere will be our Gaussian

FIGURE 24.23 A spherical Gaussian surface surrounding a sphere of charge.

surface. Because this surface surrounds the entire sphere of charge, the enclosed charge is simply $Q_{in} = Q$.

SOLVE Gauss's law is

$$\Phi_e = \oint \vec{E} \cdot d\vec{A} = \frac{Q_{in}}{\epsilon_0} = \frac{Q}{\epsilon_0}$$

To calculate the flux, notice that the electric field is everywhere perpendicular to the spherical surface. And although we don't know the electric field magnitude E, spherical symmetry dictates that E must have the same value at all points equally distant from the center of the sphere. Thus we have the simple result that the net flux through the Gaussian surface is

$$\Phi_e = EA_{sphere} = 4\pi r^2 E$$

where we used the fact that the surface area of a sphere is $A_{sphere} = 4\pi r^2$. With this result for the flux, Gauss's law is

$$4\pi r^2 E = \frac{Q}{\epsilon_0}$$

Thus the electric field at distance r outside a sphere of charge is

$$E_{outside} = \frac{1}{4\pi\epsilon_0}\frac{Q}{r^2}$$

Or in vector form, making use of the fact that $\vec{E}$ is radially outward,

$$\vec{E}_{outside} = \frac{1}{4\pi\epsilon_0}\frac{Q}{r^2}\hat{r}$$

where $\hat{r}$ is a radial unit vector.

ASSESS The field is exactly that of a point charge Q, which is what we wanted to show.

The derivation of the electric field of a sphere of charge depended crucially on a proper choice of the Gaussian surface. We would not have been able to evaluate the flux integral so simply for any other choice of surface. It's worth noting that the result of Example 24.3 can also be proven by the superposition of point-charge fields, but it requires a difficult three-dimensional integral and about a page of algebra. We obtained the answer using Gauss's law in just a few lines. Where Gauss's law works, it works *extremely* well! However, it works only in situations, such as this, with a very high degree of symmetry.

EXAMPLE 24.4 | Inside a sphere of charge

What is the electric field *inside* a uniformly charged sphere?

MODEL We haven't considered a situation like this before. To begin, we don't know if the field strength is increasing or decreasing as we move outward from the center of the sphere. But the field inside must have spherical symmetry. That is, the field must point radially inward or outward, and the field strength can depend only on r. This is sufficient information to solve the problem because it allows us to choose a Gaussian surface.

VISUALIZE FIGURE 24.24 shows a spherical Gaussian surface with radius $r \leq R$ *inside*, and *concentric with*, the sphere of charge. This surface matches the symmetry of the charge distribution, hence $\vec{E}$ is perpendicular to this surface and the field strength E has the same value at all points on the surface.

FIGURE 24.24 A spherical Gaussian surface inside a uniform sphere of charge.

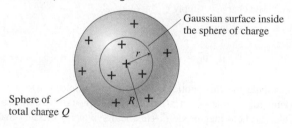

Gaussian surface inside the sphere of charge

Sphere of total charge Q

SOLVE The flux integral is identical to that of Example 24.3:

$$\Phi_e = EA_{\text{sphere}} = 4\pi r^2 E$$

Consequently, Gauss's law is

$$\Phi_e = 4\pi r^2 E = \frac{Q_{\text{in}}}{\epsilon_0}$$

The difference between this example and Example 24.3 is that Q_{in} is no longer the total charge of the sphere. Instead, Q_{in} is the amount of charge *inside* the Gaussian sphere of radius r. Because the charge distribution is *uniform*, the volume charge density is

$$\rho = \frac{Q}{V_R} = \frac{Q}{\frac{4}{3}\pi R^3}$$

The charge enclosed in a sphere of radius r is thus

$$Q_{\text{in}} = \rho V_r = \left(\frac{Q}{\frac{4}{3}\pi R^3}\right)\left(\frac{4}{3}\pi r^3\right) = \frac{r^3}{R^3}Q$$

The amount of enclosed charge increases with the cube of the distance r from the center and, as expected, $Q_{\text{in}} = Q$ if $r = R$. With this expression for Q_{in}, Gauss's law is

$$4\pi r^2 E = \frac{(r^3/R^3)Q}{\epsilon_0}$$

Thus the electric field at radius r inside a uniformly charged sphere is

$$E_{\text{inside}} = \frac{1}{4\pi\epsilon_0}\frac{Q}{R^3}r$$

The electric field strength inside the sphere increases *linearly* with the distance r from the center.

ASSESS The field inside and the field outside a sphere of charge match at the boundary of the sphere, $r = R$, where both give $E = Q/4\pi\epsilon_0 R^2$. In other words, the field strength is *continuous* as we cross the boundary of the sphere. These results are shown graphically in FIGURE 24.25.

FIGURE 24.25 The electric field strength of a uniform sphere of charge of radius R.

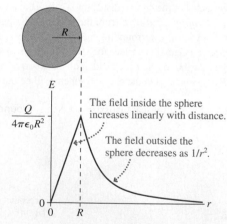

The field inside the sphere increases linearly with distance.

The field outside the sphere decreases as $1/r^2$.

EXAMPLE 24.5 | The electric field of a long, charged wire

In Chapter 23, we used superposition to find the electric field of a long, charged wire with linear charge density λ (C/m). It was not an easy derivation. Find the electric field using Gauss's law.

MODEL A long, charged wire can be modeled as an infinitely long line of charge.

VISUALIZE FIGURE 24.26 shows an infinitely long line of charge. We can use the symmetry of the situation to see that the only

FIGURE 24.26 A Gaussian surface around a charged wire.

The field is tangent to the surface on the ends. The flux is zero.

Gaussian surface

The field is perpendicular to the surface on the cylinder wall.

possible shape of the electric field is to point straight into or out from the wire, rather like the bristles on a bottle brush. The shape of the field suggests that we choose our Gaussian surface to be a cylinder of radius r and length L, centered on the wire. Because Gauss's law refers to *closed* surfaces, we must include the ends of the cylinder as part of the surface.

SOLVE Gauss's law is

$$\Phi_e = \oint \vec{E} \cdot d\vec{A} = \frac{Q_{in}}{\epsilon_0}$$

where Q_{in} is the charge *inside* the closed cylinder. We have two tasks: to evaluate the flux integral, and to determine how much charge is inside the closed surface. The wire has linear charge density λ, so

the amount of charge inside a cylinder of length L is simply

$$Q_{in} = \lambda L$$

Finding the net flux is just as straightforward. We can divide the flux through the entire closed surface into the flux through each end plus the flux through the cylindrical wall. The electric field $\vec{E}$, pointing straight out from the wire, is tangent to the end surfaces at every point. Thus the flux through these two surfaces is zero. On the wall, $\vec{E}$ is perpendicular to the surface and has the same strength E at every point. Thus

$$\Phi_e = \Phi_{top} + \Phi_{bottom} + \Phi_{wall} = 0 + 0 + EA_{cyl} = 2\pi rLE$$

where we used $A_{cyl} = 2\pi rL$ as the surface area of a cylindrical wall of radius r and length L. Once again, the proper choice of the Gaussian surface reduces the flux integral merely to finding a surface area. With these expressions for Q_{in} and Φ_e, Gauss's law is

$$\Phi_e = 2\pi rLE = \frac{Q_{in}}{\epsilon_0} = \frac{\lambda L}{\epsilon_0}$$

Thus the electric field at distance r from a long, charged wire is

$$E_{wire} = \frac{\lambda}{2\pi\epsilon_0 r}$$

ASSESS This agrees exactly with the result of the more complex derivation in Chapter 23. Notice that the result does not depend on our choice of L. A Gaussian surface is an imaginary device, not a physical object. We needed a finite-length cylinder to do the flux calculation, but the electric field of an *infinitely* long wire can't depend on the length of an imaginary cylinder.

Example 24.5, for the electric field of a long, charged wire, contains a subtle but important idea, one that often occurs when using Gauss's law. The Gaussian cylinder of length L encloses only some of the wire's charge. The pieces of the charged wire outside the cylinder are not enclosed by the Gaussian surface and consequently do not contribute anything to the net flux. Even so, *they are essential* to the use of Gauss's law because it takes the *entire* charged wire to produce an electric field with cylindrical symmetry. In other words, the charge outside the cylinder may not contribute to the flux, but it affects the *shape* of the electric field. Our ability to write $\Phi_e = EA_{cyl}$ depended on knowing that E is the same at every point on the wall of the cylinder. That would not be true for a charged wire of finite length, so we cannot use Gauss's law to find the electric field of a finite-length charged wire.

EXAMPLE 24.6 | **The electric field of a plane of charge**

Use Gauss's law to find the electric field of an infinite plane of charge with surface charge density η (C/m^2).

MODEL A uniformly charged flat electrode can be modeled as an infinite plane of charge.

VISUALIZE FIGURE 24.27 on the next page shows a uniformly charged plane with surface charge density η. We will assume that the plane extends infinitely far in all directions, although we obviously have to show "edges" in our drawing. The planar

symmetry allows the electric field to point only straight toward or away from the plane. With this in mind, choose as a Gaussian surface a cylinder with length L and faces of area A centered on the plane of charge. Although we've drawn them as circular, the shape of the faces is not relevant.

SOLVE The electric field is perpendicular to both faces of the cylinder, so the total flux through both faces is $\Phi_{faces} = 2EA$. (The fluxes add rather than cancel because the area vector $\vec{A}$ points

FIGURE 24.27 The Gaussian surface extends to both sides of a plane of charge.

Infinite plane of charge

Gaussian surface

$\vec{E}$

$d\vec{A}$

$\vec{E}$

$d\vec{A}$

Area A

L

outward on each face.) There's *no* flux through the wall of the cylinder because the field vectors are tangent to the wall. Thus the net flux is simply

$$\Phi_e = 2EA$$

The charge inside the cylinder is the charge contained in area A of the plane. This is

$$Q_{in} = \eta A$$

With these expressions for Q_{in} and Φ_e, Gauss's law is

$$\Phi_e = 2EA = \frac{Q_{in}}{\epsilon_0} = \frac{\eta A}{\epsilon_0}$$

Thus the electric field of an infinite charged plane is

$$E_{plane} = \frac{\eta}{2\epsilon_0}$$

This agrees with the result in Chapter 23.

ASSESS This is another example of a Gaussian surface enclosing only some of the charge. Most of the plane's charge is outside the Gaussian surface and does not contribute to the flux, but it does affect the shape of the field. We wouldn't have planar symmetry, with the electric field exactly perpendicular to the plane, without all the rest of the charge on the plane.

The plane of charge is an especially good example of how powerful Gauss's law can be. Finding the electric field of a plane of charge via superposition was a difficult and tedious derivation. With Gauss's law, once you see how to apply it, the problem is simple enough to solve in your head!

You might wonder, then, why we bothered with superposition at all. The reason is that Gauss's law, powerful though it may be, is effective only in a limited number of situations where the field is highly symmetric. Superposition always works, even if the derivation is messy, because superposition goes directly back to the fields of individual point charges. It's good to use Gauss's law when you can, but superposition is often the only way to attack real-world charge distributions.

STOP TO THINK 24.5 Which Gaussian surface would allow you to use Gauss's law to calculate the electric field outside a uniformly charged cube?

a. A sphere whose center coincides with the center of the charged cube
b. A cube whose center coincides with the center of the charged cube and that has parallel faces
c. Either a or b
d. Neither a nor b

24.6 Conductors in Electrostatic Equilibrium

Consider a charged conductor, such as a charged metal electrode, in electrostatic equilibrium. That is, there is no current through the conductor and the charges are all stationary. One very important conclusion is that **the electric field is zero at all points within a conductor in electrostatic equilibrium.** That is, $\vec{E}_{in} = \vec{0}$. If this weren't true, the electric field would cause the charge carriers to move and thus violate the assumption that all the charges are at rest. Let's use Gauss's law to see what else we can learn.

At the Surface of a Conductor

FIGURE 24.28 shows a Gaussian surface just barely inside the physical surface of a conductor that's in electrostatic equilibrium. The electric field is zero at all points within the conductor, hence the electric flux Φ_e through this Gaussian surface must be zero. But if $\Phi_e = 0$, Gauss's law tells us that $Q_{in} = 0$. That is, there's no net charge within this surface. There are charges—electrons and positive ions—but no *net* charge.

If there's no net charge in the interior of a conductor in electrostatic equilibrium, then **all the excess charge on a charged conductor resides on the exterior surface of the conductor.** Any charges added to a conductor quickly spread across the surface until reaching positions of electrostatic equilibrium, but there is no net charge *within* the conductor.

There may be no electric field within a charged conductor, but the presence of net charge requires an exterior electric field in the space outside the conductor. **FIGURE 24.29** shows that **the electric field right at the surface of the conductor has to be perpendicular to the surface.** To see that this is so, suppose $\vec{E}_{surface}$ had a component tangent to the surface. This component of $\vec{E}_{surface}$ would exert a force on the surface charges and cause a surface current, thus violating the assumption that all charges are at rest. The only exterior electric field consistent with electrostatic equilibrium is one that is perpendicular to the surface.

We can use Gauss's law to relate the field strength at the surface to the charge density on the surface. **FIGURE 24.30** shows a small Gaussian cylinder with faces very slightly above and below the surface of a charged conductor. The charge inside this Gaussian cylinder is ηA, where η is the surface charge density at this point on the conductor. There's a flux $\Phi = AE_{surface}$ through the outside face of this cylinder but, unlike Example 24.6 for the plane of charge, *no* flux through the inside face because $\vec{E}_{in} = \vec{0}$ within the conductor. Furthermore, there's no flux through the wall of the cylinder because $\vec{E}_{surface}$ is perpendicular to the surface. Thus the net flux is $\Phi_e = AE_{surface}$. Gauss's law is

$$\Phi_e = AE_{surface} = \frac{Q_{in}}{\epsilon_0} = \frac{\eta A}{\epsilon_0} \tag{24.19}$$

from which we can conclude that the electric field at the surface of a charged conductor is

$$\vec{E}_{surface} = \left(\frac{\eta}{\epsilon_0}, \text{ perpendicular to surface} \right) \tag{24.20}$$

In general, the surface charge density η is *not* constant on the surface of a conductor but depends on the shape of the conductor. If we can determine η, by either calculating it or measuring it, then Equation 24.20 tells us the electric field at that point on the surface. Alternatively, we can use Equation 24.20 to deduce the charge density on the conductor's surface if we know the electric field just outside the conductor.

Charges and Fields Within a Conductor

FIGURE 24.31 shows a charged conductor with a hole inside. Can there be charge on this interior surface? To find out, we place a Gaussian surface around the hole, infinitesimally close but entirely within the conductor. The electric flux Φ_e through this Gaussian surface is zero because the electric field is zero everywhere inside the conductor. Thus we must conclude that $Q_{in} = 0$. There's no net charge inside this Gaussian surface and thus no charge on the surface of the hole. Any excess charge resides on the *exterior* surface of the conductor, not on any interior surfaces.

Furthermore, because there's no electric field inside the conductor and no charge inside the hole, the electric field inside the hole must also be zero. This conclusion has an important practical application. For example, suppose we need to exclude the electric field from the region in **FIGURE 24.32a** on the next page enclosed within dashed lines. We can do so by surrounding this region with the neutral conducting box of **FIGURE 24.32b**.

FIGURE 24.28 A Gaussian surface just inside a conductor's surface.

The electric field inside is zero.

$\vec{E} = \vec{0}$

The flux through the Gaussian surface is zero. Hence all the excess charge must be on the surface.

FIGURE 24.29 The electric field at the surface of a charged conductor.

The electric field at the surface is perpendicular to the surface.

$\vec{E}$

$\vec{E} = \vec{0}$

Surface charge

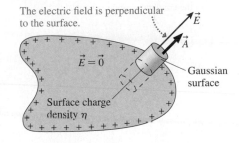

FIGURE 24.30 A Gaussian surface extending through the surface has a flux only through the outer face.

The electric field is perpendicular to the surface.

$\vec{E}$

$\vec{A}$

$\vec{E} = \vec{0}$

Gaussian surface

Surface charge density η

FIGURE 24.31 A Gaussian surface surrounding a hole inside a conductor.

A hollow completely enclosed by the conductor

$\vec{E} = \vec{0}$

The flux through the Gaussian surface is zero. There's no net charge inside, hence no charge on this interior surface.

FIGURE 24.32 The electric field can be excluded from a region of space by surrounding it with a conducting box.

(a) Parallel-plate capacitor

We want to exclude the electric field from this region.

(b) The conducting box has been polarized and has induced surface charges.

The electric field is perpendicular to all conducting surfaces.

FIGURE 24.33 A charge in the hole causes a net charge on the interior and exterior surfaces.

The flux through the Gaussian surface is zero, hence there's no *net* charge inside this surface. There must be charge $-q$ on the inside surface to balance point charge q.

Neutral conductor

$\vec{E} = \vec{0}$

Point charge q

The outer surface must have charge $+q$ so that the conductor remains neutral.

This region of space is now a hole inside a conductor, thus the interior electric field is zero. The use of a conducting box to exclude electric fields from a region of space is called **screening.** Solid metal walls are ideal, but in practice wire screen or wire mesh—sometimes called a *Faraday cage*—provides sufficient screening for all but the most sensitive applications. The price we pay is that the exterior field is now very complicated.

Finally, **FIGURE 24.33** shows a charge q inside a hole within a neutral conductor. The electric field *within* the conductor is still zero, hence the electric flux through the Gaussian surface is zero. But $\Phi_e = 0$ requires $Q_{in} = 0$. Consequently, the charge inside the hole attracts an equal charge of opposite sign, and charge $-q$ now lines the inner surface of the hole.

The conductor as a whole is neutral, so moving $-q$ to the surface of the hole must leave $+q$ behind somewhere else. Where is it? It can't be in the interior of the conductor, as we've seen, and that leaves only the exterior surface. In essence, an internal charge polarizes the conductor just as an external charge would. Net charge $-q$ moves to the inner surface and net charge $+q$ is left behind on the exterior surface.

In summary, conductors in electrostatic equilibrium have the properties described in Tactics Box 24.3.

TACTICS BOX 24.3

Finding the electric field of a conductor in electrostatic equilibrium

❶ The electric field is zero at all points within the volume of the conductor.
❷ Any excess charge resides entirely on the *exterior* surface.
❸ The external electric field at the surface of a charged conductor is perpendicular to the surface and of magnitude η/ϵ_0, where η is the surface charge density at that point.
❹ The electric field is zero inside any hole within a conductor unless there is a charge in the hole.

Exercises 20–24

EXAMPLE 24.7 | The electric field at the surface of a charged metal sphere

A 2.0-cm-diameter brass sphere has been given a charge of 2.0 nC. What is the electric field strength at the surface?

MODEL Brass is a conductor. The excess charge resides on the surface.

VISUALIZE The charge distribution has spherical symmetry. The electric field points radially outward from the surface.

SOLVE We can solve this problem in two ways. One uses the fact that a sphere, because of its complete symmetry, is the one shape for which any excess charge will spread out to a *uniform* surface charge density. Thus

$$\eta = \frac{q}{A_{\text{sphere}}} = \frac{q}{4\pi R^2} = \frac{2.0 \times 10^{-9}\,\text{C}}{4\pi (0.010\,\text{m})^2} = 1.59 \times 10^{-6}\,\text{C/m}^2$$

From Equation 24.20, we know the electric field at the surface has strength

$$E_{\text{surface}} = \frac{\eta}{\epsilon_0} = \frac{1.59 \times 10^{-6}\,\text{C/m}^2}{8.85 \times 10^{-12}\,\text{C}^2/\text{N m}^2} = 1.8 \times 10^5\,\text{N/C}$$

Alternatively, we could have used the result, obtained earlier in the chapter, that the electric field strength outside a sphere of charge Q is $E_{\text{outside}} - Q_{\text{in}}/(4\pi\epsilon_0 r^2)$. But $Q_{\text{in}} = q$ and, at the surface, $r = R$. Thus

$$E_{\text{surface}} = \frac{1}{4\pi\epsilon_0}\frac{q}{R^2} = (9.0 \times 10^9\,\text{N m}^2/\text{C}^2)\frac{2.0 \times 10^{-9}\,\text{C}}{(0.010\,\text{m})^2}$$

$$= 1.8 \times 10^5\,\text{N/C}$$

As we can see, both methods lead to the same result.

CHALLENGE EXAMPLE 24.8 | The electric field of a slab of charge

An infinite slab of charge of thickness $2a$ is centered in the xy-plane. The charge density is $\rho = \rho_0(1 - |z|/a)$. Find the electric field strengths inside and outside this slab of charge.

MODEL The charge density is not uniform. Starting at ρ_0 in the xy-plane, it decreases linearly with distance above and below the xy-plane until reaching zero at $z = \pm a$, the edges of the slab.

VISUALIZE FIGURE 24.34 shows an edge view of the slab of charge and, as Gaussian surfaces, side viwews of two cylinders with cross-section area A. By symmetry, the electric field must point away from the xy-plane; the field cannot have an x- or y-component.

FIGURE 24.34 Two cylindrical Gaussian surfaces for an infinite slab of charge.

SOLVE Gauss's law is

$$\Phi_e = \oint \vec{E} \cdot d\vec{A} = \frac{Q_{\text{in}}}{\epsilon_0}$$

With symmetry, finding the net flux is straightforward. The electric field is perpendicular to the faces of the cylinders and pointing outward, so the total flux through the faces is $\Phi_{\text{faces}} = 2EA$, where E may depend on distance z. The field is parallel to the walls of the cylinders, so $\Phi_{\text{wall}} = 0$. Thus the net flux is simply

$$\Phi_e = 2EA$$

Because the charge density is not uniform, we need to integrate to find Q_{in}, the charge *inside* the cylinder. We can slice the cylinder into small slabs of infinitesimal thickness dz and volume $dV = A\,dz$. Figure 24.34 shows one such little slab at distance z from the xy-plane. The charge in this little slab is

$$dq = \rho\,dV = \rho_0\left(1 - \frac{z}{a}\right)A\,dz$$

where we assumed that z is positive. Because the charge is symmetric about $z = 0$, we can avoid difficulties with the absolute value sign in the charge density by integrating from 0 and multiplying by 2. For the Gaussian cylinder that ends inside the slab of charge, at distance z, the total charge inside is

$$Q_{\text{in}} = \int dq = 2\int_0^z \rho_0\left(1 - \frac{z}{a}\right)A\,dz$$

$$= 2\rho_0 A\left[z\Big|_0^z - \frac{1}{2a}z^2\Big|_0^z\right]$$

$$= 2\rho_0 A z\left(1 - \frac{z}{2a}\right)$$

Gauss's law inside the slab is then

$$\Phi_e = 2E_{\text{inside}}A = \frac{Q_{\text{in}}}{\epsilon_0} = \frac{2\rho_0 A z}{\epsilon_0}\left(1 - \frac{z}{2a}\right)$$

The area A cancels, as it must because it was an arbitrary choice, leaving

$$E_{\text{inside}} = \frac{\rho_0 z}{\epsilon_0}\left(1 - \frac{z}{2a}\right)$$

The field strength is zero at $z = 0$, then increases as z increases. This expression is valid only above the xy-plane, for $z > 0$, but the field strength is symmetric on the other side.

For the Gaussian cylinder that extends outside the slab of charge, the integral for Q has to end at $z = a$. Thus

$$Q_{\text{in}} = 2\rho_0 A a\left(1 - \frac{a}{2a}\right) = \rho_0 A a$$

independent of distance z. With this, Gauss's law gives

$$E_{\text{outside}} = \frac{Q_{\text{in}}}{2\epsilon_0 A} = \frac{\rho_0 a}{2\epsilon_0}$$

This matches E_{inside} at the surface, $z = a$, so the field is continuous as it crosses the boundary.

ASSESS Outside a sphere of charge, the field is the same as that of a point charge at the center. Similarly, the field outside an infinite slab of charge should be the same as that of an infinite charged plane. We found, by integration, that the total charge in an area A of the slab is $Q = \rho_0 A a$. If we squished this charge into a plane, the surface charge density would be $\eta = Q/A = \rho_0 a$. Thus our expression for E_{outside} could be written $\eta/2\epsilon_0$, which matches the field we found in Example 24.6 for a plane of charge.

SUMMARY

The goal of Chapter 24 has been to learn about and apply Gauss's law.

GENERAL PRINCIPLES

Gauss's Law

For any *closed* surface enclosing net charge Q_{in}, the net electric flux through the surface is

$$\Phi_e = \oint \vec{E} \cdot d\vec{A} = \frac{Q_{in}}{\epsilon_0}$$

The electric flux Φ_e is the same for *any* closed surface enclosing charge Q_{in}.

To solve electric field problems with Gauss's law:

MODEL Model the charge distribution as one with symmetry.

VISUALIZE Draw a picture of the charge distribution. Draw a Gaussian surface with the same symmetry as the electric field, every part of which is either tangent to or perpendicular to the electric field.

SOLVE Apply Gauss's law and Tactics Boxes 24.1 and 24.2 to evaluate the surface integral.

ASSESS Is the result reasonable?

Symmetry

The symmetry of the electric field must match the symmetry of the charge distribution.

In practice, Φ_e is computable only if the symmetry of the Gaussian surface matches the symmetry of the charge distribution.

IMPORTANT CONCEPTS

Charge creates the electric field that is responsible for the electric flux.

Q_{in} is the sum of all enclosed charges. This charge contributes to the flux.

Gaussian surface

Charges outside the surface contribute to the electric field, but they don't contribute to the flux.

Flux is the amount of electric field passing through a surface of area A:

$$\Phi_e = \vec{E} \cdot \vec{A}$$

where $\vec{A}$ is the **area vector.**

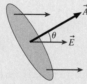

For closed surfaces:
A net flux in or out indicates that the surface encloses a net charge.

Field lines through but with no *net* flux mean that the surface encloses no *net* charge.

Surface integrals calculate the flux by summing the fluxes through many small pieces of the surface:

$$\Phi_e = \sum \vec{E} \cdot \delta\vec{A}$$

$$\rightarrow \int \vec{E} \cdot d\vec{A}$$

Two important situations:

If the electric field is everywhere tangent to the surface, then

$$\Phi_e = 0$$

If the electric field is everywhere perpendicular to the surface *and* has the same strength E at all points, then

$$\Phi_e = EA$$

APPLICATIONS

Conductors in electrostatic equilibrium

- The electric field is zero at all points within the conductor.
- Any excess charge resides entirely on the exterior surface.
- The external electric field is perpendicular to the surface and of magnitude η/ϵ_0, where η is the surface charge density.
- The electric field is zero inside any hole within a conductor unless there is a charge in the hole.

TERMS AND NOTATION

symmetric	electric flux, Φ_e	surface integral	screening
Gaussian surface	area vector, $\vec{A}$	Gauss's law	

CONCEPTUAL QUESTIONS

1. Suppose you have the uniformly charged cube in **FIGURE Q24.1**. Can you use symmetry alone to deduce the *shape* of the cube's electric field? If so, sketch and describe the field shape. If not, why not?

FIGURE Q24.1

2. **FIGURE Q24.2** shows cross sections of three-dimensional closed surfaces. They have a flat top and bottom surface above and below the plane of the page. However, the electric field is everywhere parallel to the page, so there is no flux through the top or bottom surface. The electric field is uniform over each face of the surface. For each, does the surface enclose a net positive charge, a net negative charge, or no net charge? Explain.

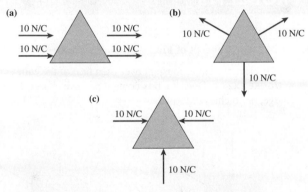

FIGURE Q24.2

3. The square and circle in **FIGURE Q24.3** are in the same uniform field. The diameter of the circle equals the edge length of the square. Is Φ_{square} larger than, smaller than, or equal to Φ_{circle}? Explain.

FIGURE Q24.3   **FIGURE Q24.4**

4. In **FIGURE Q24.4**, where the field is uniform, is the magnitude of Φ_1 larger than, smaller than, or equal to the magnitude of Φ_2? Explain.

5. What is the electric flux through each of the surfaces in **FIGURE Q24.5**? Give each answer as a multiple of q/ϵ_0.

FIGURE Q24.5

6. What is the electric flux through each of the surfaces A to E in **FIGURE Q24.6**? Give each answer as a multiple of q/ϵ_0.

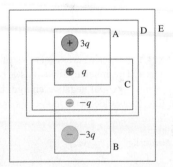

FIGURE Q24.6

7. The charged balloon in **FIGURE Q24.7** expands as it is blown up, increasing in size from the initial to final diameters shown. Do the electric field strengths at points 1, 2, and 3 increase, decrease, or stay the same? Explain your reasoning for each.

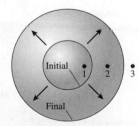

FIGURE Q24.7

8. The two spheres in **FIGURE Q24.8** on the next page surround equal charges. Three students are discussing the situation.
Student 1: The fluxes through spheres A and B are equal because they enclose equal charges.
Student 2: But the electric field on sphere B is weaker than the electric field on sphere A. The flux depends on the electric field strength, so the flux through A is larger than the flux through B.

Student 3: I thought we learned that flux was about surface area. Sphere B is larger than sphere A, so I think the flux through B is larger than the flux through A.
Which of these students, if any, do you agree with? Explain.

FIGURE Q24.8

9. The sphere and ellipsoid in **FIGURE Q24.9** surround equal charges. Four students are discussing the situation.
Student 1: The fluxes through A and B are equal because the average radius is the same.
Student 2: I agree that the fluxes are equal, but that's because they enclose equal charges.
Student 3: The electric field is not perpendicular to the surface for B, and that makes the flux through B less than the flux through A.
Student 4: I don't think that Gauss's law even applies to a situation like B, so we can't compare the fluxes through A and B.

Which of these students, if any, do you agree with? Explain.

FIGURE Q24.9

10. A small, metal sphere hangs by an insulating thread within the larger, hollow conducting sphere of **FIGURE Q24.10**. A conducting wire extends from the small sphere through, but not touching, a small hole in the hollow sphere. A charged rod is used to transfer positive charge to the protruding wire. After the charged rod has touched the wire and been removed, are the following surfaces positive, negative, or not charged? Explain.
a. The small sphere.
b. The inner surface of the hollow sphere.
c. The outer surface of the hollow sphere.

FIGURE Q24.10

EXERCISES AND PROBLEMS

Problems labeled ▨ integrate material from earlier chapters.

Exercises

Section 24.1 Symmetry

1. | **FIGURE EX24.1** shows two cross sections of two infinitely long coaxial cylinders. The inner cylinder has a positive charge, the outer cylinder has an equal negative charge. Draw this figure on your paper, then draw electric field vectors showing the shape of the electric field.

FIGURE EX24.1   **FIGURE EX24.2**

2. | **FIGURE EX24.2** shows a cross section of two concentric spheres. The inner sphere has a negative charge. The outer sphere has a positive charge larger in magnitude than the charge on the inner sphere. Draw this figure on your paper, then draw electric field vectors showing the shape of the electric field.

3. | **FIGURE EX24.3** shows a cross section of two infinite parallel planes of charge. Draw this figure on your paper, then draw electric field vectors showing the shape of the electric field.

++++++++++++++++++++

FIGURE EX24.3 ++++++++++++++++++++

Section 24.2 The Concept of Flux

4. | The electric field is constant over each face of the cube shown in **FIGURE EX24.4**. Does the box contain positive charge, negative charge, or no charge? Explain.

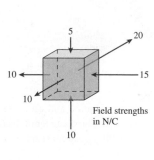

FIGURE EX24.4   **FIGURE EX24.5**

5. | The electric field is constant over each face of the cube shown in **FIGURE EX24.5**. Does the box contain positive charge, negative charge, or no charge? Explain.

6. | The cube in **FIGURE EX24.6** contains negative charge. The electric field is constant over each face of the cube. Does the missing electric field vector on the front face point in or out? What strength must this field exceed?

FIGURE EX24.6

7. | The cube in **FIGURE EX24.7** contains negative charge. The electric field is constant over each face of the cube. Does the missing electric field vector on the front face point in or out? What strength must this field exceed?

FIGURE EX24.7   **FIGURE EX24.8**

8. | The cube in **FIGURE EX24.8** contains no net charge. The electric field is constant over each face of the cube. Does the missing electric field vector on the front face point in or out? What is the field strength?

Section 24.3 Calculating Electric Flux

9. ‖ What is the electric flux through the surface shown in **FIGURE EX24.9**?

FIGURE EX24.9   **FIGURE EX24.10**

10. ‖ What is the electric flux through the surface shown in **FIGURE EX24.10**?

11. ‖ The electric flux through the surface shown in **FIGURE EX24.11** is 25 N m²/C. What is the electric field strength?

FIGURE EX24.11

12. ‖ A 2.0 cm × 3.0 cm rectangle lies in the xy-plane. What is the magnitude of the electric flux through the rectangle if
   a. $\vec{E} = (100\hat{\imath} - 200\hat{k})$ N/C?
   b. $\vec{E} = (100\hat{\imath} - 200\hat{\jmath})$ N/C?

13. ‖ A 2.0 cm × 3.0 cm rectangle lies in the xz-plane. What is the magnitude of the electric flux through the rectangle if
   a. $\vec{E} = (100\hat{\imath} - 200\hat{k})$ N/C?
   b. $\vec{E} = (100\hat{\imath} - 200\hat{\jmath})$ N/C?

14. ‖ A 3.0-cm-diameter circle lies in the xz-plane in a region where the electric field is $\vec{E} = (1500\hat{\imath} + 1500\hat{\jmath} - 1500\hat{k})$ N/C. What is the electric flux through the circle?

15. ‖ A 1.0 cm × 1.0 cm × 1.0 cm box with its edges aligned with the xyz-axes is in the electric field $\vec{E} = (350x + 150)\hat{\imath}$ N/C, where x is in meters. What is the net electric flux through the box?

16. | What is the net electric flux through the two cylinders shown in **FIGURE EX24.16**? Give your answer in terms of R and E.

FIGURE EX24.16

Section 24.4 Gauss's Law

Section 24.5 Using Gauss's Law

17. | **FIGURE EX24.17** shows three charges. Draw these charges on your paper four times. Then draw two-dimensional cross sections of three-dimensional closed surfaces through which the electric flux is (a) $2q/\epsilon_0$, (b) q/ϵ_0, (c) 0, and (d) $5q/\epsilon_0$.

FIGURE EX24.17   **FIGURE EX24.18**

18. | **FIGURE EX24.18** shows three charges. Draw these charges on your paper four times. Then draw two-dimensional cross sections of three-dimensional closed surfaces through which the electric flux is (a) $-q/\epsilon_0$, (b) q/ϵ_0, (c) $3q/\epsilon_0$, and (d) $4q/\epsilon_0$.

19. | **FIGURE EX24.19** shows three Gaussian surfaces and the electric flux through each. What are the three charges q_1, q_2, and q_3?

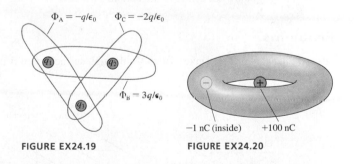

FIGURE EX24.19   **FIGURE EX24.20**

20. ‖ What is the net electric flux through the torus (i.e., doughnut shape) of **FIGURE EX24.20**?

21. ‖ What is the net electric flux through the cylinder of **FIGURE EX24.21**?

FIGURE EX24.21   +100 nC   +1 nC (inside)   −100 nC

22. || The net electric flux through an octahedron is -1000 N m^2/C. How much charge is enclosed within the octahedron?

23. || 55.3 million excess electrons are inside a closed surface. What is the net electric flux through the surface?

Section 24.6 Conductors in Electrostatic Equilibrium

24. | A spark occurs at the tip of a metal needle if the electric field strength exceeds 3.0×10^6 N/C, the field strength at which air breaks down. What is the minimum surface charge density for producing a spark?

25. || The electric field strength just above one face of a copper penny is 2000 N/C. What is the surface charge density on this face of the penny?

26. | The conducting box in FIGURE EX24.26 has been given an excess negative charge. The surface density of excess electrons at the center of the top surface is 5.0×10^{10} electrons/m^2. What are the electric field strengths E_1 to E_3 at points 1 to 3?

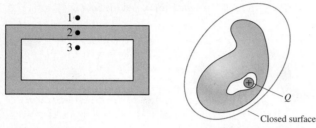

FIGURE EX24.26   FIGURE EX24.27

27. || FIGURE EX24.27 shows a hollow cavity within a neutral conductor. A point charge Q is inside the cavity. What is the net electric flux through the closed surface that surrounds the conductor?

28. | A thin, horizontal, 10-cm-diameter copper plate is charged to 3.5 nC. If the electrons are uniformly distributed on the surface, what are the strength and direction of the electric field
   a. 0.1 mm above the center of the top surface of the plate?
   b. at the plate's center of mass?
   c. 0.1 mm below the center of the bottom surface of the plate?

Problems

29. || Find the electric fluxes Φ_1 to Φ_5 through surfaces 1 to 5 in FIGURE P24.29.

FIGURE P24.29   FIGURE P24.30

30. | FIGURE P24.30 shows four sides of a 3.0 cm $\times$ 3.0 cm $\times$ 3.0 cm cube.
   a. What are the electric fluxes Φ_1 to Φ_4 through sides 1 to 4?
   b. What is the net flux through these four sides?

31. || A tetrahedron has an equilateral triangle base with 20-cm-long edges and three equilateral triangle sides. The base is parallel to the ground, and a vertical uniform electric field of strength 200 N/C passes upward through the tetrahedron.
   a. What is the electric flux through the base?
   b. What is the electric flux through each of the three sides?

32. | Charges $q_1 = -4Q$ and $q_2 = +2Q$ are located at $x = -a$ and $x = +a$, respectively. What is the net electric flux through a sphere of radius $2a$ centered (a) at the origin and (b) at $x = 2a$?

33. || A 10 nC charge is at the center of a 2.0 m $\times$ 2.0 m $\times$ 2.0 m cube. What is the electric flux through the top surface of the cube?

34. || A spherically symmetric charge distribution produces the electric field $\vec{E} = (5000r^2)\hat{r}$ N/C, where r is in m.
   a. What is the electric field strength at $r = 20$ cm?
   b. What is the electric flux through a 40-cm-diameter spherical surface that is concentric with the charge distribution?
   c. How much charge is inside this 40-cm-diameter spherical surface?

35. || A neutral conductor contains a hollow cavity in which there is a $+100$ nC point charge. A charged rod then transfers -50 nC to the conductor. Afterward, what is the charge (a) on the inner wall of the cavity, and (b) on the exterior surface of the conductor?

36. || A hollow metal sphere has inner radius a and outer radius b. The hollow sphere has charge $+2Q$. A point charge $+Q$ sits at the center of the hollow sphere.
   a. Determine the electric fields in the three regions $r \le a$, $a < r < b$, and $r \ge b$.
   b. How much charge is on the inside surface of the hollow sphere? On the exterior surface?

37. ||| A 20-cm-radius ball is uniformly charged to 80 nC.
   a. What is the ball's volume charge density (C/m^3)?
   b. How much charge is enclosed by spheres of radii 5, 10, and 20 cm?
   c. What is the electric field strength at points 5, 10, and 20 cm from the center?

38. ||| FIGURE P24.38 shows a solid metal sphere at the center of a hollow metal sphere. What is the total charge on (a) the exterior of the inner sphere, (b) the inside surface of the hollow sphere, and (c) the exterior surface of the hollow sphere?

FIGURE P24.38

39. || The earth has a vertical electric field at the surface, pointing down, that averages 100 N/C. This field is maintained by various atmospheric processes, including lightning. What is the excess charge on the surface of the earth?

40. || Figure 24.32b showed a conducting box inside a parallel-plate capacitor. The electric field inside the box is $\vec{E} = \vec{0}$. Suppose the surface charge on the exterior of the box could be frozen. Draw a picture of the electric field inside the box after the box, with its frozen charge, is removed from the capacitor.
   Hint: Superposition.

41. ‖ A hollow metal sphere has 6 cm and 10 cm inner and outer radii, respectively. The surface charge density on the inside surface is -100 nC/m². The surface charge density on the exterior surface is $+100$ nC/m². What are the strength and direction of the electric field at points 4, 8, and 12 cm from the center?

42. ‖ A positive point charge q sits at the center of a hollow spherical shell. The shell, with radius R and negligible thickness, has net charge $-2q$. Find an expression for the electric field strength (a) inside the sphere, $r < R$, and (b) outside the sphere, $r > R$. In what direction does the electric field point in each case?

43. ‖ Find the electric field inside and outside a hollow plastic ball of radius R that has charge Q uniformly distributed on its outer surface.

44. ‖ A uniformly charged ball of radius a and charge $-Q$ is at the center of a hollow metal shell with inner radius b and outer radius c. The hollow sphere has net charge $+2Q$. Determine the electric field strength in the four regions $r \le a$, $a < r < b$, $b \le r \le c$, and $r > c$.

45. | The three parallel planes of charge shown in FIGURE P24.45 have surface charge densities $-\frac{1}{2}\eta$, η, and $-\frac{1}{2}\eta$. Find the electric fields $\vec{E}_1$ to $\vec{E}_4$ in regions 1 to 4.

FIGURE P24.45

46. ‖ An infinite slab of charge of thickness $2z_0$ lies in the xy-plane between $z = -z_0$ and $z = +z_0$. The volume charge density ρ (C/m³) is a constant.
   a. Use Gauss's law to find an expression for the electric field strength inside the slab $(-z_0 \le z \le z_0)$.
   b. Find an expression for the electric field strength above the slab $(z \ge z_0)$.
   c. Draw a graph of E from $z = 0$ to $z = 3z_0$.

47. ‖ FIGURE P24.47 shows an infinitely wide conductor parallel to and distance d from an infinitely wide plane of charge with surface charge density η. What are the electric fields $\vec{E}_1$ to $\vec{E}_4$ in regions 1 to 4?

FIGURE P24.47   **FIGURE P24.48**

48. ‖ FIGURE P24.48 shows two very large slabs of metal that are parallel and distance l apart. The top and bottom of each slab has surface area A. The thickness of each slab is so small in comparison to its lateral dimensions that the surface area around the sides is negligible. Metal 1 has total charge $Q_1 = Q$ and metal 2 has total charge $Q_2 = 2Q$. Assume Q is positive. In terms of Q and A, determine
   a. The electric field strengths E_1 to E_5 in regions 1 to 5.
   b. The surface charge densities η_a to η_d on the four surfaces a to d.

49. ‖ A long, thin straight wire with linear charge density λ runs down the center of a thin, hollow metal cylinder of radius R. The cylinder has a net linear charge density 2λ. Assume λ is positive. Find expressions for the electric field strength (a) inside

the cylinder, $r < R$, and (b) outside the cylinder, $r > R$. In what direction does the electric field point in each of the cases?

50. ‖ A very long, uniformly charged cylinder has radius R and linear charge density λ. Find the cylinder's electric field strength (a) outside the cylinder, $r \ge R$, and (b) inside the cylinder, $r \le R$. (c) Show that your answers to parts a and b match at the boundary, $r = R$.

51. ‖ The electric field must be zero inside a *conductor* in electrostatic equilibrium, but not inside an insulator. It turns out that we can still apply Gauss's law to a Gaussian surface that is entirely within an insulator by replacing the right-hand side of Gauss's law, Q_{in}/ϵ_0, with Q_{in}/ϵ, where ϵ is the *permittivity* of the material. (Technically, ϵ_0 is called the *vacuum permittivity*.) Suppose that a 50 nC point charge is surrounded by a thin, 32-cm-diameter spherical rubber shell and that the electric field strength inside the rubber shell is 2500 N/C. What is the permittivity of rubber?

52. ‖ The electric field must be zero inside a *conductor* in electrostatic equilibrium, but not inside an insulator. It turns out that we can still apply Gauss's law to a Gaussian surface that is entirely within an insulator by replacing the right-hand side of Gauss's law, Q_{in}/ϵ_0, with Q_{in}/ϵ, where ϵ is the *permittivity* of the material. (Technically, ϵ_0 is called the *vacuum permittivity*.) Suppose a long, straight wire with linear charge density 250 nC/m is covered with insulation whose permittivity is $2.5\epsilon_0$. What is the electric field strength at a point inside the insulation that is 1.5 mm from the axis of the wire?

53. ‖‖ A long cylinder with radius b and volume charge density ρ has a spherical hole with radius $a < b$ centered on the axis of the cylinder. What is the electric field strength inside the hole at radial distance $r < a$ in a plane that is perpendicular to the cylinder through the center of the hole?
   Hint: Can you create this charge distribution as a superposition of charge distributions for which you can use Gauss's law to find the electric field?

54. ‖ A spherical shell has inner radius R_{in} and outer radius R_{out}. The shell contains total charge Q, uniformly distributed. The interior of the shell is empty of charge and matter.
   a. Find the electric field strength outside the shell, $r \ge R_{out}$.
   b. Find the electric field strength in the interior of the shell, $r \le R_{in}$.
   c. Find the electric field strength within the shell, $R_{in} \le r \le R_{out}$.
   d. Show that your solutions match at both the inner and outer boundaries.

55. ‖‖ An early model of the atom, proposed by Rutherford after his discovery of the atomic nucleus, had a positive point charge $+Ze$ (the nucleus) at the center of a sphere of radius R with uniformly distributed negative charge $-Ze$. Z is the atomic number, the number of protons in the nucleus and the number of electrons in the negative sphere.
   a. Show that the electric field strength inside this atom is

$$E_{in} = \frac{Ze}{4\pi\epsilon_0}\left(\frac{1}{r^2} - \frac{r}{R^3}\right)$$

   b. What is E at the surface of the atom? Is this the expected value? Explain.
   c. A uranium atom has $Z = 92$ and $R = 0.10$ nm. What is the electric field strength at $r = \frac{1}{2}R$?

56. ‖ Newton's law of gravity and Coulomb's law are both inverse-square laws. Consequently, there should be a "Gauss's law for gravity."
   CALC
   a. The electric field was defined as $\vec{E} = \vec{F}_{on\,q}/q$, and we used this to find the electric field of a point charge. Using analogous reasoning, what is the gravitational field $\vec{g}$ of a point mass?

Write your answer using the unit vector $\hat{r}$, but be careful with signs; the gravitational force between two "like masses" is attractive, not repulsive.

b. What is Gauss's law for gravity, the gravitational equivalent of Equation 24.18? Use Φ_G for the gravitational flux, $\vec{g}$ for the gravitational field, and M_{in} for the enclosed mass.

c. A spherical planet is discovered with mass M, radius R, and a mass density that varies with radius as $\rho = \rho_0(1 - r/2R)$, where ρ_0 is the density at the center. Determine ρ_0 in terms of M and R.

Hint: Divide the planet into infinitesimal shells of thickness dr, then sum (i.e., integrate) their masses.

d. Find an expression for the gravitational field strength inside the planet at distance $r < R$.

Challenge Problems

57. ‖‖ All examples of Gauss's law have used highly symmetric
CALC surfaces where the flux integral is either zero or EA. Yet we've claimed that the net $\Phi_e = Q_{in}/\epsilon_0$ is independent of the surface. This is worth checking. **FIGURE CP24.57** shows a cube of edge length L centered on a long thin wire with linear charge density λ. The flux through one face of the cube is *not* simply EA because, in this case, the electric field varies in both strength and direction. But you can calculate the flux by actually doing the flux integral.

FIGURE CP24.57

a. Consider the face parallel to the yz-plane. Define area $d\vec{A}$ as a strip of width dy and height L with the vector pointing in the x-direction. One such strip is located at position y. Use the known electric field of a wire to calculate the electric flux $d\Phi$ through this little area. Your expression should be written in terms of y, which is a variable, and various constants. It should not explicitly contain any angles.

b. Now integrate $d\Phi$ to find the total flux through this face.

c. Finally, show that the net flux through the cube is $\Phi_e = Q_{in}/\epsilon_0$.

58. ‖‖‖ An infinite cylinder of radius R has a linear charge density λ.
CALC The volume charge density (C/m^3) within the cylinder $(r \leq R)$ is $\rho(r) = r\rho_0/R$, where ρ_0 is a constant to be determined.

a. Draw a graph of ρ versus x for an x-axis that crosses the cylinder perpendicular to the cylinder axis. Let x range from $-2R$ to $2R$.

b. The charge within a small volume dV is $dq = \rho\,dV$. The integral of $\rho\,dV$ over a cylinder of length L is the total charge $Q = \lambda L$ within the cylinder. Use this fact to show that $\rho_0 = 3\lambda/2\pi R^2$.

Hint: Let dV be a cylindrical shell of length L, radius r, and thickness dr. What is the volume of such a shell?

c. Use Gauss's law to find an expression for the electric field strength E inside the cylinder, $r \leq R$, in terms of λ and R.

d. Does your expression have the expected value at the surface, $r = R$? Explain.

59. ‖‖‖ A sphere of radius R has total charge Q. The volume charge
CALC density (C/m^3) within the sphere is $\rho(r) = C/r^2$, where C is a constant to be determined.

a. The charge within a small volume dV is $dq = \rho\,dV$. The integral of $\rho\,dV$ over the entire volume of the sphere is the total charge Q. Use this fact to determine the constant C in terms of Q and R.

Hint: Let dV be a spherical shell of radius r and thickness dr. What is the volume of such a shell?

b. Use Gauss's law to find an expression for the electric field strength E inside the sphere, $r \leq R$, in terms of Q and R.

c. Does your expression have the expected value at the surface, $r = R$? Explain.

60. ‖‖‖ A sphere of radius R has total charge Q. The volume charge
CALC density (C/m^3) within the sphere is

$$\rho = \rho_0\left(1 - \frac{r}{R}\right)$$

This charge density decreases linearly from ρ_0 at the center to zero at the edge of the sphere.

a. Show that $\rho_0 = 3Q/\pi R^3$.

b. Show that the electric field inside the sphere points radially outward with magnitude

$$E = \frac{Qr}{4\pi\epsilon_0 R^3}\left(4 - 3\frac{r}{R}\right)$$

c. Show that your result of part b has the expected value at $r = R$.

61. ‖‖‖ A spherical ball of charge has radius R and total charge Q. The
CALC electric field strength inside the ball $(r \leq R)$ is $E(r) = r^4 E_{max}/R^4$.

a. What is E_{max} in terms of Q and R?

b. Find an expression for the volume charge density $\rho(r)$ inside the ball as a function of r.

c. Verify that your charge density gives the total charge Q when integrated over the volume of the ball.

25 The Electric Potential

City lights seen from space show where millions of lightbulbs are transforming electric energy into light and thermal energy.

IN THIS CHAPTER, you will learn to use the electric potential and electric potential energy.

What is electric potential energy?

Recall that potential energy is an interaction energy. Charged particles that interact via the electric force have electric potential energy U. You'll learn that there's a close analogy with gravitational potential energy.

« LOOKING BACK Section 10.1 Potential energy
« LOOKING BACK Section 10.5 Energy diagrams

What is the electric potential?

You've seen that source charges create an electric field. Source charges also create an electric potential. The electric potential V

- Exists everywhere in space.
- Is a scalar.
- Causes charges to have potential energy.
- Is measured in volts.

Electric potential

What potentials are especially important?

We'll calculate the electric potential of four important charge distributions: a point charge, a charged sphere, a ring of charge, and a parallel-plate capacitor. Finding the potential of a continuous charge distribution is similar to calculating electric fields, but easier because potential is a scalar.

Potential of a point charge

« LOOKING BACK Section 23.3 The electric field

How is potential represented?

Electric potential is a fairly abstract idea, so it will be important to visualize how the electric potential varies in space. One way of doing so is with equipotential surfaces. These are mathematical surfaces, not physical surfaces, with the same value of the potential V at every point.

How is electric potential used?

A charged particle q in an electric potential V has electric potential energy $U = qV$.

- Charged particles accelerate as they move through a potential difference.
- Mechanical energy is conserved:
$$K_f + qV_f = K_i + qV_i$$

« LOOKING BACK Section 10.4 Energy conservation

Why is energy important in electricity?

Energy allows things to happen. You want your lights to light, your computer to compute, and your music to play. All these require energy—*electric* energy. This is the first of two chapters that explore electric energy and its connection to electric forces and fields. You'll then be prepared to understand electric circuits—which are all about how energy is transformed and transferred from sources, such as batteries, to devices that utilize and dissipate the energy.

25.1 Electric Potential Energy

We started our study of electricity with electric forces and fields. But in electricity, just as in mechanics, *energy* is also a powerful idea. This chapter and the next will explore how energy is used in electricity, introduce the important concept of *electric potential,* and lay the groundwork for our upcoming study of electric circuits.

It's been many chapters since we dealt much with work and energy, but these ideas will now be *essential* to our story. Consequently, the Looking Back recommendations in the chapter preview are especially important. You will recall that a system's mechanical energy $E_{mech} = K + U$ is conserved for particles that interact with each other via *conservative forces,* where K and U are the kinetic and potential energy. That is,

$$\Delta E_{mech} = \Delta K + \Delta U = 0 \tag{25.1}$$

We need to be careful with notation because we are now using E to represent the electric field strength. To avoid confusion, we will represent mechanical energy either as the sum $K + U$ or as E_{mech}, with an explicit subscript.

NOTE Recall that for any X, the *change* in X is $\Delta X = X_{final} - X_{initial}$.

A key idea of Chapters 9 and 10 was that energy is the energy *of a system,* and clearly defining the system is crucial. Kinetic energy $K = \frac{1}{2}mv^2$ is a system's *energy of motion.* For a multiparticle system, K is the sum of the kinetic energies of each particle in the system.

Potential energy U is the *interaction energy* of the system. Suppose the particles of the system move from some initial set of positions i to final positions f. As the particles move, the action/reaction pairs of forces between the particles—the interaction forces—do work and the system's potential energy changes. In « Section 10.1 we defined the *change* in potential energy to be

$$\Delta U = -W_{interaction}(i \rightarrow f) \tag{25.2}$$

where the notation means the work done by the interaction forces as the configuration changes from i to f. This rather abstract definition will make more sense when we see specific applications.

Recall that *work* is done when a force acts on a particle as it is being displaced. In « Section 9.3 you learned that a *constant* force $\vec{F}$ does work

$$W = \vec{F} \cdot \Delta \vec{r} = F \Delta r \cos\theta \tag{25.3}$$

as the particle undergoes displacement $\Delta \vec{r}$, where θ is the angle between the two vectors. FIGURE 25.1 reminds you of the three special cases $\theta = 0°$, 90°, and 180°.

NOTE Work is *not* the oft-remembered "force times distance." Work is force times distance only in the one very special case in which the force is both constant *and* parallel to the displacement.

If the force is *not* constant, we can calculate the work by dividing the path into many small segments of length dx, finding the work done in each segment, and then summing (i.e., integrating) from the start of the path to the end. The work done in one such segment is $dW = F(x) \cos\theta\, dx$, where $F(x)$ indicates that the force is a function of position x. Thus the work done on the particle as it moves from x_i to x_f is

$$W = \int_{x_i}^{x_f} F(x) \cos\theta\, dx \tag{25.4}$$

Finally, recall that a *conservative force* is one for which the work done on a particle as it moves from position i to position f is *independent of the path followed.* We'll assert for now, and prove later, that **the electric force is a conservative force,** and thus we can define an electric potential energy.

FIGURE 25.1 The work done by a constant force.

The particle undergoes displacement $\Delta \vec{r}$.

$\theta = 0°$
$W = F\Delta r$

$\theta = 90°$
$W = 0$

$\theta = 180°$
$W = -F\Delta r$

A Gravitational Analogy

Gravity, like electricity, is a long-range force. Much as we defined the electric field $\vec{E} = \vec{F}_{\text{on }q}/q$, we can also define a *gravitational field*—the agent that exerts gravitational forces on masses—as $\vec{F}_{\text{on }m}/m$. But $\vec{F}_{\text{on }m} = m\vec{g}$ near the earth's surface; thus the familiar $\vec{g} = (9.80 \text{ N/kg, down})$ is really the gravitational field! Notice how we've written the units of $\vec{g}$ as N/kg, but you can easily show that N/kg = m/s^2. The gravitational field near the earth's surface is a *uniform* field in the downward direction.

FIGURE 25.2 shows a particle of mass m falling in the gravitational field. The gravitational force is in the same direction as the particle's displacement, so the gravitational field does a *positive* amount of work on the particle. The gravitational force is constant, hence the work done by the gravitational field is

$$W_{\text{G}} = F_{\text{G}}\,\Delta r\cos 0° = mg\left|y_{\text{f}} - y_{\text{i}}\right| = mgy_{\text{i}} - mgy_{\text{f}} \qquad (25.5)$$

We have to be careful with signs because Δr, the magnitude of the displacement vector, must be a positive number.

Now we can see how the definition of ΔU in Equation 25.2 makes sense. The *change* in gravitational potential energy is

$$\Delta U_{\text{G}} = U_{\text{f}} - U_{\text{i}} = -W_{\text{G}}(\text{i}\rightarrow\text{f}) = mgy_{\text{f}} - mgy_{\text{i}} \qquad (25.6)$$

Comparing the initial and final terms on the two sides of the equation, we see that the gravitational potential energy near the earth is the familiar quantity

$$U_{\text{G}} = U_0 + mgy \qquad (25.7)$$

where U_0 is the value of U_{G} at $y = 0$. We usually choose $U_0 = 0$, in which case $U_{\text{G}} = mgy$, but such a choice is not necessary.

FIGURE 25.2 Potential energy is transformed into kinetic energy as a particle moves in a gravitational field.

The gravitational field does work on the particle.

The net force on the particle is down. It gains kinetic energy (i.e., speeds up) as it loses potential energy.

A Uniform Electric Field

FIGURE 25.3 shows a charged particle inside a parallel-plate capacitor with electrode spacing d. This is a uniform electric field, and the situation looks very much like Figure 25.2 for a mass in a uniform gravitational field. The one difference is that $\vec{g}$ always points down whereas the positive-to-negative electric field can point in any direction. To deal with this, let's define a coordinate axis s that points *from* the negative plate, which we define to be $s = 0$, *toward* the positive plate. The electric field $\vec{E}$ then points in the negative s-direction, just as the gravitational field $\vec{g}$ points in the negative y-direction. This s-axis, which is valid no matter how the capacitor is oriented, is analogous to the y-axis used for gravitational potential energy.

A positive charge q inside the capacitor speeds up and gains kinetic energy as it "falls" toward the negative plate. Is the charge losing potential energy as it gains kinetic energy? Indeed it is, and the calculation of the potential energy is just like the calculation of gravitational potential energy. The electric field exerts a *constant* force $F - qE$ on the charge in the direction of motion; thus the work done on the charge by the electric field is

$$W_{\text{elec}} = F\,\Delta r\cos 0° = qE\left|s_{\text{f}} - s_{\text{i}}\right| = qEs_{\text{i}} - qEs_{\text{f}} \qquad (25.8)$$

where we again have to be careful with the signs because $s_{\text{f}} < s_{\text{i}}$.

The work done by the electric field causes the *electric* potential energy to change by

$$\Delta U_{\text{elec}} = U_{\text{f}} - U_{\text{i}} = -W_{\text{elec}}(\text{i}\rightarrow\text{f}) = qEs_{\text{f}} - qEs_{\text{i}} \qquad (25.9)$$

Comparing the initial and final terms on the two sides of the equation, we see that the **electric potential energy** of charge q in a uniform electric field is

$$U_{\text{elec}} = U_0 + qEs \qquad (25.10)$$

where s is measured from the negative plate and U_0 is the potential energy at the negative plate ($s = 0$). It will often be convenient to choose $U_0 = 0$, but the choice has no physical consequences because it doesn't affect ΔU_{elec}. Equation 25.10 was derived with the assumption that q is positive, but it is valid for either sign of q.

FIGURE 25.3 The electric field does work on the charged particle.

The electric field does work on the particle.

Electric field

The particle is "falling" in the direction of $\vec{E}$.

NOTE Although Equation 25.10 is sometimes called "the potential energy of charge q," it is really the potential energy of the charge + capacitor system.

FIGURE 25.4 shows positive and negative charged particles moving inside a parallel-plate capacitor. For a positive charge, U_{elec} decreases and K increases as the charge moves toward the negative plate (decreasing s). Thus a positive charge is going "downhill" if it moves in the direction of the electric field. A positive charge moving opposite the field direction is going "uphill," slowing as it transforms kinetic energy into electric potential energy.

FIGURE 25.4 A charged particle exchanges kinetic and potential energy as it moves in an electric field.

If we choose $U_0 = 0$, so that potential energy is zero at the negative plate, then a negative charged particle has *negative* potential energy. You learned in Chapter 10 that there's nothing wrong with negative potential energy—it's simply less than the potential energy at some arbitrarily chosen reference location. The more important point, from Equation 25.10, is that the potential energy *increases* (becomes less negative) as a negative charge moves toward the negative plate. A negative charge moving in the field direction is going "uphill," transforming kinetic energy into electric potential energy as it slows.

FIGURE 25.5 is an *energy diagram* for a positively charged particle in an electric field. Recall that an energy diagram is a graphical representation of how kinetic and potential energies are transformed as a particle moves. For positive q, the electric potential energy given by Equation 25.10 increases linearly from 0 at the negative plate (with $U_0 = 0$) to qEd at the positive plate. The total mechanical energy—which is under your control—is constant. If $E_{mech} < qEd$, as shown here, a positively charged particle projected from the negative plate will gradually slow (transforming kinetic energy into potential energy) until it reaches a *turning point* where $U_{elec} = E_{mech}$. But if you project the particle with greater speed, such that $E_{mech} > qEd$, it will be able to cross the gap to collide with the positive plate.

FIGURE 25.5 The energy diagram for a positively charged particle in a uniform electric field.

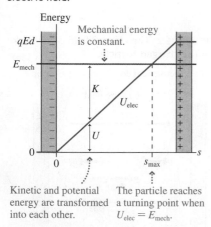

Kinetic and potential energy are transformed into each other.

The particle reaches a turning point when $U_{elec} = E_{mech}$.

EXAMPLE 25.1 | Conservation of energy

A 2.0 cm $\times$ 2.0 cm parallel-plate capacitor with a 2.0 mm spacing is charged to ± 1.0 nC. First a proton and then an electron are released from rest at the midpoint of the capacitor.

a. What is each particle's energy?

b. What is each particle's speed as it reaches the plate?

MODEL The mechanical energy is conserved. A parallel-plate capacitor has a uniform electric field.

VISUALIZE **FIGURE 25.6** is a before-and-after pictorial representation, as you learned to draw in Part II. Each particle is released from rest $(K = 0)$ and moves "downhill" toward lower potential energy. Thus the proton moves toward the negative plate, the electron toward the positive plate.

FIGURE 25.6 A proton and an electron in a capacitor.

SOLVE a. The s-axis was defined to point from the negative toward the positive plate of the capacitor. Both charged particles have $s_i = \frac{1}{2}d$, where $d = 2.0$ mm is the plate separation. If we let $U_0 = 0$, defining the negative plate as our zero-energy reference point, then the proton ($q = e$) has energy

$$E_{\text{mech p}} = K_i + U_i = 0 + eE\left(\tfrac{1}{2}d\right)$$

while the electron ($q = -e$) has

$$E_{\text{mech e}} = K_i + U_i = 0 - eE\left(\tfrac{1}{2}d\right)$$

The electric field inside the parallel-plate capacitor, from Chapter 23, is

$$E = \frac{Q}{\epsilon_0 A} = 2.82 \times 10^5 \text{ N/C}$$

Thus the particles' energies can be calculated to be

$$E_{\text{mech p}} = 4.5 \times 10^{-17} \text{ J and } E_{\text{mech e}} = -4.5 \times 10^{-17} \text{ J}$$

Notice that the electron's mechanical energy is negative.

b. Conservation of mechanical energy requires $K_f + U_f = K_i + U_i = E_{\text{mech}}$. The proton collides with the negative plate, so $U_f = 0$, and the final kinetic energy is $K_f = \frac{1}{2}m_p v_f^2 = E_{\text{mech p}}$. Thus the proton's impact speed is

$$(v_f)_p = \sqrt{\frac{2E_{\text{mech p}}}{m_p}} = 2.3 \times 10^5 \text{ m/s}$$

Similarly, the electron collides with the positive plate, where $U_f = qEd = -eEd = 2E_{\text{mech e}}$. Thus energy conservation for the electron is

$$K_f = \tfrac{1}{2}m_e v_f^2 = E_{\text{mech e}} - U_f = E_{\text{mech e}} - 2E_{\text{mech e}} = -E_{\text{mech e}}$$

We found the electron's mechanical energy to be negative, so K_f is positive. The electron reaches the positive plate with speed

$$(v_f)_e = \sqrt{\frac{-2E_{\text{mech e}}}{m_e}} = 1.0 \times 10^7 \text{ m/s}$$

ASSESS Even though both particles have mechanical energy with the same magnitude, the electron has a much greater final speed due to its much smaller mass.

STOP TO THINK 25.1 A glass rod is positively charged. The figure shows an end view of the rod. A negatively charged particle moves in a circular arc around the glass rod. Is the work done on the charged particle by the rod's electric field positive, negative, or zero?

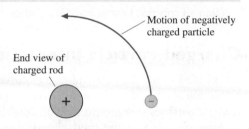

End view of charged rod

Motion of negatively charged particle

25.2 The Potential Energy of Point Charges

FIGURE 25.7a shows two charges q_1 and q_2, which we will assume to be like charges. These two charges interact, and the energy of their interaction can be found by calculating the work done by the electric field of q_1 on q_2 as q_2 moves from position x_i to position x_f. We'll assume that q_1 has been glued down and is unable to move, as shown in **FIGURE 25.7b**.

The force is entirely in the direction of motion, so $\cos\theta = 1$. Thus

$$W_{\text{elec}} = \int_{x_i}^{x_f} F_{1 \text{ on } 2} \, dx = \int_{x_i}^{x_f} \frac{Kq_1q_2}{x^2} \, dx = Kq_1q_2 \frac{-1}{x}\bigg|_{x_i}^{x_f} = -\frac{Kq_1q_2}{x_f} + \frac{Kq_1q_2}{x_i} \quad (25.11)$$

The potential energy of the two charges is related to the work done by

$$\Delta U_{\text{elec}} = U_f - U_i = -W_{\text{elec}}(i \rightarrow f) = \frac{Kq_1q_2}{x_f} - \frac{Kq_1q_2}{x_i} \quad (25.12)$$

By comparing the left and right sides of the equation we see that the potential energy of the two-point-charge system is

$$U_{\text{elec}} = \frac{Kq_1q_2}{x} \quad (25.13)$$

FIGURE 25.7 The interaction between two point charges.

(a) $\vec{F}_{2 \text{ on } 1}$ q_1 $\quad$ q_2 $\vec{F}_{1 \text{ on } 2}$

Like charges exert repulsive forces.

(b) Fixed in position q_1 $\quad$ The force changes with distance. q_2 $\quad$ $\vec{F}$

$\Delta\vec{r}$

0 $\quad$ x_i $\quad$ x_f $\quad$ x

The electric field of q_1 does work on q_2 as q_2 moves from x_i to x_f.

We could include a constant U_0, as we did in Equation 25.10, for the potential energy of a charge in a uniform electric field, but it is customary to set $U_0 = 0$.

We chose to integrate along the x-axis for convenience, but all that matters is the *distance* between the charges. Thus a general expression for the electric potential energy is

$$U_{elec} = \frac{Kq_1q_2}{r} = \frac{1}{4\pi\epsilon_0}\frac{q_1q_2}{r} \qquad \text{(two point charges)} \qquad (25.14)$$

This is explicitly the energy *of the system,* not the energy of just q_1 or q_2.

NOTE The electric potential energy of two point charges looks *almost* the same as the force between the charges. The difference is the r in the denominator of the potential energy compared to the r^2 in Coulomb's law.

Three important points need to be noted:

- The choice $U_0 = 0$ is equivalent to saying that the potential energy of two charged particles is zero only when they are infinitely far apart. This makes sense because two charged particles cease interacting only when they are infinitely far apart.
- We derived Equation 25.14 for two like charges, but it is equally valid for two opposite charges. The potential energy of two like charges is *positive* and of two opposite charges is *negative.*
- Because the electric field outside a *sphere of charge* is the same as that of a point charge at the center, Equation 25.14 is also the electric potential energy of two charged spheres. Distance r is the distance between their centers.

Charged-Particle Interactions

FIGURE 25.8a shows the potential-energy curve—a hyperbola—for two like charges as a function of the distance r between them. Also shown is the total energy line for two charged particles shot toward each other with equal but opposite momenta.

You can see that the total energy line crosses the potential-energy curve at r_{min}. This is a turning point. The two charges gradually slow down, because of the repulsive force between them, until the distance separating them is r_{min}. At this point, the kinetic energy is zero and both particles are instantaneously at rest. Both then reverse direction and move apart, speeding up as they go. r_{min} is the *distance of closest approach.*

Two opposite charges are a little trickier because of the negative energies. Negative total energies seem troubling at first, but they characterize *bound systems.* **FIGURE 25.8b** shows two oppositely charged particles shot apart from each other with equal but opposite momenta. If $E_{mech} < 0$, as shown, then their total energy line crosses the potential-energy curve at r_{max}. That is, the particles slow down, lose kinetic energy, reverse directions at *maximum separation* r_{max}, and then "fall" back together. They cannot escape from each other. Although moving in three dimensions rather than one, the electron and proton of a hydrogen atom are a realistic example of a bound system, and their mechanical energy is negative.

Two oppositely charged particles *can* escape from each other if $E_{mech} > 0$. They'll slow down, but eventually the potential energy vanishes and the particles still have kinetic energy. The threshold condition for escape is $E_{mech} = 0$, which will allow the particles to reach infinite separation $(U \to 0)$ at infinitesimally slow speed $(K \to 0)$. The initial speed that gives $E_{mech} = 0$ is called the *escape speed.*

NOTE Real particles can't be infinitely far apart, but because U_{elec} decreases with distance, there comes a point when $U_{elec} = 0$ is an excellent approximation. Two charged particles for which $U_{elec} \approx 0$ are sometimes described as "far apart" or "far away."

FIGURE 25.8 The potential-energy diagrams for two like charges and two opposite charges.

(a) Like charges

(b) Opposite charges

The Electric Force Is a Conservative Force

Potential energy can be defined only if the force is *conservative,* meaning that the work done on the particle as it moves from position i to position f is independent of the path followed between i and f. **FIGURE 25.9** demonstrates that electric force is indeed conservative.

FIGURE 25.9 The work done on q_2 is independent of the path from i to f.

Consider an alternative path for q_2 to move from i to f.

q_1 q_2
i   f

Approximate the path using circular arcs and radial lines centered on q_1.

$\vec{F}$

q_1
i   f

The electric force is a *central force.* As a result, zero work is done as q_2 moves along a circular arc because the force is perpendicular to the displacement.

q_1
i - - - - - - - - - - - - f

All the work is done along the radial line segments, which are equivalent to a straight line from i to f. This is the work that was calculated in Equation 25.11.

EXAMPLE 25.2 | Approaching a charged sphere

A proton is fired from far away at a 1.0-mm-diameter glass sphere that has been charged to $+100\ \text{nC}$. What initial speed must the proton have to just reach the surface of the glass?

MODEL Energy is conserved. The glass sphere can be modeled as a charged particle, so the potential energy is that of two point charges. The glass is so much more massive than the proton that it remains at rest as the proton moves. The proton starts "far away," which we interpret as sufficiently far to make $U_i \approx 0$.

VISUALIZE **FIGURE 25.10** shows the before-and-after pictorial representation. To "just reach" the glass sphere means that the proton comes to rest, $v_f = 0$, as it reaches $r_f = 0.50\ \text{mm}$, the *radius* of the sphere.

SOLVE Conservation of energy $K_f + U_f = K_i + U_i$ is

$$0 + \frac{Kq_p q_{\text{sphere}}}{r_f} = \tfrac{1}{2}mv_i^2 + 0$$

FIGURE 25.10 A proton approaching a glass sphere.

Before: v_i

$r_i \approx \infty$ so $U_i = 0$

$+$
R

After:

$+$

$r_f = R$
$v_f = 0$

The proton charge is $q_p = e$. With this, we can solve for the proton's initial speed:

$$v_i = \sqrt{\frac{2Keq_{\text{sphere}}}{mr_f}} = 1.86 \times 10^7\ \text{m/s}$$

EXAMPLE 25.3 | Escape speed

An interaction between two elementary particles causes an electron and a positron (a positive electron) to be shot out back to back with equal speeds. What minimum speed must each have when they are 100 fm apart in order to escape each other?

MODEL Energy is conserved. The particles end "far apart," which we interpret as sufficiently far to make $U_f \approx 0$.

VISUALIZE **FIGURE 25.11** shows the before-and-after pictorial representation. The minimum speed to escape is the speed that allows the particles to reach $r_f = \infty$ with $v_f = 0$.

FIGURE 25.11 An electron and a positron flying apart.

Before: v_i $\ominus$ $\oplus$ v_i
$r_i = 100\ \text{fm}$

$v_f = 0$ $v_f = 0$
After: $\ominus$ $\oplus$
$r_f \approx \infty$ so $U_f = 0$

SOLVE U_{elec} is the potential energy of the electron + positron system. Similarly, K is the *total* kinetic energy of the system. The electron and the positron, with equal masses and equal speeds, have equal kinetic energies. Conservation of energy $K_f + U_f = K_i + U_i$ is

$$0 + 0 + 0 = \tfrac{1}{2}mv_i^2 + \tfrac{1}{2}mv_i^2 + \frac{Kq_e q_p}{r_i} = mv_i^2 - \frac{Ke^2}{r_i}$$

Using $r_i = 100\ \text{fm} = 1.0 \times 10^{-13}\ \text{m}$, we can calculate the minimum initial speed to be

$$v_i = \sqrt{\frac{Ke^2}{mr_i}} = 5.0 \times 10^7\ \text{m/s}$$

ASSESS v_i is a little more than 10% the speed of light, just about the limit of what a "classical" calculation can predict. We would need to use the theory of relativity if v_i were any larger.

Multiple Point Charges

If more than two charges are present, their potential energy is the sum of the potential energies due to all pairs of charges:

$$U_{\text{elec}} = \sum_{i<j} \frac{Kq_iq_j}{r_{ij}} \tag{25.15}$$

where r_{ij} is the distance between q_i and q_j. The summation contains the $i < j$ restriction to ensure that each pair of charges is counted only once.

> **NOTE** For energy conservation problems, it's necessary to calculate only the potential energy for those pairs of charges for which the distance r_{ij} changes. The potential energy of any pair that doesn't move is an additive constant with no physical consequences.

EXAMPLE 25.4 | Launching an electron

Three electrons are spaced 1.0 mm apart along a vertical line. The outer two electrons are fixed in position.

a. Is the center electron at a point of stable or unstable equilibrium?

b. If the center electron is displaced horizontally by a small distance, what will its speed be when it is very far away?

MODEL Energy is conserved. The outer two electrons don't move, so we don't need to include the potential energy of their interaction.

VISUALIZE FIGURE 25.12 shows the situation.

FIGURE 25.12 Three electrons.

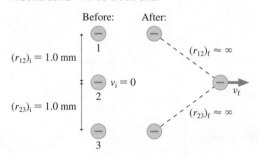

SOLVE a. The center electron is in equilibrium *exactly* in the center because the two electric forces on it balance. But if it moves a little to the right or left, no matter how little, then the horizontal components of the forces from both outer electrons will push the center electron farther away. This is an unstable equilibrium for horizontal displacements, like being on the top of a hill.

b. A small displacement will cause the electron to move away. If the displacement is only infinitesimal, the initial conditions are $(r_{12})_i = (r_{23})_i = 1.0$ mm and $v_i = 0$. "Far away" is interpreted as $r_f \rightarrow \infty$, where $U_f \approx 0$. There are now *two* terms in the potential energy, so conservation of energy $K_f + U_f = K_i + U_i$ gives

$$\tfrac{1}{2}mv_f^2 + 0 + 0 = 0 + \left[\frac{Kq_1q_2}{(r_{12})_i} + \frac{Kq_2q_3}{(r_{23})_i}\right]$$

$$= \left[\frac{Ke^2}{(r_{12})_i} + \frac{Ke^2}{(r_{23})_i}\right]$$

This is easily solved to give

$$v_f = \sqrt{\frac{2}{m}\left[\frac{Ke^2}{(r_{12})_i} + \frac{Ke^2}{(r_{23})_i}\right]} = 1000 \text{ m/s}$$

STOP TO THINK 25.2 Rank in order, from largest to smallest, the potential energies U_a to U_d of these four pairs of charges. Each + symbol represents the same amount of charge.

(a)   (b)   (c)   (d)

FIGURE 25.13 The electric field does work as a dipole rotates.

The electric forces exert a torque on the dipole.

25.3 The Potential Energy of a Dipole

The electric dipole has been our model for understanding how charged objects interact with neutral objects. In Chapter 23 we found that an electric field exerts a *torque* on a dipole. We can complete the picture by calculating the potential energy of an electric dipole in a uniform electric field.

FIGURE 25.13 shows a dipole in an electric field $\vec{E}$. Recall that the dipole moment $\vec{p}$ is a vector that points from $-q$ to q with magnitude $p = qs$. The forces $\vec{F}_+$ and $\vec{F}_-$ exert a torque on the dipole, but now we're interested in calculating the *work* done by these forces as the dipole rotates from angle ϕ_i to angle ϕ_f.

When a force component F_s acts through a small displacement ds, the force does work $dW = F_s\,ds$. If we exploit the rotational-linear motion analogy from Chapter 12, where torque τ is the analog of force and angular displacement $\Delta\phi$ is the analog of linear displacement, then a torque acting through a small angular displacement $d\phi$ does work $dW = \tau\,d\phi$. From Chapter 23, the torque on the dipole in Figure 25.13 is $\tau = -pE\sin\phi$, where the minus sign is due to the torque trying to cause a clockwise rotation. Thus the work done by the electric field on the dipole as it rotates through the small angle $d\phi$ is

$$dW_{\text{elec}} = -pE\sin\phi\,d\phi \qquad (25.16)$$

The total work done by the electric field as the dipole turns from ϕ_i to ϕ_f is

$$W_{\text{elec}} = -pE\int_{\phi_i}^{\phi_f}\sin\phi\,d\phi = pE\cos\phi_f - pE\cos\phi_i \qquad (25.17)$$

The potential energy associated with the work done on the dipole is

$$\Delta U_{\text{dipole}} = U_f - U_i = -W_{\text{elec}}(i \rightarrow f) = -pE\cos\phi_f + pE\cos\phi_i \qquad (25.18)$$

By comparing the left and right sides of Equation 25.18, we see that the potential energy of an electric dipole $\vec{p}$ in a uniform electric field $\vec{E}$ is

$$U_{\text{dipole}} = -pE\cos\phi = -\vec{p}\cdot\vec{E} \qquad (25.19)$$

FIGURE 25.14 shows the energy diagram of a dipole. The potential energy is minimum at $\phi = 0°$ where the dipole is aligned with the electric field. This is a point of stable equilibrium. A dipole exactly opposite $\vec{E}$, at $\phi = \pm 180°$, is at a point of unstable equilibrium. Any disturbance will cause it to flip around. A frictionless dipole with mechanical energy E_{mech} will oscillate back and forth between turning points on either side of $\phi = 0°$.

FIGURE 25.14 The energy of a dipole in an electric field.

EXAMPLE 25.5 | **Rotating a molecule**

The water molecule is a permanent electric dipole with dipole moment 6.2×10^{-30} C m. A water molecule is aligned in an electric field with field strength 1.0×10^7 N/C. How much energy is needed to rotate the molecule 90°?

MODEL The molecule is at the point of minimum energy. It won't spontaneously rotate 90°. However, an external force that supplies energy, such as a collision with another molecule, can cause the water molecule to rotate.

SOLVE The molecule starts at $\phi_i = 0°$ and ends at $\phi_f = 90°$. The increase in potential energy is

$$\Delta U_{\text{dipole}} = U_f - U_i = -pE\cos 90° - (-pE\cos 0°)$$
$$= pE = 6.2 \times 10^{-23}\,\text{J}$$

This is the energy needed to rotate the molecule 90°.

ASSESS ΔU_{dipole} is significantly less than $k_B T$ at room temperature. Thus collisions with other molecules can easily supply the energy to rotate the water molecules and keep them from staying aligned with the electric field.

25.4 The Electric Potential

We introduced the concept of the *electric field* in Chapter 22 because action at a distance raised concerns and difficulties. The field provides an intermediary through which two charges exert forces on each other. Charge q_1 somehow alters the space around it by creating an electric field $\vec{E}_1$. Charge q_2 then responds to the field, experiencing force $\vec{F} = q_2\vec{E}_1$.

In defining the electric field, we separated the charges that are the *source* of the field from the charge *in* the field. The force on charge q is related to the electric field of the source charges by

force on q by sources = [charge q] × [alteration of space by the source charges]

FIGURE 25.15 Source charges alter the space around them by creating an electric potential.

The potential at this point is V.

The source charges alter the space around them by creating an electric potential.

Source charges

If charge q is in the potential, the electric potential energy is $U_{q + \text{sources}} = qV$.

This battery is labeled 1.5 volts. As we'll soon see, a battery is a source of electric potential.

TABLE 25.1 Distinguishing electric potential and potential energy

The *electric potential* is a property of the source charges and, as you'll soon see, is related to the electric field. The electric potential is present whether or not a charged particle is there to experience it. Potential is measured in J/C, or V.

The *electric potential energy* is the interaction energy of a charged particle with the source charges. Potential energy is measured in J.

Let's try a similar procedure for the potential energy. The electric potential energy is due to the interaction of charge q with other charges, so let's write

potential energy of q + sources

= [charge q] × [*potential* for interaction with the source charges]

FIGURE 25.15 shows this idea schematically.

In analogy with the electric field, we will define the **electric potential** V (or, for brevity, just *the potential*) as

$$V \equiv \frac{U_{q + \text{sources}}}{q} \qquad (25.20)$$

Charge q is used as a probe to determine the electric potential, but the value of V is *independent of q*. **The electric potential, like the electric field, is a property of the source charges.** And, like the electric field, the electric potential fills the space around the source charges. It is there whether or not another charge is there to experience it.

In practice, we're usually more interested in knowing the potential energy if a charge q happens to be at a point in space where the electric potential of the source charges is V. Turning Equation 25.20 around, we see that the electric potential energy is

$$U_{q + \text{sources}} = qV \qquad (25.21)$$

Once the potential has been determined, it's very easy to find the potential energy.

The unit of electric potential is the joule per coulomb, which is called the **volt** V:

$$1 \text{ volt} = 1 \text{ V} \equiv 1 \text{ J/C}$$

This unit is named for Alessandro Volta, who invented the electric battery in the year 1800. Microvolts (μV), millivolts (mV), and kilovolts (kV) are commonly used units.

NOTE Once again, commonly used symbols are in conflict. The symbol V is widely used to represent *volume,* and now we're introducing the same symbol to represent *potential*. To make matters more confusing, V is the abbreviation for *volts*. In printed text, V for potential is italicized and V for volts is not, but you can't make such a distinction in handwritten work. This is not a pleasant state of affairs, but these are the commonly accepted symbols. It's incumbent upon you to be especially alert to the *context* in which a symbol is used.

Using the Electric Potential

The electric potential is an abstract idea, and it will take some practice to see just what it means and how it is useful. We'll use multiple representations—words, pictures, graphs, and analogies—to explain and describe the electric potential.

NOTE It is unfortunate that the terms *potential* and *potential energy* are so much alike. Despite the similar names, they are very different concepts and are not interchangeable. **TABLE 25.1** will help you to distinguish between the two.

Basically, knowing the electric potential in a region of space allows us to determine whether a charged particle speeds up or slows down as it moves through that region. **FIGURE 25.16** illustrates this idea. Here a group of source charges, which remains hidden offstage, has created an electric potential V that increases from left to right. A charged particle q, which for now we'll assume to be positive, has electric potential energy $U = qV$. If the particle moves to the right, its potential energy increases and so, by energy conservation, its kinetic energy must decrease. **A positive charge slows down as it moves into a region of higher electric potential.**

FIGURE 25.16 A charged particle speeds up or slows down as it moves through a potential difference.

- A positive charge slows down as it moves toward higher electric potential.
- The potential difference is positive.

$\Delta V > 0$

Lower potential — Direction of increasing V → Higher potential

$\Delta V < 0$

- A positive charge speeds up as it moves toward lower electric potential.
- The potential difference is negative.

It is customary to say that the particle moves through a **potential difference** $\Delta V = V_f - V_i$. The potential difference between two points is often called the **voltage.** The particle moving to the right moves through a positive potential difference ($\Delta V > 0$ because $V_f > V_i$), so we can say that a positively charged particle slows down as it moves through a positive potential difference.

The particle moving to the left in Figure 25.16 travels in the direction of decreasing electric potential—through a negative potential difference—and is losing potential energy. It speeds up as it transforms potential energy into kinetic energy. A negatively charged particle would slow down because its potential energy qV would increase as V decreases. **TABLE 25.2** summarizes these ideas.

If a particle moves through a potential difference ΔV, its electric potential energy changes by $\Delta U = q\Delta V$. We can write the conservation of energy equation in terms of the electric potential as $\Delta K + \Delta U = \Delta K + q\Delta V = 0$ or, as is often more practical,

$$K_f + qV_f = K_i + qV_i \qquad (25.22)$$

Conservation of energy is the basis of a powerful problem-solving strategy.

TABLE 25.2 Charged particles moving in an electric potential

	Electric potential	
	Increasing ($\Delta V > 0$)	Decreasing ($\Delta V < 0$)
+ charge	Slows down	Speeds up
− charge	Speeds up	Slows down

PROBLEM-SOLVING STRATEGY 25.1

Conservation of energy in charge interactions

MODEL Define the system. If possible, model it as an isolated system for which mechanical energy is conserved.

VISUALIZE Draw a before-and-after pictorial representation. Define symbols, list known values, and identify what you're trying to find.

SOLVE The mathematical representation is based on the law of conservation of mechanical energy:

$$K_f + qV_f = K_i + qV_i$$

- Is the electric potential given in the problem statement? If not, you'll need to use a known potential, such as that of a point charge, or calculate the potential using the procedure given later, in Problem-Solving Strategy 25.2.
- K_i and K_f are the sums of the kinetic energies of all moving particles.
- Some problems may need additional conservation laws, such as conservation of charge or conservation of momentum.

ASSESS Check that your result has correct units and significant figures, is reasonable, and answers the question.

Exercise 22

EXAMPLE 25.6 | **Moving through a potential difference**

A proton with a speed of 2.0×10^5 m/s enters a region of space in which there is an electric potential. What is the proton's speed after it moves through a potential difference of 100 V? What will be the final speed if the proton is replaced by an electron?

MODEL The system is the charge plus the unseen source charges creating the potential. This is an isolated system, so mechanical energy is conserved.

VISUALIZE FIGURE 25.17 is a before-and-after pictorial representation of a charged particle moving through a potential difference. A positive charge *slows down* as it moves into a region of higher potential ($K \rightarrow U$). A negative charge *speeds up* ($U \rightarrow K$).

FIGURE 25.17 A charged particle moving through a potential difference.

SOLVE The potential energy of charge q is $U = qV$. Conservation of energy, now expressed in terms of the electric potential V, is $K_f + qV_f = K_i + qV_i$, or

$$K_f = K_i - q\,\Delta V$$

where $\Delta V = V_f - V_i$ is the potential difference through which the particle moves. In terms of the speeds, energy conservation is

$$\tfrac{1}{2}mv_f^2 = \tfrac{1}{2}mv_i^2 - q\,\Delta V$$

We can solve this for the final speed:

$$v_f = \sqrt{v_i^2 - \frac{2q}{m}\Delta V}$$

For a proton, with $q = e$, the final speed is

$$(v_f)_p = \sqrt{(2.0 \times 10^5 \text{ m/s})^2 - \frac{2(1.60 \times 10^{-19} \text{ C})(100 \text{ V})}{1.67 \times 10^{-27} \text{ kg}}}$$

$$= 1.4 \times 10^5 \text{ m/s}$$

An electron, though, with $q = -e$ and a different mass, reaches speed $(v_f)_e = 5.9 \times 10^6$ m/s.

ASSESS The proton slowed down and the electron sped up, as we expected. Note that the electric potential *already existed* in space due to other charges that are not explicitly seen in the problem. The electron and proton have nothing to do with creating the potential. Instead, they *respond* to the potential by having potential energy $U = qV$.

STOP TO THINK 25.3 A proton is released from rest at point B, where the potential is 0 V. Afterward, the proton

a. Remains at rest at B.
b. Moves toward A with a steady speed.
c. Moves toward A with an increasing speed.
d. Moves toward C with a steady speed.
e. Moves toward C with an increasing speed.

$$-100\text{ V} \qquad 0\text{ V} \qquad +100\text{ V}$$
$$\text{A}\bullet \qquad \text{B}\bullet \qquad \text{C}\bullet$$

25.5 The Electric Potential Inside a Parallel-Plate Capacitor

FIGURE 25.18 A parallel-plate capacitor.

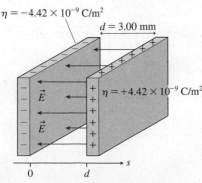

We began this chapter with the potential energy of a charge inside a parallel-plate capacitor. Now let's investigate the electric potential. FIGURE 25.18 shows two parallel electrodes, separated by distance d, with surface charge density $\pm\eta$. As a specific example, we'll let $d = 3.00$ mm and $\eta = 4.42 \times 10^{-9}$ C/m^2. The electric field inside the capacitor, as you learned in Chapter 23, is

$$\vec{E} = \left(\frac{\eta}{\epsilon_0}, \text{ from positive toward negative}\right)$$

$$= (500 \text{ N/C, from right to left})$$

(25.23)

This electric field is due to the *source charges* on the capacitor plates.

In Section 25.1, we found that the electric potential energy of a charge q in the uniform electric field of a parallel-plate capacitor is

$$U_{elec} = U_{q + sources} = qEs \qquad (25.24)$$

We've set the constant term U_0 to zero. U_{elec} is the energy of q interacting with the source charges on the capacitor plates.

Our new view of the interaction is to separate the role of charge q from the role of the source charges by defining the electric potential $V = U_{q + sources}/q$. Thus the electric potential inside a parallel-plate capacitor is

$$V = Es \qquad \text{(electric potential inside a parallel-plate capacitor)} \qquad (25.25)$$

where s is the distance from the *negative* electrode. The electric potential, like the electric field, exists at *all points* inside the capacitor. The electric potential is created by the source charges on the capacitor plates and exists whether or not charge q is inside the capacitor.

FIGURE 25.19 illustrates the important point that the electric potential increases linearly from the negative plate, where $V_- = 0$, to the positive plate, where $V_+ = Ed$. Let's define the *potential difference* ΔV_C between the two capacitor plates to be

$$\Delta V_C = V_+ - V_- = Ed \qquad (25.26)$$

In our specific example, $\Delta V_C = (500 \text{ N/C})(0.0030 \text{ m}) = 1.5 \text{ V}$. The units work out because $1.5 \text{ (N m)/C} = 1.5 \text{ J/C} = 1.5 \text{ V}$.

NOTE People who work with circuits would call ΔV_C "the voltage across the capacitor" or simply "the capacitor voltage."

Equation 25.26 has an interesting implication. Thus far, we've determined the electric field inside a capacitor by specifying the surface charge density η on the plates. Alternatively, we could specify the capacitor voltage ΔV_C (i.e., the potential difference between the capacitor plates) and then determine the electric field strength as

$$E = \frac{\Delta V_C}{d} \qquad (25.27)$$

In fact, this is how E is determined in practical applications because it's easy to measure ΔV_C with a voltmeter but difficult, in practice, to know the value of η.

Equation 25.27 implies that the units of electric field are volts per meter, or V/m. We have been using electric field units of newtons per coulomb. In fact, as you can show as a homework problem, these units are equivalent to each other. That is,

$$1 \text{ N/C} = 1 \text{ V/m}$$

NOTE Volts per meter are the electric field units used by scientists and engineers in practice. We will now adopt them as our standard electric field units.

Returning to the electric potential, we can substitute Equation 25.27 for E into Equation 25.25 for V. Thus the electric potential inside the capacitor is

$$V = Es = \frac{s}{d}\Delta V_C \qquad (25.28)$$

The potential increases linearly from $V_- = 0$ V at the negative plate ($s = 0$) to $V_+ = \Delta V_C$ at the positive plate ($s = d$).

Visualizing Electric Potential

Let's explore the electric potential inside the capacitor by looking at several different, but related, ways that the potential can be represented graphically.

FIGURE 25.19 The electric potential of a parallel-plate capacitor increases linearly from the negative to the positive plate.

Graphical representations of the electric potential inside a capacitor

A graph of potential versus s. You can see the potential increasing from 0.0 V at the negative plate to 1.5 V at the positive plate.

A three-dimensional view showing **equipotential surfaces.** These are mathematical surfaces, not physical surfaces, with the same value of V at every point. The equipotential surfaces of a capacitor are planes parallel to the capacitor plates. The capacitor plates are also equipotential surfaces.

A two-dimensional **contour map.** The capacitor plates and the equipotential surfaces are seen edge-on, so you need to imagine them extending above and below the plane of the page.

A three-dimensional **elevation graph.** The potential is graphed vertically versus the s-coordinate on one axis and a generalized "yz-coordinate" on the other axis. Viewing the right face of the elevation graph gives you the potential graph.

Potential graph   Equipotential surfaces   Contour map   Elevation graph

FIGURE 25.20 Equipotentials and electric field vectors inside a parallel-plate capacitor.

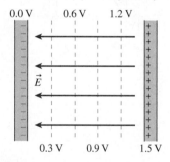

FIGURE 25.21 Using a battery to charge a capacitor to a precise value of ΔV_C.

A battery is a source of potential.

These four graphical representations show the same information from different perspectives, and the connecting lines help you see how they are related. If you think of the elevation graph as a "mountain," then the contour lines on the contour map are like the lines of a topographic map.

The potential graph and the contour map are the two representations most widely used in practice because they are easy to draw. Their limitation is that they are trying to convey three-dimensional information in a two-dimensional presentation. When you see graphs or contour maps, you need to imagine the three-dimensional equipotential surfaces or the three-dimensional elevation graph.

There's nothing special about showing equipotential surfaces or contour lines every 0.5 V. We chose these intervals because they were convenient. As an alternative, FIGURE 25.20 shows how the contour map looks if the contour lines are spaced every 0.3 V. Contour lines and equipotential surfaces are *imaginary* lines and surfaces drawn to help us visualize how the potential changes in space. Drawing the map more than one way reinforces the idea that there is an electric potential at *every* point inside the capacitor, not just at the points where we happened to draw a contour line or an equipotential surface.

Figure 25.20 also shows the electric field vectors. Notice that

- The electric field vectors are perpendicular to the equipotential surfaces.
- The electric field points in the direction of decreasing potential. In other words, the electric field points "downhill" on a graph or map of the electric potential.

Chapter 26 will present a more in-depth exploration of the connection between the electric field and the electric potential. There you will find that these observations are always true. They are not unique to the parallel-plate capacitor.

Finally, you might wonder how we can arrange a capacitor to have a surface charge density of precisely 4.42×10^{-9} C/m². Simple! As FIGURE 25.21 shows, we use wires to attach 3.00-mm-spaced capacitor plates to a 1.5 V battery. This is another topic that we'll explore in Chapter 26, but it's worth noting now that **a battery is a source of potential.** That's why batteries are labeled in volts, and it's a major reason we need to thoroughly understand the concept of potential.

EXAMPLE 25.7 | Measuring the speed of a proton

You've been assigned the task of measuring the speed of protons as they emerge from a small accelerator. To do so, you decide to measure how much voltage is needed across a parallel-plate capacitor to stop the protons. The capacitor you choose has a 2.0 mm plate separation and a small hole in one plate that you shoot the protons through. By filling the space between the plates with a low-density gas, you can see (with a microscope) a slight glow from the region where the protons collide with and excite the gas molecules. The width of the glow tells you how far the protons travel before being stopped and reversing direction. Varying the voltage across the capacitor gives the following data:

Voltage (V)	Glow width (mm)
1000	1.7
1250	1.3
1500	1.1
1750	1.0
2000	0.8

What value will you report for the speed of the protons?

MODEL The system is the proton plus the capacitor charges. This is an isolated system, so mechanical energy is conserved.

VISUALIZE FIGURE 25.22 is a before-and-after pictorial representation of the proton entering the capacitor with speed v_i and later reaching a turning point with $v_f = 0$ m/s after traveling distance s_f = glow width. For the protons to slow, the hole through which they pass has to be in the negative plate. The s-axis has $s = 0$ at this point.

FIGURE 25.22 A proton stopping in a capacitor.

SOLVE The conservation of energy equation, with the proton having charge $q = e$, is $K_f + eV_f = K_i + eV_i$. The initial potential

energy is zero, because the capacitor's electric potential is zero at $s_i = 0$, and the final kinetic energy is zero. Equation 25.28 for the potential inside the capacitor gives

$$eV_f = e\left(\frac{s_f}{d}\Delta V_C\right) = K_i = \tfrac{1}{2}mv_i^2$$

Solving for the distance traveled, you find

$$s_f = \frac{dmv_i^2}{2e}\frac{1}{\Delta V_C}$$

Thus a graph of the distance traveled versus the *inverse* of the capacitor voltage should be a straight line with zero y-intercept and slope $dmv_i^2/2e$. You can use the experimentally determined slope to find the proton speed.

FIGURE 25.23 is a graph of s_f versus $1/\Delta V_C$. It has the expected shape, and the slope of the best-fit line is seen to be 1.72 V m. The units are those of the rise-over-run. Using the slope, you calculate the proton speed:

$$v_i = \sqrt{\frac{2e}{dm}\times \text{slope}} = \sqrt{\frac{2(1.60\times 10^{-19}\,\text{C})(1.72\,\text{V m})}{(0.0020\,\text{m})(1.67\times 10^{-27}\,\text{kg})}}$$

$$= 4.1\times 10^5\,\text{m/s}$$

FIGURE 25.23 A graph of the data.

ASSESS This would be a very high speed for a macroscopic object but is quite typical of the speeds of charged particles.

In writing the electric potential inside a parallel-plate capacitor, we made the choice that $V_- = 0$ V at the negative plate. But that is not the only possible choice. FIGURE 25.24 on the next page shows three parallel-plate capacitors, each having the same capacitor voltage $\Delta V_C = V_+ - V_- = 100$ V, but each with a different choice for the location of the zero point of the electric potential. Notice the *terminal symbols* (lines with small circles at the end) showing how the potential, from a battery or a power supply, is applied to each plate; these symbols are common in electronics.

FIGURE 25.24 These three choices for $V = 0$ represent the same physical situation. These are contour maps, showing the edges of the equipotential surfaces.

The potential difference is the same.   The electric field inside is the same.

The important thing to notice is that the three contour maps in Figure 25.24 represent the *same physical situation*. The potential difference between any two points is the same in all three maps. The electric field is the same in all three. We may *prefer* one of these figures over the others, but there is no measurable physical difference between them.

STOP TO THINK 25.4 Rank in order, from largest to smallest, the potentials V_a to V_e at the points a to e.

25.6 The Electric Potential of a Point Charge

FIGURE 25.25 Measuring the electric potential of charge q.

To determine the potential of q at this point...

r

... place charge q' at the point as a probe and measure the potential energy $U_{q'+q}$.

Another important electric potential is that of a point charge. Let q in **FIGURE 25.25** be the source charge, and let a second charge q' probe the electric potential of q. The potential energy of the two point charges is

$$U_{q'+q} = \frac{1}{4\pi\epsilon_0} \frac{qq'}{r} \tag{25.29}$$

Thus, by definition, the electric potential of charge q is

$$V = \frac{U_{q'+q}}{q'} = \frac{1}{4\pi\epsilon_0} \frac{q}{r} \qquad \text{(electric potential of a point charge)} \tag{25.30}$$

The potential of Equation 25.30 extends through all of space, showing the influence of charge q, but it weakens with distance as $1/r$. This expression for V assumes that we have chosen $V = 0$ V to be at $r = \infty$. This is the most logical choice for a point charge because the influence of charge q ends at infinity.

The expression for the electric potential of charge q is similar to that for the electric field of charge q. The difference most quickly seen is that V depends on $1/r$ whereas $\vec{E}$ depends on $1/r^2$. But it is also important to notice that **the potential is a scalar** whereas the field is a vector. Thus the mathematics of using the potential are much easier than the vector mathematics using the electric field requires.

For example, the electric potential 1.0 cm from a $+1.0$ nC charge is

$$V_{1\,cm} = \frac{1}{4\pi\epsilon_0} \frac{q}{r} = (9.0 \times 10^9 \, \text{N m}^2/\text{C}^2) \frac{1.0 \times 10^{-9} \, \text{C}}{0.010 \, \text{m}} = 900 \, \text{V}$$

1 nC is typical of the electrostatic charge produced by rubbing, and you can see that such a charge creates a fairly large potential nearby. Why are we not shocked and injured

when working with the "high voltages" of such charges? The sensation of being shocked is a result of current, not potential. Some high-potential sources simply do not have the ability to generate much current. We will look at this issue in Chapter 28.

Visualizing the Potential of a Point Charge

FIGURE 25.26 shows four graphical representations of the electric potential of a point charge. These match the four representations of the electric potential inside a capacitor, and a comparison of the two is worthwhile. This figure assumes that q is positive; you may want to think about how the representations would change if q were negative.

FIGURE 25.26 Four graphical representations of the electric potential of a point charge.

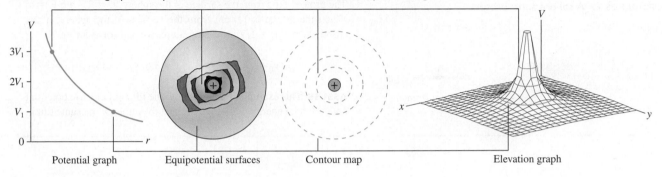

Potential graph   Equipotential surfaces   Contour map   Elevation graph

---

STOP TO THINK 25.5 Rank in order, from largest to smallest, the potential differences ΔV_{ab}, ΔV_{ac}, and ΔV_{bc} between points a and b, points a and c, and points b and c.

The Electric Potential of a Charged Sphere

In practice, you are more likely to work with a charged sphere, of radius R and total charge Q, than with a point charge. Outside a uniformly charged sphere, the electric potential is identical to that of a point charge Q at the center. That is,

$$V = \frac{1}{4\pi\epsilon_0}\frac{Q}{r} \quad \text{(sphere of charge, } r \geq R\text{)} \tag{25.31}$$

We can cast this result in a more useful form. It is customary to speak of charging an electrode, such as a sphere, "to" a certain potential, as in "Bob charged the sphere to a potential of 3000 volts." This potential, which we will call V_0, is the potential right on the surface of the sphere. We can see from Equation 25.31 that

$$V_0 = V(\text{at } r = R) = \frac{Q}{4\pi\epsilon_0 R} \tag{25.32}$$

Consequently, a sphere of radius R that is charged to potential V_0 has total charge

$$Q = 4\pi\epsilon_0 R V_0 \tag{25.33}$$

If we substitute this expression for Q into Equation 25.31, we can write the potential outside a sphere that is charged to potential V_0 as

$$V = \frac{R}{r}V_0 \quad \text{(sphere charged to potential } V_0\text{)} \tag{25.34}$$

Equation 25.34 tells us that the potential of a sphere is V_0 on the surface and decreases inversely with the distance. The potential at $r = 3R$ is $\frac{1}{3}V_0$.

A *plasma ball* consists of a small metal ball charged to a potential of about 2000 V inside a hollow glass sphere. The electric field of the high-voltage ball is sufficient to cause a gas breakdown at this pressure, creating "lightning bolts" between the ball and the glass sphere.

EXAMPLE 25.8 | **A proton and a charged sphere**

A proton is released from rest at the surface of a 1.0-cm-diameter sphere that has been charged to +1000 V.

a. What is the charge of the sphere?

b. What is the proton's speed at 1.0 cm from the sphere?

MODEL Energy is conserved. The potential outside the charged sphere is the same as the potential of a point charge at the center.

VISUALIZE FIGURE 25.27 shows the situation.

FIGURE 25.27 A sphere and a proton.

SOLVE a. The charge of the sphere is

$$Q = 4\pi\epsilon_0 R V_0 = 0.56 \times 10^{-9}\ \text{C} = 0.56\ \text{nC}$$

b. A sphere charged to $V_0 = +1000$ V is positively charged. The proton will be repelled by this charge and move away from the sphere. The conservation of energy equation $K_f + eV_f = K_i + eV_i$, with Equation 25.34 for the potential of a sphere, is

$$\tfrac{1}{2}mv_f^2 + \frac{eR}{r_f}V_0 = \tfrac{1}{2}mv_i^2 + \frac{eR}{r_i}V_0$$

The proton starts from the surface of the sphere, $r_i = R$, with $v_i = 0$. When the proton is 1.0 cm from the *surface* of the sphere, it has $r_f = 1.0$ cm $+ R = 1.5$ cm. Using these, we can solve for v_f:

$$v_f = \sqrt{\frac{2eV_0}{m}\left(1 - \frac{R}{r_f}\right)} = 3.6 \times 10^5\ \text{m/s}$$

ASSESS This example illustrates how the ideas of electric potential and potential energy work together, yet they are *not* the same thing.

25.7 The Electric Potential of Many Charges

Suppose there are many source charges $q_1, q_2, \ldots$. The electric potential V at a point in space is the sum of the potentials due to each charge:

$$V = \sum_i \frac{1}{4\pi\epsilon_0} \frac{q_i}{r_i} \tag{25.35}$$

where r_i is the distance from charge q_i to the point in space where the potential is being calculated. In other words, **the electric potential, like the electric field, obeys the principle of superposition.**

As an example, the contour map and elevation graph in FIGURE 25.28 show that the potential of an electric dipole is the sum of the potentials of the positive and negative charges. Potentials such as these have many practical applications. For example, electrical activity within the body can be monitored by measuring equipotential lines on the skin. Figure 25.28c shows that the equipotentials near the heart are a slightly distorted but recognizable electric dipole.

FIGURE 25.28 The electric potential of an electric dipole.

(a) Contour map

(b) Elevation graph

(c) Equipotentials near the heart

EXAMPLE 25.9 | **The potential of two charges**

What is the electric potential at the point indicated in **FIGURE 25.29**?

▶ **FIGURE 25.29** Finding the potential of two charges.

5.0 cm   4.0 cm

+ 3.0 cm −

+2.0 nC   −1.0 nC

MODEL The potential is the sum of the potentials due to each charge.

SOLVE The potential at the indicated point is

$$V = \frac{1}{4\pi\epsilon_0}\frac{q_1}{r_1} + \frac{1}{4\pi\epsilon_0}\frac{q_2}{r_2}$$

$$= (9.0 \times 10^9 \text{ N m}^2/\text{C}^2)\left(\frac{2.0 \times 10^{-9} \text{ C}}{0.050 \text{ m}} + \frac{-1.0 \times 10^{-9} \text{ C}}{0.040 \text{ m}}\right)$$

$$= 135 \text{ V}$$

ASSESS The potential is a *scalar*, so we found the net potential by adding two numbers. We don't need any angles or components to calculate the potential.

A Continuous Distribution of Charge

Equation 25.35 is the basis for determining the potential of a continuous distribution of charge, such as a charged rod or a charged disk. The procedure is much like the one you learned in Chapter 23 for calculating the electric field of a continuous distribution of charge, but *easier* because the potential is a scalar. We will continue to assume that the object is *uniformly charged*, meaning that the charges are evenly spaced over the object.

PROBLEM-SOLVING STRATEGY 25.2 (MP)

The electric potential of a continuous distribution of charge

MODEL Model the charge distribution as a simple shape.

VISUALIZE For the pictorial representation:

■ Draw a picture, establish a coordinate system, and identify the point P at which you want to calculate the electric potential.

■ Divide the total charge Q into small pieces of charge ΔQ, using shapes for which you *already know* how to determine V. This division is often, but not always, into point charges.

■ Identify distances that need to be calculated.

SOLVE The mathematical representation is $V = \sum V_i$.

■ Use superposition to form an algebraic expression for the potential at P. Let the (x, y, z) coordinates of the point remain as variables.

■ Replace the small charge ΔQ with an equivalent expression involving a *charge density* and a *coordinate,* such as dx. **This is the critical step in making the transition from a sum to an integral** because you need a coordinate to serve as the integration variable.

■ All distances must be expressed in terms of the coordinates.

■ Let the sum become an integral. The integration limits will depend on the coordinate system you have chosen.

ASSESS Check that your result is consistent with any limits for which you know what the potential should be.

Exercise 29

EXAMPLE 25.10 | **The potential of a ring of charge**

A thin, uniformly charged ring of radius R has total charge Q. Find the potential at distance z on the axis of the ring.

MODEL Because the ring is thin, we'll assume the charge lies along a circle of radius R.

VISUALIZE **FIGURE 25.30** on the next page illustrates the problem-solving strategy. We've chosen a coordinate system in which the ring lies in the xy-plane and point P is on the z-axis. We've then divided the ring into N small segments of charge ΔQ, each of which can be modeled as a point charge. The distance r_i between segment i and point P is

Continued

$$r_i = \sqrt{R^2 + z^2}$$

Note that r_i is a constant distance, the same for every charge segment.

FIGURE 25.30 Finding the potential of a ring of charge.

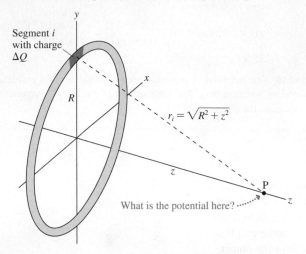

Segment i with charge ΔQ

$r_i = \sqrt{R^2 + z^2}$

What is the potential here?

SOLVE The potential V at P is the sum of the potentials due to each segment of charge:

$$V = \sum_{i=1}^{N} V_i = \sum_{i=1}^{N} \frac{1}{4\pi\epsilon_0} \frac{\Delta Q}{r_i} = \frac{1}{4\pi\epsilon_0} \frac{1}{\sqrt{R^2 + z^2}} \sum_{i=1}^{N} \Delta Q$$

We were able to bring all terms involving z to the front because z is a constant as far as the summation is concerned. Surprisingly, we don't need to convert the sum to an integral to complete this calculation. The sum of all the ΔQ charge segments around the ring is simply the ring's total charge, $\sum (\Delta Q) = Q$; hence the electric potential on the axis of a charged ring is

$$V_{\text{ring on axis}} = \frac{1}{4\pi\epsilon_0} \frac{Q}{\sqrt{R^2 + z^2}}$$

ASSESS From far away, the ring appears as a point charge Q in the distance. Thus we expect the potential of the ring to be that of a point charge when $z \gg R$. You can see that $V_{\text{ring}} \approx Q/4\pi\epsilon_0 z$ when $z \gg R$, which is, indeed, the potential of a point charge Q.

CHALLENGE EXAMPLE 25.11 | The potential of a charged disk

A thin plastic disk of radius R is uniformly coated with charge until it receives total charge Q.

a. What is the potential at distance z along the axis of the disk?

b. What is the potential energy if an electron is 1.00 cm from a 35.0-cm-diameter disk that has been charged to $+5.00$ nC?

MODEL Model the disk as a uniformly charged disk of zero thickness, radius R, and charge Q. The disk has uniform surface charge density $\eta = Q/A = Q/\pi R^2$. We can take advantage of now knowing the on-axis potential of a ring of charge.

VISUALIZE Orient the disk in the xy-plane, as shown in **FIGURE 25.31**, with point P at distance z. Then divide the disk into *rings* of equal width Δr. Ring i has radius r_i and charge ΔQ_i.

FIGURE 25.31 Finding the potential of a disk of charge.

Disk with radius R and charge Q

Ring i with charge ΔQ_i

The potential at this point is the sum of the potentials due to all the thin rings in the disk.

SOLVE a. We can use the result of Example 25.10 to write the potential at distance z of ring i as

$$V_i = \frac{1}{4\pi\epsilon_0} \frac{\Delta Q_i}{\sqrt{r_i^2 + z^2}}$$

The potential at P due to all the rings is the sum

$$V = \sum_i V_i = \frac{1}{4\pi\epsilon_0} \sum_{i=1}^{N} \frac{\Delta Q_i}{\sqrt{r_i^2 + z^2}}$$

The critical step is to relate ΔQ_i to a coordinate. Because we now have a surface, rather than a line, the charge in ring i is $\Delta Q_i = \eta \Delta A_i$, where ΔA_i is the area of ring i. We can find ΔA_i, as you've learned to do in calculus, by "unrolling" the ring to form a narrow rectangle of length $2\pi r_i$ and height Δr. Thus the area of ring i is $\Delta A_i = 2\pi r_i \Delta r$ and the charge is

$$\Delta Q_i = \eta \Delta A_i = \frac{Q}{\pi R^2} 2\pi r_i \Delta r = \frac{2Q}{R^2} r_i \Delta r$$

With this substitution, the potential at P is

$$V = \frac{1}{4\pi\epsilon_0} \sum_{i=1}^{N} \frac{2Q}{R^2} \frac{r_i \Delta r_i}{\sqrt{r_i^2 + z^2}} \rightarrow \frac{Q}{2\pi\epsilon_0 R^2} \int_0^R \frac{r\, dr}{\sqrt{r^2 + z^2}}$$

where, in the last step, we let $N \rightarrow \infty$ and the sum become an integral. This integral can be found in Appendix A, but it's not hard to evaluate with a change of variables. Let $u = r^2 + z^2$, in which case $r\, dr = \frac{1}{2} du$. Changing variables requires that we also change the integration limits. You can see that $u = z^2$ when $r = 0$, and $u = R^2 + z^2$ when $r = R$. With these changes, the on-axis potential of a charged disk is

$$V_{\text{disk on axis}} = \frac{Q}{2\pi\epsilon_0 R^2} \int_{z^2}^{R^2+z^2} \frac{\frac{1}{2} du}{u^{1/2}} = \frac{Q}{2\pi\epsilon_0 R^2} u^{1/2} \Big|_{z^2}^{R^2+z^2}$$

$$= \frac{Q}{2\pi\epsilon_0 R^2} \left(\sqrt{R^2 + z^2} - z \right)$$

b. To calculate the potential energy, we first need to determine the potential of the disk at $z = 1.00$ cm. Using $R = 0.0175$ m and $Q = 5.00$ nC, you can calculate $V = 3870$ V. The electron's charge is $q = -e = -1.60 \times 10^{-19}$ C, so the potential energy with an electron at $z = 1.00$ cm is $U = qV = -6.19 \times 10^{-16}$ J.

ASSESS Although we had to go through a number of steps, this procedure is easier than evaluating the electric field because we do not have to worry about vector components.

SUMMARY

The goals of Chapter 25 have been to use the electric potential and electric potential energy.

GENERAL PRINCIPLES

Sources of Potential

The **electric potential** V, like the electric field, is created by source charges. Two major tools for calculating the potential are:

- The potential of a point charge, $V = \dfrac{1}{4\pi\epsilon_0} \dfrac{q}{r}$

- The principle of superposition

For multiple point charges

Use superposition: $V = V_1 + V_2 + V_3 + \cdots$

For a continuous distribution of charge

MODEL Model as a simple charge distribution.

VISUALIZE Draw a pictorial representation.

- Establish a coordinate system.
- Identify where the potential will be calculated.

SOLVE Set up a sum.

- Divide the charge into point-like ΔQ.
- Find the potential due to each ΔQ.
- Use the charge density (λ or η) to replace ΔQ with an integration coordinate, then sum by integrating.

V is easier to calculate than $\vec{E}$ because potential is a scalar.

Electric Potential Energy

If charge q is placed in an electric potential V, the system's **electric potential energy** (interaction energy) is

$$U = qV$$

Point charges and dipoles

The electric potential energy of two **point charges** is

$$U_{q_1 + q_2} = \frac{Kq_1q_2}{r} = \frac{1}{4\pi\epsilon_0} \frac{q_1q_2}{r}$$

The potential energy of two opposite charges is negative.
The potential energy in an electric field of an **electric dipole** with dipole moment $\vec{p}$ is

$$U_{\text{dipole}} = -pE\cos\theta = -\vec{p}\cdot\vec{E}$$

Solving conservation of energy problems

MODEL Model as an isolated system.

VISUALIZE Draw a before-and-after representation.

SOLVE Mechanical energy is conserved.

- Mathematically $K_f + qV_f = K_i + qV_i$.
- K is the sum of the kinetic energies of all particles.
- V is the potential due to the source charges.

APPLICATIONS

Graphical representations of the potential:

Potential graph   **Equipotential surfaces**

Contour map   **Elevation graph**

Sphere of charge Q

Same as a point charge
if $r \geq R$

Parallel-plate capacitor

$V = Es$, where s is measured
from the negative plate. The
electric field inside is

$$E = \frac{\Delta V_C}{d}$$

Units

Electric potential: $1\ \text{V} = 1\ \text{J/C}$

Electric field: $1\ \text{V/m} = 1\ \text{N/C}$

TERMS AND NOTATION

electric potential energy, U	volt, V	voltage, ΔV	contour map
electric potential, V	potential difference, ΔV	equipotential surface	elevation graph

CONCEPTUAL QUESTIONS

1. a. Charge q_1 is distance r from a positive point charge Q. Charge $q_2 = q_1/3$ is distance $2r$ from Q. What is the ratio U_1/U_2 of their potential energies due to their interactions with Q?
   b. Charge q_1 is distance s from the negative plate of a parallel-plate capacitor. Charge $q_2 = q_1/3$ is distance $2s$ from the negative plate. What is the ratio U_1/U_2 of their potential energies?

2. **FIGURE Q25.2** shows the potential energy of a proton ($q = +e$) and a lead nucleus ($q = +82e$). The horizontal scale is in units of *femtometers*, where 1 fm = 10^{-15} m.
   a. A proton is fired toward a lead nucleus from very far away. How much initial kinetic energy does the proton need to reach a turning point 10 fm from the nucleus? Explain.
   b. How much kinetic energy does the proton of part a have when it is 20 fm from the nucleus and moving toward it, before the collision?

FIGURE Q25.2　　　　　　　　　**FIGURE Q25.3**

3. An electron moves along the trajectory of **FIGURE Q25.3** from i to f.
   a. Does the electric potential energy increase, decrease, or stay the same? Explain.
   b. Is the electron's speed at f greater than, less than, or equal to its speed at i? Explain.

4. Two protons are launched with the same speed from point 1 inside the parallel-plate capacitor of **FIGURE Q25.4**. Points 2 and 3 are the same distance from the negative plate.
   a. Is $\Delta U_{1\to2}$, the change in potential energy along the path $1 \to 2$, larger than, smaller than, or equal to $\Delta U_{1\to3}$?
   b. Is the proton's speed v_2 at point 2 larger than, smaller than, or equal to v_3? Explain.

FIGURE Q25.4

5. Rank in order, from most positive to most negative, the potential energies U_a to U_f of the six electric dipoles in the uniform electric field of **FIGURE Q25.5**. Explain.

FIGURE Q25.5

6. **FIGURE Q25.6** shows the electric potential along the x-axis.
   a. Draw a graph of the potential energy of a 0.1 C charged particle. Provide a numerical scale for both axes.
   b. If the charged particle is shot toward the right from $x = 1$ m with 1.0 J of kinetic energy, where is its turning point? Use your graph to explain.

FIGURE Q25.6

7. A capacitor with plates separated by distance d is charged to a potential difference ΔV_C. All wires and batteries are disconnected, then the two plates are pulled apart (with insulated handles) to a new separation of distance $2d$.
   a. Does the capacitor charge Q change as the separation increases? If so, by what factor? If not, why not?
   b. Does the electric field strength E change as the separation increases? If so, by what factor? If not, why not?
   c. Does the potential difference ΔV_C change as the separation increases? If so, by what factor? If not, why not?

8. Rank in order, from largest to smallest, the electric potentials V_a to V_e at points a to e in **FIGURE Q25.8**. Explain.

FIGURE Q25.8　　　　　　　　　**FIGURE Q25.9**

9. **FIGURE Q25.9** shows two points inside a capacitor. Let $V = 0$ V at the negative plate.
   a. What is the ratio V_2/V_1 of the electric potentials? Explain.
   b. What is the ratio E_2/E_1 of the electric field strengths?

10. **FIGURE Q25.10** shows two points near a positive point charge.
    a. What is the ratio V_2/V_1 of the electric potentials? Explain.
    b. What is the ratio E_2/E_1 of the electric field strengths?

FIGUREQ25.10

11. **FIGURE Q25.11** shows three points near two point charges. The charges have equal magnitudes. For each part, rank in order, from most positive to most negative, the potentials V_a to V_c.

FIGURE Q25.11

12. Reproduce **FIGURE Q25.12** on your paper. Then draw a dot (or dots) on the figure to show the position (or positions) at which the electric potential is zero.

FIGURE Q25.12

EXERCISES AND PROBLEMS

Problems labeled ▇ integrate material from earlier chapters.

Exercises

Section 25.1 Electric Potential Energy

1. ‖ The electric field strength is 20,000 N/C inside a parallel-plate capacitor with a 1.0 mm spacing. An electron is released from rest at the negative plate. What is the electron's speed when it reaches the positive plate?

2. ‖ The electric field strength is 50,000 N/C inside a parallel-plate capacitor with a 2.0 mm spacing. A proton is released from rest at the positive plate. What is the proton's speed when it reaches the negative plate?

3. ‖ A proton is released from rest at the positive plate of a parallel-plate capacitor. It crosses the capacitor and reaches the negative plate with a speed of 50,000 m/s. What will be the final speed of an electron released from rest at the negative plate?

4. ‖ A proton is released from rest at the positive plate of a parallel-plate capacitor. It crosses the capacitor and reaches the negative plate with a speed of 50,000 m/s. The experiment is repeated with a He⁺ ion (charge e, mass 4 u). What is the ion's speed at the negative plate?

Section 25.2 The Potential Energy of Point Charges

5. ‖ What is the potential energy of the electron-proton interactions in FIGURE EX25.5? The electrons are fixed and cannot move.

FIGURE EX25.5

6. ‖ What is the electric potential energy of the group of charges in FIGURE EX25.6?

FIGURE EX25.6   FIGURE EX25.7

7. ‖ What is the electric potential energy of the group of charges in FIGURE EX25.7?

8. ‖ Two positive point charges are 5.0 cm apart. If the electric potential energy is 72 μJ, what is the magnitude of the force between the two charges?

Section 25.3 The Potential Energy of a Dipole

9. ‖ A water molecule perpendicular to an electric field has 1.0×10^{-21} J more potential energy than a water molecule aligned with the field. The dipole moment of a water molecule is 6.2×10^{-30} C m. What is the strength of the electric field?

10. ‖ FIGURE EX25.10 shows the potential energy of an electric dipole. Consider a dipole that oscillates between ±60°.
    a. What is the dipole's mechanical energy?
    b. What is the dipole's kinetic energy when it is aligned with the electric field?

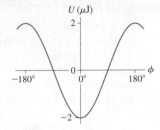

FIGURE EX25.10

Section 25.4 The Electric Potential

11. ‖ What is the speed of a proton that has been accelerated from rest through a potential difference of -1000 V?

12. ‖ What is the speed of an electron that has been accelerated from rest through a potential difference of 1000 V?

13. ‖ What potential difference is needed to accelerate an electron from rest to a speed of 2.0×10^6 m/s?

14. ‖ What potential difference is needed to accelerate a He⁺ ion (charge $+e$, mass 4 u) from rest to a speed of 2.0×10^6 m/s?

15. ‖ A proton with an initial speed of 800,000 m/s is brought to rest by an electric field.
    a. Did the proton move into a region of higher potential or lower potential?
    b. What was the potential difference that stopped the proton?

16. ‖ An electron with an initial speed of 500,000 m/s is brought to rest by an electric field.
    a. Did the electron move into a region of higher potential or lower potential?
    b. What was the potential difference that stopped the electron?

17. ‖ Through what potential difference must a proton be accelerated to reach the speed it would have by falling 100 m in vacuum?

18. ‖ In *proton beam therapy,* a high-energy beam of protons is
BIO fired at a tumor. As the protons stop in the tumor, their kinetic energy breaks apart the tumor's DNA, thus killing the tumor cells. For one patient, it is desired to deposit 0.10 J of proton energy in the tumor. To create the proton beam, protons are accelerated from rest through a 10,000 kV potential difference. What is the total charge of the protons that must be fired at the tumor?

19. ‖ A student wants to make a very small particle accelerator using a 9.0 V battery. What speed will (a) a proton and (b) an electron have after being accelerated from rest through the 9.0 V potential difference?

20. ‖ Physicists often use a different unit of energy, the *electron volt,* when dealing with energies at the atomic level. One electron volt, abbreviated eV, is defined as the amount of kinetic energy gained by an electron upon accelerating through a 1.0 V potential difference.
    a. What is 1.0 electron volt in joules?
    b. What is the speed of a proton with 5000 eV of kinetic energy?

Section 25.5 The Electric Potential Inside a Parallel-Plate Capacitor

21. | Show that 1 V/m = 1 N/C.
22. || a. What is the potential of an ordinary AA or AAA battery? (If you're not sure, find one and look at the label.)
    b. An AA battery is connected to a parallel-plate capacitor having 4.0 cm × 4.0 cm plates spaced 1.0 mm apart. How much charge does the battery supply to each plate?
23. || A 3.0-cm-diameter parallel-plate capacitor has a 2.0 mm spacing. The electric field strength inside the capacitor is 1.0×10^5 V/m.
    a. What is the potential difference across the capacitor?
    b. How much charge is on each plate?
24. || Two 2.00 cm × 2.00 cm plates that form a parallel-plate capacitor are charged to ±0.708 nC. What are the electric field strength inside and the potential difference across the capacitor if the spacing between the plates is (a) 1.00 mm and (b) 2.00 mm?
25. || Two 2.0-cm-diameter disks spaced 2.0 mm apart form a parallel-plate capacitor. The electric field between the disks is 5.0×10^5 V/m.
    a. What is the voltage across the capacitor?
    b. An electron is launched from the negative plate. It strikes the positive plate at a speed of 2.0×10^7 m/s. What was the electron's speed as it left the negative plate?
26. || In FIGURE EX25.26, a proton is fired with a speed of 200,000 m/s from the midpoint of the capacitor toward the positive plate.
    a. Show that this is insufficient speed to reach the positive plate.
    b. What is the proton's speed as it collides with the negative plate?

FIGURE EX25.26

Section 25.6 The Electric Potential of a Point Charge

27. | a. What is the electric potential at points A, B, and C in FIGURE EX25.27?
    b. What are the potential differences $\Delta V_{AB} = V_B - V_A$ and $\Delta V_{CB} = V_B - V_C$?

FIGURE EX25.27

28. || A 1.0-mm-diameter ball bearing has 2.0×10^9 excess electrons. What is the ball bearing's potential?
29. | In a semiclassical model of the hydrogen atom, the electron orbits the proton at a distance of 0.053 nm.
    a. What is the electric potential of the proton at the position of the electron?
    b. What is the electron's potential energy?

Section 25.7 The Electric Potential of Many Charges

30. | What is the electric potential at the point indicated with the dot in FIGURE EX25.30?

FIGURE EX25.30   **FIGURE EX25.31**

31. | What is the electric potential at the point indicated with the dot in FIGURE EX25.31?
32. || The electric potential at the dot in FIGURE EX25.32 is 3140 V. What is charge q?

FIGURE EX25.32

33. || A −2.0 nC charge and a +2.0 nC charge are located on the x-axis at $x = -1.0$ cm and $x = +1.0$ cm, respectively.
    a. Other than at infinity, is there a position or positions on the x-axis where the electric field is zero? If so, where?
    b. Other than at infinity, at what position or positions on the x-axis is the electric potential zero?
    c. Sketch graphs of the electric field strength and the electric potential along the x-axis.
34. || Two point charges q_a and q_b are located on the x-axis at $x = a$ and $x = b$. FIGURE EX25.34 is a graph of V, the electric potential.
    a. What are the signs of q_a and q_b?
    b. What is the ratio $|q_a/q_b|$?
    c. Draw a graph of E_x, the x-component of the electric field, as a function of x.

FIGURE EX25.34   **FIGURE EX25.35**

35. | The two halves of the rod in FIGURE EX25.35 are uniformly charged to ±Q. What is the electric potential at the point indicated by the dot?

36. || A 5.0-cm-diameter metal ball has a surface charge density of $10 \, \mu C/m^2$. How much work is required to remove one electron from this ball?

Problems

37. ||| Two point charges 2.0 cm apart have an electric potential energy $-180 \, \mu J$. The total charge is 30 nC. What are the two charges?

38. ||| A -10.0 nC point charge and a $+20.0$ nC point charge are 15.0 cm apart on the x-axis.
   a. What is the electric potential at the point on the x-axis where the electric field is zero?
   b. What is the magnitude of the electric field at the point on the x-axis, between the charges, where the electric potential is zero?

39. ||| A $+3.0$ nC charge is at $x = 0$ cm and a -1.0 nC charge is at $x = 4$ cm. At what point or points on the x-axis is the electric potential zero?

40. || A -3.0 nC charge is on the x-axis at $x = -9$ cm and a $+4.0$ nC charge is on the x-axis at $x = 16$ cm. At what point or points on the y-axis is the electric potential zero?

41. || Two small metal cubes with masses 2.0 g and 4.0 g are tied together by a 5.0-cm-long massless string and are at rest on a frictionless surface. Each is charged to $+2.0 \, \mu C$.
   a. What is the energy of this system?
   b. What is the tension in the string?
   c. The string is cut. What is the speed of each cube when they are far apart?
   Hint: There are *two* conserved quantities. Make use of both.

42. || The four 1.0 g spheres shown in FIGURE P25.42 are released simultaneously and allowed to move away from each other. What is the speed of each sphere when they are very far apart?

FIGURE P25.42   FIGURE P25.43

43. || A proton's speed as it passes point A is 50,000 m/s. It follows the trajectory shown in FIGURE P25.43. What is the proton's speed at point B?

   ||Living cells "pump" singly ionized sodium ions, Na^+, from the
BIO inside of the cell to the outside to maintain a membrane potential $\Delta V_{membrane} = V_{in} - V_{out} = -70$ mV. It is called *pumping* because work must be done to move a positive ion from the negative inside of the cell to the positive outside, and it must go on continuously because sodium ions "leak" back through the cell wall by diffusion.
   a. How much work must be done to move one sodium ion from the inside of the cell to the outside?
   b. At rest, the human body uses energy at the rate of approximately 100 W to maintain basic metabolic functions. It has been estimated that 20% of this energy is used to operate the sodium pumps of the body. Estimate—to one significant figure—the number of sodium ions pumped per second.

45. ||| An arrangement of source charges produces the electric potential $V = 5000x^2$ along the x-axis, where V is in volts and x is in meters. What is the maximum speed of a 1.0 g, 10 nC charged particle that moves in this potential with turning points at ± 8.0 cm?

46. || A proton moves along the x-axis where some arrangement of charges has produced the potential $V(x) = V_0 \sin(2\pi x / \lambda)$, where $V_0 = 5000$ V and $\lambda = 1.0$ mm.
   a. What minimum speed must the proton have at $x = 0$ to move down the axis without being reflected?
   b. What is the maximum speed reached by a proton that at $x = 0$ has the speed you calculated in part a?

47. || The electron gun in an old TV picture tube accelerates electrons between two parallel plates 1.2 cm apart with a 25 kV potential difference between them. The electrons enter through a small hole in the negative plate, accelerate, then exit through a small hole in the positive plate. Assume that the holes are small enough not to affect the electric field or potential.
   a. What is the electric field strength between the plates?
   b. With what speed does an electron exit the electron gun if its entry speed is close to zero?
   NOTE The exit speed is so fast that we really need to use the theory of relativity to compute an accurate value. Your answer to part b is in the right range but a little too big.

48. || A room with 3.0-m-high ceilings has a metal plate on the floor with $V = 0$ V and a separate metal plate on the ceiling. A 1.0 g glass ball charged to $+4.9$ nC is shot straight up at 5.0 m/s. How high does the ball go if the ceiling voltage is (a) $+3.0 \times 10^6$ V and (b) -3.0×10^6 V?

49. || A group of science and engineering students embarks on a quest to make an electrostatic projectile launcher. For their first trial, a horizontal, frictionless surface is positioned next to the 12-cm-diameter sphere of a Van de Graaff generator, and a small, 5.0 g plastic cube is placed on the surface with its center 2.0 cm from the edge of the sphere. The cube is given a positive charge, and then the Van de Graaff generator is turned on, charging the sphere to a potential of 200,000 V in a negligible amount of time. How much charge does the plastic cube need to achieve a final speed of a mere 3.0 m/s? Does this seem like a practical projectile launcher?

50. || Two 2.0 g plastic buttons each with $+50$ nC of charge are placed on a frictionless surface 2.0 cm (measured between centers) on either side of a 5.0 g button charged to $+250$ nC. All three are released simultaneously.
   a. How many interactions are there that have a potential energy?
   b. What is the final speed of each button?

51. || What is the escape speed of an electron launched from the surface of a 1.0-cm-diameter glass sphere that has been charged to 10 nC?

52. ||| An electric dipole has dipole moment p. If $r \gg s$, where s is the separation between the charges, show that the electric potential of the dipole can be written

$$V = \frac{1}{4\pi\epsilon_0} \frac{p\cos\theta}{r^2}$$

where r is the distance from the center of the dipole and θ is the angle from the dipole axis.

53. ||| Three electrons form an equilateral triangle 1.0 nm on each side. A proton is at the center of the triangle. What is the potential energy of this group of charges?

54. ||| A 2.0-mm-diameter glass bead is positively charged. The potential difference between a point 2.0 mm from the bead and a point 4.0 mm from the bead is 500 V. What is the charge on the bead?

55. || Your lab assignment for the week is to measure the amount of charge on the 6.0-cm-diameter metal sphere of a Van de Graaff generator. To do so, you're going to use a spring with spring constant 0.65 N/m to launch a small, 1.5 g bead horizontally toward the sphere. You can reliably charge the bead to 2.5 nC, and your plan is to use a video camera to measure the bead's closest approach to the edge of the sphere as you change the compression of the spring. Your data are as follows:

Compression (cm)	Closest approach (cm)
1.6	5.5
1.9	2.6
2.2	1.6
2.5	0.4

Use an appropriate graph of the data to determine the sphere's charge in nC. You can assume that the bead's motion is entirely horizontal, that the spring is so far away that the bead has no interaction with the sphere as it's launched, and that the approaching bead does not alter the charge distribution on the sphere.

56. || A proton is fired from far away toward the nucleus of an iron atom. Iron is element number 26, and the diameter of the nucleus is 9.0 fm. What initial speed does the proton need to just reach the surface of the nucleus? Assume the nucleus remains at rest.

57. ||| A proton is fired from far away toward the nucleus of a mercury atom. Mercury is element number 80, and the diameter of the nucleus is 14.0 fm. If the proton is fired at a speed of 4.0×10^7 m/s, what is its closest approach to the surface of the nucleus? Assume the nucleus remains at rest.

58. || In the form of radioactive decay known as *alpha decay,* an unstable nucleus emits a helium-atom nucleus, which is called an *alpha particle.* An alpha particle contains two protons and two neutrons, thus having mass $m = 4$ u and charge $q = 2e$. Suppose a uranium nucleus with 92 protons decays into thorium, with 90 protons, and an alpha particle. The alpha particle is initially at rest at the surface of the thorium nucleus, which is 15 fm in diameter. What is the speed of the alpha particle when it is detected in the laboratory? Assume the thorium nucleus remains at rest.

59. || One form of nuclear radiation, *beta decay,* occurs when a neutron changes into a proton, an electron, and a neutral particle called a *neutrino:* $n \rightarrow p^+ + e^- + \nu$ where ν is the symbol for a neutrino. When this change happens to a neutron within the nucleus of an atom, the proton remains behind in the nucleus while the electron and neutrino are ejected from the nucleus. The ejected electron is called a *beta particle.* One nucleus that exhibits beta decay is the isotope of hydrogen ^{3}H, called *tritium,* whose nucleus consists of one proton (making it hydrogen) and two neutrons (giving tritium an atomic mass $m = 3$ u). Tritium is radioactive, and it decays to helium: ^{3}H $\rightarrow$ ^{3}He + e$^-$ + ν.
   a. Is charge conserved in the beta decay process? Explain.
   b. Why is the final product a helium atom? Explain.
   c. The nuclei of both ^{3}H and ^{3}He have radii of 1.5×10^{-15} m. With what minimum speed must the electron be ejected if it is to escape from the nucleus and not fall back?

60. || Two 10-cm-diameter electrodes 0.50 cm apart form a parallel-plate capacitor. The electrodes are attached by metal wires to the terminals of a 15 V battery. After a long time, the capacitor is disconnected from the battery but is not discharged. What are the charge on each electrode, the electric field strength inside the capacitor, and the potential difference between the electrodes
   a. Right after the battery is disconnected?
   b. After insulating handles are used to pull the electrodes away from each other until they are 1.0 cm apart?
   c. After the original electrodes (not the modified electrodes of part b) are expanded until they are 20 cm in diameter?

61. || Two 10-cm-diameter electrodes 0.50 cm apart form a parallel-plate capacitor. The electrodes are attached by metal wires to the terminals of a 15 V battery. What are the charge on each electrode, the electric field strength inside the capacitor, and the potential difference between the electrodes
   a. While the capacitor is attached to the battery?
   b. After insulating handles are used to pull the electrodes away from each other until they are 1.0 cm apart? The electrodes remain connected to the battery during this process.
   c. After the original electrodes (not the modified electrodes of part b) are expanded until they are 20 cm in diameter while remaining connected to the battery?

62. || Electrodes of area A are spaced distance d apart to form a
CALC parallel-plate capacitor. The electrodes are charged to $\pm q$.
   a. What is the infinitesimal increase in electric potential energy dU if an infinitesimal amount of charge dq is moved from the negative electrode to the positive electrode?
   b. An uncharged capacitor can be charged to $\pm Q$ by transferring charge dq over and over and over. Use your answer to part a to show that the potential energy of a capacitor charged to $\pm Q$ is $U_{cap} = \frac{1}{2} Q \Delta V_C$.

63. || a. Find an algebraic expression for the electric field strength E_0 at the surface of a charged sphere in terms of the sphere's potential V_0 and radius R.
   b. What is the electric field strength at the surface of a 1.0-cm-diameter marble charged to 500 V?

64. || Two spherical drops of mercury each have a charge of 0.10 nC and a potential of 300 V at the surface. The two drops merge to form a single drop. What is the potential at the surface of the new drop?

65. || A Van de Graaff generator is a device for generating a large electric potential by building up charge on a hollow metal sphere. A typical classroom-demonstration model has a diameter of 30 cm.
   a. How much charge is needed on the sphere for its potential to be 500,000 V?
   b. What is the electric field strength just outside the surface of the sphere when it is charged to 500,000 V?

66. || FIGURE P25.66 shows two uniformly charged spheres. What is the potential difference between points a and b? Which point is at the higher potential?
   Hint: The potential at any point is the superposition of the potentials due to *all* charges.

FIGURE P25.66

67. || Two positive point charges q are located on the y-axis at $y = \pm \frac{1}{2}s$.

a. Find an expression for the potential along the x-axis.

b. Draw a graph of V versus x for $-\infty < x < \infty$. For comparison, use a dotted line to show the potential of a point charge $2q$ located at the origin.

68. ‖ The arrangement of charges shown in **FIGURE P25.68** is called a *linear electric quadrupole*. The positive charges are located at $y = \pm s$. Notice that the net charge is zero. Find an expression for the electric potential on the y-axis at distances $y \gg s$. Give your answer in terms of the *quadrupole moment*, $Q = 2qs^2$.

FIGURE P25.68   FIGURE P25.69

69. ‖ **FIGURE P25.69** shows a thin rod of length L and charge Q. Find
CALC an expression for the electric potential a distance x away from the center of the rod on the axis of the rod.

70. ‖‖ **FIGURE P25.69** shows a thin rod of length L and charge Q. Find
CALC an expression for the electric potential a distance z away from the center of rod on the line that bisects the rod.

71. ‖ **FIGURE P25.71** shows a thin rod with charge Q that has been bent into a semicircle of radius R. Find an expression for the electric potential at the center.

FIGURE P25.71

72. ‖ A disk with a hole has inner radius R_{in} and outer radius R_{out}.
CALC The disk is uniformly charged with total charge Q. Find an expression for the on-axis electric potential at distance z from the center of the disk. Verify that your expression has the correct behavior when $R_{in} \rightarrow 0$.

73. ‖ The wire in **FIGURE P25.73** has linear charge density λ. What is
CALC the electric potential at the center of the semicircle?

FIGURE P25.73

In Problems 74 through 76 you are given the equation(s) used to solve a problem. For each of these,

a. Write a realistic problem for which this is the correct equation(s).

b. Finish the solution of the problem.

74. $\dfrac{(9.0 \times 10^9 \ N\,m^2/C^2)q_1q_2}{0.030 \ m} = 90 \times 10^{-6} \ J$

   $q_1 + q_2 = 40 \ nC$

75. $\frac{1}{2}(1.67 \times 10^{-27} \ kg)(2.5 \times 10^6 \ m/s)^2 + 0 =$

   $\frac{1}{2}(1.67 \times 10^{-27} \ kg)v_i^2 +$

   $\dfrac{(9.0 \times 10^9 \ N\,m^2/C^2)(2.0 \times 10^{-9} \ C)(1.60 \times 10^{-19} \ C)}{0.0010 \ m}$

76. $\dfrac{(9.0 \times 10^9 \ N\,m^2/C^2)(3.0 \times 10^{-9} \ C)}{0.030 \ m} +$

   $\dfrac{(9.0 \times 10^9 \ N\,m^2/C^2)(3.0 \times 10^{-9} \ C)}{(0.030 \ m) + d} = 1200 \ V$

Challenge Problems

77. ‖‖ An electric dipole consists of 1.0 g spheres charged to $\pm 2.0 \ nC$ at the ends of a 10-cm-long massless rod. The dipole rotates on a frictionless pivot at its center. The dipole is held perpendicular to a uniform electric field with field strength 1000 V/m, then released. What is the dipole's angular velocity at the instant it is aligned with the electric field?

78. ‖‖ A proton and an alpha particle ($q = +2e$, $m = 4 \ u$) are fired directly toward each other from far away, each with an initial speed of $0.010c$. What is their distance of closest approach, as measured between their centers?

79. ‖‖ Bead A has a mass of 15 g and a charge of $-5.0 \ nC$. Bead B has a mass of 25 g and a charge of $-10.0 \ nC$. The beads are held 12 cm apart (measured between their centers) and released. What maximum speed is achieved by each bead?

80. ‖‖ Two 2.0-mm-diameter beads, C and D, are 10 mm apart, measured between their centers. Bead C has mass 1.0 g and charge 2.0 nC. Bead D has mass 2.0 g and charge $-1.0 \ nC$. If the beads are released from rest, what are the speeds v_C and v_D at the instant the beads collide?

81. ‖‖ A thin rod of length L and total charge Q has the nonuniform
CALC linear charge distribution $\lambda(x) = \lambda_0 x/L$, where x is measured from the rod's left end.

a. What is λ_0 in terms of Q and L?

b. What is the electric potential on the axis at distance d left of the rod's left end?

82. ‖‖ A hollow cylindrical shell of length L and radius R has charge
CALC Q uniformly distributed along its length. What is the electric potential at the center of the cylinder?

26 Potential and Field

These solar cells are *photovoltaic* cells, meaning that light creates a voltage— a potential difference.

IN THIS CHAPTER, you will learn how the electric potential is related to the electric field.

How are electric potential and field related?

The electric field and the electric potential are intimately connected. In fact, they are simply two different perspectives on how source charges alter the space around them.

- The electric potential can be found if you know the electric field.
- The electric field can be found if you know the electric potential.
- Electric field lines are always perpendicular to equipotential surfaces.
- The electric field points "downhill" in the direction of decreasing potential.
- The electric field is stronger where equipotentials are closer together.

« LOOKING BACK Sections 25.4– 25.6 The electric potential and its graphical representations

What are the properties of conductors?

You'll learn about the properties of conductors in electrostatic equilibrium, finding the same results as using Gauss's law:

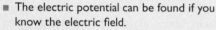

- Any excess charge is on the surface.
- The interior electric field is zero.
- The exterior electric field is perpendicular to the surface.
- The entire conductor is an equipotential.

What are sources of electric potential?

A potential difference—voltage—is created by separating positive and negative charges.

- Work must be done to separate charges. The work done per charge is called the emf of a device. Emf is measured in volts.
- We'll use a charge escalator model of a battery in which chemical reactions "lift" charges from one terminal to the other.

What is a capacitor?

Any two electrodes with equal and opposite charges form a capacitor. Their capacitance indicates their capacity for storing charge. The energy stored in a capacitor will lead us to recognize that electric energy is stored in the electric field.

« LOOKING BACK Section 23.5 Parallel-plate capacitors

How are capacitors used?

Capacitors are important circuit elements that store charge and energy.

- You'll learn to work with combinations of capacitors arranged in series and parallel.
- You'll learn that an insulator—called a dielectric—between the capacitor plates alters the capacitor in useful ways.

26.1 Connecting Potential and Field

FIGURE 26.1 shows the four key ideas of force, field, potential energy, and potential. The electric field and the electric potential were based on force and potential energy. We know, from Chapters 9 and 10, that force and potential energy are closely related. The focus of this chapter is to establish a similar relationship between the electric field and the electric potential. **The electric potential and electric field are not two distinct entities but, instead, two different perspectives or two different mathematical representations of how source charges alter the space around them.**

If this is true, we should be able to find the electric potential from the electric field. Chapter 25 introduced all the pieces we need to do so. We used the potential energy of charge q and the source charges to define the electric potential as

$$V \equiv \frac{U_{q+\text{sources}}}{q} \qquad (26.1)$$

Potential energy is defined in terms of the work done by force $\vec{F}$ on charge q as it moves from position i to position f:

$$\Delta U = -W(\text{i} \rightarrow \text{f}) = -\int_{s_i}^{s_f} F_s \, ds = -\int_i^f \vec{F} \cdot d\vec{s} \qquad (26.2)$$

But the force exerted on charge q by the electric field is $\vec{F} = q\vec{E}$. Putting these three pieces together, you can see that the charge q cancels out and the potential difference between two points in space is

$$\Delta V = V_f - V_i = -\int_{s_i}^{s_f} E_s \, ds = -\int_i^f \vec{E} \cdot d\vec{s} \qquad (26.3)$$

where s is the position along a line from point i to point f. That is, **we can find the potential difference between two points if we know the electric field.**

NOTE The minus sign tells us that the potential *decreases* along the field direction.

A graphical interpretation of Equation 26.3 is

$$V_f = V_i - (\text{area under the } E_s\text{-versus-}s \text{ curve between } s_i \text{ and } s_f) \qquad (26.4)$$

Notice, because of the minus sign in Equation 26.3, that the area is *subtracted* from V_i.

FIGURE 26.1 The four key ideas.

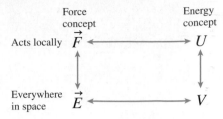

EXAMPLE 26.1 | Finding the potential

FIGURE 26.2 is a graph of E_x, the x-component of the electric field, versus position along the x-axis. Find and graph $V(x)$. Assume $V = 0$ V at $x = 0$ m.

FIGURE 26.2 Graph of E_x versus x.

MODEL The potential difference is the *negative* of the area under the curve.

VISUALIZE E_x is positive throughout this region of space, meaning that $\vec{E}$ points in the positive x-direction.

SOLVE We can see that $E_x = 1000x$ V/m, where x is in m. Thus

$$V_f = V(x) = 0 - (\text{area under the } E_x \text{ curve})$$
$$= -\tfrac{1}{2} \times \text{base} \times \text{height} = -\tfrac{1}{2}(x)(1000x) = -500x^2 \text{ V}$$

FIGURE 26.3 shows that the electric potential in this region of space is parabolic, decreasing from 0 V at $x = 0$ m to -2000 V at $x = 2$ m.

FIGURE 26.3 Graph of V versus x.

ASSESS The electric field points in the direction in which V is *decreasing*. We'll soon see that this is a general rule.

TACTICS BOX 26.1

Finding the potential from the electric field

❶ Draw a picture and identify the point at which you wish to find the potential. Call this position f.

❷ Choose the zero point of the potential, often at infinity. Call this position i.

❸ Establish a coordinate axis from i to f along which you already know or can easily determine the electric field component E_s.

❹ Carry out the integration of Equation 26.3 to find the potential.

Exercise 1

FIGURE 26.4 Finding the potential of a point charge.

❶ Identify the point at which to find the potential. This is position f at $s_f = r$.

❸ Establish a coordinate axis along which $\vec{E}$ is known.

❷ Choose a zero point of the potential. In this case, position i is at $s_i = \infty$.

❹ Integrate along the s-axis.

To see how this works, let's use the electric field of a point charge to find its electric potential. FIGURE 26.4 identifies a point P at $s_f = r$ at which we want to know the potential and calls this position f. We've chosen position i to be at $s_i = \infty$ and identified that as the zero point of the potential. The integration of Equation 26.3 is straight inward along the radial line from i to f:

$$\Delta V = V(r) - V(\infty) = -\int_\infty^r E_s \, ds = \int_r^\infty E_s \, ds \qquad (26.5)$$

The electric field is radially outward. Its s-component is

$$E_s = \frac{1}{4\pi\epsilon_0} \frac{q}{s^2}$$

Thus the potential at distance r from a point charge q is

$$V(r) = V(\infty) + \frac{q}{4\pi\epsilon_0} \int_r^\infty \frac{ds}{s^2} = V(\infty) + \frac{q}{4\pi\epsilon_0} \frac{-1}{s}\bigg|_r^\infty = 0 + \frac{1}{4\pi\epsilon_0} \frac{q}{r} \qquad (26.6)$$

We've rediscovered the potential of a point charge that you learned in Chapter 25:

$$V_{\text{point charge}} = \frac{1}{4\pi\epsilon_0} \frac{q}{r} \qquad (26.7)$$

EXAMPLE 26.2 | The potential of a parallel-plate capacitor

In Chapter 23, the electric field inside a capacitor was found to be

$$\vec{E} = \left(\frac{Q}{\epsilon_0 A}, \text{ from positive to negative} \right)$$

Find the electric potential inside the capacitor. Let $V = 0$ V at the negative plate.

MODEL The electric field inside a capacitor is a uniform field.

VISUALIZE FIGURE 26.5 shows the capacitor and establishes a point P where we want to find the potential. We've chosen an s-axis measured from the negative plate, which is the zero point of the potential.

SOLVE We'll integrate along the s-axis from $s_i = 0$ (where $V_i = 0$ V) to $s_f = s$. Notice that $\vec{E}$ points in the negative s-direction, so $E_s = -Q/\epsilon_0 A$. $Q/\epsilon_0 A$ is a constant, so

$$V(s) = V_f = V_i - \int_0^s E_s \, ds = -\left(-\frac{Q}{\epsilon_0 A} \right) \int_0^s ds = \frac{Q}{\epsilon_0 A} s = Es$$

ASSESS $V = Es$ is the capacitor potential we deduced in Chapter 25 by working directly with the potential energy. The potential

FIGURE 26.5 Finding the potential inside a capacitor.

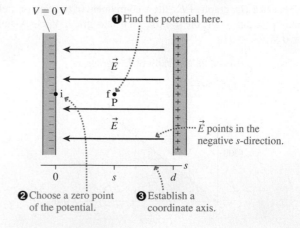

❶ Find the potential here.

$\vec{E}$ points in the negative s-direction.

❷ Choose a zero point of the potential.

❸ Establish a coordinate axis.

increases linearly from $V = 0$ at the negative plate to $V = Ed$ at the positive plate. Here we found the potential by explicitly recognizing the connection between the potential and the field.

26.2 Finding the Electric Field from the Potential

FIGURE 26.6 shows two points i and f separated by a very small distance Δs, so small that the electric field is essentially constant over this very short distance. The work done by the electric field as a charge q moves through this small distance is $W = F_s \Delta s = qE_s \Delta s$. Consequently, the potential difference between these two points is

$$\Delta V = \frac{\Delta U_{q + \text{sources}}}{q} = \frac{-W}{q} = -E_s \Delta s \qquad (26.8)$$

In terms of the potential, the component of the electric field in the s-direction is $E_s = -\Delta V/\Delta s$. In the limit $\Delta s \to 0$,

$$E_s = -\frac{dV}{ds} \qquad (26.9)$$

Now we have reversed Equation 26.3 and can find the electric field from the potential. We'll begin with examples where the field is parallel to a coordinate axis, then we'll look at what Equation 26.9 tells us about the geometry of the field and the potential.

Field Parallel to a Coordinate Axis

The derivative in Equation 26.9 gives E_s, the component of the electric field parallel to the displacement $\Delta \vec{s}$. It doesn't tell us about the electric field component perpendicular to $\Delta \vec{s}$. Thus Equation 26.9 is most useful if we can use symmetry to select a coordinate axis that is parallel to $\vec{E}$ and along which the perpendicular component of $\vec{E}$ is known to be zero.

For example, suppose we knew the potential of a point charge to be $V = q/4\pi\epsilon_0 r$ but didn't remember the electric field. Symmetry requires that the field point straight outward from the charge, with only a radial component E_r. If we choose the s-axis to be in the radial direction, parallel to $\vec{E}$, we can use Equation 26.9 to find

$$E_r = -\frac{dV}{dr} = -\frac{d}{dr}\left(\frac{q}{4\pi\epsilon_0 r}\right) = \frac{1}{4\pi\epsilon_0}\frac{q}{r^2} \qquad (26.10)$$

This is, indeed, the well-known electric field of a point charge.

Equation 26.9 is especially useful for a continuous distribution of charge because calculating V, which is a scalar, is usually much easier than calculating the vector $\vec{E}$ directly from the charge. Once V is known, $\vec{E}$ is found simply by taking a derivative.

FIGURE 26.6 The electric field does work on charge q.

A very small displacement of charge q

E_s, the component of $\vec{E}$ in the direction of motion, is essentially constant over the small distance Δs.

EXAMPLE 26.3 | **The electric field of a ring of charge**

In Chapter 25, we found the on-axis potential of a ring of radius R and charge Q to be

$$V_{\text{ring}} = \frac{1}{4\pi\epsilon_0}\frac{Q}{\sqrt{z^2 + R^2}}$$

Find the on-axis electric field of a ring of charge.

SOLVE Symmetry requires the electric field along the axis to point straight outward from the ring with only a z-component E_z. The electric field at position z is

$$E_z = -\frac{dV}{dz} = -\frac{d}{dz}\left(\frac{1}{4\pi\epsilon_0}\frac{Q}{\sqrt{z^2 + R^2}}\right)$$

$$= \frac{1}{4\pi\epsilon_0}\frac{zQ}{(z^2 + R^2)^{3/2}}$$

ASSESS This result is in perfect agreement with the electric field we found in Chapter 23, but this calculation was easier because we didn't have to deal with angles.

A geometric interpretation of Equation 26.9 is that the electric field is the negative of the *slope* of the V-versus-s graph. This interpretation should be familiar. You learned in Chapter 10 that the force on a particle is the negative of the slope of the potential-energy graph: $F = -dU/ds$. In fact, Equation 26.9 is simply $F = -dU/ds$ with both sides divided by q to yield E and V. This geometric interpretation is an important step in developing an understanding of potential.

EXAMPLE 26.4 | Finding E from the slope of V

FIGURE 26.7 is a graph of the electric potential in a region of space where $\vec{E}$ is parallel to the x-axis. Draw a graph of E_x versus x.

FIGURE 26.7 Graph of V versus position x.

MODEL The electric field is the *negative* of the slope of the potential graph.

SOLVE There are three regions of different slope:

$$0 < x < 2 \text{ cm} \quad \begin{cases} \Delta V/\Delta x = (20 \text{ V})/(0.020 \text{ m}) = 1000 \text{ V/m} \\ E_x = -1000 \text{ V/m} \end{cases}$$

$$2 < x < 4 \text{ cm} \quad \begin{cases} \Delta V/\Delta x = 0 \text{ V/m} \\ E_x = 0 \text{ V/m} \end{cases}$$

$$4 < x < 8 \text{ cm} \quad \begin{cases} \Delta V/\Delta x = (-20 \text{ V})/(0.040 \text{ m}) = -500 \text{ V/m} \\ E_x = 500 \text{ V/m} \end{cases}$$

The results are shown in **FIGURE 26.8**.

FIGURE 26.8 Graph of E_x versus position x.

The *value* of E_x is the negative of the *slope* of the potential graph.

ASSESS The electric field $\vec{E}$ points to the left (E_x is negative) for $0 < x < 2$ cm and to the right (E_x is positive) for $4 < x < 8$ cm. Notice that **the electric field is zero in a region of space where the potential is not changing.**

STOP TO THINK 26.1 Which potential graph describes the electric field at the left?

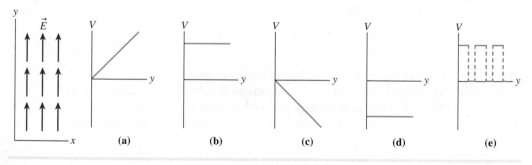

(a)   (b)   (c)   (d)   (e)

The Geometry of Potential and Field

FIGURE 26.9 The electric field at P is related to the shape of the equipotential surfaces.

Equations 26.3 for V in terms of E_s and 26.9 for E_s in terms of V have profound implications for the geometry of the potential and the field. **FIGURE 26.9** shows two equipotential surfaces, with V_+ positive relative to V_-. To learn about the electric field $\vec{E}$ at point P, allow a charge to move through the two displacements $\Delta \vec{s}_1$ and $\Delta \vec{s}_2$. Displacement $\Delta \vec{s}_1$ is *tangent* to the equipotential surface, hence a charge moving in this direction experiences *no* potential difference. According to Equation 26.9, the electric field component along a direction of *constant* potential is $E_s = -dV/ds = 0$. In other words, **the electric field component tangent to the equipotential is $E_\parallel = 0$.**

Displacement $\Delta \vec{s}_2$ is *perpendicular* to the equipotential surface. There is a potential difference along $\Delta \vec{s}_2$, hence the electric field component is

$$E_\perp = -\frac{dV}{ds} \approx -\frac{\Delta V}{\Delta s} = -\frac{V_+ - V_-}{\Delta s_2}$$

You can see that the electric field is inversely proportional to Δs_2, the spacing between the equipotential surfaces. Furthermore, because $(V_+ - V_-) > 0$, the minus sign tells us that the electric field is *opposite* in direction to $\Delta \vec{s}_2$. In other words, $\vec{E}$ is **perpendicular to the equipotential surfaces and points "downhill" in the direction of** *decreasing* **potential.**

These important ideas are summarized in **FIGURE 26.10**.

FIGURE 26.10 The geometry of the potential and the field.

1. $\vec{E}$ is everywhere perpendicular to the equipotential surfaces.

2. $\vec{E}$ points "downhill," in the direction of decreasing V.

3. The field strength is inversely proportional to the spacing Δs between the equipotential surfaces.

4. Equipotential surfaces have equal potential differences between them.

Direction of decreasing potential

Mathematically, we can calculate the individual components of $\vec{E}$ at any point by extending Equation 26.9 to three dimensions:

$$\vec{E} = E_x \,\hat{\imath} + E_y \,\hat{\jmath} + E_z \,\hat{k} = -\left(\frac{\partial V}{\partial x}\,\hat{\imath} + \frac{\partial V}{\partial y}\,\hat{\jmath} + \frac{\partial V}{\partial z}\,\hat{k}\right) \tag{26.11}$$

where $\partial V/\partial x$ is the partial derivative of V with respect to x while y and z are held constant. You may recognize from calculus that the expression in parentheses is the *gradient* of V, written ∇V. Thus, $\vec{E} = -\nabla V$. More advanced treatments of the electric field make extensive use of this mathematical relationship, but for the most part we'll limit our investigations to those we can analyze graphically.

EXAMPLE 26.5 | **Finding the electric field from the equipotential surfaces**

In **FIGURE 26.11** a 1 cm × 1 cm grid is superimposed on a contour map of the potential. Estimate the strength and direction of the electric field at points 1, 2, and 3. Show your results graphically by drawing the electric field vectors on the contour map.

FIGURE 26.11 Equipotential lines.

MODEL The electric field is perpendicular to the equipotential lines, points "downhill," and depends on the slope of the potential hill.

VISUALIZE The potential is highest on the bottom and the right. An elevation graph of the potential would look like the lower-right quarter of a bowl or a football stadium.

SOLVE Some distant but unseen source charges have created an electric field and potential. We do not need to see the source charges to relate the field to the potential. Because $E \approx -\Delta V/\Delta s$, the electric field is stronger where the equipotential lines are closer

together and weaker where they are farther apart. If Figure 26.11 were a topographic map, you would interpret the closely spaced contour lines at the bottom of the figure as a steep slope.

FIGURE 26.12 shows how measurements of Δs from the grid are combined with values of ΔV to determine $\vec{E}$. Point 3 requires an estimate of the spacing between the 0 V and the 100 V surfaces. Notice that we're using the 0 V and 100 V equipotential surfaces to determine $\vec{E}$ at a point on the 50 V equipotential.

FIGURE 26.12 The electric field at points 1 to 3.

100 V change over 4 cm is 2500 V/m.

100 V change over 2 cm is 5000 V/m.

100 V change over ≈ 2.5 cm is 4000 V/m.

ASSESS The *directions* of $\vec{E}$ are found by drawing downhill vectors perpendicular to the equipotentials. The distances between the equipotential surfaces are needed to determine the field strengths.

Kirchhoff's Loop Law

FIGURE 26.13 The potential difference between points 1 and 2 is the same along either path.

The potential difference along path 1-a-b-c-2 is
$\Delta V = 0\text{ V} + 10\text{ V}$
$\qquad + 0\text{ V} + 10\text{ V}$
$\qquad = 20\text{ V}.$

The potential difference along path 1-d-2 is
$\Delta V = 20\text{ V} + 0\text{ V} = 20\text{ V}.$

FIGURE 26.13 shows two points, 1 and 2, in a region of electric field and potential. You learned in Chapter 25 that the work done in moving a charge between points 1 and 2 is *independent of the path*. Consequently, the potential difference between points 1 and 2 along any two paths that join them is $\Delta V = 20$ V. This must be true in order for the idea of an equipotential surface to make sense.

Now consider the path 1–a–b–c–2–d–1 that ends where it started. What is the potential difference "around" this closed path? The potential increases by 20 V in moving from 1 to 2, but then decreases by 20 V in moving from 2 back to 1. Thus $\Delta V = 0$ V around the closed path.

The numbers are specific to this example, but the idea applies to any loop (i.e., a closed path) through an electric field. The situation is analogous to hiking on the side of a mountain. You may walk uphill during parts of your hike and downhill during other parts, but if you return to your starting point your *net* change of elevation is zero. So for any path that starts and ends at the same point, we can conclude that

$$\Delta V_{\text{loop}} = \sum_i (\Delta V)_i = 0 \qquad (26.12)$$

Stated in words, **the sum of all the potential differences encountered while moving around a loop or closed path is zero.** This statement is known as **Kirchhoff's loop law.**

Kirchhoff's loop law is a statement of energy conservation because a charge that moves around a loop and returns to its starting point has $\Delta U = q\,\Delta V = 0$. Kirchhoff's loop law and a second Kirchhoff's law you'll meet in the next chapter will turn out to be the two fundamental principles of circuit analysis.

STOP TO THINK 26.2 Which set of equipotential surfaces matches this electric field?

$\vec{E}$

(a) 0 V ... 50 V

(b) 0 V ... 50 V

(c) 0 V ... 50 V

(d) 50 V ... 0 V

(e) 50 V ... 0 V

(f) 50 V ... 0 V

26.3 A Conductor in Electrostatic Equilibrium

A *corona discharge* occurs at pointed metal tips where the electric field can be very strong.

The basic relationships between potential and field allow us to draw some interesting and important conclusions about conductors. Consider a conductor, such as a metal, that is in electrostatic equilibrium. The conductor may be charged, but all the charges are at rest.

You learned in Chapter 22 that any excess charges on a conductor in electrostatic equilibrium are always located on the *surface* of the conductor. Using similar reasoning, we can conclude that **the electric field is zero at any interior point of a conductor in electrostatic equilibrium**. Why? If the field were other than zero, then there would be a force $\vec{F} = q\vec{E}$ on the charge carriers and they would move, creating a current. But there are no currents in a conductor in electrostatic equilibrium, so it must be that $\vec{E} = \vec{0}$ at all interior points.

The two points inside the conductor in **FIGURE 26.14** are connected by a line that remains entirely inside the conductor. We can find the potential difference $\Delta V = V_2 - V_1$

between these points by using Equation 26.3 to integrate E_s along the line from 1 to 2. But $E_s = 0$ at all points along the line, because $\vec{E} = \vec{0}$; thus the value of the integral is zero and $\Delta V = 0$. In other words, **any two points inside a conductor in electrostatic equilibrium are at the same potential.**

When a conductor is in electrostatic equilibrium, the *entire conductor* is at the same potential. If we charge a metal sphere, then the entire sphere is at a single potential. Similarly, a charged metal rod or wire is at a single potential *if* it is in electrostatic equilibrium.

If $\vec{E} = \vec{0}$ inside a charged conductor but $\vec{E} \neq \vec{0}$ outside, what happens right at the surface? If the entire conductor is at the same potential, then the surface is an equipotential surface. You have seen that the electric field is always perpendicular to an equipotential surface, hence **the exterior electric field $\vec{E}$ of a charged conductor is perpendicular to the surface.**

We can also conclude that the electric field, and thus the surface charge density, is largest at sharp points. This follows from our earlier discovery that the field at the surface of a sphere of radius R can be written $E = V_0/R$. If we approximate the rounded corners of a conductor with sections of spheres, all of which are at the same potential V_0, the field strength will be largest at the corners with the smallest radii of curvature—the sharpest points.

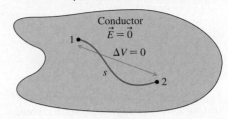

FIGURE 26.14 All points inside a conductor in electrostatic equilibrium are at the same potential.

FIGURE 26.15 Properties of a conductor in electrostatic equilibrium.

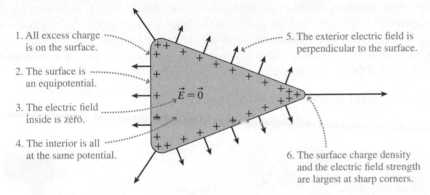

1. All excess charge is on the surface.

2. The surface is an equipotential.

3. The electric field inside is zero.

4. The interior is all at the same potential.

5. The exterior electric field is perpendicular to the surface.

6. The surface charge density and the electric field strength are largest at sharp corners.

FIGURE 26.15 summarizes what we know about conductors in electrostatic equilibrium. These are important and practical conclusions because conductors are the primary components of electrical devices.

We can use similar reasoning to estimate the electric field and potential between two charged conductors. As an example, **FIGURE 26.16** shows a negatively charged metal sphere near a flat metal plate. The surfaces of the sphere and the flat plate are equipotentials, hence the electric field must be perpendicular to both. Close to a surface, the electric field is still *nearly* perpendicular to the surface. Consequently, **an equipotential surface close to an electrode must roughly match the shape of the electrode.**

In between, the equipotential surfaces *gradually* change as they "morph" from one electrode shape to the other. It's not hard to sketch a contour map showing a plausible set of equipotential surfaces. You can then draw electric field lines (field lines are easier to draw than field vectors) that are perpendicular to the equipotentials, point "downhill," and are closer together where the contour line spacing is smaller.

FIGURE 26.16 Estimating the field and potential between two charged conductors.

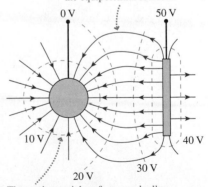

The field lines are perpendicular to the equipotential surfaces.

The equipotential surfaces gradually change from the shape of one electrode to that of the other.

STOP TO THINK 26.3 Three charged metal spheres of different radii are connected by a thin metal wire. The potential and electric field at the surface of each sphere are V and E. Which of the following is true?

a. $V_1 = V_2 = V_3$ and $E_1 = E_2 = E_3$
b. $V_1 = V_2 = V_3$ and $E_1 > E_2 > E_3$
c. $V_1 > V_2 > V_3$ and $E_1 = E_2 = E_3$
d. $V_1 > V_2 > V_3$ and $E_1 > E_2 > E_3$
e. $V_3 > V_2 > V_1$ and $E_3 = E_2 = E_1$
f. $V_3 > V_2 > V_1$ and $E_3 > E_2 > E_1$

26.4 Sources of Electric Potential

We've now studied many properties of the electric potential and seen how potential and field are connected, but we've not said much about how an electric potential is created. Simply put, **an electric potential difference is created by separating positive and negative charge.** Shuffling your feet on the carpet transfers electrons from the carpet to you, creating a potential difference between you and a doorknob that causes a spark and a shock as you touch it. Charging a capacitor by moving electrons from one plate to the other creates a potential difference across the capacitor.

As FIGURE 26.17 shows, moving charge from one electrode to another creates an electric field $\vec{E}$ pointing from the positive toward the negative electrode. As a consequence, there is a potential difference between the electrodes that is given by

$$\Delta V = V_{\text{pos}} - V_{\text{neg}} = -\int_{\text{neg}}^{\text{pos}} E_s \, ds$$

where the integral runs from any point on the negative electrode to any point on the positive. The *net* charge is zero, but pulling the positive and negative charge apart creates a potential difference.

Now electric forces try to bring positive and negative charges together, so **a nonelectrical process is needed to separate charge.** As an example, the **Van de Graaff generator** shown in FIGURE 26.18a separates charges mechanically. A moving plastic or leather belt is charged, then the charge is mechanically transported via the conveyor belt to the spherical electrode at the top of the insulating column. The charging of the belt could be done by friction, but in practice a *corona discharge* created by the strong electric field at the tip of a needle is more efficient and reliable.

FIGURE 26.18 A Van de Graaff generator.

A Van de Graaff generator has two noteworthy features:

- Charge is *mechanically* transported from the negative side to the positive side. This charge separation creates a potential difference between the spherical electrode and its surroundings.
- The electric field of the spherical electrode exerts a downward force on the positive charges moving up the belt. Consequently, *work must be done* to "lift" the positive charges. The work is done by the electric motor that runs the belt.

A classroom-demonstration Van de Graaff generator like the one shown in FIGURE 26.18b creates a potential difference of several hundred thousand volts between the upper sphere and its surroundings. The maximum potential is reached when the electric field near the sphere becomes large enough to cause a breakdown of the air. This produces a spark and temporarily discharges the sphere. A large Van de Graaff generator surrounded by vacuum can reach a potential of 20 MV or more. These generators are used to accelerate protons for nuclear physics experiments.

FIGURE 26.17 A charge separation creates a potential difference.

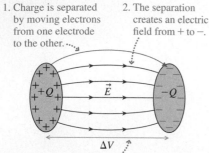

1. Charge is separated by moving electrons from one electrode to the other. ⋯⋯

2. The separation creates an electric field from + to −.

3. Because of the electric field, there's a potential difference between the electrodes.

Batteries and emf

The most common source of electric potential is a **battery,** which uses *chemical reactions* to separate charge. A battery consists of chemicals, called *electrolytes,* sandwiched between two electrodes made of different metals. Chemical reactions in the electrolytes transport ions (i.e., charges) from one electrode to the other. This chemical process pulls positive and negative charges apart, creating a potential difference between the terminals of the battery. When the chemicals are used up, the reactions cease and the battery is dead.

We can sidestep the chemistry details by introducing the **charge escalator model** of a battery.

MODEL 26.1

Charge escalator model of a battery

A battery uses chemical reactions to separate charge.

- The charge escalator "lifts" positive charges from the negative terminal to the positive terminal. This requires *work*, with the energy being supplied by the chemical reactions.
- The work done *per charge* is called the **emf** of the battery: $\mathcal{E} = W_{chem}/q$.
- The charge separation creates a potential difference ΔV_{bat} between the terminals. An *ideal battery* has $\Delta V_{bat} = \mathcal{E}$.
- Limitations: $\Delta V_{bat} < \mathcal{E}$ if current flows through the battery. In most cases, the difference is small and a battery can be considered ideal.

Charge q gains potential energy ΔU.

Emf is pronounced as the sequence of letters e-m-f. The symbol for emf is $\mathcal{E}$, a script E, and the units of emf are joules per coulomb, or volts. The rating of a battery, such as 1.5 V or 9 V, is the battery's emf.

The key idea is that **emf is work,** specifically the work done *per charge* to pull positive and negative charges apart. This work can be done by mechanical forces, chemical reactions, or—as you'll see later—magnetic forces. *Because* work is done, charges gain potential energy and their separation creates a potential difference ΔV_{bat} between the positive and negative terminals of the battery. This is called the **terminal voltage.**

In an **ideal battery,** which has no internal energy losses, the work W_{chem} done to move charge q from the negative to the positive terminal goes entirely to increasing the potential energy of the charge, and so $\Delta V_{bat} = \mathcal{E}$. In practice, the terminal voltage is slightly less than the emf when current flows through a battery—we'll discuss this in Chapter 28—but the difference usually small and in most cases we can model batteries as being ideal.

Batteries in Series

Many consumer goods, from flashlights to digital cameras, use more than one battery. Why? A particular type of battery, such as an AA or AAA battery, produces a fixed emf determined by the chemical reactions inside. The emf of one battery, often 1.5 V, is not sufficient to light a lightbulb or power a camera. But just as you can reach the third floor of a building by taking three escalators in succession, we can produce a larger potential difference by placing two or more batteries *in series*.

FIGURE 26.19 shows two batteries with the positive terminal of one literally touching the negative terminal of the next. Flashlight batteries usually are arranged like this.

FIGURE 26.19 Batteries in series.

$\Delta V_{series} = \Delta V_1 + \Delta V_2$

Other devices, such as cameras, achieve the same effect by using conducting metal wires between one battery and the next. Either way, the total potential difference of batteries in series is simply the sum of their individual terminal voltages:

$$\Delta V_{\text{series}} = \Delta V_1 + \Delta V_2 + \cdots \qquad \text{(batteries in series)} \qquad (26.13)$$

STOP TO THINK 26.4 What total potential difference is created by these three batteries?

26.5 Capacitance and Capacitors

FIGURE 26.20 Two equally but oppositely charged electrodes form a capacitor.

The separated charge has created a potential difference even though the net charge is zero.

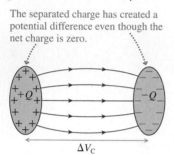

ΔV_C

FIGURE 26.20 shows two electrodes that have been charged to $\pm Q$. Their net charge is zero, but something has separated positive and negative charges. Consequently, there is a potential difference ΔV between the electrodes.

It seems plausible that ΔV is directly proportional to Q. That is, doubling the amount of charge on the electrodes will double the potential difference. We can write this as $Q = C\Delta V$, where the proportionality constant

$$C = \frac{Q}{\Delta V_C} \qquad (26.14)$$

is called the **capacitance** of the two electrodes. The two electrodes themselves form a **capacitor,** so we've written a subscript C on ΔV_C to indicate that this is the **capacitor voltage,** the potential difference between the positive and negative electrodes.

> **NOTE** You've already met the parallel-plate capacitor, but a capacitor can be formed from any two electrodes. The electrodes of a capacitor always have *equal but opposite charges* (zero net charge), and the Q appearing in equations is the magnitude (always positive) of this amount of charge.

The SI unit of capacitance is the **farad,** named in honor of Michael Faraday. One farad is defined as

$$1 \text{ farad} = 1 \text{ F} = 1 \text{ C/V}$$

One farad turns out to be an enormous amount of capacitance. Practical capacitors are usually measured in units of microfarads (μF) or picofarads ($1 \text{ pF} = 10^{-12} \text{ F}$).

Turning Equation 26.14 around, we see that the amount of charge on a capacitor that has been charged to ΔV_C is

$$Q = C\Delta V_C \qquad \text{(charge on a capacitor)} \qquad (26.15)$$

The amount of charge is determined jointly by the potential difference *and* by a property of the electrodes called capacitance. As we'll see, **capacitance depends only on the geometry of the electrodes.**

The Parallel-Plate Capacitor

Capacitors are important elements in electric circuits.

A parallel-plate capacitor consists of two flat electrodes (the plates) facing each other with a plate separation d that is small compared to the sizes of the plates. You learned in Chapter 25 that the potential difference across a parallel-plate capacitor is related to the electric field inside by $\Delta V_C = Ed$. And you know from Chapter 23 that the electric field inside a parallel-plate capacitor is

$$E = \frac{Q}{\epsilon_0 A} \qquad (26.16)$$

where A is the surface area of the plates. Combining these gives

$$Q = \frac{\epsilon_0 A}{d} \Delta V_C \qquad (26.17)$$

You can see that the charge is proportional to the potential difference, as expected. So from the definition of capacitance, Equation 26.14, we find that the capacitance of a parallel-plate capacitor is

$$C = \frac{Q}{\Delta V_C} = \frac{\epsilon_0 A}{d} \quad \text{(parallel-plate capacitor)} \qquad (26.18)$$

The capacitance is a purely *geometric* property of the electrodes, depending only on their surface area and spacing. Capacitors of other shapes will have different formulas for their capacitance, but all will depend entirely on geometry. A cylindrical capacitor is the topic of Challenge Example 26.11, and a homework problem will let you analyze a spherical capacitor.

EXAMPLE 26.6 | **Charging a capacitor**

The spacing between the plates of a 1.0 μF capacitor is 0.050 mm.

a. What is the surface area of the plates?

b. How much charge is on the plates if this capacitor is charged to 1.5 V?

MODEL Assume the capacitor is a parallel-plate capacitor.

SOLVE a. From the definition of capacitance,

$$A = \frac{dC}{\epsilon_0} = 5.65 \text{ m}^2$$

b. The charge is $Q = C \Delta V_C = 1.5 \times 10^{-6} \text{ C} = 1.5 \ \mu\text{C}$.

ASSESS The surface area needed to construct a 1.0 μF capacitor (a fairly typical value) is enormous. We'll see in Section 26.7 how the area can be reduced by inserting an insulator between the capacitor plates.

Charging a Capacitor

All well and good, but *how* does a capacitor get charged? By connecting it to a battery! **FIGURE 26.21a** shows the two plates of a capacitor shortly after two conducting wires have connected them to the two terminals of a battery. At this instant, the battery's charge escalator is moving charge from one capacitor plate to the other, and it is this work done by the battery that charges the capacitor. (The connecting wires are conductors, and you learned in Chapter 22 that charges can move through conductors as a *current.*) The capacitor voltage ΔV_C steadily increases as the charge separation continues.

FIGURE 26.21 A parallel-plate capacitor is charged by a battery.

The charge escalator moves charge from one plate to the other. ΔV_C increases as the charge separation increases.

When $\Delta V_C = \Delta V_{bat}$, the current stops and the capacitor is fully charged.

The keys on computer keyboards are capacitor switches. Pressing the key pushes two capacitor plates closer together, increasing their capacitance. A larger capacitor can hold more charge, so a momentary current carries charge from the battery (or power supply) to the capacitor. This current is sensed, and the keystroke is then recorded.

But this process cannot continue forever. The growing positive charge on the upper capacitor plate exerts a repulsive force on new charges coming up the escalator, and eventually the capacitor charge gets so large that no new charges can arrive. The capacitor in **FIGURE 26.21b** is now *fully charged*. In Chapter 28 we'll analyze how long the charging process takes, but it is typically less than a nanosecond for a capacitor connected directly to a battery with copper wires.

Once the capacitor is fully charged, with charges no longer in motion, the positive capacitor plate, the upper wire, and the positive terminal of the battery form a single conductor in electrostatic equilibrium. This is an important idea, and it wasn't true while the capacitor was charging. As you just learned, any two points in a conductor in electrostatic equilibrium are at the same potential. **Thus the positive plate of a fully charged capacitor is at the same potential as the positive terminal of the battery.**

Similarly, the negative plate of a fully charged capacitor is at the same potential as the negative terminal of the battery. Consequently, the potential difference ΔV_C between the capacitor plates exactly matches the potential difference ΔV_{bat} between the battery terminals. **A capacitor attached to a battery charges until $\Delta V_C = \Delta V_{bat}$.** Once the capacitor is charged, you can disconnect it from the battery; it will maintain this charge and potential difference until and unless something—a current—allows positive charge to move back to the negative plate. An ideal capacitor in vacuum would stay charged forever.

Combinations of Capacitors

Two or more capacitors are often joined together. **FIGURE 26.22** illustrates two basic combinations: **parallel capacitors** and **series capacitors.** Notice that a capacitor, no matter what its actual geometric shape, is represented in *circuit diagrams* by two parallel lines.

FIGURE 26.22 Parallel and series capacitors.

The circuit symbol for a capacitor is two parallel lines.

Parallel capacitors are joined top to top and bottom to bottom.

Series capacitors are joined end to end in a row.

FIGURE 26.23 Replacing two parallel capacitors with an equivalent capacitor.

(a) Parallel capacitors have the same ΔV_C.

$Q_1 = C_1 \Delta V_C \qquad Q_2 = C_2 \Delta V_C$

(b) Same ΔV_C as C_1 and C_2

$Q = Q_1 + Q_2$
Same total charge as C_1 and C_2

NOTE The terms "parallel capacitors" and "parallel-plate capacitor" do not describe the same thing. The former term describes how two or more capacitors are connected to each other, the latter describes how a particular capacitor is constructed.

As we'll show, parallel or series capacitors (or, as is sometimes said, capacitors "in parallel" or "in series") can be represented by a single **equivalent capacitance.** We'll demonstrate this first with the two parallel capacitors C_1 and C_2 of **FIGURE 26.23a**. Because the two top electrodes are connected by a conducting wire, they form a single conductor in electrostatic equilibrium. Thus the two top electrodes are at the same potential. Similarly, the two connected bottom electrodes are at the same potential. Consequently, two (or more) capacitors in parallel each have the *same* potential difference ΔV_C between the two electrodes.

The charges on the two capacitors are $Q_1 = C_1 \Delta V_C$ and $Q_2 = C_2 \Delta V_C$. Altogether, the battery's charge escalator moved total charge $Q = Q_1 + Q_2$ from the negative electrodes to the positive electrodes. Suppose, as in **FIGURE 26.23b**, we replaced the two capacitors with a single capacitor having charge $Q = Q_1 + Q_2$ and potential difference ΔV_C. This capacitor is equivalent to the original two in the sense that the battery can't tell the difference. In either case, the battery has to establish the same potential difference and move the same amount of charge.

By definition, the capacitance of this equivalent capacitor is

$$C_{eq} = \frac{Q}{\Delta V_C} = \frac{Q_1 + Q_2}{\Delta V_C} = \frac{Q_1}{\Delta V_C} + \frac{Q_2}{\Delta V_C} = C_1 + C_2 \qquad (26.19)$$

This analysis hinges on the fact that **parallel capacitors each have the same potential difference ΔV_C.** We could easily extend this analysis to more than two capacitors. If capacitors $C_1, C_2, C_3, \ldots$ are in parallel, their equivalent capacitance is

$$C_{eq} = C_1 + C_2 + C_3 + \cdots \qquad \text{(parallel capacitors)} \qquad (26.20)$$

Neither the battery nor any other part of a circuit can tell if the parallel capacitors are replaced by a single capacitor having capacitance C_{eq}.

Now consider the two series capacitors in **FIGURE 26.24a**. The center section, consisting of the bottom plate of C_1, the top plate of C_2, and the connecting wire, is electrically isolated. The battery cannot remove charge from or add charge to this section. If it starts out with no net charge, it must end up with no net charge. As a consequence, the two capacitors in series have equal charges $\pm Q$. The battery transfers Q from the bottom of C_2 to the top of C_1. This transfer polarizes the center section, as shown, but it still has $Q_{net} = 0$.

The potential differences across the two capacitors are $\Delta V_1 = Q/C_1$ and $\Delta V_2 = Q/C_2$. The total potential difference across both capacitors is $\Delta V_C = \Delta V_1 + \Delta V_2$. Suppose, as in **FIGURE 26.24b**, we replaced the two capacitors with a single capacitor having charge Q and potential difference $\Delta V_C = \Delta V_1 + \Delta V_2$. This capacitor is equivalent to the original two because the battery has to establish the same potential difference and move the same amount of charge in either case.

By definition, the capacitance of this equivalent capacitor is $C_{eq} = Q/\Delta V_C$. The inverse of the equivalent capacitance is thus

$$\frac{1}{C_{eq}} = \frac{\Delta V_C}{Q} = \frac{\Delta V_1 + \Delta V_2}{Q} = \frac{\Delta V_1}{Q} + \frac{\Delta V_2}{Q} = \frac{1}{C_1} + \frac{1}{C_2} \qquad (26.21)$$

This analysis hinges on the fact that **series capacitors each have the same charge Q.** We could easily extend this analysis to more than two capacitors. If capacitors $C_1, C_2, C_3, \ldots$ are in series, their equivalent capacitance is

$$C_{eq} = \left(\frac{1}{C_1} + \frac{1}{C_2} + \frac{1}{C_3} + \cdots \right)^{-1} \qquad \text{(series capacitors)} \qquad (26.22)$$

NOTE Be careful to avoid the common error of adding the inverses but forgetting to invert the sum.

Let's summarize the key facts before looking at a numerical example:

- Parallel capacitors all have the same potential difference ΔV_C. Series capacitors all have the same amount of charge $\pm Q$.
- The equivalent capacitance of a parallel combination of capacitors is *larger* than any single capacitor in the group. The equivalent capacitance of a series combination of capacitors is *smaller* than any single capacitor in the group.

FIGURE 26.24 Replacing two series capacitors with an equivalent capacitor.

(a) Series capacitors have the same Q.

$\Delta V_1 = Q/C_1$

No net charge on this isolated segment

$\Delta V_2 = Q/C_2$

(b) Same Q as C_1 and C_2

$\Delta V_C = \Delta V_1 + \Delta V_2$
Same total potential difference as C_1 and C_2

EXAMPLE 26.7 | A capacitor circuit

Find the charge on and the potential difference across each of the three capacitors in FIGURE 26.25.

FIGURE 26.25 A capacitor circuit.

MODEL Assume the battery is ideal, with $\Delta V_{bat} = \mathcal{E} = 12$ V. Use the results for parallel and series capacitors.

SOLVE The three capacitors are neither in parallel nor in series, but we can break them down into smaller groups that are. A useful method of *circuit analysis* is first to combine elements until reaching a single equivalent element, then to reverse the process and calculate values for each element. FIGURE 26.26 shows the analysis of this circuit. Notice that we redraw the circuit after every step. The equivalent capacitance of the 3 μF and 6 μF capacitors in series is found from

$$C_{eq} = \left(\frac{1}{3\ \mu F} + \frac{1}{6\ \mu F}\right)^{-1} = \left(\frac{2}{6} + \frac{1}{6}\right)^{-1} \mu F = 2\ \mu F$$

Once we get to the single equivalent capacitance, we find that $\Delta V_C = \Delta V_{bat} = 12$ V and $Q = C\,\Delta V_C = 24\ \mu$C. Now we can reverse direction. Capacitors in series all have the same charge, so the charge on C_1 and on C_{2+3} is $\pm 24\ \mu$C. This is enough to determine that $\Delta V_1 = 8$ V and $\Delta V_{2+3} = 4$ V. Capacitors in parallel all have the same potential difference, so $\Delta V_2 = \Delta V_3 = 4$ V. This is enough to find that $Q_2 = 20\ \mu$C and $Q_3 = 4\ \mu$C. The charge on and the potential difference across each of the three capacitors are shown in the final step of Figure 26.26.

ASSESS Notice that we had two important checks of internal consistency. $\Delta V_1 + \Delta V_{2+3} = 8$ V + 4 V add up to the 12 V we had found for the 2 μF equivalent capacitor. Then $Q_2 + Q_3 = 20\ \mu$C + 4 μC add up to the 24 μC we had found for the 6 μF equivalent capacitor. We'll do much more circuit analysis of this type in Chapter 28, but it's worth noting now that circuit analysis becomes nearly foolproof *if* you make use of these checks of internal consistency.

FIGURE 26.26 Analyzing the capacitor circuit.

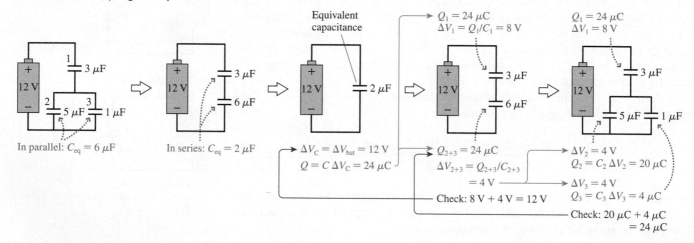

STOP TO THINK 26.5 Rank in order, from largest to smallest, the equivalent capacitance $(C_{eq})_a$ to $(C_{eq})_d$ of circuits a to d.

26.6 The Energy Stored in a Capacitor

Capacitors are important elements in electric circuits because of their ability to store energy. **FIGURE 26.27** shows a capacitor being charged. The instantaneous value of the charge on the two plates is $\pm q$, and this charge separation has established a potential difference $\Delta V = q/C$ between the two electrodes.

An additional charge dq is in the process of being transferred from the negative to the positive electrode. The battery's charge escalator must do work to lift charge dq "uphill" to a higher potential. Consequently, the potential energy of dq + capacitor increases by

$$dU = dq\,\Delta V = \frac{q\,dq}{C} \qquad (26.23)$$

NOTE Energy must be conserved. This increase in the capacitor's potential energy is provided by the battery.

The total energy transferred from the battery to the capacitor is found by integrating Equation 26.23 from the start of charging, when $q = 0$, until the end, when $q = Q$. Thus we find that the energy stored in a charged capacitor is

$$U_C = \frac{1}{C}\int_0^Q q\,dq = \frac{Q^2}{2C} \qquad (26.24)$$

In practice, it is often easier to write the stored energy in terms of the capacitor's potential difference $\Delta V_C = Q/C$. This is

$$U_C = \frac{Q^2}{2C} = \tfrac{1}{2}C(\Delta V_C)^2 \qquad (26.25)$$

The potential energy stored in a capacitor depends on the *square* of the potential difference across it. This result is reminiscent of the potential energy $U = \tfrac{1}{2}k(\Delta x)^2$ stored in a spring, and a charged capacitor really is analogous to a stretched spring. A stretched spring holds the energy until we release it, then that potential energy is transformed into kinetic energy. Likewise, a charged capacitor holds energy until we discharge it. Then the potential energy is transformed into the kinetic energy of moving charges (the current).

FIGURE 26.27 The charge escalator does work on charge dq as the capacitor is being charged.

The instantaneous charge on the plates is $\pm q$.

The charge escalator does work $dq\,\Delta V$ to move charge dq from the negative plate to the positive plate.

EXAMPLE 26.8 | **Storing energy in a capacitor**

How much energy is stored in a 220 μF camera-flash capacitor that has been charged to 330 V? What is the average power dissipation if this capacitor is discharged in 1.0 ms?

SOLVE The energy stored in the charged capacitor is

$$U_C = \tfrac{1}{2}C(\Delta V_C)^2 = \tfrac{1}{2}(220 \times 10^{-6}\,\text{F})(330\,\text{V})^2 = 12\,\text{J}$$

If this energy is released in 1.0 ms, the average power dissipation is

$$P = \frac{\Delta E}{\Delta t} = \frac{12\,\text{J}}{1.0 \times 10^{-3}\,\text{s}} = 12{,}000\,\text{W}$$

ASSESS The stored energy is equivalent to raising a 1 kg mass 1.2 m. This is a rather large amount of energy, which you can see by imagining the damage a 1 kg mass could do after falling 1.2 m. When this energy is released very quickly, which is possible in an electric circuit, it provides an *enormous* amount of power.

The usefulness of a capacitor stems from the fact that it can be charged slowly (the charging rate is usually limited by the battery's ability to transfer charge) but then can release the energy very quickly. A mechanical analogy would be using a crank to slowly stretch the spring of a catapult, then quickly releasing the energy to launch a massive rock.

The capacitor described in Example 26.8 is typical of the capacitors used in the flash units of cameras. The camera batteries charge a capacitor, then the energy stored in the capacitor is quickly discharged into a *flashlamp*. The charging process in a camera takes several seconds, which is why you can't fire a camera flash twice in quick succession.

An important medical application of capacitors is the *defibrillator*. A heart attack or a serious injury can cause the heart to enter a state known as *fibrillation* in which the heart muscles twitch randomly and cannot pump blood. A strong electric shock through the chest completely stops the heart, giving the cells that control the heart's rhythm a chance to restore the proper heartbeat. A defibrillator has a large capacitor that can store

A defibrillator, which can restore a normal heartbeat, discharges a capacitor through the patient's chest.

up to 360 J of energy. This energy is released in about 2 ms through two "paddles" pressed against the patient's chest. It takes several seconds to charge the capacitor, which is why, on television medical shows, you hear an emergency room doctor or nurse shout "Charging!"

The Energy in the Electric Field

We can "see" the potential energy of a stretched spring in the tension of the coils. If a charged capacitor is analogous to a stretched spring, where is the stored energy? It's in the electric field!

FIGURE 26.28 shows a parallel-plate capacitor in which the plates have area A and are separated by distance d. The potential difference across the capacitor is related to the electric field inside the capacitor by $\Delta V_C = Ed$. The capacitance, which we found in Equation 26.18, is $C = \epsilon_0 A/d$. Substituting these into Equation 26.25, we find that the energy stored in the capacitor is

$$U_C = \tfrac{1}{2} C(\Delta V_C)^2 = \frac{\epsilon_0 A}{2d}(Ed)^2 = \frac{\epsilon_0}{2}(Ad)E^2 \qquad (26.26)$$

The quantity Ad is the volume *inside* the capacitor, the region in which the capacitor's electric field exists. (Recall that an ideal capacitor has $\vec{E} = \vec{0}$ everywhere except between the plates.) Although we talk about "the energy stored in the capacitor," Equation 26.26 suggests that, strictly speaking, **the energy is stored in the capacitor's electric field.**

Because Ad is the volume in which the energy is stored, we can define an **energy density** u_E of the electric field:

$$u_E = \frac{\text{energy stored}}{\text{volume in which it is stored}} = \frac{U_C}{Ad} = \frac{\epsilon_0}{2}E^2 \qquad (26.27)$$

The energy density has units J/m^3. We've derived Equation 26.27 for a parallel-plate capacitor, but it turns out to be the correct expression for any electric field.

From this perspective, charging a capacitor stores energy in the capacitor's electric field as the field grows in strength. Later, when the capacitor is discharged, the energy is released as the field collapses.

We first introduced the electric field as a way to visualize how a long-range force operates. But if the field can store energy, the field must be real, not merely a pictorial device. We'll explore this idea further in Chapter 31, where we'll find that the energy transported by a light wave—the very real energy of warm sunshine—is the energy of electric and magnetic fields.

26.7 Dielectrics

FIGURE 26.29a shows a parallel-plate capacitor with the plates separated by vacuum, the perfect insulator. Suppose the capacitor is charged to voltage $(\Delta V_C)_0$, then disconnected from the battery. The charge on the plates will be $\pm Q_0$, where $Q_0 = C_0(\Delta V_C)_0$. We'll use a subscript 0 in this section to refer to a vacuum-insulated capacitor.

FIGURE 26.28 A capacitor's energy is stored in the electric field.

Capacitor plate with area A

The capacitor's energy is stored in the electric field in volume Ad between the plates.

FIGURE 26.29 Vacuum-insulated and dielectric-filled capacitors.

(a)

$(\Delta V_C)_0$

$-Q_0$ $+Q_0$

d

Capacitance C_0 in vacuum

(b)

Dielectric

ΔV_C $+Q_0$

Capacitance $C > C_0$

Now suppose, as in **FIGURE 26.29b**, an insulating material, such as oil or glass or plastic, is slipped between the capacitor plates. We'll assume for now that the insulator is of thickness d and completely fills the space. An insulator in an electric field is called a **dielectric,** for reasons that will soon become clear, so we call this a *dielectric-filled capacitor.* How does a dielectric-filled capacitor differ from the vacuum-insulated capacitor?

The charge on the capacitor plates does not change. The insulator doesn't allow charge to move through it, and the capacitor has been disconnected from the battery, so no charge can be added to or removed from either plate. That is, $Q = Q_0$. Nonetheless, measurements of the capacitor voltage with a voltmeter would find that the voltage has decreased: $\Delta V_C < (\Delta V_C)_0$. Consequently, based on the definition of capacitance, the capacitance has increased:

$$C = \frac{Q}{\Delta V_C} > \frac{Q_0}{(\Delta V_C)_0} = C_0$$

Example 26.6 found that the plate size needed to make a 1 μF capacitor is unreasonably large. It appears that we can get more capacitance *with the same plates* by filling the capacitor with an insulator.

We can utilize two tools you learned in Chapter 23, superposition and polarization, to understand the properties of dielectric-filled capacitors. Figure 23.27 showed how an insulating material becomes *polarized* in an external electric field. **FIGURE 26.30a** reproduces the basic ideas from that earlier figure. The electric dipoles in Figure 26.30a could be permanent dipoles, such as water molecules, or simply induced dipoles due to a slight charge separation in the atoms. However the dipoles originate, their alignment in the electric field—the *polarization* of the material—produces an excess positive charge on one surface, an excess negative charge on the other. The insulator as a whole is still neutral, but the external electric field separates positive and negative charge.

FIGURE 26.30b represents the polarized insulator as simply two sheets of charge with surface charge densities $\pm \eta_{\text{induced}}$. The size of η_{induced} depends both on the strength of the electric field and on the properties of the insulator. These two sheets of charge create an electric field—a situation we analyzed in Chapter 23. In essence, the two sheets of induced charge act just like the two charged plates of a parallel-plate capacitor. The **induced electric field** (keep in mind that this field is due to the insulator responding to the external electric field) is

$$\vec{E}_{\text{induced}} = \begin{cases} \left(\dfrac{\eta_{\text{induced}}}{\epsilon_0}, \text{ from positive to negative} \right) & \text{inside the insulator} \\ \vec{0} & \text{outside the insulator} \end{cases} \quad (26.28)$$

It is because an insulator in an electric field has *two* sheets of induced *electric* charge that we call it a *dielectric,* with the prefix *di,* meaning *two,* the same as in "diatomic" and "dipole."

Inserting a Dielectric into a Capacitor

FIGURE 26.31 on the next page shows what happens when you insert a dielectric into a capacitor. The capacitor plates have their own surface charge density $\eta_0 = Q_0/A$. This creates the electric field $\vec{E}_0 = (\eta_0/\epsilon_0$, from positive to negative) into which the dielectric is placed. The dielectric responds with induced surface charge density η_{induced} and the induced electric field $\vec{E}_{\text{induced}}$. Notice that $\vec{E}_{\text{induced}}$ points *opposite* to $\vec{E}_0$. By the principle of superposition, another important lesson from Chapter 23, the net electric field between the capacitor plates is the *vector* sum of these two fields:

$$\vec{E} = \vec{E}_0 + \vec{E}_{\text{induced}} = (E_0 - E_{\text{induced}}, \text{ from positive to negative}) \quad (26.29)$$

The presence of the dielectric weakens the electric field, from E_0 to $E_0 - E_{\text{induced}}$, but the field still points from the positive capacitor plate to the negative capacitor plate. The field is weakened because the induced surface charge in the dielectric acts to counter the electric field of the capacitor plates.

FIGURE 26.30 An insulator in an external electric field.

(a) The insulator is polarized.

Excess positive charge on this surface   Excess negative charge on this surface

$\vec{E}_0$ $\vec{E}_0$

(b) The polarized insulator—a dielectric—can be represented as two sheets of surface charge. This surface charge creates an electric field inside the insulator.

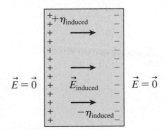

$\vec{E} = \vec{0}$ η_{induced} $\vec{E}_{\text{induced}}$ $\vec{E} = \vec{0}$ $-\eta_{\text{induced}}$

FIGURE 26.31 The consequences of filling a capacitor with a dielectric.

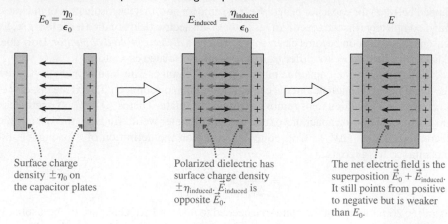

Surface charge density $\pm\eta_0$ on the capacitor plates

Polarized dielectric has surface charge density $\pm\eta_{\text{induced}} \cdot \vec{E}_{\text{induced}}$ is opposite $\vec{E}_0$.

The net electric field is the superposition $\vec{E}_0 + \vec{E}_{\text{induced}}$. It still points from positive to negative but is weaker than E_0.

Let's define the **dielectric constant** κ (Greek *kappa*) as

$$\kappa \equiv \frac{E_0}{E} \tag{26.30}$$

Equivalently, the field strength inside a dielectric in an external field is $E = E_0/\kappa$. The dielectric constant is the factor by which a dielectric *weakens* an electric field, so $\kappa \geq 1$. You can see from the definition that κ is a pure number with no units.

The dielectric constant, like density or specific heat, is a property of a material. Easily polarized materials have larger dielectric constants than materials not easily polarized. Vacuum has $\kappa = 1$ exactly, and low-pressure gases have $\kappa \approx 1$. (Air has $\kappa_{\text{air}} = 1.00$ to three significant figures, so we won't worry about the very slight effect air has on capacitors.) **TABLE 26.1** lists the dielectric constants for different materials.

The electric field inside the capacitor, although weakened, is still uniform. Consequently, the potential difference across the capacitor is

$$\Delta V_C = Ed = \frac{E_0}{\kappa}d = \frac{(\Delta V_C)_0}{\kappa} \tag{26.31}$$

where $(\Delta V_C)_0 = E_0 d$ was the voltage of the vacuum-insulated capacitor. The presence of a dielectric reduces the capacitor voltage, the observation with which we started this section. Now we see why; it is due to the polarization of the material. Further, the new capacitance is

$$C = \frac{Q}{\Delta V_C} = \frac{Q_0}{(\Delta V_C)_0/\kappa} = \kappa \frac{Q_0}{(\Delta V_C)_0} = \kappa C_0 \tag{26.32}$$

Filling a capacitor with a dielectric increases the capacitance by a factor equal to the dielectric constant. This ranges from virtually no increase for an air-filled capacitor to a capacitance 300 times larger if the capacitor is filled with strontium titanate.

We'll leave it as a homework problem to show that the induced surface charge density is

$$\eta_{\text{induced}} = \eta_0\left(1 - \frac{1}{\kappa}\right) \tag{26.33}$$

η_{induced} ranges from nearly zero when $\kappa \approx 1$ to $\approx \eta_0$ when $\kappa \gg 1$.

NOTE We assumed that the capacitor was disconnected from the battery after being charged, so Q couldn't change. If you insert a dielectric while a capacitor is attached to a battery, then it will be ΔV_C, fixed at the battery voltage, that can't change. In this case, more charge will flow from the battery until $Q = \kappa Q_0$. In both cases, the capacitance increases to $C = \kappa C_0$.

TABLE 26.1 Properties of dielectrics

Material	Dielectric constant κ	Dielectric strength $E_{\max}(10^6 \text{ V/m})$
Vacuum	1	—
Air (1 atm)	1.0006	3
Teflon	2.1	60
Polystyrene plastic	2.6	24
Mylar	3.1	7
Paper	3.7	16
Pyrex glass	4.7	14
Pure water (20°C)	80	—
Titanium dioxide	110	6
Strontium titanate	300	8

EXAMPLE 26.9 | A water-filled capacitor

A 5.0 nF parallel-plate capacitor is charged to 160 V. It is then disconnected from the battery and immersed in distilled water. What are (a) the capacitance and voltage of the water-filled capacitor and (b) the energy stored in the capacitor before and after its immersion?

MODEL Pure distilled water is a good insulator. (The conductivity of tap water is due to dissolved ions.) Thus the immersed capacitor has a dielectric between the electrodes.

SOLVE a. From Table 26.1, the dielectric constant of water is $\kappa = 80$. The presence of the dielectric increases the capacitance to

$$C = \kappa C_0 = 80 \times 5.0 \text{ nF} = 400 \text{ nF}$$

At the same time, the voltage decreases to

$$\Delta V_C = \frac{(\Delta V_C)_0}{\kappa} = \frac{160 \text{ V}}{80} = 2.0 \text{ V}$$

b. The presence of a dielectric does not alter the derivation leading to Equation 26.25 for the energy stored in a capacitor. Right after being disconnected from the battery, the stored energy was

$$(U_C)_0 = \tfrac{1}{2} C_0 (\Delta V_C)_0^2 = \tfrac{1}{2} (5.0 \times 10^{-9} \text{ F})(160 \text{ V})^2 = 6.4 \times 10^{-5} \text{ J}$$

After being immersed, the stored energy is

$$U_C = \tfrac{1}{2} C (\Delta V_C)^2 = \tfrac{1}{2} (400 \times 10^{-9} \text{ F})(2.0 \text{ V})^2 = 8.0 \times 10^{-7} \text{ J}$$

ASSESS Water, with its large dielectric constant, has a *big* effect on the capacitor. But where did the energy go? We learned in Chapter 23 that a dipole is drawn into a region of stronger electric field. The electric field inside the capacitor is much stronger than just outside the capacitor, so the polarized dielectric is actually *pulled* into the capacitor. The "lost" energy is the work the capacitor's electric field did pulling in the dielectric.

EXAMPLE 26.10 | Energy density of a defibrillator

A defibrillator unit contains a 150 μF capacitor that is charged to 2100 V. The capacitor plates are separated by a 0.050-mm-thick insulator with dielectric constant 120.

a. What is the area of the capacitor plates?

b. What are the stored energy and the energy density in the electric field when the capacitor is charged?

MODEL Model the defibrillator as a parallel-plate capacitor with a dielectric.

SOLVE a. The capacitance of a parallel-plate capacitor in a vacuum is $C_0 = \epsilon_0 A/d$. A dielectric increases the capacitance by the factor κ, to $C = \kappa C_0$, so the area of the capacitor plates is

$$A = \frac{Cd}{\kappa \epsilon_0} = \frac{(150 \times 10^{-6} \text{ F})(5.0 \times 10^{-5} \text{ m})}{120(8.85 \times 10^{-12} \text{ C}^2/\text{N m}^2)} = 7.1 \text{ m}^2$$

Although the surface area is very large, Figure 26.32 shows how very large sheets of very thin metal can be rolled up into capacitors that you hold in your hand.

b. The energy stored in the capacitor is

$$U_C = \tfrac{1}{2} C (\Delta V_C)^2 = \tfrac{1}{2}(150 \times 10^{-6} \text{ F})(2100 \text{ V})^2 = 330 \text{ J}$$

Because the dielectric has increased C by a factor of κ, the energy density of Equation 26.27 is increased by a factor of κ to $u_E = \tfrac{1}{2}\kappa\epsilon_0 E^2$. The electric field strength in the capacitor is

$$E = \frac{\Delta V_C}{d} = \frac{2100 \text{ V}}{5.0 \times 10^{-5} \text{ m}} = 4.2 \times 10^7 \text{ V/m}$$

Consequently, the energy density is

$$u_E = \tfrac{1}{2}(120)(8.85 \times 10^{-12} \text{ C}^2/\text{N m}^2)(4.2 \times 10^7 \text{ V/m})^2$$
$$= 9.4 \times 10^5 \text{ J/m}^3$$

ASSESS 330 J is a substantial amount of energy—equivalent to that of a 1 kg mass traveling at 25 m/s. And it can be delivered very quickly as the capacitor is discharged through the patient's chest.

Solid or liquid dielectrics allow a set of electrodes to have more capacitance than they would if filled with air. Not surprisingly, as **FIGURE 26.32** shows, this is important in the production of practical capacitors. In addition, dielectrics allow capacitors to be charged to higher voltages. All materials have a maximum electric field they can sustain without *breakdown*—the production of a spark. The breakdown electric field of air, as we've noted previously, is about 3×10^6 V/m. In general, a material's maximum sustainable electric field is called its **dielectric strength.** Table 26.1 includes dielectric strengths for air and the solid dielectrics. (The breakdown of water is extremely sensitive to ions and impurities in the water, so water doesn't have a well-defined dielectric strength.)

Many materials have dielectric strengths much larger than air. Teflon, for example, has a dielectric strength 20 times that of air. Consequently, a Teflon-filled capacitor can be safely charged to a voltage 20 times larger than an air-filled capacitor with the same plate separation. An air-filled capacitor with a plate separation of 0.2 mm can be charged only to 600 V, but a capacitor with a 0.2-mm-thick Teflon sheet could be charged to 12,000 V.

FIGURE 26.32 A practical capacitor.

Metal foil

Dielectric

Many real capacitors are a rolled-up sandwich of metal foils and thin, insulating dielectrics.

CHALLENGE **EXAMPLE 26.11** | The Geiger counter: A cylindrical capacitor

The radiation detector known as a *Geiger counter* consists of a 25-mm-diameter cylindrical metal tube, sealed at the ends, with a 1.0-mm-diameter wire along its axis. The wire and cylinder are separated by a low-pressure gas whose dielectric strength is 1.0×10^6 V/m.

a. What is the capacitance per unit length?

b. What is the maximum potential difference between the wire and the tube?

MODEL Model the Geiger counter as two infinitely long, concentric, conducting cylinders. Applying a potential difference between the cylinders charges them like a capacitor; indeed, they *are* a cylindrical capacitor. The charge on an infinitely long conductor is infinite, but the linear charge density λ is the charge per unit length (C/m), which is finite. So for a cylindrical capacitor we compute the capacitance per unit length $C = \lambda/\Delta V$, in F/m, rather than an absolute capacitance. To avoid breakdown of the gas, the field strength at the surface of the wire—the point of maximum field strength—must not exceed the dielectric strength.

VISUALIZE FIGURE 26.33 shows a cross section of the Geiger counter tube. We've chosen to let the outer cylinder be positive, with an inward-pointing electric field, but a negative outer cylinder would lead to the same answer since it's only the field strength that we're interested in.

FIGURE 26.33 Cross section of a Geiger counter tube.

Linear charge density $-\lambda$
Radius R_1

Linear charge density $+\lambda$
Radius R_2

$\vec{E}$ $\vec{E}$

SOLVE a. Gauss's law tells us that the electric field between the cylinders is due only to the charge on the inner cylinder. Thus $\vec{E}$ is the field of a long, charged wire—a field we found in Chapter 23 using superposition and again in Chapter 24 using Gauss's law. It is

$$\vec{E} = \left(\frac{\lambda}{2\pi\epsilon_0 r}, \text{ inward} \right)$$

where λ is the magnitude of the linear charge density. We need to connect this field to the potential difference between the wire and the outer cylinder. For that, we need to use Equation 26.3:

$$\Delta V = V_f - V_i = -\int_{s_i}^{s_f} E_s \, ds$$

We'll integrate along a radial line from $s_i = R_1$ on the surface of the inner cylinder to $s_f = R_2$ at the outer cylinder. The field component E_s is negative because the field points inward. Thus the potential difference is

$$\Delta V = -\int_{R_1}^{R_2} \left(-\frac{\lambda}{2\pi\epsilon_0 s} \right) ds = \frac{\lambda}{2\pi\epsilon_0} \int_{R_1}^{R_2} \frac{ds}{s}$$

$$= \frac{\lambda}{2\pi\epsilon_0} \ln s \Big|_{R_1}^{R_2} = \frac{\lambda}{2\pi\epsilon_0} \ln\left(\frac{R_2}{R_1} \right)$$

We see that the applied potential difference and the linear charge density are related by

$$\frac{\lambda}{2\pi\epsilon_0} = \frac{\Delta V}{\ln(R_2/R_1)}$$

Thus the capacitance per unit length is

$$C = \frac{\lambda}{\Delta V_C} = \frac{2\pi\epsilon_0}{\ln(R_2/R_1)}$$

$$= \frac{2\pi(8.85 \times 10^{-12} \text{ C}^2/\text{N m}^2)}{\ln(25)} = 17 \text{ pF/m}$$

You should convince yourself that the units of ϵ_0 are equivalent to F/m.

b. Using the above in the expression for $\vec{E}$, we find the electric field strength at distance r is

$$E = \frac{\Delta V}{r \ln(R_2/R_1)}$$

The field strength is a maximum at the surface of the wire, where it reaches

$$E_{max} = \frac{\Delta V}{R_1 \ln(R_2/R_1)}$$

The maximum applied voltage will bring E_{max} to the dielectric strength, $E_{max} = 1.0 \times 10^6$ V/m. Thus the maximum potential difference between the wire and the tube is

$$\Delta V_{max} = R_1 E_{max} \ln\left(\frac{R_2}{R_1} \right)$$

$$= (5.0 \times 10^{-4} \text{ m})(1.0 \times 10^6 \text{ V/m}) \ln(25)$$

$$= 1600 \text{ V}$$

ASSESS This is the *maximum* possible voltage, but it's not practical to operate right at the maximum. Real Geiger counters operate with typically a 1000 V potential difference to avoid an accidental breakdown of the gas. If a high-speed charged particle from a radioactive decay then happens to pass through the tube, it will collide with and ionize a number of the gas atoms. Because the tube is already very close to breakdown, the addition of these extra ions and electrons is enough to push it over the edge: A breakdown of the gas occurs, with a spark jumping across the tube. The "clicking" sounds of a Geiger counter are made by amplifying the current pulses associated with the sparks.

SUMMARY

The goal of Chapter 26 has been to learn how the electric potential is related to the electric field.

GENERAL PRINCIPLES

Connecting V and $\vec{E}$

The electric potential and the electric field are two different perspectives of how source charges alter the space around them. V and $\vec{E}$ are related by

$$\Delta V = V_f - V_i = -\int_{s_i}^{s_f} E_s \, ds$$

where s is measured from point i to point f and E_s is the component of $\vec{E}$ parallel to the line of integration.

Graphically

$\Delta V =$ the negative of the area under the E_s graph

$E_s = -\dfrac{dV}{ds}$

= the negative of the slope of the potential graph

The Geometry of Potential and Field

The electric field

- Is perpendicular to the equipotential surfaces.
- Points "downhill" in the direction of decreasing V.
- Is inversely proportional to the spacing Δs between the equipotential surfaces.

Conservation of Energy

The sum of all potential differences around a closed path is zero.

$$\sum (\Delta V)_i = 0$$

IMPORTANT CONCEPTS

A **battery** is a **source of potential**. The charge escalator in a battery uses chemical reactions to move charges from the negative terminal to the positive terminal:

$$\Delta V_{bat} = \mathcal{E}$$

where the emf $\mathcal{E}$ is the work per charge done by the charge escalator.

For a **conductor in electrostatic equilibrium**

- The interior electric field is zero.
- The exterior electric field is perpendicular to the surface.
- The surface is an equipotential.
- The interior is at the same potential as the surface.

APPLICATIONS

Capacitors

The **capacitance** of two conductors charged to $\pm Q$ is

$$C = \frac{Q}{\Delta V_C}$$

A parallel-plate capacitor has

$$C = \frac{\epsilon_0 A}{d}$$

Filling the space between the plates with a **dielectric** of dielectric constant κ increases the capacitance to $C = \kappa C_0$.

The energy stored in a capacitor is $u_C = \frac{1}{2}C(\Delta V_C)^2$.

This energy is stored in the electric field at density $u_E = \frac{1}{2}\kappa\epsilon_0 E^2$.

Combinations of capacitors

Series capacitors

$$C_{eq} = \left(\frac{1}{C_1} + \frac{1}{C_2} + \frac{1}{C_3} + \cdots \right)^{-1}$$

Parallel capacitors

$$C_{eq} = C_1 + C_2 + C_3 + \cdots$$

TERMS AND NOTATION

Kirchhoff's loop law	terminal voltage, ΔV_{bat}	farad, F	dielectric
Van de Graaff generator	ideal battery	parallel capacitors	induced electric field
battery	capacitance, C	series capacitors	dielectric constant, κ
charge escalator model	capacitor	equivalent capacitance, C_{eq}	dielectric strength
emf, $\mathcal{E}$	capacitor voltage, ΔV_C	energy density, u_E	

CONCEPTUAL QUESTIONS

1. **FIGURE Q26.1** shows the x-component of $\vec{E}$ as a function of x. Draw a graph of V versus x in this same region of space. Let $V = 0$ V at $x = 0$ m and include an appropriate vertical scale.

FIGURE Q26.1

FIGURE Q26.2

2. **FIGURE Q26.2** shows the electric potential as a function of x. Draw a graph of E_x versus x in this same region of space.

3. a. Suppose that $\vec{E} = \vec{0}$ V/m throughout some region of space. Can you conclude that $V = 0$ V in this region? Explain.
   b. Suppose that $V = 0$ V throughout some region of space. Can you conclude that $\vec{E} = \vec{0}$ V/m in this region? Explain.

4. Estimate the electric fields $\vec{E}_1$ and $\vec{E}_2$ at points 1 and 2 in **FIGURE Q26.4**. Don't forget that $\vec{E}$ is a vector.

FIGURE Q26.4

5. Estimate the electric fields $\vec{E}_1$ and $\vec{E}_2$ at points 1 and 2 in **FIGURE Q26.5**. Don't forget that $\vec{E}$ is a vector.

FIGURE Q26.5

6. An electron is released from rest at $x = 2$ m in the potential shown in **FIGURE Q26.6**. Does it move? If so, to the left or to the right? Explain.

FIGURE Q26.6

FIGURE Q26.7

7. **FIGURE Q26.7** shows an electric field diagram. Dashed lines 1 and 2 are two surfaces in space, not physical objects.
   a. Is the electric potential at point a higher than, lower than, or equal to the electric potential at point b? Explain.
   b. Rank in order, from largest to smallest, the magnitudes of the potential differences ΔV_{ab}, ΔV_{cd}, and ΔV_{ef}.
   c. Is surface 1 an equipotential surface? What about surface 2? Explain why or why not.

8. **FIGURE Q26.8** shows a negatively charged electroscope. The gold leaf stands away from the rigid metal post. Is the electric potential of the leaf higher than, lower than, or equal to the potential of the post? Explain.

FIGURE Q26.8

FIGURE Q26.9

9. The two metal spheres in **FIGURE Q26.9** are connected by a metal wire with a switch in the middle. Initially the switch is open. Sphere 1, with the larger radius, is given a positive charge. Sphere 2, with the smaller radius, is neutral. Then the switch is closed. Afterward, sphere 1 has charge Q_1, is at potential V_1, and the electric field strength at its surface is E_1. The values for sphere 2 are Q_2, V_2, and E_2.
   a. Is V_1 larger than, smaller than, or equal to V_2? Explain.
   b. Is Q_1 larger than, smaller than, or equal to Q_2? Explain.
   c. Is E_1 larger than, smaller than, or equal to E_2? Explain.

10. **FIGURE Q26.10** shows a 3 V battery with metal wires attached to each end. What are the potential differences $\Delta V_{12} = V_2 - V_1$, $\Delta V_{23} = V_3 - V_2$, $\Delta V_{34} = V_4 - V_3$, and $\Delta V_{41} = V_1 - V_4$?

FIGURE Q26.10   **FIGURE Q26.11**

11. The parallel-plate capacitor in **FIGURE Q26.11** is connected to a battery having potential difference ΔV_{bat}. Without breaking any of the connections, insulating handles are used to increase the plate separation to $2d$.

a. Does the potential difference ΔV_C change as the separation increases? If so, by what factor? If not, why not?
b. Does the capacitance change? If so, by what factor? If not, why not?
c. Does the capacitor charge Q change? If so, by what factor? If not, why not?

12. Rank in order, from largest to smallest, the potential differences $(\Delta V_C)_1$ to $(\Delta V_C)_4$ of the four capacitors in **FIGURE Q26.12**. Explain.

FIGURE Q26.12

EXERCISES AND PROBLEMS

Problems labeled ▓ integrate material from earlier chapters.

Exercises

Section 26.1 Connecting Potential and Field

1. ‖ What is the potential difference between $x_i = 10$ cm and $x_f = 30$ cm in the uniform electric field $E_x = 1000$ V/m?
2. ‖ What is the potential difference between $y_i = -5$ cm and $y_f = 5$ cm in the uniform electric field $\vec{E} = (20,000\hat{\imath} - 50,000\hat{\jmath})$ V/m?
3. ‖ **FIGURE EX26.3** is a graph of E_x. What is the potential difference between $x_i = 1.0$ m and $x_f = 3.0$ m?

FIGURE EX26.3   **FIGURE EX26.4**

4. ‖ **FIGURE EX26.4** is a graph of E_x. The potential at the origin is -50 V. What is the potential at $x = 3.0$ m?
5. ❘ a. Which point in **FIGURE EX26.5**, A or B, has a larger electric potential?
   b. What is the potential difference between A and B?

FIGURE EX26.5

Section 26.2 Finding the Electric Field from the Potential

6. ❘ Two flat, parallel electrodes 2.5 cm apart are kept at potentials of 20 V and 35 V. Estimate the electric field strength between them.

7. ❘ What are the magnitude and direction of the electric field at the dot in **FIGURE EX26.7**?

FIGURE EX26.7   **FIGURE EX26.8**

8. ❘ What are the magnitude and direction of the electric field at the dot in **FIGURE EX26.8**?
9. ❘ **FIGURE EX26.9** shows a graph of V versus x in a region of space. The potential is independent of y and z. What is E_x at (a) $x = -2$ cm, (b) $x = 0$ cm, and (c) $x = 2$ cm?

FIGURE EX26.9   **FIGURE EX26.10**

10. ❘ Determine the magnitude and direction of the electric field at points 1 and 2 in **FIGURE EX26.10**.

11. ‖ **FIGURE EX26.11** is a graph of V versus x. Draw the corresponding graph of E_x versus x.

FIGURE EX26.11   FIGURE EX26.12

12. ‖ **FIGURE EX26.12** is a graph of V versus x. Draw the corresponding graph of E_x versus x.

13. ‖ The electric potential in a region of uniform electric field is -1000 V at $x = -1.0$ m and $+1000$ V at $x = +1.0$ m. What is E_x?

14. ‖ **CALC** The electric potential along the x-axis is $V = 100x^2$ V, where x is in meters. What is E_x at (a) $x = 0$ m and (b) $x = 1$ m?

15. ‖ **CALC** The electric potential along the x-axis is $V = 100e^{-2x}$ V, where x is in meters. What is E_x at (a) $x = 1.0$ m and (b) $x = 2.0$ m?

16. ❘ What is the potential difference ΔV_{34} in **FIGURE EX26.16**?

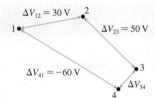

FIGURE EX26.16

Section 26.4 Sources of Electric Potential

17. ❘ How much work does the charge escalator do to move 1.0 μC of charge from the negative terminal to the positive terminal of a 1.5 V battery?

18. ❘ How much charge does a 9.0 V battery transfer from the negative to the positive terminal while doing 27 J of work?

19. ❘ How much work does the electric motor of a Van de Graaff generator do to lift a positive ion ($q = e$) if the potential of the spherical electrode is 1.0 MV?

20. ❘ Light from the sun allows a solar cell to move electrons from the positive to the negative terminal, doing 2.4×10^{-19} J of work per electron. What is the emf of this solar cell?

Section 26.5 Capacitance and Capacitors

21. ‖ Two 3.0-cm-diameter aluminum electrodes are spaced 0.50 mm apart. The electrodes are connected to a 100 V battery.
   a. What is the capacitance?
   b. What is the magnitude of the charge on each electrode?

22. ‖ What is the capacitance of the two metal spheres shown in **FIGURE EX26.22**?

FIGURE EX26.22

23. ❘ You need to construct a 100 pF capacitor for a science project. You plan to cut two $L \times L$ metal squares and insert small spacers between their corners. The thinnest spacers you have are 0.20 mm thick. What is the proper value of L?

24. ❘ A switch that connects a battery to a 10 μF capacitor is closed. Several seconds later you find that the capacitor plates are charged to ± 30 μC. What is the emf of the battery?

25. ❘ A 6 μF capacitor, a 10 μF capacitor, and a 16 μF capacitor are connected in series. What is their equivalent capacitance?

26. ❘ A 6 μF capacitor, a 10 μF capacitor, and a 16 μF capacitor are connected in parallel. What is their equivalent capacitance?

27. ❘ What is the equivalent capacitance of the three capacitors in **FIGURE EX26.27**?

FIGURE EX26.27   FIGURE EX26.28

28. ❘ What is the equivalent capacitance of the three capacitors in **FIGURE EX26.28**?

29. ❘ You need a capacitance of 50 μF, but you don't happen to have a 50 μF capacitor. You do have a 30 μF capacitor. What additional capacitor do you need to produce a total capacitance of 50 μF? Should you join the two capacitors in parallel or in series?

30. ❘ You need a capacitance of 50 μF, but you don't happen to have a 50 μF capacitor. You do have a 75 μF capacitor. What additional capacitor do you need to produce a total capacitance of 50 μF? Should you join the two capacitors in parallel or in series?

Section 26.6 The Energy Stored in a Capacitor

31. ‖ To what potential should you charge a 1.0 μF capacitor to store 1.0 J of energy?

32. ‖ 50 pJ of energy is stored in a 2.0 cm $\times$ 2.0 cm $\times$ 2.0 cm region of uniform electric field. What is the electric field strength?

33. ‖ A 2.0-cm-diameter parallel-plate capacitor with a spacing of 0.50 mm is charged to 200 V. What are (a) the total energy stored in the electric field and (b) the energy density?

34. ‖ **BIO** The 90 μF capacitor in a defibrillator unit supplies an average of 6500 W of power to the chest of the patient during a discharge lasting 5.0 ms. To what voltage is the capacitor charged?

Section 26.7 Dielectrics

35. ‖ Two 4.0 cm $\times$ 4.0 cm metal plates are separated by a 0.20-mm-thick piece of Teflon.
   a. What is the capacitance?
   b. What is the maximum potential difference between the plates?

36. ‖ Two 5.0 mm $\times$ 5.0 mm electrodes are held 0.10 mm apart and are attached to a 9.0 V battery. Without disconnecting the battery, a 0.10-mm-thick sheet of Mylar is inserted between the electrodes. What are the capacitor's potential difference, electric field, and charge (a) before and (b) after the Mylar is inserted?
   Hint: Section 26.7 considered a capacitor with *isolated* plates. What changes, and what doesn't, when the plates stay connected to the battery?

37. || A typical cell has a layer of negative charge on the inner
BIO surface of the cell wall and a layer of positive charge on the
outside surface, thus making the cell wall a capacitor. What is
the capacitance of a 50-μm-diameter cell with a 7.0-nm-thick
cell wall whose dielectric constant is 9.0? Because the
cell's diameter is much larger than the wall thickness, it is reasonable to ignore the curvature of the cell and think of it as a
parallel-plate capacitor.

Problems

38. || The electric field in a region of space is $E_x = 5000x$ V/m,
CALC where x is in meters.
   a. Graph E_x versus x over the region -1 m $\leq x \leq 1$ m.
   b. Find an expression for the potential V at position x. As a
      reference, let $V = 0$ V at the origin.
   c. Graph V versus x over the region -1 m $\leq x \leq 1$ m.
39. ||| The electric field in a region of space is $E_x = -1000x$ V/m,
CALC where x is in meters.
   a. Graph E_x versus x over the region -1 m $\leq x \leq 1$ m.
   b. What is the potential difference between $x_i = -20$ cm and
      $x_f = 30$ cm?
40. || An infinitely long cylinder of radius R has linear charge den-
CALC sity λ. The potential on the surface of the cylinder is V_0, and
the electric field outside the cylinder is $E_r = \lambda/2\pi\epsilon_0 r$. Find the
potential relative to the surface at a point that is distance r from
the axis, assuming $r > R$.
41. || **FIGURE P26.41** is an edge view of three charged metal
electrodes. Let the left electrode be the zero point of the electric
potential. What are V and $\vec{E}$ at (a) $x = 0.5$ cm, (b) $x = 1.5$ cm,
and (c) $x = 2.5$ cm?

FIGURE P26.41

42. || Use the on-axis potential of a charged disk from Chapter 25
to find the on-axis electric field of a charged disk.
43. || a. Use the methods of
      Chapter 25 to find the
      potential at distance
      x on the axis of the
      charged rod shown in
      FIGURE P26.43.
   b. Use the result of part a
      to find the electric field
      at distance x on the axis of a rod.

FIGURE P26.43

44. || It is postulated that the radial electric field of a group of
charges falls off as $E_r = C/r^n$, where C is a constant, r is the
distance from the center of the group, and n is an unknown
exponent. To test this hypothesis, you make a *field probe*
consisting of two needle tips spaced 1.00 mm apart. You
orient the needles so that a line between the tips points to the
center of the charges, then use a voltmeter to read the potential
difference between the tips. After you take measurements
at several distances from the center of the group, your data are
as follows:

Distance (cm)	Potential difference (mV)
2.0	34.7
4.0	6.6
6.0	2.1
8.0	1.2
10.0	0.6

Use an appropriate graph of the data to determine the constants
C and n.

45. || Engineers discover that the electric potential between two
CALC electrodes can be modeled as $V(x) = V_0 \ln(1 + x/d)$, where V_0 is
a constant, x is the distance from the first electrode in the direction
of the second, and d is the distance between the electrodes. What
is the electric field strength midway between the electrodes?
46. || The electric potential in a region of space is $V = (150x^2 -
CALC 200y^2)$ V, where x and y are in meters. What are the strength and
direction of the electric field at $(x, y) = (2.0$ m, 2.0 m$)$? Give the
direction as an angle cw or ccw (specify which) from the positive
x-axis.
47. || The electric potential in a region of space is $V = 200/\sqrt{x^2 + y^2}$,
CALC where x and y are in meters. What are the strength and direction of
the electric field at $(x, y) = (2.0$ m, 1.0 m$)$? Give the direction as
an angle cw or ccw (specify which) from the positive x-axis.
48. || Consider a large, thin, electrically neutral conducting plate in
CALC the xy-plane at $z = 0$ and a point charge q on the z-axis at distance
$z_{charge} = d$. What is the electric field on and above the plate?
Although the plate is neutral, electric forces from the point charge
polarize the conducting plate and cause it to have some complex
distribution of surface charge. The electric field and potential are then
a superposition of fields and potentials due to the point charge *and*
the plate's surface charge. That's complicated! However, it is shown
in more advanced classes that the field and potential outside the plate
$(z \geq 0)$ are exactly the same as the field and potential of the original
charge q plus a "mirror image" charge $-q$ located at $z_{image} = -d$.
   a. Find an expression for the electric potential in the yz-plane for
      $z \geq 0$ (i.e., in the space above the plate).
   b. We know that electric fields are perpendicular to conductors
      in electrostatic equilibrium, so the field at the surface of the
      plate has only a z-component. Find an expression for the field
      E_z on the surface of the plate $(z = 0)$ as a function of distance
      y away from the z-axis.
   c. A $+10$ nC point charge is 2.0 cm above a large conducting
      plate. What is the electric field strength at the surface of the
      plate (i) directly beneath the point charge and (ii) 2.0 cm away
      from being directly beneath the point charge?
49. || Metal sphere 1 has a positive charge of 6.0 nC. Metal sphere
2, which is twice the diameter of sphere 1, is initially uncharged.
The spheres are then connected together by a long, thin metal
wire. What are the final charges on each sphere?
50. || The metal spheres in **FIGURE P26.50** are charged to ± 300 V.
Draw this figure on your paper, then draw a plausible contour
map of the potential, showing and labeling the -300 V, -200 V,
-100 V, $\ldots$, 300 V equipotential surfaces.

FIGURE P26.50

51. || The potential at the center of a 4.0-cm-diameter copper
sphere is 500 V, relative to $V = 0$ V at infinity. How much excess
charge is on the sphere?

52. ‖ The electric potential is 40 V at point A near a uniformly charged sphere. At point B, 2.0 μm farther away from the sphere, the potential has decreased by 0.16 mV. How far is point A from the center of the sphere?

53. ‖ Two 2.0 cm $\times$ 2.0 cm metal electrodes are spaced 1.0 mm apart and connected by wires to the terminals of a 9.0 V battery.
    a. What are the charge on each electrode and the potential difference between them?
    The wires are disconnected, and insulated handles are used to pull the plates apart to a new spacing of 2.0 mm.
    b. What are the charge on each electrode and the potential difference between them?

54. ‖ Two 2.0 cm $\times$ 2.0 cm metal electrodes are spaced 1.0 mm apart and connected by wires to the terminals of a 9.0 V battery.
    a. What are the charge on each electrode and the potential difference between them?
    While the plates are still connected to the battery, insulated handles are used to pull them apart to a new spacing of 2.0 mm.
    b. What are the charge on each electrode and the potential difference between them?

55. ‖ Find expressions for the equivalent capacitance of (a) N identical capacitors C in parallel and (b) N identical capacitors C in series.

56. ‖ What are the charge on and the potential difference across each capacitor in FIGURE P26.56?

FIGURE P26.56   FIGURE P26.57

57. ‖ What are the charge on and the potential difference across each capacitor in FIGURE P26.57?

58. ‖ What are the charge on and the potential difference across each capacitor in FIGURE P26.58?

FIGURE P26.58

59. ‖ You have three 12 μF capacitors. Draw diagrams showing how you could arrange all three so that their equivalent capacitance is (a) 4.0 μF, (b) 8.0 μF, (c) 18 μF, and (d) 36 μF.

60. ‖ Six identical capacitors with capacitance C are connected as shown in FIGURE P26.60.
    a. What is the equivalent capacitance of these six capacitors?
    b. What is the potential difference between points a and b?

FIGURE P26.60

61. ‖ Initially, the switch in FIGURE P26.61 is in position A and capacitors C_2 and C_3 are uncharged. Then the switch is flipped to position B. Afterward, what are the charge on and the potential difference across each capacitor?

FIGURE P26.61   FIGURE P26.62

62. ‖ A battery with an emf of 60 V is connected to the two capacitors shown in FIGURE P26.62. Afterward, the charge on capacitor 2 is 450 μC. What is the capacitance of capacitor 2?

63. ‖ Capacitors $C_1 = 10$ μF and $C_2 = 20$ μF are each charged to 10 V, then disconnected from the battery without changing the charge on the capacitor plates. The two capacitors are then connected in parallel, with the positive plate of C_1 connected to the negative plate of C_2 and vice versa. Afterward, what are the charge on and the potential difference across each capacitor?

64. ‖ An isolated 5.0 μF parallel-plate capacitor has 4.0 mC of charge. An external force changes the distance between the electrodes until the capacitance is 2.0 μF. How much work is done by the external force?

65. ‖ An ideal parallel-plate capacitor has a uniform electric field between the plates, zero field outside. By superposition, half the field strength is due to one plate and half due to the other.
    a. The plates of a parallel-plate capacitor are oppositely charged and attract each other. Find an expression in terms of C, ΔV_C, and the plate separation d for the force one plate exerts on the other.
    b. What is the attractive force on each plate of a 100 pF capacitor with a 1.0 mm plate spacing when charged to 1000 V?

66. ‖‖ High-frequency signals are often
CALC transmitted along a *coaxial cable,* such as the one shown in FIGURE P26.66. For example, the cable TV hookup coming into your home is a coaxial cable. The signal is carried on a wire of radius R_1 while the outer conductor of radius R_2 is grounded (i.e., at $V = 0$ V). An insulating material fills the space between them, and an insulating plastic coating goes around the outside.

Coaxial cable

FIGURE P26.66

    a. Find an expression for the capacitance per meter of a coaxial cable. Assume that the insulating material between the cylinders is air.
    b. Evaluate the capacitance per meter of a cable having $R_1 = 0.50$ mm and $R_2 = 3.0$ mm.

67. ‖ The flash unit in a camera uses a 3.0 V battery to charge a capacitor. The capacitor is then discharged through a flashlamp. The discharge takes 10 μs, and the average power dissipated in the flashlamp is 10 W. What is the capacitance of the capacitor?

68. ‖ The label rubbed off one of the capacitors you are using to build a circuit. To find out its capacitance, you place it in series with a 10 μF capacitor and connect them to a 9.0 V battery. Using your voltmeter, you measure 6.0 V across the unknown capacitor. What is the unknown capacitor's capacitance?

69. || A capacitor being charged has a *current* carrying charge to and
CALC away from the plates. In the next chapter we will define current to
be dQ/dt, the rate of charge flow. What is the current to a 10 μF
capacitor whose voltage is increasing at the rate of 2.0 V/s?

70. || The current that charges a capacitor transfers energy that is
CALC stored in the capacitor's electric field. Consider a 2.0 μF capacitor,
initially uncharged, that is storing energy at a constant 200 W rate.
What is the capacitor voltage 2.0 μs after charging begins?

71. || A typical cell has a membrane potential of -70 mV, meaning
BIO that the potential inside the cell is 70 mV less than the potential
outside due to a layer of negative charge on the inner surface of
the cell wall and a layer of positive charge on the outer surface.
This effectively makes the cell wall a charged capacitor. Because
a cell's diameter is much larger than the wall thickness, it is
reasonable to ignore the curvature of the cell and think of it as a
parallel-plate capacitor. How much energy is stored in the electric
field of a 50-μm-diameter cell with a 7.0-nm-thick cell wall
whose dielectric constant is 9.0?

72. ||| A nerve cell in its resting state has a membrane potential of
BIO -70 mV, meaning that the potential inside the cell is 70 mV less
than the potential outside due to a layer of negative charge on the
inner surface of the cell wall and a layer of positive charge on the
outer surface. This effectively makes the cell wall a charged
capacitor. When the nerve cell fires, sodium ions, Na$^+$, flood
through the cell wall to briefly switch the membrane potential to
$+40$ mV. Model the central body of a nerve cell—the *soma*—as
a 50-μm-diameter sphere with a 7.0-nm-thick cell wall whose
dielectric constant is 9.0. Because a cell's diameter is much larger
than the wall thickness, it is reasonable to ignore the curvature of
the cell and think of it as a parallel-plate capacitor. How many
sodium ions enter the cell as it fires?

73. || Derive Equation 26.33 for the induced surface charge density
on the dielectric in a capacitor.

74. || A vacuum-insulated parallel-plate capacitor with plate
separation d has capacitance C_0. What is the capacitance if an
insulator with dielectric constant κ and thickness $d/2$ is slipped
between the electrodes without changing the plate separation?

In Problems 75 through 77 you are given the equation(s) used to solve
a problem. For each of these, you are to
a. Write a realistic problem for which this is the correct equation(s).
b. Finish the solution of the problem.

75. $2uz$ V/m $= -\dfrac{dV}{dz}$, where a is a constant with units of V/m^2

$V(z=0) = 10$ V

76. 400 nC $= (100$ V$)C$

$C = \dfrac{(8.85 \times 10^{-12} \text{ C}^2/\text{N m}^2)(0.10 \text{ m} \times 0.10 \text{ m})}{d}$

77. $\left(\dfrac{1}{3 \ \mu\text{F}} + \dfrac{1}{6 \ \mu\text{F}}\right)^{-1} + C = 4 \ \mu$F

Challenge Problems

78. ||| Two 5.0-cm-diameter metal disks separated by a 0.50-mm-
thick piece of Pyrex glass are charged to a potential difference of
1000 V. What are (a) the surface charge density on the disks and
(b) the surface charge density on the glass?

79. ||| An electric dipole at the origin consists of two charges $\pm q$
CALC spaced distance s apart along the y-axis.
a. Find an expression for the potential $V(x, y)$ at an arbitrary
point in the xy-plane. Your answer will be in terms of q, s,
x, and y.
b. Use the binomial approximation to simplify your result of
part a when $s \ll x$ and $s \ll y$.
c. Assuming $s \ll x$ and y, find expressions for E_x and E_y, the
components of $\vec{E}$ for a dipole.
d. What is the on-axis field $\vec{E}$? Does your result agree with
Equation 23.10?
e. What is the field $\vec{E}$ on the bisecting axis? Does your result
agree with Equation 23.11?

80. ||| Charge is uniformly distributed with charge density ρ inside
CALC a very long cylinder of radius R. Find the potential difference
between the surface and the axis of the cylinder.

81. ||| Consider a uniformly charged sphere of radius R and total
CALC charge Q. The electric field E_{out} *outside* the sphere $(r \geq R)$ is
simply that of a point charge Q. In Chapter 24, we used Gauss's
law to find that the electric field E_{in} *inside* the sphere $(r \leq R)$ is
radially outward with field strength

$$E_{in} = \frac{1}{4\pi\epsilon_0} \frac{Q}{R^3} r$$

a. The electric potential V_{out} *outside* the sphere is that of a point
charge Q. Find an expression for the electric potential V_{in} at
position r *inside* the sphere. As a reference, let $V_{in} = V_{out}$ at
the surface of the sphere.
b. What is the ratio $V_{center}/V_{surface}$?
c. Graph V versus r for $0 \leq r \leq 3R$.

82. ||| a. Find an expression for the capacitance of a *spherical
CALC capacitor*, consisting of concentric spherical shells of radii
R_1 (inner shell) and R_2 (outer shell).
b. A spherical capacitor with a 1.0 mm gap between the
shells has a capacitance of 100 pF. What are the diameters
of the two spheres?

83. ||| Each capacitor in FIGURE CP26.83 has capacitance C. What is
the equivalent capacitance between points a and b?

FIGURE CP26.83

27 Current and Resistance

A lightbulb filament is a very thin tungsten wire that is heated until it glows by passing a current through it.

IN THIS CHAPTER, you will learn how and why charge moves through a wire as a current.

What is current?
Current is the flow of charge through a conductor. We can't see charge moving, but two indicators of current are:

- A nearby compass needle is deflected.
- A wire with a current gets warm.

Current I is measured in amperes, a charge flow rate of one coulomb per second. You know this informally as "amps."

« LOOKING BACK Section 23.6 The motion of charges in electric fields

How does current flow?
We'll develop a model of conduction:

- Connecting a wire to a battery causes a nonuniform surface charge distribution.
- The surface charges create an electric field inside the wire.
- The electric field pushes the sea of electrons through the metal.
- Electrons are the charge carriers in metals, but it is customary to treat current as the motion of *positive* charges.

Current is I. Current density $J = I/A$ is the amount of current per square meter.

What law governs current?
Current is governed by Kirchhoff's junction law.

- The current is the same everywhere in a circuit with no junctions.
- The sum of currents entering a junction equals the sum leaving.

What are resistivity and resistance?
Collisions of electrons with atoms cause a conductor to *resist* the motion of charges.

- Resistivity is an electrical property of a material, such as copper.
- Resistance is a property of a specific wire or circuit element based on the material of which it is made *and* its size and shape.

What is Ohm's law?
Ohm's law says that the current flowing through a wire or circuit element depends on both the potential difference across it and the element's resistance:

$$I = \Delta V/R$$

« LOOKING BACK Section 26.4 Sources of potential

27.1 The Electron Current

We've focused thus far on situations in which charges are in static equilibrium. Now it's time to explore the *controlled motion* of charges—currents. Let's begin with a simple question: How does a capacitor get discharged? **FIGURE 27.1a** shows a charged capacitor. If, as in **FIGURE 27.1b**, we connect the two capacitor plates with a metal wire, a conductor, the plates quickly become neutral; that is, the capacitor has been *discharged*. Charge has somehow moved from one plate to the other.

FIGURE 27.1 A capacitor is discharged by a metal wire.

(a)

Isolated electrodes stay
charged indefinitely.

(b) However, a connecting
wire quickly dis-
charges the
capacitor.

The net charge of each
plate is decreasing.

(c) Current causes the
wire to get warm . . .

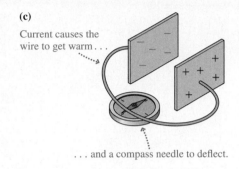

. . . and a compass needle to deflect.

In Chapter 22, we defined **current** as the motion of charges. It would seem that the capacitor is discharged by a current in the connecting wire. Let's see what else we can observe. **FIGURE 27.1c** shows that the connecting wire gets warm. If the wire is very thin in places, such as the thin filament in a lightbulb, the wire gets hot enough to glow. The current-carrying wire also deflects a compass needle, an observation we'll explore further in Chapter 29. For now, we will use "makes the wire warm" and "deflects a compass needle" as *indicators* that a current is present in a wire.

Charge Carriers

The charges that move in a conductor are called the *charge carriers*. **FIGURE 27.2** reminds you of the microscopic model of a metallic conductor that we introduced in Chapter 22. The outer electrons of metal atoms—the valence electrons—are only weakly bound to the nuclei. When the atoms come together to form a solid, the outer electrons become detached from their parent nuclei to form a fluid-like *sea of electrons* that can move through the solid. That is, **electrons are the charge carriers in metals.** Notice that the metal as a whole remains electrically neutral. This is not a perfect model because it overlooks some quantum effects, but it provides a reasonably good description of current in a metal.

> **NOTE** Electrons are the charge carriers in *metals*. Other conductors, such as ionic solutions or semiconductors, have different charge carriers. We will focus on metals because of their importance to circuits, but don't think that electrons are *always* the charge carrier.

The conduction electrons in a metal, like molecules in a gas, undergo random thermal motions, but there is no *net* motion. We can change that by pushing on the sea of electrons with an electric field, causing the entire sea of electrons to move in one direction like a gas or liquid flowing through a pipe. This net motion, which takes place at what we'll call the **drift speed** v_d, is superimposed on top of the random thermal motions of the individual electrons. The drift speed is quite small. As we'll establish later, 10^{-4} m/s is a fairly typical value for v_d.

As **FIGURE 27.3** shows, the entire sea of electrons moves from left to right at the drift speed. Suppose an observer could count the electrons as they pass through this cross section of the wire. Let's define the **electron current** i_e to be the number of electrons *per second* that pass through a cross section of a wire or other conductor. The

FIGURE 27.2 The sea of electrons is a model of electrons in a metal.

The metal as a whole is electrically neutral.

Ions (metal atoms minus valence
electrons) occupy fixed positions.

The conduction electrons are bound to the
solid as a whole, not any particular atom.
They are free to move around.

FIGURE 27.3 The electron current.

The sea of electrons flows through
a wire at the drift speed v_d.

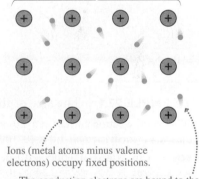

v_d

Electrons

Wire

The electron current i_e is the number
of electrons passing through this
cross section of the wire per second.

units of electron current are s^{-1}. Stated another way, the number N_e of electrons that pass through the cross section during the time interval Δt is

$$N_e = i_e \Delta t \qquad (27.1)$$

Not surprisingly, the electron current depends on the electrons' drift speed. To see how, **FIGURE 27.4** shows the sea of electrons moving through a wire at the drift speed v_d. The electrons passing through a particular cross section of the wire during the interval Δt are shaded. How many of them are there?

FIGURE 27.4 The sea of electrons moves to the right with drift speed v_d.

Wire at time t A cross section of the wire   Wire at time $t + \Delta t$ Cross-section area A

There are n_e electrons per cubic meter of wire.   The sea of electrons is moving to the right with drift speed v_d.   The sea of electrons has moved forward distance $\Delta x = v_d \Delta t$. The shaded volume is $V = A \Delta x$.

The electrons travel distance $\Delta x = v_d \Delta t$ to the right during the interval Δt, forming a cylinder of charge with volume $V = A \Delta x$. If the *number density* of conduction electrons is n_e electrons per cubic meter, then the total number of electrons in the cylinder is

$$N_e = n_e V = n_e A \Delta x = n_e A v_d \Delta t \qquad (27.2)$$

Comparing Equations 27.1 and 27.2, you can see that the electron current in the wire is

$$i_e = n_e A v_d \qquad (27.3)$$

You can increase the electron current—the number of electrons per second moving through the wire—by making them move faster, by having more of them per cubic meter, or by increasing the size of the pipe they're flowing through. That all makes sense.

In most metals, each atom contributes one valence electron to the sea of electrons. Thus the number of conduction electrons per cubic meter is the same as the number of atoms per cubic meter, a quantity that can be determined from the metal's mass density. **TABLE 27.1** gives values of the conduction-electron density n_e for several metals.

TABLE 27.1 Conduction-electron density in metals

Metal	Electron density (m^{-3})
Aluminum	6.0×10^{28}
Copper	8.5×10^{28}
Iron	8.5×10^{28}
Gold	5.9×10^{28}
Silver	5.8×10^{28}

EXAMPLE 27.1 | **The size of the electron current**

What is the electron current in a 2.0-mm-diameter copper wire if the electron drift speed is 1.0×10^{-4} m/s?

SOLVE This is a straightforward calculation. The wire's cross-section area is $A = \pi r^2 = 3.14 \times 10^{-6}$ m^2. Table 27.1 gives the electron density for copper as 8.5×10^{28} m^{-3}. Thus we find

$$i_e = n_e A v_d = 2.7 \times 10^{19} \text{ s}^{-1}$$

ASSESS This is an incredible number of electrons to pass through a section of the wire every second. The number is high not because the sea of electrons moves fast—in fact, it moves at literally a snail's pace—but because the density of electrons is so enormous. This is a fairly typical electron current.

STOP TO THINK 27.1 These four wires are made of the same metal. Rank in order, from largest to smallest, the electron currents i_a to i_d.

(a)   (b)   (c)   (d)

Discharging a Capacitor

FIGURE 27.5 shows a capacitor charged to ± 16 nC as it is being discharged by a 2.0-mm-diameter, 20-cm-long copper wire. *How long does it take* to discharge the capacitor? We've noted that a fairly typical drift speed of the electron current through a wire is 10^{-4} m/s. At this rate, it would take 2000 s, or about a half hour, for an electron to travel 20 cm.

But this isn't what happens. As far as our senses are concerned, the discharge of a capacitor is instantaneous. So what's wrong with our simple calculation?

The important point we overlooked is that the wire is *already full* of electrons. As an analogy, think of water in a hose. If the hose is already full of water, adding a drop to one end immediately (or very nearly so) pushes a drop out the other end. Likewise with the wire. As soon as the excess electrons move from the negative capacitor plate into the wire, they immediately (or very nearly so) push an equal number of electrons out the other end of the wire and onto the positive plate, thus neutralizing it. We don't have to wait for electrons to move all the way through the wire from one plate to the other. Instead, we just need to slightly rearrange the charges on the plates *and* in the wire.

Let's do a rough estimate of how much rearrangement is needed and how long the discharge takes. Using the conduction-electron density of copper in Table 27.1, we can calculate that there are 5×10^{22} conduction electrons in the wire. The negative plate in **FIGURE 27.6**, with $Q = -16$ nC, has 10^{11} excess electrons, far fewer than in the wire. In fact, the length of copper wire needed to hold 10^{11} electrons is a mere 4×10^{-13} m.

The instant the wire joins the capacitor plates together, the repulsive forces between the excess 10^{11} electrons on the negative plate cause them to push their way into the wire. As they do, 10^{11} electrons are squeezed out of the final 4×10^{-13} m of the wire and onto the positive plate. If the electrons all move together, and if they move at the typical drift speed of 10^{-4} m/s—both less than perfect assumptions but fine for making an estimate—it takes 4×10^{-9} s, or 4 ns, to move 4×10^{-13} m and discharge the capacitor. And, indeed, this is the right order of magnitude for how long the electrons take to rearrange themselves so that the capacitor plates are neutral.

STOP TO THINK 27.2 Why does the light in a room come on instantly when you flip a switch several meters away?

FIGURE 27.5 How long does it take to discharge this capacitor?

20-cm-long copper wire

FIGURE 27.6 The sea of electrons needs only a minuscule rearrangement.

1. The 10^{11} excess electrons on the negative plate move into the wire.

2.0-mm-diameter wire

2. The vast sea of electrons in the wire is pushed 4×10^{-13} m to the side in 4 ns.

3. 10^{11} electrons are pushed out of the wire and onto the positive plate. This plate is now neutral.

27.2 Creating a Current

Suppose you want to slide a book across the table to your friend. You give it a quick push to start it moving, but it begins slowing down because of friction as soon as you take your hand off. The book's kinetic energy is transformed into thermal energy, leaving the book and the table slightly warmer. The only way to keep the book moving at a constant speed is to *continue pushing it*.

As **FIGURE 27.7** shows, the sea of electrons is similar to the book. If you push the sea of electrons, you create a current of electrons moving through the conductor. But the electrons aren't moving in a vacuum. Collisions between the electrons and the atoms of the metal transform the electrons' kinetic energy into the thermal energy of the metal, making the metal warmer. (Recall that "makes the wire warm" is one of our indicators of a current.) Consequently, the sea of electrons will quickly slow down and stop *unless you continue pushing*. How do you push on electrons? With an electric field!

One of the important conclusions of Chapter 24 was that $\vec{E} = \vec{0}$ inside a conductor in electrostatic equilibrium. But a conductor with electrons moving through it is *not* in electrostatic equilibrium. **An electron current is a nonequilibrium motion of charges sustained by an internal electric field.**

Thus the quick answer to "What creates a current?" is "An electric field." But why is there an electric field in a current-carrying wire?

FIGURE 27.7 Sustaining the electron current with an electric field.

Because of friction, a steady push is needed to move the book at steady speed.

Retarding force   Sea of electrons

Because of collisions with atoms, a steady push is needed to move the sea of electrons at steady speed.

Establishing the Electric Field in a Wire

FIGURE 27.8a shows two metal wires attached to the plates of a charged capacitor. The wires are conductors, so some of the charges on the capacitor plates become spread out along the wires as a surface charge. (Remember that all excess charge on a conductor is located on the surface.)

This is an electrostatic situation, with no current and no charges in motion. Consequently—because this is always true in electrostatic equilibrium—the electric field inside the wire is zero. Symmetry requires there to be equal amounts of charge to either side of each point to make $\vec{E} = \vec{0}$ at that point; hence the surface charge density must be uniform along each wire except near the ends (where the details need not concern us). We implied this uniform density in Figure 27.8a by drawing equally spaced $+$ and $-$ symbols along the wire. Remember that a positively charged surface is a surface that is *missing* electrons.

FIGURE 27.8 The surface charge on the wires before and after they are connected.

(a) Positive plate — Uniform surface charge density

Negative plate

$\vec{E} = \vec{0}$

There is no current because electrons can't move across the gap.

$\vec{E} = \vec{0}$ at all points inside the wire.

(b) The nonuniform surface charge density creates an electric field inside the wire.

$\vec{E}$

The surface charge density now varies along the wire.

The wire is neutral at the midpoint between the capacitor plates.

Now we connect the ends of the wires together. What happens? The excess electrons on the negative wire suddenly have an opportunity to move onto the positive wire that is missing electrons. Within a *very* brief interval of time ($\approx 10^{-9}$ s), the sea of electrons shifts slightly and the surface charge is rearranged into a *nonuniform* distribution like that shown in **FIGURE 27.8b**. The surface charge near the positive and negative plates remains strongly positive and negative because of the large amount of charge on the capacitor plates, but the midpoint of the wire, halfway between the positive and negative plates, is now electrically neutral. The new surface charge density on the wire varies from positive at the positive capacitor plate through zero at the midpoint to negative at the negative plate.

This nonuniform distribution of surface charge has an *extremely* important consequence. **FIGURE 27.9** shows a section from a wire on which the surface charge density becomes more positive toward the left and more negative toward the right. Calculating the exact electric field is complicated, but we can understand the basic idea if we *model* this section of wire with four circular rings of charge.

FIGURE 27.9 A varying surface charge distribution creates an internal electric field inside the wire.

$\vec{E}_A$ points away from A and $\vec{E}_B$ points away from B, but A has more charge so the net field points to the right.

The nonuniform charge distribution creates a net field to the right at all points inside the wire.

The four rings A through D model the nonuniform charge distribution on the wire.

A   B   C   D

← More positive   More negative →

In Chapter 23, we found that the on-axis field of a ring of charge

■ Points away from a positive ring, toward a negative ring;
■ Is proportional to the amount of charge on the ring; and
■ Decreases with distance away from the ring.

The field midway between rings A and B is well approximated as $\vec{E}_{net} \approx \vec{E}_A + \vec{E}_B$. Ring A has more charge than ring B, so $\vec{E}_{net}$ points away from A.

The analysis of Figure 27.9 leads to a very important conclusion:

A *nonuniform* distribution of surface charges along a wire creates a net electric field *inside* the wire that points from the more positive end of the wire toward the more negative end of the wire. This is the internal electric field $\vec{E}$ that pushes the electron current through the wire.

Note that the surface charges are *not* the moving charges of the current. Further, the current—the moving charges—is *inside* the wire, not on the surface. In fact, as the next example shows, the electric field inside a current-carrying wire can be established with an extremely small amount of surface charge.

EXAMPLE 27.2 | **The surface charge on a current-carrying wire**

Table 23.1 in Chapter 23 gave a typical electric field strength in a current-carrying wire as 0.01 N/C or, as we would now say, 0.01 V/m. (We'll verify this value later in this chapter.) Two 2.0-mm-diameter rings are 2.0 mm apart. They are charged to $\pm Q$. What value of Q causes the electric field at the midpoint to be 0.010 V/m?

MODEL Use the on-axis electric field of a ring of charge from Chapter 23.

VISUALIZE FIGURE 27.10 shows the two rings. Both contribute equally to the field strength, so the electric field strength of the

positive ring is $E_+ = 0.0050$ V/m. The distance $z = 1.0$ mm is half the ring spacing.

SOLVE Chapter 23 found the on-axis electric field of a ring of charge Q to be

$$E_+ = \frac{1}{4\pi\epsilon_0} \frac{zQ}{(z^2 + R^2)^{3/2}}$$

Thus the charge needed to produce the desired field is

$$Q = \frac{4\pi\epsilon_0(z^2 + R^2)^{3/2}}{z} E_+$$
$$= \frac{((0.0010 \text{ m})^2 + (0.0010 \text{ m})^2)^{3/2}}{(9.0 \times 10^9 \text{ N m}^2/\text{C}^2)(0.0010 \text{ m})}(0.0050 \text{ V/m})$$
$$= 1.6 \times 10^{-18} \text{ C}$$

ASSESS The electric field of a ring of charge is largest at $z \approx R$, so these two rings are a simple but reasonable model for estimating the electric field inside a 2.0-mm-diameter wire. We find that the surface charge needed to establish the electric field is *very small*. A mere 10 electrons have to be moved from one ring to the other to charge them to $\pm 1.6 \times 10^{-18}$ C. The resulting electric field is sufficient to drive a sizable electron current through the wire.

FIGURE 27.10 The electric field of two charged rings.

STOP TO THINK 27.3 The two charged rings are a model of the surface charge distribution along a wire. Rank in order, from largest to smallest, the electron currents E_a to E_e at the midpoint between the rings.

(a)

(b)

(c)

(d)

(e)

A Model of Conduction

Electrons don't just magically move through a wire as a current. They move because an electric field inside the wire—a field created by a nonuniform surface charge density on the wire—pushes on the sea of electrons to create the electron current. The field has to *keep* pushing because the electrons continuously lose energy in collisions with the positive ions that form the structure of the solid. These collisions provide a drag force, much like friction.

We will model the conduction electrons—those electrons that make up the sea of electrons—as free particles moving through the lattice of the metal. In the absence of an electric field, the electrons, like the molecules in a gas, move randomly in all directions with a distribution of speeds. If we assume that the average thermal energy of the electrons is given by the same $\frac{3}{2}k_B T$ that applies to an ideal gas, we can calculate that the average electron speed at room temperature is $\approx 10^5$ m/s. This estimate turns out, for quantum physics reasons, to be not quite right, but it correctly indicates that the conduction electrons are moving very fast.

However, an individual electron does not travel far before colliding with an ion and being scattered to a new direction. **FIGURE 27.11a** shows that an electron bounces back and forth between collisions, but its *average* velocity is zero, and it undergoes no *net* displacement. This is similar to molecules in a container of gas.

Suppose we now turn on an electric field. **FIGURE 27.11b** shows that the steady electric force causes the electrons to move along *parabolic trajectories* between collisions. Because of the curvature of the trajectories, the negatively charged electrons begin to drift slowly in the direction opposite the electric field. The motion is similar to a ball moving in a pinball machine with a slight downward tilt. An individual electron ricochets back and forth between the ions at a high rate of speed, but now there is a slow *net* motion in the "downhill" direction. Even so, this net displacement is a *very* small effect superimposed on top of the much larger thermal motion. Figure 27.11b has greatly exaggerated the rate at which the drift would occur.

Suppose an electron just had a collision with an ion and has rebounded with velocity $\vec{v}_0$. The acceleration of the electron between collisions is

$$a_x = \frac{F}{m} = \frac{eE}{m} \tag{27.4}$$

where E is the electric field strength inside the wire and m is the mass of the electron. (We'll assume that $\vec{E}$ points in the negative x-direction.) The field causes the x-component of the electron's velocity to increase linearly with time:

$$v_x = v_{0x} + a_x \Delta t = v_{0x} + \frac{eE}{m}\Delta t \tag{27.5}$$

The electron speeds up, with increasing kinetic energy, until its next collision with an ion. The collision transfers much of the electron's kinetic energy to the ion and thus to the thermal energy of the metal. **This energy transfer is the "friction" that raises the temperature of the wire.** The electron then rebounds, in a random direction, with a new initial velocity $\vec{v}_0$, and starts the process all over.

FIGURE 27.12a shows how the velocity abruptly changes due to a collision. Notice that the acceleration (the slope of the line) is the same before and after the collision. **FIGURE 27.12b** follows an electron through a series of collisions. You can see that each collision "resets" the velocity. The primary observation we can make from Figure 27.12b is that this repeated process of speeding up and colliding gives the electron a nonzero *average* velocity. **The magnitude of the electron's average velocity, due to the electric field, is the *drift speed* v_d of the electron.**

If we observe all the electrons in the metal at one instant of time, their average velocity is

$$v_d = \overline{v_x} = \overline{v_{0x}} + \frac{eE}{m}\overline{\Delta t} \tag{27.6}$$

FIGURE 27.11 A microscopic view of a conduction electron moving through a metal.

(a) No electric field   Ions in the lattice of the metal

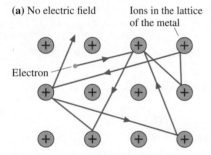

The electron has frequent collisions with ions, but it undergoes no net displacement.

(b) With an electric field   Parabolic trajectories in the electric field

A net displacement in the direction opposite to $\vec{E}$ is superimposed on the random thermal motion.

FIGURE 27.12 The electron velocity as a function of time.

(a) An electron collides with an ion.

The acceleration between collisions
(the slope of the line) is $a = eE/m$.

The electron velocity as
it collides with an ion

Collision

0

t

After the collision
the acceleration is
again eE/m.

The electron rebounds
with velocity v_{0x}.

(b) A series of collisions.

Because the acceleration is
always in the same direction,
the average velocity is *not* zero.

v_d

0

t

Δt

The mean time
between
collisions is τ.

The average
rebound
velocity is zero.

where a bar over a quantity indicates an average value. The average value of v_{0x}, the velocity with which an electron rebounds after a collision, is zero. We know this because, in the absence of an electric field, the sea of electrons moves neither right nor left.

The quantity Δt is the time between collisions, so the average value of Δt is the **mean time between collisions,** which we designate τ. The mean time between collisions, analogous to the mean free path between collisions in the kinetic theory of gases, depends on the metal's temperature but can be considered a constant in the equations below.

Thus the average speed at which the electrons are pushed along by the electric field is

$$v_d = \frac{e\tau}{m} E \qquad (27.7)$$

We can complete our model of conduction by using Equation 27.7 for v_d in the electron-current equation $i_e = n_e A v_d$. Upon doing so, we find that an electric field strength E in a wire of cross-section area A causes an electron current

$$i_e = \frac{n_e e \tau A}{m} E \qquad (27.8)$$

The electron density n_e and the mean time between collisions τ are properties of the metal.

Equation 27.8 is the main result of this model of conduction. We've found that **the electron current is directly proportional to the electric field strength.** A stronger electric field pushes the electrons faster and thus increases the electron current.

EXAMPLE 27.3 | Collisions in a copper wire

Example 27.1 found the electron current to be 2.7×10^{19} s^{-1} for a 2.0-mm-diameter copper wire in which the electron drift speed is 1.0×10^{-4} m/s. If an internal electric field of 0.020 V/m is needed to sustain this current, a typical value, how many collisions per second, on average, do electrons in copper undergo?

MODEL Use the model of conduction.

SOLVE From Equation 27.7, the mean time between collisions is

$$\tau = \frac{m v_d}{eE} = 2.8 \times 10^{-14} \text{ s}$$

The average number of collisions per second is the inverse:

$$\text{Collision rate} = \frac{1}{\tau} = 3.5 \times 10^{13} \text{ s}^{-1}$$

ASSESS This was another straightforward calculation simply to illustrate the incredibly large collision rate of conduction electrons.

27.3 Current and Current Density

We have developed the idea of a current as the motion of electrons through metals. But the properties of currents were known and used for a century before the discovery that electrons are the charge carriers in metals. We need to connect our ideas about the electron current to the conventional definition of current.

Because the coulomb is the unit of charge, and because currents are charges in motion, it seemed quite natural in the 19th century to define current as the *rate,* in coulombs per second, at which charge moves through a wire. If Q is the total amount of charge that has moved past a point in the wire, we define the current I in the wire to be the rate of charge flow:

$$I \equiv \frac{dQ}{dt} \qquad (27.9)$$

For a *steady current,* which will be our primary focus, the amount of charge delivered by current I during the time interval Δt is

$$Q = I \, \Delta t \qquad (27.10)$$

The SI unit for current is the coulomb per second, which is called the **ampere** A:

1 ampere = 1 A $\equiv$ 1 coulomb per second = 1 C/s

The current unit is named after the French scientist André Marie Ampère, who made major contributions to the study of electricity and magnetism in the early 19th century. The *amp* is an informal abbreviation of ampere. Household currents are typically ≈ 1 A. For example, the current through a 100 watt lightbulb is 0.85 A, meaning that 0.85 C of charge flow through the bulb every second. Currents in consumer electronics, such as stereos and computers, are much less. They are typically measured in milliamps $(1 \text{ mA} = 10^{-3} \text{ A})$ or microamps $(1 \, \mu\text{A} = 10^{-6} \text{ A})$.

Equation 27.10 is closely related to Equation 27.1, which said that the number of electrons delivered during a time interval Δt is $N_e = i_e \, \Delta t$. Each electron has charge of magnitude e; hence the total charge of N_e electrons is $Q = eN_e$. Consequently, the conventional current I and the electron current i_e are related by

$$I = \frac{Q}{\Delta t} = \frac{eN_e}{\Delta t} = ei_e \qquad (27.11)$$

The electron current in the copper wire of Examples 27.1 and 27.3 was 2.7×10^{19} electrons/s. To find the conventional current, multiply by e to get 4.3 C/s, or 4.3 A.

Because electrons are the charge carriers, the rate at which charge moves is e times the rate at which the electrons move.

In one sense, the current I and the electron current i_e differ by only a scale factor. The electron current i_e, the rate at which electrons move through a wire, is more *fundamental* because it looks directly at the charge carriers. The current I, the rate at which the charge of the electrons moves through the wire, is more *practical* because we can measure charge more easily than we can count electrons.

Despite the close connection between i_e and I, there's one extremely important distinction. Because currents were known and studied before it was known what the charge carriers are, **the direction of current is *defined* to be the direction in which positive charges *seem* to move.** Thus the direction of the current I is the same as that of the internal electric field $\vec{E}$. But because the charge carriers turned out to be negative, at least for a metal, **the direction of the current I in a metal is opposite the direction of motion of the electrons.**

FIGURE 27.13 The current I is opposite the direction of motion of the electrons in a metal.

The current I is in the direction that positive charges would move. It is in the direction of $\vec{E}$.

The electron current i_e is the motion of actual charge carriers. It is opposite to $\vec{E}$ and I.

The situation shown in **FIGURE 27.13** may seem disturbing, but it makes no real difference. A capacitor is discharged regardless of whether positive charges move toward the negative plate or negative charges move toward the positive plate. The primary application of current is the analysis of circuits, and in a circuit—a macroscopic device—we simply can't tell what is moving through the wires. All of our calculations will be correct and all of our circuits will work perfectly well if we choose to think of current as the flow of positive charge. The distinction is important only at the microscopic level.

The Current Density in a Wire

We found the electron current in a wire of cross-section area A to be $i_e = n_e A v_d$. Thus the current I is

$$I = e i_e = n_e e v_d A \qquad (27.12)$$

The quantity $n_e e v_d$ depends on the charge carriers and on the internal electric field that determines the drift speed, whereas A is simply a physical dimension of the wire. It will be useful to separate these quantities by defining the **current density** J in a wire as the current per square meter of cross section:

$$J = \text{current density} \equiv \frac{I}{A} = n_e e v_d \qquad (27.13)$$

The current density has units of A/m^2. A specific piece of metal, shaped into a wire with cross-section area A, carries current $I = JA$.

EXAMPLE 27.4 | Finding the electron drift speed

A 1.0 A current passes through a 1.0-mm-diameter aluminum wire. What are the current density and the drift speed of the electrons in the wire?

SOLVE We can find the drift speed from the current density. The current density is

$$J = \frac{I}{A} = \frac{I}{\pi r^2} = \frac{1.0 \text{ A}}{\pi (0.00050 \text{ m})^2} = 1.3 \times 10^6 \text{ A/m}^2$$

The electron drift speed is thus

$$v_d = \frac{J}{n_e e} = 1.3 \times 10^{-4} \text{ m/s} = 0.13 \text{ mm/s}$$

where the conduction-electron density for aluminum was taken from Table 27.1.

ASSESS We earlier used 1.0×10^{-4} m/s as a typical electron drift speed. This example shows where that value comes from.

Charge Conservation and Current

FIGURE 27.14 shows two identical lightbulbs in the wire connecting two charged capacitor plates. Both bulbs glow as the capacitor is discharged. How do you think the brightness of bulb A compares to that of bulb B? Is one brighter than the other? Or are they equally bright? Think about this before going on.

You might have predicted that B is brighter than A because the current I, which carries positive charges from plus to minus, reaches B first. In order to be glowing, B must use up some of the current, leaving less for A. Or perhaps you realized that the actual charge carriers are electrons, moving from minus to plus. The conventional current I may be mathematically equivalent, but physically it's the negative electrons rather than positive charge that actually move. Because the electron current gets to A first, you might have predicted that A is brighter than B.

In fact, both bulbs are equally bright. This is an important observation, one that demands an explanation. After all, "something" gets used up to make the bulb glow, so why don't we observe a decrease in the current? Current is the amount of charge moving through the wire per second. There are only two ways to decrease I: either decrease the amount of charge, or decrease the charge's drift speed through the wire. Electrons, the charge carriers, are charged particles. The lightbulb can't destroy electrons without violating both the law of conservation of mass and the law of conservation of charge. Thus the amount of charge (i.e., the *number* of electrons) cannot be changed by a lightbulb.

Do charges slow down after passing through the bulb? This is a little trickier, so consider the fluid analogy shown in FIGURE 27.15. Suppose the water flows into one end at a rate of 2.0 kg/s. Is it possible that the water, after turning a paddle wheel, flows out the other end at a rate of only 1.5 kg/s? That is, does turning the paddle wheel cause the water current to decrease?

We can't destroy water molecules any more than we can destroy electrons, we can't increase the density of water by pushing the molecules closer together, and there's nowhere to store extra water inside the pipe. Each drop of water entering the left end pushes a drop out the right end; hence water flows out at exactly the same rate it flows in.

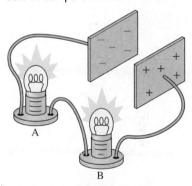

FIGURE 27.14 How does the brightness of bulb A compare to that of bulb B?

FIGURE 27.15 A current dissipates energy, but the flow is unchanged.

The amount of water leaving the turbine equals the amount entering; the number of electrons leaving the bulb equals the number entering.

Flow of electrons

The same is true for electrons in a wire. **The rate of electrons leaving a lightbulb (or any other device) is exactly the same as the rate of electrons entering the lightbulb. The current does not change.** A lightbulb doesn't "use up" current, but it *does*—like the paddlewheel in the fluid analogy—use energy. The kinetic energy of the electrons is dissipated by their collisions with the ions in the lattice of the metal (the atomic-level friction) as the electrons move through the atoms, making the wire hotter until, in the case of the lightbulb filament, it glows. The lightbulb affects the amount of current *everywhere* in the wire, a process we'll examine later in the chapter, but the current doesn't change as it passes through the bulb.

There are many issues that we'll need to look at before we can say that we understand how currents work, and we'll take them one at a time. For now, we draw a first important conclusion: **Due to conservation of charge, the current must be the same at all points in a current-carrying wire.**

FIGURE 27.16 The sum of the currents into a junction must equal the sum of the currents leaving the junction.

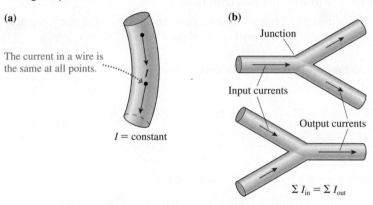

(a)

The current in a wire is the same at all points.

I

I = constant

(b)

Junction

Input currents

Output currents

$\Sigma I_{in} = \Sigma I_{out}$

FIGURE 27.16a summarizes the situation in a single wire. But what about **FIGURE 27.16b**, where two wires merge into one and another wire splits into two? A point where a wire branches is called a **junction.** The presence of a junction doesn't change our basic reasoning. We cannot create or destroy electrons in the wire, and neither can we store them in the junction. The rate at which electrons flow into one *or many* wires must be exactly balanced by the rate at which they flow out of others. For a *junction,* the law of conservation of charge requires that

$$\sum I_{in} = \sum I_{out} \tag{27.14}$$

where, as usual, the Σ symbol means summation.

This basic conservation statement—that the sum of the currents into a junction equals the sum of the currents leaving—is called **Kirchhoff's junction law.** The junction law, together with *Kirchhoff's loop law* that you met in Chapter 26, will play an important role in circuit analysis in the next chapter.

STOP TO THINK 27.4 What are the magnitude and the direction of the current in the fifth wire?

3 A   6 A

2 A   4 A

?

27.4 Conductivity and Resistivity

The current density $J = n_e e v_d$ is directly proportional to the electron drift speed v_d. We earlier used the microscopic model of conduction to find that the drift speed is $v_d = e\tau E/m$, where τ is the mean time between collisions and m is the mass of an electron. Combining these, we find the current density is

$$J = n_e e v_d = n_e e \left(\frac{e \tau E}{m} \right) = \frac{n_e e^2 \tau}{m} E \tag{27.15}$$

The quantity $n_e e^2 \tau/m$ depends *only* on the conducting material. According to Equation 27.15, a given electric field strength will generate a larger current density in a material with a larger electron density n_e or longer times τ between collisions than in materials with smaller values. In other words, such a material is a *better conductor* of current.

It makes sense, then, to define the **conductivity** σ of a material as

$$\sigma = \text{conductivity} = \frac{n_e e^2 \tau}{m} \tag{27.16}$$

Conductivity, like density, characterizes a material as a whole. All pieces of copper (at the same temperature) have the same value of σ, but the conductivity of copper is different from that of aluminum. Notice that the mean time between collisions τ can be inferred from measured values of the conductivity.

With this definition of conductivity, Equation 27.15 becomes

$$J = \sigma E \tag{27.17}$$

This is a result of fundamental importance. Equation 27.17 tells us three things:

1. Current is caused by an electric field exerting forces on the charge carriers.
2. The current density, and hence the current $I = JA$, depends linearly on the strength of the electric field. To double the current, you must double the strength of the electric field that pushes the charges along.
3. The current density also depends on the *conductivity* of the material. Different conducting materials have different conductivities because they have different values of the electron density and, especially, different values of the mean time between electron collisions with the lattice of atoms.

The value of the conductivity is affected by the structure of a metal, by any impurities, and by the temperature. As the temperature increases, so do the thermal vibrations of the lattice atoms. This makes them "bigger targets" and causes collisions to be more frequent, thus lowering τ and decreasing the conductivity. Metals conduct better at low temperatures than at high temperatures.

For many practical applications of current it will be convenient to use the inverse of the conductivity, called the **resistivity:**

$$\rho = \text{resistivity} = \frac{1}{\sigma} = \frac{m}{n_e e^2 \tau} \tag{27.18}$$

The resistivity of a material tells us how reluctantly the electrons move in response to an electric field. TABLE 27.2 gives measured values of the resistivity and conductivity for several metals and for carbon. You can see that they vary quite a bit, with copper and silver being the best two conductors.

The units of conductivity, from Equation 27.17, are those of J/E, namely $A\,C/N\,m^2$. These are clearly awkward. In the next section we will introduce a new unit called the *ohm*, symbolized by Ω (uppercase Greek omega). It will then turn out that resistivity has units of $\Omega\,m$ and conductivity has units of $\Omega^{-1}\,m^{-1}$.

This woman is measuring her percentage body fat by gripping a device that sends a small electric current through her body. Because muscle and fat have different resistivities, the amount of current allows the fat-to-muscle ratio to be determined.

TABLE 27.2 Resistivity and conductivity of conducting materials

Material	Resistivity $(\Omega\,m)$	Conductivity $(\Omega^{-1}\,m^{-1})$
Aluminum	2.8×10^{-8}	3.5×10^{7}
Copper	1.7×10^{-8}	6.0×10^{7}
Gold	2.4×10^{-8}	4.1×10^{7}
Iron	9.7×10^{-8}	1.0×10^{7}
Silver	1.6×10^{-8}	6.2×10^{7}
Tungsten	5.6×10^{-8}	1.8×10^{7}
Nichrome*	1.5×10^{-6}	6.7×10^{5}
Carbon	3.5×10^{-5}	2.9×10^{4}

*Nickel-chromium alloy used for heating wires.

EXAMPLE 27.5 | **The electric field in a wire**

A 2.0-mm-diameter aluminum wire carries a current of 800 mA. What is the electric field strength inside the wire?

SOLVE The electric field strength is

$$E = \frac{J}{\sigma} = \frac{I}{\sigma \pi r^2} = \frac{0.80 \text{ A}}{(3.5 \times 10^7 \ \Omega^{-1} \text{ m}^{-1}) \pi (0.0010 \text{ m})^2} = 0.0073 \text{ V/m}$$

where the conductivity of aluminum was taken from Table 27.2.

ASSESS This is a *very* small field in comparison with those we calculated in Chapters 22 and 23. This calculation justifies the claim in Table 23.1 that a typical electric field strength inside a current-carrying wire is ≈ 0.01 V/m. It takes *very few* surface charges on a wire to create the weak electric field necessary to push a considerable current through the wire. The reason, once again, is the enormous value of the charge-carrier density n_e. Even though the electric field is very tiny and the drift speed is agonizingly slow, a wire can carry a substantial current due to the vast number of charge carriers able to move.

Superconductors have unusual magnetic properties. Here a small permanent magnet levitates above a disk of the high-temperature superconductor $YBa_2Cu_3O_7$ that has been cooled to liquid-nitrogen temperature.

Superconductivity

In 1911, the Dutch physicist Heike Kamerlingh Onnes was studying the conductivity of metals at very low temperatures. Scientists had just recently discovered how to liquefy helium, and this opened a whole new field of *low-temperature physics*. As we noted above, metals become better conductors (i.e., they have higher conductivity and lower resistivity) at lower temperatures. But the effect is gradual. Onnes, however, found that mercury suddenly and dramatically loses *all* resistance to current when cooled below a temperature of 4.2 K. This complete loss of resistance at low temperatures is called **superconductivity.**

Later experiments established that the resistivity of a superconducting metal is not just small, it is truly zero. The electrons are moving in a frictionless environment, and charge will continue to move through a superconductor *without an electric field*. Superconductivity was not understood until the 1950s, when it was explained as being a specific quantum effect.

Superconducting wires can carry enormous currents because the wires are not heated by electrons colliding with the atoms. Very strong magnetic fields can be created with superconducting electromagnets, but applications remained limited for many decades because all known superconductors required temperatures less than 20 K. This situation changed dramatically in 1986 with the discovery of *high-temperature superconductors*. These ceramic-like materials are superconductors at temperatures as "high" as 125 K. Although $-150°C$ may not seem like a high temperature to you, the technology for producing such temperatures is simple and inexpensive. Thus many new superconductor applications are likely to appear in coming years.

STOP TO THINK 27.5 Rank in order, from largest to smallest, the current densities J_a to J_d in these four wires.

(a)   (b)   (c)   (d)

27.5 Resistance and Ohm's Law

FIGURE 27.17 shows a section of wire of length L with a potential difference $\Delta V = V_+ - V_-$ between the ends. Perhaps the two ends of the wire are connected to a battery. A potential difference represents separated positive and negative charges, and, as you saw earlier, some of these charges move onto the surface of the wire. The nonuniform charge distribution creates an electric field in the wire, and that electric field is now driving current through the wire by pushing the charge carriers.

We found in Chapter 26 that the field and the potential are closely related to each other, with the field pointing "downhill," perpendicular to the equipotential surfaces. Thus it should come as no surprise that the current through the wire is related to the potential difference between the ends of the wire.

Recall that the electric field component E_s is related to the potential by $E_s = -dV/ds$. We're interested in only the electric field strength $E = |E_s|$, so the minus sign isn't relevant. The field strength is constant inside a constant-diameter conductor; thus

$$E = \frac{\Delta V}{\Delta s} = \frac{\Delta V}{L} \tag{27.19}$$

Equation 27.19 is an important result: The electric field strength inside a constant-diameter conductor—the field that drives the current forward—is simply the potential difference between the ends of the conductor divided by its length.

Now we can use E to find the current I in the conductor. We found earlier that the current density is $J = \sigma E$, and the current in a wire of cross-section area A is related to the current density by $I = JA$. Thus

$$I = JA = A\sigma E = \frac{A}{\rho} E \tag{27.20}$$

where $\rho = 1/\sigma$ is the resistivity.

Combining Equations 27.19 and 27.20, we see that the current is

$$I = \frac{A}{\rho L} \Delta V \tag{27.21}$$

That is, **the current is directly proportional to the potential difference between the ends of a conductor.** We can cast Equation 27.21 into a more useful form if we define the **resistance** of a conductor to be

$$R = \frac{\rho L}{A} \tag{27.22}$$

The resistance is a property of a *specific* conductor because it depends on the conductor's length and diameter as well as on the resistivity of the material from which it is made.

The SI unit of resistance is the **ohm,** defined as

$$1 \text{ ohm} = 1 \ \Omega \equiv 1 \text{ V/A}$$

The ohm is the basic unit of resistance, although kilohms ($1 \text{ k}\Omega = 10^3 \ \Omega$) and meg-ohms ($1 \text{ M}\Omega = 10^6 \ \Omega$) are widely used. You can now see from Equation 27.22 why the resistivity ρ has units of Ω m while the units of conductivity σ are $\Omega^{-1} \text{m}^{-1}$.

The resistance of a wire or conductor increases as the length increases. This seems reasonable because it should be harder to push electrons through a longer wire than a shorter one. Decreasing the cross-section area also increases the resistance. This again seems reasonable because the same electric field can push more electrons through a fat wire than a skinny one.

NOTE It is important to distinguish between resistivity and resistance. *Resistivity* describes just the *material,* not any particular piece of it. *Resistance* characterizes a specific piece of the conductor with a specific geometry. The relationship between resistivity and resistance is analogous to that between mass density and mass.

FIGURE 27.17 The current I is related to the potential difference ΔV.

The potential difference creates an electric field inside the conductor and causes charges to flow through it.

Equipotential surfaces are perpendicular to the electric field.

The definition of resistance allows us to write the current through a conductor as

$$I = \frac{\Delta V}{R} \qquad \text{(Ohm's law)} \qquad (27.23)$$

In other words, establishing a potential difference ΔV between the ends of a conductor of resistance R creates an electric field (via the nonuniform distribution of charges on the surface) that, in turn, causes a current $I = \Delta V/R$ through the conductor. The smaller the resistance, the larger the current. This simple relationship between potential difference and current is known as **Ohm's law.**

NOTE Ohm's law is true for a conductor of any shape, but Equation 27.22 for the value of the resistance is valid only for a conductor with a constant cross-section area.

EXAMPLE 27.6 | **The resistivity of a leaf**

Resistivity measurements on the leaves of corn plants are a good way to assess stress and the plant's overall health. To determine resistivity, the current is measured when a voltage is applied between two electrodes placed 20 cm apart on a leaf that is 2.5 cm wide and 0.20 mm thick. The following data are obtained by using several different voltages:

Voltage (V)	Current (μA)
5.0	2.3
10.0	5.1
15.0	7.5
20.0	10.3
25.0	12.2

What is the resistivity of the leaf tissue?

MODEL Model the leaf as a bar of length $L = 0.20$ m with a rectangular cross-section area $A = (0.025 \text{ m})(2.0 \times 10^{-4} \text{ m}) = 5.0 \times 10^{-6} \text{ m}^2$. The potential difference creates an electric field inside the leaf and causes a current. The current and the potential difference are related by Ohm's law.

SOLVE We can find the leaf's resistivity ρ from its resistance R. Ohm's law

$$I = \frac{1}{R} \Delta V$$

tells us that a graph of current versus potential difference should be a straight line through the origin with slope $1/R$. The graph of the

data in **FIGURE 27.18** is as expected. Using the slope of the best-fit line, $0.50\ \mu$A/V, we find the leaf's resistance to be

$$R = \frac{1}{0.50\ \mu\text{A/V}} = 2.0 \times 10^6\ \frac{\text{V}}{\text{A}} = 2.0 \times 10^6\ \Omega$$

We can now use Equation 27.22 to find the resistivity:

$$\rho = \frac{AR}{L} = \frac{(5.0 \times 10^{-6}\ \text{m}^2)(2.0 \times 10^6\ \Omega)}{0.20\ \text{m}} = 50\ \Omega\,\text{m}$$

FIGURE 27.18 A graph of current versus potential difference.

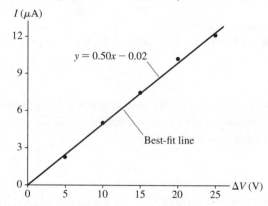

ASSESS This is a huge resistivity compared to metals, but that's not surprising; the conductivity of the salty fluids in a leaf is certainly much less than that of a metal. In fact, this value is typical of the resistivities of plant and animal tissues.

Batteries and Current

Our study of current has focused on the discharge of a capacitor because we can understand where all the charges are and how they move. By contrast, we can't easily see what's happening to the charges inside a battery. Nonetheless, current in most "real" circuits is driven by a battery rather than by a capacitor. Just like the wire discharging a capacitor, a wire connecting two battery terminals gets warm, deflects a compass needle, and makes a lightbulb glow brightly. These indicators tell us that charges flow through the wire from one terminal to the other.

The one major difference between a capacitor and a battery is the duration of the current. The current discharging a capacitor is transient, ceasing as soon as the excess charge on the capacitor plates is removed. In contrast, the current supplied by a battery is *sustained*.

We can use the charge escalator model of a battery to understand why. **FIGURE 27.19** shows the charge escalator creating a potential difference ΔV_{bat} by lifting positive charge from the negative terminal to the positive terminal. Once at the positive terminal, positive charges can move *through the wire* as current I. In essence, the charges are "falling downhill" through the wire, losing the energy they gained on the escalator. This energy transfer to the wire warms the wire.

Eventually the charges find themselves back at the negative terminal of the battery, where they can ride the escalator back up and repeat the journey. A battery, unlike a charged capacitor, has an internal source of energy (the chemical reactions) that keeps the charge escalator running. It is the charge escalator that *sustains* the current in the wire by providing a continually renewed supply of charge at the battery terminals.

An important consequence of the charge escalator model, one you learned in the previous chapter, is that **a battery is a source of potential difference.** It is true that charges flow through a wire connecting the battery terminals, but current is a *consequence* of the battery's potential difference. The battery's emf is the *cause;* current, heat, light, sound, and so on are all *effects* that happen when the battery is used in certain ways.

Distinguishing cause and effect will be vitally important for understanding how a battery functions in a circuit. The reasoning is as follows:

1. A battery is a source of potential difference ΔV_{bat}. An ideal battery has $\Delta V_{bat} = \mathcal{E}$.
2. The battery creates a potential difference $\Delta V_{wire} = \Delta V_{bat}$ between the ends of a wire.
3. The potential difference ΔV_{wire} causes an electric field $E = \Delta V_{wire}/L$ in the wire.
4. The electric field establishes a current $I = JA = \sigma AE$ in the wire.
5. The magnitude of the current is determined *jointly* by the battery and the wire's resistance R to be $I = \Delta V_{wire}/R$.

Resistors and Ohmic Materials

Circuit textbooks often write Ohm's law as $V = IR$ rather than $I = \Delta V/R$. This can be misleading until you have sufficient experience with circuit analysis. First, Ohm's law relates the current to the potential *difference* between the ends of the conductor. Engineers and circuit designers *mean* "potential difference" when they use the symbol V, but the symbol is easily misinterpreted as simply "the potential." Second, $V = IR$ or even $\Delta V = IR$ suggests that a current I causes a potential difference ΔV. As you have seen, current is a *consequence* of a potential difference; hence $I = \Delta V/R$ is a better description of cause and effect.

Despite its name, Ohm's law is *not* a law of nature. It is limited to those materials whose resistance R remains constant—or very nearly so—during use. The materials to which Ohm's law applies are called *ohmic*. **FIGURE 27.20a** shows that the current through an ohmic material is directly proportional to the potential difference. Doubling the potential difference doubles the current. Metal and other conductors are ohmic devices.

Because the resistance of metals is small, a circuit made exclusively of metal wires would have enormous currents and would quickly deplete the battery. It is useful to limit the current in a circuit with ohmic devices, called **resistors,** whose resistance is significantly larger than the metal wires. Resistors are made with poorly conducting materials, such as carbon, or by depositing very thin metal films on an insulating substrate.

FIGURE 27.19 A battery's charge escalator causes a sustained current in a wire.

The charge "falls downhill" through the wire, but a current can be sustained because of the charge escalator.

Positive terminal
$U = q\,\Delta V_{bat}$

ΔV_{bat}

Increasing U

Decreasing U

Ion flow

Current I

Negative terminal
$U = 0$

The charge escalator "lifts" charge from the negative side to the positive side. Charge q gains energy $\Delta U = q\,\Delta V_{bat}$.

FIGURE 27.20 Current-versus-potential-difference graphs for ohmic and nonohmic materials.

(a) Ohmic materials

I

The resistance is
$R = \dfrac{1}{\text{slope}}$

The current is directly proportional to the potential difference.

ΔV

(b) Nonohmic materials

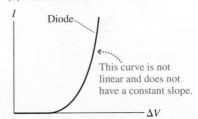

I

Diode

This curve is not linear and does not have a constant slope.

ΔV

The resistors used in circuits range from a few ohms to millions of ohms of resistance.

Some materials and devices are *nonohmic*, meaning that the current through the device is *not* directly proportional to the potential difference. For example, FIGURE 27.20b shows the *I*-versus-ΔV graph of a commonly used semiconductor device called a *diode*. Diodes do not have a well-defined resistance. Batteries, where $\Delta V = \mathcal{E}$ is determined by chemical reactions, and capacitors, where the relationship between *I* and ΔV differs from that of a resistor, are important nonohmic devices.

We can identify three important classes of ohmic circuit materials:

1. *Wires* are metals with very small resistivities ρ and thus very small resistances ($R \ll 1\ \Omega$). An **ideal wire** has $R = 0\ \Omega$; hence the potential difference between the ends of an ideal wire is $\Delta V = 0$ V *even if there is a current in it*. We will usually adopt the *ideal-wire model* of assuming that any connecting wires in a circuit are ideal.

2. *Resistors* are poor conductors with resistances usually in the range 10^1 to $10^6\ \Omega$. They are used to control the current in a circuit. Most resistors in a circuit have a specified value of *R*, such as 500 Ω. The filament in a lightbulb (a tungsten wire with a high resistance due to an extremely small cross-section area *A*) functions as a resistor as long as it is glowing, but the filament is slightly nonohmic because the value of its resistance when hot is larger than its room-temperature value.

3. *Insulators* are materials such as glass, plastic, or air. An **ideal insulator** has $R = \infty\ \Omega$; hence there is no current in an insulator even if there is a potential difference across it ($I = \Delta V/R = 0$ A). This is why insulators can be used to hold apart two conductors at different potentials. All practical insulators have $R \gg 10^9\ \Omega$ and can be treated, for our purposes, as ideal.

NOTE Ohm's law will be an important part of circuit analysis in the next chapter because resistors are essential components of almost any circuit. However, it is important that you apply Ohm's law *only* to the resistors and not to anything else.

FIGURE 27.21a shows a resistor connected to a battery with current-carrying wires. There are no junctions; hence the current *I* through the resistor is the same as the current in each wire. Because the wire's resistance is *much* less than that of the resistor, $R_{wire} \ll R_{resist}$, the potential difference $\Delta V_{wire} = IR_{wire}$ between the ends of each wire is *much* less than the potential difference $\Delta V_{resist} = IR_{resist}$ across the resistor.

If we assume ideal wires with $R_{wire} = 0\ \Omega$, then $\Delta V_{wire} = 0$ V and *all* the voltage drop occurs across the resistor. In this **ideal-wire model,** shown in FIGURE 27.21b, the wires are equipotentials, and the segments of the voltage graph corresponding to the wires are horizontal. As we begin circuit analysis in the next chapter, we will assume that all wires are ideal unless stated otherwise. Thus our analysis will be focused on the resistors.

FIGURE 27.21 The potential along a wire-resistor-wire combination.

(a) The current is constant along the wire-resistor-wire combination.

(b)

In the ideal-wire model there is no voltage drop along the wires. All the voltage drop is across the resistor: $\Delta V_{resist} = \Delta V_{bat}$.

EXAMPLE 27.7 | A battery and a resistor

What resistor would have a 15 mA current if connected across the terminals of a 9.0 V battery?

MODEL Assume the resistor is connected to the battery with ideal wires.

SOLVE Connecting the resistor to the battery with ideal wires makes $\Delta V_{resist} = \Delta V_{bat} = 9.0$ V. From Ohm's law, the resistance giving a 15 mA current is

$$R = \frac{\Delta V_{resist}}{I} = \frac{9.0\ \text{V}}{0.015\ \text{A}} = 600\ \Omega$$

ASSESS Currents of a few mA and resistances of a few hundred ohms are quite typical of real circuits.

STOP TO THINK 27.6 A wire connects the positive and negative terminals of a battery. Two identical wires connect the positive and negative terminals of an identical battery. Rank in order, from largest to smallest, the currents I_a to I_d at points a to d.

Identical batteries

Identical wires

CHALLENGE EXAMPLE 27.8 | Measuring body composition

The woman in the photo on page 753 is gripping a device that measures body fat. To illustrate how this works, **FIGURE 27.22** models an upper arm as part muscle and part fat, showing the resistivities of each. Nonconductive elements, such as skin and bone, have been ignored. This is obviously not a picture of the actual structure, but gathering all the fat tissue together and all the muscle tissue together is a model that predicts the arm's electrical character quite well.

A 0.87 mA current is recorded when a 0.60 V potential difference is applied across an upper arm having the dimensions shown in the figure. What are the percentages of muscle and fat in this person's upper arm?

FIGURE 27.22 A simple model for the resistance of an arm.

Fat tissue
25 Ω m

Muscle tissue
13 Ω m

8.0 cm

25 cm

MODEL Model the muscle and the fat as separate resistors connected to a 0.60 V battery. Assume the connecting wires to be ideal, with no "loss" of potential along the wires.

VISUALIZE FIGURE 27.23 shows the circuit, with the side-by-side muscle and fat resistors connected to the two terminals of the battery.

FIGURE 27.23 Circuit for passing current through the upper arm.

I_{total}

R_{muscle}

R_{fat}

0.60 V

I_{muscle}

I_{fat}

SOLVE The measured current of 0.87 mA is I_{total}, the current traveling from the battery to the arm and later back to the battery. This current splits at the junction between the two resistors. Kirchhoff's junction law, for the conservation of charge, requires

$$I_{total} = I_{muscle} + I_{fat}$$

The current through each resistor can be found from Ohm's law: $I = \Delta V/R$. Each resistor has $\Delta V = 0.60$ V because each is connected to the battery terminals by lossless, ideal wires, but they have different resistances.

Let the fraction of muscle tissue be x; the fraction of fat is then $1 - x$. If the cross-section area of the upper arm is $A = \pi r^2$, then the muscle resistor has $A_{muscle} = xA$ while the fat resistor has $A_{fat} = (1 - x)A$. The resistances are related to the resistivities and the geometry by

$$R_{muscle} = \frac{\rho_{muscle} L}{A_{muscle}} = \frac{\rho_{muscle} L}{x \pi r^2}$$

$$R_{fat} = \frac{\rho_{fat} L}{A_{fat}} = \frac{\rho_{fat} L}{(1 - x) \pi r^2}$$

The currents are thus

$$I_{muscle} = \frac{\Delta V}{R_{muscle}} = \frac{x \pi r^2 \Delta V}{\rho_{muscle} L} = 0.93x \text{ mA}$$

$$I_{fat} = \frac{\Delta V}{R_{fat}} = \frac{(1 - x) \pi r^2 \Delta V}{\rho_{fat} L} = 0.48(1 - x) \text{ mA}$$

The sum of these is the total current:

$$I_{total} = 0.87 \text{ mA} = 0.93x \text{ mA} + 0.48(1 - x) \text{ mA}$$

$$= (0.48 + 0.45x) \text{ mA}$$

Solving, we find $x = 0.87$. This subject's upper arm is 87% muscle tissue, 13% fat tissue.

ASSESS The percentages seem reasonable for a healthy adult. A real measurement of body fat requires a more detailed model of the human body, because the current passes through both arms and across the chest, but the principles are the same.

SUMMARY

The goal of Chapter 27 has been to learn how and why charge moves through a wire as a current.

GENERAL PRINCIPLES

Current is a nonequilibrium motion of charges sustained by an electric field. Nonuniform surface charge density creates an electric field in a wire. The electric field pushes the electron current i_e in a direction opposite to $\vec{E}$. The conventional current I is in the direction in which positive charge *seems* to move.

Conservation of Charge

The current is the same at any two points in a wire.
At a junction,

$$\sum I_{in} = \sum I_{out}$$

This is **Kirchhoff's junction law.**

Electron current

i_e = rate of electron flow

$N_e = i_e \, \Delta t$

Conventional current

I = rate of charge flow = $e i_e$

$Q = I \, \Delta t$

Current density

$J = I/A$

IMPORTANT CONCEPTS

Sea of electrons

Conduction electrons move freely around the positive ions that form the atomic lattice.

Conduction

An electric field causes a slow drift at speed v_d to be superimposed on the rapid but random thermal motions of the electrons.

Collisions of electrons with the ions transfer energy to the atoms. This makes the wire warm and lightbulbs glow. More collisions mean a higher resistivity ρ and a lower conductivity σ.

The **drift speed** is $v_d = \dfrac{e\tau}{m} E$, where τ is the mean time between collisions.

The electron current is related to the drift speed by

$$i_e = n_e A v_d$$

where n_e is the electron density.

An electric field E in a conductor causes a current density $J = n_e e v_d = \sigma E$, where the **conductivity** is

$$\sigma = \frac{n_e e^2 \tau}{m}$$

The **resistivity** is $\rho = 1/\sigma$.

APPLICATIONS

Resistors

A potential difference ΔV_{wire} between the ends of a wire creates an electric field inside the wire:

$$E_{wire} = \frac{\Delta V_{wire}}{L}$$

The electric field causes a current in the direction of decreasing potential.

The size of the current is

$$I = \frac{\Delta V_{wire}}{R}$$

where $R = \dfrac{\rho L}{A}$ is the wire's **resistance.**
This is **Ohm's law.**

TERMS AND NOTATION

current, I	ampere, A	resistivity, ρ	resistor
drift speed, v_d	current density, J	superconductivity	ideal wire
electron current, i_e	junction	resistance, R	ideal insulator
mean time between	Kirchhoff's junction law	ohm, Ω	ideal-wire model
collisions, τ	conductivity, σ	Ohm's law	

CONCEPTUAL QUESTIONS

1. Suppose a time machine has just brought you forward from 1750 (post-Newton but pre-electricity) and you've been shown the lightbulb demonstration of **FIGURE Q27.1**. Do observations or *simple* measurements you might make—measurements that must make sense to you with your 1700s knowledge—prove that something is *flowing* through the wires? Or might you advance an alternative hypothesis for why the bulb is glowing? If your answer to the first question is yes, state what observations and/or measurements are relevant and the reasoning from which you can infer that something must be flowing. If not, can you offer an alternative hypothesis about why the bulb glows that could be tested?

FIGURE Q27.1

2. Consider a lightbulb circuit such as the one in **FIGURE Q27.1**.
   a. From the simple observations and measurements you can make on this circuit, can you distinguish a current composed of positive charge carriers from a current composed of negative charge carriers? If so, describe how you can tell which it is. If not, why not?
   b. One model of current is the motion of discrete charged particles. Another model is that current is the flow of a continuous charged fluid. Do simple observations and measurements on this circuit provide evidence in favor of either one of these models? If so, describe how. If not, why not?
3. The electron drift speed in a wire is exceedingly slow—typically only a fraction of a millimeter per second. Yet when you turn on a flashlight switch, the light comes on almost instantly. Resolve this apparent paradox.
4. Is **FIGURE Q27.4** a possible surface charge distribution for a current carrying wire? If so, in which direction is the current? If not, why not?

FIGURE Q27.4

5. What is the difference between current and current density?
6. All the wires in **FIGURE Q27.6** are made of the same material and have the same diameter. Rank in order, from largest to smallest, the currents I_a to I_d. Explain.

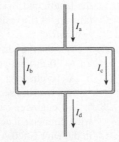

FIGURE Q27.6

7. Both batteries in **FIGURE Q27.7** are ideal and identical, and all lightbulbs are the same. Rank in order, from brightest to least bright, the brightness of bulbs a to c. Explain.

FIGURE Q27.7

8. Both batteries in **FIGURE Q27.8** are ideal and identical, and all lightbulbs are the same. Rank in order, from brightest to least bright, the brightness of bulbs a to c. Explain.

FIGURE Q27.8

9. The wire in **FIGURE Q27.9** consists of two segments of different diameters but made from the same metal. The current in segment 1 is I_1.

FIGURE Q27.9

   a. Compare the currents in the two segments. That is, is I_2 greater than, less than, or equal to I_1? Explain.
   b. Compare the current densities J_1 and J_2 in the two segments.
   c. Compare the electric field strengths E_1 and E_2 in the two segments.
   d. Compare the drift speeds $(v_d)_1$ and $(v_d)_2$ in the two segments.
10. The current in a wire is doubled. What happens to (a) the current density, (b) the conduction-electron density, (c) the mean time between collisions, and (d) the electron drift speed? Are each of these doubled, halved, or unchanged? Explain.
11. The wires in **FIGURE Q27.11** are all made of the same material. Rank in order, from largest to smallest, the resistances R_a to R_e of these wires. Explain.

FIGURE Q27.11

12. Which, if any, of these statements are true? (More than one may be true.) Explain. Assume the batteries are ideal.
   a. A battery supplies the energy to a circuit.
   b. A battery is a source of potential difference; the potential difference between the terminals of the battery is always the same.
   c. A battery is a source of current; the current leaving the battery is always the same.

EXERCISES AND PROBLEMS

Problems labeled ▓▓ integrate material from earlier chapters.

Exercises

Section 27.1 The Electron Current

1. ‖ 1.0×10^{20} electrons flow through a cross section of a 2.0-mm-diameter iron wire in 5.0 s. What is the electron drift speed?

2. ‖ The electron drift speed in a 1.0-mm-diameter gold wire is 5.0×10^{-5} m/s. How long does it take 1 mole of electrons to flow through a cross section of the wire?

3. ‖ 1.0×10^{16} electrons flow through a cross section of a silver wire in 320 μs with a drift speed of 8.0×10^{-4} m/s. What is the diameter of the wire?

4. ‖ Electrons flow through a 1.6-mm-diameter aluminum wire at 2.0×10^{-4} m/s. How many electrons move through a cross section of the wire each day?

Section 27.2 Creating a Current

5. | The electron drift speed is 2.0×10^{-4} m/s in a metal with a mean time between collisions of 5.0×10^{-14} s. What is the electric field strength?

6. ‖ a. How many conduction electrons are there in a 1.0-mm-diameter gold wire that is 10 cm long?
   b. How far must the sea of electrons in the wire move to deliver -32 nC of charge to an electrode?

7. ‖ A 2.0×10^{-3} V/m electric field creates a 3.5×10^{17} electrons/s current in a 1.0-mm-diameter aluminum wire. What are (a) the drift speed and (b) the mean time between collisions for electrons in this wire?

8. ‖ The mean time between collisions in iron is 4.2×10^{-15} s. What electron current is driven through a 1.8-mm-diameter iron wire by a 0.065 V/m electric field?

Section 27.3 Current and Current Density

9. | The wires leading to and from a 0.12-mm-diameter lightbulb filament are 1.5 mm in diameter. The wire to the filament carries a current with a current density of 4.5×10^5 A/m². What are (a) the current and (b) the current density in the filament?

10. ‖ The current in a 100 watt lightbulb is 0.85 A. The filament inside the bulb is 0.25 mm in diameter.
    a. What is the current density in the filament?
    b. What is the electron current in the filament?

11. ‖ In an integrated circuit, the current density in a 2.5-μm-thick × 75-μm-wide gold film is 7.5×10^5 A/m². How much charge flows through the film in 15 min?

12. | The current in an electric hair dryer is 10.0 A. How many electrons flow through the hair dryer in 5.0 min?

13. | When a nerve cell fires, charge is transferred across the cell
BIO membrane to change the cell's potential from negative to positive. For a typical nerve cell, 9.0 pC of charge flows in a time of 0.50 ms. What is the average current through the cell membrane?

14. ‖ The current in a 2.0 mm × 2.0 mm square aluminum wire is 2.5 A. What are (a) the current density and (b) the electron drift speed?

15. ‖ A hollow copper wire with an inner diameter of 1.0 mm and an outer diameter of 2.0 mm carries a current of 10 A. What is the current density in the wire?

16. | A car battery is rated at 90 A h, meaning that it can supply a 90 A current for 1 h before being completely discharged. If you leave your headlights on until the battery is completely dead, how much charge leaves the battery?

Section 27.4 Conductivity and Resistivity

17. | What is the mean time between collisions for electrons in (a) an aluminum wire and (b) an iron wire?

18. ‖ The electric field in a 2.0 mm × 2.0 mm square aluminum wire is 0.012 V/m. What is the current in the wire?

19. ‖ A 15-cm-long nichrome wire is connected across the terminals of a 1.5 V battery.
    a. What is the electric field inside the wire?
    b. What is the current density inside the wire?
    c. If the current in the wire is 2.0 A, what is the wire's diameter?

20. | What electric field strength is needed to create a 5.0 A current in a 2.0-mm-diameter iron wire?

21. | A 0.0075 V/m electric field creates a 3.9 mA current in a 1.0-mm-diameter wire. What material is the wire made of?

22. ‖ A 3.0-mm-diameter wire carries a 12 A current when the electric field is 0.085 V/m. What is the wire's resistivity?

23. ‖ A 0.50-mm-diameter silver wire carries a 20 mA current. What are (a) the electric field and (b) the electron drift speed in the wire?

24. | The two segments of the wire in FIGURE EX27.24 have equal diameters but different conductivities σ_1 and σ_2. Current I passes through this wire. If the conductivities have the ratio $\sigma_2/\sigma_1 = 2$, what is the ratio E_2/E_1 of the electric field strengths in the two segments of the wire?

FIGURE EX27.24

Section 27.5 Resistance and Ohm's Law

25. | A 1.5 V battery provides 0.50 A of current.
    a. At what rate (C/s) is charge lifted by the charge escalator?
    b. How much work does the charge escalator do to lift 1.0 C of charge?
    c. What is the power output of the charge escalator?

26. ‖ Wires 1 and 2 are made of the same metal. Wire 2 has twice the length and twice the diameter of wire 1. What are the ratios (a) ρ_2/ρ_1 of the resistivities and (b) R_2/R_1 of the resistances of the two wires?

27. | What is the resistance of
    a. A 2.0-m-long gold wire that is 0.20 mm in diameter?
    b. A 10-cm-long piece of carbon with a 1.0 mm × 1.0 mm square cross section?

28. ‖ An engineer cuts a 1.0-m-long, 0.33-mm-diameter piece of wire, connects it across a 1.5 V battery, and finds that the current in the wire is 8.0 A. Of what material is the wire made?

29. | The electric field inside a 30-cm-long copper wire is 5.0 mV/m. What is the potential difference between the ends of the wire?

30. ‖ a. How long must a 0.60-mm-diameter aluminum wire be to have a 0.50 A current when connected to the terminals of a 1.5 V flashlight battery?

b. What is the current if the wire is half this length?

31. ‖ The terminals of a 0.70 V watch battery are connected by a 100-m-long gold wire with a diameter of 0.10 mm. What is the current in the wire?

32. ‖ The femoral artery is the large artery that carries blood to the
BIO leg. What is the resistance of a 20-cm-long column of blood in a 1.0-cm-diameter femoral artery? The conductivity of blood is $0.63\ \Omega^{-1}\,\mathrm{m}^{-1}$.

33. | Pencil "lead" is actually carbon. What is the current if a 9.0 V potential difference is applied between the ends of a 0.70-mm-diameter, 6.0-cm-long lead from a mechanical pencil?

34. ‖‖ The resistance of a very fine aluminum wire with a $10\ \mu\mathrm{m} \times 10\ \mu\mathrm{m}$ square cross section is $1000\ \Omega$. A $1000\ \Omega$ resistor is made by wrapping this wire in a spiral around a 3.0-mm-diameter glass core. How many turns of wire are needed?

35. | FIGURE EX27.35 is a current-versus-potential-difference graph for a material. What is the material's resistance?

FIGURE EX27.35

36. ‖ A circuit calls for a 0.50-mm-diameter copper wire to be stretched between two points. You don't have any copper wire, but you do have aluminum wire in a wide variety of diameters. What diameter aluminum wire will provide the same resistance?

37. ‖ Household wiring often uses 2.0-mm-diameter copper wires. The wires can get rather long as they snake through the walls from the fuse box to the farthest corners of your house. What is the potential difference across a 20-m-long, 2.0-mm-diameter copper wire carrying an 8.0 A current?

38. ‖ An ideal battery would produce an extraordinarily large current if "shorted" by connecting the positive and negative terminals with a short wire of very low resistance. Real batteries do not. The current of a real battery is limited by the fact that the battery itself has resistance. What is the resistance of a 9.0 V battery that produces a 21 A current when shorted by a wire of negligible resistance?

Problems

39. ‖ For what electric field strength would the current in a 2.0-mm-diameter nichrome wire be the same as the current in a 1.0-mm-diameter aluminum wire in which the electric field strength is 0.0080 V/m?

40. ‖ The electron beam inside an old television picture tube is 0.40 mm in diameter and carries a current of $50\ \mu\mathrm{A}$. This electron beam impinges on the inside of the picture tube screen.

a. How many electrons strike the screen each second?

b. What is the current density in the electron beam?

c. The electrons move with a velocity of 4.0×10^7 m/s. What electric field strength is needed to accelerate electrons from rest to this velocity in a distance of 5.0 mm?

d. Each electron transfers its kinetic energy to the picture tube screen upon impact. What is the *power* delivered to the screen by the electron beam?

41. ‖ Energetic particles, such as protons, can be detected with a *silicon detector*. When a particle strikes a thin piece of silicon, it creates a large number of free electrons by ionizing silicon atoms. The electrons flow to an electrode on the surface of the detector, and this current is then amplified and detected. In one experiment, each incident proton creates, on average, 35,000 electrons; the electron current is amplified by a factor of 100; and the experimenters record an amplified current of $3.5\ \mu\mathrm{A}$. How many protons are striking the detector per second?

42. ‖ FIGURE P27.42 shows a 4.0-cm-wide plastic film being wrapped onto a 2.0-cm-diameter roller that turns at 90 rpm. The plastic has a uniform surface charge density of $-2.0\ \mathrm{nC/cm}^2$.

FIGURE P27.42

a. What is the current of the moving film?

b. How long does it take the roller to accumulate a charge of $-10\ \mu\mathrm{C}$?

43. ‖ A sculptor has asked you to help electroplate gold onto a brass statue. You know that the charge carriers in the ionic solution are singly charged gold ions, and you've calculated that you must deposit 0.50 g of gold to reach the necessary thickness. How much current do you need, in mA, to plate the statue in 3.0 hours?

44. ‖ In a classic model of the hydrogen atom, the electron moves around the proton in a circular orbit of radius 0.053 nm.

a. What is the electron's orbital frequency?

b. What is the effective current of the electron?

45. ‖ You've been asked to determine whether a new material your company has made is ohmic and, if so, to measure its electrical conductivity. Taking a $0.50\ \mathrm{mm} \times 1.0\ \mathrm{mm} \times 45\ \mathrm{mm}$ sample, you wire the ends of the long axis to a power supply and then measure the current for several different potential differences. Your data are as follows:

Voltage (V)	Current (A)
0.200	0.47
0.400	1.06
0.600	1.53
0.800	1.97

Use an appropriate graph of the data to determine whether the material is ohmic and, if so, its conductivity.

46. | The biochemistry that takes place inside cells depends on
BIO various elements, such as sodium, potassium, and calcium, that are dissolved in water as ions. These ions enter cells through narrow pores in the cell membrane known as *ion channels*. Each ion channel, which is formed from a specialized protein molecule, is selective for one type of ion. Measurements with microelectrodes have shown that a 0.30-nm-diameter potassium ion (K^+) channel carries a current of 1.8 pA.

a. How many potassium ions pass through if the ion channel opens for 1.0 ms?

b. What is the current density in the ion channel?

47. ‖ The starter motor of a car engine draws a current of 150 A from the battery. The copper wire to the motor is 5.0 mm in diameter and 1.2 m long. The starter motor runs for 0.80 s until the car engine starts.

a. How much charge passes through the starter motor?

b. How far does an electron travel along the wire while the starter motor is on?

48. || A 1.5-m-long wire is made of a metal with the same electron density as copper. The wire is connected across the terminals of a 9.0 V battery. What conductivity would the metal need for the drift velocity of electrons in the wire to be 60 mph? By what factor is this larger than the conductivity of copper?

49. || The resistivity of a metal increases slightly with increased temperature. This can be expressed as $\rho = \rho_0[1 + \alpha(T - T_0)]$, where T_0 is a reference temperature, usually 20°C, and α is the *temperature coefficient of resistivity*. For copper, $\alpha = 3.9 \times 10^{-3}$ °C^{-1}. Suppose a long, thin copper wire has a resistance of 0.25 Ω at 20°C. At what temperature, in °C, will its resistance be 0.30 Ω?

50. || Variations in the resistivity of blood can give valuable clues
BIO about changes in various properties of the blood. Suppose a medical device attaches two electrodes into a 1.5-mm-diameter vein at positions 5.0 cm apart. What is the blood resistivity if a 9.0 V potential difference causes a 230 μA current through the blood in the vein?

51. || The conducting path between the right hand and the left hand
BIO can be modeled as a 10-cm-diameter, 160-cm-long cylinder. The average resistivity of the interior of the human body is 5.0 Ω m. Dry skin has a much higher resistivity, but skin resistance can be made negligible by soaking the hands in salt water. If skin resistance is neglected, what potential difference between the hands is needed for a lethal shock of 100 mA across the chest? Your result shows that even small potential differences can produce dangerous currents when the skin is wet.

52. ||| The conductive tissues of the upper leg can be modeled as a
BIO 40-cm-long, 12-cm-diameter cylinder of muscle and fat. The resistivities of muscle and fat are 13 Ω m and 25 Ω m, respectively. One person's upper leg is 82% muscle, 18% fat. What current is measured if a 1.5 V potential difference is applied between the person's hip and knee?

53. || The resistivity of a metal increases slightly with increased
CALC temperature. This can be expressed as $\rho = \rho_0[1 + \alpha(T - T_0)]$, where T_0 is a reference temperature, usually 20°C, and α is the *temperature coefficient of resistivity*.
   a. First find an expression for the current I through a wire of length L, cross-section area A, and temperature T when connected across the terminals of an ideal battery with terminal voltage ΔV. Then, because the *change* in resistance is small, use the *binomial approximation* to simplify your expression. Your final expression should have the temperature coefficient α in the numerator.
   b. For copper, $\alpha = 3.9 \times 10^{-3}$ °C^{-1}. Suppose a 2.5-m-long, 0.40-mm-diameter copper wire is connected across the terminals of a 1.5 V ideal battery. What is the current in the wire at 20°C?
   c. What is the rate, in A/°C, at which the current changes with temperature as the wire heats up?

54. || Electrical engineers sometimes use a wire's *conductance*, $G = \sigma A/L$, instead of its resistance.
   a. Write Ohm's law in terms of conductance, starting with "$\Delta V =$".
   b. What is the conductance of a 5.4-cm-long, 0.15-mm-diameter tungsten wire?
   c. A 1.5 A current flows through the wire of part b. What is the potential difference between the ends of the wire?

55. || You need to design a 1.0 A fuse that "blows" if the current exceeds 1.0 A. The fuse material in your stockroom melts at a current density of 500 A/cm^2. What diameter wire of this material will do the job?

56. || A hollow metal cylinder has inner radius a, outer radius b, length L, and conductivity σ. The current I is *radially* outward from the inner surface to the outer surface.
   a. Find an expression for the electric field strength inside the metal as a function of the radius r from the cylinder's axis.
   b. Evaluate the electric field strength at the inner and outer surfaces of an iron cylinder if $a = 1.0$ cm, $b = 2.5$ cm, $L = 10$ cm, and $I = 25$ A.

57. || A hollow metal sphere has inner radius a, outer radius b, and conductivity σ. The current I is *radially* outward from the inner surface to the outer surface.
   a. Find an expression for the electric field strength inside the metal as a function of the radius r from the center.
   b. Evaluate the electric field strength at the inner and outer surfaces of a copper sphere if $a = 1.0$ cm, $b = 2.5$ cm, and $I = 25$ A.

58. || The total amount of charge in coulombs that has entered a
CALC wire at time t is given by the expression $Q = 4t - t^2$, where t is in seconds and $t \geq 0$.
   a. Find an expression for the current in the wire at time t.
   b. Graph I versus t for the interval $0 \leq t \leq 4$ s.

59. || The total amount of charge that has entered a wire at time t is
CALC given by the expression $Q = (20 \text{ C})(1 - e^{-t/(2.0 \text{ s})})$, where t is in seconds and $t \geq 0$.
   a. Find an expression for the current in the wire at time t.
   b. What is the maximum value of the current?
   c. Graph I versus t for the interval $0 \leq t \leq 10$ s.

60. || The current in a wire at time t is given by the expression
CALC $I = (2.0 \text{ A})e^{-t/(2.0 \, \mu s)}$, where t is in microseconds and $t \geq 0$.
   a. Find an expression for the total amount of charge (in coulombs) that has entered the wire at time t. The initial conditions are $Q = 0$ C at $t = 0$ μs.
   b. Graph Q versus t for the interval $0 \leq t \leq 10$ μs.

61. || The current supplied by a battery slowly decreases as the bat-
CALC tery runs down. Suppose that the current as a function of time is $I = (0.75 \text{ A})e^{-t/(6 \text{ h})}$. What is the total number of electrons transported from the positive electrode to the negative electrode by the charge escalator from the time the battery is first used until it is completely dead?

62. || The two wires in FIGURE P27.62 are made of the same material. What are the current and the electron drift speed in the 2.0-mm-diameter segment of the wire?

FIGURE P27.62   FIGURE P27.63

63. || What diameter should the nichrome wire in FIGURE P27.63 be in order for the electric field strength to be the same in both wires?

64. || An aluminum wire consists of the three segments shown in FIGURE P27.64. The current in the top segment is 10 A. For each of these three segments, find the
   a. Current I.
   b. Current density J.
   c. Electric field E.
   d. Drift velocity v_d.
   e. Electron current i.
Place your results in a table for easy viewing.

FIGURE P27.64

65. ‖ A wire of radius R has a current density that increases linearly
CALC with distance from the center of the wire: $J(r) = kr$, where k is a constant. Find an expression for k in terms of R and the total current I carried by the wire.

66. ‖ A 0.60-mm-diameter wire made from an alloy (a combina-
CALC tion of different metals) has a conductivity that decreases linearly with distance from the center of the wire: $\sigma(r) = \sigma_0 - cr$, with $\sigma_0 = 5.0 \times 10^7 \ \Omega^{-1}\text{m}^{-1}$ and $c = 1.2 \times 10^{11} \ \Omega^{-1}\text{m}^{-2}$. What is the resistance of a 4.0 m length of this wire?

67. ‖ A 20-cm-long hollow nichrome tube of inner diameter 2.8 mm, outer diameter 3.0 mm is connected, at its ends, to a 3.0 V battery. What is the current in the tube?

68. ‖ The batteries in FIGURE P27.68 are identical. Both resistors have equal currents. What is the resistance of the resistor on the right?

FIGURE P27.68

FIGURE P27.69

69. ‖ A 1.5 V flashlight battery is connected to a wire with a resistance of 3.0 Ω. FIGURE P27.69 shows the battery's potential difference as a function of time. What is the total charge lifted by the charge escalator?

70. ‖ Two 10-cm-diameter metal plates 1.0 cm apart are charged to ± 12.5 nC. They are suddenly connected together by a 0.224-mm-diameter copper wire stretched taut from the center of one plate to the center of the other.
   a. What is the maximum current in the wire?
   b. Does the current increase with time, decrease with time, or remain steady? Explain.
   c. What is the total amount of energy dissipated in the wire?

71. ‖ A long, round wire has resistance R. What will the wire's resistance be if you stretch it to twice its initial length?

72. ‖ You've decided to protect your house by placing a 5.0-m-tall iron lightning rod next to the house. The top is sharpened to a point and the bottom is in good contact with the ground. From your research, you've learned that lightning bolts can carry up to 50 kA of current and last up to 50 μs.

a. How much charge is delivered by a lightning bolt with these parameters?
b. You don't want the potential difference between the top and bottom of the lightning rod to exceed 100 V. What minimum diameter must the rod have?

Challenge Problems

73. ‖‖ A 5.0-mm-diameter proton beam carries a total current of
CALC 1.5 mA. The current density in the proton beam, which increases with distance from the center, is given by $J = J_{\text{edge}}(r/R)$, where R is the radius of the beam and J_{edge} is the current density at the edge.
   a. How many protons per second are delivered by this proton beam?
   b. Determine the value of J_{edge}.

74. ‖‖ FIGURE CP27.74 shows a wire that is made of two equal-diameter segments with conductivities σ_1 and σ_2. When current I passes through the wire, a thin layer of charge appears at the boundary between the segments.
   a. Find an expression for the surface charge density η on the boundary. Give your result in terms of I, σ_1, σ_2, and the wire's cross-section area A.
   b. A 1.0-mm-diameter wire made of copper and iron segments carries a 5.0 A current. How much charge accumulates at the boundary between the segments?

FIGURE CP27.74   Surface charge density η

75. ‖‖ A 300 μF capacitor is charged to 9.0 V, then connected in
CALC parallel with a 5000 Ω resistor. The capacitor will discharge because the resistor provides a conducting pathway between the capacitor plates, but much more slowly than if the plates were connected by a wire. Let $t = 0$ s be the instant the fully charged capacitor is first connected to the resistor. At what time has the capacitor voltage decreased by half, to 4.5 V?
   Hint: The current through the resistor is related to the rate at which charge is *leaving* the capacitor. Consequently, you'll need a minus sign that you might not have expected.

76. ‖‖ A thin metal cylinder of length L and radius R_1 is coaxial
CALC with a thin metal cylinder of length L and a larger radius R_2. The space between the two coaxial cylinders is filled with a material that has resistivity ρ. The two cylinders are connected to the terminals of a battery with potential difference ΔV, causing current I to flow *radially* from the inner cylinder to the outer cylinder. Find an expression for the resistance of this device.

28 Fundamentals of Circuits

A microprocessor, the heart of a powerful computer, is a complex device. Even so, a microprocessor operates on the basis of just a few fundamental physical principles.

IN THIS CHAPTER, you will learn the fundamental physical principles that govern electric circuits.

What is a circuit?

Circuits—from flashlights to computers—are the controlled motion of charges through conductors and resistors.

- This chapter focuses on DC circuits, meaning *direct current*, in which potentials and currents are constant.
- You'll learn to draw circuit diagrams.

« LOOKING BACK Section 26.4 Sources of potential

How are circuits analyzed?

Any circuit, no matter how complex, can be analyzed with Kirchhoff's two laws:

- The junction law (charge conservation) relates the currents at a junction.
- The loop law (energy conservation) relates the voltages around a closed loop.
- We'll also use Ohm's law for resistors.

How do circuits use energy?

Circuits use energy to do things, such as lighting a bulb or turning a motor. You'll learn to calculate power, the rate at which the battery supplies energy to a circuit and the rate at which a resistor dissipates energy. Many circuit elements are *rated* by their power consumption in watts.

How are resistors combined?

Resistors, like capacitors, often occur in series or in parallel. These combinations of resistors can be simplified by replacing them with a single resistor with equivalent resistance.

« LOOKING BACK Sections 27.3–27.5 Current, resistance, and Ohm's law

What is an *RC* circuit?

Capacitors are charged and discharged by current flowing through a resistor. These important circuits are called *RC* circuits. Their uses range from heart defibrillators to digital electronics. You'll learn that capacitors charge and discharge exponentially with time constant $\tau = RC$.

« LOOKING BACK Section 26.5 Capacitors

Why are circuits important?

We live in an electronic era, and electric circuits surround you: your household wiring, the ignition system in your car, your music and communication devices, and your tablets and computers. Electric circuits are one of the most important applications of physics, and in this chapter you will see how the seemingly abstract ideas of electric charge, field, and potential are the foundation for many of the things we take for granted in the 21st century.

28.1 Circuit Elements and Diagrams

The last several chapters have focused on the physics of electric forces, fields, and potentials. Now we'll put those ideas to use by looking at one of the most important applications of electricity: the controlled motion of charges in *electric circuits*. This chapter is not about circuit design—you will see that in more advanced courses—but about understanding the fundamental ideas that underlie all circuits.

FIGURE 28.1 shows an electric circuit in which a resistor and a capacitor are connected by wires to a battery. To understand the functioning of this circuit, we do not need to know whether the wires are bent or straight, or whether the battery is to the right or to the left of the resistor. The literal picture of Figure 28.1 provides many irrelevant details. It is customary when describing or analyzing circuits to use a more abstract picture called a **circuit diagram.** A circuit diagram is a *logical* picture of what is connected to what.

A circuit diagram also replaces pictures of the circuit elements with symbols. FIGURE 28.2 shows the basic symbols that we will need. The longer line at one end of the battery symbol represents the positive terminal of the battery. Notice that a lightbulb, like a wire or a resistor, has two "ends," and current passes *through* the bulb. It is often useful to think of a lightbulb as a resistor that gives off light when a current is present. A lightbulb filament is not a perfectly ohmic material, but the resistance of a *glowing* lightbulb remains reasonably constant if you don't change ΔV by much.

FIGURE 28.1 An electric circuit.

FIGURE 28.2 A library of basic symbols used for electric circuit drawings.

FIGURE 28.3 is a circuit diagram of the circuit shown in Figure 28.1. Notice how the circuit elements are labeled. The battery's emf $\mathcal{E}$ is shown beside the battery, and + and − symbols, even though somewhat redundant, are shown beside the terminals. We would use numerical values for $\mathcal{E}$, R, and C if we knew them. The wires, which in practice may bend and curve, are shown as straight-line connections between the circuit elements.

FIGURE 28.3 A circuit diagram for the circuit of Figure 28.1.

STOP TO THINK 28.1 Which of these diagrams represent the same circuit?

28.2 Kirchhoff's Laws and the Basic Circuit

FIGURE 28.4 Kirchhoff's junction law.

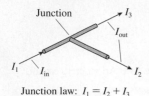

Junction law: $I_1 = I_2 + I_3$

FIGURE 28.5 Kirchhoff's loop law.

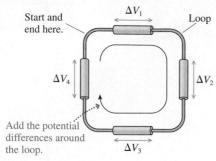

Loop law: $\Delta V_1 + \Delta V_2 + \Delta V_3 + \Delta V_4 = 0$

We are now ready to begin analyzing circuits. To analyze a circuit means to find:

1. The potential difference across each circuit component.
2. The current in each circuit component.

Because charge is conserved, the total current into the junction of **FIGURE 28.4** must equal the total current leaving the junction. That is,

$$\sum I_{\text{in}} = \sum I_{\text{out}} \qquad (28.1)$$

This statement, which you met in Chapter 27, is **Kirchhoff's junction law.**

Because energy is conserved, a charge that moves around a closed path has $\Delta U = 0$. We apply this idea to the circuit of **FIGURE 28.5** by adding all of the potential differences *around* the loop formed by the circuit. Doing so gives

$$\Delta V_{\text{loop}} = \sum (\Delta V)_i = 0 \qquad (28.2)$$

where $(\Delta V)_i$ is the potential difference of the *i*th component in the loop. This statement, introduced in Chapter 26, is **Kirchhoff's loop law.**

Kirchhoff's loop law can be true only if at least one of the $(\Delta V)_i$ is negative. To apply the loop law, we need to explicitly identify which potential differences are positive and which are negative.

TACTICS BOX 28.1

Using Kirchhoff's loop law

❶ **Draw a circuit diagram.** Label all known and unknown quantities.
❷ **Assign a direction to the current.** Draw and label a current arrow I to show your choice.

 ■ If you know the actual current direction, choose that direction.
 ■ If you don't know the actual current direction, make an arbitrary choice. All that will happen if you choose wrong is that your value for I will end up negative.

❸ **"Travel" around the loop.** Start at any point in the circuit, then go all the way around the loop in the direction you assigned to the current in step 2. As you go through each circuit element, ΔV is interpreted to mean $\Delta V = V_{\text{downstream}} - V_{\text{upstream}}$.

 ■ For an ideal battery in the negative-to-positive direction:

$$\Delta V_{\text{bat}} = +\mathcal{E}$$

Potential increases

 ■ For an ideal battery in the positive-to-negative direction:

$$\Delta V_{\text{bat}} = -\mathcal{E}$$

Potential decreases

 ■ For a resistor: $\quad \Delta V_{\text{res}} = -\Delta V_{\text{R}} = -IR$

Potential decreases

❹ **Apply the loop law:** $\sum (\Delta V)_i = 0$

Exercises 4–7

NOTE Ohm's law gives us only the *magnitude* $\Delta V_{\text{R}} = IR$ of the potential difference across a resistor. For using Kirchhoff's law, $\Delta V_{\text{res}} = V_{\text{downstream}} - V_{\text{upstream}} = -\Delta V_{\text{R}}$.

The Basic Circuit

The most basic electric circuit is a single resistor connected to the two terminals of a battery. **FIGURE 28.6a** shows a literal picture of the circuit elements and the connecting wires; **FIGURE 28.6b** is the circuit diagram. Notice that this is a **complete circuit,** forming a continuous path between the battery terminals.

The resistor might be a known resistor, such as "a 10 Ω resistor," or it might be some other resistive device, such as a lightbulb. Regardless of what the resistor is, it is called the **load.** The battery is called the **source.**

FIGURE 28.7 shows the use of Kirchhoff's loop law to analyze this circuit. Two things are worth noting:

1. This circuit has no junctions, so the current I is the same in all four sides of the circuit. Kirchhoff's junction law is not needed.
2. We're assuming the ideal-wire model, in which there are *no* potential differences along the connecting wires.

Kirchhoff's loop law, with two circuit elements, is

$$\Delta V_{\text{loop}} = \sum (\Delta V)_i = \Delta V_{\text{bat}} + \Delta V_{\text{res}}$$
$$= 0$$

(28.3)

Let's look at each of the two voltages in Equation 28.3:

1. The potential *increases* as we travel through the battery on our clockwise journey around the loop. We enter the negative terminal and, farther downstream, exit the positive terminal after having gained potential $\mathcal{E}$. Thus

$$\Delta V_{\text{bat}} = +\mathcal{E}$$

2. The potential of a conductor *decreases* in the direction of the current, which we've indicated with the $+$ and $-$ signs in Figure 28.7. Thus

$$\Delta V_{\text{res}} = V_{\text{downstream}} - V_{\text{upstream}} = -IR$$

NOTE Determining which potential differences are positive and which are negative is perhaps *the* most important step in circuit analysis.

With this information, the loop equation becomes

$$\mathcal{E} - IR = 0$$

(28.4)

We can solve the loop equation to find that the current in the circuit is

$$I = \frac{\mathcal{E}}{R}$$

(28.5)

We can then use the current to find that the magnitude of the resistor's potential difference is

$$\Delta V_R = IR = \mathcal{E}$$

(28.6)

This result should come as no surprise. The potential energy that the charges gain in the battery is subsequently lost as they "fall" through the resistor.

NOTE The current that the battery delivers depends jointly on the emf of the battery and the resistance of the load.

FIGURE 28.6 The basic circuit of a resistor connected to a battery.

FIGURE 28.7 Analysis of the basic circuit using Kirchhoff's loop law.

❶ Draw a circuit diagram.

❷ The orientation of the battery indicates a clockwise current, so assign a clockwise direction to I.

❸ Determine ΔV for each circuit element.

EXAMPLE 28.1 | Two resistors and two batteries

Analyze the circuit shown in **FIGURE 28.8**.

a. Find the current in and the potential difference across each resistor.

b. Draw a graph showing how the potential changes around the circuit, starting from $V = 0$ V at the negative terminal of the 6 V battery.

FIGURE 28.8 Circuit for Example 28.1.

MODEL Assume ideal connecting wires and ideal batteries, for which $\Delta V_{bat} = \mathcal{E}$.

VISUALIZE In **FIGURE 28.9**, we've redrawn the circuit and defined $\mathcal{E}_1, \mathcal{E}_2, R_1$, and R_2. Because there are no junctions, the current is the same through *each* component in the circuit. With some thought, we might deduce whether the current is cw or ccw, but we do not need to know in advance of our analysis. We will choose a clockwise direction and solve for the value of I. If our solution is positive, then the current really is cw. If the solution should turn out to be negative, we will know that the current is ccw.

FIGURE 28.9 Analyzing the circuit.

SOLVE a. How do we deal with *two* batteries? Can charge flow "backward" through a battery, from positive to negative? Consider the charge escalator analogy. Left to itself, a charge escalator lifts charge from lower to higher potential. But it *is* possible to run down an up escalator, as many of you have probably done. If two escalators are placed "head to head," whichever is stronger will, indeed, force the charge to run down the up escalator of the other battery. The current in a battery *can* be from positive to negative if driven in that direction by a larger emf from a second battery. Indeed, this is how rechargeable batteries are recharged.

Kirchhoff's loop law, going clockwise from the negative terminal of battery 1, is

$$\Delta V_{loop} = \sum (\Delta V)_i = \Delta V_{bat\,1} + \Delta V_{res\,1}$$
$$+ \Delta V_{bat\,2} + \Delta V_{res\,2} = 0$$

All the signs are $+$ because this is a formal statement of *adding* potential differences around the loop. Next we can evaluate each ΔV. As we go cw, the charges *gain* potential in battery 1 but *lose* potential in battery 2. Thus $\Delta V_{bat\,1} = +\mathcal{E}_1$ and $\Delta V_{bat\,2} = -\mathcal{E}_2$. There is a *loss* of potential in traveling through each resistor, because we're traversing them in the direction we assigned to the current, so $\Delta V_{res\,1} = -IR_1$ and $\Delta V_{res\,2} = -IR_2$. Thus Kirchhoff's loop law becomes

$$\sum (\Delta V)_i = \mathcal{E}_1 - IR_1 - \mathcal{E}_2 - IR_2$$
$$= \mathcal{E}_1 - \mathcal{E}_2 - I(R_1 + R_2) = 0$$

We can solve this equation to find the current in the loop:

$$I = \frac{\mathcal{E}_1 - \mathcal{E}_2}{R_1 + R_2} = \frac{6\text{ V} - 9\text{ V}}{4\ \Omega + 2\ \Omega} = -0.50\text{ A}$$

The value of I is negative; hence the actual current in this circuit is 0.50 A *counterclockwise*. You perhaps anticipated this from the orientation of the 9 V battery with its larger emf.

The potential difference across the 4 Ω resistor is

$$\Delta V_{res\,1} = -IR_1 = -(-0.50\text{ A})(4\ \Omega) = +2.0\text{ V}$$

Because the current is actually ccw, the resistor's potential *increases* in the cw direction of our travel around the loop. Similarly, $\Delta V_{res\,2} = 1.0$ V.

b. **FIGURE 28.10** shows the potential experienced by charges flowing around the circuit. The distance s is measured from the 6 V battery's negative terminal, and we have chosen to let $V = 0$ V at that point. The potential ends at the value from which it started.

FIGURE 28.10 A graphical presentation of how the potential changes around the loop.

ASSESS Notice how the potential *drops* 9 V upon passing through battery 2 in the cw direction. It then gains 1 V upon passing through R_2 to end at the starting potential.

STOP TO THINK 28.2 What is ΔV across the unspecified circuit element? Does the potential increase or decrease when traveling through this element in the direction assigned to I?

28.3 Energy and Power

The circuit of FIGURE 28.11 has two identical lightbulbs, A and B. Which is brighter? Or are they equally bright? Think about this before going on.

You might have been tempted to say that A is brighter. After all, the current gets to A first, so A might "use up" some of the current and leave less for B. But this would violate the law of conservation of charge. There are no junctions between A and B, so the current through the two bulbs must be the same. Hence the bulbs are equally bright.

What a bulb uses is not current but *energy*. The battery's charge escalator is an energy-transfer process, transferring chemical energy E_{chem} stored in the battery to the potential energy U of the charges. That energy is, in turn, transformed into the *thermal energy* of the resistors, increasing their temperature until—in the case of lightbulb filaments—they glow.

A charge gains potential energy $\Delta U = q\,\Delta V_{bat}$ as it moves up the charge escalator in the battery. For an ideal battery, with $\Delta V_{bat} = \mathcal{E}$, the battery supplies energy $\Delta U = q\mathcal{E}$ as it lifts charge q from the negative to the positive terminal.

It is useful to know the *rate* at which the battery supplies energy to the charges. Recall from Chapter 9 that the rate at which energy is transferred is *power,* measured in joules per second or *watts.* If energy $\Delta U = q\mathcal{E}$ is transferred to charge q, then the *rate* at which energy is transferred from the battery to the moving charges is

$$P_{bat} = \text{rate of energy transfer} = \frac{dU}{dt} = \frac{dq}{dt}\mathcal{E} \qquad (28.7)$$

But dq/dt, the rate at which charge moves through the battery, is the current I. Hence the power supplied by a battery, or the rate at which the battery (or any other source of emf) transfers energy to the charges passing through it, is

$$P_{bat} = I\mathcal{E} \qquad \text{(power delivered by an emf)} \qquad (28.8)$$

$I\mathcal{E}$ has units of J/s, or W. For example, a 120 V battery that generates 2 A of current is delivering 240 W of power to the circuit.

Energy Dissipation in Resistors

P_{bat} is the energy transferred per second from the battery's store of chemicals to the moving charges that make up the current. But what happens to this energy? Where does it end up? FIGURE 28.12, a section of a current-carrying resistor, reminds you of our microscopic model of conduction. The electrons accelerate in the electric field, transforming potential energy into kinetic, then collide with atoms in the lattice. The collisions transfer the electron's kinetic energy to the *thermal* energy of the lattice. The potential energy was acquired in the battery, so the entire energy-transfer process looks like

$$E_{chem} \rightarrow U \rightarrow K \rightarrow E_{th}$$

The net result is that **the battery's chemical energy is transferred to the thermal energy of the resistors,** raising their temperature.

Consider a charge q that moves all the way through a resistor with a potential difference ΔV_R between its two ends. The charge *loses potential energy* $\Delta U = -q\,\Delta V_R$,

FIGURE 28.11 Which lightbulb is brighter?

FIGURE 28.12 A current-carrying resistor dissipates energy.

The electric field causes electrons to speed up. The energy transformation is $U \rightarrow K$.

Collisions transfer energy to the lattice. The energy transformation is $K \rightarrow E_{th}$.

and, after a vast number of collisions, all that energy is transformed into thermal energy. Thus the resistor's *increase in thermal energy* due to this one charge is

$$\Delta E_{th} = q\,\Delta V_R \tag{28.9}$$

The *rate* at which energy is transferred from the current to the resistor is then

$$P_R = \frac{dE_{th}}{dt} = \frac{dq}{dt}\Delta V_R = I\,\Delta V_R \tag{28.10}$$

Power—so many joules per second—is the rate at which energy is *dissipated* by the resistor as charge flows through it. The resistor, in turn, transfers this energy to the air and to the circuit board on which it is mounted, causing the circuit and all its surroundings to heat up.

From our analysis of the basic circuit, in which a single resistor is connected to a battery, we learned that $\Delta V_R = \mathcal{E}$. That is, the potential difference across the resistor is exactly the emf supplied by the battery. But then Equations 28.8 and 28.10, for P_{bat} and P_R, are numerically equal, and we find that

$$P_R = P_{bat} \tag{28.11}$$

The answer to the question "What happens to the energy supplied by the battery?" is "The battery's chemical energy is transformed into the thermal energy of the resistor." The *rate* at which the battery supplies energy is exactly equal to the *rate* at which the resistor dissipates energy. This is, of course, exactly what we would have expected from energy conservation.

EXAMPLE 28.2 | The power of light

How much current is "drawn" by a 100 W lightbulb connected to a 120 V outlet?

MODEL Most household appliances, such as a 100 W lightbulb or a 1500 W hair dryer, have a power rating. The rating does *not* mean that these appliances *always* dissipate that much power. These appliances are intended for use at a standard household voltage of 120 V, and their rating is the power they will dissipate *if* operated with a potential difference of 120 V. Their power consumption will differ from the rating if they are operated at any other potential difference.

SOLVE Because the lightbulb is operating as intended, it will dissipate 100 W of power. Thus

$$I = \frac{P_R}{\Delta V_R} = \frac{100\ W}{120\ V} = 0.833\ A$$

ASSESS A current of 0.833 A in this lightbulb transfers 100 J/s to the thermal energy of the filament, which, in turn, dissipates 100 J/s as heat and light to its surroundings.

A resistor obeys Ohm's law, $\Delta V_R = IR$. (Remember that Ohm's law gives only the *magnitude* of ΔV_R.) This gives us two alternative ways of writing the power dissipated by a resistor. We can either substitute IR for ΔV_R or substitute $\Delta V_R/R$ for I. Thus

$$P_R = I\,\Delta V_R = I^2R = \frac{(\Delta V_R)^2}{R} \quad \text{(power dissipated by a resistor)} \tag{28.12}$$

If the same current I passes through several resistors in series, then $P_R = I^2R$ tells us that most of the power will be dissipated by the largest resistance. This is why a lightbulb filament glows but the connecting wires do not. Essentially *all* of the power supplied by the battery is dissipated by the high-resistance lightbulb filament and essentially no power is dissipated by the low-resistance wires. The filament gets very hot, but the wires do not.

EXAMPLE 28.3 | The power of sound

Most loudspeakers are designed to have a resistance of 8 Ω. If an 8 Ω loudspeaker is connected to a stereo amplifier with a rating of 100 W, what is the maximum possible current to the loudspeaker?

MODEL The rating of an amplifier is the *maximum* power it can deliver. Most of the time it delivers far less, but the maximum might be reached for brief, intense sounds like cymbal crashes.

SOLVE The loudspeaker is a resistive load. The maximum current to the loudspeaker occurs when the amplifier delivers maximum power $P_{max} = (I_{max})^2R$. Thus

$$I_{max} = \sqrt{\frac{P_{max}}{R}} = \sqrt{\frac{100\ W}{8\ \Omega}} = 3.5\ A$$

Kilowatt Hours

The energy dissipated (i.e., transformed into thermal energy) by a resistor during time Δt is $E_{th} = P_R \Delta t$. The product of watts and seconds is joules, the SI unit of energy. However, your local electric company prefers to use a different unit, the *kilowatt hour*, to measure the energy you use each month.

A load that consumes P_R kW of electricity for Δt hours has used $P_R \Delta t$ **kilowatt hours** of energy, abbreviated kWh. For example, a 4000 W electric water heater uses 40 kWh of energy in 10 hours. A 1500 W hair dryer uses 0.25 kWh of energy in 10 minutes. Despite the rather unusual name, a kilowatt hour is a unit of energy. A homework problem will let you find the conversion factor from kilowatt hours to joules.

Your monthly electric bill specifies the number of kilowatt hours you used last month. This is the amount of energy that the electric company delivered to you, via an electric current, and that you transformed into light and thermal energy inside your home. The cost of electricity varies throughout the country, but the average cost of electricity in the United States is approximately 10¢ per kWh ($0.10/kWh). Thus it costs about $4.00 to run your water heater for 10 hours, about 2.5¢ to dry your hair.

The electric meter on the side of your house or apartment records the kilowatt hours of electric energy that you use.

STOP TO THINK 28.3 Rank in order, from largest to smallest, the powers P_a to P_d dissipated in resistors a to d.

(a)
$$+ \quad \Delta V \quad -$$
$$R$$

(b)
$$+ \quad 2\Delta V \quad -$$
$$R$$

(c)
$$+ \quad \Delta V \quad -$$
$$2R$$

(d)
$$+ \quad \Delta V \quad -$$
$$\tfrac{1}{2}R$$

28.4 Series Resistors

Consider the three lightbulbs in **FIGURE 28.13**. The batteries are identical and the bulbs are identical. You learned in the previous section that B and C are equally bright, because the current is the same through both, but how does the brightness of B compare to that of A? Think about this before going on.

FIGURE 28.14a shows two resistors placed end to end between points a and b. Resistors that are aligned end to end, *with no junctions between them,* are called **series resistors** or, sometimes, resistors "in series." Because there are no junctions, the current I must be the same through each of these resistors. That is, the current out of the last resistor in a series is equal to the current into the first resistor.

The potential differences across the two resistors are $\Delta V_1 = IR_1$ and $\Delta V_2 = IR_2$. The total potential difference ΔV_{ab} between points a and b is the sum of the individual potential differences:

$$\Delta V_{ab} = \Delta V_1 + \Delta V_2 = IR_1 + IR_2 = I(R_1 + R_2) \tag{28.13}$$

FIGURE 28.13 How does the brightness of bulb B compare to that of A?

FIGURE 28.14 Replacing two series resistors with an equivalent resistor.

(a) Two resistors in series   **(b)** An equivalent resistor

Suppose, as in FIGURE 28.14b, we replaced the two resistors with a single resistor having current I and potential difference $\Delta V_{ab} = \Delta V_1 + \Delta V_2$. We can then use Ohm's law to find that the resistance R_{ab} between points a and b is

$$R_{ab} = \frac{\Delta V_{ab}}{I} = \frac{I(R_1 + R_2)}{I} = R_1 + R_2 \qquad (28.14)$$

Because the battery has to establish the same potential difference across the load and provide the same current in both cases, the two resistors R_1 and R_2 act exactly the same as a *single* resistor of value $R_1 + R_2$. We can say that the single resistor R_{ab} is *equivalent* to the two resistors in series.

There was nothing special about having only two resistors. If we have N resistors in series, their **equivalent resistance** is

$$R_{eq} = R_1 + R_2 + \cdots + R_N \qquad \text{(series resistors)} \qquad (28.15)$$

The current and the power output of the battery will be unchanged if the N series resistors are replaced by the single resistor R_{eq}. The key idea in this analysis is that **resistors in series all have the same current.**

> **NOTE** Compare this idea to what you learned in Chapter 26 about capacitors in series. The end-to-end connections are the same, but the equivalent capacitance is *not* the sum of the individual capacitances.

Now we can answer the lightbulb question posed at the beginning of this section. Suppose the resistance of each lightbulb is R. The battery drives current $I_A = \mathcal{E}/R$ through bulb A. Bulbs B and C are in series, with an equivalent resistance $R_{eq} = 2R$, but the battery has the same emf $\mathcal{E}$. Thus the current through bulbs B and C is $I_{B+C} = \mathcal{E}/R_{eq} = \mathcal{E}/2R = \frac{1}{2}I_A$. Bulb B has only half the current of bulb A, so B is dimmer.

Many people predict that A and B should be equally bright. It's the same battery, so shouldn't it provide the same current to both circuits? No! A battery is a source of emf, *not* a source of current. In other words, the battery's emf is the same no matter how the battery is used. When you buy a 1.5 V battery you're buying a device that provides a specified amount of potential difference, not a specified amount of current. The battery does provide the current to the circuit, but the *amount* of current depends on the resistance of the load. Your 1.5 V battery causes 1 A to pass through a 1.5 Ω load but only 0.1 A to pass through a 15 Ω load. As an analogy, think about a water faucet. The pressure in the water main underneath the street is a fixed and unvarying quantity set by the water company, but the amount of water coming out of a faucet depends on how far you open it. A faucet opened slightly has a "high resistance," so only a little water flows. A wide-open faucet has a "low resistance," and the water flow is large.

In summary, **a battery provides a fixed and unvarying emf (potential difference). It does *not* provide a fixed and unvarying current. The amount of current depends jointly on the battery's emf *and* the resistance of the circuit attached to the battery.**

EXAMPLE 28.4 | A series resistor circuit

a. What is the current in the circuit of FIGURE 28.15a?

b. Draw a graph of potential versus position in the circuit, going cw from $V = 0$ V at the battery's negative terminal.

MODEL The three resistors are end to end, with no junctions between them, and thus are in series. Assume ideal connecting wires and an ideal battery.

SOLVE a. The battery "acts" the same—it provides the same current at the same potential difference—if we replace the three series resistors by their equivalent resistance

$$R_{eq} = 15 \ \Omega + 4 \ \Omega + 8 \ \Omega = 27 \ \Omega$$

This is shown as an equivalent circuit in FIGURE 28.15b. Now we have a circuit with a single battery and a single resistor, for which we know the current to be

$$I = \frac{\mathcal{E}}{R_{eq}} = \frac{9 \text{ V}}{27 \ \Omega} = 0.333 \text{ A}$$

FIGURE 28.15 Analyzing a circuit with series resistors.

(a)   (b)   (c)

b. $I = 0.333$ A is the current in each of the three resistors in the original circuit. Thus the potential differences across the resistors are $\Delta V_{res\,1} = -IR_1 = -5.0$ V, $\Delta V_{res\,2} = -IR_2 = -1.3$ V, and $\Delta V_{res\,3} = -IR_3 = -2.7$ V for the 15 Ω, the 4 Ω, and the 8 Ω resistors. **FIGURE 28.15c** shows that the potential increases by 9 V due to the battery, then decreases by 9 V in three steps.

Ammeters

A device that measures the current in a circuit element is called an **ammeter.** Because charge flows *through* circuit elements, an ammeter must be placed *in series* with the circuit element whose current is to be measured.

FIGURE 28.16a shows a simple one-resistor circuit with an unknown emf $\mathcal{E}$. We can measure the current in the circuit by inserting the ammeter as shown in **FIGURE 28.16b**. Notice that we have to *break the connection* between the battery and the resistor in order to insert the ammeter. Now the current in the resistor has to first pass through the ammeter.

Because the ammeter is now in series with the resistor, the total resistance seen by the battery is $R_{eq} = 6\,\Omega + R_{ammeter}$. In order that the ammeter measure the current without changing the current, the ammeter's resistance must, in this case, be $\ll 6\,\Omega$. Indeed, an ideal ammeter has $R_{ammeter} = 0\,\Omega$ and thus has no effect on the current. Real ammeters come very close to this ideal.

The ammeter in Figure 28.16b reads 0.50 A, meaning that the current through the 6 Ω resistor is $I = 0.50$ A. Thus the resistor's potential difference is $\Delta V_R = IR = 3.0$ V. If the ammeter is ideal, with no resistance and thus no potential difference across it, then, from Kirchhoff's loop law, the battery's emf is $\mathcal{E} = \Delta V_R = 3.0$ V.

FIGURE 28.16 An ammeter measures the current in a circuit element.

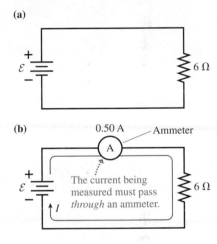

(a)

(b)

STOP TO THINK 28.4 What are the current and the potential at points a to e?

2 A → | 2 Ω | 3 Ω | 1 Ω | 4 Ω | $V = 0$ V

a   b   c   d   e

28.5 Real Batteries

Real batteries, like ideal batteries, use chemical reactions to separate charge, create a potential difference, and provide energy to the circuit. However, real batteries also provide a slight resistance to the charges on the charge escalator. They have what is called an **internal resistance,** which is symbolized by r. **FIGURE 28.17** on the next page shows both an ideal and a real battery.

From our vantage point outside a battery, we cannot see $\mathcal{E}$ and r separately. To the user, the battery provides a potential difference ΔV_{bat} called the **terminal voltage.** $\Delta V_{bat} = \mathcal{E}$ for an ideal battery, but the presence of the internal resistance affects ΔV_{bat}. Suppose the current in the battery is I. As charges travel from the negative to the

FIGURE 28.17 An ideal battery and a real battery.

Ideal battery

$\Delta V_{bat} = \mathcal{E}$

Internal resistance

Real battery

$\Delta V_{bat} = \mathcal{E} - Ir$

positive terminal, the potential increases by $\mathcal{E}$ but *decreases* by $\Delta V_{int} = -Ir$ due to the internal resistance. Thus the terminal voltage of the battery is

$$\Delta V_{bat} = \mathcal{E} - Ir \le \mathcal{E} \qquad (28.16)$$

Only when $I = 0$, meaning that the battery is not being used, is $\Delta V_{bat} = \mathcal{E}$.

FIGURE 28.18 shows a single resistor R connected to the terminals of a battery having emf $\mathcal{E}$ and internal resistance r. Resistances R and r are in series, so we can replace them, for the purpose of circuit analysis, with a single equivalent resistor $R_{eq} = R + r$. Hence the current in the circuit is

$$I = \frac{\mathcal{E}}{R_{eq}} = \frac{\mathcal{E}}{R + r} \qquad (28.17)$$

If $r \ll R$, so that the internal resistance of the battery is negligible, then $I \approx \mathcal{E}/R$, exactly the result we found before. But the current decreases significantly as r increases.

FIGURE 28.18 A single resistor connected to a real battery is in series with the battery's internal resistance, giving $R_{eq} = R + r$.

Although physically separated, the internal resistance r is electrically in series with R.

This means the two circuits are equivalent.

We can use Ohm's law to find that the potential difference across the load resistor R is

$$\Delta V_R = IR = \frac{R}{R + r} \mathcal{E} \qquad (28.18)$$

Similarly, the potential difference across the terminals of the battery is

$$\Delta V_{bat} = \mathcal{E} - Ir = \mathcal{E} - \frac{r}{R + r} \mathcal{E} = \frac{R}{R + r} \mathcal{E} \qquad (28.19)$$

The potential difference across the resistor is equal to the potential difference between the *terminals* of the battery, where the resistor is attached, *not* equal to the battery's emf. Notice that $\Delta V_{bat} = \mathcal{E}$ only if $r = 0$ (an ideal battery with no internal resistance).

EXAMPLE 28.5 | Lighting up a flashlight

A 6 Ω flashlight bulb is powered by a 3 V battery with an internal resistance of 1 Ω. What are the power dissipation of the bulb and the terminal voltage of the battery?

MODEL Assume ideal connecting wires but not an ideal battery.

VISUALIZE The circuit diagram looks like Figure 28.18. R is the resistance of the bulb's filament.

SOLVE Equation 28.17 gives us the current:

$$I = \frac{\mathcal{E}}{R + r} = \frac{3\ V}{6\ \Omega + 1\ \Omega} = 0.43\ A$$

This is 15% less than the 0.5 A an ideal battery would supply. The potential difference across the resistor is $\Delta V_R = IR = 2.6$ V, thus the power dissipation is

$$P_R = I \Delta V_R = 1.1\ W$$

The battery's terminal voltage is

$$\Delta V_{bat} = \frac{R}{R + r} \mathcal{E} = \frac{6\ \Omega}{6\ \Omega + 1\ \Omega}\ 3\ V = 2.6\ V$$

ASSESS 1 Ω is a typical internal resistance for a flashlight battery. The internal resistance causes the battery's terminal voltage to be 0.4 V less than its emf in this circuit.

A Short Circuit

In FIGURE 28.19 we've replaced the resistor with an ideal wire having $R_{wire} = 0\ \Omega$. When a connection of very low or zero resistance is made between two points in a circuit that are normally separated by a higher resistance, we have what is called a **short circuit.** The wire in Figure 28.17 is *shorting out* the battery.

If the battery were ideal, shorting it with an ideal wire $(R = 0\ \Omega)$ would cause the current to be $I = \mathcal{E}/0 = \infty$. The current, of course, cannot really become infinite. Instead, the battery's internal resistance r becomes the only resistance in the circuit. If we use $R = 0\ \Omega$ in Equation 28.17, we find that the *short-circuit current* is

$$I_{short} = \frac{\mathcal{E}}{r} \qquad (28.20)$$

A 3 V battery with 1 Ω internal resistance generates a short circuit current of 3 A. This is the *maximum possible current* that this battery can produce. Adding any external resistance R will decrease the current to a value less than 3 A.

FIGURE 28.19 The short-circuit current of a battery.

This wire is shorting out the battery.

EXAMPLE 28.6 | **A short-circuited battery**

What is the short-circuit current of a 12 V car battery with an internal resistance of 0.020 Ω? What happens to the power supplied by the battery?

SOLVE The short-circuit current is

$$I_{short} = \frac{\mathcal{E}}{r} = \frac{12\ V}{0.02\ \Omega} = 600\ A$$

Power is generated by chemical reactions in the battery and dissipated by the load resistance. But with a short-circuited battery, the load resistance is *inside* the battery! The "shorted" battery has to dissipate power $P = I^2 r = 7200\ W$ *internally.*

ASSESS This value is realistic. Car batteries are designed to drive the starter motor, which has a very small resistance and can draw a current of a few hundred amps. That is why the battery cables are so thick. A shorted car battery can produce an *enormous* amount of current. The normal response of a shorted car battery is to explode; it simply cannot dissipate this much power. Shorting a flashlight battery can make it rather hot, but your life is not in danger. Although the voltage of a car battery is relatively small, a car battery can be dangerous and should be treated with great respect.

Most of the time a battery is used under conditions in which $r \ll R$ and the internal resistance is negligible. The ideal battery model is fully justified in that case. Thus we will assume that batteries are ideal *unless stated otherwise.* But keep in mind that batteries (and other sources of emf) do have an internal resistance, and this internal resistance limits the current of the battery.

28.6 Parallel Resistors

FIGURE 28.20 is another lightbulb puzzle. Initially the switch is open. The current is the same through bulbs A and B and they are equally bright. Bulb C is not glowing. What happens to the brightness of A and B when the switch is closed? And how does the brightness of C then compare to that of A and B? Think about this before going on.

FIGURE 28.21a on the next page shows two resistors aligned side by side with their ends connected at c and d. Resistors connected *at both ends* are called **parallel resistors** or, sometimes, resistors "in parallel." The left ends of both resistors are at the same potential V_c. Likewise, the right ends are at the same potential V_d. Thus the potential *differences* ΔV_1 and ΔV_2 are the *same* and are simply ΔV_{cd}.

Kirchhoff's junction law applies at the junctions. The input current I splits into currents I_1 and I_2 at the left junction. On the right, the two currents are recombined into current I. According to the junction law,

$$I = I_1 + I_2 \qquad (28.21)$$

We can apply Ohm's law to each resistor, along with $\Delta V_1 = \Delta V_2 = \Delta V_{cd}$, to find that the current is

FIGURE 28.20 What happens to the brightness of the bulbs when the switch is closed?

Identical bulbs

FIGURE 28.21 Replacing two parallel resistors with an equivalent resistor.

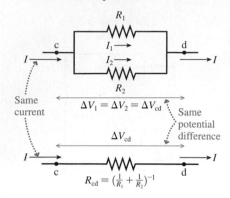

(a) Two resistors in parallel

(b) An equivalent resistor

Two identical resistors*

In series	$R_{eq} = 2R$
In parallel	$R_{eq} = \dfrac{R}{2}$

*$R_1 = R_2 = R$

$$I = \frac{\Delta V_1}{R_1} + \frac{\Delta V_2}{R_2} = \frac{\Delta V_{cd}}{R_1} + \frac{\Delta V_{cd}}{R_2} = \Delta V_{cd}\left(\frac{1}{R_1} + \frac{1}{R_2}\right) \qquad (28.22)$$

Suppose, as in **FIGURE 28.21b**, we replaced the two resistors with a single resistor having current I and potential difference ΔV_{cd}. This resistor is equivalent to the original two because the battery has to establish the same potential difference and provide the same current in either case. A second application of Ohm's law shows that the resistance between points c and d is

$$R_{cd} = \frac{\Delta V_{cd}}{I} = \left(\frac{1}{R_1} + \frac{1}{R_2}\right)^{-1} \qquad (28.23)$$

The single resistor R_{cd} draws the same current as resistors R_1 and R_2, so, as far as the battery is concerned, resistor R_{cd} is *equivalent* to the two resistors in parallel.

There is nothing special about having chosen two resistors to be in parallel. If we have N resistors in parallel, the *equivalent resistance* is

$$R_{eq} = \left(\frac{1}{R_1} + \frac{1}{R_2} + \cdots + \frac{1}{R_N}\right)^{-1} \qquad \text{(parallel resistors)} \qquad (28.24)$$

The behavior of the circuit will be unchanged if the N parallel resistors are replaced by the single resistor R_{eq}. The key idea of this analysis is that **resistors in parallel all have the same potential difference.**

NOTE Don't forget to take the inverse—the -1 exponent in Equation 28.24—after adding the inverses of all the resistances.

EXAMPLE 28.7 | **A parallel resistor circuit**

The three resistors of **FIGURE 28.22** are connected to a 9 V battery. Find the potential difference across and the current through each resistor.

FIGURE 28.22 Parallel resistor circuit of Example 28.7.

MODEL The resistors are in parallel. Assume an ideal battery and ideal connecting wires.

SOLVE The three parallel resistors can be replaced by a single equivalent resistor

$$R_{eq} = \left(\frac{1}{15\ \Omega} + \frac{1}{4\ \Omega} + \frac{1}{8\ \Omega}\right)^{-1} = (0.4417\ \Omega^{-1})^{-1} = 2.26\ \Omega$$

The equivalent circuit is shown in **FIGURE 28.23a**, from which we find the current to be

$$I = \frac{\mathcal{E}}{R_{eq}} = \frac{9\ \text{V}}{2.26\ \Omega} = 3.98\ \text{A}$$

The potential difference across R_{eq} is $\Delta V_{eq} = \mathcal{E} = 9.0$ V. Now we have to be careful. Current I divides at the junction into the smaller currents I_1, I_2, and I_3 shown in **FIGURE 28.23b**. However, the division is *not* into three equal currents. According to Ohm's law, resistor i has current $I_i = \Delta V_i / R_i$. Because the three resistors are each connected to the battery by ideal wires, as is the equivalent resistor, their potential differences are equal:

$$\Delta V_1 = \Delta V_2 = \Delta V_3 = \Delta V_{eq} = 9.0\ \text{V}$$

Thus the currents are

$$I_1 = \frac{9\ \text{V}}{15\ \Omega} = 0.60\ \text{A} \qquad I_2 = \frac{9\ \text{V}}{4\ \Omega} = 2.25\ \text{A} \qquad I_3 = \frac{9\ \text{V}}{8\ \Omega} = 1.13\ \text{A}$$

ASSESS The *sum* of the three currents is 3.98 A, as required by Kirchhoff's junction law.

FIGURE 28.23 The parallel resistors can be replaced by a single equivalent resistor.

The result of Example 28.7 seems surprising. The equivalent of a parallel combination of 15 Ω, 4 Ω, and 8 Ω was found to be 2.26 Ω. How can the equivalent of a group of resistors be *less* than any single resistance in the group? Shouldn't more resistors imply more resistance? The answer is yes for resistors in series but not for resistors in parallel. Even though a resistor is an obstacle to the flow of charge, parallel resistors provide more pathways for charge to get through. Consequently, the equivalent of several resistors in parallel is always *less* than any single resistor in the group.

Complex combinations of resistors can often be reduced to a single equivalent resistance through a step-by-step application of the series and parallel rules. The final example in this section illustrates this idea.

Summary of series and parallel resistors

	I	ΔV
Series	Same	Add
Parallel	Add	Same

EXAMPLE 28.8 | A combination of resistors

What is the equivalent resistance of the group of resistors shown in **FIGURE 28.24**?

FIGURE 28.24 A combination of resistors.

MODEL This circuit contains both series and parallel resistors.

SOLVE Reduction to a single equivalent resistance is best done in a series of steps, with the circuit being redrawn after each step. The procedure is shown in **FIGURE 28.25**. Note that the 10 Ω and 25 Ω resistors are *not* in parallel. They are connected at their top ends

but not at their bottom ends. Resistors must be connected to each other at *both* ends to be in parallel. Similarly, the 10 Ω and 45 Ω resistors are *not* in series because of the junction between them. If the original group of four resistors occurred within a larger circuit, they could be replaced with a single 15.4 Ω resistor without having any effect on the rest of the circuit.

FIGURE 28.25 The combination is reduced to a single equivalent resistor.

To return to the lightbulb question that began this section, **FIGURE 28.26** has redrawn the circuit with each bulb shown as a resistance R. Initially, before the switch is closed, bulbs A and B are in series with equivalent resistance 2R. The current from the battery is

$$I_{\text{before}} = \frac{\mathcal{E}}{2R} = \frac{1}{2}\frac{\mathcal{E}}{R}$$

This is the current in both bulbs.

Closing the switch places bulbs B and C in parallel. The equivalent resistance of two identical resistors in parallel is $R_{\text{eq}} = \frac{1}{2}R$. This equivalent resistance of B and C is in series with bulb A; hence the total resistance of the circuit is $\frac{3}{2}R$ and the current leaving the battery is

$$I_{\text{after}} = \frac{\mathcal{E}}{3R/2} = \frac{2}{3}\frac{\mathcal{E}}{R} > I_{\text{before}}$$

Closing the switch *decreases* the circuit resistance and thus *increases* the current leaving the battery.

All the charge flows through A, so A *increases* in brightness when the switch is closed. The current I_{after} then splits at the junction. Bulbs B and C have equal resistance, so the current splits equally. The current in B is $\frac{1}{3}(\mathcal{E}/R)$, which is *less* than I_{before}. Thus B *decreases* in brightness when the switch is closed. Bulb C has the same brightness as bulb B.

FIGURE 28.26 The lightbulbs of Figure 28.20 with the switch open and closed.

Voltmeters

A device that measures the potential difference across a circuit element is called a **voltmeter.** Because potential difference is measured *across* a circuit element, from

FIGURE 28.27 A voltmeter measures the potential difference across an element.

(a)

(b)

one side to the other, a voltmeter is placed in *parallel* with the circuit element whose potential difference is to be measured.

FIGURE 28.27a shows a simple circuit in which a 17 Ω resistor is connected across a 9 V battery with an unknown internal resistance. By connecting a voltmeter across the resistor, as shown in FIGURE 28.27b, we can measure the potential difference across the resistor. Unlike an ammeter, using a voltmeter does *not* require us to break the connections.

Because the voltmeter is now in parallel with the resistor, the total resistance seen by the battery is $R_{eq} = (1/17\ \Omega + 1/R_{voltmeter})^{-1}$. In order that the voltmeter measure the voltage without changing the voltage, the voltmeter's resistance must, in this case, be $\gg 17\ \Omega$. Indeed, an *ideal voltmeter* has $R_{voltmeter} = \infty\ \Omega$, and thus has no effect on the voltage. Real voltmeters come very close to this ideal, and we will always assume them to be so.

The voltmeter in Figure 28.27b reads 8.5 V. This is less than $\mathcal{E}$ because of the battery's internal resistance. Equation 28.18 found an expression for the resistor's potential difference ΔV_R. That equation is easily solved for the internal resistance r:

$$r = \frac{\mathcal{E} - \Delta V_R}{\Delta V_R} R = \frac{0.5\ \text{V}}{8.5\ \text{V}}\ 17\ \Omega = 1.0\ \Omega$$

Here a voltmeter reading was the one piece of experimental data we needed in order to determine the battery's internal resistance.

STOP TO THINK 28.5 Rank in order, from brightest to dimmest, the identical bulbs A to D.

28.7 Resistor Circuits

We can use the information in this chapter to analyze a variety of more complex but more realistic circuits.

PROBLEM-SOLVING STRATEGY 28.1	

Resistor circuits

MODEL Model wires as ideal and, where appropriate, batteries as ideal.

VISUALIZE Draw a circuit diagram. Label all known and unknown quantities.

SOLVE Base your mathematical analysis on Kirchhoff's laws and on the rules for series and parallel resistors.

- Step by step, reduce the circuit to the smallest possible number of equivalent resistors.
- Write Kirchhoff's loop law for each independent loop in the circuit.
- Determine the current through and the potential difference across the equivalent resistors.
- Rebuild the circuit, using the facts that the current is the same through all resistors in series and the potential difference is the same for all parallel resistors.

ASSESS Use two important checks as you rebuild the circuit.

- Verify that the sum of the potential differences across series resistors matches ΔV for the equivalent resistor.
- Verify that the sum of the currents through parallel resistors matches I for the equivalent resistor.

Exercise 26

EXAMPLE 28.9 | Analyzing a complex circuit

Find the current through and the potential difference across each of the four resistors in the circuit shown in FIGURE 28.28.

FIGURE 28.28 A complex resistor circuit.

MODEL Assume an ideal battery, with no internal resistance, and ideal connecting wires.

VISUALIZE Figure 28.28 shows the circuit diagram. We'll keep redrawing the diagram as we analyze the circuit.

SOLVE First, we break the circuit down, step by step, into one with a single resistor. FIGURE 28.29a shows this done in three steps. The final battery-and-resistor circuit is our basic circuit, with current

$$I = \frac{\mathcal{E}}{R} = \frac{12 \text{ V}}{400 \text{ }\Omega} = 0.030 \text{ A} = 30 \text{ mA}$$

The potential difference across the 400 Ω resistor is $\Delta V_{400} = \Delta V_{\text{bat}} = \mathcal{E} = 12$ V.

Second, we rebuild the circuit, step by step, finding the currents and potential differences at each step. FIGURE 28.29b repeats the steps of Figure 28.29a exactly, but in reverse order. The 400 Ω resistor came from two 800 Ω resistors in parallel. Because $\Delta V_{400} = 12$ V, it must be true that each $\Delta V_{800} = 12$ V. The current through each 800 Ω is then $I = \Delta V/R = 15$ mA. The checkpoint is to note that 15 mA + 15 mA = 30 mA.

The right 800 Ω resistor was formed by 240 Ω and 560 Ω in series. Because $I_{800} = 15$ mA, it must be true that $I_{240} = I_{560} = 15$ mA. The potential difference across each is $\Delta V = IR$, so $\Delta V_{240} = 3.6$ V and $\Delta V_{560} = 8.4$ V. Here the checkpoint is to note that 3.6 V + 8.4 V = 12 V = ΔV_{800}, so the potential differences add as they should.

Finally, the 240 Ω resistor came from 600 Ω and 400 Ω in parallel, so they each have the same 3.6 V potential difference as their 240 Ω equivalent. The currents are $I_{600} = 6$ mA and $I_{400} = 9$ mA. Note that 6 mA + 9 mA = 15 mA, which is our third checkpoint. We now know all currents and potential differences.

ASSESS We *checked our work* at each step of the rebuilding process by verifying that currents summed properly at junctions and that potential differences summed properly along a series of resistances. This "check as you go" procedure is extremely important. It provides you, the problem solver, with a built-in error finder that will immediately inform you if a mistake has been made.

FIGURE 28.29 The step-by-step circuit analysis.

(a) Break down the circuit.

(b) Rebuild the circuit.

EXAMPLE 28.10 | **Analyzing a two-loop circuit**

Find the current through and the potential difference across the 100 Ω resistor in the circuit of FIGURE 28.30.

FIGURE 28.30 A two-loop circuit.

FIGURE 28.31 Applying Kirchhoff's laws.

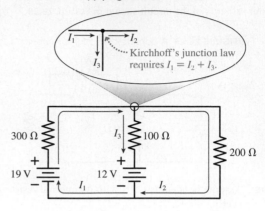

MODEL Assume ideal batteries and ideal connecting wires.

VISUALIZE Figure 28.30 shows the circuit diagram. None of the resistors are connected in series or in parallel, so this circuit cannot be reduced to a simpler circuit.

SOLVE Kirchhoff's loop law applies to *any* loop. To analyze a multi-loop problem, we need to write a loop-law equation for each loop. FIGURE 28.31 redraws the circuit and defines clockwise currents I_1 in the left loop and I_2 in the right loop. But what about the middle branch? Let's assign a downward current I_3 to the middle branch. If we apply Kirchhoff's junction law $\sum I_{in} = \sum I_{out}$ to the junction above the 100 Ω resistor, as shown in the blow-up of Figure 28.31, we see that $I_1 = I_2 + I_3$ and thus $I_3 = I_1 - I_2$. If I_3 ends up being a positive number, then the current in the middle branch really is downward. A negative I_3 will signify an upward current.

Kirchhoff's loop law for the left loop, going clockwise from the lower-left corner, is

$$\sum (\Delta V)_i = 19 \text{ V} - (300 \text{ }\Omega)I_1 - (100 \text{ }\Omega)I_3 - 12 \text{ V} = 0$$

We're traveling through the 100 Ω resistor in the direction of I_3, the "downhill" direction, so the potential decreases. The 12 V battery is traversed positive to negative, so there we have $\Delta V = -\mathcal{E} = -12$ V. For the right loop, we're going to travel "uphill" through the 100 Ω resistor, opposite to I_3, and gain potential. Thus the loop law for the right loop is

$$\sum (\Delta V)_i = 12 \text{ V} + (100 \text{ }\Omega)I_3 - (200 \text{ }\Omega)I_2 = 0$$

If we substitute $I_3 = I_1 - I_2$ and then rearrange the terms, we find that the two independent loops have given us two simultaneous equations in the two unknowns I_1 and I_2:

$$400I_1 - 100I_2 = 7$$
$$-100I_1 + 300I_2 = 12$$

We can eliminate I_2 by multiplying through the first equation by 3 and then adding the two equations. This gives $1100I_1 = 33$, from which $I_1 = 0.030$ A = 30 mA. Using this value in either of the two loop equations gives $I_2 = 0.050$ A = 50 mA. Because $I_2 > I_1$, the current through the 100 Ω resistor is $I_3 = I_1 - I_2 = -20$ mA, or, because of the minus sign, 20 mA upward. The potential difference across the 100 Ω resistor is $\Delta V_{100 \text{ }\Omega} = I_3 R = 2.0$ V, with the bottom end more positive.

ASSESS The three "legs" of the circuit are in parallel, so they must have the same potential difference across them. The left leg has $\Delta V = 19 \text{ V} - (0.030 \text{ A})(300 \text{ }\Omega) = 10$ V, the middle leg has $\Delta V = 12 \text{ V} - (0.020 \text{ A})(100 \text{ }\Omega) = 10$ V, and the right leg has $\Delta V = (0.050 \text{ A})(200 \text{ }\Omega) = 10$ V. Consistency checks such as these are very important. Had we made a numerical error in our circuit analysis, we would have caught it at this point.

28.8 Getting Grounded

The circular prong of a three-prong plug is a connection to ground.

People who work with electronics are often heard to say that something is "grounded." It always sounds quite serious, perhaps somewhat mysterious. What is it?

The circuit analysis procedures we have discussed so far deal only with potential *differences*. Although we are free to choose the zero point of potential anywhere that is convenient, our analysis of circuits has not revealed any need to establish a zero point. Potential differences are all we have needed.

Difficulties can begin to arise, however, if you want to connect two *different* circuits together. Perhaps you would like to connect your DVD to your television or your computer monitor to the computer itself. Incompatibilities can arise unless all the circuits to be connected have a *common* reference point for the potential.

You learned previously that the earth itself is a conductor. Suppose we have two circuits. If we connect *one* point of each circuit to the earth by an ideal wire, and we also agree to call the potential of the earth $V_{earth} = 0$ V, then both circuits have a common reference point. But notice something very important: *one* wire connects the circuit to the earth, but there is not a second wire returning to the circuit. That is, the wire connecting the circuit to the earth is not part of a complete circuit, so there is *no current* in this wire! Because the wire is an equipotential, it gives one point in the circuit the same potential as the earth, but it does *not* in any way change how the circuit functions. A circuit connected to the earth in this way is said to be **grounded,** and the wire is called the *ground wire.*

FIGURE 28.32a shows a fairly simple circuit with a 10 V battery and two resistors in series. The symbol beneath the circuit is the *ground symbol.* It indicates that a wire has been connected between the negative battery terminal and the earth, but the presence of the ground wire does not affect the circuit's behavior. The total resistance is $8 \, \Omega + 12 \, \Omega = 20 \, \Omega$, so the current in the loop is $I = (10 \, \text{V})/(20 \, \Omega) = 0.50 \, \text{A}$. The potential differences across the two resistors are found, using Ohm's law, to be $\Delta V_8 = 4 \, \text{V}$ and $\Delta V_{12} = 6 \, \text{V}$. These are the same values that we would find if the ground wire were *not* present. So what has grounding the circuit accomplished?

FIGURE 28.32b shows the actual potential at several points in the circuit. By definition, $V_{\text{earth}} = 0 \, \text{V}$. The negative battery terminal and the bottom of the 12 Ω resistor are connected by ideal wires to the earth, so *the* potential at these two points must also be zero. The positive terminal of the battery is 10 V more positive than the negative terminal, so $V_{\text{neg}} = 0 \, \text{V}$ implies $V_{\text{pos}} = +10 \, \text{V}$. Similarly, the fact that the potential *decreases* by 6 V as charge flows through the 12 Ω resistor now implies that *the* potential at the junction of the resistors must be +6 V. The potential difference across the 8 Ω resistor is 4 V, so the top has to be at +10 V. This agrees with the potential at the positive battery terminal, as it must because these two points are connected by an ideal wire.

All that grounding the circuit does is allow us to have *specific values* for the potential at each point in the circuit. Now we can say "The voltage at the resistor junction is 6 V," whereas before all we could say was "There is a 6 V potential difference across the 12 Ω resistor."

There is one important lesson from this: **Being grounded does not affect the circuit's behavior under normal conditions.** You cannot use "because it is grounded" to *explain* anything about a circuit's behavior.

We added "under normal conditions" because there is one exception. Most circuits are enclosed in a case of some sort that is held away from the circuit with insulators. Sometimes a circuit breaks or malfunctions in such a way that the case comes into electrical contact with the circuit. If the circuit uses high voltage, or even ordinary 120 V household voltage, anyone touching the case could be injured or killed by electrocution. To prevent this, many appliances or electrical instruments have the case itself grounded. Grounding ensures that the potential of the case will always remain at 0 V and be safe. If a malfunction occurs that connects the case to the circuit, a large current will pass through the ground wire to the earth and cause a fuse to blow. This is the *only* time the ground wire would ever have a current, and it is *not* a normal operation of the circuit.

FIGURE 28.32 A circuit that is grounded at one point.

(a)

(b)

EXAMPLE 28.11 | **A grounded circuit**

Suppose the circuit of Figure 28.32 were grounded at the junction between the two resistors instead of at the bottom. Find the potential at each corner of the circuit.

VISUALIZE FIGURE 28.33 shows the new circuit. (It is customary to draw the ground symbol so that its "point" is always down.)

SOLVE Changing the ground point does not affect the circuit's behavior. The current is still 0.50 A, and the potential differences across the two resistors are still 4 V and 6 V. All that has happened is that we have moved the $V = 0 \, \text{V}$ reference point. Because the earth has $V_{\text{earth}} = 0 \, \text{V}$, the junction itself now has a potential of 0 V. The potential decreases by 4 V as charge flows through the 8 Ω resistor.

FIGURE 28.33 Circuit of Figure 28.32 grounded at the point between the resistors.

Continued

Because it *ends* at 0 V, the potential at the top of the 8 Ω resistor must be +4 V. Similarly, the potential decreases by 6 V through the 12 Ω resistor. Because it *starts* at 0 V, the bottom of the 12 Ω resistor must be at −6 V. The negative battery terminal is at the same potential as the bottom of the 12 Ω resistor, because they are connected by a wire, so $V_{neg} = -6$ V. Finally, the potential increases by 10 V as the charge flows through the battery, so $V_{pos} = +4$ V, in agreement, as it should be, with the potential at the top of the 8 Ω resistor.

ASSESS A negative voltage means only that the potential at that point is less than the potential at some other point that we chose to call $V = 0$ V. Only potential *differences* are physically meaningful, and only potential differences enter into Ohm's law: $I = \Delta V/R$. The potential difference across the 12 Ω resistor in this example is 6 V, decreasing from top to bottom, regardless of which point we choose to call $V = 0$ V.

28.9 RC Circuits

A resistor circuit has a steady current. By adding a capacitor and a switch, we can make a circuit in which the current varies with time as the capacitor charges and discharges. Circuits with resistors and capacitors are called **RC circuits.** *RC* circuits are at the heart of timekeeping circuits in applications ranging from the intermittent windshield wipers on your car to computers and other digital electronics.

FIGURE 28.34a shows a charged capacitor, a switch, and a resistor. The capacitor has charge Q_0 and potential difference $\Delta V_0 = Q_0/C$. There is no current, so the potential difference across the resistor is zero. Then, at $t = 0$, the switch closes and the capacitor begins to discharge through the resistor.

How long does the capacitor take to discharge? How does the current through the resistor vary as a function of time? To answer these questions, **FIGURE 28.34b** shows the circuit at some point in time after the switch was closed.

Kirchhoff's loop law is valid for any circuit, not just circuits with batteries. The loop law applied to the circuit of Figure 28.34b, going around the loop cw, is

$$\Delta V_{cap} + \Delta V_{res} = \frac{Q}{C} - IR = 0 \qquad (28.25)$$

Q and I in this equation are the *instantaneous* values of the capacitor charge and the resistor current.

The current I is the rate at which charge flows through the resistor: $I = dq/dt$. But the charge flowing through the resistor is charge that was *removed* from the capacitor. That is, an infinitesimal charge dq flows through the resistor when the capacitor charge *decreases* by dQ. Thus $dq = -dQ$, and the resistor current is related to the instantaneous capacitor charge by

$$I = -\frac{dQ}{dt} \qquad (28.26)$$

Now I is positive when Q is decreasing, as we would expect. The reasoning that has led to Equation 28.26 is rather subtle but very important. You'll see the same reasoning later in other contexts.

If we substitute Equation 28.26 into Equation 28.25 and then divide by R, the loop law for the *RC* circuit becomes

$$\frac{dQ}{dt} + \frac{Q}{RC} = 0 \qquad (28.27)$$

Equation 28.27 is a first-order differential equation for the capacitor charge Q, but one that we can solve by direct integration. First, we rearrange Equation 28.27 to get all the charge terms on one side of the equation:

$$\frac{dQ}{Q} = -\frac{1}{RC} dt$$

The product RC is a constant for any particular circuit.

FIGURE 28.34 Discharging a capacitor.

(a) Before the switch closes

The switch will close at $t = 0$.

Charge Q_0
$\Delta V_0 = Q_0/C$

(b) After the switch closes

Charge Q
$\Delta V_{cap} = Q/C$

The current is reducing the charge on the capacitor.

The rear flasher on a bike helmet flashes on and off. The timing is controlled by an *RC* circuit.

The capacitor charge was Q_0 at $t = 0$ when the switch was closed. We want to integrate from these starting conditions to charge Q at a later time t. That is,

$$\int_{Q_0}^{Q} \frac{dQ}{Q} = -\frac{1}{RC} \int_0^t dt \qquad (28.28)$$

Both are well-known integrals, giving

$$\ln Q \Big|_{Q_0}^{Q} = \ln Q - \ln Q_0 = \ln\left(\frac{Q}{Q_0}\right) = -\frac{t}{RC}$$

We can solve for the capacitor charge Q by taking the exponential of both sides, then multiplying by Q_0. Doing so gives

$$Q = Q_0 e^{-t/RC} \qquad (28.29)$$

Notice that $Q = Q_0$ at $t = 0$, as expected.

The argument of an exponential function must be dimensionless, so the quantity RC must have dimensions of time. It is useful to define the **time constant** τ to be

$$\tau = RC \qquad (28.30)$$

We can then write Equation 28.29 as

$$Q = Q_0 e^{-t/\tau} \quad \text{(capacitor discharging)} \qquad (28.31)$$

The capacitor voltage, directly proportional to the charge, also decays exponentially as

$$\Delta V_C = \Delta V_0 e^{-t/\tau} \qquad (28.32)$$

The meaning of Equation 28.31 is easier to understand if we portray it graphically. **FIGURE 28.35a** shows the capacitor charge as a function of time. The charge decays exponentially, starting from Q_0 at $t = 0$ and asymptotically approaching zero as $t \to \infty$. The time constant τ is the time at which the charge has decreased to e^{-1} (about 37%) of its initial value. At time $t = 2\tau$, the charge has decreased to e^{-2} (about 13%) of its initial value. A voltage graph would have the same shape.

NOTE The *shape* of the graph of Q is always the same, regardless of the specific value of the time constant τ.

We find the resistor current by using Equation 28.26:

$$I = -\frac{dQ}{dt} = \frac{Q_0}{\tau} e^{-t/\tau} = \frac{Q_0}{RC} e^{-t/\tau} = \frac{\Delta V_0}{R} e^{-t/\tau} = I_0 e^{-t/\tau} \qquad (28.33)$$

where $I_0 = \Delta V_0/R$ is the initial current, immediately after the switch closes. **FIGURE 28.35b** is a graph of the resistor current versus t. You can see that the current undergoes the same exponential decay, with the same time constant, as the capacitor charge.

NOTE There's no specific time at which the capacitor has been discharged, because Q approaches zero asymptotically, but the charge and current have dropped to less than 1% of their initial values at $t = 5\tau$. Thus 5τ is a reasonable answer to the question "How long does it take to discharge a capacitor?"

FIGURE 28.35 The decay curves of the capacitor charge and the resistor current.

(a) Charge Q

(b) Current I

EXAMPLE 28.12 | **Measuring capacitance**

To determine the capacitance of an unmarked capacitor, you set up the circuit shown in **FIGURE 28.36**. After holding the switch in position a for several seconds, you suddenly flip it—at $t = 0$ s—to position b while monitoring the resistor voltage with a voltmeter. Your measurements are shown in the table. What is the capacitance? And what was the resistor current 5.0 s after the switch changed position?

Time (s)	Voltage (V)
0.0	9.0
2.0	5.4
4.0	2.7
6.0	1.6
8.0	1.0

FIGURE 28.36 An *RC* circuit for measuring capacitance.

Continued

MODEL The battery charges the capacitor to 9.0 V. Then, when the switch is flipped to position b, the capacitor discharges through the 25,000 Ω resistor with time constant $\tau = RC$.

SOLVE With the switch in position b, the resistor is in parallel with the capacitor and both have the same potential difference $\Delta V_R = \Delta V_C = Q/C$ at all times. The capacitor charge decays exponentially as

$$Q = Q_0 e^{-t/\tau}$$

Consequently, the resistor (and capacitor) voltage also decays exponentially:

$$\Delta V_R = \frac{Q_0}{C} e^{-t/\tau} = \Delta V_0 e^{-t/\tau}$$

where $\Delta V_0 = 9.0$ V is the potential difference at the instant the switch closes. To analyze exponential decays, we take the natural logarithm of both sides. This gives

$$\ln(\Delta V_R) = \ln(\Delta V_0) + \ln(e^{-t/\tau}) = \ln(\Delta V_0) - \frac{1}{\tau} t$$

This result tells us that a graph of $\ln(\Delta V_R)$ versus t—a *semi-log graph*—should be linear with y-intercept $\ln(\Delta V_0)$ and slope $-1/\tau$. If this turns out to be true, we can determine τ and hence C from an experimental measurement of the slope.

FIGURE 28.37 is a graph of $\ln(\Delta V_R)$ versus t. It is, indeed, linear with a negative slope. From the y-intercept of the best-fit line, we find $\Delta V_0 = e^{2.20} = 9.0$ V, as expected. This gives us confidence in our analysis. Using the slope, we find

$$\tau = -\frac{1}{\text{slope}} = -\frac{1}{-0.28 \text{ s}^{-1}} = 3.6 \text{ s}$$

With this, we can calculate

$$C = \frac{\tau}{R} = \frac{3.6 \text{ s}}{25,000 \text{ Ω}} = 1.4 \times 10^{-4} \text{ F} = 140 \text{ μF}$$

The initial current is $I_0 = (9.0 \text{ V})/(25,000 \text{ Ω}) = 360 \text{ μA}$. Current also decays exponentially with the same time constant, so the current after 5.0 s is

$$I = I_0 e^{-t/\tau} = (360 \text{ μA}) e^{-(5.0 \text{ s})/(3.6 \text{ s})} = 90 \text{ μA}$$

FIGURE 28.37 A semi-log graph of the data.

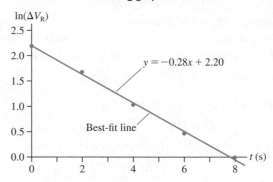

$y = -0.28x + 2.20$

Best-fit line

ASSESS The time constant of an exponential decay can be estimated as the time required to decay to one-third of the initial value. Looking at the data, we see that the voltage drops to one-third of the initial 9.0 V in just under 4 s. This is consistent with the more precise $\tau = 3.6$ s, so we have confidence in our results.

Charging a Capacitor

FIGURE 28.38 Charging a capacitor.

(a)

Switch closes at $t = 0$ s.

(b) Charge Q

FIGURE 28.38a shows a circuit that charges a capacitor. After the switch is closed, the battery's charge escalator moves charge from the bottom electrode of the capacitor to the top electrode. The resistor, by limiting the current, slows the process but doesn't stop it. The capacitor charges until $\Delta V_C = \mathcal{E}$; then the charging current ceases. The full charge of the capacitor is $Q_0 = C(\Delta V_C)_{\text{max}} = C\mathcal{E}$.

The analysis is much like that of discharging a capacitor. As a homework problem, you can show that the capacitor charge and the circuit current at time t are

$$Q = Q_0(1 - e^{-t/\tau})$$
$$I = I_0 e^{-t/\tau}$$

(capacitor charging) (28.34)

where $I_0 = \mathcal{E}/R$ and, again, $\tau = RC$. The capacitor charge's "upside-down decay" to Q_0 is shown graphically in **FIGURE 28.38b**.

STOP TO THINK 28.6 The time constant for the discharge of this capacitor is

a. 5 s.
b. 4 s.
c. 2 s.
d. 1 s.
e. The capacitor doesn't discharge because the resistors cancel each other.

CHALLENGE EXAMPLE 28.13 | Energy dissipated during a capacitor discharge

The switch in **FIGURE 28.39** has been in position a for a long time. It is suddenly switched to position b for 1.0 s, then back to a. How much energy is dissipated by the 5500 Ω resistor?

FIGURE 28.39 Circuit of a switched capacitor.

MODEL With the switch in position a, the capacitor charges through the 1200 Ω resistor with time constant $\tau_{\text{charge}} = (1200\ \Omega)$ $(2.0 \times 10^{-4}\ \text{F}) = 0.24$ s. Because the switch has been in position a for a "long time," which we interpret as being much longer than 0.24 s, we will assume that the capacitor is fully charged to 50 V when the switch is changed to position b. The capacitor then discharges through the 5500 Ω resistor until the switch is returned to position a. Assume ideal wires.

SOLVE Let $t = 0$ s be the time when the switch is moved from a to b, initiating the discharge. The battery and 1200 Ω resistor are irrelevant during the discharge, so the circuit looks like that of Figure 28.34b. The time constant is $\tau = (5500\ \Omega)(2.0 \times 10^{-4}\ \text{F}) = 1.1$ s, so the capacitor voltage decreases from 50 V at $t = 0$ s to

$$\Delta V_{\text{C}} = (50\ \text{V})e^{-(1.0\ \text{s})/(1.1\ \text{s})} = 20\ \text{V}$$

at $t = 1.0$ s.

There are two ways to determine the energy dissipated in the resistor. We learned in Section 28.3 that a resistor dissipates energy at the rate $dE/dt = P_{\text{R}} = I^2R$. The current decays exponentially as $I = I_0 \exp(-t/\tau)$, with $I_0 = \Delta V_0/R = 9.09$ mA. We can find the energy dissipated during a time T by integrating:

$$\Delta E = \int_0^T I^2 R\, dt = I_0^2 R \int_0^T e^{-2t/\tau}\, dt = \left. \tfrac{1}{2}\tau I_0^2 R e^{-2t/\tau} \right|_0^T$$

$$= \tfrac{1}{2}\tau I_0^2 R(1 - e^{-2T/\tau})$$

The 2 in the exponent appears because we squared the expression for I. Evaluating for $T = 1.0$ s, we find

$$\Delta E = \tfrac{1}{2}(1.1\ \text{s})(0.00909\ \text{A})^2(5500\ \Omega)(1 - e^{-(2.0\ \text{s})/(1.1\ \text{s})}) = 0.21\ \text{J}$$

Alternatively, we can use the known capacitor voltages at $t = 0$ s and $t = 1.0$ s and $U_{\text{C}} = \tfrac{1}{2}C(\Delta V_{\text{C}})^2$ to calculate the energy stored in the capacitor at these times:

$$U_{\text{C}}(t = 0.0\ \text{s}) = \tfrac{1}{2}(2.0 \times 10^{-4}\ \text{F})(50\ \text{V})^2 = 0.25\ \text{J}$$

$$U_{\text{C}}(t = 1.0\ \text{s}) = \tfrac{1}{2}(2.0 \times 10^{-4}\ \text{F})(20\ \text{V})^2 = 0.04\ \text{J}$$

The capacitor has lost $\Delta E = 0.21$ J of energy, and this energy was dissipated by the current through the resistor.

ASSESS Not every problem can be solved in two ways, but doing so when it's possible gives us great confidence in our result.

SUMMARY

The goal of Chapter 28 has been to learn the fundamental physical principles that govern electric circuits.

GENERAL STRATEGY

Solving Circuit Problems

MODEL Assume that wires and, where appropriate, batteries are ideal.

VISUALIZE Draw a circuit diagram. Label all quantities.

SOLVE Base the solution on Kirchhoff's laws.

• Reduce the circuit to the smallest possible number of equivalent resistors.
• Write one loop equation for each independent loop.
• Find the current and the potential difference.
• Rebuild the circuit to find I and ΔV for each resistor.

ASSESS Verify that

• The sum of potential differences across series resistors matches ΔV for the equivalent resistor.
• The sum of the currents through parallel resistors matches I for the equivalent resistor.

Kirchhoff's loop law

For a closed loop:

• Assign a direction to the current I.
• $\sum (\Delta V)_i = 0$

Kirchhoff's junction law

For a junction:

• $\sum I_{in} = \sum I_{out}$

IMPORTANT CONCEPTS

Ohm's Law

A potential difference ΔV between the ends of a conductor with resistance R creates a current

$$I = \frac{\Delta V}{R}$$

Signs of ΔV for Kirchhoff's loop law

Travel →

$\Delta V_{bat} = +\mathcal{E}$

Travel →

$\Delta V_{bat} = -\mathcal{E}$

I →

$\Delta V_{res} = -IR$

The **energy used by a circuit** is supplied by the emf $\mathcal{E}$ of the battery through the energy transformations

$$E_{chem} \rightarrow U \rightarrow K \rightarrow E_{th}$$

The battery *supplies* energy at the rate

$$P_{bat} = I\mathcal{E}$$

The resistors *dissipate* energy at the rate

$$P_R = I\Delta V_R = I^2 R = \frac{(\Delta V_R)^2}{R}$$

APPLICATIONS

Equivalent resistance

Groups of resistors can often be reduced to one equivalent resistor.

Series resistors

$$R_{eq} = R_1 + R_2 + R_3 + \cdots$$

Parallel resistors

$$R_{eq} = \left(\frac{1}{R_1} + \frac{1}{R_2} + \frac{1}{R_3} + \cdots \right)^{-1}$$

RC circuits

The charge on and current through a discharging capacitor are

$$Q = Q_0 e^{-t/\tau}$$

$$I = -\frac{dQ}{dt} = \frac{Q_0}{\tau} e^{-t/\tau} = I_0 e^{-t/\tau}$$

where $\tau = RC$ is the **time constant**.

TERMS AND NOTATION

circuit diagram	source	internal resistance, r	grounded
Kirchhoff's junction law	kilowatt hour, kWh	terminal voltage, ΔV_{bat}	*RC* circuit
Kirchhoff's loop law	series resistors	short circuit	time constant, τ
complete circuit	equivalent resistance, R_{eq}	parallel resistors	
load	ammeter	voltmeter	

CONCEPTUAL QUESTIONS

1. Rank in order, from largest to smallest, the currents I_a to I_d through the four resistors in **FIGURE Q28.1**.

FIGURE Q28.1

2. The tip of a flashlight bulb is touching the top of the 3 V battery in **FIGURE Q28.2**. Does the bulb light? Why or why not?

FIGURE Q28.2

FIGURE Q28.3

3. The wire is broken on the right side of the circuit in **FIGURE Q28.3**. What is the potential difference $V_1 - V_2$ between points 1 and 2? Explain.

4. The circuit of **FIGURE Q28.4** has two resistors, with $R_1 > R_2$. Which of the two resistors dissipates the larger amount of power? Explain.

FIGURE Q28.4

FIGURE Q28.5

5. The circuit of **FIGURE Q28.5** has two resistors, with $R_1 > R_2$. Which of the two resistors dissipates the larger amount of power? Explain.

6. Rank in order, from largest to smallest, the powers P_a to P_d dissipated by the four resistors in **FIGURE Q28.6**.

FIGURE Q28.6

7. Are the two resistors in **FIGURE Q28.7** in series or in parallel? Explain.

FIGURE Q28.7

8. A battery with internal resistance r is connected to a load resistance R. If R is increased, does the terminal voltage of the battery increase, decrease, or stay the same? Explain.

9. Initially bulbs A and B in **FIGURE Q28.9** are glowing. What happens to each bulb if the switch is closed? Does it get brighter, stay the same, get dimmer, or go out? Explain.

FIGURE Q28.9

FIGURE Q28.10

10. Bulbs A, B, and C in **FIGURE Q28.10** are identical, and all are glowing.
    a. Rank in order, from most to least, the brightnesses of the three bulbs. Explain.
    b. Suppose a wire is connected between points 1 and 2. What happens to each bulb? Does it get brighter, stay the same, get dimmer, or go out? Explain.

11. Bulbs A and B in **FIGURE Q28.11** are identical, and both are glowing. Bulb B is removed from its socket. Does the potential difference ΔV_{12} between points 1 and 2 increase, stay the same, decrease, or become zero? Explain.

FIGURE Q28.11

FIGURE Q28.12

12. Bulbs A and B in **FIGURE Q28.12** are identical, and both are glowing. What happens to each bulb when the switch is closed? Does its brightness increase, stay the same, decrease, or go out? Explain.

13. **FIGURE Q28.13** shows the voltage as a function of time of a capacitor as it is discharged (separately) through three different resistors. Rank in order, from largest to smallest, the values of the resistances R_1 to R_3.

FIGURE Q28.13

EXERCISES AND PROBLEMS

Problems labeled   integrate material from earlier chapters.

Exercises

Section 28.1 Circuit Elements and Diagrams

1. | Draw a circuit diagram for the circuit of **FIGURE EX28.1**.

FIGURE EX28.1

FIGURE EX28.2

2. | Draw a circuit diagram for the circuit of **FIGURE EX28.2**.

Section 28.2 Kirchhoff's Laws and the Basic Circuit

3. || In **FIGURE EX28.3**, what is the magnitude of the current in the wire to the right of the junction? Does the charge in this wire flow to the right or to the left?

FIGURE EX28.3

FIGURE EX28.4

4. | a. What are the magnitude and direction of the current in the 10 Ω resistor in **FIGURE EX28.4**?
   b. Draw a graph of the potential as a function of the distance traveled through the circuit, traveling cw from $V = 0$ V at the lower left corner.

5. | a. What are the magnitude and direction of the current in the 18 Ω resistor in **FIGURE EX28.5**?
   b. Draw a graph of the potential as a function of the distance traveled through the circuit, traveling cw from $V = 0$ V at the lower left corner.

FIGURE EX28.5

FIGURE EX28.6

6. | What is the magnitude of the potential difference across each resistor in **FIGURE EX28.6**?

Section 28.3 Energy and Power

7. | What is the resistance of a 1500 W (120 V) hair dryer? What is the current in the hair dryer when it is used?

8. | How much power is dissipated by each resistor in **FIGURE EX28.8**?

FIGURE EX28.8

FIGURE EX28.9

9. || A 60 W lightbulb and a 100 W lightbulb are placed in the circuit shown in **FIGURE EX28.9**. Both bulbs are glowing.
   a. Which bulb is brighter? Or are they equally bright?
   b. Calculate the power dissipated by each bulb.

10. || The five identical bulbs in **FIGURE EX28.10** are all glowing. The battery is ideal. What is the order of brightness of the bulbs, from brightest to dimmest? Some may be equal.
    A. $P = S > Q = R = T$
    B. $P = S = T > Q = R$
    C. $P > S = T > Q = R$
    D. $P > Q = R > S = T$

FIGURE EX28.10

FIGURE EX28.11

11. || The five identical bulbs in **FIGURE EX28.11** are all glowing. The battery is ideal. What is the order of brightness of the bulbs, from brightest to dimmest? Some may be equal.
    A. $P = T > Q = R = S$
    B. $P > Q = R = S > T$
    C. $P = T > Q > R = S$
    D. $P > Q > T > R = S$

12. | 1 kWh is how many joules?

13. || A standard 100 W (120 V) lightbulb contains a 7.0-cm-long tungsten filament. The high-temperature resistivity of tungsten is 9.0×10^{-7} Ω m. What is the diameter of the filament?

14. | A typical American family uses 1000 kWh of electricity a month.
    a. What is the average current in the 120 V power line to the house?
    b. On average, what is the resistance of a household?

15. | A waterbed heater uses 450 W of power. It is on 25% of the time, off 75%. What is the annual cost of electricity at a billing rate of $0.12/kWh?

Section 28.4 Series Resistors

Section 28.5 Real Batteries

16. | What is the value of resistor R in **FIGURE EX28.16**?

FIGURE EX28.16

17. | The battery in **FIGURE EX28.17** is short-circuited by an ideal ammeter having zero resistance.
    a. What is the battery's internal resistance?
    b. How much power is dissipated inside the battery?

FIGURE EX28.17

FIGURE EX28.18

18. || The circuit in **FIGURE EX28.18** is called a *voltage divider*. What value of R will make $V_{out} = V_{in}/10$?

19. || The voltage across the terminals of a 9.0 V battery is 8.5 V when the battery is connected to a 20 Ω load. What is the battery's internal resistance?

20. || Compared to an ideal battery, by what percentage does the battery's internal resistance reduce the potential difference across the 20 Ω resistor in **FIGURE EX28.20**?

FIGURE EX28.20

FIGURE EX28.21

21. | A variable resistor R is connected across the terminals of a battery. **FIGURE EX28.21** shows the current in the circuit as R is varied. What are the emf and internal resistance of the battery?

Section 28.6 Parallel Resistors

22. | Two of the three resistors in **FIGURE EX28.22** are unknown but equal. The total resistance between points a and b is 75 Ω. What is the value of R?

FIGURE EX28.22

FIGURE EX28.23

23. || What is the value of resistor R in **FIGURE EX28.23**?

24. | A metal wire of resistance R is cut into two pieces of equal length. The two pieces are connected together side by side. What is the resistance of the two connected wires?

25. | What is the equivalent resistance between points a and b in **FIGURE EX28.25**?

FIGURE EX28.25

FIGURE EX28.26

26. | What is the equivalent resistance between points a and b in **FIGURE EX28.26**?

27. | What is the equivalent resistance between points a and b in **FIGURE EX28.27**?

FIGURE EX28.27

FIGURE EX28.28

28. | What is the equivalent resistance between points a and b in **FIGURE EX28.28**?

29. || The 10 Ω resistor in **FIGURE EX28.29** is dissipating 40 W of power. How much power are the other two resistors dissipating?

FIGURE EX28.29

Section 28.8 Getting Grounded

30. || In **FIGURE EX28.30**, what is the value of the potential at points a and b?

FIGURE EX28.30

FIGURE EX28.31

31. ||| In **FIGURE EX28.31**, what is the value of the potential at points a and b?

Section 28.9 *RC* Circuits

32. | Show that the product RC has units of s.

33. | What is the time constant for the discharge of the capacitors in **FIGURE EX28.33**?

FIGURE EX28.33   **FIGURE EX28.34**

34. | What is the time constant for the discharge of the capacitors in **FIGURE EX28.34**?

35. ‖ A 10 μF capacitor initially charged to 20 μC is discharged through a 1.0 kΩ resistor. How long does it take to reduce the capacitor's charge to 10 μC?

36. | The switch in **FIGURE EX28.36** has been in position a for a long time. It is changed to position b at $t = 0$ s. What are the charge Q on the capacitor and the current I through the resistor (a) immediately after the switch is closed? (b) at $t = 50$ μs? (c) at $t = 200$ μs?

FIGURE EX28.36

37. ‖ What value resistor will discharge a 1.0 μF capacitor to 10% of its initial charge in 2.0 ms?

38. ‖ A capacitor is discharged through a 100 Ω resistor. The discharge current decreases to 25% of its initial value in 2.5 ms. What is the value of the capacitor?

Problems

39. ‖ It seems hard to justify spending $4.00 for a compact fluorescent lightbulb when an ordinary incandescent bulb costs 50¢. To see if this makes sense, compare a 60 W incandescent bulb lasting 1000 hours to a 15 W compact fluorescent bulb having a lifetime of 10,000 hours. Both bulbs produce the same amount of visible light and are interchangeable. If electricity costs $0.10/kWh, what is the total cost—purchase plus energy— to obtain 10,000 hours of light from each type of bulb? This is called the *life-cycle cost*.

40. ‖ A refrigerator has a 1000 W compressor, but the compressor runs only 20% of the time.
   a. If electricity costs $0.10/kWh, what is the monthly (30 day) cost of running the refrigerator?
   b. A more energy-efficient refrigerator with an 800 W compressor costs $100 more. If you buy the more expensive refrigerator, how many months will it take to recover your additional cost?

41. ‖ Two 75 W (120 V) lightbulbs are wired in series, then the combination is connected to a 120 V supply. How much power is dissipated by each bulb?

42. ‖ An electric eel develops a 450 V potential difference between
   BIO its head and tail. The eel can stun a fish or other prey by using this potential difference to drive a 0.80 A current pulse for 1.0 ms. What are (a) the energy delivered by this pulse and (b) the total charge that flows?

43. ‖ You have a 2.0 Ω resistor, a 3.0 Ω resistor, a 6.0 Ω resistor, and a 6.0 V battery. Draw a diagram of a circuit in which all three resistors are used and the battery delivers 9.0 W of power.

44. ‖ A 2.0-m-long, 1.0-mm-diameter wire has a variable resistivity
   CALC given by

$$\rho(x) = (2.5 \times 10^{-6})\left[1 + \left(\frac{x}{1.0\text{ m}}\right)^2\right]\Omega\text{ m}$$

where x is measured from one end of the wire. What is the current if this wire is connected to the terminals of a 9.0 V battery?

45. ‖ What is the equivalent resistance between points a and b in **FIGURE P28.45**?

FIGURE P28.45   **FIGURE P28.46**

46. ‖ What are the emf and internal resistance of the battery in **FIG-URE P28.46**?

47. ‖ A string of holiday lights can be wired in series, but all the bulbs go out if one burns out because that breaks the circuit. Most lights today are wired in series, but each bulb has a special fuse that short-circuits the bulb—making a connection around it—if it burns out, thus keeping the other lights on. Suppose a string of 50 lights is connected in this way and plugged into a 120 V outlet. By what factor does the power dissipated by each remaining bulb increase when the first bulb burns out?

48. ‖ The circuit shown in **FIGURE P28.48** is inside a 15-cm-diameter balloon filled with helium that is kept at a constant pressure of 1.2 atm. How long will it take the balloon's diameter to increase to 16 cm?

FIGURE P28.48

49. ‖ Suppose you have resistors 2.5 kΩ, 3.5 kΩ, and 4.5 kΩ and a 100 V power supply. What is the ratio of the total power delivered to the resistors if they are connected in parallel to the total power delivered if they are connected in series?

50. ‖‖ A lightbulb is in series with a 2.0 Ω resistor. The lightbulb dissipates 10 W when this series circuit is connected to a 9.0 V battery. What is the current through the lightbulb? There are two possible answers; give both of them.

51. ‖ a. Load resistor R is attached to a battery of emf $\mathcal{E}$ and
   CALC internal resistance r. For what value of the resistance R, in terms of $\mathcal{E}$ and r, will the power dissipated by the load resistor be a maximum?
   b. What is the maximum power that the load can dissipate if the battery has $\mathcal{E} = 9.0$ V and $r = 1.0$ Ω?

52. ‖ The ammeter in **FIGURE P28.52** reads 3.0 A. Find I_1, I_2, and $\mathcal{E}$.

FIGURE P28.52

53. ‖ What is the current in the 2 Ω resistor in **FIGURE P28.53**?

FIGURE P28.53   **FIGURE P28.54**

54. ‖ For an ideal battery ($r = 0$ Ω), closing the switch in **FIGURE P28.54** does not affect the brightness of bulb A. In practice, bulb A dims *just a little* when the switch closes. To see why, assume that the 1.50 V battery has an internal resistance $r = 0.50$ Ω and that the resistance of a glowing bulb is $R = 6.00$ Ω.
   a. What is the current through bulb A when the switch is open?
   b. What is the current through bulb A after the switch has closed?
   c. By what percentage does the current through A change when the switch is closed?

55. ‖ What are the battery current I_{bat} and the potential difference $V_a - V_b$ between points a and b when the switch in **FIGURE P28.55** is (a) open and (b) closed?

FIGURE P28.55   **FIGURE P28.56**

56. ‖ A battery is a voltage source, always providing the same potential difference regardless of the current. It is possible to make a *current source* that always provides the same current regardless of the potential difference. The circuit in **FIGURE P28.56** is called a *current divider*. It sends a fraction of the source current to the load. Find an expression for I_{load} in terms of R_1, R_2, and I_{source}. You can assume that the load's resistance is much less than R_2.

57. ‖ A circuit you're building needs an ammeter that goes from 0 mA to a full-scale reading of 50 mA. Unfortunately, the only ammeter in the storeroom goes from 0 μA to a full-scale reading of only 500 μA. Fortunately, you've just finished a physics class, and you realize that you can make this ammeter work by putting a resistor in parallel with it, as shown in **FIGURE P28.57**. You've measured that the resistance of the ammeter is 50.0 Ω, not the 0 Ω of an ideal ammeter.
   a. What value of R must you use so that the meter will go to full scale when the current I is 50 mA?
   b. What is the effective resistance of your ammeter?

FIGURE P28.57

58. ‖ For the circuit shown in **FIGURE P28.58**, find the current through and the potential difference across each resistor. Place your results in a table for ease of reading.

FIGURE P28.58   **FIGURE P28.59**

59. ‖ For the circuit shown in **FIGURE P28.59**, find the current through and the potential difference across each resistor. Place your results in a table for ease of reading.

60. ‖ For the circuit shown in **FIGURE P28.60**, find the current through and the potential difference across each resistor. Place your results in a table for ease of reading.

FIGURE P28.60   **FIGURE P28.61**

61. ‖ What is the current through the 20 Ω resistor in **FIGURE P28.61**?

62. ‖ For the circuit shown in **FIGURE P28.62**, find the current through and the potential difference across each resistor. Place your results in a table for ease of reading.

FIGURE P28.62   **FIGURE P28.63**

63. ‖ What is the current through the 10 Ω resistor in **FIGURE P28.63**? Is the current from left to right or right to left?

64. ‖ For what emf $\mathcal{E}$ does the 200 Ω resistor in **FIGURE P28.64** dissipate no power? Should the emf be oriented with its positive terminal at the top or at the bottom?

FIGURE P28.64

65. ‖ A 12 V car battery dies not so much because its voltage drops but because chemical reactions increase its internal resistance. A good battery connected with jumper cables can both start the engine and recharge the dead battery. Consider the automotive circuit of **FIGURE P28.65**.
   a. How much current could the good battery alone drive through the starter motor?
   b. How much current is the dead battery alone able to drive through the starter motor?
   c. With the jumper cables attached, how much current passes through the starter motor?
   d. With the jumper cables attached, how much current passes through the dead battery, and in which direction?

FIGURE P28.65

66. ‖ How much current flows through the bottom wire in **FIGURE P28.66**, and in which direction?

FIGURE P28.66

67. ‖ The capacitor in an RC circuit is discharged with a time constant of 10 ms. At what time after the discharge begins are (a) the charge on the capacitor reduced to half its initial value and (b) the energy stored in the capacitor reduced to half its initial value?

68. ‖ A circuit you're using discharges a 20 μF capacitor through an unknown resistor. After charging the capacitor, you close a switch at $t = 0$ s and then monitor the resistor current with an ammeter. Your data are as follows:

Time (s)	Current (μA)
0.5	890
1.0	640
1.5	440
2.0	270
2.5	200

Use an appropriate graph of the data to determine (a) the resistance and (b) the initial capacitor voltage.

69. ‖ A 150 μF defibrillator capacitor is charged to 1500 V. When
BIO fired through a patient's chest, it loses 95% of its charge in 40 ms. What is the resistance of the patient's chest?

70. ‖ A 50 μF capacitor that had been charged to 30 V is discharged through a resistor. **FIGURE P28.70** shows the capacitor voltage as a function of time. What is the value of the resistance?

FIGURE P28.70

71. ‖ A 0.25 μF capacitor is charged to 50 V. It is then connected in series with a 25 Ω resistor and a 100 Ω resistor and allowed to discharge completely. How much energy is dissipated by the 25 Ω resistor?

72. | A 70 μF capacitor is discharged through two parallel resistors, 15 kΩ and 25 kΩ. By what factor will the time constant of this circuit increase if the resistors are instead placed in series with each other?

73. ‖ The capacitor in **FIGURE P28.73**
CALC begins to charge after the switch closes at $t = 0$ s.
   a. What is ΔV_C a very long time after the switch has closed?
   b. What is Q_{max} in terms of $\mathcal{E}$, R, and C?
   c. In this circuit, does $I = +dQ/dt$ or $-dQ/dt$? Explain.
   d. Find an expression for the current I at time t. Graph I from $t = 0$ to $t = 5\tau$.

FIGURE P28.73

74. ‖ The capacitors in **FIGURE P28.74** are charged and the switch closes at $t = 0$ s. At what time has the current in the 8 Ω resistor decayed to half the value it had immediately after the switch was closed?

FIGURE P28.74

75. ‖ The flash on a compact camera stores energy in a 120 μF capacitor that is charged to 220 V. When the flash is fired, the capacitor is quickly discharged through a lightbulb with 5.0 Ω of resistance.
   a. Light from the flash is essentially finished after two time constants have elapsed. For how long does this flash illuminate the scene?
   b. At what rate is the lightbulb dissipating energy 250 μs after the flash is fired?
   c. What total energy is dissipated by the lightbulb?

76. ‖ Large capacitors can hold a potentially dangerous charge long
BIO after a circuit has been turned off, so it is important to make sure they are discharged before you touch them. Suppose a 120 μF capacitor from a camera flash unit retains a voltage of 150 V when an unwary student removes it from the camera. If the student accidentally touches the two terminals with his hands, and if the resistance of his body between his hands is 1.8 kΩ, for how long will the current across his chest exceed the danger level of 50 mA?

77. ‖ Digital circuits require actions to take place at precise times, so they are controlled by a *clock* that generates a steady sequence of rectangular voltage pulses. One of the most widely used integrated circuits for creating clock pulses is called a 555 timer. **FIGURE P28.77** shows how the timer's output pulses, oscillating between 0 V and 5 V, are controlled with two resistors and a capacitor. The circuit manufacturer tells users that T_H, the time the clock output spends in the high (5 V) state, is $T_H = (R_1 + R_2)C \times \ln 2$. Similarly, the time spent in the low (0 V) state is $T_L = R_2 C \times \ln 2$. You need to design a clock that

will oscillate at 10 MHz and will spend 75% of each cycle in the high state. You will be using a 500 pF capacitor. What values do you need to specify for R_1 and R_2?

FIGURE P28.77

Challenge Problems

78. ||| What power is dissipated by the $2\,\Omega$ resistor in **FIGURE CP28.78**?

FIGURE CP28.78

79. ||| You've made the finals of the Science Olympics! As one of your tasks, you're given 1.0 g of aluminum and asked to make a wire, using all the aluminum, that will dissipate 7.5 W when connected to a 1.5 V battery. What length and diameter will you choose for your wire?

80. ||| The switch in **FIGURE CP28.80** has been closed for a very long time.
   a. What is the charge on the capacitor?
   b. The switch is opened at $t = 0$ s. At what time has the charge on the capacitor decreased to 10% of its initial value?

FIGURE CP28.80

81. ||| The capacitor in Figure 28.38a begins to charge after the
CALC switch closes at $t = 0$ s. Analyze this circuit and show that $Q = Q_{max}(1 - e^{-t/\tau})$, where $Q_{max} = C\mathcal{E}$.

82. ||| The switch in Figure 28.38a closes at $t = 0$ s and, after a very
CALC long time, the capacitor is fully charged. Find expressions for (a) the total energy supplied by the battery as the capacitor is being charged, (b) total energy dissipated by the resistor as the capacitor is being charged, and (c) the energy stored in the capacitor when it is fully charged. Your expressions will be in terms of $\mathcal{E}$, R, and C. (d) Do your results for parts a to c show that energy is conserved? Explain.

83. ||| An *oscillator circuit* is important to many applications. A simple oscillator circuit can be built by adding a neon gas tube to an RC circuit, as shown in **FIGURE CP28.83**. Gas is normally a good insulator, and the resistance of the gas tube is essentially infinite when the light is off. This allows the capacitor to charge. When the capacitor voltage reaches a value V_{on}, the electric field inside the tube becomes strong enough to ionize the neon gas. Visually, the tube lights with an orange glow. Electrically, the ionization of the gas provides a very-low-resistance path through the tube. The capacitor very rapidly (we can think of it as instantaneously) discharges through the tube and the capacitor voltage drops. When the capacitor voltage has dropped to a value V_{off}, the electric field inside the tube becomes too weak to sustain the ionization and the neon light turns off. The capacitor then starts to charge again. The capacitor voltage oscillates between V_{off}, when it starts charging, and V_{on}, when the light comes on to discharge it.
   a. Show that the oscillation period is

$$T = RC \ln\left(\frac{\mathcal{E} - V_{off}}{\mathcal{E} - V_{on}}\right)$$

   b. A neon gas tube has $V_{on} = 80$ V and $V_{off} = 20$ V. What resistor value should you choose to go with a 10 μF capacitor and a 90 V battery to make a 10 Hz oscillator?

FIGURE CP28.83

29 The Magnetic Field

The aurora occurs when high-energy charged particles from the sun are steered into the upper atmosphere by the earth's magnetic field.

IN THIS CHAPTER, you will learn about magnetism and the magnetic field.

What is magnetism?

Magnetism is an interaction between *moving* charges.

- Magnetic forces, similar to electric forces, are due to the action of magnetic fields.
- A magnetic field $\vec{B}$ is created by a moving charge.
- Magnetic interactions are understood in terms of magnetic poles: north and south.
- Magnetic poles never occur in isolation. All magnets are dipoles, with two poles.
- Practical magnetic fields are created by currents—collections of moving charges.
- Magnetic materials, such as iron, occur because electrons have an inherent magnetic dipole called electron spin.

Magnetic field lines

What fields are especially important?

We will develop and use three important magnetic field models.

Long, straight wire

Current loop

Solenoid

How do charges respond to magnetic fields?

A charged particle *moving* in a magnetic field experiences a force perpendicular to both $\vec{B}$ and $\vec{v}$. The perpendicular force causes charged particles to move in circular orbits in a uniform magnetic field. This cyclotron motion has many important applications.

《 LOOKING BACK Sections 8.2–8.3 Circular motion

《 LOOKING BACK Section 12.10 The cross product

How do currents respond to magnetic fields?

Currents are moving charged particles, so:

- There's a force on a current-carrying wire in a magnetic field.
- Two parallel current-carrying wires attract or repel each other.
- There's a torque on a current loop in a magnetic field. This is how motors work.

Why is magnetism important?

Magnetism is much more important than a way to hold a shopping list on the refrigerator door. Motors and generators are based on magnetic forces. Many forms of data storage, from hard disks to the stripe on your credit card, are magnetic. Magnetic resonance imaging (MRI) is essential to modern medicine. Magnetic levitation trains are being built around the world. And the earth's magnetic field keeps the solar wind from sterilizing the surface. There would be no life and no modern technology without magnetism.

29.1 Magnetism

We began our investigation of electricity in Chapter 22 by looking at the results of simple experiments with charged rods. We'll do the same with magnetism.

Discovering magnetism

Experiment 1

If a bar magnet is taped to a piece of cork and allowed to float in a dish of water, it always turns to align itself in an approximate north-south direction. The end of a magnet that points north is called the *north-seeking pole,* or simply the **north pole.** The other end is the **south pole.**

North

South

The needle of a compass is a small magnet.

Experiment 2

If the north pole of one magnet is brought near the north pole of another magnet, they repel each other. Two south poles also repel each other, but the north pole of one magnet exerts an attractive force on the south pole of another magnet.

Experiment 3

The north pole of a bar magnet attracts one end of a compass needle and repels the other. Apparently the compass needle itself is a little bar magnet with a north pole and a south pole.

Experiment 4

Cutting a bar magnet in half produces two weaker but still complete magnets, each with a north pole and a south pole. No matter how small the magnets are cut, even down to microscopic sizes, each piece remains a complete magnet with two poles.

Experiment 5

Magnets can pick up some objects, such as paper clips, but not all. If an object is attracted to one end of a magnet, it is also attracted to the other end. Most materials, including copper (a penny), aluminum, glass, and plastic, experience no force from a magnet.

Experiment 6

A magnet does not affect an electroscope. A charged rod exerts a weak *attractive* force on *both* ends of a magnet. However, the force is the same as the force on a metal bar that isn't a magnet, so it is simply a polarization force like the ones we studied in Chapter 22. Other than polarization forces, charges have *no effects* on magnets.

No effect

What do these experiments tell us?

- First, magnetism is not the same as electricity. **Magnetic poles and electric charges share some similar behavior, but they are not the same.**
- Magnetism is a long-range force. Paper clips leap upward to a magnet.
- Magnets have two poles, called north and south poles, and thus are **magnetic dipoles.** Two like poles exert repulsive forces on each other; two opposite poles attract. The behavior is *analogous* to electric charges, but, as noted, magnetic poles and electric charges are *not* the same. Unlike charges, isolated north or south poles do not exist.
- The poles of a bar magnet can be identified by using it as a compass. The poles of other magnets, such as flat refrigerator magnets, can be identified by testing them against a bar magnet. A pole that attracts a known north pole and repels a known south pole must be a south magnetic pole.
- Materials that are attracted to a magnet are called **magnetic materials.** The most common magnetic material is iron. Magnetic materials are attracted to *both* poles of a magnet. This attraction is analogous to how neutral objects are attracted to both positively and negatively charged rods by the polarization force. The difference is that *all* neutral objects are attracted to a charged rod whereas only a few materials are attracted to a magnet.

Our goal is to develop a theory of magnetism to explain these observations.

Compasses and Geomagnetism

The north pole of a compass needle is attracted toward the geographic north pole of the earth. Apparently the earth itself is a large magnet, as shown in **FIGURE 29.1**. The

FIGURE 29.1 The earth is a large magnet.

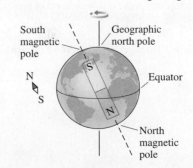

South magnetic pole

Geographic north pole

Equator

North magnetic pole

reasons for the earth's magnetism are complex, but geophysicists think that the earth's magnetic poles arise from currents in its molten iron core. Two interesting facts about the earth's magnetic field are (1) that the magnetic poles are offset slightly from the geographic poles of the earth's rotation axis, and (2) that the geographic north pole is actually a *south* magnetic pole! You should be able to use what you have learned thus far to convince yourself that this is the case.

STOP TO THINK 29.1 Does the compass needle rotate clockwise (cw), counterclockwise (ccw), or not at all?

Positive rod

Pivot

29.2 The Discovery of the Magnetic Field

As electricity began to be seriously studied in the 18th century, some scientists speculated that there might be a connection between electricity and magnetism. Interestingly, the link between electricity and magnetism was discovered *in the midst of a classroom lecture demonstration* in 1819 by the Danish scientist Hans Christian Oersted. Oersted was using a battery—a fairly recent invention—to produce a large current in a wire. By chance, a compass was sitting next to the wire, and Oersted noticed that the current caused the compass needle to turn. In other words, the compass responded as if a magnet had been brought near.

Oersted had long been interested in a possible connection between electricity and magnetism, so the significance of this serendipitous observation was immediately apparent to him. Oersted's discovery that **magnetism is caused by an electric current** is illustrated in **FIGURE 29.2**. Part c of the figure demonstrates an important **right-hand rule** that relates the orientation of the compass needles to the direction of the current.

Iron filings reveal the magnetic field around a current-carrying wire.

FIGURE 29.2 Response of compass needles to a current in a straight wire.

(a)

No current

With no current, the compass needles point north.

N

(b)

Current-carrying wire

I

With a strong current, the compass needles are tangent to a circle around the wire.

(c)

I

Right-hand rule: Point your right thumb in the direction of the current.

The compass needles are tangent to the circle . . .

. . . with the north pole in the direction your fingers are pointing.

FIGURE 29.3 The notation for vectors and currents perpendicular to the page.

× × × ×
× × × ×
Vectors into page

• • • •
• • • •
Vectors out of page

⊗
Current into page

⊙
Current out of page

Magnetism is more demanding than electricity in requiring a three-dimensional perspective of the sort shown in Figure 29.2. But since two-dimensional figures are easier to draw, we will make as much use of them as we can. Consequently, we will often need to indicate field vectors or currents that are perpendicular to the page. **FIGURE 29.3** shows the notation we will use. **FIGURE 29.4** demonstrates this notation by showing the compasses around a current that is directed into the page. To use the right-hand rule, point your right thumb in the direction of the current (into the page). Your fingers will curl cw, and that is the direction in which the north poles of the compass needles point.

The Magnetic Field

We introduced the idea of a *field* as a way to understand the long-range electric force. Although this idea appeared rather far-fetched, it turned out to be very useful.

We need a similar idea to understand the long-range force exerted by a current on a compass needle.

Let us define the **magnetic field** $\vec{B}$ as having the following properties:

1. A magnetic field is created at *all* points in space surrounding a current-carrying wire.
2. The magnetic field at each point is a vector. It has both a magnitude, which we call the *magnetic field strength B,* and a direction.
3. The magnetic field exerts forces on magnetic poles. The force on a north pole is parallel to $\vec{B}$; the force on a south pole is opposite $\vec{B}$.

FIGURE 29.5 shows a compass needle in a magnetic field. The field vectors are shown at several points, but keep in mind that the field is present at *all* points in space. A magnetic force is exerted on each of the two poles of the compass, parallel to $\vec{B}$ for the north pole and opposite $\vec{B}$ for the south pole. This pair of opposite forces exerts a torque on the needle, rotating the needle until it is parallel to the magnetic field at that point.

Notice that the north pole of the compass needle, when it reaches the equilibrium position, is in the direction of the magnetic field. Thus a compass needle can be used as a probe of the magnetic field, just as a charge was a probe of the electric field. **Magnetic forces cause a compass needle to become aligned parallel to a magnetic field, with the north pole of the compass showing the direction of the magnetic field at that point.**

Look back at the compass alignments around the current-carrying wire in Figure 29.4. Because compass needles align with the magnetic field, the magnetic field at each point must be tangent to a circle around the wire. FIGURE 29.6a shows the magnetic field by drawing field vectors. Notice that the field is weaker (shorter vectors) at greater distances from the wire.

Another way to picture the field is with the use of **magnetic field lines.** These are imaginary lines drawn through a region of space so that

- A tangent to a field line is in the direction of the magnetic field, and
- The field lines are closer together where the magnetic field strength is larger.

FIGURE 29.6b shows the magnetic field lines around a current-carrying wire. Notice that magnetic field lines form loops, with no beginning or ending point. This is in contrast to electric field lines, which stop and start on charges.

TACTICS BOX 29.1 (MP)

Right-hand rule for fields

❶ Point your *right* thumb in the direction of the current.

❷ Curl your fingers around the wire to indicate a circle.

❸ Your fingers point in the direction of the magnetic field lines around the wire.

Exercises 6–8

NOTE The magnetic field of a current-carrying wire is very different from the electric field of a charged wire. The electric field of a charged wire points radially outward (positive wire) or inward (negative wire).

Two Kinds of Magnetism?

You might be concerned that we have introduced two kinds of magnetism. We opened this chapter discussing permanent magnets and their forces. Then, without warning, we switched to the magnetic forces caused by a current. It is not at all obvious that these

FIGURE 29.4 The orientation of the compasses is given by the right-hand rule.

Current into page

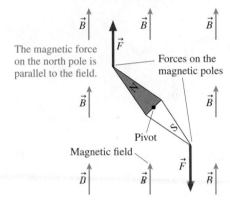

FIGURE 29.5 The magnetic field exerts forces on the poles of a compass.

The magnetic force on the north pole is parallel to the field.

Forces on the magnetic poles

Pivot

Magnetic field

FIGURE 29.6 The magnetic field around a current-carrying wire.

(a) The magnetic field *vectors* are tangent to circles around the wire, pointing in the direction given by the right-hand rule.

The field is weaker farther from the wire.

(b)

Magnetic field *lines* are circles.

forces are the same kind of magnetism as that exhibited by stationary chunks of metal called "magnets." Perhaps there are two different types of magnetic forces, one having to do with currents and the other being responsible for permanent magnets. One of the major goals for our study of magnetism is to see that these two quite different ways of producing magnetic effects are really just two different aspects of a *single* magnetic force.

STOP TO THINK 29.2   The magnetic field at position P points

a. Up.   b. Down.
c. Into the page.   d. Out of the page.

FIGURE 29.7 The magnetic field of a moving point charge.

This is the point at which we want to find $\vec{B}$.

Magnetic field of the moving point charge

$\vec{B}$

Point charge q

$\vec{r}$

θ

$\vec{v}$

Velocity of the charged particle

29.3 The Source of the Magnetic Field: Moving Charges

Figure 29.6 is a qualitative picture of the wire's magnetic field. Our first task is to turn that picture into a quantitative description. Because current in a wire generates a magnetic field, and a current is a collection of moving charges, our starting point is the idea that **moving charges are the source of the magnetic field.** FIGURE 29.7 shows a charged particle q moving with velocity $\vec{v}$. The magnetic field of this moving charge is found to be

$$\vec{B}_{\text{point charge}} = \left(\frac{\mu_0}{4\pi}\frac{qv\sin\theta}{r^2}, \text{direction given by the right-hand rule}\right) \quad (29.1)$$

where r is the distance from the charge and θ is the angle between $\vec{v}$ and $\vec{r}$.

Equation 29.1 is called the **Biot-Savart law** for a point charge (rhymes with *Leo* and *bazaar*), named for two French scientists whose investigations were motivated by Oersted's observations. It is analogous to Coulomb's law for the electric field of a point charge. Notice that the Biot-Savart law, like Coulomb's law, is an inverse-square law. However, the Biot-Savart law is somewhat more complex than Coulomb's law because the magnetic field depends on the angle θ between the charge's velocity and the line to the point where the field is evaluated.

| NOTE ► A moving charge has both a magnetic field *and* an electric field. What you know about electric fields has not changed. ◄

The SI unit of magnetic field strength is the **tesla,** abbreviated as T. The tesla is defined as

$$1 \text{ tesla} = 1 \text{ T} \equiv 1 \text{ N/A m}$$

You will see later in the chapter that this definition is based on the magnetic force on a current-carrying wire. One tesla is quite a large field; most magnetic fields are a small fraction of a tesla. TABLE 29.1 lists a few magnetic field strengths.

The constant μ_0 in Equation 29.1 is called the **permeability constant.** Its value is

$$\mu_0 = 4\pi \times 10^{-7} \text{ T m/A} = 1.257 \times 10^{-6} \text{ T m/A}$$

This constant plays a role in magnetism similar to that of the permittivity constant ϵ_0 in electricity.

The right-hand rule for finding the direction of $\vec{B}$ is similar to the rule used for a current-carrying wire: Point your right thumb in the direction of $\vec{v}$. The magnetic field vector $\vec{B}$ is perpendicular to the plane of $\vec{r}$ and $\vec{v}$, pointing in the direction in which your fingers curl. In other words, the $\vec{B}$ vectors are tangent to circles drawn about the charge's line of motion. FIGURE 29.8 shows a more complete view of the magnetic field of a moving positive charge. Notice that $\vec{B}$ is zero along the line of motion, where $\theta = 0°$ or 180°, due to the $\sin\theta$ term in Equation 29.1.

| NOTE ► The vector arrows in Figure 29.8 would have the same lengths but be reversed in direction for a negative charge. ◄

TABLE 29.1 Typical magnetic field strengths

Field source	Field strength (T)
Surface of the earth	5×10^{-5}
Refrigerator magnet	5×10^{-3}
Laboratory magnet	0.1 to 1
Superconducting magnet	10

FIGURE 29.8 Two views of the magnetic field of a moving positive charge.

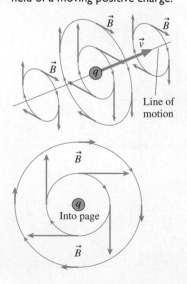

$\vec{B}$

$\vec{B}$

$\vec{v}$

$\vec{B}$

q

Line of motion

$\vec{B}$

q

Into page

$\vec{B}$

The requirement that a charge be moving to generate a magnetic field is explicit in Equation 29.1. If the speed v of the particle is zero, the magnetic field (but not the electric field!) is zero. This helps to emphasize a fundamental distinction between electric and magnetic fields: **All charges create electric fields, but only *moving* charges create magnetic fields.**

EXAMPLE 29.1 | **The magnetic field of a proton**

A proton moves with velocity $\vec{v} = 1.0 \times 10^7\,\hat{\imath}$ m/s. As it passes the origin, what is the magnetic field at the (x, y, z) positions (1 mm, 0 mm, 0 mm), (0 mm, 1 mm, 0 mm), and (1 mm, 1 mm, 0 mm)?

MODEL The magnetic field is that of a moving charged particle.

VISUALIZE FIGURE 29.9 shows the geometry. The first point is on the x-axis, directly in front of the proton, with $\theta_1 = 0°$. The second point is on the y-axis, with $\theta_2 = 90°$, and the third is in the xy-plane.

FIGURE 29.9 The magnetic field of Example 29.1.

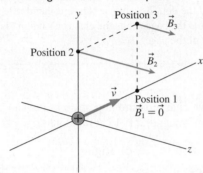

SOLVE Position 1, which is along the line of motion, has $\theta_1 = 0°$. Thus $\vec{B}_1 = \vec{0}$. Position 2 (at 0 mm, 1 mm, 0 mm) is at distance $r_2 = 1$ mm $= 0.001$ m. Equation 29.1, the Biot-Savart law, gives us the magnetic field strength at this point as

$$B = \frac{\mu_0}{4\pi} \frac{qv\sin\theta_2}{r_2^2}$$
$$= \frac{4\pi \times 10^{-7}\,\text{T m/A}}{4\pi} \frac{(1.60 \times 10^{-19}\,\text{C})(1.0 \times 10^7\,\text{m/s})\sin 90°}{(0.0010\,\text{m})^2}$$
$$= 1.60 \times 10^{-13}\,\text{T}$$

According to the right-hand rule, the field points in the positive z-direction. Thus

$$\vec{B}_2 = 1.60 \times 10^{-13}\,\hat{k}\,\text{T}$$

where $\hat{k}$ is the unit vector in the positive z-direction. The field at position 3, at (1 mm, 1 mm, 0 mm), also points in the z-direction, but it is weaker than at position 2 both because r is larger *and* because θ is smaller. From geometry we know $r_3 = \sqrt{2}$ mm $= 0.00141$ m and $\theta_3 = 45°$. Another calculation using Equation 29.1 gives

$$\vec{B}_3 = 0.57 \times 10^{-13}\,\hat{k}\,\text{T}$$

ASSESS The magnetic field of a single moving charge is *very* small.

Superposition

The Biot-Savart law is the starting point for generating all magnetic fields, just as our earlier expression for the electric field of a point charge was the starting point for generating all electric fields. Magnetic fields, like electric fields, have been found experimentally to obey the principle of superposition. If there are n moving point charges, the net magnetic field is given by the vector sum

$$\vec{B}_{\text{total}} = \vec{B}_1 + \vec{B}_2 + \cdots + \vec{B}_n \tag{29.2}$$

where each individual $\vec{B}$ is calculated with Equation 29.1. The principle of superposition will be the basis for calculating the magnetic fields of several important current distributions.

The Vector Cross Product

In « Section 22.5, we found that the electric field of a point charge can be written

$$\vec{E} = \frac{1}{4\pi\epsilon_0} \frac{q}{r^2}\hat{r}$$

where $\hat{r}$ is a *unit vector* that points from the charge to the point at which we wish to calculate the field. Unit vector $\hat{r}$ expresses the idea "away from q."

The unit vector $\hat{r}$ also allows us to write the Biot-Savart law more concisely, but we'll need to use the form of vector multiplication called the *cross product*. To remind you, FIGURE 29.10 shows two vectors, $\vec{C}$ and $\vec{D}$, with angle α between them. The **cross product** of $\vec{C}$ and $\vec{D}$ is defined to be the vector

$$\vec{C} \times \vec{D} = (CD\sin\alpha, \text{ direction given by the right-hand rule}) \tag{29.3}$$

The symbol $\times$ between the vectors is *required* to indicate a cross product.

FIGURE 29.10 The cross product $\vec{C} \times \vec{D}$ is a vector perpendicular to the plane of vectors $\vec{C}$ and $\vec{D}$.

The cross product is perpendicular to the plane. Its magnitude is $CD\sin\alpha$.

Plane of $\vec{C}$ and $\vec{D}$

FIGURE 29.11 Unit vector $\hat{r}$ defines the direction from the moving charge to the point at which we evaluate the field.

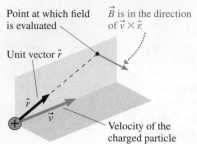

Point at which field is evaluated

$\vec{B}$ is in the direction of $\vec{v} \times \hat{r}$

Unit vector $\hat{r}$

$\hat{r}$

$\vec{v}$

Velocity of the charged particle

NOTE The cross product of two vectors and the right-hand rule used to determine the direction of the cross product were introduced in « Section 12.10 to describe torque and angular momentum. A review is worthwhile.

The Biot-Savart law, Equation 29.1, can be written in terms of the cross product as

$$\vec{B}_{\text{point charge}} = \frac{\mu_0}{4\pi} \frac{q\vec{v} \times \hat{r}}{r^2} \quad \text{(magnetic field of a point charge)} \quad (29.4)$$

where unit vector $\hat{r}$, shown in **FIGURE 29.11**, points from charge q to the point at which we want to evaluate the field. This expression for the magnetic field $\vec{B}$ has magnitude $(\mu_0/4\pi)qv \sin\theta/r^2$ (because the magnitude of $\hat{r}$ is 1) and points in the correct direction (given by the right-hand rule), so it agrees completely with Equation 29.1.

EXAMPLE 29.2 | **The magnetic field direction of a moving electron**

The electron in **FIGURE 29.12** is moving to the right. What is the direction of the electron's magnetic field at the dot?

▶ **FIGURE 29.12** A moving electron.

$\vec{v}$

$\hat{r}$

VISUALIZE Because the charge is negative, the magnetic field points *opposite* the direction of $\vec{v} \times \hat{r}$. Unit vector $\hat{r}$ points from the charge toward the dot. We can use the right-hand rule to find that $\vec{v} \times \hat{r}$ points *into* the page. Thus the electron's magnetic field at the dot points *out* of the page.

STOP TO THINK 29.3 The positive charge is moving straight out of the page. What is the direction of the magnetic field at the dot?

$\vec{v}$ out of page

a. Up   b. Down   c. Left   d. Right

29.4 The Magnetic Field of a Current

Moving charges are the source of magnetic fields, but the magnetic fields of current-carrying wires—immense numbers of charges moving together—are much more important than the feeble magnetic fields of individual charges. Real current-carrying wires, with their twists and turns, have very complex fields. However, we can once again focus on the physics by using simplified models. It turns out that three common magnetic field models are the basis for understanding a wide variety of magnetic phenomena. We present them here together as a reference; the next few sections of this chapter will be devoted to justifying and explaining these results.

MODEL 29.1

Three key magnetic fields

An infinite wire:
- Straight wires

$$B = \frac{\mu_0}{2\pi} \frac{I}{r}$$

A current loop:
- Flat coils

I N turns

$$B_{\text{center}} = \frac{\mu_0}{2} \frac{NI}{R}$$

A solenoid:
- Helical coils

Uniform field

$n = N/L$

$$B = \mu_0 nI$$

To begin, we need to write the Biot-Savart law in terms of current. FIGURE 29.13a shows a current-carrying wire. The wire as a whole is electrically neutral, but current I represents the motion of positive charge carriers through the wire. Suppose the small amount of moving charge ΔQ spans the small length Δs. The charge has velocity $\vec{v} = \Delta\vec{s}/\Delta t$, where the vector $\Delta\vec{s}$, which is parallel to $\vec{v}$, is the charge's displacement vector. If ΔQ is small enough to treat as a point charge, the magnetic field it creates at a point in space is proportional to $(\Delta Q)\vec{v}$. We can write $(\Delta Q)\vec{v}$ in terms of the wire's current I as

$$(\Delta Q)\vec{v} = \Delta Q \frac{\Delta\vec{s}}{\Delta t} = \frac{\Delta Q}{\Delta t}\Delta\vec{s} = I\Delta\vec{s} \qquad (29.5)$$

where we used the definition of current, $I = \Delta Q/\Delta t$.

If we replace $q\vec{v}$ in the Biot-Savart law with $I\Delta\vec{s}$, we find that the magnetic field of a very short segment of wire carrying current I is

$$\vec{B}_{\text{current segment}} = \frac{\mu_0}{4\pi}\frac{I\Delta\vec{s}\times\hat{r}}{r^2} \qquad (29.6)$$

(magnetic field of a very short segment of current)

Equation 29.6 is still the Biot-Savart law, only now written in terms of current rather than the motion of an individual charge. FIGURE 29.13b shows the direction of the current segment's magnetic field as determined by using the right-hand rule.

Equation 29.6 is the basis of a strategy for calculating the magnetic field of a current-carrying wire. You will recognize that it is the same basic strategy you learned for calculating the electric field of a continuous distribution of charge. The goal is to break a problem down into small steps that are individually manageable.

FIGURE 29.13 Relating the charge velocity $\vec{v}$ to the current I.

(a) Charge ΔQ in a small length Δs of a current-carrying wire

(b)

The magnetic field of the short segment of current is in the direction of $\Delta\vec{s}\times\hat{r}$.

PROBLEM-SOLVING STRATEGY 29.1

The magnetic field of a current

MODEL Model the wire as a simple shape.

VISUALIZE For the pictorial representation:
- Draw a picture, establish a coordinate system, and identify the point P at which you want to calculate the magnetic field.
- Divide the current-carrying wire into small segments for which you *already know* how to determine $\vec{B}$. This is usually, though not always, a division into very short segments of length Δs.
- Draw the magnetic field vector for one or two segments. This will help you identify distances and angles that need to be calculated.

SOLVE The mathematical representation is $\vec{B}_{\text{net}} = \Sigma\vec{B}_i$.
- Write an algebraic expression for *each* of the three components of $\vec{B}$ (unless you are sure one or more is zero) at point P. Let the (x, y, z)-coordinates of the point remain as variables.
- Express all angles and distances in terms of the coordinates.
- Let $\Delta s \rightarrow ds$ and the sum become an integral. Think carefully about the integration limits for this variable; they will depend on the boundaries of the wire and on the coordinate system you have chosen to use.

ASSESS Check that your result is consistent with any limits for which you know what the field should be.

The key idea here, as it was in Chapter 23, is that **integration is summation.** We need to add up the magnetic field contributions of a vast number of current segments, and we'll do that by letting the sum become an integral.

EXAMPLE 29.3 | **The magnetic field of a long, straight wire**

A long, straight wire carries current I in the positive x-direction. Find the magnetic field at distance r from the wire.

MODEL Model the wire as being infinitely long.

VISUALIZE FIGURE 29.14 illustrates the steps in the problem-solving strategy. We've chosen a coordinate system with point P on the y-axis. We've then divided the wire into small segments, labeled with index i, each containing a small amount ΔQ of *moving charge*. Unit vector $\hat{r}$ and angle θ_i are shown for segment i. You should use the right-hand rule to convince yourself that $\vec{B}_i$ points *out of the page*, in the positive z-direction. This is the direction no matter where segment i happens to be along the x-axis. Consequently, B_x (the component of $\vec{B}$ parallel to the wire) and B_y (the component of $\vec{B}$ straight away from the wire) are zero. The only component of $\vec{B}$ we need to evaluate is B_z, the component tangent to a circle around the wire.

FIGURE 29.14 Calculating the magnetic field of a long, straight wire carrying current I.

The z-axis is out of the page.   Divide the wire into segments.

SOLVE We can use the Biot-Savart law to find the field $(B_i)_z$ of segment i. The cross product $\Delta \vec{s}_i \times \hat{r}$ has magnitude $(\Delta x)(1)\sin\theta_i$, hence

$$(B_i)_z = \frac{\mu_0}{4\pi}\frac{I\Delta x\sin\theta_i}{r_i^2} = \frac{\mu_0}{4\pi}\frac{I\sin\theta_i}{r_i^2}\Delta x = \frac{\mu_0}{4\pi}\frac{I\sin\theta_i}{x_i^2+y^2}\Delta x$$

where we wrote the distance r_i in terms of x_i and y. We also need to express θ_i in terms of x_i and y. Because $\sin(180° - \theta) = \sin\theta$, this is

$$\sin\theta_i = \sin(180° - \theta_i) = \frac{y}{r_i} = \frac{y}{\sqrt{x_i^2+y^2}}$$

With this expression for $\sin\theta_i$, the magnetic field of segment i is

$$(B_i)_z = \frac{\mu_0}{4\pi}\frac{Iy}{(x_i^2+y^2)^{3/2}}\Delta x$$

Now we're ready to sum the magnetic fields of all the segments. The superposition is a vector sum, but in this case only the z-components are nonzero. Summing all the $(B_i)_z$ gives

$$B_{\text{wire}} = \frac{\mu_0 Iy}{4\pi}\sum_i\frac{\Delta x}{(x_i^2+y^2)^{3/2}} \rightarrow \frac{\mu_0 Iy}{4\pi}\int_{-\infty}^{\infty}\frac{dx}{(x^2+y^2)^{3/2}}$$

Only at the very last step did we convert the sum to an integral. Then our model of the wire as being infinitely long sets the integration limits at $\pm\infty$. This is a standard integral that can be found in Appendix A or with integration software. Evaluation gives

$$B_{\text{wire}} = \frac{\mu_0 Iy}{4\pi}\frac{x}{y^2(x^2+y^2)^{1/2}}\Big|_{-\infty}^{\infty} = \frac{\mu_0}{2\pi}\frac{I}{y}$$

This is the magnitude of the field. The field direction is determined by using the right-hand rule.

The coordinate system was our choice, and there's nothing special about the y-axis. The symbol y is simply the distance from the wire, which is better represented by r. With this change, the magnetic field is

$$\vec{B}_{\text{wire}} = \left(\frac{\mu_0}{2\pi}\frac{I}{r},\ \text{tangent to a circle around the wire}\ \text{in the right-hand direction}\right)$$

ASSESS FIGURE 29.15 shows the magnetic field of a current-carrying wire. Compare this to Figure 29.2 and convince yourself that the direction shown agrees with the right-hand rule.

FIGURE 29.15 The magnetic field of a long, straight wire carrying current I.

The magnetic field of an infinite, straight wire was the first of our key magnetic field models. Example 29.3 has shown that the field has magnitude

$$B_{\text{wire}} = \frac{\mu_0}{2\pi}\frac{I}{r} \quad (\text{long, straight wire}) \quad (29.7)$$

The field direction, encircling the wire, is given by the right-hand rule.

EXAMPLE 29.4 | The magnetic field strength near a heater wire

A 1.0-m-long, 1.0-mm-diameter nichrome heater wire is connected to a 12 V battery. What is the magnetic field strength 1.0 cm away from the wire?

MODEL 1 cm is much less than the 1 m length of the wire, so model the wire as infinitely long.

SOLVE The current through the wire is $I = \Delta V_{bat}/R$, where the wire's resistance R is

$$R = \frac{\rho L}{A} = \frac{\rho L}{\pi r^2} = 1.91 \ \Omega$$

The nichrome resistivity $\rho = 1.50 \times 10^{-6} \ \Omega \, m$ was taken from Table 27.2. Thus the current is $I = (12 \ V)/(1.91 \ \Omega) = 6.28 \ A$. The magnetic field strength at distance $d = 1.0 \ cm = 0.010 \ m$ from the wire is

$$B_{wire} = \frac{\mu_0}{2\pi} \frac{I}{d} = (2.0 \times 10^{-7} \ T \, m/A) \frac{6.28 \ A}{0.010 \ m}$$

$$= 1.3 \times 10^{-4} \ T$$

ASSESS The magnetic field of the wire is slightly more than twice the strength of the earth's magnetic field.

Motors, loudspeakers, metal detectors, and many other devices generate magnetic fields with *coils* of wire. The simplest coil is a single-turn circular loop of wire. A circular loop of wire with a circulating current is called a **current loop**.

EXAMPLE 29.5 | The magnetic field of a current loop

FIGURE 29.16a shows a current loop, a circular loop of wire with radius R that carries current I. Find the magnetic field of the current loop at distance z on the axis of the loop.

FIGURE 29.16 A current loop.

(a) A practical current loop

(b) An ideal current loop

MODEL Real coils need wires to bring the current in and out, but we'll model the coil as a current moving around the full circle shown in **FIGURE 29.16b**.

VISUALIZE FIGURE 29.17 shows a loop for which we've assumed that the current is circulating ccw. We've chosen a coordinate system in which the loop lies at $z = 0$ in the xy-plane. Let segment i be the segment at the top of the loop. Vector $\Delta\vec{s}_i$ is parallel to the x-axis and unit vector $\hat{r}$ is in the yz-plane, thus angle θ_i, the angle between $\Delta\vec{s}_i$ and $\hat{r}$, is 90°.

FIGURE 29.17 Calculating the magnetic field of a current loop.

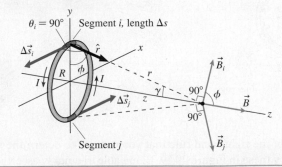

The direction of $\vec{B}_i$, the magnetic field due to the current in segment i, is given by the cross product $\Delta\vec{s}_i \times \hat{r}$. $\vec{B}_i$ must be perpendicular to $\Delta\vec{s}_i$ *and* perpendicular to $\hat{r}$. You should convince yourself that $\vec{B}_i$ in Figure 29.17 points in the correct direction. Notice that the y-component of $\vec{B}_i$ is canceled by the y-component of magnetic field $\vec{B}_j$ due to the current segment at the bottom of the loop, 180° away. In fact, *every* current segment on the loop can be paired with a segment 180° away, on the opposite side of the loop, such that the x- and y-components of $\vec{B}$ cancel and the components of $\vec{B}$ parallel to the z-axis add. In other words, the symmetry of the loop requires the on-axis magnetic field to point along the z-axis. Knowing that we need to sum only the z-components will simplify our calculation.

SOLVE We can use the Biot-Savart law to find the z-component $(B_i)_z = B_i \cos\phi$ of the magnetic field of segment i. The cross product $\Delta\vec{s}_i \times \hat{r}$ has magnitude $(\Delta s)(1) \sin 90° = \Delta s$, thus

$$(B_i)_z = \frac{\mu_0}{4\pi} \frac{I \Delta s}{r^2} \cos\phi = \frac{\mu_0 I \cos\phi}{4\pi(z^2 + R^2)} \Delta s$$

where we used $r = (z^2 + R^2)^{1/2}$. You can see, because $\phi + \gamma = 90°$, that angle ϕ is also the angle between $\hat{r}$ and the radius of the loop. Hence $\cos\phi = R/r$, and

$$(B_i)_z = \frac{\mu_0 I R}{4\pi(z^2 + R^2)^{3/2}} \Delta s$$

The final step is to sum the magnetic fields due to all the segments:

$$B_{loop} = \sum_i (B_i)_z = \frac{\mu_0 I R}{4\pi(z^2 + R^2)^{3/2}} \sum_i \Delta s$$

In this case, unlike the straight wire, none of the terms multiplying Δs depends on the position of segment i, so all these terms can be factored out of the summation. We're left with a summation that adds up the lengths of all the small segments. But this is just the total length of the wire, which is the circumference $2\pi R$. Thus the on-axis magnetic field of a current loop is

$$B_{loop} = \frac{\mu_0 I R}{4\pi(z^2 + R^2)^{3/2}} 2\pi R = \frac{\mu_0}{2} \frac{I R^2}{(z^2 + R^2)^{3/2}}$$

In practice, current often passes through a *coil* consisting of *N turns* of wire. If the turns are all very close together, so that the magnetic field of each is essentially the same, then the magnetic field of a coil is *N* times the magnetic field of a current loop. The magnetic field at the center ($z = 0$) of an *N*-turn coil, or *N*-turn current loop, is

$$B_{\text{coil center}} = \frac{\mu_0}{2} \frac{NI}{R} \qquad (N\text{-turn current loop}) \qquad (29.8)$$

This is the second of our key magnetic field models.

EXAMPLE 29.6 | **Matching the earth's magnetic field**

What current is needed in a 5-turn, 10-cm-diameter coil to cancel the earth's magnetic field at the center of the coil?

MODEL One way to create a zero-field region of space is to generate a magnetic field equal to the earth's field but pointing in the opposite direction. The vector sum of the two fields is zero.

VISUALIZE FIGURE 29.18 shows a five-turn coil of wire. The magnetic field is five times that of a single current loop.

SOLVE The earth's magnetic field, from Table 29.1, is 5×10^{-5} T. We can use Equation 29.8 to find that the current needed to generate a 5×10^{-5} T field is

$$I = \frac{2RB}{\mu_0 N} = \frac{2(0.050 \text{ m})(5.0 \times 10^{-5} \text{ T})}{5(4\pi \times 10^{-7} \text{ T m/A})} = 0.80 \text{ A}$$

FIGURE 29.18 A coil of wire.

$\vec{B}$

I

I

ASSESS A 0.80 A current is easily produced. Although there are better ways to cancel the earth's field than using a simple coil, this illustrates the idea.

29.5 Magnetic Dipoles

We were able to calculate the on-axis magnetic field of a current loop, but determining the field at off-axis points requires either numerical integrations or an experimental mapping of the field. FIGURE 29.19 shows the full magnetic field of a current loop. This is a field with *rotational symmetry,* so to picture the full three-dimensional field, imagine Figure 29.19a rotated about the axis of the loop. Figure 29.19b shows the magnetic field in the plane of the loop as seen from the right. There is a clear sense, seen in the photo of Figure 29.19c, that the magnetic field leaves the loop on one side, "flows" around the outside, then returns to the loop.

FIGURE 29.19 The magnetic field of a current loop.

(a) Cross section through the current loop   **(b)** The current loop seen from the right   **(c)** A photo of iron filings

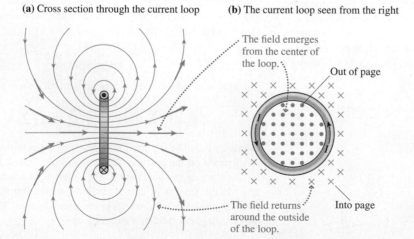

The field emerges from the center of the loop.

The field returns around the outside of the loop.

Out of page

Into page

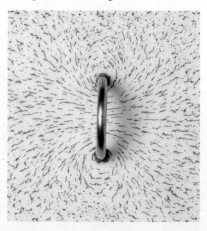

There are two versions of the right-hand rule that you can use to determine which way a loop's field points. Try these in Figure 29.19. Being able to quickly ascertain the field direction of a current loop is an important skill.

Finding the magnetic field direction of a current loop

Use either of the following methods to find the magnetic field direction:

❶ Point your right thumb in the direction of the current at any point on the loop and let your fingers curl through the center of the loop. Your fingers are then pointing in the direction in which $\vec{B}$ leaves the loop.

❷ Curl the fingers of your right hand around the loop in the direction of the current. Your thumb is then pointing in the direction in which $\vec{B}$ leaves the loop.

Exercises 18–20

A Current Loop Is a Magnetic Dipole

A current loop has two distinct sides. Bar magnets and flat refrigerator magnets also have two distinct sides or ends, so you might wonder if current loops are related to these permanent magnets. Consider the following experiments with a current loop. Notice that we're showing the magnetic field only in the plane of the loop.

Investigating current loops

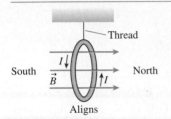

Aligns

A current loop hung by a thread aligns itself with the magnetic field pointing north.

Repels

The north pole of a permanent magnet repels the side of a current loop from which the magnetic field is emerging.

Attracts

The south pole of a permanent magnet attracts the side of a current loop from which the magnetic field is emerging.

These investigations show that **a current loop is a magnet,** just like a permanent magnet. A magnet created by a current in a coil of wire is called an **electromagnet.** An electromagnet picks up small pieces of iron, influences a compass needle, and acts in every way like a permanent magnet.

In fact, **FIGURE 29.20** shows that a flat permanent magnet and a current loop generate the same magnetic field—the field of a magnetic dipole. For both, **you can identify the north pole as the face or end** *from which* **the magnetic field emerges.** The magnetic fields of both point *into* the south pole.

FIGURE 29.20 A current loop has magnetic poles and generates the same magnetic field as a flat permanent magnet.

(a) Current loop   **(b)** Permanent magnet

Whether it's a current loop or a permanent magnet, the magnetic field emerges from the north pole.

One of the goals of this chapter is to show that magnetic forces exerted by currents and magnetic forces exerted by permanent magnets are just two different aspects of a single magnetism. This connection between permanent magnets and current loops will turn out to be a big piece of the puzzle.

The Magnetic Dipole Moment

The expression for the electric field of an electric dipole was considerably simplified when we considered the field at distances significantly larger than the size of the charge separation s. The on-axis field of an electric dipole when $z \gg s$ is

$$\vec{E}_{\text{dipole}} = \frac{1}{4\pi\epsilon_0} \frac{2\vec{p}}{z^3}$$

where the electric dipole moment $\vec{p} = (qs,$ from negative to positive charge).

The on-axis magnetic field of a current loop is

$$B_{\text{loop}} = \frac{\mu_0}{2} \frac{IR^2}{(z^2 + R^2)^{3/2}}$$

If z is much larger than the diameter of the current loop, $z \gg R$, we can make the approximation $(z^2 + R^2)^{3/2} \rightarrow z^3$. Then the loop's field is

$$B_{\text{loop}} \approx \frac{\mu_0}{2} \frac{IR^2}{z^3} = \frac{\mu_0}{4\pi} \frac{2(\pi R^2)I}{z^3} = \frac{\mu_0}{4\pi} \frac{2AI}{z^3} \qquad (29.9)$$

where $A = \pi R^2$ is the area of the loop.

A more advanced treatment of current loops shows that, if z is much larger than the size of the loop, Equation 29.9 is the on-axis magnetic field of a current loop of *any* shape, not just a circular loop. The shape of the loop affects the nearby field, but the distant field depends only on the current I and the area A enclosed within the loop. With this in mind, let's define the **magnetic dipole moment** $\vec{\mu}$ of a current loop enclosing area A to be

$$\vec{\mu} = (AI, \text{ from the south pole to the north pole})$$

The SI units of the magnetic dipole moment are $A\,m^2$.

> **NOTE** Don't confuse the magnetic dipole moment $\vec{\mu}$ with the constant μ_0 in the Biot-Savart law.

FIGURE 29.21 The magnetic dipole moment of a circular current loop.

The magnetic dipole moment is perpendicular to the loop, in the direction of the right-hand rule. The magnitude of $\vec{\mu}$ is AI.

Loop area A

The magnetic dipole moment, like the electric dipole moment, is a vector. It has the same direction as the on-axis magnetic field. Thus the right-hand rule for determining the direction of $\vec{B}$ also shows the direction of $\vec{\mu}$. **FIGURE 29.21** shows the magnetic dipole moment of a circular current loop.

Because the on-axis magnetic field of a current loop points in the same direction as $\vec{\mu}$, we can combine Equation 29.9 and the definition of $\vec{\mu}$ to write the on-axis field of a magnetic dipole as

$$\vec{B}_{\text{dipole}} = \frac{\mu_0}{4\pi} \frac{2\vec{\mu}}{z^3} \qquad \text{(on the axis of a magnetic dipole)} \qquad (29.10)$$

If you compare $\vec{B}_{\text{dipole}}$ to $\vec{E}_{\text{dipole}}$, you can see that the magnetic field of a magnetic dipole is very similar to the electric field of an electric dipole.

A permanent magnet also has a magnetic dipole moment and its on-axis magnetic field is given by Equation 29.10 when z is much larger than the size of the magnet. Equation 29.10 and laboratory measurements of the on-axis magnetic field can be used to determine a permanent magnet's dipole moment.

EXAMPLE 29.7 | **Measuring current in a superconducting loop**

You'll learn in Chapter 30 that a current can be *induced* in a closed loop of wire. If the loop happens to be made of a superconducting material, with zero resistance, the induced current will—in principle—persist forever. The current cannot be measured with an ammeter because any real ammeter has resistance that will quickly stop the current. Instead, physicists measure the persistent current in a superconducting loop by measuring its magnetic field. What is the current in a 3.0-mm-diameter superconducting loop if the axial magnetic field is 9.0 μT at a distance of 2.5 cm?

MODEL The measurements are made far enough from the loop in comparison to its radius ($z \gg R$) that we can model the loop as a magnetic dipole rather than using the exact expression for the on-axis field of a current loop.

SOLVE The axial magnetic field strength of a dipole is

$$B = \frac{\mu_0}{4\pi}\frac{2\mu}{z^3} = \frac{\mu_0}{4\pi}\frac{2\pi R^2 I}{z^3} = \frac{\mu_0 R^2 I}{2}\frac{1}{z^3}$$

where we used $\mu = AI = \pi R^2 I$ for the magnetic dipole moment of a circular loop of radius R. Thus the current is

$$I = \frac{2z^3 B}{\mu_0 R^2} = \frac{2(0.025 \text{ m})^3 (9.0 \times 10^{-6} \text{ T})}{(1.26 \times 10^{-6} \text{ T m/A})(0.0015 \text{ m})^2}$$

$$= 99 \text{ A}$$

ASSESS This would be a very large current for ordinary wire. An important property of superconducting wires is their ability to carry current that would melt an ordinary wire.

STOP TO THINK 29.4 What is the current direction in this loop? And which side of the loop is the north pole?

a. Current cw; north pole on top
b. Current cw; north pole on bottom
c. Current ccw; north pole on top
d. Current ccw; north pole on bottom

29.6 Ampère's Law and Solenoids

In principle, the Biot-Savart law can be used to calculate the magnetic field of any current distribution. In practice, the integrals are difficult to evaluate for anything other than very simple situations. We faced a similar situation for calculating electric fields, but we discovered an alternative method—Gauss's law—for calculating the electric field of charge distributions with a high degree of symmetry.

Likewise, there's an alternative method, called *Ampère's law,* for calculating the magnetic fields of current distributions with a high degree of symmetry. Whereas Gauss's law is written in terms of a surface integral, Ampère's law is based on the mathematical procedure called a *line integral.*

Line Integrals

We've flirted with the idea of a line integral ever since introducing the concept of work in Chapter 9, but now we need to take a more serious look at what a line integral represents and how it is used. **FIGURE 29.22a** shows a curved line that goes from an initial point i to a final point f.

FIGURE 29.22 Integrating along a line from i to f.

(a)

A line from i to f

(b)

The line can be divided into many small segments. The sum of all the Δs's is the length l of the line.

Suppose, as shown in **FIGURE 29.22b**, we divide the line into many small segments of length Δs. The first segment is Δs_1, the second is Δs_2, and so on. The sum of all the Δs's is the length l of the line between i and f. We can write this mathematically as

$$l = \sum_k \Delta s_k \rightarrow \int_i^f ds \qquad (29.11)$$

where, in the last step, we let $\Delta s \rightarrow ds$ and the sum become an integral.

This integral is called a **line integral.** All we've done is to subdivide a line into infinitely many infinitesimal pieces, then add them up. This is exactly what you do in calculus when you evaluate an integral such as $\int x \, dx$. In fact, an integration along the x-axis *is* a line integral, one that happens to be along a straight line. Figure 29.22 differs only in that the line is curved. **The underlying idea in both cases is that an integral is just a fancy way of doing a sum.**

The line integral of Equation 29.11 is not terribly exciting. **FIGURE 29.23a** makes things more interesting by allowing the line to pass through a magnetic field. **FIGURE 29.23b** again divides the line into small segments, but this time $\Delta \vec{s}_k$ is the displacement vector of segment k. The magnetic field at this point in space is $\vec{B}_k$.

Suppose we were to evaluate the dot product $\vec{B}_k \cdot \Delta \vec{s}_k$ at each segment, then add the values of $\vec{B}_k \cdot \Delta \vec{s}_k$ due to every segment. Doing so, and again letting the sum become an integral, we have

$$\sum_k \vec{B}_k \cdot \Delta \vec{s}_k \rightarrow \int_i^f \vec{B} \cdot d\vec{s} = \text{the line integral of } \vec{B} \text{ from i to f}$$

Once again, the integral is just a shorthand way to say: Divide the line into lots of little pieces, evaluate $\vec{B}_k \cdot \Delta \vec{s}_k$ for each piece, then add them up.

Although this process of evaluating the integral could be difficult, the only line integrals we'll need to deal with fall into two simple cases. If the magnetic field is *everywhere perpendicular* to the line, then $\vec{B} \cdot d\vec{s} = 0$ at every point along the line and the integral is zero. If the magnetic field is *everywhere tangent* to the line *and* has the same magnitude B at every point, then $\vec{B} \cdot d\vec{s} = B \, ds$ at every point and

$$\int_i^f \vec{B} \cdot d\vec{s} = \int_i^f B \, ds = B \int_i^f ds = Bl \qquad (29.12)$$

We used Equation 29.11 in the last step to integrate ds along the line.

Tactics Box 29.3 summarizes these two situations.

FIGURE 29.23 Integrating $\vec{B}$ along a line from i to f.

(a)

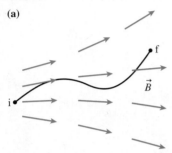

The line passes through a magnetic field.

(b)

Magnetic field at segment k

Displacement of segment k

TACTICS BOX 29.3   (MP)

Evaluating line integrals

❶ If $\vec{B}$ is everywhere perpendicular to a line, the line integral of $\vec{B}$ is

$$\int_i^f \vec{B} \cdot d\vec{s} = 0$$

❷ If $\vec{B}$ is everywhere tangent to a line of length l *and* has the same magnitude B at every point, then

$$\int_i^f \vec{B} \cdot d\vec{s} = Bl$$

Exercises 23–24

Ampère's Law

FIGURE 29.24 shows a wire carrying current I into the page and the magnetic field at distance r. The magnetic field of a current-carrying wire is everywhere tangent to a circle around the wire *and* has the same magnitude $\mu_0 I/2\pi r$ at all points on the circle. According to Tactics Box 29.3, these conditions allow us to easily evaluate the line integral of $\vec{B}$ along a circular path around the wire. Suppose we were to integrate the magnetic field *all the way around* the circle. That is, the initial point i of the integration path and the final point f will be the same point. This would be a line integral around a *closed curve*, which is denoted

$$\oint \vec{B} \cdot d\vec{s}$$

The little circle on the integral sign indicates that the integration is performed around a closed curve. The notation has changed, but the meaning has not.

Because $\vec{B}$ is tangent to the circle *and* of constant magnitude at every point on the circle, we can use Option 2 from Tactics Box 29.3 to write

$$\oint \vec{B} \cdot d\vec{s} = Bl = B(2\pi r) \tag{29.13}$$

where, in this case, the path length l is the circumference $2\pi r$ of the circle. The magnetic field strength of a current-carrying wire is $B = \mu_0 I/2\pi r$, thus

$$\oint \vec{B} \cdot d\vec{s} = \mu_0 I \tag{29.14}$$

The interesting result is that the line integral of $\vec{B}$ around the current-carrying wire is independent of the radius of the circle. Any circle, from one touching the wire to one far away, would give the same result. The integral depends only on the amount of current passing *through* the circle that we integrated around.

This is reminiscent of Gauss's law. In our investigation of Gauss's law, we started with the observation that the electric flux Φ_e through a sphere surrounding a point charge depends only on the amount of charge inside, not on the radius of the sphere. After examining several cases, we concluded that the shape of the surface wasn't relevant. The electric flux through *any* closed surface enclosing total charge Q_{in} turned out to be $\Phi_e = Q_{in}/\epsilon_0$.

Although we'll skip the details, the same type of reasoning that we used to prove Gauss's law shows that the result of Equation 29.14

- Is independent of the shape of the curve around the current.
- Is independent of where the current passes through the curve.
- Depends only on the total amount of current through the area enclosed by the integration path.

Thus whenever total current $I_{through}$ passes through an area bounded by a *closed curve*, the line integral of the magnetic field around the curve is

$$\oint \vec{B} \cdot d\vec{s} = \mu_0 I_{through} \tag{29.15}$$

This result for the magnetic field is known as **Ampère's law.**

To make practical use of Ampère's law, we need to determine which currents are positive and which are negative. The right-hand rule is once again the proper tool. If you curl your right fingers around the closed path in the direction in which you are going to integrate, then any current passing though the bounded area in the direction of your thumb is a positive current. Any current in the opposite direction is a negative current. In FIGURE 29.25, for example, currents I_2 and I_4 are positive, I_3 is negative. Thus $I_{through} = I_2 - I_3 + I_4$.

FIGURE 29.24 Integrating the magnetic field around a wire.

The integration path is a circle of radius r.

The integration starts and stops at the same point.

$\vec{B}$ is everywhere tangent to the integration path and has constant magnitude.

FIGURE 29.25 Using Ampère's law.

I_1 doesn't pass through the enclosed area.

The integration path is a closed curve.

These currents pass through the bounded area.

NOTE The integration path of Ampère's law is a mathematical curve through space. It does not have to match a physical surface or boundary, although it could if we want it to.

In one sense, Ampère's law doesn't tell us anything new. After all, we derived Ampère's law from the Biot-Savart law. But in another sense, Ampère's law is more important than the Biot-Savart law because it states a very general property about magnetic fields. We will use Ampère's law to find the magnetic fields of some important current distributions that have a high degree of symmetry.

EXAMPLE 29.8 | The magnetic field inside a current-carrying wire

A wire of radius R carries current I. Find the magnetic field *inside* the wire at distance $r < R$ from the axis.

MODEL Assume the current density is uniform over the cross section of the wire.

VISUALIZE FIGURE 29.26 shows a cross section through the wire. The wire has cylindrical symmetry, with all the charges moving parallel to the wire, so the magnetic field *must* be tangent to circles that are concentric with the wire. We don't know how the strength of the magnetic field depends on the distance from the center— that's what we're going to find—but the symmetry of the situation dictates the *shape* of the magnetic field.

FIGURE 29.26 Using Ampère's law inside a current-carrying wire.

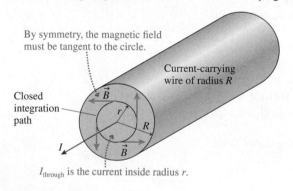

By symmetry, the magnetic field must be tangent to the circle.

Current-carrying wire of radius R

Closed integration path

$I_{through}$ is the current inside radius r.

SOLVE To find the field strength at radius r, we draw a circle of radius r. The amount of current passing through this circle is

$$I_{through} = JA_{circle} = \pi r^2 J$$

where J is the current density. Our assumption of a uniform current density allows us to use the full current I passing through a wire of radius R to find that

$$J = \frac{I}{A} = \frac{I}{\pi R^2}$$

Thus the current through the circle of radius r is

$$I_{through} = \frac{r^2}{R^2} I$$

Let's integrate $\vec{B}$ around the circumference of this circle. According to Ampère's law,

$$\oint \vec{B} \cdot d\vec{s} = \mu_0 I_{through} = \frac{\mu_0 r^2}{R^2} I$$

We know from the symmetry of the wire that $\vec{B}$ is everywhere tangent to the circle *and* has the same magnitude at all points on the circle. Consequently, the line integral of $\vec{B}$ around the circle can be evaluated using Option 2 of Tactics Box 29.3:

$$\oint \vec{B} \cdot d\vec{s} = Bl = 2\pi r B$$

where $l = 2\pi r$ is the path length. If we substitute this expression into Ampère's law, we find that

$$2\pi r B = \frac{\mu_0 r^2}{R^2} I$$

Solving for B, we find that the magnetic field strength at radius r *inside* a current-carrying wire is

$$B = \frac{\mu_0 I}{2\pi R^2} r$$

ASSESS The magnetic field *inside* a wire increases linearly with distance from the center until, at the surface of the wire, $B = \mu_0 I / 2\pi R$ matches our earlier solution for the magnetic field *outside* a current-carrying wire. This agreement at $r = R$ gives us confidence in our result. The magnetic field strength both inside and outside the wire is shown graphically in FIGURE 29.27.

FIGURE 29.27 Graphical representation of the magnetic field of a current-carrying wire.

The magnetic field increases linearly with distance inside the wire . . .

. . . and inversely with distance outside the wire.

The Magnetic Field of a Solenoid

In our study of electricity, we made extensive use of the idea of a uniform electric field: a field that is the same at every point in space. We found that two closely spaced, parallel charged plates generate a uniform electric field between them, and this was one reason we focused so much attention on the parallel-plate capacitor.

Similarly, there are many applications of magnetism for which we would like to generate a **uniform magnetic field,** a field having the same magnitude and the same direction at every point within some region of space. None of the sources we have looked at thus far produces a uniform magnetic field.

In practice, a uniform magnetic field is generated with a **solenoid.** A solenoid, shown in **FIGURE 29.28,** is a helical coil of wire with the same current I passing through each loop in the coil. Solenoids may have hundreds or thousands of coils, often called *turns,* sometimes wrapped in several layers.

FIGURE 29.28 A solenoid.

FIGURE 29.29 Using superposition to find the magnetic field of a stack of current loops.

(a) A single loop

The magnetic field vector is shown at six points.

(b) A stack of three loops

The fields of the three loops nearly cancel here. $\vec{B}_1$

$\vec{B}_2$

$\vec{B}_3$

$\vec{B}_1$

$\vec{B}_2$

$\vec{B}_3$

1   2   3

The fields reinforce each other here.

We can understand a solenoid by thinking of it as a stack of current loops. **FIGURE 29.29a** shows the magnetic field of a single current loop at three points on the axis and three points equally distant from the axis. The field directly above the loop is opposite in direction to the field inside the loop. **FIGURE 29.29b** then shows three parallel loops. We can use information from Figure 29.29b to draw the magnetic fields of each loop at the center of loop 2 and at a point above loop 2.

The superposition of the three fields at the center of loop 2 produces a *stronger* field than that of loop 2 alone. But the superposition at the point above loop 2 produces a net magnetic field that is very much weaker than the field at the center of the loop. We've used only three current loops to illustrate the idea, but these tendencies are reinforced by including more loops. With many current loops along the same axis, **the field in the center is strong and roughly parallel to the axis, whereas the field outside the loops is very close to zero.**

FIGURE 29.30a is a photo of the magnetic field of a short solenoid. You can see that the magnetic field inside the coils is nearly uniform (i.e., the field lines are nearly parallel) and the field outside is much weaker. Our goal of producing a uniform magnetic field can be achieved by increasing the number of coils until we have an *ideal solenoid* that is infinitely long and in which the coils are as close together as possible. As **FIGURE 29.30b** shows, **the magnetic field inside an ideal solenoid is uniform and parallel to the axis; the magnetic field outside is zero.** No real solenoid is ideal, but a very uniform magnetic field can be produced near the center of a tightly wound solenoid whose length is much larger than its diameter.

We can use Ampère's law to calculate the field of an ideal solenoid. **FIGURE 29.31** on the next page shows a cross section through an infinitely long solenoid. The integration path that we'll use is a rectangle of width l, enclosing N turns of the solenoid coil. Because this is a mathematical curve, not a physical boundary, there's no difficulty with letting it protrude through the wall of the solenoid wherever we wish. The solenoid's magnetic field direction, given by the right-hand rule, is left to right, so we'll integrate around this path in the ccw direction.

FIGURE 29.30 The magnetic field of a solenoid.

(a) A short solenoid

(b)

$\vec{B} = \vec{0}$

$\vec{B}$

$\vec{B} = \vec{0}$

The magnetic field is uniform inside this section of an ideal, infinitely long solenoid. The magnetic field outside the solenoid is zero.

FIGURE 29.31 A closed path inside and outside an ideal solenoid.

This is the integration path for Ampère's law. There are *N* turns inside.

$\vec{B}$ is tangent to the integration path along the bottom edge.

Each of the *N* wires enclosed by the integration path carries current *I*, so the total current passing through the rectangle is $I_{\text{through}} = NI$. Ampère's law is thus

$$\oint \vec{B} \cdot d\vec{s} = \mu_0 I_{\text{through}} = \mu_0 NI \tag{29.16}$$

The line integral around this path is the sum of the line integrals along each side. Along the bottom, where $\vec{B}$ is parallel to $d\vec{s}$ and of constant value *B*, the integral is simply *Bl*. The integral along the top is zero because the magnetic field outside an ideal solenoid is zero.

The left and right sides sample the magnetic field both inside and outside the solenoid. The magnetic field outside is zero, and the interior magnetic field is everywhere *perpendicular* to the line of integration. Consequently, as we recognized in Option 1 of Tactics Box 29.3, the line integral is zero.

Only the integral along the bottom path is nonzero, leading to

$$\oint \vec{B} \cdot d\vec{s} = Bl = \mu_0 NI$$

Thus the strength of the uniform magnetic field inside a solenoid is

$$B_{\text{solenoid}} = \frac{\mu_0 NI}{l} = \mu_0 nI \quad \text{(solenoid)} \tag{29.17}$$

where $n = N/l$ is the number of turns per unit length. Measurements that need a uniform magnetic field are often conducted inside a solenoid, which can be built quite large.

EXAMPLE 29.9 | Generating an MRI magnetic field

A 1.0-m-long MRI solenoid generates a 1.2 T magnetic field. To produce such a large field, the solenoid is wrapped with superconducting wire that can carry a 100 A current. How many turns of wire does the solenoid need?

MODEL Assume that the solenoid is ideal.

SOLVE Generating a magnetic field with a solenoid is a trade-off between current and turns of wire. A larger current requires fewer

turns, but the resistance of ordinary wires causes them to overheat if the current is too large. For a superconducting wire that can carry 100 A with no resistance, we can use Equation 29.17 to find the required number of turns:

$$N = \frac{lB}{\mu_0 I} = \frac{(1.0 \text{ m})(1.2 \text{ T})}{(4\pi \times 10^{-7} \text{ T m/A})(100 \text{ A})} = 9500 \text{ turns}$$

ASSESS The solenoid coil requires a large number of turns, but that's not surprising for generating a very strong field. If the wires are 1 mm in diameter, there would be 10 layers with approximately 1000 turns per layer.

The magnetic field of a finite-length solenoid is approximately uniform *inside* the solenoid and weak, but not zero, outside. As FIGURE 29.32 shows, the magnetic field outside the solenoid looks like that of a bar magnet. Thus **a solenoid is an electromagnet,** and you can use the right-hand rule to identify the north-pole end. A solenoid with many turns and a large current can be a very powerful magnet.

FIGURE 29.32 The magnetic fields of a finite-length solenoid and of a bar magnet.

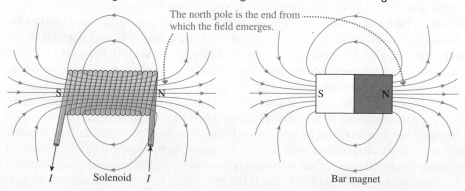

The north pole is the end from which the field emerges.

Solenoid

Bar magnet

29.7 The Magnetic Force on a Moving Charge

It's time to switch our attention from how magnetic fields are generated to how magnetic fields exert forces and torques. Oersted discovered that a current passing through a wire causes a magnetic torque to be exerted on a nearby compass needle. Upon hearing of Oersted's discovery, André-Marie Ampère, for whom the SI unit of current is named, reasoned that the current was acting like a magnet and, if this were true, that two current-carrying wires should exert magnetic forces on each other.

To find out, Ampère set up two parallel wires that could carry large currents either in the same direction or in opposite (or "antiparallel") directions. **FIGURE 29.33** shows the outcome of his experiment. Notice that, for currents, "likes" attract and "opposites" repel. This is the opposite of what would have happened had the wires been charged and thus exerting electric forces on each other. Ampère's experiment showed that **a magnetic field exerts a force on a current.**

FIGURE 29.33 The forces between parallel current-carrying wires.

"Like" currents attract.

"Opposite" currents repel.

Magnetic Force

Because a current consists of moving charges, Ampère's experiment implies that **a magnetic field exerts a force on a *moving* charge.** It turns out that the magnetic force is somewhat more complex than the electric force, depending not only on the charge's velocity but also on how the velocity vector is oriented relative to the magnetic field. Consider the following experiments:

Investigating the magnetic force on a charged particle

There is no magnetic force on a charged particle at rest.

There is no magnetic force on a charged particle moving *parallel* to a magnetic field.

As the angle α between the velocity and the magnetic field increases, the magnetic force also increases. The force is greatest when the angle is 90°. The magnetic force is always perpendicular to the plane containing $\vec{v}$ and $\vec{B}$.

Notice that the relationship among $\vec{v}$, $\vec{B}$, and $\vec{F}$ is exactly the same as the geometric relationship among vectors $\vec{C}$, $\vec{D}$, and $\vec{C} \times \vec{D}$. The magnetic force on a charge q as it moves through a magnetic field $\vec{B}$ with velocity $\vec{v}$ can be written

$$\vec{F}_{\text{on } q} = q\vec{v} \times \vec{B} = (qvB\sin\alpha, \text{ direction of right-hand rule}) \qquad (29.18)$$

where α is the angle between $\vec{v}$ and $\vec{B}$.

The right-hand rule is that of the cross product, shown in **FIGURE 29.34**. Notice that **the magnetic force on a moving charged particle is perpendicular to both $\vec{v}$ and $\vec{B}$.**

The magnetic force has several important properties:

- Only a *moving* charge experiences a magnetic force. There is no magnetic force on a charge at rest ($\vec{v} = \vec{0}$) in a magnetic field.
- There is no force on a charge moving parallel ($\alpha = 0°$) or antiparallel ($\alpha = 180°$) to a magnetic field.
- When there is a force, the force is perpendicular to *both* $\vec{v}$ and $\vec{B}$.
- The force on a negative charge is in the direction *opposite* to $\vec{v} \times \vec{B}$.
- For a charge moving perpendicular to $\vec{B}$ ($\alpha = 90°$), the magnitude of the magnetic force is $F = |q|vB$.

FIGURE 29.34 The right-hand rule for magnetic forces.

FIGURE 29.35 shows the relationship among $\vec{v}$, $\vec{B}$, and $\vec{F}$ for four moving charges. (The *source* of the magnetic field isn't shown, only the field itself.) You can see the inherent three-dimensionality of magnetism, with the force perpendicular to both $\vec{v}$ and $\vec{B}$. The magnetic force is very different from the electric force, which is parallel to the electric field.

FIGURE 29.35 Magnetic forces on moving charges.

EXAMPLE 29.10 | The magnetic force on an electron

A long wire carries a 10 A current from left to right. An electron 1.0 cm above the wire is traveling to the right at a speed of 1.0×10^7 m/s. What are the magnitude and the direction of the magnetic force on the electron?

MODEL The magnetic field is that of a long, straight wire.

VISUALIZE FIGURE 29.36 shows the current and an electron moving to the right. The right-hand rule tells us that the wire's magnetic

FIGURE 29.36 An electron moving parallel to a current-carrying wire.

field above the wire is out of the page, so the electron is moving perpendicular to the field.

SOLVE The electron charge is negative, thus the direction of the force is opposite the direction of $\vec{v} \times \vec{B}$. The right-hand rule shows that $\vec{v} \times \vec{B}$ points down, toward the wire, so $\vec{F}$ points up, away from the wire. The magnitude of the force is $|q|vB = evB$. The field is that of a long, straight wire:

$$B = \frac{\mu_0 I}{2\pi r} = 2.0 \times 10^{-4} \text{ T}$$

Thus the magnitude of the force on the electron is

$$F = evB = (1.60 \times 10^{-19} \text{ C})(1.0 \times 10^7 \text{ m/s})(2.0 \times 10^{-4} \text{ T})$$
$$= 3.2 \times 10^{-16} \text{ N}$$

The force on the electron is $\vec{F} = (3.2 \times 10^{-16} \text{ N, up})$.

ASSESS This force will cause the electron to curve away from the wire.

We can draw an interesting and important conclusion at this point. You have seen that the magnetic field is *created by* moving charges. Now you also see that magnetic forces are *exerted on* moving charges. Thus it appears that **magnetism is an interaction between moving charges.** Any two charges, whether moving or stationary, interact with each other through the electric field. In addition, two *moving* charges interact with each other through the magnetic field.

Cyclotron Motion

FIGURE 29.37 Cyclotron motion of a charged particle moving in a uniform magnetic field.

$\vec{v}$ is perpendicular to $\vec{B}$.

$\vec{B}$ into page

The magnetic force is always perpendicular to $\vec{v}$, causing the particle to move in a circle.

Many important applications of magnetism involve the motion of charged particles in a magnetic field. Older television picture tubes use magnetic fields to steer electrons through a vacuum from the electron gun to the screen. Microwave generators, which are used in applications ranging from ovens to radar, use a device called a *magnetron* in which electrons oscillate rapidly in a magnetic field.

You've just seen that there is no force on a charge that has velocity $\vec{v}$ parallel or antiparallel to a magnetic field. Consequently, **a magnetic field has no effect on a charge moving parallel or antiparallel to the field.** To understand the motion of charged particles in magnetic fields, we need to consider only motion *perpendicular* to the field.

FIGURE 29.37 shows a positive charge q moving with a velocity $\vec{v}$ in a plane that is perpendicular to a *uniform* magnetic field $\vec{B}$. According to the right-hand rule, the magnetic force on this particle is *perpendicular* to the velocity $\vec{v}$. A force that is always perpendicular to $\vec{v}$ changes the *direction* of motion, by deflecting the particle sideways, but it cannot change the particle's speed. Thus **a particle moving perpendicular to a uniform magnetic field undergoes uniform circular motion at constant speed.** This motion is called the **cyclotron motion** of a charged particle in a magnetic field.

NOTE A negative charge will orbit in the opposite direction from that shown in Figure 29.37 for a positive charge.

You've seen many analogies to cyclotron motion earlier in this text. For a mass moving in a circle at the end of a string, the tension force is always perpendicular to $\vec{v}$. For a satellite moving in a circular orbit, the gravitational force is always perpendicular to $\vec{v}$. Now, for a charged particle moving in a magnetic field, it is the magnetic force of strength $F = qvB$ that points toward the center of the circle and causes the particle to have a centripetal acceleration.

Newton's second law for circular motion, which you learned in Chapter 8, is

$$F = qvB = ma_r = \frac{mv^2}{r} \qquad (29.19)$$

Thus the radius of the cyclotron orbit is

$$r_{cyc} = \frac{mv}{qB} \qquad (29.20)$$

The inverse dependence on B indicates that the size of the orbit can be decreased by increasing the magnetic field strength.

We can also determine the frequency of the cyclotron motion. Recall from your earlier study of circular motion that the frequency of revolution f is related to the speed and radius by $f = v/2\pi r$. A rearrangement of Equation 29.20 gives the **cyclotron frequency:**

$$f_{cyc} = \frac{qB}{2\pi m} \qquad (29.21)$$

where the ratio q/m is the particle's *charge-to-mass ratio*. Notice that the cyclotron frequency depends on the charge-to-mass ratio and the magnetic field strength but *not* on the charge's speed.

Electrons undergoing circular motion in a magnetic field. You can see the electrons' path because they collide with a low-density gas that then emits light.

EXAMPLE 29.11 | The radius of cyclotron motion

In **FIGURE 29.38**, an electron is accelerated from rest through a potential difference of 500 V, then injected into a uniform magnetic field. Once in the magnetic field, it completes half a revolution in 2.0 ns. What is the radius of its orbit?

MODEL Energy is conserved as the electron is accelerated by the potential difference. The electron then undergoes cyclotron motion in the magnetic field, although it completes only half a revolution before hitting the back of the acceleration electrode.

FIGURE 29.38 An electron is accelerated, then injected into a magnetic field.

SOLVE The electron accelerates from rest ($v_i = 0$ m/s) at $V_i = 0$ V to speed v_f at $V_f = 500$ V. We can use conservation of energy $K_f + qV_f = K_i + qV_i$ to find the speed v_f with which it enters the magnetic field:

$$\tfrac{1}{2}mv_f^2 + (-e)V_f = 0 + 0$$

$$v_f = \sqrt{\frac{2eV_f}{m}} = \sqrt{\frac{2(1.60 \times 10^{-19}\,\text{C})(500\,\text{V})}{9.11 \times 10^{-31}\,\text{kg}}}$$

$$= 1.33 \times 10^7\,\text{m/s}$$

The cyclotron radius in the magnetic field is $r_{cyc} = mv/eB$, but we first need to determine the field strength. Were it not for the electrode,

Continued

the electron would undergo circular motion with period $T = 4.0$ ns. Hence the cyclotron frequency is $f = 1/T = 2.5 \times 10^8$ Hz. We can use the cyclotron frequency to determine that the magnetic field strength is

$$B = \frac{2\pi m f_{cyc}}{e} = \frac{2\pi (9.11 \times 10^{-31}\ \text{kg})(2.50 \times 10^8\ \text{Hz})}{1.60 \times 10^{-19}\ \text{C}}$$

$$= 8.94 \times 10^{-3}\ \text{T}$$

Thus the radius of the electron's orbit is

$$r_{cyc} = \frac{mv}{qB} = 8.5 \times 10^{-3}\ \text{m} = 8.5\ \text{mm}$$

ASSESS A 17-mm-diameter orbit is similar to what is seen in the photo just before this example, so this seems to be a typical size for electrons moving in modest magnetic fields.

FIGURE 29.39a shows a more general situation in which the charged particle's velocity $\vec{v}$ is neither parallel nor perpendicular to $\vec{B}$. The component of $\vec{v}$ parallel to $\vec{B}$ is not affected by the field, so the charged particle spirals around the magnetic field lines in a helical trajectory. The radius of the helix is determined by $\vec{v}_\perp$, the component of $\vec{v}$ perpendicular to $\vec{B}$.

FIGURE 29.39 In general, charged particles spiral along helical trajectories around the magnetic field lines. This motion is responsible for the earth's aurora.

(a) Charged particles spiral around the magnetic field lines.

(b) The earth's magnetic field leads particles into the atmosphere near the poles, causing the aurora.

(c) The aurora

The motion of charged particles in a magnetic field is responsible for the earth's aurora. High-energy particles and radiation streaming out from the sun, called the *solar wind,* create ions and electrons as they strike molecules high in the atmosphere. Some of these charged particles become trapped in the earth's magnetic field, creating what is known as the *Van Allen radiation belt.*

As FIGURE 29.39b shows, the electrons spiral along the magnetic field lines until the field leads them into the atmosphere. The shape of the earth's magnetic field is such that most electrons enter the atmosphere near the magnetic poles. There they collide with oxygen and nitrogen atoms, exciting the atoms and causing them to emit auroral light, as seen in FIGURE 29.39c.

STOP TO THINK 29.5 An electron moves perpendicular to a magnetic field. What is the direction of $\vec{B}$?

a. Left
b. Up
c. Into the page
d. Right
e. Down
f. Out of the page

The Cyclotron

Physicists studying the structure of the atomic nucleus and of elementary particles usually use a device called a *particle accelerator.* The first practical particle accelerator, invented in the 1930s, was the **cyclotron.** Cyclotrons remain important for many applications of nuclear physics, such as the creation of radioisotopes for medicine.

A cyclotron, shown in FIGURE 29.40, consists of an evacuated chamber within a large, uniform magnetic field. Inside the chamber are two hollow conductors shaped like the letter D and hence called "dees." The dees are made of copper, which doesn't affect the magnetic field; are open along the straight sides; and are separated by a small gap. A charged particle, typically a proton, is injected into the magnetic field from a source near the center of the cyclotron, and it begins to move in and out of the dees in a circular cyclotron orbit.

The cyclotron operates by taking advantage of the fact that the cyclotron frequency f_{cyc} of a charged particle is independent of the particle's speed. An *oscillating* potential difference ΔV is connected across the dees and adjusted until its frequency is exactly the cyclotron frequency. There is almost no electric field inside the dees (you learned in Chapter 24 that the electric field inside a hollow conductor is zero), but a strong electric field points from the positive to the negative dee in the gap between them.

Suppose the proton emerges into the gap from the positive dee. The electric field in the gap *accelerates* the proton across the gap into the negative dee, and it gains kinetic energy $e\,\Delta V$. A half cycle later, when it next emerges into the gap, the potential of the dees (whose potential difference is oscillating at f_{cyc}) will have changed sign. The proton will *again* be emerging from the positive dee and will *again* accelerate across the gap and gain kinetic energy $e\,\Delta V$.

This pattern will continue orbit after orbit. The proton's kinetic energy increases by $2e\,\Delta V$ every orbit, so after N orbits its kinetic energy is $K = 2Ne\,\Delta V$ (assuming that its initial kinetic energy was near zero). The radius of its orbit increases as it speeds up; hence the proton follows the *spiral* path shown in Figure 29.40 until it finally reaches the outer edge of the dee. It is then directed out of the cyclotron and aimed at a target. Although ΔV is modest, usually a few hundred volts, the fact that the proton can undergo many thousands of orbits before reaching the outer edge allows it to acquire a very large kinetic energy.

FIGURE 29.40 A cyclotron.

The potential ΔV oscillates at the cyclotron frequency f_{cyc}.

Proton source

$\vec{B}$

Dees

Protons exit here.

Bottom magnet

The Hall Effect

A charged particle moving through a vacuum is deflected sideways, perpendicular to $\vec{v}$, by a magnetic field. In 1879, a graduate student named Edwin Hall showed that the same is true for the charges moving through a conductor as part of a current. This phenomenon—now called the **Hall effect**—is used to gain information about the charge carriers in a conductor. It is also the basis of a widely used technique for measuring magnetic field strengths.

FIGURE 29.41a shows a magnetic field perpendicular to a flat, current-carrying conductor. You learned in Chapter 27 that the charge carriers move through a conductor at the drift speed v_d. Their motion is perpendicular to $\vec{B}$, so each charge carrier experiences a magnetic force $F_B = ev_dB$ perpendicular to both $\vec{B}$ and the current I. However, for the first time we have a situation in which it *does* matter whether the charge carriers are positive or negative.

FIGURE 29.41b, with the field out of the page, shows that positive charge carriers moving in the direction of I are pushed toward the bottom surface of the conductor. This creates an excess positive charge on the bottom surface and leaves an excess negative charge on the top. FIGURE 29.41c, where the electrons in an electron current i move opposite the direction of I, shows that electrons would be pushed toward the bottom surface. (Be sure to use the right-hand rule and the sign of the electron charge to confirm the deflections shown in these figures.) Thus the sign of the excess charge on the bottom surface is the same as the sign of the charge carriers. Experimentally, the bottom surface is negative when the conductor is a metal, and this is one more piece of evidence that the charge carriers in metals are electrons.

Electrons are deflected toward the bottom surface once the current starts flowing, but the process can't continue indefinitely. As excess charge accumulates on the top and bottom surfaces, it acts like the charge on the plates of a capacitor, creating a potential difference ΔV between the two surfaces and an electric field $E = \Delta V/w$ *inside* the conductor of width w. This electric field increases until the upward electric force $\vec{F}_E$

FIGURE 29.41 In a magnetic field, the charge carriers of a current are deflected to one side.

(a)

$\vec{B}$

The charge carriers are deflected to the side.

w

Area $A = wt$

t

I

(b) Top surface is negative.

Conventional current of positive charge carriers

$\vec{F}_E$

ΔV_H

$\vec{v}$

$\vec{E}$

I

$\vec{F}_B$ Electric field due to charge separation

(c) Electron current

Top surface is positive.

$\vec{F}_E$

i

$\vec{v}$

$\vec{E}$

ΔV_H

$\vec{F}_B$

on the charge carriers exactly balances the downward magnetic force $\vec{F}_B$. Once the forces are balanced, a steady state is reached in which the charge carriers move in the direction of the current and no additional charge is deflected to the surface.

The steady-state condition, in which $F_B = F_E$, is

$$F_B = ev_d B = F_E = eE = e\frac{\Delta V}{w} \tag{29.22}$$

Thus the steady-state potential difference between the two surfaces of the conductor, which is called the **Hall voltage** ΔV_H, is

$$\Delta V_H = wv_d B \tag{29.23}$$

You learned in Chapter 27 that the drift speed is related to the current density J by $J = nev_d$, where n is the charge-carrier density (charge carriers per m³). Thus

$$v_d = \frac{J}{ne} = \frac{I/A}{ne} = \frac{I}{wtne} \tag{29.24}$$

where $A = wt$ is the cross-section area of the conductor. If we use this expression for v_d in Equation 29.23, we find that the Hall voltage is

$$\Delta V_H = \frac{IB}{tne} \tag{29.25}$$

The Hall voltage is very small for metals in laboratory-sized magnetic fields, typically in the microvolt range. Even so, measurements of the Hall voltage in a known magnetic field are used to determine the charge-carrier density n. Interestingly, the Hall voltage is larger for *poor* conductors that have smaller charge-carrier densities. A laboratory probe for measuring magnetic field strengths, called a *Hall probe,* measures ΔV_H for a poor conductor whose charge-carrier density is known. The magnetic field is then determined from Equation 29.25.

EXAMPLE 29.12 | **Measuring the magnetic field**

A Hall probe consists of a strip of the metal bismuth that is 0.15 mm thick and 5.0 mm wide. Bismuth is a poor conductor with charge-carrier density $1.35 \times 10^{25}\,\text{m}^{-3}$. The Hall voltage on the probe is 2.5 mV when the current through it is 1.5 A. What is the strength of the magnetic field, and what is the electric field strength inside the bismuth?

VISUALIZE The bismuth strip looks like Figure 29.41a. The thickness is $t = 1.5 \times 10^{-4}\,\text{m}$ and the width is $w = 5.0 \times 10^{-3}\,\text{m}$.

SOLVE Equation 29.25 gives the Hall voltage. We can rearrange the equation to find that the magnetic field is

$$B = \frac{tne}{I}\Delta V_H$$

$$= \frac{(1.5 \times 10^{-4}\,\text{m})(1.35 \times 10^{25}\,\text{m}^{-3})(1.60 \times 10^{-19}\,\text{C})}{1.5\,\text{A}} 0.0025\,\text{V}$$

$$= 0.54\,\text{T}$$

The electric field created inside the bismuth by the excess charge on the surface is

$$E = \frac{\Delta V_H}{w} = \frac{0.0025\,\text{V}}{5.0 \times 10^{-3}\,\text{m}} = 0.50\,\text{V/m}$$

ASSESS 0.54 T is a fairly typical strength for a laboratory magnet.

29.8 Magnetic Forces on Current-Carrying Wires

Ampère's observation of magnetic forces between current-carrying wires motivated us to look at the magnetic forces on moving charges. We're now ready to apply that knowledge to Ampère's experiment. As a first step, let us find the force exerted by a uniform magnetic field on a long, straight wire carrying current I through the field. As **FIGURE 29.42a** shows, there's *no* force on a current-carrying wire *parallel* to a magnetic field. This shouldn't be surprising; it follows from the fact that there is no force on a charged particle moving parallel to $\vec{B}$.

FIGURE 29.42b shows a wire *perpendicular* to the magnetic field. By the right-hand rule, each charge in the current has a force of magnitude qvB directed to the left. Consequently, the entire length of wire within the magnetic field experiences a force to the left, perpendicular to both the current direction and the field direction.

A current is moving charge, and the magnetic force on a current-carrying wire is simply the net magnetic force on all the charge carriers in the wire. **FIGURE 29.43** shows a wire carrying current I and a segment of length l in which the charge carriers—moving with drift velocity $\vec{v}_d$—have total charge Q. Because the magnetic force is proportional to q, the *net* force on all the charge carriers in the wire is the force on the net charge: $\vec{F} = Q\vec{v}_d \times \vec{B}$. But we need to express this in terms of the current.

By definition, the current I is the amount of charge Q divided by the time t it takes the charge to flow through this segment: $I = Q/t$. The charge carriers have drift speed v_d, so they move distance l in $t = l/v_d$. Combining these equations, we have

$$I = \frac{Qv_d}{l}$$

and thus $Qv_d = Il$. If we define vector $\vec{l}$ to point in the direction of $\vec{v}_d$, the current direction, then $Q\vec{v}_d = I\vec{l}$. Substituting $I\vec{l}$ for $Q\vec{v}_d$ in the force equation, we find that the magnetic force on length l of a current-carrying wire is

$$\vec{F}_{wire} = I\vec{l} \times \vec{B} = (IlB\sin\alpha, \text{direction of right-hand rule}) \qquad (29.26)$$

where α is the angle between $\vec{l}$ (the direction of the current) and $\vec{B}$. As an aside, you can see from Equation 29.26 that the magnetic field B must have units of N/A m. This is why we defined 1 T = 1 N/A m in Section 29.3.

NOTE The familiar right-hand rule applies to a current-carrying wire. Point your right thumb in the direction of the current (parallel to $\vec{l}$) and your index finger in the direction of $\vec{B}$. Your middle finger is then pointing in the direction of the force $\vec{F}$ on the wire.

FIGURE 29.42 Magnetic force on a current-carrying wire.

(a)   (b)

There's no force on a current parallel to a magnetic field.

There is a magnetic force in the direction of the right-hand rule.

FIGURE 29.43 The force on a current is the force on the charge carriers.

EXAMPLE 29.13 | Magnetic levitation

The 0.10 T uniform magnetic field of **FIGURE 29.44** is horizontal, parallel to the floor. A straight segment of 1.0-mm-diameter copper wire, also parallel to the floor, is perpendicular to the magnetic field. What current through the wire, and in which direction, will allow the wire to "float" in the magnetic field?

FIGURE 29.44 Magnetic levitation.

MODEL The wire will float in the magnetic field if the magnetic force on the wire points upward and has magnitude mg, allowing it to balance the downward gravitational force.

SOLVE We can use the right-hand rule to determine which current direction experiences an upward force. With $\vec{B}$ pointing away from us, the direction of the current needs to be from left to right. The forces will balance when

$$F = IlB = mg = \rho(\pi r^2 l)g$$

where $\rho = 8920$ kg/m^3 is the density of copper. The length of the wire cancels, leading to

$$I = \frac{\rho\pi r^2 g}{B} = \frac{(8920 \text{ kg/m}^3)\pi(0.00050 \text{ m})^2(9.80 \text{ m/s}^2)}{0.10 \text{ T}}$$

$$= 0.69 \text{ A}$$

A 0.69 A current from left to right will levitate the wire in the magnetic field.

ASSESS A 0.69 A current is quite reasonable, but this idea is useful only if we can get the current into and out of this segment of wire. In practice, we could do so with wires that come in from below the page. These input and output wires would be parallel to $\vec{B}$ and not experience a magnetic force. Although this example is very simple, it is the basis for applications such as magnetic levitation trains.

Force Between Two Parallel Wires

Now consider Ampère's experimental arrangement of two parallel wires of length l, distance d apart. **FIGURE 29.45a** on the next page shows the currents I_1 and I_2 in the same direction; **FIGURE 29.45b** shows the currents in opposite directions. We will assume that the wires are sufficiently long to allow us to use the earlier result for the magnetic field of a long, straight wire: $B = \mu_0 I/2\pi r$.

FIGURE 29.45 Magnetic forces between parallel current-carrying wires.

(a) Currents in same direction

Magnetic field $\vec{B}_2$ created by current I_2

I_1

d

$\vec{F}_{2 \text{ on } 1}$ $\vec{F}_{2 \text{ on } 1}$

$\vec{F}_{1 \text{ on } 2}$ $\vec{F}_{1 \text{ on } 2}$

I_2

Magnetic field $\vec{B}_1$ created by current I_1

(b) Currents in opposite directions

$\vec{F}_{2 \text{ on } 1}$ $\vec{B}_2$ $\vec{F}_{2 \text{ on } 1}$

I_1

l

I_2

$\vec{F}_{1 \text{ on } 2}$ $\vec{B}_1$ $\vec{F}_{1 \text{ on } 2}$

As Figure 29.45a shows, the current I_2 in the lower wire creates a magnetic field $\vec{B}_2$ at the position of the upper wire. $\vec{B}_2$ points out of the page, perpendicular to current I_1. **It is field $\vec{B}_2$, due to the lower wire, that exerts a magnetic force on the upper wire.** Using the right-hand rule, you can see that the force on the upper wire is downward, thus attracting it toward the lower wire. The field of the lower current is not a uniform field, but it is the *same* at all points along the upper wire because the two wires are parallel. Consequently, we can use the field of a long, straight wire, with $r = d$, to determine the magnetic force exerted by the lower wire on the upper wire:

$$F_{\text{parallel wires}} = I_1 l B_2 = I_1 l \frac{\mu_0 I_2}{2\pi d} = \frac{\mu_0 l I_1 I_2}{2\pi d} \qquad (29.27)$$

As an exercise, you should convince yourself that the current in the upper wire exerts an upward-directed magnetic force on the lower wire with exactly the same magnitude. You should also convince yourself, using the right-hand rule, that the forces are repulsive and tend to push the wires apart if the two currents are in opposite directions.

Thus two parallel wires exert equal but opposite forces on each other, as required by Newton's third law. **Parallel wires carrying currents in the same direction attract each other; parallel wires carrying currents in opposite directions repel each other.**

EXAMPLE 29.14 | A current balance

Two stiff, 50-cm-long, parallel wires are connected at the ends by metal springs. Each spring has an unstretched length of 5.0 cm and a spring constant of 0.025 N/m. The wires push each other apart when a current travels around the loop. How much current is required to stretch the springs to lengths of 6.0 cm?

MODEL Two parallel wires carrying currents in opposite directions exert repulsive magnetic forces on each other.

VISUALIZE FIGURE 29.46 shows the "circuit." The springs are conductors, allowing a current to travel around the loop. In equilibrium, the repulsive magnetic forces between the wires are balanced by the restoring forces $F_{\text{Sp}} = k \Delta y$ of the springs.

SOLVE Figure 29.46 shows the forces on the lower wire. The net force is zero, hence the magnetic force is $F_B = 2F_{\text{Sp}}$. The force between the wires is given by Equation 29.27 with $I_1 = I_2 = I$:

$$F_B = \frac{\mu_0 l I^2}{2\pi d} = 2F_{\text{Sp}} = 2k \Delta y$$

FIGURE 29.46 The current-carrying wires of Example 29.14.

$\vec{F}_{\text{Sp}}$ I $\vec{F}_{\text{Sp}}$

$\vec{F}_B$

where k is the spring constant and $\Delta y = 1.0$ cm is the amount by which each spring stretches. Solving for the current, we find

$$I = \sqrt{\frac{4\pi k d \Delta y}{\mu_0 l}} = 17 \text{ A}$$

ASSESS Devices in which a magnetic force balances a mechanical force are called *current balances*. They can be used to make very accurate current measurements.

29.9 Forces and Torques on Current Loops

You have seen that a current loop is a magnetic dipole, much like a permanent magnet. We will now look at some important features of how current loops behave in magnetic fields. This discussion will be largely qualitative, but it will highlight some of the important properties of magnets and magnetic fields. We will use these ideas in the next section to make the connection between electromagnets and permanent magnets.

FIGURE 29.47a shows two current loops. Using what we just learned about the forces between parallel and antiparallel currents, you can see that **parallel current loops exert attractive magnetic forces on each other if the currents circulate in the same direction; they repel each other when the currents circulate in opposite directions.**

We can think of these forces in terms of magnetic poles. Recall that the north pole of a current loop is the side from which the magnetic field emerges, which you can determine with the right-hand rule. FIGURE 29.47b shows the north and south magnetic poles of the current loops. When the currents circulate in the same direction, a north and a south pole face each other and exert attractive forces on each other. When the currents circulate in opposite directions, the two like poles repel each other.

Here, at last, we have a real connection to the behavior of magnets that opened our discussion of magnetism—namely, that like poles repel and opposite poles attract. Now we have an *explanation* for this behavior, at least for electromagnets. **Magnetic poles attract or repel because the moving charges in one current exert attractive or repulsive magnetic forces on the moving charges in the other current.** Our tour through interacting moving charges is finally starting to show some practical results!

Now let's consider what happens to a current loop in a magnetic field. FIGURE 29.48 shows a square current loop in a uniform magnetic field along the z-axis. As we've learned, the field exerts magnetic forces on the currents in each of the four sides of the loop. Their directions are given by the right-hand rule. Forces $\vec{F}_{front}$ and $\vec{F}_{back}$ are opposite to each other and cancel. Forces $\vec{F}_{top}$ and $\vec{F}_{bottom}$ also add to give no net force, but because $\vec{F}_{top}$ and $\vec{F}_{bottom}$ don't act along the same line they will *rotate* the loop by exerting a torque on it.

Recall that torque is the magnitude of the force F multiplied by the moment arm d, the distance between the pivot point and the line of action. Both forces have the same moment arm $d = \frac{1}{2} l \sin\theta$, hence the torque on the loop—a torque exerted by the magnetic field—is

$$\tau = 2Fd = 2(IlB)\left(\tfrac{1}{2} l \sin\theta\right) = (Il^2)B\sin\theta = \mu B \sin\theta \qquad (29.28)$$

where $\mu = Il^2 = IA$ is the loop's magnetic dipole moment.

Although we derived Equation 29.28 for a square loop, the result is valid for a current loop of any shape. Notice that Equation 29.28 looks like another example of a cross product. We earlier defined the magnetic dipole moment vector $\vec{\mu}$ to be a vector perpendicular to the current loop in a direction given by the right-hand rule. Figure 29.48 shows that θ is the angle between $\vec{B}$ and $\vec{\mu}$, hence the torque on a magnetic dipole is

$$\vec{\tau} = \vec{\mu} \times \vec{B} \qquad (29.29)$$

The torque is zero when the magnetic dipole moment $\vec{\mu}$ is aligned parallel or antiparallel to the magnetic field, and is maximum when $\vec{\mu}$ is perpendicular to the field. It is this magnetic torque that causes a compass needle—a magnetic moment—to rotate until it is aligned with the magnetic field.

An Electric Motor

The torque on a current loop in a magnetic field is the basis for how an electric motor works. As FIGURE 29.49 on the next page shows, the *armature* of a motor is a coil of wire wound on an axle. When a current passes through the coil, the magnetic field exerts a torque on the armature and causes it to rotate. If the current were steady, the armature

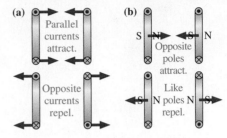

FIGURE 29.47 Two alternative but equivalent ways to view magnetic forces.

(a) Parallel currents attract. Opposite currents repel.

(b) Opposite poles attract. Like poles repel.

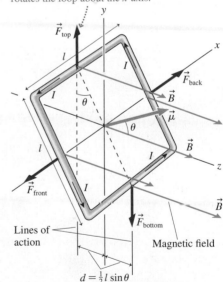

FIGURE 29.48 A uniform magnetic field exerts a torque on a current loop.

$\vec{F}_{top}$ and $\vec{F}_{bottom}$ exert a torque that rotates the loop about the x-axis.

Lines of action

$d = \frac{1}{2} l \sin\theta$

Magnetic field

would oscillate back and forth around the equilibrium position until (assuming there's some friction or damping) it stopped with the plane of the coil perpendicular to the field. To keep the motor turning, a device called a *commutator* reverses the current direction in the coils every 180°. (Notice that the commutator is split, so the positive terminal of the battery sends current into whichever wire touches the right half of the commutator.) The current reversal prevents the armature from ever reaching an equilibrium position, so the magnetic torque keeps the motor spinning as long as there is a current.

FIGURE 29.49 A simple electric motor.

The commutator reverses the current in the loop every half cycle so that the force is always upward on the left side of the loop.

STOP TO THINK 29.6   What is the current direction in the loop?

a. Out of the page at the top of the loop, into the page at the bottom
b. Out of the page at the bottom of the loop, into the page at the top

---

29.10 Magnetic Properties of Matter

Our theory has focused mostly on the magnetic properties of currents, yet our everyday experience is mostly with permanent magnets. We have seen that current loops and solenoids have magnetic poles and exhibit behaviors like those of permanent magnets, but we still lack a specific connection between electromagnets and permanent magnets. The goal of this section is to complete our understanding by developing an atomic-level view of the magnetic properties of matter.

Atomic Magnets

A plausible explanation for the magnetic properties of materials is the orbital motion of the atomic electrons. **FIGURE 29.50** shows a simple, classical model of an atom in which a negative electron orbits a positive nucleus. In this picture of the atom, the electron's motion is that of a current loop! It is a microscopic current loop, to be sure, but a current loop nonetheless. Consequently, an orbiting electron acts as a tiny magnetic dipole, with a north pole and a south pole. You can think of the magnetic dipole as an atomic-size magnet.

FIGURE 29.50 A classical orbiting electron is a tiny magnetic dipole.

However, the atoms of most elements contain many electrons. Unlike the solar system, where all of the planets orbit in the same direction, electron orbits are arranged to oppose each other: one electron moves counterclockwise for every electron that moves clockwise. Thus the magnetic moments of individual orbits tend to cancel each other and the *net* magnetic moment is either zero or very small.

The cancellation continues as the atoms are joined into molecules and the molecules into solids. When all is said and done, the net magnetic moment of any bulk matter due to the orbiting electrons is so small as to be negligible. There are various subtle magnetic effects that can be observed under laboratory conditions, but orbiting electrons cannot explain the very strong magnetic effects of a piece of iron.

The Electron Spin

The key to understanding atomic magnetism was the 1922 discovery that electrons have an *inherent magnetic moment.* Perhaps this shouldn't be surprising. An electron has a *mass,* which allows it to interact with gravitational fields, and a *charge,* which allows it to interact with electric fields. There's no reason an electron shouldn't also interact with magnetic fields, and to do so it comes with a magnetic moment.

An electron's inherent magnetic moment, shown in **FIGURE 29.51**, is often called the electron *spin* because, in a classical picture, a spinning ball of charge would have a magnetic moment. This classical picture is not a realistic portrayal of how the electron really behaves, but its inherent magnetic moment makes it seem *as if* the electron were spinning. While it may not be spinning in a literal sense, an electron really is a microscopic magnet.

We must appeal to the results of quantum physics to find out what happens in an atom with many electrons. The spin magnetic moments, like the orbital magnetic moments, tend to oppose each other as the electrons are placed into their shells, causing the net magnetic moment of a *filled* shell to be zero. However, atoms containing an odd number of electrons must have at least one valence electron with an unpaired spin. These atoms have a net magnetic moment due to the electron's spin.

But atoms with magnetic moments don't necessarily form a solid with magnetic properties. For most elements, the magnetic moments of the atoms are randomly arranged when the atoms join together to form a solid. As **FIGURE 29.52** shows, this random arrangement produces a solid whose net magnetic moment is very close to zero. This agrees with our common experience that most materials are not magnetic.

Ferromagnetism

It happens that in iron, and a few other substances, the spins interact with each other in such a way that atomic magnetic moments tend to all line up in the *same* direction, as shown in **FIGURE 29.53**. Materials that behave in this fashion are called **ferromagnetic,** with the prefix *ferro* meaning "iron-like."

In ferromagnetic materials, the individual magnetic moments add together to create a *macroscopic* magnetic dipole. The material has a north and a south magnetic pole, generates a magnetic field, and aligns parallel to an external magnetic field. In other words, it is a magnet!

Although iron is a magnetic material, a typical piece of iron is not a strong permanent magnet. You need not worry that a steel nail, which is mostly iron and is easily lifted with a magnet, will leap from your hands and pin itself against the hammer because of its own magnetism. It turns out, as shown in **FIGURE 29.54** on the next page, that a piece of iron is divided into small regions, typically less than 100 μm in size, called **magnetic domains.** The magnetic moments of all the iron atoms within each domain are perfectly aligned, so each individual domain, like Figure 29.53, is a strong magnet.

FIGURE 29.51 Magnetic moment of the electron.

The arrow represents the inherent magnetic moment of the electron.

FIGURE 29.52 The random magnetic moments of the atoms in a typical solid.

The atomic magnetic moments due to unpaired electrons point in random directions. The sample has no net magnetic moment.

FIGURE 29.53 In a ferromagnetic material, the atomic magnetic moments are aligned.

The atomic magnetic moments are aligned. The sample has north and south magnetic poles.

FIGURE 29.54 Magnetic domains in a ferromagnetic material. The net magnetic dipole is nearly zero.

Magnetic domains

Magnetic moment of a domain

The atomic moments are lined up within each domain.

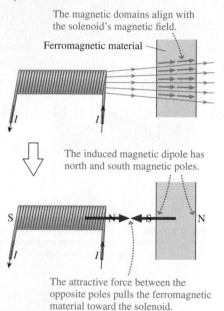

FIGURE 29.55 The magnetic field of the solenoid creates an induced magnetic dipole in the iron.

The magnetic domains align with the solenoid's magnetic field.

Ferromagnetic material

The induced magnetic dipole has north and south magnetic poles.

The attractive force between the opposite poles pulls the ferromagnetic material toward the solenoid.

Magnetic resonance imaging, or MRI, uses the magnetic properties of atoms as a noninvasive probe of the human body.

However, the various magnetic domains that form a larger solid, such as you might hold in your hand, are randomly arranged. Their magnetic dipoles largely cancel, much like the cancellation that occurs on the atomic scale for nonferromagnetic substances, so the solid as a whole has only a small magnetic moment. That is why the nail is not a strong permanent magnet.

Induced Magnetic Dipoles

If a ferromagnetic substance is subjected to an *external* magnetic field, the external field exerts a torque on the magnetic dipole of each domain. The torque causes many of the domains to rotate and become aligned with the external field, just as a compass needle aligns with a magnetic field, although internal forces between the domains generally prevent the alignment from being perfect. In addition, atomic-level forces between the spins can cause the *domain boundaries* to move. Domains that are aligned along the external field become larger at the expense of domains that are opposed to the field. These changes in the size and orientation of the domains cause the material to develop a *net magnetic dipole* that is aligned with the external field. This magnetic dipole has been *induced* by the external field, so it is called an **induced magnetic dipole.**

NOTE The induced magnetic dipole is analogous to the polarization forces and induced electric dipoles that you studied in Chapter 23.

FIGURE 29.55 shows a ferromagnetic material near the end of a solenoid. The magnetic moments of the domains align with the solenoid's field, creating an induced magnetic dipole whose south pole faces the solenoid's north pole. Consequently, the magnetic force between the poles pulls the ferromagnetic object to the electromagnet.

The fact that a magnet attracts and picks up ferromagnetic objects was one of the basic observations about magnetism with which we started the chapter. Now we have an *explanation* of how it works, based on three ideas:

1. Electrons are microscopic magnets due to their spin.
2. A ferromagnetic material in which the spins are aligned is organized into magnetic domains.
3. The individual domains align with an external magnetic field to produce an induced magnetic dipole moment for the entire object.

The object's magnetic dipole may not return to zero when the external field is removed because some domains remain "frozen" in the alignment they had in the external field. Thus a ferromagnetic object that has been in an external field may be left with a net magnetic dipole moment after the field is removed. In other words, the object has become a **permanent magnet.** A permanent magnet is simply a ferromagnetic material in which a majority of the magnetic domains are aligned with each other to produce a net magnetic dipole moment.

Whether or not a ferromagnetic material can be made into a permanent magnet depends on the internal crystalline structure of the material. *Steel* is an alloy of iron with other elements. An alloy of mostly iron with the right percentages of chromium and nickel produces *stainless steel,* which has virtually no magnetic properties at all because its particular crystalline structure is not conducive to the formation of domains. A very different steel alloy called Alnico V is made with 51% iron, 24% cobalt, 14% nickel, 8% aluminum, and 3% copper. It has extremely prominent magnetic properties and is used to make high-quality permanent magnets.

So we've come full circle. One of our initial observations about magnetism was that a permanent magnet can exert forces on some materials but not others. The *theory* of magnetism that we then proceeded to develop was about the interactions between moving charges. What moving charges had to do with permanent magnets was not obvious. But finally, by considering magnetic effects at the atomic level, we found that properties of permanent magnets and magnetic materials can be traced to the interactions of vast numbers of electron spins.

STOP TO THINK 29.7 Which magnet or magnets induced this magnetic dipole?

CHALLENGE EXAMPLE 29.15 | Designing a loudspeaker

A loudspeaker consists of a paper cone wrapped at the bottom with several turns of fine wire. As **FIGURE 29.56** shows, this coil sits in a narrow gap between the poles of a circular magnet. To produce sound, the amplifier drives a current through the coil. The magnetic field then exerts a force on this current, pushing the cone and thus pushing the air to create a sound wave. An ideal speaker would experience only forces from the magnetic field, thus responding only to the current from the amplifier. Real speakers are balanced so as to come close to this ideal unless driven very hard.

Consider a 5.5 g loudspeaker cone with a 5.0-cm-diameter, 20-turn coil having a resistance of 8.0 Ω. There is a 0.18 T field in the gap between the poles. These values are typical of the loudspeakers found in car stereo systems. What is the oscillation amplitude of this speaker if driven by a 100 Hz oscillatory voltage from the amplifier with a peak value of 12 V?

FIGURE 29.56 The coil and magnet of a loudspeaker.

MODEL Model the loudspeaker as ideal, responding only to magnetic forces. These forces cause the cone to accelerate. We'll use kinematics to relate the acceleration to the displacement.

VISUALIZE FIGURE 29.57 shows the coil in the gap between the magnet poles. Magnetic fields go from north to south poles, so the field is radially outward. Consequently, the field at all points is perpendicular to the circular current. According to the right-hand rule, the magnetic force on the current is into or out of the

FIGURE 29.57 The magnetic field in the gap, from north to south, is perpendicular to the current.

page, depending on whether the current is counterclockwise or clockwise, respectively.

SOLVE We can write the output voltage of the amplifier as $\Delta V = V_0 \cos \omega t$, where $V_0 = 12$ V is the peak voltage and $\omega = 2\pi f = 628$ rad/s is the angular frequency at 100 Hz. The voltage drives current

$$I = \frac{\Delta V}{R} = \frac{V_0 \cos \omega t}{R}$$

through the coil, where R is the coil's resistance. This causes the oscillating in-and-out force that drives the speaker cone back and forth. Even though the coil isn't a straight wire, the fact that the magnetic field is everywhere perpendicular to the current means that we can calculate the magnetic force as $F = IlB$ where l is the total length of the wire in the coil. The circumference of the coil is $\pi(0.050 \text{ m}) = 0.157$ m so 20 turns gives $l = 3.1$ m. The cone responds to the force by accelerating with $a = F/m$. Combining these pieces, we find the cone's acceleration is

$$a = \frac{IlB}{m} = \frac{V_0 lB \cos \omega t}{mR} = a_{max} \cos \omega t$$

It is straightforward to evaluate $a_{max} = 152$ m/s^2.

From kinematics, $a = dv/dt$ and $v = dx/dt$. We need to integrate twice to find the displacement. First,

$$v = \int a\, dt = a_{max} \int \cos \omega t\, dt = \frac{a_{max}}{\omega} \sin \omega t$$

The integration constant is zero because we know, from simple harmonic motion, that the average velocity is zero. Integrating again, we get

$$x = \int v\, dt = \frac{a_{max}}{\omega} \int \sin \omega t\, dt = -\frac{a_{max}}{\omega^2} \cos \omega t$$

where the integration constant is again zero if we assume the oscillation takes place around the origin. The minus sign tells us that the displacement and acceleration are out of phase. The *amplitude* of the oscillation, which we seek, is

$$A = \frac{a_{max}}{\omega^2} = \frac{152 \text{ m/s}^2}{(628 \text{ rad/s})^2} = 3.8 \times 10^{-4} \text{ m} = 0.38 \text{ mm}$$

ASSESS If you've ever placed your hand on a loudspeaker cone, you know that you can feel a slight vibration. An amplitude of 0.38 mm is consistent with this observation. The fact that the amplitude increases with the inverse square of the frequency explains why you can sometimes *see* the cone vibrating with an amplitude of several millimeters for low-frequency bass notes.

SUMMARY

The goal of Chapter 29 has been to learn about magnetism and the magnetic field.

GENERAL PRINCIPLES

At its most fundamental level, **magnetism** is an interaction between moving charges. The magnetic field of one moving charge exerts a force on another moving charge.

Magnetic Fields

The **Biot-Savart law** for a moving point charge

$$\vec{B} = \frac{\mu_0}{4\pi} \frac{q\vec{v} \times \hat{r}}{r^2}$$

Magnetic field of a current

MODEL Model wires as simple shapes.

VISUALIZE Divide the wire into short segments.

SOLVE Use superposition:

- Find the field of each segment Δs.
- Find $\vec{B}$ by summing the fields of all Δs, usually as an integral.

An alternative method for fields with a high degree of symmetry is **Ampère's law**:

$$\oint \vec{B} \cdot d\vec{s} = \mu_0 I_{\text{through}}$$

where I_{through} is the current through the area bounded by the integration path.

Magnetic Forces

The magnetic force on a moving charge is

$$\vec{F} = q\vec{v} \times \vec{B}$$

The force is perpendicular to $\vec{v}$ and $\vec{B}$.

The magnetic force on a current-carrying wire is

$$\vec{F} = I\vec{l} \times \vec{B}$$

$\vec{F} = \vec{0}$ for a charge or current moving parallel to $\vec{B}$.

The magnetic torque on a magnetic dipole is

$$\vec{\tau} = \vec{\mu} \times \vec{B}$$

APPLICATIONS

Wire

$$B = \frac{\mu_0}{2\pi} \frac{I}{r}$$

Solenoid

Uniform field

$$B = \frac{\mu_0 NI}{l}$$

Right-hand rule

Point your right thumb in the direction of I. Your fingers curl in the direction of $\vec{B}$. For a dipole, $\vec{B}$ emerges from the side that is the north pole.

Loop

$$B_{\text{center}} = \frac{\mu_0}{2} \frac{NI}{R}$$

Flat magnet

Charged-particle motion

No force if $\vec{v}$ is parallel to $\vec{B}$

Circular motion at the cyclotron frequency $f_{\text{cyc}} = qB/2\pi m$ if $\vec{v}$ is perpendicular to $\vec{B}$

Parallel wires and current loops

Parallel currents attract.
Opposite currents repel.

TERMS AND NOTATION

north pole	Biot-Savart law	line integral	Hall effect
south pole	tesla, T	Ampère's law	Hall voltage, ΔV_H
magnetic dipole	permeability constant, μ_0	uniform magnetic field	ferromagnetic
magnetic material	cross product	solenoid	magnetic domain
right-hand rule	current loop	cyclotron motion	induced magnetic dipole
magnetic field, $\vec{B}$	electromagnet	cyclotron frequency, f_{cyc}	permanent magnet
magnetic field lines	magnetic dipole moment, $\vec{\mu}$	cyclotron	

CONCEPTUAL QUESTIONS

1. The lightweight glass sphere in **FIGURE Q29.1** hangs by a thread. The north pole of a bar magnet is brought near the sphere.
   a. Suppose the sphere is electrically neutral. Is it attracted to, repelled by, or not affected by the magnet? Explain.
   b. Answer the same question if the sphere is positively charged.

FIGURE Q29.1   **FIGURE Q29.2**

2. The metal sphere in **FIGURE Q29.2** hangs by a thread. When the north pole of a magnet is brought near, the sphere is strongly attracted to the magnet. Then the magnet is reversed and its south pole is brought near the sphere. How does the sphere respond? Explain.
3. You have two electrically neutral metal cylinders that exert strong attractive forces on each other. You have no other metal objects. Can you determine if *both* of the cylinders are magnets, or if one is a magnet and the other is just a piece of iron? If so, how? If not, why not?
4. What is the current direction in the wire of **FIGURE Q29.4**? Explain.

FIGURE Q29.4   **FIGURE Q29.5**

5. What is the current direction in the wire of **FIGURE Q29.5**? Explain.
6. What is the *initial* direction of deflection for the charged particles entering the magnetic fields shown in **FIGURE Q29.6**?

FIGURE Q29.6

7. What is the *initial* direction of deflection for the charged particles entering the magnetic fields shown in **FIGURE Q29.7**?

FIGURE Q29.7

8. Determine the magnetic field direction that causes the charged particles shown in **FIGURE Q29.8** to experience the indicated magnetic force.

FIGURE Q29.8 $\vec{F}$ into page

9. Determine the magnetic field direction that causes the charged particles shown in **FIGURE Q29.9** to experience the indicated magnetic force.

FIGURE Q29.9 $\vec{F}$ out of page $\vec{v}$ into page

10. You have a horizontal cathode-ray tube (CRT) for which the controls have been adjusted such that the electron beam *should* make a single spot of light exactly in the center of the screen. You observe, however, that the spot is deflected to the right. It is possible that the CRT is broken. But as a clever scientist, you realize that your laboratory might be in either an electric or a magnetic field. Assuming that you do not have a compass, any magnets, or any charged rods, how can you use the CRT itself to determine whether the CRT is broken, is in an electric field, or is in a magnetic field? You cannot remove the CRT from the room.
11. The south pole of a bar magnet is brought toward the current loop of **FIGURE Q29.11**. Does the bar magnet attract, repel, or have no effect on the loop? Explain.

    FIGURE Q29.11

12. Give a step-by-step explanation, using both words and pictures, of how a permanent magnet can pick up a piece of nonmagnetized iron.

EXERCISES AND PROBLEMS

Problems labeled ▨ integrate material from earlier chapters.

Exercises

Section 29.3 The Source of the Magnetic Field: Moving Charges

1. | What is the magnetic field strength at points 2 to 4 in **FIGURE EX29.1**? Assume that the wires overlap closely at 2 and 3, that each point is the same distance from nearby wires, and that all other wires are too far away to contribute to the field.

FIGURE EX29.1   **FIGURE EX29.2**

2. | Points 1 and 2 in **FIGURE EX29.2** are the same distance from the wires as the point where $B = 2.0$ mT. What are the strength and direction of $\vec{B}$ at points 1 and 2?

3. | A proton moves along the x-axis with $v_x = 1.0 \times 10^7$ m/s. As it passes the origin, what are the strength and direction of the magnetic field at the (x, y, z) positions (a) (1 cm, 0 cm, 0 cm), (b) (0 cm, 1 cm, 0 cm), and (c) (0 cm, −2 cm, 0 cm)?

4. ‖ An electron moves along the z-axis with $v_z = 2.0 \times 10^7$ m/s. As it passes the origin, what are the strength and direction of the magnetic field at the (x, y, z) positions (a) (1 cm, 0 cm, 0 cm), (b) (0 cm, 0 cm, 1 cm), and (c) (0 cm, 1 cm, 1 cm)?

5. ‖ What is the magnetic field at the position of the dot in **FIGURE EX29.5**? Give your answer as a vector.

FIGURE EX29.5   **FIGURE EX29.6**

6. ‖ What is the magnetic field at the position of the dot in **FIGURE EX29.6**? Give your answer as a vector.

Section 29.4 The Magnetic Field of a Current

7. | What currents are needed to generate the magnetic field strengths of Table 29.1 at a point 1.0 cm from a long, straight wire?

8. ‖ The element niobium, which is a metal, is a superconductor (i.e., no electrical resistance) at temperatures below 9 K. However, the superconductivity is destroyed if the magnetic field at the surface of the metal reaches or exceeds 0.10 T. What is the maximum current in a straight, 3.0-mm-diameter superconducting niobium wire?

9. | Although the evidence is weak, there has been concern in recent years over possible health effects from the magnetic fields generated by electric transmission lines. A typical high-voltage transmission line is 20 m above the ground and carries a 200 A current at a potential of 110 kV.
BIO
a. What is the magnetic field strength on the ground directly under such a transmission line?
b. What percentage is this of the earth's magnetic field of 50 μT?

10. | A biophysics experiment uses a very sensitive magnetic field probe to determine the current associated with a nerve impulse traveling along an axon. If the peak field strength 1.0 mm from an axon is 8.0 pT, what is the peak current carried by the axon?
BIO

11. ‖ The magnetic field at the center of a 1.0-cm-diameter loop is 2.5 mT.
a. What is the current in the loop?
b. A long straight wire carries the same current you found in part a. At what distance from the wire is the magnetic field 2.5 mT?

12. ‖ What are the magnetic fields at points a to c in **FIGURE EX29.12**? Give your answers as vectors.

FIGURE EX29.12   **FIGURE EX29.13**

13. | A wire carries current I into the junction shown in **FIGURE EX29.13**. What is the magnetic field at the dot?

14. ‖ What are the magnetic field strength and direction at points a to c in **FIGURE EX29.14**?

FIGURE EX29.14   **FIGURE EX29.15**

15. ‖ Point A in **FIGURE EX29.15** is 2.0 mm from the wire. What is the magnetic field strength at point A? You can assume that the wire is very long and that all other wires are too far away to contribute to the magnetic field.

Section 29.5 Magnetic Dipoles

16. ‖ The on-axis magnetic field strength 10 cm from a small bar magnet is 5.0 μT.
a. What is the bar magnet's magnetic dipole moment?
b. What is the on-axis field strength 15 cm from the magnet?

17. ‖ A 100 A current circulates around a 2.0-mm-diameter super-conducting ring.
   a. What is the ring's magnetic dipole moment?
   b. What is the on-axis magnetic field strength 5.0 cm from the ring?
18. ‖‖ A small, square loop carries a 25 A current. The on-axis magnetic field strength 50 cm from the loop is 7.5 nT. What is the edge length of the square?
19. ‖ The earth's magnetic dipole moment is 8.0×10^{22} A m^2.
   a. What is the magnetic field strength on the surface of the earth at the earth's north magnetic pole? How does this compare to the value in Table 29.1? You can assume that the current loop is deep inside the earth.
   b. Astronauts discover an earth-size planet without a magnetic field. To create a magnetic field at the north pole with the same strength as earth's, they propose running a current through a wire around the equator. What size current would be needed?

Section 29.6 Ampère's Law and Solenoids

20. ‖ What is the line integral of $\vec{B}$ between points i and f in FIGURE EX29.20?

FIGURE EX29.20   FIGURE EX29.21

21. ‖ What is the line integral of $\vec{B}$ between points i and f in FIGURE EX29.21?
22. ‖ The value of the line integral of $\vec{B}$ around the closed path in FIGURE EX29.22 is 3.77×10^{-6} T m. What is I_3?

FIGURE EX29.22

23. | The value of the line integral of $\vec{B}$ around the closed path in FIGURE EX29.23 is 1.38×10^{-5} T m. What are the direction (in or out of the page) and magnitude of I_3?

FIGURE EX29.23

24. ‖ What is the line integral of $\vec{B}$ between points i and f in FIGURE EX29.24?

FIGURE EX29.24

25. ‖ Magnetic resonance imaging needs a magnetic field strength
BIO of 1.5 T. The solenoid is 1.8 m long and 75 cm in diameter. It is tightly wound with a single layer of 2.0-mm-diameter supercon-ducting wire. What size current is needed?

Section 29.7 The Magnetic Force on a Moving Charge

26. ‖ A proton moves in the magnetic field $\vec{B} = 0.50\,\hat{\imath}$ T with a speed of 1.0×10^7 m/s in the directions shown in FIGURE EX29.26. For each, what is magnetic force $\vec{F}$ on the proton? Give your answers in component form.

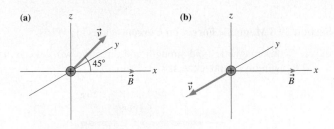

FIGURE EX29.26

27. ‖ An electron moves in the magnetic field $\vec{B} = 0.50\,\hat{\imath}$ T with a speed of 1.0×10^7 m/s in the directions shown in FIGURE EX29.27. For each, what is magnetic force $\vec{F}$ on the electron? Give your answers in component form.

FIGURE EX29.27

28. | Radio astronomers detect electromagnetic radiation at 45 MHz from an interstellar gas cloud. They suspect this radia-tion is emitted by electrons spiraling in a magnetic field. What is the magnetic field strength inside the gas cloud?
29. ‖ To five significant figures, what are the cyclotron frequencies in a 3.0000 T magnetic field of the ions (a) O_2^+, (b) N_2^+, and (c) CO^+? The atomic masses are shown in the table; the mass of the missing electron is less than 0.001 u and is not relevant at this level of precision. Although N_2^+ and CO^+ both have a nominal molecular mass of 28, they are easily distinguished by virtue of their slightly different cyclotron frequencies. Use the following constants: $1\text{ u} = 1.6605 \times 10^{-27}$ kg, $e = 1.6022 \times 10^{-19}$ C.

Atomic masses	
^{12}C	12.000
^{14}N	14.003
^{16}O	15.995

30. ‖ For your senior project, you would like to build a cyclotron that will accelerate protons to 10% of the speed of light. The largest vacuum chamber you can find is 50 cm in diameter. What magnetic field strength will you need?

31. ‖ The microwaves in a microwave oven are produced in a special tube called a *magnetron*. The electrons orbit the magnetic field at 2.4 GHz, and as they do so they emit 2.4 GHz electromagnetic waves.
    a. What is the magnetic field strength?
    b. If the maximum diameter of the electron orbit before the electron hits the wall of the tube is 2.5 cm, what is the maximum electron kinetic energy?

32. ‖ The Hall voltage across a conductor in a 55 mT magnetic field is 1.9 μV. When used with the same current in a different magnetic field, the voltage across the conductor is 2.8 μV. What is the strength of the second field?

Section 29.8 Magnetic Forces on Current-Carrying Wires

33. ‖ What magnetic field strength and direction will levitate the 2.0 g wire in **FIGURE EX29.33**?

FIGURE EX29.33

34. ‖ The two 10-cm-long parallel wires in **FIGURE EX29.34** are separated by 5.0 mm. For what value of the resistor R will the force between the two wires be 5.4×10^{-5} N?

FIGURE EX29.34

35. ‖ The right edge of the circuit in **FIGURE EX29.35** extends into a 50 mT uniform magnetic field. What are the magnitude and direction of the net force on the circuit?

FIGURE EX29.35   **FIGURE EX29.36**

36. ‖ What is the net force (magnitude and direction) on each wire in **FIGURE EX29.36**?

37. ‖ **FIGURE EX29.37** is a cross section through three long wires with linear mass density 50 g/m. They each carry equal currents in the directions shown. The lower two wires are 4.0 cm apart and are attached to a table. What current I will allow the upper wire to "float" so as to form an equilateral triangle with the lower wires?

FIGURE EX29.37

Section 29.9 Forces and Torques on Current Loops

38. ‖ A square current loop 5.0 cm on each side carries a 500 mA current. The loop is in a 1.2 T uniform magnetic field. The axis of the loop, perpendicular to the plane of the loop, is 30° away from the field direction. What is the magnitude of the torque on the current loop?

39. ‖ A small bar magnet experiences a 0.020 N m torque when the axis of the magnet is at 45° to a 0.10 T magnetic field. What is the magnitude of its magnetic dipole moment?

40. ‖ a. What is the magnitude of the torque on the current loop in **FIGURE EX29.40**?
    b. What is the loop's equilibrium orientation?

FIGURE EX29.40

Problems

41. ‖ A long wire carrying a 5.0 A current perpendicular to the *xy*-plane intersects the *x*-axis at $x = -2.0$ cm. A second, parallel wire carrying a 3.0 A current intersects the *x*-axis at $x = +2.0$ cm. At what point or points on the *x*-axis is the magnetic field zero if (a) the two currents are in the same direction and (b) the two currents are in opposite directions?

42. ‖ The two insulated wires in **FIGURE P29.42** cross at a 30° angle but do not make electrical contact. Each wire carries a 5.0 A current. Points 1 and 2 are each 4.0 cm from the intersection and equally distant from both wires. What are the magnitude and direction of the magnetic fields at points 1 and 2?

FIGURE P29.42   **FIGURE P29.43**

43. ‖ What are the strength and direction of the magnetic field at the center of the loop in **FIGURE P29.43**?

44. ‖ At what distance on the axis of a current loop is the magnetic field half the strength of the field at the center of the loop? Give your answer as a multiple of R.

45. ‖ Find an expression for the magnetic field strength at the center (point P) of the circular arc in **FIGURE P29.45**.

FIGURE P29.45   **FIGURE P29.46**

46. ‖ What are the strength and direction of the magnetic field at point P in **FIGURE P29.46**?

47. ‖ A scientist measuring the resistivity of a new metal alloy left her ammeter in another lab, but she does have a magnetic field probe. So she creates a 6.5-m-long, 2.0-mm-diameter wire of the material, connects it to a 1.5 V battery, and measures a 3.0 mT magnetic field 1.0 mm from the surface of the wire. What is the material's resistivity?

48. ‖ A 2.5-m-long, 2.0-mm-diameter aluminum wire has a potential difference of 1.5 V between its ends. Consider an electron halfway between the center of the wire and the surface that is moving parallel to the wire at the drift speed. What is the ratio of the electric force on the electron to the magnetic force on the electron?

49. ‖ Your employer asks you to build a 20-cm-long solenoid with an interior field of 5.0 mT. The specifications call for a single layer of wire, wound with the coils as close together as possible. You have two spools of wire available. Wire with a #18 gauge has a diameter of 1.02 mm and has a maximum current rating of 6 A. Wire with a #26 gauge is 0.41 mm in diameter and can carry up to 1 A. Which wire should you use, and what current will you need?

50. ‖ The magnetic field strength at the north pole of a 2.0-cm-diameter, 8-cm-long Alnico magnet is 0.10 T. To produce the same field with a solenoid of the same size, carrying a current of 2.0 A, how many turns of wire would you need?

51. ‖ The earth's magnetic field, with a magnetic dipole moment of 8.0×10^{22} A m^2, is generated by currents within the molten iron of the earth's outer core. Suppose we model the core current as a 3000-km-diameter current loop made from a 1000-km-diameter "wire." The loop diameter is measured from the centers of this very fat wire.
   a. What is the current in the current loop?
   b. What is the current density J in the current loop?
   c. To decide whether this is a large or a small current density, compare it to the current density of a 1.0 A current in a 1.0-mm-diameter wire.

52. ‖‖ Weak magnetic fields can be measured at the surface of the brain.
BIO Although the currents causing these fields are quite complicated, we can estimate their size by modeling them as a current loop around the equator of a 16-cm-diameter (the width of a typical head) sphere. What current is needed to produce a 3.0 pT field—the strength measured for one subject—at the pole of this sphere?

53. ‖ The heart produces a weak magnetic field that can be used
BIO to diagnose certain heart problems. It is a dipole field produced by a current loop in the outer layers of the heart.
   a. It is estimated that the field at the center of the heart is 90 pT. What current must circulate around an 8.0-cm-diameter loop, about the size of a human heart, to produce this field?
   b. What is the magnitude of the heart's magnetic dipole moment?

54. ‖ What is the magnetic field strength at the center of the semicircle in **FIGURE P29.54**?

FIGURE P29.54

55. ‖ The *toroid* of **FIGURE P29.55** is a coil of wire wrapped around a doughnut-shaped ring (a *torus*). Toroidal magnetic fields are used to confine fusion plasmas.
   a. From symmetry, what must be the *shape* of the magnetic field in this toroid? Explain.
   b. Consider a toroid with N closely spaced turns carrying current I. Use Ampère's law to find an expression for the magnetic field strength at a point inside the torus at distance r from the axis.
   c. Is a toroidal magnetic field a uniform field? Explain.

FIGURE P29.55

FIGURE P29.56

56. ‖ The coaxial cable shown in **FIGURE P29.56** consists of a solid inner conductor of radius R_1 surrounded by a hollow, very thin outer conductor of radius R_2. The two carry equal currents I, but

in *opposite* directions. The current density is uniformly distributed over each conductor.
   a. Find expressions for three magnetic fields: within the inner conductor, in the space between the conductors, and outside the outer conductor.
   b. Draw a graph of B versus r from $r = 0$ to $r = 2R_2$ if $R_1 = \frac{1}{3}R_2$.

57. ‖ A long, hollow wire has inner radius R_1 and outer radius R_2. The wire carries current I uniformly distributed across the area of the wire. Use Ampère's law to find an expression for the magnetic field strength in the three regions $0 < r < R_1$, $R_1 < r < R_2$, and $R_2 < r$.

58. ‖ A proton moving in a uniform magnetic field with $\vec{v}_1 = 1.00 \times 10^6 \hat{\imath}$ m/s experiences force $\vec{F}_1 = 1.20 \times 10^{-16} \hat{k}$ N. A second proton with $\vec{v}_2 = 2.00 \times 10^6 \hat{\jmath}$ m/s experiences $\vec{F}_2 = -4.16 \times 10^{-16} \hat{k}$ N in the same field. What is $\vec{B}$? Give your answer as a magnitude and an angle measured ccw from the $+x$-axis.

59. ‖ An electron travels with speed 1.0×10^7 m/s between the two parallel charged plates shown in **FIGURE P29.59**. The plates are separated by 1.0 cm and are charged by a 200 V battery. What magnetic field strength and direction will allow the electron to pass between the plates without being deflected?

FIGURE P29.59   **FIGURE P29.60**

60. ‖ An electron in a cathode-ray tube is accelerated through a potential difference of 10 kV, then passes through the 2.0-cm-wide region of uniform magnetic field in **FIGURE P29.60**. What field strength will deflect the electron by 10°?

FIGURE P29.61 $E = 1000$ V/m

61. ‖ An antiproton (same properties as a proton except that $q = -e$) is moving in the combined electric and magnetic fields of **FIGURE P29.61**. What are the magnitude and direction of the antiproton's acceleration at this instant?

62. ‖ a. A 65-cm-diameter cyclotron uses a 500 V oscillating potential difference between the dees. What is the maximum kinetic energy of a proton if the magnetic field strength is 0.75 T?
   b. How many revolutions does the proton make before leaving the cyclotron?

63. ‖ An antiproton is identical to a proton except it has the opposite charge, $-e$. To study antiprotons, they must be confined in an ultrahigh vacuum because they will annihilate—producing gamma rays—if they come into contact with the protons of ordinary matter. One way of confining antiprotons is to keep them in a magnetic field. Suppose that antiprotons are created with a speed of 1.5×10^4 m/s and then trapped in a 2.0 mT magnetic field. What minimum diameter must the vacuum chamber have to allow these antiprotons to circulate without touching the walls?

64. || **FIGURE P29.64** shows a *mass spectrometer,* an analytical instrument used to identify the various molecules in a sample by measuring their charge-to-mass ratio q/m. The sample is ionized, the positive ions are accelerated (starting from rest) through a potential difference ΔV, and they then enter a region of uniform magnetic field. The field bends the ions into circular trajectories, but after just half a circle they either strike the wall or pass through a small opening to a detector. As the accelerating voltage is slowly increased, different ions reach the detector and are measured. Consider a mass spectrometer with a 200.00 mT magnetic field and an 8.0000 cm spacing between the entrance and exit holes. To five significant figures, what accelerating potential differences ΔV are required to detect the ions (a) O_2^+, (b) N_2^+, and (c) CO^+? See Exercise 29 for atomic masses; the mass of the missing electron is less than 0.001 u and is not relevant at this level of precision. Although N_2^+ and CO^+ both have a nominal molecular mass of 28, they are easily distinguished by virtue of their slightly different accelerating voltages. Use the following constants: $1\text{ u} = 1.6605 \times 10^{-27}\text{ kg}$, $e = 1.6022 \times 10^{-19}\text{ C}$.

FIGURE P29.64

FIGURE P29.65

65. || The uniform 30 mT magnetic field in **FIGURE P29.65** points in the positive z-direction. An electron enters the region of magnetic field with a speed of 5.0×10^6 m/s and at an angle of $30°$ above the xy-plane. Find the radius r and the pitch p of the electron's spiral trajectory.

66. || Particle accelerators, such as the Large Hadron Collider, use magnetic fields to steer charged particles around a ring. Consider a proton ring with 36 identical bending magnets connected by straight segments. The protons move along a 1.0-m-long circular arc as they pass through each magnet. What magnetic field strength is needed in each magnet to steer protons around the ring with a speed of 2.5×10^7 m/s? Assume that the field is uniform inside the magnet, zero outside.

67. || A particle of charge q and mass m moves in the uniform fields $\vec{E} = E_0 \hat{k}$ and $\vec{B} = B_0 \hat{k}$. At $t = 0$, the particle has velocity $\vec{v}_0 = v_0 \hat{i}$. What is the particle's speed at a later time t?

68. || A Hall-effect probe to measure magnetic field strengths needs to be calibrated in a known magnetic field. Although it is not easy to do, magnetic fields can be precisely measured by measuring the cyclotron frequency of protons. A testing laboratory adjusts a magnetic field until the proton's cyclotron frequency is 10.0 MHz. At this field strength, the Hall voltage on the probe is 0.543 mV when the current through the probe is 0.150 mA. Later, when an unknown magnetic field is measured, the Hall voltage at the same current is 1.735 mV. What is the strength of this magnetic field?

69. ||| It is shown in more advanced courses that charged particles in circular orbits radiate electromagnetic waves, called *cyclotron radiation.* As a result, a particle undergoing cyclotron motion with speed v is actually losing kinetic energy at the rate

$$\frac{dK}{dt} = -\left(\frac{\mu_0 q^4}{6\pi c m^2}\right)B^2 v^2$$

How long does it take (a) an electron and (b) a proton to radiate away half its energy while spiraling in a 2.0 T magnetic field?

70. ||| A proton in a cyclotron gains $\Delta K = 2e\Delta V$ of kinetic energy per revolution, where ΔV is the potential between the dees. Although the energy gain comes in small pulses, the proton makes so many revolutions that it is reasonable to model the energy as increasing at the constant rate $P = dK/dt = \Delta K/T$, where T is the period of the cyclotron motion. This is *power* input because it is a rate of increase of energy.
   a. Find an expression for $r(t)$, the radius of a proton's orbit in a cyclotron, in terms of m, e, B, P, and t. Assume that $r = 0$ at $t = 0$.
   Hint: Start by finding an expression for the proton's kinetic energy in terms of r.
   b. A relatively small cyclotron is 2.0 m in diameter, uses a 0.55 T magnetic field, and has a 400 V potential difference between the dees. What is the power input to a proton, in W?
   c. How long does it take a proton to spiral from the center out to the edge?

71. || The 10-turn loop of wire shown in **FIGURE P29.71** lies in a horizontal plane, parallel to a uniform horizontal magnetic field, and carries a 2.0 A current. The loop is free to rotate about a nonmagnetic axle through the center. A 50 g mass hangs from one edge of the loop. What magnetic field strength will prevent the loop from rotating about the axle?

FIGURE P29.71   **FIGURE P29.72**

72. || The two springs in **FIGURE P29.72** each have a spring constant of 10 N/m. They are compressed by 1.0 cm when a current passes through the wire. How big is the current?

73. || **FIGURE P29.73** is an edge view of a 2.0 kg square loop, 2.5 m on each side, with its lower edge resting on a frictionless, horizontal surface. A 25 A current is flowing around the loop in the direction shown. What is the strength of a uniform, horizontal magnetic field for which the loop is in static equilibrium at the angle shown?

FIGURE P29.73

74. || Magnetic fields are sometimes measured by balancing magnetic forces against known mechanical forces. Your task is to measure the strength of a horizontal magnetic field using a 12-cm-long rigid metal rod that hangs from two nonmagnetic springs, one at each end, with spring constants 1.3 N/m. You first position the rod to be level and perpendicular to the field, whose direction you determined with a compass. You then connect the ends of the rod to wires that run parallel to the field and thus experience no forces. Finally, you measure the downward deflection of the rod, stretching the springs, as you pass current through it. Your data are as follows:

Current (A)	Deflection (mm)
1.0	4
2.0	9
3.0	12
4.0	15
5.0	21

Use an appropriate graph of the data to determine the magnetic field strength.

75. || A conducting bar of length l and mass m rests at the left end of the two frictionless rails of length d in FIGURE P29.75. A uniform magnetic field of strength B points upward.
   a. In which direction, into or out of the page, will a current through the conducting bar cause the bar to experience a force to the right?
   b. Find an expression for the bar's speed as it leaves the rails at the right end.

FIGURE P29.75   FIGURE P29.76

76. || a. In FIGURE P29.76, a long, straight, current-carrying wire of linear mass density μ is suspended by threads. A magnetic field perpendicular to the wire exerts a horizontal force that deflects the wire to an equilibrium angle θ. Find an expression for the strength and direction of the magnetic field $\vec{B}$.
   b. What $\vec{B}$ deflects a 55 g/m wire to a 12° angle when the current is 10 A?

77. ||| A wire along the x-axis carries current I in the negative x-
CALC direction through the magnetic field

$$\vec{B} = \begin{cases} B_0 \dfrac{x}{l}\hat{k} & 0 \le x \le l \\ 0 & \text{elsewhere} \end{cases}$$

   a. Draw a graph of B versus x over the interval $-\frac{3}{2}l < x < \frac{3}{2}l$.
   b. Find an expression for the net force $\vec{F}_{net}$ on the wire.
   c. Find an expression for the net torque on the wire about the point $x = 0$.

78. || A *nonuniform* magnetic field exerts a net force on a current loop of radius R. FIGURE P29.78 shows a magnetic field that is diverging from the end of a bar magnet. The magnetic field at the position of the current loop makes an angle θ with respect to the vertical.
   a. Find an expression for the net magnetic force on the current.
   b. Calculate the force if $R = 2.0$ cm, $I = 0.50$ A, $B = 200$ mT, and $\theta = 20°$.

FIGURE P29.78

Challenge Problems

79. ||| You have a 1.0-m-long copper wire. You want to make an N-turn current loop that generates a 1.0 mT magnetic field at the center when the current is 1.0 A. You must use the entire wire. What will be the diameter of your coil?

80. ||| a. Derive an expression for the magnetic field strength at
CALC   distance d from the center of a straight wire of finite length l that carries current I.
   b. Determine the field strength at the center of a current-carrying *square* loop having sides of length $2R$.
   c. Compare your answer to part b to the field at the center of a *circular* loop of diameter $2R$. Do so by computing the ratio B_{square}/B_{circle}.

81. ||| A flat, circular disk of radius R is uniformly charged with
CALC total charge Q. The disk spins at angular velocity ω about an axis through its center. What is the magnetic field strength at the center of the disk?

82. ||| A long, straight conducting wire of radius R has a nonuniform
CALC current density $J = J_0 r/R$, where J_0 is a constant. The wire carries total current I.
   a. Find an expression for J_0 in terms of I and R.
   b. Find an expression for the magnetic field strength inside the wire at radius r.
   c. At the boundary, $r = R$, does your solution match the known field outside a long, straight current-carrying wire?

83. ||| An infinitely wide flat sheet of charge flows out of the page in FIGURE CP29.83. The current per unit width along the sheet (amps per meter) is given by the linear current density J_s.
   a. What is the *shape* of the magnetic field? To answer this question, you may find it helpful to approximate the current sheet as many parallel, closely spaced current-carrying wires. Give your answer as a picture showing magnetic field vectors.
   b. Find the magnetic field strength at distance d above or below the current sheet.

FIGURE CP29.83   Linear current density J_s

30 Electromagnetic Induction

Electromagnetic induction is the physics that underlies many modern technologies, from the generation of electricity to data storage.

IN THIS CHAPTER, you will learn what electromagnetic induction is and how it is used.

What is an induced current?

A magnetic field can create a current in a loop of wire, *but only if the amount of field through the loop is changing.*

- This is called an induced current.
- The process is called **electromagnetic induction**.

« LOOKING BACK Chapter 29 Magnetic fields

What is magnetic flux?

A key idea will be the amount of magnetic field passing through a loop or coil. This is called magnetic flux. Magnetic flux depends on the strength of the magnetic field, the area of the loop, and the angle between them.

« LOOKING BACK Section 24.3 Electric flux

What is Lenz's law?

Lenz's law says that a current is induced in a closed loop if and only if the magnetic flux through the loop is changing. Simply having a flux does nothing; the flux has to change. You'll learn how to use Lenz's law to determine the direction of an induced current around a loop.

What is Faraday's law?

Faraday's law is the most important law connecting electric and magnetic fields, laying the groundwork for electromagnetic waves. Just as a battery has an emf that drives current, a loop of wire has an induced emf determined by the rate of change of magnetic flux through the loop.

« LOOKING BACK Section 26.4 Sources of potential

What is an induced field?

At its most fundamental level, Faraday's law tells us that a changing magnetic field creates an induced electric field. This is an entirely new way to create an electric field, independent of charges. It is the induced electric field that drives the induced current around a conducting loop.

How is electromagnetic induction used?

Electromagnetic induction is one of the most important applications of electricity and magnetism. Generators use electromagnetic induction to turn the mechanical energy of a spinning turbine into electric energy. Inductors are important circuit elements that rely on electromagnetic induction. All forms of telecommunication are based on electromagnetic induction. And, not least, electromagnetic induction is the basis for light and other electromagnetic waves.

30.1 Induced Currents

Oersted's 1820 discovery that a current creates a magnetic field generated enormous excitement. One question scientists hoped to answer was whether the converse of Oersted's discovery was true: that is, can a magnet be used to create a current?

The breakthrough came in 1831 when the American science teacher Joseph Henry and the English scientist Michael Faraday each discovered the process we now call *electromagnetic induction*. Faraday—whom you met in Chapter 22 as the inventor of the concept of a *field*—was the first to publish his findings, so today we study Faraday's law rather than Henry's law.

Faraday's 1831 discovery, like Oersted's, was a happy combination of an unplanned event and a mind that was ready to recognize its significance. Faraday was experimenting with two coils of wire wrapped around an iron ring, as shown in **FIGURE 30.1**. He had hoped that the magnetic field generated in the coil on the left would induce a magnetic field in the iron, and that the magnetic field in the iron might then somehow create a current in the circuit on the right.

Like all his previous attempts, this technique failed to generate a current. But Faraday happened to notice that the needle of the current meter jumped ever so slightly at the instant he closed the switch in the circuit on the left. After the switch was closed, the needle immediately returned to zero. The needle again jumped when he later opened the switch, but this time in the opposite direction. Faraday recognized that the motion of the needle indicated a current in the circuit on the right, but a momentary current only during the brief interval when the current on the left was starting or stopping.

Faraday's observations, coupled with his mental picture of field lines, led him to suggest that a current is generated only if the magnetic field through the coil is *changing*. This explains why all the previous attempts to generate a current with static magnetic fields had been unsuccessful. Faraday set out to test this hypothesis.

FIGURE 30.1 Faraday's discovery.

Closing the switch in the left circuit . . . . . . causes a momentary current in the right circuit.

No current while the switch stays closed

Opening the switch in the left circuit . . . . . . causes a momentary current in the opposite direction.

Faraday investigates electromagnetic induction

Faraday placed one coil directly above the other, without the iron ring. There was no current in the lower circuit while the switch was in the closed position, but a momentary current appeared whenever the switch was opened or closed.

He pushed a bar magnet into a coil of wire. This action caused a momentary deflection of the current-meter needle, although *holding* the magnet inside the coil had no effect. A quick withdrawal of the magnet deflected the needle in the other direction.

Must the magnet move? Faraday created a momentary current by rapidly pulling a coil of wire out of a magnetic field. Pushing the coil *into* the magnet caused the needle to deflect in the opposite direction.

Open or close switch.

Opening or closing the switch creates a momentary current.

Push or pull magnet.

Pushing the magnet into the coil or pulling it out creates a momentary current.

Push or pull coil.

Pushing the coil into the magnet or pulling it out creates a momentary current.

Faraday found that there is a current in a coil of wire if and only if the magnetic field passing through the coil is *changing*. This is an informal statement of what we'll soon call *Faraday's law*. The current in a circuit due to a changing magnetic field is called an **induced current**. An induced current is not caused by a battery; it is a completely new way to generate a current.

30.2 Motional emf

We'll start our investigation of electromagnetic induction by looking at situations in which the magnetic field is fixed while the circuit moves or changes. Consider a conductor of length l that moves with velocity $\vec{v}$ through a perpendicular uniform magnetic field $\vec{B}$, as shown in FIGURE 30.2. The charge carriers inside the wire—assumed to be positive—also move with velocity $\vec{v}$, so they each experience a magnetic force $\vec{F}_B = q\vec{v} \times \vec{B}$ of strength $F_B = qvB$. This force causes the charge carriers to move, separating the positive and negative charges. The separated charges then create an electric field inside the conductor.

FIGURE 30.2 The magnetic force on the charge carriers in a moving conductor creates an electric field inside the conductor.

Charge carriers in the conductor experience a force of magnitude $F_B = qvB$. Positive charges are free to move and drift upward.

The resulting charge separation creates an electric field in the conductor. $\vec{E}$ increases as more charge flows.

The charge flow continues until the electric and magnetic forces balance. For a positive charge carrier, the upward magnetic force $\vec{F}_B$ is equal to the downward electric force $\vec{F}_E$.

The charge carriers continue to separate until the electric force $F_E = qE$ exactly balances the magnetic force $F_B = qvB$, creating an equilibrium situation. This balance happens when the electric field strength is

$$E = vB \tag{30.1}$$

In other words, **the magnetic force on the charge carriers in a moving conductor creates an electric field $E = vB$ inside the conductor.**

The electric field, in turn, creates an electric potential difference between the two ends of the moving conductor. FIGURE 30.3a defines a coordinate system in which $\vec{E} = -vB\hat{\jmath}$. Using the connection between the electric field and the electric potential,

$$\Delta V = V_{\text{top}} - V_{\text{bottom}} = -\int_0^l E_y \, dy = -\int_0^l (-vB) \, dy = vlB \tag{30.2}$$

Thus **the motion of the wire through a magnetic field *induces* a potential difference vlB between the ends of the conductor.** The potential difference depends on the strength of the magnetic field and on the wire's speed through the field.

There's an important analogy between this potential difference and the potential difference of a battery. FIGURE 30.3b reminds you that a battery uses a nonelectric force—the charge escalator—to separate positive and negative charges. The emf $\mathcal{E}$ of the battery was defined as the work performed per charge (W/q) to separate the charges. An isolated battery, with no current, has a potential difference $\Delta V_{\text{bat}} = \mathcal{E}$. We could refer to a battery, where the charges are separated by chemical reactions, as a source of *chemical emf*.

The moving conductor develops a potential difference because of the work done by magnetic forces to separate the charges. You can think of the moving conductor as a "battery" that stays charged only as long as it keeps moving but "runs down" if it stops. The emf of the conductor is due to its motion, rather than to chemical reactions inside, so we can define the **motional emf** of a conductor moving with velocity $\vec{v}$ perpendicular to a magnetic field $\vec{B}$ to be

$$\mathcal{E} = vlB \tag{30.3}$$

FIGURE 30.3 Generating an emf.

(a) Magnetic forces separate the charges and cause a potential difference between the ends. This is a motional emf.

Electric field inside the moving conductor

$\Delta V = vlB$

(b) Chemical reactions separate the charges and cause a potential difference between the ends. This is a chemical emf.

ΔV_{bat}

Electric field inside the battery

STOP TO THINK 30.1   A square conductor moves through a uniform magnetic field. Which of the figures shows the correct charge distribution on the conductor?

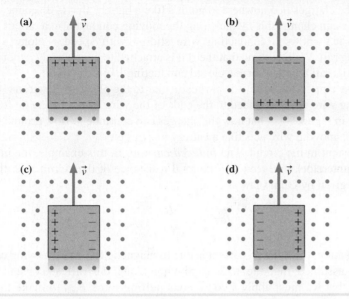

EXAMPLE 30.1 | Measuring the earth's magnetic field

It is known that the earth's magnetic field over northern Canada points straight down. The crew of a Boeing 747 aircraft flying at 260 m/s over northern Canada finds a 0.95 V potential difference between the wing tips. The wing span of a Boeing 747 is 65 m. What is the magnetic field strength there?

MODEL   The wing is a conductor moving through a magnetic field, so there is a motional emf.

SOLVE   The magnetic field is perpendicular to the velocity, so we can use Equation 30.3 to find

$$B = \frac{\mathcal{E}}{vL} = \frac{0.95 \text{ V}}{(260 \text{ m/s})(65 \text{ m})} = 5.6 \times 10^{-5} \text{ T}$$

ASSESS   Chapter 29 noted that the earth's magnetic field is roughly 5×10^{-5} T. The field is somewhat stronger than this near the magnetic poles, somewhat weaker near the equator.

EXAMPLE 30.2 | Potential difference along a rotating bar

A metal bar of length l rotates with angular velocity ω about a pivot at one end of the bar. A uniform magnetic field $\vec{B}$ is perpendicular to the plane of rotation. What is the potential difference between the ends of the bar?

VISUALIZE   FIGURE 30.4 is a pictorial representation of the bar. The magnetic forces on the charge carriers will cause the outer end to be positive with respect to the pivot.

FIGURE 30.4 Pictorial representation of a metal bar rotating in a magnetic field.

The electric field strength increases with r.

$\vec{B}$

Pivot

Angular velocity ω

The speed at distance r is $v = \omega r$.

SOLVE   Even though the bar is rotating, rather than moving in a straight line, the velocity of each charge carrier is perpendicular to $\vec{B}$. Consequently, the electric field created inside the bar is exactly that given in Equation 30.1, $E = vB$. But v, the speed of the charge carrier, now depends on its distance from the pivot. Recall that in rotational motion the tangential speed at radius r from the center of rotation is $v = \omega r$. Thus the electric field at distance r from the pivot is $E = \omega r B$. The electric field increases in strength as you move outward along the bar.

The electric field $\vec{E}$ points toward the pivot, so its radial component is $E_r = -\omega r B$. If we integrate outward from the center, the potential difference between the ends of the bar is

$$\Delta V = V_{\text{tip}} - V_{\text{pivot}} = -\int_0^l E_r \, dr$$

$$= -\int_0^l (-\omega r B) \, dr = \omega B \int_0^l r \, dr = \tfrac{1}{2}\omega l^2 B$$

ASSESS $\tfrac{1}{2}\omega l$ is the speed at the midpoint of the bar. Thus ΔV is $v_{\text{mid}} l B$, which seems reasonable.

FIGURE 30.5 A current is induced in the circuit as the wire moves through a magnetic field.

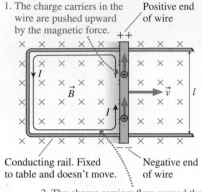

1. The charge carriers in the wire are pushed upward by the magnetic force.

Positive end of wire

Conducting rail. Fixed to table and doesn't move.

Negative end of wire

2. The charge carriers flow around the conducting loop as an induced current.

FIGURE 30.6 A pulling force is needed to move the wire to the right.

The induced current flows through the moving wire.

The magnetic force on the current-carrying wire is opposite the motion.

A pulling force to the right must balance the magnetic force to keep the wire moving at constant speed.

Induced Current in a Circuit

The moving conductor of Figure 30.2 had an emf, but it couldn't sustain a current because the charges had nowhere to go. It's like a battery that is disconnected from a circuit. We can change this by including the moving conductor in a circuit.

FIGURE 30.5 shows a conducting wire sliding with speed v along a U-shaped conducting rail. We'll assume that the rail is attached to a table and cannot move. The wire and the rail together form a closed conducting loop—a circuit.

Suppose a magnetic field $\vec{B}$ is perpendicular to the plane of the circuit. Charges in the moving wire will be pushed to the ends of the wire by the magnetic force, just as they were in Figure 30.2, but now the charges can continue to flow around the circuit. That is, the moving wire acts like a battery in a circuit.

The current in the circuit is an *induced current*. In this example, the induced current is counterclockwise (ccw). If the total resistance of the circuit is R, the induced current is given by Ohm's law as

$$I = \frac{\mathcal{E}}{R} = \frac{vlB}{R} \tag{30.4}$$

In this situation, the induced current is due to magnetic forces on moving charges.

We've assumed that the wire is moving along the rail at constant speed. It turns out that we must apply a continuous pulling force $\vec{F}_{\text{pull}}$ to make this happen. **FIGURE 30.6** shows why. The moving wire, which now carries induced current I, is in a magnetic field. You learned in Chapter 29 that a magnetic field exerts a force on a current-carrying wire. According to the right-hand rule, the magnetic force $\vec{F}_{\text{mag}}$ on the moving wire points to the left. This "magnetic drag" will cause the wire to slow down and stop *unless* we exert an equal but opposite pulling force $\vec{F}_{\text{pull}}$ to keep the wire moving.

The magnitude of the magnetic force on a current-carrying wire was found in Chapter 29 to be $F_{\text{mag}} = IlB$. Using that result, along with Equation 30.4 for the induced current, we find that the force required to pull the wire with a constant speed v is

$$F_{\text{pull}} = F_{\text{mag}} = IlB = \left(\frac{vlB}{R}\right)lB = \frac{vl^2B^2}{R} \tag{30.5}$$

STOP TO THINK 30.2 Is there an induced current in this circuit? If so, what is its direction?

Energy Considerations

The environment must do work on the wire to pull it. What happens to the energy transferred to the wire by this work? Is energy conserved as the wire moves along the rail? It will be easier to answer this question if we think about power rather than work. Power is the *rate* at which work is done on the wire. You learned in Chapter 9 that the power exerted by a force pushing or pulling an object with velocity v is $P = Fv$. The power provided to the circuit by pulling on the wire is

$$P_{\text{input}} = F_{\text{pull}}v = \frac{v^2l^2B^2}{R} \tag{30.6}$$

This is the rate at which energy is added to the circuit by the pulling force.

But the circuit also dissipates energy by transforming electric energy into the thermal energy of the wires and components, heating them up. The power dissipated by current I as it passes through resistance R is $P = I^2R$. Equation 30.4 for the induced current I gives us the power dissipated by the circuit of Figure 30.5:

$$P_{\text{dissipated}} = I^2R = \frac{v^2l^2B^2}{R} \qquad (30.7)$$

You can see that Equations 30.6 and 30.7 are identical. **The rate at which work is done on the circuit exactly balances the rate at which energy is dissipated.** Thus *energy is conserved.*

If you have to *pull* on the wire to get it to move to the right, you might think that it would spring back to the left on its own. FIGURE 30.7 shows the same circuit with the wire moving to the left. In this case, you must *push* the wire to the left to keep it moving. The magnetic force is always opposite to the wire's direction of motion.

In both Figure 30.6, where the wire is pulled, and Figure 30.7, where it is pushed, a mechanical force is used to create a current. In other words, we have a conversion of *mechanical* energy to *electric* energy. A device that converts mechanical energy to electric energy is called a **generator.** The slide-wire circuits of Figures 30.6 and 30.7 are simple examples of a generator. We will look at more practical examples of generators later in the chapter.

We can summarize our analysis as follows:

1. Pulling or pushing the wire through the magnetic field at speed v creates a motional emf $\mathcal{E}$ in the wire and induces a current $I = \mathcal{E}/R$ in the circuit.
2. To keep the wire moving at constant speed, a pulling or pushing force must balance the magnetic force on the wire. This force does work on the circuit.
3. The work done by the pulling or pushing force exactly balances the energy dissipated by the current as it passes through the resistance of the circuit.

FIGURE 30.7 A pushing force is needed to move the wire to the left.

1. The magnetic force on the charge carriers is down, so the induced current flows clockwise.

2. The magnetic force on the current-carrying wire is to the right.

EXAMPLE 30.3 | Lighting a bulb

FIGURE 30.8 shows a circuit consisting of a flashlight bulb, rated 3.0 V/1.5 W, and ideal wires with no resistance. The right wire of the circuit, which is 10 cm long, is pulled at constant speed v through a perpendicular magnetic field of strength 0.10 T.

a. What speed must the wire have to light the bulb to full brightness?

b. What force is needed to keep the wire moving?

FIGURE 30.8 Circuit of Example 30.3.

MODEL Treat the moving wire as a source of motional emf.

VISUALIZE The magnetic force on the charge carriers, $\vec{F}_B = q\vec{v} \times \vec{B}$, causes a counterclockwise (ccw) induced current.

SOLVE a. The bulb's rating of 3.0 V/1.5 W means that at full brightness it will dissipate 1.5 W at a potential difference of 3.0 V. Because the power is related to the voltage and current by $P = I\Delta V$,

the current causing full brightness is

$$I = \frac{P}{\Delta V} = \frac{1.5 \text{ W}}{3.0 \text{ V}} = 0.50 \text{ A}$$

The bulb's resistance—the total resistance of the circuit—is

$$R = \frac{\Delta V}{I} = \frac{3.0 \text{ V}}{0.50 \text{ A}} = 6.0 \text{ } \Omega$$

Equation 30.4 gives the speed needed to induce this current:

$$v = \frac{IR}{lB} = \frac{(0.50 \text{ A})(6.0 \text{ } \Omega)}{(0.10 \text{ m})(0.10 \text{ T})} = 300 \text{ m/s}$$

You can confirm from Equation 30.6 that the input power at this speed is 1.5 W.

b. From Equation 30.5, the pulling force must be

$$F_{\text{pull}} = \frac{vl^2B^2}{R} = 5.0 \times 10^{-3} \text{ N}$$

You can also obtain this result from $F_{\text{pull}} = P/v$.

ASSESS Example 30.1 showed that high speeds are needed to produce significant potential difference. Thus 300 m/s is not surprising. The pulling force is not very large, but even a small force can deliver large amounts of power $P = Fv$ when v is large.

FIGURE 30.9 Eddy currents.

Eddy currents are induced when a metal sheet is pulled through a magnetic field.

Metal sheet

The magnetic force on the eddy currents is opposite in direction to $\vec{v}$.

FIGURE 30.10 Magnetic braking system.

The electromagnets are part of the moving train car.

Rail

Eddy currents are induced in the rail. Magnetic forces between the eddy currents and the electromagnets slow the train.

Eddy Currents

These ideas have interesting implications. Consider pulling a *sheet* of metal through a magnetic field, as shown in FIGURE 30.9a. The metal, we will assume, is not a magnetic material, so it experiences no magnetic force if it is at rest. The charge carriers in the metal experience a magnetic force as the sheet is dragged between the pole tips of the magnet. A current is induced, just as in the loop of wire, but here the currents do not have wires to define their path. As a consequence, two "whirlpools" of current begin to circulate in the metal. These spread-out current whirlpools in a solid metal are called **eddy currents.**

As the eddy current passes between the pole tips, it experiences a magnetic force to the left—a retarding force. Thus **an external force is required to pull a metal through a magnetic field.** If the pulling force ceases, the retarding magnetic force quickly causes the metal to decelerate until it stops. Similarly, a force is required to push a sheet of metal *into* a magnetic field.

Eddy currents are often undesirable. The power dissipation of eddy currents can cause unwanted heating, and the magnetic forces on eddy currents mean that extra energy must be expended to move metals in magnetic fields. But eddy currents also have important useful applications. A good example is magnetic braking.

The moving train car has an electromagnet that straddles the rail, as shown in FIGURE 30.10. During normal travel, there is no current through the electromagnet and no magnetic field. To stop the car, a current is switched into the electromagnet. The current creates a strong magnetic field that passes *through* the rail, and the motion of the rail relative to the magnet induces eddy currents in the rail. The magnetic force between the electromagnet and the eddy currents acts as a braking force on the magnet and, thus, on the car. Magnetic braking systems are very efficient, and they have the added advantage that they heat the rail rather than the brakes.

STOP TO THINK 30.3 A square loop of copper wire is pulled through a region of magnetic field. Rank in order, from strongest to weakest, the pulling forces $\vec{F}_a$, $\vec{F}_b$, $\vec{F}_c$, and $\vec{F}_d$ that must be applied to keep the loop moving at constant speed.

30.3 Magnetic Flux

Faraday found that a current is induced when the amount of magnetic field passing through a coil or a loop of wire changes. And that's exactly what happens as the slide wire moves down the rail in Figure 30.5! As the circuit expands, more magnetic field passes through. It's time to define more clearly what we mean by "the amount of field passing through a loop."

Imagine holding a rectangular loop in front of the fan shown in FIGURE 30.11. The amount of air flowing *through* the loop—the *flux*—depends on the angle of the loop. The flow is maximum if the loop is perpendicular to the flow, zero if the loop is rotated to be parallel to the flow. In general, the amount of air flowing through is proportional to the *effective* area of the loop (i.e., the area facing the fan):

$$A_{\text{eff}} = ab\cos\theta = A\cos\theta \qquad (30.8)$$

where $A = ab$ is the area of the loop and θ is the tilt angle of the loop. A loop perpendicular to the flow, with $\theta = 0°$, has $A_{\text{eff}} = A$, the full area of the loop.

FIGURE 30.11 The amount of air flowing through a loop depends on the effective area of the loop.

The flux is the amount of air passing through the loop per second.

Fan

$A = ab$

$A_{\text{eff}} = ab\cos\theta$

We can apply this idea to a magnetic field passing through a loop. FIGURE 30.12 shows a loop of area $A = ab$ in a uniform magnetic field. Think of the field vectors, seen here from behind, as if they were arrows shot into the page. The density of arrows (arrows per m²) is proportional to the strength B of the magnetic field; a stronger field would be represented by arrows packed closer together. The number of arrows passing through a loop of wire depends on two factors:

1. The density of arrows, which is proportional to B, and
2. The effective area $A_{\text{eff}} = A \cos\theta$ of the loop.

The angle θ is the angle between the magnetic field and the axis of the loop. The maximum number of arrows passes through the loop when it is perpendicular to the magnetic field ($\theta = 0°$). No arrows pass through the loop if it is tilted 90°.

FIGURE 30.12 Magnetic field through a loop that is tilted at various angles.

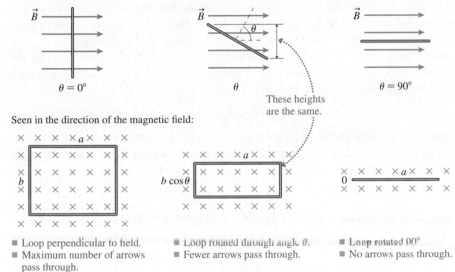

Loop seen from the side:

$\theta = 0°$ θ $\theta = 90°$

These heights are the same.

Seen in the direction of the magnetic field:

- Loop perpendicular to field.
- Maximum number of arrows pass through.

- Loop rotated through angle θ.
- Fewer arrows pass through.

- Loop rotated 90°
- No arrows pass through.

With this in mind, let's define the **magnetic flux** Φ_m as

$$\Phi_m = A_{\text{eff}}B = AB\cos\theta \qquad (30.9)$$

The magnetic flux measures the amount of magnetic field passing through a loop of area A if the loop is tilted at angle θ from the field. The SI unit of magnetic flux is the **weber.** From Equation 30.9 you can see that

$$1 \text{ weber} = 1 \text{ Wb} = 1 \text{ T m}^2$$

Equation 30.9 is reminiscent of the vector dot product: $\vec{A} \cdot \vec{B} = AB\cos\theta$. With that in mind, let's define an **area vector** $\vec{A}$ to be a vector *perpendicular* to the loop, with magnitude equal to the area A of the loop. Vector $\vec{A}$ has units of m². FIGURE 30.13a shows the area vector $\vec{A}$ for a circular loop of area A.

FIGURE 30.13b shows a magnetic field passing through a loop. The angle between vectors $\vec{A}$ and $\vec{B}$ is the same angle used in Equations 30.8 and 30.9 to define the effective area and the magnetic flux. So Equation 30.9 really is a dot product, and we can define the magnetic flux more concisely as

$$\Phi_m = \vec{A} \cdot \vec{B} \qquad (30.10)$$

Writing the flux as a dot product helps make clear how angle θ is defined: θ is the angle between the magnetic field and the axis of the loop.

FIGURE 30.13 Magnetic flux can be defined in terms of an area vector $\vec{A}$.

(a)

Loop of area A

The area vector is perpendicular to the loop. Its magnitude is the area of the loop.

(b)

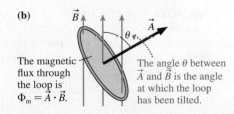

The magnetic flux through the loop is $\Phi_m = \vec{A} \cdot \vec{B}$.

The angle θ between $\vec{A}$ and $\vec{B}$ is the angle at which the loop has been tilted.

EXAMPLE 30.4 | **A circular loop in a magnetic field**

FIGURE 30.14 is an edge view of a 10-cm-diameter circular loop in a uniform 0.050 T magnetic field. What is the magnetic flux through the loop?

SOLVE Angle θ is the angle between the loop's area vector $\vec{A}$, which is perpendicular to the plane of the loop, and the magnetic field $\vec{B}$. In this case, $\theta = 60°$, not the 30° angle shown in the figure. Vector $\vec{A}$ has magnitude $A = \pi r^2 = 7.85 \times 10^{-3}$ m². Thus the magnetic flux is

$$\Phi_m = \vec{A} \cdot \vec{B} = AB\cos\theta = 2.0 \times 10^{-4} \text{ Wb}$$

FIGURE 30.14 A circular loop in a magnetic field.

Circular loop

Magnetic Flux in a Nonuniform Field

Equation 30.10 for the magnetic flux assumes that the field is uniform over the area of the loop. We can calculate the flux in a nonuniform field, one where the field strength changes from one edge of the loop to the other, but we'll need to use calculus.

FIGURE 30.15 A loop in a nonuniform magnetic field.

Loop

Small area dA. The flux through this little area is $d\Phi_m = \vec{B} \cdot d\vec{A}$.

Increasing field strength

FIGURE 30.15 shows a loop in a nonuniform magnetic field. Imagine dividing the loop into many small pieces of area dA. The infinitesimal flux $d\Phi_m$ through one such area, where the magnetic field is $\vec{B}$, is

$$d\Phi_m = \vec{B} \cdot d\vec{A} \tag{30.11}$$

The total magnetic flux through the loop is the sum of the fluxes through each of the small areas. We find that sum by integrating. Thus the total magnetic flux through the loop is

$$\Phi_m = \int_{\text{area of loop}} \vec{B} \cdot d\vec{A} \tag{30.12}$$

Equation 30.12 is a more general definition of magnetic flux. It may look rather formidable, so we'll illustrate its use with an example.

EXAMPLE 30.5 | **Magnetic flux from the current in a long straight wire**

The 1.0 cm × 4.0 cm rectangular loop of **FIGURE 30.16** is 1.0 cm away from a long, straight wire. The wire carries a current of 1.0 A. What is the magnetic flux through the loop?

FIGURE 30.16 A loop next to a current-carrying wire.

Loop

Long, straight wire

4.0 cm

I

1.0 cm   1.0 cm

MODEL Model the wire as if it were infinitely long. The magnetic field strength of a wire decreases with distance from the wire, so the field is *not* uniform over the area of the loop.

VISUALIZE Using the right-hand rule, we see that the field, as it circles the wire, is perpendicular to the plane of the loop. **FIGURE 30.17** redraws the loop with the field coming out of the page and establishes a coordinate system.

SOLVE Let the loop have dimensions a and b, as shown, with the near edge at distance c from the wire. The magnetic field varies

FIGURE 30.17 Calculating the magnetic flux through the loop.

Decreasing B

Strip of area $dA = b\,dx$ at position x. Magnetic flux through this strip is $d\Phi_m = B\,dA$.

Vector $d\vec{A}$ is coming out of the page.

with distance x from the wire, but the field is constant along a line parallel to the wire. This suggests dividing the loop into many narrow rectangular strips of length b and width dx, each forming a small area $dA = b\,dx$. The magnetic field has the same strength at all points within this small area. One such strip is shown in the figure at position x.

The area vector $d\vec{A}$ is perpendicular to the strip (coming out of the page), which makes it parallel to $\vec{B}$ ($\theta = 0°$). Thus the infinitesimal flux through this little area is

$$d\Phi_m = \vec{B} \cdot d\vec{A} = B\,dA = B(b\,dx) = \frac{\mu_0 Ib}{2\pi x}dx$$

where, from Chapter 29, we've used $B = \mu_0 I/2\pi x$ as the magnetic field at distance x from a long, straight wire. Integrating "over the area of the loop" means to integrate from the near edge of the loop at $x = c$ to the far edge at $x = c + a$. Thus

$$\Phi_m = \frac{\mu_0 Ib}{2\pi}\int_c^{c+a}\frac{dx}{x} = \frac{\mu_0 Ib}{2\pi}\ln x\Big|_c^{c+a} = \frac{\mu_0 Ib}{2\pi}\ln\left(\frac{c+a}{c}\right)$$

Evaluating for $a = c = 0.010$ m, $b = 0.040$ m, and $I = 1.0$ A gives

$$\Phi_m = 5.5 \times 10^{-9}\text{ Wb}$$

ASSESS The flux measures how much of the wire's magnetic field passes through the loop, but we had to integrate, rather than simply using Equation 30.10, because the field is stronger at the near edge of the loop than at the far edge.

30.4 Lenz's Law

We started out by looking at a situation in which a moving wire caused a loop to expand in a magnetic field. This is one way to change the magnetic flux through the loop. But Faraday found that a current can be induced by any change in the magnetic flux, no matter how it's accomplished.

For example, a momentary current is induced in the loop of **FIGURE 30.18** as the bar magnet is pushed toward the loop, increasing the flux through the loop. Pulling the magnet back out of the loop causes the current meter to deflect in the opposite direction. The conducting wires aren't moving, so this is not a motional emf. Nonetheless, the induced current is very real.

The German physicist Heinrich Lenz began to study electromagnetic induction after learning of Faraday's discovery. Three years later, in 1834, Lenz announced a rule for determining the direction of the induced current. We now call his rule **Lenz's law,** and it can be stated as follows:

Lenz's law There is an induced current in a closed, conducting loop if and only if the magnetic flux through the loop is changing. The direction of the induced current is such that the induced magnetic field opposes the *change* in the flux.

Lenz's law is rather subtle, and it takes some practice to see how to apply it.

NOTE One difficulty with Lenz's law is the term *flux*. In everyday language, the word *flux* already implies that something is changing. Think of the phrase, "The situation is in flux." Not so in physics, where *flux*, the root of the word *flow*, means "passes through." A steady magnetic field through a loop creates a steady, *un*changing magnetic flux.

Lenz's law tells us to look for situations where the flux is *changing*. This can happen in three ways.

1. The magnetic field through the loop changes (increases or decreases),
2. The loop changes in area or angle, or
3. The loop moves into or out of a magnetic field.

Lenz's law depends on the idea that an induced current generates its own magnetic field $\vec{B}_{induced}$. This is the *induced magnetic field* of Lenz's law. You learned in Chapter 29 how to use the right-hand rule to determine the direction of this induced magnetic field.

In Figure 30.18, pushing the bar magnet toward the loop causes the magnetic flux to *increase* in the downward direction. To oppose the *change* in flux, which is what Lenz's law requires, the loop itself needs to generate the *upward*-pointing magnetic field of **FIGURE 30.19**. The induced magnetic field at the center of the loop will point upward if the current is ccw. Thus pushing the north end of a bar magnet toward the loop induces a ccw current around the loop. The induced current ceases as soon as the magnet stops moving.

FIGURE 30.18 Pushing a bar magnet toward the loop induces a current.

A bar magnet pushed toward a loop increases the flux through the loop.

Which direction is the induced current?

S

In

Current meter

0

N

$\vec{B}_{magnet}$

FIGURE 30.19 The induced current is ccw.

2. The loop needs to generate an upward-pointing magnetic field to oppose the *change* in flux.

1. The flux through the loop increases downward as the magnet approaches.

S

N

$\vec{B}_{induced}$

0

Induced current

$\vec{B}_{magnet}$

3. By the right-hand rule, a ccw current is needed to induce an upward-pointing magnetic field.

Now suppose the bar magnet is pulled back away from the loop, as shown in FIGURE 30.20a. There is a downward magnetic flux through the loop, but the flux *decreases* as the magnet moves away. According to Lenz's law, the induced magnetic field of the loop *opposes this decrease*. To do so, the induced field needs to point in the *downward* direction, as shown in FIGURE 30.20b. Thus as the magnet is withdrawn, the induced current is clockwise (cw), opposite to the induced current of Figure 30.19.

FIGURE 30.20 Pulling the magnet away induces a cw current.

NOTE Notice that the magnetic field of the bar magnet is pointing downward in both Figures 30.19 and 30.20. It is not the *flux* due to the magnet that the induced current opposes, but the *change* in the flux. This is a subtle but critical distinction. If the induced current opposed the flux itself, the current in both Figures 30.19 and 30.20 would be ccw to generate an upward magnetic field. But that's not what happens. When the field of the magnet points down and is increasing, the induced current opposes the increase by generating an upward field. When the field of the magnet points down but is decreasing, the induced current opposes the decrease by generating a downward field.

Using Lenz's Law

FIGURE 30.21 shows six basic situations. The magnetic field can point either up or down through the loop. For each, the flux can either increase, hold steady, or decrease in strength. These observations form the basis for a set of rules about using Lenz's law.

FIGURE 30.21 The induced current for six different situations.

TACTICS BOX 30.1

Using Lenz's law

❶ **Determine the direction of the applied magnetic field.** The field must pass through the loop.

❷ **Determine how the flux is changing.** Is it increasing, decreasing, or staying the same?

❸ **Determine the direction of an induced magnetic field that will oppose the *change* in the flux.**

■ Increasing flux: the induced magnetic field points opposite the applied magnetic field.

■ Decreasing flux: the induced magnetic field points in the same direction as the applied magnetic field.

■ Steady flux: there is no induced magnetic field.

❹ **Determine the direction of the induced current.** Use the right-hand rule to determine the current direction in the loop that generates the induced magnetic field you found in step 3.

Exercises 10–14

Let's look at two examples.

EXAMPLE 30.6 | Lenz's law 1

FIGURE 30.22 shows two loops, one above the other. The upper loop has a battery and a switch that has been closed for a long time. How does the lower loop respond when the switch is opened in the upper loop?

MODEL We'll use the right-hand rule to find the magnetic fields of current loops.

SOLVE **FIGURE 30.23** shows the four steps of using Lenz's law. Opening the switch induces a ccw current in the lower loop. This is a momentary current, lasting only until the magnetic field of the upper loop drops to zero.

ASSESS The conclusion is consistent with Figure 30.21.

FIGURE 30.22 The two loops of Example 30.6.

FIGURE 30.23 Applying Lenz's law.

❶ By the right-hand rule, the magnetic field of the upper loop points up. It decreases rapidly after the switch is opened.

❷ The field due to the upper loop passes through the lower loop. It creates a flux through the lower loop that is up and decreasing.

❸ The induced field needs to point upward to oppose the *change* in flux.

Switch opens.

Current is dropping fast.

Induced current

❹ A ccw current induces an upward magnetic field.

$\vec{B}$

$\vec{B}_{induced}$

EXAMPLE 30.7 | Lenz's law 2

FIGURE 30.24 shows two coils wrapped side by side on a cylinder. When the switch for coil 1 is closed, does the induced current in coil 2 pass from right to left or from left to right through the current meter?

MODEL We'll use the right-hand rule to find the magnetic field of a coil.

VISUALIZE It is very important to look at the *direction* in which a coil is wound around the cylinder. Notice that the two coils in Figure 30.24 are wound in opposite directions.

FIGURE 30.24 The two solenoids of Example 30.7.

Coil 1   Coil 2

Continued

SOLVE FIGURE 30.25 shows the four steps of using Lenz's law. Closing the switch induces a current that passes from right to left through the current meter. The induced current is only momentary. It lasts only until the field from coil 1 reaches full strength and is no longer changing.

ASSESS The conclusion is consistent with Figure 30.21.

FIGURE 30.25 Applying Lenz's law.

STOP TO THINK 30.4 A current-carrying wire is pulled away from a conducting loop in the direction shown. As the wire is moving, is there a cw current around the loop, a ccw current, or no current?

30.5 Faraday's Law

Charges don't start moving spontaneously. A current requires an emf to provide the energy. We started our analysis of induced currents with circuits in which a *motional emf* can be understood in terms of magnetic forces on moving charges. But we've also seen that a current can be induced by changing the magnetic field through a stationary circuit, a circuit in which there is no motion. There *must* be an emf in this circuit, even though the mechanism for this emf is not yet clear.

The emf associated with a changing magnetic flux, regardless of what causes the change, is called an **induced emf** $\mathcal{E}$. Then, if there is a complete circuit having resistance R, a current

$$I_{\text{induced}} = \frac{\mathcal{E}}{R} \qquad (30.13)$$

is established in the wire as a *consequence* of the induced emf. The direction of the current is given by Lenz's law. The last piece of information we need is the size of the induced emf $\mathcal{E}$.

The research of Faraday and others eventually led to the discovery of the basic law of electromagnetic induction, which we now call **Faraday's law.** It states:

> **Faraday's law** An emf $\mathcal{E}$ is induced around a closed loop if the magnetic flux through the loop changes. The magnitude of the emf is
>
> $$\mathcal{E} = \left| \frac{d\Phi_{\text{m}}}{dt} \right| \qquad (30.14)$$
>
> and the direction of the emf is such as to drive an induced current in the direction given by Lenz's law.

In other words, the induced emf is the *rate of change* of the magnetic flux through the loop.

As a corollary to Faraday's law, an N-turn coil of wire in a changing magnetic field acts like N batteries in series. The induced emf of each of the coils adds, so the induced emf of the entire coil is

$$\mathcal{E}_{\text{coil}} = N \left| \frac{d\Phi_{\text{per coil}}}{dt} \right| \qquad \text{(Faraday's law for an } N\text{-turn coil)} \qquad (30.15)$$

As a first example of using Faraday's law, return to the situation of Figure 30.5, where a wire moves through a magnetic field by sliding on a U-shaped conducting rail. **FIGURE 30.26** shows the circuit again. The magnetic field $\vec{B}$ is perpendicular to the plane of the conducting loop, so $\theta = 0°$ and the magnetic flux is $\Phi = AB$, where A is the area of the loop. If the slide wire is distance x from the end, the area is $A = xl$ and the flux at that instant of time is

$$\Phi_m = AB = xlB \tag{30.16}$$

The flux through the loop increases as the wire moves. According to Faraday's law, the induced emf is

$$\mathcal{E} = \left| \frac{d\Phi_m}{dt} \right| = \frac{d}{dt}(xlB) = \frac{dx}{dt}lB = vlB \tag{30.17}$$

where the wire's velocity is $v = dx/dt$. We can now use Equation 30.13 to find that the induced current is

$$I = \frac{\mathcal{E}}{R} = \frac{vlB}{R} \tag{30.18}$$

The flux is increasing into the loop, so the induced magnetic field opposes this increase by pointing out of the loop. This requires a ccw induced current in the loop. Faraday's law leads us to the conclusion that the loop will have a ccw induced current $I = vlB/R$. This is exactly the conclusion we reached in Section 30.2, where we analyzed the situation from the perspective of magnetic forces on moving charge carriers. Faraday's law confirms what we already knew but, at least in this case, doesn't seem to offer anything new.

FIGURE 30.26 The magnetic flux through the loop increases as the slide wire moves.

Magnetic flux $\Phi_m = AB = xlB$

Induced current

Using Faraday's Law

Most electromagnetic induction problems can be solved with a four-step strategy.

PROBLEM-SOLVING STRATEGY 30.1 (MP)

Electromagnetic induction

MODEL Make simplifying assumptions about wires and magnetic fields.

VISUALIZE Draw a picture or a circuit diagram. Use Lenz's law to determine the direction of the induced current.

SOLVE The mathematical representation is based on Faraday's law

$$\mathcal{E} = \left| \frac{d\Phi_m}{dt} \right|$$

For an N-turn coil, multiply by N. The size of the induced current is $I = \mathcal{E}/R$.

ASSESS Check that your result has correct units and significant figures, is reasonable, and answers the question.

Exercise 18

EXAMPLE 30.8 | **Electromagnetic induction in a solenoid**

A 2.0-cm-diameter loop of wire with a resistance of 0.010 Ω is placed in the center of the solenoid seen in **FIGURE 30.27a** on the next page. The solenoid is 4.0 cm in diameter, 20 cm long, and wrapped with 1000 turns of wire. **FIGURE 30.27b** shows the current through the solenoid as a function of time as the solenoid is "powered up." A positive current is defined to be cw when seen from the left. Find the current in the loop as a function of time and show the result as a graph.

MODEL The solenoid's length is much greater than its diameter, so the field near the center should be nearly uniform.

VISUALIZE The magnetic field of the solenoid creates a magnetic flux through the loop of wire. The solenoid current is always positive, meaning that it is cw as seen from the left. Consequently, from the right-hand rule, the magnetic field inside the solenoid always points to the right. During the first second, while the solenoid current is

Continued

FIGURE 30.27 A loop inside a solenoid.

(a)

←————20 cm, 1000 turns————→

Positive current

4.0 cm

$\vec{B}$

2.0-cm-diameter loop

(b) Solenoid current
I_{sol} (A)

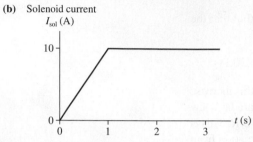

increasing, the flux through the loop is to the right and increasing. To oppose the *change* in the flux, the loop's induced magnetic field must point to the left. Thus, again using the right-hand rule, the induced current must flow ccw as seen from the left. This is a *negative* current. There's no *change* in the flux for $t > 1$ s, so the induced current is zero.

SOLVE Now we're ready to use Faraday's law to find the magnitude of the current. Because the field is uniform inside the solenoid and perpendicular to the loop ($\theta = 0°$), the flux is $\Phi_m = AB$, where $A = \pi r^2 = 3.14 \times 10^{-4}$ m^2 is the area of the loop (*not* the area of the solenoid). The field of a long solenoid of length l was found in Chapter 29 to be

$$B = \frac{\mu_0 N I_{sol}}{l}$$

The flux when the solenoid current is I_{sol} is thus

$$\Phi_m = \frac{\mu_0 A N I_{sol}}{l}$$

The changing flux creates an induced emf $\mathcal{E}$ that is given by Faraday's law:

$$\mathcal{E} = \left|\frac{d\Phi_m}{dt}\right| = \frac{\mu_0 A N}{l}\left|\frac{dI_{sol}}{dt}\right| = 2.0 \times 10^{-6}\left|\frac{dI_{sol}}{dt}\right|$$

From the slope of the graph, we find

$$\left|\frac{dI_{sol}}{dt}\right| = \begin{cases} 10 \text{ A/s} & 0.0 \text{ s} < t < 1.0 \text{ s} \\ 0 & 1.0 \text{ s} < t < 3.0 \text{ s} \end{cases}$$

Thus the induced emf is

$$\mathcal{E} = \begin{cases} 2.0 \times 10^{-5} \text{ V} & 0.0 \text{ s} < t < 1.0 \text{ s} \\ 0 \text{ V} & 1.0 \text{ s} < t < 3.0 \text{ s} \end{cases}$$

Finally, the current induced in the loop is

$$I_{loop} = \frac{\mathcal{E}}{R} = \begin{cases} -2.0 \text{ mA} & 0.0 \text{ s} < t < 1.0 \text{ s} \\ 0 \text{ mA} & 1.0 \text{ s} < t < 3.0 \text{ s} \end{cases}$$

where the negative sign comes from Lenz's law. This result is shown in FIGURE 30.28.

FIGURE 30.28 The induced current in the loop.

I_{loop} (mA)

The solenoid has a current, but it's not changing. Hence no current is induced in the loop.

There is an induced current as the flux changes.

EXAMPLE 30.9 | Current induced by an MRI machine

The body is a conductor, so rapid magnetic field changes in an MRI machine can induce currents in the body. To estimate the size of these currents, and any biological hazard they might impose, consider the "loop" of muscle tissue shown in FIGURE 30.29. This might be muscle circling the bone of your arm or thigh. Although muscle is not a great conductor—its resistivity is 1.5 Ω m—we can consider it to be a conducting loop with a rather high resistance. Suppose the magnetic field along the axis of the loop drops from

FIGURE 30.29 Edge view of a loop of muscle tissue in a magnetic field.

$\vec{B}$

8.0 cm

1.0 cm

1.6 T to 0 T in 0.30 s, which is about the largest possible rate of change for an MRI solenoid. What current will be induced?

MODEL Model the muscle as a conducting loop. Assume that B decreases linearly with time.

SOLVE The magnetic field is parallel to the axis of the loop, with $\theta = 0°$, so the magnetic flux through the loop is $\Phi_m = AB = \pi r^2 B$. The flux changes with time because B changes. According to Faraday's law, the magnitude of the induced emf is

$$\mathcal{E} = \left|\frac{d\Phi_m}{dt}\right| = \pi r^2\left|\frac{dB}{dt}\right|$$

The rate at which the magnetic field changes is

$$\frac{dB}{dt} = \frac{\Delta B}{\Delta t} = \frac{-1.60 \text{ T}}{0.30 \text{ s}} = -5.3 \text{ T/s}$$

dB/dt is negative because the field is decreasing, but all we need for Faraday's law is the absolute value. Thus

$$\mathcal{E} = \pi r^2 \left| \frac{dB}{dt} \right| = \pi (0.040 \text{ m})^2 (5.3 \text{ T/s}) = 0.027 \text{ V}$$

To find the current, we need to know the resistance of the loop. Recall, from Chapter 27, that a conductor with resistivity ρ, length L, and cross-section area A has resistance $R = \rho L/A$. The length is the circumference of the loop, calculated to be $L = 0.25$ m, and we can use the 1.0 cm diameter of the "wire" to find $A = 7.9 \times 10^{-5} \text{ m}^2$.

With these values, we can compute $R = 4700 \text{ }\Omega$. As a result, the induced current is

$$I = \frac{\mathcal{E}}{R} = \frac{0.027 \text{ V}}{4700 \text{ }\Omega} = 5.7 \times 10^{-6} \text{ A} = 5.7 \text{ }\mu\text{A}$$

ASSESS This is a very small current. Power—the rate of energy dissipation in the muscle—is

$$P = I^2 R = (5.7 \times 10^{-6} \text{ A})^2 (4700 \text{ }\Omega) = 1.5 \times 10^{-7} \text{ W}$$

The current is far too small to notice, and the tiny energy dissipation will certainly not heat the tissue.

What Does Faraday's Law Tell Us?

The induced current in the slide-wire circuit of Figure 30.26 can be understood as a motional emf due to magnetic forces on moving charges. We had not anticipated this kind of current in Chapter 29, but it takes no new laws of physics to understand it. The induced currents in Examples 30.8 and 30.9 are different. We cannot explain these induced currents on the basis of previous laws or principles. This is new physics.

Faraday recognized that all induced currents are associated with a changing magnetic flux. There are two fundamentally different ways to change the magnetic flux through a conducting loop:

1. The loop can expand, contract, or rotate, creating a motional emf.
2. The magnetic field can change.

We can see both of these if we write Faraday's law as

$$\mathcal{E} = \left| \frac{d\Phi_m}{dt} \right| = \left| \vec{B} \cdot \frac{d\vec{A}}{dt} + \vec{A} \cdot \frac{d\vec{B}}{dt} \right| \tag{30.19}$$

The first term on the right side represents a motional emf. The magnetic flux changes because the loop itself is changing. This term includes not only situations like the slide-wire circuit, where the area A changes, but also loops that rotate in a magnetic field. The physical area of a rotating loop does not change, but the area *vector* $\vec{A}$ does. The loop's motion causes magnetic forces on the charge carriers in the loop.

The second term on the right side is the new physics in Faraday's law. It says that an emf can also be created simply by changing a magnetic field, even if nothing is moving. This was the case in Examples 30.8 and 30.9. Faraday's law tells us that the induced emf is simply the rate of change of the magnetic flux through the loop, *regardless* of what causes the flux to change.

STOP TO THINK 30.5 A conducting loop is halfway into a magnetic field. Suppose the magnetic field begins to increase rapidly in strength. What happens to the loop?

a. The loop is pushed upward, toward the top of the page.
b. The loop is pushed downward, toward the bottom of the page.
c. The loop is pulled to the left, into the magnetic field.
d. The loop is pushed to the right, out of the magnetic field.
e. The tension in the wires increases but the loop does not move.

FIGURE 30.30 An induced electric field creates a current in the loop.

FIGURE 30.30 An induced electric field creates a current in the loop.

(a)

Region of increasing $\vec{B}$

Induced current

Conducting loop

(b)

Region of increasing $\vec{B}$

Induced electric field $\vec{E}$

FIGURE 30.31 Two ways to create an electric field.

A Coulomb electric field is created by charges.

$\vec{B}$ increasing or decreasing

A non-Coulomb electric field is created by a changing magnetic field.

30.6 Induced Fields

Faraday's law is a tool for calculating the strength of an induced current, but one important piece of the puzzle is still missing. What *causes* the current? That is, what *force* pushes the charges around the loop against the resistive forces of the metal? The agents that exert forces on charges are electric fields and magnetic fields. Magnetic forces are responsible for motional emfs, but magnetic forces cannot explain the current induced in a *stationary* loop by a changing magnetic field.

FIGURE 30.30a shows a conducting loop in an increasing magnetic field. According to Lenz's law, there is an induced current in the ccw direction. Something has to act on the charge carriers to make them move, so we infer that there must be an *electric* field tangent to the loop at all points. This electric field is *caused* by the changing magnetic field and is called an **induced electric field.** The induced electric field is the mechanism that creates a current inside a stationary loop when there's a changing magnetic field.

The conducting loop isn't necessary. The space in which the magnetic field is changing is filled with the pinwheel pattern of induced electric fields shown in **FIGURE 30.30b.** Charges will move if a conducting path is present, but the induced electric field is there as a direct consequence of the changing magnetic field.

But this is a rather peculiar electric field. All the electric fields we have examined until now have been created by charges. Electric field vectors pointed away from positive charges and toward negative charges. An electric field created by charges is called a **Coulomb electric field.** The induced electric field of Figure 30.30b is caused not by charges but by a changing magnetic field. It is called a **non-Coulomb electric field.**

So it appears that there are two different ways to create an electric field:

1. A Coulomb electric field is created by positive and negative charges.
2. A non-Coulomb electric field is created by a changing magnetic field.

Both exert a force $\vec{F} = q\vec{E}$ on a charge, and both create a current in a conductor. However, the origins of the fields are very different. **FIGURE 30.31** is a quick summary of the two ways to create an electric field.

We first introduced the idea of a field as a way of thinking about how two charges exert long-range forces on each other through the emptiness of space. The field may have seemed like a useful pictorial representation of charge interactions, but we had little evidence that fields are *real,* that they actually exist. Now we do. The electric field has shown up in a completely different context, independent of charges, as the explanation of the very real existence of induced currents.

The electric field is not just a pictorial representation; it is real.

Calculating the Induced Field

The induced electric field is peculiar in another way: It is nonconservative. Recall that a force is conservative if it does no net work on a particle moving around a closed path. "Uphills" are balanced by "downhills." We can associate a potential energy with a conservative force, hence we have gravitational potential energy for the conservative gravitational force and electric potential energy for the conservative electric force of charges (a Coulomb electric field).

But a charge moving around a closed path in the induced electric field of Figure 30.30 is always being pushed *in the same direction* by the electric force $\vec{F} = q\vec{E}$. There's never any negative work to balance the positive work, so the net work done in going around a closed path is not zero. Because it's nonconservative, we cannot associate an electric potential with an induced electric field. Only the Coulomb field of charges has an electric potential.

However, we can associate the induced field with the emf of Faraday's law. The emf was defined as the work required per unit charge to separate the charge. That is,

$$\mathcal{E} = \frac{W}{q} \qquad (30.20)$$

In batteries, a familiar source of emf, this work is done by chemical forces. But the emf that appears in Faraday's law arises when work is done by the force of an induced electric field.

If a charge q moves through a small displacement $d\vec{s}$, the small amount of work done by the electric field is $dW = \vec{F} \cdot d\vec{s} = q\vec{E} \cdot d\vec{s}$. The emf of Faraday's law is an emf around a *closed curve* through which the magnetic flux Φ_m is changing. The work done by the induced electric field as charge q moves around a closed curve is

$$W_{\text{closed curve}} = q \oint \vec{E} \cdot d\vec{s} \qquad (30.21)$$

where the integration symbol with the circle is the same as the one we used in Ampère's law to indicate an integral around a closed curve. If we use this work in Equation 30.20, we find that the emf around a closed loop is

$$\mathcal{E} = \frac{W_{\text{closed curve}}}{q} = \oint \vec{E} \cdot d\vec{s} \qquad (30.22)$$

If we restrict ourselves to situations such as Figure 30.30 where the loop is perpendicular to the magnetic field and only the field is changing, we can write Faraday's law as $\mathcal{E} = |d\Phi_m/dt| = A|dB/dt|$. Consequently

$$\oint \vec{E} \cdot d\vec{s} = A \left| \frac{dB}{dt} \right| \qquad (30.23)$$

Equation 30.23 is an alternative statement of Faraday's law that relates the induced electric field to the changing magnetic field.

The solenoid in **FIGURE 30.32a** provides a good example of the connection between $\vec{E}$ and $\vec{B}$. If there were a conducting loop inside the solenoid, we could use Lenz's law to determine that the direction of the induced current would be clockwise. But Faraday's law, in the form of Equation 30.23, tells us that **an induced electric field is present whether there's a conducting loop or not.** The electric field is induced simply due to the fact that $\vec{B}$ is changing.

FIGURE 30.32 The induced electric field circulates around the changing magnetic field inside a solenoid.

(a) The current through the solenoid is increasing.

Increasing current

I

$\vec{B}$ increasing

(b) The induced electric field circulates around the magnetic field lines.

$\vec{B}$ increasing

Induced $\vec{E}$

Induced $\vec{E}$

(c) Top view into the solenoid. $\vec{B}$ is coming out of the page.

$\vec{E}$ Integration curve

r $d\vec{s}$ $\vec{E}$

$\vec{E}$ $d\vec{s}$

$\vec{E}$

The shape and direction of the induced electric field have to be such that it *could* drive a current around a conducting loop, if one were present, and it has to be consistent with the cylindrical symmetry of the solenoid. The only possible choice, shown in **FIGURE 30.32b**, is an electric field that circulates clockwise around the magnetic field lines.

NOTE Circular electric field lines violate the Chapter 23 rule that electric field lines have to start and stop on charges. However, that rule applied only to Coulomb fields created by source charges. An induced electric field is a non-Coulomb field created not by source charges but by a changing magnetic field. Without source charges, induced electric field lines *must* form closed loops.

To use Faraday's law, choose a *clockwise* circle of radius r as the closed curve for evaluating the integral. FIGURE 30.32c shows that the electric field vectors are everywhere tangent to the curve, so the line integral of $\vec{E}$ is

$$\oint \vec{E} \cdot d\vec{s} = El = 2\pi rE \tag{30.24}$$

where $l = 2\pi r$ is the length of the closed curve. This is exactly like the integrals we did for Ampère's law in Chapter 29.

If we stay inside the solenoid ($r < R$), the flux passes through area $A = \pi r^2$ and Equation 30.24 becomes

$$\oint \vec{E} \cdot d\vec{s} = 2\pi rE = A\left|\frac{dB}{dt}\right| = \pi r^2 \left|\frac{dB}{dt}\right| \tag{30.25}$$

Thus the strength of the induced electric field inside the solenoid is

$$E_{\text{inside}} = \frac{r}{2}\left|\frac{dB}{dt}\right| \tag{30.26}$$

This result shows very directly that the induced electric field is created by a *changing* magnetic field. A constant $\vec{B}$, with $dB/dt = 0$, would give $E = 0$.

EXAMPLE 30.10 | An induced electric field

A 4.0-cm-diameter solenoid is wound with 2000 turns per meter. The current through the solenoid oscillates at 60 Hz with an amplitude of 2.0 A. What is the maximum strength of the induced electric field inside the solenoid?

MODEL Assume that the magnetic field inside the solenoid is uniform.

VISUALIZE The electric field lines are concentric circles around the magnetic field lines, as was shown in Figure 30.32b. They reverse direction twice every period as the current oscillates.

SOLVE You learned in Chapter 29 that the magnetic field strength inside a solenoid with n turns per meter is $B = \mu_0 nI$. In this case, the current through the solenoid is $I = I_0 \sin \omega t$, where $I_0 = 2.0$ A is the peak current and $\omega = 2\pi(60 \text{ Hz}) = 377$ rad/s. Thus the

induced electric field strength at radius r is

$$E = \frac{r}{2}\left|\frac{dB}{dt}\right| = \frac{r}{2}\frac{d}{dt}(\mu_0 nI_0 \sin \omega t) = \tfrac{1}{2}\mu_0 nr\omega I_0 \cos \omega t$$

The field strength is maximum at maximum radius ($r = R$) *and* at the instant when $\cos \omega t = 1$. That is,

$$E_{\text{max}} = \tfrac{1}{2}\mu_0 nR\omega I_0 = 0.019 \text{ V/m}$$

ASSESS This field strength, although not large, is similar to the field strength that the emf of a battery creates in a wire. Hence this induced electric field can drive a substantial induced current through a conducting loop *if* a loop is present. But the induced electric field exists inside the solenoid whether or not there is a conducting loop.

Occasionally it is useful to have a version of Faraday's law without the absolute value signs. The essence of Lenz's law is that the emf $\mathcal{E}$ opposes the *change* in Φ_m. Mathematically, this means that $\mathcal{E}$ must be opposite in sign to dB/dt. Consequently, we can write Faraday's law as

$$\mathcal{E} = \oint \vec{E} \cdot d\vec{s} = -\frac{d\Phi_m}{dt} \tag{30.27}$$

For practical applications, it's always easier to calculate just the magnitude of the emf with Faraday's law and to use Lenz's law to find the direction of the emf or the induced current. However, the mathematically rigorous version of Faraday's law in Equation 30.27 will prove to be useful when we combine it with other equations, in Chapter 31, to predict the existence of electromagnetic waves.

Maxwell's Theory of Electromagnetic Waves

In 1855, less than two years after receiving his undergraduate degree, the Scottish physicist James Clerk Maxwell presented a paper titled "On Faraday's Lines of Force." In this paper, he began to sketch out how Faraday's pictorial ideas about fields could be given a rigorous mathematical basis. Maxwell was troubled by a certain lack of symmetry. Faraday had found that a changing magnetic field creates an induced electric field, a non-Coulomb electric field not tied to charges. But what, Maxwell began to wonder, about a changing *electric* field?

To complete the symmetry, Maxwell proposed that a changing electric field creates an **induced magnetic field,** a new kind of magnetic field not tied to the existence of currents. FIGURE 30.33 shows a region of space where the *electric* field is increasing. This region of space, according to Maxwell, is filled with a pinwheel pattern of induced magnetic fields. The induced magnetic field looks like the induced electric field, with $\vec{E}$ and $\vec{B}$ interchanged, except that—for technical reasons explored in the next chapter—the induced $\vec{B}$ points the opposite way from the induced $\vec{E}$. Although there was no experimental evidence that induced magnetic fields existed, Maxwell went ahead and included them in his electromagnetic field theory. This was an inspired hunch, soon to be vindicated.

Maxwell soon realized that it might be possible to establish self-sustaining electric and magnetic fields that would be entirely independent of any charges or currents. That is, a changing electric field $\vec{E}$ creates a magnetic field $\vec{B}$, which then changes in just the right way to recreate the electric field, which then changes in just the right way to again recreate the magnetic field, and so on. The fields are continually recreated through electromagnetic induction without any reliance on charges or currents.

Maxwell was able to predict that electric and magnetic fields would be able to sustain themselves, free from charges and currents, if they took the form of an **electromagnetic wave.** The wave would have to have a very specific geometry, shown in FIGURE 30.34, in which $\vec{E}$ and $\vec{B}$ are perpendicular to each other as well as perpendicular to the direction of travel. That is, an electromagnetic wave would be a *transverse* wave.

Furthermore, Maxwell's theory predicted that the wave would travel with speed

$$v_{\text{em wave}} = \frac{1}{\sqrt{\epsilon_0 \mu_0}}$$

where ϵ_0 is the permittivity constant from Coulomb's law and μ_0 is the permeability constant from the law of Biot and Savart. Maxwell computed that an electromagnetic wave, if it existed, would travel with speed $v_{\text{em wave}} = 3.00 \times 10^8$ m/s.

We don't know Maxwell's immediate reaction, but it must have been both shock and excitement. His predicted speed for electromagnetic waves, a prediction that came directly from his theory, was none other than the speed of light! This agreement could be just a coincidence, but Maxwell didn't think so. Making a bold leap of imagination, Maxwell concluded that **light is an electromagnetic wave.**

It took 25 more years for Maxwell's predictions to be tested. In 1886, the German physicist Heinrich Hertz discovered how to generate and transmit radio waves. Two years later, in 1888, he was able to show that radio waves travel at the speed of light. Maxwell, unfortunately, did not live to see his triumph. He had died in 1879, at the age of 48.

30.7 Induced Currents: Three Applications

There are many applications of Faraday's law and induced currents in modern technology. In this section we will look at three: generators, transformers, and metal detectors.

Generators

A generator is a device that transforms mechanical energy into electric energy. FIGURE 30.35 on the next page shows a generator in which a coil of wire, perhaps spun by a windmill, rotates in a magnetic field. Both the field and the area of the loop are constant, but the magnetic flux through the loop changes continuously as the loop rotates. The induced current is removed from the rotating loop by *brushes* that press up against rotating *slip rings*.

The flux through the coil is

$$\Phi_m = \vec{A} \cdot \vec{B} = AB \cos \theta = AB \cos \omega t \qquad (30.28)$$

FIGURE 30.33 Maxwell hypothesized the existence of induced magnetic fields.

A changing magnetic field creates an induced electric field.

Region of increasing $\vec{B}$ Induced electric field $\vec{E}$

A changing electric field creates an induced magnetic field.

Region of increasing $\vec{E}$ Induced magnetic field $\vec{B}$

FIGURE 30.34 A self-sustaining electromagnetic wave.

Direction of propagation at speed $v_{\text{em wave}}$

A generator inside a hydroelectric dam uses electromagnetic induction to convert the mechanical energy of a spinning turbine into electric energy.

where ω is the angular frequency ($\omega = 2\pi f$) with which the coil rotates. The induced emf is given by Faraday's law,

$$\mathcal{E}_{coil} = -N\frac{d\Phi_m}{dt} = -ABN\frac{d}{dt}(\cos\omega t) = \omega ABN\sin\omega t \qquad (30.29)$$

where N is the number of turns on the coil. Here it's best to use the signed version of Faraday's law to see how $\mathcal{E}_{coil}$ alternates between positive and negative.

Because the emf alternates in sign, the current through resistor R alternates back and forth in direction. Hence the generator of Figure 30.35 is an alternating-current generator, producing what we call an *AC voltage*.

FIGURE 30.35 An alternating-current generator.

EXAMPLE 30.11 | An AC generator

A coil with area 2.0 m^2 rotates in a 0.010 T magnetic field at a frequency of 60 Hz. How many turns are needed to generate a peak voltage of 160 V?

SOLVE The coil's maximum voltage is found from Equation 30.29:

$$\mathcal{E}_{max} = \omega ABN = 2\pi f ABN$$

The number of turns needed to generate $\mathcal{E}_{max} = 160$ V is

$$N = \frac{\mathcal{E}_{max}}{2\pi fAB} = \frac{160\text{ V}}{2\pi(60\text{ Hz})(2.0\text{ m}^2)(0.010\text{ T})} = 21\text{ turns}$$

ASSESS A 0.010 T field is modest, so you can see that generating large voltages is not difficult with large (2 m^2) coils. Commercial generators use water flowing through a dam, rotating windmill blades, or turbines spun by expanding steam to rotate the generator coils. Work is required to rotate the coil, just as work was required to pull the slide wire in Section 30.2, because the magnetic field exerts retarding forces on the currents in the coil. Thus a generator is a device that turns motion (mechanical energy) into a current (electric energy). A generator is the opposite of a motor, which turns a current into motion.

Transformers

FIGURE 30.36 A transformer.

The magnetic field follows the iron core.

FIGURE 30.36 shows two coils wrapped on an iron core. The left coil is called the **primary coil**. It has N_1 turns and is driven by an oscillating voltage $V_1 \cos\omega t$. The magnetic field of the primary follows the iron core and passes through the right coil, which has N_2 turns and is called the **secondary coil**. The alternating current through the primary coil causes an oscillating magnetic flux through the secondary coil and, hence, an induced emf. The induced emf of the secondary coil is delivered to the load as the oscillating voltage $V_2 \cos\omega t$.

The changing magnetic field inside the iron core is inversely proportional to the number of turns on the primary coil: $B \propto 1/N_1$. (This relation is a consequence of the coil's inductance, an idea discussed in the next section.) According to Faraday's law, the emf induced in the secondary coil is directly proportional to its number of turns:

$\mathcal{E}_{\text{sec}} \propto N_2$. Combining these two proportionalities, the secondary voltage of an ideal transformer is related to the primary voltage by

$$V_2 = \frac{N_2}{N_1} V_1 \tag{30.30}$$

Depending on the ratio N_2/N_1, the voltage V_2 across the load can be *transformed* to a higher or a lower voltage than V_1. Consequently, this device is called a **transformer.** Transformers are widely used in the commercial generation and transmission of electricity. A *step-up transformer,* with $N_2 \gg N_1$, boosts the voltage of a generator up to several hundred thousand volts. Delivering power with smaller currents at higher voltages reduces losses due to the resistance of the wires. High-voltage transmission lines carry electric power to urban areas, where *step-down transformers* $(N_2 \ll N_1)$ lower the voltage to 120 V.

Transformers are essential for transporting electric energy from the power plant to cities and homes.

Metal Detectors

Metal detectors, such as those used in airports for security, seem fairly mysterious. How can they detect the presence of *any* metal—not just magnetic materials such as iron—but not detect plastic or other materials? Metal detectors work because of induced currents.

A metal detector, shown in **FIGURE 30.37**, consists of two coils: a *transmitter coil* and a *receiver coil*. A high-frequency alternating current in the transmitter coil generates an alternating magnetic field along the axis. This magnetic field creates a changing flux through the receiver coil and causes an alternating induced current. The transmitter and receiver are similar to a transformer.

Suppose a piece of metal is placed between the transmitter and the receiver. The alternating magnetic field through the metal induces eddy currents in a plane parallel to the transmitter and receiver coils. The receiver coil then responds to the *superposition* of the transmitter's magnetic field and the magnetic field of the eddy currents. Because the eddy currents attempt to prevent the flux from changing, in accordance with Lenz's law, the net field at the receiver *decreases* when a piece of metal is inserted between the coils. Electronic circuits detect the current decrease in the receiver coil and set off an alarm. Eddy currents can't flow in an insulator, so this device detects only metals.

FIGURE 30.37 A metal detector.

Induced current due to eddy currents ⋯⋯ Receiver coil

Metal

Induced current due to the transmitter coil

Transmitter coil

Eddy currents in the metal reduce the induced current in the receiver coil.

30.8 Inductors

Capacitors are useful circuit elements because they store potential energy U_C in the electric field. Similarly, a coil of wire can be a useful circuit element because it stores energy in the magnetic field. In circuits, a coil is called an **inductor** because, as you'll see, the potential difference across an inductor is an *induced* emf. An *ideal inductor* is one for which the wire forming the coil has no electric resistance. The circuit symbol for an inductor is —⟡⟡⟡⟡—.

We define the **inductance** L of a coil to be its flux-to-current ratio:

$$L = \frac{\Phi_m}{I} \tag{30.31}$$

Strictly speaking, this is called *self-inductance* because the flux we're considering is the magnetic flux the solenoid creates in itself when there is a current. The SI unit of inductance is the **henry,** named in honor of Joseph Henry, defined as

$$1 \text{ henry} = 1 \text{ H} \equiv 1 \text{ Wb/A} = 1 \text{ T}\,\text{m}^2/\text{A}$$

Practical inductances are typically millihenries (mH) or microhenries (μH).

It's not hard to find the inductance of a solenoid. In Chapter 29 we found that the magnetic field inside an ideal solenoid having N turns and length l is

$$B = \frac{\mu_0 N I}{l}$$

The magnetic flux through one turn of the coil is $\Phi_{\text{per turn}} = AB$, where A is the cross-section area of the solenoid. The total magnetic flux through all N turns is

$$\Phi_{\text{m}} = N\Phi_{\text{per turn}} = \frac{\mu_0 N^2 A}{l} I \tag{30.32}$$

Thus the inductance of the solenoid, using the definition of Equation 30.31, is

$$L_{\text{solenoid}} = \frac{\Phi_{\text{m}}}{I} = \frac{\mu_0 N^2 A}{l} \tag{30.33}$$

The inductance of a solenoid depends only on its geometry, not at all on the current. You may recall that the capacitance of two parallel plates depends only on their geometry, not at all on their potential difference.

EXAMPLE 30.12 | **The length of an inductor**

An inductor is made by tightly wrapping 0.30-mm-diameter wire around a 4.0-mm-diameter cylinder. What length cylinder has an inductance of 10 μH?

SOLVE The cross-section area of the solenoid is $A = \pi r^2$. If the wire diameter is d, the number of turns of wire on a cylinder of length l is $N = l/d$. Thus the inductance is

$$L = \frac{\mu_0 N^2 A}{l} = \frac{\mu_0 (l/d)^2 \pi r^2}{l} = \frac{\mu_0 \pi r^2 l}{d^2}$$

The length needed to give inductance $L = 1.0 \times 10^{-5}$ H is

$$l = \frac{d^2 L}{\mu_0 \pi r^2} = \frac{(0.00030 \text{ m})^2 (1.0 \times 10^{-5} \text{ H})}{(4\pi \times 10^{-7} \text{ T m/A}) \pi (0.0020 \text{ m})^2}$$

$$= 0.057 \text{ m} = 5.7 \text{ cm}$$

The Potential Difference Across an Inductor

An inductor is not very interesting when the current through it is steady. If the inductor is ideal, with $R = 0$ Ω, the potential difference due to a steady current is zero. **Inductors become important circuit elements when currents are changing.** FIGURE 30.38a shows a steady current into the left side of an inductor. The solenoid's magnetic field passes through the coils of the solenoid, establishing a flux.

In FIGURE 30.38b, the current into the solenoid is increasing. This creates an increasing flux to the left. According to Lenz's law, an induced current in the coils will oppose this increase by creating an induced magnetic field pointing to the right. This requires the induced current to be *opposite* the current into the solenoid. This induced current will carry positive charge carriers to the left until a potential difference is established across the solenoid.

You saw a similar situation in Section 30.2. The induced current in a conductor moving through a magnetic field carried positive charge carriers to the top of the wire and established a potential difference across the conductor. The induced current in the moving wire was due to magnetic forces on the moving charges. Now, in Figure 30.38b, the induced current is due to the non-Coulomb electric field induced by the changing magnetic field. Nonetheless, the outcome is the same: a potential difference across the conductor.

We can use Faraday's law to find the potential difference. The emf induced in a coil is

$$\mathcal{E}_{\text{coil}} = N \left| \frac{d\Phi_{\text{per turn}}}{dt} \right| = \left| \frac{d\Phi_{\text{m}}}{dt} \right| \tag{30.34}$$

where $\Phi_{\text{m}} = N\Phi_{\text{per turn}}$ is the total flux through all the coils. The inductance was defined such that $\Phi_{\text{m}} = LI$, so Equation 30.34 becomes

$$\mathcal{E}_{\text{coil}} = L \left| \frac{dI}{dt} \right| \tag{30.35}$$

FIGURE 30.38 Increasing the current through an inductor.

(a)

Inductor coil

$\vec{B}$

Solenoid magnetic field

Current I

(b)

The induced current is opposite the solenoid current.

The induced magnetic field opposes the change in flux.

ΔV_L

Increasing current

The induced current carries positive charge carriers to the left and establishes a potential difference across the inductor.

The induced emf is directly proportional to the *rate of change* of current through the coil. We'll consider the appropriate sign in a moment, but Equation 30.35 gives us the size of the potential difference that is developed across a coil as the current through the coil changes. Note that $\mathcal{E}_{coil} = 0$ for a steady, unchanging current.

FIGURE 30.39 shows the same inductor, but now the current (still *in* to the left side) is decreasing. To oppose the decrease in flux, the induced current is in the *same* direction as the input current. The induced current carries charge to the right and establishes a potential difference opposite that in Figure 30.38b.

NOTE Notice that the induced current does not oppose the current *through* the inductor, which is from left to right in both Figures 30.38 and 30.39. Instead, in accordance with Lenz's law, the induced current opposes the *change* in the current in the solenoid. The practical result is that it is hard to change the current through an inductor. Any effort to increase or decrease the current is met with opposition in the form of an opposing induced current. You can think of the current in an inductor as having inertia, trying to continue what it was doing without change.

Before we can use inductors in a circuit we need to establish a rule about signs that is consistent with our earlier circuit analysis. **FIGURE 30.40** first shows current I passing through a resistor. You learned in Chapter 28 that the potential difference across a resistor is $\Delta V_{res} = -\Delta V_R = -IR$, where the minus sign indicates that the potential *decreases* in the direction of the current.

We'll use the same convention for an inductor. The potential difference across an inductor, *measured along the direction of the current*, is

$$\Delta V_L = -L\frac{dI}{dt} \qquad (30.36)$$

If the current is increasing ($dI/dt > 0$), the input side of the inductor is more positive than the output side and the potential decreases in the direction of the current ($\Delta V_L < 0$). This was the situation in Figure 30.38b. If the current is decreasing ($dI/dt < 0$), the input side is more negative and the potential increases in the direction of the current ($\Delta V_L > 0$). This was the situation in Figure 30.39.

The potential difference across an inductor can be very large if the current changes very abruptly (large dI/dt). **FIGURE 30.41** shows an inductor connected across a battery. There is a large current through the inductor, limited only by the internal resistance of the battery. Suppose the switch is suddenly opened. A very large induced voltage is created across the inductor as the current rapidly drops to zero. This potential difference (plus ΔV_{bat}) appears across the gap of the switch as it is opened. A large potential difference across a small gap often creates a spark.

FIGURE 30.39 Decreasing the current through an inductor.

Decreasing current

The induced current carries positive charge carriers to the right. The potential difference is opposite that of Figure 30.38b.

FIGURE 30.40 The potential difference across a resistor and an inductor.

Resistor

$\Delta V_{res} = -IR$

The potential always decreases.

Inductor

$\Delta V_L = -L\dfrac{dI}{dt}$

The potential decreases if the current is increasing.

The potential increases if the current is decreasing.

FIGURE 30.41 Creating sparks.

Switch closed

ΔV_{bat} I

Before switch opened

Opening — Spark! The current decreases rapidly after the switch opens.

ΔV_{bat} I $\Delta V_L = -L\dfrac{dI}{dt}$ is very large.

As switch opened

Indeed, this is exactly how the spark plugs in your car work. The car's generator sends a current through the *coil,* which is a big inductor. When a switch is suddenly opened, breaking the current, the induced voltage, typically a few thousand volts, appears across the terminals of the spark plug, creating the spark that ignites the gasoline. Older cars use a *distributor* to open and close an actual switch; more recent cars have *electronic ignition* in which the mechanical switch has been replaced by a transistor.

EXAMPLE 30.13 | **Large voltage across an inductor**

A 1.0 A current passes through a 10 mH inductor coil. What potential difference is induced across the coil if the current drops to zero in 5.0 μs?

MODEL Assume this is an ideal inductor, with $R = 0 \, \Omega$, and that the current decrease is linear with time.

SOLVE The rate of current decrease is

$$\frac{dI}{dt} \approx \frac{\Delta I}{\Delta t} = \frac{-1.0 \text{ A}}{5.0 \times 10^{-6} \text{ s}} = -2.0 \times 10^5 \text{ A/s}$$

The induced voltage is

$$\Delta V_L = -L\frac{dI}{dt} \approx -(0.010 \text{ H})(-2.0 \times 10^5 \text{ A/s}) = 2000 \text{ V}$$

ASSESS Inductors may be physically small, but they can pack a punch if you try to change the current through them too quickly.

STOP TO THINK 30.6 The potential at a is higher than the potential at b. Which of the following statements about the inductor current I could be true?

$V_a > V_b$

a. I is from a to b and steady.
c. I is from a to b and decreasing.
e. I is from b to a and increasing.

b. I is from a to b and increasing.
d. I is from b to a and steady.
f. I is from b to a and decreasing.

Energy in Inductors and Magnetic Fields

Recall that electric power is $P_{elec} = I\Delta V$. As current passes through an inductor, for which $\Delta V_L = -L(dI/dt)$, the electric power is

$$P_{elec} = I\Delta V_L = -LI\frac{dI}{dt} \qquad (30.37)$$

P_{elec} is negative because a circuit with an increasing current is *losing* electric energy. That energy is being transferred to the inductor, which is *storing* energy U_L at the rate

$$\frac{dU_L}{dt} = +LI\frac{dI}{dt} \qquad (30.38)$$

where we've noted that power is the rate of change of energy.

We can find the total energy stored in an inductor by integrating Equation 30.38 from $I = 0$, where $U_L = 0$, to a final current I. Doing so gives

$$U_L = L\int_0^I I \, dI = \tfrac{1}{2}LI^2 \qquad (30.39)$$

The potential energy stored in an inductor depends on the square of the current through it. Notice the analogy with the energy $U_C = \frac{1}{2}C(\Delta V)^2$ stored in a capacitor.

In working with circuits we say that the energy is "stored in the inductor." Strictly speaking, the energy is stored in the inductor's magnetic field, analogous to how a capacitor stores energy in the electric field. We can use the inductance of a solenoid, Equation 30.33, to relate the inductor's energy to the magnetic field strength:

$$U_L = \tfrac{1}{2}LI^2 = \frac{\mu_0 N^2 A}{2l}I^2 = \frac{1}{2\mu_0}Al\left(\frac{\mu_0 NI}{l}\right)^2 \qquad (30.40)$$

We made the last rearrangement in Equation 30.40 because $\mu_0 NI/l$ is the magnetic field inside the solenoid. Thus

$$U_L = \frac{1}{2\mu_0}AlB^2 \qquad (30.41)$$

Energy in electric and magnetic fields

Electric fields	Magnetic fields
A capacitor stores energy	An inductor stores energy
$U_C = \tfrac{1}{2}C(\Delta V)^2$	$U_L = \tfrac{1}{2}LI^2$
Energy density in the field is	Energy density in the field is
$u_E = \dfrac{\epsilon_0}{2}E^2$	$u_B = \dfrac{1}{2\mu_0}B^2$

But *Al* is the volume inside the solenoid. Dividing by *Al*, the magnetic field *energy density* inside the solenoid (energy per m³) is

$$u_{\rm B} = \frac{1}{2\mu_0}B^2 \qquad (30.42)$$

We've derived this expression for energy density based on the properties of a solenoid, but it turns out to be the correct expression for the energy density anywhere there's a magnetic field. Compare this to the energy density of an electric field $u_{\rm E} = \frac{1}{2}\epsilon_0 E^2$ that we found in Chapter 26.

EXAMPLE 30.14 | **Energy stored in an inductor**

The 10 μH inductor of Example 30.12 was 5.7 cm long and 4.0 mm in diameter. Suppose it carries a 100 mA current. What are the energy stored in the inductor, the magnetic energy density, and the magnetic field strength?

SOLVE The stored energy is

$$U_{\rm L} = \tfrac{1}{2}LI^2 = \tfrac{1}{2}(1.0 \times 10^{-5}\,{\rm H})(0.10\,{\rm A})^2 = 5.0 \times 10^{-8}\,{\rm J}$$

The solenoid volume is $(\pi r^2)l = 7.16 \times 10^{-7}$ m³. Using this gives the energy density of the magnetic field:

$$u_{\rm B} = \frac{5.0 \times 10^{-8}\,{\rm J}}{7.16 \times 10^{-7}\,{\rm m}^3} = 0.070\,{\rm J/m}^3$$

From Equation 30.42, the magnetic field with this energy density is

$$B = \sqrt{2\mu_0 u_{\rm B}} = 4.2 \times 10^{-4}\,{\rm T}$$

30.9 *LC* Circuits

Telecommunication—radios, televisions, cell phones—is based on electromagnetic signals that *oscillate* at a well-defined frequency. These oscillations are generated and detected by a simple circuit consisting of an inductor and a capacitor. This is called an **LC circuit.** In this section we will learn why an *LC* circuit oscillates and determine the oscillation frequency.

FIGURE 30.42 shows a capacitor with initial charge Q_0, an inductor, and a switch. The switch has been open for a long time, so there is no current in the circuit. Then, at $t = 0$, the switch is closed. How does the circuit respond? Let's think it through qualitatively before getting into the mathematics.

As **FIGURE 30.43** shows, the inductor provides a conducting path for discharging the capacitor. However, the discharge current has to pass through the inductor, and, as we've seen, an inductor resists changes in current. Consequently, the current doesn't stop when the capacitor charge reaches zero.

FIGURE 30.42 An *LC* circuit.

FIGURE 30.43 The capacitor charge oscillates much like a block attached to a spring.

Analogy:
$v = 0$
Maximum capacitor charge is like a fully stretched spring.

Ⓐ $Q = +Q_0$ $I = 0$

The capacitor discharges until the current is a maximum.

Ⓑ $Q = 0$ Max I $\vec{B}$

Max v
Maximum current is like the block having maximum speed.

The current continues until the initial capacitor charge is restored.

The current continues until the capacitor is fully recharged with opposite polarization.

Max v

Ⓓ I $Q = 0$ Max I $\vec{B}$

Now the discharge goes in the opposite direction.

Ⓒ $Q = -Q_0$ $I = 0$

$v = 0$

A cell phone is actually a very sophisticated two-way radio that communicates with the nearest base station via high-frequency radio waves—roughly 1000 MHz. As in any radio or communications device, the transmission frequency is established by the oscillating current in an *LC* circuit.

A block attached to a stretched spring is a useful mechanical analogy. Closing the switch to discharge the capacitor is like releasing the block. The block doesn't stop when it reaches the origin; its inertia keeps it going until the spring is fully compressed. Likewise, the current continues until it has recharged the capacitor with the opposite polarization. This process repeats over and over, charging the capacitor first one way, then the other. That is, the charge and current *oscillate*.

The goal of our circuit analysis will be to find expressions showing how the capacitor charge Q and the inductor current I change with time. As always, our starting point for circuit analysis is Kirchhoff's voltage law, which says that all the potential differences around a closed loop must sum to zero. Choosing a cw direction for I, Kirchhoff's law is

$$\Delta V_C + \Delta V_L = 0 \tag{30.43}$$

The potential difference across a capacitor is $\Delta V_C = Q/C$, and we found the potential difference across an inductor in Equation 30.36. Using these, Kirchhoff's law becomes

$$\frac{Q}{C} - L\frac{dI}{dt} = 0 \tag{30.44}$$

Equation 30.44 has two unknowns, Q and I. We can eliminate one of the unknowns by finding another relation between Q and I. Current is the rate at which charge moves, $I = dq/dt$, but the charge flowing through the inductor is charge that was *removed* from the capacitor. That is, an infinitesimal charge dq flows through the inductor when the capacitor charge changes by $dQ = -dq$. Thus the current through the inductor is related to the charge on the capacitor by

$$I = -\frac{dQ}{dt} \tag{30.45}$$

Now I is positive when Q is decreasing, as we would expect. This is a subtle but important step in the reasoning.

Equations 30.44 and 30.45 are two equations in two unknowns. To solve them, we'll first take the time derivative of Equation 30.45:

$$\frac{dI}{dt} = \frac{d}{dt}\left(-\frac{dQ}{dt}\right) = -\frac{d^2Q}{dt^2} \tag{30.46}$$

We can substitute this result into Equation 30.44:

$$\frac{Q}{C} + L\frac{d^2Q}{dt^2} = 0 \tag{30.47}$$

Now we have an equation for the capacitor charge Q.

Equation 30.47 is a second-order differential equation for Q. Fortunately, it is an equation we've seen before and already know how to solve. To see this, we rewrite Equation 30.47 as

$$\frac{d^2Q}{dt^2} = -\frac{1}{LC}Q \tag{30.48}$$

Recall, from Chapter 15, that the equation of motion for an undamped mass on a spring is

$$\frac{d^2x}{dt^2} = -\frac{k}{m}x \tag{30.49}$$

Equation 30.48 is *exactly the same equation*, with x replaced by Q and k/m replaced by $1/LC$. This should be no surprise because we've already seen that a mass on a spring is a mechanical analog of the *LC* circuit.

We know the solution to Equation 30.49. It is simple harmonic motion $x(t) = x_0 \cos \omega t$ with angular frequency $\omega = \sqrt{k/m}$. Thus the solution to Equation 30.48 must be

$$Q(t) = Q_0 \cos \omega t \tag{30.50}$$

where Q_0 is the initial charge, at $t = 0$, and the angular frequency is

$$\omega = \sqrt{\frac{1}{LC}} \tag{30.51}$$

The charge on the upper plate of the capacitor oscillates back and forth between $+Q_0$ and $-Q_0$ (the opposite polarization) with period $T = 2\pi/\omega$.

As the capacitor charge oscillates, so does the current through the inductor. Using Equation 30.45 gives the current through the inductor:

$$I = -\frac{dQ}{dt} = \omega Q_0 \sin \omega t = I_{max} \sin \omega t \tag{30.52}$$

where $I_{max} = \omega Q_0$ is the maximum current.

An *LC* circuit is an *electric oscillator*, oscillating at frequency $f = \omega/2\pi$. FIGURE 30.44 shows graphs of the capacitor charge Q and the inductor current I as functions of time. Notice that Q and I are 90° out of phase. The current is zero when the capacitor is fully charged, as expected, and the charge is zero when the current is maximum.

FIGURE 30.44 The oscillations of an *LC* circuit.

EXAMPLE 30.15 | **An AM radio oscillator**

You have a 1.0 mH inductor. What capacitor should you choose to make an oscillator with a frequency of 920 kHz? (This frequency is near the center of the AM radio band.)

SOLVE The angular frequency is $\omega = 2\pi f = 5.78 \times 10^6$ rad/s. Using Equation 30.51 for ω gives the required capacitor:

$$C = \frac{1}{\omega^2 L} = 3.0 \times 10^{-11} \text{ F} = 30 \text{ pF}$$

An *LC* circuit, like a mass on a spring, wants to respond only at its natural oscillation frequency $\omega = 1/\sqrt{LC}$. In Chapter 15 we defined a strong response at the natural frequency as a *resonance*, and resonance is the basis for all telecommunications. The input circuit in radios, televisions, and cell phones is an *LC* circuit driven by the signal picked up by the antenna. This signal is the superposition of hundreds of sinusoidal waves at different frequencies, one from each transmitter in the area, but the circuit responds only to the *one* signal that matches the circuit's natural frequency. That particular signal generates a large-amplitude current that can be further amplified and decoded to become the output that you hear.

30.10 *LR* Circuits

A circuit consisting of an inductor, a resistor, and (perhaps) a battery is called an **LR circuit**. FIGURE 30.45a is an example of an *LR* circuit. We'll assume that the switch has been in position a for such a long time that the current is steady and unchanging. There's no potential difference across the inductor, because $dI/dt = 0$, so it simply acts like a piece of wire. The current flowing around the circuit is determined entirely by the battery and the resistor: $I_0 = \Delta V_{bat}/R$.

What happens if, at $t = 0$, the switch is suddenly moved to position b? With the battery no longer in the circuit, you might expect the current to stop immediately. But the inductor won't let that happen. The current will continue for some period of time as the inductor's magnetic field drops to zero. In essence, the energy stored in the inductor allows it to act like a battery for a short period of time. Our goal is to determine how the current decays after the switch is moved.

FIGURE 30.45 An *LR* circuit.

(a) The switch has been in this position for a long time. At $t = 0$ it is moved to position b.

(b)

This is the circuit with the switch in position b. The inductor prevents the current from stopping instantly.

NOTE It's important not to open switches in inductor circuits because they'll spark, as Figure 30.41 showed. The unusual switch in Figure 30.45 is designed to make the new contact just before breaking the old one.

FIGURE 30.45b shows the circuit after the switch is changed. Our starting point, once again, is Kirchhoff's voltage law. The potential differences around a closed loop must sum to zero. For this circuit, Kirchhoff's law is

$$\Delta V_{\text{res}} + \Delta V_{\text{L}} = 0 \tag{30.53}$$

The potential differences in the direction of the current are $\Delta V_{\text{res}} = -IR$ for the resistor and $\Delta V_{\text{L}} = -L(dI/dt)$ for the inductor. Substituting these into Equation 30.53 gives

$$-RI - L\frac{dI}{dt} = 0 \tag{30.54}$$

We're going to need to integrate to find the current I as a function of time. Before doing so, we rearrange Equation 30.54 to get all the current terms on one side of the equation and all the time terms on the other:

$$\frac{dI}{I} = -\frac{R}{L}dt = -\frac{dt}{(L/R)} \tag{30.55}$$

We know that the current at $t = 0$, when the switch was moved, was I_0. We want to integrate from these starting conditions to current I at the unspecified time t. That is,

$$\int_{I_0}^{I}\frac{dI}{I} = -\frac{1}{(L/R)}\int_{0}^{t} dt \tag{30.56}$$

Both are common integrals, giving

$$\ln I \Big|_{I_0}^{I} = \ln I - \ln I_0 = \ln\left(\frac{I}{I_0}\right) = -\frac{t}{(L/R)} \tag{30.57}$$

We can solve for the current I by taking the exponential of both sides, then multiplying by I_0. Doing so gives I, the current as a function of time:

$$I = I_0 e^{-t/(L/R)} \tag{30.58}$$

Notice that $I = I_0$ at $t = 0$, as expected.

The argument of the exponential function must be dimensionless, so L/R must have dimensions of time. If we define the **time constant** τ of the LR circuit to be

$$\tau = \frac{L}{R} \tag{30.59}$$

then we can write Equation 30.58 as

$$I = I_0 e^{-t/\tau} \tag{30.60}$$

The time constant is the time at which the current has decreased to e^{-1} (about 37%) of its initial value. We can see this by computing the current at the time $t = \tau$:

$$I(\text{at } t = \tau) = I_0 e^{-\tau/\tau} = e^{-1}I_0 = 0.37I_0 \tag{30.61}$$

Thus the time constant for an LR circuit functions in exactly the same way as the time constant for the RC circuit we analyzed in Chapter 29. At time $t = 2\tau$, the current has decreased to $e^{-2}I_0$, or about 13% of its initial value.

The current is graphed in **FIGURE 30.46**. You can see that the current decays exponentially. The *shape* of the graph is always the same, regardless of the specific value of the time constant τ.

FIGURE 30.46 The current decay in an LR circuit.

EXAMPLE 30.16 | Exponential decay in an *LR* circuit

The switch in FIGURE 30.47 has been in position a for a long time. It is changed to position b at $t = 0$ s.

a. What is the current in the circuit at $t = 5.0\ \mu s$?

b. At what time has the current decayed to 1% of its initial value?

FIGURE 30.47 The *LR* circuit of Example 30.16.

The switch moves from a to b at $t = 0$.

MODEL This is an *LR* circuit. We'll assume ideal wires and an ideal inductor.

VISUALIZE The two resistors will be in series after the switch is thrown.

SOLVE Before the switch is thrown, while $\Delta V_L = 0$, the current is $I_0 = (10\ \text{V})/(100\ \Omega) = 0.10\ \text{A} = 100\ \text{mA}$. This will be the initial current after the switch is thrown because the current through an inductor can't change instantaneously. The circuit resistance after the switch is thrown is $R = 200\ \Omega$, so the time constant is

$$\tau = \frac{L}{R} = \frac{2.0 \times 10^{-3}\ \text{H}}{200\ \Omega} = 1.0 \times 10^{-5}\ \text{s} = 10\ \mu s$$

a. The current at $t = 5.0\ \mu s$ is

$$I = I_0 e^{-t/\tau} = (100\ \text{mA})e^{-(5.0\mu s)/(10\mu s)} = 61\ \text{mA}$$

b. To find the time at which a particular current is reached we need to go back to Equation 30.57 and solve for t:

$$t = -\frac{L}{R}\ln\left(\frac{I}{I_0}\right) = -\tau\ln\left(\frac{I}{I_0}\right)$$

The time at which the current has decayed to 1 mA (1% of I_0) is

$$t = -(10\ \mu s)\ln\left(\frac{1\ \text{mA}}{100\ \text{mA}}\right) = 46\ \mu s$$

ASSESS For all practical purposes, the current has decayed away in $\approx 50\ \mu s$. The inductance in this circuit is not large, so a short decay time is not surprising.

STOP TO THINK 30.7 Rank in order, from largest to smallest, the time constants τ_a, τ_b, and τ_c of these three circuits.

CHALLENGE EXAMPLE 30.17 | Induction heating

Induction heating uses induced currents to heat metal objects to high temperatures for applications such as surface hardening, brazing, or even melting. To illustrate the idea, consider a copper wire formed into a 4.0 cm × 4.0 cm square loop and placed in a magnetic field—perpendicular to the plane of the loop—that oscillates with 0.010 T amplitude at a frequency of 1000 Hz. What is the wire's initial temperature rise, in °C/min?

MODEL The changing magnetic flux through the loop will induce a current that, because of the wire's resistance, will heat the wire. Eventually, when the wire gets hot, heat loss through radiation and/or convection will limit the temperature rise, but initially we can consider the temperature change due only to the heating by the current. Assume that the wire's diameter is much less than the 4.0 cm width of the loop.

FIGURE 30.48 A copper wire being heated by induction.

VISUALIZE FIGURE 30.48 shows the copper loop in the magnetic field. The wire's cross-section area A is unknown, but our assumption of a thin wire means that the loop has a well-defined area L^2.

Continued

Values of copper's resistivity, density, and specific heat were taken from tables inside the back cover of the book. We've used subscripts to distinguish between mass density ρ_{mass} and resistivity ρ_{elec}, a potentially confusing duplication of symbols.

SOLVE Power dissipation by a current, $P = I^2R$, heats the wire. As long as heat losses are negligible, we can use the heating rate and the wire's specific heat c to calculate the rate of temperature change. Our first task is to find the induced current. According to Faraday's law,

$$I = \frac{\mathcal{E}}{R} = -\frac{1}{R}\frac{d\Phi_m}{dt} = -\frac{L^2}{R}\frac{dB}{dt}$$

where R is the loop's resistance and $\Phi_m = L^2B$ is the magnetic flux through a loop of area L^2. The oscillating magnetic field can be written $B = B_0\cos\omega t$, with $B_0 = 0.010$ T and $\omega = 2\pi \times 1000$ Hz = 6280 rad/s. Thus

$$\frac{dB}{dt} = -\omega B_0\sin\omega t$$

from which we find that the induced current oscillates as

$$I = \frac{\omega B_0 L^2}{R}\sin\omega t$$

As the current oscillates, the power dissipation in the wire is

$$P = I^2R = \frac{\omega^2 B_0^2 L^4}{R}\sin^2\omega t$$

The power dissipation also oscillates, but very rapidly in comparison to a temperature rise that we expect to occur over seconds or minutes. Consequently, we are justified in replacing the oscillating P with its *average* value P_{avg}. Recall that the time average of the function $\sin^2\omega t$ is $\frac{1}{2}$, a result that can be proven by integration or justified by noticing that a graph of $\sin^2\omega t$ oscillates symmetrically between 0 and 1. Thus the average power dissipation in the wire is

$$P_{avg} = \frac{\omega^2 B_0^2 L^4}{2R}$$

Recall that power is the *rate* of energy transfer. In this case, the power dissipated in the wire is the wire's heating rate: $dQ/dt = P_{avg}$, where here Q is heat, not charge. Using $Q = mc\,\Delta T$, from thermodynamics, we can write

$$\frac{dQ}{dt} = mc\frac{dT}{dt} = P_{avg} = \frac{\omega^2 B_0^2 L^4}{2R}$$

To complete the calculation, we need the mass and resistance of the wire. The wire's total length is $4L$, and its cross-section area is A. Thus

$$m = \rho_{mass}V = 4\rho_{mass}LA$$

$$R = \frac{\rho_{elec}(4L)}{A} = \frac{4\rho_{elec}L}{A}$$

Substituting these into the heating equation, we have

$$4\rho_{mass}LAc\frac{dT}{dt} = \frac{\omega^2 B_0^2 L^3 A}{8\rho_{elec}}$$

Interestingly, the wire's cross-section area cancels. The wire's temperature initially increases at the rate

$$\frac{dT}{dt} = \frac{\omega^2 B_0^2 L^2}{32\rho_{elec}\rho_{mass}c}$$

All the terms on the right-hand side are known. Evaluating, we find

$$\frac{dT}{dt} = 3.3 \text{ K/s} = 200°\text{C/min}$$

ASSESS This is a rapid but realistic temperature rise for a small object, although the rate of increase will slow as the object begins losing heat to the environment through radiation and/or convection. Induction heating can increase an object's temperature by several hundred degrees in a few minutes.

SUMMARY

The goal of Chapter 30 has been to learn what electromagnetic induction is and how it is used.

GENERAL PRINCIPLES

Lenz's Law

There is an induced current in a closed conducting loop if and only if the magnetic flux through the loop is changing. The direction of the induced current is such that the induced magnetic field opposes the *change* in the flux.

Faraday's Law

An emf is induced around a closed loop if the magnetic flux through the loop changes.

Magnitude: $\mathcal{E} = \left| \dfrac{d\Phi_m}{dt} \right|$

Direction: As given by Lenz's law

Decreasing $\vec{B}$

Using Electromagnetic Induction

MODEL Make simplifying assumptions.

VISUALIZE Use Lenz's law to determine the direction of the **induced current**.

SOLVE The **induced emf** is

$$\mathcal{E} = \left| \dfrac{d\Phi_m}{dt} \right|$$

Multiply by N for an N-turn coil.
The size of the induced current is $I = \mathcal{E}/R$.

ASSESS Is the result reasonable?

IMPORTANT CONCEPTS

Magnetic flux

Magnetic flux measures the amount of magnetic field passing through a surface.

$$\Phi_m = \vec{A} \cdot \vec{B} = AB\cos\theta$$

Loop of area A

Three ways to change the flux

1. A loop moves into or out of a magnetic field.

2. The loop changes area or rotates.
3. The magnetic field through the loop increases or decreases.

Two ways to create an induced current

1. A **motional emf** is due to magnetic forces on moving charge carriers.

$$\mathcal{E} = vlB$$

2. An **induced electric field** is due to a changing magnetic field.

$$\oint \vec{E} \cdot d\vec{s} = -\dfrac{d\Phi_m}{dt}$$

Increasing $\vec{B}$

APPLICATIONS

Inductors

Solenoid inductance $L_{\text{solenoid}} = \dfrac{\mu_0 N^2 A}{l}$

Potential difference $\Delta V_L = -L\dfrac{dI}{dt}$

Energy stored $U_L = \frac{1}{2}LI^2$

Magnetic energy density $u_B = \dfrac{1}{2\mu_0}B^2$

LC circuit

Oscillates at $\omega = \sqrt{\dfrac{1}{LC}}$

LR circuit

Exponential change with $\tau = \dfrac{L}{R}$

TERMS AND NOTATION

electromagnetic induction	area vector, $\vec{A}$	induced magnetic field	henry, H
induced current	Lenz's law	electromagnetic wave	LC circuit
motional emf	induced emf, $\mathcal{E}$	primary coil	LR circuit
generator	Faraday's law	secondary coil	time constant, τ
eddy current	induced electric field	transformer	
magnetic flux, Φ_m	Coulomb electric field	inductor	
weber, Wb	non-Coulomb electric field	inductance, L	

CONCEPTUAL QUESTIONS

1. What is the direction of the induced current in **FIGURE Q30.1**?

FIGURE Q30.1

FIGURE Q30.2

2. You want to insert a loop of copper wire between the two permanent magnets in **FIGURE Q30.2**. Is there an attractive magnetic force that tends to *pull* the loop in, like a magnet pulls on a paper clip? Or do you need to *push* the loop in against a repulsive force? Explain.

3. A vertical, rectangular loop of copper wire is half in and half out of the horizontal magnetic field in **FIGURE Q30.3**. (The field is zero beneath the dashed line.) The loop is released and starts to fall. Is there a net magnetic force on the loop? If so, in which direction? Explain.

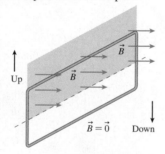

FIGURE Q30.3

4. Does the loop of wire in **FIGURE Q30.4** have a clockwise current, a counterclockwise current, or no current under the following circumstances? Explain.
   a. The magnetic field points out of the page and is increasing.
   b. The magnetic field points out of the page and is constant.
   c. The magnetic field points out of the page and is decreasing.

FIGURE Q30.4

FIGURE Q30.5

5. The two loops of wire in **FIGURE Q30.5** are stacked one above the other. Does the upper loop have a clockwise current, a counterclockwise current, or no current at the following times? Explain.
   a. Before the switch is closed.
   b. Immediately after the switch is closed.
   c. Long after the switch is closed.
   d. Immediately after the switch is reopened.

6. **FIGURE Q30.6** shows a bar magnet being pushed toward a conducting loop from below, along the axis of the loop.
   a. What is the current direction in the loop? Explain.
   b. Is there a magnetic force on the loop? If so, in which direction? Explain.
      Hint: A current loop is a magnetic dipole.
   c. Is there a force on the magnet? If so, in which direction?

FIGURE Q30.6

FIGURE Q30.7

7. A bar magnet is pushed toward a loop of wire as shown in **FIGURE Q30.7**. Is there a current in the loop? If so, in which direction? If not, why not?

8. **FIGURE Q30.8** shows a bar magnet, a coil of wire, and a current meter. Is the current through the meter right to left, left to right, or zero for the following circumstances? Explain.
   a. The magnet is inserted into the coil.
   b. The magnet is held at rest inside the coil.
   c. The magnet is withdrawn from the left side of the coil.

FIGURE Q30.8

9. Is the magnetic field strength in **FIGURE Q30.9** increasing, decreasing, or steady? Explain.

FIGURE Q30.9

10. An inductor with a 2.0 A current stores energy. At what current will the stored energy be twice as large?

11. a. Can you tell which of the inductors in **FIGURE Q30.11** has the larger current through it? If so, which one? Explain.
    b. Can you tell through which inductor the current is changing more rapidly? If so, which one? Explain.
    c. If the current enters the inductor from the bottom, can you tell if the current is increasing, decreasing, or staying the same? If so, which? Explain.

FIGURE Q30.11

12. An LC circuit oscillates at a frequency of 2000 Hz. What will the frequency be if the inductance is quadrupled?

13. Rank in order, from largest to smallest, the three time constants τ_a to τ_c for the three circuits in **FIGURE Q30.13**. Explain.

FIGURE Q30.13

14. For the circuit of **FIGURE Q30.14**:
    a. What is the battery current immediately after the switch closes? Explain.
    b. What is the battery current after the switch has been closed a long time? Explain.

FIGURE Q30.14

EXERCISES AND PROBLEMS

Problems labeled ▨ integrate material from earlier chapters.

Exercises

Section 30.2 Motional emf

1. | The earth's magnetic field strength is 5.0×10^{-5} T. How fast would you have to drive your car to create a 1.0 V motional emf along your 1.0-m-tall radio antenna? Assume that the motion of the antenna is perpendicular to $\vec{B}$.

2. | A potential difference of 0.050 V is developed across the 10-cm-long wire of **FIGURE EX30.2** as it moves through a magnetic field perpendicular to the page. What are the strength and direction (in or out) of the magnetic field?

FIGURE EX30.2

3. | A 10-cm-long wire is pulled along a U-shaped conducting rail in a perpendicular magnetic field. The total resistance of the wire and rail is 0.20 Ω. Pulling the wire at a steady speed of 4.0 m/s causes 4.0 W of power to be dissipated in the circuit.
    a. How big is the pulling force?
    b. What is the strength of the magnetic field?

Section 30.3 Magnetic Flux

4. ‖ What is the magnetic flux through the loop shown in **FIGURE EX30.4**?

FIGURE EX30.4

FIGURE EX30.5

5. ‖ **FIGURE EX30.5** shows a 10 cm × 10 cm square bent at a 90° angle. A uniform 0.050 T magnetic field points downward at a 45° angle. What is the magnetic flux through the loop?

6. ‖ An equilateral triangle 8.0 cm on a side is in a 5.0 mT uniform magnetic field. The magnetic flux through the triangle is $6.0\,\mu$Wb. What is the angle between the magnetic field and an axis perpendicular to the plane of the triangle?

7. | What is the magnetic flux through the loop shown in **FIGURE EX30.7**?

FIGURE EX30.7

8. ‖ **FIGURE EX30.8** shows a 2.0-cm-diameter solenoid passing through the center of a 6.0-cm-diameter loop. The magnetic field inside the solenoid is 0.20 T. What is the magnetic flux through the loop when it is perpendicular to the solenoid and when it is tilted at a 60° angle?

FIGURE EX30.8

Section 30.4 Lenz's Law

9. | There is a cw induced current in the conducting loop shown in **FIGURE EX30.9**. Is the magnetic field inside the loop increasing in strength, decreasing in strength, or steady?

FIGURE EX30.9

10. | A solenoid is wound as shown in **FIGURE EX30.10**.
   a. Is there an induced current as magnet 1 is moved away from the solenoid? If so, what is the current direction through resistor R?
   b. Is there an induced current as magnet 2 is moved away from the solenoid? If so, what is the current direction through resistor R?

FIGURE EX30.10

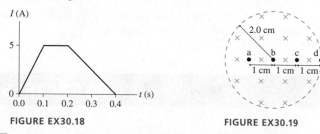

FIGURE EX30.11

11. | The metal equilateral triangle in **FIGURE EX30.11**, 20 cm on each side, is halfway into a 0.10 T magnetic field.
   a. What is the magnetic flux through the triangle?
   b. If the magnetic field strength decreases, what is the direction of the induced current in the triangle?

12. || The current in the solenoid of **FIGURE EX30.12** is increasing. The solenoid is surrounded by a conducting loop. Is there a current in the loop? If so, is the loop current cw or ccw?

FIGURE EX30.12

Section 30.5 Faraday's Law

13. | The loop in **FIGURE EX30.13** is being pushed into the 0.20 T magnetic field at 50 m/s. The resistance of the loop is 0.10 Ω. What are the direction and the magnitude of the current in the loop?

FIGURE EX30.13

14. | **FIGURE EX30.14** shows a 10-cm-diameter loop in three different magnetic fields. The loop's resistance is 0.20 Ω. For each, what are the size and direction of the induced current?

(a) B increasing at 0.50 T/s **(b)** B decreasing at 0.50 T/s **(c)** B decreasing at 0.50 T/s

FIGURE EX30.14

15. | The resistance of the loop in **FIGURE EX30.15** is 0.20 Ω. Is the magnetic field strength increasing or decreasing? At what rate (T/s)?

FIGURE EX30.15

16. || A 1000-turn coil of wire 1.0 cm in diameter is in a magnetic field that increases from 0.10 T to 0.30 T in 10 ms. The axis of the coil is parallel to the field. What is the emf of the coil?

17. || A 5.0-cm-diameter coil has 20 turns and a resistance of
CALC 0.50 Ω. A magnetic field perpendicular to the coil is $B = 0.020t + 0.010t^2$, where B is in tesla and t is in seconds.
   a. Find an expression for the induced current $I(t)$ as a function of time.
   b. Evaluate I at $t = 5$ s and $t = 10$ s.

Section 30.6 Induced Fields

18. || **FIGURE EX30.18** shows the current as a function of time through a 20-cm-long, 4.0-cm-diameter solenoid with 400 turns. Draw a graph of the induced electric field strength as a function of time at a point 1.0 cm from the axis of the solenoid.

FIGURE EX30.18

FIGURE EX30.19

19. || The magnetic field in **FIGURE EX30.19** is decreasing at the rate 0.10 T/s. What is the acceleration (magnitude and direction) of a proton initially at rest at points a to d?

20. || The magnetic field inside a 5.0-cm-diameter solenoid is 2.0 T and decreasing at 4.0 T/s. What is the electric field strength inside the solenoid at a point (a) on the axis and (b) 2.0 cm from the axis?

21. || Scientists studying an anomalous magnetic field find that it is inducing a circular electric field in a plane perpendicular to the magnetic field. The electric field strength 1.5 m from the center of the circle is 4.0 mV/m. At what rate is the magnetic field changing?

Section 30.7 Induced Currents: Three Applications

22. | Electricity is distributed from electrical substations to neighborhoods at 15,000 V. This is a 60 Hz oscillating (AC) voltage. Neighborhood transformers, seen on utility poles, step this voltage down to the 120 V that is delivered to your house.
   a. How many turns does the primary coil on the transformer have if the secondary coil has 100 turns?
   b. No energy is lost in an ideal transformer, so the output power P_{out} from the secondary coil equals the input power P_{in} to the primary coil. Suppose a neighborhood transformer delivers 250 A at 120 V. What is the current in the 15,000 V line from the substation?

23. | The charger for your electronic devices is a transformer. Suppose a 60 Hz outlet voltage of 120 V needs to be reduced to a device voltage of 3.0 V. The side of the transformer attached to the electronic device has 60 turns of wire. How many turns are on the side that plugs into the outlet?

Section 30.8 Inductors

24. | The maximum allowable potential difference across a 200 mH inductor is 400 V. You need to raise the current through the inductor from 1.0 A to 3.0 A. What is the minimum time you should allow for changing the current?

25. | What is the potential difference across a 10 mH inductor if the current through the inductor drops from 150 mA to 50 mA in 10 μs? What is the direction of this potential difference? That is, does the potential increase or decrease along the direction of the current?

26. ‖ A 100 mH inductor whose windings have a resistance of 4.0 Ω is connected across a 12 V battery having an internal resistance of 2.0 Ω. How much energy is stored in the inductor?

27. ‖ How much energy is stored in a 3.0-cm-diameter, 12-cm-long solenoid that has 200 turns of wire and carries a current of 0.80 A?

28. | MRI (magnetic resonance imaging) is a medical technique that **BIO** produces detailed "pictures" of the interior of the body. The patient is placed into a solenoid that is 40 cm in diameter and 1.0 m long. A 100 A current creates a 5.0 T magnetic field inside the solenoid. To carry such a large current, the solenoid wires are cooled with liquid helium until they become superconducting (no electric resistance).
   a. How much magnetic energy is stored in the solenoid? Assume that the magnetic field is uniform within the solenoid and quickly drops to zero outside the solenoid.
   b. How many turns of wire does the solenoid have?

Section 30.9 LC Circuits

29. ‖ An FM radio station broadcasts at a frequency of 100 MHz. What inductance should be paired with a 10 pF capacitor to build a receiver circuit for this station?

30. ‖ A 2.0 mH inductor is connected in parallel with a variable capacitor. The capacitor can be varied from 100 pF to 200 pF. What is the range of oscillation frequencies for this circuit?

31. ‖ An MRI machine needs to detect signals that oscillate at very **BIO** high frequencies. It does so with an LC circuit containing a 15 mH coil. To what value should the capacitance be set to detect a 450 MHz signal?

32. ‖ An LC circuit has a 10 mH inductor. The current has its maximum value of 0.60 A at $t = 0$ s. A short time later the capacitor reaches its maximum potential difference of 60 V. What is the value of the capacitance?

33. ‖ The switch in **FIGURE EX30.33** has been in position 1 for a long time. It is changed to position 2 at $t = 0$ s.
   a. What is the maximum current through the inductor?
   b. What is the first time at which the current is maximum?

FIGURE EX30.33

Section 30.10 LR Circuits

34. | What value of resistor R gives the circuit in **FIGURE EX30.34** a time constant of 25 μs?

FIGURE EX30.34   **FIGURE EX30.35**

35. ‖ At $t = 0$ s, the current in the circuit in **FIGURE EX30.35** is I_0. At what time is the current $\frac{1}{2}I_0$?

36. | The switch in **FIGURE EX30.36** has been open for a long time. It is closed at $t = 0$ s.
   a. What is the current through the battery immediately after the switch is closed?
   b. What is the current through the battery after the switch has been closed a long time?

FIGURE EX30.36

Problems

37. ‖ A 20 cm × 20 cm square loop has a resistance of 0.10 Ω. A **CALC** magnetic field perpendicular to the loop is $B = 4t - 2t^2$, where B is in tesla and t is in seconds. What is the current in the loop at $t = 0.0$ s, $t = 1.0$ s, and $t = 2.0$ s?

38. ‖ A 100-turn, 2.0-cm-diameter coil is at rest with its axis vertical. A uniform magnetic field 60° away from vertical increases from 0.50 T to 1.50 T in 0.60 s. What is the induced emf in the coil?

39. ‖ A 100-turn, 8.0-cm-diameter coil is made of 0.50-mm-diameter copper wire. A magnetic field is parallel to the axis of the coil. At what rate must B increase to induce a 2.0 A current in the coil?

40. ‖ A circular loop made from a flexible, conducting wire is **CALC** shrinking. Its radius as a function of time is $r = r_0 e^{-\beta t}$. The loop is perpendicular to a steady, uniform magnetic field B. Find an expression for the induced emf in the loop at time t.

41. ‖ A 10 cm × 10 cm square loop lies in the xy-plane. The **CALC** magnetic field in this region of space is $\vec{B} = (0.30t\,\hat{\imath} + 0.50t^2\,\hat{k})$ T, where t is in s. What is the emf induced in the loop at (a) $t = 0.5$ s and (b) $t = 1.0$ s?

42. ‖ A spherical balloon with a volume of 2.5 L is in a 45 mT **CALC** uniform, vertical magnetic field. A horizontal elastic but conducting wire with 2.5 Ω resistance circles the balloon at its equator. Suddenly the balloon starts expanding at 0.75 L/s. What is the current in the wire 2.0 s later?

43. ‖‖ A 3.0-cm-diameter, 10-turn coil of wire, located at $z = 0$ **CALC** in the xy-plane, carries a current of 2.5 A. A 2.0-mm-diameter conducting loop with $2.0 × 10^{-4}\,\Omega$ resistance is also in the xy-plane at the center of the coil. At $t = 0$ s, the loop begins to move along the z-axis with a constant speed of 75 m/s. What is the induced current in the conducting loop at $t = 200\,\mu$s? The diameter of the conducting loop is much smaller than that of the coil, so you can assume that the magnetic field through the loop is everywhere the on-axis field of the coil.

44. ‖‖ A 20 cm × 20 cm square loop of wire lies in the xy-plane **CALC** with its bottom edge on the x-axis. The resistance of the loop is 0.50 Ω. A magnetic field parallel to the z-axis is given by $B = 0.80y^2 t$, where B is in tesla, y in meters, and t in seconds. What is the size of the induced current in the loop at $t = 0.50$ s?

45. ‖‖ A 2.0 cm × 2.0 cm square loop of wire with resistance 0.010 Ω has one edge parallel to a long straight wire. The near edge of the loop is 1.0 cm from the wire. The current in the wire is increasing at the rate of 100 A/s. What is the current in the loop?

46. ‖ **FIGURE P30.46** shows a 4.0-cm-diameter loop with resistance 0.10 Ω around a 2.0-cm-diameter solenoid. The solenoid is 10 cm long, has 100 turns, and carries the current shown in the graph. A positive current is cw when seen from the left. Find the current in the loop at (a) $t = 0.5$ s, (b) $t = 1.5$ s, and (c) $t = 2.5$ s.

FIGURE P30.46

47. ‖‖ **FIGURE P30.47** shows a 1.0-cm-diameter loop with $R = 0.50$ Ω inside a 2.0-cm-diameter solenoid. The solenoid is 8.0 cm long, has 120 turns, and carries the current shown in the graph. A positive current is cw when seen from the left. Determine the current in the loop at $t = 0.010$ s.

FIGURE P30.47

48. ‖ **FIGURE P30.48** shows two 20-turn coils tightly wrapped on the same 2.0-cm-diameter cylinder with 1.0-mm-diameter wire. The current through coil 1 is shown in the graph. Determine the current in coil 2 at (a) $t = 0.05$ s and (b) $t = 0.25$ s. A positive current is into the page at the top of a loop. Assume that the magnetic field of coil 1 passes entirely through coil 2.

FIGURE P30.48

49. ‖ An electric generator has an 18-cm-diameter, 120-turn coil that
CALC rotates at 60 Hz in a uniform magnetic field that is perpendicular to the rotation axis. What magnetic field strength is needed to generate a peak voltage of 170 V?

50. ‖ A 40-turn, 4.0-cm-diameter coil with $R = 0.40$ Ω surrounds
CALC a 3.0-cm-diameter solenoid. The solenoid is 20 cm long and has 200 turns. The 60 Hz current through the solenoid is $I = I_0 \sin(2\pi ft)$. What is I_0 if the maximum induced current in the coil is 0.20 A?

51. ‖‖ A small, 2.0-mm-diameter circular loop with $R = 0.020$ Ω is at the center of a large 100-mm-diameter circular loop. Both loops lie in the same plane. The current in the outer loop changes from $+1.0$ A to -1.0 A in 0.10 s. What is the induced current in the inner loop?

52. ‖ A rectangular metal loop with 0.050 Ω resistance is placed next to one wire of the *RC* circuit shown in **FIGURE P30.52**. The capacitor is charged to 20 V with the polarity shown, then the switch is closed at $t = 0$ s.
    a. What is the direction of current in the loop for $t > 0$ s?
    b. What is the current in the loop at $t = 5.0$ μs? Assume that only the circuit wire next to the loop is close enough to produce a significant magnetic field.

FIGURE P30.52   **FIGURE P30.53**

53. ‖ The square loop shown in **FIGURE P30.53** moves into a 0.80 T
CALC magnetic field at a constant speed of 10 m/s. The loop has a resistance of 0.10 Ω, and it enters the field at $t = 0$ s.
    a. Find the induced current in the loop as a function of time. Give your answer as a graph of I versus t from $t = 0$ s to $t = 0.020$ s.
    b. What is the maximum current? What is the position of the loop when the current is maximum?

54. ‖ The L-shaped conductor in **FIGURE P30.54** moves at 10 m/s
CALC across and touches a stationary L-shaped conductor in a 0.10 T magnetic field. The two vertices overlap, so that the enclosed area is zero, at $t = 0$ s. The conductor has a resistance of 0.010 ohms *per meter*.
    a. What is the direction of the induced current?
    b. Find expressions for the induced emf and the induced current as functions of time.
    c. Evaluate $\mathcal{E}$ and I at $t = 0.10$ s.

FIGURE P30.54

55. ‖ A 20-cm-long, zero-resistance slide wire moves outward, on zero-resistance rails, at a steady speed of 10 m/s in a 0.10 T magnetic field. (See Figure 30.26.) On the opposite side, a 1.0 Ω carbon resistor completes the circuit by connecting the two rails. The mass of the resistor is 50 mg.
    a. What is the induced current in the circuit?
    b. How much force is needed to pull the wire at this speed?
    c. If the wire is pulled for 10 s, what is the temperature increase of the carbon? The specific heat of carbon is 710 J/kg K.

56. ‖ Your camping buddy has an idea for a light to go inside your
CALC tent. He happens to have a powerful (and heavy!) horseshoe
magnet that he bought at a surplus store. This magnet creates
a 0.20 T field between two pole tips 10 cm apart. His idea is
to build the hand-cranked generator shown in FIGURE P30.56.
He thinks you can make enough current to fully light a 1.0 Ω
lightbulb rated at 4.0 W. That's not super bright, but it should be
plenty of light for routine activities in the tent.
  a. Find an expression for the induced current as a function of
     time if you turn the crank at frequency f. Assume that the
     semicircle is at its highest point at $t = 0$ s.
  b. With what frequency will you have to turn the crank for the
     maximum current to fully light the bulb? Is this feasible?

FIGURE P30.56   FIGURE P30.57

57. ‖ The 10-cm-wide, zero-resistance slide wire shown in FIGURE
P30.57 is pushed toward the 2.0 Ω resistor at a steady speed of
0.50 m/s. The magnetic field strength is 0.50 T.
  a. How big is the pushing force?
  b. How much power does the pushing force supply to the wire?
  c. What are the direction and magnitude of the induced current?
  d. How much power is dissipated in the resistor?

58. ‖ You've decided to make the magnetic projectile launcher
shown in FIGURE P30.58 for your science project. An aluminum
bar of length l slides along metal rails through a magnetic field B.
The switch closes at $t = 0$ s, while the bar is at rest, and a battery
of emf $\mathcal{E}_{bat}$ starts a current flowing around the loop. The battery
has internal resistance r. The resistances of the rails and the bar
are effectively zero.
  a. Show that the bar reaches a terminal speed v_{term}, and find an
     expression for v_{term}.
  b. Evaluate v_{term} for $\mathcal{E}_{bat} = 1.0$ V, $r = 0.10$ Ω, $l = 6.0$ cm, and
     $B = 0.50$ T.

FIGURE P30.58

59. ‖ FIGURE P30.59 shows a U-shaped conducting rail that is
oriented vertically in a horizontal magnetic field. The rail has
no electric resistance and does not move. A slide wire with mass
m and resistance R can slide up and down without friction while
maintaining electrical contact with the rail. The slide wire is
released from rest.
  a. Show that the slide wire reaches a terminal speed v_{term}, and
     find an expression for v_{term}.
  b. Determine the value of v_{term} if $l = 20$ cm, $m = 10$ g, $R =
     0.10$ Ω, and $B = 0.50$ T.

FIGURE P30.59   FIGURE P30.60

60. ‖‖‖ Experiments to study vision often need to track the movements
BIO of a subject's eye. One way of doing so is to have the subject sit in
a magnetic field while wearing special contact lenses with a coil
of very fine wire circling the edge. A current is induced in the
coil each time the subject rotates his eye. Consider the experiment
of FIGURE P30.60 in which a 20-turn, 6.0-mm-diameter coil of
wire circles the subject's cornea while a 1.0 T magnetic field is
directed as shown. The subject begins by looking straight ahead.
What emf is induced in the coil if the subject shifts his gaze by
5° in 0.20 s?

61. ‖ A 10-turn coil of wire having a diameter of 1.0 cm and
CALC a resistance of 0.20 Ω is in a 1.0 mT magnetic field, with the
coil oriented for maximum flux. The coil is connected to an
uncharged 1.0 μF capacitor rather than to a current meter. The
coil is quickly pulled out of the magnetic field. Afterward, what
is the voltage across the capacitor?
  Hint: Use $I = dq/dt$ to relate the *net* change of flux to the amount
  of charge that flows to the capacitor.

62. ‖‖‖ The magnetic field at one place on the earth's surface is
55 μT in strength and tilted 60° down from horizontal. A 200-
turn coil having a diameter of 4.0 cm and a resistance of 2.0 Ω
is connected to a 1.0 μF capacitor rather than to a current
meter. The coil is held in a horizontal plane and the capacitor is
discharged. Then the coil is quickly rotated 180° so that the side
that had been facing up is now facing down. Afterward, what is
the voltage across the capacitor? See the Hint in Problem 61.

63. ‖ Equation 30.26 is an expression for the induced electric field
CALC inside a solenoid $(r < R)$. Find an expression for the induced
electric field outside a solenoid $(r > R)$ in which the magnetic
field is changing at the rate dB/dt.

64. ‖ A solenoid inductor has an emf of 0.20 V when the current
through it changes at the rate 10.0 A/s. A steady current of 0.10
A produces a flux of 5.0 μWb per turn. How many turns does the
inductor have?

65. ‖ One possible concern with MRI (see Exercise 28) is turning the
BIO magnetic field on or off too quickly. Bodily fluids are conductors,
and a changing magnetic field could cause electric currents to flow
through the patient. Suppose a typical patient has a maximum cross-
section area of 0.060 m². What is the smallest time interval in which
a 5.0 T magnetic field can be turned on or off if the induced emf
around the patient's body must be kept to less than 0.10 V?

66. ‖ FIGURE P30.66 shows the current through a 10 mH inductor.
Draw a graph showing the potential difference ΔV_L across the
inductor for these 6 ms.

FIGURE P30.66

67. ‖ **FIGURE P30.67** shows the potential difference across a 50 mH inductor. The current through the inductor at $t = 0$ s is 0.20 A. Draw a graph showing the current through the inductor from $t = 0$ s to $t = 40$ ms.

FIGURE P30.67

68. ‖ A 3.6 mH inductor with negligible resistance has a 1.0 A current through it. The current starts to increase at $t = 0$ s, creating a constant 5.0 mV voltage across the inductor. How much charge passes through the inductor between $t = 0$ s and $t = 5.0$ s?

69. ‖ The current through inductance L is given by $I = I_0 \sin \omega t$.
CALC a. Find an expression for the potential difference ΔV_L across the inductor.
b. The maximum voltage across the inductor is 0.20 V when $L = 50\ \mu\text{H}$ and $f = 500$ kHz. What is I_0?

70. ‖ The current through inductance L is given by $I = I_0 e^{-t/\tau}$.
CALC a. Find an expression for the potential difference ΔV_L across the inductor.
b. Evaluate ΔV_L at $t = 0$, 1.0, and 3.0 ms if $L = 20$ mH, $I_0 = 50$ mA, and $\tau = 1.0$ ms.

71. ‖ An LC circuit is built with a 20 mH inductor and an 8.0 pF capacitor. The capacitor voltage has its maximum value of 25 V at $t = 0$ s.
a. How long is it until the capacitor is first fully discharged?
b. What is the inductor current at that time?

72. ‖ An electric oscillator is made with a 0.10 μF capacitor and a 1.0 mH inductor. The capacitor is initially charged to 5.0 V. What is the maximum current through the inductor as the circuit oscillates?

73. ‖ For your final exam in electronics, you're asked to build an LC circuit that oscillates at 10 kHz. In addition, the maximum current must be 0.10 A and the maximum energy stored in the capacitor must be 1.0×10^{-5} J. What values of inductance and capacitance must you use?

74. ‖ The inductor in **FIGURE P30.74** is a 9.0-cm-long, 2.0-cm-diameter solenoid wrapped with 300 turns. What is the current in the circuit 10 μs after the switch is moved from a to b?

FIGURE P30.74

75. ‖ The 300 μF capacitor in **FIGURE P30.75** is initially charged to 100 V, the 1200 μF capacitor is uncharged, and the switches are both open.
a. What is the maximum voltage to which you can charge the 1200 μF capacitor by the proper closing and opening of the two switches?
b. How would you do it? Describe the sequence in which you would close and open switches and the times at which you would do so. The first switch is closed at $t = 0$ s.

FIGURE P30.75

76. ‖ The switch in **FIGURE P30.76** has been open for a long time. It is closed at $t = 0$ s. What is the current through the 20 Ω resistor
a. Immediately after the switch is closed?
b. After the switch has been closed a long time?
c. Immediately after the switch is reopened?

FIGURE P30.76

77. ‖ The switch in **FIGURE P30.77** has been open for a long time. It is closed at $t = 0$ s.
a. After the switch has been closed for a long time, what is the current in the circuit? Call this current I_0.
b. Find an expression for the current I as a function of time. Write your expression in terms of I_0, R, and L.
c. Sketch a current-versus-time graph from $t = 0$ s until the current is no longer changing.

FIGURE P30.77   **FIGURE P30.78**

78. ‖ To determine the inductance of an unmarked inductor, you set up the circuit shown in **FIGURE P30.78**. After moving the switch from a to b at $t = 0$ s, you monitor the resistor voltage with an oscilloscope. Your data are shown in the table:

Time (μs)	Voltage (V)
0	9.0
10	6.7
20	4.6
30	3.2
40	2.5

Use an appropriate graph of the data to determine the inductance.

79. ‖‖ 5.0 μs after the switch of **FIGURE P30.79** is moved from a to b, the magnetic energy stored in the inductor has decreased by half. What is the value of the inductance L?

FIGURE P30.79

Challenge Problems

80. ‖‖ The rectangular loop in **FIGURE CP30.80** has 0.020 Ω resis-
CALC tance. What is the induced current in the loop at this instant?

FIGURE CP30.80

81. ||| In recent years it has been possible to buy a 1.0 F capacitor. This is an enormously large amount of capacitance. Suppose you want to build a 1.0 Hz oscillator with a 1.0 F capacitor. You have a spool of 0.25-mm-diameter wire and a 4.0-cm-diameter plastic cylinder. How long must your inductor be if you wrap it with 2 layers of closely spaced turns?

82. ||| The metal wire in FIGURE CP30.82 moves with speed v parallel
CALC to a straight wire that is carrying current I. The distance between the two wires is d. Find an expression for the potential difference between the two ends of the moving wire.

FIGURE CP30.82

83. ||| Let's look at the details of eddy-current braking. A square
CALC loop, length l on each side, is shot with velocity v_0 into a uniform magnetic field B. The field is perpendicular to the plane of the loop. The loop has mass m and resistance R, and it enters the field at $t = 0$ s. Assume that the loop is moving to the right along the x-axis and that the field begins at $x = 0$ m.
   a. Find an expression for the loop's velocity as a function of time as it enters the magnetic field. You can ignore gravity, and you can assume that the back edge of the loop has not entered the field.
   b. Calculate and draw a graph of v over the interval $0 \text{ s} \leq t \leq$ 0.04 s for the case that $v_0 = 10$ m/s, $l = 10$ cm, $m = 1.0$ g, $R = 0.0010 \ \Omega$, and $B = 0.10$ T. The back edge of the loop does not reach the field during this time interval.

84. ||| An 8.0 cm × 8.0 cm square loop is halfway into a magnetic
CALC field perpendicular to the plane of the loop. The loop's mass is 10 g and its resistance is 0.010 Ω. A switch is closed at $t = 0$ s, causing the magnetic field to increase from 0 to 1.0 T in 0.010 s.
   a. What is the induced current in the square loop?
   b. With what speed is the loop "kicked" away from the magnetic field?
   Hint: What is the impulse on the loop?

85. ||| A 2.0-cm-diameter solenoid is wrapped with 1000 turns per
CALC meter. 0.50 cm from the axis, the strength of an induced electric field is 5.0×10^{-4} V/m. What is the rate dI/dt with which the current through the solenoid is changing?

86. ||| High-frequency signals are often transmitted along a *coaxial*
CALC *cable*, such as the one shown in FIGURE CP30.86. For example, the cable TV hookup coming into your home is a coaxial cable. The signal is carried on a wire of radius r_1 while the outer conductor of radius r_2 is grounded. A soft, flexible insulating material fills the space between them, and an insulating plastic coating goes around the outside.
   a. Find an expression for the inductance per meter of a coaxial cable. To do so, consider the flux through a rectangle of length l that spans the gap between the inner and outer conductors.
   b. Evaluate the inductance per meter of a cable having $r_1 = 0.50$ mm and $r_2 = 3.0$ mm.

Inner conductor radius r_1

Outer conductor

r_2

FIGURE CP30.86

31 Electromagnetic Fields and Waves

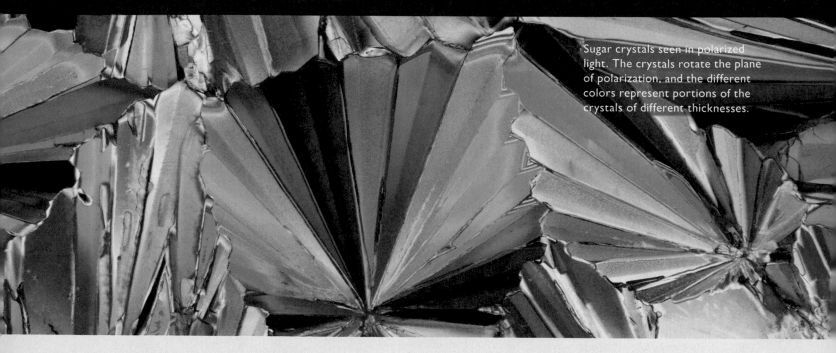

Sugar crystals seen in polarized light. The crystals rotate the plane of polarization, and the different colors represent portions of the crystals of different thicknesses.

IN THIS CHAPTER, you will study the properties of electromagnetic fields and waves.

How do fields transform?

Whether the field at a point is electric or magnetic depends, surprisingly, on your motion relative to the charges and currents. You'll learn how to transform the fields measured in one reference frame to a second reference frame moving relative to the first.

«LOOKING BACK Section 4.3 Relative motion

What is Maxwell's theory of electromagnetism?

Electricity and magnetism can be summarized in four equations for the fields, called Maxwell's equations, and one equation that tells us how charges respond to fields.

- Gauss's law: Charges create electric fields.
- Gauss's law for magnetism: There are no isolated magnetic poles.
- Faraday's law: Electric fields can also be created by changing magnetic fields.
- Ampère-Maxwell law: Magnetic fields can be created either by currents or by changing magnetic fields.

«LOOKING BACK Section 24.4 Gauss's law
«LOOKING BACK Section 29.6 Ampère's law
«LOOKING BACK Section 30.5 Faraday's law

What are electromagnetic waves?

Maxwell's equations predict the existence of self-sustaining oscillations of the electric and magnetic fields—electromagnetic waves—that travel through space without the presence of charges or currents.

- In a vacuum, all electromagnetic waves—from radio waves to x rays—travel with the *same speed* $v_{em} = 1/\sqrt{\epsilon_0 \mu_0} = c$, the speed of light.
- The fields $\vec{E}$ and $\vec{B}$ are perpendicular to each other and to the direction of travel.
- Electromagnetic waves are launched by an oscillating dipole, called an antenna.
- Electromagnetic waves transfer energy.
- Electromagnetic waves also transfer momentum and exert radiation pressure.

«LOOKING BACK Section 16.4 The wave equation

What is polarization?

An electromagnetic wave is polarized if the electric field always oscillates in the same plane—the plane of polarization. Polarizers both create and analyze polarized light. You will learn to calculate the intensity of light transmitted through a polarizer and will see that light is completely blocked by crossed polarizers. Polarization is used in many types of modern optical instrumentation.

31.1 *E* or *B*? It Depends on Your Perspective

Our story thus far has been that charges create electric fields and that moving charges, or currents, create magnetic fields. But consider **FIGURE 31.1a**, where Brittney, carrying charge q, runs past Alec with velocity $\vec{v}$. Alec sees a moving charge, and he knows that this charge creates a magnetic field. But from Brittney's perspective, the charge is at rest. Stationary charges don't create magnetic fields, so Brittney claims that the magnetic field is zero. Is there, or is there not, a magnetic field?

Or what about the situation in **FIGURE 31.1b**? Now Brittney is carrying the charge through a magnetic field that Alec has created. Alec sees a charge moving in a magnetic field, so he knows there's a force $\vec{F} = q\vec{v} \times \vec{B}$ on the charge. But for Brittney the charge is still at rest. Stationary charges don't experience magnetic forces, so Brittney claims that $\vec{F} = \vec{0}$.

Now, we may be a bit uncertain about magnetic fields, but surely there can be no disagreement over forces. After all, forces cause observable and measurable effects, so Alec and Brittney should be able to agree on whether or not the charge experiences a force. Further, if Brittney runs with constant velocity, then both Alec and Brittney are in *inertial reference frames*. You learned in Chapter 4 that these are the reference frames in which Newton's laws are valid, so we can't say that there's anything abnormal or unusual about Alec's and Brittney's observations.

This paradox has arisen because magnetic fields and forces depend on velocity, but we haven't looked at the issue of velocity *with respect to what* or velocity *as measured by whom*. The resolution of this paradox will lead us to the conclusion that $\vec{E}$ and $\vec{B}$ are not, as we've been assuming, separate and independent entities. They are closely intertwined.

FIGURE 31.1 Brittney carries a charge past Alec.

(a)

Charge q moves with velocity $\vec{v}$ relative to Alec.

(b)

Charge q moves through a magnetic field established by Alec.

Reference Frames

We introduced reference frames and relative motion in Chapter 4. To remind you, **FIGURE 31.2** shows two reference frames labeled A and B. You can think of these as the reference frames in which Alec and Brittney, respectively, are at rest. Frame B moves with velocity $\vec{v}_{BA}$ with respect to frame A. That is, an observer (Alec) at rest in A sees the origin of B (Brittney) go past with velocity $\vec{v}_{BA}$. Of course, Brittney would say that Alec has velocity $\vec{v}_{AB} = -\vec{v}_{BA}$ relative to her reference frame. We will stipulate that both reference frames are inertial reference frames, so $\vec{v}_{BA}$ is constant.

Figure 31.2 also shows a particle C. Experimenters in frame A measure the motion of the particle and find that its velocity *relative to frame A* is $\vec{v}_{CA}$. At the same instant, experimenters in B find that the particle's velocity *relative to frame B* is $\vec{v}_{CB}$. In Chapter 4, we found that $\vec{v}_{CA}$ and $\vec{v}_{CB}$ are related by

$$\vec{v}_{CA} = \vec{v}_{CB} + \vec{v}_{BA} \tag{31.1}$$

Equation 31.1, the *Galilean transformation of velocity*, tells us that the velocity of the particle relative to reference frame A is its velocity relative to frame B plus (vector addition!) the velocity of frame B relative to frame A.

Suppose the particle in Figure 31.2 is accelerating. How does its acceleration $\vec{a}_{CA}$, as measured in frame A, compare to the acceleration $\vec{a}_{CB}$ measured in frame B? We can answer this question by taking the time derivative of Equation 31.1:

$$\frac{d\vec{v}_{CA}}{dt} = \frac{d\vec{v}_{CB}}{dt} + \frac{d\vec{v}_{BA}}{dt}$$

The derivatives of $\vec{v}_{CA}$ and $\vec{v}_{CB}$ are the particle's accelerations $\vec{a}_{CA}$ and $\vec{a}_{CB}$ in frames A and B, respectively. But $\vec{v}_{BA}$ is a *constant* velocity, so $d\vec{v}_{BA}/dt = \vec{0}$. Thus the Galilean transformation of acceleration is simply

$$\vec{a}_{CA} = \vec{a}_{CB} \tag{31.2}$$

FIGURE 31.2 Reference frames A and B.

Reference frame B moves with velocity $\vec{v}_{BA}$ relative to frame A.

The velocity of this particle is measured to be $\vec{v}_{CA}$ in frame A and $\vec{v}_{CB}$ in frame B.

Brittney and Alec may measure different positions and velocities for a particle, but they *agree* on its acceleration. And if they agree on its acceleration, they must, by using Newton's second law, agree on the force acting on the particle. That is, **experimenters in all inertial reference frames agree about the force acting on a particle.**

The Transformation of Electric and Magnetic Fields

Imagine that Alec has measured the electric field $\vec{E}_A$ and the magnetic field $\vec{B}_A$ in reference frame A. Our investigations thus far give us no reason to think that Brittney's measurements of the fields will differ from Alec's. After all, it seems like the fields are just "there," waiting to be measured.

To find out if this is true, Alec establishes a region of space with a uniform magnetic field $\vec{B}_A$ but no electric field ($\vec{E}_A = \vec{0}$). Then, as shown in **FIGURE 31.3**, he shoots a positive charge q through the magnetic field. At an instant when q is moving horizontally with velocity $\vec{v}_{CA}$, Alec observes that the particle experiences force $\vec{F}_A = q\vec{v}_{CA} \times \vec{B}_A$. The direction of the force is straight up.

Suppose that Brittney, in frame B, runs alongside the charge with the same velocity: $\vec{v}_{BA} = \vec{v}_{CA}$. To her, in frame B, the charge is at rest. Nonetheless, because both experimenters must agree about forces, Brittney *must* observe the same upward force on the charge that Alec observed. But there is *no* magnetic force on a stationary charge, so how can this be?

Because Brittney sees a stationary charge being acted on by an upward force, her only possible conclusion is that there is an upward-pointing *electric field*. After all, the electric field was initially defined in terms of the force experienced by a stationary charge. If the electric field in frame B is $\vec{E}_B$, then the force on the charge is $\vec{F}_B = q\vec{E}_B$. But we know that $\vec{F}_B = \vec{F}_A$, and Alec has already measured $\vec{F}_A = q\vec{v}_{CA} \times \vec{B}_A = q\vec{v}_{BA} \times \vec{B}_A$. Thus we're led to the conclusion that

$$\vec{E}_B = \vec{v}_{BA} \times \vec{B}_A \tag{31.3}$$

As Brittney runs past Alec, she finds that at least part of Alec's magnetic field has become an electric field! **Whether a field is seen as "electric" or "magnetic" depends on the motion of the reference frame relative to the sources of the field.**

FIGURE 31.4 shows the situation from Brittney's perspective. There is a force on charge q, the same force that Alec measured in Figure 31.3, but Brittney attributes this force to an electric field rather than a magnetic field. (Brittney needs a moving charge to measure magnetic forces, so we'll need a different experiment to see whether or not there's a magnetic field in frame B.)

More generally, suppose that an experimenter in reference frame A creates both an electric field $\vec{E}_A$ and a magnetic field $\vec{B}_A$. A charge moving in A with velocity $\vec{v}_{CA}$ experiences the force $\vec{F}_A = q(\vec{E}_A + \vec{v}_{CA} \times \vec{B}_A)$ shown in **FIGURE 31.5a**. The charge is at rest in a reference frame B that moves with velocity $\vec{v}_{BA} = \vec{v}_{CA}$ so the force in B can be due only to an electric field: $\vec{F}_B = q\vec{E}_B$. Equating the forces, because experimenters in all inertial reference frames agree about forces, we find that

$$\vec{E}_B = \vec{E}_A + \vec{v}_{BA} \times \vec{B}_A \tag{31.4}$$

FIGURE 31.3 A charged particle moves through a magnetic field in reference frame A and experiences a magnetic force.

In A, the force on q is due to a magnetic field.

$\vec{F}_A = q\vec{v}_{CA} \times \vec{B}_A$

$\vec{B}_A$

q $\vec{v}_{CA}$

The situation in frame A

FIGURE 31.4 In frame B, the charge experiences an electric force.

In B, the force on q is due to an electric field.

$\vec{F}_B = q\vec{E}_B$

$\vec{B}_B = ?$

In B, there's an electric field $\vec{E}_B = \vec{v}_{BA} \times \vec{B}_A$.

q

The charge is at rest in B.

The situation in frame B

FIGURE 31.5 A charge in reference frame A experiences electric and magnetic forces. The charge experiences the same force in frame B, but it is due only to an electric field.

(a) The electric and magnetic fields in frame A

$\vec{F}_A = q(\vec{E}_A + \vec{v}_{CA} \times \vec{B}_A)$

$\vec{E}_A$

$\vec{v}_{CA}$

$\vec{B}_A$

(b) The electric field in frame B, where the charged particle is at rest

$\vec{F}_B = q\vec{E}_B$

$\vec{E}_B$

$\vec{v}_{CB} = \vec{0}$

Equation 31.4 transforms the electric and magnetic fields measured in reference frame A into the electric field measured in a frame B that moves relative to A with velocity $\vec{v}_{BA}$. FIGURE 31.5b shows the outcome. Although we used a charge as a probe to find Equation 31.4, the equation is strictly about fields in different reference frames; it makes no mention of charges.

EXAMPLE 31.1 | **Transforming the electric field**

A laboratory experimenter has created the parallel electric and magnetic fields $\vec{E} = 10{,}000\,\hat{\imath}$ V/m and $\vec{B} = 0.10\,\hat{\imath}$ T. A proton is shot into these fields with velocity $\vec{v} = 1.0 \times 10^5\,\hat{\jmath}$ m/s. What is the electric field in the proton's reference frame?

MODEL Let the laboratory be reference frame A and a frame moving with the proton be reference frame B. The relative velocity is $\vec{v}_{BA} = 1.0 \times 10^5\,\hat{\jmath}$ m/s.

VISUALIZE FIGURE 31.6 shows the geometry. The laboratory fields, now labeled A, are parallel to the *x*-axis while $\vec{v}_{BA}$ is in the *y*-direction. Thus $\vec{v}_{BA} \times \vec{B}_A$ points in the negative *z*-direction.

SOLVE $\vec{v}_{BA}$ and $\vec{B}_A$ are perpendicular, so the magnitude of $\vec{v}_{BA} \times \vec{B}_A$ is $(1.0 \times 10^5$ m/s$)$ $(0.10$ T$)(\sin 90°) = 10{,}000$ V/m. Thus the electric field in frame B, the proton's frame, is

$$\vec{E}_B = \vec{E}_A + \vec{v}_{BA} \times \vec{B}_A = (10{,}000\,\hat{\imath} - 10{,}000\,\hat{k})\ \text{V/m}$$

$$= (14{,}000\ \text{V/m}, 45°\ \text{below the }x\text{-axis})$$

ASSESS The force on the proton is the same in both reference frames. But in the proton's reference frame that force is due entirely to an electric field tilted 45° below the *x*-axis.

FIGURE 31.6 Finding electric field $\vec{E}_B$.

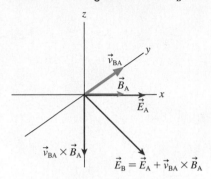

To find a transformation equation for the magnetic field, FIGURE 31.7a shows charge *q* at rest in reference frame A. Alec measures the fields of a stationary point charge:

$$\vec{E}_A = \frac{1}{4\pi\epsilon_0}\frac{q}{r^2}r \qquad \vec{B}_A = \vec{0}$$

What are the fields at this point in space as measured by Brittney in frame B? We can use Equation 31.4 to find $\vec{E}_B$. Because $\vec{B}_A = \vec{0}$, the electric field in frame B is

$$\vec{E}_B = \vec{E}_A = \frac{1}{4\pi\epsilon_0}\frac{q}{r^2}\hat{r} \tag{31.5}$$

In other words, **Coulomb's law is still valid in a frame in which the point charge is moving.**

But Brittney also measures a magnetic field $\vec{B}_B$, because, as seen in FIGURE 31.7b, charge *q* is moving in reference frame B. The magnetic field of a moving point charge is given by the Biot-Savart law:

$$\vec{B}_B = \frac{\mu_0}{4\pi}\frac{q}{r^2}\vec{v}_{CB} \times \hat{r} = -\frac{\mu_0}{4\pi}\frac{q}{r^2}\vec{v}_{BA} \times \hat{r} \tag{31.6}$$

where we used the fact that the charge's velocity in frame B is $\vec{v}_{CB} = -\vec{v}_{BA}$.

It will be useful to rewrite Equation 31.6 as

$$\vec{B}_B = -\frac{\mu_0}{4\pi}\frac{q}{r^2}\vec{v}_{BA} \times \hat{r} = -\epsilon_0\mu_0\vec{v}_{BA} \times \left(\frac{1}{4\pi\epsilon_0}\frac{q}{r^2}\hat{r}\right)$$

The expression in parentheses is simply $\vec{E}_A$, the electric field in frame A, so we have

$$\vec{B}_B = -\epsilon_0\mu_0\vec{v}_{BA} \times \vec{E}_A \tag{31.7}$$

Thus we find the remarkable idea that **the Biot-Savart law for the magnetic field of a moving point charge is nothing other than the Coulomb electric field of a stationary point charge transformed into a moving reference frame.**

We will assert without proof that if the experimenters in frame A create a magnetic field $\vec{B}_A$ in addition to the electric field $\vec{E}_A$, then the magnetic field $\vec{B}_B$ is

$$\vec{B}_B = \vec{B}_A - \epsilon_0\mu_0\vec{v}_{BA} \times \vec{E}_A \tag{31.8}$$

This is a general transformation matching Equation 31.4 for the electric field $\vec{E}_B$.

FIGURE 31.7 A charge at rest in frame A is moving in frame B.

(a) In frame A, the static charge creates an electric field but no magnetic field.

(b) In frame B, the moving charge creates both an electric and a magnetic field.

Notice something interesting. The constant μ_0 has units of T m/A; those of ϵ_0 are $C^2/N\,m^2$. By definition, 1 T = 1 N/A m and 1 A = 1 C/s. Consequently, the units of $\epsilon_0\mu_0$ turn out to be s^2/m^2. In other words, the quantity $1/\sqrt{\epsilon_0\mu_0}$, with units of m/s, is a speed. But what speed? The constants are well known from measurements of static electric and magnetic fields, so we can compute

$$\frac{1}{\sqrt{\epsilon_0\mu_0}} = \frac{1}{\sqrt{(8.85 \times 10^{-12}\,C^2/N\,m^2)(1.26 \times 10^{-6}\,T\,m/A)}} = 3.00 \times 10^8 \text{ m/s}$$

Of all the possible values you might get from evaluating $1/\sqrt{\epsilon_0\mu_0}$, what are the chances it would turn out to be c, the speed of light? It is not a random coincidence. In Section 31.5 we'll show that electric and magnetic fields can exist as a *traveling wave*, and that the wave speed is predicted by the theory to be none other than

$$v_{em} = c = \frac{1}{\sqrt{\epsilon_0\mu_0}} \tag{31.9}$$

For now, we'll go ahead and write $\epsilon_0\mu_0 = 1/c^2$. With this, our **Galilean field transformation equations** are

$$\vec{E}_B = \vec{E}_A + \vec{v}_{BA} \times \vec{B}_A$$

$$\vec{B}_B = \vec{B}_A - \frac{1}{c^2}\vec{v}_{BA} \times \vec{E}_A \tag{31.10}$$

where $\vec{v}_{BA}$ is the velocity of reference frame B relative to frame A and where, to reiterate, the fields are measured *at the same point in space* by experimenters *at rest* in each reference frame.

NOTE We'll see shortly that these equations are valid only if $v_{BA} \ll c$.

We can no longer believe that electric and magnetic fields have a separate, independent existence. Changing from one reference frame to another mixes and rearranges the fields. Different experimenters watching an event will agree on the outcome, such as the deflection of a charged particle, but they will ascribe it to different combinations of fields. Our conclusion is that **there is a single electromagnetic field that presents different faces, in terms of $\vec{E}$ and $\vec{B}$, to different viewers.**

EXAMPLE 31.2 | Two views of a magnet

The 1.0 T field of a large laboratory magnet points straight up. A rocket flies past the laboratory, parallel to the ground, at 1000 m/s. What are the fields between the magnet's pole tips as measured—very quickly!—by scientists on the rocket?

MODEL Let the laboratory be reference frame A and a frame moving with the rocket be reference frame B.

VISUALIZE FIGURE 31.8 shows the magnet and establishes the coordinate systems. The relative velocity is $\vec{v}_{BA} = 1000\,\hat{i}$ m/s.

FIGURE 31.8 The rocket and the magnet.

SOLVE The fields in the laboratory reference frame are $\vec{E}_A = \vec{0}$ and $\vec{B}_A = 1.0\,\hat{j}$ T. Transforming the fields to the rocket's reference frame gives first, for the electric field,

$$\vec{E}_B = \vec{E}_A + \vec{v}_{BA} \times \vec{B}_A = \vec{v}_{BA} \times \vec{B}_A$$

From the right-hand rule, $\vec{v}_{BA} \times \vec{B}_A$ is out of the page, in the z-direction. $\vec{v}_{BA}$ and $\vec{B}_A$ are perpendicular, so

$$\vec{E}_B = v_{BA}B_A\,\hat{k} = 1000\,\hat{k} \text{ V/m}$$

Similarly, for the magnetic field,

$$\vec{B}_B = \vec{B}_A - \frac{1}{c^2}\vec{v}_{BA} \times \vec{E}_A = \vec{B}_A = 1.0\,\hat{j} \text{ T}$$

Thus the rocket scientists measure

$$\vec{E}_B = 1000\,\hat{k} \text{ V/m} \quad \text{and} \quad \vec{B}_B = 1.0\,\hat{j} \text{ T}$$

Almost Relativity

FIGURE 31.9a shows two positive charges moving side by side through frame A with velocity $\vec{v}_{CA}$. Charge q_1 creates an electric field and a magnetic field at the position of charge q_2. These are

$$\vec{E}_A = \frac{1}{4\pi\epsilon_0} \frac{q_1}{r^2} \hat{j} \quad \text{and} \quad \vec{B}_A = \frac{\mu_0}{4\pi} \frac{q_1 v_{CA}}{r^2} \hat{k}$$

where r is the distance between the charges, and we've used $\hat{r} = \hat{j}$ and $\vec{v} \times \hat{r} = v\hat{k}$.

How are the fields seen in frame B, which moves with $\vec{v}_{BA} = \vec{v}_{CA}$ and in which the charges are at rest? From the field transformation equations,

$$\vec{B}_B = \vec{B}_A - \frac{1}{c^2}\vec{v}_{BA} \times \vec{E}_A = \frac{\mu_0}{4\pi} \frac{q_1 v_{CA}}{r^2} \hat{k} - \frac{1}{c^2}\left(v_{CA}\hat{i} \times \frac{1}{4\pi\epsilon_0}\frac{q_1}{r^2}\hat{j}\right)$$

$$= \frac{\mu_0}{4\pi} \frac{q_1 v_{CA}}{r^2}\left(1 - \frac{1}{\epsilon_0\mu_0 c^2}\right)\hat{k}$$

(31.11)

where we used $\hat{i} \times \hat{j} = \hat{k}$. But $\epsilon_0\mu_0 = 1/c^2$, so the term in parentheses is zero and thus $\vec{B}_B = \vec{0}$. This result was expected because q_1 is at rest in frame B and shouldn't create a magnetic field.

The transformation of the electric field is similar:

$$\vec{E}_B = \vec{E}_A + \vec{v}_{BA} \times \vec{B}_A = \frac{1}{4\pi\epsilon_0}\frac{q_1}{r^2}\hat{j} + v_{BA}\hat{i} \times \frac{\mu_0}{4\pi}\frac{q_1 v_{CA}}{r^2}\hat{k}$$

$$= \frac{1}{4\pi\epsilon_0}\frac{q_1}{r^2}(1 - \epsilon_0\mu_0 v_{BA}^2)\hat{j} = \frac{1}{4\pi\epsilon_0}\frac{q_1}{r^2}\left(1 - \frac{v_{BA}^2}{c^2}\right)\hat{j}$$

(31.12)

where we used $\hat{i} \times \hat{k} = -\hat{j}$, $\vec{v}_{CA} = \vec{v}_{BA}$, and $\epsilon_0\mu_0 = 1/c^2$. FIGURE 31.9b shows the charges and fields in frame B.

But now we have a problem. In frame B where the two charges are at rest and separated by distance r, the electric field due to charge q_1 should be simply

$$\vec{E}_B = \frac{1}{4\pi\epsilon_0}\frac{q_1}{r^2}\hat{j}$$

The field transformation equations have given a "wrong" result for the electric field $\vec{E}_B$.

It turns out that the field transformations of Equations 31.10, which are based on Galilean relativity, aren't quite right. We would need Einstein's relativity—a topic that we'll take up in Chapter 36—to give the correct transformations. However, the Galilean field transformations in Equations 31.10 are equivalent to the relativistically correct transformations when $v \ll c$, in which case $v^2/c^2 \ll 1$. You can see that the two expressions for $\vec{E}_B$ do, in fact, agree if v_{BA}^2/c^2 can be neglected.

Thus our use of the field transformation equations has an additional rule: Set v^2/c^2 to zero. This is an acceptable rule for speeds $v < 10^7$ m/s. Even with this limitation, our investigation has provided us with a deeper understanding of electric and magnetic fields.

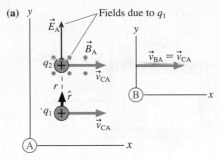

FIGURE 31.9 Two charges moving parallel to each other.

(a) Fields seen in frame A

(b) Fields seen in frame B

STOP TO THINK 31.1 The first diagram shows electric and magnetic fields in reference frame A. Which diagram shows the fields in frame B?

(a)   (b)   (c)   (d)

Faraday's Law Revisited

FIGURE 31.10 A motional emf as seen in two different reference frames.

(a) Laboratory frame A

The loop is moving to the right.

(b) Loop frame B

The induced electric field points up.

The magnetic field is moving to the left.

The transformation of electric and magnetic fields gives us new insight into Faraday's law. **FIGURE 31.10a** shows a reference frame A in which a conducting loop is moving with velocity $\vec{v}$ into a magnetic field. You learned in Chapter 30 that the magnetic field exerts a magnetic force $\vec{F}_{\mathrm{B}} = q\vec{v} \times \vec{B} = (qvB, \text{ upward})$ on the charges in the leading edge of the wire, creating an emf $\mathcal{E} = vLB$ and an induced current in the loop. We called this a *motional emf*.

How do things appear to an experimenter who is in frame B that moves with the loop at velocity $\vec{v}_{\mathrm{BA}} = \vec{v}$ and for whom the loop is at rest? We have learned the important lesson that experimenters in different inertial reference frames agree about the outcome of any experiment; hence an experimenter in frame B agrees that there is an induced current in the loop. But the charges are at rest in frame B so there cannot be any magnetic force on them. How is the emf established in frame B?

We can use the field transformations to determine that the fields in frame B are

$$\vec{E}_{\mathrm{B}} = \vec{E}_{\mathrm{A}} + \vec{v} \times \vec{B}_{\mathrm{A}} = \vec{v} \times \vec{B}$$

$$\vec{B}_{\mathrm{B}} = \vec{B}_{\mathrm{A}} - \frac{1}{c^2}\vec{v} \times \vec{E}_{\mathrm{A}} = \vec{B} \qquad (31.13)$$

where we used the fact that $\vec{E}_{\mathrm{A}} = \vec{0}$ in frame A.

An experimenter in the loop's frame sees not only a magnetic field but also the electric field $\vec{E}_{\mathrm{B}}$ shown in **FIGURE 31.10b**. The magnetic field exerts no force on the charges, because they're at rest in this frame, but the electric field does. The force on charge q is $\vec{F}_{\mathrm{E}} = q\vec{E}_{\mathrm{B}} = q\vec{v} \times \vec{B} = (qvB, \text{ upward})$. This is the same force as was measured in the laboratory frame, so it will cause the same emf and the same current. The outcome is identical, as we knew it had to be, but the experimenter in B attributes the emf to an electric field whereas the experimenter in A attributes it to a magnetic field.

Field $\vec{E}_{\mathrm{B}}$ is, in fact, the *induced electric field* of Faraday's law. Faraday's law, fundamentally, is a statement that **a changing magnetic field creates an electric field.** But only in frame B, the frame of the loop, is the magnetic field changing. Thus the induced electric field is seen in the loop's frame but not in the laboratory frame.

31.2 The Field Laws Thus Far

Let's remind ourselves where we are in terms of discovering laws about the electromagnetic field. Gauss's law, which you studied in Chapter 24, states a very general property of the electric field. It says that charges create electric fields in such a way that the electric flux of the field is the same through *any* closed surface surrounding the charges. **FIGURE 31.11** illustrates this idea by showing the field lines passing through a Gaussian surface enclosing a charge.

The mathematical statement of Gauss's law for the electric field says that for any *closed* surface enclosing total charge Q_{in}, the net electric flux through the surface is

$$(\Phi_{\mathrm{e}})_{\text{closed surface}} = \oint \vec{E} \cdot d\vec{A} = \frac{Q_{\mathrm{in}}}{\epsilon_0} \qquad (31.14)$$

The circle on the integral sign indicates that the integration is over a closed surface. Gauss's law is the first of what will turn out to be four *field equations*.

There's an analogous equation for magnetic fields, an equation we implied in Chapter 29—where we noted that isolated north or south poles do not exist—but didn't explicitly write it down. **FIGURE 31.12** shows a Gaussian surface around a magnetic dipole. Magnetic field lines form continuous curves, without starting or stopping, so every field line leaving the surface at some point must reenter it at another. Consequently, the net magnetic flux over a *closed* surface is zero.

FIGURE 31.11 A Gaussian surface enclosing a charge.

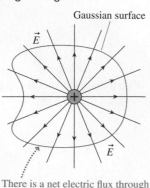

Gaussian surface

There is a net electric flux through this surface that encloses a charge.

We've shown only one surface and one magnetic field, but this conclusion turns out to be a general property of magnetic fields. Because every north pole is accompanied by a south pole, we can't enclose a "net pole" within a surface. Thus Gauss's law for magnetic fields is

$$(\Phi_m)_{\text{closed surface}} = \oint \vec{B} \cdot d\vec{A} = 0 \tag{31.15}$$

Equation 31.14 is the mathematical statement that Coulomb electric field lines start and stop on charges. Equation 31.15 is the mathematical statement that magnetic field lines form closed loops; they don't start or stop (i.e., there are no isolated magnetic poles). These two versions of Gauss's law are important statements about what types of fields can and cannot exist. They will become two of Maxwell's equations.

The third field law we've established is Faraday's law:

$$\mathcal{E} = \oint \vec{E} \cdot d\vec{s} = -\frac{d\Phi_m}{dt} \tag{31.16}$$

where the line integral of $\vec{E}$ is around the closed curve that bounds the surface through which the magnetic flux Φ_m is calculated. Equation 31.16 is the mathematical statement that an electric field can also be created by a changing magnetic field. The correct use of Faraday's law requires a convention for determining when fluxes are positive and negative. The sign convention will be given in the next section, where we discuss the fourth and last field equation—an analogous equation for magnetic fields.

31.3 The Displacement Current

We introduced Ampère's law in Chapter 29 as an alternative to the Biot-Savart law for calculating the magnetic field of a current. Whenever total current I_{through} passes through an area bounded by a closed curve, the line integral of the magnetic field around the curve is

$$\oint \vec{B} \cdot d\vec{s} = \mu_0 I_{\text{through}} \tag{31.17}$$

FIGURE 31.13 illustrates the geometry of Ampère's law. The sign of each current can be determined by using Tactics Box 31.1. In this case, $I_{\text{through}} = I_1 - I_2$.

FIGURE 31.12 There is no net flux through a Gaussian surface around a magnetic dipole.

There is no net magnetic flux through this closed surface.

FIGURE 31.13 Ampère's law relates the line integral of $\vec{B}$ around curve C to the current passing through surface S.

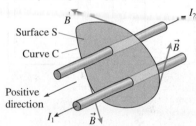

TACTICS BOX 31.1   (MP)

Determining the signs of flux and current

❶ For a surface S bounded by a closed curve C, choose either the clockwise (cw) or counterclockwise (ccw) direction around C.

❷ Curl the fingers of your *right* hand around the curve in the chosen direction, with your thumb perpendicular to the surface. Your thumb defines the positive direction.

■ A flux Φ through the surface is positive if the field is in the same direction as your thumb, negative if the field is in the opposite direction.

■ A current through the surface in the direction of your thumb is positive, in the direction opposite your thumb is negative.

Exercises 4–6

Ampère's law is the formal statement that **currents create magnetic fields**. Although Ampère's law can be used to calculate magnetic fields in situations with a high degree of symmetry, it is more important as a statement about what types of magnetic field can and cannot exist.

Something Is Missing

Nothing restricts the bounded surface of Ampère's law to being flat. It's not hard to see that any current passing through surface S_1 in FIGURE 31.14 must also pass through the curved surface S_2. To interpret Ampère's law properly, we have to say that the current $I_{through}$ is the net current passing through *any* surface S that is bounded by curve C.

FIGURE 31.14 The *net* current passing through the flat surface S_1 also passes through the curved surface S_2.

Closed curve C around wire

Surface S_1 Surface S_2

Any currents passing through S_1 must also pass through S_2.

Closed curve C around wire

S_1 S_2

Even in this case, the *net* current through S_1, namely zero, matches the net current through S_2.

FIGURE 31.15 There is no current through surface S_2 as the capacitor charges, but there is a changing electric flux.

(a)

Current I passes through surface S_1.

Cross section through a closed curve C around the wire

No current passes through surface S_2.

This is the magnetic field of the current I that is charging the capacitor.

(b)

Curve C S_2

S_1

$I = \dfrac{dQ}{dt}$ $I = \dfrac{dQ}{dt}$

The electric flux Φ_e through surface S_2 increases as the capacitor charges.

But this leads to an interesting puzzle. FIGURE 31.15a shows a capacitor being charged. Current I, from the left, brings positive charge to the left capacitor plate. The same current carries charges away from the right capacitor plate, leaving the right plate negatively charged. This is a perfectly ordinary current in a conducting wire, and you can use the right-hand rule to verify that its magnetic field is as shown.

Curve C is a closed curve encircling the wire on the left. The current passes through surface S_1, a flat surface across C, and we could use Ampère's law to find that the magnetic field is that of a straight wire. But what happens if we try to use surface S_2 to determine $I_{through}$? Ampère's law says that we can consider *any* surface bounded by curve C, and surface S_2 certainly qualifies. But *no* current passes through S_2. Charges are brought to the left plate of the capacitor and charges are removed from the right plate, but *no* charge moves across the gap between the plates. Surface S_1 has $I_{through} = I$, but surface S_2 has $I_{through} = 0$. Another dilemma!

It would appear that Ampère's law is either wrong or incomplete. Maxwell was the first to recognize the seriousness of this problem. He noted that there may be no current passing through S_2, but, as FIGURE 31.15b shows, there is an electric flux Φ_e through S_2 due to the electric field inside the capacitor. Furthermore, this flux is *changing* with time as the capacitor charges and the electric field strength grows. Faraday had discovered the significance of a changing magnetic flux, but no one had considered a changing electric flux.

The current I passes through S_1, so Ampère's law applied to S_1 gives

$$\oint \vec{B} \cdot d\vec{s} = \mu_0 I_{through} = \mu_0 I$$

We believe this result because it gives the correct magnetic field for a current-carrying wire. Now the line integral depends only on the magnetic field at points on curve C, so its value won't change if we choose a different surface S to evaluate the current. The problem is with the right side of Ampère's law, which would incorrectly give zero if applied to surface S_2. We need to modify the right side of Ampère's law to recognize that an electric flux rather than a current passes through S_2.

The electric flux between two capacitor plates of surface area A is

$$\Phi_e = EA$$

The capacitor's electric field is $E = Q/\epsilon_0 A$; hence the flux is actually independent of the plate size:

$$\Phi_e = \frac{Q}{\epsilon_0 A} A = \frac{Q}{\epsilon_0} \tag{31.18}$$

The *rate* at which the electric flux is changing is

$$\frac{d\Phi_e}{dt} = \frac{1}{\epsilon_0}\frac{dQ}{dt} = \frac{I}{\epsilon_0} \tag{31.19}$$

where we used $I = dQ/dt$. The flux is changing with time at a rate directly proportional to the charging current I.

Equation 31.19 suggests that the quantity $\epsilon_0(d\Phi_e/dt)$ is in some sense "equivalent" to current I. Maxwell called the quantity

$$I_{disp} = \epsilon_0\frac{d\Phi_e}{dt} \tag{31.20}$$

the **displacement current.** He had started with a fluid-like model of electric and magnetic fields, and the displacement current was analogous to the displacement of a fluid. The fluid model has since been abandoned, but the name lives on despite the fact that nothing is actually being displaced.

Maxwell hypothesized that the displacement current was the "missing" piece of Ampère's law, so he modified Ampère's law to read

$$\oint \vec{B} \cdot d\vec{s} = \mu_0(I_{through} + I_{disp}) = \mu_0\left(I_{through} + \epsilon_0\frac{d\Phi_e}{dt}\right) \tag{31.21}$$

Equation 31.21 is now known as the Ampère-Maxwell law. When applied to Figure 31.15b, the Ampère-Maxwell law gives

$$S_1: \quad \oint \vec{B} \cdot d\vec{s} = \mu_0\left(I_{through} + \epsilon_0\frac{d\Phi_e}{dt}\right) = \mu_0(I + 0) = \mu_0 I$$

$$S_2: \quad \oint \vec{B} \cdot d\vec{s} = \mu_0\left(I_{through} + \epsilon_0\frac{d\Phi_e}{dt}\right) = \mu_0(0 + I) = \mu_0 I$$

where, for surface S_2, we used Equation 31.19 for $d\Phi_e/dt$. Surfaces S_1 and S_2 now both give the same result for the line integral of $\vec{B} \cdot d\vec{s}$ around the closed curve C.

NOTE The displacement current I_{disp} between the capacitor plates is numerically equal to the current I in the wires to and from the capacitor, so in some sense it allows "current" to be continuous all the way through the capacitor. Nonetheless, the displacement current is *not* a flow of charge. The displacement current is equivalent to a real current in that it creates the same magnetic field, but it does so with a changing electric flux rather than a flow of charge.

The Induced Magnetic Field

Ordinary Coulomb electric fields are created by charges, but a second way to create an electric field is by having a changing magnetic field. That's Faraday's law. Ordinary magnetic fields are created by currents, but now we see that a second way to create a magnetic field is by having a changing electric field. Just as the electric field created by a changing $\vec{B}$ is called an induced electric field, the magnetic field created by a changing $\vec{E}$ is called an *induced magnetic field*.

FIGURE 31.16 shows the close analogy between induced electric fields, governed by Faraday's law, and induced magnetic fields, governed by the second term in the Ampère-Maxwell law. An increasing solenoid current causes an increasing magnetic field. The changing magnetic field, in turn, induces a circular electric field. The negative sign in Faraday's law dictates that the induced electric field direction is ccw when seen looking along the magnetic field direction.

An increasing capacitor charge causes an increasing electric field. The changing electric field, in turn, induces a circular magnetic field. But the sign of the Ampère-Maxwell law is positive, the opposite of the sign of Faraday's law, so the induced magnetic field direction is cw when you're looking along the electric field direction.

FIGURE 31.16 The close analogy between an induced electric field and an induced magnetic field.

Increasing solenoid current   Increasing $\vec{B}$

Faraday's law describes an induced electric field.

Increasing capacitor charge   Increasing $\vec{E}$

The Ampère-Maxwell law describes an induced magnetic field.

EXAMPLE 31.3 | **The fields inside a charging capacitor**

A 2.0-cm-diameter parallel-plate capacitor with a 1.0 mm spacing is being charged at the rate 0.50 C/s. What is the magnetic field strength inside the capacitor at a point 0.50 cm from the axis?

MODEL The electric field inside a parallel-plate capacitor is uniform. As the capacitor is charged, the changing electric field induces a magnetic field.

VISUALIZE FIGURE 31.17 shows the fields. The induced magnetic field lines are circles concentric with the capacitor.

FIGURE 31.17 The magnetic field strength is found by integrating around a closed curve of radius r.

The magnetic field line is a circle concentric with the capacitor. The electric flux through this circle is $\pi r^2 E$.

SOLVE The electric field of a parallel-plate capacitor is $E = Q/\epsilon_0 A = Q/\epsilon_0 \pi R^2$. The electric flux through the circle of radius r (not the full flux of the capacitor) is

$$\Phi_e = \pi r^2 E = \pi r^2 \frac{Q}{\epsilon_0 \pi R^2} = \frac{r^2}{R^2} \frac{Q}{\epsilon_0}$$

Thus the Ampère-Maxwell law is

$$\oint \vec{B} \cdot d\vec{s} = \epsilon_0 \mu_0 \frac{d\Phi_e}{dt} = \epsilon_0 \mu_0 \frac{d}{dt} \left(\frac{r^2}{R^2} \frac{Q}{\epsilon_0} \right) = \mu_0 \frac{r^2}{R^2} \frac{dQ}{dt}$$

The magnetic field is everywhere tangent to the circle of radius r, so the integral of $\vec{B} \cdot d\vec{s}$ around the circle is simply $BL = 2\pi r B$. With this value for the line integral, the Ampère-Maxwell law becomes

$$2\pi r B = \mu_0 \frac{r^2}{R^2} \frac{dQ}{dt}$$

and thus

$$B = \frac{\mu_0}{2\pi} \frac{r}{R^2} \frac{dQ}{dt} = (2.0 \times 10^{-7} \, \text{T m/A}) \frac{0.0050 \, \text{m}}{(0.010 \, \text{m})^2} (0.50 \, \text{C/s})$$
$$= 5.0 \times 10^{-6} \, \text{T}$$

ASSESS Charging a capacitor at 0.5 C/s requires a 0.5 A charging current. We've seen many previous examples in which a current-carrying wire with $I \approx 1$ A generates a nearby magnetic field of a few microtesla, so the result seems reasonable.

If a changing magnetic field can induce an electric field and a changing electric field can induce a magnetic field, what happens when both fields change simultaneously? That is the question that Maxwell was finally able to answer after he modified Ampère's law to include the displacement current, and it is the subject to which we turn next.

STOP TO THINK 31.2 The electric field in four identical capacitors is shown as a function of time. Rank in order, from largest to smallest, the magnetic field strength at the outer edge of the capacitor at time T.

31.4 Maxwell's Equations

James Clerk Maxwell was a young, mathematically brilliant Scottish physicist. In 1855, barely 24 years old, he presented a paper to the Cambridge Philosophical Society entitled "On Faraday's Lines of Force." It had been 30 years and more since the major discoveries of Oersted, Ampère, Faraday, and others, but electromagnetism remained a loose collection of facts and "rules of thumb" without a consistent theory to link these ideas together.

Maxwell's goal was to synthesize this body of knowledge and to form a *theory* of electromagnetic fields. The critical step along the way was his recognition of the need to include a displacement-current term in Ampère's law.

Maxwell's theory of electromagnetism is embodied in four equations that we today call **Maxwell's equations.** These are

$$\oint \vec{E} \cdot d\vec{A} = \frac{Q_{\text{in}}}{\epsilon_0} \qquad \text{Gauss's law}$$

$$\oint \vec{B} \cdot d\vec{A} = 0 \qquad \text{Gauss's law for magnetism}$$

$$\oint \vec{E} \cdot d\vec{s} = -\frac{d\Phi_{\text{m}}}{dt} \qquad \text{Faraday's law}$$

$$\oint \vec{B} \cdot d\vec{s} = \mu_0 I_{\text{through}} + \epsilon_0\mu_0 \frac{d\Phi_{\text{e}}}{dt} \qquad \text{Ampère-Maxwell law}$$

Maxwell's claim is that these four equations are a *complete* description of electric and magnetic fields. They tell us how fields are created by charges and currents, and also how fields can be induced by the changing of other fields. We need one more equation for completeness, an equation that tells us how matter responds to electromagnetic fields. The general force equation

$$\vec{F} = q(\vec{E} + \vec{v} \times \vec{B}) \qquad \text{(Lorentz force law)}$$

is known as the *Lorentz force law.* **Maxwell's equations for the fields, together with the Lorentz force law to tell us how matter responds to the fields, form the complete theory of electromagnetism.**

Maxwell's equations bring us to the pinnacle of classical physics. When combined with Newton's three laws of motion, his law of gravity, and the first and second laws of thermodynamics, we have all of classical physics—a total of just 11 equations.

While some physicists might quibble over whether all 11 are truly fundamental, the important point is not the exact number but how few equations we need to describe the overwhelming majority of our experience of the physical world. It seems as if we could have written them all on page 1 of this book and been finished, but it doesn't work that way. Each of these equations is the synthesis of a tremendous number of physical phenomena and conceptual developments. To know physics isn't just to know the equations, but to know what the equations *mean* and how they're used. That's why it's taken us so many chapters and so much effort to get to this point. Each equation is a shorthand way to summarize a book's worth of information!

Let's summarize the physical meaning of the five electromagnetic equations:

Classical physics

Newton's first law
Newton's second law
Newton's third law
Newton's law of gravity
Gauss's law
Gauss's law for magnetism
Faraday's law
Ampère-Maxwell law
Lorentz force law
First law of thermodynamics
Second law of thermodynamics

- **Gauss's law:** Charged particles create an electric field.
- **Faraday's law:** An electric field can also be created by a changing magnetic field.
- **Gauss's law for magnetism:** There are no isolated magnetic poles.
- **Ampère-Maxwell law, first half:** Currents create a magnetic field.
- **Ampère-Maxwell law, second half:** A magnetic field can also be created by a changing electric field.
- **Lorentz force law, first half:** An electric force is exerted on a charged particle in an electric field.
- **Lorentz force law, second half:** A magnetic force is exerted on a charge moving in a magnetic field.

These are the *fundamental ideas* of electromagnetism. Other important ideas, such as Ohm's law, Kirchhoff's laws, and Lenz's law, despite their practical importance, are not fundamental ideas. They can be derived from Maxwell's equations, sometimes with the addition of empirically based concepts such as resistance.

It's true that Maxwell's equations are mathematically more complex than Newton's laws and that their solution, for many problems of practical interest, requires advanced mathematics. Fortunately, we have the mathematical tools to get just far enough into Maxwell's equations to discover their most startling and revolutionary implication—the prediction of electromagnetic waves.

31.5 ADVANCED TOPIC Electromagnetic Waves

NOTE This optional section goes through the mathematics of showing that Maxwell's equations predict electromagnetic waves. The key results of this section are summarized at the beginning of Section 31.6, so this section may be omitted with no loss of continuity.

Maxwell developed his four equations as a mathematical summary of what was known about electricity and magnetism in the mid-19th century: the properties of *static* electric and magnetic fields plus Faraday's discovery of electromagnetic induction. Maxwell introduced the idea of *displacement current*—that a changing electric flux creates a magnetic field—on purely theoretical grounds; there was no experimental evidence at the time. But this new concept was the key to Maxwell's success because it soon allowed him to make the remarkable and totally unexpected prediction of **electromagnetic waves**—self-sustaining oscillations of the electric and magnetic fields that propagate through space without the need for charges or currents.

Our goals in this section are to show that Maxwell's equations lead to a *wave equation* for the electric and magnetic fields and to discover that all electromagnetic waves, regardless of frequency, travel through vacuum at the same speed, a speed we now call the *speed of light*. A completely general derivation of the wave equation is too mathematically advanced for this textbook, so we will make a small number of assumptions—but assumptions that will seem quite reasonable after our study of induced fields.

To begin, we will assume that electric and magnetic fields can exist independently of charges and currents in a *source-free* region of space. This is a very important assumption because it makes the statement that **fields are real entities;** they're not just cute pictures that tell us about charges and currents. The source-free Maxwell's equations, with no charges or currents, are

$$\oint \vec{E} \cdot d\vec{A} = 0 \qquad \oint \vec{E} \cdot d\vec{s} = -\frac{d\Phi_m}{dt}$$

$$\oint \vec{B} \cdot d\vec{A} = 0 \qquad \oint \vec{B} \cdot d\vec{s} = \epsilon_0\mu_0\frac{d\Phi_e}{dt} \qquad (31.22)$$

Any electromagnetic wave traveling in empty space must be consistent with these equations.

The Structure of Electromagnetic Waves

Faraday discovered that a changing magnetic field creates an induced electric field, and Maxwell's postulated displacement current says that a changing electric field creates an induced magnetic field. The idea behind electromagnetic waves, illustrated in **FIGURE 31.18**, is that the fields can exist in a self-sustaining mode *if* a changing magnetic field creates an electric field that, in turn, happens to change in just the right way to recreate the original magnetic field. Notice that it has to be an *electromagnetic* wave, with changing electric *and* magnetic fields. A purely electric or purely magnetic wave cannot exist.

You saw in Section 30.6 that an induced electric field, which can drive an induced current around a conducting loop, is *perpendicular* to the changing magnetic field.

Large radar installations like this one are used to track rockets and missiles.

FIGURE 31.18 Induced fields can be self-sustaining.

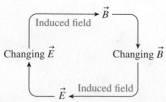

And earlier in this chapter, when we introduced the displacement current, the induced magnetic field in a charging capacitor was *perpendicular* to the changing electric field. Thus we'll make the assumption that $\vec{E}$ and $\vec{B}$ **are perpendicular to each other** in an electromagnetic wave. Furthermore—we'll justify this shortly—$\vec{E}$ and $\vec{B}$ **are each perpendicular to the direction of travel.** Thus an electromagnetic wave is a *transverse wave,* analogous to a wave on a string, rather than a sound-like longitudinal wave.

We will also assume, to keep the mathematics as simple as possible, that an electromagnetic wave can travel as a *plane wave,* which you will recall from Chapter 16 is a wave for which the fields are same *everywhere* in a plane perpendicular to the direction of travel. FIGURE 31.19a shows an electromagnetic plane wave propagating at speed v_{em} along the x-axis. $\vec{E}$ and $\vec{B}$ are perpendicular to each other, as we've assumed, *and* to the direction of travel. We've defined the y- and z-axes to be, respectively, parallel to $\vec{E}$ and $\vec{B}$. Notice how the fields are the same at every point in a yz-plane slicing the x-axis.

Because a wave is a traveling disturbance, FIGURE 31.19b shows that the fields— at one instant of time—*do* change along the x-axis. These changing fields are the disturbance that is moving down the x-axis at speed v_{em}, so $\vec{E}$ and $\vec{B}$ of a plane wave are functions of the two variables x and t. We're not assuming that the wave has any particular shape—the shape of the wave is what we want to predict from Maxwell's equations—simply that it's a transverse wave moving along the x-axis.

Now that we know something about the structure of the wave, we can start to check its consistency with Maxwell's equations. FIGURE 31.20 shows an imaginary box, a Gaussian surface, centered on the x-axis. Both electric and magnetic field vectors exist at each point in space, but the figure shows them separately for clarity. $\vec{E}$ oscillates along the y-axis, so all electric field lines enter and leave the box through the top and bottom surfaces; no electric field lines pass through the sides of the box.

FIGURE 31.19 An electromagnetic plane wave.

(a)

The fields are the same at every point in a yz-plane.

(b)

The field strengths vary with x.

FIGURE 31.20 A closed surface can be used to check Gauss's law for the electric and magnetic fields.

The net electric flux through the box is zero.

The net magnetic flux through the box is zero.

Electric field

Magnetic field

Because this is a plane wave, the magnitude of each electric field vector entering the bottom of the box is exactly matched by an electric field vector leaving the top. The electric flux through the top of the box is equal in magnitude but opposite in sign to the flux through the bottom, and the flux through the sides is zero. Thus the *net* electric flux is $\Phi_e = 0$. There is no charge inside the box, because there are no sources in this region of space, so we also have $Q_{in} = 0$. Hence the electric field of a plane wave is consistent with the first of the source-free Maxwell's equations, Gauss's law.

The exact same argument applies to the magnetic field. The net magnetic flux is $\Phi_m = 0$; thus the magnetic field is consistent with the second of Maxwell's equations.

Suppose that $\vec{E}$ or $\vec{B}$ had a component along the x-axis, the direction of travel. The fields *change* along the x-axis—that's what a traveling wave is—so it would not be possible for the flux through the right face to exactly cancel the flux through the left face at every instant of time. An x-component of either field would violate Gauss's law by creating a net flux when there are no enclosed sources. Thus our claim that an electromagnetic wave must be a transverse wave, with the fields perpendicular to the direction of travel, is a requirement of Gauss's law.

FIGURE 31.21 Applying Faraday's law.

FIGURE 31.21 Applying Faraday's law.

Faraday's Law

Gauss's law tells us that an electromagnetic wave has to be a transverse wave. What does Faraday's law have to say? Faraday's law is concerned with the changing magnetic flux through a closed curve, so let's apply Faraday's law to the narrow rectangle in the xy-plane shown in FIGURE 31.21. We'll assume that Δx is so small that $\vec{B}$ is essentially constant over the width of the rectangle.

The magnetic field $\vec{B}$ is perpendicular to the rectangle, so the magnetic flux is $\Phi_m = B_z A_{\text{rectangle}} = B_z h \Delta x$. As the wave moves, the flux *changes* at the rate

$$\frac{d\Phi_m}{dt} = \frac{d}{dt}(B_z h \Delta x) = \frac{\partial B_z}{\partial t} h \Delta x \tag{31.23}$$

The ordinary derivative dB_z/dt, which is the full rate of change of B from all possible causes, becomes a partial derivative $\partial B_z/\partial t$ in this situation because the change in magnetic flux is due entirely to the change of B_z with time and not at all to the spatial variation of B_z.

According to our sign convention, we need to go around the rectangle in a counterclockwise direction to make the flux positive. Thus we must also use a counterclockwise direction to evaluate the line integral:

$$\oint \vec{E} \cdot d\vec{s} = \int_{\text{right}} \vec{E} \cdot d\vec{s} + \int_{\text{top}} \vec{E} \cdot d\vec{s} + \int_{\text{left}} \vec{E} \cdot d\vec{s} + \int_{\text{bottom}} \vec{E} \cdot d\vec{s} \tag{31.24}$$

The electric field $\vec{E}$ points in the y-direction; hence $\vec{E} \cdot d\vec{s} = 0$ at all points on the top and bottom edges and these two integrals are zero.

Along the left edge of the loop, at position x, $\vec{E}$ has the same value at every point. Figure 31.21 shows that the direction of $\vec{E}$ is *opposite* to $d\vec{s}$; thus $\vec{E} \cdot d\vec{s} = -E_y(x)ds$. On the right edge of the loop, at position $x + \Delta x$, $\vec{E}$ is *parallel* to $d\vec{s}$, and $\vec{E} \cdot d\vec{s} = E_y(x + \Delta x)\,ds$. Thus the line integral of $\vec{E} \cdot d\vec{s}$ around the rectangle is

$$\oint \vec{E} \cdot d\vec{s} = -E_y(x)h + E_y(x + \Delta x)h = [E_y(x + \Delta x) - E_y(x)]h \tag{31.25}$$

> **NOTE** $E_y(x)$ indicates that E_y is a function of the position x. It is *not* E_y multiplied by x.

You learned in calculus that the derivative of the function $f(x)$ is

$$\frac{df}{dx} = \lim_{\Delta x \to 0}\left[\frac{f(x + \Delta x) - f(x)}{\Delta x}\right]$$

We've assumed that Δx is very small. If we now let the width of the rectangle go to zero, $\Delta x \to 0$, Equation 31.25 becomes

$$\oint \vec{E} \cdot d\vec{s} = \frac{\partial E_y}{\partial x} h \Delta x \tag{31.26}$$

We've used a partial derivative because E_y is a function of both position x and time t.

Now, using Equations 31.23 and 31.26, we can write Faraday's law as

$$\oint \vec{E} \cdot d\vec{s} = \frac{\partial E_y}{\partial x} h \Delta x = -\frac{d\Phi_m}{dt} = -\frac{\partial B_z}{\partial t} h \Delta x$$

The area $h \Delta x$ of the rectangle cancels, and we're left with

$$\frac{\partial E_y}{\partial x} = -\frac{\partial B_z}{\partial t} \tag{31.27}$$

Equation 31.27, which compares the rate at which E_y varies with position to the rate at which B_z varies with time, is a *required condition* that an electromagnetic wave must satisfy to be consistent with Maxwell's equations.

The Ampère-Maxwell Law

We have only one equation to go, but this one will now be easier. The Ampère-Maxwell law is concerned with the changing electric flux through a closed curve. **FIGURE 31.22** shows a very narrow rectangle in the xz-plane. The electric field is perpendicular to this rectangle; hence the electric flux through it is $\Phi_e = E_y A_{\text{rectangle}} = E_y l \Delta x$. This flux is changing at the rate

$$\frac{d\Phi_e}{dt} = \frac{d}{dt}(E_y l \Delta x) = \frac{\partial E_y}{\partial t} l \Delta x \tag{31.28}$$

The line integral of $\vec{B} \cdot d\vec{s}$ around this closed rectangle is calculated just like the line integral of $\vec{E} \cdot d\vec{s}$ in Figure 31.21. $\vec{B}$ is perpendicular to $d\vec{s}$ on the narrow ends, so $\vec{B} \cdot d\vec{s} = 0$. The field at *all* points on the left edge is $\vec{B}(x)$, and this field is parallel to $d\vec{s}$ to make $\vec{B} \cdot d\vec{s} = B_z(x)\,ds$. Similarly, $\vec{B} \cdot d\vec{s} = -B_z(x + \Delta x)\,ds$ at all points on the right edge, where $\vec{B}$ is opposite to $d\vec{s}$. Thus, if we let $\Delta x \to 0$,

$$\oint \vec{B} \cdot d\vec{s} = B_z(x)l - B_z(x + \Delta x)l = -[B_z(x + \Delta x) - B_z(x)]l = -\frac{\partial B_z}{\partial x} l \Delta x \tag{31.29}$$

Equations 31.28 and 31.29 can now be used in the Ampère-Maxwell law:

$$\oint \vec{B} \cdot d\vec{s} = -\frac{\partial B_z}{\partial x} l \Delta x = \epsilon_0 \mu_0 \frac{d\Phi_e}{dt} = \epsilon_0 \mu_0 \frac{\partial E_y}{\partial t} l \Delta x$$

The area of the rectangle cancels, and we're left with

$$\frac{\partial B_z}{\partial x} = -\epsilon_0 \mu_0 \frac{\partial E_y}{\partial t} \tag{31.30}$$

Equation 31.30 is a second required condition that the fields must satisfy.

The Wave Equation

In « Section 16.4, during our study of traveling waves, we derived the *wave equation:*

$$\frac{\partial^2 D}{\partial t^2} = v^2 \frac{\partial^2 D}{\partial x^2} \tag{31.31}$$

There we learned that any physical system that obeys this equation for some type of displacement D can have traveling waves that propagate along the x-axis with speed v.

If we start with Equation 31.27, the Faraday's law requirement for any electromagnetic wave, we can take the second derivative with respect to x to find

$$\frac{\partial^2 E_y}{\partial x^2} = -\frac{\partial^2 B_z}{\partial x \partial t} \tag{31.32}$$

You've learned in calculus that the order of differentiation doesn't matter, so $\partial^2 B_z/\partial x \partial t = \partial^2 B_z/\partial t \partial x$. And from Equation 31.30,

$$\frac{\partial^2 B_z}{\partial t \partial x} = -\epsilon_0 \mu_0 \frac{\partial^2 E_y}{\partial t^2} \tag{31.33}$$

Substituting Equation 31.33 into Equation 31.32 and taking the constants to the other side, we have

$$\frac{\partial^2 E_y}{\partial t^2} = \frac{1}{\epsilon_0 \mu_0} \frac{\partial^2 E_y}{\partial x^2} \qquad \text{(the wave equation for electromagnetic waves)} \tag{31.34}$$

FIGURE 31.22 Applying the Ampère-Maxwell law.

Equation 31.34 is the wave equation! And it's easy to show, by taking second derivatives of B_z rather than E_y, that the magnetic field B_z obeys exactly the same wave equation.

As we anticipated, Maxwell's equations have led to a prediction of electromagnetic waves. Referring to the general wave equation, Equation 31.31, we see that an electromagnetic wave must travel (in vacuum) with speed

$$v_{em} = \frac{1}{\sqrt{\epsilon_0 \mu_0}} \tag{31.35}$$

The constants ϵ_0 and μ_0 are known from electrostatics and magnetostatics, where they determined the size of $\vec{E}$ and $\vec{B}$ due to point charges. Thus we can calculate

$$v_{em} = \frac{1}{\sqrt{\epsilon_0 \mu_0}} = 3.00 \times 10^8 \text{ m/s} = c \tag{31.36}$$

This is a remarkable conclusion. Coulomb's law and the Biot-Savart law, in which ϵ_0 and μ_0 first appeared, have nothing to do with waves. Yet Maxwell's theory of electromagnetism ends up predicting that electric and magnetic fields can form a self-sustaining electromagnetic wave *if* that wave travels with speed $v_{em} = 1/\sqrt{\epsilon_0 \mu_0}$. No other speed will satisfy Maxwell's equations.

Laboratory measurements had already determined that light travels at 3.0×10^8 m/s, so Maxwell was entirely justified in concluding that light is an electromagnetic wave. Furthermore, we've made no assumption about the frequency of the wave, so apparently electromagnetic waves of any frequency, from radio waves to x rays, travel (in vacuum) with speed c, the speed of light.

Connecting E and B

The electric and magnetic fields of an electromagnetic wave both oscillate, but not independently of each other. The two field strengths are related. E_y and B_z both satisfy the same wave equation, so the traveling waves—just like a wave on a string—are

$$E_y = E_0 \sin(kx - \omega t) = E_0 \sin\left[2\pi\left(\frac{x}{\lambda} - ft\right)\right]$$

$$B_z = B_0 \sin(kx - \omega t) = B_0 \sin\left[2\pi\left(\frac{x}{\lambda} - ft\right)\right] \tag{31.37}$$

where E_0 and B_0 are the amplitudes of the electric and magnetic portions of the wave and, as for any sinusoidal wave, $k = 2\pi/\lambda$, $\omega = 2\pi f$, and $\lambda f = v = c$. These waves have to satisfy Equation 31.27; thus

$$\frac{\partial E_y}{\partial x} = \frac{2\pi E_0}{\lambda} \cos\left[2\pi\left(\frac{x}{\lambda} - ft\right)\right] = -\frac{\partial B_z}{\partial t} = 2\pi f B_0 \cos\left[2\pi\left(\frac{x}{\lambda} - ft\right)\right] \tag{31.38}$$

Equation 31.38 is true only if $E_0 = \lambda f B_0 = c B_0$. And since the electric and magnetic fields oscillate together, this relationship between their amplitudes has to hold true at any point on the wave. Thus **$E = cB$ at any point on the wave.**

STOP TO THINK 31.3 An electromagnetic wave is propagating in the positive x-direction. At this instant of time, what is the direction of $\vec{E}$ at the center of the rectangle?

a. In the positive x-direction
b. In the negative x-direction
c. In the positive y-direction
d. In the negative y-direction
e. In the positive z-direction
f. In the negative z-direction

31.6 Properties of Electromagnetic Waves

It had been known since the early 19th century, from experiments with interference and diffraction, that light is a wave, but no one understood what was "waving." Faraday speculated that light was somehow connected to electricity and magnetism, but Maxwell was the first to understand not only that light is an electromagnetic wave but also that electromagnetic waves can exist at any frequency, not just the frequencies of visible light.

In the previous section, we used Maxwell's equations to discover that:

1. Maxwell's equations predict the existence of sinusoidal electromagnetic waves that travel through empty space independent of any charges or currents.
2. The waves are transverse waves, with $\vec{E}$ and $\vec{B}$ perpendicular to the direction of propagation $\vec{v}_{em}$.
3. $\vec{E}$ and $\vec{B}$ are perpendicular to each other in a manner such that $\vec{E} \times \vec{B}$ is in the direction of $\vec{v}_{em}$.
4. All electromagnetic waves, regardless of frequency or wavelength, travel in vacuum at speed $v_{em} = 1/\sqrt{\epsilon_0 \mu_0} = c$, the speed of light.
5. The field strengths are related by $E = cB$ at every point on the wave.

FIGURE 31.23 illustrates many of these characteristics of electromagnetic waves. It's a useful picture, and one that you'll see in any textbook, but a picture that can be very misleading if you don't think about it carefully. First and foremost, $\vec{E}$ and $\vec{B}$ are *not* spatial vectors. That is, they don't stretch spatially in the y- or z-direction for a certain distance. Instead, these vectors are showing the field strengths and directions along a single line, the x-axis. An $\vec{E}$ vector pointing in the y-direction is saying, "At this point on the x-axis, where the tail is, this is the direction and strength of the electric field." Nothing is "reaching" to a point in space above the x-axis.

Second, we're assuming this is a *plane wave*, which, you'll recall, is a wave for which the fields are the same *everywhere* in any plane perpendicular to $\vec{v}_{em}$. Figure 31.23 shows the fields only along one line. But whatever the fields are doing at a point on the x-axis, they are doing the same thing everywhere in the yz-plane that slices the x-axis at that point. With this in mind, let's explore some other properties of electromagnetic waves.

FIGURE 31.23 A sinusoidal electromagnetic wave.

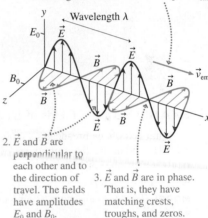

1. A sinusoidal wave with frequency f and wavelength λ travels with wave speed v_{em}.

2. $\vec{E}$ and $\vec{B}$ are perpendicular to each other and to the direction of travel. The fields have amplitudes E_0 and B_0.

3. $\vec{E}$ and $\vec{B}$ are in phase. That is, they have matching crests, troughs, and zeros.

Energy and Intensity

Waves transfer energy. Ocean waves erode beaches, sound waves set your eardrums vibrating, and light from the sun warms the earth. The energy flow of an electromagnetic wave is described by the **Poynting vector** $\vec{S}$, defined as

$$\vec{S} \equiv \frac{1}{\mu_0} \vec{E} \times \vec{B} \qquad (31.39)$$

The Poynting vector, shown in **FIGURE 31.24**, has two important properties:

1. The Poynting vector points in the direction in which an electromagnetic wave is traveling. You can see this by looking back at Figure 31.23.
2. It is straightforward to show that the units of S are W/m^2, or power (joules per second) per unit area. Thus the magnitude S of the Poynting vector measures the rate of energy transfer per unit area of the wave.

Because $\vec{E}$ and $\vec{B}$ of an electromagnetic wave are perpendicular to each other, and $E = cB$, the magnitude of the Poynting vector is

$$S = \frac{EB}{\mu_0} = \frac{E^2}{c\mu_0} = c\epsilon_0 E^2$$

The Poynting vector is a function of time, oscillating from zero to $S_{max} = E_0^2/c\mu_0$ and back to zero twice during each period of the wave's oscillation. That is, the energy

FIGURE 31.24 The Poynting vector.

The Poynting vector is in the direction of $\vec{E} \times \vec{B}$.

Wave direction

flow in an electromagnetic wave is not smooth. It "pulses" as the electric and magnetic fields oscillate in intensity. We're unaware of this pulsing because the electromagnetic waves that we can sense—light waves—have such high frequencies.

Of more interest is the *average* energy transfer, averaged over one cycle of oscillation, which is the wave's **intensity** I. In our earlier study of waves, we defined the intensity of a wave to be $I = P/A$, where P is the power (energy transferred per second) of a wave that impinges on area A. Because $E = E_0 \sin[2\pi(x/\lambda - ft)]$, and the average over one period of $\sin^2[2\pi(x/\lambda - ft)]$ is $\frac{1}{2}$, the intensity of an electromagnetic wave is

$$I = \frac{P}{A} = S_{\text{avg}} = \frac{1}{2c\mu_0}E_0^2 = \frac{c\epsilon_0}{2}E_0^2 \tag{31.40}$$

Equation 31.40 relates the intensity of an electromagnetic wave, a quantity that is easily measured, to the amplitude E_0 of the wave's electric field.

The intensity of a plane wave, with constant electric field amplitude E_0, would not change with distance. But a plane wave is an idealization; there are no true plane waves in nature. You learned in Chapter 16 that, to conserve energy, the intensity of a wave far from its source decreases with the inverse square of the distance. If a source with power P_{source} emits electromagnetic waves *uniformly* in all directions, the electromagnetic wave intensity at distance r from the source is

$$I = \frac{P_{\text{source}}}{4\pi r^2} \tag{31.41}$$

Equation 31.41 simply expresses the recognition that the energy of the wave is spread over a sphere of surface area $4\pi r^2$.

EXAMPLE 31.4 | Fields of a cell phone

A digital cell phone broadcasts a 0.60 W signal at a frequency of 1.9 GHz. What are the amplitudes of the electric and magnetic fields at a distance of 10 cm, about the distance to the center of the user's brain?

MODEL Treat the cell phone as a point source of electromagnetic waves.

SOLVE The intensity of a 0.60 W point source at a distance of 10 cm is

$$I = \frac{P_{\text{source}}}{4\pi r^2} = \frac{0.60\ \text{W}}{4\pi(0.10\ \text{m})^2} = 4.78\ \text{W/m}^2$$

We can find the electric field amplitude from the intensity:

$$E_0 = \sqrt{\frac{2I}{c\epsilon_0}} = \sqrt{\frac{2(4.78\ \text{W/m}^2)}{(3.00 \times 10^8\ \text{m/s})(8.85 \times 10^{-12}\ \text{C}^2/\text{N m}^2)}}$$
$$= 60\ \text{V/m}$$

The amplitudes of the electric and magnetic fields are related by the speed of light. This allows us to compute

$$B_0 = \frac{E_0}{c} = 2.0 \times 10^{-7}\ \text{T}$$

ASSESS The electric field amplitude is modest; the magnetic field amplitude is very small. This implies that the interaction of electromagnetic waves with matter is mostly due to the electric field.

STOP TO THINK 31.4 An electromagnetic wave is traveling in the positive y-direction. The electric field at one instant of time is shown at one position. The magnetic field at this position points

a. In the positive x-direction.
b. In the negative x-direction.
c. In the positive y-direction.
d. In the negative y-direction.
e. Toward the origin.
f. Away from the origin.

Radiation Pressure

Electromagnetic waves transfer not only energy but also momentum. An object gains momentum when it absorbs electromagnetic waves, much as a ball at rest gains momentum when struck by a ball in motion.

Suppose we shine a beam of light on an object that completely absorbs the light energy. If the object absorbs energy during a time interval Δt, its momentum changes by

$$\Delta p = \frac{\text{energy absorbed}}{c}$$

This is a consequence of Maxwell's theory, which we'll state without proof.

The momentum change implies that the light is exerting a force on the object. Newton's second law, in terms of momentum, is $F = \Delta p / \Delta t$. The radiation force due to the beam of light is

$$F = \frac{\Delta p}{\Delta t} = \frac{(\text{energy absorbed})/\Delta t}{c} = \frac{P}{c}$$

where P is the power (joules per second) of the light.

It's more interesting to consider the force exerted on an object per unit area, which is called the **radiation pressure** p_{rad}. The radiation pressure on an object that absorbs all the light is

$$p_{\text{rad}} = \frac{F}{A} = \frac{P/A}{c} = \frac{I}{c} \tag{31.42}$$

where I is the intensity of the light wave. The subscript on p_{rad} is important in this context to distinguish the radiation pressure from the momentum p.

Artist's conception of a future spacecraft powered by radiation pressure from the sun.

EXAMPLE 31.5 | **Solar sailing**

A low-cost way of sending spacecraft to other planets would be to use the radiation pressure on a solar sail. The intensity of the sun's electromagnetic radiation at distances near the earth's orbit is about 1300 W/m². What size sail would be needed to accelerate a 10,000 kg spacecraft toward Mars at 0.010 m/s²?

MODEL Assume that the solar sail is perfectly absorbing.

SOLVE The force that will create a 0.010 m/s² acceleration is $F = ma = 100$ N. We can use Equation 31.42 to find the sail

area that, by absorbing light, will receive a 100 N force from the sun:

$$A = \frac{cF}{I} = \frac{(3.00 \times 10^8 \text{ m/s})(100 \text{ N})}{1300 \text{ W/m}^2} = 2.3 \times 10^7 \text{ m}^2$$

ASSESS If the sail is a square, it would need to be 4.8 km × 4.8 km, or roughly 3 mi × 3 mi. This is large, but not entirely out of the question with thin films that can be unrolled in space. But how will the crew return from Mars?

Antennas

We've seen that an electromagnetic wave is self-sustaining, independent of charges or currents. However, charges and currents are needed at the *source* of an electromagnetic wave. We'll take a brief look at how an electromagnetic wave is generated by an antenna.

FIGURE 31.25 is the electric field of an electric dipole. If the dipole is vertical, the electric field $\vec{E}$ at points along a horizontal line is also vertical. Reversing the dipole, by switching the charges, reverses $\vec{E}$. If the charges were to oscillate back and forth,

FIGURE 31.25 An electric dipole creates an electric field that reverses direction if the dipole charges are switched.

Positive charge on top

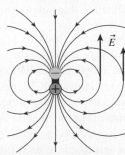

Negative charge on top

FIGURE 31.26 An antenna generates a self-sustaining electromagnetic wave.

An oscillating voltage causes the dipole to oscillate.

Antenna wire

The oscillating dipole causes an electromagnetic wave to move away from the antenna at speed $v_{em} = c$.

switching position at frequency f, then $\vec{E}$ would oscillate in a vertical plane. The changing $\vec{E}$ would then create an induced magnetic field $\vec{B}$, which could then create an $\vec{E}$, which could then create a $\vec{B}, \ldots$, and an electromagnetic wave at frequency f would radiate out into space.

This is exactly what an **antenna** does. **FIGURE 31.26** shows two metal wires attached to the terminals of an oscillating voltage source. The figure shows an instant when the top wire is negative and the bottom is positive, but these will reverse in half a cycle. The wire is basically an oscillating dipole, and it creates an oscillating electric field. The oscillating $\vec{E}$ induces an oscillating $\vec{B}$, and they take off as an electromagnetic wave at speed $v_{em} = c$. The wave does need oscillating charges as a *wave source,* but once created it is self-sustaining and independent of the source. The antenna might be destroyed, but the wave could travel billions of light years across the universe, bearing the legacy of James Clerk Maxwell.

STOP TO THINK 31.5 The amplitude of the oscillating electric field at your cell phone is 4.0 μV/m when you are 10 km east of the broadcast antenna. What is the electric field amplitude when you are 20 km east of the antenna?

a. 1.0 μV/m   b. 2.0 μV/m
c. 4.0 μV/m   d. There's not enough information to tell.

31.7 Polarization

The plane of the electric field vector $\vec{E}$ and the Poynting vector $\vec{S}$ (the direction of propagation) is called the **plane of polarization** of an electromagnetic wave. **FIGURE 31.27** shows two electromagnetic waves moving along the x-axis. The electric field in Figure 31.27a oscillates vertically, so we would say that this wave is *vertically polarized.* Similarly the wave in Figure 31.27b is *horizontally polarized.* Other polarizations are possible, such as a wave polarized 30° away from horizontal.

FIGURE 31.27 The plane of polarization is the plane in which the electric field vector oscillates.

(a) Vertical polarization

(b) Horizontal polarization

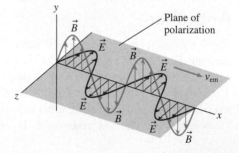

NOTE This use of the term "polarization" is completely independent of the idea of *charge polarization* that you learned about in Chapter 22.

Some wave sources, such as lasers and radio antennas, emit *polarized* electromagnetic waves with a well-defined plane of polarization. By contrast, most natural sources of electromagnetic radiation are unpolarized, emitting waves whose electric fields oscillate randomly with all possible orientations.

A few natural sources are *partially polarized,* meaning that one direction of polarization is more prominent than others. The light of the sky at right angles to the sun is partially polarized because of how the sun's light scatters from air molecules to create skylight. Bees and other insects make use of this partial polarization to navigate. Light reflected from a flat, horizontal surface, such as a road or the surface of a lake, has a predominantly horizontal polarization. This is the rationale for using polarizing sunglasses.

The most common way of artificially generating polarized visible light is to send unpolarized light through a *polarizing filter*. The first widely used polarizing filter was invented by Edwin Land in 1928, while he was still an undergraduate student. He developed an improved version, called Polaroid, in 1938. Polaroid, as shown in **FIGURE 31.28,** is a plastic sheet containing very long organic molecules known as polymers. The sheets are formed in such a way that the polymers are all aligned to form a grid, rather like the metal bars in a barbecue grill. The sheet is then chemically treated to make the polymer molecules somewhat conducting.

As a light wave travels through Polaroid, the component of the electric field oscillating parallel to the polymer grid drives the conduction electrons up and down the molecules. The electrons absorb energy from the light wave, so the parallel component of $\vec{E}$ is absorbed in the filter. But the conduction electrons can't oscillate perpendicular to the molecules, so the component of $\vec{E}$ perpendicular to the polymer grid passes through without absorption. Thus the light wave emerging from a polarizing filter is polarized perpendicular to the polymer grid. The direction of the transmitted polarization is called the *polarizer axis*.

Malus's Law

Suppose a *polarized* light wave of intensity I_0 approaches a polarizing filter. What is the intensity of the light that passes through the filter? **FIGURE 31.29** shows that an oscillating electric field can be decomposed into components parallel and perpendicular to the polarizer axis. If we call the polarizer axis the y-axis, then the incident electric field is

$$\vec{E}_{\text{incident}} = E_\perp \hat{i} + E_\parallel \hat{j} = E_0 \sin\theta \, \hat{i} + E_0 \cos\theta \, \hat{j} \qquad (31.43)$$

where θ is the angle between the incident plane of polarization and the polarizer axis.

If the polarizer is ideal, meaning that light polarized parallel to the axis is 100% transmitted and light perpendicular to the axis is 100% blocked, then the electric field of the light transmitted by the filter is

$$\vec{E}_{\text{transmitted}} = E_\parallel \hat{j} = E_0 \cos\theta \, \hat{j} \qquad (31.44)$$

Because the intensity depends on the square of the electric field amplitude, you can see that the transmitted intensity is related to the incident intensity by

$$I_{\text{transmitted}} = I_0 \cos^2\theta \qquad \text{(incident light polarized)} \qquad (31.45)$$

This result, which was discovered experimentally in 1809, is called **Malus's law.**

FIGURE 31.30a shows that Malus's law can be demonstrated with two polarizing filters. The first, called the *polarizer,* is used to produce polarized light of intensity I_0. The second, called the *analyzer,* is rotated by angle θ relative to the polarizer. As the photographs of **FIGURE 31.30b** show, the transmission of the analyzer is (ideally) 100% when $\theta = 0°$ and steadily decreases to zero when $\theta = 90°$. Two polarizing filters with perpendicular axes, called *crossed polarizers,* block all the light.

FIGURE 31.28 A polarizing filter.

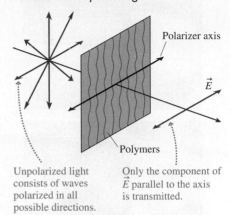

Unpolarized light consists of waves polarized in all possible directions.

Only the component of $\vec{E}$ parallel to the axis is transmitted.

FIGURE 31.29 An incident electric field can be decomposed into components parallel and perpendicular to a polarizer axis.

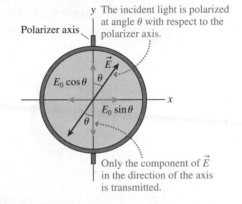

The incident light is polarized at angle θ with respect to the polarizer axis.

Only the component of $\vec{E}$ in the direction of the axis is transmitted.

FIGURE 31.30 The intensity of the transmitted light depends on the angle between the polarizing filters.

(a) Unpolarized light

Polarizer

Analyzer

$\theta = 0°$

$\theta = 45°$

$\theta = 90°$

The vertical polarizer blocks the horizontally polarized glare from the surface of the water.

Suppose the light incident on a polarizing filter is *unpolarized,* as is the light incident from the left on the polarizer in Figure 31.30a. The electric field of unpolarized light varies randomly through all possible values of θ. Because the *average* value of $\cos^2\theta$ is $\frac{1}{2}$, the intensity transmitted by a polarizing filter is

$$I_{transmitted} = \tfrac{1}{2}I_0 \qquad \text{(incident light unpolarized)} \qquad (31.46)$$

In other words, a polarizing filter passes 50% of unpolarized light and blocks 50%.

In polarizing sunglasses, the polymer grid is aligned horizontally (when the glasses are in the normal orientation) so that the glasses transmit vertically polarized light. Most natural light is unpolarized, so the glasses reduce the light intensity by 50%. But *glare*—the reflection of the sun and the skylight from roads and other horizontal surfaces—has a strong horizontal polarization. This light is almost completely blocked by the Polaroid, so the sunglasses "cut glare" without affecting the main scene you wish to see.

You can test whether your sunglasses are polarized by holding them in front of you and rotating them as you look at the glare reflecting from a horizontal surface. Polarizing sunglasses substantially reduce the glare when the glasses are "normal" but not when the glasses are 90° from normal. (You can also test them against a pair of sunglasses known to be polarizing by seeing if all light is blocked when the lenses of the two pairs are crossed.)

STOP TO THINK 31.6 Unpolarized light of equal intensity is incident on four pairs of polarizing filters. Rank in order, from largest to smallest, the intensities I_a to I_d transmitted through the second polarizer of each pair.

(a)   (b)   (c)   (d)

CHALLENGE EXAMPLE 31.6 | Light propulsion

Future space rockets might propel themselves by firing laser beams, rather than exhaust gases, out the back. The acceleration would be small, but it could continue for months or years in the vacuum of space. Consider a 1200 kg unmanned space probe powered by a 15 MW laser. After one year, how far will it have traveled and how fast will it be going?

MODEL Assume the laser efficiency is so high that it can be powered for a year with a negligible mass of fuel.

SOLVE Light waves transfer not only energy but also momentum, which is how they exert a radiation-pressure force. We found that the radiation force of a light beam of power P is

$$F = \frac{P}{c}$$

From Newton's third law, the emitted light waves must exert an equal-but-opposite reaction force on the source of the light. In this case, the emitted light exerts a force of this magnitude on the space probe to which the laser is attached. This reaction force causes the probe to accelerate at

$$a = \frac{F}{m} = \frac{P}{mc} = \frac{15 \times 10^6 \text{ W}}{(1200 \text{ kg})(3.0 \times 10^8 \text{ m/s})}$$
$$= 4.2 \times 10^{-5} \text{ m/s}^2$$

As expected, the acceleration is extremely small. But one year is a large amount of time: $\Delta t = 3.15 \times 10^7$ s. After one year of acceleration,

$$v = a\,\Delta t = 1300 \text{ m/s}$$
$$d = \tfrac{1}{2}a(\Delta t)^2 = 2.1 \times 10^{10} \text{ m}$$

The space probe will have traveled 2.1×10^{10} m and will be going 1300 m/s.

ASSESS Even after a year, the speed is not exceptionally fast—only about 2900 mph. But the probe will have traveled a substantial distance, about 25% of the distance to Mars.

SUMMARY

The goal of Chapter 31 has been to study the properties of electromagnetic fields and waves.

GENERAL PRINCIPLES

Maxwell's Equations

These equations govern electromagnetic fields:

$$\oint \vec{E} \cdot d\vec{A} = \frac{Q_{in}}{\epsilon_0} \qquad \text{Gauss's law}$$

$$\oint \vec{B} \cdot d\vec{A} = 0 \qquad \text{Gauss's law for magnetism}$$

$$\oint \vec{E} \cdot d\vec{s} = -\frac{d\Phi_m}{dt} \qquad \text{Faraday's law}$$

$$\oint \vec{B} \cdot d\vec{s} = \mu_0 I_{through} + \epsilon_0 \mu_0 \frac{d\Phi_e}{dt} \qquad \textit{Ampère-Maxwell law}$$

Maxwell's equations tell us that:

An electric field can be created by

- Charged particles
- A changing magnetic field

A magnetic field can be created by

- A current
- A changing electric field

Lorentz Force

This force law governs the interaction of charged particles with electromagnetic fields:

$$\vec{F} = q(\vec{E} + \vec{v} \times \vec{B})$$

- An electric field exerts a force on any charged particle.
- A magnetic field exerts a force on a moving charged particle.

Field Transformations

Fields measured in reference frame A to be $\vec{E}_A$ and $\vec{B}_A$ are found in frame B to be

$$\vec{E}_B = \vec{E}_A + \vec{v}_{BA} \times \vec{B}_A$$

$$\vec{B}_B = \vec{B}_A - \frac{1}{c^2}\vec{v}_{BA} \times \vec{E}_A$$

IMPORTANT CONCEPTS

Induced fields

An induced electric field is created by a changing magnetic field.

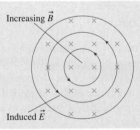

An induced magnetic field is created by a changing electric field.

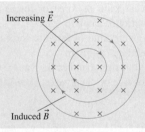

These fields can exist independently of charges and currents.

An **electromagnetic wave** is a self-sustaining electromagnetic field.

- An em wave is a transverse wave with $\vec{E}$, $\vec{B}$, and $\vec{v}_{em}$ mutually perpendicular.
- An em wave propagates with speed $v_{em} = c = 1/\sqrt{\epsilon_0 \mu_0}$.
- The electric and magnetic field strengths are related by $E = cB$.
- The **Poynting vector** $\vec{S} = (\vec{E} \times \vec{B})/\mu_0$ is the energy transfer in the direction of travel.
- The wave **intensity** is $I = P/A = (1/2c\mu_0)E_0^2 = (c\epsilon_0/2)E_0^2$.

APPLICATIONS

Polarization

The electric field and the Poynting vector define the **plane of polarization**. The intensity of polarized light transmitted through a polarizing filter is given by Malus's law:

$$I = I_0 \cos^2\theta$$

where θ is the angle between the electric field and the polarizer axis.

TERMS AND NOTATION

Galilean field transformation equations	Maxwell's equations	intensity, I	plane of polarization
displacement current	electromagnetic wave	radiation pressure, p_{rad}	Malus's law
	Poynting vector, $\vec{S}$	antenna	

CONCEPTUAL QUESTIONS

1. Andre is flying his spaceship to the left through the laboratory magnetic field of **FIGURE Q31.1**.
   a. Does Andre see a magnetic field? If so, in which direction does it point?
   b. Does Andre see an electric field? If so, in which direction does it point?

FIGURE Q31.1

FIGURE Q31.2

2. Sharon drives her rocket through the magnetic field of **FIGURE Q31.2** traveling to the right at a speed of 1000 m/s as measured by Bill. As she passes Bill, she shoots a positive charge backward at a speed of 1000 m/s relative to her.
   a. According to Bill, what kind of force or forces act on the charge? In which directions? Explain.
   b. According to Sharon, what kind of force or forces act on the charge? In which directions? Explain.

3. If you curl the fingers of your right hand as shown, are the electric fluxes in **FIGURE Q31.3** positive or negative?

FIGURE Q31.3

4. What is the current through surface S in **FIGURE Q31.4** if you curl your right fingers in the direction of the arrow?

FIGURE Q31.4

FIGURE Q31.5

5. Is the electric field strength in **FIGURE Q31.5** increasing, decreasing, or not changing? Explain.

6. Do the situations in **FIGURE Q31.6** represent possible electromagnetic waves? If not, why not?

(a)

(b)

FIGURE Q31.6

7. In what directions are the electromagnetic waves traveling in **FIGURE Q31.7**?

(a)

(b)

FIGURE Q31.7

8. The intensity of an electromagnetic wave is 10 W/m^2. What will the intensity be if:
   a. The amplitude of the electric field is doubled?
   b. The amplitude of the magnetic field is doubled?
   c. The amplitudes of both the electric and the magnetic fields are doubled?
   d. The frequency is doubled?

9. Older televisions used a *loop antenna* like the one in **FIGURE Q31.9**. How does this antenna work?

FIGURE Q31.9

10. A vertically polarized electromagnetic wave passes through the five polarizers in **FIGURE Q31.10**. Rank in order, from largest to smallest, the transmitted intensities I_a to I_e.

a   b   c   d   e

FIGURE Q31.10

EXERCISES AND PROBLEMS

Problems labeled ▓▓ integrate material from earlier chapters.

Exercises

Section 31.1 E or B? It Depends on Your Perspective

1. | **FIGURE EX31.1** shows the electric and magnetic field in frame A. A rocket in frame B travels parallel to one of the axes of the A coordinate system. Along which axis must the rocket travel, and in which direction, in order for the rocket scientists to measure (a) $B_B > B_A$, (b) $B_B = B_A$, and (c) $B_B < B_A$?

FIGURE EX31.1

2. || A rocket cruises past a laboratory at 1.00×10^6 m/s in the positive x-direction just as a proton is launched with velocity (in the laboratory frame) $\vec{v} = (1.41 \times 10^6 \hat{\imath} + 1.41 \times 10^6 \hat{\jmath})$ m/s. What are the proton's speed and its angle from the y-axis in (a) the laboratory frame and (b) the rocket frame?

3. || Laboratory scientists have created the electric and magnetic fields shown in **FIGURE EX31.3**. These fields are also seen by scientists who zoom past in a rocket traveling in the x-direction at 1.0×10^6 m/s. According to the rocket scientists, what angle does the electric field make with the axis of the rocket?

FIGURE EX31.3

4. || Scientists in the laboratory create a uniform electric field $\vec{E} = 1.0 \times 10^6 \hat{k}$ V/m in a region of space where $\vec{B} = \vec{0}$. What are the fields in the reference frame of a rocket traveling in the positive x-direction at 1.0×10^6 m/s?

5. || A rocket zooms past the earth at $v = 2.0 \times 10^6$ m/s. Scientists on the rocket have created the electric and magnetic fields shown in **FIGURE EX31.5**. What are the fields measured by an earthbound scientist?

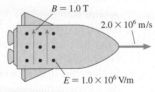

FIGURE EX31.5

Section 31.2 The Field Laws Thus Far

Section 31.3 The Displacement Current

6. || The magnetic field is uniform over each face of the box shown in **FIGURE EX31.6**. What are the magnetic field strength and direction on the front surface?

FIGURE EX31.6

7. | Show that the quantity $\epsilon_0 (d\Phi_e/dt)$ has units of current.

8. || Show that the displacement current inside a parallel-plate capacitor can be written $C(dV_C/dt)$.

9. | What capacitance, in μF, has its potential difference increasing at 1.0×10^6 V/s when the displacement current in the capacitor is 1.0 A?

10. || A 5.0-cm-diameter parallel-plate capacitor has a 0.50 mm gap. What is the displacement current in the capacitor if the potential difference across the capacitor is increasing at 500,000 V/s?

11. ||| A 10-cm-diameter parallel-plate capacitor has a 1.0 mm spacing. The electric field between the plates is increasing at the rate 1.0×10^6 V/m s. What is the magnetic field strength (a) on the axis, (b) 3.0 cm from the axis, and (c) 7.0 cm from the axis?

Section 31.5 Electromagnetic Waves

12. | What is the magnetic field amplitude of an electromagnetic wave whose electric field amplitude is 10 V/m?

13. | What is the electric field amplitude of an electromagnetic wave whose magnetic field amplitude is 2.0 mT?

14. | The magnetic field of an electromagnetic wave in a vacuum is $B_z = (3.00 \ \mu\text{T}) \sin[(1.00 \times 10^7)x - \omega t]$, where x is in m and t is in s. What are the wave's (a) wavelength, (b) frequency, and (c) electric field amplitude?

15. | The electric field of an electromagnetic wave in a vacuum is $E_y = (20.0 \ \text{V/m}) \cos[(6.28 \times 10^8)x - \omega t]$, where x is in m and t is in s. What are the wave's (a) wavelength, (b) frequency, and (c) magnetic field amplitude?

Section 31.6 Properties of Electromagnetic Waves

16. | a. What is the magnetic field amplitude of an electromagnetic wave whose electric field amplitude is 100 V/m?
    b. What is the intensity of the wave?

17. | A radio wave is traveling in the negative y-direction. What is the direction of $\vec{E}$ at a point where $\vec{B}$ is in the positive x-direction?

18. || A radio receiver can detect signals with electric field amplitudes as small as 300 μV/m. What is the intensity of the smallest detectable signal?

19. || A helium-neon laser emits a 1.0-mm-diameter laser beam with a power of 1.0 mW. What are the amplitudes of the electric and magnetic fields of the light wave?

20. || A radio antenna broadcasts a 1.0 MHz radio wave with 25 kW of power. Assume that the radiation is emitted uniformly in all directions.
    a. What is the wave's intensity 30 km from the antenna?
    b. What is the electric field amplitude at this distance?

21. || A 200 MW laser pulse is focused with a lens to a diameter of 2.0 μm.
    a. What is the laser beam's electric field amplitude at the focal point?
    b. What is the ratio of the laser beam's electric field to the electric field that keeps the electron bound to the proton of a hydrogen atom? The radius of the electron orbit is 0.053 nm.

22. | A 1000 W carbon-dioxide laser emits light with a wavelength of 10 μm into a 3.0-mm-diameter laser beam. What force does the laser beam exert on a completely absorbing target?

23. ‖ At what distance from a 10 W point source of electromagnetic waves is the magnetic field amplitude $1.0 \, \mu T$?

Section 31.7 Polarization

24. ‖ Only 25% of the intensity of a polarized light wave passes through a polarizing filter. What is the angle between the electric field and the axis of the filter?

25. | **FIGURE EX31.25** shows a vertically polarized radio wave of frequency 1.0×10^6 Hz traveling into the page. The maximum electric field strength is 1000 V/m. What are

    a. The maximum magnetic field strength?
    b. The magnetic field strength and direction at a point where $\vec{E} =$ (500 V/m, down)?

FIGURE EX31.25

Electromagnetic wave traveling into page

26. ‖ Unpolarized light with intensity $350 \, W/m^2$ passes first through a polarizing filter with its axis vertical, then through a second polarizing filter. It emerges from the second filter with intensity $131 \, W/m^2$. What is the angle from vertical of the axis of the second polarizing filter?

27. ‖ A 200 mW vertically polarized laser beam passes through a polarizing filter whose axis is 35° from horizontal. What is the power of the laser beam as it emerges from the filter?

Problems

28. ‖ What are the electric field strength and direction at the position of the electron in **FIGURE P31.28**?

FIGURE P31.28

3.2×10^{-13} N

$\vec{F}$

30°

$B = 0.10$ T

Moving out of page at 1.0×10^7 m/s

FIGURE P31.29

$E = 1.0 \times 10^6$ V/m

$v = 1.0 \times 10^7$ m/s

$\vec{v}$

$B = 0.10$ T

29. ‖ What is the force (magnitude and direction) on the proton in **FIGURE P31.29**? Give the direction as an angle cw or ccw from vertical.

30. | What electric field strength and direction will allow the proton in **FIGURE P31.30** to pass through this region of space without being deflected?

FIGURE P31.30

$B = 0.010$ T

$\vec{v}$

$v = 2.0 \times 10^7$ m/s

FIGURE P31.31

$\vec{v}$

$\vec{E}$

31. | A proton is fired with a speed of 1.0×10^6 m/s through the parallel-plate capacitor shown in **FIGURE P31.31**. The capacitor's electric field is $\vec{E} = (1.0 \times 10^5$ V/m, down).

    a. What magnetic field $\vec{B}$, both strength and direction, must be applied to allow the proton to pass through the capacitor with no change in speed or direction?
    b. Find the electric and magnetic fields in the proton's reference frame.

---

c. How does an experimenter in the proton's frame explain that the proton experiences no force as the charged plates fly by?

32. ‖ An electron travels with $\vec{v} = 5.0 \times 10^6 \hat{\imath}$ m/s through a point in space where $\vec{E} = (2.0 \times 10^5 \hat{\imath} - 2.0 \times 10^5 \hat{\jmath})$ V/m and $\vec{B} = -0.10 \hat{k}$ T. What is the force on the electron?

33. ‖ In **FIGURE P31.33**, a circular loop of radius r travels with speed v along a charged wire having linear charge density λ. The wire is at rest in the laboratory frame, and it passes through the center of the loop.

    a. What are $\vec{E}$ and $\vec{B}$ at a point on the loop as measured by a scientist in the laboratory? Include both strength and direction.
    b. What are the fields $\vec{E}$ and $\vec{B}$ at a point on the loop as measured by a scientist in the frame of the loop?
    c. Show that an experimenter in the loop's frame sees a current $I = \lambda v$ passing through the center of the loop.
    d. What electric and magnetic fields would an experimenter in the loop's frame calculate at distance r from the current of part c?
    e. Show that your fields of parts b and d are the same.

FIGURE P31.33

$\vec{v}$

Linear density λ

34. ‖ The magnetic field inside a 4.0-cm-diameter superconducting solenoid varies sinusoidally between 8.0 T and 12.0 T at a frequency of 10 Hz.

    a. What is the maximum electric field strength at a point 1.5 cm from the solenoid axis?
    b. What is the value of B at the instant E reaches its maximum value?

35. ‖ A simple series circuit consists of a 150 Ω resistor, a 25 V battery, a switch, and a 2.5 pF parallel-plate capacitor (initially uncharged) with plates 5.0 mm apart. The switch is closed at $t = 0$ s.

    a. After the switch is closed, find the maximum electric flux and the maximum displacement current through the capacitor.
    b. Find the electric flux and the displacement current at $t = 0.50$ ns.

36. ‖ A wire with conductivity σ carries current I. The current is increasing at the rate dI/dt.

    a. Show that there is a displacement current in the wire equal to $(\epsilon_0/\sigma)(dI/dt)$.
    b. Evaluate the displacement current for a copper wire in which the current is increasing at 1.0×10^6 A/s.

37. ‖‖ A 10 A current is charging a 1.0-cm-diameter parallel-plate capacitor.

    a. What is the magnetic field strength at a point 2.0 mm radially from the center of the wire leading to the capacitor?
    b. What is the magnetic field strength at a point 2.0 mm radially from the center of the capacitor?

38. ‖ **FIGURE P31.38** shows the voltage across a $0.10 \, \mu F$ capacitor. Draw a graph showing the displacement current through the capacitor as a function of time.

ΔV_C (V)

100

50

0

$t \, (\mu s)$

0 1 2 3 4

FIGURE P31.38

39. ‖ **FIGURE P31.39** shows the electric field inside a cylinder of radius $R = 3.0$ mm. The field strength is increasing with time as $E = 1.0 \times 10^8 t^2$ V/m, where t is in s. The electric field outside the cylinder is always zero, and the field inside the cylinder was zero for $t < 0$.

FIGURE P31.39

   a. Find an expression for the electric flux Φ_e through the entire cylinder as a function of time.

   b. Draw a picture showing the magnetic field lines inside and outside the cylinder. Be sure to include arrowheads showing the field's direction.

   c. Find an expression for the magnetic field strength as a function of time at a distance $r < R$ from the center. Evaluate the magnetic field strength at $r = 2.0$ mm, $t = 2.0$ s.

   d. Find an expression for the magnetic field strength as a function of time at a distance $r > R$ from the center. Evaluate the magnetic field strength at $r = 4.0$ mm, $t = 2.0$ s.

40. ‖ A long, thin superconducting wire carrying a 15 A current passes through the center of a thin, 2.0-cm-diameter ring. A uniform electric field of increasing strength also passes through the ring, parallel to the wire. The magnetic field through the ring is zero.

   a. At what rate is the electric field strength increasing?

   b. Is the electric field in the direction of the current or opposite to the current?

41. ‖ A $1.0 \ \mu$F capacitor is discharged, starting at $t = 0$ s. The displacement current through the plates is $I_{disp} = (10 \ A)\exp(-t/2.0 \ \mu s)$. What was the capacitor's initial voltage $(\Delta V_C)_0$?

42. ‖‖ At one instant, the electric and magnetic fields at one point of an electromagnetic wave are $\vec{E} = (200\,\hat{\imath} + 300\,\hat{\jmath} - 50\,\hat{k})$ V/m and $\vec{B} = B_0(7.3\,\hat{\imath} - 7.3\,\hat{\jmath} + a\,\hat{k}) \ \mu$T.

   a. What are the values of a and B_0?

   b. What is the Poynting vector at this time and position?

43. ‖ a. Show that u_E and u_B, the energy densities of the electric and magnetic fields, are equal to each other in an electromagnetic wave. In other words, show that the wave's energy is divided equally between the electric field and the magnetic field.

   b. What is the total energy density in an electromagnetic wave of intensity 1000 W/m²?

44. ‖ The intensity of sunlight reaching the earth is 1360 W/m².

   a. What is the power output of the sun?

   b. What is the intensity of sunlight reaching Mars?

45. ‖ Assume that a 7.0-cm-diameter, 100 W lightbulb radiates all its energy as a single wavelength of visible light. Estimate the electric and magnetic field strengths at the surface of the bulb.

46. ‖ The electric field of a 450 MHz radio wave has a maximum rate of change of 4.5×10^{11} (V/m)/s. What is the wave's magnetic field amplitude?

47. ‖ When the Voyager 2 spacecraft passed Neptune in 1989, it was 4.5×10^9 km from the earth. Its radio transmitter, with which it sent back data and images, broadcast with a mere 21 W of power. Assuming that the transmitter broadcast equally in all directions,

   a. What signal intensity was received on the earth?

   b. What electric field amplitude was detected?

   The received signal was somewhat stronger than your result because the spacecraft used a directional antenna, but not by much.

48. ‖ In reading the instruction manual that came with your garage-door opener, you see that the transmitter unit in your car produces a 250 mW, 350 MHz signal and that the receiver unit is supposed to respond to a radio wave at this frequency if the electric field amplitude exceeds 0.10 V/m. You wonder if this is really true. To find out, you put fresh batteries in the transmitter and start walking away from your garage while opening and closing the door. Your garage door finally fails to respond when you're 42 m away. What is the electric field amplitude at the receiver when it first fails?

49. ‖ The maximum electric field strength in air is 3.0 MV/m. Stronger electric fields ionize the air and create a spark. What is the maximum power that can be delivered by a 1.0-cm-diameter laser beam propagating through air?

50. ‖‖ A LASIK vision-correction system uses a laser that emits
BIO 10-ns-long pulses of light, each with 2.5 mJ of energy. The laser beam is focused to a 0.85-mm-diameter circle on the cornea. What is the electric field amplitude of the light wave at the cornea?

51. ‖ The intensity of sunlight reaching the earth is 1360 W/m². Assuming all the sunlight is absorbed, what is the radiation-pressure force on the earth? Give your answer (a) in newtons and (b) as a fraction of the sun's gravitational force on the earth.

52. ‖ For radio and microwaves, the depth of penetration into
BIO the human body is proportional to $\lambda^{1/2}$. If 27 MHz radio waves penetrate to a depth of 14 cm, how far do 2.4 GHz microwaves penetrate?

53. ‖ A laser beam shines straight up onto a flat, black foil of mass m.

   a. Find an expression for the laser power P needed to levitate the foil.

   b. Evaluate P for a foil with a mass of 25 μg.

54. ‖ For a science project, you would like to horizontally suspend an 8.5 by 11 inch sheet of black paper in a vertical beam of light whose dimensions exactly match the paper. If the mass of the sheet is 1.0 g, what light intensity will you need?

55. ‖ You've recently read about a chemical laser that generates a 20-cm-diameter, 25 MW laser beam. One day, after physics class, you start to wonder if you could use the radiation pressure from this laser beam to launch small payloads into orbit. To see if this might be feasible, you do a quick calculation of the acceleration of a 20-cm-diameter, 100 kg, perfectly absorbing block. What speed would such a block have if pushed *horizontally* 100 m along a frictionless track by such a laser?

56. ‖ Unpolarized light of intensity I_0 is incident on three polarizing filters. The axis of the first is vertical, that of the second is 45° from vertical, and that of the third is horizontal. What light intensity emerges from the third filter?

57. ‖ Unpolarized light of intensity I_0 is incident on two polarizing filters. The transmitted light intensity is $I_0/10$. What is the angle between the axes of the two filters?

58. ‖ Unpolarized light of intensity I_0 is incident on a stack of 7 polarizing filters, each with its axis rotated 15° cw with respect to the previous filter. What light intensity emerges from the last filter?

Challenge Problems

59. ‖‖ A cube of water 10 cm on a side is placed in a microwave beam having $E_0 = 11$ kV/m. The microwaves illuminate one face of the cube, and the water absorbs 80% of the incident energy. How long will it take to raise the water temperature by 50°C? Assume that the water has no heat loss during this time.

60. ||| An 80 kg astronaut has gone outside his space capsule to do some repair work. Unfortunately, he forgot to lock his safety tether in place, and he has drifted 5.0 m away from the capsule. Fortunately, he has a 1000 W portable laser with fresh batteries that will operate it for 1.0 h. His only chance is to accelerate himself toward the space capsule by firing the laser in the opposite direction. He has a 10-h supply of oxygen. How long will it take him to reach safety?

61. ||| An electron travels with $\vec{v} = 5.0 \times 10^6\,\hat{\imath}$ m/s through a point in space where $\vec{B} = 0.10\,\hat{\jmath}$ T. The force on the electron at this point is $\vec{F} = (9.6 \times 10^{-14}\,\hat{\imath} - 9.6 \times 10^{-14}\,\hat{k})$ N. What is the electric field?

62. ||| The radar system at an airport broadcasts 11 GHz microwaves with 150 kW of power. An approaching airplane with a 31 m² cross section is 30 km away. Assume that the radar broadcasts uniformly in all directions and that the airplane scatters microwaves uniformly in all directions. What is the electric field strength of the microwave signal received back at the airport 200 μs later?

63. ||| Large quantities of dust should have been left behind after the creation of the solar system. Larger dust particles, comparable in size to soot and sand grains, are common. They create shooting stars when they collide with the earth's atmosphere. But very small dust particles are conspicuously absent. Astronomers believe that the very small dust particles have been blown out of the solar system by the sun. By comparing the forces on dust particles, determine the diameter of the smallest dust particles that can remain in the solar system over long periods of time. Assume that the dust particles are spherical, black, and have a density of 2000 kg/m³. The sun emits electromagnetic radiation with power 3.9×10^{26} W.

64. ||| Consider current I passing through a resistor of radius r,
CALC length L, and resistance R.
   a. Determine the electric and magnetic fields at the surface of the resistor. Assume that the electric field is uniform throughout, including at the surface.
   b. Determine the strength and direction of the Poynting vector at the surface of the resistor.
   c. Show that the flux of the Poynting vector (i.e., the integral of $\vec{S} \cdot d\vec{A}$) over the surface of the resistor is $I^2 R$. Then give an interpretation of this result.

32 AC Circuits

Transmission lines carry
alternating current at
voltages as high as 500,000 V.

IN THIS CHAPTER, you will learn about and analyze AC circuits.

What is an AC circuit?

The circuits of Chapter 28, with a steady current in one direction, are called DC circuits—*direct current*. A circuit with an oscillating emf is called an AC circuit, for *alternating current*. The wires that transport electricity across the country—the grid—use alternating current.

« LOOKING BACK Chapter 28 Circuits

How do circuit elements act in an AC circuit?

Resistors in an AC circuit act as they do in a DC circuit. But you'll learn that capacitors and inductors are more useful in AC circuits than in DC circuits.

- The voltage across and the current through a capacitor or inductor are 90° out of phase. One is peaking when the other is zero, and vice versa.
- The peak voltage V and peak current I have an Ohm's-law-like relationship $V = IX$, where X, which depends on frequency, is called the reactance.
- Unlike resistors, capacitors and inductors do not dissipate energy.

« LOOKING BACK Section 26.5 Capacitors
« LOOKING BACK Section 30.8 Inductors

The current and voltage are 90° out of phase with each other.

What is a phasor?

AC voltages oscillate sinusoidally, so the mathematics of AC circuits is that of SHM. You'll learn a new way to represent oscillating quantities—as a rotating vector called a phasor. The instantaneous value of a phasor quantity is its horizontal projection.

« LOOKING BACK Chapter 15 Simple harmonic motion and resonance

What is an *RLC* circuit?

A circuit with a resistor, inductor, and capacitor in series is called an *RLC* circuit. An *RLC* circuit has a resonance—a large current over a narrow range of frequencies—that allows it to be tuned to a specific frequency. As a result, *RLC* circuits are very important in communications.

Why are AC circuits important?

AC circuits are the backbone of our technological society. Generators automatically produce an oscillating emf, AC power is easily transported over large distances, and transformers allow engineers to shift the AC voltage up or down. The circuits of radio, television, and cell phones are also AC circuits because they work with oscillating voltages and currents—at much higher frequencies than the grid, but the physical principles are the same.

FIGURE 32.1 An oscillating emf can be represented as a graph or as a phasor diagram.

(a)

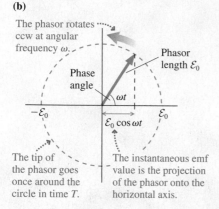

(b)

FIGURE 32.1 An oscillating emf can be represented as a graph or as a phasor diagram.

FIGURE 32.2 The correspondence between a phasor and points on a graph.

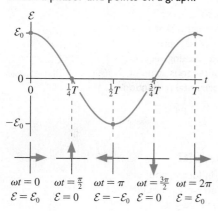

$\omega t = 0$　$\omega t = \frac{\pi}{2}$　$\omega t = \pi$　$\omega t = \frac{3\pi}{2}$　$\omega t = 2\pi$
$\mathcal{E} = \mathcal{E}_0$　$\mathcal{E} = 0$　$\mathcal{E} = -\mathcal{E}_0$　$\mathcal{E} = 0$　$\mathcal{E} = \mathcal{E}_0$

32.1 AC Sources and Phasors

One of the examples of Faraday's law cited in Chapter 30 was an electric generator. A turbine, which might be powered by expanding steam or falling water, causes a coil of wire to rotate in a magnetic field. As the coil spins, the emf and the induced current oscillate sinusoidally. The emf is alternately positive and negative, causing the charges to flow in one direction and then, a half cycle later, in the other. The oscillation frequency of the *grid* in North and South America is $f = 60$ Hz, whereas most of the rest of the world uses a 50 Hz oscillation.

The generator's peak emf—the peak voltage—is a fixed, unvarying quantity, so it might seem logical to call a generator an *alternating-voltage source*. Nonetheless, circuits powered by a sinusoidal emf are called **AC circuits,** where AC stands for *alternating current.* By contrast, the steady-current circuits you studied in Chapter 28 are called **DC circuits,** for *direct current.*

AC circuits are not limited to the use of 50 Hz or 60 Hz power-line voltages. Audio, radio, television, and telecommunication equipment all make extensive use of AC circuits, with frequencies ranging from approximately 10^2 Hz in audio circuits to approximately 10^9 Hz in cell phones. These devices use *electrical oscillators* rather than generators to produce a sinusoidal emf, but the basic principles of circuit analysis are the same.

You can think of an AC generator or oscillator as a battery whose output voltage undergoes sinusoidal oscillations. The instantaneous emf of an AC generator or oscillator, shown graphically in **FIGURE 32.1a,** can be written

$$\mathcal{E} = \mathcal{E}_0 \cos \omega t \tag{32.1}$$

where $\mathcal{E}_0$ is the peak or maximum emf and $\omega = 2\pi f$ is the angular frequency in radians per second. Recall that the units of emf are volts. As you can imagine, the mathematics of AC circuit analysis are going to be very similar to the mathematics of simple harmonic motion.

An alternative way to represent the emf and other oscillatory quantities is with the *phasor diagram* of **FIGURE 32.1b.** A **phasor** is a vector that rotates *counterclockwise* (ccw) around the origin at angular frequency ω. The length or magnitude of the phasor is the maximum value of the quantity. For example, the length of an emf phasor is $\mathcal{E}_0$. The angle ωt is the *phase angle*, an idea you learned about in Chapter 15, where we made a connection between circular motion and simple harmonic motion.

The quantity's instantaneous value, the value you would measure at time t, is the projection of the phasor onto the horizontal axis. This is also analogous to the connection between circular motion and simple harmonic motion. **FIGURE 32.2** helps you visualize the phasor rotation by showing how the phasor corresponds to the more familiar graph at several specific points in the cycle.

STOP TO THINK 32.1 The magnitude of the instantaneous value of the emf represented by this phasor is

a. Increasing.
b. Decreasing.
c. Constant.
d. It's not possible to tell without knowing t.

Resistor Circuits

In Chapter 28 you learned to analyze a circuit in terms of the current I, voltage V, and potential difference ΔV. Now, because the current and voltage are oscillating, we will use lowercase i to represent the *instantaneous* current through a circuit element and v for the circuit element's *instantaneous* voltage.

FIGURE 32.3 shows the instantaneous current i_R through a resistor R. The potential difference across the resistor, which we call the *resistor voltage* v_R, is given by Ohm's law:

$$v_R = i_R R \qquad (32.2)$$

FIGURE 32.4 shows a resistor R connected across an AC emf $\mathcal{E}$. Notice that the circuit symbol for an AC generator is —Ⓝ—. We can analyze this circuit in exactly the same way we analyzed a DC resistor circuit. Kirchhoff's loop law says that the sum of all the potential differences around a closed path is zero:

$$\sum \Delta V = \Delta V_{\text{source}} + \Delta V_{\text{res}} = \mathcal{E} - v_R = 0 \qquad (32.3)$$

The minus sign appears, just as it did in the equation for a DC circuit, because the potential *decreases* when we travel through a resistor in the direction of the current. We find from the loop law that $v_R = \mathcal{E} = \mathcal{E}_0 \cos \omega t$. This isn't surprising because the resistor is connected directly across the terminals of the emf.

The resistor voltage in an AC circuit can be written

$$v_R = V_R \cos \omega t \qquad (32.4)$$

where V_R is the peak or maximum voltage. You can see that $V_R = \mathcal{E}_0$ in the single-resistor circuit of Figure 32.4. Thus the current through the resistor is

$$i_R = \frac{v_R}{R} = \frac{V_R \cos \omega t}{R} = I_R \cos \omega t \qquad (32.5)$$

where $I_R = V_R/R$ is the peak current.

> **NOTE** Ohm's law applies to both the instantaneous *and* peak currents and voltages of a resistor.

The resistor's instantaneous current and voltage are in phase, both oscillating as $\cos \omega t$. FIGURE 32.5 shows the voltage and the current simultaneously on a graph and as a phasor diagram. The fact that the current phasor is shorter than the voltage phasor has no significance. Current and voltage are measured in different units, so you can't compare the length of one to the length of the other. Showing the two different quantities on a single graph—a tactic that can be misleading if you're not careful—illustrates that they oscillate in phase and that their phasors rotate together at the same angle and frequency.

FIGURE 32.3 Instantaneous current i_R through a resistor.

The instantaneous current in the resistor

The instantaneous resistor voltage is $v_R = i_R R$. The potential decreases in the direction of the current.

FIGURE 32.4 An AC resistor circuit.

This is the current direction when $\mathcal{E} > 0$. A half cycle later it will be in the opposite direction.

FIGURE 32.5 Graph and phasor diagrams of the resistor current and voltage.

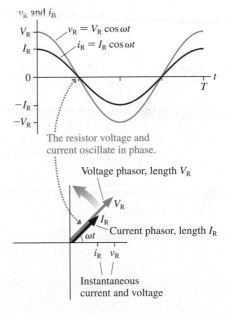

$v_R = V_R \cos \omega t$
$i_R = I_R \cos \omega t$

The resistor voltage and current oscillate in phase.

Voltage phasor, length V_R

Current phasor, length I_R

Instantaneous current and voltage

EXAMPLE 32.1 | **Finding resistor voltages**

In the circuit of FIGURE 32.6, what are (a) the peak voltage across each resistor and (b) the instantaneous resistor voltages at $t = 20$ ms?

VISUALIZE Figure 32.6 shows the circuit diagram. The two resistors are in series.

SOLVE a. The equivalent resistance of the two series resistors is $R_{\text{eq}} = 5\ \Omega +$ 15 $\Omega = 20\ \Omega$. The instantaneous current through the equivalent resistance is

$$i_R = \frac{v_R}{R_{\text{eq}}} = \frac{\mathcal{E}_0 \cos \omega t}{R_{\text{eq}}} = \frac{(100\ \text{V})\cos\left(2\pi(60\ \text{Hz})t\right)}{20\ \Omega}$$

$$= (5.0\ \text{A})\cos\left(2\pi(60\ \text{Hz})t\right)$$

FIGURE 32.6 An AC resistor circuit.

$(100\ \text{V}) \cos(2\pi(60\ \text{Hz})t)$

$5\ \Omega$

$15\ \Omega$

Continued

The peak current is $I_R = 5.0$ A, and this is also the peak current through the two resistors that form the 20 Ω equivalent resistance. Hence the peak voltage across each resistor is

$$V_R = I_R R = \begin{cases} 25 \text{ V} & 5 \text{ Ω resistor} \\ 75 \text{ V} & 15 \text{ Ω resistor} \end{cases}$$

b. The instantaneous current at $t = 0.020$ s is

$$i_R = (5.0 \text{ A}) \cos(2\pi(60 \text{ Hz})(0.020 \text{ s})) = 1.55 \text{ A}$$

The resistor voltages at this time are

$$v_R = i_R R = \begin{cases} 7.7 \text{ V} & 5 \text{ Ω resistor} \\ 23.2 \text{ V} & 15 \text{ Ω resistor} \end{cases}$$

ASSESS The sum of the instantaneous voltages, 30.9 V, is what you would find by calculating $\mathcal{E}$ at $t = 20$ ms. This self-consistency gives us confidence in the answer.

STOP TO THINK 32.2 The resistor whose voltage and current phasors are shown here has resistance R

a. $> 1 \text{ Ω}$
b. $< 1 \text{ Ω}$
c. It's not possible to tell.

32.2 Capacitor Circuits

FIGURE 32.7a shows current i_C charging a capacitor with capacitance C. The instantaneous capacitor voltage is $v_C = q/C$, where $\pm q$ is the charge on the two capacitor plates at this instant. It is useful to compare Figure 32.7a to Figure 32.3 for a resistor.

FIGURE 32.7b, where capacitance C is connected across an AC source of emf $\mathcal{E}$, is the most basic capacitor circuit. The capacitor is in parallel with the source, so the capacitor voltage equals the emf: $v_C = \mathcal{E} = \mathcal{E}_0 \cos \omega t$. It will be useful to write

$$v_C = V_C \cos \omega t \tag{32.6}$$

where V_C is the peak or maximum voltage across the capacitor. You can see that $V_C = \mathcal{E}_0$ in this single-capacitor circuit.

To find the current to and from the capacitor, we first write the charge

$$q = Cv_C = CV_C \cos \omega t \tag{32.7}$$

The current is the *rate* at which charge flows through the wires, $i_C = dq/dt$, thus

$$i_C = \frac{dq}{dt} = \frac{d}{dt}(CV_C \cos \omega t) = -\omega C V_C \sin \omega t \tag{32.8}$$

We can most easily see the relationship between the capacitor voltage and current if we use the trigonometric identity $-\sin(x) = \cos(x + \pi/2)$ to write

$$i_C = \omega C V_C \cos\left(\omega t + \frac{\pi}{2}\right) \tag{32.9}$$

In contrast to a resistor, a capacitor's current and voltage are *not* in phase. In **FIGURE 32.8a**, a graph of the instantaneous voltage v_C and current i_C, you can see that the current peaks one-quarter of a period *before* the voltage peaks. The phase angle

FIGURE 32.7 An AC capacitor circuit.

(a)

The instantaneous current to and from the capacitor

The instantaneous capacitor voltage is $v_C = q/C$. The potential decreases from $+$ to $-$.

(b)

of the current phasor on the phasor diagram of **FIGURE 32.8b** is $\pi/2$ rad—a quarter of a circle—larger than the phase angle of the voltage phasor.

We can summarize this finding:

> **The AC current of a capacitor *leads* the capacitor voltage by $\pi/2$ rad, or 90°.**

The current reaches its peak value I_C at the instant the capacitor is fully discharged and $v_C = 0$. The current is zero at the instant the capacitor is fully charged.

A simple harmonic oscillator provides a mechanical analogy of the 90° phase difference between current and voltage. You learned in Chapter 15 that the position and velocity of a simple harmonic oscillator are

$$x = A \cos \omega t$$

$$v = \frac{dx}{dt} = -\omega A \sin \omega t = -v_{max} \sin \omega t = v_{max} \cos\left(\omega t + \frac{\pi}{2}\right)$$

You can see that the velocity of an oscillator leads the position by 90° in the same way that the capacitor current leads the voltage.

Capacitive Reactance

We can use Equation 32.9 to see that the peak current to and from a capacitor is $I_C = \omega C V_C$. This relationship between the peak voltage and peak current looks much like Ohm's law for a resistor if we define the **capacitive reactance** X_C to be

$$X_C \equiv \frac{1}{\omega C} \tag{32.10}$$

With this definition,

$$I_C = \frac{V_C}{X_C} \quad \text{or} \quad V_C = I_C X_C \tag{32.11}$$

The units of reactance, like those of resistance, are ohms.

> **NOTE** Reactance relates the *peak* voltage V_C and current I_C. But reactance differs from resistance in that it does *not* relate the instantaneous capacitor voltage and current because they are out of phase. That is, $v_C \neq i_C X_C$.

A resistor's resistance R is independent of the emf frequency. In contrast, as **FIGURE 32.9** shows, a capacitor's reactance X_C depends inversely on the frequency. The reactance becomes very large at low frequencies (i.e., the capacitor is a large impediment to current). This makes sense because $\omega = 0$ would be a nonoscillating DC circuit, and we know that a steady DC current cannot pass through a capacitor. The reactance decreases as the frequency increases until, at very high frequencies, $X_C \approx 0$ and the capacitor begins to act like an ideal wire. This result has important consequences for how capacitors are used in many circuits.

FIGURE 32.8 Graph and phasor diagrams of the capacitor current and voltage.

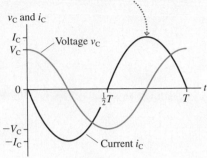

(a) i_C peaks $\frac{1}{4}T$ before v_C peaks. We say that the current *leads* the voltage by 90°.

(b) The current phasor leads the voltage phasor by 90°.

These are the instantaneous current and voltage.

FIGURE 32.9 The capacitive reactance as a function of frequency.

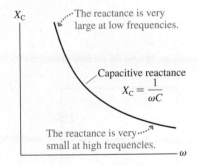

The reactance is very large at low frequencies.

Capacitive reactance $X_C = \frac{1}{\omega C}$

The reactance is very small at high frequencies.

EXAMPLE 32.2 | Capacitive reactance

What is the capacitive reactance of a 0.10 μF capacitor at 100 Hz (an audio frequency) and at 100 MHz (an FM-radio frequency)?

SOLVE At 100 Hz,

$$X_C(\text{at } 100 \text{ Hz}) = \frac{1}{\omega C} = \frac{1}{2\pi(100 \text{ Hz})(1.0 \times 10^{-7} \text{ F})} = 16,000 \ \Omega$$

Increasing the frequency by a factor of 10^6 decreases X_C by a factor of 10^6, giving

$$X_C(\text{at } 100 \text{ MHz}) = 0.016 \ \Omega$$

ASSESS A capacitor with a substantial reactance at audio frequencies has virtually no reactance at FM-radio frequencies.

EXAMPLE 32.3 | Capacitor current

A 10 μF capacitor is connected to a 1000 Hz oscillator with a peak emf of 5.0 V. What is the peak current to the capacitor?

VISUALIZE Figure 32.7b showed the circuit diagram. It is a simple one-capacitor circuit.

SOLVE The capacitive reactance at $\omega = 2\pi f = 6280$ rad/s is

$$X_C = \frac{1}{\omega C} = \frac{1}{(6280 \text{ rad/s})(10 \times 10^{-6} \text{ F})} = 16 \ \Omega$$

The peak voltage across the capacitor is $V_C = \mathcal{E}_0 = 5.0$ V; hence the peak current is

$$I_C = \frac{V_C}{X_C} = \frac{5.0 \text{ V}}{16 \ \Omega} = 0.31 \text{ A}$$

ASSESS Using reactance is just like using Ohm's law, but don't forget it applies to only the *peak* current and voltage, not the instantaneous values.

STOP TO THINK 32.3 What is the capacitive reactance of "no capacitor," just a continuous wire?

a. 0   b. ∞ c. Undefined

32.3 *RC* Filter Circuits

You learned in Chapter 28 that a resistance R causes a capacitor to be charged or discharged with time constant $\tau = RC$. We called this an *RC* circuit. Now that we've looked at resistors and capacitors individually, let's explore what happens if an *RC* circuit is driven continuously by an alternating current source.

FIGURE 32.10 shows a circuit in which a resistor R and capacitor C are in series with an emf $\mathcal{E}$ oscillating at angular frequency ω. Before launching into a formal analysis, let's try to understand qualitatively how this circuit will respond as the frequency is varied. If the frequency is very low, the capacitive reactance will be very large, and thus the peak current I_C will be very small. The peak current through the resistor is the same as the peak current to and from the capacitor (just as in DC circuits, conservation of charge requires $I_R = I_C$); hence we expect the resistor's peak voltage $V_R = I_R R$ to be very small at very low frequencies.

On the other hand, suppose the frequency is very high. Then the capacitive reactance approaches zero and the peak current, determined by the resistance alone, will be $I_R = \mathcal{E}_0/R$. The resistor's peak voltage $V_R = IR$ will approach the peak source voltage $\mathcal{E}_0$ at very high frequencies.

This reasoning leads us to expect that V_R will *increase* steadily from 0 to $\mathcal{E}_0$ as ω is increased from 0 to very high frequencies. Kirchhoff's loop law has to be obeyed, so the capacitor voltage V_C will *decrease* from $\mathcal{E}_0$ to 0 during the same change of frequency. A quantitative analysis will show us how this behavior can be used as a *filter*.

The goal of a quantitative analysis is to determine the peak current I and the two peak voltages V_R and V_C as functions of the emf amplitude $\mathcal{E}_0$ and frequency ω. Our analytic procedure is based on the fact that the instantaneous current i is the same for two circuit elements in series.

FIGURE 32.10 An *RC* circuit driven by an AC source.

Using phasors to analyze an *RC* circuit

Begin by drawing a current phasor of length *I*. This is the starting point because the series circuit elements have the same current *i*. The angle at which the phasor is drawn is not relevant.

The current and voltage of a resistor are in phase, so draw a resistor voltage phasor of length V_R parallel to the current phasor *I*. The capacitor current leads the capacitor voltage by 90°, so draw a capacitor voltage phasor of length V_C that is 90° behind [i.e., clockwise (cw) from] the current phasor.

The series resistor and capacitor are in parallel with the emf, so their *instantaneous* voltages satisfy $v_R + v_C = \mathcal{E}$. This is a *vector* addition of phasors, so draw the emf phasor as the vector sum of the two voltage phasors. The emf is $\mathcal{E} = \mathcal{E}_0 \cos \omega t$, hence the emf phasor is at angle ωt.

The length of the emf phasor, $\mathcal{E}_0$, is the hypotenuse of a right triangle formed by the resistor and capacitor phasors. Thus $\mathcal{E}_0^2 = V_R^2 + V_C^2$.

The relationship $\mathcal{E}_0^2 = V_R^2 + V_C^2$ is based on the peak values, not the instantaneous values, because the peak values are the lengths of the sides of the right triangle. The peak voltages are related to the peak current *I* via $V_R = IR$ and $V_C = IX_C$, thus

$$\mathcal{E}_0^2 = V_R^2 + V_C^2 = (IR)^2 + (IX_C)^2 = (R^2 + X_C^2)I^2$$
$$= (R^2 + 1/\omega^2 C^2)I^2 \tag{32.12}$$

Consequently, the peak current in the *RC* circuit is

$$I = \frac{\mathcal{E}_0}{\sqrt{R^2 + X_C^2}} = \frac{\mathcal{E}_0}{\sqrt{R^2 + 1/\omega^2 C^2}} \tag{32.13}$$

Knowing *I* gives us the two peak voltages:

$$V_R = IR = \frac{\mathcal{E}_0 R}{\sqrt{R^2 + X_C^2}} = \frac{\mathcal{E}_0 R}{\sqrt{R^2 + 1/\omega^2 C^2}}$$

$$V_C = IX_C = \frac{\mathcal{E}_0 X_C}{\sqrt{R^2 + X_C^2}} = \frac{\mathcal{E}_0/\omega C}{\sqrt{R^2 + 1/\omega^2 C^2}} \tag{32.14}$$

Frequency Dependence

Our goal was to see how the peak current and voltages vary as functions of the frequency ω. Equations 32.13 and 32.14 are rather complex and best interpreted by looking at graphs. **FIGURE 32.11** is a graph of V_R and V_C versus ω.

You can see that our qualitative predictions have been borne out. That is, V_R increases from 0 to $\mathcal{E}_0$ as ω is increased, while V_C decreases from $\mathcal{E}_0$ to 0. The explanation for this behavior is that the capacitive reactance X_C decreases as ω increases. For low frequencies, where $X_C \gg R$, the circuit is primarily capacitive. For high frequencies, where $X_C \ll R$, the circuit is primarily resistive.

The frequency at which $V_R = V_C$ is called the **crossover frequency** ω_c. The *crossover* frequency is easily found by setting the two expressions in Equations 32.14 equal to each other. The denominators are the same and cancel, as does $\mathcal{E}_0$, leading to

$$\omega_c = \frac{1}{RC} \tag{32.15}$$

In practice, $f_c = \omega_c/2\pi$ is also called the crossover frequency.

FIGURE 32.11 Graph of the resistor and capacitor peak voltages as functions of the emf angular frequency ω.

V_R and V_C The capacitor voltage approaches $\mathcal{E}_0$ as ω approaches 0.

The resistor voltage approaches $\mathcal{E}_0$ as ω approaches ∞.

Crossover frequency

We'll leave it as a homework problem to show that $V_R = V_C = \mathcal{E}_0/\sqrt{2}$ when $\omega = \omega_c$. This may seem surprising. After all, shouldn't V_R and V_C add up to $\mathcal{E}_0$?

No! V_R and V_C are the *peak values* of oscillating voltages, not the instantaneous values. The instantaneous values do, indeed, satisfy $v_R + v_C = \mathcal{E}$ at all instants of time. But the resistor and capacitor voltages are out of phase with each other, as the phasor diagram shows, so the two circuit elements don't reach their peak values at the same time. The peak values are related by $\mathcal{E}_0^2 = V_R^2 + V_C^2$, and you can see that $V_R = V_C = \mathcal{E}_0/\sqrt{2}$ satisfies this equation.

NOTE It's very important in AC circuit analysis to make a clear distinction between instantaneous values and peak values of voltages and currents. Relationships that are true for one set of values may not be true for the other.

Filters

FIGURE 32.12 Low-pass and high-pass filter circuits.

(a) Low-pass filter

Transmits frequencies $\omega < \omega_c$ and blocks frequencies $\omega > \omega_c$

(b) High-pass filter

Transmits frequencies $\omega > \omega_c$ and blocks frequencies $\omega < \omega_c$

FIGURE 32.12a is the circuit we've just analyzed; the only difference is that the capacitor voltage v_C is now identified as the *output voltage* v_{out}. This is a voltage you might measure or, perhaps, send to an amplifier for use elsewhere in an electronic instrument. You can see from the capacitor voltage graph in Figure 32.11 that the peak output voltage is $V_{out} \approx \mathcal{E}_0$ if $\omega \ll \omega_c$, but $V_{out} \approx 0$ if $\omega \gg \omega_c$. In other words,

- If the frequency of an input signal is well below the crossover frequency, the input signal is transmitted with little loss to the output.
- If the frequency of an input signal is well above the crossover frequency, the input signal is strongly attenuated and the output is very nearly zero.

This circuit is called a **low-pass filter.**

The circuit of **FIGURE 32.12b**, which instead uses the resistor voltage v_R for the output v_{out}, is a **high-pass filter.** The output is $V_{out} \approx 0$ if $\omega \ll \omega_c$, but $V_{out} \approx \mathcal{E}_0$ if $\omega \gg \omega_c$. That is, an input signal whose frequency is well above the crossover frequency is transmitted without loss to the output.

Filter circuits are widely used in electronics. For example, a high-pass filter designed to have $f_c = 100$ Hz would pass the audio frequencies associated with speech ($f > 200$ Hz) while blocking 60 Hz "noise" that can be picked up from power lines. Similarly, the high-frequency hiss from old vinyl records can be attenuated with a low-pass filter, allowing the lower-frequency audio signal to pass.

A simple RC filter suffers from the fact that the crossover region where $V_R \approx V_C$ is fairly broad. More sophisticated filters have a sharper transition from off ($V_{out} \approx 0$) to on ($V_{out} \approx \mathcal{E}_0$), but they're based on the same principles as the RC filter analyzed here.

EXAMPLE 32.4 | Designing a filter

For a science project, you've built a radio to listen to AM radio broadcasts at frequencies near 1 MHz. The basic circuit is an antenna, which produces a very small oscillating voltage when it absorbs the energy of an electromagnetic wave, and an amplifier. Unfortunately, your neighbor's short-wave radio broadcast at 10 MHz interferes with your reception. Having just finished physics, you decide to solve this problem by placing a filter between the antenna and the amplifier. You happen to have a 500 pF capacitor. What frequency should you select as the filter's crossover frequency? What value of resistance will you need to build this filter?

MODEL You need a low-pass filter to block signals at 10 MHz while passing the lower-frequency AM signal at 1 MHz.

VISUALIZE The circuit will look like the low-pass filter in Figure 32.12a. The oscillating voltage generated by the antenna will be the emf, and v_{out} will be sent to the amplifier.

SOLVE You might think that a crossover frequency near 5 MHz, about halfway between 1 MHz and 10 MHz, would work best. But 5 MHz is a factor of 5 higher than 1 MHz while only a factor of 2 less than 10 MHz. A crossover frequency the *same factor* above 1 MHz as it is below 10 MHz will give the best results. In practice, choosing $f_c = 3$ MHz would be sufficient. You can then use Equation 32.15 to select the proper resistor value:

$$R = \frac{1}{\omega_c C} = \frac{1}{2\pi(3 \times 10^6 \text{ Hz})(500 \times 10^{-12} \text{ F})}$$

$$= 106 \ \Omega \approx 100 \ \Omega$$

ASSESS Rounding to 100 Ω is appropriate because the crossover frequency was determined to only one significant figure. Such "sloppy design" is adequate when the two frequencies you need to distinguish are well separated.

32.4 Inductor Circuits

FIGURE 32.13a shows the instantaneous current i_L through an inductor. If the current is changing, the instantaneous inductor voltage is

$$v_L = L\frac{di_L}{dt} \tag{32.16}$$

You learned in Chapter 30 that the potential decreases in the direction of the current if the current is increasing ($di_L/dt > 0$) and increases if the current is decreasing ($di_L/dt < 0$).

FIGURE 32.13b is the simplest inductor circuit. The inductor L is connected across the AC source, so the inductor voltage equals the emf: $v_L = \mathcal{E} = \mathcal{E}_0 \cos\omega t$. We can write

$$v_L = V_L \cos\omega t \tag{32.17}$$

where V_L is the peak or maximum voltage across the inductor. You can see that $V_L = \mathcal{E}_0$ in this single-inductor circuit.

We can find the inductor current i_L by integrating Equation 32.17. First, we use Equation 32.17 to write Equation 32.16 as

$$di_L = \frac{v_L}{L}dt = \frac{V_L}{L}\cos\omega t\, dt \tag{32.18}$$

Integrating gives

$$i_L = \frac{V_L}{L}\int \cos\omega t\, dt = \frac{V_L}{\omega L}\sin\omega t = \frac{V_L}{\omega L}\cos\left(\omega t - \frac{\pi}{2}\right)$$

$$= I_L \cos\left(\omega t - \frac{\pi}{2}\right) \tag{32.19}$$

where $I_L = V_L/\omega L$ is the peak or maximum inductor current.

NOTE Mathematically, Equation 32.19 could have an integration constant i_0. An integration constant would represent a constant DC current through the inductor, but there is no DC source of potential in an AC circuit. Hence, on physical grounds, we set $i_0 = 0$ for an AC circuit.

We define the **inductive reactance,** analogous to the capacitive reactance, to be

$$X_L \equiv \omega L \tag{32.20}$$

FIGURE 32.13 Using an inductor in an AC circuit.

(a)   The instantaneous current through the inductor

The instantaneous inductor voltage is $v_L = L(di_L/dt)$.

(b)

Then the peak current $I_L = V_L/\omega L$ and the peak voltage are related by

$$I_L = \frac{V_L}{X_L} \quad \text{or} \quad V_L = I_L X_L \qquad (32.21)$$

FIGURE 32.14 The inductive reactance as a function of frequency.

FIGURE 32.14 shows that the inductive reactance increases as the frequency increases. This makes sense. Faraday's law tells us that the induced voltage across a coil increases as the time rate of change of $\vec{B}$ increases, and $\vec{B}$ is directly proportional to the inductor current. For a given peak current I_L, $\vec{B}$ changes more rapidly at higher frequencies than at lower frequencies, and thus V_L is larger at higher frequencies than at lower frequencies.

FIGURE 32.15a is a graph of the inductor voltage and current. You can see that the current peaks one-quarter of a period *after* the voltage peaks. The angle of the current phasor on the phasor diagram of FIGURE 32.15b is $\pi/2$ rad less than the angle of the voltage phasor. We can summarize this finding:

The AC current through an inductor *lags* the inductor voltage by $\pi/2$ rad, or 90°.

FIGURE 32.15 Graph and phasor diagrams of the inductor current and voltage.

EXAMPLE 32.5 | **Current and voltage of an inductor**

A 25 μH inductor is used in a circuit that oscillates at 100 kHz. The current through the inductor reaches a peak value of 20 mA at $t = 5.0$ μs. What is the peak inductor voltage, and when, closest to $t = 5.0$ μs, does it occur?

MODEL The inductor current lags the voltage by 90°, or, equivalently, the voltage reaches its peak value one-quarter period *before* the current.

VISUALIZE The circuit looks like Figure 32.15b.

SOLVE The inductive reactance at $f = 100$ kHz is

$$X_L = \omega L = 2\pi(1.0 \times 10^5 \text{ Hz})(25 \times 10^{-6} \text{ H}) = 16 \ \Omega$$

Thus the peak voltage is $V_L = I_L X_L = (20 \text{ mA})(16 \ \Omega) = 320 \text{ mV}$. The voltage peak occurs one-quarter period before the current peaks, and we know that the current peaks at $t = 5.0$ μs. The period of a 100 kHz oscillation is 10.0 μs, so the voltage peaks at

$$t = 5.0 \ \mu\text{s} - \frac{10.0 \ \mu\text{s}}{4} = 2.5 \ \mu\text{s}$$

32.5 The Series *RLC* Circuit

FIGURE 32.16 A series *RLC* circuit.

The circuit of FIGURE 32.16, where a resistor, inductor, and capacitor are in series, is called a **series *RLC* circuit.** The series *RLC* circuit has many important applications because, as you will see, it exhibits resonance behavior.

The analysis, which is very similar to our analysis of the *RC* circuit in Section 32.3, will be based on a phasor diagram. Notice that the three circuit elements are in series with each other and, together, are in parallel with the emf. We can draw two conclusions that form the basis of our analysis:

1. The instantaneous current of all three elements is the same: $i = i_R = i_L = i_C$.
2. The sum of the instantaneous voltages matches the emf: $\mathcal{E} = v_R + v_L + v_C$.

Using phasors to analyze an *RLC* circuit

Begin by drawing a current phasor of length I. This is the starting point because the series circuit elements have the same current i.

The current and voltage of a resistor are in phase, so draw a resistor voltage phasor parallel to the current phasor I. The capacitor current leads the capacitor voltage by 90°, so draw a capacitor voltage phasor that is 90° behind the current phasor. The inductor current *lags* the voltage by 90°, so draw an inductor voltage phasor 90° ahead of the current phasor.

The instantaneous voltages satisfy $\mathcal{E} = v_R + v_L + v_C$. In terms of phasors, this is a *vector* addition. We can do the addition in two steps. Because the capacitor and inductor phasors are in opposite directions, their vector sum has length $V_L - V_C$. Adding the resistor phasor, at right angles, then gives the emf phasor $\mathcal{E}$ at angle ωt.

The length $\mathcal{E}_0$ of the emf phasor is the hypotenuse of a right triangle. Thus
$$\mathcal{E}_0^2 = V_R^2 + (V_L - V_C)^2$$

If $V_L > V_C$, which we've assumed, then the instantaneous current i lags the emf by a phase angle ϕ. We can write the current, in terms of ϕ, as

$$i = I\cos(\omega t - \phi) \qquad (32.22)$$

Of course, there's no guarantee that V_L will be larger than V_C. If the opposite is true, $V_L < V_C$, the emf phasor is on the other side of the current phasor. Our analysis is still valid if we consider ϕ to be negative when i is ccw from $\mathcal{E}$. Thus ϕ can be anywhere between $-90°$ and $+90°$.

Now we can continue much as we did with the *RC* circuit. Based on the right triangle, $\mathcal{E}_0^2$ is

$$\mathcal{E}_0^2 = V_R^2 + (V_L - V_C)^2 = \left[R^2 + (X_L - X_C)^2 \right] I^2 \qquad (32.23)$$

where we wrote each of the peak voltages in terms of the peak current I and a resistance or a reactance. Consequently, the peak current in the *RLC* circuit is

$$I = \frac{\mathcal{E}_0}{\sqrt{R^2 + (X_L - X_C)^2}} = \frac{\mathcal{E}_0}{\sqrt{R^2 + (\omega L - 1/\omega C)^2}} \qquad (32.24)$$

The three peak voltages are then found from $V_R = IR$, $V_L = IX_L$, and $V_C = IX_C$.

Impedance

The denominator of Equation 32.24 is called the **impedance** Z of the circuit:

$$Z = \sqrt{R^2 + (X_L - X_C)^2} \qquad (32.25)$$

Impedance, like resistance and reactance, is measured in ohms. The circuit's peak current can be written in terms of the source emf and the circuit impedance as

$$I = \frac{\mathcal{E}_0}{Z} \qquad (32.26)$$

Equation 32.26 is a compact way to write I, but it doesn't add anything new to Equation 32.24.

Phase Angle

The current lags the emf by
$$\phi = \tan^{-1}\left(\frac{V_L - V_C}{V_R}\right) = \tan^{-1}\left(\frac{X_L - X_C}{R}\right)$$

$V_R = \mathcal{E}_0 \cos\phi$

It is often useful to know the phase angle ϕ between the emf and the current. You can see from **FIGURE 32.17** that

$$\tan\phi = \frac{V_L - V_C}{V_R} = \frac{(X_L - X_C)I}{RI}$$

The current I cancels, and we're left with

$$\phi = \tan^{-1}\left(\frac{X_L - X_C}{R}\right) \tag{32.27}$$

We can check that Equation 32.27 agrees with our analyses of single-element circuits. A resistor-only circuit has $X_L = X_C = 0$ and thus $\phi = \tan^{-1}(0) = 0$ rad. In other words, as we discovered previously, the emf and current are in phase. An AC inductor circuit has $R = X_C = 0$ and thus $\phi = \tan^{-1}(\infty) = \pi/2$ rad, agreeing with our earlier finding that the inductor current lags the voltage by 90°.

Other relationships can be found from the phasor diagram and written in terms of the phase angle. For example, we can write the peak resistor voltage as

$$V_R = \mathcal{E}_0 \cos\phi \tag{32.28}$$

Notice that the resistor voltage oscillates in phase with the emf only if $\phi = 0$ rad.

Resonance

Suppose we vary the emf frequency ω while keeping everything else constant. There is very little current at very low frequencies because the capacitive reactance $X_C = 1/\omega C$ (and thus Z) is very large. Similarly, there is very little current at very high frequencies because the inductive reactance $X_L = \omega L$ becomes very large.

If I approaches zero at very low and very high frequencies, there should be some intermediate frequency where I is a maximum. Indeed, you can see from Equation 32.24 that the denominator will be a minimum, making I a maximum, when $X_L = X_C$, or

$$\omega L = \frac{1}{\omega C} \tag{32.29}$$

The frequency ω_0 that satisfies Equation 32.29 is called the **resonance frequency:**

$$\omega_0 = \frac{1}{\sqrt{LC}} \tag{32.30}$$

This is the frequency for *maximum current* in the series *RLC* circuit. The maximum current

$$I_{max} = \frac{\mathcal{E}_0}{R} \tag{32.31}$$

is that of a purely resistive circuit because the impedance is $Z = R$ at resonance.

You'll recognize ω_0 as the oscillation frequency of the *LC* circuit we analyzed in Chapter 30. The current in an ideal *LC* circuit oscillates forever as energy is transferred back and forth between the capacitor and the inductor. This is analogous to an ideal, frictionless simple harmonic oscillator in which the energy is transformed back and forth between kinetic and potential.

Adding a resistor to the circuit is like adding damping to a mechanical oscillator. The emf is then a sinusoidal driving force, and the series *RLC* circuit is directly analogous to the driven, damped oscillator that you studied in Chapter 15. A mechanical oscillator exhibits *resonance* by having a large-amplitude response when the driving frequency matches the system's natural frequency. Equation 32.30 is the natural frequency of the series *RLC* circuit, the frequency at which the current would like to oscillate. Consequently, the circuit has a large current response when the oscillating emf matches this frequency.

FIGURE 32.18 shows the peak current I of a series *RLC* circuit as the emf frequency ω is varied. Notice how the current increases until reaching a maximum at frequency ω_0, then decreases. This is the hallmark of a resonance.

As R decreases, causing the damping to decrease, the maximum current becomes larger and the curve in Figure 32.18 becomes narrower. You saw exactly the same behavior for a driven mechanical oscillator. The emf frequency must be very close to ω_0 in order for a lightly damped system to respond, but the response at resonance is very large.

For a different perspective, **FIGURE 32.19** graphs the instantaneous emf $\mathcal{E} = \mathcal{E}_0 \cos \omega t$ and current $i = I \cos(\omega t - \phi)$ for frequencies below, at, and above ω_0. The current and the emf are in phase at resonance ($\phi = 0$ rad) because the capacitor and inductor essentially cancel each other to give a purely resistive circuit. Away from resonance, the current decreases *and* begins to get out of phase with the emf. You can see, from Equation 32.27, that the phase angle ϕ is negative when $X_L < X_C$ (i.e., the frequency is below resonance) and positive when $X_L > X_C$ (the frequency is above resonance).

Resonance circuits are widely used in radio, television, and communication equipment because of their ability to respond to one particular frequency (or very narrow range of frequencies) while suppressing others. The selectivity of a resonance circuit improves as the resistance decreases, but the inherent resistance of the wires and the inductor coil keeps R from being 0 Ω.

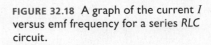

FIGURE 32.18 A graph of the current I versus emf frequency for a series *RLC* circuit.

FIGURE 32.19 Graphs of the emf $\mathcal{E}$ and the current i at frequencies below, at, and above the resonance frequency ω_0.

EXAMPLE 32.6 | Designing a radio receiver

An AM radio antenna picks up a 1000 kHz signal with a peak voltage of 5.0 mV. The tuning circuit consists of a 60 μH inductor in series with a variable capacitor. The inductor coil has a resistance of 0.25 Ω, and the resistance of the rest of the circuit is negligible.

a. To what value should the capacitor be tuned to listen to this radio station?

b. What is the peak current through the circuit at resonance?

c. A stronger station at 1050 kHz produces a 10 mV antenna signal. What is the current at this frequency when the radio is tuned to 1000 kHz?

MODEL The inductor's 0.25 Ω resistance can be modeled as a resistance in series with the inductance, hence we have a series *RLC* circuit. The antenna signal at $\omega = 2\pi \times 1000$ kHz is the emf.

VISUALIZE The circuit looks like Figure 32.16.

SOLVE a. The capacitor needs to be tuned to where it and the inductor are resonant at $\omega_0 = 2\pi \times 1000$ kHz. The appropriate value is

$$C = \frac{1}{L\omega_0^2} = \frac{1}{(60 \times 10^{-6}\,\text{H})(6.28 \times 10^6\,\text{rad/s})^2}$$
$$= 4.2 \times 10^{-10}\,\text{F} = 420\,\text{pF}$$

b. $X_L - X_C$ at resonance, so the peak current is

$$I = \frac{\mathcal{E}_0}{R} = \frac{5.0 \times 10^{-3}\,\text{V}}{0.25\,\Omega} = 0.020\,\text{A} = 20\,\text{mA}$$

c. The 1050 kHz signal is "off resonance," so we need to compute $X_L = \omega L = 396\,\Omega$ and $X_C = 1/\omega C = 361\,\Omega$ at $\omega = 2\pi \times 1050$ kHz. The peak voltage of this signal is $\mathcal{E}_0 = 10$ mV. With these values, Equation 32.24 for the peak current is

$$I = \frac{\mathcal{E}_0}{\sqrt{R^2 + (X_L - X_C)^2}} = 0.28\,\text{mA}$$

ASSESS These are realistic values for the input stage of an AM radio. You can see that the signal from the 1050 kHz station is strongly suppressed when the radio is tuned to 1000 kHz.

STOP TO THINK 32.5   A series *RLC* circuit has $V_C = 5.0$ V, $V_R = 7.0$ V, and $V_L = 9.0$ V. Is the frequency above, below, or equal to the resonance frequency?

32.6 Power in AC Circuits

A primary role of the emf is to supply energy. Some circuit devices, such as motors and lightbulbs, use the energy to perform useful tasks. Other circuit devices dissipate the energy as an increased thermal energy in the components and the surrounding air. Chapter 28 examined the topic of power in DC circuits. Now we can perform a similar analysis for AC circuits.

The emf supplies energy to a circuit at the rate

$$p_{source} = i\mathcal{E} \tag{32.32}$$

where i and $\mathcal{E}$ are the instantaneous current from and potential difference across the emf. We've used a lowercase p to indicate that this is the instantaneous power. We need to look at the power losses in individual circuit elements.

Resistors

A resistor dissipates energy at the rate

$$p_R = i_R v_R = i_R{}^2 R \tag{32.33}$$

We can use $i_R = I_R \cos \omega t$ to write the resistor's instantaneous power loss as

$$p_R = i_R{}^2 R = I_R{}^2 R \cos^2 \omega t \tag{32.34}$$

FIGURE 32.20 shows the instantaneous power graphically. You can see that, because the cosine is squared, the power oscillates twice during every cycle of the emf. The energy dissipation peaks both when $i_R = I_R$ and when $i_R = -I_R$.

In practice, we're more interested in the *average power* than in the instantaneous power. The **average power** P is the total energy dissipated per second. We can find P_R for a resistor by using the identity $\cos^2(x) = \frac{1}{2}(1 + \cos 2x)$ to write

$$P_R = I_R{}^2 R \cos^2 \omega t = I_R{}^2 R \left[\tfrac{1}{2}(1 + \cos 2\omega t)\right] = \tfrac{1}{2}I_R{}^2 R + \tfrac{1}{2}I_R{}^2 R \cos 2\omega t$$

The $\cos 2\omega t$ term oscillates positive and negative twice during each cycle of the emf. Its average, over one cycle, is zero. Thus the average power loss in a resistor is

$$P_R = \tfrac{1}{2}I_R{}^2 R \qquad \text{(average power loss in a resistor)} \tag{32.35}$$

It is useful to write Equation 32.25 as

$$P_R = \left(\frac{I_R}{\sqrt{2}}\right)^2 R = (I_{rms})^2 R \tag{32.36}$$

where the quantity

$$I_{rms} = \frac{I_R}{\sqrt{2}} \tag{32.37}$$

is called the **root-mean-square current,** or rms current, I_{rms}. Technically, an rms quantity is the square root of the average, or mean, of the quantity squared. For a sinusoidal oscillation, the rms value turns out to be the peak value divided by $\sqrt{2}$.

The rms current allows us to compare Equation 32.36 directly to the energy dissipated by a resistor in a DC circuit: $P = I^2 R$. You can see that the average power loss of a resistor in an AC circuit with $I_{rms} = 1$ A is the same as in a DC circuit with $I = 1$ A. **As far as power is concerned, an rms current is equivalent to an equal DC current.**

FIGURE 32.20 The instantaneous power loss in a resistor.

p_R is a maximum when $i_R = \pm I_R$.

The *average* power is $P_R = \frac{1}{2}I_R{}^2 R$.

p_R is zero when i_R is zero.

Similarly, we can define the root-mean-square voltage and emf:

$$V_{rms} = \frac{V_R}{\sqrt{2}} \qquad \mathcal{E}_{rms} = \frac{\mathcal{E}_0}{\sqrt{2}} \qquad (32.38)$$

The resistor's average power loss in terms of the rms quantities is

$$P_R = (I_{rms})^2 R = \frac{(V_{rms})^2}{R} = I_{rms}V_{rms} \qquad (32.39)$$

and the average power supplied by the emf is

$$P_{source} = I_{rms}\mathcal{E}_{rms} \qquad (32.40)$$

The power rating on a lightbulb is its average power at $V_{rms} = 120$ V.

The single-resistor circuit that we analyzed in Section 32.1 had $V_R = \mathcal{E}$ or, equivalently, $V_{rms} = \mathcal{E}_{rms}$. You can see from Equations 32.39 and 32.40 that the power loss in the resistor exactly matches the power supplied by the emf. This must be the case in order to conserve energy.

NOTE Voltmeters, ammeters, and other AC measuring instruments are calibrated to give the rms value. An AC voltmeter would show that the "line voltage" of an electrical outlet in the United States is 120 V. This is $\mathcal{E}_{rms}$. The peak voltage $\mathcal{E}_0$ is larger by a factor of $\sqrt{2}$, or $\mathcal{E}_0 = 170$ V. The power-line voltage is sometimes specified as "120 V/60 Hz," showing the rms voltage and the frequency.

EXAMPLE 32.7 | Lighting a bulb

A 100 W incandescent lightbulb is plugged into a 120 V/60 Hz outlet. What is the resistance of the bulb's filament? What is the peak current through the bulb?

MODEL The filament in a lightbulb acts as a resistor.

VISUALIZE FIGURE 32.21 is a simple one-resistor circuit.

FIGURE 32.21 An AC circuit with a lightbulb as a resistor.

100 W bulb

120 V/60 Hz

SOLVE A bulb labeled 100 W is designed to dissipate an average 100 W at $V_{rms} = 120$ V. We can use Equation 32.39 to find

$$R = \frac{(V_{rms})^2}{P_R} = \frac{(120 \text{ V})^2}{100 \text{ W}} = 144 \ \Omega$$

The rms current is then found from

$$I_{rms} = \frac{P_R}{V_{rms}} = \frac{100 \text{ W}}{120 \text{ V}} = 0.833 \text{ A}$$

The peak current is $I_R = \sqrt{2}I_{rms} = 1.18$ A.

ASSESS Calculations with rms values are just like the calculations for DC circuits.

Capacitors and Inductors

In Section 32.2, we found that the instantaneous current to a capacitor is $i_C = -\omega C V_C \sin \omega t$. Thus the instantaneous energy dissipation in a capacitor is

$$p_C = v_C i_C = (V_C \cos \omega t)(-\omega C V_C \sin \omega t) = -\tfrac{1}{2}\omega C V_C^2 \sin 2\omega t \qquad (32.41)$$

where we used $\sin(2x) = 2\sin(x)\cos(x)$.

FIGURE 32.22 on the next page shows Equation 32.41 graphically. Energy is transferred into the capacitor (positive power) as it is charged, but, instead of being dissipated, as it would be by a resistor, the energy is stored as potential energy in the capacitor's electric field. Then, as the capacitor discharges, this energy is given back to the circuit. Power is the rate at which energy is *removed* from the circuit, hence p is negative as the capacitor transfers energy back into the circuit.

Using a mechanical analogy, a capacitor is like an ideal, frictionless simple harmonic oscillator. Kinetic and potential energy are constantly being exchanged,

FIGURE 32.22 Energy flows into and out of a capacitor as it is charged and discharged.

but there is no dissipation because none of the energy is transformed into thermal energy. The important conclusion is that **a capacitor's average power loss is zero: $P_C = 0$.**

The same is true of an inductor. An inductor alternately stores energy in the magnetic field, as the current is increasing, then transfers energy back to the circuit as the current decreases. The instantaneous power oscillates between positive and negative, but **an inductor's average power loss is zero: $P_L = 0$.**

NOTE We're assuming ideal capacitors and inductors. Real capacitors and inductors inevitably have a small amount of resistance and dissipate a small amount of energy. However, their energy dissipation is negligible compared to that of the resistors in most practical circuits.

The Power Factor

In an *RLC* circuit, energy is supplied by the emf and dissipated by the resistor. But an *RLC* circuit is unlike a purely resistive circuit in that the current is not in phase with the potential difference of the emf.

We found in Equation 32.22 that the instantaneous current in an *RLC* circuit is $i = I\cos(\omega t - \phi)$, where ϕ is the angle by which the current lags the emf. Thus the instantaneous power supplied by the emf is

$$p_{\text{source}} = i\mathcal{E} = (I\cos(\omega t - \phi))(\mathcal{E}_0\cos\omega t) = I\mathcal{E}_0\cos\omega t\cos(\omega t - \phi) \quad (32.42)$$

We can use the expression $\cos(x - y) = \cos(x)\cos(y) + \sin(x)\sin(y)$ to write the power as

$$p_{\text{source}} = (I\mathcal{E}_0\cos\phi)\cos^2\omega t + (I\mathcal{E}_0\sin\phi)\sin\omega t\cos\omega t \quad (32.43)$$

In our analysis of the power loss in a resistor and a capacitor, we found that the average of $\cos^2\omega t$ is $\frac{1}{2}$ and the average of $\sin\omega t\cos\omega t$ is zero. Thus we can immediately write that the *average* power supplied by the emf is

$$P_{\text{source}} = \tfrac{1}{2}I\mathcal{E}_0\cos\phi = I_{\text{rms}}\mathcal{E}_{\text{rms}}\cos\phi \quad (32.44)$$

The rms values, you will recall, are $I/\sqrt{2}$ and $\mathcal{E}_0/\sqrt{2}$.

The term $\cos\phi$, called the **power factor,** arises because the current and the emf in a series *RLC* circuit are not in phase. Because the current and the emf aren't pushing and pulling together, the source delivers less energy to the circuit.

We'll leave it as a homework problem for you to show that the peak current in an *RLC* circuit can be written $I = I_{\text{max}}\cos\phi$, where $I_{\text{max}} = \mathcal{E}_0/R$ was given in Equation 32.31. In other words, the current term in Equation 32.44 is a function of the power factor. Consequently, the average power is

$$P_{\text{source}} = P_{\text{max}}\cos^2\phi \quad (32.45)$$

where $P_{\text{max}} = \frac{1}{2}I_{\text{max}}\mathcal{E}_0$ is the *maximum* power the source can deliver to the circuit.

The source delivers maximum power only when $\cos\phi = 1$. This is the case when $X_L - X_C = 0$, requiring either a purely resistive circuit or an *RLC* circuit operating at the resonance frequency ω_0. The average power loss is zero for a purely capacitive or purely inductive load with, respectively, $\phi = -90°$ or $\phi = +90°$, as found above.

Motors of various types, especially large industrial motors, use a significant fraction of the electric energy generated in industrialized nations. Motors operate most efficiently, doing the maximum work per second, when the power factor is as close to 1 as possible. But motors are inductive devices, due to their electromagnet coils, and if too many motors are attached to the electric grid, the power factor is pulled away from 1. To compensate, the electric company places large capacitors throughout the

Industrial motors use a significant fraction of the electric energy generated in the United States.

transmission system. The capacitors dissipate no energy, but they allow the electric system to deliver energy more efficiently by keeping the power factor close to 1.

Finally, we found in Equation 32.28 that the resistor's peak voltage in an *RLC* circuit is related to the emf peak voltage by $V_R = \mathcal{E}_0 \cos\phi$ or, dividing both sides by $\sqrt{2}$, $V_{rms} = \mathcal{E}_{rms} \cos\phi$. We can use this result to write the energy loss in the resistor as

$$P_R = I_{rms}V_{rms} = I_{rms}\mathcal{E}_{rms} \cos\phi \qquad (32.46)$$

But this expression is P_{source}, as we found in Equation 32.44. Thus we see that the energy supplied to an *RLC* circuit by the emf is ultimately dissipated by the resistor.

STOP TO THINK 32.6 The emf and the current in a series *RLC* circuit oscillate as shown. Which of the following (perhaps more than one) would increase the rate at which energy is supplied to the circuit?

a. Increase $\mathcal{E}_0$
b. Increase L
c. Increase C
d. Decrease $\mathcal{E}_0$
e. Decrease L
f. Decrease C

CHALLENGE EXAMPLE 32.8 | Power in an *RLC* circuit

An audio amplifier drives a series *RLC* circuit consisting of an 8.0 Ω loudspeaker, a 160 μF capacitor, and a 1.5 mH inductor. The amplifier output is 15.0 V rms at 500 Hz.

a. What power is delivered to the speaker?
b. What maximum power could the amplifier deliver, and how would the capacitor have to be changed for this to happen?

MODEL The emf and voltage of an *RLC* circuit are not in phase, and that affects the power delivered to the circuit. All the power is dissipated by the circuit's resistance, which in this case is the loudspeaker.

VISUALIZE The circuit looks like Figure 32.16.

SOLVE a. The emf delivers power $P_{source} = I_{rms}\mathcal{E}_{rms} \cos\phi$, where ϕ is the phase angle between the emf and the current. The rms current is $I_{rms} = \mathcal{E}_{rms}/Z$, where Z is the impedance. To calculate Z, we need the reactances of the capacitor and inductor, and these, in turn, depend on the frequency. At 500 Hz, the angular frequency is $\omega = 2\pi(500 \text{ Hz}) = 3140 \text{ rad/s}$. With this, we can find

$$X_C = \frac{1}{\omega C} = \frac{1}{(3140 \text{ rad/s})(160 \times 10^{-6} \text{ F})} = 1.99 \, \Omega$$

$$X_L = \omega L = (3140 \text{ rad/s})(0.0015 \text{ H}) = 4.71 \, \Omega$$

Now we can calculate the impedance:

$$Z = \sqrt{R^2 + (X_L - X_C)^2} = 8.45 \, \Omega$$

and thus

$$I_{rms} = \frac{\mathcal{E}_{rms}}{Z} = \frac{15.0 \text{ V}}{8.45 \, \Omega} = 1.78 \text{ A}$$

Lastly, we need the phase angle between the emf and the current:

$$\phi = \tan^{-1}\left(\frac{X_L - X_C}{R}\right) = 18.8°$$

The power factor is $\cos(18.8°) = 0.947$, and thus the power delivered by the emf is

$$P_{source} = I_{rms}\mathcal{E}_{rms} \cos\phi = (1.78 \text{ A})(15.0 \text{ V})(0.947) = 25 \text{ W}$$

b. Maximum power is delivered when the current is in phase with the emf, making the power factor 1.00. This occurs when $X_C = X_L$, making the impedance $Z = R = 8.0 \, \Omega$ and the current $I_{rms} = \mathcal{E}_{rms}/R = 1.88 \text{ A}$. Then

$$P_{source} = I_{rms}\mathcal{E}_{rms} \cos\phi = (1.88 \text{ A})(15.0 \text{ V})(1.00) = 28 \text{ W}$$

To deliver maximum power, we need to change the capacitance to make $X_C = X_L = 4.71 \, \Omega$. The required capacitance is

$$C = \frac{1}{(3140 \text{ rad/s})(4.71 \, \Omega)} = 68 \, \mu\text{F}$$

So delivering maximum power requires lowering the capacitance from 160 μF to 68 μF.

ASSESS Changing the capacitor not only increases the power factor, it also increases the current. Both contribute to the higher power.

SUMMARY

The goal of Chapter 32 has been to learn about and analyze AC circuits.

IMPORTANT CONCEPTS

AC circuits are driven by an emf

$$\mathcal{E} = \mathcal{E}_0 \cos \omega t$$

that oscillates with angular frequency $\omega = 2\pi f$.

Phasors can be used to represent the oscillating emf, current, and voltage.

The length of the phasor is the peak value $\mathcal{E}_0$.

The horizontal projection is the instantaneous value $\mathcal{E}$.

Basic circuit elements

Element	i and v	Resistance/ reactance	I and V	Power
Resistor	In phase	R is fixed	$V = IR$	$I_{rms}V_{rms}$
Capacitor	i leads v by 90°	$X_C = 1/\omega C$	$V = IX_C$	0
Inductor	i lags v by 90°	$X_L = \omega L$	$V = IX_L$	0

For many purposes, especially calculating power, the **root-mean-square** (rms) quantities

$$V_{rms} = V/\sqrt{2} \qquad I_{rms} = I/\sqrt{2} \qquad \mathcal{E}_{rms} = \mathcal{E}_0/\sqrt{2}$$

are equivalent to the corresponding DC quantities.

KEY SKILLS

Using phasor diagrams

- Start with a phasor (v or i) common to two or more circuit elements.

- The sum of instantaneous quantities is vector addition.

- Use the Pythagorean theorem to relate peak quantities.

For an RC circuit, shown here,

$$v_R + v_C = \mathcal{E}$$
$$V_R{}^2 + V_C{}^2 = \mathcal{E}_0{}^2$$

Instantaneous and peak quantities
The instantaneous quantities v and i vary sinusoidally. The peak quantities V and I are the maximum values of v and i. For capacitors and inductors, the peak quantities are related by $V = IX$, where X is the reactance, but this relationship does *not* apply to v and i.

Kirchhoff's loop law says that the sum of the potential differences around a loop is zero.

Charge conservation says that circuit elements in series all have the same instantaneous current i and the same peak current I.

APPLICATIONS

RC filter circuits

$$v_C = \frac{\mathcal{E}_0 X_C}{\sqrt{R^2 + X_C{}^2}}$$

$$v_C \rightarrow \mathcal{E}_0 \text{ as } \omega \rightarrow 0$$

A **low-pass filter** transmits low frequencies and blocks high frequencies.

$$v_R = \frac{\mathcal{E}_0 R}{\sqrt{R^2 + X_C{}^2}}$$

$$v_R \rightarrow \mathcal{E}_0 \text{ as } \omega \rightarrow \infty$$

A **high-pass filter** transmits high frequencies and blocks low frequencies.

Series *RLC* circuits

$I = \mathcal{E}_0/Z$ where Z is the **impedance**

$$Z = \sqrt{R^2 + (X_L - X_C)^2}$$

$$V_R = IR \qquad V_L = IX_L \qquad V_C = IX_C$$

When $\omega = \omega_0 = 1/\sqrt{LC}$ (the **resonance frequency**), the current in the circuit is a maximum $I_{max} = \mathcal{E}_0/R$.

In general, the current i lags behind $\mathcal{E}$ by the **phase angle** $\phi = \tan^{-1}\big((X_L - X_C)/R\big)$.

The power supplied by the emf is $P_{source} = I_{rms}\mathcal{E}_{rms}\cos\phi$, where $\cos\phi$ is called the **power factor**.

The power lost in the resistor is $P_R = I_{rms}V_{rms} = (I_{rms})^2 R$.

TERMS AND NOTATION

AC circuit	crossover frequency, ω_c	series RLC circuit	root-mean-square current, I_{rms}
DC circuit	low-pass filter	impedance, Z	power factor, $\cos\phi$
phasor	high-pass filter	resonance frequency, ω_0	
capacitive reactance, X_C	inductive reactance, X_L	average power, P	

CONCEPTUAL QUESTIONS

1. **FIGURE Q32.1** shows emf phasors a, b, and c.
   a. For each, what is the instantaneous value of the emf?
   b. At this instant, is the magnitude of each emf increasing, decreasing, or holding constant?

FIGURE Q32.1

2. A resistor is connected across an oscillating emf. The peak current through the resistor is 2.0 A. What is the peak current if:
   a. The resistance R is doubled?
   b. The peak emf $\mathcal{E}_0$ is doubled?
   c. The frequency ω is doubled?

3. A capacitor is connected across an oscillating emf. The peak current through the capacitor is 2.0 A. What is the peak current if:
   a. The capacitance C is doubled?
   b. The peak emf $\mathcal{E}_0$ is doubled?
   c. The frequency ω is doubled?

4. A low-pass RC filter has a crossover frequency $f_c = 200$ Hz. What is f_c if:
   a. The resistance R is doubled?
   b. The capacitance C is doubled?
   c. The peak emf $\mathcal{E}_0$ is doubled?

5. An inductor is connected across an oscillating emf. The peak current through the inductor is 2.0 A. What is the peak current if:
   a. The inductance L is doubled?
   b. The peak emf $\mathcal{E}_0$ is doubled?
   c. The frequency ω is doubled?

6. The resonance frequency of a series RLC circuit is 1000 Hz. What is the resonance frequency if:
   a. The resistance R is doubled?
   b. The inductance L is doubled?
   c. The capacitance C is doubled?
   d. The peak emf $\mathcal{E}_0$ is doubled?

7. In the series RLC circuit represented by the phasors of **FIGURE Q32.7**, is the emf frequency less than, equal to, or greater than the resonance frequency ω_0? Explain.

FIGURE Q32.7

8. The resonance frequency of a series RLC circuit is less than the emf frequency. Does the current lead or lag the emf? Explain.

9. The current in a series RLC circuit lags the emf by 20°. You cannot change the emf. What two different things could you do to the circuit that would increase the power delivered to the circuit by the emf?

10. The average power dissipated by a resistor is 4.0 W. What is P_R if:
    a. The resistance R is doubled while $\mathcal{E}_0$ is held fixed?
    b. The peak emf $\mathcal{E}_0$ is doubled while R is held fixed?
    c. Both are doubled simultaneously?

EXERCISES AND PROBLEMS

Problems labeled ▮ integrate material from earlier chapters.

Exercises

Section 32.1 AC Sources and Phasors

1. ‖ The emf phasor in **FIGURE EX32.1** is shown at $t = 2.0$ ms.
   a. What is the angular frequency ω? Assume this is the first rotation.
   b. What is the peak value of the emf?

FIGURE EX32.1

2. ‖ The emf phasor in **FIGURE EX32.2** is shown at $t = 15$ ms.
   a. What is the angular frequency ω? Assume this is the first rotation.
   b. What is the instantaneous value of the emf?

FIGURE EX32.2

3. ▮ A 110 Hz source of emf has a peak voltage of 50 V. Draw the emf phasor at $t = 3.0$ ms.

4. ‖ Draw the phasor for the emf $\mathcal{E} = (170\ \text{V})\cos\big((2\pi \times 60\ \text{Hz})t\big)$ at $t = 60$ ms.

5. ‖ **FIGURE EX32.5** shows voltage and current graphs for a resistor.
   a. What is the emf frequency f?
   b. What is the value of the resistance R?
   c. Draw the resistor's voltage and current phasors at $t = 15$ ms.

FIGURE EX32.5

6. ‖ A 200 Ω resistor is connected to an AC source with $\mathcal{E}_0 = 10$ V. What is the peak current through the resistor if the emf frequency is (a) 100 Hz? (b) 100 kHz?

Section 32.2 Capacitor Circuits

7. ‖ The peak current to and from a capacitor is 10 mA. What is the peak current if
   a. The emf frequency is doubled?
   b. The emf peak voltage is doubled (at the original frequency)?
8. ‖ A 0.30 μF capacitor is connected across an AC generator that produces a peak voltage of 10 V. What is the peak current to and from the capacitor if the emf frequency is (a) 100 Hz? (b) 100 kHz?
9. ‖ **FIGURE EX32.9** shows voltage and current graphs for a capacitor.
   a. What is the emf frequency f?
   b. What is the value of the capacitance C?

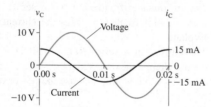

FIGURE EX32.9

10. ‖ A 20 nF capacitor is connected across an AC generator that produces a peak voltage of 5.0 V.
    a. At what frequency f is the peak current 50 mA?
    b. What is the instantaneous value of the emf at the instant when $i_C = I_C$?
11. ‖ A capacitor is connected to a 15 kHz oscillator. The peak current is 65 mA when the rms voltage is 6.0 V. What is the value of the capacitance C?
12. ‖ A capacitor has a peak current of 330 μA when the peak voltage at 250 kHz is 2.2 V.
    a. What is the capacitance?
    b. If the peak voltage is held constant, what is the peak current at 500 kHz?
13. ‖ a. Evaluate V_C in **FIGURE EX32.13** at emf frequencies 1, 3, 10, 30, and 100 kHz.
    b. Graph V_C versus frequency. Draw a smooth curve through your five points.

FIGURE EX32.13

14. ‖ a. Evaluate V_R in **FIGURE EX32.14** at emf frequencies 100, 300, 1000, 3000, and 10,000 Hz.
    b. Graph V_R versus frequency. Draw a smooth curve through your five points.

FIGURE EX32.14

Section 32.3 RC Filter Circuits

15. ‖ A high-pass RC filter is connected to an AC source with a peak voltage of 10.0 V. The peak capacitor voltage is 6.0 V. What is the resistor voltage?
16. ‖ A high-pass RC filter with a crossover frequency of 1000 Hz uses a 100 Ω resistor. What is the value of the capacitor?
17. ‖ A low-pass RC filter with a crossover frequency of 1000 Hz uses a 100 Ω resistor. What is the value of the capacitor?
18. ‖ A low-pass filter consists of a 100 μF capacitor in series with a 159 Ω resistor. The circuit is driven by an AC source with a peak voltage of 5.00 V.
    a. What is the crossover frequency f_c?
    b. What is V_C when $f = \frac{1}{2}f_c, f_c$, and $2f_c$?
19. ‖ What are V_R and V_C if the emf frequency in **FIGURE EX32.19** is 10 kHz?

FIGURE EX32.19

20. ‖ A high-pass filter consists of a 1.59 μF capacitor in series with a 100 Ω resistor. The circuit is driven by an AC source with a peak voltage of 5.00 V.
    a. What is the crossover frequency f_c?
    b. What is V_R when $f = \frac{1}{2}f_c, f_c$, and $2f_c$?
21. ‖ An electric circuit, whether it's a simple lightbulb or a complex amplifier, has two input terminals that are connected to the two output terminals of the voltage source. The impedance between the two input terminals (often a function of frequency) is the circuit's *input impedance*. Most circuits are designed to have a large input impedance. To see why, suppose you need to amplify the output of a high-pass filter that is constructed with a 1.2 kΩ resistor and a 15 μF capacitor. The amplifier you've chosen has a purely resistive input impedance. For a 60 Hz signal, what is the ratio $V_{R\,load}/V_{R\,no\,load}$ of the filter's peak voltage output with (load) and without (no load) the amplifier connected if the amplifier's input impedance is (a) 1.5 kΩ and (b) 150 kΩ?

Section 32.4 Inductor Circuits

22. ‖ A 20 mH inductor is connected across an AC generator that produces a peak voltage of 10 V. What is the peak current through the inductor if the emf frequency is (a) 100 Hz? (b) 100 kHz?
23. ‖ The peak current through an inductor is 10 mA. What is the peak current if
    a. The emf frequency is doubled?
    b. The emf peak voltage is doubled (at the original frequency)?

24. || **FIGURE EX32.24** shows voltage and current graphs for an inductor.
   a. What is the emf frequency f?
   b. What is the value of the inductance L?

FIGURE EX32.24

25. || An inductor is connected to a 15 kHz oscillator. The peak current is 65 mA when the rms voltage is 6.0 V. What is the value of the inductance L?
26. || An inductor has a peak current of 330 μA when the peak voltage at 45 MHz is 2.2 V.
   a. What is the inductance?
   b. If the peak voltage is held constant, what is the peak current at 90 MHz?

Section 32.5 The Series *RLC* Circuit

27. | A series *RLC* circuit has a 200 kHz resonance frequency. What is the resonance frequency if the capacitor value is doubled and, at the same time, the inductor value is halved?
28. | A series *RLC* circuit has a 200 kHz resonance frequency. What is the resonance frequency if
   a. The resistor value is doubled?
   b. The capacitor value is doubled?
29. || What capacitor in series with a 100 Ω resistor and a 20 mH inductor will give a resonance frequency of 1000 Hz?
30. || A series *RLC* circuit consists of a 50 Ω resistor, a 3.3 mH inductor, and a 480 nF capacitor. It is connected to an oscillator with a peak voltage of 5.0 V. Determine the impedance, the peak current, and the phase angle at frequencies (a) 3000 Hz, (b) 4000 Hz, and (c) 5000 Hz.
31. || At what frequency f do a 1.0 μF capacitor and a 1.0 μH inductor have the same reactance? What is the value of the reactance at this frequency?
32. | For the circuit of **FIGURE EX32.32**,
   a. What is the resonance frequency, in both rad/s and Hz?
   b. Find V_R and V_L at resonance.
   c. How can V_L be larger than $\mathcal{E}_0$? Explain.

FIGURE EX32.32   **FIGURE EX32.33**

33. | For the circuit of **FIGURE EX32.33**,
   a. What is the resonance frequency, in both rad/s and Hz?
   b. Find V_R and V_C at resonance.
   c. How can V_C be larger than $\mathcal{E}_0$? Explain.

Section 32.6 Power in AC Circuits

34. | The heating element of a hair drier dissipates 1500 W when connected to a 120 V/60 Hz power line. What is its resistance?
35. || A resistor dissipates 2.0 W when the rms voltage of the emf is 10.0 V. At what rms voltage will the resistor dissipate 10.0 W?
36. | For what absolute value of the phase angle does a source deliver 75% of the maximum possible power to an *RLC* circuit?
37. | The motor of an electric drill draws a 3.5 A rms current at the power-line voltage of 120 V rms. What is the motor's power if the current lags the voltage by 20°?
38. || A series *RLC* circuit attached to a 120 V/60 Hz power line draws a 2.4 A rms current with a power factor of 0.87. What is the value of the resistor?
39. || A series *RLC* circuit with a 100 Ω resistor dissipates 80 W when attached to a 120 V/60 Hz power line. What is the power factor?

Problems

40. || a. For an *RC* circuit, find an expression for the angular frequency at which $V_R = \frac{1}{2}\mathcal{E}_0$.
   b. What is V_C at this frequency?
41. || a. For an *RC* circuit, find an expression for the angular frequency at which $V_C = \frac{1}{2}\mathcal{E}_0$.
   b. What is V_R at this frequency?
42. || For an *RC* filter circuit, show that $V_R = V_C = \mathcal{E}_0/\sqrt{2}$ at $\omega = \omega_c$.
43. || A series *RC* circuit is built with a 12 kΩ resistor and a parallel-plate capacitor with 15-cm-diameter electrodes. A 12 V, 36 kHz source drives a peak current of 0.65 mA through the circuit. What is the spacing between the capacitor plates?
44. || Show that Equation 32.27 for the phase angle ϕ of a series *RLC* circuit gives the correct result for a capacitor-only circuit.
45. || a. What is the peak current supplied by the emf in **FIGURE P32.45**?
   b. What is the peak voltage across the 3.0 μF capacitor?

FIGURE P32.45

46. || You have a resistor and a capacitor of unknown values. First, you charge the capacitor and discharge it through the resistor. By monitoring the capacitor voltage on an oscilloscope, you see that the voltage decays to half its initial value in 2.5 ms. You then use the resistor and capacitor to make a low-pass filter. What is the crossover frequency f_c?
47. || **FIGURE P32.47** shows a parallel *RC* circuit.
   a. Use a phasor-diagram analysis to find expressions for the peak currents I_R and I_C.
   Hint: What do the resistor and capacitor have in common? Use that as the initial phasor.
   b. Complete the phasor analysis by finding an expression for the peak emf current I.

FIGURE P32.47

48. ‖ The small transformers that power many consumer products produce a 12.0 V rms, 60 Hz emf. Design a circuit using resistors and capacitors that uses the transformer voltage as an input and produces a 6.0 V rms output that leads the input voltage by 45°.

49. ‖ Use a phasor diagram to analyze the RL circuit of FIGURE P32.49. In particular,
    a. Find expressions for I, V_R, and V_L.
    b. What is V_R in the limits $\omega \to 0$ and $\omega \to \infty$?
    c. If the output is taken from the resistor, is this a low-pass or a high-pass filter? Explain.
    d. Find an expression for the crossover frequency ω_c.

FIGURE P32.49

50. ‖ A series RL circuit is built with a 110 Ω resistor and a 5.0-cm-long, 1.0-cm-diameter solenoid with 800 turns of wire. What is the peak magnetic flux through the solenoid if the circuit is driven by a 12 V, 5.0 kHz source?

51. ‖‖ A series RLC circuit consists of a 75 Ω resistor, a 0.12 H inductor, and a 30 μF capacitor. It is attached to a 120 V/60 Hz power line. What are (a) the peak current I, (b) the phase angle ϕ, and (c) the average power loss?

52. ‖ A series RLC circuit consists of a 25 Ω resistor, a 0.10 H inductor, and a 100 μF capacitor. It draws a 2.5 A rms current when attached to a 60 Hz source. What are (a) the emf $\mathcal{E}_{rms}$, (b) the phase angle ϕ, and (c) the average power loss?

53. ‖ A series RLC circuit consists of a 550 Ω resistor, a 2.1 mH inductor, and a 550 nF capacitor. It is connected to a 50 V rms oscillating voltage source with an adjustable frequency. An oscillating magnetic field is observed at a point 2.5 mm from the center of one of the circuit wires, which is long and straight. What is the maximum magnetic field amplitude that can be generated?

54. ‖ In FIGURE P32.54, what is the current supplied by the emf when (a) the frequency is very small and (b) the frequency is very large?

FIGURE P32.54

55. ‖ The current lags the emf by 30° in a series RLC circuit with $\mathcal{E}_0 = 10$ V and $R = 50$ Ω. What is the peak current through the circuit?

56. ‖ A series RLC circuit consists of a 50 Ω resistor, a 3.3 mH inductor, and a 480 nF capacitor. It is connected to a 5.0 kHz oscillator with a peak voltage of 5.0 V. What is the instantaneous current i when
    a. $\mathcal{E} = \mathcal{E}_0$?
    b. $\mathcal{E} = 0$ V and is decreasing?

57. ‖ A series RLC circuit consists of a 50 Ω resistor, a 3.3 mH inductor, and a 480 nF capacitor. It is connected to a 3.0 kHz oscillator with a peak voltage of 5.0 V. What is the instantaneous emf $\mathcal{E}$ when
    a. $i = I$?
    b. $i = 0$ A and is decreasing?
    c. $i = -I$?

58. ‖ Show that the power factor of a series RLC circuit is $\cos \phi = R/Z$.

59. ‖ For a series RLC circuit, show that
    a. The peak current can be written $I = I_{max} \cos \phi$.
    b. The average power can be written $P_{source} = P_{max} \cos^2 \phi$.

60. ‖ The tuning circuit in an FM radio receiver is a series RLC circuit with a 0.200 μH inductor.
    a. The receiver is tuned to a station at 104.3 MHz. What is the value of the capacitor in the tuning circuit?
    b. FM radio stations are assigned frequencies every 0.2 MHz, but two nearby stations cannot use adjacent frequencies. What is the maximum resistance the tuning circuit can have if the peak current at a frequency of 103.9 MHz, the closest frequency that can be used by a nearby station, is to be no more than 0.10% of the peak current at 104.3 MHz? The radio is still tuned to 104.3 MHz, and you can assume the two stations have equal strength.

61. ‖ A television channel is assigned the frequency range from 54 MHz to 60 MHz. A series RLC tuning circuit in a TV receiver resonates in the middle of this frequency range. The circuit uses a 16 pF capacitor.
    a. What is the value of the inductor?
    b. In order to function properly, the current throughout the frequency range must be at least 50% of the current at the resonance frequency. What is the minimum possible value of the circuit's resistance?

62. ‖ Lightbulbs labeled 40 W, 60 W, and 100 W are connected to a 120 V/60 Hz power line as shown in FIGURE P32.62. What is the rate at which energy is dissipated in each bulb?

FIGURE P32.62

63. ‖ A generator consists of a 12-cm by 16-cm rectangular loop with 500 turns of wire spinning at 60 Hz in a 25 mT uniform magnetic field. The generator output is connected to a series RC circuit consisting of a 120 Ω resistor and a 35 μF capacitor. What is the average power delivered to the circuit?

64. ‖ Commercial electricity is generated and transmitted as *three-phase electricity*. Instead of a single emf, three separate wires carry currents for the emfs $\mathcal{E}_1 = \mathcal{E}_0 \cos \omega t$, $\mathcal{E}_2 = \mathcal{E}_0 \cos(\omega t + 120°)$, and $\mathcal{E}_3 = \mathcal{E}_0 \cos(\omega t - 120°)$ over three parallel wires, each of which supplies one-third of the power. This is why the long-distance transmission lines you see in the countryside have three wires. Suppose the transmission lines into a city supply a total of 450 MW of electric power, a realistic value.
    a. What would be the current in each wire if the transmission voltage were $\mathcal{E}_0 = 120$ V rms?
    b. In fact, transformers are used to step the transmission-line voltage up to 500 kV rms. What is the current in each wire?
    c. Big transformers are expensive. Why does the electric company use step-up transformers?

65. ‖ You're the operator of a 15,000 V rms, 60 Hz electrical substation. When you get to work one day, you see that the station is delivering 6.0 MW of power with a power factor of 0.90.
    a. What is the rms current leaving the station?
    b. How much series capacitance should you add to bring the power factor up to 1.0?
    c. How much power will the station then be delivering?

66. ‖ Commercial electricity is generated and transmitted as *three-phase electricity*. Instead of a single emf $\mathcal{E} = \mathcal{E}_0 \cos \omega t$, three separate wires carry currents for the emfs $\mathcal{E}_1 = \mathcal{E}_0 \cos \omega t, \mathcal{E}_2 = \mathcal{E}_0 \cos(\omega t + 120°)$, and $\mathcal{E}_3 = \mathcal{E}_0 \cos(\omega t - 120°)$. This is why the long-distance transmission lines you see in the countryside have three parallel wires, as do many distribution lines within a city.
   a. Draw a phasor diagram showing phasors for all three phases of a three-phase emf.
   b. Show that the sum of the three phases is zero, producing what is referred to as *neutral*. In *single-phase* electricity, provided by the familiar 120 V/60 Hz electric outlets in your home, one side of the outlet is neutral, as established at a nearby electrical substation. The other, called the *hot side*, is one of the three phases. (The round opening is connected to ground.)
   c. Show that the potential difference between any two of the phases has the rms value $\sqrt{3}\,\mathcal{E}_{rms}$, where $\mathcal{E}_{rms}$ is the familiar single-phase rms voltage. Evaluate this potential difference for $\mathcal{E}_{rms} = 120$ V. Some high-power home appliances, especially electric clothes dryers and hot-water heaters, are designed to operate between two of the phases rather than between one phase and neutral. Heavy-duty industrial motors are designed to operate from all three phases, but full three-phase power is rare in residential or office use.

67. ‖ A motor attached to a 120 V/60 Hz power line draws an 8.0 A current. Its average energy dissipation is 800 W.
   a. What is the power factor?
   b. What is the rms resistor voltage?
   c. What is the motor's resistance?
   d. How much series capacitance needs to be added to increase the power factor to 1.0?

Challenge Problems

68. ‖‖ FIGURE CP32.68 shows voltage and current graphs for a series *RLC* circuit.
   a. What is the resistance R?
   b. If $L = 200\ \mu H$, what is the resonance frequency in Hz?

FIGURE CP32.68

69. ‖‖ a. Show that the average power loss in a series *RLC* circuit is
$$P_{avg} = \frac{\omega^2 (\mathcal{E}_{rms})^2 R}{\omega^2 R^2 + L^2(\omega^2 - \omega_0^2)^2}$$
   b. Prove that the energy dissipation is a maximum at $\omega = \omega_0$.

70. ‖‖ a. Show that the peak inductor voltage in a series *RLC* circuit is maximum at frequency
$$\omega_L = \left(\frac{1}{\omega_0^2} - \frac{1}{2} R^2 C^2 \right)^{-1/2}$$
   b. A series *RLC* circuit with $\mathcal{E}_0 = 10.0$ V consists of a 1.0 Ω resistor, a 1.0 μH inductor, and a 1.0 μF capacitor. What is V_L at $\omega = \omega_0$ and at $\omega = \omega_L$?

71. ‖‖ The telecommunication circuit shown in FIGURE CP32.71 has a parallel inductor and capacitor in series with a resistor.
   a. Use a phasor diagram to show that the peak current through the resistor is
$$I = \frac{\mathcal{E}_0}{\sqrt{R^2 + \left(\dfrac{1}{X_L} - \dfrac{1}{X_C} \right)^{-2}}}$$
   Hint: Start with the inductor phasor v_L.
   b. What is I in the limits $\omega \to 0$ and $\omega \to \infty$?
   c. What is the resonance frequency ω_0? What is I at this frequency?

FIGURE CP32.71

72. ‖‖ Consider the parallel *RLC* circuit shown in FIGURE CP32.72.
   a. Show that the current drawn from the emf is
$$I = \mathcal{E}_0 \sqrt{\frac{1}{R^2} + \left(\frac{1}{\omega L} - \omega C \right)^2}$$
   Hint: Start with a phasor that is common to all three circuit elements.
   b. What is I in the limits $\omega \to 0$ and $\omega \to \infty$?
   c. Find the frequency for which I is a minimum.
   d. Sketch a graph of I versus ω.

FIGURE CP32.72

Electricity and Magnetism

KEY FINDINGS   What are the overarching findings of Part VI?

- **Charge,** like mass, is a fundamental property of matter. There are two kinds of charge: positive and negative.

- Charges interact via the **electric field:**
  - Source charges create an electric field.
  - A charge in the field experiences an electric force.

- Moving charges (currents) interact with other moving charges via **magnetic fields** and forces.

- Fields are also created when other fields change:
  - A changing magnetic field induces an electric field.
  - A changing electric field induces a magnetic field.

LAWS   What laws of physics govern electricity and magnetism?

Maxwell's equations are the formal statements of the laws of electricity and magnetism, but most problem solving is based on simpler versions of these equations:

Coulomb's law	$\vec{E}_{\text{point charge}} = \dfrac{1}{4\pi\epsilon_0}\dfrac{q}{r^2}\hat{r}$
Biot-Savart law	$\vec{B}_{\text{point charge}} = \dfrac{\mu_0}{4\pi}\dfrac{q\vec{v}\times\hat{r}}{r^2}$
Lorentz force law	$\vec{F}_{\text{on } q} = q\vec{E} + q\vec{v}\times\vec{B}$
Faraday's law	$\mathcal{E} = \lvert d\Phi_{\text{m}}/dt\rvert$ $I_{\text{induced}} = \mathcal{E}/R$ in the direction of Lenz's law

Gauss's law $\oint \vec{E}\cdot d\vec{A} = \dfrac{Q_{\text{in}}}{\epsilon_0}$

Ampère's law $\oint \vec{B}\cdot d\vec{s} = \mu_0 I_{\text{through}}$

MODELS   What are the most important models of electricity and magnetism?

Charge model

- Two types of charge: positive and negative.
  - Like charges repel, opposite charges attract.
  - Charge is conserved but can be transferred.
- Two types of materials: conductors and insulators.
- Neutral objects (no net charge) can be polarized.

Electric field model

- Charges interact via the electric field.
  - Source charges create an electric field.
  - Charge q experiences $\vec{E}_{\text{on } q} = q\vec{E}$.
- Important electric field models
  - Point charge
  - Charged sphere
  - Long charged wire
  - Charged plane

Magnetic field models

- Moving charges and currents interact via the magnetic field.
  - Current creates a magnetic field.
  - Charge q experiences $\vec{F}_{\text{on } q} = q\vec{v}\times\vec{B}$.
  - Current I experiences $\vec{F}_{\text{on } I} = I\vec{l}\times\vec{B}$.
- Important magnetic field models
  - Straight, current-carrying wire
  - Current loop
  - Solenoid

Electromagnetic waves

- An **electromagnetic wave** is a self-sustaining electromagnetic field.
  - $\vec{E}$ and $\vec{B}$ are perpendicular to $\vec{v}$.
  - Wave speed $v_{\text{em}} = 1/\sqrt{\epsilon_0\mu_0} = c$.

TOOLS   What are the most important tools introduced in Part VI?

Electric potential and potential energy

- Electric interactions can also be described by an **electric potential:**

  $$V_{\text{point charge}} = q/4\pi\epsilon_0 r$$

  - A charge q in a potential has potential energy $U = qV$.
  - Mechanical energy is conserved.

- Field and potential energy are closely related.
  - The field is perpendicular to equipotential surfaces.
  - The field points "downhill" toward lower potential.

Decreasing V ⟶

Circuits

- Batteries are sources of emf.
- **Current** is the rate of flow of charge: $I = dQ/dt$.
- Circuit elements:
  - Capacitors: $Q = C\Delta V$
  - Resistors: $I = \Delta V/R$ (Ohm's law)
  - Capacitors and resistors can be combined in series or in parallel.
- **Kirchhoff's laws:**
  - Loop law: The sum of potential differences around a loop is zero.
  - Junction law: The sum of currents at a junction is zero.

Uniform fields

- A parallel-plate capacitor creates a uniform electric field.
- A solenoid creates a uniform magnetic field.

Induced currents

- **Lenz's law:** An induced electric current flows in the direction such that the induced magnetic field opposes the *change* in the magnetic flux.

Decreasing B

Induced current

OVERVIEW

The Story of Light

Optics is the area of physics concerned with the properties and applications of light, including how light interacts with matter. But what is light? This is an ancient question with no simple answer. Rather than an all-encompassing theory of light, we will develop three different *models* of light, each of which describes and explains the behavior of light within a certain range of physical situations.

The *wave model* is based on the well-known fact that light is a wave—specifically, an electromagnetic wave. This model builds on what you learned about waves in Part IV of this text. Light waves, like any wave, are diffuse, spread out through space, and exhibit superposition and interference. The wave model of light has many important applications, including

- Precision measurements.
- Optical coatings on lenses, sensors, and windows.
- Optical computing.

NOTE Although light is an electromagnetic wave, your understanding of these chapters depends on nothing more than the "waviness" of light. You can begin these chapters either before or after your study of electricity and magnetism in Part VI. The electromagnetic aspects of light waves are discussed in Chapter 31.

Another well-known fact—that light travels in straight lines—is the basis for the *ray model*. The ray model will allow us to understand

- Image formation with lenses and mirrors.
- Optical fibers.
- Optical instruments ranging from cameras to telescopes.

One of the most amazing optical instruments is your eye. We'll investigate optics of the eye and learn how glasses or contact lenses can correct some defects of vision.

The *photon model,* part of quantum physics, is mentioned here for completeness but won't be developed until Part VIII of this text. In the quantum world, light consists of tiny packets of energy—photons—that have both wave-like and particle-like properties. Photons will help us understand how atoms emit and absorb light.

For the most part, these three models are mutually exclusive; hence we'll pay close attention to establishing guidelines for when each model is valid.

These optical fibers—thin, flexible threads of glass that channel laser light much like water flowing through a pipe —are what make high-speed internet a reality.

33 Wave Optics

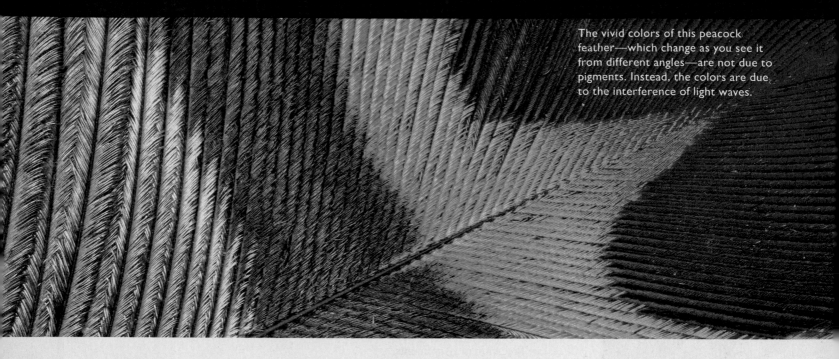

The vivid colors of this peacock feather—which change as you see it from different angles—are not due to pigments. Instead, the colors are due to the interference of light waves.

IN THIS CHAPTER, you will learn about and apply the wave model of light.

What is light?

You will learn that light has aspects of both waves and particles. We will introduce three models of light.

- The wave model of light—the subject of this chapter—describes how light waves spread out and how the superposition of multiple light waves causes interference.
- The ray model of light, in which light travels in straight lines, will explain how mirrors and lenses work. It is the subject of Chapter 34.
- The photon model of light, which will be discussed in Part VIII, is an important part of quantum physics.

One of our tasks will be to learn when each model is appropriate.

« LOOKING BACK Sections 16.5, 16.7, and 16.8 Light waves, wave fronts, phase, and intensity

Wave model

Ray model

Photon model

What is diffraction?

Diffraction is the ability of a wave to spread out after passing through a small hole or going around a corner. The diffraction of light indicates that light is a wave. One interesting finding will be that a *smaller* hole causes *more* spreading.

Does light exhibit interference?

Yes. You previously studied the thin-film interference of light reflecting from two surfaces. In this chapter we will examine the interference fringes that are seen after light passes through two narrow, closely spaced slits in an opaque screen.

« LOOKING BACK Section 17.7 Interference

Double-slit interference

What is a diffraction grating?

A diffraction grating is a periodic array of closely spaced slits or grooves. Different wavelengths are sent in different directions when light passes through a diffraction grating. Two similar wavelengths can be distinguished because the fringes of each are very narrow and precisely located.

Diffraction-grating fringes

How is interference used?

Diffraction gratings are the basis of spectroscopy, a tool for analyzing the composition of materials by the wavelengths they emit. Interferometers make precise measurements, ranging from the vibrations of wings to the movements of continents, with the controlled use of interference. And interference plays a key role in optical computers.

Interferometer fringes

33.1 Models of Light

The study of light is called **optics.** But what is light? The first Greek scientists did not make a distinction between light and vision. Light, to them, was inseparable from seeing. But gradually there arose a view that light actually "exists," that light is some sort of physical entity that is present regardless of whether or not someone is looking. But if light is a physical entity, what is it?

Newton, in addition to his pioneering work in mathematics and mechanics in the 1660s, investigated the nature of light. Newton knew that a water wave, after passing through an opening, *spreads out* to fill the space behind the opening. You can see this in FIGURE 33.1a, where plane waves, approaching from the left, spread out in circular arcs after passing through a hole in a barrier. This inexorable spreading of waves is the phenomenon called **diffraction.** Diffraction is a sure sign that whatever is passing through the hole is a wave.

In contrast, FIGURE 33.1b shows that sunlight makes sharp-edged shadows. We don't see the light spreading out in circular arcs after passing through an opening. This behavior is what you would expect if light consists of noninteracting particles traveling in straight lines. Some particles would pass through the openings to make bright areas on the floor, others would be blocked and cause the well-defined shadow. This reasoning led Newton to the conclusion that light consists of very small, light, fast particles that he called *corpuscles.*

The situation changed dramatically in 1801, when the English scientist Thomas Young announced that he had produced *interference* of light. Young's experiment, which we will analyze in the next section, quickly settled the debate in favor of a wave theory of light because interference is a distinctly wave-like phenomenon. But if light is a wave, what is waving? It was ultimately established that light is an *electromagnetic wave,* an oscillation of the electromagnetic field requiring no material medium in which to travel.

But this satisfying conclusion was soon undermined by new discoveries at the start of the 20th century. Albert Einstein's introduction of the concept of the *photon*—a wave having certain particle-like characteristics—marked the end of *classical physics* and the beginning of a new era called *quantum physics.* Equally important, Einstein's theory marked yet another shift in our age-old effort to understand light.

> **NOTE** Optics, as we will study it, was developed before it was known that light is an electromagnetic wave. This chapter requires an understanding of waves, from Chapters 16 and 17, but does not require a knowledge of electromagnetic fields. Thus you can study Part VII either before or after your study of electricity and magnetism in Part VI. Light polarization, the one aspect of optics that does require some familiarity with electromagnetic waves, is covered in Chapter 31.

Three Views

Light is a real physical entity, but the nature of light is elusive. Light is the chameleon of the physical world. Under some circumstances, light acts like particles traveling in straight lines. But change the circumstances, and light shows the same kinds of wave-like behavior as sound waves or water waves. Change the circumstances yet again, and light exhibits behavior that is neither wave-like nor particle-like but has characteristics of both.

Rather than an all-encompassing "theory of light," it will be better to develop three **models of light.** Each model successfully explains the behavior of light within a certain domain—that is, within a certain range of physical situations. Our task will be twofold:

1. To develop clear and distinct models of light.
2. To learn the conditions and circumstances for which each model is valid.

We'll begin with a brief summary of all three models.

FIGURE 33.1 Water waves spread out behind a small hole in a barrier, but light passing through an archway makes a sharp-edged shadow.

(a) Plane waves approach from the left.

Circular waves spread out on the right.

(b)

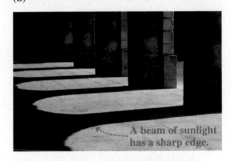

A beam of sunlight has a sharp edge.

Three models of light

The Wave Model

The wave model of light is responsible for the well-known "fact" that light is a *wave*. Indeed, under many circumstances light exhibits the same behavior as sound or water waves. Lasers and electro-optical devices are best described by the wave model of light. Some aspects of the wave model were introduced in Chapters 16 and 17, and it is the primary focus of this chapter.

The Ray Model

An equally well-known "fact" is that light travels in straight lines. These straight-line paths are called *light rays*. The properties of prisms, mirrors, and lenses are best understood in terms of light rays. Unfortunately, it's difficult to reconcile "light travels in straight lines" with "light is a wave." For the most part, waves and rays are mutually exclusive models of light. One of our important tasks will be to learn when each model is appropriate. Ray optics is the subject of Chapters 34 and 35.

The Photon Model

Modern technology is increasingly reliant on quantum physics. In the quantum world, light behaves like neither a wave nor a particle. Instead, light consists of *photons* that have both wave-like and particle-like properties. Much of the quantum theory of light is beyond the scope of this textbook, but we will take a peek at some of the important ideas in Part VIII.

FIGURE 33.2 Light, just like a water wave, does spread out behind an opening if the opening is sufficiently narrow.

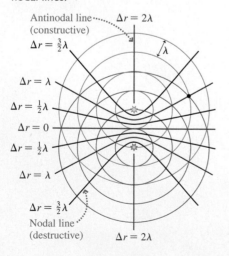

FIGURE 33.3 Constructive and destructive interference occur along antinodal and nodal lines.

33.2 The Interference of Light

Newton might have reached a different conclusion about the nature of light had he seen the experiment depicted in **FIGURE 33.2**. Here light of a single wavelength (or color) passes through a narrow slit that is only 0.1 mm wide, about twice the width of a human hair. The image shows how the light appears on a viewing screen 2 m behind the slit. If light travels in straight lines, as Newton thought, we should see a narrow strip of light, about 0.1 mm wide, with dark shadows on either side. Instead, we see a band of light extending over about 2.5 cm, a distance much wider than the aperture, with dimmer patches of light extending even farther on either side.

If you compare Figure 33.2 to the water wave of Figure 33.1a, you see that *the light is spreading out* behind the opening. The light is exhibiting diffraction, the sure sign of waviness. We will look at diffraction in more detail later in the chapter. For now, we merely need the *observation* that light does, indeed, spread out behind an opening that is sufficiently narrow. Light is acting as a wave, so we should—as Thomas Young did—be able to observe the interference of light waves.

A Brief Review of Interference

Waves obey the *principle of superposition:* If two waves overlap, their displacements add—whether it's the displacement of air molecules in a sound wave or the electric field in a light wave. You learned in **« Section 17.5–17.7** that *constructive interference* occurs when the crests of two waves overlap, adding to produce a wave with a larger amplitude and thus a greater intensity. Conversely, the overlap of the crest of one wave with the trough of another wave produces a wave with reduced amplitude (perhaps even zero) and thus a lesser intensity. This is *destructive interference.*

FIGURE 33.3 shows two in-phase sources of sinusoidal waves. The circular rings, you will recall, are the *wave crests,* and they are spaced one wavelength λ apart. Although Figure 33.3 is a snapshot, frozen in time, you should envision both sets of rings propagating outward at the wave speed as the sources oscillate.

We can measure distances simply by counting rings. For example, you can see that the dot on the right side is exactly three wavelengths from the top source and exactly four wavelengths from the bottom source, making $r_1 = 3\lambda$ and $r_2 = 4\lambda$. At this point, the *path-length difference* is $\Delta r = r_2 - r_1 = \lambda$. **Points where Δr is an integer number of wavelengths are points of constructive interference.** The waves may have traveled difference distances, but crests align with crests and troughs align with troughs to produce a wave with a larger amplitude.

If the path-length difference is a half-integer number of wavelengths, then the crest of one wave will arrive with the trough of the other. Thus **points where Δr is a half-integer number of wavelengths are points of destructive interference.** Mathematically, the conditions for interference are:

$$
\begin{array}{lll}
\text{Constructive interference:} & \Delta r = m\lambda & \\
\text{Destructive interference:} & \Delta r = \left(m + \tfrac{1}{2}\right)\lambda & \quad m = 0, 1, 2, \ldots \quad (33.1)
\end{array}
$$

Destructive interference is *perfect destructive interference*, with zero intensity, only if both waves have exactly the same amplitude.

The set of all points for which $\Delta r = m\lambda$ is called an *antinodal line*. Maximum constructive interference is happening at every point along this line. If these are sound waves, you will hear maximum loudness by standing on an antinodal line. If these are light waves, you will see maximum brightness everywhere an antinodal line touches a viewing screen.

Similarly, maximum destructive interference occurs along *nodal lines*—analogous to the *nodes* of a one-dimensional standing wave. If these are light waves, a viewing screen will be dark everywhere it intersects a nodal line.

Young's Double-Slit Experiment

Let's now see how these ideas about interference apply to light waves. FIGURE 33.4a shows an experiment in which a laser beam is aimed at an opaque screen containing two long, narrow slits that are very close together. This pair of slits is called a **double slit,** and in a typical experiment they are ≈ 0.1 mm wide and spaced ≈ 0.5 mm apart. We will assume that the laser beam illuminates both slits equally and that any light passing through the slits impinges on a viewing screen. This is the essence of Young's experiment of 1801, although he used sunlight rather than a laser.

What should we expect to see on the screen? FIGURE 33.4b is a view from above the experiment, looking down on the top ends of the slits and the top edge of the viewing screen. Because the slits are very narrow, **light spreads out behind each slit** exactly as it did in Figure 33.2. These two spreading waves overlap and interfere with each other, just as if they were sound waves emitted by two

FIGURE 33.4 A double-slit interference experiment.

(a)

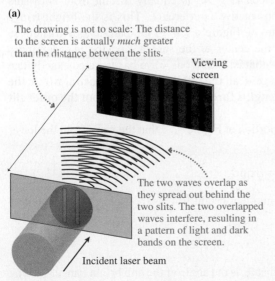

The drawing is not to scale: The distance to the screen is actually *much* greater than the distance between the slits.

Viewing screen

The two waves overlap as they spread out behind the two slits. The two overlapped waves interfere, resulting in a pattern of light and dark bands on the screen.

Incident laser beam

(b)

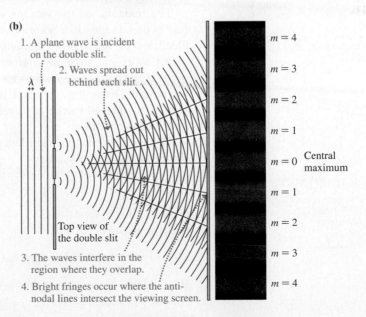

1. A plane wave is incident on the double slit.

2. Waves spread out behind each slit.

λ

Top view of the double slit

3. The waves interfere in the region where they overlap.

4. Bright fringes occur where the anti-nodal lines intersect the viewing screen.

$m = 4$

$m = 3$

$m = 2$

$m = 1$

$m = 0$ Central maximum

$m = 1$

$m = 2$

$m = 3$

$m = 4$

loudspeakers. Notice how the antinodal lines of constructive interference are just like those in Figure 33.3.

The image in Figure 33.4b shows how the screen looks. As expected, the light is intense at points where an antinodal line intersects the screen. There is no light at all at points where a nodal line intersects the screen. These alternating bright and dark bands of light, due to constructive and destructive interference, are called **interference fringes.** The fringes are numbered $m = 0, 1, 2, 3, \ldots$, going outward from the center. The brightest fringe, at the midpoint of the viewing screen, with $m = 0$, is called the **central maximum.**

STOP TO THINK 33.1   Suppose the viewing screen in Figure 33.4 is moved closer to the double slit. What happens to the interference fringes?

a. They get brighter but otherwise do not change.
b. They get brighter and closer together.
c. They get brighter and farther apart.
d. They get out of focus.
e. They fade out and disappear.

Analyzing Double-Slit Interference

Figure 33.4 showed qualitatively how interference is produced behind a double slit by the overlap of the light waves spreading out behind each slit. Now let's analyze the experiment more carefully. **FIGURE 33.5** shows a double-slit experiment in which the spacing between the two slits is d and the distance to the viewing screen is L. We will assume that L is *very* much larger than d. Consequently, we don't see the individual slits in the upper part of Figure 33.5.

Let P be a point on the screen at angle θ. Our goal is to determine whether the interference at P is constructive, destructive, or in between. The insert to Figure 33.5 shows the individual slits and the paths from these slits to point P. Because P is so far away on this scale, the two paths are virtually parallel, both at angle θ. Both slits are illuminated by the *same* wave front from the laser; hence the slits act as sources of identical, in-phase waves $(\Delta\phi_0 = 0)$. Thus the interference at point P is constructive, producing a bright fringe, if $\Delta r = m\lambda$ at that point.

The midpoint on the viewing screen at $y = 0$ is equally distant from both slits $(\Delta r = 0)$ and thus is a point of constructive interference. This is the bright fringe identified as the central maximum in Figure 33.4b. The path-length difference increases as you move away from the center of the screen, and the $m = 1$ fringes occur at the points where $\Delta r = 1\lambda$—that is, where one wave has traveled exactly one wavelength farther than the other. In general, **the mth bright fringe occurs where the wave from one slit travels m wavelengths farther than the wave from the other slit and thus $\Delta r = m\lambda$.**

You can see from the magnified portion of Figure 33.5 that the wave from the lower slit travels an extra distance

$$\Delta r = d\sin\theta \qquad (33.2)$$

If we use this in Equation 33.1, we find that bright fringes (constructive interference) occur at angles θ_m such that

$$\Delta r = d\sin\theta_m = m\lambda \qquad m = 0, 1, 2, 3, \ldots \qquad (33.3)$$

We added the subscript m to denote that θ_m is the angle of the mth bright fringe, starting with $m = 0$ at the center.

FIGURE 33.5 Geometry of the double-slit experiment.

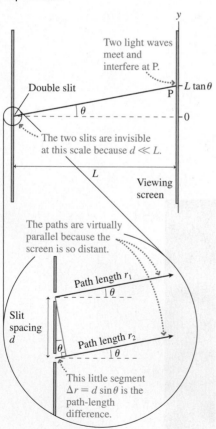

Two light waves meet and interfere at P.

Double slit

$L\tan\theta$

θ

0

The two slits are invisible at this scale because $d \ll L$.

L

Viewing screen

The paths are virtually parallel because the screen is so distant.

Path length r_1

θ

Slit spacing d

θ Path length r_2

θ

This little segment $\Delta r = d\sin\theta$ is the path-length difference.

In practice, the angle θ in a double-slit experiment is very small ($<1°$). We can use the small-angle approximation $\sin\theta \approx \theta$, where θ must be in radians, to write Equation 33.3 as

$$\theta_m = m\frac{\lambda}{d} \qquad m = 0, 1, 2, 3, \ldots \qquad \text{(angles of bright fringes)} \qquad (33.4)$$

This gives the angular positions *in radians* of the bright fringes in the interference pattern.

Positions of the Fringes

It's usually easier to measure distances rather than angles, so we can also specify point P by its position on a *y*-axis with the origin directly across from the midpoint between the slits. You can see from Figure 33.5 that

$$y = L\tan\theta \qquad (33.5)$$

Using the small-angle approximation once again, this time in the form $\tan\theta \approx \theta$, we can substitute θ_m from Equation 33.4 for $\tan\theta_m$ in Equation 33.5 to find that the *m*th bright fringe occurs at position

$$y_m = \frac{m\lambda L}{d} \qquad m = 0, 1, 2, 3, \ldots \qquad \text{(positions of bright fringes)} \qquad (33.6)$$

The interference pattern is symmetrical, so there is an *m*th bright fringe at the same distance on both sides of the center. You can see this in Figure 33.4b. As we've noted, the **$m = 1$ fringes occur at points on the screen where the light from one slit travels exactly one wavelength farther than the light from the other slit.**

> **NOTE** Equations 33.4 and 33.6 do *not* apply to the interference of sound waves from two loudspeakers. The approximations we've used (small angles, $L \gg d$) are usually not valid for the much longer wavelengths of sound waves.

Equation 33.6 predicts that **the interference pattern is a series of equally spaced bright lines** on the screen, exactly as shown in Figure 33.4b. How do we know the fringes are equally spaced? The **fringe spacing** between the *m* fringe and the $m + 1$ fringe is

$$\Delta y = y_{m+1} - y_m = \frac{(m+1)\lambda L}{d} - \frac{m\lambda L}{d} = \frac{\lambda L}{d} \qquad (33.7)$$

Because Δy is independent of *m*, *any* two adjacent bright fringes have the same spacing.

The dark fringes in the image are bands of destructive interference. They occur at positions where the path-length difference of the waves is a half-integer number of wavelengths:

$$\Delta r = \left(m + \tfrac{1}{2}\right)\lambda \qquad m = 0, 1, 2, \ldots \qquad \begin{array}{l}\text{(destructive}\\\text{interference)}\end{array} \qquad (33.8)$$

We can use Equation 33.2 for Δr and the small-angle approximation to find that the dark fringes are located at positions

$$y_m' = \left(m + \tfrac{1}{2}\right)\frac{\lambda L}{d} \qquad m = 0, 1, 2, \ldots \qquad \begin{array}{l}\text{(positions of}\\\text{dark fringes)}\end{array} \qquad (33.9)$$

We have used y_m', with a prime, to distinguish the location of the *m*th minimum from the *m*th maximum at y_m. You can see from Equation 33.9 that **the dark fringes are located exactly halfway between the bright fringes.**

As a quick example, suppose that light from a helium-neon laser ($\lambda = 633$ nm) illuminates two slits spaced 0.40 mm apart and that a viewing screen is 2.0 m behind the slits. The $m = 2$ bright fringe is located at position

$$y_2 = \frac{2\lambda L}{d} = \frac{2(633 \times 10^{-9}\ \text{m})(2.0\ \text{m})}{4.0 \times 10^{-4}\ \text{m}} = 6.3\ \text{mm}$$

Similarly, the $m = 2$ dark fringe is found at $y_2' = \left(2 + \frac{1}{2}\right)\lambda L/d = 7.9$ mm. Because the fringes are counted outward from the center, the $m = 2$ bright fringe occurs *before* the $m = 2$ dark fringe.

EXAMPLE 33.1 | **Measuring the wavelength of light**

A double-slit interference pattern is observed on a screen 1.0 m behind two slits spaced 0.30 mm apart. Ten bright fringes span a distance of 1.7 cm. What is the wavelength of the light?

MODEL It is not always obvious which fringe is the central maximum. Slight imperfections in the slits can make the interference fringe pattern less than ideal. However, you do not need to identify the $m = 0$ fringe because you can make use of the fact that the fringe spacing Δy is uniform. Ten bright fringes have *nine* spaces between them (not ten—be careful!).

VISUALIZE The interference pattern looks like the image of Figure 33.4b.

SOLVE The fringe spacing is

$$\Delta y = \frac{1.7\ \text{cm}}{9} = 1.89 \times 10^{-3}\ \text{m}$$

Using this fringe spacing in Equation 33.7, we find that the wavelength is

$$\lambda = \frac{d}{L}\Delta y = 5.7 \times 10^{-7}\ \text{m} = 570\ \text{nm}$$

It is customary to express the wavelengths of visible light in nanometers. Be sure to do this as you solve problems.

ASSESS Young's double-slit experiment not only demonstrated that light is a wave, it provided a means for measuring the wavelength. You learned in Chapter 16 that the wavelengths of visible light span the range 400–700 nm. These lengths are smaller than we can easily comprehend. A wavelength of 570 nm, which is in the middle of the visible spectrum, is only about 1% of the diameter of a human hair.

STOP TO THINK 33.2 Light of wavelength λ_1 illuminates a double slit, and interference fringes are observed on a screen behind the slits. When the wavelength is changed to λ_2, the fringes get closer together. Is λ_2 larger or smaller than λ_1?

Intensity of the Double-Slit Interference Pattern

Equations 33.6 and 33.9 locate the positions of maximum and zero intensity. To complete our analysis we need to calculate the light *intensity* at every point on the screen. In « Chapter 17, where interference was introduced, you learned that the net amplitude E of two superimposed waves is

$$E = \left| 2e \cos\left(\frac{\Delta\phi}{2}\right) \right| \tag{33.10}$$

where, for light waves, e is the electric field amplitude of each individual wave. Because the sources (i.e., the two slits) are in phase, the phase difference $\Delta\phi$ at the point where the two waves are combined is due only to the path-length difference: $\Delta\phi = 2\pi(\Delta r/\lambda)$. Using Equation 33.2 for Δr, along with the small-angle approximation and Equation 33.5 for y, we find the phase difference at position y on the screen to be

$$\Delta\phi = 2\pi\frac{\Delta r}{\lambda} = 2\pi\frac{d\sin\theta}{\lambda} \approx 2\pi\frac{d\tan\theta}{\lambda} = \frac{2\pi d}{\lambda L}y \tag{33.11}$$

Substituting Equation 33.11 into Equation 33.10, we find the wave amplitude at position y to be

$$E = \left| 2e \cos\left(\frac{\pi d}{\lambda L} y\right) \right| \tag{33.12}$$

The light *intensity,* which is what we see, is proportional to the square of the amplitude. The intensity of a single slit, with amplitude e, is $I_1 = Ce^2$, where C is a proportionality constant. For the double slit, the intensity at position y on the screen is

$$I = CE^2 = 4Ce^2 \cos^2\left(\frac{\pi d}{\lambda L} y\right) \tag{33.13}$$

Replacing Ce^2 with I_1, we see that the intensity of an *ideal* double-slit interference pattern at position y is

$$I_{\text{double}} = 4I_1 \cos^2\left(\frac{\pi d}{\lambda L} y\right) \tag{33.14}$$

We've said "ideal" because we've assumed that e, the electric field amplitude of each wave, is constant across the screen.

FIGURE 33.6a is a graph of the ideal double-slit intensity versus position y. Notice the unusual orientation of the graph, with the intensity increasing toward the *left* so that the y-axis can match the experimental layout. You can see that the intensity oscillates between dark fringes ($I_{\text{double}} = 0$) and bright fringes ($I_{\text{double}} = 4I_1$). The maxima occur at points where $y_m = m\lambda L/d$. This is what we found earlier for the positions of the bright fringes, so Equation 33.14 is consistent with our initial analysis.

One curious feature is that the light intensity at the maxima is $I = 4I_1$, four times the intensity of the light from each slit alone. You might think that two slits would make the light twice as intense as one slit, but interference leads to a different result. Mathematically, two slits make the *amplitude* twice as big at points of constructive interference ($E = 2e$), so the intensity increases by a factor of $2^2 = 4$. Physically, this is conservation of energy. The line labeled $2I_1$ in Figure 33.6a is the uniform intensity that two slits would produce *if* the waves did not interfere. Interference does not change the amount of light energy coming through the two slits, but it does redistribute the light energy on the viewing screen. You can see that the *average* intensity of the oscillating curve is $2I_1$, but the intensity of the bright fringes gets pushed up from $2I_1$ to $4I_1$ in order for the intensity of the dark fringes to drop from $2I_1$ to 0.

Equation 33.14 predicts that, ideally, all interference fringes are equally bright, but you saw in Figure 33.4b that the fringes decrease in brightness as you move away from the center. The erroneous prediction stems from our assumption that the amplitude e of the wave from each slit is constant across the screen. A more detailed calculation, which we will do in Section 33.5, must consider the varying intensity of the light that has diffracted through each of the slits. We'll find that Equation 33.14 is still correct if I_1 slowly decreases as y increases.

FIGURE 33.6b summarizes this analysis by graphing the light intensity (Equation 33.14) with I_1 slowly decreasing as y increases. Comparing this graph to the image, you can see that the wave model of light has provided an excellent description of Young's double-slit interference experiment.

FIGURE 33.6 Intensity of the interference fringes in a double-slit experiment.

(a) Ideally, the bright fringes would all have equal intensity.

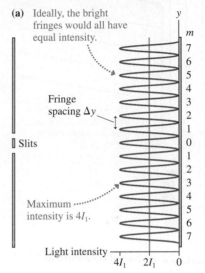

Fringe spacing Δy

Maximum intensity is $4I_1$.

(b) In reality, the fringe intensity decreases because the intensity of the light from a single slit is not uniform.

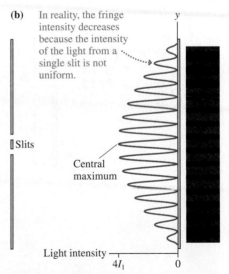

Central maximum

33.3 The Diffraction Grating

Suppose we were to replace the double slit with an opaque screen that has N closely spaced slits. When illuminated from one side, each of these slits becomes the source of a light wave that diffracts, or spreads out, behind the slit. Such a multi-slit device is called a **diffraction grating.** The light intensity pattern on a screen behind a diffraction grating is due to the interference of N overlapped waves.

FIGURE 33.7 Top view of a diffraction grating with $N = 10$ slits.

FIGURE 33.7 Top view of a diffraction grating with $N = 10$ slits.

N slits with spacing d

Spreading circular waves from each slit overlap and interfere.

λ

d

d

Toward distant screen

θ

θ $\Delta r = d \sin \theta$

θ

θ

Plane wave approaching from left

The wave from each slit travels $\Delta r = d \sin \theta$ farther than the wave from the slit above it.

FIGURE 33.8 Angles of constructive interference.

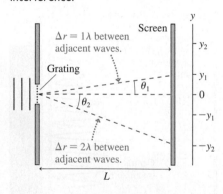

$\Delta r = 1\lambda$ between adjacent waves.

Screen

y

y_2

Grating

θ_1

y_1

0

θ_2

$-y_1$

$\Delta r = 2\lambda$ between adjacent waves.

$-y_2$

L

A microscopic side-on look at a diffraction grating.

400 nm

NOTE The terms "interference" and "diffraction" have historical roots that predate our modern understanding of wave optics, and thus their use can be confusing. Physically, both arise from the superposition of overlapped waves. *Interference* usually describes a superposition of waves coming from distinct sources—as in double-slit interference. *Diffraction* usually describes how the superposition of different portions of a single wave front causes light to bend or spread after encountering an obstacle, such as a narrow slit. Diffraction is important in a diffraction grating because light does spread out behind each slit, but the intensity pattern seen behind a diffraction grating results from the interference of light waves coming from each of the slits. It might more logically be called an *interference grating,* but it's not. Don't let the name confuse you.

FIGURE 33.7 shows a diffraction grating in which N slits are equally spaced a distance d apart. This is a top view of the grating, as we look down on the experiment, and the slits extend above and below the page. Only 10 slits are shown here, but a practical grating will have hundreds or even thousands of slits. Suppose a plane wave of wavelength λ approaches from the left. The crest of a plane wave arrives *simultaneously* at each of the slits, causing the wave emerging from each slit to be in phase with the wave emerging from every other slit. Each of these emerging waves spreads out, just like the light wave in Figure 33.2, and after a short distance they all overlap with each other and interfere.

We want to know how the interference pattern will appear on a screen behind the grating. The light wave at the screen is the superposition of N waves, from N slits, as they spread and overlap. As we did with the double slit, we'll assume that the distance L to the screen is very large in comparison with the slit spacing d; hence the path followed by the light from one slit to a point on the screen is *very nearly* parallel to the path followed by the light from neighboring slits. The paths cannot be perfectly parallel, of course, or they would never meet to interfere, but the slight deviation from perfect parallelism is too small to notice. You can see in Figure 33.7 that the wave from one slit travels distance $\Delta r = d \sin \theta$ more than the wave from the slit above it and $\Delta r = d \sin \theta$ less than the wave below it. This is the same reasoning we used in Figure 33.5 to analyze the double-slit experiment.

Figure 33.7 is a magnified view of the slits. **FIGURE 33.8** steps back to where we can see the viewing screen. If the angle θ is such that $\Delta r = d \sin \theta = m\lambda$, where m is an integer, then the light wave arriving at the screen from one slit will be *exactly in phase* with the light waves arriving from the two slits next to it. But each of those waves is in phase with waves from the slits next to them, and so on until we reach the end of the grating. In other words, N **light waves, from N different slits, will *all* be in phase with each other when they arrive at a point on the screen at angle** θ_m **such that**

$$d \sin \theta_m = m\lambda \qquad m = 0, 1, 2, 3, \ldots \qquad (33.15)$$

The screen will have bright constructive-interference fringes at the values of θ_m given by Equation 33.15. We say that the light is "diffracted at angle θ_m."

Because it's usually easier to measure distances rather than angles, the position y_m of the mth maximum is

$$y_m = L \tan \theta_m \qquad \text{(positions of bright fringes)} \qquad (33.16)$$

The integer m is called the **order** of the diffraction. For example, light diffracted at θ_2 would be the second-order diffraction. Practical gratings, with very small values for d, display only a few orders. Because d is usually very small, it is customary to characterize a grating by the number of *lines per millimeter.* Here "line" is synonymous with "slit," so the number of lines per millimeter is simply the inverse of the slit spacing d in millimeters.

NOTE The condition for constructive interference in a grating of N slits is identical to Equation 33.4 for just two slits. Equation 33.15 is simply the requirement that the path-length difference between adjacent slits, be they two or N, is $m\lambda$. But unlike the angles in double-slit interference, the angles of constructive interference from a diffraction grating are generally *not* small angles. The reason is that the slit spacing d in a diffraction grating is so small that λ/d is not a small number. Thus you *cannot* use the small-angle approximation to simplify Equations 33.15 and 33.16.

The wave amplitude at the points of constructive interference is Ne because N waves of amplitude e combine in phase. Because the intensity depends on the square of the amplitude, the intensities of the bright fringes of a diffraction grating are

$$I_{max} = N^2 I_1 \qquad (33.17)$$

where, as before, I_1 is the intensity of the wave from a single slit. You can see that the fringe intensities increase rapidly as the number of slits increases.

Not only do the fringes get brighter as N increases, they also get narrower. This is again a matter of conservation of energy. If the light waves did not interfere, the intensity from N slits would be NI_1. Interference increases the intensity of the bright fringes by an extra factor of N, so to conserve energy the width of the bright fringes must be proportional to $1/N$. For a realistic diffraction grating, with $N > 100$, the interference pattern consists of a small number of *very* bright and *very* narrow fringes while most of the screen remains dark. FIGURE 33.9a shows the interference pattern behind a diffraction grating both graphically and with a simulation of the viewing screen. A comparison with Figure 33.6b shows that the bright fringes of a diffraction grating are much sharper and more distinct than the fringes of a double slit.

Because the bright fringes are so distinct, diffraction gratings are used for measuring the wavelengths of light. Suppose the incident light consists of two slightly different wavelengths. Each wavelength will be diffracted at a slightly different angle and, if N is sufficiently large, we'll see two distinct fringes on the screen. FIGURE 33.9b illustrates this idea. By contrast, the bright fringes in a double-slit experiment are too broad to distinguish the fringes of one wavelength from those of the other.

FIGURE 33.9 The interference pattern behind a diffraction grating.

(a) Narrow, bright fringes. Most of the screen is dark.

(b) Blue light has a longer wavelength than violet, and thus diffracts more.

All wavelengths overlap at $y = 0$.

EXAMPLE 33.2 | Measuring wavelengths emitted by sodium atoms

Light from a sodium lamp passes through a diffraction grating having 1000 slits per millimeter. The interference pattern is viewed on a screen 1.000 m behind the grating. Two bright yellow fringes are visible 72.88 cm and 73.00 cm from the central maximum. What are the wavelengths of these two fringes?

VISUALIZE This is the situation shown in Figure 33.9b. The two fringes are very close together, so we expect the wavelengths to be only slightly different. No other yellow fringes are mentioned, so we will assume these two fringes are the first-order diffraction ($m = 1$).

SOLVE The distance y_m of a bright fringe from the central maximum is related to the diffraction angle by $y_m = L\tan\theta_m$. Thus the diffraction angles of these two fringes are

$$\theta_1 = \tan^{-1}\left(\frac{y_1}{L}\right) = \begin{cases} 36.08° & \text{fringe at 72.88 cm} \\ 36.13° & \text{fringe at 73.00 cm} \end{cases}$$

These angles must satisfy the interference condition $d\sin\theta_1 = \lambda$, so the wavelengths are $\lambda = d\sin\theta_1$. What is d? If a 1 mm length of the grating has 1000 slits, then the spacing from one slit to the next must be 1/1000 mm, or $d = 1.000 \times 10^{-6}$ m. Thus the wavelengths creating the two bright fringes are

$$\lambda = d\sin\theta_1 = \begin{cases} 589.0 \text{ nm} & \text{fringe at 72.88 cm} \\ 589.6 \text{ nm} & \text{fringe at 73.00 cm} \end{cases}$$

ASSESS We had data accurate to four significant figures, and all four were necessary to distinguish the two wavelengths.

The science of measuring the wavelengths of atomic and molecular emissions is called **spectroscopy.** The two sodium wavelengths in this example are called the *sodium doublet,* a name given to two closely spaced wavelengths emitted by the atoms of one element. This doublet is an identifying characteristic of sodium. Because no other element emits these two wavelengths, the doublet can be used to identify the presence of sodium in a sample of unknown composition, even if sodium is only a very minor constituent. This procedure is called *spectral analysis.*

FIGURE 33.10 Reflection gratings.

(a) Incident light — Different wavelengths diffracted at different angles

Mirror surface

Few μm

A reflection grating can be made by cutting parallel grooves in a mirror surface.

(b)

×10,000

····Parallel rods of melanin form a diffraction grating.

Reflection Gratings

We have analyzed what is called a *transmission grating,* with many parallel slits. In practice, most diffraction gratings are manufactured as *reflection gratings.* The simplest reflection grating, shown in **FIGURE 33.10a**, is a mirror with hundreds or thousands of narrow, parallel grooves cut into the surface. The grooves divide the surface into many parallel reflective stripes, each of which, when illuminated, becomes the source of a spreading wave. Thus an incident light wave is divided into *N* overlapped waves. The interference pattern is exactly the same as the interference pattern of light transmitted through *N* parallel slits.

Naturally occurring reflection gratings are responsible for some forms of color in nature. As the micrograph of **FIGURE 33.10b** shows, a peacock feather consists of nearly parallel rods of melanin. These act as a reflection grating and create the ever-changing, multicolored hues of iridescence as the angle between the grating and your eye changes. The iridescence of some insects is due to diffraction from parallel microscopic ridges on the shell.

The rainbow of colors reflected from the surface of a DVD is a similar display of interference. The surface of a DVD is smooth plastic with a mirror-like reflective coating in which millions of microscopic holes, each about 1 μm in diameter, encode digital information. From an optical perspective, the array of holes in a shiny surface is a two-dimensional version of the reflection grating shown in Figure 33.10a. Reflection gratings can be manufactured at very low cost simply by stamping holes or grooves into a reflective surface, and these are widely sold as toys and novelty items. Rainbows of color are seen as each wavelength of white light is diffracted at a unique angle.

STOP TO THINK 33.3 White light passes through a diffraction grating and forms rainbow patterns on a screen behind the grating. For each rainbow,

a. The red side is on the right, the violet side on the left.
b. The red side is on the left, the violet side on the right.
c. The red side is closest to the center of the screen, the violet side is farthest from the center.
d. The red side is farthest from the center of the screen, the violet side is closest to the center.

33.4 Single-Slit Diffraction

We opened this chapter with a photograph (Figure 33.1a) of a water wave passing through a hole in a barrier, then spreading out on the other side. You then saw an image (Figure 33.2) showing that light, after passing through a very narrow slit, also spreads out on the other side. This is called *diffraction.*

FIGURE 33.11 shows the experimental arrangement for observing the diffraction of light through a narrow slit of width *a.* Diffraction through a tall, narrow slit is known as **single-slit diffraction.** A viewing screen is placed distance *L* behind the slit, and we will assume that $L \gg a$. The light pattern on the viewing screen consists of a *central maximum* flanked by a series of weaker **secondary maxima** and dark fringes. Notice that the central maximum is significantly broader than the secondary maxima. It is also significantly brighter than the secondary maxima, although that is hard to tell here because this image has been overexposed to make the secondary maxima show up better.

FIGURE 33.11 A single-slit diffraction experiment.

Viewing screen

Central maximum — Secondary maxima

Distance *L*

Single slit of width *a*

Incident light of wavelength λ

Huygens' Principle

Our analysis of the superposition of waves from distinct sources, such as two loudspeakers or the two slits in a double-slit experiment, has tacitly assumed that the sources are *point sources,* with no measurable extent. To understand diffraction, we need to think about the propagation of an *extended* wave front. This is a problem first considered by the Dutch scientist Christiaan Huygens, a contemporary of Newton.

Huygens lived before a mathematical theory of waves had been developed, so he developed a geometrical model of wave propagation. His idea, which we now call **Huygens' principle,** has two steps:

1. Each point on a wave front is the source of a spherical *wavelet* that spreads out at the wave speed.
2. At a later time, the shape of the wave front is the line tangent to all the wavelets.

FIGURE 33.12 illustrates Huygens' principle for a plane wave and a spherical wave. As you can see, the line tangent to the wavelets of a plane wave is a plane that has propagated to the right. The line tangent to the wavelets of a spherical wave is a larger sphere.

Huygens' principle is a visual device, not a theory of waves. Nonetheless, the full mathematical theory of waves, as it developed in the 19th century, justifies Huygens' basic idea, although it is beyond the scope of this textbook to prove it.

Analyzing Single-Slit Diffraction

FIGURE 33.13a shows a wave front passing through a narrow slit of width a. According to Huygens' principle, each point on the wave front can be thought of as the source of a spherical wavelet. These wavelets overlap and interfere, producing the diffraction pattern seen on the viewing screen. The full mathematical analysis, using *every* point on the wave front, is a fairly difficult problem in calculus. We'll be satisfied with a geometrical analysis based on just a few wavelets.

FIGURE 33.13b shows the paths of several wavelets that travel straight ahead to the central point on the screen. (The screen is *very* far to the right in this magnified view of the slit.) The paths are very nearly parallel to each other, thus all the wavelets travel the same distance and arrive at the screen *in phase* with each other. The *constructive interference* between these wavelets produces the central maximum of the diffraction pattern at $\theta = 0$.

The situation is different at points away from the center. Wavelets 1 and 2 in **FIGURE 33.13c** start from points that are distance $a/2$ apart. If the angle is such that Δr_{12}, the extra distance traveled by wavelet 2, happens to be $\lambda/2$, then wavelets 1 and 2 arrive out of phase and interfere destructively. But if Δr_{12} is $\lambda/2$, then the difference Δr_{34} between paths 3 and 4 and the difference Δr_{56} between paths 5 and 6 are also $\lambda/2$.

FIGURE 33.12 Huygens' principle applied to the propagation of plane waves and spherical waves.

(a) Plane wave

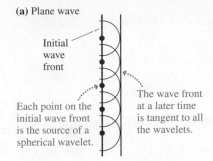

Each point on the initial wave front is the source of a spherical wavelet.

The wave front at a later time is tangent to all the wavelets.

(b) Spherical wave

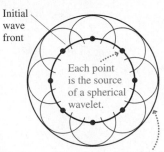

Each point is the source of a spherical wavelet.

The wave front at a later time is tangent to all the wavelets.

FIGURE 33.13 Each point on the wave front is a source of spherical wavelets. The superposition of these wavelets produces the diffraction pattern on the screen.

(a) Greatly magnified view of slit

Initial wave front

Slit width a

The wavelets from each point on the initial wave front overlap and interfere, creating a diffraction pattern on the screen.

(b)

$\theta = 0$

The wavelets going straight forward all travel the same distance to the screen. Thus they arrive in phase and interfere constructively to produce the central maximum.

(c)

Each point on the wave front is paired with another point distance $a/2$ away.

These wavelets all meet on the screen at angle θ. Wavelet 2 travels distance $\Delta r_{12} = (a/2)\sin\theta$ farther than wavelet 1.

Those pairs of wavelets also interfere destructively. The superposition of all the wavelets produces perfect destructive interference.

Figure 33.13c shows six wavelets, but our conclusion is valid for any number of wavelets. The key idea is that **every point on the wave front can be paired with another point distance $a/2$ away.** If the path-length difference is $\lambda/2$, the wavelets from these two points arrive at the screen out of phase and interfere destructively. When we sum the displacements of all N wavelets, they will—pair by pair—add to zero. The viewing screen at this position will be dark. This is the main idea of the analysis, one worth thinking about carefully.

You can see from Figure 33.13c that $\Delta r_{12} = (a/2)\sin\theta$. This path-length difference will be $\lambda/2$, the condition for destructive interference, if

$$\Delta r_{12} = \frac{a}{2}\sin\theta_1 = \frac{\lambda}{2} \tag{33.18}$$

or, equivalently, if $a\sin\theta_1 = \lambda$.

> **NOTE** Equation 33.18 cannot be satisfied if the slit width a is less than the wavelength λ. If a wave passes through an opening smaller than the wavelength, the central maximum of the diffraction pattern expands to where it *completely* fills the space behind the opening. There are no minima or dark spots at any angle. This situation is uncommon for light waves, because λ is so small, but quite common in the diffraction of sound and water waves.

We can extend this idea to find other angles of perfect destructive interference. Suppose each wavelet is paired with another wavelet from a point $a/4$ away. If Δr between these wavelets is $\lambda/2$, then all N wavelets will again cancel in pairs to give complete destructive interference. The angle θ_2 at which this occurs is found by replacing $a/2$ in Equation 33.18 with $a/4$, leading to the condition $a\sin\theta_2 = 2\lambda$. This process can be continued, and we find that the general condition for complete destructive interference is

$$a\sin\theta_p = p\lambda \qquad p = 1, 2, 3, \ldots \tag{33.19}$$

When $\theta_p \ll 1$ rad, which is almost always true for light waves, we can use the small-angle approximation to write

$$\theta_p = p\frac{\lambda}{a} \qquad p = 1, 2, 3, \ldots \qquad \text{(angles of dark fringes)} \tag{33.20}$$

Equation 33.20 gives the angles *in radians* to the dark minima in the diffraction pattern of Figure 33.11. Notice that $p = 0$ is explicitly *excluded*. $p = 0$ corresponds to the straight-ahead position at $\theta = 0$, but you saw in Figures 33.11 and 33.13b that $\theta = 0$ is the central *maximum,* not a minimum.

> **NOTE** It is perhaps surprising that Equations 33.19 and 33.20 are *mathematically* the same as the condition for the *m*th *maximum* of the double-slit interference pattern. But the physical meaning here is quite different. Equation 33.20 locates the *minima* (dark fringes) of the single-slit diffraction pattern.

FIGURE 33.14 Destructive interference by pairs leads to net destructive interference, but constructive interference by pairs does *not* necessarily lead to net constructive interference.

(a)

Each pair of vectors interferes destructively. The vector sum of all six vectors is zero.

(b)

Each pair of vectors interferes constructively. Even so, the vector sum of all six vectors is zero.

You might think that we could use this method of pairing wavelets from different points on the wave front to find the maxima in the diffraction pattern. Why not take two points on the wave front that are distance $a/2$ apart, find the angle at which their wavelets are in phase and interfere constructively, then sum over all points on the wave front? There is a subtle but important distinction. **FIGURE 33.14** shows six vector arrows. The arrows in Figure 33.14a are arranged in pairs such that the two members of each pair cancel. The sum of all six vectors is clearly the zero vector $\vec{0}$, representing destructive interference. This is the procedure we used in Figure 33.13c to arrive at Equation 33.18.

The arrows in Figure 33.14b are arranged in pairs such that the two members of each pair point in the same direction—constructive interference! Nonetheless, the sum of all six vectors is still $\vec{0}$. To have N waves interfere constructively requires more than simply having constructive interference between pairs. Each pair must also be in phase with every other pair, a condition not satisfied in Figure 33.14b. Constructive interference by pairs does *not* necessarily lead to net constructive interference. It turns out that there is no simple formula to locate the maxima of a single-slit diffraction pattern.

Optional Section 33.5 will calculate the full intensity pattern of a single slit. The results are shown graphically in FIGURE 33.15. You can see the bright central maximum at $\theta = 0$, the weaker secondary maxima, and the dark points of destructive interference at the angles given by Equation 33.20. Compare this graph to the image of Figure 33.11 and make sure you see the agreement between the two.

FIGURE 33.15 A graph of the intensity of a single-slit diffraction pattern.

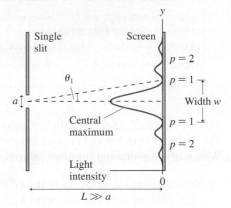

EXAMPLE 33.3 | Diffraction of a laser through a slit

Light from a helium-neon laser ($\lambda = 633$ nm) passes through a narrow slit and is seen on a screen 2.0 m behind the slit. The first minimum in the diffraction pattern is 1.2 cm from the central maximum. How wide is the slit?

MODEL A narrow slit produces a single-slit diffraction pattern. A displacement of only 1.2 cm in a distance of 200 cm means that angle θ_1 is certainly a small angle.

VISUALIZE The intensity pattern will look like Figure 33.15.

SOLVE We can use the small-angle approximation to find that the angle to the first minimum is

$$\theta_1 = \frac{1.2 \text{ cm}}{200 \text{ cm}} = 0.00600 \text{ rad} = 0.344°$$

The first minimum is at angle $\theta_1 = \lambda/a$, from which we find that the slit width is

$$a = \frac{\lambda}{\theta_1} = \frac{633 \times 10^{-9} \text{ m}}{6.00 \times 10^{-3} \text{ rad}} = 1.1 \times 10^{-4} \text{ m} = 0.11 \text{ mm}$$

ASSESS This is typical of the slit widths used to observe single-slit diffraction. You can see that the small-angle approximation is well satisfied.

The Width of a Single-Slit Diffraction Pattern

We'll find it useful, as we did for the double slit, to measure *positions* on the screen rather than angles. The position of the pth dark fringe, at angle θ_p, is $y_p = L \tan \theta_p$, where L is the distance from the slit to the viewing screen. Using Equation 33.20 for θ_p and the small-angle approximation $\tan \theta_p \approx \theta_p$, we find that the dark fringes in the single-slit diffraction pattern are located at

$$y_p = \frac{p\lambda L}{a} \qquad p = 1, 2, 3, \ldots \qquad \text{(positions of dark fringes)} \qquad (33.21)$$

A diffraction pattern is dominated by the central maximum, which is much brighter than the secondary maxima. The width w of the central maximum, shown in Figure 33.15, is defined as the distance between the two $p = 1$ minima on either side of the central maximum. Because the pattern is symmetrical, the width is simply $w = 2y_1$. This is

$$w = \frac{2\lambda L}{a} \qquad \text{(single slit)} \qquad (33.22)$$

The central maximum of this single-slit diffraction pattern appears white because it is overexposed. The width of the central maximum is clear.

The width of the central maximum is *twice* the spacing $\lambda L/a$ between the dark fringes on either side. The farther away the screen (larger L), the wider the pattern of light on it becomes. In other words, the light waves are *spreading out* behind the slit, and they fill a wider and wider region as they travel farther.

An important implication of Equation 33.22, one contrary to common sense, is that a narrower slit (smaller a) causes a *wider* diffraction pattern. **The smaller the opening you squeeze a wave through, the *more* it spreads out on the other side.**

EXAMPLE 33.4 | Determining the wavelength

Light passes through a 0.12-mm-wide slit and forms a diffraction pattern on a screen 1.00 m behind the slit. The width of the central maximum is 0.85 cm. What is the wavelength of the light?

SOLVE From Equation 33.22, the wavelength is

$$\lambda = \frac{aw}{2L} = \frac{(1.2 \times 10^{-4} \text{ m})(0.0085 \text{ m})}{2(1.00 \text{ m})}$$

$$= 5.1 \times 10^{-7} \text{ m} = 510 \text{ nm}$$

STOP TO THINK 33.4 The figure shows two single-slit diffraction patterns. The distance between the slit and the viewing screen is the same in both cases. Which of the following (perhaps more than one) could be true?

a. The slits are the same for both; $\lambda_1 > \lambda_2$.
b. The slits are the same for both; $\lambda_2 > \lambda_1$.
c. The wavelengths are the same for both; $a_1 > a_2$.
d. The wavelengths are the same for both; $a_2 > a_1$.
e. The slits and the wavelengths are the same for both; $p_1 > p_2$.
f. The slits and the wavelengths are the same for both; $p_2 > p_1$.

λ_1

λ_2

33.5 ADVANCED TOPIC A Closer Look at Diffraction

Interference and diffraction are manifestations of superposition. Mathematically, the superposition of waves at a fixed point in space (r_1 and r_2 constant) involves sums such as $e\cos(\omega t) + e\cos(\omega t + \Delta\phi)$, where e is the electric field amplitude of each wave and $\Delta\phi$ is the phase difference between them due to the fact that the waves have traveled different distances. Interestingly, we can use geometry to compute the sums that are relevant to interference and diffraction.

FIGURE 33.16a shows two vectors, each with amplitude e, rotating in the xy-plane with angular frequency ω. At any instant, the angle from the x-axis is the vector's phase, ωt or $\omega t + \Delta\phi$. A rotating *vector* that encodes amplitude and *phase* information is called a **phasor,** and Figure 33.16a is a *phasor diagram*. Notice two key features. First, the vectors rotate together, keeping a fixed angle $\Delta\phi$ between them. Second, the projections of the phasors onto the x-axis are $e\cos(\omega t)$ and $e\cos(\omega t + \Delta\phi)$, exactly what we add in a superposition calculation.

To see how this works, let's return to Young's double-slit experiment. The bright and dark interference fringes arise from the superposition of two waves, one from each slit, with a phase difference $\Delta\phi = 2\pi\,\Delta r/\lambda$ due to the path-length difference Δr. **FIGURE 33.16b** represents each wave as a phasor with amplitude e. We're interested only in their superposition, not the rapid oscillation at frequency ω, so we can draw the first phasor horizontally. If we add the phasors as vectors, using the tip-to-tail method, **the magnitude E of their vector sum is the electric field amplitude of the superposition** of the two waves.

We can use geometry and trigonometry to determine E. We have an isosceles triangle whose large angle, complementary to $\Delta\phi$, is $180° - \Delta\phi$. Consequently, the two smaller, equal angles are each $\Delta\phi/2$, and thus the base of the isosceles triangle has length $E = 2e\cos(\Delta\phi/2)$. The figure shows a triangle for which $\cos(\Delta\phi/2)$ is positive, but $\cos(\Delta\phi/2)$ can be negative at some points in the double-slit pattern. Amplitude, however, must be a positive number, so in general

$$E = \left| 2e\cos\left(\frac{\Delta\phi}{2}\right) \right| \tag{33.23}$$

FIGURE 33.16 Phasor diagrams for double-slit interference.

(a)

Phasors rotate with angular frequency ω.

The angle between the phasors, the *phase difference*, stays constant.

(b)

Vector addition of the phasors is the amplitude of the wave superposition.

Equation 33.23 is identical to Equation 33.10, which we found previously to be the amplitude of the double-slit interference pattern. The intensity of the interference pattern is proportional to E^2.

The Single Slit Revisited

Let's use phasors to find the intensity of the diffraction pattern of a single slit. FIGURE 33.17a shows a slit of width a with N point sources of Huygens' wavelets, each separated by distance a/N. (We'll soon consider the entire wave front by letting $N \rightarrow \infty$.) We need to calculate the superposition of these N wavelets at a point on a distant screen, so far away in comparison with the slit width a that the directions are all essentially parallel at angle θ.

In the double-slit experiment, two waves from slits separated by distance d had a phase difference $\Delta\phi = 2\pi d \sin\theta/\lambda$. By exactly the same reasoning, the phase difference between two adjacent wavelets separated by a/N is $\Delta\phi_{adj} = 2\pi (a/N) \sin\theta/\lambda$. This is the phase difference for *every* pair of adjacent wavelets.

FIGURE 33.17b analyzes the diffraction at $\theta = 0$, the center of the diffraction pattern. Here all the wavelets travel straight ahead, and the phase difference between adjacent wavelets is $\Delta\phi_{adj} = 0$. Consequently, the phasor diagram shows N phasors in a straight line with amplitude $E_0 = Ne$.

FIGURE 33.17c is the phasor diagram for superposition at an arbitrary point on the screen with $\theta \neq 0$. All N phasors have the same length e—so the length of the chain of phasors is still $E_0 = Ne$—but each is rotated by angle $\Delta\phi_{adj}$ with respect to the preceding phasor. The angle of the last phasor, after N rotations, is

$$\beta = N\Delta\phi_{adj} = \frac{2\pi a \sin\theta}{\lambda} \tag{33.24}$$

Notice that β is independent of N. It is the phase difference between a wavelet originating at the top edge of the slit and one originating at the bottom edge, distance a away.

E is the amplitude of the superposition of the N wavelets. To determine E, let $N \rightarrow \infty$. This makes our calculation exact, because now we're considering every point on the wave front. It also makes our calculation easier, because now the chain of phasors, with length E_0, is simply the arc of a circle.

FIGURE 33.18 shows the geometry. The triangle at the upper right is a right triangle, so $\alpha + \beta = 90°$. But α and the angle subtending the arc also add up to $90°$, so the angle subtending the arc must be β. We've divided it into two angles, each $\beta/2$, in order to create two right triangles along E. You can see that $E = 2R\sin(\beta/2)$. In addition, the arc length, if β is in radians, is $E_0 = \beta R$. Eliminating R, we find that the amplitude of the superposition is

$$E = E_0 \frac{\sin(\beta/2)}{\beta/2} = E_0 \frac{\sin(\pi a \sin\theta/\lambda)}{\pi a \sin\theta/\lambda} \tag{33.25}$$

The diffraction-pattern intensity is proportional to E^2, thus

$$I_{slit} = I_0 \left[\frac{\sin(\pi a \sin\theta/\lambda)}{\pi a \sin\theta/\lambda} \right]^2 \tag{33.26}$$

where I_0 (proportional to E_0^2) is the intensity at the center of the central maximum, $\theta = 0$. (Recall, from l'Hôpital's rule, that $\sin x/x \rightarrow 1$ as $x \rightarrow 0$.) FIGURE 33.19 is a graph of Equation 33.26. You can see that it is, indeed, exactly what we observe for single-slit diffraction—a bright central maximum flanked by weaker secondary maxima. The minima occur where the numerator of Equation 33.26 is zero. This requires $\sin\theta_p = p\lambda/a$ for $p = 1, 2, 3, \ldots$, which is exactly the result for the dark fringes that we found previously.

We usually measure positions on a screen rather than angles. For a screen at distance L, a point on the screen at distance y from the center of the pattern is at

FIGURE 33.17 Phasor diagrams for single-slit diffraction.

(a) Slit width a

N wavelets will meet at a distant screen.

The phase difference between two adjacent wavelets is $\Delta\phi_{adj}$.

(b) Center of the diffraction pattern

$E_0 = Ne$ $\theta = 0$

(c) Arbitrary point in the diffraction pattern

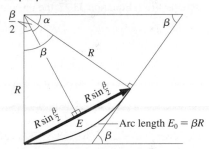

Superposition of N wavelets

Each phasor is rotated by $\Delta\phi_{adj}$.

$\beta = N\Delta\phi_{adj}$

FIGURE 33.18 Calculating the superposition.

$$\text{Arc length } E_0 = \beta R$$

FIGURE 33.19 The single-slit diffraction pattern.

Central maximum

angle $\theta = \tan^{-1}(y/L)$. For very small angles, which is also typical, the small-angle approximation $\tan\theta \approx \sin\theta \approx \theta$ allows us to write the intensity at position y as

$$I_{\text{slit}} = I_0 \left[\frac{\sin(\pi ay/\lambda L)}{\pi ay/\lambda L} \right]^2 \quad \text{(small angles)} \tag{33.27}$$

In this case the minima are at $y_p = p\lambda L/a$, also as we found previously.

You might think that the maximum would occur where the numerator in Equation 33.27 is 1. However, y also appears in the denominator, and that affects the maxima. Setting the derivative of Equation 33.27 to zero, to locate the maxima, leads to a *transcendental equation,* one that cannot be solved algebraically. It can be solved numerically, leading to the result that, for small angles, the first two maxima occur at $y_{\text{max }1} = 1.43\lambda L/a$ and $y_{\text{max }2} = 2.46\lambda L/a$.

EXAMPLE 33.5 | Single-slit intensity

Light with a wavelength of 500 nm passes through a 150-μm-wide slit and is viewed on a screen 2.5 m behind the slit. At what distance from the center of the diffraction pattern is the intensity 50% of maximum?

MODEL The slit produces a single-slit diffraction pattern. Assume that the diffraction angles are small enough to justify the small-angle approximation.

VISUALIZE Figure 33.19 showed a graph of the intensity distribution. The intensity falls to 50% of maximum before the first minimum and never returns to 50% after the first minimum.

SOLVE From Equation 33.27, the intensity at position y is

$$I = I_0 \left[\frac{\sin(\pi ay/\lambda L)}{\pi ay/\lambda L} \right]^2$$

The intensity will have fallen to 50% of maximum I_0 when

$$\frac{\sin(\pi ay/\lambda L)}{\pi ay/\lambda L} = \sqrt{\frac{1}{2}} = 0.707$$

This is a transcendental equation; there is no exact solution. However, it can easily be solved on a calculator with only a small amount of trial and error. If we let $x = \pi ay/\lambda L$, then the equation we want to solve is $\sin x/x = 0.707$, where x is in radians. Put your calculator in radian mode, guess a value of x, compute $\sin x/x$, and then use the result to make an improved guess. The first minimum, $y_1 = \lambda L/a$, has $x = \pi$ rad, and we know that the solution is less than this.

First try: $x = 1.0$ rad gives $\sin x/x = 0.841$.
Second try: $x = 1.5$ rad gives $\sin x/x = 0.665$.

With just two guesses we've narrowed the range to 1.0 rad $< x <$ 1.5 rad. It only takes about three more tries to arrive at $x = 1.39$ rad as the answer to three significant figures. Thus the intensity has dropped to 50% of maximum at

$$y = \frac{1.39}{\pi} \frac{\lambda L}{a} = 3.7 \text{ mm}$$

ASSESS Diffraction patterns seen in the laboratory are typically a centimeter or two wide. This point is within the central maximum, so ≈ 4 mm from the center is reasonable. And ≈ 4 mm from the center at a distance of 2.5 m certainly justifies our use of the small-angle approximation.

The Complete Double-Slit Intensity

Figure 33.4 showed double-slit interference occurring between two overlapping waves as they "spread out behind the two slits." The waves are spreading because light has passed through two narrow slits, and each slit is causing single-slit diffraction. What we see in double-slit interference is actually interference between two overlapping single-slit diffraction patterns. Interference produces the fringes, but the diffraction pattern—the amount of light reaching the screen—determines *how bright* the fringes are.

We earlier calculated the ideal double-slit intensity, $I_{\text{double}} = 4I_1 \cos^2(\pi dy/\lambda L)$, for two slits separated by distance d. But each slit has width a, so the double-slit pattern is *modulated* by the single-slit diffraction intensity for a slit of width a. Thus a realistic double-slit intensity, for small angles, is

$$I_{\text{double}} = I_0 \left[\frac{\sin(\pi ay/\lambda L)}{\pi ay/\lambda L} \right]^2 \cos^2(\pi dy/\lambda L) \tag{33.28}$$

The cosine term produces the fringe oscillations, but now the overall intensity is determined by the diffraction of the individual slits. If the slits are extremely narrow ($a \ll d$), which we tacitly assumed before, then the central maximum of the single-slit pattern is very broad and we see many fringes with only a slow decline in the fringe intensity. This was the case in Figure 33.6b.

But many double slits have a width a that is only slightly smaller than the slit spacing d, and this leads to a complex interplay between diffraction and interference. FIGURE 33.20 shows a double-slit interference pattern for two 0.055-mm-wide slits separated by 0.35 mm and, for comparison, the diffraction pattern of a single 0.055-mm-wide slit. Diffraction of the individual slits determines the overall brightness on the screen—it is the spreading of the light behind the slit—and within this we see the interference between light waves from the two slits. It seems there could be no better proof that light is a wave!

Notice that the $m = 7$ interference fringe is missing (and $m = 8$ is so weak as to be almost invisible). If an interference maximum falls exactly on a minimum (a zero) in the single-slit diffraction pattern, then we have what is called a **missing order.** Interference maxima occur at $y_m = m\lambda L/d$ and diffraction minima are at $y_p = p\lambda L/a$, where m and p are integers. Equating these, to find where they overlap, we see that order m is missing if

$$m_{\text{missing}} = p\frac{d}{a} \qquad p = 1, 2, 3, \ldots \qquad (33.29)$$

m has to be an integer, so the order is truly missing only for certain slit-spacing-to-width ratios d/a. In the case of Figure 33.20, $d/a = 7$ and so the $m = 7$ order is missing (it falls on the $p = 1$ diffraction minimum), as is $m = 14$. In practice, one or more interference maxima may be too weak to be seen even if they don't satisfy Equation 33.29 exactly.

33.6 Circular-Aperture Diffraction

Diffraction occurs if a wave passes through an opening of any shape. Diffraction by a single slit establishes the basic ideas of diffraction, but a common situation of practical importance is diffraction of a wave by a **circular aperture.** Circular diffraction is mathematically more complex than diffraction from a slit, and we will present results without derivation.

Consider some examples. A loudspeaker cone generates sound by the rapid oscillation of a diaphragm, but the sound wave must pass through the circular aperture defined by the outer edge of the speaker cone before it travels into the room beyond. This is diffraction by a circular aperture. Telescopes and microscopes are the reverse. Light waves from outside need to enter the instrument. To do so, they must pass through a circular lens. In fact, the performance limit of optical instruments is determined by the diffraction of the circular openings through which the waves must pass. This is an issue we'll look at in Chapter 35.

FIGURE 33.21 shows a circular aperture of diameter D. Light waves passing through this aperture spread out to generate a *circular* diffraction pattern. You should compare this to Figure 33.11 for a single slit to note the similarities and differences. The diffraction pattern still has a *central maximum,* now circular, and it is surrounded by a series of secondary bright fringes.

FIGURE 33.20 The overall intensity of a double-slit interference pattern is governed by the single-slit diffraction through each slit.

FIGURE 33.21 The diffraction of light by a circular opening.

Angle θ_1 locates the first minimum in the intensity, where there is perfect destructive interference. A mathematical analysis of circular diffraction finds

$$\theta_1 = \frac{1.22\lambda}{D} \qquad (33.30)$$

where D is the *diameter* of the circular opening. Equation 33.30 has assumed the small-angle approximation, which is almost always valid for the diffraction of light but usually is *not* valid for the diffraction of longer-wavelength sound waves.

Within the small-angle approximation, the width of the central maximum is

$$w = 2y_1 = 2L\tan\theta_1 \approx \frac{2.44\lambda L}{D} \qquad \text{(circular aperture)} \qquad (33.31)$$

The diameter of the diffraction pattern increases with distance L, showing that light spreads out behind a circular aperture, but it decreases if the size D of the aperture is increased.

EXAMPLE 33.6 | **Shining a laser through a circular hole**

Light from a helium-neon laser ($\lambda = 633$ nm) passes through a 0.50-mm-diameter hole. How far away should a viewing screen be placed to observe a diffraction pattern whose central maximum is 3.0 mm in diameter?

SOLVE Equation 33.31 gives us the appropriate screen distance:

$$L = \frac{wD}{2.44\lambda} = \frac{(3.0 \times 10^{-3}\ \text{m})(5.0 \times 10^{-4}\ \text{m})}{2.44(633 \times 10^{-9}\ \text{m})} = 0.97\ \text{m}$$

33.7 The Wave Model of Light

We opened this chapter by noting that there are three models of light, each useful within a certain range of circumstances. We are now at a point where we can establish an important condition that separates the wave model of light from the ray model of light.

When light passes through an opening of size a, the angle of the first diffraction minimum is

$$\theta_1 = \sin^{-1}\left(\frac{\lambda}{a}\right) \qquad (33.32)$$

Equation 33.32 is for a slit, but the result is very nearly the same if a is the diameter of a circular aperture. Regardless of the shape of the opening, **the factor that determines how much a wave spreads out behind an opening is the ratio λ/a, the size of the wavelength compared to the size of the opening.**

FIGURE 33.22 illustrates the difference between a wave whose wavelength is much smaller than the size of the opening and a second wave whose wavelength is comparable to the opening. A wave with $\lambda/a \approx 1$ quickly spreads to fill the region behind the opening. Light waves, because of their very short wavelength, almost always have $\lambda/a \ll 1$ and diffract to produce a slowly spreading "beam" of light.

Now we can better appreciate Newton's dilemma. With everyday-sized openings, sound and water waves have $\lambda/a \approx 1$ and diffract to fill the space behind the opening. Consequently, this is what we come to expect for the behavior of waves. We see now that light really does spread out behind an opening, but the very small λ/a ratio usually makes the diffraction pattern too small to see. Diffraction begins to be discernible only when the size of the opening is a fraction of a millimeter or less. If we wanted the diffracted light wave to *fill* the space behind the opening ($\theta_1 \approx 90°$), as a sound wave does, we would need to reduce the size of the opening to $a \approx 0.001$ mm!

FIGURE 33.23 shows light passing through a hole of diameter D. According to the ray model, light rays passing through the hole travel straight ahead to create a bright circular spot of diameter D on a viewing screen. This is the *geometric image* of the slit. In reality, diffraction causes the light to spread out behind the slit, but—and this is the important point—**we will not notice the spreading if it is less than the diameter**

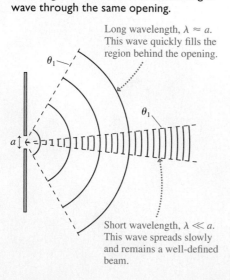

FIGURE 33.22 The diffraction of a long-wavelength wave and a short-wavelength wave through the same opening.

Long wavelength, $\lambda \approx a$. This wave quickly fills the region behind the opening.

Short wavelength, $\lambda \ll a$. This wave spreads slowly and remains a well-defined beam.

D of the geometric image. That is, we will not be aware of diffraction unless the bright spot on the screen increases in diameter.

This idea provides a reasonable criterion for when to use ray optics and when to use wave optics:

- If the spreading due to diffraction is less than the size of the opening, use the ray model and think of light as traveling in straight lines.
- If the spreading due to diffraction is greater than the size of the opening, use the wave model of light.

The crossover point between these two regimes occurs when the spreading due to diffraction is equal to the size of the opening. The central-maximum width of a circular-aperture diffraction pattern is $w = 2.44\lambda L/D$. If we equate this diffraction width to the diameter of the aperture itself, we have

$$\frac{2.44\lambda L}{D_c} = D_c \tag{33.33}$$

where the subscript c on D_c indicates that this is the crossover between the ray model and the wave model. Because we're making an estimate—the change from the ray model to the wave model is gradual, not sudden—to one significant figure, we find

$$D_c \approx \sqrt{2\lambda L} \tag{33.34}$$

This is the diameter of a circular aperture whose diffraction pattern, at distance L, has width $w \approx D$. We know that visible light has $\lambda \approx 500$ nm, and a typical distance in laboratory work is $L \approx 1$ m. For these values,

$$D_c \approx 1 \text{ mm}$$

Thus diffraction is significant, and you should use the wave model, when light passes through openings smaller than about 1 mm. The ray model, which we'll study in the next chapter, is the more appropriate model when light passes through openings larger than about 1 mm. Lenses and mirrors, in particular, are almost always larger than 1 mm, so they will be analyzed with the ray model.

We can now pull all these ideas together in a more complete presentation of the wave model of light.

FIGURE 33.23 Diffraction will be noticed only if the bright spot on the screen is wider than *D*.

If light travels in straight lines, the image on the screen is the same size as the hole. Diffraction will not be noticed unless the light spreads over a diameter larger than *D*.

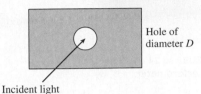

Screen

Hole of diameter *D*

Incident light

MODEL 33.1

Wave model of light

For use when diffraction is significant.

- Light is an electromagnetic wave.
  - Light travels through vacuum at speed *c*.
  - Wavelength λ and frequency *f* are related by $\lambda f = c$.
  - Most of optics depends only on the waviness of light, not on its electromagnetic properties.
- Light exhibits diffraction and interference.
  - Light spreads out after passing through an opening. The amount of spread is inversely proportional to the size of the opening.
  - Two equal-wavelength light waves interfere. Constructive and destructive interference depend on the path-length difference.
- Limitations:
  - The *ray model* is a better description in situations with no diffraction.
    Use the wave model with openings < 1 mm in size.
    Use the ray model with openings > 1 mm in size.
  - The *photon model* is a better description of extremely weak light or the light emitted in atomic transitions.

λ

Magnetic field

Electric field

$v_{em} = c$

Diffraction

Interference

33.8 Interferometers

Scientists and engineers have devised many ingenious methods for using interference to control the flow of light and to make very precise measurements with light waves. A device that makes practical use of interference is called an **interferometer.**

Interference requires two waves of *exactly* the same wavelength. One way of guaranteeing that two waves have exactly equal wavelengths is to divide one wave into two parts of smaller amplitude. Later, at a different point in space, the two parts are recombined. **Interferometers are based on the division and recombination of a single wave.**

The Michelson Interferometer

Albert Michelson, the first American scientist to receive a Nobel Prize, invented an optical interferometer, shown in **FIGURE 33.24**, in which an incoming light wave is divided by a **beam splitter,** a partially silvered mirror that reflects half the light but transmits the other half. The two waves then travel toward mirrors M_1 and M_2. Half of the wave reflected from M_1 is transmitted through the beam splitter, where it recombines with the reflected half of the wave returning from M_2. The superimposed waves travel on to a light detector, originally a human observer but now more likely an electronic photodetector.

After separating, the two waves travel distances $r_1 = 2L_1$ and $r_2 = 2L_2$ before recombining, with the factors of 2 appearing because the waves travel to the mirrors and back again. Thus the path-length difference between the two waves is

$$\Delta r = 2L_2 - 2L_1 \tag{33.35}$$

The conditions for constructive and destructive interference between the two recombined beams are the same as for double-slit interference: $\Delta r = m\lambda$ and $\Delta r = \left(m + \frac{1}{2}\lambda\right)$, respectively. Thus constructive and destructive interference occur when

$$\text{Constructive:} \quad L_2 - L_1 = m\frac{\lambda}{2}$$
$$m = 0, 1, 2, \dots \tag{33.36}$$
$$\text{Destructive:} \quad L_2 - L_1 = \left(m + \tfrac{1}{2}\right)\frac{\lambda}{2}$$

You might expect the interferometer output to be either "bright" or "dark." Instead, a viewing screen shows the pattern of circular interference fringes seen in **FIGURE 33.25**. Our analysis was for light waves that impinge on the mirrors exactly perpendicular to the surface. In an actual experiment, some of the light waves enter the interferometer at slightly different angles and, as a result, the recombined waves have slightly altered path-length differences Δr. These waves cause the alternating bright and dark fringes as you move outward from the center of the pattern. Their analysis will be left to more advanced courses in optics. Equations 33.36 are valid at the *center* of the circular pattern; thus there is a bright or dark central spot when one of the conditions in Equations 33.36 is true.

Mirror M_2 can be moved forward or backward by turning a precision screw, causing the central spot to alternate in a bright-dark-bright-dark-bright cycle that is easily seen or monitored by a photodetector. Suppose the interferometer is adjusted to produce a bright central spot. The next bright spot will appear when M_2 has moved *half* a wavelength, increasing the path-length difference by one full wavelength. The number Δm of maxima appearing as M_2 moves through distance ΔL_2 is

$$\Delta m = \frac{\Delta L_2}{\lambda/2} \tag{33.37}$$

Very precise wavelength measurements can be made by moving the mirror while counting the number of new bright spots appearing at the center of the pattern. The number Δm is counted and known exactly. The only limitation on how precisely λ

FIGURE 33.24 A Michelson interferometer.

1. The wave is divided at this point.

Mirror M_1

Mirror M_2

L_1

Source

Beam splitter

L_2 Adjustment screw

3. The detector measures the superposition of the two waves that have traveled different paths.

2. The returning waves recombine at this point.

FIGURE 33.25 Photograph of the interference fringes produced by a Michelson interferometer.

can be measured this way is the precision with which distance ΔL_2 can be measured. Unlike λ, which is microscopic, ΔL_2 is typically a few millimeters, a macroscopic distance that can be measured very accurately using precision screws, micrometers, and other techniques. Michelson's invention provided a way to transfer the precision of macroscopic distance measurements to an equal precision for the wavelength of light.

EXAMPLE 33.7 | **Measuring the wavelength of light**

An experimenter uses a Michelson interferometer to measure one of the wavelengths of light emitted by neon atoms. She slowly moves mirror M_2 until 10,000 new bright central spots have appeared. (In a modern experiment, a photodetector and computer would eliminate the possibility of experimenter error while counting.) She then measures that the mirror has moved a distance of 3.164 mm. What is the wavelength of the light?

MODEL An interferometer produces a new maximum each time L_2 increases by $\lambda/2$.

SOLVE The mirror moves $\Delta L_2 = 3.164$ mm $= 3.164 \times 10^{-3}$ m. We can use Equation 33.37 to find

$$\lambda = \frac{2\,\Delta L_2}{\Delta m} = 6.328 \times 10^{-7} \text{ m} = 632.8 \text{ nm}$$

ASSESS A measurement of ΔL_2 accurate to four significant figures allowed us to determine λ to four significant figures. This happens to be the neon wavelength that is emitted as the laser beam in a helium-neon laser.

STOP TO THINK 33.5 A Michelson interferometer using light of wavelength λ has been adjusted to produce a bright spot at the center of the interference pattern. Mirror M_1 is then moved distance λ toward the beam splitter while M_2 is moved distance λ away from the beam splitter. How many bright-dark-bright fringe shifts are seen?

a. 0   b. 1
c. 2   d. 4
e. 8   f. It's not possible to say without knowing λ.

Holography

No discussion of wave optics would be complete without mentioning holography, which has both scientific and artistic applications. The basic idea is a simple extension of interferometry.

FIGURE 33.26a shows how a **hologram** is made. A beam splitter divides a laser beam into two waves. One wave illuminates the object of interest. The light scattered by this object is a very complex wave, but it is the wave you would see if you looked at the object from the position of the film. The other wave, called the *reference beam*, is reflected directly toward the film. The scattered light and the reference beam meet at the film and interfere. The film records their interference pattern.

The interference patterns we've looked at in this chapter have been simple patterns of stripes and circles because the light waves have been well-behaved plane waves and

FIGURE 33.26 Holography is an important application of wave optics.

(a) Recording a hologram

The interference between the scattered light and the reference beam is recorded on the film.

Film — Plane waves — Reference beam — Laser — Beam splitter — Object beam — Object — The scattered light has a complex wave front.

(b) A hologram

An enlarged photo of the developed film. This is the hologram.

(c) Playing a hologram

The diffraction of the laser beam through the light and dark patches of the film reconstructs the original scattered wave.

Hologram (developed film) — An observer "sees" the object as if it were here. — Laser beam along the reference beam direction

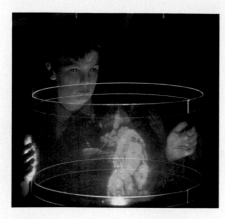

A hologram.

spherical waves. The light wave scattered by the object in Figure 33.26a is exceedingly complex. As a result, the interference pattern recorded on the film—the hologram—is a seemingly random pattern of whorls and blotches. **FIGURE 33.26b** is an enlarged photograph of a portion of a hologram. It's certainly not obvious that information is stored in this pattern, but it is.

The hologram is "played" by sending just the reference beam through it, as seen in **FIGURE 33.26c**. The reference beam diffracts through the transparent parts of the hologram, just as it would through the slits of a diffraction grating. Amazingly, the diffracted wave is *exactly the same* as the light wave that had been scattered by the object! In other words, the diffracted reference beam *reconstructs* the original scattered wave. As you look at this diffracted wave, from the far side of the hologram, you "see" the object exactly as if it were there. The view is three dimensional because, by moving your head with respect to the hologram, you can see different portions of the wave front.

CHALLENGE EXAMPLE 33.8 | Measuring the index of refraction of a gas

A Michelson interferometer uses a helium-neon laser with wavelength $\lambda_{vac} = 633$ nm. In one arm, the light passes through a 4.00-cm-thick glass cell. Initially the cell is evacuated, and the interferometer is adjusted so that the central spot is a bright fringe. The cell is then slowly filled to atmospheric pressure with a gas. As the cell fills, 43 bright-dark-bright fringe shifts are seen and counted. What is the index of refraction of the gas at this wavelength?

MODEL Adding one additional wavelength to the round trip causes one bright-dark-bright fringe shift. Changing the length of the arm is one way to add wavelengths, but not the only way. Increasing the index of refraction also adds wavelengths because light has a shorter wavelength when traveling through a material with a larger index of refraction.

VISUALIZE **FIGURE 33.27** shows a Michelson interferometer with a cell of thickness d in one arm.

FIGURE 33.27 Measuring the index of refraction.

Gas-filled cell of thickness d.
Light goes through this cell twice.

Mirror M_1

Mirror M_2

L_1

Source

Beam splitter

d

L_2

SOLVE To begin, all the air is pumped out of the cell. As light travels from the beam splitter to the mirror and back, the number of wavelengths inside the cell is

$$m_1 = \frac{2d}{\lambda_{vac}}$$

where the 2 appears because the light passes through the cell twice.

The cell is then filled with gas at 1 atm pressure. Light travels slower in the gas, $v = c/n$, and you learned in Chapter 16 that the reduction in speed decreases the wavelength to λ_{vac}/n. With the cell filled, the number of wavelengths spanning distance d is

$$m_2 = \frac{2d}{\lambda} = \frac{2d}{\lambda_{vac}/n}$$

The physical distance has not changed, but the number of wavelengths along the path has. Filling the cell has increased the path by

$$\Delta m = m_2 - m_1 = (n-1)\frac{2d}{\lambda_{vac}}$$

wavelengths. Each increase of one wavelength causes one bright-dark-bright fringe shift at the output. Solving for n, we find

$$n = 1 + \frac{\lambda_{vac}\,\Delta m}{2d} = 1 + \frac{(6.33 \times 10^{-7}\ \text{m})(43)}{2(0.0400\ \text{m})} = 1.00034$$

ASSESS This may seem like a six-significant-figure result, but there are really only two. What we're measuring is not n but $n-1$. We know the fringe count to two significant figures, and that has allowed us to compute $n - 1 = \lambda_{vac}\,\Delta m/2d = 3.4 \times 10^{-4}$.

SUMMARY

The goal of Chapter 33 has been to learn about and apply the wave model of light.

GENERAL PRINCIPLES

Huygens' principle says that each point on a wave front is the source of a spherical wavelet. The wave front at a later time is tangent to all the wavelets.

Diffraction is the spreading of a wave after it passes through an opening.

Constructive and destructive **interference** are due to the overlap of two or more waves as they spread behind openings.

IMPORTANT CONCEPTS

The **wave model** of light considers light to be a wave propagating through space. Diffraction and interference are important.

The **ray model** of light considers light to travel in straight lines like little particles. Diffraction and interference are not important.

Diffraction is important when the width of the diffraction pattern of an aperture equals or exceeds the size of the aperture. For a circular aperture, the crossover between the ray and wave models occurs for an opening of diameter $D_c \approx \sqrt{2\lambda L}$.

In practice, $D_c \approx 1$ mm for visible light. Thus

- Use the wave model when light passes through openings < 1 mm in size. Diffraction effects are usually important.
- Use the ray model when light passes through openings > 1 mm in size. Diffraction is usually not important.

APPLICATIONS

Single slit of width a.
A bright **central maximum** of width

$$w = \frac{2\lambda L}{a}$$

is flanked by weaker **secondary maxima.** Dark fringes are located at angles such that

$$a\sin\theta_p = p\lambda \qquad p = 1, 2, 3, \ldots$$

If $\lambda/a \ll 1$, then from the small-angle approximation

$$\theta_p = \frac{p\lambda}{a} \qquad y_p = \frac{p\lambda L}{a}$$

Interference due to wave-front division

Waves overlap as they spread out behind slits. Constructive interference occurs along antinodal lines. Bright fringes are seen where the antinodal lines intersect the viewing screen.

Double slit with separation d.
Equally spaced bright fringes are located at

$$\theta_m = \frac{m\lambda}{d} \qquad y_m = \frac{m\lambda L}{d} \qquad m = 0, 1, 2, \ldots$$

The **fringe spacing** is $\Delta y = \dfrac{\lambda L}{d}$

Diffraction grating with slit spacing d.
Very bright and narrow fringes are located at angles and positions

$$d\sin\theta_m = m\lambda \qquad y_m = L\tan\theta_m$$

Circular aperture of diameter D.
A bright central maximum of diameter

$$w = \frac{2.44\lambda L}{D}$$

is surrounded by circular secondary maxima. The first dark fringe is located at

$$\theta_1 = \frac{1.22\lambda}{D} \qquad y_1 = \frac{1.22\lambda L}{D}$$

For an aperture of any shape, a smaller opening causes a more rapid spreading of the wave behind the opening.

Interference due to amplitude division

An interferometer divides a wave, lets the two waves travel different paths, then recombines them. Interference is constructive if one wave travels an integer number of wavelengths more or less than the other wave. The difference can be due to an actual path-length difference or to a different index of refraction.

Michelson interferometer

The number of bright-dark-bright fringe shifts as mirror M_2 moves distance ΔL_2 is

$$\Delta m = \frac{\Delta L_2}{\lambda/2}$$

TERMS AND NOTATION

optics	double slit	spectroscopy	circular aperture
diffraction	interference fringes	single-slit diffraction	interferometer
models of light	central maximum	secondary maxima	beam splitter
wave model	fringe spacing, Δy	Huygens' principle	hologram
ray model	diffraction grating	phasor	
photon model	order, m	missing order	

CONCEPTUAL QUESTIONS

1. **FIGURE Q33.1** shows light waves passing through two closely spaced, narrow slits. The graph shows the intensity of light on a screen behind the slits. Reproduce these graph axes, including the zero and the tick marks locating the double-slit fringes, then draw a graph to show how the light-intensity pattern will appear if the right slit is blocked, allowing light to go through only the left slit. Explain your reasoning.

FIGURE Q33.1

2. In a double-slit interference experiment, which of the following actions (perhaps more than one) would cause the fringe spacing to increase? (a) Increasing the wavelength of the light. (b) Increasing the slit spacing. (c) Increasing the distance to the viewing screen. (d) Submerging the entire experiment in water.

3. **FIGURE Q33.3** shows the viewing screen in a double-slit experiment. Fringe C is the central maximum. What will happen to the fringe spacing if
   a. The wavelength of the light is decreased?
   b. The spacing between the slits is decreased?
   c. The distance to the screen is decreased?
   d. Suppose the wavelength of the light is 500 nm. How much farther is it from the dot on the screen in the center of fringe E to the left slit than it is from the dot to the right slit?

FIGURE Q33.3

4. **FIGURE Q33.3** is the interference pattern seen on a viewing screen behind 2 slits. Suppose the 2 slits were replaced by 20 slits having the same spacing d between adjacent slits.
   a. Would the number of fringes on the screen increase, decrease, or stay the same?
   b. Would the fringe spacing increase, decrease, or stay the same?
   c. Would the width of each fringe increase, decrease, or stay the same?
   d. Would the brightness of each fringe increase, decrease, or stay the same?

5. **FIGURE Q33.5** shows the light intensity on a viewing screen behind a single slit of width a. The light's wavelength is λ. Is $\lambda < a$, $\lambda = a$, $\lambda > a$, or is it not possible to tell? Explain.

FIGURE Q33.5   **FIGURE Q33.6**

6. **FIGURE Q33.6** shows the light intensity on a viewing screen behind a circular aperture. What happens to the width of the central maximum if
   a. The wavelength of the light is increased?
   b. The diameter of the aperture is increased?
   c. How will the screen appear if the aperture diameter is less than the light wavelength?

7. Narrow, bright fringes are observed on a screen behind a diffraction grating. The entire experiment is then immersed in water. Do the fringes on the screen get closer together, get farther apart, remain the same, or disappear? Explain.

8. a. Green light shines through a 100-mm-diameter hole and is observed on a screen. If the hole diameter is increased by 20%, does the circular spot of light on the screen decrease in diameter, increase in diameter, or stay the same? Explain.
   b. Green light shines through a 100-μm-diameter hole and is observed on a screen. If the hole diameter is increased by 20%, does the circular spot of light on the screen decrease in diameter, increase in diameter, or stay the same? Explain.

9. A Michelson interferometer using 800 nm light is adjusted to have a bright central spot. One mirror is then moved 200 nm forward, the other 200 nm back. Afterward, is the central spot bright, dark, or in between? Explain.

10. A Michelson interferometer is set up to display constructive interference (a bright central spot in the fringe pattern of Figure 33.25) using light of wavelength λ. If the wavelength is changed to $\lambda/2$, does the central spot remain bright, does the central spot become dark, or do the fringes disappear? Explain. Assume the fringes are viewed by a detector sensitive to both wavelengths.

EXERCISES AND PROBLEMS

Problems labeled ▓ integrate material from earlier chapters.

Exercises

Section 33.2 The Interference of Light

1. | A double slit is illuminated simultaneously with orange light of wavelength 620 nm and light of an unknown wavelength. The $m = 4$ bright fringe of the unknown wavelength overlaps the $m = 3$ bright orange fringe. What is the unknown wavelength?

2. ‖ Two narrow slits 80 μm apart are illuminated with light of wavelength 620 nm. What is the angle of the $m = 3$ bright fringe in radians? In degrees?

3. | A double-slit experiment is performed with light of wavelength 630 nm. The bright interference fringes are spaced 1.8 mm apart on the viewing screen. What will the fringe spacing be if the light is changed to a wavelength of 420 nm?

4. | Light of wavelength 550 nm illuminates a double slit, and the interference pattern is observed on a screen. At the position of the $m = 2$ bright fringe, how much farther is it to the more distant slit than to the nearer slit?

5. ‖ Light of 630 nm wavelength illuminates two slits that are 0.25 mm apart. **FIGURE EX33.5** shows the intensity pattern seen on a screen behind the slits. What is the distance to the screen?

FIGURE EX33.5

6. ‖ In a double-slit experiment, the slit separation is 200 times the wavelength of the light. What is the angular separation (in degrees) between two adjacent bright fringes?

7. ‖ Light from a sodium lamp ($\lambda = 589$ nm) illuminates two narrow slits. The fringe spacing on a screen 150 cm behind the slits is 4.0 mm. What is the spacing (in mm) between the two slits?

8. ‖ A double-slit interference pattern is created by two narrow slits spaced 0.25 mm apart. The distance between the first and the fifth minimum on a screen 60 cm behind the slits is 5.5 mm. What is the wavelength (in nm) of the light used in this experiment?

Section 33.3 The Diffraction Grating

9. | A 4.0-cm-wide diffraction grating has 2000 slits. It is illuminated by light of wavelength 550 nm. What are the angles (in degrees) of the first two diffraction orders?

10. ‖ Light of wavelength 620 nm illuminates a diffraction grating. The second-order maximum is at angle 39.5°. How many lines per millimeter does this grating have?

11. ‖ A diffraction grating produces a first-order maximum at an angle of 20.0°. What is the angle of the second-order maximum?

12. | A diffraction grating is illuminated simultaneously with red light of wavelength 660 nm and light of an unknown wavelength. The fifth-order maximum of the unknown wavelength exactly overlaps the third-order maximum of the red light. What is the unknown wavelength?

13. ‖ The two most prominent wavelengths in the light emitted by a hydrogen discharge lamp are 656 nm (red) and 486 nm (blue). Light from a hydrogen lamp illuminates a diffraction grating with 500 lines/mm, and the light is observed on a screen 1.50 m behind the grating. What is the distance between the first-order red and blue fringes?

14. ‖ A helium-neon laser ($\lambda = 633$ nm) illuminates a diffraction grating. The distance between the two $m = 1$ bright fringes is 32 cm on a screen 2.0 m behind the grating. What is the spacing between slits of the grating?

Section 33.4 Single-Slit Diffraction

15. ‖ In a single-slit experiment, the slit width is 200 times the wavelength of the light. What is the width (in mm) of the central maximum on a screen 2.0 m behind the slit?

16. ‖ A helium-neon laser ($\lambda = 633$ nm) illuminates a single slit and is observed on a screen 1.5 m behind the slit. The distance between the first and second minima in the diffraction pattern is 4.75 mm. What is the width (in mm) of the slit?

17. ‖ Light of 630 nm wavelength illuminates a single slit of width 0.15 mm. **FIGURE EX33.17** shows the intensity pattern seen on a screen behind the slit. What is the distance to the screen?

FIGURE EX33.17

18. | A 0.50-mm-wide slit is illuminated by light of wavelength 500 nm. What is the width (in mm) of the central maximum on a screen 2.0 m behind the slit?

19. ‖ You need to use your cell phone, which broadcasts an 800 MHz signal, but you're behind two massive, radio-wave-absorbing buildings that have only a 15 m space between them. What is the angular width, in degrees, of the electromagnetic wave after it emerges from between the buildings?

20. | For what slit-width-to-wavelength ratio does the first minimum of a single-slit diffraction pattern appear at (a) 30°, (b) 60°, and (c) 90°?

21. | Light from a helium neon laser ($\lambda = 633$ nm) is incident on a single slit. What is the largest slit width for which there are no minima in the diffraction pattern?

Section 33.5 A Closer Look at Diffraction

22. | A laser beam illuminates a single, narrow slit, and the diffraction pattern is observed on a screen behind the slit. The first secondary maximum is 26 mm from the center of the diffraction pattern. How far is the first minimum from the center of the diffraction pattern?

23. ‖ Two 50-μm-wide slits spaced 0.25 mm apart are illuminated by blue laser light with a wavelength of 450 nm. The interference pattern is observed on a screen 2.0 m behind the slits. How many bright fringes are seen in the *central maximum* that spans the distance between the first missing order on one side and the first missing order on the other side?

24. ‖ A laser beam with a wavelength of 480 nm illuminates two 0.12-mm-wide slits separated by 0.30 mm. The interference pattern is observed on a screen 2.3 m behind the slits. What is the light intensity, as a fraction of the maximum intensity I_0, at a point halfway between the center and the first minimum?

Section 33.6 Circular-Aperture Diffraction

25. ‖ A 0.50-mm-diameter hole is illuminated by light of wavelength 550 nm. What is the width (in mm) of the central maximum on a screen 2.0 m behind the slit?

26. ‖ Infrared light of wavelength 2.5 μm illuminates a 0.20-mm-diameter hole. What is the angle of the first dark fringe in radians? In degrees?

27. ‖ You want to photograph a circular diffraction pattern whose central maximum has a diameter of 1.0 cm. You have a helium-neon laser ($\lambda = 633$ nm) and a 0.12-mm-diameter pinhole. How far behind the pinhole should you place the screen that's to be photographed?

28. ‖ Your artist friend is designing an exhibit inspired by circular-aperture diffraction. A pinhole in a red zone is going to be illuminated with a red laser beam of wavelength 670 nm, while a pinhole in a violet zone is going to be illuminated with a violet laser beam of wavelength 410 nm. She wants all the diffraction patterns seen on a distant screen to have the same size. For this to work, what must be the ratio of the red pinhole's diameter to that of the violet pinhole?

29. ‖ Light from a helium-neon laser ($\lambda = 633$ nm) passes through a circular aperture and is observed on a screen 4.0 m behind the aperture. The width of the central maximum is 2.5 cm. What is the diameter (in mm) of the hole?

Section 33.8 Interferometers

30. ‖ A Michelson interferometer uses red light with a wavelength of 656.45 nm from a hydrogen discharge lamp. How many bright-dark-bright fringe shifts are observed if mirror M_2 is moved exactly 1 cm?

31. ‖ Moving mirror M_2 of a Michelson interferometer a distance of 100 μm causes 500 bright-dark-bright fringe shifts. What is the wavelength of the light?

32. ‖ A Michelson interferometer uses light from a sodium lamp. Sodium atoms emit light having wavelengths 589.0 nm and 589.6 nm. The interferometer is initially set up with both arms of equal length ($L_1 = L_2$), producing a bright spot at the center of the interference pattern. How far must mirror M_2 be moved so that one wavelength has produced one more new maximum than the other wavelength?

Problems

33. ‖ **FIGURE P33.33** shows the light intensity on a screen 2.5 m behind an aperture. The aperture is illuminated with light of wavelength 620 nm.
    a. Is the aperture a single slit or a double slit? Explain.
    b. If the aperture is a single slit, what is its width? If it is a double slit, what is the spacing between the slits?

Intensity

FIGURE P33.33   0 1 2 3 4 5 x (cm)

34. ‖ **FIGURE P33.34** shows the light intensity on a screen 2.5 m behind an aperture. The aperture is illuminated with light of wavelength 620 nm.
    a. Is the aperture a single slit or a double slit? Explain.
    b. If the aperture is a single slit, what is its width? If it is a double slit, what is the spacing between the slits?

Intensity

FIGURE P33.34   0 1 2 3 4 5 6 x (cm)

35. ‖ Light from a helium-neon laser ($\lambda = 633$ nm) is used to illuminate two narrow slits. The interference pattern is observed on a screen 3.0 m behind the slits. Twelve bright fringes are seen, spanning a distance of 52 mm. What is the spacing (in mm) between the slits?

36. ‖ **FIGURE P33.36** shows the light intensity on a screen behind a double slit. The slit spacing is 0.20 mm and the wavelength of the light is 620 nm. What is the distance from the slits to the screen?

Intensity (mW/m²)

12

0

FIGURE P33.36   2.0 cm x (cm)

37. ‖ **FIGURE P33.36** shows the light intensity on a screen behind a double slit. The slit spacing is 0.20 mm and the screen is 2.0 m behind the slits. What is the wavelength (in nm) of the light?

38. ‖ **FIGURE P33.36** shows the light intensity on a screen behind a double slit. Suppose one slit is covered. What will be the light intensity at the center of the screen due to the remaining slit?

39. ‖ A diffraction grating having 500 lines/mm diffracts visible light at 30°. What is the light's wavelength?

40. ‖ Helium atoms emit light at several wavelengths. Light from a helium lamp illuminates a diffraction grating and is observed on a screen 50.00 cm behind the grating. The emission at wavelength 501.5 nm creates a first-order bright fringe 21.90 cm from the central maximum. What is the wavelength of the bright fringe that is 31.60 cm from the central maximum?

41. ‖ A triple-slit experiment consists of three narrow slits, equally spaced by distance d and illuminated by light of wavelength λ. Each slit alone produces intensity I_1 on the viewing screen at distance L.
    a. Consider a point on the distant viewing screen such that the path-length difference between any two adjacent slits is λ. What is the intensity at this point?
    b. What is the intensity at a point where the path-length difference between any two adjacent slits is $\lambda/2$?

42. ‖ Because sound is a wave, it's possible to make a diffraction grating for sound from a large board of sound-absorbing material with several parallel slits cut for sound to go through. When 10 kHz sound waves pass through such a grating, listeners 10 m from the grating report "loud spots" 1.4 m on both sides of center. What is the spacing between the slits? Use 340 m/s for the speed of sound.

43. ‖ A diffraction grating with 600 lines/mm is illuminated with light of wavelength 510 nm. A very wide viewing screen is 2.0 m behind the grating.
    a. What is the distance between the two $m = 1$ bright fringes?
    b. How many bright fringes can be seen on the screen?

44. ‖ A 500 line/mm diffraction grating is illuminated by light of wavelength 510 nm. How many bright fringes are seen on a 2.0-m-wide screen located 2.0 m behind the grating?

45. ‖ White light (400–700 nm) incident on a 600 line/mm diffraction grating produces rainbows of diffracted light. What is the width of the first-order rainbow on a screen 2.0 m behind the grating?

46. ‖ A chemist identifies compounds by identifying bright lines in their spectra. She does so by heating the compounds until they glow, sending the light through a diffraction grating, and measuring the positions of first-order spectral lines on a detector 15.0 cm behind the grating. Unfortunately, she has lost the card that gives the specifications of the grating. Fortunately, she has a known compound that she can use to calibrate the grating. She heats the known compound, which emits light at a wavelength of 461 nm, and observes a spectral line 9.95 cm from the center of the diffraction pattern. What are the wavelengths emitted by compounds A and B that have spectral lines detected at positions 8.55 cm and 12.15 cm, respectively?

47. ‖‖ a. Find an expression for the positions y_1 of the first-order fringes of a diffraction grating if the line spacing is large enough for the small-angle approximation $\tan\theta \approx \sin\theta \approx \theta$ to be valid. Your expression should be in terms of d, L, and λ.
   b. Use your expression from part a to find an expression for the separation Δy on the screen of two fringes that differ in wavelength by $\Delta\lambda$.
   c. Rather than a viewing screen, modern spectrometers use detectors—similar to the one in your digital camera—that are divided into *pixels*. Consider a spectrometer with a 333 line/mm grating and a detector with 100 pixels/mm located 12 cm behind the grating. The *resolution* of a spectrometer is the smallest wavelength separation $\Delta\lambda_{min}$ that can be measured reliably. What is the resolution of this spectrometer for wavelengths near 550 nm, in the center of the visible spectrum? You can assume that the fringe due to one specific wavelength is narrow enough to illuminate only one column of pixels.

48. ‖ For your science fair project you need to design a diffraction grating that will disperse the visible spectrum (400–700 nm) over 30.0° in first order.
   a. How many lines per millimeter does your grating need?
   b. What is the first-order diffraction angle of light from a sodium lamp ($\lambda = 589$ nm)?

49. ‖ FIGURE P33.49 shows the interference pattern on a screen 1.0 m behind an 800 line/mm diffraction grating. What is the wavelength (in nm) of the light?

FIGURE P33.49

89.7 cm   89.7 cm
43.6 cm   43.6 cm

50. ‖ FIGURE P33.49 shows the interference pattern on a screen 1.0 m behind a diffraction grating. The wavelength of the light is 620 nm. How many lines per millimeter does the grating have?

51. ‖ Light from a sodium lamp ($\lambda = 589$ nm) illuminates a narrow slit and is observed on a screen 75 cm behind the slit. The distance between the first and third dark fringes is 7.5 mm. What is the width (in mm) of the slit?

52. ‖ The wings of some beetles
BIO have closely spaced parallel lines of melanin, causing the wing to act as a reflection grating. Suppose sunlight shines straight onto a beetle wing. If the melanin lines on the wing are spaced 2.0 μm apart, what is the first-order diffraction angle for green light ($\lambda = 550$ nm)?

53. | If sunlight shines straight onto a peacock feather, the feather
BIO appears bright blue when viewed from 15° on either side of the incident beam of light. The blue color is due to diffraction from parallel rods of melanin in the feather barbules, as was shown in the photograph on page 940. Other wavelengths in the incident light are diffracted at different angles, leaving only the blue light to be seen. The average wavelength of blue light is 470 nm. Assuming this to be the first-order diffraction, what is the spacing of the melanin rods in the feather?

54. ‖ You've found an unlabeled diffraction grating. Before you can use it, you need to know how many lines per mm it has. To find out, you illuminate the grating with light of several different wavelengths and then measure the distance between the two first-order bright fringes on a viewing screen 150 cm behind the grating. Your data are as follows:

Wavelength (nm)	Distance (cm)
430	109.6
480	125.4
530	139.8
580	157.2
630	174.4
680	194.8

Use the best-fit line of an appropriate graph to determine the number of lines per mm.

55. ‖‖ A diffraction grating has slit spacing d. Fringes are viewed on a screen at distance L. Find an expression for the wavelength of light that produces a first-order fringe on the viewing screen at distance L from the center of the screen.

56. ‖ FIGURE P33.56 shows the light intensity on a screen behind a single slit. The slit width is 0.20 mm and the screen is 1.5 m behind the slit. What is the wavelength (in nm) of the light?

FIGURE P33.56   0   1   2   3

57. ‖ FIGURE P33.56 shows the light intensity on a screen behind a single slit. The wavelength of the light is 600 nm and the slit width is 0.15 mm. What is the distance from the slit to the screen?

58. ‖ FIGURE P33.56 shows the light intensity on a screen behind a circular aperture. The wavelength of the light is 500 nm and the screen is 1.0 m behind the slit. What is the diameter (in mm) of the aperture?

59. ‖ A student performing a double-slit experiment is using a green laser with a wavelength of 530 nm. She is confused when the $m = 5$ maximum does not appear. She had predicted that this bright fringe would be 1.6 cm from the central maximum on a screen 1.5 m behind the slits.
    a. Explain what prevented the fifth maximum from being observed.
    b. What is the width of her slits?

60. ‖ Scientists shine a laser beam on a 35-μm-wide slit and produce a diffraction pattern on a screen 70 cm behind the slit. Careful measurements show that the intensity first falls to 25% of maximum at a distance of 7.2 mm from the center of the diffraction pattern. What is the wavelength of the laser light?
    Hint: Use the trial-and-error technique demonstrated in Example 33.5 to solve the transcendental equation.

61. ‖ Light from a helium-neon laser ($\lambda = 633$ nm) illuminates a circular aperture. It is noted that the diameter of the central maximum on a screen 50 cm behind the aperture matches the diameter of the geometric image. What is the aperture's diameter (in mm)?

62. ‖ A helium-neon laser ($\lambda = 633$ nm) is built with a glass tube of inside diameter 1.0 mm, as shown in **FIGURE P33.62**. One mirror is partially transmitting to allow the laser beam out. An electrical discharge in the tube causes it to glow like a neon light. From an optical perspective, the laser beam is a light wave that diffracts out through a 1.0-mm-diameter circular opening.
    a. Can a laser beam be *perfectly* parallel, with no spreading? Why or why not?
    b. The angle θ_1 to the first minimum is called the *divergence angle* of a laser beam. What is the divergence angle of this laser beam?
    c. What is the diameter (in mm) of the laser beam after it travels 3.0 m?
    d. What is the diameter of the laser beam after it travels 1.0 km?

FIGURE P33.62

63. ‖ One day, after pulling down your window shade, you notice that sunlight is passing through a pinhole in the shade and making a small patch of light on the far wall. Having recently studied optics in your physics class, you're not too surprised to see that the patch of light seems to be a circular diffraction pattern. It appears that the central maximum is about 1 cm across, and you estimate that the distance from the window shade to the wall is about 3 m. Estimate (a) the average wavelength of the sunlight (in nm) and (b) the diameter of the pinhole (in mm).

64. | A radar for tracking aircraft broadcasts a 12 GHz microwave beam from a 2.0-m-diameter circular radar antenna. From a wave perspective, the antenna is a circular aperture through which the microwaves diffract.
    a. What is the diameter of the radar beam at a distance of 30 km?
    b. If the antenna emits 100 kW of power, what is the average microwave intensity at 30 km?

65. ‖ Scientists use *laser range-finding* to measure the distance to the moon with great accuracy. A brief laser pulse is fired at the moon, then the time interval is measured until the "echo" is seen by a telescope. A laser beam spreads out as it travels because it diffracts through a circular exit as it leaves the laser. In order for the reflected light to be bright enough to detect, the laser spot on the moon must be no more than 1.0 km in diameter. Staying within this diameter is accomplished by using a special large-diameter laser. If $\lambda = 532$ nm, what is the minimum diameter of the circular opening from which the laser beam emerges? The earth-moon distance is 384,000 km.

66. ‖ Light of wavelength 600 nm passes though two slits separated by 0.20 mm and is observed on a screen 1.0 m behind the slits. The location of the central maximum is marked on the screen and labeled $y = 0$.
    a. At what distance, on either side of $y = 0$, are the $m = 1$ bright fringes?
    b. A very thin piece of glass is then placed in one slit. Because light travels slower in glass than in air, the wave passing through the glass is delayed by 5.0×10^{-16} s in comparison to the wave going through the other slit. What fraction of the period of the light wave is this delay?
    c. With the glass in place, what is the phase difference $\Delta \phi_0$ between the two waves as they leave the slits?
    d. The glass causes the interference fringe pattern on the screen to shift sideways. Which way does the central maximum move (toward or away from the slit with the glass) and by how far?

67. ‖ A 600 line/mm diffraction grating is in an empty aquarium tank. The index of refraction of the glass walls is $n_{glass} = 1.50$. A helium-neon laser ($\lambda = 633$ nm) is outside the aquarium. The laser beam passes through the glass wall and illuminates the diffraction grating.
    a. What is the first-order diffraction angle of the laser beam?
    b. What is the first-order diffraction angle of the laser beam after the aquarium is filled with water ($n_{water} = 1.33$)?

68. ‖ A Michelson interferometer operating at a 600 nm wavelength has a 2.00-cm-long glass cell in one arm. To begin, the air is pumped out of the cell and mirror M_2 is adjusted to produce a bright spot at the center of the interference pattern. Then a valve is opened and air is slowly admitted into the cell. The index of refraction of air at 1.00 atm pressure is 1.00028. How many bright-dark-bright fringe shifts are observed as the cell fills with air?

69. ‖‖ Optical computers require microscopic optical switches to turn signals on and off. One device for doing so, which can be implemented in an integrated circuit, is the *Mach-Zender interferometer* seen in **FIGURE P33.69**. Light from an on-chip infrared laser ($\lambda = 1.000 \ \mu$m) is split into two waves that travel equal distances around the arms of the interferometer. One arm passes through an *electro-optic crystal*, a transparent material that can change its index of refraction in response to an applied voltage. Suppose both arms are exactly the same length and the crystal's index of refraction with no applied voltage is 1.522.
    a. With no voltage applied, is the output bright (switch closed, optical signal passing through) or dark (switch open, no signal)? Explain.
    b. What is the first index of refraction of the electro-optic crystal larger than 1.522 that changes the optical switch to the state opposite the state you found in part a?

FIGURE P33.69

70. ‖ To illustrate one of the ideas of holography in a simple way, consider a diffraction grating with slit spacing d. The small-angle approximation is usually not valid for diffraction gratings, because d is only slightly larger than λ, but assume that the λ/d ratio of this grating is small enough to make the small-angle approximation valid.
   a. Use the small-angle approximation to find an expression for the fringe spacing on a screen at distance L behind the grating.
   b. Rather than a screen, suppose you place a piece of film at distance L behind the grating. The bright fringes will expose the film, but the dark spaces in between will leave the film unexposed. After being developed, the film will be a series of alternating light and dark stripes. What if you were to now "play" the film by using it as a diffraction grating? In other words, what happens if you shine the same laser through the film and look at the film's diffraction pattern on a screen at the same distance L? Demonstrate that the film's diffraction pattern is a reproduction of the original diffraction grating.

Challenge Problems

71. ‖‖ A double-slit experiment is set up using a helium-neon laser ($\lambda = 633$ nm). Then a very thin piece of glass ($n = 1.50$) is placed over one of the slits. Afterward, the central point on the screen is occupied by what had been the $m = 10$ dark fringe. How thick is the glass?

72. ‖‖ The intensity at the central maximum of a double-slit interference pattern is $4I_1$. The intensity at the first minimum is zero. At what fraction of the distance from the central maximum to the first minimum is the intensity I_1? Assume an ideal double slit.

73. ‖‖ FIGURE CP33.73 shows two nearly overlapped intensity peaks of the sort you might produce with a diffraction grating (see Figure 33.9b). As a practical matter, two peaks can just barely be resolved if their spacing Δy equals the width w of each peak, where w is measured at half of the peak's height. Two peaks closer together than w will merge into a single peak. We can use this idea to understand the *resolution* of a diffraction grating.

Two peaks are barely resolved when $\Delta y = w$.

FIGURE CP33.73

   a. In the small-angle approximation, the position of the $m = 1$ peak of a diffraction grating falls at the same location as the $m = 1$ fringe of a double slit: $y_1 = \lambda L/d$. Suppose two wavelengths differing by $\Delta\lambda$ pass through a grating at the same time. Find an expression for Δy, the separation of their first-order peaks.
   b. We noted that the widths of the bright fringes are proportional to $1/N$, where N is the number of slits in the grating. Let's hypothesize that the fringe width is $w = y_1/N$. Show that this is true for the double-slit pattern. We'll then assume it to be true as N increases.
   c. Use your results from parts a and b together with the idea that $\Delta y_{min} = w$ to find an expression for $\Delta\lambda_{min}$, the minimum wavelength separation (in first order) for which the diffraction fringes can barely be resolved.
   d. Ordinary hydrogen atoms emit red light with a wavelength of 656.45 nm. In deuterium, which is a "heavy" isotope of hydrogen, the wavelength is 656.27 nm. What is the minimum number of slits in a diffraction grating that can barely resolve these two wavelengths in the first-order diffraction pattern?

74. ‖‖‖ FIGURE CP33.74 shows light of wavelength λ incident at angle ϕ on a *reflection* grating of spacing d. We want to find the angles θ_m at which constructive interference occurs.
   a. The figure shows paths 1 and 2 along which two waves travel and interfere. Find an expression for the path-length difference $\Delta r = r_2 - r_1$.
   b. Using your result from part a, find an equation (analogous to Equation 33.15) for the angles θ_m at which diffraction occurs when the light is incident at angle ϕ. Notice that m can be a negative integer in your expression, indicating that path 2 is shorter than path 1.
   c. Show that the zeroth-order diffraction is simply a "reflection." That is, $\theta_0 = \phi$.
   d. Light of wavelength 500 nm is incident at $\phi = 40°$ on a reflection grating having 700 reflection lines/mm. Find all angles θ_m at which light is diffracted. Negative values of θ_m are interpreted as an angle left of the vertical.
   e. Draw a picture showing a *single* 500 nm light ray incident at $\phi = 40°$ and showing all the diffracted waves at the correct angles.

FIGURE CP33.74

75. ‖‖‖ The pinhole camera of FIGURE CP33.75 images distant objects by allowing only a narrow bundle of light rays to pass through the hole and strike the film. If light con-

FIGURE CP33.75

sisted of particles, you could make the image sharper and sharper (at the expense of getting dimmer and dimmer) by making the aperture smaller and smaller. In practice, diffraction of light by the circular aperture limits the maximum sharpness that can be obtained. Consider two distant points of light, such as two distant streetlights. Each will produce a circular diffraction pattern on the film. The two images can just barely be resolved if the central maximum of one image falls on the first dark fringe of the other image. (This is called Rayleigh's criterion, and we will explore its implication for optical instruments in Chapter 35.)
   a. Optimum sharpness of one image occurs when the diameter of the central maximum equals the diameter of the pinhole. What is the optimum hole size for a pinhole camera in which the film is 20 cm behind the hole? Assume $\lambda = 550$ nm, an average value for visible light.
   b. For this hole size, what is the angle α (in degrees) between two distant sources that can barely be resolved?
   c. What is the distance between two street lights 1 km away that can barely be resolved?

34 Ray Optics

The observation that light travels in straight lines—*light rays*—will help us understand the physics of lenses and prisms.

IN THIS CHAPTER, you will learn about and apply the ray model of light.

What are light rays?

A light ray is a concept, not a physical thing. It is the line along which light energy flows.

- Rays travel in straight lines. Two rays can cross without disturbing one another.
- Objects are sources of light rays.
- Reflection and refraction by mirrors and lenses create images of objects. Points to which light rays converge are called real images. Points from which light rays diverge are called virtual images.
- The eye sees an object or an image when diverging bundles of rays enter the pupil and are focused to a real image on the retina.

You'll use both graphical and mathematical techniques to analyze how light rays travel and how images are formed.

What is the law of reflection?

Light rays bounce, or reflect, off a surface.

- Specular reflection is mirror like.
- Diffuse reflection is like light reflecting from the page of this book.

The law of reflection says that the angle of reflection equals the angle of incidence. You'll learn how reflection allows images to be seen in both flat and curved mirrors.

What is refraction?

Light rays change direction at the boundary when they move from one medium to another. This is called refraction, and it is the basis for image formation by lenses. Snell's law will allow you to find the angles on both sides of the boundary.

≪ LOOKING BACK Section 16.5 Index of refraction

How do lenses form images?

Lenses form images by refraction.

- We'll start with ray tracing, a graphical method of seeing how and where images are formed.
- We'll then develop the thin-lens equation for more quantitative results.

The same methods apply to image formation by curved mirrors.

Why is optics important?

Optics is everywhere, from your smart phone camera and your car headlights to laser pointers and the optical scanners that read bar codes. Our knowledge of the microscopic world and of the cosmos comes through optical instruments. And, of course, your eye is one of the most marvelous optical devices of all. Modern optical engineering is called photonics. Photonics does draw on all three models of light, as needed, but ray optics is usually the foundation on which optical instruments are designed.

34.1 The Ray Model of Light

A flashlight makes a beam of light through the night's darkness. Sunbeams stream into a darkened room through a small hole in the shade. Laser beams are even more well defined. Our everyday experience that light travels in straight lines is the basis of the *ray model* of light.

The ray model is an oversimplification of reality but nonetheless is very useful within its range of validity. In particular, the ray model of light is valid as long as any apertures through which the light passes (lenses, mirrors, and holes) are very large compared to the wavelength of light. In that case, diffraction and other wave aspects of light are negligible and can be ignored. The analysis of Section 33.7 found that the crossover between wave optics and ray optics occurs for apertures ≈ 1 mm in diameter. Lenses and mirrors are almost always larger than 1 mm, so the ray model of light is an excellent basis for the practical optics of image formation.

To begin, let us define a **light ray** as a line in the direction along which light energy is flowing. A light ray is an abstract idea, not a physical entity or a "thing." Any narrow beam of light, such as the laser beam in **FIGURE 34.1**, is actually a bundle of many parallel light rays. You can think of a single light ray as the limiting case of a laser beam whose diameter approaches zero. Laser beams are good approximations of light rays, but any real laser beam is a bundle of many parallel rays.

Chapter 33 briefly introduced the three models of light. Here we expand on the ray model, the subject of this chapter.

FIGURE 34.1 A laser beam or beam of sunlight is a bundle of parallel light rays.

Light rays

Direction of travel

A beam of light

MODEL 34.1

Ray model of light

For use when diffraction is not significant.

■ Light rays travel in straight lines.
  • The speed of light is $v = c/n$, where n is the material's index of refraction.
  • Light rays cross without interacting.
■ Light rays travel forever unless they interact with matter.
  • At an interface between two materials, rays can be either reflected or refracted.
  • Within a material, light rays can be either scattered or absorbed.
■ An **object** is a source of light rays.
  • Rays originate at *every* point on an object.
  • Rays are sent in *all* directions.
■ The eye sees by focusing a diverging bundle of light rays.
  • Diverging rays enter the pupil and are focused on the retina.
  • Your brain perceives the object as being at the point from which the rays are diverging.
■ Limitations: Use the wave model if diffraction is significant. The ray model is usually valid if openings are larger than about 1 mm, while the wave model is more appropriate if openings are smaller than about 1 mm.

Reflection
Refraction
Absorption
Scattering

Diverging bundle of rays

Eye

Exercise 3

Objects

FIGURE 34.2 on the next page illustrates the idea that objects can be either *self-luminous,* such as the sun, flames, and lightbulbs, or *reflective*. Most objects are reflective. A tree, unless it is on fire, is seen or photographed by virtue of reflected sunlight or

FIGURE 34.2 Self-luminous and reflective objects.

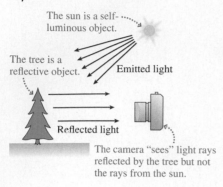

The sun is a self-luminous object.

The tree is a reflective object.

Emitted light

Reflected light

The camera "sees" light rays reflected by the tree but not the rays from the sun.

reflected skylight. People, houses, and this page in the book reflect light from self-luminous sources. In this chapter we are concerned not with how the light originates but with how it behaves after leaving the object.

Light rays from an object are emitted in all directions, but you are not *aware* of light rays unless they enter the pupil of your eye. Consequently, most light rays go completely unnoticed. For example, light rays travel from the sun to the tree in Figure 34.2, but you're not aware of these unless the tree reflects some of them into your eye. Or consider a laser beam. You've probably noticed that it's almost impossible to see a laser beam from the side unless there's dust in the air. The dust scatters a few of the light rays toward your eye, but in the absence of dust you would be completely unaware of a very powerful light beam traveling past you. **Light rays exist independently of whether you are seeing them.**

FIGURE 34.3 shows two idealized sets of light rays. The diverging rays from a **point source** are emitted in all directions. It is useful to think of each point on an object as a point source of light rays. A **parallel bundle** of rays could be a laser beam. Alternatively it could represent a *distant object,* an object such as a star so far away that the rays arriving at the observer are essentially parallel to each other.

FIGURE 34.3 Point sources and parallel bundles represent idealized objects.

Point source   Parallel bundle

FIGURE 34.4 A ray diagram simplifies the situation by showing only a few rays.

These are just a few of the infinitely many rays leaving the object.

Ray Diagrams

Rays originate from *every* point on an object and travel outward in *all* directions, but a diagram trying to show all these rays would be hopelessly messy and confusing. To simplify the picture, we usually use a **ray diagram** showing only a few rays. For example, **FIGURE 34.4** is a ray diagram showing only a few rays leaving the top and bottom points of the object and traveling to the right. These rays will be sufficient to show us how the object is imaged by lenses or mirrors.

NOTE Ray diagrams are the basis for a *pictorial representation* that we'll use throughout this chapter. Be careful not to think that a ray diagram shows all of the rays. The rays shown on the diagram are just a subset of the infinitely many rays leaving the object.

FIGURE 34.5 A camera obscura.

(a)

Image on back wall

Aperture

Darkened room

Object

Apertures

A popular form of entertainment during ancient Roman times was a visit to a **camera obscura,** Latin for "dark room." As **FIGURE 34.5a** shows, a camera obscura was a darkened room with a single, small hole to the outside world. After their eyes became dark adapted, visitors could see a dim but full-color image of the outside world displayed on the back wall of the room. However, the image was upside down! The *pinhole camera* is a miniature version of the camera obscura.

A hole through which light passes is called an **aperture**. **FIGURE 34.5b** uses the ray model of light passing through a small aperture to explain how the camera obscura works. Each point on an object emits light rays in all directions, but only a very few of these rays pass through the aperture and reach the back wall. As the figure illustrates, the geometry of the rays causes the image to be upside down.

Actually, as you may have realized, each *point* on the object illuminates a small but extended *patch* on the wall. This is because the non-zero size of the aperture—needed for the image to be bright enough to see—allows several rays from each point on the object to pass through at slightly different angles. As a result, the image is slightly blurred and out of focus. (Diffraction also becomes an issue if the hole gets

(b)

These rays don't make it through the hole.

h_o h_i

d_o d_i

The image is upside down. If the hole is sufficiently small, each point on the image corresponds to one point on the object.

too small.) We'll later discover how a modern camera, with a lens, improves on the camera obscura.

You can see from the similar triangles in Figure 34.5b that the object and image heights are related by

$$\frac{h_i}{h_o} = \frac{d_i}{d_o}$$

(34.1)

where d_o is the distance to the object and d_i is the depth of the camera obscura. Any realistic camera obscura has $d_i < d_o$; thus the image is smaller than the object.

We can apply the ray model to more complex apertures, such as the L-shaped aperture in FIGURE 34.6. The pattern of light on the screen is found by tracing all the straight-line paths—the ray trajectories—that start from the point source and pass through the aperture. We will see an enlarged L on the screen, with a sharp boundary between the image and the dark shadow.

FIGURE 34.6 Light through an aperture.

Light
Some rays are blocked by the opaque screen.
Shadow
Screen
Aperture
Point source

STOP TO THINK 34.1 A long, thin lightbulb illuminates a vertical aperture. Which pattern of light do you see on a viewing screen behind the aperture?

Screen

Light

(a) (b) (c) (d)

34.2 Reflection

Reflection of light is a familiar, everyday experience. You see your reflection in the bathroom mirror first thing every morning, reflections in your car's rearview mirror as you drive to school, and the sky reflected in puddles of standing water. Reflection from a flat, smooth surface, such as a mirror or a piece of polished metal, is called **specular reflection,** from *speculum,* the Latin word for "mirror."

FIGURE 34.7a shows a bundle of parallel light rays reflecting from a mirror-like surface. You can see that the incident and reflected rays are both in a plane that is normal, or perpendicular, to the reflective surface. A three-dimensional perspective accurately shows the relationship between the light rays and the surface, but figures such as this are hard to draw by hand. Instead, it is customary to represent reflection with the simpler pictorial representation of FIGURE 34.7b. In this figure,

- The plane of the page is the *plane of incidence,* the plane containing both incident and reflected rays. The reflective surface extends into the page.
- A *single* light ray represents the entire bundle of parallel rays. This is oversimplified, but it keeps the figure and the analysis clear.

The angle θ_i between the ray and a line perpendicular to the surface—the *normal* to the surface—is called the **angle of incidence.** Similarly, the **angle of reflection** θ_r is the angle between the reflected ray and the normal to the surface. The **law of reflection,** easily demonstrated with simple experiments, states that

1. The incident ray and the reflected ray are in the same plane normal to the surface, and
2. The angle of reflection equals the angle of incidence: $\theta_r = \theta_i$.

NOTE Optics calculations *always* use the angle measured from the normal, not the angle between the ray and the surface.

FIGURE 34.7 Specular reflection of light.

(a)
Both the incident and reflected rays lie in a plane that is perpendicular to the surface.

Reflective surface

(b)
Normal
Angle of incidence
Angle of reflection
Reflected ray
Incident ray
θ_i θ_r
Reflective surface

EXAMPLE 34.1 | Light reflecting from a mirror

A dressing mirror on a closet door is 1.50 m tall. The bottom is 0.50 m above the floor. A bare lightbulb hangs 1.00 m from the closet door, 2.50 m above the floor. How long is the streak of reflected light across the floor?

MODEL Treat the lightbulb as a point source and use the ray model of light.

FIGURE 34.8 Pictorial representation of the light rays reflecting from a mirror.

VISUALIZE FIGURE 34.8 is a pictorial representation of the light rays. We need to consider only the two rays that strike the edges of the mirror. All other reflected rays will fall between these two.

SOLVE Figure 34.8 has used the law of reflection to set the angles of reflection equal to the angles of incidence. Other angles have been identified with simple geometry. The two angles of incidence are

$$\theta_1 = \tan^{-1}\left(\frac{0.50 \text{ m}}{1.00 \text{ m}}\right) = 26.6°$$

$$\theta_2 = \tan^{-1}\left(\frac{2.00 \text{ m}}{1.00 \text{ m}}\right) = 63.4°$$

The distances to the points where the rays strike the floor are then

$$l_1 = \frac{2.00 \text{ m}}{\tan\theta_1} = 4.00 \text{ m}$$

$$l_2 = \frac{0.50 \text{ m}}{\tan\theta_2} = 0.25 \text{ m}$$

Thus the length of the light streak is $l_1 - l_2 = 3.75$ m.

Diffuse Reflection

FIGURE 34.9 Diffuse reflection from an irregular surface.

Each ray obeys the law of reflection at that point, but the irregular surface causes the reflected rays to leave in many random directions.

Magnified view of surface

Most objects are seen by virtue of their reflected light. For a "rough" surface, the law of reflection $\theta_r = \theta_i$ is obeyed at each point but the irregularities of the surface cause the reflected rays to leave in many random directions. This situation, shown in **FIGURE 34.9**, is called **diffuse reflection.** It is how you see this page, the wall, your hand, your friend, and so on.

By a "rough" surface, we mean a surface that is rough or irregular in comparison to the wavelength of light. Because visible-light wavelengths are $\approx 0.5\ \mu$m, any surface with texture, scratches, or other irregularities larger than 1 μm will cause diffuse reflection rather than specular reflection. A piece of paper may feel quite smooth to your hand, but a microscope would show that the surface consists of distinct fibers much larger than 1 μm. By contrast, the irregularities on a mirror or a piece of polished metal are much smaller than 1 μm.

The Plane Mirror

One of the most commonplace observations is that you can see yourself in a mirror. How? **FIGURE 34.10a** shows rays from point source P reflecting from a mirror. Consider the particular ray shown in **FIGURE 34.10b**. The reflected ray travels along a line that passes through point P' on the "back side" of the mirror. Because $\theta_r = \theta_i$, simple geometry dictates that P' is the same distance behind the mirror as P is in front of the mirror. That is, $s' = s$.

FIGURE 34.10 The light rays reflecting from a plane mirror.

(a)

Rays from P reflect from the mirror. Each ray obeys the law of reflection.

(b)

This reflected ray appears to have been traveling along a line that passed through point P'.

(c)   Object distance   Image distance

The reflected rays *all* diverge from P', which appears to be the source of the reflected rays. Your eye collects the bundle of diverging rays and "sees" the light coming from P'.

The location of point P' in Figure 34.10b is independent of the value of θ_i. Consequently, as FIGURE 34.10c shows, **the reflected rays all *appear* to be coming from point P'.** For a plane mirror, the distance s' to point P' is equal to the object distance s:

$$s' = s \qquad \text{(plane mirror)} \qquad (34.2)$$

If rays diverge from an object point P and interact with a mirror so that the reflected rays diverge from point P' and *appear* to come from P', then we call P' a **virtual image** of point P. The image is "virtual" in the sense that no rays actually leave P', which is in darkness behind the mirror. But as far as your eye is concerned, the light rays act exactly *as if* the light really originated at P'. So while you may say "I see P in the mirror," what you are actually seeing is the virtual image of P. Distance s' is the *image distance*.

For an extended object, such as the one in FIGURE 34.11, each point on the object from which rays strike the mirror has a corresponding image point an equal distance on the opposite side of the mirror. The eye captures and focuses diverging bundles of rays from each point of the image in order to see the full image in the mirror. Two facts are worth noting:

1. Rays from each point on the object spread out in all directions and strike *every point* on the mirror. Only a very few of these rays enter your eye, but the other rays are very real and might be seen by other observers.
2. Rays from points P and Q enter your eye after reflecting from *different* areas of the mirror. This is why you can't always see the full image of an object in a very small mirror.

FIGURE 34.11 Each point on the extended object has a corresponding image point an equal distance on the opposite side of the mirror.

The rays from P and Q that reach your eye reflect from different areas of the mirror.

Your eye intercepts only a very small fraction of all the reflected rays.

EXAMPLE 34.2 | How high is the mirror?

If your height is h, what is the shortest mirror on the wall in which you can see your full image? Where must the top of the mirror be hung?

MODEL Use the ray model of light.

VISUALIZE FIGURE 34.12 is a pictorial representation of the light rays. We need to consider only the two rays that leave your head and feet and reflect into your eye.

SOLVE Let the distance from your eyes to the top of your head be l_1 and the distance to your feet be l_2. Your height is $h = l_1 + l_2$. A light ray from the top of your head that reflects from the mirror at $\theta_r = \theta_i$ and enters your eye must, by congruent triangles, strike the mirror a distance $\frac{1}{2}l_1$ above your eyes. Similarly, a ray from your foot to your eye strikes the mirror a distance $\frac{1}{2}l_2$ below your eyes. The distance between these two points on the mirror is $\frac{1}{2}l_1 + \frac{1}{2}l_2 = \frac{1}{2}h$. A ray from anywhere else on your body can reach your eye if it strikes the mirror between these two points. Pieces of the mirror outside these two points are irrelevant, not because rays don't strike them but because the reflected rays don't reach

FIGURE 34.12 Pictorial representation of light rays from your head and feet reflecting into your eye.

your eye. Thus the shortest mirror in which you can see your full reflection is $\frac{1}{2}h$. But this will work only if the top of the mirror is hung midway between your eyes and the top of your head.

ASSESS It is interesting that the answer does not depend on how far you are from the mirror.

STOP TO THINK 34.2 Two plane mirrors form a right angle. How many images of the ball can you see in the mirrors?

a. 1
b. 2
c. 3
d. 4

Observer

FIGURE 34.13 A light beam refracts twice in passing through a glass prism.

FIGURE 34.13 A light beam refracts twice in passing through a glass prism.

34.3 Refraction

Two things happen when a light ray is incident on a smooth boundary between two transparent materials, such as the boundary between air and glass:

1. Part of the light *reflects* from the boundary, obeying the law of reflection. This is how you see reflections from pools of water or storefront windows.
2. Part of the light continues into the second medium. It is *transmitted* rather than reflected, but the transmitted ray changes direction as it crosses the boundary. The transmission of light from one medium to another, but with a change in direction, is called **refraction.**

The photograph of FIGURE 34.13 shows the refraction of a light beam as it passes through a glass prism. Notice that the ray direction changes as the light enters and leaves the glass. Our goal in this section is to understand refraction, so we will usually ignore the weak reflection and focus on the transmitted light.

> **NOTE** A transparent material through which light travels is called the *medium*. This term has to be used with caution. The material does affect the light speed, but a transparent material differs from the medium of a sound or water wave in that particles of the medium do *not* oscillate as a light wave passes through. For a light wave it is the electromagnetic field that oscillates.

FIGURE 34.14a shows the refraction of light rays in a parallel beam of light, such as a laser beam, and rays from a point source. Our analysis will be easier, however, if we focus on a single light ray. FIGURE 34.14b is a ray diagram showing the refraction of a single ray at a boundary between medium 1 and medium 2. Let the angle between the ray and the normal be θ_1 in medium 1 and θ_2 in medium 2. For the medium in which the ray is approaching the boundary, this is the *angle of incidence* as we've previously defined it. The angle on the transmitted side, *measured from the normal*, is called the **angle of refraction.** Notice that θ_1 is the angle of incidence in Figure 34.14b and the angle of refraction in FIGURE 34.14c, where the ray is traveling in the opposite direction.

FIGURE 34.14 Refraction of light rays.

(a) Refraction of parallel and point-source rays

(b) Refraction from a lower-index medium to a higher-index medium

(c) The reversed ray

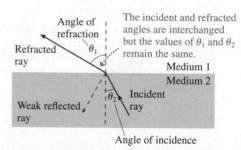

Refraction was first studied experimentally by the Arab scientist Ibn al-Haitham, in about the year 1000, and later by the Dutch scientist Willebrord Snell. **Snell's law** says that when a ray refracts between medium 1 and medium 2, having indices of refraction n_1 and n_2, the ray angles θ_1 and θ_2 in the two media are related by

$$n_1 \sin\theta_1 = n_2 \sin\theta_2 \qquad \text{(Snell's law of refraction)} \qquad (34.3)$$

Notice that Snell's law does not mention which is the incident angle and which is the refracted angle.

The Index of Refraction

To Snell and his contemporaries, n was simply an "index of the refractive power" of a transparent substance. The relationship between the index of refraction and the speed of light was not recognized until the development of a wave theory of light in the 19th century. Theory predicts, and experiment confirms, that light travels through a transparent medium, such as glass or water, at a speed *less* than its speed c in vacuum. In Section 16.5, we defined the *index of refraction n* of a transparent medium as

$$n = \frac{c}{v_{medium}} \quad (34.4)$$

where v_{medium} is the light speed in the medium. This implies, of course, that $v_{medium} = c/n$. The index of refraction of a medium is always $n > 1$ except for vacuum.

TABLE 34.1 shows measured values of n for several materials. There are many types of glass, each with a slightly different index of refraction, so we will keep things simple by accepting $n = 1.50$ as a typical value. Notice that cubic zirconia, used to make costume jewelry, has an index of refraction much higher than glass.

We can accept Snell's law as simply an empirical discovery about light. Alternatively, and perhaps surprisingly, we can use the wave model of light to justify Snell's law. The key ideas we need are:

- Wave fronts represent the crests of waves. They are spaced one wavelength apart.
- The wavelength in a medium with index of refraction n is $\lambda = \lambda_{vac}/n$, where λ_{vac} is the vacuum wavelength.
- Wave fronts are perpendicular to the wave's direction of travel.
- The wave fronts stay lined up as a wave crosses from one medium into another.

FIGURE 34.15 shows a wave crossing the boundary between two media, where we're assuming $n_2 > n_1$. **Because the wavelengths differ on opposite sides of the boundary, the wave fronts can stay lined up only if the waves in the two media are traveling in different directions.** In other words, the wave must refract at the boundary to keep the crests of the wave aligned.

To analyze Figure 34.15, consider the segment of boundary of length l between the two wave fronts. This segment is the common hypotenuse of two right triangles. From the upper triangle, which has one side of length λ_1, we see

$$l = \frac{\lambda_1}{\sin\theta_1} \quad (34.5)$$

where θ_1 is the angle of incidence. Similarly, the lower triangle, where θ_2 is the angle of refraction, gives

$$l = \frac{\lambda_2}{\sin\theta_2} \quad (34.6)$$

Equating these two expressions for l, and using $\lambda_1 = \lambda_{vac}/n_1$ and $\lambda_2 = \lambda_{vac}/n_2$, we find

$$\frac{\lambda_{vac}}{n_1 \sin\theta_1} = \frac{\lambda_{vac}}{n_2 \sin\theta_2} \quad (34.7)$$

Equation 34.7 can be true only if

$$n_1 \sin\theta_1 = n_2 \sin\theta_2 \quad (34.8)$$

which is Snell's law.

Examples of Refraction

Look back at Figure 34.14. As the ray in Figure 34.14b moves from medium 1 to medium 2, where $n_2 > n_1$, it bends closer to the normal. In Figure 34.14c, where the ray moves from medium 2 to medium 1, it bends away from the normal. This is a general conclusion that follows from Snell's law:

- When a ray is transmitted into a material with a higher index of refraction, it bends toward the normal.
- When a ray is transmitted into a material with a lower index of refraction, it bends away from the normal.

TABLE 34.1 Indices of refraction

Medium	n
Vacuum	1.00 exactly
Air (actual)	1.0003
Air (accepted)	1.00
Water	1.33
Ethyl alcohol	1.36
Oil	1.46
Glass (typical)	1.50
Polystyrene plastic	1.59
Cubic zirconia	2.18
Diamond	2.41
Silicon (infrared)	3.50

FIGURE 34.15 Snell's law is a consequence of the wave model of light.

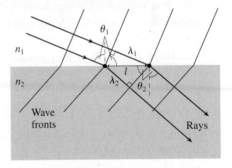

This rule becomes a central idea in a procedure for analyzing refraction problems.

TACTICS BOX 34.1

Analyzing refraction

❶ **Draw a ray diagram.** Represent the light beam with one ray.
❷ **Draw a line normal to the boundary.** Do this at each point where the ray intersects a boundary.
❸ **Show the ray bending in the correct direction.** The angle is larger on the side with the smaller index of refraction. This is the qualitative application of Snell's law.
❹ **Label angles of incidence and refraction.** Measure all angles from the normal.
❺ **Use Snell's law.** Calculate the unknown angle or unknown index of refraction.

Exercises 11–15

EXAMPLE 34.3 | Deflecting a laser beam

A laser beam is aimed at a 1.0-cm-thick sheet of glass at an angle 30° above the glass.

a. What is the laser beam's direction of travel in the glass?
b. What is its direction in the air on the other side?
c. By what distance is the laser beam displaced?

MODEL Represent the laser beam with a single ray and use the ray model of light.

VISUALIZE FIGURE 34.16 is a pictorial representation in which the first four steps of Tactics Box 34.1 are identified. Notice that the angle of incidence is $\theta_1 = 60°$, not the 30° value given in the problem.

FIGURE 34.16 The ray diagram of a laser beam passing through a sheet of glass.

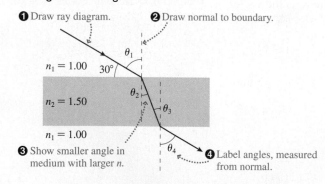

❶ Draw ray diagram. ❷ Draw normal to boundary.
θ_1
$n_1 = 1.00$ 30°
$n_2 = 1.50$ θ_2
θ_3
$n_1 = 1.00$
❸ Show smaller angle in medium with larger n. θ_4 ❹ Label angles, measured from normal.

SOLVE a. Snell's law, the final step in the Tactics Box, is $n_1 \sin\theta_1 = n_2 \sin\theta_2$. Using $\theta_1 = 60°$, we find that the direction of travel in the glass is

$$\theta_2 = \sin^{-1}\left(\frac{n_1 \sin\theta_1}{n_2}\right) = \sin^{-1}\left(\frac{\sin 60°}{1.5}\right)$$

$$= \sin^{-1}(0.577) = 35.3°$$

b. Snell's law at the second boundary is $n_2 \sin\theta_3 = n_1 \sin\theta_4$. You can see from Figure 34.16 that the interior angles are equal:

$\theta_3 = \theta_2 = 35.3°$. Thus the ray emerges back into the air traveling at angle

$$\theta_4 = \sin^{-1}\left(\frac{n_2 \sin\theta_3}{n_1}\right) = \sin^{-1}(1.5 \sin 35.3°)$$

$$= \sin^{-1}(0.867) = 60°$$

This is the same as θ_1, the original angle of incidence. The glass doesn't change the direction of the laser beam.

c. Although the exiting laser beam is parallel to the initial laser beam, it has been displaced sideways by distance d. FIGURE 34.17 shows the geometry for finding d. From trigonometry, $d = l \sin\phi$. Further, $\phi = \theta_1 - \theta_2$ and $l = t/\cos\theta_2$, where t is the thickness of the glass. Combining these gives

$$d = l \sin\phi = \frac{t}{\cos\theta_2} \sin(\theta_1 - \theta_2)$$

$$= \frac{(1.0 \text{ cm}) \sin 24.7°}{\cos 35.3°} = 0.51 \text{ cm}$$

The glass causes the laser beam to be displaced sideways by 0.51 cm.

FIGURE 34.17 The laser beam is deflected sideways by distance d.

Initial laser beam
θ_1 $\phi = \theta_1 - \theta_2$
θ_2
t l
d
Displaced laser beam d

ASSESS The laser beam exits the glass still traveling in the same direction as it entered. This is a general result for light traveling through a medium with parallel sides. Notice that the displacement d becomes zero in the limit $t \to 0$. This will be an important observation when we get to lenses.

EXAMPLE 34.4 | Measuring the index of refraction

FIGURE 34.18 shows a laser beam deflected by a 30°-60°-90° prism. What is the prism's index of refraction?

FIGURE 34.18 A prism deflects a laser beam.

FIGURE 34.19 Pictorial representation of a laser beam passing through the prism.

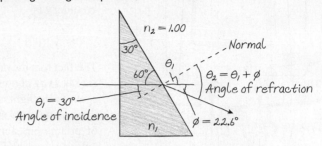

θ_1 and θ_2 are measured from the normal.

MODEL Represent the laser beam with a single ray and use the ray model of light.

VISUALIZE FIGURE 34.19 uses the steps of Tactics Box 34.1 to draw a ray diagram. The ray is incident perpendicular to the front face of the prism ($\theta_{\text{incident}} = 0°$), thus it is transmitted through the first boundary without deflection. At the second boundary it is especially important to *draw the normal to the surface* at the point of incidence and to *measure angles from the normal*.

SOLVE From the geometry of the triangle you can find that the laser's angle of incidence on the hypotenuse of the prism is $\theta_1 = 30°$,

the same as the apex angle of the prism. The ray exits the prism at angle θ_2 such that the deflection is $\phi = \theta_2 - \theta_1 = 22.6°$. Thus $\theta_2 = 52.6°$. Knowing both angles and $n_2 = 1.00$ for air, we can use Snell's law to find n_1:

$$n_1 = \frac{n_2 \sin\theta_2}{\sin\theta_1} = \frac{1.00\sin 52.6°}{\sin 30°} = 1.59$$

ASSESS Referring to the indices of refraction in Table 34.1, we see that the prism is made of plastic.

Total Internal Reflection

What would have happened in Example 34.4 if the prism angle had been 45° rather than 30°? The light rays would approach the rear surface of the prism at an angle of incidence $\theta_1 = 45°$. When we try to calculate the angle of refraction at which the ray emerges into the air, we find

$$\sin\theta_2 = \frac{n_1}{n_2}\sin\theta_1 = \frac{1.59}{1.00}\sin 45° = 1.12$$

$$\theta_2 = \sin^{-1}(1.12) = ???$$

Angle θ_2 doesn't compute because the sine of an angle can't be larger than 1. The ray is unable to refract through the boundary. Instead, 100% of the light *reflects* from the boundary back into the prism. This process is called **total internal reflection,** often abbreviated TIR. That it really happens is illustrated in FIGURE 34.20. Here three laser beams enter a prism from the left. The bottom two refract out through the right side of the prism. The blue beam, which is incident on the prism's top face, undergoes total internal reflection and then emerges through the right surface.

FIGURE 34.21 shows several rays leaving a point source in a medium with index of refraction n_1. The medium on the other side of the boundary has $n_2 < n_1$. As we've seen, crossing a boundary into a material with a lower index of refraction causes the ray to bend away from the normal. Two things happen as angle θ_1 increases. First, the refraction angle θ_2 approaches 90°. Second, the fraction of the light energy transmitted decreases while the reflected fraction increases.

A **critical angle** is reached when $\theta_2 = 90°$. Because $\sin 90° = 1$, Snell's law $n_1 \sin\theta_c = n_2 \sin 90°$ gives the critical angle of incidence as

$$\theta_c = \sin^{-1}\left(\frac{n_2}{n_1}\right) \tag{34.9}$$

FIGURE 34.20 The blue laser beam undergoes total internal reflection inside the prism.

FIGURE 34.21 Refraction and reflection of rays as the angle of incidence increases.

The angle of incidence is increasing. Transmission is getting weaker. →

$\theta_2 = 90°$

$n_2 < n_1$
n_1

$\theta_1 > \theta_c$

Critical angle when $\theta_2 = 90°$

Reflection is getting stronger. →

Total internal reflection occurs when $\theta_1 \geq \theta_c$.

FIGURE 34.22 Binoculars make use of total internal reflection.

Angles of incidence exceed the critical angle.

The refracted light vanishes at the critical angle and the reflection becomes 100% for any angle $\theta_1 \geq \theta_c$. The critical angle is well defined because of our assumption that $n_2 < n_1$. **There is no critical angle and no total internal reflection if $n_2 > n_1$.**

As a quick example, the critical angle in a typical piece of glass at the glass-air boundary is

$$\theta_{c\,glass} = \sin^{-1}\left(\frac{1.00}{1.50}\right) = 42°$$

The fact that the critical angle is less than 45° has important applications. For example, **FIGURE 34.22** shows a pair of binoculars. The lenses are much farther apart than your eyes, so the light rays need to be brought together before exiting the eyepieces. Rather than using mirrors, which get dirty and require alignment, binoculars use a pair of prisms on each side. Thus the light undergoes two total internal reflections and emerges from the eyepiece. (The actual arrangement is a little more complex than in Figure 34.22, to avoid left-right reversals, but this illustrates the basic idea.)

EXAMPLE 34.5 | **Total internal reflection**

A small lightbulb is set in the bottom of a 3.0-m-deep swimming pool. What is the diameter of the circle of light seen on the water's surface from above?

MODEL Use the ray model of light.

VISUALIZE FIGURE 34.23 is a pictorial representation. The lightbulb emits rays at all angles, but only some of the rays refract into the air and are seen from above. Rays striking the surface at greater than the critical angle undergo TIR and remain within the water. The diameter of the circle of light is the distance between the two points at which rays strike the surface at the critical angle.

SOLVE From trigonometry, the circle diameter is $D = 2h\tan\theta_c$, where h is the depth of the water. The critical angle for a water-air boundary is $\theta_c = \sin^{-1}(1.00/1.33) = 48.7°$. Thus

$$D = 2(3.0\text{ m})\tan 48.7° = 6.8\text{ m}$$

FIGURE 34.23 Pictorial representation of the rays leaving a lightbulb at the bottom of a swimming pool.

Rays at the critical angle θ_c form the edge of the circle of light seen from above.

Fiber Optics

FIGURE 34.24 Light rays are confined within an optical fiber by total internal reflection.

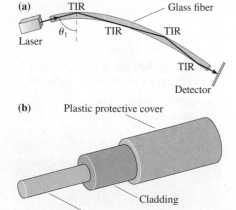

The most important modern application of total internal reflection is the transmission of light through optical fibers. **FIGURE 34.24a** shows a laser beam shining into the end of a long, narrow-diameter glass tube. The light rays pass easily from the air into the glass, but they then impinge on the inside wall of the glass tube at an angle of incidence θ_1 approaching 90°. This is well above the critical angle, so the laser beam undergoes TIR and remains inside the glass. The laser beam continues to "bounce" its way down the tube as if the light were inside a pipe. Indeed, optical fibers are sometimes called "light pipes." The rays are *below* the critical angle ($\theta_1 \approx 0$) when they finally reach the end of the fiber, thus they refract out without difficulty and can be detected.

While a simple glass tube can transmit light, a glass-air boundary is not sufficiently reliable for commercial use. Any small scratch on the side of the tube alters the rays' angle of incidence and allows leakage of light. **FIGURE 34.24b** shows the construction of a practical optical fiber. A small-diameter glass *core* is surrounded by a layer of glass *cladding*. The glasses used for the core and the cladding have $n_{core} > n_{cladding}$; thus light undergoes TIR at the core-cladding boundary and remains confined within the core. This boundary is not exposed to the environment and hence retains its integrity even under adverse conditions.

Even glass of the highest purity is not perfectly transparent. Absorption in the glass, even if very small, causes a gradual decrease in light intensity. The glass used for the core of optical fibers has a minimum absorption at a wavelength of 1.3 μm, in the infrared, so this is the laser wavelength used for long-distance signal transmission. Light at this wavelength can travel hundreds of kilometers through a fiber without significant loss.

STOP TO THINK 34.3 A light ray travels from medium 1 to medium 3 as shown. For these media,

a. $n_3 > n_1$ b. $n_3 = n_1$ c. $n_3 < n_1$
d. We can't compare n_1 to n_3 without knowing n_2.

34.4 Image Formation by Refraction at a Plane Surface

If you see a fish that appears to be swimming close to the front window of the aquarium, but then look through the side of the aquarium, you'll find that the fish is actually farther from the window than you thought. Why is this?

To begin, recall that vision works by focusing a diverging bundle of rays onto the retina. The point from which the rays diverge is where you perceive the object to be. **FIGURE 34.25a** shows how you would see a fish out of water at distance d.

Now place the fish back into the aquarium at the same distance d. For simplicity, we'll ignore the glass wall of the aquarium and consider the water-air boundary. (The thin glass of a typical window has only a very small effect on the refraction of the rays and doesn't change the conclusions.) Light rays again leave the fish, but this time they refract at the water-air boundary. Because they're going from a higher to a lower index of refraction, the rays refract *away from* the normal. **FIGURE 34.25b** shows the consequences.

A bundle of diverging rays still enters your eye, but now these rays are diverging from a closer point, at distance d'. As far as your eye and brain are concerned, it's exactly *as if* the rays really originate at distance d', and this is the location at which you see the fish. **The object appears closer than it really is because of the refraction of light at the boundary.**

We found that the rays reflected from a mirror diverge from a point that is not the object point. We called that point a *virtual image*. Similarly, if rays from an object point P refract at a boundary between two media such that the rays then diverge from a point P′ and *appear* to come from P′, we call P′ a virtual image of point P. **The virtual image of the fish is what you see.**

Let's examine this image formation a bit more carefully. **FIGURE 34.26** shows a boundary between two transparent media having indices of refraction n_1 and n_2. Point P, a source of light rays, is the object. Point P′, from which the rays *appear* to diverge, is the virtual image of P. Distance s is called the **object distance.** Our goal is to determine distance s', the **image distance. Both are measured from the boundary.**

A line perpendicular to the boundary is called the **optical axis.** Consider a ray leaving the object at angle θ_1 with respect to the optical axis. θ_1 is also the angle of incidence at the boundary, where the ray refracts into the second medium at angle θ_2. By tracing the refracted ray backward, you can see that θ_2 is also the angle between the refracted ray and the optical axis at point P′.

The distance l is common to both the incident and the refracted rays, and you can see that $l = s \tan\theta_1 = s' \tan\theta_2$. However, **it is customary in optics for virtual image distances to be negative.** (The reason will be clear when we get to image formation by lenses.) Hence we will insert a minus sign, finding that

$$s' = -\frac{\tan\theta_1}{\tan\theta_2}s \qquad (34.10)$$

FIGURE 34.25 Refraction of the light rays causes a fish in the aquarium to be seen at distance d'.

(a) A fish out of water

The rays that reach the eye are diverging from this point, the object.

(b) A fish in the aquarium

Refraction causes the rays to bend at the boundary.

Now the rays that reach the eye are diverging from this point, the image.

FIGURE 34.26 Finding the virtual image P′ of an object at P. We've assumed $n_1 > n_2$.

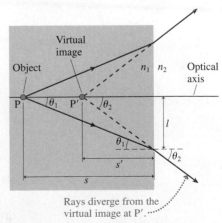

Rays diverge from the virtual image at P′.

Snell's law relates the sines of angles θ_1 and θ_2; that is,

$$\frac{\sin\theta_1}{\sin\theta_2} = \frac{n_2}{n_1} \tag{34.11}$$

In practice, the angle between any of these rays and the optical axis is very small because the size of the pupil of your eye is very much less than the distance between the object and your eye. (The angles in the figure have been greatly exaggerated.) Rays that are nearly *parallel* to the *axis* are called **paraxial rays.** The small-angle approximation $\sin\theta \approx \tan\theta \approx \theta$, where θ is in radians, can be applied to paraxial rays. Consequently,

$$\frac{\tan\theta_1}{\tan\theta_2} \approx \frac{\sin\theta_1}{\sin\theta_2} = \frac{n_2}{n_1} \tag{34.12}$$

Using this result in Equation 34.10, we find that the image distance is

$$s' = -\frac{n_2}{n_1}s \tag{34.13}$$

The minus sign tells us that we have a virtual image.

> **NOTE** The fact that the result for s' is independent of θ_1 implies that *all* paraxial rays appear to diverge from the same point P′. This property of the diverging rays is essential in order to have a well-defined image.

EXAMPLE 34.6 | An air bubble in a window

A fish and a sailor look at each other through a 5.0-cm-thick glass porthole in a submarine. There happens to be an air bubble right in the center of the glass. How far behind the surface of the glass does the air bubble appear to the fish? To the sailor?

MODEL Represent the air bubble as a point source and use the ray model of light.

VISUALIZE Paraxial light rays from the bubble refract into the air on one side and into the water on the other. The ray diagram looks like Figure 34.26.

SOLVE The index of refraction of the glass is $n_1 = 1.50$. The bubble is in the center of the window, so the object distance from either side of the window is $s = 2.5$ cm. On the water side, the image distance is

$$s' = -\frac{n_2}{n_1}s = -\frac{1.33}{1.50}(2.5\text{ cm}) = -2.2\text{ cm}$$

The minus sign indicates a virtual image. Physically, the fish sees the bubble 2.2 cm behind the surface. The image distance on the water side is

$$s' = -\frac{n_2}{n_1}s = -\frac{1.00}{1.50}(2.5\text{ cm}) = -1.7\text{ cm}$$

So the sailor sees the bubble 1.7 cm behind the surface.

ASSESS The image distance is *less* for the sailor because of the *larger* difference between the two indices of refraction.

34.5 Thin Lenses: Ray Tracing

A camera obscura or a pinhole camera forms images on a screen, but the images are faint and not perfectly focused. The ability to create a bright, well-focused image is vastly improved by using a lens. A **lens** is a transparent object that uses refraction at *curved* surfaces to form an image from diverging light rays. We will defer a mathematical analysis of lenses until the next section. First, we want to establish a pictorial method of understanding image formation. This method is called **ray tracing.**

FIGURE 34.27 shows parallel light rays entering two different lenses. The left lens, called a **converging lens,** causes the rays to refract *toward* the optical axis. The common point through which initially parallel rays pass is called the **focal point** of the lens. The distance of the focal point from the lens is called the **focal length** f of the lens. The right lens, called a **diverging lens,** refracts parallel rays *away from* the optical axis. This lens also has a focal point, but it is not as obvious.

FIGURE 34.27 Parallel light rays pass through a converging lens and a diverging lens.

Converging lens

Diverging lens

> **NOTE** A converging lens is thicker in the center than at the edges. A diverging lens is thicker at the edges than at the center.

FIGURE 34.28 clarifies the situation. In the case of a diverging lens, a backward projection of the diverging rays shows that they *appear* to have started from the same point. This is the focal point of a diverging lens, and its distance from the lens is the focal length of the lens. In the next section we'll relate the focal length to the curvature and index of refraction of the lens, but now we'll use the practical definition that **the focal length is the distance from the lens at which rays parallel to the optical axis converge or from which they diverge.**

> **NOTE** The focal length f is a property *of the lens,* independent of how the lens is used. The focal length characterizes a lens in much the same way that a mass m characterizes an object or a spring constant k characterizes a spring.

Converging Lenses

These basic observations about lenses are enough to understand image formation by a thin lens. A **thin lens** is a lens whose thickness is very small in comparison to its focal length and in comparison to the object and image distances. We'll make the approximation that the thickness of a thin lens is zero and that the lens lies in a plane called the **lens plane.** Within this approximation, **all refraction occurs as the rays cross the lens plane, and all distances are measured from the lens plane.** Fortunately, the thin-lens approximation is quite good for most practical applications of lenses.

> **NOTE** We'll *draw* lenses as if they have a thickness, because that is how we expect lenses to look, but our analysis will not depend on the shape or thickness of a lens.

FIGURE 34.29 shows three important situations of light rays passing through a thin converging lens. Part a is familiar from Figure 34.28. If the direction of each of the rays in Figure 34.29a is reversed, Snell's law tells us that each ray will exactly retrace its path and emerge from the lens parallel to the optical axis. This leads to Figure 34.29b, which is the "mirror image" of part a. Notice that the lens actually has *two* focal points, located at distances f on either side of the lens.

FIGURE 34.28 The focal lengths of converging and diverging lenses.

Focal length f

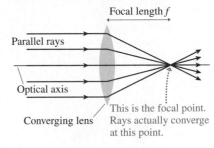

Parallel rays

Optical axis

This is the focal point.
Converging lens   Rays actually converge at this point.

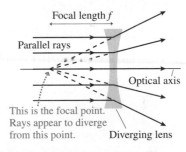

Focal length f

Parallel rays

Optical axis

This is the focal point.
Rays appear to diverge from this point.   Diverging lens

FIGURE 34.29 Three important sets of rays passing through a thin converging lens.

(a)

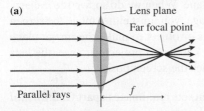

Lens plane
Far focal point
Parallel rays
f

Any ray initially parallel to the optical axis will refract through the focal point on the far side of the lens.

(b)

Lens plane
Near focal point
f
Parallel rays

Any ray passing through the near focal point emerges from the lens parallel to the optical axis.

(c)

Center of lens   Rays are not bent.

Any ray directed at the center of the lens passes through in a straight line.

Figure 34.29c shows three rays passing through the *center* of the lens. At the center, the two sides of a lens are very nearly parallel to each other. Earlier, in Example 34.3, we found that a ray passing through a piece of glass with parallel sides is *displaced* but *not bent* and that the displacement becomes zero as the thickness approaches zero. Consequently, a ray through the center of a thin lens, with zero thickness, is neither bent nor displaced but travels in a straight line.

These three situations form the basis for ray tracing.

Real Images

FIGURE 34.30 shows a lens and an object whose *object distance s* from the lens is larger than the focal length. Rays from point P on the object are refracted by the lens so as to converge at point P′ on the opposite side of the lens at *image distance s′*. If rays diverge from an object point P and interact with a lens such that the refracted rays *converge* at point P′, actually meeting at P′, then we call P′ a **real image** of point P. Contrast this with our prior definition of a *virtual image* as a point from which rays—which never meet—appear to *diverge*. For a real image, the image distance *s′* is *positive*.

FIGURE 34.30 Rays from an object point P are refracted by the lens and converge to a real image at point P′.

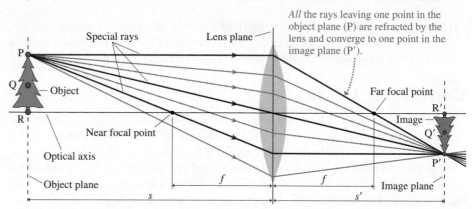

All points on the object that are in the same plane, the **object plane,** converge to image points in the **image plane.** Points Q and R in the object plane of Figure 34.30 have image points Q′ and R′ in the same plane as point P′. Once we locate *one* point in the image plane, such as P′, we know that the full image lies in the same plane.

There are two important observations to make about Figure 34.30. First, the image is upside down with respect to the object. This is called an **inverted image,** and it is a standard characteristic of real-image formation with a converging lens. Second, rays from point P *fill* the entire lens surface, and all portions of the lens contribute to the image. A larger lens will "collect" more rays and thus make a brighter image.

FIGURE 34.31 is a close-up view of the rays very near the image plane. The rays don't stop at P′ unless we place a screen in the image plane. When we do so, we see a sharp, well-focused image on the screen. To focus an image, you must either move the screen to coincide with the image plane or move the lens or object to make the image plane coincide with the screen. For example, the focus knob on a projector moves the lens forward or backward until the image plane matches the screen position.

NOTE The ability to see a real image on a screen sets real images apart from *virtual* images. But keep in mind that we need not *see* a real image in order to *have* an image. A real image exists at a point in space where the rays converge even if there's no viewing screen in the image plane.

FIGURE 34.31 A close-up look at the rays near the image plane.

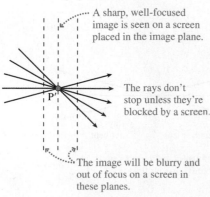

A sharp, well-focused image is seen on a screen placed in the image plane.

The rays don't stop unless they're blocked by a screen.

The image will be blurry and out of focus on a screen in these planes.

Figure 34.30 highlights three "special rays" based on the three situations of Figure 34.29. These three rays alone are sufficient to locate the image point P'. That is, we don't need to draw all the rays shown in Figure 34.30. The procedure known as *ray tracing* consists of locating the image by the use of just these three rays.

TACTICS BOX 34.2

Ray tracing for a converging lens

❶ **Draw an optical axis.** Use graph paper or a ruler! Establish an appropriate scale.

❷ **Center the lens on the axis.** Mark and label the focal points at distance f on either side.

❸ **Represent the object with an upright arrow at distance s.** It's usually best to place the base of the arrow on the axis and to draw the arrow about half the radius of the lens.

❹ **Draw the three "special rays" from the tip of the arrow.** Use a straightedge.
   a. A ray parallel to the axis refracts through the far focal point.
   b. A ray that enters the lens along a line through the near focal point emerges parallel to the axis.
   c. A ray through the center of the lens does not bend.

❺ **Extend the rays until they converge.** This is the image point. Draw the rest of the image in the image plane. If the base of the object is on the axis, then the base of the image will also be on the axis.

❻ **Measure the image distance s'.** Also, if needed, measure the image height relative to the object height.

Exercises 18–23 ✐

EXAMPLE 34.7 | Finding the image of a flower

A 4.0-cm-diameter flower is 200 cm from the 50-cm-focal-length lens of a camera. How far should the light detector be placed behind the lens to record a well-focused image? What is the diameter of the image on the detector?

MODEL The flower is in the object plane. Use ray tracing to locate the image.

VISUALIZE FIGURE 34.32 shows the ray-tracing diagram and the steps of Tactics Box 34.2. The image has been drawn in the plane where the three special rays converge. You can see *from the drawing* that the image distance is $s' \approx 67$ cm. This is where the detector needs to be placed to record a focused image.

The heights of the object and image are labeled h and h'. The ray through the center of the lens is a straight line, thus the object and image both subtend the same angle θ. Using similar triangles,

$$\frac{h'}{s'} = \frac{h}{s}$$

Solving for h' gives

$$h' = h\frac{s'}{s} = (4.0 \text{ cm})\frac{67 \text{ cm}}{200 \text{ cm}} = 1.3 \text{ cm}$$

The flower's image has a diameter of 1.3 cm.

ASSESS We've been able to learn a great deal about the image from a simple geometric procedure.

FIGURE 34.32 Ray-tracing diagram for Example 34.7.

Lateral Magnification

The image can be either larger or smaller than the object, depending on the location and focal length of the lens. But there's more to a description of the image than just its size. We also want to know its *orientation* relative to the object. That is, is the image upright or inverted? It is customary to combine size and orientation information into a single number. The **lateral magnification** m is defined as

$$m = -\frac{s'}{s} \tag{34.14}$$

You just saw in Example 34.7 that the image-to-object height ratio is $h'/h = s'/s$. Consequently, we interpret the lateral magnification m as follows:

1. A positive value of m indicates that the image is upright relative to the object. A negative value of m indicates that the image is inverted relative to the object.
2. The absolute value of m gives the size ratio of the image and object: $h'/h = |m|$.

The lateral magnification in Example 34.7 would be $m = -0.33$, indicating that the image is inverted and 33% the size of the object.

> **NOTE** The image-to-object height ratio is called *lateral* magnification to distinguish it from angular magnification, which we'll introduce in the next chapter. In practice, m is simply called "magnification" when there's no chance of confusion. Although we usually think that "to magnify" means "to make larger," in optics the magnification can be either > 1 (the image is larger than the object) or < 1 (the image is smaller than the object).

STOP TO THINK 34.4 A lens produces a sharply focused, inverted image on a screen. What will you see on the screen if the lens is removed?

a. The image will be inverted and blurry.
b. The image will be upright and sharp.
c. The image will be upright and blurry.
d. The image will be much dimmer but otherwise unchanged.
e. There will be no image at all.

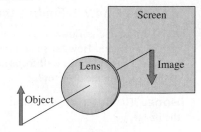

Virtual Images

The previous section considered a converging lens with the object at distance $s > f$. That is, the object was outside the focal point. What if the object is inside the focal point, at distance $s < f$? **FIGURE 34.33** shows just this situation, and we can use ray tracing to analyze it.

The special rays initially parallel to the axis and through the center of the lens present no difficulties. However, a ray through the near focal point would travel toward the left and would never reach the lens! Referring back to Figure 34.29b, you can see that the rays emerging parallel to the axis entered the lens *along a line* passing through the near focal point. It's the angle of incidence on the lens that is important, not whether the light ray actually passes through the focal point. This was the basis for the wording of step 4b in Tactics Box 34.2 and is the third special ray shown in Figure 34.33.

You can see that the three refracted rays don't converge. Instead, all three rays appear to *diverge* from point P'. This is the situation we found for rays reflecting from a mirror and for the rays refracting out of an aquarium. Point P' is a *virtual image* of

FIGURE 34.33 Rays from an object at distance $s < f$ are refracted by the lens and diverge to form a virtual image.

A ray *along a line* through the near focal point refracts parallel to the optical axis.

Virtual image — Object

Focal point

The refracted rays are diverging. They appear to come from point P'.

the object point P. Furthermore, it is an **upright image,** having the same orientation as the object.

The refracted rays, which are all to the right of the lens, *appear* to come from P′, but none of the rays were ever at that point. No image would appear on a screen placed in the image plane at P′. So what good is a virtual image?

Your eye collects and focuses bundles of diverging rays; thus, as **FIGURE 34.34a** shows, you can see a virtual image by looking *through* the lens. This is exactly what you do with a magnifying glass, producing a scene like the one in **FIGURE 34.34b**. In fact, you view a virtual image anytime you look *through* the eyepiece of an optical instrument such as a microscope or binoculars.

As before, **the image distance s′ for a virtual image is defined to be a** *negative number (s′ < 0),* indicating that the image is on the opposite side of the lens from a real image. With this choice of sign, the definition of magnification, $m = -s'/s$, is still valid. A virtual image with negative s' has $m > 0$, thus the image is upright. This agrees with the rays in Figure 34.33 and the photograph of Figure 34.34b.

> **NOTE** A lens thicker in the middle than at the edges is classified as a converging lens. The light rays from an object *can* converge to form a real image after passing through such a lens, but only if the object distance is larger than the focal length of the lens: $s > f$. If $s < f$, the rays diverge to produce a virtual image.

FIGURE 34.34 A converging lens is a magnifying glass when the object distance is less than f.

(a)

(b)

EXAMPLE 34.8 | **Magnifying a flower**

To see a flower better, a naturalist holds a 6.0-cm-focal-length magnifying glass 4.0 cm from the flower. What is the magnification?

MODEL The flower is in the object plane. Use ray tracing to locate the image.

VISUALIZE FIGURE 34.35 shows the ray-tracing diagram. The three special rays diverge from the lens, but we can use a straight-edge to extend the rays backward to the point from which they diverge. This point, the image point, is seen to be 12 cm to the left of the lens. Because this is a virtual image, the image distance is a negative $s' = -12$ cm. Thus the magnification is

$$m = -\frac{s'}{s} = -\frac{-12 \text{ cm}}{4.0 \text{ cm}} = 3.0$$

The image is three times as large as the object and, because m is positive, upright.

FIGURE 34.35 Ray-tracing diagram for Example 34.8.

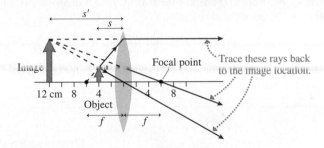

Diverging Lenses

A lens thicker at the edges than in the middle is called a *diverging lens*. **FIGURE 34.36** shows three important sets of rays passing through a diverging lens. These are based on Figures 34.27 and 34.28, where you saw that rays initially parallel to the axis diverge after passing through a diverging lens.

FIGURE 34.36 Three important sets of rays passing through a thin diverging lens.

Any ray initially parallel to the optical axis diverges along a line through the near focal point.

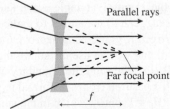

Any ray directed along a line toward the far focal point emerges from the lens parallel to the optical axis.

Any ray directed at the center of the lens passes through in a straight line.

Ray tracing follows the steps of Tactics Box 34.2 for a converging lens *except* that two of the three special rays in step 4 are different.

TACTICS BOX 34.3

Ray tracing for a diverging lens

❶–❸ **Follow steps 1 through 3 of Tactics Box 34.2.**
❹ **Draw the three "special rays" from the tip of the arrow.** Use a straightedge.
   a. A ray parallel to the axis diverges along a line through the near focal point.
   b. A ray along a line toward the far focal point emerges parallel to the axis.
   c. A ray through the center of the lens does not bend.
❺ **Trace the diverging rays backward.** The point from which they are diverging is the image point, which is always a virtual image.
❻ **Measure the image distance s'.** This will be a negative number.

Exercise 24

EXAMPLE 34.9 | Demagnifying a flower

A diverging lens with a focal length of 50 cm is placed 100 cm from a flower. Where is the image? What is its magnification?

MODEL The flower is in the object plane. Use ray tracing to locate the image.

VISUALIZE FIGURE 34.37 shows the ray-tracing diagram. The three special rays (labeled a, b, and c to match the Tactics Box) do not converge. However, they can be traced backward to an intersection ≈ 33 cm to the left of the lens. A virtual image is formed at $s' = -33$ cm with magnification

$$m = -\frac{s'}{s} = -\frac{-33 \text{ cm}}{100 \text{ cm}} = 0.33$$

The image, which can be seen by looking *through* the lens, is one-third the size of the object and upright.

FIGURE 34.37 Ray-tracing diagram for Example 34.9.

ASSESS Ray tracing with a diverging lens is somewhat trickier than with a converging lens, so this example is worth careful study.

Diverging lenses *always* make virtual images and, for this reason, are rarely used alone. However, they have important applications when used in combination with other lenses. Cameras, eyepieces, and eyeglasses often incorporate diverging lenses.

34.6 Thin Lenses: Refraction Theory

Ray tracing is a powerful visual approach for understanding image formation, but it doesn't provide precise information about the image. We need to develop a quantitative relationship between the object distance s and the image distance s'.

To begin, FIGURE 34.38 shows a *spherical* boundary between two transparent media with indices of refraction n_1 and n_2. The sphere has radius of curvature R. Consider a ray that leaves object point P at angle α and later, after refracting, reaches point P'. Figure 34.38 has exaggerated the angles to make the picture clear, but we will restrict our analysis to *paraxial rays* traveling nearly parallel to the axis. For paraxial rays, all the angles are small and we can use the small-angle approximation.

The ray from P is incident on the boundary at angle θ_1 and refracts into medium n_2 at angle θ_2, both measured from the normal to the surface at the point of incidence. Snell's law is $n_1 \sin\theta_1 = n_2 \sin\theta_2$, which in the small-angle approximation is

$$n_1\theta_1 = n_2\theta_2 \tag{34.15}$$

You can see from the geometry of Figure 34.38 that angles α, β, and ϕ are related by

$$\theta_1 = \alpha + \phi \quad \text{and} \quad \theta_2 = \phi - \beta \tag{34.16}$$

FIGURE 34.38 Image formation due to refraction at a spherical surface. The angles are exaggerated.

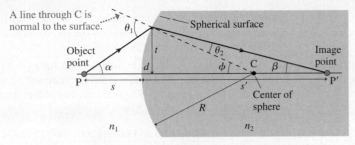

Using these expressions in Equation 34.15, we can write Snell's law as

$$n_1(\alpha + \phi) = n_2(\phi - \beta) \tag{34.17}$$

This is one important relationship between the angles.

The line of height t, from the axis to the point of incidence, is the vertical leg of three different right triangles having vertices at points P, C, and P'. Consequently,

$$\tan\alpha \approx \alpha = \frac{t}{s+d} \qquad \tan\beta \approx \beta = \frac{t}{s'-d} \qquad \tan\phi \approx \phi = \frac{t}{R-d} \tag{34.18}$$

But $d \to 0$ for paraxial rays, thus

$$\alpha = \frac{t}{s} \qquad \beta = \frac{t}{s'} \qquad \phi = \frac{t}{R} \tag{34.19}$$

This is the second important relationship that comes from Figure 34.38.

If we use Equation 34.19 in Equation 34.17, the t cancels and we find

$$\frac{n_1}{s} + \frac{n_2}{s'} = \frac{n_2 - n_1}{R} \tag{34.20}$$

Equation 34.20 is independent of angle α. Consequently, **all paraxial rays leaving point P later converge at point P'.** If an object is located at distance s from a spherical refracting surface, an image will be formed at distance s' given by Equation 34.20.

Equation 34.20 was derived for a surface that is convex toward the object point, and the image is real. However, the result is also valid for virtual images or for surfaces that are concave toward the object point as long as we adopt the *sign convention* shown in **TABLE 34.2.**

Section 34.4 considered image formation due to refraction by a plane surface. There we found (in Equation 34.13) an image distance $s' = -(n_2/n_1)s$. A plane can be thought of as a sphere in the limit $R \to \infty$, so we should be able to reach the same conclusion from Equation 34.20. Indeed, as $R \to \infty$, the term $(n_2 - n_1)/R \to 0$ and Equation 34.20 becomes $s' = -(n_2/n_1)s$.

TABLE 34.2 Sign convention for refracting surfaces

	Positive	Negative
R	Convex toward the object	Concave toward the object
s'	Real image, opposite side from object	Virtual image, same side as object

---

EXAMPLE 34.10 | Image formation inside a glass rod

One end of a 4.0-cm-diameter glass rod is shaped like a hemisphere. A small lightbulb is 6.0 cm from the end of the rod. Where is the bulb's image located?

MODEL Model the lightbulb as a point source of light and consider the paraxial rays that refract into the glass rod.

FIGURE 34.39 The curved surface refracts the light to form a real image.

VISUALIZE FIGURE 34.39 shows the situation. $n_1 = 1.00$ for air and $n_2 = 1.50$ for glass.

SOLVE The radius of the surface is half the rod diameter, so $R = 2.0$ cm. Equation 34.20 is

$$\frac{1.00}{6.0\text{ cm}} + \frac{1.50}{s'} = \frac{1.50 - 1.00}{2.0\text{ cm}} = \frac{0.50}{2.0\text{ cm}}$$

Solving for the image distance s' gives

$$\frac{1.50}{s'} = \frac{0.50}{2.0\text{ cm}} - \frac{1.00}{6.0\text{ cm}} = 0.0833\text{ cm}^{-1}$$

$$s' = \frac{1.50}{0.0833} = 18\text{ cm}$$

ASSESS This is a real image located 18 cm inside the glass rod.

EXAMPLE 34.11 | A goldfish in a bowl

A goldfish lives in a spherical fish bowl 50 cm in diameter. If the fish is 10 cm from the near edge of the bowl, where does the fish appear when viewed from the outside?

MODEL Model the fish as a point source and consider the paraxial rays that refract from the water into the air. The thin glass wall has little effect and will be ignored.

FIGURE 34.40 The curved surface of a fish bowl produces a virtual image of the fish.

VISUALIZE FIGURE 34.40 shows the rays refracting *away* from the normal as they move from the water into the air. We expect to find a virtual image at a distance less than 10 cm.

SOLVE The object is in the water, so $n_1 = 1.33$ and $n_2 = 1.00$. The inner surface is concave (you can remember "concave" because it's like looking into a cave), so $R = -25$ cm. The object distance is $s = 10$ cm. Thus Equation 34.20 is

$$\frac{1.33}{10 \text{ cm}} + \frac{1.00}{s'} = \frac{1.00 - 1.33}{-25 \text{ cm}} = \frac{0.33}{25 \text{ cm}}$$

Solving for the image distance s' gives

$$\frac{1.00}{s'} = \frac{0.33}{25 \text{ cm}} - \frac{1.33}{10 \text{ cm}} = -0.12 \text{ cm}^{-1}$$

$$s' = \frac{1.00}{-0.12 \text{ cm}^{-1}} = -8.3 \text{ cm}$$

ASSESS The image is virtual, located to the left of the boundary. A person looking into the bowl will see a fish that appears to be 8.3 cm from the edge of the bowl.

STOP TO THINK 34.5 Which of these actions will move the real image point P′ farther from the boundary? More than one may work.

a. Increase the radius of curvature R.
b. Increase the index of refraction n.
c. Increase the object distance s.
d. Decrease the radius of curvature R.
e. Decrease the index of refraction n.
f. Decrease the object distance s.

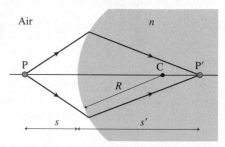

Lenses

The thin-lens approximation assumes rays refract one time, at the lens plane. In fact, as FIGURE 34.41 shows, rays refract *twice,* at spherical surfaces having radii of curvature R_1 and R_2. Let the lens have thickness t and be made of a material with index of refraction n. For simplicity, we'll assume that the lens is surrounded by air.

FIGURE 34.41 Image formation by a lens.

The object at point P is distance s_1 to the left of the lens. The first surface of the lens, of radius R_1, refracts the rays from P to create an image at point P'. We can use Equation 34.20 for a spherical surface to find the image distance s_1':

$$\frac{1}{s_1} + \frac{n}{s_1'} = \frac{n-1}{R_1} \tag{34.21}$$

where we used $n_1 = 1$ for the air and $n_2 = n$ for the lens. We'll assume that the image P' is a virtual image, but this assumption isn't essential to the outcome.

With two refracting surfaces, the image P' of the first surface becomes the object for the second surface. That is, the rays refracting at the second surface appear to have come from P'. Object distance s_2 from P' to the second surface looks like it should be $s_2 = s_1' + t$, but P' is a virtual image, so s_1' is a *negative* number. Thus the distance to the second surface is $s_2 = |s_1'| + t = t - s_1'$. We can find the image of P' by a second application of Equation 34.20, but now the rays are incident on the surface from within the lens, so $n_1 = n$ and $n_2 = 1$. Consequently,

$$\frac{n}{t - s_1'} + \frac{1}{s_2'} = \frac{1-n}{R_2} \tag{34.22}$$

For a *thin lens,* which has $t \rightarrow 0$, Equation 34.22 becomes

$$-\frac{n}{s_1'} + \frac{1}{s_2'} = \frac{1-n}{R_2} = -\frac{n-1}{R_2} \tag{34.23}$$

Our goal is to find the distance s_2' to point P'', the image produced by the lens as a whole. This goal is easily reached if we simply add Equations 34.21 and 34.23, eliminating s_1' and giving

$$\frac{1}{s_1} + \frac{1}{s_2'} = \frac{n-1}{R_1} - \frac{n-1}{R_2} = (n-1)\left(\frac{1}{R_1} - \frac{1}{R_2}\right) \tag{34.24}$$

The numerical subscripts on s_1 and s_2' no longer serve a purpose. If we replace s_1 by s, the object distance from the lens, and s_2' by s', the image distance, Equation 34.24 becomes the **thin-lens equation:**

$$\frac{1}{s} + \frac{1}{s'} = \frac{1}{f} \qquad \text{(thin-lens equation)} \tag{34.25}$$

where the *focal length* of the lens is

$$\frac{1}{f} = (n-1)\left(\frac{1}{R_1} - \frac{1}{R_2}\right) \qquad \text{(lens maker's equation)} \tag{34.26}$$

Equation 34.26 is known as the **lens maker's equation.** It allows you to determine the focal length from the shape of a thin lens and the material used to make it.

We can verify that this expression for f really is the focal length of the lens by recalling that rays initially parallel to the optical axis pass through the focal point on the far side. In fact, this was our *definition* of the focal length of a lens. Parallel rays must come from an object extremely far away, with object distance $s \rightarrow \infty$ and thus $1/s = 0$. In that case, Equation 34.25 tells us that the parallel rays will converge at distance $s' = f$ on the far side of the lens, exactly as expected.

We derived the thin-lens equation and the lens maker's equation from the specific lens geometry shown in Figure 34.41, but the results are valid for any lens as long as all quantities are given appropriate signs. The sign convention used with Equations 34.25 and 34.26 is given in TABLE 34.3.

TABLE 34.3 Sign convention for thin lenses

	Positive	Negative
R_1, R_2	Convex toward the object	Concave toward the object
f	Converging lens, thicker in center	Diverging lens, thinner in center
s'	Real image, opposite side from object	Virtual image, same side as object

NOTE For a *thick lens,* where the thickness t is not negligible, we can solve Equations 34.21 and 34.22 in sequence to find the position of the image point P''.

EXAMPLE 34.12 | Focal length of a meniscus lens

What is the focal length of the glass *meniscus lens* shown in FIGURE 34.42? Is this a converging or diverging lens?

FIGURE 34.42 A meniscus lens.

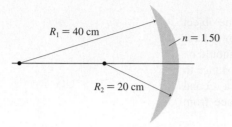

$R_1 = 40$ cm

$n = 1.50$

$R_2 = 20$ cm

SOLVE If the object is on the left, then the first surface has $R_1 = -40$ cm (concave toward the object) and the second surface has $R_2 = -20$ cm (also concave toward the object). The index of refraction of glass is $n = 1.50$, so the lens maker's equation is

$$\frac{1}{f} = (n-1)\left(\frac{1}{R_1} - \frac{1}{R_2}\right) = (1.50 - 1)\left(\frac{1}{-40 \text{ cm}} - \frac{1}{-20 \text{ cm}}\right)$$

$$= 0.0125 \text{ cm}^{-1}$$

Inverting this expression gives $f = 80$ cm. This is a converging lens, as seen both from the positive value of f and from the fact that the lens is thicker in the center.

Thin-Lens Image Formation

Although the thin-lens equation allows precise calculations, the lessons of ray tracing should not be forgotten. The most powerful tool of optical analysis is a combination of ray tracing, to gain an intuitive understanding of the ray trajectories, and the thin-lens equation.

EXAMPLE 34.13 | Designing a lens

The objective lens of a microscope uses a planoconvex glass lens with the flat side facing the specimen. A real image is formed 160 mm behind the lens when the lens is 8.0 mm from the specimen. What is the radius of the lens's curved surface?

MODEL Treat the lens as a thin lens with the specimen as the object. The lens's focal length is given by the lens maker's equation.

VISUALIZE FIGURE 34.43 clarifies the shape of the lens and defines R_2. The index of refraction was taken from Table 34.1.

FIGURE 34.43 A planoconvex microscope lens.

Image and object distances not to scale

$n = 1.50$

$s' = 160$ mm

$R_1 = \infty$

R_2

$s = 8.0$ mm

SOLVE We can use the lens maker's equation to solve for R_2 if we know the lens's focal length. Because we know both the object and image distances, we can use the thin-lens equation to find

$$\frac{1}{f} = \frac{1}{s} + \frac{1}{s'} = \frac{1}{8.0 \text{ mm}} + \frac{1}{160 \text{ mm}} = 0.131 \text{ mm}^{-1}$$

The focal length is $f = 1/(0.131 \text{ mm}^{-1}) = 7.6$ mm, but $1/f$ is all we need for the lens maker's equation. The front surface of the lens is planar, which we can consider a portion of a sphere with $R_1 \rightarrow \infty$. Consequently $1/R_1 = 0$. With this, we can solve the lens maker's equation for R_2:

$$\frac{1}{R_2} = \frac{1}{R_1} - \frac{1}{n-1}\frac{1}{f} = 0 - \left(\frac{1}{1.50 - 1}\right)(0.131 \text{ mm}^{-1})$$

$$= -0.262 \text{ mm}^{-1}$$

$$R_2 = -3.8 \text{ mm}$$

The minus sign appears because the curved surface is concave toward the object. Physically, the radius of the curved surface is 3.8 mm.

ASSESS The actual thickness of the lens has to be less than R_2, probably no more than about 1.0 mm. This thickness is significantly less than the object and image distances, so the thin-lens approximation is justified.

EXAMPLE 34.14 | A magnifying lens

A stamp collector uses a magnifying lens that sits 2.0 cm above the stamp. The magnification is 4.0. What is the focal length of the lens?

FIGURE 34.44 Pictorial representation of a magnifying lens.

MODEL A magnifying lens is a converging lens with the object distance less than the focal length $(s < f)$. Assume it is a thin lens.

VISUALIZE FIGURE 34.44 shows the lens and a ray-tracing diagram. We do not need to know the actual shape of the lens, so the figure shows a generic converging lens.

SOLVE A virtual image is upright, so $m = +4.0$. The magnification is $m = -s'/s$, thus

$$s' = -4.0s = -(4.0)(2.0 \text{ cm}) = -8.0 \text{ cm}$$

We can use s and s' in the thin-lens equation to find the focal length:

$$\frac{1}{f} = \frac{1}{s} + \frac{1}{s'} = \frac{1}{2.0 \text{ cm}} + \frac{1}{-8.0 \text{ cm}} = 0.375 \text{ cm}^{-1}$$

$$f = 2.7 \text{ cm}$$

ASSESS $f > 2$ cm, as expected.

STOP TO THINK 34.6 A lens forms a real image of a lightbulb, but the image of the bulb on a viewing screen is blurry because the screen is slightly in front of the image plane. To focus the image, should you move the lens toward the bulb or away from the bulb?

34.7 Image Formation with Spherical Mirrors

Curved mirrors—such as those used in telescopes, security and rearview mirrors, and searchlights—can be used to form images, and their images can be analyzed with ray diagrams similar to those used with lenses. We'll consider only the important case of **spherical mirrors,** whose surface is a section of a sphere.

Concave Mirrors

FIGURE 34.45 shows a **concave mirror,** a mirror in which the edges curve *toward* the light source. Rays parallel to the optical axis reflect from the surface of the mirror so as to pass through a single point on the optical axis. This is the focal point of the mirror. The focal length is the distance from the mirror surface to the focal point. A concave mirror is analogous to a converging lens, but it has only one focal point.

Let's begin by considering the case where the object's distance s from the mirror is greater than the focal length $(s > f)$, as shown in **FIGURE 34.46** on the next page. We see that the image is *real* (and inverted) because rays from the object point P converge at the image point P'. Although an infinite number of rays from P all meet at P', each ray obeying the law of reflection, you can see that three "special rays" are enough to determine the position and size of the image:

- A ray parallel to the axis reflects through the focal point.
- A ray through the focal point reflects parallel to the axis.
- A ray striking the center of the mirror reflects at an equal angle on the opposite side of the axis.

These three rays also locate the image if $s < f$, but in that case the image is *virtual* and behind the mirror. Once again, virtual images have a *negative* image distance s'.

FIGURE 34.45 The focal point and focal length of a concave mirror.

FIGURE 34.46 A real image formed by a concave mirror.

NOTE A thin lens has negligible thickness and all refraction occurs at the lens plane. Similarly, we will assume that mirrors are thin (even though drawings may show a thickness) and thus *all reflection occurs at the mirror plane.*

FIGURE 34.47 The focal point and focal length of a convex mirror.

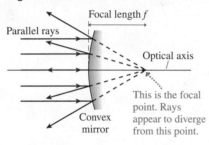

FIGURE 34.48 A city skyline is reflected in this polished sphere.

Convex Mirrors

FIGURE 34.47 shows parallel light rays approaching a mirror in which the edges curve *away from* the light source. This is called a **convex mirror.** In this case, the reflected rays appear to come from a point behind the mirror. This is the focal point for a convex mirror.

A common example of a convex mirror is a silvered ball, such as a tree ornament. You may have noticed that if you look at your reflection in such a ball, your image appears right-side-up but is quite small. As another example, **FIGURE 34.48** shows a city skyline reflected in a polished metal sphere. Let's use ray tracing to understand why the skyscrapers all appear to be so small.

FIGURE 34.49 shows an object in front of a convex mirror. In this case, the reflected rays—each obeying the law of reflection—create an upright image of reduced height behind the mirror. We see that the image is virtual because no rays actually converge at the image point P′. Instead, diverging rays *appear* to come from this point. Once again, three special rays are enough to find the image.

Convex mirrors are used for a variety of safety and monitoring applications, such as passenger-side rearview mirrors and the round mirrors used in stores to keep an eye

FIGURE 34.49 A virtual image formed by a convex mirror.

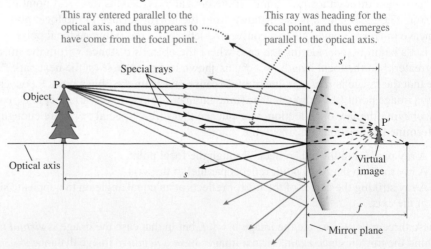

on the customers. When an object is reflected in a convex mirror, the image appears smaller than the object itself. Because the image is, in a sense, a miniature version of the object, you can *see much more of it* within the edges of the mirror than you could with an equal-sized flat mirror.

TACTICS BOX 34.4

Ray tracing for a spherical mirror

❶ **Draw an optical axis.** Use graph paper or a ruler! Establish a scale.

❷ **Center the mirror on the axis.** Mark and label the focal point at distance *f* from the mirror's surface.

❸ **Represent the object with an upright arrow at distance *s*.** It's usually best to place the base of the arrow on the axis and to draw the arrow about half the radius of the mirror.

❹ **Draw the three "special rays" from the tip of the arrow.** All reflections occur at the mirror plane.
   a. A ray parallel to the axis reflects through (concave) or away from (convex) the focal point.
   b. An incoming ray passing through (concave) or heading toward (convex) the focal point reflects parallel to the axis.
   c. A ray that strikes the center of the mirror reflects at an equal angle on the opposite side of the optical axis.

❺ **Extend the rays forward or backward until they converge.** This is the image point. Draw the rest of the image in the image plane. If the base of the object is on the axis, then the base of the image will also be on the axis.

❻ **Measure the image distance *s'*.** Also, if needed, measure the image height relative to the object height.

Exercises 28–29

EXAMPLE 34.15 | Analyzing a concave mirror

A 3.0-cm-high object is located 60 cm from a concave mirror. The mirror's focal length is 40 cm. Use ray tracing to find the position and height of the image.

MODEL Use the ray-tracing steps of Tactics Box 34.4.

VISUALIZE FIGURE 34.50 shows the steps of Tactics Box 34.4.

SOLVE We can use a ruler to find that the image position is $s' \approx 120$ cm in front of the mirror and its height is $h' \approx 6$ cm.

ASSESS The image is a *real* image because light rays converge at the image point.

FIGURE 34.50 Ray-tracing diagram for a concave mirror.

❶ Lay out the optical axis, with a scale.

❸ Draw the object as an arrow with its base on the axis.

❹ Draw the 3 special rays from the tip of the arrow.
   a. Parallel to the axis.
   b. Through the focal point.
   c. Hitting the center of the mirror.

❷ Draw the mirror and mark its focal point.

h

10 cm

h'

f = 40 cm

s = 60 cm

Mirror plane

❺ The convergence point is the tip of the image. Draw the rest of the image.

❻ Measure the image distance.

s'

The Mirror Equation

The thin-lens equation assumes lenses have negligible thickness (so a single refraction occurs in the lens plane) and the rays are nearly parallel to the optical axis (paraxial rays). If we make the same assumptions about spherical mirrors—the mirror has negligible thickness and so paraxial rays reflect at the mirror plane—then the object and image distances are related exactly as they were for thin lenses:

$$\frac{1}{s} + \frac{1}{s'} = \frac{1}{f} \qquad \text{(mirror equation)} \qquad (34.27)$$

The focal length of the mirror, as you can show as a homework problem, is related to the mirror's radius of curvature by

$$f = \frac{R}{2} \qquad (34.28)$$

TABLE 34.4 shows the sign convention used with spherical mirrors. It differs from the convention for lenses, so you'll want to carefully compare this table to Table 34.3. A concave mirror (analogous to a converging lens) has a positive focal length while a convex mirror (analogous to a diverging lens) has a negative focal length. The lateral magnification of a spherical mirror is computed exactly as for a lens:

$$m = -\frac{s'}{s} \qquad (34.29)$$

TABLE 34.4 Sign convention for spherical mirrors

	Positive	Negative
R, f	Concave toward the object	Convex toward the object
s'	Real image, same side as object	Virtual image, opposite side from object

---

EXAMPLE 34.16 | Analyzing a concave mirror

A 3.0-cm-high object is located 20 cm from a concave mirror. The mirror's radius of curvature is 80 cm. Determine the position, orientation, and height of the image.

MODEL Treat the mirror as a thin mirror.

VISUALIZE The mirror's focal length is $f = R/2 = +40$ cm, where we used the sign convention from Table 34.4. With the focal length known, the three special rays in FIGURE 34.51 show that the image is a magnified, virtual image behind the mirror.

FIGURE 34.51 Pictorial representation of Example 34.16.

SOLVE The thin-mirror equation is

$$\frac{1}{20\text{ cm}} + \frac{1}{s'} = \frac{1}{40\text{ cm}}$$

This is easily solved to give $s' = -40$ cm, in agreement with the ray tracing. The negative sign tells us this is a virtual image behind the mirror. The magnification is

$$m = -\frac{-40\text{ cm}}{20\text{ cm}} = +2.0$$

Consequently, the image is 6.0 cm tall and upright.

ASSESS This is a virtual image because light rays diverge from the image point. You could see this enlarged image by standing behind the object and looking into the mirror. In fact, this is how magnifying cosmetic mirrors work.

STOP TO THINK 34.7 A concave mirror of focal length f forms an image of the moon. Where is the image located?

a. At the mirror's surface
b. Almost exactly a distance f behind the mirror
c. Almost exactly a distance f in front of the mirror
d. At a distance behind the mirror equal to the distance of the moon in front of the mirror

CHALLENGE EXAMPLE 34.17 | Optical fiber imaging

An *endoscope* is a thin bundle of optical fibers that can be inserted through a bodily opening or small incision to view the interior of the body. As **FIGURE 34.52** shows, an *objective* lens forms a real image on the entrance face of the fiber bundle. Individual fibers, using total internal reflection, transport the light to the exit face, where it emerges. The doctor (or a TV camera) observes the object by viewing the exit face through an *eyepiece* lens.

FIGURE 34.52 An endoscope.

Consider an endoscope having a 3.0-mm-diameter objective lens with a focal length of 1.1 mm. These are typical values. The indices of refraction of the core and the cladding of the optical fibers are 1.62 and 1.50, respectively. To give maximum brightness, the objective lens is positioned so that, for an on-axis object, rays passing through the outer edge of the lens have the maximum angle of incidence for undergoing TIR in the fiber. How far should the objective lens be placed from the object the doctor wishes to view?

MODEL Represent the object as an on-axis point source and use the ray model of light.

VISUALIZE FIGURE 34.53 shows the real image being focused on the entrance face of the endoscope. Inside the fiber, rays that strike the cladding at an angle of incidence greater than the critical angle θ_c undergo TIR and stay in the fiber; rays are lost if their angle of incidence is less than θ_c. For maximum brightness, the lens is positioned so that a ray passing through the outer edge refracts into the fiber at the maximum angle of incidence θ_{max} for which TIR is possible. A smaller-diameter lens would sacrifice light-gathering power, whereas the outer rays from a larger-diameter lens would impinge on the core-cladding boundary at less than θ_c and would not undergo TIR.

FIGURE 34.53 Magnified view of the entrance of an optical fiber.

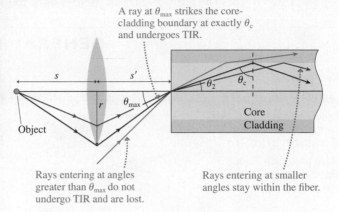

SOLVE We know the focal length of the lens. We can use the geometry of the ray at the critical angle to find the image distance s', then use the thin-lens equation to find the object distance s. The critical angle for TIR inside the fiber is

$$\theta_c = \sin^{-1}\left(\frac{n_{cladding}}{n_{core}}\right) = \sin^{-1}\left(\frac{1.50}{1.62}\right) = 67.8°$$

A ray incident on the core-cladding boundary at exactly the critical angle must have entered the fiber, at the entrance face, at angle $\theta_2 = 90° - \theta_c = 22.2°$. For optimum lens placement, this ray passed through the outer edge of the lens and was incident on the entrance face at angle θ_{max}. Snell's law at the entrance face is

$$n_{air}\sin\theta_{max} = 1.00\sin\theta_{max} = n_{core}\sin\theta_2$$

and thus

$$\theta_{max} = \sin^{-1}(1.62\sin 22.2°) = 37.7°$$

We know the lens radius, $r = 1.5\,\text{mm}$, so the distance of the lens from the fiber—the image distance s'—is

$$s' = \frac{r}{\tan\theta_{max}} = \frac{1.5\ \text{mm}}{\tan(37.7°)} = 1.9\ \text{mm}$$

Now we can use the thin-lens equation to locate the object:

$$\frac{1}{s} = \frac{1}{f} - \frac{1}{s'} = \frac{1}{1.1\ \text{mm}} - \frac{1}{1.9\ \text{mm}}$$

$$s = 2.6\ \text{mm}$$

The doctor, viewing the exit face of the fiber bundle, will see a focused image when the objective lens is 2.6 mm from the object she wishes to view.

ASSESS The object and image distances are both greater than the focal length, which is correct for forming a real image.

SUMMARY

The goals of Chapter 34 have been to learn about and apply the ray model of light.

GENERAL PRINCIPLES

Reflection

Law of reflection: $\theta_r = \theta_i$

Reflection can be **specular** (mirror-like) or **diffuse** (from rough surfaces).

Plane mirrors: A virtual image is formed at P′ with $s' = s$.

Refraction

Snell's law of refraction:

$$n_1 \sin \theta_1 = n_2 \sin \theta_2$$

Index of refraction is $n = c/v$. The ray is closer to the normal on the side with the larger index of refraction.

If $n_2 < n_1$, **total internal reflection** (TIR) occurs when the angle of incidence $\theta_1 \geq \theta_c = \sin^{-1}(n_2/n_1)$.

IMPORTANT CONCEPTS

The ray model of light

Light travels along straight lines, called **light rays,** at speed $v = c/n$.

A light ray continues forever unless an interaction with matter causes it to reflect, refract, scatter, or be absorbed.

Light rays come from **objects.** Each point on the object sends rays in all directions.

The eye sees an object (or an image) when diverging rays are collected by the pupil and focused on the retina.

▶ Ray optics is valid when lenses, mirrors, and apertures are larger than ≈ 1 mm.

Image formation

If rays diverge from P and interact with a lens or mirror so that the refracted/reflected rays *converge* at P′, then P′ is a **real image** of P.

If rays diverge from P and interact with a lens or mirror so that the refracted/reflected rays *diverge* from P′ and appear to come from P′, then P′ is a **virtual image** of P.

Spherical surface: Object and image distances are related by

$$\frac{n_1}{s} + \frac{n_2}{s'} = \frac{n_2 - n_1}{R}$$

Plane surface: $R \to \infty$, so $s' = -(n_2/n_1)s$.

APPLICATIONS

Ray tracing

3 special rays in 3 basic situations:

Converging lens   Converging lens   Diverging lens
Real image   Virtual image   Virtual image

Magnification $m = -\dfrac{s'}{s}$

m is + for an upright image, − for inverted.
The height ratio is $h'/h = |m|$.

Thin lenses

The image and object distances are related by

$$\frac{1}{s} + \frac{1}{s'} = \frac{1}{f}$$

where the focal length is given by the lens maker's equation:

$$\frac{1}{f} = (n - 1)\left(\frac{1}{R_1} - \frac{1}{R_2}\right)$$

R	+ for surface convex toward object	− for concave
f	+ for a converging lens	− for diverging
s'	+ for a real image	− for virtual

Spherical mirrors

The image and object distances are related by

$$\frac{1}{s} + \frac{1}{s'} = \frac{1}{f}$$

| R, f | + for concave mirror | − for convex |
| s' | + for a real image | − for virtual |

Focal length $f = R/2$

TERMS AND NOTATION

light ray	diffuse reflection	lens	inverted image
object	virtual image	ray tracing	lateral magnification, m
point source	refraction	converging lens	upright image
parallel bundle	angle of refraction	focal point	thin-lens equation
ray diagram	Snell's law	focal length, f	lens maker's equation
camera obscura	total internal reflection (TIR)	diverging lens	spherical mirror
aperture	critical angle, θ_c	thin lens	concave mirror
specular reflection	object distance, s	lens plane	convex mirror
angle of incidence	image distance, s'	real image	
angle of reflection	optical axis	object plane	
law of reflection	paraxial rays	image plane	

CONCEPTUAL QUESTIONS

1. Suppose you have two pinhole cameras. The first has a small round hole in the front. The second is identical except it has a square hole of the same area as the round hole in the first camera. Would the pictures taken by these two cameras, under the same conditions, be different in any obvious way? Explain.

2. You are looking at the image of a pencil in a mirror, as shown in **FIGURE Q34.2**.
   a. What happens to the image if the top half of the mirror, down to the midpoint, is covered with a piece of cardboard? Explain.
   b. What happens to the image if the bottom half of the mirror is covered with a piece of cardboard? Explain.

Midpoint

FIGURE Q34.2

3. One problem with using optical fibers for communication is that a light ray passing directly down the center of the fiber takes less time to travel from one end to the other than a ray taking a longer, zig-zag path. Thus light rays starting at the same time but traveling in slightly different directions reach the end of the fiber at different times. This problem can be solved by making the refractive index of the glass change gradually from a higher value in the center to a lower value near the edges of the fiber. Explain how this reduces the difference in travel times.

4. A light beam passing from medium 2 to medium 1 is refracted as shown in **FIGURE Q34.4**. Is n_1 larger than n_2, is n_1 smaller than n_2, or is there not enough information to tell? Explain.

FIGURE Q34.4

5. A fish in an aquarium with flat sides looks out at a hungry cat. To the fish, does the distance to the cat appear to be less than the actual distance, the same as the actual distance, or more than the actual distance? Explain.

6. Consider *one* point on an object near a lens.
   a. What is the minimum number of rays needed to locate its image point? Explain.
   b. How many rays from this point actually strike the lens and refract to the image point?

7. The object and lens in **FIGURE Q34.7** are positioned to form a well focused, inverted image on a viewing screen. Then a piece of cardboard is lowered just in front of the lens to cover the top half of the lens. Describe what you see on the screen when the cardboard is in place.

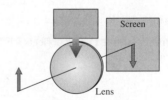

Screen

FIGURE Q34.7   Lens

8. A converging lens creates the image shown in **FIGURE Q34.8**. Is the object distance less than the focal length f, between f and $2f$, or greater than $2f$? Explain.

FIGURE Q34.8

9. A concave mirror brings the sun's rays to a focus in front of the mirror. Suppose the mirror is submerged in a swimming pool but still pointed up at the sun. Will the sun's rays be focused nearer to, farther from, or at the same distance from the mirror? Explain.

10. You see an upright, magnified image of your face when you look into a magnifying cosmetic mirror. Where is the image? Is it in front of the mirror's surface, on the mirror's surface, or behind the mirror's surface? Explain.

11. When you look at your reflection in the bowl of a spoon, it is upside down. Why?

EXERCISES AND PROBLEMS

Problems labeled ▨ integrate material from earlier chapters.

Exercises

Section 34.1 The Ray Model of Light

1. ‖ A point source of light illuminates an aperture 2.0 m away. A 12.0-cm-wide bright patch of light appears on a screen 1.0 m behind the aperture. How wide is the aperture?

2. ‖ a. How long (in ns) does it take light to travel 1.0 m in vacuum?
   b. What distance does light travel in water, glass, and cubic zirconia during the time that it travels 1.0 m in vacuum?

3. | A student has built a 15-cm-long pinhole camera for a science fair project. She wants to photograph her 180-cm-tall friend and have the image on the film be 5.0 cm high. How far should the front of the camera be from her friend?

4. ‖ A 5.0-cm-thick layer of oil is sandwiched between a 1.0-cm-thick sheet of glass and a 2.0-cm-thick sheet of polystyrene plastic. How long (in ns) does it take light incident perpendicular to the glass to pass through this 8.0-cm-thick sandwich?

Section 34.2 Reflection

5. | A light ray leaves point A in **FIGURE EX34.5**, reflects from the mirror, and reaches point B. How far below the top edge does the ray strike the mirror?

FIGURE EX34.5

6. | The mirror in **FIGURE EX34.6** deflects a horizontal laser beam by 60°. What is the angle ϕ?

FIGURE EX34.6

7. | At what angle ϕ should the laser beam in **FIGURE EX34.7** be aimed at the mirrored ceiling in order to hit the midpoint of the far wall?

FIGURE EX34.7

8. | A laser beam is incident on the left mirror in **FIGURE EX34.8**. Its initial direction is parallel to a line that bisects the mirrors. What is the angle ϕ of the reflected laser beam?

FIGURE EX34.8

9. | It is 165 cm from your eyes to your toes. You're standing 200 cm in front of a tall mirror. How far is it from your eyes to the image of your toes?

Section 34.3 Refraction

10. | A laser beam in air is incident on a liquid at an angle of 53° with respect to the normal. The laser beam's angle in the liquid is 35°. What is the liquid's index of refraction?

11. ‖ A 1.0-cm-thick layer of water stands on a horizontal slab of glass. A light ray in the air is incident on the water 60° from the normal. What is the ray's direction of travel in the glass?

12. ‖ A costume jewelry pendant made of cubic zirconia is submerged in oil. A light ray strikes one face of the zirconia crystal at an angle of incidence of 25°. Once inside, what is the ray's angle with respect to the face of the crystal?

13. ‖ An underwater diver sees the sun 50° above horizontal. How high is the sun above the horizon to a fisherman in a boat above the diver?

14. ‖ The glass core of an optical fiber has an index of refraction 1.60. The index of refraction of the cladding is 1.48. What is the maximum angle a light ray can make with the wall of the core if it is to remain inside the fiber?

15. ‖ A thin glass rod is submerged in oil. What is the critical angle for light traveling inside the rod?

16. ‖ **FIGURE EX34.16** shows a transparent hemisphere with radius R and index of refraction n. What is the maximum distance d for which a light ray parallel to the axis refracts out through the curved surface?

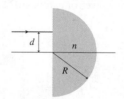

FIGURE EX34.16

Section 34.4 Image Formation by Refraction at a Plane Surface

17. | A fish in a flat-sided aquarium sees a can of fish food on the counter. To the fish's eye, the can looks to be 30 cm outside the aquarium. What is the actual distance between the can and the aquarium? (You can ignore the thin glass wall of the aquarium.)

18. | A biologist keeps a specimen of his favorite beetle embedded in a cube of polystyrene plastic. The hapless bug appears to be 2.0 cm within the plastic. What is the beetle's actual distance beneath the surface?

19. | A 150-cm-tall diver is standing completely submerged on the bottom of a swimming pool full of water. You are sitting on the end of the diving board, almost directly over her. How tall does the diver appear to be?

20. ‖ To a fish in an aquarium, the 4.00-mm-thick walls appear to be only 3.50 mm thick. What is the index of refraction of the walls?

Section 34.5 Thin Lenses: Ray Tracing

21. ‖ An object is 20 cm in front of a converging lens with a focal length of 10 cm. Use ray tracing to determine the location of the image. Is the image upright or inverted?

22. ‖ An object is 30 cm in front of a converging lens with a focal length of 5 cm. Use ray tracing to determine the location of the image. Is the image upright or inverted?

23. ‖ An object is 6 cm in front of a converging lens with a focal length of 10 cm. Use ray tracing to determine the location of the image. Is the image upright or inverted?

24. ‖ An object is 15 cm in front of a diverging lens with a focal length of −15 cm. Use ray tracing to determine the location of the image. Is the image upright or inverted?

Section 34.6 Thin Lenses: Refraction Theory

25. ‖ Find the focal length of the glass lens in **FIGURE EX34.25**.

FIGURE EX34.25   FIGURE EX34.26

26. ‖ Find the focal length of the planoconvex polystyrene plastic lens in **FIGURE EX34.26**.

27. ‖ Find the focal length of the glass lens in **FIGURE EX34.27**.

FIGURE EX34.27   FIGURE EX34.28

28. ‖ Find the focal length of the meniscus polystyrene plastic lens in **FIGURE EX34.28**.

29. ‖ An air bubble inside an 8.0-cm-diameter plastic ball is 2.0 cm from the surface. As you look at the ball with the bubble turned toward you, how far beneath the surface does the bubble appear to be?

30. ‖ A goldfish lives in a 50-cm-diameter spherical fish bowl. The fish sees a cat watching it. If the cat's face is 20 cm from the edge of the bowl, how far from the edge does the fish see it as being? (You can ignore the thin glass wall of the bowl.)

31. ‖ A 1.0-cm-tall candle flame is 60 cm from a lens with a focal length of 20 cm. What are the image distance and the height of the flame's image?

32. ‖ A 1.0-cm-tall object is 10 cm in front of a converging lens that has a 30 cm focal length.
   a. Use ray tracing to find the position and height of the image. To do this accurately, use a ruler or paper with a grid. Determine the image distance and image height by making measurements on your diagram.
   b. Calculate the image position and height. Compare with your ray-tracing answers in part a.

33. ‖ A 2.0-cm-tall object is 40 cm in front of a converging lens that has a 20 cm focal length.
   a. Use ray tracing to find the position and height of the image. To do this accurately, use a ruler or paper with a grid. Determine the image distance and image height by making measurements on your diagram.
   b. Calculate the image position and height. Compare with your ray-tracing answers in part a.

34. ‖ A 1.0-cm-tall object is 75 cm in front of a converging lens that has a 30 cm focal length.
   a. Use ray tracing to find the position and height of the image. To do this accurately, use a ruler or paper with a grid. Determine the image distance and image height by making measurements on your diagram
   b. Calculate the image position and height. Compare with your ray-tracing answers in part a.

35. ‖ A 2.0-cm-tall object is 15 cm in front of a converging lens that has a 20 cm focal length.
   a. Use ray tracing to find the position and height of the image. To do this accurately, use a ruler or paper with a grid. Determine the image distance and image height by making measurements on your diagram.
   b. Calculate the image position and height. Compare with your ray-tracing answers in part a.

36. ‖ A 1.0-cm-tall object is 60 cm in front of a diverging lens that has a −30 cm focal length.
   a. Use ray tracing to find the position and height of the image. To do this accurately, use a ruler or paper with a grid. Determine the image distance and image height by making measurements on your diagram.
   b. Calculate the image position and height. Compare with your ray-tracing answers in part a.

37. ‖ A 2.0-cm-tall object is 15 cm in front of a diverging lens that has a −20 cm focal length.
   a. Use ray tracing to find the position and height of the image. To do this accurately, use a ruler or paper with a grid. Determine the image distance and image height by making measurements on your diagram.
   b. Calculate the image position and height. Compare with your ray-tracing answers in part a.

Section 34.7 Image Formation with Spherical Mirrors

38. ‖ An object is 40 cm in front of a concave mirror with a focal length of 20 cm. Use ray tracing to locate the image. Is the image upright or inverted?

39. ‖ An object is 12 cm in front of a concave mirror with a focal length of 20 cm. Use ray tracing to locate the image. Is the image upright or inverted?

40. ‖ An object is 30 cm in front of a convex mirror with a focal length of −20 cm. Use ray tracing to locate the image. Is the image upright or inverted?

41. ‖ A 1.0-cm-tall object is 20 cm in front of a concave mirror that has a 60 cm focal length. Calculate the position and height of the image. State whether the image is in front of or behind the mirror, and whether the image is upright or inverted.

42. ‖ A 1.0-cm-tall object is 20 cm in front of a convex mirror that has a −60 cm focal length. Calculate the position and height of the image. State whether the image is in front of or behind the mirror, and whether the image is upright or inverted.

Problems

43. ‖ An advanced computer sends information to its various parts via infrared light pulses traveling through silicon fibers. To acquire data from memory, the central processing unit sends a light-pulse request to the memory unit. The memory unit processes the request, then sends a data pulse back to the central processing unit. The memory unit takes 0.5 ns to process a request. If the information has to be obtained from memory in 2.0 ns, what is the maximum distance the memory unit can be from the central processing unit?

44. ‖ A red ball is placed at point A in **FIGURE P34.44**.
    a. How many images are seen by an observer at point O?
    b. What are the (x, y) coordinates of each image?

FIGURE P34.44 O

45. ‖ The place you get your hair cut has two nearly parallel mirrors 5.0 m apart. As you sit in the chair, your head is 2.0 m from the nearer mirror. Looking toward this mirror, you first see your face and then, farther away, the back of your head. (The mirrors need to be slightly nonparallel for you to be able to see the back of your head, but you can treat them as parallel in this problem.) How far away does the back of your head appear to be? Neglect the thickness of your head.

46. ‖ A microscope is focused on a black dot. When a 1.00-cm-thick piece of plastic is placed over the dot, the microscope objective has to be raised 0.40 cm to bring the dot back into focus. What is the index of refraction of the plastic?

47. ‖ A light ray in air is incident on a transparent material whose index of refraction is n.
    a. Find an expression for the (non-zero) angle of incidence whose angle of refraction is half the angle of incidence.
    b. Evaluate your expression for light incident on glass.

48. ‖ The meter stick in **FIGURE P34.48** lies on the bottom of a 100-cm-long tank with its zero mark against the left edge. You look into the tank at a 30° angle, with your line of sight just grazing the upper left edge of the tank. What mark do you see on the meter stick if the tank is (a) empty, (b) half full of water, and (c) completely full of water?

FIGURE P34.48

49. ‖ The 80-cm-tall, 65-cm-wide tank shown in **FIGURE P34.49** is completely filled with water. The tank has marks every 10 cm along one wall, and the 0 cm mark is barely submerged. As you stand beside the opposite wall, your eye is level with the top of the water.
    a. Can you see the marks from the top of the tank (the 0 cm mark) going down, or from the bottom of the tank (the 80 cm mark) coming up? Explain.
    b. Which is the lowest or highest mark, depending on your answer to part a, that you can see?

FIGURE P34.49

50. ‖ A horizontal meter stick is centered at the bottom of a 3.0-m-deep, 3.0-m-wide pool of water. How long does the meter stick appear to be as you look at it from the edge of the pool?

51. ‖ A 4.0-m-wide swimming pool is filled to the top. The bottom of the pool becomes completely shaded in the afternoon when the sun is 20° above the horizon. How deep is the pool?

52. ‖ It's nighttime, and you've dropped your goggles into a 3.0-m-deep swimming pool. If you hold a laser pointer 1.0 m above the edge of the pool, you can illuminate the goggles if the laser beam enters the water 2.0 m from the edge. How far are the goggles from the edge of the pool?

53. ‖ An astronaut is exploring an unknown planet when she accidentally drops an oxygen canister into a 1.50-m-deep pool filled with an unknown liquid. Although she dropped the canister 21 cm from the edge, it appears to be 31 cm away when she peers in from the edge. What is the liquid's index of refraction? Assume that the planet's atmosphere is similar to earth's.

54. ‖ Shown from above in **FIGURE P34.54** is one corner of a rectangular box filled with water. A laser beam starts 10 cm from side A of the container and enters the water at position x. You can ignore the thin walls of the container.
    a. If $x = 15$ cm, does the laser beam refract back into the air through side B or reflect from side B back into the water? Determine the angle of refraction or reflection.
    b. Repeat part a for $x = 25$ cm.
    c. Find the minimum value of x for which the laser beam passes through side B and emerges into the air.

FIGURE P34.54

55. ‖ A light beam can use reflections to form a closed, N-sided polygon inside a solid, transparent cylinder if N is sufficiently large. What is the minimum possible value of N for light inside a cylinder of (a) water, (b) polystyrene plastic, and (c) cubic zirconia? Assume the cylinder is surrounded by air.

56. ‖ Optical engineers need to know the *cone of acceptance* of an optical fiber. This is the maximum angle that an entering light ray can make with the axis of the fiber if it is to be guided down the fiber. What is the cone of acceptance of an optical fiber for which the index of refraction of the core is 1.55 while that of the cladding is 1.45? You can model the fiber as a cylinder with a flat entrance face.

57. ‖‖ One of the contests at the school carnival is to throw a spear at an underwater target lying flat on the bottom of a pool. The water is 1.0 m deep. You're standing on a small stool that places your eyes 3.0 m above the bottom of the pool. As you look at the target, your gaze is 30° below horizontal. At what angle below horizontal should you throw the spear in order to hit the target? Your raised arm brings the spear point to the level of your eyes as you throw it, and over this short distance you can assume that the spear travels in a straight line rather than a parabolic trajectory.

58. ‖ There's one angle of incidence β onto a prism for which the light inside an isosceles prism travels parallel to the base and emerges at angle β.

FIGURE P34.58

    a. Find an expression for β in terms of the prism's apex angle α and index of refraction n.

    b. A laboratory measurement finds that $\beta = 52.2°$ for a prism shaped like an equilateral triangle. What is the prism's index of refraction?

59. ‖ You're visiting the shark tank at the aquarium when you see a 2.5-m-long shark that appears to be swimming straight toward you at 2.0 m/s. What is the shark's actual speed through the water? You can ignore the glass wall of the tank.

60. ‖ Paraxial light rays approach a transparent sphere parallel to an optical axis passing through the center of the sphere. The rays come to a focus on the far surface of the sphere. What is the sphere's index of refraction?

61. ‖ To determine the focal length of a lens, you place the lens in front of a small lightbulb and then adjust a viewing screen to get a sharply focused image. Varying the lens position produces the following data:

Bulb to lens (cm)	Lens to screen (cm)
20	61
22	47
24	39
26	37
28	32

Use the best-fit line of an appropriate graph to determine the focal length of the lens.

62. ‖ The illumination lights in an operating room use a concave
BIO mirror to focus an image of a bright lamp onto the surgical site. One such light uses a mirror with a 30 cm radius of curvature. If the mirror is 1.2 m from the patient, how far should the lamp be from the mirror?

63. ‖ A dentist uses a curved mirror to view the back side of teeth
BIO in the upper jaw. Suppose she wants an upright image with a magnification of 1.5 when the mirror is 1.2 cm from a tooth. Should she use a convex or a concave mirror? What focal length should it have?

64. ‖ A *keratometer* is an optical device used to measure the
BIO radius of curvature of the eye's cornea—its entrance surface. This measurement is especially important when fitting contact lenses, which must match the cornea's curvature. Most light incident on the eye is transmitted into the eye, but some light reflects from the cornea, which, due to its curvature, acts like a convex mirror. The keratometer places a small, illuminated ring of known diameter 7.5 cm in front of the eye. The optometrist, using an eyepiece, looks through the center of this ring and sees a small virtual image of the ring that appears to be behind the cornea. The optometrist uses a scale inside the eyepiece to measure the diameter of the image and calculate its magnification. Suppose the optometrist finds that the magnification for one patient is 0.049. What is the absolute value of the radius of curvature of her cornea?

65. ‖ The mirror in **FIGURE P34.65** is covered with a piece of glass. A point source of light is outside the glass. How far from the mirror is the image of this source?

FIGURE P34.65

66. ‖ A 2.0-cm-tall candle flame is 2.0 m from a wall. You happen to have a lens with a focal length of 32 cm. How many places can you put the lens to form a well-focused image of the candle flame on the wall? For each location, what are the height and orientation of the image?

67. ‖ A 25 g rubber ball is dropped from a height of 3.0 m above the center of a horizontal, concave mirror. The ball and its image coincide 0.65 s after the ball is released. What is the mirror's radius of curvature?

68. ‖ In recent years, physicists have learned to create *metamaterials*— engineered materials not found in nature—with negative indices of refraction. It's not yet possible to form a lens from a material with a negative index of refraction, but researchers are optimistic. Suppose you had a planoconvex lens (flat on one side, a 15 cm radius of curvature on the other) that is made from a metamaterial with $n = -1.25$. If you place an object 12 cm from this lens, (a) what type of image will be formed and (b) where will the image be located?

69. ‖ A lightbulb is 3.0 m from a wall. What are the focal length and the position (measured from the bulb) of a lens that will form an image on the wall that is twice the size of the lightbulb?

70. ‖ An old fashioned slide projector needs to create a 98-cm-high image of a 2.0-cm-tall slide. The screen is 300 cm from the slide.

    a. What focal length does the lens need? Assume that it is a thin lens.

    b. How far should you place the lens from the slide?

71. ‖ Some *electro-optic materials* can change their index of refraction in response to an applied voltage. Suppose a planoconvex lens (flat on one side, a 15.0 cm radius of curvature on the other), made from a material whose normal index of refraction is 1.500, is creating an image of an object that is 50.0 cm from the lens. By how much would the index of refraction need to be increased to move the image 5.0 cm closer to the lens?

72. ‖ A point source of light is distance d from the surface of a 6.0-cm-diameter glass sphere. For what value of d is there an image at the same distance d on the opposite side of the sphere?

73. ‖ A lens placed 10 cm in front of an object creates an upright image twice the height of the object. The lens is then moved along the optical axis until it creates an inverted image twice the height of the object. How far did the lens move?

74. ‖ An object is 60 cm from a screen. What are the radii of a symmetric converging plastic lens (i.e., two equally curved surfaces) that will form an image on the screen twice the height of the object?

75. ‖ A wildlife photographer with a 200-mm-focal-length tele-
CALC photo lens on his camera is taking a picture of a rhinoceros that is 100 m away. Suddenly, the rhino starts charging straight toward the photographer at a speed of 5.0 m/s. What is the speed, in μm/s, of the image of the rhinoceros? Is the image moving toward or away from the lens?

76. ‖ A concave mirror has a 40 cm radius of curvature. How far from the mirror must an object be placed to create an upright image three times the height of the object?

77. ‖‖ A 2.0-cm-tall object is placed in front of a mirror. A 1.0-cm-tall upright image is formed behind the mirror, 150 cm from the object. What is the focal length of the mirror?

78. ‖ A spherical mirror of radius R has its center at C, as shown in FIGURE P34.78. A ray parallel to the axis reflects through F, the focal point. Prove that $f = R/2$ if $\phi \ll 1$ rad.

FIGURE P34.78

Challenge Problems

79. ‖‖‖ Consider a lens having index of refraction n_2 and surfaces with radii R_1 and R_2. The lens is immersed in a fluid that has index of refraction n_1.

    a. Derive a generalized lens maker's equation to replace Equation 34.26 when the lens is surrounded by a medium other than air. That is, when $n_1 \neq 1$.

    b. A symmetric converging glass lens (i.e., two equally curved surfaces) has two surfaces with radii of 40 cm. Find the focal length of this lens in air and the focal length of this lens in water.

80. ‖‖‖ **FIGURE CP34.80** shows a light ray that travels from point A
CALC   to point B. The ray crosses the boundary at position x, making angles θ_1 and θ_2 in the two media. Suppose that you did *not* know Snell's law.

    a. Write an expression for the *time t* it takes the light ray to travel from A to B. Your expression should be in terms of the distances a, b, and w; the variable x; and the indices of refraction n_1 and n_2.

b. The time depends on x. There's one value of x for which the light travels from A to B in the shortest possible time. We'll call it x_{min}. Write an expression (but don't try to solve it!) from which x_{min} could be found.

c. Now, by using the geometry of the figure, derive Snell's law from your answer to part b.

You've proven that Snell's law is equivalent to the statement that "light traveling between two points follows the path that requires the shortest time." This interesting way of thinking about refraction is called *Fermat's principle*.

FIGURE CP34.80

81. ‖‖‖ A fortune teller's "crystal ball" (actually just glass) is 10 cm in diameter. Her secret ring is placed 6.0 cm from the edge of the ball.

    a. An image of the ring appears on the opposite side of the crystal ball. How far is the image from the center of the ball?

    b. Draw a ray diagram showing the formation of the image.

    c. The crystal ball is removed and a thin lens is placed where the center of the ball had been. If the image is still in the same position, what is the focal length of the lens?

82. ‖‖‖ Consider an object of thickness ds (parallel to the axis) in
CALC   front of a lens or mirror. The image of the object has thickness ds'. Define the *longitudinal magnification* as $M = ds'/ds$. Prove that $M = -m^2$, where m is the lateral magnification.

35 Optical Instruments

The world's greatest collection of telescopes is on the summit of Mauna Kea on the Big Island of Hawaii, towering 4200 m (13,800 ft) over the Pacific Ocean.

IN THIS CHAPTER, you will learn about some common optical instruments and their limitations.

What is an optical instrument?

Optical instruments, such as cameras, microscopes, and telescopes, are used to produce images for viewing or detection. Most use several individual lenses in combination to improve performance. You'll learn how to analyze a system with multiple lenses.

« LOOKING BACK Sections 34.5–34.6 Thin lenses

How does a camera work?

A camera uses a lens—made of several individual lenses—to project a real image onto a light-sensitive detector. The detector in a digital camera uses millions of tiny pixels.

- You'll learn about focusing and zoom.
- You'll also learn how to calculate a lens's *f*-number, which, along with shutter speed, determines the exposure.

How does vision work?

The human eye is much like a camera; the cornea and lens together focus a real image onto the retina. You'll learn about two defects of vision—myopia (nearsightedness) and hyperopia (farsightedness)—and how they can be corrected with eyeglasses.

What optical systems are used to magnify things?

Lenses and mirrors can be used to magnify objects both near and far.

- A simple magnifying glass has a low magnification of 2× or 3×.
- Microscopes use two sets of lenses to reach magnifications up to 1000×.
- Telescopes magnify distant objects.

Is color important in optics?

Color depends on the wavelength of light.

- The index of refraction is slightly wavelength dependent, so different wavelengths refract at different angles. This is the main reason we have rainbows.
- Many materials absorb or scatter some wavelengths more than others.

What is the resolution of an optical system?

Light passing through a lens undergoes diffraction, just like light passing through a circular hole. Images are not perfect points but are tiny diffraction patterns, and this limits how well two nearby objects can be resolved. You'll learn about Rayleigh's criterion for the resolution of two images.

« LOOKING BACK Section 33.6 Circular diffraction

35.1 Lenses in Combination

Only the simplest magnifiers are built with a single lens of the sort we analyzed in Chapter 34. Optical instruments, such as microscopes and cameras, are invariably built with multiple lenses. The reason, as we'll see, is to improve the image quality.

The analysis of multi-lens systems requires only one new rule: **The image of the first lens acts as the object for the second lens.** To see why this is so, **FIGURE 35.1** shows a simple telescope consisting of a large-diameter converging lens, called the *objective,* and a smaller converging lens used as the *eyepiece.* (We'll analyze telescopes more thoroughly later in the chapter.) Highlighted are the three special rays you learned to use in Chapter 34:

- A ray parallel to the optical axis refracts through the focal point.
- A ray through the focal point refracts parallel to the optical axis.
- A ray through the center of the lens is undeviated.

FIGURE 35.1 Ray-tracing diagram of a simple astronomical telescope.

The rays passing through the objective converge to a real image at P′, **but they don't stop there.** Instead, light rays *diverge* from P′ as they approach the second lens. **As far as the eyepiece is concerned, the rays are coming from P′, and thus P′ acts as the object for the second lens.** The three special rays passing through the objective lens are sufficient to locate the image P′, but these rays are generally *not* the special rays for the second lens. However, other rays converging at P′ leave at the correct angles to be the special rays for the eyepiece. That is, a new set of special rays is drawn from P′ to the second lens and used to find the final image point P″.

> **NOTE** One ray seems to "miss" the eyepiece lens, but this isn't a problem. All rays passing through the lens converge to (or diverge from) a single point, and the purpose of the special rays is to locate that point. To do so, we can let the special rays refract as they cross the *lens plane,* regardless of whether the physical lens really extends that far.

EXAMPLE 35.1 | A camera lens

The "lens" on a camera is usually a combination of two or more single lenses. Consider a camera in which light passes first through a diverging lens, with $f_1 = -120$ mm, then a converging lens, with $f_2 = 42$ mm, spaced 60 mm apart. A reasonable definition of the *effective focal length* of this lens combination is the focal length of a *single* lens that could produce an image in the same location if placed at the midpoint of the lens combination. A 10-cm-tall object is 500 mm from the first lens.

a. What are the location, size, and orientation of the image?

b. What is the effective focal length of the double-lens system used in this camera?

MODEL Each lens is a thin lens. The image of the first lens is the object for the second.

VISUALIZE The ray-tracing diagram of **FIGURE 35.2** shows the production of a real, inverted image ≈ 55 mm behind the second lens.

SOLVE a. $s_1 = 500$ mm is the object distance of the first lens. Its image, a virtual image, is found from the thin-lens equation:

$$\frac{1}{s_1'} = \frac{1}{f_1} - \frac{1}{s_1} = \frac{1}{-120 \text{ mm}} - \frac{1}{500 \text{ mm}} = -0.0103 \text{ mm}^{-1}$$

$$s_1' = -97 \text{ mm}$$

This is consistent with the ray-tracing diagram. The image of the first lens now acts as the object for the second lens. Because the lenses are 60 mm apart, the object distance is $s_2 = 97$ mm + 60 mm = 157 mm. A second application of the thin-lens equation yields

$$\frac{1}{s_2'} = \frac{1}{f_2} - \frac{1}{s_2} = \frac{1}{42 \text{ mm}} - \frac{1}{157 \text{ mm}} = 0.0174 \text{ mm}^{-1}$$

$$s_2' = 57 \text{ mm}$$

The image of the lens combination is 57 mm behind the second lens. The lateral magnifications of the two lenses are

$$m_1 = -\frac{s_1'}{s_1} = -\frac{-97 \text{ cm}}{500 \text{ cm}} = 0.194$$

$$m_2 = -\frac{s_2'}{s_2} = -\frac{57 \text{ cm}}{157 \text{ cm}} = -0.363$$

The second lens magnifies the image of the first lens, which magnifies the object, so **the total magnification is the product of the individual magnifications:**

$$m = m_1 m_2 = -0.070$$

Thus the image is 57 mm behind the second lens, inverted (m is negative), and 0.70 cm tall.

b. If a single lens midway between these two lenses produced an image in the same plane, its object and image distances would be $s = 500$ mm + 30 mm = 530 mm and $s' = 57$ mm + 30 mm = 87 mm. A final application of the thin-lens equation gives the effective focal length:

$$\frac{1}{f_{\text{eff}}} = \frac{1}{s} + \frac{1}{s'} = \frac{1}{530 \text{ mm}} + \frac{1}{87 \text{ mm}} = 0.0134 \text{ mm}^{-1}$$

$$f_{\text{eff}} = 75 \text{ mm}$$

ASSESS This combination lens would be sold as a "75 mm lens."

FIGURE 35.2 Pictorial representation of a combination lens.

The image of the first lens acts as the object for the second lens.

STOP TO THINK 35.1 The second lens in this optical instrument

a. Causes the light rays to focus closer than they would with the first lens acting alone.
b. Causes the light rays to focus farther away than they would with the first lens acting alone.
c. Inverts the image but does not change where the light rays focus.
d. Prevents the light rays from reaching a focus.

Lens 1   Lens 2

FIGURE 35.3 A camera.

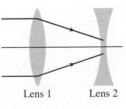

35.2 The Camera

A **camera,** shown in **FIGURE 35.3**, "takes a picture" by using a lens to form a real, inverted image on a light-sensitive detector in a light-tight box. Film was the detector of choice for well over a hundred years, but today's digital cameras use an electronic detector called a *charge-coupled device,* or CCD.

The camera "lens" is always a combination of two or more individual lenses. The simplest such lens, shown in FIGURE 35.4, consists of a converging lens and a somewhat weaker diverging lens. This combination of positive and negative lenses corrects some of the defects inherent in single lenses, as we'll discuss later in the chapter. As Example 35.1 suggested, we can model a combination lens as a single lens with an **effective focal length** (usually called simply "the focal length") f. A *zoom lens* changes the effective focal length by changing the spacing between the converging lens and the diverging lens; this is what happens when the lens barrel on your digital camera moves in and out as you use the zoom. A typical digital camera has a lens whose effective focal length can be varied from 6 mm to 18 mm, giving, as we'll see, a 3× zoom.

FIGURE 35.4 A simple camera lens is a combination lens.

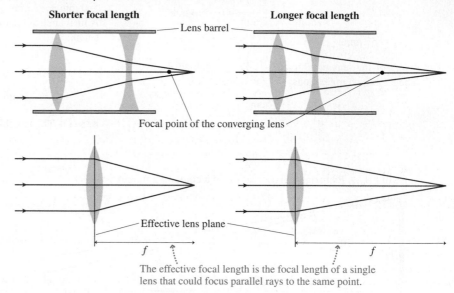

The effective focal length is the focal length of a single lens that could focus parallel rays to the same point.

A camera must carry out two important functions: focus the image on the detector and control the exposure. Cameras are focused by moving the lens forward or backward until the image is well focused on the detector. Most modern cameras do this automatically, but older cameras required manual focusing.

EXAMPLE 35.2 | Focusing a camera

Your digital camera lens, with an effective focal length of 10.0 mm, is focused on a flower 20.0 cm away. You then turn to take a picture of a distant landscape. How far, and in which direction, must the lens move to bring the landscape into focus?

MODEL Model the camera's combination lens as a single thin lens with $f = 10.0$ mm. Image and object distances are measured from the effective lens plane. Assume all the lenses in the combination move together as the camera refocuses.

SOLVE The flower is at object distance $s = 20.0$ cm $= 200$ mm. When the camera is focused, the image distance between the

effective lens plane and the detector is found by solving the thin-lens equation $1/s + 1/s' = 1/f$ to give

$$s' = \left(\frac{1}{f} - \frac{1}{s}\right)^{-1} = \left(\frac{1}{10.0 \text{ mm}} - \frac{1}{200 \text{ mm}}\right)^{-1} = 10.5 \text{ mm}$$

The distant landscape is effectively at object distance $s = \infty$, so its image distance is $s' = f = 10.0$ mm. To refocus as you shift scenes, the lens must move 0.5 mm closer to the detector.

ASSESS The required motion of the lens is very small, about the diameter of the lead used in a mechanical pencil.

Zoom Lenses

For objects more than 10 focal lengths from the lens (roughly $s > 20$ cm for a typical digital camera), the approximation $s \gg f$ (and thus $1/s \ll 1/f$) leads to $s' \approx f$. In other words, objects more than about 10 focal lengths away are essentially "at infinity," and we know that the parallel rays from an infinitely distant object are focused one focal length behind the lens. For such an object, the lateral magnification of the image is

$$m = -\frac{s'}{s} \approx -\frac{f}{s} \qquad (35.1)$$

The magnification is much less than 1, because $s \gg f$, so the image on the detector is much smaller than the object itself. This comes as no surprise. More important, **the size of the image is directly proportional to the focal length of the lens.** We saw in Figure 35.4 that the effective focal length of a combination lens is easily changed by varying the distance between the individual lenses, and this is exactly how a zoom lens works. A lens that can be varied from $f_{min} = 6$ mm to $f_{max} = 18$ mm gives magnifications spanning a factor of 3, and that is why you see it specified as a 3× zoom lens.

Controlling the Exposure

The camera also must control the amount of light reaching the detector. Too little light results in photos that are *underexposed;* too much light gives *overexposed* pictures. Both the shutter and the lens diameter help control the exposure.

The *shutter* is "opened" for a selected amount of time as the image is recorded. Older cameras used a spring-loaded mechanical shutter that literally opened and closed; digital cameras electronically control the amount of time the detector is active. Either way, the exposure—the amount of light captured by the detector—is directly proportional to the time the shutter is open. Typical exposure times range from 1/1000 s or less for a sunny scene to 1/30 s or more for dimly lit or indoor scenes. The exposure time is generally referred to as the *shutter speed.*

The amount of light passing through the lens is controlled by an adjustable **aperture,** also called an *iris* because it functions much like the iris of your eye. The aperture sets the effective diameter D of the lens. The full area of the lens is used when the aperture is fully open, but a *stopped-down* aperture allows light to pass through only the central portion of the lens.

The light intensity on the detector is directly proportional to the area of the lens; a lens with twice as much area will collect and focus twice as many light rays from the object to make an image twice as bright. The lens area is proportional to the square of its diameter, so the intensity I is proportional to D^2. The light intensity—power per square meter—is also *inversely* proportional to the area of the image. That is, the light reaching the detector is more intense if the rays collected from the object are focused into a small area than if they are spread out over a large area. The lateral size of the image is proportional to the focal length of the lens, as we saw in Equation 35.1, so the *area* of the image is proportional to f^2 and thus I is proportional to $1/f^2$. Altogether, $I \propto D^2/f^2$.

By long tradition, the light-gathering ability of a lens is specified by its **f-number,** defined as

$$f\text{-number} = \frac{f}{D} \qquad (35.2)$$

The *f*-number of a lens may be written either as *f*/4.0, to mean that the *f*-number is 4.0, or as F4.0. The instruction manuals with some digital cameras call this the *aperture value* rather than the *f*-number. A digital camera in fully automatic mode does not display shutter speed or *f*-number, but that information is displayed if you set your camera to any of the other modes. For example, the display 1/125 F5.6 means that your camera is going to achieve the correct exposure by adjusting the diameter of the lens aperture to give $f/D = 5.6$ and by opening the shutter for 1/125 s. If your lens's effective focal length is 10 mm, the diameter of the lens aperture will be

$$D = \frac{f}{f\text{-number}} = \frac{10 \text{ mm}}{5.6} = 1.8 \text{ mm}$$

An iris can change the effective diameter of a lens and thus the amount of light reaching the detector.

Focal length and f-number information is stamped on a camera lens. This lens is labeled 5.8–23.2 mm 1:2.6–5.5. The first numbers are the range of focal lengths. They span a factor of 4, so this is a 4× zoom lens. The second numbers show that the minimum f-number ranges from $f/2.6$ (for the $f = 5.8$ mm focal length) to $f/5.5$ (for the $f = 23.2$ mm focal length).

NOTE The f in f-number is not the focal length f; it's just a name. And the / in $f/4$ does not mean division; it's just a notation. These both derive from the long history of photography.

Because the aperture diameter is in the denominator of the f-number, a *larger-diameter* aperture, which gathers more light and makes a brighter image, has a *smaller* f-number. The light intensity on the detector is related to the lens's f-number by

$$I \propto \frac{D^2}{f^2} = \frac{1}{(f\text{-number})^2} \tag{35.3}$$

Historically, a lens's f-numbers could be adjusted in the sequence 2.0, 2.8, 4.0, 5.6, 8.0, 11, 16. Each differs from its neighbor by a factor of $\sqrt{2}$, so changing the lens by one "f stop" changed the light intensity by a factor of 2. A modern digital camera is able to adjust the f-number continuously.

The exposure, the total light reaching the detector while the shutter is open, depends on the product $I\Delta t_{\text{shutter}}$. A small f-number (large aperture diameter D) and short $\Delta t_{\text{shutter}}$ can produce the same exposure as a larger f-number (smaller aperture) and a longer $\Delta t_{\text{shutter}}$. It might not make any difference for taking a picture of a distant mountain, but action photography needs very short shutter times to "freeze" the action. Thus action photography requires a large-diameter lens with a small f-number.

EXAMPLE 35.3 | **Capturing the action**

Before a race, a photographer finds that she can make a perfectly exposed photo of the track while using a shutter speed of 1/250 s and a lens setting of $f/8.0$. To freeze the sprinters as they go past, she plans to use a shutter speed of 1/1000 s. To what f-number must she set her lens?

MODEL The exposure depends on $I\Delta t_{\text{shutter}}$, and the light intensity depends inversely on the square of the f-number.

SOLVE Changing the shutter speed from 1/250 s to 1/1000 s will reduce the light reaching the detector by a factor of 4. To compensate, she needs to let 4 times as much light through the lens. Because $I \propto 1/(f\text{-number})^2$, the intensity will increase by a factor of 4 if she *decreases* the f-number by a factor of 2. Thus the correct lens setting is $f/4.0$.

The Detector

FIGURE 35.5 The CCD detector used in a digital camera.

(a)   4600 × 3500 pixels

1 pixel

(b)

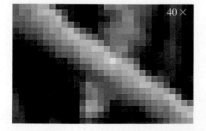

40×

For traditional cameras, the light-sensitive detector is film. Today's digital cameras use an electronic light-sensitive surface called a *charge-coupled device* or **CCD.** A CCD consists of a rectangular array of many millions of small detectors called **pixels.** When light hits one of these pixels, it generates an electric charge proportional to the light intensity. Thus an image is recorded on the CCD in terms of little packets of charge. After the CCD has been exposed, the charges are read out, the signal levels are digitized, and the picture is stored in the digital memory of the camera.

FIGURE 35.5a shows a CCD "chip" and, schematically, the magnified appearance of the pixels on its surface. To record color information, different pixels are covered by red, green, or blue filters. A pixel covered by a green filter, for instance, records only the intensity of the green light hitting it. Later, the camera's microprocessor interpolates nearby colors to give each pixel an overall true color. The pixels are so small that the picture looks "smooth" even after some enlargement, but, as you can see in **FIGURE 35.5b**, sufficient magnification reveals the individual pixels.

STOP TO THINK 35.2 A photographer has adjusted his camera for a correct exposure with a short-focal-length lens. He then decides to zoom in by increasing the focal length. To maintain a correct exposure without changing the shutter speed, the diameter of the lens aperture should

a. Be increased.   b. Be decreased.   c. Stay the same.

35.3 Vision

The human eye is a marvelous and intricate organ. If we leave the biological details to biologists and focus on the eye's optical properties, we find that it functions very much like a camera. Like a camera, the eye has refracting surfaces that focus incoming light rays, an adjustable iris to control the light intensity, and a light-sensitive detector.

FIGURE 35.6 shows the basic structure of the eye. It is roughly spherical, about 2.4 cm in diameter. The transparent **cornea,** which is somewhat more sharply curved, and the *lens* are the eye's refractive elements. The eye is filled with a clear, jellylike fluid called the *aqueous humor* (in front of the lens) and the *vitreous humor* (behind the lens). The indices of refraction of the aqueous and vitreous humors are 1.34, only slightly different from water. The lens, although not uniform, has an average index of 1.44. The **pupil,** a variable-diameter aperture in the **iris,** automatically opens and closes to control the light intensity. A fully dark-adapted eye can open to $\approx$ 8 mm, and the pupil closes down to $\approx$ 1.5 mm in bright sun. This corresponds to *f*-numbers from roughly *f*/3 to *f*/16, very similar to a camera.

FIGURE 35.6 The human eye.

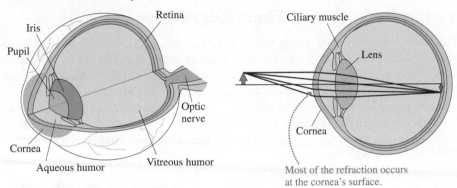

The eye's detector, the **retina,** consists of specialized light-sensitive cells called *rods* and *cones*. The rods, sensitive mostly to light and dark, are most important in very dim lighting. Color vision, which requires somewhat more light, is due to the cones, of which there are three types. **FIGURE 35.7** shows the wavelength responses of the cones. They have overlapping ranges, especially the red- and green-sensitive cones, so two or even all three cones respond to light of any particular wavelength. The relative response of the different cones is interpreted by your brain as light of a particular color. Color is a *perception*, a response of our sensory and nervous systems, not something inherent in the light itself. Other animals, with slightly different retinal cells, can see ultraviolet or infrared wavelengths that we cannot see.

FIGURE 35.7 Wavelength sensitivity of the three types of cones in the human retina.

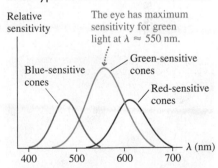

Focusing and Accommodation

The eye, like a camera, focuses light rays to an inverted image on the retina. Perhaps surprisingly, most of the refractive power of the eye is due to the cornea, not the lens. The cornea is a sharply curved, spherical surface, and you learned in Chapter 34 that images are formed by refraction at a spherical surface. The rather large difference between the index of refraction of air and that of the aqueous humor causes a significant refraction of light rays at the cornea. In contrast, there is much less difference between the indices of the lens and its surrounding fluid, so refraction at the lens surfaces is weak. The lens is important for fine-tuning, but the air-cornea boundary is responsible for the majority of the refraction.

FIGURE 35.8 Normal vision of far and near objects.

The ciliary muscles are relaxed for distant vision.

FP = ∞

NP = 25 cm

The ciliary muscles are contracted for near vision, causing the lens to curve more.

You can recognize the power of the cornea if you open your eyes underwater. Everything is very blurry! When light enters the cornea through water, rather than through air, there's almost no difference in the indices of refraction at the surface. Light rays pass through the cornea with almost no refraction, so what little focusing ability you have while underwater is due to the lens alone.

A camera focuses by moving the lens. The eye focuses by changing the focal length of the lens, a feat it accomplishes by using the *ciliary muscles* to change the curvature of the lens surface. The ciliary muscles are relaxed when you look at a distant scene. Thus the lens surface is relatively flat and the lens has its longest focal length. As you shift your gaze to a nearby object, the ciliary muscles contract and cause the lens to bulge. This process, called **accommodation,** decreases the lens's radius of curvature and thus decreases its focal length.

The farthest distance at which a relaxed eye can focus is called the eye's **far point** (FP). The far point of a normal eye is infinity; that is, the eye can focus on objects extremely far away. The closest distance at which an eye can focus, using maximum accommodation, is the eye's **near point** (NP). (Objects can be *seen* closer than the near point, but they're not sharply focused on the retina.) Both situations are shown in **FIGURE 35.8.**

Vision Defects and Their Correction

The near point of normal vision is considered to be 25 cm, but the near point of any individual changes with age. The near point of young children can be as little as 10 cm. The "normal" 25 cm near point is characteristic of young adults, but the near point of most individuals begins to move outward by age 40 or 45 and can reach 200 cm by age 60. This loss of accommodation, which arises because the lens loses flexibility, is called **presbyopia.** Even if their vision is otherwise normal, individuals with presbyopia need reading glasses to bring their near point back to 25 or 30 cm, a comfortable distance for reading.

Presbyopia is known as a *refractive error* of the eye. Two other common refractive errors are *hyperopia* and *myopia.* All three can be corrected with lenses—either eyeglasses or contact lenses—that assist the eye's focusing. Corrective lenses are prescribed not by their focal length but by their **power.** The power of a lens is the inverse of its focal length:

$$\text{Power of a lens} = P = \frac{1}{f} \tag{35.4}$$

A lens with more power (shorter focal length) causes light rays to refract through a larger angle. The SI unit of lens power is the **diopter,** abbreviated D, defined as $1\ \text{D} = 1\ \text{m}^{-1}$. Thus a lens with $f = 50\ \text{cm} = 0.50\ \text{m}$ has power $P = 2.0\ \text{D}$.

A person who is *farsighted* can see faraway objects (but even then must use some accommodation rather than a relaxed eye), but his near point is larger than 25 cm, often much larger, so he cannot focus on nearby objects. The cause of farsightedness—called **hyperopia**—is an eyeball that is too short for the refractive power of the cornea and lens. As **FIGURES 35.9a** and **b** show, no amount of accommodation allows the eye to focus on an object 25 cm away, the normal near point.

With hyperopia, the eye needs assistance to focus the rays from a near object onto the closer-than-normal retina. This assistance is obtained by adding refractive power with the positive (i.e., converging) lens shown in **FIGURE 35.9c.** To understand why this works, recall that the image of a first lens acts as the object for a second lens. The goal is to allow the person to focus on an object 25 cm away. If a corrective lens forms an upright, virtual image at the person's actual near point, that virtual image acts as an object for the eye itself and, with maximum accommodation, the eye can focus these rays onto the retina. Presbyopia, the loss of accommodation with age, is corrected in the same way.

The optometrist's prescription is −2.25 D for the right eye (top) and −2.50 D for the left (bottom), the minus sign indicating that these are diverging lenses. The optometrist doesn't write the D because the lens maker already knows that prescriptions are in diopters. Most people's eyes are not exactly the same, so each eye usually gets a different lens.

FIGURE 35.9 Hyperopia.

(a)

Retina position of normal eye

Shortened eyeball

25 cm

With maximum accommodation, the eye tries to focus the image behind the actual retina. Thus the image is blurry.

(b)

Maximum accommodation

NP > 25 cm

This is the closest point at which the eye can focus.

(c)

This is the actual object the eye wants to see.

25 cm

A converging lens forms a virtual image at the eye's near point. This image acts as the object for the eye and is what the eye actually focuses on.

Focused image

FIGURE 35.10 Myopia.

(a)

Retina position of normal eye

Elongated eyeball

Parallel rays from distant object

A fully relaxed eye focuses the image in front of the actual retina. The image is blurry.

(b)

Fully relaxed

FP < ∞

This is the farthest point at which the eye can focus.

(c)

The eye wants to see a distant object.

A diverging lens forms a virtual image at the eye's far point. This image acts as the object for the eye and is what the eye actually focuses on.

Focused image

NOTE Figures 35.9 and 35.10 show the corrective lenses as they are actually shaped—called *meniscus lenses*—rather than with our usual lens shape. Nonetheless, the lens in Figure 35.9c is a converging lens because it's thicker in the center than at the edges. The lens in Figure 35.10c is a diverging lens because it's thicker at the edges than in the center.

A person who is *nearsighted* can clearly see nearby objects when the eye is relaxed (and extremely close objects by using accommodation), but no amount of relaxation allows her to see distant objects. Nearsightedness—called **myopia**—is caused by an eyeball that is too long. As **FIGURE 35.10a** shows, rays from a distant object come to a focus in front of the retina and have begun to diverge by the time they reach the retina. The eye's far point, shown in **FIGURE 35.10b**, is less than infinity.

To correct myopia, we needed a diverging lens, as shown in **FIGURE 35.10c**, to slightly defocus the rays and move the image point back to the retina. To focus on a very distant object, the person needs a corrective lens that forms an upright, virtual image at her actual far point. That virtual image acts as an object for the eye itself and, when fully relaxed, the eye can focus these rays onto the retina.

EXAMPLE 35.4 | Correcting hyperopia

Sanjay has hyperopia. The near point of his left eye is 150 cm. What prescription lens will restore normal vision?

MODEL Normal vision will allow Sanjay to focus on an object 25 cm away. In measuring distances, we'll ignore the small space between the lens and his eye.

SOLVE Because Sanjay can see objects at 150 cm, using maximum accommodation, we want a lens that creates a virtual image at

position $s' = -150$ cm (negative because it's a virtual image) of an object held at $s = 25$ cm. From the thin-lens equation,

$$\frac{1}{f} = \frac{1}{s} + \frac{1}{s'} = \frac{1}{0.25 \text{ m}} + \frac{1}{-1.50 \text{ m}} = 3.3 \text{ m}^{-1}$$

$1/f$ is the lens power, and m^{-1} are diopters. Thus the prescription is for a lens with power $P = 3.3$ D.

ASSESS Hyperopia is always corrected with a converging lens.

EXAMPLE 35.5 | Correcting myopia

Martina has myopia. The far point of her left eye is 200 cm. What prescription lens will restore normal vision?

MODEL Normal vision will allow Martina to focus on a very distant object. In measuring distances, we'll ignore the small space between the lens and her eye.

SOLVE Because Martina can see objects at 200 cm with a fully relaxed eye, we want a lens that will create a virtual image at position

$s' = -200$ cm (negative because it's a virtual image) of a distant object at $s = \infty$ cm. From the thin-lens equation,

$$\frac{1}{f} = \frac{1}{s} + \frac{1}{s'} = \frac{1}{\infty\,\text{m}} + \frac{1}{-2.0\,\text{m}} = -0.5\,\text{m}^{-1}$$

Thus the prescription is for a lens with power $P = -0.5$ D.

ASSESS Myopia is always corrected with a diverging lens.

STOP TO THINK 35.3 You need to improvise a magnifying glass to read some very tiny print. Should you borrow the eyeglasses from your hyperopic friend or from your myopic friend?

a. The hyperopic friend
b. The myopic friend
c. Either will do.
d. Neither will work.

35.4 Optical Systems That Magnify

FIGURE 35.11 Angular size.

(a) Same object at two different distances

As the object gets closer, the angle it subtends becomes larger. Its *angular size* has increased.

Further, the size of the image on the retina gets larger. The object's *apparent size* has increased.

(b)

Near point

The camera, with its fast shutter speed, allows us to capture images of events that take place too quickly for our unaided eye to resolve. Another use of optical systems is to magnify—to see objects smaller or closer together than our eye can see.

The easiest way to magnify an object requires no extra optics at all; simply get closer! The closer you get, the bigger the object appears. Obviously the actual size of the object is unchanged as you approach it, so what exactly is getting "bigger"? Consider the green arrow in FIGURE 35.11a. We can determine the size of its image on the retina by tracing the rays that are undeviated as they pass through the center of the lens. (Here we're modeling the eye's optical system as one thin lens.) If we get closer to the arrow, now shown as red, we find the arrow makes a larger image on the retina. Our brain interprets the larger image as a larger-appearing object. The object's actual size doesn't change, but its *apparent size* gets larger as it gets closer.

Technically, we say that closer objects look larger because they subtend a larger angle θ, called the **angular size** of the object. The red arrow has a larger angular size than the green arrow, $\theta_2 > \theta_1$, so the red arrow looks larger and we can see more detail. But you can't keep increasing an object's angular size because you can't focus on the object if it's closer than your near point, which we'll take to be a normal 25 cm. FIGURE 35.11b defines the angular size θ_{NP} of an object at your near point. If the object's height is h and if we assume the small-angle approximation $\tan\theta \approx \theta$, the maximum angular size viewable by your unaided eye is

$$\theta_{NP} = \frac{h}{25\text{ cm}} \tag{35.5}$$

Suppose we view the same object, of height h, through the single converging lens in FIGURE 35.12. If the object's distance from the lens is less than the lens's focal length, we'll see an enlarged, upright image. Used in this way, the lens is called a **magnifier** or *magnifying glass*. The eye sees the virtual image subtending angle θ, and it can focus on this virtual image as long as the image distance is more than 25 cm. Within the small-angle approximation, the image subtends angle $\theta = h/s$. In practice, we usually want the image to be at distance $s' \approx \infty$ so that we can view it with a relaxed eye as a "distant object." This will be true if the object is very near the focal point: $s \approx f$. In this case, the image subtends angle

$$\theta = \frac{h}{s} \approx \frac{h}{f} \tag{35.6}$$

FIGURE 35.12 The magnifier.

The object is placed very near the focal point of the lens.

Object

Virtual image

The image of the lens acts as the object for the eye. This is what the eye focuses on.

Let's define the **angular magnification** M as

$$M = \frac{\theta}{\theta_{NP}} \tag{35.7}$$

Angular magnification is the increase in the *apparent size* of the object that you achieve by using a magnifying lens rather than simply holding the object at your near point. Substituting from Equations 35.5 and 35.6, we find the angular magnification of a magnifying glass is

$$M = \frac{25 \text{ cm}}{f} \tag{35.8}$$

The angular magnification depends on the focal length of the lens but not on the size of the object. Although it would appear we could increase angular magnification without limit by using lenses with shorter and shorter focal lengths, the inherent limitations of lenses we discuss later in the chapter limit the magnification of a simple lens to about $4\times$. Slightly more complex magnifiers with two lenses reach $20\times$, but beyond that one would use a microscope.

> **NOTE** Don't confuse angular magnification with lateral magnification. Lateral magnification m compares the height of an object to the height of its image. The lateral magnification of a magnifying glass is $\approx \infty$ because the virtual image is at $s' \approx \infty$, but that doesn't make the object seem infinitely big. Its apparent size is determined by the angle subtended on your retina, and that angle remains finite. Thus angular magnification tells us how much bigger things appear.

The Microscope

A microscope, whose major parts are shown in FIGURE 35.13, can attain a magnification of up to $1000\times$ by a *two-step* magnification process. A specimen to be observed is placed on the *stage* of the microscope, directly beneath the **objective,** a converging lens with a relatively short focal length. The objective creates a magnified real image that is further enlarged by the **eyepiece.** Both the objective and the eyepiece are complex combination lenses, but we'll model them as single thin lenses. It's common for a prism to bend the rays so that the eyepiece is at a comfortable viewing angle. However, we'll consider a simplified version of a microscope in which the light travels along a straight tube.

FIGURE 35.14 on the next page is a simple two-lens model of a microscope. The object is placed just outside the focal point of the objective, which creates a highly magnified real image with lateral magnification $m = -s'/s$. The object is so close to the focal point that $s \approx f_{obj}$ is an excellent approximation. In addition, the microscope is designed so that the image distance s' is approximately the tube length L, so $s' \approx L$. With these approximations, the lateral magnification of the objective is

$$m_{obj} = -\frac{s'}{s} \approx -\frac{L}{f_{obj}} \tag{35.9}$$

FIGURE 35.13 A microscope.

Eyepiece

Prism (bends light path so that eyepiece is at a comfortable angle)

Focus knob

Objective lens

Stage (moves up and down to focus sample)

Illuminator

FIGURE 35.14 The optics of a microscope.

The image of the objective acts as the object for the eyepiece, which functions as a simple magnifier. The angular magnification of the eyepiece is given by Equation 35.8, $M_{eye} = (25 \text{ cm})/f_{eye}$. Together, the objective and eyepiece produce a total angular magnification

$$M = m_{obj}M_{eye} \approx -\frac{L}{f_{obj}}\frac{25 \text{ cm}}{f_{eye}} \tag{35.10}$$

The minus sign shows that the image seen in a microscope is inverted.

In practice, the magnifications of the objective (without the minus sign) and the eyepiece are stamped on the barrels. A set of objectives on a rotating turret might include 10×, 20×, 40×, and 100×. When combined with a 10× eyepiece, the microscope's total angular magnification ranges from 100× to 1000×. In addition, most biological microscopes are standardized with a tube length $L = 160$ mm. Thus a 40× objective has focal length $f_{obj} = 160$ mm/40 = 4.0 mm.

EXAMPLE 35.6 | **Viewing blood cells**

A pathologist inspects a sample of 7-μm-diameter human blood cells under a microscope. She selects a 40× objective and a 10× eyepiece. What size object, viewed from 25 cm, has the same apparent size as a blood cell seen through the microscope?

MODEL Angular magnification compares the magnified angular size to the angular size seen at the near-point distance of 25 cm.

SOLVE The microscope's angular magnification is $M = -(40) \times (10) = -400$. The magnified cells will have the same apparent size as an object $400 \times 7 \ \mu m \approx 3$ mm in diameter seen from a distance of 25 cm.

ASSESS 3 mm is about the size of a capital O in this textbook, so a blood cell seen through the microscope will have about the same apparent size as an O seen from a comfortable reading distance.

STOP TO THINK 35.4 A biologist rotates the turret of a microscope to replace a 20× objective with a 10× objective. To keep the same overall magnification, the focal length of the eyepiece must be

a. Doubled.   b. Halved.   c. Kept the same.
d. The magnification cannot be kept the same if the objective is changed.

The Telescope

A microscope magnifies small, nearby objects to look large. A telescope magnifies distant objects, which might be quite large, so that we can see details that are blended together when seen by eye.

FIGURE 35.15 shows the optical layout of a simple telescope. A large-diameter objective lens (larger lenses collect more light and thus can see fainter objects) collects the parallel rays from a distant object ($s = \infty$) and forms a real, inverted image at distance $s' = f_{obj}$. Unlike a microscope, which uses a short-focal-length objective, the focal length of a telescope objective is very nearly the length of the telescope tube. Then, just as in the microscope, the eyepiece functions as a simple magnifier. The viewer observes an inverted image, but that's not a serious problem in astronomy. Telescopes for use on earth have a somewhat different design to obtain an upright image.

FIGURE 35.15 A refracting telescope.

Suppose the distant object, as seen by the objective lens, subtends angle θ_{obj}. If the image seen through the eyepiece subtends a larger angle θ_{eye}, then the angular magnification is $M = \theta_{eye}/\theta_{obj}$. We can see from the undeviated ray passing through the center of the objective lens that (using the small-angle approximation)

$$\theta_{obj} \approx -\frac{h'}{f_{obj}}$$

where the minus sign indicates the inverted image. The image of height h' acts as the object for the eyepiece, and we can see that the final image observed by the viewer subtends angle

$$\theta_{eye} = \frac{h'}{f_{eye}}$$

Consequently, the angular magnification of a telescope is

$$M = \frac{\theta_{eye}}{\theta_{obj}} = -\frac{f_{obj}}{f_{eye}} \tag{35.11}$$

The angular magnification is simply the ratio of the objective focal length to the eyepiece focal length.

Because the stars and galaxies are so distant, light-gathering power is more important to astronomers than magnification. Large light-gathering power requires a large-diameter objective lens, but large lenses are not practical; they begin to sag under their own weight. Thus **refracting telescopes,** with two lenses, are relatively small. Serious astronomy is done with a **reflecting telescope,** such as the one shown in **FIGURE 35.16**.

A large-diameter mirror (the *primary mirror*) focuses the rays to form a real image, but, for practical reasons, a *secondary mirror* reflects the rays sideways before they reach a focus. This moves the primary mirror's image out to the edge of the telescope where it can be viewed by an eyepiece on the side. None of these changes affects the overall analysis of the telescope, and its angular magnification is given by Equation 35.11 if f_{obj} is replaced by f_{pri}, the focal length of the primary mirror.

FIGURE 35.16 A reflecting telescope.

35.5 Color and Dispersion

One of the most obvious visual aspects of light is the phenomenon of color. Yet color, for all its vivid sensation, is not inherent in the light itself. Color is a *perception*, not a physical quantity. Color is associated with the wavelength of light, but the fact that we see light with a wavelength of 650 nm as "red" tells us how our visual system responds to electromagnetic waves of this wavelength. There is no "redness" associated with the light wave itself.

Most of the results of optics do not depend on color; a microscope works the same with red light and blue light. Even so, indices of refraction are slightly wavelength dependent, which will be important in the next section where we consider the resolution of optical instruments. And color in nature is an interesting subject, one worthy of a short digression.

Color

It has been known since antiquity that irregularly shaped glass and crystals cause sunlight to be broken into various colors. A common idea was that the glass or crystal somehow altered the properties of the light by *adding* color to the light. Newton suggested a different explanation. He first passed a sunbeam through a prism, producing the familiar rainbow of light. We say that the prism *disperses* the light. Newton's novel idea, shown in **FIGURE 35.17a**, was to use a second prism, inverted with respect to the first, to "reassemble" the colors. He found that the light emerging from the second prism was a beam of pure, white light.

But the emerging light beam is white only if *all* the rays are allowed to move between the two prisms. Blocking some of the rays with small obstacles, as in **FIGURE 35.17b**, causes the emerging light beam to have color. This suggests that color is associated with the light itself, not with anything that the prism is doing to the light. Newton tested this idea by inserting a small aperture between the prisms to pass only the rays of a particular color, such as green. If the prism alters the properties of light, then the second prism should change the green light to other colors. Instead, the light emerging from the second prism is unchanged from the green light entering the prism.

These and similar experiments show that

1. What we perceive as white light is a mixture of all colors. White light can be dispersed into its various colors and, equally important, mixing all the colors produces white light.
2. The index of refraction of a transparent material differs slightly for different colors of light. Glass has a slightly larger index of refraction for violet light than for green light or red light. Consequently, different colors of light refract at slightly different angles. A prism does not alter the light or add anything to the light; it simply causes the different colors that are inherent in white light to follow slightly different trajectories.

Dispersion

It was Thomas Young, with his two-slit interference experiment, who showed that different colors are associated with light of different wavelengths. The longest wavelengths are perceived as red light and the shortest as violet light. **TABLE 35.1** is a brief summary of the *visible spectrum* of light. Visible-light wavelengths are used so frequently that it is well worth committing this short table to memory.

The slight variation of index of refraction with wavelength is known as **dispersion. FIGURE 35.18** shows the *dispersion curves* of two common glasses. Notice that **n is *larger* when the wavelength is *shorter*,** thus violet light refracts more than red light.

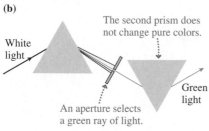

FIGURE 35.17 Newton used prisms to study color.

(a)

White light

A second prism can combine the colors back into white light.

A prism disperses white light into colors.

White light

(b)

White light

The second prism does not change pure colors.

An aperture selects a green ray of light.

Green light

TABLE 35.1 A brief summary of the visible spectrum of light

Color	Approximate wavelength
Deepest red	700 nm
Red	650 nm
Green	550 nm
Blue	450 nm
Deepest violet	400 nm

FIGURE 35.18 Dispersion curves show how the index of refraction varies with wavelength.

EXAMPLE 35.7 | **Dispersing light with a prism**

Example 34.4 found that a ray incident on a 30° prism is deflected by 22.6° if the prism's index of refraction is 1.59. Suppose this is the index of refraction of deep violet light and deep red light has an index of refraction of 1.54.

a. What is the deflection angle for deep red light?

b. If a beam of white light is dispersed by this prism, how wide is the rainbow spectrum on a screen 2.0 m away?

VISUALIZE Figure 34.19 showed the geometry. A ray of any wavelength is incident on the hypotenuse of the prism at $\theta_1 = 30°$.

SOLVE a. If $n_1 = 1.54$ for deep red light, the refraction angle is

$$\theta_2 = \sin^{-1}\left(\frac{n_1 \sin\theta_1}{n_2}\right) = \sin^{-1}\left(\frac{1.54 \sin 30°}{1.00}\right) = 50.4°$$

Example 34.4 showed that the deflection angle is $\phi = \theta_2 - \theta_1$, so deep red light is deflected by $\phi_{red} = 20.4°$. This angle is slightly smaller than the previously observed $\phi_{violet} = 22.6°$.

b. The entire spectrum is spread between $\phi_{red} = 20.4°$ and $\phi_{violet} = 22.6°$. The angular spread is

$$\delta = \phi_{violet} - \phi_{red} = 2.2° = 0.038 \text{ rad}$$

At distance r, the spectrum spans an arc length

$$s = r\delta = (2.0 \text{ m})(0.038 \text{ rad}) = 0.076 \text{ m} = 7.6 \text{ cm}$$

ASSESS The angle is so small that there's no appreciable difference between arc length and a straight line. The spectrum will be 7.6 cm wide at a distance of 2.0 m.

Rainbows

One of the most interesting sources of color in nature is the rainbow. The details get somewhat complicated, but **FIGURE 35.19a** shows that the basic cause of the rainbow is a combination of refraction, reflection, and dispersion.

Figure 35.19a might lead you to think that the top edge of a rainbow is violet. In fact, the top edge is red, and violet is on the bottom. The rays leaving the drop in Figure 35.19a are spreading apart, so they can't all reach your eye. As **FIGURE 35.19b** shows, a ray of red light reaching your eye comes from a drop *higher* in the sky than a ray of violet light. In other words, the colors you see in a rainbow refract toward your eye from different raindrops, not from the same drop. You have to look higher in the sky to see the red light than to see the violet light.

FIGURE 35.19 Light seen in a rainbow has undergone refraction + reflection + refraction in a raindrop.

(a)

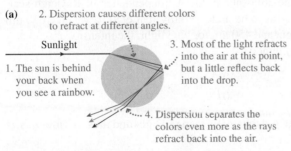

2. Dispersion causes different colors to refract at different angles.

Sunlight

1. The sun is behind your back when you see a rainbow.

3. Most of the light refracts into the air at this point, but a little reflects back into the drop.

4. Dispersion separates the colors even more as the rays refract back into the air.

(b)

Sunlight

Eye

You see a rainbow with red on the top, violet on the bottom.

Red light is refracted predominantly at 42.5°. The red light reaching your eye comes from drops higher in the sky.

Violet light is refracted predominantly at 40.8°. The violet light reaching your eye comes from drops lower in the sky.

Colored Filters and Colored Objects

White light passing through a piece of green glass emerges as green light. A possible explanation would be that the green glass *adds* "greenness" to the white light, but Newton found otherwise. Green glass is green because it *absorbs* any light that is "not green." We can think of a piece of colored glass or plastic as a *filter* that removes all wavelengths except a chosen few.

If a green filter and a red filter are overlapped, as in **FIGURE 35.20**, *no* light gets through. The green filter transmits only green light, which is then absorbed by the red filter because it is "not red."

This behavior is true not just for glass filters, which transmit light, but for *pigments* that absorb light of some wavelengths but *reflect* light at other wavelengths. For

FIGURE 35.20 No light at all passes through both a green and a red filter.

Red filter

Black where filters overlap   Green filter

FIGURE 35.21 The absorption curve of chlorophyll.

FIGURE 35.21 The absorption curve of chlorophyll.

Chlorophyll absorbs most of the red and blue/violet light for use in photosynthesis.

The green and yellow light that is not absorbed is reflected and gives plants their green color.

Sunsets are red because all the blue light has scattered as the sunlight passes through the atmosphere.

FIGURE 35.22 Rayleigh scattering by molecules in the air gives the sky and sunsets their color.

At midday the scattered light is mostly blue because molecules preferentially scatter shorter wavelengths.

Air molecules

Sun

Observer at midday

Observer at sunset

At sunset, when the light has traveled much farther through the atmosphere, the light is mostly red because the shorter wavelengths have been lost to scattering.

example, red paint contains pigments reflecting light at wavelengths near 650 nm while absorbing all other wavelengths. Pigments in paints, inks, and natural objects are responsible for most of the color we observe in the world, from the red of lipstick to the blue of a blueberry.

As an example, **FIGURE 35.21** shows the absorption curve of *chlorophyll*. Chlorophyll is essential for photosynthesis in green plants. The chemical reactions of photosynthesis are able to use red light and blue/violet light, thus chlorophyll absorbs red light and blue/violet light from sunlight and puts it to use. But green and yellow light are not absorbed. Instead, to conserve energy, these wavelengths are mostly *reflected* to give the object a greenish-yellow color. When you look at the green leaves on a tree, you're seeing the light that was reflected because it *wasn't* needed for photosynthesis.

Light Scattering: Blue Skies and Red Sunsets

In the ray model of Section 34.1 we noted that light within a medium can be scattered or absorbed. As we've now seen, the absorption of light can be wavelength dependent and can create color in objects. What are the effects of scattering?

Light can scatter from small particles that are suspended in a medium. If the particles are large compared to the wavelengths of light—even though they may be microscopic and not readily visible to the naked eye—the light essentially reflects off the particles. The law of reflection doesn't depend on wavelength, so all colors are scattered equally. White light scattered from many small particles makes the medium appear cloudy and white. Two well-known examples are clouds, where micrometer-size water droplets scatter the light, and milk, which is a colloidal suspension of microscopic droplets of fats and proteins.

A more interesting aspect of scattering occurs at the atomic level. The atoms and molecules of a transparent medium are much smaller than the wavelengths of light, so they can't scatter light simply by reflection. Instead, the oscillating electric field of the light wave interacts with the electrons in each atom in such a way that the light is scattered. This atomic-level scattering is called **Rayleigh scattering.**

Unlike the scattering by small particles, Rayleigh scattering from atoms and molecules *does* depend on the wavelength. A detailed analysis shows that the intensity of scattered light depends inversely on the fourth power of the wavelength: $I_{\text{scattered}} \propto \lambda^{-4}$. This wavelength dependence explains why the sky is blue and sunsets are red.

As sunlight travels through the atmosphere, the λ^{-4} dependence of Rayleigh scattering causes the shorter wavelengths to be preferentially scattered. If we take 650 nm as a typical wavelength for red light and 450 nm for blue light, the intensity of scattered blue light relative to scattered red light is

$$\frac{I_{\text{blue}}}{I_{\text{red}}} = \left(\frac{650}{450}\right)^4 \approx 4$$

Four times more blue light is scattered toward us than red light and thus, as **FIGURE 35.22** shows, the sky appears blue.

Because of the earth's curvature, sunlight has to travel much farther through the atmosphere when we see it at sunrise or sunset than it does during the midday hours. In fact, the path length through the atmosphere at sunset is so long that essentially all the short wavelengths have been lost due to Rayleigh scattering. Only the longer wavelengths remain—orange and red—and they make the colors of the sunset.

35.6 The Resolution of Optical Instruments

A camera *could* focus light with a single lens. A microscope objective *could* be built with a single lens. So why would anyone ever use a lens combination in place of a single lens? There are two primary reasons.

First, as you learned in the previous section, any lens has dispersion. That is, its index of refraction varies slightly with wavelength. Because the index of refraction for violet light is larger than for red light, a lens's focal length is shorter for violet light than for red light. Consequently, different colors of light come to a focus at slightly different distances from the lens. If red light is sharply focused on a viewing screen, then blue and violet wavelengths are not well focused. This imaging error, illustrated in FIGURE 35.23a, is called **chromatic aberration.**

Second, our analysis of thin lenses was based on paraxial rays traveling nearly parallel to the optical axis. A more exact analysis, taking all the rays into account, finds that rays incident on the outer edges of a spherical surface are not focused at exactly the same point as rays incident near the center. This imaging error, shown in FIGURE 35.23b, is called **spherical aberration.** Spherical aberration, which causes the image to be slightly blurred, gets worse as the lens diameter increases.

Fortunately, the chromatic and spherical aberrations of a converging lens and a diverging lens are in opposite directions. When a converging lens and a diverging lens are used in combination, their aberrations tend to cancel. A combination lens, such as the one in FIGURE 35.23c, can produce a much sharper focus than a single lens with the equivalent focal length. Consequently, most optical instruments use combination lenses rather than single lenses.

Diffraction Again

According to the ray model of light, a perfect lens (one with no aberrations) should be able to form a perfect image. But the ray model of light, though a very good model for lenses, is not an absolutely correct description of light. If we look closely, the wave aspects of light haven't entirely disappeared. In fact, the performance of optical equipment is limited by the diffraction of light.

FIGURE 35.24a shows a plane wave, with parallel light rays, being focused by a lens of diameter D. According to the ray model of light, a perfect lens would focus parallel rays to a perfect point. Notice, though, that only a piece of each wave front passes *through* the lens and gets focused. In effect, **the lens itself acts as a circular aperture** in an opaque barrier, allowing through only a portion of each wave front. Consequently, **the lens diffracts the light wave.** The diffraction is usually very small because D is usually much greater than the wavelength of the light; nonetheless, this small amount of diffraction is the limiting factor in how well the lens can focus the light.

FIGURE 35.24b separates the diffraction from the focusing by modeling the lens as an actual aperture of diameter D followed by an "ideal" diffractionless lens. You learned in Chapter 33 that a circular aperture produces a diffraction pattern with a bright central maximum surrounded by dimmer fringes. A converging lens brings this diffraction pattern to a focus in the image plane, as shown in FIGURE 35.24c. As a result, a perfect lens focuses parallel light rays not to a perfect point of light, as we expected, but to a small, circular diffraction pattern.

FIGURE 35.23 Chromatic aberration and spherical aberration prevent simple lenses from forming perfect images.

(a) Chromatic aberration

Different wavelengths focus at different points.

(b) Spherical aberration

Rays at different angles focus at different points.

(c) Correcting aberrations

High power, low dispersion

All rays meet at the same focus.

Low power, high dispersion

FIGURE 35.24 A lens both focuses and diffracts the light passing through.

(a) A lens acts as a circular aperture.

Wave fronts

Not focused

Rays are converging to the focal point.

Spherical wave fronts are converging to the focal point.

Not focused

Rays are perpendicular to the wave fronts.

(b) The aperture and focusing effects can be separated.

Wave fronts

Ideal diffractionless lens with focal length f

D

f

Circular aperture of diameter D

(c) The lens focuses the diffraction pattern in the focal plane.

Wave fronts

The first dark fringe is focused at position y_1.

θ_1

Width

w

Light intensity

Parallel rays

Light is diffracting at angle θ_1 to the first dark fringe.

f

The angle to the first minimum of a circular diffraction pattern is $\theta_1 = 1.22\lambda/D$. The ray that passes through the center of a lens is not bent, so Figure 35.24c uses this ray to show that the position of the dark fringe is $y_1 = f\tan\theta_1 \approx f\theta_1$. Thus the width of the central maximum in the focal plane is

$$w_{min} \approx 2f\theta_1 = \frac{2.44\lambda f}{D} \qquad \text{(minimum spot size)} \qquad (35.12)$$

This is the **minimum spot size** to which a lens can focus light.

Lenses are often limited by aberrations, so not all lenses can focus parallel light rays to a spot this small. A well-crafted lens, for which Equation 35.12 is the minimum spot size, is called a *diffraction-limited lens*. No optical design can overcome the spreading of light due to diffraction, and it is because of this spreading that the image point has a minimum spot size. The image of an actual object, rather than of parallel rays, becomes a mosaic of overlapping diffraction patterns, so even the most perfect lens inevitably forms an image that is slightly fuzzy.

For various reasons, it is difficult to produce a diffraction-limited lens having a focal length that is much less than its diameter. The very best microscope objectives have $f \approx 0.5D$. This implies that **the smallest diameter to which you can focus a spot of light, no matter how hard you try, is $w_{min} \approx \lambda$.** This is a fundamental limit on the performance of optical equipment. Diffraction has very real consequences!

One example of these consequences is found in the manufacturing of integrated circuits. Integrated circuits are made by creating a "mask" showing all the components and their connections. A lens images this mask onto the surface of a semiconductor wafer that has been coated with a substance called *photoresist*. Bright areas in the mask expose the photoresist, and subsequent processing steps chemically etch away the exposed areas while leaving behind areas that had been in the shadows of the mask. This process is called *photolithography*.

The power of a microprocessor and the amount of memory in a memory chip depend on how small the circuit elements can be made. Diffraction dictates that a circuit element can be no smaller than the smallest spot to which light can be focused, which is roughly the wavelength of the light. If the mask is projected with ultraviolet light having $\lambda \approx 200$ nm, then the smallest elements on a chip are about 200 nm wide. This is, in fact, just about the current limit of technology.

The size of the features in an integrated circuit is limited by the diffraction of light.

EXAMPLE 35.8 | Seeing stars

A 12-cm-diameter telescope lens has a focal length of 1.0 m. What is the diameter of the image of a star in the focal plane if the lens is diffraction limited *and* if the earth's atmosphere is not a limitation?

MODEL Stars are so far away that they appear as points in space. An ideal diffractionless lens would focus their light to arbitrarily small points. Diffraction prevents this. Model the telescope lens as a 12-cm-diameter aperture in front of an ideal lens with a 1.0 m focal length.

SOLVE The minimum spot size in the focal plane of this lens is

$$w = \frac{2.44\lambda f}{D}$$

where D is the lens diameter. What is λ? Because stars emit white light, the *longest* wavelengths spread the most and determine the size of the image that is seen. If we use $\lambda = 700$ nm as the approximate upper limit of visible wavelengths, we find $w = 1.4 \times 10^{-5}$ m = 14 μm.

ASSESS This is certainly small, and it would appear as a point to your unaided eye. Nonetheless, the spot size would be easily noticed if it were recorded on film and enlarged. Turbulence and temperature effects in the atmosphere, the causes of the "twinkling" of stars, prevent ground-based telescopes from being this good, but space-based telescopes really are diffraction limited.

Resolution

Suppose you point a telescope at two nearby stars in a galaxy far, far away. If you use the best possible detector, will you be able to distinguish separate images for the two stars, or will they blur into a single blob of light? A similar question could be asked of a microscope. Can two microscopic objects, very close together, be distinguished if sufficient magnification is used? Or is there some size limit at which their images will blur together and never be separated? These are important questions about the *resolution* of optical instruments.

Because of diffraction, the image of a distant star is not a point but a circular diffraction pattern. Our question, then, really is: How close together can two diffraction patterns be before you can no longer distinguish them? One of the major scientists of the 19th century, Lord Rayleigh, studied this problem and suggested a reasonable rule that today is called **Rayleigh's criterion.**

FIGURE 35.25 shows two distant point sources being imaged by a lens of diameter D. The angular separation between the objects, as seen from the lens, is α. Rayleigh's criterion states that

- The two objects are resolvable if $\alpha > \theta_{min}$, where $\theta_{min} = \theta_1 = 1.22\lambda/D$ is the angle of the first dark fringe in the circular diffraction pattern.
- The two objects are not resolvable if $\alpha < \theta_{min}$ because their diffraction patterns are too overlapped.
- The two objects are marginally resolvable if $\alpha = \theta_{min}$. The central maximum of one image falls exactly on top of the first dark fringe of the other image. This is the situation shown in the figure.

FIGURE 35.26 shows enlarged photographs of the images of two point sources. The images are circular diffraction patterns, not points. The two images are close but distinct where the objects are separated by $\alpha > \theta_{min}$. Two objects really were recorded in the photo at the bottom, but their separation is $\alpha < \theta_{min}$ and their images have blended together. In the middle photo, with $\alpha = \theta_{min}$, you can see that the two images are just barely resolved.

The angle

$$\theta_{min} = \frac{1.22\lambda}{D} \qquad \text{(angular resolution of a lens)} \qquad (35.13)$$

is called the **angular resolution** of a lens. The angular resolution of a telescope depends on the diameter of the objective lens (or the primary mirror) and the wavelength of the light; magnification is not a factor. Two images will remain overlapped and unresolved no matter what the magnification if their angular separation is less than θ_{min}. For visible light, where λ is pretty much fixed, the only parameter over which the astronomer has any control is the diameter of the lens or mirror of the telescope. The urge to build ever-larger telescopes is motivated, in part, by a desire to improve the angular resolution. (Another motivation is to increase the light-gathering power so as to see objects farther away.)

The performance of a microscope is also limited by the diffraction of light passing through the objective lens. Just as light cannot be focused to a spot smaller than about a wavelength, the most perfect microscope cannot resolve the features of objects separated by less than one wavelength, or roughly 500 nm. (Some clever tricks with the phase of the light can improve the resolution to about 200 nm, but diffraction is still the limiting factor.) Because atoms are approximately 0.1 nm in diameter, vastly smaller than the wavelength of visible or even ultraviolet light, there is no hope of ever seeing atoms with an optical microscope. This limitation is not simply a matter of needing a better design or more precise components; it is a fundamental limit set by the wave nature of the light with which we see.

FIGURE 35.25 Two images that are marginally resolved.

The image of each object is a small circular diffraction pattern.

Object 2

Object 1

Distant point sources

Image 1

D

α

$\alpha = \theta_{min}$

Image 2

The maximum of image 2 falls on the first dark fringe of image 1. The images are marginally resolved.

FIGURE 35.26 Enlarged photographs of the images of two point sources.

Resolved

$\alpha > \theta_{min}$

Marginally resolved

$\alpha = \theta_{min}$

Not resolved

$\alpha < \theta_{min}$

STOP TO THINK 35.5 Four diffraction-limited lenses focus plane waves of light with the same wavelength λ. Rank in order, from largest to smallest, the spot sizes w_a to w_d.

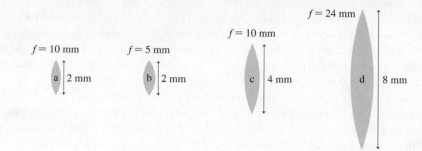

CHALLENGE EXAMPLE 35.9 | Visual acuity

The normal human eye has maximum visual acuity with a pupil diameter of about 3 mm. For larger pupils, acuity decreases due to increasing aberrations; for smaller pupils, acuity decreases due to increasing diffraction. If your pupil diameter is 2.0 mm, as it would be in bright light, what is the smallest-diameter circle that you should be able to see as a circle, rather than just an unresolved blob, on an eye chart at the standard distance of 20 ft? The index of refraction inside the eye is 1.33.

MODEL Assume that a 2.0-mm-diameter pupil is diffraction limited. Then the angular resolution is given by Rayleigh's criterion. Diffraction increases with wavelength, so the eye's acuity will be affected more by longer wavelengths than by shorter wavelengths. Consequently, assume that the light's wavelength in air is 600 nm.

VISUALIZE Let the diameter of the circle be d. **FIGURE 35.27** shows the circle at distance $s = 20$ ft $= 6.1$ m. "Seeing the circle," shown edge-on, requires resolving the top and bottom lines as distinct.

FIGURE 35.27 Viewing a circle of diameter d.

SOLVE The angular separation of the top and bottom lines of the circle is $\alpha = d/s$. Rayleigh's criterion says that a perfect lens with aperture D can just barely resolve these two lines if

$$\alpha = \frac{d}{s} = \theta_{min} = \frac{1.22\lambda_{eye}}{D} = \frac{1.22\lambda_{air}}{n_{eye}D}$$

The diffraction takes place inside the eye, where the wavelength is shortened to $\lambda_{eye} = \lambda_{air}/n_{eye}$. Thus the circle diameter that can barely be resolved with perfect vision is

$$d = \frac{1.22\lambda_{air}s}{n_{eye}D} = \frac{1.22(600 \times 10^{-9} \text{ m})(6.1 \text{ m})}{(1.33)(0.0020 \text{ m})} \approx 2 \text{ mm}$$

That's about the height of a capital O in this book, so in principle you should—in very bright light—just barely be able to recognize it as an O at 20 feet.

ASSESS On an eye chart, the O on the line for 20/20 vision—the standard of excellent vision—is about 7 mm tall, so the calculated 2 mm, although in the right range, is a bit too small. There are two reasons. First, although aberrations of the eye are reduced with a smaller pupil, they haven't vanished. And second, for a 2-mm-tall object at 20 ft, the size of the image on the retina is barely larger than the spacing between the cone cells, so the resolution of the "detector" is also a factor. Your eye is a very good optical instrument, but not perfect.

SUMMARY

The goal of Chapter 35 has been to learn about some common optical instruments and their limitations.

IMPORTANT CONCEPTS

Lens Combinations

The image of the first lens acts as the object for the second lens.

Lens **power:** $P = \dfrac{1}{f}$ diopters, $1\ D = 1\ m^{-1}$

Resolution

The **angular resolution** of a lens of diameter D is
$$\theta_{min} = 1.22\lambda/D$$

Rayleigh's criterion states that two objects separated by an angle α are marginally resolvable if $\alpha = \theta_{min}$.

APPLICATIONS

Cameras

Form a real, inverted image on a detector. The lens's *f*-**number** is
$$f\text{-number} = \frac{f}{D}$$

The light intensity on the detector is
$$I \propto \frac{1}{(f\text{-number})^2}$$

Magnifiers

For relaxed-eye viewing, the angular magnification is
$$M = \frac{25\ cm}{f}$$

For microscopes and telescopes, angular magnification, not lateral magnification, is the important characteristic. The eyepiece acts as a magnifier to view the image formed by the objective lens.

Vision

Refraction at the cornea is responsible for most of the focusing. The lens provides fine-tuning by changing its shape (**accommodation**).

In normal vision, the eye can focus from a far point (FP) at ∞ (relaxed eye) to a near point (NP) at $\approx 25\ cm$ (maximum accommodation).

- **Hyperopia** (farsightedness) is corrected with a converging lens.
- **Myopia** (nearsightedness) is corrected with a diverging lens.

Microscopes

The object is very close to the focal point of the objective. The total angular magnification is $M \approx -\dfrac{L}{f_{obj}}\dfrac{25\ cm}{f_{eye}}$.

The best possible spatial resolution of a microscope, limited by diffraction, is about one wavelength of light.

Focusing and spatial resolution

The minimum spot size to which a lens of focal length f and diameter D can focus light is limited by diffraction to
$$w_{min} = \frac{2.44\lambda f}{D}$$

With the best lenses that can be manufactured, $w_{min} \approx \lambda$.

Telescopes

The object is very far from the objective.

The total angular magnification is $M = -\dfrac{f_{obj}}{f_{eye}}$.

TERMS AND NOTATION

camera	iris	hyperopia	reflecting telescope
effective focal length, f	retina	myopia	dispersion
aperture	accommodation	angular size	Rayleigh scattering
f-number	far point	magnifier	chromatic aberration
CCD	near point	angular magnification, M	spherical aberration
pixel	presbyopia	objective	minimum spot size, w_{min}
cornea	power, P	eyepiece	Raylcigh's criterion
pupil	diopter, D	refracting telescope	angular resolution

CONCEPTUAL QUESTIONS

1. Suppose a camera's exposure is correct when the lens has a focal length of 8.0 mm. Will the picture be overexposed, underexposed, or still correct if the focal length is "zoomed" to 16.0 mm without changing the diameter of the lens aperture? Explain.

2. A camera has a circular aperture immediately behind the lens. Reducing the aperture diameter to half its initial value will
   A. Make the image blurry.
   B. Cut off the outer half of the image and leave the inner half unchanged.
   C. Make the image less bright.
   D. All the above.
   Explain your choice.

3. Suppose you wanted special glasses designed to let you see underwater without a face mask. Should the glasses use a converging or diverging lens? Explain.

4. A red card is illuminated by red light. What color will the card appear? What if it's illuminated by blue light?

5. The center of the galaxy is filled with low-density hydrogen gas that scatters light rays. An astronomer wants to take a picture of the center of the galaxy. Will the view be better using ultraviolet light, visible light, or infrared light? Explain.

6. A friend lends you the eyepiece of his microscope to use on your own microscope. He claims the spatial resolution of your microscope will be halved, since his eyepiece has the same diameter as yours but twice the magnification. Is his claim valid? Explain.

7. A diffraction-limited lens can focus light to a 10-μm-diameter spot on a screen. Do the following actions make the spot diameter larger, make it smaller, or leave it unchanged?
   A. Decreasing the wavelength of the light.
   B. Decreasing the lens diameter.
   C. Decreasing the lens focal length.
   D. Decreasing the lens-to-screen distance.

8. To focus parallel light rays to the smallest possible spot, should you use a lens with a small f-number or a large f-number? Explain.

9. An astronomer is trying to observe two distant stars. The stars are marginally resolved when she looks at them through a filter that passes green light with a wavelength near 550 nm. Which of the following actions would improve the resolution? Assume that the resolution is not limited by the atmosphere.
   A. Changing the filter to a different wavelength. If so, should she use a shorter or a longer wavelength?
   B. Using a telescope with an objective lens of the same diameter but a different focal length. If so, should she select a shorter or a longer focal length?
   C. Using a telescope with an objective lens of the same focal length but a different diameter. If so, should she select a larger or a smaller diameter?
   D. Using an eyepiece with a different magnification. If so, should she select an eyepiece with more or less magnification?

EXERCISES AND PROBLEMS

Problems labeled [blank] integrate material from earlier chapters.

Exercises

Section 35.1 Lenses in Combination

1. ‖ Two converging lenses with focal lengths of 40 cm and 20 cm are 10 cm apart. A 2.0-cm-tall object is 15 cm in front of the 40-cm-focal-length lens.
   a. Use ray tracing to find the position and height of the image. Do this accurately using a ruler or paper with a grid, then make measurements on your diagram.
   b. Calculate the image position and height. Compare with your ray-tracing answers in part a.

2. ‖ A converging lens with a focal length of 40 cm and a diverging lens with a focal length of −40 cm are 160 cm apart. A 2.0-cm-tall object is 60 cm in front of the converging lens.
   a. Use ray tracing to find the position and height of the image. Do this accurately using a ruler or paper with a grid, then make measurements on your diagram.
   b. Calculate the image position and height. Compare with your ray-tracing answers in part a.

3. ‖ A 2.0-cm-tall object is 20 cm to the left of a lens with a focal length of 10 cm. A second lens with a focal length of 15 cm is 30 cm to the right of the first lens.
   a. Use ray tracing to find the position and height of the image. Do this accurately using a ruler or paper with a grid, then make measurements on your diagram.
   b. Calculate the image position and height. Compare with your ray-tracing answers in part a.

4. ‖ A 2.0-cm-tall object is 20 cm to the left of a lens with a focal length of 10 cm. A second lens with a focal length of 5 cm is 30 cm to the right of the first lens.
   a. Use ray tracing to find the position and height of the image. Do this accurately using a ruler or paper with a grid, then make measurements on your diagram.
   b. Calculate the image position and height. Compare with your ray-tracing answers in part a.

5. ‖ A 2.0-cm-tall object is 20 cm to the left of a lens with a focal length of 10 cm. A second lens with a focal length of −5 cm is 30 cm to the right of the first lens.
   a. Use ray tracing to find the position and height of the image. Do this accurately using a ruler or paper with a grid, then make measurements on your diagram.
   b. Calculate the image position and height. Compare with your ray-tracing answers in part a.

Section 35.2 The Camera

6. | A 2.0-m-tall man is 10 m in front of a camera with a 15-mm-focal-length lens. How tall is his image on the detector?

7. | What is the f-number of a lens with a 35 mm focal length and a 7.0-mm-diameter aperture?

8. | What is the aperture diameter of a 12-mm-focal-length lens set to $f/4.0$?

9. | A camera takes a properly exposed photo at $f/5.6$ and 1/125 s. What shutter speed should be used if the lens is changed to $f/4.0$?

10. ‖ A camera takes a properly exposed photo with a 3.0-mm-diameter aperture and a shutter speed of 1/125 s. What is the appropriate aperture diameter for a 1/500 s shutter speed?

Section 35.3 Vision

11. ‖ Ramon has contact lenses with the prescription +2.0 D.
BIO a. What eye condition does Ramon have?
   b. What is his near point without the lenses?

12. | Ellen wears eyeglasses with the prescription −1.0 D.
BIO a. What eye condition does Ellen have?
   b. What is her far point without the glasses?

13. | What is the f-number of a relaxed eye with the pupil fully
BIO dilated to 8.0 mm? Model the eye as a single lens 2.4 cm in front of the retina.

Section 35.4 Optical Systems That Magnify

14. ‖ A magnifier has a magnification of 5×. How far from the lens should an object be held so that its image is seen at the near-point distance of 25 cm? Assume that your eye is immediately behind the lens.

15. ‖ A microscope has a 20 cm tube length. What focal-length objective will give total magnification 500× when used with an eyepiece having a focal length of 5.0 cm?

16. ‖ A standardized biological microscope has an 8.0-mm-focal-length objective. What focal-length eyepiece should be used to achieve a total magnification of 100×?

17. ‖ A 6.0-mm-diameter microscope objective has a focal length of 9.0 mm. What object distance gives a lateral magnification of −40?

18. ‖ A 20× telescope has a 12-cm-diameter objective lens. What minimum diameter must the eyepiece lens have to collect all the light rays from an on-axis distant source?

19. ‖ A reflecting telescope is built with a 20-cm-diameter mirror having a 1.00 m focal length. It is used with a 10× eyepiece. What are (a) the magnification and (b) the f-number of the telescope?

Section 35.5 Color and Dispersion

20. ‖ A sheet of glass has $n_{red} = 1.52$ and $n_{violet} = 1.55$. A narrow beam of white light is incident on the glass at 30°. What is the angular spread of the light inside the glass?

21. | A narrow beam of white light is incident on a sheet of quartz. The beam disperses in the quartz, with red light ($\lambda \approx 700$ nm) traveling at an angle of 26.3° with respect to the normal and violet light ($\lambda \approx 400$ nm) traveling at 25.7°. The index of refraction of quartz for red light is 1.45. What is the index of refraction of quartz for violet light?

22. ‖ A hydrogen discharge lamp emits light with two prominent wavelengths: 656 nm (red) and 486 nm (blue). The light enters a flint-glass prism perpendicular to one face and then refracts through the hypotenuse back into the air. The angle between these two faces is 35°.
   a. Use Figure 35.18 to estimate to ±0.002 the index of refraction of flint glass at these two wavelengths.
   b. What is the angle (in degrees) between the red and blue light as it leaves the prism?

23. ‖ Infrared telescopes, which use special infrared detectors, are able to peer farther into star-forming regions of the galaxy because infrared light is not scattered as strongly as is visible light by the tenuous clouds of hydrogen gas from which new stars are created. For what wavelength of light is the scattering only 1% that of light with a visible wavelength of 500 nm?

Section 35.6 The Resolution of Optical Instruments

24. ‖ A scientist needs to focus a helium-neon laser beam ($\lambda = 633$ nm) to a 10 μm diameter spot 8.0 cm behind a lens.
   a. What focal-length lens should she use?
   b. What minimum diameter must the lens have?

25. ‖ Two lightbulbs are 1.0 m apart. From what distance can these lightbulbs be marginally resolved by a small telescope with a 4.0-cm-diameter objective lens? Assume that the lens is diffraction limited and $\lambda = 600$ nm.

Problems

26. ‖ In FIGURE P35.26, parallel rays from the left are focused to a point at two locations on the optical axis. Find the position of each location, giving your answer as a distance left or right of the lens.

$f = 2.0$ cm $f = -1.5$ cm $f = 10$ cm $f = 10$ cm

5.0 cm   5.0 cm   5.0 cm

Lens   Mirror   Mirror   Lens

FIGURE P35.26   FIGURE P35.27

27. ‖ The rays leaving the two-component optical system of FIGURE P35.27 produce two distinct images of the 1.0-cm-tall object. What are the position (relative to the lens), orientation, and height of each image?

28. | A common optical instrument in a laser laboratory is a *beam expander*. One type of beam expander is shown in **FIGURE P35.28**. The parallel rays of a laser beam of width w_1 enter from the left.
   a. For what lens spacing d does a parallel laser beam exit from the right?
   b. What is the width w_2 of the exiting laser beam?

FIGURE P35.28

29. | A common optical instrument in a laser laboratory is a *beam expander*. One type of beam expander is shown in **FIGURE P35.29**. The parallel rays of a laser beam of width w_1 enter from the left.
   a. For what lens spacing d does a parallel laser beam exit from the right?
   b. What is the width w_2 of the exiting laser beam?

FIGURE P35.29

30. ‖ In **FIGURE P35.30**, what are the position, height, and orientation of the final image? Give the position as a distance to the right or left of the lens.

FIGURE P35.30

31. ‖ A 1.0-cm-tall object is 110 cm from a screen. A diverging lens with focal length −20 cm is 20 cm in front of the object. What are the focal length and distance from the screen of a second lens that will produce a well-focused, 2.0-cm-tall image on the screen?

32. ‖ A 15-cm-focal-length converging lens is 20 cm to the right of a 7.0-cm-focal-length converging lens. A 1.0-cm-tall object is distance L to the left of the first lens.
   a. For what value of L is the final image of this two-lens system halfway between the two lenses?
   b. What are the height and orientation of the final image?

33. ‖ Yang can focus on objects 150 cm away with a relaxed eye.
BIO With full accommodation, she can focus on objects 20 cm away. After her eyesight is corrected for distance vision, what will her near point be while wearing her glasses?

34. ‖‖ The cornea, a boundary between the air and the aqueous
BIO humor, has a 3.0 cm focal length when acting alone. What is its radius of curvature?

35. | The objective lens of a telescope is a symmetric glass lens with 100 cm radii of curvature. The eyepiece lens is also a symmetric glass lens. What are the radii of curvature of the eyepiece lens if the telescope's magnification is 20×?

36. ‖ Mars (6800 km diameter) is viewed through a telescope on a night when it is 1.1×10^8 km from the earth. Its angular size as seen through the eyepiece is 0.50°, the same size as the full moon seen by the naked eye. If the eyepiece focal length is 25 mm, how long is the telescope?

37. ‖ You've been asked to build a telescope from a 2.0× magnifying lens and a 5.0× magnifying lens.
   a. What is the maximum magnification you can achieve?
   b. Which lens should be used as the objective? Explain.
   c. What will be the length of your telescope?

38. | Marooned on a desert island and with a lot of time on your hands, you decide to disassemble your glasses to make a crude telescope with which you can scan the horizon for rescuers. Luckily you're farsighted, and, like most people, your two eyes have different lens prescriptions. Your left eye uses a lens of power +4.5 D, and your right eye's lens is +3.0 D.
   a. Which lens should you use for the objective and which for the eyepiece? Explain.
   b. What will be the magnification of your telescope?
   c. How far apart should the two lenses be when you focus on distant objects?

39. ‖ A microscope with a tube length of 180 mm achieves a total magnification of 800× with a 40× objective and a 20× eyepiece. The microscope is focused for viewing with a relaxed eye. How far is the sample from the objective lens?

40. ‖ Modern microscopes are more likely to use a camera than
BIO human viewing. This is accomplished by replacing the eyepiece in Figure 35.14 with a *photo-ocular* that focuses the image of the objective to a real image on the sensor of a digital camera. A typical sensor is 22.5 mm wide and consists of 5625 4.0-μm-wide pixels. Suppose a microscopist pairs a 40× objective with a 2.5× photo-ocular.
   a. What is the field of view? That is, what width on the microscope stage, in mm, fills the sensor?
   b. The photo of a cell is 120 pixels in diameter. What is the cell's actual diameter, in μm?

41. ‖ White light is incident onto a 30° prism at the 40° angle shown in **FIGURE P35.41**. Violet light emerges perpendicular to the rear face of the prism. The index of refraction of violet light in this glass is 2.0% larger than the index of refraction of red light. At what angle ϕ does red light emerge from the rear face?

FIGURE P35.41

42. ‖ A beam of white light enters a transparent material.
CALC Wavelengths for which the index of refraction is n are refracted at angle θ_2. Wavelengths for which the index of refraction is $n + \delta n$, where $\delta n \ll n$, are refracted at angle $\theta_2 + \delta\theta$.
   a. Show that the angular separation of the two wavelengths, in radians, is $\delta\theta = -(\delta n/n)\tan\theta_2$.
   b. A beam of white light is incident on a piece of glass at 30.0°. Deep violet light is refracted 0.28° more than deep red light. The index of refraction for deep red light is known to be 1.552. What is the index of refraction for deep violet light?

43. ‖ High-power lasers are used to cut and weld materials by focusing the laser beam to a very small spot. This is like using a magnifying lens to focus the sun's light to a small spot that can burn things. As an engineer, you have designed a laser cutting device in which the material to be cut is placed 5.0 cm behind the lens. You have selected a high-power laser with a wavelength of 1.06 μm. Your calculations indicate that the laser must be focused to a 5.0-μm-diameter spot in order to have sufficient power to make the cut. What is the minimum diameter of the lens you must install?

44. ‖ Once dark adapted, the pupil of your eye is approximately
BIO 7 mm in diameter. The headlights of an oncoming car are 120 cm apart. If the lens of your eye is diffraction limited, at what distance are the two headlights marginally resolved? Assume a wavelength of 600 nm and that the index of refraction inside the eye is 1.33. (Your eye is not really good enough to resolve headlights at this distance, due both to aberrations in the lens and to the size of the receptors in your retina, but it comes reasonably close.)

45. ‖ The resolution of a digital camera is limited by two factors: diffraction by the lens, a limit of any optical system, and the fact that the sensor is divided into discrete pixels. Consider a typical point-and-shoot camera that has a 20-mm-focal-length lens and a sensor with 2.5-μm-wide pixels.
   a. First, assume an ideal, diffractionless lens. At a distance of 100 m, what is the smallest distance, in cm, between two point sources of light that the camera can barely resolve? In answering this question, consider what has to happen on the sensor to show two image points rather than one. You can use $s' = f$ because $s \gg f$.
   b. You can achieve the pixel-limited resolution of part a only if the diffraction width of each image point is no greater than 1 pixel in diameter. For what lens diameter is the minimum spot size equal to the width of a pixel? Use 600 nm for the wavelength of light.
   c. What is the *f*-number of the lens for the diameter you found in part b? Your answer is a quite realistic value of the *f*-number at which a camera transitions from being pixel limited to being diffraction limited. For *f*-numbers smaller than this (larger-diameter apertures), the resolution is limited by the pixel size and does not change as you change the aperture. For *f*-numbers larger than this (smaller-diameter apertures), the resolution is limited by diffraction, and it gets worse as you "stop down" to smaller apertures.

46. ‖ The Hubble Space Telescope has a mirror diameter of 2.4 m. Suppose the telescope is used to photograph stars near the center of our galaxy, 30,000 light years away, using red light with a wavelength of 650 nm.
   a. What's the distance (in km) between two stars that are marginally resolved? The resolution of a reflecting telescope is calculated exactly the same as for a refracting telescope.
   b. For comparison, what is this distance as a multiple of the distance of Jupiter from the sun?

47. ‖ Alpha Centauri, the nearest star to our solar system, is 4.3 light years away. Assume that Alpha Centauri has a planet with an advanced civilization. Professor Dhg, at the planet's Astronomical Institute, wants to build a telescope with which he can find out whether any planets are orbiting our sun.
   a. What is the minimum diameter for an objective lens that will just barely resolve Jupiter and the sun? The radius of Jupiter's orbit is 780 million km. Assume $\lambda = 600$ nm.
   b. Building a telescope of the necessary size does not appear to be a major problem. What practical difficulties might prevent Professor Dhg's experiment from succeeding?

Challenge Problems

48. ‖‖ Your task in physics laboratory is to make a microscope from two lenses. One lens has a focal length of 2.0 cm, the other 1.0 cm. You plan to use the more powerful lens as the objective, and you want the eyepiece to be 16 cm from the objective.
   a. For viewing with a relaxed eye, how far should the sample be from the objective lens?
   b. What is the magnification of your microscope?

49. ‖‖ The lens shown in FIGURE CP35.49 is called an *achromatic doublet*, meaning that it has no chromatic aberration. The left side is flat, and all other surfaces have radii of curvature R.

FIGURE CP35.49

   a. For parallel light rays coming from the left, show that the effective focal length of this two-lens system is $f = R/(2n_2 - n_1 - 1)$, where n_1 and n_2 are, respectively, the indices of refraction of the diverging and the converging lenses. Don't forget to make the thin-lens approximation.
   b. Because of dispersion, either lens alone would focus red rays and blue rays at different points. Define Δn_1 and Δn_2 as $n_{blue} - n_{red}$ for the two lenses. What value of the ratio $\Delta n_1/\Delta n_2$ makes $f_{blue} = f_{red}$ for the two-lens system? That is, the two-lens system does *not* exhibit chromatic aberration.
   c. Indices of refraction for two types of glass are given in the table. To make an achromatic doublet, which glass should you use for the converging lens and which for the diverging lens? Explain.

	n_{blue}	n_{red}
Crown glass	1.525	1.517
Flint glass	1.632	1.616

   d. What value of R gives a focal length of 10.0 cm?

50. ‖‖ FIGURE CP35.50 shows a simple zoom lens in which the magnitudes of both focal lengths are f. If the spacing $d < f$, the image of the converging lens falls on the right side of the diverging lens. Our procedure of letting the image of the first lens act as the object of the second lens will continue to work in this case if we use a *negative* object distance for the second lens. This is called a *virtual object*. Consider an object very far to the left ($s \approx \infty$) of the converging lens. Define the effective focal length as the distance from the midpoint between the lenses to the final image.
   a. Show that the effective focal length is

$$f_{eff} = \frac{f^2 - fd + \frac{1}{2}d^2}{d}$$

   b. What is the zoom for a lens that can be adjusted from $d = \frac{1}{2}f$ to $d = \frac{1}{4}f$?

FIGURE CP35.50

PART VII

Optics

KEY FINDINGS   What are the overarching findings of Part VII?

Part VII has looked at two models of light: light waves and light rays.

- **Light waves**
  - Are electromagnetic waves.
  - Spread out after passing through openings.
  - Exhibit interference.

A third model of light, the photon model, will be introduced in Part VIII.

- **Light rays**
  - Travel in straight lines.
  - Do not interact.
  - Form images.

LAWS   What laws of physics govern optics?

Superposition/interference Constructive interference occurs where crests overlap crests, destructive interference where crests overlap troughs.

Law of reflection $\theta_r = \theta_i$

Law of refraction (Snell's law)
$n_1 \sin \theta_1 = n_2 \sin \theta_2$

Rayleigh's criterion Two objects can be resolved by a lens of diameter D if their angular separation exceeds $1.22\lambda/D$.

MODELS   What are the most important models of Part VII?

Wave model

- Light is an electromagnetic wave.
  - Light travels through vacuum at speed c.
  - Wavelength and frequency are related by $\lambda f = c$.

- Light exhibits **diffraction** and **interference.**
  - Light spreads out after passing through an opening.
  - Equal-wavelength light waves interfere. Interference depends on the path-length difference.

- The wave model is usually appropriate for openings smaller than about 1 mm.

Ray model

- Light rays travel in straight lines.
  - The speed is $v = c/n$, where n is the index of refraction.
- Light rays travel forever unless they interact with matter.
  - **Reflection** and **refraction**
  - Scattering and absorption
- An object is a source of rays.
  - Rays originate at every point.
- The eye sees by focusing a diverging bundle of rays.
- The ray model is usually appropriate for openings larger than about 1 mm.

Diverging bundle of rays

Object

Eye

TOOLS   What are the most important tools introduced in Part VII?

Diffraction

- Dark fringes in a single-slit diffraction pattern are at $\theta_p = p\lambda/a$, $p = 1, 2, \ldots$

   Slit width a

  - The **central maximum** of the diffraction has width $w = 2\lambda L/a$.
  - A circular hole has a central maximum of width $w = 2.44\lambda L/D$.

Double-slit interference

- Bright equally spaced fringes are located at
  $y_m = mL\lambda/d$ $m = 0, 1, 2, \ldots$

   Slit spacing d

Diffraction gratings

- Very narrow bright fringes are at
  $d \sin \theta_m = m\lambda$ $y_m = L \tan \theta_m$

Ray tracing

- For lenses and mirrors, three special rays locate the image.
  - Parallel to the axis
  - Through the focal point
  - Through the center

Images

- If rays converge at P′, then P′ is a **real image** and s' is positive.
- If rays diverge from P′, then P′ is a **virtual image** and s' is negative.

Thin lenses and mirrors

- The thin-lens and thin-mirror equation for focal length f is

$$\frac{1}{s} + \frac{1}{s'} = \frac{1}{f}$$

- Lateral magnification is $m = -s'/s$.

Optical instruments

- With multiple lenses, the image of one lens is the object for the next.

Vision

- **Hyperopia** occurs when the eye's near point is too far away. It is corrected with a converging lens.
- **Myopia** occurs when the eye's far point is too close. It is corrected with a diverging lens.

Resolution

- Diffraction limits optical instruments.
  - The smallest spot to which light can be focused is $w_{min} = 2.44\lambda f/D$.
  - Two objects can be resolved if their angular separation exceeds $1.22\lambda/D$.

PART
VIII Relativity and Quantum Physics

OVERVIEW

Contemporary Physics

Our journey into physics is nearing its end. We began roughly 350 years ago with Newton's discovery of the laws of motion. Parts VI and VII have brought us to the end of the 19th century, just over 100 years ago. Along the way you've learned about the motion of particles, the conservation of energy, the physics of waves, and the electromagnetic interactions that hold atoms together and generate light waves. We begin the last phase of our journey with confidence.

Newton's mechanics and Maxwell's electromagnetism were the twin pillars of science at the end of the 19th century and remain the basis for much of engineering and applied science in the 21st century. Despite the successes of these theories, a series of discoveries starting around 1900 and continuing into the first few decades of the 20th century profoundly altered our understanding of the universe at the most fundamental level.

- Einstein's theory of relativity forced scientists to completely revise their concepts of space and time. Our exploration of these fascinating ideas will end with perhaps the most famous equation in physics: Einstein's $E = mc^2$.
- Experimenters found that the classical distinction between *particles* and *waves* breaks down at the atomic level. Light sometimes acts like a particle, while electrons and even entire atoms sometimes act like waves. We will need a new theory of light and matter—quantum physics—to explain these phenomena.

These two theories form the basis for physics—and, increasingly, engineering—as it is practiced today.

The complete theory of quantum physics, as it was developed in the 1920s, describes atomic particles in terms of an entirely new concept called a *wave function*. One of our most important tasks in Part VIII will be to learn what a wave function is, what laws govern its behavior, and how to relate wave functions to experimental measurements. We will concentrate on one-dimensional models that, while not perfect, will be adequate for understanding the essential features of scanning tunneling microscopes, various semiconductor devices, radioactive decay, and other applications.

We'll complete our study of quantum physics with an introduction to atomic and nuclear physics. You will learn where the electron-shell model of chemistry comes from, how atoms emit and absorb light, what's inside the nucleus, and why some nuclei undergo radioactive decay.

The quantum world with its wave functions and probabilities can seem strange and mysterious, yet quantum physics gives the most definitive and accurate predictions of any physical theory ever devised. The contemporary perspective of quantum physics will be a fitting end to our journey.

This plot shows the instantaneous intensity of a focused laser beam.

36 Relativity

The Compact Muon Solenoid detector at the Large Hadron Collider, where protons are accelerated to 99.999999% of the speed of light.

IN THIS CHAPTER, you will learn how relativity changes our concepts of space and time.

What is an inertial reference frame?

Inertial reference frames are reference frames that move relative to each other with constant velocity.

- You'll learn to work with the positions and times of events.
- All the clocks in an inertial reference frame are synchronized.

« LOOKING BACK Section 4.3 Relative motion

What is relativity about?

Einstein's theory of relativity is based on a simple-sounding principle: The laws of physics are the same in all inertial reference frames. This leads to these conclusions:

- Light travels at the same speed c in all inertial reference frames.
- No object or information can travel faster than the speed of light.

How does relativity affect time?

Time is relative. Two reference frames moving relative to each other measure different time intervals between two events.

- Time dilation is the idea that moving clocks run slower than clocks at rest.
- We'll examine the famous twin paradox.

How does relativity affect space?

Distances are also relative. Two reference frames moving relative to each other find different distances between two events.

- Length contraction is the idea that the length of an object is less when the object is moving than when it is at rest.

How does relativity affect mass and energy?

Einstein's most famous equation, $E = mc^2$, says that mass can be transformed into energy, and energy into mass, as long as the total energy is conserved.

- Nuclear fission converts mass into energy.
- Collisions between high-speed particles create new particles from energy.

Does relativity have applications?

Abstract though it may seem, relativity is important in technologies such as medical PET scans (positron-emission tomography) and nuclear energy. And the global GPS system, a technology we use every day, functions only when the signals from precision clocks in orbiting satellites are corrected for relativistic time dilation.

36.1 Relativity: What's It All About?

What do you think of when you hear the phrase "theory of relativity"? A white-haired Einstein? $E = mc^2$? Black holes? Time travel? Perhaps you've heard that the theory of relativity is so complicated and abstract that only a handful of people in the whole world really understand it.

There is, without doubt, a certain mystique associated with relativity, an aura of the strange and exotic. The good news is that understanding the ideas of relativity is well within your grasp. Einstein's *special theory of relativity,* the portion of relativity we'll study, is not mathematically difficult at all. The challenge is conceptual because relativity questions deeply held assumptions about the nature of space and time. In fact, that's what relativity is all about—space and time.

What's Special About Special Relativity?

Einstein's first paper on relativity, in 1905, dealt exclusively with inertial reference frames, reference frames that move relative to each other with constant velocity. Ten years later, Einstein published a more encompassing theory of relativity that considered accelerated motion and its connection to gravity. The second theory, because it's more general in scope, is called *general relativity.* General relativity is the theory that describes black holes, curved spacetime, and the evolution of the universe. It is a fascinating theory but, alas, very mathematical and outside the scope of this textbook.

Motion at constant velocity is a "special case" of motion—namely, motion for which the acceleration is zero. Hence Einstein's first theory of relativity has come to be known as **special relativity.** It is special in the sense of being a restricted, special case of his more general theory, not special in the everyday sense meaning distinctive or exceptional. Special relativity, with its conclusions about time dilation and length contraction, is what we will study.

Albert Einstein (1879–1955) was one of the most influential thinkers in history.

36.2 Galilean Relativity

Relativity is the process of relating measurements in one reference frame to those in a different reference frame moving *relative to* the first. To appreciate and understand what is new in Einstein's theory, we need a firm grasp of the ideas of relativity that are embodied in Newtonian mechanics. Thus we begin with *Galilean relativity.*

Reference Frames

Suppose you're passing me as we both drive in the same direction along a freeway. My car's speedometer reads 55 mph while your speedometer shows 60 mph. Is 60 mph your "true" speed? That is certainly your speed relative to someone standing beside the road, but your speed relative to me is only 5 mph. Your speed is 120 mph relative to a driver approaching from the other direction at 60 mph.

An object does not have a "true" speed or velocity. The very definition of velocity, $v = \Delta x/\Delta t$, assumes the existence of a coordinate system in which, during some time interval Δt, the displacement Δx is measured. The best we can manage is to specify an object's velocity relative to, or with respect to, the coordinate system in which it is measured.

Let's define a **reference frame** to be a coordinate system in which experimenters equipped with meter sticks, stopwatches, and any other needed equipment make position and time measurements on moving objects. Three ideas are implicit in our definition of a reference frame:

- A reference frame extends infinitely far in all directions.
- The experimenters are at rest in the reference frame.
- The number of experimenters and the quality of their equipment are sufficient to measure positions and velocities to any level of accuracy needed.

The first two ideas are especially important. It is often convenient to say "the laboratory reference frame" or "the reference frame of the rocket." These are shorthand expressions for "a reference frame, infinite in all directions, in which the laboratory (or the rocket) and a set of experimenters happen to be at rest."

NOTE A reference frame is not the same thing as a "point of view." That is, each person or each experimenter does not have his or her own private reference frame. **All experimenters at rest relative to each other share the same reference frame.**

FIGURE 36.1 shows two reference frames called S and S'. The coordinate axes in S are x, y, z and those in S' are x', y', z'. Reference frame S' moves with velocity v relative to S or, equivalently, S moves with velocity $-v$ relative to S'. There's no implication that either reference frame is "at rest." Notice that the zero of time, when experimenters start their stopwatches, is the instant that the origins of S and S' coincide.

We will restrict our attention to *inertial reference frames,* implying that the relative velocity v is constant. You should recall from Chapter 5 that an **inertial reference frame** is a reference frame in which Newton's first law, the law of inertia, is valid. In particular, an inertial reference frame is one in which an isolated particle, one on which there are no forces, either remains at rest or moves in a straight line at constant speed.

Any reference frame moving at constant velocity with respect to an inertial reference frame is itself an inertial reference frame. Conversely, a reference frame accelerating with respect to an inertial reference frame is *not* an inertial reference frame. Our restriction to reference frames moving with respect to each other at constant velocity—with no acceleration—is the "special" part of special relativity.

NOTE An inertial reference frame is an idealization. A true inertial reference frame would need to be floating in deep space, far from any gravitational influence. In practice, an earthbound laboratory is a good approximation of an inertial reference frame because the accelerations associated with the earth's rotation and motion around the sun are too small to influence most experiments.

FIGURE 36.1 The standard reference frames S and S'.

The axes of S and S' have the same orientation.

Frame S' moves with velocity v relative to frame S, parallel to the x- and x'-axes.

The origins of S and S' coincide at $t = 0$.

Alternatively, frame S moves with velocity $-v$ relative to frame S'.

STOP TO THINK 36.1 Which of these is an inertial reference frame (or a very good approximation)?

a. Your bedroom
b. A car rolling down a steep hill
c. A train coasting along a level track
d. A rocket being launched
e. A roller coaster going over the top of a hill
f. A sky diver falling at terminal speed

The Galilean Transformations

Suppose a firecracker explodes at time t. The experimenters in reference frame S determine that the explosion happened at position x. Similarly, the experimenters in S' find that the firecracker exploded at x' in their reference frame. What is the relationship between x and x'?

FIGURE 36.2 shows the explosion and the two reference frames. You can see from the figure that $x = x' + vt$, thus

FIGURE 36.2 The position of an exploding firecracker is measured in reference frames S and S'.

At time t, the origin of S' has moved distance vt to the right. Thus $x = x' + vt$.

Firecracker

$y' = y$

Origins coincide at $t = 0$.

Distances perpendicular to the motion are not affected. Thus $y' = y$ and $z' = z$.

$$
\begin{aligned}
x &= x' + vt & x' &= x - vt \\
y &= y' \quad \text{or} \quad & y' &= y \\
z &= z' & z' &= z
\end{aligned}
\qquad (36.1)
$$

These are the *Galilean transformations of position*. If you know a position measured by the experimenters in one inertial reference frame, you can calculate the position that would be measured by experimenters in any other inertial reference frame.

Suppose the experimenters in both reference frames now track the motion of the object in **FIGURE 36.3** by measuring its position at many instants of time. The experimenters in S find that the object's velocity is $\vec{u}$. During the *same time interval* Δt, the experimenters in S' measure the velocity to be $\vec{u}'$.

> **NOTE** In this chapter, we will use v to represent the velocity of one reference frame relative to another. We will use $\vec{u}$ and $\vec{u}'$ to represent the velocities of objects with respect to reference frames S and S'.

We can find the relationship between $\vec{u}$ and $\vec{u}'$ by taking the time derivatives of Equations 36.1 and using the definition $u_x = dx/dt$:

$$u_x = \frac{dx}{dt} = \frac{dx'}{dt} + v = u'_x + v$$

$$u_y = \frac{dy}{dt} = \frac{dy'}{dt} = u'_y$$

The equation for u_z is similar. The net result is

$$
\begin{array}{lll}
u_x = u'_x + v & & u'_x = u_x - v \\
u_y = u'_y & \text{or} & u'_y = u_y \\
u_z = u'_z & & u'_z = u_z
\end{array}
\qquad (36.2)
$$

Equations 36.2 are the *Galilean transformations of velocity*. If you know the velocity of a particle in one inertial reference frame, you can find the velocity that would be measured by experimenters in any other inertial reference frame.

> **NOTE** In Section 4.3 you learned the Galilean transformation of velocity as $\vec{v}_{CB} = \vec{v}_{CA} + \vec{v}_{AB}$, where $\vec{v}_{AB}$ means "the velocity of A relative to B." Equations 36.2 are equivalent for relative motion parallel to the x-axis but are written in a more formal notation that will be useful for relativity.

FIGURE 36.3 The velocity of a moving object is measured in reference frames S and S'.

The object's velocity in frame S is $\vec{u}$.

In frame S', the velocity is $\vec{u}'$.

EXAMPLE 36.1 | The speed of sound

An airplane is flying at speed 200 m/s with respect to the ground. Sound wave 1 is approaching the plane from the front, sound wave 2 is catching up from behind. Both waves travel at 340 m/s relative to the ground. What is the speed of each wave relative to the plane?

MODEL Assume that the earth (frame S) and the airplane (frame S') are inertial reference frames. Frame S', in which the airplane is at rest, moves at $v = 200$ m/s relative to frame S.

VISUALIZE **FIGURE 36.4** shows the airplane and the sound waves.

FIGURE 36.4 Experimenters in the plane measure different speeds for the waves than do experimenters on the ground.

The plane's frame S' travels at $v = 200$ m/s relative to the ground's frame S.

Wave 2 travels at $u_2 = +340$ m/s in frame S.

Wave 1 travels at $u_1 = -340$ m/s in frame S.

SOLVE The speed of a mechanical wave, such as a sound wave or a wave on a string, is its speed *relative to its medium*. Thus the *speed of sound* is the speed of a sound wave through a reference frame in which the air is at rest. This is reference frame S, where wave 1 travels with velocity $u_1 = -340$ m/s and wave 2 travels with velocity $u_2 = +340$ m/s. Notice that the Galilean transformations use *velocities*, with appropriate signs, not just speeds.

The airplane travels to the right with reference frame S' at velocity v. We can use the Galilean transformations of velocity to find the velocities of the two sound waves in frame S':

$$u'_1 = u_1 - v = -340 \text{ m/s} - 200 \text{ m/s} = -540 \text{ m/s}$$
$$u'_2 = u_2 - v = 340 \text{ m/s} - 200 \text{ m/s} = 140 \text{ m/s}$$

ASSESS This isn't surprising. If you're driving at 50 mph, a car coming the other way at 55 mph is approaching you at 105 mph. A car coming up behind you at 55 mph is gaining on you at the rate of only 5 mph. Wave speeds behave the same. Notice that a mechanical wave appears to be stationary to a person moving at the wave speed. To a surfer, the crest of the ocean wave remains at rest under his or her feet.

The Galilean Principle of Relativity

Experimenters in reference frames S and S′ measure different values for position and velocity. What about the force on and the acceleration of the particle in FIGURE 36.5? The strength of a force can be measured with a spring scale. The experimenters in reference frames S and S′ both see the *same reading* on the scale (assume the scale has a bright digital display easily seen by all experimenters), so both conclude that the force is the same. That is, $F' = F$.

We can compare the accelerations measured in the two reference frames by taking the time derivative of the velocity transformation equation $u' = u - v$. (We'll assume, for simplicity, that the velocities and accelerations are all in the x-direction.) The relative velocity v between the two reference frames is *constant*, with $dv/dt = 0$, thus

$$a' = \frac{du'}{dt} = \frac{du}{dt} = a \qquad (36.3)$$

Experimenters in reference frames S and S′ measure different values for an object's position and velocity, but they *agree* on its acceleration.

If $F = ma$ in reference frame S, then $F' = ma'$ in reference frame S′. Stated another way, if Newton's second law is valid in one inertial reference frame, then it is valid in all inertial reference frames. Because other laws of mechanics, such as the conservation laws, follow from Newton's laws of motion, we can state this conclusion as the *Galilean principle of relativity:*

> **Galilean principle of relativity** The laws of mechanics are the same in all inertial reference frames.

The Galilean principle of relativity is easy to state, but to understand it we must understand what is and is not "the same." To take a specific example, consider the law of conservation of momentum. FIGURE 36.6a shows two particles about to collide. Their total momentum in frame S, where particle 2 is at rest, is $P_i = 9.0$ kg m/s. This is an isolated system, hence the law of conservation of momentum tells us that the momentum after the collision will be $P_f = 9.0$ kg m/s.

FIGURE 36.6b has used the velocity transformation to look at the same particles in frame S′ in which particle 1 is at rest. The initial momentum in S′ is $P_i' = -18$ kg m/s. Thus it is not the *value* of the momentum that is the same in all inertial reference frames. Instead, the Galilean principle of relativity tells us that the *law* of momentum conservation is the same in all inertial reference frames. If $P_f = P_i$ in frame S, then it must be true that $P_f' = P_i'$ in frame S′. Consequently, we can conclude that P_f' will be -18 kg m/s after the collision in S′.

FIGURE 36.5 Experimenters in both reference frames test Newton's second law by measuring the force on a particle and its acceleration.

FIGURE 36.6 Total momentum measured in two reference frames.

(a) Collision seen in frame S

(b) Collision seen in frame S′

36.3 Einstein's Principle of Relativity

The 19th century was an era of optics and electromagnetism. Thomas Young demonstrated in 1801 that light is a wave, and by midcentury scientists had devised techniques for measuring the speed of light. Faraday discovered electromagnetic induction in 1831, setting in motion a series of events leading to Maxwell's prediction, in 1864, that light waves travel with speed

$$c = \frac{1}{\sqrt{\epsilon_0 \mu_0}} = 3.00 \times 10^8 \text{ m/s}$$

This is a quite specific prediction with no wiggle room. But in what reference frame is this the speed of light? And, if light travels with speed c in some inertial reference frame S, then surely, as **FIGURE 36.7** shows, the light speed must be more or less than c in a reference frame S′ that moves with respect to S. There could even be reference frames in which light is frozen and doesn't move at all!

It was in this muddled state of affairs that a young Albert Einstein made his mark on the world. Even as a teenager, Einstein had wondered how a light wave would look to someone "surfing" the wave, traveling alongside the wave at the wave speed. You can do that with a water wave or a sound wave, but light waves seemed to present a logical difficulty. An electromagnetic wave sustains itself by virtue of the fact that a changing magnetic field induces an electric field and a changing electric field induces a magnetic field. But to someone moving with the wave, *the fields would not change.* How could there be an electromagnetic wave under these circumstances?

Several years of thinking about the connection between electromagnetism and reference frames led Einstein to the conclusion that *all* the laws of physics, not just the laws of mechanics, should obey the principle of relativity. In other words, the principle of relativity is a fundamental statement about the nature of the physical universe. Thus we can remove the restriction in the Galilean principle of relativity and state a much more general principle:

Principle of relativity All the laws of physics are the same in all inertial reference frames.

All the results of Einstein's theory of relativity flow from this one simple statement.

The Constancy of the Speed of Light

If Maxwell's equations of electromagnetism are laws of physics, and there's every reason to think they are, then, according to the principle of relativity, Maxwell's equations must be true in *every* inertial reference frame. On the surface this seems to be an innocuous statement, equivalent to saying that the law of conservation of momentum is true in every inertial reference frame. But follow the logic:

1. Maxwell's equations are true in all inertial reference frames.
2. Maxwell's equations predict that electromagnetic waves, including light, travel at speed $c = 3.00 \times 10^8$ m/s.
3. Therefore, **light travels at speed c in all inertial reference frames.**

FIGURE 36.8 shows the implications of this conclusion. *All* experimenters, regardless of how they move with respect to each other, find that *all* light waves, regardless of the source, travel in their reference frame with the *same* speed c. If Cathy's velocity toward Bill and away from Amy is $v = 0.9c$, Cathy finds, by making measurements in her reference frame, that the light from Bill approaches her at speed c, not at $c + v = 1.9c$. And the light from Amy, which left Amy at speed c, catches up from behind at speed c *relative to Cathy,* not the $c - v = 0.1c$ you would have expected.

Although this prediction goes against all shreds of common sense, the experimental evidence for it is strong. Laboratory experiments are difficult because even the highest laboratory speed is insignificant in comparison to c. In the 1930s, however, physicists R. J. Kennedy and E. M. Thorndike realized that they could use the earth itself as a laboratory. The earth's speed as it circles the sun is about 30,000 m/s. The *relative*

FIGURE 36.7 It seems as if the speed of light should differ from c in a reference frame moving through the ether.

If light travels at speed c in reference frame S, . . .

. . . then surely light travels at some other speed in a different reference frame that moves with respect to S.

FIGURE 36.8 Light travels at speed c in all inertial reference frames, regardless of how the reference frames are moving with respect to the light source.

This light wave leaves Amy at speed c relative to Amy. It approaches Cathy at speed c relative to Cathy.

$v = 0.9c$

Amy   Cathy   Bill

This light wave leaves Bill at speed c relative to Bill. It approaches Cathy at speed c relative to Cathy.

FIGURE 36.9 Experiments find that the photons travel through the laboratory with speed c, not the speed $1.99975c$ that you might expect.

A photon is emitted at speed c relative to the π meson. Measurements find that the photon's speed in the laboratory reference frame is also c.

Meson reference frame

Laboratory reference frame

velocity of the earth in January differs by 60,000 m/s from its velocity in July, when the earth is moving in the opposite direction. Kennedy and Thorndike were able to use a very sensitive and stable interferometer to show that the numerical values of the speed of light in January and July differ by less than 2 m/s.

More recent experiments have used unstable elementary particles, called π mesons, that decay into high-energy photons of light. The π mesons, created in a particle accelerator, move through the laboratory at 99.975% the speed of light, or $v = 0.99975c$, as they emit photons at speed c in the π meson's reference frame. As **FIGURE 36.9** shows, you would expect the photons to travel through the laboratory with speed $c + v = 1.99975c$. Instead, the measured speed of the photons in the laboratory was, within experimental error, 3.00×10^8 m/s.

In summary, *every* experiment designed to compare the speed of light in different reference frames has found that light travels at 3.00×10^8 m/s in every inertial reference frame, regardless of how the reference frames are moving with respect to each other.

How Can This Be?

You're in good company if you find this impossible to believe. Suppose I shot a ball forward at 50 m/s while driving past you at 30 m/s. You would certainly see the ball traveling at 80 m/s relative to you and the ground. What we're saying with regard to light is equivalent to saying that the ball travels at 50 m/s relative to my car and *at the same time* travels at 50 m/s relative to the ground, even though the car is moving across the ground at 30 m/s. It seems logically impossible.

You might think that this is merely a matter of semantics. If we can just get our definitions and use of words straight, then the mystery and confusion will disappear. Or perhaps the difficulty is a confusion between what we "see" versus what "really happens." In other words, a better analysis, one that focuses on what really happens, would find that light "really" travels at different speeds in different reference frames.

Alas, what "really happens" is that light travels at 3.00×10^8 m/s in every inertial reference frame, regardless of how the reference frames are moving with respect to each other. It's not a trick. There remains only one way to escape the logical contradictions.

The definition of velocity is $u = \Delta x/\Delta t$, the ratio of a distance traveled to the time interval in which the travel occurs. Suppose you and I both make measurements on an object as it moves, but you happen to be moving relative to me. Perhaps I'm standing on the corner, you're driving past in your car, and we're both trying to measure the velocity of a bicycle. Further, suppose we have agreed in advance to measure the position of the bicycle first as it passes the tree in **FIGURE 36.10**, then later as it passes the lamppost. Your $\Delta x'$, the bicycle's displacement, differs from my Δx because of your motion relative to me, causing you to calculate a bicycle velocity u' in your reference frame that differs from its velocity u in my reference frame. This is just the Galilean transformations showing up again.

Now let's repeat the measurements, but this time let's measure the velocity of a light wave as it travels from the tree to the lamppost. Once again, your $\Delta x'$ differs from my Δx, and the obvious conclusion is that your light speed u' differs from my light speed u. The difference will be very small if you're driving past in your car, very large if you're flying past in a rocket traveling at nearly the speed of light. Although this conclusion seems obvious, it is wrong. Experiments show that, for a light wave, we'll get the *same* values: $u' = u$.

The only way this can be true is if your Δt is not the same as my Δt. If the time it takes the light to move from the tree to the lamppost in your reference frame, a time we'll now call $\Delta t'$, differs from the time Δt it takes the light to move from the tree to the lamppost in my reference frame, then we might find that $\Delta x'/\Delta t' = \Delta x/\Delta t$. That is, $u' = u$ even though you are moving with respect to me.

FIGURE 36.10 Measuring the velocity of an object by appealing to the basic definition $u = \Delta x/\Delta t$.

Measurements made in frame S, in which the tree and lamppost are at rest. The bicycle's velocity is $u = \Delta x/\Delta t$.

Measurements made in frame S', which moves to the right relative to frame S. The bicycle's velocity is $u' = \Delta x'/\Delta t$.

Initial $t = 0$

Initial $t = 0$

Δx and $\Delta x'$ are not the same.

The tree and lamppost are moving to the left in frame S'.

Final $t = \Delta t$

Final $t = \Delta t$

We've assumed, since the beginning of this textbook, that time is simply time. It flows along like a river, and all experimenters in all reference frames simply use it. For example, suppose the tree and the lamppost both have big clocks that we both can see. Shouldn't we be able to agree on the time interval Δt the light needs to move from the tree to the lamppost?

Perhaps not. It's demonstrably true that $\Delta x' \neq \Delta x$. It's experimentally verified that $u' = u$ for light waves. Something must be wrong with *assumptions* that we've made about the nature of time. The principle of relativity has painted us into a corner, and our only way out is to reexamine our understanding of time.

36.4 Events and Measurements

To question some of our most basic assumptions about space and time requires extreme care. We need to be certain that no assumptions slip into our analysis unnoticed. Our goal is to describe the motion of a particle in a clear and precise way, making the barest minimum of assumptions.

Events

The fundamental element of relativity is called an **event.** An event is a physical activity that takes place at a definite point in space and at a definite instant of time. An exploding firecracker is an event. A collision between two particles is an event. A light wave hitting a detector is an event.

Events can be observed and measured by experimenters in different reference frames. An exploding firecracker is as clear to you as you drive by in your car as it is to me standing on the street corner. We can quantify where and when an event occurs with four numbers: the coordinates (x, y, z) and the instant of time t. These four numbers, illustrated in **FIGURE 36.11**, are called the **spacetime coordinates** of the event.

The spatial coordinates of an event measured in reference frames S and S' may differ. It now appears that the instant of time recorded in S and S' may also differ. Thus the spacetime coordinates of an event measured by experimenters in frame S are (x, y, z, t) and the spacetime coordinates of the *same event* measured by experimenters in frame S' are (x', y', z', t').

FIGURE 36.11 The location and time of an event are described by its spacetime coordinates.

An event has spacetime coordinates (x, y, z, t) in frame S and different spacetime coordinates (x', y', z', t') in frame S'.

The motion of a particle can be described as a sequence of two or more events. We introduced this idea in the preceding section when we agreed to measure the velocity of a bicycle and then of a light wave by making measurements when the object passed the tree (first event) and when the object passed the lamppost (second event).

Measurements

Events are what "really happen," but how do we learn about an event? That is, how do the experimenters in a reference frame determine the spacetime coordinates of an event? This is a problem of *measurement*.

We defined a reference frame to be a coordinate system in which experimenters can make position and time measurements. That's a good start, but now we need to be more precise as to *how* the measurements are made. Imagine that a reference frame is filled with a cubic lattice of meter sticks, as shown in FIGURE 36.12. At every intersection is a clock, and all the clocks in a reference frame are *synchronized*. We'll return in a moment to consider how to synchronize the clocks, but assume for the moment it can be done.

Now, with our meter sticks and clocks in place, we can use a two-part measurement scheme:

- The (x, y, z) coordinates of an event are determined by the intersection of the meter sticks closest to the event.
- The event's time t is the time displayed on the clock nearest the event.

You can imagine, if you wish, that each event is accompanied by a flash of light to illuminate the face of the nearest clock and make its reading known.

Several important issues need to be noted:

1. The clocks and meter sticks in each reference frame are imaginary, so they have no difficulty passing through each other.
2. Measurements of position and time made in one reference frame must use only the clocks and meter sticks in that reference frame.
3. There's nothing special about the sticks being 1 m long and the clocks 1 m apart. The lattice spacing can be altered to achieve whatever level of measurement accuracy is desired.
4. We'll assume that the experimenters in each reference frame have assistants sitting beside every clock to record the position and time of nearby events.
5. Perhaps most important, t is the time at which the event *actually happens,* not the time at which an experimenter sees the event or at which information about the event reaches an experimenter.
6. All experimenters in one reference frame agree on the spacetime coordinates of an event. In other words, **an event has a unique set of spacetime coordinates in each reference frame.**

FIGURE 36.12 The spacetime coordinates of an event are measured by a lattice of meter sticks and clocks.

The spacetime coordinates of this event are measured by the nearest meter stick intersection and the nearest clock.

Reference frame S

Reference frame S' has its own meter sticks and its own clocks.

Reference frame S'

STOP TO THINK 36.3 A carpenter is working on a house two blocks away. You notice a slight delay between seeing the carpenter's hammer hit the nail and hearing the blow. At what time does the event "hammer hits nail" occur?

a. At the instant you hear the blow
b. At the instant you see the hammer hit
c. Very slightly before you see the hammer hit
d. Very slightly after you see the hammer hit

Clock Synchronization

It's important that all the clocks in a reference frame be **synchronized,** meaning that all clocks in the reference frame have the same reading at any one instant of time. Thus we need a method of synchronization. One idea that comes to mind is to designate the clock at the origin as the *master clock*. We could then carry this clock around to every clock in the lattice, adjust that clock to match the master clock, and finally return the master clock to the origin.

This would be a perfectly good method of clock synchronization in Newtonian mechanics, where time flows along smoothly, the same for everyone. But we've been driven to reexamine the nature of time by the possibility that time is different in reference frames moving relative to each other. Because the master clock would *move,* we cannot assume that the moving master clock would keep time in the same way as the stationary clocks.

We need a synchronization method that does not require moving the clocks. Fortunately, such a method is easy to devise. Each clock is resting at the intersection of meter sticks, so by looking at the meter sticks, the assistant knows, or can calculate, exactly how far each clock is from the origin. Once the distance is known, the assistant can calculate exactly how long a light wave will take to travel from the origin to each clock. For example, light will take $1.00 \ \mu s$ to travel to a clock 300 m from the origin.

> **NOTE** It's handy for many relativity problems to know that the speed of light is $c = 300 \ \text{m}/\mu s$.

To synchronize the clocks, the assistants begin by setting each clock to display the light travel time from the origin, but they don't start the clocks. Next, as **FIGURE 36.13** shows, a light flashes at the origin and, simultaneously, the clock at the origin starts running from $t = 0$ s. The light wave spreads out in all directions at speed c. A photodetector on each clock recognizes the arrival of the light wave and, without delay, starts the clock. The clock had been preset with the light travel time, so each clock as it starts reads exactly the same as the clock at the origin. Thus all the clocks will be synchronized after the light wave has passed by.

Events and Observations

We noted above that t is the time the event *actually happens*. This is an important point, one that bears further discussion. Light waves take time to travel. Messages, whether they're transmitted by light pulses, telephone, or courier on horseback, take time to be delivered. An experimenter *observes* an event, such as an exploding firecracker, only *at a later time* when light waves reach his or her eyes. But our interest is in the event itself, not the experimenter's observation of the event. The time at which the experimenter sees the event or receives information about the event is not when the event actually occurred.

Suppose at $t = 0$ s a firecracker explodes at $x = 300$ m. The flash of light from the firecracker will reach an experimenter at the origin at $t_1 = 1.0 \ \mu s$. The sound of the explosion will reach a sightless experimenter at the origin at $t_2 = 0.88$ s. Neither of these is the time t_{event} of the explosion, although the experimenter can work backward from these times, using known wave speeds, to determine t_{event}. In this example, the spacetime coordinates of the event—the explosion—are (300 m, 0 m, 0 m, 0 s).

FIGURE 36.13 Synchronizing clocks.

1. This clock is preset to $1.00 \ \mu s$, the time it takes light to travel 300 m.

300 m

2. At $t = 0$ s, a light flashes at the origin and the origin clock starts running. A very short time later, seen here, a light wave has begun to move outward.

Wave front

3. The clock starts when the light wave reaches it. It is now synchronized with the origin clock.

EXAMPLE 36.2 | **Finding the time of an event**

Experimenter A in reference frame S stands at the origin looking in the positive x-direction. Experimenter B stands at $x = 900$ m looking in the negative x-direction. A firecracker explodes somewhere between them. Experimenter B sees the light flash at $t = 3.0 \ \mu s$.

Experimenter A sees the light flash at $t = 4.0 \ \mu s$. What are the spacetime coordinates of the explosion?

MODEL Experimenters A and B are in the same reference frame and have synchronized clocks.

Continued

VISUALIZE FIGURE 36.14 shows the two experimenters and the explosion at unknown position x.

FIGURE 36.14 The light wave reaches the experimenters at different times. Neither of these is the time at which the event actually happened.

Wave front reaches A at $t = 4.0\,\mu s$

Wave front reaches B at $t = 3.0\,\mu s$

Explosion at position x, time t

0 m   900 m

SOLVE The two experimenters observe light flashes at two different instants, but there's only one event. Light travels at 300 m/μs, so the additional 1.0 μs needed for the light to reach experimenter A implies that distance $(x - 0\text{ m})$ is 300 m longer than distance $(900\text{ m} - x)$. That is,

$$(x - 0\text{ m}) = (900\text{ m} - x) + 300\text{ m}$$

This is easily solved to give $x = 600$ m as the position coordinate of the explosion. The light takes 1.0 μs to travel 300 m to experimenter B, 2.0 μs to travel 600 m to experimenter A. The light is received at 3.0 μs and 4.0 μs, respectively; hence it was emitted by the explosion at $t = 2.0$ μs. The spacetime coordinates of the explosion are (600 m, 0 m, 0 m, 2.0 μs).

ASSESS Although the experimenters *see* the explosion at different times, they agree that the explosion actually *happened* at $t = 2.0$ μs.

Simultaneity

Two events 1 and 2 that take place at different positions x_1 and x_2 but at the *same time* $t_1 = t_2$, as measured in some reference frame, are said to be **simultaneous** in that reference frame. Simultaneity is determined by when the events actually happen, not when they are seen or observed. In general, simultaneous events are *not* seen at the same time because of the difference in light travel times from the events to an experimenter.

EXAMPLE 36.3 | **Are the explosions simultaneous?**

An experimenter in reference frame S stands at the origin looking in the positive x-direction. At $t = 3.0$ μs she sees firecracker 1 explode at $x = 600$ m. A short time later, at $t = 5.0$ μs, she sees firecracker 2 explode at $x = 1200$ m. Are the two explosions simultaneous? If not, which firecracker exploded first?

MODEL Light from both explosions travels toward the experimenter at 300 m/μs.

SOLVE The experimenter *sees* two different explosions, but perceptions of the events are not the events themselves. When did the explosions *actually* occur? Using the fact that light travels at 300 m/μs, we can see that firecracker 1 exploded at $t_1 = 1.0$ μs and firecracker 2 also exploded at $t_2 = 1.0$ μs. The events *are* simultaneous.

STOP TO THINK 36.4 A tree and a pole are 3000 m apart. Each is suddenly hit by a bolt of lightning. Mark, who is standing at rest midway between the two, sees the two lightning bolts at the same instant of time. Nancy is at rest under the tree. Define event 1 to be "lightning strikes tree" and event 2 to be "lightning strikes pole." For Nancy, does event 1 occur before, after, or at the same time as event 2?

36.5 The Relativity of Simultaneity

We've now established a means for measuring the time of an event in a reference frame, so let's begin to investigate the nature of time. The following "thought experiment" is very similar to one suggested by Einstein.

FIGURE 36.15 shows a long railroad car traveling to the right with a velocity v that may be an appreciable fraction of the speed of light. A firecracker is tied to each end of the car, just above the ground. Each firecracker is powerful enough so that, when it explodes, it will make a burn mark on the ground at the position of the explosion.

Ryan is standing on the ground, watching the railroad car go by. Peggy is standing in the exact center of the car with a special box at her feet. This box has two light detectors, one facing each way, and a signal light on top. The box works as follows:

1. If a flash of light is received at the detector facing right, as seen by Ryan, before a flash is received at the left detector, then the light on top of the box will turn green.
2. If a flash of light is received at the left detector before a flash is received at the right detector, or if two flashes arrive simultaneously, the light on top will turn red.

The firecrackers explode as the railroad car passes Ryan, and he sees the two light flashes from the explosions simultaneously. He then measures the distances to the two burn marks and finds that he was standing exactly halfway between the marks. Because light travels equal distances in equal times, Ryan concludes that the two explosions were simultaneous in his reference frame, the reference frame of the ground. Further, because he was midway between the two ends of the car, he was directly opposite Peggy when the explosions occurred.

FIGURE 36.16a shows the sequence of events in Ryan's reference frame. Light travels at speed c in all inertial reference frames, so, although the firecrackers were moving, the light waves are spheres centered on the burn marks. Ryan determines that the light wave coming from the right reaches Peggy and the box before the light wave coming from the left. Thus, according to Ryan, the signal light on top of the box turns green.

FIGURE 36.15 A railroad car traveling to the right with velocity v.

The firecrackers will make burn marks on the ground at the positions where they explode.

FIGURE 36.16 Exploding firecrackers seen in two different reference frames.

(a) The events in Ryan's frame

Explosions are simultaneous. Burn marks are equal distances from Ryan.

The waves are spheres centered on the burn marks because the light speed of both is c.

Peggy is moving to the right.

The waves reach Ryan simultaneously. The right wave has already passed Peggy and been detected. The left wave has not arrived.

(b) Are these the events in Peggy's frame?

Explosions are at the ends of the car at the instant Ryan passes Peggy.

The waves are spheres centered on the ends of the car because the light speed of both is c.

The waves reach Peggy and the light detectors simultaneously.

How do things look in Peggy's reference frame, a reference frame moving to the right at velocity v relative to the ground? As **FIGURE 36.16b** shows, Peggy sees Ryan moving to the left with speed v. Light travels at speed c in all inertial reference frames, so the light waves are spheres centered on the ends of the car. If the explosions are simultaneous, as Ryan has determined, the two light waves reach her and the box simultaneously. Thus, according to Peggy, the signal light on top of the box turns red!

Now the light on top must be either green or red. *It can't be both!* Later, after the railroad car has stopped, Ryan and Peggy can place the box in front of them. Either it has a red light or a green light. Ryan can't see one color while Peggy sees the other. Hence we have a paradox. It's impossible for Peggy and Ryan both to be right. But who is wrong, and why?

What do we know with absolute certainty?

1. Ryan detected the flashes simultaneously.
2. Ryan was halfway between the firecrackers when they exploded.
3. The light from the two explosions traveled toward Ryan at equal speeds.

The conclusion that the explosions were simultaneous in Ryan's reference frame is unassailable. The light is green.

Resolving the Paradox

Peggy, however, made an assumption. It's a perfectly ordinary assumption, one that seems sufficiently obvious that you probably didn't notice, but an assumption nonetheless. Peggy assumed that the explosions were simultaneous.

Didn't Ryan find them to be simultaneous? Indeed, he did. Suppose we call Ryan's reference frame S, the explosion on the right event R, and the explosion on the left event L. Ryan found that $t_R = t_L$. But Peggy has to use a different set of clocks, the clocks in her reference frame S′, to measure the times t'_R and t'_L at which the explosions occurred. The fact that $t_R = t_L$ in frame S does *not* allow us to conclude that $t'_R = t'_L$ in frame S′.

In fact, in frame S′ the right firecracker must explode *before* the left firecracker. Figure 36.16b, with its assumption about simultaneity, was incorrect. **FIGURE 36.17** shows the situation in Peggy's reference frame, with the right firecracker exploding first. Now the wave from the right reaches Peggy and the box first, as Ryan had concluded, and the light on top turns green.

One of the most disconcerting conclusions of relativity is that **two events occurring simultaneously in reference frame S are *not* simultaneous in any reference frame S′ moving relative to S**. This is called the **relativity of simultaneity**.

The two firecrackers *really* explode at the same instant of time in Ryan's reference frame. And the right firecracker *really* explodes first in Peggy's reference frame. It's not a matter of when they see the flashes. Our conclusion refers to the times at which the explosions actually occur.

The paradox of Peggy and Ryan contains the essence of relativity, and it's worth careful thought. First, review the logic until you're certain that there *is* a paradox, a logical impossibility. Then convince yourself that the only way to resolve the paradox is to abandon the assumption that the explosions are simultaneous in Peggy's reference frame. If you understand the paradox and its resolution, you've made a big step toward understanding what relativity is all about.

FIGURE 36.17 The real sequence of events in Peggy's reference frame.

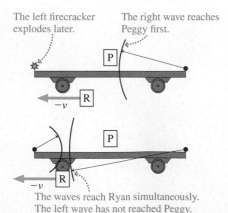

The right firecracker explodes first.

The left firecracker explodes later.

The right wave reaches Peggy first.

The waves reach Ryan simultaneously. The left wave has not reached Peggy.

STOP TO THINK 36.5 A tree and a pole are 3000 m apart. Each is hit by a bolt of lightning. Mark, who is standing at rest midway between the two, sees the two lightning bolts at the same instant of time. Nancy is flying her rocket at $v = 0.5c$ in the direction from the tree toward the pole. The lightning hits the tree just as she passes by it. Define event 1 to be "lightning strikes tree" and event 2 to be "lightning strikes pole." For Nancy, does event 1 occur before, after, or at the same time as event 2?

36.6 Time Dilation

The principle of relativity has driven us to the logical conclusion that time is not the same for two reference frames moving relative to each other. Our analysis thus far has been mostly qualitative. It's time to start developing some quantitative tools that will allow us to compare measurements in one reference frame to measurements in another reference frame.

FIGURE 36.18a shows a special clock called a *light clock*. The light clock is a box with a light source at the bottom and a mirror at the top, separated by distance h. The light source emits a very short pulse of light that travels to the mirror and reflects back to a light detector beside the source. The clock advances one "tick" each time the detector receives a light pulse, and it immediately, with no delay, causes the light source to emit the next light pulse.

Our goal is to compare two measurements of the interval between two ticks of the clock: one taken by an experimenter standing next to the clock and the other by an experimenter moving with respect to the clock. To be specific, FIGURE 36.18b shows the clock at rest in reference frame S'. We call this the **rest frame** of the clock. Reference frame S', with the clock, moves to the right with velocity v relative to reference frame S.

Relativity requires us to measure *events,* so let's define event 1 to be the emission of a light pulse and event 2 to be the detection of that light pulse. Experimenters in both reference frames are able to measure where and when these events occur *in their frame.* In frame S, the time interval $\Delta t = t_2 - t_1$ is one tick of the clock. Similarly, one tick in frame S' is $\Delta t' = t_2' - t_1'$.

To be sure we have a clear understanding of the relativity result, let's first do a classical analysis. In frame S', the clock's rest frame, the light travels straight up and down, a total distance $2h$, at speed c. The time interval is $\Delta t' = 2h/c$.

FIGURE 36.19a shows the operation of the light clock as seen in frame S. The clock is moving to the right at speed v in S, thus the mirror moves distance $\frac{1}{2}v(\Delta t)$ during the time $\frac{1}{2}(\Delta t)$ in which the light pulse moves from the source to the mirror. The distance traveled by the light during this interval is $\frac{1}{2}u_{\text{light}}(\Delta t)$, where u_{light} is the speed of light in frame S. You can see from the vector addition in FIGURE 36.19b that the speed of light in frame S is $u_{\text{light}} = (c^2 + v^2)^{1/2}$. (Remember, this is a classical analysis in which the speed of light *does* depend on the motion of the reference frame relative to the light source.)

The Pythagorean theorem applied to the right triangle in Figure 36.19a is

$$h^2 + \left(\tfrac{1}{2}v\,\Delta t\right)^2 = \left(\tfrac{1}{2}u_{\text{light}}\,\Delta t\right)^2 = \left(\tfrac{1}{2}\sqrt{c^2+v^2}\,\Delta t\right)^2$$
$$= \left(\tfrac{1}{2}c\,\Delta t\right)^2 + \left(\tfrac{1}{2}v\,\Delta t\right)^2 \tag{36.4}$$

The term $\left(\tfrac{1}{2}v\,\Delta t\right)^2$ is common to both sides and cancels. Solving for Δt gives $\Delta t = 2h/c$, identical to $\Delta t'$. In other words, a classical analysis finds that the clock ticks at exactly the same rate in both frame S and frame S'. This shouldn't be surprising. There's only one kind of time in classical physics, measured the same by all experimenters independent of their motion.

The principle of relativity changes only one thing, but that change has profound consequences. According to the principle of relativity, light travels at the same speed in *all* inertial reference frames. In frame S', the rest frame of the clock, the light simply goes straight up and back. The time of one tick,

$$\Delta t' = \frac{2h}{c} \tag{36.5}$$

is unchanged from the classical analysis.

FIGURE 36.18 The ticking of a light clock can be measured by experimenters in two different reference frames.

(a) A light clock

(b) The clock is at rest in frame S'.

FIGURE 36.19 A classical analysis of the light clock.

(a)

(b)

Light speed is the same in both frames.

Mirror

v

Light path through frame S

$\frac{1}{2}c\Delta t$

h

Emission

$\frac{1}{2}v\Delta t$

Detection

Clock moves distance $v\Delta t$.

FIGURE 36.20 shows the light clock as seen in frame S. The difference from Figure 36.19a is that the light now travels along the hypotenuse at speed c. We can again use the Pythagorean theorem to write

$$h^2 + \left(\tfrac{1}{2}v\,\Delta t\right)^2 = \left(\tfrac{1}{2}c\,\Delta t\right)^2 \tag{36.6}$$

Solving for Δt gives

$$\Delta t = \frac{2h/c}{\sqrt{1 - v^2/c^2}} = \frac{\Delta t'}{\sqrt{1 - v^2/c^2}} \tag{36.7}$$

The time interval between two ticks in frame S is *not* the same as in frame S′.

It's useful to define $\beta = v/c$, the velocity as a fraction of the speed of light. For example, a reference frame moving with $v = 2.4 \times 10^8$ m/s has $\beta = 0.80$. In terms of β, Equation 36.7 is

$$\Delta t = \frac{\Delta t'}{\sqrt{1 - \beta^2}} \tag{36.8}$$

NOTE The expression $(1 - v^2/c^2)^{1/2} = (1 - \beta^2)^{1/2}$ occurs frequently in relativity. The value of the expression is 1 when $v = 0$, and it steadily decreases to 0 as $v \to c$ (or $\beta \to 1$). The square root is an imaginary number if $v > c$, which would make Δt imaginary in Equation 36.8. Time intervals certainly have to be real numbers, suggesting that $v > c$ is not physically possible. One of the predictions of the theory of relativity, as you've undoubtedly heard, is that nothing can travel faster than the speed of light. Now you can begin to see why. We'll examine this topic more closely in Section 36.9. In the meantime, we'll require v to be less than c.

Proper Time

Frame S′ has one important distinction. It is the *one and only* inertial reference frame in which the light clock is at rest. Consequently, it is the one and only inertial reference frame in which the times of both events—the emission of the light and the detection of the light—are measured by the *same* reference-frame clock. You can see that the light pulse in Figure 36.18a starts and ends at the same position. In Figure 36.20, the emission and detection take place at different positions in frame S and must be measured by different reference-frame clocks, one at each position.

The time interval between two events that occur at the *same position* is called the **proper time** $\Delta\tau$. Only one inertial reference frame measures the proper time, and it does so with *a single clock that is present at both events*. An inertial reference frame moving with velocity $v = \beta c$ relative to the proper-time frame must use two clocks to measure the time interval: one at the position of the first event, the other at the position of the second event. The time interval in the frame where two clocks are required is

$$\Delta t = \frac{\Delta\tau}{\sqrt{1 - \beta^2}} \geq \Delta\tau \qquad \text{(time dilation)} \tag{36.9}$$

The "stretching out" of the time interval implied by Equation 36.9 is called **time dilation.** Time dilation is sometimes described by saying that "moving clocks run slow," but this statement has to be interpreted carefully. The whole point of relativity is that all inertial frames are equally valid, so there's no absolute sense in which a clock is "moving" or "at rest."

To illustrate, **FIGURE 36.21** shows two firecracker explosions—two events—that occur at different positions in the ground reference frame. Experimenters on the ground need two clocks to measure the time interval Δt. In the train reference frame,

FIGURE 36.21 The time interval between two events is measured in two different reference frames.

The ground reference frame needs two clocks, A and B, to measure the time interval Δt between events 1 and 2.

✦ Event 1

A   B

✦ Event 2

A   B

The train reference frame measures the proper time $\Delta\tau$ because one clock is present at both events.

however, a single clock is present at both events; hence the time interval measured in the train reference frame is the proper time $\Delta\tau$. You can see that $\Delta\tau < \Delta t$, so less time has elapsed in the train reference frame.

In this sense, the "moving clock," the one that is present at both events, "runs slower" than clocks that are stationary with respect to the events. More generally, **the time interval between two events is smallest in the reference frame in which the two events occur at the same position.**

NOTE Equation 36.9 was derived using a light clock because the operation of a light clock is clear and easy to analyze. But the conclusion is really about time itself. *Any clock, regardless of how it operates, behaves the same.*

EXAMPLE 36.4 | **From the sun to Saturn**

Saturn is 1.43×10^{12} m from the sun. A rocket travels along a line from the sun to Saturn at a constant speed of $0.9c$ relative to the solar system. How long does the journey take as measured by an experimenter on earth? As measured by an astronaut on the rocket?

MODEL Let the solar system be in reference frame S and the rocket be in reference frame S' that travels with velocity $v = 0.9c$ relative to S. Relativity problems must be stated in terms of *events*. Let event 1 be "the rocket and the sun coincide" (the experimenter on earth says that the rocket passes the sun; the astronaut on the rocket says that the sun passes the rocket) and event 2 be "the rocket and Saturn coincide."

FIGURE 36.22 Pictorial representation of the trip as seen in frames S and S'.

Rocket journey in frame S

Event 1

The time between these two events is Δt.

Event 2

$\Delta x = v\Delta t$

Rocket journey in frame S'

Event 1

The time between these two events is the proper time $\Delta\tau$.

Event 2

$\Delta x' = 0$

VISUALIZE FIGURE 36.22 shows the two events as seen from the two reference frames. Notice that the two events occur at the *same position* in S', the position of the rocket, and consequently can be measured by *one* clock carried on board the rocket.

SOLVE The time interval measured in the solar system reference frame, which includes the earth, is simply

$$\Delta t = \frac{\Delta x}{v} = \frac{1.43 \times 10^{12} \text{ m}}{0.9 \times (3.00 \times 10^8 \text{ m/s})} = 5300 \text{ s}$$

Relativity hasn't abandoned the basic definition $v = \Delta x/\Delta t$, although we do have to be sure that Δx and Δt are measured in just one reference frame and refer to the same two events.

How are things in the rocket's reference frame? The two events occur at the *same position* in S' and can be measured by *one* clock, the clock at the origin. Thus the time measured by the astronauts is the *proper time* $\Delta\tau$ between the two events. We can use Equation 36.9 with $\beta = 0.9$ to find

$$\Delta\tau = \sqrt{1 - \beta^2}\, \Delta t = \sqrt{1 - 0.9^2}\,(5300 \text{ s}) = 2310 \text{ s}$$

ASSESS The time interval measured between these two events by the astronauts is less than half the time interval measured by experimenters on earth. The difference has nothing to do with when earthbound astronomers *see* the rocket pass the sun and Saturn. Δt is the time interval from when the rocket actually passes the sun, as measured by a clock at the sun, until it actually passes Saturn, as measured by a synchronized clock at Saturn. The interval between *seeing* the events from earth, which would have to allow for light travel times, would be something other than 5300 s. Δt and $\Delta\tau$ are different because *time is different* in two reference frames moving relative to each other.

STOP TO THINK 36.6 Molly flies her rocket past Nick at constant velocity v. Molly and Nick both measure the time it takes the rocket, from nose to tail, to pass Nick. Which of the following is true?

a. Both Molly and Nick measure the same amount of time.
b. Molly measures a shorter time interval than Nick.
c. Nick measures a shorter time interval than Molly.

Experimental Evidence

Is there any evidence for the crazy idea that clocks moving relative to each other tell time differently? Indeed, there's plenty. An experiment in 1971 sent an atomic clock around the world on a jet plane while an identical clock remained in the laboratory. This was a difficult experiment because the traveling clock's speed was so small compared to c, but measuring the small differences between the time intervals was just barely within the capabilities of atomic clocks. It was also a more complex experiment than we've analyzed because the clock accelerated as it moved around a circle. The scientists found that, upon its return, the eastbound clock, traveling faster than the laboratory on a rotating earth, was 60 ns behind the stay-at-home clock, which was exactly as predicted by relativity.

Very detailed studies have been done on unstable particles called *muons* that are created at the top of the atmosphere, at a height of about 60 km, when high-energy cosmic rays collide with air molecules. It is well known, from laboratory studies, that stationary muons decay with a *half-life* of 1.5 μs. That is, half the muons decay within 1.5 μs, half of those remaining decay in the next 1.5 μs, and so on. The decays can be used as a clock.

The muons travel down through the atmosphere at very nearly the speed of light. The time needed to reach the ground, assuming $v \approx c$, is $\Delta t \approx (60,000 \text{ m})/ (3 \times 10^8 \text{ m/s}) = 200 \ \mu$s. This is 133 half-lives, so the fraction of muons reaching the ground should be $\approx (\frac{1}{2})^{133} = 10^{-40}$. That is, only 1 out of every 10^{40} muons should reach the ground. In fact, experiments find that about 1 in 10 muons reach the ground, an experimental result that differs by a factor of 10^{39} from our prediction!

The discrepancy is due to time dilation. In **FIGURE 36.23**, the two events "muon is created" and "muon hits ground" take place at two different places in the earth's reference frame. However, these two events occur at the *same position* in the muon's reference frame. (The muon is like the rocket in Example 36.4.) Thus the muon's internal clock measures the proper time. The time-dilated interval $\Delta t = 200 \ \mu$s in the earth's reference frame corresponds to a proper time $\Delta \tau \approx 5 \ \mu$s in the muon's reference frame. That is, in the muon's reference frame it takes only 5 μs from creation at the top of the atmosphere until the ground runs into it. This is 3.3 half-lives, so the fraction of muons reaching the ground is $(\frac{1}{2})^{3.3} = 0.1$, or 1 out of 10. We wouldn't detect muons at the ground at all if not for time dilation.

The details are beyond the scope of this textbook, but dozens of high-energy particle accelerators around the world that study quarks and other elementary particles have been designed and built on the basis of Einstein's theory of relativity. The fact that they work exactly as planned is strong testimony to the reality of time dilation.

FIGURE 36.23 We wouldn't detect muons at the ground if not for time dilation.

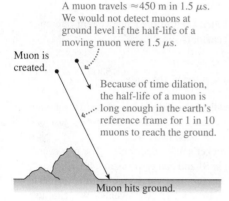

A muon travels $\approx$ 450 m in 1.5 μs. We would not detect muons at ground level if the half-life of a moving muon were 1.5 μs.

Muon is created.

Because of time dilation, the half-life of a muon is long enough in the earth's reference frame for 1 in 10 muons to reach the ground.

Muon hits ground.

The Twin Paradox

The most well-known relativity paradox is the twin paradox. George and Helen are twins. On their 25th birthday, Helen departs on a starship voyage to a distant star. Let's imagine, to be specific, that her starship accelerates almost instantly to a speed of 0.95c and that she travels to a star that is 9.5 light years (9.5 ly) from earth. Upon arriving, she discovers that the planets circling the star are inhabited by fierce aliens, so she immediately turns around and heads home at 0.95c.

A **light year,** abbreviated ly, is the distance that light travels in one year. A light year is vastly larger than the diameter of the solar system. The distance between two neighboring stars is typically a few light years. For our purpose, we can write the speed of light as $c = 1$ ly/year. That is, light travels 1 light year per year.

This value for c allows us to determine how long, according to George and his fellow earthlings, it takes Helen to travel out and back. Her total distance is 19 ly and, due to her rapid acceleration and rapid turnaround, she travels essentially the entire distance at speed $v = 0.95c = 0.95$ ly/year. Thus the time she's away, as measured by George, is

$$\Delta t_G = \frac{19 \text{ ly}}{0.95 \text{ ly/year}} = 20 \text{ years} \tag{36.10}$$

George will be 45 years old when his sister Helen returns with tales of adventure.

The global positioning system (GPS), which allows you to pinpoint your location anywhere in the world to within a few meters, uses a set of orbiting satellites. Because of their motion, the atomic clocks on these satellites keep time differently from clocks on the ground. To determine an accurate position, the software in your GPS receiver must carefully correct for time-dilation effects.

While she's away, George takes a physics class and studies Einstein's theory of relativity. He realizes that time dilation will make Helen's clocks run more slowly than his clocks, which are at rest relative to him. Her heart—a clock—will beat fewer times and the minute hand on her watch will go around fewer times. In other words, she's aging more slowly than he is. Although she is his twin, she will be younger than he is when she returns.

Calculating Helen's age is not hard. We simply have to identify Helen's clock, because it's always with Helen as she travels, as the clock that measures proper time $\Delta\tau$. From Equation 36.9,

$$\Delta t_\text{H} = \Delta\tau = \sqrt{1 - \beta^2}\,\Delta t_\text{G} = \sqrt{1 - 0.95^2}\ (20 \text{ years}) = 6.25 \text{ years} \qquad (36.11)$$

George will have just celebrated his 45th birthday as he welcomes home his 31-year-and-3-month-old twin sister.

This may be unsettling because it violates our commonsense notion of time, but it's not a paradox. There's no logical inconsistency in this outcome. So why is it called "the twin paradox"?

Helen, knowing that she had quite of bit of time to kill on her journey, brought along several physics books to read. As she learns about relativity, she begins to think about George and her friends back on earth. Relative to her, they are all moving away at 0.95c. Later they'll come rushing toward her at 0.95c. Time dilation will cause their clocks to run more slowly than her clocks, which are at rest relative to her. In other words, as FIGURE 36.24 shows, Helen concludes that people on earth are aging more slowly than she is. Alas, she will be much older than they when she returns.

Finally, the big day arrives. Helen lands back on earth and steps out of the starship. George is expecting Helen to be younger than he is. Helen is expecting George to be younger than she is.

Here's the paradox! It's logically impossible for each to be younger than the other at the time they are reunited. Where, then, is the flaw in our reasoning? It seems to be a symmetrical situation—Helen moves relative to George and George moves relative to Helen—but symmetrical reasoning has led to a conundrum.

But are the situations really symmetrical? George goes about his business day after day without noticing anything unusual. Helen, on the other hand, experiences three distinct periods during which the starship engines fire, she's crushed into her seat, and free dust particles that had been floating inside the starship are no longer, in the starship's reference frame, at rest or traveling in a straight line at constant speed. In other words, George spends the entire time in an inertial reference frame, *but Helen does not.* The situation is *not* symmetrical.

The principle of relativity applies *only* to inertial reference frames. Our discussion of time dilation was for inertial reference frames. Thus George's analysis and calculations are correct. Helen's analysis and calculations are *not* correct because she was trying to apply an inertial reference frame result while traveling in a noninertial reference frame. (Or, alternatively, Helen was in two different inertial frames while George was only in one, and thus the situation is not symmetrical.)

Helen is younger than George when she returns. This is strange, but not a paradox. It is a consequence of the fact that time flows differently in two reference frames moving relative to each other.

36.7 Length Contraction

We've seen that relativity requires us to rethink our idea of time. Now let's turn our attention to the concepts of space and distance. Consider the rocket that traveled from the sun to Saturn in Example 36.4. FIGURE 36.25a on the next page shows the rocket moving with velocity v through the solar system reference frame S. We define $L = \Delta x = x_\text{Saturn} - x_\text{sun}$ as the distance between the sun and Saturn in frame S or, more generally, the *length* of the spatial interval between two points. The rocket's speed is $v = L/\Delta t$, where Δt is the time measured in frame S for the journey from the sun to Saturn.

FIGURE 36.24 The twin paradox.

Helen is moving relative to me at 0.95c. Her clocks are running more slowly than mine, and when she returns she'll be younger than I am.

0.95c

9.5 ly

George is moving relative to me at 0.95c. His clocks are running more slowly than mine, and when I return he'll be younger than I am.

FIGURE 36.25 L and L' are the distances between the sun and Saturn in frames S and S'.

(a) Reference frame S: The solar system is stationary.

(b) Reference frame S': The rocket is stationary.

FIGURE 36.25b shows the situation in reference frame S', where the rocket is at rest. The sun and Saturn move to the left at speed $v = L'/\Delta t'$, where $\Delta t'$ is the time measured in frame S' for Saturn to travel distance L'.

Speed v is the relative speed between S and S' and is the same for experimenters in both reference frames. That is,

$$v = \frac{L}{\Delta t} = \frac{L'}{\Delta t'} \tag{36.12}$$

The time interval $\Delta t'$ measured in frame S' is the proper time $\Delta \tau$ because both events occur at the same position in frame S' and can be measured by one clock. We can use the time-dilation result, Equation 36.9, to relate $\Delta \tau$ measured by the astronauts to Δt measured by the earthbound scientists. Then Equation 36.12 becomes

$$\frac{L}{\Delta t} = \frac{L'}{\Delta \tau} = \frac{L'}{\sqrt{1 - \beta^2}\, \Delta t} \tag{36.13}$$

The Δt cancels, and the distance L' in frame S' is

$$L' = \sqrt{1 - \beta^2}\, L \tag{36.14}$$

Surprisingly, we find that **the distance between two objects in reference frame S' is *not the same* as the distance between the same two objects in reference frame S.**

Frame S, in which the distance is L, has one important distinction. It is the *one and only* inertial reference frame in which the objects are at rest. Experimenters in frame S can take all the time they need to measure L because the two objects aren't going anywhere. The distance L between two objects, or two points on one object, measured in the reference frame in which the objects are at rest is called the **proper length** ℓ. Only one inertial reference frame can measure the proper length.

We can use the proper length ℓ to write Equation 36.14 as

$$L' = \sqrt{1 - \beta^2}\, \ell \le \ell \tag{36.15}$$

The Stanford Linear Accelerator (SLAC) is a 2-mi-long electron accelerator. The accelerator's length is less than 1 m in the reference frame of the electrons.

This "shrinking" of the distance between two objects, as measured by an experiment moving with respect to the objects, is called **length contraction.** Although we derived length contraction for the distance between two distinct objects, it applies equally well to the length of any physical object that stretches between two points along the x- and x'-axes. **The length of an object is greatest in the reference frame in which the object is at rest.** The object's length is less (i.e., the length is contracted) when it is measured in any reference frame in which the object is moving.

EXAMPLE 36.5 | The distance from the sun to Saturn

In Example 36.4 a rocket traveled along a line from the sun to Saturn at a constant speed of $0.9c$ relative to the solar system. The Saturn-to-sun distance was given as 1.43×10^{12} m. What is the distance between the sun and Saturn in the rocket's reference frame?

MODEL Saturn and the sun are, at least approximately, at rest in the solar system reference frame S. Thus the given distance is the proper length ℓ.

SOLVE We can use Equation 36.15 to find the distance in the rocket's frame S':

$$L' = \sqrt{1 - \beta^2}\,\ell = \sqrt{1 - 0.9^2}\,(1.43 \times 10^{12}\text{ m})$$
$$= 0.62 \times 10^{12}\text{ m}$$

ASSESS The sun-to-Saturn distance measured by the astronauts is less than half the distance measured by experimenters on earth. L' and ℓ are different because *space is different* in two reference frames moving relative to each other.

The conclusion that space is different in reference frames moving relative to each other is a direct consequence of the fact that time is different. Experimenters in both reference frames agree on the relative velocity v, leading to Equation 36.12: $v = L/\Delta t = L'/\Delta t'$. We had already learned that $\Delta t' < \Delta t$ because of time dilation. Thus L' *has* to be less than L. That is the only way experimenters in the two reference frames can reconcile their measurements.

To be specific, the earthly experimenters in Examples 36.4 and 36.5 find that the rocket takes 5300 s to travel the 1.43×10^{12} m between the sun and Saturn. The rocket's speed is $v = L/\Delta t = 2.7 \times 10^8$ m/s $= 0.9c$. The astronauts in the rocket find that it takes only 2310 s for Saturn to reach them after the sun has passed by. But there's no conflict, because they also find that the distance is only 0.62×10^{12} m. Thus Saturn's speed toward them is $v = L'/\Delta t' = (0.62 \times 10^{12}\text{ m})/(2310\text{ s}) = 2.7 \times 10^8$ m/s $= 0.9c$.

Another Paradox?

Carmen and Dan are in their physics lab room. They each select a meter stick, lay the two side by side, and agree that the meter sticks are exactly the same length. Then, for an extra-credit project, they go outside and run past each other, in opposite directions, at a relative speed $v = 0.9c$. FIGURE 36.26 shows their experiment and a portion of their conversation.

Now, Dan's meter stick can't be both longer and shorter than Carmen's meter stick. Is this another paradox? No! Relativity allows us to compare the *same* events as they're measured in two different reference frames. This did lead to a real paradox when Peggy rolled past Ryan on the train. There the signal light on the box turns green (a single event) or it doesn't, and Peggy and Ryan have to agree about it. But the events by which Dan measures the length (in Dan's frame) of Carmen's meter stick are *not the same events* as those by which Carmen measures the length (in Carmen's frame) of Dan's meter stick.

There's no conflict between their measurements. In Dan's reference frame, Carmen's meter stick has been length contracted and is less than 1 m in length. In Carmen's reference frame, Dan's meter stick has been length contracted and is less than 1 m in length. If this weren't the case, if both agreed that one of the meter sticks was shorter than the other, then we could tell which reference frame was "really" moving and which was "really" at rest. But the principle of relativity doesn't allow us to make that distinction. Each is moving relative to the other, so each should make the same measurement for the length of the other's meter stick.

The Spacetime Interval

Forget relativity for a minute and think about ordinary geometry. FIGURE 36.27 shows two ordinary coordinate systems. They are identical except for the fact that one has been rotated relative to the other. A student using the xy-system would measure coordinates (x_1, y_1) for point 1 and (x_2, y_2) for point 2. A second student, using the $x'y'$-system, would measure (x'_1, y'_1) and (x'_2, y'_2).

FIGURE 36.26 Carmen and Dan each measure the length of the other's meter stick as they move relative to each other.

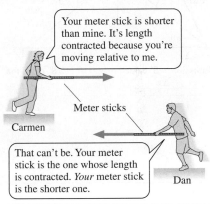

Your meter stick is shorter than mine. It's length contracted because you're moving relative to me.

Meter sticks

Carmen

That can't be. Your meter stick is the one whose length is contracted. *Your* meter stick is the shorter one.

Dan

FIGURE 36.27 Distance d is the same in both coordinate systems.

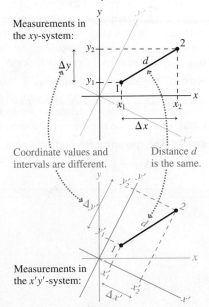

Measurements in the xy-system:

Coordinate values and intervals are different.

Distance d is the same.

Measurements in the $x'y'$-system:

The students soon find that none of their measurements agree. That is, $x_1 \neq x_1'$ and so on. Even the intervals are different: $\Delta x \neq \Delta x'$ and $\Delta y \neq \Delta y'$. Each is a perfectly valid coordinate system, giving no reason to prefer one over the other, but each yields different measurements.

Is there *anything* on which the two students can agree? Yes, there is. The distance d between points 1 and 2 is independent of the coordinates. We can state this mathematically as

$$d^2 = (\Delta x)^2 + (\Delta y)^2 = (\Delta x')^2 + (\Delta y')^2 \tag{36.16}$$

The quantity $(\Delta x)^2 + (\Delta y)^2$ is called an **invariant** in geometry because it has the same value in any Cartesian coordinate system.

Returning to relativity, is there an invariant in the spacetime coordinates, some quantity that has the *same value* in all inertial reference frames? There is, and to find it let's return to the light clock of Figure 36.20. **FIGURE 36.28** shows the light clock as seen in reference frames S′ and S″. The speed of light is the same in both frames, even though both are moving with respect to each other and with respect to the clock.

Notice that the clock's height h is common to both reference frames. Thus

$$h^2 = \left(\tfrac{1}{2}c\,\Delta t'\right)^2 - \left(\tfrac{1}{2}\Delta x'\right)^2 = \left(\tfrac{1}{2}c\,\Delta t''\right)^2 - \left(\tfrac{1}{2}\Delta x''\right)^2 \tag{36.17}$$

The factor $\tfrac{1}{2}$ cancels, allowing us to write

$$c^2(\Delta t')^2 - (\Delta x')^2 = c^2(\Delta t'')^2 - (\Delta x'')^2 \tag{36.18}$$

Let us define the **spacetime interval** s between two events to be

$$s^2 = c^2(\Delta t)^2 - (\Delta x)^2 \tag{36.19}$$

What we've shown in Equation 36.18 is that **the spacetime interval s has the same value in all inertial reference frames**. That is, the spacetime interval between two events is an invariant. It is a value that all experimenters, in all reference frames, can agree upon.

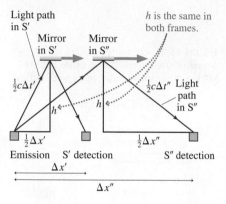

FIGURE 36.28 The light clock seen by experimenters in reference frames S′ and S″.

EXAMPLE 36.6 | Using the spacetime interval

A firecracker explodes at the origin of an inertial reference frame. Then, 2.0 μs later, a second firecracker explodes 300 m away. Astronauts in a passing rocket measure the distance between the explosions to be 200 m. According to the astronauts, how much time elapses between the two explosions?

MODEL The spacetime coordinates of two events are measured in two different inertial reference frames. Call the reference frame of the ground S and the reference frame of the rocket S′. The spacetime interval between these two events is the same in both reference frames.

SOLVE The spacetime interval (or, rather, its square) in frame S is

$$s^2 = c^2(\Delta t)^2 - (\Delta x)^2 = (600 \text{ m})^2 - (300 \text{ m})^2 = 270,000 \text{ m}^2$$

where we used $c = 300$ m/μs to determine that $c\,\Delta t = 600$ m. The spacetime interval has the same value in frame S′. Thus

$$s^2 = 270,000 \text{ m}^2 = c^2(\Delta t')^2 - (\Delta x')^2$$
$$= c^2(\Delta t')^2 - (200 \text{ m})^2$$

This is easily solved to give $\Delta t' = 1.85$ μs.

ASSESS The two events are closer together in both space and time in the rocket's reference frame than in the reference frame of the ground.

Einstein's legacy, according to popular culture, was the discovery that "everything is relative." But it's not so. Time intervals and space intervals may be relative, as were the intervals Δx and Δy in the purely geometric analogy with which we opened this section, but some things are *not* relative. In particular, the spacetime interval s between two events is not relative. It is a well-defined number, agreed on by experimenters in each and every inertial reference frame.

STOP TO THINK 36.7 Beth and Charles are at rest relative to each other. Anjay runs past at velocity v while holding a long pole parallel to his motion. Anjay, Beth, and Charles each measure the length of the pole at the instant Anjay passes Beth. Rank in order, from largest to smallest, the three lengths L_A, L_B, and L_C.

36.8 The Lorentz Transformations

The Galilean transformation $x' = x - vt$ of classical relativity lets us calculate the position x' of an event in frame S' if we know its position x in frame S. Classical relativity, of course, assumes that $t' = t$. Is there a similar transformation in relativity that would allow us to calculate an event's spacetime coordinates (x', t') in frame S' if we know their values (x, t) in frame S? Such a transformation would need to satisfy three conditions:

1. Agree with the Galilean transformations in the low-speed limit $v \ll c$.
2. Transform not only spatial coordinates but also time coordinates.
3. Ensure that the speed of light is the same in all reference frames.

We'll continue to use reference frames in the standard orientation of **FIGURE 36.29**. The motion is parallel to the x- and x'-axes, and we *define* $t = 0$ and $t' = 0$ as the instant when the origins of S and S' coincide.

The requirement that a new transformation agree with the Galilean transformation when $v \ll c$ suggests that we look for a transformation of the form

$$x' = \gamma(x - vt) \quad \text{and} \quad x = \gamma(x' + vt') \tag{36.20}$$

where γ is a dimensionless function of velocity that satisfies $\gamma \rightarrow 1$ as $v \rightarrow 0$.

To determine γ, we consider the following two events:

Event 1: A flash of light is emitted from the origin of both reference frames $(x = x' = 0)$ at the instant they coincide $(t = t' = 0)$.

Event 2: The light strikes a light detector. The spacetime coordinates of this event are (x, t) in frame S and (x', t') in frame S'.

Light travels at speed c in both reference frames, so the positions of event 2 are $x = ct$ in S and $x' = ct'$ in S'. Substituting these expressions for x and x' into Equation 36.20 gives

$$ct' = \gamma(ct - vt) = \gamma(c - v)t$$
$$ct = \gamma(ct' + vt') = \gamma(c + v)t' \tag{36.21}$$

We solve the first equation for t', by dividing by c, then substitute this result for t' into the second:

$$ct = \gamma(c + v)\frac{\gamma(c - v)t}{c} = \gamma^2(c^2 - v^2)\frac{t}{c}$$

FIGURE 36.29 The spacetime coordinates of an event are measured in inertial reference frames S and S'.

An event has spacetime coordinates (x, t) in frame S, (x', t') in frame S'.

Event

Origins coincide at $t = t' = 0$.

The t cancels, leading to

$$\gamma^2 = \frac{c^2}{c^2 - v^2} = \frac{1}{1 - v^2/c^2}$$

Thus the γ that "works" in the proposed transformation of Equation 36.20 is

$$\gamma = \frac{1}{\sqrt{1 - v^2/c^2}} = \frac{1}{\sqrt{1 - \beta^2}} \qquad (36.22)$$

You can see that $\gamma \rightarrow 1$ as $v \rightarrow 0$, as expected.

The transformation between t and t' is found by requiring that $x = x$ if you use Equation 36.20 to transform a position from S to S' and then back to S. The details will be left for a homework problem. Another homework problem will let you demonstrate that the y and z measurements made perpendicular to the relative motion are not affected by the motion. We tacitly assumed this condition in our analysis of the light clock.

The full set of equations is called the **Lorentz transformations.** They are

$$
\begin{aligned}
x' &= \gamma(x - vt) & x &= \gamma(x' + vt') \\
y' &= y & y &= y' \\
z' &= z & z &= z' \\
t' &= \gamma(t - vx/c^2) & t &= \gamma(t' + vx'/c^2)
\end{aligned}
\qquad (36.23)
$$

The Lorentz transformations transform the spacetime coordinates of *one* event. Compare these to the Galilean transformation equations in Equations 36.1.

NOTE These transformations are named after the Dutch physicist H. A. Lorentz, who derived them prior to Einstein. Lorentz was close to discovering special relativity, but he didn't recognize that our concepts of space and time have to be changed before these equations can be properly interpreted.

Using Relativity

Relativity is phrased in terms of *events;* hence relativity problems are solved by interpreting the problem statement in terms of specific events.

PROBLEM-SOLVING STRATEGY 36.1

Relativity

MODEL Frame the problem in terms of events, things that happen at a specific place and time.

VISUALIZE A pictorial representation defines the reference frames.

- Sketch the reference frames, showing their motion relative to each other.
- Show events. Identify objects that are moving with respect to the reference frames.
- Identify any proper time intervals and proper lengths. These are measured in an object's rest frame.

SOLVE The mathematical representation is based on the Lorentz transformations, but not every problem requires the full transformation equations.

- Problems about time intervals can often be solved using time dilation: $\Delta t = \gamma \Delta \tau$.
- Problems about distances can often be solved using length contraction: $L = \ell/\gamma$.

ASSESS Are the results consistent with Galilean relativity when $v \ll c$?

EXAMPLE 36.7 | Ryan and Peggy revisited

Peggy is standing in the center of a long, flat railroad car that has firecrackers tied to both ends. The car moves past Ryan, who is standing on the ground, with velocity $v = 0.8c$. Flashes from the exploding firecrackers reach him simultaneously $1.0 \ \mu s$ after the instant that Peggy passes him, and he later finds burn marks on the track 300 m to either side of where he had been standing.

a. According to Ryan, what is the distance between the two explosions, and at what times do the explosions occur relative to the time that Peggy passes him?

b. According to Peggy, what is the distance between the two explosions, and at what times do the explosions occur relative to the time that Ryan passes her?

MODEL Let the explosion on Ryan's right, the direction in which Peggy is moving, be event R. The explosion on his left is event L.

VISUALIZE Peggy and Ryan are in inertial reference frames. As FIGURE 36.30 shows, Peggy's frame S' is moving with $v = 0.8c$ relative to Ryan's frame S. We've defined the reference frames such that Peggy and Ryan are at the origins. The instant they pass, by definition, is $t = t' = 0$ s. The two events are shown in Ryan's reference frame.

FIGURE 36.30 A pictorial representation of the reference frames and events.

SOLVE a. The two burn marks tell Ryan that the distance between the explosions was $L = 600$ m. Light travels at $c = 300$ m/μs, and the burn marks are 300 m on either side of him, so Ryan can determine that each explosion took place $1.0 \ \mu s$ before he saw the flash. But this was the instant of time that Peggy passed him, so Ryan concludes that the explosions were simultaneous with each other and with Peggy's passing him. The spacetime coordinates of the two events in frame S are $(x_R, t_R) = (300 \text{ m}, 0 \ \mu s)$ and $(x_L, t_L) = (-300 \text{ m}, 0 \ \mu s)$.

b. We already know, from our qualitative analysis in Section 36.5, that the explosions are *not* simultaneous in Peggy's reference frame. Event R happens before event L in S', but we don't know how they compare to the time at which Ryan passes Peggy. We can now use the Lorentz transformations to relate the spacetime coordinates of these events as measured by Ryan to the spacetime coordinates as measured by Peggy. Using $v = 0.8c$, we find that γ is

$$\gamma = \frac{1}{\sqrt{1 - v^2/c^2}} = \frac{1}{\sqrt{1 - 0.8^2}} = 1.667$$

For event L, the Lorentz transformations are

$$x'_L = 1.667\big((-300 \text{ m}) - (0.8c)(0 \ \mu s)\big) = -500 \text{ m}$$

$$t'_L = 1.667\big((0 \ \mu s) - (0.8c)(-300 \text{ m})/c^2\big) = 1.33 \ \mu s$$

And for event R,

$$x'_R = 1.667\big((300 \text{ m}) - (0.8c)(0 \ \mu s)\big) = 500 \text{ m}$$

$$t'_R = 1.667\big((0 \ \mu s) - (0.8c)(300 \text{ m})/c^2\big) = -1.33 \ \mu s$$

According to Peggy, the two explosions occur 1000 m apart. Furthermore, the first explosion, on the right, occurs $1.33 \ \mu s$ before Ryan passes her at $t' = 0$ s. The second, on the left, occurs $1.33 \ \mu s$ after Ryan goes by.

ASSESS Events that are simultaneous in frame S are *not* simultaneous in frame S'. The results of the Lorentz transformations agree with our earlier qualitative analysis.

A follow-up discussion of Example 36.7 is worthwhile. Because Ryan moves at speed $v = 0.8c = 240$ m/μs relative to Peggy, he moves 320 m during the $1.33 \ \mu s$ between the first explosion and the instant he passes Peggy, then another 320 m before the second explosion. Gathering this information together, FIGURE 36.31 on the next page shows the sequence of events in Peggy's reference frame.

The firecrackers define the ends of the railroad car, so the 1000 m distance between the explosions in Peggy's frame is the car's length L' in frame S'. The car is at rest in frame S', hence length L' is the proper length: $\ell = 1000$ m. Ryan is measuring the length of a moving object, so he should see the car length contracted to

$$L = \sqrt{1 - \beta^2} \, \ell = \frac{\ell}{\gamma} = \frac{1000 \text{ m}}{1.667} = 600 \text{ m}$$

And, indeed, that is exactly the distance Ryan measured between the burn marks.

FIGURE 36.31 The sequence of events as seen in Peggy's reference frame.

Finally, we can calculate the spacetime interval s between the two events. According to Ryan,

$$s^2 = c^2(\Delta t^2) - (\Delta x)^2 = c^2(0\ \mu s)^2 - (600\ m)^2 = -(600\ m)^2$$

Peggy computes the spacetime interval to be

$$s^2 = c^2(\Delta t')^2 - (\Delta x')^2 = c^2(2.67\ \mu s)^2 - (1000\ m)^2 = -(600\ m)^2$$

Their calculations of the spacetime interval agree, showing that s really is an invariant, but notice that s itself is an imaginary number.

Length

We've already introduced the idea of length contraction, but we didn't precisely define just what we mean by the *length* of a moving object. The length of an object at rest is clear because we can take all the time we need to measure it with meter sticks, surveying tools, or whatever we need. But how can we give clear meaning to the length of a moving object?

A reasonable definition of an object's length is the distance $L = \Delta x = x_R - x_L$ between the right and left ends when the positions x_R and x_L are measured *at the same time t*. In other words, length is the distance spanned by the object at *one instant* of time. Measuring an object's length requires *simultaneous* measurements of two positions (i.e., two events are required); hence the result won't be known until the information from two spatially separated measurements can be brought together.

FIGURE 36.32 shows an object traveling through reference frame S with velocity v. The object is at rest in reference frame S' that travels with the object at velocity v; hence the length in frame S' is the proper length ℓ. That is, $\Delta x' = x'_R - x'_L = \ell$ in frame S'.

At time t, an experimenter (and his or her assistants) in frame S makes simultaneous measurements of the positions x_R and x_L of the ends of the object. The difference $\Delta x = x_R - x_L = L$ is the length in frame S. The Lorentz transformations of x_R and x_L are

$$x'_R = \gamma(x_R - vt)$$
$$x'_L = \gamma(x_L - vt)$$

(36.24)

where, it is important to note, t is the *same* for both because the measurements are simultaneous.

Subtracting the second equation from the first, we find

$$x'_R - x'_L = \ell = \gamma(x_R - x_L) = \gamma L = \frac{L}{\sqrt{1-\beta^2}}$$

FIGURE 36.32 The length of an object is the distance between *simultaneous* measurements of the positions of the end points.

The object is at rest in frame S'. Its length is $L' = \ell$, which can be measured at any time.

Because the object is moving in frame S, simultaneous measurements of its ends must be made to find its length L in frame S.

Solving for L, we find, in agreement with Equation 36.15, that

$$L = \sqrt{1-\beta^2}\,\ell$$

(36.25)

This analysis has accomplished two things. First, by giving a precise definition of length, we've put our length-contraction result on a firmer footing. Second, we've had good practice at relativistic reasoning using the Lorentz transformation.

NOTE Length contraction does not tell us how an object would *look*. The visual appearance of an object is determined by light waves that arrive simultaneously at the eye. These waves left points on the object at different times (i.e., *not* simultaneously) because they had to travel different distances to the eye. The analysis needed to determine an object's visual appearance is considerably more complex. Length and length contraction are concerned only with the *actual* length of the object at one instant of time.

The Binomial Approximation

You've met the binomial approximation earlier in this text and in your calculus class. The binomial approximation is useful when we need to calculate a relativistic expression for a nonrelativistic velocity $v \ll c$. Because $v^2/c^2 \ll 1$ in these cases, we can write

$$\text{If } v \ll c: \begin{cases} \sqrt{1 - \beta^2} = \left(1 - \dfrac{v^2}{c^2}\right)^{1/2} \approx 1 - \dfrac{1}{2}\dfrac{v^2}{c^2} \\[2ex] \gamma = \dfrac{1}{\sqrt{1 - \beta^2}} = \left(1 - \dfrac{v^2}{c^2}\right)^{-1/2} \approx 1 + \dfrac{1}{2}\dfrac{v^2}{c^2} \end{cases} \quad (36.26)$$

The following example illustrates the use of the binomial approximation.

The binomial approximation

If $x \ll 1$, then $(1 + x)^n \approx 1 + nx$.

EXAMPLE 36.8 | **The shrinking school bus**

An 8.0-m-long school bus drives past at 30 m/s. By how much is its length contracted?

MODEL The school bus is at rest in an inertial reference frame S' moving at velocity $v = 30$ m/s relative to the ground frame S. The given length, 8.0 m, is the proper length ℓ in frame S'.

SOLVE In frame S, the school bus is length contracted to

$$L = \sqrt{1 - \beta^2}\,\ell$$

The bus's velocity v is much less than c, so we can use the binomial approximation to write

$$L \approx \left(1 - \frac{1}{2}\frac{v^2}{c^2}\right)\ell = \ell - \frac{1}{2}\frac{v^2}{c^2}\ell$$

The *amount* of the length contraction is

$$\ell - L = \frac{1}{2}\frac{v^2}{c^2}\ell = \frac{1}{2}\left(\frac{30 \text{ m/s}}{3.0 \times 10^8 \text{ m/s}}\right)^2 (8.0 \text{ m})$$

$$= 4.0 \times 10^{-14} \text{ m} = 40 \text{ fm}$$

where 1 fm = 1 femtometer = 10^{-15} m.

ASSESS The bus "shrinks" by only slightly more than the diameter of the nucleus of an atom. It's no wonder that we're not aware of length contraction in our everyday lives. If you had tried to calculate this number exactly, your calculator would have shown $\ell - L = 0$ because the difference between ℓ and L shows up only in the 14th decimal place. A scientific calculator determines numbers to 10 or 12 decimal places, but that isn't sufficient to show the difference. The binomial approximation provides an invaluable tool for finding the very tiny difference between two numbers that are nearly identical.

The Lorentz Velocity Transformations

FIGURE 36.33 shows an object that is moving in both reference frame S and reference frame S'. Experimenters in frame S determine that the object's velocity is u, while experimenters in frame S' find it to be u'. For simplicity, we'll assume that the object moves parallel to the x- and x'-axes.

The Galilean velocity transformation $u' = u - v$ was found by taking the time derivative of the position transformation. We can do the same with the Lorentz transformation if we take the derivative with respect to the time in each frame. Velocity u' in frame S' is

$$u' = \frac{dx'}{dt'} = \frac{d[\gamma(x - vt)]}{d[\gamma(t - vx/c^2)]} \quad (36.27)$$

where we've used the Lorentz transformations for position x' and time t'.

Carrying out the differentiation gives

$$u' = \frac{\gamma(dx - v\,dt)}{\gamma(dt - v\,dx/c^2)} = \frac{dx/dt - v}{1 - v(dx/dt)/c^2} \quad (36.28)$$

But dx/dt is u, the object's velocity in frame S, leading to

$$u' = \frac{u - v}{1 - uv/c^2} \quad (36.29)$$

You can see that Equation 36.29 reduces to the Galilean transformation $u' = u - v$ when $v \ll c$, as expected.

FIGURE 36.33 The velocity of a moving object is measured to be u in frame S and u' in frame S'.

The transformation from S′ to S is found by reversing the sign of v. Altogether,

$$u' = \frac{u - v}{1 - uv/c^2} \quad \text{and} \quad u = \frac{u' + v}{1 + u'v/c^2} \tag{36.30}$$

Equations 36.30 are the Lorentz velocity transformation equations.

NOTE It is important to distinguish carefully between v, which is the relative velocity between two reference frames, and u and u', which are the velocities of an *object* as measured in the two different reference frames.

EXAMPLE 36.9 | **A really fast bullet**

A rocket flies past the earth at $0.90c$. As it goes by, the rocket fires a bullet in the forward direction at $0.95c$ with respect to the rocket. What is the bullet's speed with respect to the earth?

MODEL The rocket and the earth are inertial reference frames. Let the earth be frame S and the rocket be frame S′. The velocity of frame S′ relative to frame S is $v = 0.90c$. The bullet's velocity in frame S′ is $u' = 0.95c$.

SOLVE We can use the Lorentz velocity transformation to find

$$u = \frac{u' + v}{1 + u'v/c^2} = \frac{0.95c + 0.90c}{1 + (0.95c)(0.90c)/c^2} = 0.997c$$

The bullet's speed with respect to the earth is 99.7% of the speed of light.

NOTE Many relativistic calculations are much easier when velocities are specified as a fraction of c.

ASSESS In Newtonian mechanics, the Galilean transformation of velocity would give $u = 1.85c$. Now, despite the very high speed of the rocket and of the bullet with respect to the rocket, the bullet's speed with respect to the earth remains less than c. This is yet another indication that objects cannot travel faster than the speed of light.

Suppose the rocket in Example 36.9 fired a laser beam in the forward direction as it traveled past the earth at velocity v. The laser beam would travel away from the rocket at speed $u' = c$ in the rocket's reference frame S′. What is the laser beam's speed in the earth's frame S? According to the Lorentz velocity transformation, it must be

$$u = \frac{u' + v}{1 + u'v/c^2} = \frac{c + v}{1 + cv/c^2} = \frac{c + v}{1 + v/c} = \frac{c + v}{(c + v)/c} = c \tag{36.31}$$

Light travels at speed c in both frame S and frame S′. This important consequence of the principle of relativity is "built into" the Lorentz transformations.

36.9 Relativistic Momentum

In Newtonian mechanics, the total momentum of a system is a conserved quantity. Further, as we've seen, the law of conservation of momentum, $P_f = P_i$, is true in all inertial reference frames *if* the particle velocities in different reference frames are related by the Galilean velocity transformations.

The difficulty, of course, is that the Galilean transformations are not consistent with the principle of relativity. It is a reasonable approximation when all velocities are very much less than c, but the Galilean transformations fail dramatically as velocities approach c. We'll leave it as a homework problem to show that $P'_f \neq P'_i$ if the particle velocities in frame S′ are related to the particle velocities in frame S by the Lorentz transformations.

There are two possibilities:

1. The so-called law of conservation of momentum is not really a law of physics. It is approximately true at low velocities but fails as velocities approach the speed of light.
2. The law of conservation of momentum really is a law of physics, but the expression $p = mu$ is not the correct way to calculate momentum when the particle velocity u becomes a significant fraction of c.

Momentum conservation is such a central and important feature of mechanics that it seems unlikely to fail in relativity.

The classical momentum, for one-dimensional motion, is $p = mu = m(\Delta x/\Delta t)$. Δt is the time to move distance Δx. That seemed clear enough within a Newtonian framework, but now we've learned that experimenters in different reference frames disagree about the amount of time needed. So whose Δt should we use?

One possibility is to use the time measured *by the particle*. This is the proper time $\Delta \tau$ because the particle is at rest in its own reference frame and needs only one clock. With this in mind, let's redefine the momentum of a particle of mass m moving with velocity $u = \Delta x/\Delta t$ to be

$$p = m\frac{\Delta x}{\Delta \tau} \tag{36.32}$$

We can relate this new expression for p to the familiar Newtonian expression by using the time-dilation result $\Delta \tau = (1 - u^2/c^2)^{1/2}\Delta t$ to relate the proper time interval measured by the particle to the more practical time interval Δt measured by experimenters in frame S. With this substitution, Equation 36.32 becomes

$$p = m\frac{\Delta x}{\Delta \tau} = m\frac{\Delta x}{\sqrt{1 - u^2/c^2}\,\Delta t} = \frac{mu}{\sqrt{1 - u^2/c^2}} \tag{36.33}$$

You can see that Equation 36.33 reduces to the classical expression $p = mu$ when the particle's speed $u \ll c$. That is an important requirement, but whether this is the "correct" expression for p depends on whether the total momentum P is conserved when the velocities of a system of particles are transformed with the Lorentz velocity transformation equations. The proof is rather long and tedious, so we will assert, without actual proof, that the momentum defined in Equation 36.33 does, indeed, transform correctly. **The law of conservation of momentum is still valid in all inertial reference frames** *if the momentum of each particle is calculated with Equation 36.33.*

The factor that multiplies mu in Equation 36.33 looks much like the factor γ in the Lorentz transformation equations for x and t, but there's one very important difference. The v in the Lorentz transformation equations is the velocity of a *reference frame*. The u in Equation 36.33 is the velocity of a particle moving *in* a reference frame.

With this distinction in mind, let's define the quantity

$$\gamma_p = \frac{1}{\sqrt{1 - u^2/c^2}} \tag{36.34}$$

where the subscript p indicates that this is γ for a particle, not for a reference frame. In frame S', where the particle moves with velocity u', the corresponding expression would be called γ_p'. With this definition of γ_p, the momentum of a particle is

$$p = \gamma_p mu \tag{36.35}$$

EXAMPLE 36.10 | **Momentum of a subatomic particle**

Electrons in a particle accelerator reach a speed of $0.999c$ relative to the laboratory. One collision of an electron with a target produces a muon that moves forward with a speed of $0.95c$ relative to the laboratory. The muon mass is 1.90×10^{-28} kg. What is the muon's momentum in the laboratory frame and in the frame of the electron beam?

MODEL Let the laboratory be reference frame S. The reference frame S' of the electron beam (i.e., a reference frame in which the electrons are at rest) moves in the direction of the electrons at $v = 0.999c$. The muon velocity in frame S is $u = 0.95c$.

SOLVE γ_p for the muon in the laboratory reference frame is

$$\gamma_p = \frac{1}{\sqrt{1 - u^2/c^2}} = \frac{1}{\sqrt{1 - 0.95^2}} = 3.20$$

Thus the muon's momentum in the laboratory is

$$p = \gamma_p mu = (3.20)(1.90 \times 10^{-28}\text{ kg})(0.95 \times 3.00 \times 10^8\text{ m/s})$$
$$= 1.73 \times 10^{-19}\text{ kg m/s}$$

The momentum is a factor of 3.2 larger than the Newtonian momentum mu. To find the momentum in the electron-beam

Continued

reference frame, we must first use the velocity transformation equation to find the muon's velocity in frame S':

$$u' = \frac{u - v}{1 - uv/c^2} = \frac{0.95c - 0.999c}{1 - (0.95c)(0.999c)/c^2} = -0.962c$$

In the laboratory frame, the faster electrons are overtaking the slower muon. Hence the muon's velocity in the electron-beam frame is negative. γ_p' for the muon in frame S' is

$$\gamma_p' = \frac{1}{\sqrt{1 - u'^2/c^2}} = \frac{1}{\sqrt{1 - 0.962^2}} = 3.66$$

The muon's momentum in the electron-beam reference frame is

$$p' = \gamma_p' m u'$$

$$= (3.66)(1.90 \times 10^{-28}\,\text{kg})(-0.962 \times 3.00 \times 10^8\,\text{m/s})$$

$$= -2.01 \times 10^{-19}\,\text{kg m/s}$$

ASSESS From the laboratory perspective, the muon moves only slightly slower than the electron beam. But it turns out that the muon moves faster with respect to the electrons, although in the opposite direction, than it does with respect to the laboratory.

The Cosmic Speed Limit

FIGURE 36.34 The speed of a particle cannot reach the speed of light.

(a)

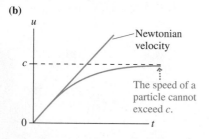

(b)

FIGURE 36.35 Assume that a causal influence can travel from A to B at a speed $u > c$.

FIGURE 36.34a is a graph of momentum versus velocity. For a Newtonian particle, with $p = mu$, the momentum is directly proportional to the velocity. The relativistic expression for momentum agrees with the Newtonian value if $u \ll c$, but p approaches ∞ as $u \rightarrow c$.

The implications of this graph become clear when we relate momentum to force. Consider a particle subjected to a constant force, such as a rocket that never runs out of fuel. If F is constant, we can see from $F = dp/dt$ that the momentum is $p = Ft$. If Newtonian physics were correct, a particle would go faster and faster as its velocity $u = p/m = (F/m)t$ increased without limit. But the relativistic result, shown in **FIGURE 36.34b**, is that the particle's velocity asymptotically approaches the speed of light $(u \rightarrow c)$ as p approaches ∞. Relativity gives a very different outcome than Newtonian mechanics.

The speed c is a "cosmic speed limit" for material particles. A force cannot accelerate a particle to a speed higher than c because the particle's momentum becomes infinitely large as the speed approaches c. The amount of effort required for each additional increment of velocity becomes larger and larger until no amount of effort can raise the velocity any higher.

Actually, at a more fundamental level, c is a speed limit for *any* kind of **causal influence.** If I throw a rock and break a window, my throw is the *cause* of the breaking window and the rock is the *causal influence.* If I shoot a laser beam at a light detector that is wired to a firecracker, the light wave is the *causal influence* that leads to the explosion. A causal influence can be any kind of particle, wave, or information that travels from A to B and allows A to be the cause of B.

For two unrelated events—a firecracker explodes in Tokyo and a balloon bursts in Paris—the relativity of simultaneity tells us that they may be simultaneous in one reference frame but not in others. Or in one reference frame the firecracker may explode before the balloon bursts but in some other reference frame the balloon may burst first. These possibilities violate our commonsense view of time, but they're not in conflict with the principle of relativity.

For two causally related events—A *causes* B—it would be nonsense for an experimenter in any reference frame to find that B occurs before A. No experimenter in any reference frame, no matter how it is moving, will find that you are born before your mother is born. If A causes B, then it must be the case that $t_A < t_B$ in *all* reference frames.

Suppose there exists some kind of causal influence that *can* travel at speed $u > c$. **FIGURE 36.35** shows a reference frame S in which event A occurs at position $x_A = 0$. The faster-than-light causal influence—perhaps some yet-to-be-discovered "z ray"—leaves A at $t_A = 0$ and travels to the point at which it will cause event B. It arrives at x_B at time $t_B = x_B/u$.

How do events A and B appear in a reference frame S' that travels at an ordinary speed $v < c$ relative to frame S? We can use the Lorentz transformations to find out.

Because $x_A = 0$ and $t_A = 0$, it's easy to see that $x'_A = 0$ and $t'_A = 0$. That is, the origins of S and S' overlap at the instant the causal influence leaves event A. More interesting is the time at which this influence reaches B in frame S'. The Lorentz time transformation for event B is

$$t'_B = \gamma\left(t_B - \frac{vx_B}{c^2}\right) = \gamma t_B\left(1 - \frac{v(x_B/t_B)}{c^2}\right) = \gamma t_B\left(1 - \frac{vu}{c^2}\right) \qquad (36.36)$$

where we first factored out t_B, then made use of the fact that $u = x_B/t_B$ in frame S.

We're assuming that $u > c$, so there exist ordinary reference frames, with $v < c$, for which $vu/c^2 > 1$. In that case, the term $(1 - vu/c^2)$ is negative and $t'_B < 0$. But if $t'_B < 0$, then event B happens *before* event A in reference frame S'. In other words, if a causal influence can travel faster than c, then there exist reference frames in which the effect happens before the cause. We know this can't happen, so our assumption $u > c$ must be wrong. **No causal influence of any kind—particle, wave, or yet-to-be-discovered z rays—can travel faster than c.**

The existence of a cosmic speed limit is one of the most interesting consequences of the theory of relativity. "Warp drive," in which a spaceship suddenly leaps to faster-than-light velocities, is simply incompatible with the theory of relativity. Rapid travel to the stars will remain in the realm of science fiction unless future scientific discoveries find flaws in Einstein's theory and open the doors to yet-undreamed-of theories. While we can't say with certainty that a scientific theory will never be overturned, there is currently not even a hint of evidence that disagrees with the special theory of relativity.

36.10 Relativistic Energy

Energy is our final topic in this chapter on relativity. Space, time, velocity, and momentum are changed by relativity, so it seems inevitable that we'll need a new view of energy.

In Newtonian mechanics, a particle's kinetic energy $K = \frac{1}{2}mu^2$ can be written in terms of its momentum $p = mu$ as $K = p^2/2m$. This suggests that a relativistic expression for energy will likely involve both the square of p and the particle's mass. We also hope that energy will be conserved in relativity, so a reasonable starting point is with the one quantity we've found that is the same in all inertial reference frames: the spacetime interval s.

Let a particle of mass m move through distance Δx during a time interval Δt, as measured in reference frame S. The spacetime interval is

$$s^2 = c^2(\Delta t)^2 - (\Delta x)^2 = \text{invariant}$$

We can turn this into an expression involving momentum if we multiply by $(m/\Delta\tau)^2$, where $\Delta\tau$ is the proper time (i.e., the time measured by the particle). Doing so gives

$$(mc)^2\left(\frac{\Delta t}{\Delta\tau}\right)^2 - \left(\frac{m\,\Delta x}{\Delta\tau}\right)^2 = (mc)^2\left(\frac{\Delta t}{\Delta\tau}\right)^2 - p^2 = \text{invariant} \qquad (36.37)$$

where we used $p = m(\Delta x/\Delta\tau)$ from Equation 36.32.

Now Δt, the time interval in frame S, is related to the proper time by the time-dilation result $\Delta t = \gamma_p\Delta\tau$. With this change, Equation 36.37 becomes

$$(\gamma_p mc)^2 - p^2 = \text{invariant}$$

Finally, for reasons that will be clear in a minute, we multiply by c^2, to get

$$(\gamma_p mc^2)^2 - (pc)^2 = \text{invariant} \qquad (36.38)$$

To say that the right side is an *invariant* means it has the same value in all inertial reference frames. We can easily determine the constant by evaluating it in the reference frame in which the particle is at rest. In that frame, where $p = 0$ and $\gamma_p = 1$, we find that

$$(\gamma_p mc^2)^2 - (pc)^2 = (mc^2)^2 \tag{36.39}$$

Let's reflect on what this means before taking the next step. The space-time interval s has the same value in all inertial reference frames. In other words, $c^2(\Delta t)^2 - (\Delta x)^2 = c^2(\Delta t')^2 - (\Delta x')^2$. Equation 36.39 was derived from the definition of the spacetime interval; hence the quantity mc^2 is also an invariant having the same value in all inertial reference frames. In other words, if experimenters in frames S and S′ both make measurements on this particle of mass m, they will find that

$$(\gamma_p mc^2)^2 - (pc)^2 = (\gamma'_p mc^2)^2 - (p'c)^2 \tag{36.40}$$

Experimenters in different reference frames measure different values for the momentum, but experimenters in all reference frames agree that momentum is a conserved quantity. Equations 36.39 and 36.40 suggest that the quantity $\gamma_p mc^2$ is also an important property of the particle, a property that changes along with p in just the right way to satisfy Equation 36.39. But what is this property?

The first clue comes from checking the units. γ_p is dimensionless and c is a velocity, so $\gamma_p mc^2$ has the same units as the classical expression $\frac{1}{2}mv^2$—namely, units of energy. For a second clue, let's examine how $\gamma_p mc^2$ behaves in the low-velocity limit $u \ll c$. We can use the binomial approximation expression for γ_p to find

$$\gamma_p mc^2 = \frac{mc^2}{\sqrt{1 - u^2/c^2}} \approx \left(1 + \frac{1}{2}\frac{u^2}{c^2}\right)mc^2 = mc^2 + \frac{1}{2}mu^2 \tag{36.41}$$

The second term, $\frac{1}{2}mu^2$, is the low-velocity expression for the kinetic energy K. This is an energy associated with motion. But the first term suggests that the concept of energy is more complex than we originally thought. It appears that **there is an inherent energy associated with mass itself.**

Rest Energy and Total Energy

With that as a possibility, subject to experimental verification, let's define the **total energy** E of a particle to be

$$E = \gamma_p mc^2 = E_0 + K = \text{rest energy} + \text{kinetic energy} \tag{36.42}$$

This total energy consists of a **rest energy**

$$E_0 = mc^2 \tag{36.43}$$

and a relativistic expression for the *kinetic energy*

$$K = (\gamma_p - 1)mc^2 = (\gamma_p - 1)E_0 \tag{36.44}$$

This expression for the kinetic energy is very nearly $\frac{1}{2}mu^2$ when $u \ll c$ but, as **FIGURE 36.36** shows, differs significantly from the classical value for very high velocities.

Equation 36.43 is, of course, Einstein's famous $E = mc^2$, perhaps the most famous equation in all of physics. Before discussing its significance, we need to tie up some loose ends. First, we can use Equations 36.42 and 36.43 to rewrite Equation 36.39 as $E^2 - (pc)^2 = E_0^2$. This is easily rearranged to give the useful result

$$E = \sqrt{E_0^2 + (pc)^2} \tag{36.45}$$

The quantity E_0 is an *invariant* with the same value mc^2 in *all* inertial reference frames.

Second, notice that we can write

$$pc = (\gamma_p mu)c = \frac{u}{c}(\gamma_p mc^2)$$

FIGURE 36.36 The relativistic kinetic energy.

The relativistic kinetic energy approaches ∞ as $u \to c$.

K

Newtonian kinetic energy

$0 \quad 0.5c \quad c \quad 1.5c$ u

The Newtonian kinetic-energy expression is valid when $u \ll c$.

But $\gamma_\text{p} mc^2$ is the total energy E and $u/c = \beta_\text{p}$, where the subscript p, as on γ_p, indicates that we're referring to the motion of a particle within a reference frame, not the motion of two reference frames relative to each other. Thus

$$pc = \beta_\text{p} E \qquad (36.46)$$

FIGURE 36.37 shows the "velocity-energy-momentum triangle," a convenient way to remember the relationships among the three quantities.

FIGURE 36.37 The velocity-energy-momentum triangle.

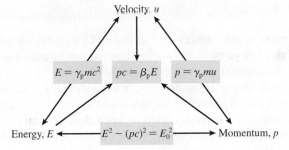

EXAMPLE 36.11 | Kinetic energy and total energy

Calculate the rest energy and the kinetic energy of (a) a 100 g ball moving with a speed of 100 m/s and (b) an electron with a speed of 0.999c.

MODEL The ball, with $u \ll c$, is a classical particle. We don't need to use the relativistic expression for its kinetic energy. The electron is highly relativistic.

SOLVE a. For the ball, with $m = 0.10$ kg,

$$E_0 = mc^2 = 9.0 \times 10^{15} \text{ J}$$
$$K = \tfrac{1}{2} mu^2 = 500 \text{ J}$$

b. For the electron, we start by calculating

$$\gamma_\text{p} = \frac{1}{(1 - u^2/c^2)^{1/2}} = 22.4$$

Then, using $m_\text{e} = 9.11 \times 10^{-31}$ kg, we find

$$E_0 - mc^2 = 8.2 \times 10^{-14} \text{ J}$$
$$K = (\gamma_\text{p} - 1)E_0 = 170 \times 10^{-14} \text{ J}$$

ASSESS The ball's kinetic energy is a typical kinetic energy. Its rest energy, by contrast, is a staggeringly large number. For a relativistic electron, on the other hand, the kinetic energy is more important than the rest energy.

STOP TO THINK 36.8 An electron moves through the lab at 99% the speed of light. The lab reference frame is S and the electron's reference frame is S'. In which reference frame is the electron's rest mass larger?

a. In frame S, the lab frame
b. In frame S', the electron's frame
c. It is the same in both frames.

Mass-Energy Equivalence

Now we're ready to explore the significance of Einstein's famous equation $E = mc^2$. FIGURE 36.38 shows two balls of clay approaching each other. They have equal masses and equal kinetic energies, and they slam together in a perfectly inelastic collision to form one large ball of clay at rest. In Newtonian mechanics, we would say that the initial energy $2K$ is dissipated by being transformed into an equal amount of thermal energy, raising the temperature of the coalesced ball of clay. But Equation 36.42, $E = E_0 + K$, doesn't say anything about thermal energy. The total energy before the

FIGURE 36.38 An inelastic collision between two balls of clay does not seem to conserve the total energy E.

The tracks of elementary particles in a bubble chamber show the creation of an electron-positron pair. The negative electron and positive positron spiral in opposite directions in the magnetic field.

FIGURE 36.39 An inelastic collision between electrons can create an electron-positron pair.

High-speed electron

Target electron e^-

e^-

Thin target material

An electron-positron pair has been created.

FIGURE 36.40 The annihilation of an electron-positron pair.

e^- e^+ An electron and a positron meet.

They annihilate.

Photon   Photon   The energy equivalent of the mass is transformed into two gamma-ray photons.

collision is $E_i = 2mc^2 + 2K$, with the factor of 2 appearing because there are two masses. It seems like the total energy after the collision, when the clay is at rest, should be $2mc^2$, but this value doesn't conserve total energy.

There's ample experimental evidence that energy is conserved, so there must be a flaw in our reasoning. The statement of energy conservation is

$$E_f = Mc^2 = E_i = 2mc^2 + 2K \tag{36.47}$$

where M is the mass of clay after the collision. But, remarkably, this requires

$$M = 2m + \frac{2K}{c^2} \tag{36.48}$$

In other words, **mass is not conserved.** The mass of clay after the collision is larger than the mass of clay before the collision. Total energy can be conserved only if kinetic energy is transformed into an "equivalent" amount of mass.

The mass increase in a collision between two balls of clay is incredibly small, far beyond any scientist's ability to detect. So how do we know if such a crazy idea is true?

FIGURE 36.39 shows an experiment that has been done countless times in the last 50 years at particle accelerators around the world. An electron that has been accelerated to $u \approx c$ is aimed at a target material. When a high-energy electron collides with an atom in the target, it can easily knock one of the electrons out of the atom. Thus we would expect to see two electrons leaving the target: the incident electron and the ejected electron. Instead, *four* particles emerge from the target: three electrons and a positron. A *positron*, or positive electron, is the antimatter version of an electron, identical to an electron in all respects other than having charge $q = +e$.

In chemical-reaction notation, the collision is

$$e^- \text{ (fast)} + e^- \text{ (at rest)} \rightarrow e^- + e^- + e^- + e^+$$

An electron and a positron have been *created*, apparently out of nothing. Mass $2m_e$ before the collision has become mass $4m_e$ after the collision. (Notice that charge has been conserved in this collision.)

Although the mass has increased, it wasn't created "out of nothing." This is an inelastic collision, just like the collision of the balls of clay, because the kinetic energy after the collision is less than before. In fact, if you measured the energies before and after the collision, you would find that the decrease in kinetic energy is exactly equal to the energy equivalent of the two particles that have been created: $\Delta K = 2m_e c^2$. The new particles have been created *out of energy!*

Particles can be created from energy, and particles can return to energy. **FIGURE 36.40** shows an electron colliding with a positron, its antimatter partner. When a particle and its antiparticle meet, they *annihilate* each other. The mass disappears, and the energy equivalent of the mass is transformed into light. In Chapter 38, you'll learn that light is *quantized*, meaning that light is emitted and absorbed in discrete chunks of energy called *photons*. For light with wavelength λ, the energy of a photon is $E_{photon} = hc/\lambda$, where $h = 6.63 \times 10^{-34}$ J s is called *Planck's constant*. Photons carry momentum as well as energy. Conserving both energy and momentum in the annihilation of an electron and a positron requires the emission in opposite directions of two photons of equal energy.

If the electron and positron are fairly slow, so that $K \ll mc^2$, then $E_i \approx E_0 = mc^2$. In that case, energy conservation requires

$$E_f = 2E_{photon} = E_i \approx 2m_e c^2 \tag{36.49}$$

Hence the wavelength of the emitted photons is

$$\lambda = \frac{hc}{m_e c^2} \approx 0.0024 \text{ nm} \tag{36.50}$$

This is an extremely short wavelength, even shorter than the wavelengths of x rays. Photons in this wavelength range are called *gamma rays*. And, indeed, the emission of 0.0024 nm gamma rays is observed in many laboratory experiments in which positrons are able to collide with electrons and thus annihilate. In recent years, with the advent of gamma-ray telescopes on satellites, astronomers have found 0.0024 nm photons coming from many places in the universe, especially galactic centers—evidence that positrons are abundant throughout the universe.

Positron-electron annihilation is also the basis of the medical procedure known as positron-emission tomography, or a PET scan. A patient ingests a very small amount of a radioactive substance that decays by the emission of positrons. This substance is taken up by certain tissues in the body, especially those tissues with a high metabolic rate. As the substance decays, the positrons immediately collide with electrons, annihilate, and create two gamma-ray photons that are emitted back to back. The gamma rays, which easily leave the body, are detected, and their trajectories are traced backward into the body. The overlap of many such trajectories shows quite clearly the tissue in which the positron emission is occurring. The results are usually shown as false-color photographs, with redder areas indicating regions of higher positron emission.

Positron-electron annihilation (a PET scan) provides a noninvasive look into the brain.

Conservation of Energy

The creation and annihilation of particles with mass, processes strictly forbidden in Newtonian mechanics, are vivid proof that neither mass nor the Newtonian definition of energy is conserved. Even so, the *total* energy—the kinetic energy *and* the energy equivalent of mass—remains a conserved quantity.

> **Law of conservation of total energy** The energy $E = \sum E_i$ of an isolated system is conserved, where $E_i = (\gamma_p)_i m_i c^2$ is the total energy of particle i.

Mass and energy are not the same thing, but, as the last few examples have shown, they are *equivalent* in the sense that mass can be transformed into energy and energy can be transformed into mass as long as the total energy is conserved.

Probably the most well-known application of the conservation of total energy is nuclear fission. The uranium isotope ^{236}U, containing 236 protons and neutrons, does not exist in nature. It can be created when a ^{235}U nucleus absorbs a neutron, increasing its atomic mass from 235 to 236. The ^{236}U nucleus quickly fragments into two smaller nuclei and several extra neutrons, a process known as **nuclear fission.** The nucleus can fragment in several ways, but one is

$$n + \,^{235}U \rightarrow \,^{236}U \rightarrow \,^{144}Ba + \,^{89}Kr + 3n$$

Ba and Kr are the atomic symbols for barium and krypton.

This reaction seems like an ordinary chemical reaction—until you check the masses. The masses of atomic isotopes are known with great precision from many decades of measurement in instruments called mass spectrometers. If you add up the masses on both sides, you find that the mass of the products is 0.185 u smaller than the mass of the initial neutron and ^{235}U, where, you will recall, 1 u = 1.66×10^{-27} kg is the atomic mass unit. In kilograms the mass loss is 3.07×10^{-28} kg.

Mass has been lost, but the energy equivalent of the mass has not. As **FIGURE 36.41** shows, the mass has been converted to kinetic energy, causing the two product nuclei and three neutrons to be ejected at very high speeds. The kinetic energy is easily calculated: $\Delta K = m_{lost}c^2 = 2.8 \times 10^{-11}$ J.

This is a very tiny amount of energy, but it is the energy released from *one* fission. The number of nuclei in a macroscopic sample of uranium is on the order of N_A, Avogadro's number. Hence the energy available if *all* the nuclei fission is enormous. This energy, of course, is the basis for both nuclear power reactors and nuclear weapons.

FIGURE 36.41 In nuclear fission, the energy equivalent of lost mass is converted into kinetic energy.

The mass of the reactants is 0.185 u more than the mass of the products.

n

^{235}U

Neutron absorbed

^{236}U

Fission

^{144}Ba ^{89}Kr

0.185 u of mass has been converted into kinetic energy.

We started this chapter with an expectation that relativity would challenge our basic notions of space and time. We end by finding that relativity changes our understanding of mass and energy. Most remarkable of all is that each and every one of these new ideas flows from one simple statement: The laws of physics are the same in all inertial reference frames.

CHALLENGE EXAMPLE 36.12 | Goths and Huns

The rockets of the Goths and the Huns are each 1000 m long in their rest frame. The rockets pass each other, virtually touching, at a relative speed of 0.8c. The Huns have a laser cannon at the rear of their rocket that fires a deadly laser beam perpendicular to the rocket's motion. The captain of the Huns wants to send a threatening message to the Goths by "firing a shot across their bow." He tells his first mate, "The Goths' rocket is length contracted to 600 m. Fire the laser cannon at the instant the tail of their rocket passes the nose of ours. The laser beam will cross 400 m in front of them."

But things are different in the Goths' reference frame. The Goth captain muses, "The Huns' rocket is length contracted to 600 m, 400 m shorter than our rocket. If they fire as the nose of their ship passes the tail of ours, the lethal laser beam will pass right through our side."

The first mate on the Huns' rocket fires as ordered. Does the laser beam blast the Goths or not?

MODEL Both rockets are inertial reference frames. Let the Huns' rocket be frame S and the Goths' rocket be frame S'. S' moves with velocity $v = 0.8c$ relative to S. We need to describe the situation in terms of events.

FIGURE 36.42 The situation seen by the Huns.

Frame S perspective

VISUALIZE Begin by considering the situation from the Huns' reference frame, as shown in **FIGURE 36.42**.

SOLVE The key to resolving the paradox is that two events simultaneous in one reference frame are not simultaneous in a different reference frame. The Huns do, indeed, see the Goths' rocket length contracted to $L_{\text{Goths}} = (1 - (0.8)^2)^{1/2}(1000 \text{ m}) = 600 \text{ m}$. Let event 1 be the tail of the Goths' rocket passing the nose of the Huns' rocket. Since we're free to define the origin of our coordinate system, we define this event to be at time $t_1 = 0$ μs and at position $x_1 = 0$ m. Then, in the Huns' reference frame, the spacetime coordinates of event 2, the firing of the laser cannon, are $(x_2, t_2) = (1000 \text{ m}, 0 \text{ μs})$. The nose of the Goths' rocket is at $x = 600$ m at $t = 0$ μs; thus the laser cannon misses the Goths by 400 m.

Now we can use the Lorentz transformations to find the spacetime coordinates of the events in the Goths' reference frame. The nose of the Huns' rocket passes the tail of the Goths' rocket at $(x_1', t_1') = (0 \text{ m}, 0 \text{ μs})$. The Huns fire their laser cannon at

$$x_2' = \gamma(x_2 - vt_2) = \tfrac{5}{3}(1000 \text{ m} - 0 \text{ m}) = 1667 \text{ m}$$

$$t_2' = \gamma\left(t_2 - \frac{vx_2}{c^2}\right) = \tfrac{5}{3}\left(0 \text{ μs} - (0.8)\frac{1000 \text{ m}}{300 \text{ m/μs}}\right) = -4.444 \text{μs}$$

where we calculated $\gamma = 5/3$ for $v = 0.8c$. Events 1 and 2 are *not* simultaneous in S'. The Huns fire the laser cannon 4.444 μs *before* the nose of their rocket reaches the tail of the Goths' rocket. The laser is fired at $x_2' = 1667$ m, missing the nose of the Goths' rocket by 667 m. **FIGURE 36.43** shows how the Goths see things.

FIGURE 36.43 The situation seen by the Goths.

Frame S' perspective

In fact, since the Huns' rocket is length contracted to 600 m, the nose of the Huns' rocket is at $x' = 1667 \text{ m} - 600 \text{ m} = 1067$ m at the instant they fire the laser cannon. At a speed of $v = 0.8c = 240$ m/μs, in 4.444 μs the nose of the Huns' rocket travels $\Delta x' = (240 \text{ m/μs})(4.444 \text{ μs}) = 1067$ m—exactly the right distance to be at the tail of the Goths' rocket at $t_1' = 0$ μs. We could also note that the 667 m "miss distance" in the Goths' frame is length contracted to $(1 - (0.8)^2)^{1/2}(667 \text{ m}) = 400$ m in the Huns' frame— exactly the amount by which the Huns think they miss the Goths' rocket.

ASSESS Thus we end up with a consistent explanation. The Huns miss the Goths' rocket because, to them, the Goths' rocket is length contracted. The Goths find that the Huns miss because event 2 (the firing of the laser cannon) occurs before event 1 (the nose of one rocket passing the tail of the other). The 400 m distance of the miss in the Huns' reference frame is the length-contracted miss distance of 667 m in the Goths' reference frame.

SUMMARY

The goal of Chapter 36 has been to learn how relativity changes our concepts of space and time.

GENERAL PRINCIPLES

Principle of Relativity

The laws of physics are the same in all inertial reference frames.

- The speed of light c is the same in all inertial reference frames.
- No particle or causal influence can travel at a speed greater than c.

Solving Relativity Problems

- Base the analysis on events.
- Time intervals can often be found using time dilation.
- Distances can often be found using length contraction.
- Use the **Lorentz transformations** for general problems.

IMPORTANT CONCEPTS

Space

Spatial measurements depend on the motion of the experimenter relative to the events. An object's length is the difference between *simultaneous* measurements of the positions of both ends.

Proper length ℓ is the length of an object measured in a reference frame in which the object is at rest. The object's length in a frame in which the object moves with velocity v is

$$L = \sqrt{1 - \beta^2}\, \ell \le \ell$$

This is called **length contraction.**

Time

Time measurements depend on the motion of the experimenter relative to the events. Events that are simultaneous in reference frame S are not simultaneous in frame S' moving relative to S.

Proper time $\Delta\tau$ is the time interval between two events measured in a reference frame in which the events occur at the same position. The time interval between the events in a frame moving with relative velocity v is

$$\Delta t = \Delta\tau / \sqrt{1 - \beta^2} \ge \Delta\tau$$

This is called **time dilation.**

Momentum

The law of conservation of momentum is valid in all inertial reference frames if the momentum of a particle with velocity u is $p = \gamma_p mu$, where

$$\gamma_p = 1/\sqrt{1 - u^2/c^2}$$

The momentum approaches ∞ as $u \to c$.

Energy

The law of conservation of energy is valid in all inertial reference frames if the energy of a particle with velocity u is $E = \gamma_p mc^2 = E_0 + K$.

Rest energy $E_0 = mc^2$

Kinetic energy $K = (\gamma_p - 1)mc^2$

Invariants

Invariants are quantities that have the same value in all inertial reference frames.

Spacetime interval: $s^2 = (c\,\Delta t)^2 - (\Delta x)^2$

Particle rest energy: $E_0^2 = (mc^2)^2 = E^2 - (pc)^2$

Mass-energy equivalence

Mass m can be transformed into energy $\Delta E = mc^2$.

Energy can be transformed into mass $m = \Delta E/c^2$.

APPLICATIONS

An **event** happens at a specific place in space and time. Spacetime coordinates are (x, t) in frame S and (x', t') in frame S'.

A **reference frame** is a coordinate system with meter sticks and clocks for measuring events.

The **Lorentz transformations** transform spacetime coordinates and velocities between reference frames S and S'.

$$x' = \gamma(x - vt) \qquad x = \gamma(x' + vt')$$
$$y' = y \qquad y = y'$$
$$z' = z \qquad z = z'$$
$$t' = \gamma(t - vx/c^2) \qquad t = \gamma(t' + vx'/c^2)$$
$$u' = \frac{u - v}{1 - uv/c^2} \qquad u = \frac{u' + v}{1 + u'v/c^2}$$

where u and u' are the x- and x'-components of an object's velocity.

$$\beta = v/c \quad \text{and} \quad \gamma = 1/\sqrt{1 - v^2/c^2} = 1/\sqrt{1 - \beta^2}$$

TERMS AND NOTATION

special relativity	spacetime coordinates,	time dilation	causal influence
reference frame	(x, y, z, t)	light year, ly	total energy, E
inertial reference frame	synchronized	proper length, ℓ	rest energy, E_0
Galilean principle of	simultaneous	length contraction	law of conservation of total
relativity	relativity of simultaneity	invariant	energy
principle of relativity	rest frame	spacetime interval, s	nuclear fission
event	proper time, $\Delta\tau$	Lorentz transformations	

CONCEPTUAL QUESTIONS

1. **FIGURE Q36.1** shows two balls. What are the speed and direction of each (a) in a reference frame that moves with ball 1 and (b) in a reference frame that moves with ball 2?

FIGURE Q36.1   **FIGURE Q36.2**

2. Teenagers Sam and Tom are playing chicken in their rockets. As **FIGURE Q36.2** shows, an experimenter on earth sees that each is traveling at $0.95c$ as he approaches the other. Sam fires a laser beam toward Tom.
   a. What is the speed of the laser beam relative to Sam?
   b. What is the speed of the laser beam relative to Tom?

3. Firecracker A is 300 m from you. Firecracker B is 600 m from you in the same direction. You see both explode at the same time. Define event 1 to be "firecracker A explodes" and event 2 to be "firecracker B explodes." Does event 1 occur before, after, or at the same time as event 2? Explain.

4. Firecrackers A and B are 600 m apart. You are standing exactly halfway between them. Your lab partner is 300 m on the other side of firecracker A. You see two flashes of light, from the two explosions, at exactly the same instant of time. Define event 1 to be "firecracker A explodes" and event 2 to be "firecracker B explodes." According to your lab partner, based on measurements he or she makes, does event 1 occur before, after, or at the same time as event 2? Explain.

5. **FIGURE Q36.5** shows Peggy standing at the center of her railroad car as it passes Ryan on the ground. Firecrackers attached to the ends of the car explode. A short time later, the flashes from the two explosions arrive at Peggy at the same time.

FIGURE Q36.5

   a. Were the explosions simultaneous in Peggy's reference frame? If not, which exploded first? Explain.
   b. Were the explosions simultaneous in Ryan's reference frame? If not, which exploded first? Explain.

6. **FIGURE Q36.6** shows a rocket traveling from left to right. At the instant it is halfway between two trees, lightning simultaneously (in the rocket's frame) hits both trees.

   a. Do the light flashes reach the rocket pilot simultaneously? If not, which reaches her first? Explain.
   b. A student was sitting on the ground halfway between the trees as the rocket passed overhead. According to the student, were the lightning strikes simultaneous? If not, which tree was hit first? Explain.

FIGURE Q36.6

7. Your friend flies from Los Angeles to New York. She carries an accurate stopwatch with her to measure the flight time. You and your assistants on the ground also measure the flight time.
   a. Identify the two events associated with this measurement.
   b. Who, if anyone, measures the proper time?
   c. Who, if anyone, measures the shorter flight time?

8. As the meter stick in **FIGURE Q36.8** flies past you, you simultaneously measure the positions of both ends and determine that $L < 1$ m.

   FIGURE Q36.8

   a. To an experimenter in frame S', the meter stick's frame, did you make your two measurements simultaneously? If not, which end did you measure first? Explain.
   b. Can experimenters in frame S' give an explanation for why your measurement is less than 1 m?

9. A 100-m-long train is heading for an 80-m-long tunnel. If the train moves sufficiently fast, is it possible, according to experimenters on the ground, for the entire train to be inside the tunnel at one instant of time? Explain.

10. Particle A has half the mass and twice the speed of particle B. Is the momentum p_A less than, greater than, or equal to p_B? Explain.

11. Event A occurs at spacetime coordinates (300 m, 2 μs).
    a. Event B occurs at spacetime coordinates (1200 m, 6 μs). Could A possibly be the cause of B? Explain.
    b. Event C occurs at spacetime coordinates (2400 m, 8 μs). Could A possibly be the cause of C? Explain.

Problems labeled [____] integrate material from earlier chapters.

Exercises

Section 36.2 Galilean Relativity

1. ‖ A firecracker explodes in reference frame S at $t = 1.0$ s. A second firecracker explodes at the same position at $t = 3.0$ s. In reference frame S′, which moves in the x-direction at speed v, the first explosion is detected at $x' = 4.0$ m and the second at $x' = -4.0$ m.
   a. What is the speed of frame S′ relative to frame S?
   b. What is the position of the two explosions in frame S?

2. ‖ At $t = 1.0$ s, a firecracker explodes at $x = 10$ m in reference frame S. Four seconds later, a second firecracker explodes at $x = 20$ m. Reference frame S′ moves in the x-direction at a speed of 5.0 m/s. What are the positions and times of these two events in frame S′?

3. | A newspaper delivery boy is riding his bicycle down the street at 5.0 m/s. He can throw a paper at a speed of 8.0 m/s. What is the paper's speed relative to the ground if he throws the paper (a) forward, (b) backward, and (c) to the side?

4. | A baseball pitcher can throw a ball with a speed of 40 m/s. He is in the back of a pickup truck that is driving away from you. He throws the ball in your direction, and it floats toward you at a lazy 10 m/s. What is the speed of the truck?

Section 36.3 Einstein's Principle of Relativity

5. | An out-of-control alien spacecraft is diving into a star at a speed of 1.0×10^8 m/s. At what speed, relative to the spacecraft, is the starlight approaching?

6. | A starship blasts past the earth at 2.0×10^8 m/s. Just after passing the earth, it fires a laser beam out the back of the starship. With what speed does the laser beam approach the earth?

Section 36.4 Events and Measurements

Section 36.5 The Relativity of Simultaneity

7. | Your job is to synchronize the clocks in a reference frame. You are going to do so by flashing a light at the origin at $t = 0$ s. To what time should the clock at $(x, y, z) = (30$ m, 40 m, 0 m) be preset?

8. ‖ Bjorn is standing at $x = 600$ m. Firecracker 1 explodes at the origin and firecracker 2 explodes at $x = 900$ m. The flashes from both explosions reach Bjorn's eye at $t = 3.0$ μs. At what time did each firecracker explode?

9. ‖ Bianca is standing at $x = 600$ m. Firecracker 1, at the origin, and firecracker 2, at $x = 900$ m, explode simultaneously. The flash from firecracker 1 reaches Bianca's eye at $t = 3.0$ μs. At what time does she see the flash from firecracker 2?

10. ‖‖ You are standing at $x = 9.0$ km and your assistant is standing at $x = 3.0$ km. Lightning bolt 1 strikes at $x = 0$ km and lightning bolt 2 strikes at $x = 12.0$ km. You see the flash from bolt 2 at $t = 10$ μs and the flash from bolt 1 at $t = 50$ μs. According to your assistant, were the lightning strikes simultaneous? If not, which occurred first, and what was the time difference between the two?

11. ‖ You are standing at $x = 9.0$ km. Lightning bolt 1 strikes at $x = 0$ km and lightning bolt 2 strikes at $x = 12.0$ km. Both flashes reach your eye at the same time. Your assistant is standing at $x = 3.0$ km. Does your assistant see the flashes at the same time? If not, which does she see first, and what is the time difference between the two?

12. ‖ You are flying your personal rocketcraft at $0.90c$ from Star A toward Star B. The distance between the stars, in the stars' reference frame, is 1.0 ly. Both stars happen to explode simultaneously in your reference frame at the instant you are exactly halfway between them. Do you see the flashes simultaneously? If not, which do you see first, and what is the time difference between the two?

Section 36.6 Time Dilation

13. ‖ A cosmic ray travels 60 km through the earth's atmosphere in 400 μs, as measured by experimenters on the ground. How long does the journey take according to the cosmic ray?

14. ‖ At what speed, as a fraction of c, does a moving clock tick at half the rate of an identical clock at rest?

15. | An astronaut travels to a star system 4.5 ly away at a speed of $0.90c$. Assume that the time needed to accelerate and decelerate is negligible.
   a. How long does the journey take according to Mission Control on earth?
   b. How long does the journey take according to the astronaut?
   c. How much time elapses between the launch and the arrival of the first radio message from the astronaut saying that she has arrived?

16. ‖ a. At what speed, as a fraction of c, must a rocket travel on a journey to and from a distant star so that the astronauts age 10 years while the Mission Control workers on earth age 120 years?
   b. As measured by Mission Control, how far away is the distant star?

17. | At what speed, as a fraction of c, would a round-trip astronaut "lose" $\frac{1}{25}$ of the elapsed time shown on her watch?

18. ‖ At what speed, in m/s, would a moving clock lose 1.0 ns in 1.0 day according to experimenters on the ground?
   Hint: Use the binomial approximation.

19. ‖ You fly 5000 km across the United States on an airliner at 250 m/s. You return two days later at the same speed.
   a. Have you aged more or less than your friends at home?
   b. By how much?
   Hint: Use the binomial approximation.

Section 36.7 Length Contraction

20. | Jill claims that her new rocket is 100 m long. As she flies past your house, you measure the rocket's length and find that it is only 80 m. What is Jill's speed, as a fraction of c?

21. | At what speed, as a fraction of c, will a moving rod have a length 60% that of an identical rod at rest?

22. ‖ A cube has a density of 2000 kg/m³ while at rest in the laboratory. What is the cube's density as measured by an experimenter in the laboratory as the cube moves through the laboratory at 90% of the speed of light in a direction perpendicular to one of its faces?

23. | A muon travels 60 km through the atmosphere at a speed of $0.9997c$. According to the muon, how thick is the atmosphere?

24. | Our Milky Way galaxy is 100,000 ly in diameter. A spaceship crossing the galaxy measures the galaxy's diameter to be a mere 1.0 ly.
    a. What is the spacecraft's speed, as a fraction of c, relative to the galaxy?
    b. How long is the crossing time as measured in the galaxy's reference frame?

25. || A human hair is about 50 μm in diameter. At what speed, in m/s, would a meter stick "shrink by a hair"?
    Hint: Use the binomial approximation.

Section 36.8 The Lorentz Transformations

26. || A rocket travels in the x-direction at speed $0.60c$ with respect to the earth. An experimenter on the rocket observes a collision between two comets and determines that the spacetime coordinates of the collision are $(x', t') = (3.0 \times 10^{10}$ m, 200 s$)$. What are the spacetime coordinates of the collision in earth's reference frame?

27. | An event has spacetime coordinates $(x, t) = (1200$ m, 2.0 μs$)$ in reference frame S. What are the event's spacetime coordinates (a) in reference frame S$'$ that moves in the positive x-direction at $0.80c$ and (b) in reference frame S$''$ that moves in the negative x-direction at $0.80c$?

28. || In the earth's reference frame, a tree is at the origin and a pole is at $x = 30$ km. Lightning strikes both the tree and the pole at $t = 10$ μs. The lightning strikes are observed by a rocket traveling in the x-direction at $0.50c$.
    a. What are the spacetime coordinates for these two events in the rocket's reference frame?
    b. Are the events simultaneous in the rocket's frame? If not, which occurs first?

29. | A rocket cruising past earth at $0.80c$ shoots a bullet out the back door, opposite the rocket's motion, at $0.90c$ relative to the rocket. What is the bullet's speed, as a fraction of c, relative to the earth?

30. || A distant quasar is found to be moving away from the earth at $0.80c$. A galaxy closer to the earth and along the same line of sight is moving away from us at $0.20c$. What is the recessional speed of the quasar, as a fraction of c, as measured by astronomers in the other galaxy?

31. || A laboratory experiment shoots an electron to the left at $0.90c$. What is the electron's speed, as a fraction of c, relative to a proton moving to the right at $0.90c$?

Section 36.9 Relativistic Momentum

32. | A proton is accelerated to $0.999c$.
    a. What is the proton's momentum?
    b. By what factor does the proton's momentum exceed its Newtonian momentum?

33. || A 1.0 g particle has momentum 400,000 kg m/s. What is the particle's speed in m/s?

34. | At what speed, as a fraction of c, is a particle's momentum twice its Newtonian value?

35. | What is the speed, as a fraction of c, of a particle whose momentum is mc?

Section 36.10 Relativistic Energy

36. | A quarter-pound hamburger with all the fixings has a mass of 200 g. The food energy of the hamburger (480 food calories) is 2 MJ.
    a. What is the energy equivalent of the mass of the hamburger?
    b. By what factor does the energy equivalent exceed the food energy?

37. || What are the rest energy, the kinetic energy, and the total energy of a 1.0 g particle with a speed of $0.80c$?

38. || At what speed, as a fraction of c, must an electron move so that its total energy is 10% more than its rest mass energy?

39. || At what speed, as a fraction of c, is a particle's kinetic energy twice its rest energy?

40. | At what speed, as a fraction of c, is a particle's total energy twice its rest energy?

41. | A modest *supernova* (the explosion of a massive star at the end of its life cycle) releases 1.5×10^{44} J of energy in a few seconds. This is enough to outshine the entire galaxy in which it occurs. Suppose a star with the mass of our sun collides with an antimatter star of equal mass, causing complete annihilation. What is the ratio of the energy released in this star-antistar collision to the energy released in the supernova?

42. || One of the important ways in which the *Higgs boson* was detected at the Large Hadron Collider was by observing a type of decay in which the Higgs—which decays too quickly to be observed directly—is immediately transformed into two photons emitted back to back. Two photons, with momenta 3.31×10^{-17} kg m/s, were detected. What is the mass of the Higgs boson? Give your answer as a multiple of the proton mass.
    Hint: The relationship between energy and momentum applies to photons if you treat a photon as a massless particle.

Problems

43. || The diameter of the solar system is 10 light hours. A spaceship crosses the solar system in 15 hours, as measured on earth. How long, in hours, does the passage take according to passengers on the spaceship?
    Hint: $c = 1$ light hour per hour.

44. | A 30-m-long rocket train car is traveling from Los Angeles to New York at $0.50c$ when a light at the center of the car flashes. When the light reaches the front of the car, it immediately rings a bell. Light reaching the back of the car immediately sounds a siren.
    a. Are the bell and siren simultaneous events for a passenger seated in the car? If not, which occurs first and by how much time?
    b. Are the bell and siren simultaneous events for a bicyclist waiting to cross the tracks? If not, which occurs first and by how much time?

45. || The star Alpha goes supernova. Ten years later and 100 ly away, as measured by astronomers in the galaxy, star Beta explodes.
    a. Is it possible that the explosion of Alpha is in any way responsible for the explosion of Beta? Explain.
    b. An alien spacecraft passing through the galaxy finds that the distance between the two explosions is 120 ly. According to the aliens, what is the time between the explosions?

46. || Two events in reference frame S occur 10 μs apart at the same point in space. The distance between the two events is 2400 m in reference frame S$'$.
    a. What is the time interval between the events in reference frame S$'$?
    b. What is the velocity of S$'$ relative to S?

47. || A starship voyages to a distant planet 10 ly away. The explorers stay 1 year, return at the same speed, and arrive back on earth 26 years, as measured on earth, after they left. Assume that the time needed to accelerate and decelerate is negligible.
    a. What is the speed of the starship?
    b. How much time has elapsed on the astronauts' chronometers?

48. | The Stanford Linear Accelerator (SLAC) accelerates electrons to $v = 0.99999997c$ in a 3.2-km-long tube. If they travel the length of the tube at full speed (they don't, because they are accelerating), how long is the tube in the electrons' reference frame?

49. ‖ On a futuristic highway, a 15-m-long rocket travels so fast that a red stoplight, with a wavelength of 700 nm, appears to the pilot to be a green light with a wavelength of 520 nm. What is the length of the rocket to an observer standing at the intersection as the rocket speeds through?

Hint: The Doppler effect for light was covered in Chapter 16.

50. ‖ In an attempt to reduce the extraordinarily long travel times for voyaging to distant stars, some people have suggested traveling at close to the speed of light. Suppose you wish to visit the red giant star Betelgeuse, which is 430 ly away, and that you want your 20,000 kg rocket to move so fast that you age only 20 years during the round trip.

a. How fast, as a fraction of c, must the rocket travel relative to earth?

b. How much energy is needed to accelerate the rocket to this speed?

c. Compare this amount of energy to the total energy used by the United States in the year 2015, which was roughly 1.0×10^{20} J.

51. ‖ The quantity dE/dv, the rate of increase of energy with speed,
CALC is the amount of additional energy a moving object needs per 1 m/s increase in speed.

a. A 25,000 kg truck is traveling at 30 m/s. How much additional energy is needed to increase its speed by 1 m/s?

b. A 25,000 kg rocket is traveling at $0.90c$. How much additional energy is needed to increase its speed by 1 m/s?

52. ‖ A rocket traveling at $0.50c$ sets out for the nearest star, Alpha Centauri, which is 4.3 ly away from earth. It will return to earth immediately after reaching Alpha Centauri. What distance will the rocket travel and how long will the journey last according to (a) stay-at-home earthlings and (b) the rocket crew? (c) Which answers are the correct ones, those in part a or those in part b?

53. ‖ The star Delta goes supernova. One year later and 2.0 ly away, as measured by astronomers in the galaxy, star Epsilon explodes. Let the explosion of Delta be at $x_D = 0$ and $t_D = 0$. The explosions are observed by three spaceships cruising through the galaxy in the direction from Delta to Epsilon at velocities $v_1 = 0.30c$, $v_2 = 0.50c$, and $v_3 = 0.70c$. All three spaceships, each at the origin of its reference frame, happen to pass Delta as it explodes.

a. What are the times of the two explosions as measured by scientists on each of the three spaceships?

b. Does one spaceship find that the explosions are simultaneous? If so, which one?

c. Does one spaceship find that Epsilon explodes before Delta? If so, which one?

d. Do your answers to parts b and c violate the idea of causality? Explain.

54. ‖ Two rockets approach each other. Each is traveling at $0.75c$ in the earth's reference frame. What is the speed, as a fraction of c, of one rocket relative to the other?

55. ‖ Two rockets, A and B, approach the earth from opposite directions at speed $0.80c$. The length of each rocket measured in its rest frame is 100 m. What is the length of rocket A as measured by the crew of rocket B?

56. ‖ A rocket fires a projectile at a speed of $0.95c$ while traveling past the earth. An earthbound scientist measures the projectile's speed to be $0.90c$. What was the rocket's speed as a fraction of c?

57. ‖ Through what potential difference must an electron be accelerated, starting from rest, to acquire a speed of $0.99c$?

58. ‖ What is the speed, in m/s, of a proton after being accelerated from rest through a 50×10^6 V potential difference?

59. ‖ The half-life of a muon at rest is 1.5 μs. Muons that have been accelerated to a very high speed and are then held in a circular storage ring have a half-life of 7.5 μs.

a. What is the speed, as a fraction of c, of the muons in the storage ring?

b. What is the total energy of a muon in the storage ring? The mass of a muon is 207 times the mass of an electron.

60. ‖ This chapter has assumed that lengths perpendicular to the direction of motion are not affected by the motion. That is, motion in the x-direction does not cause length contraction along the y- or z-axes. To find out if this is really true, consider two spray-paint nozzles attached to rods perpendicular to the x-axis. It has been confirmed that, when both rods are at rest, both nozzles are exactly 1 m above the base of the rod. One rod is placed in the S reference frame with its base on the x-axis; the other is placed in the S′ reference frame with its base on the x'-axis. The rods then swoop past each other and, as **FIGURE P36.60** shows, each paints a stripe across the other rod.

We will use proof by contradiction. Assume that objects perpendicular to the motion *are* contracted. An experimenter in frame S finds that the S′ nozzle, as it goes past, is less than 1 m above the x-axis. The principle of relativity says that an experiment carried out in two different inertial reference frames will have the same outcome in both.

a. Pursue this line of reasoning and show that you end up with a logical contradiction, two mutually incompatible situations.

b. What can you conclude from this contradiction?

Red paint nozzle   Blue paint nozzle

1.00 m in S   1.00 m in S′

FIGURE P36.60

61. ‖ Derive the Lorentz transformations for t and t'.

Hint: See the comment following Equation 36.22.

62. ‖ a. Derive a velocity transformation equation for u_y and u'_y. Assume that the reference frames are in the standard orientation with motion parallel to the x- and x'-axes.

b. A rocket passes the earth at $0.80c$. As it goes by, it launches a projectile at $0.60c$ perpendicular to the direction of motion. What is the particle's speed, as a fraction of c, in the earth's reference frame?

63. ‖ A rocket is fired from the earth to the moon at a speed of $0.990c$. Let two events be "rocket leaves earth" and "rocket hits moon."

a. In the earth's reference frame, calculate Δx, Δt, and the spacetime interval s for these events.

b. In the rocket's reference frame, calculate $\Delta x'$, $\Delta t'$, and the spacetime interval s' for these events.

c. Repeat your calculations of part a if the rocket is replaced with a laser beam.

64. ‖ Let's examine whether or not the law of conservation of momentum is true in all reference frames if we use the Newtonian definition of momentum: $p_x = mu_x$. Consider an object A of mass $3m$ at rest in reference frame S. Object A explodes into two pieces: object B, of mass m, that is shot to the left at a speed of $c/2$ and object C, of mass $2m$, that, to conserve momentum, is shot to the right at a speed of $c/4$. Suppose this explosion is observed in reference frame S′ that is moving to the right at half the speed of light.

a. Use the Lorentz velocity transformation to find the velocity and the Newtonian momentum of A in S′.

b. Use the Lorentz velocity transformation to find the velocities and the Newtonian momenta of B and C in S′.

c. What is the total final momentum in S′?

d. Newtonian momentum was conserved in frame S. Is it conserved in frame S′?

65. ‖ a. What are the momentum and total energy of a proton with speed 0.99c?

    b. What is the proton's momentum in a different reference frame in which $E' = 5.0 \times 10^{-10}$ J?

66. ‖‖ At what speed, as a fraction of c, is the kinetic energy of a particle twice its Newtonian value?

67. | A typical nuclear power plant generates electricity at the rate of 1000 MW. The efficiency of transforming thermal energy into electrical energy is $\frac{1}{3}$ and the plant runs at full capacity for 80% of the year. (Nuclear power plants are down about 20% of the time for maintenance and refueling.)

    a. How much thermal energy does the plant generate in one year?

    b. What mass of uranium is transformed into energy in one year?

68. ‖ Many science fiction spaceships are powered by antimatter reactors. Suppose a 20-m-long spaceship, with a mass of 15,000 kg when empty, carries 2000 kg of fuel: 1000 kg each of matter and antimatter. The matter and antimatter are slowly combined, and the energy of their total annihilation is used to propel the ship. After consuming all the fuel and reaching top speed, the spaceship flies past a space station that is stationary with respect to the planet from which the ship was launched. What is the length of the spaceship as measured by astronauts on the space station?

69. | The sun radiates energy at the rate 3.8×10^{26} W. The source of this energy is fusion, a nuclear reaction in which mass is transformed into energy. The mass of the sun is 2.0×10^{30} kg.

    a. How much mass does the sun lose each year?

    b. What percent is this of the sun's total mass?

    c. Fusion takes place in the core of a star, where the temperature and pressure are highest. A star like the sun can sustain fusion until it has transformed about 0.10% of its total mass into energy, then fusion ceases and the star slowly dies. Estimate the sun's lifetime, giving your answer in billions of years.

70. ‖ The radioactive element radium (Ra) decays by a process known as *alpha decay,* in which the nucleus emits a helium nucleus. (These high-speed helium nuclei were named alpha particles when radioactivity was first discovered, long before the identity of the particles was established.) The reaction is $^{226}\text{Ra} \rightarrow {}^{222}\text{Rn} + {}^{4}\text{He}$, where Rn is the element radon. The accurately measured atomic masses of the three atoms are 226.0254 u, 222.0176 u, and 4.0026 u. How much energy is released in each decay? (The energy released in radioactive decay is what makes nuclear waste "hot.")

71. ‖ The nuclear reaction that powers the sun is the fusion of four protons into a helium nucleus. The process involves several steps, but the net reaction is simply $4p \rightarrow {}^{4}\text{He} + \text{energy}$. The mass of a proton, to four significant figures, is 1.673×10^{-27} kg, and the mass of a helium nucleus is known to be 6.644×10^{-27} kg.

    a. How much energy is released in each fusion?

    b. What fraction of the initial rest mass energy is this energy?

72. ‖ Consider the inelastic collision $e^- + e^- \rightarrow e^- + e^- + e^- + e^+$ in which an electron-positron pair is produced in a head-on collision between two electrons moving in opposite directions at the same speed. This is similar to Figure 36.39, but both of the initial electrons are moving.

    a. What is the threshold kinetic energy? That is, what minimum kinetic energy must each electron have to allow this process to occur?

    b. What is the speed of an electron with this kinetic energy?

Challenge Problems

73. ‖‖ An electron moving to the right at 0.90c collides with a positron moving to the left at 0.90c. The two particles annihilate and produce two gamma-ray photons. What is the wavelength of the photons?

74. ‖‖ Two rockets are each 1000 m long in their rest frame. Rocket Orion, traveling at 0.80c relative to the earth, is overtaking rocket Sirius, which is poking along at a mere 0.60c. According to the crew on Sirius, how long does Orion take to completely pass? That is, how long is it from the instant the nose of Orion is at the tail of Sirius until the tail of Orion is at the nose of Sirius?

75. ‖‖ Some particle accelerators allow protons (p^+) and antiprotons (p^-) to circulate at equal speeds in opposite directions in a device called a *storage ring*. The particle beams cross each other at various points to cause $p^+ + p^-$ collisions. In one collision, the outcome is $p^+ + p^- \rightarrow e^+ + e^- + \gamma + \gamma$, where γ represents a high-energy gamma-ray photon. The electron and positron are ejected from the collision at 0.9999995c and the gamma-ray photon wavelengths are found to be 1.0×10^{-6} nm. What were the proton and antiproton speeds, as a fraction of c, prior to the collision?

76. ‖‖ A ball of mass m traveling at a speed of 0.80c has a perfectly inelastic collision with an identical ball at rest. If Newtonian physics were correct for these speeds, momentum conservation would tell us that a ball of mass 2m departs the collision with a speed of 0.40c. Let's do a relativistic collision analysis to determine the mass and speed of the ball after the collision.

    a. What is γ_p, written as a fraction like a/b?

    b. What is the initial total momentum? Give your answer as a fraction times mc.

    c. What is the initial total energy? Give your answer as a fraction times mc^2. Don't forget that there are two balls.

    d. Because energy can be transformed into mass, and vice versa, you cannot assume that the final mass is 2m. Instead, let the final state of the system be an unknown mass M traveling at the unknown speed u_f. You have two conservation laws. Find M and u_f.

77. ‖‖ A very fast pole vaulter lives in the country. One day, while practicing, he notices a 10.0-m-long barn with the doors open at both ends. He decides to run through the barn at 0.866c while carrying his 16.0-m-long pole. The farmer, who sees him coming, says, "Aha! This guy's pole is length contracted to 8.0 m. There will be a short interval of time when the pole is entirely inside the barn. If I'm quick, I can simultaneously close both barn doors while the pole vaulter and his pole are inside." The pole vaulter, who sees the farmer beside the barn, thinks to himself, "That farmer is crazy. The barn is length contracted and is only 5.0 m long. My 16.0-m-long pole cannot fit into a 5.0-m-long barn. If the farmer closes the doors just as the tip of my pole reaches the back door, the front door will break off the last 11.0 m of my pole."

Can the farmer close the doors without breaking the pole? Show that, when properly analyzed, the farmer and the pole vaulter agree on the outcome. Your analysis should contain both quantitative calculations and written explanation.

FIGURE CP36.77

Mathematics Review

Algebra

Using exponents:

$$a^{-x} = \frac{1}{a^x} \qquad a^x a^y = a^{(x+y)} \qquad \frac{a^x}{a^y} = a^{(x-y)} \qquad (a^x)^y = a^{xy}$$

$$a^0 = 1 \qquad a^1 = a \qquad a^{1/n} = \sqrt[n]{a}$$

Fractions:

$$\left(\frac{a}{b}\right)\left(\frac{c}{d}\right) = \frac{ac}{bd} \qquad \frac{a/b}{c/d} = \frac{ad}{bc} \qquad \frac{1}{1/a} = a$$

Logarithms:

If $a = e^x$, then $\ln(a) = x$ $\qquad\qquad$ $\ln(e^x) = x$ $\qquad\qquad$ $e^{\ln(x)} = x$

$$\ln(ab) = \ln(a) + \ln(b) \qquad\qquad \ln\left(\frac{a}{b}\right) = \ln(a) - \ln(b) \qquad\qquad \ln(a^n) = n\ln(a)$$

The expression $\ln(a + b)$ cannot be simplified.

Linear equations: The graph of the equation $y = ax + b$ is a straight line. a is the slope of the graph. b is the y-intercept.

Proportionality: To say that y is proportional to x, written $y \propto x$, means that $y = ax$, where a is a constant. Proportionality is a special case of linearity. A graph of a proportional relationship is a straight line that passes through the origin. If $y \propto x$, then

$$\frac{y_1}{y_2} = \frac{x_1}{x_2}$$

Quadratic equation: The quadratic equation $ax^2 + bx + c = 0$ has the two solutions $x = \dfrac{-b \pm \sqrt{b^2 - 4ac}}{2a}$.

Geometry and Trigonometry

Area and volume:

Rectangle
$A = ab$

Rectangular box
$V = abc$

Triangle
$A = \frac{1}{2}ab$

Right circular cylinder
$V = \pi r^2 l$

Circle
$C = 2\pi r$
$A = \pi r^2$

Sphere
$A = 4\pi r^2$
$V = \frac{4}{3}\pi r^3$

Arc length and angle:　The angle θ in radians is defined as $\theta = s/r$.

The arc length that spans angle θ is $s = r\theta$.

2π rad $= 360°$

Right triangle:　Pythagorean theorem　$c = \sqrt{a^2 + b^2}$ or $a^2 + b^2 = c^2$

$$\sin\theta = \frac{b}{c} = \frac{\text{far side}}{\text{hypotenuse}} \qquad\qquad \theta = \sin^{-1}\!\left(\frac{b}{c}\right)$$

$$\cos\theta = \frac{a}{c} = \frac{\text{adjacent side}}{\text{hypotenuse}} \qquad\qquad \theta = \cos^{-1}\!\left(\frac{a}{c}\right)$$

$$\tan\theta = \frac{b}{a} = \frac{\text{far side}}{\text{adjacent side}} \qquad\qquad \theta = \tan^{-1}\!\left(\frac{b}{a}\right)$$

General triangle:　$\alpha + \beta + \gamma = 180° = \pi$ rad

Law of cosines $c^2 = a^2 + b^2 - 2ab\cos\gamma$

Identities:

$$\tan\alpha = \frac{\sin\alpha}{\cos\alpha} \qquad\qquad \sin^2\alpha + \cos^2\alpha = 1$$

$$\sin(-\alpha) = -\sin\alpha \qquad\qquad \cos(-\alpha) = \cos\alpha$$

$$\sin(\alpha \pm \beta) = \sin\alpha\cos\beta \pm \cos\alpha\sin\beta \qquad \cos(\alpha \pm \beta) = \cos\alpha\cos\beta \mp \sin\alpha\sin\beta$$

$$\sin(2\alpha) = 2\sin\alpha\cos\alpha \qquad\qquad \cos(2\alpha) = \cos^2\alpha - \sin^2\alpha$$

$$\sin(\alpha \pm \pi/2) = \pm\cos\alpha \qquad\qquad \cos(\alpha \pm \pi/2) = \mp\sin\alpha$$

$$\sin(\alpha \pm \pi) = -\sin\alpha \qquad\qquad \cos(\alpha \pm \pi) = -\cos\alpha$$

Expansions and Approximations

Binomial expansion:　　$(1 + x)^n = 1 + nx + \dfrac{n(n-1)}{2}x^2 + \cdots$

Binomial approximation:　$(1 + x)^n \approx 1 + nx \quad$ if $\quad x \ll 1$

Trigonometric expansions:　$\sin\alpha = \alpha - \dfrac{\alpha^3}{3!} + \dfrac{\alpha^5}{5!} - \dfrac{\alpha^7}{7!} + \cdots$ for α in rad

$\cos\alpha = 1 - \dfrac{\alpha^2}{2!} + \dfrac{\alpha^4}{4!} - \dfrac{\alpha^6}{6!} + \cdots$ for α in rad

Small-angle approximation:　If $\alpha \ll 1$ rad, then $\sin\alpha \approx \tan\alpha \approx \alpha$ and $\cos\alpha \approx 1$.

The small-angle approximation is excellent for $\alpha < 5°$ (≈ 0.1 rad) and generally acceptable up to $\alpha \approx 10°$.

Calculus

The letters a and n represent constants in the following derivatives and integrals.

Derivatives

$$\frac{d}{dx}(a) = 0$$

$$\frac{d}{dx}(ax) = a$$

$$\frac{d}{dx}\left(\frac{a}{x}\right) = -\frac{a}{x^2}$$

$$\frac{d}{dx}(ax^n) = anx^{n-1}$$

$$\frac{d}{dx}\big(\ln(ax)\big) = \frac{1}{x}$$

$$\frac{d}{dx}(e^{ax}) = ae^{ax}$$

$$\frac{d}{dx}\big(\sin(ax)\big) = a\cos(ax)$$

$$\frac{d}{dx}\big(\cos(ax)\big) = -a\sin(ax)$$

Integrals

$$\int x\,dx = \frac{1}{2}x^2$$

$$\int x^2\,dx = \frac{1}{3}x^3$$

$$\int \frac{1}{x^2}\,dx = -\frac{1}{x}$$

$$\int x^n\,dx = \frac{x^{n+1}}{n+1} \qquad n \neq -1$$

$$\int \frac{dx}{x} = \ln x$$

$$\int \frac{dx}{a+x} = \ln(a+x)$$

$$\int \frac{x\,dx}{a+x} = x - a\ln(a+x)$$

$$\int \frac{dx}{\sqrt{x^2 \pm a^2}} = \ln\left(x + \sqrt{x^2 \pm a^2}\right)$$

$$\int \frac{x\,dx}{\sqrt{x^2 \pm a^2}} = \sqrt{x^2 \pm a^2}$$

$$\int \frac{dx}{x^2 + a^2} = \frac{1}{a}\tan^{-1}\left(\frac{x}{a}\right)$$

$$\int \frac{dx}{(x^2 + a^2)^2} = \frac{1}{2a^3}\tan^{-1}\left(\frac{x}{a}\right) + \frac{x}{2a^2(x^2 + a^2)}$$

$$\int \frac{dx}{(x^2 \pm a^2)^{3/2}} = \frac{\pm x}{a^2\sqrt{x^2 \pm a^2}}$$

$$\int \frac{x\,dx}{(x^2 \pm a^2)^{3/2}} = -\frac{1}{\sqrt{x^2 \pm a^2}}$$

$$\int e^{ax}\,dx = \frac{1}{a}e^{ax}$$

$$\int xe^{-x}\,dx = -(x+1)e^{-x}$$

$$\int x^2 e^{-x}\,dx = -(x^2 + 2x + 2)e^{-x}$$

$$\int \sin(ax)\,dx = -\frac{1}{a}\cos(ax)$$

$$\int \cos(ax)\,dx = \frac{1}{a}\sin(ax)$$

$$\int \sin^2(ax)\,dx = \frac{x}{2} - \frac{\sin(2ax)}{4a}$$

$$\int \cos^2(ax)\,dx = \frac{x}{2} + \frac{\sin(2ax)}{4a}$$

$$\int_0^\infty x^n e^{-ax}\,dx = \frac{n!}{a^{n+1}}$$

$$\int_0^\infty e^{-ax^2}\,dx = \frac{1}{2}\sqrt{\frac{\pi}{a}}$$

Periodic Table of Elements

Period																	
1	1 H 1.0																2 He 4.0
2	3 Li 6.9	4 Be 9.0										5 B 10.8	6 C 12.0	7 N 14.0	8 O 16.0	9 F 19.0	10 Ne 20.2
3	11 Na 23.0	12 Mg 24.3										13 Al 27.0	14 Si 28.1	15 P 31.0	16 S 32.1	17 Cl 35.5	18 Ar 39.9
4	19 K 39.1	20 Ca 40.1	21 Sc 45.0	22 Ti 47.9	23 V 50.9	24 Cr 52.0	25 Mn 54.9	26 Fe 55.8	27 Co 58.9	28 Ni 58.7	29 Cu 63.5	30 Zn 65.4	31 Ga 69.7	32 Ge 72.6	33 As 74.9	34 Se 79.0	35 Br 79.9	36 Kr 83.8
5	37 Rb 85.5	38 Sr 87.6	39 Y 88.9	40 Zr 91.2	41 Nb 92.9	42 Mo 95.9	43 Tc [98]	44 Ru 101.1	45 Rh 102.9	46 Pd 106.4	47 Ag 107.9	48 Cd 112.4	49 In 114.8	50 Sn 118.7	51 Sb 121.8	52 Te 127.6	53 I 126.9	54 Xe 131.3
6	55 Cs 132.9	56 Ba 137.3	71 Lu 175.0	72 Hf 178.5	73 Ta 180.9	74 W 183.9	75 Re 186.2	76 Os 190.2	77 Ir 192.2	78 Pt 195.1	79 Au 197.0	80 Hg 200.6	81 Tl 204.4	82 Pb 207.2	83 Bi 209.0	84 Po [209]	85 At [210]	86 Rn [222]
7	87 Fr [223]	88 Ra [226]	103 Lr [262]	104 Rf [265]	105 Db [268]	106 Sg [271]	107 Bh [272]	108 Hs [270]	109 Mt [276]	110 Ds [281]	111 Rg [280]	112 Cn [285]	113	114 Fl [289]	115	116 Lv [293]	117	118

27 Co 58.9 — Symbol

Atomic number — 27
Atomic mass — 58.9

Transition elements

Lanthanides 6	57 La 138.9	58 Ce 140.1	59 Pr 140.9	60 Nd 144.2	61 Pm 144.9	62 Sm 150.4	63 Eu 152.0	64 Gd 157.3	65 Tb 158.9	66 Dy 162.5	67 Ho 164.9	68 Er 167.3	69 Tm 168.9	70 Yb 173.0
Actinides 7	89 Ac [227]	90 Th 232.0	91 Pa 231.0	92 U 238.0	93 Np [237]	94 Pu [244]	95 Am [243]	96 Cm [247]	97 Bk [247]	98 Cf [251]	99 Es [252]	100 Fm [257]	101 Md [258]	102 No [259]

Inner transition elements

An atomic mass in brackets is that of the longest-lived isotope of an element with no stable isotopes.

Answers

Answers to Stop to Think Questions and Odd-Numbered Exercises and Problems

Chapter 1

Stop to Think Questions

1. **B.** The images of B are farther apart, so it travels a larger distance than does A during the same intervals of time.
2. **a.** Dropped ball.   **b.** Dust particle.   **c.** Descending rocket.
3. **e.** The average velocity vector is found by connecting one dot in the motion diagram to the next.
4. **b.** $\vec{v}_2 = \vec{v}_1 + \Delta\vec{v}$, and $\Delta\vec{v}$ points in the direction of $\vec{a}$.

5. **d > c > b = a.**

Exercises and Problems

1.

3.

5.

7.

9.

11. a.

   b.

13.

15.

17.

A-5

21.

x_0, v_{0x}, t_0 x_1, v_{1x}, t_1

Known

$v_{0x} = 0$ m/s $t_0 = 0$ s $x_0 = 0$ m
$a_{0x} = 1.5$ m/s^2
$v_{1x} = 7.5$ m/s

Find

x_1

23. a. 1 b. 3 c. 2 d. 5
25. a. 1.9 m b. 1.09×10^{14} s c. 2.2×10^{-4} m/s d. 5.7×10^{10} m^2
27. a. 7 m d. 100,000 m c. 30 m/s d. 0.2 m
29. a. 32,590 b. 9.0 c. 0.237 d. 4.78
31. a. 15 m
33. 32 ms
35.

Pictorial representation

x_0, v_{0x}, t_0 x_1, v_{1x}, t_1

Motion diagram

Known

$x_0 = 0$ m $x_1 = 4$ km
$v_{0x} = 300$ m/s $t_0 = 0$ s

Find

a_x

37.

| **Pictorial representation** | **Motion diagram** |

Known

$y_0 = 0$ m $v_{0y} = 10$ m/s
$t_0 = 0$ s $y_1 = 3.0$ m
$a_{0y} < 0$

Find

t_1

39. **Pictorial representation**

30°

x_1, v_1, t_1

Known

$x_0 = 0$ m $x_1 = 10$ m/s
$t_0 = 0$ s
$v_{0x} = 0$ m/s

Motion diagram

Find

v_{1x}

41. **Pictorial representation**

10°

x_0, v_{0x}, t_0

Motion diagram

Stops

Known

$x_0 = 0$ m $v_{1x} = 0$ m/s
$t_0 = 0$ s
$v_{0x} = 30$ m/s

Find

x_1

43.

Pictorial representation

David

x_{D0}, t_{D0}, v_{D0x} x_{D1}, t_{D1}, v_{D1x}

Tina

x_{T0}, t_{T0}, v_{T0x} x_{T1}, t_{T1}, v_{T1x}

Motion diagram

$\vec{a}_D = \vec{0}$

$\vec{v}_D$

$\vec{a}_T$

$\vec{v}_T$

Known

$x_{D0} = 0$ m $x_{T0} = 0$ m
$t_{D0} = 0$ s $t_{T0} = 0$ s
$v_{D0x} = 30$ m/s $v_{T0x} = 0$ m/s
$a_{D0x} = 0$ m/s^2
$a_T = 2.0$ m/s^2

Find

x_{T1}

49. a.

Coasting begins   Stop

51. a.

53. Smallest: 6.4×10^3 m², largest: 8.3×10^3 m²
55. 2.9×10^{-4} m³
57. a. 83.3 kg/m³   b. 810 kg/m³

Chapter 2

Stop to Think Questions

1. **d.** The particle starts with positive x and moves to negative x.
2. **c.** The velocity is the slope of the position graph. The slope is positive and constant until the position graph crosses the axis, then positive but decreasing, and finally zero when the position graph is horizontal.
3. **b.** A constant positive v_x corresponds to a linearly increasing x, starting from $x_i = -10$ m. The constant negative v_x then corresponds to a linearly decreasing x.
4. **a and b.** The velocity is constant while $a = 0$; it decreases linearly while a is negative. Graphs a, b, and c all have the same acceleration, but only graphs a and b have a positive initial velocity that represents a particle moving to the right.
5. **d.** The acceleration vector points downhill (negative s-direction) and has the constant value $-g \sin\theta$ throughout the motion.
6. **c.** Acceleration is the slope of the graph. The slope is zero at B. Although the graph is steepest at A, the slope at that point is negative, and so $a_A < a_B$. Only C has a positive slope, so $a_C > a_B$.

Exercises and Problems

1. a. Beth   b. 20 min
3. a. 48 mph   b. 50 mph
5. a. v_x (m/s)   b. None

7. 8.0 cm

9. a_x (m/s²)

11. a. 8.0 m   b. 2.0 m/s   c. −2.0 m/s²
13. a. 2.7 m/s²   b. 1.3×10^2 m = 4.3×10^2 feet
15. a. 360 d   b. 4.6×10^{15} m   c. 0.49
17. 2.8 m/s²
19. 216 m
21. 3.2 s
23. 5.2 cm
25. 73 m
27. 265 m
29. a. 16 m/s   b. 31 m
31. 16 m/s
33. a. 21 m   b. 26 m/s   c. 24 m/s²
35. a. 0 s and 3 s   b. 12 m and −18 m/s²; −15 m and 18 m/s²
37. a. −10 m/s   b. −20 m/s   c. 95 m/s
39. a. 2 s, 5 s   b. −3 m/s², 3 m/s²
41. a. 20 s   b. 667 m
43.

45. $v_0 = 0$

Steeper than first part

47. a. Yes   b. 35 s   c. No
49. a. 5 m   b. 22 m/s
51. 5.7 m/s
53. Yes, 10 m
55. a. 54.8 km   b. 228 s
57. 19.7 m
59. 55 cm

61. a. 2.0 h   b. 73 m
   c.   *x* (mi)

63. $v_f = \sqrt{2gh}$
65. 17 cm
67. 14 m/s
69. 0.36 m
71. 4.4 m/s²
73. *gh/d*
77. c. 17.2 m/s
79. c. 750 m
81. a. 10 s   b. 3.8 m/s²   c. 5.6%
83. 12.5 m/s
85. 4500 m/s²

Chapter 3

Stop to Think Questions

1. **c.** The graphical construction of $\vec{A}_1 + \vec{A}_2 + \vec{A}_3$ is shown in the figure.

2. **a.** The graphical construction of $2\vec{A} - \vec{B}$ is shown in the figure.

3. $C_x = -4$ cm, $C_y = 2$ cm.
4. **c.** Vector $\vec{C}$ points to the left and down, so both C_x and C_y are negative. C_x is in the numerator because it is the side opposite ϕ.

Exercises and Problems

1. a. $\vec{A} + \vec{B}$ b.

3. a. $E\sin\theta, -E\cos\theta$ b. $E\cos\phi, -E\sin\phi$
5. 8 m
7. a. 3.8 m/s, 6.5 m/s   b. –1.3 m/s², 0.80 m/s²   c. –30 N, 40 N
9. 100 m, west
11. a. 7.6, 67°   b. 4.9 m/s², 66°   c. 18 m/s, 38°   d. 4.0 m, 34°
13. a. $1\hat{i} + 3\hat{j}$
   b.

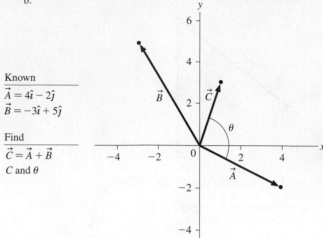

Known
$$\vec{A} = 4\hat{i} - 2\hat{j}$$
$$\vec{B} = -3\hat{i} + 5\hat{j}$$

Find
$$\vec{C} = \vec{A} + \vec{B}$$
C and θ

   c. 3.2, 72° above the +*x*-axis
15. a. $-1\hat{i} + 11\hat{j}$ c. 11, 5.2° ccw from the +*y*-axis
17. a. 2.8   b. 4.1   c. 6.1
19. $v_x = -50$ m/s, $v_y = -87$ m/s
21. $B = 2.2$ T, $\theta = 27°$
23. a. 0 m, 26 m, 160 m   b. $\vec{v}(t) = (10\hat{i} + 8.0\hat{j})\, t$ m/s
   c. 0 m/s, 26 m/s, and 6 m/s
25. $\vec{C} = 0.8\,\hat{i} - 4.5\,\hat{j}$
27. $\vec{B} = \dfrac{1}{\sqrt{2}}\hat{i} + \dfrac{1}{\sqrt{2}}\hat{j}$
29. a. $\Delta\vec{r} = (1.7\text{ cm})\hat{i} - (3.0\text{ cm})\hat{j}$ b. $\Delta\vec{r} = 0.0$ cm
31. 90 m, 46° south of west
33. $v_x = 3.7$ m/s, $v_y = 2.6$ m/s
35. a. –3.4 m/s   b. –9.4 m/s
37. 280 N, 350 N
39. 570 N, –380 N
41. 5.2 μm/s at 27° below the positive *x*-axis
43. 4.4 units at 83° below the negative *x*-axis
45. 7.3 N at 79° below the negative *x*-axis

Chapter 4

Stop to Think Questions

1. **c.** $v = 0$ requires both $v_x = 0$ and $v_y = 0$. Neither *x* nor *y* can be changing.
2. **d.** The parallel component of $\vec{a}$ is opposite $\vec{v}$ and will cause the particle to slow down. The perpendicular component of $\vec{a}$ will cause the particle to change direction downward.
3. **d.** A projectile's acceleration $\vec{a} = -g\hat{j}$ does not depend on its mass. The second marble has the same initial velocity and the same acceleration, so it follows the same trajectory and lands at the same position.
4. **f.** The plane's velocity relative to the helicopter is $\vec{v}_{PH} = \vec{v}_{PG} + \vec{v}_{GH} = \vec{v}_{PG} - \vec{v}_{HG}$, where G is the ground. The vector addition shows that $\vec{v}_{PH}$ is to the right and down with a magnitude greater than the 100 m/s of $\vec{v}_{PG}$.

$\vec{v}_{PG}$ of plane relative to ground
$-\vec{v}_{HG}$
$\vec{v}_{PH} = \vec{v}_{PG} - \vec{v}_{HG}$ of plane relative to helicopter

5. **b.** An initial cw rotation causes the particle's angular position to become increasingly negative. The speed drops to half after reversing direction, so the slope becomes positive and is half as steep as the initial slope. Turning through the same angle returns the particle to $\theta = 0°$.

6. $a_b > a_e > a_a = a_c > a_d$. Centripetal acceleration is v^2/r. Doubling r decreases a_r by a factor of 2. Doubling v increases a_r by a factor of 4. Reversing direction doesn't change a_r.

7. **c.** ω is negative because the rotation is cw. Because ω is negative but becoming *less* negative, the change $\Delta\omega$ is *positive*. So α is positive.

Exercises and Problems

3. C
5. E
7. 2.2 m/s^2
9. $(2\hat{\imath} - 8\hat{\jmath})$ m/s^2
11. 19.6 m
13. 680 m
15. $r = 16.4$ m
17. $v_x = 27.7$ m/s
19. 30 s
21. a. 42° west of north   b. 45 s
23. 10 rev
25.
27. a. 4.7 rad/s   b. 1.3 s
29. 680 km/h, 1040 mph
31. 43 m
33. a. 3.0×10^4 m/s   b. 2.0×10^{-7} rad/s   c. 6.0×10^{-3} m/s^2
35. $v = 5.7$ m/s, $a_r = 108$ m/s^2
37. a. 3.75 rad/s   b. 5.0 rad/s   c. 5.0 rad/s
39. 98 rpm
41. 47 rad/s^2
43. 38 rev
45. $\left(\frac{1}{2}bt^2 + v_{0x}\right)\hat{\imath} + (e^{-ct} + v_{0y})\hat{\jmath}$
47. a. $\dfrac{v_0^2 \sin^2\theta}{2g}$ b. 14.4 m, 28.8 m, 43.2 m
49. a. 12 m/s   b. 0.90 m
51. Clears by 1.0 m
53. 7.4 m/s
55. a. 13 m/s   b. 48°
57. 470 m/s^2
59. 4.8 m/s
61. a. 39 mi   b. 20 mph
63. No. The angular acceleration is negative.
65. a. 1.75×10^4 m/s^2 b. 4.4×10^3 m/s^2
67. 69 m/s at 21° with the vertical
69. a. -100 rad/s^2 b. 50 rev
71. 0.75 rad/s^2
73. a. $v = \sqrt{2\alpha \Delta\theta R}$ b. $a = 2\alpha \Delta\theta R$

75. 1940 rpm
77. 550 rpm
79. b. 30 m west
81. 34.3°
83. 3.8 m
85. 10°

Chapter 5

Stop to Think Questions

1. **c.**

The y-component of $\vec{F}_3$ cancels the y-component of $\vec{F}_1$.

The x-component of $\vec{F}_3$ is to the left and larger than the x-component of $\vec{F}_2$.

2. **a, b, and d.** Friction and the normal force are the only contact forces. Nothing is touching the rock to provide a "force of the kick."

3. **b.** Acceleration is proportional to force, so doubling the number of rubber bands doubles the acceleration of the original object from 2 m/s^2 to 4 m/s^2. But acceleration is also inversely proportional to mass. Doubling the mass cuts the acceleration in half, back to 2 m/s^2.

4. **d.**

First add $\vec{F}_1$ and $\vec{F}_2$.   Then add $\vec{F}_3$. This is $\vec{F}_{net}$. $\vec{a}$ is in the same direction as $\vec{F}_{net}$.

5. **c.** The acceleration vector points downward as the elevator slows. $\vec{F}_{net}$ points in the same direction as $\vec{a}$, so $\vec{F}_{net}$ also points down. This will be true if the tension is less than the gravitational force: $T < F_G$.

Exercises and Problems

1. Gravity, normal force, static friction
3. Gravity, normal force, kinetic friction
5. Gravity, drag
7. a. 2.4 m/s^2 b. 0.60 m/s^2
9. $m_1 = 0.080$ kg, $m_3 = 0.50$ kg
11. 1.5 J
13. a. 4 m/s^2 b. 2 m/s^2
15. 25 kg
17.

19.

21.

A rocket accelerates upward

23. **Force identification**   **Free-body diagram**

Normal force $\vec{n}$
Gravity $\vec{F}_G$

25. **Force identification**   **Free-body diagram**

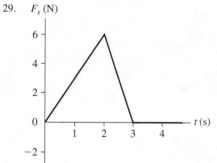

Gravity $\vec{F}_G$

Normal force $\vec{n}$
Kinetic friction $\vec{f}_k$

27.
Force identification   **Free-body diagram**

Thrust $\vec{F}_{thrust}$ Drag $\vec{F}_{drag}$

Gravity $\vec{F}_G$ Normal force $\vec{n}$

29. F_x (N)

31. a_x (m/s²)

33. a. 16 m/s²   b. 4.0 m/s²   c. 8.0 m/s²   d. 32 m/s²
35.

$\vec{F}_{net} = \vec{0}$

$\vec{a} = \vec{0}$

$\vec{v}$

37.

39.

41. a.

$F = (0.57 \pm 0.03 \text{ kg}) a$

b. Yes. 0 m/s², 0 N   c. 57 kg

43.

Air resistance (drag) $\vec{F}_{drag}$

Propulsion $\vec{F}_{thrust}$

Gravity $\vec{F}_G$

47.

49.

51.

53.

55.

57. a.

Magnified view of ball in contact with ground

First contact   Loses contact

Compressing

Expanding

Turning point

b.

Gravity $\vec{F}_G$ Normal force $\vec{n}$

c.

Chapter 6

Stop to Think Questions

1. **a.** The lander is descending and slowing. The acceleration vector points upward, and so $\vec{F}_{net}$ points upward. This can be true only if the thrust has a larger magnitude than the weight.
2. **a.** You are descending and slowing, so your acceleration vector points upward and there is a net upward force on you. The floor pushes up against your feet harder than gravity pulls down.
3. $f_b > f_c = f_d = f_e > f_a$. Situations c, d, and e are all kinetic friction, which does not depend on either velocity or acceleration. Kinetic friction is smaller than the maximum static friction that is exerted in b. $f_a = 0$ because no friction is needed to keep the object at rest.
4. **d.** The ball is shot *down* at 30 m/s, so $v_{0y} = -30$ m/s. This exceeds the terminal speed, so the upward drag force is *larger* than the downward weight force. Thus the ball *slows down* even though it is "falling." It will slow until $v_y = -15$ m/s, the terminal velocity, then maintain that velocity.

Exercises and Problems

1. 94 N, 58° below the horizontal
3. 510 N
5. 160 N
7. a. 0.0036 N   b. 0.010 N
9. $a_x = 1.5$ m/s^2, $a_y = 0$ m/s^2
11. 0 m/s^2, 4 m/s
13. a. 490 N   b. 490 N   c. 740 N   d. 240 N
15. 40 s
17. a. 540 N   b. $m = 55$ kg; $mg = 210$ N
19. a. 7.8×10^2 N   b. 1.1 kN
21. a. 3.96 N   b. 2.32 N
23. 5.90×10^{-3} m/s^2
25. Yes
27. 9 m/s
29. 10,000 N
31. 2.6×10^3 m
33. 190,000 N
35. 140 m/s
37. 4.5 m/s
39. 6400 N, 4380 N
41. a. 59 N   b. 68°   c. 79 N
43. 59 N
45. a. $v(h) = \sqrt{2\left(\dfrac{F_{thrust}}{m} - g\right)h}$ b. 54 m/s
47. 2800 kg
49. a. 16.9 m/s   b. 229 m
51. a. 3.8 m   b. 7.0 m/s
53. $T_{max} = (M + m)\mu_s g$
55. a. Yes   b. Yes
57. Stay at rest
59. a. $-5g$ b. $3g$
61. a. 0 N   b. 220 N
63. a. $\dfrac{F_0}{m}\dfrac{T}{2}$ b. $\dfrac{F_0}{m}\dfrac{T^2}{3}$

65. a. 9.4×10^{-10} N, 5.7×10^{-13} N   b. 1.8 m/s², 130 m/s²
67. a. $v_{term} = \dfrac{mg}{6\pi\eta R}$ b. 27 min
69. b. 32 m/s
71. c. 144 N
73. b. $v_x(L) = \sqrt{L\left(\dfrac{2F_0}{m} - \mu_0 g\right)}$
75. a. $a_x = (2x)g$ b. 3.9 m/s²
77. b. $v_x(x) = v_0 - \dfrac{6\pi\eta R}{m}x$ c. Not reasonable

Chapter 7

Stop to Think Questions

1. **The crate's gravitational force and the normal force are incorrectly identified as an action/reaction pair.** The normal force should be paired with a downward force of the crate on the ground. Gravity is the pull of the entire earth, so $\vec{F}_G$ should be paired with a force pulling up on the entire earth.

2. **c.** Newton's third law says that the force of A on B is *equal* and opposite to the force of B on A. This is always true. The mass of the objects isn't relevant.

3. **b.** $F_{\text{B on H}} = F_{\text{H on B}}$ and $F_{\text{A on B}} = F_{\text{B on A}}$ because these are action/reaction pairs. Box B is slowing down and therefore must have a net force to the left. So from Newton's second law we also know that $F_{\text{H on B}} > F_{\text{A on B}}$.

4. **Equal to.** Each block is hanging in equilibrium, with no net force, so the upward tension force is mg.

5. **Less than.** Block B is *accelerating* downward, so the net force on B must point down. The only forces acting on B are the tension and gravity, so $T_{\text{S on B}} < (F_G)_B$.

Exercises and Problems

1. a.   **Interaction diagram**

b. The system is the soccer ball and bowling ball.

SB = Soccer ball
BB = Bowling ball
S = Surface
EE = Entire Earth

c.   **Free-body diagrams**

Soccer ball   Bowling ball

3. a.   **Interaction diagram**

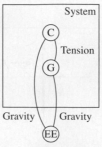

b. The system is the cable and the girder.

c.   **Free-body diagrams**

Girder   Cable

5. a.   **Interaction diagram**

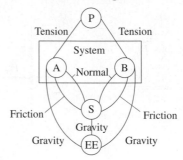

P = Pulley   S = Surface
A = Block A   B = Block B
EE = Entire Earth

b. The system is block A and block B.

c.   **Free-body diagrams**

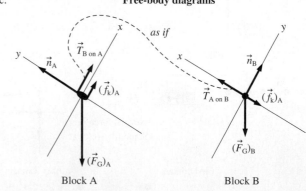

Block A   Block B

7. $m_A = 12$ kg
9. a. 6 N   b. 10 N
11. 440 N
13. 42 m
15. 6.5 m/s²

17. 590 N
19. 250 N
21. 9800 N
23. 67 N, 36°
25. 99 kg
27. a. 4 N b. 3 N
29. No
31. a. 2800 N b. 2800 N
33. Cheek, not forehead
35. 1.8 s
37. $T_1 = 100$ N, $T_2 = T_3 = T_5 = F = 50$ N, and $T_4 = 150$ N
39. a. 1.8 kg b. 1.3 m/s²
41. a. 0.67 m b. Slides back down
43. a. 8.2×10^3 N b. 4.8×10^2 N
45. 4.0×10^2 N
47. 2.4×10^{-6} N
49. $F = (m_1 + m_2)g \tan \theta$
53. 1.8 m/s²
55. 2.8 m/s²
57. 920 g

Chapter 8

Stop to Think Questions

1. **d.** The parallel component of $\vec{F}$ is opposite $\vec{v}$ and will cause the particle to slow down. The perpendicular component of $\vec{F}$ will cause the particle to change directions in a downward direction.
2. $T_d > T_b = T_e > T_c > T_a$. The center-directed force is $m\omega^2 r$. Changing r by a factor of 2 changes the tension by a factor of 2, but changing ω by a factor of 2 changes the tension by a factor of 4.
3. **b.** The car is moving in a circle, so there must be a net force toward the center of the circle. The circle is below the car, so the net force must point downward. This can be true only if $F_G > n$.
4. **c.** The ball does not have a "memory" of its previous motion. The velocity $\vec{v}$ is straight up at the instant the string breaks. The only force on the ball after the string breaks is the gravitational force, straight down. This is just like tossing a ball straight up.

Exercises and Problems

1. 39 m
3. c. 0.107°
5. 9.4 kN, static friction
7. a 3.9 m/s b. 6.2 N
9. 2.01×10^{20} N
11. 45 s
13. $v = \sqrt{\dfrac{m_2 r g}{m_1}}$
15. 6.0×10^{-3} m/s²
17. a. 24.0 h b. 0.223 m/s² c. 0 N
19. 12 m/s
21. 20 m/s
23. a. 3.77 m/s, 0.95 m/s² b. 0.90 c. 1.1
25. a. 4.9 N b. 2.9 N c. 32 N
27. 1.6 s
29. a. 2.2×10^6 N b. $-27°$
33. 0.034 N
35. 8.6 m
37. a. 165 m b. Straight line
41. a. $\omega = \sqrt{\dfrac{g}{L \sin \theta}}$ b. 72 rpm
43. 0.79 N
45. a. 2.9 m/s b. 14 N

47. a. $T = \dfrac{mgL}{\sqrt{L^2 - r^2}}$ b. $\omega = \sqrt{\dfrac{g}{\sqrt{L^2 - r^2}}}$ c. 5.0 N, 30 rpm
49. 24 rpm
51. a. 320 N, 1400 N b. 5.7 s
53. 0. 38 N
55. a. $\sqrt{gL}$ b. 5.9 mph
57. 1.4 m to the right
59. 13 N
61. a. 1.90 m/s² at 21° b. 15.7 m/s
63. a. 3.8 m/s b. 19 m/s²
65. b. 19.8 m/s
67. a. $\theta = \frac{1}{2}\tan^{-1}(mg/F)$ b. 11.5%
69. $\sqrt{gL}$

Chapter 9

Stop to Think Questions

1. **a.** Kinetic energy depends linearly on the mass but on the *square* of the velocity. A factor of 2 change in velocity is more significant than a factor of 2 change in mass.
2. **Positive.** The force (gravity) and the displacement are in the same direction. The rock gains kinetic energy.
3. **Positive.** Each puck experiences a force in the direction of motion, which is a positive amount of work. The net force is zero, but the total work is not zero because the pucks have different displacements.
4. **c.** The upward tension force is opposite the displacement, so it does negative work. The downward gravitational force is parallel to the displacement, so it does positive work.
5. **c.** $W = F(\Delta r)\cos\theta$. The 10 N force at 90° does no work at all. Because $\cos 60° = \frac{1}{2}$, the 8 N force does less work than the 6 N force.
6. **Zero.** The road does exert a forward force on the car, but the point of application does not move because the tires are not skidding on the road. The car's increasing kinetic energy is a transformation of chemical energy to kinetic energy, not a transfer of energy from the road to the car.
7. $k_a > k_b > k_c$. The spring constant is the slope of the force-versus-displacement graph.
8. **Zero.** The wall exerts a force on the right end of the spring, but the point of application does not move.
9. $P_b > P_a = P_c > P_d$. The work done is $mg\Delta y$, so the power output is $mg\Delta y/\Delta t$. Runner b does the same work as runner a, but in less time. The ratio $m/\Delta t$ is the same for runners a and c. Runner d does twice the work of a, but takes more than twice as long.

Exercises and Problems

1. The bullet
3. 2.0
5. a. 5.3 J b. 0 J
7. 0 J
9. a 1.9 J b. 26 m/s
11. a. -2 b. 0
13. 125°
15. a. -2.7 b. -20 c. 10
17. a. 0.20 J b. 3.0 m/s
19. 1.7 kJ, 1.1 kJ, -2.0 kJ
21. 8.0 m/s, 10 m/s, 11 m/s
23. $\frac{1}{3}qd^3$
25. 38 N
27. a. 3.9×10^2 N/m b. 17.5 cm
29. a. 49 N b. 1450 N/m c. 3.4 cm
31. 1360 m/s
33. 0.037
35. 540 J
37. a. $W_{net} = 176$ J b. $P = 59$ W
41. Runner: $P_{avg} = 1.2$ kW; greyhound: $P_{avg} = 2.0$ kW

43. a. -9.8×10^4 J   b. 1.1×10^5 J   c. 1.0×10^4 J
45. 2.4 m/s
47. $2h$
49. a. 2.2 m/s   b. 0.0058
51. a. 2.9 J   b. 3.6%
53. a. $Gm_1m_2\left(\dfrac{x_2 - x_1}{x_1x_2}\right)$ b. 2.1×10^5 m/s
55. 19 m/s
57. 33 N/m
59. 3.7 m/s
61. 2.5×10^5 kg/s
63. 1.4 kW
65. 1.2×10^8 ly
71. 24 W

Chapter 10

Stop to Think Questions

1. $(U_G)_c > (U_G)_b = (U_G)_d > (U_G)_a$. Gravitational potential energy depends only on height, not on speed.
2. **b.** Potential energy depends only on the vertical displacement. At the elevation of the dashed line, both have gained the same gravitational potential energy, so both have lost the same kinetic energy.
3. $v_a = v_b = v_c = v_d$. Her increase in kinetic energy depends only on the vertical distance through which she falls, not on the shape of the slide.
4. **c.** Constant speed means no change of kinetic energy. But for motion on a slope, constant speed requires friction. All the gravitational potential energy is being transformed into thermal energy.
5. **c.** U_{Sp} depends on d^2. Doubling the compression increases U_{Sp} by a factor of 4. All potential energy is transformed into kinetic energy, so K increases by a factor of 4. But K depends on v^2, so v increases by only a factor of 2.
6. $x = 6$ m. From the graph, the particle's potential energy at $x = 1$ m is $U = 3$ J. Its total energy is thus $E = K + U = 4$ J. A TE line at 4 J crosses the PE curve at $x = 6$ m.
7. **e.** Force is the negative of the slope of the potential-energy diagram. At $x = 4$ m the potential energy has risen by 4 J over a distance of 2 m, so the slope is 2 J/m = 2 N.
8. **d.** The system is losing potential energy as the weight falls. It's gaining speed, so some U is transformed into K. Energy is also being transferred out of the system to the environment via negative work done by the rope tension. The system is not isolated, so neither E_{mech} nor E_{sys} is conserved.

Exercises and Problems

1. a. $\Delta K_{tot} = 17$ J, $\Delta U_{int} = 0$ b. $\Delta K_{tot} = 12$ J, $\Delta U_{int} = 5$ J
3. $\Delta U = 2.9 \times 10^6$ J
5. a. 13 m/s   b. 14 m/s
7. 7.7 m/s
9. 1.4 m/s
11. 1.4 m/s
13. 81 m/s
15. 0.632 m
17. 2.0 m/s
19. 3.0 m/s
21. 9.7 J
23. $v_0/\sqrt{2}$
25. 3.5 m/s, 2.8 m/s, 4.5 m/s
27. a. 7.7 m/s   b. 10 m/s
29. 6.3 m/s
31. -60 N at $x = 1$ m, 15 N at $x = 4$ m
33. 2.5 N, 0.40 N, 0.16 N
35. a. 50 J   b. 50 J   c. 50 J, yes

37. Energy (J)

39. -1 J of work is done to the environment
41. 2.3 m/s
43. a. 2.2×10^4 N/m   b. 19 m/s
45. $\dfrac{5}{2}R$
47. 65 g
49. 0.54 m
51. a. 1.7 m/s   b. No
53. a. 0.51 m   b. 0.38 m
55. 1.37×10^8 m/s
57. a. $\dfrac{\pi}{3}$ and $\dfrac{2\pi}{3}$ b. $\dfrac{\pi}{3}$ is unstable, $\dfrac{2\pi}{3}$ is stable
59. a. 20 J, 40 J, 60 J, 70 J   b. 2.6 m
61. a. $-\pi B/L$ b. $-AL$ c. $-2AL + \pi B/L$
63. a. a^2b b. a^2b c. Yes
69. 80.4°
71. 11 m
73. 6.7 m

Chapter 11

Stop to Think Questions

1. **f.** The cart is initially moving in the negative x-direction, so $p_{ix} = -20$ kg m/s. After it bounces, $p_{fx} = -10$ kg m/s. Thus $\Delta p = (10 \text{ kg m/s}) - (-20 \text{ kg m/s}) = 30$ kg m/s.
2. **b.** The clay ball goes from $v_{ix} = v$ to $v_{fx} = 0$, so $J_{clay} = \Delta p_x = -mv$. The rubber ball rebounds, going from $v_{ix} = v$ to $v_{fx} = -v$ (same speed, opposite direction). Thus $J_{rubber} = \Delta p_x = -2mv$. The rubber ball has a larger momentum change, and this requires a larger impulse.
3. **Less than.** The ball's momentum $m_B v_B$ is the same in both cases. Momentum is conserved, so the *total* momentum is the same after both collisions. The ball that rebounds from C has *negative* momentum, so C must have a larger momentum than A.
4. **c.** Momentum conservation requires $(m_1 + m_2) \times v_f = m_1 v_1 + m_2 v_2$. Because $v_1 > v_2$, it must be that $(m_1 + m_2) \times v_f = m_1 v_1 + m_2 v_2 > m_1 v_2 + m_2 v_2 = (m_1 + m_2)v_2$. Thus $v_f > v_2$. Similarly, $v_2 < v_1$, so $(m_1 + m_2)v_f = m_1 v_1 + m_2 v_2 < m_1 v_1 + m_2 v_1 = (m_1 + m_2)v_1$. Thus $v_f < v_1$. The collision causes m_1 to slow down and m_2 to speed up.
5. **Right end.** The pieces started at rest, so the total momentum of the system is zero. It's an isolated system, so the total momentum after the explosion is still zero. The 6 g piece has momentum $6v$. The 4 g piece, with velocity $-2v$, has momentum $-8v$. The combined momentum of these two pieces is $-2v$. In order for P to be zero, the third piece must have a *positive* momentum $(+2v)$ and thus a positive velocity.
6. **e.** Momentum is conserved, so the total momentum of the two pieces must be the initial $2\hat{\imath}$ kg m/s. $\vec{p}_1 = 2\hat{\jmath}$ kg m/s, so $\vec{p}_2$ has to be $(2\hat{\imath} - 2\hat{\jmath})$ kg m/s.

Exercises and Problems

1. 75 m/s
3. 4.0 N s
5. 1.0×10^3 N

7. 80 m/s to the left
9. 3.0 m/s to the right
11. a. 1.5×10^4 N s   b. 30 s, 110 m/s
13. 6.0 m/s
15. 1.4 m/s
17. v_0
19. 2.0 mph
21. 7.6 cm in the direction Brutus was running
23. -1.69×10^7 m/s, 3.1×10^6 m/s
25. a. 2.6 m/s   b. 33 cm
27. a. 1.92 m/s   b. 1.90 m/s
29. 13 s
31. $(1\hat{\imath} + 5\hat{\jmath})$ kg m/s
33. 45° north of east at 1.7 m/s
35. 1800 m/s
37. 1000 m/s
39. a. 6.4 m/s   b. $F_{\text{avg}} = 3.6 \times 10^2$ N $= 612(F_G)_B$
41. 9.3×10^2 N
43. $12t$ N
45. a. 6.7×10^{-8} m/s   b. $2.2 \times 10^{-10}\%$
47. $v_0/\sqrt{2}$, 45° east of north
49. a. $v_{\text{bullet}} = \dfrac{m + M}{m}\sqrt{2\mu_k g d}$ b. 4.4×10^2 m/s
51. 4.0×10^2 m
53. 0.021 m/s
55. 25.8 cm
57. a. $v_m = \dfrac{m + M}{m}\sqrt{5Rg}$ b. $v_m = \dfrac{1}{2}\dfrac{m + M}{m}\sqrt{5Rg}$
59. 14 m
61. 16 m/s
63. 7.9 m/s
65. a. 100 g ball: 5.3 m/s to the left; 200 g ball: 1.7 m/s to the right
    b. Both balls: 0.67 m/s to the left
67. 14.0 u
69. a. -1.4×10^{-22} kg m/s   b. and c. 1.4×10^{-22} kg m/s in the direction of the electron
71. 0.85 m/s, 72° below the $+x$-axis
73. 780 m/s, 1900 m/s, 4000 m/s
79. 5.7 m/s
81. 90 m/s
83. 8 bullets

Chapter 12

Stop to Think Questions

1. **b.** The center of the hitting end is *closer* to the center of mass, so it must have *more* mass.
2. **b.** Rotational kinetic energy depends on the angular velocity and the moment of inertia. Both have the same angular velocity, but a cylindrical shell has a larger moment of inertia than a solid cylinder with the same mass and radius.
3. $I_a > I_d > I_b > I_c$. The moment of inertia is smaller when the mass is more concentrated near the rotation axis.
4. $\tau_e > \tau_a = \tau_d > \tau_b > \tau_c$. The tangential component in e is larger than 2 N.
5. $\alpha_b > \alpha_a > \alpha_c = \alpha_d$. Angular acceleration is proportional to torque and inversely proportional to the moment of inertia. The moment of inertia depends on the *square* of the radius, so angular acceleration is proportional to F/r.
6. **c.** The scale exerts an upward force at half the distance from the pivot as the weight's downward force. To exert an equal but opposite torque for static equilibrium requires twice the force.
7. **d.** There is no net torque on the bucket + rain system, so the angular momentum is conserved. The addition of mass on the outer edge of the circle increases I, so ω must decrease. Mechanical energy is not conserved because the raindrop collisions are inelastic.

Exercises and Problems

1. a. 4.2×10^2 rad/s^2 b. 8.3 rev
3. a. 1.5 m/s   b. 13 rev
5. 4.7×10^6 m
7. 8.0 cm, 5.0 cm
9. 2.4 m/s
11. a. 0.032 kg m^2 b. 16 J
13. a. 0.063 m, 0.050 m   b. 0.0082 kg m^2
15. a. (0.060 m, 0.040 m)   b. 0.0020 kg m^2
    c. 0.0013 kg m^2
17. a. 6.9 kg m^2 b. 4.1 kg m^2
19. 50 N
21. -14.7 N
23. a. $\tau = 31.0$ N m   b. $\tau = 21.9$ N m
25. 0.50 rad/s
27. a. 1.75×10^{-3} N m   b. 50 rev
29. 1.4 m
31. $F_1 = 0.75$ kN, $F_2 = 1.0$ kN
33. a. 6.4×10^2 rpm   b. 40 m/s   c. 0 m/s
35. a. 88 rad/s   b. 2/7
37. a. $\vec{A} \times \vec{B} = (17$, out of the page)   b. $\vec{C} \times \vec{D} = \vec{0}$
39. a. $\vec{D} = n\hat{\imath}$, where n could be any real number
    b. $\vec{E} = 2\hat{\jmath}$ c. $\vec{F} = 1\hat{k}$
41. $-(38$ N m$)\hat{k}$
43. $-0.025\,\hat{\imath}$ kg m^2/s
45. 91 rpm
47. 83 rpm
49. 1.8 J
51. $\dfrac{L^2}{4}\left(\dfrac{M}{3} + m_1 + \dfrac{m_2}{4}\right)$
53. $\dfrac{M}{3L}\left[(L - d)^3 + d^3\right]$
55. a. $I_{\text{disk}} = \frac{1}{2}M(R^2 + r^2)$ c. 1.4 m/s or 75%
57. 0.91 m
59. 15,300 N
61. Yes, $d_{\text{max}} = \dfrac{25}{24}L$
63. 800 N
65. a. $a = \dfrac{m_2 g}{m_1 + m_2}, T = \dfrac{m_1 m_2 g}{m_1 + m_2}$
    b. $a = \dfrac{m_2 g}{m_1 + m_2 + \frac{1}{2}m_p}, T_1 = \dfrac{m_1 m_2 g}{m_1 + m_2 + \frac{1}{2}m_p}, T_2 = \dfrac{m_2(m_1 + \frac{1}{2}m_p)g}{m_1 + m_2 + \frac{1}{2}m_p}$
67. 0.52 N
69. 4.3 m
71. $\frac{1}{3}\tan\theta$
73. a. $\sqrt{\dfrac{3}{4}Lg}$ b. $\sqrt{Lg}$
75. $h = 2.7(R - r)$
79. 50 rpm
81. 22 rpm
83. a. 137 km   b. 8.6×10^6 m/s
85. a. 2.9×10^{-5} N m   b. 7.0 m/s
87. a. kg/m^3 b. $\dfrac{12M}{L^3}$ c. $\dfrac{3}{20}ML^2$
89. $v_f = \dfrac{1}{5}v_0$ to the right

Chapter 13

Stop to Think Questions

1. **e.** The acceleration decreases inversely with the square of the distance. At height R_e, the distance from the center of the earth is $2R_e$.
2. **c.** Newton's third law requires $F_{1 \text{ on } 2} = F_{2 \text{ on } 1}$.

3. **b.** $g_{surface} = GM/R^2$. Because of the square, a radius twice as large balances a mass four times as large.

4. In absolute value, $U_e > U_a = U_b = U_d > U_c$. $|U_G|$ is proportional to $m_1 m_2/r$.

5. **a.** T^2 is proportional to r^3, or T is proportional to $r^{3/2}$. $4^{3/2} = 8$.

Exercises and Problems

1. 6.00×10^{-4}
3. a. 6.7×10^{-9} N   b. 6.8×10^{-9}
5. 9.0 N
7. 12 cm
9. a. 1.62 m/s^2 b. 5.90×10^{-3} m/s^2
11. 1.35 m/s^2
13. $0.58R_e$
15. 39 m
17. 10 km/s
19. 4.3 km/s
21. a. 1.80×10^7 m   b. 9.41 km/s
23. 33 km/s
25. $M_p = 1.5 \times 10^{25}$ kg, $M_s = 5.2 \times 10^{30}$ kg
27. 2.9×10^9 m
29. 1.0×10^{26} kg
31. 1.44 km/s, 1.72×10^7 m
33. a. 7.0 m/s   b. 12 m/s
35. a. 1.3×10^{-6} N, 83° cw from the $+y$-axis
    b. 2.3×10^{27} N, 7.5° ccw from the $-y$-axis.
37. -1.48×10^{-7} J
39. 46 kg and 104 kg
41. $2000R_e$
43. 4.2×10^5 m
45. 1.6 s
47. a. 2.3 km/s   b. 11 km/s
49. $0.414R$
51. 6.7×10^8 J
53. 1.8×10^{25} J
55. a. 0.17 m/s   b. 1.0×10^{13} kg   c. 0.84 m/s
57. $T = \left[\dfrac{4\pi^2 r^3}{G} \dfrac{1}{(M + m/4)} \right]^{1/2}$
59. a. 6.3×10^4 m/s
    b. 1.3×10^{12} m/s^2
    c. 1.3×10^{12} N
    d. 6.29×10^{-4} s
    e. 1.5×10^6 m
61. 10 yr
63. a. 14,000 km   b. 2000 km
67. 3.0×10^4 m/s
69. a. 5.8×10^{22} kg   b. 1.3×10^6 m
71. b. $v_1 = 7730$ m/s, $v_1' = 10,160$ m/s
    c. 2.17×10^{10} J
    d. $v_2' = 1600$ m/s, $v_2 = 3070$ m/s
    e. 3.43×10^9 J
    f. 2.52×10^{10} J

4. **b.** The weight of the displaced water equals the weight of the ice cube. When the ice cube melts and turns into water, that amount of water will exactly fill the volume that the ice cube is now displacing.

5. **1 cm^3/s out.** The fluid is incompressible, so the sum of what flows in must match the sum of what flows out. 13 cm^3/s is known to be flowing in, while 12 cm^3/s flows out. An additional 1 cm^3/s must flow out to achieve balance.

6. $h_b > h_d > h_c > h_a$. The liquid level is higher where the pressure is lower. The pressure is lower where the flow speed is higher. The flow speed is highest in the narrowest tube, zero in the open air.

Exercises and Problems

1. 50 mL
3. a. 8.1×10^2 kg/m^3 b. 8.4×10^2 kg/m^3
5. 2.4×10^3 kg
7. 38 cm
9. 1.2×10^5 Pa
11. 0.64 kN
13. 10.3 m
15. 3.2 m
17. 6.7×10^2 kg/m^3
19. 750 kg/m^3
21. 1.9 N
23. 2.49×10^3 kg/m^3
25. 1.1 kN
27. $1.27v_0$
29. 2.12 m/s
31. a. 0.38 N   b. 20 m/s
33. 1.0 cm
35. a. 5.1×10^7 Pa   b. -0.025 c. 1056 kg/m^3
37. 2.4 μm
39. a. 2.9×10^4 N   b. 30 players
41. 27 psi
43. 55 cm
45. a. Down by 3.0 cm   b. 3.5 cm
47. a. $F = \rho g D W L$ b. $F = \frac{1}{2}\rho g D^2 L$ c. 0.78 kN, 1.4 kN
49. 8.01%
51. $(\rho - \rho_1)/(\rho_2 - \rho_1)$
53. 0.16 kg
55. $m = 29$ g, $\rho = 1.1 \times 10^3$ kg/m^3
57. 5.5 m/s
59. 187 nm/s
61. 76 cm
63. 4.0 cm
65. a. Lower   b. 0.83 kPa   c. Blow out
67. a. $v_1 = 1.4 \times 10^2$ m/s, $v_2 = 5.8$ m/s   b. 4.5×10^{-3} m^3/s
69. a. $v_{hole}A_{hole} = \pi r^2 \sqrt{\dfrac{2gd}{1 - (r/R)^4}}$ b. 1.1 mm/min
71. 14 cm
73. $\dfrac{h}{l} = \left(1 - \dfrac{\rho_0}{\rho_f} \right)^{1/3}$
75. a. $\dfrac{2R^2 g}{9\eta}(\rho - \rho_f)$ b. $v_t\left(1 - e^{(-6t\pi\eta R/m)} \right)$ c. 3.1 s

Chapter 14

Stop to Think Questions

1. $\rho_a = \rho_b = \rho_c$. Density depends only on what the object is made of, not how big the pieces are.

2. **c.** These are all open tubes, so the liquid rises to the same height in all three despite their different shapes.

3. $F_b > F_a = F_c$. The masses in c do not add. The pressure underneath each of the two large pistons is mg/A^2, and the pressure under the small piston must be the same.

Chapter 15

Stop to Think Questions

1. **c.** $v_{max} = 2\pi A/T$. Doubling A and T leaves v_{max} unchanged.

2. **d.** Think of circular motion. At 45°, the particle is in the first quadrant (positive x) and moving to the left (negative v_x).

3. **c > b > a = d.** Energy conservation $\frac{1}{2}kA^2 = \frac{1}{2}m(v_{max})^2$ gives $v_{max} = \sqrt{k/m}A$. k or m has to be increased or decreased by a factor of 4 to have the same effect as increasing or decreasing A by a factor of 2.

4. **c.** $v_x = 0$ because the slope of the position graph is zero. The negative value of x shows that the particle is left of the equilibrium position, so the restoring force is to the right.

5. **c.** The period of a pendulum does not depend on its mass.

6. $\tau_d > \tau_b = \tau_c > \tau_a$. The time constant is the time to decay to 37% of the initial height. The time constant is independent of the initial height.

Exercises and Problems

1. a. 3.3 s   b. 0.30 Hz   c. 1.9 rad/s   d. 0.25 m   e. 0.48 m/s
3. 2.27 ms
5. a. 20 cm   b. 0.13 Hz   c. 60°
7. a. $-\frac{2}{3}\pi$ rad   b. 13.6 cm/s   c. 15.7 cm/s
9.   x (cm)

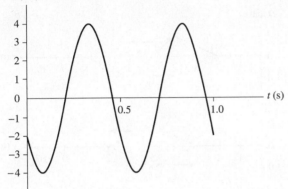

11. $x(t) = (4.0 \text{ cm})\cos\left[(8.0\pi \text{ rad/s})t - \frac{1}{2}\pi \text{ rad}\right]$
13. a. $\phi_0 = -\frac{2}{3}\pi$ rad, or $-120°$ b. $\phi = -\frac{2}{3}\pi$ rad, 0 rad, $\frac{2}{3}\pi$ rad, $\frac{4}{3}\pi$ rad
15. 5.48 N/m
17. a. 2.0 cm   b. 0.63 s   c. 5.0 N/m
    d. $-\frac{1}{4}\pi$ rad   e. 14 cm/s
    f. 20 cm/s   g. 1.00×10^{-3} J   h. 1.5 cm/s
19. 0.13 s
21. a. 25 N/m   b. 0.90 s   c. 0.70 m/s
23. 3.5 Hz
25. 36 cm
27. 33 cm
29. 3.67 m/s²
31. 54 cm
33. 5.00 s
35. 250 N/m
37. 0.0955 s
41. 1.02 m/s
43. a. 55 kg   b. 0.73 m/s
45. a. 3.2 Hz   b. 7.1 cm   c. 5.0 J
47. 1.6 Hz
49. 8.1×10^9 N/m²
51. a. 4.7×10^4 N/m   b. 1.8 Hz
53. 5.0×10^2 m/s
55. a. 7.5 m   b. 0.45 m/s
57. a. 0.84 s   b. 7.1°
59. 8.7×10^{-2} kg m²
61. a. 2.0 Hz   b. 1.2 cm
63. 7.9×10^{13} Hz
65. 5.86 m/s²
67. a. 9.5 N/m   b. 0.010 kg/s
69. 25 s
71. 21 oscillations
75. $\sqrt{\dfrac{f_1^2 f_2^2}{f_1^2 + f_2^2}}$
77. $\dfrac{1}{2\pi}\sqrt{\dfrac{5g}{7R}}$

79. 1.8 Hz
81. 0.58 s

Chapter 16

Stop to Think Questions

1. **d and e.** The wave speed depends on properties of the medium, not on how you generate the wave. For a string, $v = \sqrt{T_s/\mu}$. Increasing the tension or decreasing the linear density (lighter string) will increase the wave speed.

2. **b.** The wave is traveling to the right at 2.0 m/s, so each point on the wave passes $x = 0$ m, the point of interest, 2.0 s before reaching $x = 4.0$ m. The graph has the same shape, but everything happens 2.0 s earlier.

3. **d.** The wavelength—the distance between two crests—is seen to be 10 m. The frequency is $f = v/\lambda = (50 \text{ m/s})/(10 \text{ m}) = 5$ Hz.

4. $n_c > n_a > n_b$. $\lambda = \lambda_{vac}/n$, so a shorter wavelength corresponds to a larger index of refraction.

5. **e.** A crest and an adjacent trough are separated by $\lambda/2$. This is a phase difference of π rad.

6. **c.** Any factor-of-2 change in intensity changes the sound intensity level by 3 dB. One trumpet is $\frac{1}{4}$ the original number, so the intensity has decreased by two factors of 2.

7. **c.** Zack hears a higher frequency as he and the source approach. Amy is moving with the source, so $f_{Amy} = f_0$.

Exercises and Problems

1. 141 m/s
3. 2.0 m
5.

7.

9.

Equilibrium
$t = 0$ s

11. a. 4.2 m   b. 48 Hz
13. a. 20 Hz   b. 2.3 m   c. 46 m/s
15. $v = \sqrt{d/c}$
17. a. 3.2 mm   b. 9.3×10^{10} Hz

19. a. 6.67×10^{14} Hz b. 4.62×10^{14} Hz c. 1.44
21. a. 10 GHz b. 0.17 ms
23. a. 8.6×10^8 Hz b. 23 cm
25. a. 1.88×10^8 m/s b. 4.48×10^{14} Hz
27. a. 311 m/s b. 361 m/s
29. 40 cm
31. 2.5 m
33. 3.4×10^{-6} J
35. a. 1.1×10^{-3} W/m² b. 1.1×10^{-7} J
37. 2700 W/m², 1400 W/m², 610 W/m²
39. 5.0 W
41. a. 432 Hz b. 429 Hz
43. 86 m/s
45. a. 0.80 m b. $-\frac{\pi}{2}$ rad
    c. $D(x, t) = (2.0 \text{ mm}) \sin\left(2.5\pi x - 10\pi t - \frac{1}{2}\pi\right)$
47. 2.3 m, 1.7 m
49. 410 μs
51. 1230 km
53. 987 m/s
55. a. $-x$-direction b. 12 m/s, 5.0 Hz, 2.6 rad/m c. -1.5 cm
57. -19 m/s, 0 m/s, 19 m/s
59. 8
61. 9.4 m/s
63. 16 Hz and 2.0 cm
65. 4.0 m/s
67. a. 6.67×10^4 W b. 8.5×10^{10} W/m²
69. a. 3.10×10^9 W/m² b. 19 AU
71. 49 mW
73. 1.5
75. 19 m/s
79. Receding at 1.5×10^6 m/s
81. 0.07°C
83. 18 mJ
85. 29 s

Chapter 17

Stop to Think Questions

1. **c.** The figure shows the two waves at $t = 6$ s and their superposition. The superposition is the *point-by-point* addition of the displacements of the two individual waves.

2. **a.** The allowed standing-wave frequencies are $f_m = m(v/2L)$, so the mode number of a standing wave of frequency f is $m = 2Lf/v$. Quadrupling T_s increases the wave speed v by a factor of 2. The initial mode number was 2, so the new mode number is 1.

3. **b.** 300 Hz and 400 Hz are allowed standing waves, but they are not f_1 and f_2 because 400 Hz $\neq 2 \times 300$ Hz. Because there's a 100 Hz difference between them, these must be $f_3 = 3 \times 100$ Hz and $f_4 = 4 \times 100$ Hz, with a fundamental frequency $f_1 = 100$ Hz. Thus the second harmonic is $f_2 = 2 \times 100$ Hz $= 200$ Hz.

4. **c.** Shifting the top wave 0.5 m to the left aligns crest with crest and trough with trough.

5. **a.** $r_1 = 0.5\lambda$ and $r_2 = 2.5\lambda$, so $\Delta r = 2.0\lambda$. This is the condition for maximum constructive interference.

6. **Maximum constructive.** The path-length difference is $\Delta r = 1.0$ m $= \lambda$. For identical sources, interference is constructive when Δr is an integer multiple of λ.

7. **f.** The beat frequency is the difference between the two frequencies.

Exercises and Problems

1.

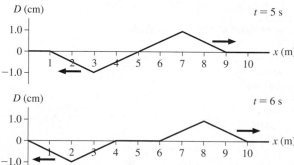

3. 4 s
5. 20 m/s
7. a. 8 b. 200 Hz
9. a. 12 Hz, 24 m/s b.

11. 12 kg

13.

m	f_m (GHz)
7	10.5
8	12.0
9	13.5
10	15.0
11	16.5
12	18.0
13	19.5

15. 400 m/s
17. 35 cm
19. 1.9 Hz
21. 1.6 kHz
23. a. 25 cm   b. 25 cm
25. 216 nm
27. a. In phase   b.

m	r_1	r_2	Δr	C/D
P	3λ	4λ	λ	C
Q	$\frac{7}{2}\lambda$	2λ	$\frac{5}{2}\lambda$	D
R	$\frac{5}{2}\lambda$	$\frac{7}{2}\lambda$	λ	C

29. 5.0 m
31. Maximum destructive interference
33. 527 Hz
35. 1.255 cm
37. 5.4 m
39. 2.4×10^{-4} N
41. 260 N
43. 8.50 m/s^2
45. $T' = \dfrac{T_0}{4}$
47. 0.54 s
49. 11 g/m
51. 13.0 cm
53. 328 m/s
55. 26 cm, 56 cm, 85 cm
57. 450 N
59. 7.9 cm
61. a. 80 cm   b. $\dfrac{3\pi}{4}$ c. 0.77a
63. a. 473 nm   b. 406 nm, 568 nm   c. Reflected light is blue and transmitted light is yellowish green.
65. a. $\lambda_C = \dfrac{2.66d}{m + \frac{1}{2}}$, $m = 0, 1, 2, 3,\ldots$
    b. 692 nm and 415 nm; this gives a purplish color.
67. 679 nm, 428 nm
69. 170 Hz
71. a. a b. 1.0 m   c. 9
73. a. 5   b. 4.6 mm
75. 7.0 m/s
77. 4.0 cm, 35 cm, 65 cm
79. 2.0 kg
81. a. $\lambda_1 = 20.0$ m, $\lambda_2 = 10.0$ m, $\lambda_3 = 6.67$ m
    b. $v_1 = 5.59$ m/s, $v_2 = 3.95$ m/s, $v_3 = 3.22$ m/s
    c. $T_1 = 3.58$ s, $T_2 = 2.53$ s, $T_3 = 2.07$ s

Chapter 18

Stop to Think Questions

1. **d.** The pressure *decreases* by 20 kPa.
2. **a.** The number of atoms depends only on the number of moles, not the substance.
3. **a.** The step size on the Kelvin scale is the same as the step size on the Celsius scale. A *change* of 10°C is a *change* of 10 K.
4. **Increase.** When an object undergoes thermal expansion, all dimensions increase by the same percentage.
5. **a.** On the water phase diagram, you can see that for a pressure just slightly below the triple-point pressure, the solid/gas transition occurs at a higher temperature than does the solid/liquid transition at high pressures. This is not true for carbon dioxide.
6. **c.** $T = pV/nR$. Pressure and volume are the same, but n differs. The number of moles in mass M is $n = M/M_{mol}$. Helium, with the smaller molar mass, has a larger number of moles and thus a lower temperature.

7. **b.** The pressure is determined entirely by the weight of the piston pressing down. Changing the temperature changes the volume of the gas, but not its pressure.
8. **b.** The temperature decreases by a factor of 4 during the isochoric process, where $p_f/p_i = \frac{1}{4}$. The temperature then increases by a factor of 2 during the isobaric expansion, where $V_f/V_i = 2$.

Exercises and Problems

1. 1900 cm^3
3. 8.3 cm
5. 1.1 mol
7. a. 6.02×10^{28} atoms/m^3 b. 3.28×10^{28} atoms/m^3
9. 2.7 cm
11. 184 K, 330 K
13. a. 171°Z   b. 671°C = 944 K
15. 12 mm
17. 1.68 L
19. a. 2   b. Unchanged
21. a. 0.050 m^3 b. 1.3 atm
23. 2.4×10^{22} molecules
25. 1.1×10^{15} particles/m^3
27. a. $T_2 = T_1$ b. $\dfrac{p_1}{2}$
29. a. 0.73 atm   b. 0.52 atm
31. a. 105,600 Pa   b. 42 cm
33. a. 93 cm^3
    b.

35. a. Isochoric   b. 1556°C, 337°C
37. a. Isothermal   b. −29°C   c. 133 cm^3
39.

41. 0.228 nm
43. 333°C
45. a. 1.3×10^{-13} b. 1.2×10^{11} molecules
47. 1.0×10^{-16} atm
49. 10 g
51. a. 2.7 m   b. 11 atm
53. 35 psi
55. 93 cm^3
57. 24 cm
59. 1.8 cm
61. a. 880 kPa   b. $T_2 = 323$°C, $T_3 = -49$°C, $T_4 = 398$°C

63. −152°C
65. a. 4.0 atm, − 73°C
    b. p (atm)

67. b. p (atm)

   c. 6 atm

69. b. p

   c. 417 cm³
71. 1.8 g
73. 4.0 × 10⁵ Pa

Chapter 19

Stop to Think Questions

1. **a.** The piston does work W on the gas. There's no heat because of the insulation, and $\Delta E_{mech} = 0$ because the gas as a whole doesn't move. Thus $\Delta E_{th} = W > 0$. The work increases the system's thermal energy and thus raises its temperature.

2. **d.** $W_A = 0$ because A is an isochoric process. $W_B = W_{1 \text{ to } 2} + W_{2 \text{ to } 3}$. $W_{2 \text{ to } 3} > W_{1 \text{ to } 2}$ because there's more area under the curve, and $W_{2 \text{ to } 3}$ is positive whereas $W_{1 \text{ to } 2}$ is negative. Thus W_B is positive.

3. **b and e.** The temperature rises in d from doing work on the gas ($\Delta E_{th} = W$), not from heat. e involves heat because there is a tempera-

ture difference. The temperature of the gas doesn't change because the heat is used to do the work of lifting a weight.

4. **c.** The temperature increases so E_{th} must increase. W is negative in an expansion, so Q must be positive and larger than $|W|$.

5. **a.** A has a smaller specific heat and thus less thermal inertia. The temperature of A will change more than the temperature of B.

6. **a.** $W_A + Q_A = W_B + Q_B$. The area under process A is larger than the area under B, so W_A is *more negative* than W_B. Q_A has to be more positive than Q_B to maintain the equality.

7. **c.** Conduction, convection, and evaporation require matter. Only radiation transfers energy through the vacuum of space.

Exercises and Problems

1. −80 J
3. 200 cm³
5. 16.6 kJ
7.

9.

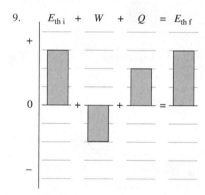

11. −700 J
13. a. 0 J   b. 3.4 kJ
15. 6.9 kJ
17. 1200 W
19. 0.55°C
21. 28°C
23. 73.5°C
25. 21%
27. a. 91 J   b. 140°C
29. a. 1.9 × 10⁻³ m³   b. 74°C
31. A: −1000 J, B: 1400 J
33. 3.14 × 10⁻⁴ m²
35. 5.8 s
37. 26 W
39. 8.7 h
41. 12 J/s
43. 87 min
45. 1.1 × 10⁵ L
47. 61 g
49. a. 5.5 kJ   b. 3.4 kJ

c. p (atm)

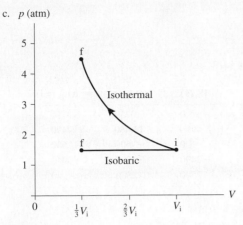

51. 5800 J
53. a. 250°C   b. 33 cm
55. a. 110 kPa   b. 24 cm
57. a. 4300 cm³, 610°C   b. 3000 J   c. 1.0 atm   d. −2200 J
    e. p (atm)

59. 15 atm
61. a. $T_{Ai} = 300$ K, $V_{Ai} = 2.5 \times 10^{-3}$ m³, $T_{Bf} = 220$ K, $V_{Bf} = 1.8 \times 10^{-3}$ m³
    b. p (atm)

63. 28°C
65. a. 5500 K   b. 0 J   c. 54 kJ   d. 20
    e. p (atm)

67. 43°C, 109°F
69. 26°C
71. 12,000 K
73. −18°C
75. b. 217°C
77. −56°C
79. 36 J
81. 150 J

Chapter 20

Stop to Think Questions

1. $\lambda_B > \lambda_A = \lambda_C > \lambda_D$. Increasing the volume makes the gas less dense, so l increases. Increasing the radius makes the targets larger, so λ decreases. The mean free path doesn't depend on the atomic mass.
2. **c.** Each v^2 increases by a factor of 16 but, after averaging, v_{rms} takes the square root.
3. **e.** Temperature is proportional to the average energy. The energy of a gas molecule is kinetic, proportional to v^2. The average energy, and thus T, increases by 4^2.
4. **b.** The bead can slide along the wire (one degree of translational motion) and rotate around the wire (one degree of rotational motion).
5. **a.** Temperature measures the average translational kinetic energy *per molecule*, not the thermal energy of the entire system.
6. **c.** With 1,000,000 molecules, it's highly unlikely that 750,000 of them would spontaneously move into one side of the box. A state with a very small probability of occurrence has a very low entropy. Having an imbalance of only 100 out of 1,000,000 is well within what you might expect for random fluctuations. This is a highly probable situation and thus one of large entropy.

Exercises and Problems

1. $\dfrac{N}{V} = 2.69 \times 10^{25}$ m⁻³
3. 0.023 Pa
5. a. 300 nm   b. 600 nm
7. 84.8
9. a. 20.0 m/s   b. 20.2 m/s
11. a. 289 K   b. 200 kPa
13. Neon
15. a. 818 m/s   b. 3600 m/s
17. −246°C
19. 2.12
21. a. 3400 J   b. 3400 J   c. 3400 J
23. a. 310 nm   b. 250 m/s   c. 2.1×10^{-22} J
25. a. 1.24×10^{-19} J   b. 1.22×10^4 m/s
27. 490 J
29. a. 0.080°C   b. 0.048°C   c. 0.040°C
31. 25°C
33. a. 62 J   b. 100 J   c. 150 J
35. 5000 J
37. 61
39. 9.6×10^{-5} m/s
41. a. 310 m   b. 2900 m
43. 5.8×10^{-22} K
45. a. $\lambda_{electron} = \dfrac{1}{\sqrt{2}\pi(N/V)r^2}$ b. 1.82×10^{-6} Pa or 1.80×10^{-11} atm
47. a. 640 m/s   b. 2.6 nm
49. 8.9×10^{25} s⁻¹
51. a. $E_{helium\,i} = 30$ J, $E_{argon\,i} = 122$ J   b. $E_{helium\,f} = 47$ J, $E_{argon\,f} = 105$ J
    c. Helium gains 16.9 J and argon loses 16.9 J.   d. 307°C
    e. $p_{helium\,f} = 3.1$ atm, $p_{argon\,f} = 3.4$ atm
53. 482 K
55. 22 kJ

59. b. $\dfrac{15n + 3}{2} p_i V_i$

61. a. 1.6 kJ   b. 2.7 kJ   c. 1550 m/s

63. a. $C_V = \dfrac{(3n_1 + 5n_2)}{2(n_1 + n_2)} R$

65. $\frac{1}{2}$

Chapter 21

Stop to Think Questions

1. $W_d > W_a = W_b > W_c$. $W_{out} = Q_H - Q_C$.
2. **b.** Energy conservation requires $Q_H = Q_C + W_{in}$. The refrigerator will exhaust more heat out the back than it removes from the front. A refrigerator with an open door will heat the room rather than cool it.
3. **c.** W_{out} = area inside triangle = 1000 J. $\eta = W_{out}/Q_H = (1000 \text{ J})/(4000 \text{ J}) = 0.25$.
4. To conserve energy, the heat Q_H exhausted to the hot reservoir needs to be $Q_H = Q_C + W_{in} = 40 \text{ J} + 10 \text{ J} = 50 \text{ J}$, not 30 J. The numbers shown here would be appropriate to a heat engine if the energy-transfer arrows were all reversed.
5. **b.** The efficiency of this engine would be $\eta = W_{out}/Q_H = 0.6$. That exceeds the Carnot efficiency $\eta_{Carnot} = 1 - T_C/T_H = 0.5$, so it is not possible.

Exercises and Problems

1. 0.33
3. a 0.27   b. 15 kJ
5. 2.0
7. a. 13 kJ/cycle   b. 50 kJ
9.

	ΔE_{th}	W_s	Q
A	–	+	0
B	+	0	+
C	0	–	–

11. 40 J
13. a. 30 J, 0.15 kJ   b. 0.21
15. a. 0.13   b. 320 J
17. 0.24
19. a. 660 W   b. 760 W
21. a. 5.0 kW   b. 1.7
23. a. Engine c violates the first law.   b. Engine b violates the second law.
25. a. 40%   b. 215°C
27. a. 40%   b. 2500 W   c. 1500 W
29. 0. 56 kJ
31. 1.7
33. a. 60 J   b. −23°C
35. a.

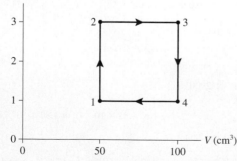

    b. 10 J   c. 0.13

37. 250 N/m
39. 28 kW
41. a. 3.6×10^6 J   b. 3.0×10^5 J

43. 8.3%
45. 6.4 J
47. 8.57 J
49. 57 g
51. a. 48 m   b. 32%
53. 37%
55. a.

	W_s (J)	Q (kJ)	ΔE_{th} (kJ)
$1 \rightarrow 2$	0	3750	3750
$2 \rightarrow 3$	4990	12,480	7490
$3 \rightarrow 4$	0	−7500	−7500
$4 \rightarrow 1$	−2490	−6230	−3740
Net	2500	2500	0

    b. 15%
57. a. 1.10 kW   b. 9.01%
59. a. 1000 cm³, 696 kPA, 522 K
    b.

	ΔE_{th} (J)	W_s (J)	Q (J)
$1 \rightarrow 2$	741.1	0	741.1
$2 \rightarrow 3$	−741.1	741.1	0
$3 \rightarrow 1$	0	−554.5	−554.5
Net	0	186.6	186.6

    c. 25%
61. a.

	p (atm)	T (K)	V (cm³)
1	1.0	406	1000
2	5.0	2030	1000
3	1.0	2030	5000

    b. 29%   c. 80%
63. a. $T_1 = 1.6$ kK, $T_2 = 2.4$ kK, $T_3 = 6.5$ kK
    b.

	ΔE_{th} (J)	W_s (J)	Q (J)
$1 \rightarrow 2$	327	−327	0
$2 \rightarrow 3$	1692	677	2369
$3 \rightarrow 1$	−2019	0	−2019
Net	0	350	350

65. b. 1.1×10^{3}°C
67. b. $Q_C = 80$ J
69. 5.3×10^4 J
71. a. 1.19   b. 227 W
73. c.

Chapter 22

Stop to Think Questions

1. **b.** Charged objects are attracted to neutral objects, so an attractive force is inconclusive. Repulsion is the only sure test.
2. $q_e(+3e) > q_a(+1e) > q_d(0) > q_b(-1e) > q_c(-2e)$.
3. **a.** The negative plastic rod will polarize the electroscope by pushing electrons down toward the leaves. This will partially neutralize the positive charge the leaves had acquired from the glass rod.

4. **b.** The two forces are an action/reaction pair, opposite in direction but *equal* in magnitude.
5. **c.** There's an electric field at *all* points, whether an $\vec{E}$ vector is shown or not. The electric field at the dot is to the right. But an electron is a negative charge, so the force of the electric field on the electron is to the left.
6. $E_b > E_a > E_d > E_c$.

Exercises and Problems

1. a Electrons removed   b. 5.0×10^{10}
3. a. Electrons transferred to the sphere   b. 3.1×10^{10}
5. $-9.6 \times 10^7 \, C$
7. 2, negative
13. a. 0.90 N   b. $0.90 \, m/s^2$
15. $-10 \, nC$
17. 0 N
19. 3.1×10^{-4} N, upward
21. $\vec{F}_{B \, on \, A} = +(7.2 \times 10^{-4} N)\hat{\jmath}, \ \vec{F}_{A \, on \, B} = -(7.2 \times 10^{-4} N)\hat{\jmath}$
23. a. $0.36 \, m/s^2$ toward glass bead   b. $0.18 \, m/s^2$ toward plastic bead
25. 8.9 kN/m
27. a. $(6.4\hat{\imath} + 1.6\hat{\jmath}) \times 10^{-17}$ N   b. $-(6.4\hat{\imath} + 1.6\hat{\jmath}) \times 10^{-17}$ N
   c. $4.0 \times 10^{10} \, m/s^2$ d. $7.3 \times 10^{13} \, m/s^2$
29. 0.11 nC
31. 12 nC
33. $-6.8 \times 10^4 \hat{\imath}$ N/C, $3.0 \times 10^4 \hat{\imath}$ N/C, $(8.1 \times 10^3 \hat{\imath} - 4.1 \times 10^4 \hat{\jmath})$ N/C
35. 1.4×10^5 C, -1.4×10^5 C
37. a. 5.0×10^2 N   b. $3.0 \times 10^{29} \, m/s^2$
39. 8.4×10^{21}
41. 4.3×10^{-3} N, 73° counterclockwise from the +x-axis
43. 2.0×10^{-4} N, 45° clockwise from the +x-axis
45. $1.1 \times 10^{-5} \hat{\jmath}$ N
47. a. -2.4 cm   b. Yes
49. -20 nC
51. $(2 - \sqrt{2}) \dfrac{KQq}{L^2}$
53. a. 2.4×10^2 N   b. Yes
55. 8.1 nC
57. 7.3 m/s
59. 33 nC
61. a. $1.1 \times 10^{18} \, m/s^2$ b. 1.0×10^{-12} N   c. 6.3×10^6 N/C   d. 69 nC
63. $-1.0 \times 10^5 \hat{\jmath}$ N/C, $(-2.9 \times 10^4 \hat{\imath} - 2.2 \times 10^4 \hat{\jmath})$ N/C, $-5.6 \times 10^4 \hat{\imath}$ N/C
65. a. $(-1.0 \, cm, 2.0 \, cm)$ b. $(3.0 \, cm, 3.0 \, cm)$ c. $(4.0 \, cm, -2.0 \, cm)$
67. $0.18 \, \mu C$
69. b. 1.0×10^3
71. b. 5.0 cm
73. $0.75 \, \mu C$
75. 41 g
77. a. $KQq \left(\dfrac{1}{(r - s/2)^2} - \dfrac{1}{(r + s/2)^2} \right) \hat{\imath}$ b. Toward Q

Chapter 23

Stop to Think Questions

1. **c.** From symmetry, the fields of the positive charges cancel. The net field is that of the negative charge, which is toward the charge.
2. $\eta_c = \eta_b = \eta_a$. All pieces of a uniformly charged surface have the same surface charge density.
3. **b, e, and h.** b and e both increase the linear charge density λ.
4. $E_a = E_b = E_c = E_d = E_e$. The field strength of a charged plane is the same at all distances from the plane. An electric field diagram shows the electric field vectors at only a few points; the field exists at all points.
5. $F_a = F_b = F_c = F_d = F_e$. The field strength inside a capacitor is the same at all points, hence the force on a charge is the same at all points. The electric field exists at all points whether or not a vector is shown at that point.

6. **c.** Parabolic trajectories require *constant* acceleration and thus a *uniform* electric field. The proton has an initial velocity component to the left, but it's being pushed back to the right.

Exercises and Problems

1. 7.6×10^3 N/C along the $+x$-axis
3. 7.6×10^3 N/C vertically downward
5. a. 2.0 nC   b. 180 N/C
7. 2.9×10^{-3} N
9. 2.3×10^5 N/C, 1.67×10^5 N/C, 2.3×10^5 N/C
11. 40 nC
13. 15 nC
15. a. 0 N/C   b. 4.1×10^3 N/C
17. 27 nC
19. 1.41×10^5 N/C
21. -0.35 pC
23. 25 nC, -25 nC
25. 0.9995 cm
27. a. 0.0023   b. 43 kN/C, up
29. 0.28 MN/C
31. $6.4 \, \mu C/m^2$
33. a. $\dfrac{1}{4\pi\epsilon_0} \dfrac{qQs}{r^3}$ b. $\dfrac{1}{4\pi\epsilon_0} \dfrac{qQs}{r^2}$
35. a. $(1.0 \times 10^5 \hat{\imath} - 3.6 \times 10^4 \hat{\jmath})$ N/C
   b. 1.1×10^5 N/C, 19.4° ccw from the +x-axis
37. a. $(-4.7 \times 10^3 \hat{\imath} + 8.6 \times 10^4 \hat{\jmath})$ N/C
   b. 8.6×10^4 N/C, 93° cw from the $+x$-axis
39. $\dfrac{1}{4\pi\epsilon_0} \left(\dfrac{q}{5\sqrt{5}a^2} \right)(3\hat{\imath} + 2\hat{\jmath})$, $-\dfrac{1}{4\pi\epsilon_0} \left(\dfrac{17q}{9a^2} \right)\hat{\imath}$, $-\dfrac{1}{4\pi\epsilon_0} \left(\dfrac{7q}{9a^2} \right)\hat{\imath}$,
   and $\dfrac{1}{4\pi\epsilon_0} \left(\dfrac{q}{5\sqrt{5}a^2} \right)(3\hat{\imath} - 2\hat{\jmath})$
41. $\dfrac{1}{4\pi\epsilon_0} \dfrac{16\lambda y}{4y^2 + d^2}$
43. a. $\dfrac{1}{4\pi\epsilon_0} \dfrac{Q}{x^2 - L^2/4}$ b. $\dfrac{1}{4\pi\epsilon_0} \dfrac{Q}{x^2}$ c. 9.8×10^4 N/C
47. a. $\dfrac{1}{4\pi\epsilon_0} \left(\dfrac{2\pi Q}{L^2} \right)\hat{\imath}$ b. 1.70×10^5 N/C
49. 1.4×10^5 N/C
51. 2.2 mm
53. 1.19×10^7 m/s
55. 1.13×10^{14} Hz
57. a. $\dfrac{\frac{4}{3}\pi r^3 \rho g + qE}{6\pi\eta r}$ b. 0.067 mm/s   c. 0.049 mm/s
59. 6.56×10^{15} Hz
63. b. 1.0 mm
65. b. $\dfrac{R}{\sqrt{3}}$
67. a.   b. $\dfrac{4Q}{L^2}$

   c. $\dfrac{8Q}{4\pi\epsilon_0 L^2} \left[1 - \dfrac{x}{\sqrt{x^2 + L^2/4}} \right]$
69. a. $\dfrac{2\eta}{4\pi\epsilon_0} \ln \left| \dfrac{2x + L}{2x - L} \right| \hat{\imath}$

c.

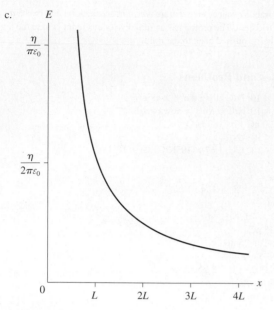

71. 8.8×10^5 N/C
73. c. 2.0×10^{12} Hz

Chapter 24

Stop to Think Questions

1. **a and d.** Symmetry requires the electric field to be unchanged if front and back are reversed, if left and right are reversed, or if the field is rotated about the wire's axis. Fields a and d both have the proper symmetry. Other factors would now need to be considered to determine the correct field.
2. **e.** The net flux is into the box.
3. **c.** There's no flux through the four sides. The flux is positive 1 N m^2/C through both the top and bottom because $\vec{E}$ and $\vec{A}$ both point outward.
4. $\Phi_b = \Phi_e > \Phi_a = \Phi_c = \Phi_d$. The flux through a closed surface depends only on the amount of enclosed charge, not the size or shape of the surface.
5. **d.** A cube doesn't have enough symmetry to use Gauss's law. The electric field of a charged cube is *not* constant over the face of a cubic Gaussian surface, so we can't evaluate the surface integral for the flux.

Exercises and Problems

1.

3.

$$\vec{E} = \vec{0} \text{ N/C}$$

5. Positive charge
7. Into the front face of the cube; field strength must exceed 5 N/C
9. -2.3 N m^2/C
11. 1.4×10^3 N/C
13. a. 0.0 N m^2/C   b. 0.12 N m^2/C
15. 3.5×10^{-4} N m^2/C

19. $+2q, +q, -3q$
21. 0.11 kN m^2/C
23. -1.00 N m^2/C
25. 17.7×10^{-9} C/m^2
27. $\dfrac{Q}{\epsilon_0}$
29. $\Phi_1 = -3.2$ kN m^2/C, $\Phi_2 = \Phi_3 = \Phi_5 = 0.0$ N m^2/C,
    $\Phi_4 = 3.2$ kN m^2/C
31. a. -3.5 N m^2/C   b. 1.2 N m^2/C
33. 0.19 kN m^2/C
35. a. -100 nC   b. $+50$ nC
37. a. 2.4×10^{-6} C/m^3 b. 1 nC, 10 nC, 80 nC
    c. 4.5 kN/C, 9.0×10^3 N/C, 1.8×10^4 N/C
39. -4.51×10^5 C
41. 2.5×10^4 N/C, outward, 0 N/C, 7.9×10^3 N/C, outward
43. 0 N/C, $\dfrac{1}{4\pi\epsilon_0} \dfrac{Q}{r^2} \hat{r}$
45. $\vec{0}$ N/C, $(\eta/2\epsilon_0)\hat{j}, -(\eta/2\epsilon_0)\hat{j}, \vec{0}$ N/C
49. a. $\dfrac{\lambda}{2\pi\epsilon_0} \dfrac{\hat{r}}{r}$ b. $\dfrac{3\lambda}{2\pi\epsilon_0} \dfrac{\hat{r}}{r}$
51. 6.2×10^{-11} C^2/N m^2
53. $\dfrac{\rho}{6\epsilon_0} r$
55. b. 0, because this is a neutral atom
    c. 4.6×10^{13} N/C
57. a. $\dfrac{\lambda L^2 dy}{4\pi\epsilon_0 [y^2 + (L/2)^2]}$ b. $\lambda L/(4\epsilon_0)Q_{in}/\epsilon_0$
59. a. $C = \dfrac{Q}{4\pi R}$ b. $\dfrac{1}{4\pi\epsilon_0} \dfrac{Q}{Rr} \hat{r}$ c. Yes
61. a. $\dfrac{Q}{4\pi\epsilon_0 R^2}$ b. $\dfrac{3Qr^3}{2\pi R^6}$

Chapter 25

Stop to Think Questions

1. **Zero.** The motion is always perpendicular to the electric force.
2. $U_b = U_d > U_a = U_c$. The potential energy depends inversely on r. The effects of doubling the charge and doubling the distance cancel each other.
3. **c.** The proton gains speed by losing potential energy. It loses potential energy by moving in the direction of decreasing electric potential.
4. $V_a = V_b > V_c > V_d = V_e$. The potential decreases steadily from the positive to the negative plate. It depends only on the distance from the positive plate.
5. $\Delta V_{ac} = \Delta V_{bc} > \Delta V_{ab}$. The potential depends only on the *distance* from the charge, not the direction. $\Delta V_{ab} = 0$ because these points are at the same distance.

Exercises and Problems

1. 2.7×10^6 m/s
3. 2.1×10^6 m/s
5. -2.2×10^{-19} J
7. -4.7×10^{-6} J
9. 1.61×10^8 N/C
11. 4.38×10^5 m/s
13. 11.4 V
15. a. Higher   b. 3340 V
17. 1.0×10^{-5} V
19. a. 4.2×10^3 m/s   b. 1.8×10^6 m/s
23. a. 200 V   b. 6.3×10^{-10} C
25. a. 1000 V   b. 7.0×10^6 m/s
27. a. 1800 V, 1800 V, 900 V   b. 0 V, 900 V
29. a. 27 V   b. -4.3×10^{-18} J
31. -1600 V

33. a. No b. Yes, at $x = 0$

c.

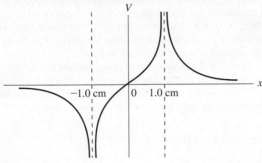

35. 0 V
37. -10 nC, 40 nC
39. 3.0 cm, 6.0 cm
41. a. 0.72 J b. 14 N c. 2.0 g cube: 22 m/s, 4.0 g cube: 11 m/s
43. 1.0×10^5 m/s
45. 2.5 cm/s
47. a. 2.1×10^6 V/m b. 9.4×10^7 m/s
49. 150 nC
51. 8.0×10^7 m/s
53. -5.1×10^{-19} J
55. 310 nC
57. 6.8 fm
59. a. Yes c. 8.21×10^8 m/s
61. a. 15 V, 3.0 kV/m, 2.1×10^{-10} C b. 15 V, 1/5 kV/m, 1.0×10^{-10} C
c. 15 V, 3.0 kV/m, 8.3×10^{-10} C
63. a. $\dfrac{V_0}{R}$ b. 100 kV/m
65. a. 8.3 μC b. 3.3×10^6 V/m
67. a. $\dfrac{2q}{4\pi\epsilon_0 x} \dfrac{1}{\sqrt{1 + s^2/4x^2}}$
b.

69. $\dfrac{Q}{4\pi\epsilon_0 L} \ln\left[(x + L/2)(x - L/2) \right]$
71. $Q/4\pi\epsilon_0 R$
73. $\dfrac{\lambda}{4\epsilon_0 R}\left(1 + \dfrac{2\ln(3)}{\pi}\right)$
75. b. 1.67×10^6 m/s
77. 0.28 rad/s

79. 0.018 m/s, 0.011 m/s
81. a. 2Q/L b. $\dfrac{K\lambda_0}{L}\left[L + d\ln\left(\dfrac{d}{L+d}\right)\right]$

Chapter 26

Stop to Think Questions

1. **c.** E_y is the negative of the slope of the V-versus-y graph. E_y is positive because $\vec{E}$ points up, so the graph has a negative slope. E_y has constant magnitude, so the slope has a constant value.
2. **c.** $\vec{E}$ points "downhill," so V must decrease from right to left. E is larger on the left than on the right, so the contour lines must be closer together on the left.
3. **b.** Because of the connecting wire, the three spheres form a single conductor in electrostatic equilibrium. Thus all points are at the same potential. The electric field of a sphere is related to the sphere's potential by $E = V/R$, so a smaller-radius sphere has a larger E.
4. **5.0 V.** The potentials add, but $\Delta V = -1.0$ V because the charge escalator goes *down* by 1.0 V.
5. $(C_{eq})_b > (C_{eq})_a = (C_{eq})_d > (C_{eq})_c$. $(C_{eq})_b = 3 \ \mu\text{F} + 3 \ \mu\text{F} = 6 \ \mu\text{F}$. The equivalent capacitance of series capacitors is less than any capacitor in the group, so $(C_{eq})_c < 3 \ \mu\text{F}$. Only d requires any real calculation. The two 4 μF capacitors are in series and are equivalent to a single 2 μF capacitor. The 2 μF equivalent capacitor is in parallel with 3 μF, so $(C_{eq})_d = 5 \ \mu\text{F}$.

Exercises and Problems

1. -0.20 kV
3. -200 V
5. a. A b. -70 V
7. $- (20\hat{\jmath})$ kV/m
9. a. -5 V/m b. 10 V/m c. -5 V/m
11.

13. -1.0 kV/m
15. a. 27 V/m b. 3.7 V/m
17. 1.5×10^{-6} J
19. 1.6×10^{-13} J
21. a. 13 pF b. 1.3 nC
23. 4.8 cm
25. 3.0 μF
27. 7.5 μF
29. 20 μF, parallel
31. 1.4 kV
33. a. 1.1×10^{-7} J b. 0.71 J/m^3
35. a. 0.15 nF b. 12 kV
37. 89 pF
39. a.

b. 25 V

41. a. $-(1.4 \times 10^7 \hat{\imath})$ V/m, 7×10^4 V   b. 0.0 V/m, 1.4×10^5 V
   c. $1.4 \times 10^7 \hat{\imath}$ V/m, 7×10^4 V

43. a. $\dfrac{Q}{4\pi\epsilon_0 L} \ln\left(\dfrac{x+L/2}{x-L/2}\right)$ b. $\dfrac{Q}{4\pi\epsilon_0} \dfrac{1}{x^2 - L^2/4}\hat{\imath}$

45. $2V_0/3d$

47. 40 V/m, 27° ccw from $+x$-axis

49. $Q_{1f} = 2$ nC, $Q_{2f} = 4$ nC

51. 1.1 nC

53. a. $\pm 3.2 \times 10^{-11}$ C, 9.0 V   b. $\pm 3.2 \times 10^{-11}$ C, 18 V

55. a. NC b. $\dfrac{C}{N}$

57. $Q_1 = 4$ μC, $\Delta V_1 = 1.0$ V; $Q_2 = 12$ μC, $\Delta V_2 = 1.0$ V; $Q_3 = 16$ μC, $\Delta V_3 = 8$ V

59. a. b. c. d.

61. $Q_1 = 0.83$ mC, $Q_2 = Q_3 = 0.67$ mC, $\Delta V_1 = 55$ V, $\Delta V_2 = 34$ V, $\Delta V_3 = 22$ V

63. $Q_1' = 33$ μC, $Q_2' = 67$ μC, $\Delta V_1' = \Delta V_2' = 3.3$ V

65. a. $\dfrac{C(\Delta V_C)^2}{2d}$ b. 5.0 mN

67. 22 μF

69. 20 μA

71. 2.4×10^{-14} J

75. b. $(10 - az^2)$ V, with z in m

77. b. 2 μF

79. a. $\dfrac{q}{4\pi\epsilon_0}\left[\dfrac{1}{\sqrt{x^2 + (y-s/2)^2}} - \dfrac{1}{\sqrt{x^2 + (y+s/2)^2}}\right]$

   b. $\dfrac{qsy}{4\pi\epsilon_0(x^2+y^2)^{3/2}}$

   c. $E_x = \dfrac{qs(3xy)}{4\pi\epsilon_0(x^2+y^2)^{5/2}}$, $E_y = -\dfrac{qs(2y^2-x^2)}{4\pi\epsilon_0(x^2+y^2)^{5/2}}$

   d. $\vec{E}_{\text{on-axis}} = \dfrac{2p}{4\pi\epsilon_0 r^3}\hat{\jmath}$, yes

   e. $\vec{E}_{\text{bisecting axis}} = -\dfrac{p}{4\pi\epsilon_0 r^3}\hat{\jmath}$, yes

81. a. $\dfrac{1}{4\pi\epsilon_0}\dfrac{Q}{R}\left[\dfrac{3}{2} - \dfrac{r^2}{2R^2}\right]$ b. 3/2

   c.

83. 2 C

Chapter 27

Stop to Think Questions

1. $i_c > i_b > i_a > i_d$. The electron current is proportional to $r^2 v_d$. Changing r by a factor of 2 has more influence than changing v_d by a factor of 2.

2. **The electrons don't have to move from the switch to the bulb, which could take hours.** Because the wire between the switch and the bulb is already full of electrons, a flow of electrons from the switch into the wire immediately causes electrons to flow from the other end of the wire into the lightbulb.

3. $E_d > E_b > E_e > E_a = E_c$. The electric field strength depends on the *difference* in the charge on the two wires. The electric fields of the rings in a and c are opposed to each other, so the net field is zero. The rings in d have the largest charge *difference*.

4. **1 A into the junction.** The total current entering the junction must equal the total current leaving the junction.

5. $J_b > J_a = J_d > J_c$. The current density $J = I/\pi r^2$ is independent of the conductivity σ, so a and d are the same. Changing r by a factor of 2 has more influence than changing I by a factor of 2.

6. $I_a = I_b = I_c = I_d$. Conservation of charge requires $I_a = I_b$. The current in each wire is $I = \Delta V_{\text{wire}}/R$. All the wires have the same resistance because they are identical, and they all have the same potential difference because each is connected directly to the battery, which is a *source of potential*.

Exercises and Problems

1. 75 μm/s

3. 0.93 mm

5. 0.023 V/m

7. a. 7.43×10^{-6} m/s   b. 2.1×10^{-14} s

9. a. 0.80 A   b. 7.0×10^7 A/m²

11. 130 C

13. 1.8 μA

15. 4.2×10^6 A/m²

17. a. 2.1×10^{-14} s   b. 4.3×10^{-15} s

19. a. 10 V/m   b. 6.7×10^6 A/m²   c. 0.62 mm

21. Nichrome

23. a. 1.64×10^{-3} V/m   b. 1.10×10^{-5} m/s

25. a. 0.50 C/s   b. 1.5 J   c. 0.75 W

27. a. 1.5 Ω   b. 3.5 Ω

29. 1.5 mV

31. 2.3 mA

33. 1.6 A

35. 50 Ω

37. 0.87 V

39. 0.10 V/m

41. 6.2×10^6

43. 23 mA

45. Yes, 2.2×10^5 Ω^{-1}m^{-1}

47. a. 120 C   b. 0.45 mm

49. 71°C

51. 100 V

53. a. $\dfrac{(\Delta V)A[1 - \alpha(T - T_0)]}{\rho_0 L}$ b. 4.4 A   c. -0.017 A/°C

55. 0.50 mm

57. a. $\dfrac{I}{4\pi\sigma r^2}$ b. $E_{\text{inner}} = 3.3 \times 10^{-4}$ V/m, $E_{\text{outer}} = 5.3 \times 10^{-5}$ V/m

59. a. $I(t) = (10 \text{ A})e^{-t/2.0\,\text{s}}$ b. 10 A
   c.

61. 1.01×10^{23}

63. 7.2 mm

65. $\dfrac{3}{2}\dfrac{I}{\pi R^3}$

67. 36 A

69. 1.80×10^3 C

71. $4R$

73. a. 9.4×10^{15} b. 115 A/m^2

75. 1.0 s

Chapter 28

Stop to Think Questions

1. **a, b, and d.** These three are the same circuit because the logic of the connections is the same. In c, the functioning of the circuit is changed by the extra wire connecting the two sides of the capacitor.

2. **ΔV increases by 2 V in the direction of I.** Kirchhoff's loop law, starting on the left side of the battery, is then $+12$ V $+ 2$ V $- 8$ V $- 6$ V $= 0$.

3. **$P_b > P_d > P_a > P_c$.** The power dissipated by a resistor is $P_R = (\Delta V_R)^2/R$. Increasing R decreases P_R; increasing ΔV_R increases P_R. But the potential has a larger effect because P_R depends on the square of ΔV_R.

4. **$I = 2$ A for all. $V_a = 20$ V, $V_b = 16$ V, $V_c = 10$ V, $V_d = 8$ V, $V_e = 0$ V.** The potential is 0 V on the right and increases by IR for each resistor going to the left.

5. **$A > B > C = D$.** All the current from the battery goes through A, so it is brightest. The current divides at the junction, but not equally. Because B is in parallel with $C + D$ but has half the resistance, twice as much current travels through B as through $C + D$. So B is dimmer than A but brighter than C and D. C and D are equal because current is the same through bulbs in series.

6. **b.** The two 2 Ω resistors are in series and equivalent to a 4 Ω resistor. Thus $\tau = RC = 4$ s.

Exercises and Problems

1.

3. 1 A to the left

5. a. 0.5 A to the right

b.

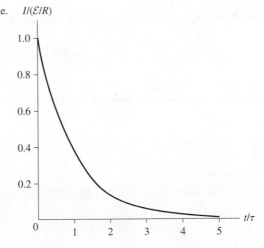

7. 9.60 Ω, 12.5 A

9. a. 60 W   b. 23 W, 14 W

11. D

13. 24 μm

15. $120

17. a. 0.65 Ω   b. 3.5 W

19. 1.2 Ω

21. 60 V, 10 Ω

23. 12 Ω

25. 14 Ω

27. 183 Ω

29. 20 W, 45 W

31. 5 V, -2 V

33. 8 ms

35. 69 ms

37. 0.87 kΩ

39. $65 for the incandescent bulb, $20 for the fluorescent tube

41. 19 W

43.

45. 7 Ω

47. 1.04

49. 9.5

51. a. $R = r$ b. 20 W

53. 3 A

55. a. 8 V   b. 0 V

57. a. 0.505 Ω   b. 0.500 Ω

59.

Resistor	Potential difference (V)	Current (A)
2 Ω	8	4
4 Ω	8	2
6 Ω	8	4/3
8 Ω	16	2
12 Ω	8	2/3

61. 2.0 A

63. 0.12 A, left to right

65. a. 200 A   b. 15 A   c. 200 A   d. 4 A

67. a. 6.9 ms   b. 3.5 ms

69. 89 Ω

71. 63 μJ

73. a. $\mathcal{E}$ b. $C\mathcal{E}$ c. $I = +dQ/dt$ d. $I = \dfrac{\mathcal{E}}{R}e^{-t/\tau}$

e.

75. a. 1.2 ms   b. 4.2 kW   c. 2.9 J

77. 140 Ω, 72 Ω

79. 2.0 m, 0.49 mm

83. b. 5.1 kΩ

Chapter 29

Stop to Think Questions

1. **Not at all.** The charge exerts weak, attractive polarization forces on both ends of the compass needle, but in this configuration the forces will balance and have no net effect.
2. **d.** Point your right thumb in the direction of the current and curl your fingers around the wire.
3. **b.** Point your right thumb out of the page, in the direction of $\vec{v}$. Your fingers are pointing down as they curl around the left side.
4. **b.** The right-hand rule gives a downward $\vec{B}$ for a clockwise current. The north pole is on the side from which the field emerges.
5. **c.** For a field pointing into the page, $\vec{v} \times \vec{B}$ is to the right. But the electron is negative, so the force is in the direction of $-(\vec{v} \times \vec{B})$.
6. **b.** Repulsion indicates that the south pole of the loop is on the right, facing the bar magnet; the north pole is on the left. Then the right-hand rule gives the current direction.
7. **a or c.** Any magnetic field to the right, whether leaving a north pole or entering a south pole, will align the magnetic domains as shown.

Exercises and Problems

1. $B_2 = 40$ mT, $B_3 = 0$ T, $B_4 = 40$ mT
3. a. 0 T   b. $1.60 \times 10^{-15}\hat{k}$ T   c. $-4.0 \times 10^{-16}\hat{k}$ T
5. $2.83 \times 10^{-16}\hat{k}$ T
7. 2.5 A, 250 A, 5000 A to 50,000 A, 500,000 A
9. a. $2.0\,\mu$T   b. 4.0%
11. a. 20 A   b. 1.6×10^{-3} m
13. $\vec{0}$ T
15. 0.12 mT
17. a. 3.1×10^{-4} A m^2 b. 5.0×10^{-7} T
19. a. 6.2×10^{-5} T   b. 6.3×10^{8} A
21. 0
23. 23.0 A, into the page
25. 2.4 kA
27. a. $8.0 \times 10^{-13}\hat{j}$ N   b. $5.7 \times 10^{-13}(-\hat{j} - \hat{k})$ N
29. a. 1.4442 MHz   b. 1.6450 MHz   c. 1.6457 MHz
31. a. 86 mT   b. 1.62×10^{-14} J
33. 0.131 T, out of page
35. 0.025 N, right
37. 240 A
39. 0.28 A m^2
41. a. $x = 0.50$ cm   b. $x = 8.0$ cm
43. 4.1×10^{-4} T, into page
45. $\dfrac{\mu_0 I \theta}{4\pi R}$
47. $2.4 \times 10^{-8}\ \Omega$ m
49. #18, 4.1 A
51. a. 1.13×10^{10} A   b. 0.014 A/m^2 c. 1.3×10^{6} A/m^2
53. a. 5.7×10^{-6} A   b. 2.9×10^{-8} A m^2
55. a. Circular   b. $\dfrac{\mu_0 NI}{2\pi r}$
57. $0;\ \dfrac{\mu_0 I}{2\pi r}\left(\dfrac{r^2 - R_1^2}{R_2^2 - R_1^2}\right);\ \dfrac{\mu_0 I}{2\pi r}$
59. 2.0 mT, into page
61. 2.4×10^{10} m/s^2, up
63. 16 cm
65. 0.82 mm, 3.0 mm
67. $\sqrt{v_0^2 + \left(\dfrac{qE_0}{m}t\right)^2}$
69. a. 0.45 s   b. 2.8×10^{9} s
71. 0.12 T
73. 0.16 T
75. a. Into the page   b. $\sqrt{\dfrac{2IlBd}{m}}$

77. a.

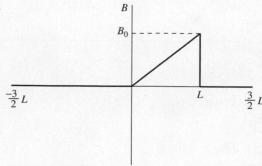

b. $\dfrac{1}{2}ILB_0\hat{j}$ c. $\dfrac{1}{3}IL^2 B_0$
79. 2.0 cm
81. $\dfrac{\mu_0 \omega Q}{2\pi R}$
83. a. Horizontal and to the left above the sheet; horizontal and to the right below the sheet   b. $\frac{1}{2}\mu_0 J_s$

Chapter 30

Stop to Think Questions

1. **d.** According to the right-hand rule, the magnetic force on a positive charge carrier is to the right.
2. **No.** The charge carriers in the wire move parallel to $\vec{B}$. There's no magnetic force on a charge moving parallel to a magnetic field.
3. $F_b = F_d > F_a = F_c$. $\vec{F}_a$ is zero because there's no field. $\vec{F}_c$ is also zero because there's no current around the loop. The charge carriers in both the right and left edges are pushed to the bottom of the loop, creating a motional emf but no current. The currents at b and d are in opposite directions, but the forces on the segments in the field are both to the left and of equal magnitude.
4. **Clockwise.** The wire's magnetic field as it passes through the loop is into the page. The flux through the loop decreases into the page as the wire moves away. To oppose this decrease, the induced magnetic field needs to point into the page.
5. **d.** The flux is increasing into the loop. To oppose this increase, the induced magnetic field needs to point out of the page. This requires a ccw induced current. Using the right-hand rule, the magnetic force on the current in the left edge of the loop is to the right, away from the field. The magnetic forces on the top and bottom segments of the loop are in opposite directions and cancel each other.
6. **b or f.** The potential decreases in the direction of increasing current and increases in the direction of decreasing current.
7. $\tau_c > \tau_a > \tau_b$. $\tau = L/R$, so smaller total resistance gives a larger time constant. The parallel resistors have total resistance $R/2$. The series resistors have total resistance $2R$.

Exercises and Problems

1. 2.0×10^{4} m/s
3. a. 1.0 N   b. 2.2 T
5. 3.5×10^{-4} Wb
7. *Bab*
9. Decreasing
11. a. 8.7×10^{-4} Wb   b. Clockwise
13. 1.6 V
15. 4.7 T/s, increasing
17. a. $1.6 \times 10^{-3}(1 + t)$ A   b. 9.4×10^{-3} A, 1.7×10^{-2} A
19. a. 4.8×10^{4} m/s^2, up   b. 0 m/s^2 c. 4.8×10^{4} m/s^2, down
    d. 9.6×10^{4} m/s^2, down
21. 5.3 mT/s
23. 2400
25. 100 V, increases

27. 9.5×10^{-5} J
29. $0.25 \ \mu$H
31. 3.8×10^{-18} F
33. a. 76 mA   b. 0.50 ms
35. 2.50×10^{-4} s
37. 1.6 A, 0.0 A, -1.6 A
39. 8.7 T/s
41. a. 5.0 mV   b. 10.0 mV
43. 44 mA
45. 44 μA
47. $-15 \ \mu$A
49. 0.15 T
51. 4.0 nA
53. a.   *I* (A)   b. 11 A

55. a. 0.20 A   b. 4.0 mN   c. 11 K
57. a. 6.3×10^{-4} N   b. 3.1×10^{-4} W   c. 1.3×10^{-2} A, ccw
    d. 3.1×10^{-4} W
59. a. $\dfrac{\mathcal{E}_{bat}}{Bl}$ b. 0.98 m/s
61. 3.9 V
63. $(R^2/2r)(dB/dt)$
65. 3.0 s
67.   *I* (A)

69. a. $-L\omega I_0 \cos\omega t$ b. 1.3 mA
71. a. 6.3×10^{-7} s   b. 0.050 mA
73. 2.0 mH, 0.13 μF
75. a. 50 V   b. Close S_1 at $t = 0$ s, open S_1 and close S_2 at $t = 0.0625$ s,
    then open S_2 at $t = 0.1875$ s
77. a. $I_0 = \Delta V_{bat}/R$ b. $I = I_0(1 - e^{-t/(L/R)})$
79. 0.72 mH
81. 0.50 m
83. a. $v_0 e^{-bt}$, where $b = l^2 B^2/(mR)$
    b.

85. 1.6×10^2 A/s

Chapter 31

Stop to Think Questions

1. **b.** $\vec{v}_{AB}$ is parallel to $\vec{B}_A$ hence $\vec{v}_{AB} \times \vec{B}_A$ is zero. Thus $\vec{E}_B = \vec{E}_A$ and points in the positive z-direction. $\vec{v}_{AB} \times \vec{E}_A$ points down, in the negative y-direction, so $-\vec{v}_{AB} \times \vec{E}_A/c^2$ points in the positive y-direction and causes $\vec{B}_B$ to be angled upward.
2. $B_c > B_a > B_d > B_b$. The induced magnetic field strength depends on the *rate dE/dt* at which the electric field is changing. Steeper slopes on the graph correspond to larger magnetic fields.
3. **e.** $\vec{E}$ is perpendicular to $\vec{B}$ and to $\vec{v}$, so it can only be along the z-axis. According to the Ampère-Maxwell law, $d\Phi_e/dt$ has the same sign as the line integral of $\vec{B} \cdot d\vec{s}$ around the closed curve. The integral is positive for a cw integration. Thus, from the right-hand rule, $\vec{E}$ is either into the page (negative z-direction) and increasing, or out of the page (positive z-direction) and decreasing. We can see from the figure that B is decreasing in strength as the wave moves from left to right, so E must also be decreasing. Thus $\vec{E}$ points along the positive z-axis.
4. **a.** The Poynting vector $\vec{S} = (\vec{E} \times \vec{B})/\mu_0$ points in the direction of travel, which is the positive y-direction. $\vec{B}$ must point in the positive x-direction in order for $\vec{E} \times \vec{B}$ to point upward.
5. **b.** The intensity along a line from the antenna decreases inversely with the square of the distance, so the intensity at 20 km is $\frac{1}{4}$ that at 10 km. But the intensity depends on the square of the electric field amplitude, or, conversely, E_0 is proportional to $I^{1/2}$. Thus E_0 at 20 km is $\frac{1}{2}$ that at 10 km.
6. $I_d > I_a > I_b = I_c$. The intensity depends on $\cos^2\theta$, where θ is the angle *between* the axes of the two filters. The filters in d have $\theta = 0°$. The two filters in both b and c are crossed ($\theta = 90°$) and transmit no light at all.

Exercises and Problems

1. a. Along the $-x$-axis   b. Along the y-axis ($+$ or $-$)
   c. Along the $+x$-axis
3. 16.3° above the x-axis
5. $-1.0 \times 10^6 \hat{k}$ V/m, $0.99998\hat{j}$ T
9. 1.0 μF
11. a. 0 T   b. 1.67×10^{-13} T   c. 1.98×10^{-13} T
13. 6.0×10^5 V/m
15. a. 10.0 nm   b. 3.00×10^{16} Hz   c. 6.67×10^{-8} T
17. $-z$-direction
19. 980 V/M, 3.3 T
21. a. 5.1×10^{11} V/m   b. 0.43
23. 16 cm
25. a. 3.33×10^{-6} T   b. 1.67×10^{-6} T
27. 66 mW
29. 2.3×10^{-13} N, 45° ccw
31. a. (0.10 T, into page)
    b. 0 V/m, (0.10 T, into page)
33. a. $\vec{E} = \dfrac{\lambda}{2\pi\epsilon_0 r}$, away from wire; $\vec{B} = \vec{0}$ T

    b. $\vec{E} = \dfrac{\lambda}{2\pi\epsilon_0 r}$, away from wire; $\vec{B} = \dfrac{1}{c^2\epsilon_0}\dfrac{v\lambda}{2\pi r}$, into page at top

    d. $\vec{E} = \dfrac{\lambda}{2\pi\epsilon_0 r}$, away from the wire; $\vec{B} = \dfrac{1}{c^2\epsilon_0}\dfrac{v\lambda}{2\pi r}$, into page at top
35. a. 7.1 V m, 0.17 A   b. 5.2 V m and 0.044 A
37. a. 1.0 mT   b. 0.160 mT

39. a. $(2.8 \times 10^3 t^2)$ V m   b.

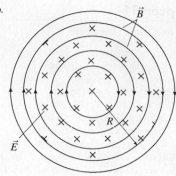

c. $(1.11 \times 10^{-9} rt)$ T, 4.4×10^{-12} T

d. $\left(1.00 \times 10^{-14} \dfrac{t}{r}\right)$ T, 5.0×10^{-12} T

41. 20 V
43. b. 6.67×10^{-6} J/m³
45. 2200 V/m, 7.4×10^{-6} T
47. a. 8.3×10^{-26} W/m²   b. 7.9×10^{-12} V/m
49. 9.4×10^7 W
51. a. 5.78×10^8 N   b. 1.64×10^{-14}
53. a. cmg b. 73.5 W
55. 0.41 m/s
57. 63°
59. 160 s
61. $(-6.0 \times 10^5 \hat{\imath} + 1.0 \times 10^5 \hat{k})$ V/m
63. 0.58 μm

Chapter 32

Stop to Think Questions

1. **a.** The instantaneous emf value is the projection down onto the horizontal axis. The emf is negative but increasing in magnitude as the phasor, which rotates ccw, approaches the horizontal axis.
2. **c.** Voltage and current are measured using different scales and units. You can't compare the length of a voltage phasor to the length of a current phasor.
3. **a.** There is "no capacitor" when the separation between the two capacitor plates becomes zero and the plates touch. Capacitance C is inversely proportional to the plate spacing d, hence $C \to \infty$ as $d \to 0$. The capacitive reactance is inversely proportional to C, so $X_C \to 0$ as $C \to \infty$.
4. $(\omega_c)_d > (\omega_c)_c = (\omega_c)_a > (\omega_c)_b$. The crossover frequency is $1/RC$.
5. **Above.** $V_L > V_C$ tells us that $X_L > X_C$. This is the condition above resonance, where X_L is increasing with ω while X_C is decreasing.
6. **a, b,** and **f.** You can always increase power by turning up the voltage. The current leads the emf, telling us that the circuit is primarily capacitive. The current can be brought into phase with the emf, thus maximizing the power, by decreasing C or increasing L.

Exercises and Problems

1. a. 1200 rad/s   c. 71 V

3.

Phasor at $t = 3.0$ ms

5. a. 25 Hz   b. 20 Ω   c.

7. a. 20 mA   b. 20 mA
9. a. 50 Hz   b. 4.8 μF
11. 81 nF
13. a. 9.95 V, 9.57 V, 7.05 V, 3.15 V, 0.990 V
    b.

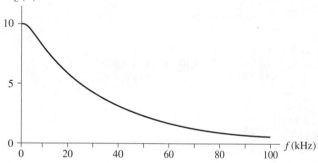

15. 8.0 V
17. 1.59 μF
19. $V_R = 6.0$ V, $V_C = 8.0$ V
21. a. 0.9770   b. 0.9998
23. a. 0.80 A   b. 0.80 mA
25. 1.4 mH
27. 200 kHz
29. 1.3 μF
31. 1.0 Ω
33. a. 5.0×10^3 Hz   b. 10 V, 32 V
35. 22 V
37. 0.40 kW
39. 0.75
41. a. $\dfrac{\sqrt{3}}{RC}$ b. $\dfrac{\sqrt{3}}{2}\mathcal{E}_0$
43. 0.50 mm
45. a. 25 mA   b. 6.7 V
47. a. $I_R = \dfrac{\mathcal{E}_0}{R}$, $I_C = \dfrac{\mathcal{E}_0}{(\omega C)^{-1}}$ b. $\mathcal{E}_0 \sqrt{(\omega C)^2 + \dfrac{1}{R^2}}$
49. a. $\mathcal{E}_0/\sqrt{R^2 + \omega^2 L^2}$, $\mathcal{E}_0 R/\sqrt{R^2 + \omega^2 L^2}$, $\mathcal{E}_0 \omega L/\sqrt{R^2 + \omega^2 L^2}$
    b. $V_R \to \mathcal{E}_0$, $V_R \to 0$ c. Low pass   d. R/L
51. a. 2.0 A   b. $-30°$ c. 150 W
53. 10 μT
55. 0.17 A
57. a. 3.6 V   b. 3.5 V   c. -3.6 V
61. a. 0.49 μH   b. 10.3 Ω
63. 24 W
65. a. 0.44 kA   b. 1.8×10^{-4} F   c. 7.4 MW
67. a. 0.83   b. 100 V   c. 13 Ω   d. 3.2×10^{-4} F
71. b. $I = \mathcal{E}_0/R$ in both cases   c. 0

Chapter 33

Stop to Think Questions

1. The antinodal lines seen in Figure 33.4b are diverging.
2. **Smaller.** Shorter-wavelength light doesn't spread as rapidly as longer-wavelength light. The fringe spacing Δy is directly proportional to the wavelength λ.
3. **d.** Larger wavelengths have larger diffraction angles. Red light has a larger wavelength than violet light, so red light is diffracted farther from the center.
4. **b or c.** The width of the central maximum, which is proportional to λ/a, has increased. This could occur either because the wavelength has increased or because the slit width has decreased.
5. **d.** Moving M_1 in by λ decreases r_1 by 2λ. Moving M_2 out by λ increases r_2 by 2λ. These two actions together change the path length by $\Delta r = 4\lambda$.

Exercises and Problems

1. 470 nm
3. 1.2 mm
5. 1.3 m
7. 0.22 mm
9. 1.6°, 3.2°
11. 43.2°
13. 14.5 cm
15. 20 mm
17. 1.2 m
19. 2.9°
21. 633 nm
23. 9
25. 5.4 mm
27. 78 cm
29. 0.25 mm
31. 400 nm
33. a. Double slit   b. 0.16 mm
35. 0.40 mm
37. 500 nm
39. 500 nm
41. a. $9I_1$ b. I_1
43. 1.3 m
45. 43 cm
47. a. $L\lambda/d$ b. $(L/d)\Delta\lambda$ c. 0.250 nm
49. 500 nm
51. 0.12 mm
53. 1.8 μm
55. $\dfrac{\sqrt{2}}{2}d$
57. 1.3 m
59. b. 50 μm
61. 0.88 mm
63. a. 550 nm   b. 0.40 mm
65. 50 cm
67. a. 22.3°   b. 16.6°
69. a. Dark   b. 1.597
71. 12.0 μm
73. a. $\Delta y = \dfrac{\Delta\lambda L}{d}$ c. $\Delta\lambda_{\min} = \dfrac{\lambda}{N}$ d. 3646 lines
75. a. 0.52 mm   b. 0.074°   c. 1.3 m

Chapter 34

Stop to Think Questions

1. **c.** The light spreads vertically as it goes through the vertical aperture. The light spreads horizontally due to different points on the horizontal lightbulb.

2. **c.** There's one image behind the vertical mirror and a second behind the horizontal mirror. A third image in the corner arises from rays that reflect twice, once off each mirror.
3. **a.** The ray travels closer to the normal in both media 1 and 3 than in medium 2, so n_1 and n_3 are both larger than n_2. The angle is smaller in medium 3 than in medium 1, so $n_3 > n_1$.
4. **e.** The rays from the object are diverging. Without a lens, the rays cannot converge to form any kind of image on the screen.
5. **a, e, or f.** Any of these will increase the angle of refraction θ_2.
6. **Away from.** You need to decrease s' to bring the image plane onto the screen. s' is decreased by increasing s.
7. **c.** A concave mirror forms a real image in front of the mirror. Because the object distance is $s \approx \infty$, the image distance is $s' \approx f$.

Exercises and Problems

1. 8.0 cm
3. 5.4 m
5. 9.0 cm
9. 433 cm
11. 35°
13. 31°
15. 76.7°
17. 23 cm
19. 113 cm
21. 20 cm behind lens, inverted
23. 15 cm in front of lens, upright
25. 30 cm
27. −40 cm
29. 1.5 cm
31. 30 cm, 0.50 cm
33. b. 40 cm, 2.0 cm, agree
35. b. −60 cm, 8.0 cm, agree
37. b. −8.6 cm, 1.1 cm, agree
39. 30 cm, behind mirror, upright
41. 30 cm, 1.5 cm, behind, upright
43. 6.4 cm
45. 10 m
47. a. $2\cos^{-1}\left(\dfrac{n}{2n_{\text{air}}}\right)$ b. 82.8°
49. b. 60 cm
51. 4.0 m
53. 1.46
55. a. 5   b. 4   c. 3
57. 35°
59. 2.7 m/s
61. 15.1 cm
63. Concave, 3.6 cm
65. 2.8 cm
67. 93 cm
69. 0.67 m, 1.0 m
71. 0.014
73. 20 cm
75. 20 μm/s away from the lens
77. −100 cm
79. a. $\dfrac{(n_2 - n_1)}{n_1}\left(\dfrac{1}{R_1} - \dfrac{1}{R_2}\right)$ b. 40 cm, 1.6 m
81. a. 24 cm
    b.

    c. 7.5 cm

Chapter 35

Stop to Think Questions

1. **b.** A diverging lens refracts rays away from the optical axis, so the rays will travel farther down the axis before converging.
2. **a.** Because the shutter speed doesn't change, the f-number must remain unchanged. The f-number is f/D, so increasing f requires increasing D.
3. **a.** A magnifier is a converging lens. Converging lenses are used to correct hyperopia.
4. **b.** If the objective magnification is halved, the eyepiece magnification must be doubled. $M_{eye} = 25 \text{ cm}/f_{eye}$, so doubling M_{eye} requires halving f_{eye}.
5. $w_a > w_d > w_b = w_c$. The spot size is proportional to f/D.

Exercises and Problems

1. b. $s_2' = 49$ cm, $h_2' = 4.6$ cm
3. b. $s_2' = -30$ cm, $h_2' = 6.0$ cm
5. b. $s_2' = -3.33$ cm, $h_2' = 0.66$ cm
7. 5.0
9. 1/250 s
11. a. Hyperopia   b. 50 cm
13. 3.0
15. 2.0 mm
17. 9.2 mm
19. a. 40   b. 5.0
21. 1.48
23. 1600 nm
25. 55 km
27. Both images are 2.0 cm tall; one upright 10 cm left of lens, the other inverted 20 cm to right of lens
29. a. $f_2 + f_1$ b. $\dfrac{f_2}{|f_1|} w_1$
31. 16 cm placed 80 cm from screen
33. 23 cm
35. 5.0 cm
37. a. $2.5\times$ b. $2.0\times$ lens   c. 17.5 cm
39. 4.6 mm
41. 1.0°
43. 2.6 cm
45. a. 1.3 cm   b. 1.2 cm   c. $f/1.7$
47. a. 3.8 cm   b. Sun is too bright
49. b. $\Delta n_2 = \frac{1}{2}\Delta n_1$ c. Crown converging, flint diverging   d. 4.18 cm

Chapter 36

Stop to Think Questions

1. **a, c, and f.** These move at constant velocity, or very nearly so. The others are accelerating.
2. **a.** $u' = u - v = -10 \text{ m/s} - 6 \text{ m/s} = -16 \text{ m/s}$. The *speed* is 16 m/s.
3. **c.** Even the light has a slight travel time. The event is the hammer hitting the nail, not your seeing the hammer hit the nail.
4. **At the same time.** Mark is halfway between the tree and the pole, so the fact that he *sees* the lightning bolts at the same time means they *happened* at the same time. It's true that Nancy *sees* event 1 before event 2, but the events actually occurred before she sees them. Mark and Nancy share a reference frame, because they are at rest relative to each other, and all experimenters in a reference frame, after correcting for any signal delays, *agree* on the spacetime coordinates of an event.

5. **After.** This is the same as the case of Peggy and Ryan. In Mark's reference frame, as in Ryan's, the events are simultaneous. Nancy *sees* event 1 first, but the time when an event is seen is not when the event actually happens. Because all experimenters in a reference frame agree on the spacetime coordinates of an event, Nancy's position in her reference frame cannot affect the order of the events. If Nancy had been passing Mark at the instant the lightning strikes occur in Mark's frame, then Nancy would be equivalent to Peggy. Event 2, like the firecracker at the front of Peggy's railroad car, occurs first in Nancy's reference frame.
6. **c.** Nick measures proper time because Nick's clock is present at both the "nose passes Nick" event and the "tail passes Nick" event. Proper time is the smallest measured time interval between two events.
7. $L_A > L_B = L_C$. Anjay measures the pole's proper length because it is at rest in his reference frame. Proper length is the longest measured length. Beth and Charles may *see* the pole differently, but they share the same reference frame and their *measurements* of the length agree.
8. **c.** The rest energy E_0 is an invariant, the same in all inertial reference frames. Thus $m = E_0/c^2$ is independent of speed.

Exercises and Problems

1. a. 4.0 m/s   b. $x_1 = 8.0$ m, $x_2 = 8.0$ m
3. a. 13 m/s   b. 3.0 m/s   c. 9.4 m/s
5. 3.0×10^8 m/s
7. 167 ns
9. $2.0\,\mu$s
11. Flash 2 is seen 40 μs after Flash 1.
15. a. 5.0 y   b. 2.2 y   c. 9.0 y
17. $0.28c$
19. a. Aged less   b. 14 ns
21. $0.80c$
23. 1.47 km
25. 3.0×10^6 m/s
27. a. 1200 m, $-2.0\,\mu$s   b. 2800 m, 8.7 μs
29. $0.36c$
31. $0.9944c$
33. 240,000,000 m/s
35. $0.707c$
37. 9.0×10^{13} J, 6.0×10^{13} J, 1.5×10^{14} J
39. $0.943c$
41. 2400
43. 11.2 h
45. a. No   b. 67.1 y
47. a. $0.80c$ b. 16 y
49. 14 m
51. a. 750 kJ   b. 8.2×10^{13} J
53. a. $t_D' = t_D'' = t_D''' = 0$ y, $t_E' = 0.42$ y, $t_E'' = 0$ y, $t_E''' = -0.56$ y   b. Yes, spaceship 2   c. Yes, spaceship 3   d. No
55. 22 m
57. 3.1×10^6 V
59. a. $0.980c$ b. 8.5×10^{-11} J
63. a. 3.84×10^8 m, 1.29 s, 5.47×10^7 m   b. 0.00 m, 0.182 s, 5.47×10^7 m   c. 3.84×10^8 m, 1.28 s, 0.00 m
65. a. 3.5×10^{-18} kg m/s, 1.1×10^{-9} J   b. 1.6×10^{-18} kg m/s
67. a. 7.6×10^{16} J   b. 0.84 kg
69. a. 1.3×10^{17} kg   b. 6.7×10^{-12}%   c. 15 billion years
71. a. 4.3×10^{-12} J   b. 0.72%
73. 1.1 pm
75. $0.85c$

Credits

Cover and title page: Thomas Vogel/Getty Images, Inc.
Page **xiii**: Stefano Garau/Shutterstock. Page **xv**: Placzkowski Piotr/ChromeOrange/Newscom. Page **xvi**: Oleksiy Maksymenko/Alamy. Page **xvii**: EpicStockMedia/Shutterstock. Page **xviii**: Dominic Harrison/Alamy. Page **xx**: Mark Axcell/Alamy. Page **xxi**: Andrew Dunn/Alamy.

CHAPTER 1
Part I Opener Page **1**: Joe Davila/U.S. Air Force. Page **2**: Kai Pfaffenbach/Reuters Pictures. Page **3** Left: Jason Edwards/National Geographic/Alamy; Center Left: Loop Images Ltd/Alamy; Center Right: Richard Megna/Fundamental Photographs; Right: RooM the Agency/Alamy. Page **4**: MarchCattle/Shutterstock. Page **8**: Rangizzz/Shutterstock. Page **9**: US Air Force. Page **11**: Courtesy of Audi/ZUMAPRESS/Newscom. Page **20**: Blaj Gabriel/Shutterstock. Page **22** Top: U.S. Department of Commerce; Bottom: Bureau International des Poids et Mesures.

CHAPTER 2
Page **32**: Oleksiy Maksymenko/AGE Fotostock America Inc. Page **38**: STS-129 Crew/NASA. Page **49**: Jim Sugar/Corbis. Page **50**: S.Tuengler - Inafrica.de/Alamy.

CHAPTER 3
Page **65**: OJO Images Ltd/Alamy. Page **69**: Pincasso/Shutterstock.

CHAPTER 4
Page **80**: Igor Ivanko/Shutterstock. Page **85**: Tony Freeman/Photo Edit Inc. Page **87**: Richard Megna/Fundamental Photographs. Page **93**: Huntstock/Getty Images.

CHAPTER 5
Page **110**: Placzkowski Piotr/ChromeOrange/Newscom. Page **111** First: Tony Freeman/Photo Edit Inc.; Second: Dmitry Kalinovsky/Shutterstock; Third: Comstock/Getty Images; Fourth: Mike Dunning/Dorling Kindersley Ltd.; Fifth: Todd Taulman/Shutterstock; Sixth: Pearson Education, Inc. Page **122**: FStop Images GmbH/Alamy. Page **123**: Olivier Le Moal /Fotolia.

CHAPTER 6
Page **131**: Tom Tracy/Alamy. Page **132**: Akira Kaede/Getty Images. Page **140**: NASA. Page **145** Top: Pearson Education, Inc.; Bottom: Marc Xavier/Fotolia. Page **157**: Richard Megna/Fundamental Photograph.

CHAPTER 7
Page **159**: C. Kurt Holter/Shutterstock. Page **161**: Todd Taulman/Shutterstock.

CHAPTER 8
Page **182**: Bernd Mellmann/Alamy. Page **185**: Doug James/Alamy. Page **191**: STS-135 Crew/NASA.

CHAPTER 9
Part II Opener Page **205**: Foxbat/Shutterstock. Page **206**: EPA/Neil Munns/Alamy. Page **207** Left: Kyodo/Newscom; Center: John Foxx/Getty Images; Right: Stacey Bates/Shutterstock. Page **208** Top Left: Kimimasa Mayama/Reuters/Corbis; Bottom Left: Murat sarica/Getty Images; Top Right: Vibrant Image Studio/Shutterstock; Bottom Right: Galaxy Picture Library/Alamy. Page **220**: Paul Harris/Paul Harris BWP Media USA/Newscom.

CHAPTER 10
Page **231**: Im Perfect Lazybones/Shutterstock.

CHAPTER 11
Page **261**: Vincent St. Thomas/Alamy. Page **262**: Jupiter Images/Getty Images. Page **268**: bl1/ZUMA Press/Newscom. Page **274**: Richard Megna/Fundamental Photographs. Page **279**: Richard Megna/Fundamental Photographs.

CHAPTER 12
Part III Opener Page **293**: NASA. Page **294**: Atomic Imagery/Getty Images. Page **304**: Polka Dot Images/Getty Images. Page **311**: Karl Thomas/Robert Harding World Imagery/Alamy. Page **314** Top: Chris Nash/Getty Images; Bottom: Richard Megna/Fundamental Photographs. Page **323**: PCN Photography/Alamy.

CHAPTER 13
Page **336**: NASA. Page **338**: bilwissedition Ltd. & Co. KG/Alamy. Page **340**: Corbis. Page **343**: NASA Earth Observing System. Page **347**: NASA Headquarters.

CHAPTER 14
Page **357**: Oleksiy Maksymenko/Alamy. Page **365**: Joseph P. Sinnot/Fundamental Photographs. Page **371**: Louise Murray/Alamy. Page **373** Top: Alix/Science Source; Bottom: TECHART Automobildesign. Page **374**: Lisa Kyle Young/Getty Images. Page **379**: Rkriminger/Shutterstock.

CHAPTER 15
Part IV Opener Page **389**: AguaSonic Acoustics/Science Source. Page **390**: George Resch/Fundamental Photographs. Page **401**: Thomas D. Rossing. Page **408**: Balan Madhavan/Balan Madhavan/Alamy. Page **412**: Martin Bough/Fundamental Photographs. Page **417**: Dr. Harold G. Craighead, Professor.

CHAPTER 16
Page **420**: EpicStockMedia/Shutterstock. Page **421**: Olga Selyutina/Shutterstock. Page **422**: Adapted from PSSC Physics, 7th edition by Haber-Schaim, Dodge, Gardner, and Shore, Published by Kendall/Hunt, 1991. Page **425**: Aflo Co., Ltd./Alamy. Page **435**: B. Benoit/Science Source. Page **436**: David Parker/Science Source. Page **446**: NOAA/AP Images. Page **448**: Image courtesy of Space Telescope Science Institute and NASA.

CHAPTER 17
Page **455**: Stefano Garau/Shutterstock. Page **457**: Renn Sminkey/Creative Digital Vision. Page **458**: Prelinger Archives. Page **466**: Brian Atkinson/Alamy. Page **473**: Eric Schrader/Pearson Education, Inc. Page **475**: Richard Megna/Fundamental Photographs.

CHAPTER 18
Part V Opener Page **489**: Pablo Paul/Alamy. Page **490**: Toshi Sasaki/Getty Images. Page **493**: Richard Megna/Fundamental Photographs. Page **494**: Richard Megna/Fundamental Photographs. Page **499**: Sean White/Design Pics Inc/Alamy.

CHAPTER 19
Page **515**: Green Angel/Alamy. Page **517**: Alexey Dudoladov/Getty Images. Page **523**: Gordon Saunders/Shutterstock. Page **528**: Stocktrek Images/Getty Images. Page **537** Left: Cn Boon/Alamy; Center Left: Gary S. Settles/Jason Listak/Science Source; Center Right: Pascal Goetgheluck/Science Source; Right: Corbis. Page **538**: Sciencephotos/Alamy. Page **539**: Image courtesy of NASA.

CHAPTER 20
Page **548**: Wjarek/iStockphoto/Getty Images. Page **562**: Randy Knight.

CHAPTER 21
Page **570**: Michael Utech/Getty Images. Page **573**: Peter Bowater/Alamy Images. Page **576**: Teresa Olson/Shutterstock. Page **580**: Malcolm Fife/Getty Images.

CHAPTER 22
Part VI Opener Page **601**: NASA Solar Dynamics Observatory. Page **602**: Lex Nichols/Alamy. Page **604**: George Resch/Fundamental Photographs. Page **617**: John Eisele/Colorado State University.

CHAPTER 23

Page **629:** Dominic Harrison/Alamy. Page **646:** Courtesy of Pacific Northwest National Laboratory.

CHAPTER 24

Page **658:** Alfred Benjamin/Science Source.

CHAPTER 25

Page **687:** NASA. Page **696:** Dave King/Dorling Kindersley Limited. Page **703:** Luc Diebold/Fotolia. Page **705:** Image provided by the Scientific Computing and Imaging Institute, University of Utah.

CHAPTER 26

Page **714:** Lightboxx/Alamy. Page **720:** Paul Silverman/Fundamental Photographs. Page **722:** AGE Fotostock America, Inc. Page **723:** Gusto/ Science Source. Page **724:** Eric Schrader/Pearson Education, Inc. Page **726:** Pushish Images/Shutterstock. Page **729:** Adam Hart-Davis/Science Source.

CHAPTER 27

Page **742:** Tibip/Fotolia. Page **750:** Iphoto/Fotolia. Page **753:** May/Science Source. Page **754:** Ktsdesign/Fotolia. Page **758:** Tim Ridley/Dorling Kindersley, Ltd.

CHAPTER 28

Page **766:** Zelenenka/Getty Images. Page **773:** Chris Ehman/Getty Images. Page **782:** Francisco Cruz/Superstock. Page **784:** Randy Knight.

CHAPTER 29

Page **796:** Shin Okamoto/Getty Images. Page **798:** John Eisele/Colorado State University. Page **806:** John Eisele/Colorado State University. Page **813:** John Eisele/Colorado State University. Page **814:** Bradwieland/Getty Images. Page **817:** Richard Megna/Fundamental Photographs. Page **818:** Neutronman/Getty Images. Page **826:** Horizon International Images Limited/Alamy. Page **827:** Andrew Lambert/Science Source.

CHAPTER 30

Page **836:** Bai Heng-Yao/Thunderbolt _TW/Getty Images. Page **855:** Jeff T. Green/Reuters. Page **857:** Miss Kanithar Aiumla-or/Shutterstock. Page **862:** Alamy.

CHAPTER 31

Page **876:** Buccaneer/Alamy. Page **888:** USACE Engineer Research and Development Center. Page **895:** NASA. Page **897** All Photos: Richard Megna/ Fundamental Photographs. Page **898** All Photos: Richard Megna/Fundamental Photographs. Page **900:** David Burton/Alamy.

CHAPTER 32

Page **905:** martinlisner/Fotolia. Page **919:** Bloomberg/Getty Images. Page **920:** Fotolia.

CHAPTER 33

Part VII Opener Page **929:** artpartner-images.com/Alamy, Page **930** Chapter Opener: Blickwinkel/Alamy; Top: Image reprinted courtesy of Dieter Zawischa; Center: Pearson Education, Inc.; Bottom: CENCO Physics/ Fundamental Photographs. Page **931** Top: Richard Megna/Fundamental Photographs; Bottom: Todd Gipstein/Getty Images. Page **938:** Pran Mukherjee et al 2009 Nanotechnology 20 325301. doi:10.1088/0957-4484/20/32/325301. Page **940** Top: Devmarya/Fotolia; Top Insert: Jian Zi. Page **943:** Ken Kay/Fundamental Photographs. Page **947** Top Right: Randy Knight; Bottom: Pearson Education, Inc. Page **950:** CENCO Physics/Fundamental Photographs. Page **951:** Image reprinted courtesy of Rod Nave. Page **952:** Philippe Plailly/Science Source. Page **957:** Alekss/Fotolia.

CHAPTER 34

Page **960** Chapter Opener: Mark Axcell/Alamy; Bottom Left: sciencephotos/ Alamy; Top Right: sciencephotos/Alamy. Page **966:** Pearson Education, Inc. Page **969:** Richard Megna/Fundamental Photographs. Page **973** All Photos: Richard Megna/Fundamental Photographs. Page **977:** Andrey Popov/Fotolia. Page **984:** Yaacov Dagan/Alamy. Page **987:** BSIP/Science Source.

CHAPTER 35

Page **995** Top: blickwinkel / Alamy Stock Photo; Bottom: Pearson Education, Inc. Page **999** All Photos: Richard Megna/Fundamental Photographs. Page **1000** Top: Underwater Photography/Canon U.S.A., Inc.; Center: NASA; Bottom: Randy Knight. Page **1002:** Tetra Images/ Alamy. Page **1005:** Tetra Images/Alamy. Page **1006:** Biophoto Associates/ Science Source. Page **1009:** Eric Schrader/Pearson Education, Inc. Page **1010:** Majeczka/Shutterstock. Page **1012:** Jeremy Burgess/Science Source. Page **1013:** Pearson Education, Inc.

CHAPTER 36

Part VIII Opener Page **1021:** Eric Toombs. Page **1022** Top: NG Images/ Alamy; Bottom: 3d brained/Shutterstock. Page **1023:** TopFoto/The Image Works. Page **1038:** U.S. Department of Defense Visual Information Center. Page **1040:** Stanford Linear Accelerator Center/Science Source. Page **1054:** LBNL/Science Source. Page **1055:** Wellcome Dept. of Cognitive Neurology/ ScienceSource.

Index

Note that pages 1063–1240 are not in the Standard Edition.

Astronomical Data

Planetary body	Mean distance from sun (m)	Period (years)	Mass (kg)	Mean radius (m)
Sun	—	—	1.99×10^{30}	6.96×10^8
Moon	3.84×10^8*	27.3 days	7.36×10^{22}	1.74×10^6
Mercury	5.79×10^{10}	0.241	3.18×10^{23}	2.43×10^6
Venus	1.08×10^{11}	0.615	4.88×10^{24}	6.06×10^6
Earth	1.50×10^{11}	1.00	5.98×10^{24}	6.37×10^6
Mars	2.28×10^{11}	1.88	6.42×10^{23}	3.37×10^6
Jupiter	7.78×10^{11}	11.9	1.90×10^{27}	6.99×10^7
Saturn	1.43×10^{12}	29.5	5.68×10^{26}	5.85×10^7
Uranus	2.87×10^{12}	84.0	8.68×10^{25}	2.33×10^7
Neptune	4.50×10^{12}	165	1.03×10^{26}	2.21×10^7

*Distance from earth

Typical Coefficients of Friction

Material	Static μ_s	Kinetic μ_k	Rolling μ_r
Rubber on dry concrete	1.00	0.80	0.02
Rubber on wet concrete	0.30	0.20	0.002
Steel on steel (dry)	0.80	0.60	0.002
Steel on steel (lubricated)	0.10	0.05	
Wood on wood	0.50	0.20	
Wood on snow	0.12	0.06	
Ice on ice	0.10	0.03	

Heats of Transformation

Substance	T_m (°C)	L_f (J/kg)	T_b (°C)	L_v (J/kg)
Water	0	3.33×10^5	100	22.6×10^5
Nitrogen (N_2)	−210	0.26×10^5	−196	1.99×10^5
Ethyl alcohol	−114	1.09×10^5	78	8.79×10^5
Mercury	−39	0.11×10^5	357	2.96×10^5
Lead	328	0.25×10^5	1750	8.58×10^5

Properties of Materials

Substance	ρ (kg/m³)	c (J/kg K)
Air at STP*	1.29	
Ethyl alcohol	790	2400
Gasoline	680	
Glycerin	1260	
Mercury	13,600	140
Oil (typical)	900	
Seawater	1030	
Water	1000	4190
Aluminum	2700	900
Copper	8920	385
Gold	19,300	129
Ice	920	2090
Iron	7870	449
Lead	11,300	128
Silicon	2330	703

*Standard temperature (0°C) and pressure (1 atm)

Coefficients of Thermal Expansion

Material	α (°C⁻¹)
Aluminum	2.3×10^{-5}
Brass	1.9×10^{-5}
Concrete	1.2×10^{-5}
Steel	1.1×10^{-5}
Invar	0.09×10^{-5}

Material	β (°C⁻¹)
Gasoline	9.6×10^{-4}
Mercury	1.8×10^{-4}
Ethyl alcohol	1.1×10^{-4}

Thermal Conductivities

Material	k (W/m K)
Diamond	2000
Silver	430
Copper	400
Aluminum	240
Iron	80
Stainless steel	14
Ice	1.7
Concrete	0.8
Glass	0.8
Styrofoam	0.035
Air (20°C, 1 atm)	0.023

Molar Specific Heats of Gases

Gas	C_P (J/mol K)	C_V (J/mol K)
Monatomic Gases		
He	20.8	12.5
Ne	20.8	12.5
Ar	20.8	12.5
Diatomic Gases		
H_2	28.7	20.4
N_2	29.1	20.8
O_2	29.2	20.9